Transient Receptor Potential Channels

ADVANCES IN EXPERIMENTAL MEDICINE AND BIOLOGY

Editorial Board:
IRUN R. COHEN, *The Weizmann Institute of Science*
ABEL LAJTHA, *N.S. Kline Institute for Psychiatric Research*
JOHN D. LAMBRIS, *University of Pennsylvania*
RODOLFO PAOLETTI, *University of Milan*

Recent Volumes in this Series

Volume 696	SOFTWARE TOOLS AND ALGORITHMS FOR BIOLOGICAL SYSTEMS Edited by Hamid R. Arabnia and Quoc-Nam Tran
Volume 697	HOT TOPICS IN INFECTION AND IMMUNITY IN CHILDREN VII Edited by Andrew Pollard, Adam Finn, and Nigel Curtis
Volume 698	BIO-FARMS FOR NUTRACEUTICALS: FUNCTIONAL FOOD AND SAFETY CONTROL BY BIOSENSORS Maria Teresa Giardi, Giuseppina Rea and Bruno Berra
Volume 699	MCR 2009: PROCEEDINGS OF THE 4TH INTERNATIONAL CONFERENCE ON MULTI-COMPONENT REACTIONS AND RELATED CHEMISTRY, EKATERINBURG, RUSSIA Maxim A. Mironov
Volume 700	REGULATION OF MICRORNAS Helge Großhans
Volume 701	OXYGEN TRANSPORT TO TISSUE XXXII Duane F. Bruley and J.C. LaManna
Volume 702	RNA EXOSOME Torben Heick Jensen
Volume 703	INFLAMMATION AND RETINAL DISEASE John D. Lambris and Anthony P. Adamis
Volume 704	TRANSIENT RECEPTOR POTENTIAL CHANNELS Md. Shahidul Islam

---

A Continuation Order Plan is available for this series. A continuation order will bring delivery of each new volume immediately upon publication. Volumes are billed only upon actual shipment. For further information please contact the publisher.

Md. Shahidul Islam
Editor

Transient Receptor Potential Channels

Editor
Md. Shahidul Islam
Karolinska Institutet
Department of Clinical Sciences
 and Education, Södersjukhuset
SE-118 83 Stockholm
Sweden

and

Uppsala University Hospital
AR Division
Uppsala, Sweden
shaisl@ki.se

Additional material to this book can be downloaded from http://extras.springer.com

ISSN 0065-2598
ISBN 978-94-007-0264-6   e-ISBN 978-94-007-0265-3
DOI 10.1007/978-94-007-0265-3
Springer Dordrecht Heidelberg London New York

© Springer Science+Business Media B.V. 2011
No part of this work may be reproduced, stored in a retrieval system, or transmitted in any form or by any means, electronic, mechanical, photocopying, microfilming, recording or otherwise, without written permission from the Publisher, with the exception of any material supplied specifically for the purpose of being entered and executed on a computer system, for exclusive use by the purchaser of the work.

Printed on acid-free paper

Springer is part of Springer Science+Business Media (www.springer.com)

*Dedicated to the living memory of my mother
Sayera Khatun*

Preface

During a conference titled "TRP channels, from sensory signaling to human disease", held at the Karolinska Institute, Stockholm, Sweden, on 26th and 27th September, 2009, I was contacted by Springer to publish the proceedings of the conference. After some discussion with some of the speakers, I understood that that was not going to happen. In stead, we were happy to publish a short meeting report [1]. I thought, the excitement and the momentum that resulted from the conference could be utilized in compiling a substantial book rather than a modest conference proceeding. The idea for a TRP book appeared very timely. This field of research has progressed fast and a few books on the TRP channels that have been published before have become outdated. My immediate concern was whether I would have enough time for editing another book. From a previous book "The Islets of Langerhans" (http://isletbook.islets.se), I knew that for completing a book, it requires a lot more time and energy than one anticipates at the onset [2]. But my real fear was whether I am the most appropriate person to edit a book on the TRP channels. After all, it is a vast and expanding field dominated by a handful of eminent electrophysiologists and biophysicists. When it comes to the TRP channels, I am at best an enthusiast and by no means an expert. I tried to adopt co-editor(s) but the ones I approached were already over committed. My other concern was whether people read books these days as they used to do in the past; books are, after all, less dynamic than journals and on-line publications. It took me some time, to overcome these perplexing thoughts, and then there was only one thing left for me to do, i.e., to take the idea of this new book on TRP channels to completion as fast and as best as possible at any cost.

During the first few weeks, it became pretty obvious to me that many scientists prefer to spend their time in publishing original papers in high-impact journals rather than in writing book chapters especially if they are not paid any remuneration for their contribution. In most academic environments, a short report in a high-impact journal counts more than an extensive and useful chapter in a book, which, often do not have any impact factors. I wrote to many scientists who have published something on TRP channels in any journal. I contacted scientists whom I knew or whom I met personally. In the end, I was rather overwhelmed that so many authors agreed to contribute a chapter in his book. The enthusiasm among the authors was noticeably high. My communication with the authors and the referees was fast, smooth, informal, and very satisfying. All authors finally submitted their respective chapters in time. The only chapter that was delayed was mine, a privilege and a problem of being the editor.

In this book one will find diverse information on the TRP channels starting from some of the essential background information to some of the cutting edge researches, from some of the most established facts to some of the most hotly debated issues of our time, and from the structural biology of the channels to the molecular basis of some human illnesses. But it is by no means an encyclopedia. The emphasis was not on making the book as complete as possible but on making the best use of the competence and interests of the authors who agreed to contribute. Some important topics are missing from the book simply because I could not persuade anyone to contribute on those topics. The authors enjoyed enormous freedom in choosing the contents of their respective chapters and in structuring the chapters as they wished. In some instances, more than one chapter was dedicated to somewhat overlapping topics to ensure that different views of different authors can be accommodated in the same book. Many authors have included their own ideas, views, and speculations which can form the basis for new testable hypotheses for future research. In this book there is something for everyone, both for the beginners and for the experts. But it is important that the readers treat the contents of this book just as starting points, question everything that they read in this book and actively find their own answers through further research.

I am grateful to all the authors and the co-authors, who, in spite of their heavy preoccupation with numerous other activities and deadlines, have worked hard to make their chapters as best as possible within the limited time that they were allotted. When I learnt from several authors that the reasons for delay of their chapters were unexpected personal or family situations or bereavement of a family member, then I paused and reflected; life is not just a bundle of papers. I would like to thank all the referees who have taken time to really read the manuscripts and to come up with very useful comments. The most important thing that I have enjoyed and I have benefited from is the reading of the comments of many referees and the authors' replies to these comments. I wish I could include some of the referees' comments in this book. In spite, of all our efforts, I am worried that the book contains many mistakes that we were not aware of. I will be grateful if readers point out such mistakes and post their comments on the website of the book: http://trpbook.islets.se. This will make the book a bit more dynamic and we all will have opportunity to learn from the mistakes.

I believe we shall all be happy, if this book can further intensify research in the field of the TRP channels in the context of understanding human physiology and pathogenesis of human diseases. Let research in this field confer some of the greatest benefits on mankind. Thanks to Karolinska Institutet that has provided the infrastructure for my academic activities over past two decades. Thanks to Melania Ruiz and Ilse Hansen for handling the practical aspects of handling the chapters and rest of the book. This editorial was written on board a high speed train that symbolizes the fast speed of research in the TRP field.

References

1. Goswami C, Islam MS (2010) Transient receptor potential channels: what is happening? reflections in the wake of the 2009 TRP meeting, Karolinska Institutet, Stockholm. Channels (Austin) 4:124–135
2. Islam MS (2010) The islets of langerhans. Springer, The Netherlands

Md. Shahidul Islam
On board X-2000 between Copenhagen and Stockholm

Contents

1. **Structural Biology of TRP Channels** 1
   Minghui Li, Yong Yu, and Jian Yang

2. **Functional and Structural Studies of TRP Channels Heterologously Expressed in Budding Yeast** 25
   Vera Moiseenkova-Bell and Theodore G. Wensel

3. **Natural Product Ligands of TRP Channels** 41
   Irina Vetter and Richard J. Lewis

4. **Synthetic Modulators of TRP Channel Activity** 87
   Christian Harteneck, Chihab Klose, and Dietmar Krautwurst

5. **Study of TRP Channels by Automated Patch Clamp Systems** . . . 107
   Morten Sunesen and Rasmus B. Jacobsen

6. **TRPC2: Of Mice But Not Men** 125
   Christoffer Löf, Tero Viitanen, Pramod Sukumaran, and Kid Törnquist

7. **TRPM1: New Trends for an Old TRP** 135
   Elena Oancea and Nadine L. Wicks

8. **The Non-selective Monovalent Cationic Channels TRPM4 and TRPM5** 147
   Romain Guinamard, Laurent Sallé, and Christophe Simard

9. **TRPM7, the Mg^{2+} Inhibited Channel and Kinase** 173
   Chris Bates-Withers, Rajan Sah, and David E. Clapham

10. **TRPM8 in Health and Disease: Cold Sensing and Beyond** 185
    Yi Liu and Ning Qin

11. **TRPML1** 209
    Grace A. Colletti and Kirill Kiselyov

12. **TRPML2 and the Evolution of Mucolipins** 221
    Emma N. Flores and Jaime García-Añoveros

13	The TRPML3 Channel: From Gene to Function	229
	Konrad Noben-Trauth	
14	TRPV5 and TRPV6 in Transcellular Ca²⁺ Transport: Regulation, Gene Duplication, and Polymorphisms in African Populations	239
	Ji-Bin Peng	
15	The TRPV5 Promoter as a Tool for Generation of Transgenic Mouse Models	277
	Marlene Vind Hofmeister, Ernst-Martin Füchtbauer, Robert Andrew Fenton, and Jeppe Praetorius	
16	TRPP Channels and Polycystins	287
	Alexis Hofherr and Michael Köttgen	
17	TRP Channels in Yeast	315
	Marta Kaleta and Christopher Palmer	
18	*C. elegans* TRP Channels	323
	Rui Xiao and X.Z. Shawn Xu	
19	Investigations of the In Vivo Requirements of Transient Receptor Potential Ion Channels Using Frog and Zebrafish Model Systems	341
	Robert A. Cornell	
20	TRP Channels in Parasites	359
	Adrian J. Wolstenholme, Sally M. Williamson, and Barbara J. Reaves	
21	Receptor Signaling Integration by TRP Channelsomes	373
	Yasuo Mori, Taketoshi Kajimoto, Akito Nakao, Nobuaki Takahashi, and Shigeki Kiyonaka	
22	Gating Mechanisms of Canonical Transient Receptor Potential Channel Proteins: Role of Phosphoinositols and Diacylglycerol	391
	Anthony P. Albert	
23	The TRPC Ion Channels: Association with Orai1 and STIM1 Proteins and Participation in Capacitative and Non-capacitative Calcium Entry	413
	Gines M. Salido, Isaac Jardín, and Juan A. Rosado	
24	Contribution of TRPC1 and Orai1 to Ca²⁺ Entry Activated by Store Depletion	435
	Kwong Tai Cheng, Hwei Ling Ong, Xibao Liu, and Indu S. Ambudkar	
25	Primary Thermosensory Events in Cells	451
	Ilya Digel	

26	**Thermo-TRP Channels: Biophysics of Polymodal Receptors** .... David Baez-Nieto, Juan Pablo Castillo, Constantino Dragicevic, Osvaldo Alvarez, and Ramon Latorre	469
27	**Complex Regulation of TRPV1 and Related Thermo-TRPs: Implications for Therapeutic Intervention** .... Rosa Planells-Cases, Pierluigi Valente, Antonio Ferrer-Montiel, Feng Qin, and Arpad Szallasi	491
28	**Voltage Sensing in Thermo-TRP Channels** .... Sebastian Brauchi and Patricio Orio	517
29	**TRP Channels as Mediators of Oxidative Stress** .... Barbara A. Miller and Wenyi Zhang	531
30	**Regulation of TRP Signalling by Ion Channel Translocation Between Cell Compartments** .... Alexander C. Cerny and Armin Huber	545
31	**Emerging Roles of Canonical TRP Channels in Neuronal Function** Sunitha Bollimuntha, Senthil Selvaraj, and Brij B. Singh	573
32	**TRP Channels and Neural Persistent Activity** .... Antonio Reboreda, Lydia Jiménez-Díaz, and Juan D. Navarro-López	595
33	**Role of TRP Channels in Pain Sensation** .... Man-Kyo Chung, Sung Jun Jung, and Seog Bae Oh	615
34	**TRPV1: A Therapy Target That Attracts the Pharmaceutical Interests** .... Rong Xia, Kim Dekermendjian, Elke Lullau, and Niek Dekker	637
35	**Expression and Function of TRP Channels in Liver Cells** ...... Grigori Y. Rychkov and Gregory J. Barritt	667
36	**Expression and Physiological Roles of TRP Channels in Smooth Muscle Cells** .... Christelle Guibert, Thomas Ducret, and Jean-Pierre Savineau	687
37	**TRPM Channels in the Vasculature** .... Alexander Zholos, Christopher Johnson, Theodor Burdyga, and Donal Melanaphy	707
38	**Molecular Expression and Functional Role of Canonical Transient Receptor Potential Channels in Airway Smooth Muscle Cells** .... Yong-Xiao Wang and Yun-Min Zheng	731
39	**TRP Channels in Skeletal Muscle: Gene Expression, Function and Implications for Disease** .... Heinrich Brinkmeier	749

40	**TRP Channels in Vascular Endothelial Cells** Ching-On Wong and Xiaoqiang Yao	759
41	**TRP Channels in the Cardiopulmonary Vasculature** Alexander Dietrich and Thomas Gudermann	781
42	**TRP Channels of Islets** Md. Shahidul Islam	811
43	**Multiple Roles for TRPs in the Taste System: Not Your Typical TRPs** Kathryn F. Medler	831
44	**Roles of Transient Receptor Potential Proteins (TRPs) in Epidermal Keratinocytes** Mitsuhiro Denda and Moe Tsutsumi	847
45	**TRP Channels in Urinary Bladder Mechanosensation** Isao Araki	861
46	**The Role of TRP Ion Channels in Testicular Function** Pradeep G. Kumar and Mohammed Shoeb	881
47	**TRP Channels in Female Reproductive Organs and Placenta** Janka Dörr and Claudia Fecher-Trost	909
48	**Oncogenic TRP Channels** V'yacheslav Lehen'kyi and Natalia Prevarskaya	929
49	**TRPV Channels in Tumor Growth and Progression** Giorgio Santoni, Valerio Farfariello, and Consuelo Amantini	947
50	**The Role of Transient Receptor Potential Channels in Respiratory Symptoms and Pathophysiology** M. Allen McAlexander and Thomas Taylor-Clark	969
51	**TRP Channels and Psychiatric Disorders** Loris A. Chahl	987
52	**Transient Receptor Potential Genes and Human Inherited Disease** Kate V. Everett	1011

Erratum to: Transient Receptor Potential Channels E1

Index 1033

Contributors

Anthony P. Albert Division of Basic Medical Sciences, St. George's University of London, London SW17 0RE, UK, aalbert@sgul.ac.uk

Osvaldo Alvarez Facultad de Ciencias, Universidad de Chile, Santiago, Chile, oalvarez@uchile.cl

Consuelo Amantini Section of Experimental Medicine, School of Pharmacy, University of Camerino, 62032 Camerino, Italy, consuelo.amantini@unicam.it

Indu S. Ambudkar Secretory Physiology Section, Molecular Physiology and Therapeutics Branch, NIDCR, NIH, Bethesda, MD 20892, USA, indu.ambudkar@nih.gov

Isao Araki Department of Urology, Interdisciplinary Graduate School of Medicine and Engineering, University of Yamanashi, Chuo, Yamanashi 409-3898, Japan; Department of Urology, Shiga University of Medical Science, Otsu, Shiga 520-2192, Japan, iaraki@yamanashi.ac.jp; iaraki@belle.shiga-med.ac.jp

Gregory J. Barritt Medical Biochemistry, School of Medicine, Flinders University, Adelaide, SA 5001, Australia, greg.barritt@flinders.edu.au

David Baez-Nieto Centro Interdisciplinario de Neurociencias de Valparaíso, Facultad de Ciencias, Universidad de Valparaíso, Valparaíso, Chile, monobolico@gmail.com

Chris Bates-Withers Department of Cardiology, Howard Hughes Medical Institute, Manton Center for Orphan Disease, Children's Hospital Boston, Boston, MA 02115, USA, cbateswithers@gmail.com

Sunitha Bollimuntha Department of Biochemistry and Molecular Biology, School of Medicine and Health Sciences, University of North Dakota, Grand Forks, ND 58201, USA, sunitha.bollimuntha@med.und.edu

Sebastian Brauchi Facultad de Medicina, Instituto de Fisiologia, Universidad Austral de Chile, Valdivia 511-0566, Chile, sbrauchi@docentes.uach.cl

Heinrich Brinkmeier Institute of Pathophysiology, University of Greifswald, D-17495 Karlsburg, Germany, heinrich.brinkmeier@uni-greifswald.de

Theodor Burdyga Department of Physiology, University of Liverpool, L69 3BX Liverpool, UK, burdyga@liverpool.ac.uk

Juan Pablo Castillo Centro Interdisciplinario de Neurociencias de Valparaíso, Facultad de Ciencias, Universidad de Valparaíso, Valparaíso, Chile; Facultad de Ciencias, Universidad de Chile, Santiago, Chile, jp.castillo@cnv.cl

Alexander C. Cerny Department of Biosensorics, Institute of Physiology, University of Hohenheim, 70599 Stuttgart, Germany, cerny@uni-hohenheim.de

Loris A. Chahl School of Biomedical Science and Pharmacy, University of Newcastle, Newcastle, NSW 2308, Australia; Schizophrenia Research Institute, Sydney, NSW, Australia, loris.chahl@newcastle.edu.au

Kwong Tai Cheng Secretory Physiology Section, Molecular Physiology and Therapeutics Branch, NIDCR, NIH, Bethesda, MD 20892, USA, chengo@mail.nih.gov

Man-Kyo Chung Department of Neural and Pain Sciences, University of Maryland Dental School, Baltimore, MD, USA, MChung@umaryland.edu

David E. Clapham Department of Cardiology, Howard Hughes Medical Institute, Manton Center for Orphan Disease, Children's Hospital Boston and Harvard University, Boston, MA 02115, USA; Department of Neurobiology, Harvard Medical School, Boston, MA 02115, USA, dclapham@enders.tch.harvard.edu

Grace A. Colletti Department of Biological Sciences, University of Pittsburgh, Pittsburgh, PA 15260, USA, gac13@pitt.edu

Robert A. Cornell Department of Anatomy and Cell Biology, Carver College of Medicine, University of Iowa, Iowa City, IA 52242, USA, robert-cornell@uiowa.edu

Kim Dekermendjian Department of Neuroscience, RA CNS and Pain Control, AstraZeneca R&D, Södertälje, Sweden, kim.dekermendjian@gmail.com

Niek Dekker DECS, Cell, Protein, and Structure Sciences, AstraZeneca R&D, Mólndal SE-43183, Sweden, Niek.dekker@astrazeneca.com

Mitsuhiro Denda Shiseido Research Center, Yokohama, Kanagawa 236-8643, Japan, mitsuhiro.denda@to.shiseido.co.jp

Alexander Dietrich Walther-Straub-Institute for Pharmacology and Toxicology, School of Medicine, Ludwig-Maximilians-University München, 80336 Munich, Germany, alexander.dietrich@lrz.uni-muenchen.de

Ilya Digel Laboratory of Cellular Biophysics, Aachen University of Applied Sciences, Juelich, Germany, digel@fh-aachen.de

Contributors

Janka Dörr Proteinfunktion Proteomics, Fachbereich Biologie, TU Kaiserslautern, D-67663 Kaiserslautern, Germany, Janka.Doerr@uniklinikum-saarland.de; jadoerr@rhrk.uni-kl.de

Constantino Dragicevic Centro Interdisciplinario de Neurociencias de Valparaíso, Facultad de Ciencias, Valparaíso, Chile, cdragicevic@gmail.com

Thomas Ducret Université Victor Segalen Bordeaux2, 33076 Bordeaux Cedex, France; INSERM U 885, 33076 Bordeaux Cedex, France, thomas.ducret@u-bordeaux2.fr

Kate V. Everett St. George's University of London, London, UK, keverett@sgul.ac.uk

Valerio Farfariello Department of Molecular Medicine, Sapienza University of Rome, 10161 Rome, Italy, valerio.farfariello@uniroma1.it

Claudia Fecher-Trost Proteinfunktion Proteomics, Fachbereich Biologie, TU Kaiserslautern, D-67663 Kaiserslautern, Germany, claudia.fecher-trost@uks.eu

Robert Andrew Fenton Department of Anatomy, Water and Salt Research Center, Aarhus University, DK-8000 Aarhus, Denmark, rofe@ana.au.dk

Antonio Ferrer-Montiel Instituto de Biología Molecular y Celular, Universidad Miguel Hernández, Elche, Spain, aferrer@umh.es

Emma N. Flores Departments of Anesthesiology, Northwestern University Institute for Neuroscience, Chicago, IL 60611, USA, enflores@nothwestern.edu

Ernst-Martin Füchtbauer Department of Molecular Biology, Aarhus University, DK-8000 Aarhus, Denmark, emf@mb.au.dk

Jaime García–Añoveros Departments of Anesthesiology, Physiology and Neurology, Northwestern University Institute for Neuroscience, Chicago, IL 60611, USA; The Hugh Knowles Center for Clinical and Basic Science in Hearing and Its Disorders, Northwestern University, Chicago, IL 60611, USA, anoveros@northwestern.edu

Thomas Gudermann Walther-Straub-Institute for Pharmacology and Toxicology, School of Medicine, Ludwig-Maximilians-University München, Munich, Germany, thomas.gudermann@lrz.uni-muenchen.de

Christelle Guibert Université Victor Segalen Bordeaux2, 33076 Bordeaux Cedex, France; INSERM U 885, 33076 Bordeaux Cedex, France, christelle.guibert@u-bordeaux2.fr

Romain Guinamard Groupe Cœur et Ischémie, EA 3212, Université de Caen, Sciences D, F-14032, Caen Cedex, France, romain.guinamard@unicaen.fr

Christian Harteneck Institute for Pharmacology and Toxicology, Interfaculty Center of Pharmacogenomics and Pharmaceutical Research (ICEPHA), Eberhard-Karls-University, Tübingen, Germany, Christian.Harteneck@uni-tuebingen.de

Alexis Hofherr Renal Division, Department of Medicine, University Medical Centre Freiburg, 79106 Freiburg, Germany; Spemann Graduate School of Biology and Medicine (SGBM), Albert-Ludwigs-University, Freiburg, Germany; Faculty of Biology, Albert-Ludwigs-University, Freiburg, Germany, alexis.hofherr@uniklinik-freiburg.de

Marlene Vind Hofmeister Department of Anatomy, Water and Salt Research Center, Aarhus University, DK-8000 Aarhus, Denmark, mvho@ana.au.dk

Armin Huber Department of Biosensorics, Institute of Physiology, University of Hohenheim, 70599 Stuttgart, Germany, armin.huber@uni-hohenheim.de

Md. Shahidul Islam Karolinska Institutet, Department of Clinical Sciences and Education, Södersjukhuset, SE-118 83 Stockholm, Sweden; Uppsala University Hospital, AR Division, Uppsala, Sweden, shahidul.islam@ki.se

Rasmus B. Jacobsen Sophion Bioscience AS, 2750 Ballerup, Denmark, rbj@sophion.com

Isaac Jardín Cell Physiology Group, Department of Physiology, University of Extremadura, Cáceres, Spain, ijp@unex.es

Lydia Jiménez-Dıaz Instituto Neurociencias F. Oloriz, Universidad de Granada, 18071 Granada, Spain; Department of Physiology, School of Medicine, Universidad de Granada, 18071 Granada, Spain, ljimdia@yahoo.es

Christopher Johnson Centre for Biomedical Science Education, School of Medicine, Dentistry and Biomedical Sciences, Queen's University of Belfast, Belfast BT9 7BL, UK, c.johnson@qub.ac.uk

Sung Jun Jung Department of Physiology College of Medicine, Hanyang University, Seoul 133-791, Republic of Korea, eurijj@naver.com

Taketoshi Kajimoto Department of Synthetic Chemistry and Biological Chemistry, Graduate School of Engineering, Kyoto University, Kyoto 615-8510, Japan, tkajimoto@sbchem.kyoto-u.ac.jp

Marta Kaleta Faculty of Life Sciences, London Metropolitan University, London N7 8DB, UK, redds@tlen.pl

Kirill Kiselyov Department of Biological Sciences, University of Pittsburgh, Pittsburgh, PA 15260, USA, kiselyov@pitt.edu

Shigeki Kiyonaka Department of Synthetic Chemistry and Biological Chemistry, Graduate School of Engineering, Kyoto University, Kyoto 615-8510, Japan, kiyonaka@sbchem.kyoto-u.ac.jp

Contributors xix

Chihab Klose Institute for Pharmacology and Toxicology, Interfaculty Center of Pharmacogenomics and Pharmaceutical Research (ICEPHA), Eberhard-Karls-University, Tübingen, Germany, chihab.klose@medizin.uni-tuebingen.de

Michael Köttgen Renal Division, Department of Medicine, University Medical Centre Freiburg, 79106 Freiburg, Germany, michael.koettgen@uniklinik-freiburg.de

Dietmar Krautwurst Deutsche Forschungsanstalt für Lebensmittelchemie, Molekulare Zellphysiology und Chemorezeption, 85354 Freising, Germany, dietmar.krautwurst@lrz.tum.de

Pradeep G. Kumar Division of Molecular Reproduction, Rajiv Gandhi Centre for Biotechnology, Thiruvananthapuram 695014, Kerela, India, kumarp@rgcb.res.in

Ramon Latorre Centro Interdisciplinario de Neurociencias de Valparaíso, Facultad de Ciencias, Universidad de Valparaíso, Valparaíso, Chile, ramon.latorre@uv.cl

V'yacheslav Lehen'kyi Department of Molecular Medicine, Sapienza University of Rome, 10161 Rome, Italy, vyacheslav.lehenkyi@univ-lille1.fn

Richard J. Lewis Institute for Molecular Bioscience, The University of Queensland, Queensland 4072, Australia, r.lewis@imb.uq.edu.au

Minghui Li Department of Biological Sciences, Columbia University, New York, NY 10027, USA, ml2311@columbia.edu

Yi Liu Johnson & Johnson Pharmaceutical Research and Development, LLC, San Diego, CA 92121, USA, yliu10@its.jnj.com

Xibao Liu Secretory Physiology Section, Molecular Physiology and Therapeutics Branch, NIDCR, NIH, Bethesda, MD 20892, USA, xiliu@mail.nih.gov

Christoffer Löf Department of Biosciences, Åbo Akademi University, 20520 Turku, Finland; Turku Graduate School of Biomedical Sciences, 20520 Turku, Finland, clof@abo.fi

Elke Lullau DECS, Cell, Protein, and Structure Sciences, AstraZeneca R&D, Mölandal SE-43183, Sweden, Elke.Lullau@astrazeneca.com

M. Allen McAlexander Respiratory Discovery Biology, GlaxoSmithKline Pharmaceuticals, King of Prussia, PA, USA, michael.a.mcalexander@gsk.com

Kathryn F. Medler Department of Biological Sciences, University at Buffalo, The State University of New York, Buffalo, NY 14260, USA, kmedler@buffalo.edu

Donal Melanaphy Centre for Vision and Vascular Science, School of Medicine, Dentistry and Biomedical Sciences, Royal Victoria Hospital, Queen's University of Belfast, Belfast BT12 6BA, UK, a.zholos@qub.ac.uk

Barbara A. Miller Departments of Pediatrics and Biochemistry and Molecular Biology, Milton S. Hershey Medical Center, Penn State Hershey Children's Hospital, Pennsylvania State University College of Medicine, Hershey, PA 17033, USA, bmiller3@psu.edu

Vera Moiseenkova-Bell Department of Pharmacology, Case Western Reserve University School of Medicine, Cleveland, OH 44106, USA, vxm102@case.edu

Yasuo Mori Department of Synthetic Chemistry and Biological Chemistry, Graduate School of Engineering, Kyoto University, Kyoto 615-8510, Japan, mori@sbchem.kyoto-u.ac.jp

Akito Nakao Department of Synthetic Chemistry and Biological Chemistry, Graduate School of Engineering, Kyoto University, Kyoto 615-8510, Japan, nakaoakito@t2005.mbox.media.kyoto-u.ac.jp

Juan D. Navarro-López Instituto Neurociencias F. Oloriz, Universidad de Granada, 18071 Granada, Spain; Department of Physiology, School of Medicine, Universidad de Granada, 18071 Granada, Spain, jdnavarro@ugr.es

Konrad Noben-Trauth Section on Neurogenetics, Laboratory of Molecular Biology, National Institute on Deafness and Other Communication Disorders, Rockville, MD 20850, USA, nobentk@nidcd.nih.gov

Elena Oancea Department of Molecular Pharmacology, Physiology and Biotechnology, Brown University, Providence, RI, USA, Elena_Oancea@brown.edu

Seog Bae Oh Department of Neurobiology and Physiology, School of Dentistry Seoul National University, Seoul 110-749, Republic of Korea, odolbae@snu.ac.kr

Hwei Ling Ong Secretory Physiology Section, Molecular Physiology and Therapeutics Branch, NIDCR, NIH, Bethesda, MD 20892, USA, ongh@mail.nih.gov

Patricio Orio Centro Interdisciplinario de Neurociencia de Valparaíso (CINV), Facultad de Ciencias, Universidad de Valparaíso, Valparaíso, Chile, patricio.orio@uv.cl

Christopher Palmer Faculty of Life Sciences, London Metropolitan University, London N7 8DB, UK, chris.palmer@londonmet.ac.uk

Ji-Bin Peng Division of Nephrology, Department of Medicine, Nephrology Research and Training Center, University of Alabama at Birmingham, Birmingham, AL 35294, USA, jpeng@uab.edu

Rosa Planells-Cases Centro de Investigación Príncipe Felipe, Valencia, Spain, rplanells@cipf.es

Jeppe Praetorius Department of Anatomy, Water and Salt Research Center, Aarhus University, DK-8000 Aarhus, Denmark, jp@ana.au.dk

Natalia Prevarskaya Inserm, U-800, Equipe labellisée par la Ligue Nationale contre le cancer, Villeneuve d'Ascq F-59655, France; Université des Sciences et Technologies de Lille (USTL), Villeneuve d'Ascq F-59655, France; Laboratoire de Physiologie Cellulaire, INSERM U1003, USTL, Villeneuve d'Ascq Cedex F-59655, France, Natacha.Prevarskaya@univ-lille1.fr

Ning Qin Johnson & Johnson Pharmaceutical Research and Development, LLC, San Diego, CA 92121, USA, NQin@its.jnj.com

Feng Qin Department of Physiology and Biophysical Sciences, State University of New York at Buffalo, Buffalo, NY, USA, qin@buffalo.edu

Barbara J. Reaves Department of Infectious Diseases, University of Georgia, Athens, GA 30602, USA, bjreaves@uga.edu

Antonio Reboreda Section of Physiology, Department of Functional Biology and Health Sciences, School of Biology, University of Vigo, Campus Lagoas-Marcosende 36310 Vigo (Pontevedra), Spain, areboreda@uvigo.es

Juan A. Rosado Cell Physiology Group, Department of Physiology, University of Extremadura, Cáceres, Spain, jarosado@unex.es

Grigori Y. Rychkov Department of Physiology, University of Adelaide, Adelaide, SA 5001, Australia, grigori.rychkov@adelaide.edu.au

Rajan Sah Department of Medicine, Cardiology, Brigham and Women's Hospital, Boston, MA 02115, USA, rsah@partners.org

Gines M. Salido Cell Physiology Group, Department of Physiology, University of Extremadura, Cáceres, Spain, gsalido@unex.es

Laurent Sallé Groupe Cœur et Ischémie, EA 3212, Université de Caen, F-14032, Caen Cedex, France, laurent.salle@unicaen.fr

Giorgio Santoni Section of Experimental Medicine, School of Pharmacy, University of Camerino, 62032 Camerino, Italy, giorgio.santoni@unicam.it

Jean-Pierre Savineau Université Victor Segalen Bordeaux2, 33076 Bordeaux Cedex, France; INSERM U 885, 33076 Bordeaux Cedex, France, jean-pierre.savineau@u-bordeaux2.fr

Senthil Selvaraj Department of Biochemistry and Molecular Biology, School of Medicine and Health Sciences, University of North Dakota, Grand Forks, ND 58201, USA, senthil.selvaraj@med.und.edu

Mohammed Shoeb Division of Molecular Reproduction, Rajiv Gandhi Centre for Biotechnology, Thiruvananthapuram, 695014 Kerela, India, shoeb@rgcb.res.in

Christophe Simard Groupe Cœur et Ischémie, EA 3212, Université de Caen, F-14032, Caen Cedex, France, christophe.simard@unicaen.fr

Brij B. Singh Department of Biochemistry and Molecular Biology, School of Medicine and Health Sciences, University of North Dakota, Grand Forks, ND 58201, USA, bsingh@medicine.nodak.edu

Pramod Sukumaran Department of Biosciences, Åbo Akademi University, 20520 Turku, Finland, psukumar@abo.fi

Morten Sunesen Sophion Bioscience AS, 2750 Ballerup, Denmark, msu@sophion.com

Arpad Szallasi Department of Pathology, Monmouth Medical Center, Long Branch, NJ 07740, USA, ASZALLASI@SBHCS.COM

Nobuaki Takahashi Department of Synthetic Chemistry and Biological Chemistry, Graduate School of Engineering, Kyoto University, Kyoto 615-8510, Japan, takahashinobuaki@t02.mbox.media.kyoto-u.ac.jp

Thomas Taylor-Clark Department of Molecular Pharmacology and Physiology, University of South Florida, Tampa, FL, USA, ttaylorc@health.usf.edu

Kid Törnquist Department of Biosciences, Åbo Akademi University, Artillerigatan 6, 20520 Turku, Finland; The Minerva Foundation Institute for Medical Research, 00290 Helsinki, Finland, ktornqvi@abo.fi

Moe Tsutsumi Shiseido Research Center, Yokohama, Kanagawa 236-8643, Japan, moe.tsutsumi@to.shiseido.co.jp

Pierluigi Valente Instituto de Biología Molecular y Celular, Universidad Miguel Hernández, Elche, Spain; Department of Neuroscience and Brain Technologies, Italian Institute of Technology – IIT, Genova, Italy, Pierluigi.Valente@iit.it

Irina Vetter Institute for Molecular Bioscience, The University of Queensland, Queensland 4072, Australia, i.vetter@uq.edu.au

Tero Viitanen The Minerva Foundation Institute for Medical Research, 00290 Helsinki, Finland, tero.viitanen@helsinki.fi

Yong-Xiao Wang Center for Cardiovascular Sciences, Albany Medical College, Albany, NY 12208, USA, wangy@mail.amc.edu

Theodore G. Wensel Verna and Marrs McLean Department of Biochemistry and Molecular Biology, Baylor College of Medicine, Houston, TX 77030, USA, twensel@bcm.edu

Nadine L. Wicks Department of Molecular Pharmacology, Physiology and Biotechnology, Brown University, Providence, RI, USA, Nadine_Wicks@brown.edu

Sally M. Williamson Department of Infectious Diseases, University of Georgia, Athens, GA 30602, USA, sallymw@uga.edu

Adrian J. Wolstenholme Department of Infectious Diseases, University of Georgia, Athens, GA 30602, USA; Center for Tropical and Emerging Global Disease, University of Georgia, Athens, GA 30602, USA, adrianw@uga.edu

Ching-On Wong Li Ka Shing Institute of Health Sciences and School of Biomedical Sciences, The Chinese University of Hong Kong, Hong Kong, China, chingon.wong@gmail.com

Rong Xia DECS, Cell, Protein, and Structure Sciences, AstraZeneca R&D, Mölndal SE-43183, Sweden, Rong.Xia@astrazeneca.com

Rui Xiao Department of Molecular and Integrative Physiology, Life Sciences Institute, University of Michigan, Ann Arbor, MI 48109, USA, rxiao@umich.edu

X.Z. Shawn Xu Department of Molecular and Integrative Physiology, Life Sciences Institute, University of Michigan, Ann Arbor, MI 48109, USA, shawnxu@umich.edu

Jian Yang Department of Biological Sciences, Columbia University, New York, NY 10027, USA, jy160@columbia.edu

Xiaoqiang Yao Li Ka Shing Institute of Health Sciences and School of Biomedical Sciences, The Chinese University of Hong Kong, Hong Kong, China, yao2068@cuhk.edu.hk

Yong Yu Department of Biological Sciences, Columbia University, New York, NY10027, USA, yy2024@columbia.edu

Wenyi Zhang Departments of Pediatrics, Milton S. Hershey Medical Center, Penn State Hershey Children's Hospital, Pennsylvania State University College of Medicine, Hershey, PA 17033, USA, wzhang1@hmc.psu.edu

Yun-Min Zheng Center for Cardiovascular Sciences, Albany Medical College, Albany, NY 12208, USA, zhengy@mail.amc.edu

Alexander Zholos Centre for Vision and Vascular Science, School of Medicine, Dentistry and Biomedical Sciences, Royal Victoria Hospital, Queen's University of Belfast, Belfast BT12 6BA, UK, a.zholos@qub.ac.uk

Chapter 1
Structural Biology of TRP Channels

Minghui Li, Yong Yu, and Jian Yang

Abstract Structural studies on TRP channels, while limited, are poised for a quickened pace and rapid expansion. As of yet, no high-resolution structure of a full length TRP channel exists, but low-resolution electron cryomicroscopy structures have been obtained for 4 TRP channels, and high-resolution NMR and X-ray crystal structures have been obtained for the cytoplasmic domains, including an atypical protein kinase domain, ankyrin repeats, coiled coil domains and a Ca^{2+}-binding domain, of 6 TRP channels. These structures enhance our understanding of TRP channel assembly and regulation. Continued technical advances in structural approaches promise a bright outlook for TRP channel structural biology.

1.1 Introduction

Full understanding of ion channel function requires high-resolution three-dimensional (3D) structures. Structural studies on ion channels entered a new phase in 1998 after the publication of the crystal structure of the bacterial K^+ channel, KcsA [1]. Since then, there has been a rapid growth in the number of ion channel structures. To date, there are ~90 crystal structures of full length or near full length ion channels, ~50 electron microscopy structures of full length or near full length ion channels, and ~130 crystal and nuclear magnetic resonance (NMR) structures of ion channel fragments. These structures have led to a quantum leap in our understanding of the molecular and biophysical mechanisms of ion channel assembly, selectivity, conduction, gating and regulation.

TRP channels constitute a distinct superfamily of ion channels and are distantly related to voltage-gated K^+, Na^+ and Ca^{2+} superfamilies. They are expressed and function in diverse organisms, including yeasts, worms, fruit flies, mice and humans. Excluding yeast TRPs, there are seven subfamilies: TRPC, TRPV, TRPM, TRPA,

J. Yang (✉)
Department of Biological Sciences, Columbia University, New York, NY 10027, USA
e-mail: jy160@columbia.edu

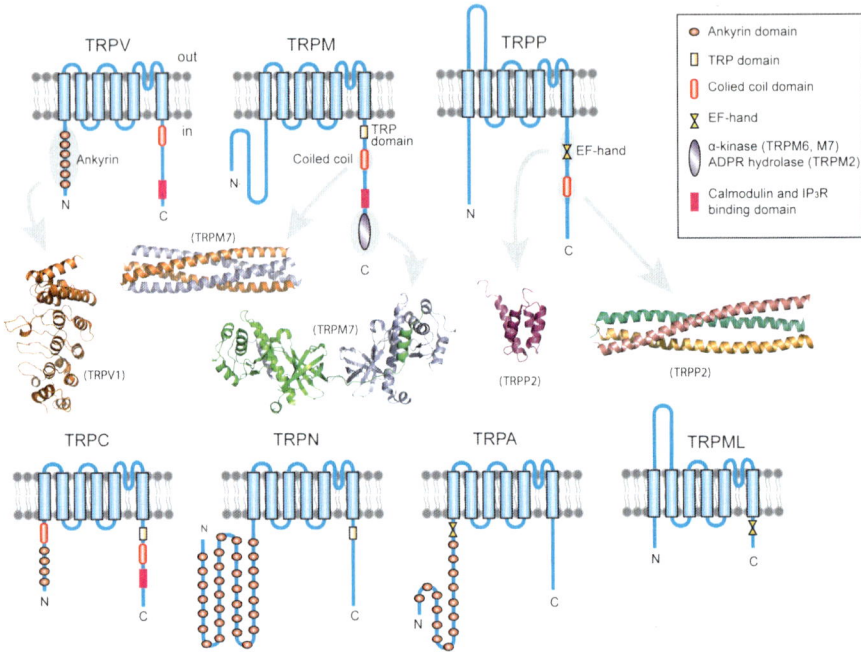

Fig. 1.1 TRP channel subfamilies and the transmembrane topology and domain organization of their subunits. Only commonly present and readily identifiable domains or motifs in the cytoplasmic N and C termini are indicated. Examples of high-resolution structures of some domains or motifs are presented

TRPN, TRPP and TRPML, with TRPN absent in mice and humans (Fig. 1.1) [2]. Each subfamily has one or more members. Mice have a total of 28 different members, and humans 27. All TRP channel subunits have six putative transmembrane segments and a pore-forming loop between the last two transmembrane segments (Fig. 1.1). The amino (N) and carboxyl (C) termini are located intracellularly and vary vastly in length (Table 1.1) and amino acid (aa) sequence. These cytoplasmic regions contain various well-recognized domains and motifs that are likely involved in channel assembly, activation and regulation through protein–protein and/or protein–ligand interactions (Fig. 1.1).

All TRP channels are cation selective, with some being highly selective for Ca^{2+} or Mg^{2+} [2]. In accord with their amino acid sequence diversity, TRP channels exhibit varied activation and modulatory mechanisms, such as stimulation of G protein coupled receptors, extracellular and intracellular ligands (including H^+, Ca^{2+} and Mg^{2+}), phosphoinositide-4,5pbisphosphate (PIP_2), temperature, and mechanical stretch [2]. To fully understand TRP channel diversity, function and regulation, it is necessary to gain structural information on different types of TRP channels.

Of the existing ion channel structures, most come from K^+ channels. This is due, in part, to their vast variety and their existence in bacteria, which make them more tractable to structural approaches, especially X-ray crystallography, because they

1 Structural Biology of TRP Channels

Table 1.1 Predicted region and length of the cytoplasmic N and C termini of TRP channel subunits and the number of low-complexity residues in these regions

Protein	N terminus			C terminus		
	Channel region	# of residues	# of low-complexity residues	Channel region	# of residues	# of low-complexity residues
TRPC1	1–316	316	14	610–759	150	0
TRPC2	1–626	626	92	918–1,172	255	82
TRPC3	1–351	351	0	671–848	178	0
TRPC4	1–327	327	38	618–977	360	21
TRPC5	1–327	327	41	622–973	352	56
TRPC6	1–404	404	22	726–931	206	7
TRPC7	1–351	351	11	671–862	192	0
TRPV1	1–433	433	0	681–839	159	0
TRPV2	1–390	390	0	645–764	120	0
TRPV3	1–438	438	56	675–790	116	0
TRPV4	1–468	468	26	716–871	156	0
TRPV5	1–326	326	0	577–729	153	22
TRPV6	1–326	326	16	577–725	149	10
TRPM1	1–760	760	84	1,053–1,533	481	27
TRPM2	1–750	750	40	1,046–1,503	458	26
TRPM3	1–716	716	59	955–1,554	600	34
TRPM4	1–687	687	43	1,041–1,214	174	34
TRPM5	1–643	643	0	975–1,158	184	0
TRPM6	1–742	742	15	1,075–2,022	948	34
TRPM7	1–756	756	15	1,102–1,864	763	13
TRPM8	1–692	692	16	977–1,104	128	24
TRPML1	1–69	69	12	518–580	63	13
TRPML2	1–61	61	0	508–566	59	0
TRPML3	1–66	66	13	503–553	51	0
TRPP2	1–224	224	99	681–968	288	87
TRPP3	1–104	104	15	558–805	248	22
TRPP5	1–33	33	0	492–613	122	12
TRPA1	1–717	717	0	962–1,119	158	0

All amino acid sequences are from humans except TRPC2, which is from mice, as human *TRPC2* is a pseudogene. Transmembrane helices were predicted using the TMHMM Server v. 2.0 at http://www.cbs.dtu.dk/services/TMHMM/. Low-complexity sequences were predicted using the program SEG [80] with the default settings.

can be more abundantly expressed, are more stable, and hence, are more amicable to purification and crystallization. TRP channels, however, are not endogenously expressed in bacteria. This is perhaps a major contributing factor in the present lack of even a single high-resolution structure of any full length TRP channel. Nevertheless, low-resolution structures have been obtained for 4 full length TRP channels by electron microscopy (EM). Meanwhile, X-ray crystallography and NMR spectroscopy have been employed effectively to garner high-resolution structures of functionally important cytosolic domains of 6 TRP channels (Table 1.2). This chapter describes the existing TRP channel structures and, when available,

Table 1.2 High-resolution structures of TRP channel fragments

Structural description	Channel region	Species	Resolution	Method	PDB code	References
TRPM7 α-kinase	1,549–1,828	Mouse	2.8 Å	X-ray crystallography	1IAJ	[27]
TRPM7 α-kinase, with AMP·PNP	1,549–1,828	Mouse	2.0 Å	X-ray crystallography	1IA9	[27]
TRPM7 α-kinase, with ADP	1,549–1,828	Mouse	2.4 Å	X-ray crystallography	1IAH	[27]
TRPV1 ankyrin repeats	101–364	Rat	2.7 Å	X-ray crystallography	2PNN	[39]
TRPV2 ankyrin repeats	75–326	Rat	1.65 Å	X-ray crystallography	2ETB	[37]
TRPV2 ankyrin repeats	69–319	Human	1.7 Å	X-ray crystallography	2F37	[40]
TRPV4 ankyrin repeats	133–382	Chicken	2.3 Å	X-ray crystallography	3JXI	[38]
TRPV6 ankyrin repeats	44–265	Mouse	1.7 Å	X-ray crystallography	2RFA	[41]
TRPM7 coiled coil	1,230–1,282	Rat	2.01 Å	X-ray crystallography	3E7K	[57]
TRPP2 coiled coil, long	833–895	Human	1.9 Å	X-ray crystallography	3HRN	[58]
TRPP2 coiled coil, short	833–872	Human	1.9 Å	X-ray crystallography	3HRO	[58]
TRPP2 E-F hand	724–796	Human		NMR	2KLE	[74]
TRPP2 E-F hand	720–797	Human		NMR	2KQ6	[75]

the mechanistic insights they provide, beginning with a brief overview of structural approaches and considerations. Advances in TRP channel structural biology have been covered in several recent reviews [3–7].

1.2 Structure-Determination Methods and Considerations

When examining the structure of a protein or a protein complex, the first and foremost concern is its resolution. At nanometer-resolutions, certain general features of the protein can be ascertained, including its shape, dimension, subunit stoichiometry and domain organization (Fig. 1.2a). At 4- to 9-Å resolutions, secondary structures can be discerned (Fig. 1.2b). At resolutions below 3.7 Å, amino acid side-chains can be visualized and assigned – the higher the resolution, the higher the precision and confidence (Fig. 1.2c). For example, aromatic side-chains can be identified at 3.5 Å, and individual atoms can be resolved at 1.5 Å [8].

Three methods are commonly used to determine protein 3D structures – electron cryomicroscopy (cryo-EM), NMR spectroscopy and X-ray crystallography. These methods have different applications, advantages and disadvantages, especially when applied to integral membrane proteins.

Cryo-EM can be used to determine the structure of proteins of various shapes, forms and sizes [9–11]. It is particularly useful for proteins that are too large or too difficult for NMR and X-ray crystallography. Moreover, cryo-EM can probe proteins in their native lipid environment. Cryo-EM can be used to visualize proteins in two-dimensional (2D) sheets or helices or in non-crystal forms. The resolution of single-particle cryo-EM, the most widely used cryo-EM method, generally ranges from 30 to ~6 Å, depending on the quality of protein preparation, protein symmetry,

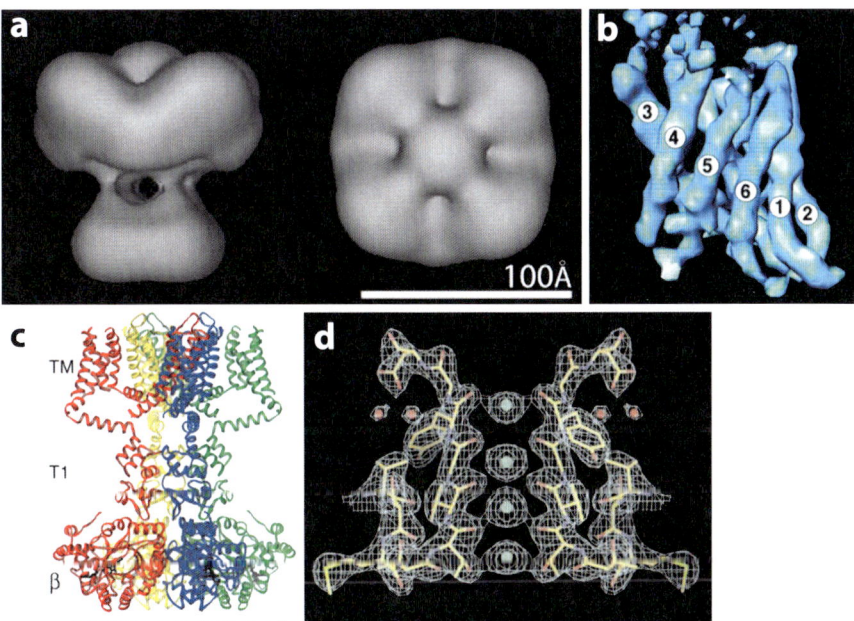

Fig. 1.2 Examples of membrane protein structures at different resolutions. (**a**) Side view (*left*) and top view (*right*) of a cryo-EM structure of the *Drosophila* Shaker K$^+$ channel at 25 Å resolution, revealing a fourfold symmetry and a two-layered architecture [76]. (**b**) Side view of the structure of a monomer of aquaporin 1 obtained by 2D cryo-EM at 6 Å resolution, revealing 6 distinct tilted rods that correspond to membrane-spanning α helices [77]. (**c**) X-ray crystal structure of the rat K$_v$1.2 channel at 2.9 Å resolution (*left*, PDB code 2A79) [78] and (**d**) The electron density map and side chain assignment of the ion selectivity filter of a rat K$_v$1.2–K$_v$2.1 chimeric channel at 2.4 Å resolution (*right*, PDB code 2R9R) [79]

sample size, data processing, and reconstruction. Near atomic resolution can be obtained for highly symmetrical complexes (see e.g., [12]). With 2D crystals, cryo-EM can achieve atomic resolution. For example, the structure of aquaporin-0 in double-layered 2D crystals has been determined at 1.9 Å [13], the highest resolution protein structure solved to date by cryo-EM.

Both NMR and X-ray crystallography allow the determination of protein structures at atomic resolutions. NMR is mainly applicable to relatively small proteins or protein fragments, usually less than 25 kDa, for structural determination, though technical advances allow proteins of up to 900 kDa to be studied [14]. Also, both soluble and membrane proteins can be examined [14, 15]. For partially or wholly unstructured proteins or protein fragments that are resistant to crystallization, NMR is often the only method for structural determination.

X-ray crystallography is by far the most widely used and most effective structure-determination method. As of March 2010, ~86% of the protein structures deposited in the Protein Data Bank and ~88% of the ion channel structures (full length and fragments) are solved by X-ray crystallography. The number of unique structures of membrane proteins solved by X-ray crystallography has been increasing exponentially, from a total of 25 in 1998 when the KcsA structure was published to 212 in 2009. Despite its power, X-ray crystallography has limitations, especially when applied to membrane proteins. Major challenges include maintaining the protein in a soluble form and in its native oligomeric state, crystallizing the protein, and achieving atomic resolution.

An important consideration in protein structure determination is the expression system. Four types of cells have been routinely employed to overexpress membrane proteins: bacteria (*Escherichia coli*), yeast (*Saccharomyces cerevisiae* and *Pichia pastoris*), insect cells (Sf9 cells), and mammalian cells (HEK293 cells and COS7 cells). Obviously, proteins that are endogenously expressed in bacteria are likely to yield better expression in *E. coli*. There are yet no well-defined guiding principles in choosing an expression system for vertebrate membrane proteins. Trial-and-error seems to be the most effective strategy.

Another key consideration is the choice of detergents. Membrane proteins are embedded in lipids and thus require detergents for solublization, purification and crystallization [16, 17]. Nonionic and zwitterionic detergents are generally less harsh on proteins than ionic detergents and have been much more successfully utilized in structural investigation. Commonly used nonionic and zwitterionic detergents include n-decyl-β-D-maltoside (DM), n-dodecyl-β-D-maltoside (DDM), lauryldimethylamine-N-oxide (LDAO), n-octyl-β-D-glucoside (OG), dodecyl octaethylene glycol ether (C12E8), and 3-[(3-cholamidopropyl)-dimethylammonio]-1-propane sulfonate (CHAPS). In general, detergent concentrations should be significantly higher than the critical micelle concentration (CMC), the concentration at which detergent monomers aggregate to form micelles. Sometimes, different detergents are used for solublization and for purification and crystallization. As with choosing the expression system, there is not a set of rules regarding detergent choice and the concentration to be used; they are largely determined empirically.

Yet another critical consideration is whether to work on full length proteins or smaller fragments. From the functional point of view, it is obviously more desirable to obtain the structure of full length proteins. With cryo-EM, this is usually achievable, even for very large proteins. This is, however, not often feasible with X-ray crystallography. To facilitate protein expression, purification, crystallization, and to improve resolution, it is often necessary to remove parts of a protein. Even with such maneuvers, it is still often unattainable to solve the structure of a membrane protein. In such cases, an alternative is to obtain the structure of the soluble domains of the protein. The extracellular and intracellular regions of membrane proteins usually contain functionally important domains and motifs, which often fold into compact and defined structures. These domains and motifs often can be independently expressed, purified and crystallized, and their structures can provide useful insights into the workings of a protein. Still, the extracellular and intracellular regions of ion channel proteins, including TRP channels, often contain low-complexity sequences (Table 1.1), which are generally detrimental to structural determination by both NMR and X-ray crystallography [18]. Thus, even when working with channel fragments, it is usually necessary to trim them further. Indeed, none of the available high-resolution structures of TRP channels comes from a full length N or C terminus (Table 1.2). Finally, it should be cautioned that the structure of an isolated protein fragment may not always represent its structure in the intact protein. Thus, the validity and usefulness of such a structure needs to be tested in the full length protein.

1.3 EM Structures

Low-resolution (15–35 Å) EM structures have been obtained for 4 TRP channels from 3 different subfamilies: TRPM2, TRPC3, TRPV1 and TRPV4 (Fig. 1.3) [19–22]. The structures of the latter 3 channels were determined by cryo-EM, but that of TRPM2 was determined by EM with negative staining. A common feature of all four structures is that they exhibit a fourfold rotational symmetry, consistent with the tetrameric subunit stoichiometry that has been demonstrated for several TRP channels by other methods [23, 24]. Strikingly, while the general structure of TRPV1 and TRPV4 is similar, that of TRPM2 and TRPC3 is markedly different (Fig. 1.3).

The structure of rat TRPV1 was determined by single particle cryo-EM at 19 Å resolution (Fig. 1.3a) [21]. The reconstructed 3D structure stands ~150 Å high and contains two interconnected regions. The small region measures ~60×60 Å, with a height of 40 Å, and accounts for ~30% of the total mass. It likely corresponds to the transmembrane portion of the channel, as suggested by its relative mass and a reasonable fit of the high-resolution structure of the transmembrane domains of the $K_v1.2$ K^+ channel into this region. The large region is shaped like a basket, with a central cavity, and is connected to the small region by 4 bridges. This region, comprising ~70% of the total mass, is ~100 Å wide and 110 Å high and

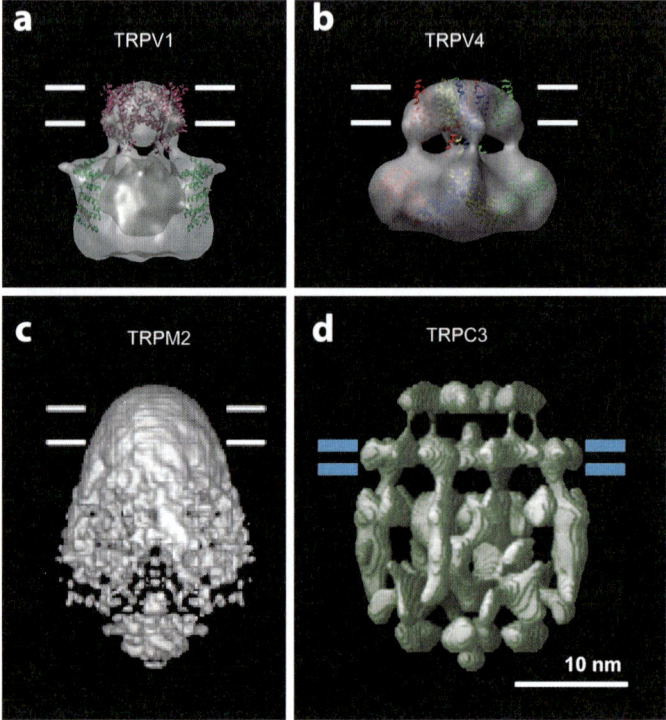

Fig. 1.3 TRP channel EM structures. (**a**) Cryo-EM structure of TRPV1 [21], superimposed with the crystal structure of the $K_v1.2$ transmembrane domains (*maroon*; PDB code 2A79) and of the ankyrin repeat domain of TRPV1 (*green*; PDB code 2PNN). (**b**) Cryo-EM structure of TRPV4 [22], superimposed with the crystal structure of Mlotik1 (*top*; PDB code 3BEH) and of the ankyrin repeat domain of TRPV1 (*bottom*). (**c**) EM structure of TRPM2 with negative staining [19]. (**d**) Cryo-EM structure of TRPC3 [20]. All structures are side-views. The *white lines* mark putative transmembrane regions, so do the *blue lines*, as presented in [20]. The resolutions of all four structures are based on the 0.5 cutoff criterion in the Fourier shell correlation

probably corresponds to cytoplasmic N and C termini. Indeed, the 6 ankyrin repeats present in the N terminus of TRPV1 can be comfortably fitted into this region in the vertical orientation. The functional importance of the vacant central chamber is unknown.

The structure of rat TRPV4, reconstructed to 35 Å resolution, is similar to that of TRPV1 and shares the two-layered general architecture (Fig. 1.3b) [22]. This is consistent with the similar size of the two channels (rat TRPV1 and TRPV4 subunits contain 838 and 871 amino acids, respectively). The small region accounts for 30% of the total volume and has a dimension of ~85 Å. The transmembrane domains of Mlotik1, a prokaryotic K^+ channel, can be largely superimposed onto this region. The large region is ~112 Å wide, and as in TRPV1, is linked to the putative transmembrane region through 4 short bridges. The N terminus of TRPV4 also contains 6 ankyrin repeats, which can be fitted into the large region, not in the vertical orientation as in TRPV1 but in a tilted orientation (Fig. 1.3b). Despite the similarities, there are some notable differences between the TRPV1 and TRPV4 structures. For example, both the small and large regions of TRPV4 are wider than their counterparts in TRPV1, but the overall height, ~130 Å, is shorter than that of TRPV1. Perhaps the most striking difference is the lack of a vacant cavity in the large region of TRPV4. This may reflect the different arrangement of the cytoplasmic N and C termini of the two channels and their interactions with distinct partners. Furthermore, due to the low resolution, smaller cavities might have not been resolved in the TRPV4 structure.

The 28 Å-resolution EM structure of human TRPM2 was obtained by negatively staining the protein samples with uranyl acetate (Fig. 1.3c) [19]. Instead of directly visualizing protein particles themselves, as in the case of cryo-EM, negative staining reveals the structure of the regions surrounding a protein. As such, this method tends to provide less structural details than single particle cryo-EM. The TRPM2 structure is composed of two main components, a bullet-shaped major component and a prism-shaped minor component. The head of the bullet presumably corresponds to the transmembrane domains, while the rest of the bullet and the prism constitute the cytoplasmic domains. The entire structure is ~250 Å high, with the base of the bullet being ~170 Å wide. The prism has a dimension of ~60 × 60 × 50 Å and is postulated to be formed by the *nu*cleoside *d*iphosphate-linked moiety X-*t*ype motif 9 *h*omology (NUDT9-H) domain located in the C-terminus.

The TRPM2 structure is significantly larger than that of TRPV1 and TRPV4. This is expected since the calculated mass of TRPM2 is >1.7 fold higher than that of TRPV1 and TRPV4. Moreover, the total mass estimated from the TRPM2 EM structure is ~30% higher than the calculated mass, presumably due to contributions from attached glycans, lipids and detergents.

Of the four existing EM structures, that of mouse TRPC3 has the highest resolution (15 Å) and is the most unique (Fig. 1.3d) [20]. The TRPC3 structure can also be divided into two components, a dense globular inner core and a sparse outer shell with a mesh-like structure containing many columns and aqueous spaces. Viewing from either the top or bottom, the outer shell columns radiate from the inner core like airport satellite terminals. These antenna-like structures are postulated to function as signal sensing modules for activators and modulators. The overall dimension of the reconstituted TRPC3 structure is ~200 × 200 × 240 Å, which is larger than that of TRPM2, even though the calculated mass of a TRPC3 tetramer is much smaller than that of a TRPM2 tetramer (388 kDa vs. 689 kDa). One reason for this disparity is the presence of many large water-filled cavities in the TRPC3 structure. It is not totally clear why the TRPC3 structure is drastically different from the TRPV1 and TRPV4 structures. It is hypothesized that the simultaneous association of TRPC3 with other protein complexes and its multi-modal activation and modulation mechanisms may underlie its expanded structure [20]. However, the use of an automated particle-selection algorithm might have caused distortions in the data analysis and structural reconstruction of TRPC3, an issue that was also discussed in a recent review [7]. This algorithm was also used for the determination of the

TRPM2 structure [19], suggesting another reason (in addition to negative staining) for the significant difference between this structure and the TRPV1 and TRPV4 structures.

1.4 NMR and X-Ray Crystal Structures

1.4.1 TRPM7 α-Kinase Domain

The mammalian TRPM subfamily has 8 members, which display distinct expression, activation, regulation and ion permeation properties [2]. TRPM7 channels and their close relatives TRPM6 channels are unique in several ways. While inhibited by intracellular Mg^{2+}, they are nevertheless highly permeable to Ca^{2+} and Mg^{2+}, and both are fusion proteins consisting of an ion channel module and an enzymatic module, with an α-kinase domain in the C terminus [2].

α-kinases belong to a family of atypical protein kinases that show little amino acid sequence similarities with classical serine, threonine or tyrosine protein kinases [25]. The substrate residues recognized by classical protein kinases are often located within β-turns, loops or irregular structures [26]. In contrast, the substrate residues targeted by α-kinases are often situated within α helices, although this is not always the case [25].

The α-kinase domain of TRPM7 is located at the distal C terminus, downstream of a coiled coil domain, whose structure was determined recently (see later). The structure of this domain, alone or in complex with AMP·PNP (an ATP analog) or ADP, was solved by X-ray crystallography at 2.8, 2.0 and 2.4 Å resolution, respectively [27]. These structures reveal that the TRPM7α-kinase domain forms a dimer, in which a 27-residue N-terminal segment of one monomer extends out and interacts extensively with the kinase domain of another monomer (Fig. 1.4a). The main component of each monomer contains two lobes (Fig. 1.4b): an N-terminal lobe consisting mainly of a stack of curved β sheets and harboring a phosphate binding P-loop, and a C-terminal lobe consisting of both α helices and β sheets and containing a glycine-rich GXA(G)XXG motif, which is conserved in the α-kinase family and is critical for catalysis. A nucleotide is bound at the interlobe cleft, where both the P-loop and the GXA(G)XXG motif project to (Fig. 1.4b). This overall architecture is highly similar to that of the PKA kinase domain (Fig. 1.4c). Strikingly, despite a lack of overall amino acid sequence conservation, many key residues involved in ATP binding and/or Mg^{2+} coordination are conserved in TRPM7α-kinase and classical kinases, including (in mouse TRPM7) K1646, D1765, Q1767 and D1775 (Fig. 1.4b).

There are, however, notable differences in the structural details of TRPM7α-kinase and classical kinases. For example, different sets of residues are engaged in contacts with the base and sugar moieties of ATP. As the ATP-binding pocket is an ideal drug target, the intricate differences between the structure of the catalytic site of TRPM7α-kinase and classical kinases might be exploited to develop

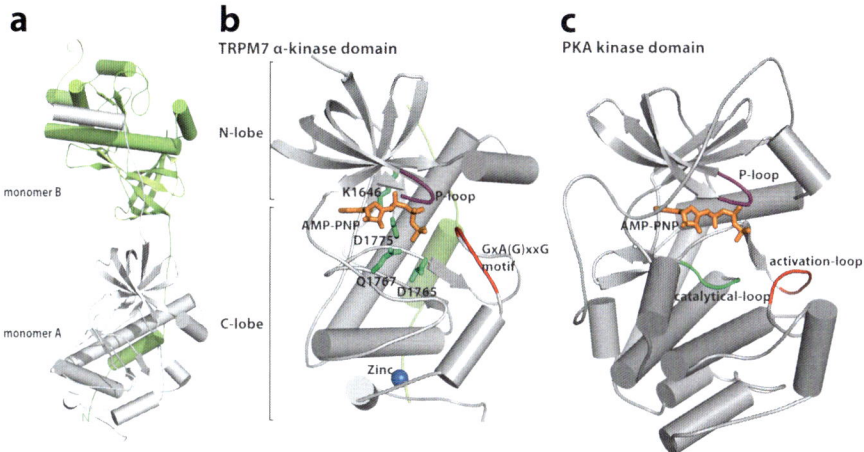

Fig. 1.4 Crystal structure of the TRPM7 α-kinase domain. (**a**) The dimeric structure of the TRPM7 α-kinase domain (PDB code 1IA9). (**b**) and (**c**) Comparison of the TRPM7 α-kinase domain (**b**, PDB code 1IA9) and PKA kinase domain (**c**, PDB code 1CDK). Regions and residues critical for catalysis are shown; they include the P-loop and the GXA(G)XXG motif in the TRPM7 α-kinase domain (**b**), and the P-loop, activation loop and catalytic loop in the PKA kinase domain (**c**) A bound zinc ion is present in the TRPM7 α-kinase domain, and a bound AMP-PNP, an ATP analog, is present in both structures

kinase-specific drugs. Furthermore, in TRPM7α-kinase, the GXA(G)XXG motif replaces the so-called activation loop present in classical kinases (Fig. 1.4), which participates in the recognition of peptide substrates. This difference may underlie, at least partly, the substrate specificity of α-kinases and classical kinases. Finally, the C-lobe of TRPM7α-kinase contains a zinc-binding module (Fig. 1.4a). Zinc coordination maybe important for the structural stability of the kinase domain.

The function of the TRPM7α-kinase domain remains controversial and incompletely understood. Initially, it was reported that replacing the final glycine in the GXA(G)XXG motif severely reduced the kinase activity and channel activity [28]. Likewise, mutating two cysteines in the zinc-binding module produced the same effect [28]. It was subsequently shown that TRPM7 channels remained active when its kinase activity was abolished by point mutations in the α-kinase domain or when the entire α-kinase domain was deleted [29, 30]. On the other hand, point mutations in the nucleotide binding pocket greatly reduced Mg^{2+}- and Mg·ATP-induced suppression of channel activity [29, 30]. Thus, it appears that the α-kinase domain is not essential for TRPM7 channel activation; instead, it contributes to channel regulation by Mg^{2+} and Mg·ATP. Currently, only a very small number of TRPM7α-kinase substrates are known, including annexin 1 and myosin II [25]. An interesting hypothesis was put forth that the opening of TRPM7 channels affects the activity of its α-kinase, thereby affecting the activity of its substrates [25]; however, this remains to be tested.

1.4.2 TRPV Ankyrin Repeats

Ankyrin repeat is one of the most common amino acid motifs and is present in numerous proteins with diverse structures and functions [4, 31, 32]. It consists of 30–34 amino acids folded into a characteristic helix-turn-helix conformation, with the two helices arranged antiparallely. In most proteins, a string of ankyrin repeats, numbering from 2 to ~30, are stacked in an array. These ankyrin repeat domains (ARDs) are generally thought to play a critical role in mediating protein–protein interactions [4, 31, 32].

The N termini of TRPA, TRPC, TRPN and TRPV contain 4 to ~30 tandem copies of ankyrin repeats. The ARDs in TRPV are highly conserved [6] and functionally important, as their deletion impairs channel activation, assembly or trafficking to the plasma membrane [33–36]. The structure of the ARD of TRPV1, TRPV2, TRPV4 and TRPV6 has been determined by X-ray crystallography at 1.6–2.7 Å resolution [37–41]. All four ARDs contain 6 ankyrin repeats and have highly similar structures (Fig. 1.5a). Each ankyrin repeat consists of 2 antiparallel α helices (named the inner and outer helix, respectively), followed by a hairpin loop (named the finger). The inner helices and fingers form a concave surface, while the outer helices form a convex surface. These surfaces likely constitute interfaces for specific protein–protein or protein–ligand interactions. The amino acid sequence and length of the finger vary among the ankyrin repeats within each ARD and among the ARDs of different TRPV subunits [6]. These differences are evident in the overlay of the 4 TRPV-ARD structures (Fig. 1.5a) and likely contribute to channel-specific functions of each ARD.

The structure of the TRPV1-ARD unexpectedly revealed a bound ATP molecule, which is cradled in the concave surface of ankyrin repeats 1–3 and interacts with residues in inner helices 1–3 and fingers 1–2 (Fig. 1.5b) [39]. Biochemical studies indicate that Ca^{2+}-calmodulin (Ca^{2+}-CaM) also binds the TRPV1-ARD. Moreover, mutations that eliminate ATP binding also prevent Ca^{2+}-CaM binding. ATP binding to the ARD sensitizes TRPV1 channels to capsaicin (an agonist) and greatly attenuates tachyphylaxis – decreasing responses to repeated applications of capsaicin. Conversely, Ca^{2+}-CaM binding to the ARD appears to play an opposing role and to be necessary for tachyphylaxis. These findings provide a structural basis for understanding the regulation of TRPV1 channels by Ca^{2+}, CaM and ATP [39]. Whether and to what extent ATP or Ca^{2+}-CaM binding alters the structure of the TRPV1-ARD remains to be investigated.

Amino acid sequence comparison and biochemical studies show that the ATP- and Ca^{2+}-CaM-binding site in the TRPV1-ARD is also present in TRPV3 and TRPV4, but not in TRPV2, TRPV5 and TRPV6 [42]. In accord with these observations, TRPV2 is insensitive to intracellular ATP, while TRPV4 is sensitized by intracellular ATP and this sensitization is eliminated by a binding site mutation. A twist is that the response of TRPV3 to agonists is *reduced* by intracellular ATP and Ca^{2+}-CaM through their interaction with the ARD [42]. Why ATP and Ca^{2+}-CaM binding to the ARD produce different modulatory effects on TRPV3 vs. TRPV1 and TRPV4 is unclear, but this example illustrates the diversified functions of the ARD.

1 Structural Biology of TRP Channels

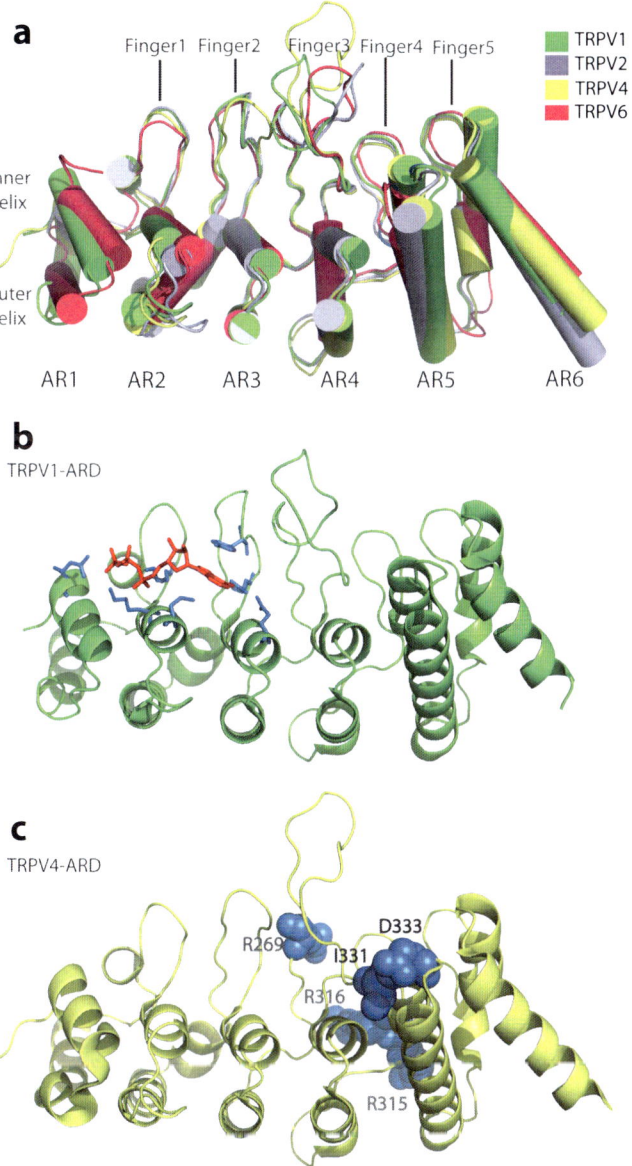

Fig. 1.5 Crystal structures of TRPV ankyrin repeat domains (ARDs). (**a**) Superposition of the structures of the ARD of TRPV1 (PDB code 2PNN), TRPV2(PDB code 2ETB), TRPV4 (PDB code 3JXI), and TRPV6 (PDB code 2RFA). AR, ankyrin repeat. (**b**) Structure of the TRPV1-ARD with a bound ATP molecule (*red*). Side-chains of residues involved in ATP binding are marked in *blue*. (**c**) Structure of the TRPV4-ARD, showing the position of the residues that cause diseases when mutated. Residues numbers correspond to human TRPV4

Recently, a number of mutations in TRPV4 has been linked to several human disorders, including spondylometaphyseal dysplasia (SMD) Kozlowski type (SMDK), metatropic dysplasia, scapuloperoneal spinal muscular atrophy (SPSMA), and Charcot–Marie–Tooth disease type 2C (CMT2C) [43]. Patients with SMDK and nonlethal metatropic dysplasia are characterized by defects in bone development, scoliosis and short stature, and those with SPSMA and CMT2C by distal and proximal muscle weakness and sensory loss. A mutation in the fifth ankyrin repeat was linked to SMDK [44]. This mutation, D333G (Fig. 1.5c), results in increased basal channel activity and increased response to the agonist 4αPDD [44]. Another mutation in the fifth ankyrin repeat, I331F, was tentatively associated with metatropic dysplasia, and its effect on TRPV4 channel activity remains to be determined [44]. Four missense mutations in the third and fourth ankyrin repeats, including R269H, R269C, R315W and R316C (Fig. 1.5c), were linked to SPSMA and CMT2C [38, 45, 46]. The effect of these mutations on TRPV4 channel activity is yet unclear, as both an increase in channel activity [38, 46] and a decrease in channel activity/surface expression [45] in heterologous systems have been reported. It is notable that all the disease-related residues are located on the concave surface formed by the inner helices and fingers. Presumably, the disease-causing mutations alter the interactions of TRPV4 with other proteins or ligands.

An intriguing proposed function of ARDs in TRP channels is that they act as the gating spring of mechanoreceptors [47, 48]. Studies based on molecular dynamics simulations [48] and single molecule measurements by atomic force microscopy [49] suggest that ARDs containing 17–24 ankyrin repeats exhibit proper elastic properties (including stiffness and extension length) that make them fit for mechanotransduction. In this regard, it is of interest to note that several TRP channels with ARDs, including TRPA1, TRPN1 and TRPV4, have been implicated or proposed in a variety of mechanosensory processes in various organisms [50].

1.4.3 TRPM7 Coiled Coil Domain

Coiled coil domains are also one of the most common amino acid motifs found in disparate proteins. They play an important role in the assembly of homomeric and heteromeric protein complexes. Coiled coils are comprised of stretches of α helices with multiple tandem copies of typically heptad (7-residue) repeats. They assemble to form oligomeric complexes, usually containing 2–5 helices, and both homomeric and heteromeric assemblies occur. The α helices in a coiled coil bundle can be parallel (i.e., with the same N and C orientation) or antiparallel (i.e., with opposite N and C orientation). The residues within the heptad, routinely designated as *a* through *g*, have characteristic features: the first and fourth residues (i.e., *a* and *d* positions) are usually non-polar (commonly Leu, Ile, Val, Phe, and Trp) and project inward to the core of the coiled coil complex; whereas the fifth and seventh residues (i.e., *e* and *g* positions) are usually charged and project outward to the outside of the coiled coil complex. The nature of the residues at *a*, *d*, *e* and *g* positions influences coiled coil stability, oligomeric state, partner selection, and helix–helix orientation, but definitive predictions on coiled coil assembly are difficult based solely on amino acid sequences. Comprehensive discussions on coiled coil structure and design can be found in recent reviews [51–54].

Coiled coil domains are predicted to be present in the cytoplasmic N and/or C termini of TRPC, TRPM, TRPP and TRPV channels and have been reported to be important for channel assembly and function [6, 55, 56]. The crystal structures of a TRPM7 and a TRPP2 C-terminal coiled coil domain have been determined at 2.8 and 1.9 Å resolution, respectively [57, 58]. In agreement with the tetrameric architecture of the full length channel, the TRPM7 coiled coil domain forms a tetramer (Fig. 1.6a). As in other coiled coil complexes, residues at the *a* and *d* positions project inward to the core of the complex and engage in extensive hydrophobic

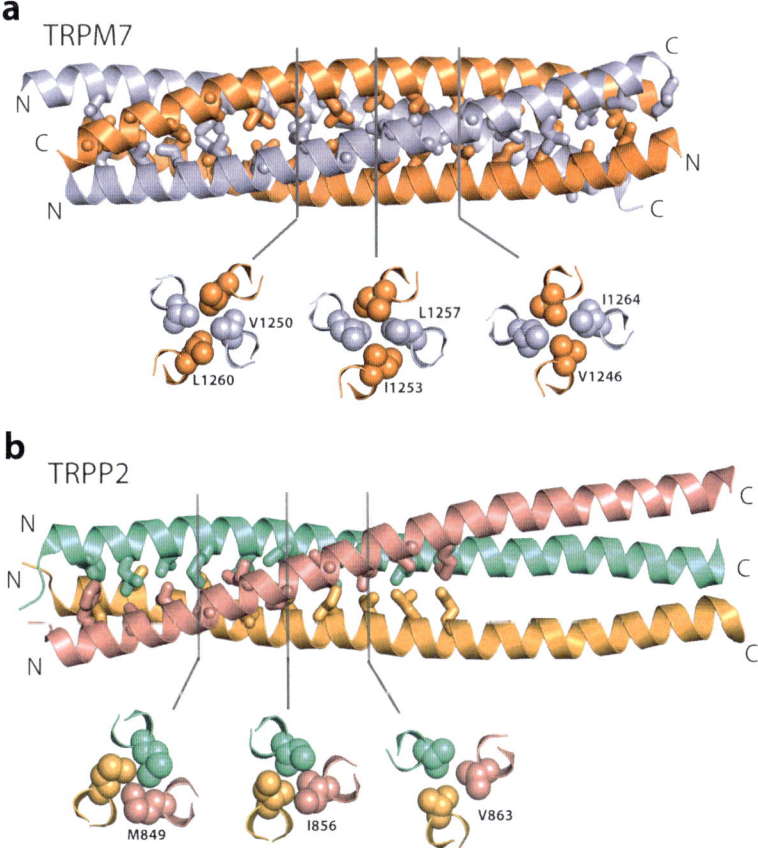

Fig. 1.6 Comparison of the crystal structure of a C-terminal coiled coil domain of TRPM7 (**a**, PDB code 3E7K) and TRPP2 (**b**, PDB code 3HRN). Side-chains of residues at the *a* and *d* positions of the heptad repeats are shown. Three cross sections, taken at the indicated positions, shows examples of the extensive van der Waals interactions among these residues

interactions. Unexpectedly, the 4 α helices are arranged antiparallely, deviating from the fourfold symmetry of the transmembrane regions. Whether these helices are also arranged in this fashion in the full length channel remains to be seen, but this arrangement fits the dimeric association of the α-kinase domain located ~230 amino acids downstream. Moreover, the bridging sequence of ~100 residues between the end of the last transmembrane segment and the coiled coil domain is sufficiently long to allow the latter to adopt a twofold symmetry. A similar symmetry mismatch has been observed in a crystal structure of a homotetrameric glutamate-gated receptor channel, where the ion channel domain exhibits a fourfold symmetry, whereas the extracellular ligand-binding domain displays a twofold symmetry [59]. As cautioned in a recent review [5], such a symmetry break complicates reconstruction of protein EM structures, where symmetry averaging is routinely used to improve resolution.

The importance of the coiled coil domain of TRPM7 in its assembly and function is not yet clear; meanwhile, deletion and mutagenesis studies on the coiled coil domains of TRPM2, TRPM4 and TRPM8, whose structures have not yet been determined, have produced varied results. In TRPM2, deleting the coiled coil domain or mutating residues at the a and d positions greatly reduced channel assembly and trafficking [60]. In TRPM4, which can be activated by depolarization and intracellular Ca^{2+}, truncation of the coiled coil domain decreased the sensitivity of the channels to Ca^{2+} and shifted the activation voltage to very positive potentials, but the mutant channels were still able to assemble and traffic properly [61]. In the case of TRPM8, one group reported that deleting the coiled coil domain severely suppressed channel activity but did not compromise assembly and trafficking [62], while another group reported that this deletion prevented TRPM8 surface expression [63]. It was also reported that a single mutation of a key a position residue, L1089P, in the coiled coil domain disrupted channel assembly [64]. Moreover, the coiled coil domain worked as a dominant negative when fused to a transmembrane segment and coexpressed with wild-type TRPM8 [63, 64]. The causes underlying the disparate findings on the function of the TRPM8 coiled coil domain are unclear, but the use of different expression systems and different cell culture temperatures are possible factors [62].

1.4.4 TRPP2 Coiled Coil Domain

TRPP2 forms homomeric channels and also interacts with members of other TRP subfamilies to form heteromeric channels, including TRPV4 and TRPC1 [65]. Furthermore, it associates with PKD1 to form a Ca^{2+}-permeable receptor/ion channel complex that is critical for kidney development and function [65–67]. PKD1 is a member of the polycystic kidney disease family of proteins. It contains 4,302 amino acids, with 11 putative transmembrane segments and a large N-terminus, which harbors several well-recognized domains and motifs involved in protein–protein and protein–carbohydrate interactions; hence, it is generally regarded as a membrane receptor and/or a mechanotransducer. TRPP2 and PKD1 assemble through the direct association of a coiled coil domain in the C terminus of both proteins [65–67]. Mutations in TRPP2 and PKD1 account for the vast majority of autosomal dominant polycystic kidney disease (ADPKD), one of the most common inherited human diseases [66, 67].

A recent study using single channel recording and atomic force microscopy imaging reported that homomeric TRPP2 channels incorporated into lipid bilayers were tetramers [68]. However, another study based on biochemical and single molecule photobleaching experiments reported that TRPP2 tended to form trimers when heterologously expressed in HEK 293 cells and *Xenopus* oocytes [58], suggesting that the association of the fourth TRPP2 subunit is weaker and that other pore-forming subunits may substitute TRPP2 to form heteromeric complexes. Biochemical studies also indicated that the TRPP2 C terminus formed a trimer complex, so did a coiled coil domain (aa 839–873) within the C terminus [58]. Consistent with these findings, the crystal structure of a TRPP2 C-terminal fragment (G833–G895) encompassing the coiled coil domain shows a 3-stranded complex (Fig. 1.6b). The 3 α helices are arranged in parallel, in accord with the propensity that opposite charges at the *e* and *g* positions favor such an orientation.

Single molecule photobleaching showed that the full length PKD1/TRPP2 complex contained 1 PKD1 and 3 TRPP2 [58]. This stoichiometry was preserved in solution for complexes formed by the coiled coil domain of both proteins and by longer C-terminal fragments. The structure of the PKD1/TRPP2 coiled coil domain complex awaits further elucidation.

Mutagenesis studies guided by the structure of the TRPP2 coiled coil domain showed that trimerization of this domain was critical for the assembly and membrane trafficking of homomeric TRPP2 trimer complexes and heteromeric PKD1/TRPP2 complexes [58]. Many ADPKD-causing mutations, including L4224P and R4227X in PKD1 and R742X, R807X, E837X, and R872X in TRPP2 [69], either alter the structure of or altogether delete the coiled coil domains. These mutations further underscore the functional importance of the TRPP2 and PKD1 coiled coil domains.

1.4.5 TRPP2 C-terminal E-F Hand

The EF-hand constitutes yet another one of the most ubiquitous amino acid motifs. It binds Ca^{2+} and through this binding regulates the activity of proteins with various structures and functions. Structurally, it is defined by a hallmark helix-loop-helix fold and it often exists in pairs [70]. In the so-called canonical EF-hand, Ca^{2+} is coordinated by 7 residues, which are usually conserved and 5 of them are located in the 9-residue loop. Ca^{2+} coordination by non-canonical (atypical) EF-hands is more diversified, as the length of the loop and the number and nature of the coordinating residues are variable. The affinity of EF-hands for Ca^{2+} ranges from nanomolar to millimolar [70].

The activities of many TRP channels are regulated by intracellular Ca^{2+} [2]. This regulation could be mediated by Ca^{2+}-binding proteins such as CaM or by the direct binding of Ca^{2+} to the channels. In this regard, it is of interest to note that several TRP channels, including TRPA1, TRPML1 and TRPP2, contain EF-hands in their cytoplasmic regions. TRPP2 channels are permeable to Ca^{2+} [65], and physiological concentrations of intracellular Ca^{2+} have been shown to regulate, in a bell-shaped concentration-dependent manner, the activity of TRPP2 channels incorporated into lipid bilayers [71, 72]. Molecular modeling predicted an EF-hand in the C-terminal region from amino acid 720–797 and yielded a structural model similar to the canonical EF-hand [73]. Subsequently, a solution structure of a slightly shorter fragment (aa 724–796) was determined by NMR [74]. This structure shows that, instead of a single EF-hand, this region contains a pair of EF-hands, one atypical and one canonical, both of which are proposed to bind Ca^{2+} [74]. However, the validity of this structure has been strongly challenged by a recent study that presents a vastly different NMR structure of the 720–797 fragment [75]. This new structure reveals a single E-F hand motif consisting of the helix α3, Ca^{2+}-binding loop, and helix α4 (Fig. 1.7). Paired with this E-F hand is a non-Ca^{2+}-binding helix-loop-helix motif comprising helices α1 and α2; this motif may have evolved from a canonical E-F hand. The 720–797 region binds Ca^{2+} noncooperatively at a single site (K_d ~214 μM) and undergoes Ca^{2+}-dependent conformational changes [73], and it shows conformational fluctuations even in the Ca^{2+}-bound state [75]. These properties make this region a prime candidate as a Ca^{2+}-sensitive regulator. However, whether and how Ca^{2+} binding to this region affect TRPP2 channel assembly and function remains to be investigated. The new structure would be very useful for guiding precision mutagenesis experiments.

It was reported that, in the absence of Ca^{2+}, a TRPP2 C-terminal fragment (aa 680–796) harboring the E-F hand and the non-canonical helix-loop-helix motif

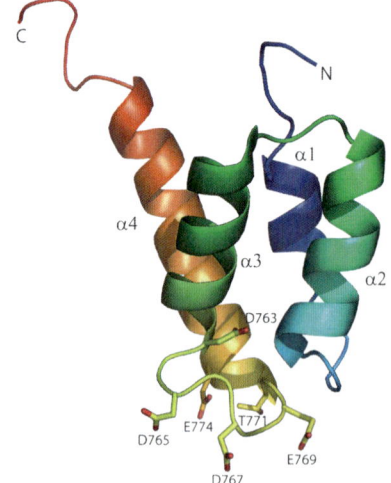

Fig. 1.7 NMR structure of an E-F hand in the TRPP2 C terminus. Side-chains of residues involved in Ca^{2+} binding are indicated. PDB code 2KQ6

forms a dimer, which dissociates upon Ca^{2+} binding [74]. In contrast, other studies reported that the 720–797 fragment was a monomer either in the absence or presence of Ca^{2+} [73, 75]. Furthermore, the reported dimeric association of the 680–796 fragment is at variance with the trimeric association of not only the downstream coiled coil domain (aa 839–873) but also a fragment (aa 723–928) encompassing both the EF-hand and the coiled coil domain [58]. Further studies are needed to reconcile this difference.

1.5 Perspectives

Structural study of TRP channels is still in its infancy. However, the few available structures have offered unique mechanistic insights on TRP channel assembly and regulation that would be otherwise difficult to obtain. As X-ray crystallography is increasingly becoming an integral part of the research arsenal in more and more research groups, crystal structures of many more TRP channel fragments will surely be solved in the coming years. It is useful to bear in mind that although protein fragments are easier to express and purify than full length proteins, they are still sometimes difficult to crystallize. In those instances, NMR would be an ideal alternative approach. Nevertheless, although structures of channel fragments undoubtedly enhance our understanding of TRP channel functions, as illustrated above, ultimately high resolution structures of full length channels are needed. With constant technical advances in membrane protein expression, screening, purification, crystallization and data processing, it is hopeful that the goal of solving high resolution structures of full length TRP channels will be achieved in the not too distant future.

Acknowledgments We thank Kathryn Abele, Ioannis Michailidis and Zafir Buraei for reading and commenting on a draft of this chapter. This work was supported by National Institutes of Health grants NS045383 and GM085234 (to J.Y.).

References

1. Doyle DA, Morais Cabral J, Pfuetzner RA, Kuo A, Gulbis JM, Cohen SL, Chait BT, MacKinnon R (1998) The structure of the potassium channel: molecular basis of K^+ conduction and selectivity. Science 280;69–77
2. Venkatachalam K, Montell C (2007) TRP channels. Annu Rev Biochem 76:387–417
3. Gaudet R (2008) TRP channels entering the structural era. J Physiol 586:3565–3575
4. Gaudet R (2008) A primer on ankyrin repeat function in TRP channels and beyond. Mol Biosyst 4:372–379
5. Gaudet R (2009) Divide and conquer: high resolution structural information on TRP channel fragments. J Gen Physiol 133:231–237
6. Latorre R, Zaelzer C, Brauchi S (2009) Structure-functional intimacies of transient receptor potential channels. Q Rev Biophys 42:201–246
7. Moiseenkova-Bell VY, Wensel TG (2009) Hot on the trail of TRP channel structure. J Gen Physiol 133:239–244
8. Blow D (2002) Outline of crystallography for biologists. Oxford University Press, Oxford

9. Jonic S, Venien-Bryan C (2009) Protein structure determination by electron cryo-microscopy. Curr Opin Pharmacol 9:636–642
10. Chiu W, Baker ML, Jiang W, Dougherty M, Schmid MF (2005) Electron cryomicroscopy of biological machines at subnanometer resolution. Structure 13:363–372
11. Frank J (2009) Single-particle reconstruction of biological macromolecules in electron microscopy – 30 years. Q Rev Biophys 42:139–158
12. Cong Y, Baker ML, Jakana J, Woolford D, Miller EJ, Reissmann S, Kumar RN, Redding-Johanson AM, Batth TS, Mukhopadhyay A, Ludtke SJ, Frydman J, Chiu W (2010) 4.0-Å resolution cryo-EM structure of the mammalian chaperonin TRiC/CCT reveals its unique subunit arrangement. Proc Natl Acad Sci USA 107:4967–4972
13. Gonen T, Cheng Y, Sliz P, Hiroaki Y, Fujiyoshi Y, Harrison SC, Walz T (2005) Lipid–protein interactions in double-layered two-dimensional AQP0 crystals. Nature 438:633–638
14. Foster MP, McElroy CA, Amero CD (2007) Solution NMR of large molecules and assemblies. Biochemistry 46:331–340
15. McDermott A (2009) Structure and dynamics of membrane proteins by magic angle spinning solid-state NMR. Annu Rev Biophys 38:385–403
16. Linke D (2009) Detergents: an overview. Methods Enzymol 463:603–617
17. Newby ZE, O'Connell JD 3rd, Gruswitz F, Hays FA, Harries WE, Harwood IM, Ho JD, Lee JK, Savage DF, Miercke LJ, Stroud RM (2009) A general protocol for the crystallization of membrane proteins for X-ray structural investigation. Nat Protoc 4:619–637
18. Bannen RM, Bingman CA, Phillips GN Jr (2007) Effect of low-complexity regions on protein structure determination. J Struct Funct Genomics 8:217–226
19. Maruyama Y, Ogura T, Mio K, Kiyonaka S, Kato K, Mori Y, Sato C (2007) Three-dimensional reconstruction using transmission electron microscopy reveals a swollen, bell-shaped structure of transient receptor potential melastatin type 2 cation channel. J Biol Chem 282:36961–36970
20. Mio K, Ogura T, Kiyonaka S, Hiroaki Y, Tanimura Y, Fujiyoshi Y, Mori Y, Sato C (2007) The TRPC3 channel has a large internal chamber surrounded by signal sensing antennas. J Mol Biol 367:373–383
21. Moiseenkova-Bell VY, Stanciu LA, Serysheva II, Tobe BJ, Wensel TG (2008) Structure of TRPV1 channel revealed by electron cryomicroscopy. Proc Natl Acad Sci USA 105: 7451–7455
22. Shigematsu H, Sokabe T, Danev R, Tominaga M, Nagayama K:A (2010) 3.5-nm structure of rat TRPV4 cation channel revealed by Zernike phase-contrast cryoelectron microscopy. J Biol Chem 285:11210–11218
23. Cheng W, Yang F, Takanishi CL, Zheng J (2007) Thermosensitive TRPV channel subunits coassemble into heteromeric channels with intermediate conductance and gating properties. J Gen Physiol 129:191–207
24. Hoenderop JG, Voets T, Hoefs S, Weidema F, Prenen J, Nilius B, Bindels RJ (2003) Homo- and heterotetrameric architecture of the epithelial Ca^{2+} channels TRPV5 and TRPV6. EMBO J 22:776–785
25. Middelbeek J, Clark K, Venselaar H, Huynen MA, van Leeuwen FN (2010) The alpha-kinase family: an exceptional branch on the protein kinase tree. Cell Mol Life Sci 67:875–890
26. Pinna LA, Ruzzene M (1996) How do protein kinases recognize their substrates? Biochim Biophys Acta 1314:191–225
27. Yamaguchi H, Matsushita M, Nairn AC, Kuriyan J (2001) Crystal structure of the atypical protein kinase domain of a TRP channel with phosphotransferase activity. Mol Cell 7: 1047–1057
28. Runnels LW, Yue L, Clapham DE (2001) TRP-PLIKa bifunctional protein with kinase and ion channel activities. Science 291:1043–1047
29. Schmitz C, Perraud AL, Johnson CO, Inabe K, Smith MK, Penner R, Kurosaki T, Fleig A, Scharenberg AM (2003) Regulation of vertebrate cellular Mg^{2+} homeostasis by TRPM7. Cell 114:191–200

30. Demeuse P, Penner R, Fleig A (2006) TRPM7 channel is regulated by magnesium nucleotides via its kinase domain. J Gen Physiol 127:421–434
31. Li J, Mahajan A, Tsai MD (2006) Ankyrin repeat: a unique motif mediating protein–protein interactions. Biochemistry 45:15168–15178
32. Mosavi LK, Cammett TJ, Desrosiers DC, Peng ZY (2004) The ankyrin repeat as molecular architecture for protein recognition. Protein Sci 13:1435–1448
33. Chang Q, Gyftogianni E, van de Graaf SF, Hoefs S, Weidema FA, Bindels RJ, Hoenderop JG (2004) Molecular determinants in TRPV5 channel assembly. J Biol Chem 279:54304–54311
34. Erler I, Hirnet D, Wissenbach U, Flockerzi V, Niemeyer BA (2004) Ca^{2+}-selective transient receptor potential V channel architecture and function require a specific ankyrin repeat. J Biol Chem 279:34456–34463
35. Jung J, Lee SY, Hwang SW, Cho H, Shin J, Kang YS, Kim S, Oh U (2002) Agonist recognition sites in the cytosolic tails of vanilloid receptor 1. J Biol Chem 277:44448–44454
36. Neeper MP, Liu Y, Hutchinson TL, Wang Y, Flores CM, Qin N (2007) Activation properties of heterologously expressed mammalian TRPV2: evidence for species dependence. J Biol Chem 282:15894–15902
37. Jin X, Touhey J, Gaudet R (2006) Structure of the N-terminal ankyrin repeat domain of the TRPV2 ion channel. J Biol Chem 281:25006–25010
38. Landoure G, Zdebik AA, Martinez TL, Burnett BG, Stanescu HC, Inada H, Shi Y, Taye AA, Kong L, Munns CH, Choo SS, Phelps CB, Paudel R, Houlden H, Ludlow CL, Caterina MJ, Gaudet R, Kleta R, Fischbeck KH, Sumner CJ (2010) Mutations in TRPV4 cause Charcot-Marie-Tooth disease type 2C. Nat Genet 42:170–174
39. Lishko PV, Procko E, Jin X, Phelps CB, Gaudet R (2007) The ankyrin repeats of TRPV1 bind multiple ligands and modulate channel sensitivity. Neuron 54:905–918
40. McCleverty CJ, Koesema E, Patapoutian A, Lesley SA, Kreusch A (2006) Crystal structure of the human TRPV2 channel ankyrin repeat domain. Protein Sci 15:2201–2206
41. Phelps CB, Huang RJ, Lishko PV, Wang RR, Gaudet R (2008) Structural analyses of the ankyrin repeat domain of TRPV6 and related TRPV ion channels. Biochemistry 47: 2476–2484
42. Phelps CB, Wang RR, Choo SS, Gaudet R (2010) Differential regulation of TRPV1, TRPV3, and TRPV4 sensitivity through a conserved binding site on the ankyrin repeat domain. J Biol Chem 285:731–740
43. Nilius B, Owsianik G (2010) Transient receptor potential channelopathies. Pflugers ArchChem 460:437–450
44. Krakow D, Vriens J, Camacho N, Luong P, Deixler H, Funari TL, Bacino CA, Irons MB, Holm IA, Sadler L, Okenfuss EB, Janssens A, Voets T, Rimoin DL, Lachman RS, Nilius B, Cohn DH (2009) Mutations in the gene encoding the calcium-permeable ion channel TRPV4 produce spondylometaphyseal dysplasia, Kozlowski type and metatropic dysplasia. Am J Hum Genet 84:307–315
45. Auer Grumbach M, Olschewski A, Papic L, Kremer H, McEntagart ME, Uhrig S, Fischer C, Frohlich E, Balint Z, Tang B, Strohmaier H, Lochmuller H, Schlotter-Weigel B, Senderek J, Krebs A, Dick KJ, Petty R, Longman C, Anderson NE, Padberg GW, Schelhaas HJ, van Ravenswaaij-Arts CM, Pieber TR, Crosby AH, Guelly C (2010) Alterations in the ankyrin domain of TRPV4 cause congenital distal SMA, scapuloperoneal SMA and HMSN2C. Nat Genet 42:160–164
46. Deng HX, Klein CJ, Yan J, Shi Y, Wu Y, Fecto F, Yau HJ, Yang Y, Zhai H, Siddique N, Hedley-Whyte ET, Delong R, Martina M, Dyck PJ, Siddique T (2010) Scapuloperoneal spinal muscular atrophy and CMT2C are allelic disorders caused by alterations in TRPV4. Nat Genet 42:165–169
47. Howard J, Bechstedt S (2004) Hypothesis: a helix of ankyrin repeats of the NOMPC-TRP ion channel is the gating spring of mechanoreceptors. Curr Biol 14:R224–R226
48. Sotomayor M, Corey DP, Schulten K (2005) In search of the hair-cell gating spring elastic properties of ankyrin and cadherin repeats. Structure 13:669–682

49. Lee G, Abdi K, Jiang Y, Michaely P, Bennett V, Marszalek PE (2006) Nanospring behaviour of ankyrin repeats. Nature 440:246–249
50. Christensen AP, Corey DP (2007) TRP channels in mechanosensation: direct or indirect activation? Nat Rev Neurosci 8:510–521
51. Lupas AN, Gruber M (2005) The structure of α-helical coiled coils. Adv Protein Chem 70:37–78
52. Parry DA, Fraser RD, Squire JM (2008) Fifty years of coiled-coils and α-helical bundles: a close relationship between sequence and structure. J Struct Biol 163:258–269
53. Woolfson DN (2005) The design of coiled-coil structures and assemblies. Adv Protein Chem 70:79–112
54. Grigoryan G, Keating AE (2008) Structural specificity in coiled–coil interactions. Curr Opin Struct Biol 18:477–483
55. Lepage PK, Boulay G (2007) Molecular determinants of TRP channel assembly. Biochem Soc Trans 35:81–83
56. Schindl R, Romanin C (2007) Assembly domains in TRP channels. Biochem Soc Trans 35:84–85
57. Fujiwara Y, Minor DL Jr (2008) X-ray crystal structure of a TRPM assembly domain reveals an antiparallel four-stranded coiled-coil. J Mol Biol 383:854–870
58. Yu Y, Ulbrich MH, Li MH, Buraei Z, Chen XZ, Ong AC, Tong L, Isacoff EY, Yang J (2009) Structural and molecular basis of the assembly of the TRPP2/PKD1 complex. Proc Natl Acad Sci USA 106:11558–11563
59. Sobolevsky AI, Rosconi MP, Gouaux E (2009) X-ray structure, symmetry and mechanism of an AMPA-subtype glutamate receptor. Nature 462:745–756
60. Mei ZZ, Xia R, Beech DJ, Jiang LH (2006) Intracellular coiled-coil domain engaged in subunit interaction and assembly of melastatin-related transient receptor potential channel 2. J Biol Chem 281:38748–38756
61. Nilius B, Prenen J, Tang J, Wang C, Owsianik G, Janssens A, Voets T, Zhu MX (2005) Regulation of the Ca^{2+} sensitivity of the nonselective cation channel TRPM4. J Biol Chem 280:6423–6433
62. Phelps CB, Gaudet R (2007) The role of the N terminus and transmembrane domain of TRPM8 in channel localization and tetramerization. J Biol Chem 282:36474–36480
63. Tsuruda PR, Julius D, Minor DL Jr (2006) Coiled coils direct assembly of a cold-activated TRP channel. Neuron 51:201–212
64. Erler I, Al-Ansary DM, Wissenbach U, Wagner TF, Flockerzi V, Niemeyer BA (2006) Trafficking and assembly of the cold-sensitive TRPM8 channel. J Biol Chem 281: 38396–38404
65. Tsiokas L (2009) Function and regulation of TRPP2 at the plasma membrane. Am J Physiol Renal Physiol 297:F1–F9
66. Harris PC, Torres VE (2009) Polycystic kidney disease. Annu Rev Med 60:321–337
67. Torres VE, Harris PC (2009) Autosomal dominant polycystic kidney disease: the last 3 years. Kidney Int 76:149–168
68. Zhang P, Luo Y, Chasan B, Gonzalez-Perrett S, Montalbetti N, Timpanaro GA, Cantero Mdel R, Ramos AJ, Goldmann WH, Zhou J, Cantiello HF (2009) The multimeric structure of polycystin-2 (TRPP2): structural-functional correlates of homo- and hetero-multimers with TRPC1. Hum Mol Genet 18:1238–1251
69. Wu G, Tian X, Nishimura S, Markowitz GS, D'Agati V, Park JH, Yao L, Li L, Geng L, Zhao H, Edelmann W, Somlo S (2002) Trans-heterozygous Pkd1 and Pkd2 mutations modify expression of polycystic kidney disease. Hum Mol Genet 11:1845–1854
70. Gifford JL, Walsh MP, Vogel HJ (2007) Structures and metal-ion-binding properties of the Ca^{2+}-binding helix-loop-helix EF-hand motifs. Biochem J 405:199–221
71. Cai Y, Anyatonwu G, Okuhara D, Lee KB, Yu Z, Onoe T, Mei CL, Qian Q, Geng L, Wiztgall R, Ehrlich BE, Somlo S (2004) Calcium dependence of polycystin-2 channel activity is modulated by phosphorylation at Ser812. J Biol Chem 279:19987–19995

72. Koulen P, Cai Y, Geng L, Maeda Y, Nishimura S, Witzgall R, Ehrlich BE, Somlo S (2002) Polycystin-2 is an intracellular calcium release channel. Nat Cell Biol 4:191–197
73. Celic A, Petri ET, Demeler B, Ehrlich BE, Boggon TJ (2008) Domain mapping of the polycystin-2 C-terminal tail using de novo molecular modeling and biophysical analysis. J Biol Chem 283:28305–28312
74. Schumann F, Hoffmeister H, Bader R, Schmidt M, Witzgall R, Kalbitzer HR (2009) Ca^{2+}-dependent conformational changes in a C-terminal cytosolic domain of polycystin-2. J Biol Chem 284:24372–24383
75. Petri ET, Celic A, Kennedy SD, Ehrlich BE, Boggon TJ, Hodsdon ME (2010) Structure of the EF-hand domain of polycystin-2 suggests a mechanism for Ca^{2+}-dependent regulation of polycystin-2 channel activity. Proc Natl Acad Sci USA 107:9176–9181
76. Sokolova O, Kolmakova-Partensky L, Grigorieff N (2001) Three-dimensional structure of a voltage-gated potassium channel at 2.5 nm resolution. Structure 9:215–220
77. Walz T, Hirai T, Murata K, Heymann JB, Mitsuoka K, Fujiyoshi Y, Smith BL, Agre P, Engel A (1997) The three-dimensional structure of aquaporin-1. Nature 387:624–627
78. Long SB, Campbell EB, Mackinnon R (2005) Crystal structure of a mammalian voltage-dependent Shaker family K+ channel. Science 309:897–903
79. Long SB, Tao X, Campbell EB, MacKinnon R (2007) Atomic structure of a voltage-dependent K^+ channel in a lipid membrane-like environment. Nature 450:376–382
80. Wootton JC, Federhen S (1993) Statistics of local complexity in amino acid sequences and sequence databases. Comput Chem 17:149–163

Chapter 2
Functional and Structural Studies of TRP Channels Heterologously Expressed in Budding Yeast

Vera Moiseenkova-Bell and Theodore G. Wensel

Abstract The transient receptor potential (TRP) superfamily is one of the largest families of cation channels. The metazoan TRP family has been subdivided into major branches: TRPC, TRPA, TRPM, TRPP, TRPV, TRPML, and TRPN, while the TRPY family is found in fungi. They are involved in many physiological processes and in the pathogenesis of various disorders. An efficient high-yield expression system for TRP channels is a necessary step towards biophysical and biochemical characterization and structural analysis of these proteins, and the budding yeast, *Saccharomyces cerevisiae* has proven to be very useful for this purpose. In addition, genetic screens in this organism can be carried out rapidly to identify amino acid residues important for function and to generate useful mutants. Here we present an overview of current developments towards understanding TRP channel function and structure using *Saccharomyces cerevisiae* as an expression system. In addition, we will summarize recent progress in understanding gating mechanisms of TRP channels using endogenously expressing TRPY channels in *S. cerevisiae,* and insights gained from genetic screens for mutants in mammalian channels. The discussion will focus particular attention of the use of cryo-electron microscopy (cryo-EM) to determine TRP channel structure, and outlines a "divide and concur" methodology for combining high resolution structures of TRP channel domains determined by X-ray crystallography with lower resolution techniques including cryo-EM and spectroscopy.

2.1 Introduction

Ion channels regulate the flow of ions across the plasma membrane in response to a variety of chemical, electrical, temperature, or mechanical signals. Determining ion channel structure is essential for understanding mechanisms of channel gating,

V. Moiseenkova-Bell (✉)
Department of Pharmacology, Case Western Reserve University School of Medicine, Cleveland, OH 44106, USA
e-mail: vxm102@case.edu

modulation, ion selectivity and permeation. TRP channels display multifunctional and polymodal behavior in their regulation and interactions with proteins, lipids, and other small molecules and ions, and with electric fields. They thus present themselves as intriguing candidates for structural analysis. Detail structural information on TRP channels will allow development of new strategies for drug design targeting these channels. Several research groups have expended considerable effort towards understanding TRP channel structure, and structure-function relationships. In this chapter, we will review the current progress in understanding TRP channel structures through structural analysis of both full-length proteins and channel fragments.

2.2 The TRP Channel Family

The TRP family of channels derive their name from a Drosophila *trp* mutant with defective vision characterized by a transient receptor potential that was reported 40 years ago [1, 2]. Twenty years later, molecular cloning and functional analysis led to the discovery that the defect lies in a gene encoding a cation channel, known as TRP in Drosophila [3]. In the last decade, subsequent investigations identified over 70 homologues to the original TRP channel in invertebrates and vertebrates. To date, a total of 27 mammalian genes belonging to the TRP family have been reported and are subdivided into six major branches [4]: TRPC (canonical), TRPA (ankyrin), TRPM (melastatin), TRPP (polycystin), TRPV (vanilloid), and TRPML (mucolipin). The TRPN (NOMP-C homologues) sub-family of proteins are not found in mammals, but they are expressed in invertebrates such as flies and worms [5], and in cold-blooded vertebrates [6, 7]. Yeast and other fungi express TRP channels known as the TRPY sub-family [8, 9].

Based on sequence comparison and structural prediction algorithms, TRP channels are related to the superfamily of voltage-gated cation channels. Typically, TRP channels are predicted to have 6 transmembrane helices (TM1-6) per subunit, with varying sizes of cytoplasmic amino- and carboxy-termini, and are thought to form tetrameric assemblies [10, 11]. Depending on the TRP family branch, the cytoplasmic amino-terminal domain contains different number of ankyrin repeats, ranging from zero to twenty nine, which have been proposed to be involved in a range of interactions (reviewed in [12]), including activating ligands, protein–protein interactions, and gating by voltage and temperature [13]. Recently it was discovered that mutations in the ankyrin domain of TRPV4 underlie autosomal dominant disorders of the peripheral nervous system [14, 15], including Charcot-Marie-Tooth disease type 2C, the most common inherited neurological disease [16]. The carboxy-termini of most contain a signature "TRP box" motive and coiled–coiled regions important in protein assembly. The majority of functionally characterized TRP channels are permeable to Ca^{2+} with the exception of TRPM4 and TRPM5, which are permeable to monovalent cations [17]. Ca^{2+} selectivity is poor for many TRP channels but TRPV5 and TRPV6 are highly permeable Ca^{2+} channels [18]. These channels function as polymodal sensors and are gated by diverse stimuli that include the binding of intracellular and extracellular messengers; changes in temperature, and chemical or mechanical stress [19].

TRP channels are widely distributed through the body, expressed in a vast number of different cell types and have numerous splice variants. TRP channels are particularly abundant in sensory receptor cells, and play a critical role in vision, touch, hearing, taste, pain and temperature sensation. The importance of determining TRP channel structure is highlighted by their emerging roles as major drug targets for the treatment of pain, inflammation, and a range of disorders [20–26], and by the association of genetic defects in TRP channels with a number of devastating diseases, ranging from the most common single-gene neurological defect to polycystic kidney disease [27], to night blindness [28–33]. The long-term hope is that understanding TRP channel structures, the structural determinants of ligand binding, and the effects of disease mutations on structure, will aid in the development of new therapeutics.

2.3 Ion Channel Structural Biology

In the past several years, considerable progress has been made in the field of membrane protein structural biology. High-resolution structures for the pharmaceutically relevant eukaryotic membrane proteins, such as G protein-coupled receptors [34–37], transporters [38] and ion channels [39–45], have been obtained and provided very valuable information about mechanism of action for these proteins. Still, membrane protein structure determination remains a difficult task. Despite extensive efforts in many laboratories, the number of solved membrane protein structures remains small because of the many challenges presented by membrane proteins. Early success in crystallization of eukaryotic membrane proteins came from the use of the native sources that provide a large amount of protein, for example, bovine retinas providing a milligram of rhodopsin per cow [34]. These problems have been especially challenging for eukaryotic ion channels among which only five high-resolution structures have been determined. Because of the low levels at which ion channels are typically expressed endogenously, structural studies of TRP channels and others require efficient heterologous systems for high level expression and purification. It is important that the cells used for expression are capable of properly folding and assembling the multiple subunits, and of producing them in stable and active form.

E. coli expression has been widely used for soluble eukaryotic proteins and for bacterial membrane proteins, as well as for soluble domains of transmembrane proteins. In the case of TRP channels, soluble fragments successfully produced in high yields from bacteria include the ankyrin repeat domains of proteins in the TRPV family [46–50], domains of TRPM7 including the C-terminal cytoplasmic coiled-coil assembly domain [51] and the α-kinase [52], and the coiled-coil region of the TRPP2 C-terminal domain [53].

Unfortunately, methods have not been found for routine expression in functional form of multi-pass eukaryotic membrane proteins such as ion channels in bacteria, likely as a result of lack of appropriate chaperones or other components of the folding and assembly machinery associated with the endoplasmic reticulum. No success has been achieved with full length TRP channels or their fragments containing transmembrane domains.

Baculovirus-mediated expression in insect cells offers another useful tool for generating recombinant membrane proteins [54]. In 2009, several new eukaryotic ion channels structures were solved using insect cell expression [42, 44, 45]. However, the high cost of this methodology represents a drawback. Expression of some TRP channels using baculovirus has been reported, including overexpression of TRPV1 [55] and TRPV4 [56]. In the case of TRPV4 sufficient protein was purified for structure determination by in insect cells showed the possibly of using this system for structural biology of TRP channels, and the TRPV4 structure was determined using electron microscopy [56].

Although even more expensive, mammalian tissue culture cells offer a native folding and assembly environment for mammalian membrane proteins. COS-1 cells have been used to express sufficient amounts of a rhodopsin mutant engineered for enhanced thermal stability [57]. Mammalian cells have been used extensively for expression and functional studies of TRP channels, but only in a few cases have TRP channels been purified from them. TRPC3 [58, 59] and TRPM2 [60] were expressed in mammalian cells, and used for structural studies by electron microscopy and single particle analysis.

Yeast is another traditional and powerful tool for the expression of eukaryotic recombinant proteins [54]. The advantages include relatively low cost, rapid cell growth and ease of producing large volume cultures. Although distant in evolution from mammals, yeast possess conserved protein folding and assembly machinery that can be exploited to produce mammalian membrane proteins in large amounts. Most commonly, *Pichia pastoris* and *Saccharomyces cerevisiae* are used for the overexpression of membrane proteins. *P. pastoris* can achieve exceptionally high cell densities, that in favorable cases can provide high levels of protein production for structural biology. The first atomic structure of a mammalian potassium channel was possible only after the methodology for the overexpression of functional channel using *P. pastoris* was published [61]. Since that published work, the *P. pastoris* system has been used to obtain structures for a number of ion channels, including aquaporin [62] and potassium channels [39, 41, 43, 63, 64]. Although attempts have been made, there has been no success reported in overexpression in functional form of TRP channels in *P. pastoris*.

2.4 TRP Channels Expression in *Saccharomyces cerevisiae* and Functional Analysis

One of the most characterized methods for the overexpression of recombinant membrane proteins has been budding yeast, *S. cerevisiae*. As eukaryotic organisms, yeast contain the machinery necessary for overexpression of eukaryotic membrane proteins, including rough endoplasmic reticulum and Golgi apparatus with associated molecules required for translocation, folding, and post-translational modifications, in addition to membrane trafficking machinery [65]. They also have the advantage of easy plasmid and genetic manipulation for protein expression. A multitude of different strains including protease-deficient strains as well as a variety of expression vectors comprising yeast episomal plasmids (Yeps) and yeast integrating plasmids (Yips) are commercially available and allow genetic manipulation [66]. In addition, yeast express an endogenous TRP channel, known as TRPY1 or YVC1 in *S. cerevisiae* [8, 67], so they have the necessary factors for folding and assembling channels of this family.

The earliest example of heterologous expression of a mammalian TRP channel in yeast was that of TRPV1 [68]. The approach was based on a method that was published for the overexpression of P-glycoprotein (MDR1) from the Al-Shawi laboratory [65]. The method is simple, and incorporates important steps that help to increase the expression of the protein by several-fold. High levels of expression of the protein were obtained using a maximally active *PMA1* promoter. Although high levels of expression of a Ca^{2+}-permeable channel, even under conditions of low activity, might be expected to be toxic to yeast; however, by transient expression produces large amounts of protein. Protein folding and stability are improved by addition of glycerol, which apparently acts as a sort of chemical chaperone. The optimal concentration of glycerol was determined for each construct, and was found to vary.

For heterologously expressed protein to be useful for structural studies, it needs to be functional. Ca^{2+} flux studies using Fura-2 in yeast confirmed that mammalian TRPV1 expressed in *S. cerevisiae* conducts Ca^{2+} in response to its well known agonist, capsaicin. A comparison of different detergents suggested that good solubilization could be obtained using either 1% egg L-α-lysophosphatidylcholine or 50 mM *n*-dodecyl-β-D-maltoside; subsequent studies (see below, and VYM and TGW unpublished observations) suggest that related detergents may be more suitable. Although TRPV1 expressed with a C-terminal His_{12} tag was purified to about 80% purity using nickel-chelate affinity chromatography, the yield was only about 1 mg per 16 L of yeast culture.

In order to obtain higher yields of functional and highly purified protein, a technique widely used for purification of rhodopsin and related visual pigments [69, 70] was used [71]. An epitope from the C-terminus of rhodopsin, recognized by a monoclonal antibody 1D4 [72], was engineered into the C-terminus of TRPV1, allowing purification to homogeneity in a single step using an affinity column of immobilized 1D4 antibody, and elution with a peptide corresponding to the epitope. For this purification, *n*-decyl-β-D-maltoside was found to be the most useful detergent among those tried.

In this preparation, some heterogeneity in subunit molecular weight was revealed by SDS PAGE. A band appearing at a lower molecular weight than full length TRPV1 was consistently observed, but in varying relative amounts. Enzymatic de-glycosylation and stringent disulphide reduction did not eliminate the heterogeneity, suggesting that the smaller fragment results from proteolysis. The smaller fragment is recognized by the C-terminal 1D4 antibody, so presumably a piece of the cytoplasmic N-terminus is removed. A construct lacking part of the N-terminal domain does not display the lower band (VYM and TGW, unpublished observations).

Gel filtration in detergent revealed that purified TRPV1 migrated predominantly as a monodisperse homotetramer. Thus the mammalian protein assembled in yeast appears to have the appropriate stoichiometry for a native TRP channel.

Although the presence of functional protein in the yeast membrane had been demonstrated, it was possible that only a small fraction of the total expressed protein was functional, or that detergent extraction disrupted the native structure and function. To address this question, the purified TRPV1 was reconstituted by dialysis in the presence of phospholipids into phospholipid bilayers under conditions that yielded, on average, less than one tetramer per vesicle. This ratio can be determined accurately by measurement of protein-to-lipid ratios in the final vesicles, and by using electron microscopy to determine the vesicle size distribution. When the reconstitution was carried out in the presence of 5 mM Ca^{2+}, and the external Ca^{2+} removed by chelate chromatography, the fraction of Ca^{2+} released by a TRPV1 agonist, resinaferatoxin, provided an estimated lower limit for the fraction of protein that was in functional form. The estimate is a lower limit because of the likelihood that more Ca^{2+} leaks out of the TRPV1-containing vesicles than out of the protein-free vesicles used as controls. This method provided a lower limit of 72% active protein, consistent with nearly all of the purified protein being in active form. This method does not distinguish among channels with cytoplasmic domains outside vs. inside the vesicles, due to the membrane permeability of the agonist. Future studies using antibodies, lectins, and/or cytoplasmic ligands such as calmodulin, will be needed to determine whether vesicle insertion happens with a preferential orientation, and to optimize conditions for achieving preferential orientation.

Other TRP channels have been expressed at high levels in *S. cerevisiae* and purified, using the same approaches as for TRPV1. These include TRPV2, TRPY1–4, TRPM8, and TRPA1 (unpublished data). They are all behaving well in this system, allowing purification of sufficient quantities of protein for functional and structural work. Thus the strategy of transient over-expression in budding yeast and epitope affinity chromatography in detergent appears to be one of general utility for members of the TRP family.

2.5 Cryo-EM Structures of TRP Channels

Single-particle EM can provide structural information for a large variety of biological molecules without the need to produce crystals [73]. Very little sample is required [74]. Single particle cryo-EM [75] is a method in which the specimen, typically a protein embedded in vitreous ice, is held at cryogenic temperatures while images are obtained by the electron microscope [76]. After completing the imaging, single-particle reconstruction methods are used to align and classify the individual particle images, and solve the structure of the protein [77–79]. Because the individual molecules (particles) are randomly oriented in the ice layer, the particle images can be classified into groups representing distinctive views of the original 3-D particle. Particle images in each group can then be aligned with each other and a consensus shape determined. Once these averaged views are obtained, a "map" of density throughout the volume of the particle can be calculated to complete the reconstruction process.

Ion channels are excellent candidates for single-particle analysis work [76]. Several channel structures have also been determined by cryo-EM, including a the voltage gated sodium channel [80], inositol triphosphate receptor [81, 82], muscle L-type voltage gated calcium channel (dihydropyridine receptor) [83, 84], muscle calcium release channel/ryanodine receptor (RyR1) [85–88], voltage-gated channel KvAP [89], large-conductance calcium- and voltage-activated potassium channel (BK) in a lipid environment [90] and others.

In the last few years, several TRP channel structures have been studied by electron microscopy [91]. These include TRPV1 [71], TRPV4 [56], and TRPC3 [59], imaged in ice, and TRPC3 [58], and TRPM2 [60] imaged in negative stain.

The structure of TRPV1 using a cryo-EM approach was solved recently using preparation affinity purified from budding yeast and tested for tetrameric structure and ligand-gated ion flux as described above [71]. Cryo-EM images and single particle reconstruction revealed that the structure is fourfold symmetric and consists of two well-defined domains (Fig. 2.1). The more compact has the right dimensions to correspond to the membrane-spanning domain, likely composed of six transmembrane segments per subunit. This domain is 40 Å in the dimension thought to be normal to the membrane surface, and about 60 Å in diameter. This domain fits well with the high-resolution structure of the voltage-gated potassium channel Kv1.2 [39]. The other domain, which contains the majority of the mass, as expected for the

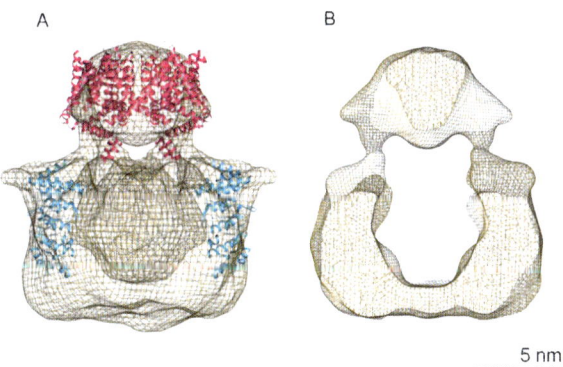

Fig. 2.1 Structure of TRPV1 determined by electron cryo-microscopy and single particle analysis [71]. On the *left* is an iso-dense surface in transparent mesh representation. Superimposed are ribbon diagrams of the *x-ray* structures of the transmembrane domain of Kv1.2 [39], 2A79 (*magenta*), and the ankyrin repeats of TRPV1 [48], 2PNN (*cyan*). On the *right* is a similar representation cut through a center plane perpendicular to the proposed plane of the membrane, showing the large space within the basket-like cytoplasmic domain of TRPV1

N- and C-terminal cytoplasmic domains, is an open, basket light structure connected by thin densities to the putative transmembrane domain. There is a region near these connecting densities, and therefore relatively near to the proposed membrane surface region, that fits well with the high-resolution structure of the ankyrin repeat domain of TRPV1 [48].

Another structure of the TRPV sub-family has been determined using electron microscopy in ice [56]. The structure of TRPV4 revealed considerable similarity to the TRPV1 structure, and contains two-distinct regions, likely corresponding to the transmembrane and cytoplasmic domains of the channel respectively (Fig. 2.2). The results from these studies provided insight into structural organization of TRPV sub-family of ion channels and can be used as an initial testable structural template for studying full-length TRPV channels at higher resolution.

As described above, there is reason for some confidence in the published structures for the TRPV sub-family, given their self-consistency and the extensive characterization of the TRPV1 preparation. From the reported structures for the TRPC and TRPM sub-families, many questions remain [91] (Fig. 2.3). Structures for TRPC3 [58] and TRPM2 [60] determined in negative stain indicated a bullet-like shape for each, with the dense bullet-head region proposed to be the channel domain with its transmembrane segments, and a more open and larger domain proposed to be the cytoplasmic regions. There are some qualitative features of these structures reminiscent of the TRPV family structures, but the detailed structures are quite distinct. Limitations on the interpretation of these structures arise from the lack of functional characterization, concerns about the presence of lipid aggregates, and the inherent limitations and artifacts associated with negative stain. A very different structure of TRPC3 was reported by the same group, based on images collected in ice [59]. The ice structure is lace-like and very open, with a very large overall volume. No regions of appropriate size and continuous density for a membrane-spanning domain are obvious.

The major strength of single-particle reconstruction method is its ability to produce structural information about proteins that are especially challenging for X-ray crystallography. These proteins include ones for which high-yield expression systems are not yet available, as well as large multi-domain proteins, which are too flexible and too large [92], and membrane proteins for which suitable crystallization

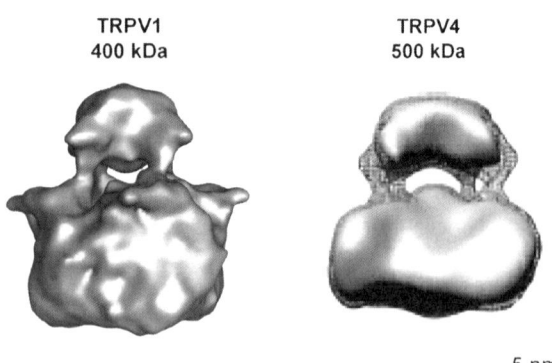

Fig. 2.2 Similarity of structures of TRPV1, *left* [71], and TRPV4, *right* [56]; (figure reproduced with permission) determined by electron cryo-microscopy

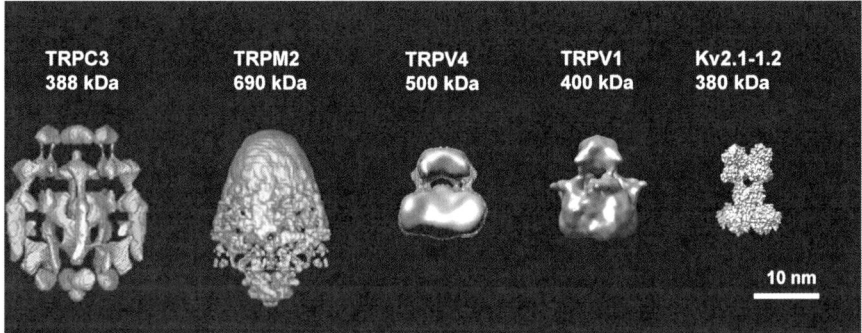

Fig. 2.3 A comparison of reported channel structures. Images reproduced by permission from [59], TRPC3 [60], TRPM2 [56, 71], TRPV1. The Kv2.1–1.2 chimera structure is from coordinates of 2R9R [41], and reproduced by permission from [91]

conditions have not yet been found. Cryo-EM allows proteins to be imaged in their native aqueous environment, and in a variety of functional states, without constrains of crystal contacts. In addition, EM data can be collected and structural information extracted as soon as purified protein is available, whereas for X-ray crystallography, considerable effort must be expended in finding suitable crystallization conditions. The major traditional disadvantage of single-particle electron microscopy has been the limitations on resolution. There have been improvements on this front in recent years, especially for large complexes and those with high symmetry, so that for the most favorable cases, near atomic resolution can be achieved, and some side chains can be visualized as well as accurate peptide backbone traces [93–95]. However, for smaller molecules, such as TRP channels, achieving atomic resolution by this method is not feasible with current methods, so that other methods such as homology modeling, mutagenesis and spectroscopy are needed to answer questions about positions of specific amino acid residues within the low resolution maps. Another limitation, which may come into play with some preparations of TRP channels, is that images of individual particles have low signal-to-noise, so it is possible to select objects other than the protein of interest, such as lipid aggregates or protein-free micelles, or to include degraded, aggregated or denatured forms of the protein in the data set. If care is not taken to avoid such "bad" particles, bizarre results can be obtained. More work will be needed, using more well-characterized TRP proteins and higher resolution cryo-EM data, to resolve the discrepancies among published TRP channel structures.

2.6 Divide and Conquer Approach: Combining X-Ray and EM Data with Computational Modeling

Given recent advances in electron microscopy and image processing [73, 74, 96], it is likely that structures of resolutions approaching 10 Å can be obtained for at least some TRP channels using electron cryo-microscopy. At this resolution, secondary elements, especially α-helices, can be identified to generate sequence-based models for understanding relationships between channel structure and function. However, in the absence of high-quality two-dimensional crystals, it is unlikely that atomic resolution structures will be obtained directly from electron microscopy, and such structures are needed to visualize such features as the conductance pore, selectivity filter, gate, and ligand-binding sites. While determining x-ray structures of full-length TRP channels remains a goal worth pursuing, success has come, and will likely continue to come, more quickly from crystallization of cytoplasmic domains of TRP channels. It should then be possible to fit these high resolution structures into lower resolution structures of complete channels determined by EM, to obtain high resolution models. For this purpose, in addition to x-ray (or perhaps NMR) structures of the domains of interest, homology models based on high resolution structures of homologous proteins (e.g. transmembrane domains of potassium channels) can also be very useful.

High resolution structures of fragments of TRP channels have included the α-kinase domain of TRPM7 [52], ankyrin repeats from TRPV subfamily of proteins [46–50], the C-terminal cytoplasmic coiled-coil domain of TRPM7 [51], and C-terminal cytoplasmic coiled-coil domain TRPP2 [53]. As an example of combining these with EM structures, the ankyrin repeat domain of TRPV1 could be readily fit into density near the membrane surface in the cytoplasmic region of the TRPV1 structure determined by cryo-EM [71]. As an example of using structures of homologous proteins for this purpose, the structure of Kv1.2, was readily fit into the transmembrane region of the same TRPV1 map.

2.7 Biochemical Studies with TRP Channels Purified from Yeast

In addition to structural analysis, purified channels can be used to measure interactions with other proteins and regulatory small molecules without the confounds arising from studies in cell membranes containing many other proteins. For example, recently, several members of TRP family have been proposed to be regulated by intracellular Ca^{2+}, and/or by calmodulin (CaM). Using calmodulin labeled with a fluorescent dye and measurements of emission anisotropy, it is possible to monitor calmodulin binding to TRPV1, TRPV2, and TRPA1 at nanomolar concentrations. Because the measurements can be carried out in real-time using T-format instrumentation, in which the light intensities for parallel and perpendicular components are detected concurrently using two independent detectors, both kinetics and thermodynamic parameters can be reliably measured (VYM and TGW, unpublished results).

2.8 Functional Studies and Genetic Screens of TRP Channels in Yeast

Once it was found that channels of other species could be expressed in functional form in budding yeast [68] it became possible to use the power and ease of yeast genetic screens to study structure-function relationships in TRP channels. This approach was initially applied to the endogenous channel of *S. cerevisiae*, TRPY1 (YVC1) [97, 98]. Heterologous expression studies revealed that TRP channels from other fungal species are also functional in *S. cerevisiae*, so they can be studied conveniently in that system [9].

TRPY1 is a large conductance (~300 pS) channel expressed in vacuolar membranes in yeast. Gain-of-function mutagenesis analysis revealed that aromatic residues on the fifth (phenylalanine) and six (tyrosine) transmembrane segments of the channel control the gating of the TRPY1 channel [98]. Alignment of all the TRP channels revealed that the phenylalanine on the intracellular base of fifth transmembrane segment of the channel is conserved and may be part of a generic gating mechanism for TRP channels. A very similar approach was used to investigate the TRPV4 gaiting mechanism, and revealed not only the importance of the main intracellular gate as reported for TRPY1, but also a new voltage-dependent gaiting mechanism for TRPV4 [99]. Genetic screens for functional changes in mammalian TRP channels have been successfully carried out in *S. cerevisiae* as aids to understanding structure-function relationships [100]. Results of this study, which used gain-or loss-of-function phenotypes, revealed that pore helix of TRPV1, TRPV2 and TRPV3 play an important role in gaiting mechanisms of these mammalian channels in addition to the intracellular gait [100].

In summary, the use of mutagenesis and the robust *S. cerevisiae* system to study functional and structure-function mechanisms of activation and gating for TRP channels will likely continue to gain popularity and produce very valuable results for understanding TRP channel biology.

2.9 Conclusion

As with any eukaryotic ion channels, studies of TRP channels at the molecular level remain challenging. However, the emergence of new tools, such as expression and genetic screens in *S. cerevisiae*, functional reconstitution protocols, cryo-electron microscopy, and x-ray crystallography of soluble fragments, will likely be combined in the next few years with the huge collection of data from functional studies in vertebrate cell membranes to provide a comprehensive picture of the structures and functional mechanisms of TRP channels.

References

1. Cosens DJ, Manning A (1969) Abnormal electroretinogram from a Drosophila mutant. Nature 224:285–287
2. Minke B, Wu C, Pak WL (1975) Induction of photoreceptor voltage noise in the dark in Drosophila mutant. Nature 258:84–87
3. Montell C, Rubin GM (1989) Molecular characterization of the Drosophila trp locus: a putative integral membrane protein required for phototransduction. Neuron 2: 1313–1323
4. Clapham DE, Montell C, Schultz G, Julius D (2003) International Union of Pharmacology. XLIII. Compendium of voltage-gated ion channels: transient receptor potential channels. Pharmacol Rev 55:591–596

5. Walker RG, Willingham AT, Zuker CS:A (2000) Drosophila mechanosensory transduction channel. Science 287:2229–2234
6. Shin JB, Adams D, Paukert M, Siba M, Sidi S, Levin M, Gillespie PG, Grunder S (2005) Xenopus TRPN1 (NOMPC) localizes to microtubule-based cilia in epithelial cells, including inner-ear hair cells. Proc Natl Acad Sci USA 102:12572–12577
7. Sidi S, Friedrich RW, Nicolson T, Nomp C (2003) TRP channel required for vertebrate sensory hair cell mechanotransduction. Science 301:96–99
8. Palmer CP, Zhou XL, Lin J, Loukin SH, Kung C, Saimi Y (2001) A TRP homolog in Saccharomyces cerevisiae forms an intracellular Ca^{2+}-permeable channel in the yeast vacuolar membrane. Proc Natl Acad Sci USA 98:7801–7805
9. Zhou XL, Loukin SH, Coria R, Kung C, Saimi Y (2005) Heterologously expressed fungal transient receptor potential channels retain mechanosensitivity in vitro and osmotic response in vivo. Eur Biophys J 34:413–422
10. Clapham DE (2003) TRP channels as cellular sensors. Nature 426:517–524
11. Montell C (2005) The TRP superfamily of cation channels. Sci STKE 2005:re3
12. Gaudet R (2008) A primer on ankyrin repeat function in TRP channels and beyond. Mol Biosyst 4:372–379
13. Lepage PK, Lussier MP, McDuff FO, Lavigne P, Boulay G (2009) The self-association of two N-terminal interaction domains plays an important role in the tetramerization of TRPC4. Cell Calcium 45:251–259
14. Auer-Grumbach M, Olschewski A, Papic L, Kremer H, McEntagart ME, Uhrig S, Fischer C, Frohlich E, Balint Z, Tang B, Strohmaier H, Lochmuller H, Schlotter-Weigel B, Senderek J, Krebs A, Dick KJ, Petty R, Longman C, Anderson NE, Padberg GW, Schelhaas HJ, van Ravenswaaij-Arts CM, Pieber TR, Crosby AH, Guelly C (2010) Alterations in the ankyrin domain of TRPV4 cause congenital distal SMA, scapuloperoneal SMA and HMSN2C. Nat Genet 42:160–164
15. Landoure G, Zdebik AA, Martinez TL, Burnett BG, Stanescu HC, Inada H, Shi Y, Taye AA, Kong L, Munns CH, Choo SS, Phelps CB, Paudel R, Houlden H, Ludlow CL, Caterina MJ, Gaudet R, Kleta R, Fischbeck KH, Sumner CJ (2010) Mutations in TRPV4 cause Charcot-Marie-Tooth disease type 2C. Nat Genet 42:170–174
16. Deng HX, Klein CJ, Yan J, Shi Y, Wu Y, Fecto F, Yau HJ, Yang Y, Zhai H, Siddique N, Hedley-Whyte ET, Delong R, Martina M, Dyck PJ, Siddique T (2010) Scapuloperoneal spinal muscular atrophy and CMT2C are allelic disorders caused by alterations in TRPV4. Nat Genet 42:165–169
17. Nilius B, Prenen J, Janssens A, Owsianik G, Wang C, Zhu MX, Voets T (2005) The selectivity filter of the cation channel TRPM4. J Biol Chem 280:22899–22906
18. Nilius B, Prenen J, Hoenderop JG, Vennekens R, Hoefs S, Weidema AF, Droogmans G, Bindels RJ (2002) Fast and slow inactivation kinetics of the Ca^{2+} channels ECaC1 and ECaC2 (TRPV5 and TRPV6). Role of the intracellular loop located between transmembrane segments 2 and 3. J Biol Chem 277:30852–30858
19. Venkatachalam K, Montell C (2007) TRP channels. Annu Rev Biochem 76:387–417
20. White JP, Cibelli M, Rei Fidalgo A, Paule CC, Noormohamed F, Urban L, Maze M, Nagy I (2010) Role of transient receptor potential and acid-sensing ion channels in peripheral inflammatory pain. Anesthesiology 112:729–741
21. Colsoul B, Nilius B, Vennekens R (2009) On the putative role of transient receptor potential cation channels in asthma. Clin Exp Allergy 39:1456–1466
22. Woudenberg-Vrenken TE, Bindels RJ, Hoenderop JG (2009) The role of transient receptor potential channels in kidney disease. Nat Rev Nephrol 5:441–449
23. Inoue R, Jian Z, Kawarabayashi Y (2009) Mechanosensitive TRP channels in cardiovascular pathophysiology. Pharmacol Ther 123:371–385
24. Cortright DN, Szallasi A (2009) TRP channels and pain. Curr Pharm Des 15:1736–1749
25. Lee LY, Gu Q (2009) Role of TRPV1 in inflammation-induced airway hypersensitivity. Curr Opin Pharmacol 9:243–249

26. Watanabe H, Murakami M, Ohba T, Ono K, Ito H (2009) The pathological role of transient receptor potential channels in heart disease. Circ J 73:419–427
27. Gallagher AR, Germino GG, Somlo S (2010) Molecular advances in autosomal dominant polycystic kidney disease. Adv Chronic Kidney Dis 17:118–130
28. Koike C, Obara T, Uriu Y, Numata T, Sanuki R, Miyata K, Koyasu T, Ueno S, Funabiki K, Tani A, Ueda H, Kondo M, Mori Y, Tachibana M, Furukawa T (2010) TRPM1 is a component of the retinal ON bipolar cell transduction channel in the mGluR6 cascade. Proc Natl Acad Sci USA 107:332–337
29. van Genderen MM, Bijveld MM, Claassen YB, Florijn RJ, Pearring JN, Meire FM, McCall MA, Riemslag FC, Gregg RG, Bergen AA, Kamermans M (2009) Mutations in TRPM1 are a common cause of complete congenital stationary night blindness. Am J Hum Genet 85:730–736
30. Shen Y, Heimel JA, Kamermans M, Peachey NS, Gregg RG, Nawy S (2009) A transient receptor potential-like channel mediates synaptic transmission in rod bipolar cells. J Neurosci 29:6088–6093
31. Morgans CW, Zhang J, Jeffrey BG, Nelson SM, Burke NS, Duvoisin RM, Brown RL (2009) TRPM1 is required for the depolarizing light response in retinal ON-bipolar cells. Proc Natl Acad Sci USA 106:19174–19178
32. Li Z, Sergouniotis PI, Michaelides M, Mackay DS, Wright GA, Devery S, Moore AT, Holder GE, Robson AG, Webster AR (2009) Recessive mutations of the gene TRPM1 abrogate ON bipolar cell function and cause complete congenital stationary night blindness in humans. Am J Hum Genet 85:711–719
33. Audo I, Kohl S, Leroy BP, Munier FL, Guillonneau X, Mohand-Said S, Bujakowska K, Nandrot EF, Lorenz B, Preising M, Kellner U, Renner AB, Bernd A, Antonio A, Moskova-Doumanova V, Lancelot ME, Poloschek CM, Drumare I, Defoort-Dhellemmes S, Wissinger B, Leveillard T, Hamel CP, Schorderet DF, De Baere E, Berger W, Jacobson SG, Zrenner E, Sahel JA, Bhattacharya SS, Zeitz C (2009) TRPM1 is mutated in patients with autosomal-recessive complete congenital stationary night blindness. Am J Hum Genet 85:720–729
34. Palczewski K, Kumasaka T, Hori T, Behnke CA, Motoshima H, Fox BA, Le Trong I, Teller DC, Okada T, Stenkamp RE, Yamamoto M, Miyano M (2000) Crystal structure of rhodopsin: a G protein-coupled receptor. Science 289:739–745
35. Park JH, Scheerer P, Hofmann KP, Choe HW, Ernst OP (2008) Crystal structure of the ligand-free G-protein-coupled receptor opsin. Nature 454:183–187
36. Jaakola VP, Griffith MT, Hanson MA, Cherezov V, Chien EY, Lane JR, Ijzerman AP, Stevens RC (2008) The 2.6 angstrom crystal structure of a human A2A adenosine receptor bound to an antagonist. Science 322:1211–1217
37. Rasmussen SG, Choi HJ, Rosenbaum DM, Kobilka TS, Thian FS, Edwards PC, Burghammer M, Ratnala VR, Sanishvili R, Fischetti RF, Schertler GF, Weis WI, Kobilka BK (2007) Crystal structure of the human beta2 adrenergic G protein coupled receptor. Nature 450:383–387
38. Shinoda T, Ogawa H, Cornelius F, Toyoshima C (2009) Crystal structure of the sodium-potassium pump at 2.4 A resolution. Nature 459:446–450
39. Long SB, Campbell EB, Mackinnon R (2005) Crystal structure of a mammalian voltage-dependent Shaker family K^+ channel. Science 309:897–903
40. Long SB, Campbell EB, Mackinnon R (2005) Voltage sensor of Kv1.2: structural basis of electromechanical coupling. Science 309:903–908
41. Long SB, Tao X, Campbell EB, MacKinnon R (2007) Atomic structure of a voltage-dependent K^+ channel in a lipid membrane-like environment. Nature 450:376–382
42. Gonzales EB, Kawate T, Gouaux E (2009) Pore architecture and ion sites in acid-sensing ion channels and P2X receptors. Nature 460:599–604
43. Tao X, Avalos JL, Chen J, MacKinnon R (2009) Crystal structure of the eukaryotic strong inward-rectifier K^+ channel Kir2.2 at 3.1 A resolution. Science 326:1668–1674

44. Sobolevsky AI, Rosconi MP, Gouaux E (2009) X-ray structure, symmetry and mechanism of an AMPA-subtype glutamate receptor. Nature 462:745–756
45. Kawate T, Michel JC, Birdsong WT, Gouaux E (2009) Crystal structure of the ATP-gated P2X(4) ion channel in the closed state. Nature 460:592–598
46. Jin X, Touhey J, Gaudet R (2006) Structure of the N-terminal ankyrin repeat domain of the TRPV2 ion channel. J Biol Chem 281:25006–25010
47. McCleverty CJ, Koesema E, Patapoutian A, Lesley SA, Kreusch A (2006) Crystal structure of the human TRPV2 channel ankyrin repeat domain. Protein Sci 15:2201–2206
48. Lishko PV, Procko E, Jin X, Phelps CB, Gaudet R (2007) The ankyrin repeats of TRPV1 bind multiple ligands and modulate channel sensitivity. Neuron 54:905–918
49. Phelps CB, Huang RJ, Lishko PV, Wang RR, Gaudet R (2008) Structural analyses of the ankyrin repeat domain of TRPV6 and related TRPV ion channels. Biochemistry 47: 2476–2484
50. Phelps CB, Wang RR, Choo SS, Gaudet R (2010) Differential regulation of TRPV1, TRPV3, and TRPV4 sensitivity through a conserved binding site on the ankyrin repeat domain. J Biol Chem 285:731–740
51. Fujiwara Y, Minor DL Jr (2008) X-ray crystal structure of a TRPM assembly domain reveals an antiparallel four-stranded coiled-coil. J Mol Biol 383:854–870
52. Yamaguchi H, Matsushita M, Nairn AC, Kuriyan J (2001) Crystal structure of the atypical protein kinase domain of a TRP channel with phosphotransferase activity. Mol Cell 7: 1047–1057
53. Yu Y, Ulbrich MH, Li MH, Buraei Z, Chen XZ, Ong AC, Tong L, Isacoff EY, Yang J (2009) Structural and molecular basis of the assembly of the TRPP2/PKD1 complex. Proc Natl Acad Sci USA 106:11558–11563
54. Brondyk WH (2009) Selecting an appropriate method for expressing a recombinant protein. Methods Enzymol 463:131–147
55. Korepanova A, Pereda-Lopez A, Solomon LR, Walter KA, Lake MR, Bianchi BR, McDonald HA, Neelands TR, Shen J, Matayoshi ED, Moreland RB, Chiu ML (2009) Expression and purification of human TRPV1 in baculovirus-infected insect cells for structural studies. Protein Expr Purif 65:38–50
56. Shigematsu H, Sokabe T, Danev R, Tominaga M, Nagayama K:A (2010) 3.5-nm structure of rat TRPV4 cation channel revealed by Zernike phase-contrast cryoelectron microscopy. J Biol Chem 285:11210–11218
57. Standfuss J, Xie G, Edwards PC, Burghammer M, Oprian DD, Schertler GF (2007) Crystal structure of a thermally stable rhodopsin mutant. J Mol Biol 372:1179–1188
58. Mio K, Ogura T, Hara Y, Mori Y, Sato C (2005) The non-selective cation-permeable channel TRPC3 is a tetrahedron with a cap on the large cytoplasmic end. Biochem Biophys Res Commun 333:768–777
59. Mio K, Ogura T, Kiyonaka S, Hiroaki Y, Tanimura Y, Fujiyoshi Y, Mori Y, Sato C (2007) The TRPC3 channel has a large internal chamber surrounded by signal sensing antennas. J Mol Biol 367:373–383
60. Maruyama Y, Ogura T, Mio K, Kiyonaka S, Kato K, Mori Y, Sato C (2007) Three-dimensional reconstruction using transmission electron microscopy reveals a swollen, bell-shaped structure of transient receptor potential melastatin type 2 cation channel. J Biol Chem 282:36961–36970
61. Parcej DN, Eckhardt-Strelau L (2003) Structural characterisation of neuronal voltage-sensitive K^+ channels heterologously expressed in Pichia pastoris. J Mol Biol 333:103–116
62. Nyblom M, Oberg F, Lindkvist-Petersson K, Hallgren K, Findlay H, Wikstrom J, Karlsson A, Hansson O, Booth PJ, Bill RM, Neutze R, Hedfalk K (2007) Exceptional overproduction of a functional human membrane protein. Protein Expr Purif 56:110–120
63. Tao X, Lee A, Limapichat W, Dougherty DA, MacKinnon R (2010) A gating charge transfer center in voltage sensors. Science 328:67–73

64. Tao X, Mackinnon R (2008) Functional analysis of Kv1.2 and paddle chimera Kv channels in planar lipid bilayers. J Mol Biol 382(1):24–33
65. Figler RA, Omote H, Nakamoto RK, Al-Shawi MK (2000) Use of chemical chaperones in the yeast Saccharomyces cerevisiae to enhance heterologous membrane protein expression: high-yield expression and purification of human P-glycoprotein. Arch Biochem Biophys 376:34–46
66. Hunte C, von Jagow G, Schagger H (2003) Membrane protein purification and crystallization: a practical guide. Academic Press, San Diego, CA
67. Denis V, Cyert MS (2002) Internal Ca^{2+} release in yeast is triggered by hypertonic shock and mediated by a TRP channel homologue. J Cell Biol 156:29–34
68. Moiseenkova VY, Hellmich HL, Christensen BN (2003) Overexpression and purification of the vanilloid receptor in yeast (Saccharomyces cerevisiae). Biochem Biophys Res Commun 310:196–201
69. Oprian DD (1993) Expression of opsin genes in COS cells. Methods Neuro. 15:301–306
70. Oprian DD, Molday RS, Kaufman RJ, Khorana HG (1987) Expression of a synthetic bovine rhodopsin gene in monkey kidney cells. Proc Natl Acad Sci USA 84:8874–8878
71. Moiseenkova-Bell VY, Stanciu LA, Serysheva II, Tobe BJ, Wensel TG (2008) Structure of TRPV1 channel revealed by electron cryomicroscopy. Proc Natl Acad Sci USA 105:7451–7455
72. Molday RS, MacKenzie D (1983) Monoclonal antibodies to rhodopsin: characterization, cross-reactivity, and application as structural probes. Biochemistry 22:653–660
73. Chiu W, Baker ML, Jiang W, Dougherty M, Schmid MF (2005) Electron cryomicroscopy of biological machines at subnanometer resolution. Structure (Camb) 13:363–372
74. Cheng Y, Walz T (2009) The advent of near-atomic resolution in single-particle electron microscopy. Annu Rev Biochem 78:723–742
75. Penczek P, Radermacher M, Frank J (1992) Three-dimensional reconstruction of single particles embedded in ice. Ultramicroscopy 40:33–53
76. Wang L, Sigworth FJ (2006) Cryo-EM and single particles. Physiology (Bethesda) 21:13–18
77. Tang G, Peng L, Baldwin PR, Mann DS, Jiang W, Rees I, Ludtke SJ (2007) EMAN2: an extensible image processing suite for electron microscopy. J Struct Biol 157:38–46
78. Hohn M, Tang G, Goodyear G, Baldwin PR, Huang Z, Penczek PA, Yang C, Glaeser RM, Adams PD, Ludtke SJ (2007) SPARX a new environment for Cryo-EM image processing. J Struct Biol 157:47–55
79. Frank J, Radermacher M, Penczek P, Zhu J, Li Y, Ladjadj M, Leith A:SPIDER (1996) WEB: processing and visualization of images in 3D electron microscopy and related fields. J Struct Biol 116:190–199
80. Sato C, Ueno Y, Asai K, Takahashi K, Sato M, Engel A, Fujiyoshi Y (2001) The voltage-sensitive sodium channel is a bell-shaped molecule with several cavities. Nature 409:1047–1051
81. Serysheva II, Bare DJ, Ludtke SJ, Kettlun CS, Chiu W, Mignery GA (2003) Structure of the type 1 inositol 1,4,5-trisphosphate receptor revealed by electron cryomicroscopy. J Biol Chem 278:21319–21322
82. Sato C, Hamada K, Ogura T, Miyazawa A, Iwasaki K, Hiroaki Y, Tani K, Terauchi A, Fujiyoshi Y, Mikoshiba K (2004) Inositol 1,4,5-trisphosphate receptor contains multiple cavities and L-shaped ligand-binding domains. J Mol Biol 336:155–164
83. Wolf M, Eberhart A, Glossmann H, Striessnig J, Grigorieff N (2003) Visualization of the domain structure of an L-type Ca^{2+} channel using electron cryo-microscopy. J Mol Biol 332:171–182
84. Serysheva II, Ludtke SJ, Baker MR, Chiu W, Hamilton SL (2002) Structure of the voltage-gated L-type Ca^{2+} channel by electron cryomicroscopy. Proc Natl Acad Sci USA 99:10370–10375

85. Sharma MR, Jeyakumar LH, Fleischer S, Wagenknecht T (2000) Three-dimensional structure of ryanodine receptor isoform three in two conformational states as visualized by cryo-electron microscopy. J Biol Chem 275:9485–9491
86. Wagenknecht T, Samso M (2002) Three-dimensional reconstruction of ryanodine receptors. Front Biosci 7:d1464–d1474
87. Serysheva II, Hamilton SL, Chiu W, Ludtke SJ (2005) Structure of Ca^{2+} release channel at 14 A resolution. J Mol Biol 345:427–431
88. Serysheva II, Ludtke SJ, Baker ML, Cong Y, Topf M, Eramian D, Sali A, Hamilton SL, Chiu W (2008) Subnanometer-resolution electron cryomicroscopy-based domain models for the cytoplasmic region of skeletal muscle RyR channel. Proc Natl Acad Sci USA 105: 9610–9615
89. Jiang QX, Wang DN, MacKinnon R (2004) Electron microscopic analysis of KvAP voltage-dependent K^+ channels in an open conformation. Nature 430:806–810
90. Wang L, Sigworth FJ (2009) Structure of the BK potassium channel in a lipid membrane from electron cryomicroscopy. Nature 461:292–295
91. Moiseenkova-Bell VY, Wensel TG (2009) Hot on the trail of TRP channel structure. J Gen Physiol 133:239–244
92. Frank J (2009) Single-particle reconstruction of biological macromolecules in electron microscopy – 30 years. Q Rev Biophys 42:139–158
93. Wolf M, Garcea RL, Grigorieff N, Harrison SC (2010) Subunit interactions in bovine papillomavirus. Proc Natl Acad Sci USA 107:6298–6303
94. Zhang X, Jin L, Fang Q, Hui WH, Zhou ZH (2010) 3.3 a cryo-EM structure of a nonenveloped virus reveals a priming mechanism for cell entry. Cell 141:472–482
95. Zhang J, Baker ML, Schroder GF, Douglas NR, Reissmann S, Jakana J, Dougherty M, Fu CJ, Levitt M, Ludtke SJ, Frydman J, Chiu W (2010) Mechanism of folding chamber closure in a group II chaperonin. Nature 463:379–383
96. Jiang W, Ludtke SJ (2005) Electron cryomicroscopy of single particles at subnanometer resolution. Curr Opin Struct Biol 15:571–577
97. Zhou X, Su Z, Anishkin A, Haynes WJ, Friske EM, Loukin SH, Kung C, Saimi Y (2007) Yeast screens show aromatic residues at the end of the sixth helix anchor transient receptor potential channel gate. Proc Natl Acad Sci USA 104:15555–15559
98. Su Z, Zhou X, Haynes WJ, Loukin SH, Anishkin A, Saimi Y, Kung C (2007) Yeast gain-of-function mutations reveal structure-function relationships conserved among different subfamilies of transient receptor potential channels. Proc Natl Acad Sci USA 104:19607–19612
99. Loukin S, Su Z, Zhou X, Kung C (2010) Forward-genetic analysis reveals multiple gating mechanisms of Trpv4. J Biol Chem 285(26):19884–19890
100. Myers BR, Bohlen CJ, Julius D (2008) A yeast genetic screen reveals a critical role for the pore helix domain in TRP channel gating. Neuron 58:362–373

Chapter 3
Natural Product Ligands of TRP Channels

Irina Vetter and Richard J. Lewis

Abstract Natural product ligands have contributed significantly to the deorphanisation of TRP ion channels. Furthermore, natural product ligands continue to provide valuable leads for the identification of ligands acting at "orphan" TRP channels. Additional naturally occurring modulators at TRP channels can be expected to be discovered in future, aiding in our understanding of not only their pharmacology and physiology, but also the therapeutic potential of this fascinating family of ion channels.

Abbreviations

$5\text{-HT}_{1A}\text{-R}$	5-hydroxytryptamine receptor 1 A
13-hydroxyoctadecadienoic acid	13-HODE
19- and 20-HETE-DA	19- and 20-hydroxyeicosa-5(Z), 8(Z), 11(Z), 14(Z)-tetraenoic acids
APHC1	Analgesic polypeptide *Heteractis crispa*
BAA	Bisandrographolide A
Ca_v	Voltage-gated Ca^{2+} channel
CFTR	Cystic fibrosis transmembrane conductance regulator
COX	Cyclo-oxygenase
DADS	Diallyl disulfide
DAS	Diallyl sulphide
DATS	Diallyl trisulfide
EGFR	Epidermal growth factor; PKCα, protein kinase C α
ERK	Extracellular Signal-Regulated Kinase
GPR55	Protein-coupled receptor 55
IP3	Inositol trisphosphate
$K_v 1.4$	Shaker-related potassium channel 1.4

R.J. Lewis (✉)
Institute for Molecular Bioscience, The University of Queensland, Queensland 4072, Australia
e-mail: r.lewis@imb.uq.edu.au

Kv2.1	Delayed rectifier potassium channel 2.1
nAChR	Nicotinic Acetylcholine receptor
N-acyl taurines	NATs
NADA	N-arachidonoyl-dopamine
Na_v	Voltage-gated sodium channel
NF-κB	Nuclear factor kappa-light-chain-enhancer of activated B cells
P2X	Ionotropic purinoreceptor
PI3/AKT	Phosphoinositide 3-kinase/protein kinase B
SERCA	Sarcoplasmic/Endoplasmic Reticulum Calcium ATPase
TRP	Transient receptor channel
Δ^9-THC	Δ^9-tetrahydrocannabinol

Natural products have been used for medicinal purposes for centuries and remain an unmatched resource to date, with more than half of currently used pharmaceuticals derived from natural origins [1]. The abundance of natural product-derived chemicals in pharmaceutical space occurs due to the often exquisite selectivity, specificity and potency of natural products compared to combinatorially designed molecules [2]. These pharmacologically desirable attributes of natural products in turn arise from millennia of evolutionary pressure and constitute a rich source of putative ligands at new and existing drug targets [2]. In addition to their use as therapeutic agents, activation or modulation by naturally occurring ligands has contributed significantly to the discovery or deorphanisation of many receptors and ion channels and has increased our understanding of their pharmacology and physiology. Natural products have undoubtedly contributed not only to the deorphanisation of several TRP channels, but indeed to the discovery of mammalian TRP channels in general, and continue to advance our understanding of this fascinating family of ion channels.

The deorphanisation of TRP channels by natural ligands is additionally unique in that many liganded TRP channels are activated by substances often encountered by humans in everyday life, and with whose sensory effects we are thus intimately familiar. Few people will not have been exposed to the culinary delight – or displeasure – of capsaicin in form of spicy chili peppers, and most are familiar with the pleasing cooling sensation elicited by menthol in oral hygiene products. While only few TRP channels have known exogenous ligands, several of these – most notably TRPV1, TRPA1 andTRPM8 – were discovered thanks to their sensitivity to naturally occurring ligands, and natural products have additionally served as scaffolds for the development of synthetic TRP modulators.

The pre-eminence of naturally occurring ligands of TRP channels highlights the role of natural products in drug discovery and receptor deorphanisation and emphasizes the possibility of future discovery of ligands for additional TRP channels, which can serve either as therapeutic agents or allow the further elucidation of the physiological role of this exciting family of ion channels.

3.1 The TRP Family

The transient receptor potential or TRP family of ion channels are cation-permeable six transmembrane homo- or heterotetrameric ion channels that can be divided into 6 mammalian subfamilies based on sequence homology. These are denoted TRPA (Ankyrin), TRPC (Canonical), TRPM (Melastatin), TRPML (Mucolipin), TRPP (Polycystin) and TRPV (Vanilloid). In nematodes, zebra fish and drosophila, an additional family known as TRPN (No mechano-potential) has also been described. In humans, 27 TRP channels have been identified to date, denoted TRPA1 (the only member of the mammalian TRPA subfamily), TRPC1, TRPC3 to TRPC7, TRPM1 to TRPM8, TRPML1 to TRPML3, TRPP2, TRPP3 and TRPP5 as well as TRPV1 to TRPV6. The function of TRP channels is diverse and ranges from thermosensation to vasoregulation, modification of growth cone morphology, detection of osmotic pressure and mechanosensation [3–6]. Several TRP channels act as polymodal sensors that are activated by physical stimuli such as temperature as well as chemical stimuli – a property that is exploited by several natural product-derived ligands. Indeed, while most TRP channels have no known exogenous ligands to date, naturally occurring chemical modulators have now been reported for most thermosensitive TRP channels, specifically TRPV1, TRPV2, TRPV3, TPV4, TRPM8 and TRPA1.

In addition to direct activation by physical stimuli such as temperature, osmotic pressure or stretch, many unliganded TRP channels are activated or modified by G-protein coupled receptors and receptor tyrosine kinases as a result of hydrolysis of phosphatidylinositol [4, 5] bisphosphate (PIP2), production of diacylglycerol (DAG), or production of inositol [1, 4, 5] trisphosphate (IP3). Given the growing number of endogenous ligands emerging as TRP channel modulators, we have summarised their affects on TRP channels in addition to the natural product-derived ligands, which are the major focus of this review.

3.2 TRPV1

TRPV1 is arguably the best known example of a TRP channel deorphanised with the help of natural ligands. The founding member of the mammalian TRP channels was discovered due to its sensitivity to capsaicin, the active ingredient from "hot" chili peppers of the genus *Capsicum* which elicits the characteristic "burning" sensation upon enjoyment of a spicy meal. These sensory qualities attributed to capsaicin by humans have implied, long before its molecular basis was understood, that capsaicin and heat might share common neuronal targets. Indeed, an elegant study published in 1997 by Julius and colleagues [7] first identified the TRPV1 using expression cloning techniques, and discovered that the TRPV1 integrates multiple pain producing stimuli and responds not only to capsaicin, but also to heat (activation threshold $> 43°$ C) and acidic pH (activation threshold $<$ pH 5.5) [8].

Nociceptive or pain sensing neurons are characterized at least in part by their responsiveness to capsaicin. It is thus not surprising that TRPV1 is predominantly expressed in nociceptive neurons, in particular small- to medium-sized DRG neurons, as well as in trigeminal and nodose ganglia [7, 9, 10]. More specifically, TRPV1 is expressed in approximately 25–50% of nociceptive C fibres, although it can also be expressed on Aδ fibres [11–16]. In addition to its predominant expression in peripheral nociceptive neurons, TRPV1 is also found in numerous brain structures such as the hypothalamus, cerebellum, cerebral cortex, medulla, hippocampus, thalamus and substantia nigra; albeit at lower levels than in DRG neurons [9, 17]. TRPV1 expression in these central structures may be involved in thermoregulation and central processing of pain [9, 17, 18]. TRPV1 immunoreactivity or mRNA has also been demonstrated in inner ear hair cells and supporting cells of the organ of Corti, keratinocytes, urothelium, bladder smooth muscle cells, cardiomyocytes, mast cells and glia [19–32]. However, TRPV1 function in non-nociceptive tissues is less well defined. Speculatively, expression of TRPV1 in these non-neuronal tissues may contribute to the development of several pathologies including over-reactive bladder and inflammatory bowel disease [4].

3.2.1 Vanilloids

Vanilloids, as the name suggests, are the prototypical vanilloid receptor TRPV1 agonists (Table 3.1). They are defined by their functional vanillyl group, and can generally belong to the class of capsaicinoids if they are structurally related to capsaicin, or the class of resiniferanoids of which resiniferatoxin is the founding member. In addition, the TRPV1 can be activated by compounds which lack a functional vanillyl moiety, in particular unsaturated dialdehydes, triprenyl phenols, polyamines as well as peptidic animal toxins [33].

3.2.1.1 Capsaicinoids

Vanilloid compounds structurally related to capsaicin are termed capsaicinoids. More than 12 pungent capsaicinoids and three major non-pungent capsinoids have been isolated from *Capsicum* species [34, 35].

Capsaicin (8-Methyl-N-vanillyl-*trans*-6-nonenamide) and dihydrocapsaicin are the major pungent constituents of *Capsicum* fruit. Sensitivity to capsaicin has long been a defining characteristic of nociceptors, and indeed capsaicin-induced Ca^{2+} responses were used to identify the TRPV1 from a rat cDNA library by expression cloning [7]. Capsaicin activates TRPV1 potently with an EC_{50} of approximately 0.04–1 μM. Interestingly, differences in capsaicin potency between heterologously expressed and endogenous TRPV1 have been attributed to the presence of FAF1, a neuronally expressed adapter protein that associates with Fas [36]. Capsaicin is thought to bind to an intracellular binding site of TRPV1, as a membrane-impermeable capsaicin analogue was able to elicit TRPV1 responses only when applied to the cytoplasmic side [37–39]. The high lipophilicity of capsaicin,

3 Natural Product Ligands of TRP Channels

Table 3.1 Natural product ligands of TRPV1

TRP subtype/ ligand	Structure/ sequence	Class	Agonist EC_{50}	Antagonist IC_{50}	Source	Alternative pharmacological target(s)	References
TRPV1							
Capsaicin		Capsaicinoid	0.04–1 μM		*Capsicum sp*		[7]
Eugenol		Capsaicinoid	~ 1mM ?		*Eugenia caryophyllata* *Cinnamomum zeylanicum* *Ocimum gratissimum* *Syzygium aromaticum*	TRPV3 TRPA1 TRPM8 Ca_v Na_v P2X	[42]
Guaiacol		Capsaicinoid	?		*Guaiacum sp*		[49]
Piperine		Capsaicinoid	38 μM		*Piper nigrum*	cytochrome P450 −3A4 P-glycoprotein	[50]
6-gingerol		Capsaicinoid	0.5–5 μM		*Zingiber officinale*	TRPA1	[53, 55]
8-gingerol		Capsaicinoid	50 μM		*Zingiber officinale*		[53, 55]
Zingerone		Capsaicinoid	0.5–5 μM		*Zingiber officinale*		[54]
6-shogaol		Capsaicinoid	0.2 μM		*Aframomum melegueta*	TRPA1	[57]
6-paradol		Capsaicinoid	0.7 μM		*Aframomum melegueta*	TRPA1	[57]
Curcumin		Capsaicinoid	?	67 nM	*Curcuma longa*	$K_v1.4$ CFTR	[58]
Resiniferatoxin		Resiniferanoid	84 pM-39 nM		*Euphorbia resinifera*		[60, 61]
Tinyatoxin		Resiniferanoid	5-30 nM				[60, 62]

Table 3.1 (continued)

TRP subtype/ ligand	Structure/ sequence	Class	Agonist EC$_{50}$	Antagonist IC$_{50}$	Source	Alternative pharmacological target(s)	References
Polygodial		α,β-unsaturated dialdehyde	5 μM		*Polygonum hydropiper*	TRPA1	[216]
Isovelleral		α,β-unsaturated dialdehyde	100 nM		*Lactarius vellereu*		[67]
Cinnamodial		α,β-unsaturated dialdehyde	0.6 μM		*Cinnamosma fragrans*		[68]
Miogadial		α,β-unsaturated dialdehyde	2 μM		*Zingiber mioga*	TRPA1	[70]
Miogatrial		α,β-unsaturated dialdehyde	6 μM		*Zingiber mioga*	TRPA1	[70]
Aframodial		α,β-unsaturated dialdehydes	12 μM		*Aframomum daniellii*		[68]
Warburganal		α,β-unsaturated dialdehyde	2.0 μM		*Warburgia ugandensis*		[67]
Merulidial		α,β-unsaturated dialdehyde	1.2 μM		*Merulius tremellosus*		[67]
Cinnamosmolide		α,β-unsaturated dialdehyde	1.5 μM		*Cinnamosma fragrans*		[68]
Cinnamolide		α,β-unsaturated dialdehyde	0.6 μM		*Cinnamosma fragrans*		[68]
Scalaradial		α,β-unsaturated dialdehyde	3.2 μM		*Cacospongia mollior*	Phospholipase A2	[68]

Table 3.1 (continued)

TRP subtype/ ligand	Structure/ sequence	Class	Agonist EC$_{50}$	Antagonist IC$_{50}$	Source	Alternative pharmacological target(s)	References
Ancistrodial		α,β-unsaturated dialdehyde	?		*Ancistrotermes cavithorax*		[68]
Drimenol		α,β-unsaturated dialdehyde	13.2 μM		*Lactarius uvidus*		[68]
Hebelomic acid F		α,β-unsaturated dialdehyde	19 μM		*Hebeloma senescens*		[68]
Isocopalendal		α,β-unsaturated dialdehyde	4.4 μM		*Spongia officinalis*		[68]
Ziniolide		α,β-unsaturated dialdehyde	?		*Xanthium catharticum*		[68]
Scutigeral		Triprenyl phenol	19 μM		*Albatrellus ovinus*		[14, 72]
Grifolin		Triprenyl phenol		26 μM	*Albatrellus confluens*	PI3/AKT	[73]
Neogrifolin		Triprenyl phenol		7 μM	*Albatrellus confluens*		[73]
Albaconol		Triprenyl phenol		17 μM	*Albatrellus confluens*	NF-κB	[73]
Cannabidiol		Cannabinoid	3 μM		*Cannabis sativa*	GPR55 5-HT$_{1A}$-R	[74]
Ginsenosides		Triterpene saponins	?	?	*Panax ginseng*		[76, 77]

Table 3.1 (continued)

TRP subtype/ ligand	Structure/ sequence	Class	Agonist EC_{50}	Antagonist IC_{50}	Source	Alternative pharmacological target(s)	References
VaTx1	SECRWFMGGCDS TLDCCKHLSCKM GLYYCAWDGTF-NH2	Vanillotoxin peptide	12 µM		*Psalmopoeus cambridgei*	Kv2.1	[79]
VaTx2	GACRWFLGGCKS TSDCCEHLSCKM GLDYCAWDGTF-NH2	Vanillotoxin peptide	3 µM		*Psalmopoeus cambridgei*	Kv2.1	[79]
VaTx3	ECRWYLGGCKED SECCEHLQCHSY WEWCLWDGSF-NH2	Vanillotoxin peptide	0.3 µM		*Psalmopoeus cambridgei*	Kv2.1	[79]
Crude venom	?	?	?		*Ornithoctonus huwena*		[79]
Agatoxin 489		Acylpolyamine		0.3 µM	*Agelenopsis aperta*	Insect glutamate receptors	[80]
Agatoxin 505		Acylpolyamine		0.3 µM	*Agelenopsis aperta*	Insect glutamate receptors	[80]
APHC1	GSICLEPKVVGPC TAYFRRFYFDSET GKCTVFIYGGCEG NGNNFETLRACR AICRA	Peptide		54 nM	*Heteractis crispa*		[81]
Leukotriene B4		Endogenous ligand	~ 30 µM		Arachidonic acid metabolite	Leukotriene receptor B1 and 2	[40]
12-HPETE		Endogenous ligand	~ 10 µM		Arachidonic acid metabolite		[40]
Anandamide		Endogenous ligand	~ 30 µM		Arachidonic acid metabolite	Cannabinoid receptors	[40, 91]
N-arachidonoyl-dopamine (NADA)		Endogenous ligand	50 nM		Arachidonic acid metabolite	T-type Ca^{2+} channels Cannabinoid receptors	[87]
19-HETE-DA		Endogenous ligand	~ 1 µM		NADA metabolite		[92]
N-oleoyl-dopamine		Endogenous ligand	36 nM		N-acyldopamine		[94]

Table 3.1 (continued)

TRP subtype/ ligand	Structure/ sequence	Class	Agonist EC_{50}	Antagonist IC_{50}	Source	Alternative pharmacological target(s)	References
N-oleoyl-ethanolamine		Endogenous ligand	~30–300 μM		N-acylethanolamine	PPAR-α	[90, 95]
13-hydroxyoctadecadienoic acid (13-HODE)		Endogenous ligand	~1 μM		Oxidized linoleic acid metabolite	PPAR-γ	[96]
Nitro-oleic acid		Endogenous ligand	? μM		Fatty acid by-product of nitric oxide and nitrite reactions		[98]
N-acyl taurine		Endogenous ligand	?		Fatty acid amides	Fatty acid amide hydrolase TRPA1	[99]
Ricinoleic acid			?		Ricinus communis		[100–102]
Thapsigargin		Sesquiterpene lactone		6 μM	Thapsia garganica	SERCA	[103]
Nicotine		Alkaloid		~1 mM	Nicotiana tabacum	nAChR TRPA1	[104]
Yohimbine		Alkaloid		25μM	Pausinystalia yohimb	α2 adrenoreceptors	[105]
Evodiamine		Alkaloid	856 nM		Evodia rutaecarpa	EGFR-PKCα-ERK	[107, 108]
Acetylsalicylic acid				1 μM	Salix sp	COX	[110, 111]
Methyl salicylate			1 mM		Gaultherias sp	TRPA1	[109]

resiniferatoxin and other vanilloids enables these compounds to readily cross the plasma membrane and activate the vanilloid binding site [37–40]. Tyrosine 511 and serine 512 in the intracellular loop linking transmembrane (TM) domains 2 and 3 appear to form part of this intracellular binding pocket [39]. This notion is supported by the observation that avian TM3, which does not facilitate hydrophobic binding, is in fact insensitive to capsaicin [38, 39, 41]. Thus, capsaicin has been an essential tool for the in-depth characterisation of TRPV1 pharmacology and continues to be the prototypical TRPV1 activator providing not only significant insights into the physiological role of TRPV1 but also culinary enjoyment for millions of people.

Eugenol is a vanilloid contained in relatively high amounts in clove oil from *Eugenia caryophyllata*, as well as cinnamon leaf oil (*Cinnamomum zeylanicum*) and oil from the clove basil *Ocimum gratissimum*. While eugenol is often referred to as a TRPV1 agonist, and indeed elicits TRPV1-mediated responses in TRPV1-expressing HEK cells and trigeminal neurons [42], recent evidence suggests that eugenol has several biological targets. In addition to TRPV1 activation, which was observed at 1 mM [42], eugenol has also been reported to activate other TRP channels including TRPV3 (3 mM) and TRPA1 (600 μM) [43, 44]. Activity at TRPM8 is less clear, with both activation and no effect reported. In addition, eugenol was recently reported to inhibit both voltage-gated calcium and sodium channels as well as purinergic P2X receptors in trigeminal ganglion neurons [45–48]. The action of eugenol at these pharmacological targets could, apart from its activity at TRPV1, contribute to the analgesic activity of eugenol.

Guaiacol is a naturally occurring vanilloid present in members of the *Guaiacum* genus of shrubs and trees as well as wood smoke and oil of cloves and can be a precursor for both eugenol and vanillin. Intrathecal guaiacol was found to inhibit formalin- and acetic acid-induced nociceptive responses in a capsazepine-sensitive manner [49], suggesting that this simple vanilloid compound acts at TRPV1.

Piperine is an alkaloid structurally related to capsaicin that occurs in plants belonging to the family of *Piperaceae* – most notably in *Piper nigrum*, commonly known as black pepper. Evidence for activation of TRPV1 by piperine stems from the observation that piperine elicits inward currents in trigeminal neurons that are blocked by the TRPV1 antagonist capsazepine [50]. In addition, whole cell patch-clamp electrophysiology using cells heterologously expressing TRPV1 showed that piperine activates TRPV1 with an EC_{50} of approximately 38 μM, making piperine a less potent TRPV1 agonist than capsaicin [51]. However, despite being less potent than capsaicin, piperine has improved efficacy at TRPV1 and also elicits enhanced desensitization and tachyphylaxis [51] through to date unclear mechanisms. Structurally, the substitution of a methylenedioxy group for the vanillyl moiety in piperine make this compound not only an agonist at TRPV1, but also a potent inhibitor of metabolizing enzymes such as cytochrome P 450-3A4 and drug transporters like p-glycoprotein [52].

Gingerols are vanillyl ketones of varying alkyl chain length contained in ginger (*Zingiber officinale*) which are transformed to zingerone, a dehydrated derivate, upon cooking. Both gingerols and zingerone activate TRPV1 in heterologous expression systems [53], with both gingerol and zingerone being more potent at rat than human TRPV1 (EC_{50} 560 nM and 4.9 µM for gingerol, respectively and 74 µM and 575 µM for zingerone, respectively) [54]. In addition, 6-gingerol was more potent than 8-gingerol in DRG neurons with EC_{50}s of 5 and 56 µM, respectively [55] and like capsaicin, gingerol-induced responses are sensitized by protons [56]. However, 6-gingerol appears to be at least somewhat non-specific, as it also displays activity at TRPA1 at a concentration of 600 µM, although an EC_{50} for gingerol at TRPA1 was not reported [44].

6-shogaol and 6-paradol, derived from the pungent oil of Melegueta pepper (*Aframomum melegueta*) and also found in ginger (*Zingiber officinale*), are agonists at TRPV1, although they were recently reported to also activate TRPA1 [57]. These hydroxy arylalkanones vanilloid compounds are slightly more potent at TRPV1 than TRPA1 with approximately tenfold higher EC_{50}s for activation of TRPA1 [57].

Curcumin, a constituent of the member of the *Zingiberaceae* family that includes tumeric, is a vanilloid compound that has been proposed to exhibit activity at TRPV1. In a mouse model of colitis, curcumin elicited protective effects that were reversed by the TRPV1 antagonist capsazepine, suggesting it acts as a TRPV1 agonist in inflamed tissue [58]. In contrast, curcumin inhibited the pro-algesic effect of capsaicin in vivo and also blocked capsaicin-induced currents in trigeminal neurons and TRPV1-expressing cells, suggesting that the vanilloid compound acts as a TRPV1 antagonist [59]. In light of these seemingly conflicting observations, the precise effect of curcumin on TRPV1 activation and sensitization remain to be determined.

3.2.1.2 Resiniferanoids

The dried latex of the cacti *Euphorbia resinifera*, *E. poissonii* and *E. unispina* yield the vanilloid resiniferatoxin [60, 61]. Resiniferatoxin and its analog tinyatoxin consist of a complex diterpene backbone to which two aromatic groups are attached, one of which is a vanillyl group [60, 62]. Interestingly, substitutions at the vanillyl moiety are relatively well tolerated, and indeed, potent non-vanilloid analogues of resiniferatoxin have been described recently [63, 64]. In contrast, resiniferanoids are very sensitive to modifications of the diterpene backbone. There is also an interesting discrepancy between the binding affinity and potency of resiniferatoxin. Resiniferatoxin binds to the TRPV1 with a K_d in the low pM range (~25–40 pM), while it its EC_{50} for activation of TRPV1 is significantly higher at approximately 1 nM [65].

3.2.2 α, β-Unsaturated Dialdehyde

While a vanillyl moiety was long believed essential for activity at TRPV1, several α,β-unsaturated dialdehydes, which lack a recognizable vanillyl motif, have now been identified as TRPV1 agonists. α,β-unsaturated dialdehydes have been isolated from a wide variety of natural sources, including plants, fungi, algae, arthropods, sponges and molluscs and are thought to form part of host defense mechanisms [66]. Perhaps for this reason, many of these compounds are relatively promiscuous at biological targets. α,β-unsaturated dialdehydes with proposed activity at TRPV1 include polygodial from the leaves of *Polygonum hydropiper*, warburganal from the bark of *Warburgia ugandensis* and *W. stuhlmannii*, isovelleral from the mushroom *Lactarius vellereu)*, merulidial from the fungus *Merulius tremellosus*, aframodial from *Aframomum daniellii*, cinnamodial, cinnamosmolide and cinnamolide from the bark of *Cinnamosma fragrans*, scalaradial from the sponge *Cacospongia mollior*, ancistrodial from the West African termite *Ancistrotermes cavithorax*, drimenol from *Lactarius uvidus*, hebelomic acid from *Hebeloma senescens*, isocopalendial from *Spongia officinalis*, fasciculol from *Naematoloma sublateritium*, ziniolide from *Xanthium catharticum*, as well as miogadial and miogatrial from flower buds of *Zingiber mioga* [67–70].

Many of these compounds, in particular aframodial, cinnamodial, cinnamosmolide, drimenol, hebelomic acid F, isocopalendial, and scalaradial, have relatively high affinity at the TRPV1 with binding affinities in the low μM range and completely displaced ^{3}H-RTX binding to rat spinal cord preparations at a concentration of 100 μM [68]. In addition, partial displacement was observed with ancistrodial, hebelomic acid A and B, fasciculol C-depsipeptide, and ziniolide [68]. Unfortunately, EC_{50} values for potency at TRPV1 were not determined for many of these compounds, with the exception of isovelleral, which induced Ca^{2+} uptake in rat DRG neurons with an EC_{50} of 100 nM, as well as cinnamodial, which evoked a biphasic concentration-response curve [67, 68] in rat DRG neurons. In addition, miogadial, miogatrial and polygodial were recently reported to activate heterologously expressed TRPV1 with EC_{50}s of 2, 6 and 5 μM, respectively, although all three compounds were found to be 10–100-fold more potent at TRPA1 [70]. Surprisingly, isovelleral was found to be inactive at concentrations up to 10 μM in a fluorescent Ca^{2+} assay at both endogenous and recombinant TRPV1 [71]. The reason(s) for this discrepancy are unclear at present time. Thus, the precise contribution of TRPV1 to biological effects of α,β-unsaturated dialdehydes, as well as their potency at TPRV1, remains to be determined systematically in both endogenous and heterologous expression systems.

3.2.3 Triprenyl Phenols

Triprenyl phenols are another class of non-vanilloid TRPV1 modulators. The first triprenyl phenol TRPV1 agonist identified was scutigeral, from the mushroom *Albatrellus ovinus*. Despite being non-pungent in human taste assays, scutigeral not only displaced ^{3}H-resiniferatoxin binding to rat DRG neurons, but also induced TRPV1-mediated Ca^{2+} influx in TRPV1-expressing cells and rat DRG neurons [14, 72]. However, like isovelleral, it was also found to be inactive in fluorescent Ca^{2+} assays at both endogenous and recombinant TRPV1 [71]. A subsequent study reported a significant decrease in the proportion of capsaicin-responsive DRG neurons after application of scutigeral, possible due to increased tachyphylaxis of the TRPV1 [73]. Additional prenyl phenols with activity at TRPV1 include grifolin, neogrifolin and albaconol (from the mushroom *Albatrellus confluens*). These compounds acted as direct TRPV1 antagonists with IC_{50} values of 26, 7 and 17 μM, respectively.

3.2.4 Cannabinoids

The marijuana plant *Cannabis sativa* has been used for medicinal purposes, including the treatment of pain, for centuries. Cannabidiol is a major nonpsychotropic constituent of *Cannabis sativa*. While virtually inactive at both CB_1 and CB_2, cannabidiol displaced ^{3}H-resiniferatoxin and elicited TRPV1-mediated Ca^{2+} responses in cells expressing human TRPV1 [74]. In addition, the in vivo analgesic activity of cannabidiol in a rat model of inflammatory pain appears to be mediated through TRPV1, as capsazepine, but not CB1 or CB2 antagonists, reversed the effect of cannabidiol [75].

3.2.5 Ginsenosides

Ginsenosides are a class of triterpene saponins found in the ginseng *Panax ginseng*. Ginsenosides were found to potentiate capsaicin-induced currents in TRPV1-expressing Xenopus oocytes, though they did not directly elicit TRPV1-mediated responses [76]. In contrast, ginsenosides inhibited TRPV1-mediated responses in rat DRG neurons [77], suggesting that ginsenosides do not act as simple TRPV1 agonists or antagonists, but rather act as mixed ligands through to-date unknown mechanisms.

3.2.6 Toxins and Peptides

Several animal-derived toxins and peptides with activity at TRPV1 have been described. Such activity is speculated to be part of host-defense mechanisms, with painful TRPV1 stimulation contributing to avoidance of venomous bites or stings [78]. The recently identified vanillotoxins VaTx1, VaTx2 and VaTx3 from the Trinidad chevron tarantula *Psalmopoeus cambridgei* exemplify such an evolutionary strategy, with all three toxins activating TRPV1 with EC_{50}s of 12, 3, and 0.3 μM, respectively [79]. Vanillotoxins belong to the group of inhibitory cysteine knot (ICK) peptides and are highly homologous to hanatoxin, which inhibits voltage-gated cation channels. Accordingly, all three peptides were demonstrated to also inhibit Kv1.2, with VaTx1 being almost equipotent at TRPV1 and Kv2.1, while VaTx3 was approximately 100-fold more selective for activation of TRPV1. Based on the similarity between vanillotoxins and huwentoxin 5 (HwTx-V), crude venom from the Chinese bird spider *Ornithoctonus huwena* was also found to activate TRPV1, although the venom component responsible for this activity was not isolated [79].

In contrast, the American funnel web spider *Agelenopsis aperta* contains high concentrations of two acylpolyamine toxins, agatoxin 489 and agatoxin 505, that inhibit rather than activate TRPV1 with an IC_{50} of 0.3 μM [80]. This inhibition is thought to occur as a result of pore block by interaction with residues in the TM5–TM6 linker and most likely reflects a mechanism common to other insect and mammalian cation channels, as agatoxin 489 and 505 also affect insect glutamate receptors [78, 80].

The first polypeptide inhibitor of TRPV1 was recently isolated from the sea anemone *Heteractis crispa* [81]. While the 56-residue polypeptide, termed analgesic polypeptide HC1 (APHC1) [81], was relatively potent at TRPV1 with an IC_{50} of 54 nM, it only elicited partial TRPV1 block [81]. Nonetheless, administration of APHC1 significantly prolonged tail-flick latency and reduced capsaicin-induced nociceptive behaviour in vivo [81].

Additional toxins that modulate TRPV1 activity rather than eliciting direct activation include gambierol and brevetoxin, marine polyethers involved in neurotoxin shellfish poisoning, which were shown to sensitize TRPV1 to activation by capsaicin, albeit at relatively high concentrations (EC_{50} 613 and 350 nM, respectively) [82]. Similarly, cnidarian venom from the anemone *Aiptasia pulchella*, the box jellyfish *Chironex fleckeri*, the bluebottle or Portugese Man O' War *Physialia physalis* and the lion's mane jellyfish *Cyanea capillata* appear to interfere with TRPV1 desensitization, an effect which could contribute to the long-lasting burning sensation after envenomation by these marine creatures [83].

3.2.7 Endogenous Ligands

The search for endogenous TRPV1 ligands has revealed that a number of endogenous compounds can act as TRPV1 agonists and may thus play a role in inflammatory and neuropathic hyperalgesia [84–90]. These so-called "endovanilloids" include 5-lipooxygenase and 12-lipooxygenase products, the most potent of which are leukotriene B4 and 12-S-HPETE; the cannabinoid receptor agonist anandamide, NADA (N-arachidonoyl-dopamine) and its ω-hydroxylated metabolites 19- and 20-HETE-DA, N-oleoyl-dopamine, as well as N-acetylethanolamines including oleoylethanolamide [40, 87, 90–95]. In recent years, the number of fatty acid derived TRPV1 ligands has been further extended with the observation that the oxidized linoleic acid metabolites 9- and 13-hydroxyoctadecadienoic acid (9- and 13-HODE) as well as nitro-oleic acid, N-arachidonoylsalsolinol and N-acyltaurines can activate TRPV1 [96–99]. However, many of these compounds are promiscuous with activity at other TRP channels, such as TRPA1 in the case of nito-oleic acid, or TRPV4 in the case of N-acyltaurines, but also non-TRP targets including the cannabinoid receptor and fatty acid amide hydrolase [99]. While the physiological role of these endogenous TRPV1 modulators remains largely unknown, additional endogenous fatty acid-derived TRPV1 ligands can be expected to be identified in future.

3.2.8 Miscellaneous Compounds

Ricinoleic acid, the main ingredient of castor oil, has been described as a capsaicin-like compound due to similarities of their biological effects. Though not pungent, some evidence suggests that ricinoleic acid has in vivo activity comparable to capsaicin in anti-inflammatory assays [100–102]. However, its direct activity at TRPV1 remains to be determined conclusively. Similarly, while thapsigargin (a sesquiterpene lactone from *Thapsia garganica*; IC_{50} at TRPV1 6 μM), nicotine (an alkaloid from *Nicotiana tabacum*; activity at TRPV1 at 1 mM) and yohimbine (from bark of the tree *Pausinystalia yohimbe* or from the root of *Rauwolfia*; IC_{50} at TRPV1 25 μM) have been reported to modulate TRPV1 activity, it is unlikely that TRPV1 is the primary pharmacological target of these compounds, as they are > 100-fold more potent at SERCA, nicotinic acetylcholine receptors and α2 adrenoreceptors, respectively [103–105]. The polyamines putrescine, spermidine and spermine are derived metabolically from the amino acids L-arginine and L-methionine and occur in many organisms. These polyamines were found to activate and sensitize endogenously or heterologously expressed TRPV1 [106]. However, the spectrum of biological activity for these compounds is varied and the physiological relevance of TRPV1 activation by polyamines remains to be determined.

Evodiamine, isolated from the dried fruits of the Chinese herb *Evodia rutaecarpa*, causes capsaicin-like effects in vivo, including capsazepine-sensitive contraction of isolated guinea pig bronchus and nocifensive behaviour, specifically increased paw licking. In addition, evodiamine elicits TRPV1-mediated Ca^{2+} influx and displaces ^{3}H-RTX binding, making it (EC_{50} of 856 nM) a moderately potent TRPV1 agonist [107, 108]. In addition, other biological effects of evodiamine include inhibition of adipogenesis via EGFR-PKCα-ERK signaling pathways as well as increased apoptosis of cancer cells [108].

Salicylates, including acetylsalicylic acid or aspirin, from the willow tree of the genus *Salix* and methylsalicylate from species of the shrub genus *Gaultheria* have been used for centuries for their analgesic and anti-inflammatory properties. While the main molecular target for acetylsalicylic acid is cyclooxygenase and methylsalicylate activates TRPA1, some evidence suggests that these compounds also act at TRPV1. Methylsalicylate was demonstrated to induce TRPV1-mediated Ca^{2+} influx and induced marked desensitization of TRPV1 with resultant inhibition of other TRPV1 stimuli such as capsaicin, anandamide and protons [109]. This activity occurred in the mM range, making this compound less potent at TRPV1 than TRPA1 [44, 109]. Similarly, acetylsalicylic acid was shown to inhibit capsaicin-induced pain, flare and allodynia after topical administration and also inhibited capsaicin-induced responses in DRG neurons with an IC_{50} of approximately 1 μM, suggesting that inhibition of TRPV1 might contribute to the analgesic activity of aspirin [110, 111].

In summary, these compounds have activity at TRPV1, which might contribute to some of their non-specific effects and needs to be taken into consideration when using these compounds.

3.3 TRPV2

TRPV2 (Table 3.2) was originally described as an analogue of TRPV1, or vanilloid-receptor-like protein 1 (VRL-1) [112]. Perhaps owing to the paucity of pharmacological modulators of TRPV2, its physiological role is relatively poorly understood. TRPV2 is generally referred to as a member of the thermosensitive TRP channels, with an activation threshold of >52° C, although intriguing species differences in the activation of heterologously expressed TRPV2 by both noxious heat and the small molecular 2-aminoethoxydiphenyl borate (2-APB) were reported recently [113]. To add further functional complexity, TRPV2 expressed in myocytes responds to changes in osmolarity and membrane stretch [114]. Accordingly, TRPV2 has been proposed to play a role in pain as well as skeletal and cardiac muscle degeneration.

3.3.1 Cannabinoids

The psychotropic cannabinoid Δ^9-tetrahydrocannabinol (THC) from the marijuana plant *Cannabis sativa* was the first reported naturally occurring ligand of TRPV2, although its potency at TRPV2 is relatively poor with EC_{50}s of 16 µM at rat and 43 µM at human TRPV2, respectively [113]. In addition, activation of TRP channels by THC is relatively non-selective as TRPA1 is also activated with similar potency [115, 116]. In contrast, the non-psychotropic constituents of *Cannabis sativa*, cannabidiol and cannabinol, were found to preferentially activate TRPV2 with at least 20-fold higher selectivity over TRPA1 and negligible effect on TRPV1 [117]. While the potency of these compounds at TRPV2 is relatively low with EC_{50}s of 3.7 µM for cannabidiol and 77 µM for cannabinol, these naturally occurring ligands may serve as scaffolds for the future development of more potent TRPV2-selective ligands.

3.3.2 Probenecid

While not a natural product, probenecid deserves a mention in this place as it appears to be the most selective TRPV2 agonist currently known. It activates TRPV2 with an EC_{50} of 32 µM with minimal effects on other TRP channels [118].

3.3.3 Endogenous Ligands

Endogenous lysophospholipids, in particular lysophosphatidylcholine and lysophosphatidylinositol have recently been described as endogenous modulators of TRPV2. Lysophospholipids caused translocation of TRPV2 to the plasma membrane and activation of TRPV2 via Gq/Go-protein and phosphatidylinositol-3,4 kinase (PI3,4 K) signaling [119].

Table 3.2 Natural product ligands of TRPV2–TRPV6

TRP subtype/ ligand	Structure/ sequence	Class	Agonist EC_{50}	Antagonist IC_{50}	Source	Alternative pharmacological target(s)	References
TRPV2							
Δ^9-tetrahydrocannabinol		Cannabinoid	16–43 µM		*Cannabis sativa*	CB1 and CB2 cannabinoid receptors TRPA1	[113]
Cannabidiol		Cannabinoid	3.7 µM		*Cannabis sativa*	TRPV1 TRPA1	[117]
Cannabinol		Cannabinoid	77 µM		*Cannabis sativa*	TRPA1	[117]
Lysophosphatidyl choline		Endogenous ligand	?		Lysophospholipid	TRPM8	[119]
Lysophosphatidyl inositol		Endogenous ligand	?		Lysophospholipid	GPR55 TRPM8	[119]
TRPV3							
Camphor		Terpenoid	6 mM		*Cinnamomum camphora*	TRPV1 TRPA1	[123]
Carvacrol		Terpenoid	? µM		*Origanum vulgare*	TRPA1 TRPM7	[43]
Thymol		Terpenoid	? µM		*Thymus vulgaris*		[43]
Incensole		Terpenoid	16 µM		*Boswellia resin*		[137]

? = not reported or unknown

Table 3.2 (continued)

TRP subtype/ ligand	Structure/ sequence	Class	Agonist EC_{50}	Antagonist IC_{50}	Source	Alternative pharmacological target(s)	References
Vanillin		Vanilloid	? mM		*Vanilla planifolia*		[43]
Farnesyl pyrophosphate		Endogenous ligand	131 nM		*Metabolite in the mevalonate pathway*	GPR92 LAP2 LAP3	[138]
TRPV4							
4α-phorbol 12,13-didecanoate		Terpenoid	200 nM		Semisynthetic from *Croton tiglium*		[163, 164]
Bisandrographoli de A		Terpenoid	750–900 nM		*Andrographis paniculata*		[146, 147]
5,6-epoxyeicosatrieno ic acid		Endogenous ligand	130 nM		*Arachidonic acid metabolite*		[166]
N-acyl taurine		Endogenous ligand	?		*Fatty acid amides*	Fatty acid amide hydrolase TRPV1	[99]

3.4 TRPV3

The third member of the vanilloid subfamily of TRP ion channels, TRPV3 (Table 3.2), is highly expressed in keratinocytes, dorsal root and trigeminal ganglia, as well as spinal cord and brain [120–122]. It responds to intermediate temperatures, with an activation threshold of between 33 and 39° C [120–122]. In contrast to TRPV1 and TRPV4, which desensitize after repeated stimulation, TRPV3 currents are increasingly sensitized upon repeated stimulation and in addition display the unique property of hysteresis, where desensitization occurs in response to slight decreases in temperature, even well above the initial threshold of activation [120–122]. Consistent with its expression patterns, TRPV3 has been suggested to play a role in thermal nociception, thermoregulation, hair growth and inflammatory skin conditions such as dermatitis, although its precise physiological roles remains to be determined [123–127].

3.4.1 Monoterpenoids and Diterpenoids

Since TRPV3 is expressed in oral and nasal epithelium as well as keratinocytes, it comes as no surprise that several naturally occurring aromatic agents activate TRPV3 [43, 128, 129]. The feeling of warmth in response to application of camphor, a naturally occurring monoterpene isolated from *Cinnamomum camphora*, as well as the resultant sensitization to heat application, was attributed to activation of TRPV3 channels in keratinocytes [123]. However, the potency of camphor is extremely poor, with an EC_{50} at TRPV3 of 6 mM [123, 128]. In addition, camphor was also found to activate TRPV1 and block TRPA1 with an IC_{50} of 0.6 mM [130]. The biosynthetic precursor of camphor (+)-borneol, also activated TRPV3 with similar potency, although borneol-evoked responses were of greater magnitude compared to camphor [128]. Subsequently, carvacrol, the major constituent of *Origanum vulgare*; thymol, an important component of *Thymus vulgaris*; and the vanilloid eugenol, the principal active constituent of the clove plant *Syzygium aromaticum*, were identified as TRPV3 activators [43]. While these compounds are slightly more potent than camphor with activity at TRPV3 in the high μM range, they nevertheless demonstrate relatively poor receptor specificity, as carvacrol also activates TRPA1 and inhibits TRPM7, while eugenol is a weak agonist at TRPV1 [43, 131]. Further illustrating the relative promiscuity of many naturally occurring TRP ligands, the monoterpene menthol, isolated from the peppermint *Mentha piperita* and generally considered a TRPM8-specific agonist, was recently found to activate TRPV3 [128, 132]. However, while activation of TRPM8 by menthol occurs with an EC_{50} of approximately 4–80 μM [133–135], its effects on other thermosensitive TRP channels, including activation of TRPV3 and inhibition of TRPA1, occur with significantly lower potency.

Citral, an isomeric mixture of the terpenoids neral and geranial and a fragrant component of lemongrass oil, lemon peel, citronella, and palmarosa grass, has recently been reported to non-selectively activate and inhibit TRPV1, TRPV3, TRPM8 and TRPA1, albeit with low potency [136].

The diterpenoid incensole acetate, a constituent of frankincense obtained from *Boswellia sp* is the most potent and the most selective TRPV3 agonist identified to date [137]. Activation of native and heterologously expressed TRPV3 by incensole acetate occurred with an EC_{50} of 16 μM, while incensole acetate did not activate TRPV2 and caused minimal responses at TRPV1 and TRPV4. In addition, incensole acetate did not affect binding to an additional 24 G protein coupled receptors, ion channels and transporters, although activity at other TRP channels remains to be confirmed [137].

3.4.2 Vanilloids

Apart from the vanilloid eugenol, which as mentioned above activates TRPV3 and TRPV1, the active component of *Vanilla planifolia*, vanillin, also weakly activates TRPV3 [43]. However, the synthetic analogue ethylvanillin elicits TRPV3 responses of approximately five to ten times greater magnitude than the native vanillin [43].

3.4.3 Endogenous Ligands

Farnesyl pyrophosphate (FPP), an intermediate metabolite in the mevalonate pathway, was recently reported to activate TRPV3 with an EC_{50} of 130 nM [138]. While no activity was observed on TRPV1, TRPV2, TRPV4, TRPA1 and TRPM8, farnesyl pyrophosphate is also known to activate the membrane receptors GPR92 and inhibits LAP2 and LAP3 receptors [138].

3.5 TRPV4

TRPV4 (Table 3.2) was initially described as an osmolarity-sensitive channel [139–142]. In addition, TRPV4 is activated by warm temperatures (> 25° C) as well as mechanical stimuli [143–147]. In contrast to TRPV3, sustained heat-evoked stimulation of TRPV4 leads to rapid desensitization [144]. The wide tissue distribution of TRPV4, including expression in DRG neurons, epidermal keratinocytes, inner ear hair cells, epithelial cells of the nephron, trachea, bile ducts and fallopian tubes as well as osteoclasts and osteoblasts, suggests that this ion channel may be involved in many physiological processes [145, 148, 149]. In vivo studies using TRPV4 knockout mice confirmed its role in central osmosensation as well as a significant contribution to mechanical and osmotic nociception [145, 150–154]. In addition, recent genetic studies suggest a role for TRPV4 in disorders of bone development, including brachyolmia, spondylometaphyseal dysplasia (SMD) and metatropic dysplasia as well as the channelopathies Scapuloperoneal spinal muscle atrophy and Charcot-Marie-Tooth disease type 2C [155–157]. The identification of novel TRPV4-specific agonists and antagonists can be expected to provide further insights into the physiological and pathophysiological role(s) of TRPV4.

3.5.1 Diterpenoids

As for TRPV3, the first selective TRPV4 activator described – the phorbol ester 4α-phorbol 12,13-didecanoate (4αPDD) – belongs to the family of tetracyclic diterpenoid compounds. Phorbol esters were first isolated from the shrub *Croton tiglium* which, along with several other plants yielding phorbol esters, belongs to the family of *Euphorbiacea* [158–160]. Of note, the placement of the OH group in the C ring of phorbols determines their activity at protein kinase C (PKC), with α isoforms inactive at PKC due to conformational shifts [161]. Although inactive at PKC [162], the semisynthetic phorbol ester 4αPDD was found to activate TRPV4 with an EC_{50} of approximately 200 nM [163, 164]. Similarly, the PKC activator 4-β-phorbol 12-myristate 13-acetate (PMA) also activated TRPV4 with a comparable EC_{50}, albeit as a partial agonist [163]. The binding site of phorbol esters at TRPV4 is formed by Y555 and S556 in the third transmembrane domain, corresponding to the Y511/S512 motif that contributes to the capsaicin binding site in TRPV1 [39, 165]. Mutation of these residues affected responses to 4αPDD but not hypotonic stimuli or the endogenous activator arachidonic acid [165].

Bisandrographolide A (BAA), derived from the bitter herbaceous plant *Andrographis paniculata*, is another diterpenoid that was identified as an activator of TRPV4 [146, 147]. While BAA is not particularly potent at TRPV4 activation (EC_{50} 750–900 nM), it appears to be relatively specific to TRPV4 as neither TRPV1, TRPV2 or TRPV3 were activated by up to 2.6 μM BAA [146]. In contrast to phorbol esters, residues involved in activation of TRPV4 by BAA were reported to include L584 and W586 [147].

3.5.2 Endogenous Ligands

The observation that arachidonic acid and its precursors anandamide and 2-arachidonoylglycerol activate TRPV4 lead to the identification of epoxyeicosatrienoic acids, in particular 5,6-epoxyeicosatrienoic acid, as an endogenous TRPV4 activator [166]. In addition, the fatty acid amide hydrolase substrates N-acyl taurines (NATs) can activate TRPV4 in addition to TRPV1 with an EC_{50} of 21 and 28 μM, respectively [99]. The physiological role of these activators remains unclear, although it is of interest that NATs were accumulated in kidney, a tissue with high TRPV4 expression.

3.6 TRPM8

The pleasant "cooling" or "fresh" sensation elicited by menthol has been exploited commercially in many products ranging from toothpaste to compression sports wear. At the molecular level, these sensory effects result from activation of the cold-activated TRPM8 by menthol (Table 3.3). Indeed, the "menthol receptor" TRPM8 was discovered – similar to TRPV1 – due to its sensitivity to a natural product that produces a distinct thermosensation upon exposure. Using bioinformatics and expression cloning techniques, the menthol receptor was initially identified in dorsal root ganglion and trigeminal neurons [133, 134], consistent with its role in thermosensation. TRPM8 is activated by cool temperatures with an upper threshold of approximately 23–27°C [133, 134]. Apart from sensory neurons, TRPM8 has also been found to be highly expressed in prostate, in particular prostate carcinoma, and liver; suggesting physiological and pathophysiological roles other than thermosensation [167, 168].

3.6.1 Monoterpenoids

Menthol is a cyclic terpene alcohol that is a major constituent of several species of mint, including the common peppermint *Mentha piperita* and the cornmint *M. arvesis*, which is the source of commercial natural menthol. The (–) optical isomer of menthol potently activates TRPM8 with an EC_{50} of approximately 4–80 μM, while (+)-menthol is slightly less potent [133, 135]. In addition, the biosynthetic menthol precursor menthone also elicited small responses in TRPM8-expressing cells, although the EC_{50} was not determined for this compound.

Other monoterpenoids that reportedly activate TRPM8 include, in order of decreasing potency and efficacy, isopulegol (from the pennyroyal *Mentha pulegium*), geraniol (a component of many essential oils, in particular rose oil), linalool (found in many plants including the family of *Lamiaceae*, to which mints belong), eucalyptol (1,8-cineole from *Eucalyptus sp.*) and hydroxy-citronellal (isolated from plants of the genus *Cymbopogon*) [135]. In contrast, the synthetic analogue of menthol, cyclohexanol, as well as the TRPV3 agonist camphor and the TRPV1 agonist eugenol had no effect at mouse TRPM8 expressed in HEK293 cells [133, 135], while one study found that eugenol activated mouse TRPM8 expressed in CHO cells [44].

3.6.2 Endogenous Ligands

In contrast to the heat-activated TRPV1, which is activated by arachidonic acid, the cold-activated TRPM8 is inhibited by arachidonic acid as well as eicosapentaenoic and docosahexaenoic acid [169]. In contrast, lysophosphatidylcholine elicited

3 Natural Product Ligands of TRP Channels 63

Table 3.3 Natural product ligands of TRPM1–TRPM8

TRP subtype/ligand	Structure/sequence	Class	Agonist EC$_{50}$	Antagonist IC$_{50}$	Source	Alternative pharmacological target(s)	References
TRPM2							
Arachidonic acid		Endogenous ligand	? μM		Polyunsaturated fatty acid		[173]
TRPM3							
Pregnenolone sulphate		Steroid	12 μM		Neurosteroid, cholesterol metabolism	Steroid receptors	[172]
D-erythro-sphingosine		Sphingolipid	~ 20 μM		Cellular sphingolipid metabolism	Ca$_v$ Ryanodine receptor	[171]
TRPM4							
Spermine		Polyamine		35 μM	Polyamine found in sperm	TRPV1 TRPM5	[177]
9-phenanthrol				~ 20 μM	Metabolite of polycyclic aromatic hydrocarbons		[178, 179]
TRPM5							
Spermine		Polyamine		35 μM	Polyamine found in sperm	TRPV1 TRPM4	[177]
TRPM8							
Menthol		Terpenoid	4 – 80 μM		*Mentha piperita* *Mentha arvesis*	TRPV3	[133, 134]
Isopulegol (Coolact P)		Terpenoid	66 μM		*Mentha pulegium*		[135]

Table 3.3 (continued)

TRP subtype/ ligand	Structure/ sequence	Class	Agonist EC_{50}	Antagonist IC_{50}	Source	Alternative pharmacological target(s)	References
Geraniol		Terpenoid	6 mM				[135]
Linalool		Terpenoid	6.7 mM		*Lamiaceae sp.*	TRPA1	[135]
Eucalyptol		Terpenoid	7.7 mM		*Eucalyptus globulus*		[135]
Hydroxy-citronellal		Terpenoid	20 mM		*Cymbopogon sp.*		[135]
Arachidonic acid		Endogenous ligand		1–3 µM	Polyunsaturated fatty acid		[169]
Eicosapentanoic acid		Endogenous ligand		2–6 µM	Polyunsaturated fatty acid		[169]
Docosahexaenoic acid		Endogenous ligand		1–2 µM	Polyunsaturated fatty acid		[169]
Lysophosphatidyl choline		Endogenous ligand	? µM		Lysophospholipid		[169]
Lysophosphatidyl inositol		Endogenous ligand	? µM		Lysophospholipid		[169]
Lysophosphatidyl serine		Endogenous ligand	? µM		Lysophospholipid		[169]

TRPM8-mediated responses in transfected CHO cells, as did the anionic lysophosphatidylinositol and lysophosphatidylserine, while the sphingolipid sphingosylphosphorylcholine also acted as a TRPM8 agonist, albeit with considerably reduced potency [169, 170]. Intriguingly, these endogenous TRPM8 modulators shifted the activation temperature of TRPM8 to more physiological temperatures, suggesting that they might be involved either in regulating the temperature sensitivity of neurons innervating tissues exposed to external temperatures, or regulating the activity of TRPM8 in neurons only exposed to physiological temperatures [169].

3.7 TRPM1 – TRPM7

For other members of the TRPM family of ion channels, naturally occurring and synthetic ligands are slowly emerging, although the study of their effects is still in its infancy. Exemplifying the newly-liganded TRPM channels, pregnenolone and other naturally occurring steroids are reported to activate TRPM3, as do sphingolipids such as sphingosine [171]. D-erythro-sphingosine is a metabolic product of de novo synthesis of cellular sphingolipids and is the only naturally occurring stereoisomer. While D-erythro-sphingosine is known to inhibit several ion channels, including voltage-gated calcium channels and skeletal muscle ryanodine receptors, it activated rather than blocked TRPM3 at a concentration of 20 μM [171]. Similarly, pregnenolone sulphate, a naturally occurring steroid, induced TRPM3-mediated Ca^{2+} influx in pancreatic β cells, with no effect on TRPM2, TRPM7, TRPM8 or TRPV4 and TRPV6 [172].

In addition to nicotinamide adenine dinucleotide, adenosine 5′-diphosphoribose or hydrogen peroxide, TRPM2, like many other TRP channels, has been reported to be modulated by arachidonic acid [173–175]. Surprisingly, while the phospholipase A2 inhibitor N-(p-amylcinnamoyl) anthranilic acid blocked TRPM2-mediated responses, this effect appeared to occur independently of phospholipase A2 activity, suggesting that mechanisms other than decreased arachidonic acid production are involved in this effect [176].

Spermine, while shown to activate and sensitize TRPV1, inhibits TRPM4 and TRPM5 [177], while 9-phenanthrol, a metabolite of polycyclic aromatic hydrocarbons, and decavanadate only affected TRPM4 [178, 179]. However, the in vivo significance of these effects remains to be determined, as these compounds are known to affect many biological systems and targets.

Similarly, carvacrol, the major constitutent of *Origanum vulgare*, is a known TRPV3 activator, but also inhibits TRPM7 with an IC_{50} of 300 μM [131]. Thus, while selectivity and specificity of TRPM ligands remains lacking, naturally occurring compounds have provided the starting point for the identification of novel pharmacological modulators to previously "orphaned" TRP channels, and may lead to the discovery and/or development of more specific and selective pharmacological modulators.

3.8 TRPA1

As for TRPV1 and TRPM8, natural products also played a significant role in the discovery as well as characterisation of TRPA1 (Table 3.4) [44, 116]. Heterologously expressed TRPA1 as well as TRPA1 endogenously expressed in DRG neurons responds to noxious cold temperatures (< 17° C) [44, 180], suggesting a crucial role in thermosensation. Interestingly, several lines of evidence support a role for TRPA1 in mechanical as well as thermal hyperalgesia, making this ion channel an attractive target for putative analgesic compounds [181–185]. Apart from activation by cold stimuli, several naturally occurring compounds owe their pungency to the activation of TRA1.

3.8.1 Isothiocyanates

The unique sensory experience often described as "pungent" or "burning" that results from consumption of mustard, wasabi and horseradish results at least in part from activation of TRPA1 by active components known as isothiocyanates [44, 116]. These compounds, in the form of allyl, benzyl, phenylethyl, isopropyl and methyl isothiocyanate are found in varying quantities in wasabi, yellow mustard, Brussels sprouts, nasturtium seeds and capers, respectively [44, 116]. Allyl isothiocyanate in particular is a relatively potent agonist at TRPA1 with an EC_{50} of approximately 1–6.5 μM, and slightly higher affinity for human than rat TRPA1 [186–188]. Interestingly, these compounds are potent electrophiles and appear to activate TRPA1, at least in part, as a result of covalent modification of the channel, rather than based on traditional lock-and-key agonist binding [189, 190]. In contrast, phospholipase C-mediated activation of TRPA1 appears to involve different mechanisms [189].

3.8.2 Thiosulfinates

Like isothiocyanates, thiosulfinates are electrophilic compounds that activate TRPA1 as a result of covalent modification of three cysteine residues in the N-terminal of the ion channel [189]. Thiosulfinates occur in plants of the genus *Allium*, to which garlic as well as onions belong. Allicin (2-propenyl 2-propene thiosulfinate) is particularly abundant in garlic and activates TRPA1 with an EC_{50} of approximately 7.5 μM [191]. However, allicin is relatively unstable in aqueous solution and rapidly yields other compounds including sulfides such as diallyl sulfide (DAS), diallyl disulfide (DADS), diallyl trisulfide (DATS) and ajoene. These compounds also modulate TRPA1 function, albeit with varying potency. DATS appears to be the most potent TRPA1 activator with an EC_{50} in the low μM range, while both DAS and DADS are considerably less potent [187, 191]. In addition, these diallyl sulfides can also activate TRPV1, which may contribute to their neurosensory effects [187]. Ajoene, a thiosulfinate derivate of allicin, appears to potentiate both allicin and allyl isothiocyanate responses rather than directly activate TRPA1 [192].

3 Natural Product Ligands of TRP Channels

Table 3.4 Natural product ligands of TRPA1

TRP subtype/ ligand	Structure/ sequence	Class	Agonist EC$_{50}$	Antagonist IC$_{50}$	Source	Alternative pharmacological target(s)	References
TRPA1							
Allyl isothiocyanate		Isothiocyanate	1–6.5 μM		*Brassica nigra* *Brassica juncea* *Brassica hirta* *Wasabia japonica* *Armoracia rusticana*		[44, 116]
Allicin		Thiosulfinate	7.5 μM		*Allium sativum*		[191]
Diallyl sulfide		Thiosulfinate	190–250 μM		*Allium sativum*	TRPV1	[187, 191]
Diallyl disulfide		Thiosulfinate	7 μM		*Allium sativum*	TRPV1	[187, 191]
Diallyl trisulfide		Thiosulfinate	0.5 μM		*Allium sativum*	TRPV1	[187, 191]
Ajoene		Thiosulfinate	0.5 μM		*Allium sativum*	Potentiation only	[192]
Cinnamaldehyde		α,β-unsaturated aldehyde	60 μM		*Cinnamomum cassia* *Cinnamomum zeylanicum*		[44]
Crotonaldehyde		α,β-unsaturated aldehyde	16 μM		Cigarette smoke		[193]
Acrolein		α,β-unsaturated aldehyde	5 μM		Cigarette smoke		[193]
4-hydroxynonenal		α,β-unsaturated aldehyde	27 μM		Peroxidated membrane phospholipids		[194]

Table 3.4 (continued)

TRP subtype/ ligand	Structure/ sequence	Class	Agonist EC_{50}	Antagonist IC_{50}	Source	Alternative pharmacological target(s)	References
Polygodial		α,β-unsaturated dialdehyde	60–600 nM		*Polygonum hydropiper*	TRPV1	[70]
Miogadial		α,β-unsaturated dialdehyde	200–400 nM		*Zingiber mioga*	TRPV1	[70]
Miogatrial		α,β-unsaturated dialdehyde	130–630 nM		*Zingiber mioga*	TRPV1	[70]
Acetaldehyde		Aldehyde	76 μM (hTRPA1)		Metabolite of ethanol		[195]
Δ^9-tetrahydrocannabinol		Cannabinoid	12 μM		*Cannabis sativa*	CB1 and CB2 cannabinoid receptors TRPV2	[116]
Cannabinol		Cannabinoid	~ 20 μM		*Cannabis sativa*	TRPV2	[116]
Hydroxy-α-sanshool		Alkylamide	69 μM		*Zanthoxylum piperitum*	TRPV1	[57]
Methyl salicylate			600 μM		*Gaultherias sp*	TRPV1	[44]
Camphor		Terpenoid		0.66 mM	*Cinnamomum camphora*	TRPV1 TRPV3	[130]
15-deoxy-$\Delta^{12,14}$-prostaglandin J_2		Endogeous ligand	? μM		Arachidonic acid metabolite	PPAR-γ	[201]

3.8.3 α, β-Unsaturated Aldehyde

Leaves and/or bark of *Cinnamomum cassia* and *Cinnamomum zeylanicum* yield a pungent oil which contains α,β-unsaturated aldehydes including cinnamaldehyde. Cinnamaldehyde activated TRPA1, but not TRPV1 or TRPM8, with an EC_{50} of approximately 60 μM [44]. Like mustard oil, cinnamaldehyde interacts with TRPA1 in a covalent manner [190]. The responsiveness of TRPA1 to unsaturated aldehydes also extends to crotonaldehyde and acrolein, which are constituents of cigarette smoke and elicit neurogenic airways inflammation through activation of TRPA1 [193]. In addition, the endogenous α,β-unsaturated aldehyde 4-hydroxynonenal as well as acetaldehyde, a metabolite of ethanol [194, 195] activate TRPA1, corroborating its role as a sensor of noxious and irritant chemicals.

3.8.4 Cannabinoids

The observation that Δ^9-THC caused G-protein-coupled receptor-independent but ruthenium red-sensitive vasorelaxation [196] led to the discovery that both Δ^9-THC and cannabinol act as TRPA1 agonists [116]. While it has been suggested that activity at TRPA1 could contribute to some of the in vivo effects of cannabinoids, activation of TRPA1 by Δ^9-THC occurred with an EC_{50} of 12 μM, which is significantly higher than the nM binding affinity of Δ^9-THC to the CB1 receptor [197]. Interestingly, Δ^9-THC appears to be able to distinguish different functional states of TRPA1, as cytosolic polyphosphates are required for activation of TRPA1 by pungent chemicals including allyl isothiocyanate, but not for activation of TRPA1 by Δ^9-THC [198]. However, in light of the activity of cannabinoids at TRPV1 in addition to TRPA1, the precise contribution of TRPA1 to the biological effects of cannabinoids remains to be determined.

3.8.5 Alkylamides

Activation of TRPV1 by hydroxy-α-sanshool, one of the many alkylamide components of the Szechuan pepper *Zanthoxylum piperitum*, has been attributed as the cause of the characteristic pungent and tingling sensation elicited by consumption of this Asian spice. However, activation of endogenously and heterologously expressed TRPA1 by hydroxy-α-sanshool as well as several derivatives has been demonstrated recently [57]. Hydroxy-α-sanshool is only poorly selective for TRPV1 and also activates TRPA1 with an EC_{50} of 69 μM [57, 199]. Like the pungent compounds described above, the response to hydroxy-α-sanshool appears to involve covalent modification of the TRPA1 ion channel.

3.8.6 Vanilloids

6-shogaol and 6-paradol, derived from the pungent oil of Melegueta pepper (*Aframomum melegueta*), activate both TRPV1 and TRPA1. Like hydroxy-α-sanshool, both hydroxy arylalkanones vanilloid compounds are slightly more potent at TRPV1 than TRPA1 with approximately tenfold higher EC_{50}s for activation of TRPA1 [57]. In addition, the weak TRPV1 and TRPV3 activator eugenol also activated both TRPM8 and TRPA1 at a concentration of 600 μM, although an EC_{50} for activation of TRPA1 was not determined [44]. Likewise, gingerol displays activity at TRPA1 in addition to TRPV1 [44].

3.8.7 Monoterpenoids

Another promiscuous TRP activator is the monoterpenoid linalool. Linalool was first described as a TRPM8 agonist, however, activity at the TRPM8 is extremely poor with an EC_{50} of 6.7 mM [135]. Recently, linalool was found to activate TRPA1 with an EC_{50} of 117 μM, making this compound somewhat selective for TRPA1 [57].

3.8.8 Methyl Salicylate

Methyl salicylate, produced by many species of wintergreen, is another non-specific TRPA1 agonist [44]. It activates TRPA1 at 600 μM, but also activates TRPV1, albeit at slightly higher concentration [44].

3.8.9 Nicotine

Nicotine is an alkaloid found in the *Solanaceae* family of plants that include tobacco (*Nicotiana tabacum*). It can cause irritation and neurogenic inflammation in both airways epithelial cells as well as after topical application While in guinea pig jugular ganglion neurons, nicotine-evoked responses were not significantly blocked by a TRPA1 antagonist [193], activation of heterologously expressed mouse TRPA1 as well as TRPA1 expressed in mouse trigeminal ganglion neurons by low μM concentrations of nicotine was reported recently [200]. In light of these seemingly conflicting results, the precise contribution of TRPA1 activation to the irritant activity of nicotine remains to be determined.

3.8.10 Endogenous Ligands

As for other TRP channels, endogenous ligands for TRPA1 are slowly emerging. The electrophilic cyclopentane prostaglandin D2 metabolite 15-deoxy-$\Delta^{12,14}$-prostaglandin J_2 (15d-PGJ_2) was recently shown to activate TRPA1, but not other thermo-TRPs, with an EC_{50} in the high μM range [201]. Similarly, nitro-oleic acid also activates TRPA1-mediated currents in rat DRG neurons, albeit this electrophilic fatty acid byproduct of nitric oxide and nitrite reactions also acted on TRPV1 [98]. While both of these endogenous metabolites are electrophilic, it remains to be determined if their mode of activation of TRPA1 occurs through covalent modification of the channel or if other mechanisms are involved.

3.9 TRPC1–TRPC6

While naturally occurring ligands have been described for all thermosensitive TRP channels, in particular TRPV1, TRPV2, TRPV3, TRPV4, TRPA1 and TRPM8, there is a relative paucity of ligands for other TRP channels. However, pharmacological modulators for some members of the TRPC (Table 3.5) family are slowly emerging.

Adenophostin A, a compound isolated from *Penicillium brevicompactum*, has been reported to activate TRPC5, though it remains unclear if this effect occurs as a result of interaction of the IP3 receptor, at which adenophostin A is a potent agonist, or if IP3 receptor independent mechanisms are involved [202, 203].

GsMTx-4, a peptide isolated from the Chilean Rose tarantula *Grammostola spatulata*, has been known for years to inhibit mechanically activated cation channels [204]. However, the molecular identity of the pharmacological target(s) for this toxin has been unclear. Several TRP channels are sensitive to various forms of mechanical stress. Both TRPV4 and TRPV2 respond osmotic cell swelling, while TRPM4 and TRPM3, as well as TRPC1, TRPC5 and TRPC6 respond to osmotic or mechanical membrane stretch. The attempt to identify the molecular target(s) of GsMTx-4 have led to reports that TRPC1, TRPC5 as well as TRPC6 are inhibited by this toxin [204, 205]. The 35 amino acid peptide blocked stretch-activated channels in rat astrocytes with relatively high potency (K_d 630 nM) and blocked TRPC6 and TRPC5-mediated responses at a concentration of 5 μM, possibly by inserting into the outer membrane leaflet and modifying boundary lipids [206, 207]. Recently, a conopeptide analogue termed NMB-1, or noxious mechanosensation blocker-1, was shown to inhibit high intensity, non-GsMTx-4-sensitive mechanoresponses in DRG neurons [208]. While the molecular target of NMB-1 was not identified, it is plausible that it also acts at mechanosensitive TRP channels. In addition to block by GsMTx-4, the acylated phloroglucinol hyperforin, a constituent of St John's Wort (*Hypericum perforatum*), activates TRPC6, as does the endogenous arachidonic acid metabolite 20-Hydroxyeicosatetraenoic acid at concentrations of > 1 μM [209, 210].

3.10 TRPP2

TRPP2 (polycystin 2) is the protein disrupted in ADPKD (autosomal dominant polycystic kidney disease). It associates with TRPP1 (polycystin-1 or PKD1) and is

Table 3.5 Natural product ligands of TRPC and TRPP channels

TRP subtype/ ligand	Structure/ sequence	Class	Agonist EC_{50}	Antagonist IC_{50}	Source	Alternative pharmacological target(s)	References
TRPC1							
GsMTx-4	YCQKWMWTCDE ERKCCEGLVCRL WCKRIINM	Peptide		< 5 μM	*Grammostola spatulata*	TRPC5	[204, 205]
TRPC5							
Adenophostin A			~ 1 nM		*Penicillium brevicompactum*	IP3 receptor	[202]
GsMTx-4	See above	Peptide		5 μM	*Grammostola spatulata*	TRPC1 TRPC6	[204, 205]
TRPC6							
GsMTx-4	See above	Peptide		5 μM	*Grammostola spatulata*	TRPC1 TRPC5	[204, 205]
Hyperforin		Phloroglucinol	1.5 μM		*Hypericum perforatum*	Serotonin reuptake	[209]
20-Hydroxyeicosatet raenoic acid		Endogenous ligand	> 1 μM ?		Arachidonic acid metabolite		[210]
TRPP2							
Triptolide		Diterpenoid	~ 100 nM		*Tripterygium wilfordii*	RNA polymerase NF-κB	[211]

translocated to the plasma membrane where it is probably activated by mechanical gating or via growth-factor initiated signaling cascades. Triptolide, from the traditional Chinese medicinal herb *Tripterygium wilfordii*, is a diterpenoid compound with proposed activity at the murine TRPP2 ion channel (Table 3.5) [211]. While activity at several other biological targets has been described for this compound, 100 nM triptolide caused TRPP2-mediated increases in Ca^{2+} and attenuated cyst formation in PKD1-knockout mice [211]. Thus, activation of TRPP2 suggests itself as a therapeutic strategy for treatment of ADPKD, although further research into the pharmacophore of triptolide and resultant activation mechanism of TRPP2 remains to be carried out.

3.11 Conclusions and Future Research

TRP channels contribute to fundamental cell function in a wide variety of tissues. Consequently, TRP channels are emerging as important drug targets involved in many pathological conditions, ranging from pain to cancer, cough, pulmonary hypertension, chronic obstructive pulmonary disease, mucolipidosis type IV, autosomal dominant polycystic kidney disease, Charcot-Marie-Tooth disease, disorders of bone development such as brachyolmia and autosomal recessive congenital stationary night blindness, with many more certain to be defined and characterized in future [212–215]. Accordingly, interest in TRP channels as novel pharmacological targets is increasing steadily.

Not only have naturally occurring substances, in particular capsaicin, menthol and allyl isothiocyanate, contributed to the deorphanisation of TRPV1, TRPM8 and TRPA1, respectively, but natural product ligands continue to provide valuable leads for the identification of ligands acting at "orphan" TRP channels. TRP channels such as TRPC5 or TRPM4, which were once considered as pure physical sensors, or modulated only by membrane lipids, are emerging as newly liganded ion channels, with natural product-derived ligands providing the first insights into the molecular pharmacology, physiology and gating mechanisms of these ion channels.

While selectivity and specificity remain lacking for many of these natural product ligands, identification of their pharmacophore through systematic structure-activity approaches can be expected to lead to the development of TRP subtype-selective agonists or antagonists, and may perhaps lead to the development of novel treatment approaches for several TRP "channelopathies". Finally, while ligands of either synthetic or natural origin remain lacking for many TRP isoforms, in light of the vast biodiversity of natural ligands, it is plausible that naturally occurring modulators for the remaining "orphan" TRP channels will be discovered in future, aiding in our understanding of not only their pharmacology and physiology, but also the therapeutic potential of this fascinating family of ion channels.

Acknowledgments This work was supported by a National Health and Medical Research Council program grant and fellowship (RJL) and a National Health and Medical Research Council Postdoctoral Training Fellowship (IV).

References

1. Newman DJ, Cragg GM, Snader KM (2003) Natural products as sources of new drugs over the period 1981–2002. J Nat Prod 66:1022–1037
2. Paterson I, Anderson EA (2005) Chemistry. The renaissance of natural products as drug candidates. Science 310:451–453
3. Greka A, Navarro B, Oancea E, Duggan A, Clapham DE (2003) TRPC5 is a regulator of hippocampal neurite length and growth cone morphology. Nat Neurosci 6:837–845
4. Nilius B, Voets T (2005) TRP channels: a TR(I)P through a world of multifunctional cation channels. Pflugers Arch 451:1–10
5. Clapham DE (2003) TRP channels as cellular sensors. Nature 426:517–524
6. Venkatachalam K, Montell C (2007) TRP channels. Annu Rev Biochem 76:387–417
7. Caterina MJ, Schumacher MA, Tominaga M, Rosen TA, Levine JD, Julius D (1997) The capsaicin receptor: a heat-activated ion channel in the pain pathway. Nature 389:816–824
8. Tominaga M, Caterina MJ, Malmberg AB, Rosen TA, Gilbert H, Skinner K, Raumann BE, Basbaum AI, Julius D (1998) The cloned capsaicin receptor integrates multiple pain-producing stimuli. Neuron 21:531–543
9. Pingle SC, Matta JA, Ahern GP (2007) Capsaicin receptor: TRPV1 a promiscuous TRP channel. Handb Exp Pharmacol 179:155–171
10. Nagy I, Santha P, Jancso G, Urban L (2004) The role of the vanilloid (capsaicin) receptor (TRPV1) in physiology and pathology. Eur J Pharmacol 500:351–369
11. Ahluwalia J, Urban L, Capogna M, Bevan S, Nagy I (2000) Cannabinoid 1 receptors are expressed in nociceptive primary sensory neurons. Neuroscience 100:685–688
12. Ichikawa H, Gouty S, Regalia J, Helke CJ, Sugimoto T (2004) Ca^{2+}/calmodulin-dependent protein kinase II in the rat cranial sensory ganglia. Brain Res 1005:36–43
13. Michael GJ, Priestley JV (1999) Differential expression of the mRNA for the vanilloid receptor subtype 1 in cells of the adult rat dorsal root and nodose ganglia and its downregulation by axotomy. J Neurosci 19:1844–1854
14. Szallasi A, Blumberg PM, Annicelli LL, Krause JE, Cortright DN (1999) The cloned rat vanilloid receptor VR1 mediates both R-type binding and C-type calcium response in dorsal root ganglion neurons. Mol Pharmacol 56:581–587
15. Guo A, Vulchanova L, Wang J, Li X, Elde R (1999) Immunocytochemical localization of the vanilloid receptor 1 (VR1): relationship to neuropeptides, the P2X3 purinoceptor and IB4 binding sites. Eur J Neurosci 11:946–958
16. Hong S, Wiley J (2005) Early painful diabetic neuropathy is associated with differential changes in the expression and function of vanilloid receptor 1. J Biol Chem 280:618–627
17. Steenland HW, Ko SW, Wu LJ, Zhuo M (2006) Hot receptors in the brain. Mol Pain 2:34
18. Gavva NR, Bannon AW, Surapaneni S, Hovland DN Jr, Lehto SG, Gore A, Juan T, Deng H, Han B, Klionsky L, Kuang R, Le A, Tamir R, Wang J, Youngblood B, Zhu D, Norman MH, Magal E, Treanor JJ, Louis JC (2007) The vanilloid receptor TRPV1 is tonically activated in vivo and involved in body temperature regulation. J Neurosci 27:3366–3374
19. Birder LA, Kanai AJ, de Groat WC, Kiss S, Nealen ML, Burke NE, Dineley KE, Watkins S, Reynolds IJ, Caterina MJ (2001) Vanilloid receptor expression suggests a sensory role for urinary bladder epithelial cells. Proc Natl Acad Sci USA 98:13396–13401
20. Dvorakova M, Kummer W (2001) Transient expression of vanilloid receptor subtype 1 in rat cardiomyocytes during development. Histochem Cell Biol 116:223–225
21. Biro T, Brodie C, Modarres S, Lewin NE, Acs P, Blumberg PM (1998) Specific vanilloid responses in C6 rat glioma cells. Brain Res Mol Brain Res 56:89–98

22. Biro T, Maurer M, Modarres S, Lewin NE, Brodie C, Acs G, Acs P, Paus R, Blumberg PM (1998) Characterization of functional vanilloid receptors expressed by mast cells. Blood 91:1332–1340
23. Takumida M, Kubo N, Ohtani M, Suzuka Y, Anniko M (2005) Transient receptor potential channels in the inner ear: presence of transient receptor potential channel subfamily 1 and 4 in the guinea pig inner ear. Acta Otolaryngol 125:929–934
24. Zheng J, Dai C, Steyger PS, Kim Y, Vass Z, Ren T, Nuttall AL (2003) Vanilloid receptors in hearing: altered cochlear sensitivity by vanilloids and expression of TRPV1 in the organ of corti. J Neurophysiol 90:444–455
25. Inoue K, Koizumi S, Fuziwara S, Denda S, Denda M (2002) Functional vanilloid receptors in cultured normal human epidermal keratinocytes. Biochem Biophys Res Commun 291: 124–129
26. Sanchez J, Krause J, Cortright D (2001) The distribution and regulation of vanilloid receptor VR1 and VR1 5′ splice variant RNA expression in rat. Neuroscience 107:373–381
27. Yiangou Y, Facer P, Ford A, Brady C, Wiseman O, Fowler C et al (2001) Capsaicin receptor VR1 and ATP-gated ion channel P2X3 in human urinary bladder. BJU Int 87:774–779
28. Avelino A, Cruz C, Nagy I, Cruz F (2002) Vanilloid receptor 1 expression in the rat urinary tract. Neuroscience 109:787–798
29. Dinh QT, Groneberg DA, Mingomataj E, Peiser C, Heppt W, Dinh S (2003) Arck Pc KBFFA: expression of substance P and vanilloid receptor (VR1) in trigeminal sensory neurons projecting to the mouse nasal mucosa. Neuropeptides 37:245–250
30. Van der Aa F, Roskams T, Blyweert W, De Ridder D (2003) Interstitial cells in the human prostate: a new therapeutic target?. Prostate 56:250–255
31. Ward S, Bayguinov J, Won K, Grundy D, Berthoud H (2003) Distribution of the vanilloid receptor (VR1) in the gastrointestinal tract. J Comp Neurol 465:121–135
32. Zahner MR, Li DP, Chen SR, Pan HL (2003) Cardiac vanilloid receptor 1-expressing afferent nerves and their role in the cardiogenic sympathetic reflex in rats. J Physiol 551: 515–523
33. Szallasi A, Blumberg PM (1999) Vanilloid (Capsaicin) receptors and mechanisms. Pharmacol Rev 51:159–212
34. Cordell GA, Araujo OE (1993) Capsaicin: identification, nomenclature, and pharmacotherapy. Ann Pharmacother 27:330–336
35. Suzuki T, Iwai K (1984) Constituents of red pepper species: chemistry, biochemistry, pharmacology, and food science of the pungent principle of capsicum species. Academic Press, Orlando, FL
36. Kim KS, Lee KW, Im JY, Yoo JY, Kim SW, Lee JK, Nestler EJ, Han PL (2006) Adenylyl cyclase type 5 (AC5) is an essential mediator of morphine action. Proc Natl Acad Sci USA 103:3908–3913
37. Jung J, Hwang SW, Kwak J, Lee SY, Kang CJ, Kim WB, Kim DOU (1999) Capsaicin binds to the intracellular domain of the capsaicin activated ion channel. J Neurosci 19: 529–538
38. Jung J, Lee SY, Hwang SW, Cho H, Shin J, Kang YS, Kim SOU (2002) Agonist recognition sites in the cytosolic tails of vanilloid receptor 1. J Biol Chem 277:44448–44454
39. Jordt SE, Julius D (2002) Molecular basis for species-specific sensitivity to "hot" chili peppers. Cell 108:421–430
40. Hwang SW, Cho H, Kwak J, Lee SY, Kang CJ, Jung J, Cho S, Min KH, Suh YG, Kim D, Oh U (2000) Direct activation of capsaicin receptors by products of lipoxygenases: endogenous capsaicin-like substances. Proc Natl Acad Sci USA 97:6155–6160
41. Marin-Burgin A, Reppenhagen S, Klusch A, Wendland JR, Petersen M (2000) Low-threshold heat response antagonized by capsazepine in chick sensory neurons, which are capsaicin-insensitive. Eur J Neurosci 12:3560–3566
42. Yang BH, Piao ZG, Kim YB, Lee CH, Lee JK, Park K, Kim JS, Oh SB (2003) Activation of vanilloid receptor 1 (VR1) by eugenol. J Dent Res 82:781–785

43. Xu H, Delling M, Jun JC, Clapham DE (2006) Oregano, thyme and clove-derived flavors and skin sensitizers activate specific TRP channels. Nat Neurosci 9:628–635
44. Bandell M, Story GM, Hwang SW, Viswanath V, Eid SR, Petrus MJ, Earley TJ, Patapoutian A (2004) Noxious cold ion channel TRPA1 is activated by pungent compounds and bradykinin. Neuron 41:849–857
45. Lee MH, Yeon KY, Park CK, Li HY, Fang Z, Kim MS, Choi SY, Lee SJ, Lee S, Park K, Lee JH, Kim JS, Oh SB (2005) Eugenol inhibits calcium currents in dental afferent neurons. J Dent Res 84:848–851
46. Park CK, Li HY, Yeon KY, Jung SJ, Choi SY, Lee SJ, Lee S, Park K, Kim JS, Oh SB (2006) Eugenol inhibits sodium currents in dental afferent neurons. J Dent Res 85:900–904
47. Chung G, Rhee JN, Jung SJ, Kim JS, Oh SB (2008) Modulation of CaV2.3 calcium channel currents by eugenol. J Dent Res 87:137–141
48. Li HY, Lee BK, Kim JS, Jung SJ, Oh SB (2008) Eugenol Inhibits ATP-induced P2X Currents in Trigeminal Ganglion Neurons. Korean J Physiol Pharmacol 12:315–321
49. Ohkubo T, Shibata M (1997) The selective capsaicin antagonist capsazepine abolishes the antinociceptive action of eugenol and guaiacol. J Dent Res 76:848–851
50. Liu L, Simon SA (1996) Similarities and differences in the currents activated by capsaicin, piperine, and zingerone in rat trigeminal ganglion cells. J Neurophysiol 76:1858–1869
51. McNamara FN, Randall A, Gunthorpe MJ (2005) Effects of piperine, the pungent component of black pepper, at the human vanilloid receptor (TRPV1). Br J Pharmacol 144:781–790
52. Bhardwaj RK, Glaeser H, Becquemont L, Klotz U, Gupta SK, Fromm MF (2002) Piperine, a major constituent of black pepper, inhibits human P-glycoprotein and CYP3A4. J Pharmacol Exp Ther 302:645–650
53. Iwasaki Y, Morita A, Iwasawa T, Kobata K, Sekiwa Y, Morimitsu Y, Kubota K, Watanabe T (2006) A nonpungent component of steamed ginger – [10]-shogaol – increases adrenaline secretion via the activation of TRPV1. Nutr Neurosci 9:169–178
54. Witte DG, Cassar SC, Masters JN, Esbenshade T, Hancock AA (2002) Use of a fluorescent imaging plate reader – based calcium assay to assess pharmacological differences between the human and rat vanilloid receptor. J Biomol Screen 7:466–475
55. Dedov VN, Tran VH, Duke CC, Connor M, Christie MJ, Mandadi S, Roufogalis BD (2002) Gingerols: a novel class of vanilloid receptor (VR1) agonists. Br J Pharmacol 137:793–798
56. Bianchi BR, Lee CH, Jarvis MF, El Kouhen R, Moreland RB, Faltynek CR, Puttfarcken PS (2006) Modulation of human TRPV1 receptor activity by extracellular protons and host cell expression system. Eur J Pharmacol 537:20–30
57. Riera CE, Menozzi-Smarrito C, Affolter M, Michlig S, Munari C, Robert F, Vogel H, Simon SA, le Coutre J (2009) Compounds from Sichuan and Melegueta peppers activate, covalently and non-covalently, TRPA1 and TRPV1 channels. Br J Pharmacol 157: 1398–1409
58. Martelli L, Ragazzi E (2007) di Mario F, Martelli M, Castagliuolo I, Dal Maschio M, Palu G, Machietto M, Scorzeto M, Vassanelli S, Brun P: a potential role for the vanilloid receptor TRPV1 in the therapeutic effect of curcumin in dinitrobenzene sulphonic acid-induced colitis in mice. Neurogastroenterol Motil 19:668–674
59. Yeon KY, Kim SA, Kim YH, Lee MK, Ahn DK, Kim HJ, Jung SJ, Oh SB (2010) Curcumin produces an antihyperalgesic effect via antagonism of TRPV1. J Dent Res 89:170–174
60. Szallasi A, Blumberg PM (1989) Resiniferatoxin, a phorbol-related diterpene, acts as an ultrapotent analog of capsaicin, the irritant constituent in red pepper. Neuroscience 30: 515–520
61. Szallasi A, Blumberg PM (1991) Characterization of vanilloid receptors in the dorsal horn of pig spinal cord. Brain Res 547:335–338
62. Szallasi A, Blumberg PM (1990) Resiniferatoxin and its analogs provide novel insights into the pharmacology of the vanilloid (capsaicin) receptor. Life Sci 47:1399–1408
63. Appendino G, Ech-Chahad A, Minassi A, Bacchiega S, De Petrocellis L, Di Marzo V (2007) Structure-activity relationships of the ultrapotent vanilloid resiniferatoxin (RTX): the homovanillyl moiety. Bioorg Med Chem Lett 17:132–135

64. Choi HK, Choi S, Lee Y, Kang DW, Ryu H, Maeng HJ, Chung SJ, Pavlyukovets VA, Pearce LV, Toth A, Tran R, Wang Y, Morgan MA, Blumberg PM, Lee J (2009) Non-vanillyl resiniferatoxin analogues as potent and metabolically stable transient receptor potential vanilloid 1 agonists. Bioorg Med Chem 17:690–698
65. Acs G, Lee J, Marquez VE, Blumberg PM (1996) Distinct structure-activity relations for stimulation of 45Ca uptake and for high affinity binding in cultured rat dorsal root ganglion neurons and dorsal root ganglion membranes. Brain Res Mol Brain Res 35:173–182
66. Jonassohn M, Anke H, Morales P, Sterner O (1995) Structure-activity relationships for unsaturated dialdehydes. 10. The generation of bioactive products by autoxidation of isovelleral and merulidial. Acta Chem Scand 49:530–535
67. Szallasi A, Jonassohn M, Acs G, Biro T, Acs P, Blumberg PM, Sterner O (1996) The stimulation of capsaicin-sensitive neurones in a vanilloid receptor-mediated fashion by pungent terpenoids possessing an unsaturated 1,4-dialdehyde moiety. Br J Pharmacol 119:283–290
68. Szallasi A, Biro T, Modarres S, Garlaschelli L, Petersen M, Klusch A, Vidari G, Jonassohn M, De Rosa S, Sterner O, Blumberg PM, Krause JE (1998) Dialdehyde sesquiterpenes and other terpenoids as vanilloids. Eur J Pharmacol 356:81–89
69. Sterner O, Szallasi A (1999) Novel natural vanilloid receptor agonists: new therapeutic targets for drug development. Trends Pharmacol Sci 20:459–465
70. Iwasaki Y, Tanabe M, Kayama Y, Abe M, Kashio M, Koizumi K, Okumura Y, Morimitsu Y, Tominaga M, Ozawa Y, Watanabe T (2009) Miogadial and miogatrial with alpha,beta-unsaturated 1,4-dialdehyde moieties – novel and potent TRPA1 agonists. Life Sci 85:60–69
71. Ralevic V, Jerman JC, Brough SJ, Davis JB, Egerton J, Smart D (2003) Pharmacology of vanilloids at recombinant and endogenous rat vanilloid receptors. Biochem Pharmacol 65:143–151
72. Szallasi A, Biro T, Szabo T, Modarres S, Petersen M, Klusch A, Blumberg PM, Krause JE, Sterner O (1999) A non-pungent triprenyl phenol of fungal origin, scutigeral, stimulates rat dorsal root ganglion neurons via interaction at vanilloid receptors. Br J Pharmacol 126:1351–1358
73. Hellwig V, Nopper R, Mauler F, Freitag J, Ji-Kai L, Zhi-Hui D, Stadler M (2003) Activities of prenylphenol derivatives from fruitbodies of Albatrellus spp. on the human and rat vanilloid receptor 1 (VR1) and characterisation of the novel natural product, confluentin. Arch Pharm (Weinheim) 336:119–126
74. Bisogno T, Hanus L, De Petrocellis L, Tchilibon S, Ponde DE, Brandi I, Moriello AS, Davis JB, Mechoulam R, Di Marzo V (2001) Molecular targets for cannabidiol and its synthetic analogues: effect on vanilloid VR1 receptors and on the cellular uptake and enzymatic hydrolysis of anandamide. Br J Pharmacol 134:845–852
75. Costa B, Giagnoni G, Franke C, Trovato AE, Colleoni M (2004) Vanilloid TRPV1 receptor mediates the antihyperalgesic effect of the nonpsychoactive cannabinoid, cannabidiol, in a rat model of acute inflammation. Br J Pharmacol 143:247–250
76. Jung SY, Choi S, Ko YS, Park CS, Oh S, Koh SR, Oh JW, Rhee MH, Nah SY (2001) Effects of ginsenosides on vanilloid receptor (VR1) channels expressed in Xenopus oocytes. Mol Cells 12:342–346
77. Hahn J, Nah SY, Nah JJ, Uhm DY, Chung S (2000) Ginsenosides inhibit capsaicin-activated channel in rat sensory neurons. Neurosci Lett 287:45–48
78. Cromer BA, McIntyre P (2008) Painful toxins acting at TRPV1. Toxicon 51:163–173
79. Siemens J, Zhou S, Piskorowski R, Nikai T, Lumpkin E, Basbaum AI, King D, Julius D (2006) Spider toxins activate the capsaicin receptor to produce inflammatory pain. Nature 444:208–212
80. Kitaguchi T, Swartz KJ (2005) An inhibitor of TRPV1 channels isolated from funnel Web spider venom. Biochemistry 44:15544–15549
81. Andreev YA, Kozlov SA, Koshelev SG, Ivanova EA, Monastyrnaya MM, Kozlovskaya EP, Grishin EV (2008) Analgesic compound from sea anemone Heteractis crispa is the first polypeptide inhibitor of vanilloid receptor 1 (TRPV1). J Biol Chem 283:23914–23921

82. Cuypers E, Yanagihara A, Rainier JD, Tytgat J (2007) TRPV1as a key determinant in ciguatera and neurotoxic shellfish poisoning. Biochem Biophys Res Commun 361: 214–217
83. Cuypers E, Yanagihara A, Karlsson E, Tytgat J (2006) Jellyfish and other cnidarian envenomations cause pain by affecting TRPV1 channels. FEBS Lett 580:5728–5732
84. Szallasi A (2002) Vanilloid (capsaicin) receptors in health and disease. Am J Clin Pathol 118:110–121
85. Hermann H, De Petrocellis L, Bisogno T, Schiano Moriello A, Lutz B, Di Marzo V (2003) Dual effect of cannabinoid CB1 receptor stimulation on a vanilloid VR1 receptor-mediated response. Cell Mol Life Sci 60:607–616
86. Di Marzo V, Blumberg PM, Szallasi A (2002) Endovanilloid signaling in pain. Curr Opin Neurobiol 12:372–379
87. Huang SM, Bisogno T, Trevisani M, Al-Hayani A, De Petrocellis L, Fezza F, Tognetto M, Petros TJ, Krey JF, Chu CJ, Miller JD, Davies SN, Geppetti P, Walker JM, Di Marzo V (2002) An endogenous capsaicin-like substance with high potency at recombinant and native vanilloid VR1 receptors. Proc Natl Acad Sci USA 99:8400–8405
88. Jerman JC, Brough SJ, Prinjha R, Harries MH, Davis JB, Smart D (2000) Characterization using FLIPR of rat vanilloid receptor (rVR1) pharmacology. Br J Pharmacol 130:916–922
89. Geppetti P, Trevisani M (2004) Activation and sensitisation of the vanilloid receptor: role in gastrointestinal inflammation and function. Br J Pharmacol 141:1313–1320
90. Movahed P, Jonsson BA, Birnir B, Wingstrand JA, Jorgensen TD, Ermund A, Sterner O, Zygmunt PM, Hogestatt ED (2005) Endogenous unsaturated C18 N-acylethanolamines are vanilloid receptor (TRPV1) agonists. J Biol Chem 280:38496–38504
91. Zygmunt PM, Petersson J, Andersson DA, Chuang H, Sorgard M, Di Marzo V, Julius D, Hogestatt ED (1999) Vanilloid receptors on sensory nerves mediate the vasodilator action of anandamide. Nature 400:452–457
92. Rimmerman N, Bradshaw HB, Basnet A, Tan B, Widlanski TS, Walker JM (2009) Microsomal omega-hydroxylated metabolites of N-arachidonoyl dopamine are active at recombinant human TRPV1 receptors. Prostaglandins Other Lipid Mediat 88:10–17
93. De Petrocellis L, Chu CJ, Moriello AS, Kellner JC, Walker JM, Di Marzo V (2004) Actions of two naturally occurring saturated N-acyldopamines on transient receptor potential vanilloid 1 (TRPV1) channels. Br J Pharmacol 143:251–256
94. Chu CJ, Huang SM, De Petrocellis L, Bisogno T, Ewing SA, Miller JD, Zipkin RE, Daddario N, Appendino G, Di Marzo V, Walker JM (2003) N-oleoyldopamine, a novel endogenous capsaicin-like lipid that produces hyperalgesia. J Biol Chem 278:13633–13639
95. Ahern GP (2003) Activation of TRPV1 by the satiety factor oleoylethanolamide. J Biol Chem 278:30429–30434
96. Patwardhan AM, Akopian AN, Ruparel NB, Diogenes A, Weintraub ST, Uhlson C, Murphy RC, Hargreaves KM (2010) Heat generates oxidized linoleic acid metabolites that activate TRPV1 and produce pain in rodents. J Clin Invest 120(5):1617–1626.
97. Patwardhan AM, Scotland PE, Akopian AN, Hargreaves KM (2009) Activation of TRPV1 in the spinal cord by oxidized linoleic acid metabolites contributes to inflammatory hyperalgesia. Proc Natl Acad Sci USA 106:18820–18824
98. Sculptoreanu A, Kullmann FA, Artim DE, Bazley FA, Schopfer F, Woodcock S, Freeman BA, de Groat W (2010) Nitro-oleic acid inhibits firing and activates TRPV1- and TRPA1-mediated inward currents in DRG neurons from adult male rats. J Pharmacol Exp Ther 333(3):883–895.
99. Saghatelian A, McKinney MK, Bandell M, Patapoutian A, Cravatt BF (2006) A FAAH-regulated class of N-acyl taurines that activates TRP ion channels. Biochemistry 45: 9007–9015
100. Vieira C, Evangelista S, Cirillo R, Terracciano R, Lippi A, Maggi CA, Manzini S (2000) Antinociceptive activity of ricinoleic acid, a capsaicin-like compound devoid of pungent properties. Eur J Pharmacol 407:109–116

101. Vieira C, Evangelista S, Cirillo R, Lippi A, Maggi CA, Manzini S (2000) Effect of ricinoleic acid in acute and subchronic experimental models of inflammation. Mediators Inflamm 9:223–228
102. Vieira C, Fetzer S, Sauer SK, Evangelista S, Averbeck B, Kress M, Reeh PW, Cirillo R, Lippi A, Maggi CA, Manzini S (2001) Pro- and anti-inflammatory actions of ricinoleic acid: similarities and differences with capsaicin. Naunyn Schmiedebergs Arch Pharmacol 364:87–95
103. Toth A, Kedei N, Szabo T, Wang Y, Blumberg PM (2002) Thapsigargin binds to and inhibits the cloned vanilloid receptor-1. Biochem Biophys Res Commun 293:777–782
104. Liu L, Zhu W, Zhang ZS, Yang T, Grant A, Oxford G, Simon SA (2004) Nicotine inhibits voltage-dependent sodium channels and sensitizes vanilloid receptors. J Neurophysiol 91:1482–1491
105. Dessaint J, Yu W, Krause JE, Yue L (2004) Yohimbine inhibits firing activities of rat dorsal root ganglion neurons by blocking Na^+ channels and vanilloid VR1 receptors. Eur J Pharmacol 485:11–20
106. Ahern GP, Wang X, Miyares RL (2006) Polyamines are potent ligands for the capsaicin receptor TRPV1. J Biol Chem 281:8991–8995
107. Pearce LV, Petukhov PA, Szabo T, Kedei N, Bizik F, Kozikowski AP, Blumberg PM (2004) Evodiamine functions as an agonist for the vanilloid receptor TRPV1. Org Biomol Chem 2:2281–2286
108. Wang T, Wang Y, Yamashita H (2009) Evodiamine inhibits adipogenesis via the EGFR-PKCalpha-ERK signaling pathway. FEBS Lett 583:3655–3659
109. Ohta T, Imagawa T, Ito S (2009) Involvement of Transient Receptor Potential Vanilloid Subtype 1 in Analgesic Action of Methylsalicylate. Mol Pharmacol 75:307–317
110. Kress M, Vyklicky L, Reeh PW (1996) Inhibition of capsaicin-induced ionic current – a new mechanism of action for aspirin-like drugs? Pflugers Arch – Eur J Physiol Suppl 431:R61 Abstract
111. Greffrath W, Kirschstein T, Nawrath H, Treede R (2002) Acetylsalicylic acid reduces heat responses in rat nociceptive primary sensory neurons – evidence for a new mechanism of action. Neurosci Lett 320:61–64
112. Caterina MJ, Rosen TA, Tominaga M, Brake AJ, Julius D (1999) A capsaicin-receptor homologue with a high threshold for noxious heat. Nature 398:436–441
113. Neeper MP, Liu Y, Hutchinson TL, Wang Y, Flores CM, Qin N (2007) Activation properties of heterologously expressed mammalian TRPV2: evidence for species dependence. J Biol Chem 282:15894–15902
114. Muraki K, Iwata Y, Katanosaka Y, Ito T, Ohya S, Shigekawa M, Imaizumi Y (2003) TRPV2 is a component of osmotically sensitive cation channels in murine aortic myocytes. Circ Res 93:829–838
115. Akopian AN, Ruparel NB, Patwardhan A, Hargreaves KM (2008) Cannabinoids desensitize capsaicin and mustard oil responses in sensory neurons via TRPA1 activation. J Neurosci 28:1064–1075
116. Jordt SE, Bautista DM, Chuang HH, McKemy DD, Zygmunt PM, Hogestatt ED, Meng ID, Julius D (2004) Mustard oils and cannabinoids excite sensory nerve fibres through the TRP channel ANKTM1. Nature 427:260–265
117. Qin N, Neeper MP, Liu Y, Hutchinson TL, Lubin ML, Flores CM (2008) TRPV2 is activated by cannabidiol and mediates CGRP release in cultured rat dorsal root ganglion neurons. J Neurosci 28:6231–6238
118. Bang S, Kim KY, Yoo S, Lee SH, Hwang SW (2007) Transient receptor potential V2 expressed in sensory neurons is activated by probenecid. Neurosci Lett 425:120–125
119. Monet M, Gkika D (2009) Lehen'kyi V, Pourtier A, Vanden Abeele F, Bidaux G, Juvin V, Rassendren F, Humez S, Prevarsakaya N: lysophospholipids stimulate prostate cancer cell migration via TRPV2 channel activation. Biochim Biophys Acta 1793:528–539

120. Peier AM, Reeve AJ, Andersson DA, Moqrich A, Earley TJ, Hergarden AC, Story GM, Colley S, Hogenesch JB, McIntyre P, Bevan S, Patapoutian A (2002) A heat-sensitive TRP channel expressed in keratinocytes. Science 296:2046–2049
121. Smith GD, Gunthorpe MJ, Kelsell RE, Hayes PD, Reilly P, Facer P, Wright JE, Jerman JC, Walhin JP, Ooi L, Egerton J, Charles KJ, Smart D, Randall AD, Anand P, Davis JB (2002) TRPV3 is a temperature-sensitive vanilloid receptor-like protein. Nature 418:186–190
122. Xu H, Ramsey IS, Kotecha SA, Moran MM, Chong JA, Lawson D, Ge P, Lilly J, Silos-Santiago I, Xie Y, DiStefano PS, Curtis R, Clapham DE (2002) TRPV3 is a calcium-permeable temperature-sensitive cation channel. Nature 418:181–186
123. Moqrich A, Hwang SW, Earley TJ, Petrus MJ, Murray AN, Spencer KS, Andahazy M, Story GM, Patapoutian A (2005) Impaired thermosensation in mice lacking TRPV3, a heat and camphor sensor in the skin. Science 307:1468–1472
124. Zimmermann K, Leffler A, Fischer MM, Messlinger K, Nau C, Reeh PW (2005) The TRPV1/2/3 activator 2-aminoethoxydiphenyl borate sensitizes native nociceptive neurons to heat in wildtype but not TRPV1 deficient mice. Neuroscience 135:1277–1284
125. Imura K, Yoshioka T, Hirasawa T, Sakata T (2009) Role of TRPV3 in immune response to development of dermatitis. J Inflamm (Lond) 6:17
126. Asakawa M, Yoshioka T, Matsutani T, Hikita I, Suzuki M, Oshima I, Tsukahara K, Arimura A, Horikawa T, Hirasawa T, Sakata T (2006) Association of a mutation in TRPV3 with defective hair growth in rodents. J Invest Dermatol 126:2664–2672
127. Imura K, Yoshioka T, Hikita I, Tsukahara K, Hirasawa T, Higashino K, Gahara Y, Arimura A, Sakata T (2007) Influence of TRPV3 mutation on hair growth cycle in mice. Biochem Biophys Res Commun 363:479–483
128. Vogt-Eisele AK, Weber K, Sherkheli MA, Vielhaber G, Panten J, Gisselmann G, Hatt H (2007) Monoterpenoid agonists of TRPV3. Br J Pharmacol 151:530–540
129. Chung MK, Guler AD, Caterina MJ (2005) Biphasic currents evoked by chemical or thermal activation of the heat-gated ion channel, TRPV3. J Biol Chem 280:15928–15941
130. Xu H, Blair NT, Clapham DE (2005) Camphor activates and strongly desensitizes the transient receptor potential vanilloid subtype 1 channel in a vanilloid-independent mechanism. J Neurosci 25:8924–8937
131. Parnas M, Peters M, Dadon D, Lev S, Vertkin I, Slutsky I, Minke B (2009) Carvacrol is a novel inhibitor of Drosophila TRPL and mammalian TRPM7 channels. Cell Calcium 45:300–309
132. Macpherson LJ, Hwang SW, Miyamoto T, Dubin AE, Patapoutian A, Story GM (2006) More than cool: promiscuous relationships of menthol and other sensory compounds. Mol Cell Neurosci 32:335–343
133. McKemy DD, Neuhausser WM, Julius D (2002) Identification of a cold receptor reveals a general role for TRP channels in thermosensation. Nature 416:52–58
134. Peier AM, Moqrich A, Hergarden AC, Reeve AJ, Andersson DA, Story GM, Earley TJ, Dragoni I, McIntyre P, Bevan S, Patapoutian A (2002) A TRP channel that senses cold stimuli and menthol. Cell 108:705–715
135. Behrendt HJ, Germann T, Gillen C, Hatt H, Jostock R (2004) Characterization of the mouse cold-menthol receptor TRPM8 and vanilloid receptor type-1 VR1 using a fluorometric imaging plate reader (FLIPR) assay. Br J Pharmacol 141:737–745
136. Stotz SC, Vriens J, Martyn D, Clardy J, Clapham DE (2008) Citral sensing by Transient [corrected] receptor potential channels in dorsal root ganglion neurons. PLoS One 3:e2082
137. Moussaieff A, Rimmerman N, Bregman T, Straiker A, Felder CC, Shoham S, Kashman Y, Huang SM, Lee H, Shohami E, Mackie K, Caterina MJ, Walker JM, Fride E, Mechoulam R (2008) Incensole acetate, an incense component, elicits psychoactivity by activating TRPV3 channels in the brain. Faseb J 22:3024–3034

138. Bang S, Yoo S, Yang TJ, Cho H, Hwang SW (2010) Farnesyl pyrophosphate is a novel pain-producing molecule via specific activation of TRPV3. J Biol Chem 285(25):19362–19371.
139. Liedtke W, Choe Y, Marti-Renom MA, Bell AM, Denis CS, Sali A, Hudspeth AJ, Friedman JM, Heller S (2000) Vanilloid receptor-related osmotically activated channel (VR-OAC), a candidate vertebrate osmoreceptor. Cell 103:525–535
140. Strotmann R, Harteneck C, Nunnenmacher K, Schultz G, Plant TD (2000) OTRPC4, a non-selective cation channel that confers sensitivity to extracellular osmolarity. Nat Cell Biol 2:695–702
141. Wissenbach U, Bodding M, Freichel M, Flockerzi V (2000) Trp12, a novel Trp related protein from kidney. FEBS Lett 485:127–134
142. Nilius B, Prenen J, Wissenbach U, Bodding M, Droogmans G (2001) Differential activation of the volume-sensitive cation channel TRP12 (OTRPC4) and volume-regulated anion currents in HEK-293 cells. Pflugers Arch 443:227–233
143. Watanabe H, Vriens J, Suh SH, Benham CD, Droogmans G, Nilius B (2002) Heat-evoked activation of TRPV4 channels in a HEK293 cell expression system and in native mouse aorta endothelial cells. J Biol Chem 277:47044–47051
144. Guler AD, Lee H, Iida T, Shimizu I, Tominaga M, Caterina M (2002) Heat-evoked activation of the ion channel, TRPV4. J Neurosci 22:6408–6414
145. Plant TD, Strotmann R (2007) Trpv4. Handb Exp Pharmacol 179(179):189–205
146. Smith PL, Maloney KN, Pothen RG, Clardy J, Clapham DE (2006) Bisandrographolide from Andrographis paniculata activates TRPV4 channels. J Biol Chem 281:29897–29904
147. Vriens J, Owsianik G, Janssens A, Voets T, Nilius B (2007) Determinants of 4 alpha-phorbol sensitivity in transmembrane domains 3 and 4 of the cation channel TRPV4. J Biol Chem 282:12796–12803
148. Chung MK, Lee H, Mizuno A, Suzuki M, Caterina MJ (2004) TRPV3 and TRPV4 mediate warmth-evoked currents in primary mouse keratinocytes. J Biol Chem 279: 21569–21575
149. Everaerts W, Nilius B, Owsianik G (2009) The vallinoid transient receptor potential channel Trpv4: from structure to disease. Prog Biophys Mol Biol 103(1):2–17
150. Suzuki M, Mizuno A, Kodaira K, Imai M (2003) Impaired pressure sensation in mice lacking TRPV4. J Biol Chem 278:22664–22668
151. Alessandri-Haber N, Dina OA, Joseph EK, Reichling D, Levine JD (2006) A transient receptor potential vanilloid 4-dependent mechanism of hyperalgesia is engaged by concerted action of inflammatory mediators. J Neurosci 26:3864–3874
152. Alessandri-Haber N, Dina OA, Yeh JJ, Parada CA, Reichling DB, Levine JD (2004) Transient receptor potential vanilloid 4 is essential in chemotherapy-induced neuropathic pain in the rat. J Neurosci 24:4444–4452
153. Alessandri-Haber N, Joseph E, Dina OA, Liedtke W, Levine JD (2005) TRPV4 mediates pain-related behavior induced by mild hypertonic stimuli in the presence of inflammatory mediator. Pain 118:70–79
154. Alessandri-Haber N, Yeh JJ, Boyd AE, Parada CA, Chen X, Reichling DB, Levine JD (2003) Hypotonicity induces TRPV4-mediated nociception in rat. Neuron 39:497–511
155. Nilius B, Owsianik G (2010) Channelopathies converge on TRPV4. Nat Genet 42:98–100
156. Auer-Grumbach M, Olschewski A, Papic L, Kremer H, McEntagart ME, Uhrig S, Fischer C, Frohlich E, Balint Z, Tang B, Strohmaier H, Lochmuller H, Schlotter-Weigel B, Senderek J, Krebs A, Dick KJ, Petty R, Longman C, Anderson NE, Padberg GW, Schelhaas HJ, van Ravenswaaij-Arts CM, Pieber TR, Crosby AH, Guelly C (2010) Alterations in the ankyrin domain of TRPV4 cause congenital distal SMA, scapuloperoneal SMA and HMSN2C. Nat Genet 42:160–164
157. Landoure G, Zdebik AA, Martinez TL, Burnett BG, Stanescu HC, Inada H, Shi Y, Taye AA, Kong L, Munns CH, Choo SS, Phelps CB, Paudel R, Houlden H, Ludlow CL, Caterina MJ, Gaudet R, Kleta R, Fischbeck KH, Sumner CJ (2010) Mutations in TRPV4 cause Charcot-Marie-Tooth disease type 2C. Nat Genet 42:170–174

158. Hecker E (1968) Cocarcinogenic principles from the seed oil of Croton tiglium and from other Euphorbiaceae. Cancer Res 28:2338–2349
159. Goel G, Makkar HP, Francis G, Becker K (2007) Phorbol esters: structure, biological activity, and toxicity in animals. Int J Toxicol 26:279–288
160. Ohuchi K, Levine L (1980) alpha-Tocopherol inhibits 12-O-tetradecanoyl-phorbol-13-acetate-stimulated deacylation of cellular lipids, prostaglandin production, and changes in cell morphology of Madin-Darby canine kidney cells. Biochim Biophys Acta 619:11–19
161. Silinsky EM, Searl TJ (2003) Phorbol esters and neurotransmitter release: more than just protein kinase C? Br J Pharmacol 138:1191–1201
162. Blumberg PM, Jaken S, Konig B, Sharkey NA, Leach KL, Jeng AY, Yeh E (1984) Mechanism of action of the phorbol ester tumor promoters: specific receptors for lipophilic ligands. Biochem Pharmacol 33:933–940
163. Watanabe H, Davis JB, Smart D, Jerman JC, Smith GD, Hayes P, Vriens J, Cairns W, Wissenbach U, Prenen J, Flockerzi V, Droogmans G, Benham CD, Nilius B (2002) Activation of TRPV4 channels (hVRL-2/mTRP12) by phorbol derivatives. J Biol Chem 277:13569–13577
164. Klausen TK, Pagani A, Minassi A, Ech-Chahad A, Prenen J, Owsianik G, Hoffmann EK, Pedersen SF, Appendino G, Nilius B (2009) Modulation of the transient receptor potential vanilloid channel TRPV4 by 4alpha-phorbol esters: a structure-activity study. J Med Chem 52:2933–2939
165. Vriens J, Watanabe H, Janssens A, Droogmans G, Voets T, Nilius B (2004) Cell swelling, heat, and chemical agonists use distinct pathways for the activation of the cation channel TRPV4. Proc Natl Acad Sci USA 101:396–401
166. Watanabe H, Vriens J, Prenen J, Droogmans G, Voets T, Nilius B (2003) Anandamide and arachidonic acid use epoxyeicosatrienoic acids to activate TRPV4 channels. Nature 424:434–438
167. Tsavaler L, Shapero MH, Morkowski S, Laus R (2001) Trp-p8, a novel prostate-specific gene, is up-regulated in prostate cancer and other malignancies and shares high homology with transient receptor potential calcium channel proteins. Cancer Res 61:3760–3769
168. Fonfria E, Murdock PR, Cusdin FS, Benham CD, Kelsell RE, McNulty S (2006) Tissue distribution profiles of the human TRPM cation channel family. J Recept Signal Transduct Res 26:159–178
169. Andersson DA, Nash M, Bevan S (2007) Modulation of the cold-activated channel TRPM8 by lysophospholipids and polyunsaturated fatty acids. J Neurosci 27:3347–3355
170. Vanden Abeele F, Zholos A, Bidaux G, Shuba Y, Thebault S, Beck B, Flourakis M, Panchin Y, Skryma R, Prevarskaya N (2006) Ca^{2+}-independent phospholipase A2-dependent gating of TRPM8 by lysophospholipids. J Biol Chem 281:40174–40182
171. Grimm C, Kraft R, Schultz G, Harteneck C (2005) Activation of the melastatin-related cation channel TRPM3 by D-erythro-sphingosine [corrected]. Mol Pharmacol 67:798–805
172. Wagner TF, Loch S, Lambert S, Straub I, Mannebach S, Mathar I, Dufer M, Lis A, Flockerzi V, Philipp SE, Oberwinkler J (2008) Transient receptor potential M3 channels are ionotropic steroid receptors in pancreatic beta cells. Nat Cell Biol 10:1421–1430
173. Hara Y, Wakamori M, Ishii M, Maeno E, Nishida M, Yoshida T, Yamada H, Shimizu S, Mori E, Kudoh J, Shimizu N, Kurose H, Okada Y, Imoto K, Mori Y (2002) LTRPC2 Ca^{2+}-permeable channel activated by changes in redox status confers susceptibility to cell death. Mol Cell 9:163–173
174. Togashi K, Hara Y, Tominaga T, Higashi T, Konishi Y, Mori Y, Tominaga M (2006) TRPM2 activation by cyclic ADP-ribose at body temperature is involved in insulin secretion. Embo J 25:1804–1815
175. Beck A, Kolisek M, Bagley LA, Fleig A, Penner R (2006) Nicotinic acid adenine dinucleotide phosphate and cyclic ADP-ribose regulate TRPM2 channels in T lymphocytes. Faseb J 20:962–964

176. Kraft R, Grimm C, Frenzel H, Harteneck C (2006) Inhibition of TRPM2 cation channels by N-(p-amylcinnamoyl)anthranilic acid. Br J Pharmacol 148:264–273
177. Ullrich ND, Voets T, Prenen J, Vennekens R, Talavera K, Droogmans G, Nilius B (2005) Comparison of functional properties of the Ca^{2+}-activated cation channels TRPM4 and TRPM5 from mice. Cell Calcium 37:267–278
178. Grand T, Demion M, Norez C, Mettey Y, Launay P, Becq F, Bois P, Guinamard R (2008) 9-phenanthrol inhibits human TRPM4 but not TRPM5 cationic channels. Br J Pharmacol 153:1697–1705
179. Nilius B, Prenen J, Janssens A, Voets T, Droogmans G (2004) Decavanadate modulates gating of TRPM4 cation channels. J Physiol 560:753–765
180. Story GM, Peier AM, Reeve AJ, Eid SR, Mosbacher J, Hricik TR, Earley TJ, Hergarden AC, Andersson DA, Hwang SW, McIntyre P, Jegla T, Bevan S, Patapoutian A (2003) ANKTM1, a TRP-like channel expressed in nociceptive neurons, is activated by cold temperatures. Cell 112:819–829
181. Viana F, Ferrer-Montiel A (2009) TRPA1 modulators in preclinical development. Expert Opin Ther Pat 19:1787–1799
182. da Costa DS, Meotti FC, Andrade EL, Leal PC, Motta EM, Calixto JB (2010) The involvement of the transient receptor potential A1 (TRPA1) in the maintenance of mechanical and cold hyperalgesia in persistent inflammation. Pain 148:431–437
183. Brierley SM, Hughes PA, Page AJ, Kwan KY, Martin CM, O'Donnell TA, Cooper NJ, Harrington AM, Adam B, Liebregts T, Holtmann G, Corey DP, Rychkov GY, Blackshaw LA (2009) The ion channel TRPA1 is required for normal mechanosensation and is modulated by algesic stimuli. Gastroenterology 137:2084–2095
184. Wei H, Hamalainen MM, Saarnilehto M, Koivisto A, Pertovaara A (2009) Attenuation of mechanical hypersensitivity by an antagonist of the TRPA1 ion channel in diabetic animals. Anesthesiology 111:147–154
185. Kwan KY, Allchorne AJ, Vollrath MA, Christensen AP, Zhang DS, Woolf CJ, Corey DP (2006) TRPA1 contributes to cold, mechanical, and chemical nociception but is not essential for hair-cell transduction. Neuron 50:277–289
186. McNamara CR, Mandel-Brehm J, Bautista DM, Siemens J, Deranian KL, Zhao M, Hayward NJ, Chong JA, Julius D, Moran MM, Fanger CM (2007) TRPA1 mediates formalin-induced pain. Proc Natl Acad Sci USA 104:13525–13530
187. Koizumi K, Iwasaki Y, Narukawa M, Iitsuka Y, Fukao T, Seki T, Ariga T, Watanabe T (2009) Diallyl sulfides in garlic activate both TRPA1 and TRPV1. Biochem Biophys Res Commun 382:545–548
188. Chen J, Kim D, Bianchi BR, Cavanaugh EJ, Faltynek CR, Kym PR, Reilly RM (2009) Pore dilation occurs in TRPA1 but not in TRPM8 channels. Mol Pain 5:3
189. Hinman A, Chuang HH, Bautista DM, Julius D (2006) TRP channel activation by reversible covalent modification. Proc Natl Acad Sci USA 103:19564–19568
190. Macpherson LJ, Dubin AE, Evans MJ, Marr F, Schultz PG, Cravatt BF, Patapoutian A (2007) Noxious compounds activate TRPA1 ion channels through covalent modification of cysteines. Nature 445:541–545
191. Bautista DM, Movahed P, Hinman A, Axelsson HE, Sterner O, Hogestatt ED, Julius D, Jordt SE, Zygmunt PM (2005) Pungent products from garlic activate the sensory ion channel TRPA1. Proc Natl Acad Sci USA 102:12248–12252
192. Yassaka RT, Inagaki H, Fujino T, Nakatani K, Kubo T (2010) Enhanced activation of the transient receptor potential channel TRPA1 by ajoene, an allicin derivative. Neurosci Res 66:99–105
193. Andre E, Campi B, Materazzi S, Trevisani M, Amadesi S, Massi D, Creminon C, Vaksman N, Nassini R, Civelli M, Baraldi PG, Poole DP, Bunnett NW, Geppetti P, Patacchini R (2008) Cigarette smoke-induced neurogenic inflammation is mediated by alpha,beta-unsaturated aldehydes and the TRPA1 receptor in rodents. J Clin Invest 118: 2574–2582

194. Trevisani M, Siemens J, Materazzi S, Bautista DM, Nassini R, Campi B, Imamachi N, Andre E, Patacchini R, Cottrell GS, Gatti R, Basbaum AI, Bunnett NW, Julius D, Geppetti P (2007) 4-Hydroxynonenal, an endogenous aldehyde, causes pain and neurogenic inflammation through activation of the irritant receptor TRPA1. Proc Natl Acad Sci USA 104:13519–13524
195. Bang S, Kim KY, Yoo S, Kim YG, Hwang SW (2007) Transient receptor potential A1 mediates acetaldehyde-evoked pain sensation. Eur J Neurosci 26:2516–2523
196. Zygmunt PM, Andersson DA, Hogestatt ED (2002) Delta 9-tetrahydrocannabinol and cannabinol activate capsaicin-sensitive sensory nerves via a CB1 and CB2 cannabinoid receptor-independent mechanism. J Neurosci 22:4720–4727
197. Howlett AC, Barth F, Bonner TI, Cabral G, Casellas P, Devane WA, Felder CC, Herkenham M, Mackie K, Martin BR, Mechoulam R, Pertwee RG (2002) International Union of Pharmacology. XXVII. Classification of cannabinoid receptors. Pharmacol Rev 54:161–202
198. Cavanaugh EJ, Simkin D, Kim D (2008) Activation of transient receptor potential A1 channels by mustard oil, tetrahydrocannabinol and Ca^{2+} reveals different functional channel states. Neuroscience 154:1467–1476
199. Koo JY, Jang Y, Cho H, Lee CH, Jang KH, Chang YH, Shin J, Oh U (2007) Hydroxy-alpha-sanshool activates TRPV1 and TRPA1 in sensory neurons. Eur J Neurosci 26: 1139–1147
200. Talavera K, Gees M, Karashima Y, Meseguer VM, Vanoirbeek JA, Damann N, Everaerts W, Benoit M, Janssens A, Vennekens R, Viana F, Nemery B, Nilius B, Voets T (2009) Nicotine activates the chemosensory cation channel TRPA1. Nat Neurosci 12:1293–1299
201. Cruz-Orengo L, Dhaka A, Heuermann RJ, Young TJ, Montana MC, Cavanaugh EJ, Kim D, Story GM (2008) Cutaneous nociception evoked by 15-delta PGJ2 via activation of ion channel TRPA1. Mol Pain 4:30
202. Venkatachalam K, Zheng F, Gill DL (2003) Regulation of canonical transient receptor potential (TRPC) channel function by diacylglycerol and protein kinase C. J. Biol. Chem 278:29031–29040
203. Kanki H, Kinoshita M, Akaike A, Satoh M, Mori Y, Kaneko S (2001) Activation of inositol 1,4,5-trisphosphate receptor is essential for the opening of mouse TRP5 channel. Mol Pharmacol 60:989–998
204. Suchyna TM, Johnson JH, Hamer K, Leykam JF, Gage DA, Clemo HF, Baumgarten CM, Sachs F (2000) Identification of a peptide toxin from Grammastola spatulata spider venom that blocks cation-selective stretch-activated channels. J Gen Physiol 115:583–598
205. Gottlieb P, Suchyna TM, Bowman CL, Sachs F (2006) The mechanosensitive TRPC1 ion channel activity is modulated by cytoskeleton and inhibited by the peptide GsMTx4. Biophys J 90:1545
206. Spassova MA, Hewavitharana T, Xu W, Sobloff J, Gill DL (2006) A common mechanism underlies stretch activation and receptor activation of TRPC6 channels. Proc Natl Acad Sci USA 103:16586–16591
207. Gomis A, Soriano S, Belmonte C, Viana F (2008) Hypoosmotic- and pressure-induced membrane stretch activate TRPC5 channels. J Physiol 586:5633–5649
208. Drew LJ, Rugiero F, Cesare P, Gale JE, Abrahamsen B, Bowden S, Heinzmann S, Robinson M, Brust A, Colless B, Lewis RJ, Wood JD (2007) High-threshold mechanosensitive ion channels blocked by a novel conopeptide mediate pressure-evoked pain. PLoS One 2:e515
209. Leuner K, Heiser JH, Derksen S, Mladenov MI, Fehske CJ, Schubert R, Gollasch M, Schneider G, Hartenecj C, Chatterjee SS, Muller WE (2010) Simple 2,4-diacylphloroglucinols as classic transient receptor potential-6 activators – identification of a novel pharmacophore. Mol Pharmacol 77:368–377
210. Basora N, Boulay G, Bilodeau L, Rousseau E, Payet MD (2003) 20-hydroxyeicosatetraenoic acid (20-HETE) activates mouse TRPC6 channels expressed in HEK293 cells. J Biol Chem 278:31709–31716

211. Leuenroth SJ, Okuhara D, Shotwell JD, Markowitz GS, Yu Z, Somlo S, Crews CM (2007) Triptolide is a traditional Chinese medicine-derived inhibitor of polycystic kidney disease. Proc Natl Acad Sci USA 104:4389–4394
212. Kiselyov K, Soyombo A, Muallem S (2007) TRPpathies. J Physiol 578:641–653
213. Nilius B (2007) TRP channels in disease. Biochim Biophys Acta 1772:805–812
214. Nilius B, Owsianik G (2010) Transient receptor potential channelopathies. Pflugers Arch 460(2):437–450.
215. Nilius B, Voets T, Peters J (2005) TRP channels in disease. Sci STKE 2005:re8
216. Andre E, Campi B, Trevisani M, Ferreira J, Malheiros A, Yunes RA, Calixto JB, Geppetti P (2006) Pharmacological characterisation of the plant sesquiterpenes polygodial and drimanial as vanilloid receptor agonists. Biochem Pharmacol 71:1248–1254

Chapter 4
Synthetic Modulators of TRP Channel Activity

Christian Harteneck, Chihab Klose, and Dietmar Krautwurst

Abstract In humans, 27 TRP channels from 6 related families contribute to a broad spectrum of cellular functions, such as thermo-, pressure-, volume-, pain- and chemosensation. Pain and inflammation-inducing compounds represent potent plant and animal defense mechanisms explaining the great variety of the naturally occurring, TRPV1-, TRPM8-, and TRPA1-activating ligands. The discovery of the first vanilloid receptor (TRPV1) and its involvement in nociception triggered the euphoria and the hope in novel therapeutic strategies treating pain, and this clear-cut indication inspired the development of TRPV1-selective ligands. On the other hand the nescience in the physiological role and putative clinical indication hampered the development of a selective drug in the case of the other TRP channels. Therefore, currently only a handful of mostly un-selective blocker is available to target TRP channels. Nevertheless, there is an ongoing quest for new, natural or synthetic ligands and modulators. In this chapter, we will give an overview on available broad-range blocker, as well as first TRP channel-selective compounds.

4.1 Introduction

The use of small molecules as drugs or tools for pharmacological interference represents an attractive measure in therapy and in basic research. In ion channel research, divalent and trivalent cations serve as tools to characterize the selectivity

---

C. Harteneck (✉)
Institute for Pharmacology and Toxicology, Interfaculty Center of Pharmacogenomics and Pharmaceutical Research (ICEPHA), Eberhard-Karls-University, Tübingen, Germany
e-mail: Christian.Harteneck@uni-tuebingen.de

and permeation properties of ion channels like the transient receptor potential (TRP) family. TRP channels represent a superfamily of calcium-permeable ion channels involved in a variety of physiological processes like hormonal signaling cascades, calcium and magnesium homeostasis, thermo-, pressure-, volume-, pain-, and chemosensation. According to their primary structure and function, mammalian TRP proteins can be classified into at least six subfamilies, the classic or canonical TRPs (TRPC), the vanilloid receptor-related TRPs (TRPV), and the melastatin-related TRPs (TRPM), the ankyrine-rich TRP (TRPA), the channels of the polycystin group (TRPP), and the members of the mucolipidin group (TRPML). Members of the TRPC are characterized by their activation due to stimulation of G-protein-coupled receptors and phospholipase C (PLC). TRPV channels (TRPV1 to TRPV4) are activated by increases in temperature and are involved in heat and pain sensation in human body, whereas TRPV4 and TRPM3 activated by extracellular hypotonic solution are involved in volume regulation [1]. Moreover, TRP channels are involved in calcium and magnesium homeostasis, since these channels permeate divalent cations. The body calcium homeostasis is maintained by the transcriptionally regulated activity of TRPV5 and TRPV6, which form channels with high calcium selectivity. The magnesium homeostasis depends on TRPM6 and TRPM7 which enable transcellular intestinal and renal magnesium uptake besides the paracellular transport mechanisms [2]. TRPM4 and TRPM5 form calcium-modulated channels selectively permeating sodium. TRPM2 represents a redox sensor being indirectly activated by hydrogen peroxide, and its closest relative TRPM8 is activated by cold temperatures, menthol and icilin, an experimentally used cooling compound [2]. The other sensor for cold temperatures of the TRP channel family is TRPA1, being activated by cold temperatures, icilin, and a broad spectrum of pain-inducing structures [3, 4]. The autosomal dominant polycystic kidney disease (ADPKD) is characterized by cysts formation in the kidney, a malformation of the renal tubular structure [5]. The cyst formation is linked to the disruption of the functional TRPP1/TRPP2 complex. TRPP1, a huge protein consisting of a complex N-terminal extracellular ligand binding and 11 transmembrane domains, and TRPP2, a calcium-permeable ion channel, form a receptor-effector complex implicated in various biological functions, such as cell proliferation, sperm fertilization and mechanosensation [5].

In the pre-genome era, organic blocker like SKF-96365, LOE908, clotrimazole and others were characterized as tools to interfere with receptor-mediated, store-operated and other Ca^{2+} entry mechanisms [6–9]. Consequently, these compounds were initially used for the characterization of mammalian TRPCs. Since some compounds are not commercially available, their effects may not be associated with distinct molecular targets anymore. Detailed analyses testing a variety of TRP channels showed that SKF-96365, clotrimazole and recently characterized compounds like 2-aminoethoxydiphenyl borate (2-APB), or N-(p-amylcinnamoyl)anthranilic acid (ACA) represent broad-spectrum TRP channel modulator (Fig. 4.1). The summary of these established structures is followed by the discussion of several new compounds described as TRP-selective tools.

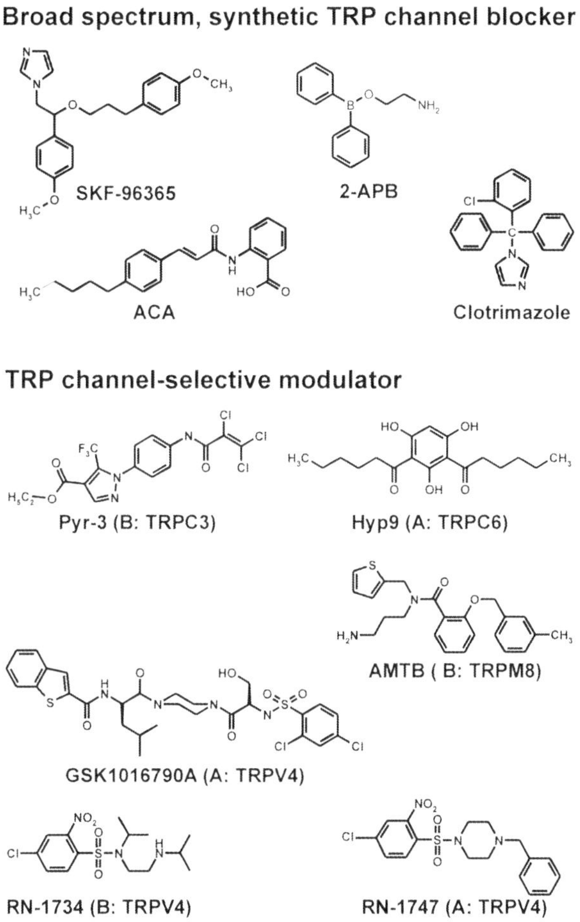

Fig. 4.1 Chemical structures of TRP channel modulator. Broad spectrum TRP channel blocker are SKF-96365: 1-{β[3-(4-methoxyphenyl)propoxy]-4-methoxyphenethyl}-1H-imidazole hydrochloride; 2-APB: 2-aminoethoxydiphenyl borate; ACA: N-(p-amylcinnamoyl)anthranilic acid; Clotrimazole: 1-[(2-Chlorophenyl)diphenyl-methyl]-1H-imidazole. The structures of several TRP-selective compounds are shown (A: activator; B: blocker). Pyr3: Ethyl-1-(4-(2,3,3-trichloroacrylamide)phenyl)-5-(trifluoromethyl)-1H-pyrazole-4-carboxylate selectively blocks TRPC3, whereas the activity of the other member of the TRPC family remains unchanged [55]. Hyp9: 2,4 Dihexanoylphloroglucinol activates TRPC6 but not the closely related TRPC3 and TRPC7 [56]. AMTB: N-(3-aminopropyl)-2-{[(3-methylphenyl)methyl]oxy}-N-(2-thienylmethyl)benzamide blocks TRPM8 expressed in the bladder, but not TRPV1 and TRPV4 also expressed in this tissue [64]. GSK1016790A: N-((1S)-1-{[4-((2S)-2-{[(2,4-Dichlorophenyl)sulfonyl]amino}-3-hydroxypropanoyl)-1-piperazinyl]carbonyl}-3-methyl-butyl)-1-benzothiophene-2-carboxamide activates human TRPV4 with an EC_{50} of 3.0 nM, whereas the EC_{50} for TRPV1 is 50 nM. TRPA1 and TRPM8 were unaffected [74]. RN-1734 blocks TRPV4 with an IC_{50} of 2.3 μM, RN-1747 activates TRPV4 with an EC_{50} of 0.77 μM whereas TRPV1, TRPV3, and TRPM8 remained unaffected [79]

4.2 Broad-Spectrum Non-natural, Synthetic TRP Channel Blocker

4.2.1 SKF-96365

SKF-96365, 1-{β[3-(4-methoxyphenyl)propoxy]-4-methoxyphenethyl}-1H-imidazole hydrochloride is an inhibitor of receptor-mediated as well as store-operated Ca^{2+} entry [8, 9]. Introduced as inhibitor of receptor-mediated Ca^{2+} entry, SKF-96365 blocked ADP-induced Ca^{2+} entry in platelets, as well as in neutrophils and endothelial cells, with IC_{50} values of around 10 μM [10]. SKF-96365 as a tool allowed to discriminate ATP- and bradykinin-induced Ca^{2+} entry mechanisms in PC-12 cells, and to characterize ATP- and N-formyl-L-methionyl-L-leucyl-L-phenylalanine (fMLP)-stimulated cation currents in HL-60 cells [6, 11]. The broad spectrum of ion channels blocked by SKF-96365 became clearer with several publications. The use of SKF-96365 side by side with LOE908, a simultaneously developed compound, showed that LOE908 selectively blocked an angiotensin-induced non-selective current, whereas SKF-96365 additionally interfered with voltage-operated dihydropyridine-sensitive currents mediated in A7r5 vascular smooth muscle cells [7]. It soon became clear that the activity of SKF-96365 is not restricted to Ca^{2+} entry mechanisms induced by hormonal stimulation of plasma membrane receptors. Refilling mechanism of intracellular stores activated by the depletion of intracellular stores by cyclopiazonic acid were also blocked by SKF-96365 [12–14]. Parallel with the identification and characterization of TRP channels, the focus of SKF-96365 applications changed away from merely describing and characterizing different Ca^{2+} entry mechanism towards the molecular characterization of cloned ion channels.

Together with the inorganic blockers, SKF-96365 was used to characterize the receptor-regulated isoforms of the TRPC family (Fig. 4.2). Here, it were mainly the diacylglycerol-regulated TRP channels, TRPC3, TRPC6, that have directly been characterized to be SKF-96365-sensitive in studies using recombinant expression systems [15–17]. Interestingly, in *Drosophila,* diacylglycerol as a PIP_2-breakdown product subsequent to receptor stimulation is ineffective in directly stimulating *Drosophila* TRPC channels. *Drosophila* TRPL as well as TRPγ are both activated by polyunsaturated fatty acids [18, 19]. Despite the vast diversity of intracellular signaling cascades and involved second messengers, TRPγ is blocked by SKF-96365 [19]. SKF-96365 was mostly used in concentrations up to 100 μM [20, 21]. To establish concentration-response relations, we studied the inhibitory effects of SKF-96365 on TRPC6, TRPM2, TRPM3 and TRPV4 stimulated by hyperforin (10 μM), hydrogen peroxide (5 mM), pregnenolone sulphate (35 μM) and 4α-phorbol-didecanoate (5 μM), respectively (Fig. 4.3). The concentration-response curves determined by Ca^{2+} imaging showed a quite distinct inhibition profile. TRPC6 was the most sensitive TRP channel within the test group with an IC_{50} value for SKF-96365 of around 2 μM. On the other end, TRPM2 emerged as the most insensitive TRP channel with an IC_{50} value of around 75 μM. Eye-catching in the set of curves is the steep slope of the TRPV4 concentration-response

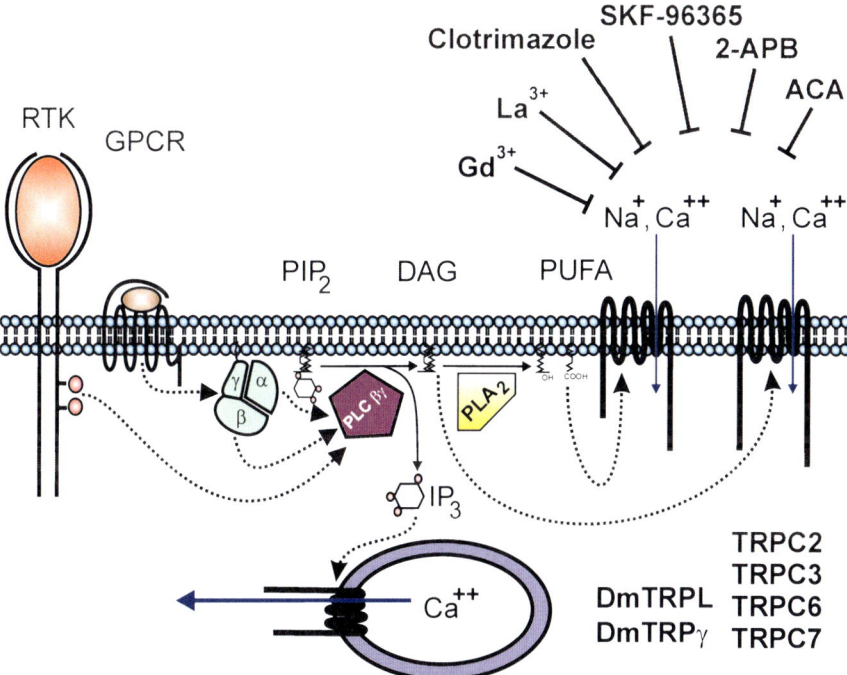

Fig. 4.2 Regulation of mammalian and fly receptor-activated TRPC channels. G-protein coupled receptor (GPCR)-mediated as well as receptor tyrosinkinase-mediated activation of phospholipase C isoforms (PLC βγ) result in the breakdown of phosphatidylinositides (PIP$_2$) leading to the formation of inositol 1,4,5-trisphosphate (IP$_3$) and diacylglycerol (DAG). IP$_3$ is the ligand of IP$_3$ receptors, which are located at the endoplasmic reticulum. Upon activation, IP$_3$ receptors mediate Ca^{2+} release from intracellular stores, whereas DAG directly activates mammalian TRPC2, TRPC3, TRPC6 and TRPC7 channels. In contrast to the mammalian signaling cascade, *Drosophila* TRPL and TRPγ are activated by polyunsaturated fatty acids (PUFA), which are generated by degradation of diacylglycerols subsequent to an activation of phospholipase A$_2$ (PLA$_2$).

curve. The nearly digital switching of TRPV4 inhibition within a narrow concentration range of less than half an order of magnitude (from 10 to 30 μM) was reproducible, and also seen in other configurations, and may imply cooperativity. An IC$_{50}$ value for SKF-96365 of around 25 μM blocking TRPV4 as well as the slope of this concentration-response relation may explain the ineffectiveness of 20 μM SKF-96365 in blocking TRPV1 [22]. In summary, SKF-96365 used in concentration up to 10 μM selectively blocks TRPC channels, whereas TRPV and TRPM channels are blocked at higher concentrations.

4.2.2 2-APB

Introduced in Ca^{2+} signaling studies in 1997, 2-minoethoxydiphenyl borate (2-APB) was initially used in many studies to probe for the involvement of inositol

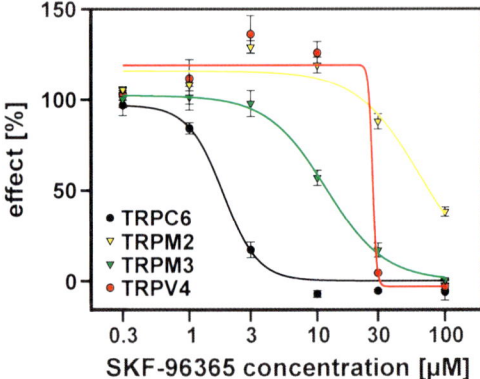

Fig. 4.3 Concentration-response relationships of SKF-96365 blocking TRP channels. Data from a representative experiment show the effect of SKF-96365 on Ca^{2+} entry in TRPC6-, TRPM2-, TRPM3-, and TRPV4-expressing cells upon hyperforin (10 μM), hydrogen peroxide (5 mM), pregnenolone sulphate (35 μM), and 4α-phorbol-didecanoate stimulation (5 μM), respectively. Cells inducibly and stably expressing TRP channels were loaded with Fluo-4, as previously described in Jörs et al. [19]. Ca^{2+} entry was measured using a FLIPR^{Tetra}. Data were analyzed offline, and were calculated from one out of at least three experiments, performed in quadruplicates per concentration and TRP channel. The concentration-response curves determined by Ca^{2+} imaging showed a quite distinct inhibition profile. The IC_{50} values of TRPC6, TRPM2, TRPM3 and TRPV4 are 2, 75, 12 and 25 μM, respectively

1,4,5-trisphosphate (IP_3) receptors in the generation of Ca^{2+} signals. As a cheap alternative to the expensive, naturally occurring xestospongin C, 2-APB was also used to study store-operated Ca^{2+} entry mechanism [23]. Whereas 2-APB was thought to block IP_3 receptors, and used as additional tool for the characterization of store-operated Ca^{2+} entry mechanisms, the molecular targets of 2-APB became clear with the identification and cloning of the stromal interactions proteins STIM1 and STIM2 and the plasma membrane channels Orai1, Orai2 and Orai3 (also known as CRACM1 to CRACM3) [24]. STIM1 and STIM2 represent the Ca^{2+} sensors of the endoplasmic reticulum (ER) interacting and regulating Orai or CRACM channels at the plasma membrane. The effects of 2-APB on Orai or CRACM channels, however, appeared to be quite complex. CRACM1 (Orai1) and CRACM2 (Orai2) were blocked, whereas 2-APB activated a current in STIM1 and CRACM3 (Orai3) co-expressing cells [25–27]. The effect on CRACM3 (Orai3) was shown to result from a 2-APB-dependent increase in pore size, thereby altering its ion selectivity [27].

Several reports about effects of 2-APB on the activity of a variety of TRP channels have finally established 2-APB as a non-selective TRP channel modulator. For instance, it has been repeatedly shown that TRPC channels are blocked by 2-APB [28–30]. Receptor-initiated signaling cascades leading to Ca^{2+} entry mediated by TRPC1, TRPC3, TRPC5, TRPC6, and TRPC7 have been described to be blocked by 2-APB. Blockade of receptor-stimulated Ca^{2+} entry by 2-APB works across species, in mammals, as well as in cockroach *Periplaneta Americana* [32].

2-APB has a stimulatory or inhibitory effect on TRP channels except TRPM2 being insensitive [30]. This is in line with the unique insensitivity of TRPM2 to gadolinium and lanthanum ions, whereas all other TRP channels are blocked or inhibited by lanthanides (see below). Nevertheless, we have to note that TRPM2 has been reported to be blocked by 2-APB by Togashi et al. [33]. Sphingosine-induced Ca^{2+} entry mediated by TRPM3 was also blocked by 2-APB, as well as the magnesium flux through TRPM7 [30, 34]. In the context of TRPM7 and its closely related homologue TRPM6, it was interesting that the quality of the 2-APB effects was concentration-dependent: In micromolar concentrations, 2-APB potentiated TRPM6, and inhibited TRPM7, whereas at millimolar concentrations TRPM6/7 heteromultimers as well as TRPM7 were potentiated [35].

Within the mammalian TRPV channels family, 2-APB mostly potentiated, or stimulated channels activity. In 2004, Chung et al. firstly showed that TRPV3 was activated and sensitized by 2-APB [36]. Later on, also TRPV1 and TRPV2 were found to be stimulated by the borate compound [37]. In the controversy, whether TRPV6 (CaT1) represents I_{CRAC}, 2-APB was used as tool to discriminate the currents under investigation: TRPV6 (CaT1), just like the other TRPV channels, was stimulated, whereas I_{CRAC} was blocked by 2-APB [38, 39]. In the case of TRPV2, the activation by 2-APB has been shown to be species-specific. Whereas mouse und rat TRPV2 were stimulated by heat above 53°C and 2-APB, human TRPV2 remained silent upon these stimulation protocols [40]. Hu et al. showed that the stimulation of TRPV3 by 2-APB was mediated by two intracellularly located amino acids [41], whereas the inhibitory effects on TRPC5 activity resulted from interaction with the extracellular side of the channel (Xhu et al. 2005). All together, this suggested that the stimulatory and inhibitory effects mediated by 2-APB are mediated through independent binding sides, and may be discussed as agonist and non-competitive allosteric modulation.

4.2.3 ACA

N (p amylcinnamoyl)anthranilic acid, commonly referred as ACA, was formerly used as a broad-spectrum inhibitor for the characterization of phospholipase A_2-mediated pathways. Phospholipase A_2 enzymes display a superfamily of structurally different enzymes classified in at least nine subfamilies by biochemical and structural properties. Phospholipase A_2 (PLA_2) enzymes release arachidonic acid as an ubiquitous intra- and extracellular mediator from phospholipids. Arachidonic acid or its subsequent metabolites, resulting from cyclooxygenase, lipoxygenase or cytochrome P_{450} enzyme reactions, modulate the activity of a great variety of proteins. Beside G-protein coupled receptors (GPCR), several TRP channels have been described to be modulated by arachidonic acid or arachidonic acid metabolites.

TRPV1, as heat receptor in our body, is involved in pain perception and is activated by endogenous ligands, known as endovanilloids [42]. Despite their common arachidonyl structure, the biosynthesis of the endovanilloids is diverse, and only partly depends on an initial PLA_2 activation. The endovanilloids are products of specifically expressed enzyme systems allowing a cell type specific TRPV1-response. By itself, arachidonic acid does not activate TRPV1. ACA partially blocked TRPV1 in a voltage-dependent manner [43].

Arachidonic acid directly does not stimulate TRPV1, but is able to trigger TRPV4-mediated Ca^{2+} entry. TRPV4, as cation channel activated by extracellular hypotonic solutions, can also be stimulated by 4α-phorbol esters, moderate heat, shear stress and arachidonic acid metabolites [1, 44]. Stimulation of arachidonic acid is in line with the stimulation mode by extracellular hypotonicity, as it has been shown, that extracellular hypotonicity results in increased intracellular arachidonic acid levels triggering a Ca^{2+} entry mechanism [45]. The hypotonicity-induced activation of TRPV4 is attenuated in the presence of inhibitors of PLA_2 enzymes, as well as inhibitors of the cytochrome P_{450} epoxygenase like miconazole [46]. Thus, it has been postulated that TRPV4 is regulated by epoxyeicosatrienoic acids [46, 47]. Beside its action as cytochrome P_{450} epoxygenase inhibitor, miconazole is an azole like econazole and clotrimazole (see below), clinically used as antifungal drug. Furthermore, azoles have in common with SKF-96365 and LOE908 a function as blocker of Ca^{2+} entry [8].

Hara et al. characterized TRPM2 as redox sensor modulated by arachidonic acid via a arachidonic acid-responsive sequence in the N-terminus [48]. To study the impact of arachidonic acid on TRPM2 endogenously expressed in microglia, we looked for tools interfering with arachidonic acid production. Beside several ineffective phospholipase A_2 inhibitors such as arachidonyl trifluoromethyl ketone (AACOCF3), and p-bromophenacyl bromide (BPB), N-(p-amylcinnamoyl)anthranilic acid (ACA) showed up as an effective interfering compound [49]. Careful characterization of its effect soon unraveled a direct blockade of TRPM2 instead of a phospholipase A_2 modulation. ACA blocks TRPM2 stimulated by hydrogen peroxide in a concentration-dependent manner with an IC_{50} value of 1.7 μM [49]. Inhibition of TRPM2-mediated whole-cell currents by ACA was voltage-independent, and accelerated at decreased pH. ACA was ineffective when applied intracellularly. Further analysis of TRP channel selectivity revealed that not only the closely related TRPM8-mediated Ca^{2+} entry is blocked by ACA but also TRPC3, TRPC6, and TRPV1-mediated currents [43]. ACA as TRPM2 blocker has been validated in further approaches characterizing natively expressed TRPM2 in neutrophils, beta-cells and hippocampal neurons [50–52].

4.2.4 Clotrimazole

Clotrimazole, 1-[(2-Chlorophenyl)diphenylmethyl]-1H-imidazole, is an azole compound being clinically used as antifungal drug. Azole compounds inhibit the fungal cytochrome P450 3A enzyme, which is responsible for the conversion of lanosterol to ergosterol, the main sterol in the fungal cell membrane. In contrast to the systemically applied azoles like fluconazole and ketoconazole, clotrimazole and econazole are used as locally applied fungistatic drugs. In the pharmacological characterization of TRP channels, econazole and clotrimazole have been firstly described as blocker of TRPM2-mediated Ca^{2+} entry and currents [53]. Like other compounds, clotrimazole has a dual effect on TRP channels. Meseguer et al. reported that clotrimazole activates TRPV1 and TRPA1, which may explain skin irritations and burning pain of skin and mucous membranes within the local antifungal therapy [54]. In order to test whether clotrimazole has an impact on additional TRP channels, we tested the effect of a single clotrimazole concentration (30 μM) on the activity of TRPM2 as a control, side by side with TRPC6, TRPM3, and TRPV4 stimulated by hydrogen peroxide, hyperforin, pregnenolone sulphate and 4α-phorbol-didecanoate, respectively (Fig. 4.4). The figure shows that 30 μM clotrimazole was effective to block all of the tested TRP channels. Based on these initial data, azole compounds like econazole and clotrimazole may be classified as broad spectrum TRP channel modulators.

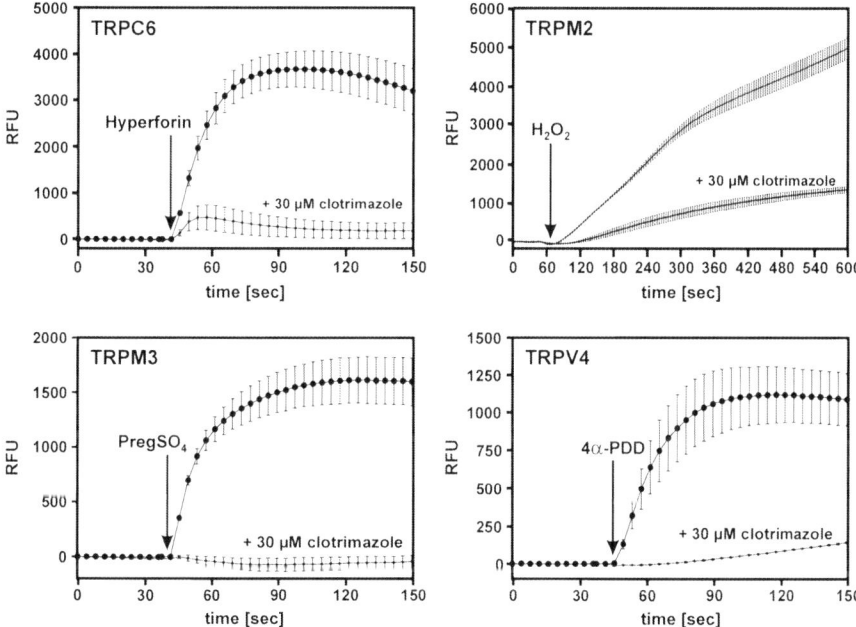

Fig. 4.4 TRPC6, TRPM2, TRPM3, and TRPV4 channels are blocked by clotrimazole. Increase in fluorescence of TRPC6-, TRPM2-, TRPM3-, and TRPV4-expressing, FLUO-4-loaded HEK293 cells upon stimulation with hyperforin (10 μM), hydrogen peroxide (5 mM), pregnenolone sulphate (35 μM), and 4α-phorbol-didecanoate (5 μM), respectively. In the absence of an inhibitor, the detectable change in fluorescence is indicative for functional expression of the different TRP channels. Agonist-induced fluorescence increase is attenuated in cells that were 10 min preincubated with 30 μM clotrimazole. The traces represent averaged data of 4 wells with 10,000 cells each. Shown is one representative experiment out of at least three similar experiments

4.3 TRP Channel-Selective Modulator

The contribution of natural occurring, secondary plant products in the deorphanization of the TRP ion channels makes the TRP channel a fascinating family of ion channels, activated by structures mediating hot, burning, sharp, cold, and spicy sensation. Simultaneously, these reports show that it is possible, in principle, to generate compounds selectively modulating individual TRP channels. The euphoria over the discovery of TRPV1 involved in nociception, the hope in novel therapeutic strategies treating pain, as well as the clear-cut indication as a therapeutic target inspired the development of TRPV1-selective ligands. Therefore, several structures modulating TRPV1 have been published. Outside the pain field, the number of TRP-selective, non-natural, synthetic modulator is very small to non-existing. This situation is based on the fact that often the TRP channels outside the pain field are not selectively linked to a clinical indication to justify focused drug screens. On the other hand, several compounds being used as tools in specific contexts were presented as TRPX-selective drugs, however the compounds were not validated for their selectivity with respect to other TRP channels. The use of such compounds bears the risk that selectivity may not be given, and that the observed effects may rather result from the modulation of other TRP channels.

4.3.1 Compounds Selectively Modulating TRPC Channels

One first step in the direction of TRPC-selective blocking tools represents the Pyr3 compound developed by the group of Yasuo Mori [55]. They showed that Pyr3, a sequel development of BTP1, selectively blocks TRPC3, whereas the activity of other TRPC channels remained unaffected. Pyr3 selectively blocked TRPC3 in electrophysiological recordings and photoaffinity labeling experiments using a Pyr3-derivative, suggesting that TRPC3 proteins directly bind Pyr3 [55]. In a subsequent physiological approach, Pyr3 potently eliminated the Ca^{2+} influx-dependent PLC translocation to the plasma membrane in DT40 B lymphocytes, and the late oscillatory phase of B cell receptor-induced Ca^{2+} response. Moreover, Pyr3 attenuated an activation of nuclear factor in activated T cells, a Ca^{2+}-dependent transcription factor and hypertrophic growth in rat neonatal cardiomyocytes, and *in vivo* pressure overload-induced cardiac hypertrophy in mice [55]. While Pyr3 represents a TRPC3-selective blocker, we recently characterized several phloroglucinol derivatives as TRPC6-selective activators [56]. Hyperforin is the naturally occurring secondary plant compound of St. John's wort, and has a polyprenylated bicyclic acylphloroglucinol structure. Pure hyperforin is not very stable when exposed to light and oxygen, which limits its clinical applications and further developments. This instability is due to the enolized β-dicarbonyl system present in the molecule. Several previous studies dealt with the chemical modification of its structure by acylation, alkylation, and oxidation, but not with a chemical simplification of hyperforin, to identify new stable and potent hyperforin analogues

[57, 58]. Based on the phloroglucinol core structure, several derivatives were tested for the ability to activate recombinantly as well as endogenously expressed TRPC6. Symmetric 2,4-diacylphloroglucinol derivatives (Hyp9) emerged as activating compounds, inducing TRPC6 activity in several experimental setups [56]. To our surprise, the new structures still appeared to be selective activators of TRPC6, whereas the activity of the closely related TRPC3 and TRPC7 channels remained unchanged [56].

4.3.2 Compounds Selectively Modulating TRPM Channels

The most prominent TRP channel of the melastatin-like TRP channel family is TRPM8 due to its function as cold sensor in our body. The natural occurring secondary plant product menthol as well as synthetic cooling compounds like icillin helped to characterize the function of TRPM8 and also TRPA1 [3, 59]. Beside icillin, several other cooling compounds of synthetic or nature identical structures have been shown to modulate TRPM8 [60–62]. However these compounds are not as selective as initially described, since they also modulate other thermo-TRPs [63]. A caveat has also been made for AMTB, a TRPM8 blocker that has been described to block overactive bladder and painful bladder syndrome in rats, which was causally linked to a TRPM8 function [64]. Due to the contribution of TRPV1 [65] and TRPV4 [66, 67] to bladder function, the TRPM8 selectivity of AMTB was tested with respect to TRPV1 and TRPV4 channel activity, and was found to be ineffective modulating both TRPV channels [64]. However, it is still unclear whether AMTB modulates the activity of other TRP channels.

In the context of TRPM-selective modulator, also nifedipine should been mentioned. Wagner et al. recently showed that nifedipine activates TRPM3 channels [68]. This effect is remarkable, since nifedipine was developed as blocker of voltage-gated calcium channels, and is normally used in electrophysiological studies characterizing native currents in order to discriminate between voltage-gated calcium channels and other calcium-permeable channels, such as TRPs.

4.3.3 Compounds Selectively Modulating TRPV Channels

With more than 500 entries in PubMed, the field of TRPV1 antagonists is overwhelming by- it would burst the space to give a detailed overview [for review see [69, 70]. The induction of pain is a potent defense mechanism in the plant and animal kingdom. For instance, snake, spider and scorpion toxins, as well as plant extracts contain TRPV1 agonistic compounds. On the other hand, the effect of pain relief of capsaicin used in traditional herbal medicine results from intracellular desensitization processes of the TRPV1 channel. Thus, it was of high interest to develop potent TRPV1 agonists as well as antagonists in order to interfere with the TRPV1-mediated pain and inflammatory processes. One of the first available antagonists was derived from a natural occurring agonist, resiniferatoxine, from *Euphorbiengummi resinifera*. The iodinated form of resiniferatoxine, iodo-resiniferatoxin, represents a TRPV1 antagonist, and, like capsazepine – another synthetic TRPV1 inhibitor, results from the pre-genome era [71, 72]. Shortly thereafter, SB-366791 is one of the first genome era TRPV1 blocker and led to other structures [73]. The euphoria, measurable by the number of publications and companies involved in the run for a TRPV1 inhibitor, is declining, as the number of individuals in clinical trials increases that display symptoms such as hyperthermia or scalding due to block of our temperature sensor [69]. In contrast to the antagonists, capsaicin and synthetic agonists induce hypothermia. The future will show whether useful TRPV1 modulating drugs will be available for therapy.

The participation of TRPV4 in bladder regulation as shown by molecular biology methods was functionally validated by demonstrating that GSK1016790A, a TRPV4-selective agonist, is able to induce bladder contractility and hyperactivity [74]. The hope of a pharmacological modulation of bladder function by GSK1016790A, however, is only one part of the coin. The other, published in Part II of this issue, is that TRPV4 plays also an important role in vascular regulation [75–77], and systemic application of GSK1016790A results in endothelial failure and circulatory collapse [78]. With this in mind, it will be interesting to learn about effects from TRPV4 antagonists. With RN-1747 and RN-1734, the first TRPV4 antagonists have been published recently, however physiological data are missing [79].

4.4 Appendix – Lanthanum and Gadolinium Ions as Modulators of TRP Channels

While divalent ions mostly are able to permeate the channel pores, the trivalent cations gadolinium and lanthanum are channel blockers. In studies on the fly phototransduction, a fly mutant with deficits in vision was characterized showing a transient receptor potential upon light activation and named *trp*. The phenotype results from inactivating mutations in *Drosophila* TRP, a Ca^{2+}-permeable ion channel involved in the light-induced signaling cascade [80]. The application of 10 μM lanthanum chloride enabled Hardie and Minke to mimic the *trp* phenotype in wild-type flies [81]. For further characterization, Ca^{2+} and barium influx measurements were used to characterize *Drosophila* TRPL, as well as heterologously expressed TRP channels in recombinant expression systems [82, 83].

The use of the trivalent cations gadolinium and lanthanum as ion channel blockers allowed the characterization of endogenous Ca^{2+} entry mechanisms in native cells, as well as the characterization of recombinant TRP channel proteins in heterologous cell systems [84]. For instance, the insensitivity of *Drosophila* TRPL to low concentration of gadolinium ions (10 μM) allowed to discriminate Ca^{2+} entry mediated by endogenously expressed TRP channels in Sf9 cells (sensitive to 10 μM gadolinium chloride) from Ca^{2+} entry mediated by the recombinantly expressed *Drosophila* TRPL channel (sensitive to 1,000 μM), which finally allowed the characterization of *Drosophila* TRPL as a receptor-regulated ion channel [84]. Like *Drosophila* TRPL, TRPγ the third member of the *Drosophila* TRPC channel family was sensitive to lanthanum and gadolinium ions (100 μM) [19].

Due to their close sequence similarity, mammalian members of the classic or canonical TRP channel family (TRPC) were identified shortly after initial functional characterization of *Drosophila* TRPs in recombinant cell systems. Despite earlier controversies on activation mechanisms, most of the mammalian TRPC members are activated by receptor-induced activation of phospholipase C isoforms, resulting in the breakdown of phosphatidylinositides, and the formation of inositol 1,4,5-trisphosphate and diacylglycerol (Fig. 4.1). Inositol 1,4,5-trisphosphate, being a ligand of inositol 1,4,5-trisphosphate receptors, which are located at the endoplasmic reticulum, induces Ca^{2+} release from intracellular stores, whereas diacylglycerols can directly activate mammalian TRPC2, TRPC3, TRPC6 and TRPC7 channels [85–87]. In contrast to the mammalian signaling cascade, *Drosophila* TRPL and TRPγ are activated by polyunsaturated fatty acids generated by phospholipase A_2 from diacylglycerols [18, 19].

Whereas most of the TRPC members are blocked by lanthanum and gadolinium ions [20, 88–91], TRPC4 and TRPC5 exclusively show a concentration-dependent stimulation or potentiation in presence of the trivalent ions [92, 93]. Neutralization by site-directed mutagenesis of negatively charged glutamic acid residues, which are situated close to the extracellular mouth of the channel pore of TRPC5, blunted this effect, and resulted in TRPC5 mutant channels that were blocked by gadolinium and lanthanum ions [94].

TRP channels of the vanilloid family (TRPV) are routinely characterized by the use of the polycationic ruthenium complex ion, ruthenium red [22, 44, 95–97]. Besides blocking ryanodine receptors, ruthenium red also interferes with the activity of TRPM6 and TRPA1 [98–100]. Gadolinium and lanthanum ions differentially modulate TRPV channels. Here, TRPV2, TRPV4, TRPV5 and TRPV6 are blocked by gadolinium and lanthanum ion, whereas TRPV1 is stimulated and potentiated by gadolinium ions [44, 101–103].

Within the melastatin-like TRP family (TRPM), information about the effects of gadolinium and lanthanum ion is still incomplete. TRPM2-mediated Ca^{2+} influx into microglia activated either by extracellular hydrogen peroxide, or by intracellular perfusion of ADP-ribose, seems to be insensitive to lanthanum ions at concentrations up to 100 μM [104]. In contrast, TRPM3 as well as TRPM8 are blocked by 100 μM gadolinium chloride [105, 106].

4.5 Outlook

The first examples of synthetic TRP-selective modulators show that not only nature is able to develop TRP channel-selective compounds – men can also develop TRP-selective compounds. Nevertheless, the diversity of compounds of animal and plant provenience provides an interesting pool of lead structures for drug development. However for a rational development, much more knowledge than currently available is necessary. For many TRP channels the physiological as well as pathophysiological role is still unclear. The availability of a first generation of "dirty" drugs will, nevertheless, facilitate a gain in knowledge. In future, deeper insights into the physiological function, together with extensive expression profiles across tissues, will enable us to develop sophisticated strategies for drug development. The diversity in function and the inspiring presence of naturally occurring structures makes TRP channels a still highly interesting group of targets.

Acknowledgments The authors were supported by the Deutsche Forschungsgemeinschaft.

References

1. Harteneck C, Reiter B (2007) TRP channels activated by extracellular hypo-osmoticity in epithelia. Biochem Soc Trans 35:91–95
2. Harteneck C (2005) Function and pharmacology of TRPM cation channels. Naunyn Schmiedebergs Arch Pharmacol 371:307–314
3. Story GM, Peier AM, Reeve AJ, Eid SR, Mosbacher J, Hricik TR, Earley TJ, Hergarden AC, Andersson DA, Hwang SW, McIntyre P, Jegla T, Bevan S, Patapoutian A (2003) ANKTM1, a TRP-like channel expressed in nociceptive neurons, is activated by cold temperatures. Cell 112:819–829
4. Caspani O, Heppenstall PA (2009) TRPA1 and cold transduction: an unresolved issue? J Gen Physiol 133:245–249
5. Zhou J (2009) Polycystins and primary cilia: primers for cell cycle progression. Annu Rev Physiol 71:83–113
6. Krautwurst D, Hescheler J, Arndts D, Losel W, Hammer R, Schultz G (1993) Novel potent inhibitor of receptor-activated nonselective cation currents in HL-60 cells. Mol Pharmacol 43:655–659
7. Krautwurst D, Degtiar VE, Schultz G, Hescheler J (1994) The isoquinoline derivative LOE 908 selectively blocks vasopressin-activated nonselective cation currents in A7r5 aortic smooth muscle cells. Naunyn Schmiedebergs Arch Pharmacol 349:301–307
8. Clementi E, Meldolesi J (1996) Pharmacological and functional properties of voltage-independent Ca^{2+} channels. Cell Calcium 19:269–279
9. Leung YM, Kwan CY (1999) Current perspectives in the pharmacological studies of store-operated Ca^{2+} entry blockers. Jpn J Pharmacol 81:253–258
10. Merritt JE, Armstrong WP, Benham CD, Hallam TJ, Jacob R, Jaxa-Chamiec A, Leigh BK, McCarthy SA, Moores KE, Rink TJ (1990) SK&F 96365, a novel inhibitor of receptor-mediated calcium entry. Biochem J 271:515–522
11. Fasolato C, Pizzo P, Pozzan T (1990) Receptor-mediated calcium influx in PC12 cells. ATP and bradykinin activate two independent pathways. J Biol Chem 265:20351–20355
12. Morgan AJ, Jacob R (1994) Ionomycin enhances Ca^{2+} influx by stimulating store-regulated cation entry and not by a direct action at the plasma membrane. Biochem J 300(Pt 3):665–672
13. Tornquist K (1993) Activation of calcium entry by cyclopiazonic acid in thyroid FRTL-5 cells. Cell Calcium 14:411–417
14. Demaurex N, Lew DP, Krause KH (1992) Cyclopiazonic acid depletes intracellular Ca^{2+} stores and activates an influx pathway for divalent cations in HL-60 cells. J Biol Chem 267:2318–2324
15. Shlykov SG, Yang M, Alcorn JL, Sanborn BM (2003) Capacitative cation entry in human myometrial cells and augmentation by hTrpC3 overexpression. Biol Reprod 69:647–655

16. Torihashi S, Fujimoto T, Trost C, Nakayama S (2002) Calcium oscillation linked to pacemaking of interstitial cells of Cajal: requirement of calcium influx and localization of TRP4 in caveolae. J Biol Chem 277:19191–19197
17. Boulay G, Zhu X, Peyton M, Jiang M, Hurst R, Stefani E, Birnbaumer L (1997) Cloning and expression of a novel mammalian homolog of Drosophila transient receptor potential (Trp) involved in calcium entry secondary to activation of receptors coupled by the Gq class of G protein. J Biol Chem 272:29672–29680
18. Chyb S, Raghu P, Hardie RC (1999) Polyunsaturated fatty acids activate the Drosophila light-sensitive channels TRP and TRPL. Nature 397:255–259
19. Jörs S, Kazanski V, Foik A, Krautwurst D, Harteneck C (2006) Receptor-induced activation of Drosophila TRP gamma by polyunsaturated fatty acids. J Biol Chem 281:29693–29702
20. Zhu X, Jiang M, Birnbaumer L (1998) Receptor-activated Ca^{2+} influx via human Trp3 stably expressed in human embryonic kidney (HEK)293 cells. Evidence for a non-capacitative Ca^{2+} entry. J Biol Chem 273:133–142
21. Jung S, Strotmann R, Schultz G, Plant TD (2002) TRPC6 is a candidate channel involved in receptor-stimulated cation currents in A7r5 smooth muscle cells. Am J Physiol Cell Physiol 282:C347–C359
22. Caterina MJ, Schumacher MA, Tominaga M, Rosen TA, Levine JD, Julius D (1997) The capsaicin receptor: a heat-activated ion channel in the pain pathway. Nature 389:816–824
23. Bootman MD, Collins TJ, Mackenzie L, Roderick HL, Berridge MJ, Peppiatt CM (2002) 2-aminoethoxydiphenyl borate (2-APB) is a reliable blocker of store-operated Ca^{2+} entry but an inconsistent inhibitor of $InsP_3$-induced Ca^{2+} release. FASEB J 16:1145–1150
24. Varnai P, Hunyady L, Balla T:STIM (2009) Orai: the long-awaited constituents of store-operated calcium entry. Trends Pharmacol Sci 30:118–128
25. Lis A, Peinelt C, Beck A, Parvez S, Monteilh-Zoller M, Fleig A, Penner R (2007) CRACM1, CRACM2, and CRACM3 are store-operated Ca^{2+} channels with distinct functional properties. Curr Biol 17:794–800
26. Zhang SL, Kozak JA, Jiang W, Yeromin AV, Chen J, Yu Y, Penna A, Shen W, Chi V, Cahalan MD (2008) Store-dependent and -independent modes regulating Ca^{2+} release-activated Ca^{2+} channel activity of human Orai1 and Orai3. J Biol Chem 283:17662–17671
27. Schindl R, Bergsmann J, Frischauf I, Derler I, Fahrner M, Muik M, Fritsch R, Groschner K, Romanin C (2008) 2-aminoethoxydiphenyl borate alters selectivity of Orai3 channels by increasing their pore size. J Biol Chem 283:20261–20267
28. Tesfai Y, Brereton HM, Barritt GJ (2001) A diacylglycerol-activated Ca^{2+} channel in PC12 cells (an adrenal chromaffin cell line) correlates with expression of the TRP-6 (transient receptor potential) protein. Biochem J 358:717–726
29. van Rossum DB, Patterson RL, Ma HT, Gill DL (2000) Ca^{2+} entry mediated by store depletion, S-nitrosylation, and TRP3 channels. Comparison of coupling and function. J Biol Chem 275:28562–28568
30. Xu SZ, Zeng F, Boulay G, Grimm C, Harteneck C, Beech DJ (2005) Block of TRPC5 channels by 2-aminoethoxydiphenyl borate: a differential, extracellular and voltage-dependent effect. Br J Pharmacol 145:405–414
31. Zagranichnaya TK, Wu X, Villereal ML (2005) Endogenous TRPC1, TRPC3, and TRPC7 proteins combine to form native store-operated channels in HEK-293 cells. J Biol Chem 280:29559–29569
32. Wicher D, Agricola HJ, Schönherr R, Heinemann SH, Derst C (2006) TRPg channels are inhibited by cAMP and contribute to pacemaking in neurosecretory insect neurons. J Biol Chem 281:3227–3236
33. Togashi K, Inada H, Tominaga M (2008) Inhibition of the transient receptor potential cation channel TRPM2 by 2-aminoethoxydiphenyl borate (2-APB). Br J Pharmacol 153:1324–1330
34. Hanano T, Hara Y, Shi J, Morita H, Umebayashi C, Mori E, Sumimoto H, Ito Y, Mori Y, Inoue R (2004) Involvement of TRPM7 in cell growth as a spontaneously activated Ca^{2+} entry pathway in human retinoblastoma cells. J Pharmacol Sci 95:403–419

35. Li M, Jiang J, Yue L (2006) Functional characterization of homo- and heteromeric channel kinases TRPM6 and TRPM7. J Gen Physiol 127:525–537
36. Chung MK, Lee H, Mizuno A, Suzuki M, Caterina MJ (2004) 2-aminoethoxydiphenyl borate activates and sensitizes the heat-gated ion channel TRPV3. J Neurosci 24:5177–5182
37. Gu Q, Lin RL, Hu HZ, Zhu MX, Lee LY (2005) 2-aminoethoxydiphenyl borate stimulates pulmonary C neurons via the activation of TRPV channels. Am J Physiol Lung Cell Mol Physiol 288:L932–L941
38. Yue L, Peng JB, Hediger MA, Clapham DE (2001) CaT1 manifests the pore properties of the calcium-release-activated calcium channel. Nature 410:705–709
39. Voets T, Prenen J, Fleig A, Vennekens R, Watanabe H, Hoenderop JG, Bindels RJ, Droogmans G, Penner R, Nilius B (2001) CaT1 and the calcium release-activated calcium channel manifest distinct pore properties. J Biol Chem 276:47767–47770
40. Neeper MP, Liu Y, Hutchinson TL, Wang Y, Flores CM, Qin N (2007) Activation properties of heterologously expressed mammalian TRPV2: evidence for species dependence. J Biol Chem 282:15894–15902
41. Hu H, Grandl J, Bandell M, Petrus M, Patapoutian A (2009) Two amino acid residues determine 2-APB sensitivity of the ion channels TRPV3 and TRPV4. Proc Natl Acad Sci USA 106:1626–1631
42. Van Der Stelt M, Di Marzo V (2004) Endovanilloids. Putative endogenous ligands of transient receptor potential vanilloid 1 channels. Eur J Biochem 271:1827–1834
43. Harteneck C, Frenzel H, Kraft R (2007) N-p-amylcinnamoyl)anthranilic acid (ACA): a phospholipase A_2 inhibitor and TRP channel blocker. Cardiovasc Drug Rev 25:61–75
44. Strotmann R, Harteneck C, Nunnenmacher K, Schultz G, Plant TD (2000) OTRPC4, a non-selective cation channel that confers sensitivity to extracellular osmolarity. Nat Cell Biol 2:695–702
45. Tinel H, Wehner F, Kinne RK (1997) Arachidonic acid as a second messenger for hypotonicity-induced calcium transients in rat IMCD cells. Pflügers Arch 433:245–253
46. Watanabe H, Vriens J, Prenen J, Droogmans G, Voets T, Nilius B (2003) Anandamide and arachidonic acid use epoxyeicosatrienoic acids to activate TRPV4 channels. Nature 424:434–438
47. Vriens J, Owsianik G, Fisslthaler B, Suzuki M, Janssens A, Voets T, Morisseau C, Hammock BD, Fleming I, Busse R, Nilius B (2005) Modulation of the Ca^{2+} permeable cation channel TRPV4 by cytochrome P_{450} epoxygenases in vascular endothelium. Circ Res 97:908–915
48. Hara Y, Wakamori M, Ishii M, Maeno E, Nishida M, Yoshida T, Yamada H, Shimizu S, Mori E, Kudoh J, Shimizu N, Kurose H, Okada Y, Imoto K, Mori Y (2002) LTRPC2 Ca^{2+}-permeable channel activated by changes in redox status confers susceptibility to cell death. Mol Cell 9:163–173
49. Kraft R, Grimm C, Frenzel H, Harteneck C (2006) Inhibition of TRPM2 cation channels by N-(p-amylcinnamoyl)anthranilic acid. Br J Pharmacol 148:264–273
50. Bari MR, Akbar S, Eweida M, Kuhn FJ, Gustafsson AJ, Lückhoff A, Islam MS (2009) H_2O_2-induced Ca^{2+} influx and its inhibition by N-(p-amylcinnamoyl) anthranilic acid in the beta-cells: involvement of TRPM2 channels. J Cell Mol Med 13:3260–3267
51. Pantaler E, Lückhoff A (2009) Inhibitors of TRP channels reveal stimulus-dependent differential activation of Ca^{2+} influx pathways in human neutrophil granulocytes. Naunyn Schmiedebergs Arch Pharmacol 380:497–507
52. Bai JZ, Lipski J (2010) Differential expression of TRPM2 and TRPV4 channels and their potential role in oxidative stress-induced cell death in organotypic hippocampal culture. Neurotoxicology 31:204–214
53. Hill K, McNulty S, Randall AD (2004) Inhibition of TRPM2 channels by the antifungal agents clotrimazole and econazole. Naunyn Schmiedebergs Arch Pharmacol 370:227–237
54. Meseguer V, Karashima Y, Talavera K, D'Hoedt D, Donovan-Rodriguez T, Viana F, Nilius B, Voets T (2008) Transient receptor potential channels in sensory neurons are targets of the antimycotic agent clotrimazole. J Neurosci 28:576–586

55. Kiyonaka S, Kato K, Nishida M, Mio K, Numaga T, Sawaguchi Y, Yoshida T, Wakamori M, Mori E, Numata T, Ishii M, Takemoto H, Ojida A, Watanabe K, Uemura A, Kurose H, Morii T, Kobayashi T, Sato Y, Sato C, Hamachi I, Mori Y (2009) Selective and direct inhibition of TRPC3 channels underlies biological activities of a pyrazole compound. Proc Natl Acad Sci USA 106:5400–5405
56. Leuner K, Heiser JH, Derksen S, Mladenov MI, Fehske CJ, Schubert R, Gollasch M, Schneider G, Harteneck C, Chatterjee SS, Müller WE (2010) Simple 2,4-diacylphloroglucinols as classic transient receptor potential-6 activators – identification of a novel pharmacophore. Mol Pharmacol 77:368–377
57. Tada M, Chiba K, Takakuwa T, Kojima E (1992) Analogues of natural phloroglucinols as antagonists against both thromboxane A2 and leukotriene D4. J Med Chem 35:1209–1212
58. Verotta L, Lovaglio E, Sterner O, Appendino G, Bombardelli E (2004) Modulation of chemoselectivity by protein additives. Remarkable effects in the oxidation of hyperforin. J Org Chem 69:7869–7874
59. McKemy DD, Neuhausser WM, Julius D (2002) Identification of a cold receptor reveals a general role for TRP channels in thermosensation. Nature 416:52–58
60. Eccles R (1994) Menthol and related cooling compounds. J Pharm Pharmacol 46:618–630
61. Behrendt HJ, Germann T, Gillen C, Hatt H, Jostock R (2004) Characterization of the mouse cold-menthol receptor TRPM8 and vanilloid receptor type-1 VR1 using a fluorometric imaging plate reader (FLIPR) assay. Br J Pharmacol 141:737–745
62. Beck B, Bidaux G, Bavencoffe A, Lemonnier L, Thebault S, Shuba Y, Barrit G, Skryma R, Prevarskaya N (2007) Prospects for prostate cancer imaging and therapy using high-affinity TRPM8 activators. Cell Calcium 41:285–294
63. Macpherson LJ, Hwang SW, Miyamoto T, Dubin AE, Patapoutian A, Story GM (2006) More than cool: promiscuous relationships of menthol and other sensory compounds. Mol Cell Neurosci 32:335–343
64. Lashinger ES, Steiginga MS, Hieble JP, Leon LA, Gardner SD, Nagilla R, Davenport EA, Hoffman BE, Laping NJ, Su X (2008) AMTB, a TRPM8 channel blocker: evidence in rats for activity in overactive bladder and painful bladder syndrome. Am J Physiol Renal Physiol 295:F803–F810
65. Birder LA, Nakamura Y, Kiss S, Nealen ML, Barrick S, Kanai AJ, Wang E, Ruiz G, De Groat WC, Apodaca G, Watkins S, Caterina MJ (2002) Altered urinary bladder function in mice lacking the vanilloid receptor TRPV1. Nat Neurosci 5:856–860
66. Gevaert T, Vriens J, Segal A, Everaerts W, Roskams T, Talavera K, Owsianik G, Liedtke W, Daelemans D, Dewachter I, Van Leuven F, Voets T, De Ridder D, Nilius B (2007) Deletion of the transient receptor potential cation channel TRPV4 impairs murine bladder voiding. J Clin Invest 117:3453–3462
67. Birder L, Kullmann FA, Lee H, Barrick S, de Groat W, Kanai A, Caterina M (2007) Activation of urothelial transient receptor potential vanilloid 4 by 4alpha phorbol 12,13 didecanoate contributes to altered bladder reflexes in the rat. J Pharmacol Exp Ther 323:227–235
68. Wagner TF, Loch S, Lambert S, Straub I, Mannebach S, Mathar I, Dufer M, Lis A, Flockerzi V, Philipp SE, Oberwinkler J (2008) Transient receptor potential M3 channels are ionotropic steroid receptors in pancreatic beta cells. Nat Cell Biol 10:1421–1430
69. Gavva NR (2008) Body-temperature maintenance as the predominant function of the vanilloid receptor TRPV1. Trends Pharmacol Sci 29:550–557
70. Szallasi A, Cortright DN, Blum CA, Eid SR (2007) The vanilloid receptor TRPV1: 10 years from channel cloning to antagonist proof-of-concept. Nat Rev Drug Discov 6:357–372
71. Walpole CS, Bevan S, Bovermann G, Boelsterli JJ, Breckenridge R, Davies JW, Hughes GA, James I, Oberer L, Winter J et al (1994) The discovery of capsazepine, the first competitive antagonist of the sensory neuron excitants capsaicin and resiniferatoxin. J Med Chem 37:1942–1954

72. Seabrook GR, Sutton KG, Jarolimek W, Hollingworth GJ, Teague S, Webb J, Clark N, Boyce S, Kerby J, Ali Z, Chou M, Middleton R, Kaczorowski G, Jones AB (2002) Functional properties of the high-affinity TRPV1 (VR1) vanilloid receptor antagonist (4-hydroxy-5-iodo-3-methoxyphenylacetate ester) iodo-resiniferatoxin. J Pharmacol Exp Ther 303:1052–1060
73. Gunthorpe MJ, Rami HK, Jerman JC, Smart D, Gill CH, Soffin EM, Luis Hannan S, Lappin SC, Egerton J, Smith GD, Worby A, Howett L, Owen D, Nasir S, Davies CH, Thompson M, Wyman PA, Randall AD, Davis JB (2004) Identification and characterisation of SB-366791, a potent and selective vanilloid receptor (VR1/TRPV1) antagonist. Neuropharmacology 46:133–149
74. Thorneloe KS, Sulpizio AC, Lin Z, Figueroa DJ, Clouse AK, McCafferty GP, Chendrimada TP, Lashinger ES, Gordon E, Evans L, Misajet BA, Demarini DJ, Nation JH, Casillas LN, Marquis RW, Votta BJ, Sheardown SA, Xu X, Brooks DP, Laping NJ, Westfall TD (2008) N-((1S)-1-{[4-((2S)-2-{[(2,4-dichlorophenyl)sulfonyl]amino}-3-hydroxypropanoyl)-1-piperazinyl]carbonyl}-3-methylbutyl)-1-benzothiophene-2-carboxamide (GSK1016790A), a novel and potent transient receptor potential vanilloid 4 channel agonist induces urinary bladder contraction and hyperactivity: part I. J Pharmacol Exp Ther 326:432–442
75. Hartmannsgruber V, Heyken WT, Kacik M, Kaistha A, Grgic I, Harteneck C, Liedtke W, Hoyer J, Kohler R (2007) Arterial response to shear stress critically depends on endothelial TRPV4 expression. PLoS One 2:e827
76. Earley S, Pauyo T, Drapp R, Tavares MJ, Liedtke W, Brayden JE (2009) TRPV4-dependent dilation of peripheral resistance arteries influences arterial pressure. Am J Physiol Heart Circ Physiol 297:H1096–H1102
77. Mendoza SA, Fang J, Gutterman DD, Wilcox DA, Bubolz AH, Li R, Suzuki M, Zhang DX (2010) TRPV4-mediated endothelial Ca^{2+} influx and vasodilation in response to shear stress. Am J Physiol Heart Circ Physiol 298:H466–H476
78. Willette RN, Bao W, Nerurkar S, Yue TL, Doe CP, Stankus G, Turner GH, Ju H, Thomas H, Fishman CE, Sulpizio A, Behm DJ, Hoffman S, Lin Z, Lozinskaya I, Casillas LN, Lin M, Trout RE, Votta BJ, Thorneloe K, Lashinger ES, Figueroa DJ, Marquis R, Xu X (2008) Systemic activation of the transient receptor potential vanilloid subtype 4 channel causes endothelial failure and circulatory collapse: part 2. J Pharmacol Exp Ther 326:443–452
79. Vincent F, Acevedo A, Nguyen MT, Dourado M, DeFalco J, Gustafson A, Spiro P, Emerling DE, Kelly MG, Duncton MA (2009) Identification and characterization of novel TRPV4 modulators. Biochem Biophys Res Commun 389:490–494
80. Montell C, Rubin GM (1989) Molecular characterization of the Drosophila trp locus: a putative integral membrane protein required for phototransduction. Neuron 2:1313–1323
81. Hardie RC, Minke B (1992) The *trp* gene is essential for a light-activated Ca^{2+} channel in *Drosophila* photoreceptors. Neuron 8:643–651
82. Vaca L, Sinkins WG, Hu Y, Kunze DL, Schilling WP (1994) Activation of recombinant trp by thapsigargin in Sf9 insect cells. Am J Physiol 267:C1501–C1505
83. Hu Y, Vaca L, Zhu X, Birnbaumer L, Kunze DL, Schilling WP (1994) Appearance of a novel Ca^{2+} influx pathway in Sf9 insect cells following expression of the transient receptor potential-like (trpl) protein of Drosophila. Biochem Biophys Res Commun 201:1050–1056
84. Harteneck C, Obukhov AG, Zobel A, Kalkbrenner F, Schultz G (1995) The *Drosophila* cation channel *trpl* expressed in insect *Sf9* cells is stimulated by agonists of G-protein-coupled receptors. FEBS Lett 358:297–300
85. Hofmann T, Obukhov AG, Schaefer M, Harteneck C, Gudermann T, Schultz G (1999) Direct activation of human TRPC6 and TRPC3 channels by diacylglycerol. Nature 397:259–263
86. Okada T, Inoue R, Yamazaki K, Maeda A, Kurosaki T, Yamakuni T, Tanaka I, Shimizu S, Ikenaka K, Imoto K, Mori Y (1999) Molecular and functional characterization of a novel

mouse transient receptor potential protein homologue TRP7. Ca^{2+}-permeable cation channel that is constitutively activated and enhanced by stimulation of G protein-coupled receptor. J Biol Chem 274:27359–27370
87. Lucas P, Ukhanov K, Leinders-Zufall T, Zufall F (2003) A diacylglycerol-gated cation channel in vomeronasal neuron dendrites is impaired in TRPC2 mutant mice: mechanism of pheromone transduction. Neuron 40:551–561
88. Okada T, Shimizu S, Wakamori M, Maeda A, Kurosaki T, Takada N, Imoto K, Mori Y (1998) Molecular cloning and functional characterization of a novel receptor-activated TRP Ca^{2+} channel from mouse brain. J Biol Chem 273:10279–10287
89. Kamouchi M, Philipp S, Flockerzi V, Wissenbach U, Mamin A, Raeymaekers L, Eggermont J, Droogmans G, Nilius B (1999) Properties of heterologously expressed hTRP3 channels in bovine pulmonary artery endothelial cells. J Physiol 518(Pt 2):345–358
90. McKay RR, Szymeczek-Seay CL, Lievremont JP, Bird GS, Zitt C, Jungling E, Luckhoff A, Putney JW Jr (2000) Cloning and expression of the human transient receptor potential 4 (TRP4) gene: localization and functional expression of human TRP4 and TRP3. Biochem J 351(Pt 3):735–746
91. Riccio A, Mattei C, Kelsell RE, Medhurst AD, Calver AR, Randall AD, Davis JB, Benham CD, Pangalos MN (2002) Cloning and functional expression of human short TRP7, a candidate protein for store-operated Ca^{2+} influx. J Biol Chem 277:12302–12309
92. Schaefer M, Plant TD, Obukhov AG, Hofmann T, Gudermann T, Schultz G (2000) Receptor-mediated regulation of the nonselective cation channels TRPC4 and TRPC5. J Biol Chem 275:17517–17526
93. Strübing C, Krapivinsky G, Krapivinsky L, Clapham DE (2001) TRPC1 and TRPC5 form a novel cation channel in mammalian brain. Neuron 29:645–655
94. Jung S, Muhle A, Schaefer M, Strotmann R, Schultz G, Plant TD (2003) Lanthanides potentiate TRPC5 currents by an action at extracellular sites close to the pore mouth. J Biol Chem 278:3562–3571
95. Caterina MJ, Rosen TA, Tominaga M, Brake AJ, Julius D (1999) A capsaicin-receptor homologue with a high threshold for noxious heat. Nature 398:436–441
96. Hoenderop JG, Vennekens R, Muller D, Prenen J, Droogmans G, Bindels RJ, Nilius B (2001) Function and expression of the epithelial Ca^{2+} channel family: comparison of mammalian ECaC1 and 2. J Physiol 537:747–761
97. Peier AM, Reeve AJ, Andersson DA, Moqrich A, Earley TJ, Hergarden AC, Story GM, Colley S, Hogenesch JB, McIntyre P, Bevan S, Patapoutian A (2002) A heat-sensitive TRP channel expressed in keratinocytes. Science 296:2046–2049
98. Voets T, Nilius B, Hoefs S, van der Kemp AW, Droogmans G, Bindels RJ, Hoenderop JG (2004) TRPM6 forms the Mg^{2+} influx channel involved in intestinal and renal Mg^{2+} absorption. J Biol Chem 279:19–25
99. Nagata K, Duggan A, Kumar G, Garcia-Anoveros J (2005) Nociceptor and hair cell transducer properties of TRPA1, a channel for pain and hearing. J Neurosci 25:4052–4061
100. Bandell M, Story GM, Hwang SW, Viswanath V, Eid SR, Petrus MJ, Earley TJ, Patapoutian A (2004) Noxious cold ion channel TRPA1 is activated by pungent compounds and bradykinin. Neuron 41:849–857
101. Tousova K, Vyklicky L, Susankova K, Benedikt J, Vlachova V (2005) Gadolinium activates and sensitizes the vanilloid receptor TRPV1 through the external protonation sites. Mol Cell Neurosci 30:207–217
102. Leffler A, Linte RM, Nau C, Reeh P, Babes A (2007) A high-threshold heat-activated channel in cultured rat dorsal root ganglion neurons resembles TRPV2 and is blocked by gadolinium. Eur J Neurosci 26:12–22
103. Nakaya K, Harbidge DG, Wangemann P, Schultz BD, Green ED, Wall SM, Marcus DC (2007) Lack of pendrin HCO$_3$-transport elevates vestibular endolymphatic [Ca^{2+}] by inhibition of acid-sensitive TRPV5 and TRPV6 channels. Am J Physiol Renal Physiol 292:F1314–F1321

104. Kraft R, Grimm C, Grosse K, Hoffmann A, Sauerbruch S, Kettenmann H, Schultz G, Harteneck C (2004) Hydrogen peroxide and ADP-ribose induce TRPM2-mediated calcium influx and cation currents in microglia. Am J Physiol Cell Physiol 286:C129–C137
105. Grimm C, Kraft R, Sauerbruch S, Schultz G, Harteneck C (2003) Molecular and functional characterization of the melastatin-related cation channel TRPM3. J Biol Chem 278:21493–21501
106. Chuang HH, Neuhausser WM, Julius D (2004) The super-cooling agent icilin reveals a mechanism of coincidence detection by a temperature-sensitive TRP channel. Neuron 43:859–869

Chapter 5
Study of TRP Channels by Automated Patch Clamp Systems

Morten Sunesen and Rasmus B. Jacobsen

Abstract Ion channels are responsible for the permeation of ions across the membrane and their central role in cellular physiology is well established. Historically, the direct study of ion channels has been considered technically challenging. As such, a significant barrier to drug discovery for ion channels has been the low throughput of high quality electrophysiological data. The emergence of automated high throughput platforms for studying ion channel kinetics and pharmacology has lowered this barrier. Ion channels are now recognized as increasingly important drug targets and a diverse range of ion channels are implicated in a variety of drug discovery and cardiac safety assessment programs. The TRP (Transient Receptor Potential) superfamily of ion channels play a crucial role in a broad range of sensory functions including vision, taste, olfaction, hearing, touch, pain and thermosensation. Many of the TRP channels are polymodal in their activation and deactivation mechanisms and even with conventional patch clamp electrophysiology, the TRP channels are considered to be a very complex target class. Here we present an update on the significant progress made on the TRP receptor assays with the available automated patch clamp systems.

5.1 Introduction

The field of patch clamping has advanced significantly since its introduction by Neher and Sakmann in the 1970s [1]. The discovery of giga-ohm patch clamp recordings, micromanipulated pipettes, and stable specialized amplifiers lead to the successful establishment of the manual patch clamp setups in both academia and industry labs. Since then a plethora of ion channels have been studied functionally and a wide range of ion channel compounds have been described pharmacologically using manual patch clamp. The most commonly used configuration, whole-cell

M. Sunesen (✉)
Sophion Bioscience AS, 2750 Ballerup, Denmark
e-mail: msu@sophion.com

patch clamp, is a direct method for measuring ion channel activity across the membrane of an entire cell. Although highly informative, the manual (conventional) whole-cell patch clamp technique is slow and requires highly skilled operators. The throughput of manual whole-cell patch clamp experiments is most often limited to examinations of one compound at any one time, and it is extremely labor intensive. As such manual patch clamp represents a major bottleneck in ion channel drug discovery and development.

Recent technological advances have led to the emergence of several automated patch clamp devices in which the classical glass microelectrode is replaced by a planar substrate, which holds the patch clamp hole. The composition of the planar substrate varies from system to system and the tightness of the seal between the cell and the planar substrate may vary according to the composition of the substrate (i.e. glass vs. plastic). However, the fundamental principles remain the same: a cell is drawn to the hole by suction, the cell membrane is ruptured (or perforated) and recordings are obtained in the whole-cell configuration. The level of automation varies a bit from platform to platform. Some systems offer full automation with onboard handling of cells, compounds and consumables, while other platforms are built as simple reader stations, where user interaction is mandatory [2].

With the current availability of devices, a broad range of ion channel currents can be measured. The automated platforms have obtained profound success in areas of electrophysiology where high quality data are needed at a low to medium high throughput. These machines are typically implemented in areas of the drug discovery cascade such as assay optimization, secondary screening, lead optimization and cardiac liability.

Automated patch clamp systems are now routinely deployed in programs with ion channels that can present significant biological and technological challenges. A number of recent publications describe the successful implementation of automated patch clamping in studies of voltage-gated sodium channels [3], voltage-gated calcium channels [4], calcium-activated potassium channels [3, 5], various fast-desensitizing ligand-gated ion channels [4, 6], two pore channels [7], hyperpolarisation activated cyclic nucleotide-gated channels [8] and calcium-release activated calcium channels [9]. This chapter describes the recent studies of a family of technically challenging ion channels, the transient receptor potential (TRP) channels on the currently available automated patch clamp devices: IonWorks (MDC), Patchliner (Nanion), PatchXpress (MDC), and QPatch (Sophion).

The TRP family of ion channels was originally named after the ion channel mutant of the *Drosophila melanogaster* fly that displayed a transient response to light rather than a sustained response [10]. TRP ion channels have six transmembrane domains and all TRP channels are permeable to cations. However, each subtype has its own cationic selectivity and activation mechanism [11]. TRP ion channels are encoded by 28 genes in mammals [12] and they are essential in a broad range of sensory physiology including thermosensation, olfaction, vision, mechanosensation, touch, taste, pain sensation and osmosensation [13]. Since their discovery TRP channels have caught profound attention as an interesting but challenging target class in both academia and industry.

The use of automated patch clamp for studying TRP channels is far from trivial and a number of biological and technological challenges must be addressed: control of intracellular calcium concentration, control of temperature, sensitivity to mechanical stimuli, and differentiation of leak current versus TRP channel permeated current.

Direct control of intracellular calcium concentrations is difficult in most automated patch clamp systems. Some devices use poreformers (e.g., amphotericin) to make holes in the membrane through which only monovalent cations can be exchanged. Manipulation of intracellular calcium levels is therefore poorly controlled on patch clamp systems using poreformers. Other automated patch clamp devices use traditional suction to establish the whole-cell configuration and a more precise control of intracellular calcium can be achieved by using well-known calcium chelator's such as BAPTA and EGTA. Only the Patchliner systems from Nanion allow intracellular solution exchange to directly study the effect of changing intracellular calcium concentrations.

All available automated patch clamp systems suffer from the same disadvantage compared to manual patch clamp setups: the lack of continuous perfusion of the liquid in the recording chamber. The technology that has been developed for automated patch clamp today falls into two categories – either an open-well system, where liquid is added/removed from the top or a laminar flow system, where liquid is drawn through flow channels in discrete bursts upon liquid applications to an inlet well (for review see [2]). Even though the exact mechanism of mechano-activation of the mechano-sensitive TRP channels remains unknown [8, 14, 15], the stop-and-go application of salines and compounds to the recording site of automated patch clamp systems could potentially affect the recordings measured. On the other hand, the stop-and-go feature may be an opportunity to address mechano-sensitivity of TRP channels. In a recent paper, the QPatch was deployed in order to discern the effects of cell swelling from those of mechanical membrane stretch of the potassium channels BK and KCNQ4 [16].

The rapid increase in temperature beyond a TRP-subtype specific temperature will activate the channel and result in an inward current [17]. As none of the available automated patch clamp devices have the capability to change the experimental temperature within seconds without a simultaneous liquid exchange around the cells, the effect of the elevated temperature might result in the activation of the channel caused by mechano-activation resulting from the stop-and-go liquid application. As illustrated in the examples below, the preferred choice when working with TRP channels on automated patch clamp devices has therefore been to either perform the experiments on the TRP channels directly as ligand-gated receptors or to perform the experiments as ligand-gated receptors with some voltage control (i.e., activation by agonist and execution of voltage ramps to detect TRP-specific current changes).

Traditionally, patch clamping has been performed with one cell at a time, however, in 2005 Molecular Devices launched the first automated patch clamp system that recorded from multiple cells residing in the same recording chamber, the IonWorks Quattro. Since then the QPatch HTX multi-hole system has been launched. The multi-hole systems have prominent advantages: the recording reliability is greatly increased, in some reports run down seems to be minimized, and the success rate of the patch clamp assay is approaching 100% [18–19]. While offering higher and more reliable throughput, the multi-hole systems have their shortcomings. The multi-hole systems lack series resistance compensation and voltage control of the individual cell. Therefore, the same high fidelity of the ion channel recordings you find in single-hole measurements cannot be expected when working with the multi-hole systems. The single-hole systems, therefore, still serve as a very useful platform for functional and biophysical characterization, assay optimization, assay validation and for assays that demands more stringent voltage control.

Automated patch clamp technologies have provided a significantly increased throughput with the same high quality as manual patch clamp recordings. This increased throughput of high quality measurements has lead to a growing number of drug discovery programs targeted towards ion channels including members of the TRP superfamily of ion channels. This review focuses on the various TRP assays that have been established on automated patch clamp systems over the last few years. Illustrated with figures from QPatch experiments from our laboratory, the respective TRPs are reviewed and discussed with special attention to the various assay conditions for automated patch clamp established by the research groups.

5.2 Results

5.2.1 TRPA1

The non-selective cation channel TRPA1 acts as a sensor for reactive chemicals in the body [20] and the active ingredients of mustard oil, wasabi and garlic have been shown to activate this channel [8, 15, 21–23]. In addition to activation by reactive chemicals, TRPA1 can also be activated by noxious cold temperatures [21, 24], and by calcium, which seems to desensitize the channel [25, 26].

In order to identify TRPA1 antagonists on PatchXpress, a ligand-gated approach was established in which the channel was activated by mustard oil (AITC) [24]. With a holding potential at –70 mV, the AITC-elicited control response was compared to the AITC signal after preincubation of the specific compounds and the concentration-dependent effect of four compounds was determined.

In a different set of experiments, TRPA1 currents were recorded on QPatch using another approach. Voltage ramps were executed during the course of the experiment and after whole-cell break-in, TRPA1 was activated by application of an agonist. In Fig. 5.1a, the same concentration of supercinnamaldehyde was applied three times and the level of the TRPA1-permeated current was measured using a blocking concentration of Ruthenium Red. Alternatively, the effective IC_{50} of TRPA1 antagonist could be determined by the cumulative addition of increasing concentrations of the antagonist (Fig. 5.1b). To evaluate the stability of the assay, the

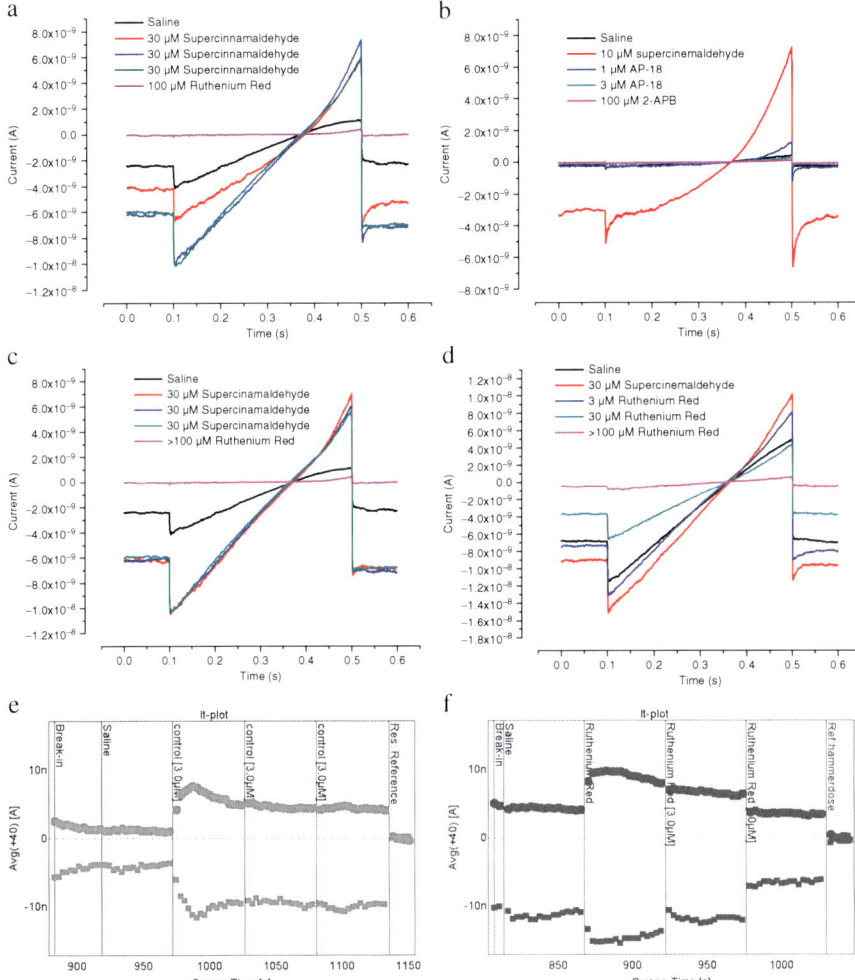

Fig. 5.1 HEK293-TRPA1 cells tested on QPatch in single- or multi-hole mode. TRPA1 currents were stimulated by a voltage ramp (−100 to +100 mV) upon supercinnamaldehyde activation. (**a**) Single-hole recording of TRPA1 current activated by 30 μM supercinnamaldehyde and blocked by 100 μM Ruthenium Red. (**b**) Single-hole recordings of TRPA1 currents activated by supercinnamaldehyde and subsequently blocked by AP-18 and 2-APB. (**c**) Multi-hole recordings of TRPA1 currents activated by 30 μM supercinnamaldehyde and blocked by Ruthenium Red. (**d**) Multi-hole recordings of TRPA1 currents activated with supercinnamaldehyde and blocked by increasing concentrations of Ruthenium Red. (**e**) Current vs. time plot demonstrating stability of the supercinnamaldehyde elicited current over time. 100 μM Ruthenium Red was used as reference to demonstrate TRPA1 specificity of the measured current. *Dots* are outward current at 40 mV, *squares* are inward current at −90 mV. (**f**) Current vs. time plot of dose response experiment, where TRPA1 currents were activated by supercinnamaldehyde and blocked by increasing concentrations of Ruthenium Red. Again, 100 μM Ruthenium Red was used as reference blocker. *Dots* are outward current at 40 mV, *squares* are inward current at −90 mV

same conditions were tested in a similar set-up on the multi-hole platform QPatch HTX with repeated application of supercinnamaldehyde (Fig. 5.1c) or increasing concentrations of Ruthenium Red (Fig. 5.1d). As illustrated in the current vs. time plots (see Fig. 5.1e, f), the TRPA1 signals remain very stable over time in the multi-hole experiments.

5.2.2 TRPCs

TRPCs were the first mammalian homolog of the *Drosophila melanogaster* TRPs to be identified [27]. Like TRPA1, TRPCs are non-selective cationic channels conducting sodium, potassium as well as calcium ions [28]. All TRPCs are activated through downstream pathways of phospholipase C (PLC) stimulation [29]. It has been suggested that certain TRPCs are functionally regulated by interactions with STIM1 [30].

5.2.2.1 TRPC1

To evaluate the functional interaction between TRPC1 and STIM1, vascular smooth muscle cells were used in electrophysiological studies on the Patchliner [31]. In whole-cell configuration the cells were clamped at a holding potential of 0 mV and during thapsigargin activation, voltage ramps were applied. In this assay, currents were evoked by applying thapsigargin extracellularly with or without an antibody directed against STIM1 or TRPC1 respectively. These experiments led the authors to conclude that TRPC1 and STIM1 interact and their activation is indeed stimulated by store-depletion but a fraction of the TRPC1 channels remain independent of STIM1 and of calcium store depletion [31].

5.2.3 TRPMs

The melastatin subfamily of TRP channels (TRPMs) forms a very diverse group of sensory ion channels that are activated by various subtype-specific mechanisms, which include sensitivity to mechanical or oxidative stress, hormones, temperature, membranous lipid composition and calcium store-depletion. TRPM ion channels are expressed in brain and in many peripheral sensory organs as well as in the immune system [32]. Several of the TRPM ion channels have been successfully tested on automated patch clamp instruments (see below).

5.2.3.1 TRPM2

A unique feature of some of the TRPM subfamilies is the presence of an enzymatic domain in the C-terminus, and hence the term *chanzyme* (Channel – Enzyme) was coined for these TRPs. TRPM2 is a chanzyme with an ADP-ribose pyrophatase domain in its C-terminus. It is a non-selective cationic channel that is activated

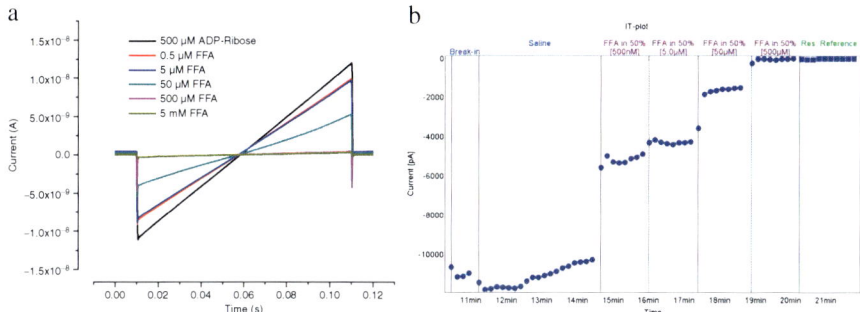

Fig. 5.2 Current activated by 500 μM ADP-Ribose and blocked by increasing concentrations of flufenamic acid (FFA) in a CHO-TRPM2 cell line. (**a**) Raw traces elicited by a 120 ms ramp from −60 to +60 mV. (**b**) Corresponding current vs. time plot. Using the QPatch assay software the average IC_{50} for FFA was found to be 38 μM, which compared well with literature values 3–30 μM [38]

by ADP-ribose, pyrimidine, nucleotides, NAD, and by oxidative stress [33–35]. TRPM2 channels are important redox sensors in the immune response [32], are involved in insulin secretion [36], and are involved in melanin synthesis and oxidative stress-induced cell death [37].

To study the effect of potential antagonists of TRPM2 on the QPatch, whole-cell currents were obtained by ADP-ribose activation of the channel. As shown in Fig. 5.2a, activation of TRPM2 results in a so-called leak current for which the leak component can be hard to discriminate from the channel permeated current. Therefore standard P/n leak subtraction can be quite problematic for these currents. Instead a known blocker of TRPM2 (5 mM flufenamic acid (FFA)), was used to measure non-TRPM2 mediated currents, which was then used as leak subtraction [18]. In the example in Fig. 5.2, FFA was also used as the test compound in a cumulative four point dose response experiment, which demonstrates the required stability of the TRPM2 current over time (See Fig. 5.2b) as well as an expected IC_{50} of FFA [38, 39].

5.2.3.2 TRPM3

TRPM3 seems to be regulated by the lipid structure of the cellular membrane. This TRP channel becomes activated by several membrane components such as sphingosine and the naturally occurring steroid pregnenolone sulphate [12, 40, 41] while being blocked by cholesterol [15]. When heterologously expressed in cultured cells the channel is constitutively open and sensitive to osmolarity of the extracellular solution [40].

TRPM3 is a calcium permeable, non-selective cationic channel but its precise function is still not well described. In hepatocytes, TRPM3 activation causes calcium influx, which results in insulin release and it has therefore been suggested that TRPM3 is involved in glucose homeostasis [41]. However, due to the lack of specific blocking agents the role of TRPM3 remains unclear.

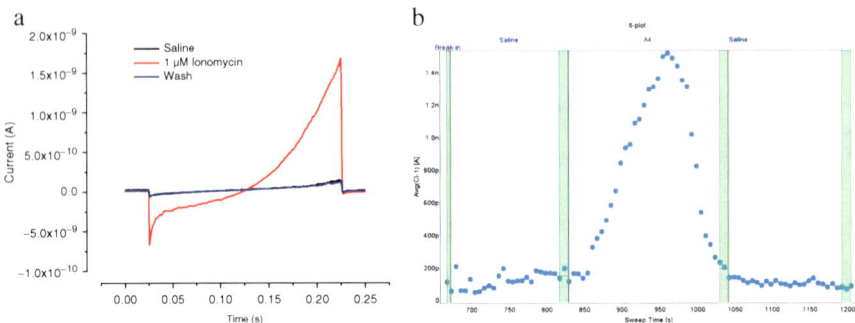

Fig. 5.3 TRPM5 current activated by application of ionomycin on CHO-TRPM5 cells. (**a**) Raw current traces obtained from a 200 ms ramp (−80 to +80 mV) before and during ionomycin (1 μM) activation and upon a wash step. (**b**) Corresponding current vs. time plot

With the use of automated patch clamp (Patchliner) Dr. Beech and coworkers recently developed specific blocking antibodies to one of the extracellular loops of TRPM3 [42]. On the Patchliner platform the HEK293-TRPM3 cells were clamped at 0 mV and by using voltage ramps from −100 to +100 mV, it was found that 50% of the pregnenolone sulphate response could be blocked by a specific polyclonal antibody directed to the E3 extracellular domain [42]. Approaches like these may in the future be extremely valuable in developing new medicines with an early preclinical prediction of drug action.

5.2.3.3 TRPM5

The TRPM5 channels are unusual to the TRP superfamily in at least three ways: the channels are activated by high concentrations of intracellular calcium, they exhibit some voltage-dependent activation, and they are together with the TRPM4 channels the only TRP channels that selectively permeate monovalent cations [43].

The poreformer ionomycin was used on the QPatch to demonstrate direct activation of TRPM5 by increasing the concentration of intracellular calcium (see Fig. 5.3a). As can be seen on the current versus time plot, a significant activation of TRPM5 is obtained with ionomycin, however, steady state of the signal is very hard to achieve (see Fig. 5.3b). While the TRPM5 assay may be a valuable tool for identifying novel TRPM5 agonists using automated patch clamp, future work will need to establish a reliable automated patch clamp assay for the identification of TRPM5 antagonists and blockers.

5.2.3.4 TRPM8

As demonstrated by its specific expression in cold sensing neurons and by several mice TRPM8 null mutants, the TRPM8 receptor is the primary sensor to innocuous cold temperatures [17, 44]. The TRPM8 channel is activated by cold temperatures and by compounds such as menthol, eucalyptol and icilin – compounds that evoke the *"cold"* sensation. Quite interestingly, it has been found that irrespective of the sensation of cold – pain or pleasant –TRPM8 plays a central role in the various cold responses [45].

The applicability of automated patch clamping on TRPM8 has been demonstrated in several studies using QPatch systems [4, 46]. In these investigations, the activity of TRPM8 was studied by voltage ramp protocols upon activation by an agonist (menthol) (Fig. 5.4a) plus or minus an antagonist (capsazepine) (Fig. 5.4b). Applications of multiple concentrations of the agonist or antagonist allowed the generation of IC_{50}/EC_{50} data from a single cell or as an average from several cells (see Fig. 5.4c, d). The results were compared to manual patch clamp data and found to be highly similar both with respect to biophysical characteristics and pharmacology [4, 46]. When setting up TRPM8 assays with automated patch clamp, remember the following: none of the existing systems have constant perfusion rather the liquid exchange around the cells happen in discrete bursts designed by the experimenter. For this ion channel the lack of constant perfusion has the consequence that two independent liquid applications are needed to reach steady state. Using this approach a reasonable match of manual and automated patch clamp data was obtained and a relatively high throughput patch clamp assay for secondary screening of TRPM8 antagonists was developed [46]. More specifically, it has been found that an average success rate beyond 60% can be achieved indicating a throughput of more than 30 individual EC_{50}s per hour (see Fig. 5.4e).

5.2.4 TRPVs

The TRPV (Vanilliod) family of receptors is like the other TRPs gating cations in general but these channels have higher a preference for permeating calcium and magnesium over sodium and potassium. Also, like other members of the TRP superfamily the TRPVs are activated by multiple apparently disparate mechanisms such as noxious heat, acidic pH, "hot taste" caused by the pungent substances in chili (capsaicin) and black pepper, garlic and camphor, the oils of clove thyme and oregano as well as mechanical stimuli [11, 13].

5.2.4.1 TRPV1

TRPV1 was originally identified and cloned based on the ability of a dorsal root ganglion cDNA fragment to confer capsaicin-evoked changes in intracellular calcium levels [47, 48]. TRPV1 channels are expressed in afferent nociceptor, pain sensing neurons, where they play a central role in transducing chemical and mechanical stimuli. As a result, TRPV1 channels have received a lot of attention in the search of blockers or antagonists for the potential prevention or treatment of pain [49].

Considering automated patch clamp, the TRPV1 channel is probably the most widely tested TRP channel. Data have appeared from nearly all machines including PatchXpress, Patchliner, QPatch, IonWorks and IonFlux. The TRPV1 assay seems to be straight forward: the TRPV1 mediated current is reliably activated by capsaicin

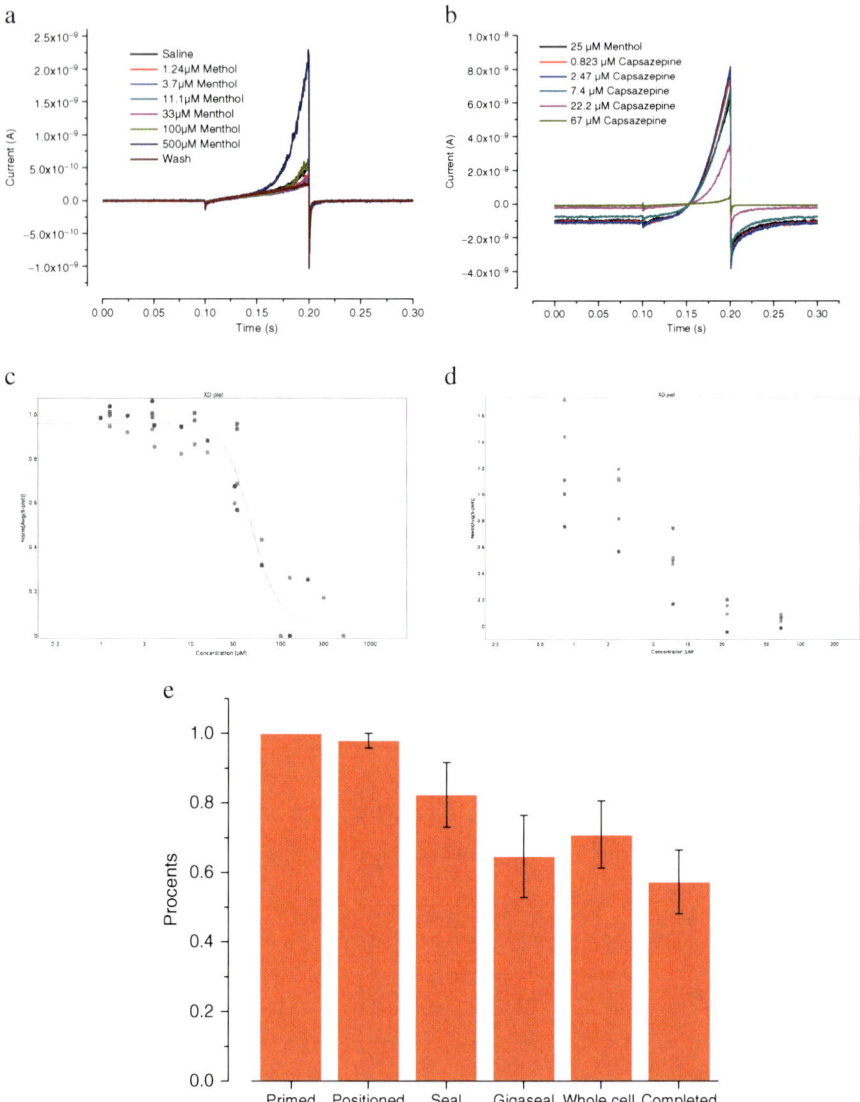

Fig. 5.4 Characterization of block of the menthol activated current of TRPM8 channels expressed in HEK293. (**a**) Raw traces were obtained from a 300 ms ramp (−80 to +80 mV) with increasing concentration of menthol. (**b**) TRPM8 currents activated with 25 μM menthol and blocked with increasing concentrations of capsazepine. (**c**) Hill fit of normalized data from 10 individual cells. Inward current measured at −60 mV. EC_{50} of menthol was estimated to be 53.30 ± 7.01 μM ($n = 11$) (reference value 101 ± 13 μM [54]). (**d**) Likewise the IC_{50} of capsazepine was estimated to be 5.06 ± 1.26 μM ($n = 7$) (reference value 18 μM [55]). (**e**) Success rates from six QPlates. Shown are percentages of sites primed, cells positioned, cells generating seals above 100 MΩ, cells generating seals above 1 GΩ, cells going whole-cell and cells completing the six point dose response experiment. Results are given as average ± SD

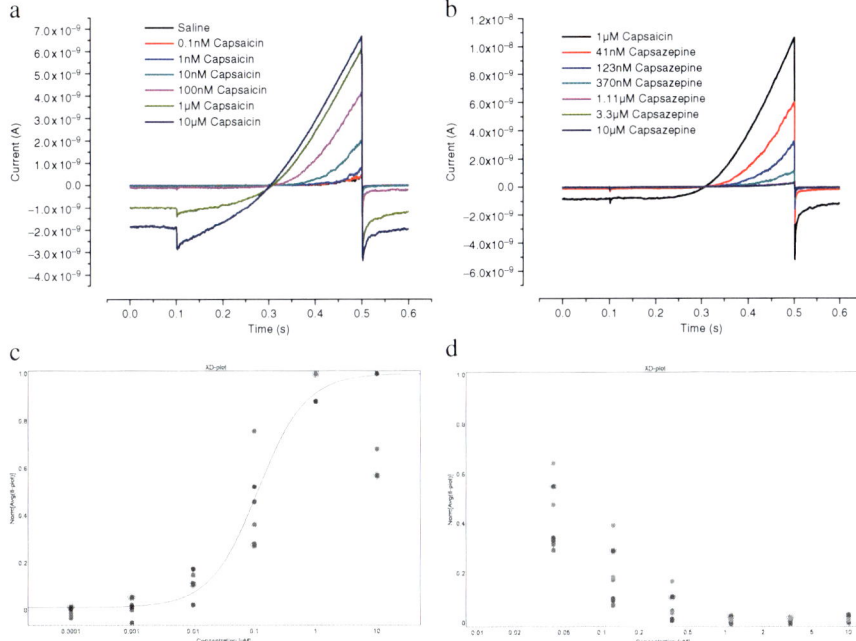

Fig. 5.5 HEK293 cells stably transfected with TRPV1. A 60 ms ramp from −100 to +100 mV was applied every 5 s. (**a**) Raw current response to increasing concentrations of capsaicin. (**b**) Raw current response to increasing concentrations of capsazepine. (**c**) Normalized group Hill fit of six individual cells. EC_{50} of capsaicin = 110.42 ± 17.55 nM. Same as previously reported [50]. (**d**) Normalized group Hill fit of seven individual cells. IC_{50} of capsazepine was 43.66 ± 7.70 nM. Which correspond well with previous found values [51]

and the capsaicin-activated current can easily be blocked with a reference compound like capsazepine. Also, there are many high quality cell lines available all exhibiting currents in the nA range.

Figure 5.5a shows some typical raw traces from a HEK cell line expressing TRPV1. A −100 to +100 mV voltage ramp protocol was used to monitor TRPV1 activation by increasing concentrations of capsaicin. In Fig. 5.5b, the same voltage protocol was used to determine the dose dependent effect of capsazepine on the capsaicin-induced current. As illustrated in Fig. 5.5c, d, the resultant dose response curves demonstrate that the obtained pharmacology is in range with previously reported values [50, 51].

5.2.4.2 TRPV4

Like many of the other TRPV subfamilies the TRPV4 channel can be activated polymodally and integrate the stimuli from a variety of sources. In addition to temperature TRPV4 is activated by various lipid messengers like endocannabinoids and some phorbol ester derivatives like 4α-PDD. Furthermore, TRPV4 is activated by hypotonicity (cell swelling) and it has been suggested that TRPV4 is involved in regulation of systemic osmotic pressure [11].

In a study of TRPV4 using PatchXpress and IonWorks, the effect of phorbol ester activation and mechanical stress was measured respectively or in combination [52]. In line with current thinking, it could be clearly demonstrated that the response to the combination of the two stimuli exceeded the effect of either stimuli on its own underscoring the polymodality of TRPV4 activation and the broad applicability of automated patch clamp in addressing polymodality of TRPs.

5.3 Conclusions

Driven by the appreciation of whole-cell patch clamp data in ion channel drug discovery and with the desire to characterize compound effects with a faster turnaround time in both drug discovery as well as in safety assessment campaigns, significant efforts have been made to develop automated patch systems [2, 53]. A few years back the automated patch clamp systems had made rather limited impact on the ion channel drug discovery programs but during the last couple of years the flexibility, versatility and reliability of automated patch clamp technologies have made a significant fingerprint on applied workflows in this arena for particular biotech and pharmaceutical companies. The number of high throughput platforms that have been fully integrated into drug discovery programs spanning hit identification, lead optimization, and liability screening has increased dramatically. With the number of programs increasing, researchers are beginning to routinely configure assays where the complexity of the assays and targets can present significant challenges to the automated patch clamp system. Most of the TRP ion channels belong to this category.

Beside the obvious increase in daily data production, a number of experimental advantages can be attributed to automated patch clamp electrophysiology compared with conventional patch clamp. As a positive side effect of automation, the limited user intervention reduces the risk of human errors and thereby increases the likelihood of obtaining reliable, objective, and reproducible data. Furthermore, in conventional patch clamp drugs are routinely applied in a constant perfusion system with flow rates around 1 ml/min. The automated systems by contrast use very small volumes (in the low μl range) thereby reducing waste of highly valuable compounds or even enabling tests at an earlier stage in the drug discovery cascade. With the relative large surface to volume ratio in the flow channels of some of the automated systems, the use of small compound volumes poses a potential problem. As illustrated above, certain sticky compounds need to be added more than once to exert its full effect on the ion channel: a fact that needs to be taking into consideration when designing experiments on automated patch clamp devices.

Being automated and with minimal user intervention, a few issues have shown to be prerequisites for optimal performance of the patch clamp systems. Most notably the cells have proven to be the key to success. Since the cell to be used in the experiment is selected by random the cell population has to be very homogeneous in particular with respect to expression as non-expressers are extremely costly. As such strict growth and harvest conditions seem to be critical in order to maintain high expression levels, high viability and to avoid clumping of the cell culture. To select the right clone with the right balance of compatibility on the system, current expression, minimal run down during the course of an experiment, and robustness over many cell culture passages, some of the higher throughput platforms like QPatch and IonWorks offer the feature of clone selection.

Generating quality cell lines can be quite cumbersome. Additionally, some TRP channels can be quite difficult to express in the traditional cell lines and expression systems. The result may be either a rather low expression profile or an uneven expression in different cells. With the introduction of the multi-hole systems the sensitivity to such factors are minimized and more stable and reliable recordings are obtainable as demonstrated on several ion channels ([19]). Furthermore, the multi-hole systems allow the usage of various transient transfection methods where the transfection efficiency does not reach 100%. Not only is the assay development time drastically reduced but novel approaches to ion channel research can be applied quickly.

The full potential of automated patch clamp in TRP research still has to be seen. The implementation of automated systems into academic institutions has been somewhat slowed down due to the capital needed to acquire and run an instrument. However, with the aid from various national stimulation packages like the US-based TARP (Troubled Asset Relief Program) there are clear signs that this will change in the next few years. With an often limited freedom and tradition to publish obtained results by groups within the pharmaceutical sector, the number of TRP channels that actually have been tested on the automated platform most probably exceed by far what has been published and reviewed in this chapter.

The availability of high quality stably expressing cell lines has also prevented the full usage of automated patch clamp in certain areas of TRP-directed drug discovery. However, in recent years a number of cell line vendors have generated a very wide panel of TRP expressing cell lines with great promise for the future.

In a meeting report from the 2009 Stockholm TRP meeting it was clearly described how substantial progress has been made in understanding the nature of the various TRP channels [15] but it was also emphasized that a lot of essential questions remain unanswered. For example by which molecular mechanisms are thermosensation and thermoactivation regulated? How is mechano-sensitivity in TRPs triggered? What is the functional consequence of TRP heteromerization or is homomerization only caused by induced fit in highly expressing heterologous expression systems? And what is the role of TRP channels in stem cells?

In the coming years, the automated patch clamp systems will most likely be instrumental in elucidating some of these questions. With the incorporation of temperature control on some of the platforms multiple mutant clones can be characterized efficiently in details that previously would have required a significant effort. Also, work is being done to allow efficient electrophysiological recordings on stem cells and primary cells on the automated platforms, which should accelerate a better understanding of TRP channels in stem cell development and differentiation. Probably most noteworthy for the contribution of automated patch clamp to the field of TRP channels is the recognition that TRP functions are essential to normal life and that the identification of new compounds that can activate or inhibit a specific TRP channel activity is of paramount importance. The increased throughput of automated patch clamp devices should enable a quicker and more competent search for candidate compounds – be it small molecules, peptides or antibodies.

Although the throughput range of automated patch clamp has not reached the need in primary screening campaigns, the diverse experimental applications, the elevated throughput and the ease of use has found significant foothold in the field of ion channel drug discovery. In the years to come, automated patch clamp is likely to remain an invaluable tool for ion channel drug discovery in general and hopefully for the development of drugs targeting the fascinating TRP ion channels in particular.

Acknowledgments The QPatch results presented in this review have been generated by many scientists at Sophion Bioscience. We would especially like to thank Rikke L. Schrøder and Hervør L. Olsen for their contribution with data: Dorthe Nielsen is acknowledged for her expert technical assistance. M. Knirke Jensen and Søren Friis are thanked for providing data and valuable critical input to this manuscript. Many thanks also to Chris Mathes for being an inspirator and scientific mentor during both data and manuscript generation.

We thank AstraZeneca, Sweden, for generously proving the TRPA1, TRPV1, and TRPM8 cell lines. The scientists at Novartis, UK, Pamela Tranter, Mark McPate and Martin Gosling are thanked for sharing their data on TRPM2.

References

1. Hamill OP, Marty A, Neher E, Sakmann B, Sigworth FJ (1981) Improved patch-clamp techniques for high-resolution current recording from cells and cell-free membrane patches. Pflugers Arch 391:85–100
2. Dunlop J, Bowlby M, Peri R, Vasilyev D, Arias R (2008) High-throughput electrophysiology: an emerging paradigm for ion-channel screening and physiology. Nat Rev Drug Discov 7:358–368
3. Korsgaard MP, Strobaek D, Christophersen P (2009) Automated planar electrode electrophysiology in drug discovery: examples of the use of QPatch in basic characterization and high content screening on Na(v), K(Ca)2.3, and K(v)11.1 channels. Comb Chem High Throughput Screen 12:51–63
4. Mathes C, Friis S, Finley M, Liu Y (2009) QPatch: the missing link between HTS and ion channel drug discovery. Comb Chem High Throughput Screen 12:78–95
5. John VH, Dale TJ, Hollands EC, Chen MX, Partington L, Downie DL, Meadows HJ, Trezise DJ (2007) Novel 384-well population patch clamp electrophysiology assays for Ca^{2+}-activated K^+ channels. J Biomol Screen 12:50–60
6. Friis S, Mathes C, Sunesen M, Bowlby MR, Dunlop J (2009) Characterization of compounds on nicotinic acetylcholine receptor alpha7 channels using higher throughput electrophysiology. J Neurosci Methods 177:142–148
7. Clark G, Todd D, Liness S, Maidment SA, Dowler S, Southan A (2005) Expression and characterization of a two pore potassium channel in HEK293 cells using different assay platforms (Abstract). Proceedings of the British Pharmacological Society at http://www.pA2online.org/abstracts/vol3Issue4abst105P.pdf

8. Lee YT, Vasilyev DV, Shan QJ, Dunlop J, Mayer S, Bowlby MR (2008) Novel pharmacological activity of loperamide and CP-339,818 on human HCN channels characterized with an automated electrophysiology assay. Eur J Pharmacol 581:97–104
9. Schroder RL, Friis S, Sunesen M, Mathes C, Willumsen NJ (2008) Automated patch-clamp technique: increased throughput in functional characterization and in pharmacological screening of small-conductance Ca^{2+} release-activated Ca^{2+} channels. J Biomol Screen 13:638–647
10. Montell C, Rubin GM (1989) Molecular characterization of the Drosophila trp locus: a putative integral membrane protein required for phototransduction. Neuron 2:1313–1323
11. Venkatachalam K, Montell C (2007) TRP channels. Annu Rev Biochem 76:387–417
12. Nilius B (2007) TRP channels in disease. Biochim Biophys Acta 1772:805–812
13. Clapham DE (2003) TRP channels as cellular sensors. Nature 426:517–524
14. Barritt G, Rychkov G (2005) TRPs as mechanosensitive channels. Nat Cell Biol 7:105–107
15. Goswami C, Islam S (2010) TRP Channels. What's happening? reflections in the wake of the 2009 TRP meeting, Karolinska Institutet, Stockholm. Channels 4:1–12
16. Hammami S, Willumsen NJ, Olsen HL, Morera FJ, Latorre R, Klaerke DA (2009) Cell volume and membrane stretch independently control K^+ channel activity. J Physiol 587:2225–2231
17. Dhaka A, Viswanath V, Patapoutian A (2006) Trp ion channels and temperature sensation. Annu Rev Neurosci 29:135–161
18. McPate M, Lilley S, Gosling M, Friis S, Jacobsen RB, Tranter P (2010) Evaluation of the QPatch HT and HTX systems as methods for ion channel screening (Abstract). Biophys J 98(3):340a–340a
19. Finkel A, Wittel A, Yang N, Handran S, Hughes J, Costantin J (2006) Population patch clamp improves data consistency and success rates in the measurement of ionic currents. J Biomol Screen 11:488–496
20. Caterina MJ (2007) Chemical biology: sticky spices. Nature 445:491–492
21. Bandell M, Story GM, Hwang SW, Viswanath V, Eid SR, Petrus MJ, Earley TJ, Patapoutian A (2004) Noxious cold ion channel TRPA1 is activated by pungent compounds and bradykinin. Neuron 41:849–857
22. Jordt SE, Bautista DM, Chuang HH, McKemy DD, Zygmunt PM, Hogestatt ED, Meng ID, Julius D (2004) Mustard oils and cannabinoids excite sensory nerve fibres through the TRP channel ANKTM1. Nature 427:260–265
23. Macpherson LJ, Geierstanger BH, Viswanath V, Bandell M, Eid SR, Hwang S, Patapoutian A (2005) The pungency of garlic: activation of TRPA1 and TRPV1 in response to allicin. Curr Biol 15:929–934
24. Klionsky L, Tamir R, Gao B, Wang W, Immke DC, Nishimura N, Gavva NR (2007) Species-specific pharmacology of Trichloro(sulfanyl)ethyl benzamides as transient receptor potential ankyrin 1 (TRPA1) antagonists. Mol Pain 3:39
25. Akopian AN, Ruparel NB, Jeske NA, Hargreaves KM (2007) Transient receptor potential TRPA1 channel desensitization in sensory neurons is agonist dependent and regulated by TRPV1-directed internalization. J Physiol 583:175–193
26. Zurborg S, Yurgionas B, Jira JA, Caspani O, Heppenstall PA (2007) Direct activation of the ion channel TRPA1 by Ca^{2+}. Nat Neurosci 10:277–279
27. Wes PD, Chevesich J, Jeromin A, Rosenberg C, Stetten G, Montell C (1995) TRPC1, a human homolog of a Drosophila store-operated channel. Proc Natl Acad Sci USA 92:9652–9656
28. Hofmann T, Obukhov AG, Schaefer M, Harteneck C, Gudermann T, Schultz G (1999) Direct activation of human TRPC6 and TRPC3 channels by diacylglycerol. Nature 397:259–263
29. DeHaven WI, Jones BF, Petranka JG, Smyth JT, Tomita T, Bird GS, Putney JW Jr (2009) TRPC channels function independently of STIM1 and Orai1. J Physiol 587:2275–2298
30. Yuan JP, Zeng W, Huang GN, Worley PF, Muallem S (2007) STIM1 heteromultimerizes TRPC channels to determine their function as store-operated channels. Nat Cell Biol 9:636–645

31. Li J, Sukumar P, Milligan CJ, Kumar B, Ma ZY, Munsch CM, Jiang LH, Porter KE, Beech DJ (2008) Interactions, functions, and independence of plasma membrane STIM1 and TRPC1 in vascular smooth muscle cells. Circ Res 103:e97–e104
32. Massullo P, Sumoza-Toledo A, Bhagat H, Partida-Sanchez S (2006) TRP channels, calcium and redox sensors during innate immune responses. Semin Cell Dev Biol 17:654–666
33. Hara Y, Wakamori M, Ishii M, Maeno E, Nishida M, Yoshida T, Yamada H, Shimizu S, Mori E, Kudoh J, Shimizu N, Kurose H, Okada Y, Imoto K, Mori Y (2002) LTRPC2 Ca^{2+}-permeable channel activated by changes in redox status confers susceptibility to cell death. Mol Cell 9:163–173
34. Kolisek M, Beck A, Fleig A, Penner R (2005) Cyclic ADP-ribose and hydrogen peroxide synergize with ADP-ribose in the activation of TRPM2 channels. Mol Cell 18:61–69
35. Sano Y, Inamura K, Miyake A, Mochizuki S, Yokoi H, Matsushime H, Furuichi K (2001) Immunocyte Ca^{2+} influx system mediated by LTRPC2. Science 293:1327–1330
36. Togashi K, Hara Y, Tominaga T, Higashi T, Konishi Y, Mori Y, Tominaga M (2006) TRPM2 activation by cyclic ADP-ribose at body temperature is involved in insulin secretion. EMBO J 25:1804–1815
37. Patel S, Docampo R (2009) In with the TRP channels: intracellular functions for TRPM1 and TRPM2. Sci Signal 2:e69
38. Hill K, McNulty S, Randall AD (2004) Inhibition of TRPM2 channels by the antifungal agents clotrimazole and econazole. Naunyn Schmiedebergs Arch Pharmacol 370:227–237
39. Naziroglu M, Brandsch C (2006) Dietary hydrogenated soybean oil affects lipid and vitamin E metabolism in rats. J Nutr Sci Vitaminol (Tokyo) 52:83–88
40. Grimm C, Kraft R, Schultz G, Harteneck C (2005) Activation of the melastatin-related cation channel TRPM3 by D-erythro-sphingosine [corrected]. Mol Pharmacol 67:798–805
41. Wagner TF, Loch S, Lambert S, Straub I, Mannebach S, Mathar I, Dufer M, Lis A, Flockerzi V, Philipp SE, Oberwinkler J (2008) Transient receptor potential M3 channels are ionotropic steroid receptors in pancreatic beta cells. Nat Cell Biol 10:1421–1430
42. Naylor J, Milligan CJ, Zeng F, Jones C, Beech DJ (2008) Production of a specific extracellular inhibitor of TRPM3 channels. Br J Pharmacol 155:567–573
43. Hofmann T, Chubanov V, Gudermann T, Montell C (2003) TRPM5 is a voltage-modulated and $Ca^{(2+)}$-activated monovalent selective cation channel. Curr Biol 13:1153–1158
44. Colburn RW, Lubin ML, Stone DJ Jr, Wang Y, Lawrence D, D'Andrea MR, Brandt MR, Liu Y, Flores CM, Qin N (2007) Attenuated cold sensitivity in TRPM8 null mice. Neuron 54:379–386
45. Takashima Y, Daniels RL, Knowlton W, Teng J, Liman ER, McKemy DD (2007) Diversity in the neural circuitry of cold sensing revealed by genetic axonal labeling of transient receptor potential melastatin 8 neurons. J.Neurosci 27:14147–14157
46. Beck EJ, Hutchinson TL, Qin N, Flores CM, Liu Y (2010) Development and validation of a secondary screening assay for TRPM8 antagonists using QPatch HT. Assay Drug Dev Technol 8:63–72
47. Tominaga M, Caterina MJ (2004) Thermosensation and pain. J Neurobiol 61:3–12
48. Caterina MJ, Schumacher MA, Tominaga M, Rosen TA, Levine JD, Julius D (1997) The capsaicin receptor: a heat-activated ion channel in the pain pathway. Nature 389:816–824
49. Patapoutian A, Tate S, Woolf CJ (2009) Transient receptor potential channels: targeting pain at the source. Nat Rev Drug Discov 8:55–68
50. Raisinghani M, Pabbidi RM, Premkumar LS (2005) Activation of transient receptor potential vanilloid 1 (TRPV1) by resiniferatoxin. J Physiol 567:771–786
51. Roberts LA, Connor M (2006) TRPV1 antagonists as a potential treatment for hyperalgesia. Recent Pat CNS Drug Discov 1:65–76
52. Jiang X (2010) Assay of TRPV4 Channel using PatchXpress 7000A and Ionworks Quattro Systems
53. Southan A, Clark G (2009) Recent advances in electrophysiology-based screening technology and the impact upon ion channel discovery research. Methods Mol Biol 565:187–208

54. Andersson DA, Chase HW, Bevan S (2004) TRPM8 activation by menthol, icilin, and cold is differentially modulated by intracellular pH. J Neurosci 24:5364–5369
55. Behrendt HJ, Germann T, Gillen C, Hatt H, Jostock R (2004) Characterization of the mouse cold-menthol receptor TRPM8 and vanilloid receptor type-1 VR1 using a fluorometric imaging plate reader (FLIPR) assay. Br J Pharmacol 141:737–745

Chapter 6
TRPC2: Of Mice But Not Men

Christoffer Löf, Tero Viitanen, Pramod Sukumaran, and Kid Törnquist

Abstract Relatively little is known in regard to the physiological significance of TRPC2 and its regulation or interaction with other calcium regulating signalling molecules. In rodents, however, the importance of TRPC2 is indisputable. In mice, transcripts for TRPC2 have been found in testis, sperm, in neurons in the vomeronasal organ, and both in the dorsal root ganglion and in the brain. In rats, TRPC2 is thought to be expressed exclusively in the vomeronasal organ. In mice, TRPC2 is of importance in regulating both sexual and social behaviour. In sperm, TRPC2 is of importance in the acrosome reaction. This review will summarize the known physiological effects of TRPC2 channels, and the regulation of the function of the channel. In addition, some new preliminary data on the role of TRPC2 in rat thyroid cells will be presented.

6.1 Introduction

Of all the different members of the TRPC – family of ion channels, the TRPC2 channel is perhaps the least investigated, and thus relatively little is known in regard to its physiological significance and regulation of interaction with other calcium regulating signalling molecules. This discrepancy stems from the fact that TRPC2 is a pseudogene in human, and thus has little importance in understanding human pathophysiology. However, in rodents the importance of TRPC2 is indisputable. In mice, transcripts for TRPC2 have been found in mice testis, sperm, in neurons in the vomeronasal organ, and both in the dorsal root ganglion and in brain. In rats, TRPC2 is thought to be expressed exclusively in the vomeronasal organ. In regard to the regulation of the acrosome reaction in sperm, TRPC2 appears to be participating in calcium signalling. Even more important, in regulating both sexual and social behaviour in mice, TRPC2 is of exceptional physiological importance. The

K. Törnquist (✉)
Department of Biosciences, Åbo Akademi University, Artillerigatan 6, 20520 Turku, Finland;
The Minerva Foundation Institute for Medical Research, 00290 Helsinki, Finland
e-mail: ktornqvi@abo.fi

knowledge regarding how the activation of TRPC2 is regulated by physiological stimuli, and information on the gating properties of the channel, is limited. Some characteristics may be deduced from comparison to other members of the TRPC family, but for understanding e.g. pheromone signalling, information regarding the regulation of TRPC2 channels is of physiological importance. This review will summarize the known physiological effects of TRPC2 channels, the known mechanisms by which TRPC2 activation and signalling is regulated, and the electrophysiological characteristics of the TRPC2 channels. For a more detailed review of the molecular biology, pharmacology and pathophysiology of TRPC2 and other members of the TRPC-family of ion channels, the reader is referred to some excellent recent reviews [1, 2].

6.2 The Regulation of TRPC2

Although TRPC2 is a pseudogene in humans, the regulation of its function converges on phospholipase C (PLC), in line with the regulation of other members of the TRPC channel family (reviewed in [3]). TRPC2 is a rather nonselective cation channel and affect cell function through its ability to mediate both Ca^{2+}-entry, to evoke calcium-mediated signalling, and Na^+-entry to collapse the plasma membrane potential. Furthermore, of all the seven members of the TRPC family, TRPC2 has the most restricted pattern of expression [1].

The first report of the TRPC2 gene was by Zhu et al. [4]. Vannier et al. [5] found two splice variants of mouse TRPC2 (TRPC2a and TRPC2b) that, when heterologously expressed in COS-M6 cells, enhanced both receptor-operated calcium entry (ROCE) and store-operated calcium entry (SOCE). However, another group found two other splice variants of mouse TRPC2, mTRPC2α and mTRPC2β, which did not enhance ROCE or SOCE when expressed in HEK-293 cells, as they failed to reach the plasma membrane and were retained in intracellular membranes [6]. A later study showed that the TRPC2α splice variant was an incompletely processed transcript of TRPC2a [7], which could explain at least in part the differences in the results. Evidence supporting the results that TRPC2 is involved in ROCE and SOCE was obtained in a study in CHO cells, where Gailly et al. [8] used a TRPC2 antisense construct and reported reductions in both ROCE and SOCE. Furthermore, inward currents were increased in HEK-293 cells transiently expressing TRPC2 after stimulation of purinergic receptors with ATP [9]. We have results that support this finding. By using shRNA for TRPC2 in rat thyroid FRTL-5 cells, we saw a reduced calcium entry when stimulating purinergic receptors [10]. In sperm, it seems like TRPC2 functions as a SOC channel, since an antibody against TRPC2 decreased both thapsigargin-induced and zona pellucida (ZP3) induced calcium entry [9].

The mechanism by which TRPC2 is activated is not yet completely understood. However, Lucas et al. [11] compared diacylglycerol (DAG) induced inward currents in wild-type and TRPC2 knockout vomeronasal sensory neurons, and found that TRPC2 knockout severely reduced the inward current. Interestingly, they reported no activation of TRPC2 in response to store depletion or increased levels of IP_3.

The involvement of TRPCs in SOCE is debatable, but it is intriguing that the master regulator of SOCE, Stim1, binds to TRPC2 and several of the TRPCs via its ezrin/radixin/moesin (ERM) domain [12].

TRPC2, like all other TRPCs, have a binding site for calmodulin (CaM) in the C-terminus [13]. In addition, the long N-terminus of TRPC2 has binding sites for CaM [7], enkurin [14] and ankyrin (see [15]). The CaM/IP$_3$ receptor-binding (CIRB) domain that is conserved in all TRPCs is also present in mouse TRPC2 [13]. CaM competes with the IP$_3$ receptor (IP$_3$R) in a calcium-dependant manner for binding to the CIRB domain [13], indicating that calcium through binding to CaM could mediate a negative feedback loop. Accordingly, currents through TRPC2 were inhibited by calcium-CaM in vomeronasal sensory neurons [16].

Different adaptor proteins have been shown to interact with TRPCs and IP$_3$Rs. One such adaptor is Homer 1, which co-immunoprecipitates with TRPC2 [17]. Homer 1 does not only function as a scaffold for TRPCs and IP$_3$Rs, it inhibits spontaneous activity of TRPC1 [17]. Furthermore, Homer 1 controls the translocation of TRPC3 to the plasma membrane in response to store depletion or stimulation of the IP$_3$R [18]. In agreement with these findings, Homer 1 knockout mice show perturbations in skeletal muscle function. In myotubes from these mice the basal current density and spontaneous cation influx was increased [19]. Another system where TRPC2 has reported functions is in erythroblasts. Stimulation of the erythropoietin (EPO) receptor activates, via PLCγ, calcium entry through TRPC2 [20]. A signaling complex between the EPO receptor, TRPC2, PLCγ, and IP$_3$R2 was shown to form in erythroblasts and in an overexpression system [21]. A complex between TRPC2 and IP$_3$R3 was also previously reported to form in the rat VNO [22]. In contrast to an earlier report [23], Chu et al. [24] showed that TRPC2 could form complexes with TRPC6 in erythroblasts.

6.3 Electrophysiological Properties of TRPC2

The notion that TRPC2 protein are highly expressed in murine vomeronasal organ (VNO) neurons suggested that it might constitute a channel involved in the transduction cascade of olfaction and opened a possibility to study the protein function in an native cellular environment [25]. VNO neurons have bipolar characteristics; single apical dendrites extend towards the surface of the sensory epithelium, which is delineated by the dendritic microvilli that contain the transduction machinery, and axons that project towards the accessory olfactory bulb (or vomeronasal bulb) [26]. These cells harbour unique sets of G-protein coupled receptors, derived from two receptor families (VR1 and VR2) suggested to be sensitive to pheromones [27–29]. Neurons having soma closer to the apical border of the sensory epithelium express receptors of type VR1 and $G_{\alpha i2}$ subunits, whereas cell bodies near the basal region contain VR2 receptors and $G_{\alpha o}$ subunits, indicating an existence of topographically and functionally segregated transduction systems [30].

VNO neurons exhibit a relatively high input resistance (several GΩ), which renders their membrane voltage (V_m) very sensitive to small fluctuations in membrane currents [31–33]. Apparently the preparation procedure may affect the resting V_m, as the isolated cells tend to be more depolarised [31–33] compared with those embedded in slices [34, 35].

When exposed to diluted urine or putative pheromones, the firing rate of VNO neurons increases robustly. The pheromone-evoked spiking depends on PLC activity and is associated with an increase in intracellular Ca^{2+} concentrations ($[Ca^{2+}]_i$) [36, 37]. The Ca^{2+} influx is initiated at distal dendrites [37] containing the sensory microvilli that show abundant TRPC2 expression [25]. Multiunit microelectrode array recordings of VNO neuroepithelium [38], or on-cell patch clamp and field potential recordings done with VNO slices [39], revealed that neurons obtained from TRPC2–/– mice are unresponsive to urine derived pheromones, confirming that TRPC2 is an essential component of olfactory transduction.

The pheromone-activated channel containing the TRPC2 protein is directly activated by DAG, the level of which is increased by PLC activity and reduced by diacylglycerol kinase (DGK) [11]. Brief stimulations of inside-out membrane patches (ripped from distal dendritic tips of VNO neurons) with the DAG analogue 1-stearoyl-2-arachidonoyl-sn-glycerol (SAG) evoked a marked increase in single-channel activity. The observed −3.3 pA single-channel current recorded at −80 mV in a symmetrical 150 mM Na^+ solution gave a slope conductance of 42 pS. The relative permeability of ions (P_{ion}) under bi-ionic conditions (as described [40]), gave P_{Ca}/P_{Na} and P_{Cs}/P_{Na} ratios of 2.7 ± 0.7 and 1.5 ± 0.3, respectively, when Na^+ was the only cation on the extracellular side. Inward whole-cell currents of VNO neurons were reduced upon replacement of Na^+ by $NMDG^+$, indicating that sodium acts as the main charge carrier. The amplitude of the TRPC2-dependent whole-cell currents was potentiated by more than twofold upon reduction of extracellular Ca^{2+} from 1 to 0.1 mM, suggesting a partial block of channels by permeating Ca^{2+} ions. Under whole-cell conditions, the TRPC2-dependent current emerged in response to DGK inhibitors, arguing for a reasonable DAG production by endogenous PLC activity. In line with this, the current evoked by DGK inhibitors was sensitive to PLC inhibitors, unlike the currents induced by application of exogenous SAG [11]. The SAG-activated currents are rapidly suppressed by a $[Ca^{2+}]_i$-and CaM-mediated feedback loop, suggesting an adaptation of the chemosensory transduction process [16], but see [36]).

A subset of VNO neurons seems to contain urine- and intracellular Ca^{2+}-activated Ca^{2+} influx routes independent of TRPC2 [16, 41–43]. The application of arachidonic acid (AA) or linolenic acid induced a slow increase in an inward current recorded at –90 mV and an increase in Ca^{2+} influx [41]. The molecular identity of this AA- and Ca^{2+}-activated channel is currently unknown. Due to the relatively high intracellular Ca^{2+} required to fully activate the channel, and due to its equal selectivity for K^+ and Na^+ and low preference for Ca^{2+}, the current is often referred to as a Ca^{2+}-activated, non-selective cation current (I_{CaNS}) [16, 42, 43]. VNO neurons exhibiting I_{CaNS} may form a TRPC2-independent transduction route serving a function of its own, or then the activation of I_{CaNS} may amplify the response initiated by TRPC2 [42, 43].

The activation of a DAG-gated and TRPC2-dependent current during the chemosensory transduction of VNO is well described [11, 38, 39]. However, the roles of other Ca^{2+} influx routes and $[Ca^{2+}]_i$-activated currents in the adaptation or amplification of pheromone-evoked signalling are less well understood. It seems that the first level of integration appears at the level of $[Ca^{2+}]_i$, the alterations of which may trigger either amplifying (I_{CaNS}) or suppressive (Ca-CaM inhibition of TRPC2) mechanisms. The common signal for both amplification and adaptation suggests that the modulation of information processing may be conducted by close interactions between TRPC2 and Ca^{2+}-sensitive ion channels and/or regulatory proteins, or by cell-specific and reciprocally activated signalling cascades. However, the options mentioned above are not mutually exclusive.

6.4 Pheromone Signal Transduction in Olfaction

In many vertebrates, olfactory cues evoked by pheromones are important in regulating both social and sexual behaviour. In terrestrial vertebrates, two classes of olfactory neurons in distinct anatomical localizations have evolved, the main olfactory epithelium (MOE) and the vomeronasal organ (VNO) (for extensive reviews, see [44] and [45]). For the detection of pheromones, the VNO seems to be of significant importance, as VNO ablation markedly changed the behaviour of the mice compared with sham-operated littermates. A breakthrough in understanding how pheromones evoke signalling in the VNO was obtained when Liman and Dulac cloned the rat TRPC2 channel, and showed that it was exclusively expressed at both mRNA and protein levels in neurons in the VNO [25]. Furthermore, these neurons project to the accessory olfactory bulb, distinctly different from the projections from the MOE [46]. When TRPC2 knockout mice were generated independently by two groups [38, 39], the importance of TRPC2 in pheromone signalling was unveiled unequivocally. The behaviour of TRPC2–/– mice differed profoundly from their wild-type and TRPC2–/+ heterozygote littermates.

The mating behaviour in TRPC2–/– males against female mice did not differ from that seen in wild-type or TRPC2–/+ males. However, when male TRPC2–/– mice encountered intruder male mice, they did not show any aggressive behaviour, instead they repeatedly tried to mount them. In addition, TRPC2–/– mice showed defects in territory marking [38, 39]. Wild type lactating female mice usually show pronounced aggressivity against intruders, but this type of behaviour was absent from female TRPC2–/– mice. Furthermore, female TRPC2–/– mice showed a deficiency in maternal behaviour, i.e. they easily abandoned their pups. In addition, female TRPC2–/– mice showed sexual behaviour similar to male mice, i.e. they tried to mount wild-type female mice in a manner similar to that seen in male mice [47].

6.5 TRPC2 as Regulator of Calcium Entry in Sperm

Observations that mRNA for TRPC2 was expressed in bovine testis, and specifically in spermatocytes, suggested that TRPC2 may have a role in sperm function

[48]. In an elegant study, Jungnickel et al. [9] showed that TRPC2 indeed was necessary for calcium entry in the sperm. In their study they showed that when sperm went through the acrosome reaction upon contact with the egg's extracelluar matrix, in particular the ZP3 glycoprotein, calcium entry dependent on TRPC2 was initiated. Furthermore, these studies clearly showed that TRPC2 was necessary for the sustained increase in calcium entry in response to ZP3. In addition, inhibiting TRPC2 also abolished the acrosome reaction. The picture is, however, complicated, as the fertility of TRPC2–/– mice is not compromised, as sperm also express both the TRPC1 and the TRPC5 channels [9]. The mechanism by which ZP3 activates TRPC2 is presently not clearly elucidated. A direct interaction between the channel and the IP_3 receptor has been suggested. In addition, TRPC2 can be activated by depleting intracellular calcium stores using thapsigargin in sperm, suggesting that TRPC2 may participate in SOCE. In HEK-cells transfected with TRPC2, the channel was activated through a receptor-mediated mechanism, possibly due to PLC-evoked breakdown of phosphatidylinositol 4,5-bisphosphate and the production of DAG [5]. However, stimulating sperm with DAG (which activates other TRPC channels) was unable to evoke the acrosome reaction [49]. Furthermore, two adaptor proteins, enkurin and junctate, may be involved in the activation and regulation of TRPC2 [14]. Thus, TRPC seems to be activated both through a receptor- and store-dependent mechanism. It is also possible that TRPC2 may be activated by slightly different mechanisms in different cell types.

6.6 TRPC2 as Mediator of Erythropoietin-Evoked Signalling

Of the few reported physiological events where TRPC2 have been reported to participate is the erythropoietin-evoked calcium signalling in murine haematopoietic cells [20]. The effect seems to be mediated through a mechanism dependent on PLCγ and IP_3-receptors [21], in a manner similar to that reported for other TRPC channels (see e.g [50]). Interestingly, TRPC2 splice variants may block calcium entry [51]. It is worth mentioning that the erythropoietin-evoked entry of calcium in human erythroid cells seems to be mediated by TRPC3 [52].

6.7 TRPC2 in Rat Thyroid Cells

We have for several years investigated calcium signalling in rat thyroid FRTL-5 cells, a well-characterized model for studying thyroid cell function. In these investigations we have recently detected a novel calcium entry mechanism dependent on a phosphatase and protein kinase A [53]. This entry mechanism was also regulated by a receptor–mediated mechanism, and was enhanced by DAG [54]. The nature of the mechanism was, however, not known. We thus performed a RT-PCR screening for

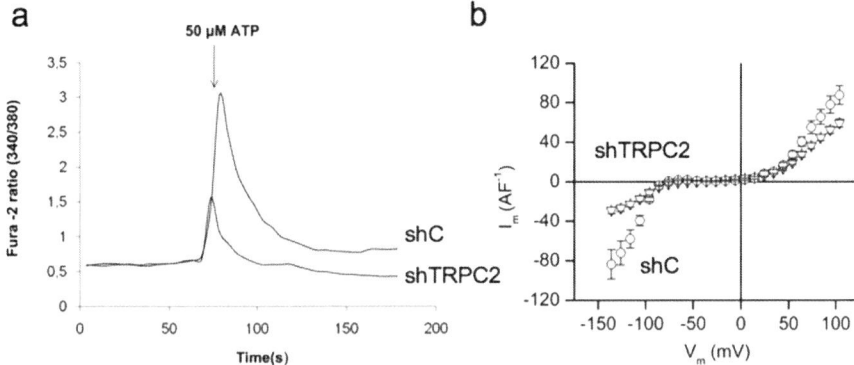

Fig. 6.1 Importance of TRPC2 for calcium signalling in rat thyroid FRTL-5 cells. (**a**). Calcium imaging studies of control cells and TRPC2 knockdown cells stimulated with ATP in calcium-containing buffer at +37°C. Each trace is representative of at least 30 separate cells. (**b**). I-V characteristics of control cells (*circle*), or TRPC2 knockdown cells (*triangle*). Whole-cell recordings were performed at +32°C in an extracellular solution containing 1.0 mM CaCl$_2$ and an intracellular solution containing 5 mM BAPTA. The voltage step protocol was initiated with a pulse to −80 mV (duration 100 ms) that was followed by a series of rectangular pulses. The series of 25 steps ascended from −120 mV to +120 mV at 20 mV increments, each pulse of 100 ms duration. Steps were applied every 2 s from a holding voltage of −0 mV. The data is given as the mean ± SEM of 4–5 separate experiments

all the TRPC channels, and found that only TRPC2 was expressed. Recent investigations have revealed that the phosphatase and PKA-regulated entry mechanism is, in fact, the result of activation of TRPC2 (Sukumaran et al. manuscript in preparation). Furthermore, knockdown of TRPC2 significantly hampered ATP-evoked calcium signalling [10] (see Fig. 6.1a). In addition, in these cells, patch clamp recordings revealed a significant change in conductances in TRPC2 knockdown cells, compared with control cells (Fig. 6.1b). Preliminary results also indicated that knockdown of TRPC2 significantly hampered proliferation of the cells (Löf et al. manuscript in preparation).

6.8 Perspective

The importance of TRPC2 in rodent social and sexual behaviour is undisputable. It will be of interest to see how future investigations will shed light on the processing of olfactory clues, and how this will interact with neural networks controlling e.g. behaviour. Furthermore, our findings that TRPC2 play a role in the rat thyroid gland physiology suggests that TRPC2 may have several other, yet undiscovered functions in rodents. This also opens up a multitude of possible interactions with different adaptor proteins and ion channels. Thus, TRPC2 may, in fact, play a much more important role in rodent physiology than hitherto believed.

References

1. Yildirim E, Birnbaumer L (2007) TRPC2: molecular biology and functional importance. Handb Exp Pharmacol 179:51–73
2. Abramowitz J, Birnbaumer L (2009) Physiology and pathophysiology of canonical transient receptor potential channels. FASEB J 23:297–328
3. Montell C (2005) The TRP superfamily of cation channels. Sci STKE www.stke org/cgi/content/full/sigtrans;2005/272/re3
4. Zhu X, Jiang M, Peyton M, Boulay G, Hurst R, Stefani E, Birnbaumer L (1996) trp, a novel mammalian gene family essential for agonist-ectivated capacitative Ca^{2+} entry. Cell 85: 661–671
5. Vannier B, Peyton M, Boulay G, Brown D, Qin N, Jiang M, Zhu X, Birnbaumer L (1999) Mouse trp2, the homologue of the human trpc2 pseudogene, encodes mTrp2, a store depletion-activated capacitative Ca^{2+} entry channel. Proc Natl Acad Sci USA 96:2060–2064
6. Hofmann T, Schaefer M, Schultz G, Gudermann T (2000) Cloning, expression and subcellular localization of two novel splice variants of mouse transient receptor potential channel 2. Biochem J 351:115–122
7. Yildirim E, Dietrich A, Birnbaumer L (2003) The mouse C-type transient receptor potential2 (TRPC2) channel: alternative splicing and calmodulin binding to its N terminus. Proc Natl Acad Sci USA 100:2220–2225
8. Gailly P, Colson-Van Schoor M (2001) Involvement of trp-2 protein in store-operated influx of calcium in fibriblasts. Cell Calcium 30:157–165
9. Jungnickel MK, Marrero H, Birnbaumer L, Lemos JR, Florman HM (2001) Trp2 regulates Ca^{2+} entry into mouse sperm triggered by egg ZP3. Nat Cell Biol 3:499–502
10. Sukumaran P, Löf C, Viitanen T, Törnquist K (2009) TRPC2 mediates calcium entry in rat thyroid FRTL-5 cells. Chem. Phys. Lipids 160(Supplement):S29
11. Lucas P, Ukhanov K, Leinders-Zufall T, Zufall F (2003) A diacylglycerol-gated cation channel in vomeronasal neuron dendrites is impaired in TRPC2 mutant mice: mechanism of pehromone transduction. Neuron 40:551–561
12. Huang GN, Zeng W, Kim JY, Yan JP, Han L, Muallem S, Worley PF (2006) STIM1 carboxyl-terminus activates native SOC I $_{(CRAC)}$ and TRPC1 channels. Nat Cell Biol 8:1003–1010
13. Tang J, Lin Y, Zhang Z, Tikunov S, Birnbaumer L, Zhu MX (2001) Identification of common binding sites for calmodulin and inositol 1,4,5-trisphosphate receptors on the carboxy termini of trp channnels. J Biol Chem 276:21303–21310
14. Sutton KA, Jungenickel MK, Wang Y, Cullen K, Lambert S, Florman HM (2004) Enkurin is a novel calmodulin and TRPC channel biinding protein in sperm. Dev Biol 274:426–435
15. Birnbaumer L, Yildirim E, Abramowitz J (2003) A comparison of the genes coding for canonical TRP channels and their M, V, and P relatives. Cell Calcium 33:419–432
16. Spehr J, Hagendorf S, Weiss J, Spehr M, Leinders-Zufall T, Zufall F (2009) Ca^{2+}-calmodulin feedback mediates sensory adaption and inhibits pheromone-sensitive ion channels in the vomeronasal organ. J Neurosci 29:2125–2135
17. Yuan JP, Kiselyov K, Shin DM, Chen J, Shcheynikov N, Schwar MK, Seeburg PH, Muallem S, Worley PF (2003) Homer binds TRPC family channels and is required for gating of TRPC1 by IP_3 receptors. Cell 114:777–789
18. Kim JY, Zeng W, Kiselyov K, Yuan JP, Dehoff MH, Mikoshiba K, Worley PF, Muallem S (2006) Homer 1 mediates store- and inositol 1,4,5-trisphosphate receptor-dependent translocation and retrieval of TRPC3 to the plasma membrane. J Biol Chem 281: 32540–32549
19. Stieber JA, Zhang ZS, Burch J, Eu JP, Zhang S, Truskey GA, Seth M, Yamaguchi N, Meissner G, Shoah R, Worley PF, Williams RS, Rosenberg PB (2008) Mice lacking Homer 1 exhibit a skeletal myopathy characterized by abnormal transient receptor potential channel activity. Mol Cell Biol 28:2637–2647

20. Chu X, Cheung JY, Barber DL, Birnbaumer L, Rothblum LI, Conrad K, Abrasonis V, Chan YM, Stahl R, Carey DJ, Miller BA (2002) Erythropoietin modulates calcium influx through TRPC2. J Biol Chem 277:34375–34382
21. Tong Q, Chu X, Cheung JY, Conrad K, Stahl R, Barber DL, Mignery G, Miller BA (2004) Erythropoietin-modulated calcium influx through TRPC2 is mediated by phospholipase Cgamma and IP3R. Am J Physiol 287:C1667–C1678
22. Brann JH, Dennis JC, Morrisin EF, Fadool DA (2002) Type-specific inositol 1,4,5-trisphosphate receptor localization in the vomeronasal organ and its interaction with a transient receptor potential channel, TRPC2. J Neurochem 83:1452–1460
23. Hofmann T, Schaefer M, Schultz G, Gudermann T (2002) Subunit composition of mammalian transient receptor potential channels in lliving cells. Proc Natl Acad Sci USA 99:7461–7466
24. Chu X, Tong Q, Cheung JY, Wozney J, Conrad K, Mazack V, Zhang W, Stahl R, Barber DL, Miller BA (2004) Interaction of TRPC2 and TRPC6 in erythropoietin modulation of calcium Influx. J Biol Chem 279:10514–10522
25. Liman ER, Corey DP, Dulac C (1999) TRP2: a candidate transduction channel for mammalian pheromone sensory signaling. Proc Natl Acad Sci USA 96:5791–5796
26. Doving KB, Trotier D (1998) Structure and function of the vomeronasal organ. J Exp Biol 21:2913–2925
27. Dulac C, Axel R (1995) A novel family of genes encoding putative pheromone receptors in mammals. Cell 83:195–206
28. Herrada G, Dulac C (1997) A novel family of putative pheromone receptors in mammals with a topographically organized and sexually dimorphic distribution. Cell 90:763–773
29. Matsunami H, Buck LB (1997) A multigene family encoding a diverse array of putative pheromone receptors in mammals. Cell 90:775–784
30. Dulac C, Torello AT (2003) Molecular detection of pheromone signals in mammals: from genes to behaviour. Nat Rev Neurosci 7:551–562
31. Liman ER, Corey DP (1996) Electrophysiological characterization of chemosensory neurons from the vomeronasal organ. J Neurosci 15:4625–4637
32. Trotier D, Doving KB, Ore K, Shalchian-Tabrizi C (1998) Scanning electron microscopy and gramicidin patch clamp recordings of microvillous receptor neurons dissociated from the rat vomeronasal organ. Chem Senses 1:49–57
33. Fieni F, Ghiaroni V, Tirindelli R, Pietra P, Biagiani A (2003) Apical and basal neuronesisolated from the mouse vomeronasal organ differ for voltage dependent currents. J Physiol 522: 425–436
34. Shimazaki R, Boccaccio A, Mazzatenta A, Pinato G, Migliore M, Menini A (2006) Electrophysiological properties and modeling of murine vomeronasal sensory neurons in acute slice preparations. Chem Senses 31:425–435
35. Ukhanov K, Leinders-Zufall T, Zufall F (2007) Patch-clamp analysis of gene-targeted vomeronasal neurons expressing a defined V1r or V2r receptor: ionic mechanisms underlying persistent firing. J Neurophysiol 98:2357–2369
36. Holy TE, Dulac C, Meister M (2000) Respnses of vomeronasal neurons to natural stimuli. Science 289:1569–1572
37. Leinders-Zufall T, Lane AP, Puche AC, Ma W, Novotny MV, Shipley MT, Zufall F (2000) Ultrasensitive pheromone detection by mammalian vomeronasal neurons. Nature 408: 792–796
38. Stowers L, Holy TE, Meister M, Dulac C, Koentges G (2002) Loss of sex discrimination and male-male aggression in mice deficient for trp2. Science 295:1493–1500
39. Leypold BG, Yu CR, Leinders-Zufall T, Kim MM, Zufall F, Axel R (2002) Altered sexual and social behaviors in trp2 mutant mice. Proc Natl Acad Sci USA 99:6376–6381
40. Lewis CA (1979) Ion-concentration dependence of the reversal potential and the single channel conductance of ion channels at the frog neuromuscular junction. J Physiol 286:417–445

41. Spehr M, Hatt H, Wetzel CH (2002) Arachidonic acid plays a role in rat vomeronasal signal transduction. J Neurosci 19:8429–8437
42. Liman ER (2003) Regulation by voltage and adenin nucleotides of a Ca^{2+}-activated cation channel from jamster vomeronasal sensory neurons. J Physiol 548:777–787
43. Zhang P, Yang C, Delay RJ (2010) Odors activate dual pathways, a TRPC2 and AA-dependent pathway, in mouse vomeronasal neurons. Am J Physiol. DOI:10.1152/ajpcell.00271.2009
44. Dulac C, Kimchi T (2007) Neural mechanisms underlying sex-specific behaviours in vertebrates. Curr Op Neurobiol 17:675–683
45. Brennan PA, Zufall F (2006) Pheromonal communication in vertebrates. Nature 444:308–315
46. Belluscio L, Koentges G, Axel R, Dulac C (1999) A map of pheromone receptor activation in the mammalian brain. Cell 97:209–220
47. Kimchi T, Xu J, Dulac C (2007) A functional circuit underlying male sexual behaviour in the female mouse brain. Nature 448:1009–1014
48. Wissenbach U, Schroth G, Philipp S, Flockerzi V (1998) Structure and mRNA expression of a bovine trp homologue related to mammalian trp2 transcripts. FEBS Lett 429:61–66
49. Stamboulian S, Moutin MJ, Treves S, Pochon N, Grunwald D, Zorzato F, De Waard M, Ronjat M, Arnoult C (2005) Junctate, an inositol 1,4,5-triphosphate receptor associated protein, is present in rodent sperm and binds TRPC2 and TRPC5 but not TRPC1 channels. Dev Biol 286:326–337
50. Kiselyov KI, Xu X, Mozhayeva GN, Kuo T, Pessah I, Mignery GA, Birnbaumer L, Muallem S (1998) Functional interaction between $InsP_3$ receptors and store-operated Htrp3 channnels. Nature 396:478–482
51. Chu X, Tong Q, Wozney J, Zhang W, Cheung JY, Conrad K, Mazack V, Stahl R, Barber DL, Miller BA (2005) Identification of an N-terminal TRPC2 splice variant which inhibits calcium influx. Cell Calcium 37:173–182
52. Tong Q, Hirschler-Laskiewicz I, Zhang W, Conrad K, Neagley DW, Barber DL, Cheung JY, Miller BA (2008) TRPC3 is the erythropoietin-regulated calcium channel in human erythroid cells. J Biol Chem 283:10385–10395
53. Gratschev D, Blom T, Björklund S, Törnquist K (2004) Phosphatase inhibition unmasks a calcium entry pathway dependent on protein kinase A in thyroid FRTL-5 cells. Comparison with store-operated calcium entry. J Biol Chem 279:49816–49824
54. Gratschev D, Löf C, Heikkilä J, Björkbom A, Sukumaran P, Hinkkanen A, Slotte JP, Törnquist K (2009) Sphingosine kinase as a regulator of calcium entry through autocrine sphingosine 1-phosphate signalling in thyroid FRTL-5 cell. Endocrinology 50:5125–5134

Chapter 7
TRPM1: New Trends for an Old TRP

Elena Oancea and Nadine L. Wicks

Abstract TRPM1, initially named Melastatin, is the founding member of the TRPM subfamily of Transient Receptor Potential (TRP) ion channels. Despite sustained efforts, the molecular properties and physiological functions of TRPM1 remained elusive until recently. New evidence has uncovered novel TRPM1 splice variants and revealed that TRPM1 is critical for a non-selective cation conductance in melanocytes and retinal bipolar cells. Functionally, TRPM1 has been shown to mediate retinal ON bipolar cell transduction and suggested to regulate melanocyte pigmentation. Notably, *TRPM1* mutations have also been associated with congenital stationary night blindness in humans. This review will summarize and discuss our present knowledge of TRPM1: its discovery, expression, regulation, and proposed functions in skin and eye.

7.1 Introduction

TRPM1 (Melastatin) was identified over a decade ago [1, 2] – a very long time in TRP years. While the rest of the TRP field has seen rapid expansion over the past decade (hundreds of TRP-related papers are published every year), in the first 10 years following its discovery, only 10 papers focusing on TRPM1 were published. In contrast, the past 2 years have seen an explosion of TRPM1 literature: 16 papers describing different aspects of TRPM1 function have been published since 2008. This surge has enhanced our knowledge of TRPM1 to the point where it is useful to review its molecular and cellular properties.

This review will summarize and discuss our present knowledge of TRPM1. We will first focus on the discovery, characterization and regulation of the *TRPM1* gene,

E. Oancea (✉)
Department of Molecular Pharmacology, Physiology and Biotechnology, Brown University, Providence, RI, USA
e-mail: Elena_Oancea@brown.edu

and will then discuss expression of TRPM1 at the tissue and cellular level. Finally, we will address proposed functions and mechanisms of action for TRPM1 in skin and eye.

7.2 Identification and Characterization of the *Melastatin* (*MLSN*) Gene

Mouse *TRPM1* was identified in a differential cDNA display between two variants of the B16 mouse melanoma cell line that differ in their ability to form metastases [1]. Expression of *Melastatin* (*MSLN*), as the gene was first named, was found to inversely correlate with melanoma progression and tumor thickness [1, 3, 4]. The identified mouse cDNA (Accession No. AF047714) encodes a 542-residue protein flanked by 5' and 3' UTRs, and represents a gene localized to chromosome 7 [2]. Shortly thereafter, human *MLSN* was identified by probing a retinal library with mouse *MSLN* [2]. Human *MSLN* (Accession No. AF071787) is localized to chromosome 15 and encodes a 1,533-residue protein with six predicted transmembrane domains and homology to the TRP family of ion channels [2]. Melastatin thus became the founding member of a new subfamily of TRP channels: TRPM. Identification of human *MLSN* suggested that the shorter mouse cDNA might represent an N-terminal splice variant of a full-length isoform. Although initial attempts to find a longer mouse *MLSN* transcript were unsuccessful [2], a full-length mouse cDNA has been recently identified (Accession No. AY180104.2) [5].

7.3 TRPM1 Splice Variants

Many TRP channels express splice variants that differ in function and tissue distribution [6] and TRPM1 is no exception. In addition to the initial N-terminal variant [1], multiple N- and C-terminal TRPM1 transcripts have been identified; these lack the putative ion channel region and some differ in expression profile from full-length transcripts [7]. Moreover, one short N-terminal variant may act as a negative regulator of TRPM1-dependent calcium influx [8], suggesting that short splice variants could have a regulatory role.

The original human TRPM1 variant [2] contains an ATG initiation codon in exon 3 and predicts a 1,533-residue protein. A single nucleotide polymorphism (C130T) in the 5' UTR of this sequence (rs4779816) generates an inframe start site within exon 2 and predicts a protein with 70 additional N-terminal amino acids, resulting in a novel 1,603-residue isoform (70+TRPM1) (Accession No. NM002420.4) [9]. Interestingly, the frequency of the C allele is significantly higher in European populations [0.217 reported in the HAPMAP database (http://hapmap.ncbi.nlm.nih.gov/) and 0.199 calculated in reference [10]] compared to Asian populations (0.033 reported by HAPMAP) and African populations (0.075 reported by HAPMAP). The higher frequency of the C allele in light-skinned individuals, the phototype most susceptible to melanoma, raises the question of whether the C130 allele confers susceptibility to melanoma and if so, by what mechanism?

Recently, two studies identified new TRPM1 splice variants that contain novel exons and result in proteins with additional N-terminal residues: 92+TRPM1 (Accession No. HM135791) [9] and 109+TRPM1 (HM135790) [9, 10]. While the exon containing the initiation site for 92+TRPM1 is conserved in humans and primates, the exon containing the start of 109+TRPM1 is highly conserved across multiple species. 92+TRPM1-encoding mRNA has been found in melanoma cell lines, as well as melanocytes, brain and retina, where its relative abundance appears significantly lower than in melanoma cell lines [9]. 109+TRPM1 transcripts have been detected in retina, brain and melanocytes [9, 10] but not in whole skin samples [10], possibly due to its expression being restricted to melanocytes, which account for less than 5% of all skin cells. As discussed below, expression of 92+TRPM1 or 109+TRPM1 in melanoma cells generates an outwardly-rectifying, non-selective cation current [9], suggesting that these isoforms may have particular physiological relevance.

A mouse *TRPM1* sequence containing a putative ion channel region homologous to that of human TRPM1 was recently identified (Accession No. AY180104) [5]. Using the initial *MSLN* cDNA [1], Koike et al. screened a mouse retinal cDNA library and isolated a 1,622-residue TRPM1 isoform highly homologous to human 109+TRPM1. Like 109+TRPM1, heterologous expression of the mouse long TRPM1 isoform generates an ionic current, albeit with a different current-voltage (IV) relationship [5].

7.4 Regulation of *TRPM1* Gene Expression

Further characterization of the *TRPM1* gene identified a promoter region containing binding domains for the microphthalmia transcription factor (MITF) [2], a helix-loop-helix leucine zipper transcription factor essential for the regulation of many melanocyte-specific genes [11]. MITF mutations in mouse result in white coat color and small eyes (microphthalmia) that often have reduced pigmentation [12]. In humans, MITF mutations lead to Waardenburg syndrome, a condition associated with hypopigmentation as well as deafness due to the absence of melanocytes in the inner ear [13]. Two studies have shown that MITF transcriptionally regulates *TRPM1* expression in melanocytes and melanoma as well as eye [14, 15], further suggesting a role for TRPM1 in melanocyte and eye physiology.

7.5 Tissue and Cellular Distribution of TRPM1

Published data agree that TRPM1 mRNA is present in melanocytes and retina [1, 2, 7, 9, 15, 16], and other studies have detected TRPM1 mRNA in brain and heart [9, 17, 18], but what is the relative abundance of TRPM1 mRNA in those tissues?

A comparative study of the distribution and expression levels of TRPM channels using quantitative reverse-transcription PCR (qRT-PCR) showed that TRPM1 mRNA levels are highest in brain and heart, but did not determine the relative abundance of TRPM1 in melanocytes and retina [17]. A similar qRT-PCR analysis showed that, across a panel of tissues and relative to other TRPM channels, TRPM1 mRNA levels are very low (at least two orders of magnitude smaller than TRPM7 in all tissues studied) [18], supporting the hypothesis that TRPM1 expression is restricted to distinct subpopulations of cells in different tissues.

Our understanding of TRPM1 at the protein level would greatly benefit from access to antibodies; however, generating specific anti-TRPM1 antibodies has proved challenging. While several anti-TRPM1 antibodies are commercially available, to date, only three of these have been used successfully to detect TRPM1 and each appears to have limited use. A polyclonal anti-human TRPM1 antibody was raised against a GST-fused N-terminal splice variant [14]; this antibody identified a 122 kDa band in melanocyte and melanoma lysates, which presumably corresponds to a full-length variant of TRPM1. However, in a separate study, the same antibody identified a ~250 kDa band in melanocytes and the relative intensity of this band was reduced in cells treated with short-hairpin (sh) RNA targeting TRPM1 [19]. While this larger band might represent an as yet unidentified splice variant, it is also possible this antibody is not highly specific for TRPM1. Recently, a second antibody raised against a C-terminal peptide of human TRPM1 stained ON bipolar cells in the retina of wild-type but not mice lacking functional TRPM1 (TRPM1–/–) [20]. A separate study reported similar results using a third antibody, raised against a C-terminal peptide of mouse TRPM1 [5]. To our knowledge, there is not yet any (published) data reflecting the ability of these antibodies to detect TRPM1 in skin or other tissues, or their utility in biochemical analysis. The C-terminal antibodies may be particularly useful for future studies, as they are more likely to detect full-length TRPM1 isoforms but not N-terminal variants lacking the putative ion channel region. Further insights into the expression, localization and function of the numerous TRPM1 isoforms might be gained by their differential detection; however, the production of isoform-specific anti-TRPM1 antibodies is not likely to happen in the near future.

The cellular localization of TRPM1 could be an important factor in further characterzing its function. In situ hybridization showed TRPM1 expression in the inner retinal layer [21], while immunostaining of retinal slices revealed TRPM1 immunoreactivity in ON bipolar cells. TRPM1 localization appears to be primarily intracellular, but imaging resolution has been thus far insufficient to determine its subcellular localization [5, 20]. A previous study using GFP-tagged TRPM1 expressed in human melanoma cells also reported primarily intracellular localization, and Total Internal Reflection Fluorescence (TIRF) analysis showed that TRPM1 localized to small and highly dynamic vesicular structures [9]. No significant differences were found in the localization, dynamics and expression level of the different isoforms studied (TRPM1, 70+TRPM1, 92+TRPM1 and 109+TRPM1) [9]. The primarily intracellular localization of TRPM1 is not uncommon within the TRP family and suggests that TRPM1 might function as an intracellular ion channel, similar to TRPML [22]. Alternatively, the amount of TRPM1 at plasma membrane could be regulated by various stimuli [23, 24]. The identity of the TRPM1-containing vesicles, and whether TRPM1 functions intracellularly, at the plasma membrane, or both, still needs be resolved, likely by a combination of electrophysiology, calcium imaging, microscopy, and biochemistry.

7.6 TRPM1 Function

7.6.1 TRPM1 Function in Skin

7.6.1.1 TRPM1 in Melanoma

Since its initial discovery, TRPM1 has been proposed to have a role in melanoma metastasis. Indeed, expression levels of TRPM1 mRNA inversely correlate with tumor aggressiveness in murine and human melanoma, suggesting that TRPM1 may function as a metastatic suppressor [1, 3, 25]. Despite intensive research, the precise contributions of TRPM1 to melanoma progression remain unclear.

Melanoma is the most severe human skin cancer, and its incidence has been steadily rising in recent decades [11]. Loss of TRPM1 is correlated with increased aggressiveness in human primary cutaneous and nodular melanomas [1, 3, 26]. TRPM1 expression levels have also been linked to tumor thickness [1, 4], and to disease-free survival in patients with localized malignant melanoma [25], emphasizing the potential utility of TRPM1 as a prognostic marker for melanoma [15]. The clinical relevance of TRPM1 prompted investigations into the mechanisms underlying its putative tumor suppressor function, which could be achieved by inhibiting cell proliferation or by promoting differentiation. Two lines of evidence suggest that TRPM1 is associated with the latter: first, *TRPM1* is transcriptionally regulated by MITF, the master transcription factor necessary for melanocyte differentiation [14, 15] (discussed above), and second, its expression is upregulated by treatment with the differentiation-inducing agent hexamethlene bisacteamide (HMBA) [7]. Evidence collected to date seems to suggest that TRPM1 downregulation in aggressive melanoma is more likely to be an effect rather than a cause of dedifferentiation: Melanoma metastasis is in most cases associated with downregulation of MITF, which results in loss of TRPM1. Nevertheless, the question of whether or not TRPM1 functions as a tumor suppressor awaits conclusive resolution.

7.6.1.2 TRPM1 in Melanocyte Pigmentation

The correlation between TRPM1 expression and melanomagenesis prompts the question: What is the role of TRPM1 in normal melanocytes? Melanocytes are present in human (but not mouse) epidermis and their primary function is to generate the pigment melanin within specialized organelles called melanosomes. Mature melanosomes are transferred to adjacent keratinocytes via a poorly defined mechanism [27], to protect DNA from UV-induced damage [28]. Melanin production in melanocytes is initiated by αMSH-dependent activation of the melanocortin-1 receptor (MC1R), a $G_{\alpha s}$-protein coupled receptor [29] that modulates expression of MITF to induce synthesis of proteins required for the production of melanin [30, 31].

A link between TRPM1 and pigmentation was first made in the Appaloosa horse: TRPM1 is markedly downregulated in the skin of horses homozygous for the spotted Appaloosa phenotype [16]. Other studies have correlated TRPM1 expression and melanin content in human melanocytes [7, 9]; TRPM1-targeted siRNA significantly reduces melanin content in these cells [9]. Accordingly, primary melanocytes from dark-skinned individuals, which have higher melanin content compared to melanocytes from light-skinned individuals, also have significantly increased levels of TRPM1 [9]. Additionally, TRPM1 expression appears to influence intracellular Ca^{2+} levels and might be regulated by UVB radiation in a p53-dependent manner [32]. Further research is needed to determine the mechanism by which TRPM1 functions in melanocytes and skin. These studies cannot take advantage of the TRPM1 –/– mouse model, since mice lack epidermal melanocytes and consequently, do not exhibit a UV-induced pigmentation response. However, further studies of human patients with TRPM1 mutations (discussed below) could identify associated skin phenotypes. Mutations resulting in TRPM1 downregulation might not result in an obvious skin phenotype in humans, but could alter the time course or degree of UV-induced pigmentation, which might lead to increased DNA damage and susceptibility to skin cancer. A humanized skin mouse model containing epidermal melanocytes has been developed [33] and exhibits an inducible skin pigmentation response [29]. Studies utilizing TRPM1–/– mice with humanized skin will provide insight into the mechanism that mediates TRPM1 function in skin pigmentation and its relationship to skin cancer.

7.6.2 TRPM1 Function in the Eye

In the Appaloosa horse, downregulation of *TRPM1* expression has been associated with depigmentation and congenital stationary night blindness (CSNB) – a visual defect that results from disrupted bipolar transmission [16]. The latter finding led to the hypothesis that TRPM1 might function in ON bipolar cells [16]. Using TRMP1–/– mice, several recent studies have identified an apparent role for TRPM1 in visual transduction [5, 20, 34]. Moreover, mutations in *TRPM1* have been linked to CSNB in humans [10, 35–37].

Vision in low light (scotopic vision) requires signaling from highly sensitive rod photoreceptors to adjacent bipolar cells in the retina. This pathway – termed the ON pathway – forms between presynaptic photoreceptor terminals and postsynaptic bipolar cells. In the dark, photoreceptors release glutamate to maintain a hyperpolarized membrane potential in adjacent ON bipolar cells. Light-induced activation of photoreceptors reduces glutamate release, which, in turn, causes ON bipolar cells to depolarize [38]. The ON bipolar cell response is mediated by a metabotropic glutamate receptor (mGluR6) coupled to $G_{\alpha o}$ [39–41]; this pathway negatively regulates a depolarizing cation channel [42]. Recent work from several groups suggests that TRPM1 is critical for the cationic current that mediates ON bipolar cell depolarization [5, 20, 34].

Electroretinograms (ERGs) of TRPM1–/– mice show abnormal bipolar cell activity and resemble the hallmark ERGs associated with CSNB: they have a normal a-wave resulting from photoreceptor activation, but lack the b-wave generated by bipolar cell depolarization [5, 20, 34]. These findings suggest that TRPM1 is important for synaptic function of ON bipolar cells and that a lack of TRPM1 is sufficient to cause CSNB. As discussed below, electrophysiological studies of native TRPM1 in bipolar cells and heterologously expressed TRPM1 support the hypothesis that mGluR6 and $G_{\alpha o}$ negatively regulate TRPM1-dependent activity. Together, these studies show that TRPM1 is necessary for the depolarizing light response of ON bipolar cells, and that TRPM1 mediates bipolar depolarization via a mechanism that involves the mGluR6/$G_{\alpha o}$ signaling pathway. While many details have yet to emerge, these studies have identified a key TRPM1-dependent step in visual transduction.

In further support of a visual role for TRPM1 is the finding that *TRPM1* mutations correlate with CSNB in humans. Mutations in two genes expressed in ON bipolar cells account for ~40% of CSNB cases: *NYX*, which encodes the membrane-tethered protein nyctalopin [43] and *GRM6*, which encodes mGluR6 [44, 45]. Four recent studies indicate that many of the remaining CSNB cases are caused by mutations in *TRPM1* [10, 35–37]. *TRPM1* genetic aberrations identified in CSNB patients of European [35, 36], South Asian [10] and Japanese [37] descent result in prematurely terminated proteins or mutations of highly conserved residues, some of which may lead to mislocalization of TRPM1 [37]. Prompted by the reported expression of TRPM1 in melanocytes and its potential role in pigmentation, Li et al. investigated whether TRPM1 mutations also result in an obvious skin phenotype in CSNB patients [10]. While all suffered from dry or scaly skin, patients appeared to have normal pigment distribution and tanning responses [10]. Broader studies of ethnic groups with darker and lighter skin tones will elucidate whether there is a correlation between CSNB and pigmentation phenotypes. Notably, a few cases of CSNB cannot be attributed to mutations in *TRPM1*, *NYX*, or *GRM6* [10], suggesting that mutations in other genes may cause the disorder. Identifying these genes will likely yield further insights into the mGluR6/TRPM1 signaling pathway

In addition to skin and eye, TRPM1 might also function in brain. TRPM1 transcripts have been detected in brain [17], but precisely where it is expressed and what role it plays is currently unknown. Based on the expression of TRPM1 in melanocytes and its role in pigmentation, it would be interesting to determine if (neuro)melanin-containing dopaminergic neurons of the substantia nigra express TRPM1. The correlation between neuromelanin content and the vulnerability of dopaminergic neurons to cell death in Parkinson's disease [46] emphasizes the need for further investigations of the localization and function of TRPM1 in brain.

7.6.3 TRPM1 is Necessary for Constitutive Currents in Melanocytes and Bipolar Cells

TRPM1 has the molecular architecture of an ion channel, but repeated attempts to measure a TRPM1-mediated ionic current have been unsuccessful. Detection of a TRPM1 current, endogenous or expressed, is further complicated by a constitutive TRPM7 current present in most cell types [6]. Recently though, a number of studies have reported TRPM1-dependent currents in human melanocytes and mouse melanoma lineage [9], in retinal bipolar cells [5, 20, 34], as well as in heterologous systems [5, 9].

In human melanocytes, an endogenous outwardly-rectifying, non-selective ionic current is abrogated by expression of TRPM1-targeted small interfering (si) RNA [9]. Two recently identified full-length TRPM1 splice variants (92+TRPM1 and 109+TRPM1) produce a similar current when expressed in human SK-Mel 19 melanoma cells (but not HEK cells or other melanoma cell lines). The amplitude of overexpressed 92+TRPM1- or 109+TRPM1-dependent currents is smaller than endogenous currents in primary melanocytes; this difference might be due to less TRPM1 at the plasma membrane, particularly if its localization depends on protein complexes absent or altered in cell lines.

The association between the spotted Appaloosa horse phenotype, reduced TRPM1, and CSNB prompted investigations of TRPM1 function in bipolar cells. Murine bipolar cells exhibit a non-selective current that is absent in TRPM1–/– mice [5, 20, 34]. The current is activated by addition of metabotropic glutamate receptor type 6 (mGluR6) antagonist [20, 34] or in response to light [5], suggesting that TRPM1 is negatively regulated by mGluR6 activity. Retinal bipolar TRPM1-dependent currents were on average smaller in amplitude than those recorded from melanocytes or melanoma cells, which might be due to different TRPM1 expression levels or regulatory mechanisms in melanocytes versus bipolar cells.

Like the TRPM1-dependent current in melanocytes, the endogenous bipolar cell current is outwardly-rectifying and non-selective. In addition to activation by mGluR6 antagonists or light, the current is also activated by two TRPV1 agonists, capsaicin and anandamide, suggesting that TRPV1 might contribute to the current [34]. However, a capsaicin-activated current persists in bipolar cells of TRPV1–/– mice, raising the possibility that TRPM1 itself is a capsaicin-sensitive ion channel [34]. The molecular identity of the channel mediating the capsaicin-activated current in TRPV1–/– mice and its relationship to TRPM1 remain to be clarified.

The regulation of TRPM1 by mGluR6 was also investigated in a heterologous system. Koike et al. recorded a non-selective current from CHO cells expressing the full-length variant of mouse TRPM1 [5]. The current densities under divalent-free conditions were small (<5pA/pF) and, unlike other reports [9, 34] the resulting IV curve lacked rectification. When TRPM1 and mGluR6 were co-expressed in CHO cells, addition of glutamate reduced the TRPM1-dependent current only in the presence of expressed $G_{\alpha o}$-protein, suggesting that mGluR6 regulates TRPM1 via $G_{\alpha o}$. Single channel recordings in CHO cells revealed a conductance of ~76pS and an open probability of 0.46. Together, these findings support the hypothesis that, in bipolar cells, TRPM1-dependent current is regulated by the mGluR6/$G_{\alpha o}$ signaling pathway. Mutational and electrophysiological analysis of putative TRPM1 pore residues will resolve whether or not TRPM1 itself constitutes a pore-forming protein.

7.7 Conclusions

Despite being the founding member of the TRPM subfamily, the physiological functions and biophysical properties of TRPM1 have been elusive. However, recent efforts have shown that TRPM1 is necessary for a non-selective cation current in primary melanocytes and retinal bipolar cells. The correlation between TRPM1 expression and melanin content in melanocytes indicates a role for TRPM1 in pigmentation, and recent evidence suggests that TRPM1 is critical for light-dependent depolarization of retinal bipolar cells. Future work will resolve the mechanisms underlying these two physiological functions and might uncover new functions for TRPM1.

References

1. Duncan LM, Deeds J, Hunter J, Shao J, Holmgren LM, Woolf EA, Tepper RI, Shyjan AW (1998) Down-regulation of the novel gene melastatin correlates with potential for melanoma metastasis. Cancer Res 58:1515–1520
2. Hunter JJ, Shao J, Smutko JS, Dussault BJ, Nagle DL, Woolf EA, Holmgren LM, Moore KJ, Shyjan AW (1998) Chromosomal localization and genomic characterization of the mouse melastatin gene (Mlsn1). Genomics 54:116–123
3. Hammock L, Cohen C, Carlson G, Murray D, Ross JS, Sheehan C, Nazir TM, Carlson JA (2006) Chromogenic in situ hybridization analysis of melastatin mRNA expression in melanomas from American Joint Committee on Cancer stage I and II patients with recurrent melanoma. J Cutan Pathol 33:599–607
4. Deeds J, Cronin F, Duncan LM (2000) Patterns of melastatin mRNA expression in melanocytic tumors. Hum Pathol 31:1346–1356
5. Koike C, Obara T, Uriu Y, Numata T, Sanuki R, Miyata K, Koyasu T, Ueno S, Funabiki K, Tani A, Ueda H, Kondo M, Mori Y, Tachibana M, Furukawa T (2010) TRPM1 is a component of the retinal ON bipolar cell transduction channel in the mGluR6 cascade. Proc Natl Acad Sci USA 107:332–337
6. Ramsey IS, Delling M, Clapham DE (2006) An introduction to TRP channels. Annu Rev Physiol 68:619–647
7. Fang D, Setaluri V (2000) Expression and Up-regulation of alternatively spliced transcripts of melastatin, a melanoma metastasis-related gene, in human melanoma cells. Biochem Biophys Res Commun 279:53–61
8. Xu XZ, Moebius F, Gill DL, Montell C (2001) Regulation of melastatin, a TRP-related protein, through interaction with a cytoplasmic isoform. Proc Natl Acad Sci USA 98: 10692–10697
9. Oancea E, Vriens J, Brauchi S, Jun J, Splawski I, Clapham DE (2009) TRPM1 forms ion channels associated with melanin content in melanocytes. Sci Signal 2:ra21
10. Li Z, Sergouniotis PI, Michaelides M, Mackay DS, Wright GA, Devery S, Moore AT, Holder GE, Robson AG, Webster AR (2009) Recessive mutations of the gene TRPM1 abrogate ON

bipolar cell function and cause complete congenital stationary night blindness in humans. Am J Hum Genet 85:711–719
11. Levy C, Khaled M, Fisher DE (2006) MITF: master regulator of melanocyte development and melanoma oncogene. Trends Mol Med 12:406–414
12. Steingrimsson E, Copeland NG, Jenkins NA (2004) Melanocytes and the microphthalmia transcription factor network. Annu Rev Genet 38:365–411
13. Price ER, Fisher DE (2001) Sensorineural deafness and pigmentation genes: melanocytes and the Mitf transcriptional network. Neuron 30:15–18
14. Zhiqi S, Soltani MH, Bhat KM, Sangha N, Fang D, Hunter JJ, Setaluri V (2004) Human melastatin 1 (TRPM1) is regulated by MITF and produces multiple polypeptide isoforms in melanocytes and melanoma. Melanoma Res 14:509–516
15. Miller AJ, Du J, Rowan S, Hershey CL, Widlund HR, Fisher DE (2004) Transcriptional regulation of the melanoma prognostic marker melastatin (TRPM1) by MITF in melanocytes and melanoma. Cancer Res 64:509–516
16. Bellone RR, Brooks SA, Sandmeyer L, Murphy BA, Forsyth G, Archer S, Bailey E, Grahn B (2008) Differential gene expression of TRPM1, the potential cause of congenital stationary night blindness and coat spotting patterns (LP) in the Appaloosa horse (Equus caballus). Genetics 179:1861–1870
17. Fonfria E, Murdock PR, Cusdin FS, Benham CD, Kelsell RE, McNulty S (2006) Tissue distribution profiles of the human TRPM cation channel family. J Recept Signal Transduct Res 26:159–178
18. Kunert-Keil C, Bisping F, Kruger J, Brinkmeier H (2006) Tissue-specific expression of TRP channel genes in the mouse and its variation in three different mouse strains. BMC Genomics 7:159
19. Devi S, Kedlaya R, Maddodi N, Bhat KM, Weber CS, Valdivia H, Setaluri V (2009) Calcium homeostasis in human melanocytes: role of transient receptor potential melastatin 1 (TRPM1) and its regulation by ultraviolet light. Am J Physiol Cell Physiol 297:C679–C687
20. Morgans CW, Zhang J, Jeffrey BG, Nelson SM, Burke NS, Duvoisin RM, Brown RL (2009) TRPM1 is required for the depolarizing light response in retinal ON-bipolar cells. Proc Natl Acad Sci USA 106:19174–19178
21. Kim DS, Ross SE, Trimarchi JM, Aach J, Greenberg ME, Cepko CL (2008) Identification of molecular markers of bipolar cells in the murine retina. J Comp Neurol 507:1795–1810
22. Dong XP, Cheng X, Mills E, Delling M, Wang F, Kurz T, Xu H (2008) The type IV mucolipidosis-associated protein TRPML1 is an endolysosomal iron release channel. Nature 455:992–996
23. Bezzerides VJ, Ramsey IS, Kotecha S, Greka A, Clapham DE (2004) Rapid vesicular translocation and insertion of TRP channels. Nat Cell Biol 6:709–720
24. Oancea E, Wolfe JT, Clapham DE (2006) Functional TRPM7 channels accumulate at the plasma membrane in response to fluid flow. Circ Res 98:245–253
25. Duncan LM, Deeds J, Cronin FE, Donovan M, Sober AJ, Kauffman M, McCarthy JJ (2001) Melastatin expression and prognosis in cutaneous malignant melanoma. J Clin Oncol 19: 568–576
26. Erickson LA, Letts GA, Shah SM, Shackelton JB, Duncan LM (2009) TRPM1 (Melastatin-1/MLSN1) mRNA expression in Spitz nevi and nodular melanomas. Mod Pathol 22:969–976
27. Boissy RE (2003) Melanosome transfer to and translocation in the keratinocyte. Exp Dermatol 12(Suppl 2):5–12
28. Lin JY, Fisher DE (2007) Melanocyte biology and skin pigmentation. Nature 445:843–850
29. D'Orazio JA, Nobuhisa T, Cui R, Arya M, Spry M, Wakamatsu K, Igras V, Kunisada T, Granter SR, Nishimura EK, Ito S, Fisher DE (2006) Topical drug rescue strategy and skin protection based on the role of Mc1r in UV-induced tanning. Nature 443:340–344
30. Bertolotto C, Abbe P, Hemesath TJ, Bille K, Fisher DE, Ortonne JP, Ballotti R (1998) Microphthalmia gene product as a signal transducer in cAMP-induced differentiation of melanocytes. J Cell Biol 142:827–835

31. Price ER, Horstmann MA, Wells AG, Weilbaecher KN, Takemoto CM, Landis MW, Fisher DE (1998) alpha-Melanocyte-stimulating hormone signaling regulates expression of microphthalmia, a gene deficient in Waardenburg syndrome. J Biol Chem 273:33042–33047
32. Maddodi N, Setaluri V (2008) Role of UV in cutaneous melanoma. Photochem Photobiol 84:528–536
33. Kunisada T, Lu SZ, Yoshida H, Nishikawa S, Mizoguchi M, Hayashi S, Tyrrell L, Williams DA, Wang X, Longley BJ (1998) Murine cutaneous mastocytosis and epidermal melanocytosis induced by keratinocyte expression of transgenic stem cell factor. J Exp Med 187:1565–1573
34. Shen Y, Heimel JA, Kamermans M, Peachey NS, Gregg RG, Nawy S (2009) A transient receptor potential-like channel mediates synaptic transmission in rod bipolar cells. J Neurosci 29:6088–6093
35. Audo I, Kohl S, Leroy BP, Munier FL, Guillonneau X, Mohand-Said S, Bujakowska K, Nandrot EF, Lorenz B, Preising M, Kellner U, Renner AB, Bernd A, Antonio A, Moskova-Doumanova V, Lancelot ME, Poloschek CM, Drumare I, Defoort-Dhellemmes S, Wissinger B, Leveillard T, Hamel CP, Schorderet DF, De Baere E, Berger W, Jacobson SG, Zrenner E, Sahel JA, Bhattacharya SS, Zeitz C (2009) TRPM1 is mutated in patients with autosomal-recessive complete congenital stationary night blindness. Am J Hum Genet 85:720–729
36. van Genderen MM, Bijveld MM, Claassen YB, Florijn RJ, Pearring JN, Meire FM, McCall MA, Riemslag FC, Gregg RG, Bergen AA, Kamermans M (2009) Mutations in TRPM1 are a common cause of complete congenital stationary night blindness. Am J Hum Genet 85:730–736
37. Nakamura M, Sanuki R, Yasuma TR, Onishi A, Nishiguchi KM, Koike C, Kadowaki M, Kondo M, Miyake Y, Furukawa T (2010) TRPM1 mutations are associated with the complete form of congenital stationary night blindness. Mol Vis 16:425–437
38. Schiller PH, Sandell JH, Maunsell JH (1986) Functions of the ON and OFF channels of the visual system. Nature 322:824–825
39. Vardi N, Matesic DF, Manning DR, Liebman PA, Sterling P (1993) Identification of a G-protein in depolarizing rod bipolar cells. Vis Neurosci 10:473–478
40. Nawy S (1999) The metabotropic receptor mGluR6 may signal through G(o), but not phosphodiesterase, in retinal bipolar cells. J Neurosci 19:2938–2944
41. Nakajima Y, Iwakabe H, Akazawa C, Nawa H, Shigemoto R, Mizuno N, Nakanishi S (1993) Molecular characterization of a novel retinal metabotropic glutamate receptor mGluR6 with a high agonist selectivity for L-2-amino-4-phosphonobutyrate. J Biol Chem 268:11868–11873
42. Shiells RA, Falk G, Naghshineh S (1981) Action of glutamate and aspartate analogues on rod horizontal and bipolar cells. Nature 294:592–594
43. Pesch K, Zeitz C, Fries JE, Munscher S, Pusch CM, Kohler K, Berger W, Wissinger B (2003) Isolation of the mouse nyctalopin gene nyx and expression studies in mouse and rat retina. Invest Ophthalmol Vis Sci 44:2260–2266
44. Dryja TP, McGee TL, Berson EL, Fishman GA, Sandberg MA, Alexander KR, Derlacki DJ, Rajagopalan AS (2005) Night blindness and abnormal cone electroretinogram ON responses in patients with mutations in the GRM6 gene encoding mGluR6. Proc Natl Acad Sci USA 102:4884–4889
45. Zeitz C, van Genderen M, Neidhardt J, Luhmann UF, Hoeben F, Forster U, Wycisk K, Matyas G, Hoyng CB, Riemslag F, Meire F, Cremers FP, Berger W (2005) Mutations in GRM6 cause autosomal recessive congenital stationary night blindness with a distinctive scotopic 15-Hz flicker electroretinogram. Invest Ophthalmol Vis Sci 46:4328–4335
46. Hirsch E, Graybiel AM, Agid YA (1988) Melanized dopaminergic neurons are differentially susceptible to degeneration in Parkinson's disease. Nature 334:345–348

Chapter 8
The Non-selective Monovalent Cationic Channels TRPM4 and TRPM5

Romain Guinamard, Laurent Sallé, and Christophe Simard

Abstract Transient Receptor Potential (TRP) proteins are non-selective cationic channels with a consistent Ca^{2+}-permeability, except for TRPM4 and TRPM5 that are not permeable to this ion. However, Ca^{2+} is a major regulator of their activity since both channels are activated by a rise in internal Ca^{2+}. Thus TRPM4 and TRPM5 are responsible for most of the Ca^{2+}-activated non-selective cationic currents (NSC_{Ca}) recorded in a large variety of tissues. Their activation induces cell-membrane depolarization that modifies the driving force for ions as well as activity of voltage gated channels and thereby strongly impacts cell physiology. In the last few years, the ubiquitously expressed TRPM4 channel has been implicated in insulin secretion, the immune response, constriction of cerebral arteries, the activity of inspiratory neurons and cardiac dysfunction. Conversely, TRPM5 whose expression is more restricted, has until now been mainly implicated in taste transduction.

Abbreviations

AMPA-R	α-amino-3-hydroxyl-5-methyl-4-isoxalone-propionate receptor
calcin.	calcineurin
DAG	diacylglycerol
ΔV	membrane potential
$\Delta \Psi$	driving force
ER	endoplasmic reticulum
G	G-protein
GPCR	G-protein coupled receptor
IP_3	inositol 1,4,5-triphosphate
IP_3R	IP_3 receptor
K_{ATP}	ATP-dependant potassium channel;

R. Guinamard (✉)
Groupe Cœur et Ischémie, EA 3212, Université de Caen, Sciences D, F-14032, Caen Cedex, France
e-mail: romain.guinamard@unicaen.fr

mGLU-R	metabotropic-glutamate receptor
PIP_2	phosphatidylinositol 4,5-biphosphate
PKC	protein kinase C
PLC	phospholipase C
RyR	ryanodine receptor
SUR	sulphonylurea receptor
TCR	T-cell receptor
TRPC6	transient receptor potential canonical 6
TRPM4	transient receptor potential melastatin 4
TRPM5	transient receptor potential melastatin 5
Tyr-K	tyrosine kinase
VDCa	voltage-dependant calcium channel
VDNa channel	voltage-dependant sodium channel
VR	vasopressin receptor

8.1 Introduction

In the community of ionic channels, non-selective cationic currents have remained orphaned while other currents have found families of proteins where they belong. However, the salvation of these currents occurred with the cloning of the Transient Receptor Potential family (TRP) of proteins at the end of the 1990s, which revealed the molecular identity for a variety of these currents. TRPs are mainly Ca^{2+}-permeable channels responsible for stretch-activated, store-operated, receptor-operated and ligand-activated Ca^{2+}-entry but also include Mg^{2+}-permeable channels. These currents are reviewed extensively in this book. The cloning of TRP channels also unmasked Ca^{2+} non-permeable channels, the so called Ca^{2+}-activated non-selective cationic channels (NSC_{Ca}) because of their sensitivity to internal Ca^{2+}. These currents are mainly supported by two members of the Transient Receptor Potential Melastatin (TRPM) family, TRPM4 and TRPM5.

NSC_{Ca} currents were described during the 20 years preceding the cloning of TRPM4 and TRPM5 in 2002 and 2000 respectively [1, 2]. Their distribution throughout a large range of tissues has been reviewed previously [3–5]. Briefly, NSC_{Ca} currents are present in (i) epithelial cells from capillary endothelium [6, 7], renal tubules [8, 9] and intestine [10]; (ii) pancreatic secretory cells [11–13]; (iii) sensory organs such as cochlear outer hair cells [14] and vomeronasal sensory neurons [15]; (iv) helix neurons [16]; and (v) cardiac cells including nodal cells [17] and cardiomyocytes [18, 19]. In addition to their calcium sensitivity and non-selective cationic permeability sequence, these channels share a single channel conductance of around 25 pS [3].

Similar to this large NSC_{Ca} distribution, the range of tissues expressing TRPM4 mRNA is very large. In humans, it is mainly expressed in heart, pancreas, placenta and prostate but is also detected in kidney, skeletal muscle, liver, intestine, thymus, spleen and hematopoietic cell lines [1, 20] (Table 8.1). In contrast, the expression of TRPM5 is mainly restricted to taste receptors with additional weak expression in the

Table 8.1 Summary of main TRPM4 and TRPM5 properties (see text for details and references)

		TRPM4	TRPM5
Structure		6 transmembrane domains	
Unitary conductance		18-25 pS	
Voltage dependance		activated by depolarization	
Permeability		$Na^+ \approx K^+ > Cs^+ > Li^+$	$Na^+ \approx K^+ \approx Cs^+ > Li^+$
		non permeable for divalent cations (Ca^{2+}, Mg^{2+})	
Ca^{2+} sensitivity (EC_{50} for activation)		Whole cell : ≈ 20 µM Single channel : ≈ 170 µM	Whole cell : ≈ 0.7 µM Single channel : ≈ 28 µM
Tissular Distribution*		pancreas, heart, prostate, renal tubule, placenta +++	taste receptors +++
		spleen, kidney, liver, skeletal muscle ++	brainstem +
		immune cells, neurons intestine, endothelium +	pancreas +
Physiological modulators	activated by	*PKC, PIP₂, heat*	*PIP₂, heat*
	Inhibited by	spermine, intracellular nucleotides (ADP > ATP > AMP >>Adenosine)	Spermine, extracellular acidification
Pharmacological modulators	activated by	decavanadate, BTP2	
	Inhibited by	flufenamic acid, quinine, quinidine, 9-phenanthrol, MPB-104	flufenamic acid, quinine, quinidine

*mRNA quantified by RT-PCR

intestinal tract, pancreas and brainstem [21]. Consequently, as will be reviewed in the following sections, TRPM4 appears to be the molecular support for most NSC_{Ca} currents and thus is implicated in a large variety of physiological processes. The processes match its tissue distribution, some examples being modulation of insulin secretion, the immune response, constriction of cerebral arteries, the firing rate of inspiratory neurons and cardiac activity. In contrast, the implication of TRPM5 is well described only in taste transduction, but additional physiological roles for this channel are emerging.

In this review, we will first describe the biophysical and regulatory properties of TRPM4 and TRPM5 channels, highlighting their differences, as summarized in Table 8.1. Then, we will describe the recent knowledge about their physiological impact.

8.2 Molecular Structure

Because the molecular properties of TRP channels have been recently reviewed [22–24], we will only focus on specificities of TRPM4 and TRPM5 channels. These channels belong to the TRPM (M for melastatin) subfamily that is composed of

8 members. While TRPM1 is not clearly identified as an ionic channel, all other members are clearly non-selective cationic channels with identified physiological functions such as the redox-sensor TRPM2, the calcium-dependent voltage regulator TRPM4, the taste sensor TRPM5, the cold sensor TRPM8 and the Mg^{2+} regulated and permeable channels TRPM6 and TRPM7 [25]. All these TRPM channels are present in humans and seem to be expressed in the plasma membrane rather than in intracellular organelles.

Several splice variants were identified for TRPM4, called TRPM4a, b and c in humans [1, 20, 26]. It is now considered that TRPM4b is the significant variant while the relevance of other variants is still unknown. In the following sections, we will refer to TRPM4b as TRPM4 because the great majority of studies were done on this variant. TRPM4 is a 1,214 amino-acid (aa) protein encoded by a gene located in chromosome 19 in humans and a 1,213 aa protein encoded by chromosome 7 in mice [1, 27]. TRPM5 is a 1,165 aa protein encoded by a gene located in chromosome 11 in humans and a 1,158 aa protein encoded by a gene in chromosome 7 in mice [2, 28].

TRPM4 and TRPM5 share 40% homology in their amino-acid sequences and less homology with all other TRPs. This confers close properties between both channels and specificities within the TRP family. Like all TRP channels, their hydropathy profile predicts six transmembrane-spanning segments (TM) (Fig. 8.1) and they are suspected to be associated as tetramers to form functional channels. Therefore, they belong to the large six TM channel family. Within the TRPM subfamily of TRP channels, members show a similar length in the N terminus but are highly variable in the C terminus. While TRPM6 and TRPM7 have the longest chain with a kinase domain, TRPM4, TRPM5 and TRPM8 exhibit a very short C terminus. However, all

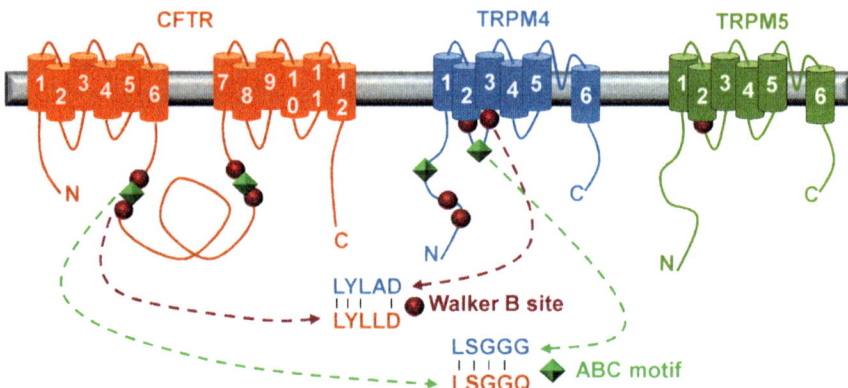

Fig. 8.1 Molecular structure of TRPM4 and TRPM5. Predicted topology of TRPM4, TRPM5 and CFTR channels. TRPM4 and TRPM5 have intracellular N and C terminal regions and six transmembrane (TM) segments [1–6] with the pore region between TM5 and TM6. While TRPM4 has two ABC motifs (*diamonds*) and four Walker B ATP-binding sites (*circles*), TRPM5 possesses only one Walker-B site at the end of TM2. Sequences of these sites are compared with their relatives in the CFTR sequence

members include a coiled-coil domain that is suspected to allow protein assembly by linking subunits, as shown for TRPM7 [29]. Although some TRPs were shown to be able to coassemblate to form heterotetramers such as TRPM6 with TRPM7 [30], no data are available for TRPM4 and TRPM5. According to their close amino-acid sequence, heterotetramer assembly should be feasible, but because they show a distinct expression profile in tissues, as detailed below, native heterotetramers of TRPM4 with TRPM5 are unlikely to occur.

A number of specific domains were reported in the TRPM4 sequence, while less data are available for TRPM5. Within its N and C terminus, TRPM4 holds several putative phosphorylation sites for PKA and PKC as well as putative calmodulin binding sites. Additionally, it possesses arginine and lysine rich stretches as putative PIP_2 binding sites [4]. In contrast to TRPM5 but similar to ATP-binding cassettes proteins (ABC proteins), TRPM4 possesses two ABC transporter-like motifs (Fig. 8.1). These 6 amino-acid sequences are known to interact with nucleotide binding domains (NBD) in ABC proteins. TRPM4 holds four NBDs that belong to the Walker B form of domains, in the vicinity of ABC transporter like-motifs (Fig. 8.1). Therefore, interaction between these sites is possible. Interestingly, TRPM5 does not contain an ABC-transporter like motif and possesses only one Walker B site that is, moreover, flanked in the second transmembrane segment and thus may not be accessible to ATP (Fig. 8.1) [31]. These structural differences might be fundamental in the function of each channel. Finally, TRPM4 and TRPM5 fold in the classical shape of 6 TM channels with the TM5 and TM6 linker forming the pore loop within which the selectivity filter is located.

In the following sections, we present evidence for the implication of several of these structures in channel biophysical properties and regulation.

8.3 Biophysical and Regulatory Properties

It must be noted that the description of the biophysical and regulatory properties of TRPM4 and TRPM5 have been mainly performed using heterologous expression in cell lines, after transient or permanent and inducible transfection. Chinese hamster ovary (CHO) and human embryonic kidney (HEK-293) cells are used by most groups investigating these channels. However, slight endogenous TRPM4-like currents have been recorded and endogenous TRPM4 mRNA was detected in these cells [1, 32]. While this point has to be kept in mind, it does not compromise studies with heterologous expression because of the enormous currents observed after cell transfection. Thus, heterologous expression relegates endogenous TRPM4 to negligible background currents [1]. Endogenous expression of TRPM4 is not surprising in a kidney cell line as TRPM4-like currents have been reported all along the renal tubule [8].

A great majority of reported studies focus on TRPM4 and so its structure is well known. Less is known about molecular determinants of TRPM5, but because its sequence, biophysical properties and regulation are close to those of TRPM4, properties described for TRPM4 might match TRPM5 in many cases.

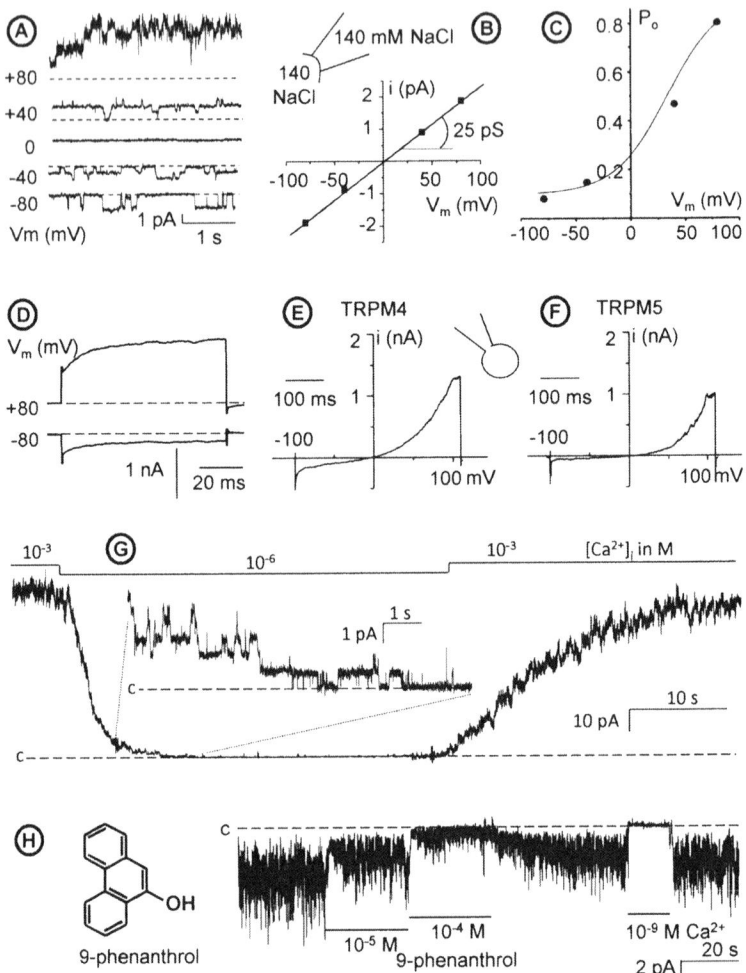

Fig. 8.2 Biophysical properties of TRPM4 and TRPM5 expressed in HEK-293 cells a, b, c: Single channel properties of TRPM4. (a) Single channel traces recorded at various voltages from an inside-out patch from a TRPM4-transfected HEK-293 cell. The pipette and bath contained in mM: 140 NaCl; 4.8 KCl; 1.2 $MgCl_2$; 1 $CaCl_2$; 10 glucose; 10 HEPES. Vm corresponds to the membrane potential. *Dashed lines* indicate the current level of closed channels. Four similar channels are present in the patch. (b) Corresponding current-voltage relationship. Points are fitted by linear regression, providing the slope of the curve corresponding to a single channel conductance of around 25 pS. (c) Voltage dependence of open probability (P_o) obtained from traces in A indicating a higher activity in the positive voltages. TRPM5 exhibits similar properties (not shown). d, e, f: Macroscopic properties of TRPM4 and TRPM5. (d) TRPM4 current tracings in the whole-cell configuration under a voltage pulse protocol from 0 to +80 or 0 to –80 mV (pipette in mM: 156 CsCl, 1 $MgCl_2$, 0.001 $CaCl_2$, 10 HEPES; bath in mM: 156 NaCl, 5 $CaCl_2$, 10 glucose, 10 HEPES). (e) TRPM4 current tracing recorded in the whole-cell condition using a ramp protocol from –100 to +100 mV (holding potential = 0 mV). Combination of the linear single channel conductance and the voltage sensitivity of the channel results in an outwardly rectifying current. (f) Same conditions as in E for a TRPM5-transfected HEK-293 cell. (g) TRPM4 calcium-sensitivity.

When expressed in HEK-293 cells, TRPM4 and TRPM5 exhibit a linear unitary current/voltage relationship characterizing a 18–25 pS single channel conductance under symmetrical ionic conditions [1, 20, 33–35]. Typical current traces for TRPM4 and TRPM5 are presented in Fig. 8.2.

8.3.1 Ionic Selectivity

TRPM4 and TRPM5 are exceptions in the TRP family because, unlike all other members, they do not conduct calcium. The permeability sequence was determined by ion substitution and showed that the TRPM4 channel poorly differentiates K^+ over Na^+ (permeability sequence: $Na^+ \sim K^+ > Cs^+ > Li^+$). Neither $NMDG^+$ nor Ca^{2+} significantly permeate through TRPM4 channels [1, 20].

Molecular determinants for this specific selectivity are partly identified. The linker between TM5 and TM6 (pore loop) is highly but not fully conserved within TRPM channels. Similar to a large variety of ionic channels, this loop contains crucial residues to control ion permeation, including several negatively charged residues. Interestingly, the neutral glutamine (Q) in position 977 in TRPM4, also present in TRPM5, is substituted by a negatively charged glutamate (E) in TRPM6 and TRPM7 and these channels are permeable to the divalent cations Ca^{2+} and Mg^{2+}. The substitution Q977E in TRPM4 confers a measurable Ca^{2+} permeability to the channel and alters the monovalent permeability sequence [36]. Reciprocally, performing the corresponding substitution E1047Q in TRPM7 results in a permeability sequence similar to TRPM4 [37]. However, TRPM2 and TRPM8, which are Ca^{2+}-permeable, also contain the Q residue, thus indicating that it is not the only residue involved in ion segregation. In that sense, swapping a 6 amino-acid sequence in TRPM4 (^{981}EDMDVA^{986}) to the corresponding sequence of TRPV6 (TIIDGP), a highly Ca^{2+}-permeable channel, confers to TRPM4 a weak but significant Ca^{2+}-permeability [36].

TRPM5 does not differentiate Na^+, K^+, and Cs^+ and displays no permeability to Ca^{2+}, Mg^{2+} and $NMDG^+$ [33–35]. This is similar to TRPM4 despite the fact that permeation of Cs^+ through TRPM4 is slightly lower than Na^+. To our knowledge, no experiment has been conducted to identify the molecular determinants of ion permeation through TRPM5.

Fig. 8.2 Single channel recording of a TRPM4-transfected HEK-293 cell in the same ionic conditions as in A with various $[Ca^{2+}]_i$. Around 70 channels are present in the patch. Changes in single channel levels can be seen in the magnification inset. In this example, all channels are closed at $[Ca^{2+}]_i = 10^{-6}$ M (holding potential = +40 mV). TRPM5 exhibits similar properties (not shown). (**h**) TRPM4 inhibition by 9-phenanthrol. *Left*: chemical structure of 9-phenanthrol. *Right*: single channel recording of a TRPM4-transfected HEK-293 cell in the same ionic conditions as in A. 9-phenanthrol at 10^{-5} or 10^{-4} M was applied during the recording. Reducing $[Ca^{2+}]_i$ from 10^{-3} to 10^{-9} M allowed the current level when all channels are closed to be revealed (c) (holding potential = −80 mV)

8.3.2 Calcium Sensitivity

In addition to their lack of Ca^{2+} permeability, TRPM4 and TRPM5, alongside TRPM2 [38], are peculiar in the TRP family because of their sensitivity to the internal calcium concentration ($[Ca^{2+}]_i$). Both are activated by a rise in $[Ca^{2+}]_i$ (Fig. 8.2). Although all authors have reported the sensitivity to $[Ca^{2+}]_i$, there are discrepancies between studies regarding the concentration for half efficiency activation ($[EC_{50}]$), probably due to differences in experimental procedures [1, 31, 33–35, 39]. Roughly, TRPM4 and TRPM5 are activated with an $[EC_{50}]$ for $[Ca^{2+}]_i$ in the range of a few μM (Table 8.1). A comparative study between both channels was performed by Ullrich et al. [31] and revealed two interesting points. Firstly, TRPM4 is activated with a 5–10 higher $[Ca^{2+}]_i$ than TRPM5 (Table 8.1). Secondly, in the whole-cell configuration, channels are about tenfold more sensitive to Ca^{2+} than in the inside-out configuration. This shift of Ca^{2+}-sensitivity in inside-out recordings might be explained by washing of intracellular regulators in this configuration. This idea is supported by progressive Ca^{2+}-desensitization of the channels following excision [35, 39]. Several experiments were performed to explain this decrease such as: (i) mutation of calmodulin to affect its interaction with Ca^{2+} reduces TRPM4 activation [39] but TRPM5 does not appear to be calmodulin sensitive [34]; (ii) mutations of ATP binding sites in TRPM4 result in higher Ca^{2+}-desensitization while ATP reverses desensitization in WT TRPM4 [39]; (iii) activation of the protein kinase C pathway increases TRPM4 Ca^{2+}-sensitivity [39] and (iv) PIP_2 partly prevents Ca^{2+}-desensitization of both TRPM4 and TRPM5 [35, 40]. Taken together, these data point to the importance of internal components in Ca^{2+}-dependent regulation of the channels, even if the molecular determinants of this sensitivity are not precisely known.

8.3.3 Voltage Sensitivity

TRPM4 and TRPM5 are voltage-dependent channels. However, like other TRP channels, they are not considered as strictly voltage-gated channels even if their voltage sensitivity may have a physiological impact, especially in excitable cells that exhibit large voltage variations. The channel's open probability increases with membrane depolarization (Fig. 8.2). Most channels are closed in negative voltages under resting conditions and opened in positive voltages, but sufficient levels of internal Ca^{2+} are required [1, 20, 33]. Despite the linear current voltage relationship observed in single channel analysis, the voltage dependence results in an outward rectifying macroscopic current (Fig. 8.2).

The molecular determinants of this voltage sensitivity remain unknown but it is tempting to extrapolate the mechanism by comparison with classical voltage-dependent channels. Indeed, in 6 TM channels that exhibit voltage dependence, the voltage sensing occurs by movement of the positively charged TM4 domain. Positively charged amino-acids (two arginines, one histidine) are presents in both TRPM4 and TRPM5 TM4 sequences, albeit in a smaller number compared to strictly voltage-gated channels. This may explain why those channels exhibit a flatter voltage-activation curve compared to classical voltage-dependent channels [20].

Moreover, it was shown that some TRPM4 and TRPM5 modulators act by shifting the voltage-activation curve. For example, decavanadate and PIP_2 shift the curve toward negative voltages, resulting in an increase in TRPM4 current [40, 41]. In a similar manner, heat activates both TRPM4 and TRPM5, not by modifying Ca^{2+}-sensitivity, but by shifting the voltage-dependent activation curve, a phenomenon observed in other thermosensitive TRP channels [42].

8.3.4 ATP Sensitivity

TRPM4 and TRPM5 exhibit a complex regulation by ATP. Both currents were first shown to increase under external ATP stimulation [1, 34]. This was attributed to phospholipase C pathway activation after P2Y receptor stimulation.

On the other hand, internal ATP was shown to inhibit TRPM4 currents with an EC_{50} around 1 μM without an effect on TRPM5 [31, 43]. TRPM4 is also sensitive to other adenine nucleotides with the highest sensitivity to ADP and the sensitivity sequence ADP > ATP > AMP >> adenosine [43]. This block is Mg^{2+} independent. The inhibition by AMP-PNP, a non hydrolysable AMP analog leads to an inhibition similar to AMP, suggesting that the effects of adenine nucleotides do not require their hydrolysis [43].

Finally, paradoxically to this TRPM4 inhibition, internal ATP prevents Ca^{2+}-desensitization of TRPM4, in favor of an increase in current [39]. Mutations of putative ATP binding domains in TRPM4 abolish this effect, indicating that ATP directly binds to the channel to express this activating effect [39].

All together, these data indicate that interaction of ATP with the channels is complex. Nevertheless, the lack of TRPM5 sensitivity to internal ATP is a powerful characteristic to differentiate TRPM4 from TRPM5. This may be used to discriminate native currents carried by one or the other protein.

8.3.5 Thermal, PIP_2 and pH Sensitivity

Within the TRP family, several members were identified early as thermal sensors. Indeed, TRPM8 is implicated in cold sensing [44, 45] while TRPV1 is heat activated [46, 47]. TRPM4 and TRPM5 are also thermosensitive channels, however, they are not considered as real thermal receptors because they do not activate without a high level of internal Ca^{2+} [42]. Heat activates both channels by shifting the voltage dependent activation curve.

Similar to temperature, phosphatidylinositol 4,5-bisphosphate (PIP_2) was shown to regulate a number of TRP channels [48] including TRPM4 and TRPM5 [35, 40]. Both channels rapidly deactivate after membrane excision during patch-clamp experiments, a process attributed to the loss of intracellular regulators. PIP_2 appeared to be one of those. It uncouples channel activity from voltage variations and increases their Ca^{2+}-sensitivity 100-fold, resulting in robust channel activity even in negative voltage ranges [40, 49]. Consequently, PIP_2 breakdown under phospholipase C stimulation would result in reduction of TRPM4 and TRPM5 activity in physiological processes. On the other hand, because TRPM4 activity is enhanced under protein kinase C (PKC) stimulation [39] and because hydrolysis of PIP_2 results in production of inositol 3 phosphate (IP_3) and diacyl glycerol, the substrate required for PKC activation, the impact of PLC stimulation on TRPM4 activity is complicated.

In contrast to TRPM4, TRPM5 is inhibited by extracellular acidification below pH 7 and is completely blocked by pH 5.9 with half maximal inhibition (IC_{50}) reached at pH 6.2 [50]. Mutation of the protonatable E830 in the TM3-TM4 linker to a neutral Q residue (E850Q) reduces TRPM5 pH-sensitivity. Similar results were obtained by performing the mutation H934N in the pore region (TM5-TM6 linker). This indicates that pH sensitivity is achieved by at least these two residues, a concept confirmed by the double mutant that is nearly insensitive to external pH [50]. Interestingly, these residues are not conserved in TRPM4 and may be responsible for the lack of external pH sensitivity of TRPM4. In addition to this inhibitory effect, acid pH also enhances TRPM5 inactivation [50]. Surprisingly, no study on internal pH sensitivity of the cloned channels has been performed. On the other hand, a bell-shaped dependence to internal pH was reported for a native TRPM4-like current in mouse renal tubules [51].

8.4 TRPM4 and TRPM5 Pharmacology

A major pitfall in the determination of the physiological implications of TRPM4 and TRPM5, as for other TRPs channels, is the poor specificity of pharmacological tools available. Thus it is difficult to specifically extract TRPM4 or TRPM5 currents from the jumble of non-selective cationic background currents. In addition, it is not easy to distinguish one from the other in native tissues. However, several molecules are known to interact with these currents.

The non-steroidal anti-inflammatory drug, Flufenamic acid, was shown to inhibit TRPM4 in the micromolar range and TRPM5 with a tenfold lower affinity [31]. It is frequently used to unmask the physiological implications of TRPM4 or TRPM5, but investigators should consider that this molecule also modulates a large variety of ionic channels. It is a common chloride channel blocker with an effect on Ca^{2+}-activated [52], swelling-activated [53] and proton-activated [54] Cl^- channels. Flufenamic acid affects several TRPs currents, inhibiting TRPC3 [55], TRPC5 [56] and TRPM2 [57] but activating TRPC6 [58] and TRPA1 [59].

The polyamine, spermine, also inhibits both TRPM4 and TRPM5 in a similar physiological 10 µM range probably by interacting with aspartate and glutamate residues flanking the pore region and present in the two channels [31, 43]. Once again, this molecule modulates other ionic channels. It potentiates kainate glutamate receptors [60], which are non-selective cationic channels, but blocks the AMPA receptor [61] and has both effects on NMDA receptors [62]. Likewise, spermine blocks the inward rectifier K$^+$ channel [63]. Within the TRP family, spermine was shown to block the magnesium inhibited TRPM7 channel [64].

The bitter compound, quinine, and its stereoisomer, quinidine, inhibit both TRPM4 and TRPM5 in the range of 100 μM [65]. TRPM5 is tenfold more sensitive. Surprisingly, the inhibition is voltage dependent and is more potent at positive potentials in TRPM4, but inversely, at negative potentials in TRPM5. Determinants involved in this discrepancy are not known. Both quinine and quinidine are also able to inhibit a variety of ionic channels including several potassium channels [66]. We found no reported effect of quinine on other TRP channels.

N-[4-3, 5-bis(trifluromethyl)pyrazol-1-yl]-4-methyl-1,2,3-thiadiazole-5-carboxamine (BTP2) is an immunosuppressive pyrazole known to reduce cytokine production and proliferation of T-cells. In these cells it was shown that BTP2 activates TRPM4 in the 10 nM range [67]. BTP2 does not only modulate TRPM4, however its effect on other channels requires 10–100-fold higher concentrations. For example, in the micromolar range, BTP2 inhibits the calcium release activated current I$_{CRAC}$ [68] and, moreover, TRPC3 and TRPC5 channels [69]. To our knowledge, no data are available for the action of BTP2 on TRPM5.

Interestingly, several molecules known to interact with ATP-dependent proteins were shown to modulate TRPM4 but not TRPM5 in accordance with the presence of ATP-binding sites and ATP-sensitivity in TRPM4 but not in TRPM5. Decavanadate increases TRPM4 activity with no effect on TRPM5 in the micromolar range by shifting the voltage-sensitivity curve [41]. Decavanadate, which holds six negative charges, usually acts by electrostatic interactions with positively charged ATP-binding sites and competes with ATP. Surprisingly, decavanadate does not have this behavior with TRPM4 since its effect does not antagonize inhibition by ATP and does not involve interactions with ATP-binding sites [41]. A six amino acid sequence in the C terminus with positively charged residues, which is absent in TRPM5, is implicated in this interaction.

Glibenclamide is a sulphonylurea that exerts an antidiabetic effect by modulating a sulphonylurea receptor on β-pancreatic cells. This receptor belongs to the ATP-binding cassette protein family (ABC transporters). It was shown that several members of this family are affected by glibenclamide such as the Cystic Fibrosis Transmembrane Conductance Regulator (CFTR) [70, 71]. Because TRPM4 holds ABC signature like motifs (Fig. 8.1) [39], we tested glibenclamide on TRPM4 currents and observed their inhibition in the 10 μM range [17]. In addition to CFTR, it has to be noted that glibenclamide inhibits other chloride channels [71–73].

We took advantage of the effect of glibenclamide to investigate the effect of other ABC transporter modulators, focusing on hydroxytricyclic compounds. The benzo[c]quinolizinium agent 5-butyl-7-chloro-6-hydroxybenzo[c]-quinolizinium chloride (MPB-104) inhibits both TRPM4 and CFTR channels [74, 75]. More interestingly, 9-hydroxyphenanthrene (9-phenanthrol) inhibits TRPM4 in the μM range (Fig. 8.2) with no effect on CFTR or TRPM5 [75]. Because no other ionic channels are known to be targeted by this last compound, it might be a powerful molecule to investigate the physiological implication of TRPM4 and to discriminate this channel from TRPM5 in native cells.

8.5 TRPM4 and TRPM5 Physiological Impact

The physiological implications of TRPM4 and TRPM5 were determined by using dominant negative mutants or siRNA and deficient mouse models. The ubiquitously expressed TRPM4 (Table 8.1) is implicated in myogenic constriction of cerebral arteries, participating in the Bayliss effect; it modulates insulin secretion in response to glucose uptake by pancreatic β-cells; it modulates the firing rate in brainstem neurons responsible for breathing rhythm; it moderates the immune response and participates in electrical activity in cardiac conductive tissue. In contrast, TRPM5, whose expression is more restricted, is implicated in fewer phenomena. In particular, it is clearly involved in taste transduction. Schematized patterns of TRPM4 and TRPM5 involvement in these phenomena are presented in Fig. 8.3.

Since a key determinant for the activation of these channels is a rise in $[Ca^{2+}]_i$, they will impact cell functions after either depletion of intracellular calcium stores or onset of calcium entry. A number of G-protein-coupled receptors may induce the activation of the PLC-signaling cascade that, through IP_3 production, leads to calcium release from the endoplasmic reticulum. Calcium entry into the cell may occur through the specific calcium-release-activated current I_{CRAC} or cationic non-specific store operated channels, after store depletion, but also through stretch activated channels and voltage gated calcium channels in excitable cells.

Activation of both TRPM4 and TRPM5 leads to a cell depolarization that modulates voltage-gated channels but also impacts the driving force for ions. Concerning Ca^{2+}, opening of channels reduces the driving force in favor of Ca^{2+} entry, thus modulating Ca^{2+} signaling [76].

8.5.1 TRPM4 Enhances Insulin Secretion by Pancreatic β-Cells

Glucose uptake in cells is enhanced by insulin after hyperglycemic stimulation. Glucose itself promotes insulin secretion by pancreatic β-cells from the islet of Langerhans. In that tissue, glucose enters the cell through a glucose transporter and is then metabolized in the mitochondria to produce ATP. ATP binds to a sulphonylurea receptor (SUR), thus inhibiting an ATP-dependent K^+ channel (K_{ATP}). This leads to a membrane depolarization that activates voltage gated Ca^{2+} channels (VDCa), enabling calcium entry. The rise in $[Ca^{2+}]_i$ releases insulin-sequestrating vesicles from the cytoskeleton. Fusion of vesicles with the plasma membrane releases insulin into the circulation [77].

TRPM4 was detected as a transcript (mRNA) and current in humans, mice, rats and hamster β-pancreatic cell lines [12, 78]. It participates in a glucose-induced depolarization that favors insulin secretion (Fig. 8.3). Indeed, use of a dominant negative mutant of TRPM4 in the rat pancreatic β-cell line, INS-1, reduces insulin secretion [12]. TRPM4 activation might be promoted by cell depolarization due to K_{ATP} channel inhibition, release of Ca^{2+} from the endoplasmic reticulum or activation by the PLC/PKC pathway. In that sense, induction of insulin secretion

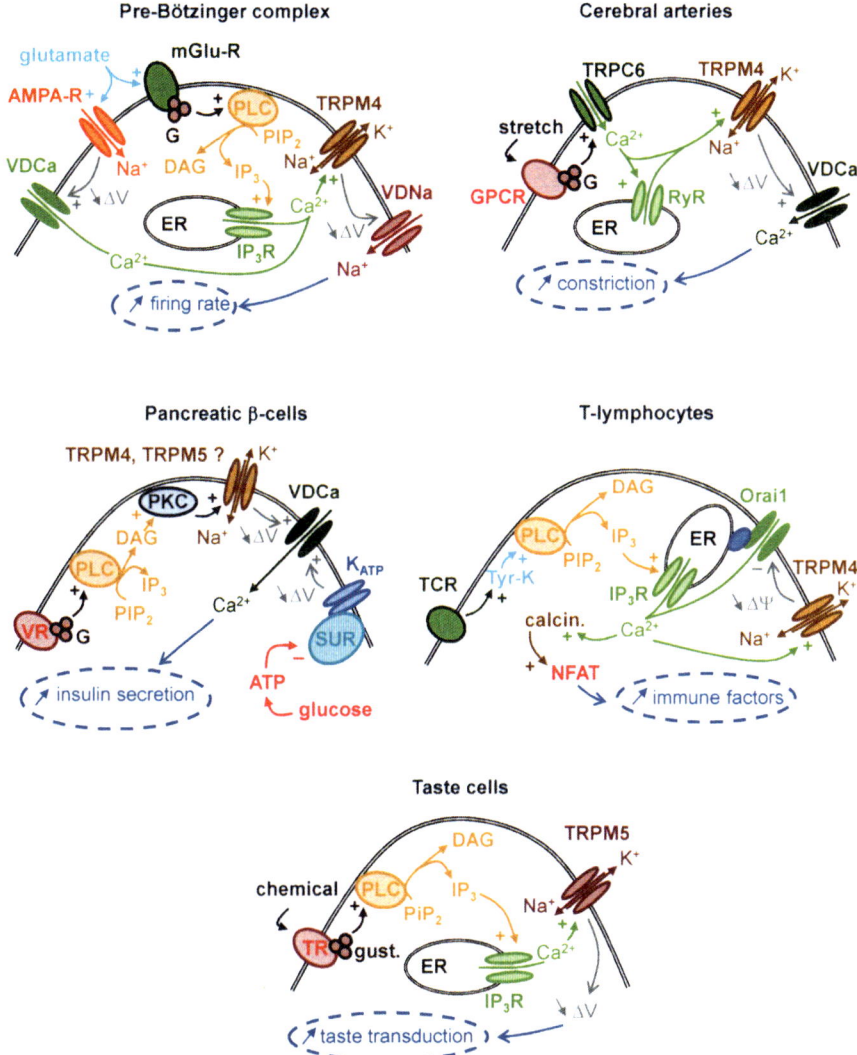

Fig. 8.3 Physiological implications of TRPM4 and TRPM5. TRPM4 is implicated in breathing rhythm regulation in the pre-Bötzinger complex, constriction of cerebral arteries, insulin secretion by pancreatic-β cells (maybe TRPM5) and T-lymphocytes. TRPM5 is involved in taste transduction in gustatory cells. For each phenomenon, the intracellular links between the external stimulus and the final physiological impact is schematized, as described in the text

by vasopressin, a Gq-coupled receptor agonist, is reduced by a dominant negative mutant of TRPM4 [78]. Insulin secretion terminates after activation of a Ca^{2+}-activated big K^+ channel at the peak of the calcium transient, inducing cell repolarization and thus closing of both VDCa and TRPM4 [79].

A recent publication casts doubt on the molecular identity of the pancreatic NSC_{Ca}. Indeed, it reports a reduction in glucose-induced secretion of insulin by pancreatic islets in TRPM5 KO mice [80]. Whether TRPM4 and TRPM5 act in different ways or whether there is confusion from the molecular characterizations made within different studies has to be determined.

8.5.2 TRPM4 Promotes the Bayliss Effect in Cerebral Arteries

Cerebral arteries respond to an increase in internal pressure by constriction, to maintain a constant cerebral blood flow. This common reaction of small arteries is named the Bayliss effect [81, 82]. As in pancreatic β-cells, a depolarization promotes VDCa opening and Ca^{2+} entry, leading to cell contraction after interactions within the contractile network.

TRPM4 is present at the mRNA and current levels in rat vascular muscle cells [83]. VDCa-activating depolarization is promoted by TRPM4, a phenomenon enhanced by PKC stimulation [84]. Therefore, reduction of its expression using an antisense oligonucleotide leads to a decrease in the Bayliss effect in rat cerebral arteries [83]. In addition to PKC, the precise pathways that link membrane stretching to TRPM4 activation are not precisely known (Fig. 8.3). This phenomenon seems to be initiated by the activation of a stretch-activated Gq-coupled receptor, namely an angiotensin receptor, leading to the opening of the calcium-permeable TRPC6. Calcium entry through TRPC6, maybe in association with Ca^{2+} release from intracellular stores, activates TRPM4 and produces cell depolarization leading to constriction [85–87].

8.5.3 TRPM4 in Immune Cells

The onset of the immune response is the capture of antigens by dendritic cells that then migrate from peripheral tissues to lymphoid organs. In these tissues, dendritic cells present antigens to T lymphocytes, producing their activation *via* a rise in $[Ca]_i$ [88, 89]. By preventing Ca^{2+} overload in both dendritic cells and T-lymphocytes, TRPM4 reduces the immune response. Indeed, in TRPM4 deficient mice, Ca^{2+} overload is increased in dendritic cells that are stimulated by bacterial infection [90]. This results in a defect in the migration of dendritic cells but does not alter their maturation. TRPM4 also acts on the activated T-cells. In these cells, activation of the antigen receptor switches on the tyrosine kinase/PLC/IP_3 pathway, resulting in depletion of calcium stores. This induces activation of the calcium-release-activated calcium current (I_{CRAC}), attributed in these cells to the Orai1 channel. The global rise in $[Ca]_i$ stimulates calcineurin that dephosphorylates the nuclear factor of activated T cells (NFAT), modulating transcription of immune factors. Use of a dominant negative mutant for TRPM4 in the Jurka T cell line increases Ca^{2+} signaling and immune factor production while stimulation of TRPM4 by

BTP2 decreases Ca^{2+} influx and immune factor release [67, 91]. In these cells, TRPM4 acts by depolarizing the cell, thus modulating the driving force for Ca^{2+} (Fig. 8.3).

Downstream to T-cells, TRPM4 is also implicated in mast-cell mediation of inflammation and the allergic response. In sensitized hosts producing specific IgE, secondary exposure to an antigen initiates mast-cell activation leading to the release of chemokines and cytokines, resulting in vasodilatation and allergic responses [92]. The intracellular signal in mast cells is driven by Ca^{2+} influx through CRAC channels [93]. In a second model of TRPM4 deficient mice, an enhanced activation of IgE-dependent mast cells was observed. This occurs because TRPM4 reduces Ca^{2+} influx by modulating membrane potential in mast cells, thus reducing the allergic response [94]. Furthermore, TRPM4 also modulates migration of activated mast cells. During inflammatory conditions, mast cells migrate to the inflammatory site from other body sites. This migration is reduced in TRPM4 knock-out mice with an impaired regulation of Ca^{2+} dependent actin cytoskeleton rearrangements [95].

It has to be noted that, until now, these modifications in immune response are the only phenotypes reported in the two models of TRPM4 deficient mice [90, 94, 95].

8.5.4 Modulation of Firing Rate of Breath Pacemaker Neurons

Rythmogenesis of breathing is initiated in the pre-Bötzinger complex (PBC) located in the brainstem. Intrinsic pacemaker activity combined with interactions within this respiratory network is necessary for inspiratory burst generation [96, 97]. In PBC, a current with the hallmarks of TRPM4 and TRPM5 was reported [98]. Its inhibition by flufenamic acid reduces the firing rate [98] while PIP_2 enhances the robustness of respiratory rhythm [99]. It is postulated that TRPM4/TRPM5 acts by depolarizing the cell, allowing it to reach the threshold for voltage-gated Na^+ channel activation, triggering action potentials. Therefore, in such excitable cells, the contribution of TRPM4/TRPM5 current does not only modulate the driving force for ions, but also directly modulates channel activity by impacting membrane potential. The initial activation of TRPM4/TRPM5 current may occur through activation of both AMPA and metabotropic glutamate receptors at the post synaptic terminals [100, 101]. While the AMPA receptor produces a depolarizing current in favor of VDCa opening, metabotropic glutamate receptors activate the PLC pathway, leading to IP_3 formation and release of Ca^{2+} from intracellular stores. All together, this produces a rise in $[Ca^{2+}]_i$ that activates the TRPM4/TRPM5 current (Fig. 8.3).

It is not clear yet whether TRPM4 or TRPM5 is responsible for this current. Indeed, both mRNAs are present in the mouse PBC [99] and no data are available from experiments with dominant negative or deficient mice from either of these channels. Nevertheless, the native current in the PBC is inhibited by flufenamic acid in the concentration range of TRPM4 and, overall, is inhibited by internal ATP [102]. That is why a contribution of TRPM4 towards this current is more likely than TRPM5.

8.5.5 TRPM4 in Cardiac Cells

The first description of a calcium-activated non-selective cationic current at the single channel level was made by Colquhoun et al. [18] in neonate rat ventricular cardiomyocytes, in the early beginnings of the patch-clamp technique. While the identity of this channel is probably not TRPM4 or TRPM5 because of its different single channel conductance, a number of studies have recently clearly demonstrated the functional expression of TRPM4 in several cardiac preparations. TRPM4 is constitutively expressed in the mouse pacemaking cardiac cells located in the sino-atrial node [17]. It is present in human and rat atrial myocytes [19, 103] but absent from ventricular myocytes [104]. However, a ventricular expression of TRPM4 is unmasked during cardiac hypertrophy [105].

Cardiac hypertrophy is an increase in ventricular mass in an adaptive process to maintain cardiac output in case of external perturbations such as hypertension [106]. In an in vitro model of primary culture of adult rat ventricular myocytes that undergo a hypertrophic process, we observed the increase in TRPM4 expression correlated with an increase in cell capacitance, a hypertrophic marker, and with a variation of the resting membrane potential through less negative values [104]. In these cells, spontaneous beating occurs after cell hypertrophy, a process initially described as an abnormal expression of the pacemaker HCN channel [107], however this may also be explained in part by the expression of TRPM4. In the in vivo model of spontaneously hypertensive rats (SHR) that develop left ventricular hypertrophy, we showed a TRPM4 expression in the ventricle at the transcript and current levels [105]. While TRPC1, TRPC3 and TRPC6 channels are involved in the development of cardiac hypertrophy, TRPM4 is suspected to support a part of the electrophysiological perturbations associated with hypertrophy [108]. The link between TRPM4 expression and arrhythmias related to hypertrophy remains unknown. It is postulated that TRPM4 may participate in a calcium-activated transient inward current (I_{ti}) responsible for arrhythmias such as delayed afterdepolarizations and in early afterdepolarizations observed in several models of cardiac hypertrophy [109, 110].

A recent report showed the first inherited disease linked to a mutation of TRPM4 [111]. A family was reported with a substitution of glutamate by lysine in position 7. Several members of the family exhibit a progressive block of the conduction through the cardiac bundle branch. Interestingly, the mutation alters SUMOylation of the channel, resulting in decreased endocytosis leading to an increased expression of the channel at the plasma membrane in a heterologous expression system (HEK-293 cells) with no effect on single channel biophysical properties. Although a stark TRPM4 mRNA expression was detected in conductive tissue, it is not established if this gain of expression occurs in cardiac tissue and how it is likely to support the conduction block. A possible link might be the alteration of the availability of the voltage dependent Na$^+$ channel (VDNa) that is known to be decreased when the resting membrane potential is depolarized [112, 113].

Thus, in cardiac cells, TRPM4 might depolarize resting membrane potentials because of its non-selective cationic selectivity, reducing the occurrence of action potentials by a negative effect on VDNa availability. On the other hand, because it is activated by membrane depolarization and a rise in [Ca^{2+}]$_i$, TRPM4 is supposed to be highly activated at the plateau of the action potential and thus may be pro-arrhythmic by promoting early afterdepolarizations. However, while NSC_{Ca} currents were firstly described in cardiac cells, their real physiological implication in the heart is still puzzling.

8.5.6 TRPM5 as a Taste Transducer

By contrast to the widely distributed TRPM4, TRPM5 expression is restricted to a small number of tissues with the highest level in taste receptor cells. According to this, the channel appears to be implicated in the bitter, sweet and amino-acid (umami) taste [114]. This taste transduction occurs through activation of G protein-coupled receptors in neuroendothelial taste receptors cells located in taste buds from the tongue and palate epithelium. Once activated, the G protein gustducin switches on the phospholipase C pathway, probably inducing hydrolysis of PIP_2 into DAG and IP_3 in favor of Ca^{2+} release from intracellular stores [114].

TRPM5 is suspected to be the subsequent link in that transduction chain (Fig. 8.3). Its activation by a rise in $[Ca^{2+}]_i$ would produce membrane depolarization that initiates an electrical response of the taste receptor. Indeed, TRPM5 expression overlaps with bitter, sweet and amino-acid receptors but also with gustducin, PLCβ2 and IP_3-receptors [21]. Moreover, knockout mice that lack a functional TRPM5 protein lose their sweet, amino-acid and bitter taste sensitivity [115]. Similar results are observed in PLCβ2 KO mice. It was also shown that these genetic modifications do not alter the salty and sour taste, suggesting that TRPM5 is not implicated in these transduction systems [115]. However, a recent report indicates that TRPM5 is implicated in divalent salt tasting in combination with TRPV1 [116]. At low concentration, divalent salts produce a hedonically positive response using the sweet and amino-acid receptors and TRPM5 pathway while, at high concentration, the activation of TRPV1 stimulates a hedonically negative pathway. In addition, the bitter/gustducin/TRPM5 pathway is implicated in the sensation of acyl-homoserine lactones by airway chemoreceptors, a molecule produced by Gram-negative pathogenic bacteria [117]. Therefore, detection of acyl-homoserine lactones should give a valuable warning before bacterial infection.

It is interesting to note that, while bitter, sweet and amino-acid taste use the same gustducin/PLC/TRPM5 transduction pathway, their membrane receptors are localized in separate taste cells within individual taste buds. That allows the fundamental discrimination between signals of nutritionally rich food (sweet and amino-acid) and toxic foods (bitter) [115, 118].

A recent study showed that KO TRPM5 mice lose their typical TRPM5 current in taste cells but a weak 30 pS NSC_{Ca} remains [119]. This current is inhibited by internal ATP and thus might be attributed to TRPM4 [119]. However, because these mice lack their sweet, amino-acid and bitter taste, this "TRPM4-like" current would not be implicated in taste transduction.

It is unlikely that TRPM5 participates in other sensory pathways such as nociceptive, thermal and mechanical sensation. Indeed, its expression profile during embryonic mouse development in the dorsal root ganglion that holds soma of neurons implicated in these sensations, reveals only a transient expression of TRPM5 that disappears in adults while a sustained increase in expression for all other TRPM members (except TRPM1) is detected [120].

8.6 Conclusion and Perspectives

The expansion of TRPM4 and TRPM5 as major physiological actors is probably far from finished. Indeed, NSC_{Ca} as well as TRPM4 and TRPM5 transcripts are present in many other tissues where a demonstration of their physiological role is still missing. For example, TRPM4 is expressed all along the renal tubule where it may influence the driving force for ions and thus modulate renal reabsorption [8]. Its expression in the cochlear outer hair cells may influence hearing [14]. TRPM5 expression in the intestinal tract may support a post-injective chemosensation [21, 121]. In this tissue it was recently shown that TRPM5 participates in the secretion of opioids into the lumen [122]. Further investigations using KO mice and specific pharmacological tools may increase the knowledge in those fields.

The pathological implication of TRPM4 is also an emerging field. In addition to its involvement in cardiac conduction block disease described above [111], recent data indicate that spinal cord injury is followed by an increase in TRPM4 expression and capillary fragmentation leading to secondary hemorrhage in control rats but not in TRPM4 mRNA antisense treated animals or in TRPM4 KO mice, suggesting that TRPM4 is implicated in this process [123].

In addition, TRPM4 and TRPM5 partner proteins may emerge. An ATP-inhibited NSC_{Ca}, suspected to be TRPM4, was shown to interact with the SUR-1 receptor to mediate cerebral edema after ischemic stroke [124]. If so, this interaction may explain, in part, the ATP and glibenclamide effect on TRPM4. Moreover, a physiological interaction between TRPC3 and TRPM4 was postulated to modulate store operated currents in HEK-293 cells [125]. Because these two channels are widely expressed in tissues, this interaction may have a strong physiological impact.

All together the data presented above show that TRPM4 and TRPM5 are fundamental proteins since nearly all cells express one of these channels and because their activation leads to variation of membrane potentials modulating the driving force for ions and activity of voltage-gated channels. Their sensitivity to $[Ca^{2+}]_i$ is a stark link to calcium signaling and thus opens a large panel of putative physiological interactions. Surprisingly, despite their expected large physiological impact, KO mice for either TRPM4 or TRPM5 are viable and, moreover, do not exhibit strongly invaliding phenotypes. The explanation may be due to a compensatory phenomenon or more probably to the fact that these channels, in particular TRPM4, may express their activity under specific conditions. In this case, the KO mouse phenotype would only emerge after challenging the animals. Even though during the last 8 years several roles for these channels have been clearly revealed, a long inquiry remains to be conducted before we can consider the matter closed.

Aknowledgments We thank Simon Patrick for constructive comments on the manuscript. Christophe Simard is a recipient of a fellowship from the French Ministère de l'Enseignement et de la Recherche.

References

1. Launay P, Fleig A, Perraud AL, Scharenberg AM, Penner R, Kinet JP (2002) TRPM4 is a Ca^{2+}-activated nonselective cation channel mediating cell membrane depolarization. Cell 109:397–407
2. Enklaar T, Esswein M, Oswald M, Hilbert K, Winterpacht A, Higgins M, Zabel B, Prawitt D (2000) Mtr1, a novel biallelically expressed gene in the center of the mouse distal chromosome 7 imprinting cluster, is a member of the Trp gene family. Genomics 67:179–187
3. Teulon J (2000) Ca^{2+}-activated nonselective cation channels. In: Endo M, Kurachi Y, Mishina M (eds) Pharmacology of ionic channel function: activators and inhibitors. Springer, Berlin, pp 625–649
4. Vennekens R, Nilius B (2007) Insights into TRPM4 function, regulation and physiological role. Handb Exp Pharmacol 179:269–285
5. Guinamard R, Demion M, Chatelier A, Bois P (2006) Calcium-activated nonselective cation channels in mammalian cardiomyocytes. Trends Cardiovasc Med 16:245–250
6. Csanády L, Adam-Vizi V (2003) $Ca(^{2+})$- and voltage-dependent gating of $Ca(^{2+})$- and ATP-sensitive cationic channels in brain capillary endothelium. Biophys J 85:313–327
7. Popp R, Gögelein H (1992) A calcium and ATP sensitive nonselective cation channel in the antiluminal membrane of rat cerebral capillary endothelial cells. Biochim Biophys Acta 1108:59–66
8. Chraïbi A, Van den Abbeele T, Guinamard R, Teulon J (1994) A ubiquitous non-selective cation channel in the mouse renal tubule with variable sensitivity to calcium. Pflugers Arch 429:90–97
9. Teulon J, Paulais M, Bouthier M (1987) A Ca2-activated cation-selective channel in the basolateral membrane of the cortical thick ascending limb of Henle's loop of the mouse. Biochim Biophys Acta 905:125–132
10. Siemer C, Gögelein H (1992) Activation of nonselective cation channels in the basolateral membrane of rat distal colon crypt cells by prostaglandin E2. Pflugers Arch 420:319–328
11. Maruyama Y, Petersen OH (1984) Single calcium-dependent cation channels in mouse pancreatic acinar cells. J Membr Biol 81:83–87
12. Cheng H, Beck A, Launay P, Gross SA, Stokes AJ, Kinet JP, Fleig A, Penner R (2007) TRPM4 controls insulin secretion in pancreatic beta-cells. Cell Calcium 41:51–61
13. Sturgess NC, Hales CN, Ashford ML (1986) Inhibition of a calcium-activated, non-selective cation channel, in a rat insulinoma cell line, by adenine derivatives. FEBS Lett 208:397–400
14. Van den Abbeele T, Tran Ba Huy P, Teulon J (1994) A calcium-activated nonselective cationic channel in the basolateral membrane of outer hair cells of the guinea-pig cochlea. Pflugers Arch 427:56–63
15. Liman ER (2003) Regulation by voltage and adenine nucleotides of a Ca^{2+}-activated cation channel from hamster vomeronasal sensory neurons. J Physiol 548:777–787
16. Partridge LD, Swandulla D (1987) Single Ca-activated cation channels in bursting neurons of Helix. Pflugers Arch 410:627–631
17. Demion M, Bois P, Launay P, Guinamard R (2007) TRPM4, a Ca^{2+}-activated nonselective cation channel in mouse sino-atrial node cells. Cardiovasc Res 73:531–538
18. Colquhoun D, Neher E, Reuter H, Stevens CF (1981) Inward current channels activated by intracellular Ca in cultured cardiac cells. Nature 294:752–754
19. Guinamard R, Chatelier A, Demion M, Potreau D, Patri S, Rahmati M, Bois P (2004) Functional characterization of a $Ca(^{2+})$-activated non-selective cation channel in human atrial cardiomyocytes. J Physiol 558:75–83

20. Nilius B, Prenen J, Droogmans G, Voets T, Vennekens R, Freichel M, Wissenbach U, Flockerzi V (2003) Voltage dependence of the Ca^{2+}-activated cation channel TRPM4. J Biol Chem 278:30813–30820
21. Pérez CA, Huang L, Rong M, Kozak JA, Preuss AK, Zhang H, Max M, Margolskee RF (2002) A transient receptor potential channel expressed in taste receptor cells. Nat Neurosci 5:1169–1176
22. Nilius B, Owsianik G, Voets T, Peters JA (2007) Transient receptor potential cation channels in disease. Physiol Rev 87:165–217
23. Ramsey IS, Delling M, Clapham DE (2006) An introduction to TRP channels. Annu Rev Physiol 68:619–647
24. Venkatachalam K, Montell C (2007) TRP channels. Annu Rev Biochem 76:387–417
25. Kraft R, Harteneck C (2005) The mammalian melastatin-related transient receptor potential cation channels: an overview. Pflugers Arch 451:204–211
26. Xu XZ, Moebius F, Gill DL, Montell C (2001) Regulation of melastatin, a TRP-related protein, through interaction with a cytoplasmic isoform. Proc Natl Acad Sci USA 98:10692–10697
27. Murakami M, Xu F, Miyoshi I, Sato E, Ono K, Iijima T (2003) Identification and characterization of the murine TRPM4 channel. Biochem Biophys Res Commun 307:522–528
28. Prawitt D, Enklaar T, Klemm G, Gärtner B, Spangenberg C, Winterpacht A, Higgins M, Pelletier J, Zabel B (2000) Identification and characterization of MTR1, a novel gene with homology to melastatin (MLSN1) and the trp gene family located in the BWS-WT2 critical region on chromosome 11p15.5 and showing allele-specific expression. Hum Mol Genet 9:203–216
29. Fujiwara Y, Minor DL Jr (2008) X-ray crystal structure of a TRPM assembly domain reveals an antiparallel four-stranded coiled-coil. J Mol Biol 383:854–870
30. Chubanov V, Waldegger S, Mederos y Schnitzler M, Vitzthum H, Sassen MC, Seyberth HW, Konrad M, Gudermann T (2004) Disruption of TRPM6/TRPM7 complex formation by a mutation in the TRPM6 gene causes hypomagnesemia with secondary hypocalcemia. Proc Natl Acad Sci USA 101:2894–2899
31. Ullrich ND, Voets T, Prenen J, Vennekens R, Talavera K, Droogmans G, Nilius B (2005) Comparison of functional properties of the Ca^{2+}-activated cation channels TRPM4 and TRPM5 from mice. Cell Calcium 37:267–278
32. Yarishkin OV, Hwang EM, Park JY, Kang D, Han J, Hong SG (2008) Endogenous TRPM4-like channel in Chinese hamster ovary (CHO) cells. Biochem Biophys Res Commun 369:712–717
33. Prawitt D, Monteilh-Zoller MK, Brixel L, Spangenberg C, Zabel B, Fleig A, Penner R (2003) TRPM5 is a transient Ca^{2+}-activated cation channel responding to rapid changes in $[Ca^{2+}]i$. Proc Natl Acad Sci USA 100:15166–15171
34. Hofmann T, Chubanov V, Gudermann T, Montell C (2003) TRPM5 is a voltage-modulated and $Ca^{(2+)}$-activated monovalent selective cation channel. Curr Biol 13:1153–1158
35. Liu D, Liman ER (2003) Intracellular Ca^{2+} and the phospholipid PIP2 regulate the taste transduction ion channel TRPM5. Proc Natl Acad Sci USA 100:15160–15165
36. Nilius B, Prenen J, Janssens A, Owsianik G, Wang C, Zhu MX, Voets T (2005) The selectivity filter of the cation channel TRPM4. J Biol Chem 280:22899–22906
37. Mederos y Schnitzler M, Wäring J, Gudermann T, Chubanov V (2008) Evolutionary determinants of divergent calcium selectivity of TRPM channels. FASEB J 22:1540–1551
38. Du J, Xie J, Yue L (2009) Intracellular calcium activates TRPM2 and its alternative spliced isoforms. Proc Natl Acad Sci USA 106:7239–7244
39. Nilius B, Prenen J, Tang J, Wang C, Owsianik G, Janssens A, Voets T, Zhu MX (2005) Regulation of the Ca^{2+} sensitivity of the nonselective cation channel TRPM4. J Biol Chem 280:6423–6433

40. Zhang Z, Okawa H, Wang Y, Liman ER (2005) Phosphatidylinositol 4,5-bisphosphate rescues TRPM4 channels from desensitization. J Biol Chem 280:39185–39192
41. Nilius B, Prenen J, Janssens A, Voets T, Droogmans G (2004) Decavanadate modulates gating of TRPM4 cation channels. J Physiol 560:753–765
42. Talavera K, Yasumatsu K, Voets T, Droogmans G, Shigemura N, Ninomiya Y, Margolskee RF, Nilius B (2005) Heat activation of TRPM5 underlies thermal sensitivity of sweet taste. Nature 438:1022–1025
43. Nilius B, Prenen J, Voets T, Droogmans G (2004) Intracellular nucleotides and polyamines inhibit the Ca^{2+}-activated cation channel TRPM4b. Pflugers Arch 448:70–75
44. Bautista DM, Siemens J, Glazer JM, Tsuruda PR, Basbaum AI, Stucky CL, Jordt SE, Julius D (2007) The menthol receptor TRPM8 is the principal detector of environmental cold. Nature 448:204–208
45. McKemy DD, Neuhausser WM, Julius D (2002) Identification of a cold receptor reveals a general role for TRP channels in thermosensation. Nature 416:52–58
46. Caterina MJ, Schumacher MA, Tominaga M, Rosen TA, Levine JD, Julius D (1997) The capsaicin receptor: a heat-activated ion channel in the pain pathway. Nature 389:816–824
47. Jordt SE, Julius D (2002) Molecular basis for species-specific sensitivity to "hot" chili peppers. Cell 108:421–430
48. Rohacs T, Nilius B (2007) Regulation of transient receptor potential (TRP) channels by phosphoinositides. Pflugers Arch 455:157–168
49. Nilius B, Mahieu F, Prenen J, Janssens A, Owsianik G, Vennekens R, Voets T (2006) The Ca^{2+}-activated cation channel TRPM4 is regulated by phosphatidylinositol 4,5-biphosphate. EMBO J 25:467–478
50. Liu D, Zhang Z, Liman ER (2005) Extracellular acid block and acid-enhanced inactivation of the Ca^{2+}-activated cation channel TRPM5 involve residues in the S3-S4 and S5-S6 extracellular domains. J Biol Chem 280:20691–20699
51. Chraïbi A, Guinamard R, Teulon J (1995) Effects of internal pH on the nonselective cation channel from the mouse collecting tubule. J Membr Biol 148:83–90
52. Greenwood IA, Large WA (1995) Comparison of the effects of fenamates on Ca-activated chloride and potassium currents in rabbit portal vein smooth muscle cells. Br J Pharmacol 116:2939–2948
53. Jin NG, Kim JK, Yang DK, Cho SJ, Kim JM, Koh EJ, Jung HC, So I, Kim KW (2003) Fundamental role of ClC-3 in volume-sensitive Cl^- channel function and cell volume regulation in AGS cells. Am J Physiol Gastrointest Liver Physiol 285:G938–G948
54. Lambert S, Oberwinkler J (2005) Characterization of a proton-activated, outwardly rectifying anion channel. J Physiol 567:191–213
55. Albert AP, Pucovsky V, Prestwich SA, Large WA (2006) TRPC3 properties of a native constitutively active Ca^{2+}-permeable cation channel in rabbit ear artery myocytes. J Physiol 571:361–369
56. Lee YM, Kim BJ, Kim HJ, Yang DK, Zhu MH, Lee KP, So I, Kim KW (2003) TRPC5 as a candidate for the nonselective cation channel activated by muscarinic stimulation in murine stomach. Am J Physiol Gastrointest Liver Physiol 284:G604–G616
57. Hill K, Benham CD, McNulty S, Randall AD (2004) Flufenamic acid is a pH-dependent antagonist of TRPM2 channels. Neuropharmacology 47:450–460
58. Inoue R, Okada T, Onoue H, Hara Y, Shimizu S, Naitoh S, Ito Y, Mori Y (2001) The transient receptor potential protein homologue TRP6 is the essential component of vascular alpha(1)-adrenoceptor-activated $Ca^{(2+)}$-permeable cation channel. Circ Res 88:325–332
59. Hu H, Tian J, Zhu Y, Wang C, Xiao R, Herz JM, Wood JD, Zhu MX (2010) Activation of TRPA1 channels by fenamate nonsteroidal anti-inflammatory drugs. Pflugers Arch 459:579–592
60. Mott DD, Washburn MS, Zhang S, Dingledine RJ (2003) Subunit-dependent modulation of kainate receptors by extracellular protons and polyamines. J Neurosci 23:1179–1188

61. Washburn MS, Dingledine R (1996) Block of alpha-amino-3-hydroxy-5-methyl-4-isoxazolepropionic acid (AMPA) receptors by polyamines and polyamine toxins. J Pharmacol Exp Ther 278:669–678
62. Jin L, Miyazaki M, Mizuno S, Takigawa M, Hirose T, Nishimura K, Toida T, Williams K, Kashiwagi K, Igarashi K (2008) The pore region of N-methyl-D-aspartate receptors differentially influences stimulation and block by spermine. J Pharmacol Exp Ther 327:68–77
63. Kurata HT, Diraviyam K, Marton LJ, Nichols CG (2008) Blocker protection by short spermine analogs: refined mapping of the spermine binding site in a Kir channel. Biophys J 95:3827–3839
64. Kerschbaum HH, Kozak JA, Cahalan MD (2003) Polyvalent cations as permeant probes of MIC and TRPM7 pores. Biophys J 84:2293–2305
65. Talavera K, Yasumatsu K, Yoshida R, Margolskee RF, Voets T, Ninomiya Y, Nilius B (2008) The taste transduction channel TRPM5 is a locus for bitter–sweet taste interactions. FASEB J 22:1343–1355
66. White NJ (2007) Cardiotoxicity of antimalarial drugs. Lancet Infect Dis 7:549–558
67. Takezawa R, Cheng H, Beck A, Ishikawa J, Launay P, Kubota H, Kinet JP, Fleig A, Yamada T, Penner R (2006) A pyrazole derivative potently inhibits lymphocyte Ca^{2+} influx and cytokine production by facilitating transient receptor potential melastatin 4 channel activity. Mol Pharmacol 69:1413–1420
68. Zitt C, Strauss B, Schwarz EC, Spaeth N, Rast G, Hatzelmann A, Hoth M (2004) Potent inhibition of Ca^{2+} release-activated Ca^{2+} channels and T-lymphocyte activation by the pyrazole derivative BTP2. J Biol Chem 279:12427–12437
69. He LP, Hewavitharana T, Soboloff J, Spassova MA, Gill DL (2005) A functional link between store-operated and TRPC channels revealed by the 3,5-bis(trifluoromethyl)pyrazole derivative, BTP2. J Biol Chem 280:10997–11006
70. Frelet A, Klein M (2006) Insight in eukaryotic ABC transporter function by mutation analysis. FEBS Lett 580:1064–1084
71. Schultz BD, DeRoos AD, Venglarik CJ, Singh AK, Frizzell RA, Bridges RJ (1996) Glibenclamide blockade of CFTR chloride channels. Am J Physiol 271:L192–L200
72. Guinamard R, Chraïbi A, Teulon J (1995) A small-conductance Cl$^-$ channel in the mouse thick ascending limb that is activated by ATP and protein kinase A. J Physiol 485:97–112
73. Demion M, Guinamard R, El Chemaly A, Rahmati M, Bois P (2006) An outwardly rectifying chloride channel in human atrial cardiomyocytes. J Cardiovasc Electrophysiol 17:60–68
74. Marivingt-Mounir C, Norez C, Dérand R, Bulteau-Pignoux L, Nguyen-Huy D, Viossat B, Morgant G, Becq F, Vierfond JM, Mettey Y, Synthesis SAR (2004) crystal structure, and biological evaluation of benzoquinoliziniums as activators of wild-type and mutant cystic fibrosis transmembrane conductance regulator channels. J Med Chem 47:962–972
75. Grand T, Demion M, Norez C, Mettey Y, Launay P, Becq F, Bois P, Guinamard R (2008) 9-phenanthrol inhibits human TRPM4 but not TRPM5 cationic channels. Br J Pharmacol 153:1697-1705
76. Fliegert R, Glassmeier G, Schmid F, Cornils K, Genisyuerek S, Harneit A, Schwarz JR, Guse AH (2007) Modulation of Ca^{2+} entry and plasma membrane potential by human TRPM4b. FEBS J 274:704–713
77. McClenaghan NH (2007) Physiological regulation of the pancreatic {beta}-cell: functional insights for understanding and therapy of diabetes. Exp Physiol 92:481–496
78. Marigo V, Courville K, Hsu WH, Feng JM, Cheng H (2009) TRPM4 impacts on Ca^{2+} signals during agonist-induced insulin secretion in pancreatic beta-cells. Mol Cell Endocrinol 299:194–203

79. Braun M, Ramracheya R, Bengtsson M, Zhang Q, Karanauskaite J, Partridge C, Johnson PR, Rorsman P (2008) Voltage-gated ion channels in human pancreatic beta-cells: electrophysiological characterization and role in insulin secretion. Diabetes 57:1618–1628
80. Colsoul B, Schraenen A, Lemaire K, Quintens R, Van Lommel L, Segal A, Owsianik G, Talavera K, Voets T, Margolskee RF, Kokrashvili Z, Gilon P, Nilius B, Schuit FC, Vennekens R (2010) Loss of high-frequency glucose-induced Ca^{2+} oscillations in pancreatic islets correlates with impaired glucose tolerance in Trpm5–/– mice. Proc Natl Acad Sci USA 107:5208–5213
81. Bayliss W (1902) On the local reactions of the arterial wall to changes of internal pressure. J Physiol 28:220–231
82. Voets T, Nilius B (2009) TRPCs, GPCRs and the Bayliss effect. EMBO J 28:4–5
83. Earley S, Waldron BJ, Brayden JE (2004) Critical role for transient receptor potential channel TRPM4 in myogenic constriction of cerebral arteries. Circ Res 95:922–929
84. Earley S, Straub SV, Brayden J, Protein Kinase C (2007) Regulates vascular myogenic tone through activation of TRPM4. Am J Physiol Heart Circ Physiol 292:2613–2622
85. Morita H, Honda A, Inoue R, Ito Y, Abe K, Nelson MT, Brayden JE (2007) Membrane stretch-induced activation of a TRPM4-like nonselective cation channel in cerebral artery myocytes. J Pharmacol Sci 103:417–426
86. Welsh DG, Morielli AD, Nelson MT, Brayden JE (2002) Transient receptor potential channels regulate myogenic tone of resistance arteries. Circ Res 90:248–250
87. Mederos y Schnitzler M, Storch U, Meibers S, Nurwakagari P, Breit A, Essin K, Gollasch M, Gudermann T (2008) Gq-coupled receptors as mechanosensors mediating myogenic vasoconstriction. EMBO J 27:3092–3103
88. Quintana A, Griesemer D, Schwarz EC, Hoth M (2005) Calcium-dependent activation of T-lymphocytes. Pflugers Arch 450:1–12
89. Luik RM, Lewis RS (2007) New insights into the molecular mechanisms of store-operated Ca^{2+} signaling in T cells. Trends Mol Med 13:103–107
90. Barbet G, Demion M, Moura IC, Serafini N, Léger T, Vrtovsnik F, Monteiro RC, Guinamard R, Kinet JP, Launay P (2008) The calcium-activated nonselective cation channel TRPM4 is essential for the migration but not the maturation of dendritic cells. Nat Immunol 9:1148–1156
91. Launay P, Cheng H, Srivatsan S, Penner R, Fleig A, Kinet JP (2004) TRPM4 regulates calcium oscillations after T cell activation. Science 306:1374–1377
92. Metz M, Grimbaldeston MA, Nakae S, Piliponsky AM, Tsai M, Galli SJ (2007) Mast cells in the promotion and limitation of chronic inflammation. Immunol Rev 217:304–328
93. Di Capite J, Parekh AB (2009) CRAC channels and Ca^{2+} signaling in mast cells. Immunol Rev 231:45–58
94. Vennekens R, Olausson J, Meissner M, Bloch W, Mathar I, Philipp SE, Schmitz F, Weissgerber P, Nilius B, Flockerzi V, Freichel M (2007) Increased IgE-dependent mast cell activation and anaphylactic responses in mice lacking the calcium-activated nonselective cation channel TRPM4. Nat Immunol 8:312–320
95. Shimizu T, Owsianik G, Freichel M, Flockerzi V, Nilius B, Vennekens R (2009) TRPM4 regulates migration of mast cells in mice. Cell Calcium 45:226–232
96. Smith JC, Ellenberger HH, Ballanyi K, Richter DW, Feldman JL (1991) Pre-Bötzinger complex: a brainstem region that may generate respiratory rhythm in mammals. Science 254:726–729
97. Ramirez JM, Viemari JC (2005) Determinants of inspiratory activity. Respir Physiol Neurobiol 147:145–157
98. Del Negro CA, Morgado-Valle C, Hayes JA, Mackay DD, Pace RW, Crowder EA, Feldman JL (2005) Sodium and calcium current-mediated pacemaker neurons and respiratory rhythm generation. J Neurosci 25:446–453

99. Crowder EA, Saha MS, Pace RW, Zhang H, Prestwich GD, Del Negro CA (2007) Phosphatidylinositol 4,5-bisphosphate regulates inspiratory burst activity in the neonatal mouse preBötzinger complex. J Physiol 582:1047–1058
100. Pace RW, Mackay DD, Feldman JL, Del Negro CA (2007) Inspiratory bursts in the pre-Bötzinger complex depend on a calcium-activated non-specific cation current linked to glutamate receptors in neonatal mice. J Physiol 582:113–125
101. Pace RW, Del Negro CA (2008) AMPA and metabotropic glutamate receptors cooperatively generate inspiratory-like depolarization in mouse respiratory neurons in vitro. Eur J Neurosci 28:2434–2442
102. Mironov SL (2008) Metabotropic glutamate receptors activate dendritic calcium waves and TRPM channels which drive rhythmic respiratory patterns in mice. J Physiol 586:2277–2291
103. Zhainazarov AB (2003) Ca^{2+}-activated nonselective cation channels in rat neonatal atrial myocytes. J Membr Biol 193:91–98
104. Guinamard R, Rahmati M, Lenfant J, Bois P (2002) Characterization of a Ca^{2+}-activated nonselective cation channel during dedifferentiation of cultured rat ventricular cardiomyocytes. J Memb Biol 188:127–135
105. Guinamard R, Demion M, Magaud C, Potreau D, Bois P (2006) Functional expression of the TRPM4 cationic current in ventricular cardiomyocytes from spontaneously hypertensive rats. Hypertension 48:587–594
106. Swynghedauw B (1999) Molecular mechanisms of myocardial remodeling. Physiol Rev 79:215–262
107. Farès N, Bois P, Lenfant J, Potreau D (1998) Characterization of a hyperpolarization-activated current in dedifferentiated adult rat ventricular cells in primary culture. J Physiol 506:73–82
108. Guinamard R, Bois P (2007) Involvement of transient receptor potential proteins in cardiac hypertrophy. Biochim Biophys Acta 1772:885–894
109. Ruocco C, Cerbai E, Failli P, Giotti A, Mugelli A (1996) Calcium-dependent electrophysiological alterations in hypertrophied rat cardiomyocytes. Biochem Biophys Res Commun 229:425–429
110. Yan GX, Rials SJ, Wu Y, Liu T, Xu X, Marinchak RA, Kowey PR (2001) Ventricular hypertrophy amplifies transmural repolarization dispersion and induces early afterdepolarization. Am J Physiol Heart Circ Physiol 281:H1968–H1975
111. Kruse M, Schulze-Bahr E, Corfield V, Beckmann A, Stallmeyer B, Kurtbay G, Ohmert I, Schulze-Bahr E, Brink P, Pongs O (2009) Impaired endocytosis of the ion channel TRPM4 is associated with human progressive familial heart block type I. J Clin Invest 119:2737–2744
112. Irvine LA, Jafri MS, Winslow RL (1999) Cardiac sodium channel Markov model with temperature dependence and recovery from inactivation. Biophys J 76:1868–1885
113. Raman IM, Bean BP (2001) Inactivation and recovery of sodium currents in cerebellar Purkinje neurons: evidence for two mechanisms. Biophys J 80:729–737
114. Liman ER (2007) TRPM5 and taste transduction. Handb Exp Pharmacol 179:287–298
115. Zhang Y, Hoon MA, Chandrashekar J, Mueller KL, Cook B, Wu D, Zuker CS, Ryba NJ (2003) Coding of sweet, bitter, and umami tastes: different receptor cells sharing similar signaling pathways. Cell 112:293–301
116. Riera CE, Vogel H, Simon SA, Damak S, le Coutre J (2009) Sensory attributes of complex tasting divalent salts are mediated by TRPM5 and TRPV1 channels. J Neurosci 29:2654–2662
117. Tizzano M, Gulbransen BD, Vandenbeuch A, Clapp TR, Herman JP, Sibhatu HM, Churchill ME, Silver WL, Kinnamon SC, Finger TE (2010) Nasal chemosensory cells use bitter taste signaling to detect irritants and bacterial signals. Proc Natl Acad Sci USA 107:3210–3215
118. Hoon MA, Adler E, Lindemeier J, Battey JF, Ryba NJ, Zuker CS (1999) Putative mammalian taste receptors: a class of taste-specific GPCRs with distinct topographic selectivity. Cell 96:541–551

119. Zhang Z, Zhao Z, Margolskee R, Liman E (2007) The transduction channel TRPM5 is gated by intracellular calcium in taste cells. J Neurosci 27:5777–5786
120. Staaf S, Franck MC, Marmigère F, Mattsson JP, Ernfors P (2010) Dynamic expression of the TRPM subgroup of ion channels in developing mouse sensory neurons. Gene Expr Patterns 10:65–74
121. Bezençon C, Fürholz A, Raymond F, Mansourian R, Métairon S, Le Coutre J, Damak S (2008) Murine intestinal cells expressing Trpm5 are mostly brush cells and express markers of neuronal and inflammatory cells. J Comp Neurol 509:514–525
122. Kokrashvili Z, Rodriguez D, Yevshayeva V, Zhou H, Margolskee RF, Mosinger B (2009) Release of endogenous opioids from duodenal enteroendocrine cells requires Trpm5. Gastroenterology 137:598–606
123. Gerzanich V, Woo SK, Vennekens R, Tsymbalyuk O, Ivanova S, Ivanov A, Geng Z, Chen Z, Nilius B, Flockerzi V, Freichel M, Simard JM (2009) De novo expression of Trpm4 initiates secondary hemorrhage in spinal cord injury. Nat Med 15:185–191
124. Simard JM, Chen M, Tarasov KV, Bhatta S, Ivanova S, Melnitchenko L, Tsymbalyuk N, West GA, Gerzanich V (2006) Newly expressed SUR1-regulated NC(Ca-ATP) channel mediates cerebral edema after ischemic stroke. Nat Med 12:433–440
125. Park JY, Hwang EM, Yarishkin O, Seo JH, Kim E, Yoo J, Yi GS, Kim DG, Park N, Ha CM, La JH, Kang D, Han J, Oh U, Hong SG (2008) TRPM4b channel suppresses store-operated Ca^{2+} entry by a novel protein–protein interaction with the TRPC3 channel. Biochem Biophys Res Commun 368:677–683

Chapter 9
TRPM7, the Mg^{2+} Inhibited Channel and Kinase

Chris Bates-Withers, Rajan Sah, and David E. Clapham

Abstract TRPM7 is a ubiquitously expressed nonselective cation channel fused to a C-terminal alpha kinase. TRPM7 current is typically small at physiological magnesium concentrations, but large outwardly rectifying currents develop in low-magnesium extracellular solution when cells are dialyzed with magnesium free solutions during whole-cell patch clamp recordings. In addition to regulation by magnesium, TRPM7 current is potentiated by low extracellular pH and inhibited by depletion of phosphatidylinositol 4,5-bisphosphate (PIP$_2$) during phospholipase C mediated signaling events. A diverse body of literature has implicated TRPM7 in fundamental cellular processes including death, survival, proliferation, cell cycle progression, magnesium homeostasis and responses to shear stress and oxidative stress. Global deletion of TRPM7 in mouse results in embryonic lethality and a thymocyte-restricted conditional knockout exhibits defective thymopoeisis, suggesting a role for TRPM7 in development and organogenesis. In disease states, TRPM7 has been linked to Guamanian amyotrophic lateral sclerosis and parkinsonian dementia (ALS/PD), various forms of neoplasia, hypertension and delayed neuronal death following cerebral ischemia.

9.1 Architecture and Expression Pattern

The transient receptor potential melastatin type 7 (TRPM7) channel-kinase, known alternatively as TRP-PLIK [1], ChaK1 [2] and LTRPC7 [3], is a 212.4 kDa, 1,863 amino acid protein consisting of a transmembrane ion channel coupled to a 39.6 kDa C-terminal α-kinase (Fig. 9.1). The complete crystal structure has not been solved, but sequence analysis predicts a channel with six transmembrane domains [1]. The TRPM7 protein has an estimated 92.2% helix content by circular dichroism

D.E. Clapham (✉)
Department of Cardiology, Howard Hughes Medical Institute, Manton Center for Orphan Disease, Children's Hospital Boston and Harvard University, Boston, MA 02115, USA;
Department of Neurobiology, Harvard Medical School, Boston, MA 02115, USA
e-mail: dclapham@enders.tch.harvard.edu

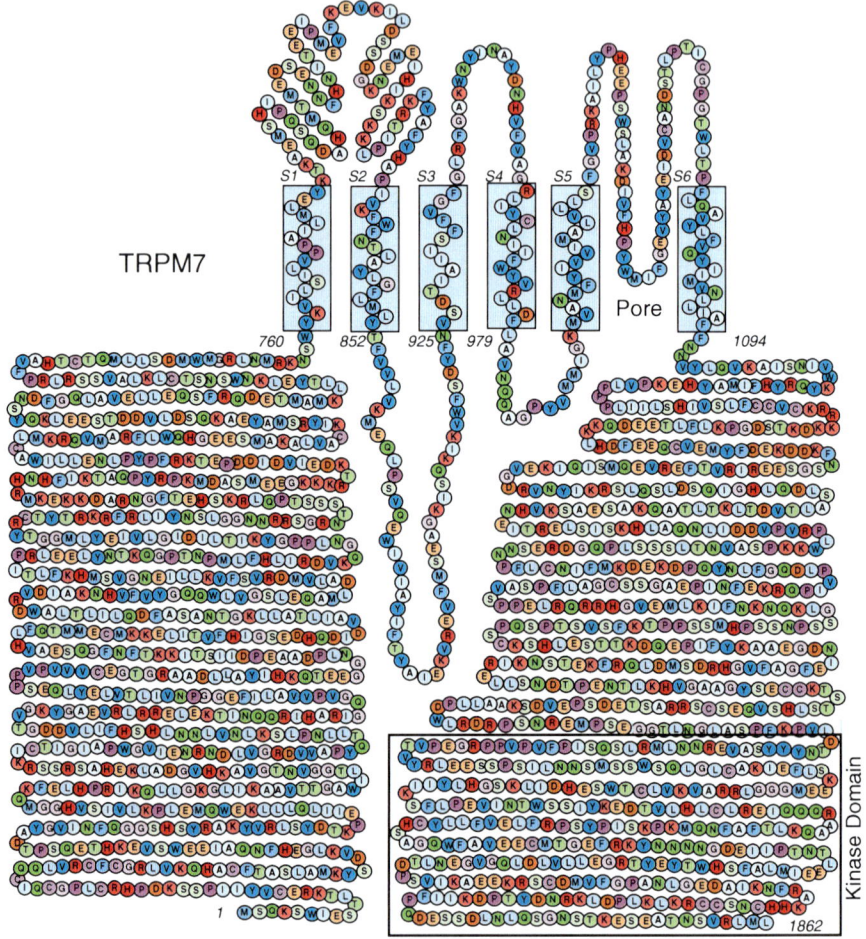

Fig. 9.1 Sequence of human TRPM7 protein showing the kinase domain at the C terminus, the predicted 6 transmembrane domains, and the putative pore domain. The channel is assumed to form as a tetramer of this subunit

spectropolarimetry [4]. The high resolution crystal structure of a portion of the rat TRPM7 C-terminus has recently been solved, revealing a coiled-coil assembly domain localized between the ion channel and kinase domains. At the coiled-coil domain, TRPM7 is predicted to form antiparallel tetramers both with itself and with the structurally similar TRPM6 [4].

TRPM7 is expressed throughout the lifespan in mouse. Expression is observed in embryonic stem cells, is concentrated in the fetal heart at embryonic day 9.5 (E9.5) and is observed ubiquitously by E14.5 [5]. In adult human, TRPM7 mRNA is most highly expressed in heart, adipose tissue and bone [6], while murine expression is strongest in heart, kidney and cerebrum. Some variability is seen across mouse strains, with greatest cardiac expression in C57/Bl mice, followed by NOD and BALB/c animals [1, 7].

TRPM7 shares 52% overall identity with TRPM6, which is implicated in defective intestinal magnesium transport in autosomal-recessive hypomagnesemia with secondary hypocalcemia [8]. One study combining electrophysiology, Mn^{2+} entry assays and FRET-based approaches proposed that TRPM6 was unable to form functional channels independently of TRPM7, and that the mutation causing hypomagnesemia with secondary hypocalcemia disrupted formation of TRPM6/TRPM7 complexes [9, 10]. The interpretation of this study is now less clear, as heterologously expressed TRPM6 without TRPM7 was subsequently demonstrated to produce functional channels with measurable current [11].

Two related neurodegenerative diseases, Guamanian amyotrophic lateral sclerosis and parkinsonian dementia (ALS/PD), have been linked epidemiologically to a missense mutation localized between the channel and kinase domains of TRPM7. Functional analysis of the heterologously expressed mutant channel in whole-cell patch clamp experiments revealed lower overall current density and slower current activation relative to heterologously expressed wild-type TRPM7 [12]. However, TRPM7 mutations do not universally account for cases of ALS/PD, as a Japanese cohort of ALS/PD patients showed no pathological TRPM7 mutations [13]. By a mechanism unrelated to ALS/PD-causing mutations, TRPM7 is also proposed to contribute to the pathogenesis of familial Alzheimer's disease, presumably through decreased currents associated with aberrant phosphoinositide metabolism [14].

9.2 Channel Properties and Regulation

The TRPM7 channel unitary single-channel conductance is 40 pS and produces a distinctive, outwardly rectifying current with a reversal potential near 0 mV in whole-cell voltage clamp recordings (Fig. 9.2) [1, 11]. The proposed permeability to divalent cations is greatest for Zn^{2+} and Ni^{2+}, followed in order by Co^{2+}, Mg^{2+}, Ca^{2+}, Mn^{2+}, Sr^{2+}, and Cd^{2+} [1, 15]. In divalent-free conditions, large inward and outward currents develop, with greatest permeability to K^+ followed by Na^+. TRPM7 current is blocked by trivalent cations [1]. TRPM7 current is typically inhibited under physiological conditions, and recordings using the perforated patch method do not produce current. The mechanism of TRPM7 activation following patch rupture in the whole cell configuration is believed to reflect either chelation of intracellular magnesium by agents in the patch pipette or gradual outward diffusion of an unknown inhibitor. The former possibility is supported by the finding that the cell-permeant magnesium chelator EDTA-AM is able to induce TRPM7 currents during recordings using the perforated patch technique [16]. TRPM7 currents were reported to be inhibited by intracellular magnesium nucleotides, particularly Mg-ATP, and early literature refers to TRPM7 currents as MagNuM (magnesium nucleotide-inhibited metal) current. However, Mg-ATP is now considered to inhibit TRPM7 currents as a source of magnesium rather than via a separate, ATP-mediated mechanism [17]. A more common designation of TRPM7 current is Magnesium Inhibitable Current (MIC: 25).

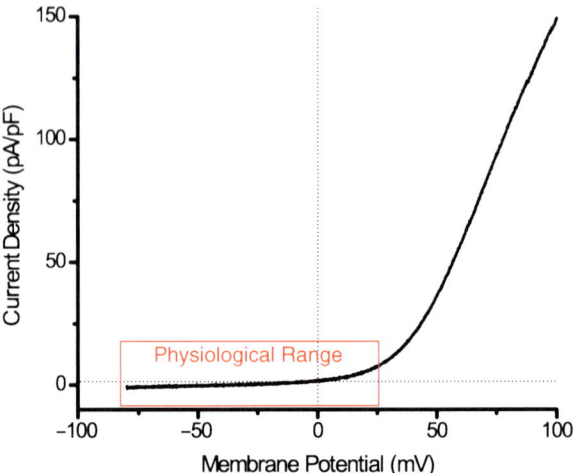

Fig. 9.2 Spontaneous, magnesium-subtracted TRPM7 current recorded in a whole-cell recording from a cardiac H9C2 cell line (cardiac) 10 min after break-in. In most cells, TRPM7 steadily increases after break-in, presumably as unknown factors wash out of the cell. In analogy to inwardly-rectifying K^+ channels, most of the current recorded is outside the physiological range. Intracellular solution, in mM: 120 Cs-aspartate, 20 CsCl, 2.5 EGTA, 2.5 EDTA, 10 HEPES, 5 Na_2ATP, 0.1 Na_2GTP, pH 7.2 Extracellular solution, in mM: 135 NaCl, 5.4 CsCl, 10 HEPES, 10 Glucose, 1 $CaCl_2$, 0.1 $CdCl_2$, pH 7.4. Extracellular solution containing 10 mM $MgCl_2$ was washed in to obtain magnesium subtracted trace. Series resistance is compensated to 75%

Inward currents through TRPM7 are typically miniscule, but can be activated in the context of decreased extracellular pH, with an ~10-fold increase in inward but not outward current observed when pH was decreased from 7.4 to 4.0 (see Table 9.1). This pH activated inward conductance is likely a monovalent current, as it is abolished by replacement of extracellular monovalents with nonpermeant N-methyl-D-glucamine [18]. Divalent conductance is not believed to be affected by pH, although the potentiation of monovalent conductance under acidic conditions is affected by external Ca^{2+} and Mg^{2+} concentration, presumably via competition between protons and divalents for binding sites in the TRPM7 pore [19]. TRPM7 has been proposed to conduct small proton currents under acidic conditions, although recording conditions used to measure proton conductances through TRPM7 do not exclude potential pH-activated chloride currents [18–20].

There is evidence for the formation of TRPM6/TRPM7 complexes as well as independently functional TRPM6 and TRPM7 channels [4, 9, 11, 21]. While TRPM7 and TRPM6 have morphologically identical I–V curves, they can be differentiated electrophysiologically via their single channel conductances and divalent permeability, pharmacologically via their responses to 2-aminophenylborate (2-APB) and biophysically via their responses to decreased extracellular pH [11]. TRPM7 and TRPM6 are considered to be functionally nonredundant, though they do interact and may functionally modulate each other, at least within some cell types [21] TRPM7 currents have also been observed under conditions used to measure

Table 9.1 Properties of TRPM6, TRPM6/TRPM7 complexes and TRPM7 (summarized from Li, Jiang and Yue, 2006)

	TRPM6	TRPM6/TRPM7	TRPM7
Cation permeation	$Zn^{2+} > Ba^{2+} > Mg^{2+} = Ca^{2+} = Mn^{2+} > Sr^{2+} > Cd^{2+} > Ni^{2+}$	$Zn^{2+} > Mg^{2+} > Mn^{2+} > Ba^{2+} = Ca^{2+} = Sr^{2+} > Cd^{2+} > Ni^{2+}$	$Ni^{2+} > Zn^{2+} > Ba^{2+} = Mg^{2+} > Ca^{2+} = Mn^{2+} = Sr^{2+} > Cd^{2+}$
Unitary single channel conductance	83.6 pS	56.6 pS	40.1 pS
$pH_{1/2}$	4.3	5.5	4.7
Increase in inward current at pH 4.0 vs 7.4	3.8-fold	5.9-fold	9.9-fold
Effect of <1 mM 2-APB	Potentiates current	Potentiates current but less than for TRPM6	Inhibits current
Effect of ≥2 mM 2-APB	Not characterized	Not characterized	Potentiates current

Ca^{2+} release-activated Ca^{2+} (CRAC) currents, however this is now believed to be a methodological artifact associated with omission of Mg^{2+} and Mg-ATP from CRAC recording solutions [22]. TRPM7 and CRAC represent different conductances with distinct biophysical properties [23–25].

TRPM7 kinase phosphorylates serine and threonine residues within alpha helices. It uses ATP but not GTP as a phosphate donor [26]. The kinase domain of TRPM7 interacts directly with multiple members of the phospholipase C (PLC) family including PLC-β_1, PLC-β_2, PLC-β_3 and PLC-γ_1. PLC activation via Gα_q-coupled muscarinic receptors produces inactivation of the TRPM7 current in a manner dependent upon phosphatidylinositol 4,5-bisphosphate (PIP$_2$) hydrolysis but independent of downstream protein kinase C activity [27, 28].

The TRPM7 kinase can phosphorylate the calcium and phospholipid binding protein annexin I [29] and the nonmuscle myosins IIA, IIB and IIC [30], but it is not known if these are phosphorylated in native conditions. Annexin I is known to mediate the anti-inflammatory effects of glucocorticoids and like TRPM7, has been implicated in apoptosis, cell growth and differentiation [29]. The nonmuscle myosins are involved in cell motility [31] and may have a role in heart tube looping during embryonic development [32]. While TRPM7 is known to undergo autophosphorylation, the physiological consequences of autophosphorylation are unknown. TRPM7 channel activity is independent of autophosphorylation and kinase activity [33]. Another proposed signaling role for TRPM7 is activation of *m*-calpain and regulation of cell adhesion. When TRPM7 is overexpressed in HEK-293T cells, cell adhesion is impaired in an *m*-calpain dependent manner [34]. Likewise, enhanced cell adhesion is observed in TRPM7-deficient cells [35]. While TRPM7 directly interacts with *m*-calpain [34], it may activate *m*-calpain by a less direct mechanism. Overexpression of TRPM7 increases levels of cellular reactive oxygen species, activating p38 mitogen-activated protein kinase (MAPK) and c-Jun N-terminal kinase (JNK). Both p38 MAPK and JNK appear necessary for TRPM7-induced changes in cell adhesion [35].

9.3 Cellular and Biological Functions of TRPM7

TRPM7 has been proposed as a mediator of cellular growth, survival and proliferation. An in vitro targeted TRPM7 deletion was performed in DT-40 cells and failed to produce viable homozygote knockout clones, except with the use of an inducible deletion system. Following induced deletion, cells exhibited impaired growth and decreased viability [3]. An early report suggested that the growth and viability phenotype observed in TRPM7 deficient DT-40 cells was rescued by supraphysiological concentrations of extracellular magnesium, leading to the hypothesis that TRPM7 may function as a mediator of cellular magnesium homeostasis [36]. A role for TRPM7 in cell survival has also been demonstrated in hepatocytes and hepatoma cells [37], mast cells [38] and gastric adenocarcinoma cells [39].

Targeted deletion of TRPM7 in mouse has produced results consistent with a role of TRPM7 in growth, survival and development, though contributions to cellular magnesium homeostasis are now in question. Global deletion of TRPM7 results in prenatal lethality, with death occurring by embryonic day 7.5 [5]. Select tissue-restricted conditional knockouts are viable and are an area of active investigation. Deletion in T lymphocytes results in developmental arrest at the double negative stage, with no appreciable differences in cellular magnesium uptake or content [5]. These studies do not preclude a local action for Mg^{2+}, and in fact, this would be a logical consequence of TRPM7's very low inward conductance. There are presently no in vivo loss-of-function models in kidney or intestine, where magnesium transport would be regulated at the organ system level. However, one study has shown increased TRPM7 expression in gut during magnesium deficiency or decreased expression during hypermagnesemia [40].

Numerous lines of evidence are consistent with a role for TRPM7 in cell proliferation, though effects are dependent upon cell type. TRPM7 is ectopically overexpressed in human breast cancer tissues, and knockdown of TRPM7 by siRNA impairs proliferation in the MCF-7 breast cancer cell line [41]. Similarly, in osteoblasts, TRPM7 knockdown by siRNA inhibited cell proliferation both under basal conditions and under stimulation with platelet-derived growth factor [42, 43]. The contribution of TRPM7 to cell proliferation and cell cycle regulation may proceed through several putative pathways. TRPM7 current densities increase nearly threefold during the G1 cell cycle phase in the RBL-2H3 mast cell line [44]. Divalent cations (specifically calcium) conducted via TRPM7 may modulate any number of calcium-dependent proteins implicated in cell-cycle progression including calpain, calmodulin, calmodulin kinase II, calcineurin or phosphoinositol-3-kinase [45]. Alternatively, phosphorylation of annexin I may contribute to cell cycle regulation.

Annexin I is known to modulate expression of cyclin D1, which in turn affects proliferation and cell cycle progression [46].

9.4 Pathological Functions in Disease Conditions

Numerous studies have implicated TRPM7 in cellular responses to physical stress and injury. TRPM7 was first identified as a potential mechanosensor when Oancea and colleagues described a modest increase in TRPM7 current and TRPM7 channel density at the plasma membrane in response to fluid flow. Oancea hypothesized that TRPM7 in vascular smooth muscle cells (VSMCs) may act as an injury sensor, becoming activated during endovascular injury if the intimal endothelial cells are disrupted, allowing exposure of VSMCs to blood flow [46]. A study published the following year also suggested that TRPM7 is activated by stretch and osmotic swelling [47–49], although TRPM7 is not particularly sensitive to these stimuli.

The known human disease correlates for TRPM7 have prompted further investigation into its roles in the nervous system, where it has been found in the synaptic terminals of acetylcholine-secreting sympathetic neurons and is believed to be involved in neurotransmitter release [50, 51]. Under a current proposed model by Michael Tymianski and colleagues, neuronal nitric oxide synthase (nNOS) produces reactive nitrogen species under ischemic conditions, activating TRPM7 current and producing cell death via calcium overload [52]. An alternative hypothesis focuses on TRPM7's reported zinc conductance, and proposes that under ischemic conditions, enhanced zinc entry via TRPM7 functions as a mechanism of toxicity and cell death [53]. Despite promising findings with knockdown of TRPM7 in an in vivo rodent model of cerebral ischemia [54], data from a prospective human trial examining TRPM7 single nucleotide polymorphisms failed to identify a connection between the TRPM7 gene and the pathogenesis of ischemic stroke [55].

To date, the role of TRPM7 in vasculature has been described much more completely than in heart. In isolated rat vascular smooth muscle cells (VSMCs), TRPM7 has been shown to mediate proinflammatory bradykinin signaling [56]. TRPM7 expression and annexin I activation are also upregulated in VSMCs in the context of bradykinin stimulation [57]. TRPM7 has been circumstantially linked to hypertension, though further functional studies are needed. In cultured VSMCs from spontaneously hypertensive rats (SHR), TRPM7 mRNA and protein are decreased relative to Wistar Kyoto control rats. In Wistar Kyoto VSMCs, treatment with angiotensin II increases TRPM7 expression, translocation of annexin I from cytosol to the plasma membrane and cellular magnesium content. All of these responses to angiotensin II are blunted in SHR VSMCs [58]. Similar results are observed in mouse and human VSMCs [59].

Functional characterization of TRPM7 in heart has focused predominantly upon cardiac fibroblasts, where it has been proposed to contribute to arrhythmogenesis and fibrosis. A recent study identified TRPM7 as the dominant route of calcium entry in human atrial fibroblasts and as a potential contributor to the pathogenesis of atrial fibrillation [60]. Under this proposed model, profibrotic TGF-β1 signals upregulate both TRPM7 expression and TRPM7 current, facilitating calcium influx into fibroblasts. This calcium influx then promotes differentiation of fibroblasts into myofibroblasts, which have extensive autocrine and paracrine activity contributing to cardiac remodeling, fibrosis and arrhythmogenic alterations in conduction. Consistent with this model, TRPM7 current density was increased in isolated atrial fibroblasts obtained from biopsies of atrial fibrillation patients [60].

In summary, TRPM7 is a protein of wide-ranging biological and physiological significance. It is a likely contributor to fundamental cellular processes including growth, survival, differentiation and death. Mouse knockout studies have now established a vital role for TRPM7 in organogenesis; a rapidly growing body of literature is now focusing on TRPM7 in disease states ranging from cerebral ischemia to cancer. A fundamental direction for future research is identification of common pathways responsible for the diverse functions of TRPM7. As the roles and processes governed by TRPM7 become more fully elucidated, TRPM7 may become a valuable therapeutic target both for developmental disorders and adult diseases.

References

1. Runnels LW, Yue L, Clapham DE (2001) TRP-PLIK, a bifunctional protein with kinase and ion channel activities. Science 291:1043–1047
2. Ryazanov AG (2002) Elongation factor-2 kinase and its newly discovered relatives. FEBS Lett 514(1):26–29
3. Nadler MJS, Hermosura MC, Inabe K, Perraud AL, Zhu Q, Stokes AJ, Kurosaki T, Kinet JP, Penner R, Scharenberg AM, Fleig A (2001) LTRPC7 is a MgATP-regulated divalent cation channel required for cell viability. Nature 411:590–595
4. Fujiwara Y, Minor DL (2008) X-ray crystal structure of a TRPM assembly domain reveals an antiparallel four-stranded coiled coil. J Mol Biol 383:854–870
5. Jin J, Desai BN, Navarro B, Donovan A, Andrews NC, Clapham DE (2008) Deletion of TRPM7 disrupts embryonic development and thymopoiesis without altering Mg^{2+} homeostasis. Science 322:756–760
6. Fonfria E, Murdock PR, Cusdin FS, Benham CD, Kelsell RE, McNulty S (2006) Tissue distribution profiles of the human TRPM7 cation channel family. J Recept Signal Transduct Res 26:159–178
7. Kunert-Keil C, Bisping F, Kruger J, Brinkmeier H (2006) Tissue-specific expression of TRP channel genes in the mouse and its variation in three different mouse strains. BMC Genomics 7:159
8. Schlingmann KP, Weber S, Peters M, Nejsum LN, Vitzthum H, Klingel K, Kratz M, Haddad E, Ristoff E, Dinour D, Syrrou M, Nielsen S, Sassen M, Waldegger S, Seyberth HW, Konrad M (2002) Hypomagnesemia with secondary hypocalcemia is caused by mutations in TRPM6, a new member of the TRPM gene family. Nat Genet 31(2):166–170
9. Chubanov V, Waldegger S, Mederos y Schnitzler M, Vitzthum H, Sassen MC, Seyberth HW, Konrad M, Gudermann T (2004) Disruption of TRPM6/TRPM7 complex formation by a mutation in the TRPM6 gene causes hypomagnesemia with secondary hypocalcemia. PNAS 101(9):2894–2899
10. Voets T, Nilius B, Hoefs S, van der Kemp AW, Droogmans G, Bindels RJ, Hoenderop JG (2004) TRPM6 forms the Mg^{2+} channel involved in intestinal and renal Mg^{2+} absorption. J Biol Chem 279(1):19–25
11. Li M, Jiang J, Yue L (2006) Functional characterization of homo- and heteromeric channel kinases TRPM6 and TRPM7. J Gen Physiol 127(5):525–537

12. Hermosura MC, Nayakanti H, Dorovkov MV, Calderon FR, Ryazanov AG, Haymer DS, Garruto RMA (2005) TRPM7 variant shows altered sensitivity to magnesium that may contribute to the pathogenesis of two guamanian neurodegenerative disorders. PNAS 102(32):11510–11515
13. Hara K, Kokubo Y, Ishiura H, Miyashita A, Kuwano R, Sasaki R, Goto J, Nishizawa M, Kuzuhara S, Tsuji S (2010) TRPM7 is not associated with amyotrophic lateral sclerosis-parkinsonism dementia complex in the Kii Peninsula of Japan. Am J Med Genet B Neuropsychiatr Genet 153B(1):310–313
14. Landman N, Jeong SY, Shin SY, Voronov SV, Serban G, Kang MS, Park MK, Di Paolo G, Chung S, Kim TW (2006) Presenilin mutations linked to familial Alzheimer's disease cause an imbalance in phosphoinositidyl 4,5-bisphosphate metabolism. PNAS 103(51):19524–19529
15. Monteilh-Zoller MK, Hermosura MC, Nadler MJS, Scharenberg AM, Penner R, Fleig A (2003) TRPM7 provides an ion channel mechanism for cellular entry of trace metal ions. J Gen Physiol 121:49–60
16. Langeslag M, Clark K, Moolenaar WH, van Leeuwen FN, Jalink K (2007) Activation of TRPM7 channels by phospholipase C-coupled receptor agonists. J Biol Chem 282(1): 232–239
17. Kozak JA, Cahalan MDMIC (2003) Channels are inhibited by internal divalent cations but not ATP. Biophys J 84:922–927
18. Jiang J, Li M, Yue L (2005) Potentiation of TRPM7 inward current by protons. J Gen Physiol 126(2):137–150
19. Cherny VV, Henderson LM, DeCoursey DE (1997) Proton and chloride currents in chinese hamster ovary cells. Membr Cell Biol 11(3):337–347
20. Numata T, Okada Y (2008) Proton conductivity through TRPM7 and its molecular determinants. J Biol Chem 283(22):15097–15103
21. Schmitz C, Dorovkov MV, Zhao X, Davenport BJ, Ryazanov AG, Perraud AL (2005) The channel-kinases TRPM6 and TRPM7 are functionally nonredundant. J Biol Chem 280(45):37763–37771
22. Bakowski D, Parekh AB (2002) Permeation through store-operated CRAC channels in divalent-free solution: potential problems and implications for putative CRAC channel genes. Cell Calcium 32(5-6):379–391
23. Kozak JA, Kerschbaum HH, Cahalan MD (2002) Distinct properties of CRAC and MIC channels in RBL cells. J Gen Physiol 120(2):221–235
24. Zakharov SI, Smani T, Dobrydneva Y, Monje F, Fichandler C, Blackmore PF, Bolotina VM (2004) Diethylstilbestrol is a potent inhibitor of store-operated channels and capacitative Ca^{2+} influx. Mol Pharmacol 66(3):702–707
25. Prakriya M, Lewis R (2002) Separation and characterization of currents through store-operated CRAC channels and Mg^{2+}-inhibited cation (MIC) channels. J Gen Physiol 119: 487–507
26. Ryazanova LV, Dorovkov MV, Ansari A, Ryazanov AG (2004) Characterization of the protein kinase activity of TRPM7/ChaK1, a protein kinase fused to the transient receptor potential ion channel. J Biol Chem 279(5):3708–3716
27. Runnels LW, Yue L, Clapham DE (2002) The TRPM7 channel is inactivated by PIP2 hydrolysis. Nat Cell Biol 4(5):329–336
28. Takezawa R, Schmitz C, Demeuse P, Scharenberg AM, Penner R, Fleig A (2004) Receptor-mediated regulation of the TRPM7 channel through its endogenous protein kinase domain. PNAS 101(16):6009–6014
29. Dorovkov MV, Ryazanov AG (2004) Phosphorylation of annexin I by TRPM7 channel-kinase. J Biol Chem 279(49):50643–50646
30. Clark K, Middelbeek J, Dorovkov MV, Figdor CG, Ryazanov AG, Lasonder E, van Leeuwen FN (2008) The α-kinases TRPM6 and TRPM7, but not eEF-2 kinase, phosphorylate the assembly domain of myosin IIA, IIB and IIC. FEBS Lett 582(20):2993–2997

31. Clark K, Langeslag M, van Leeuwen B, Ran L, Ryazanov AG, Figdor CG, Moolenaar WH, Jalink K, van Leeuwen FN (2006) TRPM7, a novel regulator of actomyosin contractility and cell adhesion. EMBO J 25:290–301
32. Lu W, Seeholzer SH, Han M, Arnold AS, Serrano M, Garita B, Philp NJ, Farthing C, Steele P, Chen J, Linask KK (2008) Cellular nonmuscle myosins NMHC-IIA and NMHC-IIB and vertebrate heart looping. Devl Dyn 237(12):3577–3590
33. Matsushita M, Kozak JA, Shimizu Y, McLachlin DT, Yamaguchi H, Wei FY, Tomizawa K, Matsui H, Chait BT, Cahalan MD, Nairn AC (2005) Channel function is dissociated from the intrinsic kinase activity and autophosphorylation of TRPM7/ChaK1. J Biol Chem 280(21):20793–20803
34. Su LT, Agapito MA, Li M, Simonson WTN, Huttenlocher A, Habas R, Yue L, Runnels LW (2006) TRPM7 regulates cell adhesion by controlling the calcium-dependent protease calpain. J Biol Chem 281(16):11260–11270
35. Su LT, Chen HC, Gonzales-Pagan O, Overton JD, Xie J, Yue L, Runnels LW (2010) TRPM7 activates m-Calpain by stress-dependent stimulation of p38 MAPK and c-Jun N-terminal kinase. J Mol Biol 396(4):858–869
36. Schmitz C, Perraud AL, Johnson CO, Inabe K, Smith MK, Penner R, Kurosaki T, Fleig A, Scharenberg AM (2003) Regulation of vertebrate cellular Mg^{2+} homeostasis by TRPM7. Cell 114:191–200
37. Mishra R, Rao V, Ta R, Shobeiri N, Hill CEA (2009) Mg^{2+} and MgATP-inhibited and Ca^{2+}/calmodulin-sensitive TRPM7 current-like current in hepatoma and hepatocytes. Am J Physiol Gastrointest Liver Physiol 297:G587–G694
38. Wykes RC, Lee M, Duffy SM, Yang W, Seward EP, Bradding P (2007) Functional transient receptor potential melastatin 7 channels are critical for human mast cell survival. J Immunol 179(6):4045–4052
39. Kim BJ, Park EJ, Lee JH, Jeon JH, Kim SJ, So I (2008) Suppression of transient receptor melastatin 7 channel induces cell death in gastric cancer. Cancer Sci 99(12):2502–2509
40. Rondon LJ, Rayssiguier Y, Dietary MA (2008) Inulin in mice stimulates Mg^{2+} absorption and modulates TRPM6 and TRPM7 expression in large intestine and kidney. Magnes Res 21(4):224–231
41. Guilbert A, Gautier M, Dhennin-Duthille I, Haren N, Sevestre H, Ouadid-Ahidouch H (2009) Evidence that TRPM7 is required for breast cancer cell proliferation. Am J Physiol Cell Physiol 297(3):C493–C502
42. Abed E, Moreau R (2007) Importance of melastatin-like transient receptor potential 7 and cations (magnesium, calcium) in human osteoblast-like cell proliferation. Cell Prolif 40(6):849–865
43. Abed E, Moreau R (2009) Importance of melastatin-like transient receptor potential 7 and magnesium in the stimulation of osteoblast proliferation and migration by platelet-derived growth factor. Am J Physiol Cell Physiol 297(2):C360–C368
44. Tani D, Monteilh-Zoller MK, Fleig A, Penner R (2007) Cell cycle-dependent regulation of store-operated I_{CRAC} and Mg^{2+}-nucleotide-regulated MagNuM (TRPM7) currents. Cell Calcium 41(3):249–260
45. Allridge LC, Bryant CE, Annexin I (2003) Regulates cell proliferation by disruption of cell morphology and inhibition of cyclin D1 expression through sustained activation of the ERK1/2 MAPK signal. Exp Cell Res 290(1):93–107
46. Oancea E, Wolfe JT, Clapham DE (2006) Functional TRPM7 channels accumulate at the plasma membrane in response to fluid flow. Circ Res 98:245–253
47. Numata T, Shimizu T, Okada Y (2002) TRPM7 is a stretch- and swelling-activated cation channel involved in volume regulation in human epithelial cells. Am J Physiol Cell Physiol 292:C460–C467
48. Numata T, Shimizu T, Okada Y (2007) Direct mechano-stress sensitivity of TRPM7 channel. Cell Physiol Biochem 19(1-4):1–8

49. Bessac BF, Fleig A (2007) TRPM7 channel is sensitive to osmotic gradients in human kidney cells. J Physiol 582(3):1073–1086
50. Krapivinsky G, Mochida S, Krapivinsky L, Cibulsky SM, Clapham DE (2006) The TRPM7 ion channel functions in cholinergic synaptic vesicles and affects transmitter release. Neuron 52(3):485–496
51. Brauchi S, Krapivinsky G, Krapivinsky L, Clapham DE (2008) TRPM7 facilitates cholinergic vesicle fusion with the plasma membrane. PNAS 105(24):8304–8308
52. Aarts M, Iihara K, Wei W, Xiong Z, Arundine M, Cerwinski W, MacDonald JF, Tymianski MA (2003) Key role for TRPM7 channels in anoxic neuronal death. Cell 115(7):863–877
53. Inoue K, Branigan D, Xiong ZG (2010) Zinc-induced neurotoxicity mediated by transient receptor potential melastatin channels. J Biolo Chem 285(10):7430–7439
54. Sun HS, Jackson MF, Martin LJ, Jansen K, Teves L, Cui H, Kiyonaka S, Mori Y, Jones M, Forder JP, Golde TE, Orser BA, Macdonald JF, Tymianski M (2009) Suppression of hippocampal TRPM7 protein prevents delayed neuronal death in brain ischemia. Nat Neurosci 12(10):1300–1307
55. Romero JR, Ridker PM, Zee RYL (2009) Gene variation of the transient receptor potential cation channel, subfamily M, member 7 (TRPM7), and risk of incident ischemic stroke: prospective, nested, case-control study. Stroke 40(9):2965–2968
56. Yogi A, Callera GE, Tostes R, Touyz RM (2009) Bradykinin regulates calpain and proinflammatory signaling through TRPM7-sensitive pathways in vascular smooth muscle cells. Am J Physiol Regulatory Integrative Comp Physiol 296:R201–R207
57. Callera GE, He Y, Yogi A, Montezano AC, Paravincini T, Yao G, Touyz RM (2009) Regulation of the novel Mg^{2+} transporter transient receptor potential melastatin 7 (TRPM7) cation channel by bradykinin in vascular smooth muscle cells. J Hypertens 27(1):155–166
58. Touyz RM, He Y, Montezano ACI, Yao G, Chubanov V, Gudermann T, Callera GE (2006) Differential regulation of transient receptor potential melastatin 6 and 7 cation channels by ANG II in vascular smooth muscle cells from spontaneously hypertensive rats. Am J Physiol Regulatory Integrative Comp Physiol 290(1):R73–R78
59. He Y, Yao G, Savoia C, Touyz RM (2005) Transient receptor potential melastatin 7 ion channels regulate magnesium homeostasis in vascular smooth muscle cells. Circ Res 96:207–215
60. Du J, Xie J, Zhang Z, Tsujikawa H, Fusco D, Silverman D, Liang B, Yue L (2010) TRPM7-mediated Ca^{2+} signals confer fibrogenesis in human atrial fibrillation. Circ Res 106(5): 992–1003

Chapter 10
TRPM8 in Health and Disease: Cold Sensing and Beyond

Yi Liu and Ning Qin

Abstract This review focuses on TRPM8, one of the ~30 members of the diverse family of transient receptor potential (TRP) ion channels. Initially identified from the prostate, TRPM8 has been studied more extensively in the sensory system and is best established as a major transducer of environmental cold temperatures. An increasing body of evidence suggests that it may also be an important player in various chronic conditions, such as inflammatory/neuropathic pain and prostate cancer. Small molecule compounds that selectively modulate TRPM8 are beginning to emerge and will be critically valuable for better understanding the role of this channel in both physiological and pathological states, on which the prospects of TRPM8 as a viable therapeutic target rest.

10.1 Introduction

TRPM8 (a.k.a. Trp-p8 or CMR1) is one of the eight members of the melastatin subfamily of transient receptor potential (TRP) ion channels. Initially cloned from the prostate [1], it has since been identified in a variety of tissues, both neuronal and non-neuronal. Much of the research has focused on its role in sensory neurons. Indeed, TRPM8 is best known for being a sensor of cold temperatures and, as such and along with a handful of other temperature-sensitive TRP channels (including TRPV1, TRPV2, TRPV3, TRPV4 and TRPA1), is dubbed a thermo-TRP. Here, we review the recent advances in our understanding of the function and physiological/pathological role of this channel in as well as beyond cold sensing. We further discuss the prospects of TRPM8 as a potential therapeutic target and drug discovery efforts targeting this channel, particularly for the treatment of cancer and pain.

Y. Liu (✉)
Johnson & Johnson Pharmaceutical Research and Development, LLC, San Diego, CA 92121, USA
e-mail: yliu10@its.jnj.com

10.2 Gene, Structure/Function and Expression

The human TRPM8 gene consists of 24 exons located on chromosome 2 (2q37) [1]. Two transcripts of the gene, 6.2 and 5.2 kb, respectively, are detected in the human prostate [1]. Similarly, two transcripts (6 and 4.5 kb, respectively) are found in rat dorsal root and trigeminal ganglia (DRG and TG) [2]. However, only a single transcript (6.3 kb) is identified in mouse DRG [3]. TRPM8 is also cloned from several other mammalian and non-mammalian species, including canine [4], chicken [5] and frog [6].

The human, rat, mouse and canine TRPM8 genes each encode a protein product of 1,104 amino acids with a predicted molecular weight of ~128 kDa. The proteins are highly conserved with 93–98% primary sequence identity (only ~75–80% between mammalian and frog/chicken TRPM8). As with other TRPs, TRPM8 contains six putative transmembrane segments (S1–S6) and a TRP domain in the C-terminus [1]. It forms a homotetrameric channel [7]. Two putative N-glycosylation sites (Asn-821 and Asn-934) are conserved across the mammalian species [4, 8]. However, only Asn-934, located in the pore region between S5 and S6, is significantly glycosylated with no obvious effects on channel function [8]. Also conserved are two cysteines (Cys-929 and Cys-940) in the pore region that flank Asn-934 and are essential for channel function [8]. In addition, there exist several putative protein kinase-A (PKA) and tyrosine kinase phosphorylation sites, of which one each (the PKA site at Thr-95 and tyrosine kinase site at Tyr-295) is conserved among all the species [4].

Evidence points to a critical role of distal S6 in the selectivity of TRPM8 for cations [9]. Similar to voltage-gated ion channels, the S4 domain, along with the S4–S5 linker, forms part of the voltage sensor (albeit with much weaker voltage sensitivity for TRPM8 than for classic voltage-gated ion channels) [10, 11]. Amino acid residues 40–86 in the N-terminus are involved in targeting the channel to the plasma membrane [12]. Interestingly, a C-terminal coiled-coil domain is shown to be required for channel tetramerization and/or expression in some [13, 14] but not other [12] expression systems. Determinants of channel sensitivity to icilin, but not to menthol or cold, map to residues in S3 (G805, which is conserved among mammalian orthologs but lacking in icilin-insensitive chicken and frog TRPM8) and the S2–S3 linker [5, 6], the same region involved in the activation of TRPV1 by capsaicin [15, 16] and of TRPV4 by 4α-phorbol 12,13-didecanoate (4α-PDD) [17], suggesting a conserved molecular mechanism of activation for these thermo-TRPs by certain chemical agonists. Residues important for menthol sensitivity are located in S2 (Y745) and the C-terminal TRP domain [18]. The C-terminus additionally plays critical roles in temperature sensitivity, gating kinetics and PIP_2 modulation [19, 20].

Since the initial identification in the prostate and tumors of the breast, colon, lung and skin origin [1], TRPM8 has also been shown to express in various other tissues, including DRG and TG [2, 3], liver [21], nodose ganglion [22–24], bladder and the male genital tract [25, 26], tongue [27], gastric fundus [28], vascular smooth muscle [29], lung [30], spinal cord [31], brain [32] and petrosal and jugular ganglia [33]. The expression of TRPM8 in sensory neurons and prostate epithelial cells is discussed in more detail in Sections 10.4.1, 10.4.2 and 10.4.3.

A truncated splice variant of TRPM8 that lacks the N-terminus and the first two transmembrane domains is expressed in human lung epithelial cells [in both endoplasmic reticulum (ER) and plasma membranes], as well as in brain, liver, kidney, testes, tongue, and colon [30]. Activation of this TRPM8 variant in human lung epithelial cells by cooling or menthol leads to increased expression of several cytokine and chemokine genes, suggesting that it may mediate inflammatory responses in the airways to cold air [34].

10.3 Biophysical/Pharmacological Properties and Modulation

TRPM8 is a cation-selective channel with high permeability to both monovalent and divalent cations and a current-voltage (I–V) relationship that exhibits strong outward rectification [2, 3]. Agonist-activated TRPM8 currents undergo pronounced Ca^{2+}-dependent desensitization (during continuous agonist application) and tachyphylaxis (upon repeated agonist applications) [2]. Despite being permeant through the channel, extracellular Ca^{2+} also blocks TRPM8 currents, an effect that is independent of channel desensitization and possibly results from Ca^{2+} inhibition of currents conducted by monovalent ions [4, 35]. Measurements of the single-channel conductance range from 21 to 83 pS under various experimental conditions [2, 35, 36].

As with TRPV1, TRPM8 is an ion channel with polymodal gating mechanisms. It can be activated by multiple types of stimuli including innocuous cool to noxious cold temperatures [2–6], chemical ligands such as menthol and icilin [2–4], and membrane depolarization [11, 36]. Effects of these stimuli on TRPM8 activation are additive [11]. It is argued that the temperature sensitivity of TRPM8 gating is a thermodynamic consequence of the difference in activation energy between voltage-dependent channel opening and closing. Cooling and chemical agonists such as menthol act to shift the voltage dependence of activation in the hyperpolarizing direction (towards more physiological potentials) [11, 37]. Conversely, antagonists exert their inhibitory effects by shifting the voltage dependence of activation towards more positive potentials [37]. As such, thermal activation of TRPM8 can be regarded as a threshold phenomenon only in the context of a given membrane potential (e.g., ~26°C at the resting membrane potential of a neuron). Other studies, however, indicate that the increase in channel open probability upon cooling is greater than what can be accounted for by a simple left-shift of voltage dependence of activation, suggesting that temperature and voltage interact allosterically to promote channel opening [36].

Unlike cold or menthol, a naturally-occurring cooling compound from mint oils, the synthetic and structurally unrelated super-cooling agent icilin requires the concurrent elevation of cytosolic Ca^{2+} to activate TRPM8 [5]. In this scenario, TRPM8 plays the role of a coincidence detector, responding only in the simultaneous presence of a pair of stimuli (icilin plus a stimulus that increases intracellular Ca^{2+}). Icilin and menthol also show differential Ca^{2+} dependence of channel desensitization [2]. Interestingly, icilin inhibits TRPM8 currents evoked by other stimuli, such as menthol and cooling, in a Ca^{2+}-independent manner and by a mechanism distinct from desensitization [38].

A variety of other agonists of TRPM8 have also been identified. These include synthetic menthol derivatives (e.g., Frescolat ML, Coolact P, Cooling agent 10 and WS-3) [39], carboxamides (e.g., WS-12, CPS-113, CPS-369), carboxylic acid (WS-30) and phosphine oxide (WS-148) derivatives [40, 41], and various naturally-occurring substances [e.g., eucalyptol [2, 39], linalool and geraniol [39], eugenol [42] and mustard oil [4]]. Other synthetic agonists are also disclosed in recent patent applications (see the end of this section and Table 10.1). The TRPM8 agonists identified to date range in potency from low nanomolar to high millimolar concentrations (Table 10.1). D-3263, a TRPM8 agonist developed by Dendreon, is in Phase I clinical trial for the treatment of cancer (see Section 10.4.3).

Lysophospholipids, end products of the Ca^{2+}-independent phospholipase A_2 (PLA_2) signaling pathway, increase TRPM8 channel open time and lower the threshold of TRPM8 activation towards physiological temperatures [43, 44]. This raises the possibility that they may act as endogenous agonists/positive modulators of TRPM8, representing a physiological mechanism for activating/sensitizing TRPM8, particularly in tissues that are not subject to significant temperature variations. In contrast, other end products of PLA_2, such as arachidonic acid, inhibit TRPM8 activation [44].

It is well documented that phosphatidylinositol 4,5-bisphosphate (PIP_2) modulates the activity of a number of TRP channels, including TRPM8 [45]. PIP_2 can activate TRPM8 directly as well as act to maintain channel activity in the presence of other stimuli [19, 46]. Evidence suggests that TRPM8-mediated Ca^{2+} influx activates Ca^{2+}-dependent phospholipase C (PLC), which down-regulates PIP_2, resulting in reduced TRPM8 activity [19, 47]. In addition, activation of protein kinase C (PKC) also desensitizes TRPM8 [48], albeit possibly via an indirect mechanism [49]. Notably, diacylglycerol (DAG), a product of PIP_2 hydrolysis by PLC, is a physiological activator of PKC. Thus, PIP_2 may be both directly and indirectly involved in Ca^{2+}-dependent desensitization of TRPM8.

Other molecules/factors that modulate TRPM8 function include: α_2A-adrenoreceptors (whose activation inhibits TRPM8 by decreasing PKA-dependent channel phosphorylation) [50], protons (which inhibit cold- and icilin-, but not menthol-activated TRPM8 currents) [51], polyphosphates (which form stable complexes with TRPM8 to maintain normal channel function) [52], and lipid rafts (in which TRPM8 is localized and the disruption of which shifts the threshold for TRPM8 activation towards warmer temperatures) [53].

A variety of TRPM8 antagonists have been identified (Table 10.2), most of which either are non-selective or lack selectivity information. Among these are a number of TRPV1 antagonists including capsazepine, *N*-(4-tertiarybutylphenyl)-4-(3-chloropyridin-2-yl)tetrahydropyrazine-1(2*H*)-carboxamide (BCTC), *N*-(4-tertiary

Table 10.1 TRPM8 agonists

Compound name/ID	Structure	Potency (EC_{50})	References
Menthol		67–101 μM	[2, 4, 51]
Icilin		125–420 nM	[2, 4, 39, 51]
Frescolat ML		3.3 μM	[39]
WS-12		39 nM; 193 nM	[40, 41]
WS-30		5.6 μM	[40]
WS-148		4.1 μM	[40]
CPS-369		3.6 μM	[40]

Table 10.1 (continued)

Compound name/ID	Structure	Potency (EC_{50})	References
Eucalyptol		3.4 mM; 7.7 mM	[2, 39]
Geraniol		5.9 mM	[39]
Eugenol		66 µM	[42]
Mustard oil (allyl isothiocyanate)		490 µM	[4]
Compound 1		3 nM	WO2009067410A1

butylphenyl)-4-(3-chloropyridin-2-yl)tetrahydropyrazine-1(2H)-(thio)carboxamide (thio-BCTC), (2R)-4-(3-chloro-2-pyridinyl)-2-methyl-N-[4-(trifluoromethyl) phenyl]-1-piperazinecarboxamide (CTPC) and N-(2-bromophenyl)-N'-{2-[ethyl(3-methylphenyl)amino]ethyl}-urea (SB-452533) [39, 54], TPRV1 agonists including capsaicin, anandamide, resiniferatoxin [55], 2-aminoethoxydiphenyl borate (2-APB) [56] and phenethyl isothiocyanate [42], TRPA1 agonists cinnamaldehyde [57] and 3′-carbamoylbiphenyl-3-yl cyclohexylcarbamate (URB597) [58], an antifungal agent clotrimazole [59], a phospholipase A_2 inhibitor N-(p-amylcinnamoyl)anthranilic acid (ACA) [60], and ethanol [54, 61]. A number of cannabinoid receptor ligands, including cannabidiol, Δ^9-tetrahydrocannabinol (Δ^9-THC) [42, 62], JWH-015 [42], cannabigerol, Δ^9-THC-acid, cannabidiol-acid, N-arachidonoyl-dopamine (NADA), 5-(4-chlorophenyl)-1-(2,4-dichloro-phenyl)-4-methyl-N-(piperidin-1-yl)-1H-pyrazole-3

-carboxamide (SR141716A) and 5-(4-chloro-3-methylphenyl)-1-[(4-methylphenyl)methyl]-*N*-[(1S, 2S, 4R)-1,3,3-trimethylbicyclo[2.2.1]hept-2-yl]-1*H*-pyrazole-3-carboxamide (SR144528) [55, 62], are also reported to inhibit TRPM8. Interestingly, some of these cannabinoids are also agonists of two other thermo-TRP channels, TPRV2 and TRPA1 [62–65].

Table 10.2 TRPM8 antagonists

Compound name/ID	Structure	Potency (IC_{50})	References
Capsazepine		18 μM	[39]
Resiniferatoxin		150 nM	[55]
Capsaicin		1.1 μM	[55]
BCTC		143 nM; 0.8 μM	[39, 54]
SB-452533		571 nM	[54]
2-APB		7.7 μM	[56]
Anandamide		150 nM	[55]

Table 10.2 (continued)

Compound name/ID	Structure	Potency (IC$_{50}$)	References
Cinnamaldehyde		1.5 mM	[57]
URB597		167 µM	[58]
Clotrimazole		200 nM	[59]
ACA		3.9 µM	[60]
Δ^9-THC		10 µM; 150 nM	[42, 62]
JWH-015		2 µM	[42]
Phenethyl isothiocyanate		17 µM	[42]

Table 10.2 (continued)

Compound name/ID	Structure	Potency (IC_{50})	References
JNJ-39267631 (compound 234)		1.6 nM	[66]; US20080027029A1
AMTB		589 nM	[67]
Compound 1d		20 nM	[68]
Compound 306		0.8 nM	US20090264474A1
Compound 150		4–12 nM	US20100048589A1

Table 10.2 (continued)

Compound name/ID	Structure	Potency (IC$_{50}$)	References
Example 4		< 5 μM	US20090082358A1
Ref. No. 1,475		100% activity at 10 μM	US20050054651A1
Example 6		< 20 nM	WO2010010435A2
Example 61		14 nM	WO2007017093A1
Example 98		0.05 nM	WO2007080109A1

JNJ-39267631, a potent and selective small molecule TRPM8 antagonist with efficacy in animal models of pain (see Section 10.4.2), inhibits menthol-, icilin- and cold-induced TRPM8 responses with low nanomolar potencies and is highly selective against TRPV1, TRPV2 and TRPA1 [66]. Another selective (against TRPV1 and TRPV4) TRPM8 antagonist, *N*-(3-aminopropyl)-2-{[(3-methylphenyl)methyl]oxy}-*N*-(2-thienylmethyl)benzamide hydrochloride salt

10 TRPM8 in Health and Disease: Cold Sensing and Beyond

Table 10.3 Radiolabeled TRPM8 ligands

Compound name/ID	Structure	Potency (EC_{50}/IC_{50})	References
Compound of formula II-A or II-B			US20100015053A1
Structure 1 (CP-129)		0.3 μM	US20070053834A1

*US20100015053A1: either ^{11}C or ^{18}F; US20070053834A1: either ^{123}I, ^{125}I, or ^{131}I.

(AMTB), is shown to attenuate the bladder micturition reflex and nociceptive reflex responses in rats [67]. A series of menthylamine derivatives are also reported to exhibit potent TRPM8 antagonist activity with good selectivity against TRPV1 and TRPA1 [68].

The past few years have seen the disclosure in various patent applications of multiple chemical classes of small molecule TRPM8 modulators, both agonists and antagonists. Many of these compounds exhibit potent activity at TRPM8 (e.g., some in the picomolar concentration range). However, the selectivity profile of these molecules is generally unknown. Examples of select chemotypes from these applications are shown in Table 10.1 (agonists) and Table 10.2 (antagonists). In addition, different series of radiolabeled small molecule TRPM8 ligands that are useful for binding and imaging studies have also been disclosed, including sulfonamide and carboxamide derivatives (Table 10.3).

10.4 Physiology/Pathophysiology and Potential as a Therapeutic Target

This section focuses on the current understanding of the role of TRPM8 in cold sensing, pain and prostate cancer, areas in which significant progress has been made in recent years.

10.4.1 TRPM8 as a Molecular Sensor of Cold Temperatures

The ability to detect (and properly respond to) environmental cold temperatures is a prime requisite for survival. How is this accomplished at the molecular level?

Evidence suggests that there likely exists more than one molecular mechanism of cold transduction. In addition to TRPM8, other candidates of potential cold transducers have been proposed, including TRPA1, background K^+ channels, voltage-gated Ca^{2+} channels, and electrogenic pumps [69]. This section primarily concerns TRPM8.

TRPM8 is expressed in a subset (~5–20%) of small- to medium-diameter sensory neurons, including TG and DRG neurons [2, 3, 27, 31, 70–72]. Expression of TRPM8 is more prevalent in TG than in DRG [2, 31, 71], consistent with the observation that a larger fraction of neurons in TG is cold sensitive than in DRG [2, 72]. TRPM8 axons innervate the skin and oral cavity, terminating in peripheral zones that contain nerve endings that mediate distinct perceptions of innocuous cool and noxious cold [70], consistent with a role in conveying a wide range of cold sensations.

Cooling activates a cationic current in DRG neurons [73–75]. Menthol potentiates the current and shifts its thermal activation threshold to warmer temperatures [73, 75]. This cold-sensitive current is blocked by extracellular Ca^{2+}. Similar properties are also observed in cells heterologously expressing TRPM8 [2–4, 35]. Heterologously expressed TRPM8 currents and cold-/menthol-sensitive currents in neurons also share similarities in thermal activation threshold, menthol sensitivity, cation permeability/selectivity, I-V relationship (outward rectification) and voltage dependence of activation [76].

When topically applied to humans, menthol produces a cooling sensation and icilin, a more potent and efficacious TRPM8 agonist than menthol [2], evokes sensations of intense cold [77].

Studies using TRPM8-knockout mice have provided critical insights into the role of TRPM8 in cold sensing [78–80]. Dissociated sensory neurons and intact sensory nerve fibers from TRPM8-null mice exhibit profoundly diminished responses to cold and menthol. In behavioral tests, mice lacking TRPM8 show severe deficits in response to non-noxious cool and, to various degrees, noxious cold temperatures, are devoid of responses to icilin (which causes the characteristic "wet-dog shakes" in wild-type mice), and display attenuated responses to acetone-induced evaporative cooling. Together with the results from neuronal preparations and heterologous TRPM8-expression systems as well as with the other observations described above, these findings demonstrate an essential and predominant role for TRPM8 in thermosensation over a wide range of cold temperatures, firmly establishing it as a principal molecular apparatus for the detection of environmental cold.

Nonetheless, TRPM8 is unlikely the only sensor involved in cold detection. TRPM8 knockout mice still display various degrees of aversion to noxious cold [78–80]. Although much reduced in number, a fraction of sensory neurons from TRPM8-null mice (some of which do not express TRPA1 either) still retain sensitivity to cold [78–80]. Other studies also describe a population of cold-sensitive neurons that do not express TRPM8 (or TRPA1) [81, 82]. A recent report suggests that the variable threshold of cold-sensitive TG neurons is determined by a balance between TRPM8 and Kv1 potassium channels [83]. These findings point to the existence of additional cold transduction molecule(s)/mechanism(s).

10.4.2 Potential of TRPM8 as a Therapeutic Target for Chronic Pain

Pain is an unpleasant sensory experience associated with hurting and soreness. While acute pain is a normal and necessary alarm/defense mechanism, chronic pain, which can be inflammatory or neuropathic in nature and serves no known beneficial function, is a pathologic and persistent expression of the nervous system. Noxious cold is overtly painful in healthy humans and can cause cold hyperalgesia in patients with neuropathic pain [84, 85]. Menthol decreases the threshold and increases the intensity of cold pain [86–88]. Non-noxious cooling can produce either cold allodynia [84, 85, 89–93] or analgesia [93–95] in patients with various neuropathies/painful conditions. The molecular mechanism underlying cold pain/analgesia is poorly understood. Might TRPM8 play a part in these processes given its role in cold sensing? Studies attempting to address the question have produced results that are both illuminating and puzzling.

Cold hypersensitivity is mediated by capsaicin-sensitive C- or Aδ-primary afferents [96–99]. TRPM8 is expressed in both C- and Aδ-fibers of sensory neurons [70]. In particular, a subset of TRPM8-expressing neurons also express TRPV1 [2, 27, 31, 70, 100], a marker of nociceptive neurons, consistent with the observation that many menthol-/cold-sensitive neurons in primary culture also respond to capsaicin [2, 101–103] and have nociceptive properties [103]. These results suggest that some TRPM8-expressing neurons may be nociceptors, thus presenting a cellular basis for a role of TRPM8 in thermal nociceptive signaling. Data from studies using topical menthol are consistent with a role of TRPM8 in innocuous cold nociception [88, 104]. Two of the three TRPM8 knockout studies also suggest involvement of the channel in nocifensive responses to noxious cold [79, 80].

A number of studies describe increased TRPM8 expression in sensory neurons after nerve injury or inflammation [105–108]. Others, however, report no change or a decrease in TRPM8 expression [109–113]. Notably, expression of TRPM8 is markedly increased in bladder specimens from patients with idiopathic detrusor overactivity and painful bladder syndrome [114]. In addition, there is an increase in the fraction of TRPM8-expressing neurons that also express TRPV1 under inflammatory conditions induced by Complete Freund's Adjuvant (CFA) [31].

Several studies indicate that nerve injury leads to cold hypersensitivity that is, at least in part, mediated by TRPM8. Cold allodynia in rodents induced by chronic constriction injury (CCI) or oxaliplatin, a chemotherapy drug, is significantly attenuated by capsazepine, an antagonist of both TRPM8 and TRPV1, but not by the selective TRPV1 antagonist 5′-iodoresiniferatoxin [106, 108], suggesting a likely TRPM8-mediated effect. Further evidence supporting this conclusion comes from JNJ-39267631, a novel, potent and selective TRPM8 antagonist, which significantly alleviates CCI-induced cold allodynia in rats [66]. JNJ-39267631 also inhibits icilin-induced wet-dog shaking, indicating an on-target effect [66]. Interestingly, JNJ-39267631 also partially reverses CFA-induced radiant heat inflammatory hyperalgesia [66]. Finally, compared to wild-type mice, TRPM8-deficient mice display a significant reduction in acetone-evoked cold hypersensitivity following

CCI and injection of CFA [79]. These results indicate that TRPM8 is critically involved in the development and maintenance of cold hypersensitivity after nerve injury and inflammation and suggest that antagonists of TRPM8 may be effective in treating cold pain hypersensitivity under conditions of nerve injury and inflammation.

Although cold hyperalgesia is observed in rats with spinal nerve ligation (SNL), it is not attenuated by intrathecal administration of TRPM8 antisense oligodeoxynucleotides [113]. It is not clear, however, whether a ~1/3 reduction of TRPM8 protein expression in rat DRG, as that produced by the antisense treatment in this study, would be adequate to significantly mitigate TRPM8-mediated cold responses. Alternatively, the noxious cold temperature (5°C) used in the study may provoke cold hypersensitivity primarily through TRPM8-independent mechanisms. It is also possible that the involvement of TRPM8 in cold hypersensitivity may be neuropathy-model dependent. Another study in patients with cold injury also concludes that the induction of cold allodynia in these patients is independent of TRPM8 (or TRPA1) [115].

Cooling, as well as relatively low doses of menthol or icilin, is also reported to attenuate thermal (noxious heat-induced) and mechanical hypersensitivity in CCI rats [107]. These effects are reversed by antisense knockdown of TRPM8, supporting a critical role of TRPM8 activation in the analgesic effects. Central mechanisms, whereby TRPM8-induced glutamate release activates inhibitory metabotropic glutamate receptors (particularly group II/III mGluRs) in the dorsal horn, are thought to mediate this analgesia. Intrathecal application of icilin similarly attenuates pain behaviors in several rodent models of inflammatory pain [107]. Furthermore, TRPM8 knockout mice lack cooling-induced analgesia normally present in wild-type mice following administration of formalin, a stimulus of acute pain followed by inflammation [80]. These results suggest that activation of TRPM8 can also mediate analgesia in certain acute/inflammatory pain states.

Consistent with these findings, TRPM8 agonists, such as menthol and cooling, are used as traditional remedies for pain relief [95, 116]. Beneficial effects of menthol or (menthol-containing) peppermint oil have also been reported in patients with postherpetic neuralgia, chemotherapy-induced neuropathic pain or preexisting cold allodynia [117–119]. Menthol has been shown to increase noxious heat-induced paw withdrawal latency in rats [120] as well as the pain threshold in the mouse hot-plate and abdominal constriction tests [121]. Eucalyptol, another TRPM8 agonist, is shown to have anti-inflammatory and analgesic properties in animal studies and used to treat rhinosinusitis and muscular pain [122]. It should be cautioned, however, that these agonists are not selective for TRPM8 and may produce effects through non-TRPM8 mechanisms [121].

Inflammation causes cold hypersensitivity via pro-inflammatory mediators such as bradykinin and prostaglandin E_2 (PGE_2). Bradykinin inhibits TRPM8 through the activation of PKC [48], and, as with PGE_2, decreases functional responses of cold- and menthol-sensitive DRG neurons to cooling [123]. Furthermore, extracellular acidification, which can result from inflammation, also inhibits TRPM8 function [51]. It is possible that these inhibitory effects on TRPM8 may, somewhat unintuitively, enhance inflammation-induced cold hypersensitivity by removing the (TRPM8-mediated) analgesic effects of cooling [123].

On the whole, studies of pain hypersensitivity in inflammatory and neuropathic states generally confirm the role of TRPM8 as a key mediator. Interestingly, both activation and inhibition of the channel appear important in attenuating pain hypersensitivity under various conditions. This raises the possibility that TRPM8 agonists as well as antagonists may have therapeutic utility in pain. However, there are critical questions about the role of TRPM8 in pain signaling that still remain unanswered. How can/does TRPM8 mediate both pain and analgesia? Why is TRPM8 activation analgesic under some conditions but painful in others? A better understanding of the mechanisms involved in these processes is critical for both definitively establishing the role of TRPM8 in pain and developing a viable therapeutic roadmap.

10.4.3 Potential of TRPM8 as a Marker and Therapeutic Target for Prostate Cancer

Prostate cancer is among the most prevalent of cancers in men. Currently available methods for screening, diagnosis and treatment have not significantly helped to reduce the rate of mortality from prostate cancer. There remains a critical need for identifying effective diagnostic markers and novel drug targets to improve the treatment and outcome of prostate cancer patients.

Since the first report linking TRPM8 with prostate cancer [1], growing interest in the role of TRPM8 in normal and cancerous prostates has contributed significantly to the understanding of how this channel works in these tissues. While the precise functions of TRPM8 in normal and malignant prostates remain to be well established, emerging evidence indicates that TRPM8 may be an important player in prostate carcinogenesis and have the potential as a therapeutic target as well as a diagnostic/prognostic marker for prostate cancer [124].

Expression of TRPM8 in the prostate (particularly in prostate epithelium) is regulated by androgen. It is elevated in relatively early-stage, androgen-sensitive prostate cancer cells [1, 21, 125, 126]. Androgen sensitivity is subsequently lost as the disease progresses to more advanced and metastatic stages, substantially lowering the level of TRPM8 expression in the process [21, 125]. Similarly, TRPM8 expression is also decreased significantly in prostate cancer tissues from patients treated with anti-androgen therapy, suggesting the possible progression to a more advanced disease state [21]. Studies show that TRPM8 levels are significantly different between malignant and non-malignant prostate tissues [127, 128] and between the urine or blood samples of patients with metastatic (but not localized) prostate cancer and those of healthy men [129]. These results suggest that TRPM8 may be potentially useful as a marker for differential diagnosis and androgen-unresponsive and metastatic prostate cancer.

The precise role of TRPM8 in prostate cancer is not well understood. However, several studies suggest that TRPM8 is likely involved in the regulation of proliferation and/or apoptosis of cancer cells. In the androgen-responsive lymph node carcinoma of the prostate (LNCaP) cell line, TRPM8 is regulated by androgen, functions as a Ca^{2+}-permeable channel, and is required for cell survival as inhibition of the channel by capsazepine or TRPM8-sepecific siRNA decreases cell viability [126]. Interestingly, menthol can also induce apoptosis [126]. These results suggest that the TRPM8 activity is tightly regulated in these cells – while it is necessary for cell survival at the resting level, overactivity can lead to cell toxicity, possibly due to intracellular Ca^{2+} and/or Na^+ overload.

TRPM8 expression is normally low in the androgen-insensitive prostate cell line PC-3. However, transfection of the channel in these cells induces cell cycle arrest and facilitates starvation-induced apoptosis [130]. Furthermore, migration is inhibited in PC-3 cells overexpressing TRPM8. These data suggest that although TRPM8 is not essential for the survival of androgen-independent prostate cancer cells, excessive TRPM8 activity negatively influences cell proliferation and migration. Notably, menthol-induced activation of TRPM8 also decreases the viability of human melanoma [131] and bladder cancer cells [132]. It should be noted, however, that conclusions based on results from established cancer cell lines should be drawn with caution, as culture conditions may alter the expression pattern in vitro such that it differs significantly from in vivo expression by cancerous tissues.

In addition to the plasma membrane, TRPM8 is also expressed in the ER membrane of prostate epithelial cells, where it appears to function as a Ca^{2+}-release channel involved in store-operated Ca^{2+} entry [126, 133, 134]. It is not clear whether the ER membrane expresses the full-length TRPM8, a recently described splice variant [135], both, or possibly other, yet-to-be-identified variants. The function and relative subcellular localization of TRPM8 are dependent on the differentiation status of prostate cells [136]. It is suggested that the balance between TRPM8 expression in the plasma and ER membranes may be important for maintaining the Ca^{2+} homeostasis in prostate epithelial cells. Tipping the balance may increase the potential for either proliferation or apoptosis. Therefore, depending on the stage and androgen-sensitivity of the targeted prostate cancer, selective modulation of either ER or plasma membrane TRPM8 may help to restore the normal cell homeostasis [136].

Although the exact nature of TRPM8's involvement in prostate cancer cell viability (e.g., pro–proliferative or anti–apoptotic) is not well understood, evidence to date suggests that it is possible to affect the viability of these cells by pharmacological manipulation of the functional expression of TRPM8, potentially offering novel therapy for treating prostate cancer. Thus, TRPM8 antagonists (and possibly agonists as well) may be effective during the androgen–responsive stages, whereas agonists may have efficacy in later–stage, hormone–refractory prostate cancers.

To this end, Dendreon has initiated a Phase 1 clinical trial of an orally bioavailable small molecule TRPM8 agonist, D–3263, in patients with advanced cancer (http://investor.dendreon.com/phoenix.zhtml?c=120739&p=irol-newsArticle&ID=1368984&highlight). Preclinical studies indicate that D-3263 selectively induces death of TRPM8–overexpressing cancer cells in vitro and significantly inhibits the growth of TRPM8–expressing and human prostate cancer xenograft tumors in animal models. In addition, D–3263 also significantly reduces the effects of androgen–induced benign prostatic hyperplasia (BPH) in animal models. (See http://investor.dendreon.com/phoenix.zhtml?c=120739&p=irol-newsArticle&ID=1368799&highlight and http://investor.dendreon.com/phoenix.zhtml?c=120739&p=irol-newsArticle&ID=1368824&highlight). These results support the potential for D-3263 as a novel investigational therapeutic for cancer, as well as possibly for BPH.

10.5 Concluding Remarks

The role of TRPM8 as a major sensor of ambient cool to cold temperatures is now well established. TRPM8 also appears to play an important part in chronic pain, causing, for instance, cold hypersensitivity or analgesia under conditions of nerve injury and inflammation. However, much still needs to be understood about the nature of TRPM8's role in various pain states in order to effectively guide drug discovery and therapeutic strategies. Studies on the role of TRPM8 in cancer, particularly prostate cancer, have also seen encouraging progress. But the precise function of TRPM8 in cancer cells remains elusive. Future studies in these and other areas will undoubtedly benefit greatly from the use of selective TRPM8 modulators, which are just beginning to emerge. TRPM8 is more broadly expressed in the body than early studies suggested. This could mean new therapeutic opportunities as well as potential challenges of on-target toxicity. A recent study, for example, suggests that inhibition of TRPM8 may have utility in treating overactive bladder and painful bladder syndromes [67]. TRPM8 is also implicated in thermoregulation, as menthol and icilin both induce hyperthermia [137–139]. In light of the role of TRPV1 in this respect [140], it will be important to evaluate the extent to which tonic activation of TRPM8 occurs in vivo and effects, if any, of selective TRPM8 antagonists on thermoregulation. The potential of TRPM8 as a viable disease target for therapeutic intervention rests on our further understanding of the role of this channel in both normal and diseased states, as well as on the ability of therapeutic molecules to achieve a fine balance between efficacy and toxicity.

References

1. Tsavaler L, Shapero MH, Morkowski S, Laus R (2001) Trp-p8, a novel prostate-specific gene, is up-regulated in prostate cancer and other malignancies and shares high homology with transient receptor potential calcium channel proteins. Cancer Res 61:3760–3769
2. McKemy DD, Neuhausser WM, Julius D (2002) Identification of a cold receptor reveals a general role for TRP channels in thermosensation. Nature 416:52–58
3. Peier AM, Moqrich A, Hergarden AC, Reeve AJ, Andersson DA, Story GM, Earley TJ, Dragoni I, McIntyre P, Bevan S, Patapoutian A (2002) A TRP channel that senses cold stimuli and menthol. Cell 108:705–715

4. Liu Y, Lubin ML, Reitz TL, Wang Y, Colburn RW, Flores CM, Qin N (2006) Molecular identification and functional characterization of a temperature-sensitive transient receptor potential channel (TRPM8) from canine. Eur J Pharmacol 530:23–32
5. Chuang HH, Neuhausser WM, Julius D (2004) The super-cooling agent icilin reveals a mechanism of coincidence detection by a temperature-sensitive TRP channel. Neuron 43:859–869
6. Myers BR, Sigal YM, Julius D (2009) Evolution of thermal response properties in a cold-activated TRP channel. PLoS One 4:e5741
7. Stewart AP, Egressy K, Lim A, Edwardson JM (2010) AFM imaging reveals the tetrameric structure of the TRPM8 channel. Biochem Biophys Res Commun 394:383–386
8. Dragoni I, Guida E, McIntyre P (2006) The cold and menthol receptor TRPM8 contains a functionally important double cysteine motif. J Biol Chem 281:37353–37360
9. Kuhn FJP, Knop G, Luckhoff A (2007) The transmembrane segment S6 determines cation versus anion selectivity of TRPM2 and TRPM8. J Biol Chem 282:27598–27609
10. Voets T, Owsianik G, Janssens A, Talavera K, Nilius B (2007) TRPM8 voltage sensor mutants reveal a mechanism for integrating thermal and chemical stimuli. Nat Chem Biol 3:174–182
11. Voets T, Droogmans G, Wissenbach U, Janssens A, Flockerzi V, Nilius B (2004) The principle of temperature-dependent gating in cold- and heat-sensitive TRP channels. Nature 430:748–754
12. Phelps CB, Gaudet R (2007) The role of the N terminus and transmembrane domain of TRPM8 in channel localization and tetramerization. J Biol Chem 282:36474–36480
13. Erler I, Al-Ansary DM, Wissenbach U, Wagner TF, Flockerzi V, Niemeyer BA (2006) Trafficking and assembly of the cold-sensitive TRPM8 channel. J Biol Chem 281: 38396–38404
14. Tsuruda PR, Julius D, Minor JDL (2006) Coiled coils direct assembly of a cold-activated TRP channel. Neuron 51:201–212
15. Jordt SE, Julius D (2002) Molecular basis for species-specific sensitivity to "hot" chili peppers. Cell 108:421–430
16. Gavva NR, Klionsky L, Qu Y, Shi L, Tamir R, Edenson S, Zhang TJ, Viswanadhan VN, Toth A, Pearce LV, Vanderah TW, Porreca F, Blumberg PM, Lile J, Sun Y, Wild K, Louis JC, Treanor JJ (2004) Molecular determinants of vanilloid sensitivity in TRPV1. J Biol Chem 279:20283–20295
17. Vriens J, Watanabe H, Janssens A, Droogmans G, Voets T, Nilius B (2004) Cell swelling, heat, and chemical agonists use distinct pathways for the activation of the cation channel TRPV4. Proc Natl Acad Sci USA 101:396–401
18. Bandell M, Dubin AE, Petrus MJ, Orth A, Mathur J, Hwang SW, Patapoutian A (2006) High-throughput random mutagenesis screen reveals TRPM8 residues specifically required for activation by menthol. Nat Neurosci 9:493–500
19. Rohacs T, Lopes CM, Michailidis I, Logothetis DE (2005) PI(4,5)P2 regulates the activation and desensitization of TRPM8 channels through the TRP domain. Nat Neurosci 8: 626–634
20. Brauchi S, Orta G, Salazar M, Rosenmann E, Latorre R (2006) A hot-sensing cold receptor: C-terminal domain determines thermosensation in transient receptor potential channels. J Neurosci 26:4835–4840
21. Henshall SM, Afar DE, Hiller J, Horvath LG, Quinn DI, Rasiah KK, Gish K, Willhite D, Kench JG, Gardiner-Garden M, Stricker PD, Scher HI, Grygiel JJ, Agus DB, Mack DH, Sutherland RL (2003) Survival analysis of genome-wide gene expression profiles of prostate cancers identifies new prognostic targets of disease relapse. Cancer Res 63:4196–4203
22. Zhang L, Jones S, Brody K, Costa M, Brookes SJ (2004) Thermosensitive transient receptor potential channels in vagal afferent neurons of the mouse. Am J Physiol Gastrointest Liver Physiol 286:G983–G991

23. Zhao H, Sprunger LK, Simasko SM (2009) Expression of transient receptor potential channels and two-pore potassium channels in subtypes of vagal afferent neurons in rat. Am J Physiol Gastrointest Liver Physiol 298:G212–G221
24. Staaf S, Franck MCM, Marmigère F, Mattsson JP, Ernfors P (2010) Dynamic expression of the TRPM subgroup of ion channels in developing mouse sensory neurons. Gene Expr Patterns 10:65–74
25. Stein RJ, Santos S, Nagatomi J, Hayashi Y, Minnery BS, Xavier M, Patel AS, Nelson JB, Futrell WJ, Yoshimura N, Chancellor MB, De Miguel F (2004) Cool (TRPM8) and hot (TRPV1) receptors in the bladder and male genital tract. J Urol 172:1175–1178
26. De Blas GA, Darszon A, Ocampo AY, Serrano CJ, Castellano LE, Hernandez-Gonzalez EO, Chirinos M, Larrea F, Beltran C, Trevino CL (2009) TRPM8, a versatile channel in human sperm. PLoS One 4:e6095
27. Abe J, Hosokawa H, Okazawa M, Kandachi M, Sawada Y, Yamanaka K, Matsumura K, Kobayashi S (2005) TRPM8 protein localization in trigeminal ganglion and taste papillae. Brain Res Mol Brain Res 136:91–98
28. Mustafa S, Oriowo M (2005) Cooling-induced contraction of the rat gastric fundus: mediation via transient receptor potential (TRP) cation channel TRPM8 receptor and Rho-kinase activation. Clin Exp Pharmacol Physiol 32:832–838
29. Yang XR, Lin MJ, McIntosh LS, Sham JS (2006) Functional expression of transient receptor potential melastatin- and vanilloid-related channels in pulmonary arterial and aortic smooth muscle. Am J Physiol Lung Cell Mol Physiol 290:L1267–L1276
30. Sabnis AS, Shadid M, Yost GS, Reilly CA (2008) Human lung epithelial cells express a functional cold-sensing TRPM8 variant. Am J Respir Cell Mol Biol 39:466–474
31. Dhaka A, Earley TJ, Watson J, Patapoutian A (2008) Visualizing cold spots: TRPM8-expressing sensory neurons and their projections. J Neurosci 28:566–575
32. Du J, Yang X, Zhang L, Zeng YM (2009) Expression of TRPM8 in the distal cerebrospinal fluid-contacting neurons in the brain mesencephalon of rats. Cerebrospinal Fluid Res 6:3
33. Hondoh A, Ishida Y, Ugawa S, Ueda T, Shibata Y, Yamada T, Shikano M, Murakami S, Shimada S (2010) Distinct expression of cold receptors (TRPM8 and TRPA1) in the rat nodose-petrosal ganglion complex. Brain Res 1319:60–69
34. Sabnis AS, Reilly CA, Veranth JM, Yost GS (2008) Increased transcription of cytokine genes in human lung epithelial cells through activation of a TRPM8 variant by cold temperatures. Am J Physiol Lung Cell Mol Physiol 295:L194–L200
35. Hui K, Guo Y, Feng Z-P (2005) Biophysical properties of menthol-activated cold receptor TRPM8 channels. Biochem Biophys Res Commun 333:374–382
36. Brauchi S, Orio P, Latorre R (2004) Clues to understanding cold sensation: thermodynamics and electrophysiological analysis of the cold receptor TRPM8. Proc Natl Acad Sci USA 101:15494–15499
37. Malkia A, Madrid R, Meseguer V, de la Pena E, Valero M, Belmonte C, Viana F (2007) Bidirectional shifts of TRPM8 channel gating by temperature and chemical agents modulate the cold sensitivity of mammalian thermoreceptors. J Physiol 581:155–174
38. Kuhn FJ, Kuhn C, Luckhoff A (2009) Inhibition of TRPM8 by icilin distinct from desensitization induced by menthol and menthol derivatives. J Biol Chem 284:4102–4111
39. Behrendt HJ, Germann T, Gillen C, Hatt H, Jostock R (2004) Characterization of the mouse cold-menthol receptor TRPM8 and vanilloid receptor type-1 VR1 using a fluorometric imaging plate reader (FLIPR) assay. Br J Pharmacol 141:737–745
40. Bödding M, Wissenbach U, Flockerzi V (2007) Characterisation of TRPM8 as a pharmacophore receptor. Cell Calcium 42:618–628
41. Beck B, Bidaux G, Bavencoffe A, Lemonnier L, Thebault S, Shuba Y, Barrit G, Skryma R, Prevarskaya N (2007) Prospects for prostate cancer imaging and therapy using high-affinity TRPM8 activators. Cell Calcium 41:285–294
42. Hutchinson T, Liu Y, Flores CM, Qin N Differential agonist and antagonist effects of cannabinoids and other chemo-sensory stimuli on thermosensitive TRP channels. In *Society for Neuroscience Meeting,* Washington, DC, 2008

43. Vanden Abeele F, Zholos A, Bidaux G, Shuba Y, Thebault S, Beck B, Flourakis M, Panchin Y, Skryma R, Prevarskaya N (2006) Ca2+-independent phospholipase A2-dependent gating of TRPM8 by lysophospholipids. J Biol Chem 281:40174–40182
44. Andersson DA, Nash M, Bevan S (2007) Modulation of the cold-activated channel TRPM8 by lysophospholipids and polyunsaturated fatty acids. J Neurosci 27:3347–3355
45. Rohacs T, Nilius B (2007) Regulation of transient receptor potential (TRP) channels by phosphoinositides. Pflugers Arch 455:157–168
46. Liu B, Qin F (2005) Functional control of cold- and menthol-sensitive TRPM8 ion channels by phosphatidylinositol 4,5-bisphosphate. J Neurosci 25:1674–1681
47. Daniels RL, Takashima Y, McKemy DD (2009) Activity of the neuronal cold sensor TRPM8 is regulated by phospholipase C via the phospholipid phosphoinositol 4,5-bisphosphate. J Biol Chem 284:1570–1582
48. Premkumar LS, Raisinghani M, Pingle SC, Long C, Pimentel F (2005) Downregulation of transient receptor potential melastatin 8 by protein kinase C-mediated dephosphorylation. J Neurosci 25:11322–11329
49. Abe J, Hosokawa H, Sawada Y, Matsumura K, Kobayashi S (2006) Ca2+-dependent PKC activation mediates menthol-induced desensitization of transient receptor potential M8. Neurosci Lett 397:140–144
50. Bavencoffe A, Gkika D, Kondratskyi A, Beck B, Borowiec AS, Bidaux G, Busserolles J, Eschalier A, Shuba Y, Skryma R, Prevarskaya N (2010) The transient receptor potential channel TRPM8 is inhibited via the alpha 2A adrenoreceptor signaling pathway. J Biol Chem 285:9410–9419
51. Andersson DA, Chase HW, Bevan S (2004) TRPM8 activation by menthol, icilin, and cold is differentially modulated by intracellular pH. J Neurosci 24:5364–5369
52. Zakharian E, Thyagarajan B, French RJ, Pavlov E, Rohacs T (2009) Inorganic polyphosphate modulates TRPM8 channels. PLoS One 4:e5404
53. Morenilla-Palao C, Pertusa M, Meseguer V, Cabedo H, Viana F (2009) Lipid raft segregation modulates TRPM8 channel activity. J Biol Chem 284:9215–9224
54. Weil A, Moore SE, Waite NJ, Randall A, Gunthorpe MJ (2005) Conservation of functional and pharmacological properties in the distantly related temperature sensors TRPV1 and TRPM8. Mol Pharmacol 68:518–527
55. De Petrocellis L, Starowicz K, Moriello AS, Vivese M, Orlando P, Di Marzo V (2007) Regulation of transient receptor potential channels of melastatin type 8 (TRPM8): effect of cAMP, cannabinoid CB1 receptors and endovanilloids. Exp Cell Res 313:1911–1920
56. Hu H-Z, Gu Q, Wang C, Colton CK, Tang J, Kinoshita-Kawada M, Lee L-Y, Wood JD, Zhu MX (2004) 2-Aminoethoxydiphenyl borate is a common activator of TRPV1, TRPV2, and TRPV3. J Biol Chem 279:35741–35748
57. Macpherson LJ, Hwang SW, Miyamoto T, Dubin AE, Patapoutian A, Story GM (2006) More than cool: promiscuous relationships of menthol and other sensory compounds. Mol Cell Neurosci 32:335–343
58. Niforatos W, Zhang XF, Lake MR, Walter KA, Neelands T, Holzman TF, Scott VE, Faltynek CR, Moreland RB, Chen J (2007) Activation of TRPA1 channels by the fatty acid amide hydrolase inhibitor 3′-carbamoylbiphenyl-3-yl cyclohexylcarbamate (URB597). Mol Pharmacol 71:1209–1216
59. Meseguer V, Karashima Y, Talavera K, D'Hoedt D, Donovan-Rodriguez T, Viana F, Nilius B, Voets T (2008) Transient receptor potential channels in sensory neurons are targets of the antimycotic agent clotrimazole. J Neurosci 28:576–586
60. Kraft R, Grimm C, Frenzel H, Harteneck C (2006) Inhibition of TRPM2 cation channels by N-(p-amylcinnamoyl)anthranilic acid. Br J Pharmacol 148:264–273
61. Benedikt J, Teisinger J, Vyklicky L, Vlachova V (2007) Ethanol inhibits cold-menthol receptor TRPM8 by modulating its interaction with membrane phosphatidylinositol 4,5-bisphosphate. J Neurochem 100:211–224
62. De Petrocellis L, Vellani V, Schiano-Moriello A, Marini P, Magherini PC, Orlando P, Di Marzo V (2008) Plant-derived cannabinoids modulate the activity of transient receptor

potential channels of ankyrin type-1 and melastatin type-8. J Pharmacol Exp Ther 325: 1007–1015
63. Neeper MP, Liu Y, Hutchinson TL, Wang Y, Flores CM, Qin N (2007) Activation properties of heterologously expressed mammalian TRPV2: evidence for species dependence. J Biol Chem 282:15894–15902
64. Qin N, Neeper MP, Liu Y, Hutchinson TL, Lubin ML, Flores CM (2008) TRPV2 is activated by cannabidiol and mediates CGRP release in cultured rat dorsal root ganglion neurons. J Neurosci 28:6231–6238
65. Jordt S-E, Bautista DM, Chuang HH, McKemy DD, Zygmunt PM, Hogestatt ED, Meng ID, Julius D (2004) Mustard oils and cannabinoids excite sensory nerve fibres through the TRP channel ANKTM1. Nature 427:260–265
66. Colburn RW, Matthews JM, Qin N, Liu Y, Hutchinson TL, Schneider CR, Stone DJJ, Lubin M, Pavlick KP, Kenigs VA, Dax SL, Brandt MR, Flores CM Small-molecule TRPM8 antagonist JNJ-39267631 reverses neuropathy-induced cold allodynia in rats. In *12th IASP World Congress on Pain,* Glasgow, UK, 2008
67. Lashinger ES, Steiginga MS, Hieble JP, Leon LA, Gardner SD, Nagilla R, Davenport EA, Hoffman BE, Laping NJ, Su X (2008) AMTB, a TRPM8 channel blocker: evidence in rats for activity in overactive bladder and painful bladder syndrome. Am J Physiol Renal Physiol 295:F803–F810
68. Ortar G, De Petrocellis L, Morera L, Moriello AS, Orlando P, Morera E, Nalli M, Di Marzo V (2010) (–)-Methylamine derivatives as potent and selective antagonists of transient receptor potential melastatin type-8 (TRPM8) channels. Bioorg Med Chem Lett 20:2729–2732
69. Foulkes T, Wood JN (2007) Mechanisms of cold pain. Channels (Austin) 1:154–160
70. Takashima Y, Daniels RL, Knowlton W, Teng J, Liman ER, McKemy DD (2007) Diversity in the neural circuitry of cold sensing revealed by genetic axonal labeling of transient receptor potential melastatin 8 neurons. J Neurosci 27:14147–14157
71. Kobayashi K, Fukuoka T, Obata K, Yamanaka H, Dai Y, Tokunaga A, Noguchi K (2005) Distinct expression of TRPM8, TRPA1, and TRPV1 mRNAs in rat primary afferent neurons with adelta/c-fibers and colocalization with trk receptors. J Comp Neurol 493:596–606
72. Nealen ML, Gold MS, Thut PD, Caterina MJ (2003) TRPM8 mRNA is expressed in a subset of cold-responsive trigeminal neurons from rat. J Neurophysiol 90:515–520
73. Reid G, Babes A, Pluteanu F (2002) A cold- and menthol-activated current in rat dorsal root ganglion neurones: properties and role in cold transduction. J Physiol 545:595–614
74. Okazawa M, Takao K, Hori A, Shiraki T, Matsumura K, Kobayashi S (2002) Ionic basis of cold receptors acting as thermostats. J Neurosci 22:3994–4001
75. Reid G, Flonta ML (2001) Physiology. Cold current in thermoreceptive neurons. Nature 413:480
76. McKemy DD (2007) TRPM8: The cold and menthol receptor. In: Liedtke WB, Heller S, (Eds.) TRP ion channel function in sensory transduction and cellular signaling cascades. CRC Press, London
77. Wei ET, Seid DA (1983) AG-3-5: a chemical producing sensations of cold. J Pharm Pharmacol 35:110–112
78. Bautista DM, Siemens J, Glazer JM, Tsuruda PR, Basbaum AI, Stucky CL, Jordt SE, Julius D (2007) The menthol receptor TRPM8 is the principal detector of environmental cold. Nature 448:204–208
79. Colburn RW, Lubin ML, Stone DJ Jr, Wang Y, Lawrence D, D'Andrea MR, Brandt MR, Liu Y, Flores CM, Qin N (2007) Attenuated cold sensitivity in TRPM8 null mice. Neuron 54:379–386
80. Dhaka A, Murray AN, Mathur J, Earley TJ, Petrus MJ, Patapoutian A (2007) TRPM8 is required for cold sensation in mice. Neuron 54:371–378
81. Munns C, AlQatari M, Koltzenburg M (2007) Many cold sensitive peripheral neurons of the mouse do not express TRPM8 or TRPA1. Cell Calcium 41:331–342

82. Babes A, Zorzon D, Reid G (2006) A novel type of cold-sensitive neuron in rat dorsal root ganglia with rapid adaptation to cooling stimuli. Eur J Neurosci 24:691–698
83. Madrid R, de la Pena E, Donovan-Rodriguez T, Belmonte C, Viana F (2009) Variable threshold of trigeminal cold-thermosensitive neurons is determined by a balance between TRPM8 and Kv1 potassium channels. J Neurosci 29:3120–3131
84. Attal N, Guirimand F, Brasseur L, Gaude V, Chauvin M, Bouhassira D (2002) Effects of IV morphine in central pain: a randomized placebo-controlled study. Neurology 58:554–563
85. Jorum E, Warncke T, Stubhaug A (2003) Cold allodynia and hyperalgesia in neuropathic pain: the effect of N-methyl-D-aspartate (NMDA) receptor antagonist ketamine – a double-blind, cross-over comparison with alfentanil and placebo. Pain 101:229–235
86. Hatem S, Attal N, Willer JC, Bouhassira D (2006) Psychophysical study of the effects of topical application of menthol in healthy volunteers. Pain 122:190–196
87. Namer B, Seifert F, Handwerker HO, Maihofner C (2005) TRPA1 and TRPM8 activation in humans: effects of cinnamaldehyde and menthol. NeuroReport 16:955–959
88. Wasner G, Schattschneider J, Binder A, Baron R (2004) Topical menthol – a human model for cold pain by activation and sensitization of C nociceptors. Brain 127:1159–1171
89. Engkvist O, Wahren LK, Wallin G, Torebjrk E, Nystrom B (1985) Effects of regional intravenous guanethidine block in posttraumatic cold intolerance in hand amputees. J Hand Surg Br 10:145–150
90. Campbell DA, Kay SP (1998) What is cold intolerance? J Hand Surg Br 23:3–5
91. Finnerup NB, Biering-Sorensen F, Johannesen IL, Terkelsen AJ, Juhl GI, Kristensen AD, Sindrup SH, Bach FW, Jensen TS (2005) Intravenous lidocaine relieves spinal cord injury pain: a randomized controlled trial. Anesthesiology 102:1023–1030
92. Leung A, Wallace MS, Ridgeway B, Yaksh T (2001) Concentration-effect relationship of intravenous alfentanil and ketamine on peripheral neurosensory thresholds, allodynia and hyperalgesia of neuropathic pain. Pain 91:177–187
93. Lindblom U, Verrillo RT (1979) Sensory functions in chronic neuralgia. J Neurol Neurosurg Psychiatry 42:422–435
94. Bini G, Cruccu G, Hagbarth KE, Schady W, Torebjork E (1984) Analgesic effect of vibration and cooling on pain induced by intraneural electrical stimulation. Pain 18:239–248
95. Sauls J (1999) Efficacy of cold for pain: fact or fallacy? Online J Knowl Synth Nurs 6:8
96. Hama AT (2002) Capsaicin-sensitive primary afferents mediate responses to cold in rats with a peripheral mononeuropathy. NeuroReport 13:461–464
97. Hao JX, Yu W, Xu XJ, Wiesenfeld-Hallin Z (1996) Capsaicin-sensitive afferents mediate chronic cold, but not mechanical, allodynia-like behavior in spinally injured rats. Brain Res 722:177–180
98. Kress M, Koltzenburg M, Reeh PW, Handwerker HO (1992) Responsiveness and functional attributes of electrically localized terminals of cutaneous C-fibers in vivo and in vitro. J Neurophysiol 68:581–595
99. LaMotte RH, Lundberg LE, Torebjork HE (1992) Pain, hyperalgesia and activity in nociceptive C units in humans after intradermal injection of capsaicin. J Physiol 448: 749–764
100. Okazawa M, Inoue W, Hori A, Hosokawa H, Matsumura K, Kobayashi S (2004) Noxious heat receptors present in cold-sensory cells in rats. Neurosci Lett 359:33–36
101. Viana F, de la Pena E, Belmonte C (2002) Specificity of cold thermotransduction is determined by differential ionic channel expression. Nat Neurosci 5:254–260
102. Hjerling-Leffler J, Alqatari M, Ernfors P, Koltzenburg M (2007) Emergence of functional sensory subtypes as defined by transient receptor potential channel expression. J Neurosci 27:2435–2443
103. Xing H, Ling J, Chen M, Gu JG (2006) Chemical and cold sensitivity of two distinct populations of TRPM8-expressing somatosensory neurons. J Neurophysiol 95: 1221–1230
104. Green BG, Schoen KL (2007) Thermal and nociceptive sensations from menthol and their suppression by dynamic contact. Behav Brain Res 176:284–291

105. Frederick J, Buck ME, Matson DJ, Cortright DN (2007) Increased TRPA1, TRPM8, and TRPV2 expression in dorsal root ganglia by nerve injury. Biochem Biophys Res Commun 358:1058–1064
106. Xing H, Chen M, Ling J, Tan W, Gu JG (2007) TRPM8 mechanism of cold allodynia after chronic nerve injury. J Neurosci 27:13680–13690
107. Proudfoot CJ, Garry EM, Cottrell DF, Rosie R, Anderson H, Robertson DC, Fleetwood-Walker SM, Mitchell R (2006) Analgesia mediated by the TRPM8 cold receptor in chronic neuropathic pain. Curr Biol 16:1591–1605
108. Gauchan P, Andoh T, Kato A, Kuraishi Y (2009) Involvement of increased expression of transient receptor potential melastatin 8 in oxaliplatin-induced cold allodynia in mice. Neurosci Lett 458:93–95
109. Caspani O, Zurborg S, Labuz D, Heppenstall PA (2009) The contribution of TRPM8 and TRPA1 channels to cold allodynia and neuropathic pain. PLoS One 4:e7383
110. Staaf S, Oerther S, Lucas G, Mattsson JP, Ernfors P (2009) Differential regulation of TRP channels in a rat model of neuropathic pain. Pain 144:187–199
111. Persson AK, Gebauer M, Jordan S, Metz-Weidmann C, Schulte AM, Schneider HC, Ding-Pfennigdorff D, Thun J, Xu XJ, Wiesenfeld-Hallin Z, Darvasi A, Fried K, Devor M (2009) Correlational analysis for identifying genes whose regulation contributes to chronic neuropathic pain. Mol Pain 5:7
112. Obata K, Katsura H, Mizushima T, Yamanaka H, Kobayashi K, Dai Y, Fukuoka T, Tokunaga A, Tominaga M, Noguchi K (2005) TRPA1 induced in sensory neurons contributes to cold hyperalgesia after inflammation and nerve injury. J Clin Invest 115:2393–2401
113. Katsura H, Obata K, Mizushima T, Yamanaka H, Kobayashi K, Dai Y, Fukuoka T, Tokunaga A, Sakagami M, Noguchi K (2006) Antisense knock down of TRPA1, but not TRPM8, alleviates cold hyperalgesia after spinal nerve ligation in rats. Exp Neurol 200:112–123
114. Mukerji G, Yiangou Y, Corcoran SL, Selmer IS, Smith GD, Benham CD, Bountra C, Agarwal SK, Anand P (2006) Cool and menthol receptor TRPM8 in human urinary bladder disorders and clinical correlations. BMC Urol 6:6
115. Namer B, Kleggetveit IP, Handwerker H, Schmelz M, Jorum E (2008) Role of TRPM8 and TRPA1 for cold allodynia in patients with cold injury. Pain 139:63–72
116. Wright A (1870) Oil of peppermint as a local anaesthetic. Lancet 2464:726
117. Colvin LA, Johnson PR, Mitchell R, Fleetwood-Walker SM, Fallon M (2008) From bench to bedside: a case of rapid reversal of bortezomib-induced neuropathic pain by the TRPM8 activator, menthol. J Clin Oncol 26:4519–4520
118. Davies SJ, Harding LM, Baranowski AP (2002) A novel treatment of postherpetic neuralgia using peppermint oil. Clin J Pain 18:200–202
119. Wasner G, Naleschinski D, Binder A, Schattschneider J, McLachlan EM, Baron R (2008) The effect of menthol on cold allodynia in patients with neuropathic pain. Pain Med 9:354–358
120. Klein AH, Sawyer CM, Carstens MI, Tsagareli M, Tsiklauri N, Carstens E (2010) Topical application of l menthol induces heat analgesia, mechanical allodynia, and a biphasic effect on cold sensitivity in rats. Behav Brain Res 212:179–186
121. Galeotti N, Di Cesare Mannelli L, Mazzanti G, Bartolini A, Ghelardini C (2002) Menthol: a natural analgesic compound. Neurosci Lett 322:145–148
122. Calixto JB, Kassuya CA, Andre E, Ferreira J (2005) Contribution of natural products to the discovery of the transient receptor potential (TRP) channels family and their functions. Pharmacol Ther 106:179–208
123. Linte RM, Ciobanu C, Reid G, Babes A (2007) Desensitization of cold- and menthol-sensitive rat dorsal root ganglion neurones by inflammatory mediators. Exp Brain Res 178:89–98
124. Zhang L, Barritt GJ (2006) TRPM8 in prostate cancer cells: a potential diagnostic and prognostic marker with a secretory function? Endocr Relat Cancer 13:27–38

125. Bidaux G, Roudbaraki M, Merle C, Crepin A, Delcourt P, Slomianny C, Thebault S, Bonnal JL, Benahmed M, Cabon F, Mauroy B, Prevarskaya N (2005) Evidence for specific TRPM8 expression in human prostate secretory epithelial cells: functional androgen receptor requirement. Endocr Relat Cancer 12:367–382
126. Zhang L, Barritt GJ (2004) Evidence that TRPM8 is an androgen-dependent Ca2+ channel required for the survival of prostate cancer cells. Cancer Res 64:8365–8373
127. Fuessel S, Sickert D, Meye A, Klenk U, Schmidt U, Schmitz M, Rost AK, Weigle B, Kiessling A, Wirth MP (2003) Multiple tumor marker analyses (PSA, hK2, PSCA, trp-p8) in primary prostate cancers using quantitative RT-PCR. Int J Oncol 23:221–228
128. Prevarskaya N, Skryma R, Bidaux G, Flourakis M, Shuba Y (2007) Ion channels in death and differentiation of prostate cancer cells. Cell Death Differ 14:1295–1304
129. Bai VU, Murthy S, Chinnakannu K, Muhletaler F, Tejwani S, Barrack ER, Kim SH, Menon M, Veer Reddy GP (2010) Androgen regulated TRPM8 expression: a potential mRNA marker for metastatic prostate cancer detection in body fluids. Int J Oncol 36:443–450
130. Yang ZH, Wang XH, Wang HP, Hu LQ (2009) Effects of TRPM8 on the proliferation and motility of prostate cancer PC-3 cells. Asian J Androl 11:157–165
131. Yamamura H, Ugawa S, Ueda T, Morita A, Shimada S (2008) TRPM8 activation suppresses cellular viability in human melanoma. Am J Physiol Cell Physiol 295:C296–C301
132. Li Q, Wang X, Yang Z, Wang B, Li S (2009) Menthol induces cell death via the TRPM8 channel in the human bladder cancer cell line T24. Oncology 77:335–341
133. Thebault S, Lemonnier L, Bidaux G, Flourakis M, Bavencoffe A, Gordienko D, Roudbaraki M, Delcourt P, Panchin Y, Shuba Y, Skryma R, Prevarskaya N (2005) Novel role of cold/menthol-sensitive transient receptor potential melastatine family member 8 (TRPM8) in the activation of store-operated channels in LNCaP human prostate cancer epithelial cells. J Biol Chem 280:39423–39435
134. Tsuzuki K, Xing H, Ling J, Gu JG (2004) Menthol-induced Ca2+ release from presynaptic Ca2+ stores potentiates sensory synaptic transmission. J Neurosci 24:762–771
135. Lis A, Wissenbach U, Philipp SE (2005) Transcriptional regulation and processing increase the functional variability of TRPM channels. Naunyn Schmiedebergs Arch Pharmacol 371:315–324
136. Bidaux G, Flourakis M, Thebault S, Zholos A, Beck B, Gkika D, Roudbaraki M, Bonnal JL, Mauroy B, Shuba Y, Skryma R, Prevarskaya N (2007) Prostate cell differentiation status determines transient receptor potential melastatin member 8 channel subcellular localization and function. J Clin Invest 117:1647–1657
137. Tajino K, Matsumura K, Kosada K, Shibakusa T, Inoue K, Fushiki T, Hosokawa H, Kobayashi S (2007) Application of menthol to the skin of whole trunk in mice induces autonomic and behavioral heat-gain responses. Am J Physiol Regul Integr Comp Physiol 293:R2128–R2135
138. Ruskin DN, Anand R, LaHoste GJ (2007) Menthol and nicotine oppositely modulate body temperature in the rat. Eur J Pharmacol 559:161–164
139. Ding Z, Gomez T, Werkheiser JL, Cowan A, Rawls SM (2008) Icilin induces a hyperthermia in rats that is dependent on nitric oxide production and NMDA receptor activation. Eur J Pharmacol 578:201–208
140. Gavva NR, Bannon AW, Surapaneni S, Hovland DN Jr, Lehto SG, Gore A, Juan T, Deng H, Han B, Klionsky L, Kuang R, Le A, Tamir R, Wang J, Youngblood B, Zhu D, Norman MH, Magal E, Treanor JJS, Louis J-C (2007) The vanilloid receptor TRPV1 is tonically activated in vivo and involved in body temperature regulation. J Neurosci 27:3366–3374

Chapter 11
TRPML1

Grace A. Colletti and Kirill Kiselyov

Abstract TRPML1 (or mucolipin 1) is the first member of the TRP family of ion channels that was found to function in the lower portions of the endocytic pathway. Mutations in the gene coding for TRPML1 (*MCOLN1*) cause the lysosomal storage disease mucolipidosis type IV (MLIV). TRPML1 localization in the lysosomes and the similarity of mucolipidosis type IV phenotype to lysosomal storage diseases whose origin has been directly linked to lysosomal dysfunction, suggest that TRPML1 activity drives some vitally important processes within the endocytic machinery. The specific aspect(s) of TRPML1 activity that make it indispensable for the proper function of the endocytic pathway as well as the specific aspect(s) of the endocytic activity that depend on TRPML1 are currently being discussed. Among the candidates are: membrane fusion within the lower portion of the endocytic pathway possibly mediated by Ca^{2+} release through TRPML1, or regulation of lysosomal ion homeostasis (pH or Fe content) by TRPML1. In addition to delineating the mechanisms of MLIV pathogenesis, identifying the role of TRPML1 in the endocytic pathway will lead to important developments in our understanding of the endocytic pathway and, due to the neurodegenerative nature of MLIV, of the integrative function of the cell. Moreover, molecular modulators of TRPML1 function may lead to novel approaches to modulating biological processes that depend on the endocytic pathway such as growth factor signaling. The present review will focus on the recent developments in identifying the TRPML1 function.

11.1 Introduction: TRPML1 Discovery and Initial Characterization

TRPML1 was cloned by three labs in 2000 as a result of the search for the genetic determinants of mucolipidosis type IV (MLIV) [1–3]. MLIV is a disease

K. Kiselyov (✉)
Department of Biological Sciences, University of Pittsburgh, Pittsburgh, PA 15260, USA
e-mail: kiselyov@pitt.edu

characterized by the buildup of storage bodies and vacuoles throughout patient tissues [4]. MLIV has a severe neurodegenerative profile, presumably due to the loss of brain tissue. Other pathological manifestations of MLIV include corneal opacity, retinal degeneration and constitutive achlorhydria, a decrease in gastric acid secretion [5–15]. The majority of TRPML1 mutations in MLIV obliterate the protein; some mutations result in improper localization or function of this channel [16–19]. The mechanisms of cell death in lysosomal storage diseases (LSDs) remain poorly understood, despite the obvious benefits promised by delineating the pathways translating dysfunction of a specific molecule to the cell-wide process of death. Some recent data suggest autophagic dysfunction and buildup of dysfunctional mitochondria as a possible pathway of cell death in LSDs, specifically in MLIV [20–22]. Similar conclusions were obtained using the *Drosophila* model of MLIV obtained by knockout of the single TRPML coding gene in *Drosophila* [23]. Various models of TRPML1 knockdown are summarized in Table 11.1. It is quite possible that some of the aspects of LSD's cellular phenotype arise from reaction of the cells, perhaps at a transcriptional level, to the insults to the endocytic pathway caused by dysfunctions in the cellular digestive machinery [24]. One possible physiological example of such a reaction is the specifically pronounced buildup of vacuoles in parietal cells, the stomach epithelial cells that produce gastric acid, which may underlie constitutive achlorhydria documented in MLIV patients [14]. Parietal involvement and achlorhydria may directly point to TRPML1 function, or may result from a specific response of specific subsets of cells to the problems in the endocytic pathway caused by the TRPML1 loss.

The gene *MCOLN1* (NG_015806, NM_020533) coding for TRPML1 is located on human chromosome 19 (19p13.2–13.3) spanning base pair positions 7,587,511–7,598,863. No splicing variants have been reported for the human gene. This gene is fairly conserved within *Vertebrata*; splice variants have been reported in mice [25]. In Drosophila, the entire TRPML family is represented by a single trpml gene (Dmel_CG8743, NM_140888, chromosome 3L, location: 76C2–76C2, base pairs 19,706,960–19,711,450). In *C. elegans*, TRPML is represented by the cup-5 gene (R13A5.1, NM_001027550, NM_001027548, NM_066263, NM_001027551, chromosome III, location: 378–564, base pairs: 7584129–7591495). Functional characterization of TRPML1 homologues in *Drosophila* [23] and *C. elegans* [26–28] yielded both consistencies and abnormalities compared with those reported in the human system. For example, a loss-of-function mutation in the *C. elegans* TRPML1 homologue *cup-5* results in the formation of large vacuoles and *Drosphila trpml* mutants had retinal degeneration consistent with the human model of MLIV [9, 29]. In contrast to reports from the human model of MLIV, *cup-5* mutants demonstrate increased endocytosis and *trpml Drosophila* mutants show normal lysosomal fusion with autophagosomes [20, 23, 26]. Similar to the reports in human MLIV fibroblasts [20, 30], the *Drosophila* mutants showed overacidification of lysosomes and autophagy defects accompanied by mitochondrial aberrations [23].

TRPML1 is a member of the TRPML subfamily of the TRP family of ion channels that are extensively reviewed in this volume. All TRPML channels function in the endocytic pathway, which sets them apart from other TRP channels; indeed

Table 11.1 The experimental systems used to report TRPML1 function

System	Observations	References
Human MLIV patients	Progressive neurodegeneration with early onset, mental retardation, corneal opacity, retinal degeneration, vacuolization of parietal cells, constitutive achlorhydria, elevated gastrin levels	[5–15]
Mouse TRPML1 knockout	Neurological defects including gait deficits and hind-limb paralysis; elevated plasma gastrin, vacuolization in parietal cells, and retinal degeneration	[53]
Drosophila trpml mutants	Neurodegeneration with accumulation of apoptotic cells in the brain, impaired synaptic transmission, impaired autophagy and mitochondrial function; lysosomal overacidification	[23]
C elegans CUP-5 mutants	Impaired endocytosis and lysosomal biogenesis	[26–28]
Human skin fibroblasts	Impaired membrane traffic	[29, 36]
	Impaired lysosomal exocytosis	[54]
Human skin fibroblasts and *Drosophila trpml* mutants	Impaired autophagy	[20–23]
Human skin fibroblasts and *Drosophila trpml* mutants	Impaired mitochondrial function	[20, 23]
Human skin fibroblasts, TRPML1 knockdown in HeLa cells, and *Drosophila trpml* mutants	Lysosomal overacidification	[23, 30, 41]
TRPML1 knockdown in HeLa cells, and *Drosophila trpml* mutants	Normal membrane fusion in the endocytic pathway	[23, 41]
Recombinant expression of TRPML1	Permeability to Ca^{2+}	[18, 33–35, 45, 46]
	Permeability to H^+	[30]
	Potentiated by H^+	[46]
	Inhibition by H^+	[18, 44]
	Permeability to Fe^{2+}	[49]

they are the only cation channels known so far to be localized and function in the cellular digestive tract. As all TRP channels, TRPML1 is presumed to have 6 transmembrane domains; its putative pore constitutes of a loop between the 5th and 6th domains. The structure-functional determinants of TRPML1 function have not been explored in detail; it does not seem to possess the full TRP box signature domain or ankyrin repeats present in some other TRP channels. Both C- and N-termini of TRPML1 contain lysosomal localization signatures [31]; the significance of a large loop connecting the 1st and 2nd domains [19] and PKA-dependent phosphorylation domains within TRPML1 [32] have not been explored in detail; some data suggest proteolytic cleavage of the loop and phosphorylation of the PKA-dependent domains as modulatory inputs.

The initial set of conclusions regarding TRPML1 function was inferred from two sources: the conductance characteristics of the currents that were associated with TRPML1 expression and measurement of membrane traffic and lysosomal enzymatic activity in fibroblasts obtained from MLIV patients (primarily the WG0909 clone). A combination of the initial reports on Ca^{2+} permeability through recombinant TRPML1 expressed in HEK293 cells [33, 34] or reconstituted into lipid bilayers [18, 35] and the delays in the lipid traffic reported in MLIV fibroblasts [29, 36] and in the *C elegans* MLIV model [26, 37] gave rise to the idea that TRPML1 is a lysosomal Ca^{2+} release channel. According to this "traffic" model (Fig. 11.1), Ca^{2+} release through TRPML1 drives the Ca^{2+} dependent step of the SNARE-mediated process of the lysosomal-endosomal fusion. Without the TRPML1-mediated Ca^{2+} release, organellar funciton within the endocytic pathway is impaired, which limits the exposure of endocytosed material to the digestive enzymes, and, therefore, digestion, with ensuing buildup of undigested material within the endocytic pathway. Several questions that remain unresolved by this model have been outlined

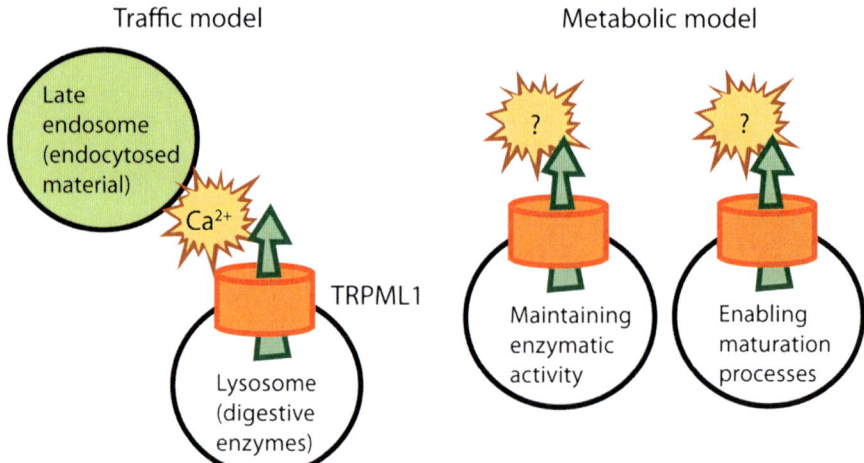

Fig. 11.1 The "traffic" and the "metabolic" models of TRPML1 function. The traffic model suggests a role for TRPML1 in the fusion or fission of vesicles within the endocytic pathway, while the "traffic" model suggests that TRPML1 regulates lysosomal ion homeostasis and, therefore, the activity of lysosomal digestive enzymes or lysosomal maturation. *Left panel*, metabolic model: Ca^{2+} release from lysosomes drives their fusion with late endosomes, and therefore, exposure of endocytosed material to digestive enzymes. In the absence of TRPML1 there is a limited exposure of endocytosed material to digestive enzymes and, therefore, limited digestion. *Right*, traffic model: ion (Fe^{2+}, H^+ etc) movement through TRPML1 established proper ionic balance inside lysosomes and promotes digestion and maturation processes

in previously published review articles [38, 39]. Briefly, a direct consequence of this model would be a general deficit in the membrane traffic within the endocytic pathway. Although such deficits have been shown in MLIV fibroblasts and in the *C elegans* model [26, 29, 36, 37], it is unclear whether they are a direct consequence of TRPML1 loss or an effect of buildup of undigested material.

The second model is built on the fact that membrane traffic delays similar to those reported in MLIV were also demonstrated in cells affected by LSDs of clearly "metabolic" origin (e.g. [40], discussed in [41]). This model (Fig. 11.1) suggests that TRPML1 regulates some aspects of lysosomal ion homeostasis and that dysregulation of the latter in the absence of TRPML1 leads to chronic or acute loss of the lysosomal digestive function. This scenario is similar to the loss of lysosomal digestive function upon downregulation of the lysosomal Cl^- channel CLC7 [42]. Recent reports from various MLIV models, discussed below, provide further evidence for "metabolic" involvement of TRPML1 and for previously unsuspected aspects of TRPML1 function. Regulation of lysosomal ion homeostasis by TRPML1 may reflect on the activity of lysosomal digestive enzymes or on processes relevant to lysosomal maturation.

Both models of TRPML1's role in the endocytic pathway offer exciting developments in our understanding of the biology of the cellular digestive system. The "traffic" model suggests the molecular identity of the first intracellular ion channel that is responsible for membrane fusion in the endocytic pathway. Furthermore, it postulates a mechanism by which the efficacy of vesicular traffic in the endocytic pathway can be modulated by signals converging on TRPML1. The "metabolic" model suggests the molecular identity of the first cationic channel that modulates the activity of lysosomal digestive enzymes. In addition to better understanding endocytic function, identifying the TRPML1 role in the endocytic tract may provide possibilities for using TRPML1 modulators to regulate cellular functions that depend on the endocytic pathway, such as parasite invasion and growth factor signaling.

Both models present several unresolved questions, primarily concerning the causal relationships between TRPML1 loss and the key aspects of the resulting phenotypes that were used to formulate the models. Specifically, the core set of evidence at the basis of the "traffic" model does not resolve the question whether the delays in lipid traffic are a direct result of TRPML1 loss or whether they are caused by the buildup of undigested material "clogging" the membrane traffic. The "metabolic" model does not answer whether the ionic dysregulations in TRPML1 deficient cells are a direct effect of TRPML1 loss or whether they reflect a cellular response to TRPML1 loss, perhaps at a transcriptional level. It is likely that future developments in this field will occur along the lines of delineating the structure/functional determinants of TRPML1 activity that make it indispensable for the proper function of the endocytic pathway, pinpointing the immediate effects of TRPML1 loss, creating a comprehensive account of TRPML1 permeability characteristics and identifying the molecular context of TRPML1 activity. Recent developments in the field outlined below already offer some exciting developments in the physiology of this interesting ion channel.

11.2 The Recent Developments

The recently published studies provide important new information on TRPML1 physiology and its role in the endocytic pathway and raise a set of exciting new questions.

In order to identify the immediate consequences of TRPML1 loss on membrane traffic in the endocytic pathway, two groups downregulated TRPML1 expression using RNAi. One group studied the effects of acute TRPML1 knockdown using siRNA [41] while the other group used a constitutively active shRNA expression system resulting in longer knockdown periods [43]. Both studies documented delayed translocation of fluorescent lipids from lysosomes to Golgi indicating that TRPML1 either modulates lipid modifications preceding lipid translocation from lysosomes to Golgi, or it regulates membrane traffic events in this step. Analysis of pre-lysosomal trafficking defects in these studies produced varied results. Using stable line shRNA knockdown of TRPML1, a delay in lysosomal delivery of lipids was shown in TRPML1-deficient murine macrophages [43], while no delay was observed 48 h after siRNA-mediated TRPML1 knockdown in HeLa cells [41]. Further studies will elucidate whether the loss of TRPML1 results in trafficking defects along later portions of the endocytic pathway. It is interesting to note that delayed lipid delivery was seen in the stable TRPML1 knockdown line similar to delays seen in MLIV fibroblast cells [41] suggesting that chronic loss of this channel could lead to compounding effects that result in trafficking delays. A detailed investigation of lipid hydrolysis and modification caused by acute TRPML1 loss will be necessary to clarify its role in lipid handling within the endocytic pathway. The acute model of TRPML1 knockdown provides an exciting new tool for studying the direct cellular responses to TRPML1 loss.

The structure-function analysis of TRPML1 based on structural comparison with its relative TRPML3 and voltage-gated ion channels revealed a region resembling the linker sequence in the voltage-gated channels that translates the movement of the S3 voltage-sensing domain to the opening and closing of the pore. Mutating this region in TRPML1 resulted in spontaneously active currents that, similar to previousuly published data, carried Ca^{2+}, but contrary to some of the previously published data [18, 44], were activated by low pH [45, 46]. Single channel recordings of TRPML1 activity confirmed its regulation by pH [47]. Although these data provide support for the role of TRPML1 in maturation and fusion of the endocytic compartments, it remains unanswered to what extent the activation mutations affected TRPML1 channel properties; similar mutations in TRPML3 essentially abolished its regulation by H^+ [48]. Nevertheless, these data provide important insights into the mechanics of TRPML1 activity and provide novel tools for exploring the TRPML1 role in the endocytic pathway.

The difficulties of inferring TRPML1 function from its permeability profiles are further compounded by the paucity of our knowledge of its permeability characteristics. It is clear that further characterizing TRPML1 permeability using the "activation" mutants or novel recording approaches, promises better understanding of its function. With this in mind, especially exciting is the recent report on current recordings using whole cell approaches and modified enlarged lysosomes that demonstrated Fe^{2+} permeability through TRPML1 [49]. It is an interesting possibility indeed that TRPML1 is a Fe^{2+} shunt that allows Fe^{2+} efflux from lysosomes that mitigates the effects of Fe^{2+} buildup, which otherwise leads, among other things, to accumulation of lipofuscin in the lysosomes. A detailed chemical and structural analysis of the storage bodies in TRPML1 deficient cells and comparison of their content with classic Fe^{2+} overload and lipofuscin buildup models will help test this interesting model.

The fourth line of research aimed at clarifying the TRPML1 role in the endocytic pathway is focused on identifying the molecular context of TRPML1 activity, its immediate working environment. As the first step towards this goal, the protein ALG-2, EF-hand-containing protein apoptosis-linked gene 2 protein suggested to be a sensor for mediating Ca^{2+} dependent vesicle fusion [50], was identified to bind to TRPML1 in a Ca^{2+} dependent manner [51]. That ALG-2 is a putative Ca^{2+} binding protein suggests that Ca^{2+} regulates TRPML1, or that the effects of Ca^{2+} currents through TRPML1 are actuated by ALG-2. Further analysis of the functional consequences of TRPML1 interaction with ALG-2 and with other proteins will delineate the molecular context of TRPML1 activity possibly helping to infer its function.

New TRPML1-deficient organisms and MLIV models provide an excellent opportunity for testing the effects of TRPML1 knock-out on the entire organism as well as probing the aspects of TRPML1 function previously reported in *vitro*, within the functional context of a working tissue. The data on the recently published murine MLIV model confirmed autophagic abnormalities reported in *in vitro* MLIV assays, judged by the increased LAMP2 stain usually associated with autophagosomes (LAMP2 is a lysosome associated membrane protein thought to play a key role in autophagy [52]), and provided a detailed analysis of the pathological manifestation of the early and late stages of MLIV development in this system [53]. Specific aspects of proposed TRPML1 function (e.g. Fe uptake, or lysosomal pH regulation) can now be probed in different tissues and correlated with phenotype severity in these tissues. Changes in gene expression in different tissues associated with TRPML1 loss can also be tested using the mouse model. The data on *Drosophila* model of MLIV reinstated dysregulation of the lysosomal pH as one of the effects of TRPML1 loss and confirmed autophagic aberrations and mitochondrial dysfunction in the absence of TRPML1 [23]. The inducible knockout systems will help illuminate the immediate consequences of TRPML1 loss within different cellular and tissue contexts, possibly helping to pinpoint the reasons for different phenotypical severity of TRPML1 loss in different tissues.

11.3 Perspectives

In addition to the immediate effects of TRPML1 loss and the molecular context of TRPML1 activity, the two central questions that remain to be resolved in order to pinpoint the TRPML1 role in the endocytic pathway are the structural determinants

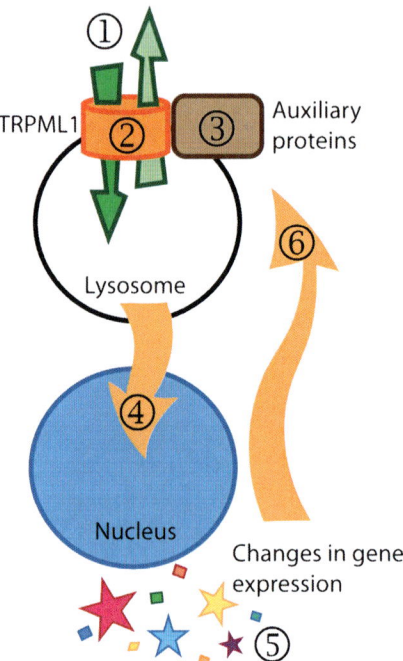

Fig. 11.2 The key paradigms and questions in TRPML1 function. This Figure delineates the Future directions and questions relevant to TRPML1 activity. *1* – a detailed account of TRPML1 permeability profile will identify its role in lysosome. *2* – structure/function analysis of TRPML1 function will pinpoint the key aspects of TRPML1 permeability that make it indispensable for the proper function of the endocytic pathway. *3* – enumerating TRPML1-interacting proteins will reveal the functional context of its role in the endocytic pathway. *4,5* – identifying the response elements that report the functional status of the endocytic pathway (*4*) resulting in changes in gene expression (*5*) in response to changes in the endocytic pathway will illuminate new aspects of the integrative function of the cell. *6* – delineating changes I gene expression associated with TRPML1 loss will identify the aspects of the endocytic function that depend on TRPML1 function directly and indirectly

of TRPML1 activity and the mechanisms of cellular "reaction" to the TRPML1 loss (Fig. 11.2). The importance of the first question lies in the difficulties that may be encountered in inferring TRPML1 function based solely on its permeability characteristics. With this in mind, it is likely that a detailed structure-function analysis of TRPML1 ion permeability focused on dissecting the structural determinants of its permeability and gating profiles followed by knock-in experiments using mutants deficient in the specific aspects of TRPML1 permeability characteristics may prove to be very useful.

Identification of the role of TRPML1 in the endocytic pathway using biological readouts of the effect of its acute downregulation on the endocytic pathway may be compounded by the cellular reaction to changes in the functional status of the endocytic pathway, which may be executed at the level of gene expression. It is, therefore, important to distinguish the direct effects of TRPML1 loss from the cellular reactions to the changes in the endocytic pathway induced by the TRPML1 loss. This can be accomplished by analyzing the changes in gene expression associated with TRPML1 downregulation and delineating the functional consequences of such changes from all of the effects of TRPML1 loss on the endocytic pathway. The initial approaches towards this goal have already yielded interesting results [24].

In summary, TRPML1 is an extremely interesting ion channel whose function we are just beginning to understand. Exploring its function will lead to better understanding of the ion channel function, the role of ion channels in the endocytic pathway and the integrative function of the cells. Applying novel physiological and molecular techniques to studying TRPML1 will lead to exciting discoveries.

Acknowledgments The authors would like to extend their apologies to those authors whose work has not been cited here due to manuscript length consideration. Work in the authors' lab is supported by the National Institutes of Health grants HD058577 and ES016782 to KK.

References

1. Bargal R et al (2000) Identification of the gene causing mucolipidosis type IV. Nat Genet 26:118
2. Bassi MT et al (2000) Cloning of the gene encoding a novel integral membrane protein, mucolipidin-and identification of the two major founder mutations causing mucolipidosis type IV. Am J Hum Genet 67:1110
3. Sun M et al (2000) Mucolipidosis type IV is caused by mutations in a gene encoding a novel transient receptor potential channel. Hum Mol Genet 9:2471
4. Bach G (2001) Mucolipidosis type IV. Mol Genet Metab 73:197
5. Tellez-Nagel I et al (1976) Mucolipidosis IV. Clinical, ultrastructural, histochemical, and chemical studies of a case, including a brain biopsy. Arch Neurol 33:828
6. Kenyon KR et al (1979) Mucolipidosis IV. Histopathology of conjunctiva, cornea, and skin. Arch Ophthalmol 97:1106
7. Goebel HH, Kohlschutter A, Lenard HG (1982) Morphologic and chemical biopsy findings in mucolipidosis IV. Clin Neuropathol 1:73
8. Zlotogora J et al (1983) A muscle disorder as presenting symptom in a child with mucolipidosis IV. Neuropediatrics 14:104
9. Abraham FA et al (1985) Retinal function in mucolipidosis IV. Ophthalmologica 191:210
10. Riedel KG et al (1985) Ocular abnormalities in mucolipidosis IV. Am J Ophthalmol 99:125
11. Weitz R, Kohn G (1988) Clinical spectrum of mucolipidosis type IV. Pediatrics 81:602
12. Chitayat D et al (1991) Mucolipidosis type IV: clinical manifestations and natural history. Am J Med Genet 41:313
13. Frei KP et al (1998) Mucolipidosis type IV: characteristic MRI findings. Neurology 51:565
14. Schiffmann R et al (1998) Constitutive achlorhydria in mucolipidosis type IV. Proc Natl Acad Sci USA 95:1207
15. Siegel H et al (1998) Electroencephalographic findings in patients with mucolipidosis type IV. Electroencephalogr Clin Neurophysiol 106:400
16. Altarescu G et al (2002) The neurogenetics of mucolipidosis type IV. Neurology 59:306
17. Bargal R et al (2002) Mucolipidosis IV: novel mutation and diverse ultrastructural spectrum in the skin. Neuropediatrics 33:199
18. Raychowdhury MK et al (2004) Molecular pathophysiology of mucolipidosis type IV: pH dysregulation of the mucolipin-1 cation channel. Hum Mol Genet 13:617
19. Kiselyov K et al (2005) TRP-ML1 is a lysosomal monovalent cation channel that undergoes proteolytic cleavage. J Biol Chem 280:43218

20. Jennings JJ Jr. et al (2006) Mitochondrial aberrations in mucolipidosis Type IV. J Biol Chem 281:39041
21. Vergarajauregui S et al (2008) Autophagic dysfunction in mucolipidosis type IV patients. Hum Mol Genet 17:2723
22. Venugopal B et al (2009) Chaperone-mediated autophagy is defective in mucolipidosis type IV. J Cell Physiol 219:344
23. Venkatachalam K et al (2008) Motor deficit in a Drosophila model of mucolipidosis type IV due to defective clearance of apoptotic cells. Cell 135:838
24. Bozzato A, Barlati S, Borsani G (2008) Gene expression profiling of mucolipidosis type IV fibroblasts reveals deregulation of genes with relevant functions in lysosome physiology. Biochim Biophys Acta 1782:250
25. Falardeau JL et al (2002) Cloning and characterization of the mouse Mcoln1 gene reveals an alternatively spliced transcript not seen in humans. BMC Genomics 3:3
26. Fares H, Greenwald I (2001) Regulation of endocytosis by CUP-5, the Caenorhabditis elegans mucolipin-1 homolog. Nat Genet 28:64
27. Hersh BM, Hartwieg E, Horvitz HR (2002) The Caenorhabditis elegans mucolipin-like gene cup-5 is essential for viability and regulates lysosomes in multiple cell types. Proc Natl Acad Sci USA 99:4355
28. Schaheen L, Patton G, Fares H (2006) Suppression of the cup-5 mucolipidosis type IV-related lysosomal dysfunction by the inactivation of an ABC transporter in C. elegans. Development 133:3939
29. Bargal R, Bach G (1997) Mucolipidosis type IV: abnormal transport of lipids to lysosomes. J Inherit Metab Dis 20:625
30. Soyombo AA et al (2006) TRP-ML1 regulates lysosomal pH and acidic lysosomal lipid hydrolytic activity. J Biol Chem 281:7294
31. Miedel MT et al (2006) Posttranslational cleavage and adaptor protein complex-dependent trafficking of mucolipin-1. J Biol Chem 281:12751
32. Vergarajauregui S et al (2008) Mucolipin 1 channel activity is regulated by protein kinase A-mediated phosphorylation. Biochem J 410:417
33. LaPlante JM et al (2002) Identification and characterization of the single channel function of human mucolipin-1 implicated in mucolipidosis type IV, a disorder affecting the lysosomal pathway. FEBS Lett 532:183
34. LaPlante JM et al (2004) Functional links between mucolipin-1 and Ca2+-dependent membrane trafficking in mucolipidosis IV. Biochem Biophys Res Commun 322:1384
35. Cantiello HF et al (2005) Cation channel activity of mucolipin-1: the effect of calcium. Pflugers Arch 451:304
36. Pryor PR et al (2006) Mucolipin-1 Is a Lysosomal Membrane Protein Required for Intracellular Lactosylceramide Traffic. Traffic 7:1388
37. Treusch S et al (2004) Caenorhabditis elegans functional orthologue of human protein h-mucolipin-1 is required for lysosome biogenesis. Proc Natl Acad Sci USA 101:4483
38. Kiselyov K, Soyombo A, Muallem S (2007) TRPpathies. J Physiol 578:641
39. Puertollano R, Kiselyov K (2009) TRPMLs: in sickness and in health. Am J Physiol Renal Physiol 296:F1245
40. Pagano RE (2003) Endocytic trafficking of glycosphingolipids in sphingolipid storage diseases. Philos Trans R Soc Lond B Biol Sci 358:885
41. Miedel MT et al (2008) Membrane traffic and turnover in TRP-ML1-deficient cells: a revised model for mucolipidosis type IV pathogenesis. J Exp Med 205:1477
42. Kasper D et al (2005) Loss of the chloride channel ClC-7 leads to lysosomal storage disease and neurodegeneration. EMBO J 24:1079
43. Thompson EG et al (2007) Lysosomal trafficking functions of mucolipin-1 in murine macrophages. BMC Cell Biol 8:54
44. Goldin E et al (2008) Isolated ocular disease is associated with decreased mucolipin-1 channel conductance. Invest Ophthalmol Vis Sci 49:3134

45. Grimm C et al (2007) A helix-breaking mutation in TRPML3 leads to constitutive activity underlying deafness in the varitint-waddler mouse. Proc Natl Acad Sci USA 104(49):19583–19588
46. Xu H et al (2007) Activating mutation in a mucolipin transient receptor potential channel leads to melanocyte loss in varitint-waddler mice. Proc Natl Acad Sci USA 104:18321
47. Kogot-Levin A et al (2009) Mucolipidosis type IV: the effect of increased lysosomal pH on the abnormal lysosomal storage. Pediatr Res 65:686
48. Kim HJ et al (2008) A novel mode of TRPML3 regulation by extracytosolic pH absent in the varitint-waddler phenotype. EMBO J 27:1197
49. Dong XP et al (2008) The type IV mucolipidosis-associated protein TRPML1 is an endolysosomal iron release channel. Nature 455:992
50. Bentley M et al (2010) Vesicular calcium regulation coat retention, fusogenicity, and size of pre-Golgi intermediates. Mol Biol Cell 21:1033
51. Vergarajauregui S, Martina JA, Puertollano R (2009) Identification of the penta-EF-hand protein ALG-2 as a Ca2+-dependent interactor of mucolipin-1. J Biol Chem 284:36357
52. Eskelinen EL (2006) Roles of LAMP-1 and LAMP-2 in lysosome biogenesis and autophagy. Mol Aspects Med 27:495
53. Venugopal B et al (2007) Neurologic, gastric, and opthalmologic pathologies in a murine model of mucolipidosis type IV. Am J Hum Genet 81:1070
54. Laplante JM et al (2006) Lysosomal exocytosis is impaired in mucolipidosis type IV. Mol Genet Metab 89:339

Chapter 12
TRPML2 and the Evolution of Mucolipins

Emma N. Flores and Jaime García-Añoveros

Abstract TRPML2, the polypeptide product of the gene *Trpml2* (aka *Mcoln2*), is a member of the TRPML or mucolipin branch of the TRP super family of ion channels. Although no known agonists have been discovered, the wild type channel gives basal currents when heterologously expressed in *Drosophila* (S2) cells and is constitutively active in mammalian cells when bearing a cell degeneration-causing, proline to alanine substitution in the fifth trans-membrane domain. TRPML2 forms channels that are inwardly rectifying and permeable to Ca^{+2}, Na^+, and Fe^{+2}. Localization studies indicate TRPML2 is present in lysosomes, late endosomes, recycling endosomes and, at a lower level, the plasma membrane. Tissue and organ distribution of TRPML2 is solely reported through RT-PCR and it is uncertain which cell types express this channel. However, various studies suggest that lymphoid cells express TRPML2. Although the function of TRPML2 is not known, distribution and channel properties suggest it could play roles in calcium release from endolysosomes, perhaps to mediate calcium-dependent events such as vesicle fusion, or to release calcium from intracellular acidic stores. However, TRPML2 may also function in the plasma membrane and its abundance in vesicles of the endocytic pathway might occur because its presence in the cell surface is regulated by endocytosis and exocytosis. An evolutionary analysis of *Trpml2* and its relatives reveals that vertebrate and invertebrate chordates have only one *Trpml* gene, that *Trpml1* and *Trpml2* are common to vertebrates, and that *Trpml3* is only found in tetrapods. Ray-finned fishes contain another isoform, which we term *Trpml4* or *Mcoln4* (and its product TRPML4). *Trpml2* is next to *Trpml3* in all tetrapod genomes except that of the frog *Xenopus tropicalis* and of the domesticated pig, which seems to lack most of the *Trpml3* gene. This close linkage across species implies that it is maintained by selective pressure and suggests that the regulation of both genes is interdependent.

J. García-Añoveros (✉)
Departments of Anesthesiology, Physiology and Neurology, Northwestern University Institute for Neuroscience, Chicago, IL 60611, USA; The Hugh Knowles Center for Clinical and Basic Science in Hearing and Its Disorders, Northwestern University, Chicago, IL 60611, USA
e-mail: anoveros@northwestern.edu

12.1 Genomics and Phylogeny

The TRPML2 protein is the product of the gene *Trpml2* (also known as *Mcoln2*, or Mucolipin 2). TRPML2 is the middle-member of a family of 3 tetrapod paralogs (close homologs from the same genome), TRPML1 to 3 (also known as Mucolipins 1 to 3), that form one of the branches of the Transient Receptor Potential ion channel superfamily. Like all TRP channel subunits, TRPML2 polypeptides contain predicted cytosolic amino-and carboxy-termini, six transmembrane domains (S1 to S6) and a pore-loop domain between S5 and S6. Like all TRPML channel subunits, TRPML2 exhibits a large extracellular or luminal loop (~210 amino acids in mouse and human TRPML2) between the first two transmembrane domains. TRPML2 is most homologous to TRPML3 and less to TRPML1. Whereas gain of function mutations in *Trpml3* (the gene for TRPML3) cause deafness and pigmentation defects in mice, and loss of function mutations in *Trpml1* cause mucolipidosis type IV in humans, *Trpml2* has not yet been implicated in any pathology. This is probably the reason for the limited knowledge regarding this channel-encoding gene.

The *Trpml2* gene maps to human chromosome I (1p22.3) and mouse chromosome 3 (3qH2) [1], very closely downstream of *Trpml3* (21 and 8 kb, respectively). Curiously, the two genes map next to each other in 27 of the 29 tetrapod genomes for which sufficient sequence information is available. The only exceptions to date are an amphibian (*Xenopus tropicalis*) genome and the domesticated pig (*Sus scrofa*) genome, in which *Trpml3* might be a pseudogene that maps to a different chromosome than *Trpml2*. Pig *Trpml3* is severely truncated, containing only the first two exons and encoding only the amino-terminal 132 residues, which lack five of the six transmembrane domains and the pore region. Sequencing the wild boar would clarify whether this loss of *Trpml3* is common to wild pigs or whether it is an effect of domestication. The conservation of the close linkage between *Trpml2* and *Trpml3* in most tetrapods suggests that a functional constraint (for example, if regulatory regions of one gene were contained within the adjacent gene) has maintained the genes together in multiple species over hundreds of millions of years.

Comparative genomics reveal that *Trpml3* orthologs (genes closer to *Trpml3* in another genome than to any other gene in their own genome) are only present in tetrapods (amphibians, reptiles, birds and mammals, but not fish) whereas *Trpml1* and *Trpml2* orthologs are present in vertebrate genomes (tetrapods as well as ray-finned fish; Figs. 12.1 and 12.2). An interpretation of these observations is that a duplication of the *Trpml2* gene took place in a common ancestor to tetrapods. One of the resulting genes maintained the *Trpml2* functions, and thus sequence similarity across vertebrates, whereas the upstream gene evolved into *Trpml3*.

Although fish contain two closely related paralogs of *Trpml1* (*a* and *b*) as well as two closely related *Trpml* paralogs, which we term *Trpml4a* and *Trpml4b* (and their protein products TRPML4a and TRPML4b) (Figs. 12.1 and 12.2). These are equally distant to *Trpml2* and *Trpml3* and are not linked to any of them. Hence fish contain two *Trpml1s*, one *Trpml2* and two *Trpml4s*, likely the result of the additional genome duplications that are thought to have occurred in the ancestor to fishes but not in the ancestor to other vertebrates.

12 TRPML2 and the Evolution of Mucolipins

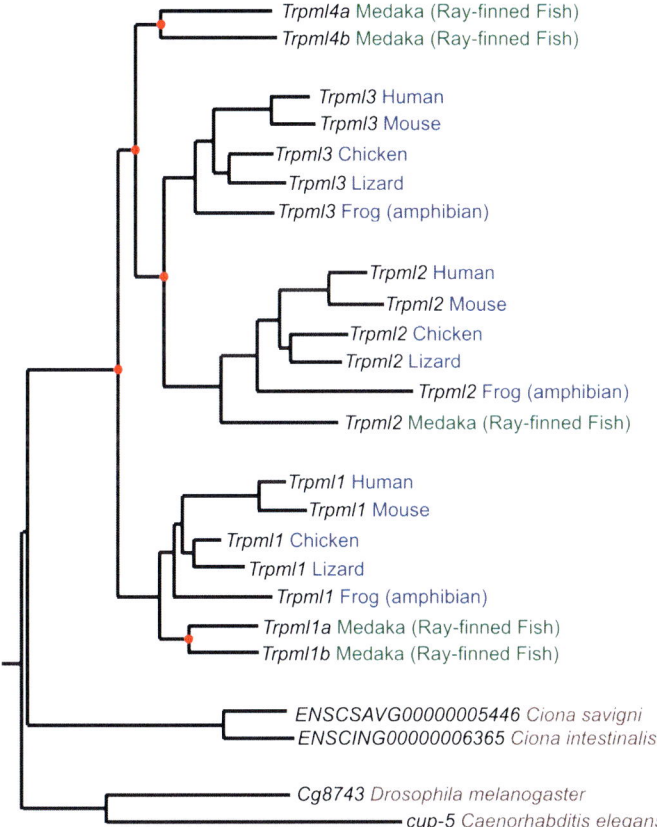

Fig. 12.1 Dendrogram of *Trpml* (aka *Mcoln*) orthologs and paralogs. The dendrogram was generated at e!Ensembl (http://uswest.ensembl.org/index.html) from the 49 available sequenced genomes that contain at least one homolog of *Trpml* gene. For simplicity of reproduction, we only list one or two species per animal class or phylum. *Red dots* indicate likely duplication events

Tunicates (urochordates, which are invertebrate chordates) as well as invertebrate protostomes (the nematode *Caenorhabditis elegans* and the insect *Drosophila*) contain one *Trpml* homolog that is equally divergent from vertebrate *Trpml1*, *2* and *3* (Figs. 12.1 and 12.2). Hence, a likely scenario is that the *Trpml* gene in the chordate ancestor to vertebrates gave rise through duplication to *Trpml1* and *Trpml2*. Then, a duplication of *Trpml2* in a vertebrate ancestor to tetrapods gave rise to two closely linked genes: one retained the functions and sequence characteristic of *Trpml2* whereas the other was modified to generate *Trpml3*. Another possibility is that *Trpml3* arose earlier but was lost in the common ancestor to ray-finned fishes (*Actinopterigy*, which is the only group of fishes from which relevant genomic sequences are available). Sequencing of other genomes, such as those of cartilaginous fishes, will clarify which of these two possibilities is correct. Whatever the evolutionary history of the *Trpml* genes, it seems likely that TRPML3 plays roles

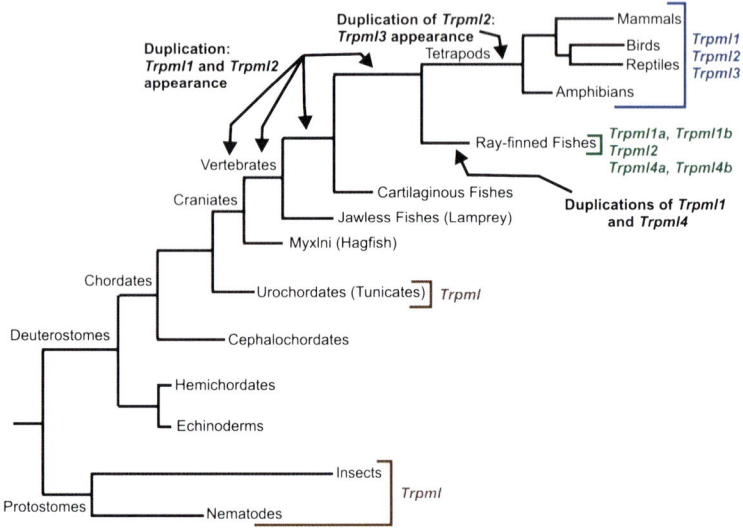

Fig. 12.2 Cladogram indicating which *Trpml* (aka *Mcoln*) genes are present in the genomes of different animal groups, and the likely evolutionary periods when *Trpml1*, *2* and *3* originated through duplication. The information regarding *Trpml* genes was inferred from a dendrogram similar but more extensive than the one displayed in Fig. 12.1

common to most tetrapods whereas TRPML1 and 2 play roles common to most vertebrates.

The mouse *Trpml2* gene generates at least two transcripts and polypeptides that result from alternative first exons: TRPML2sv (short, 2,384 bp mRNA and 538 residue protein) and TRPML2lv (long, 2,418 bp mRNA and 566 residue protein). The first exon of TRPML2sv does not code for protein, whereas the first exon of TRPML2lv encodes the additional 28 amino acids to the amino-terminus of TRPML2lv. Humans only have a long version isoform (of the same size, 566 amino acids, as mouse TRPML2lv [2]).

12.2 Tissue Distribution

The relative expression level of TRPML2 in mammals has been described solely through the use of quantitative and non-quantitative PCR. Human TRPML2 mRNA is detectable in lung, trachea, stomach, colon, mammary glands, and peripheral blood lymphocytes with lower levels in kidney, thymus, testis, and uterus. Lowest levels are seen in spleen, adipose tissue and thyroid. Mouse TRPML2 mRNA is detectable in kidney, thymus, trachea, colon, and adipose tissue with lower levels in lung, liver, pancreas, and testis [3]. As previously described, mouse *Trpml2* has two splice variants, a long variant (TRPML2lv) that contains an additional 28 amino acids when compared to a short variant (TRPML2sv). Quantitative RT-PCR analysis revealed that TRPML2sv has a higher expression level in thymus and spleen with lower levels in kidney, stomach and heart compare to TRPML2lv that was low in these and other organs [2]. All organs examined were positive for TRPML2 indicating that either *Trpml2* is ubiquitously expressed at low levels or that the technique, which is highly sensitive, detects physiologically irrelevant, "leaky" gene expression. However, the high levels of TRPML2sv expression detected in thymus and spleen suggest expression in lymphoid cells. This is further supported by RT-PCR detection of TRPML2 mRNA in B cells, T cells, primary splenocytes, mastocytoma, and myeloma cell lines [4]. Interestingly, the levels of TRPML2 (but not TRPML3) mRNA were higher in thymus, spleen and kidney of *Trpml1* KO than of *Trpml1* wild type mice [2], which suggests that the expression of both genes is coordinated in some way and that TRPML2 might be able to compensate for the lack of TRPML1 in these tissues.

Deciphering these results in order to determine biologically relevant expression versus low level mRNA transcription underscores a limitation of these and other gene expression detection methods. Another caveat of RT-PCR is that it cannot resolve individual cell expression, and thus it would miss cells that express TRPML2 at high levels but which constitute a small portion of the cells of an organ. For this reason alone the need for in situ hybridization is very important. Immunohistochemistry could also provide satisfactory information, but that would require not only antibodies that recognize the endogenous protein but also *Trpml2* KO tissues to control for non-specific immunoreactivities.

12.3 Channel Properties

Although most TRPML2 seems to localize to intracellular membranes, some heterologously expressed channels must be at the plasma membrane, because these have been detected by whole cell recordings. The initial recordings of TRPML2 [5] resorted to activating the heterologously expressed channel by introducing a proline substitution in transmembrane domain 5, mimicking the mutation in TRPML3 that leads to constitutive channel activation [5–7]. In HEK cells, calcium imaging [5] and whole cell electrophysiological recordings [2, 8], reveal that only the mutated TRPML2 channels, and not the wild type ones, produce current and calcium overload. However for unknown reasons, in *Drosophila* S2 (Schneider) cells, heterologous expression of the wild type human TRPML2 also generates currents. The currents in all these recordings have the inward rectification characteristic of TRPML3 and TRPML1 [5–7], with inward currents at negative potentials and no outward currents at positive ones. The reversal potentials and ion substitution experiments reveal that the mutated channel is permeable to Ca^{2+}, Na^+ and Fe^{2+} [9]. The latter study suggests that TRPML2, like TRPML1, may mediate transport of iron across endolysosomal membranes. There are no known agonists or antagonists of TRPML2. Because of the presence of TRPML2 in endosomes and lysosomes, characterized by low luminal pH; the effects of lowering pH have been studied on mutationally-activated TRPML2 with opposing results: increase of current by a drop from pH near 7.2 to pH4.6 [9] and 25% reduction of current by a pH drop from 7.15 to 6.0 [8]. The reasons for these different effects, or their physiological significance, are unclear.

The only single channel recordings of TRPML2 [10] were obtained from protein translated in vitro or in mammalian cells and then reconstituted into artificial lipid bilayers. These recordings indicated a single channel conductance of 7.44 pS for TRPML2 and 10.2 pS for TRPML3, and partial permeability to anions for both channels. These are uncharacteristic of other TRP channels, and disagree with the cation permeability single channel conductance of TRPML3 in mammalian cells (50 to 70 pS) [6]. Given the propensity of reconstitution into artificial lipid bilayers to generate non-physiological conditions, these channel properties await confirmation or correction by recordings from patches of biological membranes.

12.4 Subcellular Localization

Although electrophysiological recordings indicate that at least some heterologously expressed TRPML2 also exists at the plasma membrane [2, 8, 9] protein localization studies indicate that most TRPML2, similarly to TRPML1 and TRPML3, localizes to late endosomes and lysosomes [11–14]. Tagged TRPML2 heterologously expressed in HEK cells, B-lymphocytes and HeLa cells accumulated at vesicular structures that co-labeled with markers of lysosomes (Lysotracker, Lamp 3 and Rab 11) and late endosomes (CD63) but not with markers of ER (ER-YFP), Golgi (GFP-Golgin 160; GGA3-VHSGAT-GFP), early endosomes (Rab5, EEA1), late endosomes (Rab7-HA) or secretory vesicles (Rab8) [11–13]. Immunoreactivities with antibodies raised against TRPML2, which might specifically recognize endogenous TRPML2, co-localized with the lysosomal marker Lamp1 in untransfected HEK cells [14]. In addition to lysosomes, tagged TRPML2 has also been localized in HeLa cells to plasma membrane and tubular endosomes, which are recycling endosomes of the GTPase ADP-ribosylation Factor 6 (Arf6) pathway. Furthermore, TRPML2 co-localizes with markers of these endosomes (CD59 and MHCI), and inhibition of TRPML2 with shRNAs or with a dominant negative isoform of TRPML2 inhibited the recycling of CD59 from these sorting endosomes back to the plasma membrane [11]. However, inhibition of clathrin coated endocytosis by co-expression with full length AP180 (a protein that binds clathrin and causes it to be distributed into a lattice-like pattern thereby preventing clathrin-coated pit formation and clathrin-mediated endocytosis), or with the dominant-negative isoform of Dynamin (Dyn1aK44A, which is incapable of binding GTP) resulted in accumulation of TRPML2 at the plasma membrane instead of lysosomal membranes [13]. These studies suggest that TRPML2 may be internalized from the plasma membrane through clathirin-coated endocytosis but that once in endosomes it may participate in endosome recycling back to the plasma membrane through the Arf6 pathway.

12.5 Subunit Interactions

Like other TRP channels, TRPML2 may heteromultimerize with other TRP subunits, and this possibility has been examined with respect to its two closest paralogs, TRPML1 and TRPML3. FRET analysis along with co-immunoprecipitation showed that heterologously co-expressed, tagged TRPMLs can associate with one another in all combinations, so that TRPML2 likely can form homomultimers as well as heteromultiomers with TRPML1 and/or TRPML3 [3, 13]. In addition, TRPML2 co-precipitates with TRPML1/3, but with low efficiency: only 5–6% of the available TRPML2 co-immunoprecipitated with TRPML1 or TRPML3 [10, 14]. At the subcellular level, coexpression of TRPML2 and TRPML3 leads to co-localization in lysosomes positive for Lamp3 [13] and coxpression of TRPML2 and TRPML1 leads to co-localization in lysosomes positive for Lysotracker [12]. Immunoreactivities with antibodies raised to TRPML1, 2 and 3, which may detect the endogenous proteins, co-localize with one another in Lamp1-containing lysosomes of human skin fibroblasts [14]. The co-localization of TRPML2 with TRPML1 appears to be limited to lysosomes, indicating that TRPML2 may have overlapping function with TRPML1. In support, shRNA mediated knockdown of TRPML2 leads to the appearance of abnormal lysosome storage bodies reminiscent of lysosomes seen in ML-IV patients [14]. In summary, there is mounting evidence that TRPML2 interacts with TRPML1 and TRPML3 in lysosomes or late endosomes, although this interaction is very limited and only represents a small fraction of the total TRPML2 protein. Hence, although TRPML2 might heteromultimerize with TRPML1 and 3, it likely also forms channels without these subunits.

12.6 Biological Roles and Future Directions

Given the paucity of information about TRPML2, it is not surprising that its physiological function remains unknown. Based on its subcellular distribution and ionic permeability, potential functions include release of calcium from endosomes or lysosomes, either to mediate internal calcium release from acidic stores or to trigger vesicle fusions in the endocytic pathway, as well as release of iron or other metals from these organelles [15]. However, TRPML2 might also play roles in the plasma membrane, where it has also been detected, although at much lower levels than in intracellular membranes. In fact, it is conceivable that the primary function of TRPML2 takes place at the plasma membrane but that this is tightly regulated by endocytosis and exocytosis, accounting for the large amounts of TRPML2 in intracellular vesicles. Finally, the likely expression of TRPML2 in lymphoid cells and its co-localization with CD59 and the major histocompatibility complex class I (MHC-I) suggests potential roles in immune cells such as antigen processing. Of the several approaches to elucidating the physiological function of TRPML2, the most pressing are to determine which cells express it (where it is), to identify endogenous agonists or other methods of activation (what does it do for the cell), and to generate and phenotype a KO mouse (what does it do for the organism).

Note Added in Proof

Recently it has been shown that PI (3, 5) P2, a low abundance endolysosome-specific phosphoinositide (PIP) can activate TRPML2 [16].

Acknowledgments Supported by grants from NIH-NINDS (R01 NS044363), NIH-NIAMS (P30 AR057216) and *The Hugh Knowles Center for Clinical and Basic Science in Hearing and Its Disorders* to JG-A.

References

1. Di Palma F, Belyantseva IA, Kim HJ, Vogt TF, Kachar B, Noben-Trauth K (2002) Mutations in Mcoln3 associated with deafness and pigmentation defects in varitint-waddler (Va) mice. Proc Natl Acad Sci USA 99:14994–14999
2. Samie MA, Grimm C, Evans JA, Curcio-Morelli C, Heller S, Slaugenhaupt SA, Cuajungco MP (2009) The tissue-specific expression of TRPML2 (MCOLN-2) gene is influenced by the presence of TRPML1. Pflugers Arch 459:79–91
3. Grimm C, Jors S, Saldanha SA, Obukhov AG, Pan B, Oshima K, Cuajungco MP, Chase P, Hodder P, Heller S (2010) Small molecule activators of TRPML3. Chem Biol 17:135–148
4. Lindvall JM, Blomberg KE, Wennborg A, Smith CI (2005) Differential expression and molecular characterisation of Lmo7, Myo1e, Sash1, and Mcoln2 genes in Btk-defective B-cells. Cell Immunol 235:46–55
5. Grimm C, Cuajungco MP, van Aken AF, Schnee M, Jors S, Kros CJ, Ricci AJ, Heller S (2007) A helix-breaking mutation in TRPML3 leads to constitutive activity underlying deafness in the varitint-waddler mouse. Proc Natl Acad Sci USA 104:19583–19588
6. Nagata K, Zheng L, Madathany T, Castiglioni AJ, Bartles JR, Garcia-Anoveros J (2008) The varitint-waddler (Va) deafness mutation in TRPML3 generates constitutive, inward rectifying currents and causes cell degeneration. Proc Natl Acad Sci USA 105:353–358
7. Xu H, Delling M, Li L, Dong X, Clapham DE (2007) Activating mutation in a mucolipin transient receptor potential channel leads to melanocyte loss in varitint-waddler mice. Proc Natl Acad Sci USA 104:18321–18326
8. Lev S, Zeevi DA, Frumkin A, Offen-Glasner V, Bach G, Minke B (2010) Constitutive activity of the human TRPML2 channel induces cell degeneration. J Biol Chem 285:2771–2782
9. Dong XP, Cheng X, Mills E, Delling M, Wang F, Kurz T, Xu H (2008) The type IV mucolipidosis-associated protein TRPML1 is an endolysosomal iron release channel. Nature 455:992–996
10. Curcio-Morelli C, Zhang P, Venugopal B, Charles FA, Browning MF, Cantiello HF, Slaugenhaupt SA (2010) Functional multimerization of mucolipin channel proteins. J Cell Physiol 222:328–335
11. Karacsonyi C, Miguel AS, Puertollano R (2007) Mucolipin-2 localizes to the Arf6-associated pathway and regulates recycling of GPI-APs. Traffic 8:1404–1414
12. Song Y, Dayalu R, Matthews SA, Scharenberg AM (2006) TRPML cation channels regulate the specialized lysosomal compartment of vertebrate B-lymphocytes. Eur J Cell Biol 85:1253–1264
13. Venkatachalam K, Hofmann T, Montell C (2006) Lysosomal localization of TRPML3 depends on TRPML2 and the mucolipidosis-associated protein TRPML1. J Biol Chem 281: 17517–17527
14. Zeevi DA, Frumkin A, Offen-Glasner V, Kogot-Levin A, Bach G (2009) A potentially dynamic lysosomal role for the endogenous TRPML proteins. J Pathol 219:153–162
15. Cheng X, Shen D, Samie M, Xu H (2010) Mucolipins: Intracellular TRPML1-3 channels. FEBS Lett 584:2013–2021
16. Dong XP, Shen D, et al. (2010) PI (3, 5) P2 Controls Membrane Traffic by Direct Activation of Mucolipin Ca Release Channels in the Endolysosome. Nat Commun 1(4):ii–38

Chapter 13
The TRPML3 Channel: From Gene to Function

Konrad Noben-Trauth

Abstract TRPML3 is a transient receptor potential (TRP) channel that is encoded by the mucolipin 3 gene (*MCOLN3*), a member of the small mucolipin gene family. *Mcoln3* shows a broad expression pattern in embryonic and adult tissues that includes differentiated cells of skin and inner ear. Dominant mutant alleles of murine *Mcoln3* cause embryonic lethality, pigmentation defects and deafness. The TRPML3 protein features a six-transmembrane topology and functions as a Ca^{2+} permeable inward rectifying cation channel that is open at sub-physiological pH and closes as the extracytosolic pH becomes more acidic. TRPML3 localizes to the plasmamembrane and to early- and late-endosomes as well as lysosomes. Recent advances suggest that TRPML3 may regulate the acidification of early endosomes, hence playing a critical role in the endocytic pathway.

Mucolipins constitute a small gene family, mucolipin 1–3, which are named after the gene (*MCOLN1*) underlying the lysosomal storage disorder mucolipidosis type 4 [1]. In this review I will focus on different biological aspects of the *Mcoln3* gene and its protein product TRPML3.

13.1 Genomics of *Mcoln3*

TRPML3 is a transient-receptor-potential (TRP) channel that is encoded by the mucolipin-3 gene (*MCOLN3*). In humans, the gene (reference sequence NM_018298) is located on the short arm of chromosome 1 (p22.3) at base pair positions 85, 490, 242–85, 514, 169 (Genome Reference Consortium; release h37). The coding sequence of human *MCOLN3* is represented by 12 exons that give rise to an mRNA with an open reading frame of 1,659 nucleotides. *MCOLN3* orthologs are present in most vertebrate species for which a genomic sequence is available and

---

K. Noben-Trauth (✉)
Section on Neurogenetics, Laboratory of Molecular Biology, National Institute on Deafness and Other Communication Disorders, Rockville, MD 20850, USA
e-mail: nobentk@nidcd.nih.gov

homologs are also found in *Drosophila*, *C. elegans*, sea urchin and lower organisms such as Hydra, Dictyostelium, Schistosoma and Leishmania [2] (see Chapter 12).

The mouse *Mcoln3* gene (reference sequence NM_134160) is located on the distal end of chromosome 3 (qH2) at base pair position 145, 784, 755–145, 803, 610 (NCBI build37). It encodes a protein of 553 amino acids (reference sequence NP_598921) with a predicted molecular weight of 64 kDa. Human and mouse TRPML3 share 91% sequence identity with most of the discordant amino acids being located at the N-terminus and the first extracellular/luminal loop. Secondary structure analysis predicts an integral transmembrane topology with six helices, and cytosolic amino-(aa 1–63) and carboxy-(aa 502–553) termini. The protein sequence spanning the third to the sixth transmembrane domain (aa 341–501) contains an ion transport motif (PF00520) and a transient-receptor-potential-like motif (PS50272) (Fig. 13.1a and Table 13.1).

Mcoln3 shows a broad expression pattern as demonstrated by RT-PCR and is expressed in all major tissues, albeit at different levels [3]. RNA in situ hybridization on mouse inner ear sections revealed a similar expression profile showing strong expression in cochlear and vestibular sensory hair cells, marginal cells of the stria vascularis, Reissner's membrane, and weaker expression in cells of the spiral limbus and Hensen's and Claudius' cells [4]. Immunohistochemistry demonstrated that TRPML3 is expressed in native skin cells at the bulb region of hair follicles coinciding with the location of melanocyte markers (HMB45 and tyrosinase) and is also expressed in the adult mouse melanocyte cell line melan-a2 [5]. Western blot analyses of protein extracts from adult mouse inner ear and the melan-a2 cell line

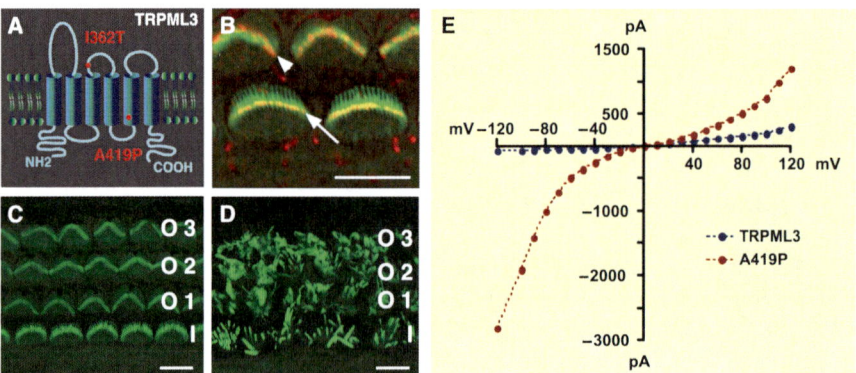

Fig. 13.1 (a) Schematic diagram of the TRPML3 channel showing the six-transmembrane topology, cytosolic amino-(NH2) and carboxy-(COOH) termini, and the location of the isoleucine to threonine (I362T) and alanine to proline (A419P) mutations (*red dots*). (b) Confocal image showing immuno labeling of TRPML3 at the base of the stereociliary hair bundle of inner (*arrow*) and outer hair cells (*arrowhead*). Stereocilia were counterstained with phalloidin (*green*). Scale bar = 5 μm. (c), (d) Confocal images of stereocilia of wildtype (c) and Va^J/Va^J (d) inner (I) and outer (O1-3) hair cells. Note the normal orientation and structure of the hair bundle in (c) and the disorganized splayed appearance in Va^J mutants. Scale bars = 5 μm. (e) Schematic diagram of the voltage/current relationship of wildtype and mutant (A419P) TRPML3. mV, milliVolt; pA, picoAmpere; from [4]

13 The TRPML3 Channel: From Gene to Function

Table 13.1 Characteristics of *Mcoln3*/TRPML3

	Human	Mouse	References
Chromosomal location	1p22.3	3qH2	[11]
Physical location	85,490–85,514 Kb	145,784–145,803 Kb	GRC
Gene symbol	*MCOLN3* (NM_018298)	*Mcoln3* (NM_134160)	[24]
Protein name	TRPML3 (NP_060768)	TRPML3 (NP_598921)	NCBI
Mutant alleles	Unknown	Varitint-waddler, *Va* and *Va^J*	[11]
Expression		Cochlea, vestibule, many other tissues	[4] [3, 5]
	Normal channel	*Mutant channel (A419P)*	
Gating mechanism	Opens at positive voltage	Constitutively active	[4–6, 17, 18]
	Removal/addition of Na^+	Inwardly rectifying	[18]
I/V function	Inwardly rectifying	Inwardly rectifying	[4, 5, 17, 18]
Conductance	50 pS	50 pS	[4]
Ion selectivity	$Ca^{2+} \gg Na^+ \geq K^+ > Cs^+$, Sr^{2+}, Ba^{2+}, and Mg^{2+}	Ca^{2+} Na^+ K^+ Mg^{2+}	[4, 5, 17, 18]
Channel inhibition	H^+ (pH 4.6)		[23]
Cellular localization	Stereocilia		[6]
	Plasma membrane, vesicles	Plasma membrane, vesicles	[21, 22]
Cellular function	Membrane trafficking,	Reduced mechanotransduction	[6]
	Endocytic pathway	Hair cell degeneration	[12, 15]
	Autophagy	Apoptosis	[4, 17]

identified a single band with a molecular weight of ~68 kDa and ~59 kDa respectively [6]. On the sub-cellular level, TRPML3 was localized to vesicular cytoplasmic structures and the plasma membrane of stereocilia of auditory sensory hair cells. High-resolution immuno-confocal and electron microscopy refined the TRPML3 location to the ankle-link region at the base of developing hair cell stereocilia (Fig. 13.1b) [6].

13.2 Normal and Mutant Alleles of *Mcoln3*

Spontaneous and induced mutant variants of mucolipin genes have been identified in different species and model organisms. Mutations in human *MCOLN1* underlie an autosomal recessive lysosomal storage disorder known as mucolipidosis type IV (OMIM 252650), which is characterized by psychomotor retardation and abnormalities affecting the cornea and photoreceptor layer in the eye [1, 7]. Induced loss-of-function mutations in the mucolipin homolog CUP-5 in *C. elegans* disrupt the biogenesis of lysosomes, leading to an accumulation of large vacuoles in many cell types, increased cell death, and embryonic lethality [8–10]. In the mouse, a spontaneous dominant gain-of-function mutation in *Mcoln3* causes hearing loss, vestibular dysfunction, pigmentation defects and embryonic lethality in the Varitint-waddler (*Va*) mutant [11–13]. No mutant allele for *Mcoln2* has yet been identified.

The *Va* mutation was first described by Cloudman and Bunker in 1945. This mutation is semidominant and heterozygotes can be distinguished from homozygotes by their lower degree of coat color variegation and greater embryonic viability. Mice of both *Va/+* and *Va/Va* genotypes are deaf and display nervous, choreic head movements together with erratic circling behavior [13]. A second spontaneous mutation occurred in the *Mcoln3* locus in a cross segregating for *Va* and coincidentally occurred *in cis*, that is, on the same parental chromosome that carried the *Va* mutation. This second site mutation, named Va^J, has an ameliorating effect on *Va* such that it partially restores embryonic viability, coat color variegation, hearing loss and choreic vestibular appearance [14].

The physiology and anatomy of the *Va* and Va^J cochlea have been studied in great detail [12, 15, 16]. The earliest phenotypic anomalies are seen in late *Va/+* embryos at gestation day 17.5, which have severely disorganized stereociliary hair bundles in the outer and inner hair cells. By postnatal day 11, most stereociliary hair bundles have become splayed in both *Va* and Va^J mutants; this is accompanied by some hair cell loss (Fig. 13.1c, d) [11]. In early postnatal and young adult mutant mice, an underpigmented stria vascularis, degeneration of outer and inner cochlear sensory hair cells, as well as degeneration of the spiral ganglion, leads to an abnormally low endocochlear potential, a loss of compound action potentials and absent auditory-evoked brain stem responses [12, 15]. All of these phenotypes are expressed in neonatal hetero- and homozygous mutants resulting in congenital deafness.

The mechanism underlying the spotted and dilute coat color in *Va* and Va^J is not precisely known, but it likely consists of a developmental defect during melanoblast differentiation and migration. A cellular defect at the level of the differentiating neural crest may also underlie the embryonic lethality in *Va* homozygotes. The embryonic lethality in Va^J homozygotes is tightly controlled by the genetic background; on the C57BL/6 J background Va^J homozygotes are lethal, whereas they survive and develop normally on a C3HeB/FeJ background [16].

A positional cloning approach aimed at identifying the *Va* locus revealed two mutations in the *Mcoln3* gene. The *Va* allele harbors a c.1,255 G to C transversion in exon 10, causing an alanine to proline substitution at amino acid position 419 (Ala419Pro). This substitution is located in the fifth predicted transmembrane domain, which is part of the ion pore region. The Va^J allele carries a c.1,085 T to C transition in exon 8, which changes the isoleucine residue at position 362 to a threonine (Ile362Thr) in the second extracellular/luminal loop. As expected from the origin and ameliorating effect of Va^J, the Va^J allele also has the Ala419Pro mutation [11].

13.3 Molecular Physiological Function of TRPML3

Heterologously-expressed wild-type TRPML3 is largely inactive at normal negative resting membrane potentials, exhibiting only small negative currents [4, 5, 17, 18], but shows a strong outward rectifying current (300 pA) when the membrane potential is shifted to positive voltages (+120 mV) [4]. TRPML3, however, can be gated at normal resting potentials by a brief depletion, then addition of Na^+ in the incubation solution, which generates a large inward rectifying current (8 nA at –100 mV) that slowly inactivates [18]. The TRPML3 channel is highly permeable to Ca^{2+}, Na^+ and K^+ and, to a lesser extent, to Cs^+, Sr^{2+}, Ba^{2+}, and Mg^{2+} [4, 5, 18]. The activity of the channel can be further modulated by decreasing the extracytosolic pH from 7.4 to 6.4, at which pH 50% of the activity is inhibited [18]. Three histidine residues (His252, His273 and His283) residing in the large first extracytosolic loop are critical for this regulation.

In contrast to wild-type TRPML3, the isoform mutated in the *Va* allele, here referred to as TRPML3(A419P), shows a strong constitutively-active inward-rectifying current at normal negative resting potentials [4, 5, 17, 18]. TRPML3(A419P) heterologously expressed in either LLC-PK1-Cl4 or HEK293 cells exhibits an inward current of ~3 nA at –120 mV that escapes regulation by extracytosolic Na^+ and does not inactivate. The A419P mutation does not alter the channel's pore properties; the mutant channel displays reversal potentials, ion permeabilities, and conductances (~50 pS) similar to that of the native channel. Interestingly, although the I362T mutation attenuates the *Va* phenotype on the organismal level, no effect was measured on the molecular level; the TRPML3(A419P/I362T) mutant showed the same physiological properties as the TRPML3(A419P) isoform (Fig. 13.1d) [4, 5, 17, 18].

Introduction of proline and glycine into an alpha helix has a destabilizing effect on the conformation for both chemical and biophysical reasons. The Ala>Pro substitution in *Va* occurs in the fifth alpha helical transmembrane domain. Amino acid substitution scans showed that A419G has a similar constitutive effect as the A419P mutation (spontaneous activity and minimal inactivation), whereas A419V had 20% of the normal inward rectifying activity, but behaved otherwise similarly to the wild-type channel. This suggests that the *Va* mutation introduces a conformational change, most likely a kink or swivel, in the transmembrane alpha helix, locking the channel in an open state [4, 5, 17, 18].

In heterologous expression systems, transient expression of TRPML3(A419P) results in cell death from permanent influx of extracellular Ca^{2+}, as demonstrated by Ca^{2+} imaging and expression of apoptotic markers such as annexin V and staining with propidium iodide. Transient expression of TRPML3(I362T) alone has no apoptotic effect. Biotinylation assays showed that surface expression of mutant TRPML3 was significantly less in TRPML3(A419P/I362T) transfected cells than in TRPML3(A419P) expressing cells [4, 5, 17, 18]. Hence, the reduced plasma membrane localization of the TRPML3(A419P/I362T) isoform in vivo could account for the attenuated *Va^J* phenotype.

The effect of the TRPML3(A419P/I362T) mutant channel in vivo was studied in detail in mouse cochlear hair cells [6]. These sensory cells express an ordered array of staggered elongated microvilli (so-called stereocilia) at their apical surface. These stereocilia respond to minute mechanical (i.e. acoustic) stimulation with nanometer-scale deflections, upon which, yet-unknown mechano-sensitive transducer channels open to initiate the signal transduction cascade. The hearing impairment in Va and Va^J mutants, the localization of TRPML3 near the base of developing stereocilia as well as the disrupted structure of the stereociliary hair bundle suggested defects in mechanosensation. Indeed, mechanical stimulation of postnatal mutant cochlear hair cells *ex vivo* revealed reduced and absent transducer currents in Va^J/+ and Va^J/Va^J respectively [6]. However, this raises the question of whether TRPML3 is directly linked to the mechanotransduction process, or whether the reduced transducer currents are secondary to the gain-of-function mutation. The lack of TRPML3 labeling in mature stereocilia seems to argue against a direct role of TRPML3 in mechanotransduction.

Consistent with the in vitro experiments, using the TRPML3(A419P/I362T) mutant, hair cells of Va^J/Va^J mutants showed an inwardly-rectifying leak current (1.6 nA at –100 mV). The demonstrated apoptotic effect of this constitutive conductance in vitro explains the hair cell and melanocyte loss in the Va^J cochlea [12, 15].

13.4 TRPML3 and Its Role in the Endocytic Pathway

The cell biological function of the TRPML proteins is best understood for TRPML1. *MCOLN1* encodes two di-leucine motifs that target TRPML1 primarily to vesicles at the late endosome-lysosome stage [19, 20]. Recent studies suggest that TRPML3 also plays a role in the endocytic pathway, although at a different point. The endocytic pathway is part of the larger intracellular vesicular traffic system, and consists of mostly uni-directional routes, highly organized and regulated, which involve early- and late endosomes, and lysosomes. The main function of this pathway is to engulf, degrade and recycle fluid, molecules and other bioparticles and, as such, is a major component of every eukaryotic cell. A series of transfection experiments using different cell lines (HeLa, HEK293, CL4 and the human epithelia cell line ARPE19) showed that heterologously-expressed TRPML3 co-localizes with markers (Hrs, EEA1, dextran) of early and late endosomes [4, 21, 22]. In addition, adenovirus-mediated overexpression of tagged TRPML3, as well as siRNA-mediated down-regulation of endogenous TRPML3, affected trafficking of the native EGF receptor, EGFR, along the endosomal pathway [21, 22]. Consistent with these findings, biochemical subcellular fractionation experiments localized transiently-expressed TRPML3 to endosomes and lysosomes [22]. Furthermore, TRPML3-overexpressing cells had higher endosomal pH compared to non-transfected cells [21].

Autophagy, or autophagocytosis, is the degradation of intra-cellular components as part of a regular recycling and clearance process or as result of cellular

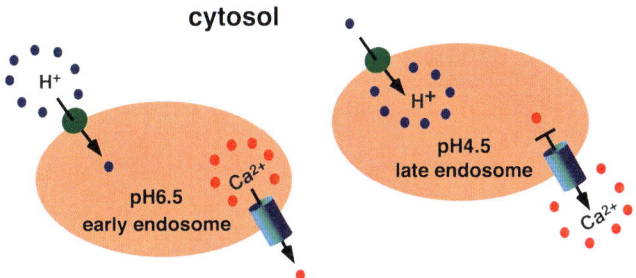

Fig. 13.2 Schematic diagram of suggested function of TRPML3 in acidification and differentiation of endosomes [21]. At pH6.5 TRPML3 releases Ca^{2+} (*red dots*) from the lumen of the early endosome into the cytosol. As H^+ (*blue dots*) enter the endosome, the lumen becomes more acidic, upon which the TRPML3 channel closes promoting differentiation into late endosomes

stress. The autophagic vesicle or autophagosome may fuse with lysosomes to form autolysosomes or with late endosomes to become amphisomes. Overexpression of TRPML3 resulted in a significant accumulation of amphisomes blocking the further maturation into autolysosomes [21].

Localization of TRPML3 to the upper and less acidic (pH 6.0) vesicular components of the endocytic pathway is consistent with the maximal activity of the TRPML3 channel being at a slightly acidic 6–6.5 pH [23]. It was proposed that TRPML3 under slightly acidic conditions, extrudes Ca^{2+} from the lumen of the vesicle into the surrounding cytosol; as Ca^{2+} exits the endosome, H^+ enters, acidifying the vesicle, thereby closing the TRPML3 channel and leading the vesicle to become a late endosome (pH 4–4.5) (Fig. 13.2) [21].

13.5 Open Questions – Future Directions

In recent years, excellent progress has been made in elucidating the molecular and cellular function of TRPML3. This work has generated exciting hypotheses and opened up promising avenues for further experimentation. On the molecular level, it will be important to identify the components that interact with TRPML3. Further precise in situ measurements of TRPML3 activity in early and late endosomes, combined with high-resolution live imaging, will provide further insights into a TRPML3 gating mechanism, and the role of TRPML3 in both in endosome differentiation and membrane trafficking. In addition, analyses of tissue-specific loss-of-function alleles of TRPML3, in combination with other members of the TRPML family, will provide first-class evidence as to the precise and indispensable role of TRPML3 in vivo, in particular during embryogenesis, in the migration and differentiation of melanocytes, and in the vestibular and cochlear development. Lastly, it will be important to identify pathological human alleles that may exist either in the form of rare and highly-penetrant alleles or as hypomorphic modifiers of *Trpml1* and/or *Trpml2*, which will aid in improved diagnostics and therapeutics.

Acknowledgments I thank Rosa Puertollano and Jamie García-Añoveros for discussions and comments on the manuscript. Supported by the Intramural Research Program at the National Institute on Deafness and Other Communication Disorders.

References

1. Bargal R, Avidan N, Ben-Asher E, Olender Z, Zeigler M, Frumkin A, Raas-Rothschild A, Glusman G, Lancet D, Bach G (2000) Identification of the gene causing mucolipidosis type IV. Nat Genet 26:118–123
2. Chenik M, Douagi F, Achour YB, Khalef NB, Ouakad M, Louzir H, Dellagi K (2005) Characterization of two different mucolipin-like genes from Leishmania major. Parasitol Res 98:5–13
3. Cuajungco MP, Samie MA (2008) The varitint-waddler mouse phenotypes and the TRPML3 ion channel mutation: cause and consequence. Pflugers Arch 457:463–473
4. Nagata K, Zheng L, Madathany T, Castiglioni AJ, Bartles JR, Garcia-Anoveros J (2008) The varitint-waddler (Va) deafness mutation in TRPML3 generates constitutive, inward rectifying currents and causes cell degeneration. Proc Natl Acad Sci USA 105:353–358
5. Xu H, Delling M, Li L, Dong X, Clapham DE (2007) Activating mutation in a mucolipin transient receptor potential channel leads to melanocyte loss in varitint-waddler mice. Proc Natl Acad Sci USA 104:18321–18326
6. van Aken AF, Atiba-Davies M, Marcotti W, Goodyear RJ, Bryant JE, Richardson GP, Noben-Trauth K, Kros CJ (2008) TRPML3 mutations cause impaired mechano-electrical transduction and depolarization by an inward-rectifier cation current in auditory hair cells of varitint-waddler mice. J Physiol 586:5403–5418
7. Sun M, Goldin E, Stahl S, Falardeau JL, Kennedy JC, Acierno JS Jr., Bove C, Kaneski CR, Nagle J, Bromley MC, Colman M, Schiffmann R, Slaugenhaupt SA (2000) Mucolipidosis type IV is caused by mutations in a gene encoding a novel transient receptor potential channel. Hum Mol Genet 9:2471–2478
8. Fares H, Greenwald I (2001) Regulation of endocytosis by CUP-5, the Caenorhabditis elegans mucolipin-1 homolog. Nat Genet 28:64–68
9. Hersh BM, Hartwieg E, Horvitz HR (2002) The Caenorhabditis elegans mucolipin-like gene cup-5 is essential for viability and regulates lysosomes in multiple cell types. Proc Natl Acad Sci USA 99:4355–4360
10. Treusch S, Knuth S, Slaugenhaupt SA, Goldin E, Grant BD, Fares H (2004) Caenorhabditis elegans functional orthologue of human protein h-mucolipin-1 is required for lysosome biogenesis. Proc Natl Acad Sci USA 101:4483–4488
11. Di Palma F, Belyantseva IA, Kim HJ, Vogt TF, Kachar B, Noben-Trauth K (2002) Mutations in *Mcoln3* associated with deafness and pigmentation defects in varitint-waddler (*Va*) mice. Proc Natl Acad Sci USA 99:14994–14999
12. Cable J, Steel KP (1998) Combined cochleo-saccular and neuroepithelial abnormalities in the Varitint-waddler-J (*Va^J*) mouse. Hear Res 123:125–136
13. Cloudman AM, Bunker LE (1945) The varitint-waddler mouse. J Hered 36:258–263
14. Lane PW (1969) VaJ – varitint-waddler-jackson. Mouse News Lett 41:32
15. Deol MS (1954) The anomalies of the labyrinth of the mutants varitint-waddler, shaker-2 and jerker in the mouse. J Genet 52:562–588
16. Kim HJ, Jackson T, Noben-Trauth K (2003) Genetic analyses of the mouse deafness mutations varitint-waddler (Va) and jerker (Espnje). J Assoc Res Otolaryngol 4:83–90
17. Grimm C, Cuajungco MP, van Aken AF, Schnee M, Jors S, Kros CJ, Ricci AJ, Heller S (2007) A helix-breaking mutation in TRPML3 leads to constitutive activity underlying deafness in the varitint-waddler mouse. Proc Natl Acad Sci USA 104:19583–19588
18. Kim HJ, Li Q, Tjon-Kon-Sang S, So I, Kiselyov K, Muallem S (2007) Gain-of-function mutation in TRPML3 causes the mouse Varitint-Waddler phenotype. J Biol Chem 282:36138–36142

19. Kiselyov K, Chen J, Rbaibi Y, Oberdick D, Tjon-Kon-Sang S, Shcheynikov N, Muallem S, Soyombo A (2005) TRP-ML1 is a lysosomal monovalent cation channel that undergoes proteolytic cleavage. J Biol Chem 280:43218–43223
20. Vergarajauregui S, Puertollano R (2006) Two di-leucine motifs regulate trafficking of mucolipin-1 to lysosomes. Traffic 7:337–353
21. Martina JA, Lelouvier B, Puertollano R (2009) The calcium channel mucolipin-3 is a novel regulator of trafficking along the endosomal pathway. Traffic 10:1143–1156
22. Kim HJ, Soyombo AA, Tjon-Kon-Sang S, So I, Muallem S (2009) The Ca(2+) channel TRPML3 regulates membrane trafficking and autophagy. Traffic 10:1157–1167
23. Kim HJ, Li Q, Tjon-Kon-Sang S, So I, Kiselyov K, Soyombo AA, Muallem S (2008) A novel mode of TRPML3 regulation by extracytosolic pH absent in the varitint-waddler phenotype. EMBO J 27:1197–1205
24. MGD (2010) Mouse Genome Informatics Project, The Jackson Laboratory, Bar Harbor, ME. World Wide Web (URL: http://www.informatics.jax.org)

Chapter 14
TRPV5 and TRPV6 in Transcellular Ca^{2+} Transport: Regulation, Gene Duplication, and Polymorphisms in African Populations

Ji-Bin Peng

Abstract TRPV5 and TRPV6 are unique members of the TRP super family. They are highly selective for Ca^{2+} ions with multiple layers of Ca^{2+}-dependent inactivation mechanisms, expressed at the apical membrane of Ca^{2+} transporting epithelia, and robustly responsive to 1,25-dihydroxivitamin D$_3$. These features are well suited for their roles as Ca^{2+} entry channels in the first step of transcellular Ca^{2+} transport pathways, which are involved in intestinal absorption, renal reabsorption of Ca^{2+}, placental transfer of Ca^{2+} to fetus, and many other processes. While TRPV6 is more broadly expressed in a variety of tissues such as esophagus, stomach, small intestine, colon, kidney, placenta, pancreas, prostate, uterus, salivary gland, and sweat gland, TRPV5 expression is relatively restricted to the distal convoluted tubule and connecting tubule of the kidney. There is only one *TRPV6*-like gene in fish and birds in comparison to both *TRPV5* and *TRPV6* genes in mammals, indicating *TRPV5* gene was likely generated from duplication of *TRPV6* gene during the evolution of mammals to meet the needs of complex renal function. TRPV5 and TRPV6 are subjected to vigorous regulations under physiological, pathological, and therapeutic conditions. The elevated TRPV6 level in malignant tumors such as prostate and breast cancers makes it a potential therapeutic target. *TRPV6*, and to a lesser extent *TRPV5*, exhibit unusually high levels of single nucleotide polymorphisms (SNPs) in African populations as compared to other populations, indicating *TRPV6* gene was under selective pressure during or after humans migrated out of Africa. The SNPs of *TRPV6* and *TRPV5* likely contribute to the Ca^{2+} conservation mechanisms in African populations.

TRPV5 and TRPV6 are unique members of the TRP super family with three major common features that distinguish them from other members: highly selective for Ca^{2+}, distributing to apical membranes in Ca^{2+} transporting epithelial cells, and

---

J.-B. Peng (✉)
Division of Nephrology, Department of Medicine, Nephrology Research and Training Center, University of Alabama at Birmingham, Birmingham, AL 35294, USA
e-mail: jpeng@uab.edu

highly responsive to 1, 25-dihydroxyvitamin D_3 (1,25$[OH]_2D_3$). These features also make them suitable for their roles as Ca^{2+} entry channels in the first step of the transcellular pathway, which is involved in many processes such as Ca^{2+} absorption in the intestine and reabsorption in the kidney. In this chapter, I will summarize the evolution of the *TRPV5* and *TRPV6* genes, the features of TRPV5 and TRPV6 in function and regulation to meet the needs for Ca^{2+} entry channels in the transcellular Ca^{2+} transport pathways, their regulation under pathophysiological and therapeutic conditions, and the unusual high allele frequency of *TRPV6* (and to a lesser extent *TRPV5*) single nucleotide polymorphisms (SNPs) in African populations. Due to the scope of this chapter, it is impossible to cover all important work. Readers are directed to relevant chapters of this series and recent reviews [1–12] for further details about various aspects of TRPV5 and TRPV6.

14.1 Identification – Results of Seeking Ca^{2+} Transporters in Transcellular Pathways

TRPV5 and TRPV6 were identified as results of searching for genes responsible for Ca^{2+} transport in the transcellular pathway in the kidney [13] and intestine [14], respectively. Until the end of the twentieth century, calbindins had remained to be the main molecules in the transcellular Ca^{2+} transport pathway with little information available about the apical entry mechanism. To tackle this problem, an expression cloning approach with *Xenopus laevis* oocytes was employed to identify Ca^{2+} transport proteins by two groups independently. Using mRNA isolated from connecting tubule and cortical collecting duct cells immunodissected from rabbit kidney, Hoenderop, Bindels and colleagues [13] constructed a cDNA library and identified a cDNA clone by functional screening of the library using radiotracer Ca^{2+} uptake assay. The resultant cDNA encodes a protein of 730 amino acids, which was named ECaC (Epithelial Ca^{2+} channel) [13]. In contrast, we chose intestinal epithelial tissues as the source to clone an intestinal Ca^{2+} transporter [14]. We observed significant increases in Ca^{2+} uptake in oocytes injected with mRNAs from cecum or duodenum from rats fed on Ca^{2+} deficient diet for 2 weeks. A 10–20% reduction in serum Ca^{2+} was observed in these rats. From the rat duodenal cDNA library, we identified a cDNA encoding a protein of 727 amino acids, which we named CaT1 (Ca^{2+} transport protein subtype 1). The two proteins share 74% amino-acid identity. ECaC and CaT1 were renamed later on as TRPV5 and TRPV6, respectively. Other alternative names include CaT2 [15] and ECaC1 [16] for TRPV5, and CaT-L [17] and ECaC2 [18] for TRPV6. Only two cloned proteins share primary sequence similarity to TRPV5 and TRPV6 at the time of cloning, the capsaicin receptor VR1 (TRPV1), a heat activated channel in the pain pathway [19] and OSM-9, a *C. elegans* protein involved in olfaction, mechanosensation, and olfactory adaptation [20]. The subsequent identification of VRL-1 (TRPV2) [21, 22], OTRPC4/VR-OAC/Trp12/VRL-2 (TRPV4) [23–26] and TRPV3 [27–29], completed the TRPV family. TRPV5 and TRPV6 share approximately 40–45% amino-acid identity with the other 4 members and are the only calcium selective channels in the family; the other four members are non-selective cation channels, they serves as sensors for heat (TRPV1-3) or osmolarity (TRPV4).

14.2 Evolution of the Genes: From One to Two

No proteins similar to TRPV members were identified in *E. coli* or yeast. Five TRPV proteins were identified in *C. elegans*, namely osm-9, ocr-1, ocr-2, ocr-3 and ocr-4 [30]. Among them, ocr-4 is closer to TRPV5 and TRPV6 (25.9 and 25.4% identity, respectively) than other TRPV proteins in *C. elegans* (20.3–23.6% identity). Ocr-4 is also closer to TRPV5 and TRPV6 among all the 6 mammalian TRPV members. This indicates that TRPV5 and TRPV6 in vertebrates were likely evolved from Ocr-4 in invertebrate.

The phylogenic tree of TRPV5 and TRPV6 from vertebrates is shown in Fig. 14.1. One interesting observation is that both TRPV5 and TRPV6 are present in mammals, but only one of the two, i.e., TRPV6, is present in birds and fish based on

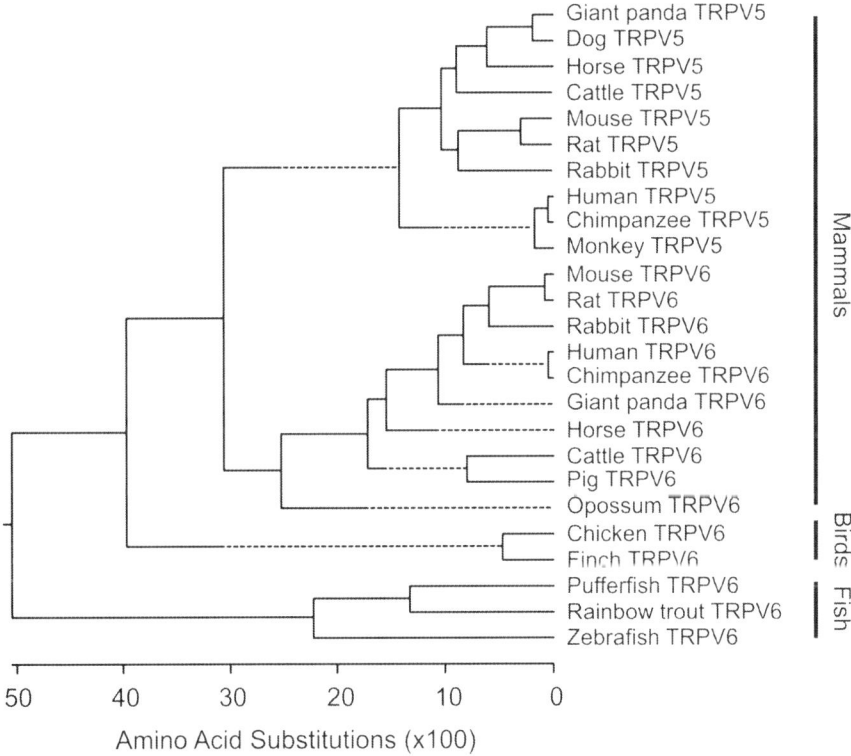

Fig. 14.1 Phylogenetic tree generated from alignment of TRPV5 and TRPV6 proteins from various species. Clustal W method was used to align the proteins with MegAlign program in DNASTAR Lasergene software (DNASTAR Inc., Madison, WI). TRPV6-like proteins are present in fish and birds, whereas both TRPV5 and TRPV6 are present in mammals

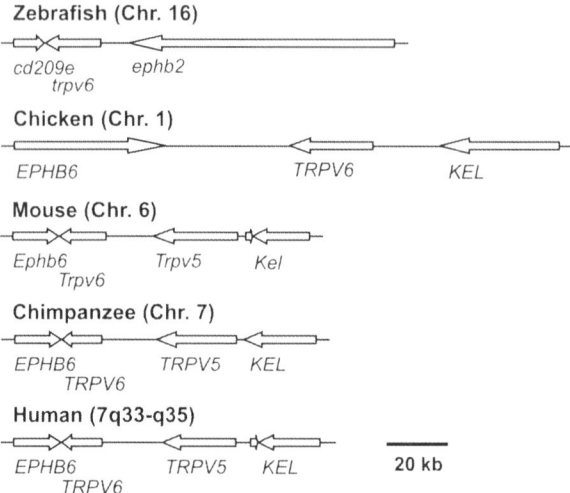

Fig. 14.2 *TRPV5* and *TRPV6* and their flanking genes in zebrafish, chicken, mouse, chimpanzee, and human genomes. Only*TRPV6* is present in zebrafish and chicken genomes, whereas both *TRPV5* and *TRPV6* are present in mouse, chimpanzee and human genomes. *Cd209e*, CD209E antigen-like protein E-like, current official gene symbol *LOC563797*; *Ephb2*, ephrin (EPH) receptor B2-like, current official gene symbol *LOC572411*; *KEL*, Kell blood group, metallo-endopeptidase; *EPHB6*, EPH receptor B6

the current available databases (Fig. 14.1). It has been noted by Qiu and Hogstrand that there is only one *TRPV5/6*-like gene in pufferfish genome [31]. This is not a single incident because the same is true in zebrafish (Fig. 14.2). Furthermore, in the chicken genome, only *TRPV6* gene is present. In contrast, both *TRPV5* and *TRPV6* genes are present in mouse, chimpanzee, and humans (Fig. 14.2). All mammals examined so far have both *TRPV5* and *TRPV6* genes in their genomes. In human and other mammal genomes, *TRPV5* and *TRPV6* genes situate side by side (Fig. 14.2). The two genes are completely conserved in the exon size in the coding region [32]. Thus, it is likely *TRPV5* gene was generated by duplication of *TRPV6* gene. As both birds and mammals are evolved from reptiles, it is likely that gene duplication event occurred during evolution of mammals from reptiles. *TRPV6* has a broader tissue expression pattern whereas *TRPV5* expression is relatively restricted to the kidney (see Sections 14.3.1 and 14.4), therefore, the difference in complex kidney function between reptiles, birds and mammals [33] is likely the driving force behind the generation of the kidney specific *TRPV5* gene.

14.3 Roles as Ca^{2+} Entry Channel in Active Intestinal/Renal Ca^{2+} Re/absorption

Transcellular Ca^{2+} transport pathway is a 3-step process: entry across the apical membrane, transfer of Ca^{2+} from the site of entry to the site of extrusion, and extrusion of Ca^{2+} across the basolateral membrane. This active pathway is involved in intestinal Ca^{2+} absorption, renal Ca^{2+} reabsorption, and many other processes. While vitamin D responsive Ca^{2+} binding protein calbindins may facilitate intracellular Ca^{2+} transfer and the Na^+-Ca^{2+} exchanger and Ca^{2+} ATPase mediate the extrusion of Ca^{2+}, TRPV5 and TRPV6 serve the role for Ca^{2+} entry as the first step of the transcellular pathway.

14.3.1 Expression in Ca^{2+} Transporting Epithelia

The major Ca^{2+} transporting epithelia reside in the intestine and kidney, where Ca^{2+} absorption and reabsorption take place, respectively. In the intestine, TRPV6 is the major Ca^{2+} entry channel whereas TRPV5 is barely detectable. Due to the high sequence similarity between TRPV5 and TRPV6, strong signals could be detected in the rat or rabbit duodenum using TRPV5 cDNA probe in Northern blot analyses [13, 15], but this is likely due to cross hybridization with TRPV6 mRNA [15]. In small intestine, TRPV6 mRNA expression decreases from duodenum to ileum in rats [14] and humans [34]; TRPV6 mRNA is also detectable in large intestine in rats and humans [14, 34]. TRPV6 protein is localized to the apical membrane of intestinal and colonic epithelial cells [35]. Even though TRPV6 mRNA is barely detectable in the ileum in mice and rats, immunostaining with an antibody against TRPV6 indicates the presence of TRPV6 protein in ileum [35]. It is worth noting that in sheep, TRPV6 mRNA and protein level are more abundant in jejunum than in duodenum [36]; higher expression of TRPV6 mRNA in proximal jejunum than duodenum was also documented in horse [37], suggesting a more important role of jejunum in active Ca^{2+} absorption in some animals. TRPV6 mRNA and protein are also weakly expressed in ovine rumen [36], where active Ca^{2+} absorption also occurs in sheep and goats [38].

In mice and rats, TRPV5 mRNA is much more abundant than that of TRPV6 in the kidney [39, 40]. TRPV5 mRNA level is approximately 10 to 20 times higher than that of TRPV6 in mouse kidney [39]. In humans, we have observed much higher mRNA level of TRPV6 than TRPV5 in the kidney [32]. Similarly, in horse kidney, TRPV6 is also much abundant than TRPV5 [37]. Thus, TRPV6 may play more significant roles in some species such as humans and horses.

TRPV5 is localized to the apical membrane of the tubular cells in the late segment of distal convoluted tubule and connecting tubule in mouse kidney [41]. In the rabbit, much TRPV5 is distributed in the connecting tubule [13]. With regards to TRPV6, the tubular distribution is less clear and controversial. Using an RT-PCR approach with microdissected mouse kidney tubules, TRPV6 signal was detected in the medullary thick ascending limb [42]. Due to its low level of expression and the lack of efficient antibodies for TRPV6, TRPV6 protein is difficult to detect in rodent kidney; however, it has been reported that TRPV6 is distributed to mouse distal convoluted tubules (DCT2), connecting tubules, and cortical and medullary collecting ducts using an immunohistochemistry approach [43]. We found a higher mRNA level of TRPV6 in outer medulla than cortex of human kidney, opposite to that of TRPV5 (unpublished observations). Using a TRPV6 specific antibody, we found TRPV6 protein is more abundant in the outer medulla, much less in the cortex, little or none in inner medulla (Zhang *et al.*, unpublished observations). This is consistent with TRPV6 protein being mostly distributed to the thick ascending limb and/or the proximal tubule in the kidney.

14.3.2 Function and Regulation as Ca^{2+} Entry Channels

TRPV5 and TRPV6 share very similar functional properties. Being identified by $^{45}Ca^{2+}$ uptake assay, TRPV5 and TRPV6 mediate Ca^{2+} influx in the absence of any exogenous ligand, altered membrane potential, mechanical stress, or other cues. This is different from other members of the TRPV family, including TRPV1-4, which are activated by ligands, heat, and/or reduction of osmolarity. At macroscopic level, this constitutive Ca^{2+} transport activity makes them act as facilitative transporters that transport solute down its concentration gradient. The Ca^{2+} uptake mediated by TRPV5 or TRPV6 is saturable, with apparent K_m values at submillimolar range, which are well suited for the luminal environment in the intestine and the distal tubule [13–15, 34].

At microscopic level, TRPV5 and TRPV6 exhibit single channel activities. The single channel activity could only be detected in the absence of extracellular divalent cation, such as Ca^{2+} and Mg^{2+}, using Na^+ or K^+ as charge carrier [44–46]. Single channel activity could not be detected with Ca^{2+} as charge carrier due to very small conductance. One common feature is that TRPV5 and TRPV6 are highly selective for Ca^{2+}, with permeability ratio between Ca^{2+} and Na^+ higher than 100:1 [45, 47]. This sets TRPV5 and TRPV6 apart from other TRPV members, which are non-selective cation channels.

TRPV5 and TRPV6 likely function as tetramers [48]. Five ankyrin repeats are present in the N-terminal region of TRPV5 and TRPV6, and it is believed that the third ankyrin repeat initiates a molecular zippering process necessary for the assembly of TRPV6 channels [49]. Regions in N- and C-terminal of TRPV5 were also found important to TRPV5 assembly [50]. However, the crystal structure of ankyrin repeats of TRPV6 indicates that ankyrin repeat domain of TRPV6 does not form a tetramer, challenging its role in channel assembly [51].

TRPV6 outer pore structure is likely similar to that of K^+ channel with a single filter preceded by a pore helix [52]. An aspartate residue at 542 (D542) from each subunit of the TRPV5/6 tetramer likely forms the filter for Ca^{2+} [53–55]. This residue is also responsible for the Mg^{2+} blockade [53, 56].

TRPV5 and TRPV6 exhibit Ca^{2+}-dependent inactivation as feedback control mechanisms via intrinsic channel structure, interaction with Ca^{2+} binding proteins, and phosphatidylinositol 4,5-bisphosphate (PIP$_2$). TRPV6 activity is tightly regulated by intracellular Ca^{2+} [57]. It has a fast inactivation mechanism compared to TRPV5 [58]. The intracellular loop between transmembrane domains 2 and 3 is the molecular determinant of this mechanism [58]. A CaM binding site is present in the C-terminal of TRPV6 [59], and it mediates a slow phase of Ca^{2+}-dependent inactivation of TRPV6 [59–61]. The interaction between Ca^{2+}/CaM and TRPV6 increases dynamically as intracellular Ca^{2+} level rises [60]. Calbindin-D$_{28k}$, a vitamin D regulated Ca^{2+} binding protein, buffers Ca^{2+} entered through TRPV5 to prevent inactivation of the channel [62]. TRPV5 activity is also regulated by a Ca^{2+} sensor 80 K-H [63]. In addition, PIP_2 increases TRPV5 and TRPV6 activity and prevents their Ca^{2+}-dependent inactivation [64–67]. The activation of phospholipase C by Ca^{2+} depletes PIP_2 and this contributes to the Ca^{2+}-dependent inactivation of the channel [65]. ATP stabilizes TRPV6 and prevents it from rundown, and this action is counteracted by PKC βII [68].

TRPV5 and TRPV6 are also regulated by extracellular cues. Both TRPV5 and TRPV6 activities increase at alkaline pH [13, 14, 69]. Glutamate residues in the pore helix and in the extracellular domain close to the pore of TRPV5 sense the pH and alter channel activity via conformational changes of the pore [70–73]. In addition, extracellular pH also regulates the dynamic of plasma membrane delivery of functional TRPV5 proteins from intracellular vesicles [74]. Furthermore, tissue kallikrein, a serine protease secreted in the connecting tubule, enhances TRPV5 plasma membrane abundance by delaying its retrieval via the PLC/DAG/PKC pathway upon activating the bradykinin 2 receptor [75]. Finally, TRPV5 and TRPV6 can be stabilized at the plasma membrane by deglycosylation via Klotho (see Section 14.3.3).

The activity, stability and trafficking of TRPV5 and TRPV6 are also regulated by associated proteins [for reviews see [7, 8]].

14.3.3 Robust Regulation by Calcitrophic Hormones

14.3.3.1 Vitamin D

TRPV6 gene is one of the genes most responsive to $1,25(OH)_2D_3$. *TRPV6* mRNA was up-regulated as early as 2 h after administration of 0.1 μM of $1,25(OH)_2D_3$ in intestinal cell line Caco-2 and a 10-fold increase was achieved by 24 h [76]. Compared to calbindin-D_{9k}, a vitamin D responsive Ca^{2+} binding protein, the response of TRPV6 to $1,25(OH)_2D_3$ was more rapid and robust [76]. In vitamin D receptor (*VDR*) knockout mice, Duodenal TRPV6 mRNA was reduced by 90%; however, TRPV5 mRNA level in the kidney appeared to be unchanged. In contrast, TRPV5 mRNA and protein abundance was decreased in vitamin D depleted rats [77]. A single injection of $1,25(OH)_2D_3$ in vitamin D deficient mice induced a ~10-fold increase of duodenal mRNA of TRPV6 which peaked at 6 h after injection, and ~three fold to fourfold increase in TRPV5 and TRPV6 mRNA in the kidney that peaked at 12 h after injection [39]. This suggests that TRPV5 and TRPV6 may be differently regulated by vitamin D in the kidney and intestine, respectively.

In humans, duodenal *TRPV6* transcript was initially found not significantly correlated with vitamin D metabolites [18]; but later studies by Walter and colleagues indicate that duodenal TRPV6 level was positively correlated with plasma $1,25(OH)_2D_3$ in a *VDR* genotype group [78], in men and likely in younger women, but not in older women [79]. This is likely due to the lower level of VDR in older women [79]. In isolated human duodenal specimens, TRPV6 transcript was very responsive to $1,25(OH)_2D_3$ [80, 81]. TRPV6 was also responsive to 25-hydroxyvitamin D [25(OH)D], both in Caco-2 cells [82] and in human duodenal specimens [81]. This is likely due to the expression of 25(OH)D 1α-hydroxylase (CYP27B1), which converts 25(OH)D to $1,25(OH)_2D_3$ [81, 82].

The most unequivocal evidence that *TRPV6* is a direct target of vitamin D action came from the identification of vitamin D responsive elements (VDREs) from *TRPV6* promoter by Pike and colleagues [83, 84]. Three active VDREs were identified in a 7 kb *TRPV6* promoter at −1.2, −2.1 and −4.3 kb from the transcriptional start site [83]. Similar sites are present in mouse *Trpv6* promoter [84]. VDR/RXR (retinoid X receptor) heterodimer binds to TRPV6 promoter, and then recruits steroid receptor coactivator 1 and RNA polymerase II; and this is accompanied by a broad change in histone 4 acetylation [83, 84]. The presence of VDREs in *TRPV6* gene was also confirmed by another study [85]. With regards to TRPV5, it is suggested that there exist VDREs in human *TRPV5* promoter [77]; however, experimental evidence is yet to be provided for these potential VDREs [86].

Finally, evidence exist to support the involvement of TRPV6 in the non-genomic action of $1,25(OH)_2D_3$ to stimulate Ca^{2+} absorption in chicken intestinal cells [87]. $1,25(OH)_2D_3$ stimulated β-glucuronidase secretion in a protein kinase A (PKA) dependent pathway. In response to β-glucuronidase administration, TRPV6 was translocated to the brush border membrane where it mediates Ca^{2+} influx. The non-genomic effect of $1,25(OH)_2D_3$ on Ca^{2+} uptake was blocked by siRNA for either β-glucuronidase or TRPV6 [87]. Thus, $1,25(OH)_2D_3$ appears to stimulate TRPV6 expression and trafficking via genomic and non-genomic pathways, respectively. TRPV6 remains the most vitamin D-responsive Ca^{2+} channel identified thus far.

14.3.3.2 Parathyroid Hormone

Parathyroid hormone (PTH) is well known to regulate renal Ca^{2+} reabsorption. The regulation of TRPV5 by PTH is twofold: expression and activity. In parathyroidectomized rats and calcimimetic compound NPS R-467 infused mice that had a lower PTH level, renal TRPV5 (calbindin-D_{28k}, and NCX1) level was decreased; and the PTH supplementation restored the level of TRPV5 and other Ca^{2+} transport proteins [88]. Inhibition of TRPV5-mediated Ca^{2+} influx by ruthenium red decreased the levels of other Ca^{2+} transport proteins, suggesting the major effect of PTH on Ca^{2+} transport machinery is to enhance TRPV5 expression, and the expression of other Ca^{2+} transport proteins depends on TRPV5-mediated Ca^{2+} influx [88].

PTH is also capable of increasing Ca^{2+} transport in the distal tubule acutely [89]. Both PKA and protein kinase C (PKC) are involved in this regulation [90]. Indeed, activation of TRPV5 by PTH via both PKA and PKC has been reported [91, 92]. PTH activated the cAMP-PKA signaling cascade and increased TRPV5 open probability via phosphorylation of threonine-709 of TRPV5 [91]. This regulation required a strong intracellular buffering of intracellular Ca^{2+}. However, the PKA phosphorylation site in TRPV5 (threonine-709) is not conserved in human TRPV5. To what extent this mechanism is applicable to humans remains to be determined.

In another study, the increase of TRPV5 activity by heterogeneously expressed PTH receptor was prevented by PKC inhibitor [92]. Mutation of PKC phosphorylation sites serine-299/serine-654 in TRPV5 abolished the regulation.

Caveolae-mediated endocytosis of TRPV5 appears to be inhibited via PKC-dependent pathway in response to PTH [92].

14.3.3.3 Calcitonin

Calcitonin affects renal Ca^{2+} reabsorption through the thick ascending limb [93, 94]. Thus, it is not surprising that this action is independent of TRPV5 because its localization to distal convoluted tubule and connecting tubule. This was confirmed using the *Trpv5* KO mice, as the effects of calcitonin on urinary excretion of Ca^{2+} (and Na^+ and K^+) was not different between wild-type (WT) and *Trpv5* KO mice [95]. To what extent TRPV6 is involved in calcitonin-mediated regulation of Ca^{2+} reabsorption is yet to be examined.

14.3.3.4 Klotho

KL gene product Klotho functions to suppress aging process [96]. The two forms of Klotho, membrane and secreted forms, function distinctively (see [97] for a recent review). While the membrane Klotho interacts with fibroblast growth factor (FGF) receptors to form a high affinity receptor for FGF23, a hormone regulating phosphate excretion, the secreted form acts as a hormonal factor that regulates the stability of TRPV5/6 [98–100] and renal outer medullary potassium (ROMK) channel [101]. Klotho is expressed in the distal convoluted tubule where it colocalizes with TRPV5 [98]. Because Klotho exhibits β-glucuronidase activity [102], it was thought that Klotho modifies TRPV5 glycan through this activity [98]. Like Klotho, β-glucuronidase also activated TRPV5 and TRPV6 but not related TRP channels TRPV4 and TRPM6 [99]. Alternatively, Klotho removed terminal sialic acids from their glycan chains of TRPV5 and exposes disaccharide galactose-N-acetylglucosamine, which bound to galactoside-binding lectin galectin-1 [100]. Galectin-1 linked TRPV5 proteins are likely resistant to endocytosis, but the underlying mechanism is not well understood. However, the removal of N-glycan by endoglycosidase-F, which completely removes *N*-glycans, also increased TRPV5 and TRPV6 activity [99]. Under this condition, galectin-1 may not be able to bind to TRPV5. Although the mechanisms are still to be clarified, the regulation of TRPV5/6 by Klotho represents a novel area of ion channel regulation. The dysregulation of TRPV5/6 may be responsible for the increased excretion of Ca^{2+} in *KL* deficient mice, in which PTH stimulated Ca^{2+} reabsorption in the connecting tubule is impaired [103].

14.3.4 Regulation Under Physiological Conditions

14.3.4.1 Low Dietary Ca^{2+}

The biosynthesis of TRPV6 has been shown to be the most responsive to the body's need for Ca^{2+} under conditions such as low Ca^{2+} diet, pregnancy, and lactation.

Restriction of dietary Ca^{2+} drastically increases duodenal TRPV6 level, and to a less extent, renal TRPV5 and TRPV6 in mice [39, 40], and in rats [104]. At least part of the effect of low Ca^{2+} diet to up-regulate TRPV6 is independent of VDR [105].

14.3.4.2 Pregnancy and Lactation

During pregnancy, duodenal *TRPV6* transcript level was up-regulated by 12-fold in either WT or *VDR* KO mice, whereas calbindin-D_{9k} and Ca^{2+}-ATPase PMCA1 were only moderately increased in WT mice [106]. This observation was confirmed with global microarray analysis in pregnant *VDR* KO mice [107]. Similarly, a 13-fold increase in duodenal *TRPV6* transcript was observed in both mouse lines during lactation [106]. The increase of prolactin during lactation causes a positive genomic effect on the *TRPV6* transcription along with paracellular claudins; however, TRPV6 appears not to be involved in the nongenomic effect of suckling induced prolactin surge on intestinal Ca^{2+} absorption [108].

14.3.4.3 Sex Hormones

TRPV5 and TRPV6 levels are also controlled by sex hormones. In estrogen receptor β KO mice, TRPV6 level was not altered, however, a 55% reduction of duodenal TRPV6 mRNA level was observed in estrogen receptor α KO mice. Pharmacological doses of estrogens increased TRPV6 mRNA by fourfold and eightfold in ovariectomized WT and *VDR* KO mice, respectively [106]. At the same level of 1,25(OH)$_2$D$_3$, female mice had a higher Ca^{2+} absorption and TRPV6 mRNA level than male mice. The response of duodenal TRPV6 mRNA to 1,25(OH)$_2$D$_3$ in female mice was twofold of that in male mice even though the VDR mRNA level was the same [109]. In ovariectomized rats, both TRPV5 mRNA and protein levels were increased by pharmacological doses of 17β-estradiol and this effect was independent of 1,25(OH)$_2$D$_3$ [110]. In aromatase deficient (ArKO) mice, deficiency in estrogen caused hypercalciuria together with decreased levels of renal TRPV5 and TRPV6 [111]. Additionally, 17β-estradiol exerted a nongenomic effect to enhance TRPV6 activity in human colonic T84 cells [112] and TRPV5 activity in rat cortical collecting duct cells [113].

We have previously observed that TRPV6 expression was negatively regulated by androgen in LNCaP cells [114]. Whether this effect of androgen on TRPV6 is related to the lowered intestinal Ca^{2+} absorption and TRPV6 level in male mice is unclear [109]. Interestingly, male mice also have a higher urinary Ca^{2+} excretion and a lower renal TRPV5 level than female mice [115]. TRPV5 level was increased in orchidectomized mice, which excreted less Ca^{2+} in their urine [115]. Males in general have a higher bone mineral density than females; therefore, a higher intestinal Ca^{2+} absorption is expected for males.

14.3.4.4 Exercise

Endurance swimming stimulated Ca^{2+} transport in the small intestine and cecum and decreased that in the proximal colon in rats [116]. The increased Ca^{2+}

absorption were associated with increased mRNA levels of VDR and Ca^{2+} transport proteins (such as TRPV6 and TRPV5) in transcellular and paracellular pathways [116]. In contrast, TRPV6 and TRPV5 mRNA levels dropped along with decreased $1,25(OH)_2D_3$ level in immobilized rats, leading to a reduction in intestinal Ca^{2+} absorption [117]. These studies suggest that TRPV6 (and to a lesser extent TRPV5) participate in exercise-induced increase in Ca^{2+} absorption in a vitamin D-dependent manner.

14.3.4.5 Aging

Decreased Ca^{2+} absorption and increased urinary Ca^{2+} excretion are associated with aging. TRPV6 (and also TRPV5) mRNA level in adult (12-month old) rats was less than half of that in young (2-month old) rats [104]. In mice, intestinal absorption decreased and urinary Ca^{2+} excretion increased with age; this was associated with decreased TRPV6 and TRPV5 in the duodenum and kidney, respectively [118]. The same is true for VDR [118]. In women, both TRPV6 and VDR transcripts were significantly decreased in older age (> 50) [79]. Thus, reduction in TRPV5 and TRPV6 may contribute to a negative Ca^{2+} balance as one ages.

In addition to the above physiological conditions, TRPV6 expression is up-regulated by short-chain fatty acids, fermentation products of indigestible oligosaccharides in rat large intestine [119]. This is likely involved in the increase of Ca^{2+} absorption by the indigestible oligosaccharides such as fructooligosaccharides [120].

14.3.5 Gene Knockout Studies

The roles of TRPV5 or TRPV6 in vitamin D-regulated Ca^{2+} transport pathways have been evaluated by gene knockout studies. The major phenotypes of *Trpv5* and *Trpv6* KO mice are listed in Table 14.1. The common features of the *Trpv5* and *Trpv6* KO mice include elevated serum $1,25(OH)_2D_3$, increased urinary Ca^{2+} excretion, and some degree of deficiency in the bone [121, 122]. Urinary Ca^{2+} excretion increased by sixfold in *Trpv5* KO mice [122]. Intestinal Ca^{2+} absorption is decreased in *Trpv6* KO mice [121], increased in *Trpv5* KO mice as duodenal TRPV6 and calbindin-D_{9k} mRNAs are increased due to an elevated $1,25(OH)_2D_3$ level [122]. However, the serum Ca^{2+} levels are largely normal, even though the PTH level is elevated in *Trpv6* KO mice [121], and in older *Trpv5* KO mice [118].

It is worth noting, however, active Ca^{2+} transport did occur in the absence of TRPV6 alone or together with calbindin-D_{9k} [123]; and the increase of intestinal Ca^{2+} absorption in response to $1,25(OH)_2D_3$ appeared to be intact in the absence of TRPV6 [124]. While Ca^{2+} balance studies are needed to verify the results, it is not surprising as transcellular Ca^{2+} transport contributes only a small fraction of total Ca^{2+} absorbed when the dietary Ca^{2+} level is adequate or high due to the short sojourn time of digested food in the duodenum [125]. The transcellular pathway plays a significant role only when Ca^{2+} in diet is low. The main driving forces for Ca^{2+} absorption through the paracellular pathway are transepithelial potential and

Table 14.1 Phenotype of Trpv5 KO and Trpv6 KO mice

Phenotype	KO mouse	Alteration	Mechanism/explanation	References
Serum1,25(OH)$_2$D$_3$	Trpv5	Elevated (2.9-fold)	Compensatory response	[122]
	Trpv6	Elevated (2.4-fold)	Compensatory response	[121]
Serum PTH	Trpv5	Elevated only in older mice	Compensatory response	[118]
	Trpv6	Elevated (3.8-fold)	Compensatory response	[121]
Intestinal Ca^{2+} absorption	Trpv5	Increased (~30% more)	Due to the increased 1,25(OH)$_2$D$_3$ and in turn TRPV6, calbindin-D$_{9k}$	[122, 126]
	Trpv6	No increase in the absence of 25(OH)D$_3$ 1α-hydroxylase	Trpv6 plays a role in Ca^{2+} absorption	[121]
Urinary Ca^{2+} excretion	Trpv5	Increased (~sixfold)	Trpv5 plays a role in Ca^{2+} reabsorption	[122]
	Trpv6	Increased (>twofold)	Trpv6 plays a role in Ca^{2+} reabsorption	[121]
Placental Ca^{2+} transport	Trpv5	Not determined		
	Trpv6	Decreased	Trpv6 is involved in the active transport of maternal Ca^{2+} to fetus	[134]
Renal function	Trpv5	Increased urine volume reduced urine pH	Activation of CaSR in the collecting duct	[122, 127]
	Trpv6	Increase urine volume	Defect in urine concentrating mechanism	[121]
Bone	Trpv5	Reduced bone thickness 9.3% decrease in BMD	Prolonged elevation of 1,25(OH)$_2$D$_3$	[122, 128] [121]
	Trpv6	Reduced bone formation/mineralization on low Ca^{2+} diet	Prolonged elevation of 1,25(OH)$_2$D$_3$	[129]
Skin	Trpv5	Not determined		
	Trpv6	Disrupted skin Ca^{2+} gradient. ~20% developed alopecia and dermatitis	Trpv6 is involved in transporting Ca^{2+} into the epidermal keratinocytes	[121]

Ca^{2+} concentration gradient. As Ca^{2+} concentration in the plasma is around 1 mM, the driving force for Ca^{2+} absorption via paracellular pathway would be decreased when dietary Ca^{2+} level is low. In contrast, there is a huge (10,000-fold normally) Ca^{2+} gradient across the brush border membrane, even when Ca^{2+} in the diet is scarce, the transcellular pathway will still function properly. When dietary Ca^{2+} is limited, $1,25(OH)_2D_3$ will be elevated in the body, which in turn will induce the transcription of TRPV6 and related Ca^{2+} transport proteins [39]. Low dietary Ca^{2+} also increases TRPV6 transcription independent of the genomic action of vitamin D [105]. Thus, TRPV6 would be an important component for Ca^{2+} absorption when dietary Ca^{2+} is restricted; when the dietary Ca^{2+} is adequate, it may play a regulatory role to fine tune the level of Ca^{2+} absorption. This role is also indispensable because $1,25(OH)_2D_3$ level is elevated in the absence of TRPV6.

14.4 Roles Beyond Intestinal/Renal Ca^{2+} Re/absorption

14.4.1 TRPV6 in Maternal-Fetal Ca^{2+} Transport

Both TRPV5 and TRPV6 are all expressed in the placenta [17, 34, 130–132], however, TRPV6 mRNA level is ~1,000 times higher than that of TRPV5 [32]. TRPV6 is expressed in the trophoblasts and syncytiotrophoblasts of human placenta [17]. In human syncyiotrophoblasts, TRPV6 and TRPV5 were detected in both apical and basal membranes [130]. Ca^{2+} transport in human syncyiotrophoblasts was insensitive to voltage and L-type Ca^{2+} channel modulators but is sensitive to TRPV5/6 blocker Mg^{2+} and ruthenium red [132]. In cultured human trophoblasts isolated from term placenta, TRPV6 and TRPV5 expression correlated with the Ca^{2+} uptake potential along the differentiation of the trophoblasts [131]. Cyclophilin B, a member of immunophilin family, was associated with TRPV6 in human syncytiotrophoblasts and increased TRPV6 activity in vitro [133].

The role of TRPV6 in maternal-fetal Ca^{2+} transport was confirmed by the *Trpv6* KO model [134]. TRPV6 mRNA and protein were detected mainly in intraplacental and the visceral layer of extraplacental yolk sac, where maternal-fetal Ca^{2+} transport takes place in mice. A 14-fold increase was observed in mouse placenta in the last 4 days of gestation to meet the need of Ca^{2+} required for fetal bone mineralization. The maternal-fetal Ca^{2+} transport was 40% lower in *Trpv6* KO fetuses than in WT. As a result, *Trpv6* KO fetuses exhibited low blood Ca^{2+} level and low ash weight [134]. Thus, TRPV6 represents an important Ca^{2+} entry route for maternal-fetal Ca^{2+} transport pathway. A significant reduction of Ca^{2+} transport along with decreases of TRPV5, TRPV6 and other Ca^{2+} transport proteins was found in primary preeclamptic syncytiotrophoblasts, suggesting that TRPV5 and TRPV6 are involved in the abnormal placental Ca^{2+} transport in preeclampsia [135].

14.4.2 TRPV6 Expression and Regulation in the Uterus

The expression of TRPV6 in uterine tissue was studied in mice [136, 137], rats [137, 138], and pigs [139]. TRPV6 mRNA was detected in rat uterine endometrium and glandular endometrium and was highly up-regulated at diestrus compared with proestrus [138]. In mice, uterine TRPV6 mRNA increased in mid- and late-pregnancy, and TRPV6 mRNA was responsive to 17β-estradiol in the luminal and glandular epithelia [136]. In rats, uterine TRPV6 was regulated by progesterone via its receptor [138], whereas TRPV6 appears to be dependent on estrogen receptors in mice [137]. Uterine and placenta TRPV6 levels changed cyclically during pregnancy, suggesting a role of TRPV6 in the embryo implantation process [137]. TRPV6 mRNA was only expressed in placenta-unattached areas of the uterus, and in the labyrinth and spongy zone of placenta. In pig uterus TRPV6 was localized mainly to luminal epithelial cells and to a lesser extent to glandular epithelial cells and chorionic membrane during pregnancy [139]. TRPV6 was higher during pregnancy than in estrous cycle [139]. These observations indicate that TRPV6 plays a role in establishing and maintaining pregnancy as well as in Ca^{2+} transfer between embryo and placenta [137, 139].

14.4.3 TRPV5 in Bone Resorption

TRPV5 was shown to be expressed in the ruffled border membrane in mouse osteoclasts [140]. In the absence of TRPV5, there were increased osteoclast numbers and osteoclast area; however, urinary bone resorption marker deoxypyridinoline was reduced [140]. This is consistent with the impaired bone resorption by osteoclasts derived from *Trpv5* KO mice in an in vitro bone marrow culture system [140]. However, bone resorption inhibitor alendronate normalized the reduced bone thickness in *Trpv5* KO mice, even though it specifically increased bone TRPV5 expression in mice [128]. This indicates that although TRPV5 may play a role in bone resorption, this process is still functional in the absence of TRPV5. Thus the defect of bone in *Trpv5* KO mice is largely due to other compensatory mechanisms as a result of urinary Ca^{2+} wasting.

14.4.4 TRPV5 and TRPV6 in Maintaining Ca^{2+} Gradient in the Inner Ear

The low Ca^{2+} concentration of mammalian endolymph in the inner ear is required for normal hearing and balance. Marcus and colleagues showed that TRPV5 and TRPV6 may play a role in the function of inner ear [141–144]. Both TRPV5 and TRPV6 were detected in primary cultures of semicircular canal duct (SCCD) epithelial cells from neonatal rats, and TRPV5 transcript was responsive to 1,25(OH)$_2$D$_3$ [144], however, the protein level was not up-regulated [141]. TRPV5 and TRPV6 were detected in native SCCD, cochlear lateral wall and stria vascularis (TRPV5 only) of adult rats along with other Ca^{2+} transport proteins [141]. TRPV5 protein was localized close to the apical membrane of strial marginal cells; and both TRPV5 and TRPV6 were found in outer and inner sulcus cells of the cochlea and in the SCCD of the vestibular system [141]. TRPV5 and TRPV6 immunostaining was detected in mouse inner ear as well, and the levels of TRPV5 and TRPV6 decreased in older mice [145]. Mutations in pendrin (Slc26a4, an anion exchanger) cause the most common form of syndromic deafness. Reduced pH and utricular endolymphatic potential and increased Ca^{2+} concentration were found in pendrin KO mice [142, 143]. The reduced pH likely blocks the activity of TRPV5 and TRPV6, whose Ca^{2+} transport activity is reduced at low pH, similar to what was observed in primary SCCD cells [143]. The elevation of endolymphatic Ca^{2+} level in pendrin KO mice may inhibit sensory transduction necessary for hearing and promote the degeneration of the sensory hair cells, which is necessary for the development of normal hearing [142]. Similar mechanisms may be responsible for loss of hearing in patients with Pendred syndrome.

14.4.5 TRPV6 in Exocrine Organs

Human TRPV6 transcripts are most abundant in exocrine organs including pancreas, prostate, mammary, salivary and sweat glands [17, 34, 35]. In pancreas, TRPV6 protein is prominently distributed to the apical membrane and granules of the secretory pole in both mice and humans [35]. In human mammary glands, TRPV6 was strongly expressed in the apical membrane of ductal epithelial cells. Apical membrane localization of TRPV6 was also observed in human prostate epithelial cells [35]. In human sweat gland, TRPV6 appears to be expressed in both apical and basolateral membranes [35]. TRPV6 mRNA is expressed in human parotid and submandibular glands, and TRPV6 protein is localized to the basolateral plasma membrane of acinar cells [146]. While TRPV6 might be involved in salivary Ca^{2+} secretion which is important in mineralization of dental enamel and exposed dentin [146], details of TRPV6 in exocrine function are largely unavailable although it is well known that Ca^{2+} plays important role in secretion. It is possible that TRPV6-mediated Ca^{2+} influx stimulates exocytosis; more likely, TRPV6 may replenish cellular Ca^{2+} by apical re-uptake of released Ca^{2+} following the secretory events, or from extracellular fluid via the basolateral membrane.

In addition, TRPV6 may play a role in Ca^{2+} induced keratinocyte differentiation [147]. Both TRPV5 and TRPV6 are expressed in retina pigment epithelium and they may mediate Ca^{2+} influx from subretinal space to regulate the Ca^{2+} level in light/dark transitions [148].

14.5 Regulation Under Pathological and Therapeutic Conditions

TRPV5 and TRPV6 undergo changes in response to diseases and therapies. Tables 14.2 and 14.3 summarize the changes in TRPV5 and TRPV6 mRNA or protein level under disease and therapeutic conditions, respectively. The regulation of TRPV6 and TRPV5 in cancers, pseudohypoaldosteronism Type II (PHA II), and under treatments of glucocorticoids and diuretics are summarized in the following.

14.5.1 TRPV6 in Cancer

TRPV6 level is elevated in prostate, breast, thyroid, colon and ovarian carcinomas [35]. In prostate cancer samples, TRPV6 mRNA level increases as tumor grade advances as defined by Gleason score [17, 114, 149]. Furthermore, TRPV6 expression also correlates with pathological stage and extraprostatic extension; and androgen-insensitive tumors exhibited decreased TRPV6 levels compared to untreated tumors [149]. Thus, TRPV6 represents a prognostic marker for advanced prostate cancer.

TRPV6 appears to be a component of the store-operated Ca^{2+} channels in prostate LNCaP cells [150, 151] while evidence exists against this [152]. Ca^{2+} store depletion induced the expression of TRPC1, TRPC3, and TRPV6 in LNCaP cells, but these channels were not sufficient to stimulate store-operated Ca^{2+} entry [153]. The store-operated current was down-regulated when LNCaP cells differentiated to androgen-insensitive, apoptotic-resistant neuroendocrine phenotype [149]. TRPV6 expression activated Ca^{2+}/NFAT pathways that enables LNCaP cells to proliferate at high rate and to become resistant to apoptosis [154]. The expression of intermediate-conductance Ca^{2+}-activated K^+ channels, which are preferentially expressed in human prostate cancer tissues, hyperpolarized the cells to drive TRPV6-mediated Ca^{2+} entry, and in turn, proliferation of cancer cells [155]. Reducing Ca^{2+} entry via blocking the K^+ channels prevented proliferation [155]. Therefore, TRPV6 represents a potential therapeutic target to treat prostate cancer.

The potential role of TRPV6 in breast cancer was recently put forward by Hediger and colleagues [156, 157]. Seven out of 12 patients exhibited 2–15-fold increases of TRPV6 mRNA in breast cancer tissues over normal tissues as control [156]. Interestingly, estrogen receptor antagonist tamoxifen not only partially blocked TRPV6 expression in breast cancer cell line T47D, but also inhibited TRPV6-mediated Ca^{2+} uptake activity in *Xenopus* oocytes [156]. In MCF-7 breast cancer cells, TRPV6 activity was blocked by tamoxifen in the presence of estrogen receptor antagonist and the effect of tamoxifen on TRPV6 remained intact in estrogen receptor-negative MDA-MB-231 cells [157]. Activation of PKC blocked the effect of tamoxifen [157]. Thus, tamoxifen-mediated inhibition of TRPV6 might be part of the therapeutic effects of tamoxifen in both estrogen sensitive and insensitive breast cancers.

In a *Citrobacter rodentium*-induced transmissible murine colonic hyperplasia model, a 10- to 20-fold increase of TRPV6 mRNA was observed during the induction of colonic hyperplasia, while VDR, calcium sensing receptor, calbindin-D_{9k} and TRPV5 mRNA levels were unaltered [158]. High Ca^{2+} diet abrogated TRPV6 expression and hyperplasia [158]. Elevated TRPV6 expression was associated with early-stage colon cancer in humans, and blocking TRPV6 expression by siRNA

Table 14.2 Alteration of TRPV5/TRPV6 expression under disease conditions

Disease conditions	Model	Tissue/cell type	Alterations	References
Arterial calcifications	TIF1alpha-deficient kidneys	Kidney	TRPV5/6 mRNA ↑	[186]
Cancer	Human specimens	Prostate, breast, thyroid, colon and ovarian carcinomas	TRPV6 mRNA/protein ↑	[17, 35, 114, 149, 156, 158]
Colonic hyperplasia	Transmissible murine colonic hyperplasia	Colon	TRPV6 mRNA/protein ↑	[158]
Crohn's disease	TNF(DeltaARE) mouse	Intestine/kidney	TRPV5/6 mRNA ↓	[187]
Familial hypomagnesemia with hypercalciuria and nephrocalcinosis	Cldn16−/− mice	Kidney	TRPV5/6 mRNA ↑	[188]
Preeclampsia	Primary culture from patients	Syncytiotrophoblasts	TRPV5/6 mRNA/protein ↓	[135]
Pseudohypoaldosteronism type II	Wnk4 D561A/+ knock-in mice	Kidney tubules	TRPV6 Protein ↓	[168]
Streptozotocin-induced diabetes mellitus	Rats	Kidney	TRPV5/6 mRNA↑ TRPV5 protein↑	[184]

Table 14.3 Alteration of TRPV5/TRPV6 expression under therapeutic conditions

Therapeutic conditions	Model	Tissue	Alterations	References
Prednisolone	Mice	Duodenum	TRPV6 mRNA↓	[172]
Dexamethasone	Mice	Duodenum	TRPV6 mRNA↓	[170, 171]
	Rats	Kidney	TRPV5 mRNA↑	[189]
Tacrolimus (FK506)	Rats	Kidney	TRPV5 mRNA↑	[189]
Hydrochlorothiazide (high dose)	Rats	Kidney	TRPV5 mRNA/Protein↓	[181]
Hydrochlorothiazide (medium dose)	Rats	Kidney	TRPV5 protein↑	[178]
Chlorothiazide (acute, low dose)	Mice	Kidney	TRPV5 mRNA↑	[180]
Furosemide	Mice	Kidney	TRPV5/6 mRNA↑	[183]
Gentamicin	Mice	Kidney	TRPV5/6 mRNA↑	[183]
Sap of acer mono	Mice	Duodenum	TRPV6 mRNA↑	[190]

inhibited colon cancer cell proliferation and induced apoptosis [158]. Thus, elevated TRPV6 expression likely contributes to colonic hyperplasia and colon cancer cell proliferation and TRPV6 represents a therapeutic target for colon cancer.

Capsaicin, an ingredient of chili pepper and TRPV1 activator, induces apoptosis in some cancer cells. Interestingly, gastric cancer cells were more susceptible to capsaicin-induced apoptosis because of the high TRPV6 expression in cancer cells [159]. Overexpression of TRPV6 in normal cells increased capsaicin-induced apoptosis and knockdown of TRPV6 in cancer cells suppressed it [159]. This suggests a strategy to selectively induce cancer cell apoptosis based on its high level TRPV6 expression. The potential roles of TRPV6 as a cancer marker and therapeutic target are still to be further explored.

14.5.2 TRPV5 and TRPV6 in Pseudohypoaldosteronism Type II

Pseudohypoaldosteronism type II (PHAII), also known as familial hyperkalemia and hypertension (FHH) or Gordon syndrome, is a genetic disorder caused by mutations in With No lysine (K) (WNK) kinases 1 and 4 [160]. Patients carrying WNK4 mutation exhibit marked sensitivity to thiazides, hypercalciuria and low bone mineral density, indicating a defect in Ca^{2+} reabsorption [161, 162]. To understand potential roles of TRPV5 and TRPV6 in this process, we investigated their regulation by WNK kinases [163–165]. In *Xenopus* oocytes expression system, WNK4 selectively up-regulated TRPV5 while exerted little effect on TRPV6 [163]. WNK1 showed no significant effect on either of them [163]. In contrast, WNK3 was capable of up-regulating both [164]. The effects of WNK3 and WNK4 on TRPV5 are at least in part via enhancing the forward trafficking of TRPV5 to the plasma membrane through the secretory pathway [164, 165]. While the disease causing mutants of WNK4 regulated TRPV5 to similar extent as WT, the presence of the thiazide-sensitive Na^+-Cl^- cotransporter, which was down-regulated by WNK4 [166, 167], blocked the effect of WNK4 on TRPV5 in a dose-dependent manner. The blocking effect of NCC was further strengthened when WT WNK4 was replaced by the PHAII-causing Q565E mutant, suggesting a reduction of TRPV5-mediated Ca^{2+} influx under this condition [163]. This is consistent with the observation of hypercalciuria and low bone mineral density in patients carrying WNK4 Q565E mutation.

In the $Wnk4^{D561A/+}$ knock-in mouse model of human PHAII, TRPV5 protein level appears to be similar to the WT littermate, but a significant increase in TRPV6 and calbindin-D_{28k} was observed, along with attenuated level of NKCC2, and increased abundance of NCC and ENaC [168]. In addition, urine Ca^{2+} excretion rate was not increased by furosemide in $Wnk4^{D561A/+}$ mice as was in WT mice [168]. These data indicate decreased Ca^{2+} reabsorption in the thick ascending loop of $Wnk4^{D561A/+}$ mice, and the enhanced level of TRPV6 and calbindin-D_{28k} represents a secondary adaptive mechanism in the distal nephron [168]. The effects of WNK4 mutation on NKCC2 and TRPV6 are consistent with the high abundance of WNK4 in the outer medulla [169]. Since TRPV6 is predominantly expressed in the thick ascending limb [42](our unpublished observation), the effect of WNK4 on TRPV6 expression (or indirectly via NKCC2) warrants further studies. The increased NCC level is also consistent with a potential decrease of TRPV5-mediated Ca^{2+} reabsorption through NCC-mediated inhibitory effect on the positive action of WNK4 on TRPV5 as noted earlier [163].

14.5.3 Glucocorticoids

Glucocorticoids are a class of hormones that are used widely as an anti-inflammatory and immunosuppressive drug. However, they have side effects such as inducing osteoporosis. Reduced intestinal Ca^{2+} absorption and renal Ca^{2+} reabsorption may be part of the mechanisms of glucocorticoid-induced osteoporosis. Regulation of TRPV5 and TRPV6 by glucocorticoids may occur through two mechanisms, expression regulation [170–172] and activity regulation via glucocorticoid inducible protein kinase SGK1 [173–176].

Oral application of prednisolone for a week in mice reduced intestinal Ca^{2+} absorption and TRPV6 mRNA level and calbindin D_{9k} mRNA and protein levels but renal TRPV5 and calbindin-D_{28k} and NCX1 mRNA levels remained unaltered [172]. In calbindin-D_{9k} and calbindin-D_{28k} KO mice, compensatory elevation of TRPV6 mRNA in the duodenum was blocked by 5-day dexamethasone treatment [171]. The renal TRPV5 mRNA in calbindin-D_{9k} KO mice was elevated while the renal TRPV6 mRNA was reduced in calbindin-D_{28k} KO mice after dexamethasone administration [171]. These regulations appear to be related to the decreased duodenal VDR level by dexamethasone [171]. The effects of dexamethasone were time dependent: at day one it increased duodenal TRPV6 mRNA and renal TRPV5 mRNA while decreased renal TRPV6 mRNA; after 5-day treatment, only TRPV6 in the duodenum was robustly reduced [170]. Thus, it appears intestinal TRPV6 is the target of glucocorticoids action and this action appears to be associated with decreased VDR level in the intestine.

The activity of TRPV5 and TRPV6 could be regulated by glucocorticoid inducible SGK kinases. In *Xenopus* oocyte system, SGK1 and SGK3 up-regulated TRPV5 activity in the presence of Na^+/H^+ exchanger (NHE) regulating factor 2 (NHERF2), and this effect could not be reproduced by active form of protein kinase B (PKB) and SGK2 [175]. The surface abundance of TRPV5 was increased by SGK1 and NHERF2 [174]. TRPV5 C-terminal tail interacts with NHERF2, and deletion of the second PDZ domain in NHERF2 abrogates the stimulating effect of SGK1/NHERF2 on TRPV5 [174]. In contrast, TRPV6 was also up-regulated by SGK1, SGK3 and PKB/Akt; however, neither NHERF1 nor NHERF2 was required for this regulation [173, 176]. The effect of SGK1 on TRPV6 was augmented in the presence of phosphatidylinositol-3-phosphate-5-kinase PIKfyve (PIP5K3), a kinase generating phosphatidylinositol 3,5-bisphosphate $[PI(3,5)P_2]$ [176]. TRPV6 expression at or close to the *Xenopus* oocytes surface was significantly increased by the co-expression of SGK1 and PIKfyve [176]. PIKfyve alone did not affect TRPV6, and ^{S318A}PIKfyve lacking the SGK1 phosphorylation site could not further increase the positive effect of SGK1 on TRPV6 [176]. Thus, PIKfyve requires its phosphorylation by SGK1 to act on TRPV6.

The effect of SGK1 on Ca^{2+} reabsorption and TRPV5 was evaluated in *Sgk1* KO mice [177]. Membrane expression of TRPV5 protein and cytosolic calbindin-D_{28k} protein were all decreased in the KO mice, suggesting that SGK1 play a role in the regulation of TRPV5 trafficking as observed in previous in vitro studies [174, 175, 177]. However, fractional excretion of Ca^{2+} was lower in the *Sgk1* KO mice than in the WT mice, despite normal $1,25(OH)_2D_3$ and PTH levels [177]. The lowered Ca^{2+} excretion is likely due to the increased reabsorption of Ca^{2+} in the thick ascending limb because NKCC2 inhibitor furosemide dissipated the difference in Ca^{2+} excretion between the KO and WT mice [177]. Whether TRPV6 is altered in the *Sgk1* KO mice is yet to be determined.

14.5.4 Diuretics

Hypocalciuria is a well known effect of thiazide diuretics, which blocks the Na^+-Cl^- cotransporter NCC. A number of studies have been undertaken to dissect the molecular mechanisms of this effect [178–181]; however, the mechanisms of thiazides-induced hypocalciuria are controversial. Application of high dose of hydrochlorothiazide (HCTZ), at 12 mg/day for a week, significantly reduced TRPV5, calbindin-D_{28k}, NCX1 and NCC mRNA and protein levels and the number of tubules expressing calbindin-D_{28k} and NCC in rat kidneys [181]. Since Na^+ repletion prevented both volume contraction and hypocalciuria in HCTZ treated rats, it is suggested volume contraction play a critical role in thiazide induced hypocalciuria [181]. At lower dose, acute application of chlorothiazide (CTZ), at either 25 or 50 mg/kg but not at 100 mg/kg, significantly increased TRPV5 expression in mouse kidneys [180]. Chronic treatment with CTZ at 25 mg/kg twice daily for 3 days caused no change in TRPV5, TRPV6 and calbindins, but salt supplement in drinking water alone or with CTZ significant increased the expression of these Ca^{2+} transport proteins [180]. This study agrees that under volume contraction, increased Ca^{2+} reabsorption via paracellular pathway in the proximal tubule plays the major role; however, without volume contraction, hypocalciuria is probably achieved through increased Ca^{2+} reabsorption in the distal convoluted tubule by up-regulating the transcellular pathway [180].

Subsequent studies with the *Trpv5* KO mice indicate that enhanced passive Ca^{2+} transport in the proximal tubule rather than active Ca^{2+} transport in distal convolution is the cause of thiazide-induced hypocalciuria [179]. HCTZ-induced hypocalciuria remained unaltered in *Trpv5* KO mice, and the reabsorption of Na^+ and Ca^{2+} was increased in the proximal tubule and the Ca^{2+} reabsorption in the distal convolution was apparently unaffected in micropuncture experiments [179]. In a recent study that evaluates the effects of HCTZ on different transporters in proximal tubule, thick ascending limb and distal convoluted tubule under high salt or high Ca^{2+} diet induced hypercalciuric rats, TRPV5, NHE3, and NKCC2 proteins were decreased in all hypercalciuric rats, whereas increased TRPV5 protein was associated with hypocalciuric effect induced by HCTZ [178]. These results indicate that thiazide-induced hypocalciuria is a result of coordinated alterations of Ca^{2+} reabsorption mechanisms in all tubular segments involved in Ca^{2+} reabsorption, including proximal tubule, thick ascending limb, and the distal convoluted and connecting tubule.

In contrast to hypocalciuric effect of thiazide diuretics, loop diuretics such as furosemide promote urinary Ca^{2+} excretion [182]. Furosemide inhibits NKCC2 in the thick ascending limb and diminishes the negative luminal potential needed to drive Ca^{2+} and Mg^{2+} reabsorption through the paracellular pathway. Lien and colleagues [183] found that acute and chronic application of furosemide increased both TRPV5 and TRPV6 levels along with that of calbindins; similar increases of TRPV5 and TRPV6 were found in hypercalciuria induced by gentamicin [183] and in streptozotocin-induced diabetes mellitus [184]. The up-regulation of Ca^{2+} transporters in these situations is likely due to the increased Ca^{2+} load to the segments [183, 184], consistent with Ca^{2+} load-dependent expression of TRPV5 in the kidney [185].

14.6 Unusual High Frequencies of SNPs in African Populations

TRPV6 and *TRPV5* have been shown to have high frequency of SNPs in African populations. By analyzing SNP data of 132 genes from 24 African–Americans and 23 European–Americans, Akey and colleagues [191] identified 4 contiguous genes under demographically robust selection, including *EPHB6*, *TRPV6*, *TRPV5*, and *KEL4*, in a 115-kb region in chromosome 7q34-35 exhibiting features of a recent selective sweep. Similarly, using data from 151 genes, Stajich and Hahn [192] found that *TRPV6* gene shows a population-specific pattern of positive selection. All these studies utilized the public available database from SeatleSNPs programs (http://pga.mbt.washington.edu/), in which approximately 200 genes from 24 African–Americans and 23 European–Americans are sequenced. The visual SNP genotype of *TRPV6* (http://pga.gs.washington.edu/data/trpv6/trpv6.prettybase.png) clearly shows that there are many rare alleles in African–Americans whereas very few in European–Americans. The same is true for *TRPV5*. Subsequent studies [193, 194] also support that *TRPV6* locus has undergone positive selection. The selective scenario put forward by Stajich and Hahn [192] is that there were many mutations in *TRPV6* locus in early humans in Africa; as early humans migrated out of Africa, a preexisting mutation in the ancestral African population became advantageous in a new environment and rose to high frequency. Akey and colleagues found that the haplotype defined by 3 nonsynonymous SNPs (C157R, M378V and M681T, Allele frequency ~0.5 in African–Americans, ~0.02 in European–Americans from SeattleSNPs) are nearly fixed for the derived alleles in non-African populations based on genotyping data of 1,064 individuals from 52 populations [195]. Hughes and colleagues [193] further provided evidence for independent, parallel selection on *TRPV6* locus in Europeans and Asians.

Which one of the 4 genes in chromosome 7q34-35 was under the selective pressure outside of Africa is unclear; however, *TRPV6* is a good candidate as suggested by Akey and colleagues [191, 195]. What drove the selection of *TRPV6* haplotye outside of Africa is unknown; however, it might be associated with the alteration of dietary Ca^{2+} availability and the change of vitamin D status in earlier humans. Akey and colleagues [191, 195] suggested that patterns of *TRPV6* sequence variation may have been influenced by availability of dairy products due to the domestication of milk-producing animals approximately 10,000 years ago, similar to positive selection for lactase persistence [196]. As early humans migrated away from the equator their vitamin D level decreased due to the reduction of ultraviolet light exposure [197]. Because *TRPV6* is a highly vitamin D-responsive gene, the alteration of vitamin D status may be a factor. Another possibility is that the derived TRPV6 allele may be advantageous in resistant to a pathogen humans encountered after migrated outside of Africa [195]. TRPV6 is expressed in the brush border membrane of intestinal epithelium, thus it is likely that TRPV6 could be a receptor for certain pathogens. Consistent with this hypothesis, M378V variation in the first extracellular loop of TRPV6 is in direct contact with digested food and potential pathogens. The 378V residue is conserved in many species. If the extracellular loop of TRPV6 interacts with a contagious pathogen, this pathogen is potentially transferable between animals and humans. Thus, the 378M mutation could potentially make TRPV6 resistant to this pathogen and provide an advantage for the 378M carriers to survive disastrous infectious diseases. In this case, the derived *TRPV6* allele may not necessarily be beneficial for Ca^{2+} absorption. Due to the increased Ca^{2+} intake from the dairy products and the increased availability of vitamin D by the development of fishery, the role of TRPV6-meidated Ca^{2+} absorption may not be as critical for human survival outside of Africa. This hypothesis is consistent with the fact that African–Americans have higher intestinal Ca^{2+} absorption compared to Caucasians [198].

The nonsynonymous SNPs in the ancestral haplotype (157R, 378V, 681T) may have an impact on TRPV6 function; however, the results from different groups are controversial [193, 199]. In one study the Ca^{2+} uptake activity of the ancestral TRPV6 protein (157R, 378V, 681T) was twofold that of the derived protein (157C, 378M, 681M) when expressed in *Xenopus* oocytes; and 378V, located to the first extracellular loop, appeared to be responsible for the increase in Ca^{2+} uptake activity [199]. In another study, however, no significant differences in channel function were identified for the ancestral TRPV6 [193]. We also failed in observing any functional difference between the two forms of TRPV6 (unpublished observations). An interesting observation is that ATP binds to TRPV6 and prevents rundown of channel activity and 2 arginine residues $R^{153}R^{154}$ close to 157R are involved in ATP binding [68]. The 157R in the ancestral TRPV6 is located to N-terminal region between ankyrin repeats 3 and 4. The S155 becomes a potential phosphorylation site ($S^{155}P^{156}R^{157}$) in the ancestral TRPV6 and phosphorylation of S155 by PKC is likely involved in the slower rundown of TRPV6 ancestral protein in the presence of ATP [68]. However, the difference is very small and may not result in significant difference in overall channel activity under normal condition.

Four nonsynonymous SNPs in *TRPV5* were identified by SeattleSNPs: three of them (A8V, A563T and L712F) were only present in African–Americans, not in European–Americans; R154H is common in both populations. In contrast to the nonsynonymous SNP variations in TRPV6 ancestral haplotype that are mostly conserved in other species, the variations in TRPV5 are newly derived as they are not commonly present in other species surveyed including chimpanzee, dog, rat and mouse, with the exceptions of 563T in dog and 8V in rat [200]. In addition, the nonsynonymous SNPs of TRPV5 are not associated with each other as are those in TRPV6. Thus, these variations more likely alter the function of TRPV5. By expression in *Xenopus* oocytes, we found that 2 of the SNPs, A563T and L712F, significantly increased TRPV5-mediated Ca^{2+} uptake by approx. 50 and 25%, respectively [200]. For A563T variant, the increased Ca^{2+} uptake activity was not associated with increased protein abundance in the plasma membrane; rather it was associated with increase apparent K_m for Ca^{2+} and increased sensitivity to extracellular Mg^{2+}, suggesting increased permeation of Ca^{2+} in the cation translocation pathway of the channel [200]. The location of A563T in the last transmembrane domain of TRPV5, 20 amino acid residues away from the D542 that forms Ca^{2+} filter in the pore, is likely close to the cation translocation path in a 3D structure.

Africans exhibit Ca^{2+} conservation mechanisms. African–Americans have lowered urinary Ca^{2+} excretion than Caucasians [198, 201, 202], and the risk of kidney stone in African–American is lower than that in Caucasians [203, 204]. African–Americans have higher bone mass [205], and lower incidence of osteoporosis related fractures than whites [206]. Because of the high allele frequencies of *TRPV5* and *TRPV6* SNPs in African populations, these SNPs may contribute to the Ca^{2+} conservation mechanisms in African populations. This is yet to be tested in population studies.

The roles of *TRPV6* or *TRPV5* SNPs have been evaluated in kidney stone disease and hypercalciuria. In a study of 170 kidney stone formers, the prevalence of the TRPV6 ancestral haplotype in stone-forming patients was significantly higher than that of non-stone-forming patients [199]. The only kidney stone patient carrying the homozygous ancestral TRPV6 haplotype exhibited more stone episodes, much higher urinary Ca^{2+} excretion and lower plasma PTH level, suggesting a potential role of ancestral TRPV6 protein in the enhanced Ca^{2+} absorption in this patient [199]. In a study involving 20 renal hypercalciuria patients, non-synonymous variation of TRPV5 (A8V, R154H and A561T) and synonymous variations were identified among these patients [207]. Although A561T is very close to A563T, which has significant effect on Ca^{2+} permeation [200], no functional difference was identified between A561T variant and control TRPV5. This result, together with the previous screening that failed to identify TRPV5 mutations in patients with autosomal dominant idiopathic hypercalciuria [208], indicates that TRPV5 may not play a primary role in hypercalciuria. However, the role of TRPV6 variations in hypercalciuria warrants further investigation.

TRPV6 expression level positively correlates with tumor grade in prostate cancer. Mortality from prostate cancer is two to three times greater in African–Americans than in Caucasians [209, 210]. It is not clear whether the high prostate cancer mortality in African–Americans is relevant to the high prevalence of ancestral *TRPV6* allele in African–Americans. However, in a recent study the frequency of *TRPV6* ancestral allele in 142 Caucasian prostate adenocarcinoma samples was not different from that in 169 Caucasian healthy control subjects [211]. The *TRPV6* ancestral allele frequency did not correlate with the onset of prostate cancer, the Gleason score, or the tumor stage [211]. Because of the low frequency of ancestral alleles in the Caucasian subjects (only one patient with homozygous *TRPV6* ancestral allele) of this study, a conclusion on a relation between *TRPV6* genotype and prostate cancer should be drawn with caution. Studies in African or African–American populations will likely provide more conclusive results.

14.7 Perspectives

Over 10 years after the identification of TRPV5 and TRPV6, a much better understanding of the apical Ca^{2+} entry mechanisms in the transcellular transport pathway has been achieved. TRPV5 and TRPV6 provide a tightly regulated selective Ca^{2+} entry in the transcellular Ca^{2+} transport pathway that is responsible for the fine tuning of Ca^{2+} absorption and reabsorption, even though this pathway does not contribute greatly to the bulk Ca^{2+} absorption and reabsorption as does the paracellular pathway. TRPV5 and TRPV6 certainly are not the sole mediator of vitamin D regulated Ca^{2+} transport; however, without either of them, the body appears to be in a negative Ca^{2+} balance as indicated by the elevated $1,25(OH)_2D_3$ and PTH levels. The roles of TRPV5 and TRPV6 in tuning Ca^{2+} homeostasis are well indicated by their robust responses to the body's need for Ca^{2+} under physiological, pathological and therapeutic conditions. Notably, TRPV6 participates in cell proliferation and apoptosis and is emerging as a potential cancer marker and therapeutic target for cancers.

The study in renal Ca^{2+} transport has been mostly focused on TRPV5 in the past 10 more years. However, *Trpv6* KO mice also lose Ca^{2+} in the urine, even though intestinal absorption of Ca^{2+} is decreased. Although TRPV6 mRNA level is much lower than TRPV5 in rodent kidney, in humans and horses, renal TRPV6 mRNA is much higher than TRPV5, indicating a more important role of TRPV6 in renal function in these species. TRPV5 and TRPV6 are not always regulated in the same way in the kidney, indicating their overlapping yet distinct roles.

While TRPV6 mediates Ca^{2+} uptake in many tissues, TRPV5 is relatively specific to the kidney. The need of a specific Ca^{2+} transporter for the complex mammalian distal tubule likely resulted in the generation of *TRPV5* by duplication of *TRPV6*. The changed source of dietary Ca^{2+} and vitamin D status during or after humans migrated out of Africa may be part of the reasons for further evolution of the *TRPV6* gene. *TRPV6* genotype is a good genetic maker in African populations. African populations could be separated into 3 groups based on their *TRPV6* genotype because of their ~50% allele frequency of the 3 *TRPV6* nonsynonymous SNPs. The involvement of *TRPV6* and *TRPV5* SNPs in the health issues related to Ca^{2+}, salt and water homeostasis in African descents will likely become a hot spot of investigation in the near future.

Acknowledgments I thank my colleagues Drs. Yi Jiang, Wei Zhang, Tao Na, Guojin Wu, and Haiyan Jing for their contributions to the research projects in our group and their participation in regular literature review. The research of our lab was supported by the National Institute of Diabetes and Digestive and Kidney Diseases (R01DK072154), and American Heart Association National Center (0430125N) and Greater Southeast Affiliate (09GRNT2160024).

References

1. Suzuki Y, Landowski CP, Hediger MA (2008) Mechanisms and regulation of epithelial Ca^{2+} absorption in health and disease. Annu Rev Physiol 70:257–271
2. Boros S, Bindels RJ, Hoenderop JG (2009, May) Active Ca^{2+} reabsorption in the connecting tubule. Pflugers Arch 458(1):99–109
3. De GT, Bindels RJ, Hoenderop JG (2008, Nov) TRPV5: an ingeniously controlled calcium channel. Kidney Int 74(10):1241–1246
4. Hoenderop JG, Bindels RJ (2008, Feb) Calciotropic and magnesiotropic TRP channels. Physiology (Bethesda) 23:32–40
5. van de Graaf SF, Bindels RJ, Hoenderop JG (2007) Physiology of epithelial Ca^{2+} and Mg^{2+} transport. Rev Physiol Biochem Pharmacol 158:77–160
6. Topala CN, Bindels RJ, Hoenderop JG (2007, July) Regulation of the epithelial calcium channel TRPV5 by extracellular factors. Curr Opin Nephrol Hypertens 16(4):319–324
7. Schoeber JP, Hoenderop JG, Bindels RJ (2007, Feb) Concerted action of associated proteins in the regulation of TRPV5 and TRPV6. Biochem Soc Trans 35(Pt 1):115–119
8. van de Graaf SF, Hoenderop JG, Bindels RJ (2006, June) Regulation of TRPV5 and TRPV6 by associated proteins. Am J Physiol Renal Physiol 290(6):F1295–F1302

9. Bodding M (2007, Mar) TRP proteins and cancer. Cell Signal 19(3):617–624
10. Perez AV, Picotto G, Carpentieri AR, Rivoira MA, Peralta Lopez ME, Tolosa de Talamoni NG (2008) Minireview on regulation of intestinal calcium absorption. Emphasis on molecular mechanisms of transcellular pathway. Digestion 77(1):22–34
11. Vriens J, Appendino G, Nilius B (2009, June) Pharmacology of vanilloid transient receptor potential cation channels. Mol Pharmacol 75(6):1262–1279
12. Vennekens R, Owsianik G, Nilius B (2008) Vanilloid transient receptor potential cation channels: an overview. Curr Pharm Des 14(1):18–31
13. Hoenderop JG, van der Kemp AW, Hartog A, van de Graaf SF, van Os CH, Willems PH, Bindels RJ (1999, Mar) Molecular identification of the apical Ca^{2+} channel in 1, 25-dihydroxyvitamin D_3-responsive epithelia. J Biol Chem 274(13):8375–8378
14. Peng JB, Chen XZ, Berger UV, Vassilev PM, Tsukaguchi H, Brown EM, Hediger MA (1999, Aug) Molecular cloning and characterization of a channel-like transporter mediating intestinal calcium absorption. J Biol Chem 274(32):22739–22746
15. Peng JB, Chen XZ, Berger UV, Vassilev PM, Brown EM, Hediger MA (2000, Sep) A rat kidney-specific calcium transporter in the distal nephron. J Biol Chem 275(36): 28186–28194
16. Muller D, Hoenderop JG, Meij IC, van den Heuvel LP, Knoers NV, den Hollander AI, Eggert P, Garcia-Nieto V, Claverie-Martin F, Bindels RJ (2000, July) Molecular cloning, tissue distribution, and chromosomal mapping of the human epithelial Ca^{2+} channel (ECAC1). Genomics 67(1):48–53
17. Wissenbach U, Niemeyer BA, Fixemer T, Schneidewind A, Trost C, Cavalie A, Reus K, Meese E, Bonkhoff H, Flockerzi V (2001, June) Expression of CaT-like, a novel calcium-selective channel, correlates with the malignancy of prostate cancer. J Biol Chem 276(22):19461–19468
18. Barley NF, Howard A, O'Callaghan D, Legon S, Walters JR (2001, Feb) Epithelial calcium transporter expression in human duodenum. Am J Physiol Gastrointest Liver Physiol 280(2):G285–G290
19. Caterina MJ, Schumacher MA, Tominaga M, Rosen TA, Levine JD, Julius. D (1997, Oct) The capsaicin receptor: a heat-activated ion channel in the pain pathway. Nature 389(6653):816–824
20. Colbert HA, Smith TL, Bargmann CI (1997, Nov) OSM-9, a novel protein with structural similarity to channels, is required for olfaction, mechanosensation, and olfactory adaptation in Caenorhabditis elegans. J Neurosci 17(21):8259–8269
21. Caterina MJ, Rosen TA, Tominaga M, Brake AJ, Julius D (1999, Apr) A capsaicin-receptor homologue with a high threshold for noxious heat. Nature 398(6726):436–441
22. Kanzaki M, Zhang YQ, Mashima H, Li L, Shibata H, Kojima I (1999, July) Translocation of a calcium-permeable cation channel induced by insulin-like growth factor-I. Nat Cell Biol 1(3):165–170
23. Strotmann R, Harteneck C, Nunnenmacher K, Schultz G, Plant TD (2000, Oct) OTRPC4, a nonselective cation channel that confers sensitivity to extracellular osmolarity. Nat Cell Biol 2(10):695–702
24. Liedtke W, Choe Y, Marti-Renom MA, Bell AM, Denis CS, Sali A, Hudspeth AJ, Friedman JM, Heller S (October 2000) Vanilloid receptor-related osmotically activated channel (VR-OAC), a candidate vertebrate osmoreceptor. Cell 103(3):525–535
25. Wissenbach U, Bodding M, Freichel M, Flockerzi V (2000, Nov) Trp12, a novel Trp related protein from kidney. FEBS Lett 485(2–3):127–134
26. Delany NS, Hurle M, Facer P, Alnadaf T, Plumpton C, Kinghorn I, See CG, Costigan M, Anand P, Woolf CJ, Crowther D, Sanseau P, Tate SN (2001, Jan) Identification and characterization of a novel human vanilloid receptor-like protein, VRL-2. Physiol Genomics 4(3):165–174
27. Smith GD, Gunthorpe MJ, Kelsell RE, Hayes PD, Reilly P, Facer P, Wright JE, Jerman JC, Walhin JP, Ooi L, Egerton J, Charles KJ, Smart D, Randall AD, Anand P, Davis JB

(2002, July) TRPV3 is a temperature-sensitive vanilloid receptor-like protein. Nature 418(6894):186–190
28. Xu H, Ramsey IS, Kotecha SA, Moran MM, Chong JA, Lawson D, Ge P, Lilly J, Silos-Santiago I, Xie Y, DiStefano PS, Curtis R, Clapham DE (2002, July) TRPV3 is a calcium-permeable temperature-sensitive cation channel. Nature 418(6894):181–186
29. Peier AM, Reeve AJ, Andersson DA, Moqrich A, Earley TJ, Hergarden AC, Story GM, Colley S, Hogenesch JB, McIntyre P, Bevan S, Patapoutian A (2002, June) A heat-sensitive TRP channel expressed in keratinocytes. Science 296(5575):2046–2049
30. Kahn-Kirby AH, Bargmann. CI (2006) TRP channels in C. elegans. Annu Rev Physiol 68:719–736
31. Qiu A, Hogstrand C (2004, Nov) Functional characterisation and genomic analysis of an epithelial calcium channel (ECaC) from pufferfish, Fugu rubripes. Gene 342(1):113–123
32. Peng JB, Brown EM, Hediger MA (2001, Aug) Structural conservation of the genes encoding CaT1, CaT2, and related cation channels. Genomics 76(1–3):99–109
33. Dantzler WH, Braun EJ (1980, Sep) Comparative nephron function in reptiles, birds, and mammals. Am J Physiol 239(3):R197–R213
34. Peng JB, Chen XZ, Berger UV, Weremowicz S, Morton CC, Vassilev PM, Brown EM, Hediger MA (2000, Nov) Human calcium transport protein CaT1. Biochem Biophys Res Commun 278(2):326–332
35. Zhuang L, Peng JB, Tou L, Takanaga H, Adam RM, Hediger MA, Freeman MR (2002, Dec) Calcium-selective ion channel, CaT1, is apically localized in gastrointestinal tract epithelia and is aberrantly expressed in human malignancies. Lab Invest 82(12):1755–1764
36. Wilkens MR, Kunert-Keil C, Brinkmeier H, Schroder B (2009, Nov) Expression of calcium channel TRPV6 in ovine epithelial tissue. Vet J 182(2):294–300
37. Rourke KM, Coe S, Kohn CW, Rosol TJ, Mendoza FJ, Toribio RE (2010, May) Cloning, comparative sequence analysis and mRNA expression of calcium transporting genes in horses. Gen Comp Endocrinol 167(1):6–10
38. Schroder B, Vossing S, Breves G (1999, Oct) In vitro studies on active calcium absorption from ovine rumen. J Comp Physiol B 169(7):487–494
39. Song Y, Peng X, Porta A, Takanaga H, Peng JB, Hediger MA, Fleet JC, Christakos S (2003, Sep) Calcium transporter 1 and epithelial calcium channel messenger ribonucleic acid are differentially regulated by 1,25 dihydroxyvitamin D3 in the intestine and kidney of mice. Endocrinology 144(9):3885–3894
40. Van Cromphaut SJ, Dewerchin M, Hoenderop JG, Stockmans I, Van HE, Kato S, Bindels RJ, Collen D, Carmeliet P, Bouillon R, Carmeliet G (2001, Nov) Duodenal calcium absorption in vitamin D receptor-knockout mice: functional and molecular aspects. Proc Natl Acad Sci USA 98(23):13324–13329
41. Loffing J, Loffing-Cueni D, Valderrabano V, Klausli L, Hebert SC, Rossier BC, Hoenderop JG, Bindels RJ, Kaissling B (2001, Dec) Distribution of transcellular calcium and sodium transport pathways along mouse distal nephron. Am J Physiol Renal Physiol 281(6):F1021–F1027
42. Suzuki M, Ishibashi K, Ooki G, Tsuruoka S, Imai M (2000, Aug) Electrophysiologic characteristics of the Ca-permeable channels, ECaC and CaT, in the kidney. Biochem Biophys Res Commun 274(2):344–349
43. Nijenhuis T, Hoenderop JG, van der Kemp AW, Bindels RJ (2003, Nov) Localization and regulation of the epithelial Ca^{2+} channel TRPV6 in the kidney. J Am Soc Nephrol 14(11):2731–2740
44. Nilius B, Vennekens R, Prenen J, Hoenderop JG, Bindels RJ, Droogmans G (2000, Sep) Whole-cell and single channel monovalent cation currents through the novel rabbit epithelial Ca^{2+} channel ECaC. J Physiol 527(Pt 2):239–248
45. Yue L, Peng JB, Hediger MA, Clapham DE (2001, Apr) CaT1 manifests the pore properties of the calcium-release-activated calcium channel. Nature 410(6829):705–709

46. Vassilev PM, Peng JB, Hediger MA, Brown EM (2001, Nov) Single-channel activities of the human epithelial Ca^{2+} transport proteins CaT1 and CaT2. J Membr Biol 184(2):113–120
47. Vennekens R, Hoenderop JG, Prenen J, Stuiver M, Willems PH, Droogmans G, Nilius B, Bindels RJ (2000, Feb) Permeation and gating properties of the novel epithelial Ca^{2+} channel. J Biol Chem 275(6):3963–3969
48. Hoenderop JG, Voets T, Hoefs S, Weidema F, Prenen J, Nilius B, Bindels RJ (2003, Feb) Homo- and heterotetrameric architecture of the epithelial Ca^{2+} channels TRPV5 and TRPV6. EMBO J 22(4):776–785
49. Erler I, Hirnet D, Wissenbach U, Flockerzi V, Niemeyer BA (2004, Aug) Ca^{2+}-selective transient receptor potential V channel architecture and function require a specific ankyrin repeat. J Biol Chem 279(33):34456–34463
50. Chang Q, Gyftogianni E, van de Graaf SF, Hoefs S, Weidema FA, Bindels RJ, Hoenderop JG (2004, Dec) Molecular determinants in TRPV5 channel assembly. J Biol Chem 279(52):54304–54311
51. Phelps CB, Huang RJ, Lishko PV, Wang RR, Gaudet R (2008, Feb) Structural analyses of the ankyrin repeat domain of TRPV6 and related TRPV ion channels. Biochemistry 47(8):2476–2484
52. Voets T, Janssens A, Droogmans G, Nilius B (2004, Apr) Outer pore architecture of a Ca^{2+}-selective TRP channel. J Biol Chem 279(15):15223–15230
53. Nilius B, Vennekens R, Prenen J, Hoenderop JG, Droogmans G, Bindels RJ (2001, Jan) The single pore residue Asp542 determines Ca^{2+} permeation and Mg^{2+} block of the epithelial Ca^{2+} channel. J Biol Chem 276(2):1020–1025
54. Dodier Y, Banderali U, Klein H, Topalak O, Dafi O, Simoes M, Bernatchez G, Sauve R, Parent L (2004, Feb) Outer pore topology of the ECaC-TRPV5 channel by cysteine scan mutagenesis. J Biol Chem 279(8):6853–6862
55. Dodier Y, Dionne F, Raybaud A, Sauve R, Parent L (2007, Dec) Topology of the selectivity filter of a TRPV channel: rapid accessibility of contiguous residues from the external medium. Am J Physiol Cell Physiol 293(6):C1962–C1970
56. Voets T, Janssens A, Prenen J, Droogmans G, Nilius B (2003, Mar) Mg^{2+}-dependent gating and strong inward rectification of the cation channel TRPV6. J Gen Physiol 121(3):245–260
57. Bodding M, Wissenbach U, Flockerzi V (2002, Sep) The recombinant human TRPV6 channel functions as Ca^{2+} sensor in human embryonic kidney and rat basophilic leukemia cells. J Biol Chem 277(39):36656–36664
58. Nilius B, Prenen J, Hoenderop JG, Vennekens R, Hoefs S, Weidema AF, Droogmans G, Bindels RJ (2002, Aug) Fast and slow inactivation kinetics of the Ca^{2+} channels ECaC1 and ECaC2 (TRPV5 and TRPV6). Role of the intracellular loop located between transmembrane segments 2 and 3. J Biol Chem 277(34):30852–30858
59. Niemeyer BA, Bergs C, Wissenbach U, Flockerzi V, Trost C (2001, Mar) Competitive regulation of CaT-like-mediated Ca^{2+} entry by protein kinase C and calmodulin. Proc Natl Acad Sci USA 98(6):3600–3605
60. Derler I, Hofbauer M, Kahr H, Fritsch R, Muik M, Kepplinger K, Hack ME, Moritz S, Schindl R, Groschner K, Romanin C (2006, Nov) Dynamic but not constitutive association of calmodulin with rat TRPV6 channels enables fine tuning of Ca^{2+}-dependent inactivation. J Physiol 577(Pt 1):31–44
61. Lambers TT, Weidema AF, Nilius B, Hoenderop JG, Bindels RJ (2004, July) Regulation of the mouse epithelial Ca^{2+} channel TRPV6 by the Ca^{2+}-sensor calmodulin. J Biol Chem 279(28):28855–28861
62. Lambers TT, Mahieu F, Oancea E, Hoofd L, de LF, Mensenkamp AR, Voets T, Nilius B, Clapham DE, Hoenderop JG, Bindels RJ (2006, July) Calbindin-D28K dynamically controls TRPV5-mediated Ca^{2+} transport. EMBO J 25(13):2978–2988
63. Gkika D, Mahieu F, Nilius B, Hoenderop JG, Bindels. RJ (2004, June) 80 K-H as a new Ca^{2+} sensor regulating the activity of the epithelial Ca^{2+} channel transient receptor potential cation channel V5 (TRPV5). J Biol Chem 279(25):26351–26357

64. Thyagarajan B, Benn BS, Christakos S, Rohacs T (2009, Mar) Phospholipase C-mediated regulation of transient receptor potential vanilloid 6 channels: implications in active intestinal Ca^{2+} transport. Mol Pharmacol 75(3):608–616
65. Thyagarajan B, Lukacs V, Rohacs T (2008, May) Hydrolysis of phosphatidylinositol 4,5-bisphosphate mediates calcium-induced inactivation of TRPV6 channels. J Biol Chem 283(22):14980–14987
66. Rohacs T, Nilius B (2007, Oct) Regulation of transient receptor potential (TRP) channels by phosphoinositides. Pflugers Arch 455(1):157–168
67. Lee J, Cha SK, Sun TJ, Huang CL (2005, Nov) PIP2 activates TRPV5 and releases its inhibition by intracellular Mg^{2+}. J Gen Physiol 126(5):439–451
68. Al-Ansary D, Bogeski I, Disteldorf BM, Becherer U, Niemeyer BA (2010, Feb) ATP modulates Ca^{2+} uptake by TRPV6 and is counteracted by isoform-specific phosphorylation. FASEB J 24(2):425–435
69. Bonny O, Rubin A, Huang CL, Frawley WH, Pak CY, Moe OW (2008, Aug) Mechanism of urinary calcium regulation by urinary magnesium and pH. J Am Soc Nephrol 19(8):1530–1537
70. Yeh BI, Yoon J, Huang CL (2006) On the role of pore helix in regulation of TRPV5 by extracellular protons. J Membr Biol 212(3):191–198
71. Yeh BI, Kim YK, Jabbar W, Huang CL (2005, Sep) Conformational changes of pore helix coupled to gating of TRPV5 by protons. EMBO J 24(18):3224–3234
72. Yeh BI, Sun TJ, Lee JZ, Chen HH, Huang CL (2003, Dec) Mechanism and molecular determinant for regulation of rabbit transient receptor potential type 5 (TRPV5) channel by extracellular pH. J Biol Chem 278(51):51044–51052
73. Cha SK, Jabbar W, Xie J, Huang CL (2007, Dec) Regulation of TRPV5 single-channel activity by intracellular pH. J Membr Biol 220(1–3):79–85
74. Lambers TT, Oancea E, de GT, Topala CN, Hoenderop JG, Bindels RJ (2007, Feb) Extracellular pH dynamically controls cell surface delivery of functional TRPV5 channels. Mol Cell Biol 27(4):1486–1494
75. Gkika D, Topala CN, Chang Q, Picard N, Thebault S, Houillier P, Hoenderop JG, Bindels RJ (2006, Oct) Tissue kallikrein stimulates Ca^{2+} reabsorption via PKC-dependent plasma membrane accumulation of TRPV5. EMBO J 25(20):4707–4716
76. Wood RJ, Tchack L, Taparia S (2001) 1,25-Dihydroxyvitamin D_3 increases the expression of the CaT1 epithelial calcium channel in the Caco-2 human intestinal cell line. BMC Physiol 1:11
77. Hoenderop JG, Muller D, van der Kemp AW, Hartog A, Suzuki M, Ishibashi K, Imai M, Sweep F, Willems PH, van Os CH, Bindels RJ (2001, July) Calcitriol controls the epithelial calcium channel in kidney. J Am Soc Nephrol 12(7):1342–1349
78. Walters JR, Barley NF, Khanji M, Rhodes-Kendler O (2004, May) Duodenal expression of the epithelial calcium transporter gene TRPV6: is there evidence for Vitamin D dependence in humans? J Steroid Biochem Mol Biol 89–90(1–5):317–319
79. Walters JR, Balesaria S, Chavele KM, Taylor V, Berry JL, Khair U, Barley NF, van Heel DA, Field J, Hayat JO, Bhattacharjee A, Jeffery R, Poulsom R (2006, Nov) Calcium channel TRPV6 expression in human duodenum: different relationships to the vitamin D system and aging in men and women. J Bone Miner Res 21(11):1770–1777
80. Walters JR, Balesaria S, Khair U, Sangha S, Banks L, Berry JL (2007, Mar) The effects of Vitamin D metabolites on expression of genes for calcium transporters in human duodenum. J Steroid Biochem Mol Biol 103(3–5):509–512
81. Balesaria S, Sangha S, Walters JR (2009, Dec) Human duodenum responses to vitamin D metabolites of TRPV6 and other genes involved in calcium absorption. Am J Physiol Gastrointest Liver Physiol 297(6):G1193–G1197
82. Taparia S, Fleet JC, Peng JB, Wang XD, Wood RJ (2006, June) 1,25-Dihydroxyvitamin D and 25-hydroxyvitamin D–mediated regulation of TRPV6 (a putative epithelial calcium channel) mRNA expression in Caco-2 cells. Eur J Nutr 45(4):196–204

83. Meyer MB, Watanuki M, Kim S, Shevde NK, Pike JW (2006, June) The human transient receptor potential vanilloid type 6 distal promoter contains multiple vitamin D receptor binding sites that mediate activation by 1,25-dihydroxyvitamin D_3 in intestinal cells. Mol Endocrinol 20(6):1447–1461
84. Meyer MB, Zella LA, Nerenz RD, Pike JW (2007, Aug) Characterizing early events associated with the activation of target genes by 1,25-dihydroxyvitamin D_3 in mouse kidney and intestine in vivo. J Biol Chem 282(31):22344–22352
85. Wang TT, Tavera-Mendoza LE, Laperriere D, Libby E, MacLeod NB, Nagai Y, Bourdeau V, Konstorum A, Lallemant B, Zhang R, Mader S, White JH (2005, Nov) Large-scale *in silico* and microarray-based identification of direct 1,25-dihydroxyvitamin D3 target genes. Mol Endocrinol 19(11):2685–2695
86. Pike JW, Zella LA, Meyer MB, Fretz JA, Kim S (2007, Dec) Molecular actions of 1,25-dihydroxyvitamin D_3 on genes involved in calcium homeostasis. J Bone Miner Res 22(Suppl 2):V16–V19
87. Khanal RC, Peters TM, Smith NM, Nemere I (2008, Nov) Membrane receptor-initiated signaling in 1,25$(OH)_2D_3$-stimulated calcium uptake in intestinal epithelial cells. J Cell Biochem 105(4):1109–1116
88. van AM, Hoenderop JG, van der Kemp AW, Friedlaender MM, van Leeuwen JP, Bindels RJ (2005, Oct) Coordinated control of renal Ca^{2+} transport proteins by parathyroid hormone. Kidney Int 68(4):1708–1721
89. Bacskai BJ, Friedman. PA (1990, Sep) Activation of latent Ca^{2+} channels in renal epithelial cells by parathyroid hormone. Nature 347(6291):388–391
90. Friedman PA, Coutermarsh BA, Kennedy SM, Gesek FA (1996, Jan) Parathyroid hormone stimulation of calcium transport is mediated by dual signaling mechanisms involving protein kinase A and protein kinase C. Endocrinology 137(1):13–20
91. de GT, Lee K, Langeslag M, Xi Q, Jalink K, Bindels RJ, Hoenderop JG (2009, Aug) Parathyroid hormone activates TRPV5 via PKA-dependent phosphorylation. J Am Soc Nephrol 20(8):1693–1704
92. Cha SK, Wu T, Huang CL (2008, May) Protein kinase C inhibits caveolae-mediated endocytosis of TRPV5. Am J Physiol Renal Physiol 294(5):F1212–F1221
93. Elalouf JM, Roinel N, de RC (1984, Feb) ADH-like effects of calcitonin on electrolyte transport by Henle's loop of rat kidney. Am J Physiol 246(2 Pt 2):F213–F220
94. Di SA, Wittner M, Nitschke R, Braitsch R, Greger R, Bailly C, Amiel C, Roinel N, de RC (1990, Oct) Effects of parathyroid hormone and calcitonin on Na^+, Cl^-, K^+, Mg^{2+} and Ca^{2+} transport in cortical and medullary thick ascending limbs of mouse kidney. Pflugers Arch 417(2):161–167
95. Hsu YJ, Dimke H, Hoenderop JG, Bindels RJ (2010, May) Calcitonin-stimulated renal Ca^{2+} reabsorption occurs independently of TRPV5. Nephrol Dial Transplant 25(5):1428–1435
96. Kuro-o M, Matsumura Y, Aizawa H, Kawaguchi H, Suga T, Utsugi T, Ohyama Y, Kurabayashi M, Kaname T, Kume E, Iwasaki H, Iida A, Shiraki-Iida T, Nishikawa S, Nagai R, Nabeshima YI (1997, Nov) Mutation of the mouse klotho gene leads to a syndrome resembling ageing. Nature 390(6655):45–51
97. Kuro-o M (2010, Jan) Klotho. Pflugers Arch 459(2):333–343
98. Chang Q, Hoefs S, van der Kemp AW, Topala CN, Bindels RJ, Hoenderop JG (2005, Oct) The beta-glucuronidase klotho hydrolyzes and activates the TRPV5 channel. Science 310(5747):490–493
99. Lu P, Boros S, Chang Q, Bindels RJ, Hoenderop JG (2008, Nov) The beta-glucuronidase klotho exclusively activates the epithelial Ca^{2+} channels TRPV5 and TRPV6. Nephrol Dial Transplant 23(11):3397–3402
100. Cha SK, Ortega B, Kurosu H, Rosenblatt KP, Kuro O, Huang CL (2008, July) Removal of sialic acid involving Klotho causes cell-surface retention of TRPV5 channel via binding to galectin-1. Proc Natl Acad Sci USA 105(28):9805–9810

101. Cha SK, Hu MC, Kurosu H, Kuro-o M, Moe O, Huang CL (2009, July) Regulation of renal outer medullary potassium channel and renal K(+) excretion by Klotho. Mol Pharmacol 76(1):38–46
102. Tohyama O, Imura A, Iwano A, Freund JN, Henrissat B, Fujimori T, Nabeshima Y (2004, Mar) Klotho is a novel beta-glucuronidase capable of hydrolyzing steroid beta-glucuronides. J Biol Chem 279(11):9777–9784
103. Tsuruoka S, Nishiki K, Ioka T, Ando H, Saito Y, Kurabayashi M, Nagai R, Fujimura A (2006, Oct) Defect in parathyroid-hormone-induced luminal calcium absorption in connecting tubules of Klotho mice. Nephrol Dial Transplant 21(10):2762–2767
104. Brown AJ, Krits I, Armbrecht HJ (2005, May) Effect of age, vitamin D, and calcium on the regulation of rat intestinal epithelial calcium channels. Arch Biochem Biophys 437(1):51–58
105. Song Y, Kato S, Fleet JC (2003, Feb) Vitamin D receptor (VDR) knockout mice reveal VDR-independent regulation of intestinal calcium absorption and ECaC2 and calbindin D9k mRNA. J Nutr 133(2):374–380
106. Van Cromphaut SJ, Rummens K, Stockmans I, Van HE, Dijcks FA, Ederveen AG, Carmeliet P, Verhaeghe J, Bouillon R, Carmeliet G (2003, Oct) Intestinal calcium transporter genes are upregulated by estrogens and the reproductive cycle through vitamin D receptor-independent mechanisms. J Bone Miner Res 18(10):1725–1736
107. Fudge NJ, Kovacs CS (2010, Mar) Pregnancy up-regulates intestinal calcium absorption and skeletal mineralization independently of the vitamin D receptor. Endocrinology 151(3):886–895
108. Charoenphandhu N, Nakkrasae LI, Kraidith K, Teerapornpuntakit J, Thongchote K, Thongon N, Krishnamra N (2009, Sep) Two-step stimulation of intestinal Ca^{2+} absorption during lactation by long-term prolactin exposure and suckling-induced prolactin surge. Am J Physiol Endocrinol Metab 297(3):E609–E619
109. Song Y, Fleet JC (2004, Aug) 1,25 dihydroxycholecalciferol-mediated calcium absorption and gene expression are higher in female than in male mice. J Nutr 134(8):1857–1861
110. van AM, Hoenderop JG, Dardenne O, St Arnaud R, van Os CH, Van Leeuwen HJ, Bindels RJ (2002, Aug) 1,25-dihydroxyvitamin D(3)-independent stimulatory effect of estrogen on the expression of ECaC1 in the kidney. J Am Soc Nephrol 13(8):2102–2109
111. Oz OK, Hajibeigi A, Howard K, Cummins CL, van AM, Bindels RJ, Word RA, Kuro-o M, Pak CY, Zerwekh JE (2007, Dec) Aromatase deficiency causes altered expression of molecules critical for calcium reabsorption in the kidneys of female mice *. J Bone Miner Res 22(12):1893–1902
112. Irnaten M, Blanchard-Gutton N, Harvey BJ (2008, Nov) Rapid effects of 17beta-estradiol on epithelial TRPV6 Ca^{2+} channel in human T84 colonic cells. Cell Calcium 44(5):441–452
113. Irnaten M, Blanchard-Gutton N, Praetorius J, Harvey BJ (2009, Aug) Rapid effects of 17beta-estradiol on TRPV5 epithelial Ca^{2+} channels in rat renal cells. Steroids 74(8):642–649
114. Peng JB, Zhuang L, Berger UV, Adam RM, Williams BJ, Brown EM, Hediger MA, Freeman MR (2001, Apr) CaT1 expression correlates with tumor grade in prostate cancer. Biochem Biophys Res Commun 282(3):729–734
115. Hsu YJ, Dimke H, Schoeber JP, Hsu SC, Lin SH, Chu P, Hoenderop JG, Bindels RJ (2010, Apr) Testosterone increases urinary calcium excretion and inhibits expression of renal calcium transport proteins. Kidney Int 77(7):601–608
116. Teerapornpuntakit J, Dorkkam N, Wongdee K, Krishnamra N, Charoenphandhu N (2009, Apr) Endurance swimming stimulates transepithelial calcium transport and alters the expression of genes related to calcium absorption in the intestine of rats. Am J Physiol Endocrinol Metab 296(4):E775–E786
117. Sato T, Yamamoto H, Sawada N, Nashiki K, Tsuji M, Nikawa T, Arai H, Morita K, Taketani Y, Takeda E (2006) Immobilization decreases duodenal calcium absorption

through a 1,25-dihydroxyvitamin D-dependent pathway. J. Bone Miner Metab 24(4): 291–299
118. van AM, Huybers S, Hoenderop JG, van der Kemp AW, van Leeuwen, JP, Bindels RJ (2006, Dec) Age-dependent alterations in Ca^{2+} homeostasis: role of TRPV5 and TRPV6. Am J Physiol Renal Physiol 291(6):F1177–F1183
119. Fukushima A, Aizaki Y, Sakuma K (2009, Jan) Short-chain fatty acids induce intestinal transient receptor potential vanilloid type 6 expression in rats and Caco-2 cells. J Nutr 139(1):20–25
120. Fukushima A, Ohta A, Sakai K, Sakuma K (2005, Dec) Expression of calbindin-D9k, VDR and Cdx-2 messenger RNA in the process by which fructooligosaccharides increase calcium absorption in rats. J Nutr Sci Vitaminol (Tokyo) 51(6):426–432
121. Bianco SD, Peng JB, Takanaga H, Suzuki Y, Crescenzi A, Kos CH, Zhuang L, Freeman MR, Gouveia CH, Wu J, Luo H, Mauro T, Brown EM, Hediger MA (2007, Feb) Marked disturbance of calcium homeostasis in mice with targeted disruption of the Trpv6 calcium channel gene. J Bone Miner Res 22(2):274–285
122. Hoenderop JG, van Leeuwen JP, van der Eerden BC, Kersten FF, van der Kemp AW, Merillat AM, Waarsing JH, Rossier BC, Vallon V, Hummler E, Bindels RJ (2003, Dec) Renal Ca^{2+} wasting, hyperabsorption, and reduced bone thickness in mice lacking TRPV5. J Clin Invest 112(12):1906–1914
123. Benn BS, Ajibade D, Porta A, Dhawan P, Hediger M, Peng JB, Jiang Y, Oh GT, Jeung EB, Lieben L, Bouillon R, Carmeliet G, Christakos S (2008, June) Active intestinal calcium transport in the absence of transient receptor potential vanilloid type 6 and calbindin-D9k. Endocrinology 149(6):3196–3205
124. Kutuzova GD, Sundersingh F, Vaughan J, Tadi BP, Ansay SE, Christakos S, Deluca HF (2008, Dec) TRPV6 is not required for 1alpha,25-dihydroxyvitamin D3-induced intestinal calcium absorption in vivo. Proc Natl Acad Sci USA 105(50):19655–19659
125. Bronner F (2009, Feb) Recent developments in intestinal calcium absorption. Nutr Rev 67(2):109–113
126. Renkema KY, Nijenhuis T, van der Eerden BC, van der Kemp AW, Weinans H, van Leeuwen JP, Bindels RJ, Hoenderop JG (2005, Nov) Hypervitaminosis D mediates compensatory Ca^{2+} hyperabsorption in TRPV5 knockout mice. J Am Soc Nephrol 16(11): 3188–3195
127. Renkema KY, Velic A, Dijkman HB, Verkaart S, van der Kemp AW, Nowik M, Timmermans K, Doucet A, Wagner CA, Bindels RJ, Hoenderop JG (2009, Aug) The calcium-sensing receptor promotes urinary acidification to prevent nephrolithiasis. J Am Soc Nephrol 20(8):1705–1713
128. Nijenhuis T, van der Eerden BC, Hoenderop JG, Weinans H, van Leeuwen JP, Bindels. RJ (2008, Nov) Bone resorption inhibitor alendronate normalizes the reduced bone thickness of TRPV5$^{-/-}$ mice. J Bone Miner Res 23(11):1815–1824
129. Lieben L, Benn BS, Ajibade D, Stockmans I, Moermans K, Hediger MA, Peng JB, Christakos S, Bouillon R, Carmeliet G (2010, Aug) Trpv6 mediates intestinal calcium absorption during calcium restriction and contributes to bone homeostasis. Bone 47(2): 301–308
130. Bernucci L, Henriquez M, Diaz P, Riquelme. G (2006, Nov) Diverse calcium channel types are present in the human placental syncytiotrophoblast basal membrane. Placenta 27 (11–12):1082–1095
131. Moreau R, Hamel A, Daoud G, Simoneau L, Lafond J (2002, Nov) Expression of calcium channels along the differentiation of cultured trophoblast cells from human term placenta. Biol Reprod 67(5):1473–1479
132. Moreau R, Daoud G, Bernatchez R, Simoneau L, Masse A, Lafond J (2002, Aug) Calcium uptake and calcium transporter expression by trophoblast cells from human term placenta. Biochim Biophys Acta 1564(2):325–332

133. Stumpf T, Zhang Q, Hirnet D, Lewandrowski U, Sickmann A, Wissenbach U, Dorr J, Lohr C, Deitmer JW, Fecher-Trost C (2008, June) The human TRPV6 channel protein is associated with cyclophilin B in human placenta. J Biol Chem 283(26):18086–18098
134. Suzuki Y, Kovacs CS, Takanaga H, Peng JB, Landowski CP, Hediger MA (2008, Aug) Calcium channel TRPV6 is involved in murine maternal-fetal calcium transport. J Bone Miner Res 23(8):1249–1256
135. Hache S, Takser L, Lebellego F, Weiler H, Leduc L, Forest JC, Giguere Y, Masse A, Barbeau B, Lafond J (2010, Feb) Alteration of calcium homeostasis in primary preeclamptic syncytiotrophoblasts: effect on calcium exchange in placenta. J Cell Mol Med [Epub ahead of print]
136. Lee GS, Jeung EB (2007, July) Uterine TRPV6 expression during the estrous cycle and pregnancy in a mouse model. Am J Physiol Endocrinol Metab 293(1):E132–E138
137. Lee BM, Lee GS, Jung EM, Choi KC, Jeung EB (2009) Uterine and placental expression of TRPV6 gene is regulated via progesterone receptor- or estrogen receptor-mediated pathways during pregnancy in rodents. Reprod Biol Endocrinol 7:49
138. Kim HJ, Lee GS, Ji YK, Choi KC, Jeung EB (2006, Aug) Differential expression of uterine calcium transporter 1 and plasma membrane Ca^{2+} ATPase 1b during rat estrous cycle. Am J Physiol Endocrinol Metab 291(2):E234–E241
139. Choi Y, Seo H, Kim M, Ka H (2009, Dec) Dynamic expression of calcium-regulatory molecules, TRPV6 and S100G, in the uterine endometrium during pregnancy in pigs. Biol Reprod 81(6):1122–1130
140. van der Eerden BC, Hoenderop JG, de Vries TJ, Schoenmaker T, Buurman CJ, Uitterlinden AG, Pols HA, Bindels RJ, van Leeuwen JP (2005, Nov) The epithelial Ca^{2+} channel TRPV5 is essential for proper osteoclastic bone resorption. Proc Natl Acad Sci USA 102(48):17507–17512
141. Yamauchi D, Nakaya K, Raveendran NN, Harbidge DG, Singh R, Wangemann P, Marcus DC (2010) Expression of epithelial calcium transport system in rat cochlea and vestibular labyrinth. BMC Physiol 10:1
142. Wangemann P, Nakaya K, Wu T, Maganti RJ, Itza EM, Sanneman JD, Harbidge DG, Billings S, Marcus DC (2007, May) Loss of cochlear. Am J Physiol Renal Physiol 292(5): F1345–F1353
143. Nakaya K, Harbidge DG, Wangemann P, Schultz BD, Green ED, Wall SM, Marcus DC (May 2007) Lack of pendrin. Am J Physiol Renal Physiol 292(5):F1314–F1321
144. Yamauchi D, Raveendran NN, Pondugula SR, Kampalli SB, Sanneman JD, Harbidge DG, Marcus DC (2005, June) Vitamin D upregulates expression of ECaC1 mRNA in semicircular canal. Biochem Biophys Res Commun 331(4):1353–1357
145. Takumida M, Ishibashi T, Hamamoto T, Hirakawa K, Anniko M (February 2009) Age-dependent changes in the expression of klotho protein, TRPV5 and TRPV6 in mouse inner ear. Acta Otolaryngol 25:1–11
146. Homann V, Kinne-Saffran E, Arnold WH, Gaengler P, Kinne RK (May 2006) Calcium transport in human salivary glands: a proposed model of calcium secretion into saliva. Histochem Cell Biol 125(5):583–591
147. Lehen'kyi V, Beck B, Polakowska R, Charveron M, Bordat P, Skryma R, Prevarskaya N (2007, Aug) TRPV6 is a Ca^{2+} entry channel essential for Ca^{2+}-induced differentiation of human keratinocytes. J Biol Chem 282(31):22582–22591
148. Kennedy BG, Torabi AJ, Kurzawa R, Echtenkamp SF, Mangini NJ (2010) Expression of transient receptor potential vanilloid channels TRPV5 and TRPV6 in retinal pigment epithelium. Mol Vis 16:665–675
149. Fixemer T, Wissenbach U, Flockerzi V, Bonkhoff H (2003, Oct) Expression of the Ca^{2+}-selective cation channel TRPV6 in human prostate cancer: a novel prognostic marker for tumor progression. Oncogene 22(49):7858–7861
150. Vanden Abeele F, Lemonnier L, Thebault S, Lepage G, Parys JB, Shuba Y, Skryma R, Prevarskaya N (2004, July) Two types of store-operated Ca^{2+} channels with different

activation modes and molecular origin in LNCaP human prostate cancer epithelial cells. J Biol Chem 279(29):30326–30337
151. Vanden Abeele F, Roudbaraki M, Shuba Y, Skryma R, Prevarskaya N (2003, Apr) Store-operated Ca^{2+} current in prostate cancer epithelial cells. Role of endogenous Ca^{2+} transporter type 1. J Biol Chem 278(17):15381–15389
152. Bodding M, Fecher-Trost C, Flockerzi V (2003, Dec) Store-operated Ca^{2+} current and TRPV6 channels in lymph node prostate cancer cells. J Biol Chem 278(51):50872–50879
153. Pigozzi D, Ducret T, Tajeddine N, Gala JL, Tombal B, Gailly P (2006, May) Calcium store contents control the expression of TRPC1, TRPC3 and TRPV6 proteins in LNCaP prostate cancer cell line. Cell Calcium 39(5):401–415
154. Lehen'kyi V, Flourakis M, Skryma R, Prevarskaya N (2007, Nov) TRPV6 channel controls prostate cancer cell proliferation via Ca^{2+}/NFAT-dependent pathways. Oncogene 26(52):7380–7385
155. Lallet-Daher H, Roudbaraki M, Bavencoffe A, Mariot P, Gackiere F, Bidaux G, Urbain R, Gosset P, Delcourt P, Fleurisse L, Slomianny C, Dewailly E, Mauroy B, Bonnal JL, Skryma R, Prevarskaya N (2009, Apr) Intermediate-conductance Ca^{2+}-activated K+ channels (IKCa1) regulate human prostate cancer cell proliferation through a close control of calcium entry. Oncogene 28(15):1792–1806
156. Bolanz KA, Hediger MA, Landowski CP (2008, Feb) The role of TRPV6 in breast carcinogenesis. Mol Cancer Ther 7(2):271–279
157. Bolanz KA, Kovacs GG, Landowski CP, Hediger MA (2009, Dec) Tamoxifen inhibits TRPV6 activity via estrogen receptor-independent pathways in TRPV6-expressing MCF-7 breast cancer cells. Mol Cancer Res 7(12):2000–2010
158. Peleg S, Sellin JH, Wang Y, Freeman MR, Umar S (2010, Sep) Suppression of aberrant transient receptor potential cation channel, subfamily V, member 6 expression in hyperproliferative colonic crypts by dietary calcium. Am J Physiol Gastrointest Liver Physiol 299(3):G593–601
159. Chow J, Norng M, Zhang J, Chai J (2007, Apr) TRPV6 mediates capsaicin-induced apoptosis in gastric cancer cells–Mechanisms behind a possible new hot cancer treatment. Biochim Biophys Acta 1773(4):565–576
160. Wilson FH, Disse-Nicodeme S, Choate KA, Ishikawa K, Nelson-Williams C, Desitter I, Gunel M, Milford DV, Lipkin GW, Achard JM, Feely MP, Dussol B, Berland Y, Unwin RJ, Mayan H, Simon DB, Farfel Z, Jeunemaitre X, Lifton RP (2001, Aug) Human hypertension caused by mutations in WNK kinases. Science 293(5532):1107–1112
161. Mayan H, Munter G, Shaharabany M, Mouallem M, Pauzner R, Holtzman EJ, Farfel Z (2004, Aug) Hypercalciuria in familial hyperkalemia and hypertension accompanies hyperkalemia and precedes hypertension: description of a large family with the Q565E WNK4 mutation. J Clin Endocrinol Metab 89(8):4025–4030
162. Mayan H, Vered I, Mouallem M, Tzadok-Witkon M, Pauzner R, Farfel Z (2002, July) Pseudohypoaldosteronism type II: marked sensitivity to thiazides, hypercalciuria, normomagnesemia, and low bone mineral density. J Clin Endocrinol Metab 87(7):3248–3254
163. Jiang Y, Ferguson WB, Peng JB (2007, Feb) WNK4 enhances TRPV5-mediated calcium transport: potential role in hypercalciuria of familial hyperkalemic hypertension caused by gene mutation of WNK4. Am J Physiol Renal Physiol 292(2):F545–F554
164. Zhang W, Na T, Peng JB (2008, Nov) WNK3 positively regulates epithelial calcium channels TRPV5 and TRPV6 via a kinase-dependent pathway. Am J Physiol Renal Physiol 295(5):F1472–F1484
165. Jiang Y, Cong P, Williams SR, Zhang W, Na T, Ma HP, Peng JB (2008, Oct) WNK4 regulates the secretory pathway via which TRPV5 is targeted to the plasma membrane. Biochem Biophys Res Commun 375(2):225–229
166. Yang CL, Angell J, Mitchell R, Ellison DH (2003, Apr) WNK kinases regulate thiazide-sensitive Na-Cl cotransport. J Clin Invest 111(7):1039–1045

167. Wilson FH, Kahle KT, Sabath E, Lalioti MD, Rapson AK, Hoover RS, Hebert SC, Gamba G, Lifton RP (2003, Jan) Molecular pathogenesis of inherited hypertension with hyperkalemia: the Na-Cl cotransporter is inhibited by wild-type but not mutant WNK4. Proc Natl Acad Sci USA 100(2):680–684
168. Yang SS, Hsu YJ, Chiga M, Rai T, Sasaki S, Uchida S, Lin SH (2010, Apr) Mechanisms for Hypercalciuria in Pseudohypoaldosteronism Type II-Causing WNK4 Knock-In Mice. Endocrinology 151(4):1829–1836
169. O'Reilly M, Marshall E, Macgillivray T, Mittal M, Xue W, Kenyon CJ, Brown RW (2006, Sep) Dietary electrolyte-driven responses in the renal WNK kinase pathway in vivo. J Am Soc Nephrol 17(9):2402–2413
170. Kim MH, Lee GS, Jung EM, Choi KC, Jeung EB (2009, July) The negative effect of dexamethasone on calcium-processing gene expressions is associated with a glucocorticoid-induced calcium-absorbing disorder. Life Sci 85(3–4):146–152
171. Kim MH, Lee GS, Jung EM, Choi KC, Oh GT, Jeung EB (2009, Jan) Dexamethasone differentially regulates renal and duodenal calcium-processing genes in calbindin-D9k and -D28k knockout mice. Exp Physiol 94(1):138–151
172. Huybers S, Naber TH, Bindels RJ, Hoenderop JG (2007, Jan) Prednisolone-induced Ca^{2+} malabsorption is caused by diminished expression of the epithelial Ca^{2+} channel TRPV6. Am J Physiol Gastrointest Liver Physiol 292(1):G92–G97
173. Bohmer C, Palmada M, Kenngott C, Lindner R, Klaus F, Laufer J, Lang F (2007, Dec) Regulation of the epithelial calcium channel TRPV6 by the serum and glucocorticoid-inducible kinase isoforms SGK1 and SGK3. FEBS Lett 581(29):5586–5590
174. Palmada M, Poppendieck S, Embark HM, van de Graaf SF, Boehmer C, Bindels RJ, Lang F (2005) Requirement of PDZ domains for the stimulation of the epithelial Ca^{2+} channel TRPV5 by the NHE regulating factor NHERF2 and the serum and glucocorticoid inducible kinase SGK1. Cell Physiol Biochem 15(1–4):175–182
175. Embark HM, Setiawan I, Poppendieck S, van de Graaf SF, Boehmer C, Palmada M, Wieder T, Gerstberger R, Cohen P, Yun CC, Bindels RJ, Lang F (2004) Regulation of the epithelial Ca^{2+} channel TRPV5 by the NHE regulating factor NHERF2 and the serum and glucocorticoid inducible kinase isoforms SGK1 and SGK3 expressed in Xenopus oocytes. Cell Physiol Biochem 14(4–6):203–212
176. Sopjani M, Kunert A, Czarkowski K, Klaus F, Laufer J, Foller M, Lang F (2010, Feb) Regulation of the Ca^{2+} channel TRPV6 by the kinases SGK1, PKB/Akt, and PIKfyve. J Membr Biol 233(1–3):35–41
177. Sandulache D, Grahammer F, Artunc F, Henke G, Hussain A, Nasir O, Mack A, Friedrich B, Vallon V, Wulff P, Kuhl D, Palmada M, Lang F (2006, July) Renal Ca^{2+} handling in sgk1 knockout mice. Pflugers Arch 452(4):444–452
178. Jang HR, Kim S, Heo NJ, Lee JH, Kim HS, Nielsen S, Jeon US, Oh YK, Na KY, Joo KW, Han JS (2009, Jan) Effects of thiazide on the expression of TRPV5, calbindin-D28K, and sodium transporters in hypercalciuric rats. J Korean Med Sci 24(Suppl):S161–S169
179. Nijenhuis T, Vallon V, van der Kemp AW, Loffing J, Hoenderop JG, Bindels RJ (2005, June) Enhanced passive Ca^{2+} reabsorption and reduced Mg^{2+} channel abundance explains thiazide-induced hypocalciuria and hypomagnesemia. J Clin Invest 115(6):1651–1658
180. Lee CT, Shang S, Lai LW, Yong KC, Lien YH (2004, Dec) Effect of thiazide on renal gene expression of apical calcium channels and calbindins. Am J Physiol Renal Physiol 287(6):F1164–F1170
181. Nijenhuis T, Hoenderop JG, Loffing J, van der Kemp AW, van Os CH, Bindels RJ (2003, Aug) Thiazide-induced hypocalciuria is accompanied by a decreased expression of Ca^{2+} transport proteins in kidney. Kidney Int 64(2):555–564
182. Friedman PA, Bushinsky DA (1999, Nov) Diuretic effects on calcium metabolism. Semin Nephrol 19(6):551–556
183. Lee CT, Chen HC, Lai LW, Yong KC, Lien YH (2007, Oct) Effects of furosemide on renal calcium handling. Am J Physiol Renal Physiol 293(4):F1231–F1237

184. Lee CT, Lien YH, Lai LW, Chen JB, Lin CR, Chen HC (2006, May) Increased renal calcium and magnesium transporter abundance in streptozotocin-induced diabetes mellitus. Kidney Int 69(10):1786–1791
185. Hoenderop JG, Dardenne O, van AM, van der Kemp AW, van Os CH, St Arnaud R, Bindels RJ (2002, Sep) Modulation of renal Ca^{2+} transport protein genes by dietary Ca^{2+} and 1,25-dihydroxyvitamin D_3 in 25-hydroxyvitamin D_3-1alpha-hydroxylase knockout mice. FASEB J 16(11):1398–1406
186. Ignat M, Teletin M, Tisserand J, Khetchoumian K, Dennefeld C, Chambon P, Losson R, Mark M (2008, Feb) Arterial calcifications and increased expression of vitamin D receptor targets in mice lacking TIF1alpha. Proc Natl Acad Sci USA 105(7):2598–2603
187. Huybers S, Apostolaki M, van der Eerden BC, Kollias G, Naber TH, Bindels RJ, Hoenderop JG (2008, June) Murine TNF(DeltaARE) Crohn's disease model displays diminished expression of intestinal Ca^{2+} transporters. Inflamm Bowel Dis 14(6):803–811
188. Will C, Breiderhoff T, Thumfart J, Stuiver M, Kopplin K, Sommer K, Gunzel D, Querfeld U, Meij IC, Shan Q, Bleich M, Willnow TE, Muller D (2010, Feb) Targeted deletion of murine Cldn16 identifies extra- and intrarenal compensatory mechanisms of Ca^{2+} and Mg^{2+} wasting. Am J Physiol Renal Physiol [Epub ahead of print]
189. Nijenhuis T, Hoenderop JG, Bindels RJ (2004, Mar) Downregulation of Ca^{2+} and Mg^{2+} transport proteins in the kidney explains tacrolimus (FK506)-induced hypercalciuria and hypomagnesemia. J Am Soc Nephrol 15(3):549–557
190. Lee GS, Byun HS, Kim MH, Lee BM, Ko SH, Jung EM, Gwak KS, Choi IG, Kang HY, Jo HJ, Lee HJ, Jeung EB (2008, Nov) The beneficial effect of the sap of Acer mono in an animal with low-calcium diet-induced osteoporosis-like symptoms. Br J Nutr 100(5):1011–1018
191. Akey JM, Eberle MA, Rieder MJ, Carlson CS, Shriver MD, Nickerson DA, Kruglyak L (2004, Oct) Population history and natural selection shape patterns of genetic variation in 132 genes. PLoS Biol 2(10):e286
192. Stajich JE, Hahn MW (2005, Jan) Disentangling the effects of demography and selection in human history. Mol Biol Evol 22(1):63–73
193. Hughes DA, Tang K, Strotmann R, Schoneberg T, Prenen J, Nilius B, Stoneking M (2008) Parallel selection on TRPV6 in human populations. PLoS One 3(2):e1686
194. Soejima M, Tachida H, Koda Y (2009, Feb) Sequence analysis of human TRPV6 suggests positive selection outside Africa. Biochem Genet 47(1–2):147–153
195. Akey JM, Swanson WJ, Madeoy J, Eberle M, Shriver MD (2006, July) TRPV6 exhibits unusual patterns of polymorphism and divergence in worldwide populations. Hum Mol Genet 15(13):2106–2113
196. Bersaglieri T, Sabeti PC, Patterson N, Vanderploeg T, Schaffner SF, Drake JA, Rhodes M, Reich DE, Hirschhorn JN (2004, June) Genetic signatures of strong recent positive selection at the lactase gene. Am J Hum Genet 74(6):1111–1120
197. Holick MF, MacLaughlin JA, Doppelt SH (1981, Feb) Regulation of cutaneous previtamin D3 photosynthesis in man: skin pigment is not an essential regulator. Science 211(4482):590–593
198. Braun M, Palacios C, Wigertz K, Jackman LA, Bryant RJ, McCabe LD, Martin BR, McCabe GP, Peacock M, Weaver CM (2007, June) Racial differences in skeletal calcium retention in adolescent girls with varied controlled calcium intakes. Am J Clin Nutr 85(6):1657–1663
199. Suzuki Y, Pasch A, Bonny O, Mohaupt MG, Hediger MA, Frey FJ (2008, June) Gain-of-function haplotype in the epithelial calcium channel TRPV6 is a risk factor for renal calcium stone formation. Hum Mol Genet 17(11):1613–1618
200. Na T, Zhang W, Jiang Y, Liang Y, Ma HP, Warnock DG, Peng JB (2009, May) The A563T variation of the renal epithelial calcium channel TRPV5 among African Americans enhances calcium influx. Am J Physiol Renal Physiol 296(5):F1042–F1051
201. Pratt JH, Manatunga AK, Peacock M (1996, Jan) A comparison of the urinary excretion of bone resorptive products in white and black children. J Lab Clin Med 127(1):67–70

202. Taylor EN, Curhan GC (2007, Feb) Differences in 24-hour urine composition between black and white women. J Am Soc Nephrol 18(2):654–659
203. Sarmina I, Spirnak JP, Resnick MI (1987, July) Urinary lithiasis in the black population: an epidemiological study and review of the literature. J Urol 138(1):14–17
204. Stamatelou KK, Francis ME, Jones CA, Nyberg LM, Curhan GC (2003, May) Time trends in reported prevalence of kidney stones in the United States: 1976–1994. Kidney Int 63(5):1817–1823
205. Bell NH, Shary J, Stevens J, Garza M, Gordon L, Edwards J (1991, July) Demonstration that bone mass is greater in black than in white children. J Bone Miner Res 6(7):719–723
206. Bohannon AD (1999, June) Osteoporosis and African American women. J Womens Health Gend Based Med 8(5):609–615
207. Renkema KY, Lee K, Topala CN, Goossens M, Houillier P, Bindels RJ, Hoenderop JG (2009, June) TRPV5 gene polymorphisms in renal hypercalciuria. Nephrol Dial Transplant 24(6):1919–1924
208. Muller D, Hoenderop JG, Vennekens R, Eggert P, Harangi F, Mehes K, Garcia-Nieto V, Claverie-Martin F, Os CH, Nilius B, Bindels JM (2002, Sep) Epithelial Ca^{2+} channel (ECAC1) in autosomal dominant idiopathic hypercalciuria. Nephrol Dial Transplant 17(9):1614–1620
209. Burks DA, Littleton RH (1992) The epidemiology of prostate cancer in black men. Henry Ford Hosp Med J 40(1–2):89–92
210. Powell IJ (1997, May) Prostate cancer and African–American men. Oncology (Williston Park) 11(5):599–605
211. Kessler T, Wissenbach U, Grobholz R, Flockerzi V (2009) TRPV6 alleles do not influence prostate cancer progression. BMC Cancer 9:380

Chapter 15
The TRPV5 Promoter as a Tool for Generation of Transgenic Mouse Models

Marlene Vind Hofmeister, Ernst-Martin Füchtbauer, Robert Andrew Fenton, and Jeppe Praetorius

Abstract The transient receptor potential vanilloid 5 (TRPV5) is a Ca^{2+} channel, which is expressed in renal late distal convoluted tubules (DCT2s) and connecting tubules (CNTs). These tubules play a major role in hormone controlled renal Ca^{2+} reabsorption, and thereby in body Ca^{2+} homeostasis, as well as urinary excretion of other electrolytes, including Na^+ and K^+. DCT2 and CNT are difficult to distinguish from the surrounding structures and thereby to study by direct functional methods. We developed a transgenic mouse model expressing enhanced green fluorescent protein (EGFP) driven by the TRPV5 promoter to identify these specific tubules. Expression of EGFP in the DCT2 and CNT allows the isolation of pure DCT2 and CNT populations for proteomic and physiological analyses. The TRPV5 promoter is also useful for generating conditional knockout mouse models in a cell-specific manner. TRPV5 promoter driven *Cre* recombinase expression will be useful for inducing DCT2 and CNT specific gene silencing of various channels, pumps, carriers, and receptors. In this chapter, we describe the strategy for developing transgenic mouse lines involving the TRPV5 promoter, provide a description of extensive validation of these mouse lines, and discuss possible uses and limitations.

15.1 Introduction

The kidney regulates the body water and electrolyte content and rids the organism of many waste products of metabolism and toxins. For instance, it plays a crucial homeostatic role in maintaining electrolyte and acid/base balance (1, reviewed in 2). Two functional units of the kidney, the nephron and the collecting duct (CD), show a complex axial differentiation into unique segments that are composed of specialized epithelia to meet the homeostatic challenges. The initial part of the nephron, the glomerulus, filters the plasma through a bed of capillaries contained within a

---

J. Praetorius (✉)
Department of Anatomy, Water and Salt Research Center, Aarhus University,
DK-8000 Aarhus, Denmark
e-mail: jp@ana.au.dk

terminal extension of the renal tubule known as the Bowman's capsule. The renal tubular system can be divided into the proximal tubule (PT), the thin limb of Henle's loop (TL), and the distal nephron consisting of the thick ascending limb of Henle's loop (TAL), the early and late distal convoluted tubules (DCT1 and DCT2, respectively), the connecting tubule (CNT), and the initial part of the cortical collecting duct (iCCD). The medullary part of the cortical collecting duct (mCCD) and the CD is not part of the nephron, as these segments originate from the ureteric bud, whereas the nephron originates from the nephrogenic blastema [3, 4].

Compelling evidence suggest that the DCT2s and CNTs are crucial for maintaining body Na^+ and Ca^{2+} homeostasis [5–8]. Regulation of Na^+ excretion is crucial for the maintenance of extracellular salt and subsequent volume homeostasis and, thus, for blood pressure control [7]. Specific hormones (e.g. aldosterone, angiotensin II, vasopressin, insulin, insulin-like growth factor I) regulate the rate of Na^+ reabsorption via the amiloride-sensitive epithelial sodium channel (ENaC) and the thiazide-sensitive NaCl cotransporter (NCC) in the DCT2 and CNT [7]. The predominant functional importance of DCT2 and CNT, in ENaC-mediated Na^+ reabsorption, is supported by the finding that mice with a collecting duct-specific gene inactivation of the α-subunit of ENaC are essentially normal regarding renal Na^+ handling, even when challenged by salt restriction, water deprivation, or potassium loading [6]. As the DCT2s and CNTs are suggested to be crucial for maintaining body Na^+ homeostasis, the expected phenotype of αENaC inactivating in the DCT2 and CNT is severe pseudohypoaldosteronism type I (PHA-1), which is associated with weight loss and dehydration, hypovolemia and hypotension, hyponatremia, hyperkalemia, and metabolic acidosis accompanied with elevated plasma aldosterone levels [6].

Ca^{2+} reabsorption is restricted to the DCT2 and CNT and occurs via the transient receptor potential vanilloid 5 (TRPV5) epithelial Ca^{2+} channels that localizes to these segments [9]. Regulation of Ca^{2+} excretion in the DCT2 and CNT (e.g. by PTH, vitamin D_3, estrogen) is crucial for the maintenance of many physiological functions, including muscle contraction, bone formation, and neuronal excitability. TRPV5 knockout mice ($TRPV5^{-/-}$) display renal Ca^{2+} wasting because of impaired Ca^{2+} reabsorption in the DCT2 and CNT [10].

TRPV5 is a member of the TRP vanilloid (TRPV) receptor family, which is composed of six members named TRPV1 to TRPV6. The TRPVs are involved in a variety of physiological processes, including sensory functions, smooth muscle proliferation, endothelial permeability, gender-specific behavior, and epithelial divalent ion transport. TRPV5 and TRPV6 show a high degree of homology (75%) and are composed of ~ 730 amino acids. The proteins contain six transmembrane domains, have a hydrophobic pore loop between transmembrane region 5 and 6, and contain large intracellular amino- and carboxyl termini. Functional TRPV5 and TRPV6 channels form a tetrameric membrane protein with four identical subunits assembled around a central aqueous pore highly selective for Ca^{2+} [11]. These channels constitute the apical and rate-limiting entry gate for active transcellular Ca^{2+} reabsorption in the DCT2 and CNT, where TRPV5 is the major isoform [5]. TRPV6 seems to be centrally involved in intestinal Ca^{2+} absorption and placental

Ca^{2+} transport [12–14]. The molecular determinant of TRPV5 and TRPV6 Ca^{2+} selectivity and permeation is a single aspartate residue present in the pore-forming region [15, 16]. TRPV4 is also expressed in DCT2 and CNT, predominantly at the basolateral membrane and is permeable to Ca^{2+} as well as Na^+ [17].

The transcellular Ca^{2+} transport through TRPV5 have been extensively studied, especially by the laboratories of Hoenderop and Bindels [18]. Since the intracellular Ca^{2+} concentration ($[Ca^{2+}]_i$) is approximately 100 nM, almost 10,000-fold less than the Ca^{2+} concentration in the extracellular fluid, and because the membrane voltage across the apical membrane is about − 70 mV, a steep electrochemical gradient favors Ca^{2+} entry across the apical plasma membrane via TRPV5. As Ca^{2+} enters the cell, it is shuttled towards the basolateral membrane bound to calbindin-$D_{28\ K}$, which has several important functions: (1) it buffers cytoplasmic Ca^{2+}, helping to keep the $[Ca^{2+}]_i$ concentration low, and maintaining a favorable gradient for Ca^{2+} influx, (2) it prevents apoptosis and protein precipitation, which are known to be induced by high levels of free Ca^{2+} in the cytoplasm, and (3) it prevent inhibition of TRPV5 activity, which has been shown to be induced by increased $[Ca^{2+}]_i$ [11, 15]. Ca^{2+} leaves the cell via extrusion to the blood compartment through NCX1 and/or PMCA1b in the basolateral membrane [19].

Regulation of TRPV5 activity in DCT2 and CNT occurs at three levels: (1) the abundance of the channel is regulated by transcriptional and translational processes in the epithelial cell, (2) TRPV5 insertion towards and retrieval from the plasma membrane provides a short-term regulatory mechanism adjusting the functional activity at the plasma membrane, and (3) once inserted in the plasma membrane, the channel activity is regulated by $[Ca^{2+}]_i$, providing negative feedback regulation [18].

15.2 TRPV5 Promoter Driven EGFP Expressing Mice

In 1981, reports indicated for the first time, that DNA from any source could be stably integrated into the mouse genome [20–22]. These methodologies (e.g. microinjection, viral infection of embryos, or manipulation of embryonic stem (ES) cells subsequently incorporated back into the embryo) are routine today, and is often provided as a commercial service.

15.2.1 Mouse Generation

A 3.6-kb fragment of the mouse TRPV5 promoter (5′ flanking region) was deduced from the Ensemble genome browser and amplified by PCR using BAC clone DNA as a template (Fig. 15.1a). *Sal*I and *Age*I restriction sites were added to the TRPV5 promoter fragment by PCR, which was subsequently inserted into the pAcGFP1-N1 vector at these restriction sites (Fig. 15.1b). The TRPV5 promoter-EGFP transgene was isolated from the vector backbone using the *Sal*I and *Afl*II restriction enzymes. To avoid vector contamination (containing bacterial DNA sequences), the

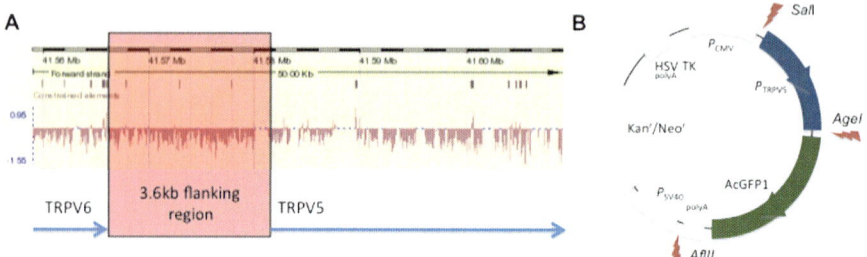

Fig. 15.1 Genomic organization of the TRPV5 gene and constructs. (**a**) The TRPV5 promoter (*orange box*) consists of a 3.6 kb region between the TRPV5 and TRPV6 genes on mouse chromosome 6. Peaks represent calculated Gerp intraspecies conservation scores. (**b**) Simplified map of the final pAcGFP1-N1 plasmid. TRPV5 was integrated into the multiple cloning sites of the plasmid at the *SalI* and *AgeI* restriction sites. The TRPV5-EGFP transgene was isolated from the vector backbone at the *SalI* and *AflII* restriction sites

transgene was further digested with the restriction enzyme *Nae*I, which cuts the vector DNA only. Purified TRPV5-EGFP DNA (Endofree Plasmid maxi, QIAGEN) was microinjected into the male pronuclei of zygotes from wild-type C57Bl/6 x DBA2 F2 hybrid mice (B6D2F2). Two cell stage embryos were retransferred into the oviduct of pseudopregnant females. Transgenic founder mice were identified by PCR as hemizygous transgenic mice, with the EGFP transgene randomly inserted into the genome. It is not possible to regulate either the expression site or the copy number of the transgene introduced into the embryo or the number of transgenes that will integrate together (usually at a single site) by this method [21, 23]. Overall, the efficiency of the methodology depends on the type of construct, the quality of the DNA preparation, the choice of the mouse strain, and the individual skill and performance of the experimenter [23].

15.2.2 Validation of the Transgenic Mouse Lines

Different founder lines with the same transgene show different patterns of transgene expression due to the regulatory influence of genomic sequences flanking the random insertion site of the transgene [24]. There are several reasons for such line-specific differences: (1) insertion into vital genes may rarely interrupt the normal expression and may be inconsequential or lethal, (2) transgenes may be silenced by the flanking heterochromatine regions, (3) insertion into chromosomal regions that act to suppress gene activity may prevent the expression of the transgene, and (4) insertion at a site that is flanked by an endogenous enhancer may stimulate gene activity at inappropriate stages or tissues. For these reasons, it was critical to analyze data from three or more founder lines as outlined in Fig. 15.2. In the case of TRPV5-EGFP transgenic mice, seven founder lines (F0) was breed with wild-type C57Bl/6 mice. Males and females from the resulting F1 generation were analyzed with respect to EGFP expression in several tissues including kidney, lung, heart,

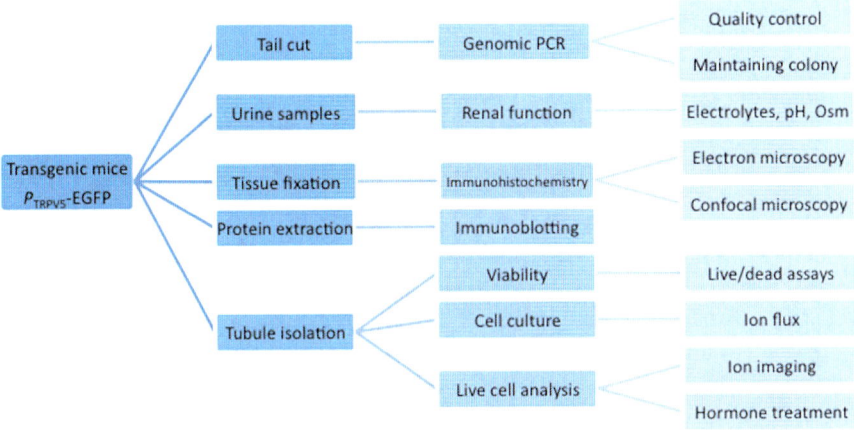

Fig. 15.2 Flow chart of the validation of the TRPV5 promoter driven EGFP expressing transgenic mouse line

liver, colon, stomach, duodenum, cecum, spleen, ovary, cerebrum, and cerebellum by immunoblotting. The EGFP localization was analyzed by immunohistochemistry (Fig. 15.3a–c). The organization of protein markers of the tubular system is shown in Fig. 15.3d. In addition, EGFP fluorescence was assessed directly in living isolated renal tubules by live cell fluorescence microscopy [25]. We chose two mouse lines as our transgenic mouse models: a line that highly expresses EGFP in many cells of the DCT2, CNT, and iCCD and a line with a low and mosaic expression pattern of EGFP in these tubules. As a consequence of the random genomic insertion of the transgene and the introduction of an exogenous polypeptide, it is crucial to ascertain that the mice did not develop a phenotype. The renal function was our main concern. Normal renal function was basically assessed by determination of urinary Ca^{2+}, Na^+, K^+, pH, and osmolality.

15.2.3 Studies in TRPV5-EGFP Mice

The endogenous expression of EGFP allows convenient identification of the DCT2 and CNT segments. To date, the transgenic mice have been used in two studies. In the first study [25], the functionality of the EGFP expressing tubules was validated by comparing intracellular Ca^{2+} responses to dDAVP and vitamin D_3 in TRPV5 promoter-driven EGFP expressing mice and non-transgenic littermates. Using the fluo-4 Ca^{2+} indicator dye, identical intracellular Ca^{2+} responses were recorded by live cell fluorescence microscopy. We used a refined enzymatic tubule isolation method to obtain long and viable segments of the relevant tubules. These tubules could be used for up to 6 h after isolation when kept on ice, as judged from their morphology and physiological responses. These initial studies provided proof of

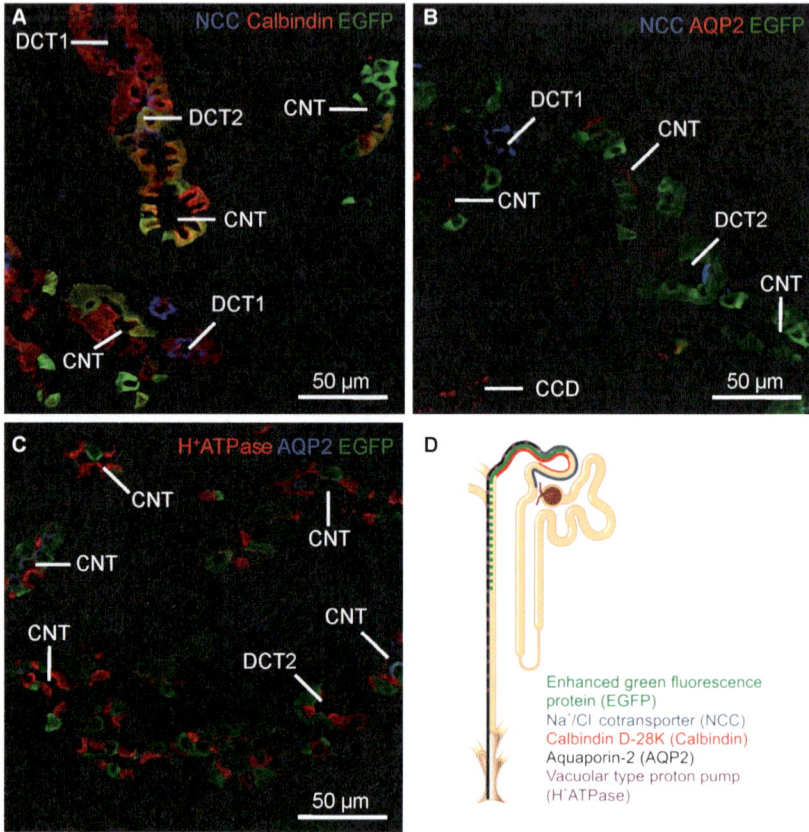

Fig. 15.3 TRPV5 promoter-driven EGFP expression in DCT2, CNT, and CCD. (**a**) Triple-immunolabeling confocal microscopical analysis of fixed kidney sections for calbindin-D_{28K}, Na^+-Cl^- cotransporter (NCC), and EGFP. (**b**) Similar immuno-fluorescence labeling for aquaporin-2 (AQP2), NCC, and EGFP. (**c**) Immuno-fluorescence labeling for vacuolar type proton pump (H^+-ATPase), AQP2, and EGFP. (**d**) Schematic representation of the renal tubular system with indication of the tubule segments and expression sites for EGFP, NCC, calbindin-D_{28K}, AQP2, and H-ATPase

concept for utilizing the TRPV5 promoter to target the DCT2 and CNT and furthermore, allowed us to initiate studies of DCT2 and CNT tubular physiology. In the second study, the mice were used to demonstrate intracellular Ca^{2+} signalling by acute administration of steroid hormones, e.g. estrogen, progesterone, and aldosterone in DCT2 and CNT (Hofmeister *et al.* manuscript). These physiologically relevant effects, which may be specific to the DCT2 and CNT, would not easily have been observed in a mixed tubule systems or in intact animals. Primary DCT2 and CNT cell cultures have been produced several times in the laboratory. The cultures form confluent monolayers with a high-resistance barrier when grown on semi-permeable supports. The cells maintain both their TRPV5 and EGFP expression for at least 2 weeks of culture. Currently, the TRPV5 activity in these cultures remains to be elucidated. To date, the TRPV5-EGFP transgenic mouse strains have been shared with three international groups in renal research.

15.2.4 Perspectives

In addition to the current studies of acute hormone effects on intracellular Ca^{2+} signalling, the TRPV5-EGFP transgenic mouse line will be of great benefit for studying Na^+, K^+, and Cl^- ion transport in the DCT2 and CNT, as well as solute transport, steroid hormone activated proteins, and protein trafficking. Both short-term and long-term hormone induced changes in protein expression in the DCT2 and CNT are of major renal physiological interest. Another possible use of the mice is in large-scale isolation of DCT2s and CNTs for proteomic analysis. Fluorescence-assisted sorting of isolated renal collecting ducts, based on EGFP expression, has elegantly been developed by others [26]. However, the technique for fluorescence-assisted DCT2 and CNT sorting remains to be established in our laboratory.

Induction of EGFP expression may be possible by treating mice, prior to isolation, with agonists of TRPV5 expression (e.g. PTH and vitamin D_3). Thus, *in vivo* regulation of renal cortical EGFP fluorescence may be monitored over time through skin windows as a putative measure for TRPV5 promoter activity during hormone treatment. As EGFP is pH sensitive, recording of intracellular pH in isolated DCT2s and CNTs would not require loading of exogenous fluorescent probes. This may be useful in studies of acute hormonal regulation of acid/base transport in the tubules. Finally, as TRPV5 is expressed in osteoclasts, the mouse line may be used as target for physiological *ex vivo* or *in vivo* studies on osteoclast cells or osteoclast culturing. In addition to *in vivo* studies, the relatively long viability of isolated DCT2s and CNTs would also facilitate studies on TRPV5-promoter function *in vitro*. For example, EGFP fluorescence could easily be measured on a plate reader as an indicator of specific transcriptional activity following agonist/antagonist treatment.

15.3 TRPV5 Promoter Driven *Cre* Recombinase Expressing Mice

15.3.1 Generation

The elaboration of the *Cre-loxP*-mediated recombination system allowed modifying the genome in specific tissues in a controlled manner [27]. We have exploited this technique and the unique tissue expression pattern of TRPV5 to develop tools for further studying the physiology and pathophysiology of the DCT2 and CNT. To develop a DCT2 and CNT *Cre* recombinase expressing transgenic mouse model, a similar strategy, as for the TRPV5-EGFP transgenic mice, was used, including an identical TRPV5 promoter fragment [25]. The TRPV5 promoter fragment was

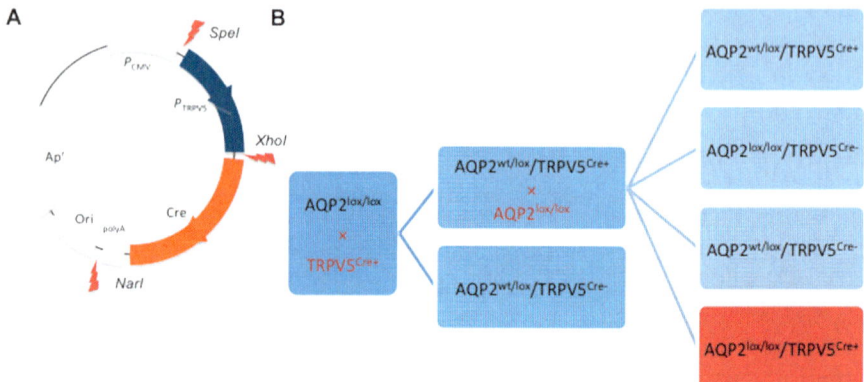

Fig. 15.4 (**a**) Map of the pBS185 plasmid containing the TRPV5 promoter fragments, including restriction sites. TRPV5 was integrated into the plasmid at the *SpeI* and *XhoI* restriction sites. The TRPV5-*Cre* transgene was isolated from the vector backbone at the *SpeI* and *NarI* restriction sites. (**b**) Flow chart of the development of DCT2/CNT-specific AQP2 knockout mice

inserted upstream of the *Cre* recombinase coding region using the *SpeI* and *XhoI* restriction sites (Fig. 15.4a). In addition, a TRPV5-*Cre* construct, insulated by a chicken lysozyme 5′ matrix attachment region (MAR) was developed. The MAR sequence (from the pSG12000 MAR backbone vector) was inserted in front of the TRPV5 in the pBS185 backbone vector, using the *NarI* and *SpeI* restriction sites. TRPV5 promoter driven *Cre* expressing transgenic mice were generated by pronuclear microinjection of purified MAR-TRPV5-*Cre* or TRPV5-*Cre* DNA into D6D2F2 zygotes with the C57Bl/6 genetic background as described for the TRPV5 promoter driven EGFP expressing mice.

15.3.2 Validation

The two *Cre* recombinase expressing transgenic mouse lines have been validated by PCR and immunoblotting using an anti-*Cre* antibody. The functional *Cre* recombinase activity is being verified by crossing the TRPV5-*Cre* or MAR-TRPV5-*Cre* mice with floxed AQP2 mice ($AQP2^{lox/lox}$, Fig. 15.4b). The outcome of this breeding included heterozygous mice ($AQP2^{wt/lox}/TRPV5^{Cre+}$ and $AQP2^{wt/lox}$). The $AQP2^{wt/lox}/TRPV5^{Cre+}$ mice are currently breeding with $AQP2^{lox/lox}$ mice. The outcome is expected to allocate according to the Mendelian inheritance generating DCT2 and CNT-specific AQP2 knockout mice ($AQP2^{lox/lox}/TRPV5^{Cre+}$) and control mice ($AQP2^{lox/lox}/TRPV5^{Cre-}$, $AQP2^{wt/lox}/TRPV5^{Cre-}$, and $AQP2^{wt/lox}/TRPV5^{Cre+}$). The site-specific deletion of AQP2 in DCT2 and CNT will be validated by immunohistochemical analysis using anti-AQP2 antibodies in combination with other tubule markers. The phenotype of these mice is not known, however, one could speculate that these mice will suffer from polyuria due to a lack of water reabsorption in the CNT.

15.3.3 Perspectives

The TRPV5 promoter driven *Cre* recombinase expressing mice can be bred with existing and future *loxP*-expressing mouse models, such as the floxed αENaC mice [6] to generate conditional gene knockouts of relevance for studying the DCT2 and CNT and osteoclast biology. This tool will be essential for studying the role of various hormone receptors, solute transporters, ion or water channels, and ion pumps at specific sites along the distal nephron.

15.4 Conclusion

The TRPV5 promoter has proved useful for driving expression of transgenes in a site- and cell-specific manner. The TRPV5 promoter driven EGFP expressing mouse lines are used routinely in studies of renal tubular physiology in our laboratory. The extended use of these mice to other kinds of experimental settings seems promising indeed. Similar optimistic prospects apply when we use the promoter to drive the *Cre* recombinase for the generation of site-specific knockout mice.

Acknowledgments The authors thank Inger Merete S. Paulsen, Christian V. Westberg, Zhila Nikrozi, and Helle Høyer for expert technical assistance. Support for this study was obtained from the Danish Medical Research Council, Karen Elise Jensens Fond, Lundbeckfonden, and Novo Nordisk Fonden. M. V. Hofmeister is supported by the Faculty of Health Sciences, University of Aarhus. The Water and Salt Research Centre at the University of Aarhus is established and supported by the Danish National Research Foundation (Danmarks Grundforskningsfond).

References

1. Reilly RF, Ellison DH (2000) Mammalian distal tubule: physiology, pathophysiology, and molecular anatomy. Physiol Rev 80:277–313
2. Boron W, Boulpaep EL (2009) Medical Physiology. Saunders Elsevier, Philadelphia, PA
3. Crayen ML, Thoenes W (1978) Architecture and cell structures in the distal nephron of the rat kidney. Cytobiologie 17:197–211
4. Madsen KM, Tisher CC (1986) Structural-functional relationships along the distal nephron. Am J Physiol 250:F1–F15
5. Hoenderop JG, Bindels R (2005) Epithelial Ca^{2+} and Mg^{2+} channels in health and disease. J Am Soc Nephrol 16:15–26
6. Rubera I, Loffing J, Palmer LG, Frindt G, Fowler-Jaeger N, Sauter D, Carroll T, McMahon A, Hummler E, Rossier BC (2003) Collecting duct-specific gene inactivation of alphaENaC in the mouse kidney does not impair sodium and potassium balance. J Clin Invest 112:554–565
7. Loffing J, Korbmacher C (2009) Regulated sodium transport in the renal connecting tubule (CNT) via the epithelial sodium channel (ENaC). Pflugers Arch 458:111–135
8. Loffing J, Loffing-Cueni D, Valderrabano V, Kläusli L, Hebert SC, Rossier BC, Hoenderop JG, Bindels RJ, Kaissling B (2001) Distribution of transcellular calcium and sodium transport pathways along mouse distal nephron. Am J Physiol Renal Physiol 281:F1021–F1027
9. Hoenderop JGJ, Müller D, Suzuki M, van Os CH, Bindels RJM (2000) Epithelial calcium channel: gate-keeper of active calcium reabsorption. Curr Opin nephrol Hypertens 9: 335–340

10. Renkema KY, Nijenhuis T, van der Eerden BC, van der Kemp AW, Weinans H, van Leeuwen JP, Bindels RJ, Hoenderop JG (2005) Hypervitaminosis D mediates compensatory Ca^{2+} hyperabsorption in TRPV5 knockout mice. J Am Soc Nephrol 16:3188–3195
11. Vennekens R, Hoenderop JG, Prenen J, Stuiver M, Willems PH, Droogmans G, Nilius B, Bindels RJ (2000) Permeation and gating properties of the novel epithelial $Ca^{(2+)}$ channel. J Biol Chem 275:3963–3969
12. Peng JB, Chen XZ, Berger UV, Vassilev PM, Tsukaguchi H, Brown EM, Hediger MA (1999) Molecular cloning and characterization of a channel-like transporter mediating intestinal calcium absorption. J Biol Chem 274:22739–22746
13. Bianco SD, Peng JB, Takanaga H, Suzuki Y, Crescenzi A, Kos CH, Zhuang L, Freeman MR, Gouveia CH, Wu J, Luo H, Mauro T, Brown EM, Hediger MA (2007) Marked disturbance of calcium homeostasis in mice with targeted disruption of the Trpv6 calcium channel gene. J Bone Miner Res 22:274–285
14. Suzuki Y, Kovacs CS, Takanaga H, Peng JB, Landowski CP, Hediger MA (2008) Calcium channel TRPV6 is involved in murine maternal-fetal calcium transport. J Bone Miner Res 23:1249–1256
15. Nilius B, Prenen J, Vennekens R, Hoenderop JG, Bindels RJ, Droogmans G (2001) Pharmacological modulation of monovalent cation currents through the epithelial Ca^{2+} channel ECaC1. Br J Pharmacol 134:453–462
16. Voets T, Janssens A, Droogmans G, Nilius B (2004) Outer pore architecture of a Ca^{2+}-selective TRP channel. J Biol Chem 279:15223–15230
17. Woudenberg-Vrenken TE, Bindels RJ, Hoenderop JG (2009) The role of transient receptor potential channels in kidney disease. Nat Rev Nephrol 5:441–449
18. Hoenderop JG, Bindels RJ (2008) Calciotropic and magnesiotropic TRP channels. Physiology (Bethesda) 23:32–40
19. Hoenderop JG, Nilius B, Bindels RJ (2002) Molecular mechanism of active Ca^{2+} reabsorption in the distal nephron. Annu Rev Physiol 64:529–549
20. Costantini F, Lacy E (1981) Introduction of a rabbit beta-globin gene into the mouse germ line. Nature 294:92–94
21. Brinster RL, Chen HY, Trumbauer M, Senear AW, Warren R, Palmiter RD (1981) Somatic expression of herpes thymidine kinase in mice following injection of a fusion gene into eggs. Cell 27:223–231
22. Gordon JW, Ruddle FH (1981) Integration and stable germ line transmission of genes injected into mouse pronuclei. Science 214:1244–1246
23. Rubera I, Hummler E, Beermann F (2009) Transgenic mice and their impact on kidney research. Pflugers Arch 458:211–222
24. Palmiter RD, Brinster RL (1986) Germ-line transformation of mice. Annu Rev Genet 20: 465–499
25. Hofmeister MV, Fenton RA, Praetorius J (2009) Fluorescence isolation of mouse late distal convoluted tubules and connecting tubules: effects of vasopressin and vitamin D3 on Ca^{2+} signaling. Am J Physiol Renal Physiol 296:F194–F203
26. Miller RL, Zhang P, Chen T, Rohrwasser A, Nelson RD (2006) Automated method for the isolation of collecting ducts. Am J Physiol Renal Physiol 291:F236–F245
27. Sauer B, Henderson N (1988) Site-specific DNA recombination in mammalian cells by the Cre recombinase of bacteriophage P1. Proc Natl Acad Sci USA 85:5166–5170

Chapter 16
TRPP Channels and Polycystins

Alexis Hofherr and Michael Köttgen

Abstract The founding member of the TRPP family, TRPP2, was identified as one of the disease genes causing autosomal dominant polycystic kidney disease (ADPKD). ADPKD is the most prevalent, potentially lethal, monogenic disorder in humans, with an average incidence of one in 400 to one in 1,000 individuals worldwide. Here we give an overview of TRPP ion channels and Polycystin-1 receptor proteins focusing on more recent studies. We include the Polycystin-1 family since these proteins are functionally linked to TRPP channels.

16.1 Introduction

The Transient Receptor Potential (TRP) channels are a family of diverse ion channels. In contrast to most ion channels, members of the TRP channel family were identified by their sequence homology rather than by ligand function or selectivity [1]. TRP channels are found in all eukaryotes, from yeast to mammals [2].

Members of the TRP channel family regulate physiological processes as diverse as thermosensation, pheromone reception, male fertility and regulation of vascular tone [3]. They share the leitmotif of permeability to cations and rather low sensitivity to membrane voltage [3]. TRP channels mediate the transmembrane flux of cations down their electrochemical gradients, thereby raising intracellular Ca^{2+} and Na^+ concentrations and depolarising the cell [4]. The depolarisation of

---

A. Hofherr (✉)
Renal Division, Department of Medicine, University Medical Centre Freiburg, 79106 Freiburg, Germany; Spemann Graduate School of Biology and Medicine (SGBM),
Albert-Ludwigs-University, Freiburg, Germany; Faculty of Biology,
Albert-Ludwigs-University, Freiburg, Germany
e-mail: alexis.hofherr@uniklinik-freiburg.de

M. Köttgen
Renal Division, Department of Medicine, University Medical Centre Freiburg,
79106 Freiburg, Germany
e-mail: michael.koettgen@uniklinik-freiburg.de

the membrane potential and Ca^{2+} serve as highly versatile signals that control many different cellular functions, like exocytosis, transcription, proliferation and apoptosis [5]. TRP channels are thought of as signal integrators and high-gain signal amplifiers, coupling the single gating event, that is channel opening, to the flux of millions of ions per second [4]. Most TRP channels appear to be responsive to multiple distinct stimuli [4]. This suggests that the physiological activity for any given TRP channel will be governed by the specifics of its cellular context, i.e. ligand concentrations, phosphorylation status, lipid environment and interacting proteins [4]. A common feature of most TRP channels is that they have a role in sensory physiology. In addition to the classical five Aristotelian senses they serve as intrinsic sensors of the cellular environment [1].

Studies on architecture of TRP channels yielded strong evidence that these channels function as tetramers [6–8]. Monomeric TRP channel subunits are predicted to have six transmembrane helices (S1–S6), with a pore loop between S5 and S6, and cytosolic amino- and carboxy-termini of varying sizes (Fig. 16.1) [1, 2]. However, to date no structural information at the atomic level is available for any full-length TRP channel [9].

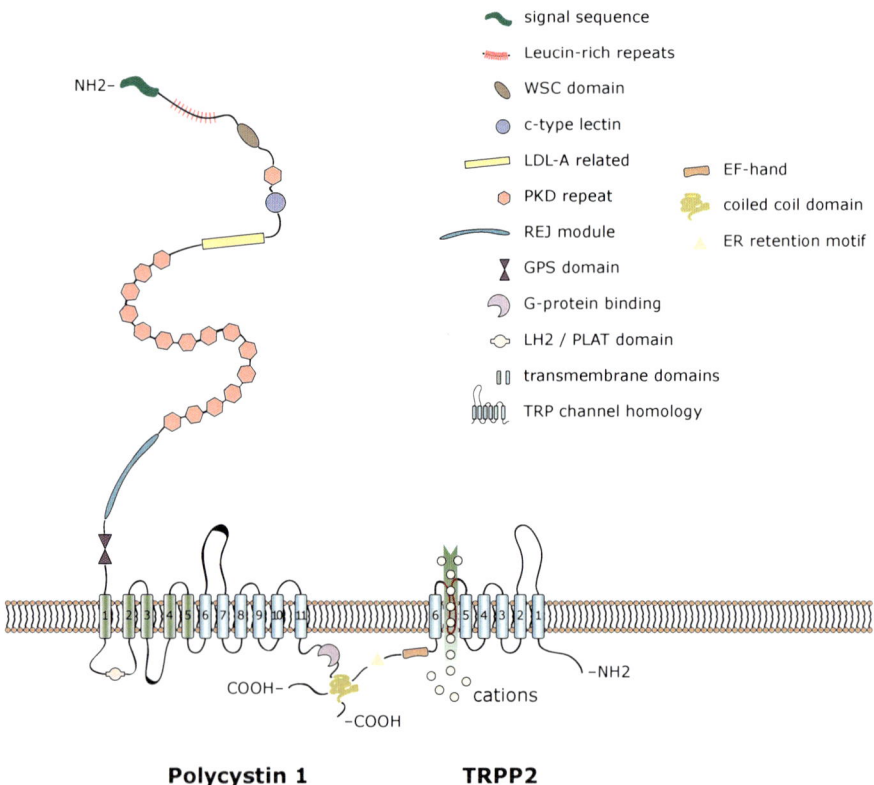

Fig. 16.1 The polycystin-1-TRPP2 receptor-channel complex

The TRP superfamily has been divided into eight subfamilies. They were named after the initially described member of each subfamily: TRPC, TRPV, TRPM, TRPA, TRPN, TRPY, TRPML and TRPP [2].

This review focuses on the TRPP subfamily. We also include the Polycystin-1 family since these proteins are physically and functionally linked to TRPP channels. The physiological role of TRPP channels as well as Polycystin-1 proteins, and their relevance to human disease, will be discussed, highlighting common themes that emerged from the study of various model systems.

16.2 Autosomal Dominant Polycystic Kidney Disease (ADPKD)

The founding member of the TRPP family, TRPP2, was identified by positional cloning as one of the disease genes causing autosomal dominant polycystic kidney disease (ADPKD) (#173,900 in Online Mendelian Inheritance of Man, http://www.ncbi.nlm.nih.gov/omim/) [10]. ADPKD is the most prevalent, potentially lethal, monogenic disorder in humans, with an average incidence of one in 400 to one in 1,000 individuals worldwide, regardless of ethnicity [11]. The disease is caused by mutations in either one of two genes, PKD1 (~85%) or PKD2 (~15%), which encode for Polycystin-1 (PC-1) and TRPP2, respectively [10, 12–14]. Disease mutations are found throughout both genes. There is rather limited information on genotype-phenotype correlation (The ADPKD Mutation Database, http://pkdb.mayo.edu) [15]. Homozygous or compound heterozygous genotypes have been shown to be embryonically lethal [16]. Individuals heterozygous for both PKD1 and PKD2 mutations usually survive to adulthood but have a more severe progression of disease [17, 18]. A significant inter- and intra-familiar phenotypic variability points to an involvement of modifier effects to the ADPKD phenotype [19]. Due to the slow but relentless growth of fluid filled renal cysts approximately 50% of all affected individuals ultimately develop fatal end-stage renal disease (ESRD) by their fourth to sixth decade of life [20–23].

Cysts in ADPKD originate from localised cystic dilatations in all nephron segments [24]. They are lined by a single layer of epithelium that is characterised by an increased cellular proliferation and decreased differentiation [25]. In 1992 the possibility of a "two hit" mechanism of cyst formation in ADPKD was first proposed [26]. This hypothesis suggested that the inactivation of both copies of a polycystic kidney disease gene is required for cyst formation: the "first hit" being the inherited germ line mutation, whereas the "second hit" is caused by acquired somatic mutations triggering cyst formation [26, 27]. Molecular genetic analysis of the cyst lining epithelia revealed that individual cysts were monoclonal in origin and in many instances resulted from somatic mutation of the copy of PKD1 or PKD2 inherited from the unaffected parent [28, 29]. Thus, a "two hit" cellular recessive mechanism appears to underpin the initiation of a majority of renal and liver cysts in ADPKD, although additional mechanisms may be involved [30, 31].

More recently, the Cre-loxP recombination system has been used to delete the mouse Pkd1 gene at different developmental stages in order to test whether the timing of Pkd1 gene inactivation impacts cyst formation and growth [32]. Deletion of Pkd1 shortly after birth resulted in rapid development of polycystic kidney disease, renal failure and death within a few weeks. Inactivation of Pkd1 in adult mice, however, resulted in overall healthy animals with mild renal cystic disease during the first months after gene disruption. Factors such as timing of gene activation, degree of inactivation of the protein product, modifier genes and the presence of other forms of renal injury are likely to contribute to the severity of ADPKD in individuals [30, 33].

Kidney pathology is the hallmark of ADPKD, but extra-renal manifestations are common, although most of them are not life-threatening [31]. Extra-renal cysts can develop in the arachnoid membrane, pancreas, seminal tract and most frequently in the liver, where they arise by excessive proliferation and dilatation of biliary ductules and peribiliary glands (Fig. 16.2) [15]. Arterial Hypertension is a common symptom of ADPKD and a major contributor to renal disease progression, as well as to cardiovascular morbidity and mortality [15, 34]. Cardiac and vascular manifestations include an increased frequency of valvular heart disease, coronary artery aneurysms, intracranial 'berry' aneurysms and chronic subdural haematoma [35–39]. Pain is a recurrent symptom in adult patients, often associated with cyst haemorrhage, kidney stones or renal infection [15, 40, 41].

It has been discussed, whether a third gene causing rare forms of ADPKD may exist [42, 43]. However, an increasing percentage of previously unmapped ADPKD patients have meanwhile been found to carry mutations in either PKD1 or PKD2.

The indistinguishable clinical manifestation of both types of ADPKD suggest that TRPP2 and PC-1 function in a common signalling pathway [44]. Consistent with this hypothesis, TRPP2 and PC-1 interact physically through their carboxy-terminal coiled-coil domains (Fig. 16.1) [45, 46].

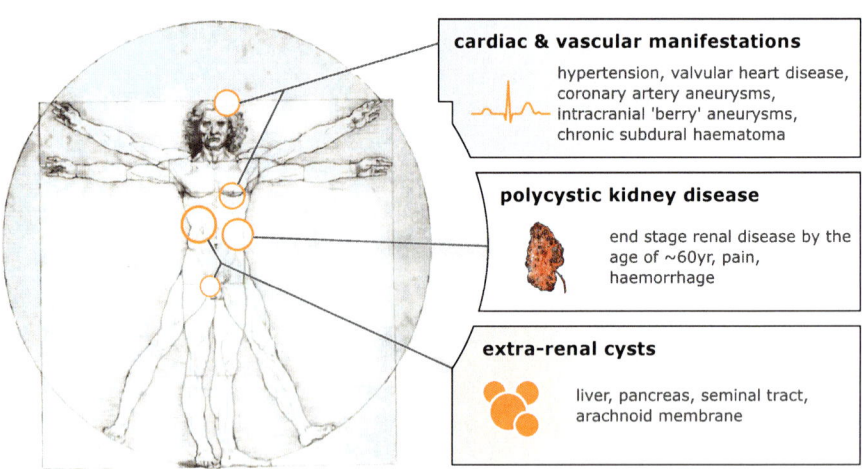

Fig. 16.2 Manifestations of autosomal dominant polycystic kidney disease

16.3 Primary Cilia and Polycystic Kidney Disease

An amazing convergence of findings from several species, including algae, nematodes, fruit flies and mice suggests that defects in structure or function of primary cilia are a possible common mechanism in the development of polycystic kidney disease [47–50]. This hypothesis is based on two observations. First, many proteins implicated in cyst formation localise to the primary cilium, and second, genetic ablation of primary cilia in model organisms cause cyst formation [47, 51–53].

The primary cilium is a solitary hair-like cellular structure emanating from the cell surface. It has been shown to be present in almost all vertebrate cells [54]. Primary cilia are sensory organelles that detect mechanical and chemical stimuli [55–58]. In the renal tubular epithelium the primary cilium is most likely to act as a sensor for the extracellular environment, as it extends from the apical cell surface into the tubule lumen [47]. Eukaryotic cilia are classified into motile cilia and nonmotile primary cilia [59]. The primary cilium develops from and is continuously anchored to the basal body, a centriole-derived microtubule organising centre [59]. Structural analysis revealed that the primary cilium has a so-called 9 + 0 axoneme, which refers to its nine peripherally located microtubule pairs and the absence of the central microtubule pair as seen in motile cilia or 9 + 2 cilia [59].

The first hint that defects in primary cilia might be involved in kidney cystogenesis came from the oak ridge polycystic kidney (orpk) mouse [51]. The orpk phenotype is characterised by multiple abnormalities including cystic kidneys and defects in left-right asymmetry, and is caused by a hypomorphic allele of the Tg737 gene [51, 60]. The discovery that Tg737 is the mouse homologue of IFT88, a protein required for the assembly of flagella in the green algae, Chlamydomonas reinhardtii, was a crucial step linking a ciliary defect to polycystic kidney disease [52]. Subsequently, it was demonstrated that orpk mice present with shortened primary cilia and that impairment of ciliary structure caused cystogenesis [53].

Several studies now highlight the common localisation of many mammalian cyst-proteins, including TRPP2 and PC-1, in primary cilia in the kidney [61, 62]. In contrast to the orpk mouse model, polycystic kidney disease patients form primary cilia that are normal in appearance. None of the human polycystic kidney disease genes appears to be required for the assembly of primary cilia. Instead, the human disease gene products seem to be essential for ciliary signalling [48, 63].

While there is a clear connection between polycystic kidney disease and dysfunction of ciliary proteins, the precise nature of this relationship is not completely understood. It is thought that PC-1 and TRPP2 act as a sensory receptor-channel complex mediating localised calcium influx at the primary cilium. Yet, the nature of the stimuli that activate the polycystin complex remains a matter of investigation. The ciliary polycystin complex has been proposed to respond to flow-dependent mechanosensory stimuli. According to this model, the loss of flow-mediated calcium influx triggers cyst formation [63, 64]. However, several studies have recently challenged this model, raising the possibility that PC1 and TRPP2 may respond to yet to be identified ligands [32, 65, 66]. Thus, it is possible that either flow- or ligand-mediated Ca^{2+} signals convey spatial information to the renal tubular epithelial cell, which is required to form and maintain a polarised epithelial tube with proper diameter [47].

It has become apparent that the primary cilium represents a nexus for multiple signalling pathways. Numerous human disorders have now been linked to defects in cilia structure or in cilia-localised proteins [67].

16.4 The TRPP Family

TRPP2 is the archetypal TRPP channel (Fig. 16.1) [1]. It was initially described as 'polycystin-2' (also referred to as PC-2, encoded by the PKD2 gene) and later classified into the Transient Receptor Potential Polycystin (TRPP) family of ion channels [1, 2, 10]. Research on TRPP2 has been largely driven by its involvement in autosomal dominant polycystic kidney disease (ADPKD).

TRPP2 is a 110 kDa protein that presents the familiar features of a TRP channel, i.e. six transmembrane segments (S1–S6) and a pore loop (between S5 and S6). The channel comprises a large loop domain separating the first two (S1 and S2) transmembrane domains [10]. Recently several studies have yielded first insights into structural requirements for assembly and function of TRPP2 channel complexes [68–70]. TRPP2 functions as a Ca^{2+}-permeable non-selective cation channel [71, 72].

Two more proteins were identified through homology cloning: TRPP3 (formerly named Polycystin L encoded by the Polycystic kidney disease 2 like 1 (PKD2L1) gene) and TRPP5 (encoded by the Polycystic kidney disease 2 like 2 (PKD2L2) gene) [73–76]. Together with TRPP2 they now comprise the TRPP family of ion channels [2]. Sequence similarity between TRPP channels is highest in transmembrane segments S1–S6, 78% between TRPP2 and TRPP3, 71% between TRPP2 and TRPP5, with rather little sequence homology in their predicted amino- and carboxy-termini [44]. All mammalian TRPP orthologues are highly conserved, with ~90% identities for TRPP2 and TRPP3, and ~80% for TRPP5 [44]. The TRPP family may be one of the most ancient TRP families, as members of this subfamily are found throughout the animal kingdom [77].

TRPP ion channels are regulated by and assemble with Polycystin-1 family proteins into receptor-channel complexes (Fig. 16.1) [45, 46, 69, 78, 79]. These protein complexes are thought to form the core of a signalling pathway that uses Ca^{2+} as a second messenger [80]. The aim of this review is to provide an overview of all TRPP channels and the corresponding Polycystin-1 family proteins.

16.5 The Polycystin-1 Family

Like TRPP2, polycystin-1 (PC-1) (also referred to as TRPP1, encoded by the PKD1 gene) was identified by positional cloning of ADPKD disease genes [12–14]. Mutations in PKD1, the gene encoding for PC-1, are causative for 85% of

ADPKD cases. Although PC-1 shares a high sequence homology in 6 of its 11 carboxy-terminal transmembrane domains (S6–S11) with TRPP2 and contains an equivalent loop domain between S6 and S7, PC-1 is not an ion channel [78, 79]. It lacks the representative TRP ion transport protein domain [44]. In contrast to the non-selective cation channel TRPP2, PC-1 is a large (~460 kDa), integral membrane protein with 11 transmembrane segments (S1–S11) (Fig. 16.1) [81, 82]. Since PC-1 does not display significant structural or functional characteristics of a TRP channel it should not be included into the TRPP channel family as has been suggested previously (TRPP1) [2]. PC-1 structurally resembles a receptor or adhesion molecule. However, the ligand or activating stimulus of PC-1 is not known. The signature features of PC-1 are the tripartite combination of a G protein-coupled receptor proteolytic site (GPS), the receptor for egg jelly (REJ) domain, and the lipoxygenase homology/polycystin, lipoxygenase, atoxin (LH2/PLAT) domain that is separated from the previous two by a transmembrane segment [44, 83–85].

Four additional PKD1-like genes were identified through homology cloning that are now members of the Polycystin-1 family [1]: Polycystic kidney disease 1 like 1 (PKD1L1) [2], Polycystic kidney disease 1 like 2 (PKD1L2) [3], Polycystic kidney disease 1 like 3 (PKD1L3) and [4] polycystic kidney disease receptor for egg jelly (PKDREJ, also referred to as polycystin receptor for egg jelly (PC-REJ)) [63, 83, 86, 87].

The Polycystin-1 family can be divided into two subsets. Both PC-1 and PKD1L1 lack an ion transport protein motif, but contain a carboxy-terminal coiled-coil domain [88]. Alternatively, the PKDREJ-like proteins, PKDREJ, PKD1L2 and PKD1L3 lack the coiled-coil domain but contain an ion transport protein motif [88].

While the PC-1 orthologues are well conserved among vertebrates, the PC-1 homologues in lower organisms (e.g. LOV-1 in C. elegans and suREJs in sea urchin) present with a non-uniform distribution of conserved amino acid residues with respect to their mammalian counterparts [44, 84, 89–91]. Like in TRPP channels, the sequence similarity is high in the transmembrane segments with low degrees of sequence similarity in the amino- and carboxy-terminal regions.

16.6 TRPP2 and PC-1

TRPP2 and PC-1 have been extensively investigated at all levels of complexity, from the single molecule to the whole organism. Here we give an overview of TRPP2 and PC-1 focusing on more recent studies. This review cannot provide a comprehensive coverage of the extensive body of work concerning all aspects of the genetics, biophysical properties, and trafficking of TRPP2 and PC-1. We refer the interested reader to the many excellent reviews covering these topics [15, 47, 63, 71, 72, 80, 92–103].

TRPP2 has been recognised as a bona fide TRP channel and is the founding member of the TRPP family [1]. The TRPP2 protein is encoded by the PKD2 gene on chromosome 4q21 and consists of 968 amino acids [10]. It presents with the familiar features of a TRP channel, i.e. six transmembrane segments and a pore loop (Fig. 16.1) [2, 10]. The detailed biophysical properties of TRPP2 are reviewed elsewhere [71, 72]. In brief, TRPP2 functions as a Ca^{2+}-permeable non-selective cation channel, with a slightly higher selectivity of Ca^{2+} over Na^+ and K^+, a high single channel conductance (100–200 pS), and a relatively high spontaneous open probability (20–40%) [71, 72]. TRPP2 is widely expressed in many tissues, including kidney, testis, intestine, pancreas, bile ducts, heart, vascular smooth muscle, and placenta [10]. In the kidney, TRPP2 has been reported in all nephron segments, but has not been detected in glomeruli [11]. Expression levels of PKD2 mRNA may be regulated on a post-transcriptional level by the RNA binding cyst-protein bicaudal C, which antagonises the repressive activity of the miR-17 microRNA family [104].

TRPP2 localises to different subcellular compartments, such as the endoplasmic reticulum (ER), the plasma membrane, the primary cilium, the centrosome and mitotic spindles in dividing cells [61, 105–111]. The functional implications of compartment-specific TRPP2 channel activity will be discussed below.

TRPP2 assembles with PC-1 into a receptor-channel complex [45, 46, 69, 78, 79]. This protein complex is thought to form the core of a signalling pathway that uses Ca^{2+} as a second messenger [80]. In contrast to TRPP2, PC-1 is a large (4,302 amino acids) glycosylated integral membrane protein with 11 transmembrane domains that structurally resembles a receptor or adhesion molecule [81, 82, 112]. The PKD1 gene encoding for PC-1 has been mapped to human chromosome 16q13 [12–14]. The protein is predicted to have an extracellular amino-terminal segment of ~3,000 amino acids and an intracellular carboxy-terminus of ~200 amino acids (Fig. 16.1). Its huge extracellular segment comprises motifs commonly involved in protein-protein and protein-carbohydrate interactions and a G-protein-coupled receptor proteolytic site (GPS domain) [85, 113]. In its short carboxy-terminus several phosphorylation sites and a putative binding sequence for heterotrimeric G proteins have been predicted [81, 114].

PC-1 has a similar expression pattern like TRPP2 [115]. Cellular compartments for which PC-1 localisation has been described include primary cilia, apical membranes, the endoplasmic reticulum as well as the adherent and desmosomal junctions [115–119].

It has been suggested that PC-1 may modulate TRPP2 channel activity in a direct and indirect manner. Delmas et al. found that PC-1 activates TRPP2 directly via conformational coupling, since antibodies directed at the PC-1 REJ domain increased whole cell currents independent of G protein and phospholipase C signaling [78]. Also, it has been reported that TRPP2 channel activity can be enhanced in response to epidermal growth factor (EGF) receptor stimulation, in a process mediated via phospholipase C (γ2), mammalian diaphanous-related formin 1 (mDia1) and phosphoinositide 3-kinase (PI3-kinase) [120, 121]. For PC-1 it was shown, that it can increase PI3-kinase activity [122]. This PC-1-induced PI3-kinase activity may regulate TRPP2, providing an indirect activation mechanism that does not require physical association between PC-1 and TRPP2.

The polycystin complex has been implicated in cellular processes as diverse as mechanosensation, cell polarity, proliferation and apoptosis [15, 47, 80, 92, 93, 103,

123, 124]. These functions are likely to be mediated in part by differential association with other proteins. Indeed, PC-1 and TRPP2 have been shown to interact with various proteins including cytoskeletal proteins, adaptor proteins, ion channels and receptors [125–139].

The last years have seen significant progress in the identification of signalling pathways through which PC-1 and TRPP2 are thought to exert their physiological functions [15, 47, 63, 103]. PC-1 has been reported to participate in various signalling pathways, including the PI3 kinase pathway, the JAK/STAT pathway, as well as mTOR, Wnt, NFAT, Id2 and PKC pathways [122, 140–150]. The fact that by now PC-1 has been implicated in that many signalling pathways raises the question whether all these pathways are relevant for PC-1 function *in vivo*. In some of these pathways PC-1 appears to require TRPP2, but it is not known how the different signalling cascades are integrated *in vivo*. Yet, in none of the described pathways a direct link to TRPP2-dependent Ca^{2+} signals has been established. To date neither the precise activation mechanisms nor the downstream effectors of the Ca^{2+} signal have been characterised. Consequently, in order to understand the physiological role and specificity of TRPP2-mediated Ca^{2+} signalling, it will be mandatory to identify Ca^{2+}-dependent downstream targets in the polycystin signalling cascade.

Based on the predicted topology of TRPP2 it was suggested to function as an ion channel [10]. Yet, it took a surprisingly long time until data on the channel function of TRPP2 was reported [79]. This is because over-expressed TRPP2 accumulates in the endoplasmic reticulum, due to an ER retention motif in the carboxy-terminal domain, making it very difficult to measure channel activity in the plasma membrane [105]. Because of the predominant localisation of over-expressed TRPP2 in the endoplasmic reticulum, several groups studied the function of the channel in this cellular compartment [105, 106, 136, 138, 151]. These studies clearly showed that TRPP2 modulates ER Ca^{2+} release and ER Ca^{2+} homeostasis. Some more recent studies suggest a role of TRPP2 in ER Ca^{2+} storage and apoptosis [123, 151]. Although there are some controversial findings with respect to the function of TRPP2 in the endoplasmic reticulum, multiple lines of evidence support a physiological role of the channel in this cellular compartment. Whether the function of TRPP2 in the endoplasmic reticulum or in the primary cilium is critical for tubular polarity and morphology remains to be determined.

Recent reports elucidated the dynamic distribution of TRPP2 to different cellular compartments. It was demonstrated that the intracellular trafficking of TRPP2 from the endoplasmic reticulum to the plasma membrane is regulated by phosphorylation and multiple interactions with adaptor proteins [100, 131, 132, 152, 153]. PC-1, glycogen synthase kinase 3 (GSK3) and Golgi- and ER-associated protein 14 (PIGEA-14) have been described to promote anterograde transport of TRPP2 along the secretory pathway, whereas the two phosphofurin acidic cluster sorting proteins 1 and 2 (PACS-2 and PACS-1) have been shown to mediate phosphorylation-dependent retention or retrieval in the endoplasmic reticulum and Golgi apparatus, respectively [79, 131, 132, 153]. In tubular epithelial cells TRPP2 has been reported to localise to cell-matrix and to cell-cell junctions within the basolateral plasma membrane [107, 109]. It has been suggested that the polycystin-1-TRPP2 receptor-channel complex may regulate processes like cell adhesion, differentiation and proliferation. However, for the reasons mentioned above the localisation of PC-1 and TRPP2 in the primary cilium has become the focus of most recent investigations.

One of the fundamental questions in the field is the activation mechanism of PC-1 and TRPP2 in the cilium. Despite intense investigations there is still no definitive answer. After identification of the polycystins in the primary cilium it was proposed that PC-1 and TRPP2 may function as ciliary mechanosensors to detect fluid flow in the renal tubule [64, 154–156]. This was based on elegant studies by Praetorius and Spring demonstrating that bending of primary cilia on MDCK cells by laminar fluid flow or by a micropipette led to an increase in the cytosolic Ca^{2+} concentration. This calcium signal was strictly dependent on the presence of cilia [55, 56]. Subsequently, Nauli et al. presented the first evidence that PC-1 and TRPP2 might be involved in flow sensing, using cells derived from Pkd1 knock-out mice [64]. In contrast to wild-type renal cells these PC-1 deficient cells did not show any increase in the cytosolic Ca^{2+} concentration in response to fluid flow [64]. Furthermore, antibodies directed against the first extracellular loop of TRPP2 (between S1 and S2) abolished the flow response in wild-type cells [64].

The flow-dependent cytosolic Ca^{2+} response appears to require PC-1 and TRPP2. But is flow-sensing sufficient to regulate tubular lumen diameter and morphology? Data from our laboratory challenge the hypothesis that the loss of flow sensing in renal epithelial cells is the critical defect leading to renal cystogenesis [65]. It has been reported that TRPP2 may form heteromeric channels with TRPC1 and TRPV4 [65, 134]. TRPP2 and TRPV4 interact physically, functionally as well as genetically and form a mechanosensitive channel complex [65]. Both channels appear to co-localise in primary cilia of renal epithelial cells. The knock-down of TRPV4 in these cells abolished flow-mediated cytosolic Ca^{2+} responses [65]. If flow-mediated Ca^{2+} signals were the critical signal regulating proper tubular morphology one would assume that TRPV4-deficient animals should develop cysts. However, zebrafish and mice lacking TRPV4 do not form renal cysts [65].

Homologues of PC-1 and TRPP2 have recently been shown to form receptor-channel complexes that are proposed to function as sour taste receptors (see below) [157, 158]. This raises the possibility that PC-1 and TRPP2 may function as a chemosensor responding to yet to be identified ligands. It is intriguing to speculate that a combined mechanism of flow-dependent delivery of a ligand is the physiological signal to activate the polycystin complex.

A new aspect to the complex ciliary function of PC-1 and TRPP2 was the recent finding that these two proteins, as well as other known cyst-proteins, are abundant in a subpopulation of urinary exosome-like vesicles [66]. *In vitro*, these PC-1 and TRPP2 containing exosome-like vesicles preferentially associate with primary cilia of kidney and biliary epithelial cells in a rapid and highly specific manner [66]. This might suggest that PC-1 and TRPP2 proteins are shed in membrane vesicles in the urine, and these vesicles interact with primary cilia. This might potentially trigger ciliary signalling, akin to mechanisms in the embryonic node [66, 159].

The kidney aside, cardiovascular manifestations are a major contributor to ADPKD morbidity and mortality (Fig. 16.2) [15, 34]. TRPP2 and PC-1 are abundantly expressed in myocytes of blood vessels, but their function in this system is incompletely understood. A recent study describes the polycystin receptor-channel complex to associate with the cytoskeletal protein filamin A and thereby regulating the activity of stretch-activated cation channels in a blood pressure-dependent manner [137]. Thus contributing to the adaptation of the vascular myogenic tone to the pressure in the lumen of blood vessels.

In summary, TRPP2 and PC-1 have been reported to have physiological functions in various cellular locations. It remains to be determined which pool and which role of TRPP2 and PC-1 are critical for the pathogenesis of ADPKD. Future experiments will have to dissect the complex interplay between different compartment-specific functions, which are further complicated by the development-dependent roles of TRPP2 and PC-1. The identification of upstream and downstream signal molecules in the polycystin signalling cascade will be one of the important next steps.

16.7 TRPP3 and PKD1L3

TRPP3 was the first TRPP2 homologue to be described. It was initially referred to as Polycystic kidney disease like, Polycystin L or PKDL [73, 74]. TRPP3 and TRPP2 show significant sequence homology, with about 50% identical and 71% homologous residues [74]. Regions of homology are clustered in the six transmembrane segments (S1-S6) and in the prominent loop domain between the first and second transmembrane segment (S1-S2) [74]. Both channel proteins contain a putative Ca^{2+} binding EF-hand-like domain and a coiled-coil domain in their carboxy-terminus [160]. TRPP3 differs from TRPP2 most significantly in the amino terminal domain where it lacks a 100 amino acid segment [74].

The multi-exon PKD2L1 gene is located on chromosome 10q31 in humans and encodes for TRPP3 (805 amino acids). TRPP3 is expressed in various tissues, like neurons (e.g. in the retina and in taste receptor cells in the tongue), testis, kidney and non-myocyte cardiac tissue (i.e. ventricular blood vessels and epicardium) [75, 157, 158, 160–162].

In contrast to TRPP2, TRPP3 was well-studied by electrophysiological methods [163]. This was possible because, unlike TRPP2, TRPP3 can be functionally expressed at the plasma membrane of *Xenopus laevis* oocytes. The reason for this may be the absence of the carboxy-terminal PACS-binding acidic cluster, which has been shown to be important for TRPP2 retention in the endoplasmic reticulum [44, 132]. Expression of human TRPP3 in *Xenopus laevis* oocytes gave rise to constitutively active, large conductance, Ca^{2+}-permeable non-selective cation channels with a single channel conductance of 137 pS [163].

The physiological function of TRPP3 *in vivo* has remained elusive. After its identification it was proposed as a candidate for a third polycystic kidney disease locus [74]. This hypothesis was supported by the fact that Pkd2l1 was shown to be deleted in the *krd* mouse model [74, 164]. But the question whether the Pkd2l1 deletion contributes to the *krd* phenotype has remained uncertain. The *krd* mouse model has a 7 cM deletion on chromosome 19 which covers Pkd2l1 and several other genes including Pax2. Heterozygous *krd* animals present with kidney and retina defects (*krd*). Importantly, patients with mutations in the PAX2 gene show eye malformations and renal hypoplasia [165]. Thus, deletion of Pax2 rather than Pkd2l1 might be causing the *krd* phenotype. To date, no linkage of TRPP3 to polycystic kidney disease has been demonstrated and there is no published data on the phenotype of Pkd2l1 knock-out mice [75, 76].

More recently, TRPP3 has been proposed to function in conjunction with PKD1L3 as a candidate sour taste receptor [157, 158]. Taste reception occurs in mammals at the apical tip of taste cells that form taste buds in the oral region [166]. Each taste bud has an onion-like shape and is composed of 50–100 taste cells that possess microvilli [166]. In mammals, taste is classified into five distinct taste modalities: bitter, sweet, umami (the taste of some L-amino acids), salty and sour [166]. The signal transduction of bitter, sweet, salty and umami taste is well characterised, but the mammalian receptors for sour taste are incompletely understood [167, 168].

Three independent research groups recently reported the circumscribed co-expression of TRPP3 and PKD1L3 in a specific subset of mouse taste receptor cells [157, 158, 169]. These cells were found in taste buds and were shown to be distinct from the taste receptor cells mediating sweet, umami and bitter taste. Genetic ablation of the TRPP3- and PKD1L3-expressing cells in mice by targeted expression of attenuated diphtheria toxin lead to a complete loss of taste response to sour taste [158]. The absence of an entire subpopulation of taste cells, however, does not necessarily establish the role of any given protein expressed by the missing cells.

Support for the hypothesis that TRPP3 in association with PKD1L3 might form a candidate sour taste receptor came from heterologous over-expression systems. Over-expressed TRPP3 appears to interact with PKD1L3, as well as PC-1, and the PC-1 like proteins PKDREJ [87, 157, 158, 160]. HEK293T cells over-expressing PKD1L3 together with TRPP3 generated an inward current upon acid stimulation [157]. Notably, neither of these proteins alone generated acid-induced currents. In a subsequent study of the same group, it was reported that PKD1L3 and TRPP3 are activated upon removal rather than addition of acid [170]. These findings complement the above-mentioned studies proposing PKD1L3 and TRPP3 might function as a sour taste receptor. But some inconsistencies with respect to the literature on sour taste transduction remain. It is well established that the proximate stimulus in sour taste transduction is the intracellular proton concentration [167, 171, 172]. HEK293T cells over-expressing PKD1L3 and TRPP3, however, were activated by protons from the extracellular space [170]. In addition, the proposed function of TRPP3 as a candidate sour receptor is at odds with the finding that TRPP3 expressed in *Xenopus laevis* oocytes or HEK293T cells is inhibited rather than activated by low pH [163, 173].

In summary, taste receptor cells involved in sour taste transduction have been identified. These cells appear to express PKD1L3 as well as TRPP3 and the association of both proteins has been proposed to form a sensory receptor-channel complex for sour taste. To test whether these two proteins indeed form the sour taste sensor *in vivo* further experimental evidence is needed. The reconstitution of the sour taste receptor with properties resembling native taste receptor cells in heterologous systems as well a sour taste transduction studies in PKD2L1 or PKD1L3 knock-out animals will be of particular interest.

Very recently Nelson et al. [174] examined acid taste responsiveness of PKD1L3 knock-out mice. PKD1L3 knock-out mice showed no reduction in taste responsiveness to acids compared to control animals. This indicates that PKD1L3 is not required for normal sour taste sensation [174, 175]. Whether or not TRPP3 has a role in sour taste remains to be elaborated.

While the physiological function of TRPP3 *in vivo* remains to be elucidated, significant progress has been made in understanding the biophysical and biochemical properties of the channel protein. High intracellular Ca^{2+} increases human TRPP3 channel activity [163]. Low extracellular pH decreases the basal human TRPP3 conductance [163, 173]. In planar lipid bilayers and mammalian cells TRPP3 channels have been shown to be voltage-dependent and inhibited by monovalent cations larger than tetraethyl ammonium, as well as by multivalent cations such as Gd^{3+}, La^{3+} and Mg^{2+} [87, 163, 176, 177, 178]. In HEK293T cells over-expressing murine TRPP3 spontaneously active channels could be observed [173]. These murine channels exhibit a high single-channel conductance of 184 pS at negative potentials and are Ca^{2+}-permeable non-selective cation channels. In contrast to previous studies on human TRPP3 no Ca^{2+}-dependent changes on murine TRPP3 current properties were observed [163, 173]. Whole-cell experiments showed voltage-dependent gating of murine TRPP3 channels: repolarisation after depolarisation caused large transient inward TRPP3 tail currents. Interestingly, murine TRPP3 activity seemed to be tonicity-dependent, because a tight correlation between cell volume and TRPP3 currents was observed: cell swelling increased currents, whereas cell shrinkage reduced them [173].

Several proteins have been suggested to interact with TRPP3, some of them linked to the cytoskeleton [176, 179]. This may be an important factor in TRPP3 regulation, in particular as several recent studies show that the association with the actin cytoskeleton is important to the sensory functions of TRP channels [2, 137, 180, 181].

Like many other TRP channels, the majority of over-expressed TRPP3 localises within cells to the endoplasmic reticulum. This is likely due to two carboxy-terminal ER retention signals [160, 176]. The association with either PC-1 [160] or PKD1L3 [157] is suggested to increase plasma membrane expression of TRPP3.

Very little is known about the putative interaction partner of TRPP3, PKD1L3. The corresponding gene has been mapped to chromosome 16q22 and codes for a large protein (1,732 amino acids) [83]. This PC-1 like protein is further distinguished by being endowed with the characteristic TRP ion transport protein motif and the lack of a carboxy-terminal coiled-coil domain [88]. Future studies are needed to analyse the physiological function of PKD1L3 and its association with TRPP3.

16.8 TRPP5

Little data is available concerning the second TRPP2 homologue, TRPP5 (624 amino acids). The gene coding for TRPP5, Polycystic kidney disease 2 like 2 (PKD2L2) was identified through homology cloning and found to reside on chromosome 5q31 in humans [75, 76].

Sequence similarity between TRPP2 and TRPP5 is 71% with rather little sequence homology in the predicted amino- and carboxy-termini [44]. Notably, TRPP5 lacks the homology to the region of TRPP2 which is involved in the interaction with PC-1 [45, 46]. The PACS-binding acidic cluster in the carboxy-terminus is also absent [44]. These two observations suggest that TRPP5 hetero-multimerisation and its trafficking might be regulated differently compared with TRPP2. But to date no conclusive functional data are available.

Unlike TRPP2, which is widely expressed, TRPP5 seems to have a restricted expression pattern [75, 76]. Northern blot analysis established the prominent presence of TRPP5 mRNA in mouse testis [75]. Additional RT-PCR studies detected smaller amounts of mRNA most eminent in the adult mouse heart and the developing mouse oocyte [75, 76, 182, 183]. Several splice variants have been described but their functional consequences remain elusive [75, 76].

Being prominently expressed in testis and to a lesser degree in oocytes it is intriguing to contemplate whether its localisation can tell us something about its physiological function. TRPP5 may function as a non-selective cation channel involved in Ca^{2+} signalling in germ line cells. It has been demonstrated that the intracellular calcium concentration plays an important role in germ cell development and function, e.g. during meiosis and fertilisation [184]. Future experiments will have to tell whether TRPP5 plays a role in these processes.

16.9 PKDREJ

Hughes et al. were the first to describe the human homologue of the sea urchin receptor for egg jelly (suREJ3), the polycystic kidney disease like 1 protein PKDREJ (polycystic kidney disease receptor for egg jelly) [185]. The corresponding gene, composed of only one single exon, encoding for the 2,253 amino acid PKDREJ protein was found on human chromosome 22q31 [75, 185].

The homology of PKDREJ with PC-1 extends over almost its entire length. The two proteins have similar hydropathy profiles and share the familiar features of the Polycystin-1 family [75, 76]. But in contrast to PC-1, PKDREJ, like suREJ3, presents with a putative ion channel pore region between the last two carboxy-terminal transmembrane segments (S10 and S11) [186]. Interestingly, *in vitro* studies revealed that PKDREJ associates with TRPP2 and TRPP3, but not with TRPP5 [87].

Northern blot analysis and RT-PCR studies suggest that the expression pattern of PKDREJ mRNA is germ cell specific [186]. The protein itself seems to be localised in mature sperm to the acrosomal region of the plasma membrane [186]. It is therefore tempting to speculate that PKDREJ, like its sea urchin homologue [84], may play a role in sperm function and fertilisation, possibly in concert with one of the TRPP ion channel proteins, that are expressed in sperm.

For mammalian sperm it was first reported in 1951 that they must spend some time in the female reproductive tract in order to acquire the ability to fertilise eggs [187–189]. In this process of capacitation sperm undergo changes in plasma membrane composition, flagellar motility and finally become competent to penetrate the oocyte's zona pellucida via the acrosome reaction [189].

To dissect the impact of PKDREJ on sperm function Sutton et al. [190] created male mice that are homozygous for a targeted mutation in the Pkdrej allele (Pkdrej$^{-/-}$). These males are fertile in unrestricted mating trials, but exhibit lower reproductive success when competing with wild-type males [190]. This seems to be the case due to a delayed sperm capacitation [190]. The ability to undergo a zona pellucida evoked acrosome reaction appears to develop more slowly in Pkdrej$^{-/-}$ sperm than in wild-type sperm [190]. These data are suggestive of PKDREJ playing a role in controlling the timing of fertilisation *in vivo* by regulating the sperm's responsiveness to signals from the oocyte's zona pellucida during capacitation.

16.10 PKD1L1

The multi exon gene coding for polycystic kidney disease 1 like 1 (PKD1L1) has been mapped to human chromosome 7p12-p13 [86].

Interestingly, PKD1L1 shares a similar degree of amino acid sequence homology to both TRPP and Polycystin-1 proteins (~38%) [86]. However, its structural properties and a phylogenetic analysis place this protein within the Polycystin-1 family [2, 86]. The protein architecture of PKD1L1 resembles the signature tripartite Polycystin-1 combination of G protein-coupled receptor proteolytic site (GPS), receptor for egg jelly (REJ) domain, and lipoxygenase homology/polycystin, lipoxygenase, atoxin (LH2/PLAT) domain [44, 83–85]. Furthermore, PKD1L1, like PC-1 lacks the characteristic ion transport protein motif present in most TRP proteins [44]. But deviant from PC-1 the REJ domain is significantly smaller, 525 vs. 111 amino acid residues, respectively, and PKD1L1 contains only two PKD domains in its large amino terminus, vs. 16 in PC-1 [86]. Future studies will have to tell how these differences in structure contribute to the physiological activity of PKD1L1.

Dot-blot analysis and RT-PCR studies show the expression of PKD1L1 prominently in Leydig cells in the testis, in the adult and foetal heart and in the mammary gland [86, 88]. In situ hybridisation experiments confirmed the abundant and specific expression of PKD1L1 in Leydig cells, a known source of testosterone in males [86]. Several splice variants have been described but their functional consequences remain elusive [86].

Given this expression pattern it was somewhat surprising when Vogel et al. presented the first data about a Pkd1l1 knock-out mouse [191]. The homozygous mutant mice present with situs inversus (30%) and a significantly reduced viability [191]. Situs inversus is a congenital condition in vertebrates characterised by a mirror-image reversal of the inner organ arrangement and is often associated with primary cilia dyskinesia [191]. Primary cilia have been described as important factors in early development, contributing to the establishment of the usual asymmetrical positioning of organs by their function in a transient structure called the embryonic node, which is of exceptional importance for the establishment of left to right asymmetry [159]. But remarkably, Pkd1l1$^{-/-}$ mice presented with no other notable phenotype that are characteristic for primary cilia dyskinesia: no respiratory tract lesions, no hydrocephalus, and no male infertility [191].

The markedly reduced viability of Pkd1l1$^{-/-}$ mice might be linked in part to the cardiac expression of PKD1L1, but future studies will have to analyse the vascular and cardiac status of these mice [191].

It is very interesting that Pkd2 as well as Pkd1l1 knock-out mice present with situs inversus. Therefore, these proteins seem to have an important developmental function within the embryonic node [191, 192]. Future studies will have to examine the role of PKD1L1 mutations as well as TRPP contribution in general to the symmetry breaking events in early vertebrate development.

16.11 PKD1L2

Polycystic kidney disease 1 like 2 (PKD1L2) is an integral membrane protein of predicted 2,460 amino acid residues characterised by the known features of the Polycystin-1 PKDREJ subfamily [83]. Like PKDREJ, but unlike PC-1, PKD1L2 presents with an ion transport domain between its last two carboxy-terminal transmembrane segments (S10-S11) and no carboxy-terminal coiled-coil domain [88]. The complex gene coding for PKD1L2 is found on human chromosome 16q23.2 [83].

Dot-blot analysis and RT-PCR studies revealed that the mRNA expression pattern of PKD1L2 is very similar to that of PKD1L1 [88]. Moderate levels of PKD1L2 mRNA have been described in the mammary gland, the central nervous system, the adult and foetal heart, as well as in testis [88]. However, compared with the marked PKD1L1 expression in testis, the amount of PKD1L2 expressed there seems to be rather low. Multiple splice variants with divergent tissue distributions have been described but their functional consequences remain elusive [88].

To date no conclusive functional data is available. A recent study proposes the causative involvement of Pkd1l2 in the neuromuscular phenotype of the ostes mouse model [193]. Homozygous$^{ostes/ostes}$ mice present with a severe growth defect, myopathy and a reduction of lipogenesis [193]. The myopathy is most likely due to an impaired peripheral synapse development and not to a loss of motor axons or demyelination [193]. Although the genomic locus of Pkd1l2 seems to lack any mutations in the ostes mouse model, the expression of PKD1L2 appears to be significantly up-regulated in these mice [193]. Immunoprecipitation and co-localisation studies of endogenous PKD1L2 proteins of $^{ostes/ostes}$ mice suggest an interaction with the soluble enzyme fatty acid synthase (FASN) [193]. This is an intriguing connection, as in humans rare cases of myopathy have been linked to the enzyme Acetyl-CoA carboxylase, which catalyses the formation of the FASN substrate malonyl-CoA [193, 194].

But overall it remains an open question whether a mutation in an yet to be identified regulatory region of Pkd1l2 gene expression may be held causative of the neuromuscular phenotype in $^{ostes/ostes}$ mice.

16.12 Conclusions

Since the identification of ADPKD disease genes 15 years ago, great progress has been made in understanding the biology of TRPP2 and Polycystin-1. The rapid expansion of knowledge in the field of PKD biology is a prime example for the power of combining genetic approaches with biochemical, cell-biological and physiological studies. The unexpected discovery of the central role of primary cilia in PKD biology supports the notion that careful and open-minded observation occasionally trumps hypothesis-driven research. The studies summarised here have laid the groundwork for important discoveries in the future. One of the most fundamental open questions is the identity of upstream and downstream players in the polycystin signalling pathway. It is hoped that a deeper mechanistic understanding will help to design rationale therapies for polycystic kidney disease.

The other members of the TRPP and polycystin-1 families are less well understood, but important first insights have been made. A common theme arising from the study of different TRPP channels appears to be the association with PC-1-like proteins. It will be exciting to learn more about the biological functions of these evolutionary conserved sensory receptor-channel complexes.

Acknowledgments The authors would like to thank Gerard Dougherty for critically reading this manuscript. We apologise to those whose work could not be cited due to space limitations. We are grateful for funding from the PKD Foundation, Deutsche Forschungsgemeinschaft and Alfried Krupp von Bohlen und Halbach Foundation as well as the Excellence Initiative of the German Federal and State Governments (GSC-4, Spemann Graduate School).

References

1. Clapham DE (2003) TRP channels as cellular sensors. Nature 426:517–524
2. Venkatachalam K, Montell C (2007) TRP channels. Annu Rev Biochem 76:387–417
3. Desai BN, Clapham DE (2005) TRP channels and mice deficient in TRP channels. Pflugers Arch 451:11–18
4. Ramsey IS, Delling M, Clapham DE (2006) An introduction to TRP channels. Annu Rev Physiol 68:619–647
5. Berridge MJ, Bootman MD, Roderick HL (2003) Calcium signalling: dynamics, homeostasis and remodelling. Nat Rev Mol Cell Biol 4:517–529
6. Jahnel R, Dreger M, Gillen C, Bender O, Kurreck J, Hucho F (2001) Biochemical characterization of the vanilloid receptor 1 expressed in a dorsal root ganglia derived cell line. Eur J Biochem 268:5489–5496
7. Kedei N, Szabo T, Lile JD, Treanor JJ, Olah Z, Iadarola MJ, Blumberg PM (2001) Analysis of the native quaternary structure of vanilloid receptor 1. J Biol Chem 276:28613–28619
8. Moiseenkova-Bell VY, Stanciu LA, Serysheva II, Tobe BJ, Wensel TG (2008) Structure of TRPV1 channel revealed by electron cryomicroscopy. Proc Natl Acad Sci USA 105:7451–7455
9. Moiseenkova-Bell VY, Wensel TG (2009) Hot on the trail of TRP channel structure. J Gen Physiol 133:239–244
10. Mochizuki T, Wu G, Hayashi T, Xenophontos SL, Veldhuisen B, Saris JJ, Reynolds DM, Cai Y, Gabow PA, Pierides A, Kimberling WJ, Breuning MH, Deltas CC, Peters DJ, Somlo S (1996) PKD2, a gene for polycystic kidney disease that encodes an integral membrane protein. Science 272:1339–1342

11. Igarashi P, Somlo S (2002) Genetics and pathogenesis of polycystic kidney disease. J Am Soc Nephrol 13:2384–2398
12. Reeders ST, Breuning MH, Davies KE, Nicholls RD, Jarman AP, Higgs DR, Pearson PL, Weatherall DJ (1985) A highly polymorphic DNA marker linked to adult polycystic kidney disease on chromosome 16. Nature 317:542–544
13. The European Polycystic Kidney Disease Consortium (1994) The polycystic kidney disease 1 gene encodes a 14 kb transcript and lies within a duplicated region on chromosome 16. Cell 77:881–894
14. The International Polycystic Kidney Disease Consortium (1995) Polycystic kidney disease: the complete structure of the PKD1 gene and its protein. Cell 81:289–298
15. Torres VE, Harris PC (2009) Autosomal dominant polycystic kidney disease: the last 3 years. Kidney Int 76:149–168
16. Paterson AD, Wang KR, Lupea D, St George-Hyslop P, Pei Y (2002) Recurrent fetal loss associated with bilineal inheritance of type 1 autosomal dominant polycystic kidney disease. Am J Kidney Dis 40:16–20
17. Wu G, Tian X, Nishimura S, Markowitz GS, D'Agati V, Park JH, Yao L, Li L, Geng L, Zhao H, Edelmann W, Somlo S (2002) Trans-heterozygous Pkd1 and Pkd2 mutations modify expression of polycystic kidney disease. Hum Mol Genet 11:1845–1854
18. Pei Y, Paterson AD, Wang KR, He N, Hefferton D, Watnick T, Germino GG, Parfrey P, Somlo S, St George-Hyslop P (2001) Bilineal disease and trans-heterozygotes in autosomal dominant polycystic kidney disease. Am J Hum Genet 68:355–363
19. Persu A, Duyme M, Pirson Y, Lens XM, Messiaen T, Breuning MH, Chauveau D, Levy M, Grünfeld J, Devuyst O (2004) Comparison between siblings and twins supports a role for modifier genes in ADPKD. Kidney Int 66:2132–2136
20. Gabow PA, Chapman AB, Johnson AM, Tangel DJ, Duley IT, Kaehny WD, Manco-Johnson M, Schrier RW (1990) Renal structure and hypertension in autosomal dominant polycystic kidney disease. Kidney Int 38:1177–1180
21. Hateboer N, v Dijk MA, Bogdanova N, Coto E, Saggar-Malik AK, San Millan JL, Torra R, Breuning M, Ravine D (1999) Comparison of phenotypes of polycystic kidney disease types 1 and 2. European PKD1-PKD2 Study Group. Lancet 353:103–107
22. King BF, Reed JE, Bergstralh EJ, Sheedy PF, Torres VE (2000) Quantification and longitudinal trends of kidney, renal cyst, and renal parenchyma volumes in autosomal dominant polycystic kidney disease. J Am Soc Nephrol 11:1505–1511
23. Fick-Brosnahan GM, Belz MM, McFann KK, Johnson AM, Schrier RW (2002) Relationship between renal volume growth and renal function in autosomal dominant polycystic kidney disease: a longitudinal study. Am J Kidney Dis 39:1127–1134
24. Baert L (1978) Hereditary polycystic kidney disease (adult form): a microdissection study of two cases at an early stage of the disease. Kidney Int 13:519–525
25. Sutters M, Germino GG (2003) Autosomal dominant polycystic kidney disease: molecular genetics and pathophysiology. J Lab Clin Med 141:91–101
26. Reeders ST (1992) Multilocus polycystic disease. Nat Genet 1:235–237
27. Pei Y (2001) A two-hit model of cystogenesis in autosomal dominant polycystic kidney disease? Trends Mol Med 7:151–156
28. Qian F, Watnick TJ, Onuchic LF, Germino GG (1996) The molecular basis of focal cyst formation in human autosomal dominant polycystic kidney disease type I. Cell 87:979–987
29. Brasier JL, Henske EP (1997) Loss of the polycystic kidney disease (PKD1) region of chromosome 16p13 in renal cyst cells supports a loss-of-function model for cyst pathogenesis. J Clin Invest 99:194–199
30. Gallagher AR, Hoffmann S, Brown N, Cedzich A, Meruvu S, Podlich D, Feng Y, Könecke V, Vries U, Hammes H, Gretz N, Witzgall R (2006) A truncated polycystin-2 protein causes polycystic kidney disease and retinal degeneration in transgenic rats. J Am Soc Nephrol 17:2719–2730

31. Pei Y, Watnick T (2010) Autosomal dominant polycystic kidney disease. Adv Chronic Kidney Dis 17:115–117
32. Piontek K, Menezes LF, Garcia-Gonzalez MA, Huso DL, Germino GG (2007) A critical developmental switch defines the kinetics of kidney cyst formation after loss of Pkd1. Nat Med 13:1490–1495
33. Gallagher AR, Germino GG, Somlo S (2010) Molecular advances in autosomal dominant polycystic kidney disease. Adv Chronic Kidney Dis 17:118–130
34. Kelleher CL, McFann KK, Johnson AM, Schrier RW (2004) Characteristics of hypertension in young adults with autosomal dominant polycystic kidney disease compared with the general U.S. population. Am J Hypertens 17:1029–1034
35. Hossack KF, Leddy CL, Johnson AM, Schrier RW, Gabow PA (1988) Echocardiographic findings in autosomal dominant polycystic kidney disease. N Engl J Med 319:907–912
36. Lumiaho A, Ikäheimo R, Miettinen R, Niemitukia L, Laitinen T, Rantala A, Lampainen E, Laakso M, Hartikainen J (2001) Mitral valve prolapse and mitral regurgitation are common in patients with polycystic kidney disease type 1. Am J Kidney Dis 38:1208–1216
37. Hadimeri H, Lamm C, Nyberg G (1998) Coronary aneurysms in patients with autosomal dominant polycystic kidney disease. J Am Soc Nephrol 9:837–841
38. Schievink WI, Huston J, Torres VE, Marsh WR (1995) Intracranial cysts in autosomal dominant polycystic kidney disease. J Neurosurg 83:1004–1007
39. Leung GKK, Fan YW (2005) Chronic subdural haematoma and arachnoid cyst in autosomal dominant polycystic kidney disease (ADPKD). J Clin Neurosci 12:817–819
40. Bajwa ZH, Gupta S, Warfield CA, Steinman TI (2001) Pain management in polycystic kidney disease. Kidney Int 60:1631–1644
41. Bajwa ZH, Sial KA, Malik AB, Steinman TI (2004) Pain patterns in patients with polycystic kidney disease. Kidney Int 66:1561–1569
42. Daoust MC, Reynolds DM, Bichet DG, Somlo S (1995) Evidence for a third genetic locus for autosomal dominant polycystic kidney disease. Genomics 25:733–736
43. Bogdanova N, Dworniczak B, Dragova D, Todorov V, Dimitrakov D, Kalinov K, Hallmayer J, Horst J, Kalaydjieva L (1995) Genetic heterogeneity of polycystic kidney disease in Bulgaria. Hum Genet 95:645–650
44. Qian F, Noben-Trauth K (2005) Cellular and molecular function of mucolipins (TRPML) and polycystin 2 (TRPP2). Pflugers Arch 451:277–285
45. Qian F, Germino FJ, Cai Y, Zhang X, Somlo S, Germino GG (1997) PKD1 interacts with PKD2 through a probable coiled-coil domain. Nat Genet 16:179–183
46. Tsiokas L, Kim E, Arnould T, Sukhatme VP, Walz G (1997) Homo- and heterodimeric interactions between the gene products of PKD1 and PKD2. Proc Natl Acad Sci USA 94:6965–6970
47. Harris PC, Torres VE (2009) Polycystic kidney disease. Annu Rev Med 60:321–337
48. Gerdes JM, Davis EE, Katsanis N (2009) The vertebrate primary cilium in development, homeostasis, and disease. Cell 137.32–45
49. Hildebrandt F, Attanasio M, Otto E (2009) Nephronophthisis:disease mechanisms of a ciliopathy. J Am Soc Nephrol 20:23–35
50. Satir P, Pedersen LB, Christensen ST (2010) The primary cilium at a glance. J Cell Sci 123:499–503
51. Yoder BK, Richards WG, Sweeney WE, Wilkinson JE, Avener ED, Woychik RP (1995) Insertional mutagenesis and molecular analysis of a new gene associated with polycystic kidney disease. Proc Assoc Am Physicians 107:314–323
52. Pazour GJ, Dickert BL, Vucica Y, Seeley ES, Rosenbaum JL, Witman GB, Cole DG (2000) Chlamydomonas IFT88 and its mouse homologue, polycystic kidney disease gene tg737, are required for assembly of cilia and flagella. J Cell Biol 151:709–718
53. Liu W, Murcia NS, Duan Y, Weinbaum S, Yoder BK, Schwiebert E, Satlin LM (2005) Mechanoregulation of intracellular Ca2+ concentration is attenuated in collecting duct of monocilium-impaired orpk mice. Am J Physiol Renal Physiol 289:F978–F988

54. Wheatley DN (1995) Primary cilia in normal and pathological tissues. Pathobiology 63: 222–238
55. Praetorius HA, Spring KR (2001) Bending the MDCK cell primary cilium increases intracellular calcium. J Membr Biol 184:71–79
56. Praetorius HA, Spring KR (2003) Removal of the MDCK cell primary cilium abolishes flow sensing. J Membr Biol 191:69–76
57. Inglis PN, Ou G, Leroux MR, Scholey JM (2007) The sensory cilia of Caenorhabditis elegans. WormBook 1–22
58. Jenkins PM, McEwen DP, Martens JR (2009) Olfactory cilia: linking sensory cilia function and human disease. Chem Senses 34:451–464
59. Praetorius HA, Spring KR (2005) A physiological view of the primary cilium. Annu Rev Physiol 67:515–529
60. Murcia NS, Richards WG, Yoder BK, Mucenski ML, Dunlap JR, Woychik RP (2000) The Oak Ridge Polycystic Kidney (orpk) disease gene is required for left-right axis determination. Development 127:2347–2355
61. Pazour GJ, San Agustin JT, Follit JA, Rosenbaum JL, Witman GB (2002) Polycystin-2 localizes to kidney cilia and the ciliary level is elevated in orpk mice with polycystic kidney disease. Curr Biol 12:R378–R380
62. Yoder BK, Hou X, Guay-Woodford LM (2002) The polycystic kidney disease proteins, polycystin-1, polycystin-2, polaris, and cystin, are co-localized in renal cilia. J Am Soc Nephrol 13:2508–2516
63. Zhou J (2009) Polycystins and primary cilia: primers for cell cycle progression. Annu Rev Physiol 71:83–113
64. Nauli SM, Alenghat FJ, Luo Y, Williams E, Vassilev P, Li X, Elia AEH, Lu W, Brown EM, Quinn SJ, Ingber DE, Zhou J (2003) Polycystins 1 and 2 mediate mechanosensation in the primary cilium of kidney cells. Nat Genet 33:129–137
65. Köttgen M, Buchholz B, Garcia-Gonzalez MA, Kotsis F, Fu X, Doerken M, Boehlke C, Steffl D, Tauber R, Wegierski T, Nitschke R, Suzuki M, Kramer-Zucker A, Germino GG, Watnick T, Prenen J, Nilius B, Kuehn EW, Walz G (2008) TRPP2 and TRPV4 form a polymodal sensory channel complex. J Cell Biol 182:437–447
66. Hogan MC, Manganelli L, Woollard JR, Masyuk AI, Masyuk TV, Tammachote R, Huang BQ, Leontovich AA, Beito TG, Madden BJ, Charlesworth MC, Torres VE, LaRusso NF, Harris PC, Ward CJ (2009) Characterization of PKD protein-positive exosome-like vesicles. J Am Soc Nephrol 20:278–288
67. Goetz SC, Anderson KV (2010) The primary cilium: a signalling centre during vertebrate development. Nat Rev Genet 11:331–344
68. Yu Y, Ulbrich MH, Li M, Buraei Z, Chen X, Ong ACM, Tong L, Isacoff EY, Yang J (2009) Structural and molecular basis of the assembly of the TRPP2/PKD1 complex. Proc Natl Acad Sci USA 106:11558–11563
69. Giamarchi A, Feng S, Rodat-Despoix L, Xu Y, Bubenshchikova E, Newby LJ, Hao J, Gaudioso C, Crest M, Lupas AN, Honoré E, Williamson MP, Obara T, Ong ACM, Delmas P (2010) A polycystin-2 (TRPP2) dimerization domain essential for the function of heteromeric polycystin complexes. EMBO J 29:1176–1191
70. Petri ET, Celic A, Kennedy SD, Ehrlich BE, Boggon TJ, Hodsdon ME (2010) Structure of the EF-hand domain of polycystin-2 suggests a mechanism for Ca^{2+}-dependent regulation of polycystin-2 channel activity. Proc Natl Acad Sci USA 107:9176–9181
71. Cantiello HF (2004) Regulation of calcium signaling by polycystin-2. Am J Physiol Renal Physiol 286:F1012–F1029
72. Delmas P, Padilla F, Osorio N, Coste B, Raoux M, Crest M (2004) Polycystins, calcium signaling, and human diseases. Biochem Biophys Res Commun 322:1374–1383
73. Wu G, Hayashi T, Park JH, Dixit M, Reynolds DM, Li L, Maeda Y, Cai Y, Coca-Prados M, Somlo S (1998) Identification of PKD2L, a human PKD2-related gene: tissue-specific expression and mapping to chromosome 10q25. Genomics 54:564–568

74. Nomura H, Turco AE, Pei Y, Kalaydjieva L, Schiavello T, Weremowicz S, Ji W, Morton CC, Meisler M, Reeders ST, Zhou J (1998) Identification of PKDL, a novel polycystic kidney disease 2-like gene whose murine homologue is deleted in mice with kidney and retinal defects. J Biol Chem 273:25967–25973
75. Veldhuisen B, Spruit L, Dauwerse HG, Breuning MH, Peters DJ (1999) Genes homologous to the autosomal dominant polycystic kidney disease genes (PKD1 and PKD2). Eur J Hum Genet 7:860–872
76. Guo L, Schreiber TH, Weremowicz S, Morton CC, Lee C, Zhou J (2000) Identification and characterization of a novel polycystin family member, polycystin-L2, in mouse and human: sequence, expression, alternative splicing, and chromosomal localization. Genomics 64:241–251
77. Palmer CP, Aydar E, Djamgoz MBA (2005) A microbial TRP-like polycystic-kidney-disease-related ion channel gene. Biochem J 387:211–219
78. Delmas P, Nauli SM, Li X, Coste B, Osorio N, Crest M, Brown DA, Zhou J (2004) Gating of the polycystin ion channel signaling complex in neurons and kidney cells. FASEB J 18:740–742
79. Hanaoka K, Qian F, Boletta A, Bhunia AK, Piontek K, Tsiokas L, Sukhatme VP, Guggino WB, Germino GG (2000) Co-assembly of polycystin-1 and -2 produces unique cation-permeable currents. Nature 408:990–994
80. Köttgen M (2007) TRPP2 and autosomal dominant polycystic kidney disease. Biochim Biophys Acta 1772:836–850
81. Hughes J, Ward CJ, Peral B, Aspinwall R, Clark K, San Millán JL, Gamble V, Harris PC (1995) The polycystic kidney disease 1 (PKD1) gene encodes a novel protein with multiple cell recognition domains. Nat Genet 10:151–160
82. Nims N, Vassmer D, Maser RL (2003) Transmembrane domain analysis of polycystin-1, the product of the polycystic kidney disease-1 (PKD1) gene: evidence for 11 membrane-spanning domains. Biochemistry 42:13035–13048
83. Li A, Tian X, Sung S, Somlo S (2003) Identification of two novel polycystic kidney disease-1-like genes in human and mouse genomes. Genomics 81:596–608
84. Moy GW, Mendoza LM, Schulz JR, Swanson WJ, Glabe CG, Vacquier VD (1996) The sea urchin sperm receptor for egg jelly is a modular protein with extensive homology to the human polycystic kidney disease protein, PKD1. J Cell Biol 133:809–817
85. Qian F, Boletta A, Bhunia AK, Xu H, Liu L, Ahrabi AK, Watnick TJ, Zhou F, Germino GG (2002) Cleavage of polycystin-1 requires the receptor for egg jelly domain and is disrupted by human autosomal-dominant polycystic kidney disease 1-associated mutations. Proc Natl Acad Sci USA 99:16981–16986
86. Yuasa T, Venugopal B, Weremowicz S, Morton CC, Guo L, Zhou J (2002) The sequence, expression, and chromosomal localization of a novel polycystic kidney disease 1-like gene, PKD1L1, in human. Genomics 79:376–386
87. Sutton KA, Jungnickel MK, Ward CJ, Harris PC, Florman HM (2006) Functional characterization of PKDREJ, a male germ cell-restricted polycystin. J Cell Physiol 209:493–500
88. Yuasa T, Takakura A, Denker BM, Venugopal B, Zhou J (2004) Polycystin-1L2 is a novel G-protein-binding protein. Genomics 84:126–138
89. Barr MM, Sternberg PW (1999) A polycystic kidney-disease gene homologue required for male mating behaviour in C. elegans. Nature 401:386–389
90. Mengerink KJ, Moy GW, Vacquier VD (2002) suREJ3, a polycystin-1 protein, is cleaved at the GPS domain and localizes to the acrosomal region of sea urchin sperm. J Biol Chem 277:943–948
91. Neill AT, Moy GW, Vacquier VD (2004) Polycystin-2 associates with the polycystin-1 homolog, suREJ3, and localizes to the acrosomal region of sea urchin spermatozoa. Mol Reprod Dev 67:472–477
92. Delmas P (2004) Polycystins: from mechanosensation to gene regulation. Cell 118:145–148

93. Delmas P (2005) Polycystins: polymodal receptor/ion-channel cellular sensors. Pflugers Arch 451:264–276
94. Anyatonwu GI, Ehrlich BE (2004) Calcium signaling and polycystin-2. Biochem Biophys Res Commun 322:1364–1373
95. Somlo S (1999) The PKD2 gene: structure, interactions, mutations, and inactivation. Adv Nephrol Necker Hosp 29:257–275
96. Somlo S, Ehrlich B (2001) Human disease: calcium signaling in polycystic kidney disease. Curr Biol 11:R356–R360
97. Giamarchi A, Padilla F, Coste B, Raoux M, Crest M, Honoré E, Delmas P (2006) The versatile nature of the calcium-permeable cation channel TRPP2. EMBO Rep 7:787–793
98. Nilius B, Voets T, Peters J (2005) TRP channels in disease. Sci STKE 2005:re8
99. Witzgall R (2005) Polycystin-2 – an intracellular or plasma membrane channel? Naunyn Schmiedebergs Arch Pharmacol 371:342–347
100. Köttgen M, Walz G (2005) Subcellular localization and trafficking of polycystins. Pflugers Arch 451:286–293
101. Woudenberg-Vrenken TE, Bindels RJM, Hoenderop JGJ (2009) The role of transient receptor potential channels in kidney disease. Nat Rev Nephrol 5:441–449
102. Tsiokas L (2009) Function and regulation of TRPP2 at the plasma membrane. Am J Physiol Renal Physiol 297:F1–F9
103. Torres VE, Harris PC, Pirson Y (2007) Autosomal dominant polycystic kidney disease. Lancet 369:1287–1301
104. Tran U, Zakin L, Schweickert A, Agrawal R, Döger R, Blum M, De Robertis EM, Wessely O (2010) The RNA-binding protein bicaudal C regulates polycystin 2 in the kidney by antagonizing miR-17 activity. Development 137:1107–1116
105. Cai Y, Maeda Y, Cedzich A, Torres VE, Wu G, Hayashi T, Mochizuki T, Park JH, Witzgall R, Somlo S (1999) Identification and characterization of polycystin-2, the PKD2 gene product. J Biol Chem 274:28557–28565
106. Koulen P, Cai Y, Geng L, Maeda Y, Nishimura S, Witzgall R, Ehrlich BE, Somlo S (2002) Polycystin-2 is an intracellular calcium release channel. Nat Cell Biol 4:191–197
107. Foggensteiner L, Bevan AP, Thomas R, Coleman N, Boulter C, Bradley J, Ibraghimov-Beskrovnaya O, Klinger K, Sandford R (2000) Cellular and subcellular distribution of polycystin-2, the protein product of the PKD2 gene. J Am Soc Nephrol 11:814–827
108. Luo Y, Vassilev PM, Li X, Kawanabe Y, Zhou J (2003) Native polycystin 2 functions as a plasma membrane Ca2+-permeable cation channel in renal epithelia. Mol Cell Biol 23:2600–2607
109. Scheffers MS, Le H, van der Bent P, Leonhard W, Prins F, Spruit L, Breuning MH, de Heer E, Peters DJM (2002) Distinct subcellular expression of endogenous polycystin-2 in the plasma membrane and Golgi apparatus of MDCK cells. Hum Mol Genet 11:59–67
110. Jurczyk A, Gromley A, Redick S, San Agustin J, Witman G, Pazour GJ, Peters DJM, Doxsey S (2004) Pericentrin forms a complex with intraflagellar transport proteins and polycystin-2 and is required for primary cilia assembly. J Cell Biol 166:637–643
111. Rundle DR, Gorbsky G, Tsiokas L (2004) PKD2 interacts and co-localizes with mDia1 to mitotic spindles of dividing cells: role of mDia1 IN PKD2 localization to mitotic spindles. J Biol Chem 279:29728–29739
112. Boletta A, Qian F, Onuchic LF, Bragonzi A, Cortese M, Deen PM, Courtoy PJ, Soria MR, Devuyst O, Monaco L, Germino GG (2001) Biochemical characterization of bona fide polycystin-1 in vitro and in vivo. Am J Kidney Dis 38:1421–1429
113. Sandford R, Sgotto B, Aparicio S, Brenner S, Vaudin M, Wilson RK, Chissoe S, Pepin K, Bateman A, Chothia C, Hughes J, Harris P (1997) Comparative analysis of the polycystic kidney disease 1 (PKD1) gene reveals an integral membrane glycoprotein with multiple evolutionary conserved domains. Hum Mol Genet 6:1483–1489
114. Parnell SC, Magenheimer BS, Maser RL, Rankin CA, Smine A, Okamoto T, Calvet JP (1998) The polycystic kidney disease-1 protein, polycystin-1, binds and activates heterotrimeric G-proteins in vitro. Biochem Biophys Res Commun 251:625–631

115. Geng L, Segal Y, Peissel B, Deng N, Pei Y, Carone F, Rennke HG, Glücksmann-Kuis AM, Schneider MC, Ericsson M, Reeders ST, Zhou J (1996) Identification and localization of polycystin, the PKD1 gene product. J Clin Invest 98:2674–2682
116. Geng L, Segal Y, Pavlova A, Barros EJ, Löhning C, Lu W, Nigam SK, Frischauf AM, Reeders ST, Zhou J (1997) Distribution and developmentally regulated expression of murine polycystin. Am J Physiol 272:F451–F459
117. Huan Y, van Adelsberg J (1999) Polycystin-1, the PKD1 gene product, is in a complex containing E-cadherin and the catenins. J Clin Invest 104:1459–1468
118. Scheffers MS, van der Bent P, Prins F, Spruit L, Breuning MH, Litvinov SV, de Heer E, Peters DJ (2000) Polycystin-1, the product of the polycystic kidney disease 1 gene, co-localizes with desmosomes in MDCK cells. Hum Mol Genet 9:2743–2750
119. Grimm DH, Cai Y, Chauvet V, Rajendran V, Zeltner R, Geng L, Avner ED, Sweeney W, Somlo S, Caplan MJ (2003) Polycystin-1 distribution is modulated by polycystin-2 expression in mammalian cells. J Biol Chem 278:36786–36793
120. Ma R, Li W, Rundle D, Kong J, Akbarali HI, Tsiokas L (2005) PKD2 functions as an epidermal growth factor-activated plasma membrane channel. Mol Cell Biol 25:8285–8298
121. Bai C, Kim S, Li W, Streets AJ, Ong ACM, Tsiokas L (2008) Activation of TRPP2 through mDia1-dependent voltage gating. EMBO J 27:1345–1356
122. Boca M, Distefano G, Qian F, Bhunia AK, Germino GG, Boletta A (2006) Polycystin-1 induces resistance to apoptosis through the phosphatidylinositol 3-kinase/Akt signaling pathway. J Am Soc Nephrol 17:637–647
123. Qian Q, Hunter LW, Li M, Marin-Padilla M, Prakash YS, Somlo S, Harris PC, Torres VE, Sieck GC (2003) Pkd2 haploinsufficiency alters intracellular calcium regulation in vascular smooth muscle cells. Hum Mol Genet 12:1875–1880
124. Aguiari G, Trimi V, Bogo M, Mangolini A, Szabadkai G, Pinton P, Witzgall R, Harris PC, Borea PA, Rizzuto R, del Senno L (2008) Novel role for polycystin-1 in modulating cell proliferation through calcium oscillations in kidney cells. Cell Prolif 41:554–573
125. Gallagher AR, Cedzich A, Gretz N, Somlo S, Witzgall R (2000) The polycystic kidney disease protein PKD2 interacts with Hax-1, a protein associated with the actin cytoskeleton. Proc Natl Acad Sci USA 97:4017–4022
126. Li Q, Dai Y, Guo L, Liu Y, Hao C, Wu G, Basora N, Michalak M, Chen XZ (2003) Polycystin-2 associates with tropomyosin-1, an actin microfilament component. J Mol Biol 325:949–962
127. Li Q, Montalbetti N, Shen PY, Dai X, Cheeseman CI, Karpinski E, Wu G, Cantiello HF, Chen X (2005) Alpha-actinin associates with polycystin-2 and regulates its channel activity. Hum Mol Genet 14:1587–1603
128. Markoff A, Bogdanova N, Knop M, Rüffer C, Kenis H, Lux P, Reutelingsperger C, Todorov V, Dworniczak B, Horst J, Gerke V (2007) Annexin A5 interacts with polycystin-1 and interferes with the polycystin-1 stimulated recruitment of E-cadherin into adherens junctions. J Mol Biol 369:954–966
129. Chen X, Li Q, Wu Y, Liang G, Lara CJ, Cantiello HF (2008) Submembraneous microtubule cytoskeleton: interaction of TRPP2 with the cell cytoskeleton. FEBS J 275:4675–4683
130. Lehtonen S, Ora A, Olkkonen VM, Geng L, Zerial M, Somlo S, Lehtonen E (2000) In vivo interaction of the adapter protein CD2-associated protein with the type 2 polycystic kidney disease protein, polycystin-2. J Biol Chem 275:32888–32893
131. Hidaka S, Könecke V, Osten L, Witzgall R (2004) PIGEA-14, a novel coiled-coil protein affecting the intracellular distribution of polycystin-2. J Biol Chem 279:35009–35016
132. Köttgen M, Benzing T, Simmen T, Tauber R, Buchholz B, Feliciangeli S, Huber TB, Schermer B, Kramer-Zucker A, Höpker K, Simmen KC, Tschucke CC, Sandford R, Kim E, Thomas G, Walz G (2005) Trafficking of TRPP2 by PACS proteins represents a novel mechanism of ion channel regulation. EMBO J. 24:705–716
133. Tsiokas L, Arnould T, Zhu C, Kim E, Walz G, Sukhatme VP (1999) Specific association of the gene product of PKD2 with the TRPC1 channel. Proc Natl Acad Sci USA 96:3934–3939

134. Bai C, Giamarchi A, Rodat-Despoix L, Padilla F, Downs T, Tsiokas L, Delmas P (2008) Formation of a new receptor-operated channel by heteromeric assembly of TRPP2 and TRPC1 subunits. EMBO Rep 9:472–479
135. Kobori T, Smith GD, Sandford R, Edwardson JM (2009) The transient receptor potential channels TRPP2 and TRPC1 form a heterotetramer with a 2:2 stoichiometry and an alternating subunit arrangement. J Biol Chem 284:35507–35513
136. Li Y, Santoso NG, Yu S, Woodward OM, Qian F, Guggino WB (2009) Polycystin-1 interacts with inositol 1,4,5-trisphosphate receptor to modulate intracellular Ca2+ signaling with implications for polycystic kidney disease. J Biol Chem 284:36431–36441
137. Sharif-Naeini R, Folgering JHA, Bichet D, Duprat F, Lauritzen I, Arhatte M, Jodar M, Dedman A, Chatelain FC, Schulte U, Retailleau K, Loufrani L, Patel A, Sachs F, Delmas P, Peters DJM, Honoré E (2009) Polycystin-1 and -2 dosage regulates pressure sensing. Cell 139:587–596
138. Sammels E, Devogelaere B, Mekahli D, Bultynck G, Missiaen L, Parys JB, Cai Y, Somlo S, Smedt H de (2010) Polycystin-2 activation by inositol 1,4,5-trisphosphate-induced Ca^{2+} release requires its direct association with the inositol 1,4,5-trisphosphate receptor in a signaling microdomain. J Biol Chem 285:18794–18805
139. Gao H, Wang Y, Wegierski T, Skouloudaki K, Pütz M, Fu X, Engel C, Boehlke C, Peng H, Kuehn EW, Kim E, Kramer-Zucker A, Walz G (2010) PRKCSH/80 K-H, the protein mutated in polycystic liver disease, protects polycystin-2/TRPP2 against HERP-mediated degradation. Hum Mol Genet 19:16–24
140. Bhunia AK, Piontek K, Boletta A, Liu L, Qian F, Xu PN, Germino FJ, Germino GG (2002) PKD1 induces p21(waf1) and regulation of the cell cycle via direct activation of the JAK-STAT signaling pathway in a process requiring PKD2. Cell 109:157–168
141. Shillingford JM, Murcia NS, Larson CH, Low SH, Hedgepeth R, Brown N, Flask CA, Novick AC, Goldfarb DA, Kramer-Zucker A, Walz G, Piontek KB, Germino GG, Weimbs T (2006) The mTOR pathway is regulated by polycystin-1, and its inhibition reverses renal cystogenesis in polycystic kidney disease. Proc Natl Acad Sci USA 103:5466–5471
142. Dere R, Wilson PD, Sandford RN, Walker CL (2010) Carboxy terminal tail of polycystin-1 regulates localization of TSC2 to repress mTOR. PLoS One 5:e9239
143. Shillingford JM, Piontek KB, Germino GG, Weimbs T (2010) Rapamycin ameliorates PKD resulting from conditional inactivation of Pkd1. J Am Soc Nephrol 21:489–497
144. Kim E, Arnould T, Sellin LK, Benzing T, Fan MJ, Grüning W, Sokol SY, Drummond I, Walz G (1999) The polycystic kidney disease 1 gene product modulates Wnt signaling. J Biol Chem 274:4947–4953
145. Lal M, Song X, Pluznick JL, Di Giovanni V, Merrick DM, Rosenblum ND, Chauvet V, Gottardi CJ, Pei Y, Caplan MJ (2008) Polycystin-1 C-terminal tail associates with beta-catenin and inhibits canonical Wnt signaling. Hum Mol Genet 17:3105–3117
146. Puri S, Magenheimer BS, Maser RL, Ryan EM, Zien CA, Walker DD, Wallace DP, Hempson SJ, Calvet JP (2004) Polycystin-1 activates the calcineurin/NFAT (nuclear factor of activated T-cells) signaling pathway. J Biol Chem 279:55455–55464
147. Li X, Luo Y, Starremans PG, McNamara CA, Pei Y, Zhou J (2005) Polycystin-1 and polycystin-2 regulate the cell cycle through the helix-loop-helix inhibitor Id2. Nat Cell Biol 7:1202–1212
148. Park J, Schutzer WE, Lindsley JN, Bagby SP, Oyama TT, Anderson S, Weiss RH (2007) p21 is decreased in polycystic kidney disease and leads to increased epithelial cell cycle progression: roscovitine augments p21 levels. BMC Nephrol 8:12
149. Nickel C, Benzing T, Sellin L, Gerke P, Karihaloo A, Liu Z, Cantley LG, Walz G (2002) The polycystin-1 C-terminal fragment triggers branching morphogenesis and migration of tubular kidney epithelial cells. J Clin Invest 109:481–489
150. Banzi M, Aguiari G, Trimi V, Mangolini A, Pinton P, Witzgall R, Rizzuto R, del Senno L (2006) Polycystin-1 promotes PKCalpha-mediated NF-kappaB activation in kidney cells. Biochem Biophys Res Commun 350:257–262

151. Wegierski T, Steffl D, Kopp C, Tauber R, Buchholz B, Nitschke R, Kuehn EW, Walz G, Köttgen M (2009) TRPP2 channels regulate apoptosis through the Ca2+ concentration in the endoplasmic reticulum. EMBO J 28:490–499
152. Geng L, Okuhara D, Yu Z, Tian X, Cai Y, Shibazaki S, Somlo S (2006) Polycystin-2 traffics to cilia independently of polycystin-1 by using an N-terminal RVxP motif. J Cell Sci 119:1383–1395
153. Streets AJ, Moon DJ, Kane ME, Obara T, Ong ACM (2006) Identification of an N-terminal glycogen synthase kinase 3 phosphorylation site which regulates the functional localization of polycystin-2 in vivo and in vitro. Hum Mol Genet 15:1465–1473
154. Nauli SM, Rossetti S, Kolb RJ, Alenghat FJ, Consugar MB, Harris PC, Ingber DE, Loghman-Adham M, Zhou J (2006) Loss of polycystin-1 in human cyst-lining epithelia leads to ciliary dysfunction. J Am Soc Nephrol 17:1015–1025
155. Kotsis F, Nitschke R, Boehlke C, Bashkurov M, Walz G, Kuehn EW (2007) Ciliary calcium signaling is modulated by kidney injury molecule-1 (Kim1). Pflugers Arch 453: 819–829
156. Xia S, Li X, Johnson T, Seidel C, Wallace DP, Li R (2010) Polycystin-dependent fluid flow sensing targets histone deacetylase 5 to prevent the development of renal cysts. Development 137:1075–1084
157. Ishimaru Y, Inada H, Kubota M, Zhuang H, Tominaga M, Matsunami H (2006) Transient receptor potential family members PKD1L3 and PKD2L1 form a candidate sour taste receptor. Proc Natl Acad Sci USA 103:12569–12574
158. Huang AL, Chen X, Hoon MA, Chandrashekar J, Guo W, Tränkner D, Ryba NJP, Zuker CS (2006) The cells and logic for mammalian sour taste detection. Nature 442:934–938
159. Vandenberg LN, Levin M (2009) Perspectives and open problems in the early phases of left-right patterning. Semin Cell Dev Biol 20:456–463
160. Murakami M, Ohba T, Xu F, Shida S, Satoh E, Ono K, Miyoshi I, Watanabe H, Ito H, Iijima T (2005) Genomic organization and functional analysis of murine PKD2L1. J Biol Chem 280:5626–5635
161. Basora N, Nomura H, Berger UV, Stayner C, Guo L, Shen X, Zhou J (2002) Tissue and cellular localization of a novel polycystic kidney disease-like gene product, polycystin-L. J Am Soc Nephrol 13:293–301
162. Volk T, Schwoerer AP, Thiessen S, Schultz J, Ehmke H (2003) A polycystin-2-like large conductance cation channel in rat left ventricular myocytes. Cardiovasc Res 58:76–88
163. Chen XZ, Vassilev PM, Basora N, Peng JB, Nomura H, Segal Y, Brown EM, Reeders ST, Hediger MA, Zhou J (1999) Polycystin-L is a calcium-regulated cation channel permeable to calcium ions. Nature 401:383–386
164. Keller SA, Jones JM, Boyle A, Barrow LL, Killen PD, Green DG, Kapousta NV, Hitchcock PF, Swank RT, Meisler MH (1994) Kidney and retinal defects (Krd), a transgene-induced mutation with a deletion of mouse chromosome 19 that includes the Pax2 locus. Genomics 23.309–320
165. Sanyanusin P, McNoe LA, Sullivan MJ, Weaver RG, Eccles MR (1995) Mutation of PAX2 in two siblings with renal-coloboma syndrome. Hum Mol Genet 4:2183–2184
166. Lindemann B (1996) Taste reception. Physiol Rev 76:718–766
167. Roper SD (2007) Signal transduction and information processing in mammalian taste buds. Pflugers Arch 454:759–776
168. Chandrashekar J, Kuhn C, Oka Y, Yarmolinsky DA, Hummler E, Ryba NJP, Zuker CS (2010) The cells and peripheral representation of sodium taste in mice. Nature 464:297–301
169. LopezJimenez ND, Cavenagh MM, Sainz E, Cruz-Ithier MA, Battey JF, Sullivan SL (2006) Two members of the TRPP family of ion channels, Pkd1l3 and Pkd2l1, are co-expressed in a subset of taste receptor cells. J Neurochem 98:68–77
170. Inada H, Kawabata F, Ishimaru Y, Fushiki T, Matsunami H, Tominaga M (2008) Off-response property of an acid-activated cation channel complex PKD1L3-PKD2L1. EMBO Rep 9:690–697

171. Lyall V, Alam RI, Phan DQ, Ereso GL, Phan TH, Malik SA, Montrose MH, Chu S, Heck GL, Feldman GM, DeSimone JA (2001) Decrease in rat taste receptor cell intracellular pH is the proximate stimulus in sour taste transduction. Am J Physiol Cell Physiol 281:C1005–C1013
172. Richter TA, Caicedo A, Roper SD (2003) Sour taste stimuli evoke Ca2+ and pH responses in mouse taste cells. J Physiol (Lond) 547:475–483
173. Shimizu T, Janssens A, Voets T, Nilius B (2009) Regulation of the murine TRPP3 channel by voltage, pH, and changes in cell volume. Pflugers Arch 457:795–807
174. Nelson TM, Lopez Jimenez ND, Tessarollo L, Inoue M, Bachmanov AA, Sullivan SL (2010) Taste Function in Mice with a Targetted Mutation of the Pkd1l3 Gene. *Chemical Senses*
175. Dotson CD (2010) The search for mechanisms underlying the sour taste evoked by acids continues. *Chemical Senses* 35:545–547
176. Li Q, Dai X, Shen PY, Wu Y, Long W, Chen CX, Hussain Z, Wang S, Chen X (2007) Direct binding of alpha-actinin enhances TRPP3 channel activity. J Neurochem 103:2391–2400
177. Dai X, Karpinski E, Chen X (2006) Permeation and inhibition of polycystin-L channel by monovalent organic cations. Biochim Biophys Acta 1758:197–205
178. Liu Y, Li Q, Tan M, Zhang Y, Karpinski E, Zhou J, Chen X (2002) Modulation of the human polycystin-L channel by voltage and divalent cations. FEBS Lett 525:71–76
179. Li Q, Liu Y, Shen PY, Dai XQ, Wang S, Smillie LB, Sandford R, Chen XZ (2003) Troponin I binds polycystin-L and inhibits its calcium-induced channel. Biochemistry 42:7618–7625
180. Montalbetti N, Li Q, González-Perrett S, Semprine J, Chen X, Cantiello HF (2005) Effect of hydro-osmotic pressure on polycystin-2 channel function in the human syncytiotrophoblast. Pflugers Arch 451:294–303
181. Christensen AP, Corey DP (2007) TRP channels in mechanosensation: direct or indirect activation? Nat Rev Neurosci 8:510–521
182. Chen Y, Zhang Z, Lv X, Wang Y, Hu Z, Sun H, Tan R, Liu Y, Bian G, Xiao Y, Li Q, Yang Q, Ai J, Feng L, Yang Y, Wei Y, Zhou Q (2008) Expression of Pkd2l2 in testis is implicated in spermatogenesis. Biol Pharm Bull 31:1496–1500
183. Taft RA, Denegre JM, Pendola FL, Eppig JJ (2002) Identification of genes encoding mouse oocyte secretory and transmembrane proteins by a signal sequence trap. Biol Reprod 67:953–960
184. Horner VL, Wolfner MF (2008) Transitioning from egg to embryo: triggers and mechanisms of egg activation. Dev Dyn 237:527–544
185. Hughes J, Ward CJ, Aspinwall R, Butler R, Harris PC (1999) Identification of a human homologue of the sea urchin receptor for egg jelly: a polycystic kidney disease-like protein. Hum Mol Genet 8:543–549
186. Butscheid Y, Chubanov V, Steger K, Meyer D, Dietrich A, Gudermann T (2006) Polycystic kidney disease and receptor for egg jelly is a plasma membrane protein of mouse sperm head. Mol Reprod Dev 73:350–360
187. Austin CR (1951) Observations on the penetration of the sperm in the mammalian egg. Aust J Sci Res B 4:581–596
188. Chang MC (1951) Fertilizing capacity of spermatozoa deposited into the fallopian tubes. Nature 168:697–698
189. Ikawa M, Inoue N, Benham AM, Okabe M (2010) Fertilization: a sperm's journey to and interaction with the oocyte. J Clin Invest 120:984–994
190. Sutton KA, Jungnickel MK, Florman HM (2008) A polycystin-1 controls postcopulatory reproductive selection in mice. Proc Natl Acad Sci USA 105:8661–8666
191. Vogel P, Read R, Hansen GM, Freay LC, Zambrowicz BP, Sands AT (2010) Situs inversus in Dpcd/Poll–/–, Nme7–/–, and Pkd1l1–/– mice. Vet Pathol 47:120–131
192. Karcher C, Fischer A, Schweickert A, Bitzer E, Horie S, Witzgall R, Blum M (2005) Lack of a laterality phenotype in Pkd1 knock-out embryos correlates with absence of polycystin-1 in nodal cilia. Differentiation 73:425–432

193. Mackenzie FE, Romero R, Williams D, Gillingwater T, Hilton H, Dick J, Riddoch-Contreras J, Wong F, Ireson L, Powles-Glover N, Riley G, Underhill P, Hough T, Arkell R, Greensmith L, Ribchester RR, Blanco G (2009) Upregulation of PKD1L2 provokes a complex neuromuscular disease in the mouse. Hum Mol Genet 18:3553–3566
194. Blom W, de Muinck Keizer SM, Scholte HR (1981) Acetyl-CoA carboxylase deficiency: an inborn error of de novo fatty acid synthesis. N Engl J Med 305:465–466

Chapter 17
TRP Channels in Yeast

Marta Kaleta and Christopher Palmer

Abstract Microbes have made numerous contributions to the study of biology and medicine. Those contributions also include many original discovery's in the study of ion channels often thought as the province of neuroscientists or cardiophysiologists. Yeast have long been used as a model organism and TRP channel genes and their transmembrane products touted as the "vanguards of the sensory system" can be identified in the genomes of many yeasts. This article aims to review the study of these TRP channels in yeast their discovery, electrophysiological properties and physiological function.

17.1 Microbial Ion Channels

Ion channel studies from a biological perspective are usually observed from a medical/human point of view. However ion channels evolved somewhat before the appearance of multicellular organisms and even the development of neurones. In other disciplines than ion channel study microbes have been utilized to advance our understanding of basic biochemical or physiological cellular processes. This became evidently possible when the patch-clamp technique was utilized to survey microbial membranes from the bacterium *Escherichia coli* [1] and the model unicellular eukaryote *Saccharomyces cerevisiae* [2]. This allowed examination of unitary ion conductances on their membranes by adaptation of techniques used for higher eukaryotic membranes and organism specific treatment of the microbial cells to remove the cell wall and enhance chances of forming reliable seals. Original contributions of microbial systems in ion channel study include discovery of the first mechanosensitive ion channel gene [3], discovery of the first vacuolar/lysosomal ion channel gene [4], discovery of the first two pore ion channel gene [5] and elucidation of the first K^+ ion channel crystal structure [6].

C. Palmer (✉)
Faculty of Life Sciences, London Metropolitan University, London N7 8DB, UK
e-mail: chris.palmer@londonmet.ac.uk

Subsequently ionic currents have been recorded from a number of microbial cells including *Paramecium* and *Stylonychia* (ciliates) [7], *Dictyostelium* (slime mold) [8], *Chlamydomonas* (a green flagellate) [9], *Neurospora* (bread mold) [10], *Uromyces* (a parasitic bean rust fungus) [11], and *Schizosaccharomyces pombe* (fission yeast) [12]. However the roles of most ion channels in eukaryotic microbes still remain to be elucidated. Fungi and protozoa are highly diverse and occupy widely different ecological niches and there is a great deal to be learned from studies of ion channels in these microorganisms not just from an evolutionary standpoint. Given the ease and utilities available when using microbial systems it is clear that the study of microbial channels will continue to open up new and novel insights into the structure, function and physiology of ion channels which will be applicable to all areas of ion channel investigation.

17.2 *S. cerevisiae* TRP Channels

Patch-clamp surveys of the plasma membrane of yeast cell reveal a 30-pS outwardly rectifying K^+ conductance, which has been identified as *TOK1* [5], and a 40-S mechanosensitive conductance [13], the identity of which is unknown. Patch clamp studies of the yeast vacuolar membrane have reported a >100-S cation conductance in the vacuolar membrane [4]. Within the yeast genome several sequences have similarity to known channels [14]. Initial survey of the yeast genome using mammalian TRP channels as a search query revealed a single TRP channel homolog. Hydrophilicity and domain prediction indicated that the ORF YOR087W encodes a protein that contains six TM domains with long N and C termini similar to the topology predicted for some members of the TRP family [5]. The most significant homology to other TRP channels is found in the predicted sixth TM domain, which forms part of the ion conduction pathway and is intimately associated with deactivation gating in cation channels [5]. The C-terminal portion contains a DDDD motif that may be involved in Ca^{2+} regulation similar to the Ca^{2+}-binding bowl in Big K^+ Ca^{2+}-activated channels. Examination of a yeast strain with this ORF deleted revealed that the known plasma membrane currents appeared to be present and intact [5].

Organeller channels have been less studied than plasma membrane channels perhaps due to the ease of access to the plasma membrane for patch clamp studies. *S. cerevisiae* vaculoar membrane derived vesicles were incorporated into artificial planar lipid membranes and a channel was described which conducts monovalent cations with broad specificity [15]. The yeast cell wall can be removed by enzyme digestion and its plasma membrane can be removed by osmotic manipulation. The exposed vacuole can directly be examined with a patch-clamp pipette. The vacuole displays an inwardly rectifying conductance of 350 pS. This channel called yeast vacuolar channel, YVC passes cations but discriminates poorly among them, and its open probability is regulated by H^+ and Ca^{2+} [5, 15, 16]. Channel gating is voltage-dependent; open probability, P., reaches maximum (−0.7) at a transmembrane voltage of −80 mV (cytoplasmic surface negative) and declines at both more negative and more positive voltages (i.e., to 0 around +80 mV). The time-averaged current-voltage curve shows strong rectification, with negative currents (positive charges flowing from vacuolar side to cytoplasmic side) much larger than positive currents [5, 15, 16]. The open probability also depends strongly on cytoplasmic Ca^{2+} concentration but, for ordinary recording conditions, is high only at unphysiologically high (21 mM) Ca^{2+}. However, reducing agents such as dithiothreitol and 2-mercaptoethanol act so that the channels can be activated by micromolar cytoplasmic Ca^{2+} [5, 15, 16].

17.3 Function of the Yvc1 Channel

Examination of yeast strains with the YOR087W ORF deleted revealed the absence of this current in the yeast vacuole which was restored when the gene was reintroduced on an inducible plasmid [5]. Although initially *yvc1Δ* had no apparent growth defects [5], cells expressing high levels of Yvc1 are extremely sensitive to the presence of $CaCl_2$ in the medium [17]. Eukaryotic cells respond to environmental changes through a variety of signal transduction mechanisms, including activation of Ca^{2+}-dependent signaling pathways. The yeast vacuole is the functional counterpart of the mammalian endoplasmic and sarcoplasmic reticulum for Ca^{2+} storage. Two transporters play complementary roles in sequestering Ca^{2+} into the vacuole: (a) Vcx1, a low-affinity Ca^{2+}/H^+ exchanger that rapidly sequesters Ca^{2+} into the vacuole; and (b) Pmc1, a high-affinity Ca^{2+} ATPase required for maintaining low cytoplasmic Ca^{2+} [18]. Using transgenic aequorin as an indicator, Denis and Cyert found that Ca^{2+} is released from the vacuole into the cytoplasm upon osmotic upshock and that Yvc1 is required for this release [17].

17.4 Mechanosensation of the Yeast Vacuolar Channel

In response to acute osmotic upshock, both the yeast cell and its vacuole(s) shrink. At the same time, Ca^{2+} is released from the vacuole into the cytoplasm [17]. Pressures at tens of millimeters of Hg (1 mmHg = 133 Pa) activate the 400-pS Yvc1 conductance in whole-vacuole recording mode as well as in the excised cytoplasmic-side-out mode. Raising the bath osmolarity activates this channel and causes vacuolar shrinkage and deformation [19]. The most convincing scheme is that on upshock, a transient osmotic force activates Yvc1 to release Ca^{2+} from the vacuole [19]. Mechanical activation of Yvc1 occurs regardless of Ca^{2+} concentration and is apparently independent of its known Ca^{2+} activation, which is proposed to be an amplification mechanism (Ca^{2+}-induced Ca^{2+} release). Note that only some 3 milliosmolar difference across the tonoplast will be sufficient to generate an osmotic force of 50 mmHg in magnitude that activates a large number of vacuolar Yvc1 [19].

17.5 Molecular Dissection of YVC1 (Also Known as TRPY1)

Yvc1 has been dissected towards understanding its detailed structure-function relationship and the molecular basis of mechanosensation. A survey of GOF mutations generated were found to be mostly those on aromatic amino acids. A mutation at the base of TM6 near the presumed gate was found to destabilize both the closed and the open state. Another aromatic mutation at the cytoplasmic end of TM5 displays constitutively open channel kinetics [20]. Removal of a dense cluster of negative charges in the C-terminal cytoplasmic domain of Yvc1 greatly diminished the Ca^{2+} activation as well as its influence on force activation [21]. These results suggest a structure-function model in which Ca^{2+} binding to the cytoplasmic domain and stretching of the membrane-embedded domain both generate gating force, reaching the gate in parallel [21].

17.6 Yvc1 Homologs in Other Yeast

Yvc1 has direct homologs in over 30 species of fungi. *Candida albicans* is a dimorphic fungus that causes opportunistic infections in immunocompromised individuals. The fungus has two forms – yeast-like form and a filamentous form [22]. One of these homologues, TRPY3 from *C. albicans* shows 45% amino acid identity with Yvc1 from *S. cerevisiae*. This *C. albicans* protein (TRPY3) expressed in *S. cerevisiae yvc1Δ* functionally replaces Yvc1 and shows unitary conductance of 380 pS – similar to 400 pS of Yvc1 [23]. It is likely that TRPY3 has a similar function to Yvc1. Indeed, Calvers and Sanders investigated mechanism of Ca^{2+} uptake and release on vacuolar membrane of *C. albicans* and found that Ca^{2+} release is regulated by two distinct mechanisms: inositol trisphosphate dependent and voltage dependent [24]. The latter is inhibited by Gd^{3+} (inhibitor of stretch-activated ion channels).

17.7 TRP Homologs in *S. pombe*

In humans autosomal dominant polycystic kidney disease is one of the most commonly inherited disorders which is characterized by the formation of large fluid filled cysts in kidneys caused by abnormal differentiation of kidney tubular epithelial cells [25]. One of the causative genes involved in this genetic disease is a TRP channel gene called TRPP2 [26]. In the primary cilium, TRPP2 has been suggested to function as a mechanosensitive channel that detects fluid flow in the renal tubule lumen, supporting the proposed role of the primary cilium as the unifying pathogenic concept for cystic kidney disease [27]. TRP homologs in the genome of *S. pombe* were discovered by BLAST analysis using the amino acid sequence of the sixth TM domain of the Yvc1 predicted protein as a probe. Several TRP channel orthologues were identified. One SPAC1F7.03 was found to possess significant homology to a TRPP2 gene called AMO in *D. melanogaster*. With most homology being identified in the fifth and sixth transmembrane domains [28]. This TRPP2 homolog in S. pombe possess six TM domains and a large predicted extracellular loop between the first and second TM domains which is similar in TM topology to all other TRPP2 channels [29]. However this channel shows minimal homology to mammalian TRPP2 channels when analysed using blast searches although convincing similarity can be viewed by analysing short stretches of the fifth and sixth TM domains as such it has been named Pkd2 (short for polycystic kidney disease). Analysis of this *pkd2* gene in *S. pombe* indicated that it was an essential gene and is involved in cell growth and cell wall synthesis and appears to be a key signalling component in the regulation of cell shape and cell wall synthesis through an interaction with a Rho1-GTPase [28]. Pkd2 was found to be expressed maximally during the exponential growth phase with both a plasma membrane and intracellular localization [30]. Under normal conditions cell surface Pkd2 was localized at the cell tip during the G(2) phase of the cell cycle, although following cell wall damage, the cell surface-expressed protein relocalized to the whole plasma membrane [30]. Pkd2 depletion affected Golgi trafficking, resulting in a build up of vesicles at the cell poles, and strongly affected plasma membrane protein delivery. Surface-localized Pkd2 was rapidly internalized internalization being dependent on Ca^{2+} and enhanced by amphipaths and inhibited by gadolinium (both modulators of mechanosensitive ion channels) [30]. The Pkd2 protein was found to be in a complex with a yeast synaptotagmin homologue and myosin V and depletion of Pkd2 severely affected the localization of glucan synthase [30]. It has been suggested that this *pkd2* gene is a candidate for the *S. pombe* mechanosensitive ion channel located in the plasma membrane although the essential nature of this gene and difficulty to express it heterologously has hindered efforts to verify this claim [Palmer, personal communication].

17.8 TRPP2 Homologs in Other Yeast

Further TRPP2 (PKD2) homologs have been identified in other strains of yeast using the sequence of the Pkd2 protein in *S. pombe* as a search tool with BLAST. Although these homologs appear to possess limited homology to mammalian TRPP2 homologs using BLAST searches, investigation with more advanced techniques reveals weak although convincing homology to the discerning eye [Palmer, unpublished]. One such TRPP2 gene has been studied in *Neurospora crassa* named the "Calcium-related Gene spray". Mutations in this gene caused slow growth and dense dichotomous branching resembling a spray of flowers. Initial investigations suggested this gene encodes a calcium controlling protein which is located to an organellar membrane [31].

17.9 TRPP2 Homologs in *S. cerevisiae*

Using short segments of the *S. pombe* Pkd2 predicted protein sequence as a homology search probe revealed further TRP homologs in the *S. cerevisiae* genome [Palmer and Kaleta, in preparation]. These TRP homologs (YAL053W, YGL139W and YPL221W) are distantly related to mammalian TRPP2 homologs but still possess a similar predicted membrane topology [Palmer and Kaleta, in preparation]. These homologs have been classified as iron transport proteins, based upon their ability to rescue iron uptake mutants [32], however to date no direct evidence of these proteins acting in iron uptake has been presented or homology to known iron sequestering proteins. Examination of the knockouts of these TRPP2 homologs reveal a role in the generation of the calcium spike in response to osmotic shock [33] suggesting they may act as mechanosensitive channels [Palmer and Kaleta, in preparation].

References

1. Martinac B, Buechner M, Delcour AH, Adler J, Kung C (1987) Pressure-sensitive ion channel in *Escherichia coli*. Proc Natl Acad Sci USA 84:2297–2301
2. Gustin MC, Martinac B, Saimi Y, Culbertson MR, Kung C (1986) Ion channels in yeast. Science 233:1195–1197
3. Sukharev SI, Blount P, Martinac B, Blattner FR, Kung C (1994) A large-conductance mechanosensitive channel in *E. coli* encoded by mscL alone. Nature 368: 265–268
4. Palmer CP, Zhou XL, Lin J, Loukin SH, Kung C, Saimi YA (2001) TRP homolog in *Saccharomyces cerevisiae* forms an intracellular Ca(2+)-permeable channel in the yeast vacuolar membrane. Proc Natl Acad Sci USA 98:7801–7805
5. Ketchum KA, Joiner WJ, Sellers AJ, Kaczmarek LK, Goldstein SA (1995) A new family of outwardly rectifying potassium channel proteins with two pore domains in tandem. Nature 376:690–695
6. Roux B, MacKinnon R (1999) The cavity and pore helices in the KcsA K+ channel: electrostatic stabilization of monovalent cations. Science 285:100–102
7. Adoutte A, Ling KY, Forte M, Ramanathan R, Nelson D, Kung C (1981) Ionic channels of Paramecium: from genetics and electrophysiology to biochemistry. J Physiol (Paris) 77: 1145–1159
8. Müller U, Malchow D, Hartung K (1986) Single ion channels in the slime mold *Dictyostelium discoideum*. Biochim Biophys Acta 857:287–290
9. Yoshimura K (1998) Mechanosensitive channels in the cell body of *Chlamydomonas*. J Membr Biol 166:149–155
10. Levina NN, Lew RR, Hyde GJ, Heath IB (1995) The roles of Ca2+ and plasma membrane ion channels in hyphal tip growth of *Neurospora crassa*. J Cell Sci 108:3405–3417
11. Zhou XL, Stumpf MA, Hoch HC, Kung C (1991) A mechanosensitive channel in whole cells and in membrane patches of the fungus *Uromyces*. Science 253:1415–1417
12. Zhou XL, Kung C (1992) A mechanosensitive ion channel in *Schizosaccharomyces pombe*. EMBO J 11:2869–2875
13. Gustin MC, Zhou XL, Martinac B, Kung C (1988) A mechanosensitive ion channel in the yeast plasma membrane. Science 242:762–765
14. Wolfe DM, Pearce DA (2006) Channeling studies in yeast: yeast as a model for channelopathies? Neuromol Med 8:279–306
15. Wada Y, Ohsumi Y, Tanifuji M, Kasai M, Anraku Y (1987) Vacuolar ion channel of the yeast, *Saccharomyces cerevisiae*. J Biol Chem 262:17260–17263
16. Bertl A, Slayman CL (1990) Cation-selective channels in the vacuolar membrane of Saccharomyces: dependence on calcium, redox state, and voltage. Proc Natl Acad Sci USA 87:7824–7828
17. Denis V, Cyert MS (2002) Internal Ca(2+) release in yeast is triggered by hypertonic shock and mediated by a TRP channel homologue. J Cell Biol 156:29–34

18. Yamaguchi T, Aharon GS, Sottosanto JB, Blumwald E (2005) Vacuolar Na+/H+ antiporter cation selectivity is regulated by calmodulin from within the vacuole in a Ca2+- and pH-dependent manner. Proc Natl Acad Sci USA 102:16107–16112
19. Zhou XL, Batiza AF, Loukin SH, Palmer CP, Kung C, Saimi Y (2003) The transient receptor potential channel on the yeast vacuole is mechanosensitive. Proc Natl Acad Sci USA 100:7105–7110
20. Zhou X, Su Z, Anishkin A, Haynes WJ, Friske EM, Loukin SH, Kung C, Saimi Y (2007) Yeast screens show aromatic residues at the end of the sixth helix anchor transient receptor potential channel gate. Proc Natl Acad Sci USA 104:15555–15559
21. Su Z, Zhou X, Loukin SH, Saimi Y, Kung C (2009) Mechanical force and cytoplasmic Ca(2+) activate yeast TRPY1 in parallel. J Membr Biol 227:141–150
22. Sabie FT, Gadd GM (1989) Involvement of a Ca2+-calmodulin interaction in the yeast-mycelial (Y-M) transition of *Candida albicans*. Mycopathologia 108:47–54
23. Zhou XL, Loukin SH, Coria R, Kung C, Saimi Y (2005) Heterologously expressed fungal transient receptor potential channels retain mechanosensitivity in vitro and osmotic response in vivo. Eur Biophys J 34:413–422
24. Calvert CM, Sanders D (1995) Inositol trisphosphate-dependent and -independent Ca2+ mobilization pathways at the vacuolar membrane of *Candida albicans*. J Biol Chem 270:7272–7280
25. Harris PC, Torres VE (2009) Polycystic kidney disease. Annu Rev Med 60:321–337
26. Tsiokas L (2009) Function and regulation of TRPP2 at the plasma membrane. Am J Physiol Renal Physiol 297:1–9
27. Zhou J (2009) Polycystins and primary cilia: primers for cell cycle progression. Annu Rev Physiol 71:83–113
28. Palmer CP, Aydar E, Djamgoz MB (2005) A microbial TRP-like polycystic-kidney-disease-related ion channel gene. Biochem J 387:211–219
29. Watnick TJ, Jin Y, Matunis E, Kernan MJ, Montell C (2003) A flagellar polycystin-2 homolog required for male fertility in Drosophila. Curr Biol 13:2179–2184
30. Aydar E, Palmer CP (2009) Polycystic kidney disease channel and synaptotagmin homologues play roles in schizosaccharomyces pombe cell wall synthesis/repair and membrane protein trafficking. J Membr Biol 229:141–152
31. Bok JW, Sone T, Silverman-Gavrila LB, Lew RR, Bowring FJ, Catcheside DE, Griffiths AJ (2001) Structure and function analysis of the calcium-related gene spray in *Neurospora crassa*. Fungal Genet Biol 32:145–158
32. Protchenko O, Rodriguez-Suarez R, Androphy R, Bussey H, Philpott CC (2006) A screen for genes of heme uptake identifies the FLC family required for import of FAD into the endoplasmic reticulum. J Biol Chem 281:21445–21457
33. Batiza AF, Schulz T, Masson PH (1996) Yeast respond to hypotonic shock with a calcium pulse. J Biol Chem 271:23357–23362

Chapter 18
C. elegans TRP Channels

Rui Xiao and X.Z. Shawn Xu

Abstract Transient receptor potential (TRP) channels represent a superfamily of cation channels found in all eukaryotes. The *C. elegans* genome encodes seventeen TRP channels covering all of the seven TRP subfamilies. Genetic analyses in *C. elegans* have implicated TRP channels in a wide spectrum of behavioral and physiological processes, ranging from sensory transduction (e.g. chemosensation, touch sensation, proprioception and osmosensation) to fertilization, drug dependence, organelle biogenesis, apoptosis, gene expression, and neurotransmitter/hormone release. Many *C. elegans* TRP channels share similar activation and regulatory mechanisms with their vertebrate counterparts. Studies in *C. elegans* have also revealed some previously unrecognized functions and regulatory mechanisms of TRP channels. *C. elegans* represents an excellent genetic model organism for the study of function and regulation of TRP channels in vivo.

18.1 Introduction

The first transient receptor potential (TRP) channel was cloned in *Drosophila* from a mutant strain with a transient instead of sustained light-evoked photoreceptor potential [1, 2]. An expanding number of TRP channels have been identified in various organisms, including *C. elegans*. Currently, the vertebrate TRP channel superfamily comprises seven subfamilies: TRPC (TRP-Canonical), TRPV (TRP-Vanilloid), TRPM (TRP-Melastatin), TRPN (TRP-NompC), TRPA (TRP-Ankyrin), TRPP (TRP-Polycystin), and TRPML (TRP-MucoLipin) [3]. The *C. elegans* genome encodes 17 TRP members that cover all of the seven TRP subfamilies. A dendrogram displaying their relationship is shown in Fig. 18.1a. *C. elegans* TRP channels are also believed to be composed of four subunits similar to their vertebrate

---

X.Z. Shawn Xu (✉)
Department of Molecular and Integrative Physiology, Life Sciences Institute,
University of Michigan, Ann Arbor, MI 48109, USA
e-mail: shawnxu@umich.edu

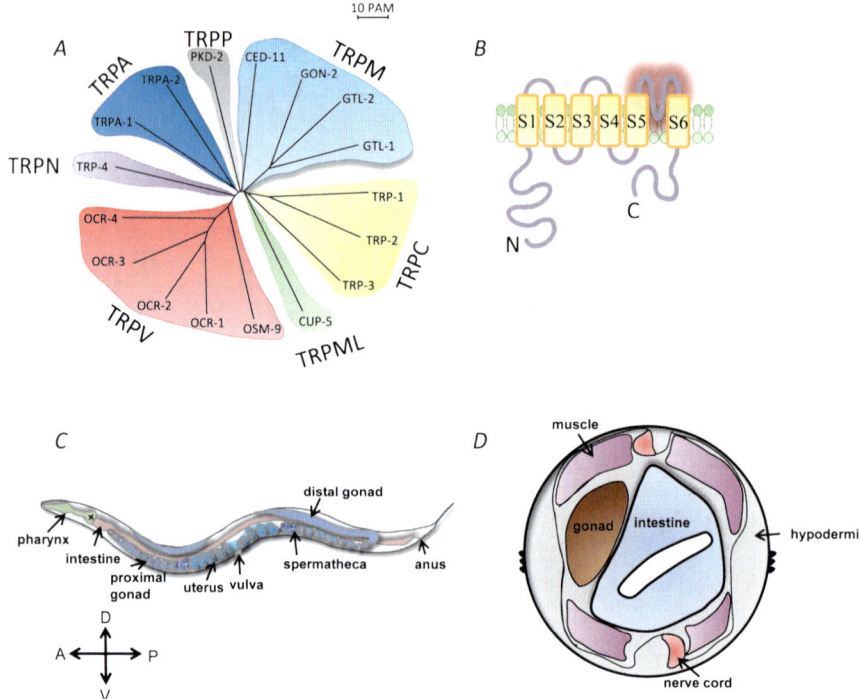

Fig. 18.1 *C. elegans* TRP channels. (**a**) A dendrogram plot of *C. elegans* TRP superfamily. The relationship between each TRP channel is indicated by the branch length in point accepted mutations (PAM) units. (**b**) Schematic representation of a TRP subunit. Extracellular Na$^+$ and Ca^{2+} flux through the putative pore between S5 and S6. (**c**) A schematic drawing of *C. elegans* (modified from www.wormatlas.com). (**d**) a schematic showing a cross section of the worm body (modified from Edwards and Wood, 1983) [4]

homologues [5, 6]. Each subunit contains six putative transmembrane segments (S1–S6) with the pore loop buried between S5 and S6 (Fig. 18.1b).

C. elegans is a powerful model organism, featuring tractable genetics and a short generation period (~3 days). The anatomy of *C. elegans* has been well characterized and reviewed by White *et al.* (Fig. 18.1c, d) [7]. The body of *C. elegans* is composed of two concentric tubes separated by a pseudocoelom (Fig. 18.1d). The outer tube comprises cuticle, hypodermis, body wall muscle, nervous system, and excretory system, and the inner tube consists of pharynx, intestine and gonad (Fig. 18.1d) [7]. Among the 959 somatic cells in adult hermaphrodite worms, 302 are neurons [8, 9]. These neurons are further classified into 118 subtypes and communicate via ~6,000 chemical synapses and ~900 gap junctions [9]. Despite its simple nervous system, *C. elegans* exhibits complex behaviors, including locomotion, mating, feeding, defecation and drug dependent behaviors, as well as learning and memory. In addition, worms are capable of detecting and reacting to a wide range of sensory cues and display a variety of sensory behaviors such as chemosensation, osmosensation, touch sensation, proprioception, and photosensation [10–12]. Importantly, many

C. elegans TRP channels are enriched in the nervous system and play important roles in behavioral control [13, 14].

Several C. elegans TRP channels were cloned by classic forward genetic screens, including *osm-9*, *gon-2* and *cup-5*. In addition, C. elegans is one of the first organisms with its whole genome sequenced [15, 16]. This has greatly facilitated the identification of other *trp* genes in C. elegans. Mutants have been isolated for every C. elegans *trp* gene. Many C. elegans TRPs exhibit high sequence homology and functional similarity to their vertebrate counterparts, and characterization of these C. elegans TRP channels has provided novel insights into the function and regulation of TRP channels in vertebrates, including humans. Here, we review the physiological functions of each worm TRP channel and also discuss their activation and regulatory mechanisms.

18.2 Functions of TRP Channels in *C. elegans*

TRPC subfamily: The C. elegans TRPC subfamily includes three members: TRP-1, TRP-2 and TRP-3. These channels were cloned based on their high sequence homology to *Drosophila* TRP [17–20]. TRP-1 and TRP-2 are expressed in motor neurons, interneurons, a few sensory neurons and muscles [19, 21]. Mutant worms lacking functional *trp-1* or *trp-2* are superficially wild type, but exhibit an interesting phenotype in nicotine-dependent behavior [21]. Similar to established rodent models, wild type worms display acute response, tolerance, withdrawal, as well as sensitization to nicotine treatment [21, 22]. These behavioral responses are nearly absent in *trp-1* and *trp-2* null worms [21]. Importantly, ectopic expression of human TRPC3 can rescue this phenotype in C. elegans, suggesting a conserved role for TRPC channels in nicotine dependence [21, 23, 24]. TRP-3, another C. elegans TRPC channel, is enriched in sperm [20, 25]. Consistent with this expression pattern, *trp-3* mutants are sterile due to a specific defect in sperm, i.e. mutant sperm cannot fuse with the egg [20]. All mammalian TRPCs are expressed in sperm, and mouse TRPC2 has been reported to be involved in triggering the acrosome reaction in sperm [26]. It is not known whether other mammalian TRPCs function in fertilization.

TRPV subfamily: The TRPV channel OSM-9 is the first characterized TRP channel in C. elegans. It was cloned from mutant strains with behavioral phenotypes in mechanosensation (nose touch response), osmosensation and chemosensation [19, 27]. OSM-9 is primarily expressed in sensory neurons such as AWA, ASH, ADL, AFD, AWC, OLQ, PHA, PHB, etc [19]. Subsequent homology analysis has identified four additional *osm-9*/capsaicin receptor related genes, *ocr-1-4* [28]. OCR channels are typically co-expressed with OSM-9, but not *vice versa* [28, 29]. In wild type worms, the chemosensory neurons AWA and AWC mediate chemotaxis to attractive odorants, while the polymodal neuron ASH is important in avoidance response to repellent odorants, high osmotic stress and nose touch [30]. AWA- and ASH-mediated sensory behaviors are severely impaired in *osm-9* and *ocr-2* mutant worms, revealing a critical role for these two TRPV channels in chemosensation, osmosensation, and mechanosensation [19, 28]. Interestingly, ectopic expression of rat TRPV4 in ASH neurons functionally rescued the avoidance response to nose touch and high osmolarity, but not repellent chemicals, suggesting an evolutionarily conserved role for TRPV channels in osmosensation and mechanosensation [31]. Several mammalian TRP channels also function as polymodal sensors in sensory physiology, especially TRPV1, TRPM8 and TRPA1 [32–35]. In this respect, *C. elegans* TRPV channels are functionally conserved as polymodal sensory detectors. In ASH and ADL neurons, OSM-9 and OCR-2 are required for sensing food shortage and population density increase [36]. In harsh environments, OSM-9 and OCR-2 are believed to directly or indirectly detect noxious chemicals in ASH and ADL neurons where they regulate social feeding behavior [36]. In the serotonergic neuron ADF, OSM-9 and OCR-2 act upstream of the serotonin synthesis gene tryptophan hydroxylase (*tph-1*), as mutations in *osm-9* or *ocr*-2 dramatically downregulate *tph-1* expression [37]. OCR channels can also function independently of OSM-9. In a subset of uterus-associated endocrine cells, OCR-1, -2 and -4 channels may form functional heteromeric channels without an involvement of OSM-9 [29]. These three TRPV channels function in uv1 endocrine cells to promote release of tyramine and other neurotransmitters to repress premature egg laying [29]. In a small set of sensory neurons, particularly ADL, OCR-2 regulates neuropeptide release [38].

TRPM subfamily: GON-2, the first TRPM channel cloned in *C. elegans*, is highly expressed in the gonad and intestine [39, 40]. GON-2 is required for normal gonadogenesis [39]. In addition, the *C. elegans* genome encodes three other TRPM channels: GTL-1, GTL-2 and CED-11. GTL-1 is co-expressed with GON-2 in the intestine, whereas GTL-2 is mainly expressed in excretory cells [40]. GON-2 and GTL-1 are essential for maintaining electrolyte homeostasis [40–42]. Consistently, mammalian TRPM6 and TRPM7 have also been reported to be important for Mg^{2+} homeostasis [43–47]. In intestine cells, GON-2 carries a Mg^{2+}-regulated outwardly rectifying current, and GTL-1 regulates the Mg^{2+} sensitivity of GON-2 [40]. GON-2 and GTL-1 may mediate the entry of some trace metals such as Ni^{2+} and Mg^{2+}, as a *gon-2;gtl-1* double mutant exhibits a significant decrease in Mg^{2+} and Ni^{2+} absorption [40]. In addition, GON-2 and GTL-1 are important for maintaining defecation rhythms in *C. elegans*. *C. elegans* defecation behavior is characterized by rhythmic cycles of posterior body wall muscle contraction (pBoc) followed by relaxation, anterior body wall muscles contraction (aBoc) and finally expulsion (Exp) [48, 49]. This defecation cycle occurs every 45 s with little variation [48]. Ca^{2+} oscillation in intestine cells is tightly coupled to defecation rhythms, and IP_3 receptor (IP_3R) is essential for this rhythmic activity [50]. Interestingly, mutant alleles of *gon-2* and *gtl-1* display abnormal defecation rhythms [40–42]. It has been proposed that these two worm TRPM channels regulate defecation rhythms through modulating IP_3Rs [41].

TRPA subfamily: Two TRPA channels (TRPA-1 and TRPA-2) are present in *C. elegans*, but only TRPA-1 has been characterized. TRPA-1 is widely expressed in multiple tissues, including many sensory neurons (e.g. ASH, AWA, AWB, ASI, ASK, PHA, PHB, PVD, PDE, IL1, and OLQ), muscle cells, excretory system, rectal gland cells, and epithelial cells [51]. In OLQ and IL1, but not ASH, TRPA-1 regulates mechanosenstive behaviors such as foraging behavior and nose-touch behavior [51]. However, it is unclear whether TRPA-1 functions as a mechanotransduction channel or indirectly modulate the signal of the transduction channel in these neurons.

TRPN subfamily: TRP-4 is the sole TRPN channel in *C. elegans*. TRP-4 is expressed in the DVA and DVC neurons, as well as the dopamine neurons CEP, ADE, and PDE [11, 52]. TRP-4 functions in mechanosensation. Specifically, it controls the body posture of the worm during locomotion through regulating the activity of the stretch-sensitive proprioceptive neuron DVA [11]. Ca^{2+} imaging shows that body stretch can elicit TRP-4-dependent calcium transients in DVA [11]. TRP-4 is also required for the function of dopamine neurons that detect mechanical attributes from the bacteria lawn on which worms feed and navigate [11, 53].

TRPP subfamily: PKD-2 is the worm homologue of mammalian PKD2/TRPP2 [54]. PKD-2 is co-localized with LOV-1/PKD1 in the male specific head neuron CEM, hook neuron HOB, and sensory ray neurons [54]. PKD-2 is required for male mating behavior. Specifically, *pkd-2* mutant male cannot efficiently locate the vulva of its mate during mating [54]. It has been proposed that PDK-2 may function as a mechanosensor and/or chemosensor in male-specific sensory neurons [54].

TRPML subfamily: CUP-5, the sole TRPML channel in *C. elegans*, was identified in genetic screens for mutant worms defective in endocytosis and apoptosis [55, 56]. It displays a high homology to human TRPML1 (mucolipin), mutations of which cause type IV mucolipidosis [57, 58]. CUP-5 is expressed in many cell types, including coelomocytes, sensory neurons, pharynx, reproductive system, and muscle [55]. Similar to mammalian TRPML1, CUP-5 is important for lysosome biogenesis and function [55, 59]. *cup-5* mutant worms accumulate large vacuoles composed of hybrids of late endosomes and lysosomes [55, 59]. The apoptosis phenotype observed in both *cup-5* mutant worms and a *Drosophila* model of type IV mucolipidosis may also result from lysosomal malfunction [56, 60].

In summary, TRP channels play diverse functions in *C. elegans*, such as sensory transduction, drug dependence, intracellular ion homeostasis, organelle biogenesis, cell fusion, apoptosis, gene expression, neurotransmitter/hormone release, etc. Given that many *C. elegans* TRP channels share similarities with their vertebrate counterparts at both the sequence and function levels, TRP channels represent an ancient, functionally conserved superfamily of cation channels.

18.3 Activation of TRP Channels in *C. elegans*

(1) *Receptor-operated activation:* Many *Drosophila* and mammalian TRP channels, particularly TRPCs, are coupled to phospholipase C (PLC) that acts downstream of GPCRs (G protein-coupled receptors) or RTKs (receptor tyrosine kinases) [3, 61]. Although PLC has a critical role in TRPC channel activation, the precise mechanism by which stimulation of PLC leads to the opening of TRPCs remains controversial. PLC hydrolyzes PIP_2 to produce IP_3 and DAG. All these three signaling molecules have been reported to activate TRP channels [62–64]. Protons, a third product of

PIP$_2$ hydrolysis, have also been shown to activate *Drosophila* TRPC channels [65]. This model proposes that protons in conjunction with the reduction of PIP$_2$ level account for the full activation of *Drospohilia* TRPC channels in vivo. However, it remains to be determined whether this model can be applied to TRPCs in other organisms.

C. elegans TRPC channels also form receptor-operated channels in vitro. When expressed in HEK293 cells, TRP-2 and TRP-3 promote receptor-operated Ca^{2+} entry that depends on the Gq-PLCβ pathway. Specifically, carbachol stimulation of endogenous muscarinic receptors induces a robust TRP-2-mediated Ca^{2+} entry following a transient release of Ca^{2+} from intracellular Ca^{2+} stores [21]. Similar results were obtained using histamine to stimulate HEK293 cells co-expressing the histamine receptor H1R [20]. Moreover, TRP-2 appears to be permeable to both Ba^{2+} and Si^{2+} in heterologous cells, a feature shared by its mammalian counterparts [21, 66].

Similar to mammalian TRPA1, *C. elegans* TRPA-1 can also be activated by stimulating Gq/11-coupled receptors. When the bradykinin 2 receptor (B2R) was co-expressed with TRPA-1 in CHO cells, bradykinin evoked a TRPA-1-dependent current that was partially blocked by Gd^{3+}, a mouse TRPA1 inhibitor [51].

In AWA and ASH neurons, the TRPV channels OSM-9 and OCR-2 are probably activated independently of PLC and Gq. Instead, the Gi-like proteins ODR-3 and GPA-3 appear to reside downstream of GPCRs and upstream of OSM-9 and OCR-2 [67]. The current model is that through an unknown mechanism, Gi-like signaling promotes the synthesis of long chain polyunsaturated fatty acids (PUFAs), which may directly or indirectly activate TRPV channels in AWA and ASH [67].

(2) *Mechanical activation*: Many mammalian TRP channels have been proposed to be mechanosensitive channels, including TRPC1, TRPC3, TRPC6, TRPV1, TPRV2, TRPV4, TRPM4, TRPM7, TRPA1, and TRPP2 [68, 69]. In addition, TRPY1 in yeast and TRPN channels in *Drosophila* and zebra fish have also been suggested to be mechanosensitive [52, 70, 71]. However, it remains enigmatic and also controversial whether these TRP channels are directly or indirectly activated by mechanical forces [68, 69].

The cellular basis of mechanosensitivity in *C. elegans* gentle touch receptor neurons has been well characterized. In adult hermaphrodites, five non-ciliated PLML/R, ALML/R and AVM neurons are responsible for sensing gentle body touch [72–74]. Three pairs of ciliated sensory neurons are responsible for detecting nose touch: ASH, FLP and OLQ [75]. The proprioceptive neuron DVA senses body stretch. Additionally, the dopamine neurons ADE, PDE and CEP detect the mechanical attributes from bacterial lawn [76]. Lastly, six IL1 cells are suggested to be mechanosensitive and involved in spontaneous foraging behaviors [77].

TRP channels are broadly involved in mechano- and osmo-sensation in *C. elegans*. In fact, OSM-9 is the first TRP family channel shown to function in mechanosensation and osmosensation [19]. Mutant worms lacking *osm-9* or *ocr-2* are insensitive to high osmotic stress and nose touch [19, 28]. However, it is unclear whether OSM-9 and OCR-2 are directly gated by forces or act downstream of an unknown transduction channel to amplify and/or modulate its signal. The fact that OSM-9 and OCR-2-dependent nose touch and osmotic responses in ASH neurons require G-protein signaling seems to support the latter possibility. *C. elegans* PKD-2 has also been implicated in mechanosensation in male mating behavior [54]. However, its role in mechanosensation could be indirect, as recent work shows that mammalian TRPP2 regulates mechanosensation not by acting as a mechanotransduction channel but rather by modulating the activity of another mechanically-gated channel of unknown molecular identity [78, 79].

TRPA-1 has also been proposed to be a mechanosensitive channel. *trpa-1* mutant worms exhibit a behavioral defect in OLQ-mediated nose touch response [51]. A Ca^{2+} imaging assay detected a reduction in the amplitude of Ca^{2+} transients in OLQ in response to the second, but not the initial, nose touch stimulus, suggesting that TRPA-1 may not be the primary mechanotransduction channel in OLQ but instead function downstream of this channel to amplify/modulate its activity. However, when expressed in CHO cells, mechanical stimulation can evoke a robust TRPA-1-dependent current, suggesting that TRPA-1 is mechanosensitive in heterologous cells [51].

TRP-4/TRPN1 has also been suggested to be a mechaosensitive channel based on its essential role for proprioception in DVA [11]. TRP-4 may also act as a mechanosensitive channel in dopamine neurons [11, 53]. However, no patch-clamp recording has been performed on TRP-4-expressing neurons. In *Drosophila* and zebrafish, direct patch-clamp recording of TRPN1/NOMPC-expressing sensory neurons has not been reported either. Therefore, it remains an open question whether TRPN channels are mechanically-gated.

(3) *Other activation modes:* Several mammalian TRP channels are sensitive to environmental and internal temperatures ranging from ~15 °C to >40 °C [33, 80]. These temperature-sensing TRP channels are called thermo-TRPs [81]. However, no TRP channel in *C. elegans* has been reported to be temperature-sensitive. In addition, many agonists from plants directly activate some mammalian TRP channels (see below), and these TRPs can thus be considered ligand-gated channels. However, most such ligands have not been tested on *C. elegans* TRPs.

Similar to TRPM6 and TRPM7, the *C. elegans* TRPM channels GON-2 and GTL-1 appear to be constitutively active [40, 41]. For these constitutively active channels, regulation of their surface expression essentially functions as a gating mechanism. Surface insertion of TRPM6 is upregulated by epidermal growth factor (EGF), a mutation of which causes TRPM6-dependent recessive hypomagnesemia [82, 83]. A similar mechanism may regulate GON-2 and GTL-1 in *C. elegans*.

Mammalian TRPA1 is directly activated by intracellular Ca^{2+} through a EF-hand Ca^{2+}-binding motif in its N-terminus [84, 85]. As this domain is conserved in *C. elegans* TRPA-1, TRPA-1 may also be directly gated by intracellular Ca^{2+}. If this is the case, TRPA-1 could function as an amplifier for other Ca^{2+}-permeable sensory channels in *C. elegans*.

18.4 Regulation of TRP Channels in *C. elegans*

As TRP channels are conserved between *C. elegans* and mammals, many of them share similar regulatory mechanisms. Additionally, genetic screens have identified some novel regulators for *C. elegans* TRP channels. Homologues of these regulators may potentially regulate mammalian TRP channels.

TRPC subfamily: Similar to *Drosophila* and mammalian TRPC channels, *C. elegans* TRPCs are coupled to the Gq-PLCβ pathway, and thus could be regulated by PIP_2, IP_3, and/or DAG. Based on sequence alignment, all three *C. elegans* TRPCs contain a calmodulin (CaM)/IP_3R-binding (CIRB) motif in their C-terminus, a feature shared by all mammalian and *Drosophila* TRPC channels [86, 87]. In addition, the CIRB motif in TRPC6 has been shown to bind several phosphoinositides (PIs) [88]. It has been proposed that CaM competes with IP_3R or/and PIs for binding to CIRB. CaM negatively regulates TRPC activity, whereas IP_3R and PIs promote channel activity [62, 88]. It is not known whether worm TRPCs are also regulated by CaM, PIs and/or IP_3R.

TRPV subfamily: A number of natural compounds isolated from plants are agonists for several mammalian TRPV channels, such as capsaicin (TRPV1), cannabinol (TRPV2), camphor (TRPV3 and TRPV1), carvacrol, thymol and eugenol (TRPV3), and bisandrographolide (TRPV4) [89–93]. However, most of these agonists have not been tested on *C. elegans* TRPV channels. One regulatory mechanism for worm TRPV channels is channel heteromerization. Genetic and cell biological studies show that distinct *C. elegans* TRPV channels may form heteromeric channels in different cell types. For example, in AWA and ASH neurons, OSM-9 and OCR-2 are co-localized and depend on each other for their ciliary localization [28]. Functionally, OSM-9 and OCR-2 both are required for chemosensation, mechanosensation and osmosensation in these neurons [28]. As many cells co-express multiple TRPV channels, the activity of these TRPV channels may be regulated by the formation of distinct heteromers. This type of regulation has also been demonstrated for many mammalian and *Drosophila* TRP channels [94–98].

A conserved acidic residue (Asp or Glu) is present in the putative pore region of all mammalian TRPV channels [99–102]. It is important for ion selectivity and also mediates channel inhibition by extracellular divalent cations [99, 102]. This acidic residue is conserved in the *C. elegans* TRPV channels OSM-9 (Asp545) and OCR-1 (Glu684). Thus, extracellular divalent cations, particularly Ca^{2+}, may directly regulate these two worm TRPVs. Given that many worm TRPV channels are highly enriched in the cilia that are directly exposed to the outside environment rich in divalent cations, these ions may represent a distinct type of sensory input directly or indirectly impinging on TRPV channels. Indeed, some heavy metals have been reported to induce avoidance behavior in worms, some of which are mediated by TRPV-expressing sensory neurons and require the TRPV channel OSM-9 [103, 104]. In addition, *C. elegans* TRPVs appear to act downstream of G proteins. Some G protein signaling regulators can directly or indirectly modulate TRPV channel activity, including GRK-2 (G protein-coupled receptor kinase-2) and

RGS-3 (regulator of G protein signaling) [105, 106]. It is not known whether these G protein singling modulators also regulate TRP channel activity in mammals.

TRPM subfamily: The worm TRPM channels GON-2 and GTL-1 are highly expressed in the intestine where they play an essential role in intracellular Ca^{2+} signaling and maintaining Mg^{2+} homeostasis [40, 41]. Whole-cell recordings of cultured embryonic intestinal cells show that GON-2 and GTL-1 are more permeable to Ca^{2+} and Mg^{2+} than to Na^+, and both divalent cations negatively regulate GON-2 and GTL-1 [41]. This inhibitory effect by divalent cations is also observed in their mammalian homologues TRPM6 and TRPM7 [107, 108]. Divalent cations bind two Glu residues (Glu1047 and Glu1052 of mouse TRPM7; Glu1030 and Asp1035 of mouse TRPM6) in the putative pore region of TRPM7 and TRPM6 [108]. Since the putative pore region, including these two residues, are very well conserved in *C. elegans* GON-2, GTL-1 and GTL-2, these worm TRPM channels may be inhibited by Ca^{2+} and Mg^{2+} in a similar manner. In a genetic screen for GON-2 suppressors, Kemp *et al*. has identified *gem-1* (*gon-2* extragenic modifier), a homolog of the SLC16 family monocarboxylate transporter, and *gem-4*, a copine family Ca^{2+}-dependent phosphatidylserine binding protein [109, 110]. It will be interesting to examine whether TRPM6 and TRPM7 are regulated by the mammalian homologues of GEM-1 and GEM-4.

TRPA subfamily: Mammalian TRPA1 is a polymodal sensor for a wide range of endogenous and environmental chemicals, noxious cold, and probably noxious mechanical forces [33, 35, 111]. Consistent with its diverse functions, TRPA1 expression and activity are under complex regulation [33, 111, 112]. However, many established agonists and blockers of mammalian TRPA1 channel are ineffective on *C. elegans* TRPA-1 [51]. This may result from the lack of the conserved cysteines in its N terminus, through which multiple structurally unrelated compounds covalently activate mammalian TRPA1 [113, 114].

TRPN subfamily: TRPN channels are found in invertebrates and lower vertebrates [14]. In *C. elegans*, TRP-4 is expressed in mechanosensory neurons, including the stretch-sensitive proprioceptive neuron DVA and dopamine neurons [11, 52]. As TRP-4 has not been well characterized at the genetic level, genetic regulators of TRP-4 remain elusive.

TRPP subfamily: The *C. elegans* TRPP channel PKD-2 is co-localized with LOV-1, the *C. elegans* homolog of mammalian PKD1 [54]. Genetic studies show that PKD-2 and LOV-1 act in the same signaling pathway [115]. Accumulating evidence supports that mammalian TRPP2 and PKD1 form a complex through physical interaction [79, 116–120]. Importantly, the coiled-coil domain in the C terminus of TRPP2, which mediates channel dimerization, is well conserved in *C. elegans* PKD-2, implying that PKD-2 may bind LOV-1 through this region [119, 120]. In addition, LOV-1 may promote the surface expression of PKD-2 by binding to the coiled-coil domain of PKD-2, a phenomenon observed in mammalian TRPP2 and PKD1 [118, 119, 121]. PKD-2 is highly expressed in the cell body and ciliary endings of male specific sensory neurons [54]. In fact, many other *C. elegans* TRP channels are also enriched in the cilia of various sensory neurons. A common mechanism may underlie the ciliary targeting of *C. elegans* TRP channels. Using PKD-2 as a model, multiple molecules regulating its ciliary localization have been identified in *C. elegans*, such as KLP-6/kinesin-3, UNC-101/AP-1, KIN-3/casein kinase II (CK2), and CIL-1 [122–124]. It will be interesting to test whether the mammalian homologues of these molecules play a similar role in ion channel trafficking.

TRPML subfamily: The *C. elegans* TRPML channel CUP-5 is required for normal endosomal and lysosomal biogenesis in worms [59]. CUP-5 activity may be regulated by PPK-3 (phosphatidylinositol phosphate kinase 3) [125]. PPK-3 phosphorylates phosphatidylinositol 3-monophosphate (PtdIns (3)P) to produce phosphatidylinositol-3,5-bisphosphate [PtdIns $(3,5)P_2$]. *ppk*-3 mutant worms display a similar phenotype to that of *cup*-5 mutants [125]. Genetic evidence reveals that *ppk*-3 and *cup*-5 function in the same pathway [125]. This suggests that PtdIns $(3,5)P_2$ may regulate and/or gate CUP-5. It will be interesting to directly test this mechanism on mammalian TRPMLs.

18.5 Perspectives

Benefiting from its simple and well characterized anatomy, facile genetics, and short generation period, *C. elegans* provides a convenient model to investigate the in vivo function and regulation of TRP channels at the whole organism level. All TRP channels in *C. elegans* have mutant strains available. Many *C. elegans* TRP channels are well correlated with their vertebrate homologues in both physiological functions and regulatory mechanisms. Insights learned from *C. elegans* TRPs could be very informative for their vertebrate counterparts. For instance, the surprising role of TRP channels in nicotine dependence was initially discovered in *C. elegans*. Interestingly, recent genome-wide association studies have implicated multiple TRPCs (TRPC4, 5 and 7) and TRPV1 in nicotine dependence and other types of drugs of abuse, suggesting that TRP channels may be broadly involved in drug dependence [21, 23, 126]. Ciliary targeting and function of TRPP2 channels were first revealed in *C. elegans* and subsequently confirmed in mammals. A functional role for TRP channels in mechanosensation and osmosensation was also first unveiled in *C. elegans* (OSM-9) and later found to be case in mammals.

Previously, electrophysiological recordings of *C. elegans* neurons were rather limited due to technical difficulties. As a result, we know very little about the biophysical properties of *C. elegans* TRP channels. Recently, conventional whole-cell recording has been successfully applied to *C. elegans* neurons to study glutamate receptors, ENaC/degenerin channels and CNG channels in vivo [12, 127–129]. Similar techniques can be utilized to characterize TRP channels in vivo. As most, if not all, worm neurons are nearly isopotential and voltage attenuation (space clamp) is not a concern [130], one can reliably record sensory channel activity by patching the soma of the sensory neuron, even though sensory channels are localized at the sensory endings distant from the soma. This permits in-depth characterization of the biophysical properties of sensory TRP channels in vivo at the whole animal level. It would be difficult to perform such experiments on mammalian DRG neurons where many sensory TRP channels are expressed. *C. elegans* thus has the potential to resolve some of the controversies surrounding the role of TRP channels in sensory transduction. For example, it is possible to address in *C. elegans* whether those "mechanosensitive" TRP channels are directly gated by mechanical forces or merely act as a downstream amplifier/modulator of another mechanically-gated transduction channel of unknown molecular identity.

Although many TRP channels in *C. elegans* have been characterized in vivo, some have not. These include *gtl-2*, *ced-11*, *trpa-2*, *ocr-3*, and *ocr-4*. Using a multidisciplinary approach combining molecular genetics, cell biology, electrophysiology, functional imaging and behavioral assays, future studies will help us gain a thorough understanding of the function and regulation of TRP channels in *C. elegans*, which may provide novel insights into their vertebrate counterparts.

Note Added in Proof

Using *C. elegans* as a model, we have recently reported TRP-4 as the first TRP family protein that acts as a pore-forming subunit of a mechanosensitive channels in vivo [131]. Thus, TRP proteins can function as mechanosensitive channels.

Acknowledgments We are grateful to members of the Xu lab, particularly Beverly Piggott, for insightful discussion and comments. Work in the lab is supported by a Pew Scholar Award and grants from the NIH (X.Z.S.X.).

References

1. Cosens DJ, Manning A (1969) Abnormal electroretinogram from a Drosophila mutant. Nature 224:285–287
2. Montell C, Rubin GM (1989) Molecular characterization of the Drosophila trp locus: a putative integral membrane protein required for phototransduction. Neuron 2: 1313–1323
3. Venkatachalam K, Montell C (2007) TRP channels. Annu Rev Biochem 76:387–417
4. Edwards MK, Wood WB (1983) Location of specific messenger RNAs in *Caenorhabditis elegans* by cytological hybridization. Dev Biol 97:375–390
5. Tsuruda PR, Julius D, Minor DL Jr (2006) Coiled coils direct assembly of a cold-activated TRP channel. Neuron 51:201–212
6. Moiseenkova-Bell VY, Stanciu LA, Serysheva II, Tobe BJ, Wensel TG (2008) Structure of TRPV1 channel revealed by electron cryomicroscopy. Proc Natl Acad Sci USA 105: 7451–7455
7. Wood WB (1988) The Nematode Caenorhabditis elegans. In: Cold Spring Harbor monograph series 17. Cold Spring Harbor Laboratory, Cold Spring Harbor, NY, p. xiii, 667 p
8. Sulston JE, Horvitz HR (1977) Post-embryonic cell lineages of the nematode, Caenorhabditis elegans. Dev Biol 56:110–156
9. White JG, Southgate E, Thomson JN, Brenner S (1986) The structure of the nervous system of the nematode C. elegans. Philos Trans R Soc Lond B Biol Sci 314:1–340
10. Hart AC Behavior. *WormBook, ed. The C. elegans Research Community, WormBook*, 2006
11. Li W, Feng Z, Sternberg PW, Xu XZ (2006) A C. elegans stretch receptor neuron revealed by a mechanosensitive TRP channel homologue. Nature 440:684–687

12. Ward A, Liu J, Feng Z, Xu XZ (2008) Light-sensitive neurons and channels mediate phototaxis in C. elegans. Nat Neurosci 11:916–922
13. Kahn-Kirby AH, Bargmann CI (2006) TRP channels in C. elegans. Annu Rev Physiol 68:719–736
14. Xiao R, Xu XZ (2009) Function and regulation of TRP family channels in C. elegans. Pflugers Arch 458:851–860
15. Sulston J, Du Z, Thomas K, Wilson R, Hillier L, Staden R, Halloran N, Green P, Thierry-Mieg J, Qiu L et al (1992) The C. elegans genome sequencing project: a beginning. Nature 356:37–41
16. Hodgkin J, Plasterk RH, Waterston RH (1995) The nematode Caenorhabditis elegans and its genome. Science 270:410–414
17. Wes PD, Chevesich J, Jeromin A, Rosenberg C, Stetten G, Montell C (1995) TRPC1, a human homolog of a Drosophila store-operated channel. Proc Natl Acad Sci USA 92:9652–9656
18. Zhu X, Chu PB, Peyton M, Birnbaumer L (1995) Molecular cloning of a widely expressed human homologue for the Drosophila trp gene. FEBS Lett 373:193–198
19. Colbert HA, Smith TL, Bargmann CI (1997) OSM-9, a novel protein with structural similarity to channels, is required for olfaction, mechanosensation, and olfactory adaptation in Caenorhabditis elegans. J Neurosci 17:8259–8269
20. Xu XZ, Sternberg PW (2003) A C. elegans sperm TRP protein required for sperm-egg interactions during fertilization. Cell 114:285–297
21. Feng Z, Li W, Ward A, Piggott BJ, Larkspur ER, Sternberg PW, Xu XZ (2006) A C. elegans model of nicotine-dependent behavior: regulation by TRP-family channels. Cell 127:621–633
22. Dwoskin LP, Crooks PA, Teng L, Green TA, Bardo MT (1999) Acute and chronic effects of nornicotine on locomotor activity in rats: altered response to nicotine. Psychopharmacology (Berl) 145:442–451
23. Caporaso N, Gu F, Chatterjee N, Sheng-Chih J, Yu K, Yeager M, Chen C, Jacobs K, Wheeler W, Landi MT, Ziegler RG, Hunter DJ, Chanock S, Hankinson S, Kraft P, Bergen AW (2009) Genome-wide and candidate gene association study of cigarette smoking behaviors. PLoS One 4:e4653
24. Drgon T, Montoya I, Johnson C, Liu QR, Walther D, Hamer D, Uhl GR (2009) Genome-wide association for nicotine dependence and smoking cessation success in NIH research volunteers. Mol Med 15:21–27
25. Reinke V, Smith HE, Nance J, Wang J, Van Doren C, Begley R, Jones SJ, Davis EB, Scherer S, Ward S, Kim SK (2000) A global profile of germline gene expression in C. elegans. Mol Cell 6:605–616
26. Jungnickel MK, Marrero H, Birnbaumer L, Lemos JR, Florman HM (2001) Trp2 regulates entry of Ca^{2+} into mouse sperm triggered by egg ZP3. Nat Cell Biol 3:499–502
27. Colbert HA, Bargmann CI (1995) Odorant-specific adaptation pathways generate olfactory plasticity in C. elegans. Neuron 14:803–812
28. Tobin D, Madsen D, Kahn-Kirby A, Peckol E, Moulder G, Barstead R, Maricq A, Bargmann C (2002) Combinatorial expression of TRPV channel proteins defines their sensory functions and subcellular localization in C. elegans neurons. Neuron 35:307–318
29. Jose AM, Bany IA, Chase DL, Koelle MR (2007) A specific subset of transient receptor potential vanilloid-type channel subunits in Caenorhabditis elegans endocrine cells function as mixed heteromers to promote neurotransmitter release. Genetics 175:93–105
30. Bargmann CI (2006) Chemosensation in C. elegans. WormBook 1–29
31. Liedtke W, Tobin DM, Bargmann CI, Friedman JM (2003) Mammalian TRPV4 (VR-OAC) directs behavioral responses to osmotic and mechanical stimuli in Caenorhabditis elegans. Proc Natl Acad Sci USA 100(Suppl 2):14531–14536
32. Clapham DE (2003) TRP channels as cellular sensors. Nature 426:517–524

33. Dhaka A, Viswanath V, Patapoutian A (2006) Trp ion channels and temperature sensation. Annu Rev Neurosci 29:135–161
34. Talavera K, Nilius B, Voets T (2008) Neuronal TRP channels: thermometers, pathfinders and life-savers. Trends Neurosci 31:287–295
35. Basbaum AI, Bautista DM, Scherrer G, Julius D (2009) Cellular and molecular mechanisms of pain. Cell 139:267–284
36. de Bono M, Tobin DM, Davis MW, Avery L, Bargmann CI (2002) Social feeding in Caenorhabditis elegans is induced by neurons that detect aversive stimuli. Nature 419: 899–903
37. Zhang S, Sokolchik I, Blanco G, Sze JY (2004) Caenorhabditis elegans TRPV ion channel regulates 5HT biosynthesis in chemosensory neurons. Development 131:1629–1638
38. Lee BH, Ashrafi K:A (2008) TRPV channel modulates C. elegans neurosecretion, larval starvation survival, and adult lifespan. PLoS Genet 4:e1000213
39. Sun AY, Lambie EJ (1997) gon-2, a gene required for gonadogenesis in Caenorhabditis elegans. Genetics 147:1077–1089
40. Teramoto T, Lambie EJ, Iwasaki K (2005) Differential regulation of TRPM channels governs electrolyte homeostasis in the C. elegans intestine. Cell Metab 1:343–354
41. Xing J, Yan X, Estevez A, Strange K (2008) Highly Ca^{2+}-selective TRPM channels regulate IP3-dependent oscillatory Ca^{2+} signaling in the C. elegans intestine. J Gen Physiol 131: 245–255
42. Kwan CS, Vazquez-Manrique RP, Ly S, Goyal K, Baylis HA (2008) TRPM channels are required for rhythmicity in the ultradian defecation rhythm of C. elegans. BMC Physiol 8:11
43. Nadler MJ, Hermosura MC, Inabe K, Perraud AL, Zhu Q, Stokes AJ, Kurosaki T, Kinet JP, Penner R, Scharenberg AM, Fleig A (2001) LTRPC7 is a Mg.ATP-regulated divalent cation channel required for cell viability. Nature 411:590–595
44. Schlingmann KP, Weber S, Peters M, Niemann Nejsum L, Vitzthum H, Klingel K, Kratz M, Haddad E, Ristoff E, Dinour D, Syrrou M, Nielsen S, Sassen M, Waldegger S, Seyberth HW, Konrad M (2002) Hypomagnesemia with secondary hypocalcemia is caused by mutations in TRPM6, a new member of the TRPM gene family. Nat Genet 31:166–170
45. Monteilh-Zoller MK, Hermosura MC, Nadler MJ, Scharenberg AM, Penner R, Fleig A (2003) TRPM7 provides an ion channel mechanism for cellular entry of trace metal ions. J Gen Physiol 121:49–60
46. Schmitz C, Perraud AL, Johnson CO, Inabe K, Smith MK, Penner R, Kurosaki T, Fleig A, Scharenberg AM (2003) Regulation of vertebrate cellular Mg^{2+} homeostasis by TRPM7. Cell 114:191–200
47. Chubanov V, Schlingmann KP, Waring J, Heinzinger J, Kaske S, Waldegger S, Mederos y Schnitzler M, Gudermann T (2007) Hypomagnesemia with secondary hypocalcemia due to a missense mutation in the putative pore-forming region of TRPM6. J Biol Chem 282: 7656–7667
48. Thomas JH (1990) Genetic analysis of defecation in Caenorhabditis elegans. Genetics 124:855–872
49. Liu DW, Thomas JH (1994) Regulation of a periodic motor program in C. elegans. J Neurosci 14:1953–1962
50. Dal Santo P, Logan MA, Chisholm AD, Jorgensen EM (1999) The inositol trisphosphate receptor regulates a 50-second behavioral rhythm in C. elegans. Cell 98:757–767
51. Kindt KS, Viswanath V, Macpherson L, Quast K, Hu H, Patapoutian A, Schafer WR (2007) Caenorhabditis elegans TRPA-1 functions in mechanosensation. Nat Neurosci 10: 568–577
52. Walker RG, Willingham AT, Zuker CS (2000) A Drosophila mechanosensory transduction channel. Science 287:2229–2234
53. Kindt KS, Quast KB, Giles AC, De S, Hendrey D, Nicastro I, Rankin CH, Schafer WR (2007) Dopamine mediates context-dependent modulation of sensory plasticity in C. elegans. Neuron 55:662–676

54. Barr MM, Sternberg PW (1999) A polycystic kidney-disease gene homologue required for male mating behaviour in C. elegans. Nature 401:386–389
55. Fares H, Greenwald I (2001) Regulation of endocytosis by CUP-5, the Caenorhabditis elegans mucolipin-1 homolog. Nat Genet 28:64–68
56. Hersh BM, Hartwieg E, Horvitz HR (2002) The Caenorhabditis elegans mucolipin-like gene cup-5 is essential for viability and regulates lysosomes in multiple cell types. Proc Natl Acad Sci USA 99:4355–4360
57. Bassi MT, Manzoni M, Monti E, Pizzo MT, Ballabio A, Borsani G (2000) Cloning of the gene encoding a novel integral membrane protein, mucolipidin-and identification of the two major founder mutations causing mucolipidosis type IV. Am J Hum Genet 67:1110–1120
58. Sun M, Goldin E, Stahl S, Falardeau JL, Kennedy JC, Acierno JS Jr, Bove C, Kaneski CR, Nagle J, Bromley MC, Colman M, Schiffmann R, Slaugenhaupt SA (2000) Mucolipidosis type IV is caused by mutations in a gene encoding a novel transient receptor potential channel. Hum Mol Genet 9:2471–2478
59. Treusch S, Knuth S, Slaugenhaupt SA, Goldin E, Grant BD, Fares H (2004) Caenorhabditis elegans functional orthologue of human protein h-mucolipin-1 is required for lysosome biogenesis. Proc Natl Acad Sci USA 101:4483–4488
60. Venkatachalam K, Long AA, Elsaesser R, Nikolaeva D, Broadie K, Montell C (2008) Motor deficit in a Drosophila model of mucolipidosis type IV due to defective clearance of apoptotic cells. Cell 135:838–851
61. Pedersen SF, Owsianik G, Nilius B (2005) TRP channels: an overview. Cell Calcium 38:233–252
62. Zhang Z, Tang J, Tikunova S, Johnson JD, Chen Z, Qin N, Dietrich A, Stefani E, Birnbaumer L, Zhu MX (2001) Activation of Trp3 by inositol 1,4,5-trisphosphate receptors through displacement of inhibitory calmodulin from a common binding domain. Proc Natl Acad Sci USA 98:3168–3173
63. Rohacs T, Lopes CM, Michailidi I, Logothetis DE (2005) PI(4,5)P2 regulates the activation and desensitization of TRPM8 channels through the TRP domain. Nat Neurosci 8:626–634
64. Hofmann T, Obukhov AG, Schaefer M, Harteneck C, Gudermann T, Schultz G (1999) Direct activation of human TRPC6 and TRPC3 channels by diacylglycerol. Nature 397:259–263
65. Huang J, Liu CH, Hughes SA, Postma M, Schwiening CJ, Hardie RC Activation of TRP channels by protons and phosphoinositide depletion in Drosophila photoreceptors. Curr Biol 20:189–197
66. Ramsey IS, Delling M, Clapham DE (2006) An introduction to TRP channels. Annu Rev Physiol 68:619–647
67. Kahn-Kirby AH, Dantzker JL, Apicella AJ, Schafer WR, Browse J, Bargmann CI, Watts JL (2004) Specific polyunsaturated fatty acids drive TRPV-dependent sensory signaling in vivo. Cell 119:889–900
68. Christensen AP, Corey DP (2007) TRP channels in mechanosensation: direct or indirect activation? Nat Rev Neurosci 8:510–521
69. Sharif-Naeini R, Dedman A, Folgering JH, Duprat F, Patel A, Nilius B, Honore E (2008) TRP channels and mechanosensory transduction: insights into the arterial myogenic response. Pflugers Arch 456:529–540
70. Zhou XL, Loukin SH, Coria R, Kung C, Saimi Y (2005) Heterologously expressed fungal transient receptor potential channels retain mechanosensitivity in vitro and osmotic response in vivo. Eur Biophys J 34:413–422
71. Sidi S, Friedrich RW, Nicolson T, Nomp C (2003) TRP channel required for vertebrate sensory hair cell mechanotransduction. Science 301:96–99
72. Goodman MB, Schwarz EM (2003) Transducing touch in Caenorhabditis elegans. Annu Rev Physiol 65:429–452
73. Bounoutas A, Chalfie M (2007) Touch sensitivity in Caenorhabditis elegans. Pflugers Arch 454:691–702
74. Chalfie M (2009) Neurosensory mechanotransduction. Nat Rev Mol Cell Biol 10:44–52

75. Kaplan JM, Horvitz HR (1993) A dual mechanosensory and chemosensory neuron in Caenorhabditis elegans. Proc Natl Acad Sci USA 90:2227–2231
76. Sawin ER, Ranganathan R, Horvitz HR (2000) C. elegans locomotory rate is modulated by the environment through a dopaminergic pathway and by experience through a serotonergic pathway. Neuron 26:619–631
77. Hart AC, Sims S, Kaplan JM (1995) Synaptic code for sensory modalities revealed by C. elegans GLR-1 glutamate receptor. Nature 378:82–85
78. Nauli SM, Alenghat FJ, Luo Y, Williams E, Vassilev P, Li X, Elia AE, Lu W, Brown EM, Quinn SJ, Ingber DE, Zhou J (2003) Polycystins 1 and 2 mediate mechanosensation in the primary cilium of kidney cells. Nat Genet 33:129–137
79. Sharif-Naeini R, Folgering JH, Bichet D, Duprat F, Lauritzen I, Arhatte M, Jodar M, Dedman A, Chatelain FC, Schulte U, Retailleau K, Loufrani L, Patel A, Sachs F, Delmas P, Peters DJ, Honore E (2009) Polycystin-1 and -2 dosage regulates pressure sensing. Cell 139:587–596
80. Jordt SE, McKemy DD, Julius D (2003) Lessons from peppers and peppermint: the molecular logic of thermosensation. Curr Opin Neurobiol 13:487–492
81. Patapoutian A, Peier AM, Story GM, Viswanath V (2003) ThermoTRP channels and beyond: mechanisms of temperature sensation. Nat Rev Neurosci 4:529–539
82. Groenestege WM, Thebault S, van der Wijst J, van den Berg D, Janssen R, Tejpar S, van den Heuvel LP, van Cutsem E, Hoenderop JG, Knoers NV, Bindels RJ (2007) Impaired basolateral sorting of pro-EGF causes isolated recessive renal hypomagnesemia. J Clin Invest 117:2260–2267
83. Thebault S, Alexander RT, Tiel Groenestege WM, Hoenderop JG, Bindels RJ (2009) EGF increases TRPM6 activity and surface expression. J Am Soc Nephrol 20:78–85
84. Zurborg S, Yurgionas B, Jira JA, Caspani O, Heppenstall PA (2007) Direct activation of the ion channel TRPA1 by Ca^{2+}. Nat Neurosci 10:277–279
85. Doerner JF, Gisselmann G, Hatt H, Wetzel CH (2007) Transient receptor potential channel A1 is directly gated by calcium ions. J Biol Chem 282:13180–13189
86. Tang J, Lin Y, Zhang Z, Tikunova S, Birnbaumer L, Zhu MX (2001) Identification of common binding sites for calmodulin and inositol 1,4,5-trisphosphate receptors on the carboxyl termini of trp channels. J Biol Chem 276:21303–21310
87. Zhu MX (2005) Multiple roles of calmodulin and other $Ca^{(2+)}$-binding proteins in the functional regulation of TRP channels. Pflugers Arch 451:105–115
88. Kwon Y, Hofmann T, Montell C (2007) Integration of phosphoinositide- and calmodulin-mediated regulation of TRPC6. Mol Cell 25:491–503
89. Caterina MJ, Schumacher MA, Tominaga M, Rosen TA, Levine JD, Julius D (1997) The capsaicin receptor: a heat-activated ion channel in the pain pathway. Nature 389:816–824
90. Moqrich A, Hwang SW, Earley TJ, Petrus MJ, Murray AN, Spencer KS, Andahazy M, Story GM, Patapoutian A (2005) Impaired thermosensation in mice lacking TRPV3, a heat and camphor sensor in the skin. Science 307:1468–1472
91. Smith PL, Maloney KN, Pothen RG, Clardy J, Clapham DE (2006) Bisandrographolide from Andrographis paniculata activates TRPV4 channels. J Biol Chem 281:29897–29904
92. Xu H, Delling M, Jun JC, Clapham DE (2006) Oregano, thyme and clove-derived flavors and skin sensitizers activate specific TRP channels. Nat Neurosci 9:628–635
93. Qin N, Neeper MP, Liu Y, Hutchinson TL, Lubin ML, Flores CM (2008) TRPV2 is activated by cannabidiol and mediates CGRP release in cultured rat dorsal root ganglion neurons. J Neurosci 28:6231–6238
94. Xu XZ, Li HS, Guggino WB, Montell C (1997) Coassembly of TRP and TRPL produces a distinct store-operated conductance. Cell 89:1155–1164
95. Strubing C, Krapivinsky G, Krapivinsky L, Clapham DE (2001) TRPC1 and TRPC5 form a novel cation channel in mammalian brain. Neuron 29:645–655
96. Hoenderop JG, Voets T, Hoefs S, Weidema F, Prenen J, Nilius B, Bindels RJ (2003) Homo- and heterotetrameric architecture of the epithelial Ca^{2+} channels TRPV5 and TRPV6. EMBO J 22:776–785

97. Chubanov V, Waldegger S, Mederos y Schnitzler M, Vitzthum H, Sassen MC, Seyberth HW, Konrad M, Gudermann T (2004) Disruption of TRPM6/TRPM7 complex formation by a mutation in the TRPM6 gene causes hypomagnesemia with secondary hypocalcemia. Proc Natl Acad Sci USA 101:2894–2899
98. Zhang P, Luo Y, Chasan B, Gonzalez-Perrett S, Montalbetti N, Timpanaro GA, Cantero Mdel R, Ramos AJ, Goldmann WH, Zhou J, Cantiello HF (2009) The multimeric structure of polycystin-2 (TRPP2): structural-functional correlates of homo- and hetero-multimers with TRPC1. Hum Mol Genet 18:1238–1251
99. Garcia-Martinez C, Morenilla-Palao C, Planells-Cases R, Merino JM, Ferrer-Montiel A (2000) Identification of an aspartic residue in the P-loop of the vanilloid receptor that modulates pore properties. J Biol Chem 275:32552–32558
100. Voets T, Prenen J, Vriens J, Watanabe H, Janssens A, Wissenbach U, Bodding M, Droogmans G, Nilius B (2002) Molecular determinants of permeation through the cation channel TRPV4. J Biol Chem 277:33704–33710
101. Chung MK, Guler AD, Caterina MJ (2005) Biphasic currents evoked by chemical or thermal activation of the heat-gated ion channel, TRPV3. J Biol Chem 280:15928–15941
102. Xiao R, Tang J, Wang C, Colton CK, Tian J, Zhu MX (2008) Calcium plays a central role in the sensitization of TRPV3 channel to repetitive stimulations. J Biol Chem 283:6162–6174
103. Sambongi Y, Nagae T, Liu Y, Yoshimizu T, Takeda K, Wada Y, Futai M (1999) Sensing of cadmium and copper ions by externally exposed ADL, ASE, and ASH neurons elicits avoidance response in Caenorhabditis elegans. NeuroReport 10:753–757
104. Hilliard MA, Apicella AJ, Kerr R, Suzuki H, Bazzicalupo P, Schafer WR (2005) In vivo imaging of C. elegans ASH neurons: cellular response and adaptation to chemical repellents. EMBO J 24:63–72
105. Fukuto HS, Ferkey DM, Apicella AJ, Lans H, Sharmeen T, Chen W, Lefkowitz RJ, Jansen G, Schafer WR, Hart AC (2004) G protein-coupled receptor kinase function is essential for chemosensation in C. elegans. Neuron 42:581–593
106. Ferkey DM, Hyde R, Haspel G, Dionne HM, Hess HA, Suzuki H, Schafer WR, Koelle MR, Hart AC (2007) C. elegans G protein regulator RGS-3 controls sensitivity to sensory stimuli. Neuron 53:39–52
107. Jiang J, Li M, Yue L (2005) Potentiation of TRPM7 inward currents by protons. J Gen Physiol 126:137–150
108. Li M, Du J, Jiang J, Ratzan W, Su LT, Runnels LW, Yue L (2007) Molecular determinants of Mg^{2+} and Ca^{2+} permeability and pH sensitivity in TRPM6 and TRPM7. J Biol Chem 282:25817–25830
109. Kemp BJ, Church DL, Hatzold J, Conradt B, Lambie EJ (2009) Gem-1 encodes an SLC16 monocarboxylate transporter-related protein that functions in parallel to the gon-2 TRPM channel during gonad development in Caenorhabditis elegans. Genetics 181:581–591
110. Church DL, Lambie EJ (2003) The promotion of gonadal cell divisions by the Caenorhabditis elegans TRPM cation channel GON-2 is antagonized by GEM-4 copine. Genetics 165:563–574
111. Patapoutian A, Tate S, Woolf CJ (2009) Transient receptor potential channels: targeting pain at the source. Nat Rev Drug Discov 8:55–68
112. Nilius B, Owsianik G, Voets T, Peters JA (2007) Transient receptor potential cation channels in disease. Physiol Rev 87:165–217
113. Macpherson LJ, Dubin AE, Evans MJ, Marr F, Schultz PG, Cravatt BF, Patapoutian A (2007) Noxious compounds activate TRPA1 ion channels through covalent modification of cysteines. Nature 445:541–545
114. Hinman A, Chuang HH, Bautista DM, Julius D (2006) TRP channel activation by reversible covalent modification. Proc Natl Acad Sci USA 103:19564–19568
115. Barr MM, DeModena J, Braun D, Nguyen CQ, Hall DH, Sternberg PW (2001) The Caenorhabditis elegans autosomal dominant polycystic kidney disease gene homologs lov-1 and pkd-2 act in the same pathway. Curr Biol 11:1341–1346

116. Qian F, Germino FJ, Cai Y, Zhang X, Somlo S, Germino GG (1997) PKD1 interacts with PKD2 through a probable coiled-coil domain. Nat Genet 16:179–183
117. Tsiokas L, Kim E, Arnould T, Sukhatme VP, Walz G (1997) Homo- and heterodimeric interactions between the gene products of PKD1 and PKD2. Proc Natl Acad Sci USA 94:6965–6970
118. Hanaoka K, Qian F, Boletta A, Bhunia AK, Piontek K, Tsiokas L, Sukhatme VP, Guggino WB, Germino GG (2000) Co-assembly of polycystin-1 and -2 produces unique cation-permeable currents. Nature 408:990–994
119. Yu Y, Ulbrich MH, Li MH, Buraei Z, Chen XZ, Ong AC, Tong L, Isacoff EY, Yang J (2009) Structural and molecular basis of the assembly of the TRPP2/PKD1 complex. Proc Natl Acad Sci USA 106:11558–11563
120. Giamarchi A, Feng S, Rodat-Despoix L, Xu Y, Bubenshchikova E, Newby LJ, Hao J, Gaudioso C, Crest M, Lupas AN, Honore E, Williamson MP, Obara T, Ong AC, Delmas P A polycystin-2 (TRPP2) dimerization domain essential for the function of heteromeric polycystin complexes. EMBO J 29:1176–1191
121. Grimm DH, Cai Y, Chauvet V, Rajendran V, Zeltner R, Geng L, Avner ED, Sweeney W, Somlo S, Caplan MJ (2003) Polycystin-1 distribution is modulated by polycystin-2 expression in mammalian cells. J Biol Chem 278:36786–36793
122. Peden EM, Barr MM (2005) The KLP-6 kinesin is required for male mating behaviors and polycystin localization in Caenorhabditis elegans. Curr Biol 15:394–404
123. Hu J, Bae YK, Knobel KM, Barr MM (2006) Casein kinase II and calcineurin modulate TRPP function and ciliary localization. Mol Biol Cell 17:2200–2211
124. Bae YK, Kim E, L"Hernault SW, Barr MM (2009) The CIL-1 PI 5-phosphatase localizes TRP Polycystins to cilia and activates sperm in C. elegans. Curr Biol 19:1599–1607
125. Nicot AS, Fares H, Payrastre B, Chisholm AD, Labouesse M, Laporte J (2006) The phosphoinositide kinase PIKfyve/Fab1p regulates terminal lysosome maturation in Caenorhabditis elegans. Mol Biol Cell 17:3062–3074
126. Bierut LJ, Madden PA, Breslau N, Johnson EO, Hatsukami D, Pomerleau OF, Swan GE, Rutter J, Bertelsen S, Fox L, Fugman D, Goate AM, Hinrichs AL, Konvicka K, Martin NG, Montgomery GW, Saccone NL, Saccone SF, Wang JC, Chase GA, Rice JP, Ballinger DG (2007) Novel genes identified in a high-density genome wide association study for nicotine dependence. Hum Mol Genet 16:24–35
127. Mellem JE, Brockie PJ, Zheng Y, Madsen DM, Maricq AV (2002) Decoding of polymodal sensory stimuli by postsynaptic glutamate receptors in C. elegans. Neuron 36:933–944
128. O'Hagan R, Chalfie M, Goodman MB (2005) The MEC-4 DEG/ENaC channel of Caenorhabditis elegans touch receptor neurons transduces mechanical signals. Nat Neurosci 8:43–50
129. Liu J, Ward A, Gao J, Dong Y, Nishio N, Inada H, Kang L, Yu Y, Ma D, Xu T, Mori I, Xie Z, Xu XZ (2010). C. elegans phototransduction requires a G protein-dependent cGMP pathway and a taste receptor homolog. Nat Neurosci 13(6):715–722
130. Goodman MB, Hall DH, Avery L, Lockery SR (1998) Active currents regulate sensitivity and dynamic range in C. elegans neurons. Neuron 20:763–772
131. Kang L, Gao J, Schafer WR, Xie Z, Xu XZS (2010) C. Elegans TRP family protein TRP-4 is a pore-forming subunit of a native mechanotransduction channel. Neuron 67:381–391

Chapter 19
Investigations of the In Vivo Requirements of Transient Receptor Potential Ion Channels Using Frog and Zebrafish Model Systems

Robert A. Cornell

Abstract Transient Receptor Potential (TRP) channels are cation channels that serve as cellular sensors on the plasma membrane, and have other less-well defined roles in intracellular compartments. The first TRP channel was identified upon molecular characterization of a fly mutant with abnormal photoreceptor function. More than 20 TRP channels have since been identified in vertebrates and invertebrate model systems, and these are divided into subfamilies based on structural similarities. The biophysical properties of TRP channels have primarily been explored in tissue culture models. The in vivo requirements for TRPs have been studied in invertebrate models like worm and flies, and also in vertebrate models, primarily mice and rats. Frog and zebrafish model systems offer certain experimental advantages relative to mammalian systems, and here a selection of papers which capitalize on these advantages to explore vertebrate TRP channel biology are reviewed. For instance, frog oocytes are useful for biochemistry and for electrophysiology, and these features were exploited in the identifcation TRPC1 as a candidate vertebrate mechanoreceptor. Also, the spinal neurons from frog embryos can be readily grown in culture. This feature was used to establish a role for TRPC1 in axon pathfinding in these neurons, and to explore how TRPC1 activity is regulated in this context. Zebrafish embryos are transparent making them well suited for in vivo imaging studies. This quality was exploited in a study in which the *trpc2* gene promoter was used to label and trace the axon pathway of a subset of olfactory sensory neurons. Another experimental advantage of zebrafish is the speed and low cost of manipulating gene expression in embryos. Using these methods, it has been shown that TRPN1 is necessary for mechanosensation in zebrafish hair cells. Frogs and fish genomes have been mined to make inferences regarding evolutionary diversification of the thermosensitive TRP channels. Finally, TRPM7 is required for early morphogenesis in mice but not in fish; the reason for this difference is unclear, but it has caused zebrafish to be favored for exploration of TRPM7's role in later

R.A. Cornell (✉)
Department of Anatomy and Cell Biology, Carver College of Medicine, University of Iowa, Iowa City, IA 52242, USA
e-mail: robert-cornell@uiowa.edu

events in embryogenesis. The special experimental attributes of frogs and zebrafish suggest that these animals will continue to play an important role as models in future explorations of TRP channel biology.

19.1 Introduction

TRP channels are cation channels that mediate receptor-operated, and possibly also store-operated, calcium entry. The TRP family was named for a fly mutant in which light fails to induce a constant receptor potential in photoreceptor cells; the mutated gene encodes a calcium channel involved in phototransduction [1–3]. While all TRPs share a basic topology, they are divided into several subfamilies – TRPA, TRPC, TRPM, TRPN, TRPP, and TRPV – based on shared structural features [4]. The function of most TRPs appears to be to participate in signal transduction in sensory cells responding to various environmental stimuli. TRPs are variously thermo, chemo, mechano and temperature sensitive [5]. There is also evidence that certain TRPs regulate cell growth and cell survival in a variety of non-excitable cell types, particularly in cells of the hematopoietic lineage [6, 7]. Consistent with this observation, mutations in TRP channels contribute to a variety of diseases [8]. The large majority of papers on the subject of TRP channel function employ cultured cells as a model system. In vivo explorations are essential if the physiology of TRP channels in a normal context is to be understood. Insights in this regard have come from analyses of the invertebrate models fly and worm (reviewed in 9). Among vertebrate models, mice and rats have been most frequently deployed in this context; however, zebrafish and frogs, although more distant from humans than mice and rats, represent attractive alternatives to mammalian models and recently have been used for explorations of TRP biology.

Frog and zebrafish share the feature of being less expensive to maintain than mice and also each offer unique experimental advantages. Frog oocytes are unusually large cells that are easily induced to express heterologous proteins. These qualities facilitate their use in electrophysiology. Another experimental feature of the frog system is that it is relatively easy to dissect spinal cords from embryos and to culture spinal neurons from them. The strength of zebrafish lies in its amenability to genetic and embryological manipulation. Zebrafish (*Danio rerio*) are hearty fresh-water fish from India, and are easy to raise in the laboratory setting or in home aquaria. In the early 1980s, George Streisinger at the University of Oregon adapted several genetic methods for use in zebrafish, including an effective means of mutagenesis [10–16]. In 1994 the first linkage map was generated, and in 1996 large-scale screens were published [17, 18]. Soon scientists around the world were using zebrafish, particularly for studies of the genetic requirements of embryonic development. They were attracted by the rapid external development, amenability to forward genetic screening, and embryo transparency in this species, with the last of these features facilitating cell-tracing studies. The experimental tool kit of zebrafish has long included means to analyze and manipulate gene expression. Recently, sophisticated methods for imaging and electrophysiology have been added. The accessibility of the zebrafish embryo makes it amenable to direct electrophysiological recordings from neurons [19], and its transparency makes it amenable to functional imaging with ion-sensitive dyes [20] and gene-encoded bioluminescent proteins [21]. There is great potential for these tools to be applied to exploration of the Transient Receptor Potential (TRP) family of ion channels. Here a selection of papers in which frogs or zebrafish embryos have been deployed to investigate the biology of TRP channels are reviewed.

19.2 TRPC1 Is a Mechanosensitive Cation Channel in Frog Oocytes

Frog oocytes have long been a favored cell type for the expression and physiological analyses of heterologous ion channels, including TRP channels (e.g., at least 6 such papers in 2009 alone, 22, 23–27). Recently Maroto sought to investigate an endogenous mechanosensitive channel in frog oocytes that turned out to be a TRP [28]. The same group had earlier determined that the *Xenopus* oocyte contains greater than three million mechanosensitive channel molecules [29]. They solubilized oocytes in detergent and identified a membrane fraction that conferred mechanosensitivity to reconstituted lysosomes; this fraction was enriched for TRPC1. Knockdown of *trpc1* reduced the oocyte's responsiveness to hypotonic swelling, and overexpression of human *TRPC1* substantially elevated this responsiveness [28]. The authors concluded that TRPC1 is a component of a vertebrate mechanosensitive ion channel, although it has been noted elsewhere that how osmotic stimuli activate channels in a cell membrane may differ substantially from how other types of mechanical force do so [30]. An important implication of this study is that the presence of at least one endogenous TRPC in *Xenopus* oocytes must be taken into account by all who would use this convenient system to assess the biophysical properties of an experimentally-overexpressed (heterologous) ion channel.

19.3 TRPC1 Is Involved in Axon Path Finding During *Xenopus* Embryogenesis

In 2005 two groups exploited the relative ease of culturing spinal neurons from *Xenopus* embryos to study the role of TRPC1 plays in these cells [31, 32]. Wang and Poo conducted patch-clamp recordings on growth cones of cultured spinal neurons and found that netrin-1 or BDNF application induced an elevation of intracellular Ca^{2+} via a TRP-like current [32]. Pharmacological inhibition of L-type voltage-dependent calcium channels only partially inhibited this current, while pharmacological inhibition of TRPC1 and antisense-mediated inhibition of *trpc1* expression fully blocked it. The growth-cone turning behavior was also inhibited by these treatments, suggesting that TRPC1 has an essential role in mediating axon path finding in frogs. Shim and colleagues found in vivo evidence to support this conclusion, showing that spinal commissural neurons fail to cross the midline in frog embryos depleted of TRPC1 expression [31].

In a 2009 paper in Neuron, Worley and colleagues explored how the event of netrin-1 binding to its receptor is biochemically coupled to the activation of TRPC1. First, they extended their earlier work to show that a peptidyl-prolyl isomerase (PPIase) called FK506-binding protein 52 (FKBP52) catalyzes *cis/trans* isomerization of regions of TRPC1 that are implicated in controlling channel opening [33]. Carrying out co-immunoprecipitation studies, they confirmed an earlier report that FKBP52 interacts with TRPC1. Amino acid substitutions in TRPC1 that blocked binding of FKBP52 usually also blocked binding of the TRPC1-regulator Homer, suggesting that FKBP52 and Homer bind the same part of TRPC1 [34]. Next, using nuclear magnetic resonance (NMR) exchange spectroscopy, the authors showed that FKBP52 and the related FKBP12 could efficiently catalyze cis-trans isomerization around Leu-Pro peptide bonds, in both amino and carboxy termini of TRPC1. In a heterologous system, they showed that ligand-induced (but not spontaneous) TRPC1-type current was significantly upregulated by co-transfection of FKBP52. Importantly this effect was dependent of the PPIase activity of FKBP52. Finally these investigators showed that inhibiting PPIase activity, by either pharmacological means or the expression of PPIase-dead FKBP52 (i.e., FKBP52-FD67DV), abolished the netrin-1-induced elevation of $[Ca^{2+}]_I$, and the turning of growth cones towards a netrin-1 source, in cultured *Xenopus* spinal neurons [33]. Importantly, overexpression of FKBP52-FD67DV, but not that of FKBP52-WT, similarly led to midline targeting errors of spinal commissural neurons in vivo [33]. The authors concluded that FKBP52 regulates TRPC1 channel activity, in a manner dependent on the PPIase activity of FKBP52, and that this regulation is necessary for normal axon guidance in the developing *Xenopus* spinal cord. Is such a mechanism found in mammals? The authors concluded their study by showing that inhibiting FKBP52 in the context of a gradient of MAG (a component of myelin) prevents TRPC1-dependent repulsion of the rat hippocampal neurons in vitro [33]. In conclusion, a study employing the tractable frog model system dissected the mechanics of TRPC1 regulation during axon path finding, and there is evidence the results are relevant to mammalian biology.

19.4 Expression of TRP Channel as a Marker of Cell-Type

Cells that express a given gene can be labeled in living zebrafish through the creation of transgenic reporter lines, using that gene's regulatory elements. A recent study made use of this feature to track a subset of olfactory sensory neurons (OSNs) that express *trpc2*. Olfactory epithelia contain two classes of OSNs, so-called ciliated OSNs and microvillous OSNs, that can be distinguished based on their characteristic gene expression. Thus, ciliated OSNs express OR-type odorant receptors, a cyclic nucleotide-gated channel A2 subunit, and an olfactory marker protein (OMP), while microvillous OSNs express V2R-type receptors and transient receptor potential channel C2 (TRPC2) [35]. Gene expression patterns indicate that these two classes of OSNs reside in different regions of the zebrafish olfactory epithelium. Interestingly, retrograde labeling studies indicated that the two classes of receptor send processes to different parts of the olfactory bulb [36, 37], but these classical approaches were not of sufficient resolution to identify the specific olfactory bulb glomeruli that are targeted by specific OSNs. Sato and colleagues constructed zebrafish transgenic lines in which regulatory elements of the *omp* gene drove expression of a fluorescent protein of one color, and regulatory elements of *trpc2* drove expression of a fluorescent protein of a different color [35]. This genetic strategy, and the transparency of the zebrafish embryo, made it possible to visualize dynamic changes in axonal projections from transgene-expressing OSNs in whole-mount preparations, and to identify the target glomeruli on histological sections at high resolution. The authors point out this method could be used to screen for and analyze mutant zebrafish in which the projection of OSN axons is impaired.

19.5 Zebrafish Functions of TRPN1

TRPN1, also known as No Mechano Potential Channel C (NOMPC), is the sole member of the TRPN subfamily. TRPN1 has been implicated in mechanotransduction in zebrafish [38]. An unusual structural feature of TRPN1 is a large number of ankyrin repeats in the amino terminus. The zebrafish ortholog was identified by Sidi and colleagues in an effort to identify proteins necessary for mechanotransduction in zebrafish hair cells; Nompc was a good candidate based on the fact that in *Drosophila*, *nompc* is expressed in touch-sensitive bristle neurons, and *nompc* homozygous mutants have reduced touch-evoked currents [39]. These authors detected a *nompc* (renamed *trpn1*) orthologue in the zebrafish, but not the mouse or human, genome (although subsequently another group has reported expression of putative *Trpn1* orthologue in mouse dorsal root ganglion neurons, 40). In situ hybridization revealed *trpn1* expression in otic hair cells in all five sensory patches of the embryo by 48 h post fertilization (hpf), which is prior to the time when sound will evoke any response in a zebrafish embryo. Interestingly, the *trpn1* transcript was not detected in the hair cells of lateral-line neuromasts. These authors had previously shown that, in several mutants that appear to be deaf, hair-cell uptake of the endocytosis marker FM1-43 does not occur [41]. Similarly, embryos injected with morpholinos targeting *trpn1* expression were deaf, and their hair cells did not accumulate FM1-43. Moreover, stimulus-evoked microphonic potentials, which are robust in control embryos, could not be detected in deaf *trpn1* knockdown embryos [38]. The authors interpreted their findings to mean that *trpn1* encodes an essential component of vertebrate hair-cell mechanotransduction, although it remains uncertain whether TRPN is itself a component of the mechanoreceptor. These findings highlight conservation of the machinery that governs the function of *Drosophila* sensory bristles and vertebrate hair cells; they also raise the question of whether TRPV1 family members are involved in the auditory response in zebrafish, as they are in flies [42].

Immunolocalization studies in *Xenopus* suggest that TRPN1 may not, in fact, be an integral component of the mechanotransducing protein itself. The apical side of hair cells in both fish and mammals features numerous stereocilia and a single, much longer kinocilium. The minimum stimulus necessary to induce an increase or decrease in the firing rate of the sensory neuron innervating the hair cell at its basal surface is deflection of the stereocilium toward or away from the kinocilium. However, a specific antibody to *Xenopus* TRPN1 was recently generated, and in *Xenopus*, anti-TRPN1 immunoreactivity was detected in the non-signal transducing kinocilium rather than in the stereocilia [43]. The authors suggested that although TRPN1 participates in hair-cell function, it may not be an integral part of the mechanosensitive molecule. The molecular composition of this elusive structure remains unresolved.

19.6 In Zebrafish as in Mouse, TRPA1 Mediates the Response to Noxious Chemicals but Not to Loud Sound or Changes in Temperature

In mice, TRPA1 is activated by cold temperatures, and also by noxious chemicals such as mustard oil [44–46]. Zebrafish possess two orthologues of mammalian *TRPA1*, i.e., *trpa1a* and *trpa1b*. Prober and colleagues found *trpa1a* to be expressed in the epibranchial sensory ganglion, whose peripheral fibers innervate viscera, and *trpa1b* to be expressed in all sensory ganglia of the cranium as well as in Rohon Beard primary sensory neurons of the trunk [47]. After transfection into HEK293 cells, both of the zebrafish TRPA1 paralogues were found to trigger calcium elevation in response to noxious chemicals such as mustard oil and acreolin, the irritants in mustard gas and vehicle exhaust, respectively. While gene targeting by homologous recombination is not yet possible in zebrafish, the authors were able to isolate null mutants of both *trpa1a* and *trpa1b,* using an undirected approach called TILLING [48]. They next used automated behavior analysis software to measure net motility in 5 day-old embryos that were untreated, treated with noxious chemicals, or bathed in noxious warm or noxious cold water. They found that all of these stimuli elicited a period of elevated movement in both wild-type and *trpa1a* homozygous mutants. By contrast, *trpa1b* homozygous mutants did not respond to noxious chemicals with elevated behavior, although they did respond like wild-type fish to changes in temperature. Noise also triggered an increase in locomotion in wild-type fish, and an equivalent response in both the *trpa1a* and *trpa1b* mutants. Furthermore, FM1-43 uptake in hair cells was the same in wild-type and mutant animals. Together these findings imply that, as in mammals, a TRPA1 orthologue that is expressed in the sensory neurons that innervate external surfaces is necessary for an appropriate response to noxious chemicals [47].

However, it is not the case that all zebrafish orthologs of mammalian chemosensitive TRP channels are chemosensitive. Cannabinoids are a class of phenolic compounds related to tetrahydrocannabinol (THC), the psychoactive compound in marijuana. Several endogenous cannabinoids are produced within the nervous system. The mammalian TRPV1 receptor is gated by multiple stimuli including heat, protons, cannabinoids, and vanilloids (another class of phenolic compounds that includes the chili pepper-derived irritant, capsaicin) [49]. Physiological studies of TRPV1 orthologues in chickens and worms showed that these receptors are gated by heat and protons but not by cannabinoids [50, 51]; ultimately these orthologues were found to lack the amino acid residues necessary for binding by the cannabinoids [52]. These findings were consistent with two mutually exclusive evolutionary scenarios: either the responsiveness of TRPV1 to cannabinoids was gained in mammals after birds diverged from mammals, or this feature was gained earlier but selectively lost in the bird lineage. Fish are basal to both mammals and birds, so the observation that TRPV1 orthologues in the fish species *D. rerio* and *T. rubripes* lack residues that are critical for ligand binding supports the first of the evolutionary scenarios just mentioned [53]. The authors additionally conclude that zebrafish TRPV1 likely serves as a receptor for heat and tissue acidity, although this remains to be confirmed experimentally. Based on these findings, the authors caution against the use of zebrafish in screens for pharmaceutical reagents that target cannabinoid receptors.

19.7 TRPC1 Regulates Angiogenesis Upstream of Map Kinase Activity

An investigation of the role of TRPC1 in angiogenesis nicely showcased the usefulness of in vivo imaging in zebrafish. In mammals, TRPCs modulate the physiology of endothelial cells [54–56]. Recent studies in zebrafish have revealed that calcium influx through TRPC1 is required for VEGF-induced elevation of endothelial-cell permeability [57]. In zebrafish embryos, *trpc1* is expressed broadly. Using the *fli:gfp* transgenic line in which the nascent vasculature expresses GFP, Yu and colleagues found that injecting a morpholino (MO) targeting *trpc1* translation inhibited the growth of intersegmental vessels (arterial progenitors) [57]. Moreover, using time-lapse imaging they showed that the filopodia of growing intersegmental vessels did not extend as rapidly in *trpc1* knockdown embryos as in controls. VEGF is known to serve as an attractant to growing vessels. Yu et al. showed that *vegfa* knockdown yields a phenotype similar to that obtained from *trpc1* knockdown, and that *trpc1* knockdown ameliorates the excessive sprouting of intersegmental vessels that occurs when VEGFA is overexpressed. Furthermore, global levels of phospho-ERK were reduced in *trpc1* morphants relative to control embryos, and an inhibitor of ERK blocked sprouting of arterial progenitors. These results are consistent with that model that in *trpc1* knockdown embryos a reduction of VEGF-stimulated ERK signaling is responsible for defects in angiogenesis.

The expression pattern of zebrafish *trpc6* fits with a conserved function of TRPC6 in maintaining vascular tone, although perhaps not in serum filtration by the kidney. In mammals, TRPC6 is expressed in kidney podocytes and in vascular smooth muscle. Consistent with these expression domains, mutations in *TRPC6* cause hereditary focal and segmental glomerulosclerosis (FSGS) [58, 59], and mice with targeted deletion of *TRPC6* have altered vaso-contractility [60, 61]. In zebrafish, *trpc6* is initially expressed ubiquitously but becomes restricted to the head, muscle of the pectoral fin, endothelial cells of the aorta, and posterior portion of the gut; notably, *trpc6* expression disappears from glomerular cells [62]. These expression data support the possibility that zebrafish TRPC6 is involved in smooth-muscle contractility in the heart vasculature, as in mammals, and possibly also in smooth-muscle contractility of the gastrointestinal tract. It remains uncertain whether a TRPC6 activity is necessary in the fish kidney. The merely transient expression of zebrafish *trpc6* in glomerular cells suggests against this possibility, however, *trpc6* may become elevated in the zebrafish kidney under pathological conditions, or an alternate TRPC family member may carry out a TRPC6-like function in zebrafish glomerular cells. Further analyses of the expression and function of zebrafish TRP genes will be necessary to resolve this question.

19.8 TRPP2 Is Necessary to Prevent Cyst Formation in Both Mammals and Zebrafish

TRPP2, more frequently called Polycystin-2 (PC2), is of clinical importance because mutations in the *TRPP2* gene cause autosomal-dominant polycystic kidney disease (ADPKD). Polycystin-1 and Polyscystin-2 form a heteromultimer that, when disrupted, causes embryonic kidney cysts, as well as defects in the vascular system, heart septum, and organ laterality [63, 64]. An insertional mutagenesis screen for zebrafish mutants with cystic kidneys yielded a *trpp2* mutant with this phenotype, and embryos injected with a MO targeting *trpp2* also have polycystic kidneys [65, 66]. Zebrafish *trpp2* mutants and *trpp2* MO-injected embryos also exhibit laterality defects due to a requirement for TRPP2 in Kupffer's vesicle, a ciliated structure homologous to the mouse node [65, 66]. In mammals, TRPP2 is localized to the apical side of epithelial cells, and there is evidence that it is a cell-surface channel responsible for initiating flow-induced calcium signaling [67]. To investigate the cellular distribution of TRPP2 in zebrafish, Obara and colleagues generated antisera to zebrafish TRPP2 [68]. Within the anterior pronephric duct, TRPP2 immunoreactivity was associated with basolateral membranes and with bundles of apical cilia. By contrast, in the posterior duct, immunostaining was excluded from the basal side of the cell but associated with apical cilia, with staining strongest in the basal portion of cilia and often absent from cilia tips. Differential subcellular localization across the pronephros implies that TRPP2 performs distinct functions depending on the cellular context in which it is expressed.

It is uncertain if the mechanism of kidney cyst formation is the same in mammals and zebrafish depleted of TRPP2. In both taxa, kidney cysts have been associated with diminished apical cilia on the luminal epithelium of the kidney [65, 69]. However, while *trpp2* MO-injected zebrafish embryos have kidney cysts, the apical cilia of kidney luminal epithelial cells are not grossly abnormal in morphology or motility, although the entire apical side of those cells appears to be expanded [68]. In such embryos the cloaca is blocked and luminal flow is obstructed [68]. The authors suggested that in zebrafish TRPP2 loss-of-function embryos, cysts form indirectly from a build-up of back pressure. The mechanisms of cyst formation in zebrafish and mammals lacking Trrp2/polycystin 2 function may be distinct, because cysts can form in patients with ADPKD even when connections to the ureter remain patent [70].

19.9 A Heteromultimer of TRPV4 and TRPP2 May Be the Mechanosensitive Complex that Detects Deflection of the Primary Cilium

Many of the proteins associated with polycystic kidney disease, including TRPP2, are localized to the primary cilium, a sensory structure that decorates renal epithelial cells and possibly all other cell types. Deflection of the primary cilium by fluid flowing over the cell elicits a transient increase in intracellular calcium that depends on the presence of TRPP2. Failure of this fluid-flow sensing function of the primary cilium is believed to be a fundamental mechanism of cystogenesis in patients with PKD; a recent study using zebrafish and other model systems challenges this view. Because TRPP2 homo-multimers lack mechanosensitivity, Köttegen and colleagues hypothesized that TRPP2 binds an auxiliary protein, specifically TRPV4, to produce a mechanosensitive channel [71]. Supporting this model, with fluorescence resonance energy transfer (FRET) and co-immunoprecipitation they showed that exogenously-expressed TRPV4 and TRPP2 physically interact. Further, they found that co-transfection of TRPV4 and TRPP2 into *Xenopus* oocytes resulted in higher swelling-induced inward currents than in oocytes transfected with TRPV4 or TRPP2 alone. Supporting a role for TRPV4 in flow sensing, there is a reduction of flow-induced calcium transients in MDCK cells transfected with a short-hairpin RNA targeting TRPV4 [71], and flow-induced potassium secretion is absent in mouse TRPV4 knockouts [72]. These data are consistent with the model that a hetero-multimer of TRPV4 and TRPP2 constitutes the mechanoreceptor that senses deflection of the primary cilium. However, mouse TRPV4 knockouts do not have cystic kidneys. Similarly, although the zebrafish orthologue is expressed in the pronephric duct [73], zebrafish injected with MOs targeting *trpv4* lack renal cysts [71]. In summary, while TRPP2 and TRPV4 localize to the cilium of renal epithelial cells and are required for flow-induced calcium transients, the lack of renal cysts in TRPV4-deficient animals indicates that renal cysts do not result from a defect in flow-induced deflection of the cilium. The authors infer that activation of the ciliary polycystin-1–TRPP2 complex by a mechanism other than fluid flow is critical for the regulation of tubular morphology. This intriguing model awaits testing.

19.10 Use of Zebrafish to Test, In Vivo, a Hypothesis Generated Based on a Tissue-Culture Model

The regulation of TRPV4 transit from the ER to the plasma membrane was the subject of a recent paper that used zebrafish for a quick in vivo test of a model developed based on tissue culture experiments. Like other membrane proteins, TRP channels are co-translationally inserted into the ER membrane, and they undergo protein folding with assistance from ER chaperones. When folding fails terminally, they are targeted to the cytoplasmic proteasome by the endoplasmic reticulum (ER)-associated protein degradation pathway [74]. The N-terminal tail of the TRP channels, and in particular the ankyrin repeats found in some TRPs, are thought to be important in regulating the folding of TRP proteins and in their trafficking to particular cellular compartments. Little is known about the cytoplasmic proteins that interact with TRPs in these domains to regulate these functions. OS-9 is a ubiquitously expressed protein that was identified in humans by virtue of being overexpressed in osteosarcomas [75]. It is homologous to a yeast protein, Yos9p, which selectively binds to misfolded or misassembled glycoproteins in the ER and targets them for destruction through the ER-associated protein degradation pathway (reviewed in 76). In 2007, Want and colleagues reported identifying OS-9 in a yeast two-hybrid screen baited with the amino terminus of TRPV4. The authors confirmed these findings by carrying out co-immunoprecipitation assays in HEK293T cells over-expressing epitope-tagged variants of OS-9 and TRPV4. Interestingly, they found that OS-9 preferentially associates with TRPV4 monomers rather than tetramers, and that OS-9 binding prevents the poly-ubiquitination of TRPV4. Depletion of OS-9 expression decreased the fraction of TRPV4 in the ER and increased the fraction present at the plasma membrane. These data supported the model that OS-9 regulates TRPV4 biogenesis by slowing its exit from the ER, and by protecting its monomers from premature poly-ubiquitination and degradation in the proteasome. As an in vivo test of this model, the authors assessed the consequences of misexpressing mouse TRPV4 in zebrafish embryos, and found that this led to widespread cell death. They hypothesized that this resulted from elevated levels of TRPV4 at the cell surface, and a consequent compromise of normal calcium homeostasis. In support of this hypothesis, they showed that forced expression of OS-9 reversed the developmental abnormalities associated with the ectopic TRPV4 expression, presumably by retaining a large fraction of it in the ER [77]. Nonetheless, further experiments are necessary to clarify the significance of the apparent interaction between OS-9 and TRPV4.

19.11 Investigation of TRP Channel Evolution During the Transition to a Warm-Blooded Lifestyle

Warm-blooded animals must constantly monitor the temperature of the environment and adjust to changes; to do so they use a subset of TRP channels that act as thermoreceptors. To ask how the repertoire of thermoreceptor TRP channels changed during the evolutionary transition from a cold-blooded to a warm-blooded lifestyle, Saito and colleagues searched for orthologues of mammalian thermoreceptor TRP channels (i.e. TRPV1, TRPV2, TRPV3, TRPV4, TRPM2, TRPM4, TRPM5, TRPM8, and TRPA1) in the draft genomes of species chosen to represent the major categories of vertebrates, i.e. birds, (chicken (*Gallus gallus*)), tetrapods (western clawed frog (*Xenopus tropicalis*)), and fishes (zebrafish (*Danio rerio*) and pufferfish (*Fugu rubripes*)) [78]. Their analyses support the evolutionary scenario that mammalian thermoreceptor TRPs were present in the common ancestor of tetrapods and ray-finned fishes, but that the repertoires of thermoreceptor TRPs have changed, i.e. duplications and deletions of family members, in the lineages leading to different vertebrate classes. An example of evidence supporting this scenario is that urochordates, which are basal to tetrapods and fish, have a single thermoreceptive TRP, TRPA1. Further, TRPM4 appears to have been lost in the lineage leading to chickens but duplicated in the one leading to zebrafish [78]. Together these analyses suggest that most vertebrate thermoreceptor TRPs emerged after the vertebrate and urochordate lineages diverged, but before the ray- and lobe-finned fish lineages diverged from the vertebrate lineage [78]. Further identification and comparison of thermoreceptor TRPs in organisms that represent key nodes in the chordate phylogenetic tree, such as cartilaginous fishes (shark), jawless fishes (lamprey), and Cephalochorda (amphioxus), will be necessary to fully illuminate the evolutionary history of thermoreceptor TRPs.

19.12 Zebrafish TRPM7 Mutants Reveal a Requirement for TRPM7 in Melanocyte Differentiation

Interestingly, loss-of-function mutations in *TRPM7* cause early embryonic lethality in mouse but not in zebrafish (Ryazanov, personal communication, 79); this situation means that zebrafish are particularly useful for investigating TRPM7 function during later stages of development. TRPM7, and also the closely related TRPM6, have the unusual structural feature of a phosphotransferase activity within their intracellular carboxy termini. They are also unusual in that they are permeable to magnesium. Mutations in the TRPM6 gene cause hypomagnesaemia with secondary hypcolcemia [80]. Mice with a targeted deletion of *TRPM7* are embryonic lethal, and when *TRPM7* is deleted specifically within thymocytes, these cells do not differentiate [79]. In zebrafish the *trpm7* gene appears to be readily targeted by ethylnitrosurea (ENU) because a large-scale ENU mutagenesis screen led to the isolation of more alleles of a mutant called *touchdown* than of any other mutant [81], and complementation tests established that *touchdown* corresponds to *trpm7* [82]. Given the ready mutability of this gene, it is not surprising that additional *trpm7* alleles were isolated in screens carried out independently in several laboratories [83–85]. Only two of the *trpm7* mutant alleles have been molecularly characterized, and in both cases, the mutation lies within the part of the gene that encodes the carboxy terminus [85]. Importantly, antisense inhibition of *trpm7* expression yields a phenotype like that seen in the case of the chemical-induced mutations, so the mutants are clearly loss of function in nature, although it is uncertain if they are true null alleles [85].

While zebrafish embryos homozygous for loss-of-function alleles of *trpm7* undergo early morphogenesis normally, a number of defects become apparent starting during the second day of development. First, embryonic melanophores are much paler than normal and eventually undergo cell death [81, 83, 84], apparently by a non-apoptotic pathway [82]. Second, 2-day-old *trpm7* mutants lack touch responsiveness and exhibit virtually no spontaneous swimming [82–84]. By 5 days of age, *trpm7* mutants have recovered touch responsiveness but remain hypomotile compared to wild-type embryos (our unpublished observations). Third, as juvenile adults, *trpm7* mutants exhibit profound defects in skeletogenesis, and are much smaller than non-mutant siblings [85]. What are the cellular bases of these defects? In early embryos, *trpm7* is expressed globally, but its expression is later restricted such that high levels are present only in the corpuscles of Stannius, structures that are adjacent to the kidney and are found only in bony fishes. Corpuscles of Stannius are thought to regulate organismal cation levels and thus to be functionally analogous to the mammalian parathyroid gland [86]. A disruption of global ion balance may explain the skeletal defects in *trpm7* mutants, but this remains to be tested. It seems unlikely that global ion balance explains the melanophore defects, because these cells have a cell-autonomous requirement for TRPM7 [83, 84]. The melanosome is the lysosome-related organelle in which melanin synthesis takes place, and in *trpm7* mutants, the melanosomes in embryonic melanophores have abnormal ultrastructure [82]. These findings suggest that TRPM7 is necessary for the normal biogenesis of melanosomes, and that in the absence of TRPM7 the intermediates of melanin synthesis (which are toxic reactive oxygen species) are released and cause cell death. In support of this model, inhibiting melanin synthesis prevents melanophore cell death in *trpm7* mutants [82]. Determining the subcellular localization of endogenous TRPM7 protein in melanophores remains an important goal. There is immunological evidence from other cell types that it is present on the plasma membrane and in intracellular compartments [87, 88]. At present the cellular basis of the absence of touch response and hypomotility in TRPM7 mutants remains unexplained. It will be interesting to determine if the cellular basis of any of the phenotypes just described is the same. Moreover, it will be important to establish how these mechanisms relate to the failure of morphogenesis and the lack of lymphocyte differentiation in mouse *TRPM7* mutants [79]. Because the zebrafish *trpm7* mutant survives early morphogenesis, it provides a convenient alternative to the mouse for these explorations.

19.13 Conclusions

A selection of papers describing the use of frog and zebrafish for the in vivo study of TRP channel function have been reviewed here. The established features of frog and zebrafish model systems, plus the emergence of zebrafish methods for live imaging [21], gene targeting [89], and pharmacological screens [90, 91], which have the potential to identify long-sought agonists and antagonists of TRP channels, will likely continue to draw TRP channel researchers to these models for years to come.

Acknowledgments I thank Christine Blaumuller for editorial comments. Work on TRPM7 in the Cornell laboratory is supported by NIH grants GM067841 and 5P30ES005605 (pilot grant).

References

1. Cosens DJ, Manning A (1969) Abnormal electroretinogram from a Drosophila mutant. Nature 224: 285–287
2. Montell C, Rubin GM (1989) Molecular characterization of the Drosophila trp locus: a putative integral membrane protein required for phototransduction. Neuron 2: 1313–1323
3. Hardie RC, Minke B (1992) The trp gene is essential for a light-activated Ca^{2+} channel in Drosophila photoreceptors. Neuron 8: 643–651
4. Venkatachalam K, Montell C (2007) TRP channels. Annu Rev Biochem 76: 387–417
5. Voets T, Talavera K, Owsianik G, Nilius B (2005) Sensing with TRP channels. Nat Chem Biol 1: 85–92
6. Nadler MJ, Hermosura MC, Inabe K, Perraud AL, Zhu Q, Stokes AJ, Kurosaki T, Kinet JP, Penner R, Scharenberg AM, Fleig A (2001) LTRPC7 is a Mg.ATP-regulated divalent cation channel required for cell viability. Nature 411: 590–595
7. Kaneko S, Kawakami S, Hara Y, Wakamori M, Itoh E, Minami T, Takada Y, Kume T, Katsuki H, Mori Y, Akaike A (2006) A critical role of TRPM2 in neuronal cell death by hydrogen peroxide. J Pharmacol Sci 101: 66–76
8. Nilius B, Owsianik G, Voets T, Peters JA (2007) Transient receptor potential cation channels in disease. Physiol Rev 87: 165–217
9. Montell C (2003) The venerable inveterate invertebrate TRP channels. Cell Calcium 33: 409–417
10. Grunwald DJ, Streisinger G (1992) Induction of mutations in the zebrafish with ultraviolet light. Genet Res 59: 93–101
11. Grunwald DJ, Streisinger G (1992) Induction of recessive lethal and specific locus mutations in the zebrafish with ethyl nitrosourea. Genet Res 59: 103–116
12. Grunwald DJ, Kimmel CB, Westerfield M, Walker C, Streisinger G (1988) A neural degeneration mutation that spares primary neurons in the zebrafish. Dev Biol 126: 115–128
13. Streisinger G, Singer F, Walker C, Knauber D, Dower N (1986) Segregation analyses and gene-centromere distances in zebrafish. Genetics 112: 311–319
14. Nawrocki L, BreMiller R, Streisinger G, Kaplan M (1985) Larval and adult visual pigments of the zebrafish, Brachydanio rerio. Vision Res 25: 1569–1576
15. Walker C, Streisinger G (1983) Induction of Mutations by gamma-Rays in Pregonial Germ Cells of Zebrafish Embryos. Genetics 103: 125–136
16. Chakrabarti S, Streisinger G, Singer F, Walker C (1983) Frequency of gamma-Ray Induced Specific Locus and Recessive Lethal Mutations in Mature Germ Cells of the Zebrafish, BRACHYDANIO RERIO. Genetics 103: 109–123
17. Postlethwait JH, Johnson SL, Midson CN, Talbot WS, Gates M, Ballinger EW, Africa D, Andrews R, Carl T, Eisen JS et al (1994) A genetic linkage map for the zebrafish. Science 264: 699–703
18. Haffter P, Granato M, Brand M, Mullins MC, Hammerschmidt M, Kane DA, Odenthal J, van Eeden FJ, Jiang YJ, Heisenberg CP, Kelsh RN, Furutani-Seiki M, Vogelsang E, Beuchle D, Schach U, Fabian C, Nusslein-Volhard C (1996) The identification of genes with unique and essential functions in the development of the zebrafish, Danio rerio. Development 123: 1–36
19. Moreno RL, Ribera AB (2009) Zebrafish motor neuron subtypes differ electrically prior to axonal outgrowth. J Neurophysiol 102: 2477–2484
20. Fetcho JR, Cox KJ (1998) O'Malley DM: Monitoring activity in neuronal populations with single-cell resolution in a behaving vertebrate. Histochem J 30: 153–167
21. Naumann EA, Kampff AR, Prober DA, Schier AF, Engert F (2010) Monitoring neural activity with bioluminescence during natural behavior. Nat Neurosci 13: 513–520

22. Liu B, Yao J, Wang Y, Li H, Qin F (2009) Proton inhibition of unitary currents of vanilloid receptors. J Gen Physiol 134: 243–258
23. Sherkheli MA, Benecke H, Doerner JF, Kletke O, Vogt-Eisele AK, Gisselmann G, Hatt H (2009) Monoterpenoids induce agonist-specific desensitization of transient receptor potential vanilloid-3 (TRPV3) ion channels. J Pharm Sci 12: 116–128
24. Na T, Zhang W, Jiang Y, Liang Y, Ma HP, Warnock DG, Peng JB (2009) The A563T variation of the renal epithelial calcium channel TRPV5 among African Americans enhances calcium influx. Am J Physiol Renal Physiol 296: F1042–F1051
25. Hu H, Grandl J, Bandell M, Petrus M, Patapoutian A (2009) Two amino acid residues determine 2-APB sensitivity of the ion channels TRPV3 and TRPV4. Proc Natl Acad Sci USA 106: 1626–1631
26. Wegierski T, Steffl D, Kopp C, Tauber R, Buchholz B, Nitschke R, Kuehn EW, Walz G, Kottgen M (2009) TRPP2 channels regulate apoptosis through the Ca^{2+} concentration in the endoplasmic reticulum. EMBO J 28: 490–499
27. Daniels RL, Takashima Y, McKemy DD (2009) Activity of the neuronal cold sensor TRPM8 is regulated by phospholipase C via the phospholipid phosphoinositol 4,5-bisphosphate. J Biol Chem 284: 1570–1582
28. Maroto R, Raso A, Wood TG, Kurosky A, Martinac B, Hamill OP (2005) TRPC1 forms the stretch-activated cation channel in vertebrate cells. Nat Cell Biol 7: 179–185
29. Zhang Y, Hamill OP (2000) Calcium-, voltage- and osmotic stress-sensitive currents in Xenopus oocytes and their relationship to single mechanically gated channels. J Physiol 523(Pt 1): 83–99
30. Stucky CL, Dubin AE, Jeske NA, Malin SA, McKemy DD, Story GM (2009) Roles of transient receptor potential channels in pain. Brain Res Rev 60: 2–23
31. Shim S, Goh EL, Ge S, Sailor K, Yuan JP, Roderick HL, Bootman MD, Worley PF, Song H, Ming GL (2005) XTRPC1-dependent chemotropic guidance of neuronal growth cones. Nat Neurosci 8: 730–735
32. Wang GX, Poo MM (2005) Requirement of TRPC channels in netrin-1-induced chemotropic turning of nerve growth cones. Nature 434: 898–904
33. Shim S, Yuan JP, Kim JY, Zeng W, Huang G, Milshteyn A, Kern D, Muallem S, Ming GL, Worley PF (2009) Peptidyl-prolyl isomerase FKBP52 controls chemotropic guidance of neuronal growth cones via regulation of TRPC1 channel opening. Neuron 64: 471–483
34. Yuan JP, Kiselyov K, Shin DM, Chen J, Shcheynikov N, Kang SH, Dehoff MH, Schwarz MK, Seeburg PH, Muallem S, Worley PF (2003) Homer binds TRPC family channels and is required for gating of TRPC1 by IP3 receptors. Cell 114: 777–789
35. Sato Y, Miyasaka N, Yoshihara Y (2005) Mutually exclusive glomerular innervation by two distinct types of olfactory sensory neurons revealed in transgenic zebrafish. J Neurosci 25: 4889–4897
36. Morita Y, Finger TE (1998) Differential projections of ciliated and microvillous olfactory receptor cells in the catfish, Ictalurus punctatus. J Comp Neurol 398: 539–550
37. Hansen A, Rolen SH, Anderson K, Morita Y, Caprio J, Finger TE (2003) Correlation between olfactory receptor cell type and function in the channel catfish. J Neurosci 23: 9328–9339
38. Sidi S, Friedrich RW, Nicolson T, Nomp C (2003) TRP channel required for vertebrate sensory hair cell mechanotransduction. Science 301: 96–99
39. Walker RG, Willingham AT, Zuker CS (2000) A Drosophila mechanosensory transduction channel. Science 287: 2229–2234
40. Zhang L, Jones S, Brody K, Costa M, Brookes SJ (2004) Thermosensitive transient receptor potential channels in vagal afferent neurons of the mouse. Am J Physiol Gastrointest Liver Physiol 286: G983–G991
41. Seiler C, Nicolson T (1999) Defective calmodulin-dependent rapid apical endocytosis in zebrafish sensory hair cell mutants. J Neurobiol 41: 424–434
42. Gopfert MC, Albert JT, Nadrowski B, Kamikouchi A (2006) Specification of auditory sensitivity by Drosophila TRP channels. Nat Neurosci 9: 999–1000

43. Shin JB, Adams D, Paukert M, Siba M, Sidi S, Levin M, Gillespie PG, Grunder S (2005) Xenopus TRPN1 (NOMPC) localizes to microtubule-based cilia in epithelial cells, including inner-ear hair cells. Proc Natl Acad Sci USA 102: 12572–12577
44. Story GM, Peier AM, Reeve AJ, Eid SR, Mosbacher J, Hricik TR, Earley TJ, Hergarden AC, Andersson DA, Hwang SW, McIntyre P, Jegla T, Bevan S, Patapoutian A (2003) ANKTM1, a TRP-like channel expressed in nociceptive neurons, is activated by cold temperatures. Cell 112: 819–829
45. Jordt SE, Bautista DM, Chuang HH, McKemy DD, Zygmunt PM, Hogestatt ED, Meng ID, Julius D (2004) Mustard oils and cannabinoids excite sensory nerve fibres through the TRP channel ANKTM1. Nature 427: 260–265
46. Bandell M, Story GM, Hwang SW, Viswanath V, Eid SR, Petrus MJ, Earley TJ, Patapoutian A (2004) Noxious cold ion channel TRPA1 is activated by pungent compounds and bradykinin. Neuron 41: 849–857
47. Prober DA, Zimmerman S, Myers BR, McDermott BM Jr, Kim SH, Caron S, Rihel J, Solnica-Krezel L, Julius D, Hudspeth AJ, Schier AF (2008) Zebrafish TRPA1 channels are required for chemosensation but not for thermosensation or mechanosensory hair cell function. J Neurosci 28: 10102–10110
48. Moens CB, Donn TM, Wolf-Saxon ER, Ma TP (2008) Reverse genetics in zebrafish by TILLING. Brief Funct Genomic Proteomic 7: 454–459
49. Damann N, Voets T, Nilius B (2008) TRPs in our senses. Curr Biol 18: R880–R889
50. Jordt SE, Julius D (2002) Molecular basis for species-specific sensitivity to "hot" chili peppers. Cell 108: 421–430
51. Tobin D, Madsen D, Kahn-Kirby A, Peckol E, Moulder G, Barstead R, Maricq A, Bargmann C (2002) Combinatorial expression of TRPV channel proteins defines their sensory functions and subcellular localization in C. elegans neurons. Neuron 35: 307–318
52. McPartland JM, Matias I, Di Marzo V, Glass M (2006) Evolutionary origins of the endo-cannabinoid system. Gene 370: 64–74
53. McPartland JM, Glass M, Matias I, Norris RW, Kilpatrick CW (2007) A shifted repertoire of endocannabinoid genes in the zebrafish (Danio rerio). Mol Genet Genomics 277: 555–570
54. Freichel M, Suh SH, Pfeifer A, Schweig U, Trost C, Weissgerber P, Biel M, Philipp S, Freise D, Droogmans G, Hofmann F, Flockerzi V, Nilius B (2001) Lack of an endothelial store-operated Ca^{2+} current impairs agonist-dependent vasorelaxation in TRP4–/– mice. Nat Cell Biol 3: 121–127
55. Jho D, Mehta D, Ahmmed G, Gao XP, Tiruppathi C, Broman M, Malik AB (2005) Angiopoietin-1 opposes VEGF-induced increase in endothelial permeability by inhibiting TRPC1-dependent Ca2 influx. Circ Res 96: 1282–1290
56. Tiruppathi C, Freichel M, Vogel SM, Paria BC, Mehta D, Flockerzi V, Malik AB (2002) Impairment of store-operated Ca^{2+} entry in TRPC4(-/-) mice interferes with increase in lung microvascular permeability. Circ Res 91: 70–76
57. Yu PC, Gu SY, Bu JW, Du JL (2010) TRPC1 is essential for in vivo angiogenesis in zebrafish. Circ Res 106: 1221–1232
58. Winn MP, Conlon PJ, Lynn KL, Farrington MK, Creazzo T, Hawkins AF, Daskalakis N, Kwan SY, Ebersviller S, Burchette JL, Pericak-Vance MA, Howell DN, Vance JM, Rosenberg PB (2005) A mutation in the TRPC6 cation channel causes familial focal segmental glomerulosclerosis. Science 308: 1801–1804
59. Reiser J, Polu KR, Moller CC, Kenlan P, Altintas MM, Wei C, Faul C, Herbert S, Villegas I, Avila-Casado C, McGee M, Sugimoto H, Brown D, Kalluri R, Mundel P, Smith PL, Clapham DE, Pollak MR (2005) TRPC6 is a glomerular slit diaphragm-associated channel required for normal renal function. Nat Genet 37: 739–744
60. Dietrich A, Mederos YSM, Gollasch M, Gross V, Storch U, Dubrovska G, Obst M, Yildirim E, Salanova B, Kalwa H, Essin K, Pinkenburg O, Luft FC, Gudermann T, Birnbaumer L (2005) Increased vascular smooth muscle contractility in TRPC6–/– mice. Mol Cell Biol 25: 6980–6989

61. Weissmann N, Dietrich A, Fuchs B, Kalwa H, Ay M, Dumitrascu R, Olschewski A, Storch U, Mederos y Schnitzler M, Ghofrani HA, Schermuly RT, Pinkenburg O, Seeger W, Grimminger F, Gudermann T (2006) Classical transient receptor potential channel 6 (TRPC6) is essential for hypoxic pulmonary vasoconstriction and alveolar gas exchange. Proc Natl Acad Sci USA 103: 19093–19098
62. Moller CC, Mangos S, Drummond IA, Reiser J (2008) Expression of trpC1 and trpC6 orthologs in zebrafish. Gene Expr Patterns 8: 291–296
63. Pennekamp P, Karcher C, Fischer A, Schweickert A, Skryabin B, Horst J, Blum M, Dworniczak B (2002) The ion channel polycystin-2 is required for left-right axis determination in mice. Curr Biol 12: 938–943
64. Wu G, Markowitz GS, Li L, D'Agati VD, Factor SM, Geng L, Tibara S, Tuchman J, Cai Y, Park JH, van Adelsberg J, Hou H Jr, Kucherlapati R, Edelmann W, Somlo S (2000) Cardiac defects and renal failure in mice with targeted mutations in Pkd2. Nat Genet 24: 75–78
65. Sun Z, Amsterdam A, Pazour GJ, Cole DG, Miller MS, Hopkins N (2004) A genetic screen in zebrafish identifies cilia genes as a principal cause of cystic kidney. Development 131: 4085–4093
66. Bisgrove BW, Snarr BS, Emrazian A, Yost HJ (2005) Polaris and Polycystin-2 in dorsal forerunner cells and Kupffer's vesicle are required for specification of the zebrafish left-right axis. Dev Biol 287: 274–288
67. Kobori T, Smith GD, Sandford R, Edwardson JM (2009) The transient receptor potential channels TRPP2 and TRPC1 form a heterotetramer with a 2:2 stoichiometry and an alternating subunit arrangement. J Biol Chem 284: 35507–35513
68. Obara T, Mangos S, Liu Y, Zhao J, Wiessner S, Kramer-Zucker AG, Olale F, Schier AF, Drummond IA (2006) Polycystin-2 immunolocalization and function in zebrafish. J Am Soc Nephrol 17: 2706–2718
69. Kramer-Zucker AG, Olale F, Haycraft CJ, Yoder BK, Schier AF, Drummond IA (2005) Cilia-driven fluid flow in the zebrafish pronephros, brain and Kupffer's vesicle is required for normal organogenesis. Development 132: 1907–1921
70. Deltas C, Papagregoriou G (2010) Cystic diseases of the kidney: molecular biology and genetics. Arch Pathol Lab Med 134: 569–582
71. Kottgen M, Buchholz B, Garcia-Gonzalez MA, Kotsis F, Fu X, Doerken M, Boehlke C, Steffl D, Tauber R, Wegierski T, Nitschke R, Suzuki M, Kramer-Zucker A, Germino GG, Watnick T, Prenen J, Nilius B, Kuehn EW, Walz G (2008) TRPP2 and TRPV4 form a polymodal sensory channel complex. J Cell Biol 182: 437–447
72. Taniguchi J, Tsuruoka S, Mizuno A, Sato J, Fujimura A, Suzuki M (2007) TRPV4 as a flow sensor in flow-dependent K^+ secretion from the cortical collecting duct. Am J Physiol Renal Physiol 292: F667–F673
73. Mangos S, Liu Y, Drummond IA (2007) Dynamic expression of the osmosensory channel trpv4 in multiple developing organs in zebrafish. Gene Expr Patterns 7: 480–484
74. Meusser B, Hirsch C, Jarosch E, Sommer T (2005) ERAD: the long road to destruction. Nat Cell Biol 7: 766–772
75. Kimura Y, Nakazawa M, Yamada M (1998) Cloning and characterization of three isoforms of OS-9 cDNA and expression of the OS-9 gene in various human tumor cell lines. J Biochem 123: 876–882
76. Cormier JH, Pearse BR, Hebert DN (2005) Yos9p: a sweet-toothed bouncer of the secretory pathway. Mol Cell 19: 717–719
77. Wang Y, Fu X, Gaiser S, Kottgen M, Kramer-Zucker A, Walz G, Wegierski T (2007) OS-9 regulates the transit and polyubiquitination of TRPV4 in the endoplasmic reticulum. J Biol Chem 282: 36561–36570
78. Saito S, Shingai R (2006) Evolution of thermoTRP ion channel homologs in vertebrates. Physiol Genomics 27: 219–230
79. Jin J, Desai BN, Navarro B, Donovan A, Andrews NC, Clapham DE (2008) Deletion of Trpm7 disrupts embryonic development and thymopoiesis without altering Mg^{2+} homeostasis. Science 322: 756–760

80. Walder RY, Landau D, Meyer P, Shalev H, Tsolia M, Borochowitz Z, Boettger MB, Beck GE, Englehardt RK, Carmi R, Sheffield VC (2002) Mutation of TRPM6 causes familial hypomagnesemia with secondary hypocalcemia. Nat Genet 31: 171–174
81. Kelsh RN, Brand M, Jiang YJ, Heisenberg CP, Lin S, Haffter P, Odenthal J, Mullins MC, van Eeden FJ, Furutani-Seiki M, Granato M, Hammerschmidt M, Kane DA, Warga RM, Beuchle D, Vogelsang L, Nusslein-Volhard C (1996) Zebrafish pigmentation mutations and the processes of neural crest development. Development 123: 369–389
82. McNeill MS, Paulsen J, Bonde G, Burnight E, Hsu MY, Cornell RA (2007) Cell death of melanophores in zebrafish trpm7 mutant embryos depends on melanin synthesis. J Invest Dermatol 127: 2020–2030
83. Arduini BL, Henion PD (2004) Melanophore sublineage-specific requirement for zebrafish touchtone during neural crest development. Mech Dev 121: 1353–1364
84. Cornell RA, Yemm E, Bonde G, Li W, d'Alencon C, Wegman L, Eisen J, Zahs A (2004) Touchtone promotes survival of embryonic melanophores in zebrafish. Mech Dev 121: 1365–1376
85. Elizondo MR, Arduini BL, Paulsen J, MacDonald EL, Sabel JL, Henion PD, Cornell RA, Parichy DM (2005) Defective skeletogenesis with kidney stone formation in dwarf zebrafish mutant for trpm7. Curr Biol 15: 667–671
86. Greenwood MP, Flik G, Wagner GF, Balment RJ (2009) The corpuscles of Stannius, calcium-sensing receptor, and stanniocalcin: responses to calcimimetics and physiological challenges. Endocrinology 150: 3002–3010
87. Krapivinsky G, Mochida S, Krapivinsky L, Cibulsky SM, Clapham DE (2006) The TRPM7 ion channel functions in cholinergic synaptic vesicles and affects transmitter release. Neuron 52: 485–496
88. Oancea E, Wolfe JT, Clapham DE (2006) Functional TRPM7 channels accumulate at the plasma membrane in response to fluid flow. Circ Res 98: 245–253
89. Ekker SC (2008) Zinc finger-based knockout punches for zebrafish genes. Zebrafish 5: 121–123
90. North TE, Goessling W, Walkley CR, Lengerke C, Kopani KR, Lord AM, Weber GJ, Bowman TV, Jang IH, Grosser T, Fitzgerald GA, Daley GQ, Orkin SH, Zon LI (2007) Prostaglandin E2 regulates vertebrate haematopoietic stem cell homeostasis. Nature 447: 1007–1011
91. Sachidanandan C, Yeh JR, Peterson QP, Peterson RT (2008) Identification of a novel retinoid by small molecule screening with zebrafish embryos. PLoS One 3: e1947

Chapter 20
TRP Channels in Parasites

Adrian J. Wolstenholme, Sally M. Williamson, and Barbara J. Reaves

Abstract A wide range of single- and multi-cellular parasites infect humans and other animals, causing some of the most prevalent and debilitating diseases on the planet. There have been virtually no published studies on the TRP channels of this diverse group of organisms. However, since many parasite genomes have been sequenced, it is simple to demonstrate that they are present in all parasitic metazoans and that sequences related to the yeast trp are present in many protozoans, including all the kinetoplastids. We compared the TRP genes of three species of animal and plant parasitic nematode to those of *C. elegans* and found that the parasitic species all had fewer such genes. These differences may reflect the phylogenetic distance between the species studied, or may be due to loss of specific gene functions following the evolution of the parasitic lifestyle. Other helminth groups, the trematodes and cestodes, seem to possess many TRPC and TRPM genes, but lack TRPV and TRPN. Most ectoparasites are insects or arachnids. We compared the TRP genes of a plant parasitic aphid and an animal parasite louse and tick with those of *Drosophila*. Again, all the parasitic species seemed to have fewer types of TRP channel, though the difference was less marked than for the nematodes. The aphid lacks TRPP and TRPML channel genes, whereas the tick lacked those encoding TRPVs. Again, these differences may reflect adaptation to parasitism, and could enable TRP channels to be targeted in the development of novel antiparasitic drugs.

20.1 Introduction

The term "parasite", though fairly easy to define in biological terms, presents a major problem for chapters such as this as it covers a wide variety of single- and multi-celled organisms which cover a large phylogenetic range. This is illustrated by Table 20.1, which describes the variety of species that specifically parasitise

---

A.J. Wolstenholme (✉)
Department of Infectious Diseases, University of Georgia, Athens, GA 30602, USA; Center for Tropical and Emerging Global Disease, University of Georgia, Athens, GA 30602, USA
e-mail: adrianw@uga.edu

Table 20.1 Some parasites infecting humans

		Example species	Disease
Endoparasites			
Protozoa	Apicomplexans	*Plasmodium spp*	Malaria
		Toxoplasma gondii	Toxoplasmosis
		Babesia spp	Babesiosis
		Cryptosporidium parva	Cryptosporidiosis
	Kinetoplastids	*Trypanosoma bruceii*	Sleeping sickness
		Trypanosoma cruzi	Chagas' disease
		Leishmania spp.	Visceral and cutaneous leishmaniasis
	Metamonada	*Giardia lamblia*	Giardiasis
		Trichonomas vaginalis	Trichomoniasis
	Ameoba	*Entamoeba histolytica*	Ameobic dysentery
Helminths	Nematoda (roundworms)	*Ascaris lumbricoides*	Ascariasis
		Wuchereria bancrofti	Lymphatic filariasis (elephantiasis)
		Necator americanus	Hookworm
	Trematoda (flatworms)	*Schistosoma spp*	Schistosomiasis (bilharzia)
		Fasciola hepatica	Fascioliasis (liver fluke)
		Clonorchis sinensis	Oriental liver fluke
	Cestodes (tapeworms)	*Taenia solium*	Cysticercosis
		Echinococcus granulosus	Hydatid disease
Ectoparasites			
	Insects	*Dermatobia hominis*	Myiasis
		Pulex irritans	Human flea
		Cimex lectularius	Bedbug
		Pediculus humanus	Head louse
	Arachnids	*Ixodes scapularis*	Deer tick (vector of lyme disease)
		Sarcoptes scabiei	Scabies

humans and other mammals. This diversity of phylogeny is matched by an equal diversity in biology and life-cycles, all of which combine to make the task of producing an authoritative chapter on TRP channels across parasitology a major challenge. However, this task can be approached via the realization that many parasites are related to well-studied model organisms, and the dramatic increase in genome sequence information means that it is possible to compare the members of the TRP family encoded by parasites with those of their model cousins and perhaps even deduce what some of the physiological roles of the channels might be. For example, several of the parasites in Table 20.1 are nematodes, and the genome of the model nematode *Caenorhabditis elegans* was not only the first to be sequenced of any metazoan [1], it has also been superbly annotated.

There is no denying, though, that there have been hardly any investigations into the functions of TRP channels in parasites. This is a pity, and a lacuna in our knowledge that should be addressed. TRP channels present attractive drug targets

[2], especially as many existing anti-parasitic compounds, especially anthelmintics, act on other ion channels [3–6], including ligand-gated ion channels and the SLO-1 potassium channels, and an improved pipeline of new drugs is always needed [7–9]. In addition, the involvement of the TRP channels in so many sensory processes implies that they will have roles in mediating the many host:parasite interactions that are essential for the completion of sometimes very complex life-cycles. Understanding these roles could dramatically improve our understanding of the molecular and cellular mechanisms that underpin the adaptations necessary for successful parasitism to evolve.

This hugely diverse grouping of parasites possesses an equal diverse collection of life-cycles. These life-cycles may be direct, involving only a single host, or indirect, requiring both a definitive host, in which sexual reproduction (if it occurs) takes place, and an intermediate host, which allows for the parasites to be dispersed and in which other developmental stages may take place. As an example, for malaria the human is a definitive host and the mosquito the intermediate host, and there are many different developmental stages within both. Even in those parasites with direct life cycles, many stages of development may take place outside the host and in the environment, for example hookworm eggs are shed into environment, then hatch and undergo several moults before reinfecting their mammalian host. A full description of the events within the various hosts is well beyond the scope of this chapter, as all parasites go through multiple developmental stages and have to find their specific niche within the host body. Many migrate around; the intestinal nematode *Ascaris lumbricoides* migrates from gut to liver to lungs and back to the gut during the various stages of its development within infected humans.

This requirement for replication within an animal host has severely limited the development of tools to study parasites. Some parasitic protozoans can be cultured in the laboratory and for these advanced molecular genetic tools such as gene knockouts have been developed [10–13]. However, no such culture methods or tools are readily available for the parasitic helminths, and this has had a severely limiting effect on research on these organisms. For the nematodes and many ectoparasites, good model organisms (*C. elegans* and *Drosophila melanogaster*, respectively) are available for comparison and, at the genomic level, these comparisons have been made facile by the amount of sequence information that is now available. A considerable amount of excellent science has been carried out on the TRP channels of these model organisms ([14–17]), but the biological relevance of these studies to parasitic species remains to be determined. Studies on other nematode gene families show that the composition of these families, and possibly their functions, differ greatly between *C. elegans* and parasitic species [18, 19] and so simple extrapolation may be misleading.

The biological diversity of parasites demands that this chapter be sub-divided, and we have chosen to do this along the lines of Table 20.1. Separate sections will discuss endoparasites, those organisms that invade their hosts and live inside them (which include the protozoan parasites and the helminthes, or "worms") and the ectoparasites, organisms that live on the outside of the host, but are dependent on them for food – many are blood feeders.

20.2 Endoparasites

20.2.1 Protozoa

A very diverse group of single-celled organisms has evolved to parasitise humans and other animals. The apicomplexans, which include the *Plasmodium* species that cause malaria, arguably the most important infectious disease on the planet [20], are an exclusively parasitic grouping that possess a unique organelle, the apicoplast, a non-photosynthetic plastid [21], and an apical complex required for entry into host cells [22]. The kinetoplastids, which cause sleeping sickness and leishmaniasis, are a group of flagellates defined by the presence of a kinetoplast, a granule within the mitochondrion that contains DNA [23]. Both of these groups are normally transmitted via an invertebrate vector, such as a mosquito or biting fly. The metamonads are anaerobic flagellates, most of which are symbiotic with their hosts, but some of which are parasitic. A few species of amoeba are also parasitic.

Two TRPML-like genes, *lmmlA* (LmjF07.0910) and *lmmlB* (LmjF26.0990), have been described from *Leishmania major*, a kinetoplastid [24]. Though *lmmlA* is constitutively expressed throughout the life-cycle, *lmmlB* expression was reported to be up-regulated in amastigotes, the form of the parasite found in the vertebrate host [24]. A simple search of TrTrypDB (http://tritrypdb.org/tritrypdb/) revealed that both of these genes are conserved throughout the kinetoplastids, with similar sequences being present in the other *Leishmania* and *Trypanosoma* species. There are no published studies on the functions of these channels.

There are no other reports of TRP channels from any of these parasites, so we undertook a simple bioinformatics search of some of the available genome sequences, using the membrane-spanning domains of the yeast *trp* sequence [25] and of *C. elegans* CUP-5 (a TRPML [26]) as probes. This search revealed sequences in *Trichomonas vaginalis* (a metamonad) with low levels of identity (~22%) to both the yeast *trp* and *C. elegans* CUP-5, and a similar sequence (24% identity to CUP-5) in *Cryptosporidium parva* (an apicomplexan). The *C. parvum* protein (cdg6_1510) possesses membrane-spanning regions and is conserved in *C. hominis* and *C. muris*. The *T. vaginalis* "hits" were less convincing, and probably do not represent true channel proteins. Though none of these proteins, including those from the kinetoplastids, has yet been shown to form a genuine ion channel, these early data do suggest that that TRP-like channels might be present in a wide variety of protozoan parasites. We found no evidence of any sequence similar to other sub-families of TRPs, except for TRPML, in these organisms, nor did we find any evidence of TRP-like channels in *Plasmodium falciparum*, *Toxoplasma gondii*, *Neospora caninum* or *Giardia lamblia*. This may be because these organisms do not have any such genes, or it might be that they are just too divergent to be easily detected – further work is clearly needed to clarify this. Apicomplexans possess most of the other components for regulation of intracellular [Ca^{2+}] found in other eukaryotes, though *P. falciparum* and *C. parva* apparently lack voltage-operated Ca^{2+} channels or PMCA [27]. Given the role of TRPML channels in intracellular events in yeast [25] and higher eukaryotes [28, 29], we would hypothesise that any such channels in protozoan parasites will be involved in the intracellular trafficking and membrane fusion events that take place in these organisms [30, 31]. One possible example of such an event is the fusion of acidocalcisomes, a dense acidic organelle, to the contractile vacuole of *T. cruzi* and related organisms that is part of the osmoregulation process [32]. Additionally, entry into and replication within host cells by the kinetoplastids is intimately associated with lysosomal-like organelles and membrane fusion events [33, 34]. These events are regulated by TRPML in uninfected cells [28] and it is possible that the related proteins encoded by these parasites function as part of the process by which they control this compartment in their hosts.

20.2.2 Helminths

20.2.2.1 Nematodes

Nematodes, or roundworms, represent about 80% of all animal species. Many of these species are parasitic, causing chronic infections and debilitating, though rarely fatal, disease. A couple of parasitic genome sequences have been published [35, 36] in addition to that of the "model worm", *C. elegans* [1], and a considerable amount of sequence information is available, primarily from the groups at Washington University, St Louis, USA (http://www.nematode.net) [37] and the Sanger Centre, Cambridge, U.K. (http://www.sanger.ac.uk/Projects/Helminths/). Many of the *C. elegans* TRP genes have been extensively studied and reviewed [17]; some of this information is summarized in Table 20.2. However, phylogenetic analysis suggest that *C. elegans* may not be typical of all nematodes [38], and previous analyses have revealed big differences between the ion channel genes of *C. elegans* and some parasitic species [19]. We searched the annotated genes of the parasitic nematode *Brugia malayi* [35] and the plant parasite *Meloidogyne incognita* [36] together with the unannotated sequence of *Trichinella spiralis* (http://genome.wustl.edu/genomes/view/trichinella_spiralis/) for TRP channel sequences; the results are summarized in Table 20.2. Compared to *C. elegans*, it seems that all three parasite species possess fewer such genes, as observed previously for the ligand-gated ion channels [19]; this may not reflect any physiological simplicity of the parasites, but rather the number of gene duplications that have occurred during *C. elegans* evolution, though considerably more comparative genomic analysis is required before any firm conclusions can be drawn. We also searched the *B. malayi* genome (http://blast.jcvi.org/er-blast/index.cgi?project=bma1) for TRPM-, TRPA- and TRPP-related sequences in case these had not been annotated; we found some evidence of TRPM-like sequences but not for TRPA or TRPP. Many of the TRP channels are involved in sensory processes, in nematodes as in other organisms, and the loss of these channels may reflect a reduction in the complexity of sensory inputs encountered by the animal parasites as they inhabit the homeostatically maintained environments of their vertebrate and, in the case of *B. malayi*, invertebrate hosts. The *T. spiralis* and

Table 20.2 TRP channels of nematode parasites compared to those in *C. elegans*. *C. elegans* TRP channels from [15]. The *Brugia malayi* and *Meloidogyne incognita* databases were searched for predicted proteins annotated as TRP-like, and then the *B. malayi*, *M. incognita* and *Trichinella spiralis* nucleotide sequence databases searched with each of the *C. elegans* protein sequences, using tblastn. The Accession Numbers are given for specific gene "hits"; where this is not possible (for example where matching sequences were found in genomic sequence that was not part of an annotated predicted gene) this is indicated by the word "present". Putative *B. malayi* and *M. incognita* TRP sequences were reciprocally blasted back against the *C. elegans* database

TRP family	*C. elegans* gene	*C. elegans* gene function	*B. malayi* sequence [35]	*M. incognita* sequence [36]	*T. spiralis* sequence
TRPC	*trp-1*	Expressed in neurons and muscle [52]	Bm1_18625	Minc09957 Minc02878	Present
	trp-2	Expressed in sensory neurons – *trp-1* and *trp-2* worms are defective in nicotine withdrawal and sensitization [53].			Present
	spe-41	Required in sperm for productive sperm-oocyte interactions during fertilization [41].			Present but divergent
TRPV	*ocr-1*	Male chemotaxis to hermaphrodite pheromones (OCR-1, OCR-2, OSM-9 heterotrimer) [54]. Promotes transmitter release (OCR-1, OCR-2, OCR-4 heterotrimer) [42]. OCR-2 and OSM-9 are expressed in sensory cilia and require each other for their correct localization [55]. OCR-3 is expressed in rectal gland cells [55].	Bm1_36495 Bm1_52135	Minc10344 Minc01718	Present Present
	ocr-2				
	ocr-3				
	ocr-4				
	osm-9				
TRPM	*gon-2*	Required for the post-embryonic mitotic cell divisions of gonadal precursor cells [56]. GON-2 and GTL-1 regulate the variability of the defecation oscillator [57, 58].Regulation of electrolyte (especially Mg^{++}) homeostasis [44, 45].	Present	Minc05592 Minc13370 Minc05513	Present
	gtl-1				
	gtl-2				
TRPA	*trpa-1*	Required for specific mechanosensory behaviors [59].			
TRPN	*trp-4*	Expressed in the CEP and ADE dopamine neurons and in two interneurons, DVA and DVC [60].	Bm1_16450 Bm1_38775	Minc14479	Present
TRPP	*lov-1*	Required for male mating behavior [39, 40].			
	pkd-2				
TRPML	*cup-5*	Required for the negative regulation of endocytosis [26].	Bm1_23790	Minc18952	Present

M. incognita genomes contain members of most of the TRP channel sub-families, but not TRPA, nor TRPP; *M. incognita* seems to have more TRPM channels than the animal parasites. The TRPP channels of *C. elegans* are required in males for successful mating and responses to hermaphrodite pheromones [39, 40]; their absence from the parasitic species may reflect differences in mating behavior between the species. This hypothesis may be supported by the absence of an obvious *spe-41* from *B. malayi* and the divergent form of this gene seen in *T. spiralis* since this gene is also involved in mating [41]. The TRPV channels of *C. elegans* function as heteromers [42], and the reduction in TRPV gene number in the parasites may reflect an expression of homomeric channels in these species.

20.2.2.2 Trematodes and Cestodes

The most important trematode, or flatworm, parasites of humans are the schistosomes, or blood flukes [20]. There are three major species of these, *S. mansoni*, *S. japomonicum* and *S. haematobium*: we searched the completed and annotated genome of *S. mansoni* [43] for TRP channel genes (Table 20.3). We also searched the incomplete and unannotated genome of the tapeworm (cestode) *Echinococcus multilocularis* (http://www.sanger.ac.uk/Projects/Echinococcus/) for TRP channel

Table 20.3 The predicted TRP channel genes of *Schistosoma mansoni* and *Echinicoccus multilocularis*. The *S. mansoni* database was searched for predicted genes annotated to encode a TRP channel (indicated as Smp_XXXXXX), then the *S. mansoni* and *E. multilocularis* nucleotide sequences searched using single members of each gene family from *C. elegans* and *D. melanogaster* and tblastn

TRP gene family	Predicted *S. mansoni* genes [43]	Predicted *E. multilocularis* genes
TRPC	Smp_151880 Smp_147860 Smp_163160 Smp_169150 Smp_116240	4 TRPC-like genes
TRPV	None	None
TRPM	Smp_000050 Smp_173720 Smp_035140 Smp_161630 Smp_161610 Smp_165170 Smp_161640 Smp_147140 Smp_130890	4 TRPM-like genes
TRPA	Smp_125690	1 TRPA-like gene
TRPN	None	None
TRPP	1 present	1 TRPP-like gene
TRPML	1 present	1 TRPML-like gene

sequences. These searches produced similar and quite remarkable results; both organisms possess multiple TRPC and TRPM genes (at least nine in *S. mansoni*), single TRPA, TRPP and TRPML genes, and apparently no TRPV or TRPN genes. To date, there is no information of the function of any of these channels; there are ESTs for several of the *S. mansoni* TRP genes, but these seem to be expressed at low levels; Sm_169150 (a TRPC) and Sm_147140 (a TRPM) are the most highly represented, with Sm_147140 having multiple ESTs from adult libraries. In mammals and nematodes, TRPM6 and TRPM7 channels are required for ion homeostasis in the gut [44–46]. Since trematodes and cestodes absorb most of their nutrients, including ions, from the blood or gut contents of the host organisms, it is tempting to speculate that one role of the parasite TRPMs might be in the uptake or release of essential ions. The apparent absence of TRPV channels from these organisms might, as with the parasitic nematodes, reflect the simpler sensory requirements of a parasitic life-style.

20.3 Ectoparasites

"Ectoparasites" is a term that makes biological but very little phylogenetic sense. For the purposes of this chapter, we have chosen to focus on a few examples of insect and arachnid parasites, the bugs, lice and ticks. Bugs, such as bedbugs, are closely related to aphids, which could be considered to be ectoparasites of plants; the genome sequence of the pea aphid, *Acyrthosiphon pisum*, has just been published [47], resulting in an analysis of the ion channel genes, including the TRP channels, present in this organism [48]. We compared the predicted TRP channel genes in *A. pisum* and *D. melanogaster* to those annotated from the human head louse, *Pediculus humanus*, and the tick, *Ixodes scapularis* (Table 20.4). We also carried out some Blast searches using the *Drosophila* peptide sequences as probes. The results, which may be incomplete, show a remarkable conservation in gene number between the various insects and ticks, with members of all seven sub-families of TRP channel present except for TRPV in the tick, and TRPP and TRPML in the aphid. This greater conservation of gene number between parasitic and non-parasitic insects, as compared to nematodes, may reflect the more limited adaptations required for ectoparasitism as opposed to endoparasitism. Ticks, which are not insects but arachnids, find their hosts by detecting their temperature and carbon dioxide "footprints" and so might be expected to conserve some of the temperature detecting channels of other arthropods. Insects detect heat via several TRP channels, including the TRPV channels [49], trpA1 [50] and pyrexia [51]; of these we found a clear homologue for only trpA1 (ISCW011428) in *I. scapularis*. TrpA1 has a role in larval thermotaxis in Drosophila [50] and so might make a good candidate for mediating the same phenomenon in ticks. Head lice move between hosts as adults and seem to have retained a larger number of temperature-sensitive TRP channels to allow them to find a new host at this stage of the life-cycle. The larvae, or nymphs, are not considered to be as infectious, which might explain the apparent loss of the larval thermotaxis gene, trpA1, in the louse.

Table 20.4 The TRP channel genes of ectoparasites compared to those of *Drosophila melanogaster*. The *D. melanogaster* gene information is taken from flybase.org. The louse and tick genes were identified by a simple text search of Genbank and by tblastn searches using the *Drosophila* protein sequences. The aphid genes are taken from [48]

TRP family	*Drosophila* gene	*Drosophila* gene function	Pea aphid genes [48]	Louse (*P. humanus*) genes	Tick (*I. scapularis*) genes
TRPC	*trp* *trpl* *trpγ*	Visual transduction [16]	Ap-trp	PHUM391770 PHUM055080 PHUM086850	ISCW016791 ISCW011394 ISCW016812
TRPV	*Nanchung* *inactive*	Sensory perception [61]	ACYPIG256024 ACYPIG776856	PHUM356850 PHUM581890	
TRPM	CG34123		ACYPIG798608	Present	ISCW013405
TRPA	*trpA1* *waterwitch* *painless* *pyrexia*	Sensory responses, including nociception [51, 62, 63]	ACYPIG433926 ACYPIG601926 ApWTRW-like ACYPIG907328	PHUM216930 PHUM546630	ISCW011428 ISCW010984
TRPN	*nompC*	Sensory perception [64]	ACYPIG949557	PHUM154320	ISCW015199
TRPP	*pkd2*	Sperm motility, larval feeding [65]	None	PHUM557350	ISCW002157
TRPML	*Trpml*	Autophagy; negative regulation of symbiont [66]	None	PHUM586230	ISCW001659

20.4 Conclusions

An immediate conclusion from this brief survey is that, for invertebrates, far more is known about the TRP channels of model than target organisms. Even from the simple bioinformatics searches that we have carried out, it is clear that the TRP channel genes can vary, especially in the protozoa and helminths, and that these differences between species do warrant further investigation. It may be premature to consider TRPs to be viable drug targets, but they are clearly worth considering and exploring further. We hope to read about such studies in the years to come.

Acknowledgments We would like to thank all of those who make their sequence information publically available. The *T. spiralis* sequence data were produced by the Genome Sequencing Center at Washington University School of Medicine in St Louis and can be obtained from http://genome.wustl.edu. Funding for the sequence characterization of the Trichinella genome is being provided by the National Human Genome Research Institute (NHGRI), National Institutes of Health (NIH). The *E. multilocularis* data were produced by the Sanger Centre, Hinxton UK and funded by the Wellcome Tust. The *P. humanus* and *I. scapularis* data were obtained from Vectorbase (www.vectorbase.org), an NIAID Bioinformatics Resource Center for Invertebrate Vectors of Human Pathogens.

References

1. C. elegans Sequencing Consortium (1998) Genome sequence of the nematode *C. elegans*: a platform for investigating biology. Science 282:2012–2018
2. Kumari S, Kumar A, Samant M, Singh N, Dube A (2008) Discovery of Novel Vaccine Candidates and Drug Targets Against Visceral Leishmaniasis Using Proteomics and Transcriptomics. Curr Drug Targets 9:938–947
3. Martin RJ, Murray I, Robertson AP, Bjorn H, Sangster N (1998) Anthelmintics and ion-channels: after a puncture, use a patch. Int J Parasitol 28:849–862
4. Wolstenholme AJ, Rogers AT (2005) Glutamate-gated chloride channels and the mode of action of the avermectin/milbemycin anthelmintics. Parasitology 131:S85–S95
5. Harder A, Holden-Dye L, Walker RJ, Wunderlich F (2005) Mechanisms of action of emodepside. Parasitol Res 97:S1–S10
6. Kaminsky R, Ducray P, Jung M, Clover R, Rufener L, Bouvier J, Schorderet Weber S, Wenger A, Wieland-Berghausen S, Goebel T, Gauvry N, Pautrat F, Skripsky T, Froelich O, Komoin-Oka C, Westlund B, Sluder A, Maser P (2008) A new class of anthelmintics effective against drug-resistant nematodes. Nature 452:176–180
7. Escalante AA, Smith DL, Kim Y (2009) The dynamics of mutations associated with anti-malarial drug resistance in *Plasmodium falciparum*. Trends Parasitol 25:557–563
8. Geary TG, Woo K, McCarthy JS, Mackenzie CD, Horton J, Prichard RK, de Silva NR, Olliaro PL, Lazdins-Helds JK, Engels DA, Bundy DA (2010) Unresolved issues in anthelmintic pharmacology for helminthiases of humans. Int J Parasitol 40:1–13
9. Wilkinson SR, Kelly JM (2009) Trypanocidal drugs: mechanisms, resistance and new targets. Expert Revs Mol Med 11:e31
10. Clayton CE (1999) Genetic manipulation of kinetoplastida. Parasitology Today 15:372–378
11. Thathy V, Menard R (2002) Gene targeting in Plasmodium berghei. Methods Mol Med 72:317–331
12. Trager W, Jensen JB (1976) Human malaria parasites in continuous culture. Science 193:673–675
13. Wu Y, Sifri CD, Lei HH, Wellems TE (1995) Transfection of *Plasmodium falciparum* within human red blood cells. Proc Natl Acad Sci USA 92:973–977
14. Benton R (2008) Chemical sensing in Drosophila. Curr Opin Neurobiol 18:357–363
15. Kahn-Kirby AH, Bargmann CI (2006) TRP channels in *C. elegans*. Ann Rev Physiol 68:719–736
16. Montell C (1999) Visual transduction in *Drosophila*. Ann Rev Cell Dev Biol 15:231–268
17. Xiao R, Xu XZ (2009) Function and regulation of TRP family channels in *C. elegans*. Pflugers Arch 458:851–860
18. Ardelli BF, Stitt LE, Tompkins JB (2010) Inventory and analysis of ATP-binding cassette (ABC) systems in *Brugia malayi*. Parasitology 137:1195–1212
19. Williamson SM, Walsh TK, Wolstenholme AJ (2007) The cys-loop ligand-gated ion channel gene family of *Brugia malayi* and *Trichinella spiralis*: a comparison with *Caenorhabditis elegans*. Inv Neurosci 7:219–226
20. Mathers CD, Ezzati M, Lopez AD (2007) Measuring the burden of neglected tropical diseases: the global burden of disease framework. PLoS Negl Trop Dis 1:e114
21. Obornik M, Janouskovec J, Chrudimsky T, Liukes J (2009) Evolution of the apicoplast and its hosts: from heterotrophy to autotrophy and back again. Int J Parasitol 39:1–12
22. Blackman MJ, Bannister LH (2001) Apical organelles of Apicomplexa: biology and isolation by subcellular fractionation. Mol Biochem Parasitol 117:11–25
23. de Souza W, Attias M, Rodrigues JC (2009) Particularities of mitochondrial structure in parasitic protists (Apicomplexa and Kinetoplastida). Int J Biochem Cell Biol 41:2069–2080
24. Chenik M, Douagi F, Ben Achour Y, Ben Khalef N, Ouakad M, Louzir H, Dellagi K (2005) Characterization of two different mucolipin-like genes from *Leishmania major*. Parasitol Res 98:5–13

25. Palmer CP, Zhou XL, Lin J, Loukin SH, Kung C, Saimi Y:A (2001) TRP homolog in *Saccharomyces cerevisiae* forms an intracellular Ca(2+)-permeable channel in the yeast vacuolar membrane. Proc Natl Acad Sci USA 98:7801–7805
26. Fares H, Greenwald I (2001) Regulation of endocytosis by CUP-5, the *Caenorhabditis elegans* mucolipin-1 homolog. Nat Genet 28:64–68
27. Nagamune K, Moreno SN, Chini EN, Sibley LD (2008) Calcium regulation and signaling in apicomplexan parasites. Subcell Biochem 47:70–81
28. Pryor PR, Reimann F, Gribble FM, Luzio JP (2006) Mucolipin-1 is a lysosomal membrane protein required for intracellular lactosylceramide traffic. Traffic 7:1388–1398
29. Martina JA, Lelouvier B, Puertollano R (2009) The calcium channel mucolipin-3 is a novel regulator of trafficking along the endosomal pathway. Traffic 10:1143–1156
30. Cesbron-Delauw MF, Gendrin C, Travier L, Ruffiot P, Mercier C (2008) Apicomplexa in mammalian cells: trafficking to the parasitophorous vacuole. Traffic 9:657–664
31. Ravindran S, Boothroyd JC (2008) Secretion of proteins into host cells by Apicomplexan parasites. Traffic 9:647–656
32. Rohloff P, Docampo R (2008) A contractile vacuole complex is involved in osmoregulation in *Trypanosoma cruzi*. Exp Parasitol 118:17–24
33. Andrade LF, Andrews NW (2004) Lysosomal fusion is essential for the retention of *Trypanosoma cruzi* inside host cells. J Exp Med 200:1135–1143
34. Wilson J, Huynh C, Kennedy KA, Ward DM, Kaplan J, Aderem A, Andrews NW (2008) Control of parasitophorous vacuole expansion by LYST/Beige restricts the intracellular growth of Leishmania amazonensis. PLoS Pathogens 4:e100179
35. Ghedin E, Wang S, Spiro D, Caler E, Zhao Q, Crabtree J, Allen JE, Delcher AL, Guiliano DB, Miranda-Saavedra D, Angiuoli SV, Creasy T, Amedeo P, Haas B, El-Sayed NM, Wortman JR, Feldblyum T, Tallon L, Schatz M, Shumway M, Koo H, Salzberg SL, Schobel S, Pertea M, Pop M, White O, Barton GJ, Carlow CKS, Crawford MJ, Daub J, Dimmic MW, Estes CF, Foster JM, Ganatra M, Gregory WF, Johnson NM, Jin J, Komuniecki R, Korf I, Kumar S, Laney S, Li B-W, Li W, Lindblom TH, Lustigman S, Ma D, Maina CV, Martin DMA, McCarter JP, McReynolds L, Mitreva M, Nutman TB, Parkinson J, Peregrin-Alvarez JM, Poole C, Ren Q, Saunders L, Sluder AE, Smith K, Stanke M, Unnasch TR, Ware J, Wei AD, Weil G, Williams DJ, Zhang Y, Williams SA, Fraser-Liggett C, Slatko B, Blaxter ML, Scott AL (2007) Draft genome of the filarial nematode parasite Brugia malayi. Science 317:1756–1760
36. Abad P, Gouzy J, Aury JM, Castagnone-Sereno P, Danchin EGJ, Deleury E, Perfus-Barbeoch L, Anthouard V, Artiguenave F, Blok VC, Caillaud MC, Coutinho PM, Dasilva C, De Luca F, Deau F, Esquibet M, Flutre T, Goldstone JV, Hamamouch N, Hewezi T, Jaillon O, Jubin C, Leonetti P, Magliano M, Maier TR, Markov GV, McVeigh P, Pesole G, Poulain J, Robinson-Rechavi M, Sallet E, Segurens B, Steinbach D, Tytgat T, Ugarte E, van Ghelder C, Veronico P, Baum TJ, Blaxter M, Bleve-Zacheo T, Davis EL, Ewbank JJ, Favery B, Grenier E, Henrissat B, Jones JT, Laudet V, Maule AG, Quesneville H, Rosso MN, Schiex T, Smant G, Weissenbach J, Wincker P (2008) Genome sequence of the metazoan plant-parasitic nematode *Meloidogyne incognita*. Nat Biotechnol 26:909–915
37. Wylie T, Martin JC, Dante M, Mitreva MD, Clifton SW, Chinwalla A, Waterston RH, Wilson RK, McCarter JP (2004) Nematode.net: a tool for navigating sequences from parasitic and free-living nematodes. Nucleic Acids Res 32:D423–D426
38. Blaxter ML, De Ley P, Garey JR, Liu LX, Scheldeman P, Vierstrate A, Vanfleteren JR, Mackey LY, Dorris M, Frisse LM, Vida JT, Thomas WK (1998) A molecular evolution framework for the phylum Nematoda. Nature 392:71–75
39. Barr MM, DeModena J, Braun D, Nguyen CQ, Hall DH, Sternberg PW (2001) The *Caenorhabditis elegans* autosomal dominant polycystic kidney disease gene homologs *lov-1* and *pkd-2* act in the same pathway. Curr Biol 11:1341–1346

40. Barr MM, Sternberg PW (1999) A polycystic kidney-disease gene homologue required for male mating behaviour in *C. elegans*. Nature 401:386–389
41. Xu XZS, Sternberg PW (2003) A *C. elegans* sperm TRP protein required for sperm-egg interactions during fertilization. Cell 114:285–297
42. Jose AM, Bany IA, Chase DL, Koelle MR (2006) A specific subset of TRPV channels in *Caenorhabditis elegans* endocrine cells function as mixed heteromers to promote neurotransmitter release. Genetics 175:93–105
43. Berriman M, Haas BJ, LoVerde PT, Wilson RA, Dillon GP, Cerqueira GC, Mashiyama ST, Al-Lazikani B, Andrade LF, Ashton PD, Aslett MA, Bartholomeu DC, Blandin G, Caffrey CR, Coghlan A, Coulson R, Day TA, Delcher A, DeMarco R, Djikeng A, Eyre T, Gamble JA, Ghedin E, Gu Y, Hertz-Fowler C, Hirai H, Hirai Y, Houston R, Ivens A, Johnston DA, Lacerda D, Macedo CD, McVeigh P, Ning ZM, Oliveira G, Overington JP, Parkhill J, Pertea M, Pierce RJ, Protasio AV, Quail MA, Rajandream MA, Rogers J, Sajid M, Salzberg SL, Stanke M, Tivey AR, White O, Williams DL, Wortman J, Wu WJ, Zamanian M, Zerlotini A, Fraser-Liggett CM, Barrell BG, El-Sayed NM (2009) The genome of the blood fluke *Schistosoma mansoni*. Nature 460:352–365
44. Teramoto T, Lambie EJ, Iwasaki K (2005) Differential regulation of TRPM channels governs electrolyte homeostasis in the *C. elegans* intestine. Cell Metab 1:343–354
45. Teramoto T, Sternick LA, Kage-Nakadai E, Sajjadi S, Siembida J, Mitani S, Iwasaki K, Lambie EJ (2010) Magnesium excretion in *C. elegans* requires the activity of the GTL-2 TRPM channel. PLoS One 5:e9589
46. Schmitz C, Perraud AL, Johnson CO, Inabe K, Smith MK, Penner R, Kurosaki T, Fleig A, Scharenberg AM (2003) Regulation of vertebrate cellular Mg^{2+} homeostasis by TRPM7. Cell 114:191–200
47. The International Aphid Genomics Consortium (2010) Genome Sequence of the Pea Aphid *Acyrthosiphon pisum*. PLoS Biol 8:e1000313
48. Dale RP, Jones AK, Tamborindeguy C, Davies TGE, Amey JS, Williamson S, Wolstenholme A, Field LM, Williamson MS, Walsh TK, Sattelle DB (2010) Identification of ion channel genes in the *Acyrthosiphon pisum* genome. Insect Mol Biol 19(Suppl 2):141–153
49. Chentsova NA, Gruntenko NE, Bogomolova EV, Adonyeva NV, Karpova EK, Rauschenbach IY (2002) Stress response in *Drosophila melanogaster* strain *inactive* with decreased tyramine and octopamine contents. J Comp Physiol B 172:643–650
50. Rosenzweig M, Brennan KM, Taylor TD, Phelps PO, Patapoutian A, Garrity PA (2005) The *Drosophila* ortholog of vertebrate TRPA1 regulates thermotaxis. Genes Dev 19:419–424
51. Lee Y, Lee Y, Lee J, Bang S, Hyun S, Kang J, Hong ST, Bae E, Kaang BK, Kim J (2005) Pyrexia is a new thermal transient receptor potential channel endowing tolerance to high temperatures in *Drosophila melanogaster*. Nat Genet 37:305–310
52. Colbert HA, Smith TL, Bargmann CI (1997) OSM-9, a novel protein with structural similarity to channels, is required for olfaction, mechanosensation, and olfactory adaptation in *Caenorhabditis elegans*. J Neurosci 17:8259–8269
53. Feng ZY, Li W, Ward A, Piggott BJ, Larkspur ER, Sternberg PW, Xu XZS (2006) A *C. elegans* model of nicotine-dependent behavior: regulation by TRP-family channels. Cell 127:621–633
54. White JQ, Nicholas TJ, Gritton J, Truong L, Davidson ER, Jorgensen EM (2007) The sensory circuitry for sexual attraction in *C. elegans* males. Curr Biol 17:1847–1857
55. Tobin D, Madsen D, Kahn-Kirby AH, Peckol E, Moulder G, Barstead R, Maricq A, Bargmann C (2002) Combinatorial expression of TRPV channel proteins defines their sensory functions and subcellular localization in *C. elegans* neurons. Neuron 35:307–318
56. West RJ, Sun AY, Church DL, Lambie EJ (2001) The *C. elegans gon-2* gene encodes a putative TRP cation channel protein required for mitotic cell progression. Gene 266:103–110
57. Kwan CS, Vazquez-Manrique RP, Ly S, Goyal K, Baylis HA (2008) TRPM channels are required for rhythmicity in the ultradian defecation rhythm of *C. elegans*. BMC Physiol 8:11
58. Xing J, Yan XH, Estevez A, Strange K (2008) Highly Ca^{2+}-selective TRPM channels regulate IP3-dependent oscillatory Ca^{2+} signaling in the *C. elegans* intestine. J Gen Physiol 131:245–255

59. Kindt KS, Viswanath V, Macpherson L, Quast K, Hu H, Patapoutian A, Schafer WR (2007) *Caenorhabditis elegans* TRPA-1 functions in mechanosensation. Nat Neurosci 10:568–577
60. Kindt KS, Quast KB, Giles AC, De S, Hendrey D, Nicastro I, Rankin CH, Schafer WR (2007) Dopamine mediates context-dependent modulation of sensory plasticity in *C. elegans*. Neuron 55:662–676
61. Kim J, Chung YD, Park DY, Choi S, Shin DW, Soh H, Lee HW, Son W, Yim J, Park CS, Kernan MJ, Kim C (2003) A TRPV family ion channel required for hearing in *Drosophila*. Nature 424:81–84
62. Hamada FN, Rosenzweig M, Kang K, Pulver SR, Ghezzi A, Jegla TJ, Garrity PA (2008) An internal thermal sensor controlling temperature preference in *Drosophila*. Nature 454: 217–220
63. Tracey WD, Wilson RI, Laurent G, Benzer S (2003) *painless*, a *Drosophila* gene essential for nociception. Cell 113:261–273
64. Gopfert MC, Robert D (2003) Motion generation by *Drosophila* mechanosensory neurons. Proc Natl Acad Sci USA 100:5514–5519
65. Gao Z, Ruden DM, Lu X (2003) PKD2 cation channel is required for directional sperm movement and male fertility. Curr Biol 13:2175–2178
66. Venkatachalam K, Long AA, Elsaesser R, Nikolaeva D, Broadie K, Montell C (2008) Motor deficit in a *Drosophila* model of mucolipidosis type IV due to defective clearance of apoptotic cells. Cell 135:838–851

Chapter 21
Receptor Signaling Integration by TRP Channelsomes

Yasuo Mori, Taketoshi Kajimoto, Akito Nakao, Nobuaki Takahashi, and Shigeki Kiyonaka

Abstract Homologues of transient receptor potential (TRP) genes encode a variety of cation channels, most of which conduct Ca^{2+} across the plasma membrane. TRP proteins interact with a variety of proteins and other biologically important factors, such as second messengers, and thereby form "channelsomes", most of which function as Ca^{2+} signalsomes. Activation mechanisms and final outputs are exquisitely incorporated in the signaling system of TRP channelsomes. In this study, we discuss the channelsomes of TRPC3, TRPC5, and TRPM2, which show unique molecular interactions and modulations of activation. Comparative studies of these specific TRP channelsomes should aid the determination of general rules that govern the formation and regulation of channelsomes and signalsomes.

21.1 Introduction

Virtually all TRP channels respond to multiple activation triggers and modulators, therefore serving as polymodal signal detectors. For example, the sensitivity of TRPV1 to noxious heat can be greatly enhanced by inflammatory agents that activate phospholipase C (PLC) signaling pathways [1]. Such integration allows TRPV1 to detect subthreshold stimuli, and provides a mechanism through which tissue injury produces thermal hypersensitivity [2]. Moreover, for TRPM8, significant and maximal activation by the super-cooling agent icilin requires intracellular Ca^{2+} [3]. The codependency of activation, wherein channel gating requires the simultaneous presence of two agonists, suggests that some TRP channels serve as coincidence detectors. As will be discussed in this chapter, another important aspect of the multimodality of the activation sensitivity of TRP channels is integration and amplification of cellular signals. When a TRP channel that is part of a specific signaling cascade is activated by downstream or upstream constituents

Y. Mori (✉)
Department of Synthetic Chemistry and Biological Chemistry, Graduate School of Engineering, Kyoto University, Kyoto 615-8510, Japan
e-mail: mori@sbchem.kyoto-u.ac.jp

(molecules/proteins/enzymes) of the cascade, in addition to the immediately upstream activation trigger, the cascade is equipped with positive feedback or feedforward loops. This mechanism ensures the fidelity of cellular responses and minimizes the variability of their magnitude, which might contribute to the synchronization of responses in neighboring cells that comprise functional domains in tissues. Thus, TRP channel research is progressing from functional identification of single molecules to analysis and integration of "channelsomes" as molecular systems. In this chapter, we discuss the molecular composition and physiological function of TRP channelsomes, focusing on our recent studies on three subtypes of TRPs: TRPC3, TRPC5, and TRPM2.

21.2 TRPC3

Invertebrate and vertebrate TRP homologues of the so-called "canonical" TRPC subfamily are characterized by channel activation induced upon stimulation of PLC-coupled receptors. TRPCs were originally hypothesized to encode store-operated Ca^{2+} channels (SOCs), and some supportive evidence for this hypothesis was obtained from cDNA expression and gene knockout experiments for various TRPCs [4, 5]. However, store-independent activation of Ca^{2+} influx and cation currents appears to be the most common function of this channel family [5–7]. Among the seven members of the vertebrate TRPCs, TRPC2, TRPC3, TRPC6, and TRPC7 have been reported to be activated by diacylglycerol (DAG) [6, 8, 9]. The physiological significance of SOCs and their roles in lymphocyte Ca^{2+} signaling has been extensively documented [10]. With regard to the physiological importance of DAG-activated cation channels, previous studies have demonstrated their function as nonselective cation channels for inducing membrane depolarization. This depolarization activates voltage-dependent channels to induce action potentials [9] and/or depolarization-induced Ca^{2+} influx, which are responsible for Ca^{2+}-dependent cellular responses, such as muscle contraction [11, 12] and activation of transcription factors/nuclear factors of activated T cells (NFAT) [13, 14]. However, in contrast to the depolarizing-function in excitable cells, the physiological significance of Ca^{2+} entry occurring directly through DAG-activated cation channels and subsequent Ca^{2+} signals remained largely unknown.

In non-excitable cells, receptor-activated Ca^{2+} signaling comprises initial transient responses followed by a Ca^{2+} entry-dependent sustained and/or oscillatory phase. We have shown that receptor-activated TRPC3 channels functionally and physically interact with PLCγ2 to enhance the second phase of Ca^{2+} signals [15]. An in vivo inositol 1,4,5-trisphosphate (IP_3) sensor revealed that in avian DT40 B lymphocytes, receptor-activated and store-operated Ca^{2+} entry greatly enhances IP_3 production, which terminates in PLCγ2-deficient cells [15, 16]. Physical association between TRPC3 Ca^{2+} channels and PLCγ2 is demonstrated by co-immunoprecipitation. This interaction requires a domain in PLCγ that includes a partial pleckstrin homology (PH) domain–a consensus lipid-binding and protein-binding sequence [17]. The partial PH domain of PLCγ interacts with a complementary partial PH-like domain in TRPC3 to elicit lipid binding and cell-surface expression of TRPC3 [18]. In DT40 cells, PLCγ2 deficiency diminishes Ca^{2+} entry-induced Ca^{2+} responses. This defect is canceled by suppressing IP_3-induced Ca^{2+} release, implying critical contributions of IP_3 and IP_3 receptors to the second Ca^{2+} phase. Furthermore, confocal visualization of PLCγ2 mutants demonstrates that Ca^{2+} entry evokes a C2 domain-mediated PLCγ2 translocation towards the plasma membrane in a lipase-independent manner to activate PLCγ2. Strikingly, Ca^{2+} entry-activated PLCγ2 maintains Ca^{2+} oscillation and extracellular signal-regulated kinase (Erk) activation. These studies reveal a novel molecular mechanism that amplifies cellular signals in the second phase of the Ca^{2+} response in B-lymphocytes. The amplification mechanism is a positive feedback cycle, in which a key event is the Ca^{2+} entry-evoked translocation towards the plasma membrane and the subsequent PLCγ2 activation/IP_3 production, which then leads to Ca^{2+} release and entry (Fig. 21.1a).

Our most recent data clearly demonstrates the important role played by TRPC3 as a DAG-activated Ca^{2+} entry channel in B-cell signaling [19]. In DT40 B lymphocytes, genetic disruption of translocation of TRPC3 proteins to the plasma membrane has revealed that native TRPC3 forms DAG-activated Ca^{2+} entry channels, but not SOCs. Upon B-cell receptor (BCR) activation, the DAG-activated Ca^{2+} influx via TRPC3 amplifies Ca^{2+} signals, and downstream NFAT activation, by controlling plasma membrane translocation of PLCγ2 and sustains protein kinase Cβ (PKCβ) translocation and activation. Similarly, another DAG-activated channel, TRPC6, signals the membrane translocation and activation of PKCβ, thereby

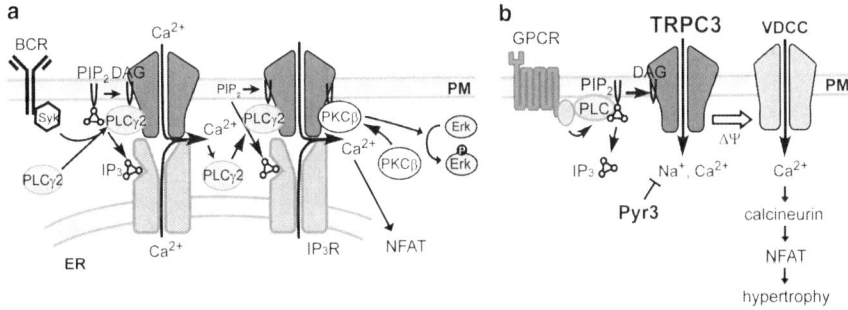

Fig. 21.1 (a) The proposed model for the role of DAG-activated Ca^{2+} influx mediated by TRPC3 in BCR signaling. TRPC3-mediated Ca^{2+} influx promotes PLCγ2 translocation from the cytosol to the PM. The targeted PLCγ2 elicits secondary production of IP_3 and DAG to amplify Ca^{2+} signaling via IP_3 receptor-mediated Ca^{2+} release and TRPC3-mediated Ca^{2+} influx, respectively. TRPC3-mediated Ca^{2+} influx is also required for sustained PKCβ membrane translocation and activation, resulting in persistent activation of Erk. PM: plasma membrane, ER: endoplasmic reticulum. (b) DAG-activated TRPC3 is essential for Ang II-induced NFAT activation and cardiac hypertrophy. Activation of TRPC3 causes cation influx responsible for depolarization of membrane potentials to a positive direction ($\Delta\psi$) and concomitantly increases the frequency of spontaneous firing due to activation of voltage-dependent Ca^{2+} channel (VDCC). TRPC3-selective inhibitor Pyr3 attenuates NFAT activation and hypertrophic growth. GPCR: G-protein coupled receptor.

inducing RhoA activation and endothelial contraction [20]. Furthermore, direct physical interaction between TRPC3 and PKCβ also regulates the stable retention of PKCβ at the plasma membrane, leading to sustained BCR-induced Erk activation [19]. These results suggest that TRPC3 has a dual function in BCR-induced signaling: it is a DAG-activated Ca^{2+} entry channel, which elicits plasma membrane translocation of PLCγ2 and PKCβ, and a scaffolding platform at the plasma membrane for PKCβ.

As summarized in Fig. 21.1a, TRPC3 mediates Ca^{2+} influx to initially amplify the production of IP_3 and DAG, the former further amplifying Ca^{2+} signals and Ca^{2+}-dependent transcription factor NFAT, and the latter sustaining PKCβ and Erk signals in B lymphocytes. It is very likely that formation of protein complexes by TRPC3 with PLCγ2, and subsequently with PKCβ, secures this process. TRPC3 might also act as a scaffold for IP_3 receptors, receptors for activated C-kinase-1 (RACK1), and SOC protein Orai1, as reported previously [21–23]. Importantly, the formation of this TRPC3 channelsome is not static, but dynamic; plasma membrane translocation and consequent interactions with TRPC3 of key molecules such as PLCγ2 and PKCβ are driven by Ca^{2+} influx. In this context, it is extremely interesting to examine association/dissociation kinetics of respective proteins on TRPC3 and relate them to the dynamics of cellular signals. Such a study should lead to an understanding of the molecular mechanism of the encoding and decoding of cellular information.

TRPC3 channels also play important roles in pathophysiological processes, such as cardiac hypertrophy induced by hypertension and mechanical load abnormalities [24]. In in vitro cultured cardiomyocytes, Ca^{2+} influx through TRPC3 and TRPC6 activation is essential for NFAT activation and cardiomyocyte hypertrophy [14] induced by Angiotensin (Ang) II receptors, that activate PLC [25]. DAG produced by Ang II-induced PLC activation directly activates TRPC3 and TRPC6, and the resulting cation (Na^+ and Ca^{2+}) influx changes the membrane potential to positive, leading to activation of voltage-dependent Ca^{2+} channels, possibly through the generation of an action potential. The increase in Ca^{2+} influx through voltage-dependent Ca^{2+} channels activates the calcineurin/NFAT pathway and hypertrophic responses in rat neonatal cardiomyocytes. In support of these findings, RNA interference (RNAi) screening for ion channels that mediate Ca^{2+}-dependent gene expression identified TRPC3 as a mediator of cardiac hypertrophy [26]. Notably, TRPC3 expression is upregulated in multiple rodent models of pathological cardiac hypertrophy [27]. In addition, TRPC3 or TRPC6 transgenic mice show increased calcineurin-NFAT activation in vivo, cardiomyopathy, and increased hypertrophy after neuroendocrine agonist or pressure overload stimulation [28, 29]. These reports suggest that the DAG-induced Ca^{2+} signaling pathway through TRPC3 and TRPC6 is essential for neuroendocrine agonist- or pressure overload stimulation-induced NFAT activation and cardiac hypertrophy.

Selective inhibitor of TRPC3 channels is extremely useful for selectively dissecting the function of the native TRPC3 channel itself and the multiple roles of channelsomes formed by TRPC3. We have reported that a pyrazole compound

(Pyr3) selectively inhibits TRPC3 channels [30]. Structure-function relationship studies of pyrazole compounds show that the trichloroacrylic amide group is important for the TRPC3 selectivity of Pyr3. Electrophysiological and photoaffinity labeling experiments reveal a direct action of Pyr3 on the TRPC3 protein. In DT40 B lymphocytes, Pyr3 potently eliminates the Ca^{2+} influx-dependent PLCγ2 translocation to the plasma membrane and the late oscillatory phase of B cell receptor-induced Ca^{2+} response. Moreover, Pyr3 attenuates the activation of NFAT, hypertrophic growth in rat neonatal cardiomyocytes, and in vivo pressure overload-induced cardiac hypertrophy in mice. These findings on the important roles of native TRPC3 channels are strikingly consistent with previous genetic studies. Thus, the TRPC3-selective inhibitor Pyr3 is a powerful tool to study the in vivo function of TRPC3 in integrating cellular signals in channelsomes.

21.3 TRPC5

TRPC5 was originally cloned from the mouse brain and functionally identified as a receptor-activated Ca^{2+}-permeable cation channel linked to PLCs [31, 32]. Although it is still controversial whether depletion of Ca^{2+} stores can activate TRPC5, a number of proteins and factors have been identified as acting as direct triggers and modulators of TRPC5 channel activation. For example, binding of intracellular Ca^{2+} and calmodulin have been implicated in TRPC5 activation and modulation [33–36], while membrane polyphosphoinositides, such as phosphatidylinositol 4,5-bisphosphate (PIP_2), exert both stimulatory and inhibitory effects in regulating TRPC5 channel activity [37]. Other proteins, including the structurally closest homologues TRPC4 and TRPC1, interact with the TRPC5 protein [38–44]. As TRPC1 interacts with both TRPC5 and caveolin-1 [38, 45], it is likely that TRPC5 forms protein complexes with caveolin-1 (the importance of this protein interaction will discussed below). An array of these proteins and factors can contribute to TRPC5 channelsomes cooperatively or independently.

Covalent modifications also mediate the action of regulatory factors. We have demonstrated that TRPC5 channels are activated by nitric oxide (NO) via cysteine S-nitrosylation [46]. TRPC1 and TRPC4 of the same subfamily and TRPV1, TRPV3, and TRPV4 of thermosensor TRPV subfamily, are very likely to be activated through comparable mechanisms. Labeling and functional assays using cysteine mutants, together with membrane sidedness in the activating action of reactive disulfides, reveal that cytoplasmically accessible Cys553 and nearby Cys558 are nitrosylation sites mediating NO sensitivity in TRPC5. The responsive TRPs harbor conserved cysteines on the same N-terminal side of the pore region. These findings have identified the structural motif for NO-sensitive activation gate in TRP channels, proposing that "NO sensors" are a new functional category of cellular receptors extending over different TRP families. In endothelial cells, native TRPC5 channels are co-localized with eNOS and are nitrosylated by NO delivered by the endothelial NO synthase (eNOS) upon G-protein-coupled ATP receptor stimulation to elicit Ca^{2+} entry.

Previous reports have provided valuable information that has allowed us to propose a model for the TRPC5 channelsome that regulates receptor-activated NO production in endothelial cells. First, as described above, TRPC1 has been described to form heterotetrameric channels with TRPC5 [45] and to form a protein complex with caveolin-1 in caveolae/lipid raft domains [38, 47], which regulate plasma membrane trafficking of TRPC1 [48]. It is therefore possible that TRPC5 forms protein complexes with caveolin-1 via TRPC1. In fact, we have found an interaction of TRPC5 with caveolin-1 and eNOS by coimmunoprecipitation experiments as well as by colocalization of TRPC5 with caveolin-1 (Mori, unpublished result). Second, among numerous caveolin-associated proteins linked to signaling cascades [49–52], interactions with three isoforms of NOS, such as eNOS, have been identified [53]. Interestingly, the inhibitory association of caveolin is disrupted by the binding of Ca^{2+}-calmodulin (Ca^{2+}-CaM) to eNOS, leading to eNOS activation [54–58]. Third, eNOS is known to be activated by different kinases, including Akt, Protein kinase A, and PKC [59–63]. On the basis of these papers and our own data, we can propose a plausible model for the TRPC5 channelsome in vascular endothelial cells (Fig. 21.2). TRPC5 proteins form protein complexes with receptors for vasodilators, G-proteins, PLCβs, and eNOS on the scaffolding protein caveolin-1 in caveolae. Upon vasodilator receptor stimulation, TRPC5 is activated through the PLCβ-mediated cascade to induce Ca^{2+} influx, which elevates the intracellular Ca^{2+} concentration ($[Ca^{2+}]_i$) and forms Ca^{2+}-CaM. This removes the inhibition of eNOS activity by associated caveolin-1 and leads to an initial NO production, which then activates TRPC5 channels on caveolin lacking receptors, G-proteins, and PLCβs for the initial eNOS activation. Ca^{2+} influx via NO-activated TRPC5 channels induces secondary activation of eNOS to amplify the production of NO, resulting in a feedback cycle of receptor-activated Ca^{2+} and NO signaling. Notably, this model is nicely summarized by Stamler and colleagues on the basis of our data in a short review [64].

Our immunolocalization studies have revealed that TRPC5 is distributed on both the apical and basal membrane in the endothelial cell layer of vascular tissue (Mori, unpublished results). Given that vasodilator receptors are distributed at the apical luminal surface of the endothelial cell layer, NO initially produced there might diffuse across the cytoplasm and activate TRPC5 located at the basolateral membrane. This could lead to an efficient propagation of Ca^{2+} signals directed toward the basal membrane. The feedback mechanism might further contribute to global $[Ca^{2+}]_i$ rises/oscillations and full activation of eNOS at the Golgi [65] of endothelial cells, leading to synchronization of neighboring smooth muscle cells during vascular relaxation as a whole at vascular tissue levels. This idea is substantiated by the fact that genetic disruption of the closest relative, TRPC4, which we demonstrated to be colocalized with TRPC5 in the endothelial cell membrane [46], impairs agonist-dependent vasorelaxation [66]. Interestingly, Greka *et al.* demonstrated the importance of TRPC5 in neurite extension [67]. As NO signals are reported to regulate neurite extension [68], the feedback mechanism might also be important in growth cone morphology. In terms of TRPVs, obtained results raise the possibility for the involvement of nitrosylation-induced Ca^{2+} entry in the exacerbation of heat or pain sensation. TRPC5 channelsomes might also be involved in the activation

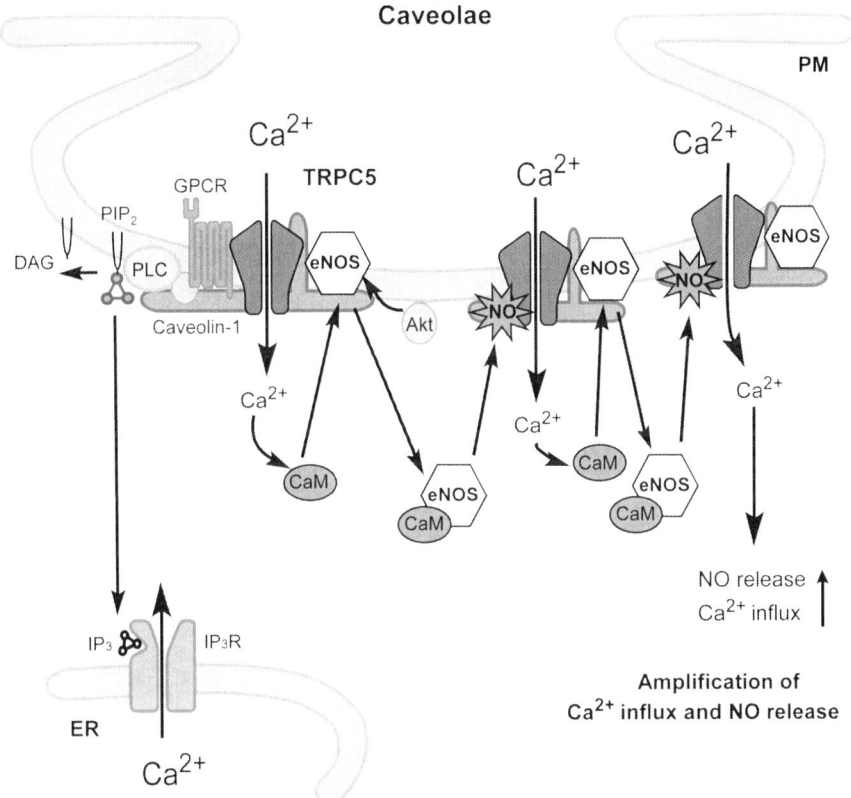

Fig. 21.2 Proposed model for TRPC5-mediated feedback cycle of receptor-activated Ca^{2+} and NO signaling in caveolae of endothelial cells. Stimulation of GPCRs (such as the ATP-activated P2Y receptor) induces Ca^{2+} influx and activation of eNOS as a consequence of binding of Ca^{2+}-CaM and release of eNOS from caveolin-1. TRPC5 undergoes eNOS-dependent S-nitrosylation after GPCR stimulation, resulting in amplified Ca^{2+} entry and secondary activation of eNOS to amplify production of NO. CaM: calmodulin.

of eNOS by shear stress [60, 69], which was thought to proceed through Ca^{2+}-independent mechanism, because membrane stretch has been reported to activate TRPC5 independently of PLC function [70]. Even in the case where initiation of NO production is evoked independently of Ca^{2+}, the secondary amplification phase might be mediated by Ca^{2+} influx via NO-activated TRPC5 channels. Thus, the positive feedback regulation of Ca^{2+} signals by NO-activated TRP channels is involved in diverse biological systems.

21.4 TRPM2

In terms of channelsome formation, TRPM2 is the most unclear among the three TRP homologs discussed in this chapter. Activation of TRPM2 is potently triggered by changes in redox status, specifically by oxidative stress. H_2O_2 application or stimulation with tumor necrosis factor-α (TNFα) evokes Ca^{2+} and/or Na^{2+} influx via native TRPM2 channels. Although we proposed NAD^+ as a mediator such sensitivity [71], it is more likely that metabolites of NAD^+, such as ADP-ribose (ADPR) and cyclic ADP-ribose (cADPR), which were reported as direct activation triggers of TRPM2 [72, 73], mediate H_2O_2-induced TRPM2 activation [74, 75]. ADPR and NAD^+ act from the intracellular side of TRPM2 [76]. In H_2O_2-induced activation, intracellular hydroxyl radicals [77] or a mechanism independent of ADPR have been also proposed [78]. Nicotine acid adenine dinucleotide phosphate (NAADP) has been reported to activate TRPM2 [79]. More recently, TRPM2 acting as a Ca^{2+} release channel has been demonstrated in the internal membrane of lysosomes [80]. In this context, it is interesting to study relationships between TRPM2 and two-pore channels (TPCs) in regulating NAADP-induced Ca^{2+} mobilization [81].

TRPM2-activating NAD^+ metabolites are produced by multiple pathways [82]. For example, a metabolic pathway in which poly(ADPR) polymerized from NAD^+ by poly(ADPR) polymerase is degraded by poly(ADPR) glycohydrolase is involved in generation of ADPR for TRPM2 channels. CD38 also produces ADPR and cADPR. Notably, poly(ADPR) production is tightly linked to DNA damage caused by oxidants and chemicals; however, a link to oxidative stress mediated via reactive oxygen species (ROS) to CD38 and other NAD^+-hydrolyzing enzymes is still elusive.

Activation of TRPM2 is modulated by various factors other than H_2O_2 and NAD^+ metabolites. Lipid metabolites are potent endogenous activators of TRP channels. Arachidonic acid is freed from a phospholipid molecule by the enzyme phospholipase A_2 (PLA_2), which cleaves off the fatty acid, but can also be generated from DAG by DAG lipase. The action of arachidonic acid on TRPM2 activation was first shown as an enhancement of its NAD^+ activation sensitivity [71]. 2'-O-acetyl-ADP-ribose (OAADPr), the unique metabolite of the Sir2 protein deacetylase reaction, binds to the NudT9-H domain and modulates TRPM2 channel activity [83]. In addition to NAD^+ and its metabolites, nucleotide phosphates show interesting modulatory action on TRP channel activation. AMP has been found to exert negative feedback regulation of TRPM2 [74]. Ca^{2+} applied from either the extracellular or the intracellular side is critical for TRPM2 channel activation [71, 84–86]. In particular, intracellular Ca^{2+} alone is likely to be sufficient for TRPM2 activation [87, 88]. Activation of TRPM2 induced by Ca^{2+} is mediated by Ca^{2+}-CaM [86, 87], whose binding site is located at the IQ-like motif in the N-terminus [89]. Tyrosine phosphorylation of TRPM2 after treatment with H_2O_2 or TNFα is important for its activation and function. Protein tyrosine phosphatase-L1 associates with TRPM2 and modulates TRPM2 phosphorylation and channel activation [90]. Electrophysiological data shows that TRPM2 can be robustly potentiated by exposure to warm temperatures, apparently via direct heat-evoked channel gating. Temperature-dependent activation of TRPM2 has a threshold of 33.6–35 °C and a temperature coefficient (Q_{10}) of 32.3–44.4 in the presence of an agonist (such as ADPR, cADPR, or NAD^+), and a threshold of 33.9 °C and a Q_{10} of 15.6 in the absence of an agonist, at –60 mV [91], implying synergistic effects of temperature and agonists on TRPM2 activation. Furthermore, external protons block human

TRPM2 with an IC_{50} of pH 5.3–6.5, by decreasing single-channel conductance, whereas internal protons inhibit TRPM2 with an IC_{50} of pH 6–6.7 [92, 93]. Three titratable residues, His958, Asp964, and Glu994, at the outer opening of the channel pore are responsible for extracellular pH sensitivity, whereas Asp933, located at the C terminus of the S4-S5 linker, is responsible for intracellular pH sensitivity.

Downstream signaling cascades of TRPM2 and their physiological roles have been slowly unraveled in cell responses, such as cell death, insulin secretion, cytokine production, and cell proliferation. The physiological role of TRPM2, based on the clear understanding of regulating signal cascades, is best studied for chemokine production in monocytes/macrophages (Fig. 21.3). TRPM2, activated by ADPR downstream of ROS such as H_2O_2, amplifies Erk signaling via Pyk2 and Ras activation and ROS-induced nuclear translocation of NF-κB via Erk, which leads to production of the inflammatory chemokine CXCL8 in human monocytes. In an inflammation model associated with ROS, *Trpm2*-knockout mice showed attenuation of production of CXCL2, a functional homolog of CXCL8, and consequent suppression of neutrophil infiltration and ulceration. Thus, ROS-evoked

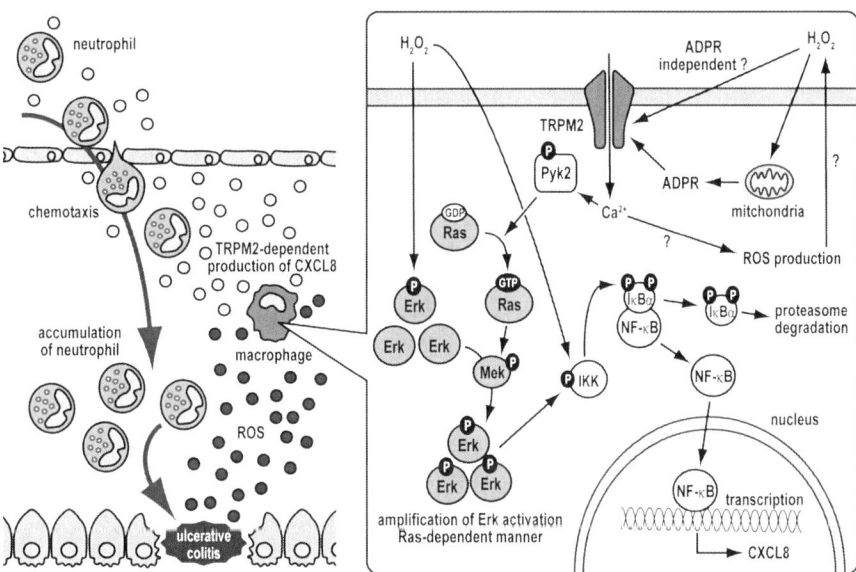

Fig. 21.3 A key role of the TRPM2 on H_2O_2-induced chemokine production in monocytes and macrophages. Chemotactic cytokines, known as the chemokines, recruits inflammatory cells to inflamed sites. For example, CXCL8 shows potent neutrophil chemotactic activity. Once monocytes adhere to endothelial cells from the bloodstream and migrate into tissues, they differentiate into macrophages. During inflammation they are the main effectors of innate immunity because of their antimicrobial activity and production of proinflammatory cytokines. Ca^{2+} influx via H_2O_2-activated TRPM2 mediates amplification of Erk activation through Pyk2-Ras phosphorylation cascade in macrophages. This results in NF-κB nuclear translocation, which leads to CXCL8 production. ADPR and protein modification processes may mediate TRPM2 activation and Ca^{2+} entry upon H_2O_2 stimulation. Analyses of protein-protein interaction are essential to establish the TRPM2 channelsome.

Ca^{2+} influx via TRPM2 might be central to inflammation, indicating that functional inhibition of TRPM2 channels might be a new therapeutic strategy for treating inflammatory diseases [94]. TRPM2 confers susceptibility to cell death induced by oxidative stress or TNFα, which is dependent on an increase in [Ca^{2+}]$_i$ [71, 90]. More recently, TRPM2-dependent lysosomal Ca^{2+} release has been suggested to induce H$_2$O$_2$-mediated pancreatic β cell death [80]. Amyloid β-peptide (Aβ) is the main component of senile plaques that characterize Alzheimer's disease and might induce neuronal death through mechanisms that include oxidative stress. As TRPM2 is involved in oxidative stress-induced cell death [71, 90], activation of TRPM2 might contribute to Aβ- and oxidative stress-induced neuronal cell death [95]. TRPM2 also plays an important role in the regulation of insulin secretion in the pancreas at body temperature [91, 96], and might also be involved in diabetes mellitus. TRPM2 is also essential for prostate cancer cell proliferation, independently of the activity of poly(ADPR) polymerases [97]. Furthermore, a heterozygous variant of *TRPM2* was found in a subset of Guamanian ALS and PD cases by epidemiological studies. Functional studies revealed that the TRPM2 variant is inactivated. The ability of TRPM2 to maintain sustained ion influx is a physiologically important function and its disruption can, under certain conditions, contribute to disease states [98]. An association has been suggested between bipolar disorder and a genetic variant of *TRPM2*, SNP rs1556314, in exon 11 [99, 100].

Channelsomes integrating multiple modes of upstream activation mechanisms and downstream cellular responses are still elusive for TRPM2. However, on the basis of the models that we proposed for chemokine production in monocytes and macrophages, analyses of protein interactions using proteomic approaches and two-hybrid screening will provide a more complete picture for several TRPM2-regulating signaling mechanisms in the near future. This should widen our understanding of Ca^{2+} signaling and the physiological functions of cation channels.

21.5 Conclusion

To fully understand TRP channelsomes, information on their three dimensional (3D) structures is essential. However, at the atomic level, 3D structures are presently available for only subfragments of a few TRPs [101–104]. 3D images using electron microscopic analyses have been revealed for several other TRP channels [105–108]. In TRPC3, the large intracellular domain represents a "nested-box" structure: a wire frame outer shell that functions as a sensor for activators and modulators, and a globular inner chamber that might modulate ion flow, as it is aligned tandemly along the central axis with the dense membrane-spanning core. The transmembrane domain demonstrates a pore-forming property. Critically, the structure implies that the TRP superfamily has evolved diversely as sensors that are specialized for various signals, rather than as simple ion-conducting apparatuses. Once the resolution of 3D images is improved, it will be extremely interesting to map signaling protein associations on the core complex of TRP channels. This information, together with analyses of the dynamics of interactions using techniques such as free resonance energy transfer assay, will provide an important basis for understanding protein-protein interactions in terms of the functions of TRP channelsomes.

References

1. Tominaga M, Wada M, Masu M (2001) Potentiation of capsaicin receptor activity by metabotropic ATP receptors as a possible mechanism for ATP-evoked pain and hyperalgesia. Proc Natl Acad Sci USA 98:6951–6956
2. Julius D, Basbaum AI (2001) Molecular mechanisms of nociception. Nature 413:203–210
3. Chuang H, Neuhausser WM, Julius D (2004) The super-cooling agent icilin reveals a mechanism of coincidence detection by a temperature-sensitive TRP channel. Neuron 43:859–869
4. Zhu X, Jiang M, Peyton M, Boulay G, Hurst R, Stefani E, Birnbaumer L (1996) *trp*, a novel mammalian gene family essential for agonist-activated capacitative Ca^{2+} entry. Cell 85:661–671
5. Parekh AB, Putney JW Jr (2005) Store-operated calcium chanels. Physiol Rev 85:757–810
6. Hofmann T, Obukhov AG, Schaefer M, Harteneck C, Gudermann T, Schultz G (1999) Direct activation of human TRPC6 and TRPC3 channels by diacylglycerol. Nature 397:259–263
7. Bird GS, Aziz O, Lievremont JP, Wedel BJ, Trebak M, Vazquez G, Putney JW Jr (2004) Mechanisms of phospholipase C-regulated calcium entry. Curr Mol Med 4:291–301
8. Okada T, Inoue R, Yamazaki K, Maeda A, Kurosaki T, Yamakuni T, Tanaka I, Shimizu S, Ikenaka K, Imoto K, Mori Y (1999) Molecular and functional characterization of a novel mouse transient receptor potential protein homologue TRP7. J Biol Chem 274: 27359–27370
9. Lucas P, Ukhanov K, Leinders-Zufall T, Zufall F (2003) A diacylglycerolgated cation channel in vomeronasal neuron dendrites is impaired in *TRPC2* mutant mice: mechanism of pheromone transduction. Neuron 40:551–561
10. Gallo EM, Canté-Barrett K, Crabtree GR (2006) Lymphocyte calcium signaling from membrane to nucleus. Nat Immunol 7:25–32
11. Inoue R, Okada T, Onoue H, Hara Y, Shimizu S, Naitoh S, Ito Y, Mori Y (2001) The transient receptor potential protein homologue TRP6 is the essential component of vascular α_1-adrenoceptor-activated Ca^{2+}-permeable cation channel. Circ Res 88:325–332
12. Welsh DG, Morielli AD, Nelson MT, Brayden JE (2002) Transient receptor potential channels regulate myogenic tone of resistance arteries. Circ Res 90:248–250
13. Thebault S, Flourakis M, Vanoverberghe K, Vandermoere F, Roudbaraki M, Lehen'kyi V, Slomianny C, Beck B, Mariot P, Bonnal JL, Mauroy B, Shuba Y, Capiod T, Skryma R, Prevarskaya N (2006) Differential role of transient receptor potential channels in Ca^{2+} entry and proliferation of prostate cancer epithelial cells. Cancer Res 66:2038–2047
14. Onohara N, Nishida M, Inoue R, Kobayashi H, Sumimoto H, Sato Y, Mori Y, Nagao T, Kurose H (2006) TRPC3 and TRPC6 are essential for angiotensin II-induced cardiac hypertrophy. EMBO J 25:5305–5316
15. Nishida M, Sugimoto K, Hara Y, Mori E, Morii T, Kurosaki T, Mori Y (2003) Amplification of receptor signaling by entry-mediated translocation and activation of PLCγ2 in B lymphocytes. EMBO J 22:4677–4688
16. Sugimoto K, Nishida M, Otsuka M, Makino K, Ohkubo K, Mori Y, Morii T (2004) Novel real-time sensors to quantitatively assess in vivo inositol 1,4,5-trisphosphate production in intact cells. Chem Biol 11:475–485
17. Patterson RL, van Rossum DB, Ford DL, Hurt KJ, Bae SS, Suh PG, Kurosaki T, Snyder SH, Gill DL (2002) Phospholipase C-γ is required for agonist-induced Ca^{2+} entry. Cell 111: 529–541

18. van Rossum DB, Patterson RL, Sharma S, Barrow RK, Kornberg M, Gill DL, Snyder SH (2005) Phospholipase Cγ1 controls surface expression of TRPC3 through an intermolecular PH domain. Nature 434:99–104
19. Numaga T, Nishida M, Kiyonaka S, Kato K, Katano M, Mori E, Kurosaki T, Inoue R, Hikida M, Putney JW Jr, Mori Y (2010) Ca^{2+} influx and protein scaffolding via TRPC3 sustain PKCβ and ERK activation in B Cells. J Cell Sci 123:927–938
20. Singh I, Knezevic N, Ahmmed GU, Kini V, Malik AB, Mehta D (2007) $G\alpha_q$-TRPC6-mediated Ca^{2+} entry induces RhoA activation and resultant endothelial cell shape change in response to thrombin. J Biol Chem 282:7833–7843
21. Boulay G, Brown DM, Qin N, Jiang M, Dietrich A, Zhu MX, Chen Z, Birnbaumer M, Mikoshiba K, Birnbaumer L (1999) Modulation of Ca^{2+} entry by polypeptides of the inositol 1,4,5-triphosphate receptor (IP_3R) that bind transient receptor potential (TRP): evidence for roles of TRP and IP_3R in store depletion-activated Ca^{2+} entry. Proc Natl Acad Sci USA 96:14955–14960
22. Bandyopadhyay BC, Ong HL, Lockwich TP, Liu X, Paria BC, Singh BB, Ambudkar IS (2008) TRPC3 controls agonist-stimulated intracellular Ca^{2+} release by mediating the interaction between inositol 1,4,5-triphosphate receptor and RACK1. J Biol Chem 283: 32821–32830
23. Woodard GE, López JJ, Jardín I, Salido GM, Rosado JA (2010) TRPC3 regulates agonist-stimulated Ca^{2+} mobilization by mediating the interaction between type I inositol 1,4,5-triphosphate receptor, RACK1, and Orai1. J Biol Chem 285:8045–8053
24. Frey N, Olson EN (2003) Cardiac hypertrophy: the good, the bad, and the ugly. Annu Rev Physiol 65:45–79
25. Molkentin JD, Dorn GW II (2001) Cytoplasmic signaling pathways that regulate cardiac hypertrophy. Annu Rev Physiol 63:391–426
26. Brenner J, Dolmetsch RE (2007) TrpC3 regulates hypertrophy-associated gene expression without affecting myocyte beating or cell size. PLoS One 2:e802
27. Bush EQ, Hood DB, Papst PJ, Chapo JA, Minobe W, Bristow MR, Olson EN, McKinsey TA (2006) Canonical transient receptor potential channels promote cardiomyocyte hypertrophy through activation of calcineurin signaling. J Biol Chem 281:33487–33496
28. Kuwahara K, Wang Y, McAnally J, Richardson JA, Bassel-Duby R, Hill JA, Olson EN (2006) TRPC6 fulfills a calcineurin signaling circuit during pathologic cardiac remodeling. J Clin Invest 116:3114–3126
29. Nakayama H, Wilkin BJ, Bodi I, Molkentin JD (2006) Calcineurin-dependent cardiomyopathy is activated by TRPC3 in the adult mouse heart. FASEB J 20:1660–1670
30. Kiyonaka S, Kato K, Nishida M, Mio K, Numaga T, Sawaguchi Y, Yoshida T, Wakamori M, Mori E, Numata T, Ishii M, Takemoto H, Ojida A, Watanabe K, Uemura A, Kurose H, Morii T, Kobayashi T, Sato Y, Sato C, Hamachi I, Mori Y (2009) Selective and direct inhibition of TRPC3 channels underlies biological activities of a pyrazole compound. Proc Natl Acad Sci USA 106:5400–5405
31. Okada T, Shimizu S, Wakamori M, Maeda A, Kurosaki T, Takada N, Imoto K, Mori Y (1998) Molecular cloning and functional characterization of a novel receptor-activated TRP Ca^{2+} channel from mouse brain. J Biol Chem 273:10279–10287
32. Philipp S, Hambrecht J, Braslavski L, Schroth G, Freichel M, Murakami M, Cavalié A, Flockerzi V (1998) A novel capacitative calcium entry channel expressed in excitable cells. EMBO J 17:4274–4282
33. Ordaz B, Tang J, Xiao R, Salgado A, Sampieri A, Zhu MX, Vaca L (2005) Calmodulin and calcium interplay in the modulation of TRPC5 channel activity. Identification of a novel C-terminal domain for calcium/calmodulin-mediated facilitation. J Biol Chem 280: 30788–30796
34. Shimizu S, Yoshida T, Wakamori M, Ishii M, Okada T, Takahashi M, Seto M, Sakurada K, Kiuchi Y, Mori Y (2006) Ca^{2+}-calmodulin-dependent myosin light chain kinase is essential for activation of TRPC5 channels expressed in HEK 293 cells. J Physiol 570:219–235

35. Blair NT, Kaczmarek JS, Clapham DE (2009) Intracellular calcium strongly potentiates agonist-activated TRPC5 channels. J Gen Physiol 133:525–546
36. Gross SA, Guzmán GA, Wissenbach U, Philipp SE, Zhu MX, Bruns D, Cavalié A (2009) TRPC5 is a Ca^{2+}-activated channel functionally coupled to Ca^{2+}-selective ion channels. J Biol Chem 284:34423–34432
37. Trebak M, Lemonnier L, DeHaven WI, Wedel BJ, Bird GS, Putney JW Jr (2009) Complex functions of phosphatidylinositol 4,5-bisphosphate in regulation of TRPC5 cation channels. Pflügers Arch 457:757–769
38. Lockwich TP, Liu X, Singh BB, Jadlowiec J, Weiland S, Ambudkar IS (2000) Assembly of Trp1 in a signaling complex associated with caveolin-scaffolding lipid raft domains. J Biol Chem 275:11934–11942
39. Tang Y, Tang J, Chen Z, Trost C, Flockerzi V, Li M, Ramesh V, Zhu MX (2000) Association of mammalian Trp4 and phospholipase C isozymes with a PDZ domain-containing protein, NHERF. J Biol Chem 275:37559–37564
40. Yuan JP, Kiselyov K, Shin DM, Chen J, Shcheynikov N, Kang SH, Dehoff MH, Schwarz MK, Seeburg PH, Muallem S, Worley PF (2003) Homer binds TRPC family channels and is required for gating of TRPC1 by IP_3 receptors. Cell 114:777–789
41. Obukhov AG, Nowycky MC (2004) TRPC5 activation kinetics are modulated by the scaffolding protein ezrin/radixin/moesin-binding phosphoprotein-50 (EBP50). J Cell Physiol 201:227–235
42. Goel M, Sinkins W, Keightley A, Kinter M, Schilling WP (2005) Proteomic analysis of TRPC5- and TRPC6-binding partners reveals interaction with the plasmalemmal Na^+/K^+-ATPase. Pflügers Arch 451:87–98
43. Schindl R, Frischauf I, Kahr H, Fritsch R, Krenn M, Derndl A, Vales E, Muik M, Derler I, Groschner K, Romanin C (2008) The first ankyrin-like repeat is the minimum indispensable key structure for functional assembly of homo- and heteromeric TRPC4/TRPC5 channels. Cell Calcium 43:260–269
44. Miehe S, Bieberstein A, Arnould I, Ihdene O, Rütten H, Strübing C (2010) The phospholipid-binding protein SESTD1 is a novel regulator of the transient receptor potential channels TRPC4 and TRPC5. J Biol Chem 285:12426–12434
45. Strübing C, Krapivinsky G, Krapivinsky L, Clapham DE (2001) TRPC1 and TRPC5 form a novel cation channel in mammalian brain. Neuron 29:645–655
46. Yoshida T, Inoue R, Morii T, Takahashi N, Yamamoto S, Hara Y, Tominaga M, Shimizu S, Sato Y, Mori Y (2006) Nitric oxide activates TRP channels by cysteine S-nitrosylation. Nat Chem Biol 2:596–607
47. Bergdahl A, Gomez MF, Dreja K, Xu SZ, Adner M, Beech DJ, Broman J, Hellstrand P, Swärd K (2003) Cholesterol depletion impairs vascular reactivity to endothelin-1 by reducing store-operated Ca^{2+} Entry dependent on TRPC1. Circ Res 93:839–847
48. Brazer SC, Singh BB, Liu X, Swaim W, Ambudkar IS (2003) Caveolin-1 contributes to assembly of store-operated Ca^{2+} influx channels by regulating plasma membrane localization of TRPC1. J Biol Chem 278:27208–27215
49. Couet J, Li S, Okamoto T, Ikezu T, Lisanti MP (1997) Identification of peptide and protein ligands for the caveolin-scaffolding domain. Implications for the interaction of caveolin with caveolae-associated proteins. J Biol Chem 272:6525–6533
50. García-Cardeña G, Martasek P, Masters BS, Skidd PM, Couet J, Li S, Lisanti MP, Sessa WC (1997) Dissecting the interaction between nitric oxide synthase (NOS) and caveolin. Functional significance of the nos caveolin binding domain in vivo. J Biol Chem 272:25437–25440
51. Sato Y, Sagami I, Shimizu T (2004) Identification of caveolin-1-interacting sites in neuronal nitric-oxide synthase. Molecular mechanism for inhibition of NO formation. J Biol Chem 279:8827–8836
52. Quest AF, Gutierrez-Pajares JL, Torres VA (2008) Caveolin-1: an ambiguous partner in cell signalling and cancer. J Cell Mol Med 12:1130–1150

53. Kone BC, Kuncewicz T, Zhang W, Yu ZY (2003) Protein interactions with nitric oxide synthases: controlling the right time, the right place, and the right amount of nitric oxide. Am J Physiol Renal Physiol 285:F178–F190
54. Michel JB, Feron O, Sacks D, Michel T (1997) Reciprocal regulation of endothelial nitric-oxide synthase by Ca^{2+}-calmodulin and caveolin. J Biol Chem 272:15583–15586
55. Michel JB, Feron O, Sase K, Prabhakar P, Michel T (1997) Caveolin versus calmodulin. Counterbalancing allosteric modulators of endothelial nitric oxide synthase. J Biol Chem 272:25907–25912
56. Ju H, Zou R, Venema VJ, Venema RC (1997) Direct interaction of endothelial nitric-oxide synthase and caveolin-1 inhibits synthase activity. J Biol Chem 272:18522–18525
57. Rizzo V, McIntosh DP, Oh P, Schnitzer JE (1998) In situ flow activates endothelial nitric oxide synthase in luminal caveolae of endothelium with rapid caveolin dissociation and calmodulin association. J Biol Chem 273:34724–34729
58. Bernatchez PN, Bauer PM, Yu J (2005) Dissecting the molecular control of endothelial NO synthase by caveolin-1 using cell-permeable peptides. Proc Natl Acad Sci USA 102:761–766
59. García-Cardeña G, Fan R, Stern DF, Liu J, Sessa WC (1996) Endothelial nitric oxide synthase is regulated by tyrosine phosphorylation and interacts with caveolin-1. J Biol Chem 271:27237–27240
60. Fulton D, Gratton JP, McCabe TJ, Fontana J, Fujio Y, Walsh K, Franke TF, Papapetropoulos A, Sessa WC (1999) Regulation of endothelium-derived nitric oxide production by the protein kinase Akt. Nature 399:597–601
61. Michell BJ, Zp C, Tiganis T, Stapleton D, Katsis F, Power DA, Sim AT, Kemp BE (2001) Coordinated Control of endothelial nitric-oxide synthase phosphorylation by protein kinase C and the cAMP-dependent Protein Kinase. J Biol Chem 276:17625–17628
62. Boo YC, Jo H (2003) Flow-dependent regulation of endothelial nitric oxide synthase: role of protein kinases. Am J Physiol Cell Physiol 285:C499–C508
63. Heijnen HF, Waaijenborg S, Crapo JD, Bowler RP, Akkerman JW, Slot JW (2004) Colocalization of eNOS and the catalytic subunit of PKA in endothelial cell junctions: a clue for regulated NO production. J Histochem Cytochem 52:1277–1285
64. Foster MW, Hess DT, Stamler JS (2006) S-nitrosylation TRiPs a calcium Switch. Nat Chem Biol 2:570–571
65. Fulton D, Fontana J, Sowa G, Gratton JP, Lin M, Li KX, Michell B, Kemp BE, Rodman D, Sessa WC (2002) Localization of endothelial nitric-oxide synthase phosphorylated on serine 1179 and nitric oxide in golgi and plasma membrane defines the existence of two pools of active enzyme. J Biol Chem 277:4277–4284
66. Freichel M, Suh SH, Pfeifer A, Schweig U, Trost C, Weissgerber P, Biel M, Philipp S, Freise D, Droogmans G, Hofmann F, Flockerzi V, Nilius B (2001) Lack of an endothelial store-operated Ca^{2+} current impairs agonist-dependent vasorelaxation in TRP4–/– mice. Nat Cell Biol 3:121–127
67. Greka A, Navarro B, Oancea E, Duggan A, Clapham DE (2003) TRPC5 is a regulator of hippocampal neurite length and growth cone morphology. Nat Neurosci 6:837–845
68. Zhang N, Beuve A, Townes-Anderson E (2005) The nitric oxide-cGMP signaling pathway differentially regulates presynaptic structural plasticity in cone and rod cells. J Neurosci 25:2761–2770
69. Boo YC, Hwang J, Sykes M, Michell BJ, Kemp BE, Lum H, Jo H (2002) Shear stress stimulates phosphorylation of eNOS at Ser^{635} by a protein kinase A-dependent mechanism. Am J Physiol Heart Circ Physiol 283:H1819–H1828
70. Gomis A, Soriano S, Belmonte C, Viana F (2008) Hypoosmotic- and pressure-induced membrane stretch activate TRPC5 channels. J Physiol 586:5633–5649
71. Hara Y, Wakamori M, Ishii M, Maeno E, Nishida M, Yoshida T, Yamada H, Shimizu S, Mori E, Kudoh J, Shimizu N, Kurose H, Okada Y, Imoto K, Mori Y (2002) LTRPC2 Ca^{2+}-permeable channel activated by changes in redox status confers susceptibility to cell death. Mol Cell 9:163–173

72. Perraud AL, Fleig A, Dunn CA, Bagley LA, Launay P, Schmitz C, Stokes AJ, Zhu Q, Bessman MJ, Penner R, Kinet JP, Scharenberg AM (2001) ADP-ribose gating of the calcium-permeable LTRPC2 channel revealed by Nudix motif homology. Nature 411: 595–599
73. Sano Y, Inamura K, Miyake A, Mochizuki S, Yokoi H, Matsushime H, Furuichi K (2001) Immunocyte Ca^{2+} influx system mediated by LTRPC2. Science 293:1327–1330
74. Kolisek M, Beck A, Fleig A, Penner R (2005) Cyclic ADP-ribose and hydrogen peroxide synergize with ADP-ribose in the activation of TRPM2 channels. Mol Cell 18:61–69
75. Perraud AL, Takanishi CL, Shen B, Kang S, Smith MK, Schmitz C, Knowles HM, Ferraris D, Li W, Zhang J, Stoddard BL, Scharenberg AM (2005) Accumulation of free ADP-ribose from mitochondria mediates oxidative stress-induced gating of TRPM2 cation channels. J Biol Chem 280:6138–6148
76. Ishii M, Shimizu S, Hagiwara T, Wajima T, Miyazaki A, Mori Y, Kiuchi Y (2006) Extracellular-added ADP-ribose increases intracellular free Ca^{2+} concentration through Ca^{2+} release from stores, but not through TRPM2-mediated Ca^{2+} entry, in rat beta-cell line RIN-5F. J Pharmacol Sci 101:174–178
77. Ishii M, Shimizu S, Hara Y, Hagiwara T, Miyazaki A, Mori Y, Kiuchi Y (2006) Intracellular-produced hydroxyl radical mediates H_2O_2-induced Ca^{2+} influx and cell death in rat beta-cell line RIN-5F. Cell Calcium 39:487–494
78. Wehage E, Eisfeld J, Heiner I, Jüngling E, Zitt C, Lückhoff A (2002) Activation of the cation channel long transient receptor potential channel 2 (LTRPC2) by hydrogen peroxide. A splice variant reveals a mode of activation independent of ADP-ribose. J Biol Chem 277:23150–23156
79. Beck A, Kolisek M, Bagley LA, Fleig A, Penner R (2006) Nicotinic acid adenine dinucleotide phosphate and cyclic ADP-ribose regulate TRPM2 channels in T lymphocytes. FASEB J 20:962–964
80. Lange I, Yamamoto S, Partida-Sanchez S, Mori Y, Fleig A, Penner R (2009) TRPM2 functions as a lysosomal Ca^{2+}-release channel in beta cells. Sci Signal 2:ra23
81. Calcraft PJ, Ruas M, Pan Z, Cheng X, Arredouani A, Hao X, Tang J, Rietdorf K, Teboul L, Chuang KT, Lin P, Xiao R, Wang C, Zhu Y, Lin Y, Wyatt CN, Parrington J, Ma J, Evans AM, Galione A, Zhu MX (2009) NAADP mobilizes calcium from acidic organelles through two-pore channels. Nature 459:596–600
82. Massullo P, Sumoza-Toledo A, Bhagat H, Partida-Sánchez S (2006) TRPM channels, calcium and redox sensors during innate immune responses. Semin Cell Dev Biol 17:654–666
83. Grubisha O, Rafty LA, Takanishi CL, Xu X, Tong L, Perraud AL, Scharenberg AM, Denu JM (2006) Metabolite of SIR2 reaction modulates TRPM2 ion channel. J Biol Chem 281:14057–14065
84. McHugh D, Flemming R, Xu SZ, Perraud AL, Beech DJ (2003) Critical intracellular Ca^{2+} dependence of transient receptor potential melastatin 2 (TRPM2) cation channel activation. J Biol Chem 278:11002–11006
85. Hill K, Tigue NJ, Kelsell RE, Benham CD, McNulty S, Schaefer M, Randall AD (2006) Characterisation of recombinant rat TRPM2 and a TRPM2-like conductance in cultured rat striatal neurones. Neuropharmacology 50:89–97
86. Starkus J, Beck A, Fleig A, Penner R (2007) Regulation of TRPM2 by extra- and intracellular calcium. J Gen Physiol 130:427–440
87. Du J, Xie J, Yue L (2009) Intracellular calcium activates TRPM2 and its alternative spliced isoforms. Proc Natl Acad Sci USA 106:7239–7244
88. Olah ME, Jackson MF, Li H, Perez Y, Sun HS, Kiyonaka S, Mori Y, Tymianski M, MacDonald JF (2009) Ca^{2+}-dependent induction of TRPM2 currents in hippocampal neurons. J Physiol 587:965–979
89. Tong Q, Zhang W, Conrad K, Mostoller K, Cheung JY, Peterson BZ, Miller BA (2006) Regulation of the transient receptor potential channel TRPM2 by the Ca^{2+} sensor calmodulin. J Biol Chem 281:9076–9085

90. Zhang W, Tong Q, Conrad K, Wozney J, Cheung JY, Miller BA (2007) Regulation of TRP channel TRPM2 by the tyrosine phosphatase PTPL1. Am J Physiol Cell Physiol 292: C1746–C1758
91. Togashi K, Hara Y, Tominaga T, Higashi T, Konishi Y, Mori Y, Tominaga M (2006) TRPM2 activation by cyclic ADP-ribose at body temperature is involved in insulin secretion. EMBO J 25:1804–1815
92. Du J, Xie J, Yue L (2009) Modulation of TRPM2 by acidic pH and the underlying mechanisms for pH sensitivity. J Gen Physiol 134:471–488
93. Starkus JG, Fleig A, Penner R (2010) The calcium-permeable non-selective cation channel TRPM2 is modulated by cellular acidification. J Physiol 588:1227–1240
94. Yamamoto S, Shimizu S, Kiyonaka S, Takahashi N, Wajima T, Hara Y, Negoro T, Hiroi T, Kiuchi Y, Okada T, Kaneko S, Lange I, Fleig A, Penner R, Nishi M, Takeshima H, Mori Y (2008) TRPM2-mediated Ca^{2+} influx induces chemokine production in monocytes that aggravates inflammatory neutrophil infiltration. Nat Med 14: 738–747
95. Fonfria E, Marshall IC, Boyfield I, Skaper SD, Hughes JP, Owen DE, Zhang W, Miller BA, Benham CD, McNulty S (2005) Amyloid beta-peptide(1-42) and hydrogen peroxide-induced toxicity are mediated by TRPM2 in rat primary striatal cultures. J Neurochem 95:715–723
96. Bari MR, Akbar S, Eweida M, Kühn FJ, Gustafsson AJ, Lückhoff A, Islam MS (2009) H_2O_2-induced Ca^{2+} influx and its inhibition by N-(p-amylcinnamoyl)anthranilic acid in the beta-cells: involvement of TRPM2 channels. J Cell Mol Med 13:3260–3267
97. Zeng X, Sikka SC, Huang L, Sun C, Xu C, Jia D, Abdel-Mageed AB, Pottle JE, Taylor JT, Li M (2009) Novel role for the transient receptor potential channel TRPM2 in prostate cancer cell proliferation. Prostate Cancer Prostatic Dis 13:195–201
98. Hermosura MC, Cui AM, Go RC, Davenport B, Shetler CM, Heizer JW, Schmitz C, Mocz G, Garruto RM, Perraud AL (2008) Altered functional properties of a TRPM2 variant in Guamanian ALS and PD. Proc Natl Acad Sci USA 105:18029–18034
99. Xu C, Macciardi F, Li PP, Yoon IS, Cooke RG, Hughes B, Parikh SV, McIntyre RS, Kennedy JL, Warsh JJ (2006) Association of the putative susceptibility gene, transient receptor potential protein melastatin type 2, with bipolar disorder. Am J Med Genet B Neuropsychiatr Genet 141B:36–43
100. Xu C, Li PP, Cooke RG, Parikh SV, Wang K, Kennedy JL, Warsh JJ (2009) TRPM2 variants and bipolar disorder risk: confirmation in a family-based association study. Bipolar Disord 11:1–10
101. Lishko PV, Procko E, Jin X, Phelps CB, Gaudet R (2007) The ankyrin repeats of TRPV1 bind multiple ligands and modulate channel sensitivity. Neuron 54:905–918
102. Fujiwara Y, Minor DL Jr (2008) X-ray crystal structure of a TRPM assembly domain reveals an antiparallel four-stranded coiled-coil. J Mol Biol 383:854–870
103. Phelps CB, Huang RJ, Lishko PV, Wang RR, Gaudet R (2008) Structural analyses of the ankyrin repeat domain of TRPV6 and related TRPV ion channels. Biochemistry 47: 2476–2484
104. Yu Y, Ulbrich MH, Li MH, Buraei Z, Chen XZ, Ong AC, Tong L, Isacoff EY, Yang J (2009) Structural and molecular basis of the assembly of the TRPP2/PKD1 complex. Proc Natl Acad Sci USA 106:11558–11563
105. Mio K, Ogura T, Hara Y, Mori Y, Sato C (2005) The non-selective cation-permeable channel TRPC3 is a tetrahedron with a cap on the large cytoplasmic end. Biochem Biophys Res Commun 333:768–777
106. Mio K, Ogura T, Kiyonaka S, Hiroaki Y, Tanimura Y, Fujiyoshi Y, Mori Y, Sato C (2007) The TRPC3 channel has a large internal chamber surrounded by signal sensing antennas. J Mol Biol 367:373–383
107. Maruyama Y, Ogura T, Mio K, Kiyonaka S, Kato K, Mori Y, Sato C (2007) Three-dimensional reconstruction using transmission electron microscopy reveals a swollen,

bell-shaped structure of transient receptor potential melastatin type 2 cation channel. J Biol Chem 282:36961–36970
108. Moiseenkova-Bell VY, Stanciu LA, Serysheva II, Tobe BJ, Wensel TG (2008) Structure of TRPV1 channel revealed by electron cryomicroscopy. Proc Natl Acad Sci USA 105: 7451–7455

Chapter 22
Gating Mechanisms of Canonical Transient Receptor Potential Channel Proteins: Role of Phosphoinositols and Diacylglycerol

Anthony P. Albert

Abstract Canonical transient receptor potential (TRPC) Ca^{2+}-permeable channels are members of the mammalian TRP super-family of cation channels, and have the closest homology to the founding members, TRP and TRPL, discovered in Drosophila photoreceptors. The TRPC subfamily is composed of 7 subunits (C1–C7, with TRPC2 a pseudogene in humans), which can all combine with one another to form homomeric and heteromeric structures. This review focuses on mechanisms involved in opening TRPC channels (i.e. gating mechanisms). It initially describes work on the involvement of phosphatidylinositol-4,5-bisphosphate (PIP_2) and diacylglycerol (DAG) in gating TRP and TRPL channels in Drosophila, and then discusses evidence that similar gating mechanisms are involved in opening mammalian TRPC channels. It concludes that there are two common activation pathways of mammalian TRPC channels. Non-TRPC1-containing channels are opened by interactions between DAG, the direct activating ligand, and PIP_2, which acts as a physiological antagonist at TRPC proteins. Competitive interactions between an excitatory effect of DAG and an inhibitory action of PIP_2 can also be modulated by IP_3 acting via an IP_3 receptor-independent mechanism. In contrast TRPC1-containing channels are gating by PIP_2, which requires PKC-dependent phosphorylation of TRPC1 proteins.

22.1 Introduction

The mammalian transient receptor potential (TRP) super-family of ion channels consists of 6 sub-families of Ca^{2+}-permeable non-selective cation channel proteins (TRPC, TRPV, TRPM, TRPA, TRPP and TRPML) [1]. These channels are expressed throughout the body, and are activated by diverse chemical and sensory stimuli (e.g. neurotransmitters, hormones, temperature and mechanical sensation).

A.P. Albert (✉)
Division of Basic Medical Sciences, St. George's University of London, London SW17 0RE, UK
e-mail: aalbert@sgul.ac.uk

In physiological conditions, opening of TRP channels at the resting membrane potential evokes an influx of Na$^+$, with most channels also having a significant permeability to Ca^{2+} ($P_{Ca}/P_{Na}=1-100$) [2, 3]. Permeability properties of TRP channels means that their activation will increase excitability and cytosolic Ca^{2+} concentration ([Ca^{2+}]$_i$). Constitutively active channels contribute to the resting membrane potential, whereas agonist-evoked channels can induce depolarisation leading to opening of voltage-dependent Ca^{2+} channels, a direct influx of Ca^{2+} and activation of Na$^+$-dependent processes [4]. Stimulation of TRP channels have been linked to different cellular functions such as contraction, secretion, synaptic and sensory potentials, proliferation, growth, motility and gene expression [4, 5]. Abnormal TRP channel activities are associated with pathophysiology, and TRP channels are considered potential therapeutic targets for treatment of disease [5].

This review focuses on the canonical transient receptor potential (TRPC) subfamily, which consists of 7 channel subunits (TRPC1-C7, with TRPC2 a pseudogene and not present in humans). It will discuss how TRPC channels are opened (i.e. gating mechanisms), with particular attention to the proposal that interactions between phosphoinositols (PI) and diacylglycerol (DAG) represent common activation mechanisms of diverse TRPC channel structures.

22.2 TRPC Subunits Produce a Diverse Group of Functional Channel Isoforms

The putative structure of TRPC channels is based on their similar homology to voltage-gated ion channels, which have 4 subunits, each containing 6 transmembrane spanning regions (S1–S6), that interact with each other to create a tetramer structure with a conducting pore located between S5 and S6 [3]. Unlike voltage-gated channels, TRPC subunits do not contain a conserved arginine residue sequence in S4, which forms a voltage sensor [1]; as such TRPC channels are considered voltage-independent ion channels. Each subunit has intracellular N- and C-termini which contain specific amino acid residues that are phosphorylated by protein kinases e.g. TRPC3 ser-712 by PKC, thr-69/70 and ser-322 by PKG, TRPC6 N-termini by pp60^{c-src} [6–11], and sequences that interact with signalling molecules and accessory proteins e.g. TRPC6 arg-852/860/864 and lys-856/859 by CaM and arg-853/861 and lys-860 by PIs [12–14]. These N- and C-termini regions are likely to be important in regulating activity of TRPC channels.

Functional TRPC channels are composed of both homomeric and heteromeric structures, with native TRPC channels predicted to mainly consist of heterotetramer compositions. TRPC heteromers were initially proposed to only occur between TRPC1, TRPC4 and TRPC5 proteins and between TRPC3, TRPC6 and TRPC7 subunits [15, 16]. It was then suggested that TRPC1 might act as linker molecule between these two groups of TRPC subtypes [17]. It is now general consensus that all TRPC subunits can associate with each other, and Table 22.1 illustrates characterised TRPC heteromers discovered so far. In addition, potential diversity of TRPC channel structures is also increased by the possible stoichiometry of TRPC

Table 22.1 Proposed functional TRPC heteromeric channel structures

Heteromers	References
C1 + C3	[18, 19]
C1 + C4	[20–22]
C1 + C5	[23–26]
C1 + C6	[21]
C1 + C3 + C7	[27]
C1 + C4 + C5	[28]
C1 + C5 + C6	[24]
C1 + C5 + C7	[24]
C1 + TRPP2	[29–31]
C3 + C4	[32]
C3 + C6	[28]
C3 + C7	[33]
C3 + TRPP2	[34]
C6 + C7	[35, 36]
C7 + TRPP2	[34]

heteromers. For example, a channel composed of two subunits has potential subunit ratios of 1:3, 2:2 and 3:1. Diversity of TRPC structures is further increased by recent work showing that TRPC1 and TRPP2 subunits associate with each other to form ion channels, which have distinct properties from those of homomeric TRPC1 and TRPP2 channels [29–31]. In addition, TRPP2 truncated mutants associate with TRPC3 and TRPC7 proteins to enhance channel activity [34]. These results indicate that TRPC proteins are also able to produce functional channels with TRP proteins from other sub-families.

22.3 Physiological Gating of TRPC Channels

TRPC channels are activated by plasmalemmal G-protein-coupled and tyrosine kinase receptors (TKRs), representing ion channels that are classically defined as receptor-operated channels (ROCs) [37, 38]. TRPC subunits are also proposed to mediate store-operated channels (SOCs), which are defined as channels activated by depletion of Ca^{2+} levels within the internal Ca^{2+} stores [3]. The SOC family also includes the recently identified Orai proteins, which are activated by stromal interaction proteins (STIM) proteins that detect reductions in Ca^{2+} levels within internal Ca^{2+} stores [39–44]. It is proposed that TRPC and Orai may combine to form SOC complexes that are modulated by STIM, although this hypothesis is unresolved and controversial [3, 45–49]. SOCs, unlike ROCs, are activated by SERCA inhibitors thapsigargin and cyclopiazonic acid (CPA), and therefore these agents are often used to selectively evoke SOC activity. However, it should be remembered that TRPC ROCs and SOCs, and Orai SOC proteins are generally activated in physiological conditions through stimulation of $G_{\alpha q}/G_{\alpha 11}$ protein-coupled receptors linked to PLCβ and TKRs linked to PLCγ. Activation of these pathways will lead to hydrolysis of PIP_2 and generation of IP_3 and DAG, to respectively activate IP_3 receptors located on the sarcoplasmic/endoplasmic reticulum to release internal Ca^{2+} stores and stimulate PKC. The release of Ca^{2+} ions from the internal Ca^{2+} stores by IP_3, and thus depletion of Ca^{2+} within these stores, is proposed be the stimulus for activation of TRPC SOCs, although these channels can also be evoked by receptor stimulation coupled to store-independent pathways [50]. Depletion of Ca^{2+} within internal Ca^{2+} stores by IP_3 is also the trigger for stimulation of STIM and gating of Orai channels. Therefore, physiologically all TRPC channels (and also Orai channels) are indeed ROCs, with channel subtypes being activated by store-dependent and -independent pathways. Moreover, receptor stimulation will only activate channels by store-dependent mechanisms if a significant depletion of internal Ca^{2+} stores is produced. Thus, properties of receptor stimulation such as circulating concentrations of agonists or frequency of nerve stimulation coupled to differential release levels of neurotransmitters may discriminate between subtypes of TRPC and Orai channels activated.

There has been considerable work on identifying whether distinct TRPC subunits mediate functional ROCs and SOCs [4]. To date, all TRPC subunits seem to be capable of being defined as ROCs or SOCs; but it is unclear if this ambiguity is due to experimental methods and measurement techniques used, e.g. these studies have mainly used over-expression systems and indirect measurements of channel activity such as Ca^{2+} measurements with fluorescent dyes. Our data from native vascular smooth muscle cells suggests that functional ROCs and SOCs have vastly different properties from each other, which indicates that they are likely to be composed of different TRPC subunits. Our proposals indicate that ROCs are composed of non-TRPC1 containing homomeric or heteromeric structures which have the general characteristics of unitary conductances between 10 and 70 pS, activation by DAG via PKC-independent mechanisms and are inhibited by PKC-dependent pathways [51]. In contrast, SOCs are TRPC1-containing channels which have a unitary conductance of 2–3 pS and are activated by PKC-dependent mechanisms [51].

There is increasing evidence that individual cells contain multiple TRPC and Orai channels, which are activated by store-dependent and -independent activation pathways. The presence of multiple types of Ca^{2+}-permeable cation channels in the same cell, all activated by stimulation of the same receptor produces problems when investigating receptor-mediated activation pathways. For example, these different ion channels are readily distinguished using single channel recording (as unitary conductances of TRPC ROCs and SOCs, and Orai channels are vastly different, see above) but not necessarily by techniques that measure macroscopic properties such as whole-cell recording or intracellular Ca^{2+} measurements with fluorescent dyes [52]. These problems must be considered when evaluating experimental conditions and techniques used for studying TRPC channel gating.

22.4 Activation of *Drosophila* TRP and TRPL Channels: Models for TRPC Gating

It is evident that TRPC subtypes compose a large group of channel isoforms, and therefore the question arises, can such a diverse group of ion channels have a common gating mechanism? To address this question it is useful to start by discussing the gating properties of TRP channels expressed in *Drosophila*.

The founding members of the mammalian TRP channel super-family, TRP and TRPL, were discovered in *Drosophila* photoreceptors and have an essential role in transduction of light into electrical activity to produce vision (see [53–56]). TRPγ channels with an unknown function have also been identified in *Drosophila* photoreceptors [57]. About 30 mammalian TRP homologues of TRP and TRPL channels have been characterised, with the TRPC channel sub-family having the closest homology to these channels in *Drosophila*. Understanding activation mechanisms of TRP and TRPL channels is therefore important for both determining invertebrate phototransduction and for providing fundamental principles of mammalian TRPC channel gating.

Upon absorption of a photon, rhodopsin is converted to metarhodospin in the photoreceptor, which stimulates $G_{\alpha q}$-proteins that activate PLC to induce hydrolysis of PIP_2 into water-soluble IP_3 and membrane-bound DAG. Stimulation of this pathway opens TRP and TRPL channels, which are respectively highly permeable ($P_{Ca}/P_{Na} = 110:1$) and non-selective ($P_{Ca}/P_{Na} = 4:1$) to Ca^{2+} [58]. Activation of TRP and TRPL channels produces changes in cell excitability and Ca^{2+} signals, which generates a visual signal. It is clear that the norpA gene, encoding PLC in Drosophila, is essential for activation of TRP and TRPL channels and photoreception [59]. An unresolved issue is the identification of molecules involved in gating of TRP and TRPL channels downstream from PLC-mediated hydrolysis of PIP_2.

IP_3 does not activate TRP or TRPL channel activity [60], and genetic removal of the one type of IP_3 receptor expressed in *Drosophila* had no effect on photoreception [61, 62]. In addition, IP_3-mediated release of Ca^{2+} is also unlikely to contribute to channel activation, as Ca^{2+} is not required for the light response or activation of TRPL channels [60, 63]. Moreover, direct application of Ca^{2+} to the cytosolic surface of inside-out patches inhibits TRPL channel activity via an open channel block mechanism [64]. These studies clearly show that IP_3, IP_3 receptors or IP_3-mediated Ca^{2+} release are not involved in activating TRP and TRPL channel activity in *Drosophila* photoreceptors.

22.4.1 Central Role for DAG in Gating Drosophila TRP and TRPL Channels

The established function of DAG is to activate PKC, but ePKC encoded by the inaC gene, is important for terminating the light response and does not affect photoreceptor stimulation in *Drosophila* [65]. It is therefore proposed that generation of DAG and/or DAG metabolites are important for gating of TRP and TRPL channels. Figure 22.1 highlights the main biochemical pathways involved in the generation and metabolism of DAG.

In a seminal study, polyunsaturated fatty acids (PUFAs, e.g. linolenic acid and arachidonic acid), were shown to activate TRP and TRPL channels in *Drosophila* and expressed TRPL channels (See Fig. 22.1) [66]. Manipulation of the DAG kinase gene, rdgA, which prevented DAG being converted into phosphatidic acid (PA) and increased basal DAG levels, also produced constitutive TRP and TRPL channel activity (see Fig. 22.1) [62]. Moreover, direct application of a cell-permeable DAG

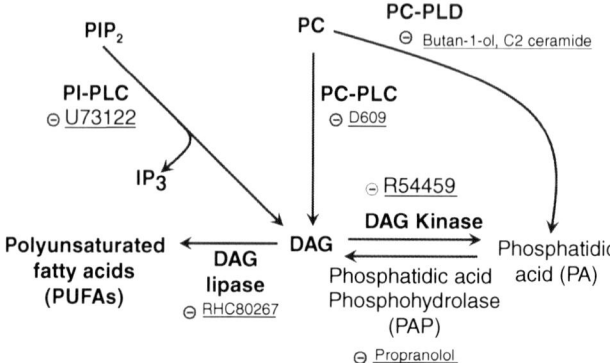

Fig. 22.1 Pathways involved in generation and metabolism of diacylglycerol (DAG). This schematic shows the biochemical pathways involved in generation of DAG from phosphatidylinositol-4,5-bisphosphate (PIP$_2$) and phosphatidylchoine (PC), and metabolism of DAG by DAG kinase and DAG lipase. Note that the underlined agents are pharmacological blockers used to inhibit enzyme activities

analogue, OAG, activated TRP and TRPL channel activity in inside-out patches from the microvilli of dissociated *Drosophila* photoreceptors [67]. Furthermore, the Lazaro gene, which encodes a phosphatidic acid phosphatase (PAP) that generates DAG from PLD-mediated hydrolysis of PC (see Fig. 22.1), is also required to induce maximum light responses in *Drosophila* photoreceptors [68].

It is apparent that generation of DAG is a pivotal event in gating of TRP and TRPL channels in *Drosophila*, although it is still not resolved if DAG itself and/or DAG metabolites directly gate these channels. There are currently three proposals, which attempt to address this problem (see Fig. 22.2). First, PIP$_2$ produces a markedly inhibition of TRPL channel activity, and this observation has led to the hypothesis that optimal activation of TRPL channel activity is produced by both hydrolysis of PIP$_2$ and generation of DAG [69]. This may explain why applications of exogenous OAG produced slow onset and small amplitude responses [67], as in these conditions the inhibitory action of PIP$_2$ had not been removed. In contrast, PIP$_2$ depletion produced decay in light responses suggesting that this phosphoinositol is also a positive modulator of channel activity [70]. However, this result may be expected if PIP$_2$ is required for generation of DAG, and DAG is the gating ligand of TRP and TRPL channels. Secondly, open channel block by Ca^{2+} has an important inhibitory action on TRPL channel activity at negative membrane potentials [64]. Recent studies have shown that PUFAs may activate TRPL channels by removing this open channel block mechanism [71]. PUFAs have also been shown to activate TRPγ channels [72]. Thirdly, the chemical reaction of PLC-mediated PIP$_2$ hydrolysis into IP$_3$ and DAG releases protons, and data indicates that the light response produces a rapid (<10 ms) local acidification of the membrane/cytosolic boundary area [73]. Consistent with this observation, reductions in pH activate TRPL channels [73], but it is still unclear whether production of protons directly gates these channels or promotes their activation by DAG and/or DAG metabolites.

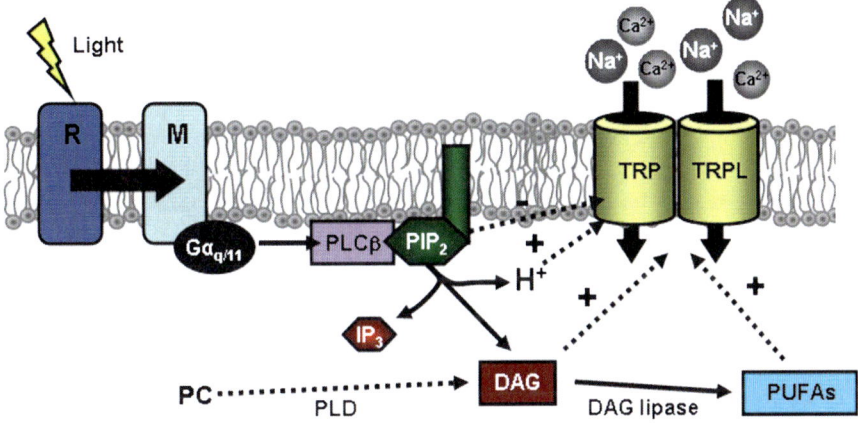

Fig. 22.2 Proposed gating mechanisms of TRP and TRPL channels in *Drosophila* Photoreceptors. Absorption of photons leads to conversion of rhodospin into metarhodospin which stimulates G-protein-mediated activation of the phosphoinositol signal transduction pathway. PIP_2 is proposed to inhibit channel activity whereas the products of PIP_2 hydrolysis; H^+, DAG and PUFAs have all been suggested to be activating ligands. Note that DAG generation from PLD-mediated PC hydrolysis is also proposed to have an excitatory action on channel activity [55, 56, 74]

Although there is considerable evidence that DAG itself and/or DAG metabolites are gating molecules of TRP and TRPL channels in *Drosophila* a number of issues still remain, such as; why does exogenous application of OAG produce much slower and small responses of TRP and TRPL channels than light stimuli, and is there a binding site for DAG on TRP and TRPL proteins? Moreover, if DAG is the gating ligand of TRP and TRPL channels, why does a mutation of the DAG lipase gene (inaE, see Fig. 22.1) produce defective light responses [74]?

22.5 Mammalian TRPC3/C6/C7 Subunits: Classical DAG-Activated Channels

This section discusses evidence, which indicates that interactions between DAG and PIP_2 are pivotal events in activating TRPC3-C7 homomers in cell lines and native channels composed of TRPC3-C7 homomers and heteromers.

22.5.1 Role of DAG

In the late 1990s, noradrenaline was shown to activate a Ca^{2+}-permeable non-selective cation conductance in portal vein vascular smooth muscle cells (VSMCs) through a sequential signal transduction pathway of α_1-adrenceptor stimulation coupled to G-proteins, PLC activation and generation of DAG [75]. Most importantly,

DAG activated this conductance via a novel PKC-independent mechanism [75]. This biochemical cascade was subsequently identified as the signature activation pathway of TRPC3, TRPC6 and TRPC7 channel subtypes [76–80], with the cation conductance in portal vein VSMCs composed of a TRPC6/C7 heteromer [36, 81]. In addition to its excitatory effect, DAG also has a pronounced inhibitory action on TRPC3, TRPC6 and TRPC7 channels through stimulation of PKC [80, 82, 83]. PKC probably decreases TRPC channel activity by reducing channel open probability by directly phosphorylating TRPC subunits [8]. Excitatory and inhibitory effects of DAG on TRPC3, TRPC6 and TRPC7 channels are proposed to be concentration-dependent, with excitatory actions of DAG occurring at <1–10 μM concentrations, and inhibitory effects via PKC activation produced with higher concentrations of DAG e.g. >10 μM [83].

Other non-PLC-mediated pathways are involved in generating DAG (see Fig. 22.1), and these pathways have also been shown to activate TRPC3 channels. In ear artery VSMCs, spontaneously active TRPC3 channel activity is produced by a signal transduction pathway involving constitutive $G_{\alpha i/o}$ protein activity coupled to PLD-mediated hydrolysis of PC, generation of PA, and conversion of PA into DAG by PAP (see Fig. 22.1) [84, 85]. PLD-mediated DAG production has also been proposed to mediate histamine-evoked TRPC3-mediated Ca^{2+} entry in HEK293 cells [86] and TRPC3-evoked excitatory postsynaptic potential in cerebellar purkinje neurones [87].

When investigating direct gating mechanisms of DAG on TRPC3, TRPC6 and TRPC7 channel activity, the potential role of DAG metabolites should be taken into account. For example, DAG lipase and DAG kinase convert DAG into respectively PUFAs and PA (see Fig. 22.1). To exclude the possibility that these DAG metabolites are involved in channel gating inhibitors of DAG lipase (RHC80267) and DAG kinase (R59949) are often used. In the presence of a receptor agonist, application of DAG lipase/DAG kinase inhibitors will prevent DAG being converted to PUFAs or PA. In addition, application of these inhibitors to un-stimulated cells is likely to increase basal levels of DAG generated by constitutive phospholipase C and/or D activity. Inhibitors of both DAG lipase and DAG have been shown to activate TRPC3, TRPC6 and TRPC7 channels [2, 76, 77, 81, 84, 85], which indicate that DAG, and not DAG metabolites, are involved in activating these TRPC channel subtypes.

22.5.2 Role of IP_3

IP_3, the other product of PLC-mediated PIP_2 hydrolysis, has been proposed to regulate opening of TRPC3, TRPC6 and TRPC7 channels. IP_3 acting via IP_3 receptors, possibly via store-dependent mechanisms, was initially reported to evoke TRPC3 [88–90] and TRPC7 channel activity [91]. These interpretations have become clouded with inconsistent data showing that TRPC3 and TRPC7 channels can be activated by either store-dependent or -independent pathways depending on channel expression levels and that channel activation can also occur in cells in which IP_3 receptors have been "knocked-out" [92, 93]. There is evidence that IP_3 acting via IP_3R1 receptors evoke TRPC3 channels in cerebral artery VSMCs [94, 95]. In other VSMCs preparations, IP_3 significantly potentiates amplitude and speed of onset of DAG-induced TRPC6/C7 and TRPC3/C7 channel activity via both IP_3 receptor-dependent and -independent mechanisms [33, 36, 96].

These results show that receptor stimulation is likely to activate TRPC3, TRPC6 and TRPC7 channel activity through generation of DAG, although they do not definitively explain how DAG gates these channels. Recent evidence, similar to a proposal for Drosophila TRPL channels (see above), indicates that interactions between PIP_2 and DAG have an important role in gating TRPC channels [36, 51, 97].

22.5.3 Interactions Between DAG and PIP_2

In mesenteric artery VSMCs, applications of high concentrations of wortmannin which inhibits PI-4- and PI-5-kinases to prevent PIP_2 synthesis, anti-PIP_2 antibodies which are thought to prevent actions of PIP_2, and poly-L-lysine which scavenges PIP_2 directly activated TRPC6 channel activity and also enhanced OAG-induced TRPC6 channel activity [97] (see Fig. 22.3). Immunoprecipitation experiments

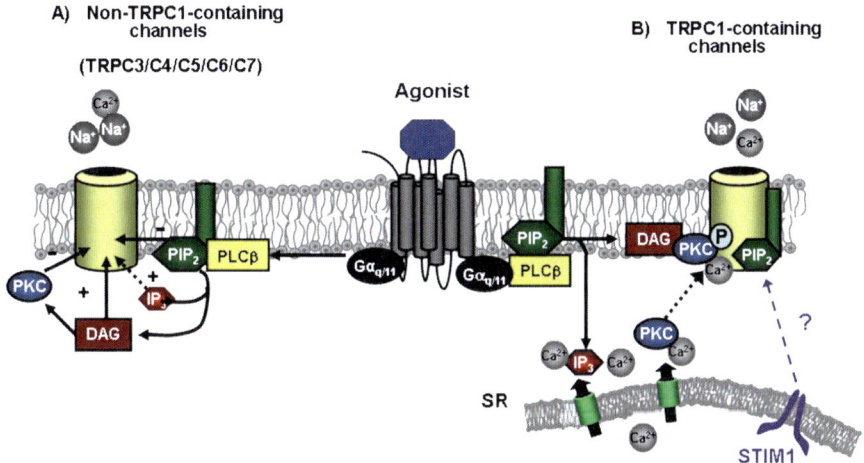

Fig. 22.3 Proposed gating mechanisms of mammalian non-TRPC1- and TRPC1-containing channels. (a) PLC-mediated hydrolysis of PIP_2 by stimulation of G-protein-coupled receptors is proposed to gate non-TRPC1-containing channels by removal of an inhibitory action of PIP_2, and generation of DAG. It is thought that PIP_2 acts as a physiological antagonist and DAG is the gating ligand. This activation model proposes the involvement of one pool of PIP_2, associated with both receptor and channel. Note that production of IP_3 can also promote channel activation, and that generation of increasing levels of DAG are thought to inhibit channel activity through stimulation of PKC. (b) Activation mechanism of TRPC1-containing channels is proposed to be via generation of DAG and stimulation of PKC-dependent phosphorylation of TRPC1 proteins, which is required for channel gating by PIP_2. In contrast to A, the model of TRPC1 gating shown in B proposes the existence of two pools of PIP_2, with one pool associated with the receptor and the other pool tethered to TRPC1 at rest [51]

showed that interactions between PIP_2 and TRPC6 were present at rest, and were prevented by pre-treatment with angiotensin II and OAG. These results indicated that optimum activation of TRPC6 channel activity by receptor stimulation required both depletion of PIP_2 and generation of DAG. It was concluded that PIP_2 acts as a physiological antagonist and that DAG is the gating ligand at TRPC6 channels, with competitive interactions between these two molecules governing opening of the channels [97].

PIP_2 also has an inhibitory action on a TRPC6/C7 heteromer in portal vein VSMCs [36]. In contrast to TRPC6 homomers in mesenteric artery VSMCs, IP_3 potentiated OAG-evoked TRPC6/C7 channel activity and overcame channel inhibition produced by PIP_2 via an IP_3 receptor-independent mechanism in portal vein VSMCs. Co-association between PIP_2 and TRPC6 was prevented by OAG but not IP_3, whereas interactions between PIP_2 and TRPC7 proteins were inhibited by IP_3 but not OAG [36]. These results indicated that synergism between DAG and IP_3 on channel activity in portal vein VSMCs may occur by these two agents maximising removal of the inhibitory action of PIP_2 from channel proteins, to allow channel gating by DAG. It was also proposed that synergistic actions of IP_3 on OAG-activated channels are mediated by TRPC7 subunits [36].

The above results indicate that PIP_2 is a common inhibitory molecule on OAG-activated TRPC channels in VSMCs. In contrast, PIP_2 has been shown to directly bind to the C-termini region of expressed TRPC6 proteins, to producing an excitatory action on channel activity through displacing an inhibitory action of CaM (13). This study also showed that PIP_3 produced similar excitatory actions to PIP_2 on TRPC6 channel activity [13], which correlated with earlier work showing PIP_3 induces TRPC6 channel activity [98]. PIP_2 as also been shown to produce a rapid activation of expressed TRPC6 channels in HEK-293 cells [99]. There is no clear explanation for why both inhibitory and excitatory effects of PIP_2 on TRPC6 channel activity have been described, although these results may represent differences in direct or indirect actions of PIP_2 and also the properties of native or expressed TRPC6 channel activity.

22.5.4 Direct vs. Indirect Actions of PIP_2

PIP_2 is a substrate for generation of IP_3/DAG and also PIP_3 by PLC and PI-3-kinase respectively. When studying the direct effect of exogenous PIP_2 on channel activity, it is important to include inhibitors of PLC and PI-3-kinase in the experimental protocols to prevent the involvement of PIP_2-related products. Presumably, there is also a physiological coupling between levels of signal transduction molecules (e.g. PIP_2, PLC etc) and functional ion channels in the cell membrane of freshly isolated cells, which may be altered when ion channel proteins are expressed in cell lines. With these latter conditions, exogenously applied PIP_2 may initially act as a substrate for generation of IP_3/DAG or PIP_3, which then activate TRPC3, TRPC6 and TRPC7 channels before direct inhibitory actions of PIP_2 are produced. Therefore, excitatory actions of PIP_2 should be monitored for several minutes to observe if these responses are transient, and if the actions of PIP_2 change over time. It should be noted that PLC and PI-3-kinase inhibitors were not included in studies showing excitatory actions of PIP_2 on TRPC6 channel activity, and that following application of PIP_2 the responses were only measured for <1 min [12, 99].

In conclusion, there is strong evidence that receptor stimulation evokes TRPC3, TRPC6 and TRPC7 channel activity by a common mechanism involving generation of DAG, which is an activating ligand. Moreover, gating of these TRPC channels by DAG also involves competitive interactions with an inhibitory effect of PIP_2 on the channel proteins. IP_3 also produces synergism with DAG on TRPC channel activity containing TRPC7 proteins, by IP_3 promoting the removal of the inhibitory action of PIP_2 from these channel subunits.

22.6 Are TRPC4 and TRPC5 Subunits DAG-Gated Channels?

Understanding the activation mechanisms of receptor-operated TRPC4 and TRPC5 channels has lagged behind our insight into gating of TRPC3, TRPC6 and TRPC7 channels. It is generally accepted that stimulation of $G_{\alpha q}/G_{\alpha 11}$ protein-coupled receptors and TKRs induce TRPC4 and TRPC5 channel activity through PLC-mediated pathways, as PLC inhibitors block receptor- and GTP-γ-S-induced channel activities [100–102].

Uncertainty comes from our lack of understanding of the mechanisms involved in activating TRPC4 and TRPC5 after stimulation of PLC. It is clear that IP_3, IP_3 receptors and alterations in $[Ca^{2+}]_i$ intracellular Ca^{2+} levels are not direct triggers for opening of TRPC4 and TRPC5 channels [100], although $[Ca^{2+}]_i$ does powerfully potentiate TRPC5 channel activity [103]. In addition, a role of DAG in gating TRPC4 and TRPC5 is also unclear. Some studies have shown that application of exogenous OAG or increasing endogenous DAG levels with DAG lipase and DAG kinase inhibitors do not evoke TRPC4 and TRPC5 channel activities, whereas other work has illustrated that OAG does stimulate TRPC5 channel activity, albeit with significantly smaller peak amplitudes than TRPC5 currents activated by receptor stimulation [104]. If there is common activation pathway of TRPC channels, it would seem likely that TRPC4 and TRPC5 channels are gated by DAG through a PKC-independent mechanism and that channel activation involves interactions between DAG and PIP_2 (similar to TRPC3/C6/C7).

Venkatachalam et al [82] provided evidence that TRPC4 and TRPC5 channel activity evoked by receptor stimulation was inhibited by exogenous application of OAG and by increasing endogenous DAG levels with DAG lipase and DAG kinase inhibitors. Moreover, these studies showed that PKC antagonists prevented these inhibitory actions of DAG on TRPC4 and TRPC5 channel activity. Moreover, activation of PKC by receptor stimulation has been shown to account for rapid the desensitisation of TRPC5 currents, through direct phosphorylation of TRPC5 proteins at the amino acid residue T972 [105]. These results indicate that adding OAG or increasing DAG levels are unlikely to activate TRPC4 or TRPC5 channel activity due to this triglyceride evoking PKC stimulation which has a marked inhibitory action on channel opening.

Recent work has indicated that PIP_2 also has an important role in activating TRPC4 and TRPC5 channels. Otsuguro et al. [102] showed that activation of the TRPC4α isoform, but not the TRPC4β isoform, was inhibited by PIP_2 through interactions between this PI and the C-terminus of the channel protein. This study concluded that hydrolysis of PIP_2 by a G-protein-coupled PLC pathway was an important event in opening TRPC4 channels, but not the gating event, as depleting PIP_2 with wortmannin or scavenging PIP_2 with poly-L-lysine did not activate TRPC4 activity. Endogenous PIP_2 has also been shown to inhibit TRPC5, with wortmannin pre-treatment inducing TRPC5 channel activity, which was blocked by application of exogenous PIP_2 [106]. In contrast, this study also showed that PIP_2 activated TRPC5 activity in inside-out patches [106]. PIP_2 has also been shown to prevent desensitisation of receptor-activated TRPC5 channel activity [107]. In a recent study, the PI binding protein SESTD1 was shown to bind to the CaM/IP_3 receptor binding region of TRPC4 and TRPC5 proteins, and was responsible sustained TRPC5-mediated Ca^{2+} entry [108]. Why some results suggest that PIP_2 is inhibitory, and others excitatory, on TRPC4 and TRPC5 channel activity is unknown. Again, it may be due to the use of expression technology in studying complex biochemical regulation of TRPC channels by PIs, or differences between direct and indirect actions of PIP_2.

There is certainly compelling evidence to indicate that PIP_2 produces a potent inhibitory action on TRPC4 and TRPC5 channel activity, that G-protein-mediated activation of PLC is an essential step in receptor stimulation of these channels, and that DAG-mediated PKC activity potently inhibits TRPC4 and TRPC5 channel activity. All these events are similar to the proposed activation mechanisms of TRPC3, TRPC6 and TRPC7 channels (see above). In light of these similarities, it is perhaps time to re-evaluate the role of DAG in activating TRPC4 and TRPC5 channels via PKC-independent mechanisms. It would be interesting to investigate whether exogenous OAG activates TRPC4 and TRPC5 channel activities following pre-treatment with PKC inhibitors. In addition, the effect of low concentrations of OAG on channel gating could be studied, in light of the proposal that low concentrations of OAG induce TRPC3 channel gating whereas high concentrations of DAG predominantly inhibit TRPC3 channel activity via PKC-dependent mechanisms [83]. Moreover, the effect of exogenous OAG on activation of TRPC4 and TRPC5 channel activity following removal of the inhibitory action of PIP_2 by wortmannin/anti-PIP_2 antibodies could be examined as shown for TRPC6 channel activity in VSMCs [97]. Lastly, it would be important to examine interactions between DAG, IP_3, PKC and PIP_2 on TRPC4 and TRPC5 proteins.

22.7 TRPC1 Containing Channels: PKC- and PIP_2-Activated Channels

TRPC1 proteins are an enigma, as few studies have been able to demonstrate that TRPC1 subunits form functional homomeric ion channels. The reason for this is not known, but it may be due by individual TRPC1 subunits being expressed at intracellular sites or that accessory subunits are also required for activation of TRPC1 channels [109]. Investigations on properties of TRPC1 channels have subsequently found that these functional channels are heteromers with TRPC1 proteins associating with other TRPC subunits (see Table 22.1). It may be expected that these TRPC1-containing heteromers would be gated by receptor stimulation via generation of DAG via a PKC-independent mechanism, as this is proposed as the common activation pathway of TRPC3-C7 subunits that are also components of these channels (see above). However, there is intriguing evidence, which indicates that receptor-mediated generation of DAG gates TRPC1-containing channels via a PKC-dependent pathway. In further contrast to the proposed activation mechanism of TRPC3-C7 channels, it has also been proposed that PIP_2 has an excitatory role on gating TRPC1-containing channels.

22.7.1 Role of PKC in Gating of TRPC1

In human glomerular mesangial cells, applications of epidermal growth factor, thapsigargin and phorbol esters (PKC activators) evoked channel activity and Ca^{2+} entry, which were abolished by PKC inhibitors [110, 111]. In portal vein, mesenteric artery and coronary artery VSMCs, vasoconstrictors, CPA and phorbol esters were also shown to activate channel activities, which were blocked by inhibition of PKC [24, 50]. In addition, it was shown that receptor-mediated PKC stimulation could activate these channels in mesangial cells and portal vein myocytes via store-independent pathways [50, 112]. In subsequent work, it was shown that TRPC1/C4 heteromers comprise the PKC-dependent channels in mesangial cells [21, 22, 113, 114], whereas the PKC-dependent channels in portal vein, mesenteric and coronary artery VSMCs were composed of TRPC1/C5/C7, TRPC1/C5 and TRPC1/C5/C6 heteromers respectively [24, 115]. These results clearly show that PKC activates TRPC1-containing channels in native cells, and importantly proved that previously defined SOCs could be activated by both store-dependent and -independent mechanisms. In human endothelial cells, direct phosphorylation of TRPC1 by PKCα was shown to be essential for whole-cell currents and Ca^{2+} entry evoked by thrombin and thapsigargin [116], with these channels being composed of TRPC1/C4 heteromers [20, 117, 118]. PKCα has also been shown to be pivotal in activating endogenous TRPC1-mediated Ca^{2+} oscillations induced by stimulation of calcium-sensing receptors stably expressed in HEK-293 cells [118], with endogenous channels in HEK-293 cells composed of TRPC1/C3/C7 heteromers [27].

22.7.2 Role of PIP_2 in Gating of TRPC1

Recent findings indicate that PIP_2 has an obligatory role in gating of TRPC1-containing channels in portal vein and coronary artery VSMCs [51, 120, 121]. In these studies, application of PIP_2 directly to the intracellular surface of inside-out patches activated TRPC1-containing channel activity indicating that PIP_2 may be a direct gating molecule. In addition, depletion of PIP_2 levels by wortmannin and application of anti-PIP_2 antibodies inhibited channel activity activated by CPA, BAPTA-AM and PDBu. It was also shown that PKC-mediated phosphorylation of TRPC1 proteins and binding of PIP_2 to TRPC1 were both required for opening of these TRPC1-containing channels. In coronary artery VSMCs, endothelin-1 (ET-1) was shown to activate TRPC1-containing channels through stimulation of ET_B receptors coupled to signal transduction pathway involving PLC, generation of DAG and activation of PKC [121]. It was concluded that PKC phosphorylated TRPC1 proteins, which enabled PIP_2 to open the channels. In contrast, stimulation of ET_A receptors activated these TRPC1-containing channels via a stimulation of PI-3-kinase and production of PIP_3, also through a PKC-dependent mechanism. It should also be noted that IP_3 facilitates opening of TRPC1-containing channels by CPA and PKC activation in portal vein [24, 122].

Interestingly, PIs are thought to also have an important role in mediating Ca^{2+} entry via interactions between STIM1 and Orai1 [123, 124]. These studies suggest that PIs may be required for Orai1 activation [123] and/or STIM1 accumulation at endoplasmic-plasma membrane junctions [124].

22.8 Summary

There is considerable evidence that TRPC3, TRPC6 and TRPC7 channels are activated by receptor-mediated generation of DAG, which directly opens these channels via a PKC-independent mechanism. Moreover, it is increasingly apparent that these PKC-independent mechanisms involves competitive interactions between the excitatory action of DAG (and sometimes IP_3) and physiological antagonism by PIP_2 on channel proteins. In light of these new ideas, the role of DAG and PIP_2 in gating of TRPC4 and TRPC5 channels should perhaps be re-evaluated. Recent work has proposed that inclusion of TRPC1 subunits within a functional TRPC heteromer structure switches the gating mechanism of TRPC channels, from being stimulated by DAG via a PKC-independent mechanism to DAG activating channel activity through a PKC-dependent pathway. Moreover, the presence of TRPC1 subunits in a TRPC heteromer also changes the actions of PIP_2 on channel activity from being inhibitory to being excitatory.

In conclusion, there does seem to be common activations pathways of TRPC channels, with non-TRPC1-containing channels directly gated by DAG, and TRPC1-containing channels gated by PIP_2, which requires PKC-dependent phosphorylation of TRPC1 proteins.

Acknowledgment Work carried out in the laboratory of the author was funded by The Wellcome Trust and the British Heart Foundation. The author would like to thank Prof WA Large and Dr SN Saleh for many helpful discussions and for reading the manuscript.

References

1. Ramsey S, Delling M, Clapham DE (2006) An introduction to TRP channels. Annu Rev Physiol 68:614–647
2. Estacion M, Sinkins WG, Jones SW, Applegate MA, Schilling WP (2006) Human TRPC6 expressed in HEK 293 cells forms non-selective cation channels with limited Ca^{2+} permeability. J Physiol 572:359–377
3. Birnbaumer L (2009) The TRPC class of ion channels: a critical review of their roles in slow, sustained increases in intracellular Ca(2+) concentrations. Annu Rev Pharmacol Toxicol 49:395–426
4. Abramowitz J, Birnbaumer L (2009) Physiology and pathophysiology of canonical transient receptor potential channels. FASEB J 23:297–328
5. Nilius B, Owsianik G, Voets T, Peters JA (2005) Transient receptor potential channels in disease. Physiol Rev 87:165–217
6. Hisatsune C, Kuroda Y, Nakamura K, Inoue T, Nakamura T, Michikawa T, Mizutani A, Mitkoshiba A (2004) Regulation of TRPC6 channel activity by tyrosine phosphorylation. J Biol Chem 279:18887–18894
7. Vazquez G, Wedel BJ, Kawasaki BT, Bird GS, Putney JW Jr (2004) Obligatory role of Src kinase in the signalling mechanisms for TRPC3 cation channels. J Biol Chem 279:40521–40528
8. Trebak M, Hempel M, Wedel BJ, Smyth JT, Bird GJ, Putney JW (2005) Negative regulation of TRPC3 channels by protein kinase C-mediated phosphorylation of serine 712. Mol Pharmacol 67:558–563
9. Takahashi S, Lin H, Geshi N, Mori Y, Kawarabayashi Y, Takami N, Mori MX, Honda A, Inoue R (2009) Nitric oxide-cGMP-protein kinase G pathway negatively regulates vascular transient receptor potential channel TRPC6. J Physiol 586:4209–4223
10. Koitabashi N, Aiba T, Hesketh GG, Rowell J, Zhang M, Takimoto E, Tomaselli GF, Kass DA (2010) Cyclic GMP/PKG-dependent inhibition of TRPC6 channel activity and expression negatively regulates cardiomyocyte NFAT activation Novel mechanism of cardiac stress modulation by PDE5 inhibition. J Mol Cell Cardiol 48:713–724
11. Nishida M, Watanabe K, Sato Y, Nakaya M, Kitajima N, Ide T, Inoue R, Kurose H (2010) Phosphorylation of TRPC6 channels at Thr69 is required for anti-hypertrophic effects of phosphodiesterase 5 inhibition. J Biol Chem 285:13244–13253
12. Zhu MX (2005) Multiple roles of calmodulin and other Ca^{2+}-binding proteins in the functional regulation of TRP channels. Pflugers Arch 451:105–115
13. Kwon Y, Hofmann T, Montell C (2007) Integration of phosphoinositide- and calmodulin-mediated regulation of TRPC6. Mol Cell 25:491–503
14. Friedlova E, Grycova L, Holakovska B, Silhan J, Janouskova H, Sulc M, Obsilova V, Obsil T, Teisinger J (2010) The interactions of the C-terminal region of the TRPC6 channel with calmodulin. Neurochem Int 56:363–366
15. Goel M, Sinkins WG, Schilling WP (2002) Selective association of TRPC channel subunits in rat brain synaptosomes. J Biol Chem 277:48303–48310
16. Hofmann T, Schaefer M, Schultz G, Gudermann T (2002) Subunit composition of mammalian transient receptor potential channel in living cells. PNAS 99:7461
17. Strubing C, Krapivinsky G, Krapivinsky L, Clapham DE (2002) Formation of novel TRPC channels by complex subunit interactions in embryonic brain. J Biol Chem 278:39014–39019
18. Wu X, Zagranichnaya TK, Gurda GT, Eves EM, Villereal MLA (2004) TRPC1/TRPC3-mediated increase in store-operated calcium entry is required for differentiation of H19-7 hippocampal neuronal cells. J Biol Chem 279:43392–43402
19. Liu X, Bandyopadhyay BC, Singh BB, Groschner K, Ambudkar IS (2005) Molecular analysis of a store-operated and 2-acetyl-sn-glycerol-sensitive non-selective

cation channel. Heteromeric assembly of TRPC1-TRPC3. J Biol Chem 280: 21600–21606

20. Antoniotti S, Fiorio Pla A, Barral S, Scalabrino O, Munaron L, Lovisolo D (2006) Interaction between TRPC channel subunits in endothelial cells. J Recept Signal Transduct Res 26: 225–240
21. Sours S, Du J, Chu S, Ding M, Zhou XJ, Ma R (2006) Expression of canonical transient receptor potential (TRPC) proteins in human glomerular mesangial cells. Am J Physiol 290:F1507–F1515
22. Sours-Brothers S, Ding M, Graham S, Ma R (2009) Interaction between TRPC1/TRPC4 assembly and STIM1 contributes to store-operated Ca^{2+} entry in mesangial cells. Exp Biol Med 234:673–682
23. Strubing C, Krapivinsky G, Krapivinsky L, Clapham DE (2001) TRPC1 and TRPC5 form a novel cation channel in mammalian brain. Neuron 29:645–655
24. Saleh SN, Albert AP, Peppiatt-Wildman CM, Large WA (2008) Diverse properties of store-operated TRPC channels activated by protein kinase C in vascular myocytes. J Physiol 586:2463–2476
25. Xu SZ, Beech DJ (2001) TrpC1 is a membrane-spanning subunit of store-operated Ca^{2+} channels in native vascular smooth muscle cells. Circ Res 88:84–87
26. Xu SZ, Boulay G, Flemming R, Beech DJ (2006) E3-targeted anti-TRPC5 antibody inhibits store-operated calcium entry in freshly isolated pial arterioles. Am J Physiol 291: H2653–H2659
27. Zagranichnaya TK, Wu X, Villereal ML (2005) Endogenous TRPC1, TRPC3, and TRPC7 proteins combine to form native store-operated channels in HEK-293 cells. J Biol Chem 280:29559–29569
28. Brownlow SL, Sage SO (2005) Transient receptor potential protein subunit assembly and membrane distribution in human platelets. Thromb Haemost 94:839–845
29. Bai CX, Giamarchi A, Rodat-Despoix L, Padilla F, Downs T, Tsiokas L, Delmas P (2009) Formation of a new receptor-operated channel by heteromeric assembly of TRPP2 and TRPC1 subunits. EMBO J 9:472–479
30. Kobori T, Smith GD, Sandford R, Edwardson JM (2009) The transient receptor potential channels TRPP2 and TRPC1 form a heterotetramer with a 2:2 stoichiometry and an alternating subunit arrangement. J Biol Chem 284:35507–35513
31. Zhang P, Luo Y, Chasan B, González-Perrett S, Montalbetti N, Timpanaro GA, Cantero Mdel R, Ramos AJ, Goldmann WH, Zhou J, Cantiello HF (2009) The multimeric structure of polycystin-2 (TRPP2): structural-functional correlates of homo- and hetero-multimers with TRPC1. Hum Mol Genet 18:1238–1251
32. Poteser M, Graziani A, Rosker C, Eder P, Derler I, Kahr H, Zhu MX, Romanin C, Groschner K (2006) TRPC3 and TRPC4 associate to form a redox-sensitive cation channel. Evidence for expression of native TRPC3-TRPC4 heteromeric channels in endothelial cells. J Biol Chem 281:13588–13595
33. Peppiatt-Wildman CM, Albert AP, Saleh SN, Large WA (2007) Endothelin-1 activates a Ca^{2+}-permeable cation channel with TRPC3 and TRPC7 properties in rabbit coronary artery myocytes. J Physiol 580:755–764
34. Miyagi K, Kiyonaka S, Yamada K, Miki T, Mori E, Kato K, Numata T, Sawaguchi Y, Numaga T, Kimura T, Kanai Y, Kawano M, Wakamori M, Nomura H, Koni I, Yamagishi M, Mori Y (2009) A pathogenic C terminus-truncated polycystin-2 mutant enhances receptor-activated Ca^{2+} entry via association with TRPC3 and TRPC7. J Biol Chem 49:34400–34412
35. Maruyama Y, Nakanishi Y, Walsh EJ, Wilson DP, Welsh DG, Cole WC (2006) Heteromultimeric TRPC6-TRPC7 channels contribute to arginine vasopressin-induced cation current of A7r5 vascular smooth muscle cells. Circ Res 98:1520–1527
36. Ju M, Shi J, Saleh SN, Albert AP, Large WA (2010) Ins(1,4,5)P$_3$ interacts with PIP$_2$ to regulate activation of TRPC6/C7 channels by diacylglycerol in native vascular myocytes. J Physiol 588:1419–1433

37. Van Breemen C, Aaronson P, Loutzenhiser R (1978) Sodium-calcium interactions in mammamlian smooth muscle. Pharmacol Rev 30:167–208
38. Bolton TB (1979) Mechanisms of transmitters and other substances on smooth muscle. Physiol Rev 59:606–718
39. Luik RM, Wu MM, Buchanan J, Lewis RS (2006) The elementary unit of store-operated Ca^{2+} entry: local activation of CRAC channels by STIM1 at ER-plasma membrane junctions. J Cell Biol 174:815–825
40. Peinelt C, Vig M, Koomoa DL, Beck A, Nadler MJ, Koblan-Huberson M, Lis A, Fleig A, Penner R, Kinet JP (2006) Amplication of CRAC current by STIM1 and CRACM1 (Orai1). Nat Cell Biol 8:771–773
41. Prakriya M, Feske S, Gwack Y, Srikanth S, Rai A, Hogan RG (2006) Orai1 is an essential pore subunit of the CRAC channel. Nature 443:230–233
42. Spassova MA, Soboloff J, He LP, Xu W, Dziadek MA, Gill DL (2006) STIM1 has a plasma membrane role in the activation of store-operated Ca^{2+} channels. PNAS 103:4040–4045
43. Yeromin AV, Zhang SL, Jiang W, Yu Y, Safrina O, Cahalan MD (2006) Molecular identification of the CRAC channel by altered ion selectivity in a mutant of Orai. Nature 443:226–229
44. Vig M, Peinelt C, Beck A, Koomoa DL, Rabah D, Koblan-Huberson M, Kraft S, Turner H, Fleig A, Penner R, Kinet JP (2006) CRACM1 is a plasma membrane protein essential for store-operated Ca^{2+} entry. Science 312:1220–1223
45. Huang GN, Zheng W, Kim JY, Yuan JP, Han L, Muallem S, Worley PF (2006) STIM1 caroxyl-terminus activates native SOC, I(crac) and TRPC1 channels. Nat Cell Biol 8: 1003–1010
46. Ong HL, Cheng KT, Liu X, Bandyopadhyay BC, Paria BC, Soboloff J, Pani B, Gwack Y, Srikanth S, Singh BB, Gill DL, Ambudkar IS (2007) Dynamic assembly of TRPC1-STIM-Orai1 ternary complex is involved in store-operated calcium influx. Evidence for similarities in store-operated and calcium release-activated calcium channel components. J Biol Chem 282:9105–9116
47. Cheng KT, Liu X, Ong HL, Ambudkar IS (2008) Functional requirement for Orai1 in store-operated TRPC1-STIM1 channels. J Biol Chem 9:12935–12940
48. Yuan JP, Zheng W, Huang GN, Worley PN, Muallem S (2007) STIM1 heteromultimerozes TRPC channels to determine their function as store-operated channels. Nat Cell Biol 9: 636–645
49. Cahalan MD (2009) STIMulating store-operated Ca^{2+} entry. Nat Cell Biol 11:669–677
50. Albert AP, Large WA (2002) Activation of store-operated channels by noradrenaline via protein kinase C in rabbit portal vein myocytes. J Physiol 544:113–125
51. Large WA, Saleh SN, Albert AP (2009) Role of phosphoinositol 4,5-bisphosphate and diacylglycerol in regulating native TRPC channel proteins in vascular smooth muscle. Cell Calcium 45:574–582
52. Albert AP, Saleh SN, Large WA (2009) Identification of canonical transient receptor potential (TRPC) channel proteins in native vascular smooth muscle cells. Curr Med Chem 16:1158–1165
53. Montell C (2005) TRP channels in *Drosophila* photoreceptor cells. J Physiol 567:45–51
54. Minke B, Parnas M (2006) Insights on TRP channels from in vivo studies in *Drosophila*. Annu Rev Physiol 68:649–684
55. Hardie RC (2007) TRP channels and lipids: from *Drosophila* to mammalian physiology. J Physiol 578:9–24
56. Katz B, Minke B (2009) *Drosophila* photoreceptors and signaling mechanisms. Front Cell Neurosci 3:1–14
57. Xu XZ, Chien F, Butler A, Salkoff L, Montell C (2000) TRPgamma, a *Drosophila* TRP-related subunit, forms a regulated cation channel with TRPL. Neuron 26:647–657
58. Hardie RC (2003) Regulation of TRPC channels by lipid second messengers. Ann Rev Physiol 65:735–759

59. Bloomquist BT, Shortridge RD, Schneuwly S, Perdew M, Montell C, Stellar H, Rubin G, Pak WL (1988) Isolation of a putative phospholipase C gene of *Drosophila*, norpA, and its role in phototransuction. Cell 54:723–733
60. Hardie RC, Raghu P (1998) Activation of heterologously expressed *Drosophila* TRPL channels: Ca^{2+} is not required and $InsP_3$ is not sufficient. Cell Calcium 24:153–163
61. Acharya JK, Jalink K, Hardy RW, Hartenstein V, Zuker CS (1997) $InsP_3$ receptor is essential for growth and differentiation but not for vision in *Drosophila*. Neuron 18:881–887
62. Raghu P, Colley NJ, Webel R, James T, Hasan G, Danin M, Slinger Z, Hardie RC (2000) Normal phototransduction in *Drosophila* photoreceptors lacking an $InsP_3$ receptor gene. Mol Cell Neurosci 15:429–445
63. Hardie RC (1995) Photolysis of caged Ca^{2+} facilitates and inactivates but does not directly excite light-sensitive channels in *Drosophila* photoreceptors. J Neurosci 15:889–902
64. Parnas M, Katz B, Minke B (2007) Open channel block by Ca^{2+} underlies the voltage dependence of *Drosophila* TRPL channel. J Gen Physiol 129:17–28
65. Hardie RC, Peretz A, Suss-Toby E, Rom-Glas A, Bishop SA, Selinger Z, Minke B (1993) Protein kinase C is required for light adaption in *Drosophila* photoreceptors. Nature 363:634–637
66. Chyb S, Raghu P, Hardie RC (1999) Polyunsaturated fatty acids activate the *Drosophila* light-sensitive channels TRP and TRPL. Nature 397:255–259
67. Raghu P, Usher K, Jonas S, Chyb S, Polyanovsky A, Hardie RC (2000) Constitutive activity of the light-sensitive channels TRP and TRPL in the *Drosophila* diacylglycerol kinase mutant, rdgA. Neuron 26:169–179
68. Delgado R, Bacigalupo J (2009) Unitary recordings of TRP and TRPL channels from isolated Drosophila retinal phostoreceptor rhabdomeres: activation by light and lipids. J Neurophysiol 101:2372–2379
69. Kwon Y, Montell C (2006) Dependence on the Lazaro phosphatidic acid phoshatase for the maximum light response. Curr Biol 16:723–729
70. Estacion M, Sinkins WG, Schilling WP (2001) Regulation of *Drosophila* transient receptor potential-like (TrpL) channels by phospholipase C-dependent mechanisms. J Physiol 530:1–19
71. Hardie RC, Raghu P, Moore S, Juusola M, Baines RA, Sweeney ST (2001) Calcium influx via TRP channels is required to maintain PIP_2 levels in *Drosophila* photoreceptors. Neuron 30:149–159
72. Parnas M, Katz B, Lev S, Tzarfaty V, Dadon D, Gordon-Shagg A, Metzner H, Yaka R, Minke B (2009) Membrane lipid modulations remove dilvant open channel block from TRP-like and NMDA channels. J Neurosci 29:2371–2383
73. Jors S, Kazanki V, Foik A, Krautwurst D, Harteneck C (2006) Receptor-induced activation of *Drosophila* TRPγ by polyunsaturated fatty acids. J Biol Chem 281:29693–29702
74. Huang J, Liu CH, Hughes SA, Postma M, Schwiening CJ, Hardie RC (2010) Activation of TRP channels by protons and phosphoinositide depletion in *Drosophila* photoreceptors. Curr Biol 20:189–197
75. Leung HT, Tseng-Crank J, Kim E, Mahapatra C, Shino S, Zhou Y, An L, Doerge RW, Pak WL (2008) DAG lipase activity is necessary for TRP channel regulation in *Drosophila* photoreceptors. Neuron 58:884–896
76. Helliwell RM, Large WA (1997) $α_1$-adrenoceptor activation of a non-selective cation current in rabbit portal vein by 1,2-diacyl-sn-glycerol. J Physiol 499:417–428
77. Hofmann T, Obukhov AG, Schaefer M, Harteneck C, Gudermann T, Schultz G (1999) Direct activation of human TRPC6 and TRPC3 channels by diacylglycerol. Nature 397:259–263
78. Okada T, Inoue R, Yamazaki K, Maeda A, Kurosaki T, Yamakuni T, Tanaka I, Shimizu S, Ikenaka K, Imoto K, Mori Y (1999) Molecular and functional characterisation of a novel mouse transient receptor potential protein homologue TRP7. Ca^{2+}-permeable cation channel that is constitutively activated and enhanced by stimulation of G-protein-coupled receptor. J Biol Chem 274:27359–27370

79. Trebak M, St J Bird G, McKay RR, Birnbaumer L, Putney JW Jr (2003) Signaling mechanism for receptor-activated canonical transient receptor potential 3 (TRPC3). J Biol Chem 278:16244–16252
80. Estacion M, Li S, Sinkins WG, Gosling M, Bahra P, Poll C, Westwick J, Schilling WP (2004) Activation of human TRPC6 channels by receptor stimulation. J Biol Chem 279: 22047–22056
81. Shi J, Mori E, Mori Y, Mori M, Li J, Ito Y, Inoue R (2004) Multiple regulation by calcium of murine homologues of transient receptor potential proteins TRPC6 and TRPC7 expressed in HEK293 cells. J Physiol 561:415–432
82. Inoue R, Okada T, Onoue H, Harea Y, Shimizu S, Naitoh S, Ito Y, Mori Y (2001) The transient receptor potential protein homologue TRP6 is the essential component of vascular α-adrenoceptor-activated Ca^{2+}-permeable cation channel. Circ Res 88:325–337
83. Ventaktachalam K, Zheng F, Gill DL (2003) Regulation of canonical transient receptor potential (TRPC) channel function by diacylglycerol and protein kinase C. J Biol Chem 278:29031–29040
84. Albert AP, Large WA (2004) Inhibitory regulation of constitutive transient receptor potential-like cation channels in rabbit ear artery myocytes. J Physiol 560: 169–180
85. Albert AP, Piper AS, Large WA (2005) Role of phospholipase D and diacylglycerol in activating constitutive TRPC-like cation channels in rabbit ear artery myocytes. J Physiol 566:769–780
86. Albert AP, Pucovsky V, Prestwich SA, Large WA (2006) TRPC3 properties of a native constitutively active Ca^{2+}-permeable cation channel in rabbit ear artery myocytes. J Physiol 571:361–369
87. Kwan HY, Wong CO, Chen ZY, Dominic Chan TW, Huang Y, Yao X (2009) Stimulation of histamine H2 receptors activates TRPC3 channels through both phospholipase C and phospholipase D. Eur J Pharmacol 602:181–187
88. Glitsch MD (2010) Activation of native TRPC3 cation channels by phospholipase D. FASEB J 24:318–325
89. Kiselyov K, Xu X, Mozhayeva G, Kuo T, Pessah I, Migery G, Zhu X, Burnbaumer L, Muallem S (1998) Functional interaction between $InsP_3$ receptors and store-operated Htrp3 channels. Nature 396:478–482
90. Ma HT, Patterson RL, van Rossum DB, Birnbaumer L, Mikoshiba K, Gill DL (2001) Requirement of the inositol trisphosphate receptor for activation of store-operated Ca^{2+} channels. Science 287:1647–1651
91. Vazquez G, Lievremont JP, St J Bird G, Putney JW Jr (2001) Human Trp3 forms both inositol trisphosphate receptor-dependent and receptor-independent store-operated cation channels in DT40 avian B lymphocytes. PNAS 98:11777–11782
92. Vazquez G, Bird GS, Mori Y, Putney JW Jr (2006) Native TRPC7 channel activation by an inositol trisphosphate receptor-dependent mechanism. J Biol Chem 281: 25250–25258
93. Vazquez G, Wedel BJ, Trebak M, St John Bird G, Putney JW Jr (2003) Expression levels of the canonical transient receptor potential 3 (TRPC3) channel determines it's mechanism of activation. J Biol Chem 278:21649–21654
94. Venkatachalam K, Ma HT, Ford DL, Gill DL (2001) Expression of functional receptor-coupled TRPC3 channels in DT40 triple receptor InsP3 knockout cells. J Biol Chem 276:33980–33985
95. Xi Q, Adebiyi A, Zhao G, Chapman KE, Waters CM, Hassied A, Jaggar JH (2008) IP_3 constricts cerebral arteries via IP_3 receptor-mediated TRPC3 channel activation and independently of sarcoplasmic reticulum Ca^{2+} release. Circ Res 102:1118–1126
96. Zhao G, Adebiyi A, Blaskova E, Xi Q, Jaggar JH (2008) Type 1 inositol 1,4,5-trisphosphate receptors mediate UTP-induced cation currents, Ca^{2+} signals, and vasoconstriction in cerebral arteries. Am J Physiol Cell Physiol 295:C1376–C1384

97. Albert AP, Large WA (2003) Synergism between inositol phosphates and diacylglycerol on native TRPC6-like channels in rabbit portal vein myocytes. J Physiol 552: 789–795
98. Albert AP, Saleh SN, Large WA (2008) Inhibition of native TRPC6 channel activity by phosphatidylinositol-4, 5-bisphosphate in mesenteric artery myocytes. J Physiol 586: 3087–3095
99. Tseng PH, Lin HP, Hu H, Wang C, Zhu MX, Chen CS (2004) The canonical transient receptor potential 6 channel as a putative phosphatidylinositol-3,4,5-trisphosphate-sensitive calcium entry system. Biochemistry 43:11701–11708
100. Lemonnier L, Trebak M, Putney JW Jr (2007) Complex regulation of the TRPC3, 6 and 7 channel subfamily by diacylglycerol and phosphatidylinositol-4, 5-bisphosphate. Cell Calcium 43:506–514
101. Schaefer M, Plant TD, Obukhov AG, Hofmann T, Gudermann T, Schultz G (2000) Receptor-mediated regulation of the nonselective cation channels TRPC4 and TRPC5. J Biol Chem 275:17517–17526
102. Plant TD, Schaefer M (2003) TRPC4 and TRPC5: receptor-operated Ca^{2+}-permeable nonselective cation channels. Cell Calcium 33:441–450
103. Otsuguro K, Tang J, Tang Y, Xiao R, Freichel M, Tsvilovskyy V, Ito S, Flockerzi V, Zhu M, Zholos AV (2008) Isoform-specific inhibition of TRPC4 channel by phosphatidylinositol 4, 5-bisphosphate. J Biol Chem 283:10026–10036
104. Blair NT, Kaczmarek JS, Clapham DE (2009) Intracellular calcium strongly potentiates agonist-activated TRPC5 channels. J Gen Physiol 133:525–546
105. Lee YM, Kim BJ, Kim HJ, Yang DK, Zhu MH, Lee KP, So I, Kim KW (2003) TRPC5 as a candidate for the nonselective cation channel activated by muscarinic stimulation in murine stomach. Am J Physiol Cell Physiol 284:G6604–G6616
106. Zhu MH, Chae M, Kim HJ, Lee YM, Kim MJ, Jin NG, Yang DK, So I, Kim KW (2005) Densensitization of canonical transient receptor potential channel 5 by protein kinase C. Am J Physiol Cell Physiol 289:C591–C600
107. Trebak M, Lemonnier L, Dehaven WI, Wedel BJ, Bird GS, Putney JW Jr. (2008) Complex functions of phosphatidylinositol-4, 5-bisphosphate on regulation of TRPC5 cation channels. Pflugers Arch 457:757–769
108. Kim BJ, Kim MT, Jeon JH, Kim SJ, So I (2008) Involvement of phosphatidylinositol 4,5-bisphosphate in the desensitization of canonical transient receptor potential 5. Biol Pharm Bull 31:1733–1738
109. Miehe S, Bieberstein A, Arnould I, Ihdene O, Rütten H, Strübing C (2010) The phospholipid-binding protein SETD1 is a novel regulator of the transient receptor potential channels TRPC4 and TRPC5. J Biol Chem 285:12426–12434
110. Beech DJ, Xu SZ, McHugh D, Flemming R (2003) TRPC1 store-operated cationic channel subunit. Cell Calcium 33:433–440
111. Ma R, Pluznick J, Kudlacek PE, Sansom SC (2001) Protein kinase C activates store-operated Ca^{2+} channels in human glomerular mesangial cells. J Biol Chem 276: 25759–25765
112. Ma R, Kudlacek PE, Sansom SC (2002) Protein kinase Calpha participates in activation of store-operated Ca^{2+} channels in human glomerular mesangial cells. Am J Physiol Cell Physiol 283:C1390–C1398
113. Li WP, Tsiokas L, Sansom SC, Ma R (2004) Epidermal growth factor activates store-operated Ca^{2+} channels through an inositol 1,4,5-trisphosphate-independent pathway in human glomerular mesangial cells. J Biol Chem 279:4570–4577
114. Du J, Sours-Brothers S, Coleman R, Ding M, Graham S, Kong DH, Ma R (2007) Canonical transient receptor potential 1 channel is involved in contractile function of glomerular mesangial cells. J Am Soc Nephrol 18:1437–1445
115. Wang X, Pluznick JL, Wei P, Padanilam BJ, Sansom SC (2004) TRPC4 forms store-operated Ca^{2+} channels in mouse mesangial cells. Am J Physiol Cell Physiol 287: C357–C364

116. Saleh SN, Albert AP, Peppiatt CM, Large WA (2006) Angiotensin II activates two cation conductances with distinct TRPC1 and TRPC6 channel properties in rabbit mesenteric artery myocytes. J Physiol 577:479–495
117. Ahmmed GU, Mehta D, Vogel S, Holinstat M, Paria BC, Tiruppathu C, Malik AB (2004) Protein kinase C alpha phosphorylates the TRPC1 channel and regulates store-operated Ca^{2+} entry in endothelial cells. J Biol Chem 279:20941–20949
118. Brough GH, Wu S, Cioffi D, Moore TM, Li M, Dean N, Stevens T (2001) Contribution of endogenously expressed Trp1 to a Ca^{2+}-selective, store-operated Ca^{2+} entry pathway. FASEB J 15:1727–1738
119. Freichel M, Suh SH, Pfeifer A, Schweig U, Trost C, Weissgerber P, Biel M, Philipp S, Freise D, Droogmans G, Hofmann F, Flockerzi V, Nilius B (2001) Lack of an endothelial store-operated Ca^{2+} current impairs agonist-dependent vasorelaxation in TRP4–/– mice. Nat Cell Biol 3:121–127
120. Rey O, Young SH, Papazyan R, Shapiro MS, Rozengurt E (2006) Requirement of the TRPC1 cation channel in the generation of transient Ca^{2+} oscillations by the calcium-sensing receptor. J Biol Chem 281:38730–38737
121. Saleh SN, Albert AP, Large WA (2009) Obligatory role for phosphatidylinositol 4,5-bisphosphate in activation of native TRPC1 store-operated channels in vascular myocytes. J Physiol 587:531–540
122. Saleh SN, Albert AP, Large WA (2009) Activation of native TRPC1/C5/C6 channels by endothelin-1 is mediated by both PIP_3 and PIP_2 in rabbit coronary artery myocytes. J Physiol 587:5361–5375
123. Liu M, Albert AP, Large WA (2005) Facilitatory effect of Ins(1,4,5)P_3 on store-operated Ca^{2+}-permeable cation channels in rabbit portal vein myocytes. J Physiol 566:161–171
124. Korzeniowski MK, Popovic MA, Szentpetery Z, Varnai P, Stojilkovic SS, Balla T (2009) Dependence of STIM1/Orai1-mediated calcium entry on plasma membrane phosphoinositides. J Biol Chem 284:21027–21035
125. Walsh CM, Chvanov M, Haynes LP, Petersen OH, Tepikin AV, Burgoyne RD (2009) Role of phosphoinositides in STIM1 dynamics and store-operated calcium entry. Biochem J 425:159–168

Chapter 23
The TRPC Ion Channels: Association with Orai1 and STIM1 Proteins and Participation in Capacitative and Non-capacitative Calcium Entry

Gines M. Salido, Isaac Jardín, and Juan A. Rosado

Abstract Transient receptor potential (TRP) proteins are involved in a large number of non-selective cation channels that are permeable to both monovalent and divalent cations. Two general classes of receptor-mediated Ca^{2+} entry has been proposed: one of then is conduced by receptor-operated Ca^{2+} channels (ROC), the second is mediated by channels activated by the emptying of intracellular Ca^{2+} stores (store-operated channels or SOC). TRP channels have been presented as sub-units of both ROC and SOC, although the precise mechanism that regulates the participation of TRP proteins in these Ca^{2+} entry mechanisms remains unclear. Recently, TRPC proteins have been shown to associate with Orai1 and STIM1 in a dynamic ternary complex regulated by the occupation of membrane receptors in several cell models, which might play an important role in the function of TRPC proteins. The present review summarizes the current knowledge concerning the association of TRP proteins with Orai and STIM proteins and how this affects the participation of TRP proteins in store-operated or receptor-operated Ca^{2+} entry.

Abbreviations

$[Ca^{2+}]_c$	cytosolic free Ca^{2+} concentration
ARC	arachidonic acid-activated
CAD	CRAC-activating domain
CMD	CRAC modulatory domain
CRAC	Ca^{2+} release-activated Ca^{2+} channel
DAG	diacylglycerol
ER	endoplasmic reticulum
FRET	Forster resonance energy transfer
IP_3	inositol 3,4,5-trisphosphate
IP_4	inositol 1,3,4,5-tetrakisphosphate
MBCD	methyl-β-cyclodextrin

J.A. Rosado (✉)
Cell Physiology Group, Department of Physiology, University of Extremadura, Cáceres, Spain
e-mail: jarosado@unex.es

OASF	Orai-activating small fragment
PIP$_2$	phosphatidilinositol 4,5-bisphosphate
PLC	phospholipase C
RACK1	receptor for activated C-kinase-1
ROCE	receptor-operated Ca^{2+} entry
SCID	severe combined immune deficiency
SERCA	sarcoplasmic/endoplasmic-reticulum Ca^{2+}-ATPase
SOC	store-operated channel
SOAR	STIM1 Orai-activating region
SOCE	store-operated calcium entry
STIM1	stromal interaction molecule 1
TRP	transient receptor potential
TG	thapsigargin.

23.1 Introduction

Regulation of the changes in cytosolic Ca^{2+} concentration is a point of convergence of many signal transduction pathways and modulates a variety of cellular functions ranging from fertilization to cell death. Changes in cytosolic free Ca^{2+} concentration ([Ca^{2+}]$_c$), also known as Ca^{2+} signals, are characterized by sudden and transitory increases in the concentration of free calcium ions [1]. Cell-generated Ca^{2+} signals require both internal and external Ca^{2+} sources. In most cell types, the major internal Ca^{2+} store is the endoplasmic reticulum (ER)/sarcoplasmic reticulum, where Ca^{2+} is stored by SERCA (sarco/endoplasmic reticulum Ca^{2+}-ATPase). Due to the finite amount of Ca^{2+} accumulated in the ER the entry of extracellular Ca^{2+} is necessary to achieve full activation of a number of cellular functions. Store-operated Ca^{2+} entry (SOCE), also known as capacitative Ca^{2+} entry, a process regulated by the filling state of the intracellular Ca^{2+} stores, is a major mechanism for Ca^{2+} entry in non-electrically excitable cells [2]. There is a body of evidence supporting an important role for SOCE in Ca^{2+} signalling and intracellular homeostasis under physiological conditions, such as supporting Ca^{2+} oscillations [3]. In addition, SOCE has been reported to be required for a number of cellular processes, including cell proliferation, muscle contraction, platelet aggregation and secretion [4, 5]. Finally, SOCE serves as a mechanism to allow ER Ca^{2+} refilling, necessary for protein synthesis and post-translational modifications [6].

The nature of the channels that conduct SOCE has been a matter of intense investigation and debate. Two types of store operated Ca^{2+} channels have been described so far, which, although show distinct biophysical properties, are activated by depletion of intracellular Ca^{2+} stores with agonists, inhibitors of SERCA and/or strong Ca^{2+} chelators. First of all, Ca^{2+} release-activated Ca^{2+}-selective (CRAC) channels have been found and extensively described on the level of whole-cell current in a variety of non-excitable cells, including mast cells, Jurkat T-lymphocytes and RBL cells [7–9]. The current through CRAC channels (I_{CRAC}) is non-voltage activated, inwardly rectifying, and selective for Ca^{2+} [10, 11]. While I_{CRAC} was the first store-operated Ca^{2+} current identified, it is not the only store-operated current, and SOCE has also been reported to include a family of Ca^{2+}-permeable channels, with different properties in different cell types known as SOC channels, which conduct the non-voltage activated, non-selective I_{SOC} current of small, but resolvable 0.7-11 pS conductance [12]. SOC channels have been found and described on single channel and whole-cell current levels in different cell types [13–16]. The mammalian homologues of the *Drosophila Transient Receptor Potential* (TRP) channels were initially presented as candidates for the conduction of SOCE and, more recently, the protein Orai1 has been proposed to form the pore of the channel mediating I_{CRAC}.

Despite intense investigations over the last two decades, the mechanisms of activation and the identity of the key molecular players conducting Ca^{2+} entry during SOCE have long remained elusive. However, in the last few years, the improvements of gene silencing protocols combined with high throughput platforms have provided important breakthroughs, especially with the identification of STIM1 (stromal interaction molecule 1) as the ER Ca^{2+} sensor and Orai1 as the pore-forming subunit of the archetypical capacitative channel, CRAC. STIM1 is a Ca^{2+}-binding protein located both in intracellular membranes, including the ER, and the plasma membrane with a single transmembrane region and an EF-hand domain in the N-terminus. STIM1 located in the ER shows the EF-hand domain in the lumen of the ER, which, by following different experimental manoeuvres, has been suggested to function as a Ca^{2+} sensor that communicates the filling state of the Ca^{2+} stores to the plasma membrane Ca^{2+} permeable channels [17, 18]. In addition, plasma membrane-resident STIM1, which shows the EF-hand domain facing the extracellular medium, has been reported to modulate the function of the capacitative channels [19, 20], probably acting as an extracellular Ca^{2+} sensor.

The involvement of Orai1 in I_{CRAC} has been identified by gene mapping in patients showing an inherited disorder called severe combined immune deficiency (SCID) syndrome attributed to loss of I_{CRAC}, which results in extreme vulnerability to infectious diseases. The ORAI1 gene located on chromosome 12 has been found to be mutated in SCID patients, and I_{CRAC} has been shown to be restored by expression of wild type Orai1 in T cells [21]. The role of Orai1 in I_{CRAC} was confirmed in a whole-genome screen of *Drosophila* S2 cells by Feske and coworkers [21], with other groups reporting similar results at the same time [22, 23]. Orai1 is a small protein with four transmembrane domains and both N- and C-terminal tails located in the cytosol. The Orai1 protein has been demonstrated to form multimeric ion channel complexes in the plasma membrane [24–29].

In addition to their involvement in receptor-operated Ca^{2+} entry (ROCE), there is now considerable evidence supporting a role for TRP proteins in the conduction of Ca^{2+} entry during SOCE. Particular attention has been paid to members of the TRPC subfamily. Using different approaches, from overexpression of specific TRP proteins to knockdown of endogenous TRPs and pharmacological studies, it has been suggested that most of the TRPC proteins can be activated by Ca^{2+} store depletion [12, 30, 31]. Among TRP proteins, the role of TRPC1 in SOCE has been extensively investigated in different cell types. TRPC1 has been reported to be involved in SOCE by antisense experiments in human salivary glands [32] and vascular endothelial cells [33]. In support of this, antibodies directed to the pore-forming region of TRPC1 have been shown to reduce SOCE in vascular smooth muscle cells and human platelets [34, 35] and TRPC1-depleted myoblasts present a largely reduced SOCE [36]. Different TRPC associations appear to give rise to channels with distinct biophysical properties. In addition, association of TRPC proteins with STIM1 and Orai1 seems to play an important role in the participation of TRPCs in different mechanisms for Ca^{2+} entry in a number of cell types, although the association between these proteins still remains controversial and further studies are required to fully understand the process.

23.2 Transient Receptor Potential (TRP) Proteins: TRPCs

TRP proteins are ion channel subunits non-selective for monovalent and divalent cations, including Na^+ and Ca^{2+} that were initially identified in the *trp* mutant of *Drosophila*. The light-sensitive current in *Drosophila* photoreceptors is conducted by two Ca^{2+}-permeable channels encoded by the *trp* and *trpl* genes [37, 38]. The *trp* mutant is characterized by transient, rather than sustained, light-sensitive depolarization due to Na^+ and Ca^{2+} influx [39]. Later on, *Drosophila* TRP channels were shown to be gated by diacylglycerol (DAG) or a metabolic byproduct, synergistically with phosphatidylinositol 4,5-bisphosphate (PIP_2) depletion [40].

The identification of mammalian homologues of *Drosophila* TRP channels raised interest in TRP proteins as candidates for Ca^{2+} entry channels. The first mammalian TRP protein, TRPC1, was identified in 1995 in human [41, 42] and mouse [43]. Since their identification, a number of TRP proteins have been found, which are grouped into seven major subfamilies: four are closely related to *Drosophila* TRP (TRPC, TRPV, TRPA and TRPM), two more distantly related subfamilies (TRPP and TRPML), and finally the TRPN group expressed so far only in fish, flies and worms [44]. The canonical TRP (TRPC) subfamily comprises seven members (TRPC1–TRPC7, which, in turn, can be divided into four groups: TRPC1, TRPC2, TRPC3/6/7 and TRPC4/5), the vanilloid TRP subfamily (TRPV) consists of six members (TRPV1–TRPV6), the TRPA (ankyrin) subfamily includes only one mammalian member, TRPA1, and the melastatin TRP subfamily (TRPM) groups eight different channels (TRPM1–TRPM8). The TRPP (polycystin) and the TRPML (mucolipin) subfamilies include three channel members each, and finally, the TRPN has no mammalian members [45].

All members of the TRP family share a common architecture: they are proteins that contain six transmembrane domains, with different cytoplasmic N- and C-termini depending on the subfamily, and a pore loop region between the transmembrane domains 5 and 6 [46]. Many TRP proteins possess long N-terminal regions with several protein–protein interaction domains known as ankyrin repeats, a coiled coil region, and a putative caveolin-binding domain. On the other hand, the C-terminus includes the TRP signature motif (EWKFAR), a proline-rich motif and different functional regions that facilitate their interaction with calmodulin or inositol 1,4,5-trisphosphate (IP$_3$) receptor [47–49]. For further information concerning the structure of TRP proteins the reader is referred to Chapter 1.

As reported above, most TRP channels are nonselective for monovalent and divalent cations with Ca^{2+}:Na$^+$ permeability ratios <10 [50]. There are a number of exceptions, such as TRPM4 and TRPM5, which are selective for monovalent cations, and TRPV5 and TRPV6, which have a Ca^{2+}:Na$^+$ permeability ratio > 100. Among the TRP channels expressed in mammals, the role of the TRPC subfamily members on agonist-evoked Ca^{2+} entry has focused much attention, and, therefore, this review present an overview of the mechanisms involved in the participation of TRPC in Ca^{2+} entry.

The TRPC members form Ca^{2+}-permeable cation channels and have been presented as candidate subunits for the channels conducting both SOCE and ROCE [30, 35, 51–53]. By means of different experimental manoeuvres, from gene inactivation to gene expression silencing using siRNA or shRNA and the use of neutralizing antibodies, all the members of the TRPC family have been reported to be activated by store depletion or to be involved in SOCE both in excitable and non-excitable cells, including TRPC1 [32, 34, 35, 54], TRPC2 [55], TRPC3 [56, 57], TRPC4 [58, 59], TRPC5 [59, 60], TRPC6 [61–63] and TRPC7 [64]. However, the participation of TRPCs in SOCE depends on special circumstances, such as the expression level. Thus, at low expression levels TRPCs are activated by depletion of the intracellular Ca^{2+} stores, while at relatively high levels of expression TRPCs are not longer sensitive to store depletion but activated by phospholipase C (PLC) or its metabolites [56]. Furthermore, it has been reported that TRPC channels might participate in SOCE or ROCE in the same cell type depending on their mode of expression. In HEK-293 cells, TRPC7 is activated by PLC-stimulating agonists and not by Ca^{2+} store discharge when transiently expressed; in contrast, stably expressed TRPC7 gating can be regulated by either Ca^{2+} stores or PLC activation [64]. Although the reason for this phenomenon has not been determined it might be attributed to the association of TRPC proteins with regulatory subunits that confer store depletion or receptor sensitivity and then participation in SOCE or ROCE. The identification of STIM1 and Orai1 as essential components of SOCE may uncover the mechanism underlying the participation of TRPC subunits in SOCE or ROCE, as reported below.

23.3 STIM and Orai Proteins

Probably one of the most significant advances occurred in the last 5 years on the intracellular Ca^{2+} homeostasis has been, together with the determination of the structure of Orai, demonstrating that STIM1 is the ER sensor that report its Ca^{2+} filling state, essential for Ca^{2+} store depletion-triggered Ca^{2+} influx across de plasma membrane. Although two single transmembrane-spanning domain stromal interaction molecules with no known catalytic activity (human STIM1 and STIM2 containing 685 and 833 amino acids, respectively, and differing primarily in the lengths of their N- and C-terminal tails) have been described, STIM1 is the most interesting for the purposes of this chapter as it was found to act not only as a sensor within the stores [17, 18, 65] but also to play a role in the plasma membrane [17, 20] to activate I_{CRAC}.

From a structural point of view, STIM1 is a Ca^{2+}-binding protein (within either the ER lumen or extracellular space) that includes a number of functional domains described in Table 23.1. Both STIM1 and STIM2 can be phosphorylated predominantly on serine and threonine residues. In addition, STIM1 contains an additional N-linked glycosylation site within the SAM domain itself [66]. STIM1 has been reported to be expressed at the cell surface, as well as in the ER membrane, while STIM2 is expressed only intracellularly, likely reflecting an ER-retention signal (KKXX) present in STIM2 but not in STIM1 [67].

Knockdown of STIM1 by siRNA or functional knockdown of STIM1 by electro-transjection of neutralizing antibodies reduces SOCE in different cell types [20, 68] and I_{CRAC} in Jurkat T cells [20]. Evidence supporting the role of STIM1 in SOCE reports that mutation of the Ca^{2+}-binding EF-hand domain of STIM1 leads to constitutive SOC channel activation, and subsequent entry of Ca^{2+} into the cytoplasm, even without any detectable change in the content of the Ca^{2+} stores [17].

It is noteworthy to mention that, in addition to its role as an ER Ca^{2+} sensor, STIM1 has been found in the plasma membrane in a number of cells, expressing the EF-hand domain in the cell surface and acting as an extracellular Ca^{2+} sensor, where it has been demonstrated to modulate the operation of CRAC and SOC channels. External application of an antibody addressed towards the STIM1 N-terminal EF-hand region has been reported to block both CRAC channels in hematopoietic cells and SOC channels in HEK293 cells [20]. In addition, external application of the anti-STIM1 antibody blocks the inhibition of SOCE induced by increasing extracellular Ca^{2+} concentrations in human platelets, revealing a role for plasma membrane-resident STIM1 in the modulation of SOCE by extracellular Ca^{2+}, probably through its interaction with Ca^{2+} channel subunits such as Orai1 [19]. The pool of STIM1 that resides in the plasma membrane has also been reported to play a key role in other mechanisms of Ca^{2+} entry different from SOCE, such as the store-independent, arachidonic acid-activated, ARC channels, which show high Ca^{2+}-selectivity and low conductance and co-exist with CRAC channels [69].

In the 2006, Vig and co-workers demonstrated that the Ca^{2+} release-activated Ca^{2+} channel protein 1 (CRACM1) is a plasma membrane protein essential for SOCE. Although overexpression of the CRACM1 did not affect CRAC currents, RNAi-mediated knockdown disrupted its activation. Also, they reported that CRACM1 could be the CRAC channel itself, a subunit of it, or a component of the CRAC signalling machinery [22]. Few months later, the same group demonstrated that STIM1 and CRACM1 interact functionally; the overexpression of both proteins greatly potentiated I_{CRAC}, suggesting that STIM1 and CRACM1 mutually limit store-operated currents and that CRACM1 may be the long-sought CRAC channel [70]. Today, CRACM1 is best known for the romantic name of Orai1 (in Greek mythology, the "Orai" are the keepers of the gates of heaven). The mammalian

Table 23.1 Main functional domains of STIM1

Region	Location	Function	References
EF-hand domain	aa 67–95	Ca^{2+} binding domain that senses ER Ca^{2+} concentration	[17, 129]
SAM motif	aa 132–200	Sterile-α motif involved in protein-protein interaction	[129]
Transmembrane region	aa 215–234	A single transmembrane segment	[130]
Coiled-coil regions	aa 238–343 aa 363–389	Include the regions involved in Orai1/CRAC channel activating domains and overlap with the ERM-like domain	[66]
ERM-like domain	aa 251–535	Ezrin-radixin-moesin (ERM)-like domain. Includes de Orai1/CRAC channel activating domain	[131, 132]
CAD	aa 342–448	Orai1/CRAC activating domain	[83]
SOAR	aa 344–442		[81]
OASF	aa 233–450/474		[82]
CCb9	aa 339–444		[84]
Homomerization domain	aa 400–474	Clusters STIM1 into regions close to the plasma membrane	[130, 133]
STIM1 inhibitory domain	aa 445–475	Inhibits the Orai1/CRAC activating domain at rest	[134]
CMD	aa 474–485	CRAC modulatory domain that induces Orai1/CRAC cannel closure	[85]
Serine/proline-rich region	aa 600–629	Localization of STIM1 into ER-PM junctions	[89, 135]
Polybasic region	aa 672–685	Involved in puncta formation	[81]

Table 23.2 Predicted functional domains of Orai1

Region	Location	Function	References
Arginine/proline-rich region	aa 3–8 aa 28–33 aa 39–47	Orai1 assembly	[130, 136]
Arginine/lysin-rich region	aa 77–88	Orai1 assembly	[130, 136]
Transmembrane regions: TM1 TM2 TM3 TM4	aa 88–105 aa 118–140 aa 175–197 aa 236–258	Four transmembrane segments	[130]
Selectivity filter	aa 106–114 E190	Pore-forming domain	[22, 23, 73]
Coiled-coil region	aa 265–294	Involved in protein-protein interactions (STIM1-Orai1 interaction)	[79]

Orai family has two additional homologs, Orai2 and Orai3. Orai proteins share no homology with any other known ion channel family or cellular proteins.

Orai1, a Ca^{2+} selective ion channel, is a 301 amino acids protein with four transmembrane domains and a number of functional regions depicted in Table 23.2. Maruyama et al. [71] have purified Orai1 in its tetrameric form and have reconstructed the three-dimensional structure from electron microscopic images, providing the first depiction of an Orai family member. According to these authors, Orai1 is a teardrop-shaped molecule 150 Å in height, 95 Å in side length, and 105 Å diagonally at the widest transmembrane region.

The structure of Orai2 and Orai3 is similar to that of their homolog Orai 1 [72, 73]. All three Orai isoforms constitute Ca^{2+} selective plasma membrane channels, whose currents have been shown to be inhibited by extracellular Ca^{2+} [74]. The three Orai isoforms can be activated by store depletion when co-expressed with STIM1 although the amplitude of the currents generated are smaller for Orai2 and Orai3, which might reflect that they interact with STIM1 with less efficiency [75, 76]. Orai isoforms show slightly different selectivity for Na^+ (being Orai3 more permeable for Na^+) and distinct sensitivity to the pharmacological agent 2-aminoethoxydiphenyl borate (2-APB) [75]. While Orai1 currents are stimulated by low concentrations of 2-APB and abolished by high 2-APB concentrations, Orai2 currents are only partially sensitive to this inhibitor and Orai3 is stimulated by 2-APB [75, 77, 78].

23.4 STIM1-Orai1-TRPC Communication

The nature of the interaction between STIM1 and the plasma membrane Ca^{2+} channel subunits is currently under intense investigation by a number of research teams in order to determine the mechanism underlying the activation of capacitative channels by STIM1. In 2008, Romanin's group demonstrated a dynamic interaction between STIM1 and Orai1 involving the C-termini of both proteins using Forster resonance energy transfer (FRET) microscopy. Interestingly, the Orai1 R91W mutant associated to SCID syndrome did not impair the interaction with STIM1 but altered the activation of Ca^{2+} currents [79]. The coiled-coil C-terminal domain of STIM1 has been reported to trigger dimerization of Orai dimers resulting in the formation of tetrameric Orai1 channels to activate I_{CRAC} [80].

Four research groups have identified in parallel that a cytoplasmic STIM1 region composed of an ezrin-radixin-moesin domain is essential for the activation of Orai1. This region has been named SOAR (STIM1 Orai-activating region) [81], OASF (Orai-activating small fragment) [82], CAD (CRAC-activating domain) [83] and CCb9 [84]. The four regions, SOAR (including the STIM1 amino acid residues 344–442), OASF (amino acids 233–450/474), CAD (amino acids 342–448) and CCb9 (amino acids 339–444), are located within STIM1 C-terminus and comprise two coiled-coil domains and an amino acid sequence that enhances interaction with Orai1, resulting in increased Ca^{2+} currents. These studies have reported several features of the Orai1-STIM1 interacting region: OASF has been reported to be able to homomerize by a novel assembly domain that occurred subsequent to the coiled-coil domains. In addition, STIM1 oligomerization has been shown to be required for CAD exposure. Furthermore, the SOAR region is able to activate all known Orai isoforms although with different conductances being greater for Orai1 than for Orai2 or Orai3 [81, 82].

In addition, a regulatory domain at aminoacids 474–485 of the cytosolic STIM1 region, containing 7 negatively charged residues, known as CMD (CRAC modulatory domain)/CDI (Ca^{2+}-dependent inactivation, reported as residues 470–491), has recently been described. This domain generates a signal that promotes Orai/CRAC channel closure in a Ca^{2+} concentration-dependent manner, a process known as fast, Ca^{2+}-dependent inactivation of the Orai channels [85–87].

The interaction of STIM1 with Orai1, following depletion of the intracellular Ca^{2+} stores, results in a conformational change in Orai1, as determined by FRET, that might be important for CRAC channel activity [88]. FRET analysis between STIM1-YFP and Orai1-CFP has revealed that STIM1 and Orai1 approach within 100 Å or less after treatment with thapsigargin to induce store depletion. Simultaneously, the interaction Orai1-Orai1 is reversibly reduced upon depletion of the stores or application of extracellular Ca^{2+}, both inducing CRAC channel activation, thus suggesting that Orai1 is subjected to a conformational rearrangement that is relevant, although not sufficient, for CRAC channel function [88].

In 2006, Huang and coworkers reported that STIM1 is able to gate TRPC1 [89]. The association of STIM1 with the TRPC proteins has been shown to be mediated by the STIM1 ERM domain [89]. More recently, the SOAR region, which has been shown to interact with Orai1 (see above), has been presented as the domain that binds to the TRPC channels [90]. The initial studies by Huang and coworkers reported that the STIM1 K-domain plays an important role in TRPC1 channel gating, although is not necessary for the interaction between STIM1 and TRPC1 [89]. Muallem's team has recently reported that STIM1 gates TRPC1 through the interaction between two conserved, negatively charged, aspartates in TRPC1 ((639)DD(640)) with the positively charged lysine residues in STIM1((684)KK(685)) located in the C-terminal polybasic region. Different charge swapping experiments confirm that STIM1 gates TRPC1 by intermolecular electrostatic interaction [91]. A similar activation mechanism has been reported for TRPC3 mediated by the negatively charged 697 and 698 aspartate residues [91]. However, STIM1 operates Orai1 by a different mechanism since the C-terminal polybasic and serine-proline rich region of STIM1 are not required for activation of Orai1 [91]. Functional association between STIM1 and TRPC1 has been reported in a number of endogenously expressing and transfected cell types, including HEK-293 cells [89, 92–95], Jurkat T cells [89], human platelets [68], salivary gland cells [96], mesangial cells [97], mouse pulmonary arterial smooth muscle cells [98], the hepatic cell line HL-7702 cells [99] and human parathyroid cells [100]. It is noteworthy to mention that association of STIM1 with TRPC1 has not been found in HEK-293 cells co-transfected with both proteins, where STIM1 overexpression has not reported an increase in the activity of different TRPCs in these cells [101], and vascular smooth muscle cells [102]. The reason of this discrepancy, which might reside on the different transfection levels or the idiosyncrasy of the cell type, is still unclear and requires further studies to fully understand.

23.5 Calcium Entry Pathways Mediated by STIM1-Orai1-TRPC Complexes

The nature of the capacitative channels, as well as the mechanisms that gate them after Ca^{2+} stores have been depleted, have been a matter of intense investigation since the identification of SOCE. One of the earliest hypotheses was formulated in sea urchin eggs, where inositol 1,3,4,5-tetrakisphosphate (IP_4) was suggested to modulate Ca^{2+} entry into the IP_3-sensitive pool by physical interaction between IP_3 and IP_4 receptors located in the plasma membrane and the ER membrane, respectively [103, 104]. The role of IP_3 receptors in SOCE has been widely investigated and a relationship between TRP proteins and IP_3 receptors has been demonstrated in different cell types, including HEK 293 cells, where exogenously expressed TRPC3 can be activated by an IP_3 receptor-dependent physical coupling mechanism [105], T3 cells stably expressing epitope-tagged TRPC3 or TRPC6, where IP_3 receptor is detected in TRP immunoprecipitates [106], and human platelets, where endogenously expressed type II IP_3 receptor has been found in TRPC1 immunoprecipitates only after depletion of the intracellular Ca^{2+} stores and independently of rises in

$[Ca^{2+}]_c$, which indicates that this interaction is capacitative in nature [107–110]. IP_3 receptors, such as the type I IP_3 receptor, have also been reported to participate in agonist-induced, probably non-capacitative, Ca^{2+} entry by interaction with TRPC3, the scaffold protein RACK1 (receptor for activated C-kinase-1), STIM1 and Orai1 [111, 112]. However, although the IP_3 receptors might play an important role in agonist-induced Ca^{2+} entry, they lack Ca^{2+} sensing capability.

With the identification of STIM1 as the ER Ca^{2+} sensor, studies concerning the communication between the ER and the plasma membrane channels focused on this protein. It is widely accepted that a functional protein-protein interaction between STIM1 and Orai1 results in the activation of SOCE. STIM1 enhances SOCE when co-expressed with Orai1 [70, 113, 114], as well as with Orai2 [113] and Orai3 [75, 78], which suggests that these combinations of proteins are sufficient to mediate the process of SOCE, although with distinct inactivation profiles and permeability properties. Special attention has been focused on the study of the interaction between STIM1 and Orai1. In HEK-293 cells, which show significant SOCE while the level of endogenous CRAC is extremely low, expression of Orai1 alone clearly reduced SOCE; however, when co-expressed with STIM1, Orai1 induces a dramatic gain in the amount of SOCE [114]. The inhibition of SOCE by Orai1 overexpression suggests that an adequate stoichiometrical relationship between STIM1 and Orai1 is necessary for this process [114]. Consistent with this, store depletion has been reported to lead to aggregation and translocation of STIM1 in close apposition to the plasma membrane in order to recruit Orai1 and assemble functional units of CRAC channels in a stoichiometric manner [115]. Studies based on electrophysiology, single-molecule fluorescence bleaching methods and FRET have demonstrated that the CRAC channels are formed by four Orai1 monomers assembled to form a tetrameric structure, which is associated to two STIM1 molecules [24, 80, 116] (Fig. 23.1). However, it remains unclear whether this is the only configuration that results in CRAC channel activation. In fact, studies in cells expressing Orai1 and STIM1 at different ratios (from 4:1 to 1:4) have reported that low Orai1:STIM1 ratios results in I_{CRAC} with strong fast Ca^{2+}-dependent inactivation, while high Orai1:STIM1 ratios produce I_{CRAC} with strong activation at negative potentials. In addition, the Orai1:STIM1 expression ratio affects Ca^{2+}, Ba^{2+} and Sr^{2+} conductance; thus suggesting that the biophysical properties of the channels formed by Orai1 depend on the stoichiometry of its interaction with STIM1 [117].

Soon after the identification of STIM1, Huang and coworkers reported that the cytosolic C-terminus of STIM1 is sufficient to activate TRPC1 channels and SOCE [89]. The association of STIM1 and TRPC1 has been reported in a number of cell types and models including human platelets endogenously expressing TRPC1 and STIM1 [68, 118, 119], rat basophilic leukemia cells [96] or HEK293 cells [120]. In addition, STIM1 has been reported to associate with other members of the TRPC family including TRPC2 [93], TRPC4, TRPC5 [92] and TRPC6 [63], although the interaction with TRPC6 has been challenged by Yuan and coworkers, suggesting that STIM1 regulates its function indirectly by promoting the heteromultimerization of TRPC6 with TRPC4 [92]. The direct or indirect association between STIM1 and TRPC6 observed in human platelets [63] and HEK293 cells [92] might be due to

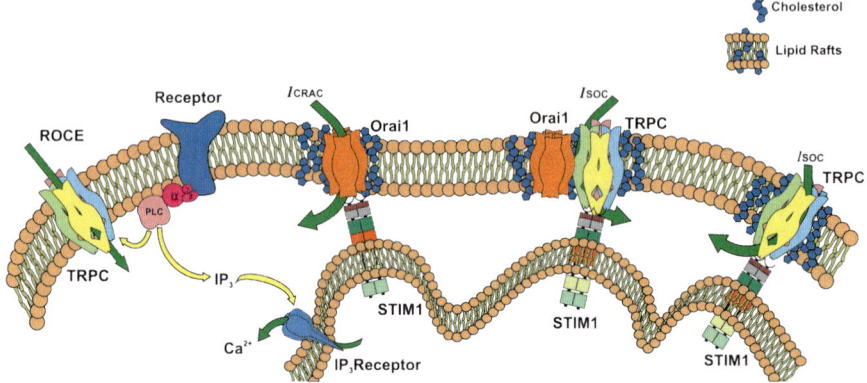

Fig. 23.1 Calcium entry into cells across the plasma membrane might occur through a variety of TRPC-dependent and – independent mechanisms. The Ca^{2+} selective capacitative current I_{CRAC} involves the activation of Orai1 forming channels by STIM1. The non-selective capacitative current I_{SOC} requires the interaction of STIM1 with either TRPC-Orai1 complexes or TRPC containing channels. Lipid raft domains have been shown to be important for capacitative channel activation. In the case of ROCE (including second messenger-operated Ca^{2+} entry) PLC metabolites activates TRPC containing channels independently of STIM1 and the plasma membrane lipid raft domains. ROCE, receptor-operated Ca^{2+} entry, TRPC, canonical transient potential receptor protein; PLC, phospholipase C; IP_3, inositol 1,4,5-trisphosphate; STIM1, stromal interaction molecule 1

the idiosyncrasy of the cells or the different expression of the proteins investigated, endogenous in human platelets, although whether the interaction between TRPC6 and STIM1 is mediated by TRPC4, or other TRPC family members, has not been investigated in these cells yet.

Interestingly, the association of STIM1 to TRPC1 has been reported to recruit TRPC1 into lipid rafts, where TRPC1 functions as a SOC channel, while in the absence of STIM1, TRPC1 interacts with other TRPC family members resulting in the formation of receptor-operated Ca^{2+} (ROC) channels (Fig. 23.1); thus providing evidence for a role of STIM1 in the regulation of TRPC1 participation in SOCE or ROCE and highlighting the role of lipid rafts in the modulation of TRPC1 channel function [121]. Lipid rafts are plasma membrane domains that contain high concentrations of cholesterol and sphingolipids. Lipid rafts recruit certain signalling molecules while excluding others. For instance, in human platelets, only TRPC1, 4 and 5 were found to associate with plasma membrane lipid rafts, while TRPC3 or TRPC6 were not found in these domains [59]. A number of studies based on the use of methyl-β-cyclodextrin (MBCD), a compound that forms soluble complexes with cholesterol and thus deplete membrane cholesterol [122], have reported that lipid raft domains are essential for the assembly of signalling complexes, although it should be taken into account that cells might restore the cholesterol level in the plasma membrane by mobilising cholesterol from intracellular cholesterol stores [123]. Therefore, MBCD might reduce intracellular cholesterol levels, and sphingolipids also participate in lipid rafts and might maintain certain raft structure [124, 125]. Lipid rafts provide a favourable environment necessary for clustering of STIM1 at ER-plasma membrane junctions upon store depletion, facilitating the Ca^{2+} store-dependent interaction between STIM1 and TRPC1 and subsequent SOCE [126]. Lipid rafts have also been reported to play a crucial role in the association between STIM1 and the plasma membrane channel subunits TRPC1 and Orai1 after depletion of the intracellular Ca^{2+} stores and is also necessary for thapsigargin-induced Ca^{2+} entry in human platelets [119].

Recent studies have presented evidence for the existence of functional interactions between Orai1 and TRPCs under the influence of STIM1, and propose that SOC channels are composed of heteromeric complexes that include TRPCs and Orai proteins [25, 26, 96] (Fig. 23.1). Knockdown of Orai1 significantly reduces I_{SOC} in human salivary gland cells [96], where TRPC1 has been demonstrated to be a major SOC channel subunit [127]. Consistent with this, Orai proteins have been reported to confer STIM1-mediated store depletion sensitivity to TRPC channels recruiting TRPC channels for the conduction of SOCE. In HEK293 cells overexpressing store-depletion insensitive TRPC3 or TRPC6, these TRPCs become sensitive to store depletion upon expression of exogenous Orai [26]. These observations suggest that the involvement of Orai proteins in SOCE might be well explained either by a model in which Orai1 are self-contained ion channels activated by STIM1, the proposed CRAC channel hypothesis [23, 70, 73, 128], or a model in which the SOC channels are formed by a combination of TRPCs and Orai proteins [26]. In the latter model, Orai proteins would communicate the information concerning the filling state of the intracellular Ca^{2+} stores from STIM, located in the ER, to TRPC proteins located in the plasma membrane. In support of a role for Orai conferring store-depletion sensitivity to TRPCs, these complexes have been found in non-transfected cells. In human platelets endogenously expressing STIM1, Orai1 and TRPC1, where electrotransjection with anti-STIM1 antibody, specific for the EF-hand domain, both prevented the interaction of STIM1 with hTRPC1 and reduced thapsigargin-evoked SOCE [68], a functional interaction between STIM1, Orai1 and TRPC1 in the activation of SOCE has been demonstrated [118]. In these cells, impairment of the interaction between STIM1 and Orai1, results in disruption of the association of STIM1 and TRPC1, and subsequently alters the behaviour of TRPC1 being no longer involved in SOCE but in ROCE mediated by DAG [118]. Similar results have been observed for TRPC6, in human platelets naturally expressing TRPC6 we have found that the participation of TRPC6 in SOCE or ROCE is regulated through its interaction with the Orai1-STIM1 complex or hTRPC3, respectively, in human platelets [63]; thus STIM1 located in the ER functions as a switch that communicates the filling state of the stores to SOC channels, involving TRPC proteins, through Orai1.

Interestingly, a number of reports have strongly suggested that Orai and TRPC proteins might form complexes that participate both in SOCE and ROCE. A study has reported that expression of Orai1, under experimental conditions that enhance SOCE, leads to the activation of ROCE. In addition, the R91W Orai1 mutant, responsible for SCID, has been shown to block both SOCE and DAG-activated ROCE into cells that, stably or transiently, express TRPC3 proteins [27]. To integrate these results with current data concerning Orai, TRPCs and STIM, it has been postulated that Orai-TRPC complexes recruited to lipid rafts mediate SOCE, whereas the same complexes mediate ROCE when they are outside of lipid rafts [27], which is consistent with previous studies reporting a role for lipid rafts in the modulation of TRPC function by STIM1 [121]. Therefore, there is a body of evidence supporting that TRPCs might be involved in the formation of ion channels responsible for ROCE or SOCE by receiving information from either PLC or STIM1-Orai, respectively [25–27, 63, 118, 120]. The activation of TRPCs by STIM1 has been challenged in a recent study, although, as reported by the authors, more complex combinations of STIM1, Orai1 and TRPCs, as described in Cheng et al. [120], Jardin et al. [63, 118] and Liao et al. [25–27] have not been addressed in that study [101]. Despite the current knowledge concerning Orai-TRPCs-STIM interactions, further studies are required to describe more accurately the molecular composition of the channels mediating SOCE and ROCE and to clarify whether the channel components are the same when Orai-TRPC complexes mediate ROCE or SOCE.

Acknowledgments This work was supported by M.E.C. grants BFU2007-60104 and BFU2010-21043-C02-01.

References

1. Petersen OH (2002) Calcium signal compartmentalization. Biol Res 35:177–182
2. Putney JW Jr (1986) A model for receptor-regulated calcium entry. Cell Calcium 7:1–12
3. Wedel B, Boyles RR, Putney JW Jr, Bird GS (2007) Role of the store-operated calcium entry proteins Stim1 and Orai1 in muscarinic cholinergic receptor-stimulated calcium oscillations in human embryonic kidney cells. J Physiol 579:679–689
4. Berridge MJ (1995) Capacitative calcium entry. Biochem J 312(Pt 1):1–11
5. Redondo PC, Harper MT, Rosado JA, Sage SO (2006) A role for cofilin in the activation of store-operated calcium entry by de novo conformational coupling in human platelets. Blood 107:973–979
6. Verkhratsky A (2005) Physiology and pathophysiology of the calcium store in the endoplasmic reticulum of neurons. Physiol Rev 85:201–279
7. Hoth M, Penner R (1992) Depletion of intracellular calcium stores activates a calcium current in mast cells. Nature 355:353–356
8. Bakowski D, Parekh AB (2000) Voltage-dependent conductance changes in the store-operated Ca2+ current ICRAC in rat basophilic leukaemia cells. J Physiol 529(Pt 2):295–306
9. Luik RM, Wang B, Prakriya M, Wu MM, Lewis RS (2008) Oligomerization of STIM1 couples ER calcium depletion to CRAC channel activation. Nature 454:538–542
10. Parekh AB, Penner R (1997) Store depletion and calcium influx. Physiol Rev 77:901–930
11. Zweifach A, Lewis RS (1993) Mitogen-regulated Ca2+ current of T lymphocytes is activated by depletion of intracellular Ca2+ stores. Proc Natl Acad Sci U S A 90:6295–6299
12. Parekh AB, Putney JW Jr (2005) Store-operated calcium channels. Physiol Rev 85:757–810
13. Trepakova ES, Gericke M, Hirakawa Y, Weisbrod RM, Cohen RA, Bolotina VM (2001) Properties of a native cation channel activated by Ca2+ store depletion in vascular smooth muscle cells. J Biol Chem 276:7782–7790
14. Smani T, Zakharov SI, Csutora P, Leno E, Trepakova ES, Bolotina VM (2004) A novel mechanism for the store-operated calcium influx pathway. Nat Cell Biol 6:113–120
15. Albert AP, Saleh SN, Peppiatt-Wildman CM, Large WA (2007) Multiple activation mechanisms of store-operated TRPC channels in smooth muscle cells. J Physiol 583:25–36

16. Guibert C, Ducret T, Savineau JP (2008) Voltage-independent calcium influx in smooth muscle. Prog Biophys Mol Biol 98:10–23
17. Zhang SL, Yu Y, Roos J, Kozak JA, Deerinck TJ, Ellisman MH, Stauderman KA, Cahalan MD (2005) STIM1 is a Ca2+ sensor that activates CRAC channels and migrates from the Ca2+ store to the plasma membrane. Nature 437:902–905
18. Roos J, DiGregorio PJ, Yeromin AV, Ohlsen K, Lioudyno M, Zhang S, Safrina O, Kozak JA, Wagner SL, Cahalan MD, Velicelebi G, Stauderman KA (2005) STIM1, an essential and conserved component of store-operated Ca2+ channel function. J Cell Biol 169:435–445
19. Jardin I, Lopez JJ, Redondo PC, Salido GM, Rosado JA (2009) Store-operated Ca2+ entry is sensitive to the extracellular Ca2+ concentration through plasma membrane STIM1. Biochim Biophys Acta 1793:1614–1622
20. Spassova MA, Soboloff J, He LP, Xu W, Dziadek MA, Gill DL (2006) STIM1 has a plasma membrane role in the activation of store-operated Ca(2+) channels. Proc Natl Acad Sci U S A 103:4040–4045
21. Feske S, Gwack Y, Prakriya M, Srikanth S, Puppel SH, Tanasa B, Hogan PG, Lewis RS, Daly M, Rao A (2006) A mutation in Orai1 causes immune deficiency by abrogating CRAC channel function. Nature 441:179–185
22. Vig M, Peinelt C, Beck A, Koomoa DL, Rabah D, Koblan-Huberson M, Kraft S, Turner H, Fleig A, Penner R, Kinet JP (2006) CRACM1 is a plasma membrane protein essential for store-operated Ca2+ entry. Science 312:1220–1223
23. Yeromin AV, Zhang SL, Jiang W, Yu Y, Safrina O, Cahalan MD (2006) Molecular identification of the CRAC channel by altered ion selectivity in a mutant of Orai. Nature 443:226–229
24. Mignen O, Thompson JL, Shuttleworth TJ (2008) Orai1 subunit stoichiometry of the mammalian CRAC channel pore. J Physiol 586:419–425
25. Liao Y, Erxleben C, Abramowitz J, Flockerzi V, Zhu MX, Armstrong DL, Birnbaumer L (2008) Functional interactions among Orai1, TRPCs, and STIM1 suggest a STIM-regulated heteromeric Orai/TRPC model for SOCE/Icrac channels. Proc Natl Acad Sci U S A 105:2895–2900
26. Liao Y, Erxleben C, Yildirim E, Abramowitz J, Armstrong DL, Birnbaumer L (2007) Orai proteins interact with TRPC channels and confer responsiveness to store depletion. Proc Natl Acad Sci U S A 104:4682–4687
27. Liao Y, Plummer NW, George MD, Abramowitz J, Zhu MX, Birnbaumer L (2009) A role for Orai in TRPC-mediated Ca2+ entry suggests that a TRPC:Orai complex may mediate store and receptor operated Ca2+ entry. Proc Natl Acad Sci U S A 106:3202–3206
28. Salido GM, Sage SO, Rosado JA (2009) TRPC channels and store-operated Ca(2+) entry. Biochim Biophys Acta 1793:223–230
29. Salido GM, Sage SO, Rosado JA (2009) Biochemical and functional properties of the store-operated Ca2+ channels. Cell Signal 21:457–461
30. Liu X, Cheng KT, Bandyopadhyay BC, Pani B, Dietrich A, Paria BC, Swaim WD, Beech D, Yildirim E, Singh BB, Birnbaumer L, Ambudkar IS (2007) Attenuation of store-operated Ca2+ current impairs salivary gland fluid secretion in TRPC1(-/-) mice. Proc Natl Acad Sci U S A 104:17542–17547
31. Cahalan MD (2009) STIMulating store-operated Ca(2+) entry. Nat Cell Biol 11:669–677
32. Liu X, Wang W, Singh BB, Lockwich T, Jadlowiec J, O'Connell B, Wellner R, Zhu MX, Ambudkar IS (2000) Trp1, a candidate protein for the store-operated Ca(2+) influx mechanism in salivary gland cells. J Biol Chem 275:3403–3411
33. Brough GH, Wu S, Cioffi D, Moore TM, Li M, Dean N, Stevens T (2001) Contribution of endogenously expressed Trp1 to a Ca2+-selective, store-operated Ca2+ entry pathway. FASEB J 15:1727–1738
34. Rosado JA, Brownlow SL, Sage SO (2002) Endogenously expressed Trp1 is involved in store-mediated Ca2+ entry by conformational coupling in human platelets. J Biol Chem 277:42157–42163

35. Xu SZ, Beech DJ (2001) TrpC1 is a membrane-spanning subunit of store-operated Ca(2+) channels in native vascular smooth muscle cells. Circ Res 88:84–87
36. Louis M, Zanou N, Van Schoor M, Gailly P (2008) TRPC1 regulates skeletal myoblast migration and differentiation. J Cell Sci 121:3951–3959
37. Hardie RC, Minke B (1992) The trp gene is essential for a light-activated Ca2+ channel in Drosophila photoreceptors. Neuron 8:643–651
38. Phillips AM, Bull A, Kelly LE (1992) Identification of a Drosophila gene encoding a calmodulin-binding protein with homology to the trp phototransduction gene. Neuron 8:631–642
39. Hardie RC, Reuss H, Lansdell SJ, Millar NS (1997) Functional equivalence of native light-sensitive channels in the Drosophila trp301 mutant and TRPL cation channels expressed in a stably transfected Drosophila cell line. Cell Calcium 21:431–440
40. Hardie RC (2003) Regulation of TRP channels via lipid second messengers. Annu Rev Physiol 65:735–759
41. Wes PD, Chevesich J, Jeromin A, Rosenberg C, Stetten G, Montell C (1995) TRPC1, a human homolog of a Drosophila store-operated channel. Proc Natl Acad Sci U S A 92: 9652–9656
42. Zhu X, Chu PB, Peyton M, Birnbaumer L (1995) Molecular cloning of a widely expressed human homologue for the Drosophila trp gene. FEBS Lett 373:193–198
43. Petersen CC, Berridge MJ, Borgese MF, Bennett DL (1995) Putative capacitative calcium entry channels: expression of Drosophila trp and evidence for the existence of vertebrate homologues. Biochem J 311(Pt 1):41–44
44. Montell C, Birnbaumer L, Flockerzi V, Bindels RJ, Bruford EA, Caterina MJ, Clapham DE, Harteneck C, Heller S, Julius D, Kojima I, Mori Y, Penner R, Prawitt D, Scharenberg AM, Schultz G, Shimizu N, Zhu MX (2002) A unified nomenclature for the superfamily of TRP cation channels. Mol Cell 9:229–231
45. Pedersen SF, Owsianik G, Nilius B (2005) TRP channels: an overview. Cell Calcium 38:233–252
46. Hoenderop JG, Voets T, Hoefs S, Weidema F, Prenen J, Nilius B, Bindels RJ (2003) Homo- and heterotetrameric architecture of the epithelial Ca2+ channels TRPV5 and TRPV6. EMBO J 22:776–785
47. Vannier B, Zhu X, Brown D, Birnbaumer L (1998) The membrane topology of human transient receptor potential 3 as inferred from glycosylation-scanning mutagenesis and epitope immunocytochemistry. J Biol Chem 273:8675–8679
48. Vazquez G, Wedel BJ, Aziz O, Trebak M, Putney JW Jr (2004) The mammalian TRPC cation channels. Biochim Biophys Acta 1742:21–36
49. Montell C, Birnbaumer L, Flockerzi V (2002) The TRP channels, a remarkably functional family. Cell 108:595–598
50. Zhu X, Jiang M, Peyton M, Boulay G, Hurst R, Stefani E, Birnbaumer L (1996) trp, a novel mammalian gene family essential for agonist-activated capacitative Ca2+ entry. Cell 85:661–671
51. Birnbaumer L, Zhu X, Jiang M, Boulay G, Peyton M, Vannier B, Brown D, Platano D, Sadeghi H, Stefani E, Birnbaumer M (1996) On the molecular basis and regulation of cellular capacitative calcium entry: roles for Trp proteins. Proc Natl Acad Sci U S A 93: 15195–15202
52. Thebault S, Zholos A, Enfissi A, Slomianny C, Dewailly E, Roudbaraki M, Parys J, Prevarskaya N (2005) Receptor-operated Ca2+ entry mediated by TRPC3/TRPC6 proteins in rat prostate smooth muscle (PS1) cell line. J Cell Physiol 204:320–328
53. Ben-Amor N, Redondo PC, Bartegi A, Pariente JA, Salido GM, Rosado JA (2006) A role for 5,6-epoxyeicosatrienoic acid in calcium entry by de novo conformational coupling in human platelets. J Physiol 570:309–323
54. Brownlow SL, Harper AG, Harper MT, Sage SO (2004) A role for hTRPC1 and lipid raft domains in store-mediated calcium entry in human platelets. Cell Calcium 35:107–113

55. Vannier B, Peyton M, Boulay G, Brown D, Qin N, Jiang M, Zhu X, Birnbaumer L (1999) Mouse trp2, the homologue of the human trpc2 pseudogene, encodes mTrp2, a store depletion-activated capacitative Ca2+ entry channel. Proc Natl Acad Sci U S A 96: 2060–2064
56. Vazquez G, Wedel BJ, Trebak M, St John Bird G, Putney JW Jr (2003) Expression level of the canonical transient receptor potential 3 (TRPC3) channel determines its mechanism of activation. J Biol Chem 278: 21649–21654
57. Yildirim E, Kawasaki BT, Birnbaumer L (2005) Molecular cloning of TRPC3a, an N-terminally extended, store-operated variant of the human C3 transient receptor potential channel. Proc Natl Acad Sci U S A 102:3307–3311
58. Philipp S, Cavalie A, Freichel M, Wissenbach U, Zimmer S, Trost C, Marquart A, Murakami M, Flockerzi V (1996) A mammalian capacitative calcium entry channel homologous to Drosophila TRP and TRPL. EMBO J 15:6166–6171
59. Brownlow SL, Sage SO (2005) Transient receptor potential protein subunit assembly and membrane distribution in human platelets. Thromb Haemost 94:839–845
60. Philipp S, Hambrecht J, Braslavski L, Schroth G, Freichel M, Murakami M, Cavalie A, Flockerzi V (1998) A novel capacitative calcium entry channel expressed in excitable cells. EMBO J 17:4274–4282
61. Jardin I, Redondo PC, Salido GM, Rosado JA (2008) Phosphatidylinositol 4,5-bisphosphate enhances store-operated calcium entry through hTRPC6 channel in human platelets. Biochim Biophys Acta 1783:84–97
62. Brechard S, Melchior C, Plancon S, Schenten V, Tschirhart EJ (2008) Store-operated Ca(2+) channels formed by TRPC1, TRPC6 and Orai1 and non-store-operated channels formed by TRPC3 are involved in the regulation of NADPH oxidase in HL-60 granulocytes. Cell Calcium 44:492–506
63. Jardin I, Gomez LJ, Salido GM, Rosado JA (2009) Dynamic interaction of hTRPC6 with the Orai1-STIM1 complex or hTRPC3 mediates its role in capacitative or non-capacitative Ca(2+) entry pathways. Biochem J 420:267–276
64. Lievremont JP, Bird GS, Putney JW Jr (2004) Canonical transient receptor potential TRPC7 can function as both a receptor- and store-operated channel in HEK-293 cells. Am J Physiol Cell Physiol 287:C1709–C1716
65. Liou J, Kim ML, Heo WD, Jones JT, Myers JW, Ferrell JE Jr, Meyer T (2005) STIM is a Ca2+ sensor essential for Ca2+-store-depletion-triggered Ca2+ influx. Curr Biol 15: 1235–1241
66. Williams RT, Manji SS, Parker NJ, Hancock MS, Van Stekelenburg L, Eid JP, Senior PV, Kazenwadel JS, Shandala T, Saint R, Smith PJ, Dziadek MA (2001) Identification and characterization of the STIM (stromal interaction molecule) gene family: coding for a novel class of transmembrane proteins. Biochem J 357:673–685
67. Soboloff J, Spassova MA, Hewavitharana T, He LP, Xu W, Johnstone LS, Dziadek MA, Gill DL (2006) STIM2 is an inhibitor of STIM1-mediated store-operated Ca2+ Entry. Curr Biol 16:1465–1470
68. Lopez JJ, Salido GM, Pariente JA, Rosado JA (2006) Interaction of STIM1 with endogenously expressed human canonical TRP1 upon depletion of intracellular Ca2+ stores. J Biol Chem 281:28254–28264
69. Mignen O, Thompson JL, Shuttleworth TJ (2009) The molecular architecture of the arachidonate-regulated Ca2+-selective ARC channel is a pentameric assembly of Orai1 and Orai3 subunits. J Physiol 587:4181–4197
70. Peinelt C, Vig M, Koomoa DL, Beck A, Nadler MJ, Koblan-Huberson M, Lis A, Fleig A, Penner R, Kinet JP (2006) Amplification of CRAC current by STIM1 and CRACM1 (Orai1). Nat Cell Biol 8:771–773
71. Maruyama Y, Ogura T, Mio K, Kato K, Kaneko T, Kiyonaka S, Mori Y, Sato C (2009) Tetrameric Orai1 Is a Teardrop-shaped Molecule with a Long, Tapered Cytoplasmic Domain. J Biol Chem 284:13676–13685

72. Cahalan MD, Zhang SL, Yeromin AV, Ohlsen K, Roos J, Stauderman KA (2007) Molecular basis of the CRAC channel. Cell Calcium 42:133–144
73. Prakriya M, Feske S, Gwack Y, Srikanth S, Rao A, Hogan PG (2006) Orai1 is an essential pore subunit of the CRAC channel. Nature 443:230–233
74. DeHaven WI, Smyth JT, Boyles RR, Putney JW Jr (2007) Calcium inhibition and calcium potentiation of Orai1, Orai2, and Orai3 calcium release-activated calcium channels. J Biol Chem 282:17548–17556
75. Lis A, Peinelt C, Beck A, Parvez S, Monteilh-Zoller M, Fleig A, Penner R (2007) CRACM1, CRACM2, and CRACM3 are store-operated Ca2+ channels with distinct functional properties. Curr Biol 17:794–800
76. Frischauf I, Schindl R, Derler I, Bergsmann J, Fahrner M, Romanin C (2008) The STIM/Orai coupling machinery. Channels (Austin) 2:261–268
77. Peinelt C, Lis A, Beck A, Fleig A, Penner R (2008) 2-Aminoethoxydiphenyl borate directly facilitates and indirectly inhibits STIM1-dependent gating of CRAC channels. J Physiol 586:3061–3073
78. Schindl R, Bergsmann J, Frischauf I, Derler I, Fahrner M, Muik M, Fritsch R, Groschner K, Romanin C (2008) 2-aminoethoxydiphenyl borate alters selectivity of Orai3 channels by increasing their pore size. J Biol Chem 283:20261–20267
79. Muik M, Frischauf I, Derler I, Fahrner M, Bergsmann J, Eder P, Schindl R, Hesch C, Polzinger B, Fritsch R, Kahr H, Madl J, Gruber H, Groschner K, Romanin C (2008) Dynamic coupling of the putative coiled-coil domain of ORAI1 with STIM1 mediates ORAI1 channel activation. J Biol Chem 283:8014–8022
80. Penna A, Demuro A, Yeromin AV, Zhang SL, Safrina O, Parker I, Cahalan MD (2008) The CRAC channel consists of a tetramer formed by Stim-induced dimerization of Orai dimers. Nature 456:116–120
81. Yuan JP, Zeng W, Dorwart MR, Choi YJ, Worley PF, Muallem S (2009) SOAR and the polybasic STIM1 domains gate and regulate Orai channels. Nat Cell Biol 11:337–343
82. Muik M, Fahrner M, Derler I, Schindl R, Bergsmann J, Frischauf I, Groschner K, Romanin C (2009) A Cytosolic Homomerization and a Modulatory Domain within STIM1 C Terminus Determine Coupling to ORAI1 Channels. J Biol Chem 284:8421–8426
83. Park CY, Hoover PJ, Mullins FM, Bachhawat P, Covington ED, Raunser S, Walz T, Garcia KC, Dolmetsch RE, Lewis RS (2009) STIM1 clusters and activates CRAC channels via direct binding of a cytosolic domain to Orai1. Cell 136:876–890
84. Kawasaki T, Lange I, Feske S (2009) A minimal regulatory domain in the C terminus of STIM1 binds to and activates ORAI1 CRAC channels. Biochem Biophys Res Commun 385:49–54
85. Derler I, Fahrner M, Muik M, Lackner B, Schindl R, Groschner K, Romanin C (2009) A CRAC modulatory domain (CMD) within STIM1 mediates fast Ca2+-dependent inactivation of ORAI1 channels. J Biol Chem 284:24933–24938
86. Lee KP, Yuan JP, Zeng W, So I, Worley PF, Muallem S (2009) Molecular determinants of fast Ca2+-dependent inactivation and gating of the Orai channels. Proc Natl Acad Sci U S A 106:14687–14692
87. Mullins FM, Park CY, Dolmetsch RE, Lewis RS (2009) STIM1 and calmodulin interact with Orai1 to induce Ca2+-dependent inactivation of CRAC channels. Proc Natl Acad Sci U S A 106:15495–15500
88. Navarro-Borelly L, Somasundaram A, Yamashita M, Ren D, Miller RJ, Prakriya M (2008) STIM1-Orai1 interactions and Orai1 conformational changes revealed by live-cell FRET microscopy. J Physiol 586:5383–5401
89. Huang GN, Zeng W, Kim JY, Yuan JP, Han L, Muallem S, Worley PF (2006) STIM1 carboxyl-terminus activates native SOC, I(crac) and TRPC1 channels. Nat Cell Biol 8:1003–1010
90. Lee KP, Yuan JP, Hong JH, So I, Worley PF, Muallem S (2009) An endoplasmic reticulum/plasma membrane junction: STIM1/Orai1/TRPCs. FEBS Lett 584:2022–2027

91. Zeng W, Yuan JP, Kim MS, Choi YJ, Huang GN, Worley PF, Muallem S (2008) STIM1 gates TRPC channels, but not Orai1, by electrostatic interaction. Mol Cell 32:439–448
92. Yuan JP, Zeng W, Huang GN, Worley PF, Muallem S (2007) STIM1 heteromultimerizes TRPC channels to determine their function as store-operated channels. Nat Cell Biol 9: 636–645
93. Worley PF, Zeng W, Huang GN, Yuan JP, Kim JY, Lee MG, Muallem S (2007) TRPC channels as STIM1-regulated store-operated channels. Cell Calcium 42:205–211
94. Sampieri A, Zepeda A, Saldaña C, Salgado A, Vaca L (2008) STIM1 converts TRPC1 from a receptor-operated to a store-operated channel: Moving TRPC1 in and out of lipid rafts. Cell Calcium 44:479–491
95. Kim MS, Zeng W, Yuan JP, Shin DM, Worley PF, Muallem S (2009) Native Store-operated Ca2+ Influx Requires the Channel Function of Orai1 and TRPC1. J Biol Chem 284: 9733–9741
96. Ong HL, Cheng KT, Liu X, Bandyopadhyay BC, Paria BC, Soboloff J, Pani B, Gwack Y, Srikanth S, Singh BB, Gill DL, Ambudkar IS (2007) Dynamic assembly of TRPC1-STIM1-Orai1 ternary complex is involved in store-operated calcium influx. Evidence for similarities in store-operated and calcium release-activated calcium channel components. J Biol Chem 282: 9105–9116
97. Sours-Brothers S, Ding M, Graham S, Ma R (2009) Interaction between TRPC1/TRPC4 assembly and STIM1 contributes to store-operated Ca2+ entry in mesangial cells. Exp Biol Med (Maywood) 234:673–682
98. Ng LC, McCormack MD, Airey JA, Singer CA, Keller PS, Shen XM, Hume JR (2009) TRPC1 and STIM1 mediate capacitative Ca2+ entry in mouse pulmonary arterial smooth muscle cells. J Physiol 587:2429–2442
99. Zhang ZY, Pan LJ, Zhang ZM (2009) Functional interactions among STIM1, Orai1 and TRPC1 on the activation of SOCs in HL-7702 cells. Amino Acids 39:195–204
100. Lu M, Branstrom R, Berglund E, Hoog A, Bjorklund P, Westin G, Larsson C, Farnebo LO, Forsberg L (2010) Expression and association of TRPC subtypes with Orai1 and STIM1 in human parathyroid. J Mol Endocrinol 44:285–294
101. DeHaven WI, Jones BF, Petranka JG, Smyth JT, Tomita T, Bird GS, Putney JW Jr (2009) TRPC channels function independently of STIM1 and Orai1. J Physiol 587:2275–2298
102. Dietrich A, Kalwa H, Storch U, Mederos y Schnitzler M, Salanova B, Pinkenburg O, Dubrovska G, Essin K, Gollasch M, Birnbaumer L, Gudermann T (2007) Pressure-induced and store-operated cation influx in vascular smooth muscle cells is independent of TRPC1. Pflugers Arch 455:465–477
103. Irvine RF, Moor RM (1987) Inositol(1,3,4,5)tetrakisphosphate-induced activation of sea urchin eggs requires the presence of inositol trisphosphate. Biochem Biophys Res Commun 146:284–290
104. Irvine RF (1992) Inositol phosphates and Ca2+ entry: toward a proliferation or a simplification? FASEB J 6:3085–3091
105. Kiselyov K, Xu X, Mozhayeva G, Kuo T, Pessah I, Mignery G, Zhu X, Birnbaumer L, Muallem S (1998) Functional interaction between InsP3 receptors and store-operated Htrp3 channels. Nature 396:478–482
106. Boulay G, Brown DM, Qin N, Jiang M, Dietrich A, Zhu MX, Chen Z, Birnbaumer M, Mikoshiba K, Birnbaumer L (1999) Modulation of Ca(2+) entry by polypeptides of the inositol 1,4, 5-trisphosphate receptor (IP3R) that bind transient receptor potential (TRP): evidence for roles of TRP and IP3R in store depletion-activated Ca(2+) entry. Proc Natl Acad Sci U S A 96:14955–14960
107. Rosado JA, Sage SO (2000) Coupling between inositol 1,4,5-trisphosphate receptors and human transient receptor potential channel 1 when intracellular Ca2+ stores are depleted. Biochem J 350(Pt 3):631–635
108. Rosado JA, Sage SO (2001) Activation of store-mediated calcium entry by secretion-like coupling between the inositol 1,4,5-trisphosphate receptor type II and human transient receptor potential (hTrp1) channels in human platelets. Biochem J 356:191–198

109. Rosado JA, Sage SO (2002) The ERK cascade, a new pathway involved in the activation of store-mediated calcium entry in human platelets. Trends Cardiovasc Med 12:229–234
110. Jardin I, Lopez JJ, Salido GM, Rosado JA (2008) Functional relevance of the de novo coupling between hTRPC1 and type II IP3 receptor in store-operated Ca2+ entry in human platelets. Cell Signal 20:737–747
111. Bandyopadhyay BC, Ong HL, Lockwich TP, Liu X, Paria BC, Singh BB, Ambudkar IS (2008) TRPC3 controls agonist-stimulated intracellular Ca2+ release by mediating the interaction between inositol 1,4,5-trisphosphate receptor and RACK1. J Biol Chem 283:32821–32830
112. Woodard GE, Lopez JJ, Jardin I, Salido GM, Rosado JA (2010) TRPC3 regulates agonist-stimulated Ca2+ mobilization by mediating the interaction between type I inositol 1,4,5-trisphosphate receptor, RACK1 and Orai1. J Biol Chem 285:8045–8053
113. Mercer JC, Dehaven WI, Smyth JT, Wedel B, Boyles RR, Bird GS, Putney JW Jr (2006) Large store-operated calcium selective currents due to co-expression of Orai1 or Orai2 with the intracellular calcium sensor, Stim1. J Biol Chem 281:24979–24990
114. Soboloff J, Spassova MA, Tang XD, Hewavitharana T, Xu W, Gill DL (2006) Orai1 and STIM reconstitute store-operated calcium channel function. J Biol Chem 281:20661–20665
115. Xu P, Lu J, Li Z, Yu X, Chen L, Xu T (2006) Aggregation of STIM1 underneath the plasma membrane induces clustering of Orai1. Biochem Biophys Res Commun 350:969–976
116. Ji W, Xu P, Li Z, Lu J, Liu L, Zhan Y, Chen Y, Hille B, Xu T, Chen L (2008) Functional stoichiometry of the unitary calcium-release-activated calcium channel. Proc Natl Acad Sci U S A 105:13668–13673
117. Scrimgeour N, Litjens T, Ma L, Barritt GJ, Rychkov GY (2009) Properties of Orai1 mediated store-operated current depend on the expression levels of STIM1 and Orai1 proteins. J Physiol 587:2903–2918
118. Jardin I, Lopez JJ, Salido GM, Rosado JA (2008) Orai1 mediates the interaction between STIM1 and hTRPC1 and regulates the mode of activation of hTRPC1-forming Ca2+ channels. J Biol Chem 283:25296–25304
119. Jardin I, Salido GM, Rosado JA (2008) Role of lipid rafts in the interaction between hTRPC1, Orai1 and STIM1. Channels (Austin) 2:401–403
120. Cheng KT, Liu X, Ong HL, Ambudkar IS (2008) Functional requirement for Orai1 in store-operated TRPC1-STIM1 channels. J Biol Chem 283:12935–12940
121. Sampieri A, Angelica Z, Carlos S, Alfonso S, Vaca L (2008) STIM1 converts TRPC1 from a receptor-operated to a store-operated channel: moving TRPC1 in and out of lipid rafts. Cell Calcium 44:479–491
122. Klein U, Gimpl G, Fahrenholz F (1995) Alteration of the myometrial plasma membrane cholesterol content with beta-cyclodextrin modulates the binding affinity of the oxytocin receptor. Biochemistry 34:13784–13793
123. Mahammad S, Parmryd I (2007) Methyl-beta-cyclodextrin does not preferentially target lipid raft cholesterol. Chem Phys Lipids 149:85
124. Aubert-Jousset E, Garmy N, Sbarra V, Fantini J, Sadoulet MO, Lombardo D (2004) The combinatorial extension method reveals a sphingolipid binding domain on pancreatic bile salt-dependent lipase: role in secretion. Structure 12:1437–1447
125. Vaca L SOCIC: The store-operated calcium influx complex. Cell Calcium doi:10.1016/j.ceca.2010.01.002
126. Pani B, Ong HL, Liu X, Rauser K, Ambudkar IS, Singh BB (2008) Lipid rafts determine clustering of STIM1 in endoplasmic reticulum-plasma membrane junctions and regulation of store-operated Ca2+ entry (SOCE). J Biol Chem 283:17333–17340
127. Liu X, Singh BB, Ambudkar IS (2003) TRPC1 is required for functional store-operated Ca2+ channels. Role of acidic amino acid residues in the S5-S6 region. J Biol Chem 278: 11337–11343
128. Lorin-Nebel C, Xing J, Yan X, Strange K (2007) CRAC channel activity in C. elegans is mediated by Orai1 and STIM1 homologues and is essential for ovulation and fertility. J Physiol 580: 67–85

129. Zheng L, Stathopulos PB, Li GY, Ikura M (2008) Biophysical characterization of the EF-hand and SAM domain containing Ca2+ sensory region of STIM1 and STIM2. Biochem Biophys Res Commun 369:240–246
130. Fahrner M, Muik M, Derler I, Schindl R, Fritsch R, Frischauf I, Romanin C (2009) Mechanistic view on domains mediating STIM1-Orai coupling. Immunol Rev 231:99–112
131. Schindl R, Muik M, Fahrner M, Derler I, Fritsch R, Bergsmann J, Romanin C (2009) Recent progress on STIM1 domains controlling Orai activation. Cell Calcium 46:227–232
132. Soboloff J, Spassova MA, Dziadek MA, Gill DL (2006) Calcium signals mediated by STIM and Orai proteins–a new paradigm in inter-organelle communication. Biochim Biophys Acta 1763:1161–1168
133. Williams RT, Senior PV, Van Stekelenburg L, Layton JE, Smith PJ, Dziadek MA (2002) Stromal interaction molecule 1 (STIM1), a transmembrane protein with growth suppressor activity, contains an extracellular SAM domain modified by N-linked glycosylation. Biochim Biophys Acta 1596:131–137
134. Feske S (2009) ORAI1 and STIM1 deficiency in human and mice: roles of store-operated Ca2+ entry in the immune system and beyond. Immunol Rev 231:189–209
135. Liou J, Fivaz M, Inoue T, Meyer T (2007) Live-cell imaging reveals sequential oligomerization and local plasma membrane targeting of stromal interaction molecule 1 after Ca2+ store depletion. Proc Natl Acad Sci U S A 104:9301–9306
136. Takahashi Y, Murakami M, Watanabe H, Hasegawa H, Ohba T, Munehisa Y, Nobori K, Ono K, Iijima T, Ito H (2007) Essential role of the N-terminus of murine Orai1 in store-operated Ca2+ entry. Biochem Biophys Res Commun 356:45–52

Chapter 24
Contribution of TRPC1 and Orai1 to Ca^{2+} Entry Activated by Store Depletion

Kwong Tai Cheng, Hwei Ling Ong, Xibao Liu, and Indu S. Ambudkar

Abstract Store-operated Ca^{2+} entry (SOCE) is activated in response to depletion of the ER-Ca^{2+} stores by the ER Ca^{2+} sensor protein, STIM1 which oligomerizes and moves to ER/PM junctional domains where it interacts with and activates channels involved in SOCE. Two types of channel activities have been described. I_{CRAC}, via Ca^{2+} release-activated Ca^{2+} (CRAC) channel, which displays high Ca^{2+} selectivity and accounts for the SOCE and cell function in T lymphocytes, mast cells, platelets, and some types of smooth muscle and endothelial cells. Orai1 has been established as the pore-forming component of CRAC channels and interaction of Orai1 with STIM1 is sufficient for generation of the CRAC channel. Store depletion also leads to activation of relatively non-selective cation currents (referred to as I_{SOC}) that contribute to SOCE in several other cell types. TRPC channels, including TRPC1, TRPC3, and TRPC4, have been proposed as possible candidate channels for this Ca^{2+} influx. TRPC1 is the best characterized channel in this regard and reported to contribute to endogenous SOCE in many cells types. TRPC1-mediated Ca^{2+} entry and cation current in cells stimulated with agonist or thapsigargin are inhibited by low [Gd^{3+}] and 10–20 μM 2APB (conditions that block SOCE). Importantly, STIM1 also associates with and gates TRPC1 via electrostatic interaction between STIM1 (^{684}KK^{685}) and TRPC1 (^{639}DD^{640}). Further, store depletion induces dynamic recruitment of a TRPC1/STIM1/Orai1 complex and knockdown of Orai1 completely abrogates TRPC1 function. Despite these findings, there has been much debate regarding the activation of TRPC1 by store depletion as well as the role of Orai1 and STIM1 in SOC channel function. This chapter summarizes recent studies and concepts regarding the contributions of Orai1 and TRPC1 to SOCE. Major unresolved questions regarding functional interaction between Orai1 and TRPC1 as well as possible mechanisms involved in the regulation of TRPC channels by store depletion will be discussed.

I.S. Ambudkar (✉)
Secretory Physiology Section, Molecular Physiology and Therapeutics Branch,
NIDCR, NIH, Bethesda, MD 20892, USA
e-mail: indu.ambudkar@nih.gov

24.1 Introduction

Store-operated calcium entry (SOCE) was first described almost 2 decades ago as a Ca^{2+} entry mechanism in the plasma membrane that is activated by the depletion of Ca^{2+} in the endoplasmic reticulum (ER)-Ca^{2+} store. Since then this Ca^{2+} entry pathway has been demonstrated to be ubiquitously present in all excitable and nonexcitable cells [1–3]. Under physiological conditions, SOCE is activated in response to stimulation of membrane receptors that lead to the hydrolysis of PIP_2, IP_3 generation, and IP_3-mediated Ca^{2+} release from the ER via activation of the IP_3 receptor. Use of the ER Ca^{2+} pump inhibitor thapsigragin (Tg) demonstrated that SOCE activation is regulated by depletion of the intracellular Ca^{2+} store rather than proximal events associated with receptor-dependent PIP_2 hydrolysis. Thus, the name store-operated calcium entry, or capacititative Ca^{2+} entry was coined. Although Ca^{2+} influx via plasma membrane store-operated calcium (SOC) channels replenishes the ER-Ca^{2+} store, it also regulates a number of critical physiological functions such as secretion, cell proliferation, endothelial cell migration, T cell activation and mast cell degranulation [2]. Identification of the mechanism(s) involved in ER-plasma membrane (PM) signaling that results in activation of SOC channels in the surface membrane, as well as the channel components themselves, has been a major challenge in this field, until very recently when some of these critical issues have been resolved.

Early studies established a close proximity of SOCE to the ER membrane. It was shown that since the ER lies very close to the PM and that Ca^{2+} entering the cells is rapidly taken up into the ER lumen via the SERCA pump [4, 5]. This functional association between ER and PM provided the basis for several models that have been proposed to explain activation of SOCE [2, 5]. The models that have garnered most attention are: (i) *conformational coupling* – a close physical association between the PM Ca^{2+} channel and an ER protein (previously proposed to be IP_3R) allows detection and relay of the luminal [Ca^{2+}] status to the surface membrane; (ii) *secretion coupling* – cortical ER is dynamically regulated so that it interacts with PM channels when luminal [Ca^{2+}] is low; (iii) *channel recruitment* – regulated trafficking and fusion of vesicles containing pre-assembled channels, and possibly accessory signaling proteins, with the PM; and (iv) *diffusible messenger* – a diffusible calcium influx factor, generated in response to store depletion, is released into the cytosol to activate the PM Ca^{2+} channel.

24.2 Characteristics of SOCE

24.2.1 *Experimental Methods for Activation and Assessment of SOCE*

Typical reagents used for activation of SOCE include agonists that stimulate PIP_2 hydrolysis and IP_3-mediated Ca^{2+} release from ER, e.g. CCh, bradykinin, angiotensin and ATP, as well as agents that lead to passive depletion of the ER Ca^{2+} store such as Tg and BHQ (SERCA inhibitors), TPEN (low affinity ER-permeant

Ca^{2+} chelator), and ionomycin. Measurements of SOCE are made either using fluorescent Ca^{2+} indicator dyes or electrophysiologically using whole cell patch clamp techniques. Agents stimulating SOCE are usually added to the bathing solution of dye-loaded cells. In the case of electrophysiological recordings, critical inclusions in the pipette solution are EGTA (to suppress store reloading and Ca^{2+}-dependent inactivation of the channel), Mg^{2+} (to suppress activation of TRPM7), or IP_3 as required (the technique has been discussed in great detail elsewhere [1, 6]). Despite the lack of very exclusive SOCE blockers, a few have been established as suitable inhibitors of SOCE. These include Gd^{3+} (1–5 µM) or 2-aminoethyldiphenyl borate (2APB, 10–20 µM). Thus, the current criteria used to identify SOCE are: (i) activation of Ca^{2+} entry and calcium currents by agonists, Tg, or IP_3 (included in the pipette solution for current measurements); and (ii) inhibition of both by 1–5 µM Gd^{3+} and 10–20 µM 2APB.

24.2.2 Characteristics of the Currents Associated with SOCE

The first store-operated current to be measured and the one studied in greatest detail, is the calcium-release-activated calcium current (I_{CRAC}) identified in T-lymphocytes and RBL cells [1]. I_{CRAC}, mediated by the CRAC channel, is a highly Ca^{2+} selective (Ca^{2+}/Na^+ of ≥ 400) and inwardly rectifying current which is greatly increased when divalent cations are removed from the external medium. The channel is also permeable to other divalents such as Mn^{2+}, Sr^{2+}, and Ba^{2+}, which are occasionally used as surrogate cations to assess SOCE. I_{CRAC} is activated relatively slowly by perfusion of the cell cytosol with EGTA (likely due to passive depletion of internal Ca^{2+} store) and faster when cells are stimulated with an agonist or Tg (see [1, 2, 6] for detailed description of I_{CRAC} properties). The CRAC channel is predicted to have very low single-channel conductance of ~15 femtosiemens (fS) [7].

Application of similar methodologies in other cell types leads to activation of Ca^{2+} entry associated with relatively non-selective cation currents [2, 4, 8–10]. These currents have been historically referred to as I_{SOC} (store-operated currents) since they are activated under the same conditions used for I_{CRAC} and inhibited by 1 µM Gd^{3+} and 10–20 µM 2APB, but are distinct from I_{CRAC} with regards to their properties. Note however that in some cells agonist stimulation also leads to activation of cation channels that are not blocked by 2APB or low Gd^{3+}. Furthermore, some channels are only activated by agonist and not by Tg, discriminating them from "store-operated" channels. Identification of SOC channels with diverse biophysical characteristics, ranging from non-selective to relatively Ca^{2+} selective, suggests the possibility that a variety of distinct channels may be involved in SOCE.

24.3 Proposed Molecular Components of SOCE

24.3.1 TRPC Channels

The long search for the channel mediating SOCE first led to the identification of mammalian transient receptor potential (TRP) channels. Members of the TRPC

(TRP Canonical) subfamily were proposed as candidates channels for SOCE based on their activation by stimuli that lead to PIP_2 hydrolysis. The TRPC subfamily consists of 7 members (TRPC1-7) that display diverse properties, modes of regulation and physiological functions. They are also suggested to be assembled as homomeric or heteromeric channels, although there is little information regarding the status of endogenous TRPC channels [11, 12]. While it is generally accepted that TRPC channels are activated downstream of agonist-stimulated PIP_2 hydrolysis, there is considerable discrepancy regarding their activation mechanisms. Many studies suggest that TRPC channels do not have an apparent contribution to endogenous CRAC channel function (exceptions to this suggestion are further discussed below). Therefore, TRPC channels activated in response to store depletion have been referred to as SOC channels, to distinguish them from CRAC channels [11–14]. TRPC1, the first mammalian TRPC protein to be identified, has been most consistently demonstrated as a SOC channel component in a variety of cell types, including keratinocytes, platelets, smooth, skeletal, and cardiac muscles, DT40, HEK293, salivary gland, neuronal, intestinal, and endothelial cells [15–26]. Mutations in the TRPC1 pore region alters the properties of I_{SOC} suggesting that TRPC1 contributes to the channel pore. Further, TRPC1 mediates sufficient Ca^{2+} entry to regulate cellular function such as K_{Ca} channel activation, cell proliferation, and gene expression. Heteromeric associations of TRPC1 with other TRPC channels could account for the diversity in channel properties and functions attributed to it (as reviewed in [8, 13]). Although overexpression of TRPC1 does not result in a substantial increase in SOCE, knockdown of TRPC1 has been associated with consistent decrease in endogenous SOCE in many cell types (noted above). We have reported severe loss of SOCE and salivary gland fluid secretion in mouse lacking TRPC1 [19], although loss of TRPC1 did not appear to affect function of platelets or cerebral artery smooth muscle cells from these mice [27, 28]. These findings suggest cell type- and tissue-specific functions for TRPC1 in SOCE. Other studies have also previously linked TRPC4 and TRPC3 to SOCE [2, 26, 29–31]. Notably, TRPC4-mediated current was highly Ca^{2+} selective and displayed inward rectification similar to that of I_{CRAC} [32].

24.3.2 Orai Channels

Orai1 is a four-transmembrane domain plasma membrane protein identified in 2006 through genome-wide RNAi screening in *Drosophila* S2 cell for CRAC channel-associated proteins and by genetic linkage analysis in severe combined immune deficiency (SCID) patients. Knockdown of Orai1 decreased CRAC channel function in S2 cells, while a single point mutation in Orai1 (R91W) was associated with loss of I_{CRAC} and defect in the Ca^{2+} signaling of T-lymphocytes obtained from SCID patients [33]. Further studies using site-directed mutagenesis revealed that Orai1 forms the pore of the CRAC channel [34]. This concept is now widely accepted as is the suggestion that Orai1 is present and is important for SOCE in all cell types. However, data on the endogenous Orai1 function in all cell types are not currently available. Two other members of the Orai family, Orai2 and Orai3, which have apparently distinct channel properties, are also present in cells. The relative levels of expression and function of these Orai channels are far from being clearly established.

24.3.3 STIM Proteins

The mechanism involved in transmitting the signal of store depletion to the channels located on the plasma membrane remained elusive until the stromal interaction molecule (STIM) was identified by the Stauderman [35] and Meyer [36] groups in 2005. STIM has two human homologs, STIM1 and STIM2, which are primarily localized in the ER membrane. They are single transmembrane protein molecules with a relatively large cytosolic C-terminal domain. The N-terminus lies within the ER lumen and has a Ca^{2+} binding site which serves as a calcium sensor in the ER [36]. The available data suggest that when the $[Ca^{2+}]$ in the ER lumen decreases, the Ca^{2+} bound to the EF hand domain of STIM1 dissociates leading to conformational changes in the protein. The net result is that STIM1 monomers oligomerize and translocate to specific ER-PM junctional regions where they aggregate into punta [37]. It is proposed that STIM1 interacts with and activates channels involved in SOCE at these locations. Both TRPC1 and Orai1 interact with and are activated by STIM1 in response to Ca^{2+} store depletion [13, 38, 39]. The role of STIM2 has not yet been conclusively established and will not be discussed in this review.

Within the past few years, the structure of STIM1 has been closely scrutinized to define the mechanism(s) involved in its association with and activation of Orai, and TRPC proteins. Mutations in the EF hand domain induce constitutive activation of SOCE [36, 37]. Further the EF hand domain, SAM motif, coiled-coil domain and proline-rich region are important in mediating the oligomerization and translocation of STIM1 to critical ER-PM junctional regions where SOCE is regulated [38–40]. The lysine-rich C-terminal end of STIM1 (referred to as the polybasic tail) has been proposed to be involved in activation of SOCE. Deletion of the polybasic tail resulted in loss of SOCE likely due to loss of STIM1 puncta formation in ER-PM junctions, although STIM1 oligomerization per se is not affected [40]. The polybasic tail region does not directly gate CRAC channels although STIM1(^{684}KK^{685}) in this domain of STIM1 is involved in gating TRPC1. The polybasic tail region contains a consensus sequence which could mediate interaction of the protein with PIP_2 in the plasma membrane [41–43]. Such an interaction could enable anchoring of STIM1 in the plasma membrane and thus facilitate its association with the channels and activation of SOCE. However, the suggested role of PIP_2 binding to and anchoring STIM1 puncta in ER-PM junctional regions has been contradicted [44]. Several other studies show that lipid raft domains (that are also enriched in PIP_2) could serve as platforms for recruitment and anchoring STIM1/channel complexes in the cell periphery [45–47]. This concept has also been questioned [48]. The interesting role of lipid rafts in SOCE will not be discussed in this chapter although this has been recently reviewed [49, 50].

24.4 STIM1/Orai1 and CRAC Channels

Studies in several cell types reveal that expression of Orai1 alone reduces endogenous SOCE while expression of STIM1 alone does not change SOCE. Co-expression of Orai1 with STIM1 induces a large increase in SOCE and generates substantial I_{CRAC} [51]. Furthermore, mutations in conserved negatively charged residues in the transmembrane region of Orai1 alter the Ca^{2+} selectivity of the CRAC channel [34, 51, 52] while expression of the Orai1 mutant associated with SCID does not lead to generation of I_{CRAC} in cells when co-expressed with STIM1. Expression of EF hand mutant of STIM1 (D76A) induces constitutive CRAC activity. Together these findings provide strong evidence that Orai1 is the essential pore-forming unit of CRAC channels and that STIM1 serves as the regulator of the channel and sensor of store depletion. Expression of Orai2 or Orai3 with STIM1 in HEK293 cells shows that both homologs are capable of constituting store-operated channels with different magnitudes and channel characteristics, leading to the speculation that native CRAC channels may involve combinations of Orai proteins [51]. Although expression of either Orai homologue leads to partial recovery of I_{CRAC} in cells from SCID patients, their exact roles in SOCE have not yet been conclusively demonstrated in any cell type.

Since Orai1 and STIM1 appear to interact within very specific regions of the cells, the recruitment of these proteins to these locations has created much interest and speculation. The currently available data have been obtained from studies with Orai1 and STIM1 tagged with fluorescent proteins. It is important to note that localization of native STIM1 and Orai1 in resting and stimulated cells has not yet been conclusively described and that introduction of a fluorescent protein tag at the N-terminus of STIM1 prevents its plasma membrane insertion. However, the currently reported data using TIRF microscopy demonstrate that STIM1 is translocated into the ER-PM junctional domains in response to store depletion. On the other hand, Orai1 is located primarily in the plasma membrane and displays a diffused pattern under resting conditions. Upon ER Ca^{2+} store depletion, Orai1 rapidly clusters and co-localizes with STIM1 puncta within ER-PM microdomains [53, 54]. Two hydrophobic amino acid residues located in the C-terminal coiled-coil motif of Orai1, L273 and L276, have been identified to be essential in Orai1-STIM1 interaction. Physical association between Orai1 and STIM1 during activation of SOCE has been further substantiated by FRET measurements [51, 55]. These data also suggest that both STIM1 and Orai1 undergo conformational changes that are associated with CRAC channel activation. Several independent studies have identified a critical Orai1-interaction domain in STIM1 [40, 56–58] that is involved in gating Orai1. Furthermore, another domain (CMD, the CRAC Modulatory Domain) within the C-terminus of STIM1 [59] appears to mediate fast Ca^{2+}-dependent inactivation of Orai1 by binding to Ca^{2+}. Interestingly, a calmodulin (CaM) binding site was recently identified in the N-terminus of Orai1 (68–91aa). Eliminating CaM binding abrogated Orai1 inactivation [59] suggesting that CaM synergistically acts with STIM1 to mediate the Ca^{2+}-dependent fast inactivation of Orai1 channel. Thus, STIM1 interaction with Orai1 accounts for several important regulatory mechanisms associated with CRAC channels including their gating in response to store depletion as well as Ca^{2+}-dependent inactivation. Interestingly, the inhibitory effect of 2-APB on SOCE has now been ascribed to the disruption of STIM1 puncta.

24.5 TRPC1/Orai1/STIM1 and SOC Channels

STIM1 also interacts with and modulates the activity of TRPC1 and other members of the TRPC channel family [60–62]. Knockdown of STIM1 in several cell types reduces endogenous TRPC1 function (Ca^{2+} entry and cation currents) stimulated by store depletion. Furthermore, co-expression of TRPC1 and STIM1 increases SOCE. STIM1 immunoprecipitates with TRPC1 and this association increases upon stimulation of cells either with an agonist or thapsigargin [61–65]. As in the case of Orai1, the D76A mutant of STIM1 also induces constitutive activation of TRPC1. Two regions of STIM1 are suggested to interact with TRPC1: the ERM (ezrin/radixin/moesin) domain [61] which is involved in the interaction with TRPC1, and the STIM1 polybasic tail (^{684}KK^{685}) residues which interacts electrostatically with the negatively charged residues in the C-terminus of TRPC1 (^{639}DD^{640}), resulting in gating of the channel [65]. Thus, STIM1 has distinct gating mechanisms for TRPC1 and Orai1.

A very important and intriguing aspect of store-dependent regulation of TRPC1 is the requirement of functional Orai1. Knocking down endogenous Orai1 or transfection of cells with functionally defective Orai1 mutants (R91W, E106Q) attenuate the increase in SOCE induced by TRPC1-STIM1 overexpression [60, 66]. Similar experimental maneuvers also abrogate endogenous SOCE in HSG cells, which is significantly dependent on endogenous TRPC1 [62]. Physical proximity and association of Orai1 and TRPC1 has been suggested in studies demonstrating the dynamic assembly of a TRPC1-STIM1-Orai1 complex in response to store depletion [60–62, 65–67]. Together, these data highlight the critical contributions of STIM1, Orai1, and TRPC1 to SOCE. While the role of STIM1 in regulation of TRPC1 has been now resolved, the exact functional interaction between TRPC1 and Orai1 is not yet known. A study in human platelets cells suggested that Orai1 mediates the communication between STIM1 and TRPC1, since in the absence of Orai1, TRPC1 functions as a store-independent channel [67]. Another study suggests that Orai1 acts as a regulatory subunit of TRPC channels based on the findings that Orai1 physically interacts with the N- and C-termini of TRPC3 and TRPC6 to transform store-insensitive channels into store-operated channels [68]. Further studies will be required to elucidate the exact contribution of Orai1 to TRPC1-SOC function.

Interesting findings also suggest that TRPC1 might be required for CRAC channel function. Low levels of expression of either Orai1 or TRPC1 with STIM1 fail to sustain SOCE or I_{CRAC}, whereas when the three proteins are expressed together at the same low levels, a normal SOCE response is obtained [66]. These findings are in contrast to previous data that knockdown of TRPC1, TRPC3 or expression of STIM1 mutants that fail to gate TRPC channels does not affect I_{CRAC} and that Orai1+STIM1-CRAC channels assemble functionally without contribution from

TRPC channels in cells such as T lymphocytes and mast cells. Furthermore, loss of SOCE in cells following knockdown of TRPC channels or in cells isolated from TRPC-knockout mice demonstrates lack of compensation of function by the residual Orai and STIM1 proteins [19]. Thus, although TRPC1 and Orai1 could be mutually dependent, they distinctly contribute to SOCE and the regulation of specific cellular functions. It is also possible that native SOCE components and signaling could vary in different cell types depending on the specific physiological functions that are regulated by SOCE.

24.6 TRPC-Generated SOC Channels – Problems and Perspectives

The currently available data support the conclusion that Orai1 and STIM1 are sufficient for the generation of CRAC channel activity following store depletion. Although the presence and function of CRAC channels has yet to be confirmed in all cell types, it has been established as the primary SOCE pathway in T lymphocytes, mast cells, and several other cell types [51, 55, 69]. The domains of interactions between STIM1 and Orai1 that are involved in key aspects of channel assembly and activation have also been revealed. On the other hand, the mechanism involved in TRPC-SOC channels has yet to be completely resolved. In this section we will primarily discuss the role of TRPC1 in SOCE as more detailed studies have been done with this channel. Scepticism regarding the relevance of TRPC1 continues despite increasing number of studies which demonstrate that TRPC1 is involved in SOCE, is regulated by STIM1, and triggers SOCE-dependent Ca^{2+} signaling mechanisms in cells, including activation of cell proliferation as well as NFAT and NFκB signaling pathways. The role of TRPC1 in SOCE has been questioned because TRPC1–/– mice do not display loss of function in all cell types [27, 28]. Further, heterologous expression of TRPC1 does not consistently result in a substantial increase in SOCE. In this context, it should be noted that mutations causing loss of Orai1 function, as in SCID patients, does not affect all the tissues [33]. Further, expression of Orai1 alone, in the absence of STIM1 does not result in an increase in SOCE [51, 55]. Thus, it is important that studies with TRPC1 be revisited keeping in mind the recently revealed regulation of the channel by STIM1. Major conundrum also arises from the findings that activation of TRPC1 channel by store depletion also requires functional Orai1 [60, 66]. Given the lack of conclusive data elucidating a possible mechanism for this, a major debate has ensued over how TRPC1 and Orai1 can simultaneously contribute to SOCE as well as the suggestion that TRPC1 and Orai1 can contribute to the same channel.

Experimental observations supporting the role for TRPC1 in SOCE and requirement of Orai1 and STIM1 for TRPC1-SOC channel function can be summarized as follows:

1. $[Ca^{2+}]$ measurements in cells stimulated by agonists, thaspigargin, BHQ, and TPEN have demonstrated activation of Ca^{2+} entry that is blocked by 1–5 μM Gd^{3+} and 10–20 μM 2APB, hallmarks of SOCE and CRAC channels.

2. Whole cell patch clamp under conditions used for CRAC channel measurements [6] such as inclusion of IP$_3$ in the pipette solution or stimulation of cells with Tg or agonist leads to generation of cation currents ranging from relatively selective (Ca:Na of 40:1 [10, 69]) to non-selective [70]. These currents are also completely blocked by low [Gd^{3+}] and 2APB.
3. Knockdown of TRPC1 or expression of TRPC1 mutants reduce SOCE and I$_{SOC}$.
4. Activation of TRPC1-SOCE regulates NFκB [71] and NFAT [72, 73].
5. STIM1 knockdown reduces endogenous TRPC1-dependent SOCE.
6. Overexpression of STIM1+TRPC1 in HSG cells and HEK293 cells leads to larger increase in SOCE than that seen with expression of TRPC1 alone. The Ca^{2+} entry is blocked completely by low [Gd^{3+}] and 2APB.
7. C-terminal residues of STIM1(^{684}KK^{685}) activate TRPC1 via electrostatic interaction. Expression of STIM1(^{684}EE^{685}) mutant exerts a dominant negative effect on SOCE, although it can gate Orai1.
8. Activation of endogenous and overexpressed TRPC1 requires functional Orai1 since knockdown of endogenous Orai1 or expression of functionally-defective mutants leads to attenuation of TRPC1-SOCE, both endogenous and that induced by expression of STIM1+TRPC1 in HSG and HEK293 cells [60].
9. Co-immunoprecipitation and TIRF experiments show that TRPC1/Orai1/STIM1 associate following store depletion.

A major problem facing the field currently is that of reconciling the data demonstrating that two distinct channels contribute to SOCE in some cells with the concept that SOCE is mediated via a single ubiquitously present channel, namely CRAC, in all cell types. Although further studies will be required to completely resolve the exact molecular mechanism that is involved in regulation of TRPC/Orai1/STIM1-channels, we will discuss currently proposed mechanisms:

- *Orai1 and TRPC1 generate distinct CRAC and SOC channels.* In this case, the larger TRPC channel activity could mask an underlying I$_{CRAC}$ activity. However, residual I$_{CRAC}$ has not been reported following reduction of I$_{SOC}$ induced by knockdown of TRPC1. Assuming (as predicted by most of the currently available data) that CRAC channels do not need TRPC proteins, detection of I$_{CRAC}$ would depend on how effectively TRPC expression has been knocked down and provided no other TRPC or cation channels are activated under these conditions.
- *TRPC1 and Orai1 interact to form one channel:* This model suggests that both channels contribute to the same channel pore. Conclusive experimental evidence supporting this suggestion is presently lacking.
- *Orai1 is a critical modulator of TRPC1 channels.* This model proposes that the pore of the SOCE channel is formed by TRPC1 and that Orai1 behaves like other tetraspanins [68] to regulate channel activity directly as a subunit of the channel. Although interesting there are not enough data available presently to support this hypothesis. Furthermore, the ion channel properties of Orai1 have been studied in great detail which provides strong evidence that Orai1 generates CRAC channel. It is also important to note that cells which display an endogenous I$_{CRAC}$

signature following store depletion do not demonstrate any apparent contribution of TRPC1 to the current.

- *Orai1 can indirectly regulate activation of TRPC channels*. This model predicts that Orai1-CRAC channel activity initiates downstream signals that lead to activation of TRP channels. An example of this type of regulation has been recently demonstrated in the case of TRPC4 and TRPC5 which are directly activated by increase in $[Ca^{2+}]_i$ that results from activation of the Orai1-CRAC channel. While these data were obtained from overexpression systems, TRPC5 activation was also seen in neuronal cells following activation of voltage-dependent calcium channels and thus is not "Orai1-specific" and more importantly not "store-depletion dependent". However, it is clear that not all TRPC channels that are activated following SOCE and show dependence on Orai1 are directly activated by Ca^{2+} (e.g. TRPC1 and TRPC3). Furthermore, TRPC1 activation in response to store depletion is determined by STIM1. Thus, other mechanisms triggered via Orai1-mediated Ca^{2+} entry such as TRPC channel recruitment, phorphorylation/dephosphorylation, or regulation by calmodulin can be proposed as possible modes for regulation of TRPCs channels. These latter would be consistent with all the reported data regarding activation of TRPC1 following store depletion. If Orai1-CRAC were the primary channel leading to activation of TRPC1, then TRPC1 would contribute to the total Ca^{2+} entry activated by store depletion. Further, TRPC1 function would be blocked under the conditions that block Orai1 function. Another possibility that cannot be presently ruled out is that direct interaction between the two channels is involved in regulation of TRPC1, although STIM1 would be the gating component of TRPC1 following store depletion. Based on the activation by STIM1, it is accurate to describe TRPC1 as a "store-dependent" channel. Further studies will be required to establish which of these possible modes of interaction between Orai1 and TRPC1 prevails in cells.

24.7 Conclusions

After more than 2 decades of intense efforts to determine the mechanisms and molecular components of SOCE, recent studies have lead to identification of novel channels and regulatory proteins for this Ca^{2+} entry pathway. More importantly, the critical physiological role of SOCE and the impact of its dysfunction of this mechanism in disease is also rapidly emerging. Exciting new studies demonstrate that SOCE is associated with highly specialized microdomains in the cells where local Ca^{2+} signaling events lead to significant regulation of cell function. There is considerable agreement that SOCE is required for a number of critical cell functions and is not restricted to the purpose of refilling of intracellular calcium stores. STIM1 appears to be a central molecule in SOCE which could serve several functions. This is demonstrated by the findings that it has multiple targets, such as Orai and TRPC channels, as well as proteins involved in protein synthesis, adenylyl cyclase, and cell adhesion. The physiological relevance of these different functions of STIM1 and whether they are all associated with SOCE needs to considered within the context of specific cell types. The complexity of the Ca^{2+} signaling mechanisms associated with SOCE is evident in T and B lymphocytes or mast cells. In these cell types, stimulation of cell surface receptors leads to the recruitment of large and distinct signaling complexes of which Orai1/STIM1-CRAC is a key component. Multiple signaling mechanisms are co-ordinately initiated within the complex. Many of these are regulated by changes in local $[Ca^{2+}]_i$ resulting from CRAC-mediated Ca^{2+} entry. In B cells $[Ca^{2+}]_i$ increases mediated by CRAC channels are modified by the concurrent activation of other channels that either directly provide Ca^{2+} or modulate CRAC channel activity by regulating membrane potential [74]. The resulting modulation of local and global Ca^{2+} signals have very significant and different outcomes on the regulation of downstream cellular functions including secretion and gene regulation. In addition to focusing on the more didactic aspects of the structure and properties of ion channels involved in SOCE, future studies should also be directed towards examining the relevance of other components that might be activated either concurrently with or downstream from SOCE activation and assess the impact of these on local and global Ca^{2+} signaling and on cell function. Such studies will provide important clues to understand and appreciate the *raison d'etre* for the complexity and diversity in SOCE.

References

1. Parekh AB, Penner R (1997) Store depletion and calcium influx. Physiol Rev 77:901–930
2. Parekh AB, Putney JW Jr. (2005) Store-operated calcium channels. Physiol Rev 85:757–810
3. Putney JW Jr. (1990) Capacitative calcium entry revisited. Cell Calcium 11:611–624
4. Ambudkar IS (2006) Ca^{2+} signaling microdomains:platforms for the assembly and regulation of TRPC channels. Trends Pharmacol Sci 27:25–32
5. Putney JW Jr., Broad LM, Braun FJ, Lievremont JP, Bird GS (2001) Mechanisms of capacitative calcium entry. J Cell Sci 114:2223–2229
6. Bird GS, DeHaven WI, Smyth JT, Putney JW Jr. (2008) Methods for studying store-operated calcium entry. Methods 46:204–212
7. Prakriya M, Lewis RS (2006) Regulation of CRAC channel activity by recruitment of silent channels to a high open-probability gating mode. J Gen Physiol 128:373–386
8. Ambudkar IS (2007) TRPC1: a core component of store-operated calcium channels. Biochem Soc Trans 35:96–100
9. Bolotina VM (2004) Store-operated channels: diversity and activation mechanisms. Sci STKE 2004:pe34
10. Brueggemann LI, Markun DR, Henderson KK, Cribbs LL, Byron KL (2006) Pharmacological and electrophysiological characterization of store-operated currents and capacitative Ca^{2+} entry in vascular smooth muscle cells. J Pharmacol Exp Ther 317:488–499
11. Montell C (2005) The TRP superfamily of cation channels. Sci STKE 2005:re3
12. Venkatachalam K, Montell C (2007) TRP channels. Annu Rev Biochem 76:387–417
13. Ambudkar IS, Ong HL, Liu X, Bandyopadhyay BC, Cheng KT (2007) TRPC1: the link between functionally distinct store-operated calcium channels. Cell Calcium 42: 213–223
14. Venkatachalam K, van Rossum DB, Patterson RL, Ma HT, Gill DL (2002) The cellular and molecular basis of store-operated calcium entry. Nat Cell Biol 4:E263–E272
15. Beech DJ (2005) Emerging functions of 10 types of TRP cationic channel in vascular smooth muscle. Clin Exp Pharmacol Physiol 32:597–603

16. Cai S, Fatherazi S, Presland RB, Belton CM, Roberts FA, Goodwin PC, Schubert MM, Izutsu KT (2006) Evidence that TRPC1 contributes to calcium-induced differentiation of human keratinocytes. Pflugers Arch 452:43–52
17. Dietrich A, Chubanov V, Kalwa H, Rost BR, Gudermann T (2006) Cation channels of the transient receptor potential superfamily: their role in physiological and pathophysiological processes of smooth muscle cells. Pharmacol Ther 112:744–760
18. Fiorio Pla A, Maric D, Brazer SC, Giacobini P, Liu X, Chang YH, Ambudkar IS, Barker JL (2005) Canonical transient receptor potential 1 plays a role in basic fibroblast growth factor (bFGF)/FGF receptor-1-induced Ca^{2+} entry and embryonic rat neural stem cell proliferation. J Neurosci 25:2687–2701
19. Liu X, Cheng KT, Bandyopadhyay BC, Pani B, Dietrich A, Paria BC, Swaim WD, Beech D, Yildrim E, Singh BB, Birnbaumer L, Ambudkar IS (2007) Attenuation of store-operated Ca^{2+} current impairs salivary gland fluid secretion in TRPC1(–/–) mice. Proc Natl Acad Sci USA 104:17542–17547
20. Liu X, Wang W, Singh BB, Lockwich T, Jadlowiec J, O'Connell B, Wellner R, Zhu MX, Ambudkar IS (2000) Trp1, a candidate protein for the store-operated Ca^{2+} influx mechanism in salivary gland cells. J Biol Chem 275:3403–3411
21. Mehta D, Ahmmed GU, Paria BC, Holinstat M, Voyno-Yasenetskaya T, Tiruppathi C, Minshall RD, Malik AB (2003) RhoA interaction with inositol 1,4,5-trisphosphate receptor and transient receptor potential channel-1 regulates Ca^{2+} entry. Role in signaling increased endothelial permeability. J Biol Chem 278:33492–33500
22. Mori Y, Wakamori M, Miyakawa T, Hermosura M, Hara Y, Nishida M, Hirose K, Mizushima A, Kurosaki M, Mori E, Gotoh K, Okada T, Fleig A, Penner R, Iino M, Kurosaki T (2002) Transient receptor potential 1 regulates capacitative Ca^{2+} entry and Ca^{2+} release from endoplasmic reticulum in B lymphocytes. J Exp Med 195:673–681
23. Rao JN, Platoshyn O, Golovina VA, Liu L, Zou T, Marasa BS, Turner DJ, Yuan JX, Wang JY (2006) TRPC1 functions as a store-operated Ca^{2+} channel in intestinal epithelial cells and regulates early mucosal restitution after wounding. Am J Physiol Gastrointest Liver Physiol 290:G782–G792
24. Tiruppathi C, Ahmmed GU, Vogel SM, Malik AB (2006) Ca^{2+} signaling, TRP channels, and endothelial permeability. Microcirculation 13:693–708
25. Vandebrouck C, Martin D, Colson-Van Schoor M, Debaix H, Gailly P (2002) Involvement of TRPC in the abnormal calcium influx observed in dystrophic (mdx) mouse skeletal muscle fibers. J Cell Biol 158:1089–1096
26. Zagranichnaya TK, Wu X, Villereal ML (2005) Endogenous TRPC1, TRPC3, and TRPC7 proteins combine to form native store-operated channels in HEK-293 cells. J Biol Chem 280:29559–29569
27. Dietrich A, Kalwa H, Storch U, Mederos Y, Schnitzler M, Salanova B, Pinkenburg O, Dubrovska G, Essin K, Gollasch M, Birnbaumer L, Gudermann T (2007) Pressure-induced and store-operated cation influx in vascular smooth muscle cells is independent of TRPC1. Pflugers Arch 455:465–477
28. Varga-Szabo D, Authi KS, Braun A, Bender M, Ambily A, Hassock SR, Gudermann T, Dietrich A, Nieswandt B (2008) Store-operated Ca^{2+} entry in platelets occurs independently of transient receptor potential (TRP) C1. Pflugers Arch 457:377–387
29. Freichel M, Vennekens R, Olausson J, Stolz S, Philipp SE, Weissgerber P, Flockerzi V (2005) Functional role of TRPC proteins in native systems: implications from knockout and knockdown studies. J Physiol 567:59–66
30. Vazquez G, Lievremont JP, St JBG, Putney JW Jr. (2001) Human Trp3 forms both inositol trisphosphate receptor-dependent and receptor-independent store-operated cation channels in DT40 avian B lymphocytes. Proc Natl Acad Sci USA 98:11777–11782
31. Wang X, Pluznick JL, Wei P, Padanilam BJ, Sansom SC (2004) TRPC4 forms store-operated Ca^{2+} channels in mouse mesangial cells. Am J Physiol Cell Physiol 287:C357–C364
32. Freichel M, Suh SH, Pfeifer A, Schweig U, Trost C, Weissgerber P, Biel M, Philipp S, Freise D, Droogmans G, Hofmann F, Flockerzi V, Nilius B (2001) Lack of an endothelial

store-operated Ca^{2+} current impairs agonist-dependent vasorelaxation in TRP4–/– mice. Nat Cell Biol 3:121–127

33. Feske S, Gwack Y, Prakriya M, Srikanth S, Puppel SH, Tanasa B, Hogan PG, Lewis RS, Daly M, Rao A (2006) A mutation in Orai1 causes immune deficiency by abrogating CRAC channel function. Nature 441:179–185
34. Prakriya M, Feske S, Gwack Y, Srikanth S, Rao A, Hogan PG (2006) Orai1 is an essential pore subunit of the CRAC channel. Nature 443:230–233
35. Roos J, DiGregorio PJ, Yeromin AV, Ohlsen K, Lioudyno M, Zhang S, Safrina O, Kozak JA, Wagner SL, Cahalan MD, Velicelebi G, Stauderman KA (2005) STIM1, an essential and conserved component of store-operated Ca^{2+} channel function. J Cell Biol 169: 435–445
36. Liou J, Kim ML, Heo WD, Jones JT, Myers JW, Ferrell JE Jr., Meyer T (2005) STIM is a Ca^{2+} sensor essential for Ca^{2+}-store-depletion-triggered Ca^{2+} influx. Curr Biol 15:1235–1241
37. Zhang SL, Yu Y, Roos J, Kozak JA, Deerinck TJ, Ellisman MH, Stauderman KA, Cahalan MD (2005) STIM1 is a Ca^{2+} sensor that activates CRAC channels and migrates from the Ca^{2+} store to the plasma membrane. Nature 437:902–905
38. Li Z, Lu J, Xu P, Xie X, Chen L, Xu T (2007) Mapping the interacting domains of STIM1 and Orai1 in Ca^{2+} release-activated Ca^{2+} channel activation. J Biol Chem 282:29448–29456
39. Worley PF, Zeng W, Huang GN, Yuan JP, Kim JY, Lee MG, Muallem S (2007) TRPC channels as STIM1-regulated store-operated channels. Cell Calcium 42:205–211
40. Park CY, Hoover PJ, Mullins FM, Bachhawat P, Covington ED, Raunser S, Walz T, Garcia KC, Dolmetsch RE, Lewis RS (2009) STIM1 clusters and activates CRAC channels via direct binding of a cytosolic domain to Orai1. Cell 136:876–890
41. Heo WD, Inoue T, Park WS, Kim ML, Park BO, Wandless TJ, Meyer T (2006) PI(3,4,5)P3 and PI(4,5)P2 lipids target proteins with polybasic clusters to the plasma membrane. Science 314:1458–1461
42. Liou J, Fivaz M, Inoue T, Meyer T (2007) Live-cell imaging reveals sequential oligomerization and local plasma membrane targeting of stromal interaction molecule 1 after Ca^{2+} store depletion. Proc Natl Acad Sci USA 104:9301–9306
43. Walsh CM, Chvanov M, Haynes LP, Petersen OH, Tepikin AV, Burgoyne RD (2009) Role of phosphoinositides in STIM1 dynamics and store-operated calcium entry. Biochem J 425: 159–168
44. Korzeniowski MK, Popovic MA, Szentpetery Z, Varnai P, Stojilkovic SS, Balla T (2009) Dependence of STIM1/Orai1-mediated calcium entry on plasma membrane phosphoinositides. J Biol Chem 284:21027–21035
45. Pani B, Ong HL, Liu X, Rauser K, Ambudkar IS, Singh BB (2008) Lipid rafts determine clustering of STIM1 in endoplasmic reticulum-plasma membrane junctions and regulation of store-operated Ca^{2+} entry (SOCE). J Biol Chem 283:17333–17340
46. Alicia S, Angelica Z, Carlos S, Alfonso S, Vaca L (2008) STIM1 converts TRPC1 from a receptor-operated to a store-operated channel: Moving TRPC1 in and out of lipid rafts. Cell Calcium 44:479–491
47. Jardin I, Salido GM, Rosado JA (2008) Role of lipid rafts in the interaction between hTRPC1, Orai1 and STIM1. Channels (Austin) 2:401–403
48. DeHaven WI, Jones BF, Petranka JG, Smyth JT, Tomita T, Bird GS, Putney JW Jr. (2009) TRPC channels function independently of STIM1 and Orai1. J Physiol 587:2275–2298
49. Ambudkar IS, Ong HL, Singh BB (2010) Molecular and functional determinants of Ca^{2+} signaling microdomains. In: Sitaramayya A (ed) Signal transduction: pathways, mechanisms and diseases. Springer, Heidelberg, pp. 237–253
50. Pani B, Singh BB (2009) Lipid rafts/caveolae as microdomains of calcium signaling. Cell Calcium 45:625–633
51. Hogan PG, Lewis RS, Rao A (2010) Molecular basis of calcium signaling in lymphocytes: STIM and ORAI. Annu Rev Immunol 28:491–533

52. Yeromin AV, Zhang SL, Jiang W, Yu Y, Safrina O, Cahalan MD (2006) Molecular identification of the CRAC channel by altered ion selectivity in a mutant of Orai. Nature 443:226–229
53. Luik RM, Wu MM, Buchanan J, Lewis RS (2006) The elementary unit of store-operated Ca^{2+} entry: local activation of CRAC channels by STIM1 at ER-plasma membrane junctions. J Cell Biol 174:815–825
54. Xu P, Lu J, Li Z, Yu X, Chen L, Xu T (2006) Aggregation of STIM1 underneath the plasma membrane induces clustering of Orai1. Biochem Biophys Res Commun 350:969–976
55. Prakriya M (2009) The molecular physiology of CRAC channels. Immunol Rev 231:88–98
56. Kawasaki T, Lange I, Feske S (2009) A minimal regulatory domain in the C terminus of STIM1 binds to and activates ORAI1 CRAC channels. Biochem Biophys Res Commun 385:49–54
57. Muik M, Fahrner M, Derler I, Schindl R, Bergsmann J, Frischauf I, Groschner K, Romanin C (2009) A cytosolic homomerization and a modulatory domain within STIM1 C terminus determine coupling to ORAI1 channels. J Biol Chem 284:8421–8426
58. Yuan JP, Zeng W, Dorwart MR, Choi YJ, Worley PF, Muallem S (2009) SOAR and the polybasic STIM1 domains gate and regulate Orai channels. Nat Cell Biol 11:337–343
59. Mullins FM, Park CY, Dolmetsch RE, Lewis RS (2009) STIM1 and calmodulin interact with Orai1 to induce Ca^{2+}-dependent inactivation of CRAC channels. Proc Natl Acad Sci USA 106:15495–15500
60. Cheng KT, Liu X, Ong HL, Ambudkar IS (2008) Functional requirement for Orai1 in store-operated TRPC1-STIM1 channels. J Biol Chem 283:12935–12940
61. Huang GN, Zeng W, Kim JY, Yuan JP, Han L, Muallem S, Worley PF (2006) STIM1 carboxyl-terminus activates native SOC, I(crac) and TRPC1 channels. Nat Cell Biol 8:1003–1010
62. Ong HL, Cheng KT, Liu X, Bandyopadhyay BC, Paria BC, Soboloff J, Pani B, Gwack Y, Srikanth S, Singh BB, Gill DL, Ambudkar IS (2007) Dynamic assembly of TRPC1-STIM1-Orai1 ternary complex is involved in store-operated calcium influx. Evidence for similarities in store-operated and calcium release-activated calcium channel components. J Biol Chem 282:9105–9116
63. Lopez JJ, Salido GM, Pariente JA, Rosado JA (2006) Interaction of STIM1 with endogenously expressed human canonical TRP1 upon depletion of intracellular Ca^{2+} stores. J Biol Chem 281:28254–28264
64. Yuan JP, Zeng W, Huang GN, Worley PF, Muallem S (2007) STIM1 heteromultimerizes TRPC channels to determine their function as store-operated channels. Nat Cell Biol 9:636–645
65. Zeng W, Yuan JP, Kim MS, Choi YJ, Huang GN, Worley PF, Muallem S (2008) STIM1 gates TRPC channels, but not Orai1, by electrostatic interaction. Mol Cell 32:439–448
66. Kim MS, Zeng W, Yuan JP, Shin DM, Worley PF, Muallem S (2009) Native store-operated Ca^{2+} influx requires the channel function of orai1 and TRPC1. J Biol Chem 284:9733–9741
67. Jardin I, Lopez JJ, Salido GM, Rosado JA (2008) Orai1 mediates the interaction between STIM1 and hTRPC1 and regulates the mode of activation of hTRPC1-forming Ca^{2+} channels. J Biol Chem 283:25296–25304
68. Liao Y, Erxleben C, Yildirim E, Abramowitz J, Armstrong DL, Birnbaumer L (2007) Orai proteins interact with TRPC channels and confer responsiveness to store depletion. Proc Natl Acad Sci USA 104:4682–4687
69. Liu X, Groschner K, Ambudkar IS (2004) Distinct Ca^{2+}-permeable cation currents are activated by internal Ca^{2+}-store depletion in RBL-2H3 cells and human salivary gland cells, HSG and HSY. J Membr Biol 200:93–104
70. Liu X, Bandyopadhyay BC, Singh BB, Groschner K, Ambudkar IS (2005) Molecular analysis of a store-operated and 2-acetyl-sn-glycerol-sensitive non-selective cation channel. Heteromeric assembly of TRPC1-TRPC3. J Biol Chem 280:21600–21606
71. Bair AM, Thippegowda PB, Freichel M, Cheng N, Ye RD, Vogel SM, Yu Y, Flockerzi V, Malik AB, Tiruppathi C (2009) Ca^{2+} entry via TRPC channels is necessary for thrombin-induced NF-kappaB activation in endothelial cells through AMP-activated protein kinase and protein kinase Cdelta. J Biol Chem 284:563–574

72. Ohba T, Watanabe H, Takahashi Y, Suzuki T, Miyoshi I, Nakayama S, Satoh E, Iino K, Sasano H, Mori Y, Kuromitsu S, Imagawa K, Saito Y, Iijima T, Ito H, Murakami M (2006) Regulatory role of neuron-restrictive silencing factor in expression of TRPC1. Biochem Biophys Res Commun 351:764–770
73. Seth M, Zhang ZS, Mao L, Graham V, Burch J, Stiber J, Tsiokas L, Winn M, Abramowitz J, Rockman HA, Birnbaumer L, Rosenberg P (2009) TRPC1 channels are critical for hypertrophic signaling in the heart. Circ Res 105:1023–1030
74. Scharenberg AM, Humphries LA, Rawlings DJ (2007) Calcium signalling and cell-fate choice in B cells. Nat Rev Immunol 7:778–789

Chapter 25
Primary Thermosensory Events in Cells

Ilya Digel

Abstract Temperature sensing is essential for the survival of living organisms. Since thermal gradients are almost everywhere, thermoreception could represent one of the oldest sensory transduction processes that evolved in organisms. There are many examples of temperature changes affecting the physiology of living cells. Almost all classes of biological macromolecules in a cell (nucleic acids, lipids, proteins) can serve as a target of the temperature-related stimuli. This review is devoted to some common features of different classes of temperature-sensing molecules as well as molecular and biological processes involved in thermosensation. Biochemical, structural and thermodynamic approaches are discussed in order to overview the existing knowledge on molecular mechanisms of thermosensation.

Abbreviations

CSP	cold shock proteins
HSP	heat shock proteins
PIP2	phosphatidylinositol bisphosphate
TRP channels	*transient receptor potential* channels
TRPA	"ankyrin" subfamily
TRPM	"melastatin" subfamily
TRPV	"vanilloid" subfamily

Temperature changes are among the main stresses experienced by organisms from bacteria to plants and animals and therefore temperature is one of the environmental cues under constant vigilance in living cells. Several problems arise from exposing a cell to a sudden change in temperature [1]: firstly, membrane fluidity changes, that affects many membrane-associated vital functions; secondly, nucleic acid topology will be affected causing shifts in processes such as transcription and translation.

I. Digel (✉)
Laboratory of Cellular Biophysics, Aachen University of Applied Sciences, Juelich, Germany
e-mail: digel@fh-aachen.de

Finally, the protein function is directly affected both from structural and catalytic points of view.

Hence, living cells need "devices" for sensing environmental temperature changes in order to adapt their biochemical processes accordingly. A successful adaptive thermotropic response cannot be performed only by corresponding changes in the rate and equilibrium of enzymatic reactions. Such a mechanism of adaptive reaction is too unspecific and uncontrollable. To cope with temperature variation, living organisms need sensing temperature alterations and translating this sensory event into a pragmatic gene response.

While such regulatory cascades may ultimately be complicated, they contain primary sensor machinery at the top of the cascade. The functional core of such machinery is usually that of a temperature-induced conformational or physicochemical change in the central constituents of the cell. Hence, a specific sensory transduction mechanism is needed, including, as a key element, *a molecular sensor*, transforming certain physical parameter (temperature) into a biologically significant signal (change in membrane permeability, specific inhibition/stimulation of gene expression, etc.). In a sense, a living organism can use structural alterations in its biomolecules as the primary thermometers or thermostats. Thus, sensory transduction is a complex biological process aimed at integrating and decoding physical and chemical stimuli performed by primary sensory molecular devices. Furthermore, sensory perception of potentially harmful stimuli can function as a warning mechanism to avert potential tissue/organ damage.

Among temperature-controlled processes in living organisms, most well-known are the expression of heat-shock and cold-shock genes [2]. Relocation of a culture of *Escherichia coli* adapted to an optimal growth to a sudden temperature increase, or decrease, by some 10–15 °C results in adaptive shock responses. Such responses involve a remodeling of bacterial gene expression, aimed at adjusting bacterial cell physiology to the new environmental demands [3, 4]. The response of prokaryotic and eukaryotic systems to heat-shock stress has been thoroughly investigated in a large number of organisms and model cell systems. Notably, all organisms from prokaryotes to higher eukaryotes respond to cold and heat shock in a similar manner. The general response of cells to temperature stress (cold or heat) is the rapid overexpression of small groups of proteins, the so-called CSPs (cold-shock proteins) or HSPs (heat shock proteins), respectively, but the initial launching mechanism is different in both cases.

In bacteria, the heat response generally invokes some 20 heat-shock proteins, mostly chaperones, whose functions are primarily to help dealing with, and alleviate, the cellular stress imposed by heat [5]. Many of these proteins participate in reconstituting and stabilizing protein structures and in removing misfolded ones. The expression of this special chaperone system, which includes the proteins DnaK, DnaJ and GrpE, is activated by the appearence of misfolded, temperature-denatured proteins. Thus, one could implicate the binding of partially unfolded proteins by chaperones as the thermosensoric event regulating expression of heat-shock proteins, where the primary sensory element is constituted by some easily denaturing proteins. This, in turn, demonstrates that even bacteria can practically utilize destructive changes in protein conformation as a means for temperature sensing.

In case of cold shock, the primary sensing event is more obscure. Various reports have now shown that when in vitro cultivation temperature is lowered, the increase of the cell membrane rigidity results in compromised membrane-associated cellular functions. Furthermore, cold stress dramatically hinders membrane-bound enzymes, slows down diffusion rates and induces cluster formation of integral membranous proteins [6]. In mammalian cells, the five known mechanisms by which cold-shock-induced changes affect gene expression are: (i) a general reduction in transcription and translation, (ii) inhibition of RNA degradation, (iii) increased transcription of specific target genes via elements in the promoter region of such genes, (iv) alternative pre-mRNA splicing, and (v) via the presence of cold-shock specific internal ribosome entry segments in mRNAs that result in the preferential and enhanced translation of such mRNAs upon cold shock [7].

It has been pointed out that cold stress exposes cells not to one but to two major stresses: those related to changes in temperature and those related to changes in dissolved oxygen concentration at decreased temperature, and it is therefore necessary to study responses to each, either independently or as part of a coordinated response. Separating the relative effects of temperature and oxygen as a result of decreased temperature is difficult and has not been extensively addressed to date. Both changes in dissolved oxygen and temperature reduction result in similar changes in cultured mammalian cells [7].

The shock response systems briefly mentioned above belong to ultimate mechanisms aimed to survival under extreme temperature conditions. However, the ability to express certain factors can be affected by reasonably small temperature changes. Less drastic changes in temperature may not induce shock responses, but can be sufficient to modulate the expression of bacterial virulence genes, for example in *Shigellae* [8] and *Yersiniae* [9]. While one might be surprised that organisms built on such minimalist approaches as bacteria respond to temperature changes, the consequence of these observations is that even bacteria actually sense temperature shifts in order to control gene expression accordingly. Investigators have now been studying the moderate temperature sensation in a variety of organisms for at least several decades or more. Recently, a number of reports have shown that exposing bacteria, yeasts or mammalian cells to sub-physiological temperatures invokes a coordinated cellular response manifesting itself as alterations in transcription, translation, metabolism, the cell cycle and the cell cytoskeleton [7, 10–13]. Nevertheless, very little is known so far about the molecular mechanisms that govern initial response on small thermal stimuli, particularly the primary sensory transduction mechanisms.

Below, we have tried to uncover some aspects of the biophysical basis of temperature sensing by biological molecular thermometers, summarizing some most general ideas concerning the primary components of temperature signal transduction.

25.1 Temperature Sensing Biomolecules

In addition to specificity and sensitivity, successful thermoresponse should be one that is reversible and controlled. Thus, complexity of thermosensing and thermoregulation on the organism level may reflect the demands to handle and fine-tune responses to an important environmental factor in a dynamic fashion. However, ultimately, it seems that basic and rather simple (bio) chemical processes are serve as primary sensory events and, for that purpose thermotropic changes in physico-chemical state of biological molecules appear highly suitable. While the information available is somewhat scant, the picture emerging shows that cells can use signals generated through changes in nucleic acid or protein conformation, or changes in membrane lipid behavior, as sensory devices. Bellow we make a short overview of temperature-sensing properties of most important groups of biological macro-molecules.

It is worthy to note that probably even water alone could serve as a primitive temperature sensor. In the middle of the twentieth century Oppenheimer and Drost-Hansen [14] reported that a number of more or less abrupt changes in the properties of water and aqueous solutions occur when the temperature is increased from 0 to 60° C. These changes or "kinks" occured within a rather narrow temperature range ($\pm 2°$ C) near 15, 30, 45, and 60° C, respectively and most probably caused by changes in the hydrogen bond network of the water. The authors argued that the temperature-induced structural changes in water and aqueous solutions exert a direct influence on biological phenomena. In a later work W. Drost-Hansen [15] suggested some mechanisms how these structural changes happening with *vicinal* (adjacent to surfaces) water can affect the behavior or activity of biological systems. It was argued that optimal conditions for a complex physiological activity (such as, for instance, growth) will occur somewhere near the middle of the interval between two consecutive kinks. This issue has been discussed in literature very controversially and has not received wide recognition.

25.2 Membrane Lipids Fluidity

The physical state of phospholipid membranes does change in response to temperature shifts in phase-transition manner [16], but the temperature-induced changes in real biological membranes are not sharp because many kinds of fatty acids and cholesterol-like molecules present, having different characteristic temperature points of phase transition. Thus, it would not be surprising if cells (even those of bacteria) could utilize the changes in membrane fluidity as a thermometer device, assisted by protein helpers, playing a role of switchers, "sharpening" the temperature response. Microorganisms counteract the membrane propensity to rigidify at lower temperature and are able to maintain a more-or-less constant degree of membrane fluidity (*homeoviscous adaptation*). The cyanobacterium *Synecocystis* responds to decreased temperature by increasing the *cis*-unsaturation of membrane-lipid fatty acids through expressing acyl-lipid desaturases [17–19]. Lipid unsaturation would then restore membrane fluidity at the lower temperature. In *B. subtilis*, this lipid modification is initiated through the activity of a so-called "two-component regulatory system" consisting of the DesK and DesR proteins [17]. Prokaryotic two-component regulatory systems usually consist of protein pairs: a sensor kinase and a regulatory protein [20].

It appears that it is a combination of membrane physical state and protein conformation that is able to sense temperature and even to translate this sensing event into proper gene expression. However, sensing of temperature through direct alteration in nucleic acid conformation might be more efficient temperature-mediated mechanism of gene expression.

25.3 RNA and DNA Thermotropic Reactions

Theoretically, RNA molecules have a strong potential as temperature sensors, in that they can form pronounced secondary and tertiary structures [21], and through their ability to form intermolecular RNA: RNA hybrids [22]. Both of these processes greatly depend on the formation of complementary base pairing, and consequently one would anticipate these to be dependent on environmental temperature. Indeed, messenger RNAs, apart from carrying their coding information for protein generation are also rapidly emerging as regulators of expression of the encoded message. With unique chemical and structural properties, sensory RNAs perform vital regulatory roles in gene expression by detecting changes in the cellular environment either alone or through interactions with small ligands [23, 24] and proteins [25, 26].

Regulatory RNA elements, "riboswitches", have been reported recently, responding to intracellular signals by conformational changes. Riboswitches are conceptually divided into two parts: an *aptamer* and an expression platform. The aptamer directly binds the small molecule, and the expression platform undergoes structural changes in response to the changes in the aptamer. The expression platform is what regulates gene expression. Riboswitches demonstrate that naturally occurring RNA can specifically response on versatile physical and chemical stimuli, a capability that many previously believed was the domain of proteins or artificially constructed RNAs [27].

RNA thermometers operate at the post transcriptional level to sense selectively the temperature and transduce a signal to the translation machinery via a conformational change. They have usually a highly structured 5′-end that shields the ribosome binding site at physiological temperatures [28–31]. Changes in temperature are manifested by the liberation of the Shine–Dalgarno (SD) sequence, thereby facilitating ribosome binding and translation initiation.

It is known that both in prokaryotic and eukaryotic cells the geometry and tension of DNA is highly dynamic and corresponds to its functional activity. In the bacterial cell, chromosome and plasmid DNAs are contained in a "twisted" superhelical conformation [32, 33], where the degree of supercoiling varies in response to changes in the ambient temperature. The expression of many genes is dependent on DNA conformation, and temperature-dependent gene regulation is mastered through changes in DNA supercoiling [3, 34, 35].

Examples of pure DNA-related temperature sensitivity are rare if ever reported. In most cases, genomic thermo-sensitivity appears to be a result of certain interplay among DNA, RNA and proteins. Some bacteria carry a DNA-plasmid which shows a controlled constant plasmid copy number at one temperature and a much higher or totally uncontrolled copy number at a different temperature The high copy number phenotype of pLO88 plasmid maintained in *Escherichia coli* (HB101) is observed only at elevated temperatures, (above 37° C), and is due to the precise position of a Tn5 insertion in DNA, but the exact mechanism remains obscure [36].

Recent experiments show [37] that artificial thermoresponsive devices may be constructed based on the temperature-dependence of the relative populations of left- and right-handed nucleic acid helical conformations. The authors reported that "upon an increase in temperature, particular sequences of DNA oligonucleotide duplexes in high salt conditions switch from a left-handed (Z) form to a right-handed (B) one, while RNA responds inversely by switching from a right- (A) to a left-handed (Z) form... Calculations revealed a complex interplay between configurational, water, and ionic entropies, which, combined with the sequence-dependence, rationalize the experimentally observed transitions from A- to Z-RNA and Z- to B-DNA in high salt concentrations and provide insight that may aid future developments of the use of nucleic acids oligomers for thermal sensing at the nanoscale in physiological conditions." [37]

The role of DNA-binding proteins has been established for plant thermosensitivity too. Kumar and Wigge [38] have revealed that eviction of the histone H2A.Z from nucleosomes performs a central role in plant thermosensory perception. Using purified nucleosomes, they showed that H2A.Z displays distinct responses to temperature in vivo, independently of transcription events.

Apparently, the temperature-induced conformational changes in DNA are mainly controlled through the presence of "nucleotid-associated" proteins, of which H–NS is the best characterized [32, 39]. In *E. coli*, creating and maintaining conformational structures in the DNA molecule are mainly regulated through the balance of two opposing topoisomerase activities, mainly those of topoisomerases II and I [40, 41]. The abovementioned examples of membrane- and nucleic acid-based temperature sensitivity imply that these systems often include proteins as a key regulatory component. Therefore, from the point of view of molecular temperature sensation, protein-based molecular "thermometers" represent an extremely interesting group.

25.4 Protein Thermometers

Many sensory pathways in living organisms use structural changes in proteins as a primary perceptive event, activating further signaling cascades. *E. coli* being exposed to an oxidative agent such as hydrogen peroxide, responds by the activation of a transcriptional regulator protein OxyR [42]. Activation of OxyR is achieved through the formation of a disulphide bound within the protein, upon which OxyR induces the expression of a set of genes adapting the bacterial cell to oxidative stress. This illustrates how it is possible both to "sense" and respond to an abrupt change in a specific environmental factor in a simple, yet elegant mode.

One would expect the organisms and cells to be similarly elegant when sensing temperature shifts. Indeed, a striking example is the temperature-controlled switching of the flagellar rotary motor of *E. coli* between the two rotational states, clockwise (CW) and counterclockwise (CCW) [43]. The molecular mechanism for switching remains unknown, but seems to be connected to the response regulator protein CheY-P. Two models of CheY-P action proposed so far explain shifting the difference in free energy between CW and CCW states in terms of (1) conformation-related differential binding [44, 45] and (2) thermodynamic changes in dissociation constants [46].

Further studies on the thermosensory transducing system in *E. coli* revealed that two major chemoreceptors, *Tar* and *Tsr*, which detect aspartate and serine, respectively, also function as thermoreceptors, together with *Trg* and *Tap* receptors [47]. Interestingly, in spite of different specificity and sensitivity, amino acid sequences of all these four chemoreceptors have a significant homology. They are transmembrane proteins having two functional domains acting as chemoreceptors: one is a ligand-binding domain located in the periplasm and the other is a signaling domain located in the cytoplasm. Thus, it is suggested that a temperature change induces a conformational change in these two receptors and that this conformational change triggers the signaling for thermoresponse. In the simplest model of thermoreception by these receptors, two conformational states of these receptors are assumed: a low-temperature state and a high-temperature state [48]. The swimming pattern of the *Trg*- and *Tap*-containing cells is determined simply by the temperature of the medium, indicating that these cells under nonadaptive conditions sense the absolute temperature as the thermal stimulus, and not the relative change in temperature.

The understanding of protein thermotropic sensory transductions in terms of their underlying molecular mechanism is fast-advancing thanks to the discovery and functional characterization of the *transient receptor potential* (TRP) channels. This protein family, first identified in *Drosophila*, is at the forefront of the sensory stem, responding to both physical and chemical stimuli and, thus having diverse functions [49, 50].

The family of TRP channels currently comprises around 30 members grouped into seven related subfamilies: TRPC, TRPV, TRPA, TRPP, TRPM, TRPN and TRPML. In higher organisms, TRPV channels are important polymodal integrators of noxious stimuli, mediating among all, thermosensation and nociception (pain sensation) [51].

To characterize thermal sensitivity of cells, molecules and processes, the Q_{10} (**temperature coefficient**) is used. Q_{10} reflects the rate of change of a biological or chemical system as a consequence of increasing the temperature by 10°C. This coefficient is used, for example, for the characterization of the nerve conduction velocity.

The Q_{10} is calculated as:

$$Q_{10} = \left(\frac{R_2}{R_1}\right)^{10/(T_2-T_1)}$$

where R is the rate of change and T is the temperature.

For biological systems, the Q_{10} value is generally between 1 and 3 but a subset of TRP channels, the thermo-TRPs, characterized by their unusually high temperature sensitivity (Q_{10} >10): TRPV1–TRPV4 are heat activated [52–54], whereas TRPM8 [54, 55] and TRPA1 [56] are activated by cold. With a Q_{10} of about 26 for TRPV1 [57] and about 24 for TRPM8 [58, 59], they far surpass the temperature dependence of the gating processes characterized by other ion channels ($Q_{10} \approx 3$) [57]. In spite of the great advances made in last years the molecular basis for regulation by temperature remains mostly obscure because of the lack of *native* structural information. Nevertheless, deeper understanding of dynamics and thermodynamics of these proteins will bring us closer to revelation of universal principles of thermal sensation.

25.5 Biophysical Aspects of Protein Thermosensitivity

It appears from the above mentioned examples of protein participation in temperature sensing events that sudden conformational changes, "structural transitions" play essential role on the primary conversion of physical stimulus into biologically relevant signal.

Phase transitions and other "critical" phenomena continue to be the subject of intensive experimental and theoretical investigation. In this context, systems consisting primarily of well characterized proteins and water can serve as particularly valuable objects of study. The importance of studies of specific phase transitions in protein/water solutions derives also from their physiological relevance to the supramolecular organization of normal tissues and to certain pathological states. For example, such phase transitions play the main role in the deformation of the erythrocyte in sickle-cell disease [25, 60] and in the cryoprecipitation of immunoglobulins in cryoglobulinemia and rheumatoid arthritis [61].

Discussions about protein stability and temperature-induced structural transitions are usually limited to the stability of the native state against denaturation. Yet the native state may include different functionally relevant conformations characterized by different Gibbs energies and therefore different stabilities (e.g. the R and T states of hemoglobin). At biological temperatures, proteins alternate between well-defined, distinct conformations. In order to those conformational states to be distinct, there must be a free-energy barrier separating them (Fig. 25.1). Adaptive alterations of protein conformation in response to signaling events might reflect corresponding changes in free-energy profile. From this point of view, temperature as a stimulus does not differ physically from, for example, a ligand-binding event. The experimental observation of distinct conformational populations by IR-spectroscopy is possible but usually requires the existence of at least two spectrally different but overlapping components of the amide I band [62].

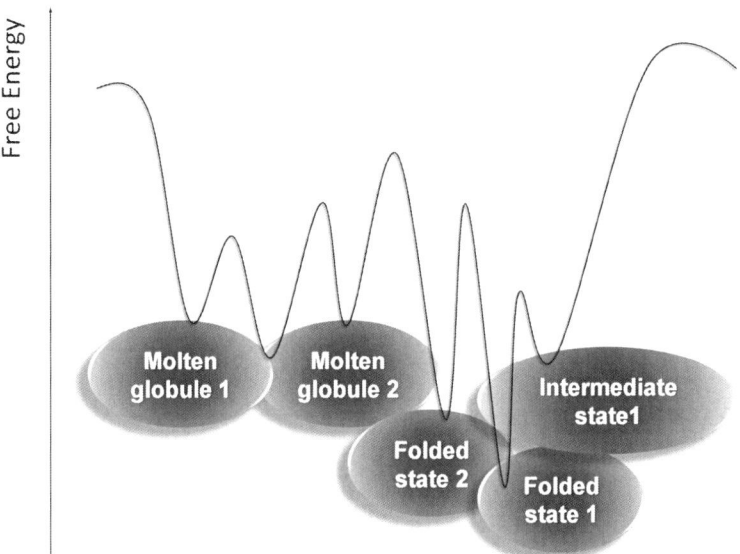

Fig. 25.1 A hypothetical model of a free energy profile of a globular protein. The threshold character of the protein reaction means that the resting states and the active states are different thermodynamic states of the system, separated by an energy barrier. Often, several protein states are thermodynamically stable and prevail in appropriate conditions. Transitions between different protein states take place in the cell in response to external stimuli

The motions involved to get from one state to another are usually much more complex than the oscillation of atoms and groups about their average positions. In proteins, because most of the forces that stabilize the native state are non-covalent, there is enough thermal energy at physiological temperature for weak interactions to break and reform frequently. Thus a protein molecule is more flexible than a molecule in which only covalent forces dictate the structure.

Recently, it became clear that natively unfolded proteins also play an important role in the cell. Dunker et al. [63] proposed to widen the notion of functional protein types in the cell: to the "classical" proteins with well defined tertiary structure, they added molten globules and proteins with unfolded conformations. Uversky [64] has suggested to supplement this list with a fourth, relatively stable protein conformation – the premolten globule, which might be called the boiling globule, as in the coordinates of the unfolding reaction it follows the globule and molten globule and precedes the completely unfolded conformation. Apparently, all these protein states are thermodynamically stable, although to different degrees.

Even when the native protein does not undergo a conformational change, it is still characterized by the occurrence of a large number of local unfolding events that give rise to many sub-states. Thus, the native state itself needs to be considered as a statistical ensemble of conformations rather than unique entity. These distinctions are very important from the functional point of view since different conformations are usually characterized by different functional properties.

The stabilizing contributions that arise from the hydrophobic effect and hydrogen bonding are largely offset by the destabilizing configurational entropy. The hydrophobic effect is strongly temperature-dependent, and is considerably weaker and perhaps even destabilizing at low temperatures than at elevated temperatures. The contribution of various interactions for a "typical" protein is reported in many works [65–69]. Apparently, the transition from stabilizing to destabilizing conditions is achieved by relatively small changes in the environment. These can be changes in temperature, pH, addition of substrates or stabilizing co-solvents. While the exact contribution of different interactions to the stability of globular proteins remains a question, our understanding seems to be refined enough to allow for the reasonable prediction of the overall folding thermodynamics applying the second law of thermodynamics for free energy changes between folded and unfolded states [68, 69]. Important to mention that both the enthalpy end entropy changes are not constant but increasing functions of temperature, and that the Gibbs (free) energy stabilization of a protein can be written as:

$$\Delta G(T) = \Delta H(T_R) + \Delta C_P(T - T_R) - T\Delta S(T_R) + \Delta C_P \ln(T/T_R)$$

where T_R is a reference temperature. ΔC_p is the heat capacity change, and $\Delta H(T_R)$ and $\Delta S(T_R)$ are the enthalpy and entropy values at that temperature, correspondingly. The temperature dependency of ΔH and ΔS is important because it transforms the Gibbs energy function from a linear into a parabolic function of temperature. This equation has only limited applicability since it does not consider the change of the solvent's entropy, which is, without doubt, an important contributor to the thermodynamics of protein behavior in solution.

For large values of ΔC_p, the Gibbs energy crosses zero point twice – one at high temperature (heat denaturation) and one at low temperature (cold denaturation). The native state is thermodynamically stable between those two temperatures and ΔG exhibits a maximum at the temperature at which $\Delta S = 0$. The peculiar shape of the free energy function of a protein does not permit a unique definition of protein stability. For example, having a higher denaturation temperature does not necessarily imply that a protein will be more stable at room temperature.

Within the context of the structural parameterization of the energetics, the free energy of protein stabilization is approximated by the equation:

$$\Delta G = \Delta G_{gen} + \Delta G_{ion} + \Delta G_{tr} + \Delta G_{other}$$

where ΔG_{gen} contains the contributions typically associated with the formation of secondary and tertiary structure (van der Waals interactions, hydrogen bonding, hydration, and conformational entropy), ΔG_{ion} comprises the electrostatic and ionization effects, and ΔG_{tr} reflects the contribution of the change in translational degrees of freedom existing in oligomeric proteins. The term ΔG_{other} includes interactions unique to specific proteins that cannot be classified in a general way (e.g. prosthetic groups, metals, and ligands) and must be treated on a case-by-case basis.

B. Nilius and co-workers have recently applied this simple thermodynamic formalism to describe the shifts in voltage dependence of protein channels due to changes in temperature [70, 71], where the probability of the opening of a channel is defined as a function of temperature, Faraday's constant, the gating charge, and the free energy difference between open and closed states of the channel.

In order to understand the nature of dynamic transitions in proteins, it is also important to consider solvent effects. Solvent can affect protein dynamics by modifying the effective characteristics of the protein surface and/or by frictional damping. Changes in the structure and internal dynamics of proteins in dependency on solvent conditions at physiological temperatures have been found by using several experimental techniques [72–74]. It follows from the works of G. Büldt, G. Artmann, J. Zaccai, A. Stadler and others that solvent affects protein dynamics differently at different temperatures and salt concentrations [74–77] Therefore, a solvent dependence of the dynamic transition might be expected. Indeed, measurements on ligand binding to myoglobin indicated that dynamic behavior of the protein is correlated with a glass transition in the surrounding solvent [78]. Recent molecular dynamics analysis of hydrated myoglobin also indicates a major solvent role in protein dynamic transition behavior [79].

One interesting aspect of thermosensation is conversion of the code conformational changes into the code of cellular signaling. In our opinion, strong methodological basis for the understanding of these events was provided by studies by the scientific schools of Dmitrii Nasonov and Gilbert Ling that have gained new appreciation over the last 20–30 years owing to advances in protein physics, and thank to series of works by Vladimir Matveev [80, 81]. The latter has postulated that when an action (for instance, temperature change) on a cell or a cell structure exceeds the threshold, (i) formation of secondary structures begins in natively unfolded proteins (or unfolded regions of proteins), while (ii) secondary structures of molten globules start to become accessible for interaction with secondary structures of other proteins and with nucleic acids. Such secondary structures induced by the external action were called by the author *centers (sites) of native aggregation* (personal communication). Thus, the first event in the activated cell is the appearance of new secondary structures able to interact selectively with each other to form tertiary, quaternary, etc. structures. Proteins whose secondary structures appear under such circumstances lose their previous inertia and become reaction-capable.

This point of view on understanding the mechanisms of cellular reactions poses the question of native and denatured protein states in a new way. According to it, in the native state the key cell proteins are inert, non-reaction-capable; they do not interact with each other or with other biopolymers. Loss of the state of inertia is partial denaturation, where new secondary structures can appear, or can be modified, or can "float up" to the surface from the hydrophobic nucleus. In all cases the secondary structures are ready to interact. Numerous intermediate protein species, corresponding to different free energy minima provide basis for native aggregation.

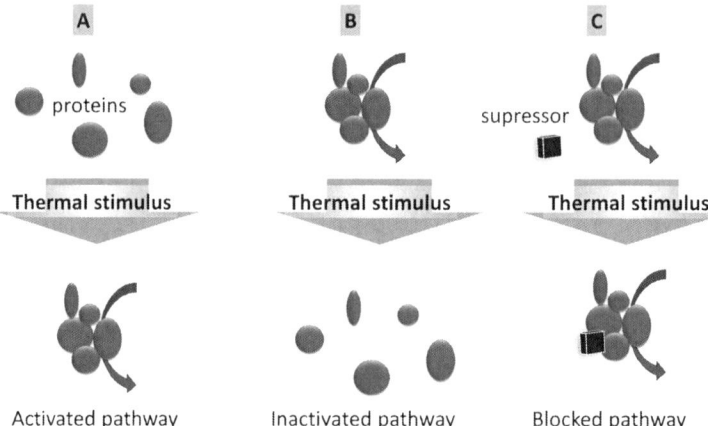

Fig. 25.2 Some possible strategies in converting protein structural information (shape, hydrophobicity, charge, domain organization) into cellular signaling events. Thermal stimulus provides energy for transferring protein molecule from the one energy state to the other, which results in changing protein surface. The induced appearance/disappearance of recognition sites on the protein surface leads to establishing inter- and intramolecular protein contacts. These native protein aggregation/disaggregation events can be interpreted as a key mechanism of signal propagation in the cell

Together with native aggregation, several other possibilities can be visualized to explain how thermo-induced conformational changes in proteins can be converted into signaling event (Fig. 25.2).

Kim et al. [82] and later Sourjik [83] studied the dynamics of the cytoplasmic domains of the *E. coli* chemotaxis receptor on interaction with repellent and attractant. It was concluded that an attractant decreases the number of secondary structures in the domain, which blocks signal transmission into the cytoplasm. A repellent produces the opposite effect: it increases the amount of secondary structures in the domain, which makes the signal function of the receptor possible. In terms of the hypothesis of native aggregation, repellent converts the domain into the excited state, enabling interactions necessary for signal transmission.

Since native aggregation results in the appearance of signaling and regulatory structures, it is obvious that as biological organization becomes more complicated during evolution, where novel mechanisms of regulation of the cell activity are needed.

25.6 Structural Features of Protein Thermometers

From the point of view of structural biophysics, thermosensation can be regarded a special case of mechanosensation and therefore many theoretical models and considerations developed for protein mechanosensors are also applicable for thermosensors. The difference between mechanosensitive channels and thermosensitive molecules is only the size and the organization of the "exciting" agents – a lot of non-coordinated events (thermal stimuli) versus a net stretch (mechanical stimuli). Therefore, nor surprisingly, many members of thermosensing TRPV family are also known as osmo- and mechano-sensors. Because mechanical stimuli are omnipresent, mechanosensation could represent one of the oldest sensory transduction processes that have evolved in living organisms. Similarly to thermal sensors, what exactly makes these channels respondent to membrane tension is unclear. The answer will not be simple because both thermal- and mechano-sensors are very diverse [84, 85]. However, there are interesting parallels in structural composition of different classes of known temperature-sensory proteins, pending comprehension.

Despite significant evolutionary distances and apparent differences of primary structure, all temperature-sensitive proteins known so far display some remarkable similarities in their tertiary/quaternary structure. The ability of a big protein TlpA responsible in *S. typhimurium* for temperature regulation of transcription undoubtedly resides in its peculiar structural design. Two-thirds of the C-terminal portion of TlpA is folded in an alpha-helical-coiled-coil structure that constitutes an oligomerization domain. The sensory capacity is consealed in this coiled-coil structure, which illustrates the means of sensing temperature through changes in protein conformation. As the temperature increases, the proportion of DNA-binding oligomers decreases, leading to a de-repression of the target gene. At moderate temperatures, the concentration of TlpA increases, shifting the balance to the formation of DNA-binding oligomers and, in part, restoring the repression potential of TlpA. Thus, TlpA undergoes a reversible conformational shift in response to temperature alteration, leading to an alteration in the oligomeric structure [48].

The coiled-coil protein structure is a very versatile and a flexible motif in mediating protein-protein interactions. In vertebrates, the thermosensitive elements of transcriptional mechanism typically contain such coiled-coil folding motifs, like those in leucine zipper family.

TRPV channel subunits have a common topology of six transmembrane segments (S1–S6) with a pore region between the fifth and sixth segment, and cytoplasmic N- and C-termini. In these thermo-TRP channels, it has been proposed that the structural rearrangement leads to a change in tension on the helical linker connecting the C-terminal domains with S6 segment. This tension on the linker provides the energy necessary to move the S6 inner helix to the open conformation [58, 59]. Indeed, partial deletions performed in the C-terminal domain of TRPV1 resulted in functional channels with attenuated heat sensitivity, whereas truncation of the whole TRPV1 C-terminal domain completely hindered channel expression [57]. Another possibility could be that temperature affects the interaction between a particular portion of the proximal C-terminal and some other region of the channel, probably an intracellular loop. Finally, it might be that independent arrangements induced by temperature on C-terminal domains directly promote gate opening [57].

Bernd Nilius' group in their study on the voltage dependence of TRP channel gating by temperature pointed out that the small gating charge of TRP channels compared to that of classical voltage-gated channels could lie at the basis of the large shifts of their voltage-dependent activation curves, and may be essential for their gating versatility [70, 71]. Thus, small changes of the free energy of activation of these channels can result in large shifts of their voltage-dependent activation curves, and concomitant gating of these channels.

In membrane, TRP channels assemble into tetramers of identical subunits [86]. Recently obtained data indicate that the homo-oligomer, modular nature of the structures involved in activation processes allow different stimuli (voltage, temperature, and agonists) to promote thermo-TRP channel opening by different interrelated mechanisms, for example, in the form of allosteric interaction [58, 59, 87].

The very interesting aspect resides in the observation that some bacterial proteins like H–NS and StpA may form not only homo-oligomers but also hetero-oligomers exactly the same way as TRPV thermosensory channels of higher animals sometimes do [32, 58]. In this context, it is important to note that the temperature-sensitive H–NS function is also associated with oligomerization and that the H–NS oligomerization domain most evidently relies on the formation of coiled-coil oligomers [33, 75].

Together with polymerization, an interesting but still pending problem is modulation of thermotropic reactions by low-molecular weight compounds. A common feature shared by many TRPM8 channels is binding of phosphatidylinositol bisphosphate (PIP2), that leads to channel activation [88]. Binding and activation by capsaicin, ADP-ribose, menthol, eucalyptol etc. are classical examples of polymodality of temperature-sensitive proteins but the field still lacks systematical study. A pool of these and related questions will be generally addressed by a quickly developing discipline, *chemical genetics*, whose subject can be defined as "a selective interaction of a small molecule with a protein that may be regarded as functionally equivalent to mutation of the protein" [89].

The molecular dynamics and organization of the temperature-sensing proteins signaling complexes are still elusive, although fast-advancing progress in this arena is uncovering the molecular identity of these elements. A series of papers published by G. Artmann and coworkers, revealed intriguing temperature-related structural transitions phenomena in hemoglobins (Hb) and myoglobins of different species [65, 90, 91]. The reported non-linearity in hemoglobin temperature behavior is determined by physiological body temperature of the given species, is strongly influenced by many small molecules (ATP, PIP2 etc.) and therefore might surprisingly imply the role of Hb as a molecular thermometer [92].

Acknowledgments The author thanks Prof. Dr. Georg Büldt (Research Center Juelich, Germany), Dr. Prof. G. Artmann (AcUAS, Aachen, Germany), Dr. A. Stadler (Research Center Juelich, Germany), Dr. V. Matveev (Institute of Cytology, Sankt-Petersburg, Russia) and Dr. G. Zaccai (ILL, Grenoble, France) for many fruitful and inspirational discussions.

References

1. Yamanaka K, Fang L, Inouye M (1998) The CspA family in Escherichia coli: multiple gene duplication for stress adaptation. Mol Microbiol 27:247–255
2. Narberhaus F, Waldminghaus T, Chowdhury S (2006) RNA thermometers. FEMS Microbiol Rev 30:3–16

3. Hurme R, Rhen M (1998) Temperature sensing in bacterial gene regulation–what it all boils down to. Mol Microbiol 30:1–6
4. Ramos L (2001) Scaling with temperature and concentration of the nonlinear rheology of a soft hexagonal phase. Phys Rev E Stat Nonlin Soft Matter Phys 64:061502
5. Lemaux PG, Herendeen SL, Bloch PL, Neidhardt FC (1978) Transient rates of synthesis of individual polypeptides in E. coli following temperature shifts. Cell 13:427–434
6. Hazel JR (1995) Thermal adaptation in biological membranes: is homeoviscous adaptation the explanation? Annu Rev Physiol 57:19–42
7. Al Fageeh MB, Smales CM (2006) Control and regulation of the cellular responses to cold shock: the responses in yeast and mammalian systems. Biochem J 397:247–259
8. Maurelli AT, Sansonetti PJ (1988) Identification of a chromosomal gene controlling temperature-regulated expression of Shigella virulence. Proc Natl Acad Sci USA 85: 2820–2824
9. Straley SC, Perry RD (1995) Environmental modulation of gene expression and pathogenesis in Yersinia. Trends Microbiol 3:310–317
10. Collins AC, Smolen A, Wayman AL, Marks MJ (1984) Ethanol and temperature effects on five membrane bound enzymes. Alcohol 1:237–246
11. Ferrer-Montiel A, Garcia-Martinez C, Morenilla-Palao C, Garcia-Sanz N, Fernandez-Carvajal A, Fernandez-Ballester G, Planells-Cases R (2004) Molecular architecture of the vanilloid receptor. Insights for drug design. Eur J Biochem 271:1820–1826
12. Ginsberg L, Gilbert DL, Gershfeld NL (1991) Membrane bilayer assembly in neural tissue of rat and squid as a critical phenomenon: influence of temperature and membrane proteins. J Membr Biol 119:65–73
13. Mandal M, Breaker RR (2004) Gene regulation by riboswitches. Nat Rev Mol Cell Biol 5:451–463
14. Oppenheimer CH, Drost-Hansen W (1960) A relationship between multiple temperature optima for biological systems and the properties of water. J Bacteriol 80:21–24
15. Drost-Hansen W (2001) Temperature effects on cell-functioning–a critical role for vicinal water. Cell Mol Biol (Noisy-le-grand) 47:865–883
16. Vigh L, Maresca B, Harwood JL (1998) Does the membrane's physical state control the expression of heat shock and other genes? Trends Biochem Sci 23:369–374
17. Aguilar PS, Cronan JE Jr., de Mendoza D (1998) A Bacillus subtilis gene induced by cold shock encodes a membrane phospholipid desaturase. J Bacteriol 180:2194–2200
18. Aguilar PS, Hernandez-Arriaga AM, Cybulski LE, Erazo AC, de Mendoza D (2001) Molecular basis of thermosensing: a two-component signal transduction thermometer in Bacillus subtilis. EMBO J 20:1681–1691
19. Suzuki T, Nakayama T, Kurihara T, Nishino T, Esaki N (2001) Cold-active lipolytic activity of psychrotrophic Acinetobacter sp. strain no. 6. J Biosci Bioeng 92:144–148
20. Dutta R, Qin L, Inouye M (1999) Histidine kinases: diversity of domain organization. Mol Microbiol 34:633–640
21. Andersen J, Forst SA, Zhao K, Inouye M, Delihas N (1989) The function of micF RNA. micF RNA is a major factor in the thermal regulation of OmpF protein in Escherichia coli. J Biol Chem 264:17961–17970
22. Lease RA, Belfort M (2000) A trans-acting RNA as a control switch in Escherichia coli: DsrA modulates function by forming alternative structures. Proc Natl Acad Sci USA 97: 9919–9924
23. Mandal M, Breaker RR (2004) Adenine riboswitches and gene activation by disruption of a transcription terminator. Nat Struct Mol Biol 11:29–35
24. Winkler WC, Breaker RR (2005) Regulation of bacterial gene expression by riboswitches. Annu Rev Microbiol 59:487–517
25. Brunel C, Romby P, Sacerdot C, de Smit M, Graffe M, Dondon J, van Duin J, Ehresmann B, Ehresmann C, Springer M (1995) Stabilised secondary structure at a ribosomal binding site enhances translational repression in E. coli. J Mol Biol 253:277–290

26. Kaempfer R (2003) RNA sensors: novel regulators of gene expression. EMBO Rep 4: 1043–1047
27. Tucker BJ, Breaker RR (2005) Riboswitches as versatile gene control elements. Curr Opin Struct Biol 15:342–348
28. Johansson J, Mandin P, Renzoni A, Chiaruttini C, Springer M, Cossart P (2002) An RNA thermosensor controls expression of virulence genes in Listeria monocytogenes. Cell 110:551–561
29. Morita MT, Tanaka Y, Kodama TS, Kyogoku Y, Yanagi H, Yura T (1999) Translational induction of heat shock transcription factor sigma32: evidence for a built-in RNA thermosensor. Genes Dev 13:655–665
30. Nocker A, Hausherr T, Balsiger S, Krstulovic NP, Hennecke H, Narberhaus F (2001) A mRNA-based thermosensor controls expression of rhizobial heat shock genes. Nucleic Acids Res 29:4800–4807
31. Yamanaka K (1999) Cold shock response in Escherichia coli. J Mol Microbiol Biotechnol 1:193–202
32. Dorman CJ (1996) Flexible response: DNA supercoiling, transcription and bacterial adaptation to environmental stress. Trends Microbiol 4:214–216
33. Dorman CJ, Hinton JC, Free A (1999) Domain organization and oligomerization among H-NS-like nucleoid-associated proteins in bacteria. Trends Microbiol 7:124–128
34. Dorman CJ (1991) DNA supercoiling and environmental regulation of gene expression in pathogenic bacteria. Infect Immun 59:745–749
35. Grau R, Gardiol D, Glikin GC, de Mendoza D (1994) DNA supercoiling and thermal regulation of unsaturated fatty acid synthesis in Bacillus subtilis. Mol Microbiol 11:933–941
36. Lupski JR, Projan SJ, Ozaki LS, Godson GN (1986) A temperature-dependent pBR322 copy number mutant resulting from a Tn5 position effect. Proc Natl Acad Sci USA 83: 7381–7385
37. Wereszczynski J, Andricioaei I (2010) Conformational and solvent entropy contributions to the thermal response of nucleic acid-based nanothermometers. J Phys Chem B 114:2076–2082
38. Kumar SV, Wigge PA (2010) H2A.Z-containing nucleosomes mediate the thermosensory response in Arabidopsis. Cell 140:136–147
39. Williams RM, Rimsky S (1997) Molecular aspects of the E. coli nucleoid protein, H-NS: a central controller of gene regulatory networks. FEMS Microbiol Lett 156:175–185
40. Drlica K (1992) Control of bacterial DNA supercoiling. Mol Microbiol 6:425–433
41. Tse-Dinh YC, Qi H, Menzel R (1997) DNA supercoiling and bacterial adaptation: thermotolerance and thermoresistance. Trends Microbiol 5:323–326
42. Carmel-Harel O, Storz G (2000) Roles of the glutathione- and thioredoxin-dependent reduction systems in the Escherichia coli and saccharomyces cerevisiae responses to oxidative stress. Annu Rev Microbiol 54:439–461
43. Turner L, Samuel AD, Stern AS, Berg HC (1999) Temperature dependence of switching of the bacterial flagellar motor by the protein CheY(13DK106YW). Biophys J 77:597–603
44. Alon U, Camarena L, Surette MG, Arcas B, Liu Y, Leibler S, Stock JB (1998) Response regulator output in bacterial chemotaxis. EMBO J 17:4238–4248
45. Monod J, Wyman J, Changeux JP (1965) On the nature of allosteric transitions: a plausible model. J Mol Biol 12:88–118
46. Scharf BE, Fahrner KA, Turner L, Berg HC (1998) Control of direction of flagellar rotation in bacterial chemotaxis. Proc Natl Acad Sci USA 95:201–206
47. Nara T, Lee L, Imae Y (1991) Thermosensing ability of Trg and Tap chemoreceptors in Escherichia coli. J Bacteriol 173:1120–1124
48. Eriksson S, Hurme R, Rhen M (2002) Low-temperature sensors in bacteria. Philos Trans R Soc Lond B Biol Sci 357:887–893
49. Harteneck C, Plant TD, Schultz G (2000) From worm to man: three subfamilies of TRP channels. Trends Neurosci 23:159–166
50. Montell C (2005) The TRP superfamily of cation channels. Sci STKE 2005:re3

51. Yang XR, Lin MJ, Sham JS (2010) Physiological functions of transient receptor potential channels in pulmonary arterial smooth muscle cells. Adv Exp Med Biol 661:109–122
52. Caterina MJ (2007) Transient receptor potential ion channels as participants in thermosensation and thermoregulation. Am J Physiol Regul Integr Comp Physiol 292:R64–R76
53. Moran MM, Xu H, Clapham DE (2004) TRP ion channels in the nervous system. Curr Opin Neurobiol 14:362–369
54. Peier AM, Reeve AJ, Andersson DA, Moqrich A, Earley TJ, Hergarden AC, Story GM, Colley S, Hogenesch JB, McIntyre P, Bevan S, Patapoutian A (2002) A heat-sensitive TRP channel expressed in keratinocytes. Science 296:2046–2049
55. McKemy DD (2005) How cold is it? TRPM8 and TRPA1 in the molecular logic of cold sensation. Mol Pain 1:16
56. Story GM, Peier AM, Reeve AJ, Eid SR, Mosbacher J, Hricik TR, Earley TJ, Hergarden AC, Andersson DA, Hwang SW, McIntyre P, Jegla T, Bevan S, Patapoutian A (2003) ANKTM1, a TRP-like channel expressed in nociceptive neurons, is activated by cold temperatures. Cell 112:819–829
57. Liu B, Hui K, Qin F (2003) Thermodynamics of heat activation of single capsaicin ion channels VR1. Biophys J 85:2988–3006
58. Brauchi S, Orio P, Latorre R (2004) Clues to understanding cold sensation: thermodynamics and electrophysiological analysis of the cold receptor TRPM8. Proc Natl Acad Sci USA 101:15494–15499
59. Brauchi S, Orta G, Salazar M, Rosenmann E, Latorre R (2006) A hot-sensing cold receptor: C-terminal domain determines thermosensation in transient receptor potential channels. J Neurosci 26:4835–4840
60. Huang HW (1976) Allosteric linkage and phase transition. Physiol Chem Phys 8:143–150
61. DeLucas LJ, Moore KM, Long MM (1999) Protein crystal growth and the International Space Station. Gravit Space Biol Bull 12:39–45
62. Leeson DT, Gai F, Rodriguez HM, Gregoret LM, Dyer RB (2000) Protein folding and unfolding on a complex energy landscape. Proc Natl Acad Sci USA 97:2527–2532
63. Dunker AK, Oldfield CJ, Meng J, Romero P, Yang JY, Chen JW, Vacic V, Obradovic Z, Uversky VN (2008) The unfoldomics decade: an update on intrinsically disordered proteins. BMC Genomics 9(Suppl 2):S1
64. Uversky VN (2010) The mysterious unfoldome: structureless, underappreciated, yet vital part of any given proteome. J Biomed Biotechnol 2010:568068
65. Artmann GM, Kelemen C, Porst D, Buldt G, Chien S (1998) Temperature transitions of protein properties in human red blood cells. Biophys J 75:3179–3183
66. Ip SH, Ackers GK (1977) Thermodynamic studies on subunit assembly in human hemoglobin. Temperature dependence of the dimer-tetramer association constants for oxygenated and unliganded hemoglobins. J Biol Chem 252:82–87
67. Valdes R Jr., Ackers GK (1977) Thermodynamic studies on subunit assembly in human hemoglobin. Calorimetric measurements on the reconstitution of oxyhemoglobin from isolated chains. J Biol Chem 252:88–91
68. Bowler BE (2007) Thermodynamics of protein denatured states. Mol Biosyst 3:88–99
69. Levy Y, Onuchic JN (2006) Water mediation in protein folding and molecular recognition. Annu Rev Biophys Biomol Struct 35:389–415
70. Nilius B, Talavera K, Owsianik G, Prenen J, Droogmans G, Voets T (2005) Gating of TRP channels: a voltage connection? J Physiol 567:35–44
71. Nilius B, Owsianik G, Voets T, Peters JA (2007) Transient receptor potential cation channels in disease. Physiol Rev 87:165–217
72. Rodger A, Marrington R, Geeves MA, Hicks M de Alwis, L, Halsall, DJ, Dafforn, TR (2006) Looking at long molecules in solution: what happens when they are subjected to Couette flow? Phys Chem Phys 8:3161–3171
73. Urry DW (1988) Entropic elastic processes in protein mechanisms. I. Elastic structure due to an inverse temperature transition and elasticity due to internal chain dynamics. J Protein Chem 7:1–34

74. Artmann GM, Digel I, Zerlin KF, Maggakis-Kelemen C, Linder P, Porst D, Kayser P, Stadler AM, Dikta G, Temiz AA (2009) Hemoglobin senses body temperature. Eur Biophys J 38: 589–600
75. Smith JC, Merzel F, Bondar AN, Tournier A, Fischer S (2004) Structure, dynamics and reactions of protein hydration water. Philos Trans R Soc Lond B Biol Sci 359:1181–1189
76. Stadler AM, Digel I, Embs JP, Unruh T, Tehei M, Zaccai G, Buldt G, Artmann GM (2009) From powder to solution: hydration dependence of human hemoglobin dynamics correlated to body temperature. Biophys J 96:5073–5081
77. Zaccai G (2004) The effect of water on protein dynamics. Philos Trans R Soc Lond B Biol Sci 359:1269–1275
78. Doster W (2010) The protein-solvent glass transition. Biochim Biophys Acta 1804:3–14
79. Muthuselvi L, Dhathathreyan A (2009) Understanding dynamics of myoglobin in heterogeneous aqueous environments using coupled water fractions. Adv Colloid Interface Sci 150:55–62
80. Matveev VV (2005) Protoreaction of protoplasm. Cell Mol Biol (Noisy-le-grand) 51:715–723
81. Matveev VV, Wheatley DN (2005) "Fathers" and "sons" of theories in cell physiology: the membrane theory. Cell Mol Biol (Noisy-le-grand) 51:797–801
82. Kim SH, Wang W, Kim KK (2002) Dynamic and clustering model of bacterial chemotaxis receptors: structural basis for signaling and high sensitivity. Proc Natl Acad Sci USA 99:11611–11615
83. Sourjik V (2004) Receptor clustering and signal processing in E. coli chemotaxis. Trends Microbiol 12:569–576
84. Hamill OP, Martinac B (2001) Molecular basis of mechanotransduction in living cells. Physiol Rev 81:685–740
85. Sachs F, Morris CE (1998) Mechanosensitive ion channels in nonspecialized cells. Rev Physiol Biochem Pharmacol 132:1–77
86. Kedei N, Szabo T, Lile JD, Treanor JJ, Olah Z, Iadarola MJ, Blumberg PM (2001) Analysis of the native quaternary structure of vanilloid receptor 1. J Biol Chem 276:28613–28619
87. McKemy DD (2007) Temperature sensing across species. Pflugers Arch 454:777–791
88. Rohacs T, Nilius B (2007) Regulation of transient receptor potential (TRP) channels by phosphoinositides. Pflugers Arch 455:157–168
89. Stockwell BR (2000) Frontiers in chemical genetics. Trends Biotechnol 18:449–455
90. Artmann GM, Burns L, Canaves JM, Temiz-Artmann A, Schmid-Schonbein GW, Chien S, Maggakis-Kelemen C (2004) Circular dichroism spectra of human hemoglobin reveal a reversible structural transition at body temperature. Eur Biophys J 33:490–496
91. Digel I, Maggakis-Kelemen C, Zerlin KF, Linder P, Kasischke N, Kayser P, Porst D, Temiz AA, Artmann GM (2006) Body temperature-related structural transitions of monotremal and human hemoglobin. Biophys J 91:3014–3021
92. Zerlin KF, Kasischke N, Digel I, Maggakis-Kelemen C, Temiz AA, Porst D, Kayser P, Linder P, Artmann GM (2007) Structural transition temperature of hemoglobins correlates with species' body temperature. Eur Biophys J 37:1–10

Chapter 26
Thermo-TRP Channels: Biophysics of Polymodal Receptors

David Baez-Nieto, Juan Pablo Castillo, Constantino Dragicevic, Osvaldo Alvarez, and Ramon Latorre

Abstract In this chapter we discuss the polymodal activation of thermo-TRP channels using as exemplars two of the best characterized members of this class of channels: TRPM8 and TRPV1. Since channel activation by temperature is the hallmark of thermo-TRP channels, we present a detailed discussion on the thermodynamics involved in the gating processes by temperature, voltage, and agonists. We also review recently published data in an effort to put together all the pieces available of the amazing puzzle of thermo-TRP channel activation. Special emphasis is made in the structural components that allow the channel-forming proteins to integrate such diverse stimuli, and in the coupling between the different sensors and the ion conduction pathway. We conclude that the present data is most economically explained by allosteric models in which temperature, voltage, and agonists act separately to modulate channel activity.

26.1 Introduction

Transient Receptor Potential (TRP) channels play important roles in sensory transduction from insects to mammals. TRP channels conform a superfamily consisting of seven subfamilies with little homology between them. The seven subfamilies include: the classical TRP subfamily (TRPC), the melastatin related subfamily (TRPM), the vanilloid-sensitive TRP subfamily (TRPV), the ankyrin subfamily (TRPA), the polycystin subfamily (TRPP), the mucolipin subfamily (TRPML), and the TRPN subfamily, named after the non mechanoreceptor potential C (nonpC) homologue [1]. Bioinformatic analyses based on primary structure of diverse sequences of TRP and topological predictions, suggest that these channels are

---

D. Baez-Nieto (✉)
Centro Interdisciplinario de Neurociencias de Valparaíso, Facultad de Ciencias, Universidad de Valparaíso, Valparaíso, Chile
e-mail: monobolico@gmail.com

tetramers, and that each monomer possess a 6 TM[1] region similar in conformation to that of the canonical voltage-activated potassium channel, Kv's [2]. The transmembrane segments in Kv channels are divided in two modules: the voltage sensor module from TM1 to TM4, and the pore module, which includes the TM5 and TM6. In Kv channels, the first module senses the membrane potential through highly conserved basic residues in TM4 (S4) that respond to changes in the membrane electric field. The pore module also has a highly conserved zone, called the selectivity filter. This selectivity filter has a signature sequence, GYGD, which is formed by a portion of the loop between TM5-TM6. This sequence is homologous amongts all potassium channels and determines the exquisite selectivity that these channels have for the K^+ ion [3, 4].

The voltage sensor domain of TRP channels, on the other hand, remains elusive, even though some published data [5] indicate that positive residues in the fourth segment in TRPM8 play a role in channel voltage sensitivity (see below). For example, it is puzzling that TRPA1, which behaves like an outward rectifier lacks positively charged residues in its TM4 [6]. The TRP pore module and the permeation properties of these channels are, however, much better understood than are the voltage sensor domain module. The pore loop in TRP channels, which is similar to Kv channels's, also has a consensus sequence[2] that differs among TRP families. Mutations in this sequence and in TM6 greatly modify the channel conduction properties [7, 8]. TRP channels are non selective cationic channels that show, in some cases, a higher relative permeability to calcium than to other ions [9, 10]. Within the TRP channels families there are outward rectifiers as well as inward rectifiers. Thus for instance, in the TRPV subfamily, TRPV1-4 are outward rectifiers, whereas TRPV5-6 are inward rectifiers, this despite the fact that homology along the TRPV family is very high. As shown in Fig. 26.1, alignment of the selectivity filter sequences shows a significant identity in the P-loop and TM6 sequences. The properties of permeation and selectivity are discussed below, including those relating to pore dilation in TRPV1 channels mediated by agonists.

Mammalian TRP channels are polymodal receptors and, as such, are targets of very diverse pungent and other chemical compounds [11–15]. Temperature-sensitive TRP channels are the best example of TRP channels that are able to sense myriad of chemical stimuli, including capsaicin (the active component of chili peppers), menthol, camphor, allicin (one of the active components of garlic), mustard oil, with the list increasing every day. Although the receptors for some of these agonists have been identified (see below), it will be a formidable task from a biophysical and structural point of view to elucidate the mechanisms by which the different sensors are coupled to the gate domains that control channel ion permeability. It is,

---

[1]TM: Transmembrane domain, helixes spanning the lipid bilayer membrane.
[2]In the TRPV subfamily the signature sequence is $TIGX_1GX_2$ (X_1 = M or L; X_2 = D or E) for TRPV1-4 and FLTXID (X = V or I) for TRPV5-6.

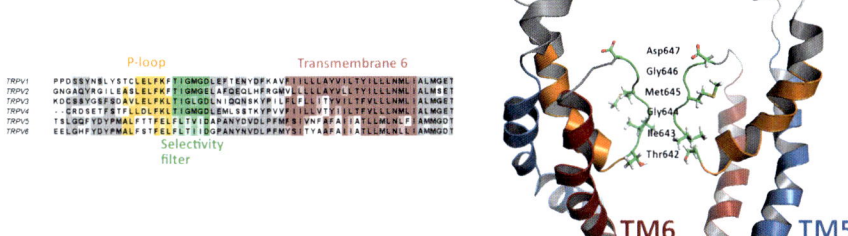

Fig. 26.1. Predicted permeation pore helix and selectivity filter of human TRPV subfamily. *Left panel* shows a partial alignment of the primary structure of members of the TRPV family including the permeation zone, selectivity filter (*light green*) and P-loop (*orange*), transmembrane domain 6 (TM6) (*light red*). The transmembrane domains were predicted by Conpred II [88], bioinformatic tool that uses a consensus prediction method. Alignment was performed by consensus using ClustalW (http://services.uniprot.org/clustalw/clustalw.html) [89] and T-coffee (http://www.ebi.ac.uk/Tools/t-coffee/index.html) [90] web servers. The sequences used in the alignment were obtained from Uniprot database. TRPV1: Q8NER1; TRPV2: Q9Y5S1; TRPV3: Q8NET8; TRPV4: Q9HBA0; TRPV5: Q9NQA5; TRPV6: Q9H1D0. *Right*. Molecular model of the human TRPV1 pore domain ([91]; coordinates drawn using PyMol molecular graphics program). The different regions are colored to match the color code used in the *left panel*. The selectivity filter residues (TIGMGD) are highlighted in sticks representation

however, the synergistic nature of channel activation that makes TRP channels such an attractive subject for the study of gating mechanisms.

If it is assumed that voltage, temperature, and agonists act separately to modulate channel activity, then TRP channel activation can be explained most economically using allosteric mechanisms for voltage, temperature, and agonist dependence of the type first proposed for hemoglobin and other allosteric proteins [16, 17]. A key element of an allosteric mechanism is the reciprocal interaction between gate and sensor. In the case of TRP channels, if temperature or voltage affect channel opening, then channel opening will also affect temperature- and voltage-sensor equilibria, with the magnitude of this effect dependent on the degree of coupling between different sensors.

The purpose of this chapter is to guide the reader through some aspects of the biophysics of TRP channels, taking as a case study one representative from the vanilloid family, TRPV1 [18], and one from the melastatin family, TRPM8 [19, 20]. Because these two ion channels are members of a class of exquisitely temperature sensitive channels, we start by discussing some basic thermodynamic principles. Due to space limitations, we will discuss in detail only the voltage and temperature dependence of these two channels trying to relate, whenever possible, the channel properties with specific protein domains. We end the chapter discussing allosteric models that provide at present a simple and plausible physical conception of the mechanical steps involved in TRP channel gating.

26.2 Energetics of Thermal Activation

When we speak of the temperature of an object, we often associate this concept with the degree of "hotness" or "coldness" of the object when we touch it. Thus our senses provide us with a qualitative indication of temperature. In mammals, nature has developed various types of molecular devices, ion channels, for sensing temperature with an exquisite sensitivity and within well defined temperature ranges. Since these devices are the main subject of the present chapter, we begin by giving some basic thermodynamic concepts.

Every thermodynamic system tends to a state of minimum energy, and proteins that form ion channels spend most of their time in stable conformations because, as depicted in Fig. 27.2, those are local minima in their energy landscape. Since temperature is an intensive measure of the average speed of atoms and molecules, everything is subject to thermal fluctuations. This is the reason why proteins are never fixed in a particular stable state. Random movement of the atoms around their equilibrium points may drive the protein to the transition state, from where it falls towards one or another conformation.

In a closed-to-open (Fig. 26.2) voltage-dependent channel transition the energy difference is given by the standard Gibbs free energy,

$$\Delta G^0 = \Delta H - T\Delta S - z\delta FV \qquad (26.1)$$

where ΔH is the difference of enthalpies between closed and open states (thermal energy absorbed by the system at constant pressure), T is absolute temperature (Kelvin scale), ΔS is difference of entropies between closed and open states (related to the number of accessible states increased by the transition), z is the gating charge

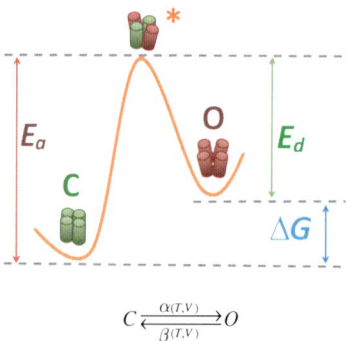

Fig 26.2. Energy barrier profile. E_a is the activation energy and E_d is the deactivation energy. ΔG represents the intrinsic configuration energy difference between the closed and the open state in the absence of an applied field. In the kinetic model C and O denotes the closed and the open state, respectively. The voltage- and temperature-dependent $\alpha(T,V)$ and $\beta(T,V)$ are the voltage and temperature-dependent forward (activation) and backwards (deactivation) rate constants, respectively

(dimensionless number of elemental charges needed to be displaced in order to open or to close the channel), F the faraday constant (the charge of one mol of electrons), V is the transmembrane electric potential, and δ is the fraction of the transmembrane potential traversed by the gating charges.

In order for the open-closed transition to be favored, ΔG^0 should be negative, so the effect of temperature depends on the sign and magnitude of ΔS and ΔH. Equation (26.1) states that the standard free energy difference between the two states is given by the sum of some intrinsic configuration energy, $E_c = \Delta H - T\Delta S$ and the work needed for the electric field to push charges around, $E_e = z\delta FV$. The negative sign in the electric term means that a positive potential stabilizes the open state.

The relative abundance of the states is governed by its standard free energy difference, according to the Boltzmann distribution law. Thus for thermal equilibrium at any potential the equilibrium constant, K, that defines the closed-open reaction is

$$K_V = \frac{\alpha}{\beta} = e^{-\frac{\Delta G^0}{RT}}, \quad (26.2)$$

where α and β represent the rates of opening and closing, respectively (see kinetic model in Fig. 26.2).

Another important effect of temperature that influences the magnitude of currents is related to ion diffusion in aqueous media. The rate at which ions enter the pore is an exponential function of temperature, so the unitary conductance of a channel also increases with temperature because the ions reach the channel more frequently. The temperature coefficient, Q_{10},[3] of unitary conductance and ion diffusion is similar, ranging from 1.2 to 1.5, and the corresponding activation energies, E_a, are ~ 4 and 7 kcal/mol (see equation that relate Q_{10} and E_a in Footnote 3).

---

[3] Temperature coefficient. Is a measure of the kinetic reaction change of a system as a consequence of increasing the temperature by 10°C. Q_{10} is a useful parameter to establish temperature dependence of a process. the Q_{10} could be easily obtained from the macroscopic currents using the following definition:

$$Q_{10} = \left(\frac{I_2}{I_1}\right)^{\frac{10}{(T_2-T_1)}},$$

where I_1 and I_2 are the currents obtained at temperatures, T_1 and T_2. Although the Q_{10} obtained by this procedure is very useful as a comparative value it lacks thermodynamic meaning. A relationship between Q_{10} and the Arrhenius activation energy (E_a) can be obtained by measuring rate constants, k, at different temperatures since $k = A\exp(-E_a/RT)$ where A is a constant called the frequency factor [28]. It is easy to show that

$$E_a = R\left(\frac{T_1 T_2}{T_2 - T_1}\right) \ln \frac{k_2}{k_1},$$

where k_1 and k_2 are the time constants measured at T_1 and T_2, respectively that can be written as:

$$E_a = -RT_1 \left(\frac{T_1 + 10}{10}\right) \ln Q_{10},$$

For thermo-TRPs such as TRPV1 and TRPM8, the thermodynamic parameters such as the overall changes in enthalpy (ΔH) and entropy (ΔS) associated with channel opening reaction have been obtained considering a two state model (Fig. 26.2; [44, 47, 50]). In this type of model, the thermodynamic meaning of the shift of the open probability (P_o) vs. voltage curve with temperature is easy to visualize since

$$P_O = \frac{1}{1 + K_V^{-1}} \quad (26.3)$$

and noticing that when $P_o = 1/2$, Eqs. (26.2) and (26.3) imply that K = 1 and $\Delta G^0 = 0$, inserting this condition in Eq. (26.1) gives that the voltage at which the probability is ½ is:

$$V_{1/2} = (\Delta H - T\Delta S)/z\delta F \quad (26.4)$$

Correa et al. [21] realized that the sign from the entropic change during channel opening can be obtained simply by measuring $V_{1/2}$ at two different temperatures, since from Eq (26.4) we have:

$$V_{1/2}(T_2) - V_{1/2}(T_1) = -\Delta S(T_2 - T_1)/z\delta F \quad (26.5)$$

Eq. (26.5) implies that a negative ΔS indicates that the closed state has greater entropy than the open state, whereas the opposite is true if a positive ΔS is obtained.

From Eqs. (26.1), (26.2) and (26.3) we have that the change in P_o with temperature, assuming that ΔH and ΔS are temperature-independent, is given by the relation:

$$\frac{\partial P_o}{\partial T} = \frac{\Delta H - z\delta FV}{\frac{RT^2}{4} \cosh^2\left(\frac{\Delta G}{2RT}\right)} \quad (26.6)$$

This function is zero when $\Delta G^0 \to \infty$ and tends to $4(\Delta H - z\delta FV)/(RT^2)$ when $\Delta G^0 \to 0$. As we discuss below, Thermo-TRP channels exhibit relatively small ΔG^0 values during the closed-opening transition; from Eq. (26.6) we see that to assure large changes in Po with temperature, the *only* value that has to be large is ΔH. This, in turn, implies that the entropy term has to be large in order to make ΔG^0 small in Eq. (26.1).

26.3 Thermal and Voltage Sensitivity of Thermo-TRP

TRP channels are expressed ubiquitously, being crucial molecular transducers and integrators of physical, chemical, and painful stimuli [22]. Though TRPs are involved in several physiological functions, a few stand out in thermosensation. A subgroup of channels contained in the TRPV, TRPM and TRPA families share an extraordinary feature: an extreme sensitivity to temperature of channel activation with $Q_{10} > 10$ implying activation energies >40 kcal/mol. This is a very large Q_{10} for channel gating, considering that, with few exceptions,[4] its value rarely exceeds $Q_{10} \sim 3$ [28] for other ion channels. Because of their very high temperature dependence these channels were dubbed thermo-TRP or thermoreceptors [12, 29]; they confer temperature sensitivity over a wide range, which extends from noxious cold (< 15°C; TRPA1)[30], to burning heat (> 52°C; TRPV2) [31]. The channels that belong to the thermosensitive subgroup so far include TRPV1-4, TRPM2, TRPM4, TRPM5, TRPM8 and TRPA1 [32]. TRPA1 is the sole member of the TRPA subfamily and is the cold sensor with the lowest threshold[5] among thermo-TRP (<15°C) [6, 33].

TRPV1 was the first thermoreceptor cloned and is mainly expressed in nociceptive fibers of trigeminal nerve and in dorsal root ganglia. It shows a temperature threshold above 42°C. In addition to high temperatures [18], it is also activated by acidic and basic pHs [34–36], by PIP_2 [37, 38], and by agonists such as capsaicin, the pungent agent of chili peppers [18, 39, 40]. On the other hand, the best characterized cold sensor is TRPM8; it is activated by temperatures below 20°C [19, 41], by menthol (mint's active compound), and by other agonists including the super cooling compound icillin [19, 42–44]. Therefore, thermo-TRP's are polymodal receptors, in the sense that they are activated by a large number of agonist and physical stimuli. In the case of voltage-dependent thermo-TRP's such as TRPM8 and TRPV1, these stimuli, by increasing channel open probability enhance channel activity in a synergistic way [20, 42, 44–49].

In thermo-TRP channels, the switch from a closed to an open state caused by a thermal stimulus has been associated with large enthalpic and entropic changes. The former implies that there is a large temperature-driven structural rearrangement involving the rupture of non covalent bonds. However, as discussed in the previous section, in order for this transition to be reversible at physiological temperatures (as it is), entropy changes must compensate the large enthalpic changes: for TRPV1, $\Delta H \sim 150$ kcal/mol, and $T\Delta S \sim 140$ kcal/mol at 25°C [50]; and for TRPM8 $\Delta H \sim -150$ kcal/mol, and $T\Delta S \sim -113$ kcal/mol at 25°C [47]. The sign of the entropy change indicates that the open state of the cold receptor is more ordered than the closed state, whereas the opposite is true for TRPV1. These results confirm the prediction of equation 6, which demands a large change in ΔH in order to have big changes in Po with temperature.

---

[4] Here we mention other ion channels with high Q_{10}. Chloride channel, ClC-0 [23], voltage gated proton channel Hv1.1 [24], mechano-activated potassium channels TRAAK and TREK-1 [25], N-type inactivation of *shaker* potassium channel [26] and Connexin 38 [27].

[5] Threshold: In general this concept should avoided in particular considering that in ion channels the average state of activation is governed by a probability distribution that does not resemble an "all or none" process typical of action potentials. In thermo-TRP, this concept must be understood as a characteristic temperature at which the channel activity is significant at a given voltage and/or agonist concentration.

26.3.1 Heat-Induced Single-Channel Activity of TRPV1

TRPV1 was the first thermo-TRP channel characterized at the single channel level [50, 51]. Patch clamp recordings of single channels showed that thermal stimuli mainly produce changes in the open probability, P_o, rather than in unitary current, accounting for the large Q_{10} value for this channel. Moreover, the single channel current was described as bursts[6] of activity with fast open-to-closed transitions. The process most affected by the thermal stimulus was found to be the burst duration ($Q_{10} \sim 32$) and, to a lesser extent, the gap length ($Q_{10} \sim 7$) [50].

Defining $V_{0.5}$ as the voltage at which the equilibrium constant K_V of Eq. (26.2) is equal 1, K_V can be rewritten as:

$$K_V = e^{\frac{z\delta F(V-V_{0.5})}{RT}}, \qquad (26.7)$$

and equation 3 becomes

$$P_O = \frac{1}{1 + e^{-\frac{z\delta F(V-V_{0.5})}{RT}}} \qquad (26.8)$$

A fit of the Po vs voltage data using Eq. (26.8) shows that thermo-TRP's are weakly voltage dependent channels with a low number of apparent charges ($z\delta \sim 0.8$ [5, 44, 47, 48] when compared to canonical voltage dependent channels like Shaker, which has a $z\delta \sim 13$ [52–54]. The source of this gating charge in Shaker has been localized to specific basic residues in TM4. However, the TM4 of thermo-TRP's has many fewer basic residues than does that of Kv channels, and the typical structure of the S4 α-helix, consisting of several basic residues separated by two neutral residues, is not present in thermo-TRP's (see Fig. 26.4). Although they have just a few basic residues in TM4, those residues are highly conserved within the TRPV family as well as within the TRPM family. As mentioned before, TRPV1 and TRPM8 are also modulated by transmembrane voltage, showing an outward rectification. The tail currents elicited at different voltages following a depolarizing prepulse show a linear behavior (i.e., they follow Ohm's Law), suggesting that the rectification is an intrinsic property of the channel rather than a fast voltage dependent blockade by divalent ions, or rectification at unitary current level [44, 47]. Since TRPs are tetramers, an apparent gating charge of 0.8 e implies that there is 0.2 e of gating charge per channel subunit, indicating that the charged particles responsible for voltage dependence do not cross the entire electric field.

---

[6]Burst: Single channel gating, where the channel activity is grouped in periods of high activity, flanked by gaps or quiescent periods.

26.3.2 Structural Determinants of Thermal Sensitivity

26.3.2.1 TRPV1

An interesting question is which part of the molecule actually senses the thermal stimulus and how this sensor is coupled to the ion conduction gate. Vlachova et al. [55] found that TRPV1 channels lacking C-terminal are non functional, but that a truncated mutant in which 72 amino acid residues of the distal part of the C-terminal were deleted, presents diminished sensitivity to capsaicin, pH and temperature activation. The truncated channel activity at 29°C was similar to that presented by the wild type TRPV1 channel at 42°C, but the Q_{10} changed from 25.6 to 4.7 [55]. However, Liu et al. [56] found that a TRPV1 channel lacking the last 88 amino acids was fully responsive to all stimuli and even exhibited the synergistic integration of the various stimuli. Apparently, the difference between the results of Vlachova et al. [55] and those of Liu et al. [56] reside in the extracellular Ca^{2+} concentration. In the presence of the divalent cation the truncated channel was unresponsive to agonist and temperature, but when the extracellular Ca^{2+} was removed the channel activity was very similar to that of the wild type channel. A Ca^{2+}-dependent phosphorylation of residue serine 502, which is a substrate of PKC[7], appears to be responsible for this Ca^{2+} effect, since phosphorylation in S502 inhibits channel activity, and somehow the C-terminal blocks accessibility to this site. When the C-terminal is removed, access to this site is unhindered and S502 can be phophorylated by PKC. In other words, truncation of the C-terminal does not specifically affect the structure of the putative temperature sensor or the coupling region between the sensor and the pore; rather, the distal part of the C-terminal plays the role of a blocking particle that prevents access by PKC to the phosphorylation site [56].

To reveal the structures involved in temperature sensitivity Brauchi et al. [57] interchanged the C-terminal domains between a hot receptor (TRPV1) and a cold receptor (TRPM8). The results showed that the C-terminal is a modular domain that confers the temperature sensitivity phenotype, since the swapping of these domains switched channel temperature sensitivity but did not alter other properties of the channel, such as menthol or capsaicin binding. Grandl et al. [58] showed, however, that a different protein region of TRPV1 is involved in thermo sensation. Using powerful mutagenesis and screening techniques used previously to identify residues critical for reduced thermal activation of TRPV3 [58a], they generated thousands of different TRPV1 mutant channels. After assaying them for changes that caused a diminished temperature activation while preserving a complete response to other stimuli, they found that only three residues are fundamental for temperature activation (Fig. 26.3) Mutations N628K, N652T and Y653T near the ion permeation pathway reduce or abolish temperature activation while maintaining the response to capsaicin, 2-APB[8] and low pH intact. Interestingly, this situation is similar to what is obtained for TRPV3, a heat activated thermo-TRP (threshold ~ 33°C). Indeed, for

---

[7]PKC: Protein Kinase C, protein that mediate S/T phosphorylations in a Ca^{+2} dependent manner.
[8]2-APB: 2-aminoethoxydiphenyl borate, agonist of TRPV1 channel

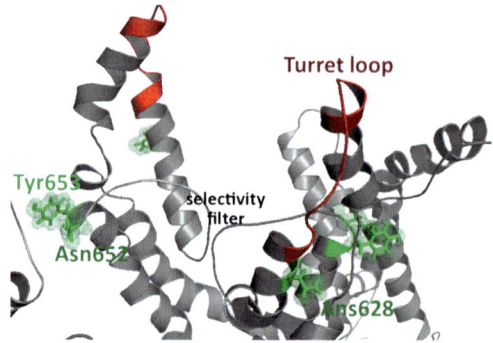

Fig. 26.3 Structural components involved in temperature activation of TRPV1. This molecular model shows the external aspect of the TRPV1 pore, highlighting the turret and the amino acids residues that have been implicated in temperature detection. The turret (*in red*) is displaced toward the pore when stimulated by heat, as reported by Yang et al. [59]. Residues Asn626, Asn652, and Phe653 (*in green*) were found to render the channel temperature-insensitive when mutated [92]. The structure shown was obtained using the TRPV1 model reported in [91]

this channel, Grandl et al. [58] also found that only three residues are fundamental for temperature activation, and these residues (Ile644, Asn647 and Tyr661) are also all located in the extracellular loop between P-loop and TM6, a region close to the ion permeation pathway of TRPV3. Mutations of these residues produced channels insensitive to temperature but able to be activated by agonist and voltage.

In another attempt to identify the specific structure responsible for activation by temperature, Yang et al. [59] replaced the entire turret loop in TRPV1 (13 residues, Fig. 26.3) with an artificial pore turret. This change produced a channel insensitive to temperature, but which showed normal activation by capsaicin and voltage. Furthermore, attaching fluorophores (fluorescein maleimide and tetramethylrodamine) to cysteines contained in the turrets and using FRET,[9] they found that this region of the channel undergoes conformational changes that suggest that the turrets move closer to each other during activation by temperature. This conclusion of Yang et al. [59] has been, however, recently challanged by Yao et al. [59a] who showed that TRPV1 channels retain normal heat response after removal of 15 amino acid residues of the turret region.

The above discussion on the temperature activation of TRPV1 has to be reconciled with previous data that attributes the feature of temperature sensitivity to the C-terminal domain [57]. One way to explain these different observations is to suppose that the sensor itself is located in the neighborhood of the ion conduction pathway, and that the structure that couples the gate with the conformational change is located in the C-terminal domain.

---

[9]FRET: **F**örster **R**esonance **E**nergy **T**ransfer, this technique uses the non radiative energy transfer between two chromophores. The efficiency of the transfer process is very sensitive to the distance and is used to establish conformational changes in proteins.

26.3.2.2 TRPM8

TRPM8 was cloned from root dorsal ganglia, DRG, and trigeminal neurons [19, 20]. In mouse DRG sensory neurons and when heterologously expressed in HEK-293 cells, TRPM8 resides mainly within lipid rafts microdomains [60]. Disrupting lipid rafts with cholesterol depleting compound methyl-beta-cyclodextrin, MCD,[10] changes the temperature threshold towards more warm temperatures, and increases the response to menthol with no change in Q_{10}, a change that can be attributable to the lack of fluidity within microdomains, interactions with cholesterol or interactions with other proteins resident in the raft. Although the possibility that membrane lipids act as the temperature-sensing molecular machinery cannot be dismissed at present, the data of Morenilla-Pelao et al. [60] indicates that in two very different lipid environments, the TRPM8 channel has essentially the same temperature sensitivity albeit different activation thresholds.

The structural determinants responsible for activation by cold temperatures remain unknown, but Brauchi et al. [57] showed that the C-terminal domain confers the phenotype of cold sensitivity. As is the case for TRPV1, TRPM8 temperature activation entails large changes in entropy and enthalpy, but with the reverse sign (see above).

26.3.3 Structural Determinants of Voltage-Dependence

Although the voltage sensor domain remains elusive, neutralization of positively charged residues in the TM4 of TRPM8 causes a decrease in its voltage dependence [5]. Voets et al. [5] neutralized all the charged residues, including that of histidine 845 contained in the TM4 segment and in the TM4-TM5 linker of the TRPM8 channel (Fig. 26.4). They found that the total apparent number of gating charges ($z\delta$ of equation 8) per TRPM8 channels is 0.85 e on average and that neutralization of R842 and K856 decreased this number to 0.7 and 0.5 e, respectively. These findings suggest that the contribution of these two charges to the total amount of gating charges/channel is not enough to explain the global voltage dependency of the channel (0.85 e). Therefore, it is possible that at least part of the total the gating charge is actually located in another position within the voltage sensor domain in TRPs (TM1-TM4). The decrease in $z\delta$ obtained following neutralization of K856 located in the TM4-TM5 linker is surprising given that, if the TRPM8 channel has a structure similar to that of the Kv channel, this linker would be located in the cell cytoplasm and therefore excluded from the electric field.

The ionic strength of the bath can also affect the voltage dependency of TRPM8. A recent study by Mahieu et al. [62] showed that an increase of extracellular cation concentration causes a partial reduction of both outward and inward currents through TRPM8. This inhibition could be explained by a shift to the right along

---

[10]MCD binds cholesterol reversibly and is commonly used to deplete membrane cholesterol acutely from both leaflets of the bilayer [61].

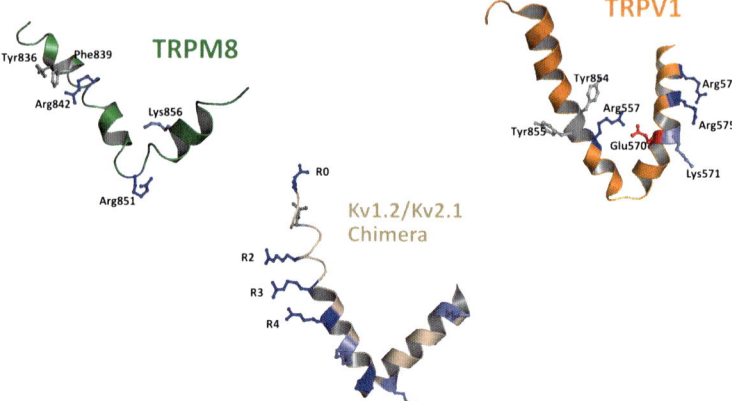

Fig. 26.4. The basic residues present in TM4 and in the TM4-TM5 linker in thermo-TRPs. The structure of the TM4 and TM4-TM5 linker of TRPM8 was obtained using the PDB of the structural model of TRPM8 developed by Pedretti et al. [69] on the basis of the crystal structure of the Kv1.2 and molecular dynamics. The structure of the TM4 and TM4-TM5 linker of the TRPV1 channel was obtained from the model reported by Brauchi et al. [91]. As a comparison, we can see that in Kv1.2 – Kv2.1 Chimera (PDB 3LNM) there are seven basic residues (arginines are shown in *blue* and lysines in *lightblue*). In addition, the position of these gating charges relative to the membrane is also different. Aromatic amino acids are shown in grey

the voltage axis in a conductance vs. voltage curve, when the Ca^{2+} concentration is increased; an increase from 0 to 20 mM in external Ca^{2+} caused a rightward shift of about 50 mV. In contrast, different concentrations of intracellular cations did not affect the magnitude of the currents, except when TRPM8 was activated by icillin, an observation in agreement with previous results of Chuang et al. [63] that demonstrated that in order to reach full efficacy, this agonist needs an elevation of cytosolic Ca^{2+}. The theoretical framework of the Gouy-Chapman-Stern equation [64–66] was used to fit the data, demonstrating that the underlying mechanism of inhibition is surface charge screening. Calculations gave an average distance between surface charges of 9–10 Å, equivalent to a total surface charge density of 0.0098–0.0126 $e/Å^2$ [62]. Whether the molecular source of these charges resides in the lipid bilayer or in the TRPM8 channel-forming protein itself is, however, not known.

26.4 TRPV1 and TRPM8 Channel Activation by Agonists

When TRPV1 was first cloned by Caterina et al. [18] it was characterized as a capsaicin receptor. Cells expressing the cloned receptor showed a calcium influx that depended on the concentration of capsaicin and other vanilloids, an effect that was potentiated by low pH [34, 67]. On the other hand, single channel results [39] showed that the increase in macroscopic currents induced by capsaicin as is the case of temperature [50], is due to an increase in burst length and shortening of the gaps with the net result of increasing *Po*.

A comparison between mammalian and avian (heat-sensitive but capsaicin-insensitive) TRPV1 channels resulted in a map of the structural determinants for capsaicin interaction. The results revealed that residues Y511 and S512, located in the transition between the second intracellular loop and the third TM segment, are required for capsaicin binding (Jordt and Julius [35]). In the rabbit (a capsaicin insensitive mammal) TRPV1 is 100 fold less sensitive to vanilloids than the rat or human TRPV1, and the key residue involved in capsaicin sensitivity is I550 [68]. The single mutation I550T is able to restore capsaicin binding and the double mutation I550T/L547M restores the high binding affinity of resiniferatoxin (RTX, an ultra potent vanilloid from the plant *Euphorbia resinifera*). In human TRPV1, M547 and T550 are located in TM4. Structural modeling suggests that capsaicin and RTX interacts with TM3 and TM4 (68).

The polymodal nature of thermo-TRP's is also present in TRPM8. It responds to cold temperatures (8–18°C) as well as different agonists such as menthol, icilin (an ultra potent super cooling agent) and eucalyptol [19, 20]. The icilin binding site is located between TM2 and TM3, in the analogous zone where capsaicin is stabilized in TRPV1 [63, 69]. Bandell et al. [43] showed that residues involved in menthol activation are Y745H, Y1005 and L1009; interestingly the first one is close to TM2 and TM3 and the last two residues are part of the TRP box[11] domain [43](cf., Brauchi et al. [57]).

26.5 Allosteric Gating in Thermo-TRP

The action of agonists like capsaicin can in principle be explained using an allosteric MWC[12] model [70]. This type of model is, however, inconsistent with the data published by Hui et al. [39], who showed that the transitions between open states are almost independent of agonist concentration. Moreover, no open state can be recorded under capsaicin-free conditions, an observation that eliminates the existence of the unbound open state that is present in the MWC model (Fig. 26.5, model 2). An important observation is that the capsaicin gating of TRPV1 involves several closed and open states, and that the durations of successive closed and open intervals are inversely related: shorter (longer) open intervals are preceded or followed by longer (shorter) closed intervals. These findings exclude kinetic models in which there is only one transition pathway between open and closed states (Fig. 26.5, model 1), but are consistent with models in which two or more closed states make direct transitions to two or more open states (Fig. 26.5, model 3), indicating that the capsaicin activation is allosterically coupled to the activation caused by other stimuli. Furthermore, high concentrations of capsaicin can activate

---

[11] TRP Box is a consensus sequence, IWKLQR, located in C-terminal of the TRP channels.

[12] MWC: Monod, Wyman and Changeux allosteric model or concerted model. This model explain transitions between symmetrical subunits that can bind agonist in an independent way, changing the subunit from relaxed to activated states.

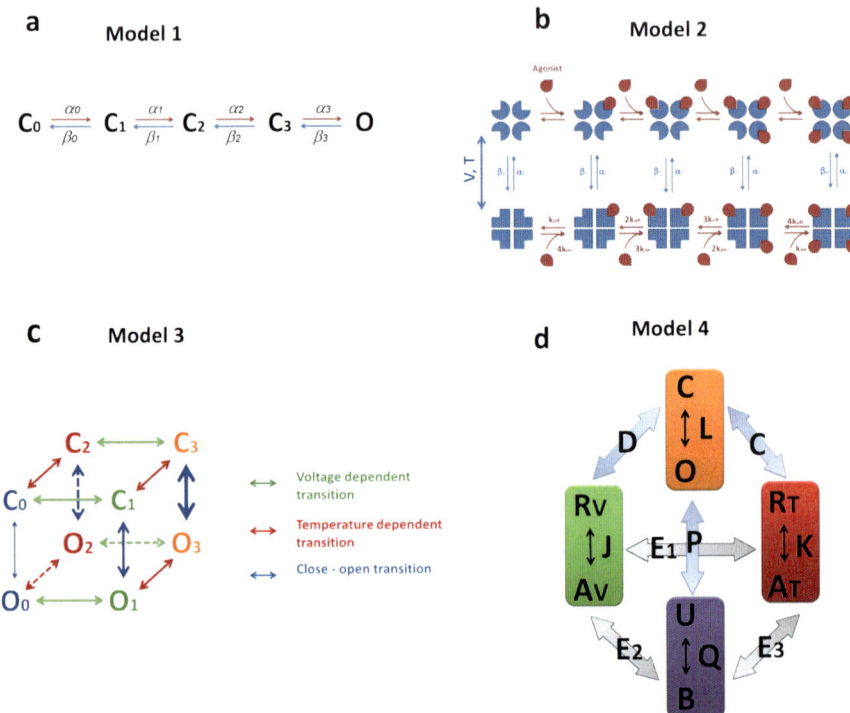

Fig. 26.5 Sequential and allosteric kinetic models. **a** A five state sequential model with four closed states, C_0, C_1, C_2 and C_3, and a single open state O. In this kind of model closed and open states are not correlated. **b** Monod-Wyman-Changeaux (MWC) model after Voets et al. [4], used to explain channel activation by agonists. In this model there are five open states (*upper tier*) and five closed states (*lower tier*). These states have 0, 1, 2, 3, or 4 bound agonist molecules (*red circles*). Open to close transitions are voltage and temperature dependent. Notice than in this model the channel can open in the absence of agonist and that the close to open transition is a one-step process. In this kind of model closed and open states are correlated. **c** Allosteric model of TRP channel activation by voltage and temperature. This model considers four closed states (*upper face of the cube*) and four open states (*lower face of the cube*). Closed to open transitions are voltage and temperature independent. Transitions from the front to the rear faces of the cube are temperature-dependent. Transitions from the left face to the right face of the cube are voltage-dependent. This model implies that temperature and voltage sensors are independent structures and that the channel can open in the absence of voltage or temperature sensor activation. In this model closed and open states are also correlated. **d** A general gating mechanism includes an allosteric interaction between channel opening (described by the equilibrium constant L), voltage sensor activation (described by the voltage-dependent equilibrium constant J), temperature sensor activation (described by the temperature-dependent equilibrium constant K) and agonist activation (described by the equilibrium constant Q). C, D, and P are allosteric factors (> 1) coupling the temperature, voltage, and agonist sensors, respectively, with pore opening. $E1$, $E2$, and $E3$ are allosteric factors that mediate the interaction between sensors. Here L is the equilibrium constant of the open to closed channel transition when the probabilities of finding the voltage, temperature and agonist sensors in their active configuration is zero. We illustrate the synergistic effect of different sensor activation with the following example: when the probability of finding all the voltage sensors in their active configuration is maximal, the closed to open equilibrium constant becomes LD but when all voltage, temperature and agonist sensors are active the closed to open equilibrium constant becomes $LDCP$ where $LDCP \gg LD$

the channel in a voltage-independent manner. This means that capsaicin drives the activation of TRPV1 through a different branch in the kinetic model that couples voltage and temperature branches allosterically (Fig. 26.5, model 4; [49]).

The integration of different type of stimuli mentioned above is not as simple as the algebraic sum of the thermal, electric, and chemical energy contributions to the process of opening the ion permeation path. As discussed below, it is likely that this integration in thermo-TRPs is driven by complex kinetic reactions between multiple closed and open states. First, as mentioned above, in single channel recordings of TRPV1 and TRPM8, the current records show bursts of activity with fast open-close transitions, and those burst are separated by quiescent periods [47, 50]. This suggests that there are at least two closed states, which account for the very short intra-burst closings and the longer inter-burst closings. As mentioned above with respect to TRPV1, rising the temperature or the capsaicin concentration makes the burst period longer and shortens the inter-burst gap with a large Q_{10}, while the faster mean open-close times have low Q_{10}. On the basis of this observation we can think that the thermal stimulus mainly affects the part of the protein responsible for destabilizing or stabilizing the long lasting closed state.

Second, in the case of TRPV1 and TRPM8, the maximum relative *Po* reached at high depolarizing voltages is lower than the maximum *Po* reached by thermal stimuli or by agonists [19, 47, 49]. This indicates that the coupling of the electric energy to the pore opening is not as efficient as the coupling of the thermal or chemical energy to the pore opening.

TRPM8 shows a delay in the activation kinetics upon depolarization, whose duration is dependent on the voltage of the applied prepulse, a characteristic also seen in the potassium conductance of the squid giant axon and calcium activated potassium channels [71, 72]. This feature demonstrates that in TRPM8, voltage gating is a process that must have at least two voltage dependent steps between various closed states, which can account for activity bursts seen in single channel recordings. Interestingly, this delay has a Q_{10} value of 1.6 [73], which implies that activation upon depolarization and "cooling" in TRPM8 must travel through different paths and could involve different structures within the protein.

Allosteric models of activation can reproduce most of the gating behavior shown by the TRPV1 and TRPM8 [47, 49, 73]. This type of model has a mechanistic interpretation that implies a structural independence between the different sensors and explains the synergistic manner in which the different stimuli operate [47, 49].

26.6 TRPV1 Channels Are Wide Pores and Show Pore Dilation in the Presence of Agonists

TRPV1 channels are permeable to large organic cations such as the polyamines spermine, spermidine and putrescine [74], the probe FMI-43 [75], the local anaesthesic QX-314 [76], the antibiotic gentamicin [77], the cations *methoxy-N-ethylquinolinium* (MEQ), and *N-(Ethoxycarbonylmethyl)-6-methoxyquinolinium*

(MQAE), [78] NMDG$^+$, and the molecular probe YO-PRO (Chung et al. [79]). Given the large size of these cations, these results indicate that the TRPV1 channel has a very wide pore ($d > 6.8$ Å; e.g., [78]) in which cationic selectivity arises as a consequence of the interaction of the permeating cations with a ring of four negative charges contained in the pore walls. In fact the members of the TRPV subfamily all show a conserved aspartate in the putative selectivity filter with the exception of TRPV2, where the aspartate is replaced by a glutamate. In TRPV1 neutralization of this aspartate decreases the strength of ruthenium red block and the relative permeability of Mg^{2+} with respect to Na^+ [80].

The pore of TRPV1 widens when the channel is activated by capsaicin. In HEK293 cells that stably expressed rat TRPV1, Chung et al. [79] measured the permeability ratio P_{NMDG}/P_{Na} as a function of time with capsaicin in the bath. On a whole cell patch clamp experiment they used a solution containing 150 mM Na^+ on the inside and 150 mM NMDG$^+$ on the outside, and applied a membrane potential of −60 mV. Upon addition of 10 μM capsaicin they observed an outward current that in a few seconds turned inward. The outward current was carried by Na^+ whereas the inward current was carried by NMDG$^+$. The biphasic nature of the current is due to a change in the P_{NMDG}/P_{Na} ratio from 0.05 to 0.28 caused by widening of the TRPV1 pore caused by by exposure to capsaicin. To evaluate the extent of the pore widening, Chung et al. [79] used the excluded volume theory, where the effective area available for diffusion by a solute of radius a in a pore of radius r is proportional to $(r - a)^2$. This theory originally derived for uncharged solutes [81] was applied to explain ion permeability ratio in the node of Ranvier [82] and to investigate pore widening in TRPV5, a highly Ca^{2+}-selective TRP channel [83]. According to the excluded volume theory the square root of the permeability ratio of a cation X and Na^+ is a linear function of the solute radius, a:

$$\sqrt{\frac{P_X}{P_{Na}}} = \frac{r - a_X}{r - a_{Na}}, \qquad (26.9)$$

where r is the pore radius. The slope of the straight line is $-1/(r - a_{Na})$ and the function becomes zero for $r = a_X$. Chung et al. [79] used three cations to evaluate the pore widening: 2-(methyl-amino)-ethanol (2-MAE $a \approx 3$ Å), trishydroxymethylaminomethane (TRIS $a \approx 2.3$ Å) and N-ethyl-D-glucamine (NMDG $a \approx 4.5$ Å). They plotted the square root of the permeability ratios measured during the initial phase, and during the steady state phase of the capsaicin elicited TRPV1 current, as a function of ion diameter. The slope of the straight line of the initial permeability ratio has a steeper slope than did the straight line measured during the final phase. This result indicates a widening of the pore according to the excluded volume theory; from the intercept with zero permeability ratio Chung et al. [79] measured a pore diameter widening from 10.1 to 12.3 Å.

Chung et al. [79] also observed the kinetics of the pore widening as measured with 2-MAE, TRIS and NMDG$^+$. Pore widening was faster the smaller the cation radius, so when conducting 2-MAE it was faster than when conducting TRIS, and with TRIS faster than NMDG$^+$ [79]. This dependence of the rate of widening as a function of solute radius was previously observed for TRPV5 [83]. The different kinetics observed as a function of solute radius indicates that pore widening is a gradual process rather than a one-step process. (In a one-step process the kinetics is independent of the solute radius and would depend only on the probability density function of the latency of the widening event). Moreover, the differences of the rate of change of the permeability ratio are orders of magnitude larger than those expected from the excluded volume theory for a gradual (exponential) change of the pore radius. This deviation from the prediction of the excluded volume theory needs to be resolved. Probably there are conformational changes of the pore induced by the permeating cation, or electrostatic effects that are not taken into account by the theory. Here, we recall a paper by Furini et al. [84], which demonstrated that increasing the radius of a water filled pore lined by fixed charges can lead to a decrease in the local concentration of ions and therefore a decrease in permeability [84].

In HEK293 cells stably expressing rat TRPV1, pore dilation as measured by YO-PRO intake occurs when the channel is activated by capsaicin and by other agonists such as piperidine, camphor, N-arachidonoyl dopamine (NADA), and resiniferatoxin (RTX). However, there is no pore dilation when TRPV1 channels are opened by heat. This observation is interesting since it demonstrates that there are at least two different molecular mechanisms for opening the channel [79].

26.6.1 Other Thermo-TRP Channels Undergo Pore Dilation

TRPA1 channels activated by allyl isothiocyanate (AITC) [85] farnesyl thiosalicylic acid (FTS) and 3'-carbamoyl biphenyl-3-yl-cyclohexylcarbamate cyclohexylcarbamic acid-3'-carbamoyl-biphenyl-3-yl ester (URB597) [86] also present pore widening. In TRM8 channels, on the other hand, there is no pore dilation when channels are activated by menthol [85]. These observations may guide further investigations into the mechanism of activation of thermo-TRP channels.

Pore widening also occurs in TRPV5, a channel that senses both intra and extracellular pH. [83]. To measure this phenomenon, organic monovalent cations of different sizes were employed. Pore radii were estimated to be 10.3, 7.5, and 5.6 Å for internal pH 9.0, 7.4, and 6.0, respectively, and 11.8, 7.5, and 6.3 Å for external pH 9.0, 7.4, and 6.0, respectively. MTSET accessibility experiments demonstrated that a rotation of the pore helix underlies the pore dilation in TRPV5. [83] This rotation of the pore helix may induce changes in the electrostatic potential in the pore (see [84]) in addition to the change in physical dimensions.

26.7 Where Is the Activation Gate?

One interesting discovery regarding the permeation pathway in thermo-TRPs was made in TRPV1 [87]. It was found that there are two constrictions along the permeation pore that could act as gates for the channel opening process. This study showed that there is a constriction near the intracellular side of the permeation pore, specifically at amino acid residue Leu681, which does not allow large molecules (such as MTSET, $a = 5.8$ Å) to pass through, whereas smaller molecules (e.g., MTSEA, $a = 3.6$ Å) can almost reach the selectivity filter. In addition, Salazar et al. [87] found that another constriction exists near the center of the ion pathway, which is formed by the Tyr671, and can impede the pass of small cations such as Ag^+ ($a = 1.15$ Å); such a constriction can, therefore, act as the relevant gate to the physiologically important permeating small cations such as K^+, Na^+ and Ca^{2+}. The above result reflects a substantial difference with the Kv channels, since in this type of channels the activation gate is located at the intracellular end of the pore, where the four TM6 ends cross each other to form a bundle [3].

As a conclusion, the notion that there is more than one gate in the TRP channels emerges. The first is assumed to be the traditional bundle crossing of TM6 seen in K^+ channels; the second would be the constriction found at the middle of the permeation pathway, and a last one would be the selectivity filter itself, revealed by the pore widening observations. However, the second one would be the most important physiologically, because it is the rate limiting step for the passage of small cations.

26.8 Coda

The *sine qua non* feature of thermo-TRP channels seems to be the ability to integrate a huge variety of chemical and physical stimuli. At present allosteric models for TRP channel gating are capable to explain much of the data presented here. It is clear, however, that the workings of these molecular machines are even more complex. Examples of this complexity are the pore dilation processes that some thermo-TRP undergo in the presence of agonist, and the existence of more than one activation gate. These features of thermo-TRP have not been taken into account in the structural models.

The link between the thermodynamics involved in the temperature activation process and the resulting conformational change in the protein is another question that remains unsolved. Although the turret of TRPV1 undergoes temperature-driven conformational changes, this effect must be generalized to explore if this region acts as a temperature sensor in all thermo-TRP channels. All the data discussed above shows that the puzzle of activation and permeation in thermo-TRP channel is far from elucidated.

Acknowledgments We thanks Dr. Giulio Vistoli for kindly providing us with the PDB of the structure of the TRPM8 channel and to Dr. Fernando González-Nilo for much help and advice in TRP channel modelling. Supported by grant from the Fondo Nacional de Investigación Científica, FONDECYT 1070049.

References

1. Montell C (2005) The TRP superfamily of cation channels. Sci STKE 2005:re3
2. Harteneck C, Plant TD, Schultz G (2000) From worm to man: three subfamilies of TRP channels. Trends Neurosci 23:159–166

3. Long SB, Campbell EB, Mackinnon R (2005) Crystal structure of a mammalian voltage-dependent Shaker family K$^+$ channel. Science 309:897–903
4. Doyle DA, Morais CJ, Pfuetzner RA, Kuo A, Gulbis JM, Cohen SL, Chait BT, Mackinnon R (1998) The structure of the potassium channel: molecular basis of K+ conduction and selectivity. Science 280:69–77
5. Voets T, Owsianik G, Janssens A, Talavera K, Nilius B (2007) TRPM8 voltage sensor mutants reveal a mechanism for integrating thermal and chemical stimuli. Nat Chem Biol 3:174–182
6. Karashima Y, Talavera K, Everaerts W, Janssens A, Kwan KY, Vennekens R, Nilius B, Voets T (2009) TRPA1 acts as a cold sensor in vitro and in vivo. Proc Natl Acad Sci USA 106: 1273–1278
7. Voets T, Prenen J, Vriens J, Watanabe H, Janssens A, Wissenbach U, Bodding M, Droogmans G, Nilius B (2002) Molecular determinants of permeation through the cation channel TRPV4. J Biol Chem 277:33704–33710
8. Owsianik G, Talavera K, Voets T, Nilius B (2006) Permeation and selectivity of TRP channels. Annu Rev Physiol 68:685–717
9. Nilius B, Vennekens R, Prenen J, Hoenderop JG, Bindels RJ, Droogmans G (2000) Whole-cell and single channel monovalent cation currents through the novel rabbit epithelial Ca^{2+} channel ECaC. J Physiol 527(Pt 2):239–248
10. Yue L, Peng JB, Hediger MA, Clapham DE (2001) CaT1 manifests the pore properties of the calcium-release-activated calcium channel. Nature 410:705–709
11. Caterina MJ (2007) Transient receptor potential ion channels as participants in thermosensation and thermoregulation. Am J Physiol Regul Integr Comp Physiol 292:R64–R76
12. Patapoutian A, Peier AM, Story GM, Viswanath V (2003) ThermoTRP channels and beyond: mechanisms of temperature sensation. Nat Rev Neurosci 4:529–539
13. Hinman A, Chuang HH, Bautista DM, Julius D (2006) TRP channel activation by reversible covalent modification. Proc Natl Acad Sci USA 103:19564–19568
14. Bandell M, Macpherson LJ, Patapoutian A (2007) From chills to chilis: mechanisms for thermosensation and chemesthesis via thermoTRPs. Curr Opin Neurobiol 17:490–497
15. Macpherson LJ, Xiao B, Kwan KY, Petrus MJ, Dubin AE, Hwang S, Cravatt B, Corey DP, Patapoutian A (2007) An ion channel essential for sensing chemical damage. J Neurosci 27:11412–11415
16. Monod J, Wyman J, Changeux JP (1965) On the nature of allosteric transitions: A plausible model. J Mol Biol 12:88–118
17. Eigen M (1968) New looks and outlooks on physical enzymology. Q Rev Biophys 1:3–33
18. Caterina MJ, Schumacher MA, Tominaga M, Rosen TA, Levine JD, Julius D (1997) The capsaicin receptor: a heat-activated ion channel in the pain pathway. Nature 389:816–824
19. McKemy DD, Neuhausser WM, Julius D (2002) Identification of a cold receptor reveals a general role for TRP channels in thermosensation. Nature 416:52–58
20. Peier AM, Moqrich A, Hergarden AC, Reeve AJ, Andersson DA, Story GM, Earley TJ, Dragoni I, McIntyre P, Bevan S, Patapoutian A (2002) A TRP channel that senses cold stimuli and menthol. Cell 108:705–715
21. Correa AM, Bezanilla F, Latorre R (1992) Gating kinetics of batrachotoxin-modified Na$^+$ channels in the squid giant axon. Voltage and temperature effects. Biophys J 61:1332–1352
22. Cortright DN, Szallasi A (2009) TRP channels and pain. Curr Pharm Des 15:1736–1749
23. Pusch M, Ludewig U, Jentsch TJ (1997) Temperature dependence of fast and slow gating relaxations of ClC-0 chloride channels. J Gen Physiol 109:105–116
24. DeCoursey TE, Cherny VV (1998) Temperature dependence of voltage-gated H$^+$ currents in human neutrophils, rat alveolar epithelial cells, and mammalian phagocytes. J Gen Physiol 112:503–522
25. Noel J, Zimmermann K, Busserolles J, Deval E, Alloui A, Diochot S, Guy N, Borsotto M, Reeh P, Eschalier A, Lazdunski M (2009) The mechano-activated K$^+$ channels TRAAK and TREK-1 control both warm and cold perception. EMBO J 28:1308–1318
26. Nobile M, Olcese R, Toro L, Stefani E (1997) Fast inactivation of Shaker K$^+$ channels is highly temperature dependent. Exp Brain Res 114:138–142

27. Ebihara L (1996) Xenopus connexin38 forms hemi-gap-junctional channels in the nonjunctional plasma membrane of Xenopus oocytes. Biophys J 71:742–748
28. Hille B (2001) Ion Channels of Excitable Membranes. Sunderlan, MA: Sinauer
29. Stucky CL, Dubin AE, Jeske NA, Malin SA, McKemy DD, Story GM (2009) Roles of transient receptor potential channels in pain. Brain Res Rev 60:2–23
30. Bandell M, Story GM, Hwang SW, Viswanath V, Eid SR, Petrus MJ, Earley TJ, Patapoutian A (2004) Noxious cold ion channel TRPA1 is activated by pungent compounds and bradykinin. Neuron 41:849–857
31. Caterina MJ, Rosen TA, Tominaga M, Brake AJ, Julius D (1999) A capsaicin-receptor homologue with a high threshold for noxious heat. Nature 398:436–441
32. Latorre R, Zaelzer C, Brauchi S (2009) Structure-functional intimacies of transient receptor potential channels. Q Rev Biophys 42:201–246
33. Sawada Y, Hosokawa H, Hori A, Matsumura K, Kobayashi S (2007) Cold sensitivity of recombinant TRPA1 channels. Brain Res 1160:39–46
34. Tominaga M, Caterina MJ, Malmberg AB, Rosen TA, Gilbert H, Skinner K, Raumann BE, Basbaum AI, Julius D (1998) The cloned capsaicin receptor integrates multiple pain-producing stimuli. Neuron 21:531–543
35. Jordt SE, Julius D (2002) Molecular basis for species-specific sensitivity to "hot" chili peppers. Cell 108:421–430
36. Dhaka A, Uzzell V, Dubin AE, Mathur J, Petrus M, Bandell M, Patapoutian A (2009) TRPV1 is activated by both acidic and basic pH. J Neurosci 29:153–158
37. Chuang HH, Prescott ED, Kong H, Shields S, Jordt SE, Basbaum AI, Chao MV, Julius D (2001) Bradykinin and nerve growth factor release the capsaicin receptor from PtdIns(4,5)P2-mediated inhibition. Nature 411:957–962
38. Yao J, Qin F (2009) Interaction with phosphoinositides confers adaptation onto the TRPV1 pain receptor. PLoS Biol 7:e46
39. Hui K, Liu B, Qin F (2003) Capsaicin activation of the pain receptor, VR1: multiple open states from both partial and full binding. Biophys J 84:2957–2968
40. Xu H, Blair NT, Clapham DE (2005) Camphor activates and strongly desensitizes the transient receptor potential vanilloid subtype 1 channel in a vanilloid-independent mechanism. J Neurosci 25:8924–8937
41. Peier AM, Moqrich A, Hergarden AC, Reeve AJ, Andersson DA, Story GM, Earley TJ, Dragoni I, McIntyre P, Bevan S, Patapoutian A (2002) A TRP channel that senses cold stimuli and menthol. Cell 108:705–715
42. Andersson DA, Chase HW, Bevan S (2004) TRPM8 activation by menthol, icilin, and cold is differentially modulated by intracellular pH. J Neurosci 24:5364–5369
43. Bandell M, Dubin AE, Petrus MJ, Orth A, Mathur J, Hwang SW, Patapoutian A (2006) High-throughput random mutagenesis screen reveals TRPM8 residues specifically required for activation by menthol. Nat Neurosci 9:493–500
44. Voets T, Droogmans G, Wissenbach U, Janssens A, Flockerzi V, Nilius B (2004) The principle of temperature-dependent gating in cold- and heat-sensitive TRP channels. Nature 430: 748–754
45. Ryu S, Liu B, Qin F (2003) Low pH potentiates both capsaicin binding and channel gating of VR1 receptors. J Gen Physiol 122:45–61
46. Bandell M, Story GM, Hwang SW, Viswanath V, Eid SR, Petrus MJ, Earley TJ, Patapoutian A (2004) Noxious cold ion channel TRPA1 is activated by pungent compounds and bradykinin. Neuron 41:849–857
47. Brauchi S, Orio P, Latorre R (2004) Clues to understanding cold sensation: thermodynamics and electrophysiological analysis of the cold receptor TRPM8. Proc Natl Acad Sci USA 101:15494–15499
48. Nilius B, Talavera K, Owsianik G, Prenen J, Droogmans G, Voets T (2005) Gating of TRP channels: a voltage connection?. J Physiol 567:35–44
49. Matta JA, Ahern GP (2007) Voltage is a partial activator of rat thermosensitive TRP channels. J Physiol 585:469–482

50. Liu B, Hui K, Qin F (2003) Thermodynamics of heat activation of single capsaicin ion channels VR1. Biophys J 85:2988–3006
51. Premkumar LS, Agarwal S, Steffen D (2002) Single-channel properties of native and cloned rat vanilloid receptors. J Physiol 545:107–117
52. Aggarwal SK, MacKinnon R (1996) Contribution of the S4 segment to gating charge in the Shaker K^+ channel. Neuron 16:1169–1177
53. Seoh SA, Sigg D, Papazian DM, Bezanilla F (1996) Voltage-sensing residues in the S2 and S4 segments of the Shaker K^+ channel. Neuron 16:1159–1167
54. Schoppa NE, Sigworth FJ (1998) Activation of shaker potassium channels. I. Characterization of voltage-dependent transitions. J Gen Physiol 111:271–294
55. Vlachova V, Teisinger J, Susankova K, Lyfenko A, Ettrich R, Vyklicky L (2003) Functional role of C-terminal cytoplasmic tail of rat vanilloid receptor 1. J Neurosci 23:1340–1350
56. Liu B, Ma W, Ryu S, Qin F (2004) Inhibitory modulation of distal C-terminal on protein kinase C-dependent phospho-regulation of rat TRPV1 receptors. J Physiol 560:627–638
57. Brauchi S, Orta G, Salazar M, Rosenmann E, Latorre R (2006) A hot-sensing cold receptor: C-terminal domain determines thermosensation in transient receptor potential channels. J Neurosci 26:4835–4840
58. Grandl J, Kim SE, Uzzell V, Bursulaya B, Petrus M, Bandell M, Patapoutian A (2010) Temperature-induced opening of TRPV1 ion channel is stabilized by the pore domain. Nat Neurosci 13:708–715
58a. Grandl J, Hu H, Bandell M, Bursulaya B, Schmidt M, Petrus M, Patapoutian A (2008) Pore region of TRPV3 ion channel is specifically required for heat activation. Nat Neurosci 11:1007–1013
59. Yang F, Cui Y, Wang K, Zheng J (2010) Thermosensitive TRP channel pore turret is part of the temperature activation pathway. Proc Natl Acad Sci USA 107:7083–7088
59a. Yao J, Liu B, Qin F (2010) Pore turret of thermal TRP channels is not essential for temperature sensing. Proc Natl Acad Sci USA 107:E125
60. Morenilla-Palao C, Pertusa M, Meseguer V, Cabedo H, Viana F (2009) Lipid raft segregation modulates TRPM8 channel activity. J Biol Chem 284:9215–9224
61. Steck TL, Ye J, Lange Y (2002) Probing red cell membrane cholesterol movement with cyclodextrin. Biophys J 83:2118–2125
62. Mahieu F, Janssens A, Gees M, Talavera K, Nilius B, Voets T (2010) Modulation of the cold-activated cation channel TRPM8 by surface charge screening. J Physiol 588:315–324
63. Chuang HH, Neuhausser WM, Julius D (2004) The super-cooling agent icilin reveals a mechanism of coincidence detection by a temperature-sensitive TRP channel. Neuron 43:859–869
64. Grahame DC (1947) The electrical double layer and the theory of electrocapillarity. Chem Rev 41:441–501
65. Gilbert DL, Ehrenstein G (1969) Effect of divalent cations on potassium conductance of squid axons: determination of surface charge. Biophys J 9:447–463
66. McLaughlin S (1989) The electrostatic properties of membranes. Annu Rev Biophys Biophys Chem 18:113–136
67. Jordt SE, Tominaga M, Julius D (2000) Acid potentiation of the capsaicin receptor determined by a key extracellular site. Proc Natl Acad Sci USA 97:8134–8139
68. Gavva NR, Klionsky L, Qu Y, Shi L, Tamir R, Edenson S, Zhang TJ, Viswanadhan VN, Toth A, Pearce LV, Vanderah TW, Porreca F, Blumberg PM, Lile J, Sun Y, Wild K, Louis JC, Treanor JJ (2004) Molecular determinants of vanilloid sensitivity in TRPV1. J Biol Chem 279:20283–20295
69. Pedretti A, Marconi C, Bettinelli I, Vistoli G (2009) Comparative modeling of the quaternary structure for the human TRPM8 channel and analysis of its binding features. Biochim Biophys Acta 1788:973–982
70. Talavera K, Nilius B, Voets T (2008) Neuronal TRP channels: thermometers, pathfinders and life-savers. Trends Neurosci 31:287–295

71. Cole KS, Moore JW (1960) Potassium ion current in the squid giant axon: dynamic characteristic. Biophys J 1:1–14
72. Horrigan FT, Cui J, Aldrich RW (1999) Allosteric voltage gating of potassium channels I. Mslo ionic currents in the absence of Ca^{2+}. J Gen Physiol 114:277–304
73. Latorre R, Brauchi S, Orta G, Zaelzer C, Vargas G (2007) Thermo TRP channels as modular proteins with allosteric gating. Cell Calcium 42:427–438
74. Ahern GP, Wang X, Miyares RL (2006) Polyamines are potent ligands for the capsaicin receptor TRPV1. J Biol Chem 281:8991–995
75. Meyers JR, MacDonald RB, Duggan A, Lenzi D, Standaert DG, Corwin JT, Corey DP (2003) Lighting up the senses: FM1-43 loading of sensory cells through nonselective ion channels. J Neurosci 23:4054–4065
76. Binshtok AM, Bean BP, Woolf CJ (2007) Inhibition of nociceptors by TRPV1-mediated entry of impermeant sodium channel blockers. Nature 449:607–610
77. Myrdal SE, Steyger PS (2005) TRPV1 regulators mediate gentamicin penetration of cultured kidney cells. Hear Res 204:156–169
78. Hellwig N, Plant TD, Janson W, Schafer M, Schultz G, Schaefer M (2004) TRPV1 acts as proton channel to induce acidification in nociceptive neurons. J Biol Chem 279:7048–7054
79. Chung MK, Guler AD, Caterina MJ (2008) TRPV1 shows dynamic ionic selectivity during agonist stimulation. Nat Neurosci 11:555–564
80. Garcia-Martinez C, Morenilla-Palao C, Planells-Cases R, Merino JM, Ferrer-Montiel A (2000) Identification of an aspartic residue in the P-loop of the vanilloid receptor that modulates pore properties. J Biol Chem 275:32552–32558
81. Renkin EM (1954) Filtration, diffusion, and molecular sieving through porous cellulose membranes. J Gen Physiol 38:225–243
82. Dwyer TM, Adams DJ, Hille B (1980) The permeability of the endplate channel to organic cations in frog muscle. J Gen Physiol 75:469–492
83. Yeh BI, Kim YK, Jabbar W, Huang CL (2005) Conformational changes of pore helix coupled to gating of TRPV5 by protons. EMBO J 24:3224–3234
84. Furini S, Zerbetto F, Cavalcanti S (2007) Role of the intracellular cavity in potassium channel conductivity. J Phys Chem B 111:13993–14000
85. Chen J, Kim D, Bianchi BR, Cavanaugh EJ, Faltynek CR, Kym PR, Reilly RM (2009) Pore dilation occurs in TRPA1 but not in TRPM8 channels. Mol Pain 5:3
86. Banke T, Chaplan S, Wickenden AD (2010) Dynamic changes in the TRPA1 selectivity filter lead to progressive but reversible pore dilation. Am J Physiol Cell Physiol 298:C1457–C1468
87. Salazar H, Jara-Oseguera A, Hernandez-Garcia E, Llorente I, Arias-Olguin II, Soriano-Garcia M, Islas LD, Rosenbaum T (2009) Structural determinants of gating in the TRPV1 channel. Nat Struct Mol Biol 16:704–710
88. Arai M, Mitsuke H, Ikeda M, Xia JX, Kikuchi T, Satake M, Shimizu T (2004) Conpred II: a consensus prediction method for obtaining transmembrane topology models with high reliability. Nucleic Acids Res 32:W390–W393
89. Thompson JD, Higgins DG, Gibson TJ (1994) CLUSTAL W: improving the sensitivity of progressive multiple sequence alignment through sequence weighting, position-specific gap penalties and weight matrix choice. Nucleic Acids Res 22:4673–4680
90. Notredame C, Higgins DG, Heringa J (2000) T-Coffee: A novel method for fast and accurate multiple sequence alignment. J Mol Biol 302:205–217
91. Brauchi S, Orta G, Mascayano C, Salazar M, Raddatz N, Urbina H, Rosenmann E, Gonzalez-Nilo F, Latorre R (2007) Dissection of the components for PIP2 activation and thermosensation in TRP channels. Proc Natl Acad Sci USA 104:10246–10251
92. Grandl J, Kim SE, Uzzell V, Bursulaya B, Petrus M, Bandell M, Patapoutian A (2010) Temperature-induced opening of TRPV1 ion channel is stabilized by the pore domain. Nat Neurosci 13:708–714

Chapter 27
Complex Regulation of TRPV1 and Related Thermo-TRPs: Implications for Therapeutic Intervention

Rosa Planells-Cases, Pierluigi Valente, Antonio Ferrer-Montiel, Feng Qin, and Arpad Szallasi

Abstract The capsaicin receptor TRPV1 (*Transient Receptor Potential, Vanilloid* family member 1), the founding member of the heat-sensitive TRP ("thermo-TRP") channel family, plays a pivotal role in pain transduction. There is mounting evidence that TRPV1 regulation is complex and is manifest at many levels, from gene expression through post-translational modification and formation of receptor heteromers to subcellular compartmentalization and association with regulatory proteins. These mechanisms are believed to be involved both in disease-related changes in TRPV1 expression, and the long-lasting refractory state, referred to as "desensitization", that follows TRPV1 agonist treatment. The signaling cascades that regulate TRPV1 and related thermo-TRP channels are only beginning to be understood. Here we review our current knowledge in this rapidly changing field. We propose that the complex regulation of TRPV1 may be exploited for therapeutic purposes, with the ultimate goal being the development of novel, innovative agents that target TRPV1 in diseased, but not healthy, tissues. Such compounds are expected to be devoid of the side-effects (e.g. hyperthermia and impaired noxious heat sensation) that plague the clinical use of existing TRPV1 antagonists.

27.1 Introduction

Acute nociceptive pain is a fundamental physiological warning system, an unpleasant sensory and emotional experience that triggers a host of protective responses to limit injury [1]. Chronic pain, by contrast, is a pathological condition which often leads to physical and mental disability [1]. Indeed, chronic pain is the most common complaint for which patients seek physician treatment. It has been estimated that nearly half billion chronic pain cases are diagnosed each year world-wide and that over 50% of individuals seeking treatment are unsatisfied with their present

A. Szallasi (✉)
Department of Pathology, Monmouth Medical Center, Long Branch, NJ 07740, USA
e-mail: ASZALLASI@SBHCS.COM

treatment options [2]. Chronic pain has a major impact on many quality-of-life measures and causes staggering losses to the society due to escalating health care costs and lost productivity (http://www.iasp-pain.org; http://www.aapainmanage.org/). However, despite a significant investment of resources by the pharmaceutical industry to identify better treatment options, the mainstay of therapy remains non-steroidal anti-inflammatory drugs (NSAIDs) and, for severe pain, opioids [3].

The difficulty in developing new effective drugs for treating chronic pain partly lies in our limited knowledge of the biochemical and molecular mechanisms involved in pain signalling. Generally speaking, pain is perceived when action potentials generated in nociceptive neurons are reaching the somatosensory cortex. Nociceptive neurons are bi-polar neurons with somata in sensory (dorsal root and trigeminal) ganglia. Pain sensation is initiated when peripheral terminals of these neurons are activated by noxious thermal, mechanical or chemical stimuli [1]. Nociceptive neurons transmit tissue injury and inflammatory information through the spinal cord to the pain processing regions of the brain (afferent function) with the aim of getting an avoidance response that prevents or minimizes damage. In addition, injury-induced activation of sensory neurons initiates the cascade of biochemical events collectively known as neurogenic inflammation (efferent function) and triggers protective reflex responses [4].

The neuronal agents that cause inflammation are bioactive peptides released in the periphery upon stimulation of thin, un-myelinated afferent C-fibres and endocrine cells in all organs [5]. Notable examples of neuropeptides released from C-fibres include substance P (SP) and calcitonin gene-related peptide (CGRP). Proinflammatory neuropeptides act both autocrinally on the nociceptor terminals and paracrinally on target cells such as mast cells, immune cells and vascular smooth muscle [6]. Importantly, nociceptors get sensitized during inflammation and this peripheral sensitization leads to profound changes in the perception of stimuli in the damaged region such as hyperalgesia (pain intensifies) and allodynia (a normally innocuous stimulus perceived as painful). In chronic conditions, this process is exacerbated due to synaptic changes at the spinal cord, a process known as central sensitization [7]. Clear analogies exist between the peripheral nociceptive sensitization and the Long Term Potentiation (LTP) in the central nervous system during memory formation and learning [8]. Sensitization is believed to play a pivotal role in the development and maintenance of chronic pathological pain conditions [9].

Over the past 15 years, significant scientific progress has been made in our understanding of the mechanisms underlying pathological pain. A key discovery was the molecular cloning of the capsaicin receptor TRPV1 (Transient Receptor Potential, Vanilloid family, member 1) [10]. In nociceptive neurons, TRPV1 functions as a molecular integrator of painful stimuli (reviewed in 11, 12). TRPV1 can be both sensitized and up-regulated during inflammation and injury [12]. Indeed, TRPV1 was suggested to play a central role in peripheral (and maybe also central) sensitization that develops after inflammation (reviewed in 12). This concept has gained strong experimental support by the absence of thermal hyperalgesia in mice whose TRPV1 gene had been deleted (or silenced) by genetic manipulation [13–15]. These findings laid the groundwork for the development of potent, small molecule TRPV1 antagonists, many of which have already entered (or even completed) clinical trials [16–19]. These studies are discussed elsewhere in the book. Here, we focus on the emerging evidence that TRPV1 is subject to a complex (in part disease state-dependant) regulation that may have important implications for drug development.

27.2 TRPV1 and Nociception

TRPV1 is highly expressed in a subset of small and medium diameter neurons of dorsal root ganglia (DRG), trigeminal ganglia and nodose ganglia, distinguished by their peculiar sensitivity to capsaicin (reviewed in 20). Albeit at much lower levels, TRPV1 is also present in various brain nuclei and non-neuronal tissues (reviewed in 21). In nociceptive neurons, TRPV1 is a major integrator of noxious stimuli, ranging from heat and voltage through exogenous chemical agents (including the natural plant products capsaicin, piperine and resiniferatoxin, snake and spider venoms and jellyfish toxins) to endogenous agonists (e.g. acidic and basic pH, anandamide, lipooxigenase metabolites) and pro-algesic substances released during tissue injury (reviewed in 12 and 22). Despite intensive research, the biological significance of TRPV1 expression in brain neurons and non-neuronal tissues remain incompletely understood.

TRPV1 is the founding member of the vanilloid family of TRP channels. Unlike most families of ion channels that have been grouped together based on their function, the 28 mammalian members of the TRP receptor superfamily are subdivided into six families based on primary amino acid structures [23]. These are the TRPC (classic or canonical), TRPV (vanilloid), TRPA (ankyrin), TRPM (melastatin-like), TRPP (polycystin), and TRPML (mucolipin) families. Two members of the TRP channel superfamily may be less than 20% identical to one another, so there is considerable diversity among family members [23]. Consistent with the diversity in primary structure is the diversity observed in functional properties. The TRPs (including TRPV1) are commonly referred to as calcium permeable non-selective cation channels, but TRPM4 and TRPM5 are sodium selective channels whereas TRPV5 and TRPV6 are inwardly rectifying calcium selective channels (reviewed in 24). TRP channels also vary in their tissue expression patterns. For example, TRPM7 is ubiquitously expressed whereas the expression of TRPA1 is highly restricted [25]. Even the sub-cellular localization of the proteins varies with family members and a number of TRP channels including TRPV1 are being re-shuffled among intracellular depots [26].

Of the currently known 28 TRP channels, seven sense hot and warm temperatures (TRPV1 to TRPV4, TRPM2, TRPM4, and TRPM5) whereas two (TRPA1 and TRPM8) are activated by cold (reviewed in 12). These channels are collectively referred to as temperature-sensitive TRP channels or "thermo-TRP" [27]. Combined, thermo-TRPs cover a wide temperature range, with extremes falling between 10°C (TRPA1) and 53°C (TRPV2). Another shared feature of these

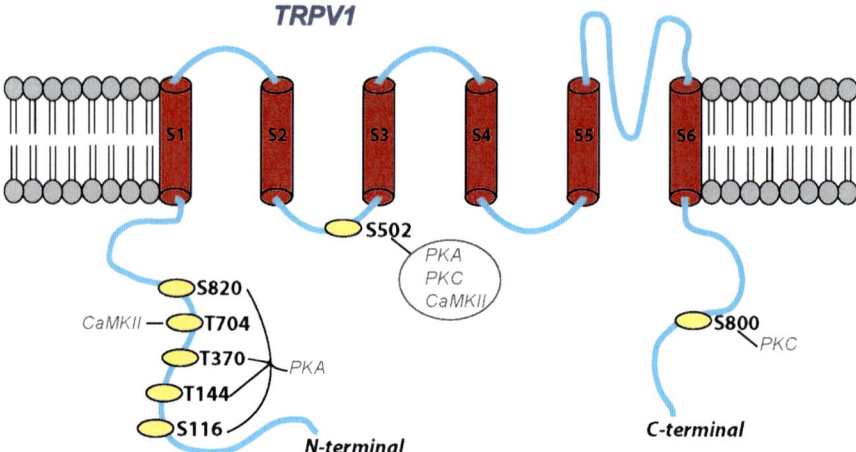

Fig. 27.1 The predicted membrane topography of the capsaicin receptor TRPV1. Please note that TRPV1 has six full serpentine trans-membrane domains (S1–S6) and a partial one between S5 and S6 which is believed to form the channel pore. The N terminus carries 5 consensus phosphorylation sites for protein kinases (PKA CaMKII) and the C terminus has one for PKC

channels is their sensitivity to a variety of natural products, many that are irritant. In fact, the TRPV1 channel was originally termed as the capsaicin receptor [10]. TRPV1 is a serpentine protein with six transmembrane domains and a loop that is believed to form a cation pore with limited selectivity for Ca^{2+} (Fig. 27.1). The human *TRPV1* gene is located on chromosome 17 and has a number of SNPs (single nucleotide polymorphisms) which have been linked to individual differences in pain sensitivity (reviewed in 28). Of note, in sensory neurons TRPV1 is co-expressed with its relatives TRPV2, TRPV3, TRPV4 and TRPA1, but not TRPM8 (reviewed in 29, 30).

In normal conditions, thermo-TRP activity is low and activators gate these channels with low efficacy and potency. Also of note, thermo-TRP function may be potentiated by the action of pro-inflammatory mediators released under pathological conditions, leading to an increment of nociceptor excitability which pivotally contributes to the inflammatory sensitization of sensory neurons (reviewed in 12, 29, 30). For instance, algogens released during inflammation potentiate the activity of TRPV1 by decreasing its threshold of temperature activation from 42 to 36°C, by enhancing its channel activity and/or by recruiting more channels to the neuronal surface (reviewed in 12, 29, 30). Therefore, TRPV1 inflammatory sensitization appears pivotal to thermal hyperalgesia and may also contribute to mechanical hyperalgesia in chronic pain states (reviewed in 12, 29, 30).

Capsaicin, the active ingredient in hot chilli peppers (*Capsicum*), causes an initial excitation of nociceptive neurons via TRPV1 activation which is perceived as itch or "hot" burning sensation (reviewed in 20). What makes capsaicin unique among naturally occurring irritant compounds is that this initial excitation is followed by a lasting refractory state in which neurons are unresponsive not only to repeated capsaicin challenge but also to unrelated stimuli (reviewed in 31). This refractory state, traditionally referred to as "desensitization", has a clear therapeutic potential [32]. The molecular mechanisms of desensitization are only beginning to be understood.

27.3 Molecular Mechanisms of TRPV1 Desensitization

The TRPV1 channel is highly permeable to extracellular calcium [10]. The functional and morphological changes of neurons induced by capsaicin presumably arise due to calcium influx through these channels (reviewed in 20). TRPV1 itself responds to calcium entry with a profound loss of its responsiveness [33]. Historically, the "desensitization" of the channel is classified into two types [31]. One is known as the "acute desensitization," referring to a decrease of channel activity during stimulation. The other is tachyphylaxis, which refers to a successive diminution of its response over repeated but brief applications of a stimulus. The separation of the two types of desensitization appears to be more phenomenology than mechanistic. They are both Ca^{2+}-dependent [33–37], and the tachyphylaxis can be at least partly explained as a result of failure of recovery from acute desensitization [36, 38]. This "desensitization" (that lasts for minutes or hours) should be distinguished from the long-lasting (weeks or even months) refractory state that develops following capsaicin administration and is thought to reflect "defunctionalisation" of the neuronal pathways [31]. However, for historical reasons, the term "desensitization" is deeply entrenched in the literature to describe this defunctionalisation [20, 31].

The desensitization of TRPV1 is triggered by Ca^{2+} influx through the channel and involves various intracellular signaling pathways (Fig. 27.2). Extensive studies done over the past decade to understand the signaling events have led to the proposition of several mechanisms (Figs. 27.1 and 27.3). In a chronological order, these include dephosphorylation of the channel by calcineurin (CN) [34, 39–44], interaction with calmodulin (CaM) [45–47], and depletion of phosphatidylinositol 4,5-bisphosphate (PIP_2) in the plasma membrane [38, 48–51]. The dephosphorylation mechanism can be further divided according to whether the phosphorylation sites are protein kinase A (PKA)-, protein kinase C (PKC)-, or CaM kinase II (CaMKII) dependent. Presently, there is still a lack of agreement on the exact signaling cascades. It is likely to involve more than one mechanism.

A new, emerging view of the desensitization is that instead of a loss of function it causes a shift of the sensitivity of the channel [51, 52]. Thus, any perturbation that could be evoked by Ca^{2+} influx and change the channel responsiveness can potentially contribute to the desensitization of the channel. Below, we provide a brief summary of the three candidate mechanisms, namely (A) dephosporylation (B) interaction with calmodulin, and (C) depletion of PIP2. Interested readers are also referred to other reviews on the subject [53–55].

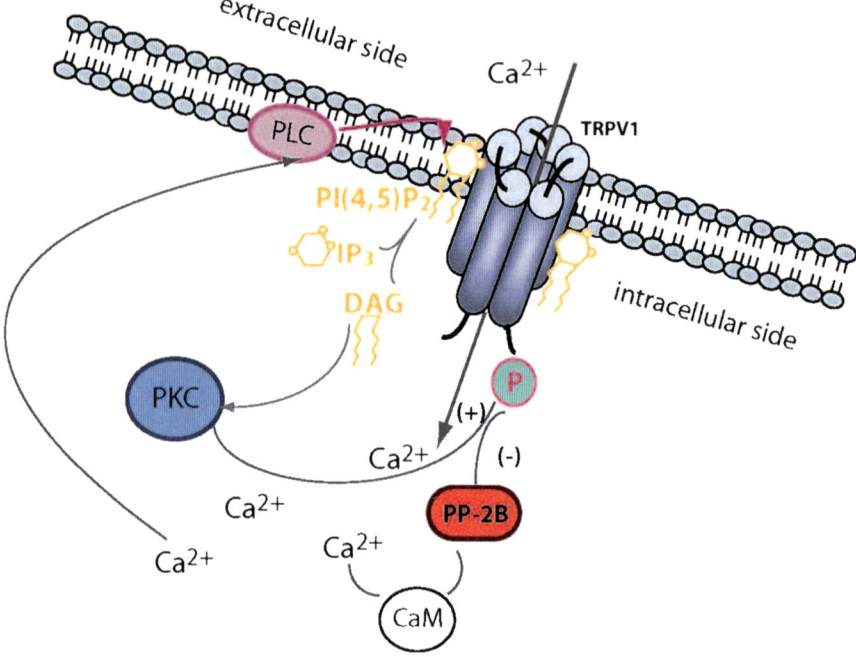

Fig. 27.2 Ca^{2+} influx through TRPV1 plays an important feed-back role by activating enzymes (PKC and PLC) that, in turn, regulate TRPV1 activity

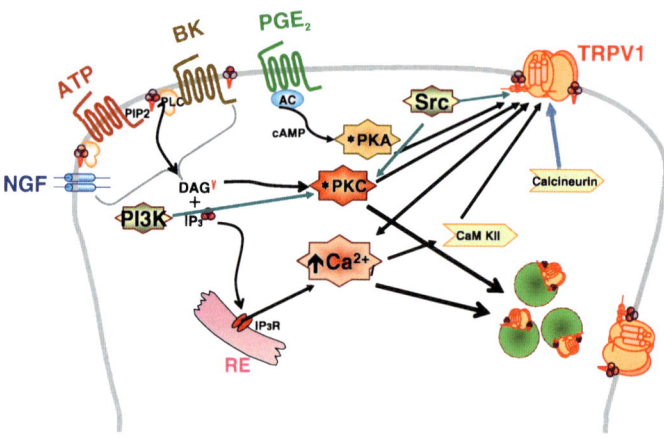

Fig. 27.3 A simplified model depicting the complex regulation of TRPV1. Pro-inflammatory mediators, acting through their receptors, activate intracellular signaling cascades via protein kinase A (PKA) and C (PKC), Src, as well as Ca^{2+}. Protein kinases phosphorylate the receptor and modify its channel activity by altering the threshold of temperature activation and the rate of desensitization, leading to higher excitability. Concomitantly, PKC and Ca^{2+} stimulate the rapid exocytotic recruitment of a vesicular population of receptor, leading to an increase of its surface expression. Taking together, pro-inflammatory mediators produce nociceptor sensitization by increasing the amount of surface expressed TRPV1 and by modifying its gating properties, leading to sensory neuron overexcitability

27.3.1 TRPV1 Dephosphorylation

The general hypothesis here is that the channel at rest is in a phosphorylated state [56]. Upon activation, the Ca^{2+} influx activates protein phosphatases that, in turn, lead to dephosphorylation of the channel and thus desensitization of its function. The key evidence for this mechanism was furnished by the pharmacological effects of interventions of protein kinases and phosphatases on the desensitization of capsaicin-evoked currents in sensory neurons [34]. The inhibition of protein kinases did not affect the desensitization of the channel, and neither did the selective inhibition of protein phosphatase 1 or 2A. Instead, only the inhibition of protein phosphatase 2B (calcineurin) was found to reduce the desensitization [34]. Experiments with cell heterologously expressing TRPV1 provide further support for this hypothesis [39, 42, 43]. In these experiments, the application of calcineurin inhibitors (e.g., cyclosporine and FK506) interfered with both tachyphylaxis and acute desensitization [42, 44].

TRPV1 is known to be versatilely regulated via phosphorylation by protein kinases. PKC was demonstrated to directly modify the channel and sensitize its responses to heat and chemical agonists [57–60]. For example, the treatment of the channel with the classic PKC activator phorbol myristoyl acetate (PMA) shifts the threshold of heat activation below 37°C or body temperature [58]. Similarly, an elevation of cellular cAMP level by stimulation of adenylylcyclase (AC) with forskolin was also found to potentiate nociceptive neurons [61–64]. Both PKA and PKC phosphorylations of TRPV1 have been implicated in the sensitization of nociceptors by proinflammatory molecules. Among them, for example, PKC phosphorylation of the channel is involved in the sensitization effect of bradykinin ([58], [65], whereas PKA phosphorylation mediates the effect of prostaglandins [62]).

The primary sequence of TRPV1 predicts a number of potential phosphorylation sites (e.g. S116, T144, T370, S502 and S800), which span over the two intracellular terminals as well as the linker domain between transmembrane segments S2 and S3 (Fig. 27.1). In vitro phosphorylation assays with isolated fragments confirm the accessibility of these sites to either PKA or PKC. There is a consensus that S502 on the S2-S3 linker is a non-selective site and may be phosphorylated by multiple kinases (e.g., PKA, PKC and CamKII) (Fig. 27.1; refs 39, 44, 46). On the other hand, the sites on the N terminus (T144, T116 and T370) appear to be preferred by PKA [39, 43], while S800 on the C terminus is specific to PKC [46]. For the dephosphorylation of the channel to be a mechanism for desensitization, the channel needs to be non-functional at the dephosphorylated state (the channel after desensitization becomes irresponsive to saturating agonist such as 1 μM capsaicin). On the other hand, the channel must not be fully phosphorylated, so that it can be further sensitized by pro-inflammatory mediators. The implication is then that only some, but not all, of these sites are phosphorylated at the normal physiological conditions.

27.3.1.1 PKA Sites

Although TRPV1 can be phosphorylated by multiple protein kinases, it has been suggested that the dephosphorylation at the PKA sites are primarily responsible for desensitization. In vitro experiments suggest that the TRPV1 channel at rest is indeed phosphorylated mainly at S116 and S502 [39]. Elevation of cAMP, which leads to activation of PKA, was shown to successively prevent the tachyphylaxis of the channel. In these experiments, the acute desensitization appeared to still exist. Thus, the effect of PKA phosphorylation seems to suggest a different mechanism for tachyphylaxis and acute desensitization. A straightforward interpretation of these findings would be that the dephosphorylation at S116 or S502 results in tachyphylaxis of the channel. However, such an interpretation became overly simplistic in lieu of the mutagenesis data. A common practice for mimicking the phosphorylated or dephosphorylated channel is to substitute the residues at the phosphorylation sites with negatively charged or neutral residues. However, the mutations of TRPV1 at the identified sites did not product all phenotypes as would be expected. The neutralization mutation (S116A) gave rise to a channel that was still functional and remained desensitized by Ca^{2+} influx instead of a non-functional mutant to mimic the desensitized channel. The mutation S116D, on the other hand, was able to blunt desensitization, suggesting that the mutant mimicked a permanently phosphorylated channel. To account for both pharmacological and mutagenic data, it was suggested that the desensitization of the channel involves multiple steps and that the dephosphorylation of the channel plays a permissive to control the process. In the other words, the phosphorylation prevents desensitization while the dephosphorylation is necessary but not sufficient to desensitize the channel.

In another study, Mohapatra and Nau [43] systematically examined all putative phosphorylation sites (S116, T144, T370, S774, S820) by point mutations. Consistent with the observations of Bhave and colleagues [39], they reported that the stimulation of PKA, but not PKC, affects tachyphylaxis. But somewhat differently, they found that the phosphorylation by PKA reduced or slowed down, but did not fully prevent tachyphylaxis [42, 43]. Accordingly, the mutation at S116 alone (S116D) only partially reduced tachyphylaxis, while mutations at other sites were also found effective, and all mutant channels retained strong acute desensitization albeit to different extents from the wild-type. Furthermore, forskolin, which stimulates production of cAMP, remained effective in inhibiting the tachyphylaxis in all mutant channels. When the phosphorylation sites were neutralized by alanine substitution, the resultant mutants were also functional, retained both strong acute desensitization (S116A, T144A and T370A were reduced) and tachyphylaxis (S6A, T144A, T370A and S820A were reduced), and remained sensitive to forskolin. In addition to S116, these authors suggest that other phosphorylation sites including T370 and T114 are also subject to dephosphorylation by calcineurin during the desensitization of TRPV1.

One consensus from these studies is that the PKA phosphorylation sites on the channel are selectively involved in desensitization and are the targets for calcineurin dephosphorylation. What is less uncertain is the exact role of the dephosphorylation at these sites in the overall desensitization process. Experimental data suggest that the inhibition of calcineurin tends to slow down, but does not prevent, both tachyphylaxis and acute desensitization [42, 44]. This failure of a complete inhibition could be attributed to the activity of additional (cyclosporine- and

FK506-independent) protein phosphatases. Alternatively, one can argue that the dephosphorylation plays a regulatory rather than an obligatory role in the desensitization of the channel. TRPV1 is a large allosteric protein, and, in principle, any allosteric effectors that may alter the gating of the channel could influence the amount of calcium influx and consequently the desensitization behaviors. Thus, the dephosphorylation of the channel by calcineurin can potentially have two separate effects, one causing the desensitization while the other regulating the desensitization through alteration of gating. These two effects are mechanistically different but hard to distinguish from the existing experimental data.

27.3.1.2 PKC Sites

By calcium imaging and patch-clamp recording, Mandadi and colleagues [40, 41] showed that the channel after desensitization could recover its capsaicin sensitivity after treatment with phorbol-12-myristate-13-acetate (PMA). PKC was further demonstrated to be responsible for this resensitization, and S502 and S800 were identified as the phosphorylation sites. This resensitization appear to be unlikely due to a reversal of the desensitization process, since it has been shown that PKC phosphorylation did not impact the tachyphylaxis of the channel while the mutants (S502A/S800A) remained sensitive to inhibition of calcineurin. Instead, it probably results from the potentiation effect of PKC phosphorylation on the gating of the channel. The channel after desensitization can still be activated by supramaximal concentrations of agonists [51, 52]. Thus, the apparent resensitization of the desensitized channel by PMA could be explained by a shift in the EC50 of agonists, which occurs also to the wild-type TRPV1.

27.3.1.3 CamKII Sites

While whole-cell patch-clamp recording is commonly used for studying the desensitization of TRPV1, Jung and co-workers exploited a different strategy in which they studied the channels in "inside-out" patches following desensitization [44]. With accessibility to the cytoplasmic side, they tested the effects of various kinases on the desensitized channels. The application of CamKII together with CaM was found to rescue the capsaicin response. Other protein kinases including PKA and PKC failed to produce any significant effect. Among five predicted CamKII motifs (T370, S502, T704, T774, S800), it was shown that double mutations at S502 and T704 (S502A/T704I) were able to blunt the capsaicin response, thereby mimicking the effect of desensitization by Ca^{2+} influx. The mutant S502A/T704I also resembled the desensitized channel with impaired binding to the high affinity agonist resiniferatoxin (RTX). Finally, a significant reduction in the phosphorylation level was observed in the mutant channels. These results led the authors to propose that the phosphorylation of the channel by CamKII at one of the two sites (S502 and T704) is a prerequisite for its function. As an implication, the dephosphorylation at these sites by calcineurin would conversely render the channel into a desensitized state.

27.3.1.4 Interaction with Calmodulin

CaM modulates Ca^{2+}-dependent inhibition (inactivation) of many ion channel receptors (e.g., CNG, NMDA, L-type Ca^{2+}, P/Q type Ca^{2+} and small conductance Ca^{2+}-activated K channels). Numazaki and colleagues first reported that CaM also mediates the desensitization of TRPV1 [46]. In their studies, it was found that typical CaM inhibitors (e.g. W-7 or calmidazoline) had no effect on the desensitization of the channel, and neither did the CaM mutants lacking Ca^{2+} binding capability exerted a dominant negative effect, both contrary to what would have been expected with a role of CaM [46]. However, biochemical assays revealed that CaM co-immunoprecipitated with the channel in the presence of Ca^{2+}. With isolated C-terminus fragments, the authors further localized the interaction domain to a region of about 35 residues at the distal C-terminal. The binding of CaM to this domain was observed in both absence and presence of Ca^{2+}, suggesting that CaM may preassociate with TRPV1. When this region was deleted, the mutant channel became resistant to tachyphylaxis and also showed reduced acute desensitization.

The interaction of CaM with TRPV1 resembles that with Ca^{2+} channels [46]. In both cases, CaM binds to the C terminal (though by different structural motifs) and the binding is present even with the apo form of CaM. It is also interesting that the functions of such interactions are resistant to conventional pharmacological inhibitions or intervention by expressions of dominant negative mutants. However, important questions remain about the contribution this CaM interaction to the desensitization of TRPV1. Mutant channels in which this region had been removed continued to desensitize albeit with slower kinetics. A much larger region of the last 88 residues of the C-terminal of the channel could be truncated while retaining strong desensitization [65]. Thus, it seems that the distal C-terminal of the channel may affect the kinetics of desensitization, but its significance to the extent of the desensitization may be limited.

Two other studies also demonstrated the interactions of CaM with TRPV1 and such interactions play an inhibitory role in the functions of the channel. However, the site of the interaction was identified on the N terminal instead of the C terminal [45, 47]. Rosenbaum and colleagues first observed that the channel activity could be significantly suppressed in inside-out patches when purified CaM was directly applied to the intracellular side of the patch [54]. Interestingly, mutant CaM was also found with a dominant negative effect, an observation at variance with previous studies [60]. Since the desensitization of TRPV1 is Ca^{2+}-dependent, these results imply that this inhibitory role of CaM may contribute to the desensitization of the channel. Indeed, Lishko and associates further demonstrated that the anti-CaM antibody reduced both acute desensitization and tachyphylaxis of heterologously expressed TRPV1 in whole-cell conditions [45]. In the crystal structure of the anykrin repeats that they obtained the CaM binding domain was mapped to a region between ANK 2 and 3 [finger 2 and inner and outer helices [66]]. The same region was also found to bind ATP and the binding had an opposite role to potentiate the channel activity. Thus, CaM and ATP compete for the same site to oppositely regulate the channel.

27.3.1.5 Depletion of PIP2

The initial clue on the involvement of PIP_2 in the desensitization of TRPV1 came from the study of its functional recovery from desensitization [38]. In typical patch-clamp recordings, the channel after desensitization became irresponsive to saturating capsaicin (~1 μM). For the channel to recover its response, ATP was found to be critical. When no ATP was included in the pipette solution, the desensitization of the channel lasted throughout experiments, showing no evidence of recovery. In the presence of ATP, the channel could fully recover in 5–10 min. Furthermore, hydrolyzable ATP was required, while non-hydrolyzable analogs competitively inhibited the recovery. The requirement for hydrolysis of ATP was consistent with the dephosphorylation mechanism, however, the concentration profile of ATP for recovery (~4 mM) was found much higher than necessary for phosphorylation by protein kinases. Furthermore, the pharmacological inhibition of either PKA or PKC did not prevent the recovery, nor did the mutations at the putative phosphorylation sites of the channel. When the lipid kinases were investigated, it was found that the inhibition of PIP2 synthesis was able to prevent the recovery of the channel from desensitization. Also in support of the involvement of PIP2, the receptor-mediated hydrolysis of PIP2 was found to similarly inhibit the recovery. Furthermore, the depletion of PIP2 from the plasma membrane was detected in live cells during the desensitization of the channel. The inward rectifier potassium channel Kir 2.1 requires PIP2 for function. When coexpressed with TRPV1, the stimulation of TRPV1 not only caused desensitization of the channel itself, but also the inactivation of the Kir channel, and in the presence of ATP, the two channels recovered concomitantly.

Simon and colleagues suggested a new mechanism for desensitization of TRPV1, which involves depletion of PIP2 on the plasma membrane resulting from Ca^{2+} influx [36]. Such a mechanism implies that a certain level of PIP2 is necessary for the function of the channel, though PIP2 was previously suggested as a tonic inhibitor for the channel [66, 67]. Further studies provided more explicit evidence in support of this mechanism [48, 49]. In particular, exogenous PIP2 was shown to directly activate the channel in excised patches. Fluorescent PIP2 binding probes also reported the depletion of PIP2 during channel activation and desensitization. PLC was further implicated as the underlying enzyme to mediate the depletion since its inhibition by U7312 could substantially suppress the extent of desensitization. Depletion of PIP2 by other approaches based on chemically inducible lipid phosphatases and voltage-activated Ci-VSP also results in decrease of channel activity. These depletion assays do not activate secondary downstream signaling pathways as receptor-mediated hydrolysis of PIP2 does. The results thus further underscore a direct effect of PIP2. With different PLC-PH domains that bind selectively to PIP2 and PIP3, Klein and co-workers demonstrated the specificity of PIP2 for the regulation of TRPV1 [48].

While the evidence for the involvement of PIP2 was increasing, it was mostly suggestive, but not conclusive, as there were few explicit studies showing a direct causal relationship between the depletion of PIP2 and the functional desensitization of the channel. Part of the problem stems from the fact that we lack specific tools for reliably inhibiting PIP2 depletion. The only pharmacological inhibitor for some isoforms of PLC, U7312, has been reported with many unwanted effects. Lukacs and colleagues showed that this inhibitor could partially reduce the desensitization of TRPV1 [47], thus providing a crucial evidence for PIP2 depletion mediating desensitization. As an alternative to the pharmacological means, Yao and co-workers resorted to physical measurements to determine [1] whether the depletion of PIP2 has kinetics fast enough to potentially modulate the channel activity and [2] whether the depletion reaches an adequate extent to significantly alter the channel function [50]. By simultaneous recording of PIP2 fluorescence and patch-clamp current, it was demonstrated that the Ca^{2+} influx through the channel led to depletion of PIP2 on a time course of a few seconds, similar to that of desensitization of currents. Furthermore, the extent of depletion reached ~60% in fluorescence changes. With a rapamycin assay that could control the extent of PIP2 depletion, it was estimated that the depletion of PIP2 at this level gave rise to ~80% of loss of channel activity. Thus the results indicate that the Ca^{2+} influx-induced depletion of PIP2 does have a significant role on the functional desensitization of the channel. Quantitatively, the desensitization shifted the capsaicin dose-response curve of the channel by ~14-fold increase in EC_{50}, while the depletion of PIP2 alone caused ~10-fold shift. Thus, in terms of changes in dose-responsiveness, the depletion of PIP2 contributed >60% to the overall desensitization induced by Ca^{2+} influx.

27.4 Physiological Functions of TRPV1 Desensitization

Desensitization is a common feature of many ion channels, and the prevention of cell toxicity due to channel overactivity has been generally cited as a physiological function. This is certainly the case for TRPV1 since it is highly permeable to Ca^{2+} and its desensitization is essential to preventing Ca^{2+} overload (reviewed in 20). However, most channels recover from desensitization immediately after the stimulus is removed. Thus, the unusual, prolonged desensitization of TRPV1 [31] prompts questions as to whether it confers additional physiological functions. The recent findings that the channel after desensitization remains functional appear to provide new insights into the question [51, 52]. Although the TRPV1 channel after desensitization becomes insensitive to normally saturating agonists, it is still responsive to agonists at the supramaximal concentrations. Both the PIP_2 depletion and the desensitization influenced only the EC_{50} of agonists while the maximal response of the channel was retained [51]. In the other words, the desensitized channel remains fully functional. Such changes on the sensitivity, but not on the dynamic range, of responses are characteristic to functional adaptation of receptors. For adaptation, the receptor needs to remain in desensitized state for a prolonged duration. Also for adaptation, the changes to the receptor should be mainly on the sensitivity rather than the dynamic response range. Adaptation is an essential feature of many sensory receptors such as those involved in vision and hearing. Thus, it is tempting to speculate that it may also occur to *some* of the pain receptors such as TRPV1. The notion of adaptation is also fit to a function of the desensitization of the channel in analgesia as induced by topical application of capsaicin [20, 31, 32].

27.5 Molecular Mechanisms of TRPV1 Sensitization

Because of its central involvement in the thermal hyperalgesia that accompanies inflammatory pain, TRPV1 has been the focus of an intense research aimed at understanding the cellular and molecular mechanisms that are implicated in its potentiation by pro-inflammatory mediators (reviewed in 12, 68, 69). These agents act upon activation of receptor-operated intracellular signaling cascades that lead to posttranslational modification of the channel, thus modifying its gating properties. Indeed, nociceptor sensitization by pro-algesic substances entails TRPV1 potentiation through phosphorylation by a series of kinases and/or by PIP_2 hydrolysis [38, 69, 70]. In this sense, one should not be surprised that TRPV1 sensitization uses complementary signalling cascades for sensitization and desensitization, namely protein phosphorylation vs. protein dephosphorylation. Thus, TRPV1 channel open probability is notably increased by the protein phosphorylation by kinases such as PKA, PKC, and CaMKII (reviewed in 71). Accordingly, it has been documented that pro-inflammatory sensitization of TRPV1 involves its chemical modification by interplay with protein kinases that lead to a decrease in its threshold of activation, and an augment of its channel activity. Noteworthy, however, TRPV1 potentiation additionally involves a rapid increase in its membrane expression level which further contributes to enhance nociceptor sensitization. Indeed, activation of PKC by pro-inflammatory mediators may potentiate TRPV1 by rapid receptor mobilization from a subcellular vesicular reservoir to the plasma membrane [72]. Therefore, it appears that sensitization of TRPV1 activity in nociceptors by algogens involves both the stimulation of the channel activity and the regulation of its levels in the cell surface.

A well studied TRPV1 potentiation mechanism implying the use of both strategies is that by nerve growth factor (NGF), which produces an acute and long lasting nociceptor sensitization by modulating TRPV1 function and expression. Release of NGF in inflamed tissues induces changes in nociceptor phenotype by activating diverse signalling cascades [8, 43]. In its acute, fast phase, NGF signals through its trkA-p75 heteromeric receptor leading to phospholipase C (PLC) activation and PIP_2 hydrolysis, with the ensuing PIP_2-inhibition release of TRPV1 [50]. Concomitantly, NGF also activates the phosphoinositide-3 kinase (PI3K)–PKCε and CaMKII signalling cascades, phosphorylating specific Ser/Thr residues on TRPV1 and thus increasing TRPV1 open probability [73, 74]. Moreover, PI3K signalling also activates Src kinase which, in turn, phosphorylates TRPV1 at Y200 inducing its translocation to the cell surface from a vesicular pool [75]. Although the mechanism of acute TRPV1 translocation evoked by Src is not well understood, it may imply PKC, since this enzyme could be also stimulated by the active signalling cascade.

In contrast, the slower and sustained inflammatory response evoked by NGF injection involves retrograde transport of the neurotrophin to the cell soma of sensory neurons, where p38 mitogen-activated protein kinase (MAPK) is activated increasing both the translation and transport of TRPV1 to the peripheral nociceptor terminal, thus contributing to enhance and maintain heat hypersensitivity [8]. It is interesting to note that the p38-mediated signalling pathway has been also involved in the inflammatory hyperalgesia evoked through activation of Bradykinin B1 receptors [76]. Several animal models of chronic pain have also shown an altered distribution of TRPV1 consisting in a shift of TRPV1 from small unmyelinated C fibers to myelinated Aβ-fibers [75, 77, 78]. Therefore, an increase of channel activity along with a change in nociceptor neurons that express the receptor gives rise to a sensitization and phenotypic change that results in the development of chronic pain associated to disease.

Similar to NGF, insulin and insulin-like growth factor-1 (IGF-1) also enhance TRPV1-mediated membrane currents through PI3K and PKC increasing channel activity and rapid receptor translocation to the cell surface [79]. In human neuroblastoma cells, both neurotrophic factors displayed long term effects augmenting the expression of the thermo-TRPs [80]. Thus, it appears that inflammatory modulation of TRPV1 surface levels is a rather pivotal mechanism of channel sensitization. However, a recent study of TRPV1 sensitization in nociceptors evaluated several pro-inflammatory mediators showing that TRPV1 membrane recruitment is not a general mechanism, thus suggesting that the strategy underlying TRPV1 sensitization may be primarily defined by the subtype of nociceptor expressing the channel [81].

Although much less information is available regarding trafficking of other thermo-TRPs, some similarities are shared with those reported for TRPV1. For instance, TRPV2 activity is also sensitized by growth factors like IGF-I by a mechanism involving surface translocation regulated by PI3K signalling [82, 83]. Nonetheless, because TRPV2 appears to associate with TRPV1 subunits [84] the increased TRPV2 expression observed during sensitization may be the consequence of TRPV1-TRPV2 heterooligomerization. Though the role of TRPV2 in thermal nociception is not clear, an upregulation has been reported in medium size diameter neurons in the nociception-induced test model of intraplantar Complete Freund's Adjuvant (CFA) injection, suggesting a role in thermal hyperalgesia during inflammation [85]. Regarding TRPA1 regulation, BK promotes TRPA1 potentiation through a PLC-dependent mechanism akin to TRPV1. Interestingly, a close functional interaction/cooperativity between both TRPV1 and TRPA1 receptors has been reported for the activation and internalization of TRPA1 [86]. Furthermore, the biophysical properties of native TRPA1 in DRGs were shown to better compare with heterologously TRPA1/TRPV1 expressing cells than those expressing [87], implying that inflammatory mobilization of TRPV1 may concomitantly recruit TRPA1 channels. Taken together, cumulative evidence substantiates the tenet that an increment in the expression of TRPV1 channels in the peripheral nociceptor terminals seems fundamental not only for the induction but also for the maintenance of a hyperalgesic inflammatory phenotype.

27.6 Modulation of TRPV1 by Signalling Complexes (Signalplex)

Ion channels do exist in membranes as isolated proteins but rather they are key components of macromolecular assemblies that act as functional protein networks. An important principle that has emerged from signal transduction research conceives the physical assembly of signalling molecules into discrete macromolecular entities termed signalling receptomes or "signalplexes" [88]. According to this notion, the biochemical steps involved in a signalling complex would require a direct physical contact between protein molecules to efficiently and specifically integrate, process, and transduce different stimuli. The assembly of a given receptor signalplex would then be governed by intracellular milieu conditions (such as intracellular Ca^{2+} concentrations ($[Ca^{2+}]_i$) or pH), ligand binding, coupling to effector molecules, as well as protein synthesis, trafficking, and post-translational modification [89]. These macromolecular organizations are composed of scaffolding, trafficking and signalling proteins. We turn now to describe the current state of the art.

27.6.1 Scaffolding Proteins

The first evidence for the formation of signalplexes important for function emerged from studies in the fly TRPC1 and TRPC2 ortholog channels. The *Drosophila* Trp signalplex is assembled in a major molecular complex protein. In fly photoreceptors, the Trp channel, its homologous TRPL, the G-protein coupled receptor rhodopsin, the scaffolding protein InaD, and the signalling proteins PLC and PKC are organized in a phototransduction signalling complex (reviewed in 90, 91). InaD is a scaffold protein with five PDZ domains; it is considered to function as a central organizer of protein complexes at the plasma membrane, capable of binding a wide diversity of signalling, cell adhesion and cytoskeletal proteins [92, 93] by typically, but not universally, binding to the most distal three residues of the C-terminal of the targeted molecules. Importantly, the association between TRPC1 and InaD has been deemed essential for correct localization of the complex in the rhabdomeres [94, 95] and to prevent degradation of unbound signalling PLC proteins. Akin to InaD, different PDZ scaffolding proteins interact selectively with other TRPC channels, like TRPC4 and TRPC5 modulating their functionality. Indeed, deletion of the PDZ-binding motif in TRPC4 strongly reduces surface expression and distribution [96].

Another scaffolding protein family that mediates associating to TRPC members is Homer, a family of proteins best known as capable of binding clusters of proteins and glutamate receptors at postsynaptic sites, and to regulate Ca^{2+} signalling. Homer 1a, for instance, has been described to increase surface expression of TRPC3, whereas Homer 1b/c, in complex with IP_3R, facilitates the exocytotic insertion and endocytosis of TRPC3 [97].

To the best of our knowledge, no PDZ-binding or Homers scaffold protein association has been clearly identified to the cytosolic domains of TRPV1. Nonetheless, immunoprecipitation experiments have shown that other non-heat sensitive TRPVs (like TRPV5 and TRPV6) interact with PDZ-proteins. For instance, TRPV5 and TRPV6 surface expression is increased in a serum- and glucocorticoid-inducible kinase (SGK1) dependent manner when coexpressed with the PDZ scaffolding protein Na^+-H^+ exchanger regulatory factor 2 (NHERF2) [98, 99]. Likewise, NHERF4 binds to TRPV5 and TRPV6 [100]. Similarly, TRPV3 appears to have a PDZ ligand at this C-end, although no interacting protein has been described, yet. Nonetheless, one should not be a surprised if in the near future proteins bearing a PDZ-binding domain were found to interact with thermo-TRPs, specially TRPV1 and TRPV3, to modulate their functionality.

In addition to their function as a backbone scaffold enabling more efficient signalling, signalplexes are likely to determine the function of individual complex elements, providing a specific protein microenvironment and/or a direct regulatory effect. For instance, AKAP79/150 (human AKAP79/rat AKAP150), a scaffolding protein that anchors PKA, PKC and calcineurin with other proteins, was found at the postsynaptic density of excitatory synapses, where it recruits neurotransmitter receptors by interaction with Membrane Associated Guanylate Kinases (MAGUK) scaffolds such as PSD-95 and SAP97 [101]. Interestingly, TRPV1 modulation by PKA and PKC and by calcineurin depends on the formation of a signalling complex with AKAP79/150 [75, 102]. Furthermore, AKAP79/150 binding to TRPV1 has been found to mediate sensitization by both bradykinin and prostaglandin PGE_2 [102]. Thus, scaffolding proteins such as AKAP appears as essential components of the TRPV1 signalplex.

27.6.2 Signalling Proteins

Several signalling proteins have been described as TRPV1 interacting proteins that could be part of a "TRPV1 receptome" that modulate nociceptor activity. As mentioned, TRPV1 associates to intracellular signalling enzymes like PKA, PKC, Src, PI3K or CaMKII kinases and also to Calcineurin 2B phosphatase [69]. Furthermore, TRPV1 may interact with the purinergic P_2X_3 receptor [103], Calmodulin (CaM) [47], the membrane protein Pirt [97], the scaffolding protein AKAP79/150 [75, 102], and with cytoskeleton proteins like tubulin [104]. All these proteins modulate channel gating by posttranslational modification, involving the phosphorylation/dephosphorylation of specific residues that, in turn, lead to a decrease in the temperatures threshold of channel activation and a potentiation of the channel activity, either by destabilizing the closed and desensitized states and/or by stabilizing the open state.

Other proteins that relevantly bind to TRPV1 are Snapin and Synaptotagmin IX, two proteins that interact with the SNARE proteins that mediate Ca^{2+}-dependent exocytosis [72]. Notably, in *Xenopus* oocytes TRPV1 could be mobilized to the cell surface upon activation of PKC signalling, indicating that this kinase modulates the receptor by direct phosphorylation and by enhancing its recruitment to the cell surface. In addition to TRPV1, SNARE complex proteins also associate with TRPM7, TRPC3 and TRPC1 and are found involved in vesicle fusion with the plasma membrane [105, 106]. TRPM7 also interacts with synapsin, a phosphoprotein present on nearly all synaptic vesicles, but no function has yet been found for this interaction.

Although the precise role of the binding of Snapin and synaptotagmin IX to TRPV1 remains elusive, it could be involved in sorting the receptor into vesicles that will be exocytose through regulated exocytosis. In support of this hypothesis, recent findings show that TRPV1 translocation by SNARE-dependent exocytosis is promoted by different components of the "Inflammatory Soup". Experiments in rat sensory neurons in culture show that membrane recruitment of TRPV1 is essential in the inflammatory sensitization of TRPV1 induced by NGF, IGF-I and ATP, and indicate that abrogation of neuronal exocytosis prevents the potentiation of TRPV1 induced by these pro-algesic agents [81]. Intriguingly, inflammation-induced TRPV1 recruitment is not a universal mechanism since pro-inflammatory mediators such as bradykinin (BK), IL-1β and artemin (ART) do not affect the surface expression of TRPV1, and their potentiation was insensitive to inhibition of regulated exocytosis. Therefore, these results indicate that TRPV1 translocation to the neuronal surface under inflammatory conditions is an important, but not general, strategy of nociceptor sensitization [81]. Furthermore, the observation that all these pro-inflammatory mediators signal towards TRPV1 by a similar intracellular pathway (i.e. activation of PKC [75]) and yet, some promote channel recruitment to the cell surface and others primarily affect channel gating is an intriguing and exciting discovery that merits further investigation.

27.6.3 Trafficking Proteins

Several proteins that regulate folding (chaperones), protein biosynthesis, surface expression and channel function have been described to associate with thermo-TRPs. For instance, Recombination Gene Activator (RGA), a protein localized in the Golgi apparatus and endoplasmic reticulum, increases cell surface levels of TRPV2. Furthermore, a role in the early biosynthesis of the TRPV2 channel and its trafficking between the endoplasmic reticulum and the different compartments of the Golgi apparatus has been suggested [107, 108]. In addition, TRPV4 binding to PACSIN3, a protein associated to synaptic vesicular trafficking and endocytosis, enhances receptor surface expression probably through endocytosis inhibition [109]. PACSIN3 modulates TRPV4 function in a stimulus specific manner [110]. Similarly, TRPV4 interacts with the microtubule-associated protein 7 (MAP7) enhancing receptor surface expression of TRPV4 [111].

Recently, using the cytosolic N-terminal domain of rat TRPV1 (Nt-TRPV1) as bait to screen a rat brain library by yeast two-hybrid technique, the $GABA_A$-Receptor Associated Protein (GABARAP), a small cytosolic protein, initially described by its ability to interact with the γ subunit of the γ-aminobutyric ($GABA_A$) receptor [112], was signalled as a TRPV1 interacting partner [113]. GABARAP has also been found to associate with other neurotransmitter receptors like the serotonergic 5-HT_3 [114], the opioid κ receptor (KOR) [115], the angiotensin (AT$_1$) receptor [116], and the Na$^+$-phosphate (NaPi-IIa) cotransporter [117]. In addition, GABARAP binds to tubulin [104], N-ethylmaleimide (NSF) [118], Calreticulin and Clathrin [119], suggesting a crucial role in intracellular trafficking. Functionally, in heterologous expression systems, GABARAP has been described to modulate GABA$_A$ receptor clustering and desensitization rates [120]. Moreover, GABARAP seems to play a key role regulating the expression of GABA$_A$ receptor in the induction of Long Term Depression (LTD) in hippocampal excitatory synapses [121], whereas in cerebellar inhibitory synapses regulates LTP (also known as Rebound Potentiation) [122].

Noteworthy, in heterologous systems GABARAP expression significantly augmented the levels of TRPV1 and its targeting to the plasma membrane (Fig. 27.4), where it appear to favor the formation of receptor clusters [113]. Paradoxically, the increase surface expression of the channels was accompanied by a significant attenuation of TRPV1 function evoked by both voltage and capsaicin in the presence of extracellular Ca^{2+}. GABARAP coexpression produced a change of the capsaicin apparent affinity and voltage-dependent activation of TRPV1 towards higher values. A plausible explanation for the discrepancy between the expression levels and channel function is that GABARAP-induced receptor clustering negatively regulates channel gating, for instance by altering channel desensitization [113].

This hypothesis is consistent with strong Ca^{2+} depencency of this inhibitory effect. A similar finding was described for the funciontal modulation of GABA$_A$ receptor by GABARAP which promotes membrane clustering and alters channel gating [120]. Paradoxically, the increased surface expression of the thermo-TRP when coexpressed with GABARAP was accompanied by a significant attenuation

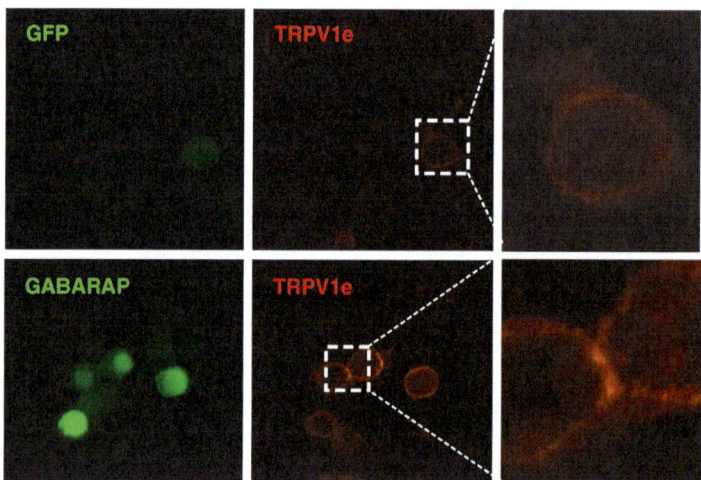

Fig. 27.4 GABA$_A$-receptor associated protein (GABARAP) induces the clustering of TRPV1 in the cell membrane. HEK293 cells were co-transfected with GFP and TRPV1 (*top panels*) or GFP-GABARAP and TRPV1 (*bottom panels*). Expression of TRPV1 was evaluated by immunocytochemistry using a polyclonal anti-TRPV1 antibody raised against the C-terminus. Surface expression was visualized by confocal microscopy

of its sensitivity to voltage and capsaicin activation in the presence of extracellular Ca^{2+}, but not in the absence of the divalent cation. A plausible explanation for the discrepancy between the higher receptor expression levels and the lower channel function could be that GABARAP-induced receptor clustering (Fig. 27.4) negatively regulates channel gating by altering the activation energy or by augmenting the agonist-induced, Ca^{2+}-dependent desensitization [113]. The strong Ca^{2+} depencency of this inhibitory effect is consitent with a modulation of receptor desensitization that is induced by intracelular Ca^{2+} levels.

It would be interesting to know what the role of GABARAP is on the nociceptive and inflammatory activity of TRPV1. It is intriguing that this thermo-TRPs is capable of interacting with both GABARAP and tubulin, which opens interesting questions as to how and where this complex is formed, especially because not all sensory neurons express GABARAP [113].

27.7 Outlook

A growing number of TRP channels are of potential therapeutic interest [123]. Clinical trials with potent, small molecule antagonists targeting the capsaicin receptor TRPV1 are on-going for diverse indications encompassing pain, chronic cough, allergic rhinitis, migraine, and overactive bladder [124]. Desensitization to capsaicin is an alternative approach to "silence" TRPV1. Indeed, site-specific capsaicin-containing patches [125] and injections [126] are already in clinical use for pain relief. A better understanding of the molecular mechanisms underlying capsaicin-desensitization may be exploited to obtain novel therapeutic agents that are devoid of the side-effects of capsaicin application. Moreover, there is mounting evidence that thermo-TRPs, and in particular TRPV1, exist as components of higher order supramolecular complexes ("signalplexes") that are essential for their spatio-temporal cellular activity. Alterations in the signalplex composition and/or dynamics may have important consequences for channels activity and nociceptor excitability. Not unexpectedly, inflammatory mediators modulate thermo-TRP channels not only [1] by directly changing their gating properties but [2] also via affecting the stability and/or dynamics of the signalling complexes. This leads to changes in TRP channel levels and/or clustering at the cell surface. We propose that a comprehensive understanding of the signalplex dynamics under nociceptive and inflammatory conditions may pave the way to innovative therapeutic strategies for pain intervention. As signalplexes may be disease state-specific, such compounds with improved therapeutic window may be synthesized that target TRPV1 in diseased, but not healthy, tissues.

Acknowledgments This work was supported by grants from el Ministerio de Ciencia e Innovación (MICINN) (BFU2009-08346 to A.F.-M. and SAF2007-63193 to R.P.-C.); the Consolider-Ingenio 2010 (MICINN, CSD2008-00005 to A.F.-M and R.P.-C.), la Fundació La Marató de TV3 (to A.F.-M and R.P.-C.).

Support for Dr. Qin was by NIH GM65994. We thank Dr. Azucena Perez Burgos for creating Figures 1, 2 and 3.

References

1. McMahon S, Koltzenburg M (eds) (2010) Wall and Melzack's textbook of pain, 5th edn. Springer, Berlin
2. Katz WA, Barkin RL (2008) Dilemmas in chronic/peristent pain mangement. Am J Ther 15:256–264
3. Noto C, Pappagallo M (2010) Current and emerging pharmacologic threrapies for pain and challenges which still lay ahead. In: Szallasi A (ed) Analgesia. Methods in Molecular Biology 617. Humana Press, Clifton, NJ
4. Geppetti P, Holzer P (eds) (1996) Neurogenic inflammation. CRC Press, Boca Raton, FL
5. Holzer P (1988) Local effector functions of capsaicin-sensitive sensory nerve endings: involvement of tachykinins, calcitonin gene-related peptide and other neuropeptides. Neuroscience 24:739–768
6. Jancsó G (ed) (2009) NeuroImmune biology. Neurogenic Inflammation in Health and Disease, Vol. 8. Elsevier, Amsterdam
7. Gudin JA (2004) Expanding our understanding of central sensitization. Medscape Neurol Neurosurg 6:1
8. Ji RR, Kohno T, Moore KA, Woolf CJ (2003) Central sensitization and LTP: do pain and memory share similar mechanisms? Trends Neurosci 26:696–705
9. Hill RG (2001) Molecular basis for the perception of pain. Neuroscientist 7:282–292
10. Caterina MJ, Schumacher MA, Tominaga M, Rosen TA, Levine JD, Julius D (1997) The capsaicin receptor: a heat-activated channels in the pain pathway. Nature 389:816–824
11. Caterina MJ, Julius D (2001) The vanilloid receptor: a molecular gateway to the pain pathway. Annu Rev Neurosci 24:487–517
12. Szallasi A, Cortright DN, Blum CA, Eid SR (2007) The vanilloid receptor TRPV1: 10 years from channel cloning to antagonist proof-of-concept. Nature Rev Drug Discov 6: 357–372
13. Caterina MJ, Leffler A, Malmberg AB, Martin W et al (2000) Impaired nociception and pain sensation in mice lacking the capsaicin receptor. Science 288:306–313
14. Davis JB, Gray J, Gunthorpe MJ, Hatcher JP et al (2000) Vanilloid receptor-1 is essential for inflammatory thermal hyperalgesia. Nature 405:183–187
15. Szabó A, Helyes Z, Sándor K, Bite A et al (2005) Role of transient receptor potential vanilloid 1 receptors in adjuvant-induced chronic arthritis: in vivo study using gene-deficient mice. J Pharmacol Exp Ther 314:111–119
16. Szallasi A, Cruz F, Geppetti P (2006) TRPV1: a therapeutic target for novel analgesic drugs? Trends Mol Med 12:545–554
17. Gunthorpe M, Szallasi A (2008) Peripheral TRPV1 receptors as targets for drug development: new molecules and mechanisms. Curr Pharm Des 14:32–41
18. Trevisani M, Szallasi A (2010) Targeting TRPV1: Challenges and issues in pain management. The Open Drug Discov J 2:37–48
19. Gomtsyan A, Faltynek CR (eds) (2010) Vanilloid receptor TRPV1 in drug discovery. Targeting Pain and Other Pathological Disorders. Wiley, Hoboken, NJ
20. Szallasi A, Blumberg PM (1999) Vanilloid (capsaicin) receptors and mechanisms. Pharmacol Rev 51:159–211
21. Szallasi A, DiMarzo V (2000) New perspectives on enigmatic vanilloid receptors. Trends Neurosci 23:491–496
22. Pingle SC, Matta JA, Ahern G (2007) Capsaicin receptor: TRPV1, a promiscuous TRP channel. Handb Exp Pharmacol 179:155–172
23. Clapham DE (2003) TRP channels as cellular sensors. Nature 426:517–524
24. Nilius B, Owsianik G (2010) Transient receptor potential channelopathies. Mol Genom Physiol 460:437–450
25. Moran MM, Xu H, Clapham DE (2004) TRP ion channels in the nervous system. Curr Opin Neurobiol 14:362–369

26. Planells-Cases R, Ferrer-Montiel A (2007) TRP channel trafficking. In: Liedtke WB, Heller S (eds) TRP ion channel function in sensory transduction and cellular signaling cascades. CRC Taylor & Francis, Boca Raton, FL
27. Huang J, Zhang X, McNaughton PA (2006) Modulation of temperature-sensitive TRP channels. Semin Cell Develop Biol 17:638–645
28. Kim H, Iadarola MJ, Dionne RA (in press) TRP polymorphism. In: Szallasi A (ed) TRP channels in health and disease – implications for diagnosis and therapy. Nova Publ, Hauppauge, NY
29. Cortright DN, Szallasi A (2009) TRP channels and pain. Curr Pharm Des 15:1736–1749
30. Patapoutian A, Tate S, Woolf CJ (2009) Transient receptor potential channels: targeting pain at the source. Nature Rev Drug Discov 8:55–68
31. Szolcsányi J (1993) Actions of capsaicin on sensory receptors. In: Wood JN (ed) Capsaicin in the study of pain. Academic Press, London
32. Knotkova H, Pappagallo M, Szallasi A (2008) Capsaicin (TRPV1 agonist) therapy for pain relief: farewell or revival? Clin J Pain 24:142–154
33. Cholewinski A, Burgess GM, Bevan S (1993) The role of calcium in capsaicin-induced desensitization in rat cultured dorsal root ganglion neurons. Neuroscience 55: 1015–1023
34. Docherty RJ, Yeats JC, Bevan S, Boddeke HW (1996) Inhibition of calcineurin inhibits the desensitization of capsaicin-evoked currents in cultured dorsal root ganglion neurones from adult rats. Pflugers Arch 431:828–837
35. Koplas PA, Rosenberg RL, Oxford GS (1997) The role of calcium in the desensitization of capsaicin responses in rat dorsal root ganglion neurons. J Neurosci 17:3525–3537
36. Liu L, Simon SA (1996) Capsaicin-induced currents with distinct desensitization and Ca^{2+} dependence in rat trigeminal ganglion cells. J Neurophysiol 75:1503–1514
37. Santicioli P, Patacchini R, Maggi CA, Meli A (1987) Exposure to calcium-free medium protects sensory fibers by capsaicin desensitization. Neurosci Lett 80:167–172
38. Liu B, Zhang C, Qin F (2005) Functional recovery from desensitization of vanilloid receptor TRPV1 requires resynthesis of phosphatidylinositol 4,5 bisphosphate. J Neurosci 25:4835–4843
39. Bhave G, Zhu W, Wang H, Brasier DJ, Oxford GS, Gereau RW (2002) cAMP-dependent protein kinase regulates desensitization of the capsaicin receptor (VR1) by direct phosphorylation. Neuron 35:721–731
40. Mandadi S, Numazaki M, Tominaga M, Bhat MB, Armati PJ, Roufogalis BD (2004) Activation of protein kinase C reverses capsaicin-induced calcium-dependent desensitization of TRPV1 ion channels. Cell Calcium 35:471–478
41. Mandadi S, Tominaga T, Numazaki M, Murayama N, Saito N, Armati PJ, Roufogalis BD, Tominaga M (2006) Increased sensitivity of desensitized TRPV1 by PMA occurs through PKCepsilon-mediated phosphorylation at S800. Pain 123:106–116
42. Mohapatra DP, Nau C (2005) Regulation of Ca^{2+} dependent desensitization in the vanilloid receptor TRPV1 by calcineurin and cAMP-dependent protein kinase. J Biol Chem 280:13424–13432
43. Mohapatra DP, Nau C (2003) Desensitization of capsaicin-activated currents in the vanilloid receptor TRPV1 is decreased by the cyclic AMP-dependent protein kinase pathway. J Biol Chem 278:50080–50090
44. Jung J, Shin JS, Lee SY, Hwang SW, Koo J, Cho H, Oh U (2004) Phosphorylation of vanilloid receptor 1 by Ca^{+2}/calmodulin-dependent kinase II regulates its vanilloid binding. J Biol Chem 279:7048–7054
45. Lishko PV, Procko E, Jin X, Phelps CB, Gaudet R (2007) The ankyrin repeats of TRPV1 bind multiple ligands and modulate channel sensitivity. Neuron 54:905–918
46. Numazaki M, Tominaga T, Takeuchi K, Murayama N, Toyooka H, Tominaga M (2003) Structural determinant of TRPV1 desensitization interacts with calmodulin. Proc Natl Acad Sci U S A 100:8002–8006

47. Lukacs V, Thyagarajan B, Varnai P, Balla A, Balla T, Rohacs T (2007) Dual regulation of TRPV1 by phosphoinositides. J Neurosci 27:7070–7080
48. Klein RM, Ufret-Vincenty CA, Hua L, Gordon SE (2008) Determinants of molecular specificity in phosphoinositide regulation. Phosphatidylinositol (4,5)-bisphosphate (PI(4,5)P2) is the endogenous lipid regulating TRPV1. J Biol Chem 283:26208–26216
49. Stein AT, Ufret-Vincenty CA, Hua L, Santana LF, Gordon SE (2006) Phosphoinositide 3-kinase binds to TRPV1 and mediates NGF-stimulated TRPV1 trafficking to the plasma membrane. J Gen Physiol 128:509–522
50. Yao JF, Qin F (2009) Interaction with phosphoinositides confers adaptation onto the TRPV1 pain receptor. PLoS Biol 7:e46
51. Novakova-Tousova K, Vyklicky L, Susankova K, Benedikt J, Samad A, Teisinger J, Vlachova V (2007) Functional changes in the vanilloid receptor subtype 1 channel during and after acute desensitization. Neuroscience 149:144–154
52. Qin F (2007) Regulation of TRP ion channels by phosphatidylinositol-4,5-bisphosphate. Handb Exp Pharmacol 179:509–525
53. Rohacs T, Thyagarajan B, Lukacs V (2008) Phospholipase C mediated modulation of TRPV1 channels. Mol Neurobiol 37:153–163
54. Rosenbaum T, Gordon-Shaag A, Munari M, Gordon SE (2004) Ca^{2+}/calmodulin modulates TRPV1 activation by capsaicin. J Gen Physiol 123:53–62
55. Vyklicky L, Novakova-Tousova K, Benedikt J, Samad A, Touska F, Vlachova V (2008) Calcium-dependent desensitization of vanilloid receptor TRPV1: a mechanism possibly involved in analgesia induced by topical application of capsaicin. Physiol Res 57(Suppl 3):S59–S68
56. Zhang X, McNaughton PA (2006) Why pain gets worse: the mechanism of heat hyperalgesia. J Gen Physiol 128:491–493
57. Cesare P, Dekker LV, Sardini A, Parker PJ, McNaughton PA (1999) Specific involvement of PKC-epsilon in sensitization of the neuronal response to painful heat. Neuron 23:617–624
58. Premkumar LS, Ahern GP (2000) Induction of vanilloid receptor channel activity by protein kinase C. Nature 408:985–990
59. Vellani V, Mapplebeck S, Moriondo A, Davis JB, McNaughton PA (2001) Protein kinase C activation potentiates gating of the vanilloid receptor VR1 by capsaicin, protons, heat and anandamide. J Physiol 534:813–825
60. Numazaki M, Tominaga T, Toyooka H, Tominaga M (2002) Direct phosphorylation of capsaicin receptor VR1 by PKCe and identification of two target serine residues. J Biol Chem 277:13375–13378
61. De Petrocellis L, Harrison S, Bisogno T, Tognetto M, Brandi I, Smith GD, Creminon C, Davis JB, Geppetti P, DiMarzo V (2001) The vanilloid receptor (VR1)-mediated effects of anandamide are potently enhanced by the cAMP-dependent protein kinase. J Neurochem 77:1660–1663
62. Lopshire JC, Nicol GD (1998) The cAMP transduction cascade mediates the prostaglandin E2 enhancement of the capsaicin-elicited current in rat sensory neurons: whole-cell and single-channel studies. J Neurosci 18:6081–6092
63. Rathee PK, Distler C, Obreja O, Neuhuber GK, Wang SY, Wang C, Nau K, Kress M (2002) PKA/AKAP/VR-1 module: A common link of Gs-mediated signaling to thermal hyperalgesia. J Neurosci 22:4740–4745
64. Liu B, Ma W, Ryu S, Qin F (2004) Inhibitory modulation of distal C-terminal on protein kinase C-dependent phospho-regulation of rat TRPV1 receptors. J Physiol 560:627–638
65. Cesare P, McNaughton P (1996) A novel heat-activated current in nociceptive neurons and its sensitization by bradykinin. Proc Natl Acad Sci USA 93:15435–15439
66. Chuang HH, Prescott ED, Kong H, Shields S, Jordt SE, Basbaum AI, Chao MV, Julius D (2001) Bradykinin and nerve growth factor release the capsaicin receptor from PtdIns(4,5)P2-mediated inhibition. Nature 411:957–962
67. Prescott ED, Julius D (2003) A modular PIP2 binding site as a determinant of capsaicin receptor sensitivity. Science 300:1284–1288

68. Di Marzo V, Blumberg PM, Szallasi A (2002) Endovanilloid signalling in pain. Curr Opin Neurobiol 12:372–379
69. Planells-Cases R, Garcia-Sanz N, Morenilla-Palao C, Ferrer-Montiel A (2005) Functional aspects and mechanisms of TRPV1 involvement in neurogenic inflammation that leads to thermal hyperalgesia. Pflugers Arch 451:151–159
70. Voets T, Nilius B (2007) Modulation of TRPs by PIPs. J Physiol 582:939–944
71. Cortright DN, Szallasi A (2004) Biochemical pharmacology of the vanilloid receptor TRPV1. An update. Eur J Biochem 271:1814–1819
72. Morenilla-Palao C, Planells-Cases R, Garcia-Sanz N, Ferrer-Montiel A (2004) Regulated exocytosis contributes to protein kinase C potentiation of vanilloid receptor activity. J Biol Che 279:25665–25672
73. Amaya F, Shimosato G, Nagano M, Ueda M et al (2004) NGF and GDNF differentially regulate TRPV1 expression that contributes to development of inflammatory thermal hyperalgesia. Eur J Neurosci 20:2303–2310
74. Bonnington JK, McNaughton PA (2003) Signalling pathways involved in the sensitisation of mouse nociceptive neurones by nerve growth factor. J Physiol 551:433–446
75. Zhang X, Huang J, McNaughton PA (2005) NGF rapidly increases membrane expression of TRPV1 heat-gated ion channels. EMBO J 24:4211–4223
76. Ganju P, Davis A, Patel S, Núñez X, Fox A (2001) p38 stress-activated protein kinase inhibitor reverses bradykinin B_1 receptor-mediated component of inflammatory hyperalgesia. Eur J Pharmacol 421:191–199
77. Hong S, Wiley JW (2005) Early painful diabetic neuropathy is associated with differential changes in the expression and function of vanilloid receptor 1. J Biol Chem 280:618–627
78. Tympanidis P, Casula MA, Yiangou Y, Terenghi G, Dowd P, Anand P (2004) Increased vanilloid receptor VR1 innervation in vulvodynia. Eur J Pain 8:129–133
79. Van Buren JJ, Bhat S, Rotello R, Pauza ME, Premkumar LS (2005) Sensitization and translocation of TRPV1 by insulin and IGF-I. Mol Pain 1:17
80. Lilja J, Laulund F, Insulin FA (2007) IGF-I up-regulate the vanilloid receptor (TRPV1) in stably TRPV1-expressing SH-SY5Y neuroblastoma cells. J Neurosci Res 85:1413–1419
81. Camprubi-Robles M, Planells-Cases R, Ferrer-Montiel A (2009) Differential contribution of SNARE-dependent exocytosis to inflammatory potentiation of TRPV1 in nociceptors. Faseb J 23:3722–3733
82. Kanzaki M, Zhang YQ, Mashima H, Li L, Shibata H, Kojima I (1999) Translocation of a calcium-permeable cation channel induced by insulin-like growth factor-I. Nat Cell Biol 1:165–170
83. Penna A, Juvin V, Chemin J, Compan V, Monet M, Rassendren FA (2006) PI3-kinase promotes TRPV2 activity independently of channel translocation to the plasma membrane. Cell Calcium 39:495–507
84. Liapi A, Wood JN (2005) Extensive co-localization and heteromultimer formation of the vanilloid receptor like protein TRPV2 and the capsaicin receptor TRPV1 in the adult rat cerebral cortex. Eur J Neurosci 22:825–834
85. Shimosato G, Amaya F, Ueda M, Tanaka Y, Decosterd I, Tanaka M (2005) Peripheral inflammation induces up-regulation of TRPV2 expression in rat DRG. Pain 119:225–232
86. Akopian AN, Ruparel NB, Jeske NA, Hargreaves KM (2007) Transient receptor potential TRPA1 channel desensitization in sensory neurons is agonist dependent and regulated by TRPV1-directed internalization. J Physiol 583:175–193
87. Salas MM, Hargreaves KM, Akopian AN (2009) TRPA1-mediated responses in trigeminal sensory neurons: interaction between TRPA1 and TRPV1. Eur J Neurosci 29:1568–1578
88. Choudhary J, Grant SG (2004) Proteomics in postgenomic neuroscience: the end of the beginning. Nat Neurosci 7:440–445
89. Sheng M, Sala C (2001) PDZ domains and the organization of supramolecular complexes. Annu Rev Neurosci 24:1–29
90. Montell C (1998) TRP trapped in fly signaling web. Curr Opin Neurobiol 8:389–397

91. Tsunoda S, Zuker CS (1999) The organization of INAD-signaling complexes by a multivalent PDZ domain protein in Drosophila photoreceptor cells ensures sensitivity and speed of signaling. Cell Calcium 26:165–171
92. Dimitratos SD, Woods DF, Stathakis DG, Bryant PJ (1999) Signaling pathways are focused at specialized regions of the plasma membrane by scaffolding proteins of the MAGUK family. Bioessays 21:912–921
93. Schillace RV, Scott JD (1999) Organization of kinases, phosphatases, and receptor signaling complexes. Clin Invest 103:761–765
94. Li HS, Montell C (2000) TRP and the PDZ protein, INAD, form the core complex required for retention of the signalplex in Drosophila photoreceptor cells. J Cell Biol 150:1411–1422
95. Tsunoda S, Sun Y, Suzuki E, Zuker C (2001) Independent anchoring and assembly mechanisms of INAD signaling complexes in Drosophila photoreceptors. J Neurosci 21:150–158
96. Mery L, Strauss B, Dufour JF, Krause KH, Hoth M (2002) The PDZ-interacting domain of TRPC4 controls its localization and surface expression in HEK293 cells. J Cell Sci 115:3497–3508
97. Kim AY, Tang Z, Liu Q, Patel KN et al (2008) Pirt, a phosphoinositide-binding protein, functions as a regulatory subunit of TRPV1. Cell 133:475–485
98. Böhmer C, Palmada M, Kenngott C, Lindner R, Klaus F, Laufer J, Lang F (2007) Regulation of the epithelial calcium channel TRPV6 by the serum and glucocorticoid-inducible kinase isoforms SGK1 and SGK3. FEBS Lett 581:5586–5590
99. Embark HM, Setiawan I, Poppendieck S, van de Graaf SF, Boehmer C, Palmada M, Wieder T, Gerstberger R, Cohen P, Yun CC, Bindels RJ, Lang F (2004) Regulation of the epithelial Ca^{2+} channel TRPV5 by the NHE regulating factor NHERF2 and the serum and glucocorticoid inducible kinase isoforms SGK1 and SGK3 expressed in Xenopus oocytes. Cell Physiol Biochem 14:203–212
100. van de Graaf SF, Hoenderop JG, van der Kemp AW, Gisler SM, Bindels RJ (2006) Interaction of the epithelial Ca^{2+} channels TRPV5 and TRPV6 with the intestine- and kidney-enriched PDZ protein NHERF4. Pflugers Arch 452:407–417
101. Dell'Acqua ML, Smith KE, Gorski JA, Horne EA, Gibson ES, Gomez LL (2006) Regulation of neuronal PKA signaling through AKAP targeting dynamics. Eur J Cell Biol 85:627–633
102. Schnizler K, Shutov LP, Van Kanegan MJ, Merrill MA, Nichols B, McKnight GS, Strack S, Hell JW, Usachev YM (2008) Protein kinase A anchoring via AKAP150 is essential for TRPV1 modulation by forskolin and prostaglandin E2 in mouse sensory neurons. J Neurosci 28:4904–4917
103. Stanchev D, Blosa M, Milius D, Gerevich Z, Rubini P, Schmalzing G, Eschrich K, Schaefer M, Wirkner K, Illes P (2009) Cross-inhibition between native and recombinant TRPV1 and P_2X_3 receptors. Pain 143:26–36
104. Goswami C, Dreger M, Jahnel R, Bogen O et al (2004) Identification and characterization of a Ca^{2+}-sensitive interaction of the vanilloid receptor TRPV1 with tubulin. J Neurochem 91:1092–1103
105. Krapivinsky G, Mochida S, Krapivinsky L, Cibulsky SM, Clapham DE (2006) The TRPM7 ion channel functions in cholinergic synaptic vesicles and affects transmitter release. Neuron 52:485–496
106. Singh BB, Lockwich TP, Bandyopadhyay BC, Liu X, Bollimuntha S, Brazer SC, Combs C, Das S, Leenders AG, Sheng ZH, Knepper MA, Ambudkar SV, Ambudkar IS (2004) VAMP2-dependent exocytosis regulates plasma membrane insertion of TRPC3 channels and contributes to agonist-stimulated Ca^{2+} influx. Mol Cell 15:635–646
107. Barnhill JC, Stokes AJ, Koblan-Huberson M, Shimoda LM, Muraguchi A, Adra CN, Turner H (2004) RGA protein associates with a TRPV ion channel during biosynthesis and trafficking. J Cell Biochem 91:808–820
108. Stokes AJ, Wakano C, Del Carmen KA, Koblan-Huberson M, Turner H (2005) Formation of a physiological complex between TRPV2 and RGA protein promotes cell surface expression of TRPV2. J Cell Biochem 94:669–683

109. Cuajungco MP, Grimm C, Oshima K, D'hoedt D, Nilius B, Mensenkamp AR, Bindels RJ, Plomann M, Heller S (2006) PACSINs bind to the TRPV4 cation channel. PACSIN3 modulates the subcellular localization of TRPV4. J Biol Chem 281:18753–18762
110. D'hoedt D, Owsianik G, Prenen J, Cuajungco MP, Grimm C, Heller S, Voets T, Nilius B (2008) Stimulus-specific modulation of the cation channel TRPV4 by PACSIN 3. J Biol Chem 283:6272–6280
111. Suzuki M, Hirao A, Mizuno A (2003) Microtubule-associated [corrected] protein 7 increases the membrane expression of transient receptor potential vanilloid 4 (TRPV4). J Biol Chem 278:51448–51453
112. Wang H, Bedford FK, Brandon NJ, Moss SJ, Olsen RW (1999) GABA(A)-receptor-associated protein links GABA(A) receptors and the cytoskeleton. Nature 397:69–72
113. Laínez S, Valente P, Ontoria-Oviedo I, Estévez-Herrera J, Camprubí-Robles M, Ferrer-Montiel A, Planells-Cases R (2010) GABAA receptor associated protein (GABARAP) modulates TRPV1 expression and channel function and desensitization. FASEB J. 19: 1745–1756
114. Sun H, Hu XQ, Emerit MB, Schoenebeck JC, Kimmel CE, Peoples RW, Miko A, Zhang L (2008) Modulation of 5-HT$_3$ receptor desensitization by the light chain of microtubule-associated protein 1B expressed in HEK 293 cells. J Physiol 586:751–762
115. Chen C, Li JG, Chen Y, Huang P, Wang Y, Liu-Chen LY (2006) GEC1 interacts with the kappa opioid receptor and enhances expression of the receptor. J Biol Chem 281:7983–7993
116. Cook JL, Re RN, deHaro DL, Abadie JM, Peters M, Alam J (2008) The trafficking protein GABARAP binds to and enhances plasma membrane expression and function of the angiotensin II type 1 receptor. Circ Res 102:1539–1547
117. Reining SC, Gisler SM, Fuster D, Moe OW et al (2009) GABARAP deficiency modulates expression of NaPi-IIa in renal brush-border membranes. Am J Physiol Renal Physiol 296:F1118–F1128
118. Kittler JT, Rostaing P, Schiavo G, Fritschy JM, Olsen R, Triller A, Moss SJ (2001) The subcellular distribution of GABARAP and its ability to interact with NSF suggest a role for this protein in the intracellular transport of GABA$_A$ receptors. Mol Cell Neurosci 18:13–25
119. Mohrluder J, Hoffmann Y, Stangler T, Hanel K, Willbold D (2007) Identification of clathrin heavy chain as a direct interaction partner for the gamma-aminobutyric acid type A receptor associated protein. Biochemistry 46:14537–14543
120. Chen L, Wang H, Vicini S, Olsen RW (2000) The gamma-aminobutyric acid type A (GABA$_A$) receptor-associated protein (GABARAP) promotes GABAA receptor clustering and modulates the channel kinetics. Proc Natl Acad Sci USA 97:11557–11562
121. Marsden KC, Beattie JB, Friedenthal J, Carroll RC (2007) NMDA receptor activation potentiates inhibitory transmission through GABA receptor-associated protein-dependent exocytosis of GABA$_A$ receptors. J Neurosci 27:14326–14337
122. Kawaguchi SY, Hirano T (2007) Sustained structural change of GABA$_A$ receptor-associated protein underlies long-term potentiation at inhibitory synapses on a cerebellar Purkinje neuron. J Neurosci 27:6788–6799
123. Moran MM, Szallasi A (eds) (2010) TRP channels as drug targets. The Open Dug Discov J 2(suppl)
124. Szallasi A (ed) (2010) TRP channels in health and disease: implications for diagnosis and therapy. Nova, Hauppauge, NY
125. Noto C, Pappagallo M, Szallasi A (2009) NGX-4010, a high-concentration capsaicin dermal patch for lasting relief of peripheral neuropathic pain. Curr Opin Invest Drugs 10:702–710
126. Remadevi R, Szallasi A (2008) Adlea (ALGRX-4975), an injectable capsaicin (TRPV1 receptor agonist) formulation for long-lasting pain relief. IDrugs 11:120–132

Chapter 28
Voltage Sensing in Thermo-TRP Channels

Sebastian Brauchi and Patricio Orio

Abstract Membrane voltage, ligand binding, mechanical force and temperature can all induce conformational changes that open ion channel pores. A key question in understanding ion channel function is how the protein domains involved in sensing stimuli (sensors) communicate with the pore to gate its opening and closing. TRP channels are considered six-transmembrane cation-permeable channels, distant relatives of voltage-gated potassium channels (Kv), which are known to be activated by membrane depolarization. Understanding the molecular nature of thermo-TRP channel gating offers a fair challenge to biophysicists. This chapter will summarize our present knowledge on the effect of voltage and temperature during thermo-TRP channel activation.

28.1 TRP Channel Family and Thermo-TRPs

Mammalian TRP channel proteins are polymodal cation channels with essential roles in cellular sensing. Other than a loose sequence homology, predicted channel architecture, and a common poor cation selectivity, there are no particular features defining the TRP family. TRP channels are grouped by homology into six subfamilies named C, M, V, A, P, and ML, for canonical, melastatin related, vanilloid binding, ankyrin repeat, polycystin, and mucolipin, respectively [1]. By integrating multiple stimuli they supply signal amplification through calcium permeation and membrane depolarization. Cooperativity intrinsic to TRP channels may result in allosteric coupling of distinct activation stimuli. A good example of the allosteric nature of TRP channels would be the case of temperature-activated TRP channels (*thermo-TRPs*) [2]. Mammalian thermo-TRPs correspond to a subgroup of 9 TRP channels which are expressed in sensory nerve endings and in skin, characterized

---

S. Brauchi (✉)
Facultad de Medicina, Instituto de Fisiologia, Universidad Austral de Chile,
Valdivia 511-0566, Chile
e-mail: sbrauchi@docentes.uach.cl

by their distinctive high temperature-dependence. Thermo-TRP channels elicit different dynamic ranges for their activation profile, are activated by different natural compounds, and are also known because of their participation in nociceptive pathways. One interesting feature of thermo-TRP channels is the presence of members from at least three different TRP families (V, M, and A). Whereas TRPV1–4, TRPM2, TRPM4, and TRPM5 are heat-activated, TRPM8 and TRPA1 are activated by cold [3]. The first member of these channels that was cloned was TRPV1 [4], and the temperature activation of TRPA1 remains controversial.

28.2 The Process of Channel Gating

Based on our understanding of canonical voltage-gated ion channels, it is predicted that membrane depolarization should induce a conformational change in the Voltage Sensor of TRP channels, where positively charged amino acids (and/or dipoles) contained within the trans-membrane region can be affected by such changes of membrane potential. However, despite this preconceived paradigm of voltage-dependence, the location of the voltage sensor in TRP channels remains undefined. Nevertheless, whatever the location and nature of the voltage sensor within the channel structure, some voltage-dependent conformational change lead to channel gating, opening the conduction pathway. Once TRP channels open, they allow the flux of cations down their electrochemical gradient.

Channel gating process can be understood as a process of distribution of particles (p_1 and p_2) in different energy levels, therefore can be described by the Boltzmann equation:

$$\frac{p_2}{p_1} = \exp\left(-\frac{u_2 - u_1}{k_b T}\right) \tag{28.1}$$

Where $u_2 - u_1$ represents the energy difference between the levels, k_b is Boltzmann's constant and T is the absolute temperature. For the case of ion channels, the Boltzmann equation will allow us to describe the distribution of the open or closed conformations at equilibrium (open probablity, P_o = open/(open + closed)). If the free energy (ΔG) for transition between the closed and open state is expressed in terms of electrical energy of an electrical particle (with charge z) responding to a change in membrane potential by moving across the electric field (V), the open probability will correspond to:

$$P_O = \frac{1}{1 + \exp\left(-\frac{zF\Delta V}{RT}\right)} \tag{28.2}$$

where F and R correspond to the Faraday constant and the gas constant respectively.

At resting membrane potentials, the closed state of voltage-sensitive channels is lower in energy than the open state, and is therefore the preferred conformation of the protein. Energy provided from external forces (e.g. agonist binding, temperature, voltage) will allow for the population to reach conformational states that otherwise would be less likely. At depolarized membrane voltages, the conformational energy for the open state is lower, allowing the stabilization of the channel in its open conformation.

TRP channels are tetramers composed by 6 transmembrane protein subunits, resembling *Shaker* type voltage-dependent potassium (Kv) channels. By analogy to Kv channels, the working hypothesis is that each TRP channel subunit may be divided into two major functional domains, a VSD-like domain (S1–S4) and a pore domain (S5–S6) that contains the actual channel gate. In Kv channels, the voltage sensing domain (VSD) is connected to the pore domain through the S4–S5 linker, which likely functions to transfer the necessary mechanical energy from VSD movement to the gated pore.

28.3 Voltage Dependence

Voltage-gated ion channels play a pivotal role in muscle contraction, neuronal excitability, and secretion. The VSD is the fundamental feature of voltage-gated ion channels for sensing transmembrane potential, and has been studied for more than 50 years [5] at the levels of both biophysics and protein structure [6, 7]. The main feature of VSD is the array of positively charged amino acids in the fourth and sometimes the second transmembrane segment [8–10]. These charges sense and induce a protein movement in response to a changing electrical field [5, 11]. For the case of classical voltage-dependent ion channels, in which VSD movements are highly coupled to gate opening, voltage-sensitivity is very strong (e.g. *Shaker* type Kv channels) and the Boltzmann distribution changes over a narrow range of voltage. Recent findings have demonstrated that a similar structural strategy for voltage sensing is used not only by voltage-dependent channels, but also in voltage-dependent enzymes [12] and voltage-dependent proton channels [13, 14].

Predicted by Hodgkin and Huxlex in 1952, the presence of the "activating particles", was first measured almost twenty years later [15, 16]. Generally speaking, gating currents correspond to a nonlinear capacitive current transient explained by the movement of a charged particle within the membrane; thus measuring the number of electrical charges associated to the gating process provides a descriptor of its voltage-dependency. Strongly voltage dependent channels such as *Shaker* K^+ channels can displace up to 12 effective electrical charges during activation of a single ion channel [17, 18]. Arginine residues located at the S4 segment of *Shaker* K^+ channels were proven to be the charges responsible for this voltage-dependence [9, 19]. For the case of *Shaker* K^+ channels the first four arginines in the S4 helix account for most of the displaced gating charge and during channel activation they move through the entire electric field. Gating current recordings are one of the important gaps present in TRP channel biophysics, however, some efforts have been made to surmount this obstacle. Equivalent gating charges can be obtained from conductance vs. voltage data using the form of the Boltzmann function described on Eq. (28.2). Using this method, the calculated number of gating charges for TRPV1 and TRPM8 is 0.6–0.8e [20, 21]. Similar weak voltage dependence ($z\delta \sim 0.4$) was also observed for TRPA1 channels [22]. Such weak voltage dependence is in contrast with the strong voltage dependence observed in Kv channels, suggesting that the inner workings of the voltage-sensing mechanism may not be exactly the same. However, regardless of mechanism, thermo-TRPs are voltage-activated channels [20, 21, 23], therefore they should contain a voltage sensor.

Because of their apparent structural similarity with Kv channels [24, 25], dogma dictates the existence of an S1–S4 voltage-sensor-module. Following this logic, Voets et al. [27] used the limiting slope analysis [18, 26] to calculate 0.85 effective gating charges associated with TRPM8 channel activation. Since TRP channels are tetramers, this finding implies that each voltage sensor only contributes about 0.2 electronic charges coupled to channel opening. The predicted S4 domain of TRPM8 channels contains 2 arginines, one of them conserved throughout the TRPM channel sub-family, and a histidine residue that, depending of local pH and pKa may contribute to the total gating charge (Fig. 28.1). Neutralization of charges contained in TRPM8s' S4 (R842) and in the S4–S5 linker (K856) reduces the total gating charge to 0.62 suggesting that these residues may participate in the voltage sensor activation process [27]. This result is somewhat unexpected since according to the crystal structure of Kv1.2 [7] the S4–S5 linker is located outside the electric field, suggesting that K856 would unlikely to contribute to the total gating charge. This result may be explained only if the tertiary structure of TRPM8 differs from Kv1.2, allowing the S4 helix to be longer (e.g. more tilted or kinked), thereby placing K856 inside the electric field. The only structural data availble for TRP channels came from cryo-EM studies. There are notorious differences between the cryo-EM structures available. While TRPVs seems similar to Kv1.2, TRPMs and TRPCs looks awkwardly bulky and bigger [3, 24]. Thus, there is no reason to dismiss the possibility that the structure of TRPM8 is different from the known structure of Kv channels.

The results obtained by Voets et al. [27] support the use of a two-state model (see *allosteric models section*), however, this model does not explain the single channel electrical activity that is characterized by bursts separated by long resting periods [23, 28]. The most simple kinetic model able to account for this type of channel gating will require more than one closed state and at least two open states [23]. If this is the case, the determination of the effective gating charge using the limiting slope may not be the best experimental approach since this method does not provide an accurate determination of the total charge when describing kinetic models containing more than one open state connected by voltage-dependent transitions [18].

Based on single channel recordings, the voltage dependence of TRP channels appears to be intrinsic to the channel-forming protein. A closer inspection of the predicted S4 segment of these channels reveals the presence of only one basic residue in TRPV1, TRPV3, and TRPV4 and possibly three in TRPM8. TRPA1 is also voltage-dependent despite the fact that it does not have a single positively charged residue in its S4. Although it is possible that the weak voltage dependence of thermo-TRPs is caused by the low density of positive charges in the S4 domain, the need for a plausible alternative model seems to be imperative because these observations strongly suggest that the actual location of the voltage sensor in TRP channels is a well kept secret.

Fig. 28.1 Topology of TRPM8 S4 helix. The model shows the localization of charged amino acids in the TRPM8 S4 segment. F839 and T848 correspond to voltage-sensing arginines in Kv channels. TRPM8 homology model was built using the crystal structure of Kv1.2 (PDB:2A79) as template. Intra- and extracellular loops were relaxed using a Monte Carlo (MC) protocol implemented in ICM. The initial minimization was followed by a short molecular dynamics simulation (1 ns). The assembly of the system and figures employed VMD (Visual Molecular Dynamics; http://www.ks.uiuc.edu/Research/vmd/)

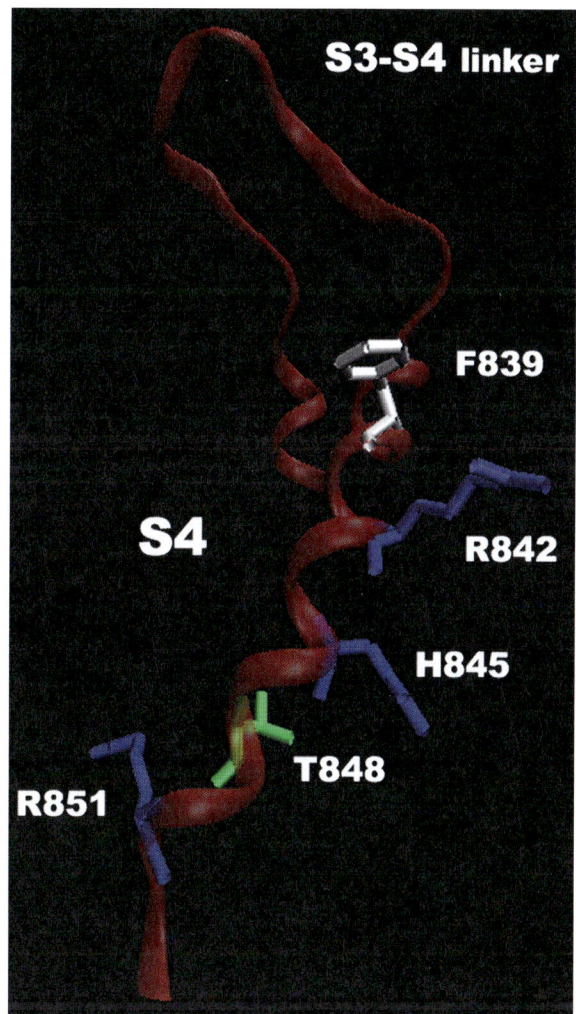

28.4 Temperature Activation

Changes in temperature affect the protein conformational landscape, therefore ion channel gating. Temperature-activated TRP channels have Q_{10}[1] values up to 30, whereas Q_{10} values for non-temperature-dependent enzymes or ion channels is about 2. Changes in ambient temperature are well correlated with changes in the firing rate of somatosensory neurons [29]. Temperature activation of thermo-TRP channels allows Na^+ and Ca^{2+} to enter cells, resulting in depolarization of sensory

---

[1] Change in enzyme activity over a change of 10°C.

neurons and triggering of action potentials. This is again a process in which the energy provided to the channel (in the form of heat) changes the distribution of open and closed states according to Eq. (28.1). Due to the stochastic nature of the gating process, the widespread concept of threshold for activation lacks meaning. However, there are temperature ranges at which different thermo-TRP channels exhibit significant changes on their open probability, from cool (24–10°C; TRPM8), to warm (> 30°C; TRPV3, TRPV4), to hot (> 40°C; TRPV1).

For the case of TRPV1 and TRPM8, temperature produces large (> 100 mV) leftward shifts of the voltage activation curve upon heating and cooling, respectively [20, 21] (Fig. 28.2). Thermodynamic parameters such as the overall changes in enthalpy (H) and entropy (S) associated with temperature- and voltage- dependent channel opening can be obtained considering a simple two-state model:

$$C \underset{k_{-1(T,V)}}{\overset{k_{1(T,V)}}{\rightleftharpoons}} O$$

Where k_1 and k_{-1} correspond to activation and deactivation rate constants, respectively. C and O can represent a collection of closed and open states, respectively, as this thermodynamical analysis is independent of the kinetic activation scheme and the number of states or transitions. It is easy to visualize the thermodynamic meaning of the observed temperature-dependent shift in the open probability vs. voltage curve (Fig. 28.2) since:

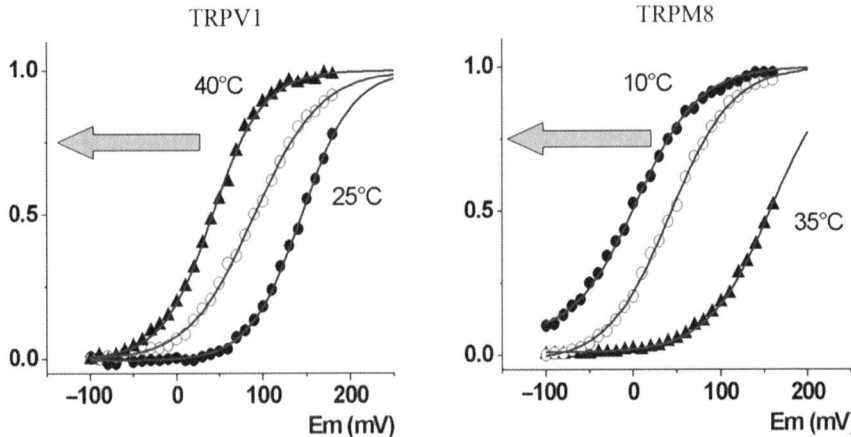

Fig. 28.2 Voltage vs normalized Conductance relationships are shifted upon heating and cooling. Plots showing the normalized conductance (G/Gmax) in function of voltage (Em) at different temperatures for cells expressing TRPV1 (*Left*) or TRPM8 (*Right*). Solid lines correspond to the best fit to Boltzmann functions. The arrow indicates the direction of curve shifting upon temperature activation

$$K_{eq} = \frac{k_1}{k_{-1}} = \exp\left(-\frac{\Delta G}{RT}\right) \tag{28.3}$$

Substitution into Eq. (28.2) results in a predicted relation between the equilibrium constant (K_{eq}) and the open probability.

$$P_O = \frac{1}{1 + K_{eq}^{-1}} \tag{28.4}$$

If we consider that ΔG corresponds to the transitional energy for the close-to-open reaction [30] according to:

$$\Delta G = \Delta H - T\Delta S - zF\Delta V \tag{28.5}$$

And assuming that for the temperature-dependent transition the term $[-zF\Delta V]$ is negligible, the enthalpy and entropy changes (ΔH and ΔS, respectively) associated with the open transition can be calculated from the slope of the ln K vs. of 1/T plot, according to the van't Hoff equation:

$$\ln K_{eq} = -\frac{\Delta H}{RT} + \frac{\Delta S}{R} \tag{28.6}$$

Voets et al. [20] have proposed that the temperature-dependent activities of TRPV1 and TRPM8 can be explained by the effect of temperature on the voltage-dependent gating. According to this, because the gating charge (z) in thermo-TRPs is small, temperature changes promote large shifts of the voltage-activation curves when compared with a channel with strong voltage dependence such as *Shaker*.

However, it can be argued that what makes thermo-TRPs temperature-dependent is not their modest gating charge, but rather the enormous entropic change and its compensation by a large change in enthalpy, rendering small values for the net free energy change (ΔG) that can be easily overcome to transit between the closed and open state. It would not be unreasonable to state that having such small voltage dependence allows TRP channels to be sensitive to other stimuli. A strongly coupled voltage-sensor would lock the channel on the closed state at non-permisive potentials, therefore, different protein rearrangements originated by the activation of alternative sensor modules (e.g. temperature or agonist binding) will not be translated on pore openings. Therefore a weak-voltage-dependence is likely necessary for the polymodal regulation observed for TRP channels. This is explained by a relation for the change in P_O with temperature presented by Latorre et al. [23]:

$$\frac{\partial P_O}{\partial T} = \frac{\Delta H - zFV}{\frac{RT^2}{4}\cosh^2\left(\frac{\Delta G}{2RT}\right)} \tag{28.7}$$

This function tends to zero when $\Delta G \to \infty$, and tends to $4(\Delta H-zFV)/(RT^2)$ when $\Delta G \to 0$. As thermo-TRP channels exhibit relatively small ΔG values during the closed-open transition, from this relation is clear that to ensure large changes in P_o, the only value that has to be large is ΔH [23]. This, in turn, implies that the entropy term has to be large in order to obtain a small change in the free energy. The absolute magnitude for the transitional enthalpy and entropy during TRP channel temperature-dependent gating has been estimated to be about 50–300 kcal/mol [20, 21, 31].

A reasonable explanation for the effect of temperature on thermo-TRP channels would be the presence of a temperature-sensing domain suffering large structural rearrangements upon temperature changes. Another case in which a highly temperature-dependent process takes place only because the large enthalpic change is compensated by a large entropic change keeping ΔG relatively small is protein denaturation [32, 33]. While heat denaturation arises from the fact that disordered conformations become accessible upon heating, cold denaturation is usually attributed to a weakening of the hydrophobic effect caused by the temperature-dependent structure of bulk water [34–36]. It has been previously proposed that the large negative entropy (about −200 kcal/mol; [21, 31]) observed during the temperature-activation process of TRPM8 channels may be linked to a net loss of hydrophobic interactions in the closed-to-open temperature-dependent transition [21, 23]. Therefore it is possible that during the process of opening, an exposure of aliphatic and aromatic groups to bulk water takes place.

Using fluorescence resonance energy transfer, in combination with electrophysiological recordings, and site directed mutagenesis, Yang et al. [31] showed that conformational rearrangements of the turret, a structure located on the external mouth of the pore domain (Fig. 28.3), are essential for temperature-dependent activation [31]. This result is somewhat supported by recently published results in which pore mutations near the turret region either ablate or severely affect temperature dependent gating [37–39]. However, the results presented by Yang et al. have been severely questioned and this controversy remains to be solved [48, 49].

28.5 Allosteric Models and Thermo-TRP Channel Activation

A contentious issue regarding voltage activation of TRMP8 and TRPV1 is the strictness of coupling between the voltage sensor and the pore gate. A strict coupling implies that whenever the voltage sensor(s) is (are) activated the channel will open, and this basic assumption underlies two-state models of channel activation. On the other hand, allosteric coupling assumes that activation of voltage sensor(s) does not lead directly to channel opening but rather to an increase of the open state probability. These two models have profound differences in their mechanistic interpretations and the prediction of channel behavior under certain conditions. Allosteric coupling has been proposed and demonstrated for several voltage-activated channels, remarkably the hyperpolarization- and cyclic nucleotide-gated (HCN) channels [43] and the high conductance voltage- and calcium-activated potassium (BK) channel [44]. A common feature of these channels, also shared by thermo-TRP channels, is the activation by voltage *and* another

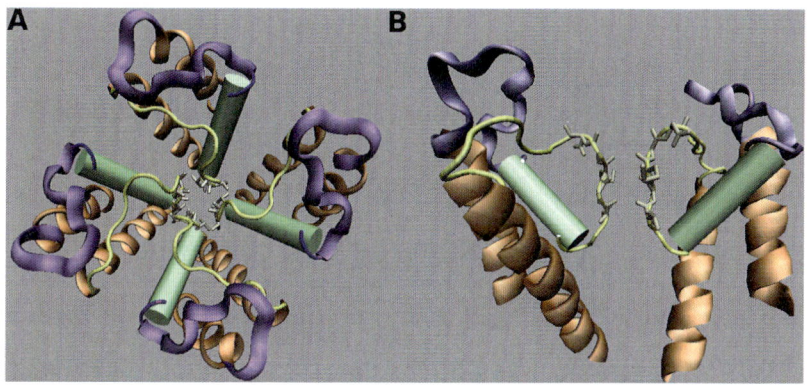

Fig. 28.3 TRP channels Pore Turret has been suggested as an structural part of the temperature activation mechanism. (**a**) An upper view of the TRPV1 pore. Purple shade ribbon highlights the location of the pore turret segment mutated by Yang et al. 2010. (**b**) A side view of the TRPV1 pore just showing two subunits, highlighting the location of the turret (purple ribbon) with respect to the selectivity filter (yellow sticks), and the pore helix (green cartoon). TRPV1's pore homology model was built using the crystal structure of Kv1.2 (PDB:2A79) as template [40]. The assembly of the system and figures employed VMD

stimulus (e.g. agonist binding). Also, their voltage dependence is much lower than that of typical voltage-activated channels, reflected in shallower conductance vs. voltage relationships.

One structural condition for a channel to be thought as allosterically activated is that its sensors have to be different protein domains. Although different domains have been proposed as part of the voltage and temperature sensors [27, 31], they are far from being well described. A modular structure has been proposed for thermo-TRP channels [2, 40], with evidence suggesting the existence of different activation domains for voltage, temperature, and PIP_2 in TRPM8. However, this is not sufficient as evidence for allosteric gating, as the classical (e.g., Shaker related) voltage-activated channels also have a modular structure [41] but their limiting slope for voltage-dependent activation suggests a strict coupling between the voltage sensors and the gate [17].

An allosteric gating scheme can be thought as two or more unconnected state transitions whose rate (and equilibrium) constants are modified depending on the state of the other equilibria. Figure 28.4a depict the simplest example of a 2-state pore allosterically gated by a 2-state voltage sensor. When the voltage sensor is in the resting state, (C_R or O_R), the channel opens with a (probably very low) equilibrium constant L. When the sensor is in the activated state (C_A or O_A), this equilibrium constant is multiplied by an allosteric factor D, thus incrementing the probability of the open state. Conversely, when the pore gate is in the open state, the equilibrium constant for voltage sensor activation (J) is multiplied by the same factor D, thus fulfilling the microscopic reversibility principle.

Additional gating mechanisms of the channel, such as inactivation or other sensors for different stimuli, can be added either as additional states or as another

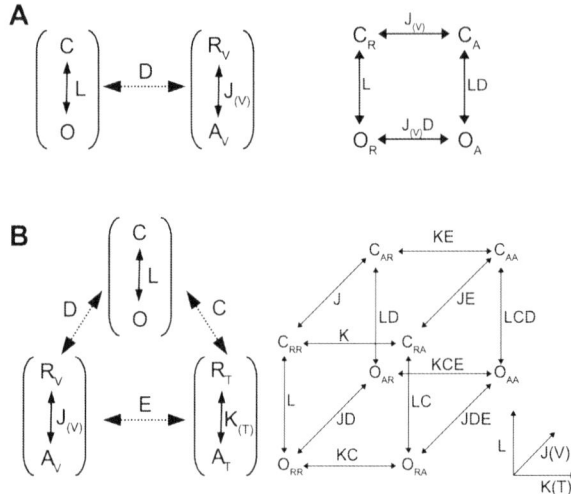

Fig. 28.4 Allosteric models for channel activation. (**a**) allosteric activation by voltage. *Left*, two independent equilibriums that interact allosterically. R_V and A_V represent the resting and activated state of the voltage sensor. C and O are the closed and open conformation of the pore gate. Only the equilibrium constant for voltage sensor activation, J, is voltage dependent. *Right*, when the combinations of the two equilibriums are considered, a 4-state diagram results. Cn and On represent closed and open conformations of the pore, respectively, with the voltage sensors in the 'n' state. The probability of being in any open state is: $P_O = \left(1 + \frac{1+J}{L(1+JD)}\right)^{-1}$ (**b**) Allosteric activation by voltage and temperature. A third equilibrium is added where R_T and A_T represent the resting and active conformations of the temperature sensor, respectively. Two new allosteric factors are introduced (C and E) to account for the possible interactions between structures. The expanded representations needs to be rendered in 3D (right). In this scheme, the probability of being in any open state is: $P_O = \left(1 + \frac{1+J+K+JKE}{L(1+JD+KC+JKCDE)}\right)^{-1}$

equilibrium that will interact allosterically. Figure 28.4b depicts the model proposed by Brauchi et al. [21] for the activation by voltage and temperature, both in its allosteric depiction and its expanded form that takes the shape of a cube. The number of states of the channel as a whole, as depicted in Fig. 28.4b, grows quickly when other sensors for stimuli such as temperature or chemical agonists are included, and soon cannot be easily depicted in 2 dimensions. For instance, Matta and Ahern [42] expanded the model to include activation by ligands and the expanded form needs four dimensions to be drawn with all the possible transitions [42]. However, when seen in its allosteric depiction (Fig. 28.4b, left), it is realized that a relatively small number of parameters are required to describe the proposed intramolecular interactions with simplicity. Also, the probability of the open state can be predicted with straightforward mathematical expressions and kinetic modeling can be simplified.

The experimental behavior of a channel predicted by an allosteric gating scheme has several features that have been described and extensively studied in BK and HCN channels [43, 44]. They include the existence of several open states evidenced as multiple mean open times or multiple exponentials in current relaxations, and the movement of gating charges between open states. However, the most notorious prediction is that there will be a minimum open probability that can be reached by hyperpolarization (given by the equilibrium constant L) and a maximum open probability upon depolarization (given by the product LD). Depending on the values of L and D this may or may not be evident from macroscopic current recordings. For instance, the BK channel shows a maximum P_O almost equal to 1 (LD≫1) and there is no minimum open probability because the equilibrium constant L has a weak voltage dependence that has been shown to be independent of the voltage sensors. Still, this phenomenon is evident only at $P_O < 10^{-6}$ [45].

When a third sensor is added to an allosteric gating scheme, its activation should modify the maximum P_O achievable by voltage sensor activation (Fig. 28.5). This led Brauchi et al. [21] to postulate an allosteric gating scheme for TRPM8, to account for recordings demonstrating that cold increases the maximum P_O. This was not noticed by Voets et al. [20] but they based their analysis and fitting only on macroscopic current recordings while Brauchi et al. measured actual P_O values with single channel recordings. Later, Matta and Ahern [42] reported the same for both TRPM8 and TRPV1 (in the case of TRPV1, heating increases maximum P_O) also with the support of single channel recordings. Moreover, Matta and Ahern showed that the minimum P_O of both channels is dramatically increased by the presence of chemical agonists (menthol for TRPM8, capsaicin and resiniferatoxin for TRPV1). The analysis presented by Voets et al. [27] for the activation of TRPM8 by agonists led them to propose an allosteric gating scheme for the effect of menthol, while retaining a 2-state mechanism for the activation by voltage (implying a strict coupling). Again, they did not determine actual values of P_O and based their conclusions on normalized macroscopic G/V curves.

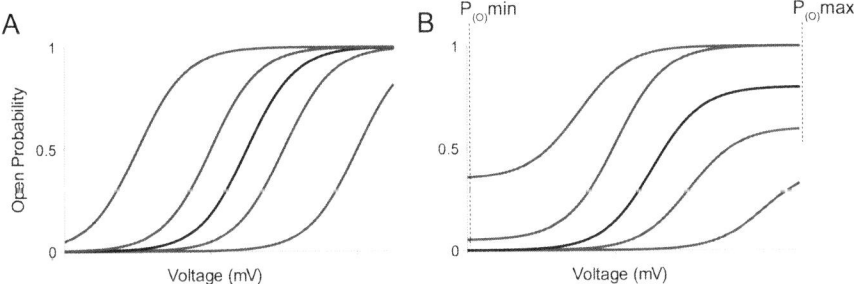

Fig. 28.5 Different behavior of strict coupling versus allosteric coupling. (**a**) A two-states model for voltage activation (strict coupling between sensor and gate) will always have maximum $P_{(O)} = 1$ and minimum $P_{(O)} = 0$. Another agonist or activator can only change the $V_{1/2}$ of the G/V relationship shifting the curve to left or right. (**b**) On the other hand, with allosteric activation by voltage minimum and maximum $P_{(O)}$ is not restricted to 0 and 1. In the model depicted in Figure 28.4b, when the temperature sensor is resting ($K_{(T)} \ll 1$) the minimum $P_{(O)}$ (when $J_{(V)} \ll 1$) is $(1+1/L)^{-1}$ and the maximum open probability that can be reached by depolarization ($J_{(V)} \gg 1$) is $(1+1/LD)^{-1}$. On the contrary, when the temperature sensor is fully active ($K_{(T)} \gg 1$) the minimum $P_{(O)}$ is $(1+1/LC)^{-1}$ and the maximum is $(1+1/LCD)^{-1}$. The values of L, C and D will determine whether these probabilities are significantly different from 0 or 1 for a macroscopic current analysis, but nevertheless they should always be measurable with single channel recordings

The debate between strict and allosteric coupling has other arguments than maximum and minimum open probabilities; Latorre et al. [2] showed that activation by voltage of TRPM8, despite being very fast, shows a brief delay after the onset of a depolarizing pulse. Moreover, this delay is shortened by a depolarizing pre-pulse. This phenomenon, known as the Cole-Moore effect, cannot be observed in a two-state scheme for voltage activation and is a clear indication of multiple closed states [46]. More evidence supporting the existence of multiple open and closed states comes from single channel dwell time analysis in the TRPV1 channel [28] and the bursting nature of the single channel activity [23].

Thus, it appears difficult to sustain a strictly coupled 2-state mechanism for voltage-activation of thermo-TRP channels. Though it has a tempting simplicity that reproduces very well the macroscopic G/V relationships and their modulation by temperature and agonists, the debate between strict and allosteric coupling of the voltage sensor and the pore gate goes beyond the best fit of the experimental data. If the purpose of the model is to reproduce the channel behavior (to be used in a conductance-based neuron model, for instance) then the choice will be the simplest model. But when it comes to draw mechanistic or structural conclusions, then the model should be challenged at every thinkable condition, even at the extremes. Evidence for allosteric coupling has arisen even in the family of Kv channels, thought to have the strictest coupling between the voltage sensor and the pore gate [47], and it would not be surprising that allosteric coupling is the norm that in classical voltage-activated channels was tuned to an almost strict coupling. However in channels activated by more than one stimuli allosteric coupling may be more convenient, because the dynamic range for the effect of each agonist can be independently tuned.

28.6 Final Words

In summary, it is clear that different stimuli act separately to modulate thermo-TRP channel activity therefore these channels act as signal integrators that sum the energies imparted by voltage, temperature, ligand binding, and pH to open the channel pore. To fulfill their role as integrators a loose pore is needed, since strict coupling between any of these stimuli to the gate will probably impede the others to open the channel pore.

Acknowledgments SB work is funded by FONDECYT grant 11070190. PO work is funded by FONDECYT 11090308 and DIPUV 51/2007 (U. de Valparaiso) grants. We thank Dr. H. Kurata for his critical reading of the manuscript.

References

1. Ramsey IS, Delling M, Clapham DE (2006) An introduction to TRP channels. Annu Rev Physiol 68:619–647
2. Latorre R, Brauchi S, Orta G, Zaelzer C, Vargas G (2007) Thermo TRP channels as modular proteins with allosteric gating. Cell Calcium 42:427–438

3. Latorre R, Zaelzer C, Brauchi S (2009) Structure-functional intimacies of transient receptor potential channels. Q Rev Biophys 42:201–246
4. Caterina MJ, Schumacher MA, Tominaga M, Rosen TA, Levine JD, Julius D (1997) The capsaicin receptor: a heat-activated ion channel in the pain pathway. Nature 389: 816–824
5. Hodgkin AL, Huxley AF (1952) A quantitative description of membrane current and itsapplication to conduction and excitation in nerve. J Physiol 117:500–544
6. Bezanilla F (2000) The voltage sensor in voltage-dependent ion channels. Physiol Rev 80:555–592
7. Long SB, Tao X, Campbell EB, MacKinnon R (2007) Atomic structure of a voltage-dependent k+ channel in a lipid membrane-like environment. Nature 450:376–382
8. Noda M, Shimizu S, Tanabe T, Takai T, Kayano T, Ikeda T, Takahashi H, Nakayama H, Kanaoka Y, Minamino N et al (1984) Primary structure of electrophorus electricus sodium channel deduced from cDNA sequence. Nature 312:121–127
9. Seoh SA, Sigg D, Papazian DM, Bezanilla F (1996) Voltage-sensing residues in the S2 and S4 segments of the shaker k+ channel. Neuron 16:1159–1167
10. Jiang Y, Lee A, Chen J, Ruta V, Cadene M, Chait BT, Mackinnon R (2003) X-ray structure of a voltage-dependent k channel. Nature 423:33–41
11. Swartz KJ (2008) Sensing voltage across lipid membranes. Nature 456:891–897
12. Murata Y, Iwasaki H, Sasaki M, Inaba K, Okamura Y (2005) Phosphoinositide phosphatase activity coupled to an intrinsic voltage sensor. Nature 435:1239–1243
13. Sasaki M, Takagi M, Okamura Y (2006) A voltage sensor-domain protein is a voltage-gated proton channel. Science 312:589–592
14. Ramsey IS, Moran MM, Chong JA, Clapham DE (2006) A voltage-gated proton-selective channel lacking the pore domain. Nature 440:1213–1216
15. Armstrong CM, Bezanilla F (1973) Currents related to movement of the gating particles of the sodium channels. Nature 242:459–461
16. Schneider MF, Chandler WK (1973) Voltage dependent charge movement of skeletal muscle: a possible step in excitation-contraction coupling. Nature 242:244–246
17. Schoppa NE, McCormack K, Tanouye MA, Sigworth FJ (1992) The size of gating charge in wild-type and mutant shaker potassium channels. Science (80) 255:1712–1715
18. Sigg D, Bezanilla F (1997) Total charge movement per channel. the relation between gating charge displacement and the voltage sensitivity of activation. J Gen Physiol 109: 27–39
19. Aggarwal SK, MacKinnon R (1996) Contribution of the S4 segment to gating charge in the shaker K+ channel. Neuron 16:1169–1177
20. Voets T, Droogmans G, Wissenbach U, Janssens A, Flockerzi V, Nilius B (2004) The principle of temperature-dependent gating in cold- and heat-sensitive TRP channels. Nature 430:748–754
21. Brauchi S, Orio P, Latorre R (2004) Clues to understanding cold sensation: thermodynamics and electrophysiological analysis of the cold receptor trpm8. Proc Natl Acad Sci USA 101:15494–15499
22. Karashima Y, Talavera K, Everaerts W, Janssens A, Kwan KY, Vennekens R, Nilius B, Voets T (2009) TRPA1 acts as a cold sensor in vitro and in vivo. Proc Natl Acad Sci USA 106:1273–1278
23. Latorre R, Vargas G, Orta G, Brauchi S (2007) Voltage and temperature gating of thermoTRP channels. In: Liedtke W, Heller S (eds) TRP ion channels function in sensory transduction and cellular signaling cascades. CRC Taylor & Francis, London, pp 287–302
24. Gaudet R (2008) P channels entering the structural era. J Physiol 586:3565–3575
25. Moiseenkova-bell VY, Stanciu LA, Serysheva II, Tobe BJ, Wensel TG (2008) Structure of TRPV1 channel revealed by electron cryomicroscopy. Proc Natl Acad Sci USA 105:7451–7455
26. Almers W (1978) Gating currents and charge movements in excitable membranes. Rev Physiol Biochem Pharmacol 82:96–190

27. Voets T, Owsianik G, Janssens A, Talavera K, Nilius B (2007) TRPM8 voltage sensor mutants reveal a mechanism for integrating thermal and chemical stimuli. Nat Chem Biol 3:174–182
28. Liu B, Hui K, Qin F (2003) Thermodynamics of heat activation of single capsaicin ion channels vr1. Biophys J 85:2988–3006
29. Iggo A (1969) Cutaneous thermoreceptors in primates and sub-primates. J Physiol 200: 403–430
30. Lecar H, Ehrenstein G, Latorre R (1975) Mechanism for channel gating in excitable bilayers. Ann NY Acad Sci 264:304–313
31. Yang F, Cui Y, Wang K, Zheng J (2010) Thermosensitive trp channel pore turret is part of the temperature activation pathway. Proc Natl Acad Sci USA 107:7083–7088
32. Privalov PL (1989) Thermodynamic problems of protein structure. Annu Rev Biophys Chem 18:47–69
33. Privalov PL. Thermodynamics of protein folding. 1997; 447–474
34. Gursky O, Atkinson D (1996) High- and low-temperature unfolding of human high-density apolipoprotein a-2. Protein Sci 5:1874–1882
35. Privalov PL, Griko YV, Venyaminov S, Kutyshenko VP (1986) Cold denaturation of myoglobin. J Mol Biol 190:487–498
36. Griko YV, Venyaminov S, Privalov PL (1989) Heat and cold denaturation of phosphoglycerate kinase (interaction of domains). FEBS Lett 244:276–278
37. Myers BR, Bohlen CJ, Julius D (2008) A yeast genetic screen reveals a critical role for the pore helix domain in TRP channel gating. Neuron 58:362–373
38. Grandl J, Hu H, Bandell M, Bursulaya B, Schmidt M, Petrus M, Patapoutian A (2008) Pore region of TRPV3 ion channel is specifically required for heat activation. Nat Neurosci 11:1007–1013
39. Grandl J, Kim SE, Uzzell V, Bursulaya B, Petrus M, Bandell M, Patapoutian A (2010) Temperature-induced opening of TRPV1 ion channel is stabilized by the pore domain. Nat Neurosci 6:708–714
40. Brauchi S, Orta G, Mascayano C, Salazar M, Raddatz N, Urbina H, Rosenmann E, Gonzalez-Nilo F, Latorre R (2007) Dissection of the components for pip2 activation and thermosensation in trp channels. Proc Natl Acad Sci USA 104:10246–10251
41. Yellen G (1998) The moving parts of voltage-gated ion channels. Quart Rev Biophys 31:239–295
42. Matta JA, Ahern GP (2007) Voltage is a partial activator of rat thermosensitive TRP channels. J Physiol 585:469–482
43. Altomare C, Bucchi A, Camatini E, Baruscotti M, Viscomi C, Moroni A, DiFrancesco D (2001) Integrated allosteric model of voltage gating of HCN channels. J Gen Physiol 117:519–532
44. Horrigan FT, Aldrich RW (2002) Coupling between voltage sensor activation, Ca2+ binding and channel opening in large conductance (BK) potassium channels. J Gen Physiol 120: 267–305
45. Horrigan FT, Cui J, Aldrich RW (1999) Allosteric voltage gating of potassium channels i. mslo ionic currents in the absence of Ca2+. J Gen Physiol 114:277–304
46. Colquhoun D, Hawkes AG (1981) On the stochastic properties of single ion channels. Proc R Soc Lond B Biol Sci 211:205–235
47. Lu Z, Klem AM, Ramu Y (2002) Coupling between voltage sensors and activation gate in voltage-gated K+ channels. J Gen Physiol 120:663–676
48. Yao J, Liu B, Qin F (2010) Pore turret of thermal TRP channels is not essential for temperature sensing. Proc Natl Acad Sci USA 107(32):E125
49. Yang F, Ciu Y, Wang K, Zheng J (2010) Reply to Yao et al.: is the pore turret just thermoTRP channels' appendix? Proc Natl Acad Sci USA 107(32):E126–E127

Chapter 29
TRP Channels as Mediators of Oxidative Stress

Barbara A. Miller and Wenyi Zhang

Abstract The transient receptor potential (TRP) protein superfamily is a diverse group of cation-permeable channels expressed in mammalian cells, which is divided into six subfamilies based on sequence identity. Three subfamilies have members with roles in oxidative stress: the TRPC subfamily characterized by receptor operated calcium entry channels; the TRPM subfamily with a number of members involved in cell proliferation and death; and the TRPV subfamily which is activated by chemical, mechanical, and physical stimuli. The TRPC members TRPC3 and TRPC4 can serve as subunits of a redox-sensitive ion channel in native aortic endothelial cells. The TRPM family member TRPM2 has a number of physiologic isoforms expressed in many cell types and responds to stimuli including oxidative stress, TNFα, and β-amyloid peptide. The important role of TRPM2 isoforms in cell proliferation and oxidant-induced cell death has been well established using divergent cell systems and techniques including overexpression, channel depletion or inhibition, and calcium chelation. TRPM7 has been shown to be involved in Ca^{2+} influx and anoxic cell death in cortical neurons. In these cells and in B cells, precise expression of TRPM7 is necessary for cell survival. TRPV1 is involved in oxidant stress-induced pain and in neuronal injury, contributing to diabetic sensory neuropathy. Future studies will likely identify additional channels involved in oxidant injury, as well as better define mechanisms through which these channels are regulated and mediate their effects. Therapeutic approaches to modulate activation of specific TRP channels are likely to have an important impact in reducing tissue damage in a number of diseases resulting from oxidant stress including ischemia/reperfusion injury and diabetes.

B.A. Miller (✉)
Departments of Pediatrics and Biochemistry and Molecular Biology, Milton S. Hershey Medical Center, Penn State Hershey Children's Hospital, Pennsylvania State University College of Medicine, Hershey, PA 17033, USA
e-mail: bmiller3@psu.edu

29.1 Introduction

The transient receptor potential (TRP) protein superfamily is a diverse group of calcium-permeable cation channels expressed in mammalian cells [1–5]. Mammalian TRP channels have been organized into six protein subfamilies designated C (canonical), V (vanilloid receptor), M (melastatin), A (ANKTM), P (polycystin), and ML (mucolipin). Mammalian isoforms have six putative transmembrane domains similar to the structure of many pore-forming subunits of voltage-gated channels. While many of them lack positively charged residues necessary for the voltage sensor and are voltage independent, some TRP channels, particularly those which are temperature sensitive, are voltage gated [6, 7]. TRP channels function as homotetramers or heterotetramers, with the pore formed by loops between the fifth and six transmembrane domains. Regulation of TRP channels includes roles for (1) extracellular signals, (2) second messengers, (3) channel subunit assembly, and (4) macromolecular complex formation. All TRP channels have multiple protein interaction motifs and regulatory domains including protein kinase A, C, and tyrosine phosphorylation sites. These channels function in many physiological processes and have roles in a number of diseases involving the cardiovascular, endocrine, neurologic, immune, respiratory, gastrointestinal, and reproductive systems as well as kidney, skeletal muscle, and bone [5, 8].

Tissue damage resulting from oxidative stress plays an important role in a number of physiological processes including aging, cancer, neurodegenerative disorders, diabetes mellitus, atherosclerosis, ischemia/reperfusion injury, and autoimmune disease [9, 10]. Oxidative stress results from a disturbance in the balance between oxidants and anti-oxidants, which may lead to tissue injury depending on severity and duration [9, 11]. Reactive oxygen species (ROS) are produced naturally during respiration by the mitochondrial electron transport chain, following activation of the arachidonic acid cascade in the cytosol, and after exposure to ionizing radiation, cytotoxic drugs, or infections which activate neutrophils or phagocytes. Free radical intermediates which are produced include ROS (superoxide anion, hydrogen peroxide, hydroxyl radical) and reactive nitrogen species (RNS; nitric oxide and its derivatives). These radicals damage cells through DNA and protein oxidation and lipid peroxidation. ROS are reduced naturally by antioxidant enzymes including catalase, superoxide dismutases, and glutathione peroxidase, and biological antioxidants include α-tocopherol and absorbic acid. A number of complex signaling events are activated in oxidative stress including enzymes such as phospholipases and protein kinases [12, 13]. Hydrogen peroxide (H_2O_2) induces apoptosis through multiple mechanisms including upregulation of Fas/Fas ligand, which activates the extrinsic cell death pathway, and activation of mitochondrial cell death pathways through modulation of the mitochondrial permeability transition pore (PT) [13–15]. H_2O_2 stimulates an increase in intracellular free calcium ($[Ca^{2+}]_i$), resulting in elevated mitochondrial matrix Ca^{2+}, which together with arachidonic acid, produced by activation of phospholipase A2, opens the mitochondrial PT pore [14, 16]. Activation of the PT pore uncouples oxidative phosphorylation, prevents ATP production, and enhances cytochrome c release into the cytosol. Cytochrome c binds to Apaf-1 (apoptotic protease activating factor 1), forming the apoptosome, activating caspase 9, followed by 3 and 7, and inactivating PARP, contributing to cell death. The rise in $[Ca^{2+}]_i$ may contribute to cell death through a number of pathways in addition to caspase cleavage, PARP inactivation and release of cytochrome c, including activation of tyrosine kinases and phosphatases, and binding of transcription factors to target genes. H_2O_2 has been proposed to mediate an increase in $[Ca^{2+}]_i$ through a number of different mechanisms including voltage-dependent calcium channels and Na^+–Ca^{2+} exchange. The role of TRP channel activation in oxidative stress will be reviewed here.

29.2 TRPC in Oxidative Stress

TRP channels were first shown to have a role in anoxic cell death in *Drosophila*. Whereas anoxia, treatment with mitochondrial uncouplers, or ATP depletion rapidly activated the *Drosophila* channels TRP and TRPL in the dark, mutation of both TRP and TRPL eliminated Ca^{2+} influx in photoreceptor cells in response to anoxia, demonstrating the role of these channels as targets of oxidative stress [17]. Furthermore, constitutive activation of these channels resulted in massive photoreceptor cell death in vivo [18].

The most closely related TRP subfamily to *Drosophila* is that of TRPC. Members are activated by stimulation of G-protein-coupled receptors and receptor tyrosine kinases with ligand, which activates phospholipase C and results in production of inositol 1,4,5-trisphosphate (IP_3) and diacylglycerol (DAG). A number of models of TRPC activation through PLC-mediated pathways have been proposed [1–3, 13, 19]. Regulation of TRPC cell surface expression is one critical component of channel activation [20, 21]. TRPC channels assemble based on structural similarities reflected in phylogenetic relationships, but specific molecular determinants for subunit assembly have not been identified [3]. While heteromeric channel formation is well established for TRPC 1/4/5 and TRPC 3/6/7 [22–24], multimerization of TRPC is complex and controversial. Different homo- and heteromeric assemblies are possible, and up to three isoforms may contribute to native pore complex formation [4, 25, 26].

Two TRPC channels, TRPC3 and TRPC4, have been shown to be important in oxidant activation of cation current in porcine endothelial cells [27, 28]. ROS cause a sustained increase in $[Ca^{2+}]_i$, resulting in protease activation, changes in the cytoskeleton, and endothelial cell dysfunction [8, 29]. TRPC3 and TRPC4 are expressed and reported to associate endogenously to form a cation-conducting pore complex in these cells. This finding was supported by several approaches including coimmunoprecipitation, FRET, and the observation that endogenous oxidant stress-mediated calcium conductance is suppressed in these cells by dominant negative TRPC3 and TRPC4 mutants [28]. Oxidant stress may also modulate TRPC function though disruption of caveolin 1-rich lipid rafts [30]. In addition, TRP channels can act as NO sensors in endothelial cells [31]. NO can activate TRPC1, TRPC4,

TRPC5, TRPV1, TRPV3, and TRPV4, inducing calcium entry into cells. The physiological significance of ROS and RNS activation of TRPC3/4 in endothelial cells is under investigation. However, because TRP channels may play an important role in oxidant-induced endothelial injury, they should be considered as potential targets to prevent oxidant stress-induced vascular damage.

29.3 TRPM in Oxidative Stress

The TRP channel TRPM subfamily is named after the first described member, TRPM1 (melastatin), a putative tumor suppressor protein [32]. TRPM1 is expressed on melanocytes, and its expression level correlates inversely with melanoma aggressiveness and the potential for metastasis, suggesting a role for this channel in cell proliferation or migration. Other members of the TRPM subfamily also have important roles in cell proliferation and survival including TRPM2 [33, 34], TRPM5 [35], TRPM7 [36], and TRPM8 [37]. Members of the TRPM subfamily share a region of high coiled coil character (CCR) in the C-terminus, which may play a role in ion channel multimerization or in recruitment of regulatory proteins [38]. The C-terminus of these channels displays considerable variability and three of these channels have unique C-terminal enzymatic domains, TRPM2, TRPM6, and TRPM7. These channels function primarily as homotetramers. However, for several TRPM channels, splice variants have been described which inhibit full-length channel function and consist only of N-terminal (TRPM1), C-terminal (TRPM2-TE), or N-terminal and truncated transmembrane domains (TRPM2-S) [34, 39–41]. A role for two of these channels, TRPM2 and TRPM7, in oxidative stress-induced cell death has been extensively studied and will be reviewed here.

29.3.1 TRPM2

TRPM2 is the second member of the TRPM subfamily to be cloned. It is expressed in many cell types including brain, hematopoietic cells, heart, vascular smooth muscle, endothelial cells, lung, endocrine system, and the gastrointestinal tract [29, 34, 42–45]. TRPM2 channels are permeable to sodium, potassium, and calcium. Extracellular signals known to activate TRPM2 include oxidative stress, TNFα, amyloid β-peptide, and concavalin A [33, 46–50]. Stimulation with these extracellular signals results in sufficient production of ADP-ribose (ADPR) to activate TRPM2 by binding to the TRPM2 COOH-terminal NUDT9-H domain, a mitochondrial ADPR hydrolase (Fig. 29.1; modified from Miller) [13, 48, 51, 52]. Cyclic adenosine diphosphoribose (cADPR) can also gate TRPM2 by itself at high concentrations and potentiates the effects of ADPR at lower concentrations [51]. ADPR may arise from a mitochondrial source [52] or via activation of poly (ADPR) polymerase (PARP) [53, 54]. PARP-1 covalently attaches ADPR polymers to proteins, which are then hydrolyzed into free ADPR by PARG [55]. Most evidence supports

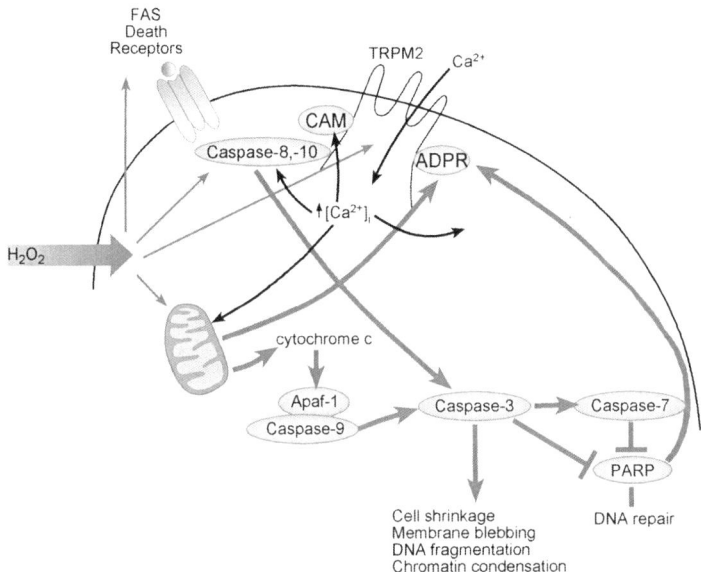

Fig. 29.1 Proposed signaling mechanisms of TRPM2 activation and induction of cell death by H_2O_2. H_2O_2 activates production of ADP-ribose (ADPR) in the mitochondria, which is released into the cytosol, and through activation of PARP/PARG. The increase in ADPR activates TRPM2 by binding to the C-terminal NUDT9-H domain. Ca^{2+} influx ensues, which enhances calmodulin (CAM) binding to TRPM2 and further channel opening. $[Ca^{2+}]_i$ rises, and in association with other oxidative stress-induced signals results in activation of extrinsic and intrinsic cell death pathways, leading to caspase-3 activation and PARP cleavage and inactivation

the conclusion that it is the binding of ADPR to the NUDT9-H domain that is critical for ADPR activation of TRPM2 rather than enzymatic activity of NUDT9-H, because ADPRase activity is low [52, 56]. While nicotinamide adenine dinucleotide (NAD) has been reported to directly induce opening of TRPM2, much evidence suggests this is secondary to conversion to or contamination by ADPR. TRPM2 currents are dependent on and positively regulated by Ca^{2+}, and have a strong requirement for Ca^{2+} at the intracellular surface of the plasma membrane [57, 58]. Low-level activation is seen at 100 nm $[Ca^{2+}]_i$ and maximal activation at 600 nm [57]. Interaction of ADPR with TRPM2 supports limited calcium entry through TRPM2, and Ca^{2+}-bound calmodulin increases. Interaction between calmodulin and an IQ-like motive in the N-terminus of TRPM2 is strengthened, providing positive feedback for TRPM2 activation leading to increased Ca^{2+} influx and enhanced $[Ca^{2+}]_i$ [59]. Recent evidence suggests that TRPM2 with mutant ADPR binding sites can still be activated by $[Ca^{2+}]_i$, and that TRPM2 may be activated under a wide range of physiological conditions through this mechanism [58].

Experimental data from a number of groups concur that oxidative stress results in Ca^{2+} influx through TRPM2 opening, and increased susceptibility to cell death [33, 34, 55, 60, 61]. The mechanisms through which the increase in $[Ca^{2+}]_i$ results in enhanced cell death were explored in the human monocytic cell line U937, in which TRPM2 isoform expression was modulated with retroviral infection (Fig. 29.1) [62]. Full length TRPM2 (TRPM2-L) activation by oxidative stress results in a significant increase in $[Ca^{2+}]_i$, and decreased cell viability. Procaspases-8, -9, -3, and -7 and PARP were cleaved, demonstrating a signaling cascade involving intrinsic (caspase-9) and extrinsic (caspase-8) cell death pathways. PARP, an important protective mechanism involved in DNA repair, was inactivated [53, 54]. These pathways have previously been linked to H_2O_2-induced apoptosis [63, 64]. This data suggests a feedback loop in which TRPM2 is activated by PARP, but TRPM2 activation in turn results in PARP cleavage and inactivation. Inhibition of the rise in $[Ca^{2+}]_i$ with the intracellular Ca^{2+} chelator BAPTA blocked caspase and PARP cleavage in TRPM-2 expressing cells, demonstrating the importance of the rise in $[Ca^{2+}]_i$ in activation of the cell death cascade.

TRPM channels function as tetramers, and subunit composition is an important factor in regulation of TRPM channel opening. Five physiological splice variants of TRPM2 have been identified: TRPM2-L (full-length or wild type), TRPM2-S (short) [34], TRPM2-ΔN [47], TRPM2-ΔC [47], and TRPM2-TE (tumor-enriched) [41] (Fig. 29.2; modified from Miller, 2006) [13]. TRPM2-S has a deletion of the entire C-terminus including four of six C-terminal transmembrane domains and the putative calcium pore [34]. TRPM2-S suppresses Ca^{2+} influx through TRPM2-L and inhibits cell death induced by oxidative stress [34, 62]. The mechanisms through which TRPM2-S inhibits TRPM2-L function are not known. However, because TRPM2-S co-associates with TRPM2-L, one hypothesis is that TRPM2-S participates in heterodimer formation, altering the tertiary structure of the TRPM2 tetramer required for ion permeability. TRPM2-ΔN has a deletion of amino acids 538-557 in the N-terminus and fails to respond to hydrogen peroxide (H_2O_2) or ADPR, suggesting that TRPM2-ΔN dominantly disrupts channel gating or assembly. TRPM2-ΔC has a deletion of amino acids 1292-1325 in the C-terminus, decreasing affinity for ADPR [47]. Cells expressing TRPM2-ΔC do not respond to ADPR but do response to H_2O_2, suggesting that oxidative stress can activate TRPM2 through mechanisms independently of ADPR [47, 51, 58]. TRPM2-TE was identified by investigators utilizing antisense technology to identify tumor suppressor genes [41]. Two TRPM2-TE transcripts were found which encode either a 218 amino acid, 25 kDa protein or a 184 amino acid, 21 kDa protein (TRPM2-TE-ΔC). These proteins are highly expressed in tumor cells including melanoma and lung, and when overexpressed with TRPM2-L, protected cells from apoptosis. Expression in malignant tissue is thought to result from hypomethylation of a specific CpG island in the TRPM2 C-terminus. Little is known about mechanisms which control differential splicing of TRPM2 isoforms, but the ratio of isoform expression may have an important impact on susceptibility to oxidative stress.

TRPM2 enhances susceptibility to oxidative stress-induced cell death in a number of cell types [33, 34, 53, 60, 62]. In heterologous expression systems, exposure to oxidative stress enhances Ca^{2+} influx and cell death in cells expressing TRPM2-L [33, 47, 52]. Inhibition of endogenous TRPM2 function by expression of the dominant negative TRPM2-S, down regulation of TRPM2-L with RNA interference, or

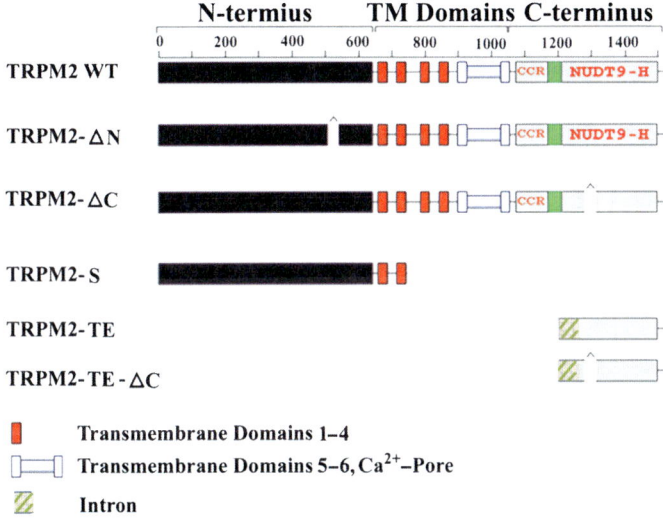

Fig. 29.2 Schematic representation of TRPM2 isoforms. Membrane spanning domains 1–4 and the putative pore region including transmembrane domains 5–6 are indicated. CCR represents the coiled coil region which may mediate protein/protein interactions. NUDT9-H represents the NUDT9 ADP-ribose hydrolase domain. TRPM2-△N has a deletion of aa 538–557 in the N-terminus. TRPM2-△C has a deletion of aa 1292–1325 in the C-terminus. TRPM2-S is missing four of six transmembrane domains and the putative calcium pore. TRPM2-TE consists of a short 22–24 kDa C-terminal fragment resulting from hypomethylation of a specific CpG island. TRPM2-TE-△C is a 184 amino acid, 21 kDa protein which differs from TRPM2-TE by a deletion of aa 1292–1325, also absent in TRPM2-△C

calcium chelation all blocked the rise in $[Ca^{2+}]_i$ induced by oxidative stress and protected cells from apoptosis [33, 34, 62].

Because of its broad expression profile, TRPM2 can modulate oxidative stress in a number of different tissues including brain, the cardiovascular system, and lymphocytes. The striatum has been shown to be highly vulnerable to ischemia/reperfusion injury [65]. TRPM2 is involved in oxidant injury to striatal cells. It may also be involved in the pathogenesis of Alzheimer's disease through activation by amyloid β-peptide, a main component of senile plaques which causes neuronal injury through generation of oxidative stress. In primary cultures of rat striatal cells which express TRPM2 endogenously, H_2O_2 or amyloid β-peptide induced an increase in $[Ca^{2+}]_i$ and cell death, which was inhibited by expression of TRPM2-S or by reduction in endogenous TRPM2 levels by RNA interference [46]. A TRPM2 variant (TRPM2P1018L) with enhanced inactivation has been identified in Guamanian amyotrophic lateral sclerosis and parkinsonism-dementia [66]. TRPM2 channels have been found to play an important role in the apoptotic component of ischemia/reperfusion injury in cardiomyocytes by inducing mitochondrial sodium and calcium overload, leading to mitochondrial membrane disruption, cytochrome c release, caspase-3 dependent chromatin condensation and myocyte death [67]. Necrotic changes were also observed which were caspase-3 independent but PARP-dependent. Inhibition of both TRPM2 and PARP totally abolished H_2O_2-induced myocyte death [67]. ROS are important regulators of the vascular barrier. H_2O_2 induces an increase in endothelial permeability through TRPM2 activated Ca^{2+} entry [68, 69]. In a concentration-dependent manner, H_2O_2 decreased transmonolayer transendothelial electrical resistance, indicating opening of interendothelial junctions. Overexpression of TRPM2-L enhanced H_2O_2-mediated Ca^{2+} entry, cationic current, and the transendothelial electrical resistance decrease, whereas these were inhibited by TRPM2 depletion with siRNA, overexpression of the dominant negative TRPM2-S, or inhibition of ADPR formation. TRPM2 may be an important target to protect against oxidant-induced endothelial barrier disruption in a number of disease processes including acute respiratory distress syndrome and ischemia/reperfusion injury [68, 69].

TRPM2 channels are widely expressed in the immune system and play an important role in immune responses to oxidative stress. CD38 is a transmembrane glycoprotein, expressed in many tissues including lymphoid and myeloid cells, which use β-NAD^+ to produce ADPR, cADPR, and nicotinic acid adenine dinucleotide phosphate ($NAADP^+$) [70]. Through TRPM2, ADPR acting in synergy with cADPR and NAADP may play a major role in CD38-dependent Ca^{2+} influx, signaling in immune cells, and cell migration in phagocytes [71, 72]. A complex issue is whether CD38 is involved in regulation of intracellular ADPR levels, since the enzymatic activity of CD38 is extracellular. ROS levels in cells increase during infection following production by neutrophils and phagocytes, or in response to environmental factors including ionizing radiation or cytotoxic drugs. Oxidants are then thought to activate PARP/PARG, resulting in production of ADPR, modulating TRPM2 opening and activating the downstream cascade. Drugs interfering with this pathway may have a potent effect on modifying the immune response.

TRPM2 is also of functional importance in diabetes through its ability to regulate oxidant-induced beta cell death [73]. Inhibition of TRPM2 function may be an important and broad approach to protect cells from death following oxidant stress. This strategy could protect a range of tissues including heart and brain from oxidative stress-induced cell death following ischemia/reperfusion injury, as well as other tissues from less acute injury associated with oxidant stress including bone marrow, pancreas, and brain (Alzheimer's). Understanding how expression of TRPM2 isoforms and channel activation are regulated is an important area of research which may result in novel approaches to modulate cell viability.

29.3.2 TRPM7

TRPM7 is a widely expressed member of the TRPM ion channel subfamily. It has a C-terminal serine/threonine kinase domain with homology to the eEF2 α-kinase family [74, 75]. TRPM7 is a divalent cation channel which is permeable to Mg^{2+}, a rare feature among ion channels [74]. TRPM7 currents are inhibited by Mg^{2+} and Zn^{2+}, and activated by low levels of MgADP. TRPM6 and TRPM7 are two ion channels involved in regulation of cellular Mg^{2+} homeostasis [76, 77].

TRPM7 has been shown to be involved in cell proliferation and cell cycle progression. Reactive oxygen/nitrogen species can activate cation conductance through TRPM7, contributing to anoxic neuronal death. Overexpression of TRPM7 in HEK cells resulted in cell swelling, detachment, and death in 48–72 h [36, 74], whereas suppression of TRPM7 expression in primary cortical neurons blocked TRPM7 currents, Ca^{2+} influx, and reactive oxygen species production, protecting cells from anoxic cell death [36]. On the other hand, targeted deletion of TRPM7 in DT-40 B was lethal. These cells exhibited Mg^{2+} deficiency, growth arrest, and death within 24 h unless rescued by increased levels of extracellular Mg^{2+} [74]. These studies, along with others in which TRPM7 expression was down regulated [78, 79], demonstrate that precise regulation of TRPM7 expression is necessary for cell survival. However, siRNA targeted to TRPM7 also reduced TRPM2 levels [36], suggesting that expression of TRPM2 and TRPM7 are interdependent. This makes it difficult to definitively distinguish the roles of TRPM2 and TRPM7 in anoxic injury. Recently, no significant evidence was found for an association between TRPM7 genetic variants and type 2 diabetes [80] or risk for ischemic stroke [81], but the single nucleotide polymorphisms studied may not have captured all of the important genetic variability in TRPM7.

29.4 TRPV in Oxidative Stress

The TRPV subfamily of TRP proteins was named because the first member, TRPV1, is activated by the inflammatory vanilloid compound capsaicin which gives spicy foods their characteristic taste. TRPV family members are involved in osmosensation, thermosensation, mechanosensation, and chemosensation [4, 5]. TRPV1 is a calcium permeable, nonselective cation channel which is gated by a number of stimuli including heat, low pH, capsaicin, and other endogenous ligands. TRPV1 can act as a signal integrator in response to multiple harmful stimuli. Repeated activation of TRPV1 has previously been shown to result in increased $[Ca^{2+}]_i$, oxidative stress, and apoptotic cell injury [82, 83]. Recently, TRPV1 activation by capsaicin was found to increase substantially following oxidative stress [84]. The sensitization is long standing, overrides receptor desensitization, and involves covalent modification of conserved cysteines [84]. Oxidation represents an independent pathway from phosphorylation, desensitization, and acidic extracellular pH, acting to increase the gain of TRPV1 [84]. Through this mechanism, oxidative stress may mediate TRPV1 responses including pain sensation during inflammation or tissue injury.

The TRPV1 channel is involved in two aspects of the pathogenesis of diabetes. TRPV1 has been shown to play a role in diabetes through its role in pancreatic beta cell death [85]. TRPV1 is also highly expressed in large sensory dorsal root ganglion (DRG) neurons. Capsaicin induced increased oxidative stress as well as cytosolic cytochrome c and activation of caspase-3 in DRG neurons isolated from diabetic rats [86]. Treatment with capsazepine, a competitive TRPV1 antagonist, markedly reduced these changes in response to capsaicin, and prevented cell injury in large DRG neurons in diabetic rats in vivo. These data suggest that increased expression and activation of TRPV1 in large DRG neurons are associated with oxidative stress and neuronal injury in early diabetic sensory neuropathy [86].

TRPC1, TRPC4, TRPV1, and TRPV4 all play an important role in endothelium-dependent vasorelaxation [8]. In addition, the TRP channels TRPC1, TRPC4, TRPV1, and TRPV4 can act as NO sensors in endothelial cells [31]. This data suggests that these channels could play an important role in diseases involving endothelial dysfunction mediated by oxidative stress.

29.5 Future Perspectives

Members of the TRP channel superfamily, particularly TRPM2, are now recognized to play important roles in oxidant stress-induced cell injury. In some cases, this is secondary to widespread tissue expression of channels which are activated indirectly by increases in oxygen or nitrogen free radicals. In certain tissues, channel expression contributes to oxidative injury through more specific activation pathways. In the near future, it is likely that additional TRP channels will be identified which contribute to oxidant injury, and the mechanisms through which they are regulated and mediate their downstream effects will be better defined. Ultimately, therapeutic approaches which modulate activation of TRP channels by oxidant stress may significantly reduce tissue damage in a number of disease processes including those resulting from ischemic injury or diabetes. Since the role of oxidant stress in malignant cell growth and chemotherapy response is increasingly recognized, TRP channel modulation may also be utilized in future targeted therapies in cancer.

References

1. Clapham DE (2003) TRP channels as cellular sensors. Nature 426:517–524
2. Birnbaumer L (2009) The TRPC class of ion channels: a critical review of their roles in slow, sustained increases in intracellular Ca($^{2+}$) concentrations. Annu Rev Pharmacol Toxicol 49:395–426
3. Montell C (2005) The TRP superfamily of cation channels. Sci STKE 2005:re3
4. Venkatachalam K, Montell C (2007) TRP channels. Annu Rev Biochem 76:387–417
5. Nilius B, Owsianik G, Voets T, Peters JA (2007) Transient receptor potential cation channels in disease. Physiol Rev 87:165–217
6. Voets T, Owsianik G, Janssens A, Talavera K, Nilius B (2007) TRPM8 voltage sensor mutants reveal a mechanism for integrating thermal and chemical stimuli. Nat Chem Biol 3:174–182
7. Nilius B, Prenen J, Droogmans G, Voets T, Vennekens R, Freichel M, Wissenbach U, Flockerzi V (2003) Voltage dependence of the Ca^{2+}-activated cation channel TRPM4. J Biol Chem 278:30813–30820
8. Watanabe H, Murakami M, Ohba T, Takahashi Y, Ito H (2008) TRP channel and cardiovascular disease. Pharmacol Ther 118:337–351
9. Langley B, Ratan RR (2004) Oxidative stress-induced death in the nervous system: cell cycle dependent or independent? J Neurosci Res 77:621–629
10. Waring P (2005) Redox active calcium ion channels and cell death. Arch Biochem Biophys 434:33–42

11. Chandra J, Samali A, Orrenius S (2000) Triggering and modulation of apoptosis by oxidative stress. Free Radic Biol Med 29:323–333
12. Gopalakrishna R, Jaken S (2000) Protein kinase C signaling and oxidative stress. Free Radic Biol Med 28:1349–1361
13. Miller BA (2006) The role of TRP channels in oxidative stress-induced cell death. J Membr Biol 209:31–41
14. Crompton M (1999) The mitochondrial permeability transition pore and its role in cell death. Biochem J 341(Pt 2):233–249
15. Duchen MR, Verkhratsky A, Muallem S (2008) Mitochondria and calcium in health and disease. Cell Calcium 44:1–5
16. Green DR, Kroemer G (2004) The pathophysiology of mitochondrial cell death. Science 305:626–629
17. Agam K, von Campenhausen M, Levy S, Ben-Ami HC, Cook B, Kirschfeld K, Minke B (2000) Metabolic stress reversibly activates the Drosophila light-sensitive channels TRP and TRPL in vivo. J Neurosci 20:5748–5755
18. Yoon J, Ben-Ami HC, Hong YS, Park S, Strong LL, Bowman J, Geng C, Baek K, Minke B, Pak WL (2000) Novel mechanism of massive photoreceptor degeneration caused by mutations in the trp gene of Drosophila. J Neurosci 20:649–659
19. Vazquez G, Wedel BJ, Trebak M, Bird StJG, Putney JW, Jr. (2003) Expression level of the canonical transient receptor potential 3 (TRPC3) channel determines its mechanism of activation. J Biol Chem 278:21649–21654
20. van Rossum DB, Patterson RL, Sharma S, Barrow RK, Kornberg M, Gill DL, Snyder SH (2005) Phospholipase Cgamma1 controls surface expression of TRPC3 through an intermolecular PH domain. Nature 434:99–104
21. Bezzerides VJ, Ramsey IS, Kotecha S, Greka A, Clapham DE (2004) Rapid vesicular translocation and insertion of TRP channels. Nat Cell Biol 6:709–720
22. Strubing C, Krapivinsky G, Krapivinsky L, Clapham DE (2001) TRPC1 and TRPC5 form a novel cation channel in mammalian brain. Neuron 29:645–655
23. Goel M, Sinkins WG, Schilling WP (2002) Selective association of TRPC channel subunits in rat brain synaptosomes. J Biol Chem 277:48303–48310
24. Hofmann T, Schaefer M, Schultz G, Gudermann T (2002) Subunit composition of mammalian transient receptor potential channels in living cells. Proc Natl Acad Sci USA 99: 7461–7466
25. Strubing C, Krapivinsky G, Krapivinsky L, Clapham DE (2003) Formation of novel TRPC channels by complex subunit interactions in embryonic brain. J Biol Chem 278:39014–39019
26. Liu X, Bandyopadhyay BC, Singh BB, Groschner K, Ambudkar IS (2005) Molecular analysis of a store-operated and 2-acetyl-sn-glycerol-sensitive non-selective cation channel. Heteromeric assembly of TRPC1-TRPC3. J Biol Chem 280:21600–21606
27. Balzer M, Lintschinger B, Groschner K (1999) Evidence for a role of Trp proteins in the oxidative stress-induced membrane conductances of porcine aortic endothelial cells. Cardiovasc Res 42:543–549
28. Poteser M, Graziani A, Rosker C, Eder P, Derler I, Kahr H, Zhu MX, Romanin C, Groschner K (2006) TRPC3 and TRPC4 associate to form a redox-sensitive cation channel. Evidence for expression of native TRPC3-TRPC4 heteromeric channels in endothelial cells. J Biol Chem 281:13588–13595
29. Yao X, Garland CJ (2005) Recent developments in vascular endothelial cell transient receptor potential channels. Circ Res 97:853–863
30. Groschner K, Rosker C, Lukas M (2004) Role of TRP channels in oxidative stress. Novartis Found Symp 258:222–230, discussion 231–225, 263–226
31. Yoshida T, Inoue R, Morii T, Takahashi N, Yamamoto S, Hara Y, Tominaga M, Shimizu S, Sato Y, Mori Y (2006) Nitric oxide activates TRP channels by cysteine S-nitrosylation. Nat Chem Biol 2:596–607

32. Duncan LM, Deeds J, Hunter J, Shao J, Holmgren LM, Woolf EA, Tepper RI, Shyjan AW (1998) Down-regulation of the novel gene melastatin correlates with potential for melanoma metastasis. Cancer Res 58:1515–1520
33. Hara Y, Wakamori M, Ishii M, Maeno E, Nishida M, Yoshida T, Yamada H, Shimizu S, Mori E, Kudoh J, Shimizu N, Kurose H, Okada Y, Imoto K, Mori Y (2002) LTRPC2 Ca^{2+}-permeable channel activated by changes in redox status confers susceptibility to cell death. Mol Cell 9:163–173
34. Zhang W, Chu X, Tong Q, Cheung JY, Conrad K, Masker K, Miller BA (2003) A novel TRPM2 isoform inhibits calcium influx and susceptibility to cell death. J Biol Chem 278:16222–16229
35. Prawitt D, Enklaar T, Klemm G, Gartner B, Spangenberg C, Winterpacht A, Higgins M, Pelletier J, Zabel B (2000) Identification and characterization of MTR1, a novel gene with homology to melastatin (MLSN1) and the trp gene family located in the BWS-WT2 critical region on chromosome 11p15.5 and showing allele-specific expression. Hum Mol Genet 9:203–216
36. Aarts M, Iihara K, Wei WL, Xiong ZG, Arundine M, Cerwinski W, MacDonald JF, Tymianski M (2003) A key role for TRPM7 channels in anoxic neuronal death. Cell 115:863–877
37. Tsavaler L, Shapero MH, Morkowski S, Laus R (2001) Trp-p8, a novel prostate-specific gene, is up-regulated in prostate cancer and other malignancies and shares high homology with transient receptor potential calcium channel proteins. Cancer Res 61:3760–3769
38. Schmitz C, Perraud AL (2005) The TRPM cation channels in the immune context. Curr Pharm Des 11:2765–2778
39. Xu XZ, Moebius F, Gill DL, Montell C (2001) Regulation of melastatin, a TRP-related protein, through interaction with a cytoplasmic isoform. Proc Natl Acad Sci USA 98:10692–10697
40. Perraud AL, Schmitz C, Scharenberg AM (2003) TRPM2 Ca^{2+} permeable cation channels: from gene to biological function. Cell Calcium 33:519–531
41. Orfanelli U, Wenke AK, Doglioni C, Russo V, Bosserhoff AK, Lavorgna G (2008) Identification of novel sense and antisense transcription at the TRPM2 locus in cancer. Cell Res 18:1128–1140
42. Perraud AL, Fleig A, Dunn CA, Bagley LA, Launay P, Schmitz C, Stokes AJ, Zhu Q, Bessman MJ, Penner R, Kinet JP, Scharenberg AM (2001) ADP-ribose gating of the calcium-permeable LTRPC2 channel revealed by Nudix motif homology. Nature 411:595–599
43. Sano Y, Inamura K, Miyake A, Mochizuki S, Yokoi H, Matsushime H, Furuichi K (2001) Immunocyte Ca^{2+} influx system mediated by LTRPC2. Science 293:1327–1330
44. Nagamine K, Kudoh J, Minoshima S, Kawasaki K, Asakawa S, Ito F, Shimizu N (1998) Molecular cloning of a novel putative Ca^{2+} channel protein (TRPC7) highly expressed in brain. Genomics 54:124–131
45. Togashi K, Hara Y, Tominaga T, Higashi T, Konishi Y, Mori Y, Tominaga M (2006) TRPM2 activation by cyclic ADP-ribose at body temperature is involved in insulin secretion. Embo J 25:1804–1815
46. Fonfria E, Marshall IC, Boyfield I, Skaper SD, Hughes JP, Owen DE, Zhang W, Miller BA, Benham CD, McNulty S (2005) Amyloid beta-peptide(1-42) and hydrogen peroxide-induced toxicity are mediated by TRPM2 in rat primary striatal cultures. J Neurochem 95:715–723
47. Wehage E, Eisfeld J, Heiner I, Jungling E, Zitt C, Luckhoff A (2002) Activation of the cation channel long transient receptor potential channel 2 (LTRPC2) by hydrogen peroxide. A splice variant reveals a mode of activation independent of ADP-ribose. J Biol Chem 277:23150–23156
48. Gasser A, Glassmeier G, Fliegert R, Langhorst MF, Meinke S, Hein D, Kruger S, Weber K, Heiner I, Oppenheimer N, Schwarz JR, Guse AH (2006) Activation of T cell calcium influx by the second messenger adp-ribose. J Biol Chem 281:2489–2496
49. Yang XR, Lin MJ, McIntosh LS, Sham JS (2006) Functional expression of transient receptor potential melastatin- and vanilloid-related channels in pulmonary arterial and aortic smooth muscle. Am J Physiol Lung Cell Mol Physiol 290:L1267–L1276

50. Wilkinson JA, Scragg JL, Boyle JP, Nilius B, Peers C, H2 O (2008) 2-stimulated Ca^{2+} influx via TRPM2 is not the sole determinant of subsequent cell death. Pflugers Arch 455: 1141–1151
51. Kolisek M, Beck A, Fleig A, Penner R (2005) Cyclic ADP-ribose and hydrogen peroxide synergize with ADP-ribose in the activation of TRPM2 channels. Mol Cell 18:61–69
52. Perraud AL, Takanishi CL, Shen B, Kang S, Smith MK, Schmitz C, Knowles HM, Ferraris D, Li W, Zhang J, Stoddard BL, Scharenberg AM (2005) Accumulation of free ADP-ribose from mitochondria mediates oxidative stress-induced gating of TRPM2 cation channels. J Biol Chem 280:6138–6148
53. Fonfria E, Marshall ICB, Benham CD, Boyfield I, Brown JD, Hill K, Hughes JP, Skaper SD, Scharenberg AM, McNulty S (2004) TRPM2 channel opening in response to oxidative stress is dependent on activation of poly (ADP-Ribose) polymerase. Br J Pharmacol 143: 186–192
54. Buelow B, Song Y, Scharenberg AM, Poly T (2008) (ADP-ribose) polymerase PARP-1 is required for oxidative stress-induced TRPM2 activation in lymphocytes. J Biol Chem 283:24571–24583
55. Kuhn FJ, Heiner I, Luckhoff A (2005) TRPM2: a calcium influx pathway regulated by oxidative stress and the novel second messenger ADP-ribose. Pflugers Arch 451:212–219
56. Kuhn FJ, Luckhoff A (2004) Sites of the NUDT9-H domain critical for ADP-ribose activation of the cation channel TRPM2. J Biol Chem 279:46431–46437
57. McHugh D, Flemming R, Xu SZ, Perraud AL, Beech DJ (2003) Critical intracellular Ca^{2+} dependence of transient receptor potential melastatin 2 (TRPM2) cation channel activation. J Biol Chem 278:11002–11006
58. Du J, Xie J, Yue L (2009) Intracellular calcium activates TRPM2 and its alternative spliced isoforms. Proc Natl Acad Sci USA 106:7239–7244
59. Tong Q, Zhang W, Conrad K, Mostoller K, Cheung JY, Peterson BZ, Miller BA (2006) Regulation of the TRP Channel TRPM2 by the Ca^{2+} sensor Calmodulin. J Biol Chem 281:9076–9085
60. McNulty S, Fonfria E (2005) The role of TRPM channels in cell death. Pflugers Arch 451:235–242
61. Kaneko S, Kawakami S, Hara Y, Wakamori M, Itoh E, Minami T, Takada Y, Kume T, Katsuki H, Mori Y, Akaike A (2006) A critical role of TRPM2 in neuronal cell death by hydrogen peroxide. J Pharmacol Sci 101:66–76
62. Zhang W, Hirschler-Laszkiewicz I, Tong Q, Conrad K, Sun SC, Penn L, Barber DL, Stahl R, Carey DJ, Cheung JY, Miller BA (2006) TRPM2 is an ion channel that modulates hematopoietic cell death through activation of caspases and PARP cleavage. Am J Physiol Cell Physiol 290:C1146–C1159
63. Ma S, Ochi H, Cui L, Zhang J, He W (2003) Hydrogen peroxide induced down-regulation of CD28 expression of Jurkat cells is associated with a change of site alpha-specific nuclear factor binding activity and the activation of caspase-3. Exp Gerontol 38:1109–1118
64. Denning TL, Takaishi H, Crowe SE, Boldogh I, Jevnikar A, Ernst PB (2002) Oxidative stress induces the expression of Fas and Fas ligand and apoptosis in murine intestinal epithelial cells. Free Radic Biol Med 33:1641–1650
65. Lipton P (1999) Ischemic cell death in brain neurons. Physiol Rev 79:1431–1568
66. Hermosura MC, Cui AM, Go RC, Davenport B, Shetler CM, Heizer JW, Schmitz C, Mocz G, Garruto RM, Perraud AL (2008) Altered functional properties of a TRPM2 variant in Guamanian ALS and PD. Proc Natl Acad Sci USA 105:18029–18034
67. Yang KT, Chang WL, Yang PC, Chien CL, Lai MS, Su MJ, Wu ML (2006) Activation of the transient receptor potential M2 channel and poly(ADP-ribose) polymerase is involved in oxidative stress-induced cardiomyocyte death. Cell Death Differ 13:1815–1826
68. Hecquet CM, Ahmmed GU, Vogel SM, Malik AB (2008) Role of TRPM2 channel in mediating H_2O_2-induced Ca^{2+} entry and endothelial hyperpermeability. Circ Res 102:347–355
69. Hecquet CM, Malik AB (2009) Role of $H_{(2)}O_{(2)}$-activated TRPM2 calcium channel in oxidant-induced endothelial injury. Thromb Haemost 101:619–625

70. Schuber F, Lund FE (2004) Structure and enzymology of ADP-ribosyl cyclases: conserved enzymes that produce multiple calcium mobilizing metabolites. Curr Mol Med 4:249–261
71. Massullo P, Sumoza-Toledo A, Bhagat H, Partida-Sanchez S (2006) TRPM channels, calcium and redox sensors during innate immune responses. Semin Cell Dev Biol 17:654–666
72. Beck A, Kolisek M, Bagley LA, Fleig A, Penner R (2006) Nicotinic acid adenine dinucleotide phosphate and cyclic ADP-ribose regulate TRPM2 channels in T lymphocytes. Faseb J 20:962–964
73. Lange I, Yamamoto S, Partida-Sanchez S, Mori Y, Fleig A, Penner R (2009) TRPM2 functions as a lysosomal Ca^{2+}-release channel in beta cells. Sci Signal 2:ra23
74. Nadler MJ, Hermosura MC, Inabe K, Perraud AL, Zhu Q, Stokes AJ, Kurosaki T, Kinet JP, Penner R, Scharenberg AM, Fleig A (2001) LTRPC7 is a Mg.ATP-regulated divalent cation channel required for cell viability. Nature 411:590–595
75. Runnels LW, Yue L, Clapham DE (2001) TRP-PLIK, a bifunctional protein with kinase and ion channel activities. Science 291:1043–1047
76. Schlingmann KP, Weber S, Peters M, Niemann Nejsum L, Vitzthum H, Klingel K, Kratz M, Haddad E, Ristoff E, Dinour D, Syrrou M, Nielsen S, Sassen M, Waldegger S, Seyberth HW, Konrad M (2002) Hypomagnesemia with secondary hypocalcemia is caused by mutations in TRPM6, a new member of the TRPM gene family. Nat Genet 31:166–170
77. Schmitz C, Perraud AL, Johnson CO, Inabe K, Smith MK, Penner R, Kurosaki T, Fleig A, Scharenberg AM (2003) Regulation of vertebrate cellular Mg^{2+} homeostasis by TRPM7. Cell 114:191–200
78. He Y, Yao G, Savoia C, Touyz RM (2005) Transient receptor potential melastatin 7 ion channels regulate magnesium homeostasis in vascular smooth muscle cells: role of angiotensin II. Circ Res 96:207–215
79. Hanano T, Hara Y, Shi J, Morita H, Umebayashi C, Mori E, Sumimoto H, Ito Y, Mori Y, Inoue R (2004) Involvement of TRPM7 in cell growth as a spontaneously activated Ca^{2+} entry pathway in human retinoblastoma cells. J Pharmacol Sci 95:403–419
80. Song Y, Hsu YH, Niu T, Manson JE, Buring JE, Liu S (2009) Common genetic variants of the ion channel transient receptor potential membrane melastatin 6 and 7 (TRPM6 and TRPM7), magnesium intake, and risk of type 2 diabetes in women. BMC Med Genet 10:4
81. Romero JR, Ridker PM, Zee RY (2009) Gene variation of the transient receptor potential cation channel, subfamily M, member 7 (TRPM7), and risk of incident ischemic stroke: prospective, nested, case-control study. Stroke 40:2965–2968
82. Shin CY, Shin J, Kim BM, Wang MH, Jang JH, Surh YJ, Oh U (2003) Essential role of mitochondrial permeability transition in vanilloid receptor 1-dependent cell death of sensory neurons. Mol Cell Neurosci 24:57–68
83. Kim SR, Lee DY, Chung ES, Oh UT, Kim SU, Jin BK (2005) Transient receptor potential vanilloid subtype 1 mediates cell death of mesencephalic dopaminergic neurons in vivo and in vitro. J Neurosci 25:662–671
84. Chuang HH, Lin S (2009) Oxidative challenges sensitize the capsaicin receptor by covalent cysteine modification. Proc Natl Acad Sci USA 106:20097–20102
85. Razavi R, Chan Y, Afifiyan FN, Liu XJ, Wan X, Yantha J, Tsui H, Tang L, Tsai S, Santamaria P, Driver JP, Serreze D, Salter MW, Dosch HM (2006) TRPV1+ sensory neurons control beta cell stress and islet inflammation in autoimmune diabetes. Cell 127:1123–1135
86. Hong S, Agresta L, Guo C, Wiley JW (2008) The TRPV1 receptor is associated with preferential stress in large dorsal root ganglion neurons in early diabetic sensory neuropathy. J Neurochem 105:1212–1222

Chapter 30
Regulation of TRP Signalling by Ion Channel Translocation Between Cell Compartments

Alexander C. Cerny and Armin Huber

Abstract The TRP (transient receptor potential) family of ion channels is a heterogeneous family of calcium permeable cation channels that is subdivided into seven subfamilies: TRPC ("Canonical"), TRPV ("Vanilloid"), TRPM ("Melastatin"), TRPA ("Ankyrin"), TRPN ("NOMPC"), TRPP ("Polycystin"), and TRPML ("Mucolipin"). TRP-mediated ion currents across the cell membrane are determined by the single channel conductance, by the fraction of activated channels, and by the total amount of TRP channels present at the plasma membrane. In many cases, the amount of TRP channels at the plasma membrane is altered in response to physiological stimuli by translocation of channels to and from the plasma membrane. Regulated translocation has been described for channels of the TRPC, TRPV, TRPM, and TRPA family and is achieved by vesicular transport of these channels along cellular exocytosis and endocytosis pathways. This review summarizes the stimuli and signalling cascades involved in the translocation of TRP channels and highlights interactions of TRP channels with proteins of the endocytosis and exocytosis machineries.

Abbreviations

4α-PDD	4α-phorbol 12,13-didecanoate
AIP4	atropine-1 interacting protein 4
AMPA-receptor	α-amino-3-hydroxy-5-methyl-4-isoxazolepropionic acid receptor
ATP	adenosine-triphosphate
BAPTA	1,2-bis(o-aminophenoxy) ethane-N,N,N',N'-tetraacetic acid
bFGF	basic fibroblast growth factor
cAMP	cyclic adenosine-monophosphate
CHO	chinese hamster ovary
DAG	diacylglycerol

A.C. Cerny (✉)
Department of Biosensorics, Institute of Physiology, University of Hohenheim, 70599 Stuttgart, Germany
e-mail: cerny@uni-hohenheim.de

DHPG	3,5-dihydroxyphenylglycine
EETs	epoxyeicosatrienoic acids
EGF	epithelial growth factor
(e)GFP	(enhanced) green fluorescent protein
ePKC	eye specific protein kinase C
ER	endoplasmatic reticulum
ERK	extracellular-signal-regulated kinase
fMLP	formyl Met-Leu-Phe
GTP	guanosine-triphosphate
HA-tag	haemagglutinin tag
HEK	human embryonic kidney
HPAEC	human pulmonary artery endothelial cells
Hrs	hepatocyte growth factor regulated tyrosine kinase substrate
IGF-1	insulin-like growth factor 1
IMCD	inner medullary collecting duct
INAD	inactivation no afterpotential D
IP_3	inositol triphosphate
IP_3R	inositol triphosphate receptor
lysoPC	lysophosphatidylcholine
NCC	thiazide-sensitive Na-Cl cotransporter
NGF	nerve growth factor
NHERF	Na/H exchanger regulatory factor
OAG	1-oleyl-2-acetyl-sn-glycerol
OS-9	osteosarcoma amplified 9
PACS	phosphofurin acidic cluster sorting protein
PACSIN	protein kinase C and casein kinase substrate in neurons
PDGF	platelet derived growth factor
PI(3) kinase	phosphatidylinositide 3 kinase
PIP	phosphatidylinositol 4-phosphate
PIP_2	phosphatidylinositol 4,5-bisphosphate
PIP_3	phosphatidylinositol 3,4,5-trisphosphate
PKA	protein kinase A
PKC	protein kinase C
PKD	polycystic kidney disease
PLC	phospholipase C
PTH	parathyroid hormone
RGA	recombination gene activator
RNF24	ring finger protein 24
SGK	serum and glucocorticoid inducible kinase
SNAP	soluble NSF attachment protein
snapin	synaptic vesicle-associated protein
SNARE	soluble NSF attachment protein receptors
STAM	signal transducing adaptor molecule
TRP	transient receptor potential
TRPL	transient receptor potential like

VAMP2	vesicle-associated membrane protein 2
WNK	with no K (lysine)
YFP	yellow fluorescent protein

30.1 Introduction

The flux of ions across the plasma membrane is mediated by ion channels that exist in an open or closed state. The ion flux is controlled by the gating, that is opening, of ion channels and also by the number of ion channels present at the plasma membrane. Indeed, regulating the number of ion channel molecules in the plasma membrane by controlled translocation of ion channels between cell compartments turned out to be an important mechanism to tune neuronal activity. A prominent example for regulated plasma membrane targeting of ion-channels are AMPA receptors, that are glutamate-regulated ion channels responsible for the primary depolarization at the postsynaptic membrane in glutamate-mediated neurotransmission [1, 2]. The translocation of AMPA-type glutamate receptors from endosomal membranes into the synapse of hippocampal neurons or of neurons in the lateral amygdale enhances the ion currents at the respective synapses, resulting in the formation of long-term potentiation, that in turn is the basis for associative learning [3, 4].

Ion channel translocation between cell compartments has also been observed for several vertebrate and invertebrate TRP channels. Including the founding member of the TRP ion channel family, a calcium-permeable cation channel required for *Drosophila* phototransduction [5–7], until today 13 TRP channels in *Drosophila*, 17 TRP channels in *C. elegans*, and 28 TRP channels in mammals have been described [8–12]. All TRP channels possess six transmembrane domains with cytosolic N- and C-termini and probably function as homo- or heterotetramers. The TRP channels are now subdivided into seven subfamilies, five belonging to the group-1 subfamilies (TRPC, TRPV, TRPM, TRPN, and TRPA) and two belonging to the more distantly related group-2 subfamilies (TRPP and TRPML) [13].

Regulated translocation of TRP channels was initially observed for the murine TRPV2 channel that translocates from an intracellular compartment to the plasma membrane upon stimulation with IGF 1 [14]. In the last years, regulated translocation of TRP channels to the plasma membrane has been demonstrated for members of the TRPC [15], TRPV [14], TRPM [16], and TRPA [17] families.

This review summarizes signalling pathways involved in regulated translocation of TRP channels between different cellular compartments and highlights the cell biological mechanisms involved in the trafficking of TRP channels. The stimuli inducing translocation of TRP channels are summarized in Table 30.1, and TRP binding proteins involved in ion channel trafficking are listed in Table 30.2. We also recommend reading excellent reviews by Cayouette and Boulay [18] and Ambudkar [19] which cover similar topics.

Table 30.1 Stimuli regulating translocation of TRP channels

Stimulus	TRP-channel	Reference
Alkalization	TRPV5+	[99]
Arginine-vasopressin	TRPC3+	[25]
Beta-Glucuronidase	TRPV6+	[103]
Carbachol	TRPC3+	[26, 27]
	TRPC6+	[31]
DHPG	TRPC4–	[29]
	TRPC5–	[29]
EETs	TRPC6+	[32]
	TRPV4+	[83]
EGF	TRPC3+	[23]
	TRPC4+	[28]
	TRPC5+	[15]
	TRPM6+	[16]
Fibronectin	TRPV1+	[70]
Fluid shear stress	TRPV4+	[83]
	TRPM7+	[104]
fMLP peptide	TRPV2+	[75]
Forskolin	TRPV1+	[71]
	TRPV2+	[77]
Head activator	TRPV2+	[74]
Insulin/IGF-1	TRPC3+	[24]
	TRPV1+	[38]
	TRPV2+	[14, 73]
Klotho	TRPV5+	[92, 93]
Light	TRPL– (Drosophila)	[45–47]
lysoPC	TRPC5+	[30]
	TRPC6+	[30]
	TRPV2+	[76]
lysoPI	TRPV2+	[76]
Mechanical stretch	TRPV2+	[78]
Methoxamine	TRPC6+	[39]
Mustard oil	TRPA1+	[17]
NGF	TRPC5+	[15]
	TRPV1+	[68, 69]
PDGF	TRPV2+	[14]
Sperm activation	TRP-3+(C. elegans)	[44]
Thrombin	TRPC1+	[22]
Tissue kallikrein	TRPV5+	[91]

+, up-regulated localization at the plasma membrane; –, down-regulated localization at the plasma membrane; DHPG, 3,5-dihydroxyphenylglycine; EETs, epoxyeicosatrienoic acids; EGF, epidermal growth factor; fMLP, formyl Met-Leu-Phe; IGF-1, insulin-like growth factor 1; lysoPC, lysophosphatidylcholine; lysoPI, lysophosphatidylinositol; NGF, nerve growth factor; PDGF, platelet derived growth factor.

Table 30.2 Proteins of vesicular trafficking pathways affecting the plasma membrane localization of TRP channels

Protein	TRP channel	Reference
AIP4	TRPC4	[42]
	TRPV4	[42]
Caveolin-1	TRPV5	[89, 90]
Dynamin	TRPV4	[84]
	TRPV5	[89, 90, 93, 94]
	TRPML3	[126]
Hrs	TRPP2	[121]
Intersectin	TRPV5	[90]
OS-9	TRPV4	[87]
PACS-1/2	TRPP2	[88]
PACSIN 3	TRPV4	[84]
Rab9	TRPC6	[43]
Rab11	TRPC6	[43]
	TRPV5	[95]
RGA	TRPV2	[77, 79]
RNF24	TRPC3-7	[41]
Snapin	TRPC6	[39]
	TRPV1	[67]
	TRPM7	[106]
STAM-1	TRPP2	[121]
Synapsin I	TRPM7	[106]
Synaptotagmin I	TRPM7	[106]
Synaptotagmin IX	TRPV1	[67]
VAMP2	TRPC3	[26]
	TRPA1	[17]

AIP4, atropine-1 interacting protein 4; Hrs, hepatocyte growth factor regulated tyrosine kinase substrate; OS-9, osteosarcoma amplified 9; PACS, phosphofurin acidic cluster sorting protein; PACSIN, protein kinase C and casein kinase substrate in neurons; RGA, recombination gene activator; RNF24, ring finger protein 24; Snapin, synaptic vesicle-associated protein; STAM, signal transducing adaptor molecule; VAMP2, vesicle-associated membrane protein 2.

30.2 Signalling Pathways Regulating Translocation of TRPC Channels

In mammals, the canonical TRP channels (TRPCs) comprise the TRP channels most closely related to the founding member *Drosophila* TRP [20, 21]. Mammalian genomes encode seven genes of the TRPC family that can be subdivided into four subfamilies: TRPC1, TRPC2 (pseudogene in humans), TRPC4/5, and TRPC3/6/7. TRPC channels are widely expressed and many cells express more than one TRPC channel. Functionally, mammalian TRPC channels have been implicated, for example, in regulation of vascular tone, kidney filtration, acrosomal reaction, and pheromone recognition [20, 21]. The analysis of TRPC channels is complicated by their tendency to form heterotetramers among the TRPC1/4/5 and among the TRPC3/6/7 channels. Regulated trafficking of vertebrate TRPCs has been reported for TRPC1 [22], TRPC3 [23–27], TRPC4 [28, 29], TRPC5 [15, 29, 30], and TRPC6 [30–32].

TRPC1 translocation from the ER to the plasma membrane without obvious requirement of other TRPC channels has been reported in human pulmonary artery endothelial cells (HPAEC) [22]. In HPAEC, application of thrombin activates a Gq signalling cascade that eventually leads to the production of inositol-triphosphate (IP_3) which triggers calcium release from the ER. This short calcium pulse is followed by a sustained calcium entry through the plasma membrane. Thrombin also activates Rho which in turn forms a complex with IP_3R and TRPC1 that translocates to the plasma membrane in an actin polymerization dependent manner within a few minutes [22]. At the plasma membrane, TRPC1 mediates sustained calcium-influx which results in increased endothelial permeability [22]. Thus, delivery of the TRPC1 channel to the plasma membrane enhances calcium entry in these cells.

In the mammalian brain, the dopamin 2 receptor colocalizes and physically interacts with TRPC1 [33]. In HEK-293 cells, overexpression of the dopamin 2 receptor increased the amount of TRPC1 at the plasma membrane, suggesting that TRP channels might be recruited to the plasma membrane by a hitchhiking mechanism [33].

The epithelial growth factor (EGF) turned out to activate the translocation of TRPC3 [23], TRPC4 [28], and TRPC5 [15] from an intracellular store to the plasma membrane, albeit using different signalling pathways. Bezzerides et al. [15] expressed TRPC5-eGFP in HEK-293 cells and observed a translocation to the plasma membrane upon EGF stimulation within minutes. Using total internal reflection microscopy (also known as evanescent field microscopy), a method to selectively visualize membrane and submembrane regions, Bezzerides et al. [15] observed the appearance of highly motile vesicles upon EGF stimulation. From these vesicles, TRPC5 finally integrates into the plasma membrane following EGF stimulation, as demonstrated by a surface biotinylation assay [15]. The translocation of the ion channel was correlated with a potentiated TRPC5 ion current upon stimulation with carbachol, a known activator of TRPC channels. Thus, increasing the number of channels at the plasma membrane potentiates the ion current in response to a gating stimulus. On the basis of experiments using pharmacological inhibitors and overexpressed dominant negative proteins, Bezzerides et al. [15] suggested that the EGF receptor activates phosphatidylinositide 3 kinase (PI(3) kinase) which converts PIP_2 into PIP_3. PIP_3 then activates the Rho GTPase Rac1 which in turn activates the PIP kinase to convert PIP into PIP_2. PIP_2 has been described to facilitate exocytosis [34], providing a possible mechanism for the increased TRPC5 surface expression in EGF stimulated cells. In addition, Bezzerides et al. [15] investigated TRPC5 translocation in cultures of rat primary hippocampal neurons endogenously expressing TRPC5. Here, application of nerve growth factor (NGF) mimics the effect of EGF in HEK-293 cells and induces TRPC5 translocation in a comparable time range. As PI(3) kinase and PIP kinase are also required for TRPC5 translocation in hippocampal neurons, the signalling cascade described above probably reflects the in vivo situation. In young rat hippocampal neurons abrogated TRPC5 activity enabled formation of significantly longer neurites and filopodia [35], while activation of TRPC5 translocation to the plasma membrane via PIP kinase stimulation in cultured rat embryonic hippocampal neurons leads to an inhibition of neurite outgrowth, suggesting a role of TRPC5 translocation in this process [15]. Interestingly, TRPC1, albeit being expressed in primary hippocampal neurons, is not translocated to the plasma membrane following stimulation with growth factors, indicating that the PI(3) kinase – Rac1 – PIP kinase pathway acts in a specific way on TRPC5 [15].

EGF also induces translocation of TRPC4 to the plasma membrane in COS-7 cells [28]. EGF enhances the calcium influx in COS-7 cells without further stimulation, suggesting that either TRPC4 is constitutively active in these cells or that a component of the EGF pathway might be involved in the activation of TRPC4 gating [28]. The signalling cascade leading to TRPC4 translocation does not involve the PI(3) kinase – Rac1 – PIP kinase pathway underlying TRPC5 translocation. Instead, TRPC4 translocation depends on Src family tyrosine kinases, with Fyn appearing to be the dominant kinase leading to the phosphorylation of Tyr-959 and Tyr-972 in the C-terminal region of the TRPC4 protein [28]. Phosphorylation of Tyr-959 and Tyr-972 results in a strengthened binding of TRPC4 to NHERF-1 (Na/H exchanger regulatory factor) [28]. The interaction of NHERF-1 with TRPC4 was shown to promote the localization of TRPC4 at the plasma membrane [36]. Mutating Tyr-959 and Tyr-972 to phenylalanine in TRPC4 disturbed the interaction with NHERF-1 and abolished translocation of the channel to the plasma membrane upon EGF stimulation [28]. In addition, TRPC4 binds to αII-spectrin in unstimulated COS-7 cells and this interaction is attenuated upon EGF stimulation [37]. A splicing variant of TRPC4 defective in spectrin binding fails to undergo plasma membrane translocation upon EGF stimulation and displays elevated basal plasma membrane localization [37]. Thus, αII-spectrin seems to retain TRPC4 in an intracellular compartment until EGF stimulation triggers the release of TRPC4 from αII-spectrin and thereby initiates translocation of the channel to the plasma membrane.

EGF was also reported to induce translocation of TRPC3 to the plasma membrane [23]. However, EGF treatment does not enhance muscarinic receptor agonist or synthetic diacylglycerol induced Ca^{2+}-influx via TRPC3 [23]. These results suggest that translocation of TRPC3 by EGF is not a major mechanism for regulating the calcium influx via TRPC3. Additionally, EGF also induces the translocation of TRPM6 to the plasma membrane (see Chapter 4) [16].

Besides EGF, various other stimuli are known to induce the translocation of TRPC channels. In murine ventricular cardiomyocytes, insulin induces a translocation of TRPC3 to the plasma membrane and potentiates TRPC3 currents gated by the DAG analog 1-oleyl-2-acetyl-sn-glycerol (OAG) [24]. Insulin-dependent translocation is not restricted to TRPC3 and has also been described for TRPV1 and TRPV2 (see Chapter 3) [14, 38]. Translocation of TRPC3 (but not TRPC6) to the plasma membrane was also observed following arginine-vasopressin stimulation in cell lines derived from the kidney inner medullary collecting duct (IMCD-3) [25].

In IMCD-3 cells, the expression level of TRPC3 correlates with the transepithelial calcium flux, suggesting a central role of TRPC3 in this process [25]. In addition, the muscarinic acetylcholine receptor agonist carbachol induces both increased calcium influx and translocation of TRPC3 from intracellular vesicles to the plasma membrane in HEK-293 cells [26, 27].

Finally, translocation of TRPC6 to the plasma membrane in HEK-293 cells was observed following either stimulation of a G-protein cascade by carbachol or stimulation of ER-store depletion by thapsigargin [31]. Thereby, surface accumulation of TRPC6 is rather fast (2 min) and parallels the increase in intracellular calcium, suggesting TRPC6 translocation to be the main mechanism to control calcium influx. In addition, epoxyeicosatrienoic acids (EETs) produced by cytochrome P450 enzymes have been shown to induce a translocation of TRPC6 in cultured human vein endothelial cells [32]. Application of EETs has no effect on basal calcium entry but causes a potentiation of the calcium influx after stimulation with the peptide hormone Bradykinin. EETs have been implicated in endothelial cell hyperpolarization suggesting that calcium influx through TRPC6 is crucial for this process [32].

Interestingly, there is a report describing translocation of TRPC channels depending on the function of another TRPC channel [30]. In bovine aortic endothelial cells, lysophosphatidylcholine (lysoPC) induces ion currents and sustained calcium entry resulting in the inhibition of cell movement [30]. In addition, lysoPC treatment initiates the translocation of both TRPC5 and TRPC6 to the plasma membrane. However, the translocation of TRPC5 is delayed compared to TRPC6. In fact, TRPC6 translocation only depends on lysoPC-stimulation, whereas TRPC5 translocation depends on both lysoPC-stimulation and calcium entry through functional TRPC6 channels.

While several reports describe translocation of vertebrate TRPC channels to the plasma membrane upon stimulation, there is only one report about translocation of vertebrate TRPC channels in the opposite direction [29]. In hippocampal neurons, the group I metabotropic glutamate receptor agonist DHPG (3,5-dihydroxyphenylglycine) induces TRPC-dependent prolonged epileptiform discharges [29]. DHPG stimulation also induces a reduction, but not a complete loss, of TRPC4 and TRPC5 at the plasma membrane (measured by surface biotinylation) paralleled by an increased intracellular amount of TRPC4 and TRPC5 (measured by chymotrypsin digest which eliminates surface protein, while intracellular protein remains unaffected). This translocation of TRPC4 and TRPC5 requires phospholipase C (PLC) activity and is thought to play a role in the desensitization of hippocampal neurons.

Translocation of TRPC channels to the plasma membrane upon stimulation of a signalling cascade is thought to include stimulated exocytosis. In line with a vesicular transport mechanism, TRP channels were found to interact with proteins of the vesicular transport/fusion machinery [26, 39]. Singh et al. [26] found that TRPC3 interacts with VAMP2, and inactivation of VAMP2 by tetanus toxin resulted in a decreased plasma membrane distribution of TRPC3 upon stimulation with the muscarinic acetylcholine receptor agonist carbachol. This finding suggests a physiological role of VAMP-mediated exocytosis in carbachol-induced TRPC3 translocation/recruitment [26]. Confocal live imaging of TRPC3-GFP revealed the existence of TRPC3 in highly mobile intracellular vesicles that are able to recover photobleached TRPC3 at the plasma membrane with a halftime of 40 s. TRPC3-GFP vesicles were lost 30–60 min after stopping the delivery of newly synthesized proteins with brefeldin A, suggesting that de novo protein synthesis is essential for maintaining the ability of TRPC3 to insert into the plasma membrane [26]. In addition, the TRPC3 channel itself contains domains facilitating exocytosis. Mutating the lipid binding TRP_2 domain attenuates the translocation of TRPC3 to the plasma membrane upon stimulation with DAG analogs [40].

In neuronal PC12 cells, TRPC6 was identified as interaction partner of snapin (synaptic vesicle-associated protein) and was found to translocate to the plasma membrane upon stimulation of the α_{1A}-adrenoceptor [39]. Interestingly, stimulation of the α_{1A}-adrenoceptor with the agonist methoxamine strengthens the formation of a snapin – TRPC6 – α_{1A}-adrenoceptor complex, while the α_{1A}-adrenoceptor antagonist prazosin weakens the formation of this complex. These findings support a model in which a tripartite complex between α_{1A}-adrenoceptor, snapin, and TRPC6 is formed to recruit TRPC6 to the plasma membrane [39].

Stimulus-dependent exocytosis of TRP channels can be regulated by proteins which retain the channels in an intracellular compartment until an activation signal releases the TRPC channels and allows them to migrate to the plasma membrane. One example of such a protein is the scaffolding protein homer1 [27]. In HEK-293 cells, carbachol treatment induces a rapid translocation of TRPC3 to the plasma membrane. Interference with the homer1b and 1c proteins leads to an increased basal concentration of TRPC3 at the plasma membrane that is unsensitive to carbachol treatment. These results suggest that homer1b and 1c function in retaining TRPC3 in an intracellular compartment during the quiescent state [27]. Another TRPC retention factor is the golgi protein RNF24 which was shown to bind to all TRPC channels and to reduce the surface expression of TRPC3-7 in HEK-293 cells upon overexpression [41]. However, RNF24 overexpression did not inhibit calcium entry through TRPC6 in both unstimulated and carbachol-stimulated conditions questioning the physiological relevance of TRPC6 retention by RNF-24 [41].

Besides exocytosis, endocytosis of TRPC channels might be regulated and affect the concentration of TRPC channels at the plasma membrane. TRPC4, but not TRPC6, has been demonstrated to undergo accelerated endocytosis upon ubiquitination by AIP4, suggesting that ubiquitin-dependent regulation of endocytosis targets selective TRPC channels [42].

At least in the case of TRPC6, Rab proteins are also involved in the trafficking of TRPC channels [43]. In HeLa cells, TRPC6 both co-immunoprecipitates with Rab9 and colocalizes with Rab9 in intracellular vesicles [43]. Introducing a dominant negative Rab9 causes an accumulation of TRPC6 at the plasma membrane [43]. Conversely, a dominant negative Rab11 eliminates TRPC6 at the plasma membrane [43]. Thus, Rab9 is required for the internalization of TRPC6, while Rab11, a recycling compartment marker, is needed for the delivery to the plasma membrane probably via a recycling pathway [43].

Translocation of TRPC channels has also been studied in *C. elegans* and *Drosophila* [44–47]. In *C. elegans*, there are three TRPC orthologs and one of it, TRP-3, is essential for male and hermaphrodite fertility and translocates from an intracellular compartment to the plasma membrane during sperm activation [44].

In *Drosophila*, three TRPC channels are expressed: TRP, TRPL, and TRPγ [48, 49]. TRP and TRPL are predominantly detected in *Drosophila* photoreceptor cells where they mediate fly phototransduction. TRPγ seems to be expressed more ubiquitously, and its role in phototransduction is not clear. The TRP protein is the major calcium channel in phototransduction, while TRPL is an unspecific cation channel. Both TRP and TRPL channels are activated by the phototransduction cascade that transduces the absorption of photons by rhodopsin via a Gq protein- and phospholipase C-mediated signalling cascade into a receptor potential. A double mutant lacking both TRP and TRPL is completely blind, indicating that the light-activated conductance in *Drosophila* photoreceptors is composed of these two channels [50]. The TRP channel is stably attached to the rhabdomere (the light-absorbing compartment of fly photoreceptor cells) by the scaffolding protein INAD that organizes a signalling complex which is believed to enhance specificity and speed of the phototransduction cascade [51–54]. There is evidence that the signalling complex containing TRP undergoes light-induced recruitment to detergent-resistant membrane rafts [55], but it constantly remains in the rhabdomeric membrane and displays no light-triggered translocation between cellular compartments.

In contrast to TRP, the TRPL protein shuttles between the rhabdomeric membrane in dark-adapted flies and an intracellular compartment in light-adapted flies (Fig. 30.1.) [45–47]. The presence of TRPL in the rhabdomere in the dark allows the photoreceptor cells to detect a wider range of light intensities, suggesting a physiological function of TRPL translocation in long-term adaptation [45]. A more detailed analysis revealed that the translocation of TRPL is in fact a two stage process: a fast (~5 min) lateral diffusion from the rhabdomer to the adjoining stalk membrane, followed by a slow (~6 h) removal of TRPL from the apical membrane [46]. Immunocytochemical studies and studies using eGFP-tagged TRPL expressed in *Drosophila* photoreceptors revealed that activation of the entire phototransduction cascade down to the opening of TRP channels (Rhodopsin, Gαq, Phospholipase Cβ, TRP) is required to induce translocation of TRPL from the plasma membrane to an internal storage compartment [46, 47]. Meyer et al. [47] showed that expression of a constitutively active TRP channel leads to localization of TRPL outside of the rhabdomere even in dark-adapted flies, suggesting that calcium influx through TRP is the activator of TRPL translocation. Calcium together with diacylglycerol is known to activate protein kinase C (PKC). In *Drosophila*, the eye specific protein kinase C (ePKC) which is part of the INAD signalling complex described above has been characterized as a protein involved in the deactivation of the light response and in light adaptation [56, 57]. Analysis of TRPL translocation in an ePKC mutant revealed that this protein kinase C is necessary for light-induced TRPL translocation [46]. It remains to be elucidated, if ePKC directly phosphorylates TRPL and

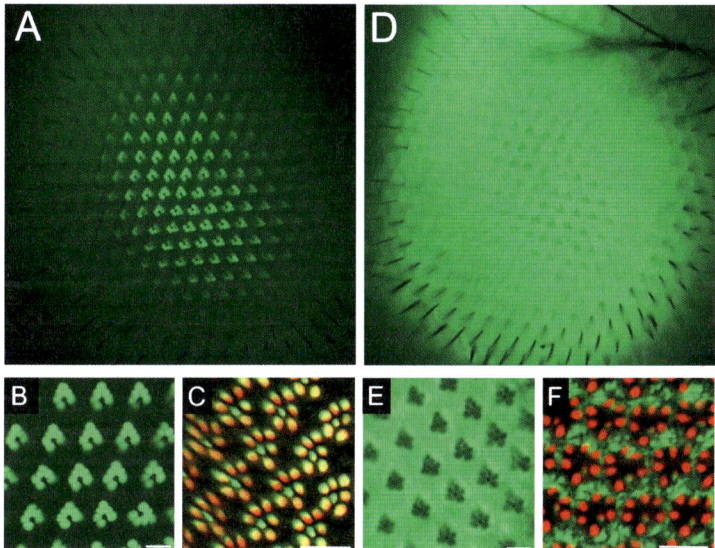

Fig. 30.1 Subcellular localization of TRPL and TRPL-eGFP in the eye of Drosophila kept in the dark or in the light. Flies were kept in the dark (**a–c**) or in *orange* light (**d–f**) for 16 h. **a**, **b** and **d**, **e** show the localization of TRPL-eGFP expressed under the control of the rhodopsin 1 promoter, as analyzed in living flies by water immersion fluorescence microscopy. **b** and **e** are higher magnifications of **a** and **d**, respectively. **c**, **f**: Immunocytochemistry of cross sections through the compound eye revealing the localization of native TRPL (*green*). The actin cytoskeleton of the rhabdomeres was labeled with phalloidin (*red*). Overlay of *green* and *red* appears *yellow*. TRPL and TRPL-eGFP are located in the rhabdomeres in dark-kept flies (**c**), but outside the rhabdomeres in flies illuminated with *orange* light (**f**). Scale bars: 10 μm

thereby induces removal of TRPL from the rhabdomere, or if ePKC exerts its function by phosphorylating other target proteins involved in the regulated translocation of TRPL.

In summary, the physiological effects of vertebrate TRPC channel translocation are not uniform. In some cases like the insulin mediated translocation of TRPC3 in cardiomyocetes, translocation of TRPC channels modulates the response to a gating stimulus and enables potentiation of the response to the same stimulus [24]. In other cases like the carbachol induced translocation of TRPC6 in HEK-293 cells, however, the translocation of the channel itself seems to determine the calcium influx without additional requirement of a gating signal [31]. Alternatively, the translocation stimulus may at the same time trigger the gating of the channel making it difficult to distinguish the relative contributions of gating and translocation to the ion current.

Enriching the amount of TRPC channels at the plasma membrane is based on multiple mechanisms: rapid stimulated exocytosis was reported for carbachol induced translocation of TRPC3 [26], while endocytosis of TRPC4 can be modified by the ubiquitin ligase AIP4 [42]. Finally, TRPC6 plasma membrane localization requires the passage of a recycling compartment [43]. However, more data needs to be collected to draw general conclusions of TRPC channel trafficking.

In *Drosophila*, calcium influx through a TRPC channel (*Drosophila* TRP) is necessary for the translocation of another TRPC channel (*Drosophila* TRPL); a situation resembling the TRPC6-dependent translocation of TRPC5 in bovine aortic endothelial cells upon stimulation with lysoPC [30]. Activation of the phototransduction cascade leads to a translocation of TRPL away from the rhabdomeric membrane towards an intracellular compartment. This translocation direction rather supports a role of TRPL internalization in light induced desensitization than in potentiating the response of a light stimulus.

30.3 Signalling Pathways Regulating Translocation of TRPV Channels

The family of TRP vanilloid (TRPV) ion channels was named after its founding member, the capsaicin sensitive ion channel TRPV1 and comprises six channels in mammals, two in *Drosophila*, and five in *C. elegans* [8–12]. The mammalian TRPV channels can be classified in two subfamilies: TRPV1-4 ("Thermo-TRPs") and TRPV5/6 [11]. TRPV1-4 are expressed in neurons and are activated by heat and numerous chemical compounds [11, 58, 59]. TRPV1 is involved in pain sensation and therefore an interesting target for pharmacological manipulation [58, 59]. Additionally, TRPV1 has been demonstrated to be necessary for synaptic plasticity in hippocampal long-term depression [58]. TRPV4 mainly functions as a mechano- and osmosensor and seems to be involved in hearing [60, 61]. In the central nervous system, TRPV4 has also been reported to contribute to the resting membrane potential [62]. TRPV5 and TRPV6 channels are not expressed in neuronal tissues [63, 64]. Instead, they are present in the kidney, intestine, pancreas, and placenta where they function in calcium reabsorption. TRPV5/6 are the channels with the highest selectivity for calcium ($pCa^{2+}/pNa^+ > 100$) among the TRP channels [63, 64].

TRPV1 is gated by numerous stimuli including heat, capsaicin, or low pH [11, 58, 59]. In *Xenopus* oocytes, TRPV1 currents evoked by capsaicin, anandamide, or protons are potentiated by PKC-mediated phosphorylation of TRPV1 [65, 66]. It was demonstrated that translocation to the plasma membrane contributes to PKC-mediated potentiation of TRPV1 ion currents [67]. The translocation of TRPV1 to the plasma membrane upon PKC stimulation requires up-regulation of exocytosis involving SNARE proteins, as it can be abolished by treatment with botulinum neurotoxin A, an inhibitor of SNAP-25 activity [67]. This result is corroborated by the finding that TRPV1 binds to and colocalizes with the two synaptic vesicle proteins snapin and synaptotagmin IX [67]. A functional role for the interaction of TRPV1 with snapin was demonstrated, as overexpression of snapin abrogated PKC-mediated translocation of TRPV1 [67]. In dorsal root ganglion cells, immunofluorescence stainings demonstrated colocalization of TRPV1 with the synaptic vesicle protein VAMP2, suggesting a vesicular localization of intracellular TRPV1 [67].

Stimulation with insulin or IGF-1 also potentiates capsaicin-gated TRPV1 currents in a receptor tyrosin kinase, PI(3) kinase, and PKC dependent manner [38]. Insulin/IGF-1-stimulation results in the translocation of the TRPV1 channel to the plasma membrane [38]. However, single channel recordings demonstrated that application of insulin also increased the open state probability of single channels [38]. Therefore, probably both increased open state probability and enhanced plasma membrane delivery contribute to the Insulin/IGF-1-induced TRPV1 potentiation.

Besides Insulin/IGF-1, treatment with NGF induced TRPV1 translocation in HEK-293 cells, involving the activity of PI(3)-kinase, PKC(δ), and Src kinase [68, 69]. In this context, a direct physical interaction between TRPV1 and the p85β subunit of the PI(3) kinase was demonstrated [69]. NGF-induced TRPV1-translocation seems to depend on Src kinase-mediated phosphorylation of Tyr200 in TRPV1, as mutagenesis of this site abolished TRPV1 translocation in HEK-293 cells treated with NGF [68]. Finally, in dorsal root ganglion cells, NGF treatment causes both translocation of endogenous TRPV1 and a potentiated capsaicin-gated calcium influx, suggesting that the results observed in HEK-293 cells represent the physiological in vivo situation [68, 69].

Endogenous TRPV1 expression is also found in trigeminal ganglia sensory neurons [70]. In these neurons, the extracellular matrix protein fibronectin induces a translocation of TRPV1 to the plasma membrane and a potentiation of capsaicin-stimulated calcium influx, resulting in an increased neuropeptide release [70]. Here again, the enhanced TRPV1 activity results from both a higher sensitivity of individual TRPV1 channels to capsaicin and an increased plasma membrane localization of TRPV1 [70]. Thereby, fibronectin-induced TRPV1 activation and translocation is dependent on Src kinase phosphorylation of TRPV1 [70].

Vetter et al. [71] reported TRPV1 translocation in HEK-293 cells upon stimulation of an adenylate cyclase/protein kinase A (PKA) pathway. Pro-inflammatory substances like prostaglandins are known to increase the cAMP level in cells and are believed to cause hyperalgesia via sensitized responses of TRPV1. Vetter et al. [71] mimicked the effect of pro-inflammatory substances with forskolin, an activator of the adenylate cyclase. Forskolin treatment of HEK-293 cells causes both a potentiated capsaicin-gated calcium entry and an enlarged expression of TRPV1 at the plasma membrane. In addition, Vetter et al. [71] combined the analysis of TRPV1 translocation with fluorescence resonance energy transfer studies, demonstrating that intracellular TRPV1 is present as monomers, while the functional channels at the plasma membrane are formed by multimeres, most likely by tetramers. Thus, translocation, of TRPV1 to the plasma membrane is tightly associated with its multimerization.

TRPV2 was the first TRP channel that has been demonstrated to undergo regulated translocation to the plasma membrane upon growth factor stimulation [14]. The translocation of TRPV2 in chinese hamster ovary (CHO) cells is rather rapid (~10 min), paralleled by an increased calcium entry and stimulated by the growth factors IGF-1 and PDGF, while other growth factors like EGF or bFGF do not exert any effect on TRPV2 translocation [14]. Blocking the PI(3)-kinase activity interferes with TRPV2 translocation, suggesting this process to be PI(3) kinase dependent [14]. However, it needs to be noted that an involvement of PI(3) kinase in TRPV2 translocation could not be reproduced by Penna et al. [72]. Instead, a role of PI(3)-kinase in gating the TRPV2 channel was proposed [72]. In MIN-6 insulinoma cells (beta cells derived tumor of the pancreas), fluorescence tagged TRPV2 colocalizes with an ER marker in resting cells and starts to overlap with a plasma membrane marker upon exposure to insulin [73]. As calcium entry is significantly stimulated by insulin in MIN-6 cells, insulin seems to enhance calcium influx by providing more TRPV2 channels at the plasma membrane [73]. In insulin-secreting beta cells, knockdown of TRPV2 results in a reduced insulin secretion upon glucose treatment, suggesting that TRPV2-mediated calcium entry is involved in a feed-forward mechanism that stimulates insulin secretion following exposure to glucose [73].

Another study pointed out that the neuropeptide "head activator" which stimulates cell proliferation triggers translocation of TRPV2 from an intracellular store to the plasma membrane in neuronal and neuroendocrine cells endogenously expressing TRPV2 [74]. Translocation of TRPV2 and increased ion currents upon stimulation with "head activator" can also be observed after expression of TRPV2 in COS-7, but not in CHO cells [74], indicating that the genetic background of different cell lines used for heterologous expression might have a major impact on the translocation of TRPV2. Using pharmacological inhibitors, Boels et al. [74] identified a pertussis-toxin sensitive G-protein, the PI(3) kinase, calcium influx, and a Ca-/calmodulin dependent kinase as components of a signalling cascade leading to the translocation of TRPV2 upon "head activator" stimulation in COS-7 cells.

A similar signalling cascade probably also acts in macrophages. Nagasawa et al. [75] reported that TRPV2 undergoes a translocation to the plasma membrane following stimulation with the chemotactic peptide formyl Met-Leu-Phe (fMLP). The translocation of TRPV2 is paralleled by an enhanced calcium influx and can be blocked by pertussis-toxin and by an inhibitor of the PI(3) kinase. Reducing TRPV2 function via a dominant negative TRPV2 results in impaired migration of macrophages, indicating that TRPV2-mediated calcium entry is required for macrophage migration [75]. Confocal microscopy using fluorescence tagged TRPV2 and immunoelectron microscopy revealed that intracellular TRPV2 is mostly present in the ER [75].

TRPV2-dependent cell migration was also observed in the prostate cancer cell line PC3 following stimulation with lysophosphatidylcholine (lysoPC) and lysophosphatidylinositol (lysoPI) [76]. Using TRPV2 transfected CHO and HEK-293 cells, Monet et al. [76] demonstrated that lysoPC treatment results both in an increased calcium influx via TRPV2 and in a translocation of TRPV2 to the plasma membrane, requiring PI(3) kinase and a pertussis-toxin sensitve G-protein [76].

Besides PI(3) kinase signalling, the adenylate cyclase activator forskolin has been reported to induce TRPV2 translocation to the plasma membrane in mast cells [77]. However, the molecular mechanism underlying cAMP-mediated TRPV2 translocation is not known so far. In myocytes, mechanical stretch is an activator of TRPV2 translocation but the signalling cascade transducing a mechanical stimulus into altered TRPV2 trafficking needs to be determined [78].

In addition, the Golgi and ER protein recombination gene activator (RGA) which binds to intracellular TRPV2 facilitates secretion of TRPV2 [79]. Overexpression of RGA increases the concentration of TRPV2 at the plasma membrane in both HEK-293 cells and a mast cell line with native TRPV2 expression [77].

Although numerous studies describe the ion influx through TRPV2 at the plasma membrane, it is noteworthy that TRPV2 was also demonstrated to act as a calcium release channel in early endosomes [80].

The TRPV4 channel is activated by osmotic swelling and requires binding to the microtubule-associated protein 7 for proper function [81]. The amount of TRPV4 channels at the plasma membrane is affected by the "With no K (Lysine)" kinases WNK1 and WNK4 which reduce plasma membrane localization of TRPV4 in HEK-293 cells [82]. In parallel, WNK1/4 reduce the calcium influx activated by hypotonic swelling or by the phorbol ester 4α-PDD [82]. However, the signalling pathway leading from WNK1/4 to TRPV4 trafficking is still elusive. In the distal nephron, WNK1/4 and TRPV4 are both highly expressed, suggesting that WNK1/4 might play a role in the regulation of TRPV4 in this tissue [82]. In cultured human endothelial cells, TRPV4 translocation from an intracellular store to the plasma membrane was observed upon fluid shear stress [83]. It has been suggested that TRPV4 translocation is activated by cytochrome P450 generated epoxyeicosatrienoic acids and participates in flow-induced vasodilatation [83].

The amount of TRPV4 channels at the plasma membrane is regulated by proteins influencing endocytosis [42, 84]. PACSIN proteins have been identified as inhibitors of dynamin mediated endocytosis [85]. PACSIN3 binds to TRPV4, and overexpression of PACSIN3 leads to a higher amount of TRPV4 at the plasma membrane in various cell lines [84]. However, analysing calcium influx in HEK-293 cells revealed that PACSIN3 inhibits rather than potentiates TRPV4 basal activity and its activation by heat and osmotic swelling [86]. Thus, plasma membrane accumulation of ion channels always needs to be analyzed for its physiological relevance, as the results described above provide an example in which increased plasma membrane concentration of TRPV4 is associated with inhibition of TRPV4-mediated current. Like TRPC4 (see Chapter 2), also TRPV4 endocytosis is stimulated upon ubiquitination by AIP4 [42].

In addition, the transport of TRPV4 to the plasma membrane is regulated by retention in the ER [87]. The ER-protein OS-9 binds to TRPV4 (and TRPV1), and overexpression of OS-9 attenuates TRPV4 surface expression [87]. TRPV4 also binds to PACS-1, a protein involved in the ER retention of TRPP2 (see Chapter 6) [88].

The TRP channels TRPV5 and TRPV6 are highly specific for calcium and play a major role in calcium homeostasis of the body [63, 64]. Calcium influx through TRPV5/6 is regulated at different levels [63, 64]: (i) The activity of single channels is modulated e.g. by a negative calcium feedback and by pH (ii) transcription of the TRPV5/6 genes is activated by vitamin D and hormones like PTH and estrogen and (iii) the amount of TRPV5/6 channels present at the plasma membrane is tightly regulated.

As already described for TRPV1 (see above), TRPV5 accumulation at the plasma membrane is stimulated upon activation of PKC [89–91]. Treatment of HEK-293 cells with the DAG analogue 1-oleoyl-acetyl-sn-glycerol (OAG) which activates PKC induces TRPV5-dependent ion currents and up-regulation of the TRPV5 content at the plasma membrane of HEK-293 cells [89]. OAG-stimulated TRPV5 accumulation is prevented by mutagenesis of PKC phosphorylation sites in TRPV5, suggesting that TRPV5 is a direct target of PKC [89]. TRPV5 dependent ion currents are also raised by interference with dynamin that is required for a number of endocytotic pathways. In addition, interference with dynamin prevented further enhancement of ion currents by OAG [89]. Thus, TRPV5 is believed to undergo constitutive dynamin-dependent endocytosis that is reduced upon TRPV5 phosphorylation by PKC, thereby leading to an accumulation of TRPV5 at the plasma membrane. In addition, caveolin-1, an essential component of lipid rafts, is also required for proper endocytosis of TRPV5 [89, 90].

Phosphorylation of TRPV5 by PKC also underlies TRPV5 accumulation at the plasma membrane upon tissue kallikrein treatment [91]. The serine protease tissue kallikrein is secreted in the renal connecting tubules and stimulates calcium entry in primary cultures of renal connecting tubules and cortical collecting duct, as well as in HEK-293 cells transfected with TRPV5. Elevation of Ca^{2+} entry by tissue kallikrein in HEK-293 cells requires the PLC/DAG/PKC pathway and is accompanied by a decreased retrieval of TRPV5 from the plasma membrane. Thus, PKC-dependent inhibition of endocytosis and subsequent accumulation of TRPV5 at the plasma membrane appears to be the major mechanism of tissue kallikrein promoted calcium influx via TRPV5 [91].

A protein specifically involved in the regulation of TRPV5 trafficking is the β-glucuronidase Klotho that has been described as anti-aging factor [92, 93]. The klotho gene is predominantly expressed in the kidney and encodes for a transmembrane protein, whose extracellular domain is cleaved and circulates in the urine [92]. Chang et al. [92] demonstrated that calcium influx through TRPV5 in HEK-293 cells can be stimulated by the biological active extracellular domain of Klotho. The increased calcium influx through TRPV5 correlates with an increased localisation of TRPV5 at the plasma membrane [92]. Introduction of a dominant negative dynamin leads to an increase of TRPV5 currents that could not be further enhanced by the application of Klotho [93]. Thus, Klotho stimulates calcium entry by accumulating TRPV5 at the plasma membrane due to inhibited constitutive endocytosis.

Accumulation of TRPV5 at the plasma membrane via inhibition of endocytosis is tightly regulated by PKC [89], tissue kallikrein [91], and klotho [92, 93]. In addition, endocytosis of TRPV5 depends on dynamin, the lipid raft protein caveolin-1, and the scaffolding protein intersectin [89, 90, 93, 94]. Endocytosis of TRPV5 can be studied in HeLa, HEK-293, and CHO cell lines, in which HA-tagged TRPV5 was traced with an anti-HA antibody in a pulse chase experiment [94]. TRPV5 disappeared quickly from the cell surface (5 min) and was found in numerous small vesicles at the periphery of the cell (5–10 min). Later (30 min), TRPV5 was predominantly localized in larger perinuclear vesicles. Internalized TRPV5 is quite stable, as almost the entire amount of surface-labelled TRPV5 protein could still be detected in the cells 3 h after internalization [94]. Furthermore, 4 h after stopping the delivery of newly synthesized proteins with brefeldin A, TRPV5 activity could still be recorded [94]. These results were confirmed by a metabolic labelling analysis, revealing that most of the TRPV5 protein is still present 8 h after synthesis [94]. Thus, internalized TRPV5 is probably not degraded but is recycled from an intracellular compartment. This hypothesis is supported by colocalization of TRPV5 with Rab11, a recycling compartment marker, and by the lack of colocalization with CD63, a marker for a degradative compartment [95]. Colocalization of Rab11 and endogenously expressed TRPV5 (and TRPV6) was also observed in subapical vesicles of connecting tubules and cortical collecting ducts kidney cells [95]. Furthermore, introducing a GTP binding-defective Rab11 prevented plasma membrane localization of TRPV5 in both *Xenopus* oocytes and primary cultures of connecting tubules/collecting ducts kidney cells, suggesting a functional role of Rab11 in the redistribution of TRPV5 to the plasma membrane [95].

TRPV5 trafficking is affected by "With no K (Lysine)" kinase 4 (WNK4). In one study, coexpression of WNK4 in HEK-293 cells was reported to result in decreased ion currents and surface abundance of TRPV5 in the basal, but not in the OAG stimulated state [90]. However, in another study, an increase of calcium uptake and TRPV5 localization at the plasma membrane upon WNK4 coexpression was reported [96]. Maybe, the difference between these studies can be explained by different expression systems, as Cha and Huang [90] performed cell surface biotinylation experiments in HEK-293 cells, while cell surface biotinylation and plasma membrane fluorescence of TRPV5-YFP in *Xenopus* oocytes were analysed by Jiang et al. [96].

In *Xenopus* oocytes, WNK3 promotes enhanced calcium uptake and upregulation of TRPV5 at the plasma membrane via enhanced exocytosis [97]. However, coexpression of the thiazide-sensitive Na-Cl cotransporter (NCC) which is coexpressed with TRPV5 in the distal convoluted tubule strongly reduced the stimulating effect of WNK3/4 on TRPV5 mediated calcium uptake [96, 97]. Therefore, involvement of NCC may weaken the physiological role of WNK3/4 stimulation in vivo.

Besides WNKs, the serum and glucocorticoid inducible kinases SGK1 and SGK3 affect the distribution of TRPV5 in *Xenopus* oocytes [98]. Coexpression of SGK1 or SGK3 induces enhanced calcium entry and increased plasma membrane abundance of the TRPV5 channel. The effect of SGK1/3 on TRPV5 recruitment to the plasma membrane requires the presence of the Na^+/H^+ exchanger regulating factor NHERF-2 [98]. NHERF binding to TRP channels has also been described for TRPC4, whose plasma membrane localization is supported by binding to NHERF-1 (see Chapter 2) [36].

In addition, the extracellular pH has a major impact on the trafficking of TRPV5 (but not TRPM7) [99]. In HEK-293 cells, total internal reflection microscopy demonstrated that alkalisation increases the number of TRPV5 containing vesicles in the membrane or submembrane region [99]. Surface biotinylation revealed that alkalisation of the extracellular milieu indeed induces a higher concentration of TRPV5 at the plasma membrane, while acidification reduces the amount of TRPV5 at the plasma membrane. Functional recovery after chemobleaching of channels at the plasma membrane shows enhanced delivery of new functional channels upon alkalization, suggesting that alkalization increases plasma membrane localization of TRPV5 by stimulating exocytotic pathways [99]. Alkalisation-induced translocation of TRPV5 to the plasma membrane is concomitant with increased calcium uptake [99]. However, it needs to be noted that extracellular pH also affects the single channel conductance of TRPV5 [100, 101].

While TRPV5 has been extensively studied, less is known about TRPV6. As is the case for TRPV5, SGK1 and SGK3 stimulate ion currents through TRPV6 and TRPV6 localisation at the plasma membrane [102]. However, in contrast to TRPV5, SGK1/3-promoted TRPV6 translocation does not require NHERF2 [98, 102]. SGK1/3 probably acts specifically on TRPV5/6, as translocation of TRPV1 is not affected by SGK1/3 in *Xenopus* oocytes [102].

In isolated chick intestinal epithelial cells, the hormonally active form of vitamin D (1,25-dihydroxycholecalciferol) stimulates calcium influx through TRPV6 and promotes TRPV6 localization at the plasma membrane [103]. Vitamin D treatment of chick intestinal epithelial cells induces increased secretion of a β-glucuronidase which in turn activates calcium uptake by enhancing the surface expression of TRPV6 [103]. Thus, β-glucuronidase-induced calcium uptake via translocation of ion channels to the plasma membrane is a mechanism shared by TRPV5 and TRPV6 [92, 93, 103].

In summary, regulated translocation is a common feature of TRPV1, TRPV2, TRPV4, TRPV5, and TRPV6. In the expression systems used, TRPV1 and TRPV4 require a gating stimulus, capsaicin or osmotic swelling, respectively, and translocation of TRPV1 or TRPV4 potentiates the ion currents induced by the gating stimulus, thereby modulating TRP signalling. Albeit regulated gating has been described in the literature, the channels TRPV2, TRPV5, and TRPV6 are constitutively active in the cell systems used, and regulated translocation of TRPV2, TRPV5, and TRPV6 seems to constitute a major mechanism controlling the calcium influx through these channels. Examples for both stimulated exocytosis [67, 97, 99] and attenuated endocytosis [42, 84, 89, 91–93] are found to increase the amount of TRPV channels at the plasma membrane.

30.4 Signalling Pathways Regulating Translocation of TRPM Channels

The TRPM channels are a family with quite diverse features, comprising eight members in vertebrates, four in *C.elegans*, and one in *Drosophila* [8–12]. Regulated translocation of TRPM channels has been described for vertebrate TRPM6 [16] and TRPM7 [104]. The TRPM6 and TRPM7 channels contain a protein kinase domain in their cytosolic C-terminal regions [10, 11, 64, 105]. TRPM6 is highly permeable for Mg^{2+} and plays an important role in Mg^{2+} homeostasis by regulating Mg^{2+} reabsorption in kidney and intestine [10, 11, 64, 105]. The activity of single TRPM6 channels is attenuated by intracellular Mg^{2+}, providing a negative feedback loop that controls Mg^{2+} homeostasis [10, 11, 64, 105]. In addition, regulated translocation to the plasma membrane is also involved in the control of TRPM6 currents [16].

Treatment with EGF which is expressed in the distal convoluted tubule where regulated transcellular Mg^{2+} reabsorption occurs was shown to increase the plasma membrane localization of TRPM6 [16]. In addition, EGF treatment stimulates ion currents of TRPM6, but not of TRPM7, in HEK-293 cells transfected with either TRPM6 or TRPM7 [16]. Thus, the increased localization of TRPM6 at the plasma membrane contributes to the enlarged ion currents [16]. EGF stimulated TRPM6 translocation does not require the kinase domain of TRPM6 and is transduced through a signalling cascade, involving src-kinases, ERK1/2, PI(3) kinase, and Rac1, while the adenylate cyclase activator forskolin inhibits the effect of EGF [16].

The TRPM7 channel is permeable for divalent cations including Mg^{2+}, Ca^{2+}, and trace metals [10, 11]. The permeability of the TRPM7 channel is regulated by various factors including Mg^{2+}, ATP, pH, or PIP_2 [10, 11]. TRPM7 has been implicated in Mg^{2+} homeostasis, neuronal cell death, and cell proliferation [10, 11, 64]. Increased ion currents through TRPM7 and translocation of TRPM7 to the plasma membrane was observed in HEK-293 and in A7R5 aortic smooth muscle cells as a response to fluid shear stress [104]. Total internal reflection microscopy images demonstrated vesicular and tubular structures of GFP-TRPM7 near the plasma membrane that are more prominent upon fluid shear stress [104]. Aortic smooth muscle cells are not exposed to fluid shear stress under physiological conditions, suggesting that increased ion currents through translocated TRPM7 is a pathophysiological response to endothelial cell injury [104].

TRPM7 also interacts with synaptic vesicle proteins [106, 107]. At least in sympathetic neurons, TRPM7 forms molecular complexes with synapsin I and synaptotagmin I, and directly interacts with synaptic vesicular snapin [106]. Interference of TRPM7's interaction with snapin is critical for neurotransmitter release through vesicle fusion [107].

30.5 Signalling Pathways Regulating Translocation of TRPA1

TRPA channels typically contain a large number of ankyrin repeats, and the TRPA family comprises one member in mammals, two in *C. elegans*, and four in *Drosophila* [8, 12]. The only vertebrate TRPA channel, TRPA1, is involved in pain perception and can be stimulated by various chemical compounds including mustard oil [10, 11, 17]. In rattlesnake, TRPA1 forms a heat-sensitive channel and constitutes the principal sensor for infrared detection in the pit organ that allows snakes to form a thermal image of their predators and prey [108].

Treatment with mustard oil induces translocation of TRPA1 to the plasma membrane via exocytosis that can be mimicked by a simultaneous stimulation of both the adenylate cyclase/PKA- and the PLC pathway [17]. Simultaneous stimulation of these two pathways increased calcium entry and potentiated the pain response of mice following stimulation with mustard oil [17]. TRPA1 translocation requires calcium influx either through TRPA1 itself or through other calcium permeable channels such as TRPV1 [17]. As TRPV1 and TRPA1 are coexpressed in sensory neurons [109], TRPA1 translocation to the plasma membrane as a consequence of TRPV1 activation might represent a mechanism to potentiate TRPA1 activity in vivo.

Translocation of TRPA1 to the plasma membrane following mustard oil treatment is achieved by stimulated exocytosis, as it depends on functional VAMP2 protein [17]. In addition, voltage-clamp techniques monitoring membrane capacitance demonstrated the occurrence of vesicle fusions after stimulation with mustard oil [17].

Note that in addition to its function at the plasma membrane, TRPA1 was found to act as an intracellular calcium channel in secretory vesicles [110].

30.6 Proteins Regulating the Localization of TRPP2 and TRPML Channels

The TRPP and TRPML channels belong to the more distantly related group-2 subfamily of TRP channels [13]. For these channels, no regulated translocation comparable to group-1 subfamily channels upon stimulation of a signalling cascade has been observed so far. However, a few proteins regulating the localization of group-2 TRP channels have been described and will be briefly discussed.

The family of TRPP proteins contains three members in vertebrates, one in *C. elegans*, and one in *Drosophila* [8–12]. The vertebrate TRPP2 channel is localized in the ER, at mitotic spindle microtubules, at the plasma membrane, and in cilia [111–115]. Mutations in the TRPP2 gene account for autosomal dominant polycystic kidney disease [111, 112].

The diverse subcellular localization of TRPP2 seems to correlate with interactions of the channel with different proteins. For example, PKD-1 is essential for the localization of TRPP2 at the plasma membrane in CHO cells [116] and sympathetic neurons [117], but not for the localization of TRPP2 in cilia [118]. Localization of TRPP2 is also determined by casein kinase 2 which phosphorylates a serine residue (S812) of TRPP2 [88]. Phosphorylated TRPP2 is bound by the phosphofurin acidic cluster sorting proteins PACS-1 and PACS-2 which retain TRPP2 at the Golgi/trans Golgi network and ER, respectively. Mutation of this casein kinase 2 phosphorylation site increased plasma membrane expression of TRPP2 and ion currents in *Xenopus* oocytes [88].

Phosphorylation of TRPP2 also occurs at its N-terminal region where phosphorylation by glycogen synthase kinase 3 is necessary for the localization of TRPP2 at the plasma membrane, but not in cilia [119].

In *C. elegans*, the TRPP2 ortholog PKD-2 forms a complex with the PKD-1 ortholog LOV-1 in the cilia of male specific sensory neurons, where both proteins are required for male mating behaviour [12]. LOV-1 interacts with casein kinase 2 which in turn phosphorylates a C-terminal amino acid of PKD-2 [120].

Phosphorylation by casein kinase 2 is antagonized by the calcium-dependent phosphatase calcineurin [120]. Mutating calcineurin phosphatase activity, as well as mimicking constitutive phosphorylation of the casein kinase 2 site (S534D) diminishes localization of PKD-2 in the cilia and results in defects in mating behaviour [120]. In addition, the expression of PKD-2 in cilia is regulated by the STAM-1/Hrs complex which binds internalized membrane proteins and sorts them for lysosomal degradation [121]. Overexpression of STAM-1 or Hrs attenuates ciliary expression of PKD-2 and disturbs male mating behaviour [121]. Lysosomal degradation by STAM/Hrs also accounts for the reduced ciliary expression of the TRPP2 S534D mutant, as this phenotype disappears in the absence of STAM-1 [121].

The results described above suggest that phosphorylation by casein kinase 2 inhibits surface expression of both vertebrate TRPP2 [88] and the *C. elegans* ortholog PKD-2 [120], albeit using different mechanisms: PACS-1/2 mediated retention in the Golgi/ER in vertebrates [88] versus STAM-1/Hrs induced targeting for degradation in *C. elegans* [121]. PACS-1 binding sites have, apart from other ion channels, also been identified in vertebrate TRPC3, TRPV4, and TRPM2 channels, and at least TRPV4 was demonstrated to bind to PACS-1 [88]. Therefore, PACS protein mediated retention in the ER might be a more general mechanism controlling cell surface abundance of ion channels.

The family of TRPML channels includes three members in vertebrates and one ortholog in each *C. elegans* and *Drosophila* [8–12]. The calcium permeable cation channel TRPML1 is an intracellular channel which is localized in lysosomes, where it functions in the fusion of late endosomes and lysosomes [122–124]. Mutations in the human TRPML1 channel account for the neurodegenerative disease mucolipidosis type IV [122–124]. Targeting of TRPML1 to the lysosome is achieved by a N-terminal and a C-terminal dileucin motive [125].

The TRPML3 channel has been demonstrated to regulate endocytosis and autophagy [126]. Inhibition of endocytosis using dominant negative dynamin triggers a plasma membrane accumulation of TRPML3, suggesting that TRPML3 trafficking between the plasma membrane and intracellular compartments is regulated by a dynamin-dependent endocytotic pathway [126]. However, the signalling cascades regulating the distribution of TRPML3 are still elusive.

30.7 Conclusion and Future Perspectives

Regulated delivery and retrieval of TRP channels to and from the plasma membrane is a mechanism to enhance or attenuate TRP-mediated ion currents, thereby modulating TRP signalling. In many cases, TRP channels are translocated from an intracellular storage compartment, where they are inactive, to the plasma membrane, where they can be activated by a variety of stimuli. However, some TRP channels (TRPV2, TRPA1 and TRPP2) function on the cell surface and in intracellular compartments. Translocation of these channels to the plasma membrane needs to be analyzed in the context, of plasma membrane versus intracellular activity.

Plasma membrane abundance of TRP channels is regulated by the interplay between exocytosis and endocytosis. In agreement with this assumption, many proteins involved in vesicular protein trafficking were found to interact with TRP channels. Most of the interaction partners of TRP channels were studied using overexpression or knockdown of candidate proteins in cell cultures. Recently, two groups performed a mutagenesis screen on the trafficking of TRPL in *Drosophila* [127] and TRPP2/PKD-2 in *C.elegans* [128] using GFP-tagged TRP channels. Maybe, applying (forward) genetic approaches to TRP channel translocation reveals new insights into this process. Only little is known about the intracellular distribution and the fate of TRP proteins after endocytosis. Therefore, future research should determine the intracellular location of TRP channels and clarify to which extent TRP channels are recycled, degraded, and synthesized de novo.

Although considerable insight into the signalling pathways triggering translocation of TRP channels could be obtained for some TRP channels, the pathways are still elusive for other TRP channels. It will be important to identify these pathways, as this may pave a way for pharmacological interference in TRP channel translocation and thus in TRP signalling.

Acknowledgments The authors are grateful to T. Oberacker, C. Oberegelsbacher, and O. Voolstra for helpful comments on the manuscript and to C. Oberegelsbacher for providing the data of Fig. 30.1. Research in the laboratory of the authors is supported by the Deutsche Forschungsgemeinschaft (Hu 839/2-5) and the German-Israeli-Foundation for Research and Development (I 1001-96.13/2008).

References

1. Greger IH, Esteban JA (2007) AMPA receptor biogenesis and trafficking. Curr Opin Neurobiol 17:289–297
2. Man HY, Ju W, Ahmadian G, Wang YT (2000) Intracellular trafficking of AMPA receptors in synaptic plasticity. Cell Mol Life Sci 57:1526–1534
3. Rumpel S, LeDoux J, Zador A, Malinow R (2005) Postsynaptic receptor trafficking underlying a form of associative learning. Science 308:83–88
4. Park M, Penick EC, Edwards JG, Kauer JA, Ehlers MD (2004) Recycling endosomes supply AMPA receptors for LTP. Science 305:1972–1975
5. Cosens DJ, Manning A (1969) Abnormal electroretinogram from a Drosophila mutant. Nature 224:285–287
6. Montell C, Rubin GM (1989) Molecular characterization of the Drosophila trp locus: a putative integral membrane protein required for phototransduction. Neuron 2:1313–1323
7. Hardie RC, Minke B: (1992) The trp gene is essential for a light-activated Ca^{2+} channel in Drosophila photoreceptors. Neuron 8:643–651
8. Montell C (2005) The TRP superfamily of cation channels. Sci STKE 2005:re3
9. Montell C (2005) Drosophila TRP channels. Pflugers Arch 451:19–28
10. Pedersen SF, Owsianik G, Nilius B (2005) TRP channels: an overview. Cell Calcium 38:233–252
11. Venkatachalam K, Montell C (2007) TRP channels. Annu Rev Biochem 76:387–417
12. Xiao R, Xu XZ: (2009) Function and regulation of TRP family channels in C. elegans. Pflugers Arch 458:851–860
13. Clapham DE, Julius D, Montell C, Schultz G: (2005) International Union of Pharmacology. XLIX. Nomenclature and structure-function relationships of transient receptor potential channels. Pharmacol Rev 57:427–450

14. Kanzaki M, Zhang YQ, Mashima H, Li L, Shibata H, Kojima I: (1999) Translocation of a calcium-permeable cation channel induced by insulin-like growth factor-I. Nat Cell Biol 1:165–170
15. Bezzerides VJ, Ramsey IS, Kotecha S, Greka A, Clapham DE: (2004) Rapid vesicular translocation and insertion of TRP channels. Nat Cell Biol 6:709–720
16. Thebault S, Alexander RT, Tiel Groenestege WM, Hoenderop JG, Bindels RJ (2009) EGF increases TRPM6 activity and surface expression. J Am Soc Nephrol 20:78–85
17. Schmidt M, Dubin AE, Petrus MJ, Earley TJ, Patapoutian A: (2009) Nociceptive signals induce trafficking of TRPA1 to the plasma membrane. Neuron 64:498–509
18. Cayouette S, Boulay G: (2007) Intracellular trafficking of TRP channels. Cell Calcium 42:225–232
19. Ambudkar IS (2007) Trafficking of TRP channels: determinants of channel function. Handb Exp Pharmacol 179:541–557
20. Birnbaumer L (2009) The TRPC class of ion channels: a critical review of their roles in slow, sustained increases in intracellular Ca(2+) concentrations. Annu Rev Pharmacol Toxicol 49:395–426
21. Kiselyov K, Patterson RL: (2009) The integrative function of TRPC channels. Front Biosci 14:45–58
22. Mehta D, Ahmmed GU, Paria BC, Holinstat M, Voyno-Yasenetskaya T, Tiruppathi C, Minshall RD, Malik AB (2003) RhoA interaction with inositol 1,4,5-trisphosphate receptor and transient receptor potential channel-1 regulates Ca^{2+} entry. Role in signaling increased endothelial permeability. J Biol Chem 278:33492–33500
23. Smyth JT, Lemonnier L, Vazquez G, Bird GS, Putney JW Jr. (2006) Dissociation of regulated trafficking of TRPC3 channels to the plasma membrane from their activation by phospholipase C. J Biol Chem 281:11712–11720
24. Fauconnier J, Lanner JT, Sultan A, Zhang SJ, Katz A, Bruton JD, Westerblad H: (2007) Insulin potentiates TRPC3-mediated cation currents in normal but not in insulin-resistant mouse cardiomyocytes. Cardiovasc Res 73:376–385
25. Goel M, Sinkins WG, Zuo CD, Hopfer U, Schilling WP: (2007) Vasopressin-induced membrane trafficking of TRPC3 and AQP2 channels in cells of the rat renal collecting duct. Am J Physiol Renal Physiol 293:F1476–F1488
26. Singh BB, Lockwich TP, Bandyopadhyay BC, Liu X, Bollimuntha S, Brazer SC, Combs C, Das S, Leenders AG, Sheng ZH, Knepper MA, Ambudkar SV, Ambudkar IS: (2004) VAMP2-dependent exocytosis regulates plasma membrane insertion of TRPC3 channels and contributes to agonist-stimulated Ca^{2+} influx. Mol Cell 15:635–646
27. Kim JY, Zeng W, Kiselyov K, Yuan JP, Dehoff MH, Mikoshiba K, Worley PF, Muallem S: (2006) Homer 1 mediates store- and inositol 1,4,5-trisphosphate receptor-dependent translocation and retrieval of TRPC3 to the plasma membrane. J Biol Chem 281:32540–32549
28. Odell AF, Scott JL, Van Helden DF: (2005) Epidermal growth factor induces tyrosine phosphorylation, membrane insertion, and activation of transient receptor potential channel 4. J Biol Chem 280:37974–37987
29. Wang M, Bianchi R, Chuang SC, Zhao W, Wong RK (2007) GroupI metabotropic glutamate receptor-dependent TRPC channel trafficking in hippocampal neurons. J Neurochem 101:411–421
30. Chaudhuri P, Colles SM, Bhat M, Van Wagoner DR, Birnbaumer L, Graham LM: (2008) Elucidation of a TRPC6-TRPC5 channel cascade that restricts endothelial cell movement. Mol Biol Cell 19:3203–3211
31. Cayouette S, Lussier MP, Mathieu EL, Bousquet SM, Boulay G: (2004) Exocytotic insertion of TRPC6 channel into the plasma membrane upon Gq protein-coupled receptor activation. J Biol Chem 279:7241–7246
32. Fleming I, Rueben A, Popp R, Fisslthaler B, Schrodt S, Sander A, Haendeler J, Falck JR, Morisseau C, Hammock BD, Busse R: (2007) Epoxyeicosatrienoic acids regulate Trp channel dependent Ca^{2+} signaling and hyperpolarization in endothelial cells. Arterioscler Thromb Vasc Biol 27:2612–2618

33. Hannan MA, Kabbani N, Paspalas CD, Levenson R: (2008) Interaction with dopamine D2 receptor enhances expression of transient receptor potential channel 1 at the cell surface. Biochim Biophys Acta 1778:974–982
34. Holz RW, Hlubek MD, Sorensen SD, Fisher SK, Balla T, Ozaki S, Prestwich GD, Stuenkel EL, Bittner MA: (2000) A pleckstrin homology domain specific for phosphatidylinositol 4, 5-bisphosphate (PtdIns-4,5-P2) and fused to green fluorescent protein identifies plasma membrane PtdIns-4,5-P2 as being important in exocytosis. J Biol Chem 275:17878–17885
35. Greka A, Navarro B, Oancea E, Duggan A, Clapham DE: (2003) TRPC5 is a regulator of hippocampal neurite length and growth cone morphology. Nat Neurosci 6:837–845
36. Mery L, Strauss B, Dufour JF, Krause KH, Hoth M (2002) The PDZ-interacting domain of TRPC4 controls its localization and surface expression in HEK293 cells. J Cell Sci 115:3497–3508
37. Odell AF, Van Helden DF, Scott JL: (2008) The spectrin cytoskeleton influences the surface expression and activation of human transient receptor potential channel 4 channels. J Biol Chem 283:4395–4407
38. Van Buren JJ, Bhat S, Rotello R, Pauza ME, Premkumar LS: (2005) Sensitization and translocation of TRPV1 by insulin and IGF-I. Mol Pain 1:17
39. Suzuki F, Morishima S, Tanaka T, Muramatsu I: (2007) Snapin, a new regulator of receptor signaling, augments alpha1A-adrenoceptor-operated calcium influx through TRPC6. J Biol Chem 282:29563–29573
40. van Rossum DB, Oberdick D, Rbaibi Y, Bhardwaj G, Barrow RK, Nikolaidis N, Snyder SH, Kiselyov K, Patterson RL: (2008) TRP_2, a lipid/trafficking domain that mediates diacylglycerol-induced vesicle fusion. J Biol Chem 283:34384–34392
41. Lussier MP, Lepage PK, Bousquet SM, Boulay G: (2008) RNF24, a new TRPC interacting protein, causes the intracellular retention of TRPC. Cell Calcium 43:432–443
42. Wegierski T, Hill K, Schaefer M, Walz G: (2006) The HECT ubiquitin ligase AIP4 regulates the cell surface expression of select TRP channels. EMBO J 25:5659–5669
43. Cayouette S, Bousquet SM, Francoeur N, Dupre E, Monet M, Gagnon H, Guedri YB, Lavoie C, Boulay G (2010) Involvement of Rab9 and Rab11 in the intracellular trafficking of TRPC6. Biochim Biophys Acta 1803:805–812
44. Xu XZ, Sternberg PW (2003) A C.elegans sperm TRP protein required for sperm-egg interactions during fertilization. Cell 114:285–297
45. Bähner M, Frechter S, Da Silva N, Minke B, Paulsen R, Huber A (2002) Light-regulated subcellular translocation of *Drosophila* TRPL channels induces long-term adaptation and modifies the light-induced current. Neuron 34:83–93
46. Cronin MA, Lieu MH, Tsunoda S: (2006) Two stages of light-dependent TRPL-channel translocation in Drosophila photoreceptors. J Cell Sci 119:2935–2944
47. Meyer NE, Joel-Almagor T, Frechter S, Minke B, Huber A: (2006) Subcellular translocation of the eGFP-tagged TRPL channel in Drosophila photoreceptors requires activation of the phototransduction cascade. J Cell Sci 119:2592–2603
48. Montell C (2005) TRP channels in Drosophila photoreceptor cells. J Physiol 567:45–51
49. Katz B, Minke B (2009) Drosophila photoreceptors and signaling mechanisms. Front Cell Neurosci 3:2 [Epub 2009 Jun 11:2]
50. Niemeyer BA, Suzuki E, Scott K, Jalink K, Zuker CS: (1996) The Drosophila light-activated conductance is composed of the two channels TRP and TRPL. Cell 85:651–659
51. Chevesich J, Kreuz AJ, Montell C: (1997) Requirement for the PDZ domain protein, INAD, for localization of the TRP store-operated channel to a signaling complex. Neuron 18:95–105
52. Shieh BH, Zhu MY: (1996) Regulation of the TRP Ca^{2+} channel by INAD in Drosophila photoreceptors. Neuron 16:991–998
53. Huber A, Sander P, Gobert A, Bahner M, Hermann R, Paulsen R: (1996) The transient receptor potential protein (Trp), a putative store-operated Ca^{2+} channel essential for

phosphoinositide-mediated photoreception, forms a signaling complex with NorpA, InaC and InaD. EMBO J 15:7036–7045
54. Tsunoda S, Sierralta J, Sun Y, Bodner R, Suzuki E, Becker A, Socolich M, Zuker CS: (1997) A multivalent PDZ-domain protein assembles signalling complexes in a G-protein-coupled cascade. Nature 388:243–249
55. Sanxaridis PD, Cronin MA, Rawat SS, Waro G, Acharya U, Tsunoda S: (2007) Light-induced recruitment of INAD-signaling complexes to detergent-resistant lipid rafts in Drosophila photoreceptors. Mol Cell Neurosci 36:36–46
56. Hardie RC, Peretz A, Suss-Toby E, Rom-Glas A, Bishop SA, Selinger Z, Minke B: (1993) Protein kinase C is required for light adaptation in Drosophila photoreceptors. Nature 363:634–637
57. Smith DP, Ranganathan R, Hardy RW, Marx J, Tsuchida T, Zuker CS: (1991) Photoreceptor deactivation and retinal degeneration mediated by a photoreceptor-specific protein kinase C. Science 254:1478–1484
58. Kauer JA, Gibson HE: (2009) Hot flash: TRPV channels in the brain. Trends Neurosci 32:215–224
59. Liu DL, Wang WT, Xing JL, Hu SJ: (2009) Research progress in transient receptor potential vanilloid 1 of sensory nervous system. Neurosci Bull 25:221–227
60. Liedtke W: (2008) Molecular mechanisms of TRPV4-mediated neural signaling. Ann N Y Acad Sci 1144:42–52
61. Plant TD, Strotmann R (2007) TRPV4. Handb Exp Pharmacol 179:189–205
62. Shibasaki K, Suzuki M, Mizuno A, Tominaga M: (2007) Effects of body temperature on neural activity in the hippocampus: regulation of resting membrane potentials by transient receptor potential vanilloid 4. J Neurosci 27:1566–1575
63. de Groot T, Bindels RJ, Hoenderop JG: (2008) TRPV5: an ingeniously controlled calcium channel. Kidney Int 74:1241–1246
64. Hoenderop JG, Bindels RJ: (2008) Calciotropic and magnesiotropic TRP channels. Physiology (Bethesda) 23:32–40
65. Premkumar LS, Ahern GP: (2000) Induction of vanilloid receptor channel activity by protein kinase C. Nature 408:985–990
66. Bhave G, Hu HJ, Glauner KS, Zhu W, Wang H, Brasier DJ, Oxford GS, Gereau RW: (2003) Protein kinase C phosphorylation sensitizes but does not activate the capsaicin receptor transient receptor potential vanilloid 1 (TRPV1). Proc Natl Acad Sci U S A 100:12480–12485
67. Morenilla-Palao C, Planells-Cases R, Garcia-Sanz N, Ferrer-Montiel A: (2004) Regulated exocytosis contributes to protein kinase C potentiation of vanilloid receptor activity. J Biol Chem 279:25665–25672
68. Zhang X, Huang J, McNaughton PA (2005) NGF rapidly increases membrane expression of TRPV1 heat-gated ion channels. EMBO J 24:4211–4223
69. Stein AT, Ufret-Vincenty CA, Hua L, Santana LF, Gordon SE: (2006) Phosphoinositide 3-kinase binds to TRPV1 and mediates NGF-stimulated TRPV1 trafficking to the plasma membrane. J Gen Physiol 128:509–522
70. Jeske NA, Patwardhan AM, Henry MA, Milam SB: (2009) Fibronectin stimulates TRPV1 translocation in primary sensory neurons. J Neurochem 108:591–600
71. Vetter I, Cheng W, Peiris M, Wyse BD, Roberts-Thomson SJ, Zheng J, Monteith GR, Cabot PJ: (2008) Rapid, opioid-sensitive mechanisms involved in transient receptor potential vanilloid 1 sensitization. J Biol Chem 283:19540–19550
72. Penna A, Juvin V, Chemin J, Compan V, Monet M, Rassendren FA: (2006) PI3-kinase promotes TRPV2 activity independently of channel translocation to the plasma membrane. Cell Calcium 39:495–507
73. Hisanaga E, Nagasawa M, Ueki K, Kulkarni RN, Mori M, Kojima I: (2009) Regulation of calcium-permeable TRPV2 channel by insulin in pancreatic beta-cells. Diabetes 58:174–184

74. Boels K, Glassmeier G, Herrmann D, Riedel IB, Hampe W, Kojima I, Schwarz JR, Schaller HC: (2001) The neuropeptide head activator induces activation and translocation of the growth-factor-regulated Ca(2+)-permeable channel GRC. J Cell Sci 114:3599–3606
75. Nagasawa M, Nakagawa Y, Tanaka S, Kojima I: (2007) Chemotactic peptide fMetLeuPhe induces translocation of the TRPV2 channel in macrophages. J Cell Physiol 210:692–702
76. Monet M, Gkika D, Lehen'kyi V, Pourtier A, Vanden AF, Bidaux G, Juvin V, Rassendren F, Humez S, Prevarsakaya N: (2009) Lysophospholipids stimulate prostate cancer cell migration via TRPV2 channel activation. Biochim Biophys Acta 1793:528–539
77. Stokes AJ, Wakano C, Del Carmen KA, Koblan-Huberson M, Turner H: (2005) Formation of a physiological complex between TRPV2 and RGA protein promotes cell surface expression of TRPV2. J Cell Biochem 94:669–683
78. Iwata Y, Katanosaka Y, Arai Y, Komamura K, Miyatake K, Shigekawa M: (2003) A novel mechanism of myocyte degeneration involving the Ca^{2+}-permeable growth factor-regulated channel. J Cell Biol 161:957–967
79. Barnhill JC, Stokes AJ, Koblan-Huberson M, Shimoda LM, Muraguchi A, Adra CN, Turner H (2004) RGA protein associates with a TRPV ion channel during biosynthesis and trafficking. J Cell Biochem 91:808–820
80. Saito M, Hanson PI, Schlesinger P: (2007) Luminal chloride-dependent activation of endosome calcium channels: patch clamp study of enlarged endosomes. J Biol Chem 282:27327–27333
81. Suzuki M, Hirao A, Mizuno A: (2003) Microtubule-associated [corrected] protein 7 increases the membrane expression of transient receptor potential vanilloid 4 (TRPV4). J Biol Chem 278:51448–51453
82. Fu Y, Subramanya A, Rozansky D, Cohen DM (2006) WNK kinases influence TRPV4 channel function and localization. Am J Physiol Renal Physiol 290:F1305–F1314
83. Loot AE, Popp R, Fisslthaler B, Vriens J, Nilius B, Fleming I: (2008) Role of cytochrome P450-dependent transient receptor potential V4 activation in flow-induced vasodilatation. Cardiovasc Res 80:445–452
84. Cuajungco MP, Grimm C, Oshima K, D'hoedt D, Nilius B, Mensenkamp AR, Bindels RJ, Plomann M, Heller S: (2006) PACSINs bind to the TRPV4 cation channel. PACSIN 3 modulates the subcellular localization of TRPV4. J Biol Chem 281:18753–18762
85. Modregger J, Ritter B, Witter B, Paulsson M, Plomann M (2000) All three PACSIN isoforms bind to endocytic proteins and inhibit endocytosis. J Cell Sci 113(Pt 24):4511–4521
86. D'hoedt D, Owsianik G, Prenen J, Cuajungco MP, Grimm C, Heller S, Voets T, Nilius B: (2008) Stimulus-specific modulation of the cation channel TRPV4 by PACSIN 3. J Biol Chem 283:6272–6280
87. Wang Y, Fu X, Gaiser S, Kottgen M, Kramer-Zucker A, Walz G, Wegierski T (2007) OS-9 regulates the transit and polyubiquitination of TRPV4 in the endoplasmic reticulum. J Biol Chem 282:36561–36570
88. Kottgen M, Benzing T, Simmen T, Tauber R, Buchholz B, Feliciangeli S, Huber TB, Schermer B, Kramer-Zucker A, Hopker K, Simmen KC, Tschucke CC, Sandford R, Kim E, Thomas G, Walz G: (2005) Trafficking of TRPP2 by PACS proteins represents a novel mechanism of ion channel regulation. EMBO J 24:705–716
89. Cha SK, Wu T, Huang CL: (2008) Protein kinase C inhibits caveolae-mediated endocytosis of TRPV5. Am J Physiol Renal Physiol 294:F1212–F1221
90. Cha SK, Huang CL: (2010) WNK4 kinase stimulates caveola-mediated endocytosis of TRPV5 amplifying the dynamic range of regulation of the channel by protein kinase C. J Biol Chem 285:6604–6611
91. Gkika D, Topala CN, Chang Q, Picard N, Thebault S, Houillier P, Hoenderop JG, Bindels RJ: (2006) Tissue kallikrein stimulates Ca(2+) reabsorption via PKC-dependent plasma membrane accumulation of TRPV5. EMBO J 25:4707–4716
92. Chang Q, Hoefs S, van der Kemp AW, Topala CN, Bindels RJ, Hoenderop JG: (2005) The beta-glucuronidase klotho hydrolyzes and activates the TRPV5 channel. Science 310: 490–493

93. Cha SK, Ortega B, Kurosu H, Rosenblatt KP, Kuro O, Huang CL: (2008) Removal of sialic acid involving Klotho causes cell-surface retention of TRPV5 channel via binding to galectin-1. Proc Natl Acad Sci U S A 105:9805–9810
94. van de Graaf SF, Rescher U, Hoenderop JG, Verkaart S, Bindels RJ, Gerke V: (2008) TRPV5 is internalized via clathrin-dependent endocytosis to enter a Ca^{2+}-controlled recycling pathway. J Biol Chem 283:4077–4086
95. van de Graaf SF, Chang Q, Mensenkamp AR, Hoenderop JG, Bindels RJ: (2006) Direct interaction with Rab11a targets the epithelial Ca^{2+} channels TRPV5 and TRPV6 to the plasma membrane. Mol Cell Biol 26:303–312
96. Jiang Y, Ferguson WB, Peng JB: (2007) WNK4 enhances TRPV5-mediated calcium transport: potential role in hypercalciuria of familial hyperkalemic hypertension caused by gene mutation of WNK4. Am J Physiol Renal Physiol 292:F545–F554
97. Zhang W, Na T, Peng JB (2008) WNK3 positively regulates epithelial calcium channels TRPV5 and TRPV6 via a kinase-dependent pathway. Am J Physiol Renal Physiol 295:F1472–F1484
98. Embark HM, Setiawan I, Poppendieck S, van de Graaf SF, Boehmer C, Palmada M, Wieder T, Gerstberger R, Cohen P, Yun CC, Bindels RJ, Lang F: (2004) Regulation of the epithelial Ca^{2+} channel TRPV5 by the NHE regulating factor NHERF2 and the serum and glucocorticoid inducible kinase isoforms SGK1 and SGK3 expressed in Xenopus oocytes. Cell Physiol Biochem 14:203–212
99. Lambers TT, Oancea E, de Groot T, Topala CN, Hoenderop JG, Bindels RJ: (2007) Extracellular pH dynamically controls cell surface delivery of functional TRPV5 channels. Mol Cell Biol 27:1486–1494
100. Yeh BI, Sun TJ, Lee JZ, Chen HH, Huang CL: (2003) Mechanism and molecular determinant for regulation of rabbit transient receptor potential type 5 (TRPV5) channel by extracellular pH. J Biol Chem 278:51044–51052
101. Yeh BI, Kim YK, Jabbar W, Huang CL: (2005) Conformational changes of pore helix coupled to gating of TRPV5 by protons. EMBO J 24:3224–3234
102. Bohmer C, Palmada M, Kenngott C, Lindner R, Klaus F, Laufer J, Lang F: (2007) Regulation of the epithelial calcium channel TRPV6 by the serum and glucocorticoid-inducible kinase isoforms SGK1 and SGK3. FEBS Lett 581:5586–5590
103. Khanal RC, Peters TM, Smith NM, Nemere I: (2008) Membrane receptor-initiated signaling in 1,25(OH)2D3-stimulated calcium uptake in intestinal epithelial cells. J Cell Biochem 105:1109–1116
104. Oancea E, Wolfe JT, Clapham DE: (2006) Functional TRPM7 channels accumulate at the plasma membrane in response to fluid flow. Circ Res 98:245–253
105. van der WJ, Hoenderop JG, Bindels RJ (2009) Epithelial Mg^{2+} channel TRPM6: insight into the molecular regulation. Magnes Res 22:127–132
106. Krapivinsky G, Mochida S, Krapivinsky L, Cibulsky SM, Clapham DE: (2006) The TRPM7 ion channel functions in cholinergic synaptic vesicles and affects transmitter release. Neuron 52:485–496
107. Brauchi S, Krapivinsky G, Krapivinsky L, Clapham DE: (2008) TRPM7 facilitates cholinergic vesicle fusion with the plasma membrane. Proc Natl Acad Sci U S A 105:8304–8308
108. Gracheva EO, Ingolia NT, Kelly YM, Cordero-Morales JF, Hollopeter G, Chesler AT, Sanchez EE, Perez JC, Weissman JS, Julius D: (2010) Molecular basis of infrared detection by snakes. Nature 464:1006–1011
109. Story GM, Peier AM, Reeve AJ, Eid SR, Mosbacher J, Hricik TR, Earley TJ, Hergarden AC, Andersson DA, Hwang SW, McIntyre P, Jegla T, Bevan S, Patapoutian A: (2003) ANKTM1, a TRP-like channel expressed in nociceptive neurons, is activated by cold temperatures. Cell 112:819–829
110. Prasad P, Yanagihara AA, Small-Howard AL, Turner H, Stokes AJ (2008) Secretogranin III directs secretory vesicle biogenesis in mast cells in a manner dependent upon interaction with chromogranin A. J Immunol 181:5024–5034
111. Giamarchi A, Padilla F, Coste B, Raoux M, Crest M, Honore E, Delmas P: (2006) The versatile nature of the calcium-permeable cation channel TRPP2. EMBO Rep 7:787–793

112. Kottgen M, Walz G: (2005) Subcellular localization and trafficking of polycystins. Pflugers Arch 451:286–293
113. Chen XZ, Li Q, Wu Y, Liang G, Lara CJ, Cantiello HF: (2008) Submembraneous microtubule cytoskeleton: interaction of TRPP2 with the cell cytoskeleton. FEBS J 275: 4675–4683
114. Tsiokas L, Kim S, Ong EC: (2007) Cell biology of polycystin-2. Cell Signal 19:444–453
115. Tsiokas L: (2009) Function and regulation of TRPP2 at the plasma membrane. Am J Physiol Renal Physiol 297:F1–F9
116. Hanaoka K, Qian F, Boletta A, Bhunia AK, Piontek K, Tsiokas L, Sukhatme VP, Guggino WB, Germino GG: (2000) Co-assembly of polycystin-1 and -2 produces unique cation-permeable currents. Nature 408:990–994
117. Delmas P, Nauli SM, Li X, Coste B, Osorio N, Crest M, Brown DA, Zhou J: (2004) Gating of the polycystin ion channel signaling complex in neurons and kidney cells. FASEB J 18: 740–742
118. Geng L, Okuhara D, Yu Z, Tian X, Cai Y, Shibazaki S, Somlo S: (2006) Polycystin-2 traffics to cilia independently of polycystin-1 by using an N-terminal RVxP motif. J Cell Sci 119:1383–1395
119. Streets AJ, Moon DJ, Kane ME, Obara T, Ong AC: (2006) Identification of an N-terminal glycogen synthase kinase 3 phosphorylation site which regulates the functional localization of polycystin-2 in vivo and in vitro. Hum Mol Genet 15:1465–1473
120. Hu J, Bae YK, Knobel KM, Barr MM: (2006) Casein kinase II and calcineurin modulate TRPP function and ciliary localization. Mol Biol Cell 17:2200–2211
121. Hu J, Wittekind SG, Barr MM (2007) STAM andHrs down-regulate ciliary TRP receptors. Mol Biol Cell 18:3277–3289
122. Dong XP, Wang X, Xu H (2010) TRP channels of intracellular membranes. J Neurochem 113:313–328
123. Puertollano R, Kiselyov K: (2009) TRPMLs: in sickness and in health. Am J Physiol Renal Physiol 296:F1245–F1254
124. Zeevi DA, Frumkin A, Bach G: (2007) TRPML and lysosomal function. Biochim Biophys Acta 1772:851–858
125. Vergarajauregui S, Puertollano R: (2006) Two di-leucine motifs regulate trafficking of mucolipin-1 to lysosomes. Traffic 7:337–353
126. Kim HJ, Soyombo AA, Tjon-Kon-Sang S, So I, Muallem S: (2009) The Ca(2+) channel TRPML3 regulates membrane trafficking and autophagy. Traffic 10:1157–1167
127. Meyer NE, Oberegelsbacher C, Dürr TD, Schäfer A, Huber A: (2008) An eGFP-based genetic screen for defects in light-triggered subcelluar translocation of the *Drosophila* photoreceptor channel TRPL. Fly 2:384–394
128. Bae YK, Lyman-Gingerich J, Barr MM, Knobel KM: (2008) Identification of genes involved in the ciliary trafficking of C. elegans PKD-2. Dev Dyn 237:2021–2029

Chapter 31
Emerging Roles of Canonical TRP Channels in Neuronal Function

Sunitha Bollimuntha, Senthil Selvaraj, and Brij B. Singh

Abstract Ca^{2+} signaling in neurons is intimately associated with the regulation of vital physiological processes including growth, survival and differentiation. In neurons, Ca^{2+} elicits two major functions. First as a charge carrier, Ca^{2+} reveals an indispensable role in information relay via membrane depolarization, exocytosis, and the release of neurotransmitters. Second on a global basis, Ca^{2+} acts as a ubiquitous intracellular messenger to modulate neuronal function. Thus, to mediate Ca^{2+}-dependent physiological events, neurons engage multiple mode of Ca^{2+} entry through a variety of Ca^{2+} permeable plasma membrane channels. Here we discuss a subset of specialized Ca^{2+}-permeable non-selective TRPC channels and summarize their physiological and pathological role in the context of excitable cells. TRPC channels are predominately expressed in neuronal cells and are activated through complex mechanisms, including second messengers and store depletion. A growing body of evidence suggests a prime contribution of TRPC channels in regulating fundamental neuronal functions. TRPC channels have been shown to be associated with neuronal development, proliferation and differentiation. In addition, TRPC channels have also been suggested to have a potential role in regulating neurosecretion, long term potentiation, and synaptic plasticity. During the past years, numerous seminal discoveries relating TRPC channels to neurons have constantly emphasized on the significant contribution of this group of ion channels in regulating neuronal function. Here we review the major groundbreaking work that has uniquely placed TRPC channels in a pivotal position for governing neuronal Ca^{2+} signaling and associated physiological responses.

31.1 Introduction

Both release of Ca^{2+} from intracellular stores as well as Ca^{2+} influx across the plasma membrane (PM) plays an important role in regulating cellular processes that range from cell division to cell death [1]. In neurons, Ca^{2+} plays a seminal

B.B. Singh (✉)
Department of Biochemistry and Molecular Biology, School of Medicine and Health Sciences, University of North Dakota, Grand Forks, ND 58201, USA
e-mail: bsingh@medicine.nodak.edu

role as a charge carrier and is an essential intracellular messenger, which could link brain function to cellular changes in humans and other multicellular organisms. Stimulation of neuronal cells using various agonists or pharmacological agents lead to an increase in intracellular Ca^{2+} ($[Ca^{2+}]_i$) [2,3]. This increase in $[Ca^{2+}]_i$ that is attributed from both release of Ca^{2+} from intracellular ER stores as well as Ca^{2+} entry across the membrane via the TRPC channels (Fig. 31.1 outlines the activation mechanism of TRPC channels). Although in most of these processes release of intracellular Ca^{2+} stores is critical, it is the influx of external Ca^{2+}, which is always essential to have a global or sustained response. Furthermore, Ca^{2+} influx followed by ER store-depletion accomplishes several critical cellular functions. First, this Ca^{2+} influx replenishes the ER Ca^{2+} stores, thereby, maintaining its ability to release Ca^{2+} upon subsequent stimuli. Second, since ER has limited Ca^{2+} capacity, Ca^{2+} influx is essential for increasing $[Ca^{2+}]_i$ levels to have a physiological response. Third, since Ca^{2+} concentrations within the ER must be maintained at sufficient levels in order for the organelle to carry out many of its fundamental functions, it could be anticipated that chronic depletion of ER Ca^{2+}, as would occur in the absence of Ca^{2+} influx via the TRPC channels, could not only influence ER-dependent processes such as protein folding and trafficking, but could also inhibit cellular functions that are dependent on increase in $[Ca^{2+}]_i$.

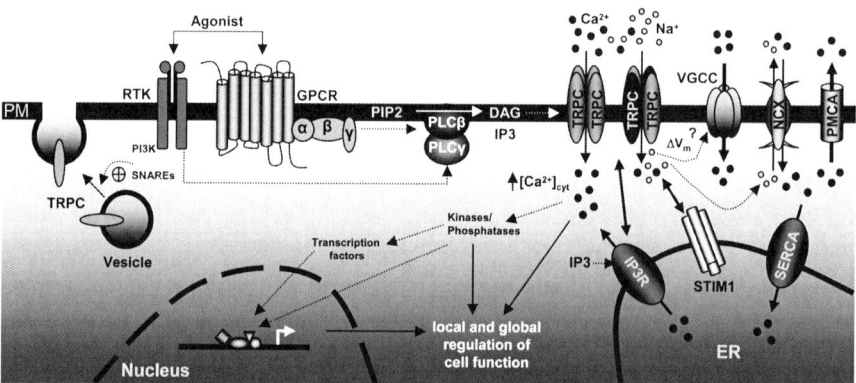

Fig. 31.1 **General mechanism of TRPC channel activation**: Agonist-mediated stimulation of RTKs or GPCRs initiate a signaling cascade leading to PLC-mediated hydrolysis of membrane bound PIP2, generating IP3 and DAG. DAG remains membrane associated, and could directly activate certain TRPC channels, whereas IP3 diffuses across the cytoplasm to activate its cognate-IP3 receptor. Activation of IP3R initiates ER Ca^{2+} store depletion. The Ca^{2+} sensor STIM1 senses the store-depletion and subsequently activates TRPC channels. IP3R also tends to interact with and activate homo/heteromeric TRPC channels. Additionally, the activation-dependant PM insertion of TRPC channels constitute a unique parameter for channel activation. PI3K and SNARE proteins are shown to positively regulate this trafficking. Thus, the resulting elevation in cytoplasmic $[Ca^{2+}]_{cyt}$, ensuing activation of TRPC channels, orchestrate cellular functions. It is also depicted that TRPC-mediated non-selective ionic conductance could depolarize membrane and presumably activate voltage-gated Ca^{2+} channels (VGCC) as well as NCX. PMCA and SERCA pumps are shown to work concertedly to maintain steady-state levels of intracellular Ca^{2+}

Ca^{2+} levels have been shown to be critical for gene regulation, muscle contraction, neurosecretion, integration of electrical signaling, neuronal excitability, synaptic plasticity, neuronal proliferation, and apoptosis-mediated neuronal loss. Although several mechanisms are known to control Ca^{2+} influx across the plasma membrane, Ca^{2+} influx could be more directly controlled either by store-depletion *per se* or by the alterations in the membrane potential which activates the voltage-gated Ca^{2+} channels. Since, Ca^{2+} regulates such diverse processes, it could not be attributed to one particular Ca^{2+} channel and factors such as amplitude, amount of cytosolic Ca^{2+}, spatial distribution of individual Ca^{2+} channels and regulators, may indeed be critical for regulating these diverse processes [2]. Furthermore, a set point for Ca^{2+} is perhaps critical to maintain normal physiological response and alterations in this Ca^{2+} set point could tilt the balance, thereby resulting in certain pathological conditions such as Alzheimer disease (AD) and Parkinson disease (PD). Although the significance of voltage-gated Ca^{2+} channels in neuronal cells is quite apparent, evidence suggesting an equally important role of the Transient receptor potential canonical (TRPC) channels is gaining momentum. Thus, the extraordinary ability of TRPC channels in regulating neuro-physiology is being discussed in the following sections.

31.2 Physiological Importance of Canonical TRP Channels in Neurons

In mammalian system, TRPC channels constitute a sub-group of the family of ion channels that comprises of 28 members (divided into TRPC (Canonical/Classical), TRPV (Vanilloid), and TRPM (Melastatin) sub-families) that are conserved and share significant homology among them [4]. A unique property of these channels is that they function as non-selective Ca^{2+} entry channels, with distinct mode of activation [4]. TRPC family contains 7 members (C1-C7) which, based on their similarities in structure-function relationships, are further divided into two sub-groups. The first group consists of C1/C4/C5 channels that are activated primarily by store-depletion. The second group comprises of C3/C6/C7 that are activated by receptor stimulation [5]. C2 is a pseudogene in humans, but has immense role in rodent behavior and pheromone sensing [6]. It is important to note that, although one can pharmacologically separate these channels in vitro, their activation in a physiological context is linked to PLC mediated signaling following stimulation of membrane G-protein coupled receptors (GPCRs) or receptor tyrosine kinases (RTKs) (Fig. 31.1). Further, since TRPC proteins are capable of forming functional channels by heteromeric interactions, the receptor- or store-dependant activation thus outlines a common feature in channel activation.

Ca^{2+} entry via the G-protein coupled mechanism has been implicated in the shaping of action potentials, synaptic transmission, and sensory transduction

[7, 8]. Additionally, changes in $[Ca^{2+}]_i$ is known to regulate the motility of many cellular structures, including the axonal growth cones [9] and dendritic filopodia of developing neurons [10]. Thus, it can be anticipated that TRPC channels may have a significant role in regulating these fundamental neuronal processes. Indeed, Ca^{2+} entry through TRPC channel has been shown to play a critical role in basic fibroblast growth factor (bFGF) induced cortical neural stem cells (NSCs) proliferation [11]. It has been shown that activation of C1, but not C3 regulate cell proliferation in neural stem cells. In contrast, Ca^{2+} influx through C1 and C3 controlled the switch between proliferation and differentiation in immortalized hippocampal H19-7 cells [12]. Similarly in mammals, C3 and C6, but not C1, have been shown to be associated with BDNF-mediated neuronal growth [13]. These distinct physiological outcomes can be explained by the differences in spatial localization and functional activation of individual TRPC channels in diverse neuronal population. Additionally, the differential distribution of membrane receptors (GPCRs/RTKs), concentration and availability of the desired agonist can further add to the complex regulation of distinct set of TRPC channels.

31.2.1 TRPC1

TRPC1 (C1), the founding member of canonical TRPs, has been shown to be expressed in both neuronal and non-neuronal tissues [14]. The first report to identify the role of C1 channels in neuronal function came from David Clapham's group. This group indicated that C1 and C5 were expressed in the hippocampus and their co-expression resulted in a novel non-selective cation channel with voltage dependence similar to NMDA receptor channels [15]. Furthermore, C1 expression was only observed in neuronal cells, with no functional activity and no localization of C1 was observed at the synapse [15]. However, another report from Paul Worley's group suggested that C1 was expressed in perisynaptic regions of the synapse and was physically associated with mGluR1 [16]. Furthermore, expression of a dominant-negative C1 – pore mutant (F561A) in cerebellar Purkinje neurons, resulted in a 49% reduction of mGluR-evoked slow excitatory postsynaptic currents (EPSCs) whereas fast transmission mediated by AMPA-type glutamate receptors remained unaffected, indicating that mGluR1 receptor activation is essential for the gating of C1. Similarly, in dopaminergic neurons C1 was functional and important for Ca^{2+} entry upon store-depletion [17]. The only explanation for these opposite results could be that in different neurons, C1 expression and function might be regulated differentially. It is possible that in hippochampal neurons C1 may not be routed to the membrane (synapse), whereas in purkinje or dopaminergic neurons it could. Also since Clapham's group did not find a functional C1 channel activity, whereas other groups were able to measure C1 function in different neurons [17, 18], suggesting that perhaps routing of C1 channels was altered in certain neuronal population. Interestingly, recently it has been shown that Caveolin1 is critical for PM routing of C1 channels [19, 20] indicating that C1 might interact with other proteins that could be critical for its localization and function. Another report indicated that C1 channels were primarily expressed on cell soma and on dendrites, whereas C5 channels were exclusively located on cell bodies [21], which support the idea that different neuronal population would have different mechanism for the expression as well as for the routing of these TRPC channels. Besides brain, C1 has also been shown to be expressed in retina [22, 23]; however its function in these tissues is not yet defined. Importantly, C1 was also found to be localized at the peripheral axons and the mechanosensory terminals, thus the involvement of C1 channels in mechanotransduction can be anticipated [24]. However, future research is needed to confirm these key findings.

Ca^{2+} entry is essential for cell proliferation and differentiation and C1 has been shown to be critical for proliferation of various non-excitable cells [5, 25], however its role in neuronal proliferation warrants elaborate investigation. Wu et al., showed for the first time that in hippochampal H19-7 cell line, C1 and C3 were essential for switching from proliferating to differentiating phenotype. Importantly, C1 was highly expressed in embryonic CNS in mammals than adult, indicating that C1 could be involved in early development and proliferation of neurons [12]. Furthermore, blocking of TRPC channels or silencing of C1 alone attenuated bFGF induced intracellular Ca^{2+} elevation and NSC's proliferation [11]. In contrast, it has been also shown that C1 is required for midline guidance of axons of commissural interneurons, and angiogenic sprouting of intersegmental vessels in the developing *Xenopus* spinal cord, but not in regeneration of adult Xenopus spinal cord, suggesting that C1 may serve as a key mediator for the Ca^{2+} influx that regulate axonal guidance during development, but inhibit axonal regeneration in adulthood [26]. Importantly, C1 homologue has also been shown to mediate BDNF-activated Ca^{2+} increase and regulate growth cone turning and extension in *Xenopus* spinal neurons [27]. Although there are slight discrepancies with regard to proliferation vs regeneration, the differences could be attributed to different neuronal populations that could have different expression of TRPC isoforms, which could form heteromultimers, and could therefore function differently in different tissues. Furthermore, stimulation of C1 via different agonists could very well have a different physiological response that could be attributed with spatial temporal resolution of Ca^{2+} signaling. Interestingly, C1, C2 and C4 had been shown to exhibit higher expression, whereas C3 and C6 expressions were decreased in NSC population [11]. These results suggest that perhaps in NSC cells, C1 could associate with C4 rather than C3 or C5, as observed above, and thus could bring about a different physiological function. Interestingly, in most of the cases the survival and proliferation induced by C1 appears to be dependent on its Ca^{2+} entry; however it remains to be seen if C1 could potentially regulate other proteins, independent of its Ca^{2+} influx ability. This could be true, since recently it was shown that in muscle cells C1 was localized in the ER/SR compartments [28] and also C1 has been shown to form large protein-protein complexes [19, 29]. Although C1 has been implicated in neuronal function, its role using animal models is not yet established. C1 knockout mice are born healthy and survive in control environment and it would be interesting to confirm the function of C1 channels in neuronal functions.

31.2.2 TRPC2

TRPC2 (C2) is a unique member of TRPC sub-family, since its expression in mammals is lost and is considered as a pseudogene in higher mammals; however C2 has been shown to be essential for small rodents and oocytes. In rodents, C2 is exclusively expressed in the vomeronasal sensory system (VNO neurons) and is localized to the dendritic tip of VNO, which are important for pheromone sensory transduction [30]. Importantly, the mouse VNO has been shown to mediate social behaviors and male mice that are deficient in C2 expression failed to display male-male aggression, and showed sexual and courtship behaviors toward both males and females [30]. C2-deficient mice also showed decreased action potential recordings of sensory responses induced by pheromonal cues in the VNO. Furthermore, another study showed that C2 knockout males increased their efforts to mount other males and demonstrated inappropriate behavior [31]. These results suggest that in rodents sensory activation of the VNO is essential for sex discrimination that ensures gender-specific behavior. Although VNO is known to be important for sensation, it could not be the only reason why these knockout mice showed behavioral issues. Importantly, it has also been shown that C2 knockout female mice had higher testosterone levels, which could be responsible for some of these behavioral issues.

C2 has also been shown to be critical for the activation of the acrosome reaction in oocyte's [32]; however since C2 knockout mice were able to reproduce normally and have similar number of offspring's, it is unlikely that C2 is the only channel that could provide Ca^{2+} needed for acrosome reaction in rodents. In contrast, C2 could still be critical in the fertilization of oocytes and more research is needed to identify if other TRPC could possibly compensate for C2 function in the fertilization of rodents. In addition, recent report also indicated that arachidonic acid generated by PLC activation in the VNO neurons activate two types of channels, a C2 channel and a separate Ca^{2+}-permeable channel [33]. Although the identification of this second channel is not known, it could be anticipated that other TRPC (mainly C3 or C6) that are expressed in these neurons could contribute to these currents. Although humans do not express C2 channels and do not have similar VNO neurons (are degenerated), but they posses a functional olfactory system and it could be suggested that perhaps another TRPC isoforms could decode olfactory cues and could be responsible for normal and abnormal social behavior in humans [34]. Thus, it would be interesting to probe this research further and establish if other TRPC isoforms are involved in abnormal human behaviors.

31.2.3 TRPC3

TRPC3 (C3) is expressed in both non-excitable and excitable cells and displays remarkable potential to form homo and hetero oligomers with different members of the TRPC family [35]. Though the expression of C3 is highest in the brain the functional role of C3 in the nervous system is not completely understood. Pioneering work on C3 in neurons revealed a role for C3 channels in BDNF signaling [36].

In pontine neurons it was shown that Trk receptors and C3 are expressed during the same developmental stages of the brain and BDNF-stimulated non-selective cationic current (I_{BDNF}) was mediated by C3. Similarly, in cerebellar granule neurons (CGNs), BDNF-induced elevation of Ca^{2+} contributes to axonal growth cone guidance [37]. The specific involvement of C3 in BDNF-induced growth cone plasticity was further demonstrated by silencing endogenous C3 or by expressing a dominant-negative C3. Furthermore, in the pyramidal neurons of rat hippocampus an involvement of C3 channels in BDNF-induced dendritic spine formation has also been reported [38]. Importantly, I_{BDNF} was blocked by antibodies directed against C3 or by C3 knockdown, which decreased BDNF, induced dendritic spine density, suggesting that C3 function as mediators of BDNF-initiated dendritic remodeling. Apart from the rapid local effects of growth cone guidance and dendritic spine morphogenesis, BDNF also has a prime role in neuronal survival. In CGNs, C3 is shown to have a protective role against serum-deprivation induced cell death, as overexpression of exogenous C3 prevented apoptosis and down-regulation of C3 induced apoptosis. This was further shown to be mediated by BDNF-induced CREB activation since the protective effect of C3 was blocked by the expression of a dominant-negative form of CREB [13]. Further, in H19-7 rat hippocampal neuronal cells C3 along with C1 was significantly increased under differentiating condition, whereas the expression of C4 and C7 were decreased [12]. Thus, this reciprocal regulation of TRPC channel expression in these neurons strongly suggests their developmentally important function.

In an interesting study performed to understand the role of Calcinuerin (CaN)-NFAT signaling pathway in neuromuscular activity, it was found that C3 channel expression was enhanced in response to neurostimulation and CaN activation. Both neurostimulation and CaN activate NFAT, but to maintain the translocation of NFAT to nucleus and modulate the gene expression a sustained Ca^{2+} influx is required. Thus, a feed back loop is established wherein the activated NFAT further enhances the expression of C3 and helps maintain the sustained flow of Ca^{2+} [39]. Additionally, recent reports from two independent studies have demonstrated a physiologically significant role of C3 channels in motor coordination [40, 41]. In one of the studies using mouse cerebellar Purkinje cells, the authors established that C3 is required for the slow synaptic potentials and inward currents evoked by group I metabotropic receptor (mGluR1) synaptic signaling. In mice lacking C3, but not C1, C4 or C6, the mGluR1 mediated slow synaptic potentials were completely absent. This abnormality in the glutamate neurotransmission in the post synaptic neurons has resulted in impaired walking behavior thus establishing a fundamental role for C3 channels in motor coordination [41]. In yet another study, to identify crucial gene products implicated in cerebral ataxia, phenotype-driven dominant mutagenesis screen was performed and an ataxic mouse mutant by the name moonwalker (Mwk) mice was identified [40]. These mice exhibited a gain-of-function mutation (T635A) and maintained sustained activation of C3 channels due to lack of negative feedback regulation by PKCγ-mediated phosphorylation. As a result diminished dendritic arborization and progressive loss of Purkinje neurons was observed. Although these studies contradict each other, they still suggest that

C3 could have a pivotal role in Purkinje neurons and future studies are needed to confirm the role of C3 in ataxia. Corroborating these results is another study with a C3 knockout mouse model, where the C3 promoter region was disrupted. These mouse exhibit atrophy and progressive paralysis suggesting an obligatory role of C3 channels in neuronal signaling, differentiation and development [42]. Importantly, C3 has been shown to associate with SNARE complex proteins [43], indicating that C3 could be important for neurosecretion.

31.2.4 TRPC4

TRPC4 (C4) is not only expressed in brain, but is also expressed in endothelium, kidney, retina, testis and adrenal glands [44]. However, the mode of activation of the channel may differ in these cell systems. C4 has shown to be co-expressed with C5 in CA1 pyramidal neurons of the hippocampus and have been shown to participate in neuronal Ca^{2+} homeostasis, however direct function of C4 was not tested in this report [45]. To elucidate the function of C4 channels, Nowycky and colleagues showed that C4 channels could provide sufficient Ca^{2+} influx to trigger a robust secretory response in voltage-clamped neurosecretory cells, indicating that C4 by itself or in combination with other TRPC channels has the potential to regulate exocytosis [46]. In addition, gamma-aminobutyric acid (GABA) release from thalamic interneurons by the activation of 5-hydroxytryptamine type 2 receptors requires Ca^{2+} entry that was critically dependent on C4, further indicating that C4 could have a role in regulating neurotransmitter release. Importantly, C4 knockout mice showed reduced GABA release from the thalamic interneurons upon 5-hydroxytryptamine stimulation, but not by acetyl-ß-methylcholine [47]. These results suggest that perhaps the F2 terminals are still functional in C4 knockout mice, and the regulation of GABA release via different mechanism may be critical for specific functions, such as the sleep–wake cycle and processing of visual information [48].

C4 expression was shown to be restricted to granule and their precursor cells in rat cerebellum, which was decreased in adults [49], suggesting that C4 is required for proper granule cell development. In contrast, in adult rat dorsal root ganglia, nerve injury mediated neurite outgrowth was dependent on C4 expression, indicating the involvement of C4 channels in axonal regeneration following nerve injury. In addition, C4 has also been shown to have a role in acute and delayed neuronal injury in focal cerebral ischemia [50]; however these studies were only based on C4 overexpression thus, further research is needed to confirm these findings. Although both these studies were performed in different neurons, they demonstrate a unique feature of C4 channels being critical for neuronal development especially during nerve injury. Although the mechanism via which C4 channels regulate neurite outgrowth is not yet identified, it could be suggested that certain kinases that are activated by increased Ca^{2+} may in turn regulate gene expression and thereby increase neurite outgrowth. Although C4 knockout mice caused impairment in store-operated Ca^{2+} currents in endothelial cells and also showed that γ-aminobutyric acid (GABA) release from thalamic interneuron's was inhibited, studies suggesting its role in neuronal cell proliferation, preferentially using the C4 knockout mice, is still lacking. Thus future research in this direction is warranted to unequivocally establish the role of C4 channels in neuronal function.

31.2.5 TRPC5

TRPC5 (C5) has been demonstrated to functionally exist as homomultimeres or in heteromultimeric assembly with other TRPC proteins. Distinct from its receptor- or store-dependant route of activation, C5 channels can also be readily activated by lipids such as lipophosphatidylcholine and sphingosine 1-phosphate, thus indicating toward a lipid-sensing ability of the channel [51]. Interestingly, the concept of rapid vesicular translocation and PM insertion has been suggested as a mode of C5 activation. EGF stimulation of hippocampal neurons is shown to target vesicle-associated C5 channels to the PM, this enables in maintaining a sustained Ca^{2+} influx and aids in neurite remodeling [52, 53]. In neuronal cell development, establishment of morphological and functional polarity is a critical feature in the formation of neural circuits and for efficient transmission of information across neurons [54]. In cultures, establishment of neuronal polarity is divided into five stages – stage 1 formation of lamellipodia, stage 2 development into neurites, stage 3 extension of one of the neurites to form axon, stage 4 dendritic arborization and stage 5 spine/synapse formation [55]. During development, every stage involved in compartmentalizing the neuron utilizes Ca^{2+} and its downstream signaling molecules to perform varied complex processes. Intracellular Ca^{2+} has diverse effects in shaping up the axons and dendrites. One of the downstream effectors of Ca^{2+} are Ca^{2+}/CaM dependent kinases (CaMK) which function by association with Ca^{2+}/CaM, an intracellular Ca^{2+} sensor. The action of CaMK in axon formation has been shown to be promoted by the Ca^{2+} influx via C5 channels, since knockdown of endogenous C5 suppresses CaMKK mediated activation of CaMKγ and thus axon formation [56]. According to a Ca^{2+} set-point hypothesis growth cone motility and neurite extensions depend on an optimal range of $[Ca^{2+}]_i$, concentrations above or below the optimal range, that alter the Ca^{2+} homeostasis, is shown to retard the growth cone protrusion [57]. On the contrary, growth cone steering induced by various extracellular guidance cues require localized Ca^{2+} influx. C5 has been shown to be a candidate protein which produces Ca^{2+} transients at resting membrane potentials [58]. In a study by Clapham's group an inverse correlation has been shown between C5 expression and hippocampal neurite length and growth cone morphology [58]. Point to note here is that C3 and XTRPC have been shown to play an opposite effect in CGNs and *Xenopus* spinal neurons respectively, wherein they both promote chemotropic turning of growth cones, suggesting different Ca^{2+} requirements for various processes.

It has been suggested that crosslinking of GM1 ganglioside associates with α5β1 integrin, which in turn initiates PLCγ and PI3K signaling cascade through auto-phosphorylation of Focal adhesion kinase (FAK) tyrosine kinase [53]. Upon activation of this signaling cascade neurite out-growth was accelerated. Further, this study identified C5 protein on cell bodies of CGNs and demonstrated C5 channels as a critical modulator of this PLC-pathway mediated neurite outgrowth. C5 knockout mice were shown to exhibit diminished innate fear levels in response to innately aversive stimuli. It has been reasoned that the lack of C5 channel potentiation by Group I mGluRs and/or CCK2 receptors and subsequent lack of membrane depolarization prevents the transmission of information to output neurons of the innate fear circuitry thus, resulting in the above mentioned fear-related behavior [59]. C5 channels were shown to have a role in generating Ca^{2+}-activated slow afterdepolarization (sADP) currents signaled by muscarinic receptors [60]. Overexpression of C5 facilitates I_{sADP} whereas expression of a pore-dead C5 dominant-negative mutant inhibits I_{sADP}. The authors argue that the afterdepolarization current generated by TRPC channels is not a result of the mere Ca^{2+} influx through the channels, but more likely due to the membrane depolarization, mediated by Na^+ entry, capable of activating voltage-dependent Ca^{2+} channels. These afterdepolarization currents allow the cell to maintain the transient input signals for a sustained period leading to long-lasting changes in excitability.

31.2.6 TRPC6

In mammals, TRPC6 (C6) is widely expressed in the cardiac neurons [61], retinal ganglion cells [62], in the neurons of olfactory epithelium [63] and in parts of brain such as – cortex, substantia niagra, hippocampus and cerebellum [49, 64–66]. Both in excitable and non-excitable cells, C6 channels largely constitute the receptor-operated, non-selective cation channels. It has been shown that receptor stimulation of PLC pathway or direct application of DAG and its analogues could specifically initiate Ca^{2+} entry [67, 68]. Importantly, this store-independent activation of C6 channels has been specifically used as readout for PLCβ activation in endocannabinoid signaling [69]. Interestingly, C6 channels could also be selectively activated by – hyperphorin [70, 71], fulfenamic acid [72] and tyrosine phosphorylation [73, 74]. Activation of Neurokinin receptors by substance-P in the noradrenergic A7 neurons has been shown to bring about a C6-specific non-selective cationic conductance, thus providing evidence for the involvement of C6 channels in nociception [75]. Regulation of TRPC channel-function has been attributed to the channel assembly into macromolecular complexes [19, 29]. In PC12 cells, activation of muscarinic GPCR has been demonstrated to organize a C6 centered channel-complex, depicting C6 in association with a multi-protein complex containing M1AChR, Calmodulin (CaM), immunophilin FKBP12, Calcineurin (CaN) and PKC [76]. Interestingly, this study also demonstrates TRPC6 to be phosphorylated by PKC and de-phosphorylated by CaN, thus regulating the channel activity. Similarly, in PC12 cells and rat cerebral cortex, a molecular complex of C6 with the GPCR – α_{1A}-adrenoceptor (α_{1A}-AR) and Snapin (a synaptic vesicle associated protein) has been reported, where following the activation of α_{1A}-AR, increased association of C6 with Snapin and α_{1A}-AR facilitates the PM recruitment of C6, thereby resulting in C6-mediated Ca^{2+} entry [77]. Additionally, in rodent rod cells, such C6 channel-complex has been shown to compartmentalize Ca^{2+} influx process distinct from that of voltage gated Ca^{2+} channels [78].

C6 channels have also been functionally associated with aggressive phenotype and growth of glioblastomas [79]. In cells derived from glioblastomas, hypoxic stress leads to Notch1-mediated upregulation of C6, which amplifies a CaN/NFAT pathway critical for cancerous growth. Further, NGF stimulation of PC12 cells results in the overexpression of C6 and is shown to be involved in the non-transferin mediated iron uptake [80]. A similar upregulation of C6-mediated Ca^{2+} influx has been reported in cultured mouse cortical astrocytes following chronic IL-1β treatment [81]. The role of C6 channels in neuronal function seems to be contextual, as these findings which indicate toward a potential involvement of C6 in neuronal toxicity and inflammation mediated neurodegeneration, there are several other significant findings that have enumerated an indispensable function of C6 in maintenance of normal neuronal physiology. C6 has been shown to co-operate with other TRP channels, in a heteromeric assembly, to regulate critical neuronal functions. The significance of heteromeric C6 channels in brain development is underscored by the finding that, in embryonic rat brain C6 physically associates with C1, C4 and C5 respectively [82]. In dorsal root ganglion (DRG) neurons, C6 co-operates with TRPC1 and TRPV4 to regulate nociception [83]. C6, in conjunction with C3, is involved in growth-cone guidance [37]. In rat cerebellar granule neurons, BDNF-induced activation of TrkB receptors resulted in a Ca^{2+} dependant growth-cone turning. BDNF-induced Ca^{2+} influx was shown to be mediated via C3/C6 channels. Since, silencing of C3, but not C1, and expression of dominant-negative constructs of C3 or C6 respectively inhibited neuronal attraction toward a BDNF gradient. Similarly, C6 together with C3 [13] and C5 [84] have been reported to collaborate in neuronal survival. In CGN, both C3 and C6, but not C1, were shown to elicit the neuroprotective effect of BDNF via the activation of CREB/ERK pathway [13]. In primary mid-brain neurons of rat, C5 and C6 co-localized with PDGF-βR and were found to regulate PDGF-mediated neuronal cell survival following HIV-1 Tat toxicity [84]. In addition to its pro-survival characteristic, C6 channels have been demonstrated to regulate dendrite development, thus influencing synaptic plasticity. In rat hippocampus the expression of C6 was detected postnatal, which peaked between postnatal day 7 through day 28 (P7-28). C6 was enriched in the synaptsomes and postsynaptic cell fractions obtained from P14 hippocampus and also significantly co-localized with PSD-95. Using immuno-electron microscopy, this study identified a predominant localization of C6 at the excitatory synapse of the hippocampal neurons. Increased expression of C6 enhanced the dendritic spine densities, whereas specific C6 silencing, but not C1, significantly reduced the dendritic spine densities. C6 silencing also abolished the effect of BDNF on spine formation. This effect of C6 channels on dendritic spine formation was largely mediated by the activation of the CaMKIV-CREB pathway. In agreement to the function of C6 in culture hippocampal neurons, overexpression of C6 in vivo also resulted in an increase in the dendritic spin densities in CA1 neurons, further emphasizing a significant function of C6 channels in the formation of excitatory synapse in rat hippocampal neurons. In addition, the C6 transgenic animals displayed an enhanced ability for learning and memory [85, 86].

31.2.7 TRPC7

TRPC7 (C7) was cloned from mouse as well as human brain [87, 88] and is thus far the last member of canonical TRPCs to be identified. Among other TRPC homologues, C7 displays the closest structural resemblance with C3 and together with C6 they constitute the receptor-activated subfamily of TRPC channels [88, 89]. In humans the highest level of C7 message is detected in the pituitary gland [66, 88]. In mouse brain, C7 is prominently expressed in the cerebellar purkinje cells in addition to hippocampal neurons, cerebral cortex, pons and the mitral layer of olfactory bulb [87]. C7 expression has also been detected in rat hippocampal neurons [12]. In spite of such a broad expression pattern of C7 in the CNS, the neuronal functions of the channel remain elusive. The existing literature, however reveal some critical observations that associate C7 channels with neuronal physiology. C7 is expressed in the rat striatal cholinergic interneurons and along with C3 has been demonstrated to develop TRPC-like currents, in response to activation of the Gαq-coupled mGluR1/5 [90], thus implying towards an involvement of C7 ion channels in regulating striatal function. In the proliferating rat H19-7 hippocampal neurons the expression of C7 was high and following differentiation the expression was dramatically reduced [12]. Although in this study C7 did not essentially contribute to SOCE, the existence of such a switch in the expression pattern of C7 and other TRPC proteins does indicate towards an internal mechanism of neurons to control physiological processes by regulating TRPC channel function. One of the cloning studies mapped the C7 gene to human chromosome 21q22.3 and surprisingly this locus is associated with bipolar disorder, thus indicating a possible involvement of C7 in the pathophysiology of bipolar disorder [91]. Interestingly, a reduced level of C7 has been reported in bipolar disorder patients. In the B-lymphoblast cells obtained from one group of bipolar patients, the expression of C7 mRNA was reduced, whereas the basal Ca^{2+} levels in these cells were high [92]. The authors speculate that this inverse relation between C7 expression and Ca^{2+} levels is probably an inbuilt cellular mechanism to restrict further Ca^{2+} influx through C7 channels. In addition, C7 expression has been reported in melanopsin-expressing retinal ganglion cells (RGCs) and inhibition of TRPC channels by non-specific inhibitors (2-APB and SKF96365) was shown to abolish light-evoked response of the photosensitive RGCs, thus emphasizing on a potential involvement of C7 channels in mammalian phototransduction [93]. In summary, reports demonstrating the role of C7 in neurons are limited however, since in native systems C7 has been shown to be equally important for receptor- and store-operated Ca^{2+} entry, it is apparent that in a neuronal context C7 would have an essential physiological role by co-operating with other TRPC channels in a heterotypic manner.

31.3 TRPC Proteins in Neurodegenerative Diseases

Neuronal cell injury is mediated via both increase and decrease of $[Ca^{2+}]_i$ concentration [94]. Changes in $[Ca^{2+}]_i$ concentration stimulate a number of intracellular events and could either trigger or inhibit cell death process [2, 3]. Importantly, disturbances in Ca^{2+} homeostasis have been implicated in many neurodegenerative diseases such as, PD, AD, and HD [95–99]. It is not surprising that disturbances in Ca^{2+} signaling pathways underlie neuronal loss, since many factors involved in neuronal function are dependent on Ca^{2+} signaling [2, 3]. However, the cellular mechanism(s) underlying neurodegeneration, due to alterations in Ca^{2+} homeostasis, remains to be elucidated [100]. Although several factors including generation of free radicals, impairment of mitochondrial function, ER stress, and apoptosis have been proposed, the role of TRPC channels is not directly tested.

Increased $[Ca^{2+}]_i$ could lead to inappropriate activation of Ca^{2+}-dependent processes, that stay inactive or operate at low Ca^{2+} levels, causing metabolic derangements leading to neuronal death [2, 3, 100]. Whereas, decrease of Ca^{2+} in the ER induces stress, which could activate cell death cascades [101], suggesting that controlled Ca^{2+} influx is critical for neuronal function and survival. Since, TRPCs are essential for replenishing and for maintaining ER Ca^{2+}, chronic depletion of ER Ca^{2+} as would occur in the absence of TRPC function, could influence ER-dependent processes such as protein folding and trafficking, the ER stress response, and apoptosis. This is most notable in neurons that respond to activation of the phosphoinositol pathway with $[Ca^{2+}]_i$ oscillations, since ablation of Ca^{2+} entry preclude the cell's ability to maintain $[Ca^{2+}]_i$ oscillations. Recently, it has been shown that TRPC1 levels were decreased in dopaminergic neurons when treated with neurotoxins such as 1-methyl, 4-phenyl pyridinium ion (MPP^+/MPTP) that cause parkinsonian syndrome [95, 102]. Importantly, loss of C1 decreased $[Ca^{2+}]_i$ levels and showed ER stress. In contrast, overexpression of C1 significantly decreased MPTP/MPP^+ mediated neuronal loss and also restored $[Ca^{2+}]_i$ levels, indicating that C1 could be critical for PD. Moreover, other TRPC channels are also expressed in substantia nigra, which could be involved in excitotoxicity [103]. Although these results are exciting, they need to be verified using PD mouse models and C1 knockout mice.

BDNF treatment is known to activate C3-mediated cationic current in pontine neurons [36]. Several studies have also reported that BDNF protect neurons from several brain insults thereby play an important role in neurodegenerative diseases such as AD, PD and HD [13, 100, 103], however the role of C3 is not yet directly established. Recent report also suggest that BDNF levels are negatively regulated with regard to the severity of AD and treatment with BDNF exerts neuroprotective effect in AD in vitro and in vivo models [104]; however the mechanism as well as the importance of C3 is not known. Interestingly, BDNF and NGF are known to modulate neurotransmitter release [105] and since BDNF activates C3 channels, a potential role of C3 in neurosecretion can be speculated. In contrast, Aβ has been shown to induce oxidative stress [100] and since C3 is important for oxidative stress (discussed below), it can potentially exacerbate AD. Thus, C3 could plausibly have both protective as well as degenerative role in the context specific manner. Moreover, Lessard et al., reported that presenilins mutations account for up to 40% of the early onset of familial AD [106] and deletion of PS1 alone or PS1 and PS2 together resulted in enhanced store-dependent Ca^{2+} entry [107]. Although it is not known if TRPCs are the major culprit in familial AD, it is still intriguing that TRPC channels could have a role in the onset/progression of AD. Overall, it seems as if SOCE is the "magical pathway" since it could not only help neuronal survival, but could also induce neuronal degeneration.

31.4 TRPC Channels and Oxidative Stress

Reactive oxygen species (ROS) over production is a widespread feature for neurodegenerative disorders including PD, AD and HD [99, 108–110]. For the past several decades, a large number of studies have been elucidated the role of Ca^{2+} homeostasis in ROS production that leads to cell death [111–114]. In contrast, Amoroso et al., reported that oxidative stress induced free radical overproduction induces cell death, but without the participation of intracellular Ca^{2+} in SH-SY5Y cells [115]. The only explanation to these discrepancies could be that Ca^{2+} induced oxidative stress is cell specific and perhaps other Ca^{2+} pumps and Ca^{2+} releasing channels and intracellular buffering system might influence ROS production [116]. So far two TRPC channels (TRPC3 and TRPC4) have been identified, which are activated under oxidative stress [117, 118]. A dominant negative TRPC3 N-terminal negatively affects the TRPC3-mediated cation current and membrane depolarization induced by the oxidants tert-butylhydroperoxide (t-BHP) in porcine aortic endothelial cells (PAEC) [119]. Also during oxidative stress the C3 and C4 forms redox-sensitive cation channel and involved in Na^+ loading and membrane depolarization in HEK293T cells [120, 121]. The mechanisms how these TRPC channels are activated by oxidative stress are not clear yet. But phospholipase c (PLC) inhibitor and tyrosine kinase inhibitor have shown to reverse the function of C3 under oxidative stress [122]. In contrast, studies performed by Crouzin et al., showed that α-tocopherol protects hippocampal neurons against Fe^{2+}-induced oxidative stress by preventing Ca^{2+} influx through TRPC-like channels [123], however the channel involved is still not yet identified and future research is needed.

31.5 Concluding Remarks

There has been intense focus on TRPC channels in the past few years, however, conclusive data regarding the exact physiological function of most of these channels in neuronal cells are still lacking. There are considerable discrepancies regarding their function, which could be due to variable expression patterns in different tissues and cells as well as their individual associations among different TRPC members. In addition their association with other accessory proteins could also be critical for their specific localization and function, which could confound the results obtained in different neuronal tissues. Overall, although the available data reveal key functional role of TRPC channels in neurons, direct evidence from individual channel is still lacking. Although individual knockout mice are available and are healthy, the issue of compensation is never resolved or investigated. Furthermore, since TRPC family members have similar activation mechanism, it could be anticipated that TRPC homologues belonging to the same sub family could compensate for the function. Nonetheless functional experiments using multiple knockout mice will be essential to decipher the role of TRPC channels in neuronal cells. Identifying specific endogenous agents that can selectively activate a given TRPC channel remains to be an outstanding challenge.

Acknowledgments We thank Dr. Biswaranjan Pani for conceptualizing the figure and helpful editorial comments. We duly acknowledge the grant support from the National Science foundation (0548733) and the National Institutes of Health (DE017102, 5P20RR017699).

References

1. Berridge MJ, Bootman MD, Roderick HL (2003) Calcium signalling: dynamics, homeostasis and remodelling. Nat Rev Mol Cell Biol 4:517–529
2. Berridge MJ, Lipp P, Bootman MD (2000) The versatility and universality of calcium signalling. Nat Rev Mol Cell Biol 1:11–21
3. Putney JW Jr. (2003) Capacitative calcium entry in the nervous system. Cell Calcium 34:339–344
4. Montell C (2005) The TRP superfamily of cation channels. Sci STKE 2005:re3
5. Minke B, Cook B (2002) TRP channel proteins and signal transduction. Physiol Rev 82:429–472
6. Lucas P, Ukhanov K, Leinders-Zufall T, Zufall F (2003) A diacylglycerol-gated cation channel in vomeronasal neuron dendrites is impaired in TRPC2 mutant mice: mechanism of pheromone transduction. Neuron 40:551–561
7. Linden R (1994) The survival of developing neurons: a review of afferent control. Neuroscience 58:671–682
8. Congar P, Leinekugel X, Ben-Ari Y, Crepel V (1997) A long-lasting calcium-activated nonselective cationic current is generated by synaptic stimulation or exogenous activation of group I metabotropic glutamate receptors in CA1 pyramidal neurons. J Neurosci 17:5366–5379
9. Gomez TM, Robles E, Poo M, Spitzer NC (2001) Filopodial calcium transients promote substrate-dependent growth cone turning. Science 291:1983–1987
10. Lohmann C, Finski A, Bonhoeffer T (2005) Local calcium transients regulate the spontaneous motility of dendritic filopodia. Nat Neurosci 8:305–312
11. Fiorio Pla A, Maric D, Brazer SC, Giacobini P, Liu X, Chang YH, Ambudkar IS, Barker JL (2005) Canonical transient receptor potential 1 plays a role in basic fibroblast growth factor (bFGF)/FGF receptor-1-induced Ca^{2+} entry and embryonic rat neural stem cell proliferation. J Neurosci 25:2687–2701
12. Wu X, Zagranichnaya TK, Gurda GT, Eves EM, Villereal ML (2004) A TRPC1/TRPC3-mediated increase in store-operated calcium entry is required for differentiation of H19-7 hippocampal neuronal cells. J Biol Chem 279:43392–43402
13. Jia Y, Zhou J, Tai Y, Wang Y (2007) TRPC channels promote cerebellar granule neuron survival. Nat Neurosci 10:559–567
14. Ambudkar IS, Bandyopadhyay BC, Liu X, Lockwich TP, Paria B, Ong HL (2006) Functional organization of TRPC-Ca^{2+} channels and regulation of calcium microdomains. Cell Calcium 40:495–504

15. Strubing C, Krapivinsky G, Krapivinsky L, Clapham DE (2001) TRPC1 and TRPC5 form a novel cation channel in mammalian brain. Neuron 29:645–655
16. Kim SJ, Kim YS, Yuan JP, Petralia RS, Worley PF, Linden DJ (2003) Activation of the TRPC1 cation channel by metabotropic glutamate receptor mGluR1. Nature 426:285–291
17. Tozzi A, Bengtson CP, Longone P, Carignani C, Fusco FR, Bernardi G, Mercuri NB (2003) Involvement of transient receptor potential-like channels in responses to mGluR-I activation in midbrain dopamine neurons. Eur J Neurosci 18:2133–2145
18. Hannan MA, Kabbani N, Paspalas CD, Levenson R (2008) Interaction with dopamine D2 receptor enhances expression of transient receptor potential channel 1 at the cell surface. Biochim Biophys Acta 1778:974–982
19. Ambudkar IS, Ong HL (2007) Organization and function of TRPC channelosomes. Pflugers Arch 455:187–200
20. Pani B, Singh BB (2009) Lipid rafts/caveolae as microdomains of calcium signaling. Cell Calcium 45:625–633
21. von Bohlen Und Halbach O, Hinz U, Unsicker K, Egorov AV (2005) Distribution of TRPC1 and TRPC5 in medial temporal lobe structures of mice. Cell Tissue Res 322:201–206
22. Bollimuntha S, Cornatzer E, Singh BB (2005) Plasma membrane localization and function of TRPC1 is dependent on its interaction with beta-tubulin in retinal epithelium cells. Vis Neurosci 22:163–170
23. Szikra T, Cusato K, Thoreson WB, Barabas P, Bartoletti TM, Krizaj D (2008) Depletion of calcium stores regulates calcium influx and signal transmission in rod photoreceptors. J Physiol 586:4859–4875
24. Glazebrook PA, Schilling WP, Kunze DL (2005) TRPC channels as signal transducers. Pflugers Arch 451:125–130
25. Beech DJ (2005) TRPC1: store-operated channel and more. Pflugers Arch 451:53–60
26. Yu PC, Gu SY, Bu JW, Du JL TRPC1 Is Essential for In Vivo Angiogenesis in Zebrafish. Circ Res 106:1221–1232
27. Shim S, Goh EL, Ge S, Sailor K, Yuan JP, Roderick HL, Bootman MD, Worley PF, Song H, Ming GL (2005) XTRPC1-dependent chemotropic guidance of neuronal growth cones. Nat Neurosci 8:730–735
28. Berbey C, Weiss N, Legrand C, Allard B (2009) Transient receptor potential canonical type 1 (TRPC1) operates as a sarcoplasmic reticulum calcium leak channel in skeletal muscle. J Biol Chem 284:36387–36394
29. Montell C (2005) TRP channels in Drosophila photoreceptor cells. J Physiol 567:45–51
30. Leypold BG, Yu CR, Leinders-Zufall T, Kim MM, Zufall F, Axel R (2002) Altered sexual and social behaviors in trp2 mutant mice. Proc Natl Acad Sci USA 99:6376–6381
31. Kimchi T, Xu J, Dulac C (2007) A functional circuit underlying male sexual behaviour in the female mouse brain. Nature 448:1009–1014
32. Jungnickel MK, Marrero H, Birnbaumer L, Lemos JR, Florman HM (2001) Trp2 regulates entry of Ca^{2+} into mouse sperm triggered by egg ZP3. Nat Cell Biol 3:499–502
33. Zhang P, Yang C, Delay RJ (2010) Odors Activate Dual Pathways, a TRPC2 and an AA-dependent Pathway, in Mouse Vomeronasal Neurons. Am J Physiol Cell Physiol 298:C1253–C1264
34. Spors H, Sobel N (2007) Male behavior by knockout. Neuron 55:689–693
35. Lintschinger B, Balzer-Geldsetzer M, Baskaran T, Graier WF, Romanin C, Zhu MX, Groschner K (2000) Coassembly of Trp1 and Trp3 proteins generates diacylglycerol- and Ca^{2+}-sensitive cation channels. J Biol Chem 275:27799–27805
36. Li HS, Xu XZ, Montell C (1999) Activation of a TRPC3-dependent cation current through the neurotrophin BDNF. Neuron 24:261–273
37. Li Y, Jia YC, Cui K, Li N, Zheng ZY, Wang YZ, Yuan XB (2005) Essential role of TRPC channels in the guidance of nerve growth cones by brain-derived neurotrophic factor. Nature 434:894–898

38. Amaral MD, Pozzo-Miller L (2007) TRPC3 channels are necessary for brain-derived neurotrophic factor to activate a nonselective cationic current and to induce dendritic spine formation. J Neurosci 27:5179–5189
39. Rosenberg P, Hawkins A, Stiber J, Shelton JM, Hutcheson K, Bassel-Duby R, Shin DM, Yan Z, Williams RS (2004) TRPC3 channels confer cellular memory of recent neuromuscular activity. Proc Natl Acad Sci USA 101:9387–9392
40. Becker EB, Oliver PL, Glitsch MD, Banks GT, Achilli F, Hardy A, Nolan PM, Fisher EM, Davies KE (2009) A point mutation in TRPC3 causes abnormal Purkinje cell development and cerebellar ataxia in moonwalker mice. Proc Natl Acad Sci USA 106:6706–6711
41. Hartmann J, Dragicevic E, Adelsberger H, Henning HA, Sumser M, Abramowitz J, Blum R, Dietrich A, Freichel M, Flockerzi V, Birnbaumer L, Konnerth A (2008) TRPC3 channels are required for synaptic transmission and motor coordination. Neuron 59:392–398
42. Rodriguez-Santiago M, Mendoza-Torres M, Jimenez-Bremont JF, Lopez-Revilla R (2007) Knockout of the trcp3 gene causes a recessive neuromotor disease in mice. Biochem Biophys Res Commun 360:874–879
43. Singh BB, Lockwich TP, Bandyopadhyay BC, Liu X, Bollimuntha S, Brazer SC, Combs C, Das S, Leenders AG, Sheng ZH, Knepper MA, Ambudkar SV, Ambudkar IS (2004) VAMP2-dependent exocytosis regulates plasma membrane insertion of TRPC3 channels and contributes to agonist-stimulated Ca^{2+} influx. Mol Cell 15:635–646
44. Freichel M, Vennekens R, Olausson J, Stolz S, Philipp SE, Weissgerber P, Flockerzi V (2005) Functional role of TRPC proteins in native systems: implications from knockout and knockdown studies. J Physiol 567:59–66
45. Philipp S, Hambrecht J, Braslavski L, Schroth G, Freichel M, Murakami M, Cavalie A, Flockerzi V (1998) A novel capacitative calcium entry channel expressed in excitable cells. EMBO J 17:4274–4282
46. Obukhov AG, Nowycky MC (2002) TRPC4 can be activated by G-protein-coupled receptors and provides sufficient $Ca(^{2+})$ to trigger exocytosis in neuroendocrine cells. J Biol Chem 277:16172–16178
47. Munsch T, Freichel M, Flockerzi V, Pape HC (2003) Contribution of transient receptor potential channels to the control of GABA release from dendrites. Proc Natl Acad Sci USA 100:16065–16070
48. Pape HC, Munsch T, Budde T (2004) Novel vistas of calcium-mediated signalling in the thalamus. Pflugers Arch 448:131–138
49. Huang WC, Young JS, Glitsch MD (2007) Changes in TRPC channel expression during postnatal development of cerebellar neurons. Cell Calcium 42:1–10
50. Gao YQ, Gao H, Zhou ZY, Lu SD, Sun FY (2004) Expression of transient receptor potential channel 4 in striatum and hippocampus of rats is increased after focal cerebral ischemia. Sheng Li Xue Bao 56:153–157
51. Beech DJ (2007) Bipolar phospholipid sensing by TRPC5 calcium channel. Biochem Soc Trans 35:101–104
52. Bezzerides VJ, Ramsey IS, Kotecha S, Greka A, Clapham DE (2004) Rapid vesicular translocation and insertion of TRP channels. Nat Cell Biol 6:709–720
53. Wu G, Lu ZH, Obukhov AG, Nowycky MC, Ledeen RW (2007) Induction of calcium influx through TRPC5 channels by cross-linking of GM1 ganglioside associated with alpha5beta1 integrin initiates neurite outgrowth. J Neurosci 27:7447–7458
54. Gomez T (2005) Neurobiology: channels for pathfinding. Nature 434:835–838
55. Barnes AP, Polleux F (2009) Establishment of axon-dendrite polarity in developing neurons. Annu Rev Neurosci 32:347–381
56. Davare MA, Fortin DA, Saneyoshi T, Nygaard S, Kaech S, Banker G, Soderling TR, Wayman GA (2009) Transient receptor potential canonical 5 channels activate Ca^{2+}/calmodulin kinase Igamma to promote axon formation in hippocampal neurons. J Neurosci 29:9794–9808
57. Gomez TM, Zheng JQ (2006) The molecular basis for calcium-dependent axon pathfinding. Nat Rev Neurosci 7:115–125

58. Hui H, McHugh D, Hannan M, Zeng F, Xu SZ, Khan SU, Levenson R, Beech DJ, Weiss JL (2006) Calcium-sensing mechanism in TRPC5 channels contributing to retardation of neurite outgrowth. J Physiol 572:165–172
59. Riccio A, Li Y, Moon J, Kim KS, Smith KS, Rudolph U, Gapon S, Yao GL, Tsvetkov E, Rodig SJ, Van't Veer A, Meloni EG, Carlezon WA Jr., Bolshakov VY, Clapham DE (2009) Essential role for TRPC5 in amygdala function and fear-related behavior. Cell 137:761–772
60. Yan HD, Villalobos C, Andrade R (2009) TRPC channels mediate a Muscarinic receptor-induced after depolarization in cerebral cortex. J Neurosci 29:10038–10046
61. Calupca MA, Locknar SA, Parsons RL (2002) TRPC6 immunoreactivity is colocalized with neuronal nitric oxide synthase in extrinsic fibers innervating guinea pig intrinsic cardiac ganglia. J Comp Neurol 450:283–291
62. Warren EJ, Allen CN, Brown RL, Robinson DW (2006) The light-activated signaling pathway in SCN-projecting rat retinal ganglion cells. Eur J Neurosci 23:2477–2487
63. Elsaesser R, Montani G, Tirindelli R, Paysan J (2005) Phosphatidyl-inositide signalling proteins in a novel class of sensory cells in the mammalian olfactory epithelium. Eur J Neurosci 21:2692–2700
64. Giampa C, DeMarch Z, Patassini S, Bernardi G, Fusco FR (2007) Immunohistochemical localization of TRPC6 in the rat substantia nigra. Neurosci Lett 424:170–174
65. Chung YH, Sun Ahn H, Kim D, Hoon Shin D, Su Kim S, Yong Kim K, Bok Lee W, Ik Cha C (2006) Immunohistochemical study on the distribution of TRPC channels in the rat hippocampus. Brain Res 1085:132–137
66. Riccio A, Medhurst AD, Mattei C, Kelsell RE, Calver AR, Randall AD, Benham CD, Pangalos MN (2002) mRNA distribution analysis of human TRPC family in CNS and peripheral tissues. Brain Res Mol Brain Res 109:95–104
67. Hofmann T, Obukhov AG, Schaefer M, Harteneck C, Gudermann T, Schultz G (1999) Direct activation of human TRPC6 and TRPC3 channels by diacylglycerol. Nature 397:259–263
68. Boulay G, Zhu X, Peyton M, Jiang M, Hurst R, Stefani E, Birnbaumer L (1997) Cloning and expression of a novel mammalian homolog of Drosophila transient receptor potential (Trp) involved in calcium entry secondary to activation of receptors coupled by the Gq class of G protein. J Biol Chem 272:29672–29680
69. Hashimotodani Y, Ohno-Shosaku T, Tsubokawa H, Ogata H, Emoto K, Maejima T, Araishi K, Shin HS, Kano M (2005) Phospholipase Cbeta serves as a coincidence detector through its Ca^{2+} dependency for triggering retrograde endocannabinoid signal. Neuron 45:257–268
70. Tu P, Gibon J, Bouron A (2010) The TRPC6 channel activator hyperforin induces the release of zinc and calcium from mitochondria. J Neurochem 112:204–213
71. Leuner K, Kazanski V, Muller M, Essin K, Henke B, Gollasch M, Harteneck C, Muller WE (2007) Hyperforin–a key constituent of St. John's wort specifically activates TRPC6 channels. FASEB J 21:4101–4111
72. Foster RR, Zadeh MA, Welsh GI, Satchell SC, Ye Y, Mathieson PW, Bates DO, Saleem MA (2009) Flufenamic acid is a tool for investigating TRPC6-mediated calcium signalling in human conditionally immortalised podocytes and HEK293 cells. Cell Calcium 45: 384–390
73. Hisatsune C, Kuroda Y, Nakamura K, Inoue T, Nakamura T, Michikawa T, Mizutani A, Mikoshiba K (2004) Regulation of TRPC6 channel activity by tyrosine phosphorylation. J Biol Chem 279:18887–18894
74. Albert AP (2004) Activation of TRPC6 channel proteins: evidence for an essential role of phosphorylation. J Physiol 561:354
75. Min MY, Shih PY, Wu YW, Lu HW, Lee ML, Yang HW (2009) Neurokinin 1 receptor activates transient receptor potential-like currents in noradrenergic A7 neurons in rats. Mol Cell Neurosci 42:56–65

76. Kim JY, Saffen D (2005) Activation of M1 muscarinic acetylcholine receptors stimulates the formation of a multiprotein complex centered on TRPC6 channels. J Biol Chem 280:32035–32047
77. Suzuki F, Morishima S, Tanaka T, Muramatsu I (2007) Snapin, a new regulator of receptor signaling, augments alpha1A-adrenoceptor-operated calcium influx through TRPC6. J Biol Chem 282:29563–29573
78. Krizaj D (2005) Compartmentalization of calcium entry pathways in mouse rods. Eur J Neurosci 22:3292–3296
79. Chigurupati S, Venkataraman R, Barrera D, Naganathan A, Madan M, Paul L, Pattisapu JV, Kyriazis GA, Sugaya K, Bushnev S, Lathia JD, Rich JN, Chan SL (2010) Receptor channel TRPC6 is a key mediator of Notch-driven glioblastoma growth and invasiveness. Cancer Res 70:418–427
80. Mwanjewe J, Grover AK (2004) Role of transient receptor potential canonical 6 (TRPC6) in non-transferrin-bound iron uptake in neuronal phenotype PC12 cells. Biochem J 378: 975–982
81. Beskina O, Miller A, Mazzocco-Spezzia A, Pulina MV, Golovina VA (2007) Mechanisms of interleukin-1beta-induced Ca^{2+} signals in mouse cortical astrocytes: roles of store- and receptor-operated Ca^{2+} entry. Am J Physiol Cell Physiol 293:C1103–C1111
82. Strubing C, Krapivinsky G, Krapivinsky L, Clapham DE (2003) Formation of novel TRPC channels by complex subunit interactions in embryonic brain. J Biol Chem 278: 39014–39019
83. Alessandri-Haber N, Dina OA, Chen X, Levine JD (2009) TRPC1 and TRPC6 channels cooperate with TRPV4 to mediate mechanical hyperalgesia and nociceptor sensitization. J Neurosci 29:6217–6228
84. Yao H, Peng F, Fan Y, Zhu X, Hu G, Buch SJ (2009) TRPC channel-mediated neuroprotection by PDGF involves Pyk2/ERK/CREB pathway. Cell Death Differ 16:1681–1693
85. Tai Y, Feng S, Ge R, Du W, Zhang X, He Z, Wang Y (2008) TRPC6 channels promote dendritic growth via the CaMKIV-CREB pathway. J Cell Sci 121:2301–2307
86. Zhou J, Du W, Zhou K, Tai Y, Yao H, Jia Y, Ding Y, Wang Y (2008) Critical role of TRPC6 channels in the formation of excitatory synapses. Nat Neurosci 11:741–743
87. Okada T, Inoue R, Yamazaki K, Maeda A, Kurosaki T, Yamakuni T, Tanaka I, Shimizu S, Ikenaka K, Imoto K, Mori Y (1999) Molecular and functional characterization of a novel mouse transient receptor potential protein homologue TRP7. $Ca^{(2+)}$-permeable cation channel that is constitutively activated and enhanced by stimulation of G protein-coupled receptor. J Biol Chem 274:27359–27370
88. Riccio A, Mattei C, Kelsell RE, Medhurst AD, Calver AR, Randall AD, Davis JB, Benham CD, Pangalos MN (2002) Cloning and functional expression of human short TRP7, a candidate protein for store-operated Ca^{2+} influx. J Biol Chem 277:12302–12309
89. Trebak M, Vazquez G, Bird GS, Putney JW Jr. (2003) The TRPC3/6/7 subfamily of cation channels. Cell Calcium 33:451–461
90. Berg AP, Sen N, Bayliss DA (2007) TrpC3/C7 and Slo2.1 are molecular targets for metabotropic glutamate receptor signaling in rat striatal cholinergic interneurons. J Neurosci 27:8845–8856
91. Nagamine K, Kudoh J, Minoshima S, Kawasaki K, Asakawa S, Ito F, Shimizu N (1998) Molecular cloning of a novel putative Ca^{2+} channel protein (TRPC7) highly expressed in brain. Genomics 54:124–131
92. Yoon IS, Li PP, Siu KP, Kennedy JL, Macciardi F, Cooke RG, Parikh SV, Warsh JJ (2001) Altered TRPC7 gene expression in bipolar-I disorder. Biol Psychiatry 50:620–626
93. Sekaran S, Lall GS, Ralphs KL, Wolstenholme AJ, Lucas RJ, Foster RG, Hankins MW (2007) 2-Aminoethoxydiphenylborane is an acute inhibitor of directly photosensitive retinal ganglion cell activity in vitro and in vivo. J Neurosci 27:3981–3986
94. Sattler R, Tymianski M (2000) Molecular mechanisms of calcium-dependent excitotoxicity. J Mol Med 78:3–13

95. Bollimuntha S, Singh BB, Shavali S, Sharma SK, Ebadi M (2005) TRPC1-mediated inhibition of 1-methyl-4-phenylpyridinium ion neurotoxicity in human SH-SY5Y neuroblastoma cells. J Biol Chem 280:2132–2140
96. Zuccato C, Cattaneo E (2007) Role of brain-derived neurotrophic factor in Huntington's disease. Prog Neurobiol 81:294–330
97. Zuccato C, Cattaneo E (2009) Brain-derived neurotrophic factor in neurodegenerative diseases. Nat Rev Neurol 5:311–322
98. O'Bryant SE, Hobson V, Hall JR, Waring SC, Chan W, Massman P, Lacritz L, Cullum CM, Diaz-Arrastia R (2009) Brain-derived neurotrophic factor levels in Alzheimer's disease. J Alzheimers Dis 17:337–341
99. Albers DS, Beal MF (2000) Mitochondrial dysfunction and oxidative stress in aging and neurodegenerative disease. J Neural Transm Suppl 59:133–154
100. Bezprozvanny I, Mattson MP (2008) Neuronal calcium mishandling and the pathogenesis of Alzheimer's disease. Trends Neurosci 31:454–463
101. Ermak G, Davies KJ (2002) Calcium and oxidative stress: from cell signaling to cell death. Mol Immunol 38:713–721
102. Selvaraj S, Watt JA, Singh BB (2009) TRPC1 inhibits apoptotic cell degeneration induced by dopaminergic neurotoxin MPTP/MPP($^+$). Cell Calcium 46:209–218
103. Selvaraj S, Sun Y, Singh BB (2010) TRPC channels and their implication in neurological diseases. CNS Neurol Disord Drug Targets 9:94–104
104. Fumagalli F, Racagni G, Riva MA (2006) The expanding role of BDNF: a therapeutic target for Alzheimer's disease? Pharmacogenomics J 6:8–15
105. Tyler WJ, Perrett SP, Pozzo-Miller LD (2002) The role of neurotrophins in neurotransmitter release. Neuroscientist 8:524–531
106. Lessard CB, Lussier MP, Cayouette S, Bourque G, Boulay G (2005) The overexpression of presenilin2 and Alzheimer's-disease-linked presenilin2 variants influences TRPC6-enhanced Ca^{2+} entry into HEK293 cells. Cell Signal 17:437–445
107. Yoo AS, Cheng I, Chung S, Grenfell TZ, Lee H, Pack-Chung E, Handler M, Shen J, Xia W, Tesco G, Saunders AJ, Ding K, Frosch MP, Tanzi RE, Kim TW (2000) Presenilin-mediated modulation of capacitative calcium entry. Neuron 27:561–572
108. Andersen JK (2004) Oxidative stress in neurodegeneration: cause or consequence? Nat Med 10(Suppl):S18–S25
109. Lin MT, Beal MF (2006) Mitochondrial dysfunction and oxidative stress in neurodegenerative diseases. Nature 443:787–795
110. Sies H, Cadenas E (1985) Oxidative stress: damage to intact cells and organs. Philos Trans R Soc Lond B Biol Sci 311:617–631
111. Boitier E, Rea R, Duchen MR (1999) Mitochondria exert a negative feedback on the propagation of intracellular Ca^{2+} waves in rat cortical astrocytes. J Cell Biol 145:795–808
112. Maciel EN, Vercesi AE, Castilho RF (2001) Oxidative stress in $Ca^{(2+)}$-induced membrane permeability transition in brain mitochondria. J Neurochem 79:1237–1245
113. Nicholls DG (2005) Mitochondria and calcium signaling. Cell Calcium 38:311–317
114. Starkov AA, Chinopoulos C, Fiskum G (2004) Mitochondrial calcium and oxidative stress as mediators of ischemic brain injury. Cell Calcium 36:257–264
115. Amoroso S, Gioielli A, Cataldi M, Di Renzo G, Annunziato L (1999) In the neuronal cell line SH-SY5Y, oxidative stress-induced free radical overproduction causes cell death without any participation of intracellular $Ca^{(2+)}$ increase. Biochim Biophys Acta 1452:151–160
116. Michaelis ML, Foster CT, Jayawickreme C (1992) Regulation of calcium levels in brain tissue from adult and aged rats. Mech Ageing Dev 62:291–306
117. Aarts MM, Tymianski M (2005) TRPMs and neuronal cell death. Pflugers Arch 451:243–249
118. Miller BA (2006) The role of TRP channels in oxidative stress-induced cell death. J Membr Biol 209:31–41

119. Balzer M, Lintschinger B, Groschner K (1999) Evidence for a role of Trp proteins in the oxidative stress-induced membrane conductances of porcine aortic endothelial cells. Cardiovasc Res 42:543–549
120. Poteser M, Graziani A, Rosker C, Eder P, Derler I, Kahr H, Zhu MX, Romanin C, Groschner K (2006) TRPC3 and TRPC4 associate to form a redox-sensitive cation channel. Evidence for expression of native TRPC3-TRPC4 heteromeric channels in endothelial cells. J Biol Chem 281:13588–13595
121. Rosker C, Graziani A, Lukas M, Eder P, Zhu MX, Romanin C, Groschner K (2004) Ca^{2+} signaling by TRPC3 involves Na^+ entry and local coupling to the Na^+/Ca^{2+} exchanger. J Biol Chem 279:13696–13704
122. Vazquez G, Wedel BJ, Kawasaki BT, Bird GS, Putney JW Jr. (2004) Obligatory role of Src kinase in the signaling mechanism for TRPC3 cation channels. J Biol Chem 279:40521–40528
123. Crouzin N, de Jesus Ferreira MC, Cohen-Solal C, Aimar RF, Vignes M, Guiramand J (2007) Alpha-tocopherol-mediated long-lasting protection against oxidative damage involves an attenuation of calcium entry through TRP-like channels in cultured hippocampal neurons. Free Radic Biol Med 42:1326–1337

Chapter 32
TRP Channels and Neural Persistent Activity

Antonio Reboreda, Lydia Jiménez-Díaz, and Juan D. Navarro-López

Abstract One of the integrative properties of the nervous system is its capability to, by transient motor commands or brief sensory stimuli, evoke persistent neuronal changes, mainly as a sustained, tonic action potential firing. This neural activity, named persistent activity, is found in a good number of brain regions and is thought to be a neural substrate for short-term storage and accumulation of sensory or motor information [1]. Examples of this persistent neural activity have been reported in prefrontal [2] and entorhinal [3] cortices, as part of the neural mechanisms involved in short-term working memory [4]. Interestingly, the general organization of the motor systems assumes the presence of bursts of short-lasting motor commands encoding movement characteristics such as velocity, duration, and amplitude, followed by a maintained tonic firing encoding the position at which the moving appendage should be maintained [5, 6]. Generation of qualitatively similar sustained discharges have also been found in spinal and supraspinal regions in relation to pain processing [7, 8]. Thus, persistent neural activity seems to be necessary for both behavioral (positions of fixation) and cognitive (working memory) processes. Persistent firing mechanisms have been proposed to involve the participation of a non-specific cationic current (CAN current) mainly mediated by activation of TRPC channels. Because the function and generation of persistent activity is still poorly understood, here we aimed to review and discuss the putative role of TRP-like channels on its generation and/or maintenance.

32.1 Introduction

Neuronal persistent activity can be defined as the ability of a neuron to maintain a sustained firing after an input is finished. Typical duration of persistent activity ranges from hundreds of milliseconds to tens of seconds or even several minutes.

A. Reboreda (✉)
Section of Physiology, Department of Functional Biology and Health Sciences, School of Biology, University of Vigo, Campus Lagoas-Marcosende 36310 Vigo (Pontevedra), Spain
e-mail: areboreda@uvigo.es

Many brain areas show persistent activity response [1] but the role of this activity is still discussed. Brain regions like amygdala, entorhinal cortex, hippocampus or prefrontal cortex (PFC) are involved in working memory and learning processes. "In vivo" experiments carried out on those areas have shown prolonged afterdischarges following a stimulus presentation [9–13] and "in vitro" experiments show persistent firing after a depolarizing pulse or synaptic stimulation, in the presence of a neuromodulator [3, 8, 14–18]. From a physiological point of view, persistent firing has been proposed to be a substrate for short-term or working memory in a variety of systems [1, 3, 9, 19, 20] and it is also involved in the coding of new long-term memory [21]. Persistent activity has also been associated to the delay phase during delayed match or nonmatch to sample task performance [9] and to the stimulus-free interval between the conditioned and unconditioned stimuli during associative learning using trace conditioning paradigms [22, 23]. More recently, after the description of spatial mapping in the medial entorhinal cortex by "grid cells" [24–26], persistent firing has become a candidate to underlie the generation of this phenomenon [20, 27, 28]. Regarding the spatial mapping and map integration, the involvement of cholinergic mediated persistent activity in postsubiculum and its participation in the head direction system has recently been proposed [29].

In the oculomotor neural integrator, persistent firing that encodes eye position is maintained in response to transient saccadic eye-velocity commands, and firing rates in the integrator fluctuate linearly with eye position. This persistent activity can also be found in cerebellar Purkinje cells indicating its putative involvement in motor control [30–32]. In fact, sustained neural activity has been proposed as an universal mechanism for the control of movements [33]. In addition, sustained depolarizations have been shown to be responsible for both sensorimotor processing (i.e. transformation of an incoming sensory signal into a motor command) [34] and somatosensory/pain processing [1, 35].

In summary, it seems that persistent activity is present in systems that require a temporal maintenance of their output (vg. learning and working memory processes) and/or a fine tuning of their output (vg. head direction system or saccadic eye movements).

32.2 Mechanisms of Persistent Activity

32.2.1 Non-specific Ca^{2+} Sensitive Cationic Current (CAN Current)

Neuronal sustained firing that is not driven by ongoing external inputs must be explained by the internal dynamics of the cell or the circuit. Indeed, it has been extensively discussed whether persistent activity emerges as a network property [36–39] or is due to a cellular intrinsic property [3, 17, 40–42]. Focusing on the intrinsic properties involved, several ionic mechanisms have been proposed to explain this phenomenon, most of them suggesting the participation of a CAN current. Activity-dependent stimulation of the CAN current can trigger a sustained depolarization, from the initial voltage level ("plateau"), that can outlast the stimulus for several minutes. If this sustained depolarization reaches the firing threshold it could generate a persistent firing. Persistent activity (plateau and persistent firing) in slice preparations is often recorded using synaptic blockers, and triggered using short depolarizing current injections (between 50 and 100 pA, 1–4 s) in the presence of muscarinic or metabotropic glutamatergic agonists [3, 14, 40, 43, 44]; however it can also be recorded without synaptic blockers by using a synaptic input to trigger the "plateau" response [3, 17, 18, 45]. Persistent activity can also be evoked without an external neuromodulator by cholinergic afferent stimulation [17], by synaptic stimulation in the presence of muscarinic blockers [45] or using step depolarizations without neuromodulators [29]. In this regard, it has also been reported that acetylcholine can be released spontaneously in slices [46] and Ca^{2+} entry can be enough to activate CAN currents [47], those mechanisms could allow the activation of the CAN current without addition of external modulators. In addition, NMDA-dependent CAN driven "plateau" depolarization have also been recorded in vitro in lamprey reticulospinal neurons after sensory stimulation [34]. In summary persistent activity has been recorded after neuronal stimulation under metabotropic activation (glutamatergic or cholinergic).

With regard to their duration, "plateau" depolarizations can be classified as self-terminating afterdepolarizations (ADP) with a burst of action potentials on top that lasts for several seconds as shown in subiculum [48] and in medial entorhinal cortex layer II (where neurons also show slow depolarizing rebounds with a burst of action potentials after the initial ADP) [16]; (see Fig. 32.1a). "Plateau" depolarization can also trigger a persistent firing that lasts for several minutes: some examples are the activity recorded in the rat postsubiculum, cingulate cortex and layer III of the medial entorhinal cortex (see Fig. 32.1b) [8, 29, 45]. Lateral entorhinal layer III persistent activity can be turned off by a second depolarizing stimulation [18]. Additionally, in medial entorhinal layer V [3] and basolateral amygdala [17], spike frequency can be graded up and down by depolarizing (Fig. 32.1c) or hyperpolarizing stimuli, and graded activity has also been reported on layer III medial entorhinal cortex in presence of atropine and using synaptic stimulation [45]. Hippocampal CA1 neurons can show a "plateau" without action potentials [19].

The duration of the "plateau" can be modified by the level of the initial depolarization or by stimulus parameters (duration and amplitude) in such a way that neurons that show long lasting "plateau" can also trigger a self terminating "plateau" by reducing duration or amplitude of the stimulus [3].

Since the CAN current properties (modulation by neurotransmitters, cation permeability and sensitivity to Ca^{2+} [17, 43, 48, 49]) are quite similar, the differences in the "plateau" response in these brain areas could be related mostly to the participation of other conductances modulated also during the "plateau" response. The presence of M-current [50], sodium persistent current [16, 51], cyclic nucleotide-gated current [52], other cationic currents [53, 54], BK current [55] and possibly other potassium and Ca^{2+} conductances [56, 57] could account for the variability of the "plateau".

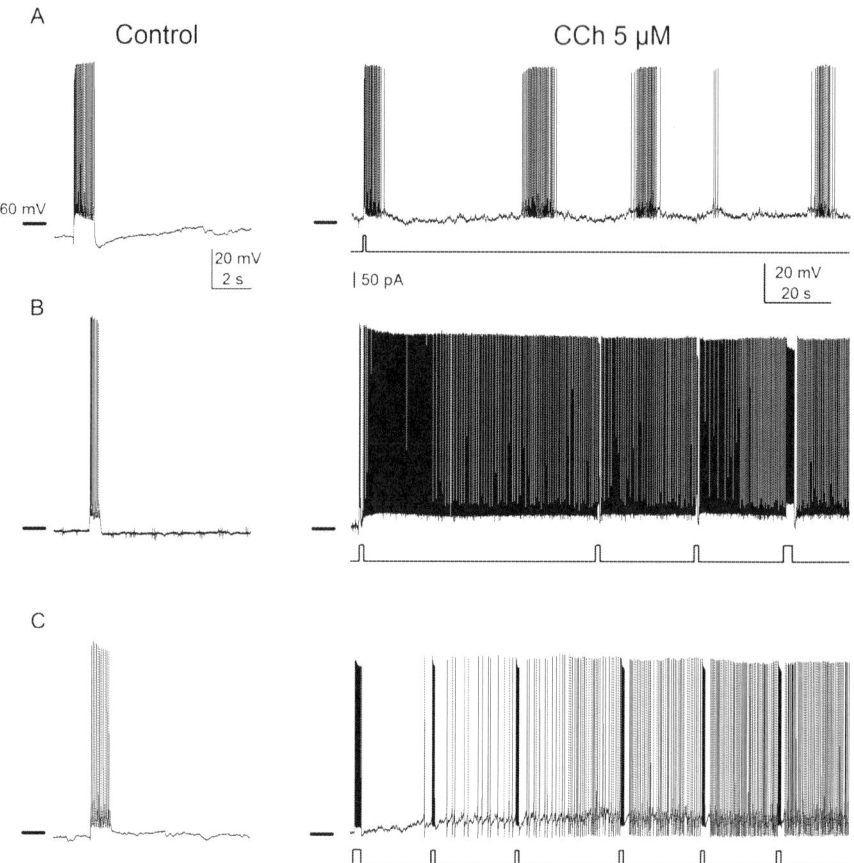

Fig. 32.1 Examples of persistent activity in medial entorhinal cortex. (**a**) Layer II medial entorhinal cortex neuron showing an ADP and rhythmic rebounds in the presence of carbachol (CCh) 5 μM. Calibrations in A also apply to (**b**) and (**c**). (**b**) Layer III medial entorhinal cortex neuron showing persistent activity during application of CCh 5 μM. Persistent activity frequency is not affected by depolarizing stimuli. (**c**) Layer V medial entorhinal cortex showing graded persistent activity in the presence of CCh 5 μM. Persistent activity frequency is graded up by depolarizing stimuli

32.3 TRP Channels and CAN Current

With the boom on transient receptor potential (TRP) channel research, the role of these channels in the generation of persistent activity has come into focus. TRPC (canonical) channels became the main candidates to underlie the CAN current, although two members of the melastatin subfamily, TRPM4 and TRPM5 have also been shown as mediators of the CAN current in brainstem areas [58]. On the other hand the presence of TRPC 4 and TRPC 5 has been demonstrated in cortical areas [59, 60] previously shown to exhibit persistent activity/CAN currents.

It has been proposed that working memory regarding to new stimulus and attentional processes requires an increase in cholinergic activity [61–63]. Cholinergic modulation through muscarinic receptors has been shown to be necessary both for working memory performance and "plateau"/persistent activity generation [64]. Application of atropine [3, 8, 65] or pirenzepine [3, 66] blocked carbachol induced persistent activity, which implies the involvement of M1 muscarinic receptors. Nevertheless "plateau" response generation has also been achieved through activation of other G-protein coupled receptors like group I metabotropic glutamate receptors (mGluRs) [8, 45, 59, 67, 68] as mGluR5 in PFC and lateral amygdala [44, 69]. Participation of glutamatergic receptors on working memory and learning has been studied on animal models by the application of different modulators and assessing the animal performance on different tasks. In this way, application of 3.3′-difluorobenzaldazine (DFB, anmGluR5 potentiator) or MPEP (mGluR5 antagonist) enhanced [70] and impaired [71] spatial related working memory tasks respectively. On the other hand, activation of group II mGluRs showed the opposite effect [72]. Experiments in slices indicate a common mechanism as it has been exposed by the occlusion of (S)-3,5-dihydroxyphenylglycine (DHPG, a potent agonist of group I mGluRs) effect, after carbachol application in cingulated cortex neurons [8]. However, persistent firing (without "plateau" depolarization by a CAN current) can be graded in frequency by step depolarizations and hyperpolarizations without external application of a neuromodulator in PFC [73], being this graded mechanism explained by modulation of the h-current. On the other hand muscarinic modulation of TRPC currents have been demonstrated in these brain areas [17, 42, 66]. The TRPC subunits proposed to intervene in the "plateau" generation were a combination mainly of the TRPC 1, 4 and 5 subunits [42, 66], although TRPC 6 could also participate [66].

32.4 TRPC Channels and G-Protein Mediated Receptors

Due to the multiple modulatory domains that a TRPC channel possesses [74, 75], one of the most controversial issues has been the mechanism of activation of the "plateau". Since receptor mediated activation seems a common mechanism, research efforts have been focused on the cascade leading to sustained depolarization, which could be a combination of structural and second messenger mechanisms. There is a wide consensus about the involvement of G-protein mediated receptors and several approaches have been used to investigate this topic. After receptor stimulation, the role of the G-protein has been studied using PLCβ-ct (phospholipase C beta1 C-terminal construct, a $G\alpha_{q-11}$ dominant negative subunit) [66], and the G-protein inhibitor GDP-β-S [8] both inhibiting the generation of the "plateau". The application of the G-protein nonhydrolyzable activator GTP-γ-S in anterior cingulated neurons was able to induce an ADP but not a complete "plateau" response, pointing out that G-protein activation is only part of the mechanism involved in the "plateau" generation [8]. Similarly, application of GTP-γ-S and GDP-β-S to mGluR evoked CAN current in CA1 pyramidal neurons was able to open and block the current respectively [67]. The G-protein activation could be explained as a necessary step for the activation cascade of the PLC, or could contribute directly to activation of TRPC channels through cytoskeletal proteins, since TRPC channels [74] and G-proteins [76] have been found to form membrane complexes in lipid rafts where Ca^{2+} is tightly regulated [77]. Lipid rafts also contain most of the accessory proteins necessary to modulate TRPC channel activation [74]. On the other hand, activation of G-protein-coupled receptors could generate a G-protein independent signal, like activation of the SRC kinase [78]. It has been shown that SRC kinases activate TRPC 3 [79, 80], TRPC 4, TRPC 5 [81] and TRPC 6 [82] channels. Nevertheless more data are required to determine SRC kinase participation in the modulation of the "plateau" activity.

32.5 TRPC Channels and PLC

Besides the dominant negative approach mentioned before, the role of the PLC has also been characterized through the application of the inhibitor U73122 and the more selective inhibitor ET-18-OCH3, both causing a reduction of the "plateau" depolarization [8, 42]. It has also been described that TRPC4 and TRPC5 have a PDZ (PSD-95/Discs-large/ZO-1) binding domain in their C-terminal region that can interact with NHERF scaffolding proteins which at the same time form complexes with PLCβ isoforms [83]. Disruption of this interaction reduces or modulates channel activation [83, 84] and using a peptide to interfere TRPC–NHERF interaction reduces the amplitude of the "plateau" [42]. Moreover, PLCβ1 knockout mice display deficits in fear conditioning and working memory-related tasks, and also show an important change in the expression of RGS4 protein, which regulates muscarinic signaling, in the hippocampus [85].

32.6 TRPC and PIP2

Since PLCβ activation will cause hydrolysis of phosphatidylinositol 4,5-bisphosphate (PIP2) and an effective reduction of PIP2 levels in the membrane [66], the influence of a change in the level of PIP_2 have been studied. It has been described that PIP2 inhibits TRPC 4 [86] and TRPC 5 [87] directly and through cytoskeleton proteins [88]. Experiments where application of DiC8-PIP_2, to maintain PIP_2 levels, reduced the "plateau" depolarization in the presence of carbachol in entorhinal cortex layer V neurons [42]. Together with the observed reduction of the PIP_2 in the membrane of PFC principal neurons during carbachol application [66], indicate that PIP_2 breakdown plays a fundamental role in "plateau" depolarization. This observation strengthens the hypothesis of TRPC 4 and 5 participation versus TRPC 3, 6, 7 and TRPM 4, 5 since maintenance of PIP_2 levels is required for the activation of those channels [89–91].

32.7 TRPC and IP$_3$

PIP$_2$ hydrolysis by the action of the PLC will render inositol trisphosphate (IP$_3$) and diacylglycerol (DAG). The roles of those second messengers have also been explored. The IP$_3$ generated will release Ca^{2+} from the intracellular stores and raise Ca^{2+} levels. TRPC channels have binding domains for Ca^{2+}/calmodulin [92–94] and other Ca^{2+} binding proteins [95], besides they are also directly modulated by Ca^{2+} [96]. Investigating the influence of Ca^{2+} from the intracellular stores proved that they are not involved in "plateau" generation since application of thapsigargin or cyclopiazonic acid, to deplete the intracellular stores, did not affect "plateau" generation [17, 97]. Nevertheless, mGluR dependent Ca^{2+} release from the intracellular stores was able to open a CAN current in hippocampal CA1 [98], opening the door for different Ca^{2+} sources modulating CAN currents in different areas. On the other hand, it is also known that all TRPC channels have a binding domain for IP$_3$ receptor (IP$_3$R) [74, 99] and they can also interact with IP$_3$R through other proteins like junctate or homer [100, 101]. However, addition of IP$_3$ to the recording pipette with strong intracellular Ca^{2+} buffering did not evoke a "plateau" depolarization in most of the medial entorhinal cortex layer V cells [42]. Nevertheless, application of heparin to block the DHPG-induced plateau was successful in PFC layer V neurons, so involvement of IP$_3$R cannot be completely ruled out [59].

32.8 TRPC and Ca^{2+} Channels

The Ca^{2+} dependency of the CAN current has been studied using different approaches. Strong buffering of intracellular Ca^{2+} close to 0 does not allow the "plateau" to be generated [8, 29, 42, 43, 48, 59, 66, 102]. Those experiments, together with the effect of the depletion of intracellular stores, point towards a different Ca^{2+} source to explain the Ca^{2+} dependency of the "plateau". In this way, application of the L-type Ca^{2+} channel blocker nifedipine [3, 8, 102], Cd^2 [55], or Co^{2+} [48], reduced the "plateau" and the CAN current. Expression of T-type Ca^{2+} and TRPC5 subunits in HEK-293 cells is enough to generate an inward current in the presence of carbachol [66]. Similarly, experiments in nominally 0 extracellular Ca^{2+} inhibited the generation of the "plateau" depolarization [8, 17, 42, 48, 55, 102]. It has been described that the same TRPC channel could be both, potentiated [103] and inhibited by different Ca^{2+} concentrations [95, 104]. Changing intracellular Ca^{2+} concentration at different levels showed that the activation of the "plateau" requires a precise control of intracellular Ca^{2+} levels; PFC neurons can trigger a "plateau" depolarization using solutions with low Ca^{2+} buffering while solutions with high buffering capabilities block the "plateau" depolarization [43]. Entorhinal layer V principal neurons are not able to generate a "plateau" below 20 nM and above 1 μM intracellular Ca^{2+} [42], the entorhinal layer II neurons showed a similar response where small increases of intracellular Ca^{2+} potentiates a CAN current that generated a "plateau" depolarization while higher intracellular Ca^{2+} levels downregulated the current [102]. It has been proposed that a tight control of intracellular Ca^{2+} levels is also important to achieve different "plateau" levels during graded persistent activity [97]. However, experiments on muscarinic graded persistent activity in lateral amygdala neurons showed that stimulation of cholinergic afferents was enough to increase the "plateau" frequency without direct stimulation, by a depolarizing step or by synaptic input, suggesting an alternative mechanism where a muscarinic second messenger could be involved in the graded response [17]. In summary, extracellular Ca^{2+} together with a precise control of intracellular Ca^{2+} concentration (probably in Ca^{2+} microdomains) [105], are both required to maintain a sustained depolarization.

32.9 TRPC and DAG

The other branch of the PLC cascade generates DAG from the hydrolysis of PIP_2. It has been described that DAG can activate TRPC 2, 3, 6 and 7 in a PKC independent way [106, 107], while it inhibits TRPC 4 and 5 through activation of PKC [108]. Moreover, activation of PKC in PFC (for example in stress situations) has been shown to reduce cognitive function in spatial working memory tasks [109]. Experiments where OAG (DAG analogous) is applied to entorhinal cortex layer V, neurons were not successful in generating a "plateau" depolarization. On the contrary, application of OAG reduced the amplitude of the "plateau" depolarization during muscarinic modulation [42] pointing out the involvement of TRPC 1, 4 and 5. In PFC layer V neurons using DHPG as an activator of the "plateau" depolarization, application of PdBU, an activator of PKC, reduces the amplitude of the depolarization reinforcing the involvement of TRPC 4 and 5 subunits in this phenomenon [59].

These experiments seemed to reduce the candidates to mainly TRPC 1, 4, and 5. However, heterologous expression of TRPC6 and a low threshold Ca^{2+} channel is also capable of generating an ADP in HEK-293 cells in the presence of carbachol. Therefore, participation of TRPC6 cannot be excluded as a mechanism to generate "plateau" depolarizations [66].

32.10 TRPC Pharmacology

Since drug selectivity for TRPC channels is very poor, different approaches have been used to investigate the mechanism of "plateau" generation relying on pharmacology only to narrow down the possible candidates. In this way, most of the studies used flufenamic acid (FFA) as a first choice to determine the presence of a CAN current [3, 54, 110]. FFA is a general cationic current blocker but with a very poor selectivity [111] and for TRPC6 acts as an enhancer [112]. Application of FFA to entorhinal cortex layer III [55], V [3], PFC [66], amygdala [17], postsubiculum [29] and cingulate cortex neurons [8] blocked the afterdischarge and the "plateau" triggered by a muscarinic agonist.

Second choice drugs are 2-APB [113] and SKF-96365 [114]. 2-APB blocks total or partially all TRPC channels [115–117], it can activate TRPV (vanilloid) channels [118] and also block the IP$_3$R (except in intact cells [115]), voltage dependent K$^+$ channels and gap junctions [119, 120]; whereas SKF-96365 affects also voltage dependent Ca^{2+} entry [114]. 2-APB blocked "plateau" potentials induced by muscarinic modulation in entorhinal cortex layer V [42], amygdala [17], and layer II/III cingulate cortex neurons [8]. SKF-96365 also blocked entorhinal cortex layer V and cingulated cortex persistent activity [8, 42], the PFC ADP is also suppressed by this drug [66].

The lack of more selective blockers requires specific design of molecular tools. In this sense the use of TRPC5 dominant negative subunits has proven to be useful in blocking the carbachol induced current in PFC [66], pointing towards the involvement of TRPC1, 4 and/or 5 subunits. Supporting this hypothesis, voltage clamp experiments in lateral amygdala [121], layer V entorhinal cortex [42], layers II/III cingulated cortex [8], and layer V PFC [122], using I-V ramps, showed a muscarinic or glutamatergic sensitive current with a phenotype similar to the one described for the TRPC 1/4, 1/5 heteromers. These heteromers are activated by G-protein coupled receptors but not by depletion of the intracellular Ca^{2+} stores [123], and are blocked by application of 2-APB, FFA and SKF-96365 [8, 42, 121]. Intracellular application of antibodies against TRPC5 also reduced the glutamate evoked current in lateral amygdala neurons while antibodies against TRPC 1 and TRPC 6 failed to do so [121].

32.11 TRP and Diseases Epilepsy

Epilepsy is the hypersynchronization of the neural activity, involving network and intrinsic mechanisms. Cholinergic activation is required for normal population dynamics [124] but it has been shown that the cholinergic system also participates in the maintenance of epileptiform discharges [125]. Experimental models of epilepsy have been achieved using muscarinic [126], but also glutamatergic [127, 128] agonists activating CAN/TRPC 4 and 5 currents through a PLCβ1-dependent mechanism [129, 130]. The application of TRPC channel blockers SKF96365 and 2-APB [130], buffering intracellular Ca^{2+} or the application of FFA [129] all blocked the sustained discharges but did not affect the onset of the discharge caused by the NMDA receptor. In turn, this receptor contributes, together with voltage dependent Ca^{2+} channels and Ca^{2+} from the intracellular stores, to the Ca^{2+} needed to activate the CAN current [129, 130]. These experimental findings indicate the participation of CAN/TRPC currents in epilepsy caused by a pathological activation of the cholinergic and/or glutamatergic neurotransmission. This activation will cause the internalization of the channels but the remaining pool in the membrane seems to be enough to maintain the activity [130].

The participation of TRPC channels in sustained discharges during epilepsy reinforces the hypothesis of their involvement in the modulation of excitability.

32.12 TRP, Pain Processing and Persistent Activity

Pain results from the complex processing of neural signals at different levels (spinal and supraspinal loci) of the central nervous system. Nociceptors – high-threshold primary sensory neurons that detect noxious stimuli-represent the beginning of the pain pathway and are the interface of the nervous system with the external and internal environments. The function of TRP channels in primary afferent nociceptors and in the peripheral mechanisms of pain sensitivity has already been extensively studied (see [131] for a review). On the other hand, the role of persistent activity and the involvement of TRPC-like channels in transient storage of information during pain processing has recently been described in supraspinal (cortical) brain regions [8]. However, although persistent neural activity has been also been documented in the spinal cord in relation to the somatosensory/pain processing [1, 35], the presence of persistent activity-related TRP channels in spinal cord neurons and its involvement in spinal pain processing remains virtually unexplored.

Deep dorsal horn neurons (DHN) integrate both innocuous and nociceptive (pain) inputs and are of fundamental importance for plasticity in information processing and transfer [7]. It has been shown both "in vitro" [7, 132, 133] and "in vivo" [134, 135] that DHN exhibit a form of persistent activity known as "wind-up" that is induced by repetitive stimulation of nociceptive primary afferents (or successive intracellular current pulses). Wind–up is an activity-dependent short-term sensitization thought to be a first step in the development of long-term spinal sensitization to pain [136, 137]. As in other areas throughout the nervous system, "plateau" potentials have been demonstrated as critical components underlying the generation of wind-up spinal persistent firing [138]. "Plateau" potentials in DHN are subjected to neuromodulation [7, 139], and are supported by two depolarizing currents in rodents: an L-type Ca^{2+} current and a Ca^{2+}-activated CAN current [138, 140]. The exact molecular identity of the channels mediating these "plateau" potentials and long-lasting persistent firing is still elusive. FFA has been the main pharmacological tool used to reveal the presence of the CAN current in DHN, thus its putative TRP-mediated nature needs to be determined with more selective blockers or alternative approaches. A member of the melastatin-like subfamily, TRPM4 channel, has been proposed to give rise to a CAN current in neonatal mouse neurons [58] and its expression in intact and injured spinal cord has been recently shown [141]. However, a role for TRPM4 in persistent firing in the anterior cingulate cortex has already been ruled out [8]. Further investigation is needed to start elucidating the role of TRP channels in the generation of persistent activity during nociceptive information processing in the spinal cord.

32.13 Future Directions

A summary of the actual proposed mechanisms for the activation of TRPC channels underlying the generation of persistent activity is shown in Fig. 32.2. Nevertheless research in the mechanism related to persistent activity still has several milestones

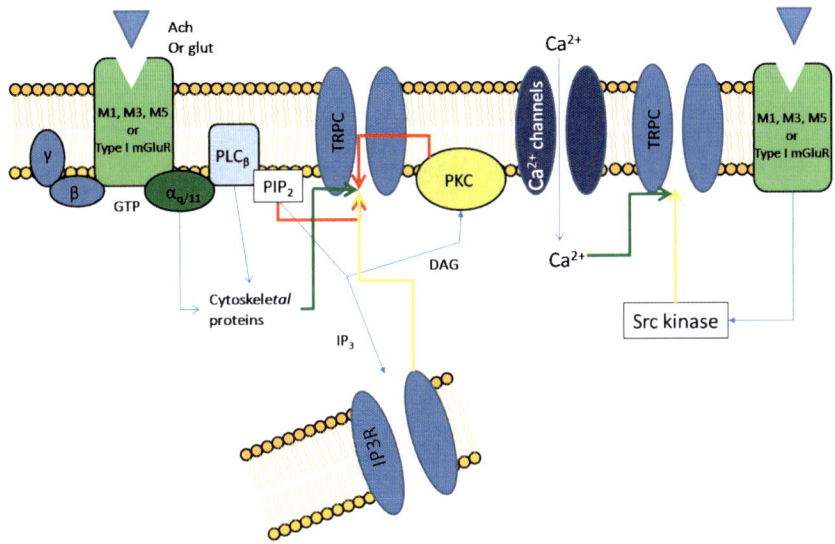

Fig. 32.2 Current and proposed mechanisms of CAN/TRPC activation during neural persistent activity. Summary of current and proposed interactions among TRPC channels involved in persistent activity and activation mechanisms. *Green lines* indicate positive modulation (*opening*), *red lines* indicate negative modulation, and *yellow lines* indicate that more data is needed

to achieve: a definitive identification and characterization of the activation and modulatory mechanisms of the TRPC subunits involved in the depolarization is needed. Besides the G-protein activation, other signaling mechanisms affecting TRPC channel in heterologous systems need to be further characterized in persistent activity preparations like the involvement of calmodulin, src kinases, IP_3R, etc. To achieve this goal and due to the complexity of these channels and the lack of selective pharmacological tools, new approaches are required. In the past few years the development and use of functional antibodies against E3, an external region of the channel, has been used successfully to block selectively the currents. This antibody can discriminate between close members of the TRPC family like TRPC 4 and TRPC 5 [142]. Another experimental approach to take advantage of is the use of RNAi against TRPC subunits. This technique was shown to be useful in several experimental models reducing TRPC expression and the associated current [143, 144]. Finally, knockout animals could also be a promising tool in TRPC research and their involvement in the generation of persistent activity. To our knowledge only behavioral and physiological studies (related to learning and memory) from the TRPC 5 knockout have been published, showing a diminished synaptic response in amygdala to mGluR I and cholecystokinin2 together with an impaired responses to G_{q-11} protein-coupled receptors. The phenotype obtained showed changes in innate and learned fear responses [145].

In summary, as illustrated in this review, TRP-mediated persistent activity has already been found in a number of areas across the CNS, supporting mainly working memory processes. The diversity of CNS regions demonstrating persistent activity is massive, suggesting its importance and pointing out possible common action mechanisms underlying this phenomenon. Therefore, experiments testing the contribution of TRP channels to persistent neural activity in other CNS areas and brain functions are new research challenges in persistent firing investigation.

Acknowledgments This work was supported by Fundación Eugenio Rodriguez Pascual. L.J-D was supported by Juan de la Cierva MICINN Programme, J.D.N-L by Ramón y Cajal MICINN Programme and A.R. was supported by a CONSOLIDER-INGENIO grant The Spanish Ion Channel Initiative (SICI) (MICINN, CSD2008-00005).

References

1. Major G, Tank D (2004) Persistent neural activity: prevalence and mechanisms. Curr Opin Neurobiol 14:675–684
2. Fuster JM (1997) Network memory. Trends Neurosci 20:451–459
3. Egorov AV, Hamam BN, Fransen E, Hasselmo ME, Alonso AA (2002) Graded persistent activity in entorhinal cortex neurons. Nature 420:173–178
4. Goldman-Rakic PS (1995) Cellular basis of working memory. Neuron 14:477–485
5. Moschovakis A (1997) The neural integrators of the mammalian saccadic system. Front Biosci 15:552–577
6. Robinson DA (1981) The use of control systems analysis in the neurophysiology of eye movements. Annu Rev Neurosci 4:463–503
7. Derjean D, Bertrand S, Le Masson G, Landry M, Morisset V, Nagy F (2003) Dynamic balance of metabotropic inputs causes dorsal horn neurons to switch functional states. Nat Neurosci 6:274–281
8. Zhang Z, Seguela P Metabotropic induction of persistent activity in layers II/III of anterior cingulate cortex. Cereb Cortex doi:10.1093/cercor/bhq1043 2010
9. Suzuki WA, Miller EK, Desimone R (1997) Object and place memory in the macaque entorhinal cortex. J Neurophysiol 78:1062–1081
10. Fuster JM, Alexander GE (1971) Neuron activity related to short-term memory. Science 173:652–654
11. Paton JJ, Belova MA, Morrison SE, Salzman CD (2006) The primate amygdala represents the positive and negative value of visual stimuli during learning. Nature 439:865–870
12. Otto T, Eichenbaum H (1992) Neuronal activity in the hippocampus during delayed non-match to sample performance in rats: evidence for hippocampal processing in recognition memory. Hippocampus 2:323–334
13. Brody CD, Hernandez A, Zainos A, Romo R (2003) Timing and neural encoding of somatosensory parametric working memory in macaque prefrontal cortex. Cereb Cortex 13:1196–1207
14. Krnjevic K, Pumain R, Renaud L (1971) The mechanism of excitation by acetylcholine in the cerebral cortex. J Physiol (Lond) 215:247–268
15. Andrade R (1991) Cell excitation enhances muscarinic cholinergic responses in rat association cortex. Brain Res 548:81–93
16. Klink R, Alonso A (1997) Muscarinic modulation of the oscillatory and repetitive firing properties of entorhinal cortex layer II neurons. J Neurophysiol 77:1813–1828
17. Egorov AV, Unsicker K (2006) von Bohlen und Halbach O: muscarinic control of graded persistent activity in lateral amygdala neurons. Eur J Neurosci 24:3183–3194
18. Tahvildari B, Fransén E, Alonso AA, Hasselmo ME (2007) Switching between "On" and "Off" states of persistent activity in lateral entorhinal layer III neurons. Hippocampus 17:257–263

19. Fraser DD, MacVicar BA (1996) Cholinergic-dependent plateau potential in hippocampal CA1 pyramidal neurons. J Neurosci 16:4113–4128
20. Hasselmo ME, Brandon MP (2008) Linking cellular mechanisms to behavior: entorhinal persistent spiking and membrane potential oscillations may underlie path integration, grid cell firing, and episodic memory. Neural Plast 2008:12
21. Schon K, Hasselmo ME, LoPresti ML, Tricarico MD, Stern CE (2004) Persistence of parahippocampal representation in the absence of stimulus input enhances long-term encoding: a functional magnetic resonance imaging study of subsequent memory after a delayed match-to-sample task. J Neurosci 24:11088–11097
22. Esclassan F, Coutureau E, Di Scala G, Marchand AR (2009) A cholinergic-dependent role for the entorhinal cortex in trace fear conditioning. J Neurosci 29:8087–8093
23. Bang SJ, Brown TH (2009) Muscarinic receptors in perirhinal cortex control trace conditioning. J Neurosci 29:4346–4350
24. Fyhn M, Molden S, Witter MP, Moser EI, Moser M-B (2004) Spatial representation in the entorhinal cortex. Science 305:1258–1264
25. Hafting T, Fyhn M, Molden S, Moser M-B, Moser EI (2005) Microstructure of a spatial map in the entorhinal cortex. Nature 436:801–806
26. Sargolini F, Fyhn M, Hafting T, McNaughton BL, Witter MP, Moser M-B, Moser EI (2006) Conjunctive representation of position, direction, and velocity in entorhinal cortex. Science 312:758–762
27. Hasselmo ME (2008) Grid cell mechanisms and function: contributions of entorhinal persistent spiking and phase resetting. Hippocampus 18:1213–1229
28. Hasselmo ME, Brandon MP, Yoshida M, Giocomo LM, Heys JG, Fransen E, Newman EL, Zilli EA (2009) A phase code for memory could arise from circuit mechanisms in entorhinal cortex. Neural Network 22:1129–1138
29. Yoshida M, Hasselmo ME (2009) Persistent firing supported by an intrinsic cellular mechanism in a component of the head direction system. J Neurosci 29:4945–4952
30. Batchelor AM, Garthwaite J (1997) Frequency detection and temporally dispersed synaptic signal association through a metabotropic glutamate receptor pathway. Nature 385:74–77
31. Tempia F, Miniaci MC, Anchisi D, Strata P (1998) Postsynaptic current mediated by metabotropic glutamate receptors in cerebellar purkinje cells. J Neurophysiol 80:520–528
32. Kim SJ, Kim YS, Yuan JP, Petralia RS, Worley PF, Linden DJ (2003) Activation of the TRPC1 cation channel by metabotropic glutamate receptor mGluR1. Nature 426:285–291
33. Delgado-García JM, Yajeya J, Navarro-López JD (2006) A cholinergic mechanism underlies persistent neural activity necessary for eye fixation. Prog Brain Res 154: 211–224, Elsevier
34. Di Prisco GV, Pearlstein E, Le Ray D, Robitaille R, Dubuc R (2000) A cellular mechanism for the transformation of a sensory input into a motor command. J Neurosci 20:8169–8176
35. Marder E (2003) Plateau properties in pain pathways. Nat Neurosci 6:210–212
36. Aksay E, Baker R, Seung HS, Tank DW (2003) Correlated discharge among cell pairs within the oculomotor horizontal velocity-to-position integrator. J Neurosci 23:10852–10858
37. Aksay E, Gamkrelidze G, Seung HS, Baker R, Tank DW (2001) In vivo intracellular recording and perturbation of persistent activity in a neural integrator. Nat Neurosci 4:184–193
38. Navarro-Lopez JD, Alvarado JC, Marquez-Ruiz J, Escudero M, Delgado-Garcia JM, Yajeya J (2004) A cholinergic synaptically triggered event participates in the generation of persistent activity necessary for eye fixation. J Neurosci 24:5109–5118
39. Navarro-Lopez JD, Delgado-Garcia JM, Yajeya J (2005) Cooperative glutamatergic and cholinergic mechanisms generate short-term modifications of synaptic effectiveness in prepositus hypoglossi neurons. J Neurosci 25:9902–9906
40. Reboreda A, Raouf R, Alonso A, Seguela P (2007) Development of cholinergic modulation and graded persistent activity in layer V of medial entorhinal cortex. J Neurophysiol 97:3937–3947

41. Russo MJ, Mugnaini E, Martina M (2007) Intrinsic properties and mechanisms of spontaneous firing in mouse cerebellar unipolar brush cells. J Physiol 581:709–724
42. Zhang Z, Reboreda A, Alonso A, Barker PA, Séguéla P (2010) TRPC channels underlie cholinergic plateau potentials and persistent activity in entorhinal cortex. Hippocampus, n/a. doi:10.1002/hipo.20755
43. Haj-Dahmane S, Andrade R (1998) Ionic mechanism of the slow afterdepolarization induced by Muscarinic receptor activation in rat prefrontal cortex. J Neurophysiol 80:1197–1210
44. Sidiropoulou K, Lu F-M, Fowler MA, Xiao R, Phillips C, Ozkan ED, Zhu MX, White FJ, Cooper DC (2009) Dopamine modulates an mGluR5-mediated depolarization underlying prefrontal persistent activity. Nat Neurosci 12:190–199
45. Yoshida M, Fransén E, Hasselmo ME (2008) mGluR-dependent persistent firing in entorhinal cortex layer III neurons. Eur J Neurosci 28:1116–1126
46. Cole AE, Nicoll RA (1983) Acetylcholine mediates a slow synaptic potential in hippocampal pyramidal cells. Science 221:1299–1301
47. Gee CE, Benquet P, Gerber U (2003) Group I metabotropic glutamate receptors activate a calcium-sensitive transient receptor potential-like conductance in rat hippocampus. J Physiol 546:655–664
48. Kawasaki H, Palmieri C, Avoli M (1999) Muscarinic receptor activation induces depolarizing plateau potentials in bursting neurons of the rat subiculum. J Neurophysiol 82:2590–2601
49. Klink R, Alonso A (1997) Ionic mechanisms of muscarinic depolarization in entorhinal cortex layer II neurons. J Neurophysiol 77:1829–1843
50. Yoshida M, Alonso A (2007) Cell-type specific modulation of intrinsic firing properties and subthreshold membrane oscillations by the M(Kv7)-current in neurons of the entorhinal cortex. J Neurophysiol 98:2779–2794
51. Cantrell AR, Ma JY, Scheuer T, Catterall WA (1996) Muscarinic modulation of sodium current by activation of protein kinase C in rat hippocampal neurons. Neuron 16:1019–1026
52. Kuzmiski JB, MacVicar BA (2001) Cyclic nucleotide-gated channels contribute to the cholinergic plateau potential in hippocampal CA1 pyramidal neurons. J Neurosci 21:8707–8714
53. Egorov AV, Angelova PR, Heinemann U, Müller W (2003) Ca^{2+}-independent muscarinic excitation of rat medial entorhinal cortex layer V neurons. Eur J Neurosci 18:3343–3351
54. Haj-Dahmane S, Andrade R (1999) Muscarinic receptors regulate two different calcium-dependent non-selective cation currents in rat prefrontal cortex. Eur J Neurosci 11:1973–1980
55. Tahvildari B, Alonso AA, Bourque CW (2008) Ionic basis of ON and OFF persistent activity in layer III lateral entorhinal cortical principal neurons. J Neurophysiol 99:2006–2011
56. Tai C, Kuzmiski JB, MacVicar BA (2006) Muscarinic enhancement of R-type calcium currents in hippocampal CA1 pyramidal neurons. J Neurosci 26:6249–6258
57. Zhang L, Han D, Carlen PL (1996) Temporal specificity of muscarinic synaptic modulation of the $Ca(^{2+})$-dependent K^+ current (ISAHP) in rat hippocampal neurones. J Physiol 496:395–405
58. Crowder EA, Saha MS, Pace RW, Zhang H, Prestwich GD, Del Negro CA (2007) Phosphatidylinositol 4,5-bisphosphate regulates inspiratory burst activity in the neonatal mouse preBÃtzinger complex. J Physiol 582:1047–1058
59. Fowler MA, Sidiropoulou K, Ozkan ED, Phillips CW, Cooper DC (2007) Corticolimbic expression of TRPC4 and TRPC5 channels in the rodent brain. PLoS One 2:e573
60. von Bohlen und Halbach O, Hinz U, Unsicker K, Egorov A (2005) Distribution of TRPC1 and TRPC5 in medial temporal lobe structures of mice. Cell Tissue Res 322:201–206
61. Hasselmo ME (2006) The role of acetylcholine in learning and memory. Curr Opin Neurobiol 16:710–715
62. Parikh V, Sarter M (2008) Cholinergic mediation of attention. Contributions of phasic and tonic increases in prefrontal cholinergic activity. Ann NY Acad Sci 1129:225–235

63. Schon K, Atri A, Hasselmo ME, Tricarico MD, LoPresti ML, Stern CE (2005) Scopolamine reduces persistent activity related to long-term encoding in the parahippocampal gyrus during delayed matching in humans. J Neurosci 25:9112–9123
64. McGaughy J, Koene RA, Eichenbaum H, Hasselmo ME (2005) Cholinergic deafferentation of the entorhinal cortex in rats impairs encoding of novel but not familiar stimuli in a delayed nonmatch-to-sample task. J Neurosci 25:10273–10281
65. Chudasama Y, Dalley JW, Nathwani F, Bouger P, Robbins TW (2004) Cholinergic modulation of visual attention and working memory: dissociable effects of basal forebrain 192-IgG-saporin lesions and intraprefrontal infusions of scopolamine. Learn Mem 11:78–86
66. Yan H-D, Villalobos C, Andrade R (2009) TRPC Channels mediate a muscarinic receptor-induced afterdepolarization in cerebral cortex. J Neurosci 29:10038–10046
67. Congar P, Leinekugel X, Ben-Ari Y, Crepel V (1997) A Long-lasting calcium-activated nonselective cationic current is generated by synaptic stimulation or exogenous activation of group I metabotropic glutamate receptors in CA1 pyramidal neurons. J Neurosci 17:5366–5379
68. Wyart C, Cocco S, Bourdieu L, Leger J-F, Herr C, Chatenay D (2005) Dynamics of excitatory synaptic components in sustained firing at low rates. J Neurophysiol 93:3370–3380
69. Rodrigues SM, Bauer EP, Farb CR, Schafe GE, LeDoux JE, The Group I (2002) Metabotropic glutamate receptor mGluR5 is required for fear memory formation and long-term potentiation in the lateral amygdala. J Neurosci 22:5219–5229
70. Balschun D, Zuschratter W, Wetzel W (2006) Allosteric enhancement of metabotropic glutamate receptor 5 function promotes spatial memory. Neuroscience 142:691–702
71. Naie K, Manahan-Vaughan D (2004) Regulation by metabotropic glutamate receptor 5 of LTP in the dentate gyrus of freely moving rats: relevance for learning and memory formation. Cereb Cortex 14:189–198
72. Gregory ML, Stech NE, Owens RW, Kalivas PW (2003) Prefrontal group II metabotropic glutamate receptor activation decreases performance on a working memory task. Ann N Y Acad Sci 1003:405–409
73. Winograd M, Destexhe A, Sanchez-Vives MV (2008) Hyperpolarization-activated graded persistent activity in the prefrontal cortex. Proc Natl Acad Sci USA 105:7298–7303
74. Ambudkar I, Ong H (2007) Organization and function of TRPC channelosomes. Pflügers Arch Eur J Physiol 455:187–200
75. Kiselyov K, Shin DM, Kim JY, Yuan JP, Muallem S (2007) TRPC channels: Interacting Proteins. Handbook of Experimental Pharmacology, Vol. 179, Part VI, 559–574, doi: 10.1007/978-3-540-34891-7_33
76. Bhatnagar A, Sheffler DJ, Kroeze WK, Compton-Toth B, Roth BL (2004) Caveolin-1 interacts with 5-HT2A serotonin receptors and profoundly modulates the signaling of selected GÎ±q-coupled protein receptors. J Biol Chem 279:34614–34623
77. Pani B, Singh BB (2009) Lipid rafts/caveolae as microdomains of calcium signaling. Cell Calcium 45:625–633
78. Heuss C, Gerber U (2000) G-protein-independent signaling by G-protein-coupled receptors. Trends Neurosci 23:469–475
79. Kawasaki BT, Liao Y, Birnbaumer L (2006) Role of Src in C3 transient receptor potential channel function and evidence for a heterogeneous makeup of receptor- and store-operated Ca^{2+} entry channels. Proc Natl Acad Sci USA 103:335–340
80. Vazquez G, Wedel BJ, Kawasaki BT, Bird GSJ, Putney JW (2004) Obligatory role of Src kinase in the signaling mechanism for TRPC3 cation channels. J Biol Chem 279: 40521–40528
81. Odell AF, Scott JL, Van Helden DF (2005) Epidermal growth factor induces tyrosine phosphorylation, membrane insertion, and activation of transient receptor potential channel 4. J Biol Chem 280:37974–37987
82. Hisatsune C, Kuroda Y, Nakamura K, Inoue T, Nakamura T, Michikawa T, Mizutani A, Mikoshiba K (2004) Regulation of TRPC6 channel activity by tyrosine phosphorylation. J Biol Chem 279:18887–18894

83. Tang Y, Tang J, Chen Z, Trost C, Flockerzi V, Li M, Ramesh V, Zhu MX (2000) Association of mammalian Trp4 and phospholipase C isozymes with a PDZ domain-containing protein, NHERF. J Biol Chem 275:37559–37564
84. Obukhov AG, Nowycky MC (2004) TRPC5 activation kinetics are modulated by the scaffolding protein ezrin/radixin/moesin-binding phosphoprotein-50 (EBP50). J Cell Physiol 201:227–235
85. McOmish CE, Burrows EL, Howard M, Hannan AJ (2008) PLC-beta1 knockout mice as a model of disrupted cortical development and plasticity: behavioral endophenotypes and dysregulation of RGS4 gene expression. Hippocampus 18:824–834
86. Otsuguro K-i, Tang J, Tang Y, Xiao R, Freichel M, Tsvilovskyy V, Ito S, Flockerzi V, Zhu MX, Zholos AV (2008) Isoform-specific inhibition of TRPC4 channel by phosphatidylinositol 4,5-bisphosphate. J Biol Chem 283:10026–10036
87. Trebak M, Lemonnier L, DeHaven W, Wedel B, Bird G, Putney J (2009) Complex functions of phosphatidylinositol 4,5-bisphosphate in regulation of TRPC5 cation channels. Pflügers Arch Eur J Physiol 457:757–769
88. Miehe S, Bieberstein A, Arnould I, Ihdene O, Rüetten H, Strubing C (2010) The phospholipid-binding protein SESTD1 is a novel regulator of the transient receptor potential channels TRPC4 and TRPC5. J Biol Chem 285(16):12426–12434
89. Lemonnier L, Trebak M, Putney JW Jr (2008) Complex regulation of the TRPC3, 6 and 7 channel subfamily by diacylglycerol and phosphatidylinositol-4,5-bisphosphate. Cell Calcium 43:506–514
90. Liu D, Liman ER (2003) Intracellular Ca^{2+} and the phospholipid PIP2 regulate the taste transduction ion channel TRPM5. Proc Natl Acad Sci USA 100:15160–15165
91. Nilius B, Mahieu F, Prenen J, Janssens A, Owsianik G, Vennekens R, Voets T (2006) The Ca^{2+}-activated cation channel TRPM4 is regulated by phosphatidylinositol 4,5-biphosphate. EMBO J 25:467–478
92. Zhang Z, Tang J, Tikunova S, Johnson JD, Chen Z, Qin N, Dietrich A, Stefani E, Birnbaumer L, Zhu MX (2001) Activation of Trp3 by inositol 1,4,5-trisphosphate receptors through displacement of inhibitory calmodulin from a common binding domain. Proc Natl Acad Sci USA 98:3168–3173
93. Shimizu S, Yoshida T, Wakamori M, Ishii M, Okada T, Takahashi M, Seto M, Sakurada K, Kiuchi Y, Mori Y (2006) Ca^{2+}-calmodulin-dependent myosin light chain kinase is essential for activation of TRPC5 channels expressed in HEK293 cells. J Physiol 570:219–235
94. Sung T, Kim M, Hong S, Jeon J-P, Kim B, Jeon J-H, Kim S, So I (2009) Functional characteristics of TRPC4 channels expressed in HEK 293 cells. Mol Cells 27:167–173
95. Kinoshita-Kawada M, Tang J, Xiao R, Kaneko S, Foskett JK, Zhu MX (2005) Inhibition of TRPC5 channels by Ca^{2+}-binding protein 1 in Xenopus oocytes. Pflügers Arch Eur J Physiol 450:345–354
96. Gross SA, Guzmán GA, Wissenbach U, Philipp SE, Zhu MX, Bruns D, Cavalié A (2009) TRPC5 Is a Ca^{2+}-activated channel functionally coupled to Ca^{2+}-selective ion channels. J Biol Chem 284:34423–34432
97. Fransén E, Tahvildari B, Egorov AV, Hasselmo ME, Alonso AA (2006) Mechanism of graded persistent cellular activity of entorhinal cortex layer V neurons. Neuron 49:735–746
98. Partridge LD, Valenzuela CF (1999) Ca^{2+} store-dependent potentiation of Ca^{2+}-activated non-selective cation channels in rat hippocampal neurones in vitro. J Physiol 521:617–627
99. Boulay G, Brown DM, Qin N, Jiang M, Dietrich A, Zhu MX, Chen Z, Birnbaumer M, Mikoshiba K, Birnbaumer L (1999) Modulation of Ca^{2+} entry by polypeptides of the inositol 1,4,5-trisphosphate receptor (IP3R) that bind transient receptor potential (TRP): evidence for roles of TRP and IP3R in store depletion-activated Ca^{2+} entry. Proc Natl Acad Sci USA 96:14955–14960
100. Stamboulian S, Moutin M-J, Treves S, Pochon N, Grunwald D, Zorzato F, De Waard M, Ronjat M, Arnoult C (2005) Junctate, an inositol 1,4,5-triphosphate receptor associated protein, is present in rodent sperm and binds TRPC2 and TRPC5 but not TRPC1 channels. Dev Biol 286:326–337

101. Yuan JP, Kiselyov K, Shin DM, Chen J, Shcheynikov N, Kang SH, Dehoff MH, Schwarz MK, Seeburg PH, Muallem S, Worley PF (2003) Homer binds TRPC family channels and is required for gating of TRPC1 by IP3 receptors. Cell 114:777–789
102. Magistretti J, Ma L, Shalinsky MH, Lin W, Klink R, Alonso A (2004) Spike patterning by Ca^{2+}-dependent regulation of a muscarinic cation current in entorhinal cortex layer II neurons. J Neurophysiol 92:1644–1657
103. Blair NT, Kaczmarek JS, Clapham DE (2009) Intracellular calcium strongly potentiates agonist-activated TRPC5 channels. J Gen Physiol 133:525–546
104. Zhu M (2005) Multiple roles of calmodulin and other Ca^{2+}-binding proteins in the functional regulation of TRP channels. Pflügers Arch Eur J Physiol 451:105–115
105. Ambudkar IS (2006) Ca^{2+} signaling microdomains:platforms for the assembly and regulation of TRPC channels. Trends Pharmacol Sci 27:25–32
106. Hofmann T, Obukhov AG, Schaefer M, Harteneck C, Gudermann T, Schultz G (1999) Direct activation of human TRPC6 and TRPC3 channels by diacylglycerol. Nature 397:259–263
107. Hardie RC (2007) TRP channels and lipids: from drosophila to mammalian physiology. J Physiol 578:9–24
108. Venkatachalam K, Zheng F, Gill DL (2003) Regulation of canonical transient receptor potential (TRPC) channel function by diacylglycerol and protein kinase C. J Biol Chem 278:29031–29040
109. Birnbaum SG, Yuan PX, Wang M, Vijayraghavan S, Bloom AK, Davis DJ, Gobeske KT, Sweatt JD, Manji HK, Arnsten AFT, Protein Kinase C (2004) Overactivity impairs prefrontal cortical regulation of working memory. Science 306:882–884
110. Partridge LD, Valenzuela CF (2000) Block of hippocampal CAN channels by flufenamate. Brain Res 867:143–148
111. Wang D, Grillner S, Wallén P (2006) Effects of flufenamic acid on fictive locomotion, plateau potentials, calcium channels and NMDA receptors in the lamprey spinal cord. Neuropharmacology 51:1038–1046
112. Inoue R, Okada T, Onoue H, Hara Y, Shimizu S, Naitoh S, Ito Y, Mori Y (2001) The transient receptor potential protein homologue TRP6 is the essential component of vascular {{alpha}}1-adrenoceptor-activated Ca^{2+}-permeable cation channel. Circ Res 88:325–332
113. Iwasaki H, Mori Y, Hara Y, Uchida K, Zhou H, Mikoshiba K (2001) 2-Aminoethoxydiphenyl borate (2-APB) inhibits capacitative calcium entry independently of the function of inositol 1,4,5-trisphosphate receptors. Receptors Channels 7:429–439
114. Merritt JE, Armstrong WP, Benham CD, Hallam TJ, Jacob R, Jaxa-Chamiec A, Leigh SAM BK, Moores KE, Rink TJ (1990) SK&F 96365, a novel inhibitor of receptor-mediated calcium entry. Biochem J 271:515–522
115. Lievremont J-P, Bird GS, Putney JW (2005) Mechanism of inhibition of TRPC cation channels by 2-aminoethoxydiphenylborane. Mol Pharmacol 68:758–762
116. Shang-Zhong X, Fanning Z, Guylain B, Christian G, Christian H, David JB (2005) Block of TRPC5 channels by 2-aminoethoxydiphenyl borate: a differential, extracellular and voltage-dependent effect. Br J Pharmacol 145:405–414
117. Poburko D, Lhote P, Szado T, Behra T, Rahimian R, McManus B, van Breemen C, Ruegg UT (2004) Basal calcium entry in vascular smooth muscle. Eur J Pharmacol 505:19–29
118. Chung M-K, Lee H, Mizuno A, Suzuki M, Caterina MJ (2004) 2-aminoethoxydiphenyl borate activates and sensitizes the heat-gated ion channel TRPV3. J Neurosci 24:5177–5182
119. Wang Y, Deshpande M, Payne R (2002) 2-Aminoethoxydiphenyl borate inhibits phototransduction and blocks voltage-gated potassium channels in limulus ventral photoreceptors. Cell Calcium 32:209–216
120. Harks EGA, Camina JP, Peters PHJ, Ypey DL, Scheenen WJJM, van Zoelen EJJ, Theuvenet APR (2003) Besides affecting intracellular calcium signaling, 2-APB reversibly blocks gap junctional coupling in confluent monolayers, thereby allowing the measurement of single-cell membrane currents in undissociated cells. FASEB J 17(8):941–943

121. Faber ESL, Sedlak P, Vidovic M, Sah P (2006) Synaptic activation of transient receptor potential channels by metabotropic glutamate receptors in the lateral amygdala. Neuroscience 137:781–794
122. Haj-Dahmane S, Andrade R (1996) Muscarinic activation of a voltage-dependent cation nonselective current in rat association cortex. J Neurosci 16:3848–3861
123. Strübing C, Krapivinsky G, Krapivinsky L, Clapham DE (2001) TRPC1 and TRPC5 form a novel cation channel in mammalian brain. Neuron 29:645–655
124. Cobb SR, Davies CH (2005) Cholinergic modulation of hippocampal cells and circuits. J Physiol 562:81–88
125. Martín ED, Ceña V, Pozo MA (2005) Cholinergic modulation of status epilepticus in the rat barrel field region of primary somatosensory cortex. Exp Neurol 196:120–125
126. Nagao T, Alonso A, Avoli M (1996) Epileptiform activity induced by pilocarpine in the rat hippocampal-entorhinal slice preparation. Neuroscience 72:399–408
127. Sayin U, Rutecki PA (2003) Group I metabotropic glutamate receptor activation produces prolonged epileptiform neuronal synchronization and alters evoked population responses in the hippocampus. Epilepsy Res 53:186–195
128. Zhao W, Bianchi R, Wang M, Wong RKS (2004) Extracellular signal-regulated kinase 1/2 is required for the induction of group I metabotropic glutamate receptor-mediated epileptiform discharges. J Neurosci 24:76–84
129. Schiller Y (2004) Activation of a calcium-activated cation current during epileptiform discharges and its possible role in sustaining seizure-like events in neocortical slices. J Neurophysiol 92:862–872
130. Wang M, Bianchi R, Chuang S-C, Zhao W, Wong RKS (2007) Group I metabotropic glutamate receptor-dependent TRPC channel trafficking in hippocampal neurons. J Neurochem 101:411–421
131. Patapoutian A, Tate S, Woolf CJ (2009) Transient receptor potential channels: targeting pain at the source. Nat Rev Drug Discov 8:55–68
132. Russo RE, Nagy F, Hounsgaard J (1998) Inhibitory control of plateau properties in dorsal horn neurones in the turtle spinal cord in vitro. J Physiol 506:795–808
133. You H-J, Mørch CD, Chen J, Arendt-Nielsen L (2003) Role of central NMDA versus non-NMDA receptor in spinal withdrawal reflex in spinal anesthetized rats under normal and hyperexcitable conditions. Brain Res 981:12–22
134. Fossat P, Sibon I, Le Masson G, Landry M, Nagy F (2007) L-type calcium channels and NMDA receptors: a determinant duo for short-term nociceptive plasticity. Eur J Neurosci 25:127–135
135. Hornby TG, Rymer WZ, Benz EN, Schmit BD (2003) Windup of flexion reflexes in chronic human spinal cord injury: a marker for neuronal plateau potentials? J Neurophysiol 89:416–426
136. Li J, Simone DA, Larson AA (1999) Windup leads to characteristics of central sensitization. Pain 79:75–82
137. Woolf C (1996) Windup and central sensitization are not equivalent. Pain 66:105–108
138. Morisset V, Frédéric N (2000) Plateau potential-dependent windup of the response to primary afferent stimuli in rat dorsal horn neurons. Eur J Neurosci 12:3087–3095
139. Russo RE, Nagy F, Hounsgaard J (1997) Modulation of plateau properties in dorsal horn neurones in a slice preparation of the turtle spinal cord. J Physiol 499:459–474
140. Morisset V, Nagy F (1999) Ionic basis for plateau potentials in deep dorsal horn neurons of the rat spinal cord. J Neurosci 19:7309–7316
141. Gerzanich V, Woo S, Vennekens R, Tsymbalyuk O, Ivanova S, Ivanov A, Geng Z, Chen Z, Nilius B, Flockerzi V, Freichel M, Simard J (2009) De novo expression of Trpm4 initiates secondary hemorrhage in spinal cord injury. Nat Med 15:185–191
142. Xu S-Z, Zeng F, Lei M, Li J, Gao B, Xiong C, Sivaprasadarao A, Beech DJ (2005) Generation of functional ion-channel tools by E3 targeting. Nat Biotech 23:1289–1293

143. Davare MA, Fortin DA, Saneyoshi T, Nygaard S, Kaech S, Banker G, Soderling TR, Wayman GA (2009) Transient receptor potential canonical 5 channels activate Ca^{2+}/calmodulin kinase I{gamma} to promote axon formation in hippocampal neurons. J Neurosci 29:9794–9808
144. Amaral MD, Pozzo-Miller L (2007) TRPC3 channels are necessary for brain-derived neurotrophic factor to activate a nonselective cationic current and to induce dendritic spine formation. J Neurosci 27:5179–5189
145. Riccio A, Li Y, Moon J, Kim K-S, Smith KS, Rudolph U, Gapon S, Yao GL, Tsvetkov E, Rodig SJ, Van't Veer A, Meloni EG, Carlezon WA Jr, Bolshakov VY, Clapham DE (2009) Essential role for TRPC5 in amygdala function and fear-related behavior. Cell 137:761–772

Chapter 33
Role of TRP Channels in Pain Sensation

Man-Kyo Chung, Sung Jun Jung, and Seog Bae Oh

Abstract It is crucial for a living organism to recognize and discern potentially harmful noxious stimuli from innocuous stimuli to avoid hazards in the environment. However, unnecessary or exaggerated nociception is at best unpleasant and often compromises the quality of life. In order to lessen the intensity of nociception or eliminate the pathological pain, it is important to understand the nature of nociception and the mechanisms of hyperalgesia or allodynia. Transient receptor potential (TRP) channels play central roles in nociception under physiological and pathological conditions including inflammation and neuropathy. In this chapter, we will highlight the enormous progress in understanding the role of TRP channels in nociception. We will mainly focus on two TRP channels (TRPV1 and TRPA1) that have been particularly implicated in transducing signals associated with pain sensation, and briefly discuss the role of TRPM8, TRPV3 and TRPV4. We will stress debatable issues that needed to be resolved and provide perspectives for the future studies.

33.1 Introduction

Pain is defined as "an unpleasant sensory and emotional experience associated with actual or potential tissue damage". Therefore, pain is in fact helpful for the maintenance of life by alerting the organism to any potentially harmful stimuli. However, there is an urgent need for treatments for persistent and unnecessary pain. For the better manipulation of such persistent pathological pain, a detailed understanding of the mechanisms of neural processes of encoding and processing noxious stimuli at the molecular level – known as "nociception" – is crucial.

Exposure to capsaicin, the hot ingredient of capsicum peppers, induces an intense burning sensation and pain. Thus, capsaicin has long been used as a useful tool studying the mechanisms of pain by evoking pain behaviors or "nocifensive" responses in various experimental animal models. The painful sensation of the initial exposure to capsaicin is followed by desensitization of nociceptors and a reduction

S.B. Oh (✉)
Department of Neurobiology and Physiology, School of Dentistry Seoul National University, Seoul 110-749, Republic of Korea
e-mail: odolbae@snu.ac.kr

of pain sensation. This paradoxical effect of capsaicin blocking nociceptive stimuli has been adopted in various over-the-counter topical capsaicin creams [1–5]. Since the molecular mechanisms of capsaicin evoking both algesic and analgesic effects had not been known, the identity of the receptor for capsaicin was the subject of intensive study.

In 1997, Caterina et al. cloned the capsaicin receptor gene from rat dorsal root ganglion (DRG), which was first named vanilloid receptor subtype 1 (VR1) after the active vanilloid moiety of capsaicin [6]. The ground-breaking discovery of the capsaicin receptor has impacted the pain research field in many ways. First, the capsaicin receptor is literally "a pain receptor". The capsaicin receptor is primarily enriched in small to medium-sized sensory ganglia neurons, which are believed to serve as nociceptors. Since the receptor is a cationic ion channel, binding of capsaicin to its receptor directly excites nociceptors by inducing the influx of cations, including Ca^{2+}. Second, the capsaicin receptor is also activated by noxious heat. It was the first known mechanism transducing the changes in the thermal energy into an electrical signal in nociceptors, which provides a mechanistic basis of thermal pain. These findings opened a new era in pain research to discover the molecular mechanisms of thermosensation and pain. Third, the capsaicin receptor was the founding member of thermosensitive transient receptor potential (TRP) channels. TRP channels were first described in 1969 in mutant *Drosophila* photoreceptors, and were given their name because this mutant lacked sustained photoreceptor activity and showed only transient changes in receptor potential [7]. The capsaicin receptor VR-1 was found to have structure that was closely related to TRP channels [6] and was renamed as TRP vanilloid subtype 1 (TRPV1). Since the initial discovery of TRPV1, almost 30 members of mammalian TRP channels have been cloned and a tremendous amount of study has been dedicated to the vanilloid receptor family and other TRP channels.

Currently, mammalian TRP channels are classified into six subfamilies: TRPC (canonical), TRPV (vanilloid), TRPM (melastatin), TRPP (polycystin), TRPML (mucolipin), and TRPA (ankyrin) [8]. Although their amino acid sequences only share homology in part, they retain a common overall structure. TRP channels are composed of six transmembrane domains, with a pore-forming region between the fifth and sixth transmembrane domains and large amino and carboxyl termini on the cytosolic side. They are tetrameric channels in which four subunits form a functional channel [4, 9]. In this chapter, we will highlight the recent findings on the role of two TRP channels in particular (TRPV1 and TRPA1) that have been implicated in pain sensation, and briefly discuss TRPM8, TRPV3 and TRPV4.

33.2 TRPV1

33.2.1 Unequivocal Role in Heat Pain and Thermal Hyperalgesia

TRPV1 can be activated not only by capsaicin but also by noxious heat; its threshold temperature (>43°C) lies close to what humans perceive as painful.

TRPV1-null mice showed diminished nocifensive responses to acute thermal stimuli arguing for the role of TRPV1 in transducing heat pain in vivo [10–12]. Surprisingly, the mice also showed impaired thermal hypersensitivity under inflammatory conditions. Compared to their wild-type counterparts, the TRPV1-null mutant animals showed attenuated thermal hypersensitivity following the injection of Complete Freund's Adjuvant (CFA) [10] and carrageenan [11]. TRPV1 is also involved in the hyperalgesia following the injection of inflammatory mediators such as bradykinin, nerve growth factor (NGF), adenosine 5′-triphosphate (ATP), or protease ([10–11, 13–16]). These studies established TRPV1 as a critical mediator of thermal hyperalgesia under inflammation and injury.

Inflammation may modulate TRPV1 in nociceptors in a number of ways; for example phosphorylation of TRPV1 protein following the activation of various kinases. Detailed studies have shown that activation of protein kinase C (PKC), protein kinase A (PKA), protein kinase D, cyclin-dependent kinase 5, phosphoinositide-3-kinase and mitogen activated protein kinase are responsible for the phosphorylation of various residues of TRPV1 that enhances the function of the receptor [16–20]. Phosphorylation of TRPV1 enhances the functional competence of the receptor by increasing the affinity to capsaicin or by reducing the temperature threshold of activation so that TRPV1 can be activated at or near body temperature. Phosphorylation of TRPV1 also increases the trafficking of TRPV1 to the plasma membrane [21–23]. Besides the strong acute functional up-regulation of TRPV1, NGF also increases the synthesis of TRPV1 upon binding trkA receptor in nociceptors [21]. Inflammatory mediators may also enhance TRPV1 activity by reversing the inhibition of TRPV1 by phosphatidylinositol bis phosphate (PIP_2) [14, 24], or releasing arachidonic acid metabolites that have been shown to act as endogenous ligands [25, 26].

TRPV1 does not directly transduce mechanical stimuli and early studies reported lack of involvement of TRPV1 in mechanical hyperalgesia under inflammation. Caternia et al. reported that mechanical hyperalgesia in TRPV1$^{-/-}$ mice 1 day after injection of CFA was similar to wild-type mice [10]. The TRPV1-null mice also exhibited normal mechanical hyperalgesia to carrageenan injection [13, 27]. Moreover, when the central terminals of primary afferents expressing TRPV1 were ablated by intrathecal injection of capsaicin, only thermosensation and not mechanosensation was profoundly diminished, arguing for the role of TRPV1 – and the afferents expressing TRPV1 – in transmitting thermal rather than mechanical nociception [28]. However, many groups have shown the involvement of TRPV1 in mechanical hypersensitivity after CFA-induced inflammation, heat injury or nerve ligation [13, 29–33]. The reason for this discrepancy is not clear but it is possible that TRPV1 may be involved in cutaneous mechanical hyperalgesia through an indirect mechanism rather than as a sensor of noxious mechanical stimulation directly. The involvement of TRPV1 in mechanical hyperalgesia is more obvious in deep tissue pain model than cutaneous pain as discussed below.

33.2.2 Role of TRPV1 in Deep Tissue Pain

Although the role of TRPV1 was primarily studied in cutaneous pain models, it is evident that TRPV1 is involved in nociception not only in skin but also in deep tissues such as musculoskeletal and visceral tissues. Pathological conditions of deep tissues produce mainly mechanical, rather than thermal, hyperalgesia and the involvement of TRPV1 has been clearly shown in a number of deep tissue pain models. The mechanosensitivity of mouse colon afferent fibers and their sensitization by inflammatory mediators require TRPV1 [34–36]. Injection of capsaicin to masseter muscle evokes acute nocifensive responses and leads to the development of masseter hypersensitivity to mechanical stimulation [37]. Injection of carrageenan or eccentric exercise of gastrocnemius muscle induces mechanical hyperalgesia that can be suppressed by TRPV1 antagonists [38]. TRPV1 also plays a critical role in joint pain in an arthritis model. TRPV1 is expressed in sensory neurons that project to the knee or ankle [39]; after the induction of arthritis in the rat the number of neurons expressing TRPV1 increased [40]. Knee joint swelling and thermal hyperalgesia was attenuated in TRPV1-null mice compared to wild type following intra-articular injection of CFA [41]. Specific antagonists against TRPV1 decreased arthritis-induced pain as assessed by hind-limb weight bearing [42, 43]. TRPV1 is also implicated in pain derived from bone cancer. Movement-induced nocifensive behavior in a bone cancer model was attenuated by specific antagonists or genetic ablation of TRPV1 [44]. Experimental bone cancer increased the percentage of TRPV1-expressing neurons in DRG [45]; it also appears that the enhanced function of TRPV1 either by acidic conditions or by chemokines (such as CCL2) and cytokines (such as TNFalpha) [46, 47] can mediate cancer induced pain.

It is not yet clear how TRPV1 is engaged in the transduction of mechanical hyperalgesia in visceral or muscle tissues, though the role of TRPV1 may be different in nociceptors projecting to the skin compared with those projecting to deep tissues. It is possible that TRPV1's function as a downstream integrator of various pronociceptive/inflammatory intracellular signals [12] is more obvious in deep tissue nociceptors and that activation of TRPV1 in this manner could sensitize nociceptors to mechanical stimuli. The mechanisms underlying the involvement of TRPV1 in mechanical hyperalgesia in deep tissues as well as in skin under injured or inflamed condition need to be further explored in the future.

33.2.3 The Role of TRPV1 in Central Terminals of Primary Afferents and Central Nervous System (CNS)

The function of TRPV1 at the peripheral terminal of primary afferents as a transducer of thermal stimuli has been the main focus of research in the field. However, expression of TRPV1 has also been demonstrated in the spinal cord, mainly at the lamina I and II of the superficial dorsal horn area [48, 49]. Immunoreactivity for TRPV1 mostly disappears following dorsal rhizotomy [48, 50], suggesting that majority of TRPV1 detected at the spinal cord is derived from the central terminals of primary afferents. There is convincing evidence that central branches of primary afferents activation of TRPV1 triggers the release of the excitatory amino acid glutamate and evokes excitatory synaptic transmission in superficial dorsal horn [51–55]. Intrathecal injection of specific TRPV1 antagonists suppressed mechanical as well as thermal hyperalgesia following inflammation or nerve constriction injury in rats [56, 57]. Moreover, the strong correlation between central nervous stem (CNS) penetrability and therapeutic efficacy of TRPV1 antagonists implies that blockade of centrally located TRPV1 is also involved in the anti-nociceptive effects of TRPV1 antagonists. These reports strongly argue that the activation of TRPV1 in the spinal cord is involved in the pathological pain state following peripheral inflammation and nerve injury.

How then can TRPV1 in the spinal cord be activated? It is unlikely that TRPV1 in the spinal cord functions as a transducer of noxious heat since the temperature in spinal cord are unlikely to rise above 43°C. In the spinal cord, the activation of TRPV1 is more likely achieved by lipid agonists as suggested in the brain [58, 59]. In the hippocampus [58], glutamate released by high-frequency stimulation of presynaptic neurons activates postsynaptic mGlu1/5 receptors, which in turn generates 12-(S)-HPETE, an endogenous agonist of TRPV1 [25]. 12-(S)-HPETE is reported to cross the synapse to activate presynaptic TRPV1, leading to long-term depression by decreasing glutamate release from the presynaptic terminal [58]. However, it is not known whether the same mechanism is involved in the activation of TRPV1 in the spinal cord.

Recently, two groups of researchers proposed two novel endogenous ligands that may potentially activate TRPV1 in spinal cord. Kim et al. [60] suggested that TRPV1 is coupled to mGluR5 in the presynaptic terminals of nociceptive primary afferents in spinal cord. Diacylglycerol (DAG) is generated following mGluR5 activation and was shown to activate TRPV1 in a membrane-deliminated manner [61]. The subsequent Ca^{2+} influx via TRPV1 may further lead to glutamate release from central terminals, thereby contributing to the modulation of nociceptive synaptic transmission in the substantia gelatinosa neurons of the spinal cord. This line of reasoning suggests that, whereas TRPV1 is a thermal and chemical sensor in the peripheral nervous system, TRPV1 in the CNS might have a more important role as a modulator of multiple G-protein-coupled receptor signals by regulating Ca^{2+} influx [60].

Another study has suggested that, metabolites of linoleic acid act as endogenous lipid ligands of TRPV1 in the spinal cord [62]. These ligands were identified from the supernatant following the depolarization of spinal cord by high potassium. Intrathecal injection of metabolites of linoleic acid induced mechanical hyperalgesia in rats, which was reversed by intrathecal administration of a selective TRPV1 antagonist. These exciting findings suggest that TRPV1 not only at peripheral terminals but also at the central terminals of primary afferents in spinal cord is involved in pathological pain (Fig. 33.1). More detailed study into the mechanisms of how lipid ligands and TRPV1 are engaged in the generation of pathological pain needs to be followed in the future.

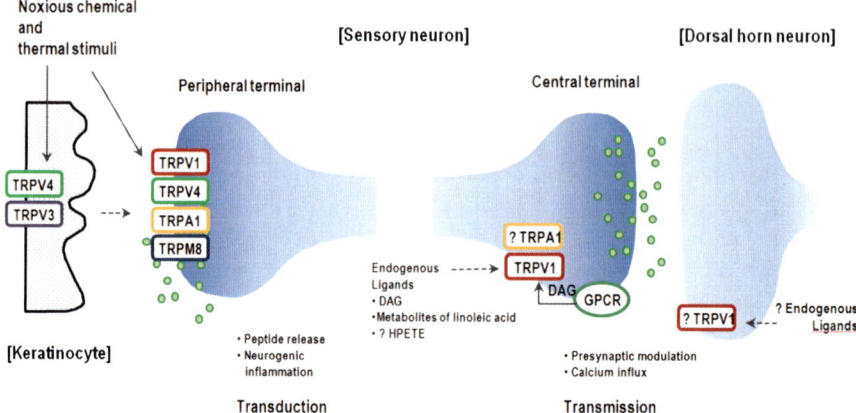

Fig. 33.1 Roles of thermosensitive TRP channels in pain. Noxious stimuli such as chemical or thermal stimuli directly activate TRP channels located in nociceptor terminals. Transduction of noxious stimuli by TRP channels generates electrical signals transmitted to the CNS and induces local release of neuropeptides producing neurogenic inflammation at peripheral tissues. TRPV3 and TRPV4 expressed in keratinocytes may also participate in transducing or modulating thermosensation and pain. TRP channels are also expressed at the central terminal of primary afferents where they modulate the release of excitatory neurotransmitter and thereby synaptic transmission. Endogenous ligands of TRP channels generated at the synaptic site may activate or enhance the activity of TRP channels at presynaptic terminals and postsynaptic neurons (modified from ref [4])

Besides the modulation of synaptic transmission at the presynaptic site, TRPV1 might also play a role in postsynaptic spinal cord neurons. Dorsal rhizotomy or resiniferatoxin (RTX)-induced degeneration of TRPV1-positive primary afferents failed to completely abolish the immunohistochemical staining of TRPV1 in spinal cord [50]. At the ultrastructural level, occasional expression of TRPV1 at postsynaptic dendrites was also reported in medullary dorsal horn neurons [63]. These reports suggest that TRPV1 is expressed not only in the central terminals of primary afferents but also in postsynaptic neurons in spinal and medullary dorsal horns. Currently it is an open question whether TRPV1 in postsynaptic site is involved in pain processing. Given the involvement of TRPV1 in long-term synaptic plasticity in hippocampus [58], it will be exciting to examine the potential role of TRPV1 in the pain modulation at the postsynaptic site (Fig. 33.1).

33.2.4 Therapeutic Approaches Targeting TRPV1

The discovery of TRPV1 and its obvious role in hyperalgesia triggered the development of novel therapeutic strategies to suppress nociception by targeting TRPV1, including the development of both agonists and antagonists of TRPV1 for the relief of pain. We will briefly discuss the rationale behind these apparently contradictory therapeutic mechanisms and also introduce a new approach utilizing TRPV1 as conduit for nociceptor-specific delivery of anesthetic molecules.

33.2.4.1 Agonists of TRPV1

Local or topical delivery of capsaicin has been used for centuries to suppress pain by desensitizing nociceptors so that subsequent noxious stimuli will not be perceived as painful. Recently, topical capsaicin application has been suggested to be effective in pain in various conditions including diabetic neuropathy, postherpetic neuralgia, and osteoarthritis [5]. However, the therapeutic application of capsaicin is limited since capsaicin is absorbed poorly by the skin and the anti-nociceptive effects are always preceded by burning sensation and pain. Newly developed analogues of capsaicin with reduced pungency in the form of patch or injectable preparation for enhancing absorbability are being tested in clinical trials [5, 64]. Another aspect of antinociceptive therapies using TRPV1 agonists is the selective ablatation of nociceptive neurons. Percutaneous, intraganglionic or intrathecal injection of resiniferatoxin can destruct cell bodies or central terminals of TRPV1-expressing neurons and, therefore, exert anti-nociceptive effects in experimental animals [65–67]. This approach may help to manage uncontrollable severe pain such as cancer-induced pain, although its efficacy and safety needs to be verified in clinical trials.

33.2.4.2 Selective Antagonists of TRPV1

A number of pharmaceutical companies have discovered small molecules that inhibit TRPV1 with high specificity, many of them showing strong efficacy in rodent models of inflammation and nerve injury [33, 42, 44, 57, 68]. Many drug candidates have progressed into clinical trials to evaluate their efficacy on various painful conditions such as dental pain, rectal pain, and migraine [64]. Although the majority of these studies are ongoing, early results from the GlaxoSmithKline developed-compound SB-705498 have shown potential therapeutic efficacy as a TRPV1 antagonist drug. SB-705498 significantly reduced capsaicin-induced flare and UV irradiation evoked hyperalgesia in healthy human subjects [69]. In order to target TRPV1 for obtaining anti-nociceptive and anti-hyperalgesic effects, it is critical to develop therapeutic agents that selectively inhibit "pathological" but not physiological activity of TRPV1. Although the role of TRPV1 for maintaining body temperature is not clearly understood, the suppression of TRPV1 by selective antagonism induced hyperthermia in experimental animals and humans [70, 71]. Along with the risk of burn injury due to the increased heat pain threshold [69], the side effect of hyperthermia is a significant barrier for the clinical application of TRPV1 antagonists. As part of an effort to develop fever-free TRPV1 antagonists, a plethora of chemicals are tested from the standpoint of modality-specific antagonism. Many TRPV1 antagonists show a differential mode of suppression of TRPV1 depending upon the type of agonist (e.g., capsaicin, proton or heat) [72]. Surprisingly, a group of antagonists which effectively block TRPV1 activation by protons show the strongest hyperthermic effects, providing a clue for developing modality-specific TRPV1 antagonists that selectively suppress pathological activity of TRPV1 only. In the future, it will be critical to understand how different modalities activate TRPV1, and how tonic or phasic activation of TRPV1 is controlled and related with functions at the systems level.

33.2.4.3 Selective Silencing of Nociceptors Through TRPV1

A new approach to analgesia targeting TRPV1 was published in 2007. Lidocaine is a widely used local anesthetic agent that works by blocking the activation of voltage-gated sodium channels. However, its use as a pain-killer is limited since lidocaine anesthetizes *all* nerve fibers, blocking signals conveying innocuous sensation, motor and autonomic control, as well as nociception. Binshtok et al. [73, 74] showed that selective silencing of nociceptive nerve fibers can be achieved by using QX314 – a hydrophilic polar derivative of lidocaine – when it is administered together with capsaicin. Upon receptor activation the pore of TRPV1 is wide enough to allow permeation of large cations; in fact, the permeability to large cations becomes greater upon strong activation of TRPV1 [75]. This phenomenon allows the delivery of membrane-impermeable nerve blocking agents such as QX314 into nociceptive neurons specifically through the pore of TRPV1.

Kim et al. demonstrated that this approach could have potential utility in selectively treating dental and orofacial pain in the trigeminal system [76]. By exploiting the jaw-opening reflex as a withdrawal reflex in the trigeminal system, and separately applying a capsaicin-QX-314 mixture either to a sensory nerve (the inferior alveolar nerve) or a motor nerve (the mylohyoid nerve), it was found that only the function of the sensory nerve was blocked (Fig. 33.2). From cellular-level experiments using whole-cell patch clamp recording, it was further shown that the blockade of sodium currents and action potentials are maintained throughout recordings and for up to at least 45 min after washout. The duration of the effect after washout of the capsaicin-QX-314 mixture suggests that the QX-314 remains in nociceptor neurons after its entry via TRPV1 channels. However, as TRPV1 is not expressed by trigeminal motor or trigeminal mesencephalic neurons QX-314 could not enter these cells; therefore, capsaicin plus QX-314 had no effect on rat trigeminal motor and proprioceptive mesencephalic neurons. These data provide additional strong evidence that selective block of pain signals can be achieved by co-application of QX-314 with TRPV1 agonists in the trigeminal system.

One drawback of this approach is the pain felt immediately following the injection of capsaicin. As an attempt to circumvent this problem, lidocaine was co-administered with the capsaicin-QX314 mixture in rats to alleviate the acute pain caused by capsaicin; the outcome was to prolong the analgesic effect and reduce acute pain [73]. Such an approach represents a new paradigm in the therapeutic targeting of nociceptive molecules. Since other nociceptive-specific channels, such as TRPA1 [77, 78] or P2X [79], are also permeable to large-sized cations, these channels could also be supplemental transporting routes for delivering QX314.

In summary, TRPV1 is a promising target for anti-nociceptive therapies in the field of pain research. Although we await a more detailed explanation, it is clear that TRPV1 is not only a transducer of noxious heat, but also a more complex gateway through which numerous factors and inflammatory mediators are integrated

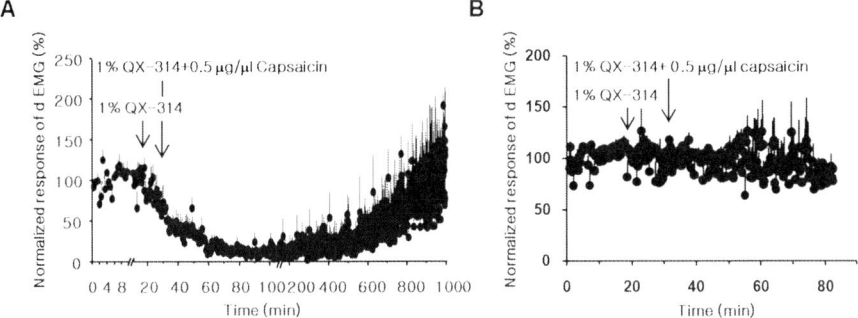

Fig. 33.2 Effects of QX-314 and capsaicin applied onto a sensory nerve (**a**) and motor nerve (**b**) tested with the jaw-opening reflex. QX-314 plus capsaicin were applied onto either the inferior alveolar nerve (the sensory nerve to the teeth) or mylohyoid nerve (the motor nerve innervating digastric muscle), while measuring dEMGs in response to electrical stimulus ($\times 2.5$ T). The normalized EMG decreased dramatically after the application of 1% QX-314 with 0.5 μg/μl capsaicin onto the inferior alveolar nerve (**a**) However, there was no significant change in dEMG amplitude after the application of 1% QX-314 with 0.5 μg/μl capsaicin onto the mylohyoid nerve (**b**) dEMG, digastric muscle electromyogram (modified from ref [76])

in the mechanisms of physiological and pathological pain. TRPV1 may also function as a modulator of synaptic plasticity in nociceptive circuits. Therefore, targeted suppression of TRPV1 at different sites could be an effective and selective analgesic strategy.

33.3 TRPA1

33.3.1 TRPA1 Senses Tissue Damages

The receptor TRPA1, formally known as ANKTM1 after its rich ankyrin repeat domains in the N-terminal domain, is activated by a remarkable variety of compounds known to induce pain or discomfort. Pungent cysteine-reactive chemicals, such as isothiocyanates (mustard oil), cinnamaldehyde (cinnamon), and allicin (garlic), activate TRPA1 through covalent modification of cysteine residues, regardless of their overall structures [4, 80, 81]. TRPA1 can also be activated by other chemicals that directly induce tissue injury, such as formalin [82]. Importantly, TRPA1 can be activated by many endogenous substances that are generated at the site of tissue injury and inflammation, for example the highly reactive aldehyde 4-hydroxy-2-nonenal (4-HNE), 15-deoxy-12,14-prostaglandin J_2 and reactive oxygen and nitrogen species [4, 83]. The activation of TRPA1 by these compounds directly excites nociceptors and thereby generates a warning signal in the organism to protect the body (Fig. 33.1).

There is growing evidence that the function and expression of TRPA1 is also modulated under inflammation or injury. TRPA1 is activated by the inflammatory mediator bradykinin [84]. Accordingly, two independently-generated lines of TRPA1 null mice showed reduced or impaired cellular and behavioral responses to bradykinin injection [85, 86]. Mechanistically, it is interesting that TRPV1 null mice also showed reduced responses to bradykinin, leaving the possibility of an interaction between TRPV1 and TRPA1 in response to bradykinin. For example, bradykinin could activate TRPV1 via phospholipase C (PLC) and the subsequent elevated intracellular Ca^{2+} may open TRPA1 [85]. Besides activation, bradykinin can also sensitize TRPA1 via PLC and PKA pathways [87]. The sensitization of TRPA1 by the activation of PLC is suggested to be mediated by the depletion of PIP_2 and a similar PLC-dependent mechanism is also involved in protease-activated receptor 2 (PAR-2)-mediated sensitization of TRPA1 [88]. However, it is still unclear how PKA sensitizes TRPA1. Recently, PLC/PKA activation in sensory neurons was shown to increase trafficking of TRPA1 to the plasma membrane and thereby sensitize nociceptors [89]. The implication of this novel mechanism of TRPA1 modulation needs to be further confirmed in inflammation and injury models.

33.3.2 The Role of TRPA1 in Cold Pain and Hyperalgesia

TRPA1 was initially found to be a cold-gated channel [90]; it is activated at temperatures below 17°C, which is lower than the temperature threshold of TRPM8 and close to the temperature known to initiate cold nociception. While TRPM8 is expressed in a different subset of neurons that do not express TRPV1, TRPA1 is specifically expressed in neurons with the nociceptive markers calcitonin gene-related peptide (CGRP) and substance P and is mostly co-expressed with TRPV1. Thus, it seemed reasonable to link TRPA1 to cold nociception. However, two groups reported that TRPA1 was irrelevant to noxious cold in vitro and in vivo [85, 91], leaving controversies as to whether TRPA1 was responsible for noxious cold sensation. Different age and sex of the animals and differences in experimental conditions might have led to mixed results in vivo [85, 86]. Moreover, a study showing that increased intracellular Ca^{2+} at cold temperature is sufficient to activate TRPA1 [92] opened up more questions about the gating of this receptor by cold stimuli. Recently, Karashima et al. clarified the cold sensitivity of TRPA1 with three sets of findings [1]: heterologously expressed TRPA1 is activated by cold in Ca^{2+}-independent manner [2] the TRPA1-null mice lacked a specific subset of cold-sensitive trigeminal ganglia (TG) neurons, and [3] the TRPA1-null mice showed impaired nocifensive behaviors to a cold plate at a temperature of 0°C [93].

Besides cold pain under normal condition, TRPA1 has also been implicated in cold hyperalgesia. Increased expression of TRPA1 is necessary to induce cold hyperalgesia after inflammation and spinal nerve ligation injury, and antisense knock-down of TRPA1 alleviates cold hyperalgesia [94–96]. However, the involvement of TRPA1 in cold allodynia following neuropathic injury is also challenged [97]. Therefore, the role of TRPA1 in cold pain and allodynia remains to be clarified.

33.3.3 The Role of TRPA1 in Mechanical Hyperalgesia

As in cold transduction, the role of TRPA1 in mechanical transduction is also controversial. Mice lacking TRPA1 do not exhibit an unequivocal deficit in noxious mechanosensation [85, 86]. TRPA1 null mice created by Kwan et al. showed a higher threshold and reduced response to suprathreshold mechanical stimuli [86], whereas the another study using an alternate TRPA1-null mouse reported no difference in threshold compared to wild-type mice [85]. *Ex vivo* skin-nerve recordings from TRPA1-null mice showed deficits in mechanical sensitivity [98]. The prolonged firing of mechanically-evoked action potentials in C-fibers, but not the initiation of firing, was inhibited by the TRPA1 antagonist HC-030031 in wild-type rats and mice; no effect of the antagonist was seen in TRPA1-null mice [99]. These results suggest the involvement of TRPA1 in mechanosensation although it is not clear whether TRPA1 functions directly as a mechanosensor.

TRPA1 may also be implicated in mechanical hyperalgesia under pathological conditions. CFA-induced mechanical hyperalgesia can be reversed by a specific TRPA1 inhibitor, AP18, in wild-type but not in TRPA1-deficient mice suggesting that TRPA1 is required for the maintenance of mechanical hyperalgesia under inflammatory conditions [100]. HC-030031 attenuates mechanical hyperalgesia following sciatic nerve ligation or CFA injection in rats [101]. Interestingly, the mechanical hyperalgesia caused by CFA was inhibited by peripheral and spinal injection of HC-030031 in mice [96]. Along with the expression of TRPA1 at the central terminals of primary afferents [102] and its modulatory role in synaptic transmission in the spinal cord [103], these results suggest the involvement of TRPA1 at central as well as peripheral terminals of primary afferents in mechanical hyperalgesia (Fig. 33.1). TRPA1 also mediates visceral pain. Behavioral responses to colonic distension were impaired in TRPA1-null mice and TRPA1 agonists caused mechanical hypersensitivity of the colon [104]. It will be important to clarify the involvement of TRPA1 in mechanical hyperalgesia for treating cutaneous and visceral mechanical hyperalgesia.

33.3.4 TRPA1 Induces Heterologous Desensitization of TRPV1

Besides pungent chemicals, TRPA1 can also be activated by cannabinoids such as Δ_9-tetrahydrocannabinol (THC) and WIN55215 [105, 106]. These compounds activate TRPA1 to mediate currents and Ca^{2+} influx in nociceptors. When these cannabinoid compounds are applied peripherally in vivo, however, they exert antinociceptive effects rather than algesic effects [107]. This seemingly paradoxical action of cannabinoids on TRPA1 and nociception turned out to be, at least in part, due to the heterologous desensitization of TRPV1 [105, 108]. Cannabinoids are weak agonists of TRPA1 with limited efficacy, and are not strong enough to induce firing of nociceptors directly. Instead, cannabinoid-induced activation of TRPA1 evokes modest influx of Ca^{2+} upon activation of TRPA1, which leads to the activation of calcineurin, a Ca^{2+}-dependent phosphatase. Activation of calcineurin is well known to desensitize TRPV1 by the dephosphorylation of the channel [109]. Since TRPA1 is mostly colocalized with TRPV1 in primary afferents, the activation of TRPA1 by cannabinoids could be an effective way to desensitize TRPV1 in nociceptors. This concept of heterologous desensitization of TRPV1 is therapeutically important because the attenuation of the function of TRPV1 can be achieved even without the preceding activation of TRPV1 and pungency. It will be exciting looking for any heterologous cross desensitization mechanisms among other TRP channels as well.

33.4 TRPM8

TRPM8 is a receptor of menthol, which evokes a cool sensation [110, 111]. The channel is activated by temperatures below 25°C and by various other chemicals including menthone, eucalyptol, spearmint, and icilin [112]. TRPM8 is mostly expressed in small-diameter DRG and TG neurons. However, unlike TRPV1, the expression of TRPM8 does not overlap with that of other common nociception-related markers [113], except for a 10–20% overlap with TRPV1 [113, 114]. Several studies of independently generated lines of TRPM8-null mice have clearly demonstrated deficient responses to moderately cool temperatures [115–117]. However, it is difficult to determine whether the TRPM8 receptor is involved in cold-evoked pain under normal conditions. Instead, TRPM8 may be involved in cold-evoked nocifensive responses under pathological conditions. In a chronic constriction injury (CCI) model, the percentage of sensory neurons expressing TRPM8-like immunoreactivity increased [118], as well as the percentage of sensory neurons showing responses to both menthol and capsaicin. Further, the amplitude of cold-evoked currents was enhanced in the same population of neurons cultured from rats with CCI [118]. When acetone was applied to the hind-paw of mice following CCI or injection of CFA, wild-type mice showed a robust paw licking behavior, while TRPM8-null mice were deficient in such behavior [117]. These reports argue for the involvement of TRPM8 in cold allodynia following CCI. However, recently the involvement of TRPM8 in cold allodynia was disputed [97] when TRPM8 mRNA was shown to decrease in rat sensory ganglia following spared nerve injury [97, 119]. Therefore, the role of TRPM8 in cold allodynia is still an open question and requires further investigation.

Interestingly, activation of TRPM8 by icilin or menthol elicits analgesia in several different pain models: neuropathic pain caused by CCI, inflammatory pain induced by CFA injection, and a peripheral demyelination model [120]. Cold temperatures exert an analgesic effect on the nocifensive behavior evoked by formalin, and the second phase of formalin-induced nocifensive behavior was impaired in TRPM8-null mice [115]. Although the mechanisms underlying such analgesic effects mediated by the activation of TRPM8 require further investigation, these findings provide the molecular rationale for applying moderately cool dressings to injured or inflamed wounds.

33.5 TRPV3 and TRPV4

TRPV3 and TRPV4 are activated at warm, ambient temperatures. TRPV3 is activated at 32–39°C and shows an increased response to higher temperature or repetitive stimuli. The TRPV3 channel protein shares 40–50% homology with TRPV1 and has certain common features such as activation and sensitization by camphor and 2-aminoethyl diphenylborinate (2-APB). Other agonists of TRPV3 include 6-tert-butyl-*m*-cresol, carvacrol, eugenol, thymol, menthol, and DPBA [83]. TRPV4 has a threshold of 27–34°C and can also be activated by hypo-osmotic conditions. TRPV4 has a variety of chemical ligands, such as phorbol esters 4-alpha-phorbol 12,13-didecanoate (4α-PDD) and phorbol 12-myristate 13-acetate (PMA), low pH, citrate, endocannabinoids, arachidonic acid metabolites, nitric oxide, and bisandrographolide A, the active compound of the Chinese herb *Andrographis paniculata* [83]. Both TRPV3 and TRPV4 are expressed most extensively in skin keratinocytes [121], providing the possibility that keratinocytes may signal to sensory neurons, instead of the direct transduction of thermal information by neurons (Fig. 33.1). TRPV3-deleted mice are deficient in responding to noxious thermal stimuli in tail immersion test and innocuous thermal selection behavior [122]. TRPV4-null mutants show partial defects in acute thermal nociception and altered thermal selection behavior [123]. TRPV4 knockout mice also show reduced thermal hyperalgesia following carrageenan-induced inflammation [124]. These reports argue that TRPV3 and TRPV4 expressed in keratinocytes may participate in thermosensation and nociception in vivo.

This argument is based upon the notion that keratinocytes communicate with sensory nerve terminal. Since no direct synapses between keratinocytes and sensory nerves have been reported, it is likely that soluble mediators are involved in such intercellular signaling. Recently, lines of transgenic mice overexpressing TRPV3 selectively in keratinocytes were generated [125]. These mice show enhanced release of prostaglandin E_2 (PGE$_2$) from keratinocytes in response to 2-APB and heat. The keratinocyte-specific TRPV3-overexpressing mice show increased escape responses to noxious heat. The enhanced thermal nociception and hyperalgesia in transgenic but not in naïve animals could be decreased by a cyclooxygenase inhibitor, ibuprofen. These results suggest that TRPV3 receptors in keratinocytes participate in thermal pain via the intercellular messenger PGE$_2$. In an independent study, ATP was suggested as a candidate molecule relaying the information from keratinocytes to sensory neurons [126] but the significance of such interaction in vivo remains to be verified.

TRPV4 is expressed in many tissues in which mechanosensitivity is critical, such as cochlear hair cells, vibrissal Merkel cells, sensory ganglia, and keratinocytes, as well as cutaneous A- and C-fiber terminals [12] The receptor can also be activated by subtle changes in osmolarity of as little as 30 milliosmoles. The distribution of TRPV4 suggests its involvement in a more general role in mechanosensation, rather than being limited to osmosensation, and may perform such a function in nociceptors as well as in keratinocytes (Fig. 33.1). Agonists of TRPV4 promote the release of substance P and CGRP from central terminals of primary afferent neurons, which suggests that TRPV4 is involved in nociception [127]. Studies conducted in TRPV4-null mice have revealed that TRPV4 is related to the development of acute inflammatory mechanical hyperalgesia [12, 128]. Both PKA and PKCε can engage TRPV4 in hyperalgesia to mechanical and osmotic changes. This process requires a certain amount of cyclic adenosine monophosphate (cAMP) produced by the concerted action of inflammatory mediators. TRPV4-null mutant mice showed marked reduction in C-fiber sensitization for mechanical and hypotonic stimuli induced by inflammatory mediators [129].

Chemotherapy-induced neuropathy can be modeled in rats by the mitotic inhibitor paclitaxel, a drug that is generally used for cancer treatment. Administration of TRPV4 antisense oligodeoxynucleotides reversed paclitaxel-induced mechanical hyperalgesia and reduced osmotic hyperalgesia [130]. Several subsequent studies discovered that the paclitaxel-induced TRPV4-mediated hyperalgesia depends on the integrin/Src tyrosine kinase pathway [131], along with PKCε and PKA [132]. Mechanical allodynia induced by direct activation of PKA or PKCε is absent in mice lacking TRPV4 and is reduced in mice treated with TRPV4 antisense oligodeoxynucleotides [12]. These results demonstrate that TRPV4 is involved in osmotic and mechanical hyperalgesia in rats under various pathological conditions.

TRPV4 is also implicated in visceral nociception. 4α-PDD-induced activation of TRPV4 in the colon caused a nociceptive response to colorectal distension in basal conditions and in PAR-2 agonist-induced hypersensitivity [133]. TRPV4 agonism increased the mechanosensory responses of colonic afferents, which was reduced in TRPV4-null mice [134]. TRPV4 and TRPV1 are co-expressed in certain DRG neurons and TRPV4 can be sensitized by PKC in DRG neurons. The co-expression of TRPV1 and TRPV4 in a subset of DRG neurons and the similar manner in which these to ion channels modulate synaptic transmission and are sensitized by PKC suggests they may play a synergistic role in nociception [135]. The above reports on the role of TRPV4 in cutaneous and visceral hyperalgesia suggest that TRPV4, along with TRPV1, may function as a molecular integrator of inflammatory signaling.

33.6 Perspectives

Since the first cloning of TRPV1 in 1997 there has been tremendous progress in our understanding for the role of TRP channels as molecular pain sensors. Two thermosensitive TRP channels in particular, TRPV1 and TRPA1, play critical roles in processing pain information. TRPV1 has definite roles for thermal nociception and thermal hyperalgesia under pathological conditions. However, its role in mechanical hyperalgesia is still debatable. TRPA1 is involved in sensing the condition of tissue damage and generating warning signals. Besides the role of TRPA1 in cold hyperalgesia, evidence of its role in mechanical hyperalgesia is accruing. It is certain that TRPV1 and TRPA1 are reliable therapeutic targets for treating pathological pain, and it is necessary to further develop pharmacological tools to selectively inhibit TRPV1 and TRPA1 without adverse side effects. From a therapeutic standpoint it is interesting that the activation of TRPM8 could elicit analgesia, providing a basis for the well known observation that a cool dressing ameliorates inflammatory pain. Further work is needed, however, to elucidate the exact role of TRPM8, as well as TPRV3/TRPV4, in nociception and pain hypersensitivity. Recent findings show that other TRP channels such as the TRPC family are widely expressed in sensory neurons [136]. It is highly likely that TRP channels other than thermo-TRP channels turn out to be involved in pain sensation in the near future. Another interesting issue is the role of TRP channels, especially TRPV1, expressed in the CNS. A better understanding in the role of TRP channels in the CNS will help to predict and screen for possible side effects and understand their possible roles as modulators of neuronal Ca^{2+} signalling and pain processing.

Acknowledgments This work was supported by a grant (R0A-2008-000-20101-0) from the National Research Laboratory Program and a grant (2009K001256) from Brain Research Center of the twenty-first century Frontier Research Program, funded by the Ministry of Education, Science and Technology, the Republic of Korea. We thank Gehoon Chung and Alexander J. Davies for helpful comments and English correction in this manuscript.

References

1. Broad LM, Mogg AJ, Beattie RE, Ogden AM, Blanco MJ, Bleakman D (2009) TRP channels as emerging targets for pain therapeutics. Expert Opin Ther Targets 13:69–81
2. Lazar J, Gharat L, Khairathkar-Joshi N, Blumberg PM, Szallasi A (2009) Screening TRPV1 antagonists for the treatment of pain: lessons learned over a decade. Expert Opin Drug Dis 4(2):159–180
3. Szallasi A, Blumberg PM (1999) Vanilloid (Capsaicin) receptors and mechanisms. Pharmacol Rev 51:159–212
4. Patapoutian A, Tate S, Woolf CJ (2009) Transient receptor potential channels: targeting pain at the source. Nat Rev Drug Discov 8:55–68
5. Knotkova H, Pappagallo M, Szallasi A (2008) Capsaicin (TRPV1 Agonist) therapy for pain relief: farewell or revival? Clin J Pain 24:142–154
6. Caterina MJ, Schumacher MA, Tominaga M, Rosen TA, Levine JD, Julius D (1997) The capsaicin receptor: a heat-activated ion channel in the pain pathway. Nature 389:816–824
7. Cosens DJ, Manning A (1969) Abnormal electroretinogram from a Drosophila mutant. Nature 224:285–287
8. Caterina MJ (2007) Transient receptor potential ion channels as participants in thermosensation and thermoregulation. Am J Physiol Regul Integr Comp Physiol 292:R64–R76
9. Clapham DE (2003) TRP channels as cellular sensors. Nature 426:517–524
10. Caterina MJ, Leffler A, Malmberg AB, Martin WJ, Trafton J, Petersen-Zeitz KR, Koltzenburg M, Basbaum AI, Julius D (2000) Impaired nociception and pain sensation in mice lacking the capsaicin receptor. Science 288(5464):306–313
11. Davis JB, Gray J, Gunthorpe MJ, Hatcher JP, Davey PT, Overend P, Harries MH, Latcham J, Clapham C, Atkinson K, Hughes SA, Rance K, Grau E, Harper AJ, Pugh PL, Rogers DC, Bingham S, Randall A, Sheardown SA (2000) Vanilloid receptor-1 is essential for inflammatory thermal hyperalgesia. Nature 405:183–187
12. Levine JD, Alessandri-Haber N (2007) TRP channels: targets for the relief of pain. Biochim Biophys Acta 1772:989–1003

13. Bölcskei K, Helyes Z, Szabó A, Sándor K, Elekes K, Németh J, Almási R, Pintér E, Petho G, Szolcsányi J (2005) Investigation of the role of TRPV1 receptors in acute and chronic nociceptive processes using gene-deficient mice. Pain 117:368–376
14. Chuang HH, Prescott ED, Kong H, Shields S, Jordt SE, Basbaum AI, Chao MV, Julius D (2001) Bradykinin and nerve growth factor release the capsaicin receptor from PtdIns(4,5)P2-mediated inhibition. Nature 411:957–962
15. Moriyama T, Iida T, Kobayashi K, Higashi T, Fukuoka T, Tsumura H, Leon C, Suzuki N, Inoue K, Gachet C, Noguchi K, Tominaga M (2003) Possible involvement of P2Y2 metabotropic receptors in ATP-induced transient receptor potential vanilloid receptor 1-mediated thermal hypersensitivity. J Neurosci 23:6058–6062
16. Amadesi S, Cottrell GS, Divino L, Chapman K, Grady EF, Bautista F, Karanjia R, Barajas-Lopez C, Vanner S, Vergnolle N, Bunnett NW (2006) Protease-activated receptor 2 sensitizes TRPV1 by protein kinase Cepsilon- and A-dependent mechanisms in rats and mice. J Physiol 575:555–571
17. Pareek TK, Keller J, Kesavapany S, Agarwal N, Kuner R, Pant HC, Iadarola MJ, Brady RO, Kulkarni AB (2007) Cyclin-dependent kinase 5 modulates nociceptive signaling through direct phosphorylation of transient receptor potential vanilloid 1. Proc Natl Acad Sci USA 104:660–665
18. Zhu W, Oxford GS (2007) Phosphoinositide-3-kinase and mitogen activated protein kinase signaling pathways mediate acute NGF sensitization of TRPV1. Mol Cell Neurosci 34:689–700
19. Zhuang Z-Y, Xu H, Clapham DE, Ji R-R (2004) Phosphatidylinositol 3-kinase activates ERK in primary sensory neurons and mediates inflammatory heat hyperalgesia through TRPV1 sensitization. J Neurosci 24:8300–8309
20. Amadesi S, Grant AD, Cottrell GS, Vaksman N, Poole DP, Rozengurt E, Bunnett NW (2009) Protein kinase D isoforms are expressed in rat and mouse primary sensory neurons and are activated by agonists of protease-activated receptor 2. J Comp Neurol 516:141–156
21. Ji RR, Samad TA, Jin SX, Schmoll R, Woolf CJ (2002) p38 MAPK activation by NGF in primary sensory neurons after inflammation increases TRPV1 levels and maintains heat hyperalgesia. Neuron 36:57–68
22. Zhang X, Huang J, McNaughton PA (2005) NGF rapidly increases membrane expression of TRPV1 heat-gated ion channels. EMBO J 24:4211–4223
23. Zhang X, Li L, McNaughton PA (2008) Proinflammatory mediators modulate the heat-activated ion channel TRPV1 via the scaffolding protein AKAP79/150. Neuron 59:450–461
24. Bhave G, Gereau RW (2004) Posttranslational mechanisms of peripheral sensitization. J Neurobiol 61:88–106
25. Hwang SW, Cho H, Kwak J, Lee SY, Kang CJ, Jung J, Cho S, Min KH, Suh YG, Kim D, Oh U (2000) Direct activation of capsaicin receptors by products of lipoxygenases: endogenous capsaicin-like substances. Proc Natl Acad Sci USA 97:6155–6160
26. Shin J, Cho H, Hwang SW, Jung J, Shin CY, Lee SY, Kim SH, Lee MG, Choi YH, Kim J, Haber NA, Reichling DB, Khasar S, Levine JD, Oh U (2002) Bradykinin-12-lipoxygenase-VR1 signaling pathway for inflammatory hyperalgesia. Proc Natl Acad Sci USA 99:10150–10155
27. Pogatzki-Zahn EM, Shimizu I, Caterina M, Raja SN (2005) Heat hyperalgesia after incision requires TRPV1 and is distinct from pure inflammatory pain. Pain 115:296–307
28. Cavanaugh DJ, Lee H, Lo L, Shields SD, Zylka MJ, Basbaum AI, Anderson DJ (2009) Distinct subsets of unmyelinated primary sensory fibers mediate behavioral responses to noxious thermal and mechanical stimuli. Proc Natl Acad Sci USA 106:9075–9080
29. Szabó A, Helyes Z, Sándor K, Bite A, Pintér E, Németh J, Bánvölgyi A, Bölcskei K, Elekes K, Szolcsányi J (2005) Role of transient receptor potential vanilloid 1 receptors in adjuvant-induced chronic arthritis: in vivo study using gene-deficient mice. J Pharmacol Exp Ther 314:111–119

30. Christoph T, Gillen C, Mika J, Grünweller A, Schäfer MK, Schiene K, Frank R, Jostock R, Bahrenberg G, Weihe E, Erdmann VA, Kurreck J (2007) Antinociceptive effect of antisense oligonucleotides against the vanilloid receptor VR1/TRPV1. Neurochem Int 50: 281–290

31. Honore P, Wismer CT, Mikusa J, Zhu CZ, Zhong C, Gauvin DM, Gomtsyan A, El Kouhen R, Lee C-H, Marsh K, Sullivan JP, Faltynek CR, Jarvis MF (2005) A-425619 [1-isoquinolin-5-yl-3-(4-trifluoromethyl-benzyl)-urea], a novel transient receptor potential type V1 receptor antagonist, relieves pathophysiological pain associated with inflammation and tissue injury in rats. J Pharmacol Exp Ther 314:410–421

32. Gavva NR, Tamir R, Qu Y, Klionsky L, Zhang TJ, Immke D, Wang J, Zhu D, Vanderah TW, Porreca F, Doherty EM, Norman MH, Wild KD, Bannon AW, Louis J-C, Treanor JJS (2005) AMG9810 [(E)-3-(4-t-butylphenyl)-N-(2,3-dihydrobenzo[b][1,4] dioxin-6-yl)acrylamide], a novel vanilloid receptor 1 (TRPV1) antagonist with antihyperalgesic properties. J Pharmacol Exp Ther 313:474–484

33. Pomonis JD, Harrison JE, Mark L, Bristol DR, Valenzano KJ, Walker K (2003) (N-4-Tertiarybutylphenyl)-4-(3-cholorphyridin-2-yl)tetrahydropyrazine-1(2H)-carbox-amide (BCTC), a novel, orally effective vanilloid receptor 1 antagonist with analgesic properties: II. in vivo characterization in rat models of inflammatory and neuropathic pain. J Pharmacol Exp Ther 306:387–393

34. Jones RC, Xu L, Gebhart GF (2005) The mechanosensitivity of mouse colon afferent fibers and their sensitization by inflammatory mediators require transient receptor potential vanilloid 1 and acid-sensing ion channel 3. J Neurosci 25:10981–10989

35. Miranda A, Nordstrom E, Mannem A, Smith C, Banerjee B, Sengupta JN (2007) The role of transient receptor potential vanilloid 1 in mechanical and chemical visceral hyperalgesia following experimental colitis. Neuroscience 148:1021–1032

36. Ravnefjord A, Brusberg M, Kang D, Bauer U, Larsson H, Lindström E, Martinez V (2009) Involvement of the transient receptor potential vanilloid 1 (TRPV1) in the development of acute visceral hyperalgesia during colorectal distension in rats. Eur J Pharmacol 611: 85–91

37. Ro JY, Lee JS, Zhang Y (2009) Activation of TRPV1 and TRPA1 leads to muscle nociception and mechanical hyperalgesia. Pain 144:270–277

38. Fujji Y, Ozaki N, Taguchi T, Mizumura K, Furukawa K, Sugiura Y (2008) TRP channels and ASICs mediate mechanical hyperalgesia in models of inflammatory muscle pain and delayed onset muscle soreness. Pain 140:292–304

39. Cho WG, Valtschanoff JG (2008) Vanilloid receptor TRPV1-positive sensory afferents in the mouse ankle and knee joints. Brain Res 1219:59–65

40. Fernihough J, Gentry C, Bevan S, Winter J (2005) Regulation of calcitonin gene-related peptide and TRPV1 in a rat model of osteoarthritis. Neurosci Lett 388:75–80

41. Keeble J, Russell F, Curtis B, Starr A, Pinter E, Brain SD (2005) Involvement of transient receptor potential vanilloid 1 in the vascular and hyperalgesic components of joint inflammation. Arthritis Rheum 52:3248–3256

42. Honore P, Wismer CT, Mikusa J, Zhu CZ, Zhong C, Gauvin DM, Gomtsyan A, El Kouhen R, Lee CH, Marsh K, Sullivan JP, Faltynek CR, Jarvis MF (2005) A-425619 [1-isoquinolin-5-yl-3-(4-trifluoromethyl-benzyl)-urea], a novel transient receptor potential type V1 receptor antagonist, relieves pathophysiological pain associated with inflammation and tissue injury in rats. J Pharmacol Exp Ther 314:410–421

43. Honore P, Chandran P, Hernandez G, Gauvin DM, Mikusa JP, Zhong C, Joshi SK, Ghilardi JR, Sevcik MA, Fryer RM, Segreti JA, Banfor PN, Marsh K, Neelands T, Bayburt E, Daanen JF, Gomtsyan A, Lee CH, Kort ME, Reilly RM, Surowy CS, Kym PR, Mantyh PW, Sullivan JP, Jarvis MF, Faltynek CR (2009) Repeated dosing of ABT-102, a potent and selective TRPV1 antagonist, enhances TRPV1-mediated analgesic activity in rodents, but attenuates antagonist-induced hyperthermia. Pain 142:27–35

44. Ghilardi JR, Rohrich H, Lindsay TH, Sevcik MA, Schwei MJ, Kubota K, Halvorson KG, Poblete J, Chaplan SR, Dubin AE, Carruthers NI, Swanson D, Kuskowski M, Flores CM, Julius D, Mantyh PW (2005) Selective blockade of the capsaicin receptor TRPV1 attenuates bone cancer pain. J Neurosci 25:3126–3131
45. Niiyama Y, Kawamata T, Yamamoto J, Omote K, Namiki A (2007) Bone cancer increases transient receptor potential vanilloid subfamily 1 expression within distinct subpopulations of dorsal root ganglion neurons. Neuroscience 148:560–572
46. Khasabova IA, Stucky CL, Harding-Rose C, Eikmeier L, Beitz AJ, Coicou LG, Hanson AE, Simone DA, Seybold VS (2007) Chemical interactions between fibrosarcoma cancer cells and sensory neurons contribute to cancer pain. J Neurosci 27:10289–10298
47. Constantin CE, Mair N, Sailer CA, Andratsch M, Xu ZZ, Blumer MJ, Scherbakov N, Davis JB, Bluethmann H, Ji RR, Kress M (2008) Endogenous tumor necrosis factor alpha (TNFalpha) requires TNF receptor type 2 to generate heat hyperalgesia in a mouse cancer model. J Neurosci 28:5072–5081
48. Guo A, Vulchanova L, Wang J, Li X, Elde R (1999) Immunocytochemical localization of the vanilloid receptor 1 (VR1): relationship to neuropeptides, the P2X3 purinoceptor and IB4 binding sites. Eur J Neurosci 11:946–958
49. Valtschanoff JG, Rustioni A, Guo A, Hwang SJ (2001) Vanilloid receptor VR1 is both presynaptic and postsynaptic in the superficial laminae of the rat dorsal horn. J Comp Neurol 436:225–235
50. Zhou H-Y, Chen S-R, Chen H, Pan H-L (2009) The glutamatergic nature of TRPV1-expressing neurons in the spinal dorsal horn. J Neurochem 108:305–318
51. Sikand P, Premkumar LS (2007) Potentiation of glutamatergic synaptic transmission by protein kinase C-mediated sensitization of TRPV1 at the first sensory synapse. J Physiol 581:631–647
52. Pan Y-Z, Pan H-L (2004) Primary afferent stimulation differentially potentiates excitatory and inhibitory inputs to spinal lamina II outer and inner neurons. J Neurophysiol 91:2413–2421
53. Yang K, Kumamoto E, Furue H, Li YQ, Yoshimura M (1999) Action of capsaicin on dorsal root-evoked synaptic transmission to substantia gelatinosa neurons in adult rat spinal cord slices. Brain Res 830:268–273
54. Yang K, Kumamoto E, Furue H, Yoshimura M (1998) Capsaicin facilitates excitatory but not inhibitory synaptic transmission in substantia gelatinosa of the rat spinal cord. Neurosci Lett 255:135–138
55. Lappin SC, Randall AD, Gunthorpe MJ, Morisset V (2006) TRPV1 antagonist, SB-366791, inhibits glutamatergic synaptic transmission in rat spinal dorsal horn following peripheral inflammation. Eur J Pharmacol 540:73–81
56. Kanai Y, Nakazato E, Fujiuchi A, Hara T, Imai A (2005) Involvement of an increased spinal TRPV1 sensitization through its up-regulation in mechanical allodynia of CCI rats. Neuropharmacology 49:977–984
57. Cui M, Honore P, Zhong C, Gauvin D, Mikusa J, Hernandez G, Chandran P, Gomtsyan A, Brown B, Bayburt EK, Marsh K, Bianchi B, McDonald H, Niforatos W, Neelands TR, Moreland RB, Decker MW, Lee CH, Sullivan JP, Faltynek CR (2006) TRPV1 receptors in the CNS play a key role in broad-spectrum analgesia of TRPV1 antagonists. J Neurosci 26:9385–9393
58. Gibson HE, Edwards JG, Page RS, Van Hook MJ, Kauer JA (2008) TRPV1 channels mediate long-term depression at synapses on hippocampal interneurons. Neuron 57:746–759
59. Kauer JA, Gibson HE (2009) Hot flash: TRPV channels in the brain. Trends Neurosci 32:215–224
60. Kim YH, Park C, Back SK, Lee CJ, Hwang S, Bae Y, Na HS, Kim JS, Jung SJ, Oh SB (2009) Membrane-delimited coupling of TRPV1 and mGluR5 on presynaptic terminals of nociceptive neurons. J Neurosci 29:10000–10009
61. Woo DH, Jung SJ, Zhu MH, Park C, Kim YH, Oh SB, Lee CJ (2008) Direct activation of transient receptor potential vanilloid 1(TRPV1) by diacylglycerol (DAG). Mol Pain 4:42

62. Patwardhan AM, Scotland PE, Akopian AN, Hargreaves KM (2009) Activation of TRPV1 in the spinal cord by oxidized linoleic acid metabolites contributes to inflammatory hyperalgesia. Proc Natl Acad Sci USA 106:18820–18824
63. Bae Y, Oh J, Hwang S, Shigenaga Y, Valtschanoff J (2004) Expression of vanilloid receptor TRPV1 in the rat trigeminal sensory nuclei. J Comp Neurol 478:62–71
64. Wong GY, Gavva NR (2009) Therapeutic potential of vanilloid receptor TRPV1 agonists and antagonists as analgesics: recent advances and setbacks. Brain Res Rev 60:267–277
65. Karai L, Brown DC, Mannes AJ, Connelly ST, Brown J, Gandal M, Wellisch OM, Neubert JK, Olah Z, Iadarola MJ (2004) Deletion of vanilloid receptor 1-expressing primary afferent neurons for pain control. J Clin Invest 113:1344–1352
66. Jeffry JA, Yu SQ, Sikand P, Parihar A, Evans MS, Premkumar LS (2009) Selective targeting of TRPV1 expressing sensory nerve terminals in the spinal cord for long lasting analgesia. PLoS One 4:e7021
67. Neubert JK, Mannes AJ, Karai LJ, Jenkins AC, Zawatski L, Abu-Asab M, Iadarola MJ (2008) Perineural resiniferatoxin selectively inhibits inflammatory hyperalgesia. Mol Pain 4:3
68. Gavva NR, Tamir R, Qu Y, Klionsky L, Zhang TJ, Immke D, Wang J, Zhu D, Vanderah TW, Porreca F, Doherty EM, Norman MH, Wild KD, Bannon AW, Louis JC, Treanor JJ (2005) AMG9810 [(E)-3-(4-t-butylphenyl)-N-(2,3-dihydrobenzo[b][1,4] dioxin-6-yl)acrylamide], a novel vanilloid receptor 1 (TRPV1) antagonist with antihyperalgesic properties. J Pharmacol Exp Ther 313:474–484
69. Chizh BA, O'Donnell MB, Napolitano A, Wang J, Brooke AC, Aylott MC, Bullman JN, Gray EJ, Lai RY, Williams PM, Appleby JM (2007) The effects of the TRPV1 antagonist SB-705498 on TRPV1 receptor-mediated activity and inflammatory hyperalgesia in humans. Pain 132:132–141
70. Gavva NR (2008) Body-temperature maintenance as the predominant function of the vanilloid receptor TRPV1. Trends Pharmacol Sci 29:550–557
71. Gavva NR, Bannon AW, Surapaneni S, Hovland DN, Lehto SG, Gore A, Juan T, Deng H, Han B, Klionsky L, Kuang R, Le A, Tamir R, Wang J, Youngblood B, Zhu D, Norman MH, Magal E, Treanor JJ, Louis JC (2007) The vanilloid receptor TRPV1 is tonically activated in vivo and involved in body temperature regulation. J Neurosci 27:3366–3374
72. Garami A, Shimansky YP, Pakai E, Oliveira DL, Gavva NR, Romanovsky AA (2010) Contributions of different modes of TRPV1 activation to TRPV1 antagonist-induced hyperthermia. J Neurosci 30:1435–1440
73. Binshtok AM, Gerner P, Oh SB, Puopolo M, Suzuki S, Roberson DP, Herbert T, Wang C-F, Kim D, Chung G, Mitani AA, Wang GK, Bean BP, Woolf CJ (2009) Coapplication of lidocaine and the permanently charged sodium channel blocker QX-314 produces a long-lasting nociceptive blockade in rodents. Anesthesiology 111:127–137
74. Binshtok AM, Bean BP, Woolf CJ (2007) Inhibition of nociceptors by TRPV1-mediated entry of impermeant sodium channel blockers. Nature 449:607–610
75. Chung M-K, Güler AD, Caterina MJ (2008) TRPV1 shows dynamic ionic selectivity during agonist stimulation. Nat Neurosci 11:555–564
76. Kim HY, Kim K, Li HY, Chung G, Park C-K, Kim JS, Jung SJ, Lee MK, Ahn DK, Hwang SJ, Kang Y, Binshtok AM, Bean BP, Woolf CJ, Oh SB (2010) Selectively targeting pain in the trigeminal system. Pain 150(1):29–40
77. Karashima Y, Prenen J, Talavera K, Janssens A, Voets T, Nilius B (2010) Agonist-induced changes in Ca^{2+} permeation through the nociceptor cation channel TRPA1. Biophys J 98:773–783
78. Chen J, Kim D, Bianchi BR, Cavanaugh EJ, Faltynek CR, Kym PR, Reilly RM (2009) Pore dilation occurs in TRPA1 but not in TRPM8 channels. Mol Pain 5:3
79. Virginio C, MacKenzie A, Rassendren FA, North RA, Surprenant A (1999) Pore dilation of neuronal P2X receptor channels. Nat Neurosci 2:315–321

80. Macpherson LJ, Dubin AE, Evans MJ, Marr F, Schultz PG, Cravatt BF, Patapoutian A (2007) Noxious compounds activate TRPA1 ion channels through covalent modification of cysteines. Nature 445:541–545
81. Hinman A, Chuang HH, Bautista DM, Julius D (2006) TRP channel activation by reversible covalent modification. Proc Natl Acad Sci USA 103:19564–19568
82. McNamara CR, Mandel-Brehm J, Bautista DM, Siemenst J, Deranian KL, Zhao M, Hayward NJ, Chong JA, Julius D, Moran MM, Fanger CM (2007) TRPA1 mediates formalin-induced pain. Proc Natl Acad Sci USA 104:13525–13530
83. Alexander SP, Mathie A, Peters JA (2008) Guide to receptors and channels (GRAC). Br J Pharmacol 153(Suppl 2):S1-S209 3rd edition
84. Bandell M, Story GM, Hwang SW, Viswanath V, Eid SR, Petrus MJ, Earley TJ, Patapoutian A (2004) Noxious cold ion channel TRPA1 is activated by pungent compounds and bradykinin. Neuron 41:849–857
85. Bautista DM, Jordt SE, Nikai T, Tsuruda PR, Read AJ, Poblete J, Yamoah EN, Basbaum AI, Julius D (2006) TRPA1 mediates the inflammatory actions of environmental irritants and proalgesic agents. Cell 124:1269–1282
86. Kwan KY, Allchorne AJ, Vollrath MA, Christensen A, Zhang DS, Woolf CJ, Corey D (2006) TRPA1 contributes to cold, mechanical, and chemical nociception but is not essential for hair-cell transduction. Neuron 50:277–289
87. Wang S, Dai Y, Fukuoka T, Yamanaka H, Kobayashi K, Obata K, Cui X, Tominaga M, Noguchi K (2008) Phospholipase C and protein kinase A mediate bradykinin sensitization of TRPA1: a molecular mechanism of inflammatory pain. Brain 131:1241–1251
88. Dai Y, Wang S, Tominaga M, Yamamoto S, Fukuoka T, Higashi T, Kobayashi K, Obata K, Yamanaka H, Noguchi K (2007) Sensitization of TRPA1 by PAR2 contributes to the sensation of inflammatory pain. J Clin Invest 117:1979–1987
89. Schmidt M, Dubin AE, Petrus MJ, Earley TJ, Patapoutian A (2009) Nociceptive signals induce trafficking of TRPA1 to the plasma membrane. Neuron 64:498–509
90. Story GM, Peier A, Reeve AJ, Eid SR, Mosbacher J, Hricik TR, Earley TJ, Hergarden AC, Andersson DA, Hwang SW, McIntyre P, Jegla T, Bevan S, Patapoutian A (2003) ANKTM1, a TRP-like channel expressed in nociceptive neurons, is activated by cold temperatures. Cell 112:819–829
91. Nagata K, Duggan A, Kumar G, García-Añoveros J (2005) Nociceptor and hair cell transducer properties of TRPA1, a channel for pain and hearing. J Neurosci 25:4052–4061
92. Zurborg S, Yurgionas B, Jira JA, Caspani O, Heppenstall PA (2007) Direct activation of the ion channel TRPA1 by Ca^{2+}. Nat Neurosci 10:277–279
93. Karashima Y, Talavera K, Everaerts W, Janssens A, Kwan KY, Vennekens R, Nilius B, Voets T (2009) TRPA1 acts as a cold sensor in vitro and in vivo. Proc Natl Acad Sci USA 106:1273–1278
94. Obata K, Katsura H, Mizushima T, Yamanaka H, Kobayashi K, Dai Y, Fukuoka T, Tokunaga A, Tominaga M, Noguchi K (2005) TRPA1 induced in sensory neurons contributes to cold hyperalgesia after inflammation and nerve injury. J Clin Invest 115: 2393–2401
95. Katsura H, Obata K, Mizushima T, Yamanaka H, Kobayashi K, Dai Y, Fukuoka T, Tokunaga A, Sakagami M, Noguchi K (2006) Antisense knock down of TRPA1, but not TRPM8, alleviates cold hyperalgesia after spinal nerve ligation in rats. Exp Neurol 200:112–123
96. da Costa DSM, Meotti FC, Andrade EL, Leal PC, Motta EM, Calixto JB (2010) The involvement of the transient receptor potential A1 (TRPA1) in the maintenance of mechanical and cold hyperalgesia in persistent inflammation. Pain 148:431–437
97. Caspani O, Zurborg S, Labuz D, Heppenstall PA (2009) The contribution of TRPM8 and TRPA1 channels to cold allodynia and neuropathic pain. PLoS One 4:e7383
98. Kwan KY, Glazer JM, Corey DP, Rice FL, Stucky CL (2009) TRPA1 modulates mechanotransduction in cutaneous sensory neurons. J Neurosci 29:4808–4819
99. Kerstein PC, del Camino D, Moran MM, Stucky CL (2009) Pharmacological blockade of TRPA1 inhibits mechanical firing in nociceptors. Mol Pain 5:19

100. Petrus M, Peier AM, Bandell M, Hwang SW, Huynh T, Olney N, Jegla T, Patapoutian A (2007) A role of TRPA1 in mechanical hyperalgesia is revealed by pharmacological inhibition. Mol Pain 3:40
101. Eid SR, Crown ED, Moore EL, Liang HA, Choong K-C, Dima S, Henze DA, Kane SA, Urban MO (2008) HC-030031, a TRPA1 selective antagonist, attenuates inflammatory- and neuropathy-induced mechanical hypersensitivity. Mol Pain 4:48
102. Kim YS, Son JY, Kim TH, Paik SK, Dai Y, Noguchi K, Ahn DK, Bae YC (2010) Expression of transient receptor potential ankyrin 1 (TRPA1) in the rat trigeminal sensory afferents and spinal dorsal horn. J Comp Neurol 518:687–698
103. Kosugi M, Nakatsuka T, Fujita T, Kuroda Y, Kumamoto E (2007) Activation of TRPA1 channel facilitates excitatory synaptic transmission in substantia gelatinosa neurons of the adult rat spinal cord. J Neurosci 27:4443–4451
104. Brierley SM, Hughes PA, Page AJ, Kwan KY, Martin CM, O'Donnell TA, Cooper NJ, Harrington AM, Adam B, Liebregts T, Holtmann G, Corey DP, Rychkov GY, Blackshaw LA (2009) The ion channel TRPA1 is required for normal mechanosensation and is modulated by algesic stimuli. Gastroenterology 137:2084–2095 e2083
105. Patwardhan AM, Jeske NA, Price TJ, Gamper N, Akopian AN, Hargreaves KM (2006) The cannabinoid WIN 55,212-2 inhibits transient receptor potential vanilloid 1 (TRPV1) and evokes peripheral antihyperalgesia via calcineurin. Proc Natl Acad Sci USA 103: 11393–11398
106. Jordt SE, Bautista DM, Chuang HH, McKemy DD, Zygmunt PM, Högestätt ED, Meng ID, Julius D (2004) Mustard oils and cannabinoids excite sensory nerve fibres through the TRP channel ANKTM1. Nature 427:260–265
107. Calignano A, La Rana G, Giuffrida A, Piomelli D (1998) Control of pain initiation by endogenous cannabinoids. Nature 394:277–281
108. Jeske NA, Patwardhan AM, Gamper N, Price TJ, Akopian AN, Hargreaves KM (2006) Cannabinoid WIN55,212-2 regulates TRPV1 phosphorylation in sensory neurons. J Biol Chem 281:32879–32890
109. Mohapatra DP, Nau C (2005) Regulation of Ca^{2+}-dependent desensitization in the vanilloid receptor TRPV1 by calcineurin and cAMP-dependent protein kinase. J Biol Chem 280:13424–13432
110. Peier A, Moqrich A, Hergarden AC, Reeve AJ, Andersson DA, Story GM, Earley TJ, Dragoni I, McIntyre P, Bevan S, Patapoutian A (2002) A TRP channel that senses cold stimuli and menthol. Cell 108:705–715
111. McKemy DD, Neuhausser WM, Julius D (2002) Identification of a cold receptor reveals a general role for TRP channels in thermosensation. Nature 416:52–58
112. Tominaga M, Caterina MJ (2004) Thermosensation and pain. J Neurobiol 61:3–12
113. Dhaka A, Earley TJ, Watson J, Patapoutian A (2008) Visualizing cold spots: TRPM8-expressing sensory neurons and their projections. J Neurosci 28:566–575
114. Takashima Y, Daniels RL, Knowlton W, Teng J, Liman ER, McKemy DD (2007) Diversity in the neural circuitry of cold sensing revealed by genetic axonal labeling of transient receptor potential melastatin 8 neurons. J Neurosci 27: 14147–14157
115. Dhaka A, Murray AN, Mathur J, Earley TJ, Petrus MJ, Patapoutian A (2007) TRPM8 is required for cold sensation in mice. Neuron 54:371–378
116. Bautista DM, Siemens J, Glazer JM, Tsuruda PR, Basbaum AI, Stucky CL, Jordt SE, Julius D (2007) The menthol receptor TRPM8 is the principal detector of environmental cold. Nature 448:204–208
117. Colburn RW, Lubin ML, Stone DJ, Wang Y, Lawrence D, D'Andrea MR, Brandt MR, Liu Y, Flores CM, Qin N (2007) Attenuated cold sensitivity in TRPM8 null mice. Neuron 54: 379–386
118. Xing H, Chen M, Ling J, Tan W, Gu JG (2007) TRPM8 mechanism of cold allodynia after chronic nerve injury. J Neurosci 27:13680–13690
119. Staaf S, Oerther S, Lucas G, Mattsson JP, Ernfors P (2009) Differential regulation of TRP channels in a rat model of neuropathic pain. Pain 144:187–199

120. Proudfoot CJ, Garry EM, Cottrell DF, Rosie R, Anderson H, Robertson DC, Fleetwood-Walker SM, Mitchell R (2006) Analgesia mediated by the TRPM8 cold receptor in chronic neuropathic pain. Curr Biol 16:1591–1605
121. Chung M-K, Lee H, Mizuno A, Suzuki M, Caterina MJ (2004) TRPV3 and TRPV4 mediate warmth-evoked currents in primary mouse keratinocytes. J Biol Chem 279:21569–21575
122. Moqrich A, Hwang SW, Earley TJ, Petrus MJ, Murray AN, Spencer KS, Andahazy M, Story GM, Patapoutian A (2005) Impaired thermosensation in mice lacking TRPV3, a heat and camphor sensor in the skin. Science 307:1468–1472
123. Lee H, Iida T, Mizuno A, Suzuki M, Caterina MJ (2005) Altered thermal selection behavior in mice lacking transient receptor potential vanilloid 4. J Neurosci 25:1304–1310
124. Todaka H, Taniguchi J, Satoh J, Mizuno A, Suzuki M (2004) Warm temperature-sensitive transient receptor potential vanilloid 4 (TRPV4) plays an essential role in thermal hyperalgesia. J Biol Chem 279:35133–35138
125. Huang SM, Lee H, Chung M-K, Park U, Yu YY, Bradshaw HB, Coulombe PA, Walker JM, Caterina MJ (2008) Overexpressed transient receptor potential vanilloid 3 ion channels in skin keratinocytes modulate pain sensitivity via prostaglandin E2. J Neurosci 28:13727–13737
126. Mandadi S, Sokabe T, Shibasaki K, Katanosaka K, Mizuno A, Moqrich A, Patapoutian A, Fukumi-Tominaga T, Mizumura K, Tominaga M (2009) TRPV3 in keratinocytes transmits temperature information to sensory neurons via ATP. Pflugers Arch 458:1093–1102
127. Grant AD, Cottrell GS, Amadesi S, Trevisani M, Nicoletti P, Materazzi S, Altier C, Cenac N, Zamponi GW, Bautista-Cruz F, Lopez CB, Joseph EK, Levine JD, Liedtke W, Vanner S, Vergnolle N, Geppetti P, Bunnett NW (2007) Protease-activated receptor 2 sensitizes the transient receptor potential vanilloid 4 ion channel to cause mechanical hyperalgesia in mice. J Physiol 578:715–733
128. Alessandri-Haber N, Dina OA, Joseph EK, Reichling D, Levine JD (2006) A transient receptor potential vanilloid 4-dependent mechanism of hyperalgesia is engaged by concerted action of inflammatory mediators. J Neurosci 26:3864–3874
129. Chen X, Alessandri-Haber N, Levine JD (2007) Marked attenuation of inflammatory mediator-induced C-fiber sensitization for mechanical and hypotonic stimuli in TRPV4–/– mice. Mol Pain 3:31
130. Alessandri-Haber N, Dina OA, Yeh JJ, Parada CA, Reichling DB, Levine JD (2004) Transient receptor potential vanilloid 4 is essential in chemotherapy-induced neuropathic pain in the rat. J Neurosci 24:4444–4452
131. Dina OA, Parada CA, Yeh J, Chen X, McCarter GC, Levine JD (2004) Integrin signaling in inflammatory and neuropathic pain in the rat. Eur J Neurosci 19:634–642
132. Dina OA, Chen X, Reichling D, Levine JD (2001) Role of protein kinase Cepsilon and protein kinase A in a model of paclitaxel-induced painful peripheral neuropathy in the rat. Neuroscience 108:507–515
133. Cenac N, Altier C, Chapman K, Liedtke W, Zamponi G, Vergnolle N (2008) Transient receptor potential vanilloid-4 has a major role in visceral hypersensitivity symptoms. Gastroenterology 135:937–946 946.e931–932
134. Brierley SM, Page AJ, Hughes PA, Adam B, Liebregts T, Cooper NJ, Holtmann G, Liedtke W, Blackshaw LA (2008) Selective role for TRPV4 ion channels in visceral sensory pathways. Gastroenterology 134:2059–2069
135. Cao D-S, Yu S-Q, Premkumar LS (2009) Modulation of transient receptor potential Vanilloid 4-mediated membrane currents and synaptic transmission by protein kinase C. Mol Pain 5:5
136. Elg S, Marmigere F, Mattsson JP, Ernfors P (2007) Cellular subtype distribution and developmental regulation of TRPC channel members in the mouse dorsal root ganglion. J Comp Neurol 503:35–46

Chapter 34
TRPV1: A Therapy Target That Attracts the Pharmaceutical Interests

Rong Xia, Kim Dekermendjian, Elke Lullau, and Niek Dekker

Abstract TRPV1 is a non-selective cation channel gated by noxious heat, vanilloids and extracellular protons, and act as an important signal integrator in sensory nociceptors. Because of its integrative signaling properties in response to inflammatory stimuli, TRPV1 antagonists are predicted to inhibit the sensation of ongoing or burning pain that is reported by patients suffering from chronic pain, therefore offering an unprecedented advantage in selectively inhibiting painful signaling from where it is initiated. In this chapter, we firstly summarize the physiological and pathological roles of TRPV1 and then describe the pharmacology of TRPV1 agonists and antagonists. Finally, we give an update and the status on TRPV1 therapies that have progressed into clinical trials.

34.1 Gene and Protein Structure of TRPV1

34.1.1 Identification of the TRPV1 as an Ion Channel

TRPV1 (formerly known as VR1) is the founding member of the subfamily of thermal-sensitive TRP channels (thermo-TRPs) expressed in sensory neurons [1]. The identification of a capsaicin receptor that mediated capsaicin induced Ca^{2+} influx and cell death from the gene products of the cDNA of the rat dorsal root ganglia is a milestone in sensory biology [1]. This finding revealed the role of TRPV1 in the somatic pain sensing, and attracted great pharmaceutical interest for TRPV1 as a novel pain target [2]. Consistent with this hypothesis, TRPV1-homozygous-null mice (knockouts) are devoid of thermal hypersensitivity that occurs in response to an acute hind-paw injection of pro-inflammatory agents [3, 4], which suggests a clinical value for TRPV1 antagonists as novel analgesic drugs.

R. Xia (✉)
DECS, Cell, Protein, and Structure Sciences, AstraZeneca R&D, Mölndal SE-43183, Sweden
e-mail: Rong.Xia@astrazeneca.com

34.1.2 From Gene to Structure

The gene of human TRPV1 is located at 17p13.2 [1]. The peptide of TRPV1 spans 839 amino acids (AA), and has a molecular weight (MW) of 95 kDa [1, 5, 6]. So far, TRPV1 has been cloned from multiple species including human [1], guinea pig [7], rabbit [8], mouse [9], zebrafish [10], and porcine [11]. In human, besides the full-length TRPV1, there are 3 variants identified so far. TRPV1b, which is expressed in trigeminal ganglion neurons and lack exon 7, forms functional ion channels active to high temperatures but unresponsive to capsaicin or protons [12]. Moreover, TRPV1b negatively regulates the TRPV1 responsiveness to capsaicin, heat and low pH [13]. An N-terminal splice variant identified in the human supraoptic nucleus (SON) is also inactive to capsaicin, but important for the intrinsic osmosensitivity of cells in the SON by controlling the release of arginine-vasopressin [14]. Furthermore, taste variants of TRPV1 (TRPV1t) has been isolated from the taste receptor cells of human and rat, which contributes to the NaCl chorda tympani response [15, 16].

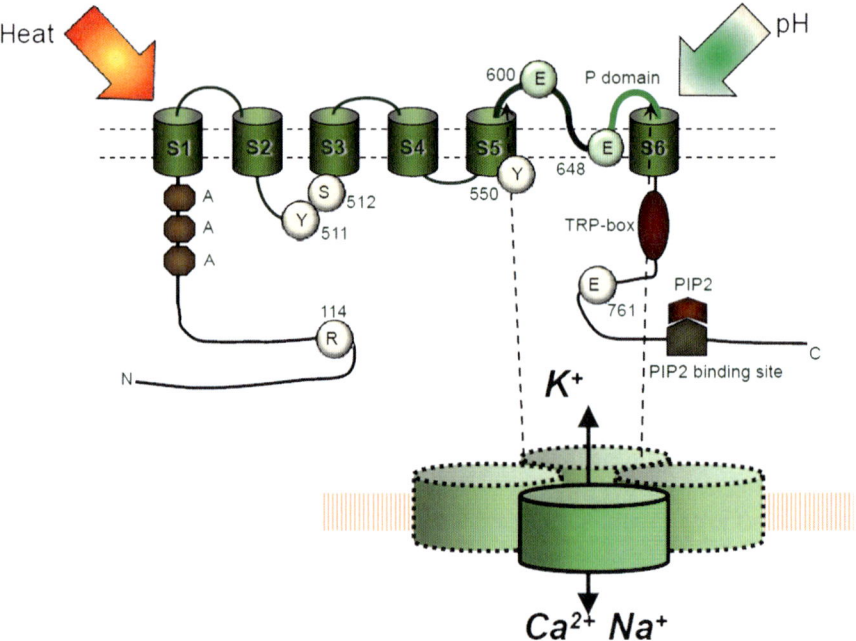

Fig. 34.1 Membrane topology of the human TRPV1 channel subunit. Six transmembrane segments (S1–S6), a P domain between S5 and S6, the TRP-box, the PIP_2 binding site, three ankyrin repeat domain (A) in the C-terminus are marked. Vanilloid-binding sites (Arg^{114}, Tyr^{511}, Ser^{512}, Tyr^{550}, Glu^{761}) and residues involved in proton mediated activation (Glu^{600} and Glu^{648}) are depicted as stippled. Four subunits assembled together on the cell membrane to form an ion channel permeable to Ca^{2+}, Na^+ and K^+

Hydrophobicity analysis of the protein reveals the presence of six putative membrane-spanning segments (TM) (Fig. 34.1), suggesting the homology to other tetrameric (6TM) ion channels (eg. Kv1.2) [17–19]. There is a re-entrant loop between the fifth and sixth transmembrane segment, which is thought to form the ion-conducting pore of the TRP channels, thus termed the pore loop (Fig. 34.1) [20–23]. Structurally, TRPV1 monomers display a hydrophilic intracellular N-terminus domain containing three conserved ankyrin repeats [24]. The ankyrin repeat domain consists of 33 amino acid residues and form a highly conserved helix-turn-helix structure, which was also reported as the calmodulin (CaM) binding site [25]. A TRP-like domain is located close to the carboxyl terminus [24]. In other TRPV channels, ankyrin repeats has been identified responsible for the protein oligomerization, whereas, in TRPV1, Glu^{684}-Arg^{721} in the TRP-like domain in the C terminus has been identified as an association domain for the tertramerization of TRPV1 [26]. the C terminus also harbors phosphoinositide and calmodulin binding domains, and phospholipase C (PKC) consensus sites [27–30].

The three-dimensional (3D) structure of TRPV1 channel have been revealed by electron cryomicroscopy (ECM) [17]. TRPV1 is 150 Å tall and displays four-fold symmetry. The whole structure consists of two major regions: a large open basketlike domain, likely corresponding to the cytoplasmic N- and C-terminal portions, and a more compact domain, corresponding to the transmembrane portion [17].

34.2 Expression Pattern and Biological Function of TRPV1

34.2.1 Tissue Distribution

The tissue distribution of TRPV1 is diverse. TRPV1 RNA was identified in dorsal root ganglion (DRG), central neuron system (CNS) (especially in cerebral cortex, hippocampus, and cerebellum), kidney, and liver [31]. A relatively higher expression was identified in DRG relative to the other sites of expression, especially mostly in subsets of small diameter DRG neurons (C and A δ). The short TRPV1 splice variant is expressed at lower levels in dorsal root ganglia but similar in other tissues compared to TRPV1, and acts as the dominant-negative regulator of TRPV1 [31]. Interestingly, TRPV1 forms agonist-sensitive channels not only in the cell membrane, but also in the endoplasmic reticulum (ER), which when activated, release Ca^{2+} from internal stores, without activating the endogenous store-operated Ca^{2+} entry [32]. Thus, TRPV1 ligands that can pass the cell membrane may have a pharmacological activity different from those who cannot. Expression of TRPV1 was also identified in skin [33, 34], in smooth muscle of the bladder [35, 36], in liver [37], in hematopoietic cells [38], and in epithelial cells in human airways [39, 40].

34.2.2 Modulation of Expression

Expression, especially surface localization of TRPV1 is rapidly up-regulated by nerve growth factor (NGF) in both native cells and HEK cells [41]. Such up-regulation is possibly due to the increased insertion of TRPV1 to the cell membrane by the Src kinase mediated phosphorylation at Tyr^{200} [41, 42]. It was found that gamma-amino butyric acid A-type (GABAA) receptor associated protein (GABARAP) associated with TRPV1 in HEK293 cells and in neurons. The GABARAP augmented TRPV1 expression and stimulated surface receptor clustering [43]. Fibronectin (FN), an extracellular matrix (ECM) molecule, is also reported to up-regulate the TRPV1 translocation to the plasma membrane in trigeminal ganglia sensory neurons [44]. Furthermore, increased expression of TRPV1 channel was observed in intrinsically aged and photoaged human skin in vivo [45]. Moreover, increased membrane expression of TRPV1 was observed at low temperature incubation (30°C).

34.2.3 Physiological and Pathological Roles of TRPV1

34.2.3.1 Pain and Temperature Sensation

The role of TRPV1 in various pain states is a subject of intense research. Over the last 15 years, evidences have been collected pointing to the critical role this ion channel plays in pain signaling (Table 34.1). The first proof of the vital role TRPV1 plays in the detection of noxious heat has been provided using knock-out mice, demonstrating reduced sensitivity to both heat and inflammatory pain [46]. Details of the role of TRPV1 in pain sensation is described in Chapter 33 by Chung et al.

Table 34.1 Summary of the involvement of TRPV1 in painful conditions

Diabetic painful neuropathy	[47–49]
Peripheral neuropathic pain	[50, 51]
Cancer pain	[52]
Rheumatoid arthritis	[53]
Osteoarthritis	[53, 54]
Chronic persistent cough	[55]
Faecal incontinence	[56]
Oesophageal reflux disease	[57, 58]
Cystitis	[59–61]

34.2.3.2 Respiratory Diseases

The expression of the TRPV1 in the mammalian respiratory tract is predominantly localized in the sensory nerves [62]. Co-immunostaining shows that TRPV1 channels are often co-localized with certain sensory neuropeptides, such as tachykinins (TKs) and calcitonin gene-related peptide (CGRP), in the same axon [40, 63]. Activation of these TRPV1-expressing vagal sensory terminals in the airways can elicit typical respiratory defense responses against inhaled irritants such as bronchoconstriction, mucus secretion, bradycardia and hypotension mediated through the efferent pathways of the autonomic nervous system, which are accompanied by rapid inspiration, airway irritation and cough [62, 64].

34.2.3.3 Vascular and Renal Hypertension

TRPV1 dysfunction is implicated in renal hypertension [65]. Sensory nerves expressing TRPV1 are abundantly distributed in the kidney, and regulate release of various neuropeptides, especially vasodilators [66, 67]. In rats with salt-sensitive hypertension caused by sensory degeneration achieved by capsaicin treatment, sodium loading resulted in impaired sodium excretion, indicating that TRPV1 might be a protective response to high salt ingestion [68, 69]. TRPV1 knockout mice showed aggravated kidney damage induced by deoxycortisosterone acetate treatment compared to the WT mice, implying the protective role of TRPV1 in protectively mediating the salt-induced renal damage [65, 69, 70].

34.2.3.4 Urinary Bladder

TRPV1 channel appears to play a major role in the afferent control of the urinary bladder. Functional TRPV1 receptors were identified in human urothelial cells, which are responsive to capsaicin and thermal stimuli [35, 71]. TRPV1 knockout mice have shown reduced frequency of bladder reflex contractions and increased bladder capacity [72]. Furthermore, the TRPV1 specific antagonists JNJ17203212 and JYL1421 can suppresses sensory reflexes of the rodent bladder [73]. Interestingly, TRPV1 is not only expressed by afferent nerves that form close contact with the bladder epithelium but also in non-neuronal cells including urothelial cells and myofibroblasts in the urinary bladder [35, 72]. Urothelial cells can release a variety of chemical mediators, such as ATP, prostaglandins and nitric oxide (NO), which can alter excitability of bladder afferent nerves. Vice versa, nociceptive afferent activation releases a number of factors, such as substance P that can activate urothelial cells [74, 75]. These data indicated the bidirectional chemical communication between urothelium and bladder nerves, where TRPV1 could play a role.

34.3 Ion Channel Properties

34.3.1 Activation

TRPV1 is a ligand-gated ion channel activated by multiple mechanisms. So far, modes of TRPV1 activation identified are heat, vanilloid compounds, voltage, and low pH.

Temperature > 43°C activate TRPV1 at the resting potential of -60 mV [1]. The activation is temperature dependant, with a Q_{10} (temperature coefficient) of approximately 27. Furthermore, low temperature inhibited the current induced by capsaicin with the Q_{10} of 2.3. Whereas, physiological temperature 37°C significantly potentiated the TRPV1 current compared to the room temperature [76]. Moreover, in the hypothalamus, TRPV1 channels are inactive at physiological temperatures. In the abdomen, TRPV1 channels are tonically activated, but not by temperature [77]. The relationship between the activation of TRPV1 by ligand and temperature is still unclear. It was proposed by Nilius et al. that both the agonist and the temperature shift the voltage dependence towards physiologically relevant potentials [78, 79].

TRPV1 was firstly identified as a vanilloid receptor (VR1). The vanilloid compounds identified to activate TRPV1 are discussed under TRPV1 agonists (Section 34.5). Vanilloid ligands interact at the intracellular valilloid binding sites. So far, 5 intracellular binding sites have been identified to form the vanilloid pocket, which are Arg^{114}, Glu^{761} [80], Tyr^{511} [81], Ser^{512} [82], and Tyr^{550} [79] (Fig. 34.1).

TRPV1 is a voltage-dependent ion channel that is activated at positive potentials and deactivated at negative potentials. In the absence of agonist or temperature stimulation, TRPV1 is active at high membrane potential with the potential for half-activation ($V_{1/2}$) around 100 mV [79]. Such high potential is not physiologically relevant for the activation of TRPV1. The sensitivity for voltage dependent activation and deactivation depends on the temperature and ligand concentration [78]. Vanilloid ligands and high temperature left-shift the voltage activation curve of TRPV1. Vice verse, the temperature threshold for TRPV1 activation is not constant, but significantly fluctuate with the membrane potential [78, 79, 83].

The last predominant activation mechanism of TRPV1 is the pH. TRPV1 is directly activated by extracellular proton at low pH (pH<6) [84]. Whereas, at higher pH (pH 6-7), protons sensitize TRPV1 to other stimuli such as heat and capsaicin. Two glutamates (Glu^{600} and Glu^{648}) located at the pore loop were identified essential for the proton regulation (Fig. 34.1) [85]. The activation of TRPV1 by proton cannot be blocked by AMG-0610, capsazepine, and SB-366791, which bind to the same binding pocket as capsaicin, indicating that proton activation does not alter the conformation of the TRPV1 capsaicin-binding pocket [86]. Moreover, divalent (Ca^{2+}, Mg^{2+} >10 mM) and polyvalent cations (Gd^{3+} <1 mM) can activate TRPV1 in a manner similar to protons. Whereas, Gd^{3+} at higher concentration (>1 mM) blocks the channel [79, 87, 88].

34.3.2 Gating Biophysics

At positive potential, the activated TRPV1 forms a pore with the single channel conductance of around 58-115 pS [1, 89]. Lower conductance (50 pS) was observed at negative potentials [1]. The single channel conductance is reduced by the sustained capsaicin, divalent cations and protons, but not by the antagonists neomycin [89] and tetrabutylammonium [90]. In whole cell recording through the ramp protocol,

TRPV1 current exhibits characteristic profound outward rectification [1, 90]. The TRP box at the C-terminus plays an important role in TRPV1 gating. Substitution of Ile^{696}, Trp^{697}, and Arg^{701} by alanine severely affected voltage- and heat-dependent activation and notably reduced the capsaicin responsiveness [91].

34.3.3 Ion Selectivity and Permeability

TRPV1 forms Ca^{2+}-permeable channels with fairly higher discrimination for divalent cations compared to monovalent cations with P_{Ca}/P_{Na} ranging between 3 and 10 [92]. When activated with capsaicin, the permeability of Ca^{2+} relative to Na$^+$ (P_{Ca}/P_{Na}) reaches 10, whereas activation by heat conferred relative low Ca^{2+} permeability with P_{Ca}/P_{Na} of 3. The permeability to monovalent cations (e.g. Na$^+$, K$^+$, Li$^+$, and Rb$^+$) is very similar. Notably, TRPV1 is also highly permeable to protons, which may underlie the intracellular acidification of nociceptive neurons after channel activation [93]. Like all the other TRPV and TRPM channels, TRPV1 contain a region in the middle of the pore loop analogous to the selectivity filter of the bacterial potassium ion channel (KcsA) [18, 94]. Residues at the pore region in TRP channels have been reported responsible to determine the ion selectivity and permeability for TRPM and TRPV channels [95–97]. Neutralization of the highly conserved negatively charged residue Asp^{646} of TRPV1 reduces the permeability for Ca^{2+} and Mg^{2+}, as well as decreases the affinity of the resultant channels to the pore blocker ruthenium red [20, 98]. Furthermore, the Tyr^{671} in the sixth transmembrane domain was also reported involved in desensitization and calcium permeability of capsaicin-activated currents. Substitution of the Tyr^{671} by positively charged lysine (K) completely abolished the Ca^{2+}-dependent desensitization, and reduced the Ca^{2+} permeability [99]. Interestingly, it was reported that activation of TRPV1 leads to increases in relative permeability to large cations and changes in Ca^{2+} permeability. Such change was time and agonist concentration dependent, and could be explained by the protein kinase C (PKC) -mediated TRPV1 phosphorylation [100]. Mutation of S800A, which impaired the phosphorylation by PKC in TRPV1, relieved the agonist-evoked ionic selectivity changes [100].

34.3.4 Modulation of TRPV1 Function

TRPV1 activity is modulated by cAMP-dependent protein kinase (PKA), PKC, Ca^{2+}/calmodulin kinase II (CaMKII) or Src kinase [79]. Capsaicin responses in TRPV1 exhibit robust potentiation by PKA [24]. Ser^{116} [24], Thr^{114}, Thr^{370} [101], and Ser^{502} [102] are important in the PKA-mediated phosphorylation and sensitization of TRPV1.

Phosphatidylinositol 4,5-bisphosphate (PIP$_2$) is found to tonically bind to the PIP$_2$ binding region in TRPV1 (Fig. 34.1) and hold it in an inhibited state. Hydrolysis of PIP$_2$ leads to TRPV1 sensitization or activation [30]. PKC-mediated phosphorylation of TRPV1 potentiates the activation of TRPV1 by capsaicine, protons and temperature. TRPV1 must be phosphorylated by CaMKII before its activation by capsaicin. In contrast, the dephosphorylation of TRPV1 by calcineurin leads to a desensitization of the receptor [103]. Two serine residues Ser^{502} and Ser^{800} have been identified essential for the PKC-mediated potentiation [29, 102]. Residues Ser^{502}, Arg^{701}, and Thr^{704} have been identified responsible for the CaMKII mediated phosphorylation of TRPV1 [104]. Furthermore, activity of TRPV1 is also regulated by N-glycosylation. Extracellular Asn^{604} has been identified as the glycosylation site of TRPV1 (Fig. 34.1) [105].

34.4 Research Toolkits

34.4.1 Current Methods for TRPV1 Production and Purification

Low-yield of functional membrane protein production has been a major limitation step in structural characterization of membrane proteins. There are few reports on successful heterologous expression and structural analysis of mammalian ion channels in bacterial compared to eukaryotic cells. So far, yeast [17], insect cells [106], and mammalian cells [107] showed to give good expression of mammalian ion channels. The protease deficient S. cerevisiae yeast strain BJ5457 has been reported to successfully expressing TRPV1 using the Yep-His-GFP vector. The fermentation was performed in 16 L scale, with addition of 5%/glycerol at 25°C and irritation of 200 rpm. The TRPV1 protein was solubilized with n-decyl-â-d-maltoside (DM) at 1% (w/v) with salt [17, 122]. High five insect cell line was reported as another successful host cell to express TRPV1. High five cells was infected with the MOI 1 at 10 L scale of fermentation, with the application of protease inhibitor cocktail at 1 ml/1 l. The cells were harvested 48 h post-infection; the protein was solubilized with FOS10 at 4% (w/v) in combination with salt and urea [120].

34.4.2 Radio Ligand Binding Assays

The basis of the receptor binding study is the binding of ligand (L) to the receptor (R) to form a ligand-receptor complex (LR), which indicates the amount of ligand that is bound to the receptor. Two constants (K_d and B_{max}) are essential to describe the binding activity of a ligand to a receptor. K_d is a measure of the affinity of a ligand (drug) for a receptor, and is described as the concentration of radioactive ligand required to occupy 50% of the receptors. B_{max} is a measure of the density of the receptor in a tissue when all of the receptors are occupied by radioactive ligand [108]. To decrease the non-specific binding background, the bovine serum albumin (BSA) at 1 mg/ml or bacitracin at 0.5 mg/ml can be routinely applied during the binding process. ^{3}H-Resiniferatoxin (PerkinElmer Life Sciences) is the most common and sensitive ligand used for TRPV1 binding assays. In mammalian and insect cell membranes, it showed a B_{max} of 0.2–15 pmol/mg and a K_d of 0.04–1 nM [166]. ^{3}H-A-778317 (Abbott Laboratories) shows a B_{max} of 0.5 pmol/mg and a K_d of 4 nM in mammalian cell membranes [106].

34.4.3 Electrophysiology: Manual and Automated Patch Clamp Recordings

As TRPV1 is a ligand- gated and voltage- potentiated ion channel selective for cations, electrophysiological recording is the most ideal and direct assay for the activity. Traditional manual patch clamp recording successfully recorded the TRPV1 current in cells expressing recombinant and native TRPV1. Commonly, a holding potential of –60 mV with the application of capsaicin at 0.1–1 μM for 1–5 s will induce currents up to thousands of pA [1, 109]. Automated patch clamp techniques have also been applied to record the currents of diversified ion channels, especially in pharmaceutical screening of compounds. Molecular Devices, Inc. have developed the automated system IonWorks. There are two versions of the system- HT and Quattro, with one hole per patch well versus 64 holes per patch well, respectively. These systems don't have the possibility to use on-line additions of compounds and voltage control at the same time and therefore not suitable for fast ligand-gated channels. Furthermore the Ionworks systems do not attain gigaohm seals. The QPatch system from Sophion gives gigaohm seals and furthermore on-line fluidics, which enables the voltage control and compound additions. These features make it suitable for both voltage- and ligand- gated channels. However, different pharmacological profiles have been identified during the recording with the IonWorks and the QPatch system. DYNAFLOW HT by Cellectricon has superior microfluidics and online control with high throughput and low cost, whereas the reliability and reproducibility of the data is in need of further evaluation [110]. The latest version of Nanions' Patchliner even includes the temperature adjusting system, which is ideal for assay of the temperature sensitive ion channels such as TRPV1, TRPA1, and TRPM8 [111]. These automated patch clamp systems have aided the discovery of TRPV1 modulators a lot and established these technologies as an important part of the screening cascade in the drug discovery process.

34.4.4 Florescence Imaging and $^{45}Ca^{2+}$ Flux Assays

TRPV1 is a cation selective ion channel permeable to Ca^{2+}, therefore suitable for Ca^{2+}-based fluorometric imaging as a rapid assay for the channel activity [112]. The Fluorometric Imaging Plate Reader (FLIPR) has been developed to perform quantitative optical detection of receptor/ion channel–mediated changes in cellular membrane potential or intracellular calcium using fluorescent indicator dyes [113]. The FLIPR platform has become critical to lead discovery programs and have been used extensively for discovery of TRPV1 antagonists. It is used not only in high throughput screening (HTS), but also for target identification, assay development, and follow-up pharmacological analysis of compound specificity and selectivity. Recently, a functional assessment of TRPV1 activity has been reported using a real-time PCR machine, based on the temperature sensitivity of TRPV1 [114]. Another assay format suitable for HTS is radioactive $^{45}Ca^{2+}$ based assays (used by Amgen and Glenmark) [115]. These assays utilizes radioactivity to trace the change in intracellular Ca^{2+} instead of florescence, and therefore less sensitive to false positives due to auto-fluorescent compounds. The drawback of the $^{45}Ca^{2+}$ flux assay is that they are not time resolved like the FLIPR assays and therefore contains no information on the Ca^{2+} signaling kinetics.

34.5 Pharmacology and Druggability

Since the cloning of TRPV1 in 1997, the interest and knowledge of TRPV1 pharmacology have expanded immensely. With the demonstration of TRPV1 being a signal integrator in sensory nociceptors responding to noxious heat, vanilloids and extracellular protons, the potential as a novel pain target was obvious and many pharmaceutical companies started drug discovery programs to find specific TRPV1 modulators, in particular antagonists.

34.5.1 Agonists of TRPV1

The vanilloid capsaicin initiated much of the research around the TRPV1 receptor and gave rise to the original name "the vanilloid receptor" or VR1 [1]. The capsaicin analogue resiniferatoxin (RTX) was shown to be much more potent on the TRPV1 receptor and have been used extensively in ligand binding assays. In addition to compounds having the vanilloid motif several endogenous TRPV1 agonists (endovanilloids) have been discovered over the years (for review and structures see [116]). It includes biogenic amines e.g., N-arachidonylethanolamine (AEA, anandamide), N-arachidonoyldopamine (NADA), N-oleoylethanolamine (OLEA), N-arachidonolylserine, various N-acyltaurines and N-acylsalsolinols [117], oxygenated eicosatetraenoic acids [118, 119] like the lipoxygenase products 5-, 12-, 15-hydroperoxyeicosatetraenoic acids (5S-, 12S-, 15S-HPETE) [120], their reduced hydroxylic analogs, prostaglandins, and leukotriene B4 [120]. Additionally components of the inflammatory soup like adenosine, ATP, and polyamines (such as spermine, spermidine, and putrescine) also activate TRPV1 [90, 121–124]. Although several TRPV1 agonists have been identified over the years, most efforts have been directed towards antagonists.

34.5.2 Antagonists of TRPV1

Long before the cloning of TRPV1 it was demonstrated that the dye ruthenium red could inhibit capsaicin induced responses in sensory neurons [125]. Later it was shown that ruthenium red is a non-competitive antagonist not only for TRPV1 but for most other TRP channels functioning as a pore blocker [116]. The first competitive antagonist for TRPV1 was the capsaicin analogue capsazepine (capz). Capz is a relative potent antagonist on human TRPV1 but demonstrated much less potency on rat TRPV1 [126, 127], illustrating the differences between species that is common for many TRPV1 antagonists. With the discovery of capz and demonstration that it was active in various preclinical pain models, many pharmaceutical companies developed TRPV1 antagonists leading to a large amount of specific antagonists [128, 129].

The TRPV1 antagonists can be broadly divided into vanilloid derived and non-vanilloid compounds based on their structure [129] and the number of TRPV1 patents has now exceeded 1,000. The most important classes of TRPV1 antagonists are summarized in Table 34.2 (for complete review with chemical structure see 129).

Most of these antagonists show high affinity to TRPV1 and are competitive antagonists binding to the same site as capsaicin and many of them have also demonstrated in-vivo effects in various pain models.

Other reported structural classes of TRPV1 antagonists are, 4-fluopyridine amides, Spiroisoxazolopyridine amides, Piperazinyl benzimidazoles, Pyridoindazolones, Piperazinyl pyrimidines, Phenylchromones, Biaryl carboxamides and Aminoquinazolines [79].

34.5.3 Clinical Status of TRPV1-Targeted Therapies

TRPV1 agonists were the first TRPV1 receptor-based therapeutics to enter clinical trials. Capsaicin creams (prescription free) have shown moderate-to-poor analgesic efficacy probably due to the low dose or poor skin absorption, while novel developed site-specific capsaicin therapy with high-dose patches and injectable preparations have been shown to provide long-lasting analgesia with rapid onset [152]. Several companies have clinical trials with TRPV1 agonist applications for pain relief.

34.5.3.1 Neurogesx Has Qutenza™ in Phase-III (NGX-4010 Dermal Patch (8% Capsaicin))

NeurogesX has successfully completed three Phase III clinical studies of Qutenza™ (NGX-4010) and the US. Food and Drug Administration (FDA) has approved Qutenza™ for the management of neuropathic pain due to postherpetic neuralgia (PHN). Qutenza™ is a synthetic trans-capsaicin and drug delivery is by a rapid-delivery patch application system [153]. NeurogesX plans to launch Qutenza™ in the United States in the first half of November 2010.

Table 34.2 Examples of TRPV1 antagonists from different structural classes

	Compounds	Comments	References
1,3-Disubstituted urea derivatives	Capsazepine 2-((4-chlorophenethylamino) methyl)-2,3,4,5-tetrahydro-1H-benzo[c]azepine-7,8-diol	IC_{50} = 420 nM ($^{45}Ca^{2+}$ uptake in rTRPV1) Inhibits voltage-activated calcium channels and nicotinic acetylcholine receptors Significantly reversed CFA-induced mechanical hyperalgesia in guinea pigs	[126, 127]
	SB-705498 1-(2-bromophenyl)-3-(1-(5-(trifluoromethyl)pyridin-2-yl)pyrrolidin-3-yl)urea	IC_{50} = 32 nM (FLIPR in rTRPV1) Phase 1: reduced capsaicin-evoked flare and acute heat-evoked pain on non-sensitized skin	[130, 131]
	A-425619 1-(isoquinolin-5-yl)-3-(4-(trifluoromethyl)benzyl)urea	IC_{50} = 5 nM (FLIPR in hTRPV1) TRPM8 IC_{50} = 8 μM; TRPA1 IC_{50} > 10 μM Inhibits CFA-induced thermal hyperalgesia (ED_{50} = 10 mg per kg)	[132–134]
	ABT-102 (R)-1-(5-tert-butyl-2,3-dihydro-1H-inden-1-yl)-3-(1H-indazol-4-yl)urea	IC_{50} = 7 nM (FLIPR in hTRPV1) Inhibits CFA-induced thermal hyperalgesia (ED_{50} = 20 μmol/kg)	[135, 136]
	BCTC N-(4-tert-butylphenyl)-4-(3-chloropyridin-2-yl)piperazine-1-carboxamide	IC_{50} = 35 nM (FLIPR in hTRPV1) TRPM8 IC_{50} = 143 nM Inhibits CFA-induced thermal and mechanical hyperalgesia (3–30 mg per kg, orally) Reduces tactile allodynia and thermal hyperalgesia in a partial nerve-ligation model	[137, 138]
Pyridyl piperazine carboxamides	JNJ-17203212 4-(3-(trifluoromethyl)pyridin-2-yl)-N-(5-(trifluoromethyl) pyridin-2-yl)piperazine-1-carboxamide	IC_{50} = 65 nM (FLIPR in hTRPV1) 1°C was increased in core body temperature in rats that took 30 mg/kg orally	[52, 139]
	1,2,3,6-tetrahydropyridyl-4-carboxamides	Attenuates nocifensive behaviours in an in vivo model of bone-cancer pain IC_{50} = 5 nM (FLIPR in hTRPV1)	[140]

34 TRPV1 as a Therapy Target

Table 34.2 (continued)

	Compounds	Comments	References
Aryl cinnamides	SB-366791 ((E)-3-(4-chlorophenyl)-N-(3-methoxyphenyl)acrylamide)	K_i = 18 nM (FLIPR assay in hTRPV1) Selectively block TRPV1 rather than TRPV4 and other TRP channels Inhibits capsaicin and heat-mediated activation of TRPV1	[141, 142]
	AMG-9810 (E)-3-(4-tert-butylphenyl)-N-(2,3-dihydrobenzo[b][1,4]dioxin-6-yl)acrylamide	IC_{50} = 25 nM ($^{45}Ca^{2+}$ uptake in hTRPV1) hIC$_{50}$ of >4 μM at TRPV3, TRPV4, TRPA1 and TRPM8 Inhibits CFA-induced thermal (30 mg/kg) and mechanical hyperalgesia (100/kg)	[143]
	AMG-517 N-4-(6-(4-((trifluoromethyl)phenyl)pyrimidine-4-carbonyl)benzo[d]thiazol-2-yl)acetamide	Initiation of Phase I clinical trials reported in September 2004 IC_{50} = 0.9 nM ($^{45}Ca^{2+}$ uptake in hTRPV1) Achieved ~40% block of CFA-induced thermal hyperalgesia at 10 mg/kg (MED 1 mg per kg)	[144]
Benzimidazol analogue	Compound 46ad 1-(5-chloro-6-(3-methyl-4-(6-(trifluoromethyl)-4-(3,4,5-tr fluorophenyl)-1H-benzo[d]imidazol-2-yl)piperazin-1-yl)pyridin-3-yl)ethane-1,2-diol	IC_{50} = 1 nM ($^{45}Ca^{2+}$ uptake in hTRPV1) Achieved significant reversal of CFA-induced thermal hyperalgesia (30 mg/kg, orally)	[145]
Quinazolinon analogue	Benzamides N-tetrahydroquinolinyl, N-quinolinyl	IC_{50} = 23 nM (FLIPR in hTRPV1) pK_b ~7 (FLIPR in hTRPV1) showed excellent potency at human, guinea pig and rat TRPV1, a favourable in vitro DMPK profile and activity in an in vivo model of inflammatory pain	[146] [147]
	N-isoquinolinyl biaryl carboxamides	pK_b ~8 (FLIPR in hTRPV1) showed excellent potency at human, guinea pig and rat TRPV1, a favourable in vitro DMPK profile and activity in an in vivo model of inflammatory pain	[147]
	N-(4-(trifluoromethyl)phenyl)-7-(3-(trifluoromethyl)pyridin-2-yl)quinazolin-4-amine	IC_{50} = 1 nM (FLIPR in hTRPV1) Achieved 80% block of carrageenan-induced thermal hyperalgesia at 3 mg/kg (MED 0.1 mg/kg)	[148]

Table 34.2 (continued)

Compounds	Comments	References
Compound 26 6-(4-chloro-3-(cyclopropylmethoxy) phenyl)-7-isopropyl-2-methylquinazolin-4(3H)-one	$IC_{50} = 50$ nM (low pH activation in hTRPV1) Achieved 60% reversal of CFA-induced mechanical hyperalgesia (30 mg/kg, orally) Achieved 57% reversal of mechanical hyperalgesia in a partial nerve-ligation model	[149]
NGD-8243/MK-2295	$IC_{50} = 0.4$-6 nM on hTRPV1 Achieved significant reversal of CFA-induced thermal hyperalgesia (3 μM plasma expousure). Initiation of Phase II trials announced in November 2006	[150]
GRC6211	$IC_{50} = 3.8$ nM (in hTRPV1) competitive and selective antagonist of the TRPV1 receptor, with greater than 2,631-fold selectivity for the TRPV1 over other vanilloid receptors. Orally available, potent A single dose of 10 mg/kg GRC-6211 in a rodent model resulted in a C_{max} of 1,550 ng/ml, a half-life of 0.72 h and a T_{max} of 0.5 h. The compound also dose-dependently reduced mechanical and neurogenic hyperalgesia in rat models. Glenmark/Eli lilly, Phase II for acute pain	[151]

34.5.3.2 Anesiva Has Adlea in Phase-III (AlgrX-4975 Injectable/Intraarticular)

Adlea is a TRPV1 agonist based on capsaicin (derived from chili peppers). Administered locally to the site of pain, Adlea has been shown to provide site-specific pain relief by binding to TRPV1 receptors, which are found predominantly on C-fiber neurons. Because Adlea acts primarily on C-fiber neurons, in clinical trials it has been shown not to have an adverse effect on normal sensation such as temperature or touch.

Adlea is intended to provide rapid, long-acting, site-specific, pain relief for the treatment of post-surgical, musculoskeletal and trauma-induced neuropathic pain and data from clinical trials in those indications demonstrate the efficacy of this compound in treating moderate to severe pain (http://www.anesiva.com/wt/page/cl_dataadlea).

34.5.3.3 Winston Laboratories Has Civamide in Phase-III (Inhaled/Injectable)

Civamide from Winston labs have been studied in various applications (http://www.winstonlabs.com/productdevelopment/civamide.asp). Civamide Cream contains Winston's proprietary TRPV1 receptor modulator. Two Phase II studies of Civamide Cream involving 178 and 151 patients and one Phase III study of 695 subjects, have been completed, demonstrating substantial relief of the signs and symptoms of osteoarthritis pain with daily use. These trials enrolled both patients who were on a stable dose of oral NSAIDs or COX-2 inhibitors (Phase III), or not on oral medications (Phase II). These studies indicate that Civamide is effective as a monotherapy as well as an adjunctive treatment.

Civamide Nasal Solution is a metered nasal spray containing Civamide that works through suppression of neuropeptides (e.g., SP and CGRP) of the trigeminal neural plexus thought to be involved in the pathogenesis of migraine and cluster headaches and PHN of the trigeminal nerve. Phase II and Phase III studies in episodic cluster headache have demonstrated that Civamide Nasal Solution can substantially reduce cluster headache frequency following a single one-week course of treatment. A U.S. Phase III study and a European Phase III study have been accomplished.

These applications are all formulations based on the natural product capsaicin (trans or cis isomers) and the route of administration is not oral due to the fact that agonists can cause pain and or erythema before desensitisation (pain relief) becomes effective. The progress of several of these clinical trials supports the concept and therapy will possibly be available for pain patients in the near future.

Several TRPV1 antagonists are currently undergoing phase I and II clinical trials for treating a variety of pain indications (Table 34.3). The first TRPV1 antagonist to reach clinical trials was SB-705498 by GSK and completed phase-I in 2005, showing it was safe in healthy volunteers and could reduce capsaicin induced flare and acute heat-evoked pain, in addition to heat-evoked pain after UV burn. SB-705498

Table 34.3 TRPV1 antagonists inclinical trials beyond phase-I

Name	Company	Clinical trials	Status
AMG 517 WO 04014871	Amgen	Dental pain	Terminated (hyperthermia)
SB-705498 WO 03022809	GlaxoSmithKline	Rectal hypersensitivity and IBS Dental pain Migraine Non-allergic rhinitis	Terminated Completed Completed Ongoing
MK-2295/NGD8243[a] WO 03062209	Merck/neurogen	Dental pain Chronic cough	Completed Ongoing
AZD1386[a] WO 07091948	AstraZeneca	Dental pain Posttraumatic neuralgia (PTN) and postherpetic neuralgia (PHN) Osteoarthritis Gastroesophageal reflux disease (GERD)	Completed Terminated Terminated Ongoing
GRC 6211[a] WO 07042906	Glenmark/lilly	Dental pain Osteoarthritis Neuropathic pain	Completed Suspended Terminated
JTS-653[a] WO 06006741	Japan tobacco	Overactive bladder	No results available

Compound name/numbers and patent references ([a]indicates that the structure given represents a patent example illustrated since the identity of the lead molecule has not been disclosed).

progressed to phase-II for migraine and dental pain but the results from these trails have not been published (http://www.clinicaltrials.gov). Currently SB-705498 is being tested as topical application against non-allergic rhinitis.

GRC-6211 (Glenmark/Lilly) was shown to be an orally available, potent (3.8 nM), competitive and selective antagonist of the VR1 receptor, with greater than 2,631-fold selectivity for the VR1 receptor over other vanilloid receptors. GRC-6211 also dose-dependently reduced mechanical and neurogenic hyperalgesia in rat models [153]. Clinical data showed GRC-6211 was well tolerated in a phase I trial in 72 healthy volunteers. Subjects received up to 200 mg of GRC-6211 in the single dose portion of the study and three doses of GRC-6211 25, 50 or 100 mg for 12 days in the multiple dose segments. GRC-6211 had a predictable safety profile after both single and multiple doses, but both the osteoarthritis and neuropathic pain clinical trials have been stopped for unknown reasons.

Partnered with Neurogen, Merck recently announced promising data for the molecule MK-2295/NGD-8243 in a dental pain proof-of-concept phase IIA trial Population pharmacokinetic (P_K) and PK/pharmacodynamic (P_K/P_D) analyses were performed using NONMEM VI based on data from five clinical trials with approximately 182 subjects treated with drug and/or placebo [154]. A 2-compartment P_K model with covariates for age and gender was developed to describe the MK-2295 P_K. A series of P_K/P_D models was developed which related MK-2295 concentration to markers of on-target and undesired activity: core body temperature (CBT, standard E_{max} with diurnal variation), capsaicin-induced dermal vasodilation (CIDV, competitive E_{max}), warmth sensation threshold (WS, standard E_{max}), and hot water bath hand withdrawal time (HWT, Weibull time to event with a standard E_{max} as the scale parameter). The EC_{50}s for CBT and CIDV, both markers for MK-2295 on-target activity, were 69.9 and 57.9 nM, respectively. The EC_{50}s for WS and HWT were 267 and 292 nM. While effects on HWT strongly suggest on-target activity, effects of MK-2295 on WS may reflect undesired activity. P_K/P_D simulation indicated that thermal sensitivity was highly correlated with other target engagement measures (CBT and CIDV) and thus it was impossible to identify a dosing regimen of MK-2295 that was predicted to be efficacious yet also devoid of risk for burn injury. Merck–Neurogen have now aligned MK-2295 with further exploratory Phase 2 studies for the treatment of cough associated with upper airway disease and are focusing on a back up compound, NGD-9611, currently in preclinical development for pain [152].

Amgen's first clinical candidate AMG-517, a highly selective TRPV1 antagonist, was shown to induce remarkable, but reversible, and generally plasma concentration-dependent hyperthermia during Phase I clinical trials [155]. Similar to what was observed in rats, dogs, and monkeys, hyperthermia was attenuated after repeated dosing of AMG-517 (at the highest dose tested) in humans during a second Phase I trial. However, AMG-517 administered after molar extraction (a surgical cause of acute pain) elicited long-lasting hyperthermia with maximal body temperature surpassing 40°C, suggesting that TRPV1 blockade elicits undesirable hyperthermia in susceptible individuals. In the SAD study, they found $t_{1/2}$ ~300 h, and established an MTD of 15 mg. where some individuals reached 39.9°C. Amgen ran a MAD with 7 days once daily dosing with 3 dose levels: 2, 5 and 10 mg. There was a dose-dependent hyperthermia, with ~1.3°C increase on day 1 in the 10 mg group, which attenuated to ~0.5°C by day 4. The proof of principal (PoP) study was dental extraction. AMG-517 was given orally 1 h after surgery. At 2 mg, 1 of 3 subjects remained at >40°C for ~ 4 days, despite treatment with NSAIDs. At 8 mg, 4 of 4 subjects reached at least 38.5°C and at 15 mg, 2 of 2 subjects remained at a temperature >39°C for a prolonged period. All of the subjects reported feeling cold and in some cases showed visible shivering before the body temperature increase occurred. The trial was stopped and further development of this compound terminated [156].

AZD-1386 from AstraZeneca have completed Phase 1 and Phase 2 clinical trials for dental pain. Furthermore AZD-1386 have been in clinical trials for Osteoarthritis (OA) and neurophatic pain, but discontinued for undisclosed reasons. AZD-1386 is currently in Phase 2 studies for the treatment of pain associated with gastro-oesophageal reflux disease (http://www.clinicaltrials.gov).

JTS-653 from Japan Tobacco that has progressed into a Phase 1 trial aligned with progression for pain and overactive bladder [152]. Additional candidates from Johnson & Johnson (JNJ-39439335, no information on molecule) clinical studies are being initiated on OA and paradoxical pain induced by a thermal grill [166].

Sanofi-Aventis has identified SAR-115740 for the treatment of chronic inflammatory and neuropathic pain. Wyeth, who partnered with Mochida Pharmaceuticals has M-68008 in preclinical development and GSK has disclosed SB-782443 as a clinical candidate [152].

Much of the information regarding these clinical candidates is not available to the public. This makes it difficult to exactly evaluate the current status, but clearly the interest in TRPV1 as a therapeutic pain target is high. From the first generation of TRPV1 antagonists entering clinical trials (Table 34.3), only SB-705498, MK-2295/NGD-8243 and AZD-1386 are still being tested.

The main reason for the considerations for TRPV1 antagonists in future pain therapy came from the first clinical findings of unwanted on-target side effects with AMG-517 (hyperthermia) and MK-2295/NGD-8243 (burn risk). TRPV1 antagonists offer a new mechanism of action for the potential treatment of a wide range of acute and chronic pain disorders. As with any novel mechanism, further preclinical and clinical work can introduce new risks and challenges such as the role of TRPV1 in core body temperature regulation and the propensity for some TRPV1 antagonist to cause hyperthermia [77, 155, 157].

34.6 Future Prospect

34.6.1 Modality Specific TRPV1 Antagonists

A majority of the first generation TRPV1 antagonists (and those in clinical trials) are full blockers of TRPV1 activation, blocking all mode of activation. Since TRPV1 is activated by various stimuli (agonists, pH, heat and voltage) that interact with different domains of the receptor, some TRPV1 antagonists have been shown to be stimulus-specific [143, 158, 159]. AMG-0610 and SB-366791 inhibit the activation of rat TRPV1 by capsaicin but not pH, whereas BCTC, A-425619, AMG-6880, AMG-7472 and AMG-9810 are TRPV1 antagonists that block both capsaicin and pH stimulated responses [86, 143, 160–162]. Furthermore, it has been shown that there are also species differences in the stimulus selectivity of TRPV1 antagonists. Both capsazepine and SB-366791 are more effective in blocking proton-induced stimulus of human TRPV1 than of rat TRPV1 [142, 143], and AMG-8562 antagonizes heat activation of human but not rat TRPV1 [158].

The role of TRPV1 in controlling body temperature have been suspected for long time, but the finding that many TRPV1 antagonists gave hyperthermia in-vivo suggests that TRPV1 is central for keeping normal body temperature regulation [77, 155, 157]. These unwanted on-target side effects lead to the search for stimulus specific TRPV1 antagonists without effect on body temperature [159, 163, 164]. Current literature suggests avoiding antagonizing pH activation could produce TRPV1 antagonists without hyperthermic effects [159].

34.6.2 Biologics: Antibodies and Toxins

The location of TRPV1 at the cell surface together with its peripheral tissue distribution makes it a good target for antagonizing antibody therapeutic intervention. For example, an E3-targeted anti-TRPC5 antibody has been successfully reported to inhibit the channel function of TRPC5 (182). A rabbit polyclonal to the extracellular prepore loop of TRPV1 has been reported to block human TRPV1 activation, whereas rabbit monoclonal antibodies did not block activation by either capsaicin or protons (183). Therefore, the E3 loop is an attractive epitope for generating functionally active anti-TRPV1 monoclonal antibodies, which prevents activation to a broad range of noxious stimuli.

Venomous animals from distinct phyla such as spiders, scorpions, snakes, cone snails, or sea anemones produce small toxic proteins that cause pain, some of which is generated by the activation of TRPV1 channels [124, 165, 166]. TRPV1 desensitization of primary sensory neurons is a powerful approach to relieve symptoms of nociceptive behavior in animal models of chronic pain [167]. Furthermore, some small toxic proteins have been identified as antagonist of TRPV1 [166], indicating another potential mechanism for pain relief using TRPV1 as a target.

34.6.3 Structure-Guided Drug Discovery

A low-resolution structure of TRPV1 has been determined by ECM and revealed the tetrameric compositon of the channel [17]. High quality mg amounts of purified TRPV1 are required for crystallization to elucidate a high resolution crystal structure. High level expression of TRPV1 has been achieved in yeast [17], insect cells

[106], and mammalian cells, but detailed analysis indicated that folding and correct plasma membrane localization was achieved in HEK cells (author's data for both stable expression in HEK293S with N-acetylglucosaminyltransferase I-negative mutation [168] and transient expression in HEK6E [169]). Purification of the channel is underway which will open the future prospect of a crystal structure of a TRP channel to guide drug discovery.

34.6.4 Concluding Remarks

TRPV1 was first cloned in 1997 and early on considered a prime target for pain therapy [5]. Several clinic trials have been stopped for unknown reasons or due to the undesirable level of hyperthermia and burn risk associated with the use of TRPV1 antagonists [170]. Ongoing clinical trials will provide the answers if TRPV1 antagonist will be useful additions to pain therapy. However, new discoveries have shown additional physiological roles of TRPV1, which is attracting the investigation of TRPV1 as the potential target for other indications such as respiratory diseases [62], urinary incontinence [171] and pancreatitis [172]. Since the molecular identity of TRPV1 was discovered 13 years ago the knowledge of this ion channel have expanded tremendously and as we now start getting clinical feedback from TRPV1 therapies it is certain that the next few years will show if it fulfill its expectations.

References

1. Caterina MJ, Schumacher MA, Tominaga M, Rosen TA, Levine JD, Julius D (1997) The capsaicin receptor: a heat-activated ion channel in the pain pathway. Nature 389:816–824
2. Cortright DN, Krause JE, Broom DC (2007) TRP channels and pain. Biochim Biophys Acta 1772:978–988
3. Breese NM, George AC, Pauers LE, Stucky CL (2005) Peripheral inflammation selectively increases TRPV1 function in IB4-positive sensory neurons from adult mouse. Pain 115: 37–49
4. Latorre R, Brauchi S, Orta G, Zaelzer C, Vargas G (2007) Thermo TRP channels as modular proteins with allosteric gating. Cell Calcium 42:427–438
5. Caterina MJ, Julius D (2001) The vanilloid receptor: a molecular gateway to the pain pathway. Annu Rev Neurosci 24:487–517
6. Kedei N, Szabo T, Lile JD, Treanor JJ, Olah Z, Iadarola MJ, Blumberg PM (2001) Analysis of the native quaternary structure of vanilloid receptor 1. J Biol Chem 276:28613–28619
7. Walker KM, Urban L, Medhurst SJ, Patel S, Panesar M, Fox AJ, McIntyre P (2003) The VR1 antagonist capsazepine reverses mechanical hyperalgesia in models of inflammatory and neuropathic pain. J Pharmacol Exp Ther 304:56–62
8. Chanda S, Sharper V, Hoberman A, Bley K (2006) Developmental toxicity study of pure trans-capsaicin in rats and rabbits. Int J Toxicol 25:205–217
9. Correll CC, Phelps PT, Anthes JC, Umland S, Greenfeder S (2004) Cloning and pharmacological characterization of mouse TRPV1. Neurosci Lett 370:55–60
10. Zimov S, Yazulla S (2004) Localization of vanilloid receptor 1 (TRPV1/VR1)-like immunoreactivity in goldfish and zebrafish retinas: restriction to photoreceptor synaptic ribbons. J Neurocytol 33:441–452

11. Ohta T, Komatsu R, Imagawa T, Otsuguro K, Ito S (2005) Molecular cloning, functional characterization of the porcine transient receptor potential V1 (pTRPV1) and pharmacological comparison with endogenous pTRPV1. Biochem Pharmacol 71:173–187
12. Lu G, Henderson D, Liu L, Reinhart PH, Simon SA (2005) TRPV1b, a functional human vanilloid receptor splice variant. Mol Pharmacol 67:1119–1127
13. Vos MH, Neelands TR, McDonald HA, Choi W, Kroeger PE, Puttfarcken PS, Faltynek CR, Moreland RB, Han P (2006) TRPV1b overexpression negatively regulates TRPV1 responsiveness to capsaicin, heat and low pH in HEK293 cells. J Neurochem 99:1088–1102
14. Sharif Naeini R, Witty MF, Seguela P, Bourque CW (2006) An N-terminal variant of Trpv1 channel is required for osmosensory transduction. Nat Neurosci 9:93–98
15. Lyall V, Heck GL, Vinnikova AK, Ghosh S, Phan TH, Alam RI, Russell OF, Malik SA, Bigbee JW, DeSimone JA (2004) The mammalian amiloride-insensitive non-specific salt taste receptor is a vanilloid receptor-1 variant. J Physiol 558:147–159
16. Lyall V, Phan TH, Mummalaneni S, Melone P, Mahavadi S, Murthy KS, DeSimone JA (2009) Regulation of the benzamil-insensitive salt taste receptor by intracellular Ca^{2+}, protein kinase C, and calcineurin. J Neurophysiol 102:1591–1605
17. Moiseenkova-Bell VY, Stanciu LA, Serysheva II, Tobe BJ, Wensel TG (2008) Structure of TRPV1 channel revealed by electron cryomicroscopy. Proc Natl Acad Sci USA 105:7451–7455
18. Gouaux E (1998) Single potassium ion seeks open channel for transmembrane travels: tales from the KcsA structure. Structure 6:1221–1226
19. Long SB, Campbell EB, Mackinnon R (2005) Crystal structure of a mammalian voltage-dependent Shaker family K+ channel. Science 309:897–903
20. Garcia-Martinez C, Morenilla-Palao C, Planells-Cases R, Merino JM, Ferrer-Montiel A (2000) Identification of an aspartic residue in the P-loop of the vanilloid receptor that modulates pore properties. J Biol Chem 275:32552–32558
21. Venkatachalam K, Montell C (2007) TRP channels. Annu Rev Biochem 76:387–417
22. Pedersen SF, Owsianik G, Nilius B (2005) TRP channels: an overview. Cell Calcium 38:233–252
23. Montell C (2005) The TRP superfamily of cation channels. Sci STKE 2005:re3
24. Bhave G, Zhu W, Wang H, Brasier DJ, Oxford GS (2002) Gereau RWt: cAMP-dependent protein kinase regulates desensitization of the capsaicin receptor (VR1) by direct phosphorylation. Neuron 35:721–731
25. Schindl R, Frischauf I, Kahr H, Fritsch R, Krenn M, Derndl A, Vales E, Muik M, Derler I, Groschner K, Romanin C (2008) The first ankyrin-like repeat is the minimum indispensable key structure for functional assembly of homo- and heteromeric TRPC4/TRPC5 channels. Cell Calcium 43:260–269
26. Garcia-Sanz N, Fernandez-Carvajal A, Morenilla-Palao C, Planells-Cases R, Fajardo-Sanchez E, Fernandez-Ballester G, Ferrer-Montiel A (2004) Identification of a tetramerization domain in the C terminus of the vanilloid receptor. J Neurosci 24:5307–5314
27. Numazaki M, Tominaga T, Toyooka H, Tominaga M (2002) Direct phosphorylation of capsaicin receptor VR1 by protein kinase Cepsilon and identification of two target serine residues. J Biol Chem 277:13375–13378
28. Numazaki M, Tominaga T, Takeuchi K, Murayama N, Toyooka H, Tominaga M (2003) Structural determinant of TRPV1 desensitization interacts with calmodulin. Proc Natl Acad Sci USA 100:8002–8006
29. Bhave G, Hu HJ, Glauner KS, Zhu W, Wang H, Brasier DJ, Oxford GS (2003) Gereau RWt: Protein kinase C phosphorylation sensitizes but does not activate the capsaicin receptor transient receptor potential vanilloid 1 (TRPV1). Proc Natl Acad Sci USA 100:12480–12485
30. Prescott ED, Julius D (2003) A modular PIP2 binding site as a determinant of capsaicin receptor sensitivity. Science 300:1284–1288
31. Sanchez JF, Krause JE, Cortright DN (2001) The distribution and regulation of vanilloid receptor VR1 and VR1 5′ splice variant RNA expression in rat. Neuroscience 107:373–381

32. Wisnoskey BJ, Sinkins WG, Schilling WP (2003) Activation of vanilloid receptor type I in the endoplasmic reticulum fails to activate store-operated Ca2+ entry. Biochem J 372: 517–528
33. Southall MD, Li T, Gharibova LS, Pei Y, Nicol GD, Travers JB (2003) Activation of epidermal vanilloid receptor-1 induces release of proinflammatory mediators in human keratinocytes. J Pharmacol Exp Ther 304:217–222
34. Liu B, Hui K, Qin F (2003) Thermodynamics of heat activation of single capsaicin ion channels VR1. Biophys J 85:2988–3006
35. Birder LA (2001) Involvement of the urinary bladder urothelium in signaling in the lower urinary tract. Proc West Pharmacol Soc 44:85–86
36. Daly D, Rong W, Chess-Williams R, Chapple C, Grundy D (2007) Bladder afferent sensitivity in wild-type and TRPV1 knockout mice. J Physiol 583:663–674
37. Avraham Y, Zolotarev O, Grigoriadis NC, Poutahidis T, Magen I, Vorobiav L, Zimmer A, Ilan Y, Mechoulam R, Berry EM (2008) Cannabinoids and capsaicin improve liver function following thioacetamide-induced acute injury in mice. Am J Gastroenterol 103:3047–3056
38. Zhang L, Taylor N, Xie Y, Ford R, Johnson J, Paulsen JE, Bates B (2005) Cloning and expression of MRG receptors in macaque, mouse, and human. Brain Res Mol Brain Res 133:187–197
39. Reilly CA, Taylor JL, Lanza DL, Carr BA, Crouch DJ, Yost GS (2003) Capsaicinoids cause inflammation and epithelial cell death through activation of vanilloid receptors. Toxicol Sci 73:170–181
40. Watanabe N, Horie S, Michael GJ, Spina D, Page CP, Priestley JV (2005) Immunohistochemical localization of vanilloid receptor subtype 1 (TRPV1) in the guinea pig respiratory system. Pulm Pharmacol Ther 18:187–197
41. Zhang X, Huang J, McNaughton PA (2005) NGF rapidly increases membrane expression of TRPV1 heat-gated ion channels. EMBO J 24:4211–4223
42. Ji RR, Samad TA, Jin SX, Schmoll R, Woolf CJ (2002) p38 MAPK activation by NGF in primary sensory neurons after inflammation increases TRPV1 levels and maintains heat hyperalgesia. Neuron 36:57–68
43. Lainez S, Valente P, Ontoria-Oviedo I, Estevez-Herrera J, Camprubi-Robles M, Ferrer-Montiel A, Planells-Cases R (2010) GABAA receptor associated protein (GABARAP) modulates TRPV1 expression and channel function and desensitization. Faseb J 24(6):1958–1970
44. Jeske NA, Patwardhan AM, Henry MA, Milam SB (2009) Fibronectin stimulates TRPV1 translocation in primary sensory neurons. J Neurochem 108:591–600
45. Lee YM, Li WH, Kim YK, Kim KH, Chung JH (2008) Heat-induced MMP-1 expression is mediated by TRPV1 through PKCalpha signaling in HaCaT cells. Exp Dermatol 17:864–870
46. Amaya F, Wang H, Costigan M, Allchorne AJ, Hatcher JP, Egerton J, Stean T, Morisset V, Grose D, Gunthorpe MJ, Chessell IP, Tate S, Green PJ, Woolf CJ (2006) The voltage-gated sodium channel Na(v)1.9 is an effector of peripheral inflammatory pain hypersensitivity. J Neurosci 26:12852–12860
47. Rashid MH, Inoue M, Bakoshi S, Ueda H (2003) Increased expression of vanilloid receptor 1 on myelinated primary afferent neurons contributes to the antihyperalgesic effect of capsaicin cream in diabetic neuropathic pain in mice. J Pharmacol Exp Ther 306:709–717
48. Kamei J, Zushida K, Morita K, Sasaki M, Tanaka S (2001) Role of vanilloid VR1 receptor in thermal allodynia and hyperalgesia in diabetic mice. Eur J Pharmacol 422:83–86
49. Hong S, Wiley JW (2005) Early painful diabetic neuropathy is associated with differential changes in the expression and function of vanilloid receptor 1. J Biol Chem 280:618–627
50. Hudson LJ, Bevan S, Wotherspoon G, Gentry C, Fox A, Winter J (2001) VR1 protein expression increases in undamaged DRG neurons after partial nerve injury. Eur J Neurosci 13:2105–2114
51. Kanai Y, Nakazato E, Fujiuchi A, Hara T, Imai A (2005) Involvement of an increased spinal TRPV1 sensitization through its up-regulation in mechanical allodynia of CCI rats. Neuropharmacology 49:977–984

52. Ghilardi JR, Rohrich H, Lindsay TH, Sevcik MA, Schwei MJ, Kubota K, Halvorson KG, Poblete J, Chaplan SR, Dubin AE, Carruthers NI, Swanson D, Kuskowski M, Flores CM, Julius D, Mantyh PW (2005) Selective blockade of the capsaicin receptor TRPV1 attenuates bone cancer pain. J Neurosci 25:3126–3131
53. Engler A, Aeschlimann A, Simmen BR, Michel BA, Gay RE, Gay S, Sprott H (2007) Expression of transient receptor potential vanilloid 1 (TRPV1) in synovial fibroblasts from patients with osteoarthritis and rheumatoid arthritis. Biochem Biophys Res Commun 359:884–888
54. Fernihough J, Gentry C, Bevan S, Winter J (2005) Regulation of calcitonin gene-related peptide and TRPV1 in a rat model of osteoarthritis. Neurosci Lett 388:75–80
55. Groneberg DA, Niimi A, Dinh QT, Cosio B, Hew M, Fischer A, Chung KF (2004) Increased expression of transient receptor potential vanilloid-1 in airway nerves of chronic cough. Am J Respir Crit Care Med 170:1276–1280
56. Chan CL, Facer P, Davis JB, Smith GD, Egerton J, Bountra C, Williams NS, Anand P (2003) Sensory fibres expressing capsaicin receptor TRPV1 in patients with rectal hypersensitivity and faecal urgency. Lancet 361:385–391
57. Banerjee B, Medda BK, Lazarova Z, Bansal N, Shaker R, Sengupta JN (2007) Effect of reflux-induced inflammation on transient receptor potential vanilloid one (TRPV1) expression in primary sensory neurons innervating the oesophagus of rats. Neurogastroenterol Motil 19:681–691
58. Matthews PJ, Aziz Q, Facer P, Davis JB, Thompson DG, Anand P (2004) Increased capsaicin receptor TRPV1 nerve fibres in the inflamed human oesophagus. Eur J Gastroenterol Hepatol 16:897–902
59. Charrua A, Cruz CD, Cruz F, Avelino A (2007) Transient receptor potential vanilloid subfamily 1 is essential for the generation of noxious bladder input and bladder overactivity in cystitis. J Urol 177:1537–1541
60. Wang ZY, Wang P, Merriam FV, Bjorling DE (2008) Lack of TRPV1 inhibits cystitis-induced increased mechanical sensitivity in mice. Pain 139:158–167
61. Dinis P, Charrua A, Avelino A, Yaqoob M, Bevan S, Nagy I, Cruz F (2004) Anandamide-evoked activation of vanilloid receptor 1 contributes to the development of bladder hyperreflexia and nociceptive transmission to spinal dorsal horn neurons in cystitis. J Neurosci 24:11253–11263
62. Jia Y, Lee LY (2007) Role of TRPV receptors in respiratory diseases. Biochim Biophys Acta 1772:915–927
63. Watanabe N, Horie S, Michael GJ, Keir S, Spina D, Page CP, Priestley JV (2006) Immunohistochemical co-localization of transient receptor potential vanilloid (TRPV)1 and sensory neuropeptides in the guinea-pig respiratory system. Neuroscience 141:1533–1543
64. Jia Y, McLeod RL, Hey JA (2005) TRPV1 receptor: a target for the treatment of pain, cough, airway disease and urinary incontinence. Drug News Perspect 18:165–171
65. Wang Y, Babankova D, Huang J, Swain GM, Wang DH (2008) Deletion of transient receptor potential vanilloid type 1 receptors exaggerates renal damage in deoxycorticosterone acetate-salt hypertension. Hypertension 52:264–270
66. Wimalawansa SJ (1996) Calcitonin gene-related peptide and its receptors: molecular genetics, physiology, pathophysiology, and therapeutic potentials. Endocr Rev 17:533–585
67. Woudenberg-Vrenken TE, Bindels RJ, Hoenderop JG (2009) The role of transient receptor potential channels in kidney disease. Nat Rev Nephrol 5:441–449
68. Kopp UC, Cicha MZ, Smith LA (2003) Dietary sodium loading increases arterial pressure in afferent renal-denervated rats. Hypertension 42:968–973
69. Wang Y, Wang DH (2006) A novel mechanism contributing to development of Dahl salt-sensitive hypertension: role of the transient receptor potential vanilloid type 1. Hypertension 47:609–614

70. Koomans HA, Blankestijn PJ, Joles JA (2004) Sympathetic hyperactivity in chronic renal failure: a wake-up call. J Am Soc Nephrol 15:524–537
71. Charrua A, Reguenga C, Cordeiro JM, Correiade-Sa P, Paule C, Nagy I, Cruz F, Avelino A (2009) Functional transient receptor potential vanilloid 1 is expressed in human urothelial cells. J Urol 182:2944–2950
72. Birder LA, Nakamura Y, Kiss S, Nealen ML, Barrick S, Kanai AJ, Wang E, Ruiz G, De Groat WC, Apodaca G, Watkins S, Caterina MJ (2002) Altered urinary bladder function in mice lacking the vanilloid receptor TRPV1. Nat Neurosci 5:856–860
73. Cefalu JS, Guillon MA, Burbach LR, Zhu QM, Hu DQ, Ho MJ, Ford AP, Nunn PA, Cockayne DA (2009) Selective pharmacological blockade of the TRPV1 receptor suppresses sensory reflexes of the rodent bladder. J Urol 182:776–785
74. Birder LA (2007) TRPs in bladder diseases. Biochim Biophys Acta 1772:879–884
75. Sadananda P, Shang F, Liu L, Mansfield KJ, Burcher E (2009) Release of ATP from rat urinary bladder mucosa: role of acid, vanilloids and stretch. Br J Pharmacol 158: 1655–1662
76. Babes A, Amuzescu B, Krause U, Scholz A, Flonta ML, Reid G (2002) Cooling inhibits capsaicin-induced currents in cultured rat dorsal root ganglion neurones. Neurosci Lett 317:131–134
77. Romanovsky AA, Almeida MC, Garami A, Steiner AA, Norman MH, Morrison SF, Nakamura K, Burmeister JJ, Nucci TB (2009) The transient receptor potential vanilloid-1 channel in thermoregulation: a thermosensor it is not. Pharmacol Rev 61:228–261
78. Nilius B, Talavera K, Owsianik G, Prenen J, Droogmans G, Voets T (2005) Gating of TRP channels: a voltage connection?. J Physiol 567:35–44
79. Pingle SC, Matta JA, Ahern GP (2007) Capsaicin receptor: TRPV1 a promiscuous TRP channel. Handb Exp Pharmacol 179:155-171
80. Vyklicky L, Lyfenko A, Kuffler DP, Vlachova V (2003) Vanilloid receptor TRPV1 is not activated by vanilloids applied intracellularly. NeuroReport 14:1061–1065
81. Woo DH, Jung SJ, Zhu MH, Park CK, Kim YH, Oh SB, Lee CJ (2008) Direct activation of transient receptor potential vanilloid 1(TRPV1) by diacylglycerol (DAG). Mol Pain 4:42
82. Sutton KG, Garrett EM, Rutter AR, Bonnert TP, Jarolimek W, Seabrook GR (2005) Functional characterisation of the S512Y mutant vanilloid human TRPV1 receptor. Br J Pharmacol 146:702–711
83. Premkumar LS, Agarwal S, Steffen D (2002) Single-channel properties of native and cloned rat vanilloid receptors. J Physiol 545:107–117
84. Holzer P (2009) Acid-sensitive ion channels and receptors. Handb Exp Pharmacol 194: 283–332
85. Jordt SE, Tominaga M, Julius D (2000) Acid potentiation of the capsaicin receptor determined by a key extracellular site. Proc Natl Acad Sci USA 97:8134–8139
86. Gavva NR, Tamir R, Klionsky L, Norman MH, Louis JC, Wild KD, Treanor JJ (2005) Proton activation does not alter antagonist interaction with the capsaicin-binding pocket of TRPV1. Mol Pharmacol 68:1524–1533
87. Tousova K, Vyklicky L, Susankova K, Benedikt J, Vlachova V (2005) Gadolinium activates and sensitizes the vanilloid receptor TRPV1 through the external protonation sites. Mol Cell Neurosci 30:207–217
88. Ahern GP, Brooks IM, Miyares RL, Wang XB (2005) Extracellular cations sensitize and gate capsaicin receptor TRPV1 modulating pain signaling. J Neurosci 25:5109–5116
89. Raisinghani M, Pabbidi RM, Premkumar LS (2005) Activation of transient receptor potential vanilloid 1 (TRPV1) by resiniferatoxin. J Physiol 567:771–786
90. Ahern GP, Wang X, Miyares RL (2006) Polyamines are potent ligands for the capsaicin receptor TRPV1. J Biol Chem 281:8991–8995
91. Valente P, Garcia-Sanz N, Gomis A, Fernandez-Carvajal A, Fernandez-Ballester G, Viana F, Belmonte C, Ferrer-Montiel A (2008) Identification of molecular determinants of channel gating in the transient receptor potential box of vanilloid receptor I. FASEB J 22:3298–3309

92. Owsianik G, Talavera K, Voets T, Nilius B (2006) Permeation and selectivity of TRP channels. Annu Rev Physiol 68:685–717
93. Hellwig N, Plant TD, Janson W, Schafer M, Schultz G, Schaefer M (2004) TRPV1 acts as proton channel to induce acidification in nociceptive neurons. J Biol Chem 279: 34553–34561
94. Voets T, Janssens A, Droogmans G, Nilius B (2004) Outer pore architecture of a Ca2+-selective TRP channel. J Biol Chem 279:15223–15230
95. Voets T, Nilius B (2003) The pore of TRP channels: trivial or neglected?. Cell Calcium 33:299–302
96. Voets T, Prenen J, Fleig A, Vennekens R, Watanabe H, Hoenderop JG, Bindels RJ, Droogmans G, Penner R, Nilius B (2001) CaT1 and the calcium release-activated calcium channel manifest distinct pore properties. J Biol Chem 276:47767–47770
97. Xia R, Mei ZZ, Mao HJ, Yang W, Dong L, Bradley H, Beech DJ, Jiang LH (2008) Identification of pore residues engaged in determining divalent cationic permeation in transient receptor potential melastatin subtype channel 2. J Biol Chem 283: 27426–27432
98. Voets T, Prenen J, Vriens J, Watanabe H, Janssens A, Wissenbach U, Bodding M, Droogmans G, Nilius B (2002) Molecular determinants of permeation through the cation channel TRPV4. J Biol Chem 277:33704–33710
99. Mohapatra DP, Nau C (2003) Desensitization of capsaicin-activated currents in the vanilloid receptor TRPV1 is decreased by the cyclic AMP-dependent protein kinase pathway. J Biol Chem 278:50080–50090
100. Chung MK, Guler AD, Caterina MJ (2008) TRPV1 shows dynamic ionic selectivity during agonist stimulation. Nat Neurosci 11:555–564
101. Mohapatra DP, Nau C (2005) Regulation of Ca2+-dependent desensitization in the vanilloid receptor TRPV1 by calcineurin and cAMP-dependent protein kinase. J Biol Chem 280:13424–13432
102. Mandadi S, Tominaga T, Numazaki M, Murayama N, Saito N, Armati PJ, Roufogalis BD, Tominaga M (2006) Increased sensitivity of desensitized TRPV1 by PMA occurs through PKCepsilon-mediated phosphorylation at S800. Pain 123:106–116
103. Jung J, Shin JS, Lee SY, Hwang SW, Koo J, Cho H, Oh U (2004) Phosphorylation of vanilloid receptor 1 by Ca2+/calmodulin-dependent kinase II regulates its vanilloid binding. J Biol Chem 279:7048–7054
104. Novakova-Tousova K, Vyklicky L, Susankova K, Benedikt J, Samad A, Teisinger J, Vlachova V (2007) Functional changes in the vanilloid receptor subtype 1 channel during and after acute desensitization. Neuroscience 149:144–154
105. Jahnel R, Bender O, Munter LM, Dreger M, Gillen C, Hucho F (2003) Dual expression of mouse and rat VRL-1 in the dorsal root ganglion derived cell line F-11 and biochemical analysis of VRL-1 after heterologous expression. Eur J Biochem 270:4264–4271
106. Korepanova A, Pereda-Lopez A, Solomon LR, Walter KA, Lake MR, Bianchi BR, McDonald HA, Neelands TR, Shen J, Matayoshi ED, Moreland RB, Chiu ML (2009) Expression and purification of human TRPV1 in baculovirus-infected insect cells for structural studies. Protein Expr Purif 65:38–50
107. Caterina MJ, Rosen TA, Tominaga M, Brake AJ, Julius D (1999) A capsaicin-receptor homologue with a high threshold for noxious heat. Nature 398:436–441
108. Bylund DB, Deupree JD, Toews ML (2004) Radioligand-binding methods for membrane preparations and intact cells. Methods Mol Biol 259:1–28
109. Salazar H, Llorente I, Jara-Oseguera A, Garcia-Villegas R, Munari M, Gordon SE, Islas LD, Rosenbaum T (2008) A single N-terminal cysteine in TRPV1 determines activation by pungent compounds from onion and garlic. Nat Neurosci 11:255–261
110. Dabrowski M (2009) Novel approaches in ion channel lead generation. In: Biophysical Society 53rd Annual Meeting, Boston, MA
111. Nanion Application Notes (2009) http://www.nanion.de/pdf/Patchliner_TRPV1.pdf

112. Smart D, Jerman JC, Gunthorpe MJ, Brough SJ, Ranson J, Cairns W, Hayes PD, Randall AD, Davis JB (2001) Characterisation using FLIPR of human vanilloid VR1 receptor pharmacology. Eur J Pharmacol 417:51–58
113. Schroeder KS, Neagle BD (1996) FLIPR: a new instrument for accurate, high throughput optical screening. J Biomol Screen 1:5
114. Reubish D, Emerling D, Defalco J, Steiger D, Victoria C, Vincent F (2009) Functional assessment of temperature-gated ion-channel activity using a real-time PCR machine. Biotechniques 47:iii–ix
115. Kym PR, Kort ME, Hutchins CW (2009) Analgesic potential of TRPV1 antagonists. Biochem Pharmacol 78:211–216
116. Vriens J, Appendino G, Nilius B (2009) Pharmacology of vanilloid transient receptor potential cation channels. Mol Pharmacol 75:1262–1279
117. Appendino G, Minassi A, Pagani A, Ech-Chahad A (2008) The role of natural products in the ligand deorphanization of TRP channels. Curr Pharm Des 14:2–17
118. Ahern GP (2003) Activation of TRPV1 by the satiety factor oleoylethanolamide. J Biol Chem 278:30429–30434
119. Wang X, Miyares RL, Ahern GP (2005) Oleoylethanolamide excites vagal sensory neurones, induces visceral pain and reduces short-term food intake in mice via capsaicin receptor TRPV1. J Physiol 564:541–547
120. Hwang SW, Oh U (2002) Hot channels in airways: pharmacology of the vanilloid receptor. Curr Opin Pharmacol 2:235–242
121. Szallasi A, Appendino G (2004) Vanilloid receptor TRPV1 antagonists as the next generation of painkillers. Are we putting the cart before the horse?. J Med Chem 47:2717–2723
122. McNamara FN, Randall A, Gunthorpe MJ (2005) Effects of piperine, the pungent component of black pepper, at the human vanilloid receptor (TRPV1. Br J Pharmacol 144:781–790
123. Xu H, Blair NT, Clapham DE (2005) Camphor activates and strongly desensitizes the transient receptor potential vanilloid subtype 1 channel in a vanilloid-independent mechanism. J Neurosci 25:8924–8937
124. Siemens J, Zhou S, Piskorowski R, Nikai T, Lumpkin EA, Basbaum AI, King D, Julius D (2006) Spider toxins activate the capsaicin receptor to produce inflammatory pain. Nature 444:208–212
125. Dray A, Bettaney J, Forster P (1990) Resiniferatoxin, a potent capsaicin-like stimulator of peripheral nociceptors in the neonatal rat tail in vitro. Br J Pharmacol 99:323–326
126. Docherty RJ, Yeats JC, Piper AS (1997) Capsazepine block of voltage-activated calcium channels in adult rat dorsal root ganglion neurones in culture. Br J Pharmacol 121: 1461–1467
127. Liu L, Simon SA (1997) Capsazepine, a vanilloid receptor antagonist, inhibits nicotinic acetylcholine receptors in rat trigeminal ganglia. Neurosci Lett 228:29–32
128. Pal M, Angaru S, Kodimuthali A, Dhingra N (2009) Vanilloid receptor antagonists: emerging class of novel anti-inflammatory agents for pain management. Curr Pharm Des 15: 1008–1026
129. Gharat LA (2008) Szallasi, Arpad: Advances in the design and therapeutic use of capsaicin receptor TRPV1 agonists and antagonists. Expert Opin Ther Pat 18:10
130. Rami HK, Thompson M, Stemp G, Fell S, Jerman JC, Stevens AJ, Smart D, Sargent B, Sanderson D, Randall AD, Gunthorpe MJ, Davis JB (2006) Discovery of SB-705498: a potent, selective and orally bioavailable TRPV1 antagonist suitable for clinical development. Bioorg Med Chem Lett 16:3287–3291
131. Chizh BA, O'Donnell MB, Napolitano A, Wang J, Brooke AC, Aylott MC, Bullman JN, Gray EJ, Lai RY, Williams PM, Appleby JM (2007) The effects of the TRPV1 antagonist SB-705498 on TRPV1 receptor-mediated activity and inflammatory hyperalgesia in humans. Pain 132:132–141
132. Honore P, Wismer CT, Mikusa J, Zhu CZ, Zhong C, Gauvin DM, Gomtsyan A, El Kouhen R, Lee CH, Marsh K, Sullivan JP, Faltynek CR, Jarvis MF (2005) A-425619 [1-isoquinolin-5-

yl-3-(4-trifluoromethyl-benzyl)-urea], a novel transient receptor potential type V1 receptor antagonist, relieves pathophysiological pain associated with inflammation and tissue injury in rats. J Pharmacol Exp Ther 314:410–421

133. El Kouhen R, Surowy CS, Bianchi BR, Neelands TR, McDonald HA, Niforatos W, Gomtsyan A, Lee CH, Honore P, Sullivan JP, Jarvis MF, Faltynek CR (2005) A-425619 [1-isoquinolin-5-yl-3-(4-trifluoromethyl-benzyl)-urea], a novel and selective transient receptor potential type V1 receptor antagonist, blocks channel activation by vanilloids, heat, and acid. J Pharmacol Exp Ther 314:400–409

134. McGaraughty S, Chu KL, Faltynek CR, Jarvis MF (2006) Systemic and site-specific effects of A-425619, a selective TRPV1 receptor antagonist, on wide dynamic range neurons in CFA-treated and uninjured rats. J Neurophysiol 95:18–25

135. Surowy CS, Neelands TR, Bianchi BR, McGaraughty S, El Kouhen R, Han P, Chu KL, McDonald HA, Vos M, Niforatos W, Bayburt EK, Gomtsyan A, Lee CH, Honore P, Sullivan JP, Jarvis MF, Faltynek CR (2008) R)-(5-tert-butyl-2,3-dihydro-1H-inden-1-yl)-3-(1H-indazol-4-yl)-urea (ABT-102) blocks polymodal activation of transient receptor potential vanilloid 1 receptors in vitro and heat-evoked firing of spinal dorsal horn neurons in vivo. J Pharmacol Exp Ther 326:879–888

136. Gomtsyan A, Bayburt EK, Schmidt RG, Surowy CS, Honore P, Marsh KC, Hannick SM, McDonald HA, Wetter JM, Sullivan JP, Jarvis MF, Faltynek CR, Lee CH (2008) Identification of (R)-1-(5-tert-butyl-2,3-dihydro-1H-inden-1-yl)-3-(1H-indazol-4-yl)urea (ABT-102) as a potent TRPV1 antagonist for pain management. J Med Chem 51:392–395

137. Valenzano KJ, Grant ER, Wu G, Hachicha M, Schmid L, Tafesse L, Sun Q, Rotshteyn Y, Francis J, Limberis J, Malik S, Whittemore ER, Hodges D (2003) N-4-tertiarybutylphenyl)-4-(3-chloropyridin-2-yl)tetrahydropyrazine -1(2H)-carbox-amide (BCTC), a novel, orally effective vanilloid receptor 1 antagonist with analgesic properties: I. in vitro characterization and pharmacokinetic properties. J Pharmacol Exp Ther 306:377–386

138. Pomonis JD, Harrison JE, Mark L, Bristol DR, Valenzano KJ, Walker K (2003) N-4-Tertiarybutylphenyl)-4-(3-cholorphyridin-2-yl)tetrahydropyrazine -1(2H)-carbox-amide (BCTC), a novel, orally effective vanilloid receptor 1 antagonist with analgesic properties: II. in vivo characterization in rat models of inflammatory and neuropathic pain. J Pharmacol Exp Ther 306:387–393

139. Swanson DM, Dubin AE, Shah C, Nasser N, Chang L, Dax SL, Jetter M, Breitenbucher JG, Liu C, Mazur C, Lord B, Gonzales L, Hoey K, Rizzolio M, Bogenstaetter M, Codd EE, Lee DH, Zhang SP, Chaplan SR, Carruthers NI (2005) Identification and biological evaluation of 4-(3-trifluoromethylpyridin-2-yl)piperazine-1-carboxylic acid (5-trifluoromethylpyridin-2-yl)amide, a high affinity TRPV1 (VR1) vanilloid receptor antagonist. J Med Chem 48:1857–1872

140. Brown BS, Keddy R, Zheng GZ, Schmidt RG, Koenig JR, McDonald HA, Bianchi BR, Honore P, Jarvis MF, Surowy CS, Polakowski JS, Marsh KC, Faltynek CR, Lee CH (2008) Tetrahydropyridine-4-carboxamides as novel, potent transient receptor potential vanilloid 1 (TRPV1) antagonists. Bioorg Med Chem 16:8516–8525

141. Lappin SC, Randall AD, Gunthorpe MJ, Morisset V (2006) TRPV1 antagonist, SB-366791, inhibits glutamatergic synaptic transmission in rat spinal dorsal horn following peripheral inflammation. Eur J Pharmacol 540:73–81

142. Gunthorpe MJ, Rami HK, Jerman JC, Smart D, Gill CH, Soffin EM, Luis Hannan S, Lappin SC, Egerton J, Smith GD, Worby A, Howett L, Owen D, Nasir S, Davies CH, Thompson M, Wyman PA, Randall AD, Davis JB (2004) Identification and characterisation of SB-366791, a potent and selective vanilloid receptor (VR1/TRPV1) antagonist. Neuropharmacology 46:133–149

143. Gavva NR, Tamir R, Qu Y, Klionsky L, Zhang TJ, Immke D, Wang J, Zhu D, Vanderah TW, Porreca F, Doherty EM, Norman MH, Wild KD, Bannon AW, Louis JC,

Treanor JJ (2005) AMG9810 [(E)-3-(4-t-butylphenyl)-N-(2,3-dihydrobenzo[b][1,4] dioxin-6-yl)acrylamide], a novel vanilloid receptor 1 (TRPV1) antagonist with antihyperalgesic properties. J Pharmacol Exp Ther 313:474–484
144. Szallasi A, Cortright DN, Blum CA, Eid SR (2007) The vanilloid receptor TRPV1: 10 years from channel cloning to antagonist proof-of-concept. Nat Rev Drug Discov 6: 357–372
145. Ognyanov VI, Balan C, Bannon AW, Bo Y, Dominguez C, Fotsch C, Gore VK, Klionsky L, Ma VV, Qian YX, Tamir R, Wang X, Xi N, Xu S, Zhu D, Gavva NR, Treanor JJ, Norman MH (2006) Design of potent, orally available antagonists of the transient receptor potential vanilloid 1. Structure-activity relationships of 2-piperazin-1-yl-1H-benzimidazoles. J Med Chem 49:3719–3742
146. Shishido Y, Jinno M, Ikeda T, Ito F, Sudo M, Makita N, Ohta A, Iki-Taki A, Ohmi T, Kanai Y, Tamura T, Shimojo M (2008) Synthesis of benzamide derivatives as TRPV1 antagonists. Bioorg Med Chem Lett 18:1072–1078
147. Westaway SM, Chung YK, Davis JB, Holland V, Jerman JC, Medhurst SJ, Rami HK, Stemp G, Stevens AJ, Thompson M, Winborn KY, Wright J (2006) N-Tetrahydroquinolinyl, N-quinolinyl and N-isoquinolinyl biaryl carboxamides as antagonists of TRPV1. Bioorg Med Chem Lett 16:4533–4536
148. Zheng X, Hodgetts KJ, Brielmann H, Hutchison A, Burkamp F, Brian Jones A, Blurton P, Clarkson R, Chandrasekhar J, Bakthavatchalam R, De Lombaert S, Crandall M, Cortright D, Blum CA (2006) From arylureas to biarylamides to aminoquinazolines: discovery of a novel, potent TRPV1 antagonist. Bioorg Med Chem Lett 16:5217–5221
149. Culshaw AJ, Bevan S, Christiansen M, Copp P, Davis A, Davis C, Dyson A, Dziadulewicz EK, Edwards L, Eggelte H, Fox A, Gentry C, Groarke A, Hallett A, Hart TW, Hughes GA, Knights S, Kotsonis P, Lee W, Lyothier I, McBryde A, McIntyre P, Paloumbis G, Panesar M, Patel S, Seiler MP, Yaqoob M, Zimmermann K (2006) Identification and biological characterization of 6-aryl-7-isopropylquinazolinones as novel TRPV1 antagonists that are effective in models of chronic pain. J Med Chem 49:471–474
150. Crutchlow M (2009) Pharmacologic Inhibition of TRPV1 Impairs Sensation of Potentially Injurious Heat in Healthy Subjects. *American Society for Clinical Pharmacology and Therapeutics (ASCPT) meeting*
151. Harrop S (2006) Pain therapeutics-SMi's seventh annual conference. IDrugs 9:548–550
152. Gunthorpe MJ, Chizh BA (2009) Clinical development of TRPV1 antagonists: targeting a pivotal point in the pain pathway. Drug Discov Today 14:56–67
153. Keith Bley Preclinical data were reported at the SMi Pain Therapeutics Meeting in London, UK, June 2006
154. Denney WS (2009) 5th Modern Drug Discovery & Development Summit. Merck & Co, Inc, San Diego, CA
155. Gavva NR, Treanor JJ, Garami A, Fang L, Surapaneni S, Akrami A, Alvarez F, Bak A, Darling M, Gore A, Jang GR, Kesslak JP, Ni L, Norman MH, Palluconi G, Rose MJ, Salfi M, Tan E, Romanovsky AA, Banfield C, Davar G (2008) Pharmacological blockade of the vanilloid receptor TRPV1 elicits marked hyperthermia in humans. Pain 136:202–210
156. Samer RE TRP1 antaonists: are they too hot to handle? *Spring pain research conference* cayman, United Kingdom 2008
157. Gavva NR (2008) Body-temperature maintenance as the predominant function of the vanilloid receptor TRPV1. Trends Pharmacol Sci 29:550–557
158. Lehto SG, Tamir R, Deng H, Klionsky L, Kuang R, Le A, Lee D, Louis JC, Magal E, Manning BH, Rubino J, Surapaneni S, Tamayo N, Wang T, Wang J, Wang J, Wang W, Youngblood B, Zhang M, Zhu D, Norman MH, Gavva NR (2008) Antihyperalgesic effects of (R,E)-N-(2-hydroxy-2,3-dihydro-1H-inden-4-yl)-3-(2-(piperidin-1-yl)-4-(tri fluoromethyl)phenyl)-acrylamide (AMG8562), a novel transient receptor potential vanilloid type 1 modulator that does not cause hyperthermia in rats. J Pharmacol Exp Ther 326:218–229

159. Garami A, Shimansky YP, Pakai E, Oliveira DL, Gavva NR, Romanovsky AA (2010) Contributions of different modes of TRPV1 activation to TRPV1 antagonist-induced hyperthermia. J Neurosci 30:1435–1440
160. Seabrook GR, Sutton KG, Jarolimek W, Hollingworth GJ, Teague S, Webb J, Clark N, Boyce S, Kerby J, Ali Z, Chou M, Middleton R, Kaczorowski G, Jones AB (2002) Functional properties of the high-affinity TRPV1 (VR1) vanilloid receptor antagonist (4-hydroxy-5-iodo-3-methoxyphenylacetate ester) iodo-resiniferatoxin. J Pharmacol Exp Ther 303: 1052–1060
161. Gavva NR, Klionsky L, Qu Y, Shi L, Tamir R, Edenson S, Zhang TJ, Viswanadhan VN, Toth A, Pearce LV, Vanderah TW, Porreca F, Blumberg PM, Lile J, Sun Y, Wild K, Louis JC, Treanor JJ (2004) Molecular determinants of vanilloid sensitivity in TRPV1. J Biol Chem 279:20283–20295
162. Neelands TR, Jarvis MF, Han P, Faltynek CR, Surowy CS (2005) Acidification of rat TRPV1 alters the kinetics of capsaicin responses. Mol Pain 1:28
163. Gavva NR, Bannon AW, Surapaneni S, Hovland DN Jr., Lehto SG, Gore A, Juan T, Deng H, Han B, Klionsky L, Kuang R, Le A, TamirR, Wang J, Youngblood B, Zhu D, Norman MH, Magal E, Treanor JJ, Louis JC (2007) The vanilloid receptor TRPV1 is tonically activated in vivo and involved in body temperature regulation. J Neurosci 27:3366–3374
164. Gavva NR, Bannon AW, Hovland DN Jr., Lehto SG, Klionsky L, Surapaneni S, Immke DC, Henley C, Arik L, Bak A, Davis J, Ernst N, Hever G, Kuang R, Shi L, Tamir R, Wang J, Wang W, Zajic G, Zhu D, Norman MH, Louis JC, Magal E, Treanor JJ (2007) Repeated administration of vanilloid receptor TRPV1 antagonists attenuates hyperthermia elicited by TRPV1 blockade. J Pharmacol Exp Ther 323:128–137
165. Cuypers E, Yanagihara A, Karlsson E, Tytgat J (2006) Jellyfish and other cnidarian envenomations cause pain by affecting TRPV1 channels. FEBS Lett 580:5728–5732
166. Andreev YA, Kozlov SA, Koshelev SG, Ivanova EA, Monastyrnaya MM, Kozlovskaya EP, Grishin EV (2008) Analgesic compound from sea anemone Heteractis crispa is the first polypeptide inhibitor of vanilloid receptor 1 (TRPV1. J Biol Chem 283:23914–23921
167. Knotkova H, Pappagallo M, Szallasi A (2008) Capsaicin (TRPV1 Agonist) therapy for pain relief: farewell or revival?. Clin J Pain 24:142–154
168. Reeves PJ, Callewaert N, Contreras R, Khorana HG (2002) Structure and function in rhodopsin: high-level expression of rhodopsin with restricted and homogeneous N-glycosylation by a tetracycline-inducible N-acetylglucosaminyltransferase I-negative HEK293S stable mammalian cell line. Proc Natl Acad Sci U S A 99:13419–13424
169. Durocher Y, Perret S, Kamen A (2002) High-level and high-throughput recombinant protein production by transient transfection of suspension-growing human 293-EBNA1 cells. Nucleic Acids Res 30:E9
170. Cortright DN, Szallasi A (2009) TRP channels and pain. Curr Pharm Des 15:1736–1749
171. Yoshimura N, Kaiho Y, Miyazato M, Yunoki T, Tai C, Chancellor MB, Tyagi P (2008) Therapeutic receptor targets for lower urinary tract dysfunction. Naunyn Schmiedebergs Arch Pharmacol 377:437–448
172. Liddle RA (2007) The role of Transient Receptor Potential Vanilloid 1 (TRPV1) channels in pancreatitis. Biochim Biophys Acta 1772:869–878

Chapter 35
Expression and Function of TRP Channels in Liver Cells

Grigori Y. Rychkov and Gregory J. Barritt

Abstract The liver plays a central role in whole body homeostasis by mediating the metabolism of carbohydrates, fats, proteins, drugs and xenobiotic compounds, and bile acid and protein secretion. Hepatocytes together with endothelial cells, Kupffer cells, smooth muscle cells, stellate and oval cells comprise the functioning liver. Many members of the TRP family of proteins are expressed in hepatocytes. However, knowledge of their cellular functions is limited. There is some evidence which suggests the involvement of TRPC1 in volume control, TRPV1 and V4 in cell migration, TRPC6 and TRPM7 in cell proliferation, and TRPPM in lysosomal Ca^{2+} release. Altered expression of some TRP proteins, including TRPC6, TRPM2 and TRPV1, in tumorigenic cell lines may play roles in the development and progression of hepatocellular carcinoma and metastatic liver cancers. It is likely that future experiments will define important roles for other TRP proteins in the cellular functions of hepatocytes and other cell types of which the liver is composed.

35.1 Introduction

The liver plays a central role in intermediary metabolism, the detoxification of endogenous and exogenous compounds, and in whole body homeostasis. The liver is responsible for the metabolism of carbohydrates, fats, proteins, and xenobiotic compounds, the synthesis and transcellular movement of bile acids and bile fluid, and the synthesis and secretion of numerous plasma proteins [1, 2]. The predominant cell type in the liver is the hepatocyte (parenchymal cell), which comprises about 70% of all cells (equivalent to 90% of the liver volume) [1, 3, 4]. In addition to hepatocytes, endothelial cells, biliary epithelial cells (cholangiocytes), hepatic stellate cells, Kupffer cells (macrophages), oval cells and vascular smooth muscle cells are

G.Y. Rychkov (✉)
Department of Physiology, University of Adelaide, Adelaide, SA 5001, Australia
e-mail: grigori.rychkov@adelaide.edu.au

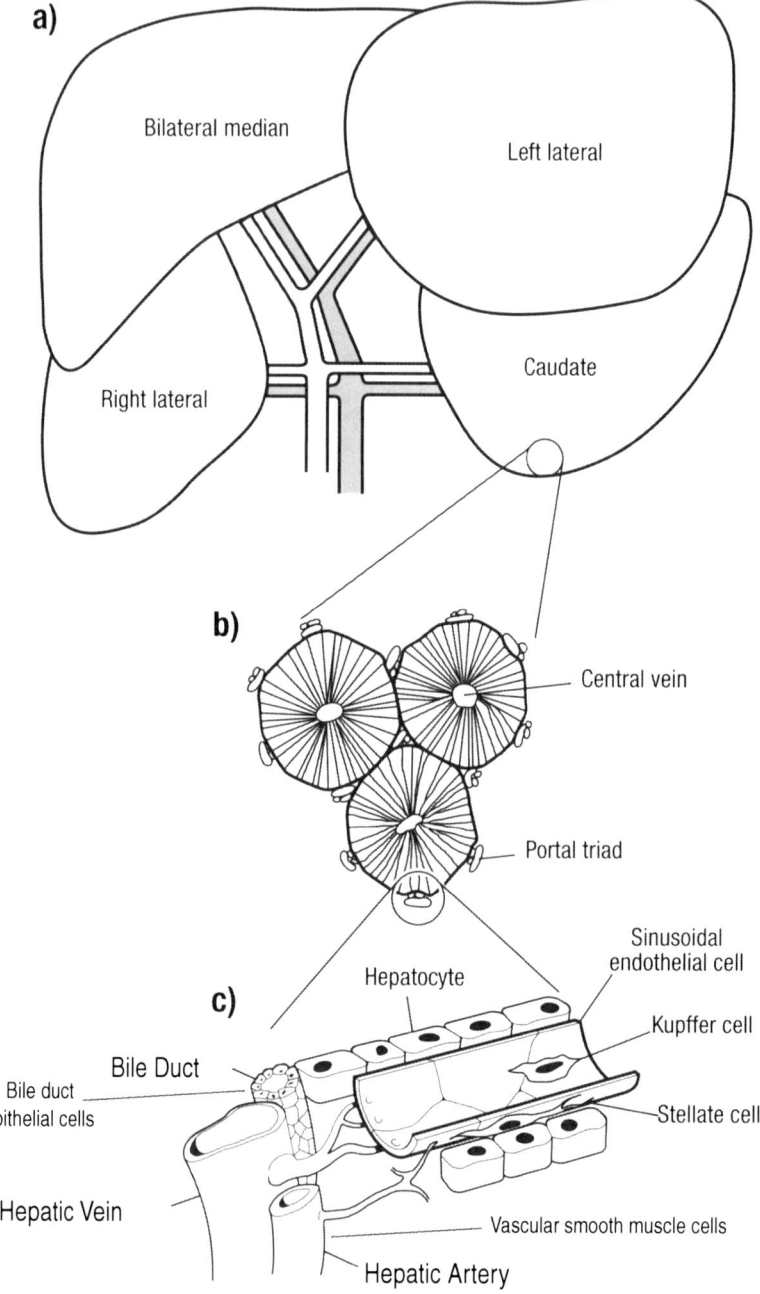

Fig. 35.1 Schematic drawing of some features of liver anatomy and the organisation of hepatocytes and other liver cell types in liver lobules. The major lobes of rat liver (**a**), the relationship between the central vein and portal triads (**b**), and the arrangement of the hepatic vein, hepatic artery, bile duct, hepatocyte plate and sinusoidal space (**c**) are shown. These cell types, in addition to hepatocytes, are shown in (**c**). Re-drawn with permission from [22]

required for normal liver physiology. Oval cells are thought to be precursors of hepatocytes and some other liver cell types. The organisation of the hepatocyte plates, blood vessels, bile ducts and these other cell types in liver is shown schematically in Fig. 35.1. Most of the metabolic, biosynthetic, biodegrative and secretory functions of the liver are carried out by hepatocytes. Consistent with the multiple functions of the liver and its complex architecture, hepatocytes are highly differentiated cells which exhibit spatial polarisation and a characteristic intracellular organization [1, 5–8].

Changes in hepatocyte $[Ca^{2+}]_{cyt}$ regulate glucose, fatty acid, amino acid and xenobiotic metabolism, bile acid secretion, protein synthesis and secretion, the movement of lysosomes and other vesicles, the cell cycle and cell proliferation, and apoptosis and necrosis [1, 2, 9–12]. Changes in the concentration of Ca^{2+} in intracellular organelles also play important regulatory roles. Thus the concentration of Ca^{2+} in the mitochondrial matrix ($[Ca^{2+}]_{mt}$) regulates the citric acid cycle and ATP synthesis [13] and apoptosis [11], the concentration of Ca^{2+} in the endoplasmic reticulum (ER) ($[Ca^{2+}]_{er}$) regulates protein synthesis and the metabolism of xenobiotic compounds [14], and the concentration of Ca^{2+} in the nucleus regulates cell proliferation [15].

The main extracellular signals which employ Ca^{2+} as an intracellular messenger in hepatocytes are hormones and local hormones [9, 16–20]. Hormones include epinephrine, norepinephrine, vasopressin, angiotensin II, glucagon and insulin. Local hormones include ATP, ADP, nitric oxide (NO), prostaglandins, serotonin, cytokines, other growth factors, neurotransmitters and extracellular Ca^{2+}. Changes in $[Ca^{2+}]_{cyt}$ can also be initiated by cell injury often mediated by the formation of reactive oxygen species [21]. Increases in $[Ca^{2+}]_{cyt}$ induced by each of these signals are due to enhanced Ca^{2+} entry across the plasma membrane and/or the release of Ca^{2+} from the ER and possibly other intracellular organelles. Ca^{2+} entry is mediated by store-operated (SOC), receptor-activated, stretch-activated and ligand-gated Ca^{2+}-permeable channels (reviewed in [22, 23]). There is no evidence for expression of voltage-operated Ca^{2+} channels (VOCCs) in hepatocytes [24–27]. The main components of the intracellular Ca^{2+} signalling system in hepatocytes are shown schematically in Fig. 35.2.

While it is difficult, or not currently possible, to assess whether any one type of Ca^{2+} entry channel plays a predominant role in hepatocyte physiology, it is likely that SOCs and the receptor-activated Ca^{2+} permeable channels are required in many processes which regulate $[Ca^{2+}]_{cyt}$. TRP channels most likely account for most of the receptor-activated Ca^{2+} permeable channels in hepatocytes, although the molecular identity and function of only a few of the channels has been reasonably well established. The aims of this review are to summarise current knowledge of the expression of TRP proteins in hepatocytes and other cell types which comprise the liver, and to discuss the functions of those TRP proteins which have so far been investigated. Changes in TRP protein expression in liver cancer and in other liver diseases will also be discussed since these observations may provide information on TRP channel function in normal liver cells, and may also be important in liver pathology.

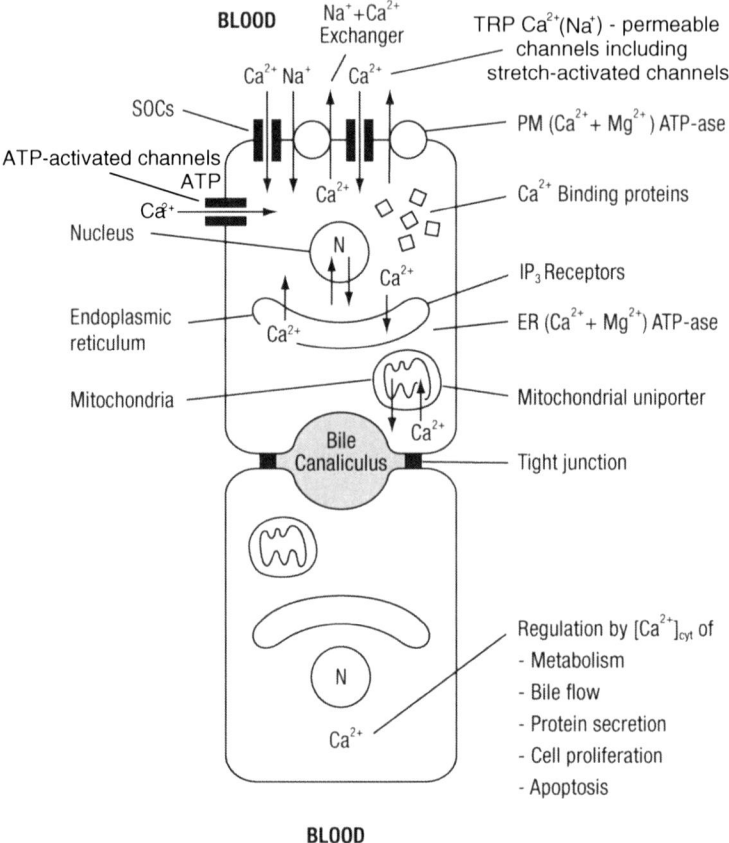

Fig. 35.2 The major elements which regulate the distribution and movement of intracellular Ca^{2+} in hepatocytes. The plasma membrane Ca^{2+} entry pathways are Ca^{2+}-selective store-operated Ca^{2+} channels (SOCs), TRP and possibly other receptor-activated Ca^{2+}-permeable channels, and ATP-activated Ca^{2+}-permeable channels [78] (ligand-gated Ca^{2+}-permeable channels). Ca^{2+} outflow across the plasma membrane is mediated by several plasma membrane (Ca^{2+} + Mg^{2+})ATP-ases [79], and by the Na^+–Ca^{2+} exchanger [80–82]. Ca^{2+} uptake by the ER is mediated by the ER (Ca^{2+} + Mg^{2+})ATP-ase (SERCAs) and Ca^{2+} outflow from the ER by types 1 and 2 IP_3 receptors [83, 84] and ryanodine receptors [85]. Ca^{2+} uptake by mitochondria is mediated by an electrogenic Ca^{2+} uniporter and Ca^{2+} outflow by Na^+/Ca^{2+} and H^+/Ca^{2+} antiporters (reviewed in [82, 86, 87]). Golgi also possess IP_3 receptors and (Ca^{2+} + Mg^{2+})ATP-ases [88]. Numerous Ca^{2+} binding proteins are present in the cytoplasmic space and in organelles. Re-drawn with permission from [22]

35.2 TRP Channels Expressed in the Liver

The expression of many members of the TRP channel family has been detected in liver tissue, in isolated hepatocytes and in immortalised liver cell lines (Table 35.1). The techniques employed have included PCR, quantitative PCR, in situ hybridisation, western blot, immunofluorescence and, in a few cases, measurement of channel

Table 35.1 TRP channels expressed in liver and liver cell lines

TRP protein	Technique employed for detection	Liver tissue, hepatocytes or liver cell line	References
TRPC1	PCR Western blot Immunofluorescence	Mouse liver Rat liver, foetal liver, hepatocytes Human liver AML12 Mouse liver cell line H4-IIE Rat liver cell line HepG2 and Huh-7 human cell lines	[23, 28, 42, 43, 90, 91]
TRPC2	PCR	Mouse liver Rat liver Rat hepatocytes AML12 mouse liver cell line H4-IIE rat liver cell line	[42, 90]
TRPC3	PCR	Mouse liver Rat liver, foetal liver, hepatocytes Human liver AML12 mouse liver cell line H4-IIE rat liver cell line	[42, 90, 91]
TRPC5	PCR	Human liver, foetal liver	[91]
TRPC6	PCR Western blot Immunofluorescence	Human liver, foetal liver HepG2 and Huh-7 human cell lines	[23, 91]
TRPC7	PCR	Mouse liver Rat liver, foetal liver, hepatocytes Human liver AML12 mouse liver cell line H4-IIE rat liver cell line HepG2 and Huh-7 human cell lines	(Chen, J L and Barritt, G J unpublished results [23, 91]);
TRPV1	PCR Patch clamp recording Intracellular fluorescent Ca^{2+} sensor	HepG2 and Huh-7 human liver cell lines	[23, 52, 53, 92]
TRPV2	PCR Dot-blot analysis	Rat liver HepG2 and Huh-7 human liver cell lines	[23, 52, 93]
TRPV3	PCR	HepG2 and Huh-7 human liver cell lines	[23, 52]
TRPV4	PCR Northern blot dot-blot analysis	Mouse liver Rat hepatocytes Human liver H4-IIE rat liver cell line HepG2 and Huh-7 human liver cell lines	[23, 52, 94, 95] (Chen, J L and Barritt, G J unpublished results)
TRPV6	PCR	Zebrafish liver	[96]

Table 35.1 (continued)

TRP protein	Technique employed for detection	Liver tissue, hepatocytes or liver cell line	References
TRPM4	Northern blot	Human liver HepG2 and Huh-7 human cell lines	[23, 97]
TRPM5	Northern blot dot-blot analysis	Human liver Huh-7 human cell line	[23, 98, 99]
TRPM6	PCR	HepG2 and Huh-7 human cell lines	[23]
TRPM7	PCR Northern blot In situ hybridisation Patch-clamp recording	Zebrafish liver Rat liver H4-IIE rat liver cell line	[22, 23, 100, 101]
TRPM8	PCR	HepG2 and Huh-7 human cell lines	[23]
PKD2	PCR Immunofluorescence	Human liver	[102–104]
TRPML1	Inhibition of function by anti-TRPML1 antibody	Rat liver	[55]

function. Since PCR can be very sensitive, the detection of mRNA encoding a given TRP protein may not necessarily indicate the expression of functional channels. Indeed, while mRNA encoding TRPC1 can readily be detected in isolated rat hepatocytes, detection of TRPC1 protein proved more difficult [28]. This may have been due partly to a low abundance of the TRPC1 protein and to a relatively low affinity of TRPC1 antibodies employed for the TRPC1 protein.

Unfortunately, results to date (summarised in Table 35.1) do not give a very clear picture of which TRP proteins are expressed in normal hepatocytes. This is because experiments have been performed with liver tissue from several species and in other cases with liver cell lines. As discussed below, immortalised liver cell lines may express some TRP proteins not expressed in their differentiated counterparts, and vice versa. Some TRP mRNA detected in liver tissue is likely expressed in non-hepatocyte cell types, such as Kupffer cells and bile duct epithelial cells. Studies of TRP channel function in hepatocytes have been somewhat limited, but there is some information about the cellular functions of TRPC1, TRPC6, TRPV1, TRPV4, TRPM7 and TRPML1 in hepatocytes and liver cell lines.

35.3 TRP Proteins and Store-Operated Ca^{2+} Entry in Hepatocytes

Intracellular Ca^{2+} measurements using fluorescent Ca^{2+} dyes have suggested that Ca^{2+} mobilising hormones that cause depletion of intracellular Ca^{2+} stores activate more than one type of Ca^{2+} channel on the hepatocyte plasma membrane [29–32].

The results of patch clamping experiments, however, have shown that there is only one type of SOC present in rat hepatocytes and in rat liver tumour cell lines. SOCs in liver cells have a high selectivity for Ca^{2+} and characteristics similar to those of the Ca^{2+} release activated Ca^{2+} (CRAC) channels in lymphocytes and mast cells [33–35]. Evidence that Orai1 constitutes the pore and stromal interaction molecule 1 (STIM1) the activator protein of the rat hepatocyte SOCs has been reported [36–38].

It has previously been proposed that a TRP protein, possibly TRPC1, TRPC3, TRPC4, TRPC6, TRPV5 and/or TRPV6, constitutes the pore of SOCs in some types of animal cells (reviewed in [23, 39–41]). Some of these TRP proteins are expressed in hepatocytes (Table 35.1). In the case of TRPC1, ectopic expression of hTRPC1 or knockdown of endogenous TRPC1 proteins using siRNA did not cause large changes in thapsigargin-stimulated Ca^{2+} entry (assessed using a fluorescent Ca^{2+} sensor and patch-clamp recording), indicating that it is unlikely that the TRPC1 peptide constitutes SOCs in liver cells [23, 42, 43]. However, a role for TRPC1 or another TRP protein in forming or modulating liver cell SOCs is not excluded. On the one hand, several studies have provided evidence that TRP polypeptides interact with STIM and/or Orai polypeptides [41, 44–47]. On the other hand, the Ca^{2+}-permeable channels formed by TRPC1 polypeptides and by most other TRP polypeptides have a relatively low selectivity for Ca^{2+} compared with Na^+ [48, 49]. As indicated above, the SOCs detected in rat liver cells have a high selectivity for Ca^{2+}. This comparison provides further evidence that it is unlikely that any of the known TRP polypeptides constitutes the Ca^{2+}-selective SOCs found in rat hepatocytes and liver cells. Nevertheless, under physiological conditions many TRP channels can be activated by intracellular messengers generated when intracellular stores are depleted, and thus may modulate the activity of SOCs and contribute to the experimentally-observed Ca^{2+} entry when intracellular stores are depleted (reviewed in [40]). Notwithstanding these observations, TRP polypeptides may contribute to SOCs in hepatocytes from some species.

Ectopic expression of a TRP protein localised, in part, in the ER has been used to investigate the role of a putative sub-compartment of the ER in SOC activation in a rat liver cell line which does not appear to express endogenous TRPV1 [50]. Ectopically-expressed TRPV1 was observed in the ER and in the plasma membrane. The amount of Ca^{2+} released from the ER by a TRPV1 agonist, measured using fura 2, was found to be the same as that released by a SERCA (ER $(Ca^{2+} + Mg^{2+})$ATP-ase) inhibitor, indicating that TRPV1 agonist-sensitive stores substantially overlap with SERCA inhibitor-sensitive stores. However, in contrast to SERCA inhibitors, TRPV1 agonists did not activate Ca^{2+} entry. Using FFP-18 to measure cytoplasmic $[Ca^{2+}]$ close to the plasma membrane it was shown that SERCA inhibitors release Ca^{2+} from regions of the ER located closer to the plasma membrane than the region from which TRPV1 agonists release Ca^{2+}. In contrast to SERCA inhibitors, TRPV1 agonists did not induce a redistribution of STIM1. TDCA (taurodeoxycholic acid), an activator of SOCs, caused the release of Ca^{2+} from the ER, which was detected by FFP-18 but not by fura-2, and a redistribution of STIM1 to puncta similar to that caused by SERCA inhibitors. It was concluded that in rat liver cells Ca^{2+} release from a small component of the ER located near the plasma membrane is required to induce STIM1 redistribution and SOC activation. Other experiments have shown that this putative subregion is enriched in type 1 IP_3 receptors [51].

35.4 Physiological Functions of TRP Channels in Hepatocytes and in Hepatocyte (Liver) Cell Lines

35.4.1 TRPV1 and TRPV4 in Cell Migration

Experiments conducted with HepG2 cells, derived from a human liver cancer cells, have provided evidence which suggests the presence of functional TRPV1 and TRPV4 channels [52, 53]. Capsaicin and RTX, known activators of TRPV1, were found to stimulate capsazepine-sensitive Ca^{2+} entry (assessed using a fluorescent Ca^{2+} sensor). It was also observed that capsaicin induces the release of Ca^{2+} from intracellular stores, suggesting that in this immortalised human liver cell line endogenous TRPV1 is located in intracellular membranes. If appropriately activated, TRPV1 could mediate intracellular Ca^{2+} release. Incubation of HepG2 cells with hepatocyte growth factor/scatter factor increased capsaicin-stimulated Ca^{2+} entry in association with cell migration, leading to the suggestion that Ca^{2+} entry through TRPV1 may be involved in the regulation of liver cell migration [52, 53]. In other experiments, Ca^{2+} entry in HepG2 cells was also found to be stimulated by 4α-phorbol-12,13-didecanoate and arachidonic acid, known activators of TRPV4, suggesting that functional TRPV4 is also expressed in this liver cell line [52, 53]. It has been suggested that both TRPV1 and TRPV4 are involved in the regulation of liver cell migration through mechanisms involving regulation of the structure of the cytoskeleton, and possibly employing the mechanosensitive properties of TRPV4 [52, 53]. Since TRPV1 and TRPV4 can be activated by arachidonic acid and some of its metabolites (reviewed in [49]), these channels may also mediate effects of local hormones on liver cells.

35.4.2 TRPC1 and Volume Control

Results of immunofluorescence experiments indicate that, in H4-IIE rat liver cells, derived from a rat liver hepatoma [54], TRPC1 is principally located in intracellular organelles with some expressed at the plasma membrane [42, 43]. In order to test the hypothesis that, in liver cells, TRPC1 serves as a non-selective cation channel which mediates Ca^{2+}- and Na^+-entry in response to physiological and pathophysiological stimuli, experiments were conducted using maitotoxin, a known activator of non-selective cation channels (reviewed in [42]). Although its effects on cells are complex, maitotoxin appears to activate a variety of intracellular signalling pathways including those involving phospholipase C, IP_3, increases in cytoplasmic Ca^{2+}

concentration and release of arachidonic acid (reviewed in [42]). Over-expression of human TRPC1 in H4-IIE cells led to a significant enhancement of maitotoxin-stimulated Ca^{2+} and Na^+ entry [42]. Moreover, knockdown of endogenous TRPC1 with siRNA led to a decrease in maitotoxin- and ATP-stimulated Ca^{2+} entry and to an enhancement of swelling in hypotonic solution and to an enhanced regulated volume decrease [43]. It was concluded that TRPC1 in a liver cell line can be activated by maitotoxin, and one possible physiological function of this TRP is in the regulation of hepatocyte volume [42, 43]. While it is possible that TRPC1 may mediate the response of liver cells to toxic agents like maitotoxin, the results obtained with maitotoxin most likely point to a role for TRPC1 in responses to hormones and other extracellular signals which lead to the formation of the same intracellular messengers as those generated artificially with maitotoxin.

35.4.3 TRPM7 and Cell Proliferation

TRPM7 is expressed in hepatocytes from several different species (Table 35.1). Mishra et al. (2009) [105] have described an outwardly rectifying non-specific cation current which is inhibited by intracellular Mg^{2+} plus MgATP and by 2-aminoethyl diphenylborate (2-APB) in rat hepatocytes and in WIF-B cells (a spatially polarised cell line derived from a rat hepatoma-human skin fibroblast cross). This current, which was also inhibited by Ca^{2+}/calmodulin inhibitors, was attributed to TPRM7. On the basis of the magnitude of the current attributed to TRPM7, the authors suggested that TRPM7 activity is lower in quiescent rat hepatocytes than in proliferating WIF-B cells. It was proposed that TRPM7 may be associated with cell proliferation.

In a separate study employing patch clamp recording, a current attributable to TRPM7 was detected in rat H4-IIE liver cells and shown to be suppressed using TRPM7 siRNA [22]. In addition to its possible roles in cell proliferation and in mediating Ca^{2+} and Na^+ entry in response to reactive oxygen species (discussed below), it is also likely that TRPM7 contributes to Mg^{2+} entry to liver cells, and may respond to changes in the concentration of intracellular ATP such as in cell injury leading to the initiation of necrosis and apoptosis.

35.4.4 TRPML1 and Lysosomal Ca^{2+} Release

TRPML1, a non-selective cation channel permeable to Ca^{2+}, Na^+ and K^+, is localised in the lysosomes in many cell types and is thought to play a role in regulating intracellular Ca^{2+} concentrations through NAAP-induced release of Ca^{2+} from lysosomes (reviewed in [55]). The results of studies by [55] have provided evidence that TRPML1 is expressed in lysosomes in rat hepatocytes. They isolated purified lysosomes, reconstituted lysosomal channel proteins into lipid bilayers, and studied NAADP-sensitive channels by lipid bilayer recording. Channel activity was inhibited by an antibody raised against the C-terminus of TRPML1 when the antibody was introduced into the *cis* (but not *trans*) solution, suggesting that the NAADP-sensitive channel is TRPML1. The authors concluded that the lysosomal TRPML1 plays a role in the regulation of intracellular Ca^{2+} homeostasis. This may also involve the interaction of TRPML1 with the recently-discovered NAAPP-sensitive two pore channel [56].

35.4.5 Other TRP Channels in Hepatocytes and in Hepatocyte (Liver) Cell Lines: Possible Functions

The cellular functions of only a few TRP proteins expressed in hepatocytes have been identified with any degree of confidence. It is likely that as more experiments are conducted, specific cellular functions will be assigned to more hepatocyte TRP proteins. Two pathological conditions associated with enhanced Ca^{2+} entry to hepatocytes, which may be mediated by TRP channels, are ischemia and reperfusion injury and drug- or xenobiotic-induced liver toxicity. Reactive oxygen species play a major role in initiating damage to hepatocytes in both of these conditions [57, 58].

When livers are subjected to ischemia reperfusion injury, an increase in total hepatocyte Ca^{2+} and in the amount of Ca^{2+} in mitochondria is observed immediately following the onset of reperfusion [59–62]. Studies with isolated hepatocytes subjected to hypoxia or anoxia show a sustained increase in $[Ca^{2+}]_{cyt}$ upon re-oxygenation [63–65]. This is due to enhanced plasma membrane Ca^{2+} entry as well as the release of Ca^{2+} from intracellular stores [64, 65]. Reactive oxygen species generated by Kupffer cells and hepatocytes during the initial stages of reperfusion are thought to be one of the main mediators of enhanced Ca^{2+} entry to hepatocytes (reviewed in [58, 66]). NO and reactive nitrogen species, generated at later stages of reperfusion, may also affect Ca^{2+} entry [66].

While the Ca^{2+} permeable channels involved in enhanced Ca^{2+} entry to hepatocytes in ischemia reperfusion injury have not been clearly identified, the results of experiments conducted with a liver cell line have shown that reactive oxygen species can activate a 16 pS Ca^{2+}-permeable non-selective cation channel [21]. While the molecular entities responsible for mediating Ca^{2+} entry to hepatocytes initiated by reactive oxygen and nitrogen species have not yet been identified, candidates are TRPM2, TRPM7 and possibly SOCs [67–72]. TRPM2 is activated by H_2O_2 and other reactive oxygen species and is thought to be responsible for Ca^{2+} entry induced by reactive oxygen species in several cell types [68, 69]. In response to toxic insults which induce the generation of reactive oxygen species, TRPM7 has been shown to mediate Ca^{2+} and Na^+ entry to neurons which, in turn, leads to cell death [67]. There is some evidence that H_2O_2 and/or reactive oxygen species can release Ca^{2+} from intracellular stores and activate SOCs [70, 72]. Other TRP channels may also be involved in responses to NO. Thus some members of the TRPC and TRPV families have been shown to be activated by NO through S-nitrosylation of cysteine residues [73]. Further experiments are required to determine which TRP channels are involved in Ca^{2+} entry to hepatocytes when this is activated by reactive oxygen and nitrogen species, and by drugs. These Ca^{2+}- and Na^+-permeable channels are likely potentially important in the events initiating liver cell death in response to pathological injury and toxic agents.

35.5 TRP Channels in Other Cell Types Present in the Liver

As described above, several cell types other than hepatocytes are essential for normal liver function. These are the endothelial, biliary epithelial, vascular smooth muscle, Kupffer oval and stellate cells (Fig. 35.1). While it is likely that TRP channels are required for the activity of each of these cell types, there is presently little knowledge about the TRP channels expressed in these cell types in the liver. Nevertheless, it can be predicted that the members of the TRP channel family expressed in liver endothelial and vascular smooth muscle cells will be similar to those expressed in these cell types in other organs, and will have similar functions.

One example of the role of TRP protein in another cell type within the liver is TRPP2. Experiments conducted with developing zebrafish embryos have provided evidence that enhanced degradation of TRPP2, most likely in epithelial cells of the biliary duct, is responsible for the development of polycystic liver disease [74]. This is a genetically-inherited autosomal dominant disease caused by mutations in either the PRKCSH or Sec63 proteins. These proteins are localised in the endoplasmic reticulum and are required for protein glycosylation, folding, and translocation. PRKCSH is thought to protect TRPP2 from degradation. It is proposed that TRPM2 is required for liver development and for adaption and regeneration of the liver following injury. Mutations in PRKCSH lead to enhanced degradation of TRPM2 and, over several decades, the symptoms of polycystic liver disease [74].

35.6 TRP Channels and Liver Cancer

Several TRP proteins exhibit altered expression in tumorigenic cells compared with their normal non-transformed counterparts, and it is thought that in some situations this altered expression of TRP proteins may play a significant role in cancer progression [75, 76]. The two most common forms of liver cancer are hepatocellular carcinoma, which involves accumulation of mutations originating in hepatocytes, and metastatic cancer which involves the migration of tumorigenic cells from the colon [77].

Studies on the role of Ca^{2+} entry channels in enhancing the proliferation of human liver tumour cells have provided evidence that TRPC6 is expressed at much greater levels in tissue from human liver tumours than it is in non-tumorigenic tissue from the same liver, or in "normal" human liver tissue and in hepatocytes isolated from "normal" human livers [23]. In normal liver tissue and hepatocytes, TRPC6 expression was found to be negligible. Experiments employing over-expression of TRPC6 or knockdown of TRPC6 using siRNA provided evidence that increased expression of TRPC6 is associated with increased thapsigargin-initiated Ca^{2+} entry and an increased rate of cell proliferation (Fig. 35.3). It was concluded that in tumorigenic human hepatocytes, TRPC6 interacts with Orai1 and STIM1 leading to enhanced SOC-mediated Ca^{2+} entry.

Liu et al. [89] have investigated the expression of TRPV2 in normal and diseased human liver tissue using quantitative PCR, western blot and immunofluorescence. They found increased expression of TRPV2 in liver tissue from patients with cirrhotic livers and moderate to well-differentiated hepatocellular tumours compared with expression of TRPM2 in liver tissue from patients with chronic hepatitis (Fig. 35.4). However, TRPV2 expression in poorly-differentiated tumours was lower than that in cirrhotic livers and in moderate to well-differentiated tumours. The authors suggested that TRPV2 may play a role in hepatocarcinogenesis and further, that TRPV2 could be evaluated further as a prognostic marker for patients with hepatocarcinoma.

Direct comparison of the expression of mRNA encoding TRP proteins in H4-IIE cells (derived from cells in a rat liver tumour [54]), with mRNA expression in hepatocytes isolated from normal rat livers showed that TRPC1, C3 and C6, TRPM4 and M7 are expressed in both rat hepatocytes and H4-IIE rat liver cells, whereas TRPV4 and TRPM2 could only be detected in rat hepatocytes (Zhang, X., Kheradpezhouh, E., Rychkov, G. and Barritt, G.J., unpublished results). In studies of human liver cancer cell lines, Boustany and colleagues detected TRPV4 mRNA in Huh-7 and HepG2 liver cell lines, but could detect no TRPM2 mRNA [23].

It is anticipated that further experiments will identify other TRP proteins which exhibit altered expression in hepatocellular carcinoma and in metastatic liver cancer, compared with that in normal liver, and will provide information on the consequences of these changes for growth and differentiation of cancer cells.

35.7 Conclusions

Many TRP proteins are expressed in hepatocytes and in the other cell types which comprise the liver. Results to date reflect experiments conducted with liver tissue, freshly-isolated hepatocytes, and immortalised liver cell lines from several species. Therefore, limited conclusions can presently be reached about the nature of the TRP channels expressed in hepatocytes and in other liver cell types in the most commonly studied species, rat and human. Results of functional studies to date have provided some evidence suggesting the involvement of TRPC1 in volume control, TRPV1 and V4 in cell migration, TRPC6 and TRPM7 in cell proliferation, and TRPPM in lysosomal Ca^{2+} release in hepatocytes and/or liver cell lines. Altered expression of some TRP proteins, including TRPC6, TRPV2 and possibly TRPV1

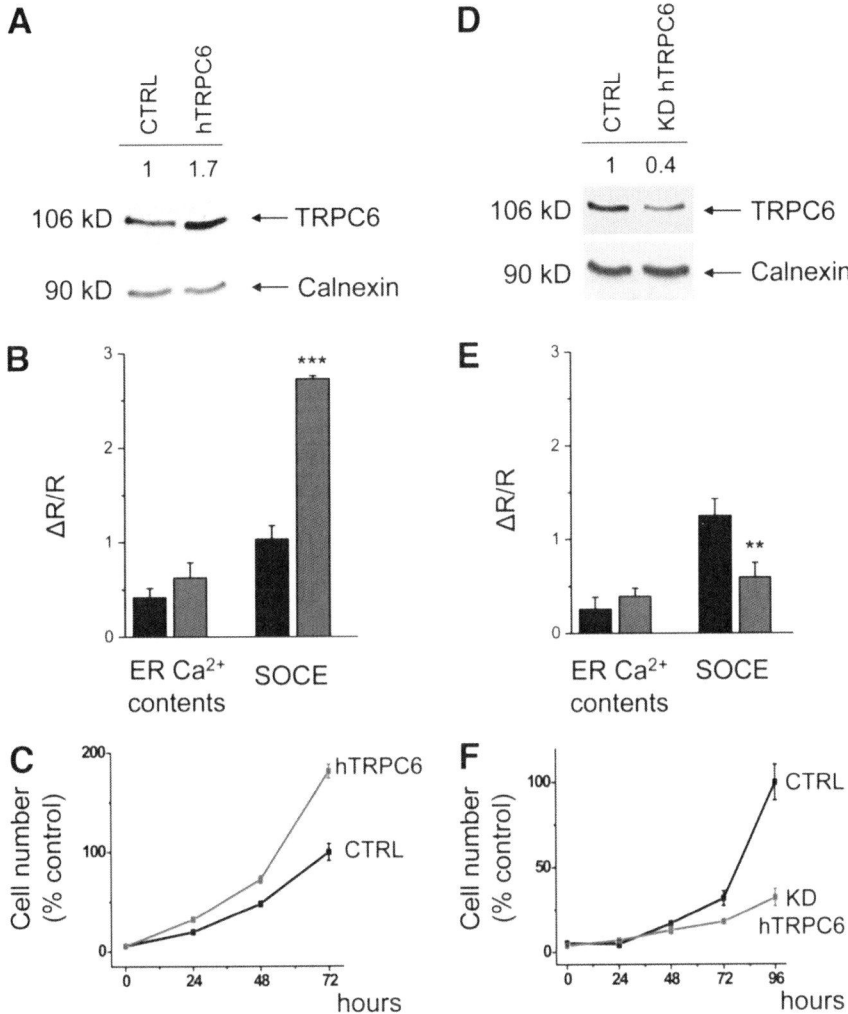

Fig. 35.3 Altered expression of TRPM6 correlates with changes in the rates of Ca^{2+} entry and cell proliferation in Huh-7 liver cells, derived from a human hepatoma. **a, d** Western blots showing the over-production (**a**) or under-production (**d**) of hTRPC6 (106 kDa) in cells over-expressing hTRPC6 (**a**) or where TRPC6 has been knocked down with siRNA (**d**). CTRL, hTRPC6 and KD hTRPC6 represent control cells, cells transfected with cDNA encoding hTRPC6, and cells transfected with siRNA targetted against hTRPC6, respectively. Calnexin was used as a reference protein. The numbers 1.7 and 0.4 represent fold increases and decreases, respectively, in TRPC6 protein expression relative to calnexin. **b, e** The amount of Ca^{2+} in the ER released by thapsigargin (ER Ca^{2+} contents) and thapsigargin-stimulated Ca^{2+} entry (SOCE) in cells over-expressing hTRPC6 (**b**) or where TRPC6 has been knocked down with siRNA (**e**). **c, f** Plots of cell number as a function of time elapsed after plating the cells over-expressing hTRPC6 (**c**) or where TRPC6 has been knocked down with siRNA (**f**). Adapted, with permission, from [23]

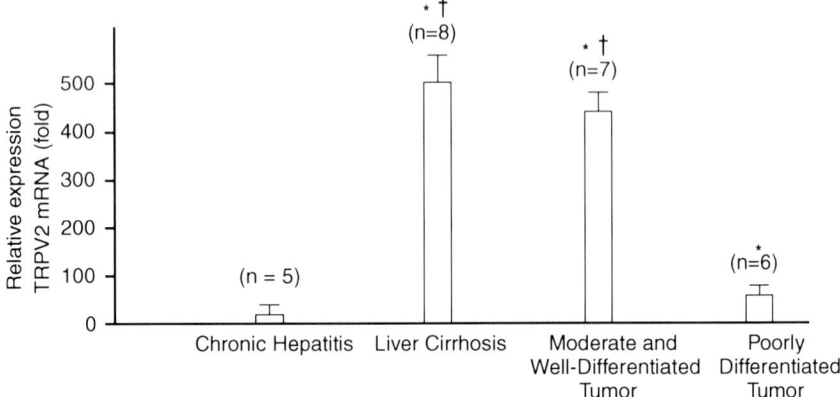

Fig. 35.4 TRPV2 expression in liver tissues at different stages of hepatocellular carcinoma (HCC). TRPV2 mRNA levels from 13 HCC tissues each paired with its non-tumour counterpart were assessed by quantitive RT-PCR. Results (mean ± SD, representative of 3 experiments) are normalised for β-actin expression. TRPV2 levels are expressed as relative fold with respect to chronic hepatitis as the control. TRPV2 mRNA relative expression in liver cirrhosis, moderately and well-differentiated tumours, and in poorly differentiated tumours is significantly increased compared with that in chronic hepatitis (*$P < 0.05$). TRPV2 mRNA relative expression in poorly differentiated liver tumours was significantly decreased compared with that in liver cirrhosis and with moderately and well-differentiated liver tumours (▲$P < 0.05$). Adapted, with permission, from [89]

and TRPM2, in tumorigenic cells may play roles in the development and progression of hepatocellular carcinoma and metastatic liver cancers. It is anticipated that future experiments will define important roles for other TRP proteins in the cellular functions of hepatocytes and other cell types of which the liver is composed.

Acknowledgments Research conducted in the authors' laboratories which has contributed to this review is supported by grants from the National Health and Medical Research Council of Australia, the Australian Research Council, and the Flinders Medical Centre Foundation of South Australia.

References

1. Boyer JL (2002) Bile formation and cholestasis. In: Schiffs diseases of the liver, 9th ed. Lippincott, Williams & Wilkins, Philadelphia, PA, pp 135–165
2. Leite MF, Nathanson MH (2001) In: The liver biology and pathobiology. Lippincott, Williams & Wilkiins, Philadelphia, PA, pp 537–554
3. Mohammed FF, Khokha R (2005) Thinking outside the cell: proteases regulate hepatocyte division. Trends Cell Biol 15: 555–563
4. Young B, Heath JW (2000) Liver and pancreas. In: Wheater's functional histology, 4th ed. Churchill Livingstone, Edinburgh, pp 274–275
5. Wakabayashi Y, Kipp H, Arias IM (2006) Transporters on demand: intracellular reservoirs and cycling of bile canalicular ABC transporters. J Biol Chem 281: 27669–27673
6. Hubbard AL, Barr VA, Scott LJ (1994) Hepatocyte surface polarity. In: The liver: biology and pathobiology. Raven Press Ltd, New York, NY, pp 189–214

7. Zegers MM, Hoekstra D (1998) Mechanisms and functional features of polarized membrane traffic in epithelial and hepatic cells. Biochem J 336:257–269
8. Berry MN, Edwards AM, Barritt G (1991) Assessment of integrity of isolated hepatocytes. In: Isolated hepatocytes – preparation, properties and applications. Elsevier, Amsterdam, pp 83–98
9. Dixon CJ, White PJ, Hall JF, Kingston S, Boarder MR (2005) Regulation of human hepatocytes by P2Y receptors: control of glycogen phosphorylase, Ca2+, and mitogen-activated protein kinases. J Pharmacol Exp Ther 313:1305–1313
10. O'Brien EM, Gomes DA, Sehgal S, Nathanson MH (2007) Hormonal regulation of nuclear permeability. J Biol Chem 282:4210–4217
11. Nieuwenhuijs VB, De Bruijn MT, Padbury RT, Barritt GJ (2006) Hepatic ischemia-reperfusion injury: roles of Ca2+ and other intracellular mediators of impaired bile flow and hepatocyte damage. Dig Dis Sci 51:1087–1102
12. Enfissi A, Prigent S, Colosetti P, Capiod T (2004) The blocking of capacitative calcium entry by 2-aminoethyl diphenylborate (2-APB) and carboxyamidotriazole (CAI) inhibits proliferation in Hep G2 and Huh-7 human hepatoma cells. Cell Calcium 36:459–467
13. Robb-Gaspers LD, Burnett P, Rutter GA, Denton RM, Rizzuto R, Thomas AP (1998) Integrating cytosolic calcium signals into mitochondrial metabolic responses. EMBO J 17:4987–5000
14. Berridge MJ (2002) The endoplasmic reticulum: a multifunctional signaling organelle. Cell Calcium 32:235–249
15. Rodrigues MA, Gomes DA, Leite MF, Grant W, Zhang L, Lam W, Cheng YC, Bennett AM, Nathanson MH (2007) Nucleoplasmic calcium is required for cell proliferation. J Biol Chem 282:17061–17068
16. Barritt GJ, Parker JC, Wadsworth JC (1981) A kinetic analysis of the effects of adrenaline on calcium distribution in isolated rat liver parenchymal cells. J Physiol 312:29–55
17. Canaff L, Petit JL, Kisiel M, Watson PH, Gascon-Barre M, Hendy GN (2001) Extracellular calcium-sensing receptor is expressed in rat hepatocytes. coupling to intracellular calcium mobilization and stimulation of bile flow. J Biol Chem 276:4070–4079
18. Lesurtel M, Graf R, Aleil B, Walther DJ, Tian Y, Jochum W, Gachet C, Bader M, Clavien PA (2006) Platelet-derived serotonin mediates liver regeneration. Science 312:104–107
19. Patel S, Robb-Gaspers LD, Stellato KA, Shon M, Thomas AP (1999) Coordination of calcium signalling by endothelial-derived nitric oxide in the intact liver. Nat Cell Biol 1:467–471
20. Schofl C, Ponczek M, Mader T, Waring M, Benecke H, von zur Muhlen A, Mix H, Cornberg M, Boker KH, Manns MP, Wagner S (1999) Regulation of cytosolic free calcium concentration by extracellular nucleotides in human hepatocytes. Am J Physiol 276:G164 G172
21. Barros LF, Stutzin A, Calixto A, Catalan M, Castro J, Hetz C, Hermosilla T (2001) Nonselective cation channels as effectors of free radical-induced rat liver cell necrosis. Hepatology 33:114–122
22. Barritt GJ, Chen J, Rychkov G (2008) Ca2+-permeable channels in the hepatocyte plasma membrane and their roles in hepatocyte physiology. Biochim Biophys Acta 1783: 651–672
23. El Boustany C, Bidaux G, Enfissi A, Delcourt P, Prevarskaya N, Capiod T (2008) Capacitative calcium entry and transient receptor potential canonical 6 expression control human hepatoma cell proliferation. Hepatology 47:2068–2077
24. Auld A, Chen J, Brereton HM, Wang YJ, Gregory RB, Barritt GJ (2000) Store-operated Ca(2+) inflow in reuber hepatoma cells is inhibited by voltage-operated Ca(2+) channel antagonists and, in contrast to freshly isolated hepatocytes, does not require a pertussis toxin-sensitive trimeric GTP-binding protein. Biochim Biophys Acta 1497:11–26

25. Brereton HM, Harland ML, Froscio M, Petronijevic T, Barritt GJ (1997) Novel variants of voltage-operated calcium channel alpha 1-subunit transcripts in a rat liver-derived cell line: deletion in the IVS4 voltage sensing region. Cell Calcium 22:39–52
26. Graf J, Haussinger D (1996) Ion transport in hepatocytes: mechanisms and correlations to cell volume, hormone actions and metabolism. J Hepatol 24 (Suppl 1):53–77
27. Sawanobori T, Takanashi H, Hiraoka M, Iida Y, Kamisaka K, Maezawa H (1989) Electrophysiological properties of isolated rat liver cells. J Cell Physiol 139:580–585
28. Ong HL, Chen J, Chataway T, Brereton H, Zhang L, Downs T, Tsiokas L, Barritt G (2002) Specific detection of the endogenous transient receptor potential (TRP)-1 protein in liver and airway smooth muscle cells using immunoprecipitation and western-blot analysis. Biochem J 364:641–648
29. Altin JG, Bygrave FL (1987) The influx of Ca2+ induced by the administration of glucagon and Ca2+-mobilizing agents to the perfused rat liver could involve at least two separate pathways. Biochem J 242:43–50
30. Ikari A, Sakai H, Takeguchi N (1997) ATP thapsigargin and cAMP increase Ca2+ in rat hepatocytes by activating three different Ca2+ influx pathways. Jpn J Physiol 47:235–239
31. Llopis J, Kass GE, Gahm A, Orrenius S (1992) Evidence for two pathways of receptor-mediated Ca2+ entry in hepatocytes. Biochem J 284:243–247
32. Striggow F, Bohnensack R (1994) Inositol 1,4,5-trisphosphate activates receptor-mediated calcium entry by two different pathways in hepatocytes. Eur J Biochem 222:229–234
33. Litjens T, Harland ML, Roberts ML, Barritt GJ, Rychkov GY (2004) Fast Ca(2+)-dependent inactivation of the store-operated Ca2+ current (ISOC) in liver cells: a role for calmodulin. J Physiol 558:85–97
34. Rychkov G, Brereton HM, Harland ML, Barritt GJ (2001) Plasma membrane Ca2+ release-activated Ca2+ channels with a high selectivity for Ca2+ identified by patch-clamp recording in rat liver cells. Hepatology 33:938–947
35. Rychkov GY, Litjens T, Roberts ML, Barritt GJ (2005) Arachidonic acid inhibits the store-operated Ca2+ current in rat liver cells. Biochem J 385:551–556
36. Litjens T, Nguyen T, Castro J, Aromataris EC, Jones L, Barritt GJ, Rychkov GY (2007) Phospholipase C-gamma1 is required for the activation of store-operated Ca2+ channels in liver cells. Biochem J 405:269–276
37. Aromataris EC, Castro J, Rychkov GY, Barritt GJ (2008) Store-operated Ca(2+) channels and stromal interaction molecule 1 (STIM1) are targets for the actions of bile acids on liver cells. Biochim Biophys Acta 1783:874–885
38. Scrimgeour N, Litjens T, Ma L, Barritt GJ, Rychkov GY (2009) Properties of Orai1 mediated store-operated current depend on the expression levels of STIM1 and Orai1 proteins. J Physiol 587:2903–2918
39. Parekh AB, Putney JW Jr. (2005) Store-operated calcium channels. Physiol Rev 85:757–810
40. Rychkov G, Barritt GJ (2007) TRPC1 Ca(2+)-permeable channels in animal cells. Handb Exp Pharmacol 179:23–52
41. Yuan JP, Zeng W, Huang GN, Worley PF, Muallem S (2007) STIM1 heteromultimerizes TRPC channels to determine their function as store-operated channels. Nat Cell Biol 9: 636–645
42. Brereton HM, Chen J, Rychkov G, Harland ML, Barritt GJ (2001) Maitotoxin activates an endogenous non-selective cation channel and is an effective initiator of the activation of the heterologously expressed hTRPC-1 (transient receptor potential) non-selective cation channel in H4-IIE liver cells. Biochim Biophys Acta 1540:107–126
43. Chen J, Barritt GJ (2003) Evidence that TRPC1 (transient receptor potential canonical 1) forms a Ca(2+)-permeable channel linked to the regulation of cell volume in liver cells obtained using small interfering RNA targeted against TRPC1. Biochem J 373:327–336
44. Huang GN, Zeng W, Kim JY, Yuan JP, Han L, Muallem S, Worley PF (2006) STIM1 carboxyl-terminus activates native SOC, I(crac) and TRPC1 channels. Nat Cell Biol 8:1003–1010

45. Liao Y, Erxleben C, Yildirim E, Abramowitz J, Armstrong DL, Birnbaumer L (2007) Orai proteins interact with TRPC channels and confer responsiveness to store depletion. Proc Natl Acad Sci USA 104:4682–4687
46. Lopez JJ, Salido GM, Pariente JA, Rosado JA (2006) Interaction of STIM1 with endogenously expressed human canonical TRP1 upon depletion of intracellular Ca2+ stores. J Biol Chem 281:28254–28264
47. Ong HL, Cheng KT, Liu X, Bandyopadhyay BC, Paria BC, Soboloff J, Pani B, Gwack Y, Srikanth S, Singh BB, Gill DL, Ambudkar IS (2007) Dynamic assembly of TRPC1-STIM1-Orai1 ternary complex is involved in store-operated calcium influx. Evidence for similarities in store-operated and calcium release-activated calcium channel components. J Biol Chem 282:9105–9116
48. Ramsey IS, Delling M, Clapham DE (2006) An introduction to TRP channels. Annu Rev Physiol 68:619–647
49. Nilius B, Owsianik G, Voets T, Peters JA (2007) Transient receptor potential cation channels in disease. Physiol Rev 87:165–217
50. Castro J, Aromataris EC, Rychkov GY, Barritt GJ (2009) A small component of the endoplasmic reticulum is required for store-operated Ca2+ channel activation in liver cells: evidence from studies using TRPV1 and taurodeoxycholic acid. Biochem J 418:553–566
51. Gregory RB, Wilcox RA, Berven LA, van Straten NC, van der Marel GA, van Boom JH, Barritt GJ (1999) Evidence for the involvement of a small subregion of the endoplasmic reticulum in the inositol trisphosphate receptor-induced activation of Ca2+ inflow in rat hepatocytes. Biochem J 341:401–408
52. Vriens J, Janssens A, Prenen J, Nilius B, Wondergem R (2004) TRPV channels and modulation by hepatocyte growth factor/scatter factor in human hepatoblastoma (HepG2) cells. Cell Calcium 36:19–28
53. Waning J, Vriens J, Owsianik G, Stuwe L, Mally S, Fabian A, Frippiat C, Nilius B, Schwab A (2007) A novel function of capsaicin-sensitive TRPV1 channels: involvement in cell migration. Cell Calcium 42:17–25
54. Reuber MD (1961) A transplantable bile-secreting hepatocellular carcinoma in the rat. J Natl Cancer Inst 26:891–899
55. Zhang F, Li PL (2007) Reconstitution and characterization of a nicotinic acid adenine dinucleotide phosphate (NAADP)-sensitive Ca2+ release channel from liver lysosomes of rats. J Biol Chem 282:25259–25269
56. Ruas M, Rietdorf K, Arredouani A, Davis LC, Lloyd-Evans E, Koegel H, Funnell TM, Morgan AJ, Ward JA, Watanabe K, Cheng XT, Churchill GC, Zhu MX, Platt FM, Wessel GM, Parrington J, Galione A (2010) Purified TPC isoforms form NAADP receptors with distinct roles for Ca2+ signalling and endolysosomal trafficking. Curr Biol 20: 703–709
57. Lee WM (2003) Drug-induced hepatotoxicity. N Engl J Med 349:474–485
58. Vardanian AJ, Busuttil RW, Kupiec-Weglinski JW (2008) Molecular mediators of liver ischemia and reperfusion injury: a brief review. Mol Med 14:337–345
59. Dhar DK, Takemoto Y, Nagasue N, Uchida M, Ono T, Nakamura T (1996) FK506 maintains cellular calcium homeostasis in ischemia-reperfusion injury of the canine liver. J Surg Res 60:142–146
60. Isozaki H, Fujii K, Nomura E, Hara H (2000) Calcium concentration in hepatocytes during liver ischaemia-reperfusion injury and the effects of diltiazem and citrate on perfused rat liver. Eur J Gastroenterol Hepatol 12:291–297
61. Kurita K, Tanabe G, Aikou T, Shimazu H (1993) Ischemic liver cell damage and calcium accumulation in rats. J Hepatol 18:196–204
62. Takemoto Y, Uchida M, Nagasue N, Ohiwa K, Kimoto T, Dhar DK, Nakamura T (1994) Changes in calcium content of the liver during hepatic ischemia-reperfusion in dogs. J Hepatol 21:743–747
63. Carini R, Castino R, De Cesaris MG, Splendore R, Demoz M, Albano E, Isidoro C (2004) Preconditioning-induced cytoprotection in hepatocytes requires Ca(2+)-dependent exocytosis of lysosomes. J Cell Sci 117:1065–1077

64. Crenesse D, Hugues M, Ferre C, Poiree JC, Benoliel J, Dolisi C, Gugenheim J (1999) Inhibition of calcium influx during hypoxia/reoxygenation in primary cultured rat hepatocytes. Pharmacology 58:160–170
65. Gasbarrini A, Borle AB, Farghali H, Bender C, Francavilla A, Van Thiel D (1992) Effect of anoxia on intracellular ATP, Na+i, Ca2+i, Mg2+i, and cytotoxicity in rat hepatocytes. J Biol Chem 267:6654–6663 [erratum appears in J Biol Chem 1992 Jun 25;267(18): 13114]
66. Glantzounis GK, Salacinski HJ, Yang W, Davidson BR, Seifalian AM (2005) The contemporary role of antioxidant therapy in attenuating liver ischemia-reperfusion injury: a review. Liver Transpl 11:1031–1047
67. Aarts M, Iihara K, Wei WL, Xiong ZG, Arundine M, Cerwinski W, MacDonald JF, Tymianski M (2003) A key role for TRPM7 channels in anoxic neuronal death. Cell 115:863–877
68. Bari MR, Akbar S, Eweida M, Kuhn FJP, Gustafsson AJ, Luckhoff A, Islam MS (2009) H_2O_2-induced Ca2+ influx and its inhibition by N-(p-amylcinnamoyl) anthranilic acid in the beta-cells: involement of TRPM2 channels. J Cell Mol Med 13: 3260–3267
69. Hara Y, Wakamori M, Ishii M, Maeno E, Nishida M, Yoshida T, Yamada H, Shimizu S, Mori E, Kudoh J, Shimizu N, Kurose H, Okada Y, Imoto K, Mori Y (2002) LTRPC2 Ca2+ permeable channel activated by changes in redox status confers susceptibility to cell death. Mol Cell 9:163–173
70. Lin MJ, Yang XR, Cao YN, Sham JS (2007) Hydrogen peroxide-induced Ca2+ mobilization in pulmonary arterial smooth muscle cells. Am J Physiol Lung Cell Mol Physiol 292: L1598–L1608
71. Miller BA (2006) The role of TRP channels in oxidative stress-induced cell death. J Membr Biol 209:31–41
72. Suzuki Y, Yoshimaru T, Inoue T, Ra C (2009) Discrete generations of intracellular hydrogen peroxide and superoxide in antigen-stimulated mast cells: reciprocal regulation of store-operated Ca2+ channel activity. Mol Immunol 46:2200–2209
73. Yoshida T, Inoue R, Morii T, Takahashi N, Yamamoto S, Hara Y, Tominaga M, Shimizu S, Sato Y, Mori Y (2006) Nitric oxide activates TRP channels by cysteine S-nitrosylation. Nat Chem Biol 2:596–607
74. Gao H, Wang Y, Wegierski T, Skouloudaki K, Putz M, Fu X, Engel C, Boehlke C, Peng H, Kuhn EW, Kim E, Kramer-Zucker A, Walz G (2010) PRKCSH/80 K-H, the protein mutated in polycystic liver disease, protects polycystin-2/TRPP2 against HERP-mediated degradation. Hum Mol Genet 19:16–24
75. Gkika D, Prevarskaya N (2009) Molecular mechanisms of TRP regulation in tumor growth and metastasis. Biochim Biophys Acta 1793:953–958
76. Prevarskaya N, Zhang L, Barritt G (2007) TRP channels in cancer. Biochim Biophys Acta 1772:937–946
77. Llovet JM, Bruix J (2008) Molecular targeted therapies in hepatocellular carcinoma. Hepatology 48:1312–1327
78. Capiod T (1998) ATP-activated cation currents in single guinea-pig hepatocytes. J Physiol 507:795–805
79. Howard A, Barley NF, Legon S, Walters JR (1994) Plasma-membrane calcium-pump isoforms in human and rat liver. Biochem J 303:275–279
80. Gasbarrini A, Borle AB, Van Thiel DH (1993) Ca2+ antagonists do not protect isolated perfused rat hepatocytes from anoxic injury. Biochim Biophys Acta 1177:1–7
81. Studer RK, Borle AB (1992) Na(+)-Ca2+ antiporter activity of rat hepatocytes. Effect of adrenalectomy on Ca2+ uptake and release from plasma membrane vesicles. Biochim Biophys Acta 1134:7–16
82. Delgado-Coello B, Trejo R, Mas-Oliva J (2006) Is there a specific role for the plasma membrane Ca^{2+}-ATPase in the hepatocyte? Mol Cell Biochem 285:1–15

83. Hirata K, Pusl T, O'Neill AF, Dranoff JA, Nathanson MH (2002) The type II inositol 1,4,5-trisphosphate receptor can trigger Ca2+ waves in rat hepatocytes. Gastroenterology 122:1088–1100
84. Wojcikiewicz RJ (1995) Type I, II, and III inositol 1,4,5-trisphosphate receptors are unequally susceptible to down-regulation and are expressed in markedly different proportions in different cell types. J Biol Chem 270:11678–11683
85. Pierobon N, Renard-Rooney DC, Gaspers LD, Thomas AP (2006) Ryanodine receptors in liver. J Biol Chem 281:34086–34095
86. Barritt GJ (2000) Calcium signalling in liver cells. In: Pochet R, Donato R, Haiech J, Heizmann CW, Gerke V (eds) Calcium: the molecular basis of calcium action in biology and medicine. Kluwer Academic Publishers, The Netherlands, pp 73–94
87. Gaspers LD, Thomas AP (2005) Calcium signaling in liver. Cell Calcium 38:329–342
88. Missiaen L, Dode L, Vanoevelen J, Raeymaekers L, Wuytack F (2007) Calcium in the Golgi apparatus. Cell Calcium 41:405–416
89. Liu G, Xie C, Sun F, Xu X, Yang Y, Zhang T, Deng Y, Wang D, Huang Z, Yang L, Huang S, Wang Q, Liu G, Zhong D, Miao X (2010) Clinical significance of transient receptor potential vanilloid 2 expression in human hepatocellular carcinoma. Cancer Genet Cytogenet 197: 54–59
90. Brereton HM, Harland ML, Auld AM, Barritt GJ (2000) Evidence that the TRP-1 protein in unlikely to account for store-operated Ca2+ inflow in Xenopus laevis oocytes. Mol Cell Biochem 214:63–74
91. Riccio A, Medhurst AD, Mattei C, Kelsell RE, Calver AR, Randall AD, Benham CD (2002) mRNA distribution analysis of human TRPC family in CNS and peripheral tissues. Brain Res Mol Brain Res 109:95–104
92. Reilly CA, Taylor JL, Lanza DL, Carr BA, Crouch DJ, Yost GS (2003) Capsaicinoids cause inflammation and epithelial cell death through activation of vanilloid receptors. Toxicol Sci 73:170–181
93. Peng JB, Hediger MA (2002) A family of calcium-permeable channels in the kidney: distinct roles in renal calcium handling. Curr Opin Nephrol Hypertens 11:555–561
94. Strotmann R, Harteneck C, Nunnenmacher K, Schultz G, Plant TD (2000) OTRPC4, a nonselective cation channel that confers sensitivity to extracellular osmolarity. Nat Cell Biol 2:695–702
95. Peng JB, Chen XZ, Berger UV, Weremowicz S, Morton CC, Vassilev PM, Brown EM (2000) Human calcium transport protein CaT1. Biochem Biophys Res Commun 278:326–332
96. Pan TC, Liao BK, Huang CJ, Lin LY, Hwang PP (2005) Epithelial Ca(2+) channel expression and Ca(2+) uptake in developing zebrafish. Am J Physiol Regul Integr Comp Physiol 289:R1202–R1211
97. Launay P, Fleig A, Perraud AL, Scharenberg AM, Penner R, Kinet JP (2002) TRPM4 is a Ca^{2+}-activated non-selective cation channel mediating cell membrane depolarization. Cell 109.397–407
98. Enklaar T, Esswein M, Oswald M, Hilbert K, Winterpacht A, Higgins M, Zabel B (2000) MTR1, a novel biallelically expressed gene in the centre of the mouse distal chromosome 7 imprinting cluster, is a member of the TRP gene family. Genomics 67:179–197
99. Prawitt D, Enklaar T, Klemm G, Gartner B, Spangenberg C, Winterpacht A, Higgins M (2000) Identification and characterization of MTR1, a novel gene with homology to melastatin (MLSN1) and the TRP gene family located in the BWS-WT2 critical region on chromosome 11p15.5 and showing allele-specific expression. Hum Mol Genet 9: 203–216
100. Runnels LW, Yue L, Clapham DE (2001) TRP-PLIK, a bifunctional protein with kinase and ion channel activities. Science 291:1043–1047
101. Elizondo MR, Arduini BL, Paulsen J, MacDonald EL, Sabel JL, Henion PD, Cornell RA, Parichy DM (2005) Defective skeletogenesis with kidney stone formation in dwarf zebrafish mutant for trpm7. Curr Biol 15:667–671

102. Ong ACM, Ward CJ, Butler RJ, Biddolph S, Bowker C, Torra R, Pei Y (1999) Coordinate expression of the autosomal dominant polycystic kidney disease proteins, polycystin-2 and polycystin-1, in normal and cystic tissue. Am J Pathol 154:1721–1729
103. Tsiokas L, Arnould T, Zhu CW, Kim E, Walz G, Sukhatme VP (1999) Specific association of the gene product of PKD2 with the TRPC1 channel. Proc Natl Acad Sci USA 96:3934–3939
104. Veldhuisen B, Spruit L, Dauwerde HG, Breuning MH, Peters DJ (1999) Genes homologous to the autosomal dominant polycystic kidney disease genes (PKD1 and PKD2). Eur J Hum Genet 7:860–872
105. Mishra R, Rao V, Ta R, Shobeiri N, Hill CE (2009) Mg^{2+} and MgATP-inhibited and Ca^{2+} colmodulin-sensitive TRPM7-like current in hepatoma and hepatocytes. Am J Physiol Gastrointest Liver Physiol 297:G687–94

Chapter 36
Expression and Physiological Roles of TRP Channels in Smooth Muscle Cells

Christelle Guibert, Thomas Ducret, and Jean-Pierre Savineau

Abstract Smooth muscles are widely distributed in mammal body through various systems such as circulatory, respiratory, gastro-intestinal and urogenital systems. The smooth muscle cell (SMC) is not only a contractile cell but is able to perform other important functions such as migration, proliferation, production of cytokines, chemokines, extracellular matrix proteins, growth factors and cell surface adhesion molecules. Thus, SMC appears today as a fascinating cell with remarkable plasticity that contributes to its roles in physiology and disease. Most of the SMC functions are dependent on a key event: the increase in intracellular calcium concentration ($[Ca^{2+}]_i$). Calcium entry from the extracellular space is a major step in the elevation of $[Ca^{2+}]_i$ in SMC and involves a variety of plasmalemmal calcium channels, among them is the superfamily of transient receptor potential (TRP) proteins. TRPC (canonical), TRPM (melastatin), TRPV (vanilloid) and TRPP (polycystin), are widely expressed in both visceral (airways, gastrointestinal tract, uterus) and vascular (systemic and pulmonary circulation) smooth muscles. Mainly, TRPC, TRPV and TRPM are implicated in a variety of physiological and pathophysiological processes such as: SMC contraction, relaxation, growth, migration and proliferation; control of blood pressure, arterial myogenic tone, pulmonary hypertension, intestinal motility, gastric acidity, uterine activity during parturition and labor. Thus it is becoming evident that TRP are major element of SMC calcium homeostasis and, thus, appear as novel drug targets for a better management of diseases originating from SMC dysfunction.

J.-P. Savineau (✉)
Université Victor Segalen Bordeaux2, 33076 Bordeaux Cedex, France; INSERM U 885, 33076 Bordeaux Cedex, France
e-mail: jean-pierre.savineau@u-bordeaux2.fr

36.1 Introduction

Smooth muscles are widely distributed in mammal body through various systems such as circulatory, respiratory, gastro-intestinal and urogenital system where they are responsible for: (i) maintaining blood pressure; (ii) controlling the passage of air in airways and food through the gut and excreting waste products, and (iii) determining the movement of sperm and eggs and the delivery of foetus. Moreover, the dysfunction of smooth muscles is implicated in various pathologies such as arterial hypertension, atherosclerosis, asthma, motility disorders of the gastro-intestinal tract, premature labour and male erectile dysfunction.

Although the contraction/relaxation cycle of smooth muscle seems to be one of its primary physiological functions [1, 2], it became obvious that smooth muscle cell (SMC) is not only a contractile cell but is able to perform other important functions such as migration, proliferation, production of cytokines, chemokines, extracellular matrix proteins, growth factors, cell surface adhesion molecules and even antigen presentation [3–6]. Thus, SMC appears today as a fascinating cell with remarkable plasticity that contributes to its roles in physiology and disease.

Most of the SMC functions are dependent on a key event: the increase in intracellular calcium concentration ($[Ca^{2+}]_i$). Calcium entry from the extracellular space is a major step in the elevation of $[Ca^{2+}]_i$ in SMC and involves a variety of plasmalemmal calcium channels. These latter can be divided into voltage-operated calcium channels (VOC) (L- and T-type calcium channels) [7–10], and voltage-independent calcium channels. Voltage-independent calcium channels involve: (1) receptor-operated channels (ROC) [11] which are regulated by agonist-receptor interaction and the downstream signal transduction proteins; (2) capacitative or store-operated channels (SOC) which are activated by the emptying of intracellular Ca^{2+} store (mainly the sarcoplasmic reticulum) [12, 13]; (3) mechanosensitive or stretch-activated channels (SAC) which are activated by SMC membrane stretch [14, 15]. In contrast to VOC, our knowledge of the functional properties, molecular structure and the role in smooth muscle function of voltage-independent calcium channels is much less advanced. However, during the ten last years, a special attention has been paid to a new family of membrane proteins: the transient receptor potential (TRP) proteins [16]. Beyond the six known subfamilies of TRP in mammals, TRPC (canonical), TRPM (melastatin), TRPV (vanilloid) TRPA (ankyrin), TRPP (polycystin), TRPML (mucolipin), all functionally characterized TRP channels are permeable to Ca^{2+} with the exceptions of TRPM4 and TRPM5, which are only permeable to monovalent cations. Most Ca^{2+}-permeable TRP channels are only poorly selective for Ca^{2+}, with permeability ratio relative to Na^+ (PCa/PNa) in the range between 0.3 and 10. Exceptions are TRPV5 and TRPV6, two highly Ca^{2+}-selective TRP channels with $PCa/PNa > 100$ [17].

There is now a growing number of studies dealing with the expression of TRP proteins in smooth muscles and the functional role of these proteins in various pathophysiological processes. The present chapter thus aims to review the expression and the physiological roles of TRP channels in a variety of smooth muscles in mammals.

36.2 Expression of TRP in SMC

36.2.1 TRPC

TRPC is a seven family member which can be subdivided into subfamilies on the basis of their amino acids sequence similarity. While TRPC1 and TRPC2 are almost unique, TRPC4 and TRPC5 share ~ 65% homology. TRPC3, – 6 and – 7 form a structural and functional subfamily sharing 70–80% homology at the amino acid level. Homo- and heteromultimerization of different TRPC proteins within a tetrameric complex has been demonstrated for TRPC1/4/5 or TRPC3/6/7 [18]. While TRPC2 is expressed in most vertebrates, it exists only as a pseudogene in humans [17]. The role of TRPC2 in smooth muscles has not yet been demonstrated so far. TRPC1, 2, 4 and 5 have been associated to a SOC therefore, they are activated by store depletion. Inversely, TRPC3, 6 and 7 are rather associated to a ROC therefore they are activated following activation of a receptor and they are sensitive to diacylglycerol (DAG) and other components of G-protein receptor signalling pathways. TRPC1 and 6 can also be activated by stretch although this is controversial (see below *TRP and myogenic tone*). All the TRPC (1–7) have been detected in SMC following RT-PCR and/or Western Blot and/or immunofluorescence analyses [19, 20]. Indeed, TRPC are ubiquitously expressed in smooth muscle including vascular smooth muscle (cerebral arteries, pulmonary arteries, caudal arteries, aorta, mesenteric arteries, renal resistance vessels, saphenous vein, ear artery, portal vein, coronary artery and pial arterioles) [19–21], airway smooth muscle, gastrointestinal tract (oesophageal body, stomach and colon) and pregnant myometrium [19, 20]. TRPC have also been found in various species including rabbit, rat, mouse, dog, guinea pig and human as well as cell lines such as the vascular cell line A7r5. However, their expressions are highly dependent on tissue and species and, sometimes, conflicting results have been observed between different studies [19]. Several reviews have extensively explored the expression of TRPC in smooth muscle [19, 20, 22] therefore we did not detail further this topic.

36.2.2 TRPV

The TRPV (Vanilloid) subfamily contains six members (TRPV1-6) which are Ca^{2+} permeable cation channels. TRPV channels are widely expressed and can be activated by a large range of stimuli, including change in cell volume, chemical ligands, extracellular protons, lipids, noxious heat [18].

TRPV1 (also known as VR1) was the first mammalian member of this family to be identified as the target for capsaicin [23]. TRPV1 channels have been characterized in both visceral and vascular smooth muscles. In airways, expression of TRPV1 mRNA was observed in human bronchial SMC [24]. TRPV1 mRNA was also detected in rat prostatic tissue and immunohistochemistry indicated that TRPV1 was highly expressed in both epithelial and SMC [25]. In vascular smooth muscles, TRPV1 mRNA and protein were characterized in human and rat pulmonary arterial smooth muscle cells (PASMC) [26, 27], in rat aorta and cultured A7r5 rat aortic SMC [28, 29]. Immunohistochemistry also revealed the expression of TRPV1 in rat skeletal muscle arteriole [28].

TRPV2 shares a 50% sequence identity with TRPV1 but is insensitive to capsaicin [30]. In visceral smooth muscles, expression of TRPV2 mRNA was observed in human bronchial SMC [24, 31] and in rat prostatic tissue where TRPV2 was the most abundantly expressed TRPV subtype [25]. In vascular smooth muscle, TRPV2 transcripts were detected in mouse aortic, mesenteric and basilar arterial myocytes using PCR amplification and, expression of TRPV2 protein was confirmed by the immunohistochemistry [32]. TRPV2 mRNA was also detected in rat aortic SMC [27] and in human and rat PASMC [26, 27].

Much less is known about TRPV3 for which mRNA have been detected only in rat aorta [27] and rat or human PASMC [26, 27].

TRPV4 shares a 40% sequence identity with TRPV1 and TRPV2 but is insensitive to capsaicin. This isoform is widely distributed in smooth muscle. In human bronchial SMC, expression of TRPV4 mRNA was observed [24, 31] and application of 4α-PDD, an agonist of TRPV4, activated a calcium influx [31]. TRPV4 mRNA and protein were also characterized in rat aortic myocytes [27], in mesenteric artery [33] and rat or human PASMC [26, 27] where a TRPV4-like signal (calcium influx or whole-cell currents) was also detected after 4α-PDD stimulation [27, 34]. Expression of TRPV4 mRNA was also observed in rat prostatic tissue [25]. In mouse urinary bladder SMC, TRPV4 mRNA was detected and protein expression demonstrated by Western blot and immunohistochemistry [35].

TRPV5 and TRPV6 are highly Ca^{2+}-selective cation channels of the apical membrane of epithelial cells. However, the expression of such TRP in smooth muscle remains uncertain since TRPV5 and TRPV6 transcripts were not detected neither in human and rat PASMC [26, 27], nor in rat aortic SMC [27].

36.2.3 TRPM

The TRPM (Melastatin) subfamily contains eight members (TRPM1-8) which exhibit highly varying cation permeability, from Ca^{2+} impermeable (TRPM4,5) to highly Ca^{2+} and Mg^{2+} permeable (TRPM6,7).

Very few studies have been performed on TRPM1,2,3 in SMC. TRPM1,2,3 mRNA were detected in rat prostatic tissue [25]. TRPM2,3 mRNA and TRPM2 protein were also characterized in rat aortic myocytes and PASMC [27]. TRPM3 (mRNA and protein) was observed in human saphenous vein, mouse aorta and femoral artery SMC [36] where pregnenolone sulphate or sphingosine, TRPM3 agonists, evoked calcium entry sensitive to nonspecific inhibitors, TRPM3 blocking antibody or knock-down of TRPM3 by RNA interference [36].

TRPM4 exhibits the highest expression (mRNA and protein) in rat and mouse cerebral arteries [37–40] but was also described in rat aortic, pulmonary arterial and prostatic SMC [25, 27].

Using conventional RT-PCR, weak signals for TRPM5 expression was identified in both rat PASMC and aortic SMC [27].

TRPM6 and TRPM7 are closely related proteins, share a 50% sequence identity and contain a C-terminal kinase domain. TRPM6 interacts with TRPM7 and may form an heterotetrameric channel important for Mg^{2+} homeostasis [41]. TRPM6,7 have been identified in human, mouse, and rat mesenteric resistance arteries [42–45] where TRPM7 was identified as a functional regulator of Mg^{2+} homeostasis and cellular growth [43]. Moreover, angiotensin II and aldosterone regulated TRPM7 acutely by inducing phosphorylation and chronically by increasing expression (mRNA and protein) [44]. Finally, bradykinin increased expression of TRPM7 and its downstream target annexin-1, and induced the rapid phosphorylation of TRPM7, such phenomena may play a role in bradykinin/B2 receptor-mediated inflammatory responses in vascular cells [42, 45]. TRPM6 and TRPM7 are also expressed in porcine carotid arteries where they contribute to Mg^{2+} homeostasis [46] and in rat prostatic, aortic, and PASMC [25, 27]. In the cultured A7r5 rat aortic cell line, it was shown that TRPM7 may act as a mechanotransducer [47].

TRPM8 is a cold-sensing Ca^{2+}-permeable channel playing a role in thermosensation. TRPM8 mRNA and protein were detected in rat aortic, femoral, mesenteric, tail, and pulmonary arteries [27, 48]. TRPM8 mRNA was also detected in rat prostatic tissue and immunohistochemistry indicated that TRPM8 was highly expressed in SMC [25]. TRPM8 mRNA was also observed in rat fundus smooth muscle [49] and in mouse bladder [50].

36.2.4 TRPP

The TRPP family is very inhomogeneous and can be divided, on structurally criteria, into PKD1-like (TRPP1-like) and PKD2-like (TRPP2-like) proteins. TRPP1 and TRPP2 are called PKD1 and PKD2 or PC1 and PC2 or polycystin 1 and polycystin 2 due to their role in polycystic kidney disease. PKD1-like members comprise TRPP1 (previously termed PKD1), PKDREJ, PKD1L1, PKD1L2, and PKD1L3. The PKD2-like members comprise PKD2 (TRPP2), PKD2L1 (TRPP3), and PKD2L2 (TRPP5) [17]. TRPP2, 3 and 5 are consistent with channels acting as calcium-activated calcium channels [17]. Most TRPP channels are recognised as sensors to mechanical stimulation. Although TRPP2 inhibits SAC activity, TRPP2 sequestration by TRPP1 restores SAC activity [51].

TRPP1 and/or TRPP2 have been detected in various vascular smooth muscles including aorta, intracranial arteries, afferent arterioles, mesenteric arteries, portal vein, hepatic arteries [19, 20, 51–53]. PKD1 and PKD2-like members have also been found in childhood tissue such as bowel from human and mouse [54–56]. TRPP1 interacts with TRPP2 explaining why they are often both expressed at the same location [57].

36.3 Physiological and Pathophysiological Roles of TRP in SMC

TRP channels expressed in SMC are implicated in variety of functions in different systems and organs. Roles of TRP in airway hyperresponsiveness, asthma and urinary bladder mechanosensation are exposed in some other chapters of the present book.

36.3.1 TRP and the Vascular System

TRP in vascular SMC underlie various cellular and integrative functions such as growth and hyperplasia, contraction and relaxation, control of myogenic tone and blood pressure and are implicated in cardiovascular diseases such as pulmonary hypertension.

36.3.1.1 TRP and Vascular SMC Growth and Hyperplasia

Vascular SMC have a remarkable plasticity allowing them to rapidly undergo, through altered gene expression, from the "differentiated" or contractile state to the "migratory" and "proliferative" state. This process, generally, requires Ca^{2+} entry which activates Ca^{2+} dependent transcriptional activities. Different studies suggest that several TRP may contribute to this process. For example, it has been shown that TRPC1 and TRPC6 expression level and SOC activity were increased in rat cerebral and human mammary arteries by short term organ culture or after vascular injury caused by balloon dilatation [58]. Similar features were also observed in mouse and rat carotid arteries with hyperplasia experimentally induced by cuff injury. Moreover, the TRPC1 specific TIE3 antibody was able to suppress the neo intimal growth of vascular SMC in organ cultured human saphenous vein obtained from patients taking coronary bypass graft surgery [36]. In proliferating pulmonary artery SMC, TRPC1, 4 and 6 are clearly upregulated and the decrease in their expression decreases SMC proliferation [59–64]. Similarly, in human coronary smooth muscle cells, TRPC1 reduction by siRNA inhibited the hypertrophic response induced by angiotensin II [65]. Moreover, the role of TRPC1 appears specific since, in proliferating PASMC, no changes in TRPC3 and voltage dependent Ca^{2+} channel β_2 subunit is observed [59]. TRPM6 and 7 are both essential for magnesium homeostasis and possess a kinase domain [66]. Although little is known regarding TRPM6, TRPM7, through its both kinase and magnesemia function, plays a major role in vascular smooth muscle viability and growth and consequently in hypertension [66].

36.3.1.2 TRP in Vascular Contractile Function in Physiological and Pathophysiological Conditions

TRP in Smooth Muscle Contraction/Relaxation Cycle

Contradictory studies have been reported concerning the role of TRPC1 as an essential component of SOC and capacitative calcium entry (CCE) associated with vascular contractility. In the one hand, overexpression of the human TRPC1 gene in rat pulmonary artery enhanced SOC-induced Ca^{2+} entry and contraction

[67]. Conversely, acute application of TIE3 antibody which specifically targets the putative outer TRPC1 channel pore inhibited CCE in pial artery [68] and endothelin-1-induced contraction of rat caudal artery [69]. In this latter preparation, TRPC1 appeared colocalised with other signalling proteins in caveolae or cholesterol-rich raft. It seems that activation of SOC and its relative importance in vascular contraction is greatly dependent on the caveolar localization of TRPC1 protein in a large signalling complex. In the other hand, SMC of TRPC1(−/−) mice isolated from thoracic aortas and cerebral arteries showed no change in SOC-induced cation influx induced by thapsigargin, inositol-1,4,5 trisphosphate, and cyclopiazonic acid compared to cells from wild-type mice [70].

TRPC6 is also largely expressed in the vascular system and its heterologous expression produced Ca^{2+} influx through nonselective cation channels activated by DAG and independently of store depletion [71, 72]. In A7r5 cells, activation of these channels and the associated Ca^{2+} influx was strongly impaired by deletion of TRPC6 with its specific antisense oligodeoxynucleotide pre-treatment or small interfering RNA (siRNA) silencing [73, 74]. Thus, TRPC6 appears as the molecular structure forming ROC and responsible for their vasoconstrictor effect. TRPC6 deficient ($TRPC6^{-/-}$) mice have a significant higher mean arterial blood pressure than wild type controls and this was associated with an upregulation of TRPC3 [71]. Interestingly, recent studies show that TRPC3 appears to be upregulated in spontaneously hypertensive rats [75] and in patients with essential hypertension [76].

TRPC3 and TRPV4 are expressed in cerebral arteries and participate to the regulation of blood flow. In intact cerebral arteries, suppression of arterial SMC TRPC3 expression with antisense oligodeoxynucleotides significantly decreased the depolarization and constriction in response to UTP [77].

In cerebral arteries TRPV4 is clearly implicated in smooth muscle hyperpolarisation and arterial dilation in response to an endothelial-derived factor [78]. TRPV4 on SMC is activated by the endothelium-derived arachidoic acid metabolite 11, 12 epoxyeicosatrienoic acid (11, 12 EET). Ca^{2+} influx through TRPV4 stimulates ryanodine receptor (RYR) in the sarcoplasmic reticulum generating Ca^{2+} sparks that signal adjacent BKCa channels to open and cause membrane hyperpolarization and vasodilation [78]. Thus TRPV4 could form a functional Ca^{2+} signalling complex with RYR and BKCa channels to induce hyperpolarization and arterial dilation. TRPV4 is also implicated in the control of vascular tone in peripheral resistances arteries and this effect contributes to the regulation of blood pressure. Indeed, TRPV4 is expressed both in endothelial cells and SMC of mesenteric arteries. [33] 11,12 EET-induced hyperpolarization and vasodilation are greatly attenuated in TRPV4 knockout (KO) mice, 50% of the effects were due to direct effects on the SMC. 11,12 EET activates a TRPV4-like current in mesenteric arterial SMC from wild type mice, but not TRPV4 KO mice. [33] TRPV4 is also expressed in arteries where it is a target for the serine/threonine kinases "With No Lysine (K) Kinases" (WNK1 and WNK4), with importance in hypertension [79]. Hypertension induced by the inhibition of NOS activity is enhanced in TRPV4 KO mice versus wild type controls [33]. TRPV4 channel activity may be an important mechanism that normally opposes systemic hypertensive stimuli in vivo. TRPV4 channels in the vasculature represent a promising new target for antihypertensive drug therapy.

Many studies have highlighted the role of TRP channels (TRPC and TRPV) in pulmonary SMC, both in control and in pathophysiological conditions (see below *Role in pulmonary artery hyperreactivity*). In addition, in rat pulmonary arterial and aortic SMC, menthol, a TRPM8 channel agonist, induced a Ca^{2+} influx [27] and application of menthol on endothelium-denuded relaxed rat vessels induced small contractions [48]. In rat skeletal muscle arterioles, capsaicin, an agonist of TRPV1, elicited constriction of isolated pressurized arterioles [28]. In freshly isolated aorta, TRPM3 agonists positively modulated contractile responses [36]. Finally, a recent study in cerebral arteries has highlighted the role of TRPM4 channels in arterial tone showing that those channels in cerebral artery SMC are regulated by Ca^{2+} release from inositol 1,4,5-trisphosphate receptor (IP_3R) on the sarcoplasmic reticulum [80].

TRP and Myogenic Tone

Altered vascular tone is responsible for elevation of intravascular pressure, and vice versa, increased intravascular pressure induces an intrinsic vasomotor mechanism termed "myogenic tone" first described by Bayliss in 1902 [81]. Indeed, change in transmural pressure is transduced by stretch-activated channels into direct Ca^{2+} entry or membrane depolarisation that activates voltage-dependent Ca^{2+} channels resulting in increased supplementary Ca^{2+} entry and elevated intracellular Ca^{2+} concentration which activates contraction through activation of calmodulin and myosin light chain kinase [82]. Currently available data suggest that at least, eleven mammalian TRP isoforms (TRPC1,5,6; TRPV1,2,4; TRPM3,4,7; TRPP2 and TRPA1) are mechanosensitive (see above and [16, 83]). In particular, in intact rat cerebral arteries, TRPC6 antisense oligodeoxynucleotides decreased TRPC6 protein expression and attenuated arterial smooth muscle depolarization and constriction caused by elevated intraluminal pressure [84]. Moreover, Inoue et al. recently reported the role of 20-HETE for mechanical potentiation of both expressed and native TRC6 channels, respectively in HEK293 and A7r5 cells, and enhanced myogenic response in rat mesenteric arteries [83]. TRPC6, as a sensor of mechanically and osmotically induced membrane stretch, has been proposed to play a key role in regulating myogenic tone in vascular tissue [85]. However, contradictory results concerning the mechanosenstivity of TRP have been obtained with TRPC6 and TRPC1 knock-out mice [70, 86]. In TRPC6−/− mice, increased myogenic contraction of cerebral arteries was observed associated with increased TRPC3 expression, although myogenic response was not altered by TRPC3 suppression in TRPC3 antisense-treated rat cerebral arteries [77]. In TRPC1−/− mice, neither pressure-induced constriction of cerebral arteries nor currents activated by hypoosmotic swelling and positive pipette pressure in SMC from cerebral arteries showed significant differences compared to wild-type cells [70].

In addition, TRPM4 appears to be necessary for the myogenic constriction in cerebral arteries. Indeed, (i) in freshly isolated cerebral artery myocytes, membrane stretch activated TRPM4-like channels with properties (unitary conductance, ion selectivity and calcium dependence) similar to those of the cloned TRPM4 [38]; (ii) myogenic contraction was attenuated in isolated rat cerebral arteries treated with TRPM4 antisense oligodeoxynucleotides [40] or in arteries resulting from in vivo TRPM4 antisense oligodeoxynucleotides infused into the cerebral spinal fluid [37]. Moreover, the control of cerebral artery myogenic tone seems to be dependent on a PKC-mediated modulation of TRPM4 activity [39]. Therefore, both TRPC6 and TRPM4 appear to be necessary for myogenic constriction. Since TRPM4 is sensitive to intracellular Ca^{2+}, increased intraluminal pressure may induce Ca^{2+} influx through TRPC6, resulting in TRPM4 activation.

Finally, TRPV2 may function as a mechano-stretch sensitive channel. In mouse aortic SMC, cell swelling induced by hypotonic solution activated a calcium influx sensitive to ruthenium red, a non-selective TRP channel blocker, that was absent in TRPV2 knock-down cells [32]. These findings suggest a potential role for TRPV2 in pressure-induced vasoconstriction.

TRP and Hypoxic Pulmonary Vasoconstriction (HPV)

Hypoxic pulmonary vasoconstriction (HPV) is an important physiological mechanism that optimises ventilation–perfusion matching and pulmonary gas exchange by diverting blood flow from poorly ventilated areas of the lung. It is generally admitted that the main trigger of HPV is the PASMC, although the endothelium plays an important facilitatory or permissive role for sustained HPV. Whilst acute hypoxia often causes a monophasic increase in PA pressure in vivo, the response in isolated PA is biphasic (also observed in some perfused lung and in vivo studies). This consists of a rapid, transient vasoconstriction lasting about 10–15 min (Phase 1) and a slowly developing vasoconstriction that is sustained as long as hypoxia is present (Phase 2) [87]. The mechanisms underlying HPV and their role in the development of chronic pulmonary hypertension remain incompletely resolved. However, direct evidence for a role of TRPC6 in HPV has been provided in a TRPC6–/– mouse model [88]. In these mice, phase 1 of HPV and the associated elevation of PASMC $[Ca^{2+}]_i$ were absent, but neither phase 2 of HPV nor chronic hypoxia-induced PH were affected. These results suggest that TRPC6 plays little role in SOC-mediated Ca entry (SOCE) activated by depletion of ryanodine-sensitive stores which have been implicated in both phenomena [87, 89]. Interestingly, hypoxia increased diacylglycerol (DAG), an activator of TRPC6, in PASMC, which could be due to ROS-mediated activation of PLC [88] and, recently it was reported that hypoxia causes translocation of TRPC6 to the membrane, a mechanism mediated by cytochrome P450-derived epoxyeicosatrienoic acids (EETs) [90].

TRP and Pulmonary Hypertension

Pulmonary arterial hypertension (PAH) is a disease of small pulmonary arteries characterized by an increase in reactivity to mediators (vasoconstriction) and a remodelling of the arterial wall due, at least in part, to proliferation of arterial SMC [91, 92]. Although molecular and cellular basis of PAH are not yet fully established, the implication of TRP in both increased reactivity of pulmonary arteries and proliferation of PASMC seems evident from results of recent studies.

Role in Pulmonary Artery Hyperreactivity

Several studies have demonstrated pulmonary artery hyperreactivity in animal PAH models as well as in human. In rats with chronic hypoxia-induced PAH (CH-PAH), the vasoreactivity to 5-HT is increased [93–95] and, since 5-HT activates a receptor-operated Ca^{2+} influx in rat intrapulmonary arteries [96], such ROC may well be involved in the hyperreactivity to 5-HT following CH-PAH. TRPC1 and TRPC6 overexpressions contribute to the increase in store-operated and receptor-operated Ca^{2+} influxes, respectively [97] and may thus participate to the elevated vascular tone in CH-PAH.

Resting $[Ca^{2+}]_i$ is usually higher in PASMC from patients or animals with PAH, except in patients with secondary PAH, than in PASMC from normotensive patients or animals [98–101]. Activation of a SOC has been implicated in the elevation of $[Ca^{2+}]_i$ in PASMC from primary pulmonary hypertensive patients. Moreover, in pulmonary arteries from patients with primary hypertension (idiopathic PAH) but not with secondary hypertension, an increased CCE and an upregulation of TRPC3 and TRPC6 was observed [59, 61, 62]. In intrapulmonary arteries from a model of normobaric CH-PAH rats, although TRPC3 and TRPC4 expression is unchanged, TRPC1 and TRPC6 have been shown to be overexpressed [97, 102]. However, another study has shown that the change in $[Ca^{2+}]_i$ associated to SOC and ROC is significantly reduced in intrapulmonary arteries from hypobaric CH-PAH rats [103]. A possible decrease in expression and/or activity of TRPC channel proteins following hypobaric CH-PAH has also been evoked [93]. In that respect, this model may be more related to secondary hypertensive patients. Altogether, SOC/ROC and TRPC regulation may vary according to animal models used (normobaric vs hypobaric CH-PAH rats). A recent study performed in rat and in TRPV4 –/– mice exposed to chronic hypoxia suggests that TRPV4 could play a significant role in the development of PAH including the modification of the pulmonary arterial vasoreactivity [29]. So far, there is no direct evidence of the involvement of TRP in the hyperreactivity associated to PAH. However, when the above studies demonstrated a concomitant increase in the role of ROC, SOC and/or CCE in the contraction and an increase in the TRPC expression, the authors consequently suggested a role of the TRPC in the hyperreactivity observed following PAH. Indeed, these conclusions were strengthened by many studies showing a link between TRPC and ROC, SOC or CCE and contraction in pulmonary artery from normal rats. For instance, downregulation of TRPC6 using an antisense oligonucleotides abolished the increase in cyclopiazonic acid-induced store-operated calcium entry associated to PDGF stimulation in rat PASMC [63] and siRNA against TRPC6 highlighted the link between ROC and TRPC6 in rat PA [97]. Using of siRNA or antisense oligonucleotides demonstrated the contribution of

TRPC1, 4 and 6 to endogenous SOC in the pulmonary vasculature [60, 63, 64, 97]. In lungs from TRPC6 –/– mice, the acute hypoxia and 11, 12 epoxyeicosatrienoic acid-induced pulmonary pressure elevation disappeared [90].

Role in SMC Proliferation

The role of TRP in PAH-induced pulmonary vascular remodelling has been addressed by several studies. In comparison to growth-arrested cells (cultured in media without serum and growth factors), proliferating human PASMC (cultured in media containing serum and growth factors) exhibit a high level of resting $[Ca^{2+}]_i$ potentially maintained by a constant Ca^{2+} influx [59]. The TRPC6 upregulation was associated to the increased proliferation in PASMC from primary hypertensive patients compared to normotensive and secondary hypertensive patients [61, 62]. In another way, the endothelin receptor blocker bosentan, which is clinically used for the treatment of patients with idiopathic PAH, was found to markedly downregulate TRPC6 expression [67]. The PDGF-mediated upregulation of TRPC6 expression is also associated with an increase in CCE and rat PASMC proliferation [63]. Interestingly, PDGF levels are higher in lung tissues of patients with severe PAH and its implication in the development of PAH has been shown. These results strengthen the role of TRPC6 in the vascular remodelling and PAH. However, it should be mentioned that pulmonary arterial remodelling and right ventricular systolic pressure were unaltered in TRPC6 –/– mice [88]. TRPC1 and TRPC3 also appeared to be involved in PASMC proliferation in PAH. Indeed, an increased TRPC3 expression has been suggested to be responsible for the higher activity of CCE and the associated PASMC proliferation in idiopathic PAH [104]. Moreover, in TRPC1 –/– mice, hypoxia-induced PAH was significantly attenuated [88].

Chronic hypoxia also up-regulated expression levels of the TRPV1 (mRNA and protein) and enhanced the proliferation of PASMC which was inhibited in a dose-dependent manner by capsazepine, a TRPV1 channel inhibitor [26]. As mentioned before, in TRPV4 –/– mice, exposed to hypoxia, PAH was attenuated [29]. Altogether, secondary and idiopathic PAH may be associated to several TRP including so far TRPC1, 3 and 6 and TRPV1 and 4. However, we can not exclude the involvement of other TRP not yet explored and further studies are needed to elucidate the exact role of the TRP in PAH.

Accordingly, agents ranging from those that downregulate TRP gene expression to specific blockers of ROC and SOC in PASMC may prove beneficial in the development of therapeutic approaches for treatment of pulmonary hypertension. It may act on pulmonary vascular resistances as well as vascular remodelling.

36.3.2 TRP and Gastro Intestinal Tract

36.3.2.1 TRP in Intestinal Motility

Parasympathetic signalling via muscarinic receptors regulates gastrointestinal smooth muscle contraction. It was well known for a long time that activation of muscarinic receptors (M2 and M3 receptors) induces a non specific cationic current named mI$_{CAT}$ which subsequently activates voltage-operated calcium channels and triggers smooth muscle contraction [105, 106]. A recent elegant work using TRPC4-, TRPC6-, and TRPC4/TRPC6-gene-deficient mice demonstrated that deletion of TRPC4 and TRPC6 impairs smooth muscle contraction and intestinal motility in vivo [107]. Patch-clamp experiments in intestinal isolated smooth muscle cells revealed that TRPC4 and TRPC6 contribute separately for 80 and 20% to mI$_{CAT}$, respectively. Thus, it clearly appears that in intestinal smooth muscle, TRPC4 and TRPC6 constitute the molecular entity of the muscarinic receptor-activated channels (ROC) mI$_{CAT}$.

Gastro intestinal motility is dependent on spontaneous electrical activity (slow waves) in interstitial cells of Cajal (ICCs) [108]. TRP channels appear implicated in this activity. In cultured ICCs from mice, TRPM7 is required for pacemaker activity in ICCs although TRPC4 is involved in electrical response induced by muscarinic stimulation [109, 110].

36.3.2.2 TRPM8 and Gatric Activity

Agonist-induced contractions of smooth muscles are modulated by several factors including temperature. Cooling induces contraction in gastrointestinal tract as in other visceral smooth muscles such as trachea [111, 112] and urinary bladder [113], by a mechanism which remains unresolved but is, at least in part, calcium-dependent. Since TRPM8 is activated by low temperatures ($<25°C$), it was tempting to assess the fact that cooling-induced contraction was TRPM8 mediated. TRPM8 is expressed in the rat gastric fundus, and capsazepine, a TRPM8 receptor antagonist, inhibited cooling-induced contraction [49]. Physiologically, The TRPM8 could be involved in the transient alteration in gastric emptying following the ingestion of cold food.

36.3.3 TRP and Uterine Contractile Activity

The regulation of uterine contractile state is of importance for the maintenance of pregnancy (quiescent uterus) and of parturition (switch to the contractile uterus) [114, 115]. It is generally admitted that extracellular calcium is essential for the generation of spontaneous and agonist-induced contractions in both rat and human myometrium [114, 116, 117]. While the role of VOC in the initiation of myometrial contraction has been recognized for a long time [118], there is increasing functional evidence that SOC-mediated Ca entry (SOCE) is also present in myometrium and that SOCE is increased during human labour [117] suggesting the presence of TRP channels in myometrial cells. Indeed, in human myometrium, RT-PCR analysis have shown TRPC1, TRPC3, TRPC4, TRPC5, TRPC6 and TRPC7 mRNA whereas Western blot analysis and immunolocalization have revealed the presence of proteins TRPC1, 3, 4 and 6 in both primary cultured human myometrial SMC [119] and immortalized pregnant human myometrial smooth muscle cells (PMH1) [120].

Moreover, SOCE activated by either oxytocin or thapsigargin is increased in PHM1 overexpressing hTRPC3 [121]. In rat myometrium, mRNA for TRPC1, TRPC2, TRPC4-C7 were expressed and TRPC4 was the most abundant [122].

Stretch due to the needs of developing foetuses is known to stimulate myometrial hyperplasia and hypertrophy in early stages of pregnancy [123]. Transduction of the stretch signal is operating via alteration of intracellular calcium concentration. From cultured human myometrial SMC submitted to 25% tonic mechanical stretch for 1–14 h, it has been shown that such a treatment induced an increase of basal calcium entry and CCE [124]. Moreover, RT-PCR and Western blot show an increased expression of TRPC3 and TRPC4 mRNA and TRPC3 protein in response to stretch [124].

Thus TRP channels could play a role in controlling myometrial intracellular Ca^{2+} concentrations and may be important transducers of agonist-mediated signals that increase at the time of parturition and labor.

36.4 Conclusion

During past 10 years, numerous studies have shown the large expression of TRP channels (TRPC, TRPV, TRPM, TRPP) in mammals SMC. These TRP are implicated in a variety of physiological and pathophysiological processes regarding cardiovascular system (control of blood pressure, myogenic tone, hypertension), gastro-intestinal system (intestinal motility, gastric acidity) and reproductive female system (uterine contractility). TRP channels may be attractive novel drug targets for disease originating from SMC dysfunction.

Acknowledgments Funding for the authors work was supported by ANR (ANR06 – Physio – 015 – 01) and the Fondation de France (2008002719).

References

1. Berridge MJ (2008) Smooth muscle cell calcium activation mechanisms. J Physiol 586:5047–5061
2. Kim HR, Appel S, Vetterkind S, Gangopadhyay SS, Morgan KG (2008) Smooth muscle signalling pathways in health and disease. J Cell Mol Med 12:2165–2180
3. Gerthoffer WT (2007) Mechanisms of vascular smooth muscle cell migration. Circ Res 100:607–621
4. House SJ, Potier M, Bisaillon J, Singer HA, Trebak M (2008) The non-excitable smooth muscle: calcium signaling and phenotypic switching during vascular disease. Pflugers Arch 456:769–785
5. Tliba O, Panettieri RA Jr. (2009) Noncontractile functions of airway smooth muscle cells in asthma. Annu Rev Physiol 71:509–535
6. Wamhoff BR, Bowles DK, Owens GK (2006) Excitation-transcription coupling in arterial smooth muscle. Circ Res 98:868–878
7. Cribbs LL (2006) T-type Ca2+ channels in vascular smooth muscle: multiple functions. Cell Calcium 40:221–230

8. Fry CH, Sui G, Wu C (2006) T-type Ca2+ channels in non-vascular smooth muscles. Cell Calcium 40:231–239
9. Liao P, Yong TF, Liang MC, Yue DT, Soong TW (2005) Splicing for alternative structures of Cav1.2 Ca2+ channels in cardiac and smooth muscles. Cardiovasc Res 68:197–203
10. Sonkusare S, Palade PT, Marsh JD, Telemaque S, Pesic A, Rusch NJ (2006) Vascular calcium channels and high blood pressure: pathophysiology and therapeutic implications. Vascul Pharmacol 44:131–142
11. Bolton TB (1979) Mechanisms of action of transmitters and other substances on smooth muscle. Physiol Rev 59:606–718
12. Parekh AB, Penner R (1997) Store depletion and calcium influx. Physiol Rev 77:901–930
13. Putney JW Jr (1990) Capacitative calcium entry revisited. Cell Calcium 11:611–624
14. Kirber MT, Walsh JV Jr., Singer JJ (1988) Stretch-activated ion channels in smooth muscle: a mechanism for the initiation of stretch-induced contraction. Pflugers Arch 412:339–345
15. Davis MJ, Donovitz JA, Hood JD (1992) Stretch-activated single-channel and whole cell currents in vascular smooth muscle cells. Am J Physiol 262:C1083–C1088
16. Guibert C, Ducret T, Savineau JP (2008) Voltage-independent calcium influx in smooth muscle. Prog Biophys Mol Biol 98:10–23
17. Nilius B, Owsianik G, Voets T, Peters JA (2007) Transient receptor potential cation channels in disease. Physiol Rev 87:165–217
18. Venkatachalam K, Montell C (2007) TRP channels. Annu Rev Biochem 76:387–417
19. Beech DJ, Muraki K, Flemming R (2004) Non-selective cationic channels of smooth muscle and the mammalian homologues of Drosophila TRP. J Physiol 559:685–706
20. Dietrich A, Chubanov V, Kalwa H, Rost BR, Gudermann T (2006) Cation channels of the transient receptor potential superfamily: their role in physiological and pathophysiological processes of smooth muscle cells. Pharmacol Ther 112:744–760
21. Large WA, Saleh SN, Albert AP (2009) Role of phosphoinositol 4,5-bisphosphate and diacylglycerol in regulating native TRPC channel proteins in vascular smooth muscle. Cell Calcium 45:574–582
22. Dietrich A, Kalwa H, Gudermann T (2010) TRPC channels in vascular cell function. Thromb Haemost 103:262–270
23. Caterina MJ, Schumacher MA, Tominaga M, Rosen TA, Levine JD, Julius D (1997) The capsaicin receptor: a heat-activated ion channel in the pain pathway. Nature 389:816–824
24. Ito S, Kume H, Naruse K, Kondo M, Takeda N, Iwata S, Hasegawa Y, Sokabe M (2008) A novel Ca2+ influx pathway activated by mechanical stretch in human airway smooth muscle cells. Am J Respir Cell Mol Biol 38:407–413
25. Wang HP, Pu XY, Wang XH (2007) Distribution profiles of transient receptor potential melastatin-related and vanilloid-related channels in prostatic tissue in rat. Asian J Androl 9:634–640
26. Wang YX, Wang J, Wang C, Liu J, Shi LP, Xu M, Wang C (2008) Functional expression of transient receptor potential vanilloid-related channels in chronically hypoxic human pulmonary arterial smooth muscle cells. J Membr Biol 223:151–159
27. Yang XR, Lin MJ, McIntosh LS, Sham JS (2006) Functional expression of transient receptor potential melastatin- and vanilloid-related channels in pulmonary arterial and aortic smooth muscle. Am J Physiol Lung Cell Mol Physiol 290:L1267–L1276
28. Kark T, Bagi Z, Lizanecz E, Pasztor ET, Erdei N, Czikora A, Papp Z, Edes I, Porszasz R, Toth A (2008) Tissue-specific regulation of microvascular diameter: opposite functional roles of neuronal and smooth muscle located vanilloid receptor-1. Mol Pharmacol 73:1405–1412
29. Yang XR, Hughes JM, Cao YN, Flavahan NA, Liedtke W, Sham JSK (2008) Upregulation of TRPV4 channels in pulmonary arteries (PAs) contribute to chronic hypoxia induced myogenic tone and pulmonary hypertension. FASEB J 22(1213):1215
30. Caterina MJ, Rosen TA, Tominaga M, Brake AJ, Julius D (1999) A capsaicin-receptor homologue with a high threshold for noxious heat. Nature 398:436–441

31. Jia Y, Wang X, Varty L, Rizzo CA, Yang R, Correll CC, Phelps PT, Egan RW, Hey JA (2004) Functional TRPV4 channels are expressed in human airway smooth muscle cells. Am J Physiol Lung Cell Mol Physiol 287:L272–L278
32. Muraki K, Iwata Y, Katanosaka Y, Ito T, Ohya S, Shigekawa M, Imaizumi Y (2003) TRPV2 is a component of osmotically sensitive cation channels in murine aortic myocytes. Circ Res 93:829–838
33. Earley S, Pauyo T, Drapp R, Tavares MJ, Liedtke W, Brayden JE (2009) TRPV4-dependent dilation of peripheral resistance arteries influences arterial pressure. Am J Physiol Heart Circ Physiol 297:H1096–H1102
34. Ducret T, Guibert C, Marthan R, Savineau JP (2008) Serotonin-induced activation of TRPV4-like current in rat intrapulmonary arterial smooth muscle cells. Cell Calcium 43:315–323
35. Thorneloe KS, Sulpizio AC, Lin Z, Figueroa DJ, Clouse AK, McCafferty GP, Chendrimada TP, Lashinger ES, Gordon E, Evans L, Misajet BA, Demarini DJ, Nation JH, Casillas LN, Marquis RW, Votta BJ, Sheardown SA, Xu X, Brooks DP, Laping NJ, Westfall TD (2008) ((N-1S)-1-{[4-((2S)-2-{[(2,4-dichlorophenyl)sulfonyl]amino}-3-hydroxypropa noyl)-1-piperazinyl]carbonyl}-3-methylbutyl)-1-benzothiophene-2-carboxamid e (GSK1016790A), a novel and potent transient receptor potential vanilloid 4 channel agonist induces urinary bladder contraction and hyperactivity: Part I. J Pharmacol Exp Ther 326:432–442
36. Kumar B, Dreja K, Shah SS, Cheong A, Xu SZ, Sukumar P, Naylor J, Forte A, Cipollaro M, McHugh D, Kingston PA, Heagerty AM, Munsch CM, Bergdahl A, Hultgardh-Nilsson A, Gomez MF, Porter KE, Hellstrand P, Beech DJ (2006) Upregulated TRPC1 channel in vascular injury in vivo and its role in human neointimal hyperplasia. Circ Res 98:557–563
37. Reading SA, Brayden JE (2007) Central role of TRPM4 channels in cerebral blood flow regulation. Stroke 38:2322–2328
38. Morita H, Honda A, Inoue R, Ito Y, Abe K, Nelson MT, Brayden JE (2007) Membrane stretch-induced activation of a TRPM4-like nonselective cation channel in cerebral artery myocytes. J Pharmacol Sci 103:417–426
39. Earley S, Straub SV, Brayden JE (2007) Protein kinase C regulates vascular myogenic tone through activation of TRPM4. Am J Physiol Heart Circ Physiol 292:H2613–H2622
40. Earley S, Waldron BJ, Brayden JE (2004) Critical role for transient receptor potential channel TRPM4 in myogenic constriction of cerebral arteries. Circ Res 95:922–929
41. Chubanov V, Waldegger S, Mederos y Schnitzler M, Vitzthum H, Sassen MC, Seyberth HW, Konrad M, Gudermann T (2004) Disruption of TRPM6/TRPM7 complex formation by a mutation in the TRPM6 gene causes hypomagnesemia with secondary hypocalcemia. Proc Natl Acad Sci USA 101:2894–2899
42. Callera GE, He Y, Yogi A, Montezano AC, Paravicini T, Yao G, Touyz RM (2009) Regulation of the novel Mg2+ transporter transient receptor potential melastatin 7 (TRPM7) cation channel by bradykinin in vascular smooth muscle cells. J Hypertens 27:155–166
43. He Y, Yao G, Savola C, Touyz RM (2005) Transient receptor potential melastatin 7 ion channels regulate magnesium homeostasis in vascular smooth muscle cells: role of angiotensin II. Circ Res 96:207–215
44. Touyz RM, He Y, Montezano AC, Yao G, Chubanov V, Gudermann T, Callera GE (2006) Differential regulation of transient receptor potential melastatin 6 and 7 cation channels by ANG II in vascular smooth muscle cells from spontaneously hypertensive rats. Am J Physiol Regul Integr Comp Physiol 290:R73–R78
45. Yogi A, Callera GE, Tostes R, Touyz RM (2009) Bradykinin regulates calpain and proinflammatory signaling through TRPM7-sensitive pathways in vascular smooth muscle cells. Am J Physiol Regul Integr Comp Physiol 296:R201–R207
46. Hamaguchi Y, Matsubara T, Amano T, Uetani T, Asano H, Iwamoto T, Furukawa K, Murohara T, Nakayama S (2008) (Na+)-independent Mg(2+) transport sensitive to 2-aminoethoxydiphenyl borate (2-APB) in vascular smooth muscle cells: involvement of TRPM-like channels. J Cell Mol Med 12:962–974

47. Oancea E, Wolfe JT, Clapham DE (2006) Functional TRPM7 channels accumulate at the plasma membrane in response to fluid flow. Circ Res 98:245–253
48. Johnson CD, Melanaphy D, Purse A, Stokesberry SA, Dickson P, Zholos AV (2009) Transient receptor potential melastatin 8 channel involvement in the regulation of vascular tone. Am J Physiol Heart Circ Physiol 296:H1868-H1877
49. Mustafa S, Oriowo M (2005) Cooling-induced contraction of the rat gastric fundus: mediation via transient receptor potential (TRP) cation channel TRPM8 receptor and Rho-kinase activation. Clin Exp Pharmacol Physiol 32:832–838
50. Kobayashi H, Yoshiyama M, Zakoji H, Takeda M, Araki I (2009) Sex differences in the expression profile of acid-sensing ion channels in the mouse urinary bladder: a possible involvement in irritative bladder symptoms. BJU Int 104:1746–1751
51. Sharif-Naeini R, Folgering JH, Bichet D, Duprat F, Lauritzen I, Arhatte M, Jodar M, Dedman A, Chatelain FC, Schulte U, Retailleau K, Loufrani L, Patel A, Sachs F, Delmas P, Peters DJ, Honore E (2009) Polycystin-1 and -2 dosage regulates pressure sensing. Cell 139:587–596
52. Boulter C, Mulroy S, Webb S, Fleming S, Brindle K, Sandford R (2001) Cardiovascular skeletal, and renal defects in mice with a targeted disruption of the Pkd1 gene. Proc Natl Acad Sci USA 98:12174–12179
53. Torres VE, Cai Y, Chen X, Wu GQ, Geng L, Cleghorn KA, Johnson CM, Somlo S (2001) Vascular expression of polycystin-2. J Am Soc Nephrol 12:1–9
54. Geng L, Segal Y, Pavlova A, Barros EJ, Lohning C, Lu W, Nigam SK, Frischauf AM, Reeders ST, Zhou J (1997) Distribution and developmentally regulated expression of murine polycystin. Am J Physiol 272:F451–F459
55. Geng L, Segal Y, Peissel B, Deng N, Pei Y, Carone F, Rennke HG, Glucksmann-Kuis AM, Schneider MC, Ericsson M, Reeders ST, Zhou J (1996) Identification and localization of polycystin, the PKD1 gene product. J Clin Invest 98:2674–2682
56. Griffin MD, Torres VE, Grande JP, Kumar R (1996) Immunolocalization of polycystin in human tissues and cultured cells. Proc Assoc Am Physicians 108:185–197
57. Giamarchi A, Padilla F, Coste B, Raoux M, Crest M, Honore E, Delmas P (2006) The versatile nature of the calcium-permeable cation channel TRPP2. EMBO Rep 7:787–793
58. Bergdahl A, Gomez MF, Wihlborg AK, Erlinge D, Eyjolfson A, Xu SZ, Beech DJ, Dreja K, Hellstrand P (2005) Plasticity of TRPC expression in arterial smooth muscle: correlation with store-operated Ca2+ entry. Am J Physiol Cell Physiol 288:C872–C880
59. Golovina VA, Platoshyn O, Bailey CL, Wang J, Limsuwan A, Sweeney M, Rubin LJ, Yuan JX (2001) Upregulated TRP and enhanced capacitative Ca(2+) entry in human pulmonary artery myocytes during proliferation. Am J Physiol Heart Circ Physiol 280:H746–H755
60. Sweeney M, Yu Y, Platoshyn O, Zhang S, McDaniel SS, Yuan JX (2002) Inhibition of endogenous TRP1 decreases capacitative Ca2+ entry and attenuates pulmonary artery smooth muscle cell proliferation. Am J Physiol Lung Cell Mol Physiol 283:L144–L155
61. Yu Y, Fantozzi I, Remillard CV, Landsberg JW, Kunichika N, Platoshyn O, Tigno DD, Thistlethwaite PA, Rubin LJ, Yuan JX (2004) Enhanced expression of transient receptor potential channels in idiopathic pulmonary arterial hypertension. Proc Natl Acad Sci USA 101:13861–13866
62. Yu Y, Keller SH, Remillard CV, Safrina O, Nicholson A, Zhang SL, Jiang W, Vangala N, Landsberg JW, Wang JY, Thistlethwaite PA, Channick RN, Robbins IM, Loyd JE, Ghofrani HA, Grimminger F, Schermuly RT, Cahalan MD, Rubin LJ, Yuan JX (2009) A functional single-nucleotide polymorphism in the TRPC6 gene promoter associated with idiopathic pulmonary arterial hypertension. Circulation 119:2313–2322
63. Yu Y, Sweeney M, Zhang S, Platoshyn O, Landsberg J, Rothman A, Yuan JX (2003) PDGF stimulates pulmonary vascular smooth muscle cell proliferation by upregulating TRPC6 expression. Am J Physiol Cell Physiol 284:C316–C330
64. Zhang S, Remillard CV, Fantozzi I, Yuan JX (2004) ATP-induced mitogenesis is mediated by cyclic AMP response element-binding protein-enhanced TRPC4 expression and activity in human pulmonary artery smooth muscle cells. Am J Physiol Cell Physiol 287:C1192–C1201

65. Takahashi Y, Watanabe H, Murakami M, Ohba T, Radovanovic M, Ono K, Iijima T, Ito H (2007) Involvement of transient receptor potential canonical 1 (TRPC1) in angiotensin II-induced vascular smooth muscle cell hypertrophy. Atherosclerosis 195:287–296
66. Touyz RM (2008) Transient receptor potential melastatin 6 and 7 channels, magnesium transport, and vascular biology: implications in hypertension. Am J Physiol Heart Circ Physiol 294:H1103–H1118
67. Kunichika N, Yu Y, Remillard CV, Platoshyn O, Zhang S, Yuan JX (2004) Overexpression of TRPC1 enhances pulmonary vasoconstriction induced by capacitative Ca2+ entry. Am J Physiol Lung Cell Mol Physiol 287:L962–L969
68. Xu SZ, Beech DJ (2001) TrpC1 is a membrane-spanning subunit of store-operated Ca(2+) channels in native vascular smooth muscle cells. Circ Res 88:84–87
69. Bergdahl A, Gomez MF, Dreja K, Xu SZ, Adner M, Beech DJ, Broman J, Hellstrand P, Sward K (2003) Cholesterol depletion impairs vascular reactivity to endothelin-1 by reducing store-operated Ca2+ entry dependent on TRPC1. Circ Res 93:839–847
70. Dietrich A, Kalwa H, Storch U, Mederos y Schnitzler M, Salanova B, Pinkenburg O, Dubrovska G, Essin K, Gollasch M, Birnbaumer L, Gudermann T (2007) Pressure-induced and store-operated cation influx in vascular smooth muscle cells is independent of TRPC1. Pflugers Arch 455:465–477
71. Dietrich A, Mederos y Schnitzler M, Kalwa H, Storch U, Gudermann T (2005) Functional characterization and physiological relevance of the TRPC3/6/7 subfamily of cation channels. Naunyn Schmiedebergs Arch Pharmacol 371:257–265
72. Hofmann T, Obukhov AG, Schaefer M, Harteneck C, Gudermann T, Schultz G (1999) Direct activation of human TRPC6 and TRPC3 channels by diacylglycerol. Nature 397: 259–263
73. Inoue R, Okada T, Onoue H, Hara Y, Shimizu S, Naitoh S, Ito Y, Mori Y (2001) The transient receptor potential protein homologue TRP6 is the essential component of vascular alpha(1)-adrenoceptor-activated Ca(2+)-permeable cation channel. Circ Res 88:325–332
74. Soboloff J, Spassova M, Xu W, He LP, Cuesta N, Gill DL (2005) Role of endogenous TRPC6 channels in Ca2+ signal generation in A7r5 smooth muscle cells. J Biol Chem 280:39786–39794
75. Liu D, Scholze A, Zhu Z, Kreutz R, Wehland-von-Trebra M, Zidek W, Tepel M (2005) Increased transient receptor potential channel TRPC3 expression in spontaneously hypertensive rats. Am J Hypertens 18:1503–1507
76. Liu D, Scholze A, Zhu Z, Krueger K, Thilo F, Burkert A, Streffer K, Holz S, Harteneck C, Zidek W, Tepel M (2006) Transient receptor potential channels in essential hypertension. J Hypertens 24:1105–1114
77. Reading SA, Earley S, Waldron BJ, Welsh DG, Brayden JE (2005) TRPC3 mediates pyrimidine receptor-induced depolarization of cerebral arteries. Am J Physiol Heart Circ Physiol 288:H2055–H2061
78. Earley S, Heppner TJ, Nelson MT, Brayden JE (2005) TRPV4 forms a novel Ca2+ signaling complex with ryanodine receptors and BKCa channels. Circ Res 97:1270–1279
79. Gamba G (2006) TRPV4: a new target for the hypertension-related kinases WNK1 and WNK4. Am J Physiol Renal Physiol 290:F
80. Gonzales AL, Amberg GC, Earley S (2010) Ca2+ release from the sarcoplasmic reticulum is required for sustained TRPM4 activity in cerebral artery smooth muscle cells. Am J Physiol Cell Physiol 299:C279-C288
81. Bayliss WM (1902) On the local reactions of the arterial wall to changes of internal pressure. J Physiol 28:220–231
82. Hill MA, Zou H, Potocnik SJ, Meininger GA, Davis MJ (2001) Invited review: arteriolar smooth muscle mechanotransduction: Ca(2+) signaling pathways underlying myogenic reactivity. J Appl Physiol 91:973–983
83. Inoue R, Jensen LJ, Jian Z, Shi J, Hai L, Lurie AI, Henriksen FH, Salomonsson M, Morita H, Kawarabayashi Y, Mori M, Mori Y, Ito Y (2009) Synergistic activation of vascular TRPC6

channel by receptor and mechanical stimulation via phospholipase C/diacylglycerol and phospholipase A2/omega-hydroxylase/20-HETE pathways. Circ Res 104:1399–1409
84. Welsh DG, Morielli AD, Nelson MT, Brayden JE (2002) Transient receptor potential channels regulate myogenic tone of resistance arteries. Circ Res 90:248–250
85. Spassova MA, Hewavitharana T, Xu W, Soboloff J, Gill DL (2006) A common mechanism underlies stretch activation and receptor activation of TRPC6 channels. Proc Natl Acad Sci USA 103:16586–16591
86. Dietrich A, Mederos YSM, Gollasch M, Gross V, Storch U, Dubrovska G, Obst M, Yildirim E, Salanova B, Kalwa H, Essin K, Pinkenburg O, Luft FC, Gudermann T, Birnbaumer L (2005) Increased vascular smooth muscle contractility in TRPC6–/– mice. Mol Cell Biol 25:6980–6989
87. Aaronson PI, Robertson TP, Knock GA, Becker S, Lewis TH, Snetkov V, Ward JP (2006) Hypoxic pulmonary vasoconstriction: mechanisms and controversies. J Physiol 570: 53–58
88. Weissmann N, Dietrich A, Fuchs B, Kalwa H, Ay M, Dumitrascu R, Olschewski A, Storch U, Mederos y Schnitzler M, Ghofrani HA, Schermuly RT, Pinkenburg O, Seeger W, Grimminger F, Gudermann T (2006) Classical transient receptor potential channel 6 (TRPC6) is essential for hypoxic pulmonary vasoconstriction and alveolar gas exchange. Proc Natl Acad Sci USA 103:19093–19098
89. Kinnear NP, Wyatt CN, Clark JH, Calcraft PJ, Fleischer S, Jeyakumar LH, Nixon GF, Evans AM (2008) Lysosomes co-localize with ryanodine receptor subtype 3 to form a trigger zone for calcium signalling by NAADP in rat pulmonary arterial smooth muscle. Cell Calcium 44:190–201
90. Keseru B, Barbosa-Sicard E, Popp R, Fisslthaler B, Dietrich A, Gudermann T, Hammock BD, Falck JR, Weissmann N, Busse R, Fleming I (2008) Epoxyeicosatrienoic acids and the soluble epoxide hydrolase are determinants of pulmonary artery pressure and the acute hypoxic pulmonary vasoconstrictor response. FASEB J 22:4306–4315
91. Humbert M, Morrell NW, Archer SL, Stenmark KR, MacLean MR, Lang IM, Christman BW, Weir EK, Eickelberg O, Voelkel NF, Rabinovitch M (2004) Cellular and molecular pathobiology of pulmonary arterial hypertension. J Am Coll Cardiol 43:13S–24S
92. Mandegar M, Fung YC, Huang W, Remillard CV, Rubin LJ, Yuan JX (2004) Cellular and molecular mechanisms of pulmonary vascular remodeling: role in the development of pulmonary hypertension. Microvasc Res 68:75–103
93. Rodat L, Savineau JP, Marthan R, Guibert C (2007) Effect of chronic hypoxia on voltage-independent calcium influx activated by 5-HT in rat intrapulmonary arteries. Pflugers Arch 454:41–51
94. Keegan A, Morecroft I, Smillie D, Hicks MN, MacLean MR (2001) Contribution of the 5-HT(1B) receptor to hypoxia-induced pulmonary hypertension: converging evidence using 5-HT(1B)-receptor knockout mice and the 5-HT(1B/1D)-receptor antagonist GR127935. Circ Res 89:1231–1239
95. MacLean MR, Sweeney G, Baird M, McCulloch KM, Houslay M, Morecroft I (1996) 5-Hydroxytryptamine receptors mediating vasoconstriction in pulmonary arteries from control and pulmonary hypertensive rats. Br J Pharmacol 119:917–930
96. Guibert C, Marthan R, Savineau JP (2004) 5-HT induces an arachidonic acid-sensitive calcium influx in rat small intrapulmonary artery. Am J Physiol Lung Cell Mol Physiol 286:L1228–L1236
97. Lin MJ, Leung GP, Zhang WM, Yang XR, Yip KP, Tse CM, Sham JS (2004) Chronic hypoxia-induced upregulation of store-operated and receptor-operated Ca2+ channels in pulmonary arterial smooth muscle cells: a novel mechanism of hypoxic pulmonary hypertension. Circ Res 95:496–505
98. Bonnet S, Belus A, Hyvelin JM, Roux E, Marthan R, Savineau JP (2001) Effect of chronic hypoxia on agonist-induced tone and calcium signaling in rat pulmonary artery. Am J Physiol Lung Cell Mol Physiol 281:L193–L201

99. Bonnet S, Dumas-de-La-Roque E, Begueret H, Marthan R, Fayon M, Dos Santos P, Savineau JP, Baulieu EE (2003) Dehydroepiandrosterone (DHEA) prevents and reverses chronic hypoxic pulmonary hypertension. Proc Natl Acad Sci USA 100:9488–9493
100. Shimoda LA, Sham JS, Shimoda TH, Sylvester JT (2000) L-type Ca(2+) channels, resting [Ca(2+)](i), and ET-1-induced responses in chronically hypoxic pulmonary myocytes. Am J Physiol Lung Cell Mol Physiol 279:L884–L894
101. Yuan JX, Aldinger AM, Juhaszova M, Wang J, Conte JV Jr., Gaine SP, Orens JB, Rubin LJ (1998) Dysfunctional voltage-gated K+ channels in pulmonary artery smooth muscle cells of patients with primary pulmonary hypertension. Circulation 98:1400–1406
102. Wang J, Weigand L, Lu W, Sylvester JT, Semenza GL, Shimoda LA (2006) Hypoxia inducible factor 1 mediates hypoxia-induced TRPC expression and elevated intracellular Ca2+ in pulmonary arterial smooth muscle cells. Circ Res 98:1528–1537
103. Jernigan NL, Broughton BR, Walker BR, Resta TC (2006) Impaired NO-dependent inhibition of store- and receptor-operated calcium entry in pulmonary vascular smooth muscle after chronic hypoxia. Am J Physiol Lung Cell Mol Physiol 290:L517–L525
104. Zhang S, Patel HH, Murray F, Remillard CV, Schach C, Thistlethwaite PA, Insel PA, Yuan JX (2007) Pulmonary artery smooth muscle cells from normal subjects and IPAH patients show divergent cAMP-mediated effects on TRPC expression and capacitative Ca2+ entry. Am J Physiol Lung Cell Mol Physiol 292:L1202–L1210
105. Inoue R, Isenberg G (1990) Acetylcholine activates nonselective cation channels in guinea pig ileum through a G protein. Am J Physiol 258:C1173–C1178
106. Zholos AV, Bolton TB (1997) Muscarinic receptor subtypes controlling the cationic current in guinea-pig ileal smooth muscle. Br J Pharmacol 122:885–893
107. Tsvilovskyy VV, Zholos AV, Aberle T, Philipp SE, Dietrich A, Zhu MX, Birnbaumer L, Freichel M, Flockerzi V (2009) Deletion of TRPC4 and TRPC6 in mice impairs smooth muscle contraction and intestinal motility in vivo. Gastroenterology 137:1415–1424
108. Sanders KM (1996) A case for interstitial cells of Cajal as pacemakers and mediators of neurotransmission in the gastrointestinal tract. Gastroenterology 111:492–515
109. Kim BJ, Lim HH, Yang DK, Jun JY, Chang IY, Park CS, So I, Stanfield PR, Kim KW (2005) Melastatin-type transient receptor potential channel 7 is required for intestinal pacemaking activity. Gastroenterology 129:1504–1517
110. Kim BJ, So I, Kim KW (2006) The relationship of TRP channels to the pacemaker activity of interstitial cells of Cajal in the gastrointestinal tract. J Smooth Muscle Res 42:1–7
111. Ortiz JL, Cortijo J, Sanz C, De Diego A, Esplugues J, Morcillo E (1991) Cooling-induced contraction of trachea isolated from normal and sensitized guinea-pigs. Naunyn Schmiedebergs Arch Pharmacol 343:418–426
112. Santacana G, Chen WY (1988) Role of Na+ and Ca++ in guinea pig trachealis contraction induced by cooling. Respiration 53:24–30
113. Mustafa SM, Thulesius O (1999) Cooling-induced bladder contraction: studies on isolated detrusor muscle preparations in the rat. Urology 53:653–657
114. Parkington HC, Tonta MA, Brennecke SP, Coleman HA (1999) Contractile activity, membrane potential, and cytoplasmic calcium in human uterine smooth muscle in the third trimester of pregnancy and during labor. Am J Obstet Gynecol 181:1445–1451
115. Wray S, Noble K (2008) Sex hormones and excitation-contraction coupling in the uterus: the effects of oestrous and hormones. J Neuroendocrinol 20:451–461
116. Tribe RM (2001) Regulation of human myometrial contractility during pregnancy and labour: are calcium homeostatic pathways important? Exp Physiol 86:247–254
117. Tribe RM, Moriarty P, Poston L (2000) Calcium homeostatic pathways change with gestation in human myometrium. Biol Reprod 63:748–755
118. Mironneau J (1973) Excitation-contraction coupling in voltage clamped uterine smooth muscle. J Physiol 233:127–141

119. Dalrymple A, Slater DM, Beech D, Poston L, Tribe RM (2002) Molecular identification and localization of Trp homologues, putative calcium channels, in pregnant human uterus. Mol Hum Reprod 8:946–951
120. Yang M, Gupta A, Shlykov SG, Corrigan R, Tsujimoto S, Sanborn BM (2002) Multiple Trp isoforms implicated in capacitative calcium entry are expressed in human pregnant myometrium and myometrial cells. Biol Reprod 67:988–994
121. Shlykov SG, Yang M, Alcorn JL, Sanborn BM (2003) Capacitative cation entry in human myometrial cells and augmentation by hTrpC3 overexpression. Biol Reprod 69:647–655
122. Babich LG, Ku CY, Young HW, Huang H, Blackburn MR, Sanborn BM (2004) Expression of capacitative calcium TrpC proteins in rat myometrium during pregnancy. Biol Reprod 70:919–924
123. Csapo A, Erdos T, De Mattos CR, Gramss E, Moscowitz C (1965) Stretch-induced uterine growth, protein synthesis and function. Nature 207:1378–1379
124. Dalrymple A, Mahn K, Poston L, Songu-Mize E, Tribe RM (2007) Mechanical stretch regulates TRPC expression and calcium entry in human myometrial smooth muscle cells. Mol Hum Reprod 13:171–179

Chapter 37
TRPM Channels in the Vasculature

Alexander Zholos, Christopher Johnson, Theodor Burdyga, and Donal Melanaphy

Abstract Recent studies show that mammalian melastatin TRPM nonselective cation channels (TRPM1-8), members of the largest and most diverse TRP subfamily, are widely expressed in the endothelium and vascular smooth muscles. When activated, these channels similarly to other TRPs permit the entry of sodium, calcium and magnesium, thus causing membrane depolarisation. Although membrane depolarisation reduces the driving force for calcium entry via TRPMs as well as other pathways for calcium entry, in smooth muscle myocytes expressing voltage-gated Ca^{2+} channels the predominant functional effect is an increase in intracellular Ca^{2+} concentration and myocyte contraction. This review focuses on several best documented aspects of vascular functions of TRPMs, including the role of TRPM2 in oxidant stress, regulation of endothelial permeability and cell death, the connection between TRPM4 and myogenic response, significance of TRPM7 for magnesium homeostasis, vessel injury and hypertension, and emerging evidence that the cold and menthol receptor TRPM8 is involved in the regulation of vascular tone.

Abbreviations

CA	constitutively active
CaM	calmodulin
CAN	Ca^{2+}-activated nonselective channel
DAG	diacylglycerol
EC	endothelial cell
GPCR	G protein-coupled receptor
$InsP_3$	inositol 1,4,5-trisphosphate
lysoPL	lysophospholipids
NSCC	non-selective cation channel

A. Zholos (✉)
Centre for Vision and Vascular Science, School of Medicine, Dentistry and Biomedical Sciences, Royal Victoria Hospital, Queen's University of Belfast, Belfast BT12 6BA, UK
e-mail: a.zholos@qub.ac.uk

PIP$_2$	phosphatidylinositol-4,5-bisphosphate
PKA	protein kinase A
PKC	protein kinase C
PLC	phospholipase C
ROC	receptor-operated channel
ROS	reactive oxygen species
SAC	stretch-activated channel
SOC	store-operated channel
SR	sarcoplasmic reticulum
TM	transmembrane domain
VGCC	voltage-gated Ca^{2+} channels
VSMC	vascular smooth muscle cell

37.1 Introduction

Vascular smooth muscle cells (VSMCs) and endothelial cells (ECs) is a structural tandem of cells comprising blood vessel wall that plays key roles in control of vascular tone under normal conditions. Disruption of normal communication between the two types of cells could result in the development of vascular diseases. VSMCs and ECs are intimately connected, both structurally, by myoendothelial gap junctions where these are present, and functionally. Locally generated endothelium-derived chemical signals play important signalling roles in the modulation of vascular tone, thus influencing blood pressure and flow. Various ion channels are largely responsible for normal responses and communication between VSMCs and ECs which underlie many, if not all, vascular functions.

Calcium-permeable ion channels, such as voltage-gated Ca^{2+} channels (VGCC) and non-voltage gated non-selective cation channels (NSCC), are the key molecules forming calcium entry pathways, which play pivotal roles in the regulation of vascular tone and permeability, angiogenesis, vascular remodelling and proliferative disorders [1–4]. Under physiological conditions, control of vascular tone is usually mediated by changes in the membrane potential. Membrane depolarisation causes Ca^{2+} entry via VGCC that may in turn trigger Ca^{2+}-induced Ca^{2+} release from the sarcoplasmic reticulum (SR) [5, 6]. The link between depolarisation-induced Ca^{2+} entry and smooth muscle contraction is termed electro-mechanical coupling. Conversely, membrane hyperpolarisation causes smooth muscle relaxation by reducing activity of VGCC. In addition, vascular tone in arteries can also be controlled by membrane potential insensitive pathways which include Ca^{2+} release from the SR and Ca^{2+} entry through receptor-operated (ROC) and/or store-operated (SOC) channels. This mechanism is known as pharmaco-mechanical coupling and can appear in the form of Ca^{2+} oscillations associated with Ca^{2+} release from the SR and/or activation of Ca^{2+} entry via ROC or SOC.

Although VGCC and Ca^{2+} release represent the two main sources of activator Ca^{2+} for smooth muscle contraction, the NSCC are very important for vascular smooth muscle responsiveness to various stimuli. In particular, NSCC are responsive to neurotransmitters, hormones, endothelial (e.g., nitric oxide), myogenic (e.g., stretch), metabolic and environmental (e.g., temperature) influences. NSCC mediate Ca^{2+} influx both directly and indirectly – via Na^+ influx causing membrane depolarisation and VGCC opening. The latter mechanism is especially important in small resistance vessels and hence in blood pressure control. In addition, significant Na^+ entry, and especially localised $[Na^+]_i$ increases, can lead to the plasmalemmal Na^+/Ca^{2+} exchanger (NCX) reversal, thus causing Na^+ removal while at the same time driving Ca^{2+} into the cells via NCX [7]. Moreover, recent evidence suggests that NSCC are critical not only for the regulation of VSMC contractile state, but also for the regulation of slow phenotypic changes, VSMC proliferation and migration – processes, which are especially important in vascular function and disease [2, 8].

The importance of NSCC for vascular smooth muscle function has been appreciated for more than 40 years. However, the molecular composition of these channels remained unclear until recently, when important clues started to rapidly emerge indicating connection between vascular NSCC and mammalian homologues of the *Drosophila* Transient Receptor Potential (*Trp*) gene [9–11]. Further molecular identification of the diverse NSCC-mediated Ca^{2+} and Na^+ entry pathways and their activation mechanisms will undoubtedly provide important information on physiological regulation of vascular homeostasis and offer new pharmacological approaches for the prevention and treatment of a wide range of vascular diseases.

Four well-characterised functional classes of NSCC include ROC, SOC, stretch-activated (SAC) and constitutively active (CA) channels [12]. One common feature of their activation is the critical involvement of $G_{q/11}$/phospholipase C (PLC) coupling that results in the breakdown of phosphatidylinositol 4,5-bisphosphate (PIP_2) and formation of inositol 1,4,5-trisphosphate ($InsP_3$) and diacylglycerol (DAG). These second messengers induce, respectively, Ca^{2+} store release and protein kinase C (PKC) activation resulting, in particular, in SOC and ROC activation. Naturally, in the past decade research on molecular delineation of vascular NSCC has mainly focused on the roles of "classical" TRP (TRPC), which are also commonly activated secondary to receptor-stimulated PIP_2 hydrolysis [11]. This has revealed numerous key roles of vascular TRPC channels [1 4, 8, 12 15]. However, recent surveys of TRP expression in the vasculature show that members of the other two major TRP subfamilies, vanilloid (TRPV) and melastatin (TRPM), are also widely expressed in both VSMCs and ECs [3, 4]. Although their functional significance is just beginning to be elucidated, it is already becoming clear that TRPV and TRPM channels may play pivotal roles in several specialised functions. This review summarises emerging vascular functions of TRPMs. Among them TRPM2/4/7/8 appear to be NSCCs of most notable importance in myogenic response, oxidant stress, regulation of endothelial permeability, magnesium homeostasis and thermal behaviour of blood vessels.

37.2 TRPM Channels

37.2.1 Channel Structure

Members of the TRPM subfamily belong to Group 1 which also includes TRPC, TRPV, TRPA and TRPN channels [11, 16]. Interestingly, several TRPM members (TRPM1/5/8) were identified by analysis of gene expression in cancer cells [17]. Although these channels are clearly involved in cancer development, their role in vascular angiogenesis has not been investigated.

Eight members of the TRPM subfamily are characterised by relatively long N and C termini (hence they are termed long TRPs), the latter containing unusual enzyme domains in TRPM2/6/7. Similar to other TRP channels, the cytoplasmic N- and C-terminals in TRPMs are separated by six putative transmembrane domains (TM) with the pore-forming region between TM5 and TM6. The distal part of TM6 determines cation versus anion selectivity, at least in TRPM2 and TRPM8 channels [18]. Similar to TRPC channels, there is a TRP box in the C-terminus, but their N-terminus instead of ankyrin repeats found in TRPCs and TRPVs has a common large TRPM homology domain. The C-terminus coiled-coil domain is necessary for TRPM channel assembly and sufficient for tetrameric formation [19]. While TRPM8 functions as a homomeric complex [19], other TRPMs (e.g. TRPM6 and TRPM7) can either function as homotetramers or as TRPM6/7 heterotetramers [20]. This is similar to known diversity of subunit assembly in TRPC and TRPV channels which show certain preferred interactions (e.g., TRPC1/4/5, TRPC3/6/7, TRPV5/6), or preferential homotetramerisation (e.g., TRPC2, TRPV1) [21]. Structural similarities within the family suggests TRPM subdivision into 4 groups: TRPM1/3, TRPM6/7, TRPM4/5 and TRPM2/8 [16, 22, 23] (Table 37.1).

Numerous studies of TRPMs heterologously expressed in HEK293 or CHO cells revealed significant diversity of their activation mechanisms and their involvement in processes ranging from detection of cold, taste, osmolarity, redox state and pH to control of Mg^{2+} homeostasis and cell proliferation or death. These studies paved the way to our understanding of TRPM roles in native cells, including VSMCs and ECs, and will be briefly summarised below. In addition, several recent reviews provide in-depth coverage of structural aspects, biophysical and pharmacological properties, interacting molecules, tissue expression profiles, activation mechanisms and functional roles of recombinant and native TRPMs [17, 22–27].

37.2.2 Biophysical Properties

All TRPMs function as NSCCs formed either as homo- or heteromultimers as was demonstrated by direct patch-clamp measurements of cation currents in mammalian cells lines heterologously expressing individual TRPM subtypes. Single channel measurements revealed channel conductances in the range from 15–25 pS (TRPM4/5) to 80–120 pS (TRPM3/8) [17, 22, 24].

Table 37.1 Expression and properties of vascular melastatin TRPM channels. Known modes of activation/regulation, inhibitors and possible vascular functions for each TRPM subtype are also summarised

Structural homology	Channel (Hs AA)	VSMC expression	EC expression	Activation and regulation	Inhibitors	Vascular function
	TRPM1 (1.602 AA)		PCR [130]	CA? Transloocation?	La^{3+}, Gd^{3+}	?
	TRPM3 (1.555 AA)	PCR [109]	PCR [130]	CA, hypoosmolarity, D-erythrosphingosine, dihydrosphingosine	La^{3+}, Gd^{3+}, 2-APB, SKF-96365	Mechanotransduction? Myogenic tone?
	TRPM6 (2.022 AA)	PCR [101, 109, 131] WB [101]	PCR [130] -PCR [118]	CA, translocation. 2-APB, decrease in [Mg^{2+}]$_i$, H$^+$, V-dep.	Ruthenium red. Mg^{2+}, Ca^{2+}	Mg^{2+} homeostasis, hypertension
	TRPM7 (1.86= AA)	PCR [3, 100-102, 109, 123, 131] WB [99-102] IC [102]	PCR [118, 130]	CA, Mg-ATP, cAMP-PKA, PIP$_2$, H$^+$, V-dep	La^{3+}, Gd^{3+}, Mg^{2+}, spermine, 2-APB, PIP$_2$ breakdown?	Mg^{2+} homeostasis, hyper-tension, mechanotransduction. Angiogenesis? Vessel injury?
	TRPM4 (1.203 AA)	PCR [3, 109, 116, 117]	PCR [121, 130] NB [58] IC, ISH [121] -PCR [130]	[Ca^{2+}]$_i$, decavanadate. PIP$_2$, heat. BTP2. V-dep	ATP, ATP^{4-}, ADP, AMP, La^{3+}, Gd^{3+}, FA, clotrimazole, polyamines	Myogenic tone CAN? Vessel injury?
	TRPM5 (1.165 AA)	PCR [109] -PCR [116]		[Ca^{2+}]$_i$, PIP$_2$, heat, V-dep.	H$^+$, FA, polyamines, ATP	?
	TRPM2 (1.503 AA)	PCR [3, 109] WB [109]	PCR [4, 44] -PCR[130] WB [4, 44]	ADP-ribose, cADPR, H$_2$O$_2$, NAD [Ca^{2+}]$_i$, heat	2-APB, FA, econazole, clotrimazole, miconazole, ACA	Oxidant stress and cell death, endothelial permeability Vessel injury? Atherosclerosis?
	TRPM8 (1.104 AA)	PCR [3, 109, 110] WB [109, 110] IC [110]	PCR [130]	Cold, menthol, Ca^{2+} store depletion, PIP$_2$, lysoPL, V-dep.	Clotrimazole, AMTB, 2-APB, BCTC, thio-BCTC, capsazepine, ACA	Regulation of vascular tone Atherosclerosis? SOC?

Abbreviations: 2-APB, 2-aminoethoxydiphenyl borate; ACA, N-(p-amylcinnamoyl)anthranilic acid; AMTB, N-(3-aminopropyl)-2-{[(3-methylphenyl)methyl]oxy}-N-(2-thienylmethyl)benzamide hydrochloride salt; BCTC, N-(4-tert-butylphenyl)-4-(3-chloropyridin-2-yl)piperazine-1-carboxamide; BTP2, 4-methy-4′-[3,5-bis(trifluoromethyl)-1H-pyrazol-1-yl]-1,2,3-thiadiazole-5-carboxanilide; FA, flufenamic acid; Hs AA, number of amino acid residues in human protein; IC, immunocytochemistry; NB, Northern blots; PCR, RT-PCR (-PCR: not detected by RT-PCR); V-dep., voltage-dependent activation, WB, Western blots.

Regarding the ion selectivity, with the exception of TRPM4/5 channels which are permeable only to monovalent cations (Na^+, K^+, Cs^+) TRPMs permeate divalent cations including Mg^{2+} (TRPM6/7) and, to a variable degree, Ca^{2+}. Among them, TRPM8 has the highest $P_{Ca}:P_{Na}$ selectivity ratio of 3.3 [28]. Thus, under physiological gradients sodium is expected to carry most of the inward current through TRPMs.

TRPM2 and TRPM3 show almost linear current-voltage (I-V) relation, but other TRPMs produce outwardly rectifying I-V curves suggesting that membrane depolarisation increases their open probability. The voltage dependence in some cases is very strong, especially if compared to other weakly voltage-sensitive TRPs, but not as strong as in classical voltage-gated channels, such as VGCC or K^+ channels.

These biophysical properties of TRPMs suggest that under physiological conditions these channels can universally induce membrane depolarisation via Na^+ influx with additional Ca^{2+} influx, either directly via TRPM1-3/6–8 or indirectly via VGCC if these are expressed, such as in arterial VSMC. Furthermore, activation by membrane depolarisation (TRPM1/4–8) provides a mechanism for self-potentiation of channel gating. Indeed, functional activation of several TRPMs involves a considerable, many tens of millivolts, shifts of the activation range towards more negative potentials, which by itself can open these channels. For example, activation of TRPM8 by menthol is known to lower its voltage activation threshold [28–33]. Various phospholipids, most notably PIP_2, are also important regulators of the voltage dependence of TRPs, including TRPM4/5/7/8 channels [34–36].

37.2.3 Activation Mechanisms and Functional Roles

37.2.3.1 TRPM1

Very little is still known about the activation properties and function of the founding member of the TRPM subfamily. Originally, TRPM1 was characterised as gene down-regulated in melanoma metastasis [37]. Recent studies indicate that the main function of TRPM1 can be intracellular and critical to normal melanocyte pigmentation [38]. When expressed in HEK293 cells, TRPM1 mediates Ca^{2+} influx and appears to be constitutively active, whereas its short isoform suppresses the activity of the full-length channel by reducing its membrane expression [39]. Expression of TRPM1 results in an outwardly rectifying currents sensitive to La^{3+}. TRPM1 channels are slightly more permeable to Na^+ than to Ca^{2+}, and they are not responsive to receptor agonists, protein kinase A activation, external pH or osmolarity [38].

37.2.3.2 TRPM2

TRPM2 is one of the three TRPMs with enzymatic activity. In this case, there is a Nudix domain in the C-terminus with nucleotide pyrophosphatase activity [40]. This domain is important for ADP-ribose (ADPR)-mediated gating of the channel. TRPM2 is activated by reactive oxygen species (ROS, for which H_2O_2 is often used as an experimental approach), ADPR, NAD$^+$ and intracellular Ca^{2+}: the latter increases its sensitivity to ADPR. Heat (>35°C) also potentiates TRPM2 activity by acting in synergy with cADPR [41]. Thus, TRPM2 is a multifunctional channel which plays central roles in oxidative and nitrosative stress and cell death [17, 23, 26, 40, 42–48]. Interaction of TRPM2 with the silent information regulator 2 (Sir2) also contributes to its role in cell death [49].

37.2.3.3 TRPM3

TRPM3 is primarily expressed in kidney and shows constitutive activity, which can be further potentiated by hypotonic solution [50] or Ca^{2+} store depletion [51]. Thus, its major role in renal Ca^{2+} homeostasis has been postulated. The effects of hypotonicity on TRPM3 are likely to be mediated by cell swelling suggesting that TRPM3 can function as a volume- and/or mechano-sensory channel. Intracellular Mg^{2+} inhibits TRPM3 [52], which is also a feature of TRPM6/7 regulation.

37.2.3.4 TRPM4 and TRPM5

Ca^{2+}-activated NSCC (CAN) channels are widely expressed in various cell types [53], including cardiovascular system [54, 55], where they play important roles in resting membrane potential control, rhythmical electrical activity and regulation of Ca^{2+} oscillations. Identification of TRPM4b with the distinct properties of a CAN channel opened up a good prospect for uncovering the molecular identity of CANs [56, 57]. Membrane depolarisation alone is insufficient to open TRPM4, but it strongly modulates channel activity in a Ca^{2+}-dependent manner [58]. TRPM4 sensitivity to [Ca^{2+}]$_i$ is also controlled by ATP, PKC-dependent phosphorylation and calmodulin (CaM) binding to C-terminal CaM domains [59]. Phosphorylation of the channel at two C-terminal serine residues enhanced its intracellular Ca^{2+} sensitivity about 4-fold [59], which can account for PKC-dependent modulation of TRPM4-mediated myogenic response [60]. Alternatively, the voltage range of channel activation (often characterised by the potential of half-maximal activation, V$_{1/2}$) is strongly dependent on [Ca^{2+}]$_i$, calmodulin, PIP$_2$ concentration, channel phosphorylation and temperature [61]. Interestingly, TRPM4 appears to be responsive to Ca^{2+} influx rather than Ca^{2+} release [24].

The related TRPM5 channel shows several important similarities as well as differences with TRPM4 [62]. TRPM5 is directly activated by [Ca^{2+}]$_i$ rise, especially in response to InsP$_3$-induced Ca^{2+} release, and shows interesting sensing properties of the rate, rather than steady-state levels, of [Ca^{2+}]$_i$ change, which emphasises its role in coupling Ca^{2+} release events to electrical activity [24, 63]. TRPM5 is highly expressed in the taste buds of the tongue where it has a key role in taste transduction [64–67]. Similarly to TRPM4, PIP$_2$ and CaM binding regulate Ca^{2+} and voltage sensitivity of these channels [68], while heat (15–35°C) shifts the V$_{1/2}$ towards more negative potentials [66].

37.2.3.5 TRPM6 and TRPM7

These related channels are two other "chanzymes" in the TRPM subfamily which possess similar C-terminal atypical protein kinase domain. This domain is essential for channel function [69]. The channels are not only Mg^{2+} permeable but they are also inhibited by intracellular Mg^{2+}, which suggest their importance for normal Mg^{2+} homeostasis [70, 71]. TRPM7 is a widely expressed channel, while TRPM6 is restricted mainly to colon and kidney. Non-functional mutant TRPM6 channels cause hypomagnesemia with secondary hypocalcemia (HSH) due to impaired renal and intestinal Mg^{2+} homeostasis. Homomeric TRPM6 and TRPM7 channels can be activated at reduced pH with similar $pH_{1/2}$ values of about 4.5 while the $pH_{1/2}$ of heteromeric TRPM6/7 is increased to 5.5 [20]. It was proposed that TRPM7 channel has mechanosensitive properties and is involved in volume regulation in epithelial cells [72]. Alternatively, functional TRPM7 channels translocate to the plasma membrane in response to fluid flow [73].

37.2.3.6 TRPM8

TRPM8 has recently emerged as a primary neuronal cold and menthol receptor [28, 74–77] as well as a promising prognostic marker and therapeutic target in prostate cancer [78–80]. In sensory neurones, cold is the physiological stimulus for TRPM8 activation, and therefore in our past research we addressed the apparently paradoxical expression of TRPM8 in tissues not exposed to any essential temperature variations, like the prostate from which TRPM8 (Trp-p8) was originally cloned [79]. The novel biochemical pathway for TRPM8 activation at physiological temperature involves Ca^{2+} store depletion, $iPLA_2$ activation and synthesis of lysophospholipids (lysoPLs) [81, 82]. These pathways are critically important for SOC activation in VSMC and agonist-induced vasoconstriction [83, 84]. Moreover, we found that TRPM8 can be expressed both in the plasma membrane (classical TRPM8 denoted $_{PM}$TRPM8) and in the endoplasmic reticulum membrane (classical as well as shorter isoform denoted $_{ER}$TRPM8). TRPM8 can thus mediate both Ca^{2+} entry and Ca^{2+} release [78, 85].

Biophysical and pharmacological properties of TRPM8 have been extensively characterised [30, 31, 86–90]. Notably, TRPM8 shows considerable $V_{1/2}$ shifting behaviour during its activation by cold and menthol. PIP_2 is a critical common determinant of TRPM8 activation by both stimuli [91–93].

37.2.4 Roles of G-Protein Coupled Receptors (GPCR) and Ca^{2+} Store Depletion in TRPM Regulation

TRPM regulation by various GPCRs and Ca^{2+} store depletion is especially relevant to our understanding of the functions of these channels in the vasculature where Ca^{2+} homeostasis is strongly influenced by various receptor agonists acting on VSMCs and ECs and evoking ROC and SOC currents. PIP_2-binding domains and PIP$_2$ roles in regulating TRPM4/5/7/8 are well understood, suggesting important roles of G$_{q/11}$/PLC coupled receptors in the functional control of these channels [92–98].

PIP$_2$ potentiates TRPM4/5/7/8 by various mechanisms, that include direct gating, negative shift of V$_{1/2}$, and increase in Ca^{2+} sensitivity. Accordingly, activation of M$_1$ (G$_{q/11}$/PLC-coupled) muscarinic receptor subtype was shown to potently inhibit TRPM4 activity [96]. This was also the case for negative regulation of TRPM8 by the NGF receptor trkA, which stimulates PLCγ and hence PIP$_2$ hydrolysis [92]. However, TRPM5 channel was stimulated by acetylcholine in HEK293 M$_1$-expressing cells contransfected with the chimeric G protein G16z44. This activation was only observed without intracellular Ca^{2+} buffering showing that physiological rise in [Ca^{2+}]$_i$ can activate TRPM5 despite parallel PIP$_2$ depletion [95]. Interestingly, TRPM7 shows dual regulation by GPCRs. Activation of the G$_{q/11}$-coupled M$_1$ muscarinic receptor or the epidermal growth factor (EGF) receptor inhibited heterologously expressed TRPM7 via PIP$_2$ depletion, while TRPM7 currents in ventricular fibroblasts were not modulated by angiotensin II or bradykinin, but were inhibited by another G$_{q/11}$-coupled receptor, the lysophosphatidic acid (LPA) receptor [97].

Importantly, several key vasoactive agonists, including angiotensin II, bradykinin and aldosterone, have been shown to influence TRPM6/7 expression and activity in primary rat, mouse and human VSMCs [99–102]. In addition, at least two TRPM members, TRPM3 and TRPM8, can be activated by Ca^{2+} store depletion [51, 81]. Both studies found that HEK293 overexpressing these channels showed enhanced Ca^{2+} entry upon external Ca^{2+} readmission following a period of thapsigargin treatment in Ca^{2+}-free solution to deplete the calcium stores. In addition, Abeele et al. [81] showed enhanced TRPM8 activity in cell-attached patches induced by thapsigargin. Although these studies showed the possibility that some TRPMs can contribute to store-dependent Ca^{2+} entry, it should be noted that TRPM8 is primarily activated by cold while TRPM3 is a constitutively active channel. Thus, both channels do not necessarily require calcium store depletion for their activation.

37.2.5 Pharmacological Properties of TRPM Channels

For most TRPM channels both agonists and antagonists are available aiding investigation of their roles in native cells (Table 37.1). However, selective and potent TRPM ligands are generally lacking, a common problem in TRP research. For example, 2-APB which is considered as a general inhibitor of TRP channels, also inhibits TRPM2/3/7/8 channels. Other blockers, such as SKF-96365, flufenamic acid, ruthenium red and spermine also inhibit other TRP and non-TRP channels. Nevertheless, there are several interesting and useful TRPM ligands.

TRPM3 was the first TRP channel found to be activated by D-erythro-sphingosine (but not by sphingosine-1-phosphate, S1P) [103]. TRPM3 activation by the neurosteroid pregnelone sulphate allowed its recent identification as an essential component of an ionotropic steroid receptor in pancreatic β cells [104]. The action of pregnelone sulphate seems very specific to TRPM3 as other TRPs (TRPM2/7/8 and TRPV1/4/6) are insensitive to this steroid. Decavanadate activates TRPM4 but not related TRPM5, while ATP^{4-} allows further pharmacological differentiation between TRPM4 and TRPM5 [105, 106]. In the case of TRPM8 its most potent known antagonist clotrimazole also shows characteristic voltage dependency. This blocker can be an especially useful pharmacological tool to discriminate between TRPM8 and another cold-, menthol- and icilin-activated channel, TRPA1, as it has the opposite, activating effect on TRPA1 ([107]. Another recently discovered antagonist AMTB also shows high affinity for TRPM8 [108]). As already described, TRPM6/7 channels are inhibited by Mg^{2+} or Ca^{2+}, but in contrast intracellular Ca^{2+} potentiates activity of TRM2/4/5. Furthermore, 2-APB at micromolar concentrations maximally activates TRPM6 but inhibits the related TRPM7 [20]. Thus, although it may be sometimes possible to discriminate pharmacologically between even closely related TRPM members, given significant lack of knowledge of the action of putative TRPM ligands on all other ion channels expressed in the tissue the future and imminent work in the field of vascular TRPM channels would require the use of combined molecular and pharmacological approaches.

37.3 TRPM Channels in Vascular Function and Disease

The diverse mechanisms of channel activation discussed above are paralleled by diverse vascular functions of TRPM channels, which became increasingly recognised during the last five years. These range from mechanosensory transduction and regulation of the arterial myogenic response to magnesium transport in hypertension, and from cold sensitivity to vascular injury.

37.3.1 TRPM Expression in Endothelial and Vascular Smooth Muscle Cells

TRPM channels are widely expressed in VSMCs and ECs obtained from various vascular beds which suggests their wide-spread physiological roles in the vasculature (Table 37.1 and references therein). Several studies detected TRPM mRNA expression using RT-PCR analysis, the most sensitive and selective technique. With the exception of TRPM1 and TRPM5, all other TRPM genes appear to be ubiquitously expressed in vascular tissues. It should be noted, however, that most studies performed on ECs used cultured cells, but cell culture conditions can alter gene expression [4]. In addition, there is lack of information on the relative expression levels of different TRPMs in ECs. In contrast, many expression studies in VSMCs used freshly isolated cells which represent the physiological phenotype most closely. One study examined gene expression profile of TRPMs in pulmonary artery and aortic smooth muscles and found consistent rank order

TRPM8 > TRPM4 > TRPM7 > TRPM3 = TRPM2 > TRPM5 = TRPM6 [109]. In addition, we showed TRPM8 expression in rat aorta, tail, femoral and mesenteric arteries by semiquantitative PCR, Western blotting and immunocytochemistry [110]. There is evidence that TRPM2/6/7 are also expressed in VSMC at the protein levels (Table 37.1).

37.3.2 Specific Vascular Functions of TRPM Channels

37.3.2.1 TRPM2 Role in Oxidant Stress and Endothelial Permeability

NSCCs have been implicated in cell death associated with ischemia/hypoxia, oxidative stress and Ca^{2+} overload [111]. Although several TRPs can be activated by oxidative stress (e.g. TRPC3/C4/M2/M7/A1) TRPM2/7 appear the leading candidates involved in these processes. TRPM2 channel senses and responds to oxidative stress while the ubiquitously expressed TRPM7 directly supports cell viability. TRPM2 channels are likely to be involved in a range of pathophysiological processes in oxidant-induced vascular injury, cerebral ischemia and stroke [43, 111]. Oxidative and nitrosative stress causes accumulation of ADPR in the cytosol that activates TRPM2 resulting in Ca^{2+} influx and increased susceptibility to cell death. Conversely, inhibition of TRPM2 enhances cell survival [112].

In endothelial cells, TRPM2 activation by H_2O_2 causes Ca^{2+} entry thus increasing endothelial permeability [43, 44, 113]. These studies showed that in human pulmonary artery ECs H_2O_2-induced cation current, Ca^{2+} signals and trans-endothelial resistance decrease were all blocked by siRNA, anti-TRPM2 antibodies, as well as by the expression of the dominant negative short TRPM2 isoform. In monocytes, ROS-evoked Ca^{2+} entry via TRPM2 is a key trigger of chemokine production in inflammation. These processes are attenuated by the *Trpm2* gene disruption in mice [114]. The functional roles of TRPM2 channels expressed in VSMCs are not known, but they may be involved in vessel injury or atherosclerosis development, which is associated with mitochondrial dysfunction, ROS production and inflammation.

37.3.2.2 TRPM4 and Myogenic Tone

Both vessel constriction in response to an increase of transmural pressure and vessel dilation after a decrease of transmural pressure are referred to as the "myogenic response". Pressure-induced myogenic constriction is of central importance to the local regulation of blood flow in small resistance arteries and arterioles. It also provides the basal level of vascular "myogenic" tone that can be further modulated by local endothelial, metabolic and neural factors. The initial event in this process is stretch-induced VSMC depolarisation which then opens VGCC channels leading to Ca^{2+} influx and myogenic response. Several types of ion channels have been proposed to couple a change in intraluminal pressure to membrane potential change including activation of one or more types of NSCC [115]. Although TRPM3 can be directly or indirectly activated by membrane stretch, it was the

TRPM4 channel that was actually implicated in myogenic constriction. TRPM4 can offer new insights in the molecular nature of myogenic tone control by Ca^{2+} and PKC [60, 116, 117]. Using the antisense technology to suppress TRPM4 expression in cerebral arteries these studies found reduction of pressure- and PKC-induced VSMC depolarisation as well as pressure-induced vasoconstriction. In addition, single channel studies provided direct support for the hypothesis that TRPM4 channel in cerebral artery myocytes functions as SAC [117]. It is still unclear how stretch activates TRPM4, although the latter study suggested the involvement of ryanodine receptors. In this study, caffeine application activated and then inhibited SAC activity, which can be explained by the already described potentiating effect of the released Ca^{2+} on TRPM4 activity, rather than by Ca^{2+} store depletion as seen in "classical" SOC activation. More importantly, many other TRPs including TRPC1/C3-6/V1/V2/V4/M7/P1/P2 have been implicated in vascular mechanotransduction [113, 118, 119]. In different arteries, these TRP subtypes are found both in VSMCs and ECs and both cell types are known to express mechanosensitive channels which may differ in single channel conductance, selectivity and pharmacological properties [118]. Thus, the relative importance of TRPM4 in myogenic response in other arteries remains to be established. With the use of similar antisense strategy, an essential role of TRPC6 in the depolarising response to pressure and in the regulation of myogenic tone of cerebral arteries has been earlier postulated by Welsh et al. [120]. TRPC6 is a stretch-sensitive Ca^{2+}-permeable channel activated downstream of the PLC/DAG cascade, but it is negatively regulated by PKC. Earley et al. [60] proposed that both "myogenic" channels, TRPM4 and TRPC6, exist in intracellular microdomains, whereby TRPC6 provides the initial Ca^{2+} signal in response to pressure/stretch while TRPM4, PKC and VGCC act in concert to amplify this signal. This hypothesis is very attractive since with the addition of TRPC6 not only PKC dependence (TRPM4-mediated) but also other known molecular links between pressure and membrane depolarisation such as PLC/DAG (TRPC6-mediated) could be interpreted as an ion channel effect.

Functional properties of TRPM4/5 channels also make them excellent candidates for various poorly understood cardiovascular CAN channels [57], but any insight here is still missing and TRPM5 is either not expressed or expressed only at a very low level in VSMCs. Remarkably, in a model of spinal cord injury downregulation of TRPM4 expression with the antisense approach or TRPM4 knockout eliminated secondary hemorrhage and preserved capillary structural integrity [121]. These novel findings suggest that TRPM4, together with TRPM2/7, may play important roles in vessel injury.

37.3.2.3 TRPM7 Role in Magnesium Homeostasis and Hypertension

TRPM6 and TRPM7 channels regulate Mg^{2+} homeostasis as was initially recognised in epithelial and neuronal cells. However, these channels, and especially TRPM7, are widely expressed in different cell types whereby they may play similar roles. In VSMCs, Mg^{2+} plays many important roles in cell growth, remodelling associated with vascular injury, atherosclerosis and hypertension, and it also causes vasodilation and inhibits agonist-induced vasoconstriction. Using a range of pharmacological and molecular biology approaches He et al. [100] showed that membrane expression of TRPM7 in VSMCs is regulated by angiotensin II and aldosterone and that this channel contributes to the control of basal $[Mg^{2+}]_i$ and cell growth. In addition, bradykinin was shown to regulate VSMC $[Mg^{2+}]_i$ and cell migration via TRPM7 [99]. Moreover, in spontaneously hypertensive rats expression of TRPM7, but not TRPM6, was reduced and this was associated with lower $[Mg^{2+}]_i$ [101]. Interestingly, angiotensin II failed to increase TRPM7 expression in VSMCs from spontaneously hypertensive rats.

In ECs, downregulation of TRPM7 caused enhanced growth and proliferation, as well as increased expression of nitric oxide synthase and nitric oxide production [122]. This implies that endothelial TRPM7 reduces NO-dependent vasodilation. These findings are also consistent with dysregulation of TRPM7 and endothelial dysfunction in inherited hypomagnesimia which is associated with hypertension [123]. Taken together with the already discussed TRPM7 role in mechanotransduction, there is now considerable evidence implicating TRPM7 in diverse vascular functions and disease processes [124]. Remarkably, in VSMCs, but not in ECs, fluid flow increases TRPM7 current as the channel is translocated to the plasma membrane further supporting its role in cellular response to vessel injury [73].

37.3.2.4 TRPM8 Role in Thermal Behaviour of Blood Vessels

The first study to show TRPM8 expression (at the mRNA and protein level) in vascular tissue was conducted by Yang et al. [109]. TRPM8 was found in both rat aorta and pulmonary artery. Their studies also suggested a functional role for the channel as they were able to elicit increases in $[Ca^{2+}]_i$ by exposure of isolated myocytes to 300 μM menthol, which is thought to be a specific agonist for TRPM8 channels in the micromolar range. Calcium responses depended on external Ca^{2+}, but VGCCs were not involved. However, these studies gave no indication as to how this channel might actually contribute to actual function in whole vessels or vessels *in vivo*, and no information as to possible mechanisms by which activation of the channel might result in functional consequences. Our laboratory has picked up this gauntlet to conduct the first assessment of a functional role for TRPM8 channels in a number of arterial vessels. We used a number of approaches with this intention, ranging from examining expression at the molecular level in several rat vessel types to assessing possible contributions to control of human forearm blood flow. Conventional semi-quantitative PCR revealed expression of TRPM8 receptor mRNA in rat aortae, tail, femoral and mesenteric arteries [110]. We originally showed protein expression by immunocytochemistry in tail artery myocytes using the Alamone anti-TRPM8 antibody that has an epitope in the pore region of the channel (ACC049). We have now further confirmed these findings using two different Abcam anti-TRPM8 antibodies which have different epitopes in the C- and N-termini of TRPM8 (ab63073 and ab3243). In each case, the cold receptor appears to be present on or at least near the plasma membrane and also in regions throughout the cytosol, potentially

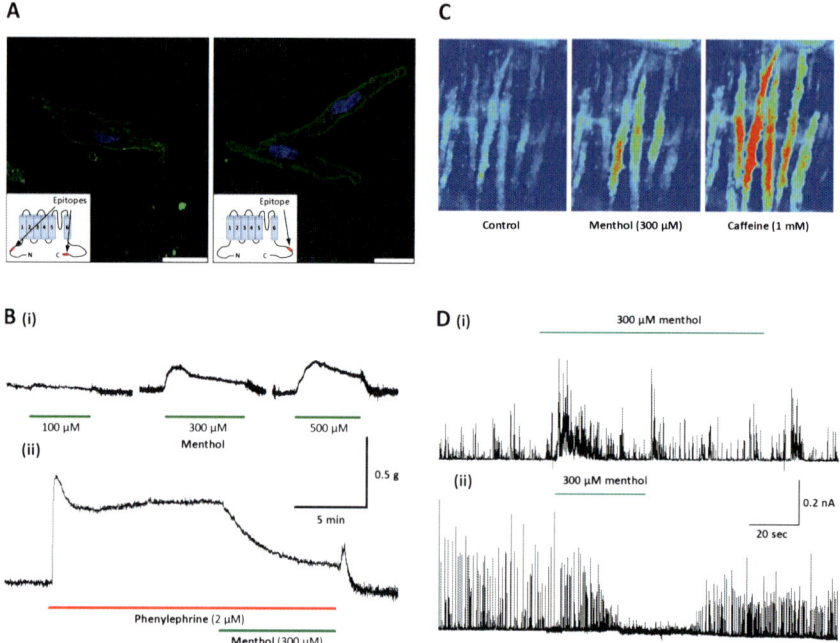

Fig. 37.1 Functional expression of TRPM8 in rat tail artery VSMC. (**a**) Immunocytochemically stained vascular myocytes labeled for TRPM8 using two different antibodies that react with different epitopes (see *inset cartoon*). The fluorescence profiles of the cells indicate that TRPM8 channels are localized on or near to the plasma membrane as well as within the cytoplasm (Scale bar = 25 μm). (**b**) Raw trace from vascular ring isometric contraction experiment. Vascular rings taken from rat tail artery were mounted in an organ bath via two stainless steel hooks. With one hook fastened to the base of the organ bath and the other attached to a force transducer a baseline tension was applied. (*i*) Incrementing concentrations of the TRPM8 agonist, menthol, evoked a dose dependent vasoconstriction. (*ii*) Application of TRPM8 agonists, menthol and icilin, to vascular rings pre-contracted with either high-K^+ or phenylephrine, resulted in profound vasodilatations. (**c**) Confocal calcium imaging of rat tail artery vascular myocytes in situ. Sections of rat tail artery, loaded with a calcium sensitive dye (Fluo-4 AM, Invitrogen), were imaged using a laser scanning confocal microscope with an argon 488 nm laser. Application of menthol and the ryanodine-sensitive calcium-release channel agonist, caffeine, induced robust increases in $[Ca^{2+}]_i$ in myocytes. (**d**) Rat tail artery vascular myocytes, under voltage clamp conditions, display STOCs that reflect BK_{Ca} channel activation caused by localized spontaneous calcium release events. Upon application of menthol, there is an initial acceleration of STOCs discharge (*i, ii*), suggesting an increased frequency of elementary calcium release events from the sarcoplasmic reticulum. (ii) In some cells this initial increase in STOCs frequency was followed by a decrease in STOCs amplitude until their activity almost ceased, possibly due to store depletion

on the SR (Figs. 37.1a and 37.2). Work in progress supports the latter by showing co-localisation of TRPM8 and Ca^{2+} store release channels.

In isometric contraction studies in rat tail artery, the effects of TRPM8 agonists appeared to depend on the state of vasomotor tone of the artery preparations. In relaxed vessel, menthol caused small dose-dependent contractions (Fig. 37.1b(i)).

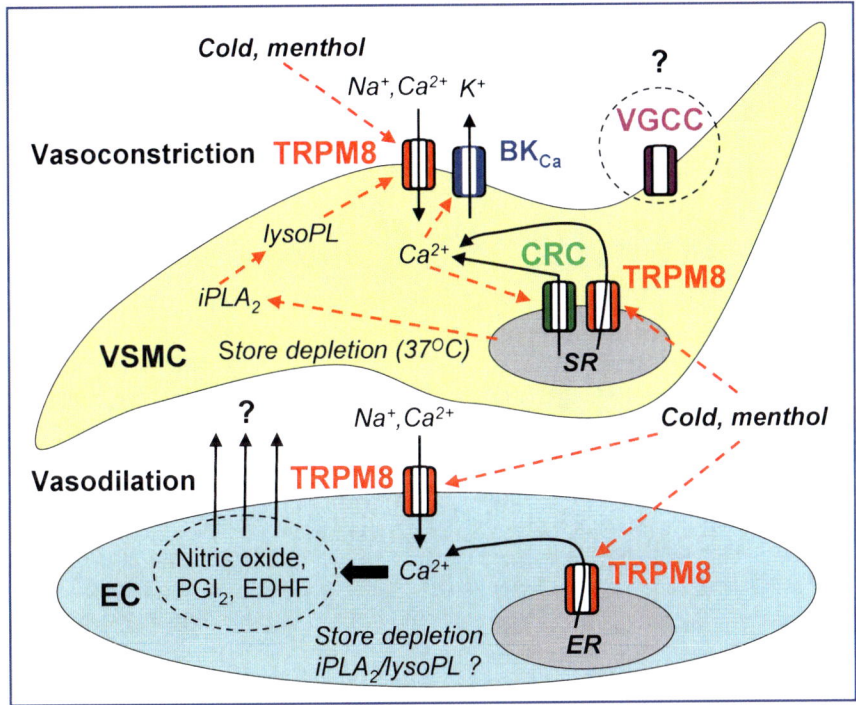

Fig. 37.2 Proposed mechanism that can explain both vasoconstrictor and vasodilator influences of TRPM8 activation. According to the model, TRPM8 is closely associated with plasmalemmal BK_{Ca} channels as well Ca^{2+}-release channels and thus may support a number of processes, including direct membrane depolarisation and Ca^{2+} influx, indirect membrane hyperpolarisation via BK_{Ca} activation (e.g. STOC discharge), Ca^{2+} release and Ca^{2+} store depletion (with secondary recruitment of the $iPLA_2$/lysoPL pathway for channel activation), and possibly vasodilation through the increase of $[Ca^{2+}]_i$ in ECs and release of vasoactive substances. It is likely that similar possibilities exist for other TRPMs with dual expression (Table 37.1) but further research is needed to establish the relative contributions of TRPMs expressed in VSMC and EC to various vascular functions

This may be explained in a straightforward way by increased $[Ca^{2+}]_i$ induced by TRPM8 channel activation. Menthol-induced $[Ca^{2+}]_i$ rise was visualised using confocal microscopy in Fluo 4 loaded myocytes, both in isolated cells and cells *in situ* (Fig. 37.1c). Calcium fluorescence experiments, both in single myocytes and in whole vessels, have shown that TRPM8 agonists can induce large, rapid increased intracellular calcium associated with myocyte or vessel contraction. These signals were somewhat smaller compared to caffeine-induced $[Ca^{2+}]_i$ increments, and they were also observed, albeit smaller in amplitude, under Ca^{2+}-free conditions suggesting that both Ca^{2+} entry and Ca^{2+} release contribute to TRPM8-mediated Ca^{2+} signalling (Fig. 37.2).

Remarkably, in contractile studies the greatest effects were seen in vessels pre-contracted with either KCl (60 mM) or phenylephrine (2 μM), in which case both menthol and icilin induced a profound relaxation (Fig. 37.1b(ii)). A similar effect was observed on responses evoked by electrical field stimulation of sympathetic nerves, which were markedly reduced by menthol [110]. These actions of TRPM8 agonists appear to be directly on smooth muscle as they were observed in vessels with the endothelium removed. However, these relaxations were more significant with intact endothelium suggesting that endothelial TRPM8 may contribute to the regulation of vascular tone as proposed in the model of Fig. 37.2.

Dilatory effects of menthol have been reported in a number of other smooth muscle preparations [125, 126] and would seem contradictory to TRPM8-induced increases in $[Ca^{2+}]_i$ resulting in vessel contraction. On the other hand, the presence of TRPM8 receptors on the SR might allow store depletion that gives rise to vasodilatation, and becomes apparent when there is already considerable store mobilization from the prior induction of contraction (Fig. 37.2). Patch clamp recordings have provided support for this explanation. Rat tail artery myocytes display spontaneous transient outwards currents (STOCs) that reflect BK_{Ca} channel activation caused by localized spontaneous calcium release events. STOCs frequency may be increased in the presence of menthol followed by a decrease in STOCs amplitude until their activity almost ceases due to store depletion (see two examples in Fig. 37.1d), which would be consistent with the observed vessel vasodilation. Furthermore, the menthol-induced relaxations were significantly reduced by (S)-BEL suggesting the involvement of the iPLA$_2\beta$ pathway in these responses [110]. It is also possible that some of the actions of menthol and icilin can involve another cold sensitive channel, TRPA1. TRPA1 is expressed in ECs, but probably not in VSMCs, and activation of TRPA1 causes vasodilation mediated by Ca^{2+}-activated K^+ channels [127]. Delineation of specific roles of these two cold-sensitive channels in vasodilation clearly needs further studies with the use of more selective molecular biology approaches [128].

Extension of these studies to humans has provided some similar observations, as menthol applied to the skin of the forearm causes a profound increase in local cutaneous blood flow [110]. The interpretation of this observation is less straightforward, however, as control of human forearm cutaneous circulation is more complicated than control of vessel tone in isolated rat tail artery. This is reflected by the observation that the menthol-induced dilatation was reduced by up to 50% by L-NAME or atropine, belying possible contributions to the response by sympathetic cholinergic innervation and vascular endothelium in this vascular target [129].

In conclusion, there is now good evidence that TRPM8 channels are functional on VSMCs from rats and possibly humans. The precise mechanisms by which they act and their contribution to vascular tone in normal conditions and in disease states remain to be determined using a combination of molecular biology approaches, pharmacological tools and antibodies. Interestingly, we have found that vasodilator responses to menthol in tail artery are significantly potentiated in 20 week old streptozotocin-diabetic rats (Johnson et al., unpublished observations). Understanding these problems and their interaction with TRPM8 channels on sensory nerves is particularly intriguing for the cutaneous circulation. A more complete understanding may be of immense benefit for treatment of certain cutaneous conditions that are likely to be influenced by a combination of factors such as temperature, neural sensory and vascular innervation, endothelial factors and smooth muscle properties, including vascular problems associated with diabetes, Reynaud's phenomenon and complex regional pain syndrome.

37.4 Conclusions

In recent years, multifunctional TRPM channels emerged as important players regulating vascular function and hence potential new targets for treating vascular disease. With the use of pharmacological tools, biophysical analysis, molecular biology approaches such as antisense or siRNA technologies and genetic studies of hereditary disorders and knock-out mouse models in many cases it became possible to link several vascular physiological responses to specific TRPM channels. Each of these approaches has its limitations, but with potential caveats in mind there have been several novel and unexpected developments in this area, including the identification of mechanosensory roles of vascular TRPM4/7 channels and vascular predominance and functionality of the cold receptor TRPM8. Numerous other vascular functions of TRPMs can be envisaged based on their known properties as indicated by the question marks in Table 37.1, but these are awaiting further investigations.

The major challenges remain in understanding of TRPM roles in different blood vessels, and also of the mechanisms of cross-talk between the same type of channel expressed in VSMC and EC. The overall picture so far is that the vascular "TRP pool" is considerable and TRPM channels make important contributions, especially on the side of communication between blood vessels and local tissue environment, but individual TRP members may be expressed at low levels and decoding their specific roles represents a significant challenge to researchers.

Acknowledgments Research in the authors' laboratories is funded by BHF and NIH. We thank Mrs Tetyana Zholos for her assistance in the preparation of this review.

References

1. Beech DJ (2005) Emerging functions of 10 types of TRP cationic channel in vascular smooth muscle. Clin Exp Pharmacol Physiol 32:597–603
2. Beech DJ (2007) Ion channel switching and activation in smooth-muscle cells of occlusive vascular diseases. Biochem Soc Trans 35:890–894
3. Inoue R, Jensen LJ, Shi J, Morita H, Nishida M, Honda A, Ito Y (2006) Transient receptor potential channels in cardiovascular function and disease. Circ Res 99:119–131
4. Yao X, Garland CJ (2005) Recent developments in vascular endothelial cell transient receptor potential channels. Circ Res 97:853–863
5. Kotlikoff MI (2003) Calcium-induced calcium release in smooth muscle: the case for loose coupling. Prog Biophys Mol Biol 83:171–191
6. Wray S, Burdyga T (2010) Sarcoplasmic reticulum function in smooth muscle. Physiol Rev 90:113–178
7. Poburko D, Liao CH, Lemos VS, Lin E, Maruyama Y, Cole WC, van Breemen C (2007) Transient receptor potential channel 6 mediated, localized cytosolic [Na$^+$] transients drive

Na^+/Ca^{2+} exchanger mediated Ca^{2+} entry in purinergically stimulated aorta smooth muscle cells. Circ Res 101:1030–1038
8. House SJ, Potier M, Bisaillon J, Singer HA, Trebak M (2008) The non-excitable smooth muscle: calcium signaling and phenotypic switching during vascular disease. Pflugers Arch 456:769–785
9. Clapham DE, Julius D, Montell C, Schultz G (2005) International union of pharmacology. XLIX. Nomenclature and structure-function relationships of transient receptor potential channels. Pharmacol Rev 57:427–450
10. Ramsey IS, Delling M, Clapham DE (2006) An introduction to TRP channels. Annu Rev Physiol 68:619–647
11. Venkatachalam K, Montell C (2007) TRP Channels. Annu Rev Biochem 76:387–417
12. Albert AP, Large WA (2006) Signal transduction pathways and gating mechanisms of native TRP-like cation channels in vascular myocytes. J Physiol 570:45–51
13. Beech DJ, Muraki K, Flemming R (2004) Non-selective cationic channels of smooth muscle and the mammalian homologues of *Drosophila* TRP. J Physiol 559:685–706
14. Albert AP, Saleh SN, Peppiatt-Wildman CM, Large WA (2007) Multiple activation mechanisms of store-operated TRPC channels in smooth muscle cells. J Physiol 583:25–36
15. Tiruppathi C, Ahmmed GU, Vogel SM, Malik AB (2006) Ca^{2+} signaling, TRP channels, and endothelial permeability. Microcirc 13:693–708
16. Montell C (2005) The TRP superfamily of cation channels. Sci STKE 2005:re3
17. Kraft R, Harteneck C (2005) The mammalian melastatin-related transient receptor potential cation channels: an overview. Pflugers Arch 451:204–211
18. Kuhn FJP, Knop G, Luckhoff A (2007) The transmembrane segment S6 determines cation versus anion selectivity of TRPM2 and TRPM8. J Biol Chem 282:27598–27609
19. Tsuruda PR, Julius D, Minor DL Jr. (2006) Coiled coils direct assembly of a cold-activated TRP channel. Neuron 51:201–212
20. Li M, Jiang J, Yue L (2006) Functional characterization of homo- and heteromeric channel kinases TRPM6 and TRPM7. J Gen Physiol 127:525–537
21. Lepage PK, Boulay G (2007) Molecular determinants of TRP channel assembly. Biochem Soc Trans 35:81–83
22. Fleig A, Penner R (2004b) The TRPM ion channel subfamily: molecular, biophysical and functional features. Trends Pharmacol Sci 25:633–639
23. Harteneck C (2005) Function and pharmacology of TRPM cation channels. N-S Arch Pharmacol 371:307–314
24. Fleig A, Penner R (2004a) Emerging roles of TRPM channels. Novartis Found Symp 258:248–258
25. Fonfria E, Murdock PR, Cusdin FS, Benham CD, Kelsell RE, McNulty S (2006) Tissue distribution profiles of the human TRPM cation channel family. J Rec Signal Transduct 26:159–178
26. McNulty S, Fonfria E (2005) The role of TRPM channels in cell death. Pflugers Arch 451:235–242
27. Zholos A (2010) Pharmacology of transient receptor potential melastatin channels in the vasculature. Br J Pharmacol 159:1559–1571
28. McKemy DD, Neuhausser WM, Julius D (2002) Identification of a cold receptor reveals a general role for TRP channels in thermosensation. Nature 416:52–58
29. Brauchi S, Orio P, Latorre R (2004) Clues to understanding cold sensation: thermodynamics and electrophysiological analysis of the cold receptor TRPM8. Proc Natl Acad Sci USA 101:15494–15499
30. Malkia A, Madrid R, Meseguer V, de la Pe E, Valero M, Belmonte C, Viana F (2007) Bidirectional shifts of TRPM8 channel gating by temperature and chemical agents modulate the cold sensitivity of mammalian thermoreceptors. J Physiol 581:155–174
31. Matta JA, Ahern GP (2007) Voltage is a partial activator of rat thermosensitive TRP channels. J Physiol 585:469–482

32. Reid G, Flonta ML (2002) Ion channels activated by cold and menthol in cultured rat dorsal root ganglion neurones. Neurosci Lett 324:164–168
33. Voets T, Droogmans G, Wissenbach U, Janssens A, Flockerzi V, Nilius B (2004b) The principle of temperature-dependent gating in cold- and heat-sensitive TRP channels. Nature 430:748–754
34. Nilius B, Mahieu F, Karashima Y, Voets T (2007) Regulation of TRP channels: a voltage-lipid connection. Biochem Soc Trans 35:105–108
35. Rohacs T, Nilius B (2007) Regulation of transient receptor potential (TRP) channels by phosphoinositides. Pflugers Arch 455:157–168
36. Rohacs T (2009) Phosphoinositide regulation of non-canonical transient receptor potential channels. Cell Calcium 45:554–565
37. Duncan LM, Deeds J, Hunter J, Shao J, Holmgren LM, Woolf EA, Tepper RI, Shyjan AW (1998) Down-regulation of the novel gene melastatin correlates with potential for melanoma metastasis. Cancer Res 58:1515–1520
38. Oancea E, Vriens J, Brauchi S, Jun J, Splawski I, Clapham DE (2009) TRPM1 forms ion channels associated with melanin content in melanocytes. Sci STKE 2:ra21
39. Xu X-Z S, Moebius F, Gill DL, Montell C (2001) Regulation of melastatin, a TRP-related protein, through interaction with a cytoplasmic isoform. Proc Natl Acad Sci USA 98:10692–10697
40. Perraud AL, Fleig A, Dunn CA, Bagley LA, Launay P, Schmitz C, Stokes AJ, Zhu Q, Bessman MJ, Penner R, Kinet JP, Scharenberg AM (2001) ADP-ribose gating of the calcium-permeable LTRPC2 channel revealed by Nudix motif homology. Nature 411: 595–599
41. Togashi K, Hara Y, Tominaga T, Higashi T, Konishi Y, Mori Y, Tominaga M (2006) TRPM2 activation by cyclic ADP-ribose at body temperature is involved in insulin secretion. EMBO J 25:1804–1815
42. Perraud AL, Takanishi CL, Shen B, Kang S, Smith MK, Schmitz C, Knowles HM, Ferraris D, Li W, Zhang J, Stoddard BL, Scharenberg AM (2005) Accumulation of free ADP-ribose from mitochondria mediates oxidative stress-induced gating of TRPM2 cation channels. J Biol Chem 280:6138–6148
43. Hecquet CM, Malik AB (2009) Role of H_2O_2-activated TRPM2 calcium channel in oxidant-induced endothelial injury. Thromb Haemost 101:619–625
44. Hecquet CM, Ahmmed GU, Vogel SM, Malik AB (2008) Role of TRPM2 channel in mediating H_2O_2-induced Ca^{2+} entry and endothelial hyperpermeability. Circ Res 102:347–355
45. Kaneko S, Kawakami S, Hara Y, Wakamori M, Itoh E, Minami T, Takada Y, Kume T, Katsuki H, Mori Y, Akaike A (2006) A critical role of TRPM2 in neuronal cell death by hydrogen peroxide. J Pharmacol Sci 101:66–76
46. Kuhn FJ, Heiner I, Luckhoff A (2005) TRPM2: a calcium influx pathway regulated by oxidative stress and the novel second messenger ADP-ribose. Pflugers Arch 451:212–219
47. Scharenberg A (2005) TRPM2 and TRPM7: channel/enzyme fusions to generate novel intracellular sensors. Pflugers Arch 451:220–227
48. Zhang W, Hirschler-Laszkiewicz I, Tong Q, Conrad K, Sun SC, Penn L, Barber DL, Stahl R, Carey DJ, Cheung JY, Miller BA (2006) TRPM2 is an ion channel that modulates hematopoietic cell death through activation of caspases and PARP cleavage. Am J Physiol 290:C1146–C1159
49. Grubisha O, Rafty LA, Takanishi CL, Xu X, Tong L, Perraud AL, Scharenberg AM, Denu JM (2006) Metabolite of SIR2 reaction modulates TRPM2 ion channel. J Biol Chem 281:14057–14065
50. Grimm C, Kraft R, Sauerbruch S, Schultz G, & Harteneck C (2003) Molecular and functional characterization of the melastatin-related cation channel TRPM3. J Biol Chem 278: 21493–21501
51. Lee N, Chen J, Sun L, Wu S, Gray KR, Rich A, Huang M, Lin JH, Feder JN, Janovitz EB, Levesque PC, Blanar MA (2003) Expression and characterization of human transient receptor potential melastatin 3 (hTRPM3). J Biol Chem 278:20890–20897

52. Oberwinkler J, Lis A, Giehl KM, Flockerzi V, Philipp SE (2005) Alternative splicing switches the divalent cation selectivity of TRPM3 channels. J Biol Chem 280:22540–22548
53. Petersen OH (2002) Cation channels: Homing in on the elusive CAN channels. Curr Biol 12:R520–R522
54. Colquhoun D, Neher E, Reuter H, Stevens CF (1981) Inward current channels activated by intracellular Ca in cultured cardiac cells. Nature 294:752–754
55. Suh S, Watanabe H, Droogmans G, Nilius B (2002) ATP and nitric oxide modulate a Ca^{2+}-activated non-selective cation current in macrovascular endothelial cells. Pflugers Arch 444:438–445
56. Launay P, Fleig A, Perraud AL, Scharenberg AM, Penner R, Kinet JP (2002) TRPM4 is a Ca^{2+}-activated nonselective cation channel mediating cell membrane depolarization. Cell 109:397–407
57. Nilius B, Vennekens R (2006) From cardiac cation channels to the molecular dissection of the transient receptor potential channel TRPM4. Pflugers Arch 453:313–321
58. Nilius B, Prenen J, Droogmans G, Voets T, Vennekens R, Freichel M, Wissenbach U, Flockerzi V (2003) Voltage dependence of the Ca^{2+}-activated cation channel TRPM4. J Biol Chem 278:30813–30820
59. Nilius B, Prenen J, Tang J, Wang C, Owsianik G, Janssens A, Voets T, Zhu MX (2005a) Regulation of the Ca^{2+} sensitivity of the nonselective cation channel TRPM4. J Biol Chem 280:6423–6433
60. Earley S, Straub SV, Brayden JE (2007) Protein kinase C regulates vascular myogenic tone through activation of TRPM4. Am J Physiol 292:H2613–H2622
61. Vennekens R, Nilius B (2007) Insights into TRPM4 function, regulation and physiological role. Handb Exp Pharmacol 179:269–285
62. Ullrich ND, Voets T, Prenen J, Vennekens R, Talavera K, Droogmans G, Nilius B (2005) Comparison of functional properties of the Ca^{2+}-activated cation channels TRPM4 and TRPM5 from mice. Cell Calcium 37:267–278
63. Prawitt D, Monteilh-Zoller MK, Brixel L, Spangenberg C, Zabel B, Fleig A, Penner R (2003) TRPM5 is a transient Ca^{2+}-activated cation channel responding to rapid changes in $[Ca^{2+}]_i$. Proc Natl Acad Sci USA 100:15166–15171
64. Ohkuri T, Yasumatsu K, Horio N, Jyotaki M, Margolskee RF, Ninomiya Y (2009) Multiple sweet receptors and transduction pathways revealed in knockout mice by temperature dependence and gurmarin sensitivity. Am J Physiol 296:R960–R971
65. Perez CA, Huang L, Rong M, Kozak JA, Preuss AK, Zhang H, Max M, Margolskee RF (2002) A transient receptor potential channel expressed in taste receptor cells. Nat Neurosci 5:1169–1176
66. Talavera K, Yasumatsu K, Voets T, Droogmans G, Shigemura N, Ninomiya Y, Margolskee RF, Nilius B (2005) Heat activation of TRPM5 underlies thermal sensitivity of sweet taste. Nature 438:1022–1025
67. Zhang Z, Zhao Z, Margolskee R, Liman E (2007) The transduction channel TRPM5 is gated by intracellular calcium in taste cells. J Neurosci 27:5777–5786
68. Nilius B, Talavera K, Owsianik G, Prenen J, Droogmans G, Voets T (2005b) Gating of TRP channels: a voltage connection? J Physiol 567:35–44
69. Runnels LW, Yue L, Clapham DE (2001) TRP-PLIK, a bifunctional protein with kinase and ion channel activities. Science 291:1043–1047
70. Nadler MJ, Hermosura MC, Inabe K, Perraud AL, Zhu Q, Stokes AJ, Kurosaki T, Kinet JP, Penner R, Scharenberg AM, Fleig A (2001) LTRPC7 is a MgATP-regulated divalent cation channel required for cell viability. Nature 411:590–595
71. Voets T, Nilius B, Hoefs S, van der Kemp AW, Droogmans G, Bindels RJ, Hoenderop JG (2004a) TRPM6 forms the Mg^{2+} influx channel involved in intestinal and renal Mg^{2+} absorption. J Biol Chem 279:19–25
72. Numata T, Shimizu T, Okada Y (2007) TRPM7 is a stretch- and swelling-activated cation channel involved in volume regulation in human epithelial cells. Am J Physiol 292: C460–C467

73. Oancea E, Wolfe JT, Clapham DE (2006) Functional TRPM7 channels accumulate at the plasma membrane in response to fluid flow. Circ Res 98:245–253
74. Bautista DM, Siemens J, Glazer JM, Tsuruda PR, Basbaum AI, Stucky CL, Jordt SE, Julius D (2007) The menthol receptor TRPM8 is the principal detector of environmental cold. Nature 448:204–208
75. Colburn RW, Lubin ML, Stone J, Wang Y, Lawrence D, D'Andrea MR, Brandt MR, Liu Y, Flores CM, Qin N (2007) Attenuated cold sensitivity in TRPM8 null mice. Neuron 54: 379–386
76. Dhaka A, Murray AN, Mathur J, Earley TJ, Petrus MJ, Patapoutian A (2007) TRPM8 is required for cold sensation in mice. Neuron 54:371–378
77. Peier AM, Moqrich A, Hergarden AC, Reeve AJ, Andersson DA, Story GM, Earley TJ, Dragoni I, McIntyre P, Bevan S, Patapoutian A (2002) A TRP channel that senses cold stimuli and menthol. Cell 108:705–715
78. Bidaux G, Flourakis M, Thebault S, Zholos A, Beck B, Gkika D, Roudbaraki M, Bonnal JL, Mauroy B, Shuba Y, Skryma R, Prevarskaya N (2007) Prostate cell differentiation status determines transient receptor potential melastatin member 8 channel subcellular localization and function. J Clin Invest 117:1647–1657
79. Tsavaler L, Shapero MH, Morkowski S, Laus R (2001) Trp-p8, a novel prostate-specific gene, is up-regulated in prostate cancer and other malignancies and shares high homology with Transient Receptor Potential calcium channel proteins. Cancer Res 61:3760–3769
80. Zhang L, Barritt GJ (2006) TRPM8 in prostate cancer cells: a potential diagnostic and prognostic marker with a secretory function? Endocr-Relat Cancer 13:27–38
81. Abeele FV, Zholos A, Bidaux G, Shuba Y, Thebault S, Beck B, Flourakis M, Panchin Y, Skryma R, Prevarskaya N (2006) Ca^{2+}-independent phospholipase A_2-dependent gating of TRPM8 by lysophospholipids. J Biol Chem 281:40174–40182
82. Andersson DA, Nash M, Bevan S (2007) Modulation of the cold-activated channel TRPM8 by lysophospholipids and polyunsaturated fatty acids. J Neurosci 27:3347–3355
83. Smani T, Zakharov SI, Csutora P, Leno E, Trepakova ES, Bolotina VM (2004) A novel mechanism for the store-operated calcium influx pathway. Nat Cell Biol 6:113–120
84. Park KM, Trucillo M, Serban N, Cohen RA, Bolotina VM (2008) Role of $iPLA_2$ and store-operated channels in agonist-induced Ca^{2+} influx and constriction in cerebral, mesenteric, and carotid arteries. Am J Physiol 294:H1183–H1187
85. Thebault S, Lemonnier L, Bidaux G, Flourakis M, Bavencoffe A, Gordienko D, Roudbaraki M, Delcourt P, Panchin Y, Shuba Y, Skryma R, Prevarskaya N (2005) Novel role of cold/menthol-sensitive Transient Receptor Potential melastatine family member 8 (TRPM8) in the activation of store-operated channels in LNCaP human prostate cancer epithelial cells. J Biol Chem 280:39423–39435
86. Andersson DA, Chase HWN, Bevan S (2004) TRPM8 activation by menthol, icilin, and cold is differentially modulated by intracellular pH. J Neurosci 24:5364–5369
87. Bodding M, Wissenbach U, Flockerzi V (2007) Characterisation of TRPM8 as a pharmacophore receptor. Cell Calcium 42:618–628
88. Hui K, Guo Y, Feng ZP (2005) Biophysical properties of menthol-activated cold receptor TRPM8 channels. Biochem Biophys Res Commun 333:374–382
89. Latorre R, Brauchi S, Orta G, Zaelzer C, Vargas G (2007) ThermoTRP channels as modular proteins with allosteric gating. Cell Calcium 42:427–438
90. Reid G (2005) ThermoTRP channels and cold sensing: what are they really up to? Pflugers Arch 451:250–263
91. Brauchi S, Orta G, Mascayano C, Salazar M, Raddatz N, Urbina H, Rosenmann E, Gonzalez-Nilo F, Latorre R (2007) Dissection of the components for PIP_2 activation and thermosensation in TRP channels. Proc Natl Acad Sci USA 104:10246–10251
92. Liu B, Qin F (2005) Functional control of cold- and menthol-sensitive TRPM8 ion channels by phosphatidylinositol 4,5-bisphosphate. J Neurosci 25:1674–1681
93. Rohacs T, Lopes CM, Michailidis I, Logothetis DE (2005) $PI(4,5)P_2$ regulates the activation and desensitization of TRPM8 channels through the TRP domain. Nat Neurosci 8:626–634

94. Daniels RL, Takashima Y, McKemy DD (2009) Activity of the neuronal cold sensor TRPM8 is regulated by phospholipase C via the phospholipid phosphoinositol 4,5-bisphosphate. J Biol Chem 284:1570–1582
95. Liu D, Liman ER (2003) Intracellular Ca^{2+} and the phospholipid PIP_2 regulate the taste transduction ion channel TRPM5. Proc Natl Acad Sci USA 100:15160–15165
96. Nilius B, Mahieu FF, Prenen J, Janssens A, Owsianik G, Vennekens RF, Voets T (2006) The Ca^{2+}-activated cation channel TRPM4 is regulated by phosphatidylinositol 4,5-biphosphate. EMBO J 25:467–478
97. Runnels LW, Yue L, Clapham DE (2002) The TRPM7 channel is inactivated by PIP_2 hydrolysis. Nat Cell Biol 4:329–336
98. Zhang Z, Okawa H, Wang Y, Liman ER (2005) Phosphatidylinositol 4,5-bisphosphate rescues TRPM4 channels from desensitization. J Biol Chem 280:39185–39192
99. Callera GE, He Y, Yogi A, Montezano AC, Paravicini T, Yao G, Touyz RM (2009) Regulation of the novel Mg^{2+} transporter transient receptor potential melastatin 7 (TRPM7) cation channel by bradykinin in vascular smooth muscle cells. J Hypertens 27:155–166
100. He Y, Yao G, Savoia C, Touyz RM (2005) Transient receptor potential melastatin 7 ion channels regulate magnesium homeostasis in vascular smooth muscle cells: role of angiotensin II. Circ Res 96:207–215
101. Touyz RM, He Y, Montezano ACI, Yao G, Chubanov V, Gudermann T, Callera GE (2006) Differential regulation of transient receptor potential melastatin 6 and 7 cation channels by ANG II in vascular smooth muscle cells from spontaneously hypertensive rats. Am J Physiol 290:R73–R78
102. Yogi A, Callera GE, Tostes R, Touyz RM (2009) Bradykinin regulates calpain and proinflammatory signaling through TRPM7-sensitive pathways in vascular smooth muscle cells. Am J Physiol 296:R201–R207
103. Grimm C, Kraft R, Schultz G, Harteneck C (2005) Activation of the melastatin-related cation channel TRPM3 by D-erythro-sphingosine. Mol Pharmacol 67:798–805
104. Wagner TFJ, Loch S, Lambert S, Straub I, Mannebach S, Mathar I, Dufer M, Lis A, Flockerzi V, Philipp SE, Oberwinkler J (2008) Transient receptor potential M3 channels are ionotropic steroid receptors in pancreatic β cells. Nat Cell Biol 10:1421–1430
105. Nilius B, Prenen J, Voets T, Droogmans G (2004a) Intracellular nucleotides and polyamines inhibit the Ca^{2+}-activated cation channel TRPM4b. Pflugers Arch 448:70–75
106. Nilius B, Prenen J, Janssens A, Voets T, Droogmans G (2004b) Decavanadate modulates gating of TRPM4 cation channels. J Physiol 560:753–765
107. Meseguer V, Karashima Y, Talavera K, D'Hoedt D, Donovan-Rodriguez T, Viana F, Nilius B, Voets T (2008) Transient Receptor Potential channels in sensory neurons are targets of the antimycotic agent clotrimazole. J Neurosci 28:576–586
108. Lashinger ESR, Steiginga MS, Hieble JP, Leon LA, Gardner SD, Nagilla R, Davenport EA, Hoffman BE, Laping NJ, Su X (2008) AMTB, a TRPM8 channel blocker: evidence in rats for activity in overactive bladder and painful bladder syndrome. Am J Physiol 295: F803–F810
109. Yang XR, Lin MJ, McIntosh LS, Sham JSK (2006) Functional expression of transient receptor potential melastatin- and vanilloid-related channels in pulmonary arterial and aortic smooth muscle. Am J Physiol 290:L1267–L1276
110. Johnson CD, Melanaphy D, Purse A, Stokesberry SA, Dickson P, Zholos AV (2009) Transient receptor potential melastatin 8 channel involvement in the regulation of vascular tone. Am J Physiol 296:H1868–H1877
111. Simard JM, Tarasov KV, Gerzanich V (2007) Non-selective cation channels, transient receptor potential channels and ischemic stroke. Biochim Biophys Acta 1772:947–957
112. Miller BA (2004) Inhibition of TRPM2 function by PARP inhibitors protects cells from oxidative stress-induced death. Br J Pharmacol 143:515–516
113. Di A, Malik AB (2010) TRP channels and the control of vascular function. Curr Opin Pharmacol 10:127–132

114. Yamamoto S, Shimizu S, Kiyonaka S, Takahashi N, Wajima T, Hara Y, Negoro T, Hiroi T, Kiuchi Y, Okada T, Kaneko S, Lange I, Fleig A, Penner R, Nishi M, Takeshima H, Mori Y (2008) TRPM2-mediated Ca^{2+} influx induces chemokine production in monocytes that aggravates inflammatory neutrophil infiltration. Nat Med 14:738–747
115. Hill MA, Davis MJ (2007) Coupling a change in intraluminal pressure to vascular smooth muscle depolarization: still stretching for an explanation. Am J Physiol 292:H2570–H2572
116. Earley S, Waldron BJ, Brayden JE (2004) Critical role for Transient Receptor Potential channel TRPM4 in myogenic constriction of cerebral arteries. Circ Res 95:922–929
117. Morita H, Honda A, Inoue R, Ito Y, Abe K, Nelson MT, Brayden JE (2007) Membrane stretch-induced activation of a TRPM4-like nonselective cation channel in cerebral artery myocytes. J Pharmacol Sci 103:417–426
118. Inoue R, Jian Z, Kawarabayashi Y (2009) Mechanosensitive TRP channels in cardiovascular pathophysiology. Pharmacol Ther 123:371–385
119. Sharif-Naeini R, Dedman A, Folgering JH, Duprat F, Patel A, Nilius B, Honore E (2008) TRP channels and mechanosensory transduction: insights into the arterial myogenic response. Pflugers Arch 456:529–540
120. Welsh DG, Morielli AD, Nelson MT, Brayden JE (2002) Transient receptor potential channels regulate myogenic tone of resistance arteries. Circ Res 90:248–250
121. Gerzanich V, Woo SK, Vennekens R, Tsymbalyuk O, Ivanova S, Ivanov A, Geng Z, Chen Z, Nilius B, Flockerzi V, Freichel M, Simard JM (2009) De novo expression of Trpm4 initiates secondary hemorrhage in spinal cord injury. Nat Med 15:185–191
122. Inoue K, Xiong ZG (2009) Silencing TRPM7 promotes growth/proliferation and nitric oxide production of vascular endothelial cells via the ERK pathway. Cardiovasc Res 83:547–557
122. Paravicini TM, Yogi A, Mazur A, Touyz RM (2009a) Dysregulation of vascular TRPM7 and annexin-1 is associated with endothelial dysfunction in inherited hypomagnesemia. Hypertension 53:423–429
124. Touyz RM (2008) Transient receptor potential melastatin 6 and 7 channels, magnesium transport, and vascular biology: implications in hypertension. Am J Physiol 294:H1103–H1118
125. Hawthorn M, Ferrante J, Luchowski E, Rutledge A, Wei XY, Triggle DJ (1988) The actions of peppermint oil and menthol on calcium channel dependent processes in intestinal, neuronal and cardiac preparations. Aliment Pharmacol Ther 2:101–118
126. Wright CE, Laude EA, Grattan TJ, Morice AH (1997) Capsaicin and neurokinin A-induced bronchoconstriction in the anaesthetised guinea-pig: evidence for a direct action of menthol on isolated bronchial smooth muscle. Br J Pharmacol 121:1645–1650
127. Earley S, Gonzales AL, Crnich R (2009) Endothelium-dependent cerebral artery dilation mediated by TRPA1 and Ca^{2+}-activated K^+ channels. Circ Res 104:987–994
128. Sherkheli MA (2009) Letter to the editor: "Is menthol- or icilin-induced vasodilation mediated by the activation of TRPM8?" Am J Physiol 297:H887
129. Kellogg DL Jr., Pergola PE, Piest KL, Kosiba WA, Crandall CG, Grossmann M, Johnson JM (1995) Cutaneous active vasodilation in humans is mediated by cholinergic nerve cotransmission. Circ Res 77:1222–1228
130. Fantozzi I, Zhang S, Platoshyn O, Remillard CV, Cowling RT, Yuan JXJ (2003) Hypoxia increases AP-1 binding activity by enhancing capacitative Ca^{2+} entry in human pulmonary artery endothelial cells. Am J Physiol 285:L1233–L1245
131. Hamaguchi Y, Matsubara T, Amano T, Uetani T, Asano H, Iwamoto T, Furukawa K, Murohara T, Nakayama S (2008) Na^+-independent Mg^{2+} transport sensitive to 2-aminoethoxydiphenyl borate (2-APB) in vascular smooth muscle cells: involvement of TRPM-like channels. J Cell Mol Med 12:962–974

Chapter 38
Molecular Expression and Functional Role of Canonical Transient Receptor Potential Channels in Airway Smooth Muscle Cells

Yong-Xiao Wang and Yun-Min Zheng

Abstract Multiple canonical or classic transient receptor potential (TRPC) molecules are expressed in animal and human airway smooth muscle cells (SMCs). TRPC3, but not TRPC1, is a major molecular component of native non-selective cation channels (NSCCs) to contribute to the resting $[Ca^{2+}]_i$ and muscarinic increase in $[Ca^{2+}]_i$ in freshly isolated airway SMCs. TRPC3-encoded NSCCs are significantly increased in expression and activity in airway SMCs from ovalbumin-sensitized/challenged "asthmatic" mice, whereas TRPC1-encoded channel activity, but not its expression, is largely augmented. The upregulated TRPC3- and TRPC1-encoded NSCC activity both mediate "asthmatic" membrane depolarization in airway SMCs. Supportively, tumor necrosis factor-α (TNFα), an important asthma mediator, increases TRPC3 expression, and TRPC3 gene silencing inhibits TNFα-mediated augmentation of acetylcholine-evoked increase in $[Ca^{2+}]_i$ in passaged airway SMCs. In contrast, TRPC6 gene silencing has no effect on 1-oleoyl-2-acetyl-sn-glycerol (OAG)-evoked increase in $[Ca^{2+}]_i$ in primary isolated cells. These findings provide compelling information indicating that TRPC3-encoded NSCCs are important for physiological and pathological cellular responses in airway SMCs. However, continual studies are necessary to further determine whether, which, and how TRPC-encoded channels are involved in cellular responses in normal and diseased (e.g., asthmatic) airway SMCs.

Y.-X. Wang (✉)
Center for Cardiovascular Sciences, Albany Medical College, Albany, NY 12208, USA
e-mail: wangy@mail.amc.edu

Y.-M. Zheng
Center for Cardiovascular Sciences, Albany Medical College, Albany, NY 12208, USA
e-mail: zhengy@mail.amc.edu

38.1 Introduction

The canonical or classic transient receptor potential (TRPC) channels are encoded by genes that are most closely related to the *trp* gene that was originally identified in *Drosophila*. Photoreceptor cells in *Drosophila* produce a transient receptor potential, consisted of an initial rapid spike followed by a sustained phase in response to light. These two phases of a receptor potential are mediated by TRP-encoded and TRP-like channels. The first mammalian TRP gene was cloned from the human brain using an expressed sequence tag, categorized into the TRPC family based on primary amino acid sequence, and thus termed TRPC1. Currently, the TRPC family is known to consist of seven members designated TRPC1-7. In addition to the TRPC family, many other mammalian TRP molecules have been discovered and classified into the following five different subfamilies based on their activation stimuli and homology: TRPV (V for vanilloid), TRPM (from the tumor suppressor melastatin), TRPP (P for polycystin), TRPML (ML for mucolipin), and TRPA (A for ankyrin).

TRPC molecules may encode nonselective cation channels (NSCCs), playing important roles in physiological and pathological cellular responses in various types of cells [1, 2]. NSCCs are present in airway smooth muscle cells (SMCs). The activity of these channels is significantly increased in response to agonists to initiate and maintain intracellular Ca^{2+} signaling, which may play an important role in various cellular responses including contraction, proliferation, migration, and gene expression in airway SMCs. We and other investigators have started to explore which TRPC molecules may encode NSCCs and show functional importance in airway SMCs. Studies have yielded interesting data, greatly improving our current knowledge in the field. In this chapter, we highlight the important findings of studies on TRPC-encoded NSCCs in airway SMCs and discuss the fundamental questions that remain to be addressed as well.

38.2 Multiple TRPC Molecules are Expressed in Airway SMCs

An earlier study using a standard *reverse transcription*-polymerase chain reaction method has shown that TRPC1, TRPC2, TRPC3, TRPC4, TRPC5 and TRPC6 mRNAs are expressed in primary isolated guinea-pig airway SMCs [3]. However, another study has reported that TRPC1, TRPC3, TRPC4 and TRPC6, but not TRPC2 and TRPC5 mRNAs are detected in passaged human airway SMCs [4]. It has also been found that TRPC1, TRPC3, TRPC4, TRPC5 and TRPC6, but not TRPC2 mRNAs are present in passaged human airway SMCs [5]. Using Western blot analysis or immunofluorescence staining, previous studies further reveal expression of TRPC1, TRPC3 and TRPC6 proteins in primary isolated guinea pig airway SM cells and tissues [3, 6], TRPC1, TRPC3 and TRPC4 proteins in freshly isolated porcine airway SM tissues [7], TRPC6 protein in primary isolated guinea pig and passaged human airway SMCs [4, 8], and TRPC1, TRPC3,

Table 38.1 Profiles of molecular expression and functional roles of TRPC members in passaged, primary cultured (un-passaged) and freshly isolated airway SMCs

	TRPC1	TRPC2	TRPC3	TRPC4	TRPC5	TRPC6	TRPC7
Passaged	Guinea pig Human [3–6, 66]	Guinea pig (reference [3])	Guinea pig Human [3–5]	Guinea pig Human [3–5]	Guinea pig Human [3, 5]	Guinea pig Human [3–5]	N/A
Primary cultured	Mouse [9]	N/A	Mouse [9]	N/A	N/A	Guinea pig [8]	N/A
Freshly isolated	Mouse [9]	N/A	Mouse [9]	N/A	N/A	N/A	N/A

N/A, no data available.

TRPC4, TRPC5 and TRPC6 proteins in passaged human airway SMCs [5]. Our recent studies unveil that TRPC1 and TRPC3 mRNAs and proteins are expressed in freshly isolated and primary mouse airway SMCs [9]. As summarized in Table 38.1, mRNAs and proteins of TRPC1 and TRPC3, but not other members, are expressed in freshly isolated, primary isolated and passaged airway SMCs.

38.3 TRPC3 Is an Important Molecular Component of Native Constitutively-Active NSCCs in Airway SMCs

Using the cell-attached single channel recording, a number of researchers have found that native constitutively-active channels are present in freshly isolated bovine and human airway SMCs [10–13]. These channels are blocked by the substitution of extracellular Na^+ with the non-permeable cation N-methyl-D-glucamine (indicating that the channels are Na^+-permeable), but unaffected by inhibition of Ca^{2+}-activated or voltage-activated K^+ channels with iberiotoxin, tetraethylammonium or 4-aminopyridine. Substitution of extracellular Na^+ with Ca^{2+} does not change the channel slope conductance, but rather shifts its reversal potential. Thus, the identified constitutively-active single channels in bovine and human AMSCs are the NSCCs, which are permeable to both Na^+ and Ca^{2+}, with a higher permeability to the former ion than the later.

We have further characterized NSCCs in freshly isolated mouse airway SMCs using the excised inside-out single channel recording [9]. These channels are constitutively active and can maintain their activity for up to 20 min without a time-dependent decrease. The relationship between the single channel current amplitudes and membrane potentials shows that the channels have 15, 37 and 57 pS conductance states. All three conductance states have the same reverse potential of ~0 mV. Cell-attached single channel recordings have also observed the ~15 pS NSCCs in bovine and human airway SMCs [10–13]. The inability to detect the two other conductances may be due to the different modes of single channel recordings; as such, the activity of native constitutively-active NSCCs is more readily recorded using the excised, rather than cell-attached recording, mode. The channel activity in freshly isolated mouse airway SMCs is significantly higher at positive than negative potentials, suggesting that the native NSCCs exhibit outward rectification in airway SMCs. A diacylglycerol analogue, 1-oleoyl-2-acetyl-sn-glycerol (OAG), significantly increases the channel activity. The channel activity is augmented by elevating extracellular Ca^{2+} concentration and inhibited by reducing extracellular Ca^{2+} concentration. Thus, constitutively-active NSCCs in airway SMCs possess diacylglycerol- and Ca^{2+}-gated properties, which are very similar to native constitutively-active NSCCs in vascular SMCs [14, 15].

Our recent study also reveals that application of specific TRPC3 antibodies blocks the activity of native single NSCCs by ~80% in freshly isolated airway SMCs [9]. TRPC3 gene silencing by specific siRNAs inhibits the single channel activity to a very similar extent in primary isolated cells. These patch clamp and genetic studies for the first time provide compelling evidence to reveal that native constitutively-active NSCCs are mainly encoded by TRPC3 in airway SMCs. These findings, together with the fact that the three conductance states of the channels exhibit the same reversal potentials, also suggest that TRPC3 may form homomeric and/or heteromeric NSCCs that have the different conductance states. TRPC1 is not involved in the formation of these channels, as its antibodies and siRNAs have no effect on the channel activity. More recently, we have found that application of specific TRPC4 antibodies does not produce an inhibitory effect (Xiao and Wang, unpublished data). As shown in Table 38.2, each of the TRPC channels has different biophysical and pharmacological properties. The properties that individual channels manifest may be determined by intracellular regulatory molecules and experimental conditions. Thus, further work is necessary to determine the biophysical and pharmacological characteristics of TRPC3 and other potential TRPC channels in airway SMCs.

38.4 TRPC3-Encoded NSCCs are Important for Controlling the Resting Membrane Potential and $[Ca^{2+}]_i$ in Airway SMCs

The resting membrane potential (V_m) in airway SMCs is between -40 and -50 mV, similar to other types of SMCs, significantly less negative than a K^+ equilibrium potential of ~-85 mV [16, 17]. Consistent with these previous reports, we have also found that the resting V_m is -44 mV in freshly isolated airway SMCs [9]. More importantly, our data reveal that specific TRPC3 antibodies and gene silencing both result in a pronounced V_m hyperpolarization by ~14 mV. Conversely, TRPC1 antibodies and gene silencing have no significant effect on the resting V_m. Thus, TRPC3-encoded NSCCs are important for controlling the resting V_m in airway SMCs, consistent with the central role of TRPC3 in mediating the activity of constitutively-active NSCCs. It is interesting to point out that constitutively-active NSCCs contribute to the less negative V_m than the K^+ equilibrium potential of

Table 38.2 Biophysical and pharmacological properties of expressed TRPC channels in cell lines (modified from references [1, 20, 92])

	Conductance (pS)	Ion selectivity (PCa/PNa)	Constitutive activity	Activation	Inhibition	Binding partners
TRPC1	16, 24	1.0, Nonselective	–	PLC, OAG, stretch, store depletion	La^{3+}, Gd^{3+} 2-APB, SKF-96365	TRPC3, TRPC4, TRPC5, IP_3R, CaM, Homer, Stim1
TRPC2	42	2.7 Nonselective	N/A	PLC, OAG, store depletion	La^{3+}, 2-APB, SKF-96365	IP_3R, CaM
TRPC3	42, 66	1.5, 1.6	+	PLC, OAG, Src kinase, IP_3 $[Ca^{2+}]_i$, store depletion	FFA, La^{3+}, Gd^{3+} 2-APB, SKF-96365	TRPC6, TRPC7, IP_3R, CaM, RyR, NCX1
TRPC4	30, 42	1.1, 7.7	–	PLC, GTPγS, receptor-operation, La^{3+}, store depletion,	La^{3+} 2-APB, SKF-96365	TRPC1, TRPC5, IP_3R, CaM
TRPC5	38, 47, 66	1.8, 9.0	+	PLC, OAG, store depletion, Src kinase, IP_3 $[Ca^{2+}]_i$, PI3K, MLCK	FFA 2-APB, SKF-96365	TRPC1, TRPC4, CaM, MLCK
TRPC6	28, 35, 37	5.0	–	PLC, OAG, store depletion, Src, 20-HETE	FFA, La^{3+}, Gd^{3+} 2-APB, SKF-96365	TRPC3, TRPC7, CaM, FKBP12, Fyn kinase
TRPC7	25, 50, 75	2	+	PLC, OAG, store depletion, 20-HETE	La^{3+}, Gd^{3+}, 2-APB, SKF-96365	TRPC3, TRPC6, CaM, FKBP12

2-APB, 2-aminoethoxydiphenyl borate; CaM, calmodulin; FKBP12, FK506 binding protein 12; 20-HETE, 20-hydroxyeicosatetraenoic acid; IP_3R, inositol 1,4,5-trisphosphate receptors; MLCK, myosin light chain kinase; NCX1, Na^+/Ca^{2+} exchanger-1; OAG, 1-oleoyl-2-acetyl-sn-glycerol; PI3K, phosphoinositide-3-kinase; PLC, phospholipase C; RyR, ryanodine receptor.

~–85 mV in coronary and pulmonary artery SMCs [18, 19]. These data, together with the fact that TRPC3 is a major molecular constituent of native constitutively-active NSCCs in vascular SMCs [20], suggest that TRPC3-formed NSCCs are most likely to regulate the resting V_m as well in vascular SMCs. However, this view needs confirmation through further experimentation. In addition, whether other TRPC members are involved in the control of the less negative V_m in airway and vascular myocytes remain to be determined.

Comparable to the effect on the resting membrane potential, we have found that siRNA-mediated TRPC3 gene silencing significantly lowers the resting $[Ca^{2+}]_i$ in primary isolated airway SMCs [9]. However, TRPC1 gene silencing does not alter the resting $[Ca^{2+}]_i$. These findings provide evidence that TRPC3, but not TRPC1-encoded NSCCs, play an important role in the control of the resting $[Ca^{2+}]_i$ in airway SMCs.

38.5 TRPC3-Encoded NSCCs May Mediate Agonist-Induced Increase in $[Ca^{2+}]_i$ in Airway SMCs

Whole-cell patch clamp studies demonstrate that muscarinic agonists acetylcholine and methacholine activate NSCCs in freshly isolated canine, equine, guinea-pig and swine airway SMCs [21–25]. Simultaneous measurements of membrane currents and $[Ca^{2+}]_i$ reveal that activation of NSCCs during muscarinic stimulation is always accompanied by a sustained increase in $[Ca^{2+}]_i$ due to extracellular Ca^{2+} influx [22–25]. The sustained increase in $[Ca^{2+}]_i$ induced by methacholine and other agonists are largely inhibited or abolished by the general NSCC blockers Ni^{2+}, Cd^{2+}, La^{3+}, Gd^{3+} and SKF-96365 [12, 23, 25–32]. Thus, functional NSCCs mediate agonist-induced Ca^{2+} influx and associated increase in $[Ca^{2+}]_i$ in airway SMCs.

We and other investigators have started to indentify which of the TRPC-encoded NSCCs are responsible for agonist-induced increase in $[Ca^{2+}]_i$. Our recent study reveals that TRPC3 gene silencing inhibits methacholine-evoked increase in $[Ca^{2+}]_i$ in primary isolated airway SMCs. Differently, TRPC1 gene silencing does not produce an effect. Similar to our findings, a previous report has shown that TRPC3 gene silencing blocks acetylcholine- and tumor necrosis factor-α (TNFα)-induced increase in $[Ca^{2+}]_i$ as well in passaged human airway SMCs [5]. Moreover, OAG, which is known to be a putative activator for TRPC-encoded channels [1, 2], causes a significant increase in $[Ca^{2+}]_i$ in primary isolated guinea pig airway SMCs; however, TRPC6 gene silencing does not affect OAG-evoked response [8]. Collectively, TRPC3-encoded NSCCs are important for agonist-induced Ca^{2+} responses in airway myocytes. Relative to Ca^{2+} release from the sarcoplasmic reticulum (SR, a major intracellular Ca^{2+} store), extracellular Ca^{2+} influx through TRPC3-encoded NSCCs makes a smaller contribution to agonist-induced initial increase in $[Ca^{2+}]_i$. On the other hand, this TRPC3-mediated Ca^{2+} signaling is persistent during agonist stimulation. As such, this persistent Ca^{2+} signaling may be essential for maintaining cell contraction and other cellular responses, as well as refilling intracellular Ca^{2+} stores to start a new response.

Our previous studies have shown that the muscarinic activation of NSCCs in airway SMCs is mediated by a pertussis toxin-sensitive, Gi/o protein-coupled M_2 receptor signaling pathway [22]. It is noted that the same signaling pathway is involved in the muscarinic activation of expressed TRPC4 channels in HEK293 cells [33]. Furthermore, Flockerzi and his colleagues have recently reported that NSCCs in intestinal myocytes, which show similar biophysical properties and activation mechanisms to those in airway SMCs, are mainly mediated by TRPC4 [34]. Hence, it would be interesting to conduct further studies to determine the role of TRPC-encoded NSCCSs in mediating agonist-induced increase in $[Ca^{2+}]_i$ in airway SMCs.

38.6 TRPC-Encoded NSCCs Cause Ca^{2+} Influx Through Themselves Directly, Na^+/Ca^{2+} Exchanger-1 and/or L-type Ca^{2+} Channels in Airway SMCs

A previous study using the whole-cell patch clamp recording with cation substitution has shown that extracellular Ca^{2+} can pass through NSCCs in freshly isolated airway SMCs during muscarinic stimulation [21]. In addition, our simultaneous measurements of membrane currents and $[Ca^{2+}]_i$ further reveal that under physiological conditions, a direct Ca^{2+} permeability of NSCCs contributes to 14% of the associated muscarinic Ca^{2+} influx in freshly isolated airway SMCs [23]. Thus, functional NSCCs mediate agonist-induced increase in $[Ca^{2+}]_i$ in airway SMCs by directly permitting Ca^{2+} influx through the opening channels, as diagrammed in Fig. 38.1.

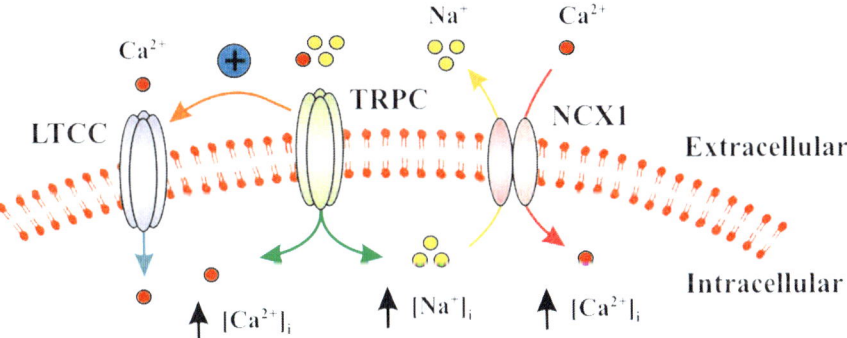

Fig. 38.1 A schematic diagram illustrates the three possible pathways for extracellular Ca^{2+} influx mediated by canonical transient receptor potential (TRPC)-encoded non-selective cation channels (NSCCs) in airway smooth muscle cells. First, TRPC-encoded NSCCs show a direct Ca^{2+} permeability and thus lead to Ca^{2+} influx through the opening channels. Second, TRPC channels are functionally coupled to Na^+/Ca^{2+} exchanger-1 (NCX1), by which substantial Na^+ entry through TRPC-encoded NSCCs makes NCX1 operate in the reverse mode, causing Ca^{2+} influx to extrude Na^+ ions out of the cell. Finally, activation of TRPC-encoded NSCCs may result in membrane epolarization and then opening of L-type Ca^{2+} channels (LTCCs), leading to Ca^{2+} influx

The plasmalemmal Na^+/Ca^{2+} exchangers play an important role in the control of intracellular Ca^{2+} homeostasis in many types of cells. These exchangers can operate in the forward mode (also termed the Ca^{2+} removal mode) to remove one Ca^{2+} ion out of the cell in exchange for three Na^+ ions entering the cell, or in the reverse mode (Ca^{2+} entry mode) to bring one Ca^{2+} ion into the cell in exchange for three Na^+ ions leaving the cell. The ion transport direction and activity of Na^+/Ca^{2+} exchangers depend on the electrochemical gradient of both Ca^{2+} and Na^+ ions as well as membrane potential [35, 36]. TRPC-encoded NSCCs exhibit a high permeability to Na^+ and may thus result in substantial Na^+ entry, which makes Na^+/Ca^{2+} exchangers operate in the reverse mode, leading to Ca^{2+} influx (Fig. 38.1). In agreement with this view, an earlier study has found that reducing extracellular Na^+ from 137 to 5 mM or application of the Na^+/Ca^{2+} exchanger inhibitor KB-R9743 almost completely blocks TRPC3-mediated extracellular Ca^{2+} influx in HEK293 cells [37]. Extracellular Ca^{2+} influx is also inhibited by dominant negative suppression of TRPC3 channel function. Low density membrane fractionation and co-immunoprecipitation experiments reveal the co-localization of TRPC3 and Na^+/Ca^{2+} exchanger-1, and glutathione *S*-transferase pull-down experiments demonstrate that Na^+/Ca^{2+} exchanger-1 binds to TRPC3 C-terminus. Moreover, several reports have shown that the Na^+/Ca^{2+} exchanger blocker KB-R7943 and the NSCC inhibitor SKF-96365 both largely inhibit or abolish agonist-evoked increase in $[Ca^{2+}]_i$ in freshly isolated airway SMCs and muscle contraction in freshly isolated airway strips [28–32, 38]. These results suggest that Na^+/Ca^{2+} exchangers are important for the role of TRPC-encoded NSCCs in mediating Ca^{2+} influx and associated cellular responses in airway SMCs.

Significant Na^+ entry through TRPC-encoded NSCCs may also cause membrane depolarization and then activation of L-type Ca^{2+} channels, contributing to Ca^{2+} influx (Fig. 38.1). Supportively, the whole-cell patch clamp recording in combination with measurement of $[Ca^{2+}]_i$ has shown the presence of L-type Ca^{2+} channels and associated Ca^{2+} influx in freshly isolated bovine, canine, equine, guinea pig and human airway SMCs [39–46]. A number of studies have found that a sustained increase in $[Ca^{2+}]_i$ and associated contraction following application of muscarinic and other agonists are inhibited by the L-type Ca^{2+} channel blockers nifedipine and verapamil in bovine, canine and mouse airway SMCs [29, 30, 38, 47–51]. In contrast, it has been reported that agonist-induced contraction is not significantly affected by L-type Ca^{2+} channel blockers in airway SMCs [51–55]. It should also be noted that clinical studies indicate that L-type Ca^{2+} channel blockers are ineffective in the treatment of asthma [56–63]. Furthermore, our recent work demonstrates that membrane depolarization causes a direct activation of G protein-coupled receptors, leading to Ca^{2+} release and contraction without the involvement of L-type Ca^{2+} channel in airway SMCs [46]. Thus, the significance of L-type Ca^{2+} channels in the role of TRPC-encoded NSCCs in regulating agonist-induced Ca^{2+} influx needs to be further evaluated in airway SMCs.

38.7 TRPC-Encoded NSCC-Mediated Ca^{2+} Influx Occurs Due to Store-Operated Ca^{2+} Entry, Receptor-Operated Ca^{2+} Entry, or Both Processes in Airway SMCs

Agonist-induced, NSCC-mediated Ca^{2+} influx exhibits two distinct signaling processes: store-operated Ca^{2+} entry (also called capacitance Ca^{2+} entry) and receptor-operated Ca^{2+} entry. Stimulation of muscarinic receptors or other G-protein coupled receptors result in activation of phospholipase C and generation of inositol 1,4,5-trisphosphate (IP_3). IP_3 binds to and activates its receptors, inducing Ca^{2+} release from the SR, which plays an important role in cellular responses. On the other hand, IP_3-mediated Ca^{2+} release leads to a decrease in the SR Ca^{2+} and subsequently store-operated Ca^{2+} entry. The NSCCs mediating this ubiquitous Ca^{2+} entry process are proposed to be essential for agonist-induced sustained increase in $[Ca^{2+}]_i$ and for the refilling of the SR Ca^{2+} store in airway SMCs [17, 64, 65]. It has been shown that after depletion of the SR Ca^{2+} store with thapsigargin or cyclopiazonic acid in the absence of extracellular Ca^{2+}, readmission of extracellular Ca^{2+} results in store-operated Ca^{2+} entry in primary isolated and passaged airway SMCs [5, 13, 66–70]. Whole-cell patch clamp recordings indicate that cyclopiazonic acid generates store-operated NSCC currents [13, 66, 70, 71]. A number of studies have reported that store-operated Ca^{2+} entry is significantly increased by muscarinic and other agonists [5, 7, 66, 68]. Moreover, it has been found that incubation of TNFα for hours significantly increases store-operated Ca^{2+} entry in passaged human airway SMCs, and the effect of TNFα is attenuated by TRPC3 gene silencing [5]. These data indicate the direct activation of store-operated NSCCs and associated Ca^{2+} entry in airway SMCs during agonist stimulation, which can be mediated by TRPC3.

Activation of phospholipase C following stimulation of muscarinic receptors and other G-protein-coupled receptors can also produce diacylglycerol and its derivates (e.g., arachidonic acid). These products lead to NSCC-mediated receptor-operated Ca^{2+} entry independent of depletion of the SR Ca^{2+} in many types of cells [1, 2]. Application of OAG, a diacylglycerol analogue, causes an increase in $[Ca^{2+}]_i$ in primary isolated guinea pig airway SMCs [8]. These results suggest that receptor-operated NSCCs and associated Ca^{2+} influx exist, contributing to agonist-induced increase in $[Ca^{2+}]_i$ and contraction in airway SMCs. In support, methacholine or histamine can cause a further increase in $[Ca^{2+}]_i$ and tension in freshly isolated guinea pig tracheal rings under the condition that store-operated Ca^{2+} entry is fully activated by thapsigargin [69]. We and others have discovered that TRPC3 gene silencing significantly reduces the muscarinic increase in $[Ca^{2+}]_i$ in primary isolated mouse and passaged humans airway SMCs [5, 9]. It has also been reported that TRPC6 gene silencing has no effect on OAG-induced increase in $[Ca^{2+}]_i$ in primary isolated guinea pig airway myocytes [8]. Collectively, TRPC3-, but not TRPC6-encoded, NSCCs are likely to be involved in receptor-operated Ca^{2+} entry following stimulation of G-protein-coupled receptors.

38.8 TRPC3-Encoded Native Constitutively-Active NSCCs are Significantly Upregulated in Expression and Activity in Asthmatic Airway SMCs

Asthma is a chronic lung disease characterized with two distinct features: airway hyperresponsiveness and inflammation. Both asthmatic airway hyperresponsiveness and inflammation are closely associated with an abnormal increase in Ca^{2+} signaling in airway SMCs. Simultaneous measurements of $[Ca^{2+}]_i$ and contraction have shown that the increased contraction following exposure to spasomogens is nicely mirrored by an increased rise in $[Ca^{2+}]_i$ in freshly isolated asthmatic rabbit airway muscle strips [72]. The potentiation of spasomogen-induced increase in $[Ca^{2+}]_i$ has also been found in freshly isolated hyperresponsive rat airway SMCs [73–75]. Moreover, the removal of extracellular Ca^{2+} and inhibition of an increase in $[Ca^{2+}]_i$ with various pharmacological agents attenuate or prevent asthmatic airway muscle hyperresponsiveness in animals [72, 76, 77]. There is compelling evidence that an abnormal alteration in $[Ca^{2+}]_i$ occurs in asthmatic human airway SMCs as well [5, 68, 78–83]. In addition, clinical studies demonstrate that drugs that interrupt intracellular Ca^{2+} mobilization can reverse airway muscle hyperresponsiveness [84–86]. Furthermore, the actions of glucocorticoids, β-adrenergic agonists and cholinergic blockers, the most effective clinical drugs in the treatment of asthma, are related to their inhibition of abnormal increases in $[Ca^{2+}]_i$ in freshly isolated and passaged airway SMCs during asthmatic stimuli [64, 65, 87–90]. Collectively, it is reasonable to state that the enhanced increase in $[Ca^{2+}]_i$ in airway SMCs is an important converging pathway for asthmatic airway hyperresponsiveness and inflammation as well.

To explore the potentially significant contribution of TRPC-encoded NSCCs in the development of asthma, White et al. have shown that TRPC3 mRNA and protein expression are significantly increased in passaged human airway SMCs following treatment with TNFα, an important asthma mediator [5]. These authors have further found that TRPC3 gene silencing inhibits TNFα-induced Ca^{2+} influx and associated increase in $[Ca^{2+}]_i$, as well as TNFα-mediated augmentation of acetylcholine-evoked increase in $[Ca^{2+}]_i$. All these findings suggest that TRPC3-encoded NSCCs may be upregulated in the molecular expression and function role, contributing to the asthmatic cellular responses in airway SMCs. However, there is a lack of direct experimental evidence; thus, we have begun to determine the potential changes in TRPC molecule expression and functional activities in asthmatic airway SMCs. Using ovalbumin-sensitized/challenged "asthmatic" mice, we have shown that TRPC1 protein expression level is not changed in freshly isolated "asthmatic" airway SM tissues [9]. On the other hand, TRPC3 protein expression level is increased by threefolds.

Comparable to airway muscle hyperresponsiveness in vitro and in vivo, the activity of native NSCCs is increased by fourfolds in freshly isolated "asthmatic" airway SMCs [9]. The increased channel activity is completely blocked by specific TRPC1 antibodies. Similar to a previous report detailing "asthmatic" rat airway

SMCs [16], our data unveil that membrane depolarization is found in freshly isolated "asthmatic" mouse airway SMCs. The "asthmatic" membrane depolarization is prevented by TRPC1 antibodies. In spite of having no clear explanation for these interesting but puzzling results, we would speculate that TRPC1-encoded NSCCs, which are dormant in normal cells, can be activated by an unknown mechanism in "asthmatic" cells. We have also revealed that the "asthmatic" increase in the activity of constitutively-active NSCCs and depolarization in membrane potential are both blocked by TRPC3 antibodies [9]. Thus, TRPC3-encoded native constitutively-active NSCCs are upregulated in both the molecular expression and functional activity, contributing to membrane depolarization and hyperresponsiveness in "asthmatic" airway SMCs, whereas TRPC1-encoded NSCCs are only functionally upregulated to mediate the increased cellular responses in "asthmatic" airway SMCs.

38.9 Perspective

We and other investigators have started to explore the molecular expression and potential functional roles of TRPC-encoded NSCCs in airway SMCs. A number of reports have shown that TRPC1-6 mRNAs and proteins are expressed in primary isolated and passaged airway SMCs [3–9]. Hitherto, there are only three studies to explore the functional importance of these channels in airway SMCs [5, 8, 9]. One of these three available functional studies has found that TNFα, an important asthma mediator, significantly increases TRPC3 mRNA and protein expression, and TRPC3 gene silencing inhibits TNFα-mediated augmentation of acetylcholine-evoked increase in $[Ca^{2+}]_i$ in passaged airway SMCs [5]. Another study indicates that TRPC6 gene silencing has no effect on OAG-evoked increase in $[Ca^{2+}]_i$ in primary isolated cells [8]. Finally, our recent patch clamp study, for the first time, provides evidence that TRPC3, but not TRPC1, is a major molecular component of native constitutively-active NSCCs in freshly isolated cells [9]. Further research using specific antibodies and gene silencing reveal that TRPC3-, but not TRPC1-encoded NSCCs show important functional roles in mediating the resting V_m and $[Ca^{2+}]_i$ as well as muscarinic increase in $[Ca^{2+}]_i$ in primary isolated cells. We have also found that TRPC3 protein expression and encoded NSCC activity are both significantly increased in freshly isolated "asthmatic" airway SMCs. Interestingly, TRPC1 protein expression is unaltered, but its encoded channel activity is largely augmented. The upregulated TRPC3 and TRPC1-encoded NSCCs mediate membrane depolarization in "asthmatic" airway SMCs.

The findings described above provide novel and interesting information indicating that TRPC-encoded NSCCs may possibly contribute to asthmatic airway hyperresponsiveness and inflammation. However, due to limited studies, many fundamental questions remain unanswered. For instance, it is unknown whether other TPRC molecules, in addition to the identified TRPC3 [9], are also involved in the formation of constitutively-active NSCCs in airway SMCs. TRPC3 can be assembled with TRPC1 or TRPC6 to form heterotetrameric channels in vascular SMCs and other cell types [1, 2]. TRPC1 antibodies and gene silencing both have no effect on the activity of constitutively-active NSCCs in airway SMCs [9], indicating the lack of the involvement of this molecule. Regardless of this, future research is needed to investigate the possibility of TRPC6 or other TRPC members to form functional channels in airway SMCs.

TRPC-encoded NSCCs play an important role in the control of extracellular Ca^{2+} influx, contributing to the resting $[Ca^{2+}]_i$ and agonist-induced increase in $[Ca^{2+}]_i$ in airway SMCs [9]. Studies in other types of cells have shown that TRPC-encoded channels in mediating Ca^{2+} influx is determined by three distinct routes: the channels themselves (due to a direct Ca^{2+} permeability), Na^+/Ca^{2+} exchange-1 and L-type Ca^{2+} channels [1, 2]. However, there is no report to investigate which of these Ca^{2+} influx routes and to what extent they may participate in individual TRPC-encoded NSCC-mediated Ca^{2+} influx in airway SMCs. These important questions should be addressed in the near future. As TRPC-encoded NSCC-mediated Ca^{2+} influx occurs via two different signaling processes: store-operated and receptor-operated Ca^{2+} entry, it is also worthwhile to conduct a series of experiments to identify which of these processes is responsible for the role of individual TRPC-encoded NSCCs.

It should be pointed out that extensive attention has been paid to studies on Ca^{2+} signaling derived from TRPC-encoded NSCCs in many types of cells [1, 2], but not in airway SMCs. The low enthusiasm for studies on TRPC-encoded NSCCs in airway SMCs is possibly due to the following two main reasons. First, scientists in the field may question the functional importance of TRPC-encoded Ca^{2+} influx. However, current available information, even though it is limited, suggests that TRPC channels are potentially important for physiological and pathological cellular responses. Second, one may consider that much work has been done in other types of cells. The fact is that each type of organs and tissues are different in the structural, functional and molecular nature. For instance, the airway and vasculature – "the two preparations are quite different"; as such, "we should not look for short-cuts through the field of cardiovascular research" [91]. Thus, continual studies to determine whether, which, and how TRPC-encoded channels are involved in cellular responses in normal and diseased (e.g., asthmatic) airway SMCs are very necessary, which will yield novel and important findings to greatly enhance our understanding of the functional roles, regulatory mechanisms and signaling processes for TRPC-encoded NSCCs and also aid in the creation of effective therapeutic targets for the treatment of asthma and other respiratory diseases.

Acknowledgments Our work presented in this article was supported by NIH R01HL071000, and AHA Scientist Development Grant 0630236 N and Established Investigator Award 0340160 N.

References

1. Nilius B, Owsianik G, Voets T, Peters JA (2007) Transient receptor potential cation channels in disease. Physiol Rev 87:165–217
2. Abramowitz J, Birnbaumer L (2009) Physiology and pathophysiology of canonical transient receptor potential channels. FASEB J 23:297–328

3. Ong HL, Brereton HM, Harland ML, Barritt GJ (2003) Evidence for the expression of transient receptor potential proteins in guinea pig airway smooth muscle cells. Respirology 8:23–32
4. Corteling RL, Li S, Giddings J, Westwick J, Poll C, Hall IP (2004) Expression of transient receptor potential, C6 and related transient receptor potential family members in human airway smooth muscle and lung tissue. Am J Respir Cell Mol Biol 30:145–154
5. White TA, Xue A, Chini EN, Thompson M, Sieck GC, Wylam ME (2006) Role of transient receptor potential C3 in, TNF-alpha-enhanced calcium influx in human airway myocytes. Am J Respir Cell Mol Biol 35:243–251
6. Ong HL, Chen J, Chataway T et al (2002) Specific detection of the endogenous transient receptor potential (TRP)-1 protein in liver and airway smooth muscle cells using immunoprecipitation and Western-blot analysis. Biochem J 364:641–648
7. Ay B, Prakash YS, Pabelick CM, Sieck GC (2004) Store-operated Ca^{2+} entry in porcine airway smooth muscle. Am J Physiol Lung Cell Mol Physiol 286:L909–L917
8. Godin N, Rousseau E (2007) TRPC6 silencing in primary airway smooth muscle cells inhibits protein expression without affecting OAG-induced calcium entry. Mol Cell Biochem 296:193–201
9. Xiao JH, Zheng YM, Liao B, Wang YX (2010) Functional role of canonical transient receptor potential 1 and canonical transient receptor potential 3 in normal and asthmatic airway smooth muscle cells. Am J Resoir Cell Mol Biol 43:17–25
10. Snetkov VA, Pandya H, Hirst SJ, Ward JP (1998) Potassium channels in human fetal airway smooth muscle cells. Pediatr Res 43:548–554
11. Snetkov VA, Ward JP (1999) Ion currents in smooth muscle cells from human small bronchioles: presence of an inward rectifier K^+ current and three types of large conductance K^+ channel. Exp Physiol 84:835–8460
12. Snetkov VA, Hapgood KJ, McVicker CG, Lee TH, Ward JP (2001) Mechanisms of leukotriene D4-induced constriction in human small bronchioles. Br J Pharmacol 133:243–252
13. Helli PB, Janssen LJ (2008) Properties of a store-operated nonselective cation channel in airway smooth muscle. Eur Respir J 32:1529–1538
14. Albert AP, Piper AS, Large WA (2003) Properties of a constitutively active Ca^{2+}-permeable non-selective cation channel in rabbit ear artery myocytes. J Physiol 549:143–156
15. Albert AP, Large WA (2001) Comparison of spontaneous and noradrenaline-evoked non-selective cation channels in rabbit portal vein myocytes. J Physiol 530:457–468
16. Liu XS, Xu YJ (2005) Potassium channels in airway smooth muscle and airway hyperreactivity in asthma. Chin Med J (Engl) 118:574–580
17. Hirota S, Helli P, Janssen LJ (2007) Ionic mechanisms and Ca^{2+} handling in airway smooth muscle. Eur Respir J 30:114–133
18. Bae YM, Park MK, Lee SH, Ho WK, Earm YE (1999) Contribution of, Ca^{2+}-activated, K^+ channels and non-selective cation channels to membrane potential of pulmonary arterial smooth muscle cells of the rabbit. J Physiol (Lond) 514:747–758
19. Terasawa K, Nakajima T, Iida H et al (2002) Nonselective cation currents regulate membrane potential of rabbit coronary arterial cell: modulation by lysophosphatidylcholine. Circulation 106:3111–3119
20. Albert AP, Pucovsky V, Prestwich SA, Large WA (2006) TRPC3 properties of a native constitutively active Ca^{2+}-permeable cation channel in rabbit ear artery myocytes. J Physiol 571:361–369
21. Janssen LJ, Sims SM (1992) Acetylcholine activates non-selective cation and chloride conductances in canine and guinea-pig tracheal myocytes. J Physiol 453:197–218
22. Wang YX, Fleischmann BK, Kotlikoff MI (1997) M2 receptor activation of nonselective cation channels in smooth muscle cells: calcium and Gi/Go requirements. Am J Physiol 273:C500–C508
23. Fleischmann BK, Wang YX, Kotlikoff MI (1997) Muscarinic activation and calcium permeation of nonselective cation currents in airway myocytes. Am J Physiol 272:C341–C349

24. Wang YX, Kotlikoff MI (2000) Signalling pathway for histamine activation of non-selective cation channels in equine tracheal myocytes. J Physiol (Lond) 523:131–138
25. Yamashita T, Kokubun S (1999) Nonselective cationic currents activated by acetylcholine in swine tracheal smooth muscle cells. Can J Physiol Pharmacol 77:796–805
26. Murray RK, Kotlikoff MI (1991) Receptor-activated calcium influx in human airway smooth muscle cells. J Physiol 435:123–144
27. Gorenne I, Labat C, Gascard JP, Norel X, Nashashibi N, Brink C (1998) Leukotriene D4 contractions in human airways are blocked by SK&F 96365, an inhibitor of receptor-mediated calcium entry. J Pharmacol Exp Ther 284:549–552
28. Parvez O, Voss AM, de KM, Roth-Kleiner M, Belik J (2006) Bronchial muscle peristaltic activity in the fetal rat. Pediatr Res 59:756–761
29. Dai JM, Kuo KH, Leo JM, van BC, Lee CH (2006) Mechanism of ACh-induced asynchronous calcium waves and tonic contraction in porcine tracheal muscle bundle. Am J Physiol Lung Cell Mol Physiol 290:L459–L469
30. Dai JM, Kuo KH, Leo JM, Pare PD, van BC, Lee CH (2007) Acetylcholine-induced asynchronous calcium waves in intact human bronchial muscle bundle. Am J Respir Cell Mol Biol 36:600–608
31. Hirota S, Janssen LJ (2007) Store-refilling involves both L-type calcium channels and reverse-mode sodium-calcium exchange in airway smooth muscle. Eur Respir J 30:269–278
32. Hirota S, Pertens E, Janssen LJ (2007) The reverse mode of the Na^+/Ca^{2+} exchanger provides a source of Ca^{2+} for store refilling following agonist-induced Ca^{2+} mobilization. Am J Physiol Lung Cell Mol Physiol 292:L438–L447
33. Jeon JP, Lee KP, Park EJ et al (2008) The specific activation of, TRPC4 by Gi protein subtype. Biochem Biophys Res Commun 377:538–543
34. Tsvilovskyy VV, Zholos AV, Aberle T et al (2009) Deletion of TRPC4 and TRPC6 in mice impairs smooth muscle contraction and intestinal motility in vivo. Gastroenterology 137:1415–1424
35. Blaustein MP, Lederer WJ (1999) Sodium/Calcium exchange: its physiological implications. Physiol Rev 79:763–854
36. Zheng YM, Wang YX (2007) Sodium-calcium exchanger in pulmonary artery smooth muscle cells. Ann NY Acad Sci 1099:427–435
37. Rosker C, Graziani A, Lukas M et al (2004) Ca^{2+} signaling by TRPC3 involves Na^+ entry and local coupling to the Na^+/Ca^{2+} exchanger. J Biol Chem 279:13696–13704
38. Algara-Suarez P, Romero-Mendez C, Chrones T et al (2007) Functional coupling between the Na^+/Ca^{2+} exchanger and nonselective cation channels during histamine stimulation in guinea pig tracheal smooth muscle. Am J Physiol Lung Cell Mol Physiol 293:L191–L198
39. Kotlikoff MI (1988) Calcium currents in isolated canine airway smooth muscle cells. Am J Physiol 254:C793–C801
40. Fleischmann BK, Murray RK, Kotlikoff MI (1994) Voltage window for sustained elevation of cytosolic calcium in smooth muscle cells. Proc Natl Acad Sci USA 91:11914–11918
41. Fleischmann BK, Wang YX, Pring M, Kotlikoff MI (1996) Voltage-dependent calcium currents and cytosolic calcium in equine airway myocytes. J Physiol (Lond) 492:347–358
42. Welling A, Felbel J, Peper K, Hofmann F (1992) Hormonal regulation of calcium current in freshly isolated airway smooth muscle cells. Am J Physiol 262:L351–L359
43. Hisada T, Kurachi Y, Sugimoto T (1990) Properties of membrane currents in isolated smooth muscle cells from guinea-pig trachea. Pflugers Arch 416:151–161
44. Marthan R, Martin C, Amedee T, Mironneau J (1989) Calcium channel currents in isolated smooth muscle cells from human bronchus. J Appl Physiol 66:1706–1714
45. Janssen LJ (1997) T-type and L-type Ca^{2+} currents in canine bronchial smooth muscle: characterization and physiological roles. Am J Physiol 272:C1757–C1765
46. Liu QH, Zheng YM, Korde AS et al (2009) Membrane depolarization causes a direct activation of G protein-coupled receptors leading to local Ca^{2+} release in smooth muscle. Proc Natl Acad Sci USA 106:11418–11423

47. Semenov I, Wang B, Herlihy JT, Brenner R (2006) BK channel beta1-subunit regulation of calcium handling and constriction in tracheal smooth muscle. Am J Physiol Lung Cell Mol Physiol 291:L802–L810
48. Tao L, Huang Y, Bourreau JP (2000) Control of the mode of excitation-contraction coupling by Ca^{2+} stores in bovine trachealis muscle. Am J Physiol Lung Cell Mol Physiol 279: L722–L732
49. Qian Y, Bourreau JP (1999) Two distinct pathways for refilling Ca^{2+} stores in permeabilized bovine trachealis muscle. Life Sci 64:2049–2059
50. Kannan MS, Davis C, Ladenius AR, Kannan L (1987) Agonist interactions at the calcium pools in skinned and unskinned canine tracheal smooth muscle. Can J Physiol Pharmacol 65:1780–1787
51. Farley JM, Miles PR (1978) The sources of calcium for acetylcholine-induced contractions of dog tracheal smooth muscle. J Pharmacol Exp Ther 207:340–346
52. Bourreau JP, Abela AP, Kwan CY, Daniel EE (1991) Acetylcholine Ca^{2+} stores refilling directly involves a dihydropyridine-sensitive channel in dog trachea. Am J Physiol 261:C497–C505
53. Bourreau JP, Kwan CY, Daniel EE (1993) Distinct pathways to refill ACh-sensitive internal Ca^{2+} stores in canine airway smooth muscle. Am J Physiol 265:C28–C35
54. Daniel EE, Jury J, Serio R, Jager LP (1991) Role of depolarization and calcium in contractions of canine trachealis from endogenous or exogenous acetylcholine. Can J Physiol Pharmacol 69:518–525
55. Perez JF, Sanderson MJ (2005) The frequency of calcium oscillations induced by 5-HT ACH, and KCl determine the contraction of smooth muscle cells of intrapulmonary bronchioles. J Gen Physiol 125:535–553
56. Barnes PJ (1985) Clinical studies with calcium antagonists in asthma. Br J Clin Pharmacol 20(Suppl 2):289S–298S
57. Fish JE (1984) Calcium channel antagonists in the treatment of asthma. J Asthma 21:407–418
58. Gordon EH, Wong SC, Klaustermeyer WB (1987) Comparison of nifedipine with a new calcium channel blocker, flordipine, in exercise-induced asthma. J Asthma 24:261–265
59. Hoppe M, Harman E, Hendeles L (1992) The effect of inhaled gallopamil, a potent calcium channel blocker, on the late-phase response in subjects with allergic asthma. J Allergy Clin Immunol 89:688–695
60. Middleton E Jr (1985) Calcium antagonists and asthma. J Allergy Clin Immunol 76:341–346
61. Middleton E Jr (1985) The treatment of asthma–beyond bronchodilators. N Engl Reg Allergy Proc 6:235–237
62. Riska H, Stenius-Aarniala B, Sovijarvi AR (1987) Comparison of the effects of an angiotensin converting enzyme inhibitor and a calcium channel blocker on blood pressure and respiratory function in patients with hypertension and asthma. J Cardiovasc Pharmacol 10(Suppl 10):S79–S81
63. Sly PD, Olinsky A, Landau LI (1986) Does nifedipine affect the diurnal variation of asthma in children? Pediatr Pulmonol 2:206–210
64. Jude JA, Wylam ME, Walseth TF, Kannan MS (2008) Calcium signaling in airway smooth muscle. Proc Am Thorac Soc 5:15–22
65. Sanderson MJ, Delmotte P, Bai Y, Perez-Zogbhi JF (2008) Regulation of airway smooth muscle cell contractility by Ca^{2+} signaling and sensitivity. Proc Am Thorac Soc 5:23–31
66. Sweeney M, McDaniel SS, Platoshyn O et al (2002) Role of capacitative Ca^{2+} entry in bronchial contraction and remodeling. J Appl Physiol 92:1594–1602
67. Ay B, Iyanoye A, Sieck GC, Prakash YS, Pabelick CM (2006) Cyclic nucleotide regulation of store-operated Ca^{2+} influx in airway smooth muscle. Am J Physiol Lung Cell Mol Physiol 290:L278–L283
68. Sieck GC, White TA, Thompson MA, Pabelick CM, Wylam ME, Prakash YS (2008) Regulation of store-operated Ca^{2+} entry by CD38 in human airway smooth muscle. Am J Physiol Lung Cell Mol Physiol 294:L378–L385

69. Ito S, Kume H, Yamaki K et al (2002) Regulation of capacitative and noncapacitative receptor-operated Ca^{2+} entry by rho-kinase in tracheal smooth muscle. Am J Respir Cell Mol Biol 26:491–498
70. Helli PB, Pertens E, Janssen LJ (2005) Cyclopiazonic acid activates a Ca^{2+}-permeable, nonselective cation conductance in porcine and bovine tracheal smooth muscle. J Appl Physiol 99:1759–1768
71. Peel SE, Liu B, Hall IP (2008) ORAI and store-operated calcium influx in human airway smooth muscle cells. Am J Respir Cell Mol Biol 38:744–749
72. Ali S, Metzger WJ, Mustafa SJ (1996) Simultaneous measurement of cyclopentyladenosine-induced contraction and intracellular calcium in bronchial rings from allergic rabbits and it's antagonism. J Pharmacol Exp Ther 278:639–644
73. Tao FC, Tolloczko B, Eidelman DH, Martin JG (1999) Enhanced Ca^{2+} mobilization in airway smooth muscle contributes to airway hyperresponsiveness in an inbred strain of rat. Am J Respir Crit Care Med 160:446–453
74. Tao FC, Tolloczko B, Mitchell CA, Powell WS, Martin JG (2000) Inositol (1,4,5)-trisphosphate metabolism and enhanced calcium mobilization in airway smooth muscle of hyperresponsive rats [In Process Citation]. Am J Respir Cell Mol Biol 23:514–520
75. Zacour ME, Tolloczko B, Martin JG (2000) Calcium and growth responses of hyperresponsive airway smooth muscle to different isoforms of platelet-derived growth factor (PDGF). Can J Physiol Pharmacol 78:867–873
76. Perpina M, Palau M, Cortijo J, Fornas E, Ortiz JL, Morcillo E (1989) Sources of calcium for the contraction induced by various agonists in trachealis muscle from normal and sensitized guinea pigs. Respiration 55:105–112
77. McCaig D, De Jonckheere S (1993) Effects of two Ca^{2+} modulators in normal and albumin-sensitized guinea- pig trachea. Eur J Pharmacol 249:53–63
78. Akasaka K, Konno K, Ono Y, Mue S, Abe C (1975) Electromyographic study of bronchial smooth muslce in bronchial asthma. Tohoku J Exp Med 117:55–59
79. Amrani Y, Panettieri RA Jr, Frossard N, Bronner C (1996) Activation of the TNF alpha-p55 receptor induces myocyte proliferation and modulates agonist-evoked calcium transients in cultured human tracheal smooth muscle cells. Am J Respir Cell Mol Biol 15:55–63
80. Amrani Y, Krymskaya V, Maki C, Panettieri RA Jr (1997) Mechanisms underlying TNF-alpha effects on agonist-mediated calcium homeostasis in human airway smooth muscle cells. Am J Physiol 273:L1020–L1028
81. Amrani Y, Panettieri RA Jr (1998) Cytokines induce airway smooth muscle cell hyperresponsiveness to contractile agonists. Thorax 53:713–716
82. Deshpande DA, Dogan S, Walseth TF et al (2004) Modulation of calcium signaling by interleukin-13 in human airway smooth muscle: role of CD38/cyclic adenosine diphosphate ribose pathway. Am J Respir Cell Mol Biol 31:36–42
83. Kang BN, Deshpande DA, Tirumurugaan KG, Panettieri RA, Walseth TF, Kannan MS (2005) Adenoviral mediated anti-sense CD38 attenuates TNF-alpha-induced changes in calcium homeostasis of human airway smooth muscle cells. Can J Physiol Pharmacol 83:799–804
84. Pellegrino R, Violante B, Crimi E, Brusasco V (1991) Time course and calcium dependence of sustained bronchoconstriction induced by deep inhalation in asthma. Am Rev Respir Dis 144:1262–1266
85. Ahmed T, Danta I (1992) Modification of histamine- and methacholine-induced bronchoconstriction by calcium antagonist gallopamil in asthmatics. Respiration 59:332–338
86. Quast U (1992) Potassium channel openers: pharmacological and clinical aspects. Fundam Clin Pharmacol 6:279–293
87. Hakonarson H, Grunstein MM (1998) Regulation of second messengers associated with airway smooth muscle contraction and relaxation. Am J Respir Crit Care Med 158:S115–S122
88. Hirst SJ (2000) Airway smooth muscle as a target in asthma. Clin Exp Allergy 30(Suppl 1):54–59

89. An SS, Bai TR, Bates JH et al (2007) Airway smooth muscle dynamics: a common pathway of airway obstruction in asthma. Eur Respir J 29:834–860
90. Berend N, Salome CM, King GG (2008) Mechanisms of airway hyperresponsiveness in asthma. Respirology 13:624–631
91. Janssen LJ (2009) Asthma therapy: how far have we come, why did we fail and where should we go next? Eur Respir J 33:11–20
92. Ramsey IS, Delling M, Clapham DE (2006) An introduction to TRP channels. Annu Rev Physiol 68:619–647

Chapter 39
TRP Channels in Skeletal Muscle: Gene Expression, Function and Implications for Disease

Heinrich Brinkmeier

Abstract Besides the well known voltage-gated Ca^{2+} channels skeletal muscle fibres contain several non-voltage gated Ca^{2+} conducting cation channels. They have been physiologically characterized as stretch activated, store operated and Ca^{2+} leak channels. TRP channels are good candidates to account for these sarcolemmal channels and Ca^{2+} influx pathways or at least contribute to the responsible macromolecular complexes. Several members of the TRPC, TRPV and TRPM subfamilies of TRP channels are expressed in skeletal muscle as shown by RT-PCR, Western blot and immunohistochemistry. The most prominent and consistently found are TRPC1, C3, C4 and C6, TRPV2 and V4 as well as TRPM4 and M7. However, the precise function of individual channels is largely unknown. Linking physiologically characterized channels of the muscle fibre membrane to TRP channel proteins has been a major challenge during the last years. It has been successful only in a few cases and is complicated by the fact that some channels have dual functions in cultured, immature muscle cells and adult fibres. The best characterized TRP channel in skeletal muscle is TRPC1, a small-conductance channel of the sarcolemma. It is needed for Ca^{2+} homeostasis during sustained contractile muscle activity. In addition to certain physiological functions TRP channels seem to be involved in the pathomechanisms of muscle disorders. There is a broad body of evidence that dysregulation of Ca^{2+} conducting channels plays a key role in the pathomechanism of Duchenne muscular dystrophy. Lack of the cytoskeletal protein dystrophin or δ-sarcoglycan, seems to disturb the function of one or several Ca^{2+} channels of the muscle fibre membrane, leading to pathological dystrophic changes. Almost 10 different TRP channels have been detected in skeletal muscle. They seem to be involved in muscle development, Ca^{2+} homeostasis, Ca^{2+} signalling and in disease progression of certain muscle disorders. However, we are still at the beginning of understanding the impact of TRP channel functions in skeletal muscle.

H. Brinkmeier (✉)
Institute of Pathophysiology, University of Greifswald, D-17495 Karlsburg, Germany
e-mail: heinrich.brinkmeier@uni-greifswald.de

39.1 Calcium Entry Pathways in the Sarcolemma of Skeletal Muscle Fibres

Besides the well known voltage-gated Ca^{2+} channels a variety of voltage-independent Ca^{2+} conducting cation channels has been detected in skeletal muscle fibres. These cation channels have been classified as stretch-activated channels (SACs) [1], store operated channels (SOCs) [2–5] and Ca^{2+} leak channels [6]. The latter ones have a certain open probability at rest that is not modulated by mechanical stress applied to the muscle fibre membrane. In addition, some authors show the presence of stretch-inactivated channels in skeletal muscle [7]. SACs and stretch-inactivated channels have been characterized by single channel recordings and the simultaneous application of suction to membrane patches [1, 7, 8], while SOCs have been preferentially studied by Ca^{2+} imaging, i.e. the use fluorescent Ca^{2+} indicators [4, 5, 9]. An alternative method to characterize divalent cation influx is the so called Mn^{2+} quench technique. Mn^{2+}, when applied extracellularly, may flow through ion channels or membrane lesions, but it is not extruded from the cytoplasm by Ca^{2+} pumps and exchangers. Intracellularly accumulating Mn^{2+} binds to Fura-2 and quenches its fluorescence. Muscle fibres loaded with Fura-2 show a fluorescence decline by 1–5% per min (manganese quench) depending on the applied Mn^{2+} concentration [2, 10, 11]. Since Mn^{2+} influx can be inhibited by ion channel blockers, such as Gd^{3+}, amiloride [11] and the 3,5-bis(trifluoromethyl)pyrazole derivative, BTP2 [2], it has to be assumed that the resting permeability of muscle fibres to divalent cations is caused by cation channel activity. Under physiological conditions Ca^{2+} is the responsible ion to account for divalent cation ion permeability of the sarcolemma. A recent pharmacological study suggested that store operated Ca^{2+} entry, stretch activated and background Ca^{2+} entry might be caused by the same population of ion channels or channel complexes sharing common constituents [8].

39.2 Linking Physiologically Detected Cation Channels to TRP Channel Proteins

TRP channels comprise a superfamily of 27 cation channels in humans. The channels show a great diversity in activation and ion selectivity. The superfamily can be divided into 6 subfamilies called TRPC (6), TRPV (6), TRPM (8), TRPA (1), TRPP (3) and TRPML (3) [12]. In skeletal muscle, only members of the TRPC, TRPV and TRPM subfamilies have been detected and characterized. TRP channels are candidates to account for the above mentioned sarcolemmal currents and Ca^{2+} influx mechanisms or at least contribute to the postulated ion channel complexes. However, to prove the contribution of a distinct TRP channel, several lines of evidence have to be in agreement to obtain a conclusive view. First, the respective TRP channel has to be expressed in skeletal muscle. Second, it has to be verified that the channel protein is localized in the muscle fibre membrane, at least under certain physiological conditions. Third, specific blockers, gene disruption or RNA targeting of TRP channels should inhibit or modify ion currents associated with SOCs, SACs or leak channels.

39.3 Gene Expression of TRP Channels in Mammalian Skeletal Muscle

The majority of data on TRP channel expression in skeletal muscle has been obtained with mouse and rat muscle tissue and with muscle cultures of mouse, rat and human origin. The first report dealing with the presence and role of TRP channels in skeletal muscle is from Vandebrouck and coworkers [13]. The authors showed the presence of TRPC channel transcripts 1, 2, 3, 4, and 6 by RT-PCR and confirmed protein expression in the sarcolemma of TRPC1, C4 and C6. TRPC3 showed an intracellular localization while TRPC2 was hardly detectable on the protein level. It should be added that the localization studies on TRP channels were performed with enzymatically isolated muscle fibres of the flexor digitorum brevis (FDB). The fibres had been allowed to adhere to culture dishes for > 2 h before experimentation. Thus, the localization of the channel proteins within the cells might have been influenced by the procedure. Further studies widely confirmed the presence of all TRPC channel proteins in mouse skeletal muscle, except TRPC2 and C7 [14–17]. However, comparisons of TRPC channel expression between different skeletal muscles and mouse strains revealed variations by a factor of 5–10 [15]. Therefore, one should be careful to generalize data obtained with single muscles from certain mouse strains, such as the frequently used flexor digitorum brevis of the C57BL/10 mouse strain. Nevertheless several studies confirmed the presence of at least TRPC1, C3, C4 and C6 in mouse skeletal muscle [13–15, 17] by PCR, Western blot and immunohistochemical staining. TRPC1, C4 and C6 showed a sarcolemmal localization while TRPC3 was preferentially found in the intracellular compartment, both, in isolated fibres [13] and muscle cross sections [14].

In muscle cultures TRPC1 channel expression has been shown at all stages, i.e. in mononucleated myoblasts and in differentiated, multinucleated myotube cultures [18, 19]. Interestingly, TRPC1 was found to be transiently up-regulated during early myoblast differentiation and fusion. Among the TRPC channels only TRPC3 has been intensively studied so far in cultured muscle. TRPC3 transcript and protein have been detected in chick [20] and mouse muscle cultures showing a sharp upregulation during the early phase of myotube differentiation [16]. In the latter study also TRPC1, C4 and C6 were detected by Western blot. Knockdown of TRPC3 in these cultures caused an up-regulation of TRPC1 while C4 and C6 protein levels remained unchanged.

The data obtained so far suggest that TRPC1, C3, C4 and C6 are expressed in skeletal muscle, since they have been detected by PCR techniques, Western blot and/or immunohistochemistry. However, several of the above mentioned data and studies have to be regarded with caution, since at least some of the commercially available antibodies are obviously not sufficiently characterized. Recently, the presence of TRPC4 in skeletal muscle was questioned because a frequently used anti-TRPC4 antibody (or at least the used batch) recognized a band of the correct molecular weight, both in wildtype and TRPC4 deficient animals. A well characterized antibody designed by the authors [21] proved TRPC4 expression in brain and showed absence or very low expression in mouse skeletal muscle.

Only few data do exist about gene expression of members of TRPV and TRPM subfamilies in skeletal muscle. The best characterized is the TRPV2 protein. It has been originally described as a insulin like growth factor (IGF-1) regulated Ca^{2+} conducting channel in mouse, hamster and human muscle [22]. IGF-1 application caused translocation of TRPV2 to the plasma membrane in cultured mouse and hamster myotubes. The presence of TRPV2 transcripts was independently confirmed, and Western blot and immunohistochemistry revealed in addition the presence of the TRPV4 protein mouse skeletal muscle [14].

TRPM channels have been little studied in skeletal muscle. Transcripts coding for TRPM4 and TRPM7 have been consistently found in mouse tibial muscles and diaphragm [14] and also in human skeletal muscle tissue [23]. Though quantitative analyses of target genes that are merely based on RT-PCR give little insights in protein expression and function, it looks that TPM4 and M7 are the dominating members of the TRPM subfamily in skeletal muscle. Their transcript levels were 10–100 fold higher than those of TRPM2, M3 and M6. TRPM1, M5, and M8 were hardly detectable. In addition, by immunohistochemical staining, the TRPM7 protein has been localized in the sarcolemma of mouse muscle. Western blot analysis supported the finding of TRPM7 protein expression in mouse muscle [14].

39.4 Localization and Function of TRP Channels

39.4.1 The TRPC Subfamily

The most frequent and intensely studied TRP channel in skeletal muscle is TRPC1. Vandebrouck and coworkers [13] presented already in 2002 good evidence that the TRPC1 protein is localized in the sarcolemma of mouse muscle fibres. They further showed that antisense-oligonucleotides directed against TRPC1 reduced the corresponding band detectable by Western blot. Single channel recordings revealed the reduction of an 8 pS mechanosensitive Ca^{2+} conducting channel, in agreement with the view that TRPC1 contributes to or forms mechanosensitive channels in the sarcolemma [24]. In a recent comprehensive study the same group continued their investigations on TRPC1 and included data about TRPC1 deficient mice [25]. The major outcome of this study was that TRPC1 is needed for Ca^{2+} homeostasis during sustained contractile muscle activity. Muscle fatigue induced by repetitive trains of high frequency electrical stimulation was remarkably faster in TRPC1 deficient mice. Further, a small conductance-channel (13 pS) was absent in TRPC1 knockout mice. Interestingly, the channel was not mechanosensitive. Lack of TRPC1 neither had much impact on resting Ca^{2+} influx nor was the store operated Ca^{2+} influx impaired. Thus, gating of the channel remained unclear.

An unexpected alternative function of TRPC1 function in skeletal muscle was suggested by Berbey and coworkers [26]. The group studied TRPC1-YFP overexpressing muscle fibres and concluded that TRPC1 functions as Ca^{2+} leak channel in the sarcoplasmic reticulum. However, these results have to be re-investigated, since again a problem with a commercially available antibody was revealed later [25]. Other results support the view of TRPC1 as a sarcolemmal channel. Co-immunoprecipitation and co-localization experiments suggested an association of TRPC1, TRPC4 and α-syntrophin in cultured myotubes [9, 17]. The data indicate that TRPC1 is associated via syntrophin to the complex of dystrophin-associated glycoproteins (DAG) and that TRPC1 forms a Ca^{2+} channel or at least contributes to a Ca^{2+} influx channel.

Besides connecting to the DAG complex, TRPC1 co-localizes with the scaffolding protein Homer 1 in skeletal muscle. Mice lacking Homer 1 exhibit a myopathy characterized by decreased muscle fiber cross-sectional area and decreased skeletal muscle force generation. The myopathy was explained by increased Ca^{2+} influx through overactivity of dysregulated TRPC1 channels during muscle development [27]. The authors concluded that Homer 1 anchors TRPC1 in the costamer and thereby normalises TRPC1 channel function and Ca^{2+} influx.

Another channel of the TRPC family, TRPC3, is upregulated during muscle differentiation and functionally coupled to the type 1 ryanodine receptor (RyR1) in cultured mouse myotubes [16]. Knockdown of TRPC3 by small interference RNAs did not impair muscle differentiation and neither SOCE nor excitation-contraction coupled Ca^{2+} entry were impaired. TRPC3 was shown to interact with the triadic proteins TRPC1, junctophilin-2, homer, mitsugumin-29, calreticulin and calmodulin, indicating that all these proteins may form a macromolecular complex together with RyR1. In a continuative study the authors suggested that TRPC1 may be the protein linking TRPC3 to RyR1 [28]. A further function of TRPC3 was discovered by Rosenberg et al. who showed coupling of TRPC3 to the calcineurin-NFAT (nuclear factor of activated T cells) signal transduction pathway. This pathway couples Ca^{2+} influx from extracellular to gene regulation in cultured muscle [29]. The authors obtained evidence that the latter pathway is also functional in adult slow twitch muscle connecting neuromuscular activity and Ca^{2+} influx to NFAT-dependent gene regulation [29].

However, in adult muscle localization and function of TRPC3 seems to be different from that in myotubes. Both, in resting isolated fibres [13] and in muscle cross sections [14] TRPC3 appeared to be intracellularly localized. In a recent study it was shown that TRPC3 is co-localized with the glucose transporter GLUT4 in the proximity of the transverse tubule system in adult mouse muscle. Upon stimulation with insulin, a TRPC3 related Ca^{2+} influx was shown to modulate insulin-mediated glucose uptake [30]. In conclusion, localization and function of TRPC3 seem to change during muscle development. In myotubes characterized by developing T tubules TRPC3 could serve as a RyR1 binding protein in the early tubules, a function that is later taken over by the skeletal muscle calcium channel $Ca_v1.2$. In adult muscle TRPC3 seems to fulfil specific functions in Ca^{2+} dependent signal transduction and regulation of insulin dependent glucose transport [30].

To complete the TRPC family and the possible function of its members in skeletal muscle it should be noted that TRPC6 has been detected in mouse muscle on mRNA and protein level [13, 14]. The sarcolemmal location of TRPC6 was confirmed by immunohistochemistry. Since functional studies have not been performed the physiological role of TRPC6 is unclear. In heart muscle TRPC6 is coupled to the calcineurin-NFAT signalling pathway and contributes to cardiac hypertrophy and remodelling in response to mechanical stress [31].

39.4.2 The TRPV Subfamily

The TRPV subfamily is represented by at least two members in skeletal muscle, TRPV2 and V4. TRPV2 was originally described as a growth factor-regulated channel occurring at increased levels in dystrophic muscle cells [22]. Mechanical stretch and application of IGF-1 lead to translocation of the channel to the surface membrane of cultured muscle cells. Cardiac-specific overexpression of TRPV2 caused a cardiomyopathy due to Ca^{2+} overload. Interestingly, the authors found expression of TRPV2 in all tested preparations, muscle cultures from normal and dystrophin deficient mice, normal and δ-sarcoglycan deficient hamsters and in muscle tissue of mouse, hamster and human origin. Though the physiological role of TRPV2 in normal muscle has still not been sufficiently clarified, increased activity and/or overexpression of the channel seems to cause muscle fibre and myocyte degeneration. This view has recently been supported by an opposite approach. Iwata and colleagues showed that muscle-specific knockdown of functional TRPV2 and consequential inhibition of TRPV2-related Ca^{2+} influx ameliorates muscular dystrophy in mdx mice [32]. Furthermore, mdx muscle is in contrast to normal muscle very sensitive to eccentric contractions. Isolated mdx muscles respond with damage and irreversible loss of force when exposed to appropriate eccentric stimulation protocols. This increased sensitivity was considerably reduced by the muscle-specific knockdown of functional TRPV2 channels [33]. The observations point to a role of TRPV2 for the pathophysiology of dystrophin deficient muscle.

The physiological role of TRPV2 in skeletal muscle may be related to signal cascades. TRPV2 activity was shown to cause phosphorylation of Ca^{2+}/calmodulin dependent protein kinase II. Thus, TRPV2 activity in normal muscle may be connected to Ca^{2+} dependent remodeling processes during muscle injury and/or exercise [32].

Little is known about the function of TRPV4 in skeletal muscle. At least, it was found to be expressed at about the same level as other above mentioned TRP channels. The TRPV4 protein was further detected in the sarcolemma of mouse muscle fibres by immunofluorescence staining [14]. Recently, preliminary results indicated that TRPV4 is functional in isolated mouse muscle fibres, since application of the TRPV4 activator 4α-PDD caused an increase of resting Ca^{2+} influx by 60% [34]. Physiological role and endogenous activator or activation process remain to be clarified for TRPV4 in skeletal muscle.

39.4.3 The TRPM Subfamily

The functions of TRPM channels have not yet been studied in skeletal muscle. TRPM4, a Ca^{2+} activated cation channel that is permeable to Na^+ and K^+ [35], may modulate cellular excitability. TRPM7 is permeable to Mg^{2+}, Ca^{2+} and trace metals ions, such as Zn^{2+} and Ni^{2+} [36] and is activated by low intracellular Mg^{2+}. Since TRPM7 is important for cellular Mg^{2+} homeostasis during cell growth and proliferation [37], one could assume that it is also significant for myoblast proliferation and early muscle development. In contrast to other TRP channels, TRPM7 is a bifunctional protein that contains both, an ion channel and a protein kinase domain [37, 38]. Alternatively to its function for Mg^{2+} homeostasis, the kinase domain of TRPM7 may be important for muscle development and/or function. There is so far no evidence that TRPM7 or TRPM4 contribute to store operated Ca^{2+} entry, stretch activated or background Ca^{2+} entry in muscle fibres.

39.5 Implications for Disease

A driving force behind the recent research in the field of TRP channels in skeletal muscle is the idea that TRP channels could be involved in the pathomechanism of muscle diseases. Duchenne muscular dystrophy, the most severe and frequent muscular dystrophy [39] is in the focus. There is a broad body of evidence that non-voltage gated Ca^{2+} channels contribute to an imbalance of sub-sarcolemmal Ca^{2+} and to abnormal Ca^{2+} signalling in dystrophin deficient muscle [40–43]. It is assumed that the abnormalities finally lead to necrosis of muscle fibres, muscle damage, fibrosis and progressive muscle weakness. In addition, a mechanical role of dystrophin seems to be obvious since it connects the membrane cytoskeleton of muscle fibers via a complex of dystrophin associated glycoproteins to the extracellular matrix [44]. Nevertheless, many data indicate that dysregulation of Ca^{2+} influx and Ca^{2+} handling significantly contribute to the pathomechanism of Duchenne muscular dystrophy. Recent data on TRP channel location and function support this hypothesis.

The association of TRPC1 and TRPC4 to α-syntrophin [9, 17] in normal muscle highlighted a link between TRP channels and dystrophin. Lack of dystrophin and α-syntrophin could cause alteration in TRPC1 and/or TRPC4 localization and function. It has not become finally clear whether decoupled TRPC1 and/or TRPC4 channels generate abnormal Ca^{2+} influx, but the studies proved the basic principle of a connection of TRP channels and the membrane cytoskeleton in muscle. An intact membrane cytoskeleton seems to be a prerequisite for normal functioning of non-voltage gated Ca^{2+} channels of the sarcolemma and intact muscular Ca^{2+} signalling. A further example for the interaction of a sarcolemmal channel with a cytoskeletal scaffolding protein is the connection between TRPC1 and Homer [27]. Homer 1 localizes TRPC1 in the costamer and lack of Homer 1 causes a Ca^{2+} dependent myopathy due to TRPC1 overactivity. Obviously in Homer 1 deficient mouse muscle TRPC1 is not degraded, but functional and dysregulated.

The second TRP channel that seems to be involved in the pathomechanism of muscular dystrophies is TRPV2. Mechanical stress and IGF-1 cause an increased occurrence of TRPV2 in the sarcolemma of dystrophin-deficient muscle [22]. Cardiac-specific overexpression of TRPV2 caused a cardiomyopathy due to Ca^{2+} overload. These data clearly indicate a pathogenic role of overexpressed or overactive TRPV2 channels. In contrast, the muscle-specific knockdown of functional TRPV2 ameliorates muscular dystrophy in mdx mice [32] and reduces the sensitivity of dystrophin-deficient muscle to mechanical stress, i.e. eccentric contractions [33]. The observations clearly point out the role of the TRPV2 channel for the pathophysiology of dystrophin-deficient muscle. Interestingly TRPV2 dysregulation was also observed in the δ-sarcoglycan deficient hamster [22, 32]. The latter observation rises expectations that the presumed Ca^{2+} related pathomechanism is not restricted to dystrophinopathies. Future research will have to elucidate whether muscular dystrophies characterized by loss of function of dystrophin associated glycoproteins are likewise related to abnormal TRP channel function.

References

1. Yeung EW, Whitehead NP, Suchyna TM, Gottlieb PA, Sachs F, Allen DG (2005) Effects of stretch-activated channel blockers on $[Ca^{2+}]_i$ and muscle damage in the mdx mouse. J Physiol 562:367–380
2. Boittin FX, Petermann O, Hirn C, Mittaud P, Dorchies OM, Roulet E, Rüegg UT (2006) Ca^{2+}-independent phospholipase A2 enhances store-operated Ca^{2+} entry in dystrophic skeletal muscle fibers. J Cell Sci 119:3733–3742
3. Dirksen RT (2009) Checking your SOCCs and feet: the molecular mechanisms of Ca^{2+} entry in skeletal muscle. J Physiol 587:3139–3147
4. Edwards JN, Murphy RM, Cully TR, von Wegner F, Friedrich O, Launikonis BS (2010) Ultra-rapid activation and deactivation of store-operated Ca^{2+} entry in skeletal muscle. Cell Calcium 47:458–467
5. Launikonis BS, Rios E (2007) Store-operated Ca^{2+} entry during intracellular Ca2+ release in mammalian skeletal muscle. J Physiol 583:81–97
6. Alderton JM, Steinhardt RA (2000) How calcium influx through calcium leak channels is responsible for the elevated levels of calcium-dependent proteolysis in dystrophic myotubes. Trends Cardiovasc Med 10:268–272
7. Franco-Obregon A, Lansman JB (2002) Changes in mechanosensitive channel gating following mechanical stimulation in skeletal muscle myotubes from the mdx mouse. J Physiol 539:391–407
8. Ducret T, Vandebrouck C, Cao ML, Lebacq J, Gailly P (2006) Functional role of store-operated and stretch-activated channels in murine adult skeletal muscle fibres. J Physiol 575:913–924

9. Vandebrouck A, Sabourin J, Rivet J, Balghi H, Sebille S, Kitzis A, Raymond G, Cognard C, Bourmeyster N, Constantin B (2007) Regulation of capacitative calcium entries by alpha1-syntrophin: association of TRPC1 with dystrophin complex and the PDZ domain of alpha1-syntrophin. FASEB J 21:608–617
10. De Backer F, Vandebrouck C, Gailly P, Gillis JM (2002) Long-term study of Ca^{2+} homeostasis and of survival in collagenase-isolated muscle fibres from normal and mdx mice. J Physiol 542:855–865
11. Tutdibi O, Brinkmeier H, Rüdel R, Föhr KJ (1999) Increased calcium entry into dystrophin-deficient muscle fibres of MDX and ADR-MDX mice is reduced by ion channel blockers. J Physiol 515:859–868
12. Venkatachalam K, Montell C (2007) TRP channels. Annu Rev Biochem 76:387–417
13. Vandebrouck C, Martin D, Colson-Van Schoor M, Debaix H, Gailly P (2002) Involvement of TRPC in the abnormal calcium influx observed in dystrophic (mdx) mouse skeletal muscle fibers. J Cell Biol 158:1089–1096
14. Krüger J, Kunert-Keil C, Bisping F, Brinkmeier H (2008) Transient receptor potential cation channels in normal and dystrophic mdx muscle. Neuromuscul Disord 18:501–513
15. Kunert-Keil C, Bisping F, Krüger J, Brinkmeier H (2006) Tissue-specific expression of TRP channel genes in the mouse and its variation in three different mouse strains. BMC Genomics 7:159
16. Lee EH, Cherednichenko G, Pessah IN, Allen PD (2006) Functional coupling between TRPC3 and RyR1 regulates the expressions of key triadic proteins. J Biol Chem 281:10042–10048
17. Sabourin J, Lamiche C, Vandebrouck A, Magaud C, Rivet J, Cognard C, Bourmeyster N, Constantin B (2009) Regulation of TRPC1 and TRPC4 cation channels requires an alpha1-syntrophin-dependent complex in skeletal mouse myotubes. J Biol Chem 284:36248–36261
18. Formigli L, Sassoli C, Squecco R, Bini F, Martinesi M, Chellini F, Luciani G, Sbrana F, Zecchi-Orlandini S, Francini F, Meacci E (2009) Regulation of transient receptor potential canonical channel 1 (TRPC1) by sphingosine 1-phosphate in C2C12 myoblasts and its relevance for a role of mechanotransduction in skeletal muscle differentiation. J Cell Sci 122:1322–1333
19. Louis M, Zanou N, Van Schoor M, Gailly P (2008) TRPC1 regulates skeletal myoblast migration and differentiation. J Cell Sci 121:3951–3959
20. Santillan G, Katz S, Vazquez G, Boland RL (2004) TRPC3-like protein and vitamin D receptor mediate 1alpha,25(OH)2D3-induced SOC influx in muscle cells. Int J Biochem Cell Biol 36:1910–1918
21. Flockerzi V, Jung C, Aberle T, Meissner M, Freichel M, Philipp SE, Nastainczyk W, Maurer P, Zimmermann R (2005) Specific detection and semi-quantitative analysis of TRPC4 protein expression by antibodies. Pflügers Arch 451:81–86
22. Iwata Y, Katanosaka Y, Arai Y, Komamura K, Miyatake K, Shigekawa M (2003) A novel mechanism of myocyte degeneration involving the Ca^{2+} permeable growth factor regulated channel. J Cell Biol 161:957–967
23. Fonfria E, Murdock PR, Cusdin FS, Benham CD, Kelsell RE, McNulty S (2006) Tissue distribution profiles of the human TRPM cation channel family. J Recept Signal Transduct Res 26:159–178
24. Franco-Obregon A Jr., Lansman JB (1994) Mechanosensitive ion channels in skeletal muscle from normal and dystrophic mice. J Physiol 481:299–309
25. Zanou N, Shapovalov G, Louis M, Tajeddine N, Gallo C, Van Schoor M, Anguish I, Cao ML, Schakman O, Dietrich A, Lebacq J, Ruegg U, Roulet E, Birnbaumer L, Gailly P (2010) Role of TRPC1 channel in skeletal muscle function. Am J Physiol Cell Physiol 298:C149–C162
26. Berbey C, Weiss N, Legrand C, Allard B (2009) Transient receptor potential canonical type 1 (TRPC1) operates as a sarcoplasmic reticulum calcium leak channel in skeletal muscle. J Biol Chem 284:36387–36394

27. Stiber JA, Zhang ZS, Burch J, Eu JP, Zhang S, Truskey GA, Seth M, Yamaguchi N, Meissner G, Shah R, Worley PF, Williams RS, Rosenberg PB (2008) Mice lacking homer 1 exhibit a skeletal myopathy characterized by abnormal TRP channel activity. Mol Cell Biol 28(8):2637–2647
28. Turner PR, Westwood T, Regen CM, Steinhardt RA (1988) Increased protein degradation results from elevated free calcium levels found in muscle from mdx mice. Nature 335:735–738
29. Rosenberg P, Hawkins A, Stiber J, Shelton JM, Hutcheson K, Bassel-Duby R, Shin DM, Yan Z, Williams RS (2004) TRPC3 channels confer cellular memory of recent neuromuscular activity. Proc Natl Acad Sci USA 101:9387–9392
30. Lanner JT, Bruton JD, Assefaw-Redda Y, Andronache Z, Zhang SJ, Severa D, Zhang ZB, Melzer W, Zhang SL, Katz A, Westerblad H (2009) Knockdown of TRPC3 with siRNA coupled to carbon nanotubes results in decreased insulin-mediated glucose uptake in adult skeletal muscle cells. FASEB J 23:1728–1738
31. Kuwahara K, Wang Y, McAnally J, Richardson JA, Bassel-Duby R, Hill JA, Olson EN (2006) TRPC6 fulfills a calcineurin signaling circuit during pathologic cardiac remodeling. J Clin Invest 116:3114–3126
32. Iwata Y, Katanosaka Y, Arai Y, Shigekawa M, Wakabayashi S (2009) Dominant-negative inhibition of Ca^{2+} influx via TRPV2 ameliorates muscular dystrophy in animal models. Hum Mol Genet 18:824–834
33. Zanou N, Iwata Y, Schakman O, Lebacq J, Wakabayashi S, Gailly P (2009) Essential role of TRPV2 ion channel in the sensitivity of dystrophic muscle to eccentric contractions. FEBS Lett 583:3600–3604
34. Lange T, Pritschow B, Kasch J, Kunert-Keil C, Brinkmeier H (2009) Functional TRPV4 channels are expressed in mouse skeletal muscle an can modulate background Ca^{2+} entry and muscle fatigue. Acta Physiol Scand 195:23
35. Vennekens R, Nilius B (2007) Insights into TRPM4 function, regulation and physiological role. Handb Exp Pharmacol 179:269–285
36. Monteilh-Zoller MK, Hermosura MC, Nadler MJ, Scharenberg AM, Penner R, Fleig A (2003) TRPM7 provides an ion channel mechanism for cellular entry of trace metal ions. J Gen Physiol 121:49–60
37. Penner R, Fleig A (2007) The Mg^{2+} and Mg^{2+}-nucleotide-regulated channel-kinase TRPM7. Handb Exp Pharmacol 179:313–328
38. Wolf FI (2004) TRPM7: channeling the future of cellular magnesium homeostasis? Sci STKE 2004:pe23
39. Emery AEH, Emery MHL (1995) Refining the clinical picture. In: Emery AEH, Emery MHL (ed) The history of a genetic disease: Duchenne muscular dystrophy or Meryonss´ disease. The Royal Society of Medicine Press Limited, London, 89–99
40. Constantin B, Sebille S, Cognard C (2006) New insights in the regulation of calcium transfers by muscle dystrophin-based cytoskeleton: implications in DMD. J Muscle Res Cell Motil 27:375–386
41. Hopf FW, Turner PR, Steinhardt RA (2007) Calcium misregulation and the pathogenesis of muscular dystrophy. Subcell Biochem 45:429–464
42. Allen DG, Gervasio OL, Yeung EW, Whitehead NP (2010) Calcium and the damage pathways in muscular dystrophy. Can J Physiol Pharmacol 88:83–91
43. Berchtold MW, Brinkmeier H, Müntener M (2000) Calcium ion in skeletal muscle: its crucial role for muscle function, plasticity, and disease. Physiol Rev 80:1215–1265
44. Michele DE, Campbell KP (2003) Dystrophin-glycoprotein complex: post-translational processing and dystroglycan function. J Biol Chem 278:15457–15460

Chapter 40
TRP Channels in Vascular Endothelial Cells

Ching-On Wong and Xiaoqiang Yao

Abstract Endothelial cells regulate multiple vascular functions, such as vascular tone, permeability, remodeling, and angiogenesis. It is known for long that cytosolic Ca^{2+} level ($[Ca^{2+}]_i$) and membrane potential of endothelial cells are crucial factors to initiate the signal transduction cascades, leading to diverse vascular functions. Among the various kinds of endothelial ion channels that regulate ion homeostasis, transient receptor potential (TRP) channels emerge as the prime mediators for a diverse range of vascular signaling. The characteristics of TRP channels, including subunit heteromultimerization, diverse ion selectivity, and multiple modes of activation, permit their versatile functional roles in vasculatures. Substantial amount of evidence demonstrates that many TRP channels in endothelial cells participate in physiological and pathophysiological processes of vascular system. In this article, we summarize the recent findings of TRP research in endothelial cells, aiming at providing up-to-date information to the researchers in this rapidly growing field.

40.1 Introduction

Endothelial cells form a continuous lining on the blood vessel lumen. The structure not only serves as a barrier between circulation and vascular tissue, but also is a platform for signal transductions that mediate multiple vascular processes. The endothelial cells "sense" the hemodynamic mechanical force generated by blood flow, and they also "sense" the change of chemical components in blood. These circulating components include hormones, growth factors and cytokines. In addition, endothelial cells synthesize vasoactive factors to modulate vascular tone and alter the diameter of the blood vessels. They also interact with the circulating cells to modulate the processes of blood clotting, thrombosis and inflammation; change their own contractile state to regulate the permeability of the endothelial barrier; and initiate the repair of vascular injury and the growth of new blood vessel. In

---

C.-O. Wong (✉)
Li Ka Shing Institute of Health Sciences and School of Biomedical Sciences, The Chinese University of Hong Kong, Hong Kong, China
e-mail: chingon.wong@gmail.com

short, endothelial cells are the major regulator of vascular homeostasis. A healthy vasculature relies on proper endothelial functions.

A variety of ion channels is expressed in endothelial cells. These endothelial ion channels play a critical role in multiple vascular processes [1]. K^+ channels, Na^+ channels and Cl^- channels regulate the membrane potential of the endothelial cells. Myo-endothelial gap junction allows the change in membrane potential to spread to the smooth muscle layer to modulate contractility of smooth muscles. Ca^{2+}-permeable channels regulate the cytosolic Ca^{2+} concentration ($[Ca^{2+}]_i$), which has profound effects on endothelial cell functions. Spatial and temporal changes in endothelial $[Ca^{2+}]_i$ regulate the synthesis of vasoactive factors, contractile state of endothelial cells, endothelial gene transcription and membrane potential. Furthermore, different ion channels may be expressed in different types of endothelial cells to meet the functional needs of different vasculatures.

Transient receptor potential (TRP) channels belong to a superfamily of cation channel proteins comprising seven subfamilies. Based on sequence similarity, the 28 (mouse) or 27 (human) TRP isoforms are classified into six subfamilies: the canonical TRPs (TRPC); melastatin TRPs (TRPM); vanilloid receptor TRPs (TRPV); ankyrin TRP (TRPA), polycystins (TRPP) and mucolipins (TRPML). Most of the characterized TRP channels are non-selective Ca^{2+}-permeable channels, except for TRPM4 and TRPM5, which conduct monovalent cations only, and TRPV5 and TRPV6, which are Ca^{2+}-selective. Native TRP channels are often formed by selective tetramerization of subunits, and display a high degree of diversity in cation selectivity, channel gating and physiological functions. Such characteristic allows native TRP channels to play versatile roles in vascular system. Several previous reviews have highlighted the properties and functions of TRP channels in endothelial cells and in other systems [2–5]. However, endothelial TRP research has been growing rapidly in recent years. This article aims at providing link between TRP channels and specific endothelial functions and giving up-to-date information on recent development.

40.2 Expression of TRP Isoforms in Endothelial Cells

The blooming of vascular TRP research in recent years provides more information on the TRP expression in endothelial cells. Tables 40.1, 40.2 and 40.3 summarize the expression profile of TRP isoforms in endothelial cells of different vasculatures. Most mammalian TRP isoforms are detected in endothelial cells, including: TRPC1-7; TRPV1/2/4; TRPP1/2; TRPA1; and TRPM1-8 except TRPM5. Note that the expression patterns of TRP isoforms vary in different vascular beds and in different species. For instance, TRPC3 expression was detected in bovine aortic endothelial cells and human pulmonary artery endothelial cells, but not in bovine pulmonary endothelial cells (Table 40.1). However, in some cases there are conflicting reports as to whether a specific TRP isoform is expressed in certain type(s) of endothelial cells [18, 44]. Such discrepancy in results could be due to the difference in cell culture condition and/or methods used for detecting expression [25].

Table 40.1 Expression profile of TRPC isoforms in vascular endothelial cells (EC)

	TRPC1	TRPC3	TRPC4	TRPC5	TRPC6	TRPC7
Bovine aortic EC [6–9]	+(RT, WB)	+(WB)/−(RT)	+(RT, WB)	+(IC, WB)/−(RT)	+(IC,WB)/−(RT)	−(RT)
Bovine corneal EC [10]			+(WB)			
Bovine pulmonary artery EC	+(RT)	−(RT)	+(RT)		+(WB)	
Ewe uterine artery EC [11]		+(WB)				
Human cerebral artery EC	+(RT, ISH, IHC)	+(RT, ISH, IHC)	+(RT, ISH, IHC)	+(RT, ISH, IHC)	+(RT, ISH, IHC)	+(RT, ISH)
Human coronary artery EC	+(RT, ISH, IHC)	+(RT, ISH, IHC)	+(RT, ISH, IHC)	+(RT, ISH, IHC)	+(RT, ISH, IHC)	+(RT, ISH)
Human dermal microvascular EC	+(RT)	+(RT)	+(RT)	−(RT)	+(RT)	+(RT)
Human intact mesenteric artery EC	+(RT)	+(RT)	−(RT)			
Human microvascular EC [12]	+(RT)				+(RT, WB)	+(RT)
Human preglomerular arteriole EC [13]		+(IHC)				
Human pulmonary artery EC [4, 15]	+(RT, NB, WB, IC)	+(RT)	+(RT, WB)	−(RT)	+(RT, WB)/−(RT)	+(RT)/−(RT)
Human renal artery EC [16]		+(WB, IHC, RT)				
Human umbilical vein EC [12, 17–20]	+(RT,WB)	+(RT,WB)/−(RT)	+(RT,WB)/−(RT)	+(RT)/−(RT)	+(RT,WB)/−(RT)	+(RT)/−(RT)
Mouse aortic EC [8]			+(NB, WB)		+(WB)	
Mouse cerebral microvessel EC [21]	+(RT)		+(RT)			
Mouse pulmonary artery EC	+(RT)	+(RT)	+(RT)	−(RT)	+(RT)	+(RT)
Porcine aortic EC [22]		+(RT, WB)	+(WB)			
Rat coronary artery EC [23]	+(RT, IHC)					
Rat pulmonary artery EC	+(RT)	+(RT)	−(RT)	+(RT)	−(RT)	
Rat splenic sinus EC	+(IHC)					
Zebrafish aortic EC [24]					+(ISH)	

+, signal detected; −, signal not detected. Methods of detection are indicated in brackets: IC, immunostaining of cultured cells; IHC, immunohistochemistry; ISH, in situ hybridization; NB, Northern blot; RT, RT-PCR; WB, Western blot. Updated from Yao and Garland 2003 [25]. References published from 2005 to early 2010 are cited.

Table 40.2 Expression profile of TRPV, -P, -A isoforms in vascular endothelial cells (EC)

	TRPV1	TRPV2	TRPV3	TRPV4	TRPV5	TRPV6	TRPP1	TRPP2	TRPA1
Bovine adrenal cortex capillary EC [26]				+(RT, WB)					
Human cerebral EC [27]	+(RT, IC)			+(IHC)					
Human dermal microvascular EC [26]				+(RT, WB)					
Human pulmonary artery EC [27]	+(RT)			+(RT, IHC)	−(RT)	−(RT)			
Human renal artery EC							+(IHC)	+(IHC)	
Human renal glomerular EC								+(IC)	
Human renal interlobar artery EC [28]				+(IHC)					
Human submucosal vascular EC [27]				+(IC)					
Human umbilical vein EC [28, 29]							+(RT, IC)	+(RT, WB)/−(WB)	
Mouse aortic EC [30–32]				+(RT, NB, IHC, WB)			+(IHC)	+(RT, IC, WB)	
Mouse cerebral microvascular EC [21, 33]				+(RT, WB)					
Mouse femoral artery EC [28]		+(RT)						+(IHC)	
Mouse mesenteric artery EC [34, 35]				+(RT, IHC)					
Porcine aortic EC [36, 37]				+(RT, IC)					
Rat carotid artery EC [38]				+(RT)					
Rat cerebral artery EC [39, 40]				+(RT)					+(RT, IC)
Rat femoral artery EC [37]				+(IHC)					
Zebrafish ventricular EC [41]				+(ISH)					

+, signal detected; −, signal not detected. Methods of detection are indicated in brackets: IC, immunostaining of cultured cells; IHC, immunohistochemistry; ISH, in situ hybridization; NB, Northern blot; RT, RT –PCR; WB, Western blot. Updated from Yao and Garland 2003 [25]. References published from 2005 to early 2010 are cited.

Table 40.3 Expression profile of TRPM isoforms in vascular endothelial cells (EC)

	TRPM1	TRPM2	TRPM3	TRPM4	TRPM5	TRPM6	TRPM7	TRPM8
Human pulmonary artery EC [42]	+(RT)	−(RT)	+(RT)	+(RT)	−(RT)	+(RT)	+(RT)	+(RT)
Human umbilical vein EC [43]							−(RT)	+(RT)
Mouse aortic EC				+(NB)				
Mouse cerebral microvascular EC [21]		+(RT, WB)	+(RT)	+(RT)			+(RT)	
Mouse heart microvascular EC		+(RT, WB)						

+, signal detected; −, signal not detected. Methods of detection are indicated in brackets: IC, immunostaining of cultured cells; IHC, immunohistochemistry; ISH, in situ hybridization; NB, Northern blot; RT, RT-PCR; WB, Western blot. Updated from Yao and Garland 2003 [25]. References published from 2005 to early 2010 are cited.

A growing body of evidence is affirming that native TRP channels are often composed of heteromeric subunits. Subunit multimerizations were often found between the members within one TRP subfamily, for example TRPC1 with other TRPC isoforms. However, heteromerization across different subfamilies has also been reported for several TRP isoforms. The reported heteromeric TRP channels in endothelial cells include: TRPC3-TRPC4 [22]; TRPC1-TRPC4 [6]; TRPC1-TRPV4 [45] and TRPP1-TRPP2 [28]. Many other subunit combinations have been found in other organs/cell types, thus it can be expected that more heteromeric TRP channels will be uncovered in endothelial cells in future. Because heteromeric TRP channels may have distinct channel properties, including current-voltage relationships, relative permeability ratio of Na^+/Ca^{2+}, unitary conductance, and gating behavior, that are different from those of homomeric ones, uncovering the subunit composition of these heteromeric TRP channels could provide key information on structural, functional and therapeutic roles of these channels in endothelial cells.

40.3 Functional Role of Endothelial TRP Channels

40.3.1 Ca^{2+} Influx and Electrogenesis

Endothelial cells express a plethora of ion channels that contribute to the electrogenesis and Ca^{2+}-signaling upon receptor activation and/or mechanical stimulation [1]. Vasoactive agents elicit transient Ca^{2+} release from Ca^{2+} stores, followed by a sustained $[Ca^{2+}]_i$ rise due to Ca^{2+} entry via Ca^{2+}-permeable channels in the plasma membrane. These Ca^{2+}-permeable channels are sensitized by the messengers produced after receptor activation (e.g. diacylglycerol, arachidonic acids and derivatives) or Ca^{2+}-store release/depletion [46]. TRP channels are a major class of Ca^{2+}-permeable channels in endothelial cells, where they mediate Ca^{2+} influx upon agonist-stimulation. ATP, angiotensin II, thrombin and bradykinin have been shown to stimulate Ca^{2+} influx through TRPC1/3/4/5/6 [7, 15, 20, 25]. Endocannabinoids and epoxyeicosatrienoic acids (EETs) are potent vasodilatory agents acting on endothelium [47], and are capable of stimulating TRPV1/4 [31, 32, 48]. One controversial aspect of TRP channels is whether they mediate the Ca^{2+} influx pathway upon store depletion, the so-called store-operated Ca^{2+} influx (SOC) pathway. SOC pathway has been documented in endothelial cells, and is related to various endothelial functions [49]. Numerous studies have suggested that TRPC1,3-5,7 may mediate SOC in endothelial cells at least under certain conditions [50]. Indeed, TRPC1- and TRPC4-mediated SOC in endothelial cells has been suggested to play key functional role in control of vascular tone and permeability [44, 49, 51, 52]. In this regard, a scaffold protein caveolin-1 appears to be an important protein that links TRPC1 to IP_3 receptors to regulate SOC [14, 53]. In contrast, however, a recent report showed that suppressing the protein expression of TRPC1 and TRPC4 by siRNAs did not alter SOC in endothelial cells, casting a doubt on the role of TRPC1 and TRPC4 in SOC [17, 54, 55]. The same study also provided evidence that STIM1 and Orai1 may be the key SOC components in vascular endothelial cells [17]. One compromise could be that TRPCs may interact with STIM1 and Orai1 to participate in SOC in endothelial cells [56]. Nevertheless, it is clear that more works are necessary to resolve the issue.

A rise in $[Ca^{2+}]_i$ through TRP channels initiates multiple signaling cascades, leading to diverse cellular responses. Some responses are fast, occurring within seconds to minutes, including release of vasodilators such as nitric oxide (NO) and endothelium-derived hyperpolarizing factor (EDHF). They may quickly lead to end responses such as vascular relaxation, as we will discuss later. Other responses are relative slow, occurring in minutes to several hours, involving transcription factor activation and gene transcription. In this regard, it has been shown that thrombin and tumor necrosis factor-α can stimulate Ca^{2+} influx via TRPC1, leading to the activation of nuclear factor kappa-B [19, 57, 58]. Hypoxia may also induce Ca^{2+} influx through TRP channels, resulting in enhanced binding of activating protein 1 transcription factors to DNA [59].

Activation of TRP channels not only alters endothelial $[Ca^{2+}]_i$ homeostasis, but also modulate the membrane potential of endothelial cells. Following agonist stimulation, endothelial cells often display an initial membrane hyperpolarization, followed by a sustained depolarization mediated by non-selective cation channels [1]. A feature of TRP channels is the non-selective cation conductivity. The influx of cations (e.g. Na^+, Ca^{2+}) through TRP channels modulates the membrane potentials of endothelial cells in two different ways. Firstly, influx of Na^+ and Ca^{2+} results in membrane depolarization. In this regard, TRPM4 and TRPM5 are of particular interest, because these two TRP isoforms are highly selective for monovalent cation. Endothelial cells express TRPM4 [59], but the precise functional role of TRPM4 in endothelial cells is still undefined. In vascular smooth muscle cells, pressure-induced activation of TRPM4 results in membrane depolarization, which subsequently activates voltage-gated Ca^{2+} channels, leading to $[Ca^{2+}]_i$ rise and smooth muscle contraction [60]. However, unlike vascular smooth muscle cells, endothelial cells do not possess functional voltage-gated Ca^{2+} channels., TRPM4-mediated membrane depolarization in endothelial cells reduces the driving force of Ca^{2+} influx, and furthermore, Na^+ influx through TRPM4 is expected to stimulate Ca^{2+} extrusion via plasma membrane Na^+/Ca^{2+} exchanger, both of which results in a reduction in $[Ca^{2+}]_i$ in endothelial cells. Secondly, apart from their direct depolarization effect, activity of TRP channels could indirectly cause membrane hyperpolarization. This is because Ca^{2+} entry through TRP channels may activate neighboring Ca^{2+}-sensitive K^+ channels, resulting in membrane hyperpolarization. Recently, Earley et al. found a functional coupling between TRPA1 and two Ca^{2+}-sensitive K^+ channels, intermediate conductance and small conductance Ca^{2+}-sensitive K^+ channels (IK_{Ca} and SK_{Ca}, respectively) in endothelial cells [39]. They suggested that a localized $[Ca^{2+}]_i$ rise via TRPA1 activates SK_{Ca} and IK_{Ca} in endothelial cells, resulting in membrane hyperpolarization, which is conducted to the nearby myocytes. Furthermore, activity of SK_{Ca} and IK_{Ca} also cause K^+ efflux, which activates inwardly-rectifying K^+ channels in smooth muscle. Both effects contribute to smooth muscle hyperpolarization and subsequent vascular relaxation [39]. Given the fact that many TRPs are expressed in endothelial cells, it will be interesting to explore whether other TRPs may also function in a similar way. In this regard, in vascular smooth muscle cells two other TRP isoforms TRPC1 [61] and TRPV4 [62] are found to be coupled to the large conductance Ca^{2+}-sensitive K^+ channels (BK_{Ca}). Activity of these two TRPs stimulates BK_{Ca} to result in membrane hyperpolarization of smooth muscle cells [61, 62].

40.4 Control of Vascular Tone

One of the major functions provided by the endothelial cells is the control of vascular tone. Endothelium generates multiple factors to regulate vascular tone. Two important factors are NO and EDHF, both of which act on smooth muscle cells to cause relaxation. The production of these vasodilators is regulated by $[Ca^{2+}]_i$ and/or membrane potentials of endothelial cells. $[Ca^{2+}]_i$ elevation in endothelial cells, either by vasoactive agents or mechanical forces, activates NO synthase to enhance NO production. NO diffuses into nearby smooth muscle cells where it stimulates guanylate cyclases to enhance cyclic GMP (cGMP) production. cGMP then triggers downstream events, resulting in smooth muscle relaxation. The term EDHF describes a diverse array of chemicals that can hyperpolarize smooth muscle cells but with different chemical structures. These include EETs generated from cytochrome P_{450}, K^+ ions and H_2O_2 [63]. $[Ca^{2+}]_i$ is a key factor that can regulate the production of EETs and H_2O_2 [64, 65]. $[Ca^{2+}]_i$ rise also stimulates the activity of IK_{Ca} and SK_{Ca}, both of which are critical mediators of EDHF signaling [66].

Studies show that multiple TRP isoforms contribute to Ca^{2+} influx in endothelial cells (see above). Some TRPs, including TRPC4, -V1, -V4, and -A1, have been shown to play a role in agonist-induced vascular tone control. The most convincing evidence comes from the knock-out animal studies. Endothelial cells from the *trpc4–/–* and *trpv4–/–* knock-out mice showed diminished $[Ca^{2+}]_i$ rise in response to agonist [32, 51]. As a result, endothelium-dependent relaxation to agonists was impaired in these knock-out animals [32, 51]. These studies conclusively demonstrate a functional role of TRPC4 and TRPV4 in vascular tone control. In another study, stimulation of TRPA1 by allyl isothiocyanate (a mustard oil component) was found to induce an endothelium-dependent relaxation of rat cerebral artery, the effect of which was blocked by a selective TRPA1 inhibitor HC-030031. In this case TRPA1 activates SK_{Ca} and IK_{Ca} in endothelial cells, resulting in an EDHF-mediated vascular relaxation [39]. With the use of selective pharmacological activators and/or inhibitors, several studies also demonstrated that the activity of TRPV1 in endothelial cells causes endothelium-dependent relaxation [67, 68]. TRPV1 activity evokes Ca^{2+} entry in endothelial cells, which enhances NO production, resulting in endothelium-dependent vascular dilation [67, 68]. Endothelial TRPC3 is also implicated in the vascular tone regulation. TRPC3 expression in the endothelium of preglomerular arterioles was found to be upregulated in patients with malignant hypertension [13]. Beside their roles in agonist-induced vascular tone control, several TRPs are involved in flow-induced vascular dilation. This will be discussed later.

40.5 Control of Vascular Permeability

The endothelium is a continuous monolayer of cells tightly adhered to each other and to the basement. The structure forms a semi-permeable dynamic barrier between vascular lumen and the underlying tissues. The permeability of this structure is governed by paracellular pathway and transcellular pathway. The paracellular pathway relies on the integrity of the cell-cell junctions, which is maintained by the balance between the adhesive force at junctions and the contractile forces within the endothelial cells. Any shift in such force balance, for example by increasing cytoskeletal contractile force, changes the permeability of the interendothelial gap, and thus alters the barrier function. Aberration of endothelial barrier function leads to an abnormal filtration of blood components across the endothelium, resulting in tissue dysfunction, for example leakage and edema.

Pro-inflammatory agents (such as histamine and thrombin), cytokines, growth factors and reactive oxygen species (ROS) can increase vascular permeability. These agents elicit a rise in $[Ca^{2+}]_i$, which signals the reorganization of cytoskeletal proteins and disassembly of transmembrane adhesive complexes, leading to an increased vascular permeability. Many TRPs conduct Ca^{2+} and they also interact with cytoskeletal proteins, therefore it is likely that they may play a role in the control of vascular permeability [69–71]. Current research suggests that TRPC1, -C4, -C6, -V4, -M2 and -M4 may participate in the vascular permeability control.

In a mouse model of spinal cord injury, TRPM4 expression is upregulated in the capillary endothelium after injury [72]. Destruction in capillary integrity contributes to post-trauma secondary hemorrhage, which further injures the neighboring tissues. TRPM4 is apparently linked to the post-trauma capillary dysfunction since knock-down or knock-out of TRPM4 expression greatly reduced the secondary hemorrhage by preserving the capillary integrity [72]. Although the mechanism by which TRPM4 destroys the capillary barrier is still unknown, the study suggests a possibility of targeting TRP channels as a potential treatment in post-trauma spinal cord injury.

By measuring the transendothelial cell electrical resistance across confluent monolayer of primary endothelial cells, several studies have identified an essential role of TRPC1 and TRPC4 in endothelial permeability control [42, 44, 52]. Inflammatory mediator thrombin stimulates TRPC1 and -C4 in endothelial cells to induce a $[Ca^{2+}]_i$ rise, which in turn triggers a signaling cascade, resulting in cytoskeletal stress fiber formation and endothelial hyperpermeability [44, 52]. In *trpc4–/–* mice, thrombin-induced $[Ca^{2+}]_i$ rise and endothelial hyperpermeability are drastically reduced [52]. Studies also found a role of TRPM2 and TRPC6 in vascular permeability control. Exposure of endothelial cell monolayers to H_2O_2, which is an ROS, stimulates TRPM2, resulting in endothelial hyperpermeability [42]. In another study, Pocock et al. found a correlation between TRPC6 activity and the vascular endothelial growth factor (VEGF)-induced vascular permeability in frog mesenteric microvessels [73].

TRPV4 channels also influence vascular permeability most prominently in the lung capillary endothelial barrier. TRPV4 activators, 4α-Phorbol 12,13-Didecanoate (4α-PDD) and EETs, produce blebs or breaks in the endothelial and epithelial layers of the alveolar wall, resulting in an increased lung endothelial permeability and alveolar septal barrier disruption [74]. Increased activity of TRPV4 channels is also believed to be the culprit for acute vascular permeability increase and lung injury following high pressure ventilation. Lung distention causes rapid Ca^{2+} entry in lung tissues, which was absent in *trpv4–/–* animals [75]. The high pressure ventilation-induced lung filtration and edema incidence were greatly reduced in *trpv4–/–* lungs [75]. Collectively, these studies provide strong evidence that activation of TRPV4 causes vascular hyperpermeability in lung, contributing to alveolar septal barrier disruption and acute lung injury.

40.6 Angiogenesis and Blood Vessel Formation

Formation of blood vessel relies on two processes: (1) vasculogenesis, in which new vascular cells are differentiated from endothelial precursor cells called angioblast; (2) angiogenesis, which involves the formation of new branch of vessels from pre-existing vessels. Vasculogenesis occurs mainly during embryogenesis. On the other hand, angiogenesis is responsible for the development of blood vessels in later stage of embryogenesis and in adulthood. In adult, endothelial cells are generally quiescent and do not initiate angiogenesis, except for menstruation, wound healing and pathological conditions. Angiogenesis is coordinated by various vascular growth factors, which act on endothelial cells to initiate the sprouting of the vessel through extracellular matrix degradation, directional endothelial cell migration and proliferation. VEGF is crucial in promoting the proliferation and differentiation of endothelial cells in angiogenesis. Several other factors, including fibroblast growth factor, platelet-derived growth factor, transforming growth factor beta and mechanical signals, are also important for the remodeling and patterning of the developing vasculature. Endothelial TRP channels are associated with angiogenic process by their ability to affect $[Ca^{2+}]_i$ and cytoskeleton reorganization [1, 71], the two requisites for the initial sprouting of the vessels by motile endothelial cells. Up to the present, TRPC1, -C4, -C6, -V4 and -M7 in endothelial cells have been suggested to be involved in angiogenesis.

Two independent studies have found that dominant-negative pore mutant of TRPC6 was capable of inhibiting VEGF-induced cationic inward currents and $[Ca^{2+}]_i$ rise in human umbilical vein endothelial cells (HUVECs) and human microvascular endothelial cells (HMVECs). Disruption of TRPC6 function also inhibited endothelial cell proliferation by arresting the cell cycle at G2/M phase [18]. Importantly, the VEGF-induced cell migration, sprouting and capillary tube formation was also inhibited by the dominant-negative construct of TRPC6 [12, 18]. These studies point out a correlation between TRPC6 function and endothelial motility and proliferation.

TRPV4 is also linked to vessel formation. TRPV4 mediates endothelial cell orientation in response to cyclic mechanical strain produced by pulsatile blood flow [26], suggesting its function in endothelial cell motility. A recent study suggested an important role of TRPV4 in collateral artery formation after arterial occlusion [37]. Ligature of femoral artery causes elevated fluid shear stress, which causes an upregulation in TRPV4 protein expression and induces the formation of new collateral arteries [37]. Although the detailed mechanism is not known, TRPV4 seems to be the link between hemodynamic force and new vessel formation.

TRPM7 activity is positively correlated with cell proliferation in most tissues [76]. However, the opposite may be true for the endothelial cells. A recent study found that knockdown of TRPM7 with siRNA or inhibition of TRPM7 function with 2-APB results in an enhanced growth/proliferation of HUVECs via the ERK-dependent pathway [43], suggesting an inhibitory rather than stimulatory effect of TRPM7 to endothelial cells proliferation [43].

TRPC4 is also associated with endothelial cell proliferation. In HMVECs [77], epidermal growth factor recruited TRPC4 into the plasma membrane in subconfluent endothelial cells [77]. Interestingly, after the cells reached confluence and establish mature cell-cell contact, growth factor simulation caused retrieval of TRPC4 from plasma membrane to cytosol, and this process appears to be beta-catenin-dependent [77]. It is possible that TRPC4-mediated Ca^{2+} influx may be a potential key player in endothelial phenotype switching from a proliferating to a quiescent phenotype [77].

TRPC1 may play a role in in vivo angiogenesis during zebrafish development [78]. After introduction of morpholino construct to knock down TRPC1 expression in zebrafish embryos, angiogenic sprouting of intersegmental vessels was disrupted [78]. Importantly, the phenotypes were rescued by TRPC1 mRNA [78]. TRPC1 function was found to be associated with the tip cell migration and proliferation in the sprouting vessels [78]. Furthermore, by showing that TRPC1 mediates the VEGF-activated signaling of MEK-ERK cascade, this study also provided a possible mechanistic view of how TRPC1 regulates angiogenesis [78].

40.7 Oxidative Stress

ROS plays important roles in vascular pathophysiology as well as normal physiology [79]. Major cellular ROS includes superoxide ($O_2 \cdot^-$), hydroxyl radicals ($OH \cdot$), hydrogen peroxide (H_2O_2), and peroxynitrite ($ONOO^-$). They are the products of normal oxygen consuming metabolic process in the body. ROS is produced mostly in circulating cells, including neutrophils and macrophage, and also in vascular cells. The enzymatic sources of ROS include nicotinamide adenine dinucleotide phosphate (NADPH)-oxidase, xanthine oxidase, uncoupled NO synthase and mitochondrial respiration system. Cells also possess anti-oxidant systems, including superoxide dismutase, glutathione and thioredoxin, that serve to scavenge ROS. In normal physiological status, a balance between ROS production and ROS scavenging is maintained. However, excessive production of ROS occurs during pathological conditions of cell injury, inflammation, ischemia/reperfusion and hyperglycemia. Due to the reactive nature of the ROS, ROS causes redox modifications of lipids and proteins, resulting in cell damage. ROS also interferes with the vasodilator pathways including NO, prostacyclin and EDHF. ROS-induced vascular damage is often preceded by an alteration of endothelial Ca^{2+} homeostasis, which could result from ROS action on plasma membrane ion channels such as TRP and BK_{Ca}, or on endoplasmic reticulum receptors/pumps such as Ca^{2+}-ATPase, inositol trisphosphate receptor (IP_3R) and ryanodine receptors. There is a suggestion that TRP channels in endothelial cells could serve as the oxidative stress sensor to amplify the vascular effects brought by ROS. In this regard, an attractive strategy is to develop selective TRP antagonists, including pharmacological agents, blocking antibodies, and/or siRNA, into therapeutic tools for the treatment of ROS-related cardiovascular disease such as atherosclerosis and ischemic heart disease.

TRPC3 and TRPC4 are redox sensitive channels activated by oxidative stress [80]. In porcine aortic endothelial cells (PAECs), oxidant tert-butylhydroperoxide (tBHP) activates a cationic current. Expression of an N-terminal fragment of TRPC3 abolished the oxidant-induced cationic current, suggesting that this channel is related to TRPC3 [80]. A follow-up study by the same group provided evidence that this redox-sensitive channel is actually a heteromeric channel composed of TRPC3 and TRPC4 [22]. Immunoprecipitation and fluorescence resonance energy transfer experiments demonstrated that TRPC3 proteins physically associated with TRPC4 proteins in close proximity in PAECs [22]. tBHP or cholesterol oxidase induced cationic currents in PAECs, which was suppressed by applications of extracellular antibody targeting TRPC3 or TRPC4 respectively [22]. Sensitivity of the native channels to cholesterol oxidase suggests that the TRPC3-TRPC4 complex is capable of sensing the oxidation of membrane lipids, rather than the redox modification of the channel proteins.

TRPM2 is another oxidative stress-activated channel in vascular endothelial cells. Exposure of endothelial cell monolayers to sublytic concentrations of H_2O_2 induced a cationic current and a Ca^{2+} entry, both of which were inhibited by TRPM2-siRNA and a TRPM2-specific blocking antibody [42]. Overexpression of the full-length TRPM2 enhanced H_2O_2-mediated Ca^{2+} entry and cationic currents, whereas a dominant-negative splice variant of TRPM2 diminished the H_2O_2 responses [42]. These studies provided strong evidence that H_2O_2 can activate TRPM2 in endothelial cells. Further studies indicate that activation of TRPM2 by H_2O_2 results in endothelial hyperpermeability [42]. Mechanistically, H_2O_2 may activate TRPM2 via enhancing the formation of ADP-ribose, which then activates TRPM2 [42]. Alternatively, H_2O_2 may directly act on TRPM2 to stimulate channel activity [81, 82]. Therefore, TRPM2 may act as an endothelial H_2O_2 sensor, and mediates vascular barrier dysfunction during inflammation or ischemia/reperfusion.

Several other TRP isoforms, including TRPM7 and -A1 are also activated by oxidative stress. TRPM7 mediates the anoxic Ca^{2+} uptake and anoxic cell death in cortical neurons [83]. TRPA1 is activated by hypochlorite and H_2O_2, and serves as the oxidant sensor in airway sensory neurons [84]. TRPM7 and TRPA1 are expressed in endothelial cells. However, functional role of these channels in oxidative stress of endothelial cells remained to be resolved.

40.8 Mechanosensing

Hemodynamic blood flow exerts a frictional force or shear stress on the endothelium. Shear force is parallel to the direction of blood flow. A high physiological shear stress is beneficial to vessel health. It promotes the production and release of endothelial vasodilatory factors such as NO and prostacyclin; inhibits leukocyte binding and chemoattractant protein expression on endothelial surface thus preventing the formation of artherosclerotic lesions; and exerts anti-apoptotic effect on endothelial cells [85]. In contrast, low shear stress and a reversal in the direction of flow may lead to endothelial dysfunction. In the area of vessel bifurcation, where flow is turbulent and shear force is low, atherosclerotic lesion is more likely to occur [85].

Although the primary sensor(s) that can detect the flow shear force remain unknown, many components have been suggested to be involved in the early stage of flow sensing. These include membrane-bound G proteins, cytoskeleton, focal adhesion molecules, cell-cell junction, membrane microviscosity and mechanosensitive ion channels [86]. Mechanosensitive ion channels are attractive candidates as mechanosensors and mechanotransducers. These channels open in response to shear stress, and the opening of these channels results in change in membrane potentials and/or cytosolic ionic concentration. Several TRP channels have been shown to be sensitive to flow shear forces. Activation of TRP channels in response to flow shear force causes a $[Ca^{2+}]_i$ rise in endothelial cells, leading to the production and release of vasodilators from endothelial cells and subsequent vascular dilation.

There are compelling evidences linking TRPV4 channels to flow-induced Ca^{2+} influx in endothelial cells and subsequent vascular dilation. Initial studies using knock-out animal suggested that TRPV4 may be a mechanosensitive channel playing a role in pressure sensation [87]. However, it was later found that TRPV4 is not activated by direct membrane stretch, but rather, is activated by enzymatic signaling following cell swelling [88, 89]. Despite of the controversy, several independent studies did demonstrate a functional role of endothelial TRPV4 in flow-induced vascular dilation. Kohler et al. first demonstrated that 4α-PDD, a TRPV4 activator, and shear stress both induce relaxation of carotid artery, the effect of which was blocked by ruthenium red, an inhibitor to TRPV4 [38]. Later, Loot et al. found that flow-induced endothelium-dependent dilation of carotid artery was greatly reduced in *trpv4–/–* mice [29]. Interestingly, the TRPV4-mediated vasodilatory effect was dependent on cytochrome P_{450} (CYP) epoxygenase activity [29]. CYP epoxygenase catalyses the formation of EETs from arachidonic acid, which is produced by the cell swelling-sensitive phospholipase A. Therefore it is likely that TRPV4 activation is downstream of a mechanosensitive pathway that leads to the production of EETs, which then activate TRPV4. Activation of TRPV4 enhances the production of NO and EDHF in endothelial cells, leading to the subsequent vascular dilation [29, 38, 45]. Recent studies from our group indicate that TRPV4 and TRPC1 form heteromeric channels in endothelial cells. It is TRPV4-C1 hetermeric channel rather than TRPV4 homomeric channel that actually mediates flow-induced endothelial Ca^{2+} influx and subsequent vascular dilation [45]. Inhibition of either TRPV4 or TRPC1 each could diminish the flow-induced endothelial Ca^{2+} influx and flow-induced dilation. Flow-induced Ca^{2+} rise in endothelial cells is long known to be sensitive to inhibition by NO-cGMP-protein kinase G signaling pathway [90, 91]. It is now clear that protein kinase G acts on TRPV4-C1 heteromer to exert its inhibition on flow-induced Ca^{2+} influx in endothelial cells [45]. Collectively, the present evidences indicate the TRPV4 activity is crucial to flow-induced endothelium-dependent vasodilation. The mechanism may involve upstream mechanosensitive signaling pathway(s) and heteromultimerization with TRPC1.

TRPP1-TRPP2 complex could also mediate flow-induced Ca^{2+} rise in endothelial cells. Mutations in *pkd1* and *pkd2* genes, which code for TRPP1 and TRPP2 respectively, cause autosomal dominant polycystic kidney disease (ADPKD). Earlier study has demonstrated the function of TRPP1 and TRPP2 in transducing the shear stress into $[Ca^{2+}]_i$ in epithelial cilia [92]. Recently, a ciliary structure has also been identified in cultured endothelial cells and the endothelia of isolated artery [28, 30]. This ciliary structure contains TRPP2 and TRPP1 proteins [28]. Importantly, flow-induced endothelial $[Ca^{2+}]_i$ rise and NO production is drastically reduced by silencing TRPP2 or TRPP1 expression [28, 30]. The flow-induced endothelial responses were also absent in cells derived from *pkd2–/–* mice or ADPKD patients [28]. These data suggest an important role of TRPP1 and TRPP2 in flow-induced endothelial Ca^{2+} rise and NO production. However, the mechanistic details of how TRPP1 and TRPP2 respond to flow shear stress remains unclear. Early study from

Nauli et al. suggested that TRPP1 may serve as a sensory molecule that transduces the flow stimuli to TRPP2, which in turn enables Ca^{2+} influx [92]. A recent study, however, suggests that TRPP2 functions to modulate the activity of a yet unidentified stretch-activated channel [93]. A third possibility is that TRPP2 may interact with TRPV4, which is flow-sensitive [94, 95]. Further studies are apparently needed to clarify which mechanism is correct, or if all these mechanisms operate independently or coordinate with each other to mediate flow responses.

In addition to flow shear force, vascular endothelial cells are also exposed to circumferential stretch due to pulsatile blood flow. The circumferential stretch results from transmural pressure, which acts perpendicularly on the vascular wall. Several TRP channels could be activated directly or indirectly by transmural pressure. These channels, which include TRPV2, -M4, -M7 and -C6, are all expressed in vascular endothelial cells [96]. However, up to the present there is no concrete evidence linking these channels to endothelial pressure sensing.

40.9 TRP Channels, Endothelial Dysfunction, Diseases

As discussed above, endothelial TRP channels perform multiple functions. Thus it is not surprising that malfunction of TRPs may result in endothelial dysfunction and contribute to vascular diseases. Indeed, the dysfunction and dysregulation of endothelial TRP channels are suggested to result in excessive oxidative damage, decreased NO availability, the disruption of the endothelial barrier, and the progression of angiogenesis-associated diseases [3].

Malfunction of endothelial cell TRPP1 and TRPP2 may contribute to ADPKD. ADPKD patients present an impaired endothelium-dependent relaxation to acetylcholine in resistance vessels [97]. In animal studies, TRPP1 heterozygous mice ($TRPP1^{+/-}$) also display impairment in endothelium-dependent vascular relaxation. It has been suggested that the malfunction of endothelial TRPP1 and -P2 impairs the complex regulation of endothelial NO synthase, which results in endothelial dysfunction and contributes to the progression of ADPKD [98].

Dysfunction of endothelial TRPC3, -M7, and -V4 is also associated with vascular diseases. Activity of endothelial TRPC3 may contribute to hypertension. In patients with malignant hypertension, TRPC3 expression in the endothelium of preglomerular arterioles was found to be upregulated [13]. It is possible that excessive Ca^{2+} influx through TRPC3 stimulates endothelial cells to release vascular constrictors, resulting in an increased vascular tone and hypertension [13]. TRPV4 plays a crucial role in regulating airway and lung epithelial-endothelial barrier function. Activation of TRPV4 disrupts alveolar walls by increasing lung epithelial-endothelial permeability [74]. Increased activity of TRPV4 channels is also believed to be the culprit for acute vascular permeability increase and lung injury following high pressure ventilation [75]. These functions of TRPV4 implicate a possible role of TRPV4 in the pathogenesis of chronic obstructive pulmonary disease. Indeed, TRPV4 gene polymorphisms have been found to be associated with chronic obstructive pulmonary disease [99]. In mice with inherited hypomagnesemia, TRPM7 dysregulation is found to be associated with endothelial dysfunction with reduced endothelial NO synthase expression [100].

40.10 Conclusive Remarks and Goal for Future

Efforts made by vascular biologists and TRP researchers in recent years have provided extensive information on the functional relevance of TRP channels in endothelium (Fig. 40.1 and Table 40.4). Several TRP isoforms apparently stand out to be critical mediators for certain endothelial functions. Due to the cell surface expression and the versatile functional roles, TRP channels have long been proposed as therapeutic targets. In particular, the endothelial TRP channels localized on the luminal side are susceptible to be modulated by circulating molecules.

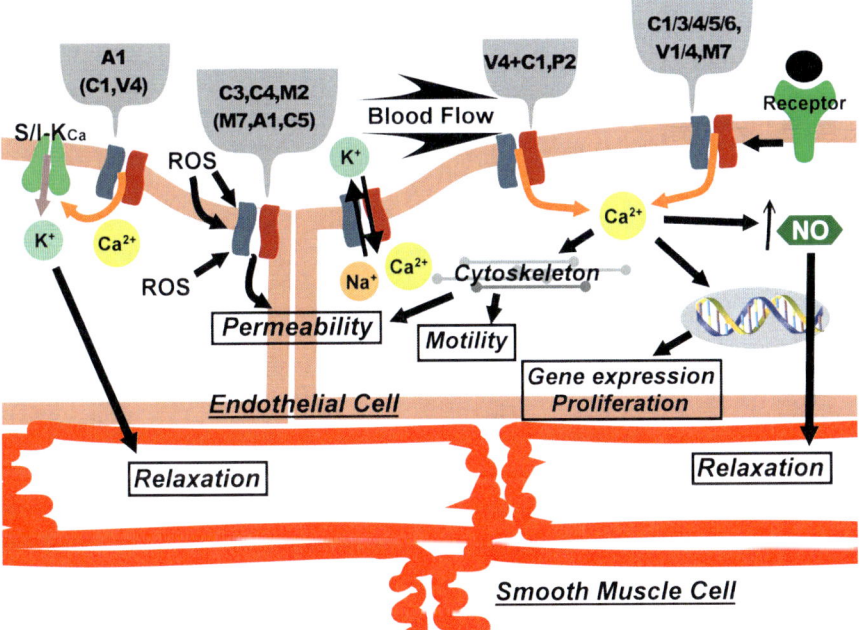

Fig. 40.1 TRP channels in endothelial cells. The schematic diagram summarizes the functions of TRP channels in endothelial cells. Native TRP channels in endothelial cells consist of either homomeric or heteromeric isoforms. Most of them have non-selective conductance to K^+, Na^+ and Ca^{2+}. TRP channels act to modulate membrane potential (by Na^+ influx or Ca^{2+}-activation of K^+ channels), stimulate the production of vasodilatory substance (NO, EET, K^+), sense ROS, control vascular permeability, modulate cytoskeleton remodeling and cell motility, regulate gene transcription and cell proliferation (by Ca^{2+}-signaling cascades). (TRP isoforms are grouped according to their mode of activation and related endothelial function; Candidate isoforms are marked in bracket)

Table 40.4 Functional roles of TRP channels in endothelial cells

	Functional roles in endothelial cell
TRPC1	Receptor agonist-stimulated Ca^{2+} influx; thrombin-induced hyperpermeability; flow-induced vasodilation (heteromeric with TRPV4) [7, 20, 45]
TRPC3	Receptor agonist-stimulated Ca^{2+} influx; sensor of lipid oxidation (heteromeric with TRPC4); upregulated expression in hypertensive patients [13, 22, 90]
TRPC4	Receptor agonist-stimulated Ca^{2+} influx and vasodilation; thrombin-induced hyperpermeability; sensor of lipid oxidation (heteromeric with TRPC3); endothelial cell proliferation [22, 51, 52, 77]
TRPC5	Receptor agonist-, lipophilic molecule- stimulated Ca^{2+} influx [8, 9, 25]
TRPC6	Receptor agonist-stimulated Ca^{2+} influx; VEGF-induced hyperpermeability; VEGF-induced endothelial cell proliferation [12, 15, 18, 73]
TRPM2	H_2O_2-induced Ca^{2+} influx and hyperpermeability [42]
TRPM7	Inhibit endothelial cell proliferation [43]
TRPV1	Arachidonic acid metabolite-evoked Ca^{2+} influx; nitric oxide production and vasodilation [48, 67, 68]
TRPV4	Arachidonic acid metabolite-evoked Ca^{2+} influx; flow-induced vasodilation; angiogenesis; high pressure-induced alveolar hyperpermeability; gene polymorphisms in chronic obstructive pulmonary disease [29, 32, 34, 37, 38, 45, 74, 75, 99]
TRPP2	Flow-induced Ca^{2+} influx and NO production; loss of function impairs endothelium-dependent vasdilation in ADPKD [28, 97]
TRPA1	Endothelium-dependent vasodilation by coupling with SKCa and IKCa [39]

By stimulating or inhibiting the activity of specific TRP channels, one may be able to alter the localized vascular processes. For example, 4α-PDD, the TRPV4 activator, is capable of inducing endothelial-dependent vascular dilation [38]. Targeting TRPC1 by a strategically derived antibody was found to reduce the neointimal growth in human vein [101].

However, for future development of endothelial TRP fields, there are several outstanding issues need to be addressed. (1) Currently, only very few specific and potent stimulators/inhibitors of TRP channels are available. Future development of TRP isoform-specific pharmacological agents, specific blocking antibodies, siRNA technology, and conditional or tissue-specific knock-out animals, should greatly facilitate TRP research and help the development of therapeutic tools targeting TRP channels. (2) It becomes more and more apparent that TRP channels in native cells exist and function as heteromers, and that one TRP isoform may assemble with different isoforms to form multiple heteromeric channels, each of which serves different function. In this regard, we need to develop a reliable new technology to determine the subunit composition and stoichiometry of each heteromeric channel in native cells, and we also need to quantify the relative amount of different heteromeric channels. This is especially important, because different heteromeric channels have different properties and functions. (3) Many of current research on endothelial TRP rely on the data generated from cultured endothelial cells. Due to phenotypic drifting of endothelial function during cell culture and passage, the obtained results may not always reflect *in vivo* situation. In future, it will be desirable to study TRP functions using intact vascular tissues. (4) Information on expression pattern and function profile of different TRPs in different vasculatures is still scarce. This is one of the limiting factors that have not drawn sufficient attention. Note that an established TRP function in endothelial cells of one particular vascular bed could not readily be applicable to another vascular bed, due to vasculature-dependent differential expression, change in subunit assembly patterns, and/or altered coupling of TRP with other vasculature-dependent proteins. (5) Future establishment of more animal models and more data from patient samples may also shed light on TRP function in pathophysiology, and provide important insight into the therapeutic potential of targeting endothelial TRP channels. Taken together, on the voyage to applying the knowledge from basic TRP channel research, there is spacious avenue ahead of us.

Acknowledgments We thank the support from Hong Kong RGC Grants (CUHK477307, CUHK477408 and CUHK479109), Focused Investment Scheme of CUHK and Li Ka Shing Institute of Health Sciences.

References

1. Nilius B, Droogmans G (2001) Ion channels and their functional role in vascular endothelium. Physiol Rev 81:1415–1459
2. Nilius B, Droogmans G, Wondergem R (2003) Transient receptor potential channels in endothelium: solving the calcium entry puzzle? Endothelium 10:5–15
3. Kwan HY, Huang Y, Yao X (2007) TRP channels in endothelial function and dysfunction. Biochim Biophys Acta 1772:907–914
4. Nilius B (2007) TRP channels in disease. Biochim Biophys Acta 1772:805–812
5. Venkatachalam K, Montell C (2007) TRP channels. Annu Rev Biochem 76:387–417
6. Antoniotti S, Fiorio Pla A, Barral S, Scalabrino O, Munaron L, Lovisolo D (2006) Interaction between TRPC channel subunits in endothelial cells. J Recept Signal Transduct Res 26: 225–240
7. Bishara NB, Ding H (2010) Glucose enhances expression of TRPC1 and calcium entry in endothelial cells. Am J Physiol Heart Circ Physiol 298:H171–H178
8. Chaudhuri P, Colles SM, Bhat M, Van Wagoner DR, Birnbaumer L, Graham LM (2008) Elucidation of a TRPC6-TRPC5 channel cascade that restricts endothelial cell movement. Mol Biol Cell 19:3203–3211
9. Wong CO, Huang Y, Yao X (2010) Genistein potentiates activity of the cation channel TRPC5 independently of tyrosine kinases. Br J Pharmacol 159:1486–1496
10. Xie Q, Zhang Y, Cai Sun X, Zhai C, Bonanno JA (2005) Expression and functional evaluation of transient receptor potential channel 4 in bovine corneal endothelial cells. Exp Eye Res 81:5–14
11. Gifford SM, Yi FX, Bird IM (2006) Pregnancy-enhanced store-operated Ca2+ channel function in uterine artery endothelial cells is associated with enhanced agonist-specific transient receptor potential channel 3-inositol 1,4,5-trisphosphate receptor 2 interaction. J Endocrinol 190:385–395
12. Hamdollah Zadeh MA, Glass CA, Magnussen A, Hancox JC, Bates DO (2008) VEGF-mediated elevated intracellular calcium and angiogenesis in human microvascular endothelial cells in vitro are inhibited by dominant negative TRPC6. Microcirculation 15:605–614
13. Thilo F, Loddenkemper C, Berg E, Zidek W, Tepel M (2009) Increased TRPC3 expression in vascular endothelium of patients with malignant hypertension. Mod Pathol 22: 426–430

14. Kwiatek AM, Minshall RD, Cool DR, Skidgel RA, Malik AB, Tiruppathi C (2006) Caveolin-1 regulates store-operated Ca2+ influx by binding of its scaffolding domain to transient receptor potential channel-1 in endothelial cells. Mol Pharmacol 70:1174–1183
15. Singh I, Knezevic N, Ahmmed GU, Kini V, Malik AB, Mehta D (2007) Galphaq-TRPC6-mediated Ca2+ entry induces RhoA activation and resultant endothelial cell shape change in response to thrombin. J Biol Chem 282:7833–7843
16. Thilo F, Baumunk D, Krause H, Schrader M, Miller K, Loddenkemper C, Zakrzewicz A, Krueger K, Zidek W, Tepel M (2009) Transient receptor potential canonical type 3 channels and blood pressure in humans. J Hypertens 27:1217–1223
17. Abdullaev IF, Bisaillon JM, Potier M, Gonzalez JC, Motiani RK, Trebak M (2008) Stim1 and Orai1 mediate CRAC currents and store-operated calcium entry important for endothelial cell proliferation. Circ Res 103:1289–1299
18. Ge R, Tai Y, Sun Y, Zhou K, Yang S, Cheng T, Zou Q, Shen F, Wang Y (2009) Critical role of TRPC6 channels in VEGF-mediated angiogenesis. Cancer Lett 283:43–51
19. Paria BC, Bair AM, Xue J, Yu Y, Malik AB, Tiruppathi C (2006) Ca2+ influx induced by protease-activated receptor-1 activates a feed-forward mechanism of TRPC1 expression via nuclear factor-kappaB activation in endothelial cells. J Biol Chem 281:20715–20727
20. Yang LX, Guo RW, Liu B, Wang XM, Qi F, Guo CM, Shi YK, Wang H (2009) Role of TRPC1 and NF-kappaB in mediating angiotensin II-induced Ca2+ entry and endothelial hyperpermeability. Peptides 30:1368–1373
21. Brown RC, Wu L, Hicks K, O'Neil RG (2008) Regulation of blood-brain barrier permeability by transient receptor potential type C and type v calcium-permeable channels. Microcirculation 15:359–371
22. Poteser M, Graziani A, Rosker C, Eder P, Derler I, Kahr H, Zhu MX, Romanin C, Groschner K (2006) TRPC3 and TRPC4 associate to form a redox-sensitive cation channel. Evidence for expression of native TRPC3-TRPC4 heteromeric channels in endothelial cells. J Biol Chem 281:13588–13595
23. Huang H, Wang W, Liu P, Jiang Y, Zhao Y, Wei H, Niu W (2009) TRPC1 expression and distribution in rat hearts. Eur J Histochem 53:217–223
24. Moller CC, Mangos S, Drummond IA, Reiser J (2008) Expression of trpC1 and trpC6 orthologs in zebrafish. Gene Expr Patterns 8:291–296
25. Yao X, Garland CJ (2005) Recent developments in vascular endothelial cell transient receptor potential channels. Circ Res 97:853–863
26. Thodeti CK, Matthews B, Ravi A, Mammoto A, Ghosh K, Bracha AL, Ingber DE (2009) TRPV4 channels mediate cyclic strain-induced endothelial cell reorientation through integrin-to-integrin signaling. Circ Res 104:1123–1130
27. Willette RN, Bao W, Nerurkar S, Yue TL, Doe CP, Stankus G, Turner GH, Ju H, Thomas H, Fishman CE, Sulpizio A, Behm DJ, Hoffman S, Lin Z, Lozinskaya I, Casillas LN, Lin M, Trout RE, Votta BJ, Thorneloe K, Lashinger ES, Figueroa DJ, Marquis R, Xu X (2008) Systemic activation of the transient receptor potential vanilloid subtype 4 channel causes endothelial failure and circulatory collapse: Part 2. J Pharmacol Exp Ther 326:443–452
28. AbouAlaiwi WA, Takahashi M, Mell BR, Jones TJ, Ratnam S, Kolb RJ, Nauli SM (2009) Ciliary polycystin-2 is a mechanosensitive calcium channel involved in nitric oxide signaling cascades. Circ Res 104:860–869
29. Loot AE, Popp R, Fisslthaler B, Vriens J, Nilius B, Fleming I (2008) Role of cytochrome P450-dependent transient receptor potential V4 activation in flow-induced vasodilatation. Cardiovasc Res 80:445–452
30. Nauli SM, Kawanabe Y, Kaminski JJ, Pearce WJ, Ingber DE, Zhou J (2008) Endothelial cilia are fluid shear sensors that regulate calcium signaling and nitric oxide production through polycystin-1. Circulation 117:1161–1171
31. Vriens J, Owsianik G, Fisslthaler B, Suzuki M, Janssens A, Voets T, Morisseau C, Hammock BD, Fleming I, Busse R, Nilius B (2005) Modulation of the Ca2 permeable cation channel TRPV4 by cytochrome P450 epoxygenases in vascular endothelium. Circ Res 97:908–915

32. Zhang DX, Mendoza SA, Bubolz AH, Mizuno A, Ge ZD, Li R, Warltier DC, Suzuki M, Gutterman DD (2009) Transient receptor potential vanilloid type 4-deficient mice exhibit impaired endothelium-dependent relaxation induced by acetylcholine in vitro and in vivo. Hypertension 53:532–538
33. Ma YY, Huo HR, Li CH, Zhao BS, Li LF, Sui F, Guo SY, Jiang TL (2008) Effects of cinnamaldehyde on PGE2 release and TRPV4 expression in mouse cerebral microvascular endothelial cells induced by interleukin-1beta. Biol Pharm Bull 31:426–430
34. Mendoza SA, Fang J, Gutterman DD, Wilcox DA, Bubolz AH, Li R, Suzuki M, Zhang DX (2010) TRPV4-mediated endothelial Ca2+ influx and vasodilation in response to shear stress. Am J Physiol Heart Circ Physiol 298:H466–H476
35. Gao F, Wang DH (2010) Hypotension induced by activation of the transient receptor potential vanilloid 4 channels: role of Ca2+-activated K+ channels and sensory nerves. J Hypertens 28:102–110
36. Fian R, Grasser E, Treiber F, Schmidt R, Niederl P, Rosker C (2007) The contribution of TRPV4-mediated calcium signaling to calcium homeostasis in endothelial cells. J Recept Signal Transduct Res 27:113–124
37. Troidl C, Troidl K, Schierling W, Cai WJ, Nef H, Mollmann H, Kostin S, Schimanski S, Hammer L, Elsasser A, Schmitz-Rixen T, Schaper W (2009) Trpv4 induces collateral vessel growth during regeneration of the arterial circulation. J Cell Mol Med 13:2613–2621
38. Kohler R, Heyken WT, Heinau P, Schubert R, Si H, Kacik M, Busch C, Grgic I, Maier T, Hoyer J (2006) Evidence for a functional role of endothelial transient receptor potential V4 in shear stress-induced vasodilatation. Arterioscler Thromb Vasc Biol 26:1495–1502
39. Earley S, Gonzales AL, Crnich R (2009) Endothelium-dependent cerebral artery dilation mediated by TRPA1 and Ca2+-Activated, K+ channels. Circ Res 104:987–994
40. Marrelli SP, O'Neil RG, Brown RC, Bryan RM Jr. (2007) PLA2 and TRPV4 channels regulate endothelial calcium in cerebral arteries. Am J Physiol Heart Circ Physiol 292:H1390–H1397
41. Mangos S, Liu Y, Drummond IA (2007) Dynamic expression of the osmosensory channel trpv4 in multiple developing organs in zebrafish. Gene Expr Patterns 7:480–484
42. Hecquet CM, Ahmmed GU, Vogel SM, Malik AB (2008) Role of TRPM2 channel in mediating H_2O_2-induced Ca2+ entry and endothelial hyperpermeability. Circ Res 102:347–355
43. Inoue K, Xiong ZG (2009) Silencing, TRPM7 promotes growth/proliferation and nitric oxide production of vascular endothelial cells via the ERK pathway. Cardiovasc Res 83:547–557
44. Paria BC, Vogel SM, Ahmmed GU, Alamgir S, Shroff J, Malik AB, Tiruppathi C (2004) Tumor necrosis factor-alpha-induced TRPC1 expression amplifies store-operated Ca2+ influx and endothelial permeability. Am J Physiol Lung Cell Mol Physiol 287:L1303–L1313
45. Ma X, Qiu S, Luo J, Ma Y, Ngai CY, Shen B, Wong CO, Huang Y, Yao X (2010) Functional role of vanilloid transient receptor potential 4-canonical transient receptor potential 1 complex in flow-induced Ca2+ influx. Arterioscler Thromb Vasc Biol 30:851–858
46. Berridge MJ, Bootman MD, Roderick HL (2003) Calcium signalling: dynamics, homeostasis and remodelling. Nat Rev Mol Cell Biol 4:517–529
47. Spector AA (2009) Arachidonic acid cytochrome P450 epoxygenase pathway. J Lipid Res 50 (Suppl):S52-S56
48. Ross RA (2003) Anandamide and vanilloid TRPV1 receptors. Br J Pharmacol 140:790–801
49. Cioffi DL, Wu S, Stevens T (2003) On the endothelial cell I(SOC). Cell Calcium 33:323–336
50. Salido GM, Sage SO, Rosado JA (2009) TRPC channels and store-operated Ca(2+) entry. Biochim Biophys Acta 1793:223–230
51. Freichel M, Suh SH, Pfeifer A, Schweig U, Trost C, Weissgerber P, Biel M, Philipp S, Freise D, Droogmans G, Hofmann F, Flockerzi V, Nilius B (2001) Lack of an endothelial store-operated Ca2+ current impairs agonist-dependent vasorelaxation in TRP4–/– mice. Nat Cell Biol 3:121–127

52. Tiruppathi C, Freichel M, Vogel SM, Paria BC, Mehta D, Flockerzi V, Malik AB (2002) Impairment of store-operated Ca2+ entry in TRPC4(−/−) mice interferes with increase in lung microvascular permeability. Circ Res 91:70–76
53. Sundivakkam PC, Kwiatek AM, Sharma TT, Minshall RD, Malik AB, Tiruppathi C (2009) Caveolin-1 scaffold domain interacts with TRPC1 and IP3R3 to regulate Ca2+ store release-induced Ca2+ entry in endothelial cells. Am J Physiol Cell Physiol 296:C403–C413
54. Trebak M (2009) STIM1/Orai1, ICRAC, and endothelial SOC. Circ Res 104:e56–57
55. Beech DJ (2009) Harmony and discord in endothelial calcium entry. Circ Res 104:e22–23
56. Cahalan MD (2009) STIMulating store-operated Ca(2+) entry. Nat Cell Biol 11:669–677
57. Bair AM, Thippegowda PB, Freichel M, Cheng N, Ye RD, Vogel SM, Yu Y, Flockerzi V, Malik AB, Tiruppathi C (2009) Ca2+ entry via TRPC channels is necessary for thrombin-induced NF-kappaB activation in endothelial cells through AMP-activated protein kinase and protein kinase Cdelta. J Biol Chem 284:563–574
58. Thippegowda PB, Singh V, Sundivakkam PC, Xue J, Malik AB, Tiruppathi C (2010) Ca2+ influx via TRPC channels induces NF-kappaB-dependent A20 expression to prevent thrombin-induced apoptosis in endothelial cells. Am J Physiol Cell Physiol 298:C656–C664
59. Fantozzi I, Zhang S, Platoshyn O, Remillard CV, Cowling RT, Yuan JX (2003) Hypoxia increases AP-1 binding activity by enhancing capacitative Ca2+ entry in human pulmonary artery endothelial cells. Am J Physiol Lung Cell Mol Physiol 285:L1233–L1245
60. Earley S, Waldron BJ, Brayden JE (2004) Critical role for transient receptor potential channel TRPM4 in myogenic constriction of cerebral arteries. Circ Res 95:922–929
61. Kwan HY, Shen B, Ma X, Kwok YC, Huang Y, Man YB, Yu S, Yao X (2009) TRPC1 associates with BK(Ca) channel to form a signal complex in vascular smooth muscle cells. Circ Res 104:670–678
62. Earley S, Heppner TJ, Nelson MT, Brayden JE (2005) TRPV4 forms a novel Ca2+ signaling complex with ryanodine receptors and BKCa channels. Circ Res 97:1270–1279
63. Feletou M, Vanhoutte PM (2009) EDHF: an update. Clin Sci (Lond) 117:139–155
64. Liu C, Ngai CY, Huang Y, Ko WH, Wu M, He GW, Garland CJ, Dora KA, Yao X (2006) Depletion of intracellular Ca2+ stores enhances flow-induced vascular dilatation in rat small mesenteric artery. Br J Pharmacol 147:506–515
65. Jenkins CM, Cedars A, Gross RW (2009) Eicosanoid signalling pathways in the heart. Cardiovasc Res 82:240–249
66. Grgic I, Kaistha BP, Hoyer J, Kohler R (2009) Endothelial, Ca+-activated K+ channels in normal and impaired EDHF-dilator responses–relevance to cardiovascular pathologies and drug discovery. Br J Pharmacol 157:509–526
67. Poblete IM, Orliac ML, Briones R, Adler-Graschinsky E, Huidobro-Toro JP (2005) Anandamide elicits an acute release of nitric oxide through endothelial TRPV1 receptor activation in the rat arterial mesenteric bed. J Physiol 568:539–551
68. Bratz IN, Dick GM, Tune JD, Edwards JM, Neeb ZP, Dincer UD, Sturek M (2008) Impaired capsaicin-induced relaxation of coronary arteries in a porcine model of the metabolic syndrome. Am J Physiol Heart Circ Physiol 294:H2489–H2496
69. Cioffi DL, Lowe K, Alvarez DF, Barry C, Stevens T (2009) TRPing on the lung endothelium: calcium channels that regulate barrier function. Antioxid Redox Signal 11:765–776
70. Tiruppathi C, Ahmmed GU, Vogel SM, Malik AB (2006) Ca2+ signaling, TRP channels, and endothelial permeability. Microcirculation 13:693–708
71. Clark K, Middelbeek J, van Leeuwen FN (2008) Interplay between TRP channels and the cytoskeleton in health and disease. Eur J Cell Biol 87:631–640
72. Gerzanich V, Woo SK, Vennekens R, Tsymbalyuk O, Ivanova S, Ivanov A, Geng Z, Chen Z, Nilius B, Flockerzi V, Freichel M, Simard JM (2009) De novo expression of Trpm4 initiates secondary hemorrhage in spinal cord injury. Nat Med 15:185–191
73. Pocock TM, Foster RR, Bates DO (2004) Evidence of a role for TRPC channels in VEGF-mediated increased vascular permeability in vivo. Am J Physiol Heart Circ Physiol 286:H1015–H1026

74. Alvarez DF, King JA, Weber D, Addison E, Liedtke W, Townsley MI (2006) Transient receptor potential vanilloid 4-mediated disruption of the alveolar septal barrier: a novel mechanism of acute lung injury. Circ Res 99:988–995
75. Hamanaka K, Jian MY, Weber DS, Alvarez DF, Townsley MI, Al-Mehdi AB, King JA, Liedtke W, Parker JC (2007) TRPV4 initiates the acute calcium-dependent permeability increase during ventilator-induced lung injury in isolated mouse lungs. Am J Physiol Lung Cell Mol Physiol 293:L923–L932
76. He Y, Yao G, Savoia C, Touyz RM (2005) Transient receptor potential melastatin 7 ion channels regulate magnesium homeostasis in vascular smooth muscle cells: role of angiotensin II. Circ Res 96:207–215
77. Graziani A, Poteser M, Heupel WM, Schleifer H, Krenn M, Drenckhahn D, Romanin C, Baumgartner W, Groschner K (2010) Cell-cell contact formation governs Ca2+ signaling by TRPC4 in the vascular endothelium: evidence for a regulatory TRPC4-beta-catenin interaction. J Biol Chem 285:4213–4223
78. Yu PC, Gu SY, Bu JW, Du JL (2010) TRPC1 is essential for in vivo angiogenesis in zebrafish. Circ Res 106:1221–1232
79. Droge W (2002) Free radicals in the physiological control of cell function. Physiol Rev 82:47–95
80. Balzer M, Lintschinger B, Groschner K (1999) Evidence for a role of Trp proteins in the oxidative stress-induced membrane conductances of porcine aortic endothelial cells. Cardiovasc Res 42:543–549
81. Naziroglu M (2007) New molecular mechanisms on the activation of TRPM2 channels by oxidative stress and ADP-ribose. Neurochem Res 32:1990–2001
82. Naziroglu M, Luckhoff A (2008) A calcium influx pathway regulated separately by oxidative stress and ADP-Ribose in TRPM2 channels: single channel events. Neurochem Res 33:1256–1262
83. Aarts M, Iihara K, Wei WL, Xiong ZG, Arundine M, Cerwinski W, MacDonald JF, Tymianski M (2003) A key role for TRPM7 channels in anoxic neuronal death. Cell 115:863–877
84. Bessac BF, Sivula M, von Hehn CA, Escalera J, Cohn L, Jordt SE (2008) TRPA1 is a major oxidant sensor in murine airway sensory neurons. J Clin Invest 118:1899–1910
85. Resnick N, Yahav H, Shay-Salit A, Shushy M, Schubert S, Zilberman LC, Wofovitz E (2003) Fluid shear stress and the vascular endothelium: for better and for worse. Prog Biophys Mol Biol 81:177–199
86. White CR, Frangos JA (2007) The shear stress of it all: the cell membrane and mechanochemical transduction. Philos Trans R Soc Lond B Biol Sci 362:1459–1467
87. Suzuki M, Mizuno A, Kodaira K, Imai M (2003) Impaired pressure sensation in mice lacking TRPV4. J Biol Chem 278:22664–22668
88. Strotmann R, Harteneck C, Nunnenmacher K, Schultz G, Plant TD (2000) OTRPC4, a non-selective cation channel that confers sensitivity to extracellular osmolarity. Nat Cell Biol 2:695–702
89. Vriens J, Watanabe H, Janssens A, Droogmans G, Voets T, Nilius B (2004) Cell swelling, heat, and chemical agonists use distinct pathways for the activation of the cation channel TRPV4. Proc Natl Acad Sci USA 101:396–401
90. Yao X, Kwan HY, Chan FL, Chan NW, Huang Y (2000) A protein kinase G-sensitive channel mediates flow-induced Ca(2+) entry into vascular endothelial cells. FASEB J 14:932–938
91. Ohata H, Ikeuchi T, Kamada A, Yamamoto M, Momose K (2001) Lysophosphatidic acid positively regulates the fluid flow-induced local Ca(2+) influx in bovine aortic endothelial cells. Circ Res 88:925–932
92. Nauli SM, Alenghat FJ, Luo Y, Williams E, Vassilev P, Li X, Elia AE, Lu W, Brown EM, Quinn SJ, Ingber DE, Zhou J (2003) Polycystins 1 and 2 mediate mechanosensation in the primary cilium of kidney cells. Nat Genet 33:129–137

93. Sharif-Naeini R, Folgering JH, Bichet D, Duprat F, Lauritzen I, Arhatte M, Jodar M, Dedman A, Chatelain FC, Schulte U, Retailleau K, Loufrani L, Patel A, Sachs F, Delmas P, Peters DJ, Honore E (2009) Polycystin-1 and -2 dosage regulates pressure sensing. Cell 139:587–596
94. Kottgen M, Buchholz B, Garcia-Gonzalez MA, Kotsis F, Fu X, Doerken M, Boehlke C, Steffl D, Tauber R, Wegierski T, Nitschke R, Suzuki M, Kramer-Zucker A, Germino GG, Watnick T, Prenen J, Nilius B, Kuehn EW, Walz G (2008) TRPP2 and TRPV4 form a polymodal sensory channel complex. J Cell Biol 182:437–447
95. Tsiokas L, Arnould T, Zhu C, Kim E, Walz G, Sukhatme VP (1999) Specific association of the gene product of PKD2 with the TRPC1 channel. Proc Natl Acad Sci USA 96:3934–3939
96. Pedersen SF, Nilius B (2007) Transient receptor potential channels in mechanosensing and cell volume regulation. Methods Enzymol 428:183–207
97. Wang D, Iversen J, Strandgaard S (2000) Endothelium-dependent relaxation of small resistance vessels is impaired in patients with autosomal dominant polycystic kidney disease. J Am Soc Nephrol 11:1371–1376
98. Devuyst O, Persu A, Vo-Cong MT (2003) Autosomal dominant polycystic kidney disease: modifier genes and endothelial dysfunction. Nephrol Dial Transplant 18:2211–2215
99. Zhu G, Gulsvik A, Bakke P, Ghatta S, Anderson W, Lomas DA, Silverman EK, Pillai SG (2009) Association of TRPV4 gene polymorphisms with chronic obstructive pulmonary disease. Hum Mol Genet 18:2053–2062
100. Paravicini TM, Yogi A, Mazur A, Touyz RM (2009) Dysregulation of vascular TRPM7 and annexin-1 is associated with endothelial dysfunction in inherited hypomagnesemia. Hypertension 53:423–429
101. Kumar B, Dreja K, Shah SS, Cheong A, Xu SZ, Sukumar P, Naylor J, Forte A, Cipollaro M, McHugh D, Kingston PA, Heagerty AM, Munsch CM, Bergdahl A, Hultgardh-Nilsson A, Gomez MF, Porter KE, Hellstrand P, Beech DJ (2006) Upregulated, TRPC1 channel in vascular injury in vivo and its role in human neointimal hyperplasia. Circ Res 98:557–563

Chapter 41
TRP Channels in the Cardiopulmonary Vasculature

Alexander Dietrich and Thomas Gudermann

Abstract Transient receptor potential (TRP) channels are expressed in almost every human tissue, including the heart and the vasculature. They play unique roles not only in physiological functions but, if over-expressed, also in pathophysiological disease states. Cardiovascular diseases are the leading cause of death in the industrialized countries. Therefore, TRP channels are attractive drug targets for more effective pharmacological treatments of these diseases. This review focuses on three major cell types of the cardiovascular system: cardiomyocytes as well as smooth muscle cells and endothelial cells from the systemic and pulmonary circulation. TRP channels initiate multiple signals in all three cell types (e.g. contraction, migration) and are involved in gene transcription leading to cell proliferation or cell death. Identification of their genes has significantly improved our knowledge of multiple signal transduction pathways in these cells. Some TRP channels are important cellular sensors and are mostly permeable to Ca^{2+}, while most other TRP channels are receptor activated and allow for the entry of Na^+, Ca^{2+} and Mg^{2+}. Physiological functions of TRPA, TRPC, TRPM, TRPP and TRPV channels in the cardiovascular system, dissected by down-regulating channel activity in isolated tissues or by the analysis of gene-deficient mouse models, are reviewed. The involvement of TRPs as homomeric or heteromeric channels in pathophysiological processes in the cardiovascular system like heart failure, cardiac hypertrophy, hypertension as well as edema formation by increased endothelial permeability will be discussed.

41.1 Introduction

The cardiovascular system includes the heart as well as the blood circulation trough the vasculature. The action of the heart provides sufficient blood flow to peripheral tissues and organs based on their metabolic demand as well as to the pulmonary

A. Dietrich (✉)
Walther-Straub-Institute for Pharmacology and Toxicology, School of Medicine,
Ludwig-Maximilians-University München, 80336 Munich, Germany
e-mail: alexander.dietrich@lrz.uni-muenchen.de

circulation which supplies the blood circulation with oxygen and removes carbon dioxide. This review will focus on three major cell types of the cardiovascular system: cardiomyocytes, as well as smooth muscle cells and endothelial cells from the systemic and pulmonary circulation. Ion channels not only initiate contraction, but also induce cell proliferation and cell death via regulation of gene transcription. Therefore, the identification of TRP channel genes has significantly improved our knowledge of multiple signal transduction pathways in these cells. Moreover, evidence for their role in slowly progressing remodelling processes of the cardiovascular system manifested as cardiac hypertrophy, cardiomyopathy, arteriosclerosis, pulmonary hypertension, and other proliferative/degenerative disorders has also accumulated in recent years.

The superfamily of **t**ransient **r**eceptor **p**otential (**TRP**) cation channels is composed of proteins that were initially identified in the *Drosophila* eye, where they are involved in the photoreceptor signal transduction pathway (see [1] for a recent review). The first mammalian family of *classical* or *canonical* TRP cation channels (TR**PC**) to be discovered is highly related to the *Drosophila* TRPs. In the following years other TRP families like TR**PM** (for *melastatin*), TR**PML** (for *mucolipidins*), TR**PV** (for *vanilloid* receptor), TR**PP** (for **P**KD proteins) and TR**PA** (for *ankyrin rich proteins*) were described. This review will focus on the TRPA, TRPC, TRPM, TRPP and TRPV families (see Fig. 41.1) whose members are expressed in the heart, the endothelium and vascular smooth muscle cells. Their functional properties as well as evidence for their involvement in physiological and pathophysiological processes of the heart, of the endothelium and in smooth muscle cells will be discussed below in alphabetical order. Other members of the TRP superfamily and their proposed physiological function are summarized elsewhere [1, 2].

41.1.1 TRPA Channels

The TRPA subfamily consists only of a single member, TRPA1. A structural characteristic of this channel is its high number of 14 N-terminal ankyrin repeats. These motifs which are also present in other TRP families are believed to be important for membrane localization, channel tetramerization and intramolecular interactions (reviewed in [3]). Electrophilic compounds activate TRPA1, including substances found in pungent foods such as garlic and mustard oil. TRPA1 seems to have a central role in the pain response to volatile irritants like tear gas. Expression of TRPA1 was demonstrated in endothelial cells [4] and its function there will be discussed in the designated paragraph below.

41.1.2 TRPC Channels

The TRPC family consists of seven family members which can be subdivided into subfamilies on the basis of their amino acid homology. While TRPC1 and TRPC2 are almost unique, TRPC4 and TRPC5 share ~65% homology. TRPC3, 6 and 7 form a structural and functional subfamily sharing 70–80% homology and their direct activation by diacylglycerol (DAG) [5, 6]. Accordingly, TRPC3/6/7 are gated by agonist-induced receptor activation and subsequent phospholipase C activation resulting in DAG production from phosphatidylinositol 4,5-bisphosphate (PIP_2) and mediate receptor operated cation entry (**ROCE**) (see Fig. 41.2) [7]. For the other TRPC channels (TRPC1, 4, 5) as well as for TRPC3 and TRPC7 store-operated cation entry (**SOCE**) mechanisms are discussed. These types of channels are proposed to be regulated by the filling status of intracellular Ca^{2+} stores. Inositol-1,4,5 trisphosphate (IP_3)-induced store depletion through IP_3 receptors localized at the membrane of the endoplasmatic reticulum is mimicked by application of thapsigargin or cyclopiazonic acid (CPA) which inhibit ATP-driven SERCA (for sarcoplasmic/endoplasmatic reticulum **Ca^{2+}**) pumps resulting in Ca^{2+} leakage from intracellular stores and subsequent Ca^{2+} entry from extracellular compartments. The underlying mechanisms of this signalling pathway were not known [8]. Recently however, a major breakthrough in understanding SOC was reported by the identification of the calcium sensor STIM and the Orai channels [9]. STIM1 which is predominantly located in the ER is a single transmembrane domain protein containing two N-terminal Ca^{2+}-binding EF hands in the ER lumen and is, thus, ideally situated for detecting calcium levels in the ER. Upon Ca^{2+} release reduced calcium levels induce STIM redistribution to punctae and the opening of store-operated channels the so called Orai proteins in the plasma membrane [9]. Although Orai channels can form homomultimers as pore forming units, it was also suggested that Orai molecules might interact with TRPC channels when heterologously expressed in HEK293 cells to form heteromeric store-operated channels [10]. Moreover, evidence for the interaction of the calcium sensor STIM1 with TRPC proteins was presented for the same heterologous expression system (see Fig. 41.2) [11]. However until now, we were not able to verify these interactions in native tissues like aortic smooth muscle cells and blood platelets [12–14].

Members of the TRPC family share common structures: they contain three to four N-terminal ankyrin repeats [3], six transmembrane-spanning domains (S1-S6), a putative pore region between S5 and S6 and several protein binding domains. TRPC monomers assemble to a functional channel homo- or heterotetramer. Different TRPC monomers heteromultimerize within the confines of TRPC subfamilies [15]. In addition, complexes containing three different TRPC monomers (TRPC1 and TPRC4 or 5 and TRPC3 or 6) were detected in embryonic brain [16]. In these combinations TRPC1 serves as linker or adaptor protein joining monomers of the TRPC4/5 subfamily with the TRPC3/6/7 subfamily in one heteromeric channel complex [16]. Evidence for the importance of the highly conserved pore region for TRPC function stems from in vitro mutagenesis studies resulting not only in the complete loss of channel activity, but also in a dominant-negative (dn) phenotype of the mutated channel monomer for the functional homo- or heteromeric channel tetramer [15].

All members of the TRPC subfamily are molecular correlates of non-selective cation channels identified in the cell types of the cardiopulmonary vasculature [17, 18] except TRPC2 which is a pseudogene in humans in contrast to mice where it is functional only in testis and in the murine specific vomeronasal organ [19]. In general TRPC channels are non-selective Ca^{2+}-permeable cation channels, although the selectivity ratio $P_{Ca^{2+}}/P_{Na^+}$ varies significantly from non selective (TRPC1) to slightly selective (range of 1–3: TRPC3, TRPC4, TRPC5, TRPC7) to more selective (range of 5: TRPC6) [17], if over-expressed in HEK 293 cells. Studying TRPC channels in vitro in heterologous overexpression systems, however, has given rise to a substantial body of controversial data concerning channel properties and regulation. The lack of specific channel blockers and the use of TRPC antibodies which are often unspecific [20–23] have complicated the in situ identification and characterization of TRPC channels in cardiovascular tissues. Therefore, gene-deficient mouse models and specific down-regulation of TRPC proteins appear to be straightforward strategies for conclusive *in vivo* and cellular studies.

41.1.3 The TRPM Family

The mammalian TRPM family consists of 8 genes, TRPM1-8. Three members of the TRPM family, i.e. TRPM2, TRPM6 and TRPM7, differ from other ion channels because they harbour enzyme domains in their C-termini. The NUDT9 domain in TRPM2 was shown to have ADP-ribose pyrophosphatase activity, while the C-termini of TRPM7 and TRPM6 contain a serine/threonine protein kinase domain resembling that of elongation factor 2 (eEF-2) kinase and other α-kinases [24–26].

TRPM4 and TRPM5 are mainly permeable to monovalent cations, are gated by increases in intracellular Ca^{2+} and exhibit a pronounced voltage modulation [27–33]. Intriguingly, mice with a genetically disrupted TRPM5 gene display impaired sweet, bitter and umami taste perception [28], but the precise role of TRPM5 in taste receptor cells is still under debate [29–33].

TRPM7 was shown to be a constitutively active cation channel, which is inhibited by intracellular free Mg^{2+} ($[Mg^{2+}]_i$) and Mg·ATP ($[Mg·ATP]_i$) [34, 35]. TRPM7 is permeable to a broad range of divalent cations and, in contrast to other TRPs, is slightly more permeable to Mg^{2+} than to Ca^{2+} [36]. Recently, annexin A1 and myosin II were identified as physiological substrates of the TRPM7 kinase domain [37]. The TRPM7-dependent phosphorylation of myosin II may play a role in the regulation of actomyosin contractility and cell adhesion [38], while the physiological role of TRPM7-mediated phosphorylation of annexin A1 is still unknown. The kinase domain of TRPM7 also displays autophosphorylation activity [39, 40], but kinase activity appears not to be essential for TRPM7-mediated currents [40–42].

Another TRPM channel subunit coupled to a kinase domain, TRPM6, was found to be an essential component of the physiological cellular machinery governing Mg^{2+} homeostasis. Loss-of-function mutations in the human TRPM6 gene result in hypomagnesemia with secondary hypocalcemia (HSH) [43–45]. HSH is an autosomal recessive disorder characterized by low serum Mg^{2+} levels due to defective intestinal absorption and/or renal Mg^{2+} wasting. Supplementation with high Mg^{2+} doses overcomes the Mg^{2+} deficiency of HSH patients [25, 46].

TRPM7 also seems to play a pivotal role of for body Mg^{2+} and Ca^{2+} homeostasis of vertebrates. It was demonstrated that DT40 chicken lymphocytes lacking an intact TRPM7 gene are not viable [41]. However, supplementation of the cell culture medium with high levels of Mg^{2+} restores the viability of mutant DT40 cells. Importantly, mammalian TRPM7 (but not the homologous TRPM6) was able to rescue the mutant DT40 cell line cultured in normal, physiological Mg^{2+} concentrations [41, 42].

There is substantial evidence [42, 47] that TRPM6 functions only as a subunit of heterooligomeric channel complexes formed by TRPM7 and TRPM6: TRPM6 potentiates TRPM7 channel activity. In line with these observations, human TRPM6 was not able to rescue the DT40 cell line containing a loss-of-function mutation in the TRPM7 gene [42].

Expression of TRPM4, 5 and 7 in cardiomyocytes was demonstrated recently and all TRPM channels are expressed in smooth muscle cells. In endothelial cells only TRPM5 is not detectable by reverse-transcription (RT) PCR (reviewed in [48]).

41.1.4 The TRPP Family

The polycystin family is structurally divided into two groups, the polycystic kidney disease 1-like proteins and the polycystic disease 2-like proteins. The latter ones are also called TRPP2-like proteins with TRPP2, TRPP3 and TRPP5 as family members. They are cation-permeable channels, usually with $P_{Ca}^{2+}/P_{Na}^{+} > 1$ and exhibit significant homology to other TRPs, although they neither contain a TRP domain nor ankyrin repeats [49, 50]. Their predicted topology is characterized by a large extracellular loop between the first and second transmembrane domain. TRPP2 was originally discovered as a gene mutated in autosomal dominant polycystic kidney disease (ADPKD) [51]. Both TRPP1 and TRPP2 are widely expressed, and a significant body of evidence supports the notion that TRPP1 and TRPP2 are physically coupled and act as a signaling complex [52]. TRPP1 may be required for the localization of TRPP2 at the plasma membrane [52] and suppresses the constitutive channel activity of TRPP2 [50]. Yet, it should be noted that TRPP2 might also have roles independently of TRPP1, as TRPP2 expression and TRPP2-like activity has been detected in left ventricular myocytes in the absence of TRPP1 [53].

Both TRPPs are expressed in heart and in human coronary arterial smooth muscle cells, while other smooth muscle cells express only TRPP2. TRPP1 is the only TRPP channel detected in umbilical vein endothelial cells (reviewed in [48])

41.1.5 TRPV Channels

The TRPV family consists of six members. The vanilloid receptor 1, TRPV1, was the first mammalian member of this family to be identified [54], and the one which has been studied most intensively. TRPV1 is predominately expressed in sensory neurons and is believed to play a crucial role in temperature sensing and nociception

[55]. Initially, TRPV1 and TRPV2 were mainly associated with sensory processes. However, TRPV2 mRNA can be detected in a wide range of tissues and organs including smooth muscle cells of blood vessels. Like TRPV2, TRPV4 is activated by heat and is expressed in neurons (reviewed in [56]). However, TRPV4 mRNA is also widely expressed in tissues like kidney, bladder, keratinocytes as well as in smooth muscle cells and the vascular endothelium [56]. A possible formation of TRPV2/4 multiprotein complexes in smooth muscle cells has not been analyzed yet. By FRET-analysis and co-immunoprecipitation experiments of heterologously overexpressed proteins, no interaction of TRPV2 and 4, but instead of TRPV1 and 2 as well as TRPV5 and 6 was detected [57]. For these reasons the relevance of these channels and their exact function in smooth muscle cells and the vascular endothelium remain to be elucidated.

41.2 TRP Expression in the Heart: Physiological Function and Pathophysiological Implications

Cardiac output is influenced by altered size and functionality of the heart resulting from inotropic and chronotropic changes. The involvement of TRP channels in physiological and pathophysiological processes like cardiac hypertrophy or impairment of heart function by arrhythmia or cardiomyopathy are summarized in this part of the review.

There is mounting evidence for the role of Ca^{2+} influx in slowly progressive remodelling processes of the heart such as *cardiac hypertrophy* induced by pressure overload such as chronic hypertension and aortic stenosis. Intracellular Ca^{2+} acts to induce the hypertrophic response, although the source of the Ca^{2+} responsible for this is still elusive. Previous studies, however, have convincingly demonstrated the importance of the Ca^{2+}/calmodulin-dependent serine/threonine phosphatase calcineurin in mediating cardiac hypertrophy and progressive heart failure (reviewed in [58]). Calcineurin dephosphorylates transcription factors of the nuclear factor of the activated T-cell (**NFAT**) family, thus allowing for their translocation into the nucleus, resulting in the activation of hypertrophic response genes. The calcineurin/NFAT pathway was first discovered in lymphocytes. Most interestingly, it was demonstrated in these cells that NFAT remains in the nucleus only in response to prolonged, low-amplitude Ca^{2+} signals and is insensitive to a transient high-amplitude Ca^{2+} influx (reviewed in [59]). Large increases in Ca^{2+} which are fully reversible in milliseconds are initiated by voltage-gated calcium channels with a subsequent release of Ca^{2+} from the sarcoplasmatic reticulum (SR), the internal calcium store of myocytes. Therefore, voltage-gated calcium channels would be inadequate to activate calcineurin. Hypertrophic factors like endothelin 1 (ET1) and angiotensin II (Ang II) however, bind to G-protein-coupled receptors which initiate ROC by activating TRPC3/6/7 cation channels resulting in prolonged Ca^{2+} increases with lower amplitudes. Along these lines, several studies reported the involvement of TRPC channels, especially TRPC1, 3 and 6, in cardiac hypertrophy (reviewed in

[18]). TRPC3 is up-regulated in several animal models of cardiac hypertrophy and promotes the hypertrophic response [60] as well as hypertrophy-associated gene expression [61]. Most interestingly, a pharmacological approach to reduce cardiac hypertrophy by inhibiting TRPC3 activity was demonstrated recently. A pyrazole compound (Pyr3) selectively and directly inhibits TRPC3 channels [62]. Moreover, Pyr3 also inhibited activation of NFAT and hypertrophic growth in rat neonatal cardiomyocytes and *in vivo* pressure overload-induced cardiac hypertrophy in mice [62]. Because of its high specifity, Pyr3 may represent a novel therapeutic strategy for preventing cardiac hypertrophy without influencing normal cardiac contraction.

TRPC6 may be also involved in cardiac hypertrophy (reviewed in [18]). Knockdown of TRPC6 reduced hypertrophic signalling induced by ET-1, while transgenic mice over-expressing TRPC6 showed increased NFAT-dependent expression of the β-myosin heavy chain, a sensitive marker for cardiomyopathy (see Fig. 41.1 [63]). Interestingly, both TRPC3 and TRPC6 mediate AngII-induced

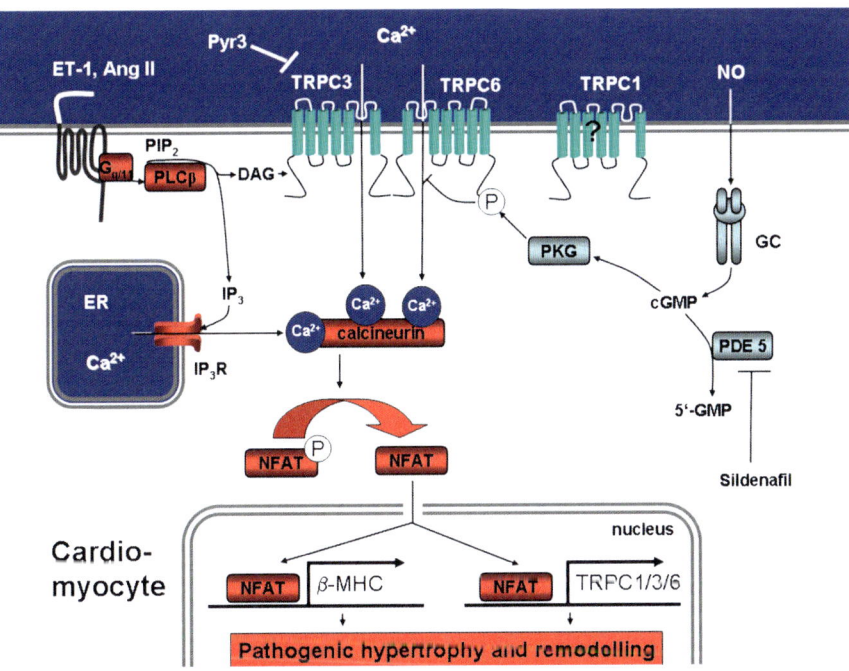

Fig. 41.1 Overview of the implication of TRPC channel activity for cardiac hypertrophy and remodelling in the heart (modified from [18]). Pharmacological inhibition of TRPC channel function by Pyr3 (a novel pyrazole compound) [62] and sildenafil [69] are indicated. Blue areas indicate high Ca^{2+} concentration (1–2 mM), white areas low Ca^{2+} concentrations (50–200 nM). AngII: angiotensin II; cGMP: cyclic guanosine monophosphate; β-MHC: β myosin heavy chain; DAG: diacylglycerol; ET-1: endothelin-1; ER: endoplasmic reticulum; $G_{q/11}$: G protein type q and 11; GC: guanylyl cyclase; IP_3: inositol-1,4,5 trisphosophate; IP_3-R: IP_3 receptor; NFAT: nuclear factor of activated T-cells; PDE5: phosphodiesterase5; PIP_2: phosphatidylinositol 4,5-bisphosphate; PKG: cGMP activated protein kinase; PLC: phospholipase C. See text for details

Ca^{2+} influx in rat neonatal cardiomyocytes and knock-down of either TRPC3 or TRPC6 channels completely suppresses AngII-induced hypertrophy [64]. Thus, TRPC3 and TRPC6 may form heteromeric channel complexes in cardiomyocytes similarly to smooth muscle cells [65] as described below. Both channels are also targets for phosphorylation by protein kinases. While tyrosine phosphorylation stimulates TRPC3 channels [66], serine/threonine (Ser/Thr) phosphorylation by cGMP-activated protein kinase G (PKG) inhibits channel activity. The amino acids Thr11 and Ser263 in TRPC3 [67] as well as Thr69 in TRPC6 [68] are the targets for this negative regulation by the NO-cGMP-PKG pathway. Very recently, it was demonstrated that phosphorylation of TRPC6 might be essential for the anti-hypertrophic effects of phosphodiesterase 5 (PDE5) inhibition [69]. Specific inhibitors of the cGMP-dependent PDE5 like sildenafil prevented TRPC6 mediated Ca^{2+} influx in cardiomyocytes and induced TRPC6 phosphorylation in mouse hearts (see Fig. 41.1 [69]). Therefore, the described antihypertrophic effects after chronic treatment with sildenafil in mice [70, 71] and in patients with systolic heart failure [72] may be due to TRPC6 phosphorylation rather than to PKG-dependent phosphorylation of regulator of G protein signalling 2 (RGS2) as reported earlier [73]. Very recently, TRPC6 phosphorylation at Thr69 was also made responsible for the antihypertrophic effects of artrial (ANP) and brain natriuretic peptides (BNP) [74].

The proposed role of TRPC1 in cardiac hypertrophy is somewhat more difficult to evaluate. Because heterologous expression of TRPC1 has often not produced a functional correlate clearly distinguishable from background signals, TRPC1 may not act physiologically on its own (reviewed in [75]). TRPC1 might rather contribute to functional ion channels composed of heterotetramers complexed with other TRPC isoforms. Along these lines, interactions with other TRP channels appear to be necessary to quantitatively translocate TRPC1 to the plasma membrane as shown by TRPC1-TRPC4 coexpression studies [15]. Nevertheless, TRPC1 seems to be important in heart remodelling processes, because it was reported that gene silencing of TRPC1 using small interference RNA (siRNA) attenuated cardiac hypertrophy [76]. Moreover, not only the TRPC3 and TRPC6 genes [60, 63] but also the TRPC1 gene has conserved NFAT consensus sites in their promoters [76]. Therefore, activated NFAT might stimulate TRPC channel expression through a positive feedback mechanism explaining the long term development of cardiac hypertrophy. Recently, it was reported that a TRPC1-deficient mouse model [12] was protected from pressure overload and lacks a significant hypertrophic response [77]. Moreover, changes of gene expression profiles during the remodelling processes in wild-type mice were not detected in TRPC1–/– mice [77]. Therefore, many different TRPC channels seem to be involved in these processes. Along these lines, in a very recent analysis each of three cardiac transgenic mouse lines expressing dominant-negative (dn) TRPC3, TRPC4 or TRPC6 show attenuated cardiac hypertrophy by neuroendocrine agonist infusion or pressure overload [78]. Most interestingly, dn TRPC4 was able to suppress the activity of the TRPC3/6/7 subfamily [78]. These data favour heterotrimeric complexes of TRPC1 and TRPC4 or TRPC5 as well as TRPC3 or TRPC6 or TRPC7 in the heart like they were originally described in embryonic brain microsomes [16]. However, further studies will be necessary to identify the real molecular make-up of the heteromeric TRPC channel complex in cardiomyocytes.

Increased TRPM4 levels were also demonstrated in cardiac hypertrophy, using freshly isolated ventricular myocytes from spontaneously hypertensive rats. An ion current reflecting salient TRPM4 characteristics was identified by the patch-clamp technique and was easily recordable in cardiomyocytes of spontaneous hypertensive rats but not in myocytes from normal Wistar-Kyoto rats [79].

Compensated cardiac hypertrophy may progress to *heart failure*, in which apoptopic cells will be more abundant than in non-failing hearts. Therefore, myocardial apoptosis is a critical process in heart failure emphasizing inhibition of apoptosis as a promising therapeutic option. Apoptosis is induced by various stimuli including oxidative stress, pro-inflamatory cytokines, catecholamines and AngII [80]. The involvement of four different TRP channels was identified by studies in animal models. TRPC7 seems to link the activation of angiotensin type 1 (AT_1) receptors to myocardial apoptosis [81] and TRPC3 over-expression following ischemia-reperfusion increases apoptosis in adult mouse cardiomyocytes [82]. In contrast to these results, TRPC6 activation might suppress heart failure via inhibition of myofibroblast differentiation. It was reported that TRPC6 over-expression inhibited ET-1-induced myofibroblast formation which resulted in decreasing amounts of cardiac fibroblasts [83]. These fibroblasts serve as mediators of inflammatory and fibrotic myocardial remodelling which also induces heart failure and artrial fibrillation [84].

TRPM2 and its poly(ADP-ribose) polymerase-dependent activation are also involved in oxidative stress-induced cardiomyocyte death [85]. A predominant feature of TRPM2 is the so called Nudix box, a consensus region for pyrophosphatases localized in the cytoplasmic C-terminal tail of the channel protein. This domain confers a unique activation mechanism, gating by adenosine 5′diphosphoribose (ADPR), on TRPM2 [86] Oxidative stress, mimicked by application of H_2O_2, induces TRPM2 currents and increases in $[Ca^{2+}]_i$ in various cell types heterologously expressing TRPM2 [87]. H_2O_2 activates the mitochondrial production of ADPR and may also result in the activation of poly-ADPR polymerases (PARPs). PARP enzymes catalyze the breakdown of NAD into nicotinamide and ADPR [88] Subsequently, ADPR may activate TRPM2 by binding to the C-terminal Nudix domain, inducing large cation currents These currents result in a Na^+ and Ca^{2+} overload leading to mitochondrial membrane disruption, cytochrome c release and caspase 3-dependent chromatin condensation/fragmentation as essential steps for cardiac apoptosis.

As outlined above, mutation in either the TRPP1 or the TRPP2 gene lead to autosomal dominant polycystic kidney disease (PKD), characterized by progressive development of fluid filled cysts in the kidney but also *cardiac valvular abnormalities*. Most interestingly, TRPP2-deficient mice have structural defects in cardiac septation [89]. These abnormalities together with a stable expression pattern of TRPP2 in the heart [90] as well as the identification of a cardiac TRPP2-like large conductance current [53] point to an important role of TRPP2 in the development of interventricular and interatrial septa. Deletion of the tolloid-like 1 gene, which codes for a bone morphogenesis protein-1 – related metalloprotease, produces a cardiac-restricted phenotype [91] similar to the TRPP2–/– phenotype in the heart. Therefore, tolloid-like 1 might regulate heart morphogenesis through a signal transduction cascade involving TRPP2. However, a detailed TRPP2 signalling pathways remains elusive. Moreover, it is important to emphasize that TRPP2 function in the heart seems to be TRPP1-independent, because TRPP2 induces left-right asymmetry development during embryogenesis independent of TRPP1 [92].

Cardiac arrhytmia resulting from elevated intracellular Ca^{2+} levels $[Ca^{2+}]_i$, have been attributed to the activation of a transient inward current. Ca^{2+}-activated non selective calcium currents (CAN) of 25 pS equally permeable to Na^+ and K^+ but not permeable to Ca^{2+} were identified by the patch clamp technique in cardiac cells [93]. Channel characteristics of TRPM4 match these criteria very closely [94] and TRPM4 seems to be expressed to a greater degree in the artrial myocardium than in the ventricular myocardium of the heart [95]. Most interestingly, a mutation in the gene coding for TRPM4 was recently identified in patients with human progressive familial heart block type I [96]. Mutant TRPM4 channels show constitutive SUMOylation which prevents channel internalization and induces enhanced channel expression at the cell membrane [96]. However, elucidating the detailed mechanistic link between increased Na^+ and Ca^{2+} influx and progressive familial heart block type I, a defect in the conduction system, is an important task for the future.

Duchenne muscular dystrophy (DMD) patients with defects in the dystrophin gene develop progressive skeletal muscular degeneration. Some DMD patients also suffer from a *dilated cardiomyopathy*. It is not known why the absence of functional dystrophin induces cardiomyocyte degeneration leading to cardiomyopathy, but changes in Ca^{2+} handling are involved in the pathogenic process. Most interestingly, TRPV2 expression is elevated in the sarcolemma of cardiac muscle in dystrophic human patients and animal models and TRPV2 is activated in response to myocyte stretch [97]. The resulting enhanced Ca^{2+} influx induces cell damage. Moreover, transgenic mice over-expressing TRPV2 develop cardiomyopathy due to Ca^{2+} overload [97]. Therefore, TRPV2 seems to be a key player in the pathogensis of myocyte degeneration caused by disruption of the dystrophin-glycoprotein complex.

Chest pain is a characteristic symptom of *myocardial ischemia* but its underlying mechanism is poorly understood. Recently, new evidence was provided that iodoresiniferatoxin, a selective antagonist of TRPV1 channels attenuates both bradykinin- and ischemia-induced firing of cardiac spinal nerves [98]. Therefore, TRPV1 located on the cardiac sensory nerve might function as a molecular sensor to detect tissue ischemia and activate cardiac nociceptors. Moreover, TRPV1 activation causes the release of substance P, calcitonin gene-related peptide (CGRP) and other neurokines from sensory nerve terminals, protecting the heart from ischemic injury [98]. Along these lines, TRPV1–/– mice show impaired postischemic recovery of various hemodynamic parameters [99]. Myocardial ischemia also generates 12-lipoxygenase-derived eicosanoids, which protect against ischemia/reperfusion injury (IRI) via activation of TRPV1 in sensory C-fibres through subsequent release of substance P and CGRP [100]. Most interestingly, N-oleoyldopamine, a novel capsaicin-like lipid, protects the heart against IRI via the activation of TRPV1 [101]. Further clinical studies are required to elucidate whether pharmacological activation of TRPV1 has beneficial effects on the outcome of myocardial ischemia.

41.3 TRP Expression and Function in Endothelial Cells of the Systemic and Pulmonary Circulation

Endothelial cells are a specialized type of epithelial cells which reduce blood flow turbulence and regulate the transport of liquids across the semi-permeable vascular endothelial barrier. Vascular inflammation induces changes in endothelial cell shape and consequently increases in endothelial permeability which is induced by gaps between endothelial cells after Ca^{2+} influx. Other important functions of endothelial cells include angiogenesis, endothelial-dependent vasodilatations of smooth muscle cells and an important role in vascular remodelling processes.

Several reports demonstrated the involvement of Ca^{2+} entry through TRPC1, 4 and 6 channels in the disruption of the *barrier function* in pulmonary arteries [102–104]. The TRPC1-mediated signal may be associated with Rho activation or PKCalpha phosphorylation of TRPC1 as well as mechanisms involving nuclear factor κB (NF-κB) signalling [105, 106]. Deletion of TRPC4 in TRPC4–/– mice inhibits increases in lung vascular permeability to about 50% [103]. Cultures of lung endothelial cells from TRPC4–/– mice fail to respond to either thrombin or a PAR-1 agonist peptide with enhanced Ca^{2+} influx [103]. Moreover, isolated perfused TRPC4-deficient lungs showed a reduced PAR-1 receptor-mediated increase in vascular permeability [103]. The abnormal Ca^{2+} influx in TRPC4–/– endothelial cells was associated with a lack of thrombin-mediated actin-stress fiber formation as well as a reduced cellular retraction response [103]. The inability of thrombin to induce actin-stress fiber formation may be a direct result of TRPC4 interaction with protein 4.1, an endothelial cytoskeletal protein [107]. In addition, the intracellular C-terminal region of TRPC4 has been shown to interact with cytoskeletal αII- and βV-spectrin and to regulate plasma membrane expression of TRPC4 [108].

TRPC6- [109] and TRPC1/4 [107]-mediated Ca^{2+} entry were proposed to induce the resultant changes in endothelial cell shape in response to inflammatory agonists (e.g. thrombin and bradykinin). Vascular endothelial growth factor (VEGF) also increases vascular permeability by stimulating endothelial Ca^{2+} entry. In human microvascular endothelial cells, the VEGF-induced cation current has characteristics similar to those of VEGF-mediated TRPC currents in cells heterologously expressing VEGFR2 and TRPC3 or TRPC6 [110], while another group also implicated TRPC1 in VEGF-induced vascular hyper-permeability [111].

Recent evidence support a role of TRPM4 channels in a secondary spinal hemorrhage induced by spinal cord injury in a rat model [112]. Although TRPM4 expression was increased only in capillary endothelium distant from the point of injury, suppression of TRPM4 expression or block of channel activity greatly reduced the secondary capillary pathology associated with spinal cord injury. It is hypothesized that a large Na$^+$ influx in endothelial cells, due to increased TRPM4 channel density results in osmotically driven capillary damage [112].

TRPV4 channels are also involved in regulation of permeability of the pulmonary vascular endothelium, because the TRPV4 activator 4α-PDD increased the filtration coefficient (Kf), a measure of lung endothelial permeability, in perfused rat lungs of wild-type but not TRPV4–/– mice [113]. This response was blocked by ruthenium red an inhibitor of TRPV4 channel activity [113].

Recently, TRPA1 channels were identified in endothelial cells of cerebral arteries and an important role in *endothelium derived vasodilatation* was suggested [4]. These channels appear to be concentrated in region of the endothelial cell membrane that span the internal elastic lamina and are in close proximity to smooth muscle cells which are the effector cells. Stimulation of TRPA1 by allyl isothiocyanate (AITC), an TRPA1 agonist found in garlic and moustard oil, causes endothelium-dependent smooth muscle cell hyperpolarization and subsequent vasodilatation of cerebral arteries [4]. This mechanism that requires small and intermediate-conductance Ca^{2+} activated K$^+$ channels would explain the cardiovascular benefits of a garlic and moustard oil rich diet.

Other agonists regulate production of vasoactive compounds e.g. nitric oxide (NO) via TRPC channel-mediated Ca^{2+} entry. For instance, in primary aortic endothelial cells of TRPC4–/– mice, acetylcholine-induced Ca^{2+} entry is reduced markedly, resulting in a significant decrease in endothelium-dependent NO-mediated vasorelaxation of blood vessels [114]. These findings indicate that TRPC4 is part of the Ca^{2+} influx signal transduction pathway regulating vascular tone. However, TRPC5 might also be involved in this process, because treatment of bovine aortic endothelial cells with siRNA against TRPC5 prevented NO-induced Ca^{2+} entry [115]. This study also demonstrated that NO caused cysteine S-nitrosylation of TRPC5 on two cysteine residues in the S5-S6 loop region which are conserved in TRPC1 and TRPC4. The authors proposed a positive feed back loop in which Ca^{2+} influx-mediated by NO-induced nitrosylation of TRPC channels leads to increased endothelial nitric oxide synthase (eNOS) activity which increases NO production resulting in enhanced smooth muscle relaxation [115]. The described data clearly contradict other publications which describe NO as an inhibitor of Ca^{2+} entry in endothelial cells [116, 117]. Along these lines, inhibition of TRPC3 activity by PKG-dependent phosphorylation was demonstrated to protect endothelial cells from the detrimental effect of excessive NO and Ca^{2+} [67, 118]. Because TRPC3–/– [119] and TRPC5–/– mice [120] are now available, it is essential to verify the molecular mechanism in TRPC-deficient endothelial cells in comparison to wild-type cells.

In human umbilical vein endothelial cells, 11,12-epoxygenase-derived epoxyeicosanotrienoic acids (EETs) are some of the predominant mechanically produced metabolites from arachidonic acid (AA) by phospholipase A$_2$ (PLA$_2$) and cyclooxygenase (COX). EETs are found to facilitate the translocation of the TRPC6 protein to caveolin-1-rich cell membrane areas (see Fig. 41.2). This event leads to enhanced Ca^{2+} influx in endothelial cells and prolongation of the membrane hyperpolarization in response to bradykin [121]. 11,12-EETs from endothelial cells can also directly

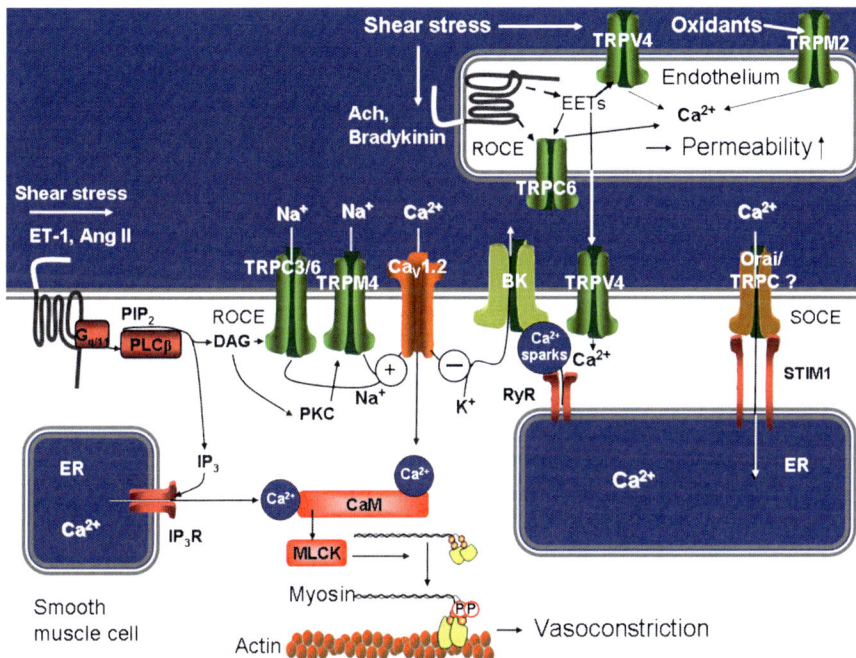

Fig. 41.2 Overview of TRP channel functions involved in inducing increased permeability of vascular endothelium as well as constriction of vascular smooth muscle cells. +: membrane depolarisation by Na^+ accumulation inducing activation of $Ca_V 1.2$; –: membrane hyperpolarization by K^+ efflux inhibiting $Ca_V 1.2$; Ang II: angiotensin II; CaM: calmodulin; $Ca_V 1.2$: voltage gated calcium channel; DAG: diacylglycerol; EETS: epoxygenase-derived epoxyeicosanotrienoic acids; ET-1: endothelin-1; ER: endoplasmic reticulum; $G_{q/11}$: G protein type q and 11; IP_3: inositol-1,4,5 trisphosophate; IP_3-R: IP_3 receptor; MLCK: myosin-light chain kinase; PIP_2: phosphatidylinositol 4,5-bisphosphate; PKC: proteinkinase C; PLC: phospholipase C. See text for details

activate large conductance K_{Ca} (BK_{Ca}) channels in vascular smooth muscle cells to induce hyperpolarisation and vasorelaxation (reviewed in [122]).

An important role of TRPM7 channels in the regulation of NO-mediated vasodilatation is suggested by a study using a mouse model of hypomagnesemia with low levels of Mg^{2+} in their erythrocytes (MgL) [123]. Aortas from these mice express higher levels of TRPM7 than mice with high Mg^{2+} concentrations (MgH). Cultured endothelial cells from MgL mice exhibit diminished eNOS expression and Ach-induced vasodilatation of isolated mesenteric arteries was decreased compared to MgH arteries. Silencing of TRPM7 using siRNA also increased expression of nitric oxide synthase and nitric oxide production [123]. Two studies demonstrate TRPM8 channel function in endothelial cells. Icillin a TRPM8 agonist induces Ca^{2+} influx in immortalized human corneal endothelial cells [124] and menthol another activator of TRPM8 dilates rat tail arteries, an effect which is partially dependent on the endothelium [125].

TRPV1 channels are expressed in endothelial cells of pig coronary arteries and regulate endothelium dependent dilation in response to the TRPV1 activator capsaicin [126]. Interestingly, this response was impaired in obese pigs [126]. However, the exact signalling mechanisms by which TRPM7 inhibits and TRPM8 as well as TRPV1 increase endothelial vasodilatation needs to be elucidated.

More mechanistic insight of TRPV4 function in the endothelium derived vasodilatation is provided by the analysis of a TRPV4–/– mouse model. Application of the TRPV4 activators 4α-PDD, heating to 43°C and 5,6-EETs induced currents with biophysical properties of TRPV4 in mouse aortic endothelial cells (MAEC) (see Fig. 41.2). These activators failed to increase Ca^{2+} influx in MAEC from TRPV4–/– mice [127]. Changes in Ca^{2+} influx in response to AA and hypotonicity which also activate TRPV4 channels were attenuated. However, TRPV4 activation by 4α-PDD and heat were not altered by a pharmacological block of the EETs pathway [127] indicating a clearly distinct pathway of these agents to TRPV4 stimulation. TRPV4 expression and TRPV4-like currents were also detected in rat carotid artery endothelial cells [128]. Most interestingly, application of 4α-PDD resulted in dilation of preconstricted carotid and gracilis arteries which was blocked by the TRPV4 inhibitor ruthenium red. While 4α-PDD-induced dilation was largely decreased by inhibition of NOS and COX in aortic endothelial cells, 4α-PDD-induced dilation of cells from gracilis arteries was insensitive to NOS and COX inhibition [128]. A similar scenario prevails in mesenteric arteries, another resistance vessel where endothelium-dependent vasodilation is impaired in TRPV4–/– mice [129]. These data provide the first evidence that TRPV4 activation mechanisms might be different in conduit versus resistance vessels. Moreover, shear-stress induced vasodilatation is also blocked by ruthenium red and shear-stress induced vasodilatation is absent in carotid arteries of TRPV4–/– mice [130]. A decreased endothelium derived hyperpolarization in vessels of caveolin1 –/– mice is also dependent on TRPV4, because TRPV4–mediated Ca^{2+} increase is diminished in endothelial cells of these mice. Therefore, caveolin expression is most probably important for membrane localization and activation of TRPV4 channels [131].

The above mentioned study on MgL mice also indicates an important role of TRPM7 in suppressing *endothelial cell proliferation* [123]. SiRNA-mediated downregulation of TRPM7 increased phosphorylation of extracellular signal-regulated kinase (ERK) and enhanced the proliferation of these cells.

Hypoxia sensed by endothelial cells activates transcription factors leading to the production of growth factors, which stimulate smooth muscle cell proliferation, resulting in vascular remodelling and hypertension [132]. One of the transcription factors involved in this process is AP-1, which regulates Ca^{2+}-sensitive genes. Culture of human pulmonary arterial endothelial cells under hypoxic conditions results in increased TRPC4 mRNA and protein expression [133]. This is accompanied by enhanced binding of AP-1 to a number of AP-1-responsive genes involved in proliferation. Expression of siRNA against TRPC4 in endothelial cells prevented hypoxia-induced increases in TRPC4 expression and AP-1 binding [133]. Thus, TRPC4 appears to be involved in mediating some aspects of hypoxia-induced gene expression and cell proliferation.

TRPC3 channels have been suggested to serve as redox sensors which monitor *oxidative stress in endothelial cells*. Porcine aortic endothelial cells express TRPC3 and display oxidant-induced cation currents with properties similar to those of TRPC3 [134, 135]. Expression of the N-terminus of TRPC3, which acts as a dominant negative TRPC3 fragment, abolished the oxidant-induced cation current and reduced membrane depolarisation in endothelial cells [134]. However, TRPC3 may not be acting alone in this process, because it was demonstrated recently that heteromeric channels composed of TRPC3 and TRPC4 form redox-sensitive channels both in native endothelial cells and when heterologously expressed in HEK293 cells [136]. Expression of dominant negative constructs for either TRPC3 or TRPC4 was able to suppress oxidant-induced channel activity in both systems [136]. It may be interesting to test these hypotheses by comparing primary endothelial cells from TRPC3–/– and TRPC3/4–/– double deficient mice [119] with these of wild-type mice.

Functional TRPM2 channels are present in primary cultures of human pulmonary artery endothelial cells, where the channel mediates Ca^{2+} influx in response to hydrogen peroxide (H_2O_2) [137]. Small interfering RNA (siRNA)-mediated down-regulation of TRPM2 expression reduces H_2O_2-induced increases in the permeability of the endothelium, suggesting that TRPM2 is important for regulation of the pulmonary endothelial barrier during oxidative stress. Most intriguingly, the short variant of TRPM2 (TRPM2 s), which lacks the pore domain and acts as a dominant-negative form by inhibiting the formation of functional homotetrameric channels, was also able to significantly diminish the increase in endothelial permeability [137]. As both forms of TRPM2 are expressed in HPAE cells, the control of the relative expression levels is an enticing potential regulatory mechanism of TRPM2 activity in these cells (reviewed in [138]).

Endothelial cell migration plays an important role in vascular remodelling and regeneration. A TRPC5/6 activation cascade has been shown to take part in the regulation of endothelial cell migration [139]. Treatment of bovine aortic endothelial cells with lysophosphatidylcholine (lsyoPC) which inhibits cell migration by blocking non-voltage-gated Ca^{2+} channels [140] induced translocation of TRPC5 and 6 to the plasma membrane. Treatment of endothelial cells with siRNA against either TRPC5 or TRPC6 reduced lysoPC-induced increases in $[Ca^{2+}]_i$ and cell migration. Similar results were obtained in TRPC6 deficient endothelial cells, which also showed a lack of lysoPC-induced translocation of TRPC5 to the plasma membrane. However, this report is the only one demonstrating the existence of TRPC5/6 heteromeric channels while other reports excluded this heteromultimerization pattern [15, 16, 141].

TRPC6 channels may be important in growth factor induced *angiogenesis*. As outlined above a current was recorded in primary cultures of human microvascular endothelial cells in response to VEGF [110]. The biophysical properties of this current were reconstituted in Chinese hamster ovary (CHO) cells expressing the VEGF receptor, TRPC6 and TRPC3, but not TRPC3 or TRPC6 [110]. These finding strongly suggest that a TRPC3/6 heteromeric channel is responsible for VEGF-induced migration, sprouting, and proliferation as well as other functions necessary for VEGF-induced angiogenesis. Similar results were obtained in human umbilical cord endothelial cells (HUVECs) [142]. A VEGF-induced TRPC6-like cation current was absent in cells expressing a dominant negative form of TRPC6. TRPC6 specific siRNA suppressed VEGF – but not fibroblast growth factor-induced proliferation and capillary tube formation in HUVEC cultures [142]. These data are especially interesting, because antiangiogenic compounds like VEGF antibodies are used in anticancer therapy. Therefore, specific inhibitors of TRPC6 activity may also be successful in endothelial cells to stop cancer-related angiogenesis.

In summary, TRP channels support endothelial functions like vascular regeneration, increased permeability and endothelium-derived NO-mediated vasorelaxation of smooth muscle cells.

41.4 Expression and Pathological Overexpression of TRPC Channels in Smooth Muscle Cells of the Systemic Vasculature

Smooth muscle cells do not only provide structural integrity for the vessel but also precise regulation of vascular tone and blood pressure. However, a striking feature of smooth muscle cells is their cellular heterogeneity in different vessels. Moreover, in response to various environmental stimuli, including growth factors, cytokines, mechanical influences, and various inflammatory mediators smooth muscle cells change their phenotype. The quiescent contractile vascular smooth muscle cells undergo transcriptional changes resulting in both the downregulation of contractile proteins and concurrent up-regulation of proteins supporting a proliferative, so-called "synthetic" phenotype. This switch from the contractile to the proliferative and migratory phenotype is believed to have evolved as a mechanism of vascular repair during injury or vascular adaptation (reviewed in [143]).

Smooth muscle cell proliferation is an important step in vascular remodelling and evidence has been accumulated that TRP channels are important for these processes. TRPC1 activity is associated with pulmonary artery smooth muscle cell proliferation [144] and TRPC1 expression is increased during vascular hyperplasia associated with vascular occlusive disease [145]. Smooth muscle cell proliferation results in neointima formation. Most interestingly it was reported that TRPC1 expression is also up-regulated in neointima from mice, rats, pigs and humans [145] and in vitro treatment of neointima with a TRPC1-antibody inhibits cell proliferation [145]. However, we were not able to reproduce these data by a comparison of neointima formation in wild-type and TRPC1–/– mice (R. Köhler and A. Dietrich unpublished observation). Studies in human pulmonary arterial smooth muscle cells demonstrated that enhanced proliferation was associated with an increase in TRPC4 expression, and treatment of these cells with siRNA against TRPC4 inhibited ATP-induced proliferation [146]. Activation of a heteromeric TRPC1/5 complex in pial arterioles does not constrict smooth muscle cells but rather induced cell migration and proliferation [147]. Most interestingly, TRPC3 and TRPC6 up-regulation in smooth muscle cells was also observed in patients with idiopathic pulmonary arterial hypertension (IPAH) which is characterized by excessive proliferation of smooth muscle cells in the pulmonary artery [148]. SiRNAs directed against TRPC6 markedly attenuated proliferation of these cells in vitro [148]. However, TRPC6–/– mice are not protected from chronic pulmonary hypertension and vascular remodelling [149]. ATP-induced mitogenesis was also accompanied by an increased TRPC4 expression. This TRPC4 upregulation was preceded by increases in phosphorylated CREB, and was prevented by blockage of purinergic receptors, as well as inhibition of CREB phosphorylation in human pulmonary artery smooth muscle cells [146].

There is also evidence that TRPM7 is a functionally important regulator of Mg^{2+} hemostasis and growth in vascular smooth muscle cells. If TRPM7 expression is suppressed by specific siRNAs, basal intracellular Mg^{2+} levels are reduced and Ang II-induced proliferation of vascular smooth muscle cells is inhibited [150].

Patients with polcystic kidney disease suffer from aneurism resulting from thinning of the arterial wall with ensuing rupture and internal bleeding [151]. Both, TRPP1 and TRPP2 are expressed in vascular smooth muscle cells, but their exact function in smooth muscle cell proliferation is still elusive.

In rat pulmonary artery smooth muscle cells serotonin induced Ca^{2+}-influx and cation currents were mimicked by the TRPV4 activator 4α-PDD and blocked by the TRPV4 inhibitor ruthenium red. The TRPV4-mediated Ca^{2+} influx induced a mitogenic response suggesting a role for TRPV4 in proliferation of pulmonary vascular smooth muscle cells [152].

Migration of vascular smooth muscle cells is another important component of neointima formation. Sphingosine-1-phosphate (S1P) binding to G protein-coupled receptors (GPCR) stimulates smooth muscle cell migration and induces Ca^{2+} entry in cultured human saphenous vein SMC [153]. Both primary SMC from human saphenous veins and human saphenous veins express TRPC1 and TRPC5, and treatment of these cells with a dominant negative construct of TRPC5 or a TRPC5-antibody inhibited S1P-induced smooth muscle cell migration [153].

Vascular smooth muscle *cell contraction* is an important regulator of blood pressure. Treatment of cultured arteries with an antibody to TRPC1 inhibited SOCE-induced contractions [154, 155], and adenovirus-mediated overexpression of human TRPC1 in rat pulmonary artery rings led to enhanced SOCE and enhanced SOCE-mediated contraction [156]. However, in aortic smooth muscle cells from TRPC1-deficient mice, the amount of SOCE was indistinguishable from wild-type cells [12]. Down-regulation of STIM1, however, completely abolished SOCE in these cells [12]. The latter data clearly indicate that SOCE is completely independent of TRPC1 and no functionally relevant STIM1-TRPC1 interaction exists, at least in aortic smooth muscle cells.

Smooth muscle cell contraction is activated by α1-adrenergic agonists and can be blocked by suppressing TRPC6 expression [157]. Activation of TRPC6 was also shown in a A7r5 smooth muscle cell line stimulated by vasopression (AVP) [158] another important vasoconstrictor. In a TRPC6–/– mouse model however, we observed airway smooth muscle hyperreactivity after exposure to bronchoconstrictors and higher agonist-induced contractility in isolated tracheal [159] and aortic rings [65] prepared from these mice. Futhermore, systemic blood pressure was elevated [65]. These effects are explained by *in vivo* replacement of TRPC6 by TRPC3-type channels which are closely related, but constitutively active [65, 159], resulting in enhanced basal and agonist-induced cation entry into smooth muscle cells which increases smooth muscle contractility [65, 159]. Moreover, smooth muscle cells from TRPC6–/– aorta or cerebral arteries are more depolarized and demonstrate enhanced spontaneous and agonist-induced Ca^{2+} entry [65]. TRPC6 has also been shown to be involved in the regulation of myogenic tone in vascular smooth muscle cells of small resistance vessels. The elevation of intravascular pressure in these vessels increases myogenic tone, a phenomenon already described by Arthur Bayliss in 1902. Treatment of rat cerebral arteries with antisense oligonucleotides directed against TRPC6 inhibited vasoconstriction-induced by elevating intravascular pressure, and suppressed pressure-induced depolarization of VSM cells [160]. However, TRPC6-deficient mice displayed an earlier onset of the so-called Bayliss effect, excluding an exclusive role of TRPC6 for this effect. Along these lines, UTP-induced depolarization of cerebral arteries and subsequent contraction of smooth muscle cells is inhibited by treatment with antisense oligonucleotides for TRPC3 [161], demonstrating again an important role of TRPC3 in these cells. Changes in TRPC3 expression has also been proposed to be a factor involved in essential hypertension, because spontaneously hypertensive rats express abnormally high levels of TRPC3 compared to wild-type control animals [162, 163].

In contrast to the systemic vasculature, the pulmonary circulation responds to hypoxia by constricting pulmonary arteries and diverting blood flow to the well-ventilated areas of the lung to ensure maximal oxygenation of the venous blood. This alveolar hypoxia-mediated vasoconstrictive phenomenon is known as acute hypoxic pulmonary vasoconstriction (HPV) and was first described by von Euler and Liljestrand in 1946. Sustained pulmonary vasoconstriction (chronic HPV) is often accompanied by vascular remodelling, i.e. the muscularization of smaller arteries and arterioles due to SMC proliferation and migration. In severe forms of pulmonary hypertension such as idiopathic and familial PAH, pulmonary artery remodelling, resulting from intimal fibrosis and medial hypertrophy is extensive. As outlined above, up-regulation of TRPC3 and TRPC6 was observed in smooth muscle cells of IPAH patients [148] and in a rat model of pulmonary hypertension, TRPC1 and 6 up-regulation was shown to require the expression of hypoxia-inducible factor 1 (HIF-1) [164, 165]. For these reasons, we analyzed pulmonary arterial pressure (PAP) in isolated lungs during acute (<20 min) and prolonged (60–160 min) hypoxia. Much to our surprise, acute HPV was completely absent in TRPC6–/– mice, while the vasoconstriction after prolonged hypoxia was not significantly different in TRPC6–/– mice compared to wild-type mice [149]. Moreover, the lack of acute HPV in TRPC6–/– mice has profound physiological relevance because partial occlusion of alveolar ventilation provoked severe hypoxemia in TRPC6–/–, but not in wild-type mice [149]. However, vascular as well as cardiac remodelling after chronic hypoxia of 3 weeks was not significantly different in TRPC6–/– mice compared to control mice [149]. Most interestingly, TRPC3, which is up-regulated in TRPC6 deficient mice, is only expressed in large, but not in small precapillary pulmonary arterial smooth muscle cells [149, 166, 167]. For this reason, we analyzed Ca^{2+} influx in small precapillary pulmonary arterial smooth muscle cells (PASMC) from TRPC6-deficient and wild-type mice after priming the cells with endothelin-1. Hypoxic incubation of wild-type PASMC resulted in an increase in $[Ca^{2+}]_i$, which was completely absent in PASMC from TRPC6–/– mice [149]. Most interestingly, Ca^{2+} influx was completely dependent on extracellular Ca^{2+} [149], excluding a proposed contribution of STIM1 and Orai, as well as store-operated TRPC channels [168]. It is notable that the entry of Ca^{2+}-ions in response to hypoxia is mostly carried by voltage-gated Ca^{2+} channels, because nicardipine, a potent blocker of these channels, almost completely inhibited acute HPV in isolated lungs and Ca^{2+} influx in wild-type PASMC [149]. These data support a model in which Na^+ influx through TRPC6 channels leads to membrane depolarization and activation of voltage-gated Ca^{2+} channels (see Fig. 41.2) [169–172].

TRP channels might also be involved in vasodilatation of smooth muscle cells. TRPM7 expression levels are lower in cultured mesenteric artery smooth muscle cells from spontaneous hypertensive rats compared to normotensive WKY rats [173] and low intracellular Mg^{2+} levels are associated with reduced vasodilator function, vascular remodelling and elevated blood pressure [174, 175]. Thus patients with hypertension might benefit from TRPM7 mediated Mg^{2+} influx in smooth muscle cells. EETs produced in endothelial cells described above activate TRPV4 channels in isolated cerebral arteries (see Fig. 41.2) [176]. Ca^{2+} influx through these channels may induce the release Ca^{2+} from ryanodine-sensitive receptors located on the sarcoplasmic reticulum. These so called Ca^{2+} sparks then activate nearby Ca^{2+} sensitive K^+ (BK) channels and increase K^+ efflux leading to smooth muscle hyperpolarization and vasodilatation (see Fig. 41.2) [176]. For this mechanism TRPV4 has to form close signalling complexes with ryanodine receptors and BK channels near to the SR.

Stretch activated channels (SAC) which respond to mechanical stress as well as cell swelling are thought to be responsible for the pressure-induced vasoconstriction e.g. the Bayliss effect described above. Characteristics of these channels were already described in vascular smooth muscle cells (reviewed in [177]). Two TRPC channels TRPC1 and TRPC6 were proposed to be activated by mechanical stimulation [178, 179]. Overexpression of human TRPC1 protein in *Xenopus* oocytes or the human cell line CHO-K1 remarkably enhanced, whereas microinjection of antisense TRPC1 copy RNA (cRNA) into the oocytes greatly reduced mechanosensitive channel activity [178]. However, in smooth muscle cells from isolated cerebral arteries, we could not find any differences in mechanically-activated cation currents and pressure-induced vasoconstriction between TRPC1-deficient and wild-type mice [12]. In addition, another report found that the mechanosensitive channel activity was not significantly different between non transfected and TRPC1-transfected CHO-K cells, and concluded that the observed mechanically stimulated current may reflect an endogenous channel activity present in the expression system [180]. Similar conflicting evidence has been obtained regarding the apparent mechanosensitivity of TRPC6 channels. While one report proposed a common bilayer-dependent mechanism for mechanical activation of expressed TRPC6 channels [179], Gottlieb et al. found no TRPC6-mediated mechanosensitive current in a heterologous expression system [180]. Moreover, a careful analysis of TRPC6 mechanosensitivity suggest that mechanical activation of TRPC6 channels is mediated by accumulation of DAG through mechanosensitive GPCR-activation, rather than by direct stimulation of TRPC6 by mechanical stimuli [181]. Along these lines, another indirect mechanisms result in stimulation of TRPC6-activity after mechanical stretching of smooth muscle cells. Mechanical stress can activate phospholipase A_2 (PLA_2) to produce arachidonic acid which can be converted into EETs (reviewed in [177]) and mechanically produced 11,12-EETs are able to promote translocation of TRPC6 channels to the plasma membrane leading to increased acute hypoxic pulmonary vasoconstriction which is absent in TRPC6-deficient smooth muscle cells [182].

Several reports indicate that TRPM4 channels contribute to stretch induced constriction of cerebral artery myocytes [183–185]. TRPM4 like currents which were also activated by intracellular Ca^{2+} and PKC activation (see Fig. 41.2) were identified in these studies. These currents as well as smooth muscle constriction were inhibited by down-regulation of TRPM4 by TRPM4 antisense oligonucleotides in living rats [184]. Shear stress also caused rapid accumulation of TRPM7 in cultured aortic smooth muscle cells (A7r5 cells) [186]. An intriguing hypothesis by the authors is that TRPM7 up-regulation occurs only in areas where endothelial damage by shear stress was induced. Increased Mg^{2+} influx by TRPM7 might then alter vascular myocyte function [186].

Cell swelling induced by hypoosmotic bath solutions increase cation currents in murine aortic myocytes which are suppressed by treatment of the cells with TRPV2 antisense oligonucleotides [187]. Along these lines, a mechanosensitive function of TRPV2 channels in a variety of tissues is discussed [187]. However, until now indirect mechanisms for mechanosensitive activation of TRPM7 and TRPV2 like described above for TRPC1 and TRPC6 can not be excluded.

41.5 Conclusion

As outlined above, TRP channels are involved in numerous physiological and pathophysiological processes in cardiomyocytes, smooth muscle and endothelial cells of the cardiovascular system. Thus, specific drugs designed to target TRP channels may be important therapeutic agents for cardiac hypertrophy, hypertension and vascular remodelling in the near future.

References

1. Venkatachalam K, Montell C (2007) TRP channels. Annu Rev Biochem 76:387–417
2. Nilius B, Owsianik G (2010) Transient receptor potential channelopathies. Pflugers Arch 460(2):437–450

3. Latorre R, Zaelzer C, Brauchi S (2009) Structure-functional intimacies of transient receptor potential channels. Q Rev Biophys 42:201–246
4. Earley S, Gonzales AL, Crnich R (2009) Endothelium-dependent cerebral artery dilation mediated by TRPA1 and Ca^{2+}-Activated K+ channels. Circ Res 104:987–994
5. Hofmann T, Obukhov AG, Schaefer M, Harteneck C, Gudermann T, Schultz G (1999) Direct activation of human TRPC6 and TRPC3 channels by diacylglycerol. Nature 397:259–263
6. Okada T, Inoue R, Yamazaki K, Maeda A, Kurosaki T, Yamakuni T, Tanaka I, Shimizu S, Ikenaka K, Imoto K, Mori Y (1999) Molecular and functional characterization of a novel mouse transient receptor potential protein homologue TRP7. Ca(2+)-permeable cation channel that is constitutively activated and enhanced by stimulation of G protein-coupled receptor. J Biol Chem 274:27359–27370
7. Dietrich A, Mederos y Schnitzler M, Kalwa H, Storch U, Gudermann T (2005) Functional characterization and physiological relevance of the TRPC3/6/7 subfamily of cation channels. Naunyn Schmiedebergs Arch Pharmacol 371:257–265
8. Dietrich A, Kalwa H, Rost BR, Gudermann T (2005) The diacylgylcerol-sensitive TRPC3/6/7 subfamily of cation channels: functional characterization and physiological relevance. Pflugers Arch 451:72–80
9. Cahalan MD (2009) STIMulating store-operated Ca(2+) entry. Nat Cell Biol 11:669–677
10. Liao Y, Plummer NW, George MD, Abramowitz J, Zhu MX, Birnbaumer L (2009) A role for Orai in TRPC-mediated Ca^{2+} entry suggests that a TRPC:Orai complex may mediate store and receptor operated Ca^{2+} entry. Proc Natl Acad Sci USA 106:3202–3206
11. Yuan JP, Kim MS, Zeng W, Shin DM, Huang G, Worley PF, Muallem S (2009) TRPC channels as STIM1-regulated SOCs. Channels (Austin) 3:221–225
12. Dietrich A, Kalwa H, Storch U, Mederos y Schnitzler M, Salanova B, Pinkenburg O, Dubrovska G, Essin K, Gollasch M, Birnbaumer L, Gudermann T (2007) Pressure-induced and store-operated cation influx in vascular smooth muscle cells is independent of TRPC1. Pflugers Arch 455:465–477
13. Varga-Szabo D, Authi KS, Braun A, Bender M, Ambily A, Hassock SR, Gudermann T, Dietrich A, Nieswandt B (2008) Store-operated $Ca(^{2+})$ entry in platelets occurs independently of transient receptor potential (TRP) C1. Pflugers Arch 457:377–387
14. Dietrich A, Kalwa H, Gudermann T (2010) TRPC channels in vascular cell function. Thromb Haemost 103:262–270
15. Hofmann T, Schaefer M, Schultz G, Gudermann T (2002) Subunit composition of mammalian transient receptor potential channels in living cells. Proc Natl Acad Sci USA 99:7461–7466
16. Strubing C, Krapivinsky G, Krapivinsky L, Clapham DE (2003) Formation of novel TRPC channels by complex subunit interactions in embryonic brain. J Biol Chem 278:39014–39019
17. Dietrich A, Chubanov V, Kalwa H, Rost BR, Gudermann T (2006) Cation channels of the transient receptor potential superfamily: their role in physiological and pathophysiological processes of smooth muscle cells. Pharmacol Ther 112:744–760
18. Dietrich A, Kalwa H, Fuchs B, Grimminger F, Weissmann N, Gudermann T (2007) In vivo TRPC functions in the cardiopulmonary vasculature. Cell Calcium 42:233–244
19. Yildirim E, Birnbaumer L (2007) TRPC2: molecular biology and functional importance. Handb Exp Pharmacol 179:53–75
20. Ong HL, Chen J, Chataway T, Brereton H, Zhang L, Downs T, Tsiokas L, Barritt G (2002) Specific detection of the endogenous transient receptor potential (TRP)-1 protein in liver and airway smooth muscle cells using immunoprecipitation and Western-blot analysis. Biochem J 364:641–648
21. Sage SO, Brownlow SL, Rosado JA (2002) TRP channels and calcium entry in human platelets. Blood 100:4245–4246 author reply 4246–4247

22. Flockerzi V, Jung C, Aberle T, Meissner M, Freichel M, Philipp SE, Nastainczyk W, Maurer P, Zimmermann R (2005) Specific detection and semi-quantitative analysis of TRPC4 protein expression by antibodies. Pflugers Arch 451:81–86
23. Tajeddine N, Zanou N, Van Schoor M, Lebacq J, Gailly P (2010) TRPC1: subcellular localization? J Biol Chem 285:le1; author reply le2
24. Perraud AL, Schmitz C, Scharenberg AM (2003) TRPM2 Ca^{2+} permeable cation channels: from gene to biological function. Cell Calcium 33:519–531
25. Chubanov V, Mederos y Schnitzler M, Waring J, Plank A, Gudermann T (2005) Emerging roles of TRPM6/TRPM7 channel kinase signal transduction complexes. Naunyn Schmiedebergs Arch Pharmacol 371:334–341
26. Scharenberg AM (2005) TRPM2 and TRPM7: channel/enzyme fusions to generate novel intracellular sensors. Pflugers Arch 451:220–227
27. Launay P, Fleig A, Perraud AL, Scharenberg AM, Penner R, Kinet JP (2002) TRPM4 is a Ca^{2+}-activated nonselective cation channel mediating cell membrane depolarization. Cell 109:397–407
28. Zhang Y, Hoon MA, Chandrashekar J, Mueller KL, Cook B, Wu D, Zuker CS, Ryba NJ (2003) Coding of sweet, bitter, and umami tastes: different receptor cells sharing similar signaling pathways. Cell 112:293–301
29. Hofmann T, Chubanov V, Gudermann T, Montell C (2003) TRPM5 is a voltage-modulated and $Ca^{(2+)}$-activated monovalent selective cation channel. Curr Biol 13: 1153–1158
30. Liu D, Liman ER (2003) Intracellular Ca^{2+} and the phospholipid PIP2 regulate the taste transduction ion channel TRPM5. Proc Natl Acad Sci USA 100: 15160–15165
31. Perez CA, Margolskee RF, Kinnamon SC, Ogura T (2003) Making sense with TRP channels: store-operated calcium entry and the ion channel Trpm5 in taste receptor cells. Cell Calcium 33:541–549
32. Prawitt D, Monteilh-Zoller MK, Brixel L, Spangenberg C, Zabel B, Fleig A, Penner R (2003) TRPM5 is a transient Ca^{2+}-activated cation channel responding to rapid changes in $[Ca^{2+}]i$. Proc Natl Acad Sci USA 100:15166–15171
33. Damak S, Rong M, Yasumatsu K, Kokrashvili Z, Perez CA, Shigemura N, Yoshida R, Mosinger B Jr., Glendinning JI, Ninomiya Y, Margolskee RF (2006) Trpm5 null mice respond to bitter, sweet, and umami compounds. Chem Senses 31:253–264
34. Nadler MJ, Hermosura MC, Inabe K, Perraud AL, Zhu Q, Stokes AJ, Kurosaki T, Kinet JP, Penner R, Scharenberg AM, Fleig A (2001) LTRPC7 is a Mg.ATP-regulated divalent cation channel required for cell viability. Nature 411:590–595
35. Runnels LW, Yue L, Clapham DE (2001) TRP-PLIK, a bifunctional protein with kinase and ion channel activities. Science 291:1043–1047
36. Monteilh-Zoller MK, Hermosura MC, Nadler MJ, Scharenberg AM, Penner R, Fleig A (2003) TRPM7 provides an ion channel mechanism for cellular entry of trace metal ions. J Gen Physiol 121:49–60
37. Dorovkov MV, Ryazanov AG (2004) Phosphorylation of annexin I by TRPM7 channel-kinase. J Biol Chem 279:50643–50646
38. Clark K, Langeslag M, van Leeuwen B, Ran L, Ryazanov AG, Figdor CG, Moolenaar WH, Jalink K, van Leeuwen FN (2006) TRPM7, a novel regulator of actomyosin contractility and cell adhesion. EMBO J 25:290–301
39. Ryazanova LV, Dorovkov MV, Ansari A, Ryazanov AG (2004) Characterization of the protein kinase activity of TRPM7/ChaK1, a protein kinase fused to the transient receptor potential ion channel. J Biol Chem 279:3708–3716
40. Matsushita M, Kozak JA, Shimizu Y, McLachlin DT, Yamaguchi H, Wei FY, Tomizawa K, Matsui H, Chait BT, Cahalan MD, Nairn AC (2005) Channel function is dissociated from the intrinsic kinase activity and autophosphorylation of TRPM7/ChaK1. J Biol Chem 280:20793–20803

41. Schmitz C, Perraud AL, Johnson CO, Inabe K, Smith MK, Penner R, Kurosaki T, Fleig A, Scharenberg AM (2003) Regulation of vertebrate cellular Mg^{2+} homeostasis by TRPM7. Cell 114:191–200
42. Schmitz C, Dorovkov MV, Zhao X, Davenport BJ, Ryazanov AG, Perraud AL (2005) The channel kinases TRPM6 and TRPM7 are functionally nonredundant. J Biol Chem 280:37763–37771
43. Schlingmann KP, Sassen MC, Weber S, Pechmann U, Kusch K, Pelken L, Lotan D, Syrrou M, Prebble JJ, Cole DE, Metzger DL, Rahman S, Tajima T, Shu SG, Waldegger S, Seyberth HW, Konrad M (2005) Novel TRPM6 Mutations in 21 Families with Primary Hypomagnesemia and Secondary Hypocalcemia. J Am Soc Nephrol 16:3061–3069
44. Schlingmann KP, Weber S, Peters M, Niemann Nejsum L, Vitzthum H, Klingel K, Kratz M, Haddad E, Ristoff E, Dinour D, Syrrou M, Nielsen S, Sassen M, Waldegger S, Seyberth HW, Konrad M (2002) Hypomagnesemia with secondary hypocalcemia is caused by mutations in TRPM6, a new member of the TRPM gene family. Nat Genet 31:166–170
45. Walder RY, Landau D, Meyer P, Shalev H, Tsolia M, Borochowitz Z, Boettger MB, Beck GE, Englehardt RK, Carmi R, Sheffield VC (2002) Mutation of TRPM6 causes familial hypomagnesemia with secondary hypocalcemia. Nat Genet 31:171–174
46. Schlingmann KP, Waldegger S, Konrad M, Chubanov V, Gudermann T (2007) TRPM6 and TRPM7–Gatekeepers of human magnesium metabolism. Biochim Biophys Acta 1772:813–821
47. Chubanov V, Waldegger S, Mederos y Schnitzler M, Vitzthum H, Sassen MC, Seyberth HW, Konrad M, Gudermann T (2004) Disruption of TRPM6/TRPM7 complex formation by a mutation in the TRPM6 gene causes hypomagnesemia with secondary hypocalcemia. Proc Natl Acad Sci USA 101:2894–2899
48. Watanabe H, Murakami M, Ohba T, Takahashi Y, Ito H (2008) TRP channel and cardiovascular disease. Pharmacol Ther 118:337–351
49. Delmas P, Nauli SM, Li X, Coste B, Osorio N, Crest M, Brown DA, Zhou J (2004) Gating of the polycystin ion channel signaling complex in neurons and kidney cells. FASEB J 18:740–742
50. Delmas P (2005) Polycystins: polymodal receptor/ion-channel cellular sensors. Pflugers Arch 451:264–276
51. Mochizuki T, Wu G, Hayashi T, Xenophontos SL, Veldhuisen B, Saris JJ, Reynolds DM, Cai Y, Gabow PA, Pierides A, Kimberling WJ, Breuning MH, Deltas CC, Peters DJ, Somlo S (1996) PKD2, a gene for polycystic kidney disease that encodes an integral membrane protein. Science 272:1339–1342
52. Hanaoka K, Qian F, Boletta A, Bhunia AK, Piontek K, Tsiokas L, Sukhatme VP, Guggino WB, Germino GG (2000) Co-assembly of polycystin-1 and -2 produces unique cation-permeable currents. Nature 408:990–994
53. Volk T, Schwoerer AP, Thiessen S, Schultz JH, Ehmke H (2003) A polycystin-2-like large conductance cation channel in rat left ventricular myocytes. Cardiovasc Res 58:76–88
54. Caterina MJ, Schumacher MA, Tominaga M, Rosen TA, Levine JD, Julius D (1997) The capsaicin receptor: a heat-activated ion channel in the pain pathway. Nature 389:816–824
55. Caterina MJ, Leffler A, Malmberg AB, Martin WJ, Trafton J, Petersen Zeitz KR, Koltzenburg M, Basbaum AI, Julius D (2000) Impaired nociception and pain sensation in mice lacking the capsaicin receptor. Science 288:306–313
56. Nilius B, Vriens J, Prenen J, Droogmans G, Voets T (2004) TRPV4 calcium entry channel: a paradigm for gating diversity. Am J Physiol Cell Physiol 286:C195–C205
57. Schaefer M (2005) Homo- and heteromeric assembly of TRP channel subunits. Pflugers Arch 451:35–42
58. Molkentin JD, Dorn GW (2001) 2nd: Cytoplasmic signaling pathways that regulate cardiac hypertrophy. Annu Rev Physiol 63:391–426
59. Crabtree GR (1999) Generic signals and specific outcomes: signaling through Ca^{2+}, calcineurin, and NF-AT. Cell 96:611–614

60. Bush EW, Hood DB, Papst PJ, Chapo JA, Minobe W, Bristow MR, Olson EN, McKinsey TA (2006) Canonical transient receptor potential channels promote cardiomyocyte hypertrophy through activation of calcineurin signaling. J Biol Chem 281:33487–33496
61. Brenner JS, Dolmetsch RE (2007) TrpC3 regulates hypertrophy-associated gene expression without affecting myocyte beating or cell size. PLoS One 2:e802
62. Kiyonaka S, Kato K, Nishida M, Mio K, Numaga T, Sawaguchi Y, Yoshida T, Wakamori M, Mori E, Numata T, Ishii M, Takemoto H, Ojida A, Watanabe K, Uemura A, Kurose H, Morii T, Kobayashi T, Sato Y, Sato C, Hamachi I, Mori Y (2009) Selective and direct inhibition of TRPC3 channels underlies biological activities of a pyrazole compound. Proc Natl Acad Sci USA 106:5400–5405
63. Kuwahara K, Wang Y, McAnally J, Richardson JA, Bassel-Duby R, Hill JA, Olson EN (2006) TRPC6 fulfills a calcineurin signaling circuit during pathologic cardiac remodeling. J Clin Invest 116:3114–3126
64. Onohara N, Nishida M, Inoue R, Kobayashi H, Sumimoto H, Sato Y, Mori Y, Nagao T, Kurose H (2006) TRPC3 and TRPC6 are essential for angiotensin II-induced cardiac hypertrophy. EMBO J 25:5305–5316
65. Dietrich A, Mederos YSM, Gollasch M, Gross V, Storch U, Dubrovska G, Obst M, Yildirim E, Salanova B, Kalwa H, Essin K, Pinkenburg O, Luft FC, Gudermann T, Birnbaumer L (2005) Increased vascular smooth muscle contractility in TRPC6–/– mice. Mol Cell Biol 25:6980–6989
66. Kawasaki BT, Liao Y, Birnbaumer L (2006) Role of Src in C3 transient receptor potential channel function and evidence for a heterogeneous makeup of receptor- and store-operated Ca^{2+} entry channels. Proc Natl Acad Sci USA 103:335–340
67. Kwan HY, Huang Y, Yao X (2004) Regulation of canonical transient receptor potential isoform 3 (TRPC3) channel by protein kinase G. Proc Natl Acad Sci USA 101: 2625–2630
68. Takahashi S, Lin H, Geshi N, Mori Y, Kawarabayashi Y, Takami N, Mori MX, Honda A, Inoue R (2008) Nitric oxide-cGMP-protein kinase G pathway negatively regulates vascular transient receptor potential channel TRPC6. J Physiol 586:4209–4223
69. Nishida M, Watanabe K, Sato Y, Nakaya M, Kitajima N, Ide T, Inoue R, Kurose H Phosphorylation of TRPC6 channels at Thr69 is required for anti-hypertrophic effects of phosphodiesterase 5 inhibition. J Biol Chem. 2010 Apr 23; 285(17):13244–53
70. Takimoto E, Champion HC, Li M, Belardi D, Ren S, Rodriguez ER, Bedja D, Gabrielson KL, Wang Y, Kass DA (2005) Chronic inhibition of cyclic GMP phosphodiesterase 5A prevents and reverses cardiac hypertrophy. Nat Med 11:214–222
71. Hsu S, Nagayama T, Koitabashi N, Zhang M, Zhou L, Bedja D, Gabrielson KL, Molkentin JD, Kass DA, Takimoto E (2009) Phosphodiesterase 5 inhibition blocks pressure overload-induced cardiac hypertrophy independent of the calcineurin pathway. Cardiovasc Res 81:301–309
72. Lewis GD, Lachmann J, Camuso J, Lepore JJ, Shin J, Martinovic ME, Systrom DM, Bloch KD, Semigran MJ (2007) Sildenafil improves exercise hemodynamics and oxygen uptake in patients with systolic heart failure. Circulation 115:59–66
73. Takimoto E, Koitabashi N, Hsu S, Ketner EA, Zhang M, Nagayama T, Bedja D, Gabrielson KL, Blanton R, Siderovski DP, Mendelsohn ME, Kass DA (2009) Regulator of G protein signaling 2 mediates cardiac compensation to pressure overload and antihypertrophic effects of PDE5 inhibition in mice. J Clin Invest 119:408–420
74. Kinoshita H, Kuwahara K, Nishida M, Jiang Z, Rong X, Kiyonaka S, Kuwabara Y, Kurose H, Inoue R, Mori Y, Li Y, Nakagawa Y, Usami S, Fujiwara M, Yamada Y, Minami T, Ueshima K, Nakao K Inhibition of TRPC6 channel activity contributes to the antihypertrophic effects of natriuretic peptides-guanylyl cyclase-a signaling in the heart. Circ Res. 2010 Jun 25; 106(12):1849–60. Epub 2010 May 6
75. Beech DJ, Muraki K, Flemming R (2004) Non-selective cationic channels of smooth muscle and the mammalian homologues of Drosophila TRP. J Physiol 559:685–706

76. Ohba T, Watanabe H, Murakami M, Takahashi Y, Iino K, Kuromitsu S, Mori Y, Ono K, Iijima T, Ito H (2007) Upregulation of TRPC1 in the development of cardiac hypertrophy. J Mol Cell Cardiol 42:498–507
77. Seth M, Zhang ZS, Mao L, Graham V, Burch J, Stiber J, Tsiokas L, Winn M, Abramowitz J, Rockman HA, Birnbaumer L, Rosenberg P (2009) TRPC1 channels are critical for hypertrophic signaling in the heart. Circ Res 105:1023–1030
78. Wu X, Eder P, Chang B, Molkentin JD TRPC channels are necessary mediators of pathologic cardiac hypertrophy. Proc Natl Acad Sci USA. 2010 Apr 13; 107(15):7000–5. Epub 2010 Mar 29
79. Guinamard R, Demion M, Chatelier A, Bois P (2006) Calcium-activated nonselective cation channels in mammalian cardiomyocytes. Trends Cardiovasc Med 16:245–250
80. Kang PM, Izumo S (2003) Apoptosis in heart: basic mechanisms and implications in cardiovascular diseases. Trends Mol Med 9:177–182
81. Satoh S, Tanaka H, Ueda Y, Oyama J, Sugano M, Sumimoto H, Mori Y, Makino N (2007) Transient receptor potential (TRP) protein 7 acts as a G protein-activated Ca^{2+} channel mediating angiotensin II-induced myocardial apoptosis. Mol Cell Biochem 294:205–215
82. Shan D, Marchase RB, Chatham JC (2008) Overexpression of TRPC3 increases apoptosis but not necrosis in response to ischemia-reperfusion in adult mouse cardiomyocytes. Am J Physiol Cell Physiol 294:C833–C841
83. Nishida M, Onohara N, Sato Y, Suda R, Ogushi M, Tanabe S, Inoue R, Mori Y, Kurose H (2007) Galpha12/13-mediated up-regulation of TRPC6 negatively regulates endothelin-1-induced cardiac myofibroblast formation and collagen synthesis through nuclear factor of activated T cells activation. J Biol Chem 282:23117–23128
84. Brown RD, Ambler SK, Mitchell MD, Long CS (2005) The cardiac fibroblast: therapeutic target in myocardial remodeling and failure. Annu Rev Pharmacol Toxicol 45:657–687
85. Yang KT, Chang WL, Yang PC, Chien CL, Lai MS, Su MJ, Wu ML (2006) Activation of the transient receptor potential M2 channel and poly(ADP-ribose) polymerase is involved in oxidative stress-induced cardiomyocyte death. Cell Death Differ 13:1815–1826
86. Eisfeld J, Luckhoff A (2007) Trpm2. Handb Exp Pharmacol 179:237–252
87. Hara Y, Wakamori M, Ishii M, Maeno E, Nishida M, Yoshida T, Yamada H, Shimizu S, Mori E, Kudoh J, Shimizu N, Kurose H, Okada Y, Imoto K, Mori Y (2002) LTRPC2 Ca^{2+}-permeable channel activated by changes in redox status confers susceptibility to cell death. Mol Cell 9:163–173
88. Miller BA (2006) The role of TRP channels in oxidative stress-induced cell death. J Membr Biol 209:31–41
89. Wu G, Markowitz GS, Li L, D'Agati VD, Factor SM, Geng L, Tibara S, Tuchman J, Cai Y, Park JH, van Adelsberg J, Hou H Jr, Kucherlapati R, Edelmann W, Somlo S (2000) Cardiac defects and renal failure in mice with targeted mutations in Pkd2. Nat Genet 24:75–78
90. Chauvet V, Qian F, Boute N, Cai Y, Phakdeekitacharoen B, Onuchic LF, Attie-Bitach T, Guicharnaud L, Devuyst O, Germino GG, Gubler MC (2002) Expression of PKD1 and PKD2 transcripts and proteins in human embryo and during normal kidney development. Am J Pathol 160:973–983
91. Clark TG, Conway SJ, Scott IC, Labosky PA, Winnier G, Bundy J, Hogan BL, Greenspan DS (1999) The mammalian Tolloid-like 1 gene, Tll1, is necessary for normal septation and positioning of the heart. Development 126:2631–2642
92. Pennekamp P, Karcher C, Fischer A, Schweickert A, Skryabin B, Horst J, Blum M, Dworniczak B (2002) The ion channel polycystin-2 is required for left-right axis determination in mice. Curr Biol 12:938–943
93. Colquhoun D, Neher E, Reuter H, Stevens CF (1981) Inward current channels activated by intracellular Ca in cultured cardiac cells. Nature 294:752–754
94. Guinamard R, Chatelier A, Demion M, Potreau D, Patri S, Rahmati M, Bois P (2004) Functional characterization of a $Ca^{(2+)}$-activated non-selective cation channel in human atrial cardiomyocytes. J Physiol 558:75–83

95. Nilius B, Mahieu F, Prenen J, Janssens A, Owsianik G, Vennekens R, Voets T (2006) The Ca^{2+}-activated cation channel TRPM4 is regulated by phosphatidylinositol 4,5-biphosphate. EMBO J 25:467–478
96. Kruse M, Schulze-Bahr E, Corfield V, Beckmann A, Stallmeyer B, Kurtbay G, Ohmert I, Brink P, Pongs O (2009) Impaired endocytosis of the ion channel TRPM4 is associated with human progressive familial heart block type I. J Clin Invest 119:2737–2744
97. Iwata Y, Katanosaka Y, Arai Y, Komamura K, Miyatake K, Shigekawa M (2003) A novel mechanism of myocyte degeneration involving the Ca^{2+}-permeable growth factor-regulated channel. J Cell Biol 161:957–967
98. Pan HL, Chen SR (2004) Sensing tissue ischemia: another new function for capsaicin receptors? Circulation 110:1826–1831
99. Wang L, Wang DH (2005) TRPV1 gene knockout impairs postischemic recovery in isolated perfused heart in mice. Circulation 112:3617–3623
100. Sexton A, McDonald M, Cayla C, Thiemermann C, Ahluwalia A (2007) 12-Lipoxygenase-derived eicosanoids protect against myocardial ischemia/reperfusion injury via activation of neuronal TRPV1. FASEB J 21:2695–2703
101. Zhong B, Wang DH (2008) N-oleoyldopamine, a novel endogenous capsaicin-like lipid, protects the heart against ischemia-reperfusion injury via activation of TRPV1. Am J Physiol Heart Circ Physiol 295:H728–H735
102. Paria BC, Malik AB, Kwiatek AM, Rahman A, May MJ, Ghosh S, Tiruppathi C (2003) Tumor necrosis factor-alpha induces nuclear factor-kappaB-dependent TRPC1 expression in endothelial cells. J Biol Chem 278:37195–37203
103. Tiruppathi C, Freichel M, Vogel SM, Paria BC, Mehta D, Flockerzi V, Malik AB (2002) Impairment of store-operated Ca^{2+} entry in TRPC4(–/–) mice interferes with increase in lung microvascular permeability. Circ Res 91:70–76
104. Singh I, Knezevic N, Ahmmed GU, Kini V, Malik AB, Mehta D (2007) Galphaq-TRPC6-mediated Ca^{2+} entry induces RhoA activation and resultant endothelial cell shape change in response to thrombin. J Biol Chem 282:7833–7843
105. Paria BC, Vogel SM, Ahmmed GU, Alamgir S, Shroff J, Malik AB, Tiruppathi C (2004) Tumor necrosis factor-alpha-induced TRPC1 expression amplifies store-operated Ca^{2+} influx and endothelial permeability. Am J Physiol Lung Cell Mol Physiol 287:L1303–L1313
106. Paria BC, Bair AM, Xue J, Yu Y, Malik AB, Tiruppathi C (2006) Ca^{2+} influx induced by protease-activated receptor-1 activates a feed-forward mechanism of TRPC1 expression via nuclear factor-kappaB activation in endothelial cells. J Biol Chem 281:20715–20727
107. Cioffi DL, Stevens T (2006) Regulation of endothelial cell barrier function by store-operated calcium entry. Microcirculation 13:709–723
108. Odell AF, Van Helden DF, Scott JL (2008) The spectrin cytoskeleton influences the surface expression and activation of human transient receptor potential channel 4 channels. J Biol Chem 283:4395–4407
109. Singh I, Knezevic N, Ahmmed GU, Kini V, Malik AB, Mehta D (2006) Galpha q-TRPC6-mediated Ca^{2+} entry induces RhoA activation and resultant endothelial cell shape change in response to thrombin. J Biol Chem. 2007 Mar 16; 282(11):7833–43. Epub 2006 Dec 29
110. Cheng HW, James AF, Foster RR, Hancox JC, Bates DO (2006) VEGF activates receptor-operated cation channels in human microvascular endothelial cells. Arterioscler Thromb Vasc Biol 26:1768–1776
111. Jho D, Mehta D, Ahmmed G, Gao XP, Tiruppathi C, Broman M, Malik AB (2005) Angiopoietin-1 opposes VEGF-induced increase in endothelial permeability by inhibiting TRPC1-dependent Ca2 influx. Circ Res 96:1282–1290
112. Gerzanich V, Woo SK, Vennekens R, Tsymbalyuk O, Ivanova S, Ivanov A, Geng Z, Chen Z, Nilius B, Flockerzi V, Freichel M, Simard JM (2009) De novo expression of Trpm4 initiates secondary hemorrhage in spinal cord injury. Nat Med 15:185–191
113. Alvarez DF, King JA, Weber D, Addison E, Liedtke W, Townsley MI (2006) Transient receptor potential vanilloid 4-mediated disruption of the alveolar septal barrier: a novel mechanism of acute lung injury. Circ Res 99:988–995

114. Freichel M, Suh SH, Pfeifer A, Schweig U, Trost C, Weissgerber P, Biel M, Philipp S, Freise D, Droogmans G, Hofmann F, Flockerzi V, Nilius B (2001) Lack of an endothelial store-operated Ca^{2+} current impairs agonist-dependent vasorelaxation in TRP4–/– mice. Nat Cell Biol 3:121–127
115. Yoshida T, Inoue R, Morii T, Takahashi N, Yamamoto S, Hara Y, Tominaga M, Shimizu S, Sato Y, Mori Y (2006) Nitric oxide activates TRP channels by cysteine S-nitrosylation. Nat Chem Biol 2:596–607
116. Dedkova EN, Blatter LA (2002) Nitric oxide inhibits capacitative Ca^{2+} entry and enhances endoplasmic reticulum Ca^{2+} uptake in bovine vascular endothelial cells. J Physiol 539:77–91
117. Takeuchi K, Watanabe H, Tran QK, Ozeki M, Sumi D, Hayashi T, Iguchi A, Ignarro LJ, Ohashi K, Hayashi H (2004) Nitric oxide: inhibitory effects on endothelial cell calcium signaling, prostaglandin I2 production and nitric oxide synthase expression. Cardiovasc Res 62:194–201
118. Kwan HY, Huang Y, Yao X (2000) Store-operated calcium entry in vascular endothelial cells is inhibited by cGMP via a protein kinase G-dependent mechanism. J Biol Chem 275: 6758–6763
119. Hartmann J, Dragicevic E, Adelsberger H, Henning HA, Sumser M, Abramowitz J, Blum R, Dietrich A, Freichel M, Flockerzi V, Birnbaumer L, Konnerth A (2008) TRPC3 channels are required for synaptic transmission and motor coordination. Neuron 59:392–398
120. Riccio A, Li Y, Moon J, Kim KS, Smith KS, Rudolph U, Gapon S, Yao GL, Tsvetkov E, Rodig SJ, Vant Veer A, Meloni EG, Carlezon WA Jr., Bolshakov VY, Clapham DE (2009) Essential role for TRPC5 in amygdala function and fear-related behavior. Cell 137:761–772
121. Fleming I, Rueben A, Popp R, Fisslthaler B, Schrodt S, Sander A, Haendeler J, Falck JR, Morisseau C, Hammock BD, Busse R (2007) Epoxyeicosatrienoic acids regulate Trp channel dependent Ca^{2+} signaling and hyperpolarization in endothelial cells. Arterioscler Thromb Vasc Biol 27:2612–2618
122. Busse R, Fleming I (2006) Vascular endothelium and blood flow. Handb Exp Pharmacol 176:43–78
123. Inoue K, Xiong ZG (2009) Silencing TRPM7 promotes growth/proliferation and nitric oxide production of vascular endothelial cells via the ERK pathway. Cardiovasc Res 83:547–557
124. Mergler S, Pleyer U (2007) The human corneal endothelium: new insights into electrophysiology and ion channels. Prog Retin Eye Res 26:359–378
125. Johnson CD, Melanaphy D, Purse A, Stokesberry SA, Dickson P, Zholos AV (2009) Transient receptor potential melastatin 8 channel involvement in the regulation of vascular tone. Am J Physiol Heart Circ Physiol 296:H1868–H1877
126. Bratz IN, Dick GM, Tune JD, Edwards JM, Neeb ZP, Dincer UD, Sturek M (2008) Impaired capsaicin-induced relaxation of coronary arteries in a porcine model of the metabolic syndrome. Am J Physiol Heart Circ Physiol 294:H2489–H2496
127. Vriens J, Owsianik G, Fisslthaler B, Suzuki M, Janssens A, Voets T, Morisseau C, Hammock BD, Fleming I, Busse R, Nilius B (2005) Modulation of the Ca2 permeable cation channel TRPV4 by cytochrome P450 epoxygenases in vascular endothelium. Circ Res 97:908–915
128. Kohler R, Heyken WT, Heinau P, Schubert R, Si H, Kacik M, Busch C, Grgic I, Maier T, Hoyer J (2006) Evidence for a functional role of endothelial transient receptor potential V4 in shear stress-induced vasodilatation. Arterioscler Thromb Vasc Biol 26: 1495–1502
129. Earley S, Pauyo T, Drapp R, Tavares MJ, Liedtke W, Brayden JE (2009) TRPV4-dependent dilation of peripheral resistance arteries influences arterial pressure. Am J Physiol Heart Circ Physiol 297:H1096–H1102
130. Hartmannsgruber V, Heyken WT, Kacik M, Kaistha A, Grgic I, Harteneck C, Liedtke W, Hoyer J, Kohler R (2007) Arterial response to shear stress critically depends on endothelial TRPV4 expression. PLoS One 2:e827

131. Saliez J, Bouzin C, Rath G, Ghisdal P, Desjardins F, Rezzani R, Rodella LF, Vriens J, Nilius B, Feron O, Balligand JL, Dessy C (2008) Role of caveolar compartmentation in endothelium-derived hyperpolarizing factor-mediated relaxation: Ca^{2+} signals and gap junction function are regulated by caveolin in endothelial cells. Circulation 117:1065–1074
132. Stenmark KR, Fagan KA, Frid MG (2006) Hypoxia-induced pulmonary vascular remodeling: cellular and molecular mechanisms. Circ Res 99:675–691
133. Fantozzi I, Zhang S, Platoshyn O, Remillard CV, Cowling RT, Yuan JX (2003) Hypoxia increases AP-1 binding activity by enhancing capacitative Ca^{2+} entry in human pulmonary artery endothelial cells. Am J Physiol Lung Cell Mol Physiol 285:L1233–L1245
134. Balzer M, Lintschinger B, Groschner K (1999) Evidence for a role of Trp proteins in the oxidative stress-induced membrane conductances of porcine aortic endothelial cells. Cardiovasc Res 42:543–549
135. Groschner K, Rosker C, Lukas M (2004) Role of TRP channels in oxidative stress. Novartis Found Symp 258:222–230 discussion 231–225, 263–226
136. Poteser M, Graziani A, Rosker C, Eder P, Derler I, Kahr H, Zhu MX, Romanin C, Groschner K (2006) TRPC3 and TRPC4 associate to form a redox-sensitive cation channel. Evidence for expression of native TRPC3-TRPC4 heteromeric channels in endothelial cells. J Biol Chem 281:13588–13595
137. Hecquet CM, Ahmmed GU, Vogel SM, Malik AB (2008) Role of TRPM2 channel in mediating H_2O_2-induced Ca^{2+} entry and endothelial hyperpermeability. Circ Res 102:347–355
138. Dietrich A, Gudermann T (2008) Another TRP to endothelial dysfunction: TRPM2 and endothelial permeability. Circ Res 102:275–277
139. Chaudhuri P, Colles SM, Damron DS, Graham LM (2003) Lysophosphatidylcholine inhibits endothelial cell migration by increasing intracellular calcium and activating calpain. Arterioscler Thromb Vasc Biol 23:218–223
140. Chaudhuri P, Colles SM, Bhat M, Van Wagoner DR, Birnbaumer L, Graham LM (2008) Elucidation of a TRPC6-TRPC5 channel cascade that restricts endothelial cell movement. Mol Biol Cell 19:3203–3211
141. Goel M, Sinkins WG, Schilling WP (2002) Selective association of TRPC channel subunits in rat brain synaptosomes. J Biol Chem 277:48303–48310
142. Ge R, Tai Y, Sun Y, Zhou K, Yang S, Cheng T, Zou Q, Shen F, Wang Y (2009) Critical role of TRPC6 channels in VEGF-mediated angiogenesis. Cancer Lett 283:43–51
143. House SJ, Potier M, Bisaillon J, Singer HA, Trebak M (2008) The non-excitable smooth muscle: calcium signaling and phenotypic switching during vascular disease. Pflugers Arch 456:769–785
144. Sweeney M, Yu Y, Platoshyn O, Zhang S, McDaniel SS, Yuan JX (2002) Inhibition of endogenous TRP1 decreases capacitative Ca^{2+} entry and attenuates pulmonary artery smooth muscle cell proliferation. Am J Physiol Lung Cell Mol Physiol 283:L144–L155
145. Kumar B, Dreja K, Shah SS, Cheong A, Xu SZ, Sukumar P, Naylor J, Forte A, Cipollaro M, McHugh D, Kingston PA, Heagerty AM, Munsch CM, Bergdahl A, Hultgardh-Nilsson A, Gomez MF, Porter KE, Hellstrand P, Beech DJ (2006) Upregulated TRPC1 channel in vascular injury in vivo and its role in human neointimal hyperplasia. Circ Res 98:557–563
146. Zhang S, Remillard CV, Fantozzi I, Yuan JX (2004) ATP-induced mitogenesis is mediated by cyclic AMP response element-binding protein-enhanced TRPC4 expression and activity in human pulmonary artery smooth muscle cells. Am J Physiol Cell Physiol 287:C1192–C1201
147. Flemming PK, Dedman AM, Xu SZ, Li J, Zeng F, Naylor J, Benham CD, Bateson AN, Muraki K, Beech DJ (2006) Sensing of lysophospholipids by TRPC5 calcium channel. J Biol Chem 281:4977–4982
148. Yu Y, Fantozzi I, Remillard CV, Landsberg JW, Kunichika N, Platoshyn O, Tigno DD, Thistlethwaite PA, Rubin LJ, Yuan JX (2004) Enhanced expression of transient receptor potential channels in idiopathic pulmonary arterial hypertension. Proc Natl Acad Sci USA 101:13861–13866
149. Weissmann N, Dietrich A, Fuchs B, Kalwa H, Ay M, Dumitrascu R, Olschewski A, Storch U, Mederos y Schnitzler M, Ghofrani HA, Schermuly RT, Pinkenburg O, Seeger W,

Grimminger F, Gudermann T (2006) Classical transient receptor potential channel 6 (TRPC6) is essential for hypoxic pulmonary vasoconstriction and alveolar gas exchange. Proc Natl Acad Sci USA 103:19093–19098
150. He Y, Yao G, Savoia C, Touyz RM (2005) Transient receptor potential melastatin 7 ion channels regulate magnesium homeostasis in vascular smooth muscle cells: role of angiotensin II. Circ Res 96:207–215
151. Schievink WI, Torres VE, Piepgras DG, Wiebers DO (1992) Saccular intracranial aneurysms in autosomal dominant polycystic kidney disease. J Am Soc Nephrol 3:88–95
152. Ducret T, Guibert C, Marthan R, Savineau JP (2008) Serotonin-induced activation of TRPV4-like current in rat intrapulmonary arterial smooth muscle cells. Cell Calcium 43:315–323
153. Xu SZ, Muraki K, Zeng F, Li J, Sukumar P, Shah S, Dedman AM, Flemming PK, McHugh D, Naylor J, Cheong A, Bateson AN, Munsch CM, Porter KE, Beech DJ (2006) A sphingosine-1-phosphate-activated calcium channel controlling vascular smooth muscle cell motility. Circ Res 98:1381–1389
154. Bergdahl A, Gomez MF, Dreja K, Xu SZ, Adner M, Beech DJ, Broman J, Hellstrand P, Sward K (2003) Cholesterol depletion impairs vascular reactivity to endothelin-1 by reducing store-operated Ca^{2+} entry dependent on TRPC1. Circ Res 93:839–847
155. Bergdahl A, Gomez MF, Wihlborg AK, Erlinge D, Eyjolfson A, Xu SZ, Beech DJ, Dreja K, Hellstrand P (2005) Plasticity of TRPC expression in arterial smooth muscle: correlation with store-operated Ca^{2+} entry. Am J Physiol Cell Physiol 288:872–C880
156. Kunichika N, Yu Y, Remillard CV, Platoshyn O, Zhang S, Yuan JX (2004) Overexpression of TRPC1 enhances pulmonary vasoconstriction induced by capacitative Ca^{2+} entry. Am J Physiol Lung Cell Mol Physiol 287:L962–L969
157. Inoue R, Okada T, Onoue H, Hara Y, Shimizu S, Naitoh S, Ito Y, Mori Y (2001) The transient receptor potential protein homologue TRP6 is the essential component of vascular alpha(1)-adrenoceptor-activated $Ca^{(2+)}$-permeable cation channel. Circ Res 88: 325–332
158. Jung S, Strotmann R, Schultz G, Plant TD (2002) TRPC6 is a candidate channel involved in receptor-stimulated cation currents in A7r5 smooth muscle cells. Am J Physiol Cell Physiol 282:C347–C359
159. Sel S, Rost BR, Yildirim AO, Sel B, Kalwa H, Fehrenbach H, Renz H, Gudermann T, Dietrich A (2008) Loss of classical transient receptor potential 6 channel reduces allergic airway response. Clin Exp Allergy 38:1548–1558
160. Welsh DG, Morielli AD, Nelson MT, Brayden JE (2002) Transient receptor potential channels regulate myogenic tone of resistance arteries. Circ Res 90: 248–250
161. Reading SA, Earley S, Waldron BJ, Welsh DG, Brayden JE (2005) TRPC3 mediates pyrimidine receptor-induced depolarization of cerebral arteries. Am J Physiol Heart Circ Physiol 288:H2055–H2061
162. Liu D, Scholze A, Zhu Z, Kreutz R, Wehland-von-Trebra M, Zidek W, Tepel M (2005) Increased transient receptor potential channel TRPC3 expression in spontaneously hypertensive rats. Am J Hypertens 18:1503–1507
163. Liu D, Scholze A, Zhu Z, Krueger K, Thilo F, Burkert A, Streffer K, Holz S, Harteneck C, Zidek W, Tepel M (2006) Transient receptor potential channels in essential hypertension. J Hypertens 24:1105–1114
164. Wang J, Weigand L, Lu W, Sylvester JT, Semenza GL, Shimoda LA (2006) Hypoxia inducible factor 1 mediates hypoxia-induced TRPC expression and elevated intracellular Ca^{2+} in pulmonary arterial smooth muscle cells. Circ Res 98:1528–1537
165. Wang J, Shimoda LA, Weigand L, Wang W, Sun D, Sylvester JT (2005) Acute hypoxia increases intracellular $[Ca^{2+}]$ in pulmonary arterial smooth muscle by enhancing capacitative Ca^{2+} entry. Am J Physiol Lung Cell Mol Physiol 288:L1059–L1069
166. Walker RL, Hume JR, Horowitz B (2001) Differential expression and alternative splicing of TRP channel genes in smooth muscles. Am J Physiol Cell Physiol 280:C1184–C1192

167. Wang J, Shimoda LA, Sylvester JT (2004) Capacitative calcium entry and TRPC channel proteins are expressed in rat distal pulmonary arterial smooth muscle. Am J Physiol Lung Cell Mol Physiol 286:L848–L858
168. Yuan JP, Kim MS, Zeng W, Shin DM, Huang G, Worley PF, Muallem S (2009) TRPC channels as STIM1-regulated SOCs. Channels (Austin) 3:221–225
169. Estacion M, Sinkins WG, Jones SW, Applegate MA, Schilling WP (2006) TRPC6 forms non-selective cation channels with limited Ca^{2+} Permeability. J Physiol 572(2):359–377
170. Gudermann T, Mederos y Schnitzler M, Dietrich A (2004) Receptor-operated cation entry–more than esoteric terminology? Sci STKE 2004:pe35
171. Soboloff J, Spassova M, Xu W, He LP, Cuesta N, Gill DL (2005) Role of endogenous TRPC6 channels in Ca^{2+} signal generation in A7r5 smooth muscle cells. J Biol Chem 280:3 9786–39794
172. Dietrich A, Gudermann T (2007) Trpc6. Handb Exp Pharmacol 179:125–141
173. Touyz RM, He Y, Montezano AC, Yao G, Chubanov V, Gudermann T, Callera GE (2005) Differential Regulation of TRPM6/7 Cation Channels by Ang II in Vascular Smooth Muscle Cells from Spontaneously Hypertensive Rats. Am J Physiol Regul Integr Comp Physiol. 2006 Jan; 290(1):R73–8. Epub 2005 Aug 18
174. Northcott CA, Watts SW (2004) Low [Mg^{2+}]e enhances arterial spontaneous tone via phosphatidylinositol 3-kinase in DOCA-salt hypertension. Hypertension 43:125–129
175. Quamme GA, Dai LJ, Rabkin SW (1993) Dynamics of intracellular free Mg^{2+} changes in a vascular smooth muscle cell line. Am J Physiol 265:H281–H288
176. Earley S, Heppner TJ, Nelson MT, Brayden JE (2005) TRPV4 forms a novel Ca2+ signaling complex with ryanodine receptors and BKCa channels. Circ Res 97:1270–1279
177. Inoue R, Jian Z, Kawarabayashi Y (2009) Mechanosensitive TRP channels in cardiovascular pathophysiology. Pharmacol Ther. 2009 Sep; 123(3):371–85. Epub 2009 Jun 6. Review
178. Maroto R, Raso A, Wood TG, Kurosky A, Martinac B, Hamill OP (2005) TRPC1 forms the stretch-activated cation channel in vertebrate cells. Nat Cell Biol 7:179–185
179. Spassova MA, Hewavitharana T, Xu W, Soboloff J, Gill DL (2006) A common mechanism underlies stretch activation and receptor activation of TRPC6 channels. Proc Natl Acad Sci USA 103:16586–16591
180. Gottlieb P, Folgering J, Maroto R, Raso A, Wood TG, Kurosky A, Bowman C, Bichet D, Patel A, Sachs F, Martinac B, Hamill OP, Honore E (2008) Revisiting TRPC1 and TRPC6 mechanosensitivity. Pflugers Arch 455:1097–1103
181. Mederos y Schnitzler M, Storch U, Meibers S, Nurwakagari P, Breit A, Essin K, Gollasch M, Gudermann T (2008) Gq-coupled receptors as mechanosensors mediating myogenic vasoconstriction. EMBO J 27:3092–3103
182. Keseru B, Barbosa-Sicard E, Popp R, Fisslthaler B, Dietrich A, Gudermann T, Hammock BD, Falck JR, Weissmann N, Busse R, Fleming I (2008) Epoxyeicosatrienoic acids and the soluble epoxide hydrolase are determinants of pulmonary artery pressure and the acute hypoxic pulmonary vasoconstrictor response. FASEB J 22:4306–4315
183. Earley S, Straub SV, Brayden JE (2007) Protein kinase C regulates vascular myogenic tone through activation of TRPM4. Am J Physiol Heart Circ Physiol 292:H2613–H2622
184. Reading SA, Brayden JE (2007) Central role of TRPM4 channels in cerebral blood flow regulation. Stroke 38:2322–2328
185. Morita H, Honda A, Inoue R, Ito Y, Abe K, Nelson MT, Brayden JE (2007) Membrane stretch-induced activation of a TRPM4-like nonselective cation channel in cerebral artery myocytes. J Pharmacol Sci 103:417–426
186. Oancea E, Wolfe JT, Clapham DE (2006) Functional TRPM7 channels accumulate at the plasma membrane in response to fluid flow. Circ Res 98:245–253
187. O'Neil RG, Heller S (2005) The mechanosensitive nature of TRPV channels. Pflugers Arch 451:193–203

Chapter 42
TRP Channels of Islets

Md. Shahidul Islam

Abstract In the normal human body pancreatic β-cells spend most of the time in a READY mode rather than in an OFF mode. When in the READY mode, normal β-cells can be easily SWITCHED ON by a variety of apparently trivial stimuli. In the READY mode β-cells are highly excitable because of their high input resistance. A variety of small depolarizing currents mediated through a variety of cation channels triggered by a variety of chemical and physical stimuli can SWITCH ON the cells. Several polymodal ion channels belonging to the transient receptor potential (TRP) family may mediate the depolarizing currents necessary to shift the β-cells from the READY mode to the ON mode. Thanks to the TRP channels, we now know that the Ca^{2+}-activated monovalent cation selective channel described by Sturgess et al. in 1986 (FEBS Lett 208:397–400) is TRPM4, and that the H_2O_2-activate non-selective cation channel described by Herson and Ashford, in 1997 (J Physiol 501:59–66) is TRPM2. Glucose metabolism generates heat which appears to be a second messenger sensed by the temperature-sensitive TRP channels like the TRPM2 channel. Global knock-out of TRPM5 channel impairs insulin secretion in mice. Other TRPs that may be involved in the regulation of β-cell function include TRPC1, TRPC4, TRPM3, TRPV2 and TRPV4. Future research needs to be intensified to study the molecular regulation of the TRP channels of islets, and to elucidate their roles in the regulation of human β-cell function, in the context of pathogenesis of human islet failure.

42.1 Introduction

The pancreatic β-cells are sensors for glucose, amino acids, fatty acids as well as for incretins, other hormones, neurotransmitters and growth factors [1]. Many ion channels, G-protein coupled receptors, and other plasma membrane and intracellular receptors are involved in triggering different sets of interconnected signaling events

---

M.S. Islam (✉)
Karolinska Institutet, Department of Clinical Sciences and Education, Södersjukhuset, SE-118 83 Stockholm, Sweden; Uppsala University Hospital, AR Division, Uppsala, Sweden
e-mail: shahidul.islam@ki.se

that regulate exocytosis of insulin in response to a variety of stimuli. From in vitro experiments where glucose has been used as the only active agent, it is known that two important events occur when glucose stimulates insulin secretion. These are depolarization of the membrane potential and an increase in the cytoplasmic free Ca^{2+} concentration ($[Ca^{2+}]_i$), often in the form of oscillations [2]. A handful of ion channels are known to be involved in mediating these two interdependent events [3]. Among them, the most extensively studied one is the ATP-sensitive K^+ (K_{ATP}) channel. β-cells and some cells in the hypothalamus are the only cells in the body that use ATP as a cytoplasmic second messenger; the K_{ATP} channels being the sensors for this unique second messenger. These channels are targets for some widely prescribed anti-diabetic drugs. Mutations of K_{ATP} channel can cause congenital hyperinsulinism or neonatal diabetes in human [4]. Not surprisingly, K_{ATP} channels hugely dominate the field of islet research; cellular processes are defined in terms of this channel, and are branded either "K_{ATP}-channel-dependent" or "K_{ATP}-channel-*in*dependent". This scenario, including the use of the term "consensus model" leaves little room to accommodate the TRP channels in the existing model(s) of stimulus secretion coupling in β-cells.

It is worthwhile to consider a slightly different conceptual scenario where one could possibly implicate the TRP channels in the regulation of insulin secretion. Human β-cells secrete insulin all the time both during the fasting state and during the fed state. Insulin can cause potentially life-threatening hypoglycemia. As a safety device, nature has equipped these cells with K_{ATP} channels which can rapidly switch OFF insulin secretion to a great extent, when hypoglycemia is imminent. In most in vitro experiments, where relatively simple physiological solutions containing < 3 mM glucose are used as the sole nutrient, β-cells are in OFF mode [2]. In the OFF mode, the cytoplasmic free Ca^{2+} concentration ($[Ca^{2+}]_i$) is stable around 50–100 nM [2]. In vivo, β-cells are unlikely to be in the OFF mode even after overnight fasting because they are continuously bathed in many nutrients and hormones in addition to about 5 mM glucose. In fact, β-cells secrete insulin, albeit at low level, even during hypoglycemia, which is corrected by counter-regulatory hormones and gluconeogenesis. In their native environment, β-cells probably spend most of the time at least in a READY mode rather than in the OFF mode. In the READY mode, the degree of readiness or excitability of β-cells depends upon the input-resistance of the plasma membrane which in turn depends mostly on the degree of closure of the K_{ATP} channels and the magnitude of the background depolarizing currents. When in the READY mode, normal β-cells can be easily switched ON by a variety of apparently trivial stimuli acting alone or in combination. A variety of small depolarizing currents mediated through a variety of cation channels triggered by a variety of chemical and physical stimuli can switch ON these cells. The molecular basis of such depolarizing currents remained elusive for more than two decades. After the discovery of the TRP channels, their roles in β-cell was not extensively studied partly because of the scenario described in the previous paragraph, and partly because of the lack of specific pharmacological tools, specific antibodies and lack of obvious phenotypes in knock-out mouse models.

TRP channels of β-cells have been described in a recent review [2]. In this chapter, I shall sum up the background information and elaborate some of the recent developments in the field and give my own views and speculations. One note of caution is that research in cell and molecular biology is neither randomized, nor double-blinded, and seldom adequately controlled. The huge pressure on scientists for publication, and lack of adequate regulatory mechanisms for controlling accuracy of data make it inevitable that some enthusiasts can reveal what is suggestive and hide what is vital. Nevertheless, the review is based on a handful of published papers on the topic assuming that all published reports are of high scientific quality.

It is becoming increasingly evident that β-cells have many TRP channels (Table 42.1), and that some of these channels can contribute to membrane depolarization, Ca^{2+} signaling, insulin secretion and cell survival. I shall start with the TRPM4 and TRPM5 because these two are pretty hot at the time of this writing.

Table 42.1 TRP channels of insulin-secreting cells

Channel	Cell type	Methods	Reference
TRPC1	MIN6, mouse islet	RT-PCR, NB	[5, 6]
	INS-1, rat β-cells	RT-PCR	[7]
TRPC2	MIN6, mouse islet	RT-PCR	[6]
TRPC3	Mouse islets	RT-PCR	[6]
TRPC4	MIN6, βTC3, INS-1		
	rat β-cells	RT-PCR [7], NB [6]	[6, 7]
TRPC5	βTC3	RT-PCR	[6]
TRPC6	MIN6	RT-PCR	[6]
TRPM2	Human islets	RT-PCR [8], WB [9]	[8, 9]
	INS-1E	EP	[9]
	RIN-5F	EP, IF [10], WB [10]	[10, 11]
	CRI-G1	EP, RT-PCR	[12]
	HIT-T15	EP	[13]
	Mouse β-cells	IF [11], EP [14]	[11, 14]
	Rat β-cells	IF	[11]
TRPM3	INS-1, mouse islets	EP, RT-PCR, NB, WB	[15]
	Mouse β-cells	EP	[15]
TRPM4	INS-1, RINm5F	RT-PCR, IP, EP	[16]
	HIT-T15		
	MIN6, H11-115, RINm5F		
	β-TC3, INR1G9	EP	[17]
	Human β-cells	IF	[17]
TRPM5	MIN6, INS-1		
	Human islets	RT-PCR	[18, 19]
	Mouse β-cells	RT-PCR, EP, IF	[19]
TRPV1	INS-1, RINm5F, rat islets	RT-PCR, WB, IF	[20]
TRPV2	MIN6	RT-PCR, IB	[21]
	Mouse β-cells	IF	[21]
TRPV4	MIN6	RT-PCR, Ca^{2+} imaging	[22]

EP, electrophysiology; IF, immunofluorescence; IB, immunoblot; NB, Northern blot; WB, Western blot

42.2 TRPM4 and TRPM5

TRPM4 and TRPM5 have been reviewed in Chapter 8 by Romain Guinamard, Laurent Sallé, and Christophe Simard, in this book. These channels are activated by an elevated $[Ca^{2+}]_i$, and later on desensitized to further activation by Ca^{2+}. They are permeable to monovalent cations but almost impermeable to the divalent ones. It is speculated that TRPM4 and/or TRPM5, and possibly heterotetramers of TRPM4/TRPM5 may account for some components of the Ca^{2+} activated non-selective cation currents (Ca-NS) described in β-cells in some earlier reports [6, 23–25].

TRPM4 is inhibited by adenine nucleotides whereas TRPM5 is not, allowing one to distinguish between these two currents. The earliest indication for the presence of a TRPM4-like current in β-cells can be tracked back to the studies of Sturgess et al., reported in 1986 [25]. They described in CRI-G1 rat insulinoma cells a ~ 25 pS nonselective cation current (NSCC) activated by Ca^{2+} and inhibited by adenine nucleotides. The potencies of different adenine derivatives for the inhibition of this current (AMP > ADP > ATP > adenosine) are slightly different from those reported for the cloned TRPM4 over-expressed in heterologous cells (ADP > ATP > AMP >> adenosine) [26]. More than a decade later, Leech and Habener described in HIT-T15 cells, a ~ 25–30 pS Ca^{2+}-activated NSCC that was inhibited by ATP. Their results suggest the possibility that they were perhaps dealing with the TRPM4 current [24]. However, the linear current voltage (*I/V*) relationship of their current is not typical for TRPM4 or TRPM5, which usually shows activation at positive membrane potentials [27].

TRPM4 currents have now been described in a variety of rodent insulinoma and glucagonoma cells. In these cells, an elevated $[Ca^{2+}]_i$ activates the channel with an EC_{50} of ~ 0.57–1.25 μM [17]. Open probability of TRPM4 increases also upon membrane depolarization but only if $[Ca^{2+}]_i$ is sufficiently high. The TRPM4 current activated by an increase of $[Ca^{2+}]_i$ is biphasic where the first phase develops within seconds and the second phase develops slowly. Development of the second phase is thought to be due to incorporation of TRPM4 channels into the plasma membrane as a result of exocytosis. Inhibition of endogenous TRPM4 by a truncated dominant negative construct of TRPM4 reduces the magnitude of Ca^{2+} signal and insulin secretion in response to glucose or agonists of receptors coupled to phosphoinositide-specific phospholipase C (PI-PLC) [17]. So far, TRPM4 current has not been described in primary β-cells. However, TRPM4 protein can be detected by immunofluorescence in human β-cells [17]. Several splice variants of TRPM4 are known and of them TRPM4b (often referred to simply as TRPM4) is the one most studied. In TRPM4 knockout mice insulin secretion is normal possibly because of the presence of the related TRPM5 channel [28].

Amino acid sequence of TRPM4 shows two stretches that look like signature motifs of ATP binding cassette proteins (ABC transporters), and four nucleotide binding domains (NBD). Consistent with this, TRPM4 is inhibited by glibenclamide at least in some tissues [29]. The adenine nucleotide sensitive Ca^{2+}-activated NSCC described by Sturgess et al., is not inhibited by tolbutamide or glibenclamide [30]. As mentioned earlier, TRPM4 is directly inhibited by cytoplasmic ATP and other adenine nucleotides without requiring Mg^{2+} or hydrolysis of the nucleotides. However, ATP also inhibits Ca^{2+}-desensitization of TRPM4 [31]. TRPM4 has several putative phosphorylation sites for PKA and PKC. It also has arginine lysine rich sequences that are putative binding sites for phosphatidylinositol-4,5-bisphosphate (PIP2). Two well known regulators of TRPM4 are PKC and PIP2. The Ca^{2+}-sensitivity of TRPM4 is enhanced by PKC phosphorylation [31]. This is interesting given that different PKC isoforms regulate diverse functions of β-cells [32]. PIP2 shifts the voltage activation curve of TRPM4 towards negative voltages and also prevents Ca^{2+}-mediated desensitization of the channel; depletion of PIP2 inhibits the channel [33]. This mode of regulation may be relevant for β-cells since glucose, by way of altering the cytoplasmic ATP-to-ADP ratio, alters the PIP2 concentration in the plasma membrane in an oscillatory fashion [34].

Compared to TRPM4 which is expressed in a wide variety of tissues, the expression of TRPM5 is more restricted [35]. The latter channel is abundant in taste bud, intestine, pancreas and pituitary [35]. In fact, TRPM5 is best known for its role in taste signaling [36]. TRPM5 mRNA is detected in MIN6 cells [18] [but see [37]], INS-1 cells, purified mouse β-cells and human islets [18, 19]. By immunofluorescence, TRPM5 protein is detected in the β-cells of wild type mice and not in those of the knock-out mice [19].

Human TRPM4 and TRPM5 channels are activated by $[Ca^{2+}]_i$ with nearly similar EC_{50} values (~ 840–885 nM) [18]. Mouse TRPM4 is much less sensitive to $[Ca^{2+}]_i$ (EC_{50} of ~ 2,000 nM) compared to the mouse TRPM5, which is highly sensitive to $[Ca^{2+}]_i$ (EC_{50} of ~ 700 nM) [38]. Human TRPM5 is rapidly inhibited by high $[Ca^{2+}]_i$ (IC_{50} of 1 μM) (18), whereas mouse TRPM5 is not inhibited by high $[Ca^{2+}]_i$ (18). Activation of human TRPM5 depends more on the rate of rise of $[Ca^{2+}]_i$ rather than the magnitude of $[Ca^{2+}]_i$ [18]. In β-cells Ca^{2+} induced Ca^{2+} release (CICR) causes fast and transient increase of $[Ca^{2+}]_i$ [39, 40], and my speculation is that CICR is particularly suitable for activation of TRPM5 in human β-cells. Activation of mouse TRPM5 and TRPM4 depends on the magnitude of $[Ca^{2+}]_i$ and not on the rate of increase of $[Ca^{2+}]_i$ [38].

42.2.1 Role of TRPM4 and TRPM5 in Stimulus-Secretion Coupling in β-Cells

TRPM5 turns out to be essential for stimulation of insulin secretion by glucose in mouse. Knock-out of the channel reduces [19] or abolishes [41] glucose-induced insulin secretion. Colsoul et al., demonstrate that fast oscillations in membrane potential and corresponding fast oscillations in $[Ca^{2+}]_i$ that are induced by glucose in normal mouse islets are almost completely lost when TRPM5 is knocked-out [19]. It appears that glucose alone can successfully shift β-cells of TRPM5 knock-out mouse from the OFF mode to the READY mode, but fails to shift to the ON mode. In the READY mode, β-cells of TRPM5 knock-out mouse display slow oscillations in membrane potential and corresponding slow oscillations in $[Ca^{2+}]_i$, and the cells secrete insulin at low level [19]. Apparently, shifting from the READY mode to the ON mode requires, in this instance, functional TRPM5 channels. Furthermore, it is apparent that even small depolarizing currents are enough to shift the β-cells from the READY mode to the ON mode. For instance, TRPM5 current described by Colsoul et al., is of small magnitude (~ 20 pA at –80 mV under maximal stimulation by Ca^{2+}) [19]. Perhaps one electrophysiological fingerprint of the ON mode in mouse islets is the fast oscillations in membrane potential and corresponding fast oscillations in the $[Ca^{2+}]_i$.

Potential factors that can link stimulation of β-cells by glucose to the activation of the TRPM5 channels include 1. glucose-induced increase in $[Ca^{2+}]_i$; 2. glucose-induced increase in the membrane potential [42]; 3. glucose-induced increase in the concentration of cytoplasmic arachidonic acid [43, 44]; and 4. glucose-induced increase in the concentration of PIP2 [34, 36]. It is noteworthy that mouse β-cells express the G-protein coupled heterodimeric sweet taste receptor T1R2/T1R3, and that even noncaloric sweet agents can stimulate insulin secretion in a Na^+-dependent manner [37]. Glucose may possibly directly activate these sweet taste receptors in the β-cells and thereby activate the TRPM5 channel as is the case in the taste cells [36].

TRPM5 takes part in mediating insulin secretion even when glucose is used as the sole agonist. It is however tempting to speculate that both TRPM4 and TRPM5 may also participate in the regulation of insulin secretion by different incretins, neurotransmitters and hormones. For instance, GLP-1 increases cytoplasmic $[Na^+]$ [45] by activating a Ca^{2+}-activated non-selective cation current mainly carried by Na^+ and blocked by La^{3+} [23]. It has been reported that GLP-1 activates a current that is mimicked by mitotoxin [46]. However, it is not known whether maitotoxin activates any TRP channels. There are some indirect evidence that mitotoxin activates TRPC1 [47] and β-cells have TRPC1 [5]. On the other hand, it has been demonstrated that maitotoxin binds to the plasma membrane Ca^{2+} ATPase and converts it into a Ca^{2+}-permeable non-selective cation channel [48].

In normal human body, insulin secretion ensues almost immediately in connection with the ingestion or even anticipation of food. Such preabsorptive phase of insulin secretion is subdivided into a cephalic phase and an enteric phase. In reality, these phases of insulin secretion merge with one another to a variable extent. One can speculate that in the cephalic phase of insulin secretion $[Ca^{2+}]_i$ increases in β-cells because of the actions of neurotransmitters like acetylcholine released from the vagus nerve. Such increase of $[Ca^{2+}]_i$ in turn activates TRPM4 and TRPM5 leading to membrane depolarization. In this scenario, TRPM4 and TRPM5 may play a key role in shifting the β-cells from the READY mode to the ON mode. Thus, TRPM4 and TRPM5 provide a distinct mechanism for coupling $[Ca^{2+}]_i$ to membrane depolarization. It is possible that an increase of $[Ca^{2+}]_i$ by a variety of mechanisms, for instance, by CICR, IP3, NAADP, or cADPR may increase or sustain depolarization through the activation of TRPM4 and TRPM5 channels. It may be noted that TRPM5 affects insulin secretion not just by way of membrane depolarization. For instance, islets of TRPM5 knock-out mice fail to secret insulin even when they are persistently depolarized by arginine, suggesting that a distinct type of oscillations in membrane potential i.e. the fast oscillations are necessary for optimal insulin secretion [41].

42.3 TRPM2 (Formerly Called LTRPC2)

42.3.1 TRPM2 and β-Cells

In 1994, Reale et al., described a non-selective Ca^{2+} permeable cation channel activated by β-NAD^+ and inhibited by AMP in CRI-G1 insulinoma cells [49]. Later on, the same group reported that this channel is activated by H_2O_2 [50]. The currents described in these papers have some key features of the TRPM2 current e.g. linear current-voltage relationship, extremely long single channel open times, and the requirement of intracellular Ca^{2+} for activation of the current by H_2O_2 [50, 51]. Today we know that TRPM2 channels are present not only in the rodent insulinoma cells but also in the primary β-cells [14]. The most potent and specific endogenous activator of TRPM2 is cytoplasmic ADP ribose (ADPR) that binds to the NUDT9-H domain which acts as an enzyme (ADPR hydrolase) [52]. ADP ribose formed by degradation of NAD^+ by poly(ADP ribose) polymerase activates TRPM2. H_2O_2 does not have any direct effect on the TRPM2 channel; in stead it activates TRPM2 by increasing the concentration of cytoplasmic free ADPR [53]. In primary mouse β-cells, ADPR activates a plasma membrane current characteristic of TRPM2, whereas it fails to do so in the β-cells from the TRPM2 knock out mouse [14]. It is likely that human β-cells also express TRPM2. In these cells, H_2O_2 activates Ca^{2+} entry through the plasma membrane [9]. By Western blot two isoforms of TRPM2 protein can be detected in the membrane fraction obtained from whole human islets [9]. These are the full length or the long form of the channel (TRPM2-L), and a short form of the channel (TRPM2-S) where the four C-terminal transmembrane domains, the putative pore region and the entire C-terminus are truncated [54].

To study the distribution of TRPM2 in different cells of islets, we performed immunohistochemistry of formalin fixed, paraffin embedded human pancreas. We used two antibodies: 1. an affinity purified rabbit polyclonal IgG directed against an epitope on the N-terminal part of TRPM2 (anti-TRPM2-N) (BL 970. Cat. no. A300-414A, Bethyl laboratories Inc., USA). 2. an affinity purified rabbit polyclonal IgG directed against the C-terminal part of the TRPM2 (anti TRPM2-C) (BL969, Cat. no.A300-413A, Bethyl laboratories Inc.). The immunogen for anti-TRPM2-N was the peptide ILKELSKEEEDTDSSEEMLA, which represents the amino acids 658–677 of human TRPM2 encoded within exon 13. The immunogen for anti-TRPM2-C was the peptide KAAEEPDAEPGGRKKTEEPGDS, which represents amino acids 1,216–1,237 of human TRPM2 encoded within exon 25. However, from the Western blot and from immunohistochemistry, we concluded that these antibodies are not suitable for use in immunohistochemistry.

42.3.2 Role of TRPM2 Channel in Stimulus-Secretion Coupling in β-Cells

Like TRPM4 and TRPM5, TRPM2 is also a Ca^{2+}-activated non-selective cation channel [13, 55], but unlike TRPM4 and TRPM5, TRPM2 is permeable to Ca^{2+}.

TRPM2 current is mainly carried by Na^+. The permeability ratio $p_{Ca}:p_{Cs}$, as estimated from shifts in reversal potentials is low (~0.54 in CRI-G1 insulinoma cells) [12]. Nevertheless, activation of the TRPM2 channel increases $[Ca^{2+}]_i$ [9]. Surprisingly, the permeability of TRPM2 for Ca^{2+} increases ($p_{Ca}:p_{Na} = 5.83$), when the current is activated by heat [11]. Activation of human TRPM2 by ADPR requires Ca^{2+}. Ca^{2+} activates all isoforms of TRPM2, including the ones that are unable to bind ADPR [13]. The Ca^{2+} binding sites are located in a restricted space intracellularly but in the immediate vicinity of the pore region [56]. Thus extracellular Ca^{2+} may be the major source of Ca^{2+} for the activation of TRPM2. In the presence of stable low micromolar concentration of cytoplasmic ADPR, TRPM2 essentially behaves like a Ca^{2+}-activated channel [56, 57]. Perhaps this mode of regulation enables TRPM2 mediate its physiological function in stimulus-secretion coupling in the β-cells. Extracellular Ca^{2+} that enters through the TRPM2 channel can activate the channel by binding to the activation sites that lie near the pore entrance of the channel [56]. This is a potential mechanism for prolongation of activation of TRPM2 in a self sustained manner [56]. An increase of $[Ca^{2+}]_i$ upon stimulation of PI-PLC linked receptors is also able to activate TRPM2. For instance, activation of muscarinic acetylcholine receptor in hamster insulinoma HIT T15 cells activates TRPM2 channel [13]. Thus, TRPM2 provides another mechanism for linking $[Ca^{2+}]_i$ increase to membrane depolarization as has been described for TRPM4 and TRPM5 in a previous paragraph.

From whole cell studies it appears that TRPM2 channel is activated by many agents some of which act synergistically. Some of these agents act directly on the TRPM2 channel and others act indirectly. Effects of ADPR, Ca^{2+}, NAADP and nicotinic acid adenine dinucleotide (NAAD) are direct and their effects can be demonstrated in inside-out patches. NAADP is a low affinity partial agonist and is thus unlikely to be a physiologically relevant activator of TRPM2 [53]. TRPM2 could possibly provide a mechanism whereby mitochondrial reactive oxygen species (ROS) couple glucose metabolism to insulin secretion [58, 59]. However, in intact cells TRPM2 is activated by ROS only indirectly by increasing the concentration of cytoplasmic free ADPR as a result of oxidative/nitrosative stress [53]. It was reported that one isofom of TRPM2 e.g. TRPM2-ΔC is not activated by ADPR, but is activated by H_2O_2 suggesting that ROS can directly activate the channel [60]. However, a later study could not reproduce this finding [61]. "Activation" of TRPM2 by β-NAD^+ and cyclic ADP ribose (cADPR) is actually due to the contamination of these substances with ADPR [53, 57]. β-NAD^+ may have some direct effects comparable to the effects of NAAD. In intact cells, high concentrations of cADPR potentiate ADPR-incuced activation of TRPM2 [62] but such high concentrations of cADPR do not occur in cells. According to Togashi et all cADPR couples glucose metabolism to the activation of TRPM2. They have used high concentrations of cADPR (up to 100 μM). Such concentrations of cADPR have nothing to do with physiology [11]. They used cADPR from Sigma and did not comment on the purity of cADPR [11]. It should be noted that some lots of cADPR may contain as much as 25–50% ADPR as contaminant [53, 57, 62]. I suspect that many effects of cADPR published in many papers are due to the use of impure cADPR.

Other modulators of TRPM2 activity include arachidonic acid, PKA, and heat. In β-cells, arachidonic acid is produced upon glucose metabolism and it has been demonstrated that arachidonic acid can stimulate TRPM2 via a specific binding domain [10]. TRPM2 channel activity is potentiated by PKA phosphorylation [11]. Consistent with this, stimulation of insulin secretion by GLP-1 analogue Exendin-4 is inhibited by siTRPM2 [11]. GLP-1 also produces NAADP and cADPR but as mentioned before, these agents can positively modulate TRPM2 channel activity only when they are used in high concentrations [63]. TRPM2 knock-out mice have impaired insulin secretion in response to glucose and GLP-1[64]. In human, at least three TRPM2 gene variants are associated with reduced insulin secretion as estimated by homeostatic model assessment [65].

It is possible that various insulinotropic agents may engage TRPM2 by producing a multitude of factors like Ca^{2+}, heat, ROS, arachidonic acid, cADPR, and NAADP. The channel, thus, may act as a coincidence detector and even though increases in any of these factors alone are small or absent, the net effect of such increases on the TRPM2 channel can be large. This is however a pure speculation.

ADPR-induced activation of TRPM2 is inhibited by AMP (IC_{50} ~ 10 μM) and cADPR-induced activation of TRPM2 in intact cells is inhibited by 8-Bromo-cADPR [62]. However effects of these inhibitors are likely to be indirect [53]. It is not known whether inhibition by AMP is via the AMP-kinase. Among the pharmacological inhibitors of TRPM2, we have found N-(p-amylcinnamoyl) anthranilic acid (ACA) a useful one but it does inhibit other TRP channels e.g. TRPM8 and TRPC6 to a variable extent [66]. ACA is also an inhibitor of phospholipase A2 [9]. It is unclear whether the inhibitory effect of ACA on TRPM2 is a direct one or an indirect one via phospholipase A2.

42.3.3 Heat as a Physical Second Messenger

The ionic flux through many channels is sensitive to temperature (typical Q_{10} values 1.2–4), probably because of temperature-dependence of diffusion. However, TRPM2 is specialized to detect heat (Q_{10} — 15.6), the temperature threshold and the temperatures for optimal activity being ~ 34°C and ~ 37°C respectively [11]. In the presence of ADPR, the Q_{10} value is as high as 44 [11]. At normal body temperature TRPM2 of β-cells is constitutively active possibly contributing to the well known background depolarizing current. It is plausible that nutrient metabolism increases local temperature of individual β-cells [67] leading to further activation of the TRPM2 channel leading to membrane depolarization. In this scenario, heat could be seen as a physical second messenger for stimulus-secretion coupling in the β-cells. Thus, TRPM2 provides a molecular basis for the well known steep temperature dependence of insulin secretion, which is not the case for secretion of other hormones like glucagon from the α-cells or catecholamines from the chromaffin cells [68].

42.3.4 TRPM2 and β-Cells Death

Being a Ca^{2+} permeable redox sensitive channel, TRPM2 confers susceptibility to cell death caused by oxidative stress and consequent perturbation of Ca^{2+} homeostasis. Under conditions of oxidative stress, the concentration of cytoplasmic ADPR is high. Such high concentration of ADPR, together with high concentration of $[Ca^{2+}]_i$ synergize to cause massive Ca^{2+} influx and Ca^{2+} overload. Thus, in RIN-5F cells, H_2O_2 and TNFα -induce cell death which can be suppressed by treatment with antisense TRPM2 [10]. Alloxan-induced oxidative stress can cause β-cell death by many mechanisms and one of these may be the activation of TRPM2 [50]. One short isoform of TRPM2 (TRPM2-S) does not form a channel; in stead, it inhibits the activity of the TRPM2-L and thereby inhibit cell death [54, 69]. It is possible that the ratio of these two isoforms of TRPM2 channels in human β-cell may determine the degree of susceptibility or resistance of these cells to oxidative stress. Human β-cells are resistant to alloxan and such resistance has been attributed to the fact that human β-cells express only low level of the glucose transporter GLUT2 which transport alloxan into the β-cells [70–72]. However, since TRPM2 mediates alloxan-induced Ca^{2+} overload [50], it can be speculated that resistance of human β-cells to alloxan could partly be attributed to the inhibitory action of the TRPM2-S isoform expressed in these cells.

42.3.5 TRPM2 as an Intracellular Ca^{2+} Release Channel

Almost all mammalian TRP channels are also located on the intracellular membranes in addition to the plasma membrane [73]. Thus it is not so surprising that, in β-cells TRPM2 is also an intracellular Ca^{2+}-release channel. It is located on the acidic Ca^{2+} stores like the lysosomes [14]. Thus, TRPM2 increases $[Ca^{2+}]_i$ not just by activating the channels located on the plasma membrane but also by release of Ca^{2+} through the TRPM2 channels located on the lysosomal Ca^{2+} stores [14]. Mobilization of Ca^{2+} from the lysosomes is important for externalization of phosphatidylserine, a hallmark of apoptotic cells [74]. It has been speculated that TRPM2 may provide a mechanism for mediating apoptosis and elimination of those β-cells that have been severely damaged by oxidative stress [75].

42.4 TRPM3

TRPM3 has many isoforms, some having different functional properties [76]. TRPM3 channels display some constitutive activity [77, 78], and may thus partly account for the background depolarizing currents in the β-cells. Some sphingolipids e.g. D-*erythro*-sphingosine, dihydro-D-*erythro*-sphingosine and *N,N*-dimethyl-D-*erythro*-sphingosine but not sphingosine-1-phosphate and ceramides activate TRPM3 channel [79]. Interestingly, TRPM3 of β-cells is directly activated by supraphysiological concentrations ($EC5_0 = 23$ μM) of the steroid pregnenolone sulphate (PS), which makes it a useful pharmacological tool for the study of this

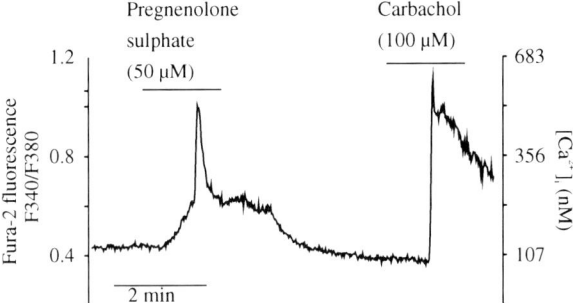

Fig. 42.1 Pregnenolone sulphate, an activator of TRPM3 increases [Ca^{2+}]$_i$ in human β-cells. [Ca^{2+}]$_i$ was measured by microfluometry from fura-2-loaded single human β-cells identified by their size and appearance. Cells plated on coverslips were perfused by modified Kreb's-Ringer bicarbonate buffer containing 3 mM glucose, at 37° C. Pregnenolone sulphate (*PS*) and carbachol were added during the times indicated by the horizontal bars. PS increased [Ca^{2+}]$_i$ which returned to the base line on wash out of the substance. The trace is representative of experiments repeated at least four times

channel [15]. Another low affinity pharmacological activator of TRPM3 is nifedipine (EC$_{50}$ = ~ 32 μM), which is commonly used as a potent inhibitor of L-type voltage-gated Ca^{2+} channels.

In in vitro experiments, activation of TRPM3 by high concentration of PS (e.g. 36 μM) increases [Ca^{2+}]$_i$ in mouse islet cells and augments glucose-stimulated insulin secretion. In such experiments, at least 5 μM PS is required to detect TRPM3 channel activity. We have found that PS increases [Ca^{2+}]$_i$ also in human β-cells (Fig. 42.1) suggesting that human β-cells also have functional TRPM3 channels. The concentrations of PS used in these in vitro experiments are too high to be physiologically relevant for regulation of insulin secretion. However, in some conditions e.g. in mid-pregnancy [80], hyperthyroidism [81], anxiety-depressive disorder [82], and 21-hydroxylate deficiency [83], concentration of PS in the plasma is elevated (normal plasma concentration of PS in adult human is ~ 0.2–0.4 μM). It is possible that in such conditions, modest increases in the concentration of PS synergize with other unknown activators of TRPM3 and may thus render the β-cells more excitable.

Another interesting feature of TRPM3 channel is that it conducts Zn^{2+} ions [84]. Zn^{2+} can pass through TRPM3 channels even when concentration of extracellular Zn^{2+} is as low as 10 μM and concentrations of Ca^{2+} and Mg^{2+} are 1–2 mM. Thus TRPM3, like ZnT-8, constitutes one of the pathways for Zn^{2+} transport across the plasma membrane of β-cells [84].

42.5 TRPV4 (Other Names OTRPC4, VR-OAC, VRL-2, TRP-12)

According to the Human Protein Atlas (version 6.0, updated 2010–03–28) published by the Swedish Human Proteome Resource (HPR) program (www.hpr.se), TRPV4 is abundant in the islets, and is almost completely absent in the pancreatic acinar cells. However, reliability of immunohistochemistry depends on control experiments done to elucidate the specificity of the antibody used. One important control experiment that should be employed in IHC is neutralization of the antibody by the corresponding peptide antigen. Rather surprisingly, this control is not included in the protocols used by HPR because it can not be adapted to their high throughput systems. An examination of the TRPV4 antibody developed by HPR (Product number HPA007150, marketed by Atlas Antibodies under category "Prestige Antibodies") shows that it stains mostly in the nuclei and that it does not detect any band in the Western blot (information obtained from HPR website). Thus, IHC results obtained by this antibody are unreliable.

Nevertheless, TRPV4 has been described at least in a mouse insulinoma cell line where it acts as a stretch-activated plasma membrane channel. In MIN6 cells aggregated human islet amyloid polypeptide (hIAPP) interacts with the plasma membrane and increases $[Ca^{2+}]_i$ by activating TRPV4, an observation consistent with the role of TRV4 as a mechanosensitive channel [22]. Increase of $[Ca^{2+}]_i$ by hIAPP and consequent increase in ER stress response are suppressed by siRNA against TRPV4 [22].

TRPV4 can be activated by diverse physical stimuli including warm temperatures (> 25–34°C), volume changes [85, 86], osmolality, stretch, and mechanical stimuli [87]. As pointed out by Ilya Digel in Chapter 25 in this book "The difference between mechanosensitive channels and thermosensitive molecules is only the size and the organization of the "exciting" agents – a lot of non-coordinated events (thermal stimuli) versus a net stretch (mechanical stimuli). Therefore, not surprisingly, many members of thermosensing TRPV family are also known as osmo- and mechanosensors". Stimulation of β-cells by glucose generates heat [67] and increases volume of β-cells [88]. It is, thus, possible that such changes may act as physical second messengers to activate the TRPV4 channel of β-cells. TRPV4 may thus be involved in $[Ca^{2+}]_i$ increase in β-cells activated by hypotonicity, volume increase and mechanical stretch [89].

Endogenous chemical agonists of TRPV4 include arachidonic acid, anandamide, and 5,6- or 14,15-epoxyeicosatrienoic acids. Phorbol derivatives 4α-Phorbol 12,13-didecanoate (4α-PDD) and Phorbol 12-myristate 13-acetate activate the channel directly without involving protein kinase C [90]. 4α-PDD is a commonly used pharmacological tool for activation of TRPV4. Other potent agonists for TRPV4 are bisandrographolide A [91] and GSK1016790A [92].

42.6 TRPV2

TRPV2 protein has been demonstrated in mouse insulinoma MIN6 cells and cultured mouse β-cells. In these cells insulin stimulates translocation of TRPV2 from the cytoplasm to the plasma membrane as demonstrated by use of TRPV2 tagged with GFP or c-Myc [21]. Such action of insulin on the TRPV2 protein enhances Ca^{2+} entry, insulin secretion and β-cell growth. These effects are inhibited by tranilast, a pharmacological inhibitor of TRPV2 or by konock-down of TRPV2 by molecular techniques. Importantly, glucose, by stimulating insulin secretion promotes insulin-induced translocation and insertion of TRPV2 to the plasma membrane, providing a positive feed-back mechanism [21]. It should be noted that TRPV2 current has not yet been demonstrated in any of the insulin secreting cells. Thus, it is not entirely clear whether TRPV2 protein exerts its function by acting as a channel or it acts as a signaling protein independent of its function as an ion channel. It is known that TRPV2 may display some spontaneous basal activity. In β-cells, such constitutive activity of TRPV2 may account partly for the background depolarizing current. Like TRPV4, TRPV2 is also activated by heat, mechanical stretch and osmotic swelling which may possibly couple glucose metabolism to the activation of the channel.

42.7 TRPV1

Rat insulinoma cell lines RIN and INS-1 express TRPV1 [20]. In these cells capsaicin, an activator of TRPV1 increases $[Ca^{2+}]_i$ and the increase can be inhibited by capsazepine. In primary β-cells, capsaicin does not increase $[Ca^{2+}]_i$. TRPV1 immunoreactivity has been described in primary β-cells of Sprague Dawley rats by one group [20], but not in those of Zucker diabetic rats [93] or NOD mice [94]. TRPV1 channels are present in some nerve fibers in mouse islets [94] but we have not observed any TRPV1 positive nerve fiber in human islets.

42.8 TRPC1 and TRPC4

From Human Protein Atlas (version 6.0) it appears that TRPC1 is expressed at high level in the islets and numerous other cells and tissues (www.hpr.se). I find it hard to interpret these data since the antibody (HPA021130, distributed by Atlas Antibodies as one of the "Prestige Antibodies") does not detect the expected protein band in Western blot (according to the information posted in the HPR website). As mentioned earlier, an important control experiment i.e. the effect of neutralization of the antibody by the corresponding antigen is not included in the protocol used by the Human Proteome Research program. Nevertheless, TRPC1 mRNA has been demonstrated in mouse islets, MIN6 cells, rat β-cells and INS-1 cells [5, 7]. The level of expression of TRPC1 in rat primary β-cells is higher than that in the rat insulinoma INS-1 cells [7]. In MIN6 cells, four splice variants of TRPC1 have been identified. Of these, the β variant is the most abundant one in MIN6 cells and probably also in mouse islets [5]. Different insulinoma cell lines differ in terms of the level of expression of different TRPC channels. Thus, TRPC1 is highly expressed in MIN6 cell and TRPC4 is highly expressed in β-TC3 cells [5, 6]. TRPC4 is also expressed at high level in INS-1 cells and rat β-cells [7]. TRPC4 has at least two major splice variants: the full length TRPC4α and a shorter TRPC4β which lacks 84 amino acids in the C-terminus. In rat β-cells TRPC4β is the main isoform whereas in rat insulinoma INS-1 cells TRPC4α is the main isoform [7]. TRPC5 which is closely related to TRPC4 is not expressed in mouse islets [6].

TRPC1 and TRPC4 are non-selective cation channels with almost equal permeability for Ca^{2+} and Na^+. Since these channels are expressed in many cells, it appears less likely that they perform any β-cell-specific function. Nevertheless, these channels are there in the β-cells and one needs to speculate what their roles could be in these cells. It is possible that low level of spontaneous activity of these channels constitutes part of the background depolarizing current in the β-cells. More, importantly, both of these channels are molecular candidates for the non-selective cation current activated by Gq/PI-PLC-coupled receptors or by depletion of the ER Ca^{2+} store. TRPC1 is the most well characterized TRP channel that, according to many investigators, mediates store-operated Ca^{2+} entry (SOCE) in many cells. Such SOCE has been described in electrically excitable cells like β-cells [95], and it is possible that channels like TRPC1 or TRPC4 or their heteromers mediate SOCE in β-cells. However, there are many unresolved issues as to whether the TRPC channels are regulated by the ER Ca^{2+} sensor STIM1 or just by mechanisms dependent on PI-PLC activation [96] (see Chapter 24 by Kwong Tai Cheng, Hwei Ling Ong, Xibao Liu, and Indu S. Ambudkar, in this book). One hypothesis is that hydrolysis of PIP2 by PI-PLC induces shortening of certain membrane lipids and thus alters the lipid packing at the inner surface of the membrane microdomain [see fig 2 of reference [97] for explanation]. Changes in the membrane curvature and in the lipid-TRP channel interaction induce conformational changes and consequent opening of the TRP channels. Such changes in the membrane curvature can also alter the "open channel block" by forcefully removing the metal ions from the channels and this may lead to further opening of the channels [97, 98].

42.9 Perspectives

The presence of at least 8 TRP channels in the β-cells call for a revision of the current views on the mechanisms of insulin secretion in response to various insulin secretagogues. The background depolarizing currents in the β-cells could be mediated by these 8 redundant TRP channels. Consequently knock out of one or other of these channels may not cause any major change in the background depolarizing current and may not lead to readily identifiable phenotypes. When β-cells are in the READY mode, small increases in the inward depolarizing currents mediated through one or other TRP channels may shift the β-cells to the ON mode. The coupling factors between insulin-secretagogues and the activation of the TRP channels could be diverse physical second messengers like heat, swelling, stretch, changes in the curvature of the plasma membrane microdomains, and chemical factors like arachidonic acid, cAMP, PIP_2, Ca^{2+} and other as yet unknown ones. It will be challenging to investigate the quantitative contribution of different second messengers and different TRP channels in β-cells under different physiological and pathological conditions.

Acknowledgments I would like to thank Romain Guinamard and Frank Kühn for useful discussions. I am grateful to Stockholm County Council, Forskningscentrum, Landes Bioscience, and Engelbrechts kliniken. Financial support was obtained from the Swedish Research Council

References

1. Newsholme P, Gaudel C, McClenaghan NH (2010) Nutrient regulation of insulin secretion and beta-cell functional integrity. Adv Exp Med Biol 654:91–114
2. Islam MS (2010) Calcium signaling in the islets. Adv Exp Med Biol 654:235–259
3. Drews G, Krippeit-Drews P, Dufer M (2010) Electrophysiology of islet cells. Adv Exp Med Biol 654:115–163
4. Clark R, Proks P (2010) ATP-sensitive potassium channels in health and disease. Adv Exp Med Biol 654:165–192
5. Sakura H, Ashcroft FM (1997) Identification of four trp1 gene variants murine pancreatic beta-cells. Diabetologia 40:528–532
6. Roe MW, Worley JF III, Qian F, Tamarina N, Mittal AA, Dralyuk F, Blair NT, Mertz RJ, Philipson LH, Dukes ID (1998) Characterization of a Ca^{2+} release-activated nonselective cation current regulating membrane potential and $[Ca^{2+}]_i$ oscillations in transgenically derived beta-cells. J BiolChem 273:10402–10410
7. Li F, Zhang ZM (2009) Comparative identification of Ca^{2+} channel expression in INS-1 and rat pancreatic beta cells. World J Gastroenterol 15:3046–3050
8. Qian F, Huang P, Ma L, Kuznetsov A, Tamarina N, Philipson LH (2002) TRP genes: candidates for nonselective cation channels and store-operated channels in insulin-secreting cells. Diabetes 51:S183–S189
9. Bari MR, Akbar S, Eweida M, Kuhn FJ, Gustafsson AJ, Lückhoff A, Islam MS (2009) H_2O_2-induced Ca^{2+} influx and its inhibition by N-(p-amylcinnamoyl) anthranilic acid in the beta-cells: involvement of TRPM2 channels. J Cell Mol Med 13:3260–3267
10. Hara Y, Wakamori M, Ishii M, Maeno E, Nishida M, Yoshida T, Yamada H, Shimizu S, Mori E, Kudoh J, Shimizu N, Kurose H, Okada Y, Imoto K, Mori Y (2002) LTRPC2 Ca^{2+}-permeable channel activated by changes in redox status confers susceptibility to cell death. Mol Cell 9:163–173
11. Togashi K, Hara Y, Tominaga T, Higashi T, Konishi Y, Mori Y, Tominaga M (2006) TRPM2 activation by cyclic ADP-ribose at body temperature is involved in insulin secretion. EMBO J 25:1804–1815
12. Inamura K, Sano Y, Mochizuki S, Yokoi H, Miyake A, Nozawa K, Kitada C, Matsushime H, Furuichi K (2003) Response to ADP-Ribose by Activation of TRPM2 in the CRI-G1 Insulinoma Cell Line. J Membr Biol 191:201–207
13. Du J, Xie J, Yue L (2009) Intracellular calcium activates TRPM2 and its alternative spliced isoforms. Proc Natl Acad Sci USA 106.7239–7244
14. Lange I, Yamamoto S, Partida-Sanchez S, Mori Y, Fleig A, Penner R (2009) TRPM2 functions as a lysosomal Ca^{2+}-release channel in beta cells. Sci Signal 2:ra23
15. Wagner TF, Loch S, Lambert S, Straub I, Mannebach S, Mathar I, Dufer M, Lis A, Flockerzi V, Philipp SE, Oberwinkler J (2008) Transient receptor potential M3 channels are ionotropic steroid receptors in pancreatic beta cells. Nat Cell Biol 10:1421–1430
16. Cheng H, Beck A, Launay P, Gross SA, Stokes AJ, Kinet JP, Fleig A, Penner R (2007) TRPM4 controls insulin secretion in pancreatic beta-cells. Cell Calcium 41:51–61
17. Marigo V, Courville K, Hsu WH, Feng JM, Cheng H (2009) TRPM4 impacts on Ca^{2+} signals during agonist-induced insulin secretion in pancreatic beta-cells. Mol Cell Endocrinol 299:194–203
18. Prawitt D, Monteilh-Zoller MK, Brixel L, Spangenberg C, Zabel B, Fleig A, Penner R (2003) TRPM5 is a transient Ca^{2+}-activated cation channel responding to rapid changes in $[Ca^{2+}]_i$. Proc Natl Acad Sci USA 100:15166–15171

19. Colsoul B, Schraenen A, Lemaire K, Quintens R, Van Lommel L, Segal A, Owsianik G, Talavera K, Voets T, Margolskee RF, Kokrashvili Z, Gilon P, Nilius B, Schuit FC, Vennekens R (2010) Loss of high-frequency glucose-induced Ca^{2+} oscillations in pancreatic islets correlates with impaired glucose tolerance in Trpm5-/- mice. Proc Natl Acad Sci USA 107:5208–5213
20. Akiba Y, Kato S, Katsube KI, Nakamura M, Takeuchi K, Ishii H, Hibi T (2004) Transient receptor potential vanilloid subfamily 1 expressed in pancreatic islet beta cells modulates insulin secretion in rats. Biochem Biophys Res Commun 321:219–225
21. Hisanaga E, Nagasawa M, Ueki K, Kulkarni RN, Mori M, Kojima I (2009) Regulation of calcium-permeable TRPV2 channel by insulin in pancreatic beta-cells. Diabetes 58:174–184
22. Casas S, Novials A, Reimann F, Gomis R, Gribble FM (2008) Calcium elevation in mouse pancreatic beta cells evoked by extracellular human islet amyloid polypeptide involves activation of the mechanosensitive ion channel TRPV4. Diabetologia 51:2252–2262
23. Holz GG, Leech CA, Habener JF (1995) Activation of a cAMP-regulated Ca^{2+}-signaling pathway in pancreatic beta-cells by the insulinotropic hormone glucagon-like peptide-1. J BiolChem 270:17749–17757
24. Leech CA, Habener JF (1998) A role for Ca^{2+}-sensitive nonselective cation channels in regulating the membrane potential of pancreatic beta-cells. Diabetes 47:1066–1073
25. Sturgess NC, Hales CN, Ashford ML (1986) Inhibition of a calcium-activated, non-selective cation channel, in a rat insulinoma cell line, by adenine derivatives. FEBS Lett 208:397–400
26. Nilius B, Prenen J, Voets T, Droogmans G (2004) Intracellular nucleotides and polyamines inhibit the Ca^{2+}-activated cation channel TRPM4b. Pflugers Arch 448:70–75
27. Nilius B, Mahieu F, Karashima Y, Voets T (2007) Regulation of TRP channels: a voltage-lipid connection. Biochem Soc Trans 35:105–108
28. Vennekens R, Olausson J, Meissner M, Bloch W, Mathar I, Philipp SE, Schmitz F, Weissgerber P, Nilius B, Flockerzi V, Freichel M (2007) Increased IgE-dependent mast cell activation and anaphylactic responses in mice lacking the calcium-activated nonselective cation channel TRPM4. Nat Immunol 8:312–320
29. Demion M, Bois P, Launay P, Guinamard R (2007) TRPM4, a Ca^{2+}-activated nonselective cation channel in mouse sino-atrial node cells. Cardiovasc Res 73:531–538
30. Sturgess NC, Kozlowski RZ, Carrington CA, Hales CN, Ashford ML (1988) Effects of sulphonylureas and diazoxide on insulin secretion and nucleotide-sensitive channels in an insulin-secreting cell line. Br J Pharmacol 95:83–94
31. Nilius B, Prenen J, Tang J, Wang C, Owsianik G, Janssens A, Voets T, Zhu MX (2005) Regulation of the Ca^{2+} sensitivity of the nonselective cation channel TRPM4. J Biol Chem 280:6423–6433
32. Biden TJ, Schmitz-Peiffer C, Burchfield JG, Gurisik E, Cantley J, Mitchell CJ, Carpenter L (2008) The diverse roles of protein kinase C in pancreatic beta-cell function. Biochem Soc Trans 36:916–919
33. Nilius B, Mahieu F, Prenen J, Janssens A, Owsianik G, Vennekens R, Voets T (2006) The Ca^{2+}-activated cation channel TRPM4 is regulated by phosphatidylinositol 4,5-biphosphate. EMBO J 25:467–478
34. Thore S, Wuttke A, Tengholm A (2007) Rapid turnover of phosphatidylinositol-4,5-bisphosphate in insulin-secreting cells mediated by Ca^{2+} and the ATP-to-ADP ratio. Diabetes 56:818–826
35. Fonfria E, Murdock PR, Cusdin FS, Benham CD, Kelsell RE, McNulty S (2006) Tissue distribution profiles of the human TRPM cation channel family. J Recept Signal Transduct Res 26:159–178
36. Liu D, Liman ER (2003) Intracellular Ca^{2+} and the phospholipid PIP2 regulate the taste transduction ion channel TRPM5. Proc Natl Acad Sci USA 100:15160–15165
37. Nakagawa Y, Nagasawa M, Yamada S, Hara A, Mogami H, Nikolaev VO, Lohse MJ, Shigemura N, Ninomiya Y, Kojima I (2009) Sweet taste receptor expressed in pancreatic

beta-cells activates the calcium and cyclic AMP signaling systems and stimulates insulin secretion. PLoS One 4:e5106
38. Ullrich ND, Voets T, Prenen J, Vennekens R, Talavera K, Droogmans G, Nilius B (2005) Comparison of functional properties of the Ca^{2+}-activated cation channels TRPM4 and TRPM5 from mice. Cell Calcium 37:267–278
39. Kang G, Chepurny OG, Rindler MJ, Collis L, Chepurny Z, Li WH, Harbeck M, Roe MW, Holz GG (2005) A cAMP and Ca^{2+} coincidence detector in support of Ca^{2+}-induced Ca^{2+} release in mouse pancreatic beta cells. J Physiol 566:173–188
40. Dyachok O, Gylfe E (2004) Ca^{2+}-induced Ca^{2+} release via inositol 1,4,5-trisphosphate receptors is amplified by protein kinase A and triggers exocytosis in pancreatic beta-cells. J Biol Chem 279:45455–45461
41. Brixel LR, Monteilh-Zoller MK, Ingenbrandt CS, Fleig A, Penner R, Enklaar T, Zabel BU, Prawitt D (2010) TRPM5 regulates glucose-stimulated insulin secretion. Pflugers Arch 460:69–76
42. Hofmann T, Chubanov V, Gudermann T, Montell C (2003) TRPM5 is a voltage-modulated and Ca^{2+}-activated monovalent selective cation channel. Curr Biol 13:1153–1158
43. Wolf BA, Turk J, Sherman WR, McDaniel ML (1986) Intracellular Ca^{2+} mobilization by arachidonic acid. Comparison with myo-inositol 1,4,5-trisphosphate in isolated pancreatic islets. J BiolChem 261:3501–3511
44. Oike H, Wakamori M, Mori Y, Nakanishi H, Taguchi R, Misaka T, Matsumoto I, Abe K (2006) Arachidonic acid can function as a signaling modulator by activating the TRPM5 cation channel in taste receptor cells. Biochim Biophys Acta 1761:1078–1084
45. Miura Y, Matsui H (2003) Glucagon-like peptide-1 induces a cAMP-dependent increase of [Na+]i associated with insulin secretion in pancreatic beta-cells. Am J Physiol Endocrinol Metab 285:E1001–E1009
46. Leech CA, Habener JF (1997) Insulinotropic glucagon-like peptide-1-mediated activation of non-selective cation currents in insulinoma cells is mimicked by maitotoxin. J BiolChem 272:17987–17993
47. Brereton HM, Chen J, Rychkov G, Harland ML, Barritt GJ (2001) Maitotoxin activates an endogenous non-selective cation channel and is an effective initiator of the activation of the heterologously expressed hTRPC-1 (transient receptor potential) non-selective cation channel in H4-IIE liver cells. Biochim Biophys Acta 1540:107–126
48. Sinkins WG, Estacion M, Prasad V, Goel M, Shull GE, Kunze DL, Schilling WP (2009) Maitotoxin converts the plasmalemmal Ca^{2+} pump into a Ca^{2+}-permeable nonselective cation channel. Am J Physiol Cell Physiol 297:C1533–C1543
49. Reale V, Hales CN, Ashford ML (1994) The effects of pyridine nucleotides on the activity of a calcium-activated nonselective cation channel in the rat insulinoma cell line, CRI-G1. J Membr Biol 142:299–307
50. Herson PS, Ashford ML (1997) Activation of a novel non-selective cation channel by alloxan and H_2O_2 in the rat insulin-secreting cell line CRI-G1. J Physiol 501(Pt 1):59–66
51. Eisfeld J, Luckhoff A (2007) TRPM2. Handb.Exp.Pharmacol 179:237–252
52. Perraud AL, Fleig A, Dunn CA, Bagley LA, Launay P, Schmitz C, Stokes AJ, Zhu Q, Bessman MJ, Penner R, Kinet JP, Scharenberg AM (2001) ADP-ribose gating of the calcium-permeable LTRPC2 channel revealed by Nudix motif homology. Nature 411:595–599
53. Toth B, Csanady L (2010) Identification of direct and indirect effectors of the transient receptor potential melastatin 2 (TRPM2) cation channel. J Biol Chem 285:30091–30102
54. Zhang W, Chu X, Tong Q, Cheung JY, Conrad K, Masker K, Miller BA (2003) A Novel TRPM2 Isoform Inhibits Calcium Influx and Susceptibility to Cell Death. J BiolChem 278:16222–16229
55. McHugh D, Flemming R, Xu SZ, Perraud AL, Beech DJ (2003) Critical intracellular Ca^{2+} dependence of transient receptor potential melastatin 2 (TRPM2) cation channel activation. J Biol Chem 278:11002–11006

56. Csanady L, Torocsik B (2009) Four Ca^{2+} ions activate TRPM2 channels by binding in deep crevices near the pore but intracellularly of the gate. J Gen Physiol 133:189–203
57. Heiner I, Eisfeld J, Warnstedt M, Radukina N, Jungling E, Lückhoff A (2006) Endogenous ADP-ribose enables calcium-regulated cation currents through TRPM2 channels in neutrophil granulocytes. Biochem J 398:225–232
58. Leloup C, Tourrel-Cuzin C, Magnan C, Karaca M, Castel J, Carneiro L, Colombani AL, Ktorza A, Casteilla L, Penicaud L (2009) Mitochondrial reactive oxygen species are obligatory signals for glucose-induced insulin secretion. Diabetes 58:673–681
59. Pi J, Bai Y, Zhang Q, Wong V, Floering LM, Daniel K, Reece JM, Deeney JT, Andersen ME, Corkey BE, Collins S (2007) Reactive oxygen species as a signal in glucose-stimulated insulin secretion. Diabetes 56:1783–1791
60. Wehage E, Eisfeld J, Heiner I, Jungling E, Zitt C, Lückhoff A (2002) Activation of the Cation Channel Long Transient Receptor Potential Channel 2 (LTRPC2) by Hydrogen Peroxide. A splice variant reveals a mode of activation independent of ADP-ribose. J BiolChem 277:23150–23156
61. Perraud AL, Takanishi CL, Shen B, Kang S, Smith MK, Schmitz C, Knowles HM, Ferraris D, Li W, Zhang J, Stoddard BL, Scharenberg AM (2005) Accumulation of free ADP-ribose from mitochondria mediates oxidative stress-induced gating of TRPM2 cation channels. J BiolChem 280:6138–6148
62. Lange I, Penner R, Fleig A, Beck A (2008) Synergistic regulation of endogenous TRPM2 channels by adenine dinucleotides in primary human neutrophils. Cell Calcium 44:604–615
63. Kim BJ, Park KH, Yim CY, Takasawa S, Okamoto H, Im MJ, Kim UH (2008) Generation of nicotinic acid adenine dinucleotide phosphate and cyclic ADP-ribose by glucagon-like peptide-1 evokes Ca^{2+} signal that is essential for insulin secretion in mouse pancreatic islets. Diabetes 57:868–878
64. Uchida K, Dezaki K, Dambindorj B, Inada H, Shiuchi T, Mori Y, Yada T, Minokoshi Y, Tominaga M (2010) Lack of TRPM2 impaired insulin secretion and glucose metabolisms in mice. Diabetes doi: 10.2337/db10-0276
65. Romero JR, Germer S, Castonguay AJ, Barton NS, Martin M, Zee RY (2010) Gene variation of the transient receptor potential cation channel, subfamily M, member 2 (TRPM2) and type 2 diabetes mellitus: a case-control study. Clin Chim Acta 411:1437–1440
66. Kraft R, Grimm C, Frenzel H, Harteneck C (2006) Inhibition of TRPM2 cation channels by N-(p-amylcinnamoyl)anthranilic acid. Br J Pharmacol 148:264–273
67. Silva-Alves JM, Mares-Guia TR, Oliveira JS, Costa-Silva C, Bretz P, Araujo S, Ferreira E, Coimbra C, Sogayar MC, Reis R, Mares-Guia ML, Santoro MM (2008) Glucose-induced heat production, insulin secretion and lactate production in isolated Wistar rat pancreatic islets. Thermochim Acta 474:67–71
68. Ohta M, Nelson D, Nelson J, Meglasson MD, Erecinska M (1990) Oxygen and temperature dependence of stimulated insulin secretion in isolated rat islets of Langerhans. J Biol Chem 265:17525–17532
69. Fonfria E, Marshall ICB, Boyfield I, Skaper SD, Hughes JP, Owen DE, Zhang W, Miller BA, Benham CD, McNulty S (2005) Amyloid beta-peptide(1–42) and hydrogen peroxide-induced toxicity are mediated by TRPM2 in rat primary striatal cultures. J Neurochem 95:715–723
70. De Vos A, Heimberg H, Quartier E, Huypens P, Bouwens L, Pipeleers D, Schuit F (1995) Human and rat beta cells differ in glucose transporter but not in glucokinase gene expression. J Clin Invest 96:2489–2495
71. Eizirik DL, Pipeleers DG, Ling Z, Welsh N, Hellerstrom C, Andersson A (1994) Major species differences between humans and rodents in the susceptibility to pancreatic beta-cell injury. Proc Natl Acad Sci USA 91:9253–9256
72. Elsner M, Tiedge M, Lenzen S (2003) Mechanism underlying resistance of human pancreatic beta cells against toxicity of streptozotocin and alloxan. Diabetologia 46:1713–1714
73. Dong XP, Wang X, Xu H (2010) TRP channels of intracellular membranes. J Neurochem 113:313–328

74. Mirnikjoo B, Balasubramanian K, Schroit AJ (2009) Mobilization of lysosomal calcium regulates the externalization of phosphatidylserine during apoptosis. J Biol Chem 284:6918–6923
75. Scharenberg A (2009) TRPM2 and pancreatic beta cell responses to oxidative stress. Islets 1:165–166
76. Oberwinkler J, Lis A, Giehl KM, Flockerzi V, Philipp SE (2005) Alternative splicing switches the divalent cation selectivity of TRPM3 channels. J Biol Chem 280:22540–22548
77. Grimm C, Kraft R, Sauerbruch S, Schultz G, Harteneck C (2003) Molecular and functional characterization of the melastatin-related cation channel TRPM3. J Biol Chem 278:21493–21501
78. Lee N, Chen J, Sun L, Wu S, Gray KR, Rich A, Huang M, Lin JH, Feder JN, Janovitz EB, Levesque PC, Blanar MA (2003) Expression and characterization of human transient receptor potential melastatin 3 (hTRPM3). J Biol Chem 278:20890–20897
79. Grimm C, Kraft R, Schultz G, Harteneck C (2005) Activation of the melastatin-related cation channel TRPM3 [corrected] by D-erythro-sphingosine. Mol Pharmacol 67:798–805
80. Bicikova M, Klak J, Hill M, Zizka Z, Hampl R, Calda P (2002) Two neuroactive steroids in midpregnancy as measured in maternal and fetal sera and in amniotic fluid. Steroids 67: 399–402
81. Tagawa N, Tamanaka J, Fujinami A, Kobayashi Y, Takano T, Fukata S, Kuma K, Tada H, Amino N (2000) Serum dehydroepiandrosterone, dehydroepiandrosterone sulfate, and pregnenolone sulfate concentrations in patients with hyperthyroidism and hypothyroidism. Clin Chem 46:523–528
82. Bicikova M, Tallova J, Hill M, Krausova Z, Hampl R (2000) Serum concentrations of some neuroactive steroids in women suffering from mixed anxiety-depressive disorder. Neurochem Res 25:1623–1627
83. de Peretti E, Forest MG, Loras B, Morel Y, David M, Francois R, Bertrand J (1986) Usefulness of plasma pregnenolone sulfate in testing pituitary-adrenal function in children. Acta Endocrinol Suppl (Copenh) 279:259–263
84. Wagner TF, Drews A, Loch S, Mohr F, Philipp SE, Lambert S, Oberwinkler J (2010) TRPM3 channels provide a regulated influx pathway for zinc in pancreatic beta cells. Pflugers Arch 460:755–765
85. Nilius B, Prenen J, Wissenbach U, Bodding M, Droogmans G (2001) Differential activation of the volume-sensitive cation channel TRP12 (OTRPC4) and volume-regulated anion currents in HEK-293 cells. Pflugers Arch 443:227–233
86. Becker D, Blase C, Bereiter-Hahn J, Jendrach M (2005) TRPV4 exhibits a functional role in cell-volume regulation. J Cell Sci 118:2435–2440
87. Phan MN, Leddy HA, Votta BJ, Kumar S, Levy DS, Lipshutz DB, Lee SH, Liedtke W, Guilak F (2009) Functional characterization of TRPV4 as an osmotically sensitive ion channel in porcine articular chondrocytes. Arthritis Rheum 60:3028–3037
88. Milcy IIE, Sheader EA, Brown PD, Best L (1997) Glucose-induced swelling in rat pancreatic beta-cells. J Physiol 504(Pt 1):191–198
89. Grapengiesser E, Gylfe E, Dansk H, Hellman B (2003) Stretch activation of Ca^{2+} transients in pancreatic beta cells by mobilization of intracellular stores. Pancreas 26:82–86
90. Watanabe H, Davis JB, Smart D, Jerman JC, Smith GD, Hayes P, Vriens J, Cairns W, Wissenbach U, Prenen J, Flockerzi V, Droogmans G, Benham CD, Nilius B (2002) Activation of TRPV4 channels (hVRL-2/mTRP12) by phorbol derivatives. J BiolChem 277: 13569–13577
91. Smith PL, Maloney KN, Pothen RG, Clardy J, Clapham DE (2006) Bisandrographolide from Andrographis paniculata activates TRPV4 channels. J Biol Chem 281:29897–29904
92. Thorneloe KS, Sulpizio AC, Lin Z, Figueroa DJ, Clouse AK, McCafferty GP, Chendrimada TP, Lashinger ES, Gordon E, Evans L, Misajet BA, Demarini DJ, Nation JH, Casillas LN, Marquis RW, Votta BJ, Sheardown SA, Xu X, Brooks DP, Laping NJ, Westfall TD (2008) N-((1S)-1-{[4-((2S)-2-{[(2,4-dichlorophenyl)sulfonyl]amino}-3-hydroxypropa noyl)-1-piperazinyl]carbonyl}-3-methylbutyl)-1-benzothiophene-2-carboxamid e (GSK1016790A),

a novel and potent transient receptor potential vanilloid 4 channel agonist induces urinary bladder contraction and hyperactivity: Part I. J Pharmacol Exp Ther 326:432–442
93. Gram DX, Ahren B, Nagy I, Olsen UB, Brand CL, Sundler F, Tabanera R, Svendsen O, Carr RD, Santha P, Wierup N, Hansen AJ (2007) Capsaicin-sensitive sensory fibers in the islets of Langerhans contribute to defective insulin secretion in Zucker diabetic rat, an animal model for some aspects of human type 2 diabetes. Eur J Neurosci 25:213–223
94. Razavi R, Chan Y, Afifiyan FN, Liu XJ, Wan X, Yantha J, Tsui H, Tang L, Tsai S, Santamaria P, Driver JP, Serreze D, Salter MW, Dosch HM (2006) TRPV1(+) Sensory Neurons Control beta Cell Stress and Islet Inflammation in Autoimmune Diabetes. Cell 127:1123–1135
95. Dyachok O, Gylfe E (2001) Store-operated influx of Ca^{2+} in pancreatic beta-cells exhibits graded dependence on the filling of the endoplasmic reticulum. J Cell Sci 114:2179–2186
96. Dehaven WI, Jones BF, Petranka JG, Smyth JT, Tomita T, Bird GS, Putney JW Jr (2009) TRPC channels function independently of STIM1 and Orai1. J Physiol 587:2275–2298
97. Goswami C, Islam MS (2010) Transient receptor potential channels: what is happening? Reflections in the wake of the 2009 TRP meeting, karolinska institutet, stockholm. Channels (Austin) 4:124–135
98. Parnas M, Katz B, Lev S, Tzarfaty V, Dadon D, Gordon-Shaag A, Metzner H, Yaka R, Minke B (2009) Membrane lipid modulations remove divalent open channel block from TRP-like and NMDA channels. J Neurosci 29:2371–2383

Chapter 43
Multiple Roles for TRPs in the Taste System: Not Your Typical TRPs

Kathryn F. Medler

Abstract The peripheral taste system is contained within taste buds located in the oral cavity. These taste buds are comprised of a heterogeneous group of taste receptor cells that use multiple signaling pathways to transduce chemical taste stimuli into an output signal that is sent to the brain. Salty and sour taste involve the detection of charged ions that directly interact with receptors to cause cell depolarization while bitter, sweet and umami taste stimuli activate G-protein coupled receptors and their second messenger pathways. The roles of TRP channels in these different signaling pathways are not well characterized and to date, only three TRP channels have been identified in taste receptor cells. This book chapter discusses the current understanding of how the three known TRP channels function in peripheral taste cell signaling: TRPM5, TRPV1, and the heterodimer PKD1L3/PKD2L1.

43.1 Introduction

All organisms, including unicellular organisms, detect and respond to chemicals found in their external environment. Chemical sensory systems, which are comprised of olfaction and taste in vertebrates and many invertebrates, play important roles in feeding and social interactions. Specifically, the sense of taste is used to determine whether potential food items will be ingested or rejected and is critical for an organism's survival. The "chemical detectors" in the taste system are the taste receptor cells which are specialized neuroepithelial cells that are housed in taste buds found on the tongue and palate in the oral cavity. Taste buds contain 50–150 taste receptors cells that extend apical processes into the oral cavity to detect chemical stimuli, which the taste receptor cells convert into output signals

K.F. Medler (✉)
Department of Biological Sciences, University at Buffalo, The State University of New York, Buffalo, NY 14260, USA
e-mail: kmedler@buffalo.edu

that are transmitted to the central nervous system. These cells share characteristics with both neurons and epithelial cells. Like neurons, taste cells can fire action potentials, express voltage-gated channels, and form conventional synapses with afferent neurons. Like epithelial cells, taste receptor cells express several types of epithelial proteins and have a limited lifespan of about 2 weeks. As a result of this turnover, gustatory neurons must disassociate from dying cells and form new synaptic connections with newly-matured taste cells throughout the organism's life [1–7].

In mammals, taste buds are localized in specialized bumps or protrusions called papillae. There are three types of papillae on the tongue: circumvallate, foliate and fungiform, with additional taste buds located on the soft palate and scattered throughout the oral cavity. Multiple cranial nerves innervate these taste buds which vary by location. A branch of the facial nerve innervates the fungiform papillae and the anterior portion of the foliate papillae while the glossopharyngeal nerve innervates the circumvallate papillae and the posterior portion of the foliate. A separate branch of the facial nerve and the vagus nerve also innervate the taste buds scattered throughout the oral cavity as well as those located in the palate [8–9].

Taste stimuli are very diverse in their structure and characteristics. There are five known taste qualities: salty, sour, bitter, sweet, and umami (the detection of amino acids, primarily glutamate). The first two qualities, salty and sour, involve the detection of small ions, primarily Na^+ and H^+. These ions can interact directly with apically located ion channels to cause cell depolarization. The other three taste qualities involve the detection of more chemically complex molecules that primarily interact with apically located G-protein coupled receptors (GPCRs). As a result, there are multiple signaling pathways and many channels that contribute to the detection of taste stimuli [1–2, 10].

It is not well understood how the detection of chemicals in the oral cavity is translated into a taste perception in the brain. Within taste buds there are different taste receptor cell populations that use distinct signaling mechanisms to generate an output signal in response to a particular taste stimuli [11]. Sour and salty taste stimuli are charged molecules that interact directly with ion channels and cause taste cells to depolarize. In sour sensitive cells, this depolarization causes calcium influx through voltage-gated calcium channels (VGCCs) [12] and these cells belong to the population of taste cells that have conventional, chemical synapses (Fig. 43.1a). This population of taste cells expresses synaptic type proteins like the presynaptic protein SNAP-25 and use vesicular release of neurotransmitter to communicate to the afferent nerves [13–17]. The signal transduction processes for salty taste stimuli are less clear since these stimuli do not appear to activate taste cells that express the machinery necessary for neurotransmitter release through conventional synapses [18].

There is a separate taste receptor cell population that expresses GPCRs which detect bitter, sweet or umami stimuli. These tastants bind to taste GPCRs to activate phospholipase C (PLCβ2) and trigger the release of calcium from intracellular stores [1, 19]. Taste cells using the PLCβ2 signaling pathway do not have chemical synapses [11, 20–24] and rely entirely on calcium release from internal stores to

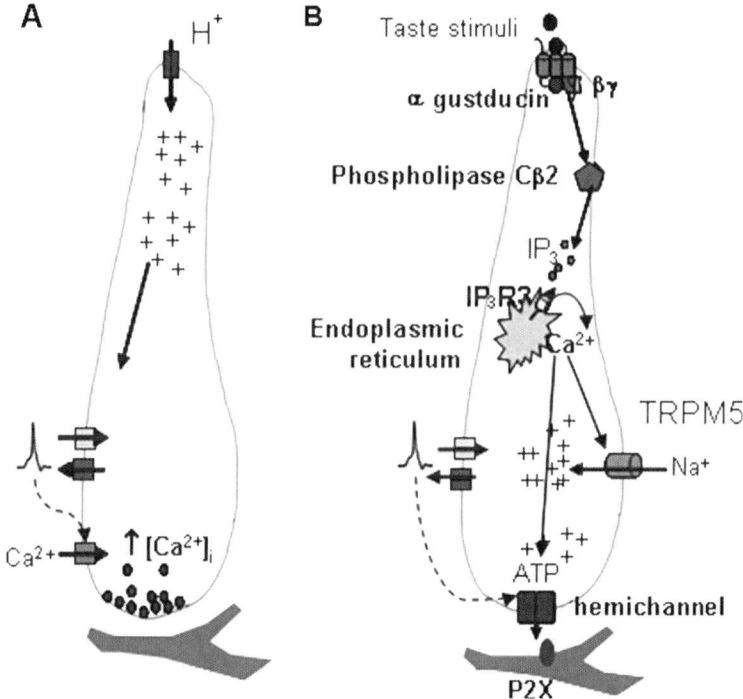

Fig. 43.1 Current model of the signaling pathways involved in transducing some ionic (**a**) and chemically complex taste stimuli (**b**) in taste cells. (**a**) Ionic stimuli interact with channels to cause cell depolarization and Ca^{2+} influx through VGCCs. This results in neurotransmitter release onto afferent gustatory neurons (*gray*). (**b**) Bitter, sweet or umami taste stimuli bind to GPCRs to activate phospholipase C (PLCβ2). This causes Ca^{2+} release from internal stores that activates the TRPM5 channel and ATP release through a hemichannel. ATP activates the P2X receptors on the gustatory neurons

cause neurotransmitter release through hemichannels [25–26] (Fig. 43.1b). It is currently unclear how the same cellular signaling pathway can generate output signals that result in distinct taste perceptions in the brain. This overlap between signaling pathways activated by different tastants suggests that there is some peripheral processing of taste signals within the taste receptor cells which shapes the final output signal that is sent to the brain. Therefore, there are functional differences between taste cells that correlate with the type of taste stimuli detected by the cell [11, 14, 20, 24, 27–28].

Surprisingly, little is known about the role of TRP channels in the transduction of taste-evoked signals in taste cells. To date, only three TRP channels (TRPM5, PKD1L3/PKD2L1, and TRPV1t) have been identified in these cells and their relative functions are not generally well characterized. Of the three, TRPM5 is the best studied and has the most definitive role in the transduction of taste stimuli [29–32]. A heterodimer of PKD1L3/PKD2L1 has been proposed as a sour receptor, but it has not yet been demonstrated as the sour receptor in vivo [33–35]. A taste variant of

TRPV1 has also been identified in taste cells [36]. This tTRPV1 was initially identified as a potential salty receptor, but the data is conflicting about its role in salt transduction [37–38] and it may serve additional roles in these cells [39–40]. This chapter will focus on what is currently known about these three TRP channels in taste receptor cells and their unique roles in taste transduction.

43.2 TRP-Melastatin 5 (TRPM5)

The best characterized TRP channel in taste transduction is TRPM5, a member of the TRPM subfamily. This TRPM family also includes TRPM8 which functions as a cold receptor [41] as well as TRPM2, TRPM6, and TRPM7 which all contain a C-terminal enzyme domain [42]. Compared to other TRPs, TRPM4 and TRPM5 have several unique characteristics. Both TRPM4 and TRPM5 are voltage-modulated, selective for monovalent cations, and are activated by increases in intracellular calcium [29–30, 32]. The voltage dependence of these channels allows for their activity to be regulated by changes in membrane potential [30, 32]. These unique characteristics have allowed TRPM5 to have an important role in the transduction of multiple taste stimuli.

The taste stimuli that evoke bitter, sweet and umami sensation are large and interact with apically located GPCRs. There are two identified families of GPCRs that act as taste receptors, the T2Rs that detect bitter stimuli and the T1Rs that detect sweet and umami stimuli [43–49]. There is evidence that other receptors also contribute to the detection of these taste stimuli, but their roles are not as well defined [50–55]. What has been well-established is that upon activation of the taste GPCRs, the $\beta\gamma$ subunits from the activated heterotrimeric G protein stimulate the activity of a PLC signaling pathway which produces inositol trisphosphate (IP_3) and diacylglycerol (DAG) [56–58]. The IP_3 binds to an IP_3 receptor found on the endoplasmic reticulum and causes the release of calcium from these intracellular stores into the cytosol. This increase in intracellular calcium activates TRPM5 [29–30, 32, 59], which opens to allow a sodium influx and cell depolarization. It is postulated that this cell depolarization, either independently or in conjunction with an increase in intracellular calcium, causes the opening of a hemichannel and ATP release which acts as a neurotransmitter for these cells [25–26, 60–62]. There is still some debate as to the exact nature of these hemichannels and whether an increase of intracellular calcium is required for the normal transmission of these taste signals [25–26, 60, 62], but in all cases, TRPM5 is necessary for normal taste transduction to occur [20, 63].

Since TRPM5 has such a pivotal role in the transduction of multiple types of taste stimuli, there is great interest in understanding its function. Results from heterologous expression studies have at times, been conflicting [20, 29–30, 32, 64] and it is not clear what the functional significance is of some of the findings. In taste cells, TRPM5 is half activated by 8 μM calcium and is desensitized by prolonged exposure to intracellular calcium [59]. However, addition of phosphatidylinositol-4,5-bisphosphate (PIP_2) causes a partial reversal of this desensitization and restores the TRPM5 channel's sensitivity to calcium. This effect only occurs after the TRPM5 currents had been desensitized, which may indicate that the desensitization is due to the loss of bound PIP$_2$ from the TRPM5 channel [30]. One study has also reported that arachidonic acid can activate TRPM5 in heterologous cells. While arachidonic acid is present in taste cells, its functional relationship with TRPM5 is still unknown [65].

TRPM5 has also been shown to be a highly temperature sensitive, heat-activated channel. Inward currents through this channel in the peripheral taste cells are greatly increased between 15 and 35°C (Fig. 43.2). A corresponding increase in the subsequent gustatory nerve response to sweet compounds was also measured in wild type animals but not TRPM5-KO mice. This temperature sensitivity in the TRPM5 may underlie the mechanism of how temperature influences taste perceptions [66]. As shown in Fig. 43.2, when TRPM5 was activated by high intracellular calcium, an increase in temperature caused a corresponding increase in the magnitude of the TRPM5 current. Increasing temperatures induced a shift in the activation curve and

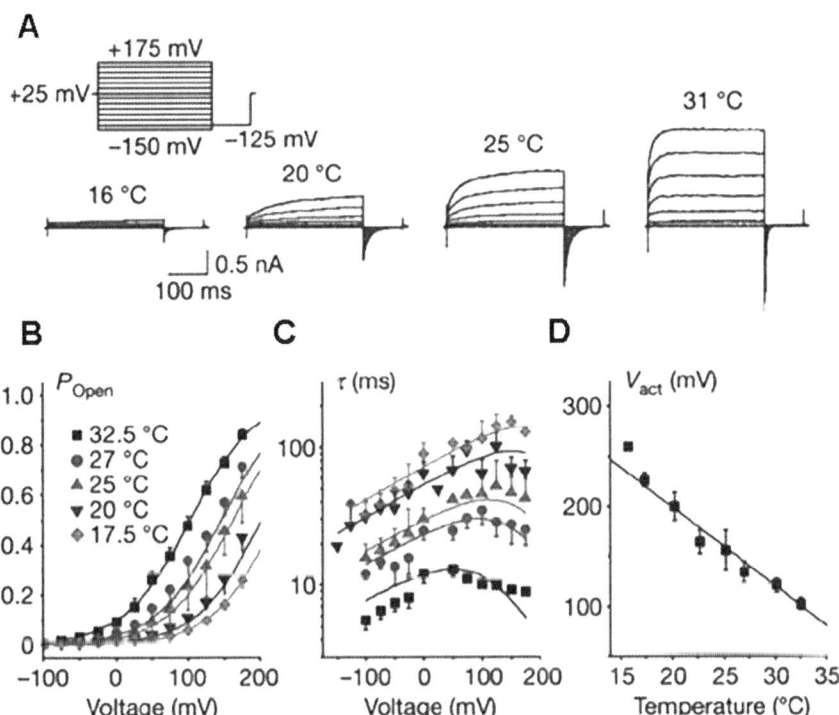

Fig. 43.2 Temperature dependence of TRPM5 in a high intracellular calcium concentration. (**a**) Current traces elicited at different temperatures in response to the indicated voltage protocol at an intracellular Ca^{2+} concentration of 100 μM. (**b**), (**c**) average activation curves and voltage dependence of the time constant of current relaxation at different temperatures ($n = 3$–5). (**d**) Temperature dependence of V_{act}. Continuous lines represent the fit of the two-state gating model in (**b**) and (**c**). Data are the mean ±s.e.m. This figure was adapted by permission from Macmillan Publishers, Ltd: nature (2005) [66]

also affected the time constant of the response at every membrane potential tested [66]. These data clearly show that modulating the activity of TRPM5 with temperature has an effect on the transmission of taste signals, particularly sweet taste stimuli.

Interestingly, TRPM5 has also been localized in olfactory neurons, chemosensory cells in the vomeronasal organ and as solitary chemoreceptive cells in the respiratory and gastrointestinal tract [67–72]. Within these cells, the physiological role of TRPM5 has not been well-defined, though studies have shown that the TRPM5 expressing cells respond to odorants in olfactory neurons and to pheromones in vomeronasal neurons [67–69]. The solitary chemoreceptive cells expressing TRPM5 in the respiratory tract respond to bitter ligands as well as irritants, including acyl-homoserine lactone which is a bacterial indicator. In TRPM5-KO mice, the responsiveness of these chemoreceptive cells to those stimuli was lost and the authors postulate that these TRPM5-expressing chemoreceptive cells may have important roles in the trigeminal system [71]. These studies reveal a broad role for TRPM5 as an intrinsic signaling component in chemosensory organs in addition to its role in taste transduction.

43.3 Polycystic Kidney Disease-1-Like 3/Polycystic Kidney Disease-2-Like 1 (PKD1L3/PKD2L1) TRPP Channels

TRPP channels comprise one of the subfamilies within the group two TRPs that are distantly related to the group one TRPs which contain the five main subfamilies of TRPs. Members of the TRPP subfamily are non-selective cation channels that are permeable to both sodium and calcium, are widely expressed from yeast to mammals, and when mutated, are responsible for polycystic kidney disease [42]. Within the taste system, a heterodimer of PKD1L3/PKD2L1 is expressed in a subpopulation of taste cells [33–34, 73] and strong evidence suggests that this heterodimer is involved in the detection of sour stimuli [33, 35, 74].

The search for the sour receptor in the taste system has been a long one. Multiple candidate receptors have been proposed [75–79], but none have been definitively identified as the sour receptor. This is attributed to the fact that protons which comprise acidic stimuli, will modulate most ionic conductances and therefore almost all taste cells respond to acids with a change in membrane conductance and intracellular acidification [80–81]. However, it is not known if all of these taste cells communicate that information to the nervous system. Some taste cells also respond to acidic stimuli by increasing intracellular calcium, likely via calcium influx through VGCCs [12] and it is assumed that these taste cells are generating the acid signals to the nervous system. It is also possible that multiple receptors contribute to the detection of sour stimuli and that this system redundancy has made it more difficult to determine which receptors are primarily responsible for the detection of sour stimuli.

Multiple laboratories have identified the PKD1L3/PKD2L1 heterodimer as a candidate sour receptor [33–35, 74]. PKD1L3 and PKD2L1 are both highly expressed in the peripheral taste system [33–34] and have been localized in the sub-population of taste cells that express VGCCs and synaptic proteins [73]. Diphtheria toxin A fragment (DTA)-mediated ablation of PKD2L1 expressing taste cells demonstrated that these cells are needed for the detection of sour stimuli in vivo. Specifically, the ablation of PKD2L1-expressing taste cells eliminated the gustatory nerve responses to multiple acidic stimuli without significantly impacting the nerve responses to other types of taste stimuli [35]. These results indicate that the PKD2L1-expressing taste cells are the primary sour sensitive taste cells and support the hypothesis that the PKD1L3/PKD2L1 heterodimer is involved in sour transduction.

However, some evidence suggests that additional receptors may be involved in sour detection. When the PKD1L3/PKD2L1 heterodimer is expressed in human embryonic kidney cells (HEK) 293T cells, both PKD1L3 and PKD2L1 are needed to form a functional channel at the cell surface [33]. When either isoform is expressed alone, they do not localize at the cell surface in the heterologous cells. In taste cells, only PKD2L1 is expressed in fungiform papillae [33, 35] and yet these taste cells are very sensitive to sour stimuli [82] which indicates that additional mechanisms for sour taste transduction likely exist. Supporting a role for both the PKDs and ASIC channels in human sour taste, Huque et al. [83] recently reported that in patients with deficits in their ability to detect sour taste stimuli, there was an absence mRNA transcripts for multiple ASIC channels as well as PKD1L3 and PKD2L1. These transcripts were all readily amplified in normal humans that detected sour stimuli. In addition, immunocytochemical analysis of PKD1L3 found that PKD1L3 was expressed in a population of cells outside the taste buds while ASICs and PKD2L1 were localized in the fungiform taste cells [83]. These human data for the PKDs parallel what has been reported in mice which suggests that the detection and transduction of sour stimuli are conserved processes within the mammalian taste system. Therefore, this PKD heterodimer appears to have an important role in sour taste transduction, but its mode of action as well as the determination if these channels are acting as the sour receptor *per se*, have not been definitely identified.

The functional properties of PKD1L3/PKD2L1 in HEK293 cells have been studied using calcium imaging and patch clamp analysis. When co-expressed, the PKD1L3/PKD2L1 heterodimer responded specifically to sour stimuli and not to other tastants (see Fig. 43.3) with little activity when expressed alone [33]. Patch-clamp analysis of these channels in HEK293 cells found that sour stimuli generated a strong current with rapid inactivation, while calcium imaging revealed a large cytosolic calcium increase in response to sour stimuli. Compared to stimulus onset, there was a delay of almost 20 s before the stimulus responses were recorded for both increases in intracellular calcium and membrane currents. It may be that intracellular acidification is needed to activate these channels and that the channels are not directly gated by protons [33].

Another study in HEK293 cells reported that when the acidic stimulus was strong (pH < 3.0), the PKD1L3/PKD2L1 heterodimer is activated after the acid stimulus is removed. The authors concluded that the heterodimer was activated by stimulus application but was not gated open until the stimulus was removed. At that time, the HEK293 cells had a large increase in intracellular calcium as well as a large

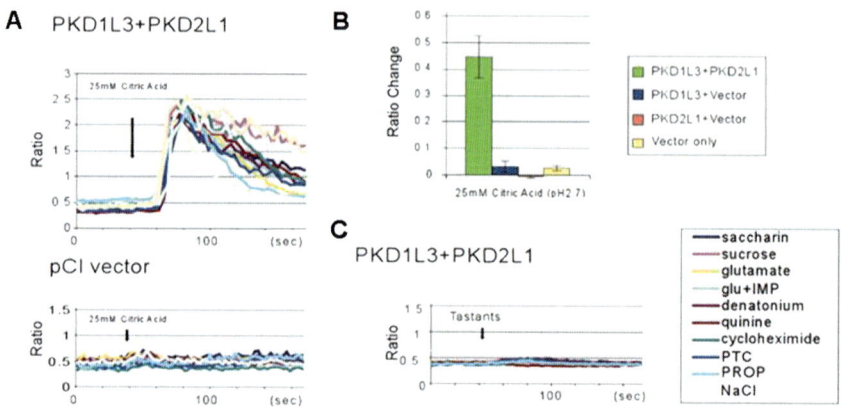

Fig. 43.3 PKD1L3 and PKD2L1 are stimulated by sour tastants. (**a**) Kinetics of the Fluo-4/Fura red ratio changes for 10 representative cells. (**b**) PKD1L3 and PKD2L1 respond to citric acid. In contrast, cells expressing PKD1L3 or PKD2L1 alone, or neither, do not respond. Fluorescent ratio of the entire field (≈200 cells) before and after stimulation (±SEM) are shown ($n = 3$–5). (**c**) PKD1L3 and PKD2L1 do not respond to sweet, bitter, umami, and salty chemicals. Averages of the kinetics of 12–31 representative cells for each ligand are shown. These cells responded to 25 mM citric acid (data not shown). glu+IMP, l-glutamate plus inosine monophosphate; PTC, phenylthiocarbamide; PROP, 6-n-propylthiouracil. This data was adapted with permission from [33], copyright (2006) National Academy of Sciences, USA

inward current that quickly inactivated [84]. This "off-response" might also explain the delay in the acid-induced responses seen in the earlier study that was reporting the initial characterization of the PKD1L3/PKD2L1 heterodimer in HEK293 cells [33]. It will be interesting to see if future studies determine if the PKD1L3/PKD2L1 heterodimer has similar properties in taste cells and clearly define the role of these TRPP channels in sour taste transduction.

43.4 Vanilloid Receptor 1 (TRPV1)

The TRPV receptors have well established roles in pain perception and are activated by diverse stimuli, including heat, protons, lipids, phosphorylation, and depletion of intracellular calcium stores. The first identified TRPV receptor, TRPV1, is also activated by vanilloids, including capsaicin [85]. These particular characteristics made TRPV1 an attractive candidate taste receptor because of the relationship of some of its ligands to the taste system, but early studies reported conflicting findings. Initial immunohistochemical analysis of rat taste buds reported that VR1 was not localized in taste receptor cells but was highly expressed in nerve fibers that were concentrated around the taste papillae [86]. Conversely, physiological studies on rat taste receptor cells reported that capsaicin inhibited potassium currents in these cells at nanomolar concentrations and affected their intracellular calcium levels at micromolar concentrations. These data suggest that capsaicin, likely working through the

VR1 receptor, modifies gustatory sensation, however the authors did not link this capsaicin response to a particular taste quality [87].

Other studies indicated that TRPV1 may be acting as a salt receptor [36, 88–89]. Salt taste transduction relies on two different components: an amiloride-sensitive component and an amiloride-insensitive component. Amiloride is a diuretic that blocks the resting sodium conductance in the epithelial sodium channel (ENaC), a member of the degenerin channel family that is widely expressed in epithelial cells. It is well established that ENaC is the receptor responsible for the salt sensation that is blocked by amiloride but the channel responsible for the amiloride-insensitive component of the salt response has not been identified [90]. Studies from Lyall and DeSimone provide evidence that a splice variant of the TRPV1 channel may be the amiloride-insensitive channel that is responsible for the other component of salt taste [36, 88–89]. These studies were based on afferent gustatory nerve recordings which found that amiloride-insensitive salt responses were blocked by TRPV1 antagonists and activated by vanilloids and temperature. However, there were functional differences between the identified channel and TRPV1 that led them to conclude that this channel was a splice variant of TRPV1 (labeled TRPV1t) [36]. Follow up studies in the TRPV1 knock out mouse produced conflicting data. Gustatory nerve recordings from the TRPV1 knock out mouse found no functional amiloride-insensitive salt response and the sensitivity to vanilloids and temperature were also abolished which support a critical role for this channel, or a splice variant of this channel, in the detection of salt taste [88]. However, behavioral studies with the TRPV1 knock out mouse determined that these mice have no impairment in their ability to detect salt, even when amiloride was added to block the amiloride-sensitive component of the salt response [37–38]. These data suggest that while TRPV1t appears to play a role in salt detection, it is not solely responsible for the amiloride-insensitive component of the taste response.

There are several additional studies that indicate other potential roles for TRPV1 in the modulation of signaling in the peripheral taste receptor cells. One recent behavioral study demonstrated that TRPV1 contributes to the detection of complex taste divalent salts such as iron, zinc, copper and magnesium. For some of these salts, TRPV1 appears to contribute to the aversive response that normally occurs at high concentrations in wild type mice, since the TRPV1 knock out mice were less sensitive to the aversive effect. However, since these studies were not performed on taste receptor cells directly, it is not clear where the TRPV1 was having its effect [91]

Two other studies have demonstrated that TRPV1 is one of several channels that contribute to the maintenance of intracellular calcium levels in taste receptor cells even in the absence of cell stimulation [39–40]. In order to prevent non-specific activation of calcium-dependent processes, regulating cytosolic calcium levels is critical for all cells including taste cells. However, little is currently known about cytosolic calcium regulation and the calcium clearance mechanisms that contribute to this process in taste receptor cells. Calcium regulation is especially important in taste cells, because unlike most cells, taste cells are exposed to a variable external environment. While most of the taste receptor cell is housed in a normal extracellular milieu, the apical ends of these cells are in physical contact with the external environment in order to detect taste stimuli. Generally, this external environment has a very low ionic content unless potential nutrients are present, at which time the external environment will change depending on what is being consumed. This variable external environment appears to have created a need for taste receptor cells to have constitutively activated channels that allow calcium influx to occur as needed. These studies [39–40] demonstrated that when external calcium concentrations were altered, the resting calcium levels in the taste cells had a corresponding change. In addition, inhibiting either mitochondrial calcium uptake or calcium efflux through sodium-calcium exchangers caused an elevation in cytosolic calcium even if the taste cells had not been stimulated. In other cell types, inhibiting these calcium clearance mechanisms generally has little effect in the absence of an imposed calcium load [92–94]. In taste cells, this elevated cytosolic calcium levels were maintained as long as external calcium was present and these calcium clearance mechanisms were inhibited. Applying TRPV1 antagonists significantly reduced or abolished the

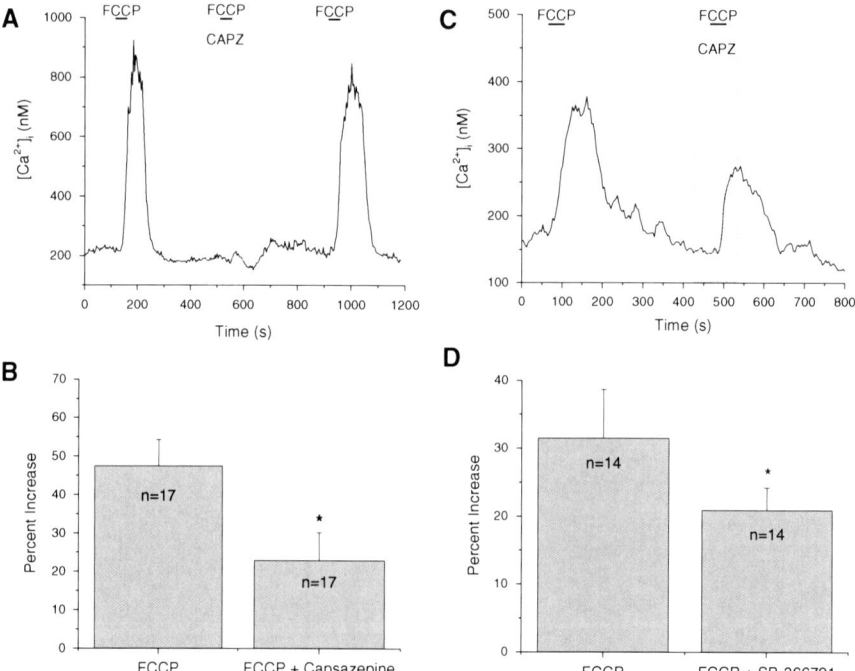

Fig. 43.4 TRPV1 antagonists inhibit FCCP-induced calcium increases from the mitochondria. (**a**) Application of 10 μM capsazepine (CAPZ) abolished the FCCP-dependent cytosolic calcium increases in some mouse taste cells. (**b**) In other taste cells, capsazepine inhibited the rise in intracellular calcium when mitochondria were disabled with FCCP, but did not completely block the response. (**c**) Capsazepine significantly inhibited the peak of the FCCP-induced calcium response in taste cells ($n = 17, p = 0.019$). (**d**) Application of 1 μM SB366791, another TRPV1 antagonist, significantly reduced the amplitude of the FCCP-induced calcium elevation compared to control ($n = 14, p = 0.03$). This data is reprinted from [40] and is being used with permission

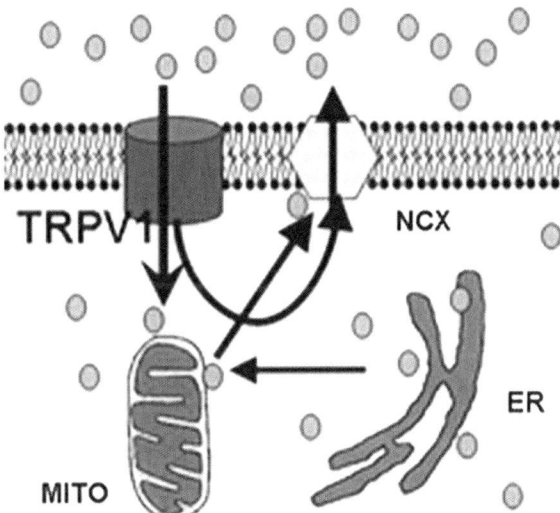

Fig. 43.5 Model summarizing the proposed regulation of cytosolic calcium in taste cells in the absence of cell stimulation. TRPV1 or TRPV1t channels contribute to a constitutive calcium influx across the plasma membrane that is regulated by mitochondria (MITO) and sodium/calcium exchangers (NCX). Mitochondria also regulate calcium leak from internal calcium stores in the endoplasmic reticulum (ER). Excess internal calcium is routinely extruded by sodium/calcium exchangers on the plasma membrane. Circles represent calcium ions

elevated cytosolic calcium levels in many of the taste receptor cells [39–40] which indicates that TRPV1, or probably TRPV1t, makes a significant contribution to this constitutive calcium influx in some taste cells (see Figs. 43.4 and 43.5). While TRPV1 has well defined roles in pain perception [85], in taste cells TRPV1 appears to have evolved into a constitutively active form, presumably TRPV1t, that is important in maintaining appropriate cytosolic calcium levels in these cells. If TRPV1 is absent in taste receptor cells, there may be a significant effect on the resting calcium levels that would affect the cells' ability to respond appropriately to taste stimuli, including salt stimuli. Future studies are needed to more precisely define how this constitutively active TRPV channel affects taste transduction.

43.5 Conclusions

Clearly, TRP channels play important roles in the transduction of taste stimuli in the peripheral taste system. Of the TRPs identified to date, none act as the characteristic non-selective cation channel that reside in the plasma membrane and pass calcium and sodium when the cell is stimulated. Instead, TRPM5 is activated by voltage and intracellular calcium, but is monovalent selective. PKD2L1 forms a heterodimer with PKD1L3 in some sour sensitive taste cells, but not others, while responding to cell acidification and finally, TRPV1t appears to contribute to maintaining the taste cell in a physiological state that allows it to respond appropriately to external stimuli. Each of these TRP channels appears to have important roles in the normal functioning of the peripheral taste system, but their physiological roles are not completely clear. It is possible, even likely, that other currently unidentified TRPs are also contributing to taste transduction. Their identification and physiological function within the taste system await future studies.

References

1. Lindemann B (1996 Jul) Taste reception. Physiol Rev 76(3):718–766
2. Lindemann B (2001 Sep 13) Receptors and transduction in taste. Nature 413(6852):219–225
3. Farbman AI (1980 Jul) Renewal of taste bud cells in rat circumvallate papillae. Cell Tissue Kinet 13(4):349–357
4. Finger TE, Simon SA (2000) Cell biology of taste epithelium. In: Finger TE, Silver WL, Restrepo D (eds) The neurobiology of taste and smell. Wiley-Liss, New York, NY, pp 287–314
5. Herness MS, Sun XD (1995 Jul) Voltage-dependent sodium currents recorded from dissociated rat taste cells. J Membr Biol 146(1):73–84
6. Chen Y, Sun XD, Herness S (1996 Feb) Characteristics of action potentials and their underlying outward currents in rat taste receptor cells. J Neurophysiol 75(2):820–831
7. Behe P, DeSimone JA, Avenet P, Lindemann B (1990 Nov) Membrane currents in taste cells of the rat fungiform papilla. Evidence for two types of Ca currents and inhibition of K currents by Saccharin. J Gen Physiol 96(5):1061–1084
8. Smith DV, Davis BJ (2000) Neural representation of taste. In: Finger TE, Silver WL, Restrepo D (eds) The neurobiology of taste and smell. Wiley-Liss, New York, NY, pp 353–394
9. Bryant BP, Silver WL (2000) Chemesthesis: the common chemical sense. In: Finger TE, Silver WL, Restrepo D (eds) The neurobiology of taste and smell. Wiley-Liss, New York, NY, pp 73–100
10. Medler K (2008) Signaling mechanisms controlling taste cell function. Crit Rev Eukaryot Gene Expr 18(2):125–137
11. DeFazio RA, Dvoryanchikov G, Maruyama Y, Kim JW, Pereira E, Roper SD, Chaudhari N (2006 Apr 12) Separate populations of receptor cells and presynaptic cells in mouse taste buds. J Neurosci 26(15):3971–3980
12. Richter TA, Caicedo A, Roper SD (2003 Mar 1) Sour taste stimuli evoke Ca2+ and pH responses in mouse taste cells. J Physiol 547(Pt 2):475–483
13. Yang R, Crowley HH, Rock ME, Kinnamon JC (2000 Aug 21) Taste cells with synapses in rat circumvallate papillae display SNAP-25-like immunoreactivity. J Comp Neurol 424(2):205–215
14. Medler KF, Margolskee RF, Kinnamon SC (2003 Apr 1) Electrophysiological characterization of voltage-gated currents in defined taste cell types of mice. J Neurosci 23(7):2608–2617
15. Royer SM, Kinnamon JC (1988 Apr 1) Ultrastructure of mouse foliate taste buds: synaptic and nonsynaptic interactions between taste cells and nerve fibers. J Comp Neurol 270(1):11–24
16. Yang R, Stoick CL, Kinnamon JC (2004 Mar 22) Synaptobrevin-2-like immunoreactivity is associated with vesicles at synapses in rat circumvallate taste buds. J Comp Neurol 471(1):59–71
17. Yee CL, Yang R, Bottger B, Finger TE, Kinnamon JC (2001 Nov 5) "Type III" cells of rat taste buds: immunohistochemical and ultrastructural studies of neuron-specific enolase, protein gene product 9.5, and serotonin. J Comp Neurol 440(1):97–108
18. Vandenbeuch A, Clapp TR, Kinnamon SC (2008 Jan 2) Amiloride-sensitive channels in type I fungiform taste cells in mouse. BMC Neurosci 9(1):1

19. Akabas MH, Dodd J, Al-Awqati Q (1988 Nov 18) A bitter substance induces a rise in intracellular calcium in a subpopulation of rat taste cells. Science 242(4881):1047–1050
20. Zhang Y, Hoon MA, Chandrashekar J, Mueller KL, Cook B, Wu D, Zuker CS, Ryba NJ (2003 Feb 7) Coding of sweet, bitter, and umami tastes: different receptor cells sharing similar signaling pathways. Cell 112(3):293–301
21. Yang R, Tabata S, Crowley HH, Margolskee RF, Kinnamon JC (2000 Sep 11) Ultrastructural localization of gustducin immunoreactivity in microvilli of type II taste cells in the rat. J Comp Neurol 425(1):139–151
22. Clapp TR, Stone LM, Margolskee RF, Kinnamon SC (2001) Immunocytochemical evidence for co-expression of Type III IP3 receptor with signaling components of bitter taste transduction. BMC Neurosci 2:6
23. Clapp TR, Yang R, Stoick CL, Kinnamon SC, Kinnamon JC (2004 Jan 12) Morphologic characterization of rat taste receptor cells that express components of the phospholipase C signaling pathway. J Comp Neurol 468(3):311–321
24. Clapp TR, Medler KF, Damak S, Margolskee RF, Kinnamon SC (2006) Mouse taste cells with G protein-coupled taste receptors lack voltage-gated calcium channels and SNAP-25. BMC Biol 4:7
25. Huang YJ, Maruyama Y, Dvoryanchikov G, Pereira E, Chaudhari N, Roper SD (2007 Apr 10) The role of pannexin 1 hemichannels in ATP release and cell-cell communication in mouse taste buds. Proc Natl Acad Sci USA 104(15):6436–6441
26. Romanov RA, Rogachevskaja OA, Bystrova MF, Jiang P, Margolskee RF, Kolesnikov SS (2007 Feb 7) Afferent neurotransmission mediated by hemichannels in mammalian taste cells. Embo J 26(3):657–667
27. Roper SD (2006 Jul) Cell communication in taste buds. Cell Mol Life Sci 63(13):1494–1500
28. Roper SD (2007 Aug) Signal transduction and information processing in mammalian taste buds. Pflugers Arch 454(5):759–776
29. Hofmann T, Chubanov V, Gudermann T, Montell C (2003 Jul 1) TRPM5 is a voltage-modulated and Ca(2+)-activated monovalent selective cation channel. Curr Biol 13(13):1153–1158
30. Liu D, Liman ER (2003 Dec 9) Intracellular Ca2+ and the phospholipid PIP2 regulate the taste transduction ion channel TRPM5. Proc Natl Acad Sci USA 100(25):15160–15165
31. Perez CA, Margolskee RF, Kinnamon SC, Ogura T (2003 May-Jun) Making sense with TRP channels: store-operated calcium entry and the ion channel Trpm5 in taste receptor cells. Cell Calcium 33(5–6):541–549
32. Prawitt D, Monteilh-Zoller MK, Brixel L, Spangenberg C, Zabel B, Fleig A, Penner R (2003 Dec 9) TRPM5 is a transient Ca2+-activated cation channel responding to rapid changes in [Ca2+]i. Proc Natl Acad Sci USA 100(25):15166–15171
33. Ishimaru Y, Inada H, Kubota M, Zhuang H, Tominaga M, Matsunami H (2006 Aug 15) Transient receptor potential family members PKD1L3 and PKD2L1 form a candidate sour taste receptor. Proc Natl Acad Sci USA 103(33):12569–12574
34. LopezJimenez ND, Cavenagh MM, Sainz E, Cruz-Ithier MA, Battey JF, Sullivan SL (2006 Jul) Two members of the TRPP family of ion channels, Pkd1l3 and Pkd2l1, are co-expressed in a subset of taste receptor cells. J Neurochem 98(1):68–77
35. Huang AL, Chen X, Hoon MA, Chandrashekar J, Guo W, Trankner D, Ryba NJ, Zuker CS (2006 Aug 24) The cells and logic for mammalian sour taste detection. Nature 442(7105):934–938
36. Lyall V, Heck GL, Vinnikova AK, Ghosh S, Phan TH, Alam RI, Russell OF, Malik SA, Bigbee JW, DeSimone JA (2004 Jul 1) The mammalian amiloride-insensitive non-specific salt taste receptor is a vanilloid receptor-1 variant. J Physiol 558(Pt 1):147–159
37. Ruiz C, Gutknecht S, Delay E, Kinnamon S (2006 Nov) Detection of NaCl and KCl in TRPV1 knockout mice. Chem Senses 31(9):813–820

38. Treesukosol Y, Lyall V, Heck GL, Desimone JA, Spector ACA (2007 Jan 18) Psychophysical and electrophysiological analysis of salt taste in Trpv1 null mice. Am J Physiol Regul Integr Comp Physiol 292(5):R1799–R1809
39. Laskowski AI, Medler KF (2009 Jul 6) Sodium/calcium exchangers contribute to the regulation of cytosolic calcium levels in mouse taste cells. J Physiol 587(16):4077–4089
40. Hacker K, Medler KF (2008 Oct) Mitochondrial calcium buffering contributes to the maintenance of Basal calcium levels in mouse taste cells. J Neurophysiol 100(4): 2177–2191
41. Peier AM, Moqrich A, Hergarden AC, Reeve AJ, Andersson DA, Story GM, Earley TJ, Dragoni I, McIntyre P, Bevan S, Patapoutian AA (2002 Mar 8) TRP channel that senses cold stimuli and menthol. Cell 108(5):705–715
42. Venkatachalam K, Montell C (2007) TRP channels. Annu Rev Biochem 76:387–417
43. Adler E, Hoon MA, Mueller KL, Chandrashekar J, Ryba NJ, Zuker CS (2000 Mar 17) A novel family of mammalian taste receptors. Cell 100(6):693–702
44. Chandrashekar J, Hoon MA, Ryba NJ, Zuker CS (2006 Nov 16) The receptors and cells for mammalian taste. Nature 444(7117):288–294
45. Chandrashekar J, Mueller KL, Hoon MA, Adler E, Feng L, Guo W, Zuker CS, Ryba NJ (2000 Mar 17) T2Rs function as bitter taste receptors. Cell 100(6):703–711
46. Montmayeur JP, Liberles SD, Matsunami H, Buck LB (2001 May) A candidate taste receptor gene near a sweet taste locus. Nat Neurosci 4(5):492–498
47. Montmayeur JP, Matsunami H (2002 Aug) Receptors for bitter and sweet taste. Curr Opin Neurobiol 12(4):366–371
48. Mueller KL, Hoon MA, Erlenbach I, Chandrashekar J, Zuker CS, Ryba NJ (2005 Mar 10) The receptors and coding logic for bitter taste. Nature 434(7030):225–229
49. Li X, Staszewski L, Xu H, Durick K, Zoller M, Adler E (2002 Apr 2) Human receptors for sweet and umami taste. Proc Natl Acad Sci USA 99(7):4692–4696
50. Kinnamon SC, Vandenbeuch A (2009 Jul) Receptors and transduction of umami taste stimuli. Ann NY Acad Sci 1170:55–59
51. Yoshida R, Yasumatsu K, Shirosaki S, Jyotaki M, Horio N, Murata Y, Shigemura N, Nakashima K, Ninomiya Y (2009 Jul) Multiple receptor systems for umami taste in mice. Ann NY Acad Sci 1170:51–54
52. Ninomiya Y, Beauchamp GK (2009 Jul) Symposium overview: Umami reception in the oral cavity: receptors and transduction. Ann NY Acad Sci 1170:39–40
53. Chaudhari N, Pereira E, Roper SD (2009 Sep) Taste receptors for umami: the case for multiple receptors. Am J Clin Nutr 90(3):738S–42S
54. Kinnamon SC (2009 Sep) Umami taste transduction mechanisms. Am J Clin Nutr 90(3):753S–5S
55. Temussi PA (2009 Jun) Sweet, bitter and umami receptors: a complex relationship. Trends Biochem Sci 34(6):296–302
56. Huang L, Shanker YG, Dubauskaite J, Zheng JZ, Yan W, Rosenzweig S, Spielman AI, Max M, Margolskee RF (1999 Dec) Ggamma13 colocalizes with gustducin in taste receptor cells and mediates IP3 responses to bitter denatonium. Nat Neurosci 2(12):1055–1062
57. Ming D, Ninomiya Y, Margolskee RF (1999 Aug 17) Blocking taste receptor activation of gustducin inhibits gustatory responses to bitter compounds. Proc Natl Acad Sci USA 96(17):9903–9908
58. Yan W, Sunavala G, Rosenzweig S, Dasso M, Brand JG, Spielman AI (2001 Apr) Bitter taste transduced by PLC-beta(2)-dependent rise in IP(3) and alpha-gustducin-dependent fall in cyclic nucleotides. Am J Physiol Cell Physiol 280(4):C742–C751
59. Zhang Z, Zhao Z, Margolskee R, Liman E (2007 May 23) The transduction channel TRPM5 is gated by intracellular calcium in taste cells. J Neurosci 27(21): 5777–5786
60. Romanov RA, Rogachevskaja OA, Khokhlov AA, Kolesnikov SS (2008 Dec) Voltage dependence of ATP secretion in mammalian taste cells. J Gen Physiol 132(6):731–744

61. Finger TE, Danilova V, Barrows J, Bartel DL, Vigers AJ, Stone L, Hellekant G, Kinnamon SC (2005 Dec 2) ATP signaling is crucial for communication from taste buds to gustatory nerves. Science 310(5753):1495–1499
62. Dando R, Roper SD (2009 Dec 15) Cell-to-cell communication in intact taste buds through ATP signalling from pannexin 1 gap junction hemichannels. J Physiol 587(Pt 24): 5899–5906
63. Damak S, Rong M, Yasumatsu K, Kokrashvili Z, Perez CA, Shigemura N, Yoshida R, Mosinger B Jr, Glendinning JI, Ninomiya Y, Margolskee RF (2006 Mar) Trpm5 null mice respond to bitter, sweet, and umami compounds. Chem Senses 31(3):253–264
64. Perez CA, Huang L, Rong M, Kozak JA, Preuss AK, Zhang H, Max M, Margolskee RF (2002 Nov) A transient receptor potential channel expressed in taste receptor cells. Nat Neurosci 5(11):1169–1176
65. Oike H, Wakamori M, Mori Y, Nakanishi H, Taguchi R, Misaka T, Matsumoto I, Abe K (2006 Sep) Arachidonic acid can function as a signaling modulator by activating the TRPM5 cation channel in taste receptor cells. Biochim Biophys Acta 1761(9):1078–1084
66. Talavera K, Yasumatsu K, Voets T, Droogmans G, Shigemura N, Ninomiya Y, Margolskee RF, Nilius B (2005 Dec 15) Heat activation of TRPM5 underlies thermal sensitivity of sweet taste. Nature 438(7070):1022–1025
67. Lin W, Ezekwe EA Jr., Zhao Z, Liman ER, Restrepo D (2008) TRPM5-expressing microvillous cells in the main olfactory epithelium. BMC Neurosci 9:114
68. Lin W, Ogura T, Margolskee RF, Finger TE, Restrepo D (2008 Mar) TRPM5-expressing solitary chemosensory cells respond to odorous irritants. J Neurophysiol 99(3): 1451–1460
69. Lin W, Margolskee R, Donnert G, Hell SW, Restrepo D (2007 Feb 13) Olfactory neurons expressing transient receptor potential channel M5 (TRPM5) are involved in sensing semiochemicals. Proc Natl Acad Sci USA 104(7):2471–2476
70. Kaske S, Krasteva G, Konig P, Kummer W, Hofmann T, Gudermann T, Chubanov V (2007) TRPM5, a taste-signaling transient receptor potential ion-channel, is a ubiquitous signaling component in chemosensory cells. BMC Neurosci 8:49
71. Tizzano M, Gulbransen BD, Vandenbeuch A, Clapp TR, Herman JP, Sibhatu HM, Churchill ME, Silver WL, Kinnamon SC, Finger TE (2010 Feb 16) Nasal chemosensory cells use bitter taste signaling to detect irritants and bacterial signals. Proc Natl Acad Sci USA 107(7): 3210–3215
72. Bezencon C, le Coutre J, Damak S (2007 Jan) Taste-signaling proteins are coexpressed in solitary intestinal epithelial cells. Chem Senses 32(1):41–49
73. Kataoka S, Yang R, Ishimaru Y, Matsunami H, Sevigny J, Kinnamon JC, Finger TE (2008 Mar) The candidate sour taste receptor, PKD2L1, is expressed by type III taste cells in the mouse. Chem Senses 33(3):243–254
74. Ishii S, Misaka T, Kishi M, Kaga T, Ishimaru Y, Abe K (2009 Jul 31) Acetic acid activates PKD1L3-PKD2L1 channel a candidate sour taste receptor. Biochem Biophys Res Commun 385(3):346–350
75. Ugawa S, Minami Y, Guo W, Saishin Y, Takatsuji K, Yamamoto T, Tohyama M, Shimada S (1998 Oct 8) Receptor that leaves a sour taste in the mouth. Nature 395(6702):555–556
76. Gilbertson DM, Gilbertson TA (1994 Oct) Amiloride reduces the aversiveness of acids in preference tests. Physiol Behav 56(4):649–654
77. Miyamoto T, Fujiyama R, Okada Y, Sato T (1998 Oct) Sour transduction involves activation of NPPB-sensitive conductance in mouse taste cells. J Neurophysiol 80(4):1852–1859
78. Stevens DR, Seifert R, Bufe B, Muller F, Kremmer E, Gauss R, Meyerhof W, Kaupp UB, Lindemann B (2001 Oct 11) Hyperpolarization-activated channels HCN1 and HCN4 mediate responses to sour stimuli. Nature 413(6856):631–635
79. Ugawa S, Yamamoto T, Ueda T, Ishida Y, Inagaki A, Nishigaki M, Shimada S (2003 May 1) Amiloride-insensitive currents of the acid-sensing ion channel-2a (ASIC2a)/ASIC2b heteromeric sour-taste receptor channel. J Neurosci 23(9):3616–3622

80. Lin W, Ogura T, Kinnamon SC (2002 Jul) Acid-activated cation currents in rat vallate taste receptor cells. J Neurophysiol 88(1):133–141
81. Lyall V, Alam RI, Phan DQ, Ereso GL, Phan TH, Malik SA, Montrose MH, Chu S, Heck GL, Feldman GM, DeSimone JA (2001 Sep) Decrease in rat taste receptor cell intracellular pH is the proximate stimulus in sour taste transduction. Am J Physiol Cell Physiol 281(3): C1005–C1013
82. Yoshida R, Miyauchi A, Yasuo T, Jyotaki M, Murata Y, Yasumatsu K, Shigemura N, Yanagawa Y, Obata K, Ueno H, Margolskee RF, Ninomiya Y (2009 Sep 15) Discrimination of taste qualities among mouse fungiform taste bud cells. J Physiol 587(Pt 18):4425–4439
83. Huque T, Cowart BJ, Dankulich-Nagrudny L, Pribitkin EA, Bayley DL, Spielman AI, Feldman RS, Mackler SA, Brand JG (2009) Sour ageusia in two individuals implicates ion channels of the ASIC and PKD families in human sour taste perception at the anterior tongue. PLoS One 4(10):e7347
84. Inada H, Kawabata F, Ishimaru Y, Fushiki T, Matsunami H, Tominaga M (2008 Jun 6) Off-response property of an acid-activated cation channel complex PKD1L3-PKD2L1. EMBO Rep 9(7):690–697
85. Gunthorpe MJ, Benham CD, Randall A, Davis JB (2002 Apr) The diversity in the vanilloid (TRPV) receptor family of ion channels. Trends Pharmacol Sci 23(4):183–191
86. Ishida Y, Ugawa S, Ueda T, Murakami S, Shimada S (2002 Oct 30) Vanilloid receptor subtype-1 (VR1) is specifically localized to taste papillae. Brain Res Mol Brain Res 107(1):17–22
87. Park K, Brown PD, Kim YB, Kim JS (2003 Feb 7) Capsaicin modulates K+ currents from dissociated rat taste receptor cells. Brain Res 962(1–2):135–143
88. Lyall V, Heck GL, Vinnikova AK, Ghosh S, Phan TH, Desimone JA (2005 Jan) A novel vanilloid receptor-1 (VR-1) variant mammalian salt taste receptor. Chem Senses 30(Suppl 1): i42–i43
89. Lyall V, Heck GL, Phan TH, Mummalaneni S, Malik SA, Vinnikova AK, Desimone JA (2005 Jun) Ethanol modulates the VR-1 variant amiloride-insensitive salt taste receptor. II. Effect on chorda tympani salt responses. J Gen Physiol 125(6):587–600
90. Lindemann B (1997 Sep) Sodium taste. Curr Opin Nephrol Hypertens 6(5):425–429
91. Riera CE, Vogel H, Simon SA, Damak S, le Coutre J (2009 Feb 25) Sensory attributes of complex tasting divalent salts are mediated by TRPM5 and TRPV1 channels. J Neurosci 29(8):2654–2662
92. Lawrie AM, Rizzuto R, Pozzan T (1996 May 3) Simpson AWM. A role for calcium influx in the regulation of mitochondrial calcium in endothelial cells. J Biol Chem 271(18): 10753–10759
93. Mironov SL, Ivannikov MV, Johansson M (2005 Jan 7) [Ca2+]i signaling between mitochondria and endoplasmic reticulum in neurons is regulated by microtubules. From mitochondrial permeability transition pore to Ca2+-induced Ca2+ release. J Biol Chem 280(1):715–721
94. White RJ, Reynolds IJ (1995 Feb) Mitochondria and Na+/Ca2+ exchange buffer glutamate-induced calcium loads in cultured cortical neurons. J Neurosci 15(2):1318–1328

Chapter 44
Roles of Transient Receptor Potential Proteins (TRPs) in Epidermal Keratinocytes

Mitsuhiro Denda and Moe Tsutsumi

Abstract Epidermal keratinocytes are the epithelial cells of mammalian skin. At the basal layer of the epidermis, these cells proliferate strongly, and as they move towards the skin surface, differentiation proceeds. At the uppermost layer of the epidermis, keratinocytes undergo apoptosis and die, forming a thin, water-impermeable layer called the stratum corneum. Peripheral blood vessels do not reach the epidermis, but peripheral nerve fibers do penetrate into it. Until recently, it was considered that the main role of epidermal keratinocytes was to construct and maintain the water-impermeable barrier function. However, since the functional existence of TRPV1, which is activated by heat and low pH, in epidermal keratinocytes was identified, our understanding of the role of keratinocytes has changed enormously. It has been found that many TRP channels are expressed in epidermal keratinocytes, and play important roles in differentiation, proliferation and barrier homeostasis. Moreover, because TRP channels expressed in keratinocytes have the ability to sense a variety of environmental factors, such as temperature, mechanical stress, osmotic stress and chemical stimuli, epidermal keratinocytes might form a key part of the sensory system of the skin. The present review deals with the potential roles of TRP channels expressed in epidermal keratinocytes and focuses on the concept of the epidermis as an active interface between the body and the environment.

Abbreviations

TRP   transient receptor potential

44.1 Expression of TRPs in Epidermal Keratinocytes

The transient receptor potential protein (TRP) superfamily was first cloned from the visual system of *Drosophila* [1]. In mammals, attention was initially focused on thermo-sensitive TRPs. Julius and his co-workers found TRPV1 (VR1) as

M. Denda (✉)
Shiseido Research Center, Yokohama, Kanagawa 236-8643, Japan
e-mail: mitsuhiro.denda@to.shiseido.co.jp

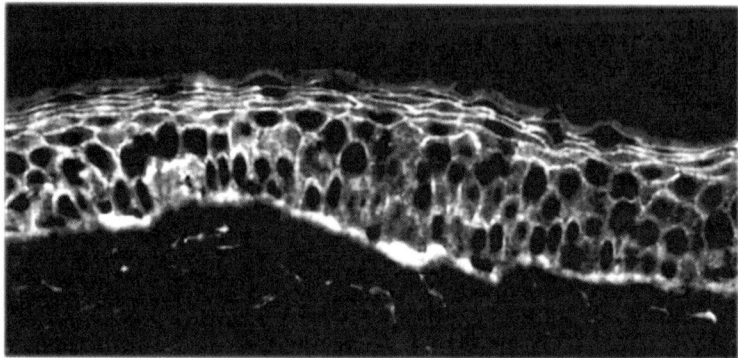

Fig. 44.1 Immunostaining of human epidermis by anti-TRPV1 serum. TRPV1 is expressed in membrane of keratinocytes throughout the epidermis

a polymodal detector of pain-producing heat (>43°C) or chemicals, such as capsaicin and protons, in primary afferent neurons [2]. Several reports described the expression of TRPV1 in the peripheral nervous system. However, we first discovered that TRPV1 is also expressed in human epidermal keratinocytes (Fig. 44.1) [3], and then demonstrated the functional existence of TRPV1 in human cultured keratinocytes [4]. Subsequently, expression of TRPV3 [5] and TRPV4 [6], both of which are activated by high temperature (around 30°C), in keratinocytes was reported [76].

Recently, immunostaining with TRPA1 antibody indicated the presence of TRPA1 in human epidermis [7]. TRPA1 is activated by cold (<17°C) [8]. We demonstrated that intracellular calcium was elevated in cultured human keratinocytes at 17~22°C (Fig. 44.2) [9], and the elevation was more marked in non-differentiated cells than in differentiated cells. Application of ruthenium red (a non-selective TRP blocker) and HC030031 (a specific antagonist of TRPA1) reduced the elevation. These results suggest the existence of TRPA1 as a functional cold-sensitive calcium channel in human epidermal keratinocytes [9]. Because some keratinocytes showed a response at around 22°C, another cold-sensitive receptor, such as TRPM8, might also exist in epidermal keratinocytes [83].

Not only thermo-sensitive TRPs, but also other types of TRPs have been found in keratinocytes. TRPC1, C5, C6, and C7 are expressed in gingival keratinocytes [10]. TRPC4 was also found in epidermal keratinocytes [11]. As described below, TRPC receptors play important roles in the calcium dynamics and differentiation of keratinocytes.

Although a variety of TRP expressions in the epidermal keratinocytes have been reported, the role of them has not clarified perfectly. In following sections, specific roles of some of TRPs will be introduced.

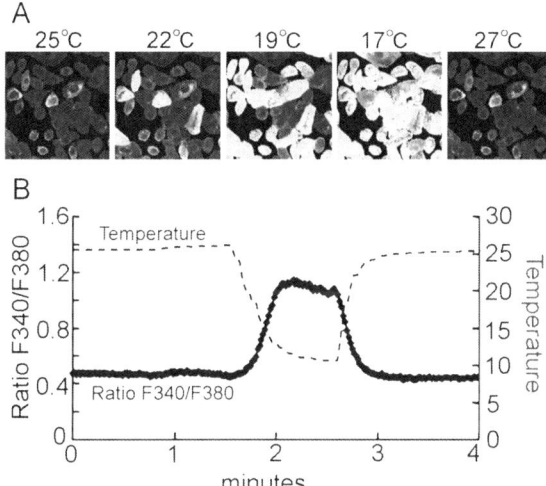

Fig. 44.2 (a) Calcium imaging of cultured keratinocytes. Intracellular calcium level is displayed in ratiometric color (F340/F380). White numbers on the *upper right* of each figure indicate the bath temperature. *White color* indicated the highest concentration of intracellular calcium. (b) Representative profiles of intracellular calcium and temperature in undifferentiated keratinocytes. During the experiments, medium at 25°C was continuously flowed. When we decreased the temperature, cold medium (around 10°C) was added. For measurement of the medium temperature during the experiment, we used a Data, Logger Thermometer (MT-309, Mother, Tool, Ueda, Japan) with an extra-fine thermoelectric couple (diameter 0.5 mm, T35 type, Sakaguchi, Dennetsu, Tokuo, Japan). Data were collected using the supplied software (TestLink, Mother, Tool, Ueda, Japan). When the temperature decreased below 14°C, artifact was appeared because of the deform of the camber

44.2 Role of TRPs in Keratinocyte Differentiation

Calcium ion dynamics in epidermal keratinocytes is strongly associated with cutaneous homeostasis. The concentration of calcium is highest in the uppermost region of the epidermis (the epidermal granular layer) in healthy normal skin, and the gradient disappears immediately after barrier disruption [12, 78]. Abnormal calcium gradients in the epidermis have been observed in a variety of skin diseases [13]. Modulation of epidermal calcium was found to coordinately regulate events late in epidermal differentiation that together lead to barrier formation [14]. Calcium induces terminal differentiation of keratinocytes [15], formation of the cornified envelope (Nemes and Steinert [16]), and also epidermal lipid synthesis [17]. Menon et al. demonstrated that alteration of calcium gradation in the epidermis affects exocytosis of the epidermal lamellar bodies [18].

Several, TRP receptors influence calcium dynamics and differentiation of keratinocytes. TRPC subtypes 1, 4 and 6 and TRPV6 were reported to be associated with keratinocyte differentiation. Human gingival keratinocytes in which TRPC1 was knocked down with siRNA showed a decrease of involucrin and morphologically abnormal differentiation at high calcium concentration [19]. Knockdown of TRPC1 and TRPC4 in HaCaT human keratinocytes and primary keratinocytes also interfered with calcium-induced keratinocyte differentiation [11]. Because, TRPC1 and TRPC4 are not expressed in basal carcinoma cells, these receptors might play a role in skin pathology [11]. Another report demonstrated that knockdown of TRPC1 or TRPC4 inhibited store-operated calcium entry. TRPC expressed in intracellular calcium stores might be associated with keratinocyte differentiation [20].

Application of hyperforin, a specific TRPC6 activator, induced influx of calcium and differentiation of keratinocytes, and knockdown of TRPC6 blocked hyperforin-induced differentiation [21]. Application of triterpenes induced calcium influx and differentiation of keratinocytes, and increased the expression of TRPC6 [22]. Because triterpenes inhibit actinic keratoses [23], TRPC6 might be a potential target for treatment of skin cancer.

1,25-Dihydroxyvitamin, D3, an active metabolite of vitamin D, has a role in keratinocyte differentiation. Application of 1,25-dihydroxyvitamin D3 increased TRPV6 mRNA and protein in keratinocytes. Knockdown of TRPV6 with siRNA blocked calcium-induced keratinocyte differentiation [24]. Thus, TRPV6 might mediate the action of 1,25-dihydroxyvitamin D3. Another report indicated that TRPV6 knockout mice had fewer and thinner layers of the stratum corneum [25]. TRPV6 might play an important role for constract stratum corneum.

Abnormal differentiation of keratinocytes is observed in atopic dermatitis, psoriasis, and contact dermatitis. Thus, study of the TRP channels mentioned above might have implications for the treatment of these skin diseases.

44.3 Role of TRPs in Cutaneous Inflammation and Epidermal Proliferation

Various studies have demonstrated that TRPV1 and TRPV3 are associated with skin inflammation. Activation of TRPV1 by the agonist capsaicin induced dose-dependent increase of cyclooxygenase-2, interleukin-8, and prostaglandin 2 expression in HaCaT keratinocytes and this was blocked by capsazepine, a specific antagonist of TRPV1 [26]. Matrix metalloproteinase-1 (MMP-1) induces collagen degradation and inflammatory responses. Activation of TRPV1 by heat induced MMP-1 mRNA and protein in a temperature-dependent manner in human keratinocytes [27]. The response was blocked by capsazepin. Moreover, knockdown of TRPV1 decreased MMP-1 expression in keratinocytes. Thus, TRPV1 might be associated with skin inflammatory responses.

TRPV1 is also activated by low pH. We recently demonstrated that application of glycolic acid induced epidermal proliferation and this response was blocked by capsazepine in a skin equivalent model [28], We also observed ATP release in a glycolic acid-treated skin equivalent model, and this was blocked by capsazepine. Thus, we speculated that TRPV1 and purinergic receptor play a role in the proliferative response induced by low pH.

Ultraviolet (UV) radiation appears to induce the expression of TRPV1 in human epidermal keratinocytes [29]; acute UV radiation increased TRPV1 in human skin, and moreover, TRPV1 expression in sun-exposed skin was significantly higher than that in sun-protected skin. They also demonstrated that UV radiation induced matrix metalloproteinase-1 in cultured HaCaT cells [30]. Interestingly, TRPV1 expression increased with not only photo-aging, but also intrinsic aging [29]. Another study demonstrated that expression of TRPV1 was significantly increased in epidermal keratinocytes of skin from a distal small nerve fibre disease patient [31]. Although no mechanisms were identified, TRPV1 might potentially be associated with a variety of skin pathologies, including painful neuropathies.

Other work has suggested that TRPV3 is also involved in inflammatory responses, pain, and itching sensation [32]. Asakawa et al. showed that mutation at Gly573 in the TRPV3 gene is associated with defective hair growth and dermatitis in mice [33]. It has been speculated that TRPV3^{Gly573Ser} mutation is a cause of pruritus and dermatitis [32].

We reported that topical application of unsaturated fatty acid induced abnormal differentiation and barrier dysfunction [34]. We also found that application of unsaturated fatty acid to cultured human keratinocytes induced elevation of intracellular calcium. Hu et al. showed that elevation of intracellular calcium was induced by unsaturated fatty acids in TRPV3-expressing oocytes [35]. TRPV3 might be associated with abnormal keratinization induced by unsaturated fatty acids, such as comedogenesis.

TRPC1 may play a role in Darier's disease [36]. Keratinocytes of patients with Darier's disease show up-regulation of TRPC1 and enhanced proliferation. Up-regulation was also observed in sarco(endo)plasmic reticulum Ca^{2+} ATPase 2 (SERCA2) knockout mice. Knockdown of SERCA2 in HaCaT cells increased both TRPC1 expression and thapsigargin-induced apoptosis. The authors speculated that over-expression of TRPC1 in keratinocytes of Darier's disease patients might induce abnormal keratosis.

Calcium dynamics in epidermal keratinocytes plays a key role in epidermal homeostasis. Some, TRPs act as calcium-permeable channels. Thus, abnormal expression or mutation of TRP family members might be associated with various kinds of dermatitis involving abnormal differentiation or barrier dysfunction.

44.4 Role of TRPs in Epidermal Permeability Barrier Homeostasis

One of the most important roles of the skin in terrestrial mammals is to generate a water-impermeable barrier in order to prevent excessive transcutaneous water loss. A decline in barrier function often parallels increased severity of clinical symptomatology [37]. Stratum corneum, which forms the water-impermeable barrier, is composed of two components, i.e., protein-rich nonviable cells and intercellular lipid domains [37]. When the stratum corneum barrier is damaged, a series of homeostatic processes act to restore the barrier [37]. At the first stage of the barrier repair process, exocytosis of lipid-containing granules, lamellar bodies, is accelerated and the interior lipid is secreted into the intercellular domain between the stratum granulosum and stratum corneum, forming a water-impermeable membrane [37].

We have shown that several TRPs in epidermal keratinocytes are associated with epidermal permeability barrier homeostasis. We first demonstrated that TRPV1 and TRPV4 were involved in the epidermal barrier homeostasis [38]. To evaluate the influence of these receptors on epidermal permeability barrier homeostasis, we kept both hairless mouse skin and human skin at various temperatures immediately after tape stripping. At temperatures from 36 to 40°C, the barrier recovery was accelerated in both cases compared with that at 34°C. At 42°C, the barrier recovery was delayed compared with the un-occluded area. Topical application of 4α-phorbol 12,13-didecanone, an activator of TRPV4, accelerated the barrier recovery, while ruthenium red, a blocker of TRP, blocked the acceleration. Topical application of capsaicin, an activator of TRPV1, delayed the barrier recovery, while capsazepin, an antagonist of TRPV1, blocked this delay. Application of TRPV3 activators, 2-aminoethoxydiphenyl borate and camphor, did not affect the barrier recovery rate. Since, TRPV4 is activated at about 35°C and above, while TRPV1 is activated at about 42°C and above, these results suggest that TRPV1 and TRPV4 both play important roles in skin permeability barrier homeostasis. An earlier report suggested the existence of a water flux sensor in the epidermis [39], and as TRPV4 is known to be activated by osmotic pressure [40], our results indicate that TRPV4 might be this sensor of water flux from the surface of the epidermis.

We recently demonstrated that TRPA1 was also involved in epidermal barrier homeostasis [41]. TRPA1, which is activated at temperatures lower than 17°C, is expressed in human epidermis [7]. Thus, we hypothesized that activation of TRPA1 would also influence epidermal permeability barrier homeostasis. To test this idea, we examined the effects of agonists of TRPA1 and brief cold exposure on the barrier recovery rate after barrier disruption. Topical application of a TRPA1 agonist, allyl isothiocyanate or cinnamaldehyde, accelerated the barrier recovery after tape stripping. The effect of both agonists was blocked by HC030031, an antagonist of TRPA1. Brief exposure (1 min) to cold (10–15°C) also accelerated the barrier recovery and this acceleration was also blocked by HC030031. Electron-microscopic studies indicated that brief cold exposure accelerated lamellar body secretion between stratum corneum and stratum granulosum, while pretreatment with HC030031 inhibited the secretion. These results suggest that TRPA1 is associated with epidermal permeability barrier homeostasis. Interestingly, topical application of bradykinin accelerated the barrier repair, and this acceleration was blocked by HC030031. Moreover, application of HC030031 delayed the barrier recovery [41]. An earlier report demonstrated that bradykinin activates TRPA1 [42]. Thus, TRPA1 might act as a sensor of barrier insult, serving to trigger the barrier recovery process.

Epidermal barrier dysfunction is observed in a variety of skin diseases, such as atopic dermatitis, psoriasis, and contact dermatitis, and barrier disruption induces an inflammatory response that might aggravate dermatitis [43]. On the other hand, acceleration of barrier recovery was shown to improve epidermal hyperplasia [44, 45]. Therefore, acceleration of the barrier recovery by regulation of TRP channels could have implications for clinical dermatology.

44.5 TRPs in Hair and Sebaceous Gland

Expression of TRPV1 was reported in hair follicle and outer root sheath keratinocytes (46). In organ culture of human hair, activation of TRPV1 inhibited hair shaft elongation, proliferation and apoptosis (46). Moreover, a comparison of the hair follicle cycle in TRPV1 knockout and wild-type mice showed that the knockout mice exhibited a delay in spontaneous involution of hair follicles [47]. These results indicate that TRPV1 in the hair follicle might play a role in the hair cycle.

TRPV3 is also associated with hair growth. Mutation of TRPV3 was found in two spontaneous hairless rodents, DS-Nh mice and WBN/Kob-Ht rat [33]. Moreover, DS-Nh mice have an abnormal hair growth cycle [48].

TRPV1 is also expressed in human sebaceous gland [49]. Application of a TRPV1 agonist, capsaicin, inhibited basal lipid synthesis of sebocytes without affecting cell viability or apoptosis. By using TRPV1-knockdown cells, evidence was obtained that those effects were mediated by TRPV1. Thus, TRPV1 might be associated with the response of sebaceous glands to environmental stimuli, such as temperature or pH.

44.6 Keratinocyte TRPs as a Cutaneous Sensory System

It has long been recognized that peripheral nerve fibers play a major role in cutaneous thermo-sensation and mechanical sensation. However, until the existence of TRPV1 in epidermal keratinocytes was discovered [3, 4], nobody imagined that keratinocytes might also be involved in the thermo-sensory system. Now, not only TRPV1, but also TRPV3, V4 and A1 have been reported to be present in epidermal keratinocytes [5, 6, 9]. That is, at least 4 thermo-sensitive TRPs exist in the epidermal keratinocytes and they cover the temperature range from 17 to 43°C. Thus, keratinocytes may be the main thermo-sensory cells for skin surface temperature [80].

These, TRP are activated by not only temperature, but also other physical or chemical stimuli. TRPV1 is activated by protons (pH< 5.4) and capsaicin [50]. TRPV3 in epidermal keratinocytes is activated by oregano, thyme, and clove-derived flavors, such as carvacrol, eugenol, and thymol [51]. Interestingly, TRPV3 in epidermal keratinocytes is activated by camphor, whereas TRPV3 in sensory neurons is not [52]. Thus, TRPV3 in epidermal keratinocytes might be a sensor of these herbal extracts.

TRPV4 is activated by osmotic stimuli [40, 53]. We previously demonstrated that environmental humidity influences many aspects of epidermal homeostasis [54, 2000]. Thus, TRPV4 might be a sensor of environmental humidity.

We recently demonstrated that mechanical stress induced elevation of intracellular calcium in cultured keratinocytes [55]. This elevation was blocked by a non specific TRP antagonist, ruthenium red. Previous reports suggested that TRPA1 and TRPV4 might be activated by mechanical stress [56, 57]. Thus, these TRPs in keratinocytes might play a role as a cutaneous sensory system for external mechanical stimuli.

Even if epidermal keratinocytes have a sensory function for a variety of environment factors, the signals from the environment must be passed to the nervous system. Various interactions between keratinocytes and the peripheral nervous system have been suggested. Koizumi and his co-workers demonstrated that mechanical stress induced calcium propagation in cultured keratinocytes and this in turn activated co-cultured nerve cells [58]. Because the effect was blocked by application of an ATP-degrading enzyme, they concluded that ATP played a role in signal transmission from keratinocytes to the nervous system. Prostaglandin, E(2) may also act as a messenger from keratinocytes to the nervous system [59]. We showed that glutamate, dopamine and nitric oxide were released from keratinocytes immediately after insult of the stratum corneum [60, 61, 62]. These molecules might also be candidate messengers from keratinocytes to the nervous system. Because unmyelinated nerve fibers reach the uppermost layer of the epidermis [63], direct communication between keratinocytes and nerve fibers might occur in the epidermis. Potentially, a variety of communication modes between epidermal keratinocytes and the nervous system might operate in the epidermis as a part of the cutaneous sensation system.

Although the relationship between TRPs on keratinocytes and nerve systems has not clarified, studies of communication between keratinocytes and peripheral nerve fibers would be important for understanding the mechanisms of itching or sensitive skin.

44.7 Epidermal Keratinocytes as the Interface Between Body and Environment

The traditional image of the epidermis used to be that it is the uppermost layer of skin which manifests the water-impermeable barrier function. However, in the past decade, that picture has changed dramatically. As described above, several TRPs activated by various environmental factors are expressed in the epidermal keratinocytes [79, 38]. For example, we first discovered the expression of pain receptor TRPV1 in keratinocytes [3, 4]. Then, we found that another pain receptor, P2X3, which is activated by ATP [64], is also expressed in keratinocytes [68]. These receptors might be associated with pain or itching of the skin.

Epidermal permeability barrier homeostasis is influenced by temperature, and TRPV1, TRPV4, and TRPA1 are involved in this phenomenon [38, 41]. Not only temperature, but also environmental humidity, visible light radiation, and inaudible sound all influence barrier homeostasis [28, 54, 81]. TRPV4 is activated by osmotic pressure [40]. Thus, it might act as a humidity sensor in the epidermis. We have reported the expression of photoreceptor proteins (opsin and rhodopsin), which are expressed in the retina, in epidermal keratinocytes [65]. Thus, the epidermis could potentially recognize visible light or color. It was also shown that TRPA1 might play a role in acoustic sensation [66]. We recently demonstrated that 10–30 kHz sound influenced epidermal barrier homeostasis [67]. Thus, TRPA1 in epidermal keratinocytes might recognize sound. Indeed, a variety of environmental factors might be detected by epidermal keratinocytes.

Epidermal keratinocytes are derived from ectoderm during the early stage of development. The nervous system, including the brain, and all sensory systems are also ectoderm-derived organs. We have shown that a variety of neurotransmitter receptors, which are known to have a role in information processing in the brain, are also expressed in epidermal keratinocytes [68–71, 61, 62, 82]. Thus, we hypothesized that there might be an information processing function in the epidermis [70, 38].

Overall, these findings suggest that epidermal keratinocytes might have the ability to sense a variety of environmental factors, and this information from the

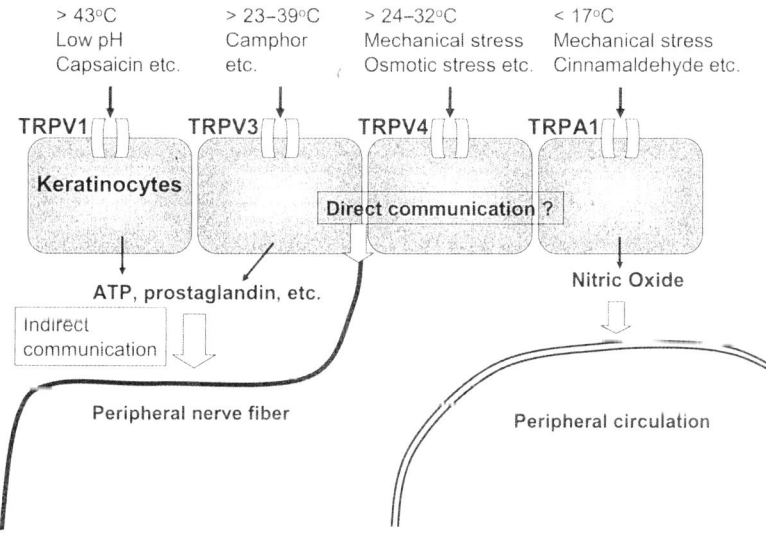

Fig. 44.3 Schematic illustration of the putative skin surface sensory system. Epidermal keratinocytes form the forefront of the sensory system of the skin. Various environmental factors are sensed by a range of TRP receptors, and signals are transferred directly or indirectly to peripheral nerve fibers or peripheral circulation

environment might be processed in the epidermis and passed to the nervous system. As mentioned above, epidermal keratinocytes generate and release a variety of neurotransmitters, including ATP, nitric oxide (NO), glutamate, and dopamine (Denda et al. 2002) [60–62, 68], which might interact with the peripheral nerve system or peripheral circulation. For example, we demonstrated that NO released from epidermal keratinocytes enlarged cutaneous lymphatic vessels and accelerated blood flow in the skin [72]. Other studies have suggested that epidermal keratinocytes produce β-endorphin, CGRP, and substance-P [73–75], which might influence whole body physiology or the nervous system (Fig. 44.3).

In summary, TRPs expressed in epidermal keratinocytes have been shown to have important roles in modulating skin structure and mediating sensation. They might be involved in whole body homeostasis in response to environmental stimuli. Consequently, the study of keratinocyte TRPs may have implications not only for treatment of skin diseases, but also for our basic understanding of human interaction with the environment.

References

1. Friel DD (1996) TRP: its role in phototransduction and store-operated Ca^{2+} entry. Cell 85:617–619
2. Caterina MJ, Schumacher MA, Tominaga M et al (2007) The capsaicin receptor: a heat-activated ion channel in the pain pathway. Nature 389:816–824
3. Denda M, Fuziwara, Inoue K et al (2001) Immunoreactivity of VR1 on epidermal keratinocyte of human skin. Biochem Biophys Res Commun 285:1250–1252
4. Inoue K, Koizumi S, Fuziwara S et al (2002) Functional vanilloid receptors in cultured normal human keratinocytes. Biochem Biophys Res Commun 291:124–129
5. Peier AM, Reeve AJ, Andersson DA, Moqrich A, Earley TJ, Hergarden AC, Story GM, Colley S, Hogenesch JB, McIntyre P, Bevan S, Patapoutian A (2002) A heat-sensitive TRP channel expressed in keratinocytes. Science 296:2046–2049
6. Chung MK, Lee H, Caterina MJ (2003) Warm temperatures activate TRPV4 in mouse 308 keratinocytes. J Biol Chem 278:32037–32046
7. Atoyan R, Shander D, Botchkarvera NV (2009) Non-neural expression of transient receptor potential type A1 (TRPA1) in human skin. J Invest Dermatol 129:2096–2099
8. Story GM, Peier AM, Reeve AJ et al (2003) ANKTM1, a TRP-like channel expressed in nociceptive neurons, is activated by cold temperatures. Cell 112:819–829
9. Tsutsumi M, Denda S, Ikeyama K et al (2010) Exposure to low temperature induces elevation of intracellular calcium in cultured human keratinocytes. J Invest Dermatol 130:1945–1948
10. Cai S, Fatherazi S, Presland RB et al (2005) TRPC channel expression during calcium-induced differentiation of human gingival keratinocytes. J Dermatol Sci 40:21–28
11. Beck B, Lehen'kyi V, Roudbaraki M et al (1998) TRPC channels determine human keratinocyte differentiation: new insight into basal cell carcinoma. Cell Calcium 43:492–505
12. Mauro T, Bench G, Sidderas-Haddad E et al (1998) Acute barrier perturbation abolishes the Ca^{2+} and K^+ gradients in murine epidermis: quantitative measurement using PIXE. J Invest Dermatol 111:1198–1201
13. Forslind B, Werner-Linde Y, Lindberg M et al (1999) Elemental analysis mirrors epidermal differentiation. Acta Derm Venereol 79:12–17
14. Elias PM, Ahn SK, Denda M et al (2002) Modulations in epidermal calcium regulate the expression of differentiation-specific proteins. J Invest Dermatol 119:1128–1136

15. Watt FM (1989) Terminal differentiation of epidermal keratinocytes. Curr Opin Cell Biol 1:1107–1115
16. Nemes Z, Steinert PM (1999) Bricks and mortar of the epidermal barrier. Exp Mol Med 31:5–19
17. Watanabe R, Wu K, Paul P et al (1998) Up-regulation of glucosylceramide synthase expression and activity during human keratinocyte differentiation. J Biol Chem 273:9651–9655
18. Menon GK, Price LF, Bommannan B et al (1994) Selective obliteration of the epidermal calcium gradient leads to enhanced lamellar body secretion. J Invest Dermatol 102: 789–795
19. Cai S, Fatherazi S, Presland RB et al (2006) Evidence that TRPC1 contributes to calcium-induced differentiation of human keratinocytes. Pflugers Arch 452:43–52
20. Tu CL, Chang W, Bikle DD (2005) Phospholipase cgamma1 is required for activation of store-operated channels in human keratinocytes. J Invest Dermatol 124:187–197
21. Müller M, Essin K, Hill K et al (2008) Specific, TRPC6 channel activation, a novel approach to stimulate keratinocyte differentiation. J Biol Chem 283:33942–33954
22. Woelfle U, Laszczyk MN, Kraus M et al (2010) Triterpenes promote keratinocyte differentiation in vitro, *ex vivo* and in vivo: a role for the transient receptor potential canonical (subtype) 6. J Invest Dermatol 130:113–123
23. Huyke C, Reuter J, Rödig M et al (2009) Treatment of actinic keratoses with a novel betulin-based oleogel. A prospective randomized comparative pilot study. J Dtsch Dermatol Ges 7:128–133
24. Lehen'kyi V, Beck B, Polakowska R et al (2007) TRPV6 is a Ca^{2+} entry channel essential for Ca^{2+}-induced differentiation of human keratinocytes. J Biol Chem 282:22582–22591
25. Bianco SD, Peng JB, Takanaga H, Suzuki Y, Crescenzi A, Kos CH, Zhuang L, Freeman MR, Gouveia CH, Wu J, Luo H, Mauro T, Brown EM, Hediger MA (2007) Marked disturbance of calcium homeostasis in mice with targeted disruption of the Trpv6 calcium channel gene. J Bone Miner Res 22:274–285
26. Southall MD, Li T, Gharibova LS et al (2003) Activation of epidermal vanilloid receptor-1 induces release of proinflammatory mediators in human keratinocytes. J Pharmacol Exp Ther 304:217–222
27. Li WH, Lee YM, Kim JY et al (2007) Transient receptor potential vanilloid-1 mediates heat-shock-induced matrix metalloproteinase-1 expression in human epidermal keratinocytes. J Invest Dermatol 127:2328–2335
28. Denda S, Denda M, Inoue K et al (2010) Glycolic acid induces keratinocyte proliferation in a skin equivalent model via TRPV1 activation. J Dermatol Sci 57:108–113
29. Lee YM, Kim YK, Chung JH (2009a) Increased expression of TRPV1 channel in intrinsically aged and photoaged human skin in vivo. Exp Dermatol 18:431–436
30. Lee YM, Kim YK, Kim KH et al (2009b) A novel role for the TRPV1 channel in UV-induced matrix metalloproteinase (MMP)-1 expression in HaCaT cells. J Cell Physiol 219: 766–775
31. Wilder-Smith EP, Ong WY, Guo Y et al (2007) Epidermal transient receptor potential vanilloid 1 in idiopathic small nerve fibre disease, diabetic neuropathy and healthy human subjects. Histopathology 51:674–680
32. Yoshioka T, Imura K, Asakawa M et al (2009) Impact of the Gly573Ser substitution in TRPV3 on the development of allergic and pruritic dermatitis in mice. J Invest Dermatol 129: 714–722
33. Asakawa M, Yoshioka T, Matsutani T et al (2006) Association of a mutation in TRPV3 with defective hair growth in rodents. J Invest Dermatol 126:2664–2672
34. Katsuta Y, Iida T, Inomata S et al (2005) Unsaturated fatty acids induce calcium influx into keratinocytes and cause abnormal differentiation of epidermis. J Invest Dermatol 124: 1008–1013
35. Hu HZ, Xiao R, Wang C et al (2006) Potentiation of TRPV3 channel function by unsaturated fatty acids. J Cell Physiol 208:201–212

36. Pani B, Cornatzer E, Cornatzer W et al (2006) Up-regulation of transient receptor potential canonical 1 (TRPC1) following sarco(endo)plasmic reticulum Ca^{2+} ATPase 2 gene silencing promotes cell survival: a potential role for TRPC1 in Darier's disease. Mol Biol Cell 17: 4446–4458
37. Elias PM, Feingold KR (2001) Coordinate regulation of epidermal differentiation and barrier homeostasis. Skin Pharmacol Appl Skin Physiol 14(Suppl 1):28–34
38. Denda M, Sokabe T, Tominaga T (2007) Effects of skin surface temperature on epidermal permeability barrier homeostasis. J Invest Dermatol 127:654–659
39. Grubauer G, Elias PM, Feingold KR (1989) Transepidermal water loss: the signal for recovery of barrier structure and function. J Lipid Res 30:323–333
40. Liedtke W (2007) Role of TRPV ion channels in sensory transduction of osmotic stimuli in mammals. Exp Physiol 92:507–512
41. Denda M, Tsutsumi M, Goto M et al (2010) Topical application of TRPA1 agonists and brief cold exposure accelerate skin permeability barrier recovery. J Invest Dermatol 130:1942–1945
42. Bandell M, Story GM, Hwang SW et al (2004) Noxious cold ion channel TRPA1 is activated by pungent compounds and bradykinin. Neuron 41:849–857
43. Grice KA (1980) Transepidermal water loss in pathologic skin. In: Jarrett A (ed) The, Physiology and Pathophysiology of the Skin. Academic Press, London, UK, pp 2147–2155
44. Denda M, Kitamura K, Elias PM, Feingold KR (1997) Trans-4-(aminomethyl)cyclohexane carboxylic acid (t-AMCHA), an anti-fibrinolytic agent, accelerates barrier recovery and prevents the epidermal hyperplasia induced by epidermal injury in hairless mice and humans. J Invest Dermatol 109:84–90
45. Fuziwara S, Ogawa K, Aso D et al (2004) Barium sulfate with a negative ζ potential accelerates skin permeable barrier recovery and prevents epidermal hyperplasia indueced by barrier disruption. Br J Dermarol 151:557–564
46. Bodó E, Bíró T, Telek A et al (2005) A hot new twist to hair biology: involvement of vanilloid receptor-1 (VR1/TRPV1) signaling in human hair growth control. Am J Pathol 166: 985–998
47. Bíró T, Bodó E, Telek A et al (2006) Hair cycle control by vanilloid receptor-1 (TRPV1): evidence from TRPV1 knockout mice. J Invest Dermatol 126:1909–1912
48. Imura K, Yoshioka T, Hikita I et al (2007) Influence of TRPV3 mutation on hair growth cycle in mice. Biochem Biophys Res Commun 363:479–483
49. Tóth BI, Géczy T, Griger Z et al (2009) Transient receptor potential vanilloid-1 signaling as a regulator of human sebocyte biology. J Invest Dermatol 129:329–339
50. Tominaga M, Caterina MJ, Malmberg AB et al (1998) The cloned capsaicin receptor integrates multiple pain-producing stimuli. Neuron 21:531–543
51. Xu H, Delling M, Jun JC et al (2006) Oregano, thyme and clove-derived flavors and skin sensitizers activate specific TRP channels. Nat Neurosci 9:628–635
52. Moqrich A, Hwang SW, Earley TJ et al (2005) Impaired thermosensation in mice lacking TRPV3, a heat and camphor sensor in the skin. Science 307:1468–1472
53. Becker D, Blase C, Bereiter-Hahn J et al (2005) TRPV4 exhibits a functional role in cell-volume regulation. J Cell Sci 118(Pt 11):2435–2440
54. Denda M, Sato J, Masuda Y et al (1998) Exposure to a dry environment enhances epidermal permeability barrier function. J Invest Dermatol 111:858–863
55. Goto M, Ikeyama, Tsutsumi KM et al (2010) Calcium ion propagation in cultured keratinocytes and other cells in skin in response to hydraulic pressure stimulation. J Cell Physiol 224:229–233
56. Liedtke W (2008) Molecular mechanisms of TRPV4-mediated neural signaling. Ann NY Acad Sci 1144:42–52
57. Story GM, Gereau RW 4th (2006) Numbing the senses: role of TRPA1 in mechanical and cold sensation. Neuron 50:177–180

58. Koizumi S, Fijishita K, Inoue K et al (2004) Ca^{2+} waves in keratinocytes are transmitted to sensory neurons: the involvement of extracellular ATP and P2Y2 receptor activation. Biochem J 380:329–338
59. Huang SM, Lee H, Chung MK et al (2008) Overexpressed transient receptor potential vanilloid 3 ion channels in skin keratinocytes modulate pain sensitivity via prostaglandin E2. J Neurosci 28:13727–13737
60. Ikeyama K, Fuziwara S, Denda M (1719) Topical application of neuronal nitric oxide synthase inhibitor accelerates cutaneous barrier recovery and prevents epidermal hyperplasia induced by barrier disruption. J Invest Dermatol 2007:127:1713–1719
61. Fuziwara S, Inoue K, Denda M (2003) NMDA-type glutamate receptor is associated with cutaneous barrier homeostasis. J Invest Dermatol 120:1023–1029
62. Fuziwara S, Suzuki A, Inoue K et al (2005) Dopamine, D2-like receptor agonists accelerate barrier repair and inhibit the epidermal hyperplasia induced by barrier disruption. J Invest Dermatol 125:783–789
63. Navarro X, Verdu E, Wendelschafer-Crabb G et al (1995) Innervation of cutaneous structures in the mouse hind paw: a confocal microscopy immmunohistochemical study. J Neurosci Res 41:111–120
64. Cockayne DA, Hamilton SG, Zhu QM et al (2000) Urinary bladder hyporeflexia and reduced pain-related behavior in P2X3-deficient mice. Nature 407:1011–1015
65. Tsutsumi M, Ikeyama K, Denda S et al (2009) Expressions of rod and cone photoreceptor-like proteins in human epidermis. Exp Dermatol 18:567–570
66. Corey DP, García-Añoveros J, Holt JR et al (2004) TRPA1 is a candidate for the mechanosensitive transduction channel of vertebrate hair cells. Nature 432:723–730
67. Denda M, Nakatani M (2010) Acceleration of permeability barrier recovery by exposure of skin to 10–30 kilohertz sound. Br J Dermatol 162:503–507
68. Denda M, Inoue K, Fuziwara S et al (2002a) P2X purinergic receptor antagonist accelerates skin barrier repair and prevents epidermal hyperplasia induced by skin barrier disruption. J Invest Dermatol 119:1034–1040
69. Denda M, Inoue K, Inomata S et al (2002b) GABA (A) receptor agonists accelerate cutaneous barrier recovery and prevent epidermal hyperplasia induced by barrier disruption. J Invest Dermatol 119:1041–1047
70. Denda M (2003a) Epidermis of the skin as a self-organizing electrochemical sensor. In: Nakata S (eds) Chemical, Analysis Based on Nonlinearity. NOVA, New York, NY, pp 132–138
71. Denda M, Fuziwara S, Inoue K (2003b) Beta-2-adrenergic receptor antagonist accelerates skin barrier recovery and reduces epidermal hyperplasia induced by barrier disruption. J Invest Dermatol 121:142–148
72. Ikeyama K, Denda S, Tsutsumi M et al (2010) Neuronal nitric oxide synthase in epidermis is involved in cutaneous circulatory response to mechanical stimulation. J Invest Dermatol 130:1158–1166
73. Zanello SB, Jackson DM, Holick MF (1999) An immunocytochemical approach to the study of beta-endorphin production in human keratinocytes using confocal microscopy. Ann N Y Acad Sci 885:85–99
74. Seike M, Ikeda M, Morimoto A et al (2002) Increased synthesis of calcitonin gene-related peptide stimulates keratinocyte proliferation in murine UVB-irradiated skin. J Dermatol Sci 28:135–143
75. Bae S, Matsunaga Y, Tanaka Y et al (1999) Autocrine induction of substance P mRNA and peptide in cultured normal human keratinocytes. Biochem Biophys Res Commun 263:327–333
76. Chung MK, Lee H, Mizuno A et al (2004) TRPV3 and TRPV4 mediate warmth-evoked currents in primary mouse keratinocytes. J Biol Chem 279:21569–21575
77. Denda M (2000a) Influence of dry environment on epidermal function. J Dermatol Sci 24(Suppl 1):S22–S28
78. Denda M, Hosoi J, Ashida Y (2000b) Visual imaging of ion distribution in human epidermis. Biochem Biophys Res Commun 272:134–137

79. Denda M, Fuziwara S, Inoue K (2003c) Influx of calcium and chloride ions into epidermal keratinocytes regulates exocytosis of epidermal lamellar bodies and skin permeability barrier homeostasis. J Invest Dermatol 121:362–367
80. Denda M, Nakatani M, Ikeyama K et al (2007a) Epidermal keratinocytes as the forefront of the sensory system. Exp Dermatol 16:157–161
81. Denda M, Fuziwara S (2008) Visible radiation affects epidermal permeability barrier recovery: selective effects of red and blue light. J Invest Dermatol 128:1335–1336
82. Denda M, Fuziwara S, Inoue K (2004) Association of cyclic adenosine monophosphate with permeability barrier homeostasis of murine skin. J Invest Dermatol 122:140–146
83. Denda M, Tsutsumi M, Denda S (2010) Topical application of TRPM8 agonists accelerates skin permeability barrier recovery and reduces epidermal proliferation induced by barrier insult: Therde of cold-sensitive TRP receptors in epidermal permeability barrier homeostasis. Exp Dermatol 19:791–795

Chapter 45
TRP Channels in Urinary Bladder Mechanosensation

Isao Araki

45.1 Introduction

The lower urinary tract has two main functions: storage and periodic elimination of urine [1]. These functions are controlled by complex neural circuits of reflex pathways located in the brain, the spinal cord and the periphery. The lower urinary tract constantly sends mechanosensory information to the central nervous system via afferent pathway. These signals generate sensation and trigger voiding responses. Alterations in afferent activity may lead to lower urinary tract dysfunction. The increase of afferent excitability is one of mechanisms for overactive bladder syndrome and for painful bladder syndrome [1, 2]. However, the precise mechanisms by which mechanical stimuli excite bladder afferents remain unclear.

For mechanosensory transduction in the physiological condition, the presence of mechanosensitive molecules may be basically essential in the peripheral sensory systems, although other receptor molecules could modulate mechanosensory transduction especially in pathological conditions [3]. For many years, it has been postulated that the core components of mechanosensors are specific ion channels that could convert mechanical energy rapidly into an electric signal, in contrast to sensory receptors for odors and most tastes [4]. Thus, mechanosensitive ion channels in the peripheral sensory systems have attracted particular interests as candidates for mechanosensor in mechanosensory transduction systems.

The transient receptor potential (TRP) superfamily of cation channels, comprising seven subfamilies on the basis of sequence similarity, is an emerging group of channel proteins implicated in a wide variety of mechanical transduction processes in diverse organs [5, 6]. In the urinary bladder, there is an increasing body

---

I. Araki (✉)
Department of Urology, Interdisciplinary Graduate School of Medicine and Engineering,
University of Yamanashi, Chuo, Yamanashi 409-3898, Japan; Department of Urology, Shiga
University of Medical Science, Otsu, Shiga 520-2192, Japan
e-mails: iaraki@yamanashi.ac.jp; iaraki@belle.shiga-med.ac.jp

of evidence for implication of TRP channels in bladder function and disease [7–9]. This article reviews the possible roles of TRP channel family in mechanosensation of the urinary bladder.

45.2 Mechanosensory Machinery in the Urinary Bladder

The mechanisms underlying the activation of bladder afferent pathway are not completely understood. The peripheral sensory systems of the urinary bladder include sensory nerve endings, urothelial cells and others whose locations are suitable for perceiving mechanical and chemical stimuli.

Two main types of mechanisms have been proposed to operate during mechanotransduction in afferent systems: indirect chemical and direct physical transduction mechanisms [6, 10, 11]. An indirect, chemical transduction mechanism relies on activation of afferents by mediators released from non-neuronal cells by mechanical stimulation [10]. A direct, physical transduction is due to direct activation of mechano-gated ion channels in the afferent nerve endings without involvement of extracellular mediators [6, 11].

45.2.1 Sensory Nerve Endings

Sensations from the bladder are conveyed by pelvic and hypogastric nerve afferent fibers [1]. The pelvic nerves are the principal pathway for afferent input related to micturition. The bladder afferent fibers are thinly myelinated or unmyelinated (Aδ and C fibers). In the cat, it has been proposed that Aδ afferent fibers are mechanosensitive and involved in physiological micturition reflex, whereas C-fiber afferents are mechanically insensitive and participate in nociception under painful pathological conditions (silent nociceptors) [12, 13]. However, this scheme in the cat can not be generalized in different species. In the rat, both Aδ and C fibers respond to bladder distension [14, 15]. Among stretch-sensitive bladder afferents, low threshold and high threshold afferents have been identified in in vivo and in vitro preparations. Low threshold fibers are considered to be involved in physiological control of micturition, while high threshold afferents are associated with painful sensations. Both Aδ and C fibers are included in both low and high threshold types. There are no relationship between the conduction velocities of individual mechanoreceptors and their response thresholds [14, 15].

Afferent fibers are abundant within the muscle and in the suburothelial layers [16]. Recent experiments in vitro have identified several distinct functional classes of bladder afferents in the rat and mouse [14, 15, 17]. In addition to the muscle afferents including Aδ and C fibers, the suburothelial nerve plexus primarily comprising C fibers could be mechanosensitive even in physiological condition. Zagorodnyuk et al. [18] showed that the guinea pig bladder is innervated by at least four classes of extrinsic sensory neurons identified by the location of their receptive fields (mucosal and muscle afferents), their function (mechanoreceptors, chemoreceptors, nociceptors) and the magnitude of their responses (low and high responders). These afferents include stretch-sensitive 1) muscle mechano-afferents and 2) muscle-mucosal mechano-afferents, and stretch-insensitive 3) mucosal mechano-afferents and 4) mucosal chemo-afferents. Removal of the mucosa (urothelium and lamina propria) reduces the stretch-induced firing of muscle-mucosal mechanoreceptors, but not that of muscle mechanoreceptors in the guinea-pig [18]. Further, stretch-evoked firing of bladder afferents in the rat and guinea-pig is not affected by the non-selective P2 purinoreceptor antagonists and Ca^{2+}-free solution for blocking synaptic transmission [19, 20]. These results suggest that mechanosensitivity arises by a direct physical mechanism at the nerve endings rather than by a chemical mediator released from the urothelium. The density of suburothelial presumptive sensory nerves in the bladder wall is increased in women with idiopathic detrusor overactivity, compared with asymptomatic women [21]. Mechanosensory information of the bladder could be conveyed by direct activation of mechanosensitive channels in sensory nerve endings in response to mechanical stimuli. In dorsal root ganglia (DRG), sensory neurons express a variety of receptors and channels including mechanosensitive channels belonging to TRP channel family [22, 23]. At present, however, only a part of these mechanosensitive channels have been demonstrated to be located in the suburothelial nerve plexus (see below) (Table 45.1 and Fig. 45.1).

Immunocytochemical studies have revealed that numerous peptides, including substance P, neurokinines, calcitonin gene-related peptide (CGRP), vasoactive intestinal polypeptide, enkephalins and cholecystokinin are localized in bladder afferents [24]. The release of these peptides from sensory nerve endings may be involved in local regulation of sensory nerve excitability, transmitter release from urothelial cells and muscle cell activity [25–27]. Activation of TRP channels in capsaicin-sensitive sensory terminals induces the contraction of isolated bladder

Table 45.1 TRP channels and mechanosensitive ion channels related to bladder mechanosensory transduction

	Mechano-/thermo-/chemo-sensitivity	Localization in the bladder	Agonist/antagonist (commercially available)
TRPV1	−/+/+	DRG (Sen), Ur, SM? ICC?	Capsaicin, resiniferatoxin/capsazepine
TRPV2	+/+ (noxious)/−	DRG (Sen?), Ur, SM? ICC?	Tetrahydrocannabinol
TRPV4	+/+/+	DRG, Ur, SM?	Phorbol esters (4αPDD)
TRPM8	−/+/+	DRG (Sen), Ur	Menthol, icilin
TRPA1	+/+?/+	DRG (Sen), Ur, SM?	Mustard oil, cinnamaldehype
ENaC	+/+/−	DRG (Sen), Ur	/Amiloride, benzamil
ASIC	+?/+/+	DRG (Sen), Ur, SM	/(Amiloride)
TREK-1	+/+/+	DRG, Ur, SM	/L-methionine, methioninol

DRG, dorsal root ganglia; Sen, bladder sensory endings; Ur, urothelium; SM, detrusor smooth muscle; ICC, interstitial cells of Cajal.

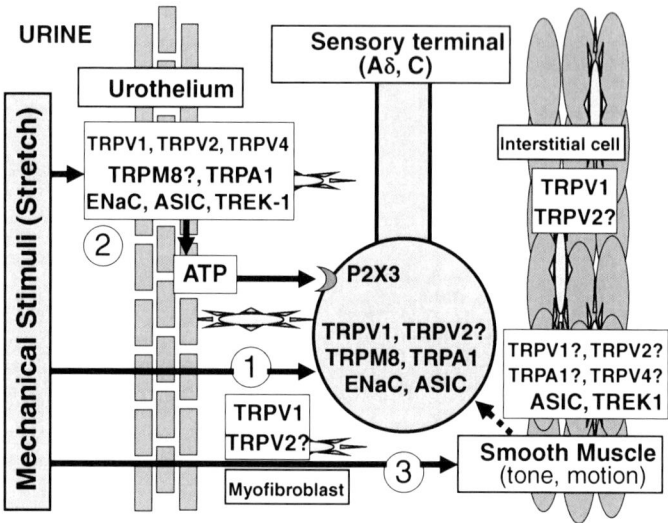

Fig. 45.1 Candidates for mechanosensor in the urinary bladder. Bladder filling with urine stretches the bladder wall. This mechanosensory information is conveyed to the central nervous system via the bladder afferent excitation. Several mechanisms have been proposed to operate during mechanotransduction in the bladder afferent systems. TRP channels, especially mechanosensitive TRP channels, are possibly implicated in these mechanosensory transduction mechanisms. (*1*) A direct activation of mechanosensitive TRP channels at afferent nerve endings. (*2*) An indirect excitation of bladder afferents by chemical mediators, such as ATP, released from non-neuronal cells (e.g. urothelial cells) via activation of mechanosensitive TRP channels. Alternatively, these mediators may play a crucial role in setting the afferent excitability and pathophysiology of the bladder function. (*3*) Tone and spontaneous focal contraction (micromotion) of smooth muscle are influenced by bladder wall stretch possibly via activation of mechanosensitive TRP channels. These muscular activities may regulate the basal level of afferent excitability rather than excite afferents directly. Interstitial cells of Cajal in the suburothelium and detrusor muscle might be implicated in mechanosensory transduction mechanisms as an amplifier or a modulator. Spontaneous or chemically-induced neuromediator releases from efferent or afferent nerve terminals could modulate the afferent excitability

strip via neuropeptide release from C-fiber terminals [28]. In women with idiopathic detrusor overactivity, the densities of CGRP- and substance P-containing nerves are increased in the suburothelial nerve plexus [24].

45.2.2 Urothelium

Recently, the bladder epithelium has been shown to play an important role in mechanosensory transduction [29]. In response to mechanical and chemical stimuli, various neuromediators, such as adenosine triphosphate (ATP), acetylcholine, nitric oxide, prostaglandins and nerve growth factor, are released from urothelial cells [30–32]. ATP is abundant in the cell cytoplasm and can be released extracellularly by several mechanisms including vesicular exocytosis, transporters such as a member of ATP-binding cassette transporter superfamily, or anion-selective channels such as maxi-anion channel [33]. $P2X_3$ is expressed at suburothelial nerve plexus, and exogenous ATP activates several types of bladder afferents and sensitizes their mechanosensory responses [19, 34, 35]. Purinergic antagonists reduce distension-induced firing of bladder afferents [34, 35]. Transgenic mice ($P2X_3$, $P2X_2$ and $P2X_2/P2X_3$) show reduced urinary bladder reflexes and decreased pelvic afferent firing in response to bladder distension [35, 36]. These findings indicate that ATP released from bladder epithelial cells in response to distention acts on $P2X_3$ and $P2X_{2/3}$ receptors located in the subepithelial afferent nerve plexus [36, 37]. In addition, there are suggestions that receptors for acetylcholine and prostaglandins are present in bladder afferent endings, although these studies were primarily based on indirect in vivo experiments [38, 39]. Muscarinic receptors (M2 and M3) have been supposed to be involved in sensory transduction [40], however those presence is still ambiguous in the suburothelial nerve endings. M2 and M3 immunoreactivities have been shown in the suburothelial nerve bundle [41], while this nerve bundle was located near the detrusor layer and its sensory origin was not defined. However, the reliability of antisera against muscarinic receptors has been recently questioned [42, 43].

It has been shown that urothelial cells express various receptors and channels including receptors for bradykinin (B1 and B2), prostaglandin (EP1 and EP3), substance P (NK1), CGRP, neurotrophins (TrkA), purines (P2X and P2Y), noradrenalin (α1 and β3), acetylcholine (muscarinic and nicotinic), protease-activated receptors, pituitary adenylate cyclase-activating peptide and several members of TRP channel and degenerin/epithelial Na^+ channel (ENaC) families (see below; Table 45.1 and Fig. 45.1) [44, 45]. Activation of these receptors and channels could leads to ATP release from the urothelial cells [46–49]. It has been reported that stretch-evoked ATP release from the urothelium is reduced by blocking α1 adrenoreceptor [50, 51], muscarinic receptors (M2 and M3) [52] or TRP or degenerin/ENaC channel family (see below).

45.2.3 Detrusor Smooth Muscle and Interstitial Cells of Cajal

Smooth muscles in the bladder show spontaneous contractile activity during the storage phase [53]. These contractions, so called "micromotions", are localized and can be multifocal in separate areas of bladder wall. It is still unclear what exactly underlies this spontaneous activity, although the involvements of interstitial cells [54, 55], intramural ganglia [54, 56], gap junction [57] and prostanoids [55] are under consideration. There is spontaneous and TTX-resistant release of acetylcholine from autonomic nerve endings that affects bladder tone and contractility [56]. The autonomous contractile activity in the isolated bladder increases as the bladder is filled [58]. Furthermore, it has been suggested that the stretches resulting from local phasic activity of bladder wall generate bladder afferent discharges [59, 60]. Thus, spontaneous focal contraction of the detrusor muscle could be implicated in the bladder afferent transduction.

The bladder has interstitial cells in the detrusor (interstitial cells of Cajal: ICC), although a human study could not distinguish such cells from fibroblasts [61]. Their functional role in the detrusor is still unexplained, but similar cells generate pacemaker activity responsible for phasic or tonic muscular contraction in the gut [62] and possibly in the urethra [63]. In the detrusor, only a small percentage of ICC show spontaneous Ca^{2+} transients [64], and the frequency and duration of these transients are quite different from those in the smooth muscle [57]. Thus, ICC may modulate the spread of action potentials along the muscle bundles rather than being the pacemaker of spontaneous activity.

It has been proposed that myofibroblasts beneath urothelium (ICC in the lamina propria: ICC-LP) act as a relay or an amplifier in the sensory response to bladder-wall stretch [65]. Isolated myofibroblasts display spontaneous transients of membrane potential and intracellular Ca^{2+} [66], and response to exogenous application of ATP via P2Y receptor [67]. Additionally, muscarinic receptors (M2 and M3) are expressed in presumptive myofibroblasts [41, 68], although application of the cholinergic agonist, carbachol failed to evoke a response [66]. Suburothelial myofibroblasts have rich connexin 43 gap junctions between adjacent cells [69] and show a close apposition to unmyelinated nerve fibers [70]. These findings suggest that myofibroblasts link the urothelial ATP signaling to afferent excitation [71]. However, this possibility is unwarranted at present because a functional role of bladder myofibroblasts has not yet been demonstrated.

45.3 TRP Channels in the Urinary Bladder

45.3.1 TRPV1

TRPV1 is primarily expressed in small to medium-sized primary sensory neurons, the majority of which synthesize neuropeptides such as substance P and CGRP [72, 73]. In the mammalian urinary bladder, the structures in which TRPV1 (vanilloid receptor type 1: VR-1) is expressed include the bladder sensory fibers, urothelial cells, ICC (myofibroblasts) and probably smooth muscle (Table 45.1 and Fig.45.1) [72–76]. Before TRPV2 is identified as a separate channel, VRL-1 (TRPV2) was thought to be a variant of VR-1 (TRPV1). Thus, a part of old reports need great care for interpretation and to be revisited (ex., ref. #75). Furthermore, evidence of TRPV1 expression in non-neuronal cells has been recently questioned with the observation of nonspecific cellular TRPV1-immunoreactivity in bladders from TRPV1 knockout mice [77]. TRPV1 regulates pain perception and bladder reflex by modulating sensory activity. Agonists of TRPV1, capsaicin and resiniferatoxin, have been used for therapeutic purposes of bladder dysfunction [78].

TRPV1-immunoreactive nerve fibers form varicose plexuses in the subepithelial layer and the surface of the smooth muscle of the bladder wall [79, 80]. A role of TRPV1 is well established in nociception as an integrator for thermal and chemical noxious stimuli [81–83], although acidic stimuli are capable of sensitizing TRPV1-independent mechanisms of bladder sensation [31, 84]. Activation of

TRPV1 in peripheral nerve endings promotes the depolarization and the release of neuropeptides, such as substance P and CGRP [85, 86]. Capsazepine, a TRPV1 antagonist, decreases the frequency of reflex voiding in cyclophosphamide inflamed rat bladder [82]. Cyclophospamide- or acrolein-induced cystitis leads to bladder mechanical hyperactivity in wild-type mice, but not in TRPV1 KO mice [87]. After bladder inflammation with lipopolysaccharide, the frequency of reflex voiding and the number of *c-fos* expressing spinal neurons by innocuous bladder distension increase in mice. These changes in the inflamed bladder are suppressed by a potent TRPV1 antagonist and not observed in the TRPV1-deficient mice [83, 88]. Recently a growing body of evidence has led to the emergence of TRPV1 as a prominent participator in normal sensory transduction. Compared with wild-type, mice lacking TRPV1 has been shown to increase the frequency of non-voiding bladder contraction and enhance bladder capacity under anesthesia [31]. In TRPV1-deficient mice, sensitivity of low threshold bladder afferents to distension is reduced [89] and spinal *c-fos* response to distension is abolished [31]. These results suggest that TRPV1 is implicated in mechanosensitivity of the bladder. However, TRPV1 is not considered to be mechanically gated and a TRPV1 antagonist, capsazepine, had no effect on bladder reflex activity of normal mice [82]. Stroking-induced activity of mucosal high-responding afferents in the guinea-pig is not influenced by capsazepine in vitro [20].

TRPV1 may be also functional in the urothelium. In excised bladder strips and cultured urothelial cells from mice lacking TRPV1, hypoosmolality-evoked ATP and NO releases are diminished [31]. Calcium influx and ATP release increase in human urothelial cells when exposed to vanilloid compounds [90]. However, recent studies questioned the functional expression of TRPV1 in the mouse and guinea-pig urothelial cells [91–94]. Capsaicin induced neither Ca^{2+} influx nor current in cultured urothelial cells [91–93].

The functional meaning of TRPV1 in ICC and smooth muscle is completely unclear at the moment, even if the immunolabelling is specific [77].

45.3.2 TRPM8

TRPM8, a cool receptor, expressed in the urothelium and suburothelial sensory fiber might be implicated in the bladder cooling reflex (Table 45.1 and Fig. 45.1) [95–97]. Intravesical infusion of menthol, a widely used TRPM8 activator, facilitates the micturition reflex in conscious rats [98]. Intravenous administration of a TRPM8 channel blocker decreases the frequency of isovolumetric bladder contractions without reducing the contraction amplitude in anesthetized rats [99]. These findings suggest that TRPM8 play a role in the bladder afferent pathways. The number of TRPM8-immunoreactive suburothelial nerve is increased in patients with idiopathic detrusor overactivity, and the relative density of TRPM8-immunoreactive nerve fibers significantly correlates with symptoms of overactive bladder and painful bladder syndrome [97]. However, there is a report showing that TRPM8-immunoreactivity is detected in only a small proportion of the DRG sensory neurons innervating the urinary bladder (1.2%) [100]. Furthermore, TRPM8-deficient mice still have menthol-sensitive neuron [101, 102] and menthol may cause TRPM8-independent release of Ca^{2+} from intracellular stores [103]. Also, in the rat and mouse urothelium, the expression level of TRPM8 is much lower than those of other TRP channels [93, 94, 104]. The expression of TRPM8 mRNA is not changed in the human bladder mucosa with bladder outlet obstruction and detrusor overactivity [105]. TRPM8 does not show any mechanosensitivity originally and its endogenous activators await further studies.

45.3.3 TRPA1

Recently, it was reported that the mechanosensitive channels belonging to TRP channel family, TRPA1, are expressed in the rodent and human bladders [104–107]. TRPA1 is a potential candidate for mechanosensor and/or nociceptor responding to chemical and thermal stimuli [5, 6]. TRPA1 shares similar channel properties with the mechanosensitive channels of the hair cells [104]. In the DRG and trigeminal ganglia, TRPA1 is expressed in small sensory neurons (C or Aδ) with nociceptive markers, CGRP and substance P, and coexpressed with TRPV1 [108]. TRPA1 agonist causes nociceptive reaction and heat hyperalgesia in the skin [109]. TRPA1 knockout mice displayed behavioral deficits in response to mechanical, cold and chemical stimuli [110], while a conflicting observation was made in another independently developed knockout mice [111]. TRPA1 has been reported to be expressed in the epithelium, suburothelial sensory C-fibers and probably muscle layer of the urinary bladder (Table 45.1 and Fig.45.1) [104–107]. Nagata et al. [105] first reported that TRPA1 is expressed at sensory nerve terminals beneath the mouse urinary bladder mucosa. TRPA1 agonists cause a concentration-dependent contraction of isolated rat bladder strips through sensory fiber stimulation [112]. Intravesical infusion of TRPA1 agonists causes bladder hyper-reflexia through C-fiber afferent pathway during continuous infusion cystometrograms in rats [106, 107]. In the patient with bladder outlet obstruction, the expression of TRPA1 is upregulated in the bladder mucosa [104]. Our preliminary study indicates that TRPA1 is the most abundant TRP channels expressed in the human bladder mucosa (unpublished data). Thus, TRPA1 in the bladder epithelia and/or sensory nerve endings might be involved in the bladder sensory transduction and the induction process of overactive bladder by bladder outlet obstruction.

45.3.4 TRPV4

TRPV4 is originally postulated to serve as a mechano- or osmosensor [5, 6]. Recent studies using mice lacking TRPV4 revealed the involvement of TRPV4 in sensing mechanical pressure, osmolality, and warmth in vivo [113, 114]. TRPV4 is abundantly expressed in the rodent and human bladder epithelium [115, 116] and the most abundant TRP channel expressed in the mouse cultured urothelial cells [93, 94]. TRPV4 knockout mice manifest an incontinence phenotype in spontaneous voiding pattern and exhibit a lower frequency of voiding contraction in continuous filling cystometry under anesthesia [115, 117]. Intravesical infusion of a TRPV4 agonist induces bladder overactivity in the conscious mice, an effect that is undetected in TRPV4-deficient mice [117]. In rodent cultured urothelial cells, TRPV4 agonist promotes Ca^{2+} influx and current [91–94], and enhances ATP release [116]. In cultured urothelial cells from TRPV4-deficient mice, stretch-evoked Ca^{2+} influx and ATP release decrease [94]. These findings indicate a critical role of TRPV4 in physiological bladder function and mechanosensory transduction.

A study showed the expression of TRPV4 channel in the mouse bladder smooth muscle and an agonist-induced contraction of bladder strips [117], whereas other did not observe the alteration of rat bladder strip contractility by a widely-used agonist, 4α-phorbol 12,13-didecanoate [116].

45.3.5 TRPV2 and Other Mechanosensitive TRP Channels

TRPV2 mediates hypotonic swelling and stretch-induced Ca^{2+} influx [118]. TRPV2 has been reported to express in the urothelium, smooth muscle and suburothelial sensory nerve endings in the rodent bladder [74, 75, 93] and urothelium in the human bladder [119]. Before TRPV2 is identified as a separate channel, VRL-1 (TRPV2) was thought to be a variant of VR-1 (TRPV1). Thus, a great care is needed for interpretation of old reports [74, 75]. A known agonist for TRPV2, tetrahydrocannabinol, induced Ca^{2+} influx and current in mouse urothelial cells [93]. A role of TRPV2 in bladder function is yet to be explored.

TRPC1, 5, 6, TRPM3, 4, 7, TRPP1, 2 and TRPML3 are proposed to have mechanosensitivity in mammals [120], but their expressions in the lower urinary tract have not been reported so far.

45.4 Other Mechanosensitive Channels in the Urinary Bladder

45.4.1 Degenerin/Epithelial Na^+ Channel Family

The degenerin/ENaC family represents a new class of cation channels that was discovered at the early 1990s [22, 121, 122]. This cationic channel family is characterized by amiloride-sensitivity, and is either constitutively active or activated by mechanical stimuli, and/or by ligands such as peptides or protons. Recent studies on these channels have implicated them in various mechanosensory modalities, such as baroreceptors and cutaneous sensory structures [123–125]. It has been reported that the ENaC is expressed in the mammalian urinary tract epithelia (renal pelvis, ureter and urinary bladder) [126–129] and suburothelial nerve fibers (unpublished data). In rabbit bladder epithelium, ENaC is mechano-sensitive, having the ability to change their sodium transport properties following changes in hydrostatic pressure [130]. The basal ATP release from the rabbit bladder epithelium is altered by amiloride, a blocker of ENaC [30]. In cultured urothelial cells, amiloride and Gd^{3+}, a non-specific blocker of mechanosensitive channels, suppresses ATP release by a hypotonic stimulus [131]. Furthermore, intravesical infusion of amiloride reduces the frequency of reflex voiding during bladder filling in anesthetized rats, and stretch-evoked ATP release from bladder strips is largely diminished by amiloride [129]. These indicate that ENaC expressed in the bladder epithelium is implicated in the mechanosensory transduction by controlling stretch-evoked ATP release. The ENaC expression in the bladder epithelium has a remarkable species difference between the rat and human [128, 129]. In the human bladder, the expression level of ENaC is extremely low, but is markedly up-regulated in obstructed bladders [128]. The expression level of ENaC mRNA correlates significantly with storage symptom score. The over-expression of ENaC in the human obstructed bladder might be associated with the induction of detrusor overactivity by bladder outlet obstruction.

Acid-sensing ion channels (ASICs), an H^+-gated subgroup of the degenerin/ENaC family, are encoded by three different subunit genes, ASIC1, ASIC2 and ASIC3, and the subunits form homo- and hetero-multimeric channels, which differ in their pH sensitivity and other pharmacological properties [122]. In the central and peripheral nervous system, ASICs have emerged as key receptors for extracellular protons, and recent studies suggest diverse roles for these channels in the physiology of mechanosensation and the pathophysiology of acid-evoked pain [132, 133]. ASICs, especially ASIC1 and ASIC2, are abundantly expressed in the urothelium and detrusor muscle of mouse bladder [134]. ASIC1 is a dominant subunit in the bladder mucosa, and both ASIC1 and ASIC2 are expressed in the bladder muscle. ASIC2 and ASIC3 are expressed in suburothelial nerve plexus of the rat [135]. The expressions of ASIC2 and ASIC3 in the urothelium and suburothelial nerve plexus increase in cyclophospamide-induced cystitis, while ASIC1 expression is not altered. Recent studies in the rat suggested that acid-induced Ca^{2+} influx and ATP release in the urothelium are partly attributed to ASIC activation [136, 137]. Roles of each ASIC subunit in sensory function are not well understood at present. The experiments on gastrointestinal sensation indicate that the disruption of ASIC1 or ASIC3 increases or decreases the mechanical sensitivity, respectively, while disrupting ASIC2 has varied effects [138–140].

45.4.2 Two-Pore-Domain K^+ Channels

The two-pore-domain K^+ channels (K2P or KCNK) are highly expressed in the central and peripheral nervous system and non-neuronal tissues, and they provide a wide variety of important functions including responses to mechanical stretch, temperature and pH [141]. TREK-1 (KCNK2), a mechanosensitive subfamily of K2P channels, has a higher expression level (12-fold) in the human bladder myocytes relative to the aorta [142]. Its opener produces a relaxation of KCl-induced contraction in rat bladder strips, but no effect on aortic strips. The expression of TREK-1 protein is decreased in the smooth muscle of mouse bladder with bladder outlet obstruction [143]. Systemic administration of a TREK-1 blocker induces an increase in premature detrusor contraction (non-voiding contraction) during bladder filling in anesthetized mice [143]. Our preliminary study indicates that TREK-1 is also expressed in the bladder epithelium (unpublished data).

45.5 Summary

Mechanosensory transduction may be mediated by a number of factors that together enhance the electrical activity to excite afferents. Our study on TRPV4 [94] suggested that multiple mechanosensitive channels with different thresholds to be activated contribute the bladder response to mechanical stimuli. Mechanosensitive TRP channels in the sensory nerve ending, urothelium and smooth muscle are possible candidates for the bladder mechanosensor and/or mechanotransducer (Fig.45.1). TRP channels without mechanosensitivity like TRPV1 could contribute to the mechanosensory transduction indirectly by tuning the general level of afferent excitability and play a crucial role during pathological conditions, such as overactive bladder and painful bladder syndrome.

Mechanosensitive TRP channels in the urothelium may function as a mechanotransducer by releasing ATP, which acts on P2X3 receptor at suburothelial sensory nerve endings, in response to bladder distension. Thus, P2X3 receptor is a possible candidate for bladder mechanosensor. However, recent studies assessing bladder afferent sensitivity without urothelial influence suggest that bladder mechanosensitivity arises by a direct physical mechanism at the nerve terminal rather than via release of neuromediators from the urothelium [18–20]. Thus, mechanosensitive TRP channels in the sensory nerve terminals are potent candidates for bladder mechanosensor. Mechanosensitive TRP channels are also implicated in the mediator release from the urothelium and possibly in the regulation of smooth muscle tone and micromotion in response to mechanical stretch. These urothelial and muscular actions may play an important role in setting the excitability of bladder afferents and be a key factor in pathophysiology of the bladder function. Pharmacological interventions targeting TRP channels may provide a new strategy for the treatment of bladder dysfunction.

References

1. de Groat WC (2006) Integrative control of the lower urinary tract: preclinical perspective. Br J Pharmacol 147:S25–S40
2. Yoshimura N, Seki S, Chancellor MB, de Groat WC, Ueda T (2002) Targeting afferent hyperexcitability for therapy of the painful bladder syndrome. Urology 59(suppl 5A):61–67
3. Lewin GR, Lu Y, Park TJ (2004) A plethora of painful molecules. Curr Opin Neurobiol 14:443–449

4. Gillespie PG, Walker RG (2001) Molecular basis of mechanosensory transduction. Nature 413:194–202
5. Voets T, Talavera K, Owsianik G, Nilius B (2005) Sensing with TRP channels. Nat Chem Biol 1:85–92
6. Christensen AP, Corey DP (2007) TRP channels in mechanosensation: direct or indirect activation? Nat Rev Neurosci 8:510–521
7. Birder LA (2007) TRPs in bladder diseases. Biochim Biophys Acta 1772:879–884
8. Araki I, Du S, Kobayashi H, Sawada F, Mochizuki T, Zakoji H, Takeda M (2008) Roles of mechanosensitive ion channels in bladder sensory transduction and overactive bladder. Int J Urol 15:681–687
9. Everaerts W, Gevaert T, Nilius B, De Ridder D (2008) On the origin of bladder sensing: Tr(i)ps in urology. Neurourol Urodyn 27:264–273
10. Burnstock G (2001) Purine-mediated signalling in pain and visceral perception. Trends Pharmacol Sci 22:182–188
11. Tsunozaki M, Bautista DM (2009) Mammalian somatosensory mechanotransduction. Curr Opin Neurobiol 19:362–369
12. Häbler HJ, Jänig W, Koltzenburg M (1990) Activation of unmyelinated afferent fibres by mechanical stimuli and inflammation of the urinary bladder in the cat. J Physiol 425:545–562
13. Häbler HJ, Jänig W, Koltzenburg M (1993) Myelinated primary afferents of the sacral spinal cord responding to slow filling and distension of the cat urinary bladder. J Physiol 463: 449–460
14. Sengupta JN, Gebhart GF (1994) Mechanosensitive properties of pelvic nerve afferent fibers innervating the urinary bladder of the rat. J Neurophysiol 72:2420–2430
15. Shea VK, Cai R, Crepps B, Mason JL, Perl ER (2000) Sensory fibers of pelvic nerve innervating the rat's urinary bladder. J Neurophysiol 84:1924–1933
16. Gabella G, Davis C (1998) Distribution of afferent axons in the bladder of rats. J Neurocytol 27:141–155
17. Xu L, Gebhart GF (2008) Characterization of mouse lumbar splanchnic and pelvic nerve urinary bladder mechanosensory afferents. J Neurophysiol 99:244–253
18. Zagorodnyuk VP, Gibbins IL, Costa M, Brookes SJH, Gregory SJ (2007) Properties of the major classes of mechanoreceptors in the guinea pig bladder. J Physiol 585(1):147–163
19. Yu Y, de Groat WC (2008) Sensitization of pelvic afferent nerves in the in vitro rat urinary bladder-pelvic nerve preparation by purinergic agonists and cyclophosphamide pretreatment. Am J Physiol Renal Physiol 294:F1146–F1156
20. Zagorodnyuk VP, Brookes SJH, Spencer NJ, Gregory S (2009) Mechanotransduction and chemosensitivity of two major classes of bladder afferents with endings in the vicinity to the urothelium. J Physiol 587(14):3523–3538
21. Moore KH, Gilpin SA, Dixon JS, Richmond DH, Sutherst JR (1992) Increase in presumptive sensory nerves of the urinary bladder in idiopathic detrusor instability. Brit J Urol 70: 370–372
22. Welsh MJ, Price MP, Xie J (2002) Biochemical basis of touch perception: mechanosensory function of degenerin/epithelial Na$^+$ channels. J Biol Chem 277:2369–2372
23. Moran MM, Xu H, Clapham DE (2004) TRP ion channels in the nervous system. Curr Opin Neurobiol 14:362–369
24. Smet PJ, Moore KH, Jonavicius J (1997) Distribution and colocalization of calcitonin gene-related peptide, tachykinins, and vasoactive intestinal peptide in normal and idiopathic unstable human urinary bladder. Lab Invest 77:37–49
25. Maggi CA, Meli A (1988) The sensory-efferent function of capsaicin-sensitive sensory neurons. Gen Pharmacol 19:1–43
26. Meini S, Maggi CA (1994) Evidence for a capsaicin-sensitive, tachykinin-mediated, component in the NANC contraction of the rat urinary bladder to nerve stimulation. Br J Pharmacol 112:1123–1131
27. Andersson KE (2002) Bladder activation: afferent mechanisms. Urology 59(suppl 5A): 43–50

28. Patacchini R, Santicioli P, Giuliani S, Maggi CA (2005) Pharmacological investigation of hydrogen sulfide (H2S) contractile activity in rat detrusor muscle. Eur J Pharmacol 509: 171–177
29. Birder LA (2005) More than just a barrier: urothelium as a drug target for urinary bladder pain. Am J Physiol Renal Physiol 289:F489–F495
30. Ferguson DR, Kennedy I, Burton TJ (1997) ATP is released from rabbit urinary bladder epithelial cells by hydrostatic pressure changes-a possible sensory mechanism? J Physiol 505:503–511
31. Birder LA, Nakamura Y, Kiss S, Nealen ML, Barrick S, Kanai AJ, Wang E, Ruiz G, de Groat WC, Apodaca G, Watkins S, Caterina MJ (2002) Altered urinary bladder function in mice lacking the vanilloid receptor TRPV1. Nat Neurosci 5:856–860
32. Yoshida M, Inadome A, Maeda Y, Satoji Y, Masunaga K, Sugiyama Y, Murakami S (2006) Non-neuronal cholinergic system in human bladder urothelium. Urology 67:425–430
33. Sabirov RZ, Okada Y (2004) ATP-conducting maxi-anion channel: a new player in stress-sensory transduction. Jpn J Physiol 54:7–14
34. Rong W, Spyer KM, Burnstock G (2002) Activation and sensitisation of low and high threshold afferent fibres mediated by P2X receptors in the mouse urinary bladder. J Physiol 541:591–600
35. Vlaskovska M, Kasakov L, Rong W, Bodin P, Bardini M, Cockayne DA, Ford APDW, Burnstock G (2001) P2X 3 knock-out mice reveal a major sensory role for urothelially released ATP. J Neurosci 21:5670–5677
36. Cockayne DA, Dunn PM, Zhaong Y, Rong W, Hamilton SG, Knight GE, Ruan HZ, Ma B, Yip P, Nunn P, McMahon SB, Burnstock G, Ford APDW (2005) P2X2 knockout mice and P2X2/P2X3 double knockout mice reveal a role for the P2X2 receptor subunit in mediating multiple sensory effects of ATP. J Physiol 567(2):621–639
37. Burnstock G (2009) Purinergic mechanosensory transduction and visceral pain. Mol Pain 5:69
38. Hedlund P, Streng T, Lee T, Andersson KL (2007) Effects of tolterodine on afferent neurotransmission in normal and resiniferatoxin treated conscious rats. J Urol 178:326–331
39. Iijima K, De Wachter S, Wyndaele JJ (2007) Effects of the M3 receptor selective muscarinic antagonist darifenacin on bladder afferent activity of the rat pelvic nerve. Eur Urol 52: 842–849
40. Bernardini N, Sauer SK, Haberberger R, Fischer MJM, Reeh PW (2001) Excitatory nicotinic and desensitizing muscarinic (M2) effects on C-nociceptors in isolated rat skin. J Neurosci 21:3295–3302
41. Mukerji G, Yiangou Y, Grogono J, Underwood J, Agarwal SK, Khullar V, Anand P (2006) Localization of M_2 and M_3 muscarinic receptors in human bladder disorders and their clinical correlations. J Urol 176:367–373
42. Pradidarcheep W, Stallen J, Labruyere WT, Dabhoiwala NF, Michel MC, Lamers WH (2008) Lack of specificity of commercially available antisera: better specifications needed. J Histochem Cytochem 56:1099–1111
43. Jositsch G, Papadakis T, Haberberger RV, Wolff M, Wess J, Kummer W (2009) Suitability of muscarinic acetylcholine receptor antibodies for immunohistochemistry evaluated on tissue sections of receptor gene-deficient mice. Naunyn Schmiedebergs Arch Pharmacol 379: 389–395
44. Yoshimura N, Kaiho Y, Miyazato M, Yunoki T, Tai C, Chancellor MB, Tyagi P (2007) Therapeutic receptor targets for lower urinary tract dysfunction. Naunyn Schmiedebergs Arch Pharmacol 377:437–448
45. Birder LA (2010) Urothelial signaling. Auton Neurosci 153:33–40
46. Kullman FA, Artim DE, Birder LA, de Groat WC (2008) Activation of muscarinic receptors in rat bladder sensory pathways alters reflex bladder activity. J Neurosci 28:1977–1987
47. Chopra B, Gever J, Barrick SR, Hanna-Mitchell AT, Beckel JM, Ford APDW, Birder LA (2008) Expression and function of rat urothelial P2Y receptors. Am J Physiol Renal Physiol 294:F821–F829

48. Wang X, Momota Y, Yanase H, Narumiya S, Maruyama T, Urothelium KM (2008) EP1 receptor facilitates the micturition reflex in mice. Biomed Res 29:105–111
49. Girard BM, Wolf-Johnston A, Braas KM, Birder LA, May V, Vizzard MA (2008) PACAP-mediated ATP release from rat urothelium and regulation of PACAP/VIP and receptor mRNA in micturition pathways after cyclophosphamide (CYP)-induced cystitis. J Mol Neurosci 36:310–320
50. Sun Y, MaLossi J, Jacobs SC, Chai TC (2002) Effect of doxazosin on stretch-activated adenosine triphosphate release in bladder urothelial cells from patients with benign prostatic hyperplasia. Urology 60:351–356
51. Ishihama H, Momota Y, Yanase H, Wang X, de Groat WC, Kawatani M (2006) Activation of alpha1D adrenergic receptors in the rat urothelium facilitates the micturition reflex. J Urol 175:358–364
52. Yoshida M, Masunaga K, Nagata T, Maeda Y, Miyamoto Y, Kudoh J, Homma Y (2009) Attenuation of non-neuronal adenosine triphosphate release from human bladder mucosa by antimuscarinic agents. LUTS 1:88–92
53. Brading AF (2006) Spontaneous activity of lower urinary tract smooth muscles: correlation between ion channels and tissue function. J Physiol 570:13–22
54. Gillespie JI (2004) The autonomous bladder: a view of the origin of bladder overactivity and sensory urge. BJU Int 93:478–483
55. Collins C, Klausner AP, Herrick B, Koo HP, Miner AS, Henderson SC, Ratz PH (2009) Potential for control of detrusor smooth muscle spontaneous rhythmic contraction by cyclooxygenase products released by interstitial cells of Cajal. J Cell Mol Med 13: 3236–3250
56. Zagorodnyuk VP, Gregory S, Costa M, Brookes SJH, Tramontana M, Giuliani S, Maggi CA (2009) Spontaneous release of acetylcholine from autonomic nerves in the bladder. Br J Pharmacol 157:607–619
57. Hashitani H, Yanai Y, Suzuki H (2004) Role of interstitial cells and gap junctions in the transmission of spontaneous Ca^{2+} signals in detrusor smooth muscle of the guinea-pig urinary bladder. J Physiol 559(2):567–581
58. Drake MJ, Harvey IJ, Gillespie JI (2003) Autonomous activity in the isolated guinea pig bladder. Exp Physiol 88:19–30
59. Gillespie JI (2005) A developing view of the origins of urgency: the importance of animal models. BJU Int 96(suppl 1):22–28
60. McCarthy CJ, Zabbarova IV, Brumovsky PR, Gebhart GF, Kanai AJ (2009) Spontaneous contractions evoke afferent nerve firing in mouse bladders with detrusor overactivity. J Urol 181:1459–1466
61. Drake MJ, Hedlund P, Andersson KE, Brading AF, Hussain I, Fowler C, Landon DN (2003) Morphology, phenotype and ultrastructure of fibroblastic cells from normal and neuropathic human detrusor: absence of myofibroblast characteristics. J Urol 169:1573–1576
62. Sanders KM (1996) A case for interstitial cells of Cajal as pacemakers and mediators of neurotransmission in the gastrointestinal tract. Gastroenterology 111:492–515
63. Sergeant GP, Hollywood MA, McCloskey KD, Thornbury KD, McHale NG (2000) Specialised pacemaking cells in the rabbit urethra. J Physiol 526:359–366
64. McCloskey KD, Gurney AM (2002) Kit positive cells in the guinea pig bladder. J Urol 168:832–836
65. Fry CH, Sui GP, Kanai AJ, Wu C (2007) The function of suburothelial myofibroblasts in the bladder. Neurourol Urodyn 26:914–919
66. Sui GP, Fry CH (2004) Electrical characteristics of suburothelial cells isolated from the human bladder. J Urol 171:938–943
67. Wu C, Sui GP, Fry CH (2004) Purinergic regulation of guinea pig suburothelial myofibroblasts. J Physiol 559:231–243
68. Grol S, Essers PBM, van Koeverringe GA, Martinez-Martinez P, de Vente J, Gillespie JI (2009) M3 muscarinic receptor expression on suburothelial interstitial cells. BJU Int 104:398–405

69. Sui GP, Rothery S, Dupont E, Fry CH, Severs NJ (2002) Gap junctions and connexin expression in human suburothelial interstitial cells. BJU Int 90:118–129
70. Wiseman OJ, Fowler CJ, Landon DN (2003) The role of the human bladder lamina propria myofibroblast. BJU Int 91:89–93
71. Sui GP, Wu C, Fry CH (2008) Modulation of bladder myofibroblast activity: implications for bladder function. Am J Physiol Renal Physiol 295:F688–F697
72. Avelino A, Cruz F (2006) TRPV1 (vanilloid receptor) in the urinary tract: expression, function and clinical applications. Naunyn Schmiedebergs Arch Pharmacol 373:287–299
73. Hwang AJ, Oh JM, Valtschanoff JG (2005) Expression of the vanilloid receptor TRPV1 in rat dorsal root ganglion neurons supports different roles of the receptor in visceral and cutaneous afferents. Brain Res 1047:261–266
74. Birder LA, Kanai AJ, de Groat WC, Kiss S, Nealen ML, Burke NE, Dineley KE, Watkins A, Reynolds IJ, Caterina MJ (2001) Vanilloid receptor expression suggests a sensory role for urinary bladder epithelial cells. Proc Natl Acad Sci USA 98:13396–13401
75. Ost D, Roskams T, Van Der Aa F, De Ridder D (2002) Topography of the vanilloid receptor in the human bladder: more than just the nerve fibers. J Urol 168:293–297
76. Lazzeri M, Vannucchi MG, Zardo C, Spinelli M, Beneforti P, Turini D, Faussone-Pellegrini MS (2004) Immunohistochemical evidence of vanilloid receptor 1 in normal human urinary bladder. Eur Urol 46:792–798
77. Everaerts W, Sepulvera MR, Gavaert T, Roskams T, Nillius B, De Ridder D (2009) Where is TRPV1 expressed in the bladder, do see the real channel? Naunyn Schmiedebergs Arch Pharmacol 379:421–425
78. Chancellor MB, de Groat WC (1999) Intravesical capsaicin and resiniferatoxin therapy: spicing up the ways to treat the overactive bladder. J Urol 162:3–11
79. Tominaga M, Caterina MJ, Malmberg AB, Rosen TA, Gilbert H, Skinner K, Raumann BE, Basbaum AI, Julius D (1998) The cloned capsaicin receptor integrates multiple pain-producing stimuli. Neuron 21:531–543
80. Avelino A, Cruz C, Nagy I, Cruz F (2002) Vanilloid receptor 1 expression in the rat urinary tract. Neuroscience 109:787–798
81. Vizzard MA (2000) Alterations in spinal cord Fos protein expression induced by bladder stimulation following cystitis. Am J Physiol Regul Integr Comp Physiol 278:R1027–R1039
82. Dinis P, Charrus A, Avelino A, Yaqoob M, Bevan S, Nagy I, Cruz F (2004) Anandamide-evoked activation of vanilloid receptor 1 contributes to the development of bladder hyper-reflexia and nociceptive transmission to spinal dorsal horn neurons in cystitis. J Neurosci 24:11253–11263
83. Charrua A, Cruz CD, Narayanan S, Gharat L, Gullapalli S, Cruz F, Avelino A (2009) GRC-6211, a new oral specific TRPV1 antagonist, decreases bladder overactivity and noxious bladder input in cystitis animal models. J Urol 181:379–386
84. Yoshiyama M, Araki I, Kobayashi H, Zakoji H, Takeda M (2010) Functional roles of TRPV1 channels in lower urinary tract irritated by acetic acid: in-vivo evaluations on the sex difference in decerebrate unanesthetized mice. Am J Physiol Renal Physiol doi: 10.1152/ajprenal.00685.2009
85. Zeihofer HU, Kress M, Swandulla D (1997) Functional Ca^{2+} currents through capsaicin- and proton-activated ion channels in rat dorsal root ganglion neurons. J Physiol 503:67–78
86. Rigoni M, Trevisani M, Gazzieri D, Nadaletto R, Tognetto M, Creminon C, Davis JB, Campi B, Amadesi S, Geppetti P, Harrison S (2003) Neurogrnic responses mediated by vanilloid receptor-1 (TRPV1) are blocked by the high affinity antagonist, iodo-resiniferatoxin. Br J Pharmacol 138:977–985
87. Wang ZY, Wang P, Merriam FV, Bjorling DE (2008) Lack of TRPV1 inhibits cystitis-induced increased mechanical sensitivity in mice. Pain 139:158–167
88. Charrua A, Cruz CD, Cruz F, Avelino A (2007) Transient receptor potential vanilloid sub-family 1 is essential for the generation of noxious bladder input and bladder overactivity in cystitis. J Urol 177:1537–1541

89. Daly D, Rong W, Chess-Williams R, Chapple C, Grundy D (2007) Bladder afferent sensitivity in wild-type and TRPV1 knockout mice. J Physiol 583(2):663–674
90. Charrua A, Reguenga C, Cordeiro JM, Correiade-Sa P, Paule C, Nagy I, Cruz F, Avelino A (2009) Functional transient receptor potential vanilloid 1 is expressed in human urothelial cells. J Urol 182:2944–2950
91. Yamada T, Ugawa S, Ueda T, Ishida Y, Kajita K, Shimada S (2009) Differential localizations of the transient receptor potential channels TRPV4 and TRPV1 in the mouse urinary bladder. J Histochem Cytochem 57:277–287
92. Xu X, Gordon E, Lin Z, Lozinskaya IM, Chen Y, Thorneloe KS (2009) Functional TRPV4 channels and an absence of capsaicin-evoked currents in freshly–isolated, guinea-pig urothelial cells. Channels 3:15–160
93. Everaerts W, Vriens J, Owsianik G, Appendino G, Voets T, De Ridder D, Nilius B Functional characterisation of transient receptor potential channels in mouse urothelial cells. Am J Physiol Renal Physiol doi: 10.1152/ajprenal.00599.2009 in press
94. Mochizuki T, Sokabe T, Araki I, Fujishita K, Shibasaki K, Uchida K, Koizumi S, Takeda M, Tominaga M (2009) The TRPV4 cation channel mediates stretch-evoked Ca^{2+} influx and ATP release in primary urothelial cell cultures. J Biol Chem 284:21257–21264
95. Stein RJ, Santos S, Nagatomi J, Hayashi Y, Minnery BS, Xavier M, Patel AS, Nelson JB, Futrell WJ, Yoshimura N, Chancellor MB, De Miguel F (2004) Cool (TRPM8) and hot (TRPV1) receptors in the bladder and male genital tract. J Urol 172:1175–1178
96. Tsukimi Y, Mizuyachi K, Yamasaki T, Niki T, Hayashi F (2005) Cold response of the bladder in guinea pig: involvement of transient receptor potential channel, TRPM8. Urology 65: 406–410
97. Mukerji G, Yiangou Y, Corcoran SL, Selmer IS, Smith GD, Benham CD, Bountra C, Agarwal SK, Anand P (2006) Cool and menthol receptor TRPM8 in human urinary bladder disorders and clinical correlations. BMC Urol 6:6
98. Nomoto Y, Yoshida A, Ikeda S, Kamikawa Y, Harada K, Ohwatashi A, Kawahira K (2008) Effect of menthol on detrusor smooth-muscle contraction and micturition reflex in rats. Urology 72:701–705
99. Lashinger ESR, Steiginga MS, Hieble JP, Leon LA, Gardner SD, Nagilla R, Davenport EA, Hoffman BE, Laping NJ, Su X (2008) AMTB, a TRPM8 channel blocker: evidence in rats for activity in overactive bladder and painful bladder syndrome. Am J Physiol Renal Physiol 295:F803–F810
100. Hayashi T, Kondo T, Ishimatsu M, Yamada S, Nakamura K, Matsuoka K, Akasu T (2009) Expression of the TRPM8-immunoreactivity in dorsal root ganglion neurons innervating the rat urinary bladder. Neurosci Res 65:245–251
101. Dhaka A, Murray AN, Mathur J, Earley TJ, Petrus MJ, Patapoutian A (2007) TRPM8 is required for cold sensation in mice. Neuron 54:371–378
102. Colbum RW, Lubin ML, Stone DJ Jr, Wang Y, Lawrence D, D'Andrea MR, Brandt MR, Liu Y, Flores CM, Qin N (2007) Attenuated cold sensitivity in TRPM8 null mice. Neuron 54:379–386
103. Mahieu F, Owsianik G, Verbert L, Janssens A, De Smedt H, Nillius B, Voets T (2007) TRPM8-independent menthol-induced Ca2+ release from endoplasmic reticulum and Golgi. J Biol Chem 282:3325–3336
104. Du S, Araki I, Kobayashi H, Zakoji H, Sawada N, Takeda M (2008) Differential expression profile of cold (TRPA1) and cool (TRPM8) receptors in human urogenital organs. Urology 72:450–455
105. Nagata K, Duggan A, Kumar G, Garcia-Anoveros J (2005) Nociceptor and hair cell transducer properties of TRPA1, a channel for pain and hearing. J Neurosci 25:4052–4061
106. Du S, Araki I, Yoshiyama M, Nomura T, Takeda M (2007) Transient receptor potential channel A1 involved in sensory transduction of rat urinary bladder through C-fiber pathway. Urology 70:826–831

107. Streng T, Axelsson HE, Hedlund P, Andersson DA, Jordt SE, Bevan S, Andersson KE, Hogestatt ED, Zygmunt PM (2008) Distribution and function of the hydrogen sulfide-sensitive TRPA1 ion channels in rat urinary bladder. Eur Urol 53:391–400
108. Story GM, Peier AM, Reeve AJ, Eid SR, Mosbacher J, Hricik TR, Earley TJ, Hergarden AC, Andersson DA, Hwang SW, McIntyre P, Jegla T, Bevan S, Patapoutian A (2003) ANKTM1, a TRP-like channel expressed in nociceptive neurons, is activated by cold temperatures. Cell 112:819–829
109. Bandell M, Story GM, Hwang SW, Viswanath V, Eid SR, Petrus MJ, Earley TJ, Patapoutian A (2004) Noxious cold ion channel TRPA1 is activated by pungent compounds and bradykinin. Neuron 41:849–857
110. Kwan KY, Allchorne AJ, Vollrath MA, Christensen AP, Zhang DS, Woolf CJ, Corey DP (2006) TRPA1 contributes to cold, mechanical, and chemical nociception but is not essential for hair-cell transduction. Neuron 50:277–289
111. Bautista DM, Jordt S-E, Nikai T, Tsuruda PR, Read AJ, Poblete J, Yamoah EN, Basbaum AI, Julius D (2006) TRPA1 mediates the inflammatory actions of environmental irritants and proalgesic agents. Cell 124:1269–1282
112. Andrade EL, Ferreira J, Andre E, Calixto JB (2006) Contractile mechanisms coupled to TRPA1 receptor activation in rat urinary bladder. Biochem Pharmacol 72:104–114
113. Suzuki M, Mizuno A, Kodaira K, Imai M (2003) Impaired pressure sensation in mice lacking TRPV4. J Biol Chem 278:22664–22668
114. Liedtke W, Friedman JM (2003) Abnormal osmotic regulation in $trp4^{-/-}$ mice. Proc Natl Acad Sci USA 100:13698–13703
115. Gevaert T, Vriens J, Segal A, Everaerts W, Roskams T, Talavera K, Owsianik G, Liedtke W, Daelemans D, Dewachter I, Van Leuven F, Voets T, De Ridder D, Nilius B (2007) Deletion of the transient receptor potential cation channel TRPV4 impairs murine bladder voiding. J Clin Invest 117:3453–3462
116. Birder L, Kullmann FA, Lee H, Barrick S, de Groat W, Kanai A, Caterina M (2007) Activation of urothelial transient receptor potential vanilloid 4 by 4αphorbol 12,13-didecanoate contributes to altered bladder reflexes in the rat. J Pharmacol Exp Ther 323:227–235
117. Thorneloe KS, Sulpizio AC, Lin Z, Figueroa DJ, Clouse AK, McCafferty GP, Chendrimada TP, Lashinger ESR, Gordon E, Evans L, Misajet BA, DeMarini DJ, Nation JH, Casillas LN, Marquis RW, Votta BJ, Sheardown SA, Xu X, Brooks DP, Laping NJ, Westfall TD (2008) N-((1S)-1-{[4-((2S)-2-{[(2,4-Dichlorophenyl)sulfonyl]amino}-3-hydroxypropanoyl)-1-piperazinyl}-3-methylbutyl)-1-benzothiophene-2-carboxamide (GSK1016790A), a novel and potent transient receptor potential vanilloid 4 channel agonist induces urinary bladder contraction and hyperactivity: part I. J Pharmacol Exp Ther 326:432–442
118. Muraki K, Iwata Y, Katanosaka Y, Ito T, Ohya S, Shigekawa M, Imaizumi Y (2003) TRPV2 is a component of osmotically sensitive cation channels in murine aortic myocytes. Circ Res 93:829–838
119. Caprodossi S, Lucciarini R, Amantini C, Nabissi M, Canesin G, Ballarini P, Di Spilimbergo A, Cardarelli MA, Servi L, Mammana G, Santoni G (2008) Transient receptor potential vanilloid type 2 (TRPV2) expression in normal urothelium and in urothelial carcinoma of human bladder: correlation with the pathologic stage. Eur Urol 54:612–620
120. Pedersen SF, Nilius B (2007) Transient receptor potential channels in mechanosensing and cell volume regulation. Methods Enzymol 428:183–207
121. de la Rosa DA, Canessa CM, Fyfe GK, Zhang P (2000) Structure and regulation of amiloride-sensitive sodium channels. Annu Rev Physiol 62:573–594
122. Kellenberger S, Schild L (2002) Epithelial sodium channel/degenerin family of ion channels: a variety of functions for a shared structure. Physiol Rev 82:735–767

123. Drummond HA, Price MP, Welsh MJ, Abboud FM (1998) A molecular component of the arterial baroreceptor mechanotransducer. Neuron 21:1435–1441
124. Drummond HA, Abboud FM, Welsh MJ (2000) Localization β and γ subunits of ENaC in sensory nerve endings in the rat foot pad. Brain Res 884:1–12
125. Fricke B, Lints R, Stewart G, Drummond H, Dodt G, Driscoll M, von During M (2000) Epithelial Na^+ channels and stomatin are expressed in rat trigeminal mechanosensory neurons. Cell Tissue Res 299:327–334
126. Kopp UC, Matsushita K, Sigmund RD, Smith LA, Watanabe S, Stokes JB (1792) Amiloride-sensitive Na^+ channels in pelvic uroepithelium involved in renal sensory receptor activation. Am J Physiol Regul Integr Comp Physiol 1998(275):R1780–R1792
127. Smith PR, Mackler SA, Weiser PC, Brooker DR, Ahn YJ, Harte BJ, McNulty KA, Kleyman TR (1998) Expression and localization of epithelial sodium channel in mammalian urinary bladder. Am J Physiol Renal Physiol 274:F91–F96
128. Araki I, Du S, Kamiyama M, Mikami Y, Matsushita K, Komuro M, Furuya Y, Takeda M (2004) Overexpression of epithelial sodium channels in epithelium of human urinary bladder with outlet obstruction. Urology 64:1255–1260
129. Du S, Araki I, Mikami Y, Zakoji H, Beppu M, Yoshiyama M, Takeda M (2007) Amiloride-sensitive ion channels in urinary bladder epithelium involved in mechanosensory transduction by modulating stretch-evoked adenosine triphosphate release. Urology 69:590–595
130. Ferguson DR (1999) Urothelial function. BJU Int 84:235–242
131. Birder LA, Barrick SR, Roppolo JR, Kanai AJ, de Groat WC, Kiss S, Buffington CA (2003) Feline interstitial cystitis results in mechanical hypersensitivity and altered ATP release from bladder urothelium. Am J Physiol Renal Physiol 285:F423–F429
132. Wemmie JA, Price MP, Welsh MJ (2006) Acid-sensing ion channels: advances, questions and therapeutic opportunities. Trends Neurosci 29:578–586
133. Lingueglia E (2007) Acid-sensing ion channels in sensory perception. J Biol Chem 282:17325–17329
134. Kobayashi H, Yoshiyama M, Zakoji H, Takeda M, Araki I (2009) Sex differences in expression profile of acid-sensing ion channels in the mouse urinary bladder: a possible involvement in irritative bladder symptoms. BJU Int 104:1746–1751
135. Corrow K, Girard BM, Vizzard MA Expression and response of acid-sensing ion channels (ASICs) in urinary bladder to cyclophosphamide (CYP)-induced cystitis. Am J Physiol Renal Physiol in press; doi: 10.1152/ajprenal.00618.2009
136. Sadananda P, Shang F, Liu L, Mansfield KJ, Burcher E (2009) Release of ATP from rat urinary bladder mucosa: role of acid, vanilloids and stretch. Br J Pharmacol 158: 1655–1662
137. Kullmann FA, Shah MA, Birder LA, de Groat WC (2009) Functional TRP and ASIC-like channels in cultured urothelial cells from the rat. Am J Physiol Renal Physiol 296: F892–F901
138. Page AJ, Brierley SM, Martin CM, Martinez-Salgado C, Wemmie JA, Brennan TJ, Symonds E, Omari T, Lewin GR, Welsh MJ, Blackshaw LA (2004) The ion channel ASIC1 contributes to visceral but not cutaneous mechanoreceptor function. Gastroenterology 127:1739–1747
139. Page AJ, Brierley SM, Martin CM, Price M, Symonds E, Butler R, Wemmie JA, Blackshaw LA (2005) Different contributions of ASIC channels 1a, 2, and 3 in gastrointestinal mechanosensory function. Gut 54:1408–1415
140. Jones III RCW, Xu L, Gebhart GF (2005) The mechanosensitivity of mouse colon afferent fibers and their sensitization by inflammatory mediators require transient receptor potential vanilloid 1 and acid-sensing ion channel 3. J Neurosci 25:10981–10989
141. Sanders KM, Koh SD (2006) Two-pore-domain potassium channels in smooth muscles: new components of myogenic regulation. J Physiol 570:37–43

142. Tertyshnikova S, Knox RJ, Plym MJ, Thalody G, Griffin C, Neelands T, Harden DG, Signor L, Weaver D, Myers RA, Lodge NJ (2005) JBL-1249[(5,6,7,8-tetrahydro-naphthalen-1-yl)-[2-(1H-tetrazol-5-yl)-phenyl]-amine]: a putative potassium channel opener with bladder-relaxant properties. J Pharmacol Exp Ther 313:250–259
143. Baker SA, Hatton WJ, Han J, Hennig GW, Britton FC, Koh SD (2010) Role of TREK-1 potassium channel in bladder overactivity after partial bladder outlet obstruction in mouse. J Urol 183:793–800

Chapter 46
The Role of TRP Ion Channels in Testicular Function

Pradeep G. Kumar and Mohammed Shoeb

Abstract Transient receptor potential (TRP) proteins are homologues of Drosophila transient receptor potential ion channels first identified in the photo receptors and reported to be involved in calcium entry following calcium store depletion during photo transduction. TRP is a large super family divided in several families including the TRPC (Canonical) family, the TRPV (Vanilloid) family, the TRPM (Melastatin) family, the TRPP (Polycystin) family, the TRPML (Mucolipin) family, the TRPA (Ankyrin) family, and the TRPN (NOMPC) family. TRP proteins are six transmembrane ion channels and act as components of multimeric complexes which allow cation entry either after internal calcium depletion or in response to receptor stimulation. TRP ion channels have been reported to act as molecular sensors of environment. Trp genes are expressed in a wide range of tissues including testis. In addition to this TRP proteins have also been detected in mature sperm from a number of species including humans. TRP may be involved in regulating calcium dependent functions of sperm including motility, capacitation, and acrosome reaction. Here we review the available information about TRP proteins reported in the sperm, as well as in other cells/tissue systems.

46.1 Introduction

Spermatozoa are terminally differentiated cells and depend on the cues from their external environment in the testis, epididymis and vas deferens to complete maturation and achieve fertilization potential. In species with external fertilization, the spermatozoa require specific chemo-attractants released in water to locate and reach eggs. In higher (land dwelling) vertebrates with internal fertilization, the spermatozoa undergo two important changes in the female reproductive tract, viz., capacitation and acrosome reaction [1], before they are capable of fertilizing the oocyte.

P.G. Kumar (✉)
Division of Molecular Reproduction, Rajiv Gandhi Centre for Biotechnology, Thiruvananthapuram 695014, Kerela, India
e-mail: kumarp@rgcb.res.in

Capacitation is a complex event, which involves alterations in metabolism, intracellular ion concentrations, membrane fluidity [2, 3], membrane hyper polarization, intracellular pH [4], cAMP concentration [5–7] and protein tyrosine phosphorylation [8–10] leading to the sperm hyper activation and development of the ability to undergo acrosomal exocytosis. On coming in contact with zona pellucida the spermatozoa undergo an exocytosis releasing contents of its acrosome which dissolve the zona pellucida and enable it to reach the oocyte.

Several extrinsic factors regulate sperm development, and the role of ion fluxes has been recognized as an important factor in programming the physiology of spermatozoa. Calcium, sodium, potassium, bicarbonate and chloride are reported to be necessary for sperm motility, capacitation, acrosome reaction and sperm-egg fusion [11–20].

The role of calcium in initiation of motility [21–29], capacitation [30, 31] and acrosome reaction [32–34] in spermatozoa is well documented. A number of calcium channels have been reported in the sperm membrane including calcium conductance channels [35] which are receptor operated and second messenger operated [36], voltage dependent calcium selective channels [37], cyclic nucleotide-gated (CNG) calcium channels [38], T-type calcium channels [39], L-type calcium channels [40], store operated calcium channels [41], CatSper1, CatSper2, CatSper3 and CatSper4 [42], low threshold voltage-gated Ca(2+) (Ca(V)) channels [43] and homologues of the drosophila Transient receptor potential (TRP) proteins.

TRP proteins are six transmembrane ion channels and have been proposed to act as molecular sensors of the environment. TRP proteins are assembled in multicomponent signalling complexes and function as store operated or receptor operated cation channels [44]. As molecular sensors of external environment [45, 46], the TRP ion channels may play an important role in sperm maturation and development of its fertilization potential. The expression of these proteins have been reported mostly in the spermatocytes [47, 48] indicating their possible presence in mature spermatozoa. TRPs reported on mammalian spermatozoa include TRPC1, TRPC3, TRPC4, TRPC6 [47, 49], TRPC2 [48, 50], TRPP, TRPV and TRPM (Table 46.1). In this chapter, we focus on these TRP channels expressed in testis and their role in sperm function.

46.2 TRPC Family

TRPC1: TRPC1 is the archetypal member of TRPC family, showing 40% similarity to drosophila TRP protein [51]. TRPC1 is a constitutively active non selective cation channel in SF9 cells [52]. It co-assembles with itself or with other members of the TRPC family to form either store operated [53, 54] or receptor operated [55, 56] cation channels. In differentiating hippocampal neurons TRPC1/TRPC3 work in tandem to increase store operated calcium entry [57]. In HSY cells TRPC1 and TRPC3 co-assemble, via N-terminal interactions, to form a heteromeric store-operated non-selective cation channel [58]. TRPC1/TRPC4 complexes constitute the functional subunits of store operated channels (SOC) in glomerular mesangial cells [59]. TRPC1 and TRPC5 are subunits of a heteromeric neuronal channel [60].

Table 46.1 Expression of TRP genes in mammalian testis

TRP family	Member	Species	Technique	References
TRP C	trpC-1	H, M	RT-PCR, NB	[47–50]
	trpC-2	B, M	RT-PCR, NB	[51, 52]
	trpC-3	H, M	RT-PCR	[47, 48]
	trpC-4	M	RT-PCR	[48]
	trpC-5	M	RT-PCR	[48]
	trpC-6	H, M	RT-PCR	[47, 48]
	trpC-7	H, M	RT-PCR, NB	[47, 48, 53]
TRP M	trpM-3	H, R, M	RT-PCR, NB	[54–57]
	trpM-4	R, M	RT-PCR	[56–58]
	trpM-5	M	RT-PCR	[58]
	trpM-7	R	RT-PCR	[56]
	trpM-8	R	RT-PCR	[56, 57]
TRP P	trpP-3	M	IF	[59]
	trpP-5	M	NB, RT-PCR	[60, 61]
TRP V	trpV-1	M	RT-PCR	[62, 63]
	trpV-4	R	RT-PCR	[57]
	trpV-5	R	RT-PCR	[56, 57]
	trpV-6	R	RT-PCR	[57]

M, mouse; R, rat; H, human; B, bovine; RT-PCR, reverse transcriptase polymerase chain reaction; NB, Northern blotting and IF, immunofluorescence.

TRPC1 forms a unique heteromeric channels with TRPP2 in cilium of kidney cells with properties distinct from that of TRPP2 or TRPC1 alone and has implications in mechanosensation and cilium-based Ca(2+) signalling [61]. In HEK293 cells endogenous TRPC1/TRPC3/TRPC7 participate in forming heteromeric store-operated channels [62]. TRPC1 forms complexes with STIM1 (Stromal interaction molecule1) and Orai1 in human salivary gland cells and this dynamic assembly of TRPC1-STIM1-Orai1 forms internal Ca2+ store depletion activated SOC channel [63]. STIM1 binds TRPC1, TRPC4 and TRPC5 and determines their function as SOCs [64]. In rabbit coronary artery TRPC1 is component of Endothelin ET(A) and ET(B) receptors activated Protein kinase C (PKC) dependent cation channels [65]. TRPC1 is a component of stretch operated mechno-sensitive cation channel in frog oocytes [66]. TRPC1 is definitely a component of the store dependent cation entry mechanism as in TRPC1 knockout mice store operated calcium entry is significantly reduced [67]. TRPC1 is a multifunctional protein able to form intracellular calcium release channels when expressed alone, and plasma membrane channels when co-expressed with TRPC4 or TRPC5, but not TRPC3 or TRPC6. Both endoplasmic reticular and plasma membrane forms of the channel are activated upon addition of agonists coupled to the Inositol triphosphate IP [3] cascade [68].

Mouse TRPC1 (NM_011643.2, NP_035773.1) is a 765 amino acid protein. SMART analysis of mTRPC1 shows that it has N-terminus ankyrin protein-protein interaction motifs followed by TRP_2 motif of unknown function. There are 6 transmembrane helices of which the 4 towards C terminus form ion transport channels (Fig. 46.1). TRPC1 has been reported in the mid-piece of the mouse sperm flagellum with punctuate pattern [47] and in flagellum of human sperm [49]. TRPC1

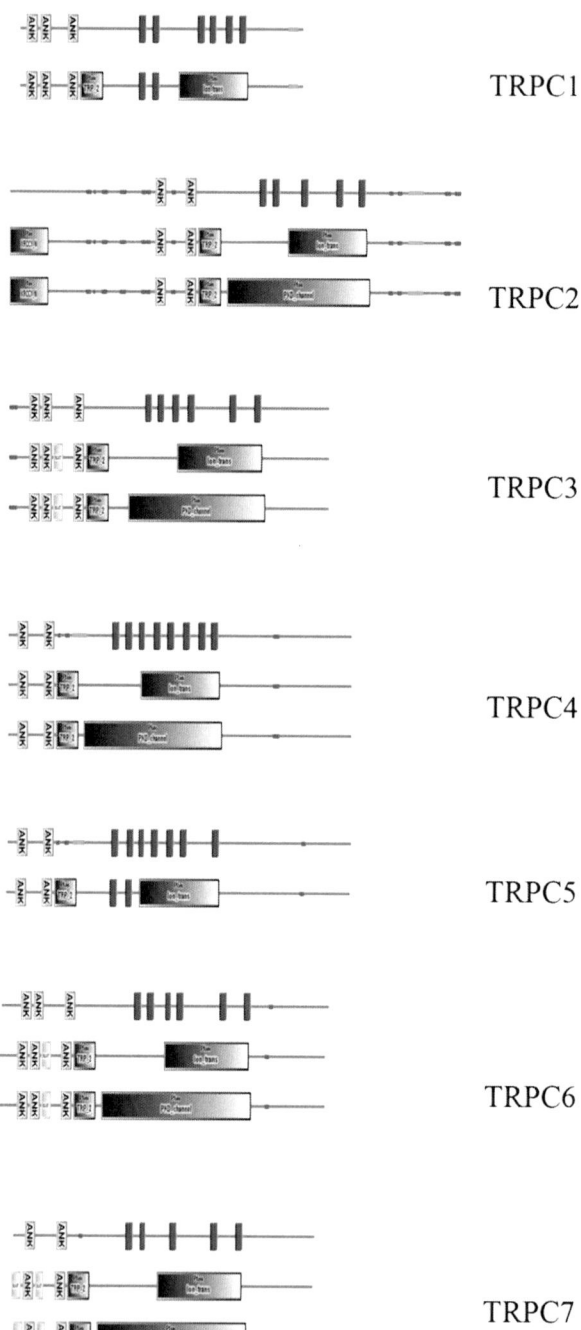

Fig. 46.1 (continued)

has been reported in the acrosomal domain of mouse sperm head as well [69]. We have studied the presence of TRPC1 in goat sperm and detected it in the acrosome head domain as well as in the mid-piece and proximal tail region (unpublished). Although no study has been done on the functional role of TRPC1 in the spermatozoa, its localisation on the mid-piece or tail of the sperm indicates a possible role in mediating calcium controlled functions like motility [47], and its presence in the acrosome domain of the sperm head indicates a possible role in acrosome reaction and sperm-oocyte fusion.

TRPC2: In mouse four variants of TRPC2 have been reported including TRPC2-14, TRPC2-17 [70], mTRPC2α and mTRPC2β [71]. While mTRPC2α is detected in many tissues including testis, mTRPC2β is reported only in the vomero-nasal organ [71]. TRPC2 is activated by mechanisms which are store depletion dependent [70] and independent [72]. In the microvilli of rat vomero-nasal receptor cells TRPC2 co-localizes with G-protein subunits [73] and type-III IP3R [74], indicating a possibility of interaction of GPCR/PLC mediated receptor activated mechanism of calcium entry through TRPC2 channels. TRPC2 interacts with Inositol trisphosphate receptor IP3R, PLC-gamma and Epo-receptor in erythroid cells and causes an increase in calcium influx in response to Epo (Erythropoietin) stimulation [75] involving PLC-gamma [76]. In Neutrophiles TRPC2 forms non selective cation channels regulated by ADP-ribose and the redox state [77]. In humans TRPC2 is a pseudo-gene [70].

Mouse TRPC2α (NM_011644.2, NP_035774.2) is a 1220 amino acid protein. It has two N-terminal ankyrin repeats followed by TRP_2 domain and five transmembrane domains of which the 3 towards C-terminus form ion transport channel (Fig. 46.1). TRPC2 is localised on anterior head in mouse spermatozoa and is an essential component of the ZP3-activated cation channel that drives acrosome reaction [50]. Enkurin acts as an adaptor protein interacting with TRPC1, TRPC2 and TRPC5 in mouse spermatozoa and localizes a Ca2+ sensitive signal transduction machinery to a Ca2+-permeable ion channel [78]. TRPC2 has IP3-R binding domains and junctate, an IP3R associated protein also interacts with TRPC2 at its C-terminus. Junctate binds to TRPC2 independently of the calcium concentration and the junctate binding site does not overlap with the common IP3R/calmodulin binding sites. Thus junctate may participate in gating of TRPC2 in mouse sperm [69]. TRPC2 knockout mice are fertile although TRPC2 antibody shows a decrease

Fig. 46.1 Simple Modular Architecture Research Tool (SMART) analysis of domain organizations of various mouse TRPCs expressed in mouse testis. Transmembrane segments as predicted by the *TMHMM2* program ( ), coiled coil regions determined by the *Coils2* program ( ), segments of low compositional complexity determined by the *SEG* program ( ), disordered regions detected by *DisEMBL* ( ). SMART domains and PFAM domains are represented as blocks with respective abbreviated labels. Multiple cartoons for the same protein are generated to represent overlapping domains. ANK, ankyrin; XRCC1_N, XRCC N-terminal domain; Ion_trans, Ion transport domain; PKD_channel, cation channel region of PKD1 and PKD2 proteins

in acrosome reaction [79, 80]. TRPC2 was reported from bovine testis where its mRNA message was localised in spermatocytes and not in Leydig or Sertoli cells suggesting that bTRP2 may contribute to the formation of ion channels in sperm cells [48].

TRPC3: TRPC3 forms a constitutionally active calcium entry channel in resting conditions [80–82]. The basal constitutive activity of TRPC3 depends on its mono-glycosylated state [83]. TRPC3 is activated via membrane receptor activation, leading to phospolipase mediated diacylglycerol (DAG) production [84–87]. Phospholipase C-generated diacylglycerol has been shown to have a dual role in regulation of TRPC3 activity. It serves as a signal for TRPC3 activation as well as a signal for negative feedback via protein kinase C-mediated phosphorylation [88]. TRPC3 can also be activated by calcium store depletion pathway [89, 90]. In fact the coupling hypothesis of TRPC activation involving calcium store depletion by InsP3 binding to InsP3R and its interaction/activation with calcium entry channels on plasma membrane has been studied by using TRPC3 as the representative of TRPC family member [91]. By expression of HTRPC-1 and HTRPC-3 antisense constructs in HEK293 cells [92] and by suppression of TRPC3 protein levels using siRNA in A431 cells, TRPC3 has been reported to form store-operated channels [93]. In pancreas of TRPC3–/– mice receptor-stimulated and SOC-mediated Ca^{2+} influx is reduced by about 50%, indicating that TRPC3 functions as an SOC [94]. A novel human variant of TRPC3 which has a 73-aa N-terminal extension, and referred to as hTRPC3a mediates both receptor operated as well as store operated calcium entry [95].

TRPC3 expression levels affect its functional activation. In DT40 cells at low level of expression TRPC3 forms store operated channels and at high levels of expression TRPC3 form receptor operated channels [87, 96–98]. A short splice variant of TRPC3 named TRP3sv is reported in rat expressed predominantly in heart, kidney, and liver and detected in very low quantity in testis also. TRP3sv has been shown to be activated by membrane depolarisation and acts as a calcium activated non-selective cation channel [99].

TRPC3 co-assembles in heteromeric complexes with other members of TRPC family. With TRPC1 it forms Phospholipase-C and intra-cellular calcium regulated calcium channels [100] and heteromeric store-operated non-selective cation channel [58]. TRPC3 and TRPC4 are able to form redox sensitive cation channels [101], A key mechanism involved in redox activation of TRPC3 appears to be ROS-induced promotion of protein tyrosine phosphorylation and stimulation of phospolipase C activity [102]. By Fluorescence resonance energy transfer (FRET) studies it is shown that diacylglycerol-regulated TRPC3, TRPC6 and TRPC7 channels are capable of forming heteromeric complexes [103].

In HEK293 cells TRPC3 has been shown to associate with Na^+/Ca^{2+} exchanger (NCX) and Ca^{2+} entry into TRPC3-expressing cells involves reversed mode Na^+/Ca^{2+} exchange [104]. TRPC3 is the only member of TRPC family present in rat ventricular myocytes and it co-localises with sodium-potassium ATPase pump and sodium-calcium exchange pump, forming a signalling complex [105]. PLC mediated increase in membrane association of TRPC3-NCX1 complex is key component of calcium homeostasis in cardiac myocytes [106]. Inhibition of sodium/calcium exchange channel also inhibits TRPC3 [107].

TRPC3 activity is regulated by a number of molecules. Receptor activation leading to phospholipase mediated DAG generation is a major pathway of TRPC3 mediated calcium entry. Calcium entry in the T-cells is mediated by TRPC3 via activation of T-cell receptor and is a key triggering event for the T-cell-associated immune response [108]. Bradykinin and 1-oleoyl-2-acetyl-*sn*-glycerol (OAG) stimulated an apical calcium influx in polarised MDCK cells, indicating a receptor-coupled $[Ca^{2+}]i$ increase mediated by TRPC3/TRPC6 [109]. TRPC3 is involved in flow- and bradykinin-induced vasodilation in rat small mesenteric arteries probably by mediating the Ca(2+) influx into endothelial cells [110]. Vascular endothelial growth factor (VEGF) activated heterologously expressed TRPC3/6 channels through VEGFR2 in human microvascular endothelial cells (HMVECs) [111]. In a subset of CD133$^+$ stem cells TRPC3 plays a key role in VEGF-induced Ca^{2+} entry as a key Ca^{2+} entry channel and acts as an essential determinant of cell fate in CD133$^+$ progenitor-derived colonies [101]. In HEK 293 cells histamine could activate TRPC3 via histamine H [2] receptors, and both phospholipase C and phospholipase D participate in this process [112]. In rat cerebellar slices, metabotropic glutamate receptors activate TRPC3 channels through the small GTP-binding protein Rho and subsequent phospholipase D stimulation [113].

TRPC3 associates with inositol triphosphate receptor and this plays an important role in gating of TRPC3 channels. TRPC3 has a common calcium/calmodulin and IP3R binding site. Displacement of CaM from this binding site by activated IP3R is an important step in activation of TRPC3 [114]. The C terminus IP3 receptor, calmodulin binding (CIRB) region of TRPC3, has been suggested to be involved in its targeting to the plasma membrane and deletion of 20 amino acids corresponding to this highly conserved CIRB region results in the loss of diacylglycerol and agonist-mediated TRPC3 channel activation in HEK293 cells [115]. In resting cells TRPC3 exists in a complex with IP [3]-R and scaffold protein Homer H1-b/c located in part at plasma membrane and in part at intra-cellular membrane. Binding of IP [3] to the IP [3]Rs dissociates the interaction between IP [3]Rs and H1 but not between H1 and TRPC3. TRPC3 translocates to the plasma membrane upon receptor activation and store depletion. In HEK293, dissociating the H1b/c-IP [3]R complex with H1a results in TRPC3 translocation to the PM, where it is spontaneously active. Thus assembly of the TRPC3-H1b/c-IP [3]Rs complexes by H1b/c mediates both the translocation of TRPC3-containing vesicles to the PM and gating of TRPC3 by IP [3]Rs [116]. Junctate, an integral calcium binding protein of endo (sarco) plasmic reticulum membrane forms a supramolecular complex with the InsP [3]R, TRPC3. The full-length protein modulates both agonist-induced and store depletion-induced calcium entry, whereas its NH [2] terminus affects receptor-activated calcium entry by TRPC3 [117]. However in InsP [3]Rs knockout DT40 cells TRPC3 is shown to be activated independently of InsP [3]Rs through DAG production resulting from receptor-mediated activation of either PLC-gamma or PLC-beta [98]. In mouse muscle myotubules junctophilin-2(JP-2) has been shown to interact with TRPC3 by glutamate in the F1-2 region (E227).

This substantial binding between JP2 and TRPC3 suggests that JP2 can be a regulatory protein of TRPC3 and/or TRPC3-mediated Ca(2+) homeostasis in skeletal muscle [118].

TRPC3 also interacts with Orai-1 (a plasma membrane calcium channel) and STIM1 (a single spanning membrane protein which functions as a sensor of ER luminal calcium). TRPC3 and TRPC1 have been reported to interact with STIM1 by the electrostatic interaction of negatively charged aspartate residues of TRPCs with positively charged residues of the poly basic domain of STIM1 [119], while another study reported that STIM1 indirectly binds to TRPC3 via it heteromerization with TRPC1 [64]. Orai interacts with TRPC3 and TRPC6 and confers sensitivity to store depletion and it has been suggested that Orai proteins, by interacting with TRPCs, act as regulatory subunits that confer STIM1-mediated store depletion sensitivity to these channels [120]. TRPC3 interacts with receptor for activated C-kinase-1 (RACK1) and this not only determines plasma membrane localization of the channel but also the interaction of IP [3]R with RACK1 and IP [3]-dependent intracellular Ca^{2+} release [121]. Agonist stimulation results in the formation of an Orai1-STIM1-TRPC3-RACK1-type I IP [3]R complex, where TRPC3 plays a central role. This Ca^{2+} signalling complex might be important for agonist induced both Ca^{2+} release and entry [122].

Phosphorylation of TRPC3 has been reported to regulate its activity. TRPC3 channels could be directly phosphorylated by Protein kinase G (PKG) at position T11 and S263, and this phosphorylation abolishes the store-operated Ca^{2+} influx mediated by TRPC3 channels [123]. TRPC3 activity is inhibited by Protein kinase C by a Protein kinase G dependent mechanism, Protein kinase G (PKG) inactivates TRPC3 by direct phosphorylation on Thr-11 and Ser-263 of the TRPC3 proteins, and protein kinase C (PKC) inactivates TRPC3 by phosphorylation on Ser-712 [124]. Receptor tyrosine kinase Src phosphorylates TRPC3 at Y226, which is essential for TRPC3 activation [125] during receptor and diacylglycerol activation of TRPC3 [126]. Interactions of TRPC3 with cytoskeleton [127], SNARE proteins [128], partial PH domains of phospolipase C [129] and plasma membrane cholesterol [130] also play an important role in regulating TRPC3 activity.

Mouse TRPC3 (NM_019510.2, NP_062383.2) is an 866 amino acid protein which has three N-terminal ankyrin repeats followed by TRP_2 domain and six transmembrane domains forming the ion transport channel (Fig. 46.1). In the mouse spermatozoa, TRPC3 has been localised in the post acrosomal head and sperm tail [47, 78]. In human spermatozoa, TRPC3 is localised on the acrosomal region and mid-piece by confocal microscopy and by electron microscopy it is shown to be present on the plasma membrane of sperm head and flagella [49]. In goat spermatozoa, we detected TRPC3 in the principal acrosomal domain and sperm mid-piece (unpublished). The presence of TRPC3 in sperm head may suggest its role in the calcium entry during acrosome reaction and its presence in the sperm mid-piece indicate its role in sperm motility [49].

TRPC4: TRPC4 acts as a store operated cation channel in Xenopus oocyte [131], adrenal cortex [132] and mice endothelial cells [133]. Deletion of TRPC4 in mouse caused impairment in store-operated Ca^{2+} current and Ca^{2+} store release activated Ca^{2+} influx in aortic and lung endothelial cells (LEC) [134]. In human keratinocytes knockdown of the level of endogenous TRPC1 or TRPC4 inhibited store-operated calcium entry, indicating they are part of the native SOC [135]. TRPC4 is a component of SOC in human corneal epithelial cells [136].

In HEK cells heterologously expressing TRPC4 and TRPC5 (mTRPC4/5) form nonselective cation channels that integrate signalling pathways from G-protein-coupled receptors and receptor tyrosine kinases independently of store depletion [137]. TRPC4 forms a receptor operated cation channel rather than store operated channels in HEK293 cells [138] and neuronal cells [139]. In visceral smooth muscle cells Galphai is involved specifically in the activation of TRPC4 [140].

Two isoforms of the human TRPC4 protein, referred to as alpha-hTRP4 and beta-hTRP4 are reported [141]. In mouse mesangial cells TRPC4-alpha may form the homo-tetrameric SOC channels [142]. The shorter variant TRPC-4 beta forms receptor regulated cation channels when expressed in HEK 293 cells [137]. A new variant of TRPC-4 designated as rTRPC4gamma is detected from rat heart and brain. When rTRPC4gamma cDNA was transiently transfected to HEK-293 cells, thapsigargin (TG)-induced Ca2+ entry was suppressed significantly, suggesting that rTRPC4gamma may play a modulatory role in Capacitative calcium entry (CCE) channel activity [143].

TRPC4 associates with phospolipase C-beta1 and NHERF (Na^+/H^+ exchanger regulatory factor) which is a two PDZ domain-containing protein that associates with the actin cytoskeleton via interactions with members of ezrin/radixin/moesin family. Thus, store-operated channels involving TRPC4 can form signalling complexes with phospolipase C isozymes via interactions with NHERF and thereby linking the lipase and the channels to the actin cytoskeleton. The interaction with the PDZ protein may constitute an important mechanism for distribution and regulation of TRPC4 store-operated channels [144]. TRPC4 interacts with IP [3]Rs. The IP [3]R-binding domain of TRPC4 also interacts with calmodulin (CaM) in a Ca^{2+}-dependent manner and both CaM and IP [3]Rs play important roles in controlling the gating of Trp-based channels [145]. In glomerular mesangial cells, the TRPC1/TRPC4 complexes constitute the functional subunits of SOC and that the interaction between STIM1 and TRPC4 may be the mechanism for the activation of the channels [59]. STIM1 binds TRPC4 and TRPC5 and determines their function as SOCs [64].

Mouse TRPC4 (NM_016984.2) is a 930 amino acid protein with two N-terminal Ankyrin repeats followed by coiled coil region and TRP_2 domain and 8 predicated transmembrane helices. The 6 transmembrane helices towards C terminus form ion transport domain of the protein (Fig. 46.1). TRPC4 has not been detected at protein level in mice. In human spermatozoa, weak TRPC4 fluorescence has been detected on head and a strong signal on mid piece and tail [49]. The presence of TRPC4 on sperm head suggests its role in sperm acrosome reaction and its presence in mid-piece may suggest its role in motility.

TRPC6: TRPC6 is a closely related isoform of TRPC3 and is activated by GPCR coupled generation of second messengers like DAG in a non selective membrane delimited fashion [84]. TRPC6 act as a receptor operated cation channel in COS cells [146] and PC12D cells [147]. In A7r5 smooth muscle cells TRPC6 forms receptor operated cation channels alone [148] or as heteromultimers with TRPC-7 [149]. In human platelets TRPC6 is reported to be involved in receptor activated cation entry [150], but with TRPC1 it also forms store-operated Ca2+ entry (SOCE) channels [151, 152]. In platelets and megakaryocytic cell lines TRPC6 and TRPC1 have been reported as the non-selective cation channels stimulated by the physiological agonist ADP [153]. In primary erythroid cells TRPC2 and TRPC6 interact to modulate calcium influx stimulated by erythropoietin [154]. However in neutrophiles TRPC6 is reported to induce store operated calcium entry after Platelet activating factor (PAF) treatment [155]. TRPC6 is detected in neuronal cells [156], where it forms DAG activated cation channels [157].

TRPC6 forms receptor activated cation channel in rabbit portal vein myocytes [158], in human airway smooth muscle (HASM) [159] and it also forms a mechanosensitve channels in addition to receptor operated non selective cation channel in vascular smooth muscle cells [160].

TRPC6 engages in functional interaction with plasmalemmal Na(+)/K(+)-ATPase (NKA) pump in brain and kidney [161], L-type Ca2+ channel in A7r5 smooth muscle cells [162], sodium calcium exchanger (NCX) in smooth muscle cells [163] and Large-conductance (BK (Ca) type) Ca (2+)-activated K(+) channels [164].

STIM1 does not bind to TRPC6 directly but mediates heteromerization of TRPC6 with TRPC4 [64]. Ca2+-store depletion by TG (thapsigargin) + ionomycin enhances the interaction between hTRPC6 and the CCE proteins Orai1 and STIM1. In contrast, stimulation with the diacylglycerol analogue OAG (1-oleoyl-2-acetyl-sn-glycerol) displaces hTRPC6 from Orai1 and STIM1 and enhances the association between hTRPC6 and hTRPC3. Thus hTRPC6 participates both in Capacitative and non capacitative calcium entry through its interaction with the Orai1-STIM1 complex or hTRPC3 respectively [165].

TRPC6 activity is regulated by arachidonic acid metabolite, 20-HETE which increases the open probability of TRPC6 channels by approximately threefold [166]. Ca(2+)-CaM binds to TRPC6 and CaM is involved in the modulation of receptor operated calcium entry by TRPC6 [167]. Phosphoinositide modulates TRPC6 activity by directly interacting with it, which disrupts the association of calmodulin (CaM) with TRPC6 [168]. In COS-7 cells epidermal growth factor receptor induced rapid tyrosine phosphorylation of TRPC6 by Src family PTKs and this is a novel regulatory mechanism of TRPC6 channel activity [169]. Flufenamic acid (FFA) activates TRPC6 [Ca(2+)](i) signalling in both podocyte cell line ciPod and HEK293 cells independently of TRPC3 and TRPC7, and independently of properties of the fenamate family [170]. High glucose increases TRPC6 channel protein expression on the platelet surface and calcium influx which is mediated by a phophatidylinositol 3-kinase-dependent pathway [171].

An exocytotic mechanism has been reported to increase the membrane localisation of TRPC-6 thereby modulating its activity [172]. RNF24, a new membrane RING-H2 protein, interacts with the ankyrin-like repeat domain of TRPC6 in the Golgi apparatus and affects TRPC intracellular trafficking [173].

In rabbit mesenteric artery myocytes constitutively produced PIP [2] exerts a powerful inhibitory action on native TRPC6 channels [174]. TRPC6 channels can be negatively regulated by the NO-cGMP-PKG pathway, probably via T69 phosphorylation of the N-terminal. This mechanism may be physiologically important in vascular tissues where nitric oxide (NO) is constantly released from vascular endothelial cells or nitrergic nerve [175].

TRPC6 channels play an essential role in regulation of myogenic tone via inducing depolarization current, constriction (myogenic tone) of small arteries and arterioles, and this response is a key element in blood flow regulation [176]. TRPC6 channel activity at the slit diaphragm is essential for proper regulation of podocyte structure and function in kidney [177]. Human and mouse erythrocytes express TRPC6 cation channels which participate in cation leak and Ca(2+)-induced suicidal death [178]. In neutrophils macrophage-inflammatory protein (MIP-2) stimulation of Gq-protein-coupled chemokine receptors activates TRPC6 channels effecting cytoskeleton rearrangement and migration of neutrophils [179]. TrpC-6 plays an important role in immunological function in Th2 cells, eosinophils, mast cells and B cells [180]. TRPC6 is an obligatory component of cation channels required for the VEGF-mediated increase in cytosolic calcium and subsequent downstream signalling that leads to processes associated with angiogenesis [181]. In HaCaT cells and in primary cultures of human keratinocytes hyperforin triggers differentiation and inhibits proliferation by inducing Ca(2+) influx via TRPC6 channels [182].

TRPC6 has been reported to have a critical role in progression and development of gastric cancer [183], oesophageal squamous cells cancer [184] and prostate cancer [185]. Mutations in TRPC6 have been reported to cause focal segmental glomerulosclerosis (FSGS) [186]. In metabolic syndrome, abnormally elevated adrenal TRPC-1, TRPC-5 and TRPC-6 expression may underlie increased plasma epinephrine and heart rate [187].

Mouse TRPC6 (NM_013838.2, NP_038866.2) is an 886 amino acid protein which has three N-terminal ankyrin repeats followed by TRP_2 domain. There are 6 transmembrane alpha helices which form part of the ion transport region (Fig. 46.1). In mouse sperm TRPC6 has been localised in the post-acrosomal head region and sperm flagellum [47]. In human sperm TRPC6 has been localised in mid-piece and sperm tail in addition to this a diffuse fluorescence has also been detected on sperm head [49]. In goat sperm we have detected TRPC6 in sperm head as well as mid-piece. The presence of TRPC6 in sperm head suggests its role in acrosome reaction and its presence in the mid-piece and tail suggests its role in the calcium controlled sperm motility and hyper activation.

46.3 TRPP Family

Polycystic kidney disease (PKD) gene encode two families of proteins: polycystin 1 (PKD1)-like and polycystin 2 (PKD2)-like proteins [44, 188–190]. The PKD2-like subgroup contains three homologous proteins – PKD2, PKD2L1 and PKD2L2 now referred to as TRPP2, TRPP3 and TRPP5. These proteins are like TRP proteins having 6 transmembrane domains and so included into TRPP (TRP polycystin) family of TRP super family. The PKD1-like subgroup contains five homologous proteins, all with an 11-transmembrane topology, and – by virtue of their structure – are not considered as members of the TRP superfamily [44, 191].

Drosophila PKD2 has been reported to be a Ca (2+)-activated, nonselective cation channel associated with the tail and the acrosome-containing head region of mature spermatozoa. Targeted disruption of Pkd2 results in male sterility without affecting spermatogenesis. The mutant sperm are motile but fail to swim into the storage organs in the female and hence are unable to fertilize [192]. Drosophila polycystin-2 homologue (amo), present at the distal tip of sperm tail, is necessary for the sperm entry into female sperm storage organ before fertilization. In flies with mutated *amo* normal motile sperm were present but they failed to enter female sperm storage organ resulting in sterility [193]. In sea urchin sperm TRPP-2 homologue sea urchin polycystin-2 (suPC2) associates with sea urchin polycystin-1 homolog suREJ3 and is located on the acrosomal domain of sperm head [194] and forms a receptor for egg jelly [195, 196]. Thus TRPP-2 (suPC2) and polycystin-1 homolog suREJ3 form a receptor/ion channel complex facilitating sperm binding to the egg coat, subsequent cation entry and acrosome reaction in sea-urchin.

Not much is known about the role of TRPP proteins in mammalian sperm function. Humans with autosomal polycystic kidney disease due to defects in polycystin (PKD) genes have necrospermia and immotile sperm along with seminal vesicle cysts, and ejaculatory duct cysts [197]. Therefore TRPP proteins may have a role in sperm motility in mammals including humans. Pkdrej (a member of polycystin-1 gene family though not a TRPP member) is shown to confer the ability to undergo a zona pellucida–evoked acrosome reaction, as the sperm from mutant Pkdrej (tm/tm) mice develop the ability to undergo a zona pellucida-evoked acrosome reaction more slowly than sperm from wild-type males, there by conferring the wild type males a selective advantage in case of sequential mating condition [198].

46.4 TRPV Family

Transient receptor potential vanilloid receptor-1 (TRPV1) is a ligand-gated, nonselective cation channel. TRPV1 can be activated by vanilloid compounds such as capsaicin, protons, products of the arachidonic metabolism, and cations [199–203]. TRPV1 activity can also be influenced by ambient temperature; a slight warming above room temperature potentiates the responsiveness of TRPV1 to its chemical agonists, and at temperatures above 42°C, it is activated in the absence of exogenous chemical ligands [204].

In mammals, six members of TRPV family TRPV 1-6 have been reported. Of these, TRPV1 and TRPV5 have been reported in the spermatozoa. In testis TRPV1 expression has been shown to increase in meiotic spermatocytes [205], and it plays an important role in preventing spermatogonia from undergoing massive cell death under heat stress [206].

TRPV1 is activated by an endo-cannabinoid anandamide (arachidonoylethanolamide, AEA) [207]. In boar spermatozoa, it co-localises with cannabinoid receptor CB1R and fatty acid amide hydrolase (FAAH) (which hydrolyses anandamide into ethanolamine and arachidonic acid) on post-acrosomal head domain and mid-piece. After capacitation, TRPV1 relocalizes to the anterior region of the sperm head [208]. TRPV1 has been reported on the post-acrosomal region of human spermatozoa as well, and is important in the progesterone induced sperm oocyte fusion [209]. TRPV1 plays a role in stabilization of the plasma membranes in capacitated spermatozoa before the sperm-zona pellucida interaction and the zona pellucida mediated true acrosome reaction [210]. In addition to this TRPV1 is also reported in the sertoli cells where it regulates an acid sensing chloride channel (ASCC) necessary for maintaining the acidic milieu for spermatogenesis [211]. However TRPV1 knockout mice are normal and fertile [206, 212]. TRPV5 is reported in the spermatogenic cells and spermatozoa of rat and have been proposed to function as cation channel during spermatogenesis [213].

46.5 TRPM

Transient receptor potential melastatin (TRPM) is a family of TRP super family closely related to the TRPC and TRPV. Members of the TRPM family share close to 20% amino acid identity with TRPC over a 325 amino acid segment that includes the C-terminal five transmembrane domains and a highly conserved 25-residue TRP domain [214]. TRPM member melastatin (MLSN), is encoded by a gene that was down-regulated in mouse melanoma tumor-cell lines and has been suggested to be a tumor-suppressor gene [215, 216]. TRPM family has eight members TRPM1 to TRPM8. TRPM4, TRPM7 and TRPM8 are reported in spermatozoa.

TRPM4: TRPM4 is a calcium-activated nonselective (CAN) cation channel which conducts monovalent cations such as Na^+ and K^+ without significant permeation of Ca^{2+}. TRPM has two variants TRPM4a and TRPM4b. TRPM4b is activated following receptor-mediated Ca^{2+} mobilization, representing a regulatory mechanism that controls the magnitude of Ca^{2+} influx by modulating the membrane potential [217, 218]. TRPM4 has been detected in spermatogenic cells and spermatozoa in rat testis [213].

TRPM7: TRPM7 (ChaK1, TRP-PLIK, LTRPC7) is a ubiquitous, calcium-permeant ion channel that is unique in being both an ion channel and a serine/threonine kinase [219, 220]. TRPM7 is a constitutively active divalent cation-selective ion channel, whose basal activity is regulated by intracellular levels of Mg(2+) and MgATP. TRPM7 activity is up- and down-regulated through its endogenous kinase in a cAMP- and PKA-dependent manner [221]. In addition to calcium and magnesium, it is also permeable to trace metal ions such as Zn(2+), Fe(2+), Cu(2+), Mn(2+), and Co(2+) into vertebrate cells [222]. TRPM7 has also been detected in the spermatogenic cells and spermatozoa of rat testis and may be involved in spermatogensis [213].

TRPM8: TRPM8 is cold, menthol sensitive receptor in somatosensory neurons [223], which is also detected in the prostate [224]. TRPM8 channels are permeable to Ca^{2+} and account for increases in $[Ca^{2+}]_i$ in response to cold and menthol [225]. TRPM8 is a channel activated by cold temperatures, voltage, and menthol. Temperature and voltage has been reported to interact allosterically to enhance channel opening [226].

TRPM-8 is present in the human sperm. It is localised on sperm head and flagellum. Activation of TRPM8 either by temperature or menthol induced [Ca(2+)]i increases in human sperm and is suggested to be involved in cell signalling events such as thermotaxis or chemotaxis [227].

46.6 Conclusions

Among the members of TRP super-family, TRPC family proteins are more widely studied and reported in the spermatozoa. The ankyrin protein interaction domain, CRIB domain and the coiled coil motifs in TRP proteins enable them to not only form homomeric and heteromeric complexes with other TRPC proteins, but also to interact with a wide variety of other proteins like Inositol receptor [69, 76, 117], phospolipase C [76], calcium calmodulin [114], orai [63, 122], stim-1 [59, 122], enkurin [78], junctate [69], homer [116], rack [121], NHERF [144] and a number of other ion channels including plasmalemmal Na(+)/K(+)-ATPase (NKA) pump [161], L-type Ca2+ channel [162], sodium calcium exchanger (NCX) [163], and large-conductance (BK (Ca) type) Ca (2+)-activated K(+) channels [164]. These protein-protein interactions provide the TRPCs a wide variety of activation/gating and functional regulation mechanisms in different cell/tissue systems.

In spermatozoa, TRPC ion channels may function as classical store operated calcium channels and as cation channels responding to different maturation promoting factors in testis, epididymis and vas deferens and factors which promote capacitation and acrosome reaction in the female reproductive tract. The ubiquitous localization of multiple TRPC members on acrosomal head domain, the mid-piece and/or the tail of spermatozoa indicates that TRPCs may form heteromers and form store operated as well as receptor operated ion channels. However we do not know how these ion channels interact with other regulatory proteins like inositiol triphosphate receptors, calmodulin, orai, stim-1 etc.

Data from our and other labs has revealed the presence of membrane raft micro domains in sperm membrane [47, 228]. These are highly dynamic, nanoscale, cholesterol, glycosphingolipid enriched ordered membrane micro domains which function as platforms for forming cell signalling complexes by bringing together different components of the signalling complex thereby regulating in space and time different cell signalling pathways. In addition to this these micro domains also regulate intracellular trafficking of lipids and proteins [229, 230]. TRPC proteins have been reported in the membrane rafts along with other interacting/regulatory proteins like caveolin, G-alpha subunit, inositol triphosphate receptor and phospolipase C. Caveolar membrane raft microdomains are important for formation of the calcium signalling complex involving TRPCs and other regulatory proteins [231]. Caveolin (Cav1) has an important role in the assembly of SOCE channel(s) [232]. Caveolin scaffold domain (NH [2]-terminal residues 82-101; CSD) interacts with transient receptor potential canonical channel 1 (TRPC1) and inositol 1,4,5-trisphosphate receptor 3 (IP [3]R3) to regulate Ca(2+) entry [127, 233–235]. Caveolin acts as a negative regulator of TRPC-1 function in store operated calcium entry and it has been suggested that activation of TRPC1-SOC by STIM1 mediates release of the channel from Caveolin [236].

In sperm membrane also TRPCs have been detected in caveolin enriched membrane microdomains [47]. Membrane raft micro-domains may be important for assembly of the TRP proteins and other interacting proteins of the calcium signalling complex in sperm. After detection of TRP ion channels in the sperm the next question will be to identify their regulatory/interacting proteins in spermatozoa during stages of sperm maturation, capacitation and acrosome reaction. Use of conventional electrophysiological techniques has been difficult in case of mature spermatozoa due to their small size [39] but high resolution scanning patch clamp has made it possible to study ion channels in sperm [237, 238] and can be useful to study TRP ion channels function in mature spermatozoa. As spermatozoa are terminally differentiated cells with no transcriptional activity, protein level manipulation by using si-RNA technique is also not applicable to study functions of these ion channels in mature spermatozoa. Use of specific inhibitory peptides/antibodies against the TRP proteins and their interacting partners may help in understanding their functional role in spermatozoa. In addition, knockout mouse models can also help in understanding the importance of TRP proteins in sperm function.

Acknowledgments This work was supported by Grant Nos. SP/SO/B-47/2000 (Department of Science and Technology, New Delhi) and BT/PR5512/BRB/10/394/2004 (Department of Biotechnology, New Delhi) to PGK. MS received Senior Research Fellowship (3/1/2/13/07-RHN) from Indian Council of Medical Research, New Delhi.

References

1. Yanagimachi R (1994) Mammalian fertilization. In: Knobil E, Neill JD (ed) The physiology of reproduction. Raven Press, New York, NY, pp. 189–317
2. Abou-haila A, Tulsiani DR (2009) Signal transduction pathways that regulate sperm capacitation and the acrosome reaction. Arch Biochem Biophys 485:72–81
3. Wolf DE, Hagopian SS, Ishijima S (1986) Changes in sperm plasma membrane lipid diffusibility after hyperactivation during in vitro capacitation in the mouse. J Cell Biol 102:1372–1377
4. Zeng Y, Clark EN, Florman HM (1995) Sperm membrane potential: hyperpolarization during capacitation regulates zona pellucida-dependent acrosomal secretion. Dev Biol 171:554–563
5. Visconti PE, Moore GD, Bailey JL, Leclerc P, Connors SA, Pan D, Olds-Clarke P, Kopf GS (1995) Capacitation of mouse spermatozoa. II. Protein tyrosine phosphorylation and capacitation are regulated by a cAMP-dependent pathway. Development 121:1139–1150
6. White DR, Aitken RJ (1989) Relationship between calcium, cyclic AMP, ATP, and intracellular pH and the capacity of hamster spermatozoa to express hyperactivated motility. Gamete Res 22:163–177

7. Fraser LR (1990) Adenosine and its analogues, possibly acting at A2 receptors, stimulate mouse sperm fertilizing ability during early stages of capacitation. J Reprod Fertil 89: 467–476
8. Visconti PE, Bailey JL, Moore GD, Pan D, Olds-Clarke P, Kopf GS (1995) Capacitation of mouse spermatozoa. I. Correlation between the capacitation state and protein tyrosine phosphorylation. Development 121:1129–1137
9. Naz RK, Ahmad K, Kumar R (1991) Role of membrane phosphotyrosine proteins in human spermatozoal function. J Cell Sci 99:157–165
10. Naz RK, Rajesh PB (2004) Role of tyrosine phosphorylation in sperm capacitation/acrosome reaction. Reprod Biol Endocrinol 2:75
11. Babcock DF, Pfeiffer DR (1987) Independent elevation of cytosolic [Ca2+] and pH of mammalian sperm by voltage-dependent and pH-sensitive mechanisms. J Biol Chem 262:15041–15047
12. Babcock DF, Bosma MM, Battaglia DE, Darszon A (1992) Early persistent activation of sperm K+ channels by the egg peptide speract. Proc Natl Acad Sci USA 89:6001–6005
13. Garcia-Soto J, Darszon A (1985) High pH-induced acrosome reaction and Ca2+ uptake in sea urchin sperm suspended in Na+-free seawater. Dev Biol 110:338–345
14. Labarca P, Santi C, Zapata O, Morales E (1996) Beltr'an,C, Li'evano,A, Darszon, A: a cAMP regulated K+-selective channel from the sea urchin sperm plasma membrane. Dev Biol 174:271–280
15. Lee HC, Garbers DL (1986) Modulation of the voltage-sensitive Na+/H+ exchange in sea urchin spermatozoa through membrane potential changes induced by the egg peptide speract. J Biol Chem 261:16026–16032
16. Silvestroni L, Menditto A (1989) Calcium uptake in human spermatozoa: characterization and mechanisms. Arch Androl 23:87–96
17. Spira B, Breitbart H (1992) The role of anion channels in the mechanism of acrosome reaction in bull spermatozoa. Biochim Biophys Acta 1109:65–73
18. Wistrom CA, Meizel S (1993) Evidence suggesting involvement of a unique human sperm steroid receptor/Cl– channel complex in the progesterone-initiated acrosome reaction. Dev Biol 159:679–690
19. Wong PY, Lee WM (1983) Potassium movement during sodium-induced motility initiation in the rat caudal epididymal spermatozoa. Biol Reprod 28:206–212
20. Florman HM, Corron ME, Kim TD, Babcock DF (1992) Activation of voltage-dependent calcium channels of mammalian sperm is required for zona pellucida-induced acrosomal exocytosis. Dev Biol 152:304–314
21. Breitbart H, Rubinstein S, Nass-Arden L (1985) The role of calcium and Ca2+-ATPase in maintaining motility in ram spermatozoa. J Biol Chem 260:11548–11553
22. Chinoy NJ, Verma RJ, Patel KG (1983) Effect of calcium on sperm motility of cauda epididymis in vitro. Acta Eur Fertil 14:421–423
23. Davis BK (1978) Effect of calcium on motility and fertilization by rat spermatozoa in vitro. Proc Soc Exp Biol Med 157:54–56
24. Gorus FK, Finsy R, Pipeleers DG (1982) Effect of temperature, nutrients, calcium, and cAMP on motility of human spermatozoa. Am J Physiol 242:C304–C311
25. Hoskins DD, Brandt H, Acott TS (1978) Initiation of sperm motility in the mammalian epididymis. Fed Proc 37:2534–2542
26. Iwasa F, Shimizu H, Mohri H (1981) Effects of Ca2+ and Mg2+ on motility of sea urchin spermatozoa. Experientia 37:861–862
27. Morton BE, Sagadraca R, Fraser C (1978) Sperm motility within the mammalian epididymis: species variation and correlation with free calcium levels in epididymal plasma. Fertil Steril 29:695–698
28. Nelson L, Gardner ME, Young MJ (1982) Regulation of calcium distribution in bovine sperm cells: cytochemical evidence for motility control mechanisms. Cell Motil 2:225–242
29. Lee WM, Tsang AY, Wong PY (1981) Effects of divalent and lanthanide ions on motility initiation in rat caudal epididymal spermatozoa. Br J Pharmacol 73:633–638

30. Anand SR, Atreja SK, Chauhan MS, Behl R (1989) In vitro capacitation of goat spermatozoa. Indian J Exp Biol 27:921–924
31. Singh JP, Babcock DF, Lardy HA (1978) Increased calcium-ion influx is a component of capacitation of spermatozoa. Biochem J 172:549–556
32. Hong CY, Chiang BN, Turner P (1984) Calcium ion is the key regulator of human sperm function. Lancet 2:1449–1451
33. Summers RG, Talbot P, Keough EM, Hylander BL, Franklin LE (1976) Ionophore A23187 induces acrosome reactions in sea urchin and guinea pig spermatozoa. J Exp Zool 196: 381–385
34. Thomas P, Meizel S (1988) An influx of extracellular calcium is required for initiation of the human sperm acrosome reaction induced by human follicular fluid. Gamete Res 20:397–411
35. Cox T, Peterson RN (1989) Identification of calcium conducting channels in isolated boar sperm plasma membranes. Biochem Biophys Res Commun 161:162–168
36. Guerrero A, Darszon A (1989) Evidence for the activation of two different Ca2+ channels during the egg jelly-induced acrosome reaction of sea urchin sperm. J Biol Chem 264:19593–19599
37. Beltran C, Darszon A, Labarca P, Lievano A (1994) A high-conductance voltage-dependent multistate Ca2+ channel found in sea urchin and mouse spermatozoa. FEBS Lett 338:23–26
38. Weyand I, Godde M, Frings S, Weiner J, Muller F, Altenhofen W, Hatt H, Kaupp UB (1994) Cloning and functional expression of a cyclic-nucleotide-gated channel from mammalian sperm. Nature 368:859–863
39. Lievano A, Santi CM, Serrano CJ, Trevino CL, Bellve AR, Hernandez-Cruz A, Darszon A (1996) T-type Ca2+ channels and alpha1E expression in spermatogenic cells, and their possible relevance to the sperm acrosome reaction. FEBS Lett 388:150–154
40. Goodwin LO, Leeds NB, Hurley I, Mandel FS, Pergolizzi RG, Benoff S (1997) Isolation and characterization of the primary structure of testis-specific L-type calcium channel: implications for contraception. Mol Hum Reprod 3:255–268
41. O'Toole CM, Arnoult C, Darszon A, Steinhardt RA, Florman HM (2000) Ca(2+) entry through store-operated channels in mouse sperm is initiated by egg ZP3 and drives the acrosome reaction. Mol Biol Cell 11:1571–1584
42. Lobley A, Pierron V, Reynolds L, Allen L, Michalovich D (2003) Identification of human and mouse CatSper3 and CatSper4 genes: characterisation of a common interaction domain and evidence for expression in testis. Reprod Biol Endocrinol 1:53
43. Trevino CL, Felix R, Castellano LE, Gutierrez C, Rodriguez D, Pacheco J, Lopez-Gonzalez I, Gomora JC, Tsutsumi V, Hernandez-Cruz A, Fiordelisio T, Scaling AL, Darszon A (2004) Expression and differential cell distribution of low-threshold Ca(2+) channels in mammalian male germ cells and sperm. FEBS Lett 563:87–92
44. Delmas P (2004) Assembly and gating of TRPC channels in signalling microdomains. Novartis Found Symp 258:75–89 (discussion 89–102, 263-6.:75–89)
45. Nilius B (2007) TRP channels in disease. Biochim Biophys Acta 1772:805–812
46. Voets T, Talavera K, Owsianik G, Nilius B (2005) Sensing with TRP channels. Nat Chem Biol 1:85–92
47. Trevino CL, Serrano CJ, Beltran C, Felix R, Darszon A (2001) Identification of mouse trp homologs and lipid rafts from spermatogenic cells and sperm. FEBS Lett 509:119–125
48. Wissenbach U, Schroth G, Philipp S, Flockerzi V (1998) Structure and mRNA expression of a bovine trp homologue related to mammalian trp2 transcripts. FEBS Lett 429: 61–66
49. Castellano LE, Trevino CL, Rodriguez D, Serrano CJ, Pacheco J, Tsutsumi V, Felix R, Darszon A (2003) Transient receptor potential (TRPC) channels in human sperm: expression, cellular localization and involvement in the regulation of flagellar motility. FEBS Lett 541:69–74
50. Jungnickel MK, Marrero H, Birnbaumer L, Lemos JR, Florman HM (2001) Trp2 regulates entry of Ca2+ into mouse sperm triggered by egg ZP3. Nat Cell Biol 3:499–502

51. Wes PD, Chevesich J, Jeromin A, Rosenberg C, Stetten G, Montell C (1995) TRPC1, a human homolog of a Drosophila store-operated channel. Proc Natl Acad Sci USA 92:9652–9656
52. Sinkins WG, Estacion M, Schilling WP (1998) Functional expression of TrpC1: a human homologue of the Drosophila Trp channel. Biochem J 331:331–339
53. Rao JN, Platoshyn O, Golovina VA, Liu L, Zou T, Marasa BS, Turner DJ, Yuan JX, Wang JY (2006) TRPC1 functions as a store-operated Ca2+ channel in intestinal epithelial cells and regulates early mucosal restitution after wounding. Am J Physiol Gastrointest Liver Physiol 290:G782–G792
54. Xu SZ, Beech DJ (2001) TrpC1 is a membrane-spanning subunit of store-operated Ca(2+) channels in native vascular smooth muscle cells. Circ Res 88:84–87
55. Antoniotti S, Lovisolo D, Fiorio PA, Munaron L (2002) Expression and functional role of bTRPC1 channels in native endothelial cells. FEBS Lett 510:189–195
56. Fiorio PA, Maric D, Brazer SC, Giacobini P, Liu X, Chang YH, Ambudkar IS, Barker JL (2005) Canonical transient receptor potential 1 plays a role in basic fibroblast growth factor (bFGF)/FGF receptor-1-induced Ca2+ entry and embryonic rat neural stem cell proliferation. J Neurosci 25:2687–2701
57. Wu X, Zagranichnaya TK, Gurda GT, Eves EM, Villereal ML (2004) A TRPC1/TRPC3-mediated increase in store-operated calcium entry is required for differentiation of H19-7 hippocampal neuronal cells. J Biol Chem 279:43392–43402
58. Liu X, Bandyopadhyay BC, Singh BB, Groschner K, Ambudkar IS (2005) Molecular analysis of a store-operated and 2-acetyl-sn-glycerol-sensitive non-selective cation channel. Heteromeric assembly of TRPC1-TRPC3. J Biol Chem 280:21600–21606
59. Sours-Brothers S, Ding M, Graham S, Ma R (2009) Interaction between TRPC1/TRPC4 assembly and STIM1 contributes to store-operated Ca2+ entry in mesangial cells. Exp Biol Med (Maywood) 234:673–682
60. Strubing C, Krapivinsky G, Krapivinsky L, Clapham DE (2001) TRPC1 and TRPC5 form a novel cation channel in mammalian brain. Neuron 29:645–655
61. Bai CX, Giamarchi A, Rodat-Despoix L, Padilla F, Downs T, Tsiokas L, Delmas P (2008) Formation of a new receptor-operated channel by heteromeric assembly of TRPP2 and TRPC1 subunits. EMBO Rep 9:472–479
62. Zagranichnaya TK, Wu X, Villereal ML (2005) Endogenous TRPC1, TRPC3, and TRPC7 proteins combine to form native store-operated channels in HEK-293 cells. J Biol Chem 280:29559–29569
63. Ong HL, Cheng KT, Liu X, Bandyopadhyay BC, Paria BC, Soboloff J, Pani B, Gwack Y, Srikanth S, Singh BB, Gill DL, Ambudkar IS (2007) Dynamic assembly of TRPC1-STIM1-Orai1 ternary complex is involved in store-operated calcium influx. Evidence for similarities in store-operated and calcium release-activated calcium channel components. J Biol Chem 282:9105–9116
64. Yuan JP, Zeng W, Huang GN, Worley PF, Muallem S (2007) STIM1 heteromultimerizes TRPC channels to determine their function as store-operated channels. Nat Cell Biol 9:636–645
65. Saleh SN, Albert AP, Large WA (2009) Activation of native TRPC1/C5/C6 channels by endothelin-1 is mediated by both PIP(3) and PIP(2) in rabbit coronary artery myocytes. J Physiol 587:5361–5375
66. Maroto R, Raso A, Wood TG, Kurosky A, Martinac B, Hamill OP (2005) TRPC1 forms the stretch-activated cation channel in vertebrate cells. Nat Cell Biol 7:179–185
67. Liu X, Cheng KT, Bandyopadhyay BC, Pani B, Dietrich A, Paria BC, Swaim WD, Beech D, Yildrim E, Singh BB, Birnbaumer L, Ambudkar IS (2007) Attenuation of store-operated Ca2+ current impairs salivary gland fluid secretion in TRPC1(-/-) mice. Proc Natl Acad Sci USA 104:17542–17547
68. Alfonso S, Benito O, Alicia S, Angelica Z, Patricia G, Diana K, Vaca L (2008) Regulation of the cellular localization and function of human transient receptor potential channel 1 by other members of the TRPC family. Cell Calcium 43:375–387

69. Stamboulian S, Moutin MJ, Treves S, Pochon N, Grunwald D, Zorzato F, De Waard M, Ronjat M, Arnoult C (2005) Junctate, an inositol 1,4,5-triphosphate receptor associated protein, is present in rodent sperm and binds TRPC2 and TRPC5 but not TRPC1 channels. Dev Biol 286:326–337
70. Vannier B, Peyton M, Boulay G, Brown D, Qin N, Jiang M, Zhu X, Birnbaumer L (1999) Mouse trp2, the homologue of the human trpc2 pseudogene, encodes mTrp2, a store depletion-activated capacitative Ca2+ entry channel. Proc Natl Acad Sci USA 96:2060–2064
71. Hofmann T, Schaefer M, Schultz G, Gudermann T (2000) Cloning, expression and subcellular localization of two novel splice variants of mouse transient receptor potential channel 2. Biochem J 351:115–122
72. Liman ER, Corey DP, Dulac C (1999) TRP2: a candidate transduction channel for mammalian pheromone sensory signaling. Proc Natl Acad Sci USA 96:5791–5796
73. Menco BP, Carr VM, Ezeh PI, Liman ER, Yankova MP (2001) Ultrastructural localization of G-proteins and the channel protein TRP2 to microvilli of rat vomeronasal receptor cells. J Comp Neurol 438:468–489
74. Brann JH, Dennis JC, Morrison EE, Fadool DA (2002) Type-specific inositol 1,4,5-trisphosphate receptor localization in the vomeronasal organ and its interaction with a transient receptor potential channel, TRPC2. J Neurochem 83:1452–1460
75. Chu X, Cheung JY, Barber DL, Birnbaumer L, Rothblum LI, Conrad K, Abrasonis V, Chan YM, Stahl R, Carey DJ, Miller BA (2002) Erythropoietin modulates calcium influx through TRPC2. J Biol Chem 277:34375–34382
76. Tong Q, Chu X, Cheung JY, Conrad K, Stahl R, Barber DL, Mignery G, Miller BA (2004) Erythropoietin-modulated calcium influx through TRPC2 is mediated by phospholipase Cgamma and IP3R. Am J Physiol Cell Physiol 287:C1667–C1678
77. Heiner I, Eisfeld J, Halaszovich CR, Wehage E, Jungling E, Zitt C, Luckhoff A (2003) Expression profile of the transient receptor potential (TRP) family in neutrophil granulocytes: evidence for currents through long TRP channel 2 induced by ADP-ribose and NAD. Biochem J 371:1045–1053
78. Sutton KA, Jungnickel MK, Wang Y, Cullen K, Lambert S, Florman HM (2004) Enkurin is a novel calmodulin and TRPC channel binding protein in sperm. Dev Biol 274:426–435
79. Leypold BG, Yu CR, Leinders-Zufall T, Kim MM, Zufall F, Axel R (2002) Altered sexual and social behaviors in trp2 mutant mice. Proc Natl Acad Sci USA 99:6376–6381
80. Stowers L, Holy TE, Meister M, Dulac C, Koentges G (2002) Loss of sex discrimination and male-male aggression in mice deficient for TRP2. Science 295:1493–1500
81. Hurst RS, Zhu X, Boulay G, Birnbaumer L, Stefani E (1998) Ionic currents underlying HTRP3 mediated agonist-dependent Ca2+ influx in stably transfected HEK293 cells. FEBS Lett 422:333–338
82. Zhu X, Jiang M, Peyton M, Boulay G, Hurst R, Stefani E, Birnbaumer L (1996) trp, a novel mammalian gene family essential for agonist-activated capacitative Ca2+ entry. Cell 85:661–671
83. Dietrich A, Schnitzler M, Emmel J, Kalwa H, Hofmann T, Gudermann T (2003) N-linked protein glycosylation is a major determinant for basal TRPC3 and TRPC6 channel activity. J Biol Chem 278:47842–47852
84. Hofmann T, Obukhov AG, Schaefer M, Harteneck C, Gudermann T, Schultz G (1999) Direct activation of human TRPC6 and TRPC3 channels by diacylglycerol. Nature 397:259–263
85. Kamouchi M, Philipp S, Flockerzi V, Wissenbach U, Mamin A, Raeymaekers L, Eggermont J, Droogmans G, Nilius B (1999) Properties of heterologously expressed hTRP3 channels in bovine pulmonary artery endothelial cells. J Physiol 518(Pt 2):345–358
86. Li HS, Xu XZ, Montell C (1999) Activation of a TRPC3-dependent cation current through the neurotrophin BDNF. Neuron 24:261–273
87. Trebak M, St JB, McKay RR, Birnbaumer L, Putney JW Jr (2003) Signaling mechanism for receptor-activated canonical transient receptor potential 3 (TRPC3) channels. J Biol Chem 278:16244–16252

88. Trebak M, Hempel N, Wedel BJ, Smyth JT, Bird GS, Putney JW Jr (2005) Negative regulation of TRPC3 channels by protein kinase C-mediated phosphorylation of serine 712. Mol Pharmacol 67:558–563
89. Thebault S, Zholos A, Enfissi A, Slomianny C, Dewailly E, Roudbaraki M, Parys J, Prevarskaya N (2005) Receptor-operated Ca2+ entry mediated by TRPC3/TRPC6 proteins in rat prostate smooth muscle (PS1) cell line. J Cell Physiol 204:320–328
90. Vazquez G, Lievremont JP, St JB, Putney JW Jr (2001) Human Trp3 forms both inositol trisphosphate receptor-dependent and receptor-independent store-operated cation channels in DT40 avian B lymphocytes. Proc Natl Acad Sci USA 98:11777–11782
91. Kiselyov K, Xu X, Mozhayeva G, Kuo T, Pessah I, Mignery G, Zhu X, Birnbaumer L, Muallem S (1998) Functional interaction between InsP3 receptors and store-operated Htrp3 channels. Nature 396:478–482
92. Wu X, Babnigg G, Villereal ML (2000) Functional significance of human trp1 and trp3 in store-operated Ca(2+) entry in HEK-293 cells. Am J Physiol Cell Physiol 278: C526–C536
93. Kaznacheyeva E, Glushankova L, Bugaj V, Zimina O, Skopin A, Alexeenko V, Tsiokas L, Bezprozvanny I, Mozhayeva GN (2007) Suppression of TRPC3 leads to disappearance of store-operated channels and formation of a new type of store-independent channels in A431 cells. J Biol Chem 282:23655–23662
94. Kim MS, Hong JH, Li Q, Shin DM, Abramowitz J, Birnbaumer L, Muallem S (2009) Deletion of TRPC3 in mice reduces store-operated Ca2+ influx and the severity of acute pancreatitis. Gastroenterology 137:1509–1517
95. Yildirim E, Kawasaki BT, Birnbaumer L (2005) Molecular cloning of TRPC3a, an N-terminally extended, store-operated variant of the human C3 transient receptor potential channel. Proc Natl Acad Sci USA 102:3307–3311
96. Trebak M, Vazquez G, Bird GS, Putney JW Jr (2003) The TRPC3/6/7 subfamily of cation channels. Cell Calcium 33:451–461
97. Vazquez G, Wedel BJ, Trebak M, St John BG, Putney JW Jr (2003) Expression level of the canonical transient receptor potential 3 (TRPC3) channel determines its mechanism of activation. J Biol Chem 278:21649–21654
98. Venkatachalam K, Ma HT, Ford DL, Gill DL (2001) Expression of functional receptor-coupled TRPC3 channels in DT40 triple receptor InsP3 knockout cells. J Biol Chem 276:33980–33985
99. Ohki G, Miyoshi T, Murata M, Ishibashi K, Imai M, Suzuki M (2000) A calcium-activated cation current by an alternatively spliced form of Trp3 in the heart. J Biol Chem 275: 39055–39060
100. Lintschinger B, Balzer-Geldsetzer M, Baskaran T, Graier WF, Romanin C, Zhu MX, Groschner K (2000) Coassembly of Trp1 and Trp3 proteins generates diacylglycerol- and Ca2+-sensitive cation channels. J Biol Chem 275:27799–27805
101. Poteser M, Graziani A, Rosker C, Eder P, Derler I, Kahr H, Zhu MX, Romanin C, Groschner K (2006) TRPC3 and TRPC4 associate to form a redox-sensitive cation channel. Evidence for expression of native TRPC3-TRPC4 heteromeric channels in endothelial cells. J Biol Chem 281:13588–13595
102. Groschner K, Rosker C, Lukas M (2004) Role of TRP channels in oxidative stress. Novartis Found Symp 258:222–230 (discussion 231-5, 263-6.:222-230)
103. Amiri H, Schultz G, Schaefer M (2003) FRET-based analysis of TRPC subunit stoichiometry. Cell Calcium 33:463–470
104. Rosker C, Graziani A, Lukas M, Eder P, Zhu MX, Romanin C, Groschner K (2004) Ca(2+) signaling by TRPC3 involves Na(+) entry and local coupling to the Na(+)/Ca(2+) exchanger. J Biol Chem 279:13696–13704
105. Goel M, Zuo CD, Sinkins WG, Schilling WP (2007) TRPC3 channels colocalize with Na+/Ca2+ exchanger and Na+ pump in axial component of transverse-axial tubular system of rat ventricle. Am J Physiol Heart Circ Physiol 292:H874–H883

106. Eder P, Probst D, Rosker C, Poteser M, Wolinski H, Kohlwein SD, Romanin C, Groschner K (2007) Phospholipase C-dependent control of cardiac calcium homeostasis involves a TRPC3-NCX1 signaling complex. Cardiovasc Res 73:111–119
107. Kraft R (2007) The Na+/Ca2+ exchange inhibitor KB-R7943 potently blocks TRPC channels. Biochem Biophys Res Commun 361:230–236
108. Philipp S, Strauss B, Hirnet D, Wissenbach U, Mery L, Flockerzi V, Hoth M (2003) TRPC3 mediates T-cell receptor-dependent calcium entry in human T-lymphocytes. J Biol Chem 278:26629–26638
109. Bandyopadhyay BC, Swaim WD, Liu X, Redman RS, Patterson RL, Ambudkar IS (2005) Apical localization of a functional TRPC3/TRPC6-Ca2+-signaling complex in polarized epithelial cells. Role in apical Ca2+ influx. J Biol Chem 280:12908–12916
110. Liu CL, Huang Y, Ngai CY, Leung YK, Yao XQ (2006) TRPC3 is involved in flow- and bradykinin-induced vasodilation in rat small mesenteric arteries. Acta Pharmacol Sin 27:981–990
111. Cheng HW, James AF, Foster RR, Hancox JC, Bates DO (2006) VEGF activates receptor-operated cation channels in human microvascular endothelial cells. Arterioscler Thromb Vasc Biol 26:1768–1776
112. Kwan HY, Wong CO, Chen ZY, Dominic Chan TW, Huang Y, Yao X (2009) Stimulation of histamine H2 receptors activates TRPC3 channels through both phospholipase C and phospholipase D. Eur J Pharmacol 602:181–187
113. Glitsch MD (2010) Activation of native TRPC3 cation channels by phospholipase D. FASEB J 24:318–325
114. Zhang Z, Tang J, Tikunova S, Johnson JD, Chen Z, Qin N, Dietrich A, Stefani E, Birnbaumer L, Zhu MX (2001) Activation of Trp3 by inositol 1,4,5-trisphosphate receptors through displacement of inhibitory calmodulin from a common binding domain. Proc Natl Acad Sci USA 98:3168–3173
115. Wedel BJ, Vazquez G, McKay RR, St JB, Putney JW Jr (2003) A calmodulin/inositol 1,4,5-trisphosphate (IP3) receptor-binding region targets TRPC3 to the plasma membrane in a calmodulin/IP3 receptor-independent process. J Biol Chem 278:25758–25765
116. Kim JY, Zeng W, Kiselyov K, Yuan JP, Dehoff MH, Mikoshiba K, Worley PF, Muallem S (2006) Homer 1 mediates store- and inositol 1,4,5-trisphosphate receptor-dependent translocation and retrieval of TRPC3 to the plasma membrane. J Biol Chem 281:32540–32549
117. Treves S, Franzini-Armstrong C, Moccagatta L, Arnoult C, Grasso C, Schrum A, Ducreux S, Zhu MX, Mikoshiba K, Girard T, Smida-Rezgui S, Ronjat M, Zorzato F (2004) Junctate is a key element in calcium entry induced by activation of InsP3 receptors and/or calcium store depletion. J Cell Biol 166:537–548
118. Woo JS, Hwang JH, Ko JK, Kim dH, Ma J, Lee EH (2009) Glutamate at position 227 of junctophilin-2 is involved in binding to TRPC3. Mol Cell Biochem 328:25–32
119. Zeng W, Yuan JP, Kim MS, Choi YJ, Huang GN, Worley PF, Muallem S (2008) STIM1 gates TRPC channels, but not Orai1, by electrostatic interaction. Mol Cell 32:439–448
120. Liao Y, Erxleben C, Yildirim E, Abramowitz J, Armstrong DL, Birnbaumer L (2007) Orai proteins interact with TRPC channels and confer responsiveness to store depletion. Proc Natl Acad Sci USA 104:4682–4687
121. Bandyopadhyay BC, Ong HL, Lockwich TP, Liu X, Paria BC, Singh BB, Ambudkar IS (2008) TRPC3 controls agonist-stimulated intracellular Ca2+ release by mediating the interaction between inositol 1,4,5-trisphosphate receptor and RACK1. J Biol Chem 283:32821–32830
122. Woodard GE, Lopez JJ, Jardin I, Salido GM, Rosado JA (2010) TRPC3 regulates agonist-stimulated Ca2+ mobilization by mediating the interaction between type I inositol 1,4,5-trisphosphate receptor, RACK1 and Orai1. J Biol Chem 285:8045–8053
123. Kwan HY, Huang Y, Yao X (2004) Regulation of canonical transient receptor potential isoform 3 (TRPC3) channel by protein kinase G. Proc Natl Acad Sci USA 101:2625–2630
124. Kwan HY, Huang Y, Yao X (2006) Protein kinase C can inhibit TRPC3 channels indirectly via stimulating protein kinase G. J Cell Physiol 207:315–321

125. Kawasaki BT, Liao Y, Birnbaumer L (2006) Role of Src in C3 transient receptor potential channel function and evidence for a heterogeneous makeup of receptor- and store-operated Ca2+ entry channels. Proc Natl Acad Sci USA 103:335–340
126. Vazquez G, Wedel BJ, Kawasaki BT, Bird GS, Putney JW Jr (2004) Obligatory role of Src kinase in the signaling mechanism for TRPC3 cation channels. J Biol Chem 279:40521–40528
127. Lockwich T, Singh BB, Liu X, Ambudkar IS (2001) Stabilization of cortical actin induces internalization of transient receptor potential 3 (Trp3)-associated caveolar Ca2+ signaling complex and loss of Ca2+ influx without disruption of Trp3-inositol trisphosphate receptor association. J Biol Chem 276:42401–42408
128. Singh BB, Lockwich TP, Bandyopadhyay BC, Liu X, Bollimuntha S, Brazer SC, Combs C, Das S, Leenders AG, Sheng ZH, Knepper MA, Ambudkar SV, Ambudkar IS (2004) VAMP2-dependent exocytosis regulates plasma membrane insertion of TRPC3 channels and contributes to agonist-stimulated Ca2+ influx. Mol Cell 15:635–646
129. van Rossum DB, Patterson RL, Sharma S, Barrow RK, Kornberg M, Gill DL, Snyder SH (2005) Phospholipase Cgamma1 controls surface expression of TRPC3 through an intermolecular PH domain. Nature 434:99–104
130. Graziani A, Rosker C, Kohlwein SD, Zhu MX, Romanin C, Sattler W, Groschner K, Poteser M (2006) Cellular cholesterol controls TRPC3 function: evidence from a novel dominant-negative knockdown strategy. Biochem J 396:147–155
131. Tomita Y, Kaneko S, Funayama M, Kondo H, Satoh M, Akaike A (1998) Intracellular Ca2+ store-operated influx of Ca2+ through TRP-R, a rat homolog of TRP, expressed in Xenopus oocytes. Neurosci Lett 248:195–198
132. Philipp S, Trost C, Warnat J, Rautmann J, Himmerkus N, Schroth G, Kretz O, Nastainczyk W, Cavalie A, Hoth M, Flockerzi V (2000) TRP4 (CCE1) protein is part of native calcium release-activated Ca2+-like channels in adrenal cells. J Biol Chem 275:23965–23972
133. Freichel M, Suh SH, Pfeifer A, Schweig U, Trost C, Weissgerber P, Biel M, Philipp S, Freise D, Droogmans G, Hofmann F, Flockerzi V, Nilius B (2001) Lack of an endothelial store-operated Ca2+ current impairs agonist-dependent vasorelaxation in TRP4−/− mice. Nat Cell Biol 3:121–127
134. Tiruppathi C, Minshall RD, Paria BC, Vogel SM, Malik AB (2002) Role of Ca2+ signaling in the regulation of endothelial permeability. Vascul Pharmacol 39:173–185
135. Tu CL, Chang W, Bikle DD (2005) Phospholipase cgamma1 is required for activation of store-operated channels in human keratinocytes. J Invest Dermatol 124:187–197
136. Yang H, Mergler S, Sun X, Wang Z, Lu L, Bonanno JA, Pleyer U, Reinach PS (2005) TRPC4 knockdown suppresses epidermal growth factor-induced store-operated channel activation and growth in human corneal epithelial cells. J Biol Chem 280:32230–32237
137. Schaefer M, Plant TD, Obukhov AG, Hofmann T, Gudermann T, Schultz G (2000) Receptor-mediated regulation of the nonselective cation channels TRPC4 and TRPC5. J Biol Chem 275:17517–17526
138. Wu X, Babnigg G, Zagranichnaya T, Villereal ML (2002) The role of endogenous human Trp4 in regulating carbachol-induced calcium oscillations in HEK-293 cells. J Biol Chem 277:13597–13608
139. Obukhov AG, Nowycky MC (2002) TRPC4 can be activated by G-protein-coupled receptors and provides sufficient Ca(2+) to trigger exocytosis in neuroendocrine cells. J Biol Chem 277:16172–16178
140. Jeon JP, Lee KP, Park EJ, Sung TS, Kim BJ, Jeon JH, So I (2008) The specific activation of TRPC4 by Gi protein subtype. Biochem Biophys Res Commun 377:538–543
141. Mery L, Magnino F, Schmidt K, Krause KH, Dufour JF (2001) Alternative splice variants of hTrp4 differentially interact with the C-terminal portion of the inositol 1,4,5-trisphosphate receptors. FEBS Lett 487:377–383
142. Wang X, Pluznick JL, Wei P, Padanilam BJ, Sansom SC (2004) TRPC4 forms store-operated Ca2+ channels in mouse mesangial cells. Am J Physiol Cell Physiol 287:C357–C364

143. Satoh E, Ono K, Xu F, Iijima T (2002) Cloning and functional expression of a novel splice variant of rat TRPC4. Circ J 66:954–958
144. Tang Y, Tang J, Chen Z, Trost C, Flockerzi V, Li M, Ramesh V, Zhu MX (2000) Association of mammalian trp4 and phospholipase C isozymes with a PDZ domain-containing protein, NHERF. J Biol Chem 275:37559–37564
145. Tang J, Lin Y, Zhang Z, Tikunova S, Birnbaumer L, Zhu MX (2001) Identification of common binding sites for calmodulin and inositol 1,4,5-trisphosphate receptors on the carboxyl termini of trp channels. J Biol Chem 276:21303–21310
146. Boulay G, Zhu X, Peyton M, Jiang M, Hurst R, Stefani E, Birnbaumer L (1997) Cloning and expression of a novel mammalian homolog of Drosophila transient receptor potential (Trp) involved in calcium entry secondary to activation of receptors coupled by the Gq class of G protein. J Biol Chem 272:29672–29680
147. Zhang L, Guo F, Kim JY, Saffen D (2006) Muscarinic acetylcholine receptors activate TRPC6 channels in PC12D cells via Ca2+ store-independent mechanisms. J Biochem 139:459–470
148. Jung S, Strotmann R, Schultz G, Plant TD (2002) TRPC6 is a candidate channel involved in receptor-stimulated cation currents in A7r5 smooth muscle cells. Am J Physiol Cell Physiol 282:C347–C359
149. Maruyama Y, Nakanishi Y, Walsh EJ, Wilson DP, Welsh DG, Cole WC (2006) Heteromultimeric TRPC6-TRPC7 channels contribute to arginine vasopressin-induced cation current of A7r5 vascular smooth muscle cells. Circ Res 98:1520–1527
150. Hassock SR, Zhu MX, Trost C, Flockerzi V, Authi KS (2002) Expression and role of TRPC proteins in human platelets: evidence that TRPC6 forms the store-independent calcium entry channel. Blood 100:2801–2811
151. Jardin I, Redondo PC, Salido GM, Rosado JA (2008) Phosphatidylinositol 4,5-bisphosphate enhances store-operated calcium entry through hTRPC6 channel in human platelets. Biochim Biophys Acta 1783:84–97
152. Redondo PC, Jardin I, Lopez JJ, Salido GM, Rosado JA (2008) Intracellular Ca2+ store depletion induces the formation of macromolecular complexes involving hTRPC1, hTRPC6, the type II IP3 receptor and SERCA3 in human platelets. Biochim Biophys Acta 1783: 1163–1176
153. Carter RN, Tolhurst G, Walmsley G, Vizuete-Forster M, Miller N, Mahaut-Smith MP (2006) Molecular and electrophysiological characterization of transient receptor potential ion channels in the primary murine megakaryocyte. J Physiol 576:151–162
154. Chu X, Tong Q, Cheung JY, Wozney J, Conrad K, Mazack V, Zhang W, Stahl R, Barber DL, Miller BA (2004) Interaction of TRPC2 and TRPC6 in erythropoietin modulation of calcium influx. J Biol Chem 279:10514–10522
155. McMeekin SR, Dransfield I, Rossi AG, Haslett C, Walker TR (2006) E-selectin permits communication between PAF receptors and TRPC channels in human neutrophils. Blood 107:4938–4945
156. Sergeeva OA, Korotkova TM, Scherer A, Brown RE, Haas HL (2003) Co-expression of non-selective cation channels of the transient receptor potential canonical family in central aminergic neurones. J Neurochem 85:1547–1552
157. Nasman J, Bart G, Larsson K, Louhivuori L, Peltonen H, Akerman KE (2006) The orexin OX1 receptor regulates Ca2+ entry via diacylglycerol-activated channels in differentiated neuroblastoma cells. J Neurosci 26:10658–10666
158. Albert AP, Large WA (2003) Synergism between inositol phosphates and diacylglycerol on native TRPC6-like channels in rabbit portal vein myocytes. J Physiol 552:789–795
159. Corteling RL, Li S, Giddings J, Westwick J, Poll C, Hall IP (2004) Expression of transient receptor potential C6 and related transient receptor potential family members in human airway smooth muscle and lung tissue. Am J Respir Cell Mol Biol 30:145–154
160. Spassova MA, Hewavitharana T, Xu W, Soboloff J, Gill DL (2006) A common mechanism underlies stretch activation and receptor activation of TRPC6 channels. Proc Natl Acad Sci USA 103:16586–16591

161. Goel M, Sinkins W, Keightley A, Kinter M, Schilling WP (2005) Proteomic analysis of TRPC5- and TRPC6-binding partners reveals interaction with the plasmalemmal Na(+)/K(+)-ATPase. Pflugers Arch 451:87–98
162. Soboloff J, Spassova M, Xu W, He LP, Cuesta N, Gill DL (2005) Role of endogenous TRPC6 channels in Ca2+ signal generation in A7r5 smooth muscle cells. J Biol Chem 280: 39786–39794
163. Lemos VS, Poburko D, Liao CH, Cole WC, van Breemen C (2007) Na+ entry via TRPC6 causes Ca2+ entry via NCX reversal in ATP stimulated smooth muscle cells. Biochem Biophys Res Commun 352:130–134
164. Kim EY, Alvarez-Baron CP, Dryer SE (2009) Canonical transient receptor potential channel (TRPC)3 and TRPC6 associate with large-conductance Ca2+-activated K+ (BKCa) channels: role in BKCa trafficking to the surface of cultured podocytes. Mol Pharmacol 75:466–477
165. Jardin I, Gomez LJ, Salido GM, Rosado JA (2009) Dynamic interaction of hTRPC6 with the Orai1-STIM1 complex or hTRPC3 mediates its role in capacitative or non-capacitative Ca(2+) entry pathways. Biochem J 420:267–276
166. Basora N, Boulay G, Bilodeau L, Rousseau E, Payet MD (2003) 20-hydroxyeicosatetraenoic acid (20-HETE) activates mouse TRPC6 channels expressed in HEK293 cells. J Biol Chem 278:31709–31716
167. Boulay G (2002) Ca(2+)-calmodulin regulates receptor-operated Ca(2+) entry activity of TRPC6 in HEK-293 cells. Cell Calcium 32:201–207
168. Tseng PH, Lin HP, Hu H, Wang C, Zhu MX, Chen CS (2004) The canonical transient receptor potential 6 channel as a putative phosphatidylinositol 3,4,5-trisphosphate-sensitive calcium entry system. Biochemistry 43:11701–11708
169. Hisatsune C, Kuroda Y, Nakamura K, Inoue T, Nakamura T, Michikawa T, Mizutani A, Mikoshiba K (2004) Regulation of TRPC6 channel activity by tyrosine phosphorylation. J Biol Chem 279:18887–18894
170. Foster RR, Zadeh MA, Welsh GI, Satchell SC, Ye Y, Mathieson PW, Bates DO, Saleem MA (2009) Flufenamic acid is a tool for investigating TRPC6-mediated calcium signalling in human conditionally immortalised podocytes and HEK293 cells. Cell Calcium 45:384–390
171. Liu D, Maier A, Scholze A, Rauch U, Boltzen U, Zhao Z, Zhu Z, Tepel M (2008) High glucose enhances transient receptor potential channel canonical type 6-dependent calcium influx in human platelets via phosphatidylinositol 3-kinase-dependent pathway. Arterioscler Thromb Vasc Biol 28:746–751
172. Cayouette S, Lussier MP, Mathieu EL, Bousquet SM, Boulay G (2004) Exocytotic insertion of TRPC6 channel into the plasma membrane upon Gq protein-coupled receptor activation. J Biol Chem 279:7241–7246
173. Lussier MP, Lepage PK, Bousquet SM, Boulay G (2008) RNF24, a new TRPC interacting protein, causes the intracellular retention of TRPC. Cell Calcium 43:432–443
174. Albert AP, Saleh SN, Large WA (2008) Inhibition of native TRPC6 channel activity by phosphatidylinositol 4,5-bisphosphate in mesenteric artery myocytes. J Physiol 586: 3087–3095
175. Takahashi S, Lin H, Geshi N, Mori Y, Kawarabayashi Y, Takami N, Mori MX, Honda A, Inoue R (2008) Nitric oxide-cGMP-protein kinase G pathway negatively regulates vascular transient receptor potential channel TRPC6. J Physiol 586:4209–4223
176. Welsh DG, Morielli AD, Nelson MT, Brayden JE (2002) Transient receptor potential channels regulate myogenic tone of resistance arteries. Circ Res 90:248–250
177. Reiser J, Polu KR, Moller CC, Kenlan P, Altintas MM, Wei C, Faul C, Herbert S, Villegas I, Avila-Casado C, McGee M, Sugimoto H, Brown D, Kalluri R, Mundel P, Smith PL, Clapham DE, Pollak MR (2005) TRPC6 is a glomerular slit diaphragm-associated channel required for normal renal function. Nat Genet 37:739–744
178. Foller M, Kasinathan RS, Koka S, Lang C, Shumilina E, Birnbaumer L, Lang F, Huber SM (2008) TRPC6 contributes to the Ca(2+) leak of human erythrocytes. Cell Physiol Biochem 21:183–192

179. Damann N, Owsianik G, Li S, Poll C, Nilius B (2009) The calcium-conducting ion channel transient receptor potential canonical 6 is involved in macrophage inflammatory protein-2-induced migration of mouse neutrophils. Acta Physiol (Oxf) 195:3–11
180. Sel S, Rost BR, Yildirim AO, Sel B, Kalwa H, Fehrenbach H, Renz H, Gudermann T, Dietrich A (2008) Loss of classical transient receptor potential 6 channel reduces allergic airway response. Clin Exp Allergy 38:1548–1558
181. Hamdollah Zadeh MA, Glass CA, Magnussen A, Hancox JC, Bates DO (2008) VEGF-mediated elevated intracellular calcium and angiogenesis in human microvascular endothelial cells in vitro are inhibited by dominant negative TRPC6. Microcirculation 15:605–614
182. Muller M, Essin K, Hill K, Beschmann H, Rubant S, Schempp CM, Gollasch M, Boehncke WH, Harteneck C, Muller WE, Leuner K (2008) Specific TRPC6 channel activation, a novel approach to stimulate keratinocyte differentiation. J Biol Chem 283:33942–33954
183. Cai R, Ding X, Zhou K, Shi Y, Ge R, Ren G, Jin Y, Wang Y (2009) Blockade of TRPC6 channels induced G2/M phase arrest and suppressed growth in human gastric cancer cells. Int J Cancer 125:2281–2287
184. Shi Y, Ding X, He ZH, Zhou KC, Wang Q, Wang YZ (2009) Critical role of TRPC6 channels in G2 phase transition and the development of human oesophageal cancer. Gut 58:1443–1450
185. Yue D, Wang Y, Xiao JY, Wang P, Ren CS (2009) Expression of TRPC6 in benign and malignant human prostate tissues. Asian J Androl 11:541–547
186. Santin S, Ars E, Rossetti S, Salido E, Silva I, Garcia-Maset R, Gimenez I, Ruiz P, Mendizabal S, Luciano NJ, Pena A, Camacho JA, Fraga G, Cobo MA, Bernis C, Ortiz A, de Pablos AL, Sanchez-Moreno A, Pintos G, Mirapeix E, Fernandez-Llama P, Ballarin J, Torra R, Zamora I, Lopez-Hellin J, Madrid A, Ventura C, Vilalta R, Espinosa L, Garcia C, Melgosa M, Navarro M, Gimenez A, Cots JV, Alexandra S, Caramelo C, Egido J, San Jose MD, de la CF, Sala P, Raspall F, Vila A, Daza AM, Vazquez M, Ecija JL, Espinosa M, Justa ML, Poveda R, Aparicio C, Rosell J, Muley R, Montenegro J, Gonzalez D, Hidalgo E, de Frutos DB, Trillo E, Gracia S, los Rios FJ (2009) TRPC6 mutational analysis in a large cohort of patients with focal segmental glomerulosclerosis. Nephrol Dial Transplant 24:3089–3096
187. Hu G, Oboukhova EA, Kumar S, Sturek M, Obukhov AG (2009) Canonical transient receptor potential channels expression is elevated in a porcine model of metabolic syndrome. Mol Endocrinol 23:689–699
188. Igarashi P, Somlo S (2002) Genetics and pathogenesis of polycystic kidney disease. J Am Soc Nephrol 13:2384–2398
189. Nauli SM, Zhou J (2004) Polycystins and mechanosensation in renal and nodal cilia. Bioessays 26:844–856
190. Delmas P, Nauli SM, Li X, Coste B, Osorio N, Crest M, Brown DA, Zhou J (2004) Gating of the polycystin ion channel signaling complex in neurons and kidney cells. FASEB J 18:740–742
191. Giamarchi A, Padilla F, Coste B, Raoux M, Crest M, Honore E, Delmas P (2006) The versatile nature of the calcium-permeable cation channel TRPP2. EMBO Rep 7:787–793
192. Gao Z, Ruden DM, Lu X (2003) PKD2 cation channel is required for directional sperm movement and male fertility. Curr Biol 13:2175–2178
193. Watnick TJ, Jin Y, Matunis E, Kernan MJ, Montell C (2003) A flagellar polycystin-2 homolog required for male fertility in Drosophila. Curr Biol 13:2179–2184
194. Neill AT, Moy GW, Vacquier VD (2004) Polycystin-2 associates with the polycystin-1 homolog, suREJ3, and localizes to the acrosomal region of sea urchin spermatozoa. Mol Reprod Dev 67:472–477
195. Hughes J, Ward CJ, Aspinwall R, Butler R, Harris PC (1999) Identification of a human homologue of the sea urchin receptor for egg jelly: a polycystic kidney disease-like protein. Hum Mol Genet 8:543–549

196. Mengerink KJ, Moy GW, Vacquier VD (2002) suREJ3, a polycystin-1 protein, is cleaved at the GPS domain and localizes to the acrosomal region of sea urchin sperm. J Biol Chem 277:943–948
197. Vora N, Perrone R, Bianchi DW (2008) Reproductive issues for adults with autosomal dominant polycystic kidney disease. Am J Kidney Dis 51:307–318
198. Sutton KA, Jungnickel MK, Florman HM (2008) A polycystin-1 controls postcopulatory reproductive selection in mice. Proc Natl Acad Sci USA 105:8661–8666
199. Ahern GP, Brooks IM, Miyares RL, Wang XB (2005) Extracellular cations sensitize and gate capsaicin receptor TRPV1 modulating pain signaling. J Neurosci 25:5109–5116
200. Cortright DN, Szallasi A (2004) Biochemical pharmacology of the vanilloid receptor TRPV1. An update. Eur J Biochem 271:1814–1819
201. Ferrer-Montiel A, Garcia-Martinez C, Morenilla-Palao C, Garcia-Sanz N, Fernandez-Carvajal A, Fernandez-Ballester G, Planells-Cases R (2004) Molecular architecture of the vanilloid receptor. Insights for drug design. Eur J Biochem 271:1820–1826
202. Neubert JK, Karai L, Jun JH, Kim HS, Olah Z, Iadarola MJ (2003) Peripherally induced resiniferatoxin analgesia. Pain 104:219–228
203. Tominaga M, Caterina MJ, Malmberg AB, Rosen TA, Gilbert H, Skinner K, Raumann BE, Basbaum AI, Julius D (1998) The cloned capsaicin receptor integrates multiple pain-producing stimuli. Neuron 21:531–543
204. Dhaka A, Viswanath V, Patapoutian A (2006) Trp ion channels and temperature sensation. Annu Rev Neurosci 29:135–161
205. Grimaldi P, Orlando P, Di Siena S, Lolicato F, Petrosino S, Bisogno T, Geremia R, De Petrocellis L, Di Marzo V (2009) The endocannabinoid system and pivotal role of the CB2 receptor in mouse spermatogenesis. Proc Natl Acad Sci USA 106:11131–11136
206. Mizrak SC, Dissel-Emiliani FM (2008) Transient receptor potential vanilloid receptor-1 confers heat resistance to male germ cells. Fertil Steril 90:1290–1293
207. De Petrocellis L, Bisogno T, Maccarrone M, Davis JB, Finazzi-Agro A, Di Marzo V (2001) The activity of anandamide at vanilloid VR1 receptors requires facilitated transport across the cell membrane and is limited by intracellular metabolism. J Biol Chem 276:12856–12863
208. Bernabo N, Pistilli MG, Gloria A, Di Pancrazio C, Falasca G, Barboni B, Mattioli M (2008) Factors affecting TRPV1 receptor immunolocalization in boar spermatozoa capacitated in vitro. Vet Res Commun 32(Suppl 1):S103–S105
209. Francavilla F, Battista N, Barbonetti A, Vassallo MR, Rapino C, Antonangelo C, Pasquariello N, Catanzaro G, Barboni B, Maccarrone M (2009) Characterization of the endocannabinoid system in human spermatozoa and involvement of transient receptor potential vanilloid 1 receptor in their fertilizing ability. Endocrinology 150:4692–4700
210. Maccarrone M, Barboni B, Paradisi A, Bernabo N, Gasperi V, Pistilli MG, Fezza F, Lucidi P, Mattioli M (2005) Characterization of the endocannabinoid system in boar spermatozoa and implications for sperm capacitation and acrosome reaction. J Cell Sci 118:4393–4404
211. Auzanneau C, Norez C, Antigny F, Thoreau V, Jougla C, Cantereau A, Becq F, Vandebrouck C (2008) Transient receptor potential vanilloid 1 (TRPV1) channels in cultured rat Sertoli cells regulate an acid sensing chloride channel. Biochem Pharmacol 75:476–483
212. Caterina MJ, Leffler A, Malmberg AB, Martin WJ, Trafton J, Petersen-Zeitz KR, Koltzenburg M, Basbaum AI, Julius D (2000) Impaired nociception and pain sensation in mice lacking the capsaicin receptor. Science 288:306–313
213. Li S, Wang X, Ye H, Gao W, Pu X, Yang Z (2010) Distribution profiles of transient receptor potential melastatin- and vanilloid-related channels in rat spermatogenic cells and sperm. Mol Biol Rep 37:1287–1293
214. Montell C (2001) Physiology, phylogeny, and functions of the TRP superfamily of cation channels. Sci STKE 2001:re1

215. Duncan LM, Deeds J, Hunter J, Shao J, Holmgren LM, Woolf EA, Tepper RI, Shyjan AW (1998) Down-regulation of the novel gene melastatin correlates with potential for melanoma metastasis. Cancer Res 58:1515–1520
216. Hunter JJ, Shao J, Smutko JS, Dussault BJ, Nagle DL, Woolf EA, Holmgren LM, Moore KJ, Shyjan AW (1998) Chromosomal localization and genomic characterization of the mouse melastatin gene (Mlsn1). Genomics 54:116–123
217. Launay P, Fleig A, Perraud AL, Scharenberg AM, Penner R, Kinet JP (2002) TRPM4 is a Ca2+-activated nonselective cation channel mediating cell membrane depolarization. Cell 109:397–407
218. Xu XZ, Moebius F, Gill DL, Montell C (2001) Regulation of melastatin, a TRP-related protein, through interaction with a cytoplasmic isoform. Proc Natl Acad Sci USA 98:10692–10697
219. Deeds J, Cronin F, Duncan LM (2000) Patterns of melastatin mRNA expression in melanocytic tumors. Hum Pathol 31:1346–1356
220. Runnels LW, Yue L, Clapham DE (2001) TRP-PLIKa bifunctional protein with kinase and ion channel activities. Science 291:1043–1047
221. Takezawa R, Schmitz C, Demeuse P, Scharenberg AM, Penner R, Fleig A (2004) Receptor-mediated regulation of the TRPM7 channel through its endogenous protein kinase domain. Proc Natl Acad Sci USA 101:6009–6014
222. Monteilh-Zoller MK, Hermosura MC, Nadler MJ, Scharenberg AM, Penner R, Fleig A (2003) TRPM7 provides an ion channel mechanism for cellular entry of trace metal ions. J Gen Physiol 121:49–60
223. McKemy DD, Neuhausser WM, Julius D (2002) Identification of a cold receptor reveals a general role for TRP channels in thermosensation. Nature 416:52–58
224. Tsavaler L, Shapero MH, Morkowski S, Laus R (2001) Trp-p8, a novel prostate-specific gene, is up-regulated in prostate cancer and other malignancies and shares high homology with transient receptor potential calcium channel proteins. Cancer Res 61: 3760–3769
225. Voets T, Droogmans G, Wissenbach U, Janssens A, Flockerzi V, Nilius B (2004) The principle of temperature-dependent gating in cold- and heat-sensitive TRP channels. Nature 430:748–754
226. Brauchi S, Orio P, Latorre R (2004) Clues to understanding cold sensation: thermodynamics and electrophysiological analysis of the cold receptor TRPM8. Proc Natl Acad Sci USA 101:15494–15499
227. De Blas GA, Darszon A, Ocampo AY, Serrano CJ, Castellano LE, Hernandez-Gonzalez EO, Chirinos M, Larrea F, Beltran C, Trevino CL (2009) TRPM8, a versatile channel in human sperm. PLoS One 4:e6095
228. Shoeb M, Laloraya M, Kumar PG (2010) Progesterone-induced reorganisation of NOX-2 components in membrane rafts is critical for sperm functioning in Capra hircus. *Andrologia* 42:6 (in press)
229. Harder T, Simons K (1997) Caveolae, DIGs, and the dynamics of sphingolipid-cholesterol microdomains. Curr Opin Cell Biol 9:534–542
230. Helms JB, Zurzolo C (2004) Lipids as targeting signals: lipid rafts and intracellular trafficking. Traffic 5:247–254
231. Ambudkar IS, Brazer SC, Liu X, Lockwich T, Singh B (2004) Plasma membrane localization of TRPC channels: role of caveolar lipid rafts. Novartis Found Symp 258:63–70 (discussion 70-4, 98-102, 263-6.:63-70)
232. Brazer SC, Singh BB, Liu X, Swaim W, Ambudkar IS (2003) Caveolin-1 contributes to assembly of store-operated Ca2+ influx channels by regulating plasma membrane localization of TRPC1. J Biol Chem 278:27208–27215
233. Gervasio OL, Whitehead NP, Yeung EW, Phillips WD, Allen DG (2008) TRPC1 binds to caveolin-3 and is regulated by Src kinase – role in Duchenne muscular dystrophy. J Cell Sci 121:2246–2255

234. Lockwich TP, Liu X, Singh BB, Jadlowiec J, Weiland S, Ambudkar IS (2000) Assembly of Trp1 in a signaling complex associated with caveolin-scaffolding lipid raft domains. J Biol Chem 275:11934–11942
235. Sundivakkam PC, Kwiatek AM, Sharma TT, Minshall RD, Malik AB, Tiruppathi C (2009) Caveolin-1 scaffold domain interacts with TRPC1 and IP3R3 to regulate Ca2+ store release-induced Ca2+ entry in endothelial cells. Am J Physiol Cell Physiol 296:C403–C413
236. Pani B, Ong HL, Brazer SC, Liu X, Rauser K, Singh BB, Ambudkar IS (2009) Activation of TRPC1 by STIM1 in ER-PM microdomains involves release of the channel from its scaffold caveolin-1. Proc Natl Acad Sci USA 106:20087–20092
237. Gu Y, Gorelik J, Spohr HA, Shevchuk A, Lab MJ, Harding SE, Vodyanoy I, Klenerman D, Korchev YE (2002) High-resolution scanning patch-clamp: new insights into cell function. FASEB J 16:748–750
238. Jimenez-Gonzalez MC, Gu Y, Kirkman-Brown J, Barratt CL, Publicover S (2007) Patch-clamp 'mapping' of ion channel activity in human sperm reveals regionalisation and co-localisation into mixed clusters. J Cell Physiol 213:801–808

Chapter 47
TRP Channels in Female Reproductive Organs and Placenta

Janka Dörr and Claudia Fecher-Trost

Abstract TRP channel proteins are widely expressed in female reproductive organs. Based on studies detecting TRP transcripts and proteins in different parts of the female reproductive organs and placenta they are supposed to be involved in the transport of the oocyte or the blastocyte through the oviduct, implantation of the blastocyte, development of the placenta and transport processes across the feto-maternal barrier. Furthermore uterus contractility and physiological processes during labour and in mammary glands seem to be dependant on TRP channel expression.

47.1 Introduction

In animals of sexual reproduction, the appropriate communication between mature male and female gametes determines the generation of a new individual. Sexual reproduction is a matter of life or death. Ion channels are involved in the composition of the propagable female uterine endo- and myometrium, and ovary, given under tight hormonal control the prerequisite for maturation of the egg cell and implantation of blastocyte. In the case of a successful fertilisation, the mammalian embryo cannot develop without the placenta, which is early formed by specialised cells (trophoblasts, endoderm, and extraembryonic mesoderm). These cells allow the attachment of the embryo to the endometrium of the uterus and the development of vascular connections necessary for mineral and nutrient exchange between mother and foetus. In haemochorial mammalians as humans and rodents, foetal trophoblast cells invade the maternal uterus, which develops into the decidua. Maternal uterine blood vessels are infiltrated by trophoblast cells which causes the rupture of blood vessels and release of blood into the intervillous space. The outer layer of

---

J. Dörr (✉)
Proteinfunktion Proteomics, Fachbereich Biologie, TU Kaiserslautern,
D-67663 Kaiserslautern, Germany
e-mail: Janka.Doerr@uniklinikum-saarland.de; jadoerr@rhrk.uni-kl.de

the chorionic villi (syncytiotrophoblast) is thereby exposed to the maternal blood [1]. In humans the multinuclear syncytiotrophoblast (STB) represents a specialised polarised endothelial cell layer, which forms the barrier for the exchange between the mother and foetus. The apical, microvillous surface of the STB is the mother-facing membrane, which is in direct contact to the maternal blood in the intervillous space, while the basolateral surface of the STB is opposed to the foetal environment. The STB cells cover a tree like structure, the chorionic villi. In mice, STBs form a labyrinth-like structure which serves the same purpose [2]. In contrast to humans, the yolk sac (YS) in rodents is composed of visceral (vYS) and parietal (pYS) layers and provides an additional maternal-foetal epithelial interface for the exchange of ions and nutrients [3]. The bidirectional transepithelial transport across the polarised STB monolayer is regulated by the polarised expression of transporters in the apical and basolateral membrane [4]. The cell monolayer also protects the developing foetus against the attack of the maternal immune system.

Several members of different TRP subfamilies have been reported in these specialised cell layer and the underlying trophoblast cells of the placenta or localise in ovary and the smooth muscle cells of the pregnant or non pregnant uterus. Present insights which and how TRP channels participate in female reproductive physiological processes were based on electrophysiological recordings of reconstituted giant liposomes derived from placenta cells or cultured placenta/uterine cells, RT-PCR studies, calcium uptake measurements, heterologous expression, biochemical interaction studies and generation of different TRP deficient animal models.

47.2 The TRPC Family

The family of the canonical TRP proteins is expressed in a variety of cell types. Therefore it is not surprisingly, that TRPCs are also expressed in female reproductive organs, mainly in the myometrium representing the smooth muscle cells of the uterine wall, and even they are expressed in placenta (Table 47.1 and Fig. 47.1, Fig. 47.2). Beside the responsibility of the myometrium for giving a secure environment during pregnancy for the foetus it is also responsible for the uterine contractions during labour. For initiating muscle contraction a rise of intracellular free calcium is very important, also in the uterine myometrium [5]. Although TRPC channels are rather nonselective cation channels they are all Ca^{2+} permeable and hence, may play a role in regulating intracellular calcium concentrations ($[Ca^{2+}]_i$). In vertebrate genomes there are coding regions for seven TRPCs. TRPC2 is probably not a functional ion channel in humans because its gene exhibits several in frame stop codons, but if it is a pseudogene, a gene without function, is still questionable. First evidence for the expression of TRPC channels in myometrium came from McKay et al. [6] detecting TRPC4 mRNA in placenta and uterus. In several studies [7–12] TRPC mRNA and protein expression in myometrium was investigated, where they might play a role in capacitative calcium entry and constitute store-operated calcium entry (SOCE) channels which trigger the contractile activity of

Table 47.1 TRPC mRNA and protein expression in reproductive organs and placenta of different mammalian species

	Human reproductive organ		Human placenta		Rat reproductive organ		Porcine reproductive organ	
	RNA	Protein	RNA	Protein	RNA	Protein	RNA	Protein
TRPC1	$+^1$	$+^{3,4}$	$+^1$	$-^3$	$+^1$	$+^3$	n.d.	n.d.
TRPC2	Pseudogene		Pseudogene		$\pm^1$	−	n.d.	n.d.
TRPC3	$+^1$	$+^{3,4}$	$+^1$	$+^{3,4}$	−	−	$+^1$	n.d.
TRPC4	$+^{1,2}$	$+^{3,4}$	$+^{1,2}$	$+^4$	$+^1$	$+^3$	n.d.	n.d.
TRPC5	$\pm^1$	−	$+^1$	n.d.	$+^1$	$+^3$	n.d.	n.d.
TRPC6	$+^1$	$+^{(3),4}$	$+^1$	$+^{3,4}$	$+^1$	$+^3$	n.d.	n.d.
TRPC7	$\pm^1$	−	n.d.	n.d.	$\pm^1$	−	n.d.	n.d.

Superior numbers specify the technique used for detection of mRNA/proteins: 1, RT-PCR; 2, Northern Blot; 3, Western Blot/immunoprecipitation with adjacent western blot; 4, immunohistochemistry; n.d., not determined.

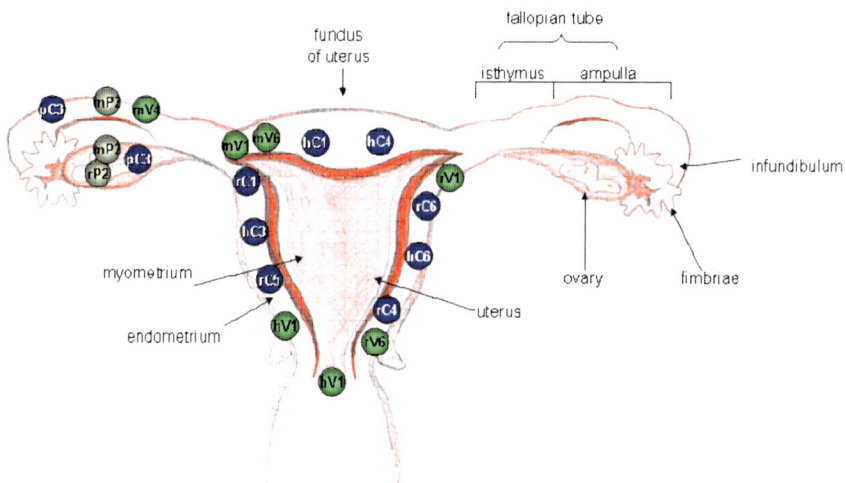

Fig. 47.1 TRP proteins present in female reproductive organs. This illustration of the female reproductive organs shows all TRP proteins of the different subfamilies detected by Western Blots or immunolocalization in uterus, the fallopian tubes and in ovary. Most of them are expressed in the myometrium of the uterus (human TRPC1, C3, C4, C6, rat TRPC1, C4, C5, C6, human TRPV1, mouse TRPV1, V6 and rat TRPV1 and V6). Along the fallopian tubes, porcine TRPC3, mouse TRPV4 and mouse TRPP2 are expressed. In ovary, porcine TRPC3 as well as mouse and rat TRPP2 are present. *Blue circles* represent the TRPC family, *green circles* the TRPV family and *yellow circles* stand for the TRPP family (h, human; m, mouse; p, porcine; r, rat)

the uterus during labour. Expression of TRPC1, C3, C4, C6 and C7 mRNA in pregnant myometrium and in different myometrial cell lines was detected (7;8;10). In different immortalised or primary myometrial cells additionally TRPC5 mRNA was found by a quantitative real-time PCR method [10]. Concordant with all other studies TRPC1 and C4 are the most abundant mRNAs followed by TRPC6, and the least mRNA is found for TRPC3, C5 and C7 (TRPC1 ≈ TRPC4 > TRPC6 >> TRPC3 >> TRPC5 > TRPC7). On protein level (Western Blot or immunolocalisation) TRPC1, C3, C4 and also a weak signal of TRPC6 were visualised in both, tissues samples and the cells (7;8;10). In contrast to the investigations on humans, in non-pregnant rat myometrium mRNAs coding for TRPC 1-7 are described with the exception of TRPC3 which was not detectable (TRPC4 >> TRPC1, C5, C6 >> TRPC2, C7) [11]. Western Blot analysis of rat myometrium supported the identified mRNA expression profile on protein levels for TRPC1, 4, 5 and 6 in the non-pregnant and pregnant status. Due to the fact that the TRPC channels expressed in myometrium might be involved in uterus contractions during labour some studies deal with the question whether the expression level is changed, especially upregulated, with the onset of labour. For rat myometrium neither significant mRNA or protein level changes throughout pregnancy nor with the onset of labour are identified. Even slight decreases in the mRNA levels of TRPC5 and C6 were observed in pregnancy compared to non-pregnant myometrium but no changes between day 13 and 21 of gestation or in labour were detected [11]. In one study of human myometrium no changes of mRNA and protein concentrations were reported for TRPCs before and after the onset of labour. TRPC4 mRNA actually decreased after the onset of labour in fundal myometrium [10]. In contrast to that unchanged mRNA levels for TRPC3 and TRPC4 were reported for non-pregnant myometrium and in term myometrium before and during labour phase. Also not even a decrease for TRPC4 mRNA could be detected [9]. For TRPC1, a significant increase in mRNA expression was revealed already at term (not being in labour) and the levels remained constant during labour, when additionally TRPC6 and C7 mRNA expression was up-regulated. Unlike the TRPC mRNA expression in non-pregnant status, only TRPC3 protein was found – in little amounts – expressed before pregnancy but not all the others TRPC proteins. However, whereas the expression of the TRPC isoforms (TRPC1, TRPC3, TRPC4, TRPC6) was at least minimal in term myometrium (not in labour) it was significantly increased with the onset of labour. These results support a role for TRPC proteins in adaptation of the uterus to pregnancy, particularly affecting the contractile activity. Especially the TRPC3 protein might be involved in this process because interleukin 1-β, a cytokine implicated in contraction [13, 14] and initiating uterine contractions in primates [15], causes an upregulation of the TRPC3 protein expression in primary cultured HMSM cells, whereas all other TRPC channel proteins and also mRNA levels remain unaffected. Moreover, in term pregnancy mechanical stretch of human myometrial cells in culture was observed to upregulate beside TRPC4 mRNA also TRPC3 mRNA and protein levels [16]. Mechanical stretch is assumed to be a stimulus for uterine growth and, in this case more interesting, for contraction during pregnancy ([16] and references therein), considering the reported results maybe particular in labour. Further this is supported by the proof of capacitative calcium entry in human myometrial cells (PHM1) induced by thapsigargin, oxytocin and OAG (1-oleoyl-2-acetyl-*sn*-glycerol), which can be enhanced by overexpression of human TRPC3. But if a capacitative stimulus is lacking, the overexpression of TRPC3 did not increase calcium entry. Concerning increases in calcium entry elicited by a special signal, possible roles for TRPC4 and TRPC6 are examined in more detail. Signal-regulated Ca^{2+} entry (SRCE) is defined as receptor-, store- or OAG-mediated increase in $[Ca^{2+}]_i$ dependant on extracellular Ca^{2+} and as reviewed by Sanborn [17] all major stimulants of SRCE ascribed to TRPC channels have been observed in myometrium. Upon a reduced expression of TRPC4 by RNA interference knockdown strategy in human myometrial cells (PHM1-41) and primary uterine smooth muscle cells, specifically the GPCR (G protein coupled receptor)-stimulated calcium entry is attenuated but not the thapsigargin (store depletion)-, or OAG-stimulated SRCE. As GPCR stimulants oxytocin, ATP and Prostaglandin F (PGF)2α were tested [18]. In a similar study a knockdown of TRPC6 resulted only in a reduced OAG-mediated increase in calcium concentrations in myometrial cells, whereas oxytocin or thapsigargin responses were not altered [19]. Further studies implicated an enhanced sodium entry by OAG activated TRPC6 channels, shown by inhibition of OAG-induced SRCE by removal of extracellular Na^+. The influx of sodium in turn might activate voltage-dependant Ca^{2+} entry channels. Also the OAG-mediated increase in $[Ca^{2+}]_i$ is partially inhibited by nifedipine treatment of PHM1-41 cells, suggesting an involvement of L-type calcium channels (blocked by nifedipine) in the OAG effect. Receptor (vasopressin + oxytocin)- and store-operated SRCE were not effected by nifedipine. In conclusion, several signals might affect specific types of SRCE in part triggered by different TRPC channels in human myometrium.

TRPC channels are not only expressed in the myometrium, also expression of these channels in term human placenta was observed [20] (Fig. 47.2). Previous studies detected, if any, only faint signals for TRPC1 mRNA in placenta [21, 22] but to a much greater intense in ovary consistent with all the studies mentioned above. In this more detailed analysis of human placenta, mRNAs for TRPC1, TRPC3, TRPC4, TRPC5 and TRPC6 were found in first trimester, second trimester and term placenta by RT-PCR. The TRPC5 expression is very low in all three stages but from the authors explained by the fact that normally its expression is associated with brain and neural tissue [23] and placental tissue is not innervated. In Western Blots only TRPC3 and TRPC6 were detected specifically in term but not in first trimester placenta plasma membranes. TRPC1 proteins were identified neither in first trimester nor in term placenta. Using immunhistochemical studies TRPC3 and TRPC4 proteins were mainly detected in cytotrophoblast cells in first trimester tissue and also with lower amount in syncytiotrophoblast basal plasma and microvillous membranes of term placenta. TRPC4 KO mice do not show reduced fertility [24] and the role of TRPC4 in placenta is not known so far. TRPC6 staining is observed in syncytiotrophoblast of both, first trimester and term samples with a much greater intense in term tissue [20]. In conclusion these results show that placental expression of TRPC channels increase at the end of pregnancy and reaches a maximum in labour phase, just as observed for myometrial expression. Furthermore in the same study it is demonstrated that a store-operated calcium entry exists especially in term placenta constituting a possible pathway for calcium entry into the syncytiotrophoblast. Based on the expression of TRPC channels in placenta and

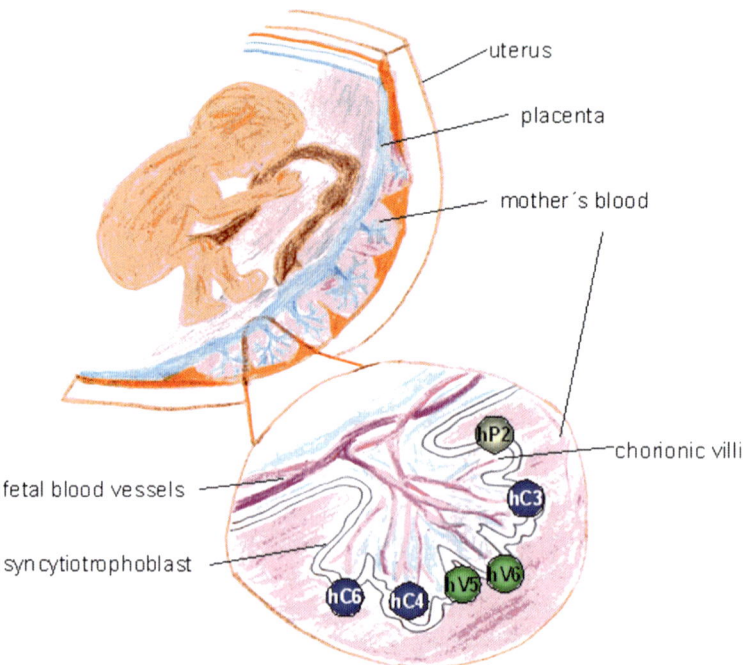

Fig. 47.2 TRP proteins present in human placenta. The figure depicts the human developing foetus (within the uterus), who is connected with the umbilical cord to the placenta. The enlarged part shows the detailed structure of a chorionic villous tree bathing in maternal blood and surrounded by the syncytiotrophoblast (STB). The STB is a special cell layer building a barrier between the mother and foetus and is important for nutrient exchange during pregnancy. Nearly all found TRP proteins are expressed in this STB layer, like human TRPV5, TRPV6, TRPP2 and TRPC6. TRPC3 and 4 are also weakly expressed in syncytiotrophoblast but in a higher extend in the underlying cytotrophoblast. *Circles* with different colours represent the different TRP families (*blue*, TRPC; *green*, TRPV; *yellow*, TRPP; h, human)

the observed store-operated calcium entry, the same mechanisms as described for myometrium might be responsible for calcium entry in the placenta. But in this field further studies are necessary to gain a more detailed insight.

47.3 The TRPV Family

47.3.1 TRPV1

The vanilloid receptor TRPV1 is the key molecule in sensory neurons, who was originally identified as a sensor for uncomfortable warm ($\geq 43°C$) [25] and is activated by many chemicals like capsaicin, endocannabinoids and anandamide [26]. During the last few years TRPV1 was also examined in nerve fibers innervating mammalian reproductive organs [27–30]. Rat uterine cervical afferents in the hypogastric nerve express TRPV1 [28], and estrogen amplifies responses to painful stimuli of the uterine cervix [31]. Another study investigated the TRPV1 innervation in the human uterus during pregnancy and labour [29]. Whereas an almost complete disappearance of TRPV1 positive nerve fibers was observed in the pregnant uterus, the cervical innervation remained high throughout pregnancy and labour and seems to be responsible for the pain during cervical ripening (Table 47.2 and Fig. 47.1).

A recent study investigated the expression of TRPV1 and the anandamide-binding receptors (CB1, CB2) in pregnant rats during placenta development [30]. The endocannabinoid system has been suggested as a key pathway involved in regulating blastocyst maturation, oviductal transport, implantation and maintenance of early pregnancy and successful implantation and progression of pregnancy is characterised by low levels of the endocannabinoids anandamide and 2-arachidonylglycerol (2-AG) [30]. TRPV1, CB1 and CB2 proteins were expressed in decidualised cells of the uterus and in the placenta. Interestingly, CB1 and CB2 were also expressed in smooth muscle cells of maternal blood vessels and in endovascular trophoblast cells, whereas TRPV1 was mainly expressed in uterine natural killer (uNK) cells and in the longitudinal muscle layer throughout pregnancy. The authors suggest that the localisation of the three proteins support a role of the endocannabinoid system during the decidualisation and development of the placenta. In summary these data suggest that TRPV1 expression in nerve fibers may be involved in different pain mechanisms associated with cervical ripening, labour or cervical cancer and that the TRPV1 expression in the uterus may be involved in successful implantation of the blastocyte.

47.3.2 TRPV2/TRPV3

The TRPV2 channel shares a 50% sequence homology to TRPV1, is insensitive to capsaicin, low pH and is activated by noxious heat ($\geq 52°C$) [32]. TRPV3, the last cloned member of the TRPV family is thermosensitive in the physiological temperature range of 22–40°C [33–35].

Transcripts of TRPV2/TRPV3 have been reported in mice ovary [36] (see Table 47.2) but their functional role is to date unknown. In contrast to members of the TRPC and TRPM subfamilies, the number of expressed TRPV transcripts seems to be lower than of TRPC transcripts in reproductive organs [36]

47.3.3 TRPV4

The vertebrate TRPV4 channel, which displays a 45% sequence identity with TRPV1 has been proposed as an osmo- and mechanosensitive channel [37–40] and responds to mechanical stress, acidic pH, endogenous ligands like arachidonic acid,

Table 47.2 TRPV mRNA and protein expression in reproductive organs and placenta of different mammalian species

	Human reproductive organ		Human placenta		Mouse reproductive organ		Mouse placenta		Rat reproductive organ		Rat placenta		Rabbit placenta	
	RNA	Protein	RNA	Protein	RNA	Protein	RNA	Protein	RNA	Protein	RNA	Protein	RNA	Protein
TRPV1	n.d.	$+^5$	n.d.	n.d.	$-^2$	n.d.	n.d.	n.d.	n.d.	$+^5$	$+^2$	$+^{4,5}$	n.d.	n.d.
TRPV2	n.d.	n.d.	n.d.	n.d.	$+^2$	n.d.	n.d.	n.d.	n.d.	n.d.	n.d.	n.d.	n.d.	n.d.
TRPV3	n.d.	n.d.	n.d.	n.d.	$+^2$	n.d.	n.d.	n.d.	n.d.	n.d.	n.d.	n.d.	n.d.	n.d.
TRPV4	n.d.	n.d.	n.d.	n.d.	$+^2$	$+^5$	n.d.	n.d.	n.d.	n.d.	n.d.	n.d.	n.d.	n.d.
TRPV5	n.d.	n.d.	$+^{1,2,3}$	$+^{4,5}$	$+^2$	n.d.	$+^1$	n.d.	n.d.	n.d.	n.d.	n.d.	$+^2$	n.d.
TRPV6	$-^2$	n.d.	$+^{1,2,3}$	$+^{4,5}$	$+^{2,1}$	$+^5$	$+^{1,3}$	$+^5$	$+^1$	$+^5$	$+^1$	$+^5$	n.d.	n.d.

Superior numbers specify the technique used for detection of mRNA/proteins: 1, RT-PCR; 2, Northern Blot; 3, in situ hybridization; 4, Western Blot/immunoprecipitation with adjacent western blot; 5, immunohistochemistry; n.d., not determined.

heat and synthetic 4α-phorbols [41]. TRPV4 transcripts and protein have been identified in native ciliated epithelial cells of hamster and mouse female reproductive organs [42–44] (Table 47.2). Here TRPV4 seems to be a key molecule involved in the mechanotransduction process of hamster oviductal ciliated cells. TRPV4 localizes to a sub-population of motile cilia on the epithelial cells of the ampulla and isthmus with high intensity in proximal invaginations of the epithelial folds (Fig. 47.1). TRPV4 channels were stimulated by changes in mucus viscosity, which leads to increased intracellular calcium levels [42]. This stimulation requires also the activation of a phospholipase A2 dependant pathway. The coupling of fluid viscosity and TRPV4 channel activation leads to the generation of the calcium signal, which is required for the autoregulation of the cilliary beat frequency. The authors assume that a dysfunction of TRPV4 may result in reduced or failure of oocytes and embryos transport through the oviduct and therefore is a potential matter for infertility. However a reduced fertility or an apparent defect in reproduction has not been observed in mice lacking the TRPV4 gene [45, 46].

47.3.4 TRPV5/6

TRPV5 and TRPV6 channels are distinct from the above mentioned TRPV family members, because they are not sensitive for heat or mechanical stimuli and they are the most calcium selective cation channels within the TRP family ($P_{ca}:P_{Na}>100$). TRPV5 (ECaC, CaT2) and TRPV6 (ECaC2, CaT1, CaT-like), the presumptive key players in intestinal and renal calcium absorption/reabsorption are prominently expressed in small intestine and kidney [47, 48]. Moreover both channels are coexpressed in some other tissues with exocrine function and they are supposed to form functional heterotetramers [49–53]. The calcium absorbing processes in kidney and intestine are relatively well understood [54]. In contrast to that the uterine calcium transport and the participating proteins are not properly discovered so far. Both TRPV channel proteins are identified in different cells of the placenta [52, 55–58] and the uterus [59]. The expression of TRPV5/6 and/or calcium absorption/reabsorption in duodenum, kidney and bone is under tight hormonal control by parathyroid hormone (PTH) [60, 61], Vitamin D3 [62–64], progesterone and estrogens [59, 65–68] but may be regulated by diverse mechanisms in different species during estrous and pregnancy. In mice TRPV6 transcript levels fluctuated during the estrous cycle in the uterus, increased at the time of estrous to a maximum [59] and TRPV6 expression was induced after 17β-estradiol treatment with prominent localization in the luminal and glandular epithelium of the uterus. Uterine TRPV6 expression increased in mid- and late pregnancy in mice with highest expression in the labyrinth and spongy zone of the murine placenta and in the foetal yolk sac membranes. In contrast to that, in pregnant rats *trpv6* transcription appeared to be solely controlled by progesterone [66]. A recent study demonstrates that Vitamin D receptor (VDR) knock out mice adapt to pregnancy by increasing TRPV6 expression 13.5 fold in the duodenum and by up-regulating intestinal $^{45}Ca^{2+}$ absorption. These data imply that in pregnant VDR$^{(-/-)}$ mice the up-regulation of intestinal calcium entry and skeletal mineralisation is independent of vitamin D receptor activity [69].

Maternal calcium transfer to the fetus is crucial for infant calcium homeostasis, growth and bone development. Calcium transport increases in response to the foetus demand during the last trimester, where nearly 80% of Ca^{2+}, equivalent to ~25 g, are transferred to the human foetus [70]. Placental calcium transfer in humans mainly occurs through the syncytiotrophoblast (STB). The expression of human TRPV6 and TRPV5 transcripts in cultured human STBs has been demonstrated, and STBs show a dose-dependant inhibition of Ca^{2+} uptake after ruthenium red and magnesium treatment, two known blocker of TRPV5 and 6 [56, 71]. Another study showed that calcium itself inhibits the activity of heterologously expressed human TRPV6 channel proteins. This effect was mediated through a calcium/calmodulin/PKC dependant regulation mechanism [72]. The presence of the glycosylated mature TRPV6 protein was shown in the apical membrane of STBs of human term placenta [56, 57]. Endogenously expressed TRPV6 channel proteins were enriched out of human term placenta together with Annexin A2 and cyclophilin B, a peptidyl-cis-trans-isomerase. Cyclophilin B induced an activation of TRPV6 mediated calcium entry in *Xenopus laevis* oocytes, which was reversed by the immunosuppressive drug cyclosporine A [57].

Murine foetuses with a double knock-out of *trpv6* and *eph*b6 show a marked disturbance in calcium homeostasis with reduced transplacental $^{45}Ca^{2+}$ transport, lower weight and a reduced bone density. The adult KO mice were less fertile [68, 73]: males deficient in TRPV6 rarely impregnated females and females with a TRPV6 KO took longer to get pregnant than control female mice. In the case of becoming pregnant the TRPV6 deficient females have a smaller number of pups. These observations could be the first evidence that *trpv6* gene expression is essential for male and female fertility. However the underlying molecular mechanism how TRPV6 modulates fertility of both sexes is not known.

Wildtype mice show increasing TRPV6 mRNA levels during the last 5 days of gestation corresponding to the increasing need of Ca^{2+} during foetal bone formation. In contrast to humans the transcellular Ca^{2+} transport in mice is mediated through both, labyrinth and yolk sac and immunolocalisation studies show that the mouse TRPV6 protein is mainly expressed in the membranes of the intra- and extraplacental yolk sac [68].

In contrast to TRPV6 knockouts, mice lacking TRPV5 (TRPV5$^{-/-}$) are fertile. They display profound renal calcium wasting (hypercalciuria) and a compensatory intestinal calcium hyperabsorption [74, 75]. The knockout mice exhibit significant disturbances in bone structure, including reduced femoral cortical and trabecular bone thickness [74]. Another study of the same group demonstrates that TRPV5 is essential for osteoclastic bone resorption: TRPV5$^{(-/-)}$ mice have an increase in the number and size of osteoclasts, but these cells have a significant reduced bone resorption [75].

In summary during pregnancy these both TRPV channels seem to be strongly involved in the transepithelial calcium transport from the mother to the infant against the concentration gradient.

47.4 The TRPM Family

TRPM1 (melastatin) was identified in melanomic cell lines and its down-regulation correlates with the metastatic potential of melanomas [76]. The eight family members of the TRPM family, TRPM1-8, share about 20% amino acid identities to TRPC channels over transmembrane domains 2–6 and contain similar to TRPC and TRPN channels the TRP domain [77]. A special feature share TRPM2, 6 and 7 channels in that they are chanzymes with an atypical proteinkinase domain at the C-terminus. Among mammals only in mice an incidence of mRNA transcripts for TRPM4 and TRPM7 has been reported [36]. Low levels of TRPM4 and TRPM7 transcripts were detected in the uterus and ovary but the physiological processes these channels might be involved in murine reproductive organs are to date not known (Table 47.3).

47.5 TRPP Family

The founding member of the polycystin subfamily is TRPP2 that is encoded by the gene PKD2 (polycystin-2) which is mutated in the autosomal dominant polycystic kidney disease (ADPKD). The classification as a TRP channel arises from its topology as a 6-membrane spanner with intracellular N- and C-termini [78]. The members of the polycystin family are one of the most distantly related within the TRP superfamily and can be divided into two subfamilies, the polycystin 1 (PKD-1)-like and polycystin 2 (PKD-2)-like channels. Proteins of the PKD-1 group contain 11 transmembrane domains with only little homology to the rest of the TRP superfamily and are therefore not regarded as TRP proteins. Within the PKD-2-like group there are three mammalian homologues TRPP2 (PKD2), TRPP3 (PKD2L1) and TRPP5 (PKD2L2) [79]. For TRPP3 there is only evidence on RNA level that it might be expressed in human placenta and uterus [80] but TRPP2 as a non-specific cation channel is described in detail in female reproductive organs and placenta, respectively (see Table 47.4; Fig. 47.1). In 1999 polycytins PKD-1 and 2 were

Table 47.3 TRPM mRNA and protein expression in reproductive organs and placenta of different mammalian species

	Mouse reproductive organ	
	RNA	Protein
TRPM1	n.d.	n.d.
TRPM2	n.d.	n.d.
TRPM3	n.d.	n.d.
TRPM4	+[1]	n.d.
TRPM5	n.d.	n.d.
TRPM6	n.d.	n.d.
TRPM7	+[1]	n.d.
TRPM8	n.d.	n.d.

Superior numbers specify the technique used for detection of mRNA/proteins: 1, RT-PCR; n.d., not determined.

Table 47.4 TRPP mRNA and protein expression in reproductive organs and placenta of different mammalian species

	Human reproductive organ		Human placenta		Mouse reproductive organ		Mouse placenta		Rat reproductive organ	
	RNA	Protein	RNA	Protein	RNA	Protein	RNA	Protein	RNA	protein
TRPP2	n.d.	n.d.	$+^{1,2}$	$+^{4,5}$	n.d.	$+^{4,5}$	n.d.	Indirect via KO	n.d.	$+^{4,5}$
TRPP3	$+^3$	n.d.	$+^{2,3}$	n.d.	n.d.	n.d.	n.d.	n.d.	n.d.	n.d.

Superior numbers specify the technique used for detection of mRNA/proteins: 1, RT-PCR; 2, Northern Blot; 3, RNA dot blot; 4, Western Blot/immunoprecipitation with adjacent western blot; 5, immunohistochemistry; n.d., not determined.

described for the first time in the fallopian tubes and oocytes of the female reproductive organs and in different cells of the placenta [81]. A more detailed distribution of TRPP2 in the syncytiotrophoblast of the human placenta is presented in another study where the presence of mRNA and the protein (by Western Blot) is ensured. More precisely, TRPP2 is suggested to be located at the apical membrane of the syncytiotrophoblast by immunolocalisation [82] (Fig. 47.2). The channel itself is characterised as an nonselective cation channel, that is more permeable to divalent than to monovalent cations (Na^+ and K^+) with a high perm-selectivity to Ca^{2+} and is strongly voltage dependant [82, 83]. In vesicles of human syncytiotrophoblasts multiple subconductance states were described. Inhibition of the channel activity is induced by La^{3+}, Gd^{3+} and amiloride, a known inhibitor of some epithelial cation channels that are Na^+-permeable [84, 85]. But TRPP2 is also impeded by a low pH, via an H^+ ion regulatory site in the cytoplasmic part of the protein [83] and by reactive oxygen species (ROS) that completely inhibit channel activity by targeting not only the membrane proteins but also the surrounding membrane lipids [86]. As a cation channel TRPP2 may be important for Ca^{2+} transport in placenta and several studies deal with the question of the functional role of the channel in human syncytiotrophoblasts. A strong influence of the cytoskeletal components alpha-actinin and tubulin on channel activity is described. Each one colocalises with the polycystin in the human syncytiotrophoblast, illustrated by immunfluorescence [87, 88]. For alpha-actinin even a physical interaction was determined by co-immunoprecipitation. Both, tubulin and alpha-actinin effect a stimulation of the channel activity. However, monomeric alpha-actinin leads to inhibition of the channel whereas the actin-disrupting agent cytochalasin D has a stimulatory effect on TRPP2 which is also achieved by addition of the actin-severing protein gelsolin, but here only in presence of Ca^{2+} [89]. Similar results were obtained for tubulin, where the microtubule stabiliser taxol increases the activity and by contrast colchizine as a disrupter of microtubules causes an inhibition of cannel activity [88]. Therefore it is likely that the interaction of TRPP2 with the dynamic cytoskeletal structures plays an important role concerning the regulation and control of ion transport in the placenta. Another study shows that changes in environmental signals such as hydrostatic pressure and osmolarity also regulate TRPP2 channel activity in reconstituted human syncytiotrophoplast vesicles and gives new insights how changes in physical forces are sensed by mechanism involving the actin cytoskeleton [90]. TRPP2 is not only described in human placenta and reproductive organ but also in rodents. For mouse there are several *Pkd* knockout models described [91, 92] in which all homozygous knockouts were lethal at an embryogenic stage between E12.5 and parturition. Using $Pkd2^{-/-}$ mice, Allen and co-workers [93] analysed a possible contribution of TRPP2 in placenta for in utero lethality of embryos. What they observed was that the labyrinth layer of the placenta has fewer fetal vessels and the networks built by them are disorganized and showed reduced branching, so this might be an alternate explanation for the fetal demise on the contrary to the earlier described cardiac defects. Additional to the presence of TRPP2 in placenta it plays also a role in murine reproductive organs [43]. Protein localisation is observed in primary cilia of granulosa cells of antral follicles in ovary by immunfluorescence and confirmed by western blot analysis. In cultured granulosa cells without primary cilia the protein is localised in cytoplasma, when these cells are growth-arrested and form a primary cilia the protein is predominantly present in cilia. TRPP2 is also found in motile cilia of the oviduct in adult mice by western blot and immunfluorescence analysis. Simulation of ovulation in oviductal cells by gonadotropins increased the level of TRPP2 in motile cilia and at the apical surface especially of ciliated cells but not in non-ciliated goblet cells. By virtue of their results the authors suggest an important role of polycystins in granulosa cell differentiation just as in development and maturation of ovarian follicles. Accordingly rat TRPP2 was also found with punctate expression in cytoplasma of granulosa cells of corpora lutea but not in developing follicles in rat ovary [94]. Surprisingly, the proteins in rat ovary were more heavily glycosylated than in other organs such as kidney and salivary gland.

47.6 Mammary Glands

After labour the mammary glands are subjected to extensive calcium loads during lactation to support the requirements of milk calcium enrichment. Mammary glands represent accessory female reproductive organs and play a underpart in reproduction processes. They are therefore not further considered in this review article. But nonetheless it should be mentioned here that the expression of TRP proteins is reported in mammary glands or derived cell lines for TRPV1-6 and TRPC3 and C6 transcripts/proteins [95, 96]. These expression profile was clearly different from studies with breast carcinoma sections or derived malign cell lines where increased levels of TRPC3, TRPM8, TRPV4, TRPV6 and TRPC6 are reported [97–101].

47.7 Conclusion

Based on the different studies TRP channels seem to be strongly involved in the transport of the oocyte or the blastocyte through the oviduct, implantation of the blastocyte, development of the placenta, transport processes across the feto-maternal barrier, uterus contractility and physiological processes during labour.

Many of the here reviewed experiments were performed on the basis of transcript/messenger RNA measurements which do not fully substitute for protein expression data. In studies where the protein expression was investigated with antibodies against TRP proteins to detect their expression in cell culture or native tissue the data should be regarded with caution because the antibodies used have not always been rigorously validated. Unfortunately commercially available antibodies against TRP proteins fail in many cases even the most fundamental tests of activity and/or specificity. We are now only beginning to understand how TRP channels may be involved in reproductive processes.

Considering the finding that TRP proteins are widely expressed in female reproductive organs it is surprising that no female infertility phenotype or prenatal lethality has been described so far, although for most of the *trp* genes knock-out animals are available (reviewed in [102]). One might argue that a strong redundancy in function of the various TRP proteins does exist which may compensate for the loss of function in knock-outs or that the expression levels of TRP proteins vary. Future investigations of TRP protein expression levels and TRP interacting proteins in primary tissues still awaits careful validation by appropriate antibodies and/or mass spectrometry approaches.

References

1. Moffett A, Loke C (2006) Immunology of placentation in eutherian mammals. Nat Rev Immunol 6:584–594
2. Rossant J, Cross JC (2001) Placental development: lessons from mouse mutants. Nat Rev Genet 2:538–548
3. Jollie WP (1990) Development, morphology, and function of the yolk-sac placenta of laboratory rodents. Teratology 41:361–381
4. Enders AC, Blankenship TN (1999) Comparative placental structure. Adv Drug Deliv Rev 38:3–15
5. Sanborn BM (2001) Hormones and calcium: mechanisms controlling uterine smooth muscle contractile activity. The Litchfield Lecture. Exp Physiol 86:223–237
6. McKay RR, Szymeczek-Seay CL, Lievremont JP, Bird GS, Zitt C, Jungling E, Luckhoff A, Putney JW Jr (2000) Cloning and expression of the human transient receptor potential 4 (TRP4) gene: localization and functional expression of human TRP4 and TRP3. Biochem J 351(Pt 3):735–746
7. Yang M, Gupta A, Shlykov SG, Corrigan R, Tsujimoto S, Sanborn BM (2002) Multiple Trp isoforms implicated in capacitative calcium entry are expressed in human pregnant myometrium and myometrial cells. Biol Reprod 67:988–994
8. Dalrymple A, Slater DM, Beech D, Poston L, Tribe RM (2002) Molecular identification and localization of Trp homologues, putative calcium channels, in pregnant human uterus. Mol Hum Reprod 8:946–951

9. Dalrymple A, Slater DM, Poston L, Tribe RM (2004) Physiological induction of transient receptor potential canonical proteins, calcium entry channels, in human myometrium: influence of pregnancy, labor, and interleukin-1 beta. J Clin Endocrinol Metab 89:1291–1300
10. Ku CY, Babich L, Word RA, Zhong M, Ulloa A, Monga M, Sanborn BM (2006) Expression of transient receptor channel proteins in human fundal myometrium in pregnancy. J Soc Gynecol Investig 13:217–225
11. Babich LG, Ku CY, Young HW, Huang H, Blackburn MR, Sanborn BM (2004) Expression of capacitative calcium TrpC proteins in rat myometrium during pregnancy. Biol Reprod 70:919–924
12. Machaty Z, Ramsoondar JJ, Bonk AJ, Bondioli KR, Prather RS (2002) Capacitative calcium entry mechanism in porcine oocytes. Biol Reprod 66:667–674
13. Young A, Thomson AJ, Ledingham M, Jordan F, Greer IA, Norman JE (2002) Immunolocalization of proinflammatory cytokines in myometrium, cervix, and fetal membranes during human parturition at term. Biol Reprod 66:445–449
14. Osman I, Young A, Ledingham MA, Thomson AJ, Jordan F, Greer IA, Norman JE (2003) Leukocyte density and pro-inflammatory cytokine expression in human fetal membranes, decidua, cervix and myometrium before and during labour at term. Mol Hum Reprod 9: 41–45
15. Baggia S, Gravett MG, Witkin SS, Haluska GJ, Novy MJ (1996) Interleukin-1 beta intra-amniotic infusion induces tumor necrosis factor-alpha, prostaglandin production, and preterm contractions in pregnant rhesus monkeys. J Soc Gynecol Investig 3:121–126
16. Dalrymple A, Mahn K, Poston L, Songu-Mize E, Tribe RM (2007) Mechanical stretch regulates TRPC expression and calcium entry in human myometrial smooth muscle cells. Mol Hum Reprod 13:171–179
17. Sanborn BM (2007) Hormonal signaling and signal pathway crosstalk in the control of myometrial calcium dynamics. Semin Cell Dev Biol 18:305–314
18. Ulloa A, Gonzales AL, Zhong M, Kim YS, Cantlon J, Clay C, Ku CY, Earley S, Sanborn BM (2009) Reduction in TRPC4 expression specifically attenuates G-protein coupled receptor-stimulated increases in intracellular calcium in human myometrial cells. Cell Calcium 46:73–84
19. Chung D, Kim YS, Phillips JN, Ulloa A, Ku CY, Galan HL, Sanborn BM (2010) Attenuation of canonical transient receptor potential-like channel 6 expression specifically reduces the diacylglycerol-mediated increase in intracellular calcium in human myometrial cells. Endocrinology 151:406–416
20. Clarson LH, Roberts VH, Hamark B, Elliott AC, Powell T (2003) Store-operated Ca2+ entry in first trimester and term human placenta. J Physiol 550:515–528
21. Wes PD, Chevesich J, Jeromin A, Rosenberg C, Stetten G, Montell C (1995) TRPC1, a human homolog of a Drosophila store-operated channel. Proc Natl Acad Sci USA 92: 9652–9656
22. Zhu X, Chu PB, Peyton M, Birnbaumer L (1995) Molecular cloning of a widely expressed human homologue for the Drosophila trp gene. FEBS Lett 373:193–198
23. Sossey-Alaoui K, Lyon JA, Jones L, Abidi FE, Hartung AJ, Hane B, Schwartz CE, Stevenson RE, Srivastava AK (1999) Molecular cloning and characterization of TRPC5 (HTRP5), the human homologue of a mouse brain receptor-activated capacitative Ca2+ entry channel. Genomics 60:330–340
24. Freichel M, Suh SH, Pfeifer A, Schweig U, Trost C, Weissgerber P, Biel M, Philipp S, Freise D, Droogmans G, Hofmann F, Flockerzi V, Nilius B (2001) Lack of an endothelial store-operated Ca2+ current impairs agonist-dependent vasorelaxation in TRP4−/− mice. Nat Cell Biol 3:121–127
25. Caterina MJ, Schumacher MA, Tominaga M, Rosen TA, Levine JD, Julius D (1997) The capsaicin receptor: a heat-activated ion channel in the pain pathway. Nature 389:816–824
26. Zygmunt PM, Petersson J, Andersson DA, Chuang H, Sorgard M, Di Marzo V, Julius D, Hogestatt ED (1999) Vanilloid receptors on sensory nerves mediate the vasodilator action of anandamide. Nature 400:452–457

27. Peng HY, Chang HM, Lee SD, Huang PC, Chen GD, Lai CH, Lai CY, Chiu CH, Tung KC, Lin TB (2008) TRPV1 mediates the uterine capsaicin-induced NMDA NR2B-dependent cross-organ reflex sensitization in anesthetized rats. Am J Physiol Renal Physiol 295:F1324–F1335
28. Tong C, Conklin D, Clyne BB, Stanislaus JD, Eisenach JC (2006) Uterine cervical afferents in thoracolumbar dorsal root ganglia express transient receptor potential vanilloid type 1 channel and calcitonin gene-related peptide, but not P2X3 receptor and somatostatin. Anesthesiology 104:651–657
29. Tingaker BK, Ekman-Ordeberg G, Facer P, Irestedt L, Anand P (2008) Influence of pregnancy and labor on the occurrence of nerve fibers expressing the capsaicin receptor TRPV1 in human corpus and cervix uteri. Reprod Biol Endocrinol 6:8
30. Fonseca BM, Correia-da-Silva G, Taylor AH, Konje JC, Bell SC, Teixeira NA (2009) Spatio-temporal expression patterns of anandamide-binding receptors in rat implantation sites: evidence for a role of the endocannabinoid system during the period of placental development. Reprod Biol Endocrinol 7:121
31. Yan T, Liu B, Du D, Eisenach JC, Tong C (2007) Estrogen amplifies pain responses to uterine cervical distension in rats by altering transient receptor potential-1 function. Anesth Analg 104:1246–1250
32. Caterina MJ, Rosen TA, Tominaga M, Brake AJ, Julius D (1999) A capsaicin-receptor homologue with a high threshold for noxious heat. Nature 398:436–441
33. Peier AM, Reeve AJ, Andersson DA, Moqrich A, Earley TJ, Hergarden AC, Story GM, Colley S, Hogenesch JB, McIntyre P, Bevan S, Patapoutian A (2002) A heat-sensitive TRP channel expressed in keratinocytes. Science 296:2046–2049
34. Smith GD, Gunthorpe MJ, Kelsell RE, Hayes PD, Reilly P, Facer P, Wright JE, Jerman JC, Walhin JP, Ooi L, Egerton J, Charles KJ, Smart D, Randall AD, Anand P, Davis JB (2002) TRPV3 is a temperature-sensitive vanilloid receptor-like protein. Nature 418:186–190
35. Xu H, Ramsey IS, Kotecha SA, Moran MM, Chong JA, Lawson D, Ge P, Lilly J, Silos-Santiago I, Xie Y, DiStefano PS, Curtis R, Clapham DE (2002) TRPV3 is a calcium-permeable temperature-sensitive cation channel. Nature 418:181–186
36. Kunert-Keil C, Bisping F, Kruger J, Brinkmeier H (2006) Tissue-specific expression of TRP channel genes in the mouse and its variation in three different mouse strains. BMC Genomics 7:159
37. Liedtke W, Choe Y, Marti-Renom MA, Bell AM, Denis CS, Sali A, Hudspeth AJ, Friedman JM, Heller S (2000) Vanilloid receptor-related osmotically activated channel (VR-OAC), a candidate vertebrate osmoreceptor. Cell 103:525–535
38. Strotmann R, Harteneck C, Nunnenmacher K, Schultz G, Plant TD (2000) OTRPC4, a nonselective cation channel that confers sensitivity to extracellular osmolarity. Nat Cell Biol 2:695–702
39. Wissenbach U, Bodding M, Freichel M, Flockerzi V (2000) Trp12, a novel Trp related protein from kidney. FEBS Lett 485:127–134
40. Nilius B, Prenen J, Wissenbach U, Bodding M, Droogmans G (2001) Differential activation of the volume-sensitive cation channel TRP12 (OTRPC4) and volume-regulated anion currents in HEK-293 cells. Pflugers Arch 443:227–233
41. Nilius B, Vriens J, Prenen J, Droogmans G, Voets T (2004) TRPV4 calcium entry channel: a paradigm for gating diversity. Am J Physiol Cell Physiol 286:C195–C205
42. Andrade YN, Fernandes J, Vazquez E, Fernandez-Fernandez JM, Arniges M, Sanchez TM, Villalon M, Valverde MA (2005) TRPV4 channel is involved in the coupling of fluid viscosity changes to epithelial ciliary activity. J Cell Biol 168:869–874
43. Teilmann SC, Byskov AG, Pedersen PA, Wheatley DN, Pazour GJ, Christensen ST (2005) Localization of transient receptor potential ion channels in primary and motile cilia of the female murine reproductive organs. Mol Reprod Dev 71:444–452
44. Andrade YN, Fernandes J, Lorenzo IM, Arniges M, Valverde MA (2007) The TRPV4 channel in ciliated epithelia. In: TRP ion channel function in sensory transduction and cellular signaling cascades. CRC Press, Taylor & Francis Group, Boca Raton, FL

45. Liedtke W, Friedman JM (2003) Abnormal osmotic regulation in trpv4–/– mice. Proc Natl Acad Sci USA 100:13698–13703
46. Guilak F, Leddy HA, Liedtke W (2010) Transient receptor potential vanilloid 4. Ann NY Acad Sci 1192:404–409
47. Hoenderop JG, van der Kemp AW, Hartog A, van de Graaf SF, van Os CH, Willems PH, Bindels RJ (1999) Molecular identification of the apical Ca2+ channel in 1, 25-dihydroxyvitamin D3-responsive epithelia. J Biol Chem 274:8375–8378
48. Peng JB, Chen XZ, Berger UV, Vassilev PM, Tsukaguchi H, Brown EM, Hediger MA (1999) Molecular cloning and characterization of a channel-like transporter mediating intestinal calcium absorption. J Biol Chem 274:22739–22746
49. den Dekker E, Hoenderop JG, Nilius B, Bindels RJ (2003) The epithelial calcium channels, TRPV5 & TRPV6: from identification towards regulation. Cell Calcium 33:497–507
50. Hoenderop JG, Voets T, Hoefs S, Weidema F, Prenen J, Nilius B, Bindels RJ (2003) Homo- and heterotetrameric architecture of the epithelial Ca2+ channels TRPV5 and TRPV6. EMBO J 22:776–785
51. Barley NF, Howard A, O'Callaghan D, Legon S, Walters JR (2001) Epithelial calcium transporter expression in human duodenum. Am J Physiol Gastrointest Liver Physiol 280:G285–G290
52. Wissenbach U, Niemeyer BA, Fixemer T, Schneidewind A, Trost C, Cavalie A, Reus K, Meese E, Bonkhoff H, Flockerzi V (2001) Expression of CaT-like, a novel calcium-selective channel, correlates with the malignancy of prostate cancer. J Biol Chem 276:19461–19468
53. Hirnet D, Olausson J, Fecher-Trost C, Bodding M, Nastainczyk W, Wissenbach U, Flockerzi V, Freichel M (2003) The TRPV6 gene, cDNA and protein. Cell Calcium 33:509–518
54. Hoenderop JG, Nilius B, Bindels RJ (2003) Epithelial calcium channels: from identification to function and regulation. Pflugers Arch 446:304–308
55. Peng JB, Chen XZ, Berger UV, Weremowicz S, Morton CC, Vassilev PM, Brown EM, Hediger MA (2000) Human calcium transport protein CaT1. Biochem Biophys Res Commun 278:326–332
56. Bernucci L, Henriquez M, Diaz P, Riquelme G (2006) Diverse calcium channel types are present in the human placental syncytiotrophoblast basal membrane. Placenta 27:1082–1095
57. Stumpf T, Zhang Q, Hirnet D, Lewandrowski U, Sickmann A, Wissenbach U, Dorr J, Lohr C, Deitmer JW, Fecher-Trost C (2008) The human TRPV6 channel protein is associated with cyclophilin B in human placenta. J Biol Chem 283:18086–18098
58. Moreau R, Hamel A, Daoud G, Simoneau L, Lafond J (2002) Expression of calcium channels along the differentiation of cultured trophoblast cells from human term placenta. Biol Reprod 67:1473–1479
59. Lee GS, Jeung EB (2007) Uterine TRPV6 expression during the estrous cycle and pregnancy in a mouse model. Am J Physiol Endocrinol Metab 293:E132–E138
60. van Abel M, Hoenderop JG, van der Kemp AW, Friedlaender MM, van Leeuwen JP, Bindels RJ (2005) Coordinated control of renal Ca(2+) transport proteins by parathyroid hormone. Kidney Int 68:1708–1721
61. de Groot T, Lee K, Langeslag M, Xi Q, Jalink K, Bindels RJ, Hoenderop JG (2009) Parathyroid hormone activates TRPV5 via PKA-dependent phosphorylation. J Am Soc Nephrol 20:1693–1704
62. Hoenderop JG, Muller D, van der Kemp AW, Hartog A, Suzuki M, Ishibashi K, Imai M, Sweep F, Willems PH, van Os CH, Bindels RJ (2001) Calcitriol controls the epithelial calcium channel in kidney. J Am Soc Nephrol 12:1342–1349
63. Brown AJ, Krits I, Armbrecht HJ (2005) Effect of age, vitamin D, and calcium on the regulation of rat intestinal epithelial calcium channels. Arch Biochem Biophys 437:51–58
64. Muller D, Hoenderop JG, Merkx GF, van Os CH, Bindels RJ (2000) Gene structure and chromosomal mapping of human epithelial calcium channel. Biochem Biophys Res Commun 275:47–52

65. Van Cromphaut SJ, Rummens K, Stockmans I, Van HE, Dijcks FA, Ederveen AG, Carmeliet P, Verhaeghe J, Bouillon R, Carmeliet G (2003) Intestinal calcium transporter genes are upregulated by estrogens and the reproductive cycle through vitamin D receptor-independent mechanisms. J Bone Miner Res 18:1725–1736
66. Lee BM, Lee GS, Jung EM, Choi KC, Jeung EB (2009) Uterine and placental expression of TRPV6 gene is regulated via progesterone receptor- or estrogen receptor-mediated pathways during pregnancy in rodents. Reprod Biol Endocrinol 7:49
67. Kim HJ, Lee GS, Ji YK, Choi KC, Jeung EB (2006) Differential expression of uterine calcium transporter 1 and plasma membrane Ca2+ ATPase 1b during rat estrous cycle. Am J Physiol Endocrinol Metab 291:E234–E241
68. Suzuki Y, Kovacs CS, Takanaga H, Peng JB, Landowski CP, Hediger MA (2008) Calcium channel TRPV6 is involved in murine maternal-fetal calcium transport. J Bone Miner Res 23:1249–1256
69. Fudge NJ, Kovacs CS (2010) Pregnancy up-regulates intestinal calcium absorption and skeletal mineralization independently of the vitamin D receptor. Endocrinology 151:886–895
70. Gertner JM, Coustan DR, Kliger AS, Mallette LE, Ravin N, Broadus AE (1986) Pregnancy as state of physiologic absorptive hypercalciuria. Am J Med 81:451–456
71. Moreau R, Daoud G, Bernatchez R, Simoneau L, Masse A, Lafond J (2002) Calcium uptake and calcium transporter expression by trophoblast cells from human term placenta. Biochim Biophys Acta 1564:325–332
72. Niemeyer BA, Bergs C, Wissenbach U, Flockerzi V, Trost C (2001) Competitive regulation of CaT-like-mediated Ca2+ entry by protein kinase C and calmodulin. Proc Natl Acad Sci USA 98:3600–3605
73. Bianco SD, Peng JB, Takanaga H, Suzuki Y, Crescenzi A, Kos CH, Zhuang L, Freeman MR, Gouveia CH, Wu J, Luo H, Mauro T, Brown EM, Hediger MA (2007) Marked disturbance of calcium homeostasis in mice with targeted disruption of the Trpv6 calcium channel gene. J Bone Miner Res 22:274–285
74. Hoenderop JG, van Leeuwen JP, van der Eerden BC, Kersten FF, van der Kemp AW, Merillat AM, Waarsing JH, Rossier BC, Vallon V, Hummler E, Bindels RJ (2003) Renal Ca2+ wasting, hyperabsorption, and reduced bone thickness in mice lacking TRPV5. J Clin Invest 112:1906–1914
75. van der Eerden BC, Hoenderop JG, de Vries TJ, Schoenmaker T, Buurman CJ, Uitterlinden AG, Pols HA, Bindels RJ, van Leeuwen JP (2005) The epithelial Ca2+ channel TRPV5 is essential for proper osteoclastic bone resorption. Proc Natl Acad Sci USA 102:17507–17512
76. Duncan LM, Deeds J, Hunter J, Shao J, Holmgren LM, Woolf EA, Tepper RI, Shyjan AW (1998) Down-regulation of the novel gene melastatin correlates with potential for melanoma metastasis. Cancer Res 58:1515–1520
77. Montell C (2005) The TRP superfamily of cation channels. Sci STKE 2005:re3
78. Mochizuki T, Wu G, Hayashi T, Xenophontos SL, Veldhuisen B, Saris JJ, Reynolds DM, Cai Y, Gabow PA, Pierides A, Kimberling WJ, Breuning MH, Deltas CC, Peters DJ, Somlo S (1996) PKD2, a gene for polycystic kidney disease that encodes an integral membrane protein. Science 272:1339–1342
79. Clapham DE, Julius D, Montell C, Schultz G (2005) International union of pharmacology. XLIX. Nomenclature and structure-function relationships of transient receptor potential channels. Pharmacol Rev 57:427–450
80. Nomura H, Turco AE, Pei Y, Kalaydjieva L, Schiavello T, Weremowicz S, Ji W, Morton CC, Meisler M, Reeders ST, Zhou J (1998) Identification of PKDL, a novel polycystic kidney disease 2-like gene whose murine homologue is deleted in mice with kidney and retinal defects. J Biol Chem 273:25967–25973
81. Ong AC, Ward CJ, Butler RJ, Biddolph S, Bowker C, Torra R, Pei Y, Harris PC (1999) Coordinate expression of the autosomal dominant polycystic kidney disease

proteins, polycystin-2 and polycystin-1, in normal and cystic tissue. Am J Pathol 154: 1721–1729
82. Gonzalez-Perrett S, Kim K, Ibarra C, Damiano AE, Zotta E, Batelli M, Harris PC, Reisin IL, Arnaout MA, Cantiello HF (2001) Polycystin-2, the protein mutated in autosomal dominant polycystic kidney disease (ADPKD), is a Ca2+-permeable nonselective cation channel. Proc Natl Acad Sci USA 98:1182–1187
83. Gonzalez-Perrett S, Batelli M, Kim K, Essafi M, Timpanaro G, Moltabetti N, Reisin IL, Arnaout MA, Cantiello HF (2002) Voltage dependence and pH regulation of human polycystin-2-mediated cation channel activity. J Biol Chem 277:24959–24966
84. Light DB, McCann FV, Keller TM, Stanton BA (1988) Amiloride-sensitive cation channel in apical membrane of inner medullary collecting duct. Am J Physiol 255:F278–F286
85. Cantiello HF, Stow JL, Prat AG, Ausiello DA (1991) Actin filaments regulate epithelial Na+ channel activity. Am J Physiol 261:C882–C888
86. Montalbetti N, Cantero MR, Dalghi MG, Cantiello HF (2008) Reactive oxygen species inhibit polycystin-2 (TRPP2) cation channel activity in term human syncytiotrophoblast. Placenta 29:510–518
87. Li Q, Montalbetti N, Shen PY, Dai XQ, Cheeseman CI, Karpinski E, Wu G, Cantiello HF, Chen XZ (2005) Alpha-actinin associates with polycystin-2 and regulates its channel activity. Hum Mol Genet 14:1587–1603
88. Montalbetti N, Li Q, Wu Y, Chen XZ, Cantiello HF (2007) Polycystin-2 cation channel function in the human syncytiotrophoblast is regulated by microtubular structures. J Physiol 579:717–728
89. Montalbetti N, Li Q, Timpanaro GA, Gonzalez-Perrett S, Dai XQ, Chen XZ, Cantiello HF (2005) Cytoskeletal regulation of calcium-permeable cation channels in the human syncytiotrophoblast: role of gelsolin. J Physiol 566:309–325
90. Montalbetti N, Li Q, Gonzalez-Perrett S, Semprine J, Chen XZ, Cantiello HF (2005) Effect of hydro-osmotic pressure on polycystin-2 channel function in the human syncytiotrophoblast. Pflugers Arch 451:294–303
91. Wu G, Markowitz GS, Li L, D'Agati VD, Factor SM, Geng L, Tibara S, Tuchman J, Cai Y, Park JH, van AJ, Hou H Jr, Kucherlapati R, Edelmann W, Somlo S (2000) Cardiac defects and renal failure in mice with targeted mutations in Pkd2. Nat Genet 24:75–78
92. Pennekamp P, Karcher C, Fischer A, Schweickert A, Skryabin B, Horst J, Blum M, Dworniczak B (2002) The ion channel polycystin-2 is required for left-right axis determination in mice. Curr Biol 12:938–943
93. Allen E, Piontek KB, Garrett-Mayer E, Garcia-Gonzalez M, Gorelick KL, Germino GG (2006) Loss of polycystin-1 or polycystin-2 results in dysregulated apolipoprotein expression in murine tissues via alterations in nuclear hormone receptors. Hum Mol Genet 15:11–21
94. Obermuller N, Gallagher AR, Cai Y, Gassler N, Gretz N, Somlo S, Witzgall R (1999) The rat pkd2 protein assumes distinct subcellular distributions in different organs. Am J Physiol 277:F914–F925
95. Reiter B, Kraft R, Gunzel D, Zeissig S, Schulzke JD, Fromm M, Hartenech C (2006) TRPV4-mediated regulation of epithelial permeability. FASEB J 20:1802–1812
96. Guilbert A, Dhennin-Duthille I, Hiani YE, Haren N, Khorsi H, Sevestre H, Ahidouch A, Ouadid-Ahidouch H (2008) Expression of TRPC6 channels in human epithelial breast cancer cells. BMC Cancer 8:125
97. Prevarskaya N, Zhang L, Barritt G (2007) TRP channels in cancer. Biochim Biophys Acta 1772:937–946
98. Bodding M (2007) TRP proteins and cancer. Cell Signal 19:617–624
99. Bolanz KA, Hediger MA, Landowski CP (2008) The role of TRPV6 in breast carcinogenesis. Mol Cancer Ther 7:271–279
100. Zhuang L, Peng JB, Tou L, Takanaga H, Adam RM, Hediger MA, Freeman MR (2002) Calcium-selective ion channel, CaT1, is apically localized in gastrointestinal

tract epithelia and is aberrantly expressed in human malignancies. Lab Invest 82: 1755–1764
101. Tsavaler L, Shapero MH, Morkowski S, Laus R (2001) Trp-p8, a novel prostate-specific gene, is up-regulated in prostate cancer and other malignancies and shares high homology with transient receptor potential calcium channel proteins. Cancer Res 61:3760–3769
102. Freichel M, Flockerzi V (2007) Biological functions of TRPs unravelled by spontaneous mutations and transgenic animals. Biochem Soc Trans 35:120–123

Chapter 48
Oncogenic TRP Channels

V'yacheslav Lehen'kyi and Natalia Prevarskaya

Abstract Ion channels and notably TRP channels play a crucial role in a variety of physiological functions and in addition these channels have been also shown associated with several diseases including cancer. The process of cancer initiation and progression involves the altered expression of one or more of TRP proteins, depending on the nature of the cancer. The most clearly described role in pathogenesis has been evidenced for TRPM8, TRPV6 and TRPM1 channels. The increased expression of some other channels, such as TRPV1, TRPC1, TRPC6, TRPM4, and TRPM5 has also been demonstrated in some cancers. Further investigations are required to precise the role of TRP channels in cancer development and/or progression and to specifically develop further knowledge of TRP proteins as discriminative markers and prospective targets for pharmaceutical intervention in treating cancer.

48.1 Introduction

The transition from normal to cancerous phenotype is caused by the accumulation of mutations in certain key signalling proteins, encoded by oncogenes and tumour suppressor genes, and the formation and selection by evolution of those cells which can more aggressively compete in their local environment and, in the case of metastatic cells, in the environments of other organs. Among the signalling pathways altered in tumorigenesis those which enhance cell proliferation and inhibit apoptosis are some of the most important. Many proteins in cancer cells exhibit increased or decreased expression compared to their levels of expression in normal cells. Some of these proteins, encoded by oncogenes and tumour suppressor genes, play key roles in tumorigenesis and in the development of metastases, while others, most likely including those involved in intracellular Ca^{2+} homeostasis are associated with

---

N. Prevarskaya (✉)
Inserm, U-800, Equipe labellisée par la Ligue Nationale contre le cancer, Villeneuve d'Ascq F-59655, France; Université des Sciences et Technologies de Lille (USTL), Villeneuve d'Ascq F-59655, France; Laboratoire de Physiologie Cellulaire, INSERM U1003, USTL, Villeneuve d'Ascq Cedex F-59655, France
e-mail: Natacha.Prevarskaya@univ-lillel.fr

cancer progression but are not causative in further development of the tumour and/or malignant cells (reviewed in [1]).

Several members of the TRP family of cationic-permeable channels (reviewed in [2] and [3]) show altered expression in cancer cells. The most studied are TRPM8, TRPV6, TRPM1 and TRPV1. To date, most changes involving TRP proteins do not involve mutations in the TRP gene but rather increased or decreased levels of expression of the normal (wild type) TRP protein, depending on the stage of the cancer. It is not yet possible to say whether these changes in TRP expression are central steps in the progression of the cancer or are secondary to other changes, although the latter is most likely.

There are over 100 distinct types of cancer, which are defined by specific genotypes, and much research effort is directed towards identifying new genes with altered expression in specific cancers. However, as defined by Hanahan and Weinberg [4], the vast catalog of cancer cell genotypes can be characterized in terms of six essential pathophysiological phenotypes that define malignant growth: (1) self-sufficiency in growth signals, (2) insensitivity to antigrowth signals, (3) evasion of programmed cell death (apoptosis), (4) limitless replicative potential, (5) sustained angiogenesis and (6) tissue invasion and metastasis.

A number of recent reviews have discussed the roles of TRP channels in cancer from two principal standpoints: examining how specific these channels are involved in certain cancer-related cellular behaviors such as proliferation, apoptosis, migration or angiogenesis [5, 6], or examining specific expression and functional profiles of various channels that are characteristic of certain human cancers [7, 8]. In this review, we discuss how the six pathophysiological hallmarks of cancer, defined by Hanahan and Weinberg [4], depend on various channels from TRP superfamily.

48.2 Role of TRP Channels in Cancer Development and Progression

The mammalian TRP channels are encoded by at least 28 genes (reviewed by [2, 9, 10]). Most of these proteins display a putative topology of six transmembrane domains with a pore loop between the fifth and sixth segment. Both C-and N-termini are presumably located intracellularly. Very similar primary structures are typical for other ion channels such as voltage-gated channels and the cyclic-nucleotide-gated channels. Amino acid sequences flanking the pore are highly conserved within the TRP superfamily. This is especially the case for about 25 amino acid residues immediately C-terminally to the 6th transmembrane domain which is called TRP domain. However, the overall sequence identity of the mammalian TRP proteins is as low as 20%. They all form cation channels most likely by homo- and/or hetero-multimerisation to tetramers. TRP proteins are almost ubiquitously expressed, of most are splice variants known. Based on homology and partially also on channel function, TRP channels can be divided into three main subfamilies: the canonical or classical TRP proteins (TRPC), the vanilloid receptor related TRP proteins (TRPV) and the melastatin related TRP proteins (TRPM). The mucolipins (TRPML), the polycystins (TRPP) and the ankyrin transmembrane protein 1 (TRPA1) are more distantly related to the founding member TRP from Drosophila. TRPN proteins have not been detected in mammals but in Caenorhabditis elegans, the fruit fly and zebra fish.

TRP proteins have diverse functional properties and profound effects on a variety of physiological and pathological conditions. Only TRP proteins which are associated with cancer will be addressed below.

48.2.1 TRPC Channels and Cancer

The first data demonstrating the expression of TRPC channels in tumor cells have appeared in the last decade showing the expression of TRPC1 channel in H4-IIE rat liver hepatoma cell line [11], and TRPC2 in murine B16 melanoma cell line [12]. Initially, the function of TRPC1 channel has been ascribed to the regulation of cell volume in H4-IIE cells [13]. The further investigation using the human prostate cancer model showed the expression of both TRPC1 and TRPC3 channel and their function [14]. The TRPC1 channel was shown to be the most likely molecular candidate for the formation of prostate-specific endogenous store-operated channel (SOC) [15]. Differentiation of LNCaP cells to an androgen-insensitive, apoptotic-resistant neuroendocrine phenotype downregulates SOC current. The authors conclude that prostate-specific SOCs are important determinants in the transition to androgen-independent prostate cancer [15]. The role of TRPC1 in human neuroblastoma (SH-SY5Y) cells has also been studied [16]. Activation of TRPC1 by thapsigargin or carbachol decreased neurotoxicity, which was partially dependent on external $Ca^{(2+)}$. The authors concluded that TRPC1 protects SH-SY5Y neuronal cells against apoptosis which is an important hallmark of cancer development and progression. In the cannabinoid-2 receptor-expressing human peripheral blood-acute lymphoid leukemia (HPB-ALL) human T cell line small interference RNA knockdown of TRPC1 attenuated the Delta9-tetrahydrocannabinol-evoked elevation of $[Ca^{2+}]i$. The results suggest that Delta9-THC-induced elevation in $[Ca^{2+}]i$ is attributable entirely to extracellular calcium influx, which is independent of $[Ca^{2+}]i$ store depletion, and is mediated, at least partially, through the DAG-sensitive TRPC1 channels [17]. In the other work, the high level expression of the functional TRPC2 and induction of protective immunity by adoptive transfer of TRPC2 gene modified dendritic cells were demonstrated [18].

The other study showed that alpha1-AR signaling required the coupled activation of TRPC6 channels and NFAT to promote proliferation of human prostate cancer (PCa) cells [19]. In androgen responsive prostate cancer cell line LNCaP, TRPC1 and TRPC4 have been reported to function as SOCs, while TRPC3 might be involved in the response to agonist stimulation [20]. The upregulation of TRPC1 and TRPC3 depends on the store contents level and involves the activation of the $Ca^{(2+)}$/calmodulin/calcineurin/NFAT pathway participating in proliferation [20]. Human SH-SY5Y neuroblastoma cells treated with salsolinol, an endogenous neurotoxin, showed approximately 60% reduction in TRPC1 protein levels

[21]. Confocal microscopy also showed that SH-SY5Y cells treated with salsolinol had a significant decrease in the plasma membrane staining of the TRPC1 protein. Interestingly, overexpression of TRPC1 increases TRPC1 protein levels and also protected SH-SY5Y neuroblastoma cells against salsolinol-mediated cytotoxicity [21]. In breast cancer MCF-7 cells, calcium receptor is functionally coupled to $Ca^{(2+)}$-permeable cationic TRPCs, for which TRPC1 and TRPC6 are the most likely candidates for the highly selective $Ca^{(2+)}$ current which have been shown to be involved in cell proliferation [22]. TRPC1 is required for the ERK1/2 phosphorylation, $Ca^{(2+)}$ entry and the CaR-proliferative effect in human breast cancer cell line MCF-7 [23, 24].

TRPC1/TRPC4-mediated calcium entry and endoplasmic reticulum $Ca^{(2+)}$ content increased significantly in differentiated keratinocytes [25]. However, the failure of BCC cells to differentiate was related to a lack of TRPC channel expression and calcium entry. The data demonstrate that TRPC1 and TRPC4 channels are key elements in keratinocyte $Ca^{(2+)}$ homeostasis and differentiation and may therefore be responsible for skin pathologies. In the other case TRPC4 loss in renal cell carcinoma leads to impaired $Ca^{(2+)}$ intake, misfolding, retrograde transport and diminished secretion of antiangiogenic TSP1, thus enabling angiogenic switch during renal carcinoma progression [26].

The consistent expression of four channel family members (TRPC-1, -3, -5, -6) in glioma cell lines and acute patient-derived tissues has been shown [27]. These channels gave rise to small, non-voltage-dependent cation currents that were blocked by the TRPC inhibitors GdCl [3], 2-APB, or SKF96365. The authors conclude that nuclear atypia and enlarged cells are histopathological hallmarks for glioblastoma multiforme, the highest grade glioma, suggesting that a defect in TRPC channel function may contribute to cellular abnormalities in these tumours [27]. TRPC6 channels were shown to be strongly expressed and functional in breast cancer epithelial cells [28]. Moreover, the overexpression of these channels appears without any correlation with tumour grade, ER expression and lymph node metastasis. The findings support the idea that TRPC6 may have a role in breast carcinogenesis. Furthermore, TRPC6 was shown to be very weakly expressed in isolated hepatocytes from healthy patients and expressed more strongly in tumoral samples from the liver of a cancer patient controlling cell proliferation and strongly supporting a role for these calcium channels in liver oncogenesis [29].

Ovarian cancer (OC) is the leading cause of death from gynecological malignancy [30]. However, the mechanism by which OC develops remains largely unknown. The authors report that TRPC3 channels promote human OC growth. Downregulating TRPC3 expression in SKOV3 cells, a human OC cell line, led to reduction of proliferation, suppression in epidermal growth factor-induced $Ca^{(2+)}$ influx, dephosphorylation of Cdc2 and CaMKIIalpha and prolonged progression through M phase of these cells suggesting that the increased activity of TRPC3 channels is necessary for the development of OCs [30].

TRPC6 channel has been reported to have a critical role in human gastric cancer development [31]. Expression of TRPC6 was greatly upregulated in human gastric cancer epithelial cells compared with that in normal gastric epithelial cells.

Moreover, inhibition of TRPC6 suppressed the formation of gastric tumors in nude mice. These results suggest that $Ca^{(2+)}$ elevation regulated by TRPC6 channels is essential for G2/M phase transition and for the development of gastric cancers [31]. TRPC6 channels play also a critical role in the development of oesophageal squamous cell carcinoma [32]. Therefore the $[Ca^{(2+)}](i)$ elevation regulated by TRPC6 channels is essential for G2 phase progression and oesophageal squamous cell carcinoma development.

The expression of TRPC6 protein in benign and malignant human prostate tissues and in prostate cancer cell lines and the association with the stage, grade and androgen responsiveness of the tumors was investigated [33]. Tumors of higher stage tended to have a higher frequency of TRPC6 protein staining, but the difference was not significant among T2, T3 and T4. TRPC6 expression difference between androgen-independent (AI) tumors and androgen-dependent (AD) tumors was not statistically significant. TRPC6 was also observed in prostate cancer cell lines. In summary, TRPC6 is detected in benign and malignant human prostate tissues and prostate cancer cell lines and is associated with the histological grade, Gleason score and extraprostatic extension of prostate cancer [33].

Therefore, TRPC family of channels may be considered as potentially oncogenic family though not all of the members are directly involved in oncogenesis.

48.2.2 TRPV Channels and Cancer

TRPV6 channel is expressed in the intestine, kidney, placenta and pancreas [34] and [35, 36]. One of the biological functions of TRPV6 is to mediate transcellular Ca^{2+} movement across epithelial cells of the gut, kidney and placenta [35, 36], and [37].

When compared with normal tissue or cells, the expression of TRPV6 mRNA (measured using semi-quantitative or real time quantitative RT-PCR) and/or expression of the TRPV6 protein (measured by immunofluorescence) is substantially increased in prostate cancer tissue, and in human carcinomas of the colon, breast, thyroid, and ovary [38, 39, 40], and [34]. Increased expression of TRPV6 mRNA compared with that in normal cells is also observed in the LNCaP and PC-3 prostate cancer cell lines [40], in SW480 colorectal cancer cell lines [40] and in the K-562 chronic myelogenous leukaemia cell line [36]. Increased expression of TRPV6 is not observed in pancreatic carcinoma and some other cancers [34] indicating that TRPV6 is not a general marker for tumorigenic cells [41] and [34].

More extensive studies of TRPV6 expression have been made in prostate cancer. The expression of TRPV6 mRNA is very low or not detectable in healthy and benign human prostate tissue [38] and [34]. Studies with prostate cancer tissue obtained from biopsies or resections show substantial expression of TRPV6 mRNA which increases with the degree of aggressiveness of the cancer, assessed by the Gleason score (grading of the pathological stage) and the degree of metastasis outside the prostate [38] and [40]. These observations have led to the suggestion that the level of expression of TRPV6 could be used as a marker to predict the clinical outcome of prostate cancer [38] and [41].

As mentioned above, TRPV6 is expressed in the LNCaP human prostate cancer cell line. Prevarskaya's team have recently shown that TRPV6 is directly involved in the control of proliferation in LNCaP cells by decreasing: (i) proliferation rate; (ii) cells accumulation into S-phase of the cell cycle; and (iii) proliferating cell nuclear antigen (PCNA) expression [42]. This team has demonstrated that Ca^{2+} uptake into LNCaP cells is mediated by TRPV6, with the subsequent downstream activation of nuclear factor of activated T cells (NFAT). The possible role of androgens in the regulation of TRPV6 mRNA expression remains unclear. Previous studies have shown that the androgen receptor agonist dihydrotestosterone inhibits TRPV6 expression while the androgen receptor antagonist becalutamide increases TRPV6 expression [40, 15], and [43]. However, TRPV6 expression has not been identified in androgen-insensitive prostate cancer cell lines DU-145 and PC-3 [38] and, moreover, Lehenkyi et al. have shown that TRPV6 expression in LNCaP cells is regulated by androgen receptor, however in a ligand-independent fashion [42]. To date, little is known about whether the observed increased expression of TRPV6 mRNA and protein in prostate cancer cells is associated with increased Ca^{2+} and Na+ entry through functional TRPV6 channels, and what the physiological and pathological consequences might be [41].

The very close to TRPV6 channel, TRPV5 channel was originally cloned from rabbit kidney, and belongs to the "apical calcium channels" family like TRPV6 channel. Unlike the other four members of TRPV family that are non selective cation channel, TRPV5 together with TRPV6, are remarkable Ca^{2+}-selective channels which serve as apical calcium entry mechanisms in absortive and secretory tissues [44], however no data have been provided regarding the potential involvement of TRPV5 in renal cancers. From the other hand, endogenous TRPV5 mRNA expression was observed using RT-PCR analysis in human Jurkat T leukemia cell line [45] and in K562 erythroleukemia cell line [46]. K562 cells co-express TRPV5 and TRPV6 calcium channels that form a functional homotetrameric structure by interacting each other. Based on the evidence that TRPV5/V6 expression levels are strongly controlled by 1,25-dihydroxyvitamin D3 (1,25-(OH)2D3) [46] that exerts antiproliferative effects and induces K562 cell differentiation, it has been suggested that TRPV5/TRPV6 channels regulate leukemia cell differentiation in a 1,25-(OH)2D3-dependent manner.

The very recent discovery of glycosidase klotho (a mammalian senescence suppression protein that shows homology with other glycosidases) in the renal epithelium, showed that TRPV5 expression is regulated by the latter [47, 48]. Klotho increases cell surface expression of TRPV5 by removing terminal sialic acids from their glycan chains. Removal of sialic acids exposes the disaccharide galactose-N-acetylglucosamine, a ligand for the beta-galactoside-binding lectin Galectin 1 (GAL1), involved in cell cycle progression, and surface binding to GAL1 leads to accumulation of functional TRPV5 on the plasma membrane [49]. As GAL1 overexpression is associated with neoplastic transformation, and differentiation of K562 cells induces externalisation and binding of GAL1 on the cell surface Klotho/TRPV5/GAL1 pathway has been suggested to be involved in myeloid transformation [50].

The expression of other TRPV family channels like TRPV1 channel in cancer has been reported. However, as compared to TRPV6 channel its expression and function as a potential oncogene is different than that of TRPV6. TRPV1 mRNA and protein expression was shown in normal astrocytes and glioma cells and tissues [51–53]. Its expression inversely correlated with glioma grading and a marked loss of TRPV1 expression in the majority of grade IV glioblastoma tissues was observed. TRPV1 activation by CPS induced apoptosis of glioma cells, by rising the Ca^{2+} influx, p38MAPK activation, inducing mitochondrial permeability transmembrane pore opening and caspase-3 activation [51].

TRPV1 mRNA and protein are significantly expressed in human pancreatic cancer [53]. In human hepatoblastoma cells the hepatocyte growth factor/scatter factor (HGF/SF) promoted cell migration by stimulating the activity of TRPV1. This action of TRPV1 in hepatoblastoma cells may be in part mediated by increasing intracellular Ca^{2+} and promoting tumor invasion by triggering a signaling cascade leading to cell locomotion and acquisition of a migratory phenotype [54–56]. Interestingly, Miao et al. have recently demonstrated in hepatocarcinoma patients that high TRPV1 expression is associated with increased disease-free survival [57].

TRPV1 expression has been also reported in human cervical cancer cell lines and tissues, and the role of TRPV1-dependent tumor cell apoptosis has been demonstrated [58]. In addition, TRPV1 stimulation completely reverted the cannabidiol (CBD)-mediated inhibitory effect on human cervical cancer cell invasion [59].

TRPV1 is highly and specifically expressed in both premalignant (leukoplasia) and lowgrade papillary skin carcinoma with its expression being substantially absent in invasive carcinoma [60]. TRPV1 overexpression in HEK293 cells resulted in decreased EGFR protein expression, and higher EGFR levels were observed in the skin of TRPV1 deficient mice (TRPV1-/-) as compared to wild-type control animals [61]. Changes in the TRPV1 expression can also occur during the development of human urothelial cancer [52]. Lazzeri and colleagues have demonstrated that transitional cell carcinoma (TCC) show a progressive decrease in TRPV1 expression as the tumor stage increases [62].

TRPV1 expression has been also demonstrated on the plasma membrane of rat pheochromocytomaderived PC12 cell line [63]. CPS exposure triggered Ca^{2+} influx, which enhanced mitochondrial Ca^{2+} accumulation and promoted NO generation by NO-synthase, events that have been associated with tumor progression [64].

A functional TRPV1 channel is also expressed in human prostate cancer cells (PC3 and LNCaP) and in prostate hyperplasic tissue [65]. CPS inhibited tumor growth in vivo, in a xenograft human prostate PC3 cancer model [66]. However, in androgen-responsive LNCaP prostate cancer cells, CPS was found to stimulate TRPV1-dependent cell proliferation. The increased TRPV1 mRNA and protein expression was found in human prostate cancer tissues as compared to prostate hyperplasic and healthy donors, and this increase correlated with the degree of malignancy [67]. In addition, CPS induced in TRPV1-independent manner a growth inhibition and apoptosis of PC3 prostate cancer cells through ROS generation, mitochondrial inner transmembrane potential dissipation and caspase-3 activation. The constitutive activity of TRPV2 is also critical for castration-resistant prostate cancer development and progression in vivo. The progression of prostate cancer to the castration-resistant phenotype is characterized by de novo expression of TRPV2, and higher levels of TRPV2 transcripts are found in patients with metastatic cancer (stage M1) as compared to primary solid tumors (stage T2a and T2b) [68]. In addition, induction of MMP9 and cathepsin B expression was observed upon transfection of TRPV2 in the LNCaP cell line, and knockout of TRPV2 by siRNA reduced the growth and the invasive properties of PC3 cells by progressive downregulation of MMP2, MMP9 and cathepsin B expression in a xenograft tumor model [68]. The other work from the same group shows the Ca^{2+}-dependent activation and translocation of TRPV2 by lysophospholipids (LPL), such as lysophosphatidylcoline and lysophosphatidylinositol resulted in increased PC3 prostate tumor cell migration, and involved Gq/Go-protein and PI3,4-K pathways. TRPV2 knockdown in PC3 cells or inhibition of PI3,4-K pathway, abolished the stimulatory effects of LPL on both calcium entry and migration [69]. TRPV2-regulated cytosolic calcium levels promoted PC3 migration by induction of key proteases, namely MMP2, MMP9 and cathepsin B [68].

TRPV2 mRNA is expressed in benign astrocyte tissues, and its expression progressively declines in high-grade glioma tissues as histological grade increased [70]. TRPV2 siRNA increases glioblastoma proliferation and survival to Fas/CD95-induced apoptosis in an ERK-dependent manner [70]. In the same way TRPV2 mRNA and protein up-regulation inversely correlated with the degree of tumor differentiation, indicating a potential role for TRPV2 in hepatocarcinogenesis [68].

Thus, TRPV2 could represent a prospective prognostic marker and potential therapeutic target for novel interventions aimed at improving the life expectancy of prostate cancer patients.

TRPV3 is a calcium-permeable temperature-sensitive cation channel [71] which is expressed in tongue, nasal epithelium and epidermal keratinocytes [72, 73]. TRPV3 seems to be involved in prostate cancer progression [74]. TRPV3 expression was shown to be stimulated by androgens and more strongly expressed in castrate-resistant C4-2b prostate cancer cells. In addition, immunohistochemistry and RT-PCR analyses revealed TRPV3 mRNA and protein expression in rat prostatic tissues, although at lower levels with respect to other TRPV channels (TRPV2 and TRPV4) [75]. As to the other type of cancer the expression of TRPV3 channel has been shown in colorectal cancer development where TRPV3 expression was positively associated with higher colorectal cancer risk [76].

Recent findings demonstared TRPV4 mRNA expression in primary cultured cortical astrocytes. TRPV4 protein was abundant in astrocytes of the superficial layers of the neocortex and in astrocyte end feet facing pia and blood vessels [77]. Activated outwardly rectifying cation current of TRPV4 promoted $[Ca^{2+}]I$ elevation, suggesting that astrocyte TRPV4 channel is functional [77]. Functional expression of TRPV4 has been described in HepG2 cells hepatoblastoma cells [78], as this receptor channel can regulate both Ca^{2+} influx and Ca^{2+} release channels. A functional interaction between TRPV4 and F-actin in sensing hypotonicity and the onset of regulatory volume decrease has also been demonstrated [79]. TRPV4 transcripts and protein were shown to be abundantly expressed in mouse basal urothelial cells [80], as well as TRPV4-mediated strech-evoked increase of $[Ca^{2+}]i$ and ATP release in primary isolated urothelial cell cultures [81].

Thus TRPV family represents a very different but rather oncogenic family of TRP channels with the pathological function including but not limited to proliferation, apoptosis, migration and invasion.

48.2.3 TRPM Channels and Cancer

TRPM1 (initially called melastatin) was discovered in B-16 mouse melanoma cell lines as a result of differential display analysis [82]. Heterologous expression of TRPM1 in HEK293 cells has been shown to lead to an increase in the cytoplasmic free Ca^{2+} concentration ($[Ca^{2+}]cyt$) at the normal extracellular Ca^{2+} concentration ($[Ca^{2+}]ext$), and a corresponding increase or decrease in $[Ca^{2+}]cyt$ when $[Ca^{2+}]ext$ was increased or decreased compared to the absence of such changes in control HEK cells [83]. This suggests that when heterologously expressed, TRPM1 is constitutively active (active in the absence of addition of any known activator). Ca^{2+} entry through TRPM1 is inhibited by La3+ and Gd3+ [83]. While it has been suggested that TRPM1 may be involved in the regulation of cell proliferation [41], the cellular and biological roles of TRPM1 are not well understood.

Studies of TRPM1 expression in B-16 mouse melanoma cell lines have shown that poorly metastatic variants, such as B-16-F1, express higher levels of TRPM1 mRNA compared to highly metastatic variants, such as B-16-F10 [84]. In specimens of human melanoma tissue, Duncan and colleagues, using in situ hybridisation, found high levels of TRPM1 expression in benign tissue (nevi), decreased expression in primary melanomas, and no detectable mRNA in metastatic melanomas [82]. TRPM1 mRNA expression appears to correlate with the state of melanocytic tumour progression, thickness of the tumour, and the degree of aggression, with low or undetectable TRPM1 mRNA expression in aggressive tumours [85, 86].

These observations have led to the suggestion that the trpm1 gene is a tumour suppressor [85]. It has also been suggested that the level of expression of TRPM1 mRNA may be a prognostic marker for the development of melanoma metastasis, and hence, may permit the prediction of the development of non-metastatic or metastatic melanomas [82] and [41].

Studies of the regulation of expression of TRPM1 in melanocytes and melanoma cells have provided evidence that TRPM1 expression is linked to the state of cell differentiation, and is regulated by the microphthalmia transcription factor (MIRF). Thus, the treatment of human pigmented metastatic melanoma cells with hexamethylene bisaretamide, an inducer of cell differentiation, led to an increase in TRPM1 mRNA expression [85]. Other studies have shown that the promoter region of the TRPM1 gene contains binding sites for microphthalmia, indicating that TRPM1 expression is likely to be regulated by MIRF. This transcription factor also regulates expression of a number of other melanoma cell proteins [87, 88].

While remaining at moderate levels in a normal prostate, TRPM8 expression strongly increases in prostate cancer. For this reason it has been proposed to be a pro-oncogenic actor in prostate cancer cells [89]. Other non-prostatic primary human tumours (breast, colon, lung, and skin) also become highly enriched in TRPM8, although it is virtually undetectable in corresponding normal tissues [89]. Thus, even this initial information strongly pointed to much broader roles of TRPM8 beyond cold sensation, especially in the prostate and during carcinogenesis. The role of TRPM8 in organs not exposed to ambient temperatures, and especially in the prostate gland, remains a gnawing mystery. However, the data accumulated in the last 3 years allows the formulation of several hypotheses.

TRPM8 could function as a cold sensor in the prostate [90], but may also be involved in other functions such as ion and protein secretion, regulation of proliferation and/or apoptosis in prostate epithelial cells (for review see [91]). Nevertheless, the secretion of citric acid, prostate-specific antigen (PSA), acid phosphatase, several enzymes, lipids and other products is the major function of apical epithelial prostate cells. Therefore, considering the specific TRPM8 expression in these cells, the potential role of this channel in secretion has been suggested [92]. Furthermore, the recent study of Mergler et al. has confirmed this hypothesis on neuroendocrine pancreatic tumour cells [93]. The authors have shown that TRPM8 activation increases the secretion of neurotensin in these cells.

In normal prostate, trpm8 gene expression seems to be directly controlled by androgen receptors [94] positioning it as a primary androgen-response gene [92]. Single-cell RT-PCR and immunohistochemical experiments conducted on primary human prostate cancer cells have shown that TRPM8 is mainly expressed in androgen-dependent, apical secretory epithelial cells, and that its expression becomes down-regulated in cells losing the androgen receptor activity and regressing to the basal epithelial phenotype [92]. Mature prostate epithelial cells are non-proliferative cells which are highly sensitive to apoptotic stimuli. (Due to the specific regulation of the expression of genes belonging to the Bcl-2 family, anti-apoptotic Bcl-2 gene expression is repressed whereas pro-apoptotic Bax gene expression is stimulated by androgen receptors [96, 97]).

In prostate cancer tumours, a significant difference in the expression level of TRPM8 mRNA between malignant and non-malignant tissue specimens has been detected [95]. This was comparable to the currently used prostate cancer marker, PSA, thus, qualifying TRPM8 as its potential competitor in prostate cancer diagnosis and staging. A significant difference in TRPM8 expression between human benign prostate hyperplasia and prostate cancer tissues is also obvious at protein level [7]. According to Tsavaler's hypothesis defining trpm8 as an oncogene [89], TRPM8 over-expression and over activity in circumscribed, androgen-dependent prostate cancer may be correlated to the higher rate of growth of these cells compared to normal ones [96]. During the transition to androgen independence, TRPM8 is lost in a xenograft model of prostate cancer and also in prostate cancer tissue from patients treated preoperatively with anti-androgen therapy, suggesting that its loss may be associated with a more advanced form of the disease [97]. It has been demonstrated that LNCaP cells resistant to anti-androgen bicalutamide treatment displayed a reduced doubling time [7]. This is correlated with a decreased expression of mRNA encoding the androgen receptor, TRPM8 and the Proliferating Cell Nuclear Antigen (PCNA), while anti-apoptotic Bcl-2 mRNA expression is increased. All these data reinforce the putative pro-proliferative role of TRPM8 in androgen-dependent prostate cancer cells. Finally, it has also been shown that both pharmacological activation of TRPM8 and siRNA-mediated TRPM8 silencing in LNCaP cells can decrease the cell viability [94], probably by perturbing the TRPM8-dependent intracellular Ca^{2+} homeostasis. However, it is still not clear whether TRPM8 involvement in cell viability is carried out through a pro-proliferative and/or an anti-apoptotic mechanism.

Suguro et al. have studied the CD5+ subgroup of diffuse large B-cell lymphomas which comprise about 30% of non-Hodgkin's lymphomas [100]. CD5+ is associated with poor prognosis. TRPM4 mRNA expression was found to be increased in CD5+ lymphomas along with increased or decreased expression of a number of other genes. It was suggested that TRPM4 is one of several genes which constitute a "CD5 signature" which can be used as a prognostic clinical marker.

Indirect genetic evidence suggests that altered expression of TRPM5 may be associated with tumorigenesis. Alterations in region 11p15.5 of the human chromosome are known to be associated with Beckwith–Wiedemann syndrome and with a predisposition to neoplasias including Wilms' tumours, rhaboid tumours, and rhabdomyosarcomas (reviewed in [98] In a study of candidate genes in the 11p15.5 region, Prawitt et al. identified TRPM5 as one of the genes in this region of DNA [76]). Moreover, TRPM5 mRNA was found to be expressed in a large proportion of Wilms' tumours and rhabdomyosarcomas [98].

Yet another member of the TRPM subfamilily, redox state- and Mg^{2+}-sensitive TRPM7, was recently reported to be specifically overexpressed in large-size breast tumors [99], and its silencing in the MCF-7 breast cancer cell line reduced cell proliferation and basal intracellular Ca^{2+} concentration ($[Ca^{2+}]i$), suggesting that TRPM7 is involved in the proliferative potential of breast cancer cells, most likely by regulating Ca^{2+} influx.

Thus, TRPM subfamily of channels represents also potentially oncogenic members involved in cancer initiation and progression.

48.3 Conclusions

It seems evident from the discussion above that some of the TRP channels like TRPM8, TRPM1, and TRPV6 are highly expressed in cancer cells and the amount of protein expressed changes with progression from normal through tumorigenic to metastatic cells. The expression of some other TRP channels, including TRPC1, TRPC6, TRPM5 and TRPV1 is also increased in cancer tissues. So far, there is no evidence to indicate that the expression of a given TRP protein is consistently increased or decreased in many different cancers, although further studies may show this is the case for some TRP proteins, for example, TRPM8. As regards the TRP channels as the oncogenes, it has been suggested that TRPM8, TRPV6 and TRPV1 may be oncogenes and TRPM1 a tumour suppressor gene, but further experiments are required to test these hypotheses. Many of the physiological and pathophysiological roles of the TRP proteins which are expressed at high levels in cancer tissue are still to be elucidated. This task is presently made more difficult because, for some TRP channels, little is known of their functions in normal cells. The roles of TRP channels in cancer progression may involve changes in intracellular Ca^{2+} and Na^+, although effects on Ca^{2+} would be complex since the expression of some TRP proteins increases while that of others decreases as cancer progresses [41]. Another likely possibility is that some of the effects of TRP proteins in cancer cells are exerted by their interactions with other intracellular proteins [41]. Elucidation of the cellular functions of TRP channels in tumorigenesis and in cancer cells might be achieved by ablation of a given TRP protein, for example, by use of a non-functional TRP mutant or a knockout model [41]. TRPM8, TRPV6, TRPM1 and TRPV1 are potential diagnostic markers for the prognosis of tumour development especially the degree of tumour aggression, and are potential targets for pharmaceutical interventions. Such interventions may involve use of the TRP protein as a recognition site for antibody-mediated delivery of a toxic pay load, or possibly more directly through selective activation of the TRP channel to induce sustained Ca^{2+} and Na^+ entry, and subsequent necrosis and apoptosis.

It is predicted that there will be considerable further progress in understanding changes in TRP channel expression in cancer cells in the next few years as knowledge of both the molecular events involved in cancer progression and the physiological functions of TRP channels increases. One of the main challenges will be gaining an understanding of whether the involvement of a given TRP channel is part of the causative mechanisms involved in cancer progression or is associated in a secondary manner with these mechanisms and changes.

References

1. Prevarskaya N, Skryma R, Shuba Y (2004) Ca^{2+} homeostasis in apoptotic resistance of prostate cancer cells (Translated from eng). Biochem Biophys Res Commun 322(4):1326–1335 (in eng)
2. Ramsey IS, Delling M, Clapham DE (2006) An introduction to TRP channels (Translated from eng). Annu Rev Physiol 68:619–647 (in eng)
3. Nilius B, Owsianik G, Voets T, Peters JA (2007) Transient receptor potential cation channels in disease (Translated from eng). Physiol Rev 87(1):165–217 (in eng)
4. Hanahan D, Weinberg RA (2000) The hallmarks of cancer (Translated from eng). Cell 100(1):57–70 (in eng)
5. Kunzelmann K (2005) Ion channels and cancer (Translated from eng). J Membr Biol 205(3):159–173 (in eng)
6. Prevarskaya N, Zhang L, Barritt G (2007) TRP channels in cancer (Translated from eng). Biochim Biophys Acta 1772(8):937–946 (in eng)
7. Prevarskaya N, Skryma R, Bidaux G, Flourakis M, Shuba Y (2007) Ion channels in death and differentiation of prostate cancer cells (Translated from eng). Cell Death Differ 14(7):1295–1304 (in eng)
8. McFerrin MB, Sontheimer H (2006) A role for ion channels in glioma cell invasion (Translated from Eng). Neuron Glia Biol 2(1):39–49 (in Eng)
9. Pedersen SF, Owsianik G, Nilius B (2005) TRP channels: an overview (Translated from eng). Cell Calcium 38(3–4):233–252 (in eng)

10. Clapham DE, Julius D, Montell C, Schultz G (2005) International, Union of Pharmacology. XLIX. Nomenclature and structure-function relationships of transient receptor potential channels (Translated from eng). Pharmacol Rev 57(4):427–450 (in eng)
11. Ong HL et al (2002) Specific detection of the endogenous transient receptor potential (TRP)-1 protein in liver and airway smooth muscle cells using immunoprecipitation and Western-blot analysis (Translated from eng). Biochem J 364(Pt 3):641–648 (in eng)
12. Steitz J, Bruck J, Knop J, Tuting T (2001) Adenovirus-transduced dendritic cells stimulate cellular immunity to melanoma via a CD4(+) T cell-dependent mechanism (Translated from eng). Gene Ther 8(16):1255–1263 (in eng)
13. Chen J, Barritt GJ (2003) Evidence that TRPC1 (transient receptor potential canonical 1) forms a $Ca^{(2+)}$-permeable channel linked to the regulation of cell volume in liver cells obtained using small interfering RNA targeted against TRPC1 (Translated from eng). Biochem J 373(Pt 2):327–336 (in eng)
14. Sydorenko V et al (2003) Receptor-coupled, DAG-gated Ca^{2+}-permeable cationic channels in LNCaP human prostate cancer epithelial cells (Translated from eng). J Physiol 548(Pt 3):823–836 (in eng)
15. Vanden Abeele F et al (2003) Store-operated Ca^{2+} channels in prostate cancer epithelial cells: function, regulation, and role in carcinogenesis (Translated from eng). Cell Calcium 33 (5–6):357–373 (in eng)
16. Bollimuntha S, Singh BB, Shavali S, Sharma SK, Ebadi M (2005) TRPC1-mediated inhibition of 1-methyl-4-phenylpyridinium ion neurotoxicity in human SH-SY5Y neuroblastoma cells (Translated from eng). J Biol Chem 280(3):2132–2140 (in eng)
17. Rao GK, Kaminski NE (2006) Induction of intracellular calcium elevation by Delta9-tetrahydrocannabinol in T cells involves TRPC1 channels (Translated from eng). J Leukoc Biol 79(1):202–213 (in eng)
18. Metharom P, Ellem KA, Wei MQ (2005) Gene transfer to dendritic cells induced a protective immunity against melanoma (Translated from eng). Cell Mol Immunol 2(4):281–288 (in eng)
19. Thebault S et al (2006) Differential role of transient receptor potential channels in Ca^{2+} entry and proliferation of prostate cancer epithelial cells (Translated from eng). Cancer Res 66(4):2038–2047 (in eng)
20. Pigozzi D et al (2006) Calcium store contents control the expression of TRPC1, TRPC3 and TRPV6 proteins in LNCaP prostate cancer cell line (Translated from eng). Cell Calcium 39(5):401–415 (in eng)
21. Bollimuntha S, Ebadi M, Singh BB (2006) TRPC1 protects human SH-SY5Y cells against salsolinol-induced cytotoxicity by inhibiting apoptosis (Translated from eng). Brain Res 1099(1):141–149 (in eng)
22. El Hiani Y et al (2006) Calcium-sensing receptor stimulation induces nonselective cation channel activation in breast cancer cells (Translated from eng). J Membr Biol 211(2):127–137 (in eng)
23. El Hiani Y et al (2009) Extracellular signal-regulated kinases 1 and 2 and TRPC1 channels are required for calcium-sensing receptor stimulated MCF-7 breast cancer cell proliferation (Translated from eng). Cell Physiol Biochem 23(4–6):335–346 (in eng)
24. El Hiani Y, Lehen'kyi V, Ouadid-Ahidouch H, Ahidouch A (2009) Activation of the calcium-sensing receptor by high calcium induced breast cancer cell proliferation and TRPC1 cation channel over-expression potentially through EGFR pathways (Translated from eng). Arch Biochem Biophys 486(1):58–63 (in eng)
25. Beck B et al (2008) TRPC channels determine human keratinocyte differentiation: new insight into basal cell carcinoma (Translated from eng). Cell Calcium 43(5):492–505 (in eng)
26. Veliceasa D et al (2007) Transient potential receptor channel 4 controls thrombospondin-1 secretion and angiogenesis in renal cell carcinoma (Translated from eng). FEBS J 274(24):6365–6377 (in eng)

27. Bomben VC, Sontheimer HW (2008) Inhibition of transient receptor potential canonical channels impairs cytokinesis in human malignant gliomas (Translated from eng). Cell Prolif 41(1):98–121 (in eng)
28. Guilbert A et al (2008) Expression of TRPC6 channels in human epithelial breast cancer cells (Translated from eng). BMC Cancer 8:125 (in eng)
29. El Boustany C et al (2008) Capacitative calcium entry and transient receptor potential canonical 6 expression control human hepatoma cell proliferation (Translated from eng). Hepatology 47(6):2068–2077 (in eng)
30. Yang SL, Cao Q, Zhou KC, Feng YJ, Wang YZ (2009) Transient receptor potential channel C3 contributes to the progression of human ovarian cancer (Translated from eng). Oncogene 28(10):1320–1328 (in eng)
31. Cai R et al (2009) Blockade of TRPC6 channels induced G2/M phase arrest and suppressed growth in human gastric cancer cells (Translated from eng). Int J Cancer 125(10):2281–2287 (in eng)
32. Shi Y et al (2009) Critical role of TRPC6 channels in G2 phase transition and the development of human oesophageal cancer (Translated from eng). Gut 58(11):1443–1450 (in eng)
33. Yue D, Wang Y, Xiao JY, Wang P, Ren CS (2009) Expression of TRPC6 in benign and malignant human prostate tissues (Translated from eng). Asian J Androl 11(5):541–547 (in eng)
34. Wissenbach U et al (2001) Expression of CaT-like, a novel calcium-selective channel, correlates with the malignancy of prostate cancer (Translated from eng). J Biol Chem 276(22):19461–19468 (in eng)
35. Peng JB et al (1999) Molecular cloning and characterization of a channel-like transporter mediating intestinal calcium absorption (Translated from eng). J Biol Chem 274(32):22739–22746 (in eng)
36. Peng JB et al (2000) Human calcium transport protein CaT1 (Translated from eng). Biochem Biophys Res Commun 278(2):326–332 (in eng)
37. Bianco SD et al (2007) Marked disturbance of calcium homeostasis in mice with targeted disruption of the Trpv6 calcium channel gene (Translated from eng). J Bone Miner Res 22(2):274–285 (in eng)
38. Fixemer T, Wissenbach U, Flockerzi V, Bonkhoff H (2003) Expression of the Ca^{2+}-selective cation channel TRPV6 in human prostate cancer: a novel prognostic marker for tumor progression (Translated from eng). Oncogene 22(49):7858–7861 (in eng)
39. Zhuang L et al (2002) Calcium-selective ion channel, CaT1, is apically localized in gastrointestinal tract epithelia and is aberrantly expressed in human malignancies (Translated from eng). Lab Invest 82(12):1755–1764 (in eng)
40. Peng JB et al (2001) CaT1 expression correlates with tumor grade in prostate cancer (Translated from eng). Biochem Biophys Res Commun 282(3):729–734 (in eng)
41. Bodding M (2007) TRP proteins and cancer (Translated from eng). Cell Signal 19(3):617–624 (in eng)
42. Lehen'kyi V, Flourakis M, Skryma R, Prevarskaya N (2007) TRPV6 channel controls prostate cancer cell proliferation via $Ca^{(2+)}$/NFAT-dependent pathways (Translated from eng). Oncogene 26(52):7380–7385 (in eng)
43. Bodding M, Fecher-Trost C, Flockerzi V (2003) Store-operated Ca^{2+} current and TRPV6 channels in lymph node prostate cancer cells (Translated from eng). J Biol Chem 278(51):50872–50879 (in eng)
44. Nijenhuis T, Hoenderop JG, Nilius B, Bindels RJ (2003) (Patho)physiological implications of the novel epithelial Ca^{2+} channels TRPV5 and TRPV6 (Translated from eng). Pflugers Arch 446(4):401–409 (in eng)
45. Vasil'eva IO, Neguliaev IuA, Marakhova II, Semenova SB (2008) TRPV5 and TRPV6 calcium channels in human T cells (Translated from rus). Tsitologiia 50(11):953–957 (in rus)

46. Semenova SB, Vassilieva IO, Fomina AF, Runov AL, Negulyaev YA (2009) Endogenous expression of TRPV5 and TRPV6 calcium channels in human leukemia K562 cells (Translated from eng). Am J Physiol Cell Physiol 296(5):C1098–C1104 (in eng)
47. Kuro-o M et al (1997) Mutation of the mouse klotho gene leads to a syndrome resembling ageing (Translated from eng). Nature 390(6655):45–51 (in eng)
48. Ito S, Fujimori T, Hayashizaki Y, Nabeshima Y (2002) Identification of a novel mouse membrane-bound family 1 glycosidase-like protein, which carries an atypical active site structure (Translated from eng). Biochim Biophys Acta 1576(3):341–345 (in eng)
49. Chang Q et al (2005) The beta-glucuronidase klotho hydrolyzes and activates the TRPV5 channel (Translated from eng). Science 310(5747):490–493 (in eng)
50. Lutomski D et al (1997) Externalization and binding of galectin-1 on cell surface of K562 cells upon erythroid differentiation (Translated from eng). Glycobiology 7(8):1193–1199 (in eng)
51. Amantini C et al (2007) Capsaicin-induced apoptosis of glioma cells is mediated by TRPV1 vanilloid receptor and requires p38 MAPK activation (Translated from eng). J Neurochem 102(3):977–990 (in eng)
52. Amantini C et al (2009) Triggering of transient receptor potential vanilloid type 1 (TRPV1) by capsaicin induces Fas/CD95-mediated apoptosis of urothelial cancer cells in an ATM-dependent manner (Translated from eng). Carcinogenesis 30(8):1320–1329 (in eng)
53. Hartel M et al (2006) Vanilloids in pancreatic cancer: potential for chemotherapy and pain management (Translated from eng). Gut 55(4):519–528 (in eng)
54. Zhang YW, Vande Woude GF (2003) HGF/SF-met signaling in the control of branching morphogenesis and invasion (Translated from eng). J Cell Biochem 88(2):408–417 (in eng)
55. Waning J et al (2007) A novel function of capsaicin-sensitive TRPV1 channels: involvement in cell migration (Translated from eng). Cell Calcium 42(1):17–25 (in eng)
56. Wondergem R, Ecay TW, Mahieu F, Owsianik G, Nilius B (2008) HGF/SF and menthol increase human glioblastoma cell calcium and migration (Translated from eng). Biochem Biophys Res Commun 372(1):210–215 (in eng)
57. Miao X et al (2008) High expression of vanilloid receptor-1 is associated with better prognosis of patients with hepatocellular carcinoma (Translated from eng). Cancer Genet Cytogenet 186(1):25–32 (in eng)
58. Contassot E, Tenan M, Schnuriger V, Pelte MF, Dietrich PY (2004) Arachidonyl ethanolamide induces apoptosis of uterine cervix cancer cells via aberrantly expressed vanilloid receptor-1 (Translated from eng). Gynecol Oncol 93(1):182–188 (in eng)
59. Ramer R, Merkord J, Rohde H, Hinz B (2010) Cannabidiol inhibits cancer cell invasion via upregulation of tissue inhibitor of matrix metalloproteinases-1 (Translated from eng). Biochem Pharmacol 79(7):955–966 (in eng)
60. Marincsak R et al (2009) Increased expression of TRPV1 in squamous cell carcinoma of the human tongue (Translated from eng). Oral Dis 15(5):328–335 (in eng)
61. Bode AM et al (2009) Transient receptor potential type vanilloid 1 suppresses skin carcinogenesis (Translated from eng). Cancer Res 69(3):905–913 (in eng)
62. Lazzeri M et al (2005) Transient receptor potential vanilloid type 1 (TRPV1) expression changes from normal urothelium to transitional cell carcinoma of human bladder (Translated from eng). Eur Urol 48(4):691–698 (in eng)
63. Qiao S, Li W, Tsubouchi R, Murakami K, Yoshino M (2004) Role of vanilloid receptors in the capsaicin-mediated induction of iNOS in PC12 cells (Translated from eng). Neurochem Res 29(4):687–693 (in eng)
64. Hosseini A et al (2006) Enhanced formation of nitric oxide in bladder carcinoma in situ and in BCG treated bladder cancer (Translated from eng). Nitric Oxide 15(4):337–343 (in eng)
65. Sanchez MG et al (2005) Expression of the transient receptor potential vanilloid 1 (TRPV1) in LNCaP and PC-3 prostate cancer cells and in human prostate tissue (Translated from eng). Eur J Pharmacol 515(1–3):20–27 (in eng)

66. Sanchez AM, Sanchez MG, Malagarie-Cazenave S, Olea N, Diaz-Laviada I (2006) Induction of apoptosis in prostate tumor PC-3 cells and inhibition of xenograft prostate tumor growth by the vanilloid capsaicin (Translated from eng). Apoptosis 11(1):89–99 (in eng)
67. Czifra G et al (2009) Increased expressions of cannabinoid receptor-1 and transient receptor potential vanilloid-1 in human prostate carcinoma (Translated from eng). J Cancer Res Clin Oncol 135(4):507–514 (in eng)
68. Monet M et al (2010) Role of cationic channel TRPV2 in promoting prostate cancer migration and progression to androgen resistance (Translated from eng). Cancer Res 70(3):1225–1235 (in eng)
69. Monet M et al (2009) Lysophospholipids stimulate prostate cancer cell migration via TRPV2 channel activation (Translated from eng). Biochim Biophys Acta 1793(3):528–539 (in eng)
70. Nabissi M et al (2010) TRPV2 channel negatively controls glioma cell proliferation and resistance to Fas-induced apoptosis in ERK-dependent manner (Translated from eng). Carcinogenesis 31(5):794–803 (in eng)
71. Xu H et al (2002) TRPV3 is a calcium-permeable temperature-sensitive cation channel (Translated from eng). Nature 418(6894):181–186 (in eng)
72. Mandadi S et al (2009) TRPV3 in keratinocytes transmits temperature information to sensory neurons via ATP (Translated from eng). Pflugers Arch 458(6):1093–1102 (in eng)
73. Ueda T, Yamada T, Ugawa S, Ishida Y, Shimada S (2009) TRPV3, a thermosensitive channel is expressed in mouse distal colon epithelium (Translated from eng). Biochem Biophys Res Commun 383(1):130–134 (in eng)
74. Jariwala U et al (2007) Identification of novel androgen receptor target genes in prostate cancer (Translated from eng). Mol Cancer 6:39 (in eng)
75. Wang HP, Pu XY, Wang XH (2007) Distribution profiles of transient receptor potential melastatin-related and vanilloid-related channels in prostatic tissue in rat (Translated from eng). Asian J Androl 9(5):634–640 (in eng)
76. Hoeft B et al (2010) Polymorphisms in fatty-acid-metabolism-related genes are associated with colorectal cancer risk (Translated from eng). Carcinogenesis 31(3):466–472 (in eng)
77. Benfenati V et al (2007) Expression and functional characterization of transient receptor potential vanilloid-related channel 4 (TRPV4) in rat cortical astrocytes (Translated from eng). Neuroscience 148(4):876–892 (in eng)
78. Vriens J, Janssens A, Prenen J, Nilius B, Wondergem R (2004) TRPV channels and modulation by hepatocyte growth factor/scatter factor in human hepatoblastoma (HepG2) cells (Translated from eng). Cell Calcium 36(1):19–28 (in eng)
79. Becker D, Bereiter-Hahn J, Jendrach M (2009) Functional interaction of the cation channel transient receptor potential vanilloid 4 (TRPV4) and actin in volume regulation (Translated from eng). Eur J Cell Biol 88(3):141–152 (in eng)
80. Yamada T et al (2009) Differential localizations of the transient receptor potential channels TRPV4 and TRPV1 in the mouse urinary bladder (Translated from eng). J Histochem Cytochem 57(3):277–287 (in eng)
81. Everaerts W et al (2010) Functional characterization of transient receptor potential channels in mouse urothelial cells (Translated from eng). Am J Physiol Renal Physiol 298(3):F692–F701 (in eng)
82. Duncan LM et al (1998) Down-regulation of the novel gene melastatin correlates with potential for melanoma metastasis (Translated from eng). Cancer Res 58(7):1515–1520 (in eng)
83. Xu XZ, Moebius F, Gill DL, Montell C (2001) Regulation of melastatin, a TRP-related protein, through interaction with a cytoplasmic isoform (Translated from eng). Proc Natl Acad Sci USA 98(19):10692–10697 (in eng)
84. Nilius B (2007) TRP channels in disease (Translated from eng). Biochim Biophys Acta 1772(8):805–812 (in eng)

85. Fang D, Setaluri V (2000) Expression and Up-regulation of alternatively spliced transcripts of melastatin, a melanoma metastasis-related gene, in human melanoma cells (Translated from eng). Biochem Biophys Res Commun 279(1):53–61 (in eng)
86. Deeds J, Cronin F, Duncan LM (2000) Patterns of melastatin mRNA expression in melanocytic tumors (Translated from eng). Hum Pathol 31(11):1346–1356 (in eng)
87. Miller AJ et al (2004) Transcriptional regulation of the melanoma prognostic marker melastatin (TRPM1) by MITF in melanocytes and melanoma (Translated from eng). Cancer Res 64(2):509–516 (in eng)
88. Zhiqi S et al (2004) Human melastatin 1 (TRPM1) is regulated by MITF and produces multiple polypeptide isoforms in melanocytes and melanoma (Translated from eng). Melanoma Res 14(6):509–516 (in eng)
89. Tsavaler L, Shapero MH, Morkowski S, Laus R (2001) Trp-p8, a novel prostate-specific gene, is up-regulated in prostate cancer and other malignancies and shares high homology with transient receptor potential calcium channel proteins (Translated from eng). Cancer Res 61(9):3760–3769 (in eng)
90. Stein RJ et al (2004) Cool (TRPM8) and hot (TRPV1) receptors in the bladder and male genital tract (Translated from eng). J Urol 172(3):1175–1178 (in eng)
91. Zhang L, Barritt GJ (2006) TRPM8 in prostate cancer cells: a potential diagnostic and prognostic marker with a secretory function? (Translated from eng). Endocr Relat Cancer 13(1):27–38 (in eng)
92. Bidaux G et al (2005) Evidence for specific TRPM8 expression in human prostate secretory epithelial cells: functional androgen receptor requirement (Translated from eng). Endocr Relat Cancer 12(2):367–382 (in eng)
93. Mergler S et al (2007) Transient receptor potential channel TRPM8 agonists stimulate calcium influx and neurotensin secretion in neuroendocrine tumor cells (Translated from eng). Neuroendocrinology 85(2):81–92 (in eng)
94. Zhang L, Barritt GJ (2004) Evidence that TRPM8 is an androgen-dependent Ca^{2+} channel required for the survival of prostate cancer cells (Translated from eng). Cancer Res 64(22):8365–8373 (in eng)
95. Fuessel S et al (2003) Multiple tumor marker analyses (PSA, hK2, PSCA, trp-p8) in primary prostate cancers using quantitative RT-PCR (Translated from eng). Int J Oncol 23(1):221–228 (in eng)
96. Kiessling A et al (2003) Identification of an HLA-A*0201-restricted T-cell epitope derived from the prostate cancer-associated protein trp-p8 (Translated from eng). Prostate 56(4):270–279 (in eng)
97. Henshall SM et al (2003) Survival analysis of genome-wide gene expression profiles of prostate cancers identifies new prognostic targets of disease relapse (Translated from eng). Cancer Res 63(14):4196–4203 (in eng)
98. Prawitt D et al (2000) Identification and characterization of MTR1, a novel gene with homology to melastatin (MLSN1) and the trp gene family located in the BWS-WT2 critical region on chromosome 11p15.5 and showing allele-specific expression (Translated from eng). Hum Mol Genet 9(2):203–216 (in eng)
99. Guilbert A et al (2009) Evidence that TRPM7 is required for breast cancer cell proliferation (Translated from eng). Am J Physiol Cell Physiol 297(3):C493–C502 (in eng)
100. Suguro M, Tagawa H, Kagami Y, Okamoto M, Ohshima K, Shiku H, Morishima Y, Nakamura S, Seto M. Expression profiling analysis of the CD5+ diffuse large B-cell lymphoma subgroup: development of a CD5 signature. Cancer Sci. 2006 Sep; 97(9):868–74. Epub 2006 Jul 13.

Chapter 49
TRPV Channels in Tumor Growth and Progression

Giorgio Santoni, Valerio Farfariello, and Consuelo Amantini

Abstract Transient receptor potential (TRP) channels affect several physiological and pathological processes. In particular, TRP channels have been recently involved in the triggering of enhanced proliferation, aberrant differentiation, and resistance to apoptotic cell death leading to the uncontrolled tumor invasion. About thirty TRPs have been identified to date, and are classified in seven different families: TRPC (Canonical), TRPV (Vanilloid), TRPM (Melastatin), TRPML (Mucolipin), TRPP (Polycystin), and TRPA (Ankyrin transmembrane protein) and TRPN (NomPC-like). Among these channel families, the TRPC, TRPM, and TRPV families have been mainly correlated with malignant growth and progression. The aim of this review is to summarize data reported so far on the expression and the functional role of TRPV channels during cancer growth and progression. TRPV channels have been found to regulate cancer cell proliferation, apoptosis, angiogenesis, migration and invasion during tumor progression, and depending on the stage of the cancer, up- and down-regulation of TRPV mRNA and protein expression have been reported. These changes may have cancer promoting effects by increasing the expression of constitutively active TRPV channels in the plasma membrane of cancer cells by enhancing Ca^{2+}-dependent proliferative response; in addition, an altered expression of TRPV channels may also offer a survival advantage, such as resistance of cancer cells to apoptotic-induced cell death. However, recently, a role of TRPV gene mutations in cancer development, and a relationship between the expression of specific TRPV gene single nucleotide polymorphisms and increased cancer risk have been reported. We are only at the beginning, a more deep studies on the physiopathology role of TRPV channels are required to understand the functional activity of these channels in cancer, to assess which TRPV proteins are associated with the development and progression of cancer and to develop further knowledge of TRPV proteins as valuable diagnostic and/or prognostic markers, as well as targets for pharmaceutical intervention and targeting in cancer.

G. Santoni (✉)
Section of Experimental Medicine, School of Pharmacy, University of Camerino, 62032 Camerino, Italy
e-mail: giorgio.santoni@unicam.it

49.1 Introduction

The processes involved in the transformation of normal cells to tumorigenic cells and in tumor progression are complex and only partly understood [1, 2]. Tumorigenesis involves the transition from normal cells to hyperplastic, metaplastic, dysplastic, neoplastic and then to metastatic cells.

It is well established that over a period of time many tumors become more aggressive and acquire greater malignant potential. This phenomenon is referred as "tumor progression". Careful clinical and experimental studies reveal that an increasingly malignant phenotype (e.g. accelerated growth, invasiveness, and ability to form metastasis) is often acquired in an incremental fashion. Tumor progression is related to the sequential appearance of cell subpopulations that differ for several phenotypic attributes, such as invasiveness, rate of growth, genetic instability, hormonal responsiveness, and susceptibility to anti-neoplastic drugs [3].

Invasion and metastasis are biologic hallmarks of malignant tumors. Tumor cells detached from the primary mass, enter blood vessels or lymphatics and produce a secondary growth at a distant site. The metastatic cascade results in invasion of extracellular matrix and vascular dissemination and homing of tumor cells. Invasion of extracellular matrix is not merely due to passive growth pressure but requires active enzymatic degradation of extracellular matrix components. Tumor cells themselves secrete proteolytic enzymes (metalloproteinases, MMPs) or induce host cells to elaborate proteases [4–6].

At molecular level, tumor progression and the associated heterogeneity is likely the result of multiple mutations in certain key signaling proteins, along with the formation and selection of subclones capable of competing more aggressively in the local microenvironment, and in the case of metastatic cells, in the environment of other organs.

Among these proteins, the Transient receptor potential (TRP) channels have been identified to profoundly affect a variety of physiological and pathological processes [7, 8]. In the last years, the extent to which TRP channels are associated with cancer has been increasingly clarified, and the involvement of these receptors in triggering enhanced proliferation, aberrant differentiation, and impaired ability to die, thus leading to the uncontrolled cancer expansion and invasion have been recognized [9–15]. Approximately thirty TRPs have been identified to date, and are classified in seven different families: TRPC (Canonical), TRPV (Vanilloid), TRPM (Melastatin), TRPML (Mucolipin), TRPP (Polycystin), and TRPA (Ankyrin transmembrane protein) and TRPN (NomPC-like) [16]. The expression levels and activity of members of the TRPC, TRPM, and TRPV families have been correlated with malignant growth and progression [17–23].

TRPV channels may regulate cancer cell proliferation, apoptosis, angiogenesis, migration and invasion during tumor progression at different levels [1]: by acting as Ca^{2+} entry pathways in the plasma membrane [24–26]; by regulating the expression and the activity of cell-surface glycoproteins [3, 27, 28]; by regulating the binding, trafficking and functional activity of several growth factors [4, 29]; by interacting with specific G protein-coupled receptors (GPCRs) at the plasma membrane [5, 30];

by directly or indirectly triggering different intracytoplasmatic signaling pathways [6, 21, 23, 31]; by regulating at transcriptional and translational levels the production/release of different cytokines, chemokines, growth factors and neuropeptides [32–34]. Further studies are required to assess which TRPV proteins are associated with the development and progression of cancer, which role TRPV proteins play in this process, and to develop further knowledge of TRPV proteins as valuable diagnostic and/or prognostic markers, as well as targets for pharmaceutical intervention and targeting in cancer.

Depending on the stage of the cancer, either increased or decreased expression of TRPV mRNA and protein levels have been reported. These changes may have cancer promoting effects by increasing the expression of constitutively active TRPV channels in the plasma membrane of cancer cells, and thus enhancing Ca^{2+}-dependent proliferative response. Alternatively, altered expression of TRPV channels may offer a survival advantage, such as resistance of cancer cells to apoptotic-induced cell death. Moreover, a role of TRPV gene mutations in cancer development, and a relationship between the expression of specific TRPV gene single nucleotide polymorphisms (SNPs) and increased cancer risk have been recently reported. Thus, by using tagging SNPs, two synonymous-coding *TRPV3* SNPs (rs11078458 and rs65002729), were positively associated with higher colon cancer (CRC) risk [35].

TRPV2 overexpression was evidenced by microarray analysis in N-RAS-mutated melanoma cell lines [36], and multiple myeloma patients [37]. In addition, it has been described *TRPV2* down-regulation in acute myeloid leukaemia and myelodysplastic syndrome (AML/MDS) patients [38], and aberrant TRPV2 expression in B lymphocytes from Mantle cell lymphoma (MCL) cell lines and tissues [39].

The aim of this review is to summarize the observations reported so far on the expression and activity of the different TRPVs that are associated with cancer growth and progression.

49.2 TRPV1

Although changes in the expression and activity of TRPV1 frequently occur in human cancer cells, mainly of epithelial origin, at present the role of TRPV1 in cancer growth and progression and the molecular events and intracellular signaling pathways activated by TRPV1, are only partially understood (Tables 49.1 and 49.2).

TRPV1 is highly and specifically expressed in both premalignant (leukoplasia) [40] and low-grade papillary skin carcinoma, whereas its expression is substantially absent in invasive carcinoma. Recently, TRPV1 has been found to exhibit tumor suppressive activity on skin carcinogenesis in mice because of its ability to down-regulate epidermal growth factor receptor (EGFR) expression; conversely, loss of TRPV1 expression resulted in marked increase in papilloma development. TRPV1 by interacting with EGFR through its terminal cytosolic domain, facilitates

Table 49.1 Expression profiles of TRPVs in cancer progression. Data reported are related to TRPV channels evaluated at mRNA and/or protein levels in human cancer tissues. n.c., no changes; n.d., not detected; (f), full length; (s), short splice variant

Channel	Tissue	Expression in tumor tissues respect to normal controls	Modulation in tumor progression	References
TRPV1	Bladder	↓	↓	[41, 44]
	Glioma	↑	↓	[21]
	Prostate	↑	↑	[59]
	Tongue	↑	n.c.	[40]
	Hepatocellular carcinoma	↓	↓	[55]
	Skin	↓	n.d.	[29]
	Pancreas	↑	n.c.	[50]
TRPV2	Bladder	↑	↑(f) ↓(s)	[22]
	Glioma	↓	↓	[23]
	Prostate	↑	↑	[71]
	Hepatocellular carcinoma	↑	↓	[72]
TRPV6	Breast	↑	n.d.	[109]
	Prostate	↑	↑	[102]
	Ovarian	↑	n.d.	[108]
	Thyroid	↑	n.d.	
	Colon	↑	n.d.	

Cbl-mediated EGFR ubiquitination and subsequently its degradation via the lysosomal pathway. In addition, ectopic TRPV1 expression in HEK293 cells resulted in decreased EGFR protein expression, and higher EGFR levels were observed in the skin of TRPV1 deficient mice (TRPV1$^{-/-}$) as compared to wild-type control animals [29].

Changes in the TRPV1 expression can also occur during the development of human urothelial cancer. Lazzeri and colleagues have demonstrated that transitional cell carcinoma (TCC) show a progressive decrease in TRPV1 expression as the tumor stage increases [41]. In accordance with Lazzeri's data, we found [31] that TRPV1 was highly expressed in low-grade human papillary urothelial carcinoma, whereas its expression was strongly reduced in high-grade and stage invasive TCC. Treatment of low-grade RT4 urothelial cancer cells with a specific TRPV1 agonist, capsaicin (CPS) induced a TRPV1-dependent G0/G1 cell cycle arrest and apoptosis. These events were associated with the transcription of pro-apoptotic genes including Fas/CD95, Bcl-2 and caspases, and with the activation of the DNA damage response pathway. Moreover, stimulation of TRPV1 by CPS significantly increased Fas/CD95 protein expression and more importantly induced a TRPV1-dependent redistribution and clustering of Fas/CD95 that colocalized with the vanilloid receptor, suggesting that Fas/CD95ligand-independent TRPV1-mediated Fas/CD95 clustering results in death-inducing signaling complex formation and triggering of apoptotic signaling through both the extrinsic and

Table 49.2 Pathophysiological functions TRPV channels in cancers

Channels	Proposed functions	Cancer	Roles in cancer
TRPV1	Sensing spice peppers pain sensation noxious chemical and thermal stimuli sensor bladder distension mediate proton influx hyperalgesia	Papillary skin carcinoma[a]	Inhibition of tumor development
		Papillary urothelial carcinoma[b]	Cell cycle arrest and apoptosis
		Glioma[b]	Apoptosis
		Hepatoblastoma[b]	Migration
		Cervical carcinoma[b]	Apoptosis
		Prostate carcinoma[b]	Proliferation
TRPV2	High-temperature heat responses mechanosensing	Glioma[b]	Inhibition of proliferation and cell survival
		Prostate carcinoma[b]	Migration
TRPV3	Temperature sensation vasoregulation	Prostate carcinoma[b]	Androgen-resistance
TRPV4	Osmotic sensitivity mechanosensitivity	Hepatoblastoma[b]	Calcium influx
TRPV5	Ca^{2+} reabsorption in intestine and kidney	Myeloid leukemia[b]	Differentiation
TRPV6	Ca^{2+} reabsorption in intestine	Prostate carcinoma[b]	Inhibition of proliferation
		Breast adenocarcinoma[b]	Proliferation
		Gastric cancer[b]	Apoptosis

[a]Mouse.
[b]Human cell lines.

intrinsic mitochondrial-dependent pathways [31]. In accordance with Amantini's findings, previous evidence demonstrated that TRPV1 N-terminus binds to Fas-associated factor-1, a Fas/CD95-associated protein [42], showing regulatory functions in TRPV1-dependent CPS-mediated apoptosis [43]. Moreover, by the use of the specific ATM inhibitor KU55933, we found that CPS activates the ATM kinase involved in p53 Ser15, Ser20 and Ser392 phosphorylation. ATM activation is involved in Fas/CD95 up-regulation and co-clustering with TRPV1 as well as in urothelial cancer cell growth and apoptosis (Fig. 49.1).

Recently, we have also assessed the role of TRPV1 mRNA down-regulation as a negative prognostic factor in patients with bladder cancer [44]. By univariate analysis, cumulative survival curves calculated according to the Kaplan-Meier method for the canonic prognostic parameters such as tumor grade and high stage (pT4), lymph nodes and distant diagnosed metastasis, reached significance. Notably, the reduction of TRPV1 mRNA expression was associated with a shorter survival of urothelial cancer patients ($P = 0.008$) (Table 49.3, A). On multivariate Cox regression analysis, TRPV1 mRNA expression retained its significance as an independent risk

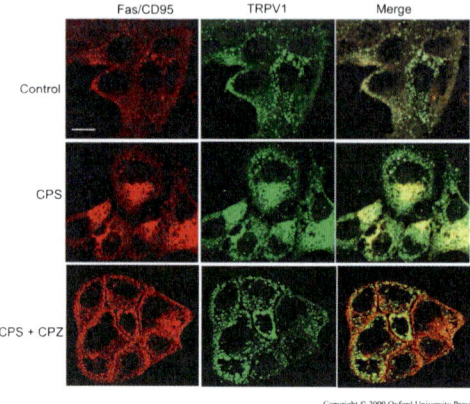

Fig. 49.1 CPS increases Fas/CD95 expression and clustering in TRPV1-dependent manner in RT4 cells. The immunocytochemical localization of Fas/CD95 and TRPV1 in RT4 cells treated for 24 h with 100 μM of CPS alone or in combination with 10 μM of CPZ was analyzed by confocal microscopy using an anti-Fas/CD95 mAb and a goat anti-TRPV1 Ab followed by respective secondary Abs. Control sample indicates DMSO vehicle. Data are representative of three different experiments. Bar = 10 μm

factor (Table 49.3, B), also in a subgroup of patients without diagnosed metastasis (M0). These findings may be particularly important in the stratification of urothelial cancer patients with higher risk of tumor progression for the choice of therapy options [44].

The knowledge of the mechanism controlling TRPV1 expression would be of importance for a better understanding of urothelial cancer growth and progression. Moreover, as TRPV1 agonists such as CPS and RTX are widely employed in the treatment of detrusor overactivity or interstitial cystitis [45–48], the comprehension of the molecular mechanisms underlying their pro-apoptotic activity would be clinically relevant to extend the use of these agents also to the therapy of superficial urothelial malignancies.

TRPV1 mRNA and protein expression was also evidenced in normal astrocytes and glioma cells and tissues [21, 49]. Its expression inversely correlated with glioma grading, with a marked loss of TRPV1 expression in the majority of grade IV glioblastoma tissues. TRPV1 activation by CPS induced apoptosis of U373MG glioma cells, and involved rise of Ca^{2+} influx, p38MAPK activation, mitochondrial permeability transmembrane pore opening and transmembrane potential dissipation, and caspase-3 activation [21].

Human pancreatic cancer, too, significantly expressed increased levels of TRPV1 mRNA and protein. However, resinferatoxin (RTX), a potent TRPV1 agonist, induced apoptosis by targeting mitochondrial respiration, and decreased pancreatic cancer cell growth in a TRPV1-independent manner [50].

In human hepatoblastoma cells, hepatocyte growth factor/scatter factor (HGF/SF) promotes cell migration by stimulating the activity of TRPV1. TRPV1

Table 49.3 (**A**) Univariate Kaplan-Meier survival analysis of clinico-pathological parameters and TRPV1 mRNA expression (log-rank test). (**B**) Multivariate Cox proportional hazard regression analysis of clinico-pathological parameters and TRPV1 mRNA expression in relation to survival rates

A

	All patients (n = 62)		N° events (n = 20)	M0 patients (n = 50)	N° events (n = 12)
	Log-rank test		p value	Log-rank test	p value
Age ($\leq$70/>70)	0.29		0.588	0.80	0.372
Tumor stage (pTa-1/pT2-4)	2.21		0.137	0.57	0.449
Tumor grade (G1/G2/G3)	10.47		0.001	8.15	0.004
TRPV1 (low/ moderate/ high)	9.69		0.008	6.20	0.045
Lymph node stage (N0/N1)	10.65		0.001	-	-
Metastasis (M0/M1)	5.76		0.016	not included	

B

	Relative risk (95% CI)					
	All patients (n = 62)		pTa/T1 (n = 24)		pT2/T3/T4 (n = 38)	
TRPV1 (low/ moderate/ high)	0.26 (0.10-0.64)	$p = 0.004$	0.61 (0.07-5.03)	$p = 0.65$	0.30 (0.10-0.88)	$p = 0.030$
Tumor grade (G1/G2/G3)	1.52 (0.74-3.12)	$p = 0.26$	2.83 (0.29-27.44)	$p = 0.37$	2.85 (0.81-10.12)	$p = 0.011$
			Relative risk (95% CI)			
			G2/G3 pT2/T3 N0M0 (n = 27)			
TRPV1 (low/ moderate/ high)			0.19 (0.03-1.00)			$p = 0.05$
Tumor grade (G2/G3)			5.65 (0.68-47.0)			$p = 0.11$
Tumer stage (T2/T3)			0.43 (0.09-1.96)			$p = 0.28$

may in part mediate HGF/SF effects in hepatoblastoma cells by increasing intracellular Ca^{2+} and promoting tumor invasion by triggering a signaling cascade leading to cell locomotion and acquisition of a migratory phenotype [51–54]. Moreover, Miao et al. have recently demonstrated in hepatocarcinoma patients that high TRPV1 expression is associated with increased disease-free survival [55].

TRPV1 expression has been also reported in human cervical cancer cell lines and tissues, and the endocannabinod anandamide (AEA) induced TRPV1-dependent tumor cell apoptosis [56]. In addition, TRPV1 stimulation completely reverted the cannabidiol (CBD)-mediated inhibitory effect on human cervical cancer cell invasion by blocking CBD-induced increase of TIMP-1 MMP inhibitor both at mRNA and protein levels, and ERK1/2 and p38MAPK activation [57].

A functional TRPV1 channel is also expressed in human prostate cancer cells (PC3 and LNCaP) and in prostate hyperplasic tissue [58]. Moreover, increased

TRPV1 mRNA and protein expression was found in human prostate cancer tissues as compared to prostate hyperplastic and healthy donors, and this increase correlated with degree of malignancy [59]. CPS induced a growth inhibition and apoptosis of PC3 prostate cancer cells, but in TRPV1-independent manner, through ROS generation, mitochondrial inner transmembrane potential dissipation and caspase-3 activation. Moreover, CPS or the specific antagonist capsazepin inhibited tumor growth in vivo, in a xenograft human prostate PC3 cancer model [60]. By contrast, in androgen-responsive LNCaP prostate cancer cells, CPS was found to stimulate TRPV1-dependent cell proliferation. CPS effects were attributable to decreased ceramide levels and to activation of Akt/PKB and ERK pathways, and were associated with increased androgen receptor expression [61].

TRPV1 expression has been also demonstrated on the plasma membrane of rat pheochromocytoma-derived PC12 cell line [62]. PC12 stimulation by CPS resulted in TRPV1-dependent nitric oxide synthase (iNOS) expression. CPS exposure triggered Ca^{2+} influx, which in turn enhanced mitochondrial Ca^{2+} accumulation and promoted NO generation, events that have been associated with tumor progression [63].

49.3 TRPV2

Because of its regulation by growth factor signaling (i.e. platelet-derived growth factor, PDGF; Insulin-like growth factor-I, IGF-I), TRPV2 is alternatively called "growth receptor channel" (GRC) [64]. In particular, a cross-talk between TRPV2 and IGF-I receptor has been described. IGF-I can enhance Ca^{2+} influx in TRPV2-transfected cells by promoting its translocation from intracellular pools to the plasma membrane [64]. In addition, TRPV2 is over-expressed in urothelial cancer cells together with IGF-I and its receptor [65]. Moreover, increased proliferation and survival of human bladder smooth-muscle cells induced by mechanical stress was associated with increased IGF-I levels [66]. Therefore, it can be hypothesized that TRPV2 plays an important role in controlling urothelial cancer cell growth and progression through the modulation of IGF-1/IGF-1R pathway.

TRPV2 expression and activity can also depend on the presence of alternative splicing variants, resulting in multiple channel proteins. We have recently demonstrated [22] two different TRPV2 transcripts in normal human urothelial cells and bladder tissue specimens: the full length TRPV2 and a short-splice variant, sTRPV2. Analysis of *TRPV2* gene and protein expression in superficial and invasive tumors has evidenced that TRPV2 mRNA increased gradually as tumor grade and stage increased (Table 49.1), while sTRPV2 expression gradually decreased. The differences observed in the short/full TRPV2 form ratio at different stages of neoplastic disease, suggest that sTRPV2 is lost as an early event in bladder carcinogenesis, whereas enhanced expression of full-length TRPV2 in high-stage muscle-invasive urothelial cancer rather represents a secondary event. Similarly, a different sTRPV2 lacking the pore-forming region and the fifth and sixth transmembrane domains, was characterized in human macrophages [67]. As demonstrated for TRPV1 channel, these naturally occurring alternative splice variants could act as dominant-negative mutants by forming heterodimers with TRPV2, thus inhibiting its trafficking and translocation to the plasma membrane [68].

We have recently reported that TRPV2 negatively controls glioblastoma survival and proliferation, as well as resistance to Fas/CD95-induced apoptosis in an ERK-dependent manner. TRPV2 mRNA was expressed in benign astrocyte tissues, and its expression progressively declined in high-grade glioma tissues as histological grade increased (Tables 49.1 and 49.2) [23]. Silencing of TRPV2 by RNA interference (siRNA) in U87MG glioma cells, down-regulated Fas/CD95 and procaspase-8 expression, and up-regulated CCNE1, CDK2, E2F1, Raf-1 and Bcl-X_L mRNA expression. Moreover, TRPV2 siRNA increased glioblastoma proliferation and survival to Fas/CD95-induced apoptosis in an ERK-dependent manner [23]. Inhibition of ERK activation by treatment of the siRNA-TRPV2 U87MG cells with the specific MEK-1 inhibitor PD98059, reduced Bcl-X_L protein levels, promoted Fas/CD95 expression and restored Akt/PKB pathway activation leading to reduced cell survival and proliferation and increased sensitivity to Fas/CD95-induced apoptosis [23]. These events are consistent with previous evidence showing that PI3K pharmacological inhibitors, inhibited calcium overload and cell death in TRPV2-transfected mouse cells [69]. Consistently, TRPV2 transfection of the primary MZC glioblastoma cells also reduced glioma viability and increased spontaneous and Fas/CD95-induced apoptosis, by inducing Fas/CD95 expression [23].

A role for TRPV2 in the invasive capability of prostate cancer cells has been recently reported. Ca^{2+}-dependent activation and translocation of TRPV2 by lysophospholipids (LPL), such as lysophosphatidylcoline and lysophosphatidylinositol resulted in increased PC3 prostate tumor cell migration (Table 49.2), and involved Gq/Go-protein and PI3,4-K pathways. Silencing of *TRPV2* gene in PC3 cells or inhibition of PI3,4-K pathway, abolished the stimulatory effects of LPL on both calcium entry and migration [70]. TRPV2-regulated cytosolic calcium levels promoted PC3 migration by induction of key proteases, namely MMP2, MMP9 and cathepsin B.

The constitutive activity of TRPV2 is also critical for castration-resistant prostate cancer development and progression in vivo. Indeed, the progression of prostate cancer to the castration-resistant phenotype is characterized by de novo expression of TRPV2, and higher levels of TRPV2 transcripts are found in patients with metastatic cancer (stage M1) as compared to primary solid tumors (stage T2a and T2b) (Table 49.1) [71]. In addition, induction of MMP9 and cathepsin B expression was observed upon transfection of TRPV2 in the LNCaP cell line, and knockout of TRPV2 by siRNA reduced the growth and the invasive properties of PC3 cells by progressive down-regulation of MMP2, MMP9 and cathepsin B expression in a xenograft tumor model [71].

Thus, TRPV2 could represent a prospective prognostic marker and potential therapeutic target for novel interventions aimed at improving the life expectancy of prostate cancer patients.

Finally, increased TRPV2 mRNA and protein expression was observed in human hepatocellular carcinoma (Table 49.1). Clinico-pathological assessment suggested a significant association between TRPV2 expression and portal vein invasion and histo-pathological differentiation. Moreover TRPV2 mRNA and protein up-regulation inversely correlated with the degree of tumor differentiation, indicating a potential role for TRPV2 in hepatocarcinogenesis [72].

49.4 TRPV3

TRPV3 is a calcium-permeable temperature-sensitive cation channel [73] expressed in tongue, nasal epithelium and epidermal keratinocytes [74, 75]. Recently, it has been reported that TRPV3 is abundantly and selectively expressed, both at mRNA and protein levels, in the superficial enterocytes of the distal colon, where it participates to a variety of cellular functions including immune defense, cell volume regulation, cell proliferation and mitogenesis, apoptosis and epithelial ion and water transport in surrounding cells [75].

TRPV3 is activated by a number of chemical substances and its activity is strongly potentiated by phospholipase C, arachidonic acid (AA) and other unsaturated fatty acid [76–78]. The fatty acid-induced TRPV3 potentiation is PKC-independent, does not require AA metabolism, but is mimicked by non-metabolizable analogs of AA [79]. Carvacrol, a TRPV3 agonist, increases the cytosolic Ca^{2+} concentration, and evokes ATP-release in a subset of primary colonic epithelial cells [75].

Some recent studies suggest a role for TRPV3 in colorectal cancer development, as by using tagging SNPs, two synonymous-coding *TRPV3* SNPs (rs11078458 and rs65002729), were positively associated with higher colorectal cancer risk, with the later SNP significantly associated with higher risk in women [35].

In addition to a potential role in distal colon epithelium pathophysiology, TRPV3 seems to be involved in prostate cancer progression [80]. By ChIP display, TRPV3 expression was shown to be stimulated by androgens, but also more strongly expressed in castrate-resistant C4-2b prostate cancer cells. In addition, immunohistochemistry and RT-PCR analyses revealed TRPV3 mRNA and protein expression in rat prostatic tissues, although at lower levels with respect to other TRPV channels (TRPV2 and TRPV4) [81]. Although TRPV3 has been suggested to represent a new target gene for androgen receptors in prostate cancer development and progression, further studies are needed to better understand its contribution in the regulation of human prostate tissue physiology and transformation.

Finally, as shown in Fig. 49.2a, we detected by PCR, TRPV3 mRNA expression in a number of urothelial (RT4, TCCSUP, J82 and EJ), and glioblastoma cancer cell lines (U373, T98G and U87MG) (Santoni et al., unpublished results) (Fig. 49.2a, b), but the relevance of TRPV3 expression in these cancers has not been addressed so far.

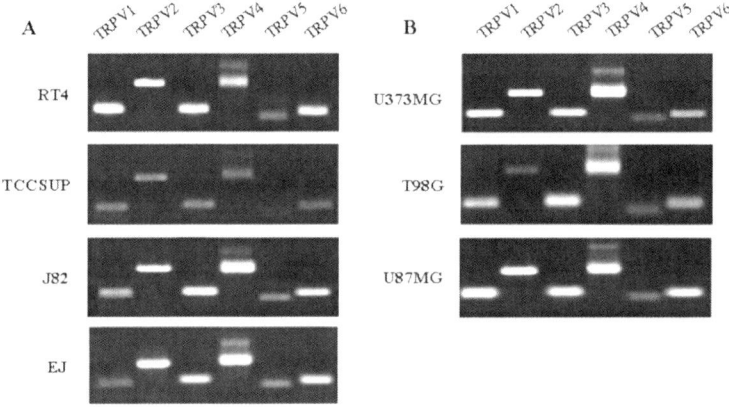

Fig. 49.2 TRPV mRNA expression as determined by reverse transcription–PCR. (**a**) TRPVs mRNA expression in urothelial carcinoma cell lines. (**b**) TRPVs mRNA expression in glioma cell lines

49.5 TRPV4

Functional expression of TRPV4 has been described in HepG2 cells hepatoblastoma cells [54], as this receptor channel can regulate both Ca^{2+} influx and Ca^{2+} release channels. In addition, based on the evidence that HGF/SF enhanced Ca^{2+} entry via TRPV1 activation and that its action was limited to migrating cells, activation of both TRPV4 and TRPV1 channels by HGF/SF has been suggested as the first step of a signaling cascade that gives rise to cell locomotion and migratory phenotype by the reorganization of actin cytoskeleton [82]. In this regard, a functional interaction between TRPV4 and F-actin in sensing hypotonicity and the onset of regulatory volume decrease has been demonstrated [83].

Recent findings evidenced TRPV4 mRNA expression in primary cultured cortical astrocytes. TRPV4 protein was enriched in astrocytic processes of the superficial layers of the neocortex and in astrocyte end feet facing pia and blood vessels. The selective TRPV4 agonist 4-alpha-phorbol 12,13-didecanoate (4alphaPDD) activated an outwardly rectifying cation current and promoted $[Ca^{2+}]i$ elevation, suggesting that astrocyte TRPV4 channel was functional [84].

Consistent with its role in other tissues, it is conceivable that TRPV4 may be involved in the astroglial control of brain volume homeostasis; whether TRPV4 is also involved in glioblastoma invasion and metastatic spread is under investigation. In this regard, we have preliminary evidence on TRPV4 mRNA expression on a number of glioblastoma cell lines (Fig. 49.2b) (Santoni et al., unpublished observations).

TRPV4 transcripts and protein are also abundantly expressed in mouse basal urothelial cells [85, 86], and 4alphaPDD induces TRPV4-mediated strech-evoked increase of $[Ca^{2+}]i$ and ATP release in primary isolated urothelial cell cultures

[87]. In addition, two different TRPV4 transcript variants, variant 1 that encodes the longer isoform (a), and variant 2 that lacks an alternate in-frame exon and encodes a shorter isoform (b) lacking an internal segment, have been identified in human urothelial cancer cell lines (Fig. 49.2a) (Santoni et al., unpublished results), but the significance of different TRPV4 isoform expression is presently unknown.

49.6 TRPV5

This receptor originally cloned from rabbit kidney, belongs to the "apical calcium channels" family. Unlike the other four members of TRPV family that are non selective cation channel, TRPV5 together with TRPV6, are remarkable Ca^{2+}-selective channels which serve as apical calcium entry mechanisms in absortive and secretory tissues [88]. TRPV5 is most prominent in the distal convoluted and connecting tubules of the kidney [89].

Van Abel et al. [90] have demonstrated that 17β-estradiol increases renal TRPV5 mRNA and protein expression, as well as Ca^{2+} levels [88, 91]; however no data have been provided regarding the potential involvement of TRPV5 in renal cancers. Instead, endogenous TRPV5 mRNA expression was evidenced by RT-PCR analysis in human Jurkat T leukemia cell line [92] and in K562 erythroleukemia cell line [93]. K562 cells co-express TRPV5 and TRPV6 calcium channels that form a functional homotetrameric structure by interacting each other. Based on the evidence that TRPV5/V6 expression levels are strongly controlled by 1,25-dihydroxyvitamin D3 (1,25-(OH)2D3) [93] that exerts anti-proliferative effects and induces K562 cell differentiation, it has been suggested that TRPV5/TRPV6 channels regulate leukemia cell differentiation in a 1,25-(OH)2D3-dependent manner.

In the renal epithelium, TRPV5 expression is regulated by klotho, a mammalian senescence suppression protein that shows homology with glycosidases [94–96].

Klotho increases cell surface expression of TRPV5 by removing terminal sialic acids from their glycan chains. Removal of sialic acids exposes the disaccharide galactose-N-acetylglucosamine, a ligand for the beta-galactoside-binding lectin Galectin 1 (GAL1), involved in cell cycle progression, and surface binding to GAL1 leads to accumulation of functional TRPV5 on the plasma membrane [27, 28]. As GAL1 overexpression is associated with neoplastic transformation, and differentiation of K562 cells induces externalisation and binding of GAL1 on the cell surface [97], Klotho/TRPV5/GAL1 pathway may be suggested to be involved in myeloid transformation.

The regulation of TRPV5 by klotho represents a novel mechanism regulating the activity of cell-surface glycoproteins in cancer. Understanding klotho protein function, and its interaction with TRPV5 protein in leukemias may provide new insights in the knowledge of senescence process and vitamin D metabolism in cancer.

49.7 TRPV6

TRPV6 channel is involved in the control of growth and progression (Tables 49.1 and 49.2) of hormone-dependent tumors, such as prostate and breast cancers [15, 98].

TRPV6 channel is regulated by androgens and seems to play a key role in the progression of prostate cancer [24, 99, 100]. TRPV6 expression in healthy and benign human prostate tissues is very low or even undetectable, and increases with disease aggressiveness and the extent of metastasis [101–103]. Androgen treatment of the human LNCaP cell line that constitutively expresses the *TRPV6* gene, causes a strong reduction of TRPV6 mRNA levels while bicalutamide, an oral non-steroidal anti-androgen resulted in a time-dependent TRPV6 up-regulation [104], indicating that TRPV6 is negatively regulated by androgens. TRPV6 is directly involved in the control of LNCaP proliferation by decreasing the proliferation rate, cell accumulation in the S-phase of cell cycle, and PCNA expression. Ca^{2+} entry in LNCaP cells is mediated by TRPV6 with subsequent downstream activation of the nuclear factor of activated T-cell transcription factor (NFAT), and has been involved in resistance to apoptosis [105]. TRPV6 is not a direct androgen receptor-responsive gene, however in these cells, androgen receptor is required to maintain TRPV6 expression [105, 106].

Finally, two different allelic variants of the human *TRPV6* gene, TRPV6a and TRPV6b have been identified and the TRPV6b genotype is predominant. However, the expression of TRPV6b variant does not correlate with the onset of prostate cancer, Gleason score and tumor stage, whereas TRPV6 transcript levels correlate with a more aggressive behavior of prostate cancer [107].

In regard to breast cancer, immunohistological analysis of mammary adenocarcinoma tissues showed a clear enhancement of TRPV6 expression as compared to normal tissue [108]. Short-term estrogen stimulation slightly reduced RNA transcription of TRPV6, while it was enhanced by long-term treatment [109]. In addition, tamoxifen, a selective anti-estrogen-receptor (ER) drug reduces TRPV6 transcription [109]. Limited ER signaling apparently leads to lower TRPV6 expression. These observations suggest that ER regulates TRPV6 expression. Ca^{2+} entry via TRPV6 can increase the rate of Ca^{2+}-dependent cell proliferation and thus can be directly linked to modulation of tumor cell proliferation [110]. It is worth noting that recent results demonstrate a non-genomic effect of estrogens on Ca^{2+} influx via TRPV6, whereas 17β-estradiol can exert a rapid and specific action on the channel [111]. This suggests that the biosynthesis and accumulation of estrogens in breast carcinoma [112] may have a direct action on TRPV6 other than the transcriptional regulation of the channel, rapidly promoting Ca^{2+}-dependent proliferation. The effects of tamoxifen on breast cancer cell viability were enhanced when TRPV6 expression was silenced by means of siRNA, and similarly to LNCaP prostate cells, silencing of TRPV6 in breast cancer cells caused cell cycle arrest. In MCF-7 estrogen-negative breast cancer cells, tamoxifen decreased the basal intracellular calcium concentration in a PKC-dependent manner, suggesting that the therapeutic effect of tamoxifen and protein kinase C inhibitors used in breast cancer therapy, involve TRPV6-mediated calcium entry [113].

TRPV6 is also expressed more abundantly in gastric cancer cells than in normal gastric epithelial cells. This channel has been found to mediate CPS-induced apoptosis in gastric cancer cells by increasing mitochondrial permeability through activation of Bax and p53, in a JNK-dependent manner. The relative expression of TRPV6 in normal and gastric cancer cells determines their susceptibility to cell death following CPS treatment. Thus, overexpression of TRPV6 in gastric cancer cells increased CPS-induced apoptosis and knockdown of TRPV6 suppressed this action and induced cell survival [114].

Finally, as evidenced in Fig. 49.2a, we also detected TRPV6 mRNA expression in different urothelial cancers, as well as in glioblastoma cell lines by RT-PCR, (Fig. 49.2b) (Santoni et al., unpublished results), but the relevance of TRPV6 mRNA expression has not been addressed so far.

49.8 Conclusions and Perspectives

Target therapy is considerably changing the treatment and prognosis of cancer. Progressive understanding of the molecular mechanisms that regulate the establishment and progression of different tumors is leading to ever more specific and efficacious pharmacological approaches [115].

In this picture, ion channels, and in particular TRPV channels, represent an unexpected, but very promising player [8]. The expression and activity of different TRPV channels mark and regulates specific stages of cancer progression. Their contribution to the neoplastic phenotype ranges from control of cell proliferation and apoptosis, to regulation of invasiveness and metastatic spread.

As increasingly recognized, some of these roles can be attributed to signaling mechanisms other than ion flow [116].

Evidences are particularly numerous for TRPV channels. Their expression is altered in some human primary tumors, and they frequently exert pleiotropic effects on the tumor cell behaviour by regulating membrane potential, TRPV can control Ca^{2+} fluxes and thus the cell cycle machinery, or cancer cell volume. TRPV channels are also implicated in late stages of cancer disease, by stimulating angiogenesis, modulating tumor cell-matrix interaction and regulating cell motility and invasion capability [116].

Altered TRPV channel expression can be exploited for diagnostic purposes or for addressing traceable or cytotoxic compounds to specific neoplastic tissues [117]. Of more importance, recent evidence indicates that blocking channel activity impairs the growth of some tumors, both in vivo and in vitro. This open a new field for medical chemistry, biology and biotechnology studies, allowing the design and the development of many anti-cancer innovative tools, not only agonists and antagonists, but also blocking antibodies, antisense oligonucleotides, small interfering RNAs, peptide toxins and a large variety of small organic compounds.

We are only at the beginnig, a more deep studies on the physiopathology of TRPV channels are required to further understand the functional activity of these receptors in tumor growth and progression

References

1. Gupta GP, Massagué J (2006) Cancer metastasis: building a framework. Cell 127: 679–695
2. Weinberg RA (2006) Multi-step tumorigenesis. In: The biology of cancer. Garland Sciences, New York, NY, Chapter 11
3. Pusztai L, Lewis CE, Lorenzen J, McGee JO (1993) Growth factors: regulation of normal and neoplastic growth. J Pathol 169:191–201
4. Pluda JM (1997) Tumor-associated angiogenesis: mechanisms, clinical implications, and therapeutic strategies. Semin Oncol 24:203–218
5. Price JT, Bonovich MT, Kohn EC (1997) The biochemistry of cancer dissemination. Crit Rev Biochem Mol Biol 32:175–253
6. Liotta LA, Kohn EC (1997) Invasion and metastasis. In: Holland JF, Bast RC Jr, Morton DL, Frei III E, Kufe DW, Weichselbaum RR, (eds) Cancer Medicine, 4th ed. Williams & Wilkins, Baltimore, MD, p. 165
7. Kiselyov K, Soyombo A, Muallem S (2007) TRPpathies. J Physiol 578:641–653
8. Nilius B, Owsianik G, Voets T, Peters JA (2007) Transient receptor potential cation channels in disease. Physiol Rev 87:165–217
9. Xu XZ, Moebius F, Gill DL, Montell C (2001) Regulation of melastatin, a TRP-related protein, through interaction with a cytoplasmic isoform. Proc Natl Acad Sci USA 98: 10692–10697
10. Wisnoskey BJ, Sinkins WG, Schilling WP (2003) Activation of vanilloid receptor type 1 in the endoplasmic reticulum fails to activate store-operated Ca^{2+} entry. Biochem J 372: 517–528
11. Thebault S, Lemonnier L, Bidaux G, Flourakis M, Bavencoffe A, Gordienko D, Roudbaraki M, Delcourt P, Panchin Y, Shuba Y, Skryma R, Prevarskaya N (2005) Novel role of cold/menthol-sensitive transient receptor potential melastatine family member 8 (TRPM8) in the activation of store-operated channels in LNCaP human prostate cancer epithelial cells. J Biol Chem 280:39423–39435
12. Xin H, Tanaka H, Yamaguchi M, Takemori S, Nakamura A, Kohama K (2005) Vanilloid receptor expressed in the sarcoplasmic reticulum of rat skeletal muscle. Biochem Biophys Res Commun 332:756–762
13. Bidaux G, Flourakis M, Thebault S, Zholos A, Beck B, Gkika D, Roudbaraki M, Bonnal JL, Mauroy B, Shuba Y, Skryma R, Prevarskaya N (2007) Prostate cell differentiation status determines transient receptor potential melastatin member 8 channel subcellular localization and function. J Clin Invest 117:1647–1657
14. Prevarskaya N, Zhang L, Barritt G (2007) TRP channels in cancer. Biochim Biophys Acta 1772:937–946
15. Gkika D, Prevarskaya N (2009) Molecular mechanisms of TRP regulation in tumor growth and metastasis. Biochim Biophys Acta 1793:953–958
16. Montell C (2003) Thermosensation: hot findings make TRPNs very cool. Curr Biol 13:R476–R478
17. Duncan LM, Deeds J, Hunter J, Shao J, Holmgren LM, Woolf EA, Tepper RI, Shyjan AW (1998) Down-regulation of the novel gene melastatin correlates with potential for melanoma metastasis. Cancer Res 58:1515–1520
18. Tsavaler L, Shapero MH, Morkowski S, Laus R (2001) Trp-p8, a novel prostate-specific gene, is up-regulated in prostate cancer and other malignancies and shares high homology with transient receptor potential calcium channel proteins. Cancer Res 61:3760–3769

19. Wissenbach U, Niemeyer BA, Fixemer T, Schneidewind A, Trost C, Cavalie A, Reus K, Meese E, Bonkhoff H, Flockerzi V (2001) Expression of CaT-like, a novel calcium-selective channel, correlates with the malignancy of prostate cancer. J Biol Chem 276:19461–19468
20. Thebault S, Flourakis M, Vanoverberghe K, Vandermoere F, Roudbaraki M, Lehen'kyi V, Slomianny C, Beck B, Mariot P, Bonnal JL, Mauroy B, Shuba Y, Capiod T, Skryma R, Prevarskaya N (2006) Differential role of transient receptor potential channels in Ca^{2+} entry and proliferation of prostate cancer epithelial cells. Cancer Res 66:2038–2047
21. Amantini C, Mosca M, Nabissi M, Lucciarini R, Caprodossi S, Arcella A, Giangaspero F, Santoni G (2007) Capsaicin-induced apoptosis of glioma cells is mediated by TRPV1 vanilloid receptor and requires p38 MAPK activation. J Neurochem 102:977–990
22. Caprodossi S, Lucciarini R, Amantini C, Nabissi M, Canesin G, Ballarini P, Di Spilimbergo A, Cardarelli MA, Servi L, Mammana G, Santoni G (2008) Transient receptor potential vanilloid type 2 (TRPV2) expression in normal urothelium and in urothelial carcinoma of human bladder: correlation with the pathologic stage. Eur Urol 54:612–620
23. Nabissi M, Morelli MB, Amantini C, Farfariello V, Ricci-Vitiani L, Caprodossi S, Arcella A, Santoni M, Giangaspero F, De Maria R, Santoni G (2010) TRPV2 channel negatively controls glioma cell proliferation and resistance to Fas-induced apoptosis in ERK-dependent manner. Carcinogenesis 31:794–803
24. Prevarskaya N, Skryma R, Bidaux G, Flourakis M, Shuba Y (2007) Ion channels in death and differentiation of prostate cancer cells. Cell Death Differ 14:1295–1304
25. Flourakis M, Prevarskaya N (2009) Insights into Ca^{2+} homeostasis of advanced prostate cancer cells. Biochim Biophys Acta 1793:1105–1109
26. Lallet-Daher H, Roudbaraki M, Bavencoffe A, Mariot P, Gackière F, Bidaux G, Urbain R, Gosset P, Delcourt P, Fleurisse L, Slomianny C, Dewailly E, Mauroy B, Bonnal JL, Skryma R, Prevarskaya N (2009) Intermediate-conductance Ca^{2+}-activated K^+ channels (IKCa1) regulate human prostate cancer cell proliferation through a close control of calcium entry. Oncogene 28:1792–1806
27. Chang Q, Hoefs S, van der Kemp AW, Topala CN, Bindels RJ, Hoenderop JG (2005) The beta-glucuronidase klotho hydrolyzes and activates the TRPV5 channel. Science 310: 490–493
28. Cha SK, Ortega B, Kurosu H, Rosenblatt KP, Kuro-O M, Huang CL (2008) Removal of sialic acid involving Klotho causes cell-surface retention of TRPV5 channel via binding to galectin-1. Proc Natl Acad Sci USA 105:9805–9810
29. Bode AM, Cho YY, Zheng D, Zhu F, Ericson ME, Ma WY, Yao K, Dong Z (2009) Transient receptor potential type vanilloid 1 suppresses skin carcinogenesis. Cancer Res 69:905–913
30. Zhang N, Oppenheim JJ (2005) Crosstalk between chemokines and neuronal receptors bridges immune and nervous systems. J Leukoc Biol 78:1210–1214
31. Amantini C, Ballarini P, Caprodossi S, Nabissi M, Morelli MB, Lucciarini R, Cardarelli MA, Mammana G, Santoni G (2009) Triggering of transient receptor potential vanilloid type 1 (TRPV1) by capsaicin induces Fas/CD95-mediated apoptosis of urothelial cancer cells in an ATM-dependent manner. Carcinogenesis 30:1320–1329
32. Reilly CA, Johansen ME, Lanza DL, Lee J, Lim JO, Yost GS (2005) Calcium-dependent and independent mechanisms of capsaicin receptor (TRPV1)-mediated cytokine production and cell death in human bronchial epithelial cells. J Biochem Mol Toxicol 19:266–275
33. Zhang F, Yang H, Wang Z, Mergler S, Liu H, Kawakita T, Tachado SD, Pan Z, Capó-Aponte JE, Pleyer U, Koziel H, Kao WW, Reinach PS (2007) Transient receptor potential vanilloid 1 activation induces inflammatory cytokine release in corneal epithelium through MAPK signaling. J Cell Physiol 213:730–739
34. Kaur R, O'Shaughnessy CT, Jarvie EM, Winchester WJ, McLean PG (2009) Characterization of a calcitonin gene-related peptide release assay in rat isolated distal colon. Arch Pharm Res 32:1775–1781
35. Hoeft B, Linseisen J, Beckmann L, Müller-Decker K, Canzian F, Hüsing A, Kaaks R, Vogel U, Jakobsen MU, Overvad K, Hansen RD, Knüppel S, Boeing H, Trichopoulou A, Yvoni K, Trichopoulos D, Berrino F, Palli D, Panico S, Tumino R, Bueno-de-Mesquita HB, van

Duijnhoven FJ, van Gils CH, Peeters PH, Dumeaux V, Lund E, Huerta Castaño JM, Muñoz X, Rodriguez L, Barricarte A, Manjer J, Jirström K, Van Guelpen B, Hallmans G, Spencer EA, Crowe FL, Khaw KT, Wareham N, Morois S, Boutron-Ruault MC, Clavel-Chapelon F, Chajes V, Jenab M, Boffetta P, Vineis P, Mouw T, Norat T, Riboli E, Nieters A (2009) Polymorphisms in fatty acid metabolism-related genes are associated with colorectal cancer risk. Carcinogenesis 31:466–472
36. Bloethner S, Chen B, Hemminki K, Müller-Berghaus J, Ugurel S, Schadendorf D, Kumar R (2005) Effect of common B-RAF and N-RAS mutations on global gene expression in melanoma cell lines. Carcinogenesis 26:1224–1232
37. Fabris S, Todoerti K, Mosca L, Agnelli L, Intini D, Lionetti M, Guerneri S, Lambertenghi-Deliliers G, Bertoni F, Neri A (2007) Molecular and transcriptional characterization of the novel 17p11.2-p12 amplicon in multiple myeloma. Genes Chromosomes Cancer 46:1109–1118
38. Zatkova A, Merk S, Wendehack M, Bilban M, Muzik EM, Muradyan A, Haferlach C, Haferlach T, Wimmer K, Fonatsch C, Ullmann R (2009) AML/MDS with 11q/MLL amplification show characteristic gene expression signature and interplay of DNA copy number changes. Genes Chromosomes Cancer 48:510–520
39. Boyd RS, Jukes-Jones R, Walewska R, Brown D, Dyer MJ, Cain K (2009) Protein profiling of plasma membranes defines aberrant signaling pathways in mantle cell lymphoma. Mol Cell Proteomics 8:1501–1515
40. Marincsák R, Tóth BI, Czifra G, Márton I, Rédl P, Tar I, Tóth L, Kovács L, Bíró T (2009) Increased expression of TRPV1 in squamous cell carcinoma of the human tongue. Oral Dis 15:328–335
41. Lazzeri M, Vannucchi MG, Spinelli M, Bizzoco E, Beneforti P, Turini D, Faussone-Pellegrini MS (2005) Transient receptor potential vanilloid type 1 (TRPV1) expression changes from normal urothelium to transitional cell carcinoma of human bladder. Eur Urol 48:691–698
42. Kim S, Kang C, Shin CY, Hwang SW, Yang YD, Shim WS, Park MY, Kim E, Kim M, Kim BM, Cho H, Shin Y, Oh U (2006) TRPV1 recapitulates native capsaicin receptor in sensory neurons in association with Fas-associated factor 1. J Neurosci 26:2403–2412
43. Ghosh AK, Basu S (2010) Fas-associated factor 1 is a negative regulator in capsaicin induced cancer cell apoptosis. Cancer Lett 287:142–149
44. Kalogris C, Caprodossi S, Amantini C, Lambertucci F, Nabissi M, Farfariello V, Filosa A, Emiliozzi MC, Mammana G, Santoni G Expression of transient receptor potential vanilloid-1 (TRPV1) in urothelial cancers of human bladder: relation to clinicopathological and molecular parameters. Histopathology. DOI: 10.1111/j.1365-2559.2010.03683.x
45. Birder LA (2007) TRPs in bladder diseases. Biochem Biophys Acta 772:879–884
46. Cruz CD, Charrua A, Vieira E, Valente J, Avelino A, Cruz F (2008) Intrathecal delivery of resiniferatoxin (RTX) reduces detrusor overactivity and spinal expression of TRPV1 in spinal cord injured animals. Exp Neurol 214:301–308
47. Apostolidis A, Brady CM, Yiangou Y, Davis J, Fowler CJ, Anand P (2005) Capsaicin receptor TRPV1 in urothelium of neurogenic human bladders and effect of intravesical resiniferatoxin. Urology 65:400–405
48. Liu HT, Kuo HC (2007) Increased expression of transient receptor potential vanilloid sub-family 1 in the bladder predicts the response to intravesical instillations of resiniferatoxin in patients with refractory idiopathic detrusor overactivity. BJU Int 100:1086–1090
49. Contassot E, Wilmotte R, Tenan M, Belkouch MC, Schnüriger V, de Tribolet N, Burkhardt K, Dietrich PY (2004) Arachidonylethanolamide induces apoptosis of human glioma cells through vanilloid receptor-1. J Neuropathol Exp Neurol 63:956–963
50. Hartel M, di Mola FF, Selvaggi F, Mascetta G, Wente MN, Felix K, Giese NA, Hinz U, Di Sebastiano P, Büchler MW, Friess H (2006) Vanilloids in pancreatic cancer: potential for chemotherapy and pain management. Gut 55:519–528
51. Zhang YW, Vande Woude GF (2003) HGF/SF-met signaling in the control of branching morphogenesis and invasion. J Cell Biochem 88:408–417

52. Waning J, Vriens J, Owsianik G, Stüwe L, Mally S, Fabian A, Frippiat C, Nilius B, Schwab A (2007) A novel function of capsaicin-sensitive TRPV1 channels: involvement in cell migration. Cell Calcium 42:17–25
53. Wondergem R, Ecay TW, Mahieu F, Owsianik G, Nilius B (2008) HGF/SF and menthol increase human glioblastoma cell calcium and migration. Biochem Biophys Res Commun 372:210–215
54. Vriens J, Janssens A, Prenen J, Nilius B, Wondergem R (2004) TRPV channels and modulation by hepatocyte growth factor/scatter factor in human hepatoblastoma (HepG2) cells. Cell Calcium 36:19–28
55. Miao X, Liu G, Xu X, Xie C, Sun F, Yang Y, Zhang T, Hua S, Fan W, Li Q, Huang S, Wang Q, Liu G, Zhong D (2008) High expression of vanilloid receptor-1 is associated with better prognosis of patients with hepatocellular carcinoma. Cancer Genet Cytogenet 186:25–32
56. Contassot E, Tenan M, Schnüriger V, Pelte MF, Dietrich PY (2004) Arachidonyl ethanolamide induces apoptosis of uterine cervix cancer cells via aberrantly expressed vanilloid receptor-1. Gynecol Oncol 93:182–188
57. Ramer R, Merkord J, Rohde H, Hinz B (2010) Cannabidiol inhibits cancer cell invasion via upregulation of tissue inhibitor of matrix metalloproteinases-1. Biochem Pharmacol 79: 955–966
58. Sanchez MG, Sanchez AM, Collado B, Malagarie-Cazenave S, Olea N, Carmena MJ, Prieto JC, Diaz-Laviada I (2005) Expression of the transient receptor potential vanilloid 1 (TRPV1) in LNCaP and PC-3 prostate cancer cells and in human prostate tissue. Eur J Pharmacol 515:20–27
59. Czifra G, Varga A, Nyeste K, Marincsák R, Tóth BI, Kovács I, Kovács L, Bíró T (2009) Increased expressions of cannabinoid receptor-1 and transient receptor potential vanilloid-1 in human prostate carcinoma. J Cancer Res Clin Oncol 135:507–514
60. Sánchez AM, Sánchez MG, Malagarie-Cazenave S, Olea N, Díaz-Laviada I (2006) Induction of apoptosis in prostate tumor PC-3 cells and inhibition of xenograft prostate tumor growth by the vanilloid capsaicin. Apoptosis 11:89–99
61. Malagarie-Cazenave S, Olea-Herrero N, Vara D, Díaz-Laviada I (2009) Capsaicin, a component of red peppers, induces expression of androgen receptor via PI3K and MAPK pathways in prostate LNCaP cells. FEBS Lett 583:141–147
62. Qiao S, Li W, Tsubouchi R, Murakami K, Yoshino M (2004) Role of vanilloid receptors in the capsaicin-mediated induction of iNOS in PC12 cells. Neurochem Res 29:687–689
63. Hosseini A, Koskela LR, Ehrén I, Aguilar-Santelises M, Sirsjö A, Wiklund NP (2006) Enhanced formation of nitric oxide in bladder carcinoma in situ and in BCG treated bladder cancer. Nitric Oxide 15:337–343
64. Kanzaki M, Zhang YQ, Mashima H, Li L, Shibata H, Kojima I (1999) Translocation of a calcium-permeable cation channel induced by insulin-like growth factor-I. Nat Cell Biol 1:165–170
65. Rochester MA, Patel N, Turney BW, Davies DR, Roberts IS, Crew J, Protheroe A, Macaulay VM (2007) The type 1 insulin-like growth factor receptor is over-expressed in bladder cancer. BJU Int 100:1396–1401
66. Zhao H, Grossman HB, Spitz MR, Lerner SP, Zhang K, Wu X (2003) Plasma levels of insulin-like growth factor-1 and binding protein-3, and their association with bladder cancer risk. J Urol 169:714–717
67. Nagasawa M, Nakagawa Y, Tanaka S, Kojima I (2007) Chemotactic peptide fMetLeuPhe induces translocation of the TRPV2 channel in macrophages. J Cell Physiol 210: 692–702
68. Wang C, Hu HZ, Colton CK, Wood JD, Zhu MX (2004) An alternative splicing product of the murine trpv1 gene dominant negatively modulates the activity of TRPV1 channels. J Biol Chem 279:37423–37430
69. Penna A, Juvin V, Chemin J, Compan V, Monet M, Rassendren FA (2006) PI3-kinase promotes TRPV2 activity independently of channel translocation to the plasma membrane. Cell Calcium 39:495–507

70. Monet M, Gkika D, Lehen'kyi V, Pourtier A, Vanden Abeele F, Bidaux G, Juvin V, Rassendren F, Humez S, Prevarsakaya N (2009) Lysophospholipids stimulate prostate cancer cell migration via TRPV2 channel activation. Biochim Biophys Acta 1793: 528–539
71. Monet M, Lehen'kyi V, Gackiere F, Firlej V, Vandenberghe M, Roudbaraki M, Gkika D, Pourtier A, Bidaux G, Slomianny C, Delcourt P, Rassendren F, Bergerat JP, Ceraline J, Cabon F, Humez S, Prevarskaya N (2010) Role of cationic channel TRPV2 in promoting prostate cancer migration and progression to androgen resistance. Cancer Res 70: 1225–1235
72. Liu G, Xie C, Sun F, Xu X, Yang Y, Zhang T, Deng Y, Wang D, Huang Z, Yang L, Huang S, Wang Q, Liu G, Zhong D, Miao X (2010) Clinical significance of transient receptor potential vanilloid 2 expression in human hepatocellular carcinoma. Cancer Genet Cytogenet 197: 54–59
73. Xu H, Ramsey IS, Kotecha SA, Moran MM, Chong JA, Lawson D, Ge P, Lilly J, Silos-Santiago I, Xie Y, DiStefano PS, Curtis R, Clapham DE (2002) TRPV3 is a calcium-permeable temperature-sensitive cation channel. Nature 418:181–186
74. Mandadi S, Sokabe T, Shibasaki K, Katanosaka K, Mizuno A, Moqrich A, Patapoutian A, Fukumi-Tominaga T, Mizumura K, Tominaga M (2009) TRPV3 in keratinocytes transmits temperature information to sensory neurons via ATP. Pflugers Arch 458:1093–1102
75. Ueda T, Yamada T, Ugawa S, Ishida Y, Shimada S (2009) TRPV3, a thermosensitive channel is expressed in mouse distal colon epithelium. Biochem Biophys Res Commun 383:130–134
76. Lee H, Caterina MJ (2005) TRPV channels as thermosensory receptors in epithelial cells. Pflugers Arch 451:160–167
77. Caterina MJ (2007) Transient receptor potential ion channels as participants in thermosensation and thermoregulation. Am J Physiol Regul Integr Comp Physiol 292:R64–R76
78. Levine JD, Alessandri-Haber N (2007) TRP channels: targets for the relief of pain. Biochim Biophys Acta 1772:989–1003
79. Hu HZ, Xiao R, Wang C, Gao N, Colton CK, Wood JD, Zhu MX (2006) Potentiation of TRPV3 channel function by unsaturated fatty acids. J Cell Physiol 208:201–212
80. Jariwala U, Prescott J, Jia L, Barski A, Pregizer S, Cogan JP, Arasheben A, Tilley WD, Scher HI, Gerald WL, Buchanan G, Coetzee GA, Frenkel B (2007) Identification of novel androgen receptor target genes in prostate cancer. Mol Cancer 6:39
81. Wang HP, Pu XY, Wang XH (2007) Distribution profiles of transient receptor potential melastatin-related and vanilloid-related channels in prostatic tissue in rat. Asian J Androl 9:634–640
82. Potter DA, Tirnauer JS, Janssen R, Croall DE, Hughes CN, Fiacco KA, Mier JW, Maki M, Herman IM (1998) Calpain regulates actin remodeling during cell spreading. J Cell Biol 141:647–662
83. Becker D, Bereiter-Hahn J, Jendrach M (2009) Functional interaction of the cation channel transient receptor potential vanilloid 4 (TRPV4) and actin in volume regulation. Eur J Cell Biol 88:141–152
84. Benfenati V, Amiry-Moghaddam M, Caprini M, Mylonakou MN, Rapisarda C, Ottersen OP, Ferroni S (2007) Expression and functional characterization of transient receptor potential vanilloid-related channel 4 (TRPV4) in rat cortical astrocytes. Neuroscience 148:876–892
85. Yamada T, Ugawa S, Ueda T, Ishida Y, Kajita K, Shimada S (2009) Differential localizations of the transient receptor potential channels TRPV4 and TRPV1 in the mouse urinary bladder. J Histochem Cytochem 57:277–287
86. Everaerts W, Vriens J, Owsianik G, Appendino G, Voets T, De Ridder D, Nilius B (2010) Functional characterization of transient receptor potential channels in mouse urothelial cells. Am J Physiol Renal Physiol 298:F692–F701
87. Mochizuki T, Sokabe T, Araki I, Fujishita K, Shibasaki K, Uchida K, Naruse K, Koizumi S, Takeda M, Tominaga M (2009) The TRPV4 cation channel mediates stretch-evoked Ca^{2+} influx and ATP release in primary urothelial cell cultures. J Biol Chem 284: 21257–21264

88. Nijenhuis T, Hoenderop JG, Nilius B, Bindels RJ (2003) (Patho)physiological implications of the novel epithelial Ca^{2+} channels TRPV5 and TRPV6. Pflugers Arch 446:401–409
89. Peng JB, Brown EM, Hediger MA (2003) Epithelial Ca^{2+} entry channels: transcellular Ca^{2+} transport and beyond. J Physiol 551:729–740
90. van Abel M, Hoenderop JG, van der Kemp AW, van Leeuwen JP, Bindels RJ (2003) Regulation of the epithelial Ca^{2+} channels in small intestine as studied by quantitative mRNA detection. Am J Physiol Gastrointest Liver Physiol 285:G78–G85
91. Irnaten M, Blanchard-Gutton N, Praetorius J, Harvey BJ (2009) Rapid effects of 17beta-estradiol on TRPV5 epithelial Ca^{2+} channels in rat renal cells. Steroids 74:642–649
92. Vasil'eva IO, Neguliaev IuA, Marakhova II, Semenova SB (2008) TRPV5 and TRPV6 calcium channels in human T cells. Tsitologiia 50:953–957
93. Semenova SB, Vassilieva IO, Fomina AF, Runov AL, Negulyaev YA (2009) Endogenous expression of TRPV5 and TRPV6 calcium channels in human leukemia K562 cells. Am J Physiol Cell Physiol 296:1098–1104
94. Kuro-o M, Matsumura Y, Aizawa H, Kawaguchi H, Suga T, Utsugi T, Ohyama Y, Kurabayashi M, Kaname T, Kume E, Iwasaki H, Iida A, Shiraki-Iida T, Nishikawa S, Nagai R, Nabeshima YI (1997) Mutation of the mouse klotho gene leads to a syndrome resembling ageing. Nature 390:45–51
95. Ito S, Fujimori T, Hayashizaki Y, Nabeshima Y (2002) Identification of a novel mouse membrane-bound family 1 glycosidase-like protein, which carries an atypical active site structure. Biochim Biophys Acta 1576:341–345
96. Kuro-o M (2008) Klotho as a regulator of oxidative stress and senescence. Biol Chem 389:233–241
97. Lutomski D, Fouillit M, Bourin P, Mellottée D, Denize N, Pontet M, Bladier D, Caron M, Joubert-Caron R (1997) Externalization and binding of galectin-1 on cell surface of K562 cells upon erythroid differentiation. Glycobiology 7:1193–1199
98. Wissenbach U, Niemeyer BA (2007) TRPV6. Handb Exp Pharmacol 179:221–234
99. Bödding M (2007) TRP proteins and cancer. Cell Signal 19:617–624
100. Prevarskaya N, Flourakis M, Bidaux G, Thebault S, Skryma R (2007) Differential role of TRP channels in prostate cancer. Biochem Soc Trans 35:133–135
101. Peng JB, Chen XZ, Berger UV, Vassilev PM, Tsukaguchi H, Brown EM, Hediger MA (1999) Molecular cloning and characterization of a channel-like transporter mediating intestinal calcium absorption. J Biol Chem 274:22739–22746
102. Fixemer T, Wissenbach U, Flockerzi V, Bonkhoff H (2003) Expression of the Ca^{2+}-selective cation channel TRPV6 in human prostate cancer: a novel prognostic marker for tumor progression. Oncogene 22:7858–7861
103. Wissenbach U, Niemeyer B, Himmerkus N, Fixemer T, Bonkhoff H, Flockerzi V (2004) TRPV6 and prostate cancer: cancer growth beyond the prostate correlates with increased TRPV6 Ca^{2+} channel expression. Biochem Biophys Res Commun 322:1359–1363
104. Peng JB, Zhuang L, Berger UV, Adam RM, Williams BJ, Brown EM, Hediger MA, Freeman MR (2001) CaT1 expression correlates with tumor grade in prostate cancer. Biochem Biophys Res Commun 282:729–734
105. Lehen'kyi V, Flourakis M, Skryma R, Prevarskaya N (2007) TRPV6 channel controls prostate cancer cell proliferation via $Ca^{(2+)}$/NFAT-dependent pathways. Oncogene 26:7380–7385
106. Bödding M, Fecher-Trost C, Flockerzi V (2003) Store-operated Ca^{2+} current and TRPV6 channels in lymph node prostate cancer cells. J Biol Chem 278:50872–50879
107. Kessler T, Wissenbach U, Grobholz R, Flockerzi V (2009) TRPV6 alleles do not influence prostate cancer progression. BMC Cancer 9:380
108. Zhuang L, Peng JB, Tou L, Takanaga H, Adam RM, Hediger MA, Freeman MR (2002) Calcium-selective ion channel, CaT1, is apically localized in gastrointestinal tract epithelia and is aberrantly expressed in human malignancies. Lab Invest 82:1755–1764

109. Bolanz KA, Hediger MA, Landowski CP (2008) The role of TRPV6 in breast carcinogenesis. Mol Cancer Ther 7:271–279
110. Schwarz EC, Wissenbach U, Niemeyer BA, Strauss B, Philipp SE, Flockerzi V, Hoth M (2006) TRPV6 potentiates calcium-dependent cell proliferation. Cell Calcium 39:163–173
111. Irnaten M, Blanchard-Gutton N, Harvey BJ (2008) Rapid effects of 17beta-estradiol on epithelial TRPV6 Ca^{2+} channel in human T84 colonic cells. Cell Calcium 44:441–452
112. Chetrite GS, Cortes-Prieto J, Philippe JC, Wright F, Pasqualini JR (2000) Comparison of estrogen concentrations, estrone sulfatase and aromatase activities in normal, and in cancerous, human breast tissues. J Steroid Biochem Mol Biol 72:23–27
113. Bolanz KA, Kovacs GG, Landowski CP, Hediger MA (2009) Tamoxifen inhibits TRPV6 activity via estrogen receptor-independent pathways in TRPV6-expressing MCF-7 breast cancer cells. Mol Cancer Res 7:2000–2010
114. Chow J, Norng M, Zhang J, Chai J (2007) TRPV6 mediates capsaicin-induced apoptosis in gastric cancer cells – Mechanisms behind a possible new "hot" cancer treatment. Biochim Biophys Acta 1773:565–576
115. Arcangeli A, Crociani O, Lastraioli E, Masi A, Pillozzi S, Becchetti A (2009) Targeting ion channels in cancer: a novel frontier in antineoplastic therapy. Curr Med Chem 16:66–93
116. Kunzelmann K (2005) Ion channels and cancer. J Membr Biol 205:159–173
117. Schönherr R (2005) Clinical relevance of ion channels for diagnosis and therapy of cancer. J Membr Biol 205:175–184

Chapter 50
The Role of Transient Receptor Potential Channels in Respiratory Symptoms and Pathophysiology

M. Allen McAlexander and Thomas Taylor-Clark

Abstract The Transient Receptor Potential channels constitute a superfamily of ion channels that is unmatched in its functional diversity. Recent research employing pharmacological and genetic methods has demonstrated that these channels are widely distributed within the respiratory tract, where they may mechanistically link noxious irritant exposures and inflammation to heightened airway reflex sensitivity, pathological remodeling and airflow limitation. Herein, we summarize the state of the art in this rapidly expanding area, emphasizing the known roles of Transient Receptor Potential channels in airway sensory nerves in addition to highlighting their roles in non-excitable cells.

The heterogeneous collection of cells in mammalian respiratory tracts must constantly survey the external environment and react accordingly in order to preserve gas exchange, the essential process whereby oxygen from inspired air is used to fuel mitochondrial respiration and CO_2, a primary waste product of this aerobic metabolism, is eliminated. Maintaining this homeostatic balance to match the metabolic state of the host organism would be a challenge without further perturbation; however, the respiratory passages are constantly exposed to a wide and varied range of irritants, allergens, pollutants and infectious agents. Complicating matters further, the "dead space," or surface area of the conducting airways where active gas exchange does not take place, can become occluded by airway smooth muscle constriction and the trapping of cellular debris in viscous mucus.

The multi-faceted nature of the lung thus offers many sources of vulnerability, although it also presents many opportunities for therapeutic intervention, provided that specific molecular entities can be identified that can dissociate necessary protective events from deleterious pathological consequences. In this regard, the Transient Receptor Potential (TRP) channels are exceptionally noteworthy, as these non-selective cation channels serve as environmental sensors capable of transducing extracellular physical or chemical stimuli into biological responses via membrane

M.A. McAlexander (✉)
Respiratory Discovery Biology, GlaxoSmithKline Pharmaceuticals, King of Prussia, PA, USA
e-mail: michael.a.mcalexander@gsk.com

depolarization and/or increases in intracellular calcium concentration. Indeed, many members of the mammalian TRP super family of ion channels are functionally expressed in the respiratory tract, where they contribute to processes as varied as mucociliary clearance, perception of noxious chemicals and initiation of irritant dilution/expulsion/avoidance reflexes, regulation of alveolar septal barrier permeability, and airway smooth muscle constriction. At least one human genetic study has also linked multiple SNPs of one TRP channel, the TRPV4 ion channel, to COPD [1]. Thus, evidence is mounting from many fronts that members of the TRP superfamily may be critical links between exogenous irritant perturbations, pathological tissue remodeling, and generation of noxious sensations accompanied by reflex sensitization in the respiratory tract.

50.1 Sensory Neuronal TRP Channels as Initiators and Regulators of Respiratory Symptoms

Although chronic respiratory diseases have markedly different underlying etiologies, timecourses, and degrees of severity, affected patients tend to present with a relatively small and stereotypical set of symptoms including cough, dyspnea (also known as "air hunger"), sputum production, throat irritation, wheeze, chest tightness, and pain upon inspiration. While the development of these symptoms likely mimics disease development in being the product of multifactorial interactions between the pathological milieu in susceptible individuals and environmental factors such as exposure to hazardous pollutants, allergens and infectious agents, they can be evoked acutely in humans by inhalation of irritants such as capsaicin [2] and ATP [3] that activate sensory nerves. Thus, future treatments that reduce activation of respiratory sensory neurons hold considerable promise as therapies for the alleviation of symptoms caused by diverse means. Intriguingly, primary sensory neurons contain a specialized array of ion channels – including members of the Transient Receptor Potential superfamily – that depolarize nerve terminals to initiate reflex-provoking electrical impulses in response to irritant and inflammatory stimuli.

50.1.1 Transient Receptor Potential Vanilloid 1: A Capsaicin-Sensitive Polymodal Respiratory Irritant Sensor

Capsaicin, the active principle responsible for the pungency of hot peppers, produces noxious sensations accompanied by dilution and expulsion reflexes such as plasma extravasation, cough and sneeze when applied to respiratory mucosa of mammals including humans [2]. Owing to these irritant properties, capsaicin is frequently utilized in pepper sprays which cause incapacitating irritation in humans.

Capsaicin aerosols have been used in human experimental studies to provoke cough [4], and have more recently been used to measure "cough receptor sensitivity," defined as the threshold concentration of acidic or capsaicin aerosol necessary to produce either two or five coughs [5]. This test is useful in demonstrating how sensitive individuals are to graded stimulation of a known sensory irritant mechanism. Of note, sensitivity to this challenge is increased under multiple pathological conditions, varying from idiopathic pulmonary fibrosis [6] to upper airway/nasal disturbances such as allergic rhinitis and rhinosinusitis [7]. Curiously, an exception to this general rule is that capsaicin cough threshold is increased in cigarette smokers due to unknown mechanisms, and this is reversible upon smoking cessation [8].

The molecular entity responsible for capsaicin's irritant properties proved elusive for decades until the capsaicin-sensitive ion channel Vanilloid Receptor-1 (VR1; now known as TRPV1 for Transient Receptor Potential Vanilloid 1) was cloned and characterized [9]. Studies in TRPV1-deficient mice demonstrated that capsaicin caused minimal sensory neuron activation [10] or behavioral avoidance responses in these animals [11], confirming TRPV1 as the critical target of capsaicin in vivo. Not only has this concept met no significant challenges during the ensuing decade, it has been supported by subsequent development of selective, orally-active blockers of the TRPV1 channel that abolish capsaicin-induced pain behaviors [12, 13] and expanded to include TRPV1 as the critical target of capsaicin in sensory neurons within the respiratory tract [14, 15].

Since capsaicin aerosols robustly and reproducibly evoke coughing in volunteers and no other molecular target of capsaicin beyond TRPV1 has been identified, respiratory TRPV1 activation would appear to be sufficient to cause coughing in humans. The intriguing question of whether TRPV1 activation facilitates pathological coughing in human respiratory diseases must remain unanswered until selective TRPV1 blockers suitable for dosing in humans become available. That said, TRPV1 is highly unusual in its ability to respond to a broad variety of noxious physical, chemical and inflammatory stimuli, including heat [9], acidic pH [16], basic pH [17], bioactive lipids [18, 19] and inflammatory mediators such as nerve growth factor, bradykinin and histamine [20–22]. TRPV1 responses can also be enhanced both by kinase activation and oxidative stress [23]. Thus, while the specific endogenous stimuli causing disease-associated respiratory symptoms including cough are elusive at present, the polymodal nature of TRPV1 makes it likely that many if not all of these stimuli act through this capsaicin-sensitive channel.

Available data suggest: TRPV1 activation causes respiratory symptoms including coughing in humans, TRPV1 is activated by inflammatory and oxidant stimuli known to be present in diseased human airways, and that the inflammatory and oxidant conditions seen in disease states could sensitize TRPV1 such that it may be opened by innocuous stimuli. Consistent with this hypothesis, respiratory reflex responses to capsaicin are enhanced in allergically inflamed mouse airways [24] and in rats exposed to the hazardous pollutant acrolein [24, 25]. These animal data are consistent with the aforementioned human data that cough thresholds to capsaicin are decreased in various disease states, suggesting that increased reflex sensitivity to inhaled irritants such as TRPV1 agonists in diseased airways is a phenomenon that likely translates into multiple patient populations.

50.1.2 Transient Receptor Potential Ankyrin 1: A Unique, Polymodal Sensor of Noxious Irritants

TRPA1 was initially characterized as a noxious cold sensor selectively expressed in a subset of capsaicin-sensitive sensory neurons [26]. Since this time, its role as a direct sensor of noxious cold has become a source of considerable controversy, while its role as a sensor for an unprecedented array of hazardous irritants and mediators of oxidative tissue damage and as a contributor to human pain states [27] has become increasingly clear [28–35]. TRPA1's unique promiscuity is due at least in part to reactive cysteine residues in its cytosolic N terminus, since many channel agonists (e.g., acrolein, maleimides and allyl isothiocyanate) act as electrophiles that can covalently react with free thiols and the ability of these agonists to activate the channel is greatly impaired in mutants lacking these critical cytosolic residues [36, 37].

Many of the aforementioned noxious chemicals that activate TRPA1 are known to cause marked to severe respiratory symptoms in humans, including acrolein, chlorine, formaldehyde, toluene diisocyanate, ozone, and tear gases. While potent, selective, safe and efficacious blockers of TRPA1 are not yet available to confirm that these agents produce their irritant effects in humans via TRPA1 activation, genetic disruption or block of the channel protects animals from respiratory reflexes and irritation behaviors including cough [38] evoked by these chemicals [33–35, 39, 40].

TRPA1 is not activated solely by reactive molecules, as many non-electrophilic molecules such as Δ9-THC (Delta-9 tetrahydrocannabinol) and menthol activate the channel via an apparently direct means and inflammatory mediators such as bradykinin (BK) lead to channel opening downstream of their canonical G_q signaling pathways [31, 32]. Of note, gaseous anesthetics used routinely during surgical procedures activate TRPA1 [41] and may paradoxically contribute to post-operative pain hypersensitivity. This finding provides the most parsimonious explanation for the observations that many human volunteers consider the respiratory irritation caused by inhalation of desflurane or isoflurane literally intolerable, as roughly half of the volunteers could not complete a 60s exposure period, whereas no volunteers pre-maturely abandoned their 60s exposure to sevoflurane [42], a gaseous anesthetic that does not activate TRPA1 [41]. Consistent with this notion, the selective TRPA1 blocker HC-030031 protects guinea pigs from increases in airway resistance caused by desflurane inhalation [43].

In general, for an ion channel, kinase or receptor to be a viable target for pharmaceutical inhibition, the molecule must exhibit intrinsic activity driven by endogenous factors. TRPA1 would present an interesting possible exception to this concept, given that it is activated by such a wide range of hazardous air pollutants and cigarette smoke constituents; however, TRPA1 is also activated by multiple endogenously occurring agonists. These include lipid peroxidation products such as 4-hydroxynonenal (4-HNE) and 4-oxononenal (4-ONE) [29, 30, 40, 44, 45], products of nitrative stress such as nitro-oleate [46], and reactive electrophilic metabolites of prostaglandin D_2 and E_2 [30]. Other molecules that may act as endogenous agonists of TRPA1 through direct and/or indirect means include bradykinin [31], PAR2 agonists [47], Zn^{2+} [48, 49], and nitric oxide [50]. Moreover, intracellular Ca^{2+} elevations appear to directly activate TRPA1 [51], which raises the possibility that TRPA1 could amplify the ability of any anatomically-colocalized G_q-coupled receptor or Ca^{2+}-permeable ion channel to depolarize nociceptors.

In the airways of small laboratory animals, activation of TRPA1 in nociceptor terminals causes nerve depolarization that leads to both action potential discharge and local neuropeptide release. This nociceptor activation culminates in respiratory reflexes in vivo; most notably, the stereotypical drop in respiratory rate without concomitant changes in tidal volume known as respiratory sensory irritation. Such behaviors are readily observed in conscious mice following inhalation of aerosols or intra-nasal application of irritants including TRPA1 agonists [33, 34]. Nicotine, the addictive principle in tobacco plants that can cause irritation when applied to the human respiratory mucosa, acts through TRPA1 in addition to nicotinic acetylcholine receptors to activate sensory neurons, although TRPA1's contribution to nicotine-induced respiratory irritation reflexes is controversial. TRPA1 knockout mice do not exhibit increases in the derived ventilatory parameter PenH when challenged with intra-nasal nicotine in unrestrained plethysmography experiments [52], suggesting they are protected from nicotine's irritating and/or bronchospastic effects. However, when respiratory rates are measured directly in nose-out whole-body plethysmography experiments, TRPA1-deficient mice exhibit marked respiratory rate decreases indicative of respiratory sensory irritation when exposed to similar intra-nasal nicotine application [34]. The latter results suggest TRPA1-deficient mice can perceive intra-nasal nicotine as noxious, although the former study offers evidence that TRPA1 can influence sensory perception of nicotine in some manner. Clearly, further study is required to fully characterize the relative roles of TRPA1 and nicotinic receptors in nicotine-evoked nocifensive reflexes.

Importantly, TRPA1 deletion or small molecule blockade protects mice from inflammation and airway hyperreactivity accompanying an ovalbumin model of allergic asthma [53]. This protection was not mimicked by deletion of the polymodal nocisensor and capsaicin receptor TRPV1, but it would likely be mimicked by capsaicin desensitization, which broadly defunctionalizes TRPV1-expressing nociceptors. Since many TRPA1 agonists are reactive, hazardous irritants that cause covalent modifications including crosslinking of biomolecules, further studies to elucidate the role of TRPA1 in the in vivo responses to these toxins are eagerly awaited.

50.1.3 Transient Receptor Potential Melastatin 8: The Cold and Menthol Receptor

The most extensively-studied member of the TRPM family in the respiratory tract is TRPM8, a sensor for cold and menthol that is expressed on primary sensory nerves [54, 55]. This ion channel is presumed to be the molecular entity responsible for the respiratory symptom relief afforded by menthol-containing lozenges and related products, although menthol also exerts pronounced effects on TRPA1 [56, 57], and cold air is more frequently associated with the onset of respiratory symptoms than their relief [58]. While TRPM8 has not been firmly established as the molecular entity responsible for the respiratory effects of menthol in the human airways, studies have reported that menthol application can produce sensations of enhanced nasal patency in the absence of alterations in objective measures of airflow [59, 60]. Additionally, menthol prevents constriction of guinea pig airway smooth muscle, although the mechanism appears to be pharmacologically distinct from TRPM8 [61]. Beneficial effects of menthol on respiratory sensations have not yet been tied directly to TRPM8 through selective pharmacological tools or genetic disruption in animal studies, but TRPM8 agonism is one likely mechanism.

While TRPM8's response to cold and menthol is well-accepted, it may also respond to endogenous ligands. Indeed, independent studies have now shown that TRPM8 can be gated by lysophospholipids [62, 63]. Lysophospholipids are produced in high concentrations following segmental antigen provocation in the airways of allergic subjects [64], yet their contribution to the associated inflammation, airflow limitation and related symptoms remains unknown. Whether lysophospholipid-sensitive TRP channels serve as ionotropic receptors for these mediators in the context of allergic disease is a topic worthy of future study.

50.2 TRP Channels in Airway Cells of Non-Neuronal Origin

Perhaps owing to the pronounced effects of capsaicin in the human airways and the comparatively widespread availability of tool probes for exploring TRPV1 function, research on TRP channels in the airways was initially skewed primarily toward sensory neurobiology. Further research has, however, revealed that TRPA, TRPC, TRPM and TRPV channels serve diverse purposes in non-excitable cell types of multiple origins.

50.2.1 TRPA1 as a Calcium-Permeable Ion Channel in Human Pulmonary Fibroblasts

Since its first characterization as an ion channel, TRPA1 expression has been thought to be predominantly limited to a subset of TRPV1-positive nociceptors [26]. More recent results have demonstrated that multiple known TRPA1 activators cause Ca^{2+} influx and outwardly-rectifying, non-selective cation currents in the human embryonic pulmonary fibroblast line WI-38 that are blocked by standard TRPA1 blockers [65]. This study provides compelling evidence that pulmonary fibroblast TRPA1 has electrophysiological and pharmacological properties nearly identical to the more extensively-characterized neuronal TRPA1, although it is worth noting that these findings were from an embryonic cell line and the question of whether TRPA1 plays a physiological or pathophysiological role in adult pulmonary fibroblasts remains unanswered at present. If TRPA1 is functional in adult pulmonary fibroblasts, it could conceivably contribute to aberrant Ca^{2+} influx that may alter collagen synthesis, differentiation and/or migration, impairing wound healing in the airways and facilitating chronic airflow limitation.

50.2.2 TRPC Channels Regulate Intracellular Calcium Homeostasis and Inflammation

Possible roles of TRPC channels in the respiratory tract will be reviewed comprehensively elsewhere within this volume, but it is worth noting here that TRPC channels are widely expressed and can modulate both Ca^{2+} entry through the plasma membrane and Ca^{2+} release from the ER. Various members of the TRPC family have been implicated in the function of airway and vascular smooth muscle [66–69], migration of neutrophils [70] and platelet activating factor (PAF)-induced Ca^{2+} mobilization in human alveolar macrophages [71]. Furthermore, TRPC6 knockout mice have reduced BAL inflammatory cell counts and T_h2 cytokine (IL-5 and IL-13) levels in response to ovalbumin sensitization and challenge [72]. Thus, TRPC channels may contribute to airway inflammation and/or airflow limitation via multiple mechanisms and effector cells.

50.2.3 Wide-Ranging Functions of TRPM Channels in Airway Structural and Immune Cells

The role of TRPM channels beyond TRPM8 in the airways has not been extensively explored to-date. There are, nonetheless, reasons to believe that TRPM channels could contribute to respiratory disease. For example, TRPM2, the oxidant-sensitive channel/kinase hybrid (sometimes referred to as a "chanzyme"), can positively regulate neutrophil bacteria-killing ability [73] and this sort of activity may help the lung efficiently clear infectious pathogens. Any potential benefits of evoking TRPM2 activation would have to be weighed against the known deleterious effects of TRPM2 opening, including enhancement of endothelial permeability [74], monocyte-dependent inflammatory chemokine production [75], and oxidant-mediated cytotoxicity [76]. At present, the available data suggest that TRPM2 may be a downstream effector of multiple pathological features of oxidant stress within the airways, including excess neutrophil infiltration and endothelial barrier disruption.

TRPM3 channels have received heightened attention of late following their identification as a novel class of ionotropic steroid receptor that can be activated by pregnenolone [77]. Decades of evidence support the contention that nuclear glucocorticoid receptors are the primary biological targets of steroids used as antiinflammatory therapies, although the possibility exists that if they activate TRPM3, this ion channel may modulate their effects.

Mice lacking TRPM4 channels exhibited increased Ca^{2+} influx accompanied by increased release of aminergic (histamine) and lipid (cysteinyl leukotriene) mediators and the cytokine TNF-α upon engagement of Fcε-RI, the high-affinity receptor for IgE [78]. The authors went on to establish that TRPM4 acts as a calcium-activated non-selective cation channel that can cause membrane depolarization and limit the driving force for calcium entry into mast cells. Consistent with these in vitro findings, cutaneous anaphylaxis responses were also exaggerated in TRPM4-deficient mice, confirming that removal of TRPM4 exaggerates mast cell-driven responses to antigen in vivo. Further experiments have shown that TRPM4 can specifically influence the migration of mast cells [79] and dendritic cells [80]. Together, these data suggest that TRPM4 may influence the immune response to inhaled allergens in allergic asthma at many levels. In addition, TRPM4 has been implicated in trauma-associated injury, as removing TRPM4 by genetic ablation in mice or antisense knockdown in rats limited capillary fragmentation associated with secondary hemorrhage and improved performance in behavioral assays after spinal cord injury [81]. If similar processes occur in the vulnerable alveolar capillary beds, disrupting TRPM4 function may provide a novel means to rescue gas exchange during pulmonary hemorrhage associated with events such as trauma, severe infection or chemical toxicity.

The TRPM7 ion channel is nearly ubiquitously expressed and was initially considered to be an essential regulator of cellular Mg^{2+} homeostasis, as TRPM7 deficiency is embryonic lethal [82]. Consistent with this hypothesis, RNA-mediated knockdown of TRPM7 is toxic to human lung mast cells [83]. More intense scrutiny has revealed that TRPM7 is dispensable for Mg^{2+} homeostasis in adult cells and that its deletion is lethal because of its critical role in organogenesis [82]. Further investigations of possible roles of TRPM7 in the respiratory tract are warranted, in light of its wide expression profile, distinct biology, and the observation that it is a necessary downstream mediator of TGF-beta1-mediated fibrogenesis in human atrial fibroblasts [84].

As mentioned earlier in this chapter, TRPM8 is most widely-studied for its role in transduction of cold stimuli by sensory nerves. Studies from three independent laboratories produced only subtly different results and uniformly agreed that mice lacking TRPM8 are essentially unable to respond to non-noxious cold stimuli in a variety of behavioral assays [85–87]. If this unique role of TRPM8 translates to humans, it suggests that the channel could play an important role in regulating autonomic outflow, as cold exposure causes autonomic changes that result in heat generation/preservation reflexes such as shivering and subcutaneous vasoconstriction. While the involvement of TRPM8 in human diseases has not yet been rigorously tested, the efficacy of anti-muscarinics as bronchodilators provides clear evidence that unknown factors lead to net autonomic dysfunction in respiratory diseases such as COPD.

50.2.4 TRPV Channels: Beyond the Capsaicin Receptor

TRPV1 has been most exhaustively studied as a sensor of noxious stimuli in nociceptor neurons, although its expression and function have been characterized in a limited number of other contexts. Notably, multiple airway epithelial cell lines express TRPV1 and their exposure to particulate matter can lead to Ca^{2+} influx, cytokine release and apoptosis that are inhibited by the TRPV1 blocker capsazepine [88, 89]. These findings suggest that non-neuronal TRPV1 may contribute to airway inflammation and pathology caused by particulate matter, an established cause of respiratory disease development and exacerbations.

Following the discovery of TRPV1 as the ion channel target of capsaicin, other TRPV channels were identified using similar methods. The first of these, TRPV2, was described as a TRPV1-related channel present in sensory neurons and sensitive to noxious heat [90]. Studies since this initial characterization have failed to prove that TRPV2 plays an obligatory role in pain-related behaviors or nociceptor activation in response to noxious thermal stimuli [91]. Indeed, the human TRPV2 ortholog is not gated by noxious heat in the range that activates the rodent channels when they are expressed in heterologous expression systems [92].

In the absence of convincing evidence that TRPV2 serves as a nociceptor transducer in vivo, investigators have sought other possible physiological roles for the channel. In addition to sensory neurons, TRPV2 is also expressed in airway epithelium and parasympathetic ganglia [93], as well as a subset of motor neurons [94]. Its function and endogenous activator(s) in these cellular contexts has not yet been well-established, but both rodent and human orthologs of the channel respond to cannabinoids [92]. More recently, TRPV2 was identified as a channel causing Ca^{2+} influx and non-selective cation currents downstream of formyl peptide receptors in TtT/M87 macrophages [95]. A separate lab was unable to reproduce these findings in TRPV2-deficient mice, although they demonstrated that TRPV2 is expressed in multiple macrophage populations (including alveolar macrophages) in vivo, plays a critical role in binding and phagocytosis of opsonized particles, and is necessary for optimal migration of peritoneal macrophages toward chemotactic cues such as the chemokine MCP1 [96]. Likely owing to these macrophage functional deficits, TRPV2-deficient mice could not clear *Listeria monocytogenes* as well as wild-types, which resulted in higher liver and spleen bacterial burdens and more rapid mortality [96].

The TRPV3 channel is gated by heat in the warm range and is present in keratinocytes within the skin [97–99]. The activation of TRPV3 in keratinocytes can lead to the production of mediators that act on sensory nerves and modulate aversion behaviors in vivo [100]. Mouse models also suggest that over-active TRPV3 facilitates development of dermatitis-like skin lesions accompanied by scratching behaviors [101–103]. More recently, TRPV3 was established as a downstream mediator of epithelial growth factor receptor (EGFR) signaling in skin barrier formation [104]. If TRPV3 plays a similar role in airway epithelial cells, its activity could lead to alterations in phenomena as varied as wound healing, barrier strength, inflammatory mediator release and mucus secretion. Additionally, TRPV3 mutations leading to over-active channels in mice are associated with increased levels of nerve growth factor (NGF), which is elevated in patients with allergic diseases including asthma [105].

TRPV4 is widely expressed in multiple cell types including airway epithelium, vascular endothelium and smooth muscle. Studies to date, based almost exclusively on knockout mice, suggest a role for TRPV4 in osmoregulation and in bladder function [106–110] and in osteoclast differentiation [111]. More recently, TRPV4 mutations that uniformly result in channels that are overactive when expressed in the plasma membrane have been linked to multiple human genetic disorders that highlight a role of TRPV4 in the musculoskeletal system [112–116]. Multiple TRPV4 SNPs have also been linked to COPD [1], although since these changes fall almost exclusively within introns, it is not yet clear what impact they have on TRPV4 expression and function.

Perhaps owing to the large variety of cells that express the channel, TRPV4's functions in the respiratory tract have remained elusive. The first critical insight into the function of TRPV4 in the lung was the demonstration that chemical activation of TRPV4 increases the lung filtration coefficient (K_f), a value representative of the lung's permeability to water [117]. This study also employed electron microscopy to demonstrate that TRPV4 activation specifically disrupts the alveolar septal barrier, unlike thapsigargin, which elevates intracellular Ca^{2+} via a distinct mechanism and leads to K_f increases by disrupting the endothelial barrier of larger, extra-alveolar vessels [117]. The alveolar capillary endothelium has been known for some time to be unusually vulnerable to insult, and its compromise can lead to pulmonary edema and life-threatening impairment of gas exchange. The identification of TRPV4 as a molecular entity that could contribute to this vulnerability is exciting, as it raises the possibility that decreasing TRPV4's channel activity could be a treatment strategy for diseases where gas exchange is impaired by pulmonary edema of multiple origins. Since this groundbreaking study, TRPV4 has also been implicated in alveolar septal barrier disruption in response to high vascular pressures [118], which can be seen during cardiovascular compromise and to high tidal volume ventilation [119], a known cause of lung injury in animals and humans. This ability of TRPV4 activation to disrupt the alveolar barrier has also been demonstrated in vivo, where intravenous administration of the TRPV4 activator GSK1016790A causes rapid circulatory collapse accompanied by severe pulmonary hemorrhage [120].

TRPV4 is also expressed in human airway smooth muscle [121], and hypotonic solutions or 4alpha-phorbol 12,13-didecanoate (4αPDD), which are both known TRPV4 openers, cause Ca^{2+} influx into cultured human airway smooth muscle cells that is blocked by the pan-TRPV blocker ruthenium red but not the TRPV1 blocker capsazepine or the L-type Ca^{2+} channel blocker verapamil [121]. Combined with the mounting body of data that TRPV4 activation increases lung endothelial permeability, these studies suggest that TRPV4 may contribute to respiratory pathophysiology through distinct mechanisms in the lungs and conducting airways. The airway smooth muscle data are also compelling since a subset of COPD-associated TRPV4 SNPs was specifically associated with a reduced FEV1/FVC measurement [1], which is representative of airway obstruction.

50.2.5 Summary

Understanding of the influence of the Transient Receptor Potential channels in biological systems continues to grow at a rapid pace. Due to their diversity of function, expression, and sensitivity to chemical and physical stimuli, TRP channels are uniquely positioned as polymodal sensors that can couple stressors to alterations in the effector functions of structural, immune and neuronal cells. The study of TRP channels is still in its relative infancy, and already many members of this diverse superfamily have been implicated in respiratory noxious chemical sensation, inflammation and pathological remodeling. Over the next several years, further examination of these channels will hopefully yield both novel biological insights and benefits to human health.

References

1. Zhu G, Investigators ICGN, Gulsvik A, Bakke P, Ghatta S, Anderson W, Lomas DA, Silverman EK, Pillai SG (2009) Association of TRPV4 gene polymorphisms with chronic obstructive pulmonary disease. Hum Mol Genet 18:2053–2062
2. Fuller RW (1991) Pharmacology of inhaled capsaicin in humans. Respir Med 85:31–34
3. Basoglu OK, Pelleg A, Essilfie-Quaye S, Brindicci C, Barnes PJ, Kharitonov SA (2005) Effects of Aerosolized Adenosine 5′-Triphosphate-Triphosphate vs Adenosine 5′-Monophosphate on Dyspnea and Airway Caliber in Healthy Nonsmokers and Patients With Asthma*. Chest 128:1905–1909
4. Fuller RW, Dixon CM, Barnes PJ (1985) Bronchoconstrictor response to inhaled capsaicin in humans. J Appl Physiol 58:1080–1084
5. Pecova R, Javorkova N, Kudlicka J, Tatar M (2007) Tussigenic agents in the measurement of cough reflex sensitivity. J Physiol Pharmacol 58(Suppl 5):531–538
6. Hope-Gill BDM, Hilldrup S, Davies C, Newton RP, Harrison NK (2003) A Study of the Cough Reflex in Idiopathic Pulmonary Fibrosis. Am J Respir Crit Care Med 168: 995–1002
7. Tatar M, Plevkova J, Brozmanova M, Pecova R, Kollarik M (2009) Mechanisms of the cough associated with rhinosinusitis. Pulm Pharmacol Ther 22:121–126
8. Sitkauskiene B, Stravinskaite K, Sakalauskas R, Dicpinigaitis PV (2007) Changes in cough reflex sensitivity after cessation and resumption of cigarette smoking. Pulm Pharmacol Ther 20:240–243
9. Caterina MJ, Schumacher MA, Tominaga M, Rosen TA, Levine JD, Julius D (1997) The capsaicin receptor: a heat-activated ion channel in the pain pathway. Nature 389:816–824
10. Davis JB, Gray J, Gunthorpe MJ, Hatcher JP, Davey PT, Overend P, Harries MH, Latcham J, Clapham C, Atkinson K, Hughes SA, Rance K, Grau E, Harper AJ, Pugh PL, Rogers DC, Bingham S, Randall A, Sheardown SA (2000) Vanilloid receptor-1 is essential for inflammatory thermal hyperalgesia. Nature 405:183–187
11. Caterina MJ, Leffler A, Malmberg AB, Martin WJ, Trafton J, Petersen-Zeitz KR, Koltzenburg M, Basbaum AI, Julius D (2000) Impaired Nociception and Pain Sensation in Mice Lacking the Capsaicin Receptor. Science 288:306–313
12. Gavva NR, Tamir R, Qu Y, Klionsky L, Zhang TJ, Immke D, Wang J, Zhu D, Vanderah TW, Porreca F, Doherty EM, Norman MH, Wild KD, Bannon AW, Louis JC, Treanor JJS (2005) AMG 9810 [(E)-3-(4-t-Butylphenyl)-N-(2,3-dihydrobenzo[b][1,4] dioxin-6-yl)acrylamide], a Novel Vanilloid Receptor 1 (TRPV1) Antagonist with Antihyperalgesic Properties. J Pharmacol Exp Ther 313:474–484
13. Gavva NR, Bannon AW, Hovland DN Jr, Lehto SG, Klionsky L, Surapaneni S, Immke DC, Henley C, Arik L, Bak A, Davis J, Ernst N, Hever G, Kuang R, Shi L, Tamir R, Wang J, Wang W, Zajic G, Zhu D, Norman MH, Louis JC, Magal E, Treanor JJ (2007) Repeated administration of vanilloid receptor TRPV1 antagonists attenuates hyperthermia elicited by TRPV1 blockade. J Pharmacol Exp Ther 323:128–137

14. Symanowicz PT, Gianutsos G, Morris JB (2004) Lack of role for the vanilloid receptor in response to several inspired irritant air pollutants in the C57Bl/6J mouse. Neurosci Lett 362:150–153
15. Kollarik M, Undem BJ (2004) Activation of bronchopulmonary vagal afferent nerves with bradykinin, acid and vanilloid receptor agonists in wild-type and TRPV1-/- mice. J Physiol (Lond) 555:115–123
16. Tominaga M, Caterina MJ, Malmberg AB, Rosen TA, Gilbert H, Skinner K, Raumann BE, Basbaum AI, Julius D (1998) The Cloned Capsaicin Receptor Integrates Multiple Pain-Producing Stimuli. Neuron 21:531–543
17. Dhaka A, Uzzell V, Dubin AE, Mathur J, Petrus M, Bandell M, Patapoutian A (2009) TRPV1 Is Activated by Both Acidic and Basic pH. J Neurosci 29:153–158
18. Ahern GP (2003) Activation of TRPV1 by the Satiety Factor Oleoylethanolamide. J Biol Chem 278:30429–30434
19. Patwardhan AM, Scotland PE, Akopian AN, Hargreaves KM (2009) Activation of TRPV1 in the spinal cord by oxidized linoleic acid metabolites contributes to inflammatory hyperalgesia. PNAS 106:18820–18824
20. Chuang Hh, Prescott ED, Kong H, Shields S, Jordt SE, Basbaum AI, Chao MV, Julius D (2001) Bradykinin and nerve growth factor release the capsaicin receptor from PtdIns(4,5)P2-mediated inhibition. Nature 411:957–962
21. Shim WS, Tak MH, Lee MH, Kim M, Kim M, Koo JY, Lee CH, Kim M, Oh U (2007) TRPV1 Mediates Histamine-Induced Itching via the Activation of Phospholipase A2 and 12-Lipoxygenase. J Neurosci 27:2331–2337
22. Ji RR, Samad TA, Jin SX, Schmoll R, Woolf CJ (2002) p38 MAPK Activation by NGF in Primary Sensory Neurons after Inflammation Increases TRPV1 Levels and Maintains Heat Hyperalgesia. Neuron 36:57–68
23. Chuang Hh, Lin S (2009) Oxidative challenges sensitize the capsaicin receptor by covalent cysteine modification. PNAS 106:20097–20102
24. Braun A, Lommatzsch M, Neuhaus-Steinmetz U, Quarcoo D, Glaab T, McGregor GP, Fischer A, Renz H (2004) Brain-derived neurotrophic factor (BDNF) contributes to neuronal dysfunction in a model of allergic airway inflammation. Br J Pharmacol 141: 431–440
25. Hazari MS, Rowan WH, Winsett DW, Ledbetter AD, Haykal-Coates N, Watkinson WP, Costa DL Potentiation of pulmonary reflex response to capsaicin 24 h following whole-body acrolein exposure is mediated by TRPV1. Respir Physiol Neurobiol (in press, Corrected Proof)
26. Story GM, Peier AM, Reeve AJ, Eid SR, Mosbacher J, Hricik TR, Earley TJ, Hergarden AC, Andersson DA, Hwang SW (2003) ANKTM1, a TRP-like Channel Expressed in Nociceptive Neurons, Is Activated by Cold Temperatures. Cell 112:819–829
27. Kremeyer B, Lopera F, Cox JJ, Momin A, Rugiero F, Marsh S, Woods CG, Jones NG, Paterson KJ, Fricker FR, Villegas A, Acosta N, Pineda-Trujillo NG, Ramírez JD, Zea J, Burley MW, Bedoya G, Bennett DLH, Wood JN, Ruiz-Linares A (2010) A Gain-of-Function Mutation in TRPA1 Causes Familial Episodic Pain Syndrome. Neuron 66: 671–680
28. Bautista DM, Jordt SE, Nikai T, Tsuruda PR, Read AJ, Poblete J, Yamoah EN, Basbaum AI, Julius D (2006) TRPA1 Mediates the Inflammatory Actions of Environmental Irritants and Proalgesic Agents. Cell 124:1269–1282
29. Trevisani M, Siemens J, Materazzi S, Bautista DM, Nassini R, Campi B, Imamachi N, Andre E, Patacchini R, Cottrell GS, Gatti R, Basbaum AI, Bunnett NW, Julius D, Geppetti P (2007) 4-Hydroxynonenal, an endogenous aldehyde, causes pain and neurogenic inflammation through activation of the irritant receptor TRPA1. PNAS 104:13519–13524
30. Taylor-Clark TE, Undem BJ, MacGlashan DW Jr, Ghatta S, Carr MJ, McAlexander MA (2008) Prostaglandin-Induced Activation of Nociceptive Neurons via Direct Interaction with Transient Receptor Potential A1 (TRPA1). Mol Pharmacol 73:274–281

31. Bandell M, Story GM, Hwang SW, Viswanath V, Eid SR, Petrus MJ, Earley TJ, Patapoutian A (2004) Noxious Cold Ion Channel TRPA1 Is Activated by Pungent Compounds and Bradykinin. Neuron 41:849–857
32. Jordt SE, Bautista DM, Chuang Hh, McKemy DD, Zygmunt PM, Hogestatt ED, Meng ID, Julius D (2004) Mustard oils and cannabinoids excite sensory nerve fibres through the TRP channel ANKTM1. Nature 427:260–265
33. Bessac BF, Sivula M, von Hehn CA, Escalera J, Cohn L, Jordt SE (2008) TRPA1 is a major oxidant sensor in murine airway sensory neurons. J Clin Invest 118:1899–1910
34. Taylor-Clark TE, Kiros F, Carr MJ, McAlexander MA (2009) Transient Receptor Potential Ankyrin 1 Mediates Toluene Diisocyanate-Evoked Respiratory Irritation. Am J Respir Cell Mol Biol 40:756–762
35. McNamara CR, Mandel-Brehm J, Bautista DM, Siemens J, Deranian KL, Zhao M, Hayward NJ, Chong JA, Julius D, Moran MM, Fanger CM (2007) TRPA1 mediates formalin-induced pain. PNAS 104:13525–13530
36. Hinman A, Chuang Hh, Bautista DM, Julius D (2006) TRP channel activation by reversible covalent modification. PNAS 103:19564–19568
37. Macpherson LJ, Dubin AE, Evans MJ, Marr F, Schultz PG, Cravatt BF, Patapoutian A (2007) Noxious compounds activate TRPA1 ion channels through covalent modification of cysteines. Nature 445:541–545
38. Birrell MA, Belvisi MG, Grace M, Sadofsky L, Faruqi S, Hele DJ, Maher SA, Freund-Michel V, Morice AH (2009) TRPA1 Agonists Evoke Coughing in Guinea Pig and Human Volunteers. Am J Respir Crit Care Med 180:1042–1047
39. Bessac BF, Sivula M, von Hehn CA, Caceres AI, Escalera J, Jordt SE (2009) Transient receptor potential ankyrin 1 antagonists block the noxious effects of toxic industrial isocyanates and tear gases. FASEB J 23:1102–1114
40. Macpherson LJ, Xiao B, Kwan KY, Petrus MJ, Dubin AE, Hwang S, Cravatt B, Corey DP, Patapoutian A (2007) An Ion Channel Essential for Sensing Chemical Damage. J Neurosci 27:11412–11415
41. Matta JA, Cornett PM, Miyares RL, Abe K, Sahibzada N, Ahern GP (2008) General anesthetics activate a nociceptive ion channel to enhance pain and inflammation. PNAS 105:8784–8789
42. TerRiet MF, DeSouza GJA, Jacobs JS, Young D, Lewis MC, Herrington C, Gold MI (2000) Which is most pungent: isoflurane, sevoflurane or desflurane? Br J Anaesth 85:305–307
43. Satoh JI, Yamakage M (2009) Desflurane induces airway contraction mainly by activating transient receptor potential A1 of sensory C-fibers. J Anesth 23:620–623
44. Taylor-Clark TE, McAlexander MA, Nassenstein C, Sheardown SA, Wilson S, Thornton J, Carr MJ, Undem BJ (2008) Relative contributions of TRPA1 and TRPV1 channels in the activation of vagal bronchopulmonary C-fibres by the endogenous autacoid 4-oxononenal. J Physiol 586:3447–3459
45. Andersson DA, Gentry C, Moss S, Bevan S (2008) Transient Receptor Potential A1 Is a Sensory Receptor for Multiple Products of Oxidative Stress. J Neurosci 28:2485–2494
46. Taylor-Clark TE, Ghatta S, Bettner W, Undem BJ (2009) Nitrooleic Acid, an Endogenous Product of Nitrative Stress, Activates Nociceptive Sensory Nerves via the Direct Activation of TRPA1. Mol Pharmacol 75:820–829
47. Dai Y, Wang S, Tominaga M, Yamamoto S, Fukuoka T, Higashi T, Kobayashi K, Obata K, Yamanaka H, Noguchi K (2007) Sensitization of TRPA1 by PAR2 contributes to the sensation of inflammatory pain. J Clin Invest 117:1979–1987
48. Andersson DA, Gentry C, Moss S, Bevan S (2009) Clioquinol and pyrithione activate TRPA1 by increasing intracellular Zn^{2+}. PNAS 106:8374–8379
49. Hu H, Bandell M, Petrus MJ, Zhu MX, Patapoutian A (2009) Zinc activates damage-sensing TRPA1 ion channels. Nat Chem Biol 5:183–190
50. Miyamoto T, Dubin AE, Petrus MJ, Patapoutian A (2009) TRPV1 and TRPA1 Mediate Peripheral Nitric Oxide-Induced Nociception in Mice. PLoS One 4:e7596

51. Doerner JF, Gisselmann G, Hatt H, Wetzel CH (2007) Transient receptor potential channel A1 is directly gated by calcium ions. J Biol Chem 282:13180–13189, M607849200
52. Talavera K, Gees M, Karashima Y, Meseguer VM, Vanoirbeek JAJ, Damann N, Everaerts W, Benoit M, Janssens A, Vennekens R, Viana F, Nemery B, Nilius B, Voets T (2009) Nicotine activates the chemosensory cation channel TRPA1. Nat Neurosci 12:1293–1299
53. Caceres AI, Brackmann M, Elia MD, Bessac BF, del Camino D, D'Amours M, Witek JS, Fanger CM, Chong JA, Hayward NJ, Homer RJ, Cohn L, Huang X, Moran MM, Jordt SE (2009) A sensory neuronal ion channel essential for airway inflammation and hyperreactivity in asthma. PNAS 106:9099–9104
54. McKemy DD, Neuhausser WM, Julius D (2002) Identification of a cold receptor reveals a general role for TRP channels in thermosensation. Nature 416:52–58
55. Peier AM, Moqrich A, Hergarden AC, Reeve AJ, Andersson DA, Story GM, Earley TJ, Dragoni I, McIntyre P, Bevan S, Patapoutian A (2002) A TRP Channel that Senses Cold Stimuli and Menthol. Cell 108:705–715
56. Karashima Y, Damann N, Prenen J, Talavera K, Segal A, Voets T, Nilius B (2007) Bimodal Action of Menthol on the Transient Receptor Potential Channel TRPA1. J Neurosci 27:9874–9884
57. Macpherson LJ, Hwang SW, Miyamoto T, Dubin AE, Patapoutian A, Story GM (2006) More than cool: Promiscuous relationships of menthol and other sensory compounds. Mol Cell Neurosci 32:335–343
58. Lumme A, Haahtela T, Ounap J, Rytila P, Obase Y, Helenius M, Remes V, Helenius I (2003) Airway inflammation, bronchial hyperresponsiveness and asthma in elite ice hockey players. Eur Respir J 22:113–117
59. Naito K, Komori M, Kondo Y, Takeuchi M, Iwata S (1997) The effect of l-menthol stimulation of the major palatine nerve on subjective and objective nasal patency. Auris Nasus Larynx 24:159–162
60. Eccles R (2003) Menthol: Effects on nasal sensation of airflow and the drive to breathe. Curr Allergy Asthma Rep 3:210–214
61. Ito S, Kume H, Shiraki A, Kondo M, Makino Y, Kamiya K, Hasegawa Y (2008) Inhibition by the cold receptor agonists menthol and icilin of airway smooth muscle contraction. Pulm Pharmacol Ther 21:812–817
62. Andersson DA, Nash M, Bevan S (2007) Modulation of the Cold-Activated Channel TRPM8 by Lysophospholipids and Polyunsaturated Fatty Acids. J Neurosci 27:3347–3355
63. Vanden Abeele F, Zholos A, Bidaux G, Shuba Y, Thebault S, Beck B, Flourakis M, Panchin Y, Skryma R, Prevarskaya N (2006) iPLA2-dependent gating of TRPM8 by lysophospholipids. J Biol Chem 281:40174–40182, M605779200
64. Chilton FH, Averill FJ, Hubbard WC, Fonteh AN, Triggiani M, Liu MC (1996) Antigen-induced generation of lyso-phospholipids in human airways. J Exp Med 183:2235–2245
65. Hu H, Tian J, Zhu Y, Wang C, Xiao R, Herz J, Wood J, Zhu M (2010) Activation of TRPA1 channels by fenamate nonsteroidal anti-inflammatory drugs. Pflugers Arch 459:579–592
66. White TA, Xue A, Chini EN, Thompson M, Sieck GC, Wylam ME (2006) Role of Transient Receptor Potential C3 in TNF-{alpha}-Enhanced Calcium Influx in Human Airway Myocytes. Am J Respir Cell Mol Biol 35:243–251
67. Corteling RL, Li S, Giddings J, Westwick J, Poll C, Hall IP (2004) Expression of Transient Receptor Potential C6 and Related Transient Receptor Potential Family Members in Human Airway Smooth Muscle and Lung Tissue. Am J Respir Cell Mol Biol 30:145–154
68. Yu Y, Sweeney M, Zhang S, Platoshyn O, Landsberg J, Rothman A, Yuan JXJ (2003) PDGF stimulates pulmonary vascular smooth muscle cell proliferation by upregulating TRPC6 expression. Am J Physiol Cell Physiol 284:C316–C330
69. Lin MJ, Leung GPH, Zhang WM, Yang XR, Yip KP, Tse CM, Sham JSK (2004) Chronic Hypoxia-Induced Upregulation of Store-Operated and Receptor-Operated Ca^{2+} Channels in Pulmonary Arterial Smooth Muscle Cells: A Novel Mechanism of Hypoxic Pulmonary Hypertension. Circ Res 95:496–505

70. Damann N, Owsianik G, Li S, Poll C, Nilius B (2009) The calcium-conducting ion channel transient receptor potential canonical 6 is involved in macrophage inflammatory protein-2-induced migration of mouse neutrophils. Acta Physiol 195:3–11
71. Finney-Hayward TK, Popa O, Bahra P, Li S, Poll CT, Gosling M, Nicholson AG, Russell REK, Kon OM, Jarai G, Westwick J, Barnes PJ, Donnelly LE (2010) Expression of TRPC6 channels in human lung macrophages. Am J Respir Cell Mol Biol 43(6): 296–304
72. Sel S, Rost BR, Yildirim AO, Sel B, Kalwa H, Fehrenbach H, Renz H, Gudermann T, Dietrich A (2008) Loss of classical transient receptor potential 6 channel reduces allergic airway response. Clin Exp Allergy 38(9):1548–1558
73. Hong CW, Kim TK, Ham HY, Nam JS, Kim YH, Zheng H, Pang B, Min TK, Jung JS, Lee SN, Cho HJ, Kim EJ, Hong IH, Kang TC, Lee J, Oh SB, Jung SJ, Kim SJ, Song DK (2010) Lysophosphatidylcholine Increases Neutrophil Bactericidal Activity by Enhancement of Azurophil Granule-Phagosome Fusion via Glycine{middle dot}GlyR{alpha}2/TRPM2/p38 MAPK Signaling. J Immunol 184:4401–4413
74. Hecquet CM, Ahmmed GU, Vogel SM, Malik AB (2007) Role of TRPM2 Channel in Mediating H_2O_2-Induced Ca^{2+} Entry and Endothelial Hyperpermeability. Circ Res 102(3):347–355
75. Yamamoto S, Shimizu S, Kiyonaka S, Takahashi N, Wajima T, Hara Y, Negoro T, Hiroi T, Kiuchi Y, Okada T, Kaneko S, Lange I, Fleig A, Penner R, Nishi M, Takeshima H, Mori Y (2008) TRPM2-mediated Ca^{2+} influx induces chemokine production in monocytes that aggravates inflammatory neutrophil infiltration. Nat Med 14:738–747
76. Ishii M, Shimizu S, Hara Y, Hagiwara T, Miyazaki A, Mori Y, Kiuchi Y (2006) Intracellular-produced hydroxyl radical mediates H_2O_2-induced Ca^{2+} influx and cell death in rat [beta]-cell line RIN-5F. Cell Calcium 39:487–494
77. Wagner TFJ, Loch S, Lambert S, Straub I, Mannebach S, Mathar I, Dufer M, Lis A, Flockerzi V, Philipp SE, Oberwinkler J (2008) Transient receptor potential M3 channels are ionotropic steroid receptors in pancreatic [beta] cells. Nat Cell Biol 10:1421–1430
78. Vennekens R, Olausson J, Meissner M, Bloch W, Mathar I, Philipp SE, Schmitz F, Weissgerber P, Nilius B, Flockerzi V, Freichel M (2007) Increased IgE-dependent mast cell activation and anaphylactic responses in mice lacking the calcium-activated nonselective cation channel TRPM4. Nat Immunol 8:312–320
79. Shimizu T, Owsianik G, Freichel M, Flockerzi V, Nilius B, Vennekens R (2009) TRPM4 regulates migration of mast cells in mice. Cell Calcium 45:226–232
80. Barbet G, Demion M, Moura IC, Serafini N, Leger T, Vrtovsnik F, Monteiro RC, Guinamard R, Kinet JP, Launay P (2008) The calcium-activated nonselective cation channel TRPM4 is essential for the migration but not the maturation of dendritic cells. Nat Immunol 9:1148–1156
81. Gerzanich V, Woo SK, Vennekens R, Tsymbalyuk O, Ivanova S, Ivanov A, Geng Z, Chen Z, Nilius B, Flockerzi V, Freichel M, Simard JM (2009) De novo expression of Trpm4 initiates secondary hemorrhage in spinal cord injury. Nat Med 15:185–191
82. Jin J, Desai BN, Navarro B, Donovan A, Andrews NC, Clapham DE (2008) Deletion of Trpm7 Disrupts Embryonic Development and Thymopoiesis Without Altering Mg^{2+} Homeostasis. Science 322:756–760
83. Wykes RCE, Lee M, Duffy SM, Yang W, Seward EP, Bradding P (2007) Functional Transient Receptor Potential Melastatin 7 Channels Are Critical for Human Mast Cell Survival. J Immunol 179:4045–4052
84. Du J, Xie J, Zhang Z, Tsujikawa H, Fusco D, Silverman D, Liang B, Yue L (2010) TRPM7-Mediated Ca^{2+} Signals Confer Fibrogenesis in Human Atrial Fibrillation. Circ Res 106:992–1003
85. Colburn RW, Lubin ML, Stone J, Wang Y, Lawrence D, D'Andrea MR, Brandt MR, Liu Y, Flores CM, Qin N (2007) Attenuated Cold Sensitivity in TRPM8 Null Mice. Neuron 54: 379–386

86. Dhaka A, Murray AN, Mathur J, Earley TJ, Petrus MJ, Patapoutian A (2007) TRPM8 Is Required for Cold Sensation in Mice. Neuron 54:371–378
87. Bautista DM, Siemens J, Glazer JM, Tsuruda PR, Basbaum AI, Stucky CL, Jordt SE, Julius D (2007) The menthol receptor TRPM8 is the principal detector of environmental cold. Nature 448:204–208
88. Agopyan N, Bhatti T, Yu S, Simon SA (2003) Vanilloid receptor activation by 2- and 10-[mu]m particles induces responses leading to apoptosis in human airway epithelial cells. Toxicol Appl Pharmacol 192:21–35
89. Agopyan N, Head J, Yu S, Simon SA (2004) TRPV1 receptors mediate particulate matter-induced apoptosis. Am J Physiol Lung Cell Mol Physiol 286:L563–L572
90. Caterina MJ, Rosen TA, Tominaga M, Brake AJ, Julius D (1999) A capsaicin-receptor homologue with a high threshold for noxious heat. Nature 398:436–441
91. Woodbury CJ, Zwick M, Wang S, Lawson JJ, Caterina MJ, Koltzenburg M, Albers KM, Koerber HR, Davis BM (2004) Nociceptors Lacking TRPV1 and TRPV2 Have Normal Heat Responses. J Neurosci 24:6410–6415
92. Neeper MP, Liu Y, Hutchinson TL, Wang Y, Flores CM, Qin N (2007) Activation Properties of Heterologously Expressed Mammalian TRPV2. J Biol Chem 282:15894–15902
93. Yamamoto Y, Taniguchi K (2005) Immunolocalization of VR1 and VRL1 in rat larynx. Auton Neurosci 117:62–65
94. Lewinter RD, Scherrer G, Basbaum AI (2008) Dense transient receptor potential cation channel, vanilloid family, type 2 (TRPV2) immunoreactivity defines a subset of motoneurons in the dorsal lateral nucleus of the spinal cord, the nucleus ambiguus and the trigeminal motor nucleus in rat. Neuroscience 151:164–173
95. Nagasawa M, Nakagawa Y, Tanaka S, Kojima I (2007) Chemotactic peptide fMetLeuPhe induces translocation of the TRPV2 channel in macrophages. J Cell Physiol 210(3):692–702
96. Link TM, Park U, Vonakis BM, Raben DM, Soloski MJ, Caterina MJ (2010) TRPV2 has a pivotal role in macrophage particle binding and phagocytosis. Nat Immunol 11:232–239
97. Xu H, Ramsey IS, Kotecha SA, Moran MM, Chong JA, Lawson D, Ge P, Lilly J, Silos-Santiago I, Xie Y, DiStefano PS, Curtis R, Clapham DE (2002) TRPV3 is a calcium-permeable temperature-sensitive cation channel. Nature 418:181–186
98. Smith GD, Gunthorpe MJ, Kelsell RE, Hayes PD, Reilly P, Facer P, Wright JE, Jerman JC, Walhin JP, Ooi L, Egerton J, Charles KJ, Smart D, Randall AD, Anand P, Davis JB (2002) TRPV3 is a temperature-sensitive vanilloid receptor-like protein. Nature 418:186–190
99. Chung MK, Lee H, Mizuno A, Suzuki M, Caterina MJ (2004) TRPV3 and TRPV4 Mediate Warmth-evoked Currents in Primary Mouse Keratinocytes. J Biol Chem 279:21569–21575
100. Huang SM, Lee H, Chung MK, Park U, Yu YY, Bradshaw HB, Coulombe PA, Walker JM, Caterina MJ (2008) Overexpressed Transient Receptor Potential Vanilloid 3 Ion Channels in Skin Keratinocytes Modulate Pain Sensitivity via Prostaglandin E2. J Neurosci 28:13727–13737
101. Xiao R, Tian J, Tang J, Zhu MX (2008) The TRPV3 mutation associated with the hairless phenotype in rodents is constitutively active. Cell Calcium 43:334–343
102. Imura K, Yoshioka T, Hirasawa T, Sakata T (2009) Role of TRPV3 in immune response to development of dermatitis. J Inflamm 6:17
103. Yoshioka T, Imura K, Asakawa M, Suzuki M, Oshima I, Hirasawa T, Sakata T, Horikawa T, Arimura A (2009) Impact of the Gly573Ser Substitution in TRPV3 on the Development of Allergic and Pruritic Dermatitis in Mice. J Invest Dermatol 129:714–722
104. Cheng X, Jin J, Hu L, Shen D, Dong Xp, Samie MA, Knoff J, Eisinger B, Liu Ml, Huang SM, Caterina MJ, Dempsey P, Michael LE, Dlugosz AA, Andrews NC, Clapham DE, Xu H (2010) TRP Channel Regulates EGFR Signaling in Hair Morphogenesis and Skin Barrier Formation. Cell 141:331–343
105. Bonini S, Lambiase A, Bonini S, Angelucci F, Magrini L, Manni L, Aloe L (1996) Circulating nerve growth factor levels are increased in humans with allergic diseases and asthma. Proc Natl Acad Sci USA 93:10955–10960

106. Liedtke W, Friedman JM (2003) Abnormal osmotic regulation in trpv4–/– mice. PNAS 100:13698–13703
107. Suzuki M, Mizuno A, Kodaira K, Imai M (2003) Impaired Pressure Sensation in Mice Lacking TRPV4. J Biol Chem 278:22664–22668
108. Birder L, Kullmann FA, Lee H, Barrick S, de Groat W, Kanai A, Caterina M (2007) Activation of Urothelial-TRPV4 by 4{alpha}PDD contributes to altered bladder reflexes in the rat. J Pharmacol Exp Ther 323(1):227–235
109. Thorneloe KS, Sulpizio AC, Lin Z, Figueroa DJ, Clouse AK, McCafferty GP, Chendrimada TP, Lashinger ESR, Gordon E, Evans L, Misajet BA, DeMarini DJ, Nation JH, Casillas LN, Marquis RW, Votta BJ, Sheardown SA, Xu X, Brooks DP, Laping NJ, Westfall TD (2008) N-((1S)-1-{[4-((2S)-2-{[(2,4-Dichlorophenyl)sulfonyl]amino}-3-hydroxypropanoyl)-1-piperazinyl]carbonyl}-3-methylbutyl)-1-benzothiophene-2-carboxamide (GSK1016790A), a Novel and Potent Transient Receptor Potential Vanilloid 4 Channel Agonist Induces Urinary Bladder Contraction and Hyperactivity: Part I. J Pharmacol Exp Ther 326:432–442
110. Gevaert T, Vriens J, Segal A, Everaerts W, Roskams T, Talavera K, Owsianik G, Liedtke W, Daelemans D, Dewachter I, Van Leuven F, Voets T, De Ridder D, Nilius B (2007) Deletion of the transient receptor potential cation channel TRPV4 impairs murine bladder voiding. J Clin Invest 117:3453–3462
111. Masuyama R, Vriens J, Voets T, Karashima Y, Owsianik G, Vennekens R, Lieben L, Torrekens S, Moermans K, Vanden Bosch A, Bouillon R, Nilius B, Carmeliet G (2008) TRPV4-Mediated Calcium Influx Regulates Terminal Differentiation of Osteoclasts. Cell Metab 8:257–265
112. Rock MJ, Prenen J, Funari VA, Funari TL, Merriman B, Nelson SF, Lachman RS, Wilcox WR, Reyno S, Quadrelli R, Vaglio A, Owsianik G, Janssens A, Voets T, Ikegawa S, Nagai T, Rimoin DL, Nilius B, Cohn DH (2008) Gain-of-function mutations in TRPV4 cause autosomal dominant brachyolmia. Nat Genet 40:999–1003
113. Deng HX, Klein CJ, Yan J, Shi Y, Wu Y, Fecto F, Yau HJ, Yang Y, Zhai H, Siddique N, Hedley-Whyte ET, DeLong R, Martina M, Dyck PJ, Siddique T (2009) Scapuloperoneal spinal muscular atrophy and CMT2C are allelic disorders caused by alterations in TRPV4. Nat Genet 42:165–169, advance online publication
114. Krakow D, Vriens J, Camacho N, Luong P, Deixler H, Funari TL, Bacino CA, Irons MB, Holm IA, Sadler L, Okenfuss EB, Janssens A, Voets T, Rimoin DL, Lachman RS, Nilius B, Cohn DH (2009) Mutations in the Gene Encoding the Calcium-Permeable Ion Channel TRPV4 Produce Spondylometaphyseal Dysplasia, Kozlowski Type and Metatropic Dysplasia. Am J Hum Genet 84:307–315
115. uer-Grumbach M, Olschewski A, Papic L, Kremer H, McEntagart ME, Uhrig S, Fischer C, Frohlich E, Balint Z, Tang B, Strohmaier H, Lochmuller H, Schlotter-Weigel B, Senderek J, Krebs A, Dick KJ, Petty R, Longman C, Anderson NE, Padberg GW, Schelhaas HJ, van Ravenswaaij-Arts CMA, Pieber TR, Crosby AH, Guelly C (2010) Alterations in the ankyrin domain of TRPV4 cause congenital distal SMA, scapuloperoneal SMA and HMSN2C. Nat Genet 42:160–164
116. Landoure G, Zdebik AA, Martinez TL, Burnett BG, Stanescu HC, Inada H, Shi Y, Taye AA, Kong L, Munns CH, Choo SS, Phelps CB, Paudel R, Houlden H, Ludlow CL, Caterina MJ, Gaudet R, Kleta R, Fischbeck KH, Sumner CJ (2010) Mutations in TRPV4 cause Charcot-Marie-Tooth disease type 2C. Nat Genet 42:170–174
117. Alvarez DF, King JA, Weber D, Addison E, Liedtke W, Townsley MI (2006) Transient Receptor Potential Vanilloid 4-Mediated Disruption of the Alveolar Septal Barrier. A Novel Mechanism of Acute Lung Injury. Circ Res 99:988–995
118. Jian MY, King JA, Al-Mehdi AB, Liedtke W, Townsley MI (2008) High Vascular Pressure-Induced Lung Injury Requires P450 Epoxygenase-Dependent Activation of TRPV4. Am J Respir Cell Mol Biol 38:386–392
119. Hamanaka K, Jian MY, Weber DS, Alvarez DF, Townsley MI, Al Mehdi AB, King JA, Liedtke W, Parker JC (2007) TRPV4 initiates the acute calcium-dependent permeability

increase during ventilator-induced lung injury in isolated mouse lungs. Am J Physiol Lung Cell Mol Physiol 293:L923–L932
120. Willette RN, Bao W, Nerurkar S, Yue Tl, Doe CP, Stankus G, Turner GH, Ju H, Thomas H, Fishman CE, Sulpizio A, Behm DJ, Hoffman S, Lin Z, Lozinskaya I, Casillas LN, Lin M, Trout REL, Votta BJ, Thorneloe K, Lashinger ESR, Figueroa DJ, Marquis R, Xu X (2008) Systemic Activation of the Transient Receptor Potential Vanilloid Subtype 4 Channel Causes Endothelial Failure and Circulatory Collapse: Part 2. J Pharmacol Exp Ther 326:443–452
121. Jia Y, Wang X, Varty L, Rizzo CA, Yang R, Correll CC, Phelps PT, Egan RW, Hey JA (2004) Functional TRPV4 channels are expressed in human airway smooth muscle cells. Am J Physiol Lung Cell Mol Physiol 287:L272–L278

Chapter 51
TRP Channels and Psychiatric Disorders

Loris A. Chahl

Abstract Depression and schizophrenia are major psychiatric disorders that cause much human suffering. Current treatments have major limitations and new drug targets are eagerly sought. Study of transient receptor potential (TRP) channels in these disorders is at an early stage and the potential of agents that activate or inhibit these channels remains speculative. The findings that TRPC6 channels promote dendritic growth and are selectively activated by hyperforin, the key constituent of St John's wort, suggest that TRPC6 channels might prove to be a new target for antidepressant drug development. There is now considerable evidence that TRPV1 antagonists have anxiolytic activity but there is no direct evidence that they have antidepressant activity. There is also no direct evidence that TRP channels play a role in schizophrenia. However, the findings that TRPC channels are involved in neuronal development and fundamental synaptic mechanisms, and that TRPV1 channels play a role in central dopaminergic and cannabinoid mechanisms is suggestive of potential roles of these channels in schizophrenia. Investigation of TRP channels in psychiatric disorders holds the promise of yielding further understanding of the aetiology of psychiatric disorders and the development of new drug treatments.

51.1 Introduction

Depression and schizophrenia are psychiatric disorders that cause great human suffering and make a major contribution to the burden of disease worldwide. Thus any new drug target warrants investigation for its potential to provide improved treatment options for these disorders. Transient receptor potential (TRP) channels mediate the flux of cations across cell membranes and play an important role in

L.A. Chahl (✉)
School of Biomedical Science and Pharmacy, University of Newcastle, Newcastle,
NSW 2308, Australia; Schizophrenia Research Institute, Sydney, NSW, Australia
e-mail: loris.chahl@newcastle.edu.au

cell sensing and signaling. There are six subfamilies of TRP channels, TRPA, TRPC, TRPM, TRPML, TRPP and TRPV [1, 2]. The structure of TRP channels and their structure-function relationships have been extensively reviewed elsewhere [3]. Although understanding of the roles of transient receptor potential (TRP) channels in brain function is currently limited, there is increasing evidence that they have actions that might be of consequence to psychiatric disorders. This chapter reviews the actions of TRP channels that have possible relevance to psychiatric disorders.

51.2 TRP Channels in the Brain

Several TRP channels are present in the central nervous system (CNS) and pharmacological studies have shown that they have actions in the CNS which may be of relevance to psychiatric disorders. The TRP channels that have been most extensively investigated for actions in the CNS are the TRPC and TRPV channels.

51.2.1 TRPC Channels

The canonical TRP (TRPC) channels have been classified into three groups based on sequence and functional similarities, TRPC1/4/5, TRPC3/6/7 and TRPC2 [4]. The human TRPC2 gene encodes a non-functional truncated protein [5]. All TRPC channels are highly expressed in human brain each with a discrete pattern of distribution [6]. The human TRPC1, TRPC3 and TRPC5 mRNAs are widely expressed across most brain regions, with TRPC5 exhibiting the most CNS-specific expression having ten-fold higher levels in the CNS than in the periphery [6]. Within neurons, specific cellular localization has also been reported. In dopaminergic neurons of the rat substantia nigra TRPC1 was found to be localized to dendritic processes of dopaminergic neurons [7], TRPC6 was localized to proximal dendrites [8], and TRPC5 to neuronal nuclei [9]. TRPC3 was mainly found in oligodendrocytes [10]. During development mouse brain exhibited high densities of TRPC4 mRNA in the septal, cerebellar and hippocampal areas, whereas in the adult brain high densities of TRPC4 occurred in the olfactory bulb, the CA1 to CA3 areas of the hippocampus and in the dentate gyrus [11].

The functional roles of TRPC channels in the developing brain have been reviewed recently [12]. TRPC1 channels promote basic fibroblast growth factor (bFGF)-induced stem cell proliferation, and are required for proper growth cone turning responses to gradients of Netrin-1. TRPC3 and TRPC6 promote cerebellum granule cell survival in response to brain-derived neurotrophic factor (BDNF), mediate BDNF-induced chemoattraction involved in axon guidance, and BDNF-induced spine formation. TRPC5 and TRPC6 have opposing effects to shape neuronal morphogenesis, TRPC5 inhibiting neurite outgrowth and TRPC6 promoting dendritic growth. TRPC6 is highly expressed in the rat brain during the period of maximal dendritic growth and promotes dendritic growth through a Ca^{2+}/calmodulin-dependent kinase IV (CaMKIV) and cyclic AMP-response-element binding protein (CREB) and has been proposed to play a critical role in early brain development when neuronal activity is low [13]. TRPC6 has also been shown to be required for excitatory synaptogenesis [12].

Interestingly, TRPC1, TRPC4 and TRPC6 knock-out mice have apparently normal brain development [12, 14]. However, TRPC3 knock-out mice have defective motor coordination [15]. Recently, TRPC5 knock-out mice have been shown to have diminished fear levels in response to innately aversive stimuli. TRPC5 channels, activated via G protein-coupled neuronal receptors in the amygdala, were found to play an essential role in fear-related behaviour [16]. It is possible that redundancy in TRPC channel mechanisms might require knock-out of several TRPC channels in order to fully elucidate their importance in brain development.

51.2.2 TRPV Channels

The TRPV subfamily may be divided into two groups, TRPV1-4 and TRPV5-6. TRPV1-4 channels are expressed in the nervous system, the highest levels being found in primary afferent neurons. They share the property of thermosensitivity with distinct thresholds ranging from moderate to noxious heat [17], and thus are known as "thermo-TRPs" [18–20]. Other thermo-TRP channels include TRPM8 and TRPA1 which are cold activated [21, 22].

Much research has focused on the role of thermo-TRPs in sensory transduction, but there has been increasing interest in the roles of several TRP channels in the CNS. The distribution and possible roles of TRPV channels in the brain has recently been reviewed [23]. The TRPV channel that has been most extensively studied in the CNS is the TRPV1 channel, previously known as the capsaicin receptor or vanilloid receptor 1 (VR1). The TRPV1 channel was first identified in sensory neurons [18, 24] but is now known to be widely distributed in both the central and peripheral nervous systems. The TRPV1 channel shares with other TRP channels the characteristic of polymodal activation and is activated by exogenous plant products such as capsaicin, endogenous substances including diacylglycerol (DAG), phosphoinositides, eicosanoids and anandamide, purine nucleotides, inorganic ions (Ca^{2+} and Mg^{2+}), as well as pH change and physical stimuli such as heat and mechanical stimuli [24–26].

Studies in rat and mouse brain using immunohistochemistry, autoradiography using the selective TRPV1 channel ligand [^{3}H]resiniferatoxin (RTX), and ribonuclease protection assays, have demonstrated that TRPV1 is widely expressed throughout the brain, including the cortex, the CA1-CA3 subfields of Ammon's horn and the dentate gyrus of the hippocampus, basal ganglia, cerebellum and olfactory bulb as well as in the mesencephalon and hindbrain [27–29], albeit at lower levels than in dorsal root ganglia [30]. Confirmation that the central [^{3}H]RTX binding sites were TRPV1 channels was obtained in studies on wild type and TRPV-knock out (KO) mice [31]. In the cortex and hippocampus high expression of TRPV1 was found in cell bodies and post-synaptic dendritic spines of neurons, and also on astrocytes and pericytes [32]. There is less known about the distribution of TRPV1 in human brain. However, TRPV1 receptors have been found in the third and fifth layers of the human parietal cortex [27].

Studies in rat brain have found extensive co-localization of TRPV1 with TRPV2 [33] whereas in the basal ganglia TRPV1 immunoreactivity was found in dopaminergic neurons as shown by its co-localization with tyrosine hydroxylase [27]. Extensive co-localization of TRPV1 with cannabinoid CB_1 receptors has also been found in several regions of mouse brain, including the pyramidal neurons of the hippocampus, basal ganglia, Purkinje cells and ventral periaqueductal (PAG) grey neurons [29]. This extensive co-localization supports the pharmacological studies showing a functional relationship between TRPV1 channels and CB_1 receptors. In human brain TRPV1 channels have been shown to be co-localized with CB_1 and CB_2 receptors on cerebromicrovascular endothelial cells [34].

The discovery of TRPV1 channels in the brain led to a search for endogenous ligands or endovanilloids. A surprising finding was that the endocannabinoid, anandamide, which is a CB_1 receptor agonist [35], had affinity for TRPV1 channels [36, 37]. There is now substantial evidence that anandamide functions as an endovanilloid [see 38]. Other endovanilloids that have since been identified include N-acyldopamines and some lipoxygenase metabolites of arachidonic acid, the hydroperoxyeicosatetraenoic acids (HPETEs) [23, 38].

Studies in TRPV1-KO mice have shown that TRPV1 channels play important roles in several pathophysiological processes involving sensory mechanisms including thermal hyperalgesia [39, 40], joint inflammation [41], nociception due to bone metastases [42], cough, airways disease and urinary incontinence [43]. Several TRPV1 antagonists have been developed and their clinical efficacy for pain and a number of other conditions is currently being evaluated [44]. The increasing evidence that TRPV1 channels have functional roles in CNS processes such as synaptic transmission and synaptic plasticity [23, 45], has raised the possibility that the central actions of these drugs might be found to be useful therapeutically.

51.3 Depression and Anxiety

Depression is a disorder of affect and is the most common psychiatric condition. The major symptoms include depressed mood, anhedonia, difficulty in concentrating, and abnormalities in appetite and sleep. Bipolar disorder is a disorder in which depression is interspersed with episodes of mania. The aetiology of depression is poorly understood, but both genetic factors and stressful life events appear to be involved. The observation that agents which enhanced central noradrenergic or serotonergic transmission had antidepressant action gave rise to the classical monoamine hypothesis which proposed that depression results from decreased central monoamine function. At present the first line therapy for depression is treatment with drugs such as selective serotonin reuptake inhibitors that enhance monoaminergic transmission. These drugs have the limitation of a long therapeutic delay whereby the antidepressant effect takes weeks to develop despite an immediate effect on monoamine levels, indicating that neuroplastic changes in neural circuits are involved in depression [46]. Recognition that processes in addition to monoaminergic transmission may be involved in depression has lead to active investigation of alternative treatment options. This search has resulted in several potential new targets for drug treatment including growth factors such as brain-derived neurotrophic factor (BDNF) [47], and receptors for glucocorticoid and corticotrophin-releasing factor, for neuropeptides such as substance P, and for glutamate, in particular *N*-methyl-D-aspartate (NMDA) receptors [46].

Although understanding of the circuitry involved in depression is incomplete, several brain regions appear to be involved. Studies on post-mortem brain as well as investigations using neuroimaging techniques have revealed structural changes including reduced grey matter volume in the prefrontal cortex and hippocampus of depressed patients [48–50]. Functional studies such as functional magnetic resonance imaging (fMRI) and positron emission tomography (PET) have implicated the amygdala and subgenual cingulate cortex in dysphoria [48, 51]. Furthermore, deep brain stimulation of the subgenual cingulate cortex [52], or the nucleus accumbens [53], a region known to be involved in reward, has been reported to be effective in patients with refractory depression.

Depression and anxiety commonly co-occur and anxiety is present in a majority of patients with depression. Nevertheless, anxiety is not considered a core symptom of depression [54]. Although there is overlap in efficacy of antidepressant drugs and anti-anxiety drugs, antidepressant drugs do not all have the same efficacy in the treatment of anxiety. Furthermore, the preclinical tests in animals for potential anti-anxiety and antidepressant drugs also differ. Tests for anti-anxiety agents include tests of conditioned and unconditioned fear, whereas common animal tests for antidepressant activity include social defeat and learned helplessness, and responses to inescapable stress and to rewarding stimuli [46].

51.3.1 TRPC Channels, Depression and Anxiety

The proposal that neuroplastic changes in neural circuits are involved in depression raises the possibility that TRPC channels might play a role in the disorder. Of considerable interest has been the demonstration that hyperforin, the key constituent of St. John's wort that is responsible for its antidepressant action, produced a highly selective activation of TRPC6 channels [55]. Indeed, hyperforin is the first selective activator of a TRPC channel protein to be discovered. Synthetic antidepressants inhibit the neuronal uptake of serotonin and noradrenaline directly by inhibiting the transporter proteins. However, hyperforin was found to reduce monoamine uptake indirectly by elevating the intracellular sodium concentration, thereby decreasing the sodium gradient as the driving force of the monoamine transporters. Since depression is associated with loss of grey matter volume (see above), and activation of TRPC6 channels promotes dendritic growth [13], TRPC6 channels might prove to be a valuable new target for development of antidepressant drugs.

51.3.2 TRPV Channels, Depression and Anxiety

The finding that systemic administration of the TRPV1 antagonist, capsazepine, induced an anxiolytic effect in rats and the TRPV1 agonist, olvanil, produced anxiogenic effects [56], suggested that TRPV1 channel activation might be anxiogenic. This proposal was strengthened by studies using TRPV1-knock out (KO) mice. These animals showed no difference from wild type mice in general activity but exhibited reduced unconditioned fear responses in the light-dark test and elevated plus maze, and reduced conditioned fear responses following auditory fear conditioning [57].

Studies on possible sites of action of the TRPV1 antagonists as anxiolytic agents have implicated the ventral hippocampus. Rats infused with capsazepine into the ventral hippocampus exhibited less inhibitory avoidance in the elevated plus maze test than controls [58]. Another region that has been implicated is the periaqueductal grey (PAG), a midbrain region that has been extensively studied for its role in responses to noxious and aversive stimuli [59]. The suggestion that the dorsolateral PAG might be a site of the anxiolytic action of TRPV1 antagonists arose from the observations that local injection of anandamide into the dorsolateral PAG induced an anxiolytic effect mediated by CB_1 receptors [60], and that several neurons in the ventrolateral PAG co-expressed cannabinoid CB_1 receptors and TRPV1 channels [61]. Local injection of the TRPV1 antagonist, capsazepine, into the dorsolateral PAG of rats was subsequently shown to produce an anxiolytic effect in the elevated plus-maze and Vogel conflict tests [62]. Interestingly, the classic TRPV1 agonist, capsaicin, was also anxiolytic, which has been suggested to be due to rapid desensitization of the channels, a well-known effect of this agent on peripheral sensory TRPV1 channels [63, 64]. Recently it has been shown that N-arachidonoyl-serotonin, a drug with a dual action to block TRPV1 channels and the endocannabinoid-inactivating enzyme, fatty acid amide hydrolase (FAAH), thereby indirectly activating CB_1 receptors, was a more effective anxiolytic than selective TRPV1 blockers [65].

Stress is considered to play a major role in the aetiology of anxiety and depression. Animal studies have shown that exposure of rats to chronic, mild, unpredictable stressors for several weeks induces behavioural changes that resemble the core symptoms of depression in humans, including reduced sensitivity to rewarding stimuli, decreased sexual activity, reduced grooming, sleep and neuroendocrine disturbances [66]. This animal model is considered to be a valid model of depression since these effects were reversed by antidepressant drugs but not by antipsychotic drugs [66, 67]. There is considerable evidence that stress induces alterations in synaptic plasticity in the hippocampus that produce changes in learning and memory [68]. Synaptic plasticity has been most extensively studied by investigation of the responses of pyramidal cells in the CA1 region of the rodent hippocampus to Schaffer collateral-commissural pathway stimulation. Long-lasting enhanced synaptic responses induced by bursts of rapidly repeated stimuli, known as long-term potentiation (LTP), and reduced synaptic responses induced by low frequency stimulation, known as long-term depression (LTD), are well-characterized forms of synaptic plasticity. Hippocampal LTD has been reported to mediate the impairment of spatial memory retrieval induced by acute stress [69]. Support for the proposal that LTD might be involved in the pathophysiology of depression [70] has been obtained in studies on rats subjected to chronic mild stress. In these animals LTD was facilitated but LTP was unaffected [67]. Chronic treatment with the antidepressant, fluvoxamine, during the stress protocol, prevented the facilitation of LTD and enhanced LTP [67]. Other candidate antidepressant drugs might therefore be expected to reduce LTD and facilitate LTP.

The observation that TRPV1-KO mice showed less contextual fear memory than wild type mice indicating that TRPV1 channels played a role in the synaptic plasticity mechanisms involved in learning and memory [57], prompted an investigation of synaptic plasticity in CA1 neurons in brain slices of the dorsal hippocampus of TRPV1-KO mice. It was found that LTP was impaired in these mice [57]. In rats it was shown that the TRPV1 agonists, capsaicin and resiniferatoxin, facilitated LTP but suppressed LTD [71] by a TRPV1 channel mediated mechanism since the selective TRPV1 antagonists, capsazepine and SB366791 inhibited these actions [72]. In the same study it was shown that acute stress induced by placement of rats on an elevated platform, suppressed LTP, enhanced LTD, and impaired spatial memory, effects that were inhibited by capsaicin [71]. If LTD, depression and anxiety were correlated, TRPV1 agonists would be expected to be anxiolytic and antidepressant. However, the behavioural studies discussed above have shown that TRPV1 antagonists are anxiolytic. The difficulties in attempting to correlate changes in synaptic plasticity in brain slices to behavioural changes in animals and disorders of affect in humans, have been discussed [73].

TRPV1 channels have been shown to be involved in synaptic plasticity at other synapses. A study by Gibson et al. [74] reported that TRPV1 channel activation triggers interneuron LTD at synapses on hippocampal interneurons in rats. Interestingly, interneuron LTD was absent in TRPV1-KO mice and TRPV1 agonists were unable to elicit LTD in these mice, indicating that TRPV1 channels were necessary for this form of interneuron synaptic plasticity. The implications of these findings for the future of TRPV1-targeted therapeutics has been discussed [75]. TRPV1 channels have also been shown to control synaptic plasticity in the developing superior colliculus, a model of synaptic refinement. Antagonism of TRPV1 prevented the induction of LTD but not its reversal if already established [76]. These findings contribute to the emerging concept that brain TRPV1 channels are functionally active and contribute to long-term regulation of synaptic strength during brain development [76].

There is considerable evidence that TRPV1 antagonists are effective anxiolytic agents but there is no direct experimental evidence that TRPV1 channels play a role in depression. However, the overlap in efficacy of antidepressant drugs and anti-anxiety drugs suggest that investigation of their possible antidepressant action is warranted. There is still much to be learned about the role of TRP channels in affective disorders. Nevertheless, there is now sufficient evidence to indicate that TRPV1 channels should be considered as putative therapeutic targets for the treatment of affective disorders [45].

Less is known about other TRPV channels in brain function. TRPV2 channels have a restricted distribution within the hypothalamic paraventricular, suprachiasmatic and supraoptic nuclei of the primate brain [77], where they are preferentially localized to oxytocinergic and vasopressinergic neurons. This distribution is suggestive of a role for TRPV2 channels in disorders of the hypothalamic-pituitary-adrenal axis, including anxiety and depression. Interestingly, incensole acetate, a component of incense, is a potent TRPV3 channel agonist and produces anxiolytic and antidepressant like behavioural effects in wild type mice but not in TRPV3-KO animals [78]. These findings suggest that the role of TRPV2 and TRPV3 channels in affective states is also worthy of further investigation.

51.4 Schizophrenia

Schizophrenia is a severe, debilitating psychiatric disorder that carries a life-time risk of the order of 1%. Typically the overt signs and symptoms of schizophrenia manifest in early adulthood. The complexity of the disorder is reflected in the varying symptomatology. The disorder may be characterized by the classical positive symptoms of hallucinations and thought disorder or by negative symptoms of social withdrawal, poverty of speech and anhedonia. However, it is now recognized that schizophrenia is primarily a disorder of cognition with impaired cognition, particularly in visual memory and working memory, being a core symptom of the disorder [79, 80].

Although the cause of schizophrenia remains unknown, it is generally accepted that it is a neurodevelopmental disorder and that both genetic and environmental factors are involved. Several genes have been implicated as susceptibility genes for schizophrenia, including neuregulin 1 (NRG1), catechol-O-methyltransferase (COMT), dysbindin, disrupted in schizophrenia 1 (DISC1), and regulator of G-protein signaling (RGS) protein-4 (RGS-4), with the strongest evidence being for NRG1 [81–86]. The association of NRG1 with schizophrenia first reported by Stefansson et al. [87], has been confirmed in several studies including two meta-analyses [88–90]. Although the conserved regions of the NRG1 gene and the gene of its receptor, erbB4, are unchanged, altered expression levels of specific isoforms of the genes have been found in schizophrenia [91–94].

Structural changes have been found in studies on post-mortem human brain from subjects with schizophrenia. The brains of subjects with schizophrenia are reduced in volume [95–97], have larger ventricles and thinner cortices, particularly in the prefrontal and temporal regions, compared with those of healthy individuals [98, 99]. Selemon et al. [100, 101] observed that neuronal density was increased in the prefrontal cortex of subjects with schizophrenia. They proposed that the symptoms of schizophrenia resulted from reduced cortical connectivity rather than a reduction in neuron numbers, a proposal that is now known as the "reduced neuropil hypothesis" [102]. Other neuropathological findings of reduced interneuronal space [103], mean cell spacing abnormalities [104], and reduced neuronal size [105] support this proposal. Recently Pajonk et al [106] found hippocampal plasticity in response to aerobic exercise in normal subjects as well as subjects with schizophrenia suggesting that some of the structural brain changes in schizophrenia may be partially reversible.

The observation that antipsychotic drugs shared the common property of dopamine D_2 receptor antagonism [107] led to the "dopamine hypothesis" of schizophrenia which dominated the field for many years. In particular, the dopaminergic mesolimbic pathway that projects from the ventral tegmental area (VTA) to the nucleus accumbens and ventral striatum, and the mesocortical pathway that projects from the VTA to the cortex were considered to be involved in the symptomatology of schizophrenia [108]. Although the classical dopamine D_2 receptor antagonists, are effective in treating the positive symptoms of schizophrenia, they produce serious Parkinsonian–like side effects, presumably due to action on the nigrostriatal pathway which is involved in movement control, and they lack efficacy in treating the negative symptoms and cognitive impairment of schizophrenia. The newer atypical antipsychotic drugs produce fewer motor side effects than the older drugs and have some efficacy in treating the negative symptoms, but none of the current drugs is effective in treating the core symptom of cognitive impairment.

Lack of understanding of the aetiology of schizophrenia and the limitations of current therapies have led to an ongoing search for new drug targets. A number of neurotransmitter systems have been found to be affected in schizophrenia, including D_1 receptors, the serotonin, gamma-aminobutyric acid (GABA), glutamate and cholinergic systems [109]. In recent years the "hypoglutamatergic hypothesis" of schizophrenia has received considerable support [110–112]. The hypothesis arose from the observation that the NMDA antagonists, phencyclidine (PCP) and ketamine, induced psychotic symptoms and cognitive dysfunction in normal humans. Mutant mice with reduced levels of the NMDA receptor subunit, NR1, exhibited hyperactivity and impaired social behaviours that were reduced by antipsychotic drugs [113]. Interestingly, NMDA receptor hypofunction has been reported to produce a hyperdopaminergic state in humans [114] which might explain the role of dopamine. A link between NMDA receptor hypofunction and enhanced NRG1-erbB4 signaling was also observed in post-mortem prefrontal cortex of subjects with schizophrenia [115].

51.4.1 TRPC Channels and Schizophrenia

A diversity of TRPC heteromers has been found in mammalian brain. Of interest to schizophrenia, which is considered to be a neurodevelopmental disorder, is the finding that several novel TRPC heteromers are present in developing brain.

These novel TRPC heteromers have been proposed to play specific roles in developing brain, particularly as voltage-dependent Ca^{2+} entry channels emerge later than TRPC channels during development [116]. The possibility that formation of particular TRPC heteromers in developing human brain might predispose to schizophrenia warrants investigation.

TRPC channels play key roles in neurite extension and growth cone guidance [117]. Following growth factor receptor stimulation [118], homomeric TRPC5 channels are rapidly delivered to the plasma membrane and have been shown to control neurite length and growth cone morphology of cultured mouse hippocampal neurons by regulating Ca^{2+} influx [117]. The mechanisms governing the inhibitory role of TRPC5 in neuronal outgrowth have been proposed to involve a protein complex between neuronal calcium sensor-1 (NCS-1) protein and TRPC5 [119]. In addition Ca^{2+} influx through TRPC3 channels has been shown to have a selective effect to control growth cone guidance [120]. The reduced neuropil found in postmortem human brain of subjects with schizophrenia has been proposed to be due to reduced neuronal dendritic arborization [102]. Therefore, aberrant TRPC3 and TRPC5 channel function might play a role in the pathogenesis of schizophrenia.

Studies in post-mortem human brain have shown a reduction in transcripts encoding synaptic vesicle proteins and synaptic plasma membrane proteins involved in the presynaptic release of neurotransmitters [121–123]. There is increasing evidence that certain TRP channels play critical roles in fundamental synaptic mechanisms. A critical functional role of agonist-activated TRPC4 channels in the release of GABA from dendrites has been proposed [124]. A critical functional role of TRPM7 channels in neurotransmitter release in sympathetic neurons has also been proposed [125]. The discovery that TRP channels play critical functional roles in fundamental mechanisms of neurotransmitter release suggests their possible involvement in the pathogenesis of schizophrenia.

51.4.2 TRPV Channels and Schizophrenia

Of particular relevance to schizophrenia was the observation that many dopaminergic cells in the mesencephalon are TRPV1-immunopositive [27, 126]. Activation of TRPV1 channels in rat midbrain VTA slices by capsaicin increased the rate of firing of dopamine neurons in a concentration dependent manner suggesting an action in reward mechanisms and possibly schizophrenia. The response was inhibited by the selective TRPV1 receptor antagonist, iodoresiniferatoxin and also by ionotropic glutamate antagonists indicating that a glutamatergic mechanism was involved [127]. In vivo experiments in rats showed that noxious tail stimulation and microinjection of capsaicin into the VTA transiently increased dopamine release in the nucleus accumbens, a response that was also blocked by iodoresiniferatoxin [127]. TRPV1 channel activation has been found to facilitate glutamate transmission in several other brain regions including the striatum [128], paraventricular nucleus [129], brainstem descending antinociceptive pathways [130], dorsolateral PAG [131] and spinal cord [132].

Capsaicin administered systemically to rats suppressed spontaneous locomotion, an effect that was inhibited by the specific TRPV1 antagonist, capsazepine [133]. This observation suggested that TRPV1 channels might be involved in the control of movement. It had previously been reported that the inhibitory effect of anandamide on motor behaviour was inhibited by capsazepine and presumably mediated by TRPV1 channels [134]. On the other hand the endogenous endovanilloid/endocannabinoid, N-arachidonoyl-dopamine (NADA), had a "tuning" effect on dopaminergic neurons of the substantia nigra since it exerted opposing actions, increasing glutamatergic transmission onto dopamine neurons by activating presynaptic TRPV1 channels or reducing transmission by activating presynaptic CB_1 receptors [126]. In the presence of a selective blocker of the putative endocannabinoid membrane transporter, the balance between excitatory and inhibitory effects was altered in favour of CB_1 mediated inhibition, confirming that NADA, like capsaicin and anandamide, must be taken up by cells to interact with the binding site on TRPV1 channels which is in the cytosolic domain [135].

A model of the hyperdopaminergic state is produced in mice by knockout of the dopamine transporter (DAT). These animals exhibit hyperactivity and have markedly reduced anandamide levels in the striatum [136]. The hyperactivity was reduced by administration of indirect endocannabinoids acting via TRPV1 channels and not on CB_1 receptors [136]. Anandamide has also been found to produce a disruption of the operant attention task in rats by a vanilloid-dependent mechanism which might involve modulation of dopaminergic neurotransmission [137]. Thus in the CNS, as in the peripheral sensory system [138], TRPV1 channels apparently function as ionotropic cannabinoid receptors.

Activation of TRPV1 channels and CB_1 receptors has been shown to mediate cell death of mesencephalic dopaminergic neurons in vivo and in vitro [139]. The neurotoxicity was mediated by the lipoxygenase product, 12(S)-hydroperoxyeicosatetraenoic acid (12(S)-HPETE) and involved a functional interaction between TRPV1 channels and CB_1 receptors in which activation of CB_1 receptors in rat mesencephalic neurons was associated with biosynthesis of 12(S)-HPETE, which stimulated TRPV1 activity, calcium entry, mitochondrial damage and neuronal death [140]. Neuroprotective effects of capsaicin and the CB_1 antagonist, rimonabant, have been described in a model of transient global ischemia in gerbils, effects which have been proposed to be mediated by TRPV1 desensitization (capsaicin) or antagonism (rimonabant) [141, 142]. Whether TRPV1 channels and CB_1 receptors mediate neurotoxic or neuroprotective effects remains controversial [143]. Nevertheless, the relative roles of the neurotoxic and neuroprotective actions of TRPV1 channel activation are clearly of major importance for future drug development. The significance of the neurotoxic or neuroprotective effects of TRPV1 channel activation in psychiatric and neurological disorders remains to be explored.

An association between adolescent cannabis use and first episode psychosis in susceptible individuals is now accepted [144–147], although the mechanisms involved are unclear. Complex interactions between TRPV1, cannabinoid and dopaminergic mechanisms may be involved. Cross-talk between CB_1 and D_2 receptors by heterodimer formation has been identified [148] and cooperativity between

TRPV1 channels and CB_1 receptor mechanisms has been observed, for example, in mediating mesencephalic dopaminergic neuronal cell death described above. Further understanding of the functional interrelationships between TRPV1 channels and CB_1 receptors in different brain regions, and of the relative efficacy of exogenous cannabinoid agonists on TRPV1 channels and CB_1 receptors, may shed light on the role of cannabis use in precipitation of psychosis. Genetic studies have suggested an association between microsatellite and single nucleotide polymorphism (SNP) genetic markers in the *CNR1* gene that encodes the CB_1 receptor, and the incidence of schizophrenia [149]. However polymorphisms that influence the full-length *CNR1* transcript expression have not been reported. In a Japanese population SNPs in the *CNR2* gene that encodes the CB_2 receptor have recently been found that reduce the function of the CB_2 receptor, and it has been suggested that this increases susceptibility to schizophrenia [150]. Replication studies are necessary to confirm whether low CB_2 receptor function plays a role in susceptibility to schizophrenia.

Abnormal daily body temperature ranges and an impaired ability to compensate for temperature stress have been reported in subjects with schizophrenia [151–153]. These effects are apparently not the result of treatment with antipsychotic drugs as they have been observed in drug-free subjects [154]. Thermoregulation involves both central and peripheral mechanisms. In the CNS a complex regulatory thermostat mechanism involving dopaminergic and serotonergic mechanisms has been described [153]. The preoptic area of the anterior hypothalamus and midbrain dopaminergic mechanisms have been shown to play important roles in thermoregulation [155], with dopamine D_2 receptors being mainly involved in the maintenance of body temperature in euthermia [156].

A striking similarity has been found between the thermoregulatory deficit in animals following capsaicin desensitization and that observed in schizophrenia, suggesting that TRPV1 channel function might be abnormal in schizophrenia. Injection of capsaicin into the preoptic area caused hypothermia and enhanced frequency of spontaneous glutamatergic excitatory postsynaptic currents and GABAergic inhibitory postsynaptic currents in medial preoptic neurons [157]. Desensitization of TRPV1 channels by repeated capsaicin administration induced impaired ability to thermoregulate against heat [158]. Capsaicin induced its central effects on thermoregulation via action on TRPV1 channels [40] since injection of capsaicin into the anterior hypothalamus of TRPV1-KO mice did not produce a change in body temperature. The effect of TRPV1 channel ligands on thermoregulation is an important consideration in the development of drugs that target TRPV1 channels, and is a possible limitation to their use for disorders where brain penetration is required.

Subjects with schizophrenia and their relatives have been reported to be less sensitive to pain than normal subjects [159–162]. The hypoalgesia was independent of antipsychotic drug treatment [163]. Pain is a sensory and emotional experience. Nociception is the physiological component of the pain response and is transduced by primary afferent neurons of the peripheral nervous system. The perception of a nociceptive stimulus as pain and the emotional responses it evokes involve higher CNS mechanisms. The primary afferent neurons of the somatosensory system involved in nociception, the majority of which are unmyelinated [164], contain neuropeptides such as the tachykinin substance P, and express TRPV1 channels [63, 165]. These neurons are glutamatergic and contact spinal neurons that co-express tachykinin NK_1 receptors and ionotropic (NMDA or alpha-amino-3-hydroxy-5-methyl-4-isoxazolepropionic acid (AMPA)) glutamate receptors [166]. Activation of TRPV1 channels by agonists such as capsaicin results in nociception by activation of spinal neurons, and neurogenic flare and inflammation by release from their peripheral terminals of neuropeptides, including substance P and calcitonin gene related peptide (CGRP) [63, 165]. Studies in TRPV1-KO mice have shown that TRPV1 channels are not essential for normal nociceptive heat responses but they play an essential role in thermal hyperalgesia and neurogenic inflammation [167, 168]. Thus the TRPV1 channel is now an important target for the development of novel anti-nociceptive and anti-inflammatory drugs.

Not only have subjects with schizophrenia been reported to be less sensitive to pain but they and their relatives have also been reported to have reduced flare responses to niacin (nicotinic acid) and methylnicotinate [169, 170]. These observations considered together raised the possibility that capsaicin-sensitive primary afferent neurons might be involved in the pathogenesis of schizophrenia. The question arose as to whether the structural and functional changes in the CNS that give rise to schizophrenia could result from a deficit in input via a subset of primary afferent neurons throughout development. Although such a possibility at first seems remote, other studies in developmental neurobiology have shown that neonatal somatosensory deprivation such as that induced by whisker trimming in the mouse whisker barrel model, results in reduced synaptic density in the barrel cortex [171]. Thus it is possible that input deficit in other somatosensory systems could result in reduced synaptic density and "reduced neuropil" such as that seen in the cortex of subjects with schizophrenia [102]. However, the role of generalized somatosensory deficits on brain development is a relatively unexplored area.

It has been known for many years that treatment of neonatal rats with capsaicin produces life-long loss of a high proportion of TRPV1-expressing primary afferent neurons [172]. Newson et al. [173] used this property of capsaicin to test the proposal that somatosensory deficit induces brain changes similar to those found in schizophrenia. At 5–7 weeks the rats treated as neonates with capsaicin had increased locomotor activity in a novel environment and the male rats had reduced brain weight. The capsaicin treated rats also had reduced hippocampal and coronal cross-sectional area, reduced cortical thickness and increased neuronal density in several cortical areas [173]. These brain changes in neonatal capsaicin treated rats were reminiscent of those found in post-mortem human brain of subjects with schizophrenia. Furthermore, the effects of neonatal capsaicin treatment in rats were long-lasting since they were maintained into adulthood (11–12 weeks) (Newson et al. unpublished). Nevertheless, it cannot be concluded that deficit in the peripheral somatosensory system was responsible for the brain changes [174]. Although neonatal capsaicin treatment has been reported not to affect TRPV1 receptor mRNA expression in rat brain [27], capsaicin administration to 10 day old and adult rats has been found to produce degenerating terminals in many brain areas not previously known to receive primary afferent input [175, 176]. Recently it has been found that neonatal capsaicin treatment induced changes in central cholinergic, monoaminergic and cannabinoid systems in the adult animal [177]. Capsaicin injection in neonates produces a brief period of hypoxia which has been proposed to cause long-lasting changes in rat brain [73]. Therefore, further studies are required to determine whether the changes observed by Newson et al. [173] and Zavitsanou et al. [177] in rat brain following neonatal capsaicin treatment resulted from loss of TRPV1 channel function in the peripheral or central nervous system or from some other toxic effect. Investigations in TRPV1-KO mice might provide more definitive evidence on the possible role of TRPV1 channels in schizophrenia.

A greater understanding of the role of TRPV1 channels in the brain is required before drugs targeting these channels are proposed for use in schizophrenia. Nevertheless, results indicating that TRPV1 channels play a role in dopaminergic and cannabinoid mechanisms in the brain suggest that their role in schizophrenia is worthy of evaluation. In particular, the role of TRPV1 channels in psychosis precipitated by cannabis use requires investigation. Whether TRPV1 channels will prove to be a useful drug target for treatment of schizophrenia remains to be elucidated.

51.5 Conclusion

The brain mechanisms involved in psychiatric disorders are poorly understood and remain a major challenge to neuroscience. Current drug treatments for psychiatric disorders have major limitations and thus there is a constant search for new targets for drug development. TRP channels are involved in CNS processes that suggest they might be promising new targets for drug development but there is insufficient understanding of their actions in the CNS to make firm conclusions about their potential roles in psychiatric disorders. Current evidence suggests that TRPC6 channels might be a drug target for the development of a new class of antidepressant agent and TRPV1 channel antagonists might be useful anti-anxiety agents. There is currently a lack of direct evidence linking TRP channels with schizophrenia. Nevertheless, the links between schizophrenia and cannabis use, the interactions of cannabinoids with TRPV1 channels, and the disorders of thermoregulation and sensation observed in subjects with schizophrenia, are tantalizing and suggestive of a potential role of TRPV channels in schizophrenia. Future research on the central actions of TRP channels holds the promise of increased understanding of the aetiology of psychiatric disorders, and the development of new therapeutic approaches to the treatment of these disorders.

References

1. Montell C, Birnbaumer L, Flockerzi V, Bindels RJ, Bruford EA, Caterina MJ, Clapham DE, Harteneck C, Heller S, Julius D, Kojima I, Mori Y, Penner R, Prawitt D, Scharenberg AM, Schultz G, Shimizu N, Zhu MX (2002) A unified nomenclature for the superfamily of TRP cation channels. Mol Cell 9:229–231

2. Clapham DE, Montell C, Schultz G, Julius D (2003) International Union of Pharmacology. XLIII. Compendium of voltage-gated ion channels: transient receptor potential channels. Pharmacol Rev 55:591–596
3. Owsianik G, D'hoedt D, Voets T, Nilius B (2006) Structure-function relationship of the TRP channel superfamily. Rev Physiol Biochem Pharmacol 156:61–90
4. Clapham DE (2003) TRP channels as cellular sensors. Nature 426:517–524
5. Liman ER, Corey DP, Dulac C (1999) TRP2: a candidate transduction channel for mammalian pheromone sensory signaling. Proc Natl Acad Sci USA 96:5791–5796
6. Riccio A, Medhurst AD, Mattei C, Kelsell RE, Calver AR, Randall AD, Benham CD, Pangalos MN (2002) mRNA distribution analysis of human TRPC family in CNS and peripheral tissues. Brain Res Mol Brain Res 109:95–104
7. Martorana A, Giampa C, De March Z, Viscomi MT, Patassini S, Sancesario G, Bernardi G, Fusco FR (2006) Distribution of TRPC1 receptors in dendrites of rat substantia nigra: a confocal and electron microscopy study. Eur J Neurosci 24:732–738
8. Giampa C, De March Z, Patassini S, Bernardi G, Fusco FR (2007) Immunohistochemical localization of TRPC6 in the rat substantia nigra. Neurosci Lett 424:170–174
9. De March Z, Giampa C, Patassini S, Bernardi G, Fusco FR (2006) Cellular localization of TRPC5 in the substantia nigra of rat. Neurosci Lett 402:35–39
10. Fusco FR, Martorana A, Giampa C, De March Z, Vacca F, Tozzi A, Longone P, Piccirilli S, Paolucci S, Sancesario G, Mercuri NB, Bernardi G (2004) Cellular localization of TRPC3 channel in rat brain; preferential distribution to oligodendrocytes. Neurosci Lett 365: 137–142
11. Zechel S, Werner S, von Bohlen und Halbach O (2007) Distribution of TRPC4 in developing and adult murine brain. Cell Tissue Res 328:651–656
12. Tai Y, Feng S, Du W, Wang Y (2009) Functional roles of TRPC channels in the developing brain. Pflügers Arch Eur J Physiol 458:283–289
13. Tai Y, Feng S, Ge R, Du W, Zhang X, He Z, Wang Y (2008) TRPC6 channels promote dendritic growth via the CaMKIV-CREB pathway. J Cell Sci 121:2301–2307
14. Talavera K, Nilius B, Voets T (2008) Neuronal TRP channels: thermometers, pathfinders and life-savers. Trends Neurosci 31:287–295
15. Hartmann J, Dragicevic E, Adelsberger H, Henning HA, Sumser M, Abramowitz J, Blum R, Dietrich A, Freichel M, Flockerzi V, Birnbaumer L, Konnerth A (2008) TRPC3 channels are required for synaptic transmission and motor coordination. Neuron 59:392–398
16. Riccio A, Li Y, Moon J, Kim K-S, Smith KS, Rudolph U, Gapon S, Yao GL, Tsvetkov E, Rodig SJ, Veer AV, Meloni EG, Carlezon WA, Bolshakov VY, Clapham DE (2009) Essential role for TRPC5 in amygdala function and fear-related behaviour. Cell 137:761–772
17. Dhaka A, Viswanath V, Patapoutian A (2006) TRP ion channels and temperature sensation. Annu Rev Neurosci 29:135–161
18. Caterina MJ, Schumacher MA, Tominaga M, Rosen TA, Levine JD, Julius D (1997) The capsaicin receptor: a heat-activated ion channel in the pain pathway. Nature 389:816–824
19. Todaka H, Taniguchi J, Satoh J, Mizuno A, Suzuki M (2004) Warm temperature-sensitive transient receptor potential vanilloid 4 (TRPV4) plays an essential role in thermal hyperalgesia. J Biol Chem 279:35133–35138
20. Moqrich A, Hwang SW, Earley TJ, Petrus MJ, Murray AN, Spencer KS, Andahazy M, Story GM, Patapoutian A (2005) Impaired thermosensation in mice lacking TRPV3, a heat and camphor sensor in the skin. Science 307:1468–1472
21. Peier AM, Moqrich A, Hergarden AC, Reeve AJ, Andersson DA, Story GM, Earley TJ, Dragoni I, McIntyre P, Bevan S, Patapoutian AA (2002) TRP channel that senses cold stimuli and menthol. Cell 108:705–715
22. Story GM, Peier AM, Reeve AJ, Eid SR, Mosbacher J, Hricik TR, Earley TJ, Hergarden AC, Andersson DA, Hwang SW, McIntyre P, Jegla T, Bevan S, Patapoutian A (2003) ANKTM1, a TRP-like channel expressed in nociceptive neurons, is activated by cold temperatures. Cell 112:819–829

23. Kauer JA, Gibson HE (2009) Hot flash: TRPV channels in the brain. Trends Neurosci 32:215–224
24. Tominaga M, Caterina MJ, Malmberg AB, Rosen TA, Gilbert H, Skinner K, Raumann BE, Basbaum AI, Julius D (1998) The cloned capsaicin receptor integrates multiple pain-producing stimuli. Neuron 21:531–543
25. Tominaga M, Tominaga T (2005) Structure and function of TRPV1. Pflugers Arch Eur J Physiol 451:143–150
26. Szallasi A, Cortright DN, Blum CA, Eid SR (2007) The vanilloid receptor TRPV1: 10 years from channel cloning to antagonist proof-of-concept. Nat Rev Drug Discov 6:357–372
27. Mezey E, Toth ZE, Cortright DN, Arzubi MK, Krause JE, Elde R, Guo A, Blumberg PM, Szallasi A (2000) Distribution of mRNA for vanilloid receptor subtype 1 (VR1), and the VR1-like immunoreactivity in the central nervous system of the rat and human. Proc Natl Acad Sci USA 97:3655–3660
28. Szabo T, Biro T, Gonzalez AF, Palkovits M, Blumberg PM (2002) Pharmacological characterization of vanilloid receptor located in the brain. Brain Res Mol Brain Res 98:51–57
29. Cristino L, de Petrocellis L, Pryce G, Baker D, Guglielmotti V, Di Marzo V (2006) Immunohistochemical localization of cannabinoid type1 and vanilloid transient receptor potential vanilloid type 1 receptors in mouse brain. Neuroscience 139:1405–1415
30. Sanchez JF, Krause JE, Cortwright DN (2001) The distribution and regulation of vanilloid receptor VR1 and VR1 5' splice variant RNA expression in rat. Neuroscience 107:373–381
31. Roberts JC, Davis JB, Benham CD (2004) (3H)Resiniferatoxin autoradiography in the CNS of wild-type and TRPV1 null mice defines TRPV1 (VR-1) protein distribution. Brain Res 995:176–183
32. Toth A, Boczan J, Kedei N, Lizanecz E, Bagi Z, Papp Z, Edes I, Csiba L, Blumberg PM (2005) Expression and distribution of vanilloid receptor 1 (TRPV1) in the adult rat brain. Brain Res Mol Brain Res 135:162–168
33. Liapi A, Wood JN (2005) Extensive co-localization and heteromultimer formation of the vanilloid receptor-like protein TRPV2 and the capsaicin receptor TRPV1 in the adult rat cerebral cortex. Eur J Neurosci 22:825–834
34. Golech SA, McCarron RM, Chen Y, Bembry J, Lenz F, Mechoulam R, Shohami E, Spatz M (2004) Human brain endothelium: coexpression and function of vanilloid and endocannabinoid receptors. Brain Res Mol Brain Res 132:87–92
35. Devane WA, Hanus L, Breuer A, Pertwee RG, Stevenson LA, Griffin G, Gibson D, Mandelbaum A, Etinger A, Mechoulam R (1992) Isolation and structure of a brain constituent that binds to the cannabinoid receptor. Science 258:1946–1949
36. Smart D, Gunthorpe MJ, Jerman JC, Nasir S, Gray J, Muir AI, Chambers JK, Randall AD, Davis JB (2000) The endogenous lipid anandamide is a full agonist at the human vanilloid receptor (hVR1). Br J Pharmacol 129:227–230
37. Zygmunt PM, Petersson J, Andersson DA, Chuang H, Sorgard M, Di Marzo V, Julius D, Hogestatt ED (1999) Vanilloid receptors on sensory nerves mediate the vasodilator action of anandamide. Nature 400:452–457
38. Starowicz K, Nigam S, Di Marzo V (2007a) Biochemistry and pharmacology of endovanilloids. Pharmacol Ther 114:13–33
39. Davis JB, Gray J, Gunthorpe MJ, Hatcher JP, Davey PT, Overend P, Harries MH, Latcham J, Clapham C, Atkinson K, Hughes SA, Rance K, Grau E, Harper AJ, Pugh PL, Rogers DC, Bingham S, Randall A, Sheardown SA (2000) Vanilloid receptor-1 is essential for inflammatory thermal hyperalgesia. Nature 405:183–187
40. Caterina MJ, Leffler A, Malmberg AB, Martin WJ, Trafton J, Petersen-Zeitz KR, Koltzenburg M, Basbaum AI, Julius D (2000) Impaired nociception and pain sensation in mice lacking the capsaicin receptor. Science 288:306–313
41. Keeble J, Russell F, Curtis B, Starr A, Pinter E, Brain SD (2005) Involvement of transient receptor potential vanilloid 1 in the vascular and hyperalgesic components of joint inflammation. Arthritis Rheum 52:3248–3256

42. Ghilardi JR, Rohrich H, Lindsay TH, Sevcik MA, Schwei MJ, Kubota K, Halvorson KG, Poblete J, Chaplan SR, Dubin AE, Carruthers NI, Swanson D, Kuskowski M, Flores CM, Julius D, Mantyh PW (2005) Selective blockade of the capsaicin receptor TRPV1 attenuates bone cancer pain. J Neurosci 25:3126–3131
43. Jia Y, McLeod RL, Hey JA (2005) TRPV1 receptor: a target for the treatment of pain, cough, airway disease and urinary incontinence. Drug News Perspect 18:165–171
44. Gunthorpe MJ, Szallasi A (2008) Peripheral TRPV1 receptors as targets for drug development: new molecules and mechanisms. Curr Pharm Des 14:32–41
45. Starowicz K, Cristino L, Di Marzo V (2008) TRPV1 receptors in the central nervous system: potential for previously unforeseen therapeutic applications. Curr Pharm Des 14:42–54
46. Krishnan V, Nestler EJ (2008) The molecular neurobiology of depression. Nature 455: 894–902
47. Yulug B, Ozan E, Gönül SAS, Kilic E (2009) Brain-derived neurotrophic factor, stress and depression: A minireview. Brain Res Bull 78:267–269
48. Drevets WC (2001) Neuroimaging and neuropathological studies of depression: implications for the cognitive-emotional features of mood disorders. Curr Opin Neurobiol 11:240–249
49. Sheline YI (2003) Neuroimaging studies of mood disorder effects on the brain. Biol Psychiatry 54:338–352
50. Harrison PJ (2002) The neuropathology of primary mood disorder. Brain 125:1428–1449
51. Ressler KJ, Mayberg HS (2007) Targeting abnormal neural circuits in mood and anxiety disorders: from the laboratory to the clinic. Nat Neurosci 10:1116–1124
52. Mayberg HS, Lozano AM, Voon V, McNeely HE, Seminowicz D, Hamani C, Schwalb JM, Kennedy SH (2005) Deep brain stimulation for treatment-resistant depression. Neuron 45:651–660
53. Schlaepfer TE, Cohen MX, Frick C, Kosel M, Brodesser D, Axmacher N, Joe AY, Kreft M, Lenartz D, Sturm V (2008) Deep brain stimulation to reward circuitry alleviates anhedonia in refractory major depression. Neuropsychopharmacology 33:368–377
54. Kennedy SH (2008) Core symptoms of major depressive disorder: relevance to diagnosis and treatment. Dialogues Clin Neurosci 10:271–277
55. Leuner K, Kazanski V, Muller M, Essin K, Henke B, Gollasch M, Harteneck C, Muller WE (2007) Hyperforin – a key constituent of St John's wort specifically activates TRPC6 channels. FASEB J 21:4101–4111
56. Kasckow JW, Mulchahey JJ, Geracioti TD (2004) Effects of the vanilloid agonist olvanil and antagonist capsazepine on rat behaviours. Biol Psychiatry 28:291–295
57. Marsch R, Foeller E, Rammes G, Bunck M, Kössl M, Holsboer F, Zieglgänsberger W, Landgraf R, Lutz B, Wotjak CT (2007) Reduced anxiety, conditioned fear, and hippocampal long-term potentiation in transient receptor potential vanilloid type 1 receptor-deficient mice. J Neurosci 27:832–839
58. Santos CJPA, Stern CAJ, Bertoglio LJ (2008) Attenuation of anxiety-related behaviour after the antagonism of transient receptor potential vanilloid type 1 channels in the rat ventral hippocampus. Behav Pharmacol 19:357–360
59. Bandler R, Keay KA, Floyd N, Price J (2000) Central circuits mediating patterned autonomic activity during active vs. passive emotional coping. Brain Res Bull 53:95–104
60. Moreira FA, Aguiar DC, Guimarães FS (2007) Anxiolytic-like effect of cannabinoids injected into the rat dorsolateral periaqueductal gray. Neuropharmacology 52:958–965
61. Maione S, Bisogno T, De Novellis V, Palazzo E, Cristino L, Valenti M, Petrosino S, Guglielmotti V, Rossi F, Di Marzo V (2006) Elevation of endocannabinoid levels in the ventrolateral periaqueductal grey through inhibition of fatty acid amide hydrolase affects descending nociceptive pathways via both cannabinoid type 1 and transient receptor potential type-1 receptors. J Pharmacol Exp Ther 316:969–982
62. Terzian ALB, Aguiar DC, Guimarães FS, Moreira FA (2009) Modulation of anxiety-like behaviour by transient receptor potential vanilloid type 1 (TRPV1) channels located in the dorsolateral periaqueductal gray. Eur Neuropsychopharmacol 19:188–195

63. Szallasi A, Blumberg PM (1999) Vanilloid (capsaicin) receptors and mechanisms. Pharmacol Rev 51:159–211
64. Szallasi A, Di Marzo V (2000) New perspectives on enigmatic vanilloid receptors. Trends Neurosci 23:491–497
65. Micale V, Cristino L, Tamburella A, Petrosino S, Leggio GM, Drago F, Di Marzo V (2009) Anxiolytic effects in mice of a dual blocker of fatty acid amide hydrolase and transient receptor potential vanilloid type-1 channels. Neuropsychopharmacology 34:593–606
66. Willner P (2005) Chronic mild stress (CMS) revisited: consistency and behavioural-neurobiological concordance in the effects of CMS. Neuropsychobiology 52:90–110
67. Holderbach R, Clark K, Moreau J-L, Bischofberger J, Normann C (2007) Enhanced long-term synaptic depression in an animal model of depression. Biol Psychiatry 62:92–100
68. Howland JG, Wang YT (2008) Synaptic plasticity in learning and memory: Stress effects in the hippocampus. Prog Brain Res 169:145–158
69. Wong TP, Howland JG, Robillard JM, Ge Y, Yu W, Titterness AK, Brebner K, Liu L, Weinberg J, Christie BR, Phillips AG, Wang YT (2007) Hippocampal long-term depression mediates acute stress-induced spatial memory retrieval impairment. Proc Natl Acad Sci USA 104:11471–11476
70. Xu L, Anwyl R, Rowan MJ (1997) Behavioral stress facilitates the induction of long-term depression in the hippocampus. Nature 387:497–500
71. Li H-B, Mao R-R, Zhang J-C, Yang Y, Cao J, Xu L (2008) Antistress effect of TRPV1 channel on synaptic plasticity and spatial memory. Biol Psychiatry 64:286–292
72. Gunthorpe MJ, Rami HK, Jerman JC, Smart D, Gill CH, Soffin EM, Luis Hannan S, Lappin SC, Egerton J, Smith GD, Worby A, Howett L, Owen D, Nasir S, Davies CH, Thompson M, Wyman PA, Randall AD, Davis JB (2004) Identification and characterisation of SB-366791, a potent and selective vanilloid receptor (VR1/TRPV1) antagonist. Neuropharmacology 46:133–149
73. Di Marzo V, Gobbi G, Szallasi A (2008) Brain TRPV1: a depressing TR(i)P down memory lane? Trends Pharmacol Sci 29:594–600
74. Gibson HE, Edwards JG, Page RS, Van Hook MJ, Kauer JA (2008) TRPV1 channels mediate long-term depression at synapses on hippocampal interneurons. Neuron 57:746–759
75. Alter BJ, Gereau RW (2008) Hotheaded: TRPV1 as mediator of hippocampal synaptic plasticity. Neuron 57:629–631
76. Maione S, Cristino L, Migliozzi AL, Georgiou AL, Starowicz K, Salt TE, Di Marzo V (2009) TRPV1 channels control synaptic plasticity in the developing superior colliculus. J Physiol 587:2521–2535
77. Wainwright A, Rutter AR, Seabrook GR, Reilly K, Oliver KR (2004) Discrete expression of TRPV2 within the hypothalamo-neurohypophysial system: implications for regulatory activity within the hypothalamic-pituitary-adrenal axis. J Comp Neurol 474:24–42
78. Moussaieff A, Rimmerman N, Bregman T, Straiker A, Felder CC, Shoham S, Kashman Y, Huang SM, Lee H, Shohami E, Mackie K, Caterina MJ, Walker JM, Fride E, Mechoulam R (2008) Incensole acetate, an incense component, elicits psychoactivity by activating TRPV3 channels in the brain. FASEB J 22:3024–3034
79. First MB, Frances A, Pincus HA (2002) DSM-IV-TR. handbook of differential diagnosis. American Psychiatric Association Publishing, Washington, DC
80. Freedman R (2003) Schizophrenia. N Engl J Med 349:1738–1749
81. Schwab SG, Wildenauer DB (2009) Update on key previously proposed candidate genes for schizophrenia. Curr Opin Psychiatry 22:147–153
82. Mei L, Xiong WC (2008) Neuregulin 1 in neural development, synaptic plasticity and schizophrenia. Nat Rev Neurosci 9:437–452
83. Harrison PJ (2007) Schizophrenia susceptibility genes and neurodevelopment. Biol Psychiatry 61:1119–1120
84. O'Tuathaigh CMP, Babovic D, O'Meara G, Clifford JJ, Croke DT, Waddington JL (2007) Susceptibility genes for schizophrenia: characterization of mutant mouse models at the level of phenotypic behaviour. Neurosci Biobehav Rev 31:60–78

85. Chen J, Lipska BK, Weinberger DR (2006) Genetic mouse models of schizophrenia: from hypothesis-based to susceptibility gene-based models. Biol Psychiatry 59:1180–1188
86. Ross CA, Margolis RL, Reading SAJ, Pletnikov M, Coyle JT (2006) Neurobiology of schizophrenia. Neuron 52:139–153
87. Stefansson H, Sigurdsson E, Steinthorsdottir V, Bjornsdottir S, Sigmundsson T, Ghosh S, Brynjolfsson J, Gunnarsdottir S, Ivarsson O, Chou TT, Hjaltason O, Birgisdottir B, Jonsson H, Gudnadottir VG, Gudmundsdottir E, Bjornsson A, Ingvarsson B, Ingason A, Sigfusson S, Hardardottir H, Harvey RP, Lai D, Zhou M, Brunner D, Mutel V, Gonzalo A, Lemke G, Sainz J, Johannesson G, Andresson T, Gudbjartsson D, Manolescu A, Frigge ML, Gurney ME, Kong A, Gulcher JR, Petursson H, Stefansson K (2002) Neuregulin 1 and susceptibility to schizophrenia. Am J Hum Genet 71:877–892
88. Li D, Collier DA, He L (2006) Meta-analysis shows strong positive association of the neuregulin 1 (NRG1) gene with schizophrenia. Hum Mol Genet 15:1995–2002
89. Munafò MR, Thiselton DL, Clark TG, Flint J (2006) Association of the NRG1 gene and schizophrenia: a meta-analysis. Mol Psychiatry 11:539–546
90. Munafò MR, Attwood AS, Flint J (2008) Neuregulin 1 genotype and schizophrenia. Schizophrenia Bull 34:9–12
91. Hashimoto R, Straub RE, Weickert CS, Hyde TM, Kleinman JE, Weinberger DR (2004) Expression analysis of neuregulin-1 in the dorsolateral prefrontal cortex in schizophrenia. Mol Psychiatry 9:299–307
92. Law AJ, Lipska BK, Weickert CS, Hyde TM, Straub RE, Hashimoto R, Harrison PJ, Kleinman JE, Weinberger DR (2006) Neuregulin 1 transcripts are differentially expressed in schizophrenia and regulated by 5' SNPs associated with the disease. Proc Natl Acad Sci USA 103:6747–6752
93. Law AJ, Kleinman JE, Weinberger DR, Weickert CS (2007) Disease-associated intronic variants in the ErbB4 gene are related to altered ErbB4 splice-variant expression in the brain in schizophrenia. Hum Mol Genet 16:129–141
94. Silberberg G, Darvasi A, Pinkas-Kramarski R, Navon R (2006) The involvement of ErbB4 with schizophrenia: association and expression studies. Am J Med Genet 141B:142–148
95. Schlaepfer TE, Harris GJ, Tien AY, Peng LW, Lee S, Federman EB, Chase GA, Barta PE, Pearlson GD (1994) Decreased regional cortical gray matter volume in schizophrenia. Am J Psychiatry 151:842–848
96. Selemon LD, Kleinman JE, Herman MM, Goldman-Rakic PS (2002) Smaller frontal gray matter volume in post-mortem schizophrenic brains. Am J Psychiatry 159:1983–1991
97. McDonald C, Grech A, Toulopoulou T, Schulze K, Chapple B, Sham P, Walshe M, Sharma T, Sigmundsson T, Chintis X, Murray RM (2002) Brain volumes in familial and non-familial schizophrenic probands and their unaffected relatives. Am J Med Genet (Neuropsychiatr Genet) 114:616–625
98. Shenton ME, Kikinis R, Jolesz FA, Pollak SD, Lemay M, Wible CG, Hokama H, Martin J, Metcalf D, Coleman M, McCarley RW (1992) Abnormalities of the left temporal lobe and thought disorder in schizophrenia. A quantitative magnetic resonance imaging study. New Engl J Med 327:604–612
99. McCarley RW, Wible CG, Frumin M, Hirayasu Y, Levitt JJ, Fischer IA, Shenton ME (1999) MRI anatomy of schizophrenia. Biol Psychiatry 45:1099–1119
100. Selemon LD, Rajkowska G, Goldman-Rakic PS (1995) Abnormally high neuronal density in the schizophrenic cortex. A morphometric analysis of prefrontal area 9 and occipital area 17. Arch Gen Psychiatry 52:805–820
101. Selemon LD, Rajkowska G, Goldman-Rakic PS (1998) Elevated neuronal density in prefrontal area 46 in brains from schizophrenic patients: application of a 3-dimensional, stereologic counting method. J Comp Neurol 392:402–412
102. Selemon LD, Goldman-Rakic PS (1999) The reduced neuropil hypothesis: a circuit based model of schizophrenia. Biol Psychiatry 45:17–25
103. Buxhoeveden D, Roy E, Switala A (2000) Reduced interneuronal space in schizophrenia. Biol Psychiatry 47:681–683

104. Casanova MF, De Zeeuw L, Switala A, Kreczmanski P, Korr H, Ulfig N, Heinsen H, Steinbusch HWM, Schmitz C (2005) Mean cell spacing abnormalities in the neocortex of patients with schizophrenia. Psychiatry Res 133:1–12
105. Rajkowska G, Selemon LD, Goldman-Rakic PS (1998) Neuronal and glial somal size in the prefrontal cortex: a post-mortem morphometric study of schizophrenia and Huntington disease. Arch Gen Psychiatry 55:215–224
106. Pajonk F-G, Wobrock T, Gruber O, Scherk H, Berner D, Kaizl I, Kierer A, Müller S, Oest M, Meyer T, Backens M, Schneider-Axmann T, Thornton AE, Honer WG, Falkai P (2010) Hippocampal plasticity in response to exercise in schizophrenia. Arch Gen Psychiatry 67:133–143
107. Seeman P (1992) Dopamine receptor sequences. Therapeutic levels of neuroleptics occupy D2 receptors, clozapine occupies D4. Neuropsychopharmacology 7:261–284
108. Marsden CA (2006) Dopamine: the rewarding years. Br J Pharmacol 147:S136–S144
109. Lewis DA, Gonzalez-Burgos G (2006) Pathophysiologically based treatment interventions in schizophrenia. Nat Med 12:1016–1022
110. Tsai G, Coyle JT (2002) Glutamatergic mechanisms in schizophrenia. Annu Rev Pharmacol Toxicol 42:165–179
111. Krystal JH, D'Souza DC, Mathalon D, Perry E, Belger A, Hoffman R (2003) NMDA receptor antagonist effects, cortical glutamatergic function, and schizophrenia: toward a paradigm shift in medication development. Psychopharmacology (Berl) 169:215–233
112. du Bois TM, Huang X-F (2007) Early brain development disruption from NMDA receptor hypofunction: relevance to schizophrenia. Brain Res Rev 53:260–270
113. Mohn AR, Gainetdinov RR, Caron MG, Koller BH (1999) Mice with reduced NMDA receptor expression display behaviors related to schizophrenia. Cell 98:427–436
114. Vollenweider FX, Vontobel P, Oye I, Hell D, Leenders KL (2000) Effects of (S)-ketamine on striatal dopamine: a [^{11}C]raclopride PET study of a model psychosis in humans. J Psychiatric Res 34:35–43
115. Hahn CG, Wang HY, Cho DS, Talbot K, Gur RE, Berrettini WH, Bakshi K, Kamins J, Borgmann-Winter KE, Siegel SJ, Gallop RJ, Arnold SE (2006) Altered neuregulin 1-erbB4 signaling contributes to NMDA receptor hypofunction in schizophrenia. Nat Med 12:824–828
116. Strubing C, Krapivinsky G, Krapivinsky L, Clapham DE (2003) Formation of novel TRPC channels by complex subunit interactions in embryonic brain. J Biol Chem 278:39014–39019
117. Greka A, Navarro B, Oancea E, Duggan A, Clapham DE (2003) TRPC5 is a regulator of hippocampal neurite length and growth cone morphology. Nat Neurosci 6:837–845
118. Bezzerides VJ, Ramsey IS, Kotecha S, Greka A, Clapham DE (2004) Rapid vesicular translocation and insertion of TRP channels. Nat Cell Biol 6:709–720
119. Hui H, McHugh D, Hannan M, Zeng F, Xu S-Z, Khan S-U-H, Levenson R, Beech DJ, Weiss JL (2006) Calcium-sensing mechanism in TRPC5 channels contributing to retardation of neurite outgrowth. J Physiol 572:165–172
120. Li Y, Jia YC, Cui K, Li N, Zheng ZY, Wang YZ, Yuan XB (2005) Essential role of TRPC channels in the guidance of nerve growth cones by brain-derived neurotrophic factor. Nature 434:894–898
121. Mirnics K, Middleton FA, Marquez A, Lewis DA, Levitt P (2000) Molecular characterization of schizophrenia viewed by microarray analysis of gene expression in prefrontal cortex. Neuron 28:53–67
122. Vawter MP, Crook JM, Hyde TM, Kleinman JE, Weinberger DR, Becker KG, Freed WJ (2002) Microarray analysis of gene expression in the prefrontal cortex in schizophrenia: A preliminary study. Schizophrenia Res 58:11–20
123. Hemby SE, Ginsberg SD, Brunk B, Arnold SE, Trojanowski JQ, Erberwine JH (2002) Gene expression profile for schizophrenia: discrete neuron transcription patterns in the entorhinal cortex. Arch Gen Psychiatry 59:631–640

124. Munsch T, Freichel M, Flockerzi V, Pape HC (2003) Contribution of transient receptor potential channels to the control of GABA release from dendrites. Proc Natl Acad Sci USA 100:16065–16070
125. Krapivinsky G, Mochida S, Krapivinsky L, Cibulsky SM, Clapham DE (2006) The TRPM7 ion channel functions in cholinergic synaptic vesicles and affects transmitter release. Neuron 52:485–496
126. Marinelli S, Di Marzo V, Florenzano F, Fezza F, Viscomi MT, van der Stelt M, Bernardi G, Molinari M, Maccarrone M, Mercuri NBN- (2007) Arachidonoyl-dopamine tunes synaptic transmission onto dopaminergic neurons by activating both cannabinoid and vanilloid receptors. Neuropsychopharmacology 32:298–308
127. Marinelli S, Pascucci T, Bernardi G, Puglisi-Allegra S, Mercuri NB (2005) Activation of TRPV1 in the VTA excites dopaminergic neurons and increases chemical- and noxious-induced dopamine release in the nucleus accumbens. Neuropsychopharmacology 30: 864–870
128. Musella A, De Chiara V, Rossi S, Prosperetti C, Bernardi G, Maccarrone M, Centonze D (2009) TRPV1 channels facilitate glutamate transmission in the striatum. Mol Cell Neurosci 40:89–97
129. Li DP, Chen SR, Pan HL (2004) VR1 receptor activation induces glutamate release and postsynaptic firing in the paraventricular nucleus. J Neurophysiol 92:1807–1816
130. Starowicz K, Maione S, Cristino L, Palazzo E, Marabese I, Rossi F, de Novellis V, Di Marzo V (2007b) Tonic endovanilloid facilitation of glutamate release in brainstem descending antinociceptive pathways. J Neurosci 27:13739–13749
131. Xing J, Li J (2007) TRPV1 receptor mediates glutamatergic synaptic input to dorsolateral periaqueductal gray (dl-PAG) neurons. J Neurophysiol 97:503–511
132. Yang K, Kumamoto E, Furue H, Yoshimura M (1998) Capsaicin facilitates excitatory but not inhibitory synaptic transmission in substantia gelatinosa of the rat spinal cord. Neurosci Lett 255:135–138
133. Lee J, Di Marzo V, Brotchie JM (2006) A role for vanilloid receptor 1 (TRPV1) and endocannabinoid signaling in the regulation of spontaneous and L-DOPA induced locomotion in normal and reserpine-treated rats. Neuropharmacology 51:557–565
134. de Lago E, de Miguel R, Lastres-Becker I, Ramos JA, Fernandez-Ruiz J (2004) Involvement of vanilloid-like receptors in the effects of anandamide on motor behavior and nigrostriatal dopaminergic activity: in vivo and in vitro evidence. Brain Res 1007:152–159
135. Jordt SE, Julius D (2002) Molecular basis for species-specific sensitivity to 'hot' chili peppers. Cell 108:421–430
136. Tzavara ET, Li DL, Moutsimilli L, Bisogno T, Di Marzo V, Phebus LA, Nomikos GG, Giros B (2006) Endocannabinoids activate transient receptor potential vanilloid 1 receptors to reduce hyperdopaminergia-related hyperactivity: therapeutic implications. Biol Psychiatry 59:508–515
137. Panlilio LV, Mazzola C, Medalie J, Hahn B, Justinova Z, Drago F, Cadet JL, Yasar S, Goldberg SR (2009) Anandamide-induced behavioural disruption through a vanilloid-dependent mechanism in rats. Psychopharmacology 203:529–538
138. Akopian AN, Ruparel NB, Jeske NA, Patwardhan A, Hargreaves KM (2008) Role of ionotropic cannabinoid receptors in peripheral antinociception and antihyperalgesia. Trends Pharmacol Sci 30:79–84
139. Kim SR, Lee DY, Chung ES, Oh UT, Kim SU, Jin BK (2005) Transient receptor potential vanilloid subtype 1 mediates cell death of mesencephalic dopaminergic neurons in vivo and in vitro. J Neurosci 25:662–671
140. Kim SR, Bok E, Chung YC, Chung ES, Jin BK (2008) Interactions between CB_1 receptors and TRPV1 channels mediated by 12-HPETE are cytotoxic to mesencephalic dopaminergic neurons. Br J Pharmacol 155:253–264
141. Pegorini S, Braida D, Verzoni C, Guerini-Rocco C, Consalez GG, Croci L, Sala M (2005) Capsaicin exhibits neuroprotective effects in a model of transient global ischemia in Mongolian gerbils. Br J Pharmacol 144:727–735

142. Pegorini S, Zani A, Braida D, Guerinin-Rocco C, Sala M (2006) Vanilloid VR1 receptor is involved in rimonabant-induced neuroprotection. Br J Pharmacol 147:552–559
143. Kim SR, Chung YC, Chung ES, Park KW, Won SY, Bok E, Park ES, Jin BK (2007) Roles of transient receptor potential vanilloid subtype 1 and cannabinoid type 1 receptors in the brain: neuroprotection versus neurotoxicity. Mol Neurobiol 35:245–254
144. Degenhardt L, Hall W (2006) Is cannabis use a contributory cause of psychosis? Canad. J Psychiatry 51:556–565
145. Laviolette SR, Grace AA (2006) The roles of cannabinoid and dopamine receptor systems in neural emotional learning circuits: implications for schizophrenia and addiction. Cell Mol Life Sci 63:1597–1613
146. Sundram S (2006) Cannabis and neurodevelopment: implications for psychiatric disorders. Human Psychopharmacol 21:245–254
147. Malone DT, Hill MN, Rubino T (2010) Adolescent cannabis use and psychosis: epidemiology and neurodevelopmental models. Brit J Pharmacol 160:511–522
148. Harkany T, Guzman M, Galve-Roperh I, Berghuis P, Devi LA, Mackie K (2007) The emerging functions of endocannabinoid signaling during CNS development. Trends Pharmacol Sci 28:83–92
149. Chavarria-Siles I, Contreras-Rojas J, Hare E, Walss-Bass C, Quezada P, Dassori A, Contreras S, Medina N, Ramirez M, Salazar R, Raventos H, Escamilla MA (2008) Cannabinoid receptor 1 gene (CNR1) and susceptibility to a quantitative phenotype for hebephrenic schizophrenia. Am J Med Genet B Neuropsychiatr Genet 147:279–284
150. Ishiguro H, Horiuchi Y, Ishikawa M, Koga M, Imai K, Suzuki Y, Morikawa M, Inada T, Watanabe Y, Takahashi M, Someya T, Ujike H, Iwata N, Ozaki N, Onaivi ES, Kunugi H, Sasaki T, Itokawa M, Arai M, Niizato K, Iritani S, Naka I, Ohashi J, Kakita A, Takahashi H, Nawa H, Arinami T (2010) Brain cannabinoid CB2 receptor in schizophrenia. Biol Psychiatry 67:974–982
151. Hermesh H, Shiloh R, Epstein Y, Manaim H, Weizman A, Munitz H (2000) Heat intolerance in patients with chronic schizophrenia maintained with antipsychotic drugs. Am J Psychiatry 157:1327–1329
152. Chong TWH, Castle DJ (2004) Layer upon layer: thermoregulation in schizophrenia. Schizophrenia Res 69:149–157
153. Schwartz PJ, Erk SD (2004) Regulation of central dopamine-2 receptor sensitivity by a proportional control thermostat in humans. Psychiatry Res 127:19–26
154. Shiloh R, Weizman A, Epstein Y, Rosenberg SL, Valevski A, Dorfman-Etrog P, Wiezer N, Katz N, Munitz H, Hermesh H (2001) Abnormal thermoregulation in drug-free male schizophrenia patients. Eur Neuropsychopharmacol 11:285–288
155. Lee TF, Mora F, Myers RD (1985) Dopamine and thermoregulation: an evaluation with special reference to dopaminergic pathways. Neurosci Biobehav Rev 9:589–598
156. Barros RCH, Branco LGS, Carnio EC (2004) Evidence for thermoregulation by dopamine D1 and D2 receptors in the anteroventral preoptic region during normoxia and hypoxia. Brain Res 1030:165–171
157. Karlsson U, Sundgren-Andersson AK, Johansson S, Krupp JJ (2005) Capsaicin augments synaptic transmission in the rat medial preoptic nucleus. Brain Res 1043:1–11
158. Szolcsanyi J (2004) Forty years in capsaicin research for sensory pharmacology and physiology. Neuropeptides 38:377–384
159. Kudoh A, Ishihara H, Matsuki A (2000) Current perception thresholds and postoperative pain in schizophrenic patients. Reg Anesth Pain Med 25:475–479
160. Blumensohn R, Ringler D, Eli I (2002) Pain perception in patients with schizophrenia. J Nerv Mental Dis 190:481–483
161. Hooley JM, Delgado ML (2001) Pain insensitivity in the relatives of schizophrenia patients. Schizophrenia Res 47:265–273
162. Bonnot O, Anderson GM, Cohen D, Willer JC, Tordjman S (2009) Are patients with Schizophrenia insensitive to pain? a reconsideration of the question. Clin J Pain 25:244–252

163. Potvin S, Marchand S (2008) Hypoalgesia in schizophrenia is independent of antipsychotic drugs: A systematic quantitative review of experimental studies. Pain 138:70–78
164. Ren K, Williams GM, Ruda MA, Dubner R (1994) Inflammation and hyperalgesia in rats neonatally treated with capsaicin: effects on two classes of nociceptive neurons in the superficial dorsal horn. Pain 59:287–300
165. Holzer P (1991) Capsaicin: cellular targets, mechanisms of action, and selectivity for thin sensory neurons. Pharmacol Rev 43:143–201
166. Hwang SJ, Burette A, Rustioni A, Valtschanoff JG (2004) Vanilloid receptor VR1-positive primary afferents are glutamatergic and contact spinal neurons that co-express neurokinin receptor NK_1 and glutamate receptors. J Neurocytology 33:321–329
167. Bolcskei K, Helyes Z, Szabo A, Sandor K, Elekes K, Nemeth J, Almasi R, Pinter E, Petho G, Szolcsanyi J (2005) Investigation of the role of TRPV1 receptors in acute and chronic nociceptive processes using gene-deficient mice. Pain 117:368–376
168. Zimmermann K, Leffler A, Fischer MMJ, Messlinger K, Nau C, Reeh PW (2005) The TRPV1/2/3 activator 2-aminoethoxydiphenyl borate sensitizes native nociceptive neurons to heat in wildtype but not TRPV1 deficient mice. Neuroscience 135:1277–1284
169. Waldo MC (1999) Co-distribution of sensory gating and impaired niacin flush response in the parents of schizophrenics. Schizophrenia Res 40:49–53
170. Messamore E, Hoffman WE, Janowsky A (2003) The niacin skin flush abnormality in schizophrenia: a quantitative dose-response study. Schizophrenia Res 62:251–258
171. Sadaka Y, Weinfeld E, Lev DL, White EL (2003) Changes in mouse barrel synapses consequent to sensory deprivation from birth. J Comp Neurol 457:75–86
172. Jansco G, Király E, Janscó-Gábor A (1977) Pharmacologically induced selective degeneration of chemosensitive primary sensory neurons. Nature (London) 270:741–743
173. Newson P, Lynch-Frame A, Roach R, Bennett S, Carr V, Chahl LA (2005) Intrinsic sensory deprivation induced by neonatal capsaicin treatment induces changes in rat brain and behaviour of possible relevance to schizophrenia. Brit J Pharmacol 146:408–418
174. Chahl LA (2007) TRP's: links to schizophrenia? Biochim Biophys Acta 1772:968–977
175. Ritter S, Dinh TT (1988) Capsaicin-induced neuronal degeneration: silver impregnation of cell bodies, axons, and terminals in the central nervous system of the adult rat. J Comp Neurol 271:79–90
176. Ritter S, Dinh TT (1990) Capsaicin-induced neuronal degeneration in the brain and retina of preweanling rats. J Comp Neurol 296:447–461
177. Zavitsanou K, Dalton VS, Wang H, Newson P, Chahl LA (2010) Receptor changes in brain tissue of rats treated as neonates with capsaicin. J Chem Neuroanat 39:248–255

Chapter 52
Transient Receptor Potential Genes and Human Inherited Disease

Kate V. Everett

Abstract Transient receptor potential (TRP) genes have been implicated in a wide array of human disorders, from cancers to bipolar disorder. The extraordinary range of diseases in whose pathogenesis they may play a role exemplifies the equally broad range of functions of the TRP proteins. TRP proteins primarily form homomeric or heteromeric channels in the cell membrane but there may also be intracellular non-channel functions for TRPs. Mutations in TRP genes have been causally associated with at least 12 hereditary human diseases. This chapter aims to summarise those associations and focuses on the following diseases: focal segmental glomerulosclerosis; polycystic kidney disease; brachyolmia; spondylometaphyseal dysplasia; metatropic dysplasia; hereditary motor and sensory neuropathy; spinal muscular atrophy; congenital stationary night blindness; progressive familial heart block; hypomagnesaemia; and mucolipidosis. There appears to be very little to connect these disorders except the involvement of a TRP gene but by understanding more about the genes involved in diseases, we understand more about disease biology and about the function of those genes causally associated. This feedback loop of information will serve to enhance our knowledge of disease and elucidate basic gene and protein function of the TRPs.

52.1 Introduction

The mammalian transient receptor potential (TRP) superfamily of ion channels consists of 28 cation selective ion channels encoded by 28 genes. TRP channels contain six transmembrane domains flanked by cytoplasmic N and C termini. They are broadly expressed with diverse functions and activated by a wide variety of stimuli [1, 2]. In mammals, they are grouped into six subfamilies according to their amino acid sequence homology (TRPC, TRPV, TRPM, TRPA, TRPP, and TRPML). All except the TRPA genes, so far, have been implicated in human disease. The

K.V. Everett (✉)
St. George's University of London, London, UK
e-mail: keverett@sgul.ac.uk

major hereditary disorders are channelopathies but there is evidence for a role in systemic disorders as well. This chapter aims to summarise the spectra of mutations associated with these hereditary TRP-related genetic diseases in humans.

52.2 The Canonical TRPs

The canonical transient receptor potential (TRPC) cation channel family has seven members (*TRPC1-7*) with widespread expression. TRPC channels mediate cation entry in response to phospholipase C activation. They are important for the increase of intracellular Ca^{2+} concentration in response to activation of G-coupled protein receptors and receptor tyrosine kinases. They have a variety of functions, and their dysfunction has been implicated in hypertrophic phenotypes [3, 4].

52.2.1 TRPC6

The transient receptor potential cation channel, subfamily 3, member 6 encoded by *TRPC6* on chromosome 11q21-q22 has been implicated in cardiac hypertrophy [3], idiopathic pulmonary hypertension [4, 5] and infantile hypertrophic pyloric stenosis [6]. However, the only definitive evidence for disease involvement is in focal segmental glomerulosclerosis (FSGS). FSGS is a disease that attacks the glomeruli of the kidneys causing severe scarring. Initially the scarring is localised (focal), affecting a subset of the glomeruli, and is specific to a section of the glomerular tuft (i.e. is segmental). FSGS causes inefficient filtering of the blood resulting in the following symptoms: oedema; hypoalbuminaemia; and proteinuria. Ultimately, the damage to the kidney can be so severe that end stage renal disease is reached. Inherited forms of FSGS demonstrate both recessive and dominant patterns of inheritance. Five disease loci have been identified: FSGS1 on chromosome 19q13 (MIM #603278); FSGS2 on chromosome 11q21-q22 (MIM #603965); FSGS3 on chromosome 6p12 (MIM #607832); FSGS4 on chromosome 22q12 (MIM #612551); and FSGS5 on chromosome 14q32.33 (MIM #613237). FSGS2 is an autosomal dominant form of the condition attributable to mutations in *TRPC6*. Genome-wide linkage analysis of a large family of British ancestry with 13 confirmed cases in which FSGS segregated as a dominant disorder initially mapped the locus to chromosome 11q [7]. Haplotype analysis reduced the minimal disease interval to a 2.1 cM area containing > 40 genes. Expression analysis shows that TRPC6 is highly expressed in the podocyte cells of the glomeruli, and is a key component of the glomerular slit diaphragm which spans the space between adjacent podocytes. Furthermore, TRPC6 channel activity at the slit diaphragm is essential for proper regulation of podocyte structure and function [8].

FSGS2 demonstrates allelic heterogeneity; at this time eight different mutations have been identified. Seven different missense mutations have been identified (Table 52.1) each co-segregating with the disease in the heterozygous state [8, 9]. Five of these mutations are predicted to be gain-of-function mutations, enhancing

Table 52.1 Summary of FSGS2 mutations in *TRPC6*

Mutation	Protein location	Ethnicity of family	Functional effect	References
P112Q	1st Ankyrin repeat	British origin	Increased current amplitude Alters cellular localisation of protein	[9]
M132T	2nd Ankyrin repeat	Turkish	Increased current amplitude	[10]
N143S	2nd Ankyrin repeat	African American		[8]
S270T	4th Ankyrin repeat	Colombian		[8]
K874X	C terminus	Polish		[8]
Q889K	C terminus	Chinese	Increased current amplitude	[11]
R895C	C terminus	Mexican	Increased current amplitude	[8]
E897K	C terminus	Irish/German	Increased current amplitude	[8]

Amino acid positions based on protein sequence NP_004612.2.

TRPC6-mediated calcium signals. Four of these five missense mutations are fully penetrant, being present in all affected individuals and absent from all controls tested. The P112Q mutation alters a highly conserved residue located in the first ankyrin repeat domain of the TRPC6 protein [9]. Not only does this lead to enhanced TRPC6-mediated calcium signals in response to agonists such as angiotensin II, but the P112Q TRPC6 mutant also has an altered intracellular distribution with an increased proportion localised to the plasma membrane. A fully penetrant mutation in the second ankyrin repeat domain (M132T) leads to an increase in channel currents and impaired channel inactivation resulting, in one case, in an unusually early onset case in childhood [10]. Q889K alters a highly conserved residue in the C-terminus leading to increase amplitude [11]. The final fully penetrant mutation, R895C, alters a highly conserved residue in the C terminus and leads to an increase in current amplitude. The same is true for E897K but this mutation demonstrates incomplete penetrance suggesting that in this case the mechanism by which this effect is elicited must be different. One possible result of the increased current amplitude observed is excess activation of the nuclear factor of activated T cell (NFAT) transcription factors, a downstream signalling target of TRPC6. P112Q, R895C and E897K lead to constitutive transcriptional activation of NFAT-responsive reporters. This ability to activate NFAT is dominant over wild type TRPC6 [12]. This suggests that targets of NFAT will need to be investigated as possible contributors to TRPC6-induced FSGS2.

Two missense mutations (N143S and S270T) and a nonsense mutation (K874X) do not result in a change in current amplitude but because they cosegregate with disease phenotype and alter highly conserved residues or lead to a premature stop codon, it has been proposed that they cause FSGS2 via another mechanism such as altered interaction with other slit diaphragm proteins [8].

52.3 The Vanilloid TRPs

The first transient receptor potential gene identified in this family, TRPV1, was shown to be sensitive to capaiscin (8-Methyl-*N-vanillyl-trans*-6-nonenamide) hence the name, vanilloid TRPs. Functional TRPV channels are tetrameric, made either of four identical subunits or of two or more different subunits. TRPV channels respond to physical and chemical stimuli such as temperature, pH and mechanical stress.

52.3.1 TRPV4

The transient receptor potential cation channel, subfamily V, member 4 gene on chromosome 12q24.1 encodes the TRPV4 protein which is thought to form a channel which can be activated by hypotonic swelling, temperature and acidic pH [13–15]. Mutations have been identified in four different heritable human disorders: brachyolmia; spondylometaphyseal dysplasia; metatropic dysplasia; and hereditary motor and sensory neuropathy type IIC.

52.3.1.1 Skeletal Disorders

Mutations in *TRPV4* are associated with three skeletal disorders ranging in severity from the relatively mild brachyolmia, through spondylometaphyseal dysplasia to the more severe metatropic dysplasia (Table 52.2).

52.3.1.1.1 Type 3 Brachyolmia

Brachyolmia is a clinically heterogeneous group of skeletal dysplasias characterised by a short trunk, scoliosis and mild short stature. Type 1 and Type 2 brachyolmia are autosomal recessive disorders. Type 3 brachyolmia (MIM #113500) is an autosomal dominant form with severe scoliosis and flattened irregular cervical vertebrae. The gene for Type 3 brachyolmia was initially mapped to chromosome 12q24.1-q24.2. Analysis of microarray data demonstrated that *TRPV4* had a tenfold higher cartilage-selectivity than any other gene in this region indicating that it was the most plausible candidate gene for brachyolmia [16]. Two different missense mutations have been identified in the *TRPV4* gene in two families which segregate in the heterozygous state with disease. These mutations were not found in controls. Both of these mutations (R616Q and V620I) change a conserved residue in the fifth transmembrane region of the TRPV4 protein. Expression of the R616Q and V620I mutations in human embryonic kidney (HEK) cells yielded considerably larger constitutive current before agonist application. Thus, both missense mutations conferred a gain-of-function phenotype by significantly increasing the fraction of constitutively open channels and by potentiating agonist activation. The mechanism by which this may lead to the skeletal problems associated with brachyolmia is not clear but TRPV4 has been shown to have a role in chondrogenesis via regulation of the SOX9 pathway [17]. SOX9 is a transcription factor which regulates the activation of genes encoding cartilage-specific extracellular matrix

Table 52.2 Summary of TRPV4 mutations in human heritable disease

Mutation	Protein location	Human phenotype	References
R269C	4th Ankyrin repeat	HMSN2C	[26]
R269H	4th Ankyrin repeat	Congenital distal SMA HMSN2C	[24–26]
R315W	5th Ankyrin repeat	Congenital distal SMA SPSMA HMSN2C	[24]
R316C	5th Ankyrin repeat	HMSN2C SPSMA	[24, 25]
I331F	5th Ankyrin repeat	Metatropic dysplasia	[19]
D333G	5th Ankyrin repeat	SMDK	[19]
D333_E337delinsE	5th Ankyrin repeat	Metatropic dysplasia	[23]
R594H	Cytoplasmic S4 domain	SMDK	[19]
R616Q	5th TM domain	Type 3 brachyolmia	[16]
F617L	5th TM domain	Metatropic dysplasia	[23]
V620I	5th TM domain	Type 3 brachyolmia	[16]
A716S	Cytoplasmic S6 domain	SMDK	[19]
E797K	Cytoplasmic C terminus	Metatropic dysplasia	[23]
P799L	Cytoplasmic C terminus	Metatropic dysplasia	[19, 23]

Amino acid positions based on protein sequence NP_067638.3; TM, transmembrane; HMSN2C, hereditary motor and sensory neuropathy type IIC; SMA, spinal muscular atrophy; SPSMA, scapuloperoneal spinal muscular atrophy; SMDK, Kozlowski type spondylometaphyseal dysplasia.

molecules. Increased Ca^{2+} influx followed by TRPV4 activation is essential for enhanced SOX9-dependant promoter activity. A constitutively active TRPV4 channel may result in dysregulation of the SOX9 pathway ultimately affecting bone formation.

Recent work in zebrafish supports a role for TRPV4 in skeletal formation as overexpression of mouse TRPV4 in zebrafish leads to marked shortening and curvature of the axial bones [18].

52.3.1.1.2 Kozlowski Type Spondylometaphyseal Dysplasia

Spondylometaphyseal dysplasias are characterized by abnormalities in the vertebrae and the metaphyses of the long bones (e.g. femur, tibia). Kozlowski type spondylometaphyseal dysplasia (SMDK; MIM #184252) has the specific characteristics of short stature and significant kyphoscoliosis (curvature of the spine in both a coronal and sagittal plane). This kyphoscoliosis leads to progressive deformity. Inheritance is autosomal dominant.

The similarity of SMDK to brachyolmia led to the screening of *TRPV4* as a candidate gene for SMDK. Heterozygous mutations in *TRPV4* have been identified in six probands with SMDK. Four had an R594H mutation in the cytoplasmic S4 domain, which was associated with increased intracellular calcium ion concentration and activity [19]. On two occasions, this mutation was shown to be de novo; in the other two families, information on the parents was not reported. It was not found in more than 200 control chromosomes. Another mutation (D333G) led to increased basal intracellular Ca^{2+} concentration and intracellular Ca^{2+} activity and in this case was shown to be inherited from the proband's affected mother.

52.3.1.1.3 Metatropic Dysplasia

Metatropic dysplasia (MIM #156530) is a spondylometaphyseal dysplasia which presents with a variable clinical phenotype ranging from mild to lethal. Typically patients present with shortened limbs and a long, narrow trunk with progressive kyphoscoliosis often needing surgical correction. Affected individuals also have a distinct facies with a prominent forehead and squared-off jaw. Inheritance may be autosomal recessive or autosomal dominant [20–22].

The dominant non-lethal form of metatropic dysplasia has been associated with *TRPV4* mutations. Heterozygous mutations have been identified in seven patients (see Table 52.2). Four patients exhibited de novo mutations (I331F in one patient and P799L in three patients) affecting conserved residues in the fifth ankyrin domain and the carboxyl-terminal cytoplasmic tail respectively [19]. These patients all exhibited moderately severe metatropic dysplasia [19, 23]. Two cases presenting with a mild form had point mutations resulting in amino acid substitutions in the carboxyl-terminal tail (E797K) and the fifth transmembrane region (F617L). An in-frame deletion was identified in a third patient with relatively mild metatropic dysplasia. It was not possible to determine whether these mutations were de novo or inherited from the parents but they were not found to be present in over 200 control chromosomes indicating that they are not likely to be polymorphisms.

In summary, there are a range of mutations in *TRPV4* which cause skeletal disorders. There does not seem to be any relationship between location of mutation and phenotype: for example, mutations in the fifth ankyrin domain can cause metatropic dysplasia or SMDK; mutations in the fifth transmembrane domain can cause Type 3 brachyolmia and metatropic dysplasia. This suggests that genetic modifiers may exist which dictate the form and severity of skeletal dysplasia.

52.3.1.2 Neuromuscular Disorders

A large five-generation family in which 10 individuals were affected with hereditary motor and sensory neuropathy type IIC (HMSN2C; MIM #606071), congenital spinal muscular atrophy (MIM #600175), or scapuloperoneal spinal muscular atrophy (SPSMA; MIM #181405) was recently described [24]. HMSN2C, SPSMA and congenital SMA are subtypes of sensory neuropathies and spinal muscular atrophies characterised by proximal and distal muscle weakness and wasting. All 10 affected individuals were found to have the same heterozygous mutation R315W. This R315W mutation was also identified in an unrelated family in which six members had HMSN2C. A second mutation, also altering a residue in the fifth ankyrin domain (R316C), was found in two small families with HMSN2C [24]. Neither of these two missense mutations was found in more than 300 controls. This mutation has also been found in a family with SPSMA [25]. One family with congenital distal SMA had an altered residue in the fourth ankyrin domain (R269H) which was not found in more than 150 controls. Two families with HMSN2C have the same mutation whilst another has an alteration of the same residue; R269C [25, 26]. These results emphasize the spectrum of mutations found in *TRPV4* and the range of effects they can have. Functional analyses of these mutations are contradictory. Expression of mutant TRPV4 protein has been shown to fail to localize correctly to the plasma membrane. Furthermore, mutant TRPV4 channels demonstrated a reduced response to hypo-osmotic stimulation or stimulation with the TRPV4-specific agonist, 4αPDD [24]. The cytoplasmic ankyrin domains are crucial for channel maturation and function [27–29] and these experiments would indicate that mutations in these domains affect both these processes. However, independent experiments by other groups suggest that normal trafficking is maintained and instead mutant TRPV4 channels show increased calcium channel activity consistent with a gain of function mutation [25, 26].

52.4 The Melastatin TRPs

There are eight mammalian members of the melastatin TRPS. Unlike TRPC and TRPV proteins, they lack ankyrin domains but still contain the conserved TRP domain in the N-terminus. The TRPM subfamily of TRP genes are named after the founding member, melastatin, so named because it was found to be downregulated in a highly metastatic mouse melanoma cell line [30].

52.4.1 TRPM1

The protein encoded by *TRPM1*, melastatin, is primarily found in melanin-producing cells but its exact function is unknown. However, it has been shown to be expressed in cells responsible for transmitting signals from the retina to the ganglions.

52.4.1.1 Congenital Stationary Night Blindness

Appaloosa horses have a distinctive "leopard-spotted" coat pattern attributable to a single incomplete dominant locus (LP). Homozygosity not only affects pigmentation but also leads to congenital stationary night blindness (CSNB) in the horse. This condition principally affects the rod receptors of the eye and diminishes night vision. In severely affected horses, vision in normal light may also be affected. Expression analysis of genes in the LP region in horses demonstrated that *TRPM1* had only 0.05% the level of expression in the retina of homozygous Appaloosa horses in comparison to non-Appaloosa horses [31]. *TRPM1* was also downregulated in both pigmented and non-pigmented skin of homozygous Appaloosa horses. Re-sequencing identified a single 173 kb haplotype across TRPM1 significantly associated with both leopard-spotted pigmentation and CSNB reinforcing the relationship between TRPM1 and these traits [32].

In humans, recessive and X-linked forms of CSNB have been identified. Patients with "complete" CSNB have total loss of rod pathway function due to postsynaptic defects in depolarising or ON bipolar cell signalling. Bipolar cells transmit signals from the rod cells to the ganglions. In the absence of light, rod photoreceptors release glutamate which activates the glutamate receptor, mGluR, on the ON bipolar cell. This inhibits the activity of the ON bipolar cell by closing the TRPM1 channel via an intracellular cascade preventing Ca^{2+} influx. CSNB1C (MIM #613216) is a form of complete CSNB, inherited in an autosomal recessive manner and shown to be caused by mutations in *TRPM1*. The evidence of downregulation of *TRPM1* in Appaloosa horses had made this gene an excellent functional candidate for CSNB1C in humans. Autozygosity mapping of a large South Asian family with three consanguineous marriages and five affected individuals identified a large region of homozygosity on chromosome 15q [33]. This region contains *TRPM1* making this gene a positional as well as functional candidate. Screening identified a single homozygous mutation in the affected mother of the proband whilst the unaffected father was shown to be heterozygous for this change (IVS16+2T>C). This mutation was predicted to alter splicing although this was not proven experimentally. Other family members were not screened therefore segregation of this mutation with disease has not been proven definitively. In a Caucasian non-consanguineous family with a single affected individual with CSNB1C, the proband was shown to be heterozygous for two mutations in *TRPM1* both of which are predicted to result in protein truncation (G138fs and Y1035X). Analysis of the mother showed that these mutations were in *trans* (i.e. on different homologous chromosomes). In a second non-consanguineous Caucasian family with two affected sibs, the proband was a compound heterozygote for two missense mutations in *TRPM1* (R74C and I1002F). Other family members were not available for segregation analysis but these four mutations were not found in 384 control chromosomes.

Three other homozygous mutations have been identified: one missense mutation (R14W) which segregated with disease and was found to be heterozygous in the parents and unaffected sibs; a premature truncation mutation (W856X); and a large deletion of exons two to seven (Y72-K365del) for which both parents were heterozygous and was absent from an unaffected sib. None of these mutations were found in more than 200 control chromosomes.

Twelve other patients have been shown to be compound heterozygotes for a range of mutations, including missense, frameshift and deletions (see Table 52.3). None of these have been found in controls [34–36]. There is evidently a high level of allelic heterogeneity in CSNB1B, without any clear pattern with regards to location of mutation. Mutations affect residues throughout TRPM1: the intracellular N-terminus; the transmembrane domains; and the C-terminus. There is not any noticeable genotype-phenotype relationship. However, it would seem likely that

Table 52.3 Summary of mutations in TRPM1 associated with CSNB1B

	Mutation	Protein location	Ethnicity of family	References
Homozygous mutations	c.-27C>T (5′ UTR expression defect)		Flemish-Belgian	[34]
	IVS16+2T>C (splice defect)		South Asian	[33]
	Y72-K365del	N terminus	European	[36]
	W856X	2nd TM domain	Turkish	[34]
Compound heterozygotes	Q11X	N terminus	German	[34]
	L99P	N terminus		
	N28MfsX62	N terminus	European	[36]
	G534R	N terminus		
	Y72C	N terminus	French	[34]
	E1032X	Extracellular S5 domain		
	Y72C	N terminus	Italian	[34]
	IVS4-3C>G (splice defect)			
	R74C	N terminus	European	[36]
	L364R	N terminus		
	R74C	N terminus	White European	[33]
	I1002F			
	K99P	N terminus	European	[36]
	P611H			
	G138fs	N terminus	White European	[33]
	Y1035X	Extracellular S5 domain		
	P399P (splice defect)	N terminus	German	[34]
	Q1164RfsX31	C terminus		
	R473P (+ maternal deletion)	N terminus	Portuguese-French	[34]
	R624C	N terminus	Japanese	[35]
	S882X	Cytosolic S3 domain		
	R624C	N terminus	Japanese	[35]
	IVS8+3_6delAAGT (splice defect)			
	F1075S	6th TM domain	Japanese	[35]
	IVS2-3C>G (splice defect)			
	N1278N	C terminus	German	[34]
	IVS20+3G>A (splice defect)			

Amino acid positions based on protein sequence NP_002411.3; TM, transmembrane.

missense mutations may affect the functional gating properties of TRPM1; whilst splice-site defects, nonsense mutations and large deletions could result in a small dysfunctional form of TRPM1 being transcribed which either does not localise correctly and/or prevents full-length TRPM1 localising correctly [37].

52.4.2 TRPM4

The transient receptor potential cation channel, subfamily M, member 4 gene on chromosome 19q13.32 encodes a cation channel which, unlike other TRP channels, is not activated by Ca^{2+}, but is equally permeable to K^+ and Na^+. It has been shown to be highly expressed in the Purkinje fibres of the heart and plays an important role in the human heart conductance system.

52.4.2.1 Progressive Familial Heart Block, Type 1B

Progressive familial heart block, Type 1B (PFHB1B, MIM #604559) is an autosomal dominant disorder shown to be caused in a subset of cases by mutations in *TRPM4*. It affects the heart conduction system ultimately leading to a full block of the conductance system. Complete block of the conductance system is a common indicator for pacemaker implantation.

A large South African kindred with multiple affected individuals across six generations was used to map the locus for PFHB1B to chromosome 19q13.33 [38, 39]. Screening of *TRPM4* identified a heterozygous missense mutation (E7K) which segregated with the disease and was not found in controls [39]. Cells expressing mutated TRPM4 result in elevated channel density at the cell surface. It has been proposed that elevated expression of the $TRPM4^{E7K}$ channel increases membrane leak conductance, disabling action potential propagation down the Purkinje fibres. This explains the block of the conductance system seen in PFHB1B.

52.4.3 TRPM6

The transient receptor potential cation channel, subfamily M, member 6 gene on chromosome 9q22 encodes multiple mRNA isoforms of TRPM6. It forms homomeric complexes or heteromeric complexes with TRPM7 and is primarily expressed in the intestinal epithelium and kidney. TRPM6 is crucial in Mg^{2+} homeostasis.

52.4.3.1 Hypomagnesaemia with Secondary Hypocalcaemia

Familial hypomagnesaemia with secondary hypocalcaemia (HSH, MIM #602014) is an autosomal recessive disorder resulting in electrolyte abnormalities in the neonatal period. Severe magnesium deficiency leads to secondary hypocalcaemia (very low serum calcium levels) as a result of parathyroid failure. Untreated individuals suffer from muscle spasms and seizures which can lead to neurological damage or even death. Treatment is with life-long high dosage magnesium supplements. The primary defect appears to be in intestinal magnesium absorption potentially coupled with renal magnesium leak.

HSH was initially mapped to chromosome 9q21.13 using three large inbred Bedouin families from Israel [40]. Sequencing of genes in this region identified a homozygous mutation in *TRPM6* segregating in all affected individuals in these families and not found in 156 control chromosomes. This mutation altered a splice donor site in intron 16 (IVS16+1G>A) predicted to cause a splicing defect. Since the identification of this mutation, 22 other putative causal mutations have been identified, demonstrating the extensive allelic heterogeneity of HSH (Table 52.4). All except one of these mutations (IVS20+1G>A) are family-specific, and segregate with disease, and none have been found in control chromosomes. Five other homozygous intronic mutations have been found each of which segregates with the disease and are not found in control chromosomes. Each has been predicted to affect splicing but this has not been proven experimentally [41–43].

Five missense mutations predicted to result in premature truncation of the TRPM6 protein have been found in the homozygous state. Four are likely to lead to premature termination of the gene product prior to the first transmembrane domain (E157X, R484X, S590X and R736fsX737). The fifth is predicted to truncate the protein after the fourth transmembrane domain (I944fsX959) [41–43]. In theory, the probable functional effect of these truncation mutations seems more likely to be causal than the intronic variants but this has not been shown experimentally.

A single synonymous change (W1925W) has been identified in a Turkish family which is predicted to result in the loss of a splice site although this has not be proven [43].

Four large deletions affecting various combinations of exons have been identified. The homozygous deletion of exon 21 is predicted to lead to a frameshift and premature truncation of the protein at the sixth transmembrane domain. The deletions of exons 22–23 and 25–27 would also result in a frameshift and premature termination of the protein. However, the deletion of exons 32 and 33 is in frame and may allow a shorter but functional, or partially functional, protein to be generated. The transmembrane region, N terminus and TRP domain are all intact.

Compound heterozygosity has been identified in four patients with HSH: one having two frameshift mutations leading to premature stop codons; two having the combination of a missense mutation and an intronic mutation predicted to alter splicing; and one having a frameshift mutation leading to premature termination combined with an intronic splice site mutation [41, 43, 44]. The likely effect of two premature truncation mutations seems likely to be a lack of functional protein although this has not been demonstrated. However, experiments in *Xenopus* oocytes have shown that the S141L missense mutation seen in one compound heterozygote leads to intracellular retention of TRPM6. Correct localisation of TRPM6 to the cell membrane requires the presence of TRPM7 to allow the formation of heteromeric complexes. A lack of TRPM7 prevents channel formation [45]. The molecular basis of disease causality of the S141L mutation seems to be to prevent TRPM6 and TRPM7 forming a functional channel at the cell membrane resulting in HSH. The functional effect of the second missense mutation seen in a compound heterozygote (P1017R) has also been investigated. Unlike the S141L mutation which prevents heteromeric assembly of TRPM6 and TRPM7, TRPM6^{P1017R} correctly assembles

Table 52.4 Mutations in TRPM6 associated with HSH

	Mutation	Protein location	Ethnicity of family	References
Homozygous mutations	IVS8+5G>C (splice defect)		Israeli-Arab	[42]
	IVS11+1G>A (splice defect)		Egyptian	[43]
	IVS16+1G>A (splice defect)		Bedouin	[42]
	IVS18+1G>A (splice defect)		Turkish	[41]
	IVS20+1G>A (splice defect)		Turkish	[43]
	IVS23-68A>G (splice defect)		Israeli-Arab	[42]
	E157X	N terminus	Turkish	[43]
	R484X	N terminus	Greek-Arab	[42]
	S590X	N terminus	Turkish	[41, 43]
	R736fsX737	N terminus	Albanian	[43]
	I944fsX959	4th TM domain	Romanian	[43]
	W1925W (splice defect)	C terminus	Turkish	[43]
	I890_M973del	Cytoplasmic S2 domain to 4th TM domain	Greek	[43]
	T974_N1070delfsX	5th TM domain to C terminus	Indian	[43]
	K1135_N1596delfsX	C terminus	Pakistani	[43]
	K1695_A1734del	C terminus	Australian	[43]
Compound heterozygotes	S141L IVS25-1G>A (splice defect)	N terminus	Swedish	[41, 43]
	H427fsX429 E1260fsX1283	N terminus C terminus	Israeli	[41, 43]
	P599fsX609 IVS30+2T>C (splice defect)	N terminus	Japanese	[43]
	P1017R IVS10-1G>A (splice defect)	Pore-forming S5 domain	German	[44]

Amino acid positions based on protein sequence NP_060132.3; TM, transmembrane.

with TRPM7 to form channels at the cell membrane but their activity is suppressed [44]. The control of Mg^{2+} flux is thus impaired causing HSH.

A single patient with one heterozygous mutation predicted to result in a truncated protein has also been proposed as a causal variant but this seems unlikely given the recessive mode of inheritance of HSH [42].

52.5 The Polycystin TRPs

This particular subfamily of TRP genes have undergone a number of name changes but the accepted nomenclature now lists three different TRPP genes: *TRPP1* (or *PKD2*); *TRPP2* (*PKD2L1* or PKD2-like 1); and *TRPP3* (*PKD2L2* or PKD2-like 2) [46]. As the names suggest, these genes have long been associated with polycystic kidney disease. Adult polycystic kidney disease (ADPKD; MIM #173900) is an autosomal dominant disorder primarily characterised by renal cysts. Extrarenal features include liver cysts, occasionally pancreatic cysts and, in about 5% of cases, intracranial aneurysm, which is a significant cause of death. ADPKD affects approximately one in 1,000 people worldwide and progresses to end stage renal failure in the majority of patients. Genetic heterogeneity is recognized, with one locus (PKD1), responsible for approximately 80% of cases. The *PKD1* gene proved problematic to identify, as have mutations within it, due to the high CpG content and high level of duplication seen (exon one to intron 34 is duplicated elsewhere in the genome). *PKD1*, initially called *PBP* (polycystic breakpoint, and encoding polycystin-1), was identified using a family in which a mother and her daughter were both affected with ADPKD. They both had a balanced translocation between chromosomes 16 and 22 which was found to disrupt the *PBP* gene on chromosome 16 [47]. Many insertion/deletions, nonsense, missense and splice site mutations have now been identified across the gene which affect the resulting transcript and are associated with ADPKD. The remaining 20% of cases are caused by at least two other loci, *PKD2* (TRPP1) and *PKD3* (gene currently unknown).

52.5.1 TRPP1=PKD2

The *TRPP1* gene on chromosome 4q21-q23 encodes polycystin-2, and is involved in Ca^{2+} transport and Ca^{2+} signalling in renal epithelial cells. Polycystin-1 and polycystin-2 are thought to physically interact via their C-termini to form a complex at the renal primary cilium which can respond to mechanical stimuli [48, 49]. However, there is also evidence that polycystin-2 can function independently and may be important in the regulation of calcium signalling in the endoplasmic reticulum. The role of polycystin 2 in the cilium may be of particular interest as many ciliopathies display renal cysts as part of their phenotype and PKD may fall within this spectrum (see Patel et al., 2009 for review [50]). ADPKD due to mutations in *TRPP1* seems to be clinically less severe than that due to *PKD1* mutations although it is not understood why this may be.

Many mutations in *TRPP1* leading to ADPKD have now been identified as can be seen in Table 52.5. The vast majority are nonsense and frameshift mutations predicted to lead to a truncated protein. Where it has been possible to look at family members, all mutations have segregated with disease and have not been present in controls (where tested; see Table 52.5 for references). There does not seem to be a genotype-phenotype relationship between mutation location or type and clinical presentation of ADPKD and mutations are found in all 15 exons

Table 52.5 List of major mutations in TRPP1 (PKD2) which cause ADPKD

Mutation	Protein location	Ethnicity of family	References
P68fsX22	N terminus	Cypriot	[51, 67]
L231fsX37	1st TM domain		
L180fsX32	N terminus	Dutch	[68]
c.596-12_599del (splice defect)	1st TM domain	Icelandic	[68]
W201fsX9	N terminus	Spanish	[69]
IVS2+1G>A (splice defect)		Belgian	[68]
IVS2-2A>G (splice defect)		Dutch	[68]
R306X	Extracellular S1 domain	Bulgarian	[68]
R320X	Extracellular S1 domain	Dutch	[70]
R325fsX15	Extracellular S1 domain	Dutch	[68]
W380X	Extracellular S1 domain	Spanish	[71]
T398fsX8	Extracellular S1 domain	English	[68]
T398fsX52	Extracellular S1 domain	English	[68]
Q405X	Extracellular S1 domain	Cypriot	[71]
W414G	Extracellular S1 domain	Australian	[68]
W455fsX5	Extracellular S1 domain	Spanish	[68]
R464X	Extracellular S1 domain	Spanish	[69]
F482fsX31	2nd TM domain	Spanish Dutch	[68, 70]
E494X	Cytoplasmic S2 domain	Australian	[68]
D511V	3rd TM domain	Spanish	[70]
Q555X	4th TM domain	Spanish	[69]
F567fsX15	4th TM domain	Spanish	[69]
IVS7-3C>G (splice defect)		Spanish	[68]
A594fsX15	Cytoplasmic S4 domain	Dutch	[68]
N645fsX	Extracellular S5 domain		[72]
M675fsX5	6th TM domain	Dutch	[70]
L717fsX18	C terminus	Spanish	[69]
D723X	C terminus	Canadian	[73]
L736X[a]	C terminus	Canadian	[74]
R742X	C terminus	Cypriot	[71]
IVS11-2A>G		Danish	[68]
Y762X	C terminus	Dutch Bulgarian	[68]
R807X	C terminus	Spanish	[69]
E837X	C terminus	Spanish	[69]
R872X	C terminus	Icelandic	[70]
R872_E890fsX8 (aberrant splicing and early truncation)	C terminus		
D886G (aberrant splicing)	C terminus	US	[70]

[a]Family segregating with a PKD1 mutation as well; amino acid positions based on protein sequence NP_000288.1; TM, transmembrane.

of *TRPP1* except exons 3 and 15. The fact that disease presentation is similar regardless of mutation location indicates that all are likely to be inactivating mutations.

It has been postulated that rather than being a straightforward autosomal dominant condition, ADPKD may instead be explained by a two-hit hypothesis whereby a germline mutation followed by a somatic mutation is required for disease. This hypothesis would explain the adult onset of ADPKD and is supported by experimental evidence in which somatic mutations are found at a high level in renal and hepatic cysts in patients with ADPKD [51]. Furthermore, it has been shown that mutations in *PKD1* and *TRPP1* can exist in a "transheterozygous" state in ADPKD patients, such that the mutation in *PKD1* is germline whereas that in *TRPP1* is somatic and vice versa [52–54]. This represents a fascinating mechanism by which disease is caused by the combination of an inherited mutation in one gene and an acquired mutation in another related gene. Mouse models of PKD also provide support for the two-hit hypothesis and interaction between *PKD1* and *TRPP1* [55, 56]. Mice heterozygous for an unstable allele of *Pkd2*, designed to be able to undergo somatic rearrangement, develop renal and pancreatic cysts, ultimately suffering renal failure and early death. Mice who are doubly heterozygous for *Pkd1* and *Pkd2* null alleles suffered from a greater severity of cyst formation than would be predicted from a simple additive model, suggestive of a more complex epistatic interaction [56].

In summary, ADPKD is an important disease due to its prevalence and morbidity. *TRPP1* contributes significantly to ADPKD manifestation and many mutations have been identified although their actual functional effect remains unclear. The potential role of the polycystin-1/polycystin-2 complex as a mechanosensor in primary cilium presents the intriguing possibility that ADPKD is a ciliopathy which presents, at the cellular level, via a two-hit mechanism in accordance with Knudson's model as seen in carcinogenesis.

52.6 The Mucolipin TRPs

The TRPML subfamily has three members: *TRPML1*; *TRPML2*; and *TRPML3*. They share 75% amino acid similarity and have certain highly characteristic features, such as a short cytosolic tail (see Puertollano and Kiselyov, 2009 for review [57]). The founding member is *TRPML1*, a 14 exon gene which encodes mucolipin-1, a protein highly expressed in the endocytic pathway and thought to be involved in membrane trafficking and degradation.

52.6.1 TRPML1

The transient receptor potential cation channel, mucolipin subfamily, member 1 gene on chromosome 19p13.3-p13.2 used to be known as *MCOLN1* because it encodes the protein mucolipin-1. However, the similarity in general structure and sequence to other members of the TRP superfamily led to the name change and inclusion within this group [58, 59]. It has been shown to be involved in the pathogenesis of a particular form of mucolipidosis. The mucolipidoses are inherited lysosomal storage disorders characterised by an inability to process certain carbohydrates and lipids. These accumulate in the lysosomes of the cells causing distortion and swelling. This cellular dysfunction leads to a range of symptoms dependent upon the location of the cell. There are four types of mucolipidosis, three of which are due to enzymatic dysfunction leading to a block in the catabolic pathway and excessive lysosomal storage. Mucolipidosis type IV (MLIV, MIM #252650), however, appears to affect the endocytic pathway and trafficking leading to a build-up of material in the lysosomes, rather than being due to a lack of enzyme [60, 61]. MLIV is a progressive neurodevelopmental disorder usually presenting in the 1st year of life with mental retardation, corneal opacity, retinal degeneration and delayed developmental milestones. Typically, sufferers rarely learn to walk unaided and communication is normally only through sign language, if at all. Occasional cases of atypical MLIV occur in which symptoms are mild, but these cases are rare. The gene for MLIV was initially localised to chromosome 19p13.2-p13.3 using a set of 26 Ashkenazi Jewish families in which two major founder haplotypes were identified and found to segregate with disease suggesting the existence of at least two founder mutations for MLIV [62]. The majority of cases of MLIV have been observed in the Ashkenazim population [63]. Re-sequencing of the region excluded all genes except the newly identified *MCOLN1* (*TRPML1*) [58, 59]. Sequence analysis identified two relatively common mutations (IVS3-2A>G and g.511_6943del) which occur either in the homozygous state or as compound heterozygotes in affected individuals and accord to the two major disease haplotypes previously defined. In the Ashkenazim population, ~60% of MLIV sufferers are homozygous for the IVS-3A>G splice site mutation which prevents correct splicing of exon 4 and results in truncated unstable mRNA species. Approximately 33% are compound heterozygotes for the two mutations whilst just one individual has so far been found to be homozygous for the g.511_6943del mutation [58, 59, 64, 65]. This large genomic deletion results in a transcript missing all of exons one to six and part of exon seven (c.1_788del). Patients with any combination of these mutations show similar clinical severity. *TRPML1* expression has been shown to be decreased in patients with these genotypes [64]. A range of other mutations have now been identified in both Ashkenazi and in some non-Jewish families (Table 52.6). Although most cause the severe phenotype associated with MLIV, a recent report of a compound heterozygote bearing the D362Y and G568fsX9 mutations and presenting with an extremely mild form of the condition, suggests that a degree of genotype-phenotype correlation may be seen with certain mutations.

An excellent mouse model provides strong support for *TRPML1* as the MLIV gene. Knockout Mcoln1−/− mice present with retinal degeneration and gait deficits strongly reminiscent of MLIV in humans [66]. Furthermore, they have raised plasma gastrin levels in comparison to wildtype mice and inclusion bodies in the brain.

Table 52.6 Mutations in TRPML1 causing MLIV

Mutation	Protein location	References
c.1_788del	Extracellular S1 domain	[58, 59, 64, 75]
Q79X	1st TM domain	[76]
F101fsX	Extracellular S1 domain	[76]
R102X	Extracellular S1 domain	[63, 64]
L106P	Extracellular S1 domain	[63]
IVS3-2A>G (aberrant splicing and early truncation)		[58, 59, 63, 64, 75]
T158fsX	Extracellular S1 domain	[64]
R172X	Extracellular S1 domain	[64]
T232P	Extracellular S1 domain	[75]
R322X	Cytosolic S2 domain	[58, 75]
D362Y	3rd TM domain	[63, 64, 77, 78]
R403C	4th TM domain	[76]
Y404fsX	4th TM domain	[64, 75]
F408del	4th TM domain	[63, 64, 75, 77]
V446L	5th TM domain	[64]
L447P	5th TM domain	[63]
c.1444insGCCCTGCTGCG	Extracellular s5 domain	[75]
F454_N569del	Extracellular S5 domain	[64]
S456L	Extracellular S5 domain	[79]
F465L	Extracellular S5 domain	[75]
N469S (aberrant splicing)	Extracellular S5 domain	[63]
G568fsX9	C terminus	[78]

Amino acid positions based on protein sequence NP_065394.1; TM, transmembrane; c., coding sequence position using AF287269.1.

Raised serum gastrin levels are a clinical indicator for MLIV in humans as are inclusion bodies in the skin or conjunctivae.

52.7 Summary

The transient receptor potential superfamily of genes and their encoded proteins are proving to be extremely important in human disease. This review has focused, for brevity's sake, on those human disorders which (a) are truly heritable, and (b) for which there is conclusive evidence for a role for TRP genes. The disorders caused by mutations in TRP genes range from skeletal dysplasias to kidney diseases. They have been implicated as ciliopathies, endosymal trafficking disorders, and channelopathies. This range implies much, not only about disease pathogenesis but also about TRP functionality. Although TRP genes are generally considered to encode non-selectively permeable cation channels, it is clear that they may not always locate to the cell membrane as functional channels but may have other intracellular roles yet to be fully elucidated. As we understand more about the way TRP genes cause disease, and link genotype and phenotype more closely, the potential for improved therapy for the debilitating conditions described here has to improve.

References

1. Huang CL (2004) The transient receptor potential superfamily of ion channels. J Am Soc Nephrol 15:1690–1699
2. Montell C (2005) The TRP superfamily of cation channels. Sci STKE 2005:re3
3. Bush EW, Hood DB, Papst PJ, Chapo JA, Minobe W, Bristow MR, Olson EN, McKinsey TA (2006) Canonical transient receptor potential channels promote cardiomyocyte hypertrophy through activation of calcineurin signaling. J Biol Chem 281: 33487–33496
4. Yu Y, Fantozzi I, Remillard CV, Landsberg JW, Kunichika N, Platoshyn O, Tigno DD, Thistlethwaite PA, Rubin LJ, Yuan JX (2004) Enhanced expression of transient receptor potential channels in idiopathic pulmonary arterial hypertension. Proc Natl Acad Sci USA 101:13861–13866
5. Yu Y, Keller SH, Remillard CV, Safrina O, Nicholson A, Zhang SL, Jiang W, Vangala N, Landsberg JW, Wang JY, Thistlethwaite PA, Channick RN, Robbins IM, Loyd JE, Ghofrani HA, Grimminger F, Schermuly RT, Cahalan MD, Rubin LJ, Yuan JX (2009) A functional single-nucleotide polymorphism in the TRPC6 gene promoter associated with idiopathic pulmonary arterial hypertension. Circulation 119:2313–2322
6. Everett KV, Chioza BA, Georgoula C, Reece A, Gardiner RM, Chung EM (2009) Infantile hypertrophic pyloric stenosis: evaluation of three positional candidate genes, TRPC1, TRPC5 and TRPC6, by association analysis and re-sequencing. Hum Genetics Vol 126, number 6, pp 819–831 (2009)
7. Winn MP, Conlon PJ, Lynn KL, Howell DN, Slotterbeck BD, Smith AH, Graham FL, Bembe M, Quarles LD, Pericak-Vance MA, Vance JM (1999) Linkage of a gene causing familial focal segmental glomerulosclerosis to chromosome 11 and further evidence of genetic heterogeneity. Genomics 58:113–120
8. Reiser J, Polu KR, Moller CC, Kenlan P, Altintas MM, Wei C, Faul C, Herbert S, Villegas I, vila-Casado C, McGee M, Sugimoto H, Brown D, Kalluri R, Mundel P, Smith PL, Clapham DE, Pollak MR (2005) TRPC6 is a glomerular slit diaphragm-associated channel required for normal renal function. Nat Genet 37:739–744
9. Winn MP, Conlon PJ, Lynn KL, Farrington MK, Creazzo T, Hawkins AF, Daskalakis N, Kwan SY, Ebersviller S, Burchette JL, Pericak-Vance MA, Howell DN, Vance JM, Rosenberg PB (2005) A mutation in the TRPC6 cation channel causes familial focal segmental glomerulosclerosis. Science 308:1801–1804
10. Heeringa SF, Moller CC, Du J, Yue L, Hinkes B, Chernin G, Vlangos CN, Hoyer PF, Reiser J, Hildebrandt F (2009) A novel TRPC6 mutation that causes childhood FSGS. PLoS One 4:e7771
11. Zhu B, Chen N, Wang ZH, Pan XX, Ren H, Zhang W, Wang WM (2009) Identification and functional analysis of a novel TRPC6 mutation associated with late onset familial focal segmental glomerulosclerosis in Chinese patients. Mutat Res 664:84–90
12. Schlondorff J, Del CD, Carrasquillo R, Lacey V, Pollak MR (2009) TRPC6 mutations associated with focal segmental glomerulosclerosis cause constitutive activation of NFAT-dependent transcription. Am J Physiol Cell Physiol 296:C558–C569
13. Xu H, Zhao H, Tian W, Yoshida K, Roullet JB, Cohen DM (2003) Regulation of a transient receptor potential (TRP) channel by tyrosine phosphorylation. SRC family kinase-dependent tyrosine phosphorylation of TRPV4 on TYR-253 mediates its response to hypotonic stress. J Biol Chem 278:11520–11527
14. Chen X, essandri-Haber N, Levine JD (2007) Marked attenuation of inflammatory mediator-induced C-fiber sensitization for mechanical and hypotonic stimuli in TRPV4–/– mice. Mol Pain 3:31
15. Loukin SH, Su Z, Kung C (2009) Hypotonic shocks activate rat TRPV4 in yeast in the absence of polyunsaturated fatty acids. FEBS Lett 583:754–758
16. Rock MJ, Prenen J, Funari VA, Funari TL, Merriman B, Nelson SF, Lachman RS, Wilcox WR, Reyno S, Quadrelli R, Vaglio A, Owsianik G, Janssens A, Voets T, Ikegawa S, Nagai

T, Rimoin DL, Nilius B, Cohn DH (2008) Gain-of-function mutations in TRPV4 cause autosomal dominant brachyolmia. Nat Genet 40:999–1003
17. Muramatsu S, Wakabayashi M, Ohno T, Amano K, Ooishi R, Sugahara T, Shiojiri S, Tashiro K, Suzuki Y, Nishimura R, Kuhara S, Sugano S, Yoneda T, Matsuda A (2007) Functional gene screening system identified TRPV4 as a regulator of chondrogenic differentiation. J Biol Chem 282:32158–32167
18. Wang Y, Fu X, Gaiser S, Kottgen M, Kramer-Zucker A, Walz G, Wegierski T (2007) OS-9 regulates the transit and polyubiquitination of TRPV4 in the endoplasmic reticulum. J Biol Chem 282:36561–36570
19. Krakow D, Vriens J, Camacho N, Luong P, Deixler H, Funari TL, Bacino CA, Irons MB, Holm IA, Sadler L, Okenfuss EB, Janssens A, Voets T, Rimoin DL, Lachman RS, Nilius B, Cohn DH (2009) Mutations in the gene encoding the calcium-permeable ion channel TRPV4 produce spondylometaphyseal dysplasia, Kozlowski type and metatropic dysplasia. Am J Hum Genet 84:307–315
20. Beck M, Roubicek M, Rogers JG, Naumoff P, Spranger J (1983) Heterogeneity of metatropic dysplasia. Eur J Pediatr 140:231–237
21. Genevieve D, Le MM, Feingold J, Munnich A, Maroteaux P, Cormier-Daire V (2008) Revisiting metatropic dysplasia: presentation of a series of 19 novel patients and review of the literature. Am J Med Genet A 146A:992–996
22. Kannu P, Aftimos S, Mayne V, Donnan L, Savarirayan R (2007) Metatropic dysplasia: clinical and radiographic findings in 11 patients demonstrating long-term natural history. Am J Med Genet A 143A:2512–2522
23. Camacho N, Krakow D, Johnykutty S, Katzman PJ, Pepkowitz S, Vriens J, Nilius B, Boyce BF, Cohn DH (2010) Dominant TRPV4 mutations in nonlethal and lethal metatropic dysplasia. Am J Med Genet A 152A:1169–1177
24. Auer-Grumbach M, Olschewski A, Papic L, Kremer H, McEntagart ME, Uhrig S, Fischer C, Frohlich E, Balint Z, Tang B, Strohmaier H, Lochmuller H, Schlotter-Weigel B, Senderek J, Krebs A, Dick KJ, Petty R, Longman C, Anderson NE, Padberg GW, Schelhaas HJ, van Ravenswaaij-Arts CM, Pieber TR, Crosby AH, Guelly C (2010) Alterations in the ankyrin domain of TRPV4 cause congenital distal SMA, scapuloperoneal SMA and HMSN2C. Nat Genet 42:160–164
25. Deng HX, Klein CJ, Yan J, Shi Y, Wu Y, Fecto F, Yau HJ, Yang Y, Zhai H, Siddique N, Hedley-Whyte ET, Delong R, Martina M, Dyck PJ, Siddique T (2010) Scapuloperoneal spinal muscular atrophy and CMT2C are allelic disorders caused by alterations in TRPV4. Nat Genet 42:165–169
26. Landoure G, Zdebik AA, Martinez TL, Burnett BG, Stanescu HC, Inada H, Shi Y, Taye AA, Kong L, Munns CH, Choo SS, Phelps CB, Paudel R, Houlden H, Ludlow CL, Caterina MJ, Gaudet R, Kleta R, Fischbeck KH, Sumner CJ (2010) Mutations in TRPV4 cause Charcot-Marie-Tooth disease type 2C. Nat Genet 42:170–174
27. Hellwig N, Albrecht N, Harteneck C, Schultz G, Schaefer M (2005) Homo- and heteromeric assembly of TRPV channel subunits. J Cell Sci 118:917–928
28. Arniges M, Fernandez-Fernandez JM, Albrecht N, Schaefer M, Valverde MA (2006) Human TRPV4 channel splice variants revealed a key role of ankyrin domains in multimerization and trafficking. J Biol Chem 281:1580–1586
29. Cuajungco MP, Grimm C, Oshima K, D'hoedt D, Nilius B, Mensenkamp AR, Bindels RJ, Plomann M, Heller S (2006) PACSINs bind to the TRPV4 cation channel. PACSIN 3 modulates the subcellular localization of TRPV4. J Biol Chem 281:18753–18762
30. Duncan LM, Deeds J, Hunter J, Shao J, Holmgren LM, Woolf EA, Tepper RI, Shyjan AW (1998) Down-regulation of the novel gene melastatin correlates with potential for melanoma metastasis. Cancer Res 58:1515–1520
31. Bellone RR, Brooks SA, Sandmeyer L, Murphy BA, Forsyth G, Archer S, Bailey E, Grahn B (2008) Differential gene expression of TRPM1, the potential cause of congenital stationary night blindness and coat spotting patterns (LP) in the Appaloosa horse (Equus caballus). Genetics 179:1861–1870

32. Bellone RR, Forsyth G, Leeb T, Archer S, Sigurdsson S, Imsland F, Mauceli E, Engensteiner M, Bailey E, Sandmeyer L, Grahn B, Lindblad-Toh K, Wade CM (2010) Fine-mapping and mutation analysis of TRPM1: a candidate gene for leopard complex (LP) spotting and congenital stationary night blindness in horses. Brief Funct Genomics 9:193–207
33. Li Z, Sergouniotis PI, Michaelides M, Mackay DS, Wright GA, Devery S, Moore AT, Holder GE, Robson AG, Webster AR (2009) Recessive mutations of the gene TRPM1 abrogate ON bipolar cell function and cause complete congenital stationary night blindness in humans. Am J Hum Genet 85:711–719
34. Audo I, Kohl S, Leroy BP, Munier FL, Guillonneau X, Mohand-Said S, Bujakowska K, Nandrot EF, Lorenz B, Preising M, Kellner U, Renner AB, Bernd A, Antonio A, Moskova-Doumanova V, Lancelot ME, Poloschek CM, Drumare I, foort-Dhellemmes S, Wissinger B, Leveillard T, Hamel CP, Schorderet DF, De BE, Berger W, Jacobson SG, Zrenner E, Sahel JA, Bhattacharya SS, Zeitz C (2009) TRPM1 is mutated in patients with autosomal-recessive complete congenital stationary night blindness. Am J Hum Genet 85:720–729
35. Nakamura M, Sanuki R, Yasuma TR, Onishi A, Nishiguchi KM, Koike C, Kadowaki M, Kondo M, Miyake Y, Furukawa T (2010) TRPM1 mutations are associated with the complete form of congenital stationary night blindness. Mol Vis 16:425–437
36. van Genderen MM, Bijveld MM, Claassen YB, Florijn RJ, Pearring JN, Meire FM, McCall MA, Riemslag FC, Gregg RG, Bergen AA, Kamermans M (2009) Mutations in TRPM1 are a common cause of complete congenital stationary night blindness. Am J Hum Genet 85: 730–736
37. Xu XZ, Moebius F, Gill DL, Montell C (2001) Regulation of melastatin, a TRP-related protein, through interaction with a cytoplasmic isoform. Proc Natl Acad Sci USA 98:10692–10697
38. Brink PA, Ferreira A, Moolman JC, Weymar HW, van der Merwe PL, Corfield VA (1995) Gene for progressive familial heart block type I maps to chromosome 19q13. Circulation 91:1633–1640
39. Kruse M, Schulze-Bahr E, Corfield V, Beckmann A, Stallmeyer B, Kurtbay G, Ohmert I, Schulze-Bahr E, Brink P, Pongs O (2009) Impaired endocytosis of the ion channel TRPM4 is associated with human progressive familial heart block type I. J Clin Invest 119: 2737–2744
40. Walder RY, Shalev H, Brennan TM, Carmi R, Elbedour K, Scott DA, Hanauer A, Mark AL, Patil S, Stone EM, Sheffield VC (1997) Familial hypomagnesemia maps to chromosome 9q, not to the X chromosome: genetic linkage mapping and analysis of a balanced translocation breakpoint. Hum Mol Genet 6:1491–1497
41. Schlingmann KP, Weber S, Peters M, Niemann NL, Vitzthum H, Klingel K, Kratz M, Haddad E, Ristoff E, Dinour D, Syrrou M, Nielsen S, Sassen M, Waldegger S, Seyberth HW, Konrad M (2002) Hypomagnesemia with secondary hypocalcemia is caused by mutations in TRPM6, a new member of the TRPM gene family. Nat Genet 31:166–170
42. Walder RY, Landau D, Meyer P, Shalev H, Tsolia M, Borochowitz Z, Boettger MB, Beck GE, Englehardt RK, Carmi R, Sheffield VC (2002) Mutation of TRPM6 causes familial hypomagnesemia with secondary hypocalcemia. Nat Genet 31:171–174
43. Schlingmann KP, Sassen MC, Weber S, Pechmann U, Kusch K, Pelken L, Lotan D, Syrrou M, Prebble JJ, Cole DE, Metzger DL, Rahman S, Tajima T, Shu SG, Waldegger S, Seyberth HW, Konrad M (2005) Novel TRPM6 mutations in 21 families with primary hypomagnesemia and secondary hypocalcemia. J Am Soc Nephrol 16:3061–3069
44. Chubanov V, Schlingmann KP, Waring J, Heinzinger J, Kaske S, Waldegger S, Schnitzler M, Gudermann T (2007) Hypomagnesemia with secondary hypocalcemia due to a missense mutation in the putative pore-forming region of TRPM6. J Biol Chem 282:7656–7667
45. Chubanov V, Waldegger S, Schnitzler M, Vitzthum H, Sassen MC, Seyberth HW, Konrad M, Gudermann T (2004) Disruption of TRPM6/TRPM7 complex formation by a mutation in the TRPM6 gene causes hypomagnesemia with secondary hypocalcemia. Proc Natl Acad Sci USA 101:2894–2899

46. Clapham DE, Nilius B, Owsianik G (2009) Transient Receptor Potential Channels: TRPP1. [article online], 2009, Available from http://www.iuphar-db.org/DATABASE/ObjectDisplay Forward?objectId=504.
47. The European Polycystic Kidney Disease Consortium (1994) The polycystic kidney disease 1 gene encodes a 14 kb transcript and lies within a duplicated region on chromosome 16. Cell 77:881–894
48. Qian F, Germino FJ, Cai Y, Zhang X, Somlo S, Germino GG (1997) PKD1 interacts with PKD2 through a probable coiled-coil domain. Nat Genet 16:179–183
49. Sharif-Naeini R, Folgering JH, Bichet D, Duprat F, Lauritzen I, Arhatte M, Jodar M, Dedman A, Chatelain FC, Schulte U, Retailleau K, Loufrani L, Patel A, Sachs F, Delmas P, Peters DJ, Honore E (2009) Polycystin-1 and -2 dosage regulates pressure sensing. Cell 139: 587–596
50. Patel V, Chowdhury R, Igarashi P (2009) Advances in the pathogenesis and treatment of polycystic kidney disease. Curr Opin Nephrol Hypertens 18:99–106
51. Koptides M, Hadjimichael C, Koupepidou P, Pierides A, Constantinou DC (1999) Germinal and somatic mutations in the PKD2 gene of renal cysts in autosomal dominant polycystic kidney disease. Hum Mol Genet 8:509–513
52. Koptides M, Mean R, Demetriou K, Constantinides R, Pierides A, Harris PC, Deltas CC (2000) Screening of the PKD1 duplicated region reveals multiple single nucleotide polymorphisms and a de novo mutation in Hellenic polycystic kidney disease families. Hum Mutat 16:176
53. Pei Y (2001) A "two-hit" model of cystogenesis in autosomal dominant polycystic kidney disease? Trends Mol Med 7:151–156
54. Watnick T, He N, Wang K, Liang Y, Parfrey P, Hefferton D, St George-Hyslop P, Germino G, Pei Y (2000) Mutations of PKD1 in ADPKD2 cysts suggest a pathogenic effect of trans-heterozygous mutations. Nat Genet 25:143–144
55. Wu G, Somlo S (2000) Molecular genetics and mechanism of autosomal dominant polycystic kidney disease. Mol Genet Metab 69:1–15
56. Wu G, Tian X, Nishimura S, Markowitz GS, D'Agati V, Park JH, Yao L, Li L, Geng L, Zhao H, Edelmann W, Somlo S (2002) Trans-heterozygous Pkd1 and Pkd2 mutations modify expression of polycystic kidney disease. Hum Mol Genet 11:1845–1854
57. Puertollano R, Kiselyov K (2009) TRPMLs: in sickness and in health. Am J Physiol Renal Physiol 296:F1245–F1254
58. Bargal R, Avidan N, Ben-Asher E, Olender Z, Zeigler M, Frumkin A, Raas-Rothschild A, Glusman G, Lancet D, Bach G (2000) Identification of the gene causing mucolipidosis type IV. Nat Genet 26:118–123
59. Bassi MT, Manzoni M, Monti E, Pizzo MT, Ballabio A, Borsani G (2000) Cloning of the gene encoding a novel integral membrane protein, mucolipidin-and identification of the two major founder mutations causing mucolipidosis type IV. Am J Hum Genet 67: 1110–1120
60. LaPlante JM, Ye CP, Quinn SJ, Goldin E, Brown EM, Slaugenhaupt SA, Vassilev PM (2004) Functional links between mucolipin-1 and Ca2+-dependent membrane trafficking in mucolipidosis IV. Biochem Biophys Res Commun 322:1384–1391
61. Vergarajauregui S, Puertollano R (2008) Mucolipidosis type IV: the importance of functional lysosomes for efficient autophagy. Autophagy 4:832–834
62. Slaugenhaupt SA, Acierno JS Jr, Helbling LA, Bove C, Goldin E, Bach G, Schiffmann R, Gusella JF (1999) Mapping of the mucolipidosis type IV gene to chromosome 19p and definition of founder haplotypes. Am J Hum Genet 65:773–778
63. Altarescu G, Sun M, Moore DF, Smith JA, Wiggs EA, Solomon BI, Patronas NJ, Frei KP, Gupta S, Kaneski CR, Quarrell OW, Slaugenhaupt SA, Goldin E, Schiffmann R (2002) The neurogenetics of mucolipidosis type IV. Neurology 59:306–313
64. Sun M, Goldin E, Stahl S, Falardeau JL, Kennedy JC, Acierno JS Jr, Bove C, Kaneski CR, Nagle J, Bromley MC, Colman M, Schiffmann R, Slaugenhaupt SA (2000) Mucolipidosis

type IV is caused by mutations in a gene encoding a novel transient receptor potential channel. Hum Mol Genet 9:2471–2478
65. Wang ZH, Zeng B, Pastores GM, Raksadawan N, Ong E, Kolodny EH (2001) Rapid detection of the two common mutations in Ashkenazi Jewish patients with mucolipidosis type IV. Genet Test 5:87–92
66. Venugopal B, Browning MF, Curcio-Morelli C, Varro A, Michaud N, Nanthakumar N, Walkley SU, Pickel J, Slaugenhaupt SA (2007) Neurologic, gastric, and opthalmologic pathologies in a murine model of mucolipidosis type IV. Am J Hum Genet 81: 1070–1083
67. Xenophontos S, Constantinides R, Hayashi T, Mochizuki T, Somlo S, Pierides A, Deltas CC (1997) A translation frameshift mutation induced by a cytosine insertion in the polycystic kidney disease 2 gene (PDK2). Hum Mol Genet 6:949–952
68. Veldhuisen B, Saris JJ, de HS, Hayashi T, Reynolds DM, Mochizuki T, Elles R, Fossdal R, Bogdanova N, van Dijk MA, Coto E, Ravine D, Norby S, Verellen-Dumoulin C, Breuning MH, Somlo S, Peters DJ (1997) A spectrum of mutations in the second gene for autosomal dominant polycystic kidney disease (PKD2). Am J Hum Genet 61:547–555
69. Viribay M, Hayashi T, Telleria D, Mochizuki T, Reynolds DM, Alonso R, Lens XM, Moreno F, Harris PC, Somlo S, San Millan JL (1997) Novel stop and frameshifting mutations in the autosomal dominant polycystic kidney disease 2 (PKD2) gene. Hum Genet 101:229–234
70. Reynolds DM, Hayashi T, Cai Y, Veldhuisen B, Watnick TJ, Lens XM, Mochizuki T, Qian F, Maeda Y, Li L, Fossdal R, Coto E, Wu G, Breuning MH, Germino GG, Peters DJ, Somlo S (1999) Aberrant splicing in the PKD2 gene as a cause of polycystic kidney disease. J Am Soc Nephrol 10:2342–2351
71. Mochizuki T, Wu G, Hayashi T, Xenophontos SL, Veldhuisen B, Saris JJ, Reynolds DM, Cai Y, Gabow PA, Pierides A, Kimberling WJ, Breuning MH, Deltas CC, Peters DJ, Somlo S (1996) PKD2, a gene for polycystic kidney disease that encodes an integral membrane protein. Science 272:1339–1342
72. Bergmann C, Bruchle NO, Frank V, Rehder H, Zerres K (2008) Perinatal deaths in a family with autosomal dominant polycystic kidney disease and a PKD2 mutation. N Engl J Med 359:318–319
73. Pei Y, Wang K, Kasenda M, Paterson AD, Liang Y, Huang E, Lian J, Rogovea E, Somlo S, St George-Hyslop P (1998) A novel frameshift mutation induced by an adenosine insertion in the polycystic kidney disease 2 (PKD2) gene. Kidney Int 53:1127–1132
74. Pei Y, Paterson AD, Wang KR, He N, Hefferton D, Watnick T, Germino GG, Parfrey P, Somlo S, St George-Hyslop P (2001) Bilineal disease and trans-heterozygotes in autosomal dominant polycystic kidney disease. Am J Hum Genet 68:355–363
75. Bargal R, Avidan N, Olender T, Ben AE, Zeigler M, Raas-Rothschild A, Frumkin A, Ben-Yoseph O, Friedlender Y, Lancet D, Bach G (2001) Mucolipidosis type IV: novel MCOLN1 mutations in Jewish and non-Jewish patients and the frequency of the disease in the Ashkenazi Jewish population. Hum Mutat 17:397–402
76. Bach G, Webb MB, Bargal R, Zeigler M, Ekstein J (2005) The frequency of mucolipidosis type IV in the Ashkenazi Jewish population and the identification of 3 novel MCOLN1 mutations. Hum Mutat 26:591
77. Raychowdhury MK, Gonzalez-Perrett S, Montalbetti N, Timpanaro GA, Chasan B, Goldmann WH, Stahl S, Cooney A, Goldin E, Cantiello HF (2004) Molecular pathophysiology of mucolipidosis type IV: pH dysregulation of the mucolipin-1 cation channel. Hum Mol Genet 13:617–627
78. Dobrovolny R, Liskova P, Ledvinova J, Poupetova H, Asfaw B, Filipec M, Jirsova K, Kraus J, Elleder M, Mucolipidosis IV (2007) Report of a case with ocular restricted phenotype caused by leaky splice mutation. Am J Ophthalmol 143:663–671
79. Tuysuz B, Goldin E, Metin B, Korkmaz B, Yalcinkaya C (2009) Mucolipidosis type IV in a Turkish boy associated with a novel MCOLN1 mutation. Brain Dev 31:702–705

ERRATUM

Erratum to: Transient Receptor Potential Channels

Md. Shahidul Islam

Karolinska Institutet, Department of Clinical Sciences and Education,
Södersjukhuset, SE-118 83 Stockholm, Sweden
and
Uppsala University Hospital, AR Division, Uppsala, Sweden
shaisl@ki.se

M.S. Islam (ed.), *Transient Receptor Potential Channels*, Advances in Experimental Medicine
and Biology 704, DOI 10.1007/978-94-007-0265-3,© Springer Science+Business Media B.V. 2011

DOI 10.1007/978-94-007-0265-3_53

The author has found some errors in this book which should read as follows:

1. In Contributors, page xviii

   (a) The email address of the author **Marta Kaleta** is incorrect. The correct e-mail address is "m.kaleta@londonmet.ac.uk"

2. In Contributors, page xix

   (a) Affiliation of the author **"V'yacheslav Lehen'kyi"** has been updated incorrectly. The correct affiliation should read as below:

   V'yacheslav Lehen'kyi Laboratoire de Physiologie Cellulaire, INSERM U1003, Equipe labellisée par la Ligue Nationale contre le cancer, Université des Sciences et Technologies de Lille (USTL), Villeneuve d'Ascq F-59655, France, vyacheslav.lehenkyi@univ-lille1.fr"

3. In Chapter 32, page 595

   (a) One of the corresponding author name and his affiliation was missed. The corresponding author name and his affiliation should read as below:

   Juan D. Navarro-López (✉)
   Department of Physiology, School of Medicine,
   University of Granada, 18071 Granada, Spain
   e-mail: jdnavarro@ugr.es

The online version of the book can be found at
http://dx.doi.org/10.1007/978-94-007-0265-3

Index

Note: The letters 'f' and 't' following the locators refers to figures and tables cited in the text.

A
AA, *see* Arachidonic acid (AA)
aa, *see* Amino acid (aa)
AACOCF3, *see* Arachidonyl trifluoromethyl ketone (AACOCF3)
α1A-adrenoceptor antagonist prazosin, 553
aa 723–928 fragment, 19
aa 839–873 fragment, 17, 19
Aβ, *see* Amyloid β-peptide (Aβ)
ABC protein, *see* ATP-binding cassette protein (ABC transporter)
Abeele, F. V., 715
Abbott Laboratories, 645
ACA, *see* N-(p-amylcinnamoyl)anthranilic acid (ACA)
Acetaldehyde, 68t, 69
Acetic acid-induced nociceptive response, 50
Acetylcholine-secreting sympathetic neuron, 179
Acetylsalicylic acid, 49t, 55
Acrolein, 67t, 69, 867, 971–972
Acrosome reaction, 125, 130, 301, 325, 378, 881–882, 885, 888–889, 891–895
Actinopterigy, 223
Activation gate, 485–486
 Leu681, amino acid residue, 485–486
 Tyr671, ion pathway, 486
Acute myeloid leukaemia and myelodysplastic syndrome (AML/MDS), 949
Acylated phloroglucinol hyperforin, 71
Acylation, 96
Acyl-homoserine lactone, 163, 836

Acyl-lipid desaturas, 455
Adenophostin A, 71, 72t
Adenosine 5′-diphosphoribose, 65, 789
Adlea, 651
ADP, 4t, 10, 155, 597, 598f, 599, 602, 669, 711t, 814–815, 890
ADPKD, *see* Autosomal dominant polycystic kidney disease (ADPKD)
ADPKD-causing mutations
 PKD1
  L4224P, 17
  R4227X, 17
 TRPP2
  E837X, 17
  R742X, 17
  R807X, 17
  R872X, 17
ADPR, *see* ADP-ribose (ADPR)
ADP-ribose (ADPR), 99, 112–113, 113f, 380–382, 464, 534–536, 535f, 537f, 538, 711t, 712–713, 717, 770, 784, 789, 816–820, 885
β-Adrenergic agonists, 740
α2 Adrenoceptor, 49t, 55, 188
Adult polycystic kidney disease (ADPKD), 1023
Aframomum daniellii (aframodial), 46t, 52
Aframomum melegueta (Melegueta pepper), 45t, 51, 70
Agelenopsis aperta (American funnel web spider), 48t, 54
Agonist-dependent vasorelaxation, 378
Ahern, G. P., 526–527
Aiptasia pulchella (anemone), 54

Airway cells of non-neuronal origin, TRP channels in, 974
  airway structural and immune cells, 975–976
  capsaicin receptor and, 976–978
  intracellular calcium homeostasis and inflammation, 975
  TRPA1, 974–975
Airway chemoreceptor, 163
Airway irritation, 641
Airway SMC, molecular expression and functional role
  active NSCCs
    cell-attached single channel recordings, 733
    homomeric/heteromeric, 734
  multiple TRPC molecules
    guinea-pig airway SMCs, 732
    human airway SMCs, 732
    porcine airway SM tissues, 732
  TRPC3-encoded NSCCs
    agonist-induced increase in [Ca^{2+}]i, 736–737
    expression and activity in asthmatic airway SMCs, 740–741
    mediated Ca^{2+} influx, 739
    Na$^+$/Ca^{2+} exchanger-1 and/or L-type Ca^{2+} channels, 737–738, 737f
    phospholipase C activation, 739
    resting [Ca^{2+}]i lowered, 736
    V$_m$ hyperpolarization, 734
  TRPC3/TRPC6 gene silencing, 739
Airway smooth muscle hyperreactivity, 797–798
AITC
  elicited control response *vs.* signal, comparison, 110
Ajoene, 66, 67t
Aka Mcoln2, *see* TRPML2
Akey, J. M., 259–260
Akt/PKB, 954–955
Alarm/defense mechanism, 197
Albatrellus confluens (neogri-folin/albaconol/grifolin), 47t, 52–53
Albatrellus ovinus (scutigeral), 47t, 52
Albert, A. P., 391–404
ALG-2 protein, 215
Alkylation, 96
Allen, E., 921
Allergic asthma, 973, 976
Allicin (garlic), 66, 67t, 110, 115, 470, 623, 782, 792

Allicin (2-propenyl 2-propene thiosulfinate), 66
Allium, 66, 67t
Allosteric coupling, 517, 524, 527f, 528
Allosteric gating in thermoTRP, 481–483
  allosteric MWC model, 481, 482f
  capsaicin gating of TRPV1, 481
  coupling of electric energy, 483
  single channel recordings, 483
  voltage gating, 483
Allosteric gating scheme, 525–527
Allyl isothiocyanate (AITC), 66, 67t, 69, 73, 110, 190t, 485, 766, 792, 852, 972
Alpha-amino-3-hydroxy-5-methyl-4-isoxazolepropionic acid (AMPA), 156, 161, 547, 576, 999
Al-Shawi laboratory, 29
ALS/PD, *see* Guamanian amyotrophic lateral sclerosis and parkinsonian dementia (ALS/PD)
Alvarez, O., 469–486
Alveolar capillary endothelium, 978
Alzheimer, 175, 382, 537–538, 575
Alzheimer's disease, 175, 382, 537
Amantini, C., 947–961
Ambudkar, I. S., 435–445, 547, 824
AMG-517, 653–654
Amgen, 646, 652–653
Amide I band, 458
Amiloride, 278, 750, 839, 863t, 869–870, 920
Amino acid (aa), 2, 17, 19, 150, 230, 419t–420t, 537f
  184, 536, 537f
  218, 536
  538–557, 536, 537f
  1292–1325, 536, 537f
2-Aminoethoxydiphenyl borate (2-APB), 56, 88, 89f, 91–93, 190, 420, 477, 711t, 735t, 852
2-Aminoethyl diphenylborinate (2-APB), 627
Amino (N) termini, 2f, 2, 3t, 8–9, 12, 15, 211, 316, 392, 415, 440, 463, 547, 710, 719, 919, 930, 1011
2-Aminophenylborate (2-APB), 176
4-Aminopyridine, 733
Aminoquinazolines, 647
Amoroso, S., 586
AMPA receptor, 156, 161, 547, 576, 999
Amplification mechanism, 317, 375
AMP–PNP (an ATP analog), 10, 11f, 155

Index 1035

AMTB (TRPM8 blocker), 89f, 97, 193t, 195, 711t, 716
Amyloid β-peptide (Aβ), 382, 534, 537
Analgesia, 197–199, 201, 503, 622, 626, 629, 647
Analgesic polypeptide HC1 (APHC1), 48t, 54
Anandamide, 48t, 54–55, 61, 142, 190, 191t, 493, 556, 646, 822, 893, 914–915, 953, 989–990, 992, 997
Anaphylaxis, 976
Ancistrodial, 47t, 52
Ancistrotermes cavithorax (ancistrodial), 47t, 52
Androgen-dependent (AD), 933, 938–939
Androgen-independent (AI), 200, 931, 933
Androgen-insensitive neuroendocrine phenotype, 254
Androgen sensitivity, 199–200
Andrographis paniculata (bitter herbaceous plant), 61
Andrographis paniculata (Chinese herb), 627
Angiogenesis, 347–348, 708, 710, 711t, 767–769, 772, 774t, 791, 795–796, 891, 930, 948, 960
Angiotensin (Ang) II receptor, 160, 179, 278, 376, 400, 669, 691–692, 715, 719, 764, 787f, 789, 793f, 1013
Anionic lysophosphatidylinositol, 65
Ankyrine-rich TRP (TRPA), 12, 26, 43, 88, 289, 323, 326–327, 331, 342, 363, 364t, 365, 365t, 367t, 416, 457, 469, 474–475, 493, 547, 563, 616, 688, 710, 732, 750, 760, 782, 948, 974, 988, 1011
Ankyrin repeat domain (ARD), 12, 14
  X-ray crystallography, 12
Ankyrin repeats, 8f, 12, 13f, 14, 27, 32, 34, 244, 517, 623, 638f, 639, 1013t, 1013, 1015t
Annexin A1, 11, 177–179, 233, 691, 784, 918
Anterior hypothalamus, 998
Anti-aging factor, *see* Kallikrein
Anti-androgen therapy, 199, 938
Anti-anxiety drugs, 991
Antidepressant drugs, 991–994
Anti-estrogen-receptor (ER), 959
Anti-inflammatory assay, 55
Antioxidant enzymes, 532
Aortic smooth muscle cells (A7r5 cells), 563, 693–694, 783, 797, 800
Aortic stenosis, 786

2-APB, *see* 2-Aminoethoxydiphenyl borate (2-APB); 2-Aminoethyl diphenylborinate (2-APB)
2-APB action, 91–93, 111f, 176, 177t, 190, 191t, 420, 441, 477, 584, 603, 627, 675, 711t, 715–716, 735t, 768, 932
APHC1, *see* Analgesic polypeptide HC1 (APHC1)
Apical calcium channels, 934, 958
Apoptosis, 55, 177, 200, 215, 231t, 254, 256, 263, 279, 288, 294–295, 327, 532, 536–537, 575, 579, 585, 669, 675, 789, 820, 851, 853, 929–931, 935–938, 940, 948, 950–956, 959–960, 977
Apoptotic-induced cell death, 947, 949
Apoptotic-resistant neuroendocrine phenotype, 254, 931
Appaloosa horse phenotype, 140, 142, 1017–1018
AP180 protein, 226
Aquaporin, 5f, 6, 28, 282f
Arachidonic acid (AA), 48t, 58t, 61–62, 63t, 64t, 65, 68t, 71, 72t, 94, 128, 188, 380, 395, 418, 532, 578, 617, 627, 674–675, 739, 763–764, 771, 774t, 792, 800, 816, 819, 822, 824, 835, 890, 893, 915–916, 956, 990
Arachidonic acid metabolites, 48t, 58t, 68t, 71, 72t, 94, 617, 627, 774t, 890
2-Arachidonoylglycerol, 61
Arachidonyl trifluoromethyl ketone (AACOCF3), 94
Araki, I., 861–871
A7R5 aortic smooth muscle cells, 563
A7r5 cells, *see* Aortic smooth muscle cells (A7r5 cells)
ARD, *see* Ankyrin repeat domain (ARD)
Arf6, *see* GTPase ADP-ribosylation Factor 6 (Arf6) pathway
Arg701, 643–644
Aromatase deficient (ArKO) mice, 248
ARPE19 cell line, 234
Arrhenius, 473
Arrhenius activation energy, 473
Arrhythmogenic alteration, 180
Arteriosclerosis, 782
Arthritis, 458, 618, 640t
Artmann, G., 461, 464
A7r5 vascular smooth muscle cell, 90
Aryl cinnamides, 649t

Asakawa, M., 851
Ashford, M. L., 811
Asian spice, see Zanthoxylum piperitum (Szechuan pepper)
Asn604, 644
Aspirin, 55
Asp964 residue, 381
Asthma, 688, 691–692, 740–741, 973, 976–977
Astrocytes, 936, 952, 955, 957
A563T, 261–262
ATG initiation codon, 136
Atomic force microscopy, 14, 17
Atomic resolution, 6, 33–34
ATP binding, 10, 12, 13f, 150f, 154–155, 157, 261
ATP-binding cassette protein (ABC transporter), 151, 157, 814
ATP-dependent K^+ channel (K_{ATP}), 158, 812
ATP-evoked calcium signalling, 131
$ATP^+ Mg^{2+}$, 437
ATP synthesis, 669
Atropine, 549t, 597, 599, 722
Automated patch clamp Anno 2010
  results, 110–120
    HEK293-TRPA1 cell testing on QPatch, 111f
    TRPA1, 110–112
    TRPCs, 112–115
    TRPMs, 112–115
    TRPVs, 115–118
    See also individual entries
Autophagic dysfunction, 210
Autophagic vesicle, 235
Autophagocytosis, see Autophagy
Autophagosome, see Autophagic vesicle
Autophagy, 210, 211t, 215, 231t, 234, 367t
Autosomal dominant polycystic kidney disease (ADPKD), 17, 71, 73, 88, 289–290, 771, 919
  arterial hypertension, 290
  cardiac and vascular manifestations, 290, 290f
  Cre-loxP recombination system deletion of Pkd1, 290
  cysts in
    "two hit" mechanism, 289
  extra-renal cysts, 290
  fatal ESRD, 289
  PKD1 or PKD2 genes, 289
  TRPP2
    family, 289
    and PC-1 function, signalling pathway, 290
A8V, 261–262
Axonal growth cones, 576, 579
AZD-1386 (AstraZeneca), 652t, 654

B

BAA, see Bisandrographolide A (BAA)
Bacterial K^+ channel (KcsA), 1, 6, 643
Baez-Nieto, D., 469–486
Bandell, M., 481
BAPTA-AM, 404
Barritt, G. J., 667–680
Basic fibroblast growth factor (bFGF), 557, 576–577, 988
Bates-Withers, C., 173–180
Bax, 938, 960
Bayliss, A., 798
Bayliss effect, 158, 160, 798–799
Bayliss, W., 158
Bayliss, W. M., 694
B-cell receptor (BCR), 375–376, 375f
B-cell signaling, 375
Bcl-2, 938–939, 950
BCR, see B-cell receptor (BCR)
Beckwith–Wiedemann syndrome, 939
Bedouin, 1020–1021
Beech, D. J., 114
Benign prostatic hyperplasia (BPH), 201
Benzimidazol analogue, 649t
Berbey, C., 753
Bernd Nilius, 463
Beta-cell, 94, 538–539, 558
Beta-galactoside-binding lectin Galectin 1 (GAL1), 934, 958
Bezzerides, V. J., 550
bFGF, see Basic fibroblast growth factor (bFGF)
Bhave, G., 498
Biaryl carboxamides, 647, 649t
Biliary epithelial cells (cholangiocytes), 298, 667
Bindels, R., 240, 279
Binshtok, A. M., 622
(Bio) chemical process, 452, 454
Bioinformatics, 62, 362, 367, 469, 471f
Biological antioxidants
  absorbic acid, 532
  α-tocopherol, 532
Biosynthetic menthol precursor, see Menthone
Bisandrographolide A (BAA), 61, 330, 627, 822
3,5-Bis(trifluoromethyl)pyrazole, 750

Bitter/gustducin/TRPM5 pathway, 61, 157, 163, 298, 784, 832, 833f, 834, 836, 838f
BK channels, 524, 527, 799
Bladder cancer cell, 200
Bladder micturition reflex, 195
Blastocyst, 915
Blaumuller C., 353
B-lymphocyte, 226, 375–376, 445, 949
B16 mouse melanoma cell line, 136
BNP, *see* Brain natriuretic peptides (BNP)
Boels, K., 558
Boiling globule, 459
Bollimuntha, S., 573–587
Boltzmann distribution law, 473
Boltzmann equation, 518
Bone cancer, 618, 648t
Borate compound, 93
Borneol-evoked response, 59
Boswellia sp., 57t, 60
Botulinum neurotoxin A, 556
Boulay, G., 547
Boustany, C., 678
Bovine serum albumin (BSA), 644
Bovine testis, 129, 886
Bowman's capsule, 278
BPB, *see* p-bromophenacyl bromide (BPB)
BPH, *see* Benign prostatic hyperplasia (BPH)
Brachyolmia, 61, 73, 1014–1016
Brachyolmia, type 3, 1014–1015
 chromosome 12q24.1-q24.2, 1014
 R616Q/V620I mutations in HEK, expression of, 1014
 SOX9 pathway, 1014–1015
 TRPV4 gene, mutations in, 1014
Bradycardia, 641
Bradykinin (BK), 31, 109, 179, 198, 328, 436, 497, 504, 506–507, 524, 552, 597, 617, 624, 691, 715, 719, 764, 790–791, 799, 852, 865, 887, 890, 894, 971–973
Brain, 537, 596, 988–990
Brain-derived neurotrophic factor (BDNF), 343, 576–577, 579, 583, 585, 988, 991
Brain natriuretic peptides (BNP), 788
Brain, TRP channels in
 TRPC channels, 988–989
 TRPV channels, 989–990
Brauchi, S., 477, 479–481, 517–528
Brefeldin A, 553, 561
Brevetoxin, 54
Brinkmeier, H., 749–756

Broad-spectrum inhibitor, *see* N-(p-amylcinnamoyl)anthranilic acid (ACA)
Bronchoconstriction, 641
Bronchodilators, 976
Brussels sprouts, 66
BSA, *see* Bovine serum albumin (BSA)
B. subtilis, 455
BTP2, *see* N-[4-3,5-bis(trifluromethyl)pyrazol-1-yl]-4-methyl-1,2,3-thiadiazole-5-carboxamine (BTP2)
Budding Yeast
 biochemical studies with purification, 34
 cryo-EM structures, 30–33
  electron cryo-microscopy/single particle analysis, TRPV1 structure, 31f
  reported channel structures, comparison, 33f
  TRPV1 structures, similarity, 32f
 divide and conquer approach, 33–34
 functional studies and genetic screens, 34–35
 ion channel structural biology, 27–28
  eukaryotic membrane proteins, 27
 Saccharomyces cerevisiae, expression/functional analysis, 28–30
 TRP channel family, 26–27
  cytoplasmic amino-terminal domain, 26
  distribution/determination, significance, 27
  heterologous expression studies, 35
  TRPA (ankyrin), 26
  TRPC (canonical), 26
  TRPML (mucolipin), 26
  TRPM (melastatin), 26
  TRPP (polycystin), 26
  TRPV (vanilloid), 26
  voltage-gated cation channel, relation, 26
Büldt, Dr. G., 461, 464
Bunker, L. E., 232
Buraei, Z., 19
Burdyga, T., 707–723
5-Butyl-7-chloro-6-hydroxybenzo[c]-quinolizinium chloride (MPB-104), 157

C
Ca^{2+}-activated CAN current, 604
Ca^{2+}-activated NSCC (CAN) channels, 713, 814

Ca^{2+}-bound calmodulin, 535
Ca^{2+}-calmodulin (Ca^{2+}-CaM), 12, 16, 378, 379f, 380
Ca^{2+}/calmodulindependent kinase IV (CaMKIV), 583, 989
Ca^{2+}/calmodulin inhibitors, 675
Ca^{2+}/calmodulin phosphorylation, 754
Ca^{2+}-CaM, see Ca^{2+}-calmodulin (Ca^{2+}-CaM)
Caco-2 cell, 245–246
Cacospongia mollior (scalaradial), 46t, 52
Ca^{2+}-dependent cellular response, 374
Ca^{2+}-dependent inactivation (CDI), 245, 421, 423, 437, 440–441
Ca^{2+}-desensitization, 154–155, 815
cADPR, see Cyclic adenosine diphosphoribose (cADPR)
Caenorhabditis elegans, 223, 360, 931
Ca^{2+} entry-evoked translocation, 375
Ca^{2+} entry mechanism, 88, 90, 92, 94, 98, 262, 436
$[Ca^{2+}]_i$, see Intracellular Ca^{2+} concentration ($[Ca^{2+}]_i$)
Ca^{2+} induced Ca^{2+} release (CICR), 815–816
Ca^{2+} influx, 52, 55, 65, 96, 99, 128–129, 161, 188, 244, 246, 253, 257, 279, 374–378, 375f, 379f, 382, 416–417, 436, 495, 496f, 497–499, 501–502, 533, 535f, 535–536, 538–539, 573–577, 579–586, 619, 625, 637, 693–697, 709, 712–713, 717, 721f, 737–740, 737f, 742, 750, 753–755, 763–765, 768, 771–772, 774t, 786, 788, 791–795, 797, 799, 820, 833f, 867, 869–870, 886, 888–889, 893, 935–936, 939, 952, 954, 957, 959, 974–975, 977–978, 996, 1015, 1018
Calbindin-D_{28k}, 244, 246, 257–258
Calcineurin-NFAT signalling pathway, 376, 754
Calcitonin gene related peptide (CGRP), 492, 624, 627, 640, 651, 790, 856, 863–868, 999
Calcium ($[Ca^{2+}]_i$), 128–129, 153f, 154, 158, 161–163, 279, 344, 378, 382, 392, 401, 436, 439, 444–445, 505, 532–533, 535f, 536–537, 539, 574, 576, 581, 585, 688, 695–697, 711t, 713, 715, 719–722, 720f, 721f, 734–742, 735t, 760, 763–768, 771, 790, 795, 799, 812, 814–816, 818, 820–823, 821f, 887, 894, 910, 913, 931, 936–937, 939, 957
Calcium chelators
 BAPTA, 109
 EGTA, 109
Calcium homeostasis, 88, 295, 350, 559, 687, 887, 918, 975
Calcium-permeable ion channels
 non-voltage gated non-selective cation channels (NSCC), 708
 non-voltage gated NSCC, 708
 voltage-gated Ca^{2+} channels (VGCC), 708
"Calcium-related gene spray," 319
Calcium-release-activated current (I_{CRAC}), 93, 157–158, 160, 414–415, 418, 421, 423, 424f, 437–438, 440–441, 443
Calcium store depletion, 112, 715, 886
Calmodulin (CaM), 30, 34, 127, 151, 154, 178, 330, 377, 379, 417, 440, 444, 495, 500, 506, 535f, 558, 582, 601, 605, 639, 675, 694, 713, 735t, 753–754, 793f, 885, 887, 889–890, 894, 918, 931
Calmodulin (CaM)/IP3R-binding (CIRB) motif, 330
Calvert, C. M., 318
CaM, see Calmodulin (CaM)
CaM dependent kinases (CaMK), 581
CaM/IP_3 receptor-binding (CIRB) domain, 127
CaMK, see CaM dependent kinases (CaMK)
cAMP, see Cyclic adenosine monophosphate (cAMP)
Camphor, 57t, 59, 62, 68t, 115, 330, 470, 485, 627, 852–853
Camphor (+)-borneol, 59
Cancer pain, 640t, 648t
Cannabinoids
 Δ^9-tetrahydrocannabinol (THC), 625
 WIN55215, 625
Cannabinol, 56, 57t, 68t, 69, 330
Cannabis sativa
 CB_1 antagonist, 53
 CB_2 antagonist, 53
 non-psychotropic constituents, 56
  cannabidiol, 56
  cannabinol, 56
Canonical TRPs, 99, 416, 549, 573–587, 782, 910, 988, 1012
Capacitative Ca^{2+} entry, see Store-operated Ca^{2+} entry (SOCE)

Ca^{2+} permeability, 153–154, 212, 643, 737, 737f, 742
Ca^{2+}-permeable channels, 29, 148, 153, 200, 415–416, 578, 643, 669, 670f, 673, 691, 718, 760, 763–764
Capers, 66
Capsaicin (CPS), 470, 539, 615–616, 914–915, 950, 970, 974, 977, 997–1000
  receptor, 976–978
  -sensitive polymodal respiratory irritant sensor, 970–971
Capsaicin-induced dermal vasodilation (CIDV), 653
Capsaicin injection to masseter muscle
  acute nocifensive responses, 618
  masseter hypersensitivity development, 618
Capsazepine, 50–51, 53, 55, 98, 115, 116f, 117, 117, 188, 191t, 197, 200, 539, 642, 647, 648t, 655, 674, 697–698, 711t, 823, 840f, 850, 863t, 867, 977–978, 992–993, 997
Capsicum, 43
  capsaicin (8-Methyl-N-vanillyl-*trans*-6-nonenamide), 44
  dihydrocapsaicin, 44
Carbachol, 328, 548t, 550, 552–553, 555, 598–603, 821f, 866, 931
Carboxamide
  derivatives
    CPS-113, 188
    CPS-369, 188
    WS-12, 188
Carboxyl (C) termini, 2, 2f, 3t, 8–9, 15, 278, 316, 392, 400, 416, 421, 440–441, 463, 547, 616, 639, 710, 784, 919, 1011, 1016, 1023
Carboxylic acid
  derivative
    WS-30, 188
Cardiac hypertrophy, 754, 782
  arrhythmias, 162
    delayed afterdepolarization, 162
    early afterdepolarization, 162–163
Cardiac remodeling, 180
Cardiomyopathy, 376, 754, 756, 782, 786–787, 790
Cardiopulmonary vasculature
  endothelial cells of systemic and pulmonary circulation
    angiogenesis, 795
    cardiovascular benefits, 792
    cysteine S-nitrosylation of TRPC5, 792
    11,12-EETs, 792–793
    endothelial cell migration, 795
    icillin, 793
    increases endothelial permeability, 791, 795
    MgL mice exhibit diminished eNOS expression, 793
    mouse model of hypomagnesemia, 793
    permeability of pulmonary vascular endothelium regulation, 792
    role in endothelium derived vasodilatation, 792
    treatment with siRNA, 795
    TRPA1 identified in endothelial cells, 792
    TRPC3 activity, inhibition, 792
    TRPC4 in TRPC4–/– mice, deletion, 791
    TRPC6 protein, translocation, 792, 793f
    VEGF increases vascular permeability, 791
  heart, physiological function and pathophysiological implications
    in adult mouse cardiomyocytes, 789
    apoptosis, 789
    arrhythmia, 786
    cardiac arrhytmia, 790
    cardiac hypertrophy, 786
    cardiomyopathy, 786
    chest pain, 790
    dilated cardiomyopathy, 790
    Duchenne muscular dystrophy (DMD), 790
    hypertrophic factors, 786
    12-lipoxygenase-derived eicosanoids, 790
    NO-cGMP-PKG pathway, 788
    PARP activation, 789
    in rat neonatal cardiomyocytes, 787
    sildenafil in mice, 788
    tolloid-like 1 gene deletion, 789–790
    TRPC3 channels, inhibited, 787
    TRPC1-deficient mouse model, 788
    TRPC1 in cardiac hypertrophy, 788
    TRPC6 phosphorylation in mouse hearts, 787f, 788
    TRPM4 levels increased, 789
  SMC of systemic vasculature
    activated by α1-adrenergic agonists, 797

Cardiopulmonary vasculature (*cont.*)
   alveolar hypoxia-mediated vaso-
      constrictive phenomenon,
      798
   ATP-induced mitogenesis, 797
   Bayliss effect, 798
   blocked by suppressing TRPC6
      expression, 797
   Ca^{2+} influx, 797
   cell swelling, 799
   heteromeric TRPC1/5 complex,
      activation, 796
   intracellular Ca^{2+} and PKC activation,
      800
   Mg^{2+} influx, 799
   migration, 797
   neointima formation, 796
   normotensiveWKY rats, 799
   PLA2 activated, 800
   proliferation, 795–796
   in rat pulmonary artery smooth muscle
      cells, 797
   regulation of myogenic tone, 798
   STIM1, down-regulation, 797
   synthetic phenotype, 796
   TRPC1 activity and expression, 796
   TRPC3 and TRPC6, up-regulation, 798
   TRPC6–/– mice and wild-type mice,
      comparison, 798
   in TRPC6–/– mouse model, 797
   TRPM4 antisense oligonucleotides in
      living rats, 800
   TRPA channels
      channel tetramerization, 782
      intramolecular interactions, 782
      membrane localization, 782
      TRPA1 activated, 782
   TRPC channels
      calcium sensor STIM, 783
      DAG production, 783, 787f
      detected in embryonic brain, 783
      Orai channels, 783
   TRPM family
      ADP-ribose pyrophosphatase activity,
         784
      C-termini, 784
      mice with genetically disrupted
         TRPM5, 784
      NUDT9 domain, 784
      pronounced voltage modulation, 784
      TRPM4, 5 and 7 in cardiomyocytes,
         expression, 785
      TRPM6 gene in HSH, 785
   TRPP family
      detected in umbilical vein endothelial
         cells, 785
      in heart, 785
      in human coronary arterial smooth
         muscle cells, 785
   TRPV channels
      activated by heat, 786
      in bladder, 786
      in keratinocytes, 786
      in kidney, 786
      in smooth muscle cells, 786
Cardiovascular system, 537, 699, 781–782
Ca^{2+} release-activated Ca^{2+} channel protein 1
   (CRACM1), 92, 418
Ca^{2+} release activated Ca^{2+} (CRAC), 161,
   177, 414–415, 418, 419t, 421,
   423, 425, 437–445, 673
Ca^{2+} release-activated Ca^{2+}-selective (CRAC)
   channels, 414–415, 934, 958
Carrageenan, 617–618, 627, 649t
Carrageenan injection to gastrocnemius muscle
   mechanical hyperalgesia, 618
Carvacrol, 57t, 59, 65, 330, 627, 853, 956
Case-by-case basis, 460
Ca^{2+} sensor, 92, 245, 415, 418, 423, 574f, 581,
   671t, 673–674, 824
Ca^{2+} signals, 92, 291, 295, 374–376, 378–379,
   395, 414, 445, 717, 786
Caspases, 950
   -3 activation, 952
Castillo, J. P., 469–486
CaT1, *see* Ca^{2+} transport protein subtype 1
   (CaT1)
Catalase, 532
Catechol-O-methyltransferase (COMT), 994
Caterina, M. J., 480
Cathepsin B, 936, 955
Ca^{2+} transportation
   function, 244–245
   intestinal absorption, 243
   regulation, 244–245
   renal reabsorption, 243
   transcellular pathways identification,
      240–241
   TRPV5, epithelia, 243–244
   See also TRPV5, TRPV6
Ca^{2+} transport protein subtype 1 (CaT1), 93,
   240, 917
Caveolin-1, 377–379, 549t, 560, 764, 792
Caveolin 1-rich lipid rafts, 533
Cayouette, S., 547
CB_1 receptor, 69, 990, 992, 997–998

Index 1041

CB$_2$ receptor, 990, 998
CCI, *see* Chronic constriction injury (CCI)
CCR, *see* Coiled coil character (CCR)
CD38, 380, 538
cDNA fragment, 115
cDNA library, 44, 137, 240
C12E8, *see* Dodecyl octaethylene glycol ether (C12E8)
C. elegans, 210, 213, 230–232, 240–241, 293, 323–333, 361–363, 364t, 365t, 365, 547, 548t, 554, 556, 562–566
C. elegans, activation of TRP channels in, 323–333, 324f
  mechanical activation, 328–329
    mammalian TRP channels, 328
    mechano/osmo-sensation, 328–329
    mechanosensitivity, 328
    TRPA-1, 329
    TRP-4/TRPN1, 329
  other activation modes, 329
    mammalian TRPA1, 329
    thermo-TRPs, 329
    TRPM channels, 329
  receptor-operated activation, 327–328
    PIP$_2$ hydrolysis, 328
    TRPA-1, 328
    TRPC channels, 328
    TRPV channels, 328
C. elegans TRP channels, 323–333, 324f
  activation of TRP channels in *C. elegans*
    mechanical activation, 328–329
    other activation modes, 329
    receptor-operated activation, 327–328
  functions of TRP channels in *C. elegans*
    TRPA subfamily, 326–327
    TRPC subfamily, 325
    TRPML subfamily, 327
    TRPM subfamily, 326
    TRPN subfamily, 327
    TRPP subfamily, 327
    TRPV subfamily, 325–326
  regulation of TRP channels in *C. elegans*
    TRPA subfamily, 331
    TRPC subfamily, 330
    TRPML subfamily, 331–332
    TRPM subfamily, 331
    TRPN subfamily, 331
    TRPP subfamily, 331–332
    TRPV subfamily, 330–331
Cell-cycle progression, calcium-dependent proteins
  calcineurin, 178

calmodulin, 178
calmodulin kinase II, 178
calpain, 178
phosphoinositol 3-kinase, 178
Cell proliferation, 88, 139, 178, 200, 256, 263, 381–382, 414, 436, 442, 534, 539, 558, 563, 576–577, 581, 669, 675, 678, 679f, 710, 768, 773f, 774t, 781, 794, 796–797, 929, 932, 935, 937, 939, 948, 954, 956, 959–960, 988
Cellular magnesium homeostasis, 178
Centers (sites) of native aggregation, 461
Central nervous system (CNS), 302, 492, 556, 577, 584, 604–606, 618–620, 629, 639, 832, 861, 864f, 988–990, 997–1000
Cerebellar purkinje cells, 579, 584, 596, 990
Cerebellar purkinje neurone, 398, 576, 579–580
Cerebral ataxia, 579
Cerebral cortex, 44, 582, 584, 639
Cerny, A. C., 545–566
Cervical cancer cell, 935, 953
CFA, *see* Complete Freund's Adjuvant (CFA)
CFA-induced radiant heat inflammatory hyperalgesia, 197
CFTR, *see* Cystic Fibrosis Transmembrane Conductance Regulator (CFTR)
CGRP, *see* Calcitonin gene related peptide (CGRP)
Cha, S. K., 561
Chahl, L. A., 987–1000
Changeux, J. P., 481
Chang, Q., 560
Channel gating process, 518–519
Channel recruitment, 436, 444, 507
Chanzyme (Channel – Enzyme), 112, 975
  TRPM6 and TRPM7, 714
Chaperone
  expression proteins
    DnaJ, 452
    DnaK, 452
    GrpE, 452
CHAPS, *see* 3-[(3-Cholamidopropyl)-dimethylammonio]-1-propane sulfonate (CHAPS)
Charcot–Marie–Tooth disease type 2C (CMT2C), 14, 26, 61
Chelate chromatography, 30
Chemical genetics, 464

Chemokine gene, 161, 187, 381, 381f, 382, 618, 688, 717, 891, 949, 975, 977
Chemokine MCP1, 977
Chemokines (CCL2), 161, 187, 381, 381f, 382, 618, 688, 717, 891, 949, 975, 977
Cheng, K. T., 426, 435–445, 824
Chest tightness, 970
CheY-P (response regulator protein), 457
Chicken *(Gallus gallus)*, 350
Chinese hamster ovary (CHO) cell, 62, 65, 126, 142, 151, 328–329, 557–558, 560, 564, 710, 795, 799
Chironex fleckeri (box jellyfish), 54
5-(4-Chloro-3-methylphenyl)-1-[(4-methylphenyl)methyl]-N-[(1S, 2S, 4R)-1,3,3-trimethylbicyclo[2.2.1]hept-2-yl]-1H-pyrazole-3-carboxamide (SR144528), 191
Chlorothiazide (CTZ), 256, 258–259
CHO cell, *see* Chinese hamster ovary (CHO) cell
3-[(3-Cholamidopropyl)-dimethylammonio]-1-propane sulfonate (CHAPS), 6
Cholinergic blockers, 740
Chromosome 2, 186
Chromosome 7, 136, 150
Chromosome 19, 150, 210, 297
Chromosome III, 210
Chromosome 3L, 210
Chronic constriction injury (CCI), 197–198, 626
Chronic hypertension, 786
Chronic hypoxia-induced PAH (CH-PAH), 696
Chronic obstructive pulmonary disease (COPD), 73, 772, 774t, 970, 976, 978
Chronic persistent cough, 640t
Chronic respiratory diseases, 970
Chuang, H. H., 480
Chu, X., 127
Chung, G., 629
Chung, M. K., 93, 484, 615–629, 640
Chymotrypsin, 552
Cilia (primary) and polycystic kidney disease, 291–292
  cyst formation, 291
  defects in primary cilium
    chlamydomonas reinhardtii, 291

oak ridge polycystic kidney (orpk) mouse, 291
Tg737 gene, 291
features of primary cilium, 291
  9 + 0 axoneme, 291
  eukaryotic cilia, 291
  TRPP2 and PC-1, 291
Ciliated OSNs, 344
  cyclic nucleotide-gated channel A2 subunit, 344
  olfactory marker protein (OMP), 344
  OR-type odorant receptors, 344
1,8-Cineole, 62
Cinnamaldehyde (cinnamon), 67t, 69, 110, 111, 111f, 190, 192t, 623, 852
Cinnamomum camphora, 57t, 59, 68t
Cinnamomum cassia, 67t, 69
Cinnamomum zeylanicum, 45t, 50, 67t, 69
Cinnamosma fragrans (cinnamodial/cinnamosmolide/cinnamolid), 46t, 52
CIRB, *see* CaM/IP$_3$ receptor-binding (CIRB) domain
Circular dichroism spectropolarimetry, 173–174
Circumferential stretch, 772
Citral
  fragrant component, 59
    citronella, 59
    lemongrass oil, 59
    lemon peel, 59
    palmarosa grass, 59
  isomeric mixture, 59
    geranial, 59
    terpenoids neral, 59
Citrate, 627
Citric acid cycle, 669
Citrobacter rodentium-induced transmissible murine colonic hyperplasia model, 254
Civamide Cream, 651
Civamide Nasal Solution, 651
c-Jun N-terminal kinase (JNK), 178, 960
Clapham, D., 576
Clapham, D. E., 173–180, 576, 581
Classic or canonical TRP (TRPC), 88, 99, 493
Clathirin-coated endocytosis, 226
Claudius' cell, 230
CL4 cell line, 233–234
CLC7 (lysosomal Cl$^-$ channel), 213
Clogging, 213
Clotrimazole, 88, 89f, 94–95, 95f, 190, 192t, 711t, 716

Clotrimazole, 1-[(2-Chlorophenyl)
diphenylmethyl]-1H-imidazole,
94
Cloudman, A. M., 232
CMC, see Critical micelle concentration
(CMC)
CMD (CRAC modulatory domain), 419t, 421,
440
CMT2C, see Charcot–Marie–Tooth disease
type 2C (CMT2C)
CNR1 and *CNR2* gene, 998
Cochlear hair cells, 234, 627
Coda, 486
  pore dilation processes, 486
  sine qua non feature of thermoTRP
  channels, 486
Coiled coil character (CCR), 534, 537f
Coiled–coiled region, 26
Coiled-coil folding motif, 463
Cold allodynia, 197–198, 624, 626
Cold and menthol receptor, 714, 973–974
Cold hyperalgesia, 197–198, 624, 628
Cold-shock protein (CSP), 452
Cole-Moore effect, 528
Colletti, G. A., 209–217
Colquhoun, D., 162
Colsoul, B., 815–816
Complete Freund's Adjuvant (CFA), 197–198,
504, 617–618, 625–626,
648t–650t
Complex kidney function, 242
Concavalin A, 534
Conformational coupling, 294, 436
Conformational entropy, 460
Congenital stationary night blindness (CSNB),
140–141, 1017–1020
  in Appaloosa horses, 1017–1018
  in Caucasian non-consanguineous family,
  1018
  chromosome 15q, homozygosity on, 1018
  homozygous mutations
    allelic heterogeneity in CSNB1B,
    1018–1019, 1019t
    deletion of exons (Y72-K365del), 1018
    missense mutation (R14W), 1018
    truncation mutation (W856X), 1018
  in humans, 1018
  screening, 1018
Connecting tubule (CNT), 240, 243, 245, 247,
259, 278–279, 281–285, 282f,
284f

Conopeptide analogue, see Noxious
mechanosensation blocker-1
(NMB-1)
Constitutively active (CA) channels, 329, 392,
709, 715, 733
Core body temperature (CBT), 648t, 653–654
Cornell, R. A., 341–352
Corpuscles of Stannius, 352
Correa, A. M., 474
COS-1 cell, 28
COS-7 cell, 6, 551, 558, 890
COS-M6 cell, 126
Cough, 73, 509, 640t, 641, 652t, 653, 970–972
CPA, see Cyclopiazonic acid (CPA)
CPS, 935, 950–952, 952f, 954, 960
CRAC, see Ca^{2+} release activated Ca^{2+}
(CRAC)
Cre recombinase expressing mice, TRPV5
promoter driven
  generation, 283–284
    of DCT2 and CNT, 283
    map of pBS185 plasmid, 284f
    MAR and *NarI/SpeI* restriction sites,
    284
    SpeI and *XhoI* restriction sites, 284,
    284f
  validation and breeding, 284
    DCT2/CNT-specific AQP2 knockout
    mice and control mice, 284
    TRPV5-*Cre* or MAR-TRPV5-*Cre* mice
    with floxed AQP2 mice, 284
Critical micelle concentration (CMC), 6
Crotonaldehyde, 67t, 69
Croton tiglium, 58t, 61
Crouzin, N., 586
cryo-EM, see Electron cryo-microscopy
(cryo-EM)
CSNB, see Congenital stationary night
blindness (CSNB)
CSP, see Cold-shock protein (CSP)
C-terminal His_{12}, 29
C-terminal (TRPM2-TE), 534, 536, 537f
C-terminus, 534
  gating kinetics, 186
  PIP_2 modulation, 186
  temperature sensitivity, 186
CTZ, see Chlorothiazide (CTZ)
Cup-5 gene, 210
Curcumin, 45t, 51
Current-voltage (I–V) relationship, 137, 152f,
176, 187, 196, 317, 712, 763,
817
Cutaneous A- and C-fiber terminals, 627

CXCL2, 381
CXCL8, 381, 381f
Cyanea capillata (lion's mane jellyfish), 54
Cyanobacterium *Synecocystis*, 454
Cyclic adenosine diphosphoribose (cADPR), 380, 534, 538, 711t, 713, 816, 818–819
Cyclic adenosine monophosphate (cAMP), 246, 497–498, 557–558, 628, 643, 711t, 824, 882, 893
Cyclic AMP-response-element binding protein (CREB), 579, 583, 797, 989
Cyclic terpene alcohol, *see* Menthol
Cyclin D1, 179
Cyclooxygenase inhibitor, *see* Ibuprofen
Cyclooxygenase reaction, 55, 93, 792, 850
Cyclopiazonic acid (CPA), 90, 393, 403–404, 601, 693, 696, 739, 783
Cymbopogon (hydroxy-citronellal), 62, 64t
CYP, *see* Cytochrome P450 (CYP)
Cys553, 377
Cys558, 377
Cysteines
　Cys-929, 186
　Cys-940, 186
Cysteine S-nitrosylation, 377
Cyst formation in mammals/zebrafish, 348–349
　polycystin-2 (PC2), 348
　trpp2 MO-injected zebrafish embryos, 348
　zebrafish *trpp2* mutants, 348
Cystic Fibrosis Transmembrane Conductance Regulator (CFTR), 45t, 150f, 157
Cystitis, 640t, 870, 952
Cytochrome P450 3A enzyme, 95
Cytochrome P_{450} (CYP), 771
Cytochrome P_{450} enzyme reaction, 93
Cytokine gene, 157, 161, 187, 381, 381f, 618, 669, 688, 759, 766, 789, 796, 912, 949, 975–977
Cytokine TNF-α, 618, 976
Cytosolic Ca^{2+} concentration, 296, 392, 414, 760, 956
Cytrochrome P 450-3A4, 50

D
DADS, *see* Diallyl disulfide (DADS)
DAG, *see* Diacylglycerol (DAG)
DAS, *see* Diallyl sulfide (DAS)
DATS, *see* Diallyl trisulfide (DATS)
Davies, A. J., 629
ΔC_p, *see* Heat capacity change (ΔC_p)
DCT2, *see* Distal convoluted tubule (DCT2)
DDM, *see* N-dodecyl-β-D-maltoside (DDM)
Decavanadate, 65, 155, 157, 711t, 716
Dekermendjian, K., 637–656
Dekker, N., 637–656
Delmas, P., 294
Denda, M., 847–856
Dendreon, 188, 200
Dendritic cell, 160, 931, 976
Dendritic filopodia, 576
Dendritic microvilli, 127
De novo synthesis, 65
Dental and orofacial pain, 622
Dental pain, 621, 652t, 653–654
15-Deoxy-$\Delta^{12,14}$-prostaglandin J_2 (15d-PGJ$_2$), 68, 70, 623
Depression and anxiety, 990–994
　TRPC channels, 991–992
　TRPV channels, 992–994
D-erythro-sphingosine, 65, 820
Desensitization, 50, 54–55, 59–60, 97, 154–155, 187–188, 495–503, 508–509, 539, 552, 556, 615, 625–626, 643–644, 655, 815, 834–835, 973, 992, 997–998
DeSimone, J. A., 839
DesK protein, 455
DesR protein, 455
Destabilizing configurational entropy, 460
Detergents, use
　crystallization, 6
　purification, 6
　solublization, 6
ΔG_{gen}, 460
ΔG_{ion}, 460
DGK, *see* Diacylglycerol kinase (DGK)
D333G mutation, 14, 1015t, 1016
ΔG_{other}, 460
ΔG_{tr}, 460
DHN, *see* Dorsal horn neurons (DHN)
DHPG
　activator of PKC, 602
　activator of "plateau" depolarization, 602
　amplitude of depolarization, reduced, 602
　application of PdBU, 602
Diabetic neuropathy, 621
Diabetic painful neuropathy, 642t
Diacylglycerol (DAG), 43, 91f, 99, 126, 163, 188, 374, 533, 553, 619, 783
　activated channel, 397–400
　　IP$_3$, role, 398
　　PIP$_2$, direct *vs.* indirect actions, 399–400

and PIP$_2$ interactions, 398–399
role, 397–398
biochemical cascade, 398
Ca^{2+} open channel block, 396
excitatory and inhibitory effects, 398
gated channel, 401–402
　TRPC4 and TRPC5 channel activation, PIP$_2$ role, 401
generation and metabolism pathways, 396f
physiological coupling, 400
PIP$_2$, inhibitory action, 400
TRP and TRPL channel gating, pivotal generation, 396
　non-PLC-mediated pathways, 398
　TRPC3-C7 homomer, 397
TRPC4 and TRPC5
　exogenous OAG inhibition, 401
　PKC activity inhibition, 402
Diacylglycerol kinase (DGK), 128
2,4-Diacylphloroglucinol derivative (Hyp9), 89f, 97
Diallyl disulfide (DADS), 66, 67t
Diallyl sulfide (DAS), 66, 67t
Diallyl trisulfide (DATS), 66, 67t
Dictyostelium, 230, 316
Dietrich, A., 781–800
Diffusible messenger, 436
Digel, I., 451–464, 822
1,25-Dihydroxycholecalciferol, 562
Diphtheria toxin A fragment (DTA), 837
Disrupted bipolar transmission, 140
Disrupted in schizophrenia 1 (DISC1), 994
Distal (early/late) convoluted tubules (DCT1/DCT2), 243, 247, 259, 278, 561, 563
1,3-Disubstituted urea derivatives, 648t
Diterpenoid incensole acetate, 60
Divalent cations, 88, 153, 175, 178, 187, 244, 330, 642
1D4 (monoclonal antibody), 29
DNA-binding oligomer, 463
Docosahexaenoic acid, 62, 64t
Dodecyl octaethylene glycol ether (C12E8), 6
Dopamine transporter (DAT), 997
Dörr, J., 909–922
Dorsal horn neurons (DHN), 604, 620
Dorsal rhizotomy, 618, 620
Dorsal root ganglia (DRG), 44, 51–53, 55, 60, 66, 71, 186, 196, 198, 332, 475, 479, 493, 539–540, 580, 583, 616, 618, 626, 628, 637, 639, 863, 863t, 867–868, 989

Dot-blot analysis and RT-PCR studies, 305–306
Dougherty, G., 303
3-D particle, 31
DPBA, 627
15d-PGJ$_2$, see 15-Deoxy-$\Delta^{12,14}$-prostaglandin J$_2$ (15d-PGJ$_2$)
Dragicevic, C., 469–486
DRG, see Dorsal root ganglia (DRG)
Drosophila, 90, 230
　TRP, 98
　TRPγ, 90, 99
　TRPL, 90, 99
Drosophila melanogaster, 108, 112, 361, 367t
Drosophila phototransduction, 547
Drosophila S2 (Schneider) cell, 221, 225, 415, 438
Drosophila trp mutant, 26
Drost-Hansen, W., 454
Drug discovery cascade, machine implementation
　assay optimization, 108
　cardiac liability, 108
　lead optimization, 108
　secondary screening, 108
DT40 B lymphocyte, 96, 374–375, 377
DT-40 cell, 178, 374, 438, 539, 785, 886–887
Ducret, T., 687–699
Dulac, C., 129
Duncan, L. M., 937
Dunker, A. K., 459
Dye ruthenium red, 69, 99, 110, 111f, 246, 251, 484, 643, 647, 695, 711t, 715, 771, 792, 794, 797, 848, 852, 854, 918, 978
Dynamin-dependent endocytotic pathway, 565
Dynamin (Dyn1aK44A), 226, 549t, 559
Dysbindin, 994
Dysfunctional mitochondria, 210
Dyspnea air hunger, 970

E
E9.5, see Embryonic day 9.5 (E9.5)
Earley, S., 718, 765
EC$_{50}$, see Half efficiency activation (EC$_{50}$)
ECaC, see Epithelial Ca^{2+} channel (ECaC)
E. coli, see Escherichia coli (E. coli)
E. coli chemotaxis receptor, 462
Econazole, 94–95, 711t
Ectoparasites, 366
　bugs (bedbugs), 366
　head lice, 366
　pea aphid, Acyrthosiphon pisum, 366

Ectoparasites (*cont.*)
  ticks, 366
    trpA1 (ISCW011428) in *I. scapularis*, 366
    TRP channel genes/*D. melanogaster*,
      comparison, 366, 367t
Ectopic TRPV1 expression, 950
EDTA-AM (cell-permeant magnesium
  chelator), 175
Edwards, M. K., 324
EET, *see* Epoxyeicosatrienoic acid (EET)
E3 extracellular domain, 114
EF-hands
  Ca^{2+} binding *vs.* non-Ca^{2+}-binding, 18
  Ca^{2+} coordination, 17
    affinity, 17
    canonical *vs.* non-canonical (atypical),
      17–18
    NMR structure in TRPP2 C terminus, 18f
    TRPA1, 18
    TRPML1, 18
    TRPP2, 18
EGF, *see* Epithelial growth factor (EGF)
EGFP expressing mice, TRPV5 promoter
  driven
  mouse generation, 279–280
    genomic organization of TRPV5 gene
      and constructs, 280f
    restriction enzyme *Nae*I, 280
    *Sal*I and *Age*I restriction sites, 279
    TRPV5-EGFP DNA, 280
    wild-type C57Bl/6 x DBA2 F2 hybrid
      mice (B6D2F2), 280
  studies in TRPV5-EGFP mice, 281–283
    EGFP expressing tubules, functionality
      of, 281
    intracellular Ca^{2+} signalling, 282
  validation of transgenic mouse lines,
    280–281
    differences of transgene expression,
      280
    EGFP expression in DCT2/CNT/CCD,
      280–281, 282f
    flow chart of, 281f
    TRPV5-EGFP transgenic mice, founder
      lines (F0), 280
EGF receptor, 234, 294, 550, 715, 795
EGFR-PKCα-ERK signaling pathway, 55
Eicosapentaenoic acid, 62
EJ (urothelial cell line), 956
El Boustany, C., 678
Electro-mechanical coupling, 708
Electron cryo-microscopy (cryo-EM), 5–7, 9,
  33

2D crystal, 6
2D sheet, 5
Electron microscopy (EM), 3, 28, 30, 32, 34
  structures, features, 34
    conductance pore, 34
    gate, 34
    ligand-binding site, 34
    selectivity filter, 34
  TRP channel structures
    imaged in ice, 31
    imaged in negative stain, 31
    TRPC3, 31
    TRPM2, 31
    TRPV1, 31
    TRPV4, 31
Electrophilic cyclopentane prostaglandin
  D2 metabolite, *see* 15-Deoxy-
  $\Delta^{12,14}$-prostaglandin J_2
  (15d-PGJ_2)
Electrophysiological and photoaffinity labeling
  experiment, 377
Electrophysiological recording, 96, 119,
  225–226, 332, 343, 437, 524,
  645, 910
Electrophysiology, 50, 107, 118, 139, 175,
  342, 423, 645, 813
Electroretinogram (ERG), 141
Elke, L., 637–656
EM, *see* Electron microscopy (EM)
Embryo implantation process, 252
Embryonic day 9.5 (E9.5), 174
Embryonic stem (ES) cells, 174, 279
Emerging roles in neuronal function
  activation mechanism, 574f
  Ca^{2+} influx, 574
  physiological importance, TRPC
    TRPC1, 576–578
    TRPC2, 578
    TRPC3, 578–582
    TRPC4, 582–583
    TRPC5, 581–582
    TRPC6, 582–584
    TRPC7, 584
  TRPC channels and oxidative stress
    participation in SH-SY5Y cells, 586
  TRPC proteins in neurodegenerative
    diseases
    BDNF treatment, 585
    $[Ca^{2+}]_i$ concentration, changes, 585
    neuroprotective effect, 585
    neurotoxins, 585
    phosphoinositol pathway, activation,
      585

E830 mutation, 156
Endocannabinod anandamide, 953
Endocannabinoid-inactivating enzyme, 992
Endocannabinoids, 117–118, 582, 627, 764, 914–915, 990, 992, 997
Endoparasites
    helminths, 361–366
    protozoa, 362–363
Endoplasmic reticulum (ER), 28–29, 91f, 92, 99, 158, 187, 294–295, 297, 299, 350, 375f, 393, 414, 436
Endoplasmic reticulum (ER)-associated protein degradation pathway, 350
Endoplasmic reticulum (ER) ($[Ca^{2+}]_{er}$)
    metabolism of xenobiotic compounds, 669
    protein synthesis, 669
Endothelial $[Ca^{2+}]_i$ homeostasis, 764
Endothelial cell, 90, 533, 667
    migration, 795
    dysfunction, 533
Endothelial NO synthase (eNOS), 377–379, 379f, 793
Endothelin-1 (ET-1), 404, 697, 787f–788f, 787, 789, 799
Endovanilloid, 94, 990
    5-lipooxygenase and 12-lipooxygenase products, 54
        leukotriene B4, 54
        12-S-HPETE, 54
End-stage renal disease (ESRD), 289
Enolized α-dicarbonyl system, 96
eNOS, see Endothelial NO synthase (eNOS)
Enthalpy value ($\Delta H(T_R)$), 460
Entropy value ($\Delta S(T_R)$), 460
Envenomation, 54
Enzymatic de-glycosylation, 29
Epidermal keratinocytes
    TRP expression, 847–849
        Calcium imaging, 849f
        environmental factors, 854–856
        immunostaining of human epidermis, 848f
        putative skin surface sensory system., 848f
    TRP's role in, 849–850
        low pH, activation, 850–851
        skin inflammation, 850
        Ultraviolet (UV) radiation, 851
Epidermal permeability
    TRP's role
        barrier homeostasis, 851–853
Epilepsy, 603

Epinephrine, 669, 891
Epithelial Ca^{2+} channel (ECaC), 240, 917
Epithelial growth factor (EGF), 550
    receptor, 294, 329, 548t, 715, 890, 932, 949, 977
Epithelial sodium channel (ENaC), 257, 278, 332, 839, 863t, 864f, 865, 869–870
Epitope affinity chromatography, 30
ePKC, see Eye specific protein kinase C (ePKC)
EPO, see Erythropoietin (EPO) receptor
E. poissonii, 51
Epoxyeicosatrienoic acid (EET), 61, 94, 548t, 552, 559, 693, 695, 697, 764, 822
E1047Q substitution, 153
ER, see Endoplasmic reticulum (ER)
ERG, see Electroretinogram (ERG)
Erk, see Extracellular signal-regulated kinase (ERK)
ERM, see Ezrin/radixin/moesin (ERM) domain
ER marker, 558
ER-PM microdomains, 440
Erythropoietin (EPO) receptor, 127, 130–131, 885, 890
Erythropoietin-evoked calcium signalling, 130
Escherichia coli (E. coli), 6, 241, 452
    chemoreceptors
        Tar, 457
        Tsr, 457
    HB101, 456
    temperature-controlled switching
        clockwise (CW), 457
        counterclockwise (CCW), 457
    thermoreceptors
        Tap, 457
        Trg, 457
ET-1, see Endothelin-1 (ET-1)
ET_A receptor, 404
ET_B receptor, 404
Eucalyptol, 192, 630
    treatment
        muscular pain, 198
        rhinosinusitis, 198
Eucalyptus sp. (eucalyptol), 62
Eugenia caryophyllata, 45t, 50
Eugenol, 45t, 50, 59–60, 62, 70, 188, 190t, 330, 627, 853
Eukaryotic proteins
    advantages, 28–29
    Baculovirus-mediated expression, 28
    crystallization, 27

Eukaryotic proteins (*cont.*)
　E. coli expression, 27
　Golgi apparatus, 29
　insect cell expression, 28
　protease-deficient strain, 29
　recombinant protein, expression, 28
　rough endoplasmic reticulum, 29
　vector expression, 29
E. unispina, 51
Euphorbiacea, 61
Euphorbia resinifera, 45t, 51, 481
Euphorbiengummi resinifera (resiniferatoxine), 98
Everett, K. V., 1011–1027
Evodiamine
　biological effects, 55
　　adipogenesis inhibition, 55
　　cancer cells, increased apoptosis, 55
　　capsaicin-like effect, 55
　　capsazepine-sensitive contraction, 55
Evodia rutaecarpa (Chinese herb), 55
Evolution, TRP channel, 350–351
　chordate phylogenetic tree, 351
　mammalian thermoreceptors, 350
　urochordates, 351
　warm-blooded animals, 351
Exogenous OAG, 396, 401
Expression and function in liver cells
　ectopic expression, 673
　in other cell types present in liver
　　liver endothelial cells, 677
　　polycystic liver disease development, 677
　　PRKCSH mutations, 677
　　vascular smooth muscle cells, 677
　physiological functions
　　TRPC1 and volume control, 674–675
　　TRPM7 and cell proliferation, 675
　　TRPML1 and lysosomal Ca^{2+} release, 675–676
　　TRPV1 and TRPV4 in cell migration, 674
　store-operated Ca^{2+} entry in hepatocytes
　　ectopic expression, 673
　　with SERCA inhibitors, 673
　　SOCs in liver cells, 673
　TRP channels and liver cancer
　　with chronic hepatitis, 678
　　with cirrhotic livers, 678
　　mRNA expression, 678, 679f
　　thapsigargin-initiated Ca^{2+} entry increased, 678
　TRP channels expressed in liver, 671t–672t

Expression cloning approach/technique, 43–44, 62, 240
Expression pattern and biological function, TRPV1
　modulation of expression
　　extracellular matrix (ECM) molecule, 640
　　fibronectin, 640
　　HEK cells, 640
　　increased expression, 640
　　native cells, 640
　physiological and pathological roles
　　deoxycortisosterone acetate treatment, 641
　　elicit respiratory defense responses, 641
　　pain and temperature sensation, 641
　　respiratory diseases, 640–641
　　urinary bladder, 641
　　vascular and renal hypertension, 641
　tissue distribution
　　epithelial cells in human airways, 639
　　hematopoietic cells, 639
　　liver, 639
　　skin, 639
　　smooth muscle of bladder, 639
Extracellular hypotonicity, 94
Extracellular signal-regulated kinase (ERK), 347, 375–376, 381, 381f, 769, 794, 936, 954–955
Eye specific protein kinase C (ePKC), 395, 554–555
Ezrin/radixin/moesin (ERM) domain, 127, 419t, 422, 441

F
Faecal incontinence, 640t
FAK, *see* Focal adhesion kinase (FAK)
Familial hyperkalemia and hypertension (FHH), 256
Faraday's constant, 461, 473, 518
Farfariello, V., 947–961
Farnesyl pyrophosphate (FPP), 58t, 60
　activation, 60
　　GPR92 membrane receptor, 60
　inhibition, 60
　　LAP2 receptor, 60
　　LAP3 receptor, 60
Farnesyl thiosalicylic acid (FTS), 485
Fas-associated factor-1, 951
Fas/CD95, 950, 951, 952f
　-associated protein, 951
　-induced apoptosis, 955
Fasciculol C-depsipeptide, 52

Fatty acid amide hydrolase (FAAH), 893, 992
Fatty acid synthase (FASN), 302
Fecher-Trost, C., 909–922
Fenton, R. A., 277–285
Ferrer-Montiel, A., 491–509
Feske, S., 415
FFA, see Flufenamic acid (FFA)
FGF, see Fibroblast growth factor (FGF)
FHH, see Familial hyperkalemia and hypertension (FHH)
Fibroblast growth factor (FGF), 247, 576, 768, 796, 988
Fibrosis, 180, 755, 798, 971
Field potential recording, 128
FK506-binding protein 52 (FKBP52), 344
FLIPR, see Fluorometric Imaging Plate Reader (FLIPR)
Flockerzi, V., 737
Flores, E. N., 221–228
Fluconazole, 95
Flufenamic acid, 156
    activation
        TRPA1, 156
        TRPC6, 156
    common chloride channel blocker
        Ca^{2+}-activated Cl^- channel, 156
        proton-activated Cl^- channel, 156
        swelling-activated Cl^- channel, 156
    inhibition
        TRPC3, 156
        TRPC5, 156
        TRPM2, 156
    PBC inhibition, firing rate reduction, 161
Flufenamic acid (FFA), 113f, 113, 156, 161, 602, 711t, 715, 735t, 890
Fluorescence resonance energy transfer (FRET), 175, 227, 349, 421, 423, 440, 478, 524, 533, 557, 769, 786, 886
Fluorescent Ca^{2+} assay, 52
Fluorometric Imaging Plate Reader (FLIPR), 645–646, 648t
fMLP, see N-formyl-Lmethionyl-L-leucyl-L-phenylalanine (fMLP)
Focal adhesion kinase (FAK), 581–582
Focal cerebral ischemia, 580
Focal segmental glomerulosclerosis (FSGS), 347, 1012
    FSGS3 on chromosome 6p12, 1012
    FSGS1 on chromosome 19q13, 1012
    FSGS4 on chromosome 22q12, 1012

    FSGS5 on chromosome 14q32.33, 1012
    FSGS2 on chromosome 11q21-q22, 1012–1013, 1013f
Formalin-induced nociceptive response, 50
Forskolin, 497–498, 548t, 557–558, 563
Fourfold rotational symmetry, 7
FPP, see Farnesyl pyrophosphate (FPP)
680–796 Fragment, 19
720–797 Fragment, 19
Frankincense, 60
Frictional damping, 461
Friis, S., 120
Frog/zebrafish model systems, TRP ion channels using, 341–352
    cyst formation in mammals/zebrafish, 348–349
    expression of TRP channel, 344–345
    Map kinase activity, TRPC1 regulation, 347–348
    RPN1, zebrafish functions of, 345–346
    tissue-culture model hypothesis, 349–350
    TRPA1, noxious chemicals, 346–347
    TRP channel evolution, 350–351
    TRPC1 in frog oocytes, 341
    TRPM7 in embryonic development, 351–352
    TRPV4/TRPP2, heteromultimer of, 349
    Xenopus embryogenesis, TRPC1 in, 343–344
FRTL-5 cell, 126, 130, 131f
Fructooligosaccharide, 249
Füchtbauer, E.-M., 277–285
Functional magnetic resonance imaging (fMRI), 991
Functional studies, Budding Yeast, 25–35
Functional TRPV1 channel, 953–954
Function of temperature, 461
Fura-2, 29, 673, 750, 821f
Furini, S., 485
Future prospect, TRPV1
    antibodies and toxins, biologics
        E3 loop, 655
        venomous animals, 655
    modality specific TRPV1 antagonists
        A-425619, 655
        AMG-0610, 655
        AMG-6880, 655
        AMG-7472, 655
        AMG-8562, 655
        AMG-9810, 655
        BCTC, 655
        SB-366791, 655

Future prospect, TRPV1 (cont.)
  structure-guided drug discovery
    insect cells, 655
    mammalian cells, 656
    yeast, 655

G
GABAergic inhibitory postsynaptic currents, 998
Gadolinium ion, 93, 98–100
Gailly, P., 126
Gain-of-function mutagenesis analysis, 35
Gain-of-function mutation, 232, 234, 1012
Gain-or loss-of-function phenotype, 35
Gαi2 subunit, 127
Galectin 1 (GAL1), 247, 934, 958
Galectin-1 linked TRPV5 protein, 247
Gambierol, 54
Gamma-aminobutyric acid (GABA), 580, 995
García-Añoveros, J., 221–228
Garland, C. J., 761–763
Gastric adenocarcinoma cell, 178
Gastro intestinal tract and TRP
  in intestinal motility, 697–698
  TRP in intestinal motility, 697–698
  TRPM8 and gatric activity, 698
Gating currents, 519
Gating mechanism
  diverse group of functional channel isoforms, 392–393
    amino acid residues, 392
    heteromeric structures, 392
    homomeric structures, 392
    proposed functional heteromeric channel structures, 393t
    voltage-gated channels, 392
  Drosophila photoreceptors, 395, 397f
    IP$_3$-mediated release of Ca^{2+}, 395
  Drosophila TRP channels, activation, 394–397
    DAG, central role, 395–397
  Drosophila TRPL channels, activation, 394–397
    DAG, central role, 395–397
    of mammalian non-TRPC1- and TRPC1-containing channels, 399f
  photon absorption, 395
  physiological channels, 393–394
    SOC protein, activation, 393
    subunits as ROCs and SOCs, 394
  PIP$_2$-activated channels, 402–404
    Ca^{2+} entry mediation, PI role, 404
    role, 402–404
    TRPC1-containing channel activity, 403
  PKC-activated channels, 402–404
    epidermal growth factor, applications, 403
    role, 403
  See also Diacylglycerol (DAG)
Gaultheria, 49t, 55, 68t
G1 cell cycle phase, 178, 950
Gα_q-coupled muscarinic receptor, 177
Gd^{3+}, 299, 318, 328, 435, 442–443, 642, 711t, 735t–736t, 750, 870, 920, 937
Gel filtration, 30
Generic gating mechanism, 35
Genes (TRP) and human inherited disease, 1011–1027
  canonical TRPs, 1012
  congenital stationary night blindness, 1017–1019
  HSH, 1020–1022
  polycystin TRPs, 1023
  Kozlowski type spondylometaphyseal dysplasia, 1015–1016
  metatropic dysplasia, 1016
  neuromuscular disorders
    melastatin TRP, 1017
  progressive familial heart block, type 1B, 1020
  skeletal disorders, 1014
  TRPC6, 1012–1013
    vanilloid TRPs, 1013
  TRPM1, 1017
  TRPM4, 1020
  TRPM6, 1020
  TRPML1, 1025–1027
  TRPP1=PKD2, 1023–1025
    mucolipin TRPs, 1025
  TRPV4, 1014
  type 3 brachyolmia, 1014–1015
Geraniol, 62, 64t, 188, 190t
G$_\alpha$θ/G$_\alpha$11 protein-coupled receptor, 393, 401
Gibbs energy, 458, 460, 472
  zero point crossing
    high temperature (heat denaturation), 460
    low temperature (cold denaturation), 460
Gibson, H. E., 993
Giga-ohm patch clamp recording, 107
Gilbert Ling, 461
Gingerol, 50–51, 70
6-Gingerol, 45t, 51

8-Gingerol, 45t, 51
GlaxoSmithKline, 621, 652t
Gleason score, 254, 262, 933, 959
Glenmark, 646, 650t, 652t, 653
Glibenclamide, 157, 164, 814
Glioblastoma, grade IV, 935, 952
Glioma cells, 932, 935, 952, 955, 957
Glomerulus/glomeruli, 277, 294, 345, 1012
Glucagon, 669, 819
Glucocorticoid-induced osteoporosis mechanism, 257
Glucocorticoids, 177, 254, 257–258, 506, 561, 740, 975, 991
β-Glucuronidase, 246–247, 548t, 562
β-Glucuronidase Klotho, 560
Glu994 residue, 381
Glutathione peroxidase, 532
Glutathione S-transferase, 738
Glycine-rich GXA(G)XXG motif, 10–11
Glycogen synthase kinase 3 (GSK3), 295, 564
Glycosidase klotho, 934
Golgi, 29, 214, 226, 295, 319, 378, 507, 559, 564–565, 670f, 890
Golgi protein RNF24, 549t, 553, 890
González-Nilo, Dr. F., 486
Gordon syndrome, 256
Gottlieb, P., 800
Gouy-Chapman-Stern equation, 480
GPCR, see G-protein coupled receptor (GPCR)
$G_{\alpha o}$-protein, 140, 142
G-protein-coupled PLC pathway, 402
G protein-coupled receptor proteolytic site (GPS), 293–294, 301
G-protein coupled receptor (GPCR), 327, 948–949
  and Ca^{2+} store depletion, roles of, 713–714
    PIP_2-binding domains, 714–715
    TRPM4/5/7/8, 715
    vasoactive agonists (angiotensin II/bradykinin/aldosterone), 715
G protein gustducin switch, 163
G-protein inhibitor GDP-β-S, 599
Gq-coupled receptor agonist, see Vasopressin
Gq/Go-protein, 56, 936, 955
Gq/11/phospholipase C (PLC) coupling, 709, 715
Grammostola spatulata (Chilean Rose tarantula), 71, 72t
Gram-negative pathogenic bacteria, 163
Grandl, J., 477–478
GRC-6211, 650t, 653
Greka, A., 378
G805 residue, 186

Grid cells, 596
Grimm, C., 713
GRM6 gene, 141
Growth cone morphology, 43, 378, 581, 996
Growth cone steering, 581
Growth receptor channel (GRC), 231, 650t, 652t, 653, 954
GSK1016790A (TRPV4-selective agonist), 89f, 98, 822, 978
GsMTx-4 peptide, 71, 72t
GST-fused N-terminal splice variant, 138
$G_{\alpha o}$ subunit, 127
GTPase ADP-ribosylation Factor 6 (Arf6) pathway, 226
Guaiacol, 45t, 50
Guaiacum, 45t, 50
Guamanian ALS and PD case, 382
Guamanian amyotrophic lateral sclerosis, 175, 537
Guamanian amyotrophic lateral sclerosis and parkinsonian dementia (ALS/PD), 175, 537
Gudermann, T., 781–800
Guibert, C., 687–699
Guinamard, R., 814
Guinea pig airway SM cells, 732, 736, 739
Guinea pig airway smooth muscle, 974

H
3H-A-778317, 645
Habener, J. F., 814
Hahn, M. W., 260
Half efficiency activation (EC_{50}), 44, 45t–49t, 50–56, 57t–58t, 59–62, 63t–64t, 66, 67t–68t, 69–71, 72t, 89f, 115, 116f–117f, 154–155, 189t–190t, 195t, 499, 502, 653, 814–815, 820–821
Half maximal inhibition (IC_{50}), 55, 59, 65, 91, 94, 156
Hallmark helix-loop-helix fold, 17
Hanahan, D., 930
HAPMAP database, 136
Hara, Y., 94
Hardie, R. C., 99
Harteneck, C., 87–100
Hb, see Hemoglobin (Hb)
HC-030031, 625, 766, 972
H19-7 cells, 576–577, 579, 584
HCN channels, 162, 524, 526
HCTZ, see Hydrochlorothiazide (HCTZ)
Heart tube looping, 177
Heat capacity change (ΔC_p), 460

Heat sensitive TRPs, *see* Thermo TRP channels
Heat shock protein (HSP), 452
Hebeloma senescens (hebelomic acid), 47t, 52
Hebelomic acid A, 52
Hebelomic acid B, 52
Hediger, M. A., 254
HEK cell, 134, 229–230
HEK-293 cell, *see* Heterologous expression system (HEK-293 cell); Human embryonic kidney (HEK-293) cell
HEK293 cell line, 234
HEK-293T cell, 177, 586
HEK293-TRPM3 cell, 114
HeLa cell, 211t, 214, 226, 553
HeLa cell line, 234
Helix-loop-helix leucine zipper transcription factor, 137
Helminths
   nematodes, 363–365
   trematodes and cestodes, 365–366
Hemoglobin (Hb), 458, 464, 471
Hensen's cell, 230
Hepatic stellate cells, 667
Hepatocyte, 113, 178, 549t, 667, 668f, 669–670, 670f, 671t–672t, 672–678, 932, 935, 952
Hepatocyte $[Ca^{2+}]_{cyt}$, changes
   amino acid, 669
   apoptosis and necrosis, 669
   bile acid secretion, 669
   cell cycle and cell proliferation, 669
   fatty acid, 669
   glucose, 669
   lysosomes movement, 669
   protein synthesis and secretion, 669
   xenobiotic metabolism, 669
Hepatocyte growth factor/scatter factor (HGF/SF), 952–953
Hepatoma cell, 178, 931
HepG2 cell, 674, 678, 936, 957
HepG2 cells hepatoblastoma cells, 957
Hereditary motor and sensory neuropathy type IIC (HMSN2C), 1015t, 1016–1017
Herson, P. S., 811
19-HETE-DA, 48t, 54
20-HETE-DA, 54
Heteractis crispa, 48t, 54
Heterologous expression system (HEK-293 cell), 50, 52, 119, 162, 233, 508, 536, 783, 800, 977
Hexamethlene bisacteamide (HMBA), 139

He, Y., 719
HIF-1, *see* Hypoxia-inducible factor 1 (HIF-1)
H4-IIE rat liver cells, 671t–672t, 674, 678, 931
Hippocampal neuron, 94, 344, 547, 550–552, 579, 581, 583–584, 586, 882, 996
His958 residue, 381
Histidine residues
   His252, 233
   His273, 233
   His283, 233
Histone H2A.Z, 456
HMBA, *see* Hexamethlene bisacteamide (HMBA)
HMVEC, *see* Human microvascular endothelial cell (HMVEC)
H934N mutation, 156
H–NS (bacterial protein), 456, 464
H_2O_2 application, 379–380
9-HODE, *see* 9-Hydroxyoctadecadienoic acid (9-HODE)
13-HODE, *see* 13-Hydroxyoctadecadienoic acid (13-HODE)
Hodgkin, A. L., 519
Hoenderop, J. G., 240, 279
Hofherr, A., 287–303
Hofmeister, M. V., 277–285
Hogstrand, C., 242
Homeoviscous adaptation, 454
Homer 1, 127, 505, 753, 756
Homology modeling method, 33
Homo-oligomer, 464
Homotetrameric channel, 186, 795
Homotetrameric glutamate-gated receptor channel, 16
Homotetramers, 16, 30, 186, 532, 534, 710, 795, 934, 958
Homozygote knockout clone, 178
Hormone–refractory prostate cancer, 200
Horseradish, 66
Host-defense mechanism, 53
Hot receptor (TRPV1), 477
Hot water bath hand withdrawal time (HWT), 653
Høyer, H., 285
HPAEC, *see* Human pulmonary artery endothelial cells (HPAEC)
HPV, *see* Hypoxic pulmonary vasoconstriction (HPV)
^{3}H-resiniferatoxin, 52–53, 644–645
[^{3}H]resiniferatoxin (RTX), 989
^{3}H-RTX binding, 52, 55

HSH, see Hypomagnesaemia with secondary hypocalcaemia (HSH)
HSP, see Heat shock protein (HSP)
Huang, C. L., 561
Huang, G. N., 421–423
Hu, H., 93
Hu, H. Z., 851
Huber, A., 545–566
Hughes, D. A., 260
Hughes, J., 300
Huh-7, 671t–672t, 678, 679f
Hui, K., 481
Human embryonic kidney (HEK) cell, 837, 1014
Human embryonic kidney (HEK-293) cell, 151, 153
Human glomerular mesangial cell, 403
Human islet amyloid polypeptide (hIAPP), 822
Human lung epithelial cell, 187
Human melanoma, 138–139, 200, 937
Human microvascular endothelial cell (HMVEC), 768, 791, 795, 887
Human pancreatic cancer, 935, 952
Human peripheral bloodacute lymphoid leukemia (HPB-ALL), 931
Human prostate cancer xenograft tumor, 201
Human Protein Atlas, 821, 823
Human pulmonary artery endothelial cells (HPAEC), 550, 760, 795
Human umbilical vein endothelial cell (HUVEC), 768, 792
Human urothelial cancer, 935, 950, 958
Huque, T., 837
HUVEC, see Human umbilical vein endothelial cell (HUVEC)
Huwentoxin 5 (HwTx-V), 53
Huxley, A. F., 519
HwTx-V, see Huwentoxin 5 (HwTx-V)
Hydra, 230
Hydration, 460
Hydrochlorothiazide (HCTZ), 256t, 258
Hydrogen bonding, 460
Hydrogen peroxide (H_2O_2), 65, 88, 90, 92, 94–95, 95f, 99, 456–457, 532, 536, 769, 795
Hydroperoxyeicosatetraenoic acids (HPETE), 646, 990, 997
Hydrophobic amino acid residues, 440
Hydrophobic effect, 460, 524
Hydrophobic nucleus, 461

Hydroxy-α-sanshool, 68t, 69–70
20-Hydroxyeicosatetraenoic acid, 71, 735
4-Hydroxynonenal (4-HNE), 69, 623, 972
9-Hydroxyoctadecadienoic acid (9-HODE), 54
13-Hydroxyoctadecadienoic acid (13-HODE), 49t, 54
9-Hydroxyphenanthrene (9-Phenanthrol), 157
25-Hydroxyvitamin D (25(OH)D), 245
Hyperalgesia, 54, 66, 197–198, 492, 494, 503–504, 557, 616–621, 624–625, 627–628, 648t–650t, 868, 951t, 990, 999
Hypercalciuria, 248, 256t, 257, 259, 262, 918
Hyperforin, 71, 72t, 90, 92f, 95, 95f, 96, 850, 891, 991
Hypericum perforatum (St John's Wort), 71
Hyperphorin, 582
Hyperthermia, 98, 201, 621, 652t, 653–656
Hypertrophic process, 162
Hypoalgesia, 998
Hypocalciuria effect, 258–259
Hypoglutamatergic hypothesis, 995
Hypomagnesaemia with secondary hypocalcaemia (HSH), 784, 1020–1022
  compound heterozygosity, 1021
  deletion of exons, 1021
  magnesium deficiency, 1020
  mapped to chromosome 9q21.13, 1020
  missense mutations in TRPM6 protein, 1021, 1022t
  polycystin TRPs, 1023
    adult polycystic kidney disease (ADPKD), 1023
    PKD1, *PBP* (polycystic breakpoint), 1023
    TRPP1 (or *PKD2*), 1023
    TRPP2 (*PKD2L1* or PKD2-like 1), 1023
    TRPP3 (*PKD2L2* or PKD2-like 2), 1023
  S141L mutation, 1021
  W1925W in Turkish family, 1021
Hypomagnesemia, 175, 255t, 329, 714, 773, 784, 793
Hypotension, 278, 641
Hypoxia-inducible factor 1 (HIF-1), 798
Hypoxic pulmonary vasoconstriction (HPV), 695, 798
  acute hypoxia, 695
  diacylglycerol (DAG), 695
  role of TRPC6 in HPV, 695
  SOC-mediated Ca entry (SOCE), 695

Hypoxic pulmonary vasoconstriction (*cont.*)
  and TRP, 695
    acute hypoxia, 695
    diacylglycerol (DAG), 695
    role of TRPC6 in HPV, 695
    SOC-mediated Ca entry (SOCE), 695
Hysteresis, 59

I

Iberiotoxin, 733
Ibuprofen, 627
IC_{50}, *see* Half maximal inhibition (IC_{50})
I_{CaNS}, *see* Non-selective cation current (I_{CaNS})
Icilin, 114–115, 186–188, 189t, 194, 196–198, 201, 373, 626, 716, 720, 722, 863t
Icillin, 88, 97, 475, 480–481, 793
ICK, *see* Inhibitory cysteine knot (ICK)
I_{CRAC}, *see* Calcium-release-activated current (I_{CRAC})
Idiopathic detrusor overactivity, 197, 863–864, 867
Idiopathic pulmonary arterial hypertension (IPAH), 797
I331F mutation, 14
IGF-1/IGF-1R pathway, 954
IMCD-3, *see* Inner medullary collecting duct (IMCD-3)
Immunohistochemistry approach, 243
Immunoprecipitation, 227, 302, 344, 349–350, 374, 378, 399–400, 443, 505, 533, 738, 753, 786, 911t, 916t, 920t
InaC gene, 395
INAD, 554
Incensole acetate, 60, 994
Inducible deletion system, 178
Inhibitory cysteine knot (ICK), 53
Initial part of the cortical collecting duct (iCCD), 278, 281
Inner medullary collecting duct (IMCD-3), 551–552
Inositol triphosphate (IP_3), 31, 43, 156, 550, 883, 887, 894
Inositol 1,4,5-trisphosphate (IP_3), 91f, 92, 99, 374–377, 417, 424f, 533, 694, 709, 735t, 739, 895
Inoue, R., 694
Insulin, 113, 149, 158–160, 278, 381–382, 504, 548t, 551, 555, 557–558, 669, 752–754, 812, 813t, 814–819, 821–824, 954
Insulin/IGF-1, 548t, 557

Integral transmembrane topology, 230
Integrin/Src tyrosine kinase pathway, 628
Intersectin, 549t, 560
Interstitial cells of Cajal (ICCs), 698, 863t, 864f, 865–866
Intestinal motility, TRP in, 697–698
  gastro intestinal motility, 698
  parasympathetic signalling, 697
  TRPC4-, TRPC6-, and TRPC4/TRPC6-gene-deficient mice, 698
  TRPC4-/TRPC6-/TRPC4/TRPC6-gene-deficient mice, 698
Intracellular Ca^{2+} concentration ($[Ca^{2+}]_i$), 128, 279, 378, 505, 601–602, 675, 694, 699, 835f, 939, 1012, 1016
Intracellular desensitization process, 97
Intracellular signalling pathways
  cytoplasmic Ca^{2+} increases, 674–675
  IP3, 674–675
  phospholipase C, 674–675
Intra-nasal nicotine, 973
Intrinsic pacemaker activity, 161
In vitro cultivation temperature, 453
In vitro cultured cardiomyocyte, 376
In vitro model, 162
In vivo model, 162
In vivo rodent model, 179
Iodoresiniferatoxin, 97–98, 197, 790, 996
Ion channel properties
  activation
    heat, 641
    low pH, 641
    vanilloid compounds, 641
    voltage, 641
  function modulation
    N-glycosylation, 644
    PIP2 hydrolysis, 643
  gating biophysics
    Ile696, Trp697, and Arg701, substitution, 642
    lower conductance (50 pS), 642
  ion selectivity and permeability
    activated with capsaicin, 643
    Asp646, 643
    neutralization, 643
    S800A mutation, 643
    Tyr671, 643
Ion channels, 25, 27, 960
  chemical signals, 25
  cryo-EM determinnation channels, 31
    inositol triphosphate receptor, 31
    large-conductance calcium-channel, 31

muscle calcium release channel/ryanodine receptor (RyR1), 31
muscle L-type voltage gated calcium channel (dihydropyridine receptor), 31
voltage-activated potassium channel (BK), 31
voltage-gated channel KvAP, 31
voltage gated sodium channel, 31
determination mechanisms
  channel gating, 25
  ion selectivity, 26
  modulation, 26
  permeation, 26
electrical signals, 25
mechanical signals, 25
selectivity and permeation, characterization tools, 87–88
  divalent cation, 87
  trivalent cation, 87
single-particle analysis work, 31
temperature signals, 25
Ion channel translocation, 545–565
Ion flux, 31, 547, 882
IonWorks (MDC), 108, 645
IonWorks Quattro, 109
IP$_3$, see Inositol triphosphate (IP$_3$); Inositol 1,4,5-trisphosphate (IP$_3$)
IPAH, see Idiopathic pulmonary arterial hypertension (IPAH)
IP$_3$R, 609
  See also IP$_3$ receptor (IP$_3$R)
IP$_3$ receptor-independent mechanism, 400
IP$_3$ receptor (IP$_3$R), 71, 72t, 91f, 92, 127, 130, 163, 326, 375f, 376, 395, 398, 400–402, 417, 422–423, 436, 574f, 601, 670, 674, 764, 783f–784f, 834, 887
IR-spectroscopy, 458
Ischemia, 179–180, 531, 537–538, 580, 676, 717, 769–770, 789–790, 997
Ischemia/reperfusion injury, 532, 537–538, 790
Ischemic stroke, 164, 179, 539
Islam, M. S., 811–825
Islet of Langerhans, 158
Islets, TRP channels of, 811–824
  of insulin-secreting cells, 813t
  TRPC1 and TRPC4, 823–824
  TRPM3, 820–821
  TRPM4 and TRPM5, 817–818
    in stimulus-secretion coupling in β-cells, 815–816
  TRPM2 (LTRPC2)
    and β-cells, 817
    and β-cells death, 820
    heat, 819
    intracellular Ca^{2+} release channel, 820
    in stimulus-secretion coupling in β-cells, 817–818
  TRPV1, 823
  TRPV2, 822–823
  TRPV4 (OTRPC4/VR-OAC/VRL-2/TRP-12), 821–822
Isothiocyanate, 66
  forms, 66
    allyl, 66
    benzyl, 66
    isopropyl, 66
    methyl, 66
    phenylethyl, 66
Isothiocyanates (mustard oil), 623
I$_{ti}$, see Transient inward current (I$_{ti}$)
IV, see Current-voltage (I–V) relationship
I–V curve, 180, 712
Iwata, Y., 754

J

J82 (urothelial cell line), 956
Jacobsen, R. B., 107–120
Jardín, I., 412–425
Jaw-opening reflex, 622, 623f
Jiang, J., 177
Jiang, Y., 263, 561
Jiménez-Díaz, L., 595–606
Jing, H., 263
JNJ17203212, 641
JNJ-39267631, 193f, 194, 197
JNK, see c-Jun N-terminal kinase (JNK)
Johnson, C., 707–723
Jordt, S. E., 481
JTS-653 (Japan Tobacco), 652f, 654
Julius, D., 43, 481, 847
Jung, J., 499
Jung, S. J., 615–629
Jungnickel, M. K., 130
Jurka T cell line, 160
JYL1421, 641

K

Kajimoto, T., 373–382
Kainate glutamate receptor, 56
Kaleta, M., 315–320
Kallikrein, 245, 548t, 560
Karashima, Y., 624

K_{ATP}, see ATP-dependent K^+ channel (K_{ATP})
K_{ATP} channels, 158, 812
KB-R7943, 738
KB-R9743, 738
21 kDa protein, 536, 537f
25 kDa protein, 536
K^+ efflux, 765, 793f, 799
Kelvin scale, 472
Kemp, B. J., 331
Keratinocyte, 44, 59–60, 139, 250t, 253, 438, 620f, 627, 786, 847–856, 891, 932, 936, 956, 977
K562 erythroleukemia cell, 934, 958
Ketoconazole, 95
Kheradpezhouh, E., 678
Kim, S. H., 460
Kim, Y. H, 619
Kim, H. Y., 622
α-Kinase, 4t, 10–11, 27, 34, 173, 178, 258, 294, 375, 398–400, 404, 423, 538, 558–559, 565, 617, 735t, 784, 819, 890, 893
Kinetic model, 472f, 471, 481, 482f, 520, 525
Kiselyov, K., 209–217, 1025
Kiyonaka, S., 373–383
Klein, R. M., 501
Klose, C., 87
Klotho (*KL* gene product), 247, 958
Knee joint swelling, 618
Knock-in experiment, 216
Knudson's model, 1025
Kohler, R., 771, 796
Koike, C., 137, 142
Koizumi, S., 854
KO mouse/mice, 164, 227, 250t
  functional TRPM5 protein, lacking, 163
Köttgen, M., 297–301, 351
Kozlowski type spondylometaphyseal dysplasia, 1015–1016
  kyphoscoliosis, 1015
  mutation (D333G), 1015
  R594H mutation, 1015
Krautwurst, D., 87–99
KU55933, 951
Kühn, F., 825
Kumar, P. G., 881–895
Kumar, S. V., 456
Kupffer cells, 672, 676
Kupffer cells (macrophages), 667
Kurata, Dr. H., 528
K_v1.2, 5f, 7, 8f, 31f, 34, 53, 480f, 520, 521f, 525f, 639

K_v channels, 470, 476, 479, 486, 519–520, 521f, 528
K_v1.2 K^+ channel, 7
K_v1 potassium channel, 196
K_v's, canonical voltage-activated potassium channel, 468
  pore module (TM5 and TM6), 470
  voltage sensor module (TM1 to TM4), 470
Kwan, K. Y., 625
Kyphoscoliosis, 1015–1016

L
L273, 442
L276, 442
Lactarius uvidus (drimenol), 47t, 52
Lactarius vellereu (isovelleral), 46t, 52
L-α-lysophosphatidylcholine, 29
Lamiaceae, 62, 64t
LAMP2 stain, 215
Lanthanum ion, 93, 98–99
Large, Prof. W. A., 404
L-arginine, 55
Latorre, R., 467–484, 521, 525
Lauryldimethylamine-N-oxide (LDAO), 6
Lazaro gene, 394
Lazzeri, M., 935, 950
LDAO, see Lauryldimethylamine-N-oxide (LDAO)
"Leaky" gene expression, 225
Leech, C. A., 814
Lehen'kyi, V., 929–940
Lehenkyi, V., 934
Leishmania, 230, 358t, 360
Lessard, C. B., 586
Leucine zipper, 137, 463
Lewis, R. J., 41–73
L712F, 261–262
Liao, Y., 424
Lidocaine, 624
Lien, 260
Ligand-gated approach, see PatchXpress (MDC)
Light-induced signaling cascade, 98
Liljestrand, G., 798
Li, M., 1–19, 177
Liman, E. R., 129
Linalool, 62, 64t, 70, 188
Linear current voltage *(I/V)* relationship, 814
Ling, G., 459
Linolenic acid, 128, 395
Lipid aggregate, 32–33
Lipid bilayer recording, 675
Lipid hydrolysis, 214

Lipid rafts, 188, 376, 422f, 423–424, 437, 477, 533, 560, 600
Lipofuscin buildup model, 215
Lipophosphatidylcholine, 581
Lipoxygenase homology/polycystin, lipoxygenase, atoxin (LH2/PLAT) domain, 293, 301
Lipoxygenase reaction, 93, 293, 301, 790, 990, 997
Lishko, P.V., 498
Listeria monocytogenes, 977
Liu, X., 433–443, 824
Liu, Y., 185–200
Liver, 62, 148, 186–187, 224, 281, 289–290, 358t, 359, 639, 667–680, 886, 931–932, 977, 1023
Li, Z., 141
L-methionine, 55, 863t
LNCaP, 200, 248, 254, 931, 933–935, 938, 953–955, 959, *see* Lymph node carcinoma of the prostate (LNCaP) cell line
Local antifungal therapy, 95
Lock-and-key agonist binding, 66
LOE908, 88, 90, 94
Löf, C., 125–131
Long-term depression (LTD), 506, 556, 993
Long-term potentiation (LTP), 490, 556, 547, 993
Loot, 771
Loss-of-function mutation, 210, 231–232, 349, 786
LOV-1, 293, 326, 329, 362t, 564
L1089P residue, 16
LSD, *see* Lysosomal storage disease (LSD)
L-type Ca^{2+} channel blocker verapamil, 978
L-type Ca^{2+} channels, 251, 601, 737–738, 742, 890, 894, 978
L type Ca^{2+} current, 604
Lucas, P., 126
Lukacs, V., 500
Lung filtration coefficient (K_f), 978
Lyall, V., 839
Lymph node carcinoma of the prostate (LNCaP) cell line, 200, 936, 955, 959
Lymphocytes, 96, 159f, 160, 178, 224, 226, 350, 372–375, 412, 435–436, 440, 443, 537, 673, 786, 788, 949
LysoPC, *see* Lysophosphatidylcholine (lysoPC)

Lysophosphatidylcholine (lysoPC), 56, 62, 552, 558, 795, 955
Lysophosphatidylinositol (lysoPI), 56, 65, 548t, 558, 936, 955
Lysophosphatidylserine, 65
Lysophospholipid (LPL/lysoPLs), 188, 714, 935, 955, 974
LysoPI, *see* Lysophosphatidylinositol (lysoPI)
Lysosomal-endosomal fusion, 212
Lysosomal ion homeostasis, 212f, 213
Lysosomal storage disease (LSD), 210, 213

M
M-68008, 654
Magnesium homeostasis, 88, 178, 692, 709, 718–719
Magnesium Inhibitable Current (MIC), 175
Magnesium nucleotide-inhibited metal (MagNuM), 175
MagNuM, *see* Magnesium nucleotide-inhibited metal (MagNuM)
Mahieu, F., 479
Main olfactory epithelium (MOE), 129
Maitotoxin, 674–675, 816
Major histocompatibility complex class I (MHC-I), 227
Maleimides, 478, 972
Malignant tumors, 948
Mammalian cell, 6, 28, 226, 299, 453, 532, 644–645, 656, 710
Mammalian diaphanous-related formin 1 (mDia1), 294
Mammalian respiratory tracts, 640, 969
Mammalian thermoTRPs, 517
Mammalian TRP channels
  ANKTM (A), 532
  canonical (C), 532
  melastatin (M), 532
  mucolipin (ML), 532
  polycystin (P), 532
  vanilloid receptor (V), 532
Mammary glands, 224, 253, 301–302, 921
Mandadi, S., 499
Mantle cell lymphoma (MCL), 949
MAPK, *see* Mitogen-activated protein kinase (MAPK)
Map kinase activity, TRPC1 regulation
  expression pattern of zebrafish *trpc6*, 347–348
  fli:gfp transgenic line, 347
  MO targeting *trpc1* translation, 347
  VEGF, 347
  in zebrafish embryos, 347

Marcus, D. C., 252
Marine polyether, 54
Marker of cell-type, TRP channel, 344–345
   olfactory sensory neuron (OSN), 344
      ciliated OSNs, 344
      microvillous OSNs, 344
      zebrafish transgenic lines, 345
Maroto, R., 343
Maruyama, Y., 420
M. arvesis (cornmint), 62
Mast cell, 44, 161, 178, 436, 442, 445, 492, 558–559, 673, 891, 976
Mast-cell activation, 165
Mathes, C., 120
Matrix attachment region (MAR), 284,
Matta, J. A., 526–527
Matveev, V. V., 461, 464
McAlexander, M. A., 969–979
M-calpain, 177–178
MCF-7 breast cancer cell, 178, 254, 939
MCF-7 breast cancer cell line, 178, 939
McKay, R. R., 910
MCOLN3
   mutant alleles, 231t, 231–232
      Va/+ and *Va/Va, of mice*, 232
      Va mutation, 232
   normal alleles, 231–232
   See also Mucolipin-3 gene *(MCOLN3)*; TRPML3
Mcoln4, see Trpml4
MCOLN1 (NG_015806, NM_020533) coding, 210
MC1R, *see* Melanocortin-1 receptor (MC1R)
MDA-MB-231 cell, 254
Mechanical (acoustic) stimulation, 234
Mechanical allodynia, 628
Mechanical membrane stretch, 71, 109
Mechano-activation, 109
Mechanosensation, 43, 71, 88, 108, 234, 240, 294–295, 317–319, 325–330, 332, 462–463, 539, 617, 625, 627, 692, 861–871, 883
Mechano-sensor, 463, 713
Mechanotransducer, 17, 691, 770, 871
Mechanotransduction, 14, 231t, 234, 327, 329, 345, 711t, 718, 862, 864f, 917
Medial entorhinal cortex layer V cells, 601
Mediators of oxidative stress
   TRPC
      diacylglycerol (DAG), 533
      inositol 1,4,5-trisphosphate (IP$_3$), 533
      NO sensors, 533
      TRPC3/4, activation, 534
   TRPM
      TRPM2, 534–538
      TRPM7, 538–539
   TRPV
      capsaicin, 539
      caspase-3 in DRG neurons, activation, 539
Medler, K. F., 831–842
Medullary part of the cortical collecting duct (mCCD), 278
Melanaphy, D., 707–723
Melanocortin-1 receptor (MC1R), 139
Melanosome, 139, 352
Melastatin (MSLN), 136
Melastatin-related TRP (TRPM), 88, 99
Membrane depolarization, 154, 158, 162–163, 187, 374, 518–519, 582, 586, 738, 741, 764–765, 799, 813–814, 816, 818–819
Membrane receptor, 17, 60, 90, 163, 436, 576, 886
Membrane voltage (V$_m$), 127–128
Mentha piperita (peppermint), 59, 62
Mentha pulegium (isopulegol), 62
Menthol, 62, 88, 97, 186–188, 189t, 198, 470, 627
   synthetic derivatives
      Coolact P, 188
      Cooling agent 10, 188
      Frescolat ML, 188
      WS-3, 188
   as traditional methods, 198
Menthone, 62, 626
Merck, 652t, 653
Mergler, S., 938
Merulius tremellosus (merulidial), 46t, 52
Meseguer, V., 95
"Metabolic" model, 212f, 213
Metabotropic glutamate receptor (mGluR6), 141–143, 161, 198, 552, 599, 887
Metalloproteinases (MMP), 948
Metatropic dysplasia, 14, 61, 1016
   characteristics, 1016
   point mutations, 1016
   TRPV4 mutations, 1016
Methacholine, 736, 739
1-(β[3-(4-Methoxyphenyl)propoxy]-4-methoxyphenethyl)-1H-imidazole hydrochloride, 89f, 90
Methyl-β-cyclodextrin (MBCD), 479
   use of, 424

Methylnicotinate, 999
Methylsalicylate, 55
Meyer, N. E., 554
Meyer, T., 439
mGluR6, see Metabotropic glutamate receptor (mGluR6)
MHC-I, see Major histocompatibility complex class I (MHC-I)
Miao, X., 935, 953
MIC, see Magnesium Inhibitable Current (MIC)
Mice lacking TRPM4 channels, 975
Michailidis, I., 19
Miconazole, 94, 711t
Microbial ion channels, 315–316
Micromanipulated pipette, 107
Microphthalmia transcription factor (MITF), 137, 139–140, 937
Microsatellite, 998
Microvillous OSNs, 344
 TRPC2, 344
 V2R-type receptors, 344
Midbrain dopaminergic mechanisms, 998
Migraine, 509, 621, 651, 652t, 653
Miller, B. A., 531–540
MIN-6 insulinoma, 558
Minke, B., 98
Mishra, R., 675
MITF, see Microphthalmia transcription factor (MITF)
Mitochondrial matrix ($[Ca^{2+}]_{mt}$)
 apoptosis, 669
 ATP synthesis, 669
 citric acid cycle, 669
Mitogen-activated protein kinase (MAPK), 178, 504
Mitral layer of olfactory bulb, 584
MK-2295/NGD-8243, 653–654
MLIV, see Mucolipidosis type IV (MLIV)
MLIV fibroblast cell, 214
MMP2, 936, 955
MMP9, 936, 955
Mn^{2+} entry assay, 175
Mochida Pharmaceuticals, 654
MOE, see Main olfactory epithelium (MOE)
Mohapatra, D. P., 498
Moiseenkova-Bell, V., 25–35
Molecular dissection of YVC1 (TRPY1), 318
Monet, M., 558
Monod, J., 481–482
Monoterpenoid linalool, 70

Monovalent cations, 26, 109, 114, 147–165, 299, 316, 417, 485, 643, 688, 712, 760, 764, 784, 814, 834, 893, 920
Moonwalker (Mwk) mice, 579
Morenilla-Pelao, C., 479
Mori, Y., 96, 373–383
Morpholino (MO), 345
Mouse
 TRPC2–/–, mating behaviour, 129
 TRPC2, splice variants
  mTRPC2α, 126
  mTRPC2β, 126
  TRPC2a, 126
  TRPC2b, 126
Mouse colon afferent fibers, mechanosensitivity, 618
Movement-induced nocifensive behavior, 618
MSLN, see Melastatin (MSLN)
Muallem, S., 422
Mucolipidin group (TRPML), 88
Mucolipidosis type IV (MLIV), 209, 222, 1026, 1027t
 achlorhydria, 210
 Drosophila model, 210
 pathological manifestations
  constitutive achlorhydria, 210
  corneal opacity, 210
  gastric acid secretion, decrease in, 210
  retinal degeneration, 210
Mucolipin-3 gene (MCOLN3), 229–230, 231t, 231–232
Mucolipins 1 to3, see TRPML1; TRPML2; TRPML3
Mucus secretion, 641, 977
Multiunit microelectrode array, 128
Multivariate Cox regression, 951–952
Murine haematopoietic cell, 130
Muscle fatigue, 752
Muscular dystrophy in mdx mice, 754, 756
Mustard oil, 69, 110, 188, 190t, 346, 548t, 563–564, 623, 766, 782, 863t
Mutagenesis method, 16–18, 33, 35, 99, 342, 348, 351, 438, 477, 498, 524, 557, 560, 566, 579, 783
Myocardial ischemia, 790–791
Myo-endothelial gap junction, 760
Myofibroblasts, 180, 641, 789, 864f, 866
Myogenic tone and TRP, 694–695
 calmodulin and myosin light chain kinase, 694
 mammalian TRP isoforms, 694

Myogenic tone and TRP (cont.)
  TRPC6 antisense oligodeoxynucleotides, 694
  TRPC6–/– mice, 694
  TRPM4-like channels, 694–695
  TRPV2, 695
Myoglobin, 461, 464
Myosin II, 11, 784

N

NAADP, see Nicotine acid adenine dinucleotide phosphate (NAADP)
NAADP$^+$, see Nicotinic acid adenine dinucleotide phosphate (NAADP$^+$)
Na$^+$/Ca^{2+} exchange-1, 742
Na$^+$/Ca^{2+} exchanger (NCX), 709
Na-Cl cotransporter (NCC), 282f, 561
N-acyldopamines, 48t, 990
N-acylsalsolinols, 646
N-acyltaurine (NAT), 54, 61, 646
NAD, see Nicotinamide adenine dinucleotide (NAD)
β-NAD$^+$, 538, 817–818
NADA, see N-arachidonoyl-dopamine (NADA)
Naematoloma sublateritium (fasciculol), 52
Nagasawa, M., 558
Nagata, K., 868
Na$^+$/H$^+$ exchanger (NHE) regulating factor 2 (NHERF2), 258, 506, 561–562
Nakao, A., 373–383
N-(3-aminopropyl)-2-([(3-methylphenyl)methyl]oxy)-*N*-(2-thienylmethyl)benzamide hydrochloride salt (AMTB), 194–195, 711t
N-arachidonolylserine, 646
N-arachidonoyl-dopamine (NADA), 48t, 54, 190, 485, 646, 997
N-arachidonoylsalsolinol, 54
N-arachidonylethanolamine (AEA, anandamide), 646
Nasonov, D., 461
Nasturtium seeds, 66
Na, T., 263
NAT, see N-acyltaurine (NAT)
Natural product ligands
  deorphanisation, 42
  diverse function
    growth cone morphology, modification, 43
  mechanosensation, 43
  osmotic pressure, detection, 43
  thermosensation to vasoregulation, 43
mammalian subfamilies
  TRPA (Ankyrin), 43
  TRPC (Canonical), 43
  TRPML (Mucolipin), 43
  TRPM (Melastatin), 43
  TRPN (No mechano-potential), 43
  TRPP (Polycystin), 43
  TRPV (Vanilloid), 43
pharmacological attributes, 42
as polymodal sensors, 43
role and emphasis, 42
as therapeutic agents, 42
TRPA1, 66–71
  α, β-unsaturated aldehyde, 69
  alkylamides, 69
  cannabinoids, 69
  endogenous ligands, 70–71
  isothiocyanates, 66
  methyl salicylate, 70
  monoterpenoids, 70
  nicotine, 70
  thiosulfinates, 66–68
  vanilloids, 70
TRPC and TRPP channels, 72t
TRPC1–TRPC6, 71
TRP family, 43
TRPM8, 62–65
  endogenous ligands, 62–65
  menthol receptor, 62
  monoterpenoids, 62
  prostate, expression in, 62
TRPM1–TRPM7, 65
TRPM1–TRPM8, 63t–64t
TRPP2, 71–73
TRPV1, 4355, 45t–49t, 67t–68t
  α, β-unsaturated dialdehyde, 51–52
  brain structures, 44
  cannabinoids, 53
  compounds, affinities, 52
  endogenous ligands, 54
  functional vanillyl group, 44
  ginsenosides, 53
  identification, 43
  immunoreactivity or mRNA, 44
  miscellaneous compounds, 55
  nociceptive or pain sensing neurons, characterization, 44
  toxins and peptides, 53–54
  triprenyl phenols, 52–53
  vanilloids, 44–51

TRPV2, 56–58
    cannabinoids, 56
    endogenous ligands, 56–58
    probenecid, 56
TRPV3, 59–60
    dermatitis, 59
    endogenous ligands, 60
    hair growth, 59
    monoterpenoids and diterpenoids, 59–60
    thermal nociception, 59
    thermoregulation, 59
    vanilloids, 60
TRPV4, 60–61
    diterpenoids, 61
    endogenous ligands, 61
    phorbol esters, binding site formation, 61
TRPV2–TRPV6, 57t–58t
use, 42
Nau, C., 498
Nauli, S. M., 296, 772
Navarro-López, J. D., 595–606
NBD, see Nucleotide binding domain (NBD)
N-[4-3,5-bis(trifluromethyl)pyrazol-1-yl]-4-methyl-1,2,3-thiadiazole-5-carboxamine (BTP2), 157, 161
NCC, see Na-Cl cotransporter (NCC)
N-dodecyl-β-D-maltoside (DDM), 6, 29
Negatively charged glutamate, 153
Neher, E., 107
Nematodes
    Brugia malayi, 363
    Meloidogyne incognita (plant parasite), 363
    "model worm," C. elegans, 363
    or roundworms, 363
    TRP channels/C. elegans, comparison, 364t–365t
    T. spiralis and M. incognita genomes, 363, 364t, 365
Nemes, Z., 849
Nephron and the collecting duct (CD), 277
    Bowman's capsule, 278
    glomerulus, 277
    renal tubular system, 278
Nerve growth factor (NGF), 500, 548t, 550, 617, 640, 864, 971, 977
Netrin-1, 343–344, 988
Neural persistent activity
    mechanisms

non-specific Ca^{2+} sensitive cationic current (CAN Current), 596–598
TRP and diseases epilepsy
    CAN/TRPC 4 and 5 currents, activated, 603, 605f
    experimental models, 603
TRPC and Ca^{2+} channels
    intracellular Ca^{2+} levels, 601
    L-type Ca^{2+} channel blocker nifedipine, 601
TRPC and DAG
    application of OAG, 602
    DHPG, 602
TRPC and IP3
    medial entorhinal cortex layer V cells, 601
    thapsigargin/cyclopiazonic acid, application, 601
TRPC and PIP2
    inhibits TRPC 4/5, 600
    role in "plateau" depolarization, 600
TRPC channels and G-protein mediated receptors
    G-protein inhibitor GDP-β-S, 599
    G-protein nonhydrolyzable activator GTP-γ-S, 599
TRPC channels and PLC
    ET-18-OCH3, 600
    TRPC–NHERF interaction, 600
    U73122, 600
TRP channels and CAN current
    mediators, 598
    M1 muscarinic receptors, involvement, 599
    (S)-3,5-dihydroxyphenylglycine effect, occlusion, 599
TRPC pharmacology
    2-APB, 603
    carbachol induced current, blockage, 603
    FFA application, 602
    heteromers activated, 603
    SKF-96365, 603
TRP/pain processing/persistent activity
    nociceptors, 604
    "plateau" potentials, 604
    wind–up, 604
Neural stem cell (NSC), 576–577
Neuregulin 1 (NRG1), 994
NeurogesX, 647
(neuro)melanin-containing dopaminergic neuron, 141

Neuromuscular disorders
  4αPDD agonist, 1017
  congenital spinal muscular atrophy, 1016
  HMSN2C, 1016–1017
  melastatin TRP, 1017
  R315W/R316C/R269H mutation, 1017
  SPSMA, 1016
Neuronal calcium sensor-1 (NCS-1), 996
Neuronal cell death, 382, 563, 998
Neuronal nitric oxide synthase (nNOS), 179
Neurotoxins
  1-methyl, 4-phenyl pyridinium ion (MPP$^+$/MPTP), 585
Neurotoxin shellfish poisoning, 54
Neutral glutamine, 153
Neutralization, 99, 479, 484, 498, 520, 643, 822–823
Neutrophil, 90, 95, 381f, 532, 538, 769, 885, 890–891, 975
Newson, P., 999–1000
NFAT, see Nuclear factor of activated T cell (NFAT)
N-formyl-Lmethionyl-L-leucyl-L-phenylalanine (fMLP), 90, 548t, 558
NGD-8243/MK-2295, 650t
NGF, see Bradykinin (BK); Nerve growth factor (NGF)
N-glycosylation sites
  Asn-821, 186
  Asn-934, 186
NHERF-1, 551, 561
NHERF2, see Na$^+$/H$^+$ exchanger (NHE) regulating factor 2 (NHERF2)
Niacin (nicotinic acid), 999
Nickel-chelate affinity chromatography, 29
Nicotiana tabacum (nicotine), 49, 55, 70
Nicotinamide adenine dinucleotide (NAD), 65, 113, 380, 535, 538, 711t, 713, 769, 789, 817–818
Nicotine, 49t, 55, 70, 325, 332, 364t, 380, 973
Nicotine acid adenine dinucleotide phosphate (NAADP), 380, 538, 676, 816, 818–819
Nicotinic acid adenine dinucleotide phosphate (NAADP$^+$), 538
Nielsen, D., 120
Nifedipine, 97, 601, 738, 821, 913
Night blindness, 27, 73, 140, 1017–1019
Nikrozi, Z., 285
Nilius, B., 461, 463,
Nitric oxide (NO), 49t, 71, 179, 377, 532, 627, 641, 669, 709, 719, 764, 774t, 793, 854, 856, 864, 891, 954, 973
Nitro-oleic acid, 49t, 54, 71
NMB-1, see Noxious mechanosensation blocker-1 (NMB-1)
N-methyl-D-aspartate (NMDA), 999
  receptors, 991
N-methyl-D-glucamine, 176, 733
NMR, see Nuclear magnetic resonance (NMR)
NMR spectroscopy, 3, 5–7
nNOS, see Neuronal nitric oxide synthase (nNOS)
NO, see Nitric oxide (NO)
Noben-Trauth, K., 229–236
NO-cGMP-protein kinase G signaling pathway, 771
Nociception, 59–60, 96, 197, 367t, 457, 493–495, 504, 582–583, 615, 617–618, 620, 624, 625–629, 785, 862, 866, 998
Nociceptive reflex, 195
Nociceptors, 44, 115, 197, 492, 494, 496f, 497, 503–504, 506–507, 509, 604, 615–618, 620f, 621–627, 646, 790, 862–863, 868, 973–974, 976–977
N-octyl-β-D-glucoside (OG), 6
Node of Ranvier, 484
N-oleoyl-dopamine, 54, 790
N-oleoylethanolamine (OLEA), 646
No Mechano Potential Channel C (NOMPC), 345
Nongenomic effect of suckling, 248
Nonionic and zwitterionic detergents, 6
  See also individual entries
Nonmuscle myosin IIA, 177
Nonmuscle myosin IIB, 177
Nonmuscle myosin IIC, 177
Non-selective cation channels (NSCC), 708
  Ca^{2+} influx and Na$^+$ influx, 709
  "classical" TRP (TRPC), 709
  functional classes of, 709
    CA channels, 709
    constitutively active (CA) channels, 709
    ROC and SOC, 709
    SAC channels, 709
    stretch activated (SAC) channels, 709
  Gq/11/phospholipase C (PLC) coupling, 709
  Na$^+$/Ca^{2+} exchanger (NCX), 709
  NCX, 709
  PLC coupling, 709
  for vascular smooth muscle function, 709

vascular smooth muscle responsiveness, 709
Non-selective cation current (I_{CaNS}), 128–129, 137, 437, 814, 816, 824, 974, 977
Nonselective cation current (NSCC), 708–709, 713, 717, 736, 738–739, 741–742, 814
Non-selective cationic channels (NSC_{Ca}), 148
    presence, 148
Non-TRPM8 mechanism, 198
Noradrenaline, 397, 991
Norepinephrine, 669
NorpA gene, 395
Northern blot analyses, 243
NO sensor, 377, 463, 533, 540, 713
Novel therapeutic strategy, 787
Nowycky, M. C., 580
Noxious mechanosensation blocker-1 (NMB-1), 71
N-(p-amylcinnamoyl)anthranilic acid (ACA), 65, 88, 89f, 94–95, 190, 711t, 819
NPS R-467 (calcimimetic compound), 246
NSC, *see* Neural stem cell (NSC)
NSC_{Ca}, *see* Non-selective cationic channels (NSC_{Ca})
NSCC blockers
    Cd^{2+}, 736
    Gd^{3+}, 736
    La^{3+}, 736
    Ni^{2+}, 736
    SKF-96365, 736
N-terminal and truncated transmembrane domains (TRPM2-S), 534, 536–538, 770, 817, 820
14 N-terminal ankyrin, 782
N-terminal extracellular ligand binding, 88
N-terminal (TRPM1), 534
N-(4-tertiarybutylphenyl)-4-(3-chloropyridin-2-yl)tetrahydropyrazine-1(2*H*)-carboxamide (BCTC), 188
N-(4-tertiary-carboxamide (SR141716A), 188–191
Nuclear factor kappa-B, 764
Nuclear factor of activated T cell (NFAT), 160, 374, 375f, 376–377
    transcription factors, 1013
Nuclear magnetic resonance (NMR), 1, 3, 4t, 5–7, 10–19, 34, 344

*Nu*cleoside *d*iphosphate-linked moiety X-*t*ype motif 9 *h*omology (NUDT9-H), 9, 380, 534–535, 547f, 817
Nucleotid-associated proteins, 456
Nucleotide binding domain (NBD), 151, 814
Nudix box, 789
NUDT9-H, *see Nu*cleoside *d*iphosphate-linked moiety X-*t*ype motif 9 *h*omology (NUDT9-H)
NudT9-H domain, 9, 380, 534, 535f, 817
Numazaki, M., 500
NYX gene, 141

O

*O*AADPr, *see* 2'-*O*acetyl-ADP-ribose (*O*AADPr)
2'-*O*acetyl-ADP-ribose (*O*AADPr), 380
OAG, *see* 1-Oleoyl-2-acetyl-sn-glycerol (OAG) effect
OAG-induced TRPC6 channel activity, 399
Oak ridge polycystic kidney (orpk) mouse, 291
Oancea, E., 135–143, 180
Obara, T., 348
Oberacker, T., 566
Oberegelsbacher, C., 566
OASF (Orai-activating small fragment), 421t, 421
Ocimum gratissimum, 45t, 50
Oesophageal reflux disease, 640t, 652t, 656
OG, *see* N-octyl-β-D-glucoside (OG)
25(OH)D, *see* 25-Hydroxyvitamin D (25(OH)D)
Ohm's Law, 476
Oh, S. B., 615–629
1-Oleoyl-2-acetyl-sn-glycerol (OAG) effect, 564, 894, 912
Oleoylethanolamide, 54
Olfaction, 108, 127, 129, 240, 831
Olfactory adaptation, 240
Olfactory marker protein (OMP), 344
Olfactory sensory neurons (OSNs)
    ciliated OSNs, 344
        cyclic nucleotide-gated channel A2 subunit, 344
        OMP, 344
        OR-type odorant receptors, 344
    microvillous OSNs, 344
        TRPC2, 344
        V2R-type receptors, 344
ON bipolar cell, 138, 140–141, 1018
On-cell patch clamp, 128

Oncogenic TRP channels, 929–940
    cancer cell genotypes characterization, 930
    role in cancer development and progression, 930–939
        TRPC channels and cancer, 931–933
        TRPM channels and cancer, 937–939
        TRPV channels and cancer, 933–927
Ong, H. L., 435–445, 824
ON pathway, 140
On-target effect, 201
Open channel block mechanism, 395–396
Oppenheimer, C. H., 454
Orai1, 160, 404
    Ca^{2+} selective ion channel, 420
    functional domains of, 420t
    Orai2 and Orai3, structure of, 420
    Orai1 currents, 420
Orai protein, 393, 417–420, 425, 440, 783, 888
Orai1 (SOC protein), 376, 393
Orio, P., 517–528
Organic blocker
    clotrimazole, 88
    LOE908, 88
    SKF-96365, 88
Origanum vulgare (carvacrol), 57t, 59, 65
Ornithoctonus huwena (Chinese bird spider), 48t, 53
Orphaned TRP channel, 65
Osmolarity-sensitive channel, 60
Osmo-sensor, 463
Osmotic cell swelling, 71
Osmotic membrane stretch, 71
OSM-9 protein, 240–241, 325–326, 328–330, 330, 364t
Osteoarthritis, 621, 640t, 651, 652t, 653–654
Otsuguro, K., 402
Oval cells, 669, 669
Ovarian cancer (OC), 936
Over-expression system, 298, 394
Oxaliplatin (chemotherapy drug), 197
Oxidation, 96, 532, 539, 770, 774t
Oxidative stress, 112–113, 379–380, 382, 457, 531–540, 585–586, 717, 769–770, 789, 795, 820, 971
4-Oxononenal (4-ONE), 972
Oxygenated eicosatetraenoic acids, 646
OxyR (transcriptional regulator protein), 457

P

P53, 140, 951, 960
PA, *see* Phosphatidic acid (PA)
Pacemaker HCN channel, 162
Paclitaxel, 628

PACSIN proteins, 549t, 559
PAEC, *see* Porcine aortic endothelial cells (PAEC)
Painful bladder syndrome, 97, 201, 867, 871
Pain upon inspiration, 970
Pajonk, F. G., 995
Palmer, C., 315–320
Panax ginseng, 47t, 53
Pancreatic β-cell, 65, 160, 382, 716
β-Pancreatic cell, 157
β-Pancreatic cell line, 158
Pani, Dr. B., 587
Piggott, B., 333
PAP, *see* Phosphatidic acid phosphatase (PAP)
Paracellular pathway, 249, 251, 259–260, 766
Para-cellular transport mechanism, 88
6-Paradol, 45t, 51, 70
Parasites, TRP channels in, 359–367
    ectoparasites, 366–368
    endoparasites
        helminths, 363–366
        protozoa, 2–363
    parasites infecting humans, 360t
Parathyroid hormone (PTH), 246–248, 250t, 251, 258, 278, 283, 559, 917
Parkinsonian syndrome, 585
Parkinsonism-dementia, 537
Parkinson's disease, 141
PARP, *see* Poly (ADPR) polymerase (PARP)
Patch-clamp electrophysiology, 50
Patch-clamp technique, 162, 315, 789
Patchliner (Nanion), 108–109, 112
PatchXpress (MDC), 108–110, 118, 122
Patel, V., 1023
Pathological cardiac hypertrophy, 376
Pathophysiological functions TRPV channels in cancers, 951t
Paulsen, I. M. S., 285
Pausinystalia yohimbe (yohimbine), 55
PBC, *see* Pre-Bötzinger complex (PBC)
PBP (polycystic breakpoint), 1027
p-bromophenacyl bromide (BPB), 94
PC-3 cell, 200
PC-12 cell, 90, 582
PC3 (prostate cancer) cells, 954, 955
PDBu, 404, 602
4α-PDD, *see* 4α-Phorbol 12,13-didecanoate (4α-PDD)
4αPDD agonist, 14, 1017
Pedretti, A., 480
Pendred syndrome, 253
Peng, J.-B., 239–263
Penicillium brevicompactum, 71, 72t

Penna, A., 558
Peptidyl-prolyl isomerase (PPIase), 344
Perforated patch technique, 175
Periaqueductal grey (PAG), 990, 992, 996
Peripheral demyelination model, 626
Peripheral neuropathic pain, 640t
Periplaneta Americana (cockroach), 93
Permeability of ion (P_{ion}), 128
P-glycoprotein (MDR1), 29, 50
PH, *see* Pleckstrin homology (PH)
PHA II, *see* Pseudohypoaldosteronism type II (PHA II)
Pharmacology and druggability
    agonists
        biogenic amines, 646
        resiniferatoxin (RTX), 646
    antagonists
        Aminoquinazolines, 647
        Biaryl carboxamides, 647
        dye ruthenium red, 647
        4-fluopyridine amides, 647
        non-vanilloid compounds, 647
        Phenylchromones, 647
        Piperazinyl benzimidazoles, 647
        Piperazinyl pyrimidines, 647
        pore blocker, 647
        Pyridoindazolones, 647
        spiroisoxazolopyridine amides, 647
        vanilloid derived, 647
Pharmaco-mechanical coupling, 708
Phase I clinical trial
    D-3263, 188, 200–201
9-Phenanthrol, 63t, 65, 153f, 157
Phencyclidine (PCP), 995
Phenylalanine, 35, 551
Phenylchromones, 647
Pheromone-evoked spiking, 128
4α-Phorbol 12,13-didecanoate (4α-PDD), 14, 58t, 61, 90, 95, 118, 186, 549, 627–628, 674, 690, 755, 767, 771, 774, 792, 794, 797, 822, 863t, 869, 957, 978, 1017
Phorbol ester, 61
    binding site formation at TRPV4, 61
        S556, 61
        Y555, 61
    TRPV4, activation residues, 61
        L584, 61
        W586, 61
Phorbol esters 4-alpha-phorbol 12,13-didecanoate (4α-PDD), 627

4-β-Phorbol 12-myristate 13-acetate (PMA), 61, 499, 627, 822
Phosphatidic acid (PA), 395, 396f, 715
Phosphatidic acid phosphatase (PAP), 395–396
Phosphatidylinositide, 99
Phosphatidylinositide 3 kinase (PI(3) kinase), 550
Phosphatidylinositol 4,5-bisphosphate, 130, 159, 177, 188, 244, 377, 396f, 416, 495, 600, 643, 709, 783, 787f, 815, 834
Phosphatidylinositol bisphosphate (PIP_2), 2, 43, 90, 91f, 151, 154–156, 161, 163, 177, 188, 244, 327–328, 330, 378, 393, 395–404, 416, 436, 436–437, 464, 475, 495, 501–503, 525, 550, 574f, 600, 602, 617, 624, 638f, 643, 709, 711f, 712–715, 787f, 815–816, 824, 834–836
Phosphatidylinositol 3,5-bisphosphate (PI(3,5)P_2), 258, 332
Phosphatidylinositol 4,5-bisphosphate (PIP_2), 130, 154, 177, 188, 244, 377, 396f, 416, 495, 600, 643, 709, 783f–784f, 815, 834–835
Phosphatidylinositol-3,4 kinase (PI3,4 K), 56
Phosphatidylinositol-3-phosphate-5-kinase PIKfyve (PIP5K3), 258
Phosphine oxide
    derivative
        WS-148, 188
Phosphofurin acidic cluster sorting proteins
    PACS-1, 559
    PACS-2, 564
Phosphofurin acidic cluster sorting proteins 1 and 2 (PACS-2 and PACS-1), 295
Phosphoinositide-4,5-bisphosphate (PIP_2), 2
Phosphoinositide 3-kinase (PI3-kinase), 294, 617, 735f
Phosphoinositide-specific phospholipase C (PI-PLC), 396f, 814, 818, 824
Phospholipase A_2 (PLA_2), 93, 532
    activity, 65
    signaling pathway, 188
Phospholipase C (PLC), 88, 99, 112, 128, 327, 373, 956
    members, interaction, 177
        PLC-β_1, 177
        PLC-β_2, 177
        PLC-β_3, 177
        PLC-γ_1, 177

Phospholipase C (PLC) activity, 88, 112, 126, 177, 188, 327, 373, 417, 503, 554, 586, 624
Phospholipases, 532
Photoaffinity labeling experiment, 96, 377
Photoreceptor cells, 342, 533, 554, 732
Physialia physalis (bluebottle or Portugese Man O' War), 54
Physiological mechanism, 188, 695
Pichia pastoris (P. pastoris), 6, 28
PIGEA-14, 295
PI3,4 K, *see* Phosphatidylinositol-3,4 kinase (PI3,4 K)
Pike, J. W., 246
PI(3) kinase, *see* Phosphatidylinositide 3 kinase (PI(3) kinase)
PI-3-kinase, 400–401, 404, 557
PI-4-kinase, 399
PI-5-kinase, 399
P_{ion}, *see* permeability of ion (P_{ion})
PIP_2, *see* Phosphatidylinositol bisphosphate (PIP_2); Phosphatidylinositol 4,5-bisphosphate (PIP_2)
$PI(3,5)P_2$, *see* Phosphatidylinositol 3,5-bisphosphate ($PI(3,5)P_2$)
Piperaceae, 50
Piperazinyl benzimidazoles, 647
Piperazinyl pyrimidines, 647
Piperine, 45t, 50, 493
Piper nigrum (black pepper), 45t, 50
PIP5K3, *see* Phosphatidylinositol-3-phosphate-5-kinase PIKfyve (PIP5K3)
Pirenzepine, 599
PKA, *see* Protein kinase A (PKA)
PKA-regulated entry mechanism, 131
PKB, *see* Protein kinase B (PKB)
PKC, *see* Protein kinase C (PKC)
PKCβ, *see* Protein kinase Cβ (PKCβ)
PKC(δ), 557
PKC-dependent mechanism, 394, 402, 404
PKC-independent mechanism, 394, 398, 401–402
PKC phosphorylation, 246, 497, 499, 560, 815
PKD1, 16–17, 71
   knockout mice, 73
      attenuated cyst formation, 73
PKD-1, 564, 919–920
PKD2, 289–290, 293–294, 302, 319, 327, 672t, 61, 885, 891–892, 919, 921, 1023–1025
PKD2L1, 292, 297–298, 691, 833, 836–838, 841, 891, 919, 1023

PKD2L2, 292, 299, 691, 891, 919, 1023
PKD1L3/PKD2L1 TRPP channels, 836–840
   ASIC channels, 837
   DTA, 837
   HEK 293T cells, 837
      functional properties, 837
      PKD1L3/PKD2L1 heterodimer, 838
   peripheral taste system, 837
   sour receptor in taste system, identification of, 836
   stimulated by sour tastants, 838f
PKD1L1 (Polycystic kidney disease 1 like 1), 301–302
   Dot-blot analysis and RT-PCR studies, 301
   in testis's Leydig cells, 301
   Pkd1l1$^{-/-}$ mice, 301
   primary cilia dyskinesia, 301
   protein architecture of, 301
PKD1L2 (Polycystic kidney disease 1 like 2), 301
   Dot-blot analysis and RT-PCR studies, 302
   homozygous$^{ostes/ostes}$ mice, 302
   Polycystin-1 PKDREJ subfamily, 302
PKD2 (polycystin-2) gene, 919
PKDREJ, 300–301
   homology of, 300
   for mammalian sperm, 300–301
   northern blot analysis and RT-PCR studies, 300
   sea urchin receptor for egg jelly (suREJ3), 300
PLA_2, *see* Phospholipase A_2 (PLA_2)
Placenta, 148, 251–252, 294, 556, 909–921, 933
Planar substrate, 108
Planells-Cases, R., 491–509
Plant thermosensory perception, 456
Plasma membrane channel
   Orai1 (CRACM1), 92
   Orai2 (CRACM2), 92
   Orai3 (CRACM3), 92
Platelet activating factor (PAF), 890, 975
Platelet-derived growth factor, 178, 768, 954
PLC, *see* Phospholipase C (PLC) activity
PLCγ2-deficient cell, 374
PLC-signaling cascade, 158
Pleckstrin homology (PH), 374
PMA, *see* 4-β-Phorbol 12-myristate 13-acetate (PMA)
PMA1 promoter, 29
P/n leak subtraction, 113
Pocock, 767
Polcystic kidney disease, 797

Poly(ADPR) glycohydrolase, 380
Poly (ADPR) polymerase (PARP), 534
Polyamine, 44, 55, 63t, 156, 483, 646, 711t
Polybasic tail, 439, 441
Polycystic kidney disease, 17, 71–73, 88, 289–293, 298–303, 318–319, 348, 564, 691, 771, 785, 789, 836–838, 891–892, 919, 1023
Polycystic kidney disease-1-like 3 (PKD1L3) TRPP channels, 836–838
Polycystic kidney disease-2-like 1(PKD2L1) TRPP channels, 846–838
Polycystic kidney disease (PKD), 17, 27, 71–73, 88, 289–293, 297–303, 318–319, 348–349, 564, 691, 771, 785, 789, 836–838, 891–892, 919, 1023
Polycystin-1, 71–72, 288f, 289, 292–293, 300–303, 348–349, 892, 1023, 1025
Polycystin 2, 71, 292, 348–349, 691, 891–892, 919, 1023
Polycystin group (TRPP), 88
Polycystin-1 (PC-1), 292–293
 features of, 293
  GPS, 293
  LH2/PLAT domain, 293
  REJ domain, 293
 mutations in PKD1, 292–293
 PC-1 and PKD1L1, 293
 PKD1-like genes, 293
  PKD1L1, 293
  PKD1L2, 293
  PKD1L3, 293
  PKDREJ (PC-REJ), 293
 PKDREJ-like proteins
  PKDREJ/PKD1L2/PKD1L3, 293
 receptor or adhesion molecule, 293
 TRPP1 encoded by PKD1 gene, 292
Polycystin 2 (PC 2), 292, 348, 602, 883t, 892, 919, 1023
Polycystin TRPs, 1023
 adult polycystic kidney disease (ADPKD), 1023
 PKD1, *PBP* (polycystic breakpoint), 1023
 TRPP1 (or *PKD2*), 1023
 TRPP2 (*PKD2L1* or PKD2-like 1), 1023
 TRPP3 (*PKD2L2* or PKD2-like 2), 1023
Polygonum hydropiper (polygodial), 46t, 52, 68
Poly-L-lysine, 399, 402
Polymodal gating mechanism, 187
Polymodal receptors, 469–486

Polymodal sensor of noxious irritants, 972–973
Polyunsaturated fatty acid (PUFA), 63t–64t, 328, 395–396
Polyvalent cations, 642
Pons, 584
Poo, M. M., 343
Porcine airway SM tissues, 732
Porcine aortic endothelial cells (PAEC), 586, 769, 795
Pore turret, 478, 525, 525f
Pore widening, 484–485
Positional cloning approach, 232
Positron emission tomography (PET), 991
Postherpetic neuralgia, 198, 621, 647, 652f
Potassium channel, 28
 BK, 109
 KCNQ4, 109
Potent defensemechanism, 98
Praetorius, J., 277–285, 296
Pre-Bötzinger complex (PBC), 159f, 161
Pregnant human myometrial smooth muscle cells (PMH1), 698–693
Pregnenolone, 63t, 65, 90, 92f, 95, 96f, 114, 690, 820, 821f, 975
Pregnenolone sulphate, 63t, 65, 90, 92f, 95, 96f, 114, 690, 820, 821f
Premolten globule, 459
Prenatal lethality, 178, 922
Pressure overload-induced cardiac hypertrophy, 96, 787
Prevarskaya, N., 929–940
Prevarskaya's team, 934
Primary thermosensory events in cells
 CSP-induced change mechanisms, 452
 DNA thermotropic reactions, 455–456
  genomic thermo-sensitivity, 456
  oligonucleotide duplexes, 456
  pLO88 plasmid, 456
  transitions, 456
  "twisted" superhelical conformation, 455
 membrane lipids fluidity, 454–455
  thermometers, role, 454
 protein thermometers, 456–458
  OxyR activation, 457
  phase transitions, 458
  structural transitions, 458
  thermal sensitivity, characterization, 457
 protein thermometers, structural features, 462–464
  polymerization, 464
  tertiary/quaternary structure, 463

Primary thermosensory events in cells (*cont.*)
  thermotropic reactions, modulation, 464
  TRPV1 C-terminal domain, 463
  protein thermosensitivity, biophysical aspects, 458–462
    enthalpy end entropy changes, 460
    free energy profile of globular protein, hypothetical model, 459f
    hypothesis of native aggregation, 462
    natively unfolded proteins, 459
    protein structural information conversion, strategies, 462f
  RNA thermotropic reactions, 455–456
    hybrid (intermolecular forms), 455
    regulatory elements, 455
    secondary and tertiary structures, 455
    thermometers operation, 455
  temperature change, effects, 451–452
    physico-chemical change, 452
    temperature-induced conformational change, 452
  temperature-controlled process
    cold-shock gene, expression, 452
    heat-shock gene, expression, 452
  temperature sensing biomolecules, 454
  TRP subfamilies
    TRPA, 457
    TRPC, 457
    TRPM, 457
    TRPML, 457
    TRPN, 457
    TRPP, 457
    TRPV, 457
PRKCSH, 677
Pro-algesic effect, 51
Prober, D. A., 346
Profibrotic TGF-β1 signal, 179–180
Progressive familial heart block, type 1B (PFHB1B), 1020
Pro-inflammatory agents
  histamine, 766
  thrombin, 766
Proinflammatory bradykinin signaling, 179
Proliferating Cell Nuclear Antigen (PCNA), 934, 939, 959
Proliferation, role of TRP in SMC, 697
  chronic hypoxia, 697
  secondary and idiopathic PAH, 697
  TRPC1 and TRPC3, 697
  TRPC6 upregulation, 697
Prostaglandin E_2 (PGE_2), 198, 506, 627

Prostaglandins, 68t, 70–71, 198, 497, 506, 557, 623, 627, 641, 646, 669, 850, 854, 864–865, 913, 972
Prostate-specific antigen (PSA), 938
Protein Data Bank, 6
Protein 3D structure
  determination methods
    electron cryomicroscopy (cryo-EM), 5
    NMR spectroscopy, 5
    X-ray crystallography, 5
Protein expression, 7, 19, 29, 198, 283, 669, 679f, 690, 694, 719, 740–741, 751–752, 764, 768, 770, 794, 890, 910, 911t, 912, 916t, 919t–920t, 935–936, 950, 952, 954, 956, 958, 1014
Protein-free micelles, 33
Protein kinase A (PKA), 10, 11f, 130–131, 151, 186, 188, 212, 246–247, 378, 494f, 495, 496f, 497–498, 499, 501, 503, 520, 536, 557, 563, 617, 624, 628, 643, 712, 711t, 815, 819, 893
Protein kinase B (PKB), 258, 649t, 954–955
Protein kinase Cβ (PKCβ), 375–376
Protein kinase C (PKC), 61, 158, 188, 246, 495, 554, 709, 883, 888
Protein kinases, 10, 61, 130, 154, 156, 177–178, 186, 188, 246–247, 258, 375, 378, 392, 477, 494f, 495, 496f, 497, 499, 501, 503–504, 532, 554, 557, 562, 617, 643, 709, 712, 714, 754–755, 771, 784, 787f, 788, 822, 883, 886, 888, 960
Protein–ligand interaction, 2, 12
Protein–protein interaction, 12, 16, 26, 381f, 416, 419t–420t, 423, 463, 883, 894
Proteins (TRP), TRPCs, 416–417
  Ca^{2+}:Na^+ permeability ratios, 417
  canonical TRP (TRPC) subfamily, 416
  cytoplasmic N- and C-termini, 416
  expression levels, 417
  siRNA or shRNA, 417
  in SOCE or ROCE, 417
  subfamilies of TRP, 416
  TRPC1 (mammalian TRP protein), 416
  trp mutant of *Drosophila*, 416
  TRP signature motif (EWKFAR), 416
Proteomic approach, 382
Protozoa, 362–363
  apicomplexans (*Plasmodium* species), 362

Index 1069

Cryptosporidium parva protein, 362
  kinetoplastids, 362
  metamonads, 362
  mosquito or biting fly, 362
  TRP-like channels, 362
  TRPML-like genes, 362
    lmmlA (LmjF07.0910), 362
    lmmlB (LmjF26.0990), 362
  TrTrypDB, 362
Proximal tubule (PT), 259, 278
Psalmopoeus cambridgei, 53
  VaTx1, 53
  VaTx2, 53
  VaTx3, 53
P53 Ser15, 951
Pseudohypoaldosteronism type II (PHA II), 254, 255t, 256
Pseudohypoaldosteronism type I (PHA 1), 278
18–25 pS single channel conductance, 153
Psychiatric disorders, 987–988, 1000
  depression and anxiety, 990–991
    TRPC channels, 991–992
    TRPV channels, 992–993
  schizophrenia, 994–995
    TRPC channels and, 995–996
    TRPV channels and, 996–997
  TRP channels in brain
    TRPC channels, 988–989
    TRPV channels, 989–990
PTH, *see* Parathyroid hormone (PTH)
PT pore, activation, 532
PubMed, 97
Puertollano, R., 236, 1025
PUFA, *see* Polyunsaturated fatty acid (PUFA)
Pufferfish *(Fugu rubripes)*, 351
Pulmonary arterial smooth muscle cells (PASMC), 422, 690–691, 695–696, 796, 799
Pulmonary artery hyperreactivity, 696–697
  CH-PAH in rats, 696
  chronic hypoxia-induced PAH (CH-PAH), in rats with, 696
  model of normobaric CH-PAH rats, 696
  primary hypertension (idiopathic PAH), 696
  resting [Ca^{2+}]i, 696
  siRNA or antisense oligonucleotides, 696
  studies in rat and in TRPV4 –/– mice, 696
Pulmonary hypertension, 782
Pulmonary hypertension and TRP, 695–696
  pulmonary arterial hypertension (PAH), 695
Purinergic P2X receptor, 50

Putrescine, 55, 483, 646
Pyrazole compound (Pyr3), 89f, 96–97, 375f, 376–377, 787f, 787
Pyr3 compound, 96–97
Pyr3-derivative, 96
Pyridoindazolones, 647
Pyridyl piperazine carboxamides, 648t

Q
Q977E substitution, 153
Qin, F., 491–509
Qin, N., 185–201
Qiu, A., 242
QPatch HTX multi-hole system, 109, 111
QPatch (Sophion), 108–111, 111f, 113f, 115–116, 118, 119, 120, 645
qRT-PCR, *see* Quantitative reverse-transcription PCR (qRT-PCR)
Q_{10} (temperature coefficient), 380, 457–458, 473, 475–477, 479, 483, 521, 642, 819
Quantitative reverse-transcription PCR (qRT-PCR), 138
Quinazolinon analogue, 649t
QutenzaTM, 647
QX314, 622

R
Rab9, 549t, 553
Rab11, 549t, 553, 561
RACK1, *see* Receptors for activated C-kinase-1 (RACK1)
Rapid inspiration, 641
Rauwolfia, 55
RBL-2H3mast cell line, 178
R269C mutation, 14, 1015t, 1017
R316C mutation, 14, 1015t, 1017
R59949 (DAG kinase), 395, 396f, 398, 401
rdgA gene, 395
Reactive nitrogen species (RNS), 179, 532, 534, 676
Reactive oxygen species (ROS), 178, 352, 380–381, 532–533, 539–540, 586, 669, 675–676, 695, 717, 766–767, 769, 773f, 818–819, 886, 920, 935, 954
Reale, V., 817
Reaves, B. J., 359–367
Reboreda, A., 595–606
Receptor for egg jelly (REJ) domain, 293, 300–301, 892
Receptor-mediated mechanism, 130
Receptor-operated Ca^{2+} entry (ROCE), 328, 415, 424f, 739, 742

Receptor operated cation entry (ROCE), 783
Receptor-operated channel (ROC), 328, 393, 688, 708
Receptors for activated C-kinase-1 (RACK1), 376, 423, 888
Receptor signaling integration
  TRPC3, 374–377
    DAG-activated Ca^{2+} influx role, proposed model, 375f
    partial PH-like domain, 374–375
    pathophysiological process, 376
    SOC encoding, 374
  TRPC5, 377–379
    activation proteins and factors, 377
    covalent modifications, 377
    endothelial cell layer, distribution, 378
    feedback cycle, proposed model, 379f
    labeling and functional assay, 377
    in neurite extension, 378
  TRPM2, 379–382
    downstream signaling cascades, 381
    electrophysiological data, 380
    on H_2O_2-induced chemokine production in monocytes and macrophages, 381f
    H_2O_2-induced TRPM2 activation, 380
    insulin secretion, 382
    NAD^+ metabolites, 380
    oxidative stress, activation, 379
    sustained ion influx, maintenance, 382
    tyrosine phosphorylation, 380
Receptor tyrosine kinase (RTK), 43, 327, 535, 574f, 576–587, 888–889, 1012
Recombination gene activator (RGA), 507, 549, 559
Rectal pain, 621
Redox sensor, 88, 94, 113, 150, 795
Refilling mechanism, 90
Regulator of Gprotein signaling (RGS), 331, 994
Renal outer medullary potassium (ROMK), 247
Reperfusion injury, 532, 537–538, 676, 790
Reproductive (female) organs and placenta, TRP channels in, 909–922
  mammary glands, 921
  TRPC family, 911–914
  TRPM family, 919
  TRPP family, 919–921
  TRP proteins in
    female reproductive organs, 911f
    human placenta, 914f
  TRPV family, 914–918

TRPV1, 914–915
TRPV2/TRPV3, 915
TRPV4, 915–917
TRPV5/6, 917–918
Research toolkits, TRPV1
  clinical status
    anesiva has Adlea in phase-III, 651
    clinical trials beyond phase-I, 652t
    Merck–Neurogen alignment on MK-2295, 653
    neurogesx has QutenzaTM in phase-III, 647–650
    Winston Laboratories has Civamide in phase-III, 651–654
  electrophysiology
    automated patch clamp techniques, 645
    IonWorks, 645
    manual patch clamp recording, 645
    QPatch system, 645
  florescence imaging and $45Ca^{2+}$ flux assays
    $45Ca^{2+}$ based assays, 646
    FLIPR platform, 646
  production and purification, current methods
    fermentation, 644
    insect cells, 644
    mammalian cells, 644
    S. cerevisiae yeast strain BJ5457, 644
    yeast, 644
  radio ligand binding assays
    3H-A-778317 (Abbott Laboratories), 645
    3H-Resiniferatoxin (PerkinElmer Life Sciences), 644–646
    Kd and Bmax, 644
Resiniferatoxin (RTX), 44, 45t, 50–53, 55, 98, 190, 191t, 481, 485, 493, 499, 527, 620–621, 644, 646, 674, 863t, 866, 989, 952, 989, 996
Respiratory sensory irritation, 973
Respiratory symptoms and pathophysiology, 969–970
  sensory neuronal TRP channels as initiators and regulators, 970
    TRPA1, 972–973
    TRPM8, 975–976
    TRPV1, 971–972
  TRP channels in airway cells of non-neuronal origin, 974
    airway structural and immune cells, 975–976
    capsaicin receptor and, 976–978

intracellular calcium homeostasis and inflammation, 975
TRPA1, 974–975
Reverse transcription–PCR, 957f
Reynaud's phenomenon, 723
RGA, see Recombination gene activator (RGA)
RGS-4, 994
R154H, 261–262
Rhabdomyosarcomas, 939
Rhaboid tumours, 939
RHC80267 (DAG lipase), 396f, 398
Rheumatoid arthritis, 458, 640t
R269H mutation, 14, 1015t, 1017
Rhodopsin mutant, 28
"Riboswitches," 455
    aptamer, 455
    expression platform, 455
Ricinoleic acid, 49f, 55
RN-1734, 89, 98
RN-1747, 89f, 98
RNAi, see RNA interference (RNAi)
RNA interference (RNAi), 376, 536–537, 690, 913, 955
RNS, see Reactive nitrogen species (RNS)
ROC, see Receptor-operated channel (ROC)
ROCE, see Receptor operated cation entry (ROCE)
Rohon Beard primary sensory neurons, 348
Role in pain sensation
    perspectives
        thermal hyperalgesia, 628
        thermal nociception, 628
    TRPA1
        in cold pain and hyperalgesia, 624
        induces heterologous desensitization of TRPV1, 625–626
        in mechanical hyperalgesia, 625
        tissue damages, 623–624
    TRPM8
        analgesia in different pain models, 626
        CFA injection, 626
        formalin-induced nocifensive behavior, 626
        receptor of menthol, 626
    TRPV1
        in central terminals of primary afferents and CNS, 618–620
        decreased arthritis-induced pain, 618
        in deep tissue pain, 618
        in heat pain and thermal hyperalgesia, 616–617
        therapeutic approaches, 620–623

TRPV3 and TRPV4
    activated by hypo-osmotic conditions, 627
    acute inflammatory mechanical hyperalgesia development, 628
    agonists, 627
    chemical ligands, 627
    inhibitor paclitaxel, 628
    prostaglandin E2 (PGE2) release, 627
    TRPV4 in visceral nociception, 628
Romanin, C., 421
ROMK, see Renal outer medullary potassium (ROMK)
ROS, see Reactive oxygen species (ROS)
Rosado, J. A., 413–426
Rosenbaum, T., 500
Rosenberg, P., 753
RPN1, zebrafish functions of, 345–346
    immunolocalization studies in *Xenopus*, 345–346
    TRPN1 (NOMPC), 345
Rs11078458, 949, 956
Rs65002729, 949, 956
RT4 (urothelial cell line), 950, 952f, 956
RTK, see Receptor tyrosine kinase (RTK)
RT-PCR screening, 130
RTX, 952, see Resiniferatoxin (RTX)
Ruthenium red, 69, 99, 110–111, 111f, 246, 251, 483, 643, 647, 695, 711t, 715, 771, 792, 794, 797, 848, 852, 856, 918, 978
Ruthenium red-sensitive vasorelaxation, 69
R315W mutation, 14, 1016
Rythmogenesis, 161
Ryazanov, A. G., 351
Rychkov, G. Y., 667–680

S

S556, 61
SAC, see Stretch activated channels (SAC)
Saccharomyces cerevisiae (S. cerevisiae), 6, 28–30, 35
SAG, see 1-Stearoyl-2-arachidonoyl-*sn*-glycerol (SAG)
Sah, R., 173–180
Saito, S., 350
Sakmann, B., 107
Salazar, H., 486
Saleh, S. N., 404
Salicylates, 49t, 55, 68t, 70
Salido, G. M., 413–426
Salix, 49t, 55
Sallé, L., 147–165, 814

Sanders, K. M., 318
Sanofi-Aventis, 654
Santoni, G., 947–961
SAR-115740, 654
Sarcoplasmic reticulum (SR), 317, 414, 688, 693–694, 708, 720, 736, 753, 799
Sato, Y., 345
Savineau, J. P., 687–699
SB-705498, 621, 648t, 651, 652f, 653–654
SB-782443, 654
SB-366791 (TRPV1 blocker), 98, 642, 649t, 655
Scaffold protein caveolin-1, 764
Scalding, 98
Scapuloperoneal spinal muscular atrophy (SPSMA), 14, 61, 1015t, 1016–1017
SCCD, *see* Semicircular canal duct (SCCD)
S. cerevisiae
    TRP channels, 316–316
        channel gating, 316–317
        mammalian, 316
        patch-clamp surveys, 316
        yeast strain, 316
        yeast vacuolar channel (YVC), 316
    TRPP2 homologs in, 319–320
Schistosoma, 230, 360t, 365t
Schizophrenia, 994–995
    TRPC channels and, 995–996
    TRPV channels and, 996–1000
Schrøder, R. L., 120
Scotopic vision, 140
SD, *see* Shine–Dalgarno (SD) sequence
SDS PAGE, 29
Sea squirts (invertebrate chordates), 351
SeatleSNPs program, 260–261
Second law of thermodynamics, 460
Sec63 proteins, 677
Secretion coupling, 436, 812, 815–816
Selectivity filter, 5f, 34, 151, 420t, 470, 471f, 484, 486, 525f, 643
Selemon, L. D., 994
Selvaraj, S., 573–587
Semicircular canal duct (SCCD), 252
Sensory ganglia, 346, 616, 626–627
Sensory neuronal TRP channels as initiators and regulators, 970
    TRPA1, 974–975
    TRPM8, 975–976
    TRPV1, 971–972
Sensory transduction process/mechanism, 451–452, 463

Ser20, 951
Ser502, 643–644
Ser800, 644
SERCA inhibitors, 393, 436, 673
SERCA pump, 436, 574f
SERCA (sarco/endoplasmic reticulum Ca^{2+}-ATPase), 414
Serine, 10, 50, 64t, 177, 245–246, 418, 419t, 422, 457, 477, 538, 560, 564, 644, 693, 713, 786, 788, 893
Serotonin, 72t, 326, 669, 797, 990–992, 995
Ser392 phosphorylation, 951
SESTD1 (PI binding protein), 402
Severe combined immune deficiency (SCID) syndrome, 415, 438
Sf9 cell, 6, 98, 882
sh, *see* Short-hairpin (sh)
Shaker K^+ channels, 5f, 519
Sham-operated littermate, 129
Shawn Xu, X. Z., 323–333
Shear force, 770, 772
Shigellae, 453
Shim, S., 343–344
Shine–Dalgarno (SD) sequence, 455
Shoeb, M., 881–895
6-Shogaol, 45t, 51, 70
Short-hairpin (sh), 138, 349
SHR, *see* Spontaneously hypertensive rat (SHR)
ShRNA, 126, 214, 226–227, 417
SH-SY5Y cells, 586, 931–932
Sickle-cell disease, 458
Sidi, S., 345
Signal-regulated Ca^{2+} entry (SRCE), 913
Sildenafil, 787f, 798–789
Simard, C., 147–164, 814
Simon, P., 165
Simon, S. A., 501
Singh, B. B., 552, 573–587
Single channel analysis, 154
Single molecule photobleaching, 17
Single nucleotide polymorphisms (SNP), 136, 179, 239–240, 494, 539, 947, 998
Single particle cryo-EM, 7, 9, 30
Single-particle electron microscopy, 33
Single-particle reconstruction method, 30, 32
SiRNA inhibited cell proliferation, 178
Sir2 protein deacetylase reaction, 380
Six putative transmembrane segments (S1–S6), 2, 186, 324
Skeletal disorders, 1014, 1016

Skeletal muscle, gene expression/function/implications
  calcium entry pathways in sarcolemma
    ion channel blockers, inhibition, 750
    Mn^{2+} quench technique, 750
  gene expression
    detected in chick and mouse muscle cultures, 751
    flexor digitorum brevis isolated, 751
    transcripts coding, 752
    TRPV2 translocation, 752
  implications for disease
    cardiac-specific overexpression of TRPV2, 756
    cytoskeletal scaffolding protein, 756
    Duchenne muscular dystrophy, 755
    fibrosis, 755
    muscle damage, 755
    muscular dystrophy in mdx mice, 756
    necrosis of muscle fibres, 755
    progressive muscle weakness, 755
    α-syntrophin and dystrophin, 755
  linking physiologically detected cation channels, 750–751
  localization and function
    Ca^{2+} leak channel in sarcoplasmic reticulum, 753
    glucose transporter GLUT4, 753
    interaction with triadic proteins, 753
    myopathy, 753
    TRPC subfamily, 752–754
    TRPM subfamily, 755
    TRPV subfamily, 754–755
Skeletal muscle ryanodine receptor, 65
SKF-96365, 738
  action, 603
  characterization
    ATP-stimulated cation current, 90
    fMLP-stimulated cation current, 90
    receptor regulated isoform, 90
  discrimination
    ATP-induced Ca^{2+} entry mechanism, 90
    bradykinin-induced Ca^{2+} entry mechanism, 90
SK-Mel 19 melanoma cell, 142
SMD, see Spondylometaphyseal dysplasia (SMD)
SMDK, see Spondylometaphyseal dysplasia Kozlowski type (SMDK)
Smooth muscle cells (SMC), TRP channels in, 687–699
  expression of
    TRPC, 689
    TRPM, 690–691
    TRPP, 691
    TRPV, 689–690
  physiological/pathophysiological roles of
    TRP and gastro intestinal tract, 697–698
    TRP and uterine contractile activity, 698–699
    TRP and vascular system, 692
  proliferation, 796
  relaxation cycle, TRP in, 692–694
    human TRPC1 gene in rat pulmonary artery, 692
    rat pulmonary arterial and aortic SMC, menthol in, 694
    rat skeletal muscle arterioles, capsaicin in, 694
    ryanodine receptor (RYR), 693
    SMC of TRPC1(–/–) mice, 693
    TRPC6, 693
    TRPC3 and TRPV4, 693
    "With No Lysine (K) Kinases" (WNK1 and WNK4), 693
SNAP-25, 556, 832
Snapin, 506–507, 549t, 553, 556, 563, 582–583
Snapin – TRPC6 – α1A-adrenoceptor complex, 553
SNARE-mediated process, 212
SNARE proteins, 506, 556, 574f, 888
SNL, see Spinal nerve ligation (SNL)
SNP rs1556314, 382
SOAR (STIM1 Orai-activating region), 419t, 422–423
SOC, see store-operated Ca^{2+} channel (SOC); store operated channel (SOC)
SOCE, see Store-operated calcium (Ca^{2+}) entry (SOCE)
Solanaceae, 70
Somatosensory systems, 998–999
Sourjik, V, 462
S1P, see Sphingosine-1-phosphate (S1P)
Spearmint, 626
αII-Spectrin, 551
Spermatocyte, 129, 882, 886, 892
Sperm fertilization, 88
Spermidine, 55, 483, 646
Spermine, 55, 63t, 65, 156–157, 483, 646, 711t, 715
Sphingolipid sphingosylphosphorylcholine, 65
Sphingosine-1-phosphate (S1P), 581, 715, 797, 820
Spinal nerve ligation (SNL), 198, 624
Spiroisoxazolopyridine amides, 647

S. pombe, TRP homologs in, 318–319
  ADPKD, 348
  BLAST analysis, 318
  pkd2 gene analysis, 319
  TRPP2, 318
Spondylometaphyseal dysplasia Kozlowski type (SMDK), 14, 1015t, 1016–1017
Spondylometaphyseal dysplasia (SMD), 14, 61, 1015–1016
Spongia officinalis (isocopalendial), 47t, 52
Spontaneously hypertensive rat (SHR), 162, 179, 693, 719, 789, 798
Spontaneous transient outwards currents (STOCs), 722
Sprouting and capillary tube formation, 768
SPSMA, see Scapuloperoneal spinal muscular atrophy (SPSMA)
Sputum production, 970
S4 (R842), 520
Src kinases, 503, 557, 563, 600, 605, 640, 643, 735t
S4–S5 linker, 186, 381, 519–520
Stable specialized amplifier, 107
Stadler, A., 461
Stadler, Dr. A. M., 461
Stajich, J. E., 260
STAM-1/Hrs, 565
Stamler, J. S., 378
Stark TRPM4 mRNA expression, 162
Starvation-induced apoptosis, 200
State of inertia, 461
Stauderman, K. A., 439
1-Stearoyl-2-arachidonoyl-sn-glycerol (SAG), 128
Steinert, P. M., 849
STIM, see Stromal interaction protein (STIM)
Stim and Orai proteins, 417–420
  Ca^{2+} homeostasis, 417
  CRAC channels, 418
  CRACM1, 418
  ER Ca^{2+} sensor, 418
  knockdown of STIM1 by siRNA, 418
  Orai1 functional domains, 420t
  STIM1 functional domains, 419t
  STIM1 and STIM2, 417
Stim1-Orai1-TRPC
  communication, 421–422
    aspartates in TRPC1 ((639)DD(640)), 422
    CAD (CRAC-activating domain), 421
    CCb9, 421
    CMD/CDI, 421
    FRET microscopy, 421
    lysine residues in STIM1((684)KK(685)), 421
    OASF (Orai-activating small fragment), 421
    SOAR (STIM1 Orai-activating region), 421
    STIM1 and TRPC1, functional association, 422
    STIM1 ERM domain, 422
    STIM1 K-domain, 422
  complexes, calcium entry pathways, 422–426, 424f
    CRAC channels, 423
    in HEK-293 cells, 423
    in human platelets, 425
    human platelets, expression of, 423
    lipid rafts, role of, 424–425
    methyl-β-cyclodextrin (MBCD), use of, 424
    protein-protein interaction, 423
    role of IP_3 receptors, 422–423
    in sea urchin eggs, 422
    SOC channel subunit, 425
    in SOCE and ROCE, 425–426
St. John... wort, 97, 991
Stop-and-go liquid application, 109
Store-dependent channel, 444
Store-dependent mechanism, 130, 394, 398
Store-operated calcium (Ca^{2+}) entry (SOCE), 90, 92, 126–127, 130, 200, 254, 342, 374, 414–415, 422–426, 436–445, 584, 586, 679f, 695, 739, 850, 890, 910, 914–915
Store-operated channels (SOC), 126, 376, 393, 414–415, 418, 424–425, 436–438, 441–443, 669, 674–675, 678, 688, 692–693, 695–698, 708–709, 711t, 714, 718, 750, 764, 783, 882–883, 886, 889, 895, 931
StpA (bacterial protein), 464
Streisinger, G., 342
Streptozotocin-induced diabetes mellitus, 259
Stretch activated channels (SAC), 71, 158, 670f, 688, 691, 694, 709, 718, 750, 772, 799
Stretch and osmotic swelling, 179, 823
Stringent disulphide reduction, 29
Stromal interaction protein, 92–93
  STIM1, 92
  STIM2, 93

Stromal interaction protein (STIM), 393–394, 417–420, 425–426, 439, 673, 783, 894
Structural biology
  with ARD
    TRPA1, 14
    TRPN1, 14
    TRPV4, 14
  coiled coil bundle
    antiparallel α helices, 14
    coiled coil stability, 15
    helix–helix orientation, 15
    heptad residues, characteristic features, 14–15
    homomeric vs. heteromeric protein complex, assembly, 14
    hydrophobic interaction, 15
    oligomeric state, 15
    parallel α helices, 14
    partner selection, 15
    TRPM7, C-terminal coiled coil domain, crystal structure comparison, 16f
    twofold symmetry, 16
  EM structures, 7–10, 8f
  fragments, high-resolution structures, 4t
  methods and considerations, 4–7
    detergents, choice, 6
    expression system, cells, 6
    full length proteins or smaller fragments, working, 7
    membrane protein structures at different resolutions, 5f
  NMR and X-Ray crystal structures, 10–19
    gating spring of mechanoreceptors, 14
    TRPM7 α–kinase domain, 10–12
    TRPM7 coiled coil domain, 14–16
    TRPP2 coiled coil domain, 16–17
    TRPP2 C-terminal E-F hand, 17–19
    TRPV ankyrin repeat, 12–14
    TRPV ankyrin repeat domain (ARD), 13f
  perspectives, 19
  subfamilies, 1–2
    cytoplasmic N and C termini, 3t
    TRPA, 1
    TRPC, 1
    TRPM, 1
    TRPML, 2
    TRPN, 2
    TRPP, 2
    TRPV, 1
  TRP channel subfamilies and transmembrane topology, 2f

Structural studies, Budding Yeast, 25–35
Sturgess, N. C., 814
S. typhimurium, 463
Substance P, 492, 582, 624, 627, 641, 790, 856, 863–874, 991, 999
Suguro, M., 939
Sukumaran, P., 125–131
Sulphonylurea receptor (SUR), 157–158, 164
SUMOylation, 162, 790
Sunesen, M., 107–120
Supercinnamaldehyde, 110–111, 111f
Superoxide dismutases, 532, 769
SUR, see Sulphonylurea receptor (SUR)
Surface biotinylation assay, 550, 552, 561
Sus scrofa, 222
Sutton, K. A., 301
S3 voltage-sensing domain, 214
Swedish Human Proteome Resource (HPR) program, 821
Swimming pattern, 457
Syncytiotrophoblast (STB), 251–252, 255t, 910, 913–914, 914f, 918, 922
Synthetic antidepressants, 991
Synthetic modulator
  broad-spectrum non-natural, synthetic blocker, 90–95
    ACA, 94–95
    2-APB, 92–93
    clotrimazole, 95
    mammalian and fly receptor-activated TRPC channel, regulation, 91f
    SKF-96365, 90–92
    SKF-96365 blocking, concentration-response relationship, 92f
  chemical structures, 89f
  lanthanum and gadolinium ions as modulators, 98–100
  selective, 95–98
    clotrimazole effects on TRPC6, TRPM2, TRPM3, and TRPV4 channels, 95f
    compounds modulating TRPC channels, 96–97
    compounds modulating TRPM channels, 97
    compounds modulating TRPV channels, 97–98
Syzygium aromaticum (vanilloid eugenol), 45, 59–60
Szallasi, A., 491–509

T
Tachykinin substance P, 999
Tachykinin (TK), 640, 999

Tachyphylaxis, 12, 50, 52, 187, 495, 497–500
Takahashi, N., 373–383
Taste buds, 831–832
  detection of chemicals in oral cavity, 832
  like neurons and epithelial cells, 832
  in mammals
    papillae (circumval-late/foliate/fungiform), 832
    PLCβ2 signaling pathway, 832–833
    50–150 taste receptors cells, 831
    taste stimuli, detection of
      ionic and chemically complex taste stimuli, 833f
      salty/sour/bitter/sweet/umami, 832
    TRPM5, 833
    tTRPV1, 834
    voltage-gated calcium channels (VGCCs), 832
Taste system, TRPs in, 831–842
  PKD1L3/PKD2L1 TRPP channels, 836–838
  TRP-melastatin 5 (TRPM5), 834–836
  vanilloid receptor 1 (TRPV1), 838–842
Taurodeoxycholic acid (TDCA), 673
Taylor-Clark, T., 969–978
TCCSUP (urothelial cell line), 956
T-cell, 157, 160–161, 787f, 790, 887, 959
TDCA, see TAURODEOXYCHOLIC acid (TDCA)
Temperature-activated TRP channels (ThermoTRPs), 517
Temperature -sensitive TRP channels, see Thermo TRP channels
Tert-butylhydroperoxide (t-BHP), 586, 769
6-Tert-butyl-m-cresol, 627
Testicular function
  PKD, 891–892
  spermatozoa, 881–882, 885, 888–889, 892–893
  TRPC1, 883–884
    SOCE channel assembly, 890, 895
  TRPC2 and, 885–886
  TRPC3 and, 886–888
  TRPC4 and, 888–889
  TRPC6 and, 889–890
  TRPM family, 893–894
    TRPM4, 893
    TRPM7, 893
    TRPM8, 894
  TRPV family, 892–893
    TRPV 1-6, 893
Tetraethylammonium, 733

Δ^9-Tetrahydrocannabinol (THC), 56, 69, 190, 192t, 346, 625, 931, 972
Tetrameric subunit stoichiometry, 7
Tetramers, 9, 15, 17, 30, 150, 174, 174f, 244–245, 288, 350, 392, 464, 470, 476, 519–520, 536, 557, 783, 930
Tetraspanins, 443
T-format instrumentation, 34
  light intensities, 34
    kinetics parameter, 34
    thermodynamic parameter, 34
TG, see Trigeminal ganglia (TG) neuron
T98G, 956
Thapsia garganica (thapsigargin), 49t, 55, 112–113, 126, 130, 393, 403, 421, 425, 441, 552, 601, 673, 678, 679f, 693, 699, 715, 739, 783, 851, 889–890, 912–913, 931, 978
THC, see Δ^9-Tetrahydrocannabinol (THC)
Therapeutic intervention, implications of, 491–509
  See also TRP channels, complex regulation of TRPV1; Thermo TRP channels
Thermal activation, energetics of, 472–477
  Boltzmann distribution law, 473
  closed-to-open voltage-dependent channel transition, 472–473
  Gibbs free energy, 472
  energy barrier profile, 472f
  open-closed transition, 473
  parameters, thermodynamic, 474
  temperature coefficient, 473
  thermodynamic system, 472
Thermal and voltage sensitivity of thermoTRP
  enthalpic and entropic changes, 475
  heat-induced single-channel activity of TRPV1, 476
    Ohm's Law, 476
    patch clamp recordings, 476
    single channel current (bursts), 476
    TM4 of thermoTRP's, 476
    $V_{0.5}$, definition, 476
  Q10 for channel gating, 475
  structural determinants of thermal sensitivity
    TRPM8, 479–480
    TRPV1, 478–479
  structural determinants of voltage-dependence, 479–480
  thermosensation, 474

threshold, 475
TRPM8 (cold sensor), 475
TRPV1 (thermoreceptor), 475
TRPV/TRPM/TRPA families, 474–475
Thermal hyperalgesia, 66, 492, 494, 503–504, 616–619, 627–628, 648t–650t, 990, 999
Thermal hypersensitivity, 373, 617, 637
Thermal nociceptive signaling, 197
Thermal sensitivity, structural determinants of
  TRPM8, 478
    DRG sensory neurons, in mouse, 478
    membrane lipids, 479
    methyl-beta-cyclodextrin (MCD), 479
  TRPV1, 477–479
    activation by capsaicin and voltage, 478
    Ca^{+2} concentration, 477
    cold receptor (TRPM8), 477
    FRET, 478
    hot receptor (TRPV1), 477
    lacking C-terminal, 477
    temperature activation, 478
    temperature activation of, 478f
Thermal sensors, 155, 463
Thermodynamic formalism, 461
Thermo-induced conformational change, 462
Thermoregulation, 44, 59, 201, 454, 998
Thermosensation, 43, 62, 66, 108, 119, 196, 287, 457, 461–462, 474, 539, 616–617, 620f, 627, 691
Thermosensing, 454, 463, 822
ThermoTRP channels
  activation gate, 485–486
  allosteric gating in thermoTRP, 481–483
  allosteric models and activation
    allosteric coupling, 525
    allosteric gating scheme, 525
    Cole-Moore effect, 528
    PO values, 528
    strict coupling vs allosteric coupling, 527f
  biophysics of polymodal receptors, 469–486
  channel gating process
    Boltzmann equation, 518
    conformational change, 518
    Kv channels, 519
    opened and closed state, 518–519
  Coda, 486
  energetics of thermal activation, 472–473
  family
    allosteric nature, example, 517
    features, 518
    mammalian thermoTRPs, 517
    subfamilies grouped by homology, 517
  in normal conditions, 494
  temperature activation
    entry of Na^+ and Ca^{2+}, 521–522
    heat and cold denaturation, 524
    polymodal regulation, 523
    pore turret, 525f
    protein denaturation, 524
    Q10 values, 521
    thermodynamic parameters, 522
    van't Hoff equation, 523
    voltage vs normalized conductance relationships, 522f
  thermal and voltage sensitivity
    heat-induced single-channel activity of TRPV1, 476
    structural determinants of thermal sensitivity, 477–479
    structural determinants of voltage-dependence, 479–480
  trafficking proteins, 507–509
  TRPV1 trafficking, 503
  TRPV1 and TRPM8 channel activation by agonists, 480–481
  TRPV1 channels, 483–485
    thermoTRP channels undergo pore dilation, 485
  voltage dependence
    Arginine residues at S4 segment, 519
    changes on open probability, 523
    gating current recordings, 519
    Hodgkin and Huxley prediction in 1952, 519
    Shaker K^+ channels, 519
    S4–S5 linker (K856), 520
    S1–S4 voltage-sensor-module, 520
    TRPM8 channels, S4 domain, 520
    TRPM8 S4 helix, topology, 521f
    two-state model, 520
    voltage-gated ion channels, features, 518
    weak, 520
  voltage sensing, 517–528
  See also Thermo TRP channels; *specific TRPs*
Thiazide-induced hypocalciuria, 259
Thick ascending limb of Henle's loop (TAL), 278
Thin limb of Henle's loop (TL), 278
Thr-95, 186
Thr704, 644

Three-dimensional (3D) structure, 1, 5, 7, 261, 420, 639
Threonine, 10, 177, 230f, 232, 246, 418, 538, 693, 786, 788, 893
Threshold phenomenon, 187
Throat irritation, 970
Thymol, 57t, 59, 330, 627, 853
Thymus vulgaris (thymol), 57t, 59, 330, 627, 857
TIRF, *see* Total Internal Reflection Fluorescence (TIRF) analysis
TIRF microscopy, 440
Tissue-culture model in zebrafish, *in vivo*, 349–350
  ER-associated protein degradation pathway, 350
  OS-9 expression, 350
Tissue damage, 532, 615, 623–624, 628, 972
TKR, *see* Tyrosine kinase receptor (TKR)
TK, *see* Tachykinin (TK)
TlpA protein, 463
T lymphocytes, 159f, 160, 178, 414, 437–438, 442
TM, *see* Transmembrane-spanning segment (TM)
TM1-6, *see* 6 Transmembrane helices (TM1-6)
TM3-TM4 linker, 156
TM5-TM6 linker, 54, 156
TNFα, *see* Tumor necrosis factor-α (TNFα)
Togashi, K., 93, 818
Topoisomerase I, 456
Topoisomerase II, 456
Törnquist, K., 125–131
Total Internal Reflection Fluorescence (TIRF) analysis, 138
TPC, *see* Two-pore channel (TPC)
"Traffic" model, 212–213, 212f
Transcellular Ca^{2+} transport pathway, 242
Transcriptional mechanism, 463
Transforming growth factor, 768
Transient inward current (I_{ti}), 162, 790
Transient receptor potential channel C2 (TRPC2), 3t, 91f, 99, 125–131, 325, 341, 344–345, 374, 392, 416–417, 423, 505, 549, 578, 671t, 689, 699, 710, 732, 733t, 735t, 751, 782, 783, 813t, 882, 885–886, 890, 910, 911t, 912, 931, 988
Transient receptor potential vanilloid 5 (TRPV5), 278
  in DCT2 and CNT, regulation, 279

intracellular Ca^{2+} concentration ($[Ca^{2+}]i$) is, 279
Transitional cell carcinoma (TCC), 935, 950
6 Transmembrane helices (TM1-6), 26, 889
Transmembrane-spanning segment (TM), 50, 150–152, 150f, 154, 316, 318–319, 481, 639, 710, 1015t, 1019t, 1022t, 1024t, 1027t
Transmural pressure, 694, 717, 772
Tranter, P., 120
Trematodes and cestodes
  S. mansoni/*S. japomonicum*/*S. haematobium*, 365
  tapeworm (cestode) *Echinococcus multilocularis*, 365
  TRP channel genes of *Echinococcus multilocularis*, 365t
  TRP channel genes of *Schistosoma mansoni*, 365t
  TRPM6 and TRPM7 channels, 366
Trial-and-error strategy, 6
Trigeminal ganglia (TG) neuron, 50–51, 70, 186
Tripterygium wilfordii (Chinese medicinal herb), 72t, 73
Triptolide, 72t, 73
TRPA, *see* Ankyrine-rich TRP (TRPA)
TRPA (ankyrin transmembrane protein), 694, 947
  functions of TRP channels in *C. elegans*, 325–327
    in multiple tissues, 326
    TRPA-1/TRPA-2 channels, 326
  regulation of TRP channels in *C. elegans*, 330
    mammalian TRPA1, 331
  subfamily, 325
TRPA1, 520, 972–973, 974–975
  activation ingredients
    garlic, 110
    mustard oil, 110
    wasabi, 110
  activators, 974
  agonist application, 110
  antagonists, identification, 110
  blockers, 974
  calcium influx requirement, 564
  in cold pain and hyperalgesia
    increased expression, 624
    increased intracellular Ca^{2+}, 624
  epidermal permeability barrier homeostasis, 855

Index 1079

induces heterologous desensitization of TRPV1
  activated by cannabinoids, 625
  heterologous desensitization, 626
knockout mice, 975
in mechanical hyperalgesia
  CFA-induced mechanical hyperalgesia, 625
  Ex vivo skin-nerve recordings, 625
  mechanical hypersensitivity of colon, 625
  mediates visceral pain, 625
mediates noxious chemicals in zebrafish, 346–347
  cannabinoids, 346
  TILLING, 346
  trpa1a and *trpa1b*, 346
  TRPV1 orthologues in chickens and worms, 347
tissue damages
  activation, 620f
  bradykinin, 624
  formalin, 623
  PLC/PKA activation, 624
  pungent cysteine-reactive chemicals, 623
treatment with mustard oil, 563
voltage-clamp techniques, 564
TRPC, *see* Classic or canonical TRP (TRPC)
TRPC (canonical) channels, 688, 689, 948
and cancer, 933–937
  in $Ca^{(2+)}$ homeostasis and differentiation, 932
  DAG-sensitive TRPC1 channels, 931
  GAL1 overexpression, 934
  human gastric cancer, 932
  LNCaP cells differentiation, 933
  ovarian cancer, 932
  SH-SY5Y neuroblastoma cells, human, 931–932
  TRPC1 expression in rat liver hepatoma cell line, 931
  TRPC2 expression in murine B16 melanoma cell line, 931
  TRPC6 prostate cancer cell lines, 933
  TRPC6 channels and NFAT, coupled activation, 931
in *C. elegans* and *Drosophila*, 554
DHPG stimulation, 552
EGF receptor, 550
endocytosis, 553
family
  GPCR stimulants, 913
  OAG activated TRPC6 channels, 913
  PHM1 induction, 912
  TRPC mRNA and protein expression, 911t
  TRPC1/C3/C4/C6/C7 mRNA, expression of, 911, 912
functions of TRP channels in *C. elegans*, 325
  mammalian, 325
  trp-1 and *trp-2* null worms, 325
  TRP-1/TRP-2/TRP-3, 325
heteromers, 996
homo-/hetero-multimerization, 689
ion channels, 413–426
  Stim and Orai proteins, 417–420
  Stim1-Orai1-TRPC communication, 421–422
  Stim1-Orai1-TRPC complexes, 422–426
  TRP proteins, TRPCs, 416–417
IP_3 production, 550
PIP kinase stimulation, 551
protein kinase C activated, 554
regulation of TRP channels in *C. elegans*, 330
  CIRB motif, 330
  Gq-PLCβ pathway, 330
  RT-PCR/Western Blot/immunofluorescence, 690
stimulus-dependent exocytosis, 553
αII-spectrin, 551
subfamilies (TRPC1-7), 689
TRPC1, 549
TRPC2 (pseudogene in humans), 549
TRPC3/6/7, 549, 550
TRPC4 translocation
  EGF pathway, 551
TRPC5 translocation
  EGF effect in HEK-293, 550
TRPC6 translocation
  α1A-adrenoceptor stimulation, 553
  EETs application, 552
  ER-store depletion, stimulation, 552
  G-protein cascade, stimulation, 552
  Tyr-959 and Tyr-972, phosphorylation, 551
  VAMP-mediated exocytosis role, 552–553
TRPC1 and TRPC4, 823–824
  in frog oocytes, 341
  Human Proteome Research program, 823
  intracellular Ca^{2+} elevation, 577
  involvement in mechanotransduction, 577
  mechanosensitive cation channel, 343

TRPC1 and TRPC4 (cont.)
  mGluR1 receptor activation, 576
  NSC's proliferation, 577
  pore region, mutations, 438
  vs. STIM1, functional interaction, 112–113
  store-depletion stimulation, 112
  TRPC4, 823–824
  TRPC1 mRNA, 823
  Western blot, 822
  Xenopus oocyte, 343
TRPC1 and Orai1 to Ca^{2+} entry, contribution
  proposed molecular components
    Orai channels, 438–439
    STIM proteins, 438
    TRPC channels, 437–438
  SOCE characteristics
    activation and assessment, experimental methods, 436–437
    currents associated, 437
  STIM1/Orai1 and CRAC channels
  TRPC-generated SOC channels
    [Ca^{2+}] measurements in cells, 442
    co-immunoprecipitation and TIRF experiments, 443
    C-terminal residues of STIM1, 443
    experimental observations, 442
    major conundrum, 442
    mechanisms, 443–444
    scepticism, 442
    STIM1 knockdown, 443
    STIM1+TRPC1 overexpression, 443
    TRPC1 activation, 443
    TRPC1 knockdown, 443
    TRPC1-SOCE activation, 443
    whole cell patch clamp, 443
  TRPC1/Orai1/STIM1 and SOC channels
    C-terminus of TRPC1 (639DD640), 440
    D76A mutant, 441
    STIM1 knockdown, 441
    STIM1 polybasic tail (684KK685), 441
TRPC2
  acrosome reaction activation, 578
  action potential recordings, decreased, 578
  activation
    receptor-dependent mechanism, 130
    store-dependent mechanism, 130
  arachidonic acid generated, 578
  as calcium entry regulator in sperm, 129–130
  CaM binding sites
    ankyrin, 127
    enkurin, 127
  electrophysiological properties, 127–129
    VNO neurons, 127
  erythroblasts, 127
  as erythropoietin-evoked signalling mediator, 130
  knockdown cell, 131
  males against female, mating, see Mouse
  in mice, 127
  olfaction, pheromone signal transduction, 129
  as pseudogene in human, 125–126
  in rat thyroid cells, 130–131
    calcium signalling, 130f
    novel calcium entry mechanism, 130
  regulation, 126–127
    Ca^{2+}-entry, 126
    calcium-mediated signalling, 126
    inward current, reduction, 126
    Na^+-entry, 126
  ROCE vs. SOCE, 126
  sensory activation of VNO, 578
  as SOC channel, 126
  splice variants, see Mouse
  whole-cell condition, 128
TRPC3, 7, 9
  atrophy and progressive paralysis exhibited, 580
  BDNF-induced growth cone plasticity, involvement in, 579
  dendritic spine formation, 579
  dimension, 9
  disparity, 9
  feed back loop established, 579
  knock-out mice, 989
  modulation mechanism, 9
  multi-modal activation, 9
  NFAT activated, 579
  overexpression, 579
  structure, components, 9
    dense globular inner core, 9
    sparse outer shell, 9
TRPC3–/–, 795
TRPC4, 898
  Ca^{2+} and Na^+ in β-cells, 824
  hydrolysis of PIP2, 824
  5-hydroxytryptamine, activated, 580
  in INS-1 cells and rat β-cells, 823
  involvement in axonal regeneration, 580
  neurite outgrowth, increased, 580
  role in acute and delayed neuronal injury, 580
  sites
    adrenal glands, 580

Index 1081

endothelium, 580
kidney, 580
retina, 580
testis, 580
store-operated Ca^{2+} entry (SOCE), 824
TRPC4α and TRPC4β, 823–824
TRPC5
activated by lipids, 581
CaMK action, 581
knock-out mice, 989
–/– mice, 794
neuronal polarity establishment, stages
dendritic arborization, 581
development into neurites, 581
extension of neurites to form axon, 581
lamellipodia formation, 581
spine/synapse formation, 581
overexpression, 582
TRPC6, 989, 1012–1013
association with multiprotein complex
Calcinurin (CaN), 582
Calmodulin (CaM), 582
immunophilin FKBP12, 582
M1AChR, 582
PKC, 582
Ca^{2+} entry initiated, 582
on chromosome 11q21-q22, 1012
dendritic spine formation, effects, 583
expressed in
brain, 582
cardiac neurons, 582
neurons of olfactory epithelium, 582
retinal ganglion cells, 582
FSGS, see Focal segmental glomerulosclerosis (FSGS)
knock-out mice, 975, 989
neurokinin receptors, activation, 582
NFAT transcription factors, 1013
potential involvement, 583
P112Q/R895C/E897K, mutations in, 1013
vanilloid TRPs, 1014
TRPC7
in bipolar disorder patients, 584
expressed in
cerebellar purkinje cells, 584
cerebral cortex, 584
hippocampal neurons, 584
mitral layer of olfactory bulb, 584
pons, 584
in mouse brain, 584
TRP channels, 287, 930–931
complex regulation of TRPV1, 491–492, 496f

depolarisation, 287
monomeric TRP channel subunits, 288
polycystin-1-TRPP2 receptor-channel complex, 288f
studies on architecture of, 288
subfamilies
ankyrin transmembrane protein 1 (TRPA1), 931
canonical/classical TRP proteins (TRPC), 930
melastatin related TRP proteins (TRPM), 930
mucolipins (TRPML), 930
polycystins (TRPP), 931
vanilloid receptor related TRP proteins (TRPV), 930
of thermo TRPs, 491–492
TRP domain, 930
TRP channel family
subfamilies
ankyrin repeat (A), 517
canonical (C), 517
melastatin related (M), 517
mucolipin (ML), 517
polycystin (P), 517
vanilloid binding (V), 517
TRPγ, 554
TRPM (melastatin), 688, 690–691, 947
atypical proteinkinase domain/TRPM2/6/7 channels, 919
cladogram, different animals, 224f
conventional RT-PCR, 690
EGF treatment, 563
functions of TRP channels in C. elegans, 325
C. elegans defecation cycle, 324
GON-2, 325
GTL-1/GTL-2/CED-11, 326
heterotetramers, 151
ionic channel TRPM1, 150
NSC$_{Ca}$, physiological functions
calcium-dependent voltage regulator TRPM4, 150
cold sensor TRPM8, 150
Mg^{2+} regulated and permeable channels, TRPM6 and TRPM7, 150
redox-sensor TRPM2, 150
taste sensor TRPM5, 150
permeability regulated by, 563
regulation of TRP channels in C. elegans, 330
GEM-1 and GEM-4, 331

TRPM (melastatin) (cont.)
    Glu residues, 331
    GON-2 and GTL-1, 331
  subfamilies (TRPM1-8), 690
  TRPM1, 1017
    abnormal bipolar cell activity, 141
    access to antibodies, 138
    in brain, 141
    C allele frequency rate, 136
    CSNB correlation, 140
    in eye, 140–141
    function, 140–143
    in situ hybridization, 138
    intracellular localization, 138–139
    ion channel, molecular architecture, 142
    melanocytes and bipolar cells, necessity, 142–143
  melanin production, 140
  melanoma metastasis, 139
  melastatin (MLSN) gene, identification/characterization, 136
    MITF mutations, 137
    –/– mouse model, 140, 142
    and pigmentation, 140
    qRT-PCR analysis, 138
    regulation by mGluR6, 142–143
    in skin, 139–140
    splice variants, 136–137
    tissue and cellular distribution, 137–139
    92+TRPM1, 137
    109+TRPM1, 137
  TRPM2, activation features
    ADP-ribose, production, 112, 534
    ADPR interaction with TRPM2, 535
    amino acids 1292-1325 deletion, TRPM2-ΔX, 536
    amino acids 538-557 deletion, TRPM2-ΔN, 536
    amyloid â-peptide, activation, 537
    bullet-shaped major component, 9
    CD38, 538
    current activation, 113f
    in diabetes, 538
    endogenous TRPM2 function, inhibition, 536–537
    exposure to oxidative stress, 536
    extracellular signals, 534
    increase in $[Ca^{2+}]_i$, 535
    inhibition results, 538
    insulin secretion, 113
    isoforms, schematic representation, 537f
    leak current, 113–114
    melanin synthesis, 113
    NAD, 113
    necrotic changes, 538
    nucleotide, 113
    overexpression, 538
    oxidant injury to striatal cells, 537
    oxidative stress-induced cell death, 113
    oxidative stress modulated, 112, 536
    and PARP, inhibition, 538
    physiological splice variants, 536
    prism-shaped minor component, 9
    proposed signaling mechanisms, 535f
    pyrimidine, 113
  TRPM3, activation
    glucose homeostasis, 113
    hepatocytes, activation, 113
    pregnenolone sulphate (PS), 114, 820, 821f
    sphingolipids, 820
    sphingosine, 113
    Zn^{2+} ions, 821
  TRPM mRNA and protein expression, 919t
  TRPM4 and myogenic tone, 717–718
    antisense technology, 718
    Ca^{2+} on TRPM4 activity, 718
    "myogenic" channels (TRPM4/TRPC6), 718
    "myogenic response," 717–718
    role of TRPC6, 718
    stretch-induced VSMC depolarisation, 717
  TRPM4/5 channels, 718
  and TRPM5
    amino acid sequence, 814
    ATP selectivity, 155
    ABC transporters, 814
    automated patch clamp, 112
    Bayliss effect in cerebral arteries, TRPM4 promotion, 160
    biophysical and regulatory properties, 151–156, 152f–153f
    calcium selectivity, 154
    $[Ca^{2+}]i$ concentration, 814
    chemoreceptive cells, 836
    CICR, 815
    CRI-Gl rat insulinoma cells, 814
    diacylglycerol (DAG), 834
    heterologous expression, 151
    expression in humans, 148–149
    firing rate of breath pacemaker neurons, modulation, 161
    in humans and mouse, 815

Index   1083

inositol trisphosphate (IP$_3$), 834
insulin secretion, TRPM4 enhancement, 158–160, 816
ionic selectivity, 153
ionomycin, 114, 114f
linear current voltage (I/V) relationship, 814
  fast/slow oscillations in [Ca^{2+}]i, 815–816
  by glucose, 816
  GPCRs
  main properties, 149t
  molecular structure, 149–151, 150f
  NBD, 814
  PIP2, 834–835
  PLC signaling pathway, 834
  rise in [Ca^{2+}]i, 158
  T1Rs, sweet and umami stimuli, 834
  T2Rs, bitter stimuli, 834
  temperature sensitivity, 835, 835f
  thermal, PIP$_2$ and pH selectivity, 154–156
  TRPM subfamily, 834
  voltage selectivity, 154–155
PI-PLC, 814
PKC and PIP2, 815
putative calmodulin binding sites, 151
putative phosphorylation sites, 151
in stimulus-secretion coupling in β-cells, 815–816
TRPM4 in cardiac cells, 162–165
TRPM4 in immune cells, 160–161
TRPM5 as taste transducer, 163–164
TRPM5 knock-out mouse, 815
TRPM5 mRNA, 815
TRPM5 activation, 156
unusual features, 112
TRPM6, 175, 177t
TRPM7, activated by low levels of MgADP, 538
  alternatives, 173
  architecture and expression pattern, 173–175
  autophosphorylation, 177
  BALB/c animal, 174
  cardiac expression, 174
  cations, permeability, 175
  C57/Bl mice, 174
  channel properties and regulation, 175–178
  cell survival, 178
  cellular and biological functions, 178–179

cell swelling, 539
ChaK1, 173
classical kinases, 10
coiled-coil region, 27
controversial function, 11
crystal structure, 11f
C-terminal cytoplasmic coiled-coil assembly domain, 27
C-terminal lobe, 10
current densities, 178
D1765, 10
D1775, 10
death, 539
detachment, 539
expression, observation, 174
functional characterization, 179–180
growth arrest, 539
helix content, estimate, 173
inhibited by Mg^{2+} and Zn^{2+}, 538
K1646, 10
α-kinase, 27
α-kinase domain, 10–11
LTRPC7, 173
NOD, 174
targeted deletion vs. global deletion, 178
TRP-PLIK, 173
Mg^{2+} deficiency, 539
monomer components, 10
nonmuscle myosins, involvement, 177
N-terminal lobe, 10
overexpression
pathological functions in disease conditions, 179–180
pharmacology, 156–157
phosphorylation, 177
physiological impact, 158–164, 159f
PIP$_2$ hydrolysis, 177
PLC members, interaction, see Phospholipase C (PLC) activity
p38 MAPK and JNK, 178
potential mechanosensor, 179
Q1767, 10
role in anoxic injury, 539
spontaneous, magnesium-subtracted current, 176f
suppression in primary cortical neurons, 539
tetrameric architecture, 15
TRPM6/TRPM7, formation, 176
TRPM8 activation, 534, 691, 973–974, 974, 976
  agonists, 191t–192t

TRPM (melastatin) (cont.)
   antagonists, 191t–194t
   automated patch clamping, 115
   biophysical/pharmacological properties
     and modulation, 187–195
   activation features, 187
   α2A-adrenoreceptors, 188
   agonists, identification, 188
   Ca^{2+}-dependent desensitization, 187
   for chronic pain, 197–203
   cold detection, 196
   cold hypersensitivity, 197–198
   cold sensation compounds, 115
   cold temperatures, 114
   ER membrane, expression, 200
   eucalyptol, 114
   functions, 188
   gene, structure/function and expression,
     186–187
   icilin, 114
   -independent mechanism, 198
   KO mice, role in cold sensing, 196
   lipid rafts, 188
   menthol, 114
   menthol block, characterization,
     116f–117f
   menthol/cooling compound, effects,
     197–198
   modulation factors, 188
   as molecular sensor of cold
     temperatures, 195–196
   –overexpressing cancer cell, 201
   polyphosphates, 188
   protons, 188
   physiology/pathophysiology/therapeutic
     target potential, 194–201
   for prostate cancer, 199–200
   QPatch systems, 115
   sensory neurons, expression, 197
   TG and DRG neurons, expression, 198
   voltage ramp protocols, 115
   zinc-binding module, 11
TRPM channels and cancer, 937–938
   HEK293 cells, 937
   heterologous expression, 937
   microphthalmia transcription factor
     (MIRF), 937
   TRPM8 expression, 937–938
     cold sensor in prostate, 938
     immunohistochemical experiment, 938
     single-cell RT-PCR experiment, 938
   TRPM5 expression in
     Wilms'/rhaboid tumours and
       rhabdomyosarcomas, 939
   TRPM4 expression in CD5+ lymphomas,
     939
   trpm1 gene, tumour suppressor, 937
   TRPM1 (melastatin), 937
   TRPM7, Mg^{2+}-sensitive, 939
     expressed in breast tumors, 939
   trpm8 oncogene, 938
TRPM channels in vasculature, 707–723
   activation mechanisms/functional roles
     TRPM1, 712
     TRPM2, 712–713
     TRPM3, 713
     TRPM8, 714
     TRPM4 and TRPM5, 713
     TRPM6 and TRPM7, 714
   biophysical properties, 710–712
     ion selectivity, 712
     linear current-voltage (I-V) relation,
       712
     membrane depolarisation, 712
     patch-clamp measurements of cation
       currents, 710
   channel structure, 710
     HEK293 or CHO cells, 710
     homotetramerisation, 710
     long TRPs, 710
     N-terminus and C-terminus, 710
     transmembrane (TM) domains, 710
     TRPC/TRPV/TRPA/TRPN channels,
       710
     TRPM6/7 heterotetramers, 710
     TRPM roles in native cells
       (VSMC/ECs), 710
     vascular melastatin TRPM channels,
       expression/properties of, 711t
   GPCR/Ca^{2+} store depletion, roles of,
     714–715
   pharmacological properties, 715–716
     inhibitor of TRP channels, 715
     TRPM3, 715–716
     TRPM8, 716
     TRPM6/7 channels, 716
   vascular function/disease, 716–723
TRPM2 (LTRPC2)
   and β-cells, 817
     anti-TRPM2-N and anti TRPM2-C, 817
     cytoplasmic ADP ribose (ADPR), 817
     long form of channel (TRPM2-L), 817
     short form of channel (TRPM2-S), 817
     Western blot, 817
   and β-cells death, 820

Index 1085

    alloxan-induced oxidative stress, 820
    short isoform of TRPM2 (TRPM2-S), 820
heat, 819
intracellular Ca^{2+} release channel, 820
in stimulus-secretion coupling in β-cells, 817–819
    activation of TRPM2 channel, 818
    ADPR/Ca^{2+}/NAADP/NAAD, effects of, 818
    Ca^{2+}-activated channel, 818
    cADPR, high concentrations of, 818
    insulinotropic agents, 819
    modulators of TRPM2 activity, 819
    N-(p-amylcinnamoyl) anthranilic acid (ACA), 819
    reactive oxygen species (ROS), 818
TRPM2 in oxidant stress/endothelial permeability, 717
    oxidative and nitrosative stress, 717
    TRPM2 activation by H_2O_2, 717
    TRPM2/7 channels, 716
TRPM2-S (short), 536
TRPM2-TE (tumor-enriched), 536
TRPM7 in embryonic development, functions of, 351–352
    endogenous TRPM7 protein in melanophores, 352
    loss-of-function alleles of *trpm7*, 351
    melanin synthesis, 352
    mutations in TRPM6 gene, 351
    scale ENU mutagenesis screen, 351
TRPM7 in magnesium homeostasis/hypertension, 718–719
    in ECs, 719
    in VSMCs, 719
TRPM8 in thermal behaviour of blood vessels, 719–723
    calcium responses, 719
    effects of TRPM8 agonists, 720
    functional expression in rat tail artery VSMC, 720f
    menthol
        cutaneous circulation, 722
        dilatory effects of, 722
        effects in vessels precontraction, 721
        -induced $[Ca^{2+}]i$ rise, 720–721
        patch clamp recordings, 722
        STOCs, 722
    protein expression
        Abcam anti-TRPM8 antibodies, 719
        Alamone anti-TRPM8 antibody, 719
    TRPM8 receptor mRNA in rat, 719
    vasoconstrictor/vasodilator influences, 721f
TRPML, *see* Mucolipidin group (TRPML)
TRPML (mucolipin), 323, 688, 948
    functions of TRP channels in *C. elegans*, 325
        CUP-5, 325
        TRPML1, 325
    PPK-3 (phosphatidylinositol phosphate kinase 3), 330
    regulation of TRP channels in *C. elegans*, 330
        CUP-5 activity, 330
TRPML1, 222, 1026–1027
    active shRNA expression system, 216
    acute knockdown, effects, 216
    ALG-2 protein, role, 215
    c.1_788del (genomic deletion), 1026
    on chromosome 19p13.3-p13.2, 1026
    -deficient murine macrophage, 216
    dendograms of, 223f
    discovery and initial characterization, 209–213
    endocytic pathway functions, 210–211, 216
    experimental systems used to report, 211t
    Fe^{2+} permeability, 215
    function, specific aspects, 215
    genome sequencing, 223
    IVS3-2A_G and g.511_6943del mutations, 1026
    key paradigms and questions, 216f
    Knockout Mcoln1–/– mice, 1027
    MLIV, genetic deteminants, 209–210
    mucolipidosis type IV (MLIV), 1026, 1027t
    mucolipin subfamily, 1026
    perspectives, 219–221
    PKA-dependent phosphorylation, 212
    pre-lysosomal trafficking defects, analysis, 214
    recent developments, 214–215
        consequences, identification groups, 214
    structure-function analysis, 214
    "traffic" and "metabolic" models, 212–213, 212f
        endocytic pathway, 213
TRPML2
    biological roles of, 227
    CD59, recycling, 226–227
    channel formation, 221
    channel properties, 225–226
    early endosome marker
        EEA1, 226

TRPML2 (cont.)
  Rab5, 226
  endocytosis regulation, 221, 226–227
  endosome marker
    CD59, 226
    MHCI, 226
  ER marker
    ER-YFP, 226
  exocytosis regulation, 221, 227
  genome sequencing, 223
  genomics, 222–224
  Golgi marker
    GFP-Golgin 160, 226
    GGA3-VHSGAT-GFP, 226
  late endosome marker
    CD63, 226
    Rab7-HA, 226
  lysosome marker
    Lamp 3, 226
    Lysotracker, 226
    Rab 11, 226
  in vertebrates, 221
  phylogeny, 222–224
  plasma mebrane, heterologous expression, 226–227
  secretory vesicle marker
    Rab8, 226
  subcellular localization, 226
  subunit interactions, 227
  tissue distribution, 224–225
TRPML3
  characteristics of, 231t
  comparative genomics, 222
  disease causes, 222
  endocytic pathway roles of, 234–235
  endosome acidification and differentiation, 235f
  endosome marker
    dextran, 234
    EEA1, 234
    Hrs, 233
  genome sequencing, 223
  genomics, 239–231
  heterologous expression systems, 233
  homologs, 230
  immunohistochemistry, 230
  melanocyte marker
    HMB45, 230
    tyrosinase, 230
  molecular physiological function of, 233–234
  orthologs, 229–230
  RT-PCR expressions, 230
  schematic diagram, 230f
  in vitro experiments, 234
  in vivo experiments, 234
  Western blot analyses, 230
Trpml4, 222
TRPN (NomPC-like), 948
TRPN1 (NOMPC), 341
TRPN (NOMP-C homologues), 26
TRPN (TRP-NompC) subfamily, 323
  functions of TRP channels in *C. elegans*, 325
    mechanosensation, 327
    TRP-4, 326
  regulation of TRP channels in *C. elegans*, 330
TRPP, *see* Polycystin group (TRPP)
TRPP (Polycystin), 287–303, 688, 691, 948
  ADPKD, 287–290
  in humans, 892
  family, 292, 919–921
    co-immunoprecipitation, 920
    colchizine, 920–921
    cytochalasin D, 920
    ion channels, 292
    mammalian homologues
  functions of TRP channels in *C. elegans*, 325
    PKD-2, 325
  gene encoding (PKD2L2), 299–300
  mRNA in mouse testis/oocytes, 300
  mRNA and protein expression, 920t
  PKD1L1, 301–302
  PKD1L2, 302
  PKD1-like (TRPP1-like) proteins, 691
  PKDREJ, 300–301
  polycystin-1 family, 292–293
  primary cilia and PKD, 295–296
  regulation of TRP channels in *C. elegans*, 331–332
    mammalian TRPP2 and PKD1, 331
    PKD-2 and LOV-1, 331
  TRPP1 and TRPP2, 691
    PC1 and PC2, 691
    PKD1 and PKD2, 691
    polycystin 1 and polycystin 2, 691
  TRPP2 and PC-1, 293–297
  TRPP3 and PKD1L3, 297–299
    TRPP2 (PKD2), TRPP3 (PKD2L1) and TRPP5 (PKD2L2), 919
    6-membrane spanner, 919
    ovulation simulation, 921
    Pkd knockout models in mouse, 921

Index

polycytins PKD-1 and 2 in fallopian
tubes and oocytes, 920
reactive oxygen species (ROS), 920
TRPP2
in ADPKD, 292
encoded by PKD2L2, 292
features, 292
PC 2 encoded by PKD2 gene, 292
regulation, 921
six transmembrane segments (S1–S6), 292
TRPP3
polycystin L encoded by PKD2L1, 292
TRPP5, 300
TRPP1=PKD2, 1023–1025
ADPKD and mutations, 1024t, 1025
on chromosome 4q21-q23, 1023
Knudson's model, 1025
mucolipin TRPs, 1025
TRPML1, 1025
TRPML subfamily, 1025
polycystin-1 and polycystin-2, 1023
TRPP2, 891–892
ADPKD morbidity and mortality, 296–297
anterograde transport, 295
chemosensor, 296
in cilium, activation of, 296
cyst-proteins, 296
features of, 293–294
flow-dependent cytosolic Ca^{2+} response, 296
functions, 294
HEK 293 cells, 17
heteromeric complex, 17
intra-cellular trafficking, 295
PC-1, 293–296
EGF receptor stimulation, 294
expression and localisation, 294
induced PI3-kinase activity, 294
signalling pathways, 295
3-stranded complex, 17
subcellular compartments, 294
in tisssues, 294
topology of, 295
and TRPML channels
casein kinase 2 phosphorylation site, mutation, 564
lysosomal degradation, 565
phosphorylation at N-terminal region, 564
PKD-1, 564
STAM-1 or Hrs, overexpression, 565
TRPML1 mutation, 565

Xenopus oocytes, 17
TRPP2 C-terminal fragment (G833–G895), 17
TRPP2 homologs in other yeast, 319
TRPP3 and PKD1L3, 297–299, 892
HEK293T cells, over-expression of, 302
human TRPP3 channel activity, 303
krd mouse model, 297–298
multi-exon PKD2L1 gene, 297
murine TRPP3, 299
Pkd2l1 deletion, 297
polycystin L or PKDL, 297
regions of homology, 297
sour taste receptor, 298
Xenopus laevis oocytes, 297
TRP receptor supefamilies
subdivsions of, 493
TRP signalling, regulation
ion channel trafficking, 549t
signalling pathways regulating translocation
TRPA1, 563–564
TRPC channels, 549–556
TRPM channels, 562–563
TRPP2 and TRPML channels, proteins regulating, 564–565
TRPV channels, 556–562
TRP channels, translocation, 548t
TRPP2 and TRPML channels, proteins regulating, 564–565
TRPV, *see* Vanilloid receptor-related TRP (TRPV)
TRPV (vanilloid), 689–690, 948, 989–990, 992–994, 996–1000
in cancer progression
expression profiles of, 950
channels in tumor growth, 948–949
TRPV1, 949–954
TRPV2, 954–956
TRPV3, 956–957
TRPV4, 957–958
TRPV5, 958
TRPV6, 959–960
subfamilies (TRPV1-6), 690
TRPV1 mRNA, 689
TRPV1 (VR1), 689
TRPV2 mRNA, 689
TRPV4 mRNA, 689
TRPV5 and TRPV6, 689
TRPV1, 7–9, 12–13, 518, 823, 949–954, 970–971, 973, 977–978, 989–990, 992–994, 996–1000
activation, 957
Ca^{2+} influx, as activating enzymes, 494f

TRPV1 (cont.)
    capsaicin receptor, 492f
    in central terminals of primary afferents and CNS
        immunoreactivity, 618
        inflammation/nerve constriction injury in rats, 619
        12-(S)-HPETE, 619
    co-expressions of, 492
    cryo-EM approach, 31
    data appearance
        IonFlux, 115
        IonWorks, 115
        Patchliner, 115
        PatchXpress, 115
        QPatch, 115
    decreased arthritis-induced pain, 618
    in deep tissue pain
        arthritis-induced pain decreased, 618
        capsaicin injection, 618
        carrageenan injection, 618
        chemokines (CCL2), 618
        cytokines (TNFalpha), 618
        mechanical hyperalgesia, transduction of, 618
        mouse colon afferent fibers, mechanosensitivity, 618
    deficient mice (TRPV1$^{-/-}$), 950
    -dependent G0/G1, 950
    -dependent nitric oxide synthase (iNOS), 954
    -dependent tumor cell apoptosis, 953
    dephosphorylation
        PIP$_2$ depletion, 499–500
        PKA sites, 495–497
        PKC sites, 497
    desensitization of, 493
        molecular mechanisms, 501–505
        physiological functions, 500–501
    epidermal permeability barrier homeostasis, 855
    expression, 954
        nociceptor, 115
        pain sensing neuron, 1115
    expression pattern and biological function
        modulation of expression, 640
        physiological and pathological roles, 640–641
        tissue distribution, 639
    functional protein, presence, 30
    future prospect
        antibodies and toxins, biologics, 655
        modality specific TRPV1 antagonists, 654–655
        structure-guided drug discovery, 655–656
    GABAA-receptor associated protein (GABARAP), in cell membrane, 508f
    gene and protein structure
        from gene to structure, 638–639
        identification as ion channel, 657
    in heat pain and thermal hyperalgesia, 616–617
    ion channel properties
        activation, 641–642
        function modulation, 643–644
        gating biophysics, 642–643
        ion selectivity and permeability, 643
    membrane topology, 638f
    molecular mechanisms, 493–494
    "defunctionalisation" of, 493
    monodisperse homotetramer, 30
    mRNA, 952, 954
    mRNA expression, 951
    and nociception
        capsicum example, 492–493
            dorsal root ganglia (DRG), 491
    pharmacology and druggability
        agonists, 646
        antagonists, 647, 648t–650t
    preferential orientation, 30
    protein-to-lipid ratios, measurement, 30
    research toolkits
        clinical status, 647–648
        electrophysiology, 645
        florescence imaging and 45Ca^{2+} flux assays, 645–646
        production and purification, current methods, 644
        radio ligand binding assays, 644–645
    resinaferatoxin (agonist), 30
    signalling modulation, 505–509
    proteins
        SNARE complex proteins, 506–507
        scaffolding proteins, 505–506
        trafficking proteins, 507–509
    stable transfection with HEK293 cell, 117f
    therapeutic approaches
        agonists, 621
        selective antagonists, 621–622
        selective silencing of nociceptors, 622–623
    therapeutic intervention, 489–491
TRPV2, 822–823, 948, 954–956, 978

insulin secretion, 823
in mouse insulinoma MIN6 cells and
    β-cells, 822
mRNA, 955–956
siRNA, 955
TRPV3, 12, 520, 956, 977
TRPV4, 7–9, 12–13, 520, 957–958, 978, 1014
    activation
        4α-PDD, 118
        endocannabinoids, 117
        hypotonicity (cell swelling), 118
        polymodality, 118
        stimuli integration, 118
    activator GSK1016790A, 978
    epidermal permeability barrier homeostasis, 855
    human disorders, mutation linked, 14
    See also individual entries
    IonWorks, 119
    mRNA expression, 957
    PatchXpress, 118
    protein, 957
    systemic osmotic pressure, regulation, 118
TRPV5, 26, 249, 958
    in bone resorption, 252–253
    calcitonin regulation, 247
    Ca^{2+} transporting, 239–241
    gene knockout studies, 249–251
    genes evolution, 240–242
    in inner ear function, 253
    kidney function and, 242
    Klotho functions, 247–248
    pathological and therapeutic conditions
        in cancer, 254–257
        in diuretics, 259–260
        glucocorticoids, 258–259
        in Pseudohypoaldosteronism type II (PHAII), 257
    phylogenic tree of, 241f
    PTH, 246–247
    SNP data, African population, 260–263
        African–Americans *vs.* European-Americans, 260–263
    in vertebrates, 241
TRPV6, 12, 26, 278–279, 959–960
    in birds and mammals, 242
    Ca^{2+} transporting, 240–241
    in exocrine organs, 253–254
    gene knockout studies, 249–251
    genes evolution, 241–243
    in inner ear function, 253
    intestinal Ca^{2+} absorption, 278–279
    Klotho functions, 247–248

maternal-fetal Ca^{2+} transport, 251–252
mRNA, 960
non-genomic actions of, 246
pathological and therapeutic conditions
    in cancer, 254–257
    in diuretics, 259–260
    glucocorticoids, 258–259
    in Pseudohypoaldosteronism type II (PHAII), 257
phylogenic tree of, 241f
physiological condtions, regulations
    aging, 249
    exercise, 249
    lactation, 248
    low dietary Ca^{2+}, 248
    pregnancy, 248
    sex hormones, 248–249
SNP data, African population, 260–263
in uterine, 252
vitamin D regulation
    in humans, 246
    in mouse, 245–246
TRPV1 and TRPM8 channel activation by agonists, 480–481
    mammalian and avian TRPV1 channels, comparison, 480–481
    polymodal thermoTRP's, 481
    single mutation I550T, 481
TRP vanilloid subtype 1 (TRPV1), *see* Capsaicin (CPS)
TRPV6b, 959
TRPV channels
    functions of TRP channels in *C. elegans*, 325–326
        AWA/ASH-mediated sensory behaviors, 325
        OCR channels, 326
        OSM-9, 325
        as polymodal sensors, 326
    mRNA and protein expression, 916t
    NCC involvement, 561
    regulation of TRP channels in *C. elegans*, 330–332
        Asp or Glu, 330
        bisandrographolide (TRPV4), 330
        camphor (TRPV3 and TRPV1), 330
        cannabinol (TRPV2), 330
        capsaicin (TRPV1), 330
        carvacrol, thymol and eugenol (TRPV3), 330
        GRK-2 (G protein-coupled receptor kinase-2), 330
        OSM-9 and OCR-2, 330

TRPV channels (cont.)
    RGS-3 (regulator of G protein signaling), 331
    sites
        intestine, 556
        kidney, 556
        pancreas, 556
        placenta, 556
    subfamilies (TRPV1-6), 690
    TRPV1, 689, 914–915
        activated by capsaicin, endocannabinoids and anandamide, 914
        anandamide binding receptors (CB1, CB2), 915
        cervical ripening, 914
        cAMP level in cells increased, 497
        enhanced activity, 497
        human pancreatic cancer, 935
        human prostate cancer cells, 935
        pheochromocytoma derived PC12 cell line, 935
        premalignant (leukoplasia), 935
        Src kinase phosphorylation, 557
        translocation, 557
    TRPV1-4 ("Thermo-TRPs"), 554
    TRPV2 translocation
        -dependent cell migration, 558
        RGA overexpression, 559
    TRPV2 expression, 936
        cathepsin B expression, 936
        MMP9 induction, 936
    TRPV2/3 transcription, 915
    TRPV3
        calcium-permeable temperature-sensitive, 935
    TRPV4, 690
        activated by cytochrome P450, 559
        binding with PACS-1, 559
        cilliary beat frequency autoregulation, 917
        dysfunction results in infertility, 917
        expression, 936
        mechanotransduction process, 917
        osmo- and mechanosensitive channel, 915
        PACSIN3 overexpression, 559
        phospholipase A2 dependant pathway activation, 917
        WNK1 and WNK4 role, 559
    TRPV5, 556, 559
        Ca^{2+} elevation, 560
        dynamin-dependent endocytosis, 560
        endocytosis, 560
        phosphorylation, 560
        Rab11 colocalization, 561
        serine protease tissue, see Kallikrein
        SGK1/3 effects, 561
        surface biotinylation, 561
        TRPV5 trafficking, 560
    TRPV5/6, 690
        Annexin A2, 918
        17β-estradiol treatment, 917
        calcium selective cation channels, 917
        Cyclophilin B induced activation, 918
        double knock-out of *trpv6* and *eph*b6, 918
        endogenous expression, 917–918
        functional heterotetramers, 917
        heterologous expression, 918
        mice lacking TRPV5 are fertile, 918
        osteoclastic bone resorption, TRPV5, 918
        ruthenium red and magnesium treatment, 918
        TRPV5/6 expression, hormonal control, 917
        trpv6 transcription by progesterone, 917
        Xenopus laevis oocytes, 918
    TRPV1 channels, 481–483
        in HEK293 cells, 483
        node of Ranvier, 482
        organic cations, 481
        permeability ratio P_{NMDG}/P_{Na}, 482
        pore widening, 482–483
        thermoTRP channels undergo pore dilation, 483
    and cancer, 933–937
        expression in LNCaP cells, 934
        homotetrameric structure, 934
        human Jurkat T leukemia cell line, 934
        K562 erythroleukemia cell line, 934
        prostate cancer clinical outcome prediction, 933
    TRPV1-mediated pain and inflammatory process, 98
        in cutaneous sensory system, 853–854
        in hair, 853
        in human epidermis, 848
        in peripheral nervous system., 848
        in sebaceous gland, 853
        See also Epidermal keratinocytes
TRPV2 gene, 954–955
TRPV4 concentration-response curve, 91

TRPV4 knockout (KO) mice, 60, 990, 993, 998–1000
  bone development, disorders, 60–61
  channelopathies, 61
  in vivo studies
    central osmosensation, 60
    mechanical and osmotic nociception, contribution, 60
TRPV4 (OTRPC4/VR-OAC/VRL-2/TRP-12), 821–822
  antibody development by HPR, 822
  human islet amyloid polypeptide (hIAPP), 822
  immunohistochemistry, 822
  osmo and mechanosensors, 822
  phorbol derivatives, 822
  SNPs, 978
TRPV4/TRPP2, heteromultimer of, 349
  co-immunoprecipitation, 349
  deflection of primary cilium, 349
  fluorescence resonance energy transfer (FRET), 349
  polycystic kidney disease, 349
  role for TRPV4 in flow sensing, 349
TRPV5 promoter for generation of transgenic mouse models, 277–285
  Cre recombinase expressing mice
    generation, 283–284
    perspectives, 285
    validation, 284
  EGFP expressing mice
    mouse generation, 279–280
    perspectives, 283
    studies in TRPV5-EGFP mice, 281–283
    validation, 280–281
TRPV6 gene, 242, 246, 245–246, 260, 263, 280, 918, 959
TRPY sub-family, 26
TRPY1 (YVC1), 35
  gating control, aromatic residues, 35
Tsavaler's hypothesis, 938
Tsavaler, L., 938
Tsutsumi, M., 847–856
TtT/M87 macrophages, 977
Tumor cells, 536, 893, 931, 935–946, 948, 953, 955, 959–960
Tumorigenesis, 929, 939–940, 948
Tumorigenic cells, 677–678, 933, 948
Tumor necrosis factor-α (TNFα), 380, 382, 534, 618, 736, 739–741, 764, 820, 976
Tumor progression, 935, 948, 950t, 952, 954
Tumor suppressor function, 139

cell proliferation, inhibition, 139
differentiation, promotion, 139
Two-component regulatory system, 455
Two-hybrid screening, 382
Two-pore channel (TPC), 380
Two-state model, 520, 522, 524
Type 2 diabetes, 539
Tyr-295, 186
Tymianski, M. A., 179
Tyrosine kinase phosphorylation, 186
Tyrosine kinase/PLC/IP$_3$ pathway, 160
Tyrosine kinase receptor (TKR), 392, 393
Tyrosine phosphatase-L1, 380
Tyrosine phosphorylation, 380, 532, 582, 788, 882, 886, 890

U
Ullrich, N. D., 154
U87MG glioma cells, 955–956
U373MG glioma cells, 952, 956
Upstream activation mechanisms, 382
Urinary bladder mechanisms
  degenerin/ENaC family, 869–870
    acid-sensing ion channels (ASICs), 870
    in rabbit bladder epithelium, 870
  mechano sensitive channels, 864f
  mechanosensory transduction, 863t
  sensory nerve endings, 861–864
  Smooth muscles, detrusor, 865–866
  TRP channels in
    P2X3 receptor, 871
    TRPA1, 868
    TRPM8, 867–868
    TRPV1, 866–867
    TRPV2, 867
    TRPV4, 868–869
  two-pore-domain K$^+$ channels (K2P or KCNK), 870–871
  urothelium, 864–865
Urochordates, 223, 351
Urothelial cells, 641, 862–867, 864f, 869–870, 937, 954, 957
Uterine contractile activity and TRP, 698–699
  myometrial hyperplasia and hypertrophy, 699
  PMH1, 698–699
  RT-PCR analysis in human myometrium, 698
  SOC-mediated Ca entry (SOCE), 698
  VOC, role of, 698
  Western blot analysis and immunolocalization, 698
Uterine natural killer (uNK) cells, 915

UVB radiation, 140
Uversky, V. N., 459

V

Valente, P., 489–507
Varitint-waddler *(Va* and *Va^J)* mutant
  anatomy, 232
  cellular defect, 232
  physiology, 232
Van Abel, M., 958
Vandebrouck, C., 751–752
Van der Waals interaction, 16f, 460
Vanilla planifolia (vanillin), 50, 58t, 60
Vanilloid compounds
  Arg114, 638f, 642
  Glu761, 638f, 642
  Ser512, 638f, 642
  Tyr511, 638f, 642
  Tyr550, 638f, 642
Vanilloid receptor, 44, 532, 616, 642, 646,
    650t, 653, 760, 782, 787,
    838–841, 866, 892, 914, 950,
    971, 989
Vanilloid-receptor-like protein 1 (VRL-1), 56,
    240, 866, 869, 930
Vanilloid receptor-related TRP (TRPV), 1,
    12–15, 26–27, 32, 34, 43,
    88, 91, 93, 97–99, 115, 117,
    240–241, 242, 278, 289, 321,
    323–325–326, 330, 342, 364t,
    365, 365t–366t, 375, 389,
    414, 457, 463–464, 469–470,
    471f, 474, 484, 493, 539–540,
    547, 556–562, 575, 603,
    616, 639, 643, 676, 688–690,
    694, 709–710, 732, 750, 752,
    754–755, 760, 762t, 782, 786,
    822, 838, 841, 882, 883t,
    892–893, 911f, 914–920,
    914f, 933–937, 947–961, 974,
    976–978, 988–990
Vanilloid receptor subtype 1 (VR1), 127, 240,
    616, 637, 642, 646, 653, 689,
    838–839, 842, 971, 989
Vanilloid receptor 1 (TRPV1), 838–841
  calcium regulation, 839
  capsaicin, 838–839
  cytosolic calcium in taste cells, 839–840,
    840f
  epithelial sodium channel (ENaC), 839
  FCCP-induced calcium, 840f
  knock out mouse, 839
  salt taste transduction, 839

  amiloride-insensitive component, 839
  amiloride-sensitive omponent, 839
  signaling in peripheral taste receptor cells,
    839
Vanillotoxin, 48t, 53
Vannier, B., 126
Vascular endothelial cells
  angiogenesis and blood vessel formation
    femoral artery ligature, 768
    HUVECs and HMVECs, rise in, 768
    remodeling and patterning, 768
    siRNA or inhibition of TRPM7, 768
    TRPV4 in collateral artery formation,
      768
    vasculogenesis, 767
  endothelial TRP fields, future development,
    774
  functional roles, 774t
  Ca^{2+} influx and electrogenesis,
    763–765
  mechanosensing
    anti-apoptotic effect, 770
    ciliary structure, 771
    CYP epoxygenase activity, 771
    exposed to circumferential stretch, 772
    flow-induced Ca^{2+} influx, inhibition,
      771
    hemodynamic blood flow, 770
    ion channels, 770
    mechanosensitive signaling pathway(s),
      771
    TRP channels, activation, 770–771
    TRPC1 heteromultimerization, 771
    TRPV4 in pressure sensation, 771
    vasodilatory effect, 771
  oxidative stress
    hydrogen peroxide (H2O2), 769
    hydroxyl radicals (OH·), 769
    lipids and proteins, redox modifications,
      769
    peroxynitrite (ONOO–), 769
    superoxide (O2–), 769
  TRP channels/endothelial dysfunc-
      tion/diseases
    angiogenesis-associated diseases,
      progression, 772
    chronic obstructive pulmonary disease,
      772
    endothelial barrier disruption, 772
    endothelial dysfunction, 773
    endothelial NO synthase expression,
      reduced, 773

Index 1093

endothelium-dependent relaxation impaired, 772
excessive oxidative damage, 772
high pressure ventilation, 772
hypertension, 772
lung epithelial-endothelial permeability, 772
lung injury, 772
in mice with inherited hypomagnesemia, 773
NO availability decreased, 772
TRPC3 dysfunction, 772
TRPM7 dysregulation, 773
TRPP1 and TRPP2, malfunction, 772
TRPV4 activation, 772
vascular permeability increase, 772
TRP isoforms, expression of
bovine aortic endothelial cells, 760
human pulmonary artery endothelial cells, 760
reported heteromeric TRP channels, 763
vascular permeability, control of
cytoskeletal stress fiber formation, 767
endothelial barrier function aberration, 766
endothelial hyperpermeability, 767
in lung capillary endothelial barrier, 767
lung distention, 767
lung filtration and edema incidence, 767
in mouse model of spinal cord injury, 766–767
paracellular pathway, 766
post-trauma capillary dysfunction, 767
pro-inflammatory agents, 766
transcellular pathway, 766
vascular tone, control of
$[Ca^{2+}]i$ elevation, 765
guanylate cyclases stimulation, 765
knock-out animals, impairment in, 766
SKCa and IKCa activated, 766
TRPA1 stimulation, 766
Vascular endothelial growth factor (VEGF), 347, 763–769, 791, 795–795, 887, 891
Vascular function and disease, 716–723

specific functions of TRPM channels, 717–723
TRPM4 and myogenic tone, 717–718
TRPM7 in magnesium homeostasis/hypertension, 718–719
TRPM2 in oxidant stress/endothelial permeability, 717
TRPM8 in thermal behaviour of blood vessels, 719–723
TRPM expression in endothelial/VSMC, 716–717
gene expression, studies of, 716
TRPM8 expression, 717
TRPM mRNA expression (RT-PCR analysis), 716
Vascular hyperplasia, 796
Vascular occlusive disease, 796
Vascular smooth muscle cell (VSMC), 90, 112, 183, 394, 397, 399–40, 416, 422, 668, 677, 708–712, 711t, 714–721, 720f–721f, 764–765, 782, 784f, 793, 796–799, 890
Vascular system and TRP, 691
in vascular contractile function, 692–698
pulmonary artery hyperreactivity, 696–696
SMC proliferation, 397
TRP and HPV, 695
TRP and myogenic tone, 694
TRP and pulmonary hypertension, 697
TRP in smooth muscle contraction/relaxation cycle, 692–696
and vascular SMC growth and hyperplasia, 692
TRPC1 specific TIE3 antibody, 692
TRPC1/TRPC6 expression level and SOC activity, 692
TRPM6 and 7, 692
Vasculogenesis, 767
Vasopressin, 159, 278, 548, 551, 638, 669, 913, 994
VDCa-activating depolarization, 160
VDNa, see Voltage dependent Na^+ channel (VDNa)
VDRE, see Vitamin D responsive element (VDRE)
VEGF, see Vascular endothelial growth factor (VEGF)
VEGF antibodies, 796
VEGF-induced cell migration, 796
Venkatachalam, K., 401
Ventral tegmental area (VTA), 995–996
Verapamil, 738, 978

Vertebrata, 2140
Vertebrate TRP channel superfamily
  TRPA (TRP-Ankyrin), 322
  TRPC (TRP-Canonical), 322
  TRPML (TRP-MucoLipin), 322
  TRPM (TRP-Melastatin), 322
  TRPN (TRP-NompC), 322
  TRPP (TRP-Polycystin), 323
  TRPV (TRP-Vanilloid), 322
Vetter, I., 41–73, 576
VGCC, *see* Voltage-gated Ca^{2+} channels (VGCC)
Vibrissal Merkel cells, 627
Vitamin D receptor (VDR), 918
  knock out mice, 989
Vitamin D responsive element (VDRE), 246
Viitanen, T., 125–131
Vistoli, Dr. G., 486
Vlachova, V., 477
V_m, *see* Membrane voltage (V_m)
VNO, *see* Vomeronasal organ (VNO)
VOCCs, *see* Voltage-operated Ca^{2+} channel (VOCC)
Voets, T., 479, 482, 520, 523, 527
Vogel, P., 301
Voltage-dependence, structural determinants of, 479–480
  Gouy-Chapman-Stern equation, 480
  TM4-TM5 linker, 479, 480f
Voltage-dependent gaiting mechanism, 35
Voltage dependent Na^+ channel (VDNa), 162
Voltage-gated Ca^{2+} channels (VGCC), 158–164, 196, 574–575, 574f, 707–708, 712, 717–719, 750, 765, 795, 799, 824, 832, 833f, 836–837
Voltage-gated ion channels, 168, 214, 390, 518–519
Voltage-operated Ca^{2+} channel (VOCC), 362, 669
Voltage-operated calcium channel (VOC), 688, 698
Voltage sensong domain (VSD), 519
Vomeronasal organ (VNO), 127–129
  input resistance, 127
von Euler, U. S., 802
Voolstra, O., 566
VR1, *see* Vanilloid receptor subtype 1 (VR1)
VRL-1, *see* Vanilloid-receptor-like protein 1 (VRL-1)
VR1 receptors, 127, 653, 839
VR2 receptors, 127
VSD, *see* Voltage sensong domain (VSD)
VSMC, *see* Vascular smooth muscle cell (VSMC)
VSM cells, 798

W

Waardenburg syndrome, 137
Wagner, T. F., 97
Walker B form of domain, 151
Walter, K. A., 245
Wang, G. H., 343
Wang, Y.-X., 731–742
Warburgia ugandensis (warburganal), 46t, 52
Warmth sensation threshold (WS), 188, 189t, 653
Wasabi, 66, 67t, 110
Weinberg, R. A, 930
Welsh, D. G., 718
Wensel, T. G., 25–35
Westberg, C. V., 285
Western clawed frog *(Xenopus tropicalis)*, 221–222, 351
Wet-dog shake, 196–197
White, J. G., 324, 740
Whole-cell patch clamp method, 107–108, 112
  recent technological advances, 108
Wicks, N. L., 135–143
Wigge, P. A., 456
Wild-type (WT), 16, 98, 126, 129, 138, 175, 196–197, 233, 247, 252, 257, 259, 280, 301, 303, 344, 346, 352, 498–499, 618, 625–646, 641, 693–694, 789, 792–793, 795–799, 853, 867, 892, 930, 950
Williamson, S. M., 359–368
Wilms' tumours, 939
Wind-up activity, 604
Wistar Kyoto control rat, 179
Wistar-Kyoto rats, 789
Wistar Kyoto VSMC, 179
With No lysine (K) (WNK), 256
  kinase 1, 256
  kinase 4, 256
WNK, *see* With No lysine (K) (WNK)
WNK4, 255t, 257, 561
Wnk4$^{D561A/+}$ knock-in mouse model, 257
Wolstenholme, A. J., 359–368
Wong, C. O., 789–774
Wood, W. B., 324
Wortmannin, 399, 402, 404
W. stuhlmannii (warburganal), 46t, 52
WT, *see* Wild-type (WT)
Wu, G., 263

Wu, X., 577
Wyeth, 654
Wyman, J., 481–482

X

Xanthium catharticum (ziniolide), 47t, 52
Xenobiotic-induced liver toxicity, 676
Xenopus embryogenesis, TRPC1 in axon path during, 343–344
    L-type voltage-dependent calcium channels, 343
    netrin-1, 344
    NMR exchange spectroscopy, 344
    patch-clamp recordings, 343
    PPIase activity, 344
Xenopus laevis oocytes, 240, 297–298, 918
Xenopus oocytes, 17, 53, 254, 256, 258, 261, 343, 349, 506, 556, 561–562, 564, 799, 888, 1021
Xenopus tropicalis, 221–222, 351
Xestospongin C, 92
Xhu, L., 93
Xia, R., 637–656
Xiao, R., 323–333
X-ray crystallography, 2–3, 5–7, 10, 19, 32–33
    TRPV1, ARD structure, 12
    TRPV2, ARD structure, 12
    TRPV4, ARD structure, 12
    TRPV6, ARD structure, 12
Xu, X. Z. S., 323–333

Y

Y555, 61
Yang, F., 478, 524–525
Yang, J., 1–19
Yang, X. R., 719
Yao, J. F., 502
Yao, X., 759–775
Yasuo Mori, 96, 373–383
Yeast, 241
    TRP channels in, 315–320
        function of the Yvc1 channel, 317
        mechanosensation of, 317
        microbial ion channels, 315–316
        molecular dissection of YVC1 (TRPY1), 318
        S. cerevisiae TRP channels, 316–317
        TRP homologs in *S. pombe*, 318–319
        TRPP2 homologs in other yeast, 319
        TRPP2 homologs in *S. cerevisiae*, 319–320
        Yvc1 homologs in other yeast, 318
        vacuolar channel, mechanosensation of, 317
        Yvc1 homologs in other, 318
Yeast episomal plasmids (Yeps), 29
Yeast integrating plasmids (Yips), 29
Yellow mustard, 66
Yeps, *see* Yeast episomal plasmids (Yeps)
Yersiniae, 453
Yips, *see* Yeast integrating plasmids (Yips)
Y745 residues, 186, 481
Y511/S512 motif, 61
Yuan, X. J., 423
Yu, P. C., 347
Yu, Y., 1–19
Yvc1 channel, function of, 317

Z

Zaccai, Dr. G., 461, 462
Zaccai, J., 461, 464
Zanthoxylum piperitum (Szechuan pepper), 68t, 69
Zavitsanou, K., 1000
Zebrafish *(Danio rerio)*, 244, 296, 329, 341–352, 638, 671t–672t, 677, 761t–762t, 768, 1015
Zhang, W., 244, 264, 531–540
Zhang, X., 678
Zhang, Z., 244
Zheng, Y.-M., 731–742
Zholos, A., 707–723
Zholos, T., 723
Zhu, M. X., 126
Zingerone, 45t, 50
Zingiberaceae, 51
Zingiber mioga (miogadial/miogatrial), 46t, 52, 68t
Zingiber officinale, 45t, 50–51
Ziniolide, 47t, 52
Zona pellucida (ZP3), 126, 130, 301, 882, 892–893
ZP3, *see* Zona pellucida (ZP3)
ZP3 glycoprotein, 130
ZP3-induced calcium entry, 126
Zucker, K., 823